The Handbook of PHOTONICS

SECOND EDITION

The Handbook of PHOTONICS

SECOND EDITION

Edited by
Mool C. Gupta
John Ballato

CRC Press is an imprint of the
Taylor & Francis Group, an **informa** business

CRC Press
Taylor & Francis Group
6000 Broken Sound Parkway NW, Suite 300
Boca Raton, FL 33487-2742

First issued in paperback 2019

ISBN-13: 978-0-8493-3095-7 (hbk)
ISBN-13: 978-0-367-38963-5 (pbk)

Library of Congress Cataloging-in-Publication Data

The Handbook of photonics / Mool C. Gupta and John Ballato, editors. -- 2nd ed.
p. cm.
"A CRC title."
Includes bibliographical references and index.
ISBN-13: 978-0-8493-3095-7 (alk. paper)
ISBN-10: 0-8493-3095-5 (alk. paper)
1. Photonics--Handbooks, manuals, etc. I. Gupta, M. C. (Mool Chand). II. Ballato, John. III. Title.

TA1520.H37 2006
621.36--dc22 2006024648

Visit the Taylor & Francis Web site at
http://www.taylorandfrancis.com

and the CRC Press Web site at
http://www.crcpress.com

Preface

Since the publication of the first edition of *The Handbook of Photonics* in 1997 significant growth has taken place in the field of photonics. In the first edition, Professor Mool C. Gupta served as the sole editor in chief. In this new updated second edition, both Professor Gupta and Professor John Ballato have combined their knowledge and experiences to strengthen this handbook. The second edition of the *Handbook of Photonics* continues to serve as a reference source for introductory material and a collection of published data for research and teaching in this field. The subject of photonics continues to grow rapidly and plays a key role in many segments of industry and academic research. The handbook provides a comprehensive source of information for scientists, engineers and students working in this field, with an extensive list of references for further details.

The book is divided into three general sections, photonic materials, devices, and systems, respectively. An introductory chapter, new to this edition, provides an overview of photonics technology, innovation and economic development. The section devoted to photonic materials contains information pertaining to semiconductors, inorganic and organic materials. The devices section includes information on miniature photovoltaic devices, optoelectronic devices, miniature solid-state lasers, optical modulators, optical fibers, ion-exchanged glass waveguides, design methodologies for guided wave devices, optical and holographic data storage, binary optics, and electronic displays. The photonic systems section contains information on optical communications and digital computing. Although the second edition contains several new chapters, it has not been possible to cover every aspect of photonics in a single volume. We have however tried to cover many of the important areas with the latest information and a large collection of data in a single source.

Professor Mool C. Gupta:
I would like to sincerely thank my wife, Rita, children, Nita, Varun and Ashish, my parents and other family members for their continued support and for taking on additional tasks during the updating of this edition. I would also like to thank the National Science Foundation (NSF) Industry University Cooperative Research Center (I/UCRC) program and its industrial members for their support in establishing a Laser and Plasma Applications Center at the University of Virginia. In addition my thanks go to the Department of Electrical and Computer Engineering at the University of Virginia, National Institute of Aerospace and NASA Langley Research Center, all for providing me with the honor of a distinguished Langley professorship.

Professor John Ballato:
I wish to thank my wife, Heidi, and son, Elijah Louis, for their constant and unwavering support. By the time this second edition is published we will also have another child who has already given us great joy. Additional gratitude is deeply owed to my parents, Arthur and Margaret, and my brother and sister-in-law, Tom and Dawn. Friends and colleagues at Clemson University, the School of Materials Science and

Engineering, and the Center for Optical Materials Science and Engineering Technologies (COMSET) are also gratefully acknowledged.

Additional thanks go to all of the authors, editorial board and Taylor & Francis for their input and share of hard work.

Editors

Mool C. Gupta is currently Langley distinguished professor in the department of electrical and computer engineering and director of the National Science Foundation (NSF) Industry/University Cooperative Research (I/UCRC) Center for Lasers and Plasmas at the University of Virginia. As a Langley professor he has collaborations and research facilities at NASA Langley Research Center, National Institute of Aerospace and University of Virginia. He actively participates and represents the University of Virginia as a Langley professor in a seven-university consortium that formed the National Institute of Aerospace. He is the founding director of the NSF I/UCRC and Commonwealth of Virginia centers for Plasma and Photon Processing. Prior to this he was the founding director of the Applied Research Center, program director for materials science and engineering and a research professor in the department of electrical and computer engineering at Old Dominion University. Professor Gupta was a senior scientist and a group leader at the research laboratories of Eastman Kodak Company for 17 years. At Eastman Kodak Company he made contributions in the field of optical data storage, integrated optics and optical frequency conversion. There he represented and led the Kodak team in a consortium of large corporations and multi university partnership for five years in the area of integrated optics based optical read/write head. Before joining Kodak he was senior scientist at the Jet Propulsion Laboratory, California Institute of Technology, Pasadena, California. His educational experiences include: senior research fellow at California Institute of Technology, Pasadena, California (78–79), postdoctoral fellow at Cornell University, Ithaca, New York, (76–78) and PhD in physics from Washington State University, Pullman, Washington (1973). Other professional activities include: Materials Research Society (MRS) short course instructor for optoelectronic materials, processes and devices for over six years, adjunct professor in the department of materials science and engineering at Cornell University for over eight years, conference chair for 1996 SPIE Conference on Nonlinear Frequency Conversion. He is editor-in-chief for *CRC Handbook of Photonics* first edition and editor for *SPIE Proceedings* Vol. 2700 (1996). He has over 85 research publications and 25 patents and was inducted in Kodak's Inventors Gallery. He has taught courses at the University of Virginia, Cornell University, University of Rochester, MRS and SPIE meetings. Professor Gupta has been an invited and plenary speaker at various conferences and universities. His research interests include photon processing of materials, thin films, nanomaterials, devices and sensors. He has been principal investigator for NSF, DARPA, DOE, NASA, AFOSR, ARO, Navy and for various industries supported projects.

John Ballato, associate professor of materials science and engineering at Clemson University, Clemson, South Carolina, directs COMSET, the center for optical materials science and engineering technologies, a South Carolina research center of economic excellence.

Dr. Ballato earned a BS in ceramic science and engineering (1993) and a PhD in ceramic and materials engineering (1997) from Rutgers, The State University of New Jersey. His research interests include the optical properties of materials, optical fiber fabrication, photonic bandgap materials and structures, and rare-earth doping for active and passive applications including lasers, optical amplifiers, and isolators. Dr. Ballato has published over 120 archival scientific papers, holds 10 U.S. and 5 international patents, has given in excess of 110 invited lectures/colloquia, and has co-organized 20 national and international conferences and symposia.

Dr. Ballato has received the Kreidl Award (ACerS 1997), the Byar's Prize for Excellence in Teaching (Clemson University, 2001), the Robert Lansing Hardy Award (TMS 2003) "in recognition of exceptional promise for a successful career in materials science," the Robert L. Coble Award (ACerS 2004) "in recognition of the nation's outstanding young ceramic scientist who is conducting research in academia, industry or a government-funded laboratory," and the Schwartzwalder-PACE Award (NICE 2004) "in recognition of the nation's outstanding young ceramic engineer whose achievements have been significant to the profession and the general welfare of the American people." Dr. Ballato also has been a traveling lecturer for the United States Advisory Committee to the International Commission on Optics (2002). Dr. Ballato was chosen as a liberty fellow (2005). The liberty fellowship is a statewide leadership effort to identify 20 emerging South Carolina leaders each year. The fellowship is a two-year program aimed at identifying the future leaders of South Carolina and giving them special executive management and leadership training. Dr. Ballato's mentor is the Honorable David Wilkins; U.S. Ambassador to Canada. In 2006, Dr. Ballato was selected as the distinguished young alumni for Rutgers University's graduate school and also was selected as the faculty representative on the Clemson University board of trustees.

Dr. Ballato has been a principal investigator (PI) on over $25,000,000 worth of sponsored programs, co-PI on an additional $1,250,000 worth of sponsored programs, and PI on a $10,000,000 endowed chair in optical fiber from the State of South Carolina. Dr. Ballato is an associate editor of the *Journal of the American Ceramic Society*, and is an active participant on the "Optoelectronic Glasses," technical committee for the International Commission on Glass, as well as the Ferroelectrics Standards Committee of the IEEE Ultrasonics, Ferroelectrics, and Frequency Control Society. His Erdös number is 4 and Einstein number is 5.

Contributors

Ilesanmi Adesida
University of Illinois
Urbana, Illinois

John Ballato
Clemson University
Clemson, South Carolina

Ivan Biaggio
Lehigh University
Bethlehem, Pennsylvania

Robert W. Boyd
University of Rochester
Rochester, New York

Charles F. Brucker
Eastman Kodak Company
Rochester, New York

James J. Coleman
University of Illinois
Urbana, Illinois

Larry Dalton
University of Washington
Seattle, Washington

Volkmar Dierolf
Lehigh University
Bethlehem, Pennsylvania

Seydi Doğan
Atatürk University
Erzurum, Turkey

Michael W. Farn
Palo Alto, California

Venkatraman Gopalan
Pennsylvania State University
University Park, Pennsylvania

Mool C. Gupta
University of Virginia
Charlottesville, Virginia

J. Harrison
Spectra-Physics
Semiconductor Lasers
Tucson, Arizona

Carlo Infante
CBI Technology Consultants
Scottsdale, Arizona

Jürgen Jahns
University of Hagen
Hagen, Germany

Alex K.-Y. Jen
University of Washington
Seattle, Washington

John N. Lee
United States Naval Research Laboratory
Washington, District of Columbia

Hong Luo
State University of New York
Buffalo, New York

Terry W. McDaniel
IBM Corporation
White Plains, New York

Hadis Morkoç
Virginia Commonwealth University
Richmond, Virginia

Daniel A. Nolan
Corning, Inc.
Corning, New York

Sergei S. Orlov
Stanford University
Stanford, California

Zhongqi Pan
University of Louisiana at Lafayette
Lafayette, Louisiana

Athos Petrou
State University of New York
Buffalo, New York

Richard Pouder
Appalachian State University
Boone, North Carolina

Bryce S. Richards
Heriot-Watt University
Edinburgh, United Kingdom

Avi Shalav
University of New South Wales
New South Wales, Australia

Kenneth L. Schepler
United States Air Force Research Laboratory
Wright Patterson Air Force Base
Dayton, Ohio

Caron St. John
Clemson University
Clemson, North Carolina

Andrew J. Stentz
AT&T Bell Laboratories
Murray Hill, New Jersey

Philip Sullivan
University of Washington
Seattle, Washington

Ali Teke
Balıkesir University
Balıkesir, Turkey

David Welford
Endeavour Laser Technologies
Hathorne, Massachusetts

Brian West
University of Toronto
Toronto, Ontario, Canada

Gary W. Wicks
University of Rochester
Rochester, New York

Alan E. Willner
University of Southern California
Los Angeles, California

G. L. Yip
McGill University
Montreal, Quebec, Canada

John J. Zayhowski
Massachusetts Institute of Technology
Lexington, Massachusetts

Editorial Board

Contents

PART V Devices

PART VI Systems

I

Introduction

1

Photonics Technology, Innovation, and Economic Development

Caron St. John
Richard Pouder

1.1 Introduction

Optics refers to the diverse body of science and technology that seeks to explain, generate, transmit, manipulate, detect, and use light. The use of light for practical purposes has a long history. Mirrors have been in use for thousands of years. In the early 1600s, the first lenses were ground for telescopes and microscopes. During the early 1700s, Isaac Newton framed the fundamental principles of reflection and refraction. Optics continues to be an evolving field, with several significant technological advances and application opportunities that have positioned it for substantial growth in the future (Saleh and Teich 1991).

Most experts agree that the development of the first laser in 1960 marked the beginning of the modern era of optics or, as it is increasingly known, photonics. During this modern era, photonic technologies have been used in a wide variety of applications including high-speed transmission, data and image storage, process and inspection monitoring, lithography, microscopy, medical procedures, calibration, spectroscopy, surveillance technologies, laser weaponry, image projection, and—importantly—lighting. The industries that have benefited from photonics technologies include virtually every industry of significance: telecommunications, semiconductors and computers, manufacturing, medicine, entertainment, and defense.

Because of its pervasiveness, optics plays an important role in the global economy. There are thousands of optics-related companies in the United States, with annual revenues of over $50 billion

(National Research Council 1998). Worldwide, optics-related companies generated revenues of over $200 billion in 2003 (OIDA 2004). These optoelectronic companies play an important role in generating revenues and jobs in the U.S. economy and in the world. The Optoelectronics Industry Development Association estimates that optoelectronics businesses, research labs, and universities in North America provide jobs for roughly 150,000 people (OIDA 2005).

The primary impact of optics technology on the economy is through its role as an enabler of innovative downstream products, such as the devices and systems noted above. As shown in Figure 1.1, the full supply chain for optics-related industries encompasses many of the largest sectors in the economy. Because of this important enabling role, it is virtually impossible to estimate the full economic impact of optics technologies, but it is well over a trillion dollars (OIDA 2005). As reported by the National Academy of Science in *Harnessing Light* (1998), an investment of a few hundred million dollars in optical-fiber technology has enabled a trillion-dollar worldwide communications revolution. Just six laser transmitters are used in a transatlantic telephone transmission system that can carry 40 million simultaneous conversations. Advances in optics technology allowed these phenomenal achievements, which have also enabled an order of magnitude increase in the downstream economic development impact.

The purpose of this chapter is to frame photonics technology within its economic development context. Technologies generally impact the economy through job creation, new firm formation, and introduction of fresh technologies that enable the development of innovative products which benefit firms and society at large. First, we will define economic development and the role of technology in stimulating economic development. Then, we will discuss the special role of enabling technologies, of which optics and photonic technologies are examples, and the significance of entrepreneurial activity as a

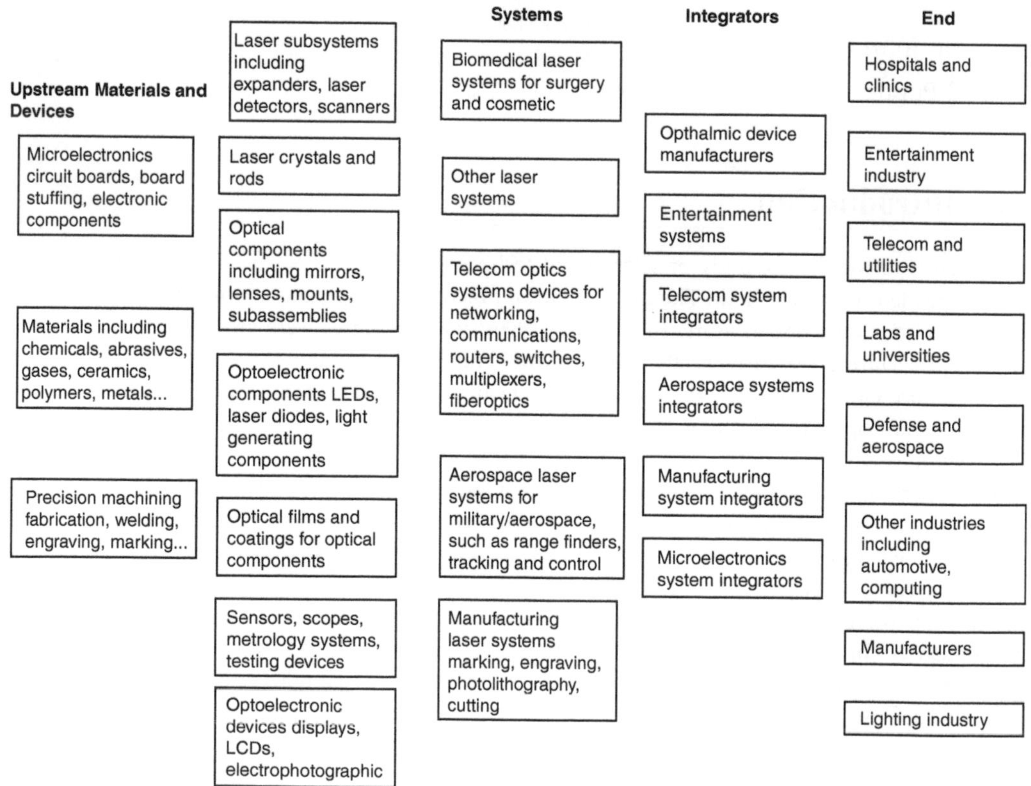

FIGURE 1.1 Optics industries' supply chains.

mechanism for introducing new technologies. We will conclude with a profile of a sample of photonics-related companies in the U.S. to demonstrate economic development impact.

1.2 Technology and Economic Development: An Introduction

New technology and innovation play an essential role in economic development by creating opportunities for new products, new businesses, and whole new industries, all of which stimulate job growth. Furthermore, new technology and innovation provide opportunities for cost reductions and productivity increases that work to improve competitiveness and increase the standard of living in the economy. By definition, successful technology introductions and new innovations are primary drivers of job growth, wealth increases, and improvements in the standard of living. Experts widely credit the U.S. economy's ability to develop original products and new businesses from fresh technologies and inventions as the key driver of the remarkable U.S. economic growth experienced during the last century (Porter and van Opstal 2001).

1.2.1 Definitions and Concepts

Invention involves the creation of something new—a new material, product, or process. Innovation involves putting the invention to use, which then provides an opportunity for economic impact and societal benefit. In discussing the role of photonics technology in driving innovation and economic development, it is important to first define some terms, discuss how technology influences innovation in a general sense, and describe how innovation and new product development tend to track predictable patterns over time.

Incremental innovations provide changes to existing products allowing an extension of product lines, creating opportunities for cost reductions, and reinforcing the capabilities and competitive positions of existing firms within an industry. Within optics, improvements in the performance of optical fibers and refinements in the design of communications systems are both examples of incremental innovations. On the other hand, a *radical innovation* is based on a different set of engineering and design principles, results in wholly new product classes, opens up new markets, and may spawn entirely new industries (Henderson and Clark 1990). Within optics, both the laser and optical fibers are radical innovations that gave rise to whole new markets and industries.

Descriptions of modular and architectural innovations capture the relationship between the subcomponents and the overall system. A *modular innovation* involves changing a core technology within an overall system, but not the relationships among subsystems. An example is the replacement of analog dialing devices with digital devices (Henderson and Clark 1990). The telephone itself, with all of its other subcomponents, still works in the same basic way and interacts with other telephones, over the same telephone lines, in the same way. Similarly, an optical switch as a replacement for an electronic switch is a significant modular innovation—because the components continues to interact with each other in the same fashion and system itself continues to operate as originally designed. In contrast, an *architectural innovation* involves a reconfiguration of existing components in a new way, often brought about by an improvement in a component technology (Henderson and Clark 1990). In the late 1970s-early 1980s, the personal computer was an example of an architectural innovation, building on known technologies within the traditional computer industry, but capitalizing on a window of opportunity created by advances in miniaturization.

Innovations may also be classified according to the degree to which they build on the existing capabilities of firms that are already in the industry. A *competence-enhancing innovation* is one that relies on the same scientific and engineering base as that of the existing firms in the industry. Competence-enhancing innovations, which may originate as incremental or modular innovations, often extend the life cycle of the technological base for the existing firms in the industry. Examples of competence-enhancing innovations are those many new initiatives that have led to the continuous miniaturization of silicon-based microprocessors. A *competence-destroying innovation* is one that draws from different scientific and engineering competences; thereby eliminating any advantage existing firms might have in adopting and deploying the technology. An example of such an innovation is the transition of photography from a largely chemical process to a wholly digital process, with implications for the design of cameras, the need for film, all leading to disruption of business models and displacement of existing competitors. An innovation that is competence-destroying may originate as either a radical innovation, such as the development of the laser, or an architectural innovation, such as the application of digital technology to photography or laser technology to printers.

In general, incremental innovations are part of the normal R&D activity of most firms, fueling growth and sustaining the competitive positions of existing firms in an industry. Incremental innovations virtually always are competence-enhancing for the companies in the industry, allowing growth opportunities and improved competitive positions. Radical innovations, on the other hand, are often introduced into an industry by new firms rather than by existing firms. They are often competence destroying and disrupt the established order among existing competitors. Yet, the revenue generated by the new technology is largely a substitution of revenue previously generated by the old technology, as with the replacement of old cameras with digital cameras. Often there is no real net job creation in the economy as a whole—just a shifting of employment from companies with the older technology to companies with the newer technology. In some cases however, a radical innovation can lead to development of whole new industries—creating opportunities for new products and new firms.

1.2.2 Technology as an Enabler of Higher Value Products

As noted earlier, the primary way that many new technologies impact economic development is through their role as enablers of new higher value products. An *enabling technology* opens up opportunities for innovation in downstream markets. A *general purpose technology* is a special case of an enabling technology that finds widespread application in a range of products and industry sectors, in ways that alter the innovative opportunities and performance of downstream products (Bresnahan and Trajtenberg 1995; Helpman 1998). The laser, internal combustion engine, electricity, and semiconductors are all examples of general purpose enabling technologies. To qualify as a general purpose technology, a technology must embody three characteristics: (1) pervasiveness—it is used in many downstream applications, (2) dynamism—it can be adapted and improved over time, and (3) innovational complementarity—innovation in the general purpose technology increases opportunities for further innovation in downstream applications (Bresnahan and Trajtenberg 1995).

1.2.3 Patterns of Innovation over Time

The relationships between advances in technology and the advances in downstream enabled products and applications generally follow typical patterns over time. Furthermore, the patterns are usually associated with particular economic development outcomes. Therefore, for any class of technologies or class of products, an understanding of the particular life cycle stage and the drivers of transitions between stages can shed light on current and future economic impact.

Innovations are often discussed in terms of s-curves or life cycles that capture patterns over time. A *technology s-curve* captures the pattern of performance improvement over time for a particular technology. Performance improvement is a function of the amount of money and effort invested by researchers in advancing the capability of the technology. For example, one performance parameter in the semiconductor industry is the number of transistors on a chip. In the early 1970s, it was possible to include 2000 transistors per chip but now it is possible to add tens of millions of transistors per chip (Betz 2003). As shown in Figure 1.2, there are often substantial gains in performance during the early years of a technology, slowing, and then reaching the natural performance limit of the technology. As advances are made in performance, new applications of the technology are possible—which stimulates growth and creates new downstream markets or allows cost reductions and improved efficiencies. As an example, for

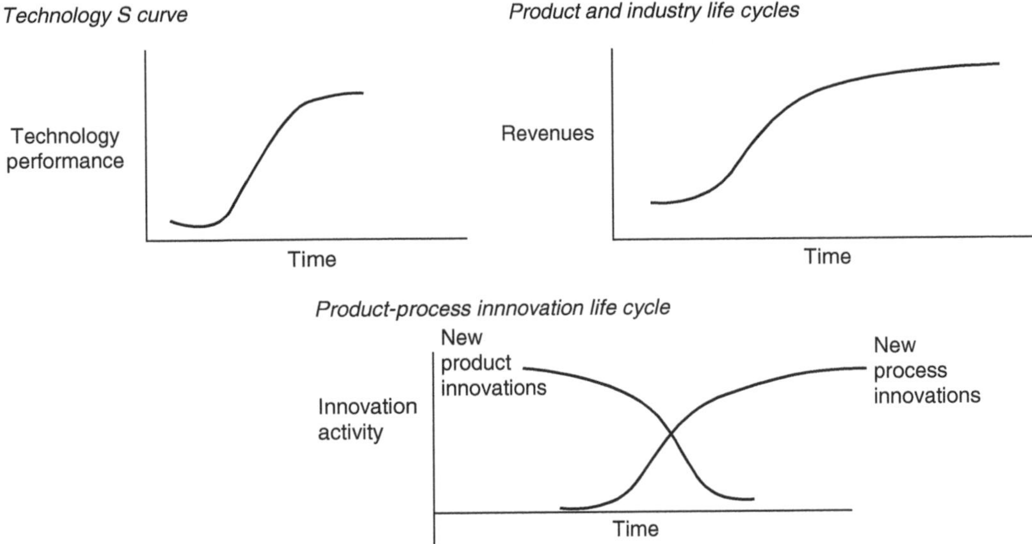

FIGURE 1.2 Technology and innovation life cycles.

semiconductors the increased processing ability in smaller and smaller configurations has enabled thousands of completely new products and sharply reduced the cost structure across several industries.

When a completely original product is introduced to the market, it is very common for revenues to grow slowly initially, followed by more rapid growth as the product is accepted and new applications are identified. Eventually, the product fully penetrates all available markets, demand is driven largely by replacements, and revenue increases come more slowly. This revenue growth pattern is captured in an s-shaped product life cycle, also shown in Figure 1.2. The product life cycle curve is similar in shape to that of the technology s-curve, but captures market acceptance of the innovation rather than the improvements in the capability of the underlying technology. An *industry life cycle* curve is a summation of all new products and their variations and extensions over time—and reflects the overall grow and maturity of the industry.

During the early stage of the industry life cycle, most of the innovations offered by firms in the industry are incremental product innovations. The frequent new product introductions help increase industry revenues, find new applications, and create new markets. As the industry matures, the emphasis on new product innovations shifts to incremental new *process* innovations to reduce costs and improve productivity. This pattern of innovative behavior over time can be captured with the *product-process innovation life cycle* (Figure 1.2), developed by Abernathy and Utterback (1978). Often the transition from a focus on product innovation to a focus on process innovation is triggered by the emergence of a dominant product design that most customers prefer (Abernathy and Utterback 1978). This dominant design then becomes the standard product configuration that serves the needs of most of the market, which provides a disincentive for major new product innovations. During the later process-focused stage, firms are very reluctant to introduce new products because they can require changes in the production system and disrupt process efficiencies. As a consequence, once an industry progresses to the process-innovation stage, firms are very reluctant to develop and adopt fresh product innovations.

These patterns of innovative behavior over time have implications for the level and type of economic impact. In general, during the early stages of a product life cycle, when the focus of innovative efforts is on new product development, successes fuel employment growth and investment. Both of which provide a positive economic development impact. New products are introduced, new plants are constructed, new markets penetrated. The companies that participate in the industry during this stage may growth quite rapidly by hiring new employees and making significant capital investments.

Over time, however, as the firms in the industry transition to a mature stage, they turn their attention to a more stable product offering as well as process and productivity improvements. At that transition point, it is common to see a reduction or stabilization in the number of employees and far fewer investments in new plants. This process occurs even while investments in productivity-enhancing new equipment continue. The slowing of employment growth is a result of productivity improvements that allow higher levels of output with fewer employees. Consequently, judgments about the economic development impact of a particular technology, measured by jobs, wages, and wealth, should consider the stage of technology maturation, the life cycle stage of enabled products, and overall drivers of industry growth.

1.2.4 The Role of Entrepreneurial Activity

When we think of technology and innovation, we often assume that the large well-funded corporations are generating most of the innovations in the economy. In fact, they are generating most of the incremental, competence-enhancing innovations. Yet, interestingly, entrepreneurial firms are more likely to introduce significant new technologies, radical innovations, into the market. Research has demonstrated that small technical firms, on average, patent more technically significant inventions than larger firms, participate in innovations that are more leading edge, and play an essential role in advancing new technologies in the economy (CHI 2003). Furthermore, entrepreneurial firms are responsible for the vast majority of new job growth. For example, between 1990 and 2000, the U.S. economy created approximately 15 million new jobs. Over one-third of those jobs were created by the approximately 6.4 million new firms that were formed during those same years. In spite of the economic slowdown precipitated by the 911 terrorist attacks and the wars that followed, the U.S. economy generated over a million new jobs in 2004. Together, new and expanding small firms account for well over 75% of all new job creation in the country—a phenomenon that has been observed in the U.S. for the last three decades. Therefore, the number of new firms formed in a particular technology category is an important indicator of its current and future economic development impact.

1.3 The Economic Impact of Optics Technologies

As described in the previous section, the economic impact of technologies can be described in terms of the downstream innovations and applications that are enabled, the number of firms and number of workers employed in the industry, and the prevalence of entrepreneurial activity, i.e., the number of new firms that are created to commercialize new innovations. Cast against this backdrop, we will now discuss each of the aforementioned topics as they relate to photonics technologies.

1.3.1 Optics-Photonics as an Enabler

As noted earlier, optics realizes its greatest economic impact in its role as an enabling technology. Using the criteria of pervasiveness, dynamism, and complementarity set forth in the previous section, a laser may be classified as a general purpose enabling technology. It has found applications in manufacturing, health care, information technology, entertainment, and communications. Consistent with the idea of a general purpose enabling technology, it has experienced substantial improvements over time, and has enhanced opportunities for innovation in many downstream applications.

In 2003, the value of optoelectronic components was $63 billion and the value of the largest volume products that were enabled by an optical technology was $140 billion (OIDA 2004). Optoelectronic component revenues are dominated by flat panel displays, which account for roughly 70% of the dollar value of all optoelectronic component sales. Other components include LEDs, storage media, fiber cable, image sensors, and diode lasers—each of which generates revenues in the $3–4 billion dollar range annually (OIDA 2004). As would be expected by the component sales patterns, the products enabled by optical components are dominated by displays, including notebook computers, LCD monitors and televisions, digital cameras, and camera phones. Other enabled products include optical storage devices and fiber network equipment. These enabled products have found extensive application in information technology, telecommunications, health care, life sciences, optical sensing, lighting, energy, and manufacturing, with current and future implications for economic development activity.

1.3.2 Communication and Information Technology

In the late 1990s, optical fiber for communications was installed worldwide at a rate of 1000 meters every second (National Research Council 1998). Between the late 1980s and the late 1990s, the percentage of U.S. transcontinental calls carried over fiber-optic cables increased from 10 to 90% (National Research Council 1998). As of 2004, over half of U.S. residential Internet users accessed the Web via a broadband connection, which illustrates the degree to which broadband technologies have rapidly penetrated the market (OIDA 2005). Historically, optical fiber for the telecommunications industry dominated sales of optics and related products in the communication and information technology sectors. Recently, telecommunications applications have dropped to 7% of total enabled products (OIDA 2004). This decrease reflects the telecommunications industry slump worldwide, the widespread penetration of established markets, as well as the extraordinary growth of flat panel displays. Whereas, in the past, demand for optoelectronics in the communications and information technology sector was driven by communications needs, it is now driven by the need for optical displays in notebook computers, monitors, televisions, PDAs, and camera phones.

For the existing array of communications and information technology applications, the industry is now moving into the latter phase of the growth stage of the product life cycle. Much of the extraordinary growth of the last decade has been in replacing older communications, display, and storage technologies with optical technologies. For example, converting from copper wire to fiber optic cable, moving from magnetic to optical storage, replacing CRT monitors with FPD monitors. Once the installed base is replaced with optical technologies, future growth will come from new network installations, new products, and replacements and repairs of existing products and networks. This demand pattern will sustain a large *volume* of demand in these existing applications, but calculated over a larger base or denominator, will result in a slower *growth rate*. Although there will continue to be substantial growth opportunities in the future, particularly into developing countries, once optical technologies have replaced the traditional technologies used in these applications, the growth rate will be increasingly dependent upon the underlying replacement growth rate of the enabled products themselves. Significant new innovations in transmission, switching, data storage, and displays will provide new growth opportunities as well.

1.3.3 Healthcare and the Life Sciences

Within health care and life sciences, optics has enabled both new therapies and new diagnostic techniques, which have reduced the invasiveness of procedures and the resulting risk of infection, pain and suffering. For example, lasers have been used in procedures ranging from open heart surgery, to ophthalmic surgery, to knee repairs, to dermatological treatments. Optic technologies have been used in imaging, monitoring, and diagnostic techniques during surgical procedures. As in all optical applications, the value of the optical components included in the medical equipment and devices is very small—but the economic benefit in terms of new medical devices and improved health and well-being is significant.

One of the most promising applications under study is the use of light to study tissue and blood, in vivo, to detect, diagnose, and treat diseases noninvasively. Biophotonics, which is defined as the use of light to image, detect, and manipulate biological materials, is a substantial growth field within the optical components industry. Applications of biophotonics include bioimaging, biosensing, optical diagnostics, light based therapies, tissue engineering, and light manipulation of cells (Prasad 2003). Medical and diagnostic applications of photonics technology hold great promise for the future and are likely to lead to many opportunities for new and existing firms.

1.3.4 Optical Sensing, Lighting, and Energy

Innovative optics technologies promise to cut the energy costs associated with lighting by as much as 20% (National Research Council 1998). This outcome results in significant economic implications for industrial and residential customers. Furthermore, new sensors are improving security systems and industrial process monitoring and control systems. Night vision systems, which have been so successful on the battlefield and with law enforcement, are being incorporated into automobiles. This use will increase growth opportunities substantially. One of the fastest growing optics-enabled industries in recent years has been digital cameras and camera phones. Innovations within this area will likely result in new product applications and new market opportunities that will continue to drive sales growth for some time to come.

1.3.5 Other Industrial Applications

Optical technologies are applied to production problems in a diverse set of industries to perform production processing, such as laser welding and machining, and to control production, including sensors and vision systems (National Research Council 1998). One of the most significant production processing applications is in the photolithography used in making semiconductor chips. The equipment used for photolithography is a $1 billion industry, which then enables a $200 billion electronics industry (National Research Council 1998). Other production processing technologies make use of lasers for welding, sintering, and curing of epoxy resins. Continued, significant improvements in optical lithography, precision laser machining, and in MEMS systems are expected to continue to advanced production processing across a variety of industries. The companies that provide optical components for production applications are highly specialized and innovative, and are likely to continue aggressive new product development. This fact will allow continued process improvements downstream.

1.3.6 Firms and Employment in the Photonics Industry

The Optoelectronics Industry Development Association estimates that optoelectronics businesses, research labs, and universities in North America provide jobs for roughly 150,000 people (OIDA 2005). The *CorpTech Directory of High Technology Companies* lists approximately 2500 firms or business units involved in photonics-related products or services in the U.S., classified into the following product groups: acousto-optics, camera and related equipment, display, fiber optics, laser, optoelectronic devices, optics, photonics services, and other photonics equipment(Table 1.1). As shown in Table 1.2, most states

TABLE 1.1 Number of Firms in Specific Photonic Product Categories

Product Category	Number of Firms
Acousto-optics	20
Camera and related equipment	174
Display	234
Fiber optics	353
Laser	262
Optoelectronic devices	428
Optics	678
Photonics services	511
Other photonics equipment	4

Source: One Source Information Services. 2003. *CorpTech Directory of High Technology Companies*. One Source Information Services, Omaha, NE. With permission.

TABLE 1.2 Photonics Companies in Each State

State	Number of Photonics-Related Companies and Divisions
Alabama	19
Arizona	57
California	534
Colorado	44
Connecticut	74
Delaware	4
District of Columbia	4
Florida	95
Georgia	25
Iowa	6
Idaho	3
Illinois	106
Indiana	21
Kentucky	16
Kansas	10
Louisiana	1
Massachusetts	219
Maryland	61
Maine	219
Michigan	38
Minnesota	54
Missouri	20
Mississippi	5
Montana	3
Nebraska	6
North Carolina	20
New Hampshire	35
New Jersey	140
New Mexico	5
Nevada	8
New York	190
Ohio	68
Oklahoma	12
Oregon	33
Pennsylvania	111
Rhode Island	18
South Carolina	7
South Dakota	2
Tennessee	11
Texas	93
Utah	14
Vermont	46
Washington	31
West Virginia	4
Wisconsin	30
Wyoming	1

Source: One Source Information Services. 2003. *CorpTech Directory of High Technology Companies*. One Source Information Services, Omaha, NE. With permission.

have at least a few photonics-related firms, but as confirmed in Table 1.3 and Figure 1.3, the firms and business units tend to concentrate in a handful of states, which in turn focuses the employment opportunities in those areas.

As described in the early sections of this chapter, an important indicator of future economic growth is the number of new ventures that are formed to commercialize new photonics technologies. The highly

TABLE 1.3 States with the Most Photonics Companies

States with Most Photonics Companies		States with Most Photonic Start-Ups Since 1990	
California	534	California	142
Massachusetts	219	Massachusetts	54
New York	190	New York	40
New Jersey	140	New Jersey	27
Pennsylvania	111	Florida	26
Illinois	106	Texas	23
Florida	95	Pennsylvania	17
Texas	93	Arizona	15
Connecticut	74	Colorado	12
Ohio	68	Connecticut	11

Source: One Source Information Services. 2003. *CorpTech Directory of High Technology Companies*. One Source Information Services, Omaha, NE. With permission.

innovative activities of new and small firms work to create new products and new industry segments, reinvigorating the industry life cycle and sustaining growth overall. As shown in Table 1.3, hundreds of new photonics firms have been formed since 1990 and they also tend to concentrate in a handful of states. Most of the established photonics-related companies are located in California and Massachusetts, and most of the new ventures are created there as well.

The rate at which firms are being formed within a particular state is also of interest. The proportion of firms within a state formed since 1990 has been greatest in the following states: Arizona (31%), North Carolina (30%), Florida (27%), California (27%), Colorado (24%), Massachusetts (25%), and Texas (21%). Among these states, California, Massachusetts, and Texas have historically had large numbers of photonics firms. These states' firms cover the full range of photonics technologies, making products that act as enablers to a wide range of downstream markets. On the other hand, Arizona, North Carolina,

FIGURE 1.3 Dark, greater than 60; Medium, 21–60; Light, 20 or fewer. (From One Source Information Services. 2003. *CorpTech Directory of High Technology Companies*. One Source Information Services, Omaha, NE.)

Florida, and Colorado, which are emerging as new centers of growth in photonics-related firms, tend to focus on a few photonics technologies and make products that are enablers to a narrower range of applications. For example, a high proportion of recently created firms in Florida produce optical and laser systems that act as enablers of downstream technologies in aerospace and astronomy applications, consistent with the needs of local industry in that region. Recent start-ups in North Carolina focus on optics and fiber optics and enable applications in data communication and telecommunications systems, an industry that is well represented in that region.

Among states historically having large numbers of photonics firms, the proportion of total firms formed since 1990 has been lowest in the following states: Illinois (12%), Connecticut (15%), Pennsylvania (15%), and Ohio (17%). These statistics suggest that the industries that purchase products from these photonic firms are well served by the existing array of companies—and new firm formations as a mechanism for introducing new products into the market is less important. It also implies that the industries in the aforementioned states that are enabled by photonics technologies are at a mature stage of their industry life cycle, and in less need of innovative new technologies.

In some regions these patterns of geographic concentration may be partly explained by the industries that are enabled by photonics technologies. States having a large concentration of photonics firms and a high proportion of new firms created since 1990 tend to be in proximity to firms that are in the industries that they enable (i.e., their customers). For example, photonics firms in California are near firms that use optical and display products for movie production technologies. Other photonics firms in California are near semiconductor manufacturers, another customer group of the photonics industry.

The geographical clustering of entrepreneurial activity in photonics is not surprising. Clustering is very common in technology-based industries because the skills and resources needed to support new ventures are highly specialized and spin-offs must locate near those resources. Once a region develops the resources to support a particular type of industry or technology, it is likely that similar firms will form or locate in that same region near universities or established firms. Tucson and Tampa–Orlando in the United States and Ottawa in Canada have particularly vibrant photonics clusters that are benefiting from strong local university research programs and aggressive regional economic development groups.

1.4 Concluding Remarks

Economic development is about enhancing the factors of production including land, labor, capital, and technology for the purpose of creating jobs and wealth. New jobs provide opportunities for a growing population and help bid up the wages of existing jobs, which, when accompanied by productivity increases, help raise the standard of living. As wages, numbers of jobs, and wealth increase, it becomes possible to make further investments in education, infrastructure, and quality of life, which stimulates a virtuous cycle of growth and development.

Photonics technology has played an important role in economic development by driving innovations that have led to significant societal advancements in communications, computing, medicine, national security, and defense. Researchers have contributed to both incremental and radical innovations, which have spurred growth for existing firms and provided opportunities for new start-ups. Whereas North American fiber optic cable companies dominated the industry in the 1990s, display technologies, largely centered in Asia, are dominant in the mid 2000s. As a large collection of enabling technologies, the viability of the photonics industry is tied very much to the health of the industries that it supports. When the telecommunications industry falls on hard times, the photonics technologies targeted toward those applications are hurt as well.

In the future, new economic opportunity will come from continued incremental and modular innovations in materials, components, devices, and systems that build on growth in worldwide telecommunications, computing, medical, and security demand. The U.S. optics industry is strongest in developing high performance specialty niche products, which play an important role in enabling innovation in downstream applications. Although these niche products are unlikely to result in large numbers of new jobs, they can be the basis for forming new entrepreneurial ventures that provide some job stimulus in their regions. Research on new firm formation and innovation demonstrates that significant (maybe radical) innovations are more likely to originate in regions having a higher proportion of new firms. Such areas are also likely to be centers of early-stage industries—those with products at earlier stages of their product life cycles. Therefore, once a state or region develops the infrastructure to support the growth and development of new, innovative firms, it is likely that the momentum will continue to spur regional economic development. When specific products become commodity-like, however, they are likely to move offshore. This pattern of early-stage innovation located in the higher wage, early-adopter nations—followed by movement to lower wage, late adopter nations during the commodity stage—is very likely to continue. This pattern has implications for job growth—and the sustainability of job levels over time.

References

Abernathy, W. J. and Utterback, J. M. 1978. Patterns of industrial innovation. *Technology Review*, 80, 7, 40–47.

Betz, F. 2003. *Managing Technological Innovation: Competitive Advantage from Change, 2nd Ed.*, Wiley, Hoboken, NJ.

Bresnahan, T. and Trajtenberg, M. 1995. General purpose technologies: Engines of growth. *Journal of Econometrics*, 65, 83–108.

CHI Research, Inc. 2003. *Small Serial Innovators: The Small Firm Contribution to Technical Change.* U.S. Small Business Administration, Office of Advocacy, Washington, DC.

One Source Information Services. 2003. *CorpTech Directory of High Technology Companies.* One Source Information Services, Omaha, NE.

Helpman, E. 1998. *General Purpose Technologies and Economic Growth.* MIT Press, Cambridge, MA.

Henderson, R. M. and Clark, K. B. 1990. Architectural innovation: The reconfiguration of existing product technologies and the failure of established firms. *Administrative Science Quarterly*, 35, 9–30.

National Research Council. 1998. *Harnessing Light: Optical Science and Engineering for the 21st Century.* National Academy Press, Washington, DC.

OIDA (Opto-electronic Industry Development Association). 2004. *Worldwide Optoelectronics Markets-2003.* OIDA, Washington, DC.

OIDA (Opto-electronics Industry Development Association). 2005. *Worldwide Optoelectronics Markets-2004.* OIDA, Washington, DC.

Porter, M. and van Opstal, D. 2000. *U.S. Competitiveness 2001: Strengths, Vulnerabilities, and Long Term Priorities.* Council on Competitiveness, Washington, DC.

Prasad, P. N. 2003. *Introduction to Biophotonics.* Wiley, Hoboken, NJ.

Saleh, B. E. A. and Teich, M. C. 1991. *Fundamentals of Photonics.* Wiley, New York.

II

Materials—Semiconductor

2

III–V Semiconductor Materials

Gary W. Wicks

2.1 Introduction

The purpose of this chapter is to provide a source of data on the properties of III–V semiconductors that are relevant to optoelectronic applications. The family of III–V materials discussed in this chapter consists of compounds of the column III elements, aluminum, gallium, and indium, and the column V elements, phosphorus, arsenic, and antimony. The III–V nitrides and bismuthides are excluded from consideration here. The three column III and three column V elements can be grouped into 9 binary, 18 ternary and 15 quaternary compounds. Not all of these possible combinations have been successfully grown, and only about a quarter of them have important optoelectronic applications. This subset of the III–V family that has important optoelectronic applications is the material covered in this chapter.

This chapter organizes the III–V family into material systems, i.e., sets of materials that are compatible with each other and can be used to construct useful epitaxial heterostructures. The important III–V material systems are listed in Table 2.1.

2.2 Fundamentals

2.2.1 Optical Processes and Relationships among Constants

The main optical processes of importance in optoelectronic applications are reflection, waveguiding, diffraction, absorption, emission, and electrooptic and nonlinear optical effects. The main aspects of reflection, waveguiding, diffraction, absorption, and emission are expressed with the refractive index (n), absorption coefficient (α), and emission energy or direct bandgap energy (E_Γ), respectively, and are tabulated in this chapter. In material systems that have an indirect bandgap (E_X or E_L) that can become smaller than E_Γ, the indirect bandgap is also tabulated. Only those direct gap materials, i.e., materials with E_Γ smaller than E_X or E_L, can be efficient light emitters. Electrooptic effects and optical nonlinearities are discussed in other chapters of this handbook.

TABLE 2.1 III–V Material Systems with Important Optoelectronic Applications

Material System	Substrate	Lattice-Matched Members	Important Strained Members	Main Optoelectronic Applications
AlGaAs	GaAs	GaAs	$Ga_{1-x}In_xAs$ $0 \leq x \leq 0.25$	Emitters and modulators: 0.75 μm $\leq \lambda \leq$ 1.1 μm
		$Al_xGa_{1-x}As$ $0 \leq x \leq 1$		Detectors: 0.4 μm $\leq \lambda \leq$ 1.1 μm
		AlAs		Saturable absorbers: $\lambda \sim$ 0.8–0.9 μm
GaInAsP/InP	InP	$Ga_{0.47}In_{0.53}As$	$Ga_{1-x}In_xAs$ $0.4 \leq x \leq 06$	Optoelectronic devices at $\lambda = 1.3$ μm and 1.55 μm
		$Ga_xIn_{1-x}As_yP_{1-y}$ $x = 0.47y$; $0 \leq y \leq 1$	$InAs_xP_{1-x}$ $0 \leq x \leq 0.2$	
		InP		
AlGaInAs/InP	InP	$Ga_{0.47}In_{0.53}As$	$Ga_{1-x}In_xAs$ $0.4 \leq x \leq 0.6$	Optoelectronic devices at $\lambda = 1.3$ μm and 1.55 μm
		$(Al_xGa_{1-x})_{0.47}In_{0.53}As$ $0 \leq x \leq 1$		
		$Al_{0.48}In_{0.52}As$		
AlGaInP	GaAs	GaAs	$Ga_{1-x}In_xAs$ $0 \leq x \leq 0.25$	Red emitters
		$Ga_{0.5}In_{0.5}P$	$Ga_{1-x}In_xP$ $0.4 \leq x \leq 0.6$	
		$(Al_xGa_{1-x})_{0.5}In_{0.5}P$ $0 \leq x \leq 1$		
		$Al_{0.5}In_{0.5}P$		
AlGaAsSb/GaInAsSb/GaSb	GaSb	GaSb		Emitters and detectors: $\lambda \sim$ 2–3 μm
		$Al_xGa_{1-x}As_ySb_{1-y}$ $x = 12y$; $0 \leq x \leq 1$		
		$Ga_{1-x}In_xAs_{1-y}Sb_y$ $x = 1.1y$; $0 \leq x \leq 1$		
GaAsP	GaAs or GaP	GaAs (on GaAs substrates); GaP (on GaP substrates)	GaAsP	Visible LED's

Other optical properties more relevant to solid state physics than to optoelectronic applications will not be discussed here. Examples of optical data that are not included in this chapter are critical point energies above the fundamental bandgap, Raman phonon energies, mid-IR properties, and low temperature (4 K) data. These data can be found in other standard references (Madelung 1982).

In many cases, low temperature data on these materials is easier to find and is more complete. Room temperature data, however, are more relevant to optoelectronic applications. Thus, this chapter provides room temperature data on the spectral and material composition dependence of the optical constants and emission energies. Not discussed here are the effects of doping and optical intensity on the optical constants and emission energy.

Data is provided here on the optical constants, n and α, and bandgaps. Rather than listing the optical constants n and α, some references list n and k (the extinction coefficient), or ε_1 and ε_2 (the real and imaginary parts of the dielectric function). These pairs of optical constants are all interrelated. Any of the three pairs of optical constants can be determined from either of the other two pairs. The relationships among these optical constants are given in many common references (Pankove 1975). For most common optoelectronic applications, the most convenient form of the optical constants is n and α, thus, it is this pair of optical constants that is tabulated here.

2.2.2 Quantum Wells and Strained Materials

The optical properties of a semiconductor are altered by quantum size effects when at least one of the dimensions of the material is less than a few hundred Å. At present, the main structure in which quantum size effects are important are quantum wells—structures consisting of a thin well material sandwiched between two layers of a barrier material. Bulk material, i.e., material with large enough dimensions so that quantum size effects are negligible, are simpler cases to tabulate than quantum wells The optical characteristics of bulk material depend mainly on photon energy and material composition, and are comprehensively documented here. The optical properties of quantum wells depend on photon energy—two materials compositions (well and barrier materials), and the thicknesses of the two materials. The additional parameters affecting the quantum wells make it impractical to completely catalogue optical properties for all possible quantum well structures. Only general trends for quantum wells are discussed here.

The present documentation of materials is mainly limited to bulk materials that lattice match an available substrate. Lattice mismatch produces strain. If the strain exceeds a critical value, it will be relieved via the production of dislocations or cracks, drastically degrading material quality. The critical amount of strain is determined by the magnitude of the lattice mismatch and the thickness of the layer. Thick layers need to be accurately lattice matched but thin layers, such as those in quantum wells, can be quite severely mismatched without suffering a degradation in quality. By relaxing the lattice matching requirement, strained quantum wells present the device designer with an additional adjustable parameter, an expanded range of usable material compositions. Additionally, in many cases the alteration of the bandstructure resulting from strain can be used to advantage. For these reasons strained quantum wells, incorporated into epitaxial structures with thicker nominally lattice matched layers, are becoming increasingly used in such optoelectronic devices as diode lasers. Although not exhaustively catalogued here, expanded ranges of bandgaps attainable with strained quantum wells are mentioned here, where relevant.

2.3 Data on Specific Systems

2.3.1 AlGaAs Material System

2.3.1.1 General Characteristics and Optoelectronic Applications

The materials technology and optical characteristics of the $Al_xGa_{1-x}As$ material system is the best understood of all the III–V materials. A primary reason for the large interest in this material system is the fact that it is a *ternary* material the lattice matches an available substrate, GaAs, over its whole composition range. Other material systems must to resort to the more difficult *quaternary* materials to have continuous composition ranges that lattice match substrates. The only other ternaries that similarly lattice match substrates are AlGaSb and AlGaP, but these materials are less useful than AlGaAs. AlGaSb is either indirect or nearly indirect over its whole composition range, and AlGaP is an indirect gap material over its whole composition range AlGaAs is *direct* over nearly half of its composition range.

The AlGaAs material system is the prototype III–V semiconductor where new physics and device concepts are usually initially explored before being moved into other, more difficult material systems. AlGaAs has widespread applications in electronics and optoelectronics. Electronics applications include advanced devices, such as high electronic mobility transistors (HEMTs), heterojunction bipolar transistors (HBTs) and resonant tunneling diodes. Optoelectronic applications include diode lasers, solar cells and high speed photodetectors, optical modulators, and saturable absorbers. AlGaAs, like many of the III–V systems, has a certain relationship between electronic and optical properties that is extremely useful for optoelectronic applications; namely, the compositions of AlGaAs with larger bandgap energies also have smaller refractive indices and lower optical absorption. The relationship between bandgap and refractive index enables double heterostructure waveguides to confine injected electrons and holes together with the optical mode, an extremely useful characteristic for laser diodes. The relationship between bandgap and transparency allows the construction of high bandgap, transparent windows to let light in or out of a heterostructure, while limiting access of electrons to non-radiative recombination centers at surfaces. In a typical optoelectronic application, the active region of the device would consist of GaAs (or strained GaInAs); cladding/electron confinement layers of waveguides and transparent window layers of photodetectors employ AlGaAs.

2.3.1.2 Bandgaps, Refractive Index, and Absorption Coefficient

The ternary, $Al_xGa_{1-x}As$, is a direct gap material (like GaAs) over the gallium rich portion of its composition range and indirect (like AlAs) over the aluminum rich portion of its composition range. Several bandgap-composition curves have been published. One of the most accurate results gives the direct–indirect crossover composition at room temperature to be at a composition of $x_c = 0.43$ and a bandgap energy near 2 eV (Chang et al. 1991). In its direct gap range, dependence of the composition, x, on the T = 300 K bandgap is given by (Chang et al. 1991)

$$x = 0.702(E(\text{eV}) - 1.424), \quad \text{for } x \leq 0.43.$$

Rearranging this expression, the dependence of the direct gap on composition is obtained:

$$E_\Gamma(\text{eV}) = 1.424 + 1.4245x, \quad \text{for } x < 0.43.$$

A second reference (Guzzi et al. 1992) cites a value of $x_c = 0.40$ and gives expressions for the direct and indirect room temperature bandgaps over the whole composition range of the ternary. This information is given in the following expressions and in Figure 2.1:

$$E_\Gamma(\text{eV}) = 1.423 + 1.36x + 0.22x^2$$

and

$$E_X(\text{eV}) = 1.906 + 0.207x + 0.55x^2 \quad \text{for } 0 \leq x \leq 1.$$

The above two expressions for $E_\Gamma(x)$ agree within 7 meV for the direct gap compositions of $Al_xGa_{1-x}As$, i.e., x values less than 0.40.

The compositional dependence of the refractive index spectra and absorption spectra (or equivalently, n and k spectra or ε_1 and ε_2 spectra) have been measured (Marple 1964; Sell et al. 1974; Aspnes et al. 1986) and modeled (Adachi 1988; Jenkins 1990) by several authors. A fairly accurate model is that of

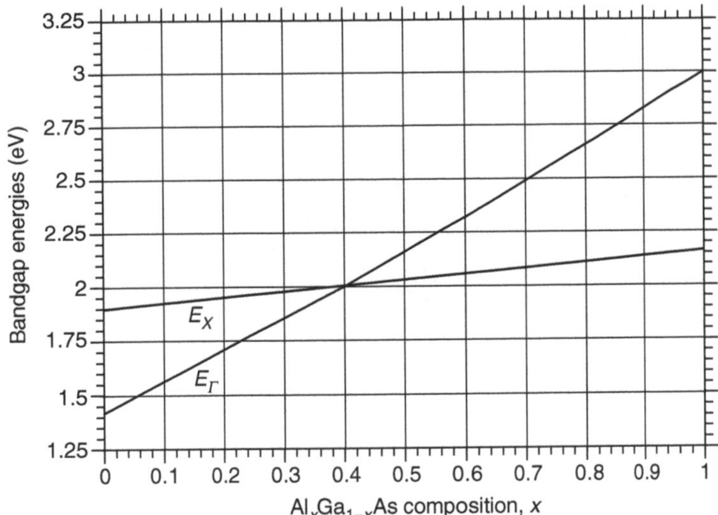

FIGURE 2.1 Room temperature values of the lowest direct (E_Γ) and indirect (E_x) bandgaps vs. $Al_xGa_{1-x}As$ composition. (From Guzzi, M., Grilli, E., Oggioni, S., et al. 1992. *Phys. Rev. B.*, 45, 10951. With permission.)

Jenkins (Jenkins 1990) which is an extension of the model of Adachi (1988). Figure 2.2 and Figure 2.3, displaying the compositional dependence of the refractive index and absorption spectra, respectively, are based mainly on the Jenkins model.

2.3.1.3 Additional Data

2.3.1.3.1 Strained Quantum Wells

The long wavelength limit of emitters and detectors that employ lattice-matched AlGaAs is set by the GaAs bandgap wavelength at $\lambda \sim 870$ nm. The use of strained, pseudomorphic $Ga_{1-x}In_xAs$ quantum

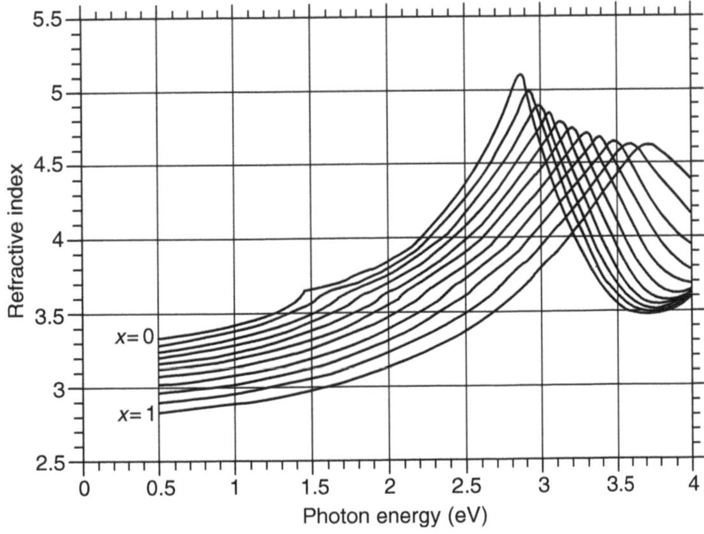

FIGURE 2.2 Compositional dependence of the room temperature refractive index spectrum of $Al_xGa_{1-x}As$. The eleven curves represent compositions from $x=0$ to $x=1$ in equal increments of $\Delta x=0.1$.

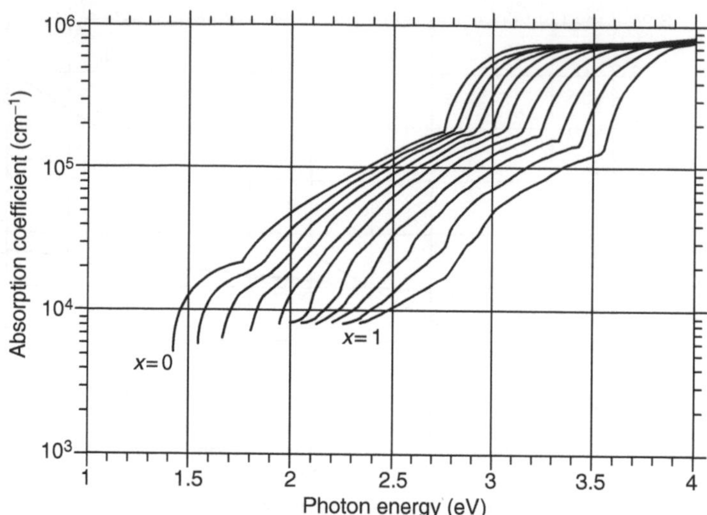

FIGURE 2.3 Compositional dependence of the room temperature above-bandgap absorption spectrum of $Al_xGa_{1-x}As$. The eleven curves represent compositions from $x=0$ to $x=1$ in equal increments of $\Delta x=0.1$.

wells extends the wavelength to beyond 1 μm (Feketa et al. 1986; Choi and Wang 1990), thereby encompassing the important wavelengths of $\lambda=980$ nm (for pumping erbium-doped fiber amplifiers) and $\lambda=1.06$ μm (for YAG replacements). Typical InAs mole fractions for these longer wavelength applications are as large as $x \sim 0.2$–0.25.

2.3.1.3.2 *Thermal Effects*

The variation of the bandgap energy of GaAs with temperature can be fit with the Varshni expression (Thurmond 1975)

$$E_g(T) = 1.519 - \frac{5.405 \times 10^{-4} T^2}{T + 204} \text{eV}, \quad 0 < T(\text{K}) < 1000$$

Near room temperature, this expression indicates that the GaAs bandgap shrinks as the material is heated at a rate of 0.45 meV per degree:

$$\frac{dE_g}{dT} = -4.5 \times 10^{-4} \frac{\text{eV}}{\text{K}}, \quad T \sim 300\text{K}.$$

Equivalently, near room temperature the heating-induced lengthening of the GaAs bandgap wavelength is given by

$$\frac{d\lambda_g}{dT} = 2.8 \frac{\text{Å}}{\text{K}}, \quad T \sim 300 \text{ K}.$$

A second report (Kirillov and Merz 1983) on the rate of bandgap change with temperature gives values 30% larger than those cited above.

The thermal variation of the refractive index of GaAs is given by (Cardona 1961a)

$$\frac{1}{n}\frac{dn}{dT} = 1.1 \times 10^{-4}\,\mathrm{K}^{-1}.$$

2.3.2 GaInAsP/InP Material System

2.3.2.1 General Characteristics and Optoelectronic Applications

The general $Ga_xIn_{1-x}As_yP_{1-y}$ quaternary has compositions specified by the parameters x and y that can independently take on any values between 0 and 1. The most important of these compositions are those that lattice match InP, denoted here as GaInAsP/InP. The GaInAsP/InP quaternary can be viewed as a mixture of two simpler materials, $Ga_{0.47}In_{0.53}As$ and InP, each of which lattice matches InP:

$$Ga_xIn_{1-x}As_yP_{1-y} = (Ga_{0.47}In_{0.53}As)_y(InP)_{1-y}.$$

By comparing the Ga subscript on both sides of the above relationship, it is easily seen that the $Ga_xIn_{1-x}As_yP_{1-y}$ compositions that lattice match InP are specified by

$$x = 0.47y \quad (0, \leq y \leq 1).$$

The lowest bandgap, highest refractive index mixture of these two materials is the endpoint composition, $Ga_{0.47}In_{0.53}As$; the highest bandgap, lowest refractive index mixture of these two materials is the other endpoint composition, InP. All other compositions of GaInAsP/InP have bandgaps and refractive indices (and most other properties) intermediate to those of $Ga_{0.47}In_{0.53}As$ and InP.

Similar to many of the III–V systems, GaInAsP/InP materials with higher bandgaps have lower refractive indices. As discussed earlier (see Section 2.3.1.1) this useful characteristic enables optical and electrical confinement in the same layer of a double heterostructure. Important applications of the GaInAsP/InP material system involve optoelectronic devices for optical communications systems based on 1.3 or 1.55 μm wavelengths. Diode lasers operating at either of these two wavelengths typically consist of InP cladding/electron confinement layers with GaInAsP cores and may or may not contain $Ga_{0.47}In_{0.53}As$ quantum wells. Infrared detectors with $Ga_{0.47}In_{0.53}As$ active layers cover wavelengths shorter than the bandgap wavelength of $Ga_{0.47}In_{0.53}As$, namely 1.65 μm, thus including the important communications wavelengths of 1.3 and 1.55 μm.

There are two other material systems closely related to GaInAsP/InP. AlGaInAs/InP, another quaternary that lattice matches InP, spans a similar range of bandgaps and refractive indices. This system is covered in the section on AlGaInAs/InP. The $Ga_xIn_{1-x}As_yP_{1-y}$ quaternary can also lattice match GaAs (with $x=0.48y+0.52$; $0 \leq y \leq 1$). However, it is not especially important because its range of bandgaps and optical constants are also covered by the well-developed AlGaAs/GaAs material system.

2.3.2.2 Direct Bandgap, Refractive Index, and Absorption Coefficient

As discussed above, the compositions of the quaternary $Ga_xIn_{1-x}As_yP_{1-y}$/InP can be pictured as mixtures of InP and $Ga_{0.47}In_{0.53}As$. Because both of these constituents of $Ga_xIn_{1-x}As_yP_{1-y}$/InP are direct gap, the quaternary itself is direct gap over the whole range of compositions that lattice matches InP. The room temperature direct bandgap energy vs. composition is given by Nahory et al. 1978.

$$E_g(y) = 1.35 - 0.72y + 0.12y^2(\mathrm{eV}) \quad \text{for } x = 0.47y;\ 0 \leq y \leq 1.$$

This relationship is depicted in Figure 2.4.

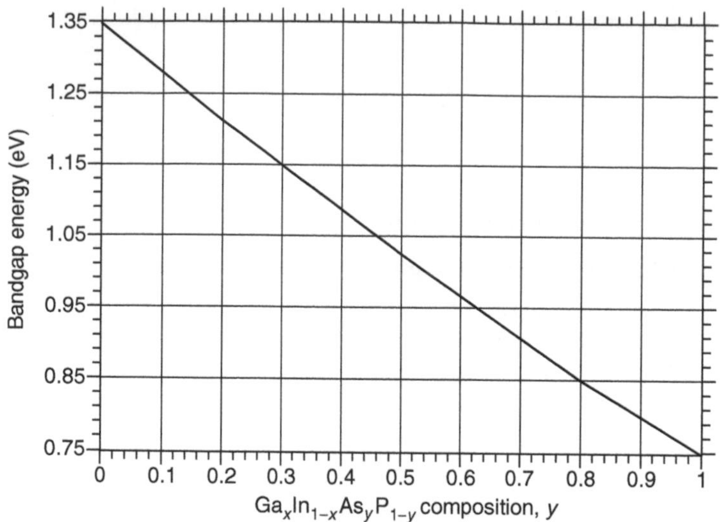

FIGURE 2.4 Room temperature bandgap of Ga_xIn_{1-x} As_yP_{1-y} for compositions lattice matched to InP, i.e., $x=0.47y$ and $0\leq y\leq 1$.

Fairly complete refractive index data (Chandra et al. 1981) and models (Jensen and Torabi 1983) exist for photon energies below the bandgap. The modeled spectral dependence of the room temperature refractive index is shown in Figure 2.5.

Data on above-bandgap absorption of GaInAsP is scarce. An absorption spectrum of one composition of GaInAsP, with a bandgap wavelength of $\lambda_g=1.31$ μm, has been reported (Yokouchi et al. 1992) and is graphed in Figure 2.6, along with that of the two end point compositions, InP, and of $Ga_{0.47}In_{0.53}As$ (Backer et al. 1988) (Newman 1958; Cardona 1961b, 1965; Bacher et al. 1988; Petti and Turner 1965).

A simple model of above-gap absorption, valid within a few tenths of an eV of the fundamental gap, uses the expression

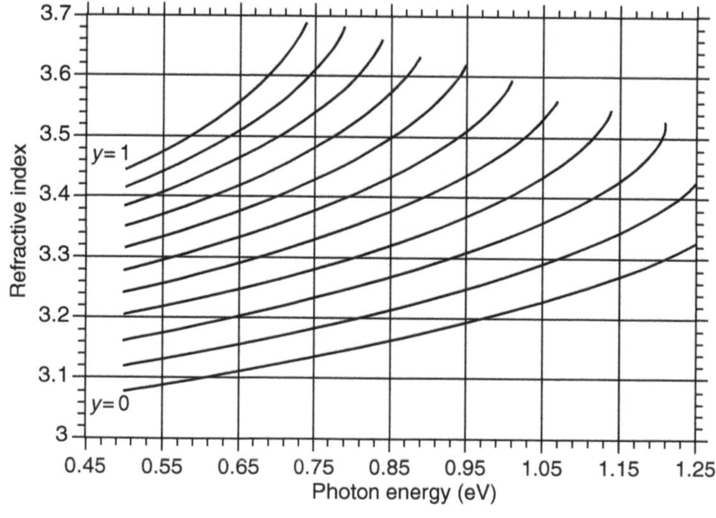

FIGURE 2.5 Room temperature refractive index of $Ga_xIn_{1-x}As_yP_{1-y}$ for compositions lattice matched to InP, i.e., $x=0.47y$ and $0\leq y\leq 1$.

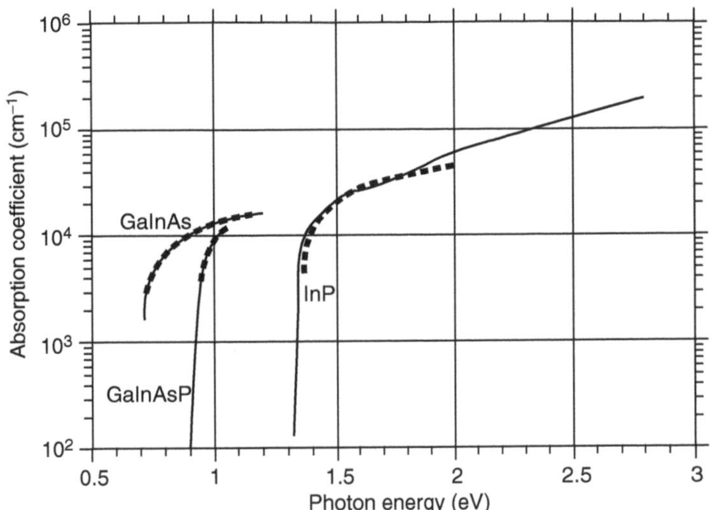

FIGURE 2.6 Room temperature absorption spectra of $Ga_{0.47}In_{0.53}As$, GaInAsP/InP ($\lambda_g = 1.3$ μm), and InP. The solid line represents the experimental data. The dashed line is the fit to the above gap absorption with the expression, $\alpha = \text{const} \times (E - E_g)^{1/2}$, as discussed in the text.

$$\alpha(E) = \text{constant} \times (E - E_\Gamma)^{1/2}.$$

The resulting fits of the three absorption spectra in Figure 2.6 are

$$\alpha_{\text{GaInAs}} = 22900(\text{cm}^{-1}\ \text{eV}^{-1/2})(E - 0.705\ \text{eV})^{1/2},$$

$$\alpha_{\text{GaInAsP}} = 36600(\text{cm}^{-1}\ \text{eV}^{-1/2})(E - 0.938\ \text{eV})^{1/2},$$

$$\alpha_{\text{InP}} = 54500(\text{cm}^{-1}\ \text{eV}^{-1/2})(E - 1.354\ \text{eV})^{1/2}.$$

In the absence of other absorption data, absorption spectra of compositions near that with $\lambda_g = 1.31$ μm can be estimated by inserting the appropriate direct bandgap into the fit expression from Figure 2.6 and converting bandgap wavelength to bandgap energy:

$$E_g = \frac{1.2398\ \mu\text{m eV}}{\lambda_g}.$$

Note that the α fit for the sample labeled $\lambda_g = 1.31$ μm in Figure 2.6 indicates that its bandgap energy is approximately 0.938 eV, corresponding to a bandgap wavelength closer to 1.32 μm than the cited value of 1.31 μm.

2.3.2.3 Additional Data

2.3.2.3.1 Thermal Effects

The heating-induced shrinkage of the bandgap has been reported for the two endpoint compositions and one intermediate composition of GaInAsP/InP. The Varshni expression for the full temperature dependence of the $Ga_{0.47}In_{0.53}As$ bandgap (Yu and Kuphal 1984) is

$$E_g(T) = 0.814 - \frac{4.906 \times 10^{-4} T^2}{T + 301} \text{eV} \quad (T \text{ in degrees Kelvin}).$$

Evaluating the derivative of this expression at room temperature gives

$$\frac{dE_g}{dT} = -3.7 \times 10^{-4} \frac{\text{eV}}{\text{K}} \quad (Ga_{0.47}In_{0.53}As).$$

The rate of bandgap shift with temperature has been measured (Turner, Reese, and Pettit 1964) for InP:

$$\frac{dE_g}{dT} = -2.9 \times 10^{-4} \frac{\text{eV}}{\text{K}} \quad (\text{InP}).$$

A second source (Kirillov and Merz 1983) gives this temperature coefficient of InP as -5.3×10^{-4} eV/K.

Finally, the temperature coefficient of the bandgap of a quaternary composition, $Ga_{0.28}In_{0.72}As_{0.59}P_{0.415}$ has been reported (Madelon and Dore 1981):

$$\frac{dE_g}{dT} = -3.9 \times 10^{-4} \frac{\text{eV}}{\text{K}} \quad (Ga_{0.28}In_{0.72}As_{0.59}P_{0.41}).$$

2.3.3 AlGaInAs/InP Material System

2.3.3.1 General Characteristics and Optoelectronic Applications

The range of compositions of the general quaternary, $Al_xGa_yIn_{1-x-y}As$, can produce lattice constants as large as that of InAs ($x=y=0$) or as small as that of GaAs ($x+y \approx 1$). The most useful members of this quaternary are the compositions that produce a lattice constant equal to that of InP. The InP-lattice-matched compositions of this quaternary are denoted here as AlGaInAs/InP. The AlGaInAs/InP quaternary can be pictured as consisting of mixtures of two ternaries, $Al_{0.48}In_{0.52}As$ and $Ga_{0.47}In_{0.53}As$, each of which lattices matches InP. Because the InAs mole fraction varies only from 0.52 to 0.53 (or, more precisely, from 0.523 to 0.532) over this whole composition range, it is a good approximation to take it to be constant at 0.53. This approximation leads to writing the compositions of AlGaInAs that lattice match InP as

$$(Al_xGa_{1-x})_{0.47}In_{0.53}As, \quad 0 \leq x \leq 1.$$

AlGaInAs/InP and GaInAsP/InP have very similar properties. Both lattice match InP. Both quaternaries have the same low bandgap end point composition, namely, the ternary $Ga_{0.47}In_{0.53}As$, with a room temperature bandgap of 0.75 eV (Figure 2.7). The upper limits of the two quaternaries' bandgaps are similar. The high bandgap member of AlGaInAs/InP is $Al_{0.48}In_{0.52}As$ with a bandgap of 1.45 eV; the high bandgap member of GaInAsP/InP is InP with a bandgap of 1.35 eV. The choice between the two material systems is usually determined by the epitaxial growth technique. AlGaInAs is the easier system for molecular beam epitaxy (MBE), whereas GaInAsP is easier for the other epitaxial techniques.

Applications of AlGaInAs/InP are similar to those of GaInAsP/InP (see Section 2.3.2.1).

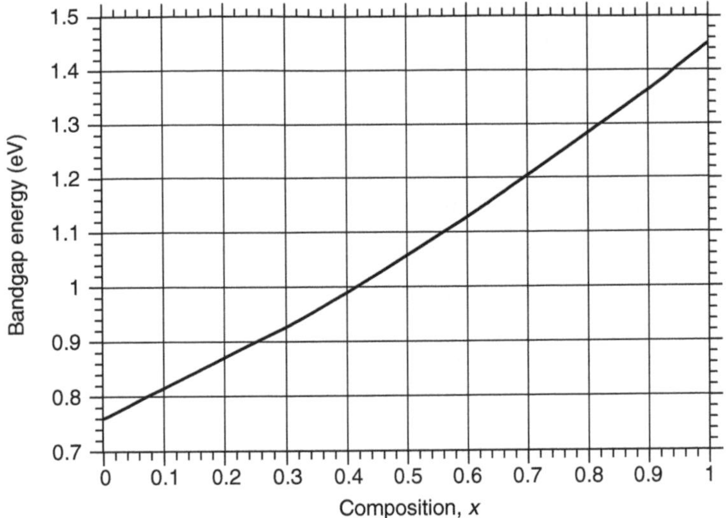

FIGURE 2.7 Compositional dependence of the room temperature bandgap of $(Al_xGa_{1-x})_{0.47}In_{0.53}As$.

2.3.3.2 Direct Bandgap, Refractive Index, and Absorption Coefficient

The quaternary is a direct bandgap material over the whole range of compositions that lattice match InP. Its room temperature bandgap (Olego et al. 1982) is given by:

$$E_g = 0.76 + 0.49x + 0.20x^2$$

The room temperature refractive index at photon energies below the bandgap has been measured for compositions of $x \geq 0.3$ (Mondry et al. 1992) and $x = 0$ (Chandra et al. 1981), and is graphed in Figure 2.8.

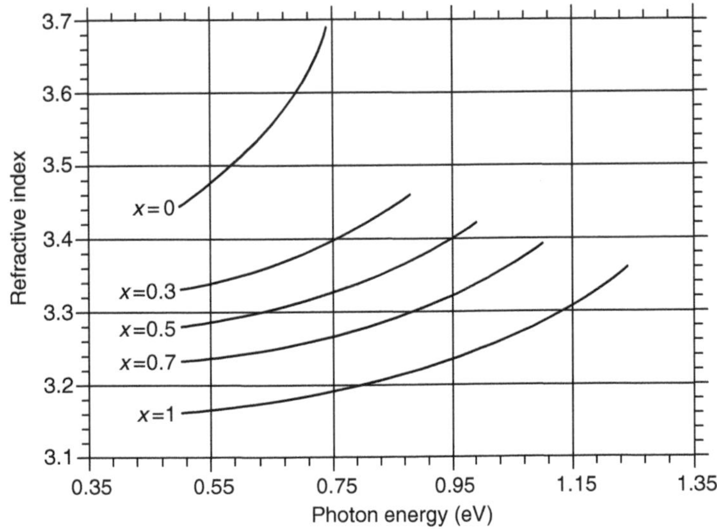

FIGURE 2.8 Compositional dependence of the room temperature refractive index spectrum of $(Al_xGa_{1-x})_{0.47}In_{0.53}As$.

Absorption data is scarce for the AlGaInAs quaternary, except for the binary end point, $Ga_{0.47}In_{0.53}As$. Because this ternary is common to the GaInAsP/InP material system, its absorption spectrum is displayed along with that of GaInAsP and InP in Figure 2.6. The absorption spectra of other compositions of the AlGaInAs quaternary can be estimated by comparison with materials of similar direct bandgaps in the GaInAsP/InP system.

2.3.4 AlGaInP Material System

2.3.4.1 General Characteristics and Optoelectronic Applications

The range of compositions of the general quaternary AlGaInP can produce lattice constants as large as that of InP or as small as that of GaP. Compositions which produce an intermediate lattice constant equal to that of GaAs are the most important members of this quaternary. The GaAs-lattice-matched compositions of this quaternary are denoted here as AlGaInP/GaAs. The AlGaInP/GaAs quaternary can be thought of as consisting of mixtures of two ternaries, $Al_{0.52}In_{0.48}P$ and $Ga_{0.52}In_{0.48}P$, each of which lattice match GaAs. Because the InP mole fractions of these two ternaries are equal (to at least two significant digits) at 0.52, their mixture (the compositions of AlGaInP that lattice match GaAs) can be written as

$$(Al_xGa_{1-x})_{0.52}In_{0.48}P, \quad 0 \leq x \leq 1$$

The main application of AlGaInP/GaAs is in the construction of red diode lasers. As discussed in Section 2.3.1.1 on AlGaAs Material System (General Characteristics and Optoelectronic Applications) the bandgap and refractive index trends are such that electrical and optical confinement are possible in a single layer of a double heterostructure. Cladding/electron confinement layers are constructed of the aluminum rich compositions, often $Al_{0.52}In_{0.48}P$. Core layers consist of low aluminum compositions, e.g., $(Al_{0.2}Ga_{0.8})_{0.52}In_{0.48}P$. The active quantum well often consists of (lattice matched) $Ga_{0.52}In_{0.48}P$ or strained $Ga_{1-x}In_xP$ with InP mole fractions, x, of 0.6–0.65.

2.3.4.2 Bandgaps, Refractive Index, and Absorption Coefficient

The most important member of the quaternary $(Al_xGa_{1-x})_{0.52}In_{0.48}P$ is the ternary end point, $Ga_{0.52}In_{0.48}P$. Certain MOCVD epitaxial conditions cause this ternary to grow in an ordered CuPt-type structure (Kondow and Minagawa 1988). Other MOCVD conditions as well as MBE growth (Wicks et al. 1991) produce a random ternary. The bandgap of the ordered structure is 90 meV lower than that of the random ternary. The possibility of ordering and its effects on the bandgap energy across the composition range of the quaternary $(Al_xGa_{1-x})_{0.52}In_{0.48}P$ are not known. The data reported here apply to the random alloys.

The lower bandgap compositions of the quaternary are direct, like $Ga_{0.52}In_{0.48}P$; the higher bandgap compositions are indirect, like $Al_{0.52}In_{0.48}P$. The dependence of the bandgap energies on composition can be approximated as linear interpolations between the two ternary end points. The room temperature direct gap, E_Γ, of the (random) quaternary has been reported (Kato et al. 1994):

$$E_\Gamma(\mathrm{eV}) = 1.89 + 0.64x$$

The room temperature indirect gap involving the X minimum of conduction band and Γ maximum of the valence band, E_X, can be approximated by a linear interpolation between the values reported for $Ga_{0.52}In_{0.48}P$ (Nelson and Holonyak 1976) and $Al_{0.52}In_{0.48}P$ (Onton and Chicotka 1970):

$$E_X(\mathrm{eV}) = 2.25 + 0.09x$$

The direct/indirect crossover is at the composition $x_c \sim 0.65$, and a room temperature bandgap of $E_g^c \sim 2.3$ eV. The compositional variations of these two bandgaps are graphed in Figure 2.9.

The refractive index of $(Al_xGa_{1-x})_{0.52}In_{0.48}P$ has been experimentally measured and modeled (Kato et al. 1994). The room temperature indexes are graphed in Figure 2.10.

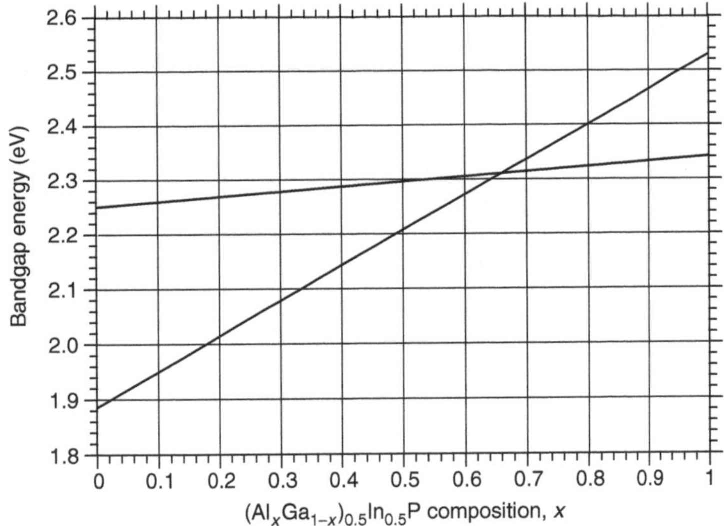

FIGURE 2.9 Room temperature bandgaps of $(Al_xGa_{1-x})_{0.52}In_{0.48}P$. Data applies to random alloy materials.

The model used to generate the refractive index data of $(Al_xGa_{1-x})_{0.52}In_{0.48}P$ in Figure 2.10 can also be applied to generate absorption coefficient data (Kato et al. 1994). The modeled above bandgap absorption coefficients of the quaternary are exhibited in Figure 2.11.

2.3.5 GaSb Material System

2.3.5.1 General Characteristics and Optoelectronic Applications

GaSb material system is more complicated than the others listed in this chapter in that it of two quaternaries, AlGaAsSb and GaInAsSb, each of which lattice matches GaSb substrates. Each quaternary can be envisioned as a mixture of GaSb and a GaSb-lattice-matched ternary:

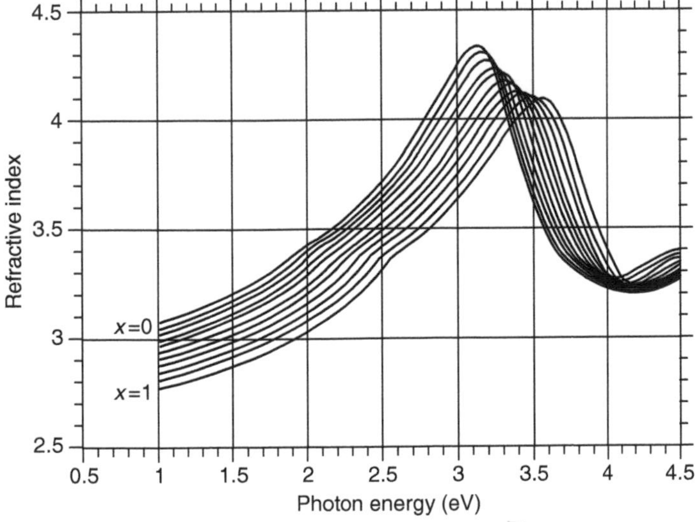

FIGURE 2.10 Modeled room temperature refractive index spectra of $(Al_xGa_{1-x})_{0.52}In_{0.48}P$. (From Kato, H., Adachi, S., and Nakanishi, H. et al. 1994. *Jpn J. Appl Phys.*, 33, 186. With permission.)

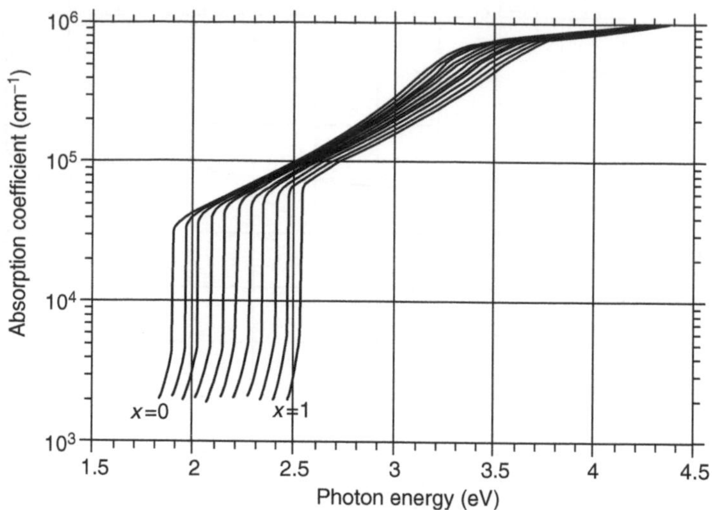

FIGURE 2.11 Modeled room temperature absorption spectra of $(Al_xGa_{1-x})_{0.52}In_{0.48}P$.

$$Al_xGa_{1-x}As_ySb_{1-y} = (GaSb)_{1-x}(AlAs_{0.083}Sb_{0.917})_x$$

and

$$Ga_{1-x}In_xAs_ySb_{1-y} = (GaSb)_{1-x}(InAs_{0.911}Sb_{0.089})_x$$

The alloying of GaSb with $AlAs_{0.083}Sb_{0.917}$ increases the AlGaAsSb's bandgap above that of GaSb, whereas the alloying of GaSb with $InAs_{0.911}Sb_{0.089}$ decreases GaInAsSb's bandgap (at least the direct bandgap, E_Γ) below that of GaSb (see Figure 2.12). Equating the As subscript on both sides of the above expressions gives the GaSb lattice matching conditions, $y=0.083x$ and $y=0.911x$, for $Al_xGa_{1-x}As_ySb_{1-y}$ and $Ga_{1-x}In_xAs_ySb_{1-y}$, respectively.

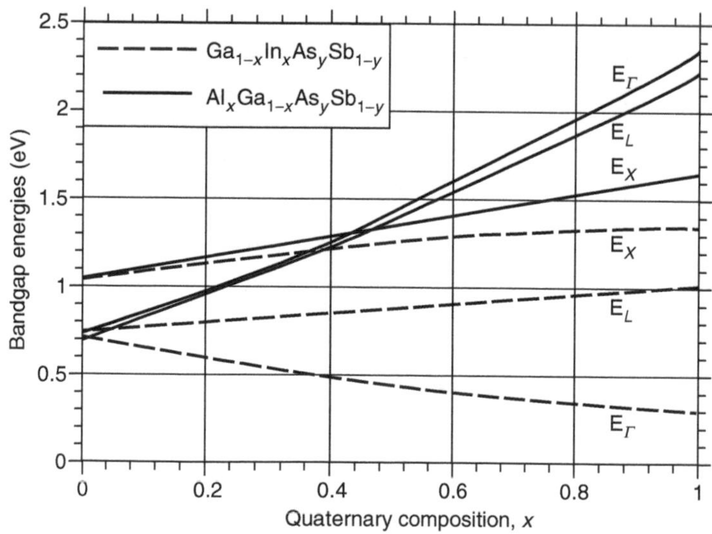

FIGURE 2.12 Modeled room temperature bandgaps vs. compositions of $Al_xGa_{1-x}As_ySb_{1-y}$ and $Ga_{1-x}In_xAs_ySb_{1-y}$ lattice matched to GaSb. Note that the $x=0$ binary endpoint, GaSb, is common to both quaternaries. (From Adachi, S. 1987. *J. Appl. Phys.*, 61, 4869. With permission.)

Common optoelectronic applications of these GaSb-based materials involve detectors and emitters operating in the 2–3 μm wavelength region. In such applications, GaInAsSb is often the active layer of the laser or detector, and AlGaAsSb is the cladding/electron confinement layer of the laser or the window/cap layer of the detector.

Because the bandgaps of GaSb and InAs differ by only 0.6%, AlGaAsSb and GaInAsSb can also be lattice matched to InAs substrates. The range of bandgaps of GaInAsSb/InAs is smaller than that of GaInAsSb/GaSb, and the former material is generally used more rarely than the latter. Additionally AlGaAsSb and GaInAsSb can lattice match InP, but are rarely employed because of competition from the GaInAsP/InP and AlGaInAs/InP systems.

2.3.5.2 Bandgaps, Refractive Indices, and Absorption Coefficients

GaSb is just barely a direct bandgap material, with the conduction band minimum at the I-point only a few tens of meV above that at the Γ point. Because this energy difference is on the order of k_BT at room temperature, a sizable fraction of the conduction band electrons populate indirect minima, making the material unsuitable as the active layer for an efficient room temperature laser. This difficulty is rapidly alleviated by a rising E_L and a falling E_Γ as the composition moves away from GaSb in the GaInAsSb/GaSb quaternary system. The GaInAsSb/GaSb quaternary remains direct over its whole composition range, spanning direct gaps from that of GaSb, $E_\Gamma=0.72$ eV, down to that of $InAs_{0.911}Sb_{0.089}$, $E_\Gamma \sim 0.3$ eV.

The AlGaAsSb/GaSb quaternary is indirect or nearly indirect over its whole bandgap range. Its bandgap ranges from that of the barely direct gap of GaSb, $E_\Gamma=0.72$ eV, to that of $AlAs_{0.08}Sb_{0.92}$, $E_X \sim 1.6$ eV. Moving away from the GaSb end point, the bandgap of the AlGaAsSb/GaSb quaternary rises from that of GaSb, becoming indirect around 1 eV when E_Γ crosses E_L at an aluminum composition just above $x=0.2$. At yet higher aluminum concentrations, $x \sim 0.45$, the indirect nature changes as E_X becomes the lowest lying bandgap.

The compositional dependence of the bandgaps of AlGaAsSb/GaSb and GaInAsSb/GaSb have been modeled (Adachi 1987) and are graphed in Figure 2.12.

Although the modeling is somewhat involved, the results of the modeling can be fit with simple quadratic expressions with errors of typically only a few meV. The fits are listed in Table 2.2.

The room temperature refractive index of AlGaAsSb/GaSb has been measured (Alibert et al. 1991) and modeled (Adachi 1987; Alibert et al. 1991) at room temperature. The two models agree reasonably well with each other and with available experimental data. We use the first (Adachi 1987) of these two models here, as it also applies to GaInAsSb/GaSb. The room temperature below-gap refractive indices of these two quaternaries that lattice match GaSb are graphed in Figure 2.13 and Figure 2.14.

As do most of the III–V materials, the larger bandgap AlGaAsSb/GaSb quaternaries have smaller refractive indices.

The behavior of the refractive index of GaInAsSb/GaSb is anomalous in that the larger bandgap materials have larger refractive indices, at least at below gap photon energies. The cause of this anomaly is, perhaps, related to the aspect of the band structure that causes both the X and L bandgaps to have a compositional trend is opposite to that of the fundamental gap at Γ (see Figure 2.12). A practical consequence of this index anomaly affects the design of a typical diode laser structure (double

TABLE 2.2 Quadratic Fits to Modeled Room Temperature Bandgaps of AlGaAsSb/GaSb and GaInAsSb/GaSb

	$Al_xGa_{1-x}As_ySb_{1-y}$ $0 \le x \le 1$; $y=0.083x$	$Ga_{1-y}In_yAs_ySb_{1-x}$ $0 \le x \le 1$; $y=0.911x$
E_Γ(eV)	$0.717 + 1.233x + 0.414x^2$	$0.721 - 0.670x + 0.0252x^2$
E_X(eV)	$1.051 + 0.612x - 0.0937x^2$	$1.057 + 0.537x - 0.0242x^2$
E_L(eV)	$0.756 + 1.075x + 0.0404x^2$	$0.760 + 0.261x - 0.00190x^2$

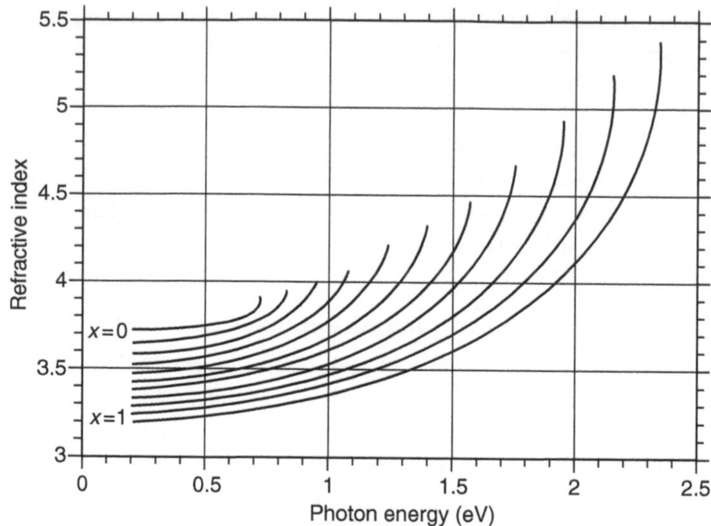

FIGURE 2.13 Modeled room temperature refractive index of AlGaAsSb/GaSb. Below-bandgap photon energies are displayed. (From Adachi, S. 1987. *J. Appl. Phys.*, 61, 4869. With permission.)

heterostructure or quantum well heterostructure). In most III–V material systems the cladding layer of the waveguide (lower refractive index) also functions as an electron confinement layer (larger bandgap). This scheme will not work in the GaInAsSb system, because the trend toward high bandgaps is not accompanied by the trend toward low refractive indices. Fortunately, the problem can be solved by the constructing laser cladding/electron confinement layers of AlGaAsSb, which have a higher bandgap and lower refractive index than those of the GaInAsSb core and quantum well.

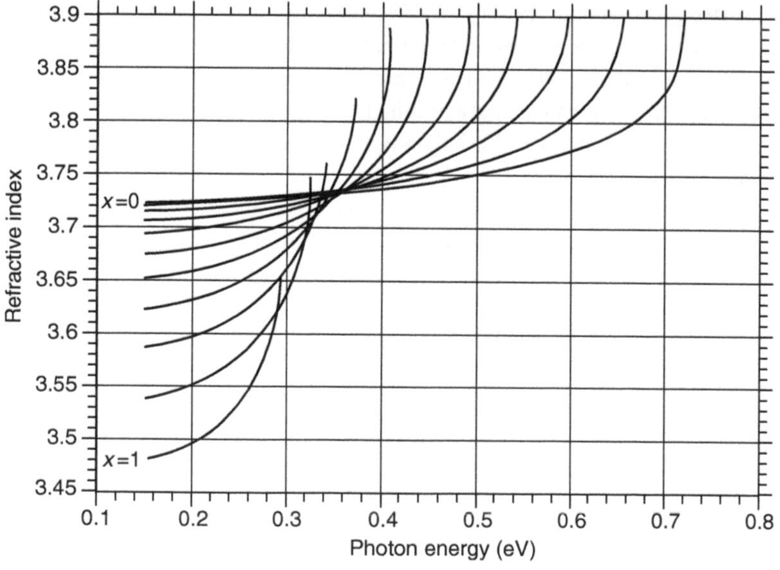

FIGURE 2.14 Modeled room temperature refractive index in GaInAsSb/GaSb. Below bandgap photon energies are displayed.

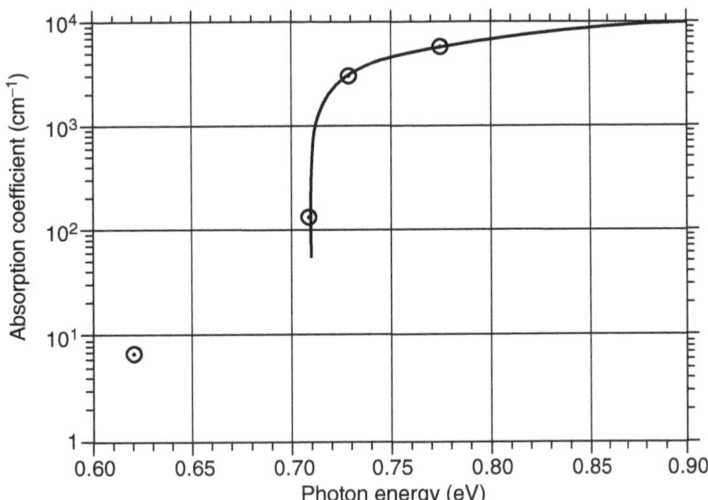

FIGURE 2.15 Room temperature absorption spectrum of GaSb. The circles are experimental data points, the line is a fit to the three highest energy data points, as discussed in the text.

Very little has been published on the room temperature absorption of the GaSb-based materials. A few experimental data points exist (Becker et al. 1961) only for GaSb itself, as shown in Figure 2.15.

Fitting the simple expression, $\alpha \propto (E - E_g)^{1/2}$ to the three above gap data points of Figure 2.15 gives the approximation:

$$\alpha_{\text{GaSb}} = 22600(\text{cm}^{-1}\ \text{eV}^{-1/2})(E - 0.71\ \text{eV})^{1/2}.$$

In the absence of more complete data, an estimate of the absorption coefficient of other materials in the AlGaAsSb/GaSb or GaInAsSb/GaSb systems could be made by replacing the 0.71 eV in the above expression with the relevant E_Γ bandgap.

2.3.6 GaAsP Material System

2.3.6.1 General Characteristics and Optoelectronic Applications

$GaAs_xP_{1-x}$ has a band structure similar to that of $Al_xGa_{1-x}As$ in that it is a direct gap material in its GaAs-rich composition range and crosses over to an indirect bandgap in its GaAs-deficient compositions. It has long been the dominant material for visible light emitting diodes, although this is in the process of changing as newer, higher performance materials become available. The GaAsP approach to visible LEDs involves simple epitaxial growth techniques, thus, minimizing cost, but relatively low performance results because of lattice matching and band structure factors. Only the two binary end points, GaAs and GaP, lattice match available substrates; the general quaternary is grown mismatched on either GaAs or GaP substrates, resulting in dislocations. The direct, lower bandgap compositions are used for red LEDs; the higher compositions (GaP) are used for yellow–green LEDs despite their indirect bandgap. Nitrogen doping of the indirect gap GaAsP compositions increases their radiative efficiencies to modest levels (10^{-4}–10^{-3}) (Campbell et al. 1974). Strong electron confinement by the nitrogen isoelectronic impurity relaxes the k-selection rules, enabling light emission from the indirect material. Emission efficiencies, however, are still at least an order of magnitude lower than that of direct material.

2.3.6.2 Bandgaps, Electroluminescence Energies, and Absorption Spectra

The direct bandgaps of a series of 11 samples of $GaAs_{1-x}P_x$ were measured by Thompson et al. (1966) in an early work. In this paper, the data is fit with an expression produce is a value for the GaAs bandgap that is 15 meV or so larger than the presently accepted value. Although the GaAs bandgap produced by

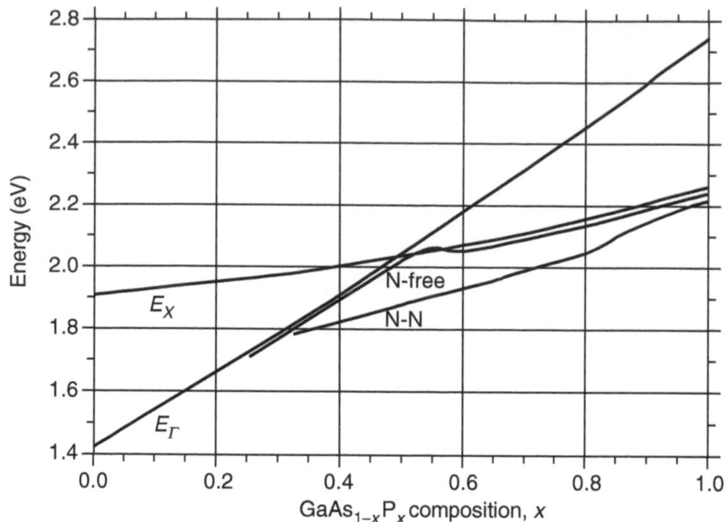

FIGURE 2.16 Room temperature energies of E_Γ and E_X bandgaps of GaAsP. Also displayed are the electroluminescence peak energies of nitrogen-free GaAsP and the electroluminescence peak involving nitrogen–nitrogen pairs (N–N) of nitrogen doped GaAsP.

the original fit appears to be slightly too large, the original data itself does produce the correct value of the GaAs bandgap. The error in the fit appears to result from errors in the measurements of the compositions of the ternary samples. A more reliable fitting expression can be obtained from the original data by weighting the binary end points more heavily than the ternary points, because there can be no error in binary composition measurements. This fitting procedure produces the following result:

$$E_g(\text{eV}) = 1.428 + 1.125x + 1.952x^2.$$

The crossover of the direct and indirect bandgaps of $GaAs_xP_{1-x}$ is near a composition of $x=0.5$ and a bandgap of 2 eV. At compositions greater than $x=0.5$, the X indirect bandgap is the lowest. The Γ

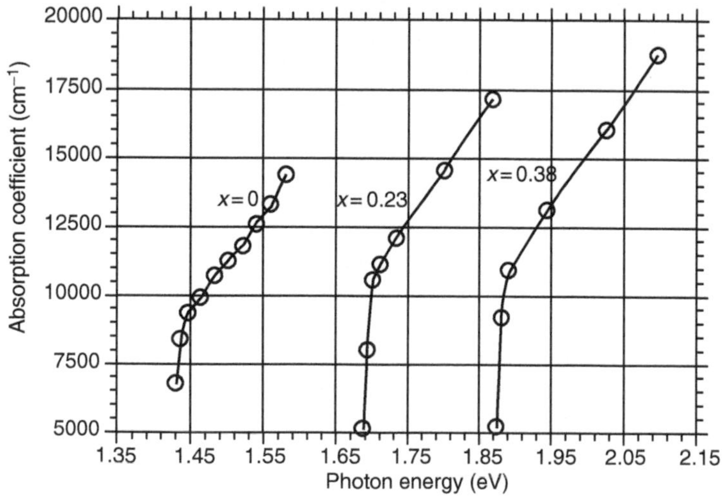

FIGURE 2.17 Compositional dependence of the room temperature absorption spectra near the fundamental absorption edge of $GaAs_{1-x}Px$.

(direct) and the X bandgaps are displayed as functions of composition in Figure 2.16. Also displayed in the figure are the peak electroluminescence energies (Craford and Holonyak 1976) of N-doped and N-free GaAsP, useful information in the analysis of LEDs.

Optical absorption spectra are available for only a few compositions (Hasehawa et al. 1984) and are displayed in Figure 2.17.

References

Adachi, S. 1987. *Appl. Phys.*, 61, 4869.

Adachi, S. 1988. *Phys. Rev. B.*, 38, 12345.

Alibert, C., Skouri, M., Joullie, A., Benouna, M., and Sadiq, S. J. 1991. *J. Appl. Phys.*, 69, 3208.

Aspnes, D. E., Kelso, S. M., Logan, R. A., and Bhat, R. J. 1986. *J. Appl. Phys.*, 60, 754.

Bacher, F. R., Blakemore, J. S., Ebner, J. I., and Arthur, J. S. 1988. *Phys. Rev. B*, 37, 2551.

Becker, W. M., Ramdas, A. K., and Fan, H. Y. 1961. *J. Appl. Phys. Suppl.*, 32, 2094.

Campbell, J. C., Holonyak, N. Jr., Craford, M. G., and Keune, D. L. 1974. *J. Appl. Phys.*, 45, 4543.

Cardona, M. 1961a. In *Proceedings of International Conference on Semiconductor Physics, Prague, 1960*, p. 388.

Cardona, M. 1961b. *J. Appl. Phys.*, 32, 958.

Cardona, M. 1965. *J. Appl. Phys.*, 36, 2181.

Chandra, P., Coldren, L. A., and Strege, K. E. 1981. *Electron. Lett.*, 17, 6.

Chang, K. H., Lee, C. P., Wu, J. S., Liu, D. G., Liou, D. C., Wang, M. H., Chen, L. J., and Marais, M. A. 1991. *J. Appl. Phys.*, 70, 4877.

Choi, H. K. and Wang, C. A. 1990. *Appl. Phys. Lett.*, 57, 321.

Craford, M. G. and Holonyak, N. Jr., 1976. In *Optical Properties of Solids New Developments*, B.O. Seraphin, ed., p. 1659. The Netherlands, North-Holland, Amsterdam, (Chap. 5).

Feketa, D., Chan, K. T., Ballantyne, J. M., and Eastman, L. F. 1986. *Appl. Phys. Lett.*, 49, 1659.

Guzzi, M., Grilli, E., Oggioni, S., Stachli, J. I., Bosio, C., and Pavesi, L. 1992. *Phys. Rev. B.*, 45, 10951.

Hasehawa, S., Tanaka, A., and Sukegawa, I. J. 1984. *Appl. Phys.*, 55, 3188.

Jenkins, D. W. 1990. *J. Appl. Phys.*, 68, 1848.

Jensen, B. and Torabi, A. 1983. *J. Appl. Phys.*, 54, 3623.

Kato, H., Adachi, S., Nakanishi, H., and Ohtsuka, K. 1994. *Jpn J. Appl. Phys.*, 33, 186.

Kirillov, D. and Merz, J. L. 1983. *J. Appl. Phys.*, 54, 4104.

Kondow, M. and Minagawa, S. 1988. *J. Appl. Phys.*, 64, 793.

Madelon, R. and Dore, M. 1981. *Solid State Commun.*, 39, 639.

Madelung, O. ed. 1982. *Landholt-Bornstein, New Series, Group III, Vol. 17*, p. 1241. Springer, Berlin, Germany.

Marple, D. T. F. 1964. *J. Appl. Phys.*, 35, 1241.

Mondry, M. J., Babic, D. I., Bowers, J. E., and Coldren, L. A. 1992. *IEEE Phontonics Technol. Lett.*, 4, 627.

Nahory, R. R., Pollack, M. A., Johnston, W. D. Jr., and Barnes, R. L. 1978. *Appl. Phys. Lett.*, 33, 659.

Nelson, R. J. and Holonyak, N. Jr., 1976. *J. Phys. Chem. Solids.*, 37, 629.

Newman, R. 1958. *Phys. Rev.*, 111, 1518.

Olego, D., Chang, T. Y., Silberg, E., Caridi, E. A., and Pinczuk, P. 1982. *Appl. Phys. Lett.*, 41, 476.

Onton, A. and Chicotka, R. J. 1970. *J. Appl. Phys.*, 41, 4205.

Pankove, J. L. 1975. *Optical Processes in Semiconductors*. Dover Publications, New York.

Pettit, G. D. and Turner, W. J. 1965. *J. Appl. Phys.*, 36, 2081.

Sell, D. D., Casey, H. C., and Weeht, K. W. 1974. *J. Appl. Phys.*, 45, 2650.

Thompson, A. G., Cardona, M., Shaklee, K. L., and Woolley, J. C. 1966. *Phys. Rev.*, 146, 601.

Thurmond, C. D. 1975. *J. Electrochem. Soc.*, 122, 1133.

Turner, W. J., Reese, W. E., and Pettit, G. D. 1964. *Phys. Rev.*, 136, A1467.

Wicks, G. W., Koch, M. W., Varriano, J. A., Johnson, F. G., Wie, C. R., Kim, H. M., and Colombo, P. 1991. *Appl. Phys. Lett.*, 59, 342.

Yokouchi, N., Uchida, T., Miyamoto, F., Inaba, Y., Koyama, F., and Iga, K. 1992. *Jpn J. Appl. Phys.*,1255.

Yu, P. W. and Kuphal, E. 1984. *Solid State Commun.*, 49, 907.

3

Optical Properties and Optoelectronic Applications of II–VI Semiconductor Heterostructures

Hong Luo

Athos Petrou

3.1 Introduction

Semiconductor heterostructures of II–VI compounds offer a wide range of properties which can be exploited for fundamental studies and for device applications. We will review recent studies addressing these issues, emphasizing ZnSe-based structures for blue–green light emitting devices. Other systems, such as type-II and diluted magnetic semiconductor heterostructures and narrow-gap type-III heterostructures, will be discussed.

Semiconductor applications have long been dominated by electronic devices, i.e., transistor-based integrated circuits. However, optoelectronic devices and photonic applications are becoming increasingly important—for example, optical data storage and communication, laser printing, flat panel displays, and so on. II–VI semiconductors are of interest, primarily, because of their wide range of optoelectronic properties. This is exemplified by the wide spectrum of energy gaps E_g of these materials, ranging from 0

TABLE 3.1 Energy Gaps for II–VI Compounds for Liquid Helium (LHe), Liquid Nitrogen (LN_2), and Room Temperature (RT) (Mn-Based Chalcogenides Are Included)

Compound	T	E_g (eV)
ZnS	LHe	3.84
	LN_2	3.78
	RT	3.68
ZnSe	LHe	2.82
	LN_2	2.80
	RT	2.71
ZnTe	LHe	2.40
	LN_2	2.38
	RT	2.28
CdS	LHe	2.56
	LN_2	2.55
	RT	2.45
CdSe	LHe	1.83
	LN_2	1.82
	RT	1.75
CdTe	LHe	1.606
	LN_2	1.596
	RT	1.528
HgS	LHe	< 0
	LN_2	~ 0
	RT	> 0
HgSe	LHe	−0.27
	LN_2	−0.205
	RT	+0.10
HgTe	LHe	−0.300
	LN_2	−0.252
	RT	−0.146
MnS	LHe	~3.6
	LN_2	~3.5
	RT	~3.4
MnSe	LHe	3.3
	LN_2	3.14
	RT	2.9
MnTe	LHe	3.2
	LN_2	3.05
	RT	2.9

to over 3.8 eV, as listed in Table 3.1. Studies of most II–VI semiconductors can be traced back to the 1960s, concentrating on the fundamental properties of bulk materials. Most of the excitement in these materials, however, came from heterostructures after the introduction of molecular beam epitaxy (MBE), as demonstrated in the area of blue–green laser diodes (Haase et al. 1991). Considering the current interests in these materials and their application prospects, we will focus on heterostructures of II–VI semiconductors.

II–VI semiconductor heterostructures are unique in many ways. First, the spectrum of energy gaps allows the exploration of optical and optoelectronic phenomena and device applications covering wavelengths ranging from the far infrared to the ultraviolet. Furthermore, II–VI semiconductor heterostructures involve a wide range of band offsets, which adds variety and flexibility to band gap engineering. The II–VIs also have stronger polarity compared to other well studied semiconductors, e.g., III–Vs, Si and Ge, providing testing ground for various effects related to lattice distortion and vibration. Magnetic ions, such as Mn^{++} and Fe^{++}, can also be easily incorporated into II–VIs, which have been traditionally referred to diluted magnetic semiconductors (although a concentration of magnetic ions as large as 100% can be achieved with MBE).

In this paper, we will review various optical and optoelectronic studies involving II–VI semiconductor heterostructures and the development of II–VI optoelectronic devices. In light of the recent rapid progress in the fabrication of ZnSe-based blue–green laser diodes and their importance to the future of II–VI materials, we will focus our attention on the progress and remaining problems in this area.

3.2 Overview

Modern growth techniques, such as MBE and metal-organic chemical vapor phase deposition (MOCVD), have had a major impact on the evolution of the general field of II–VI semiconductors in several important ways. First, it was demonstrated that the II–VIs can be grown in monolithic multilayer structures, making it possible to form a variety of quantum well systems involving type-I, type-II and type-III (a zero-gap and open-gap combination) band alignments. Because of the various lattice constants of the II–VIs, both lattice matched systems and strained layer, structures have been achieved. Second, it was demonstrated that epitaxy can be used to form new phases which do not exist naturally in the bulk. All II–VI compounds except HgS are tetrahedrally coordinated in their stable bulk phases, crystallizing either in the cubic (zincblende) or in hexagonal (wurtzite) structures. It was demonstrated that epitaxy allows one to grow zincblende phases of CdSe (Samarth et al. 1989), MnTe (Durbin et al. 1989), MnSe (Kolodziejski et al. 1986) and FeSe (Jonker et al. 1988)—all of which have electronic and structural properties closely related to the common II–VIs. These new phases can, therefore, be alloyed and combined with the zinc blende II–VIs, thus, dramatically extending the scope of the physical phenomena that can be investigated, as will be discussed below.

Lattice constants and band gaps are essential as guides in designing various heterostructures. In Figure 3.1, we show the relationship between energy gaps and lattice parameters of II–VI compounds. As evident in the figure, the II–VIs cover a wide range of energy gaps. It should be pointed out that II–VIs resemble III–V compounds in many ways. A useful guide can be obtained from a portion of the Periodic Table containing these elements, with their tetrahedral radii r_i in tetrahedrally bonded compounds, shown in Table 3.2. We note that, for any zinc blende or diamond-structure crystal, the lattice constant is given with reasonable accuracy by $(4/\sqrt{13})(r_i + r_j)$, where r_i and r_j are the tetrahedral radii of nearest-neighbor atoms (e.g., of Zn and Se in ZnSe). Thus, Table 3.2 also provides a very useful and compact guide for selecting lattice matched systems not only between II–VI compounds themselves, but between

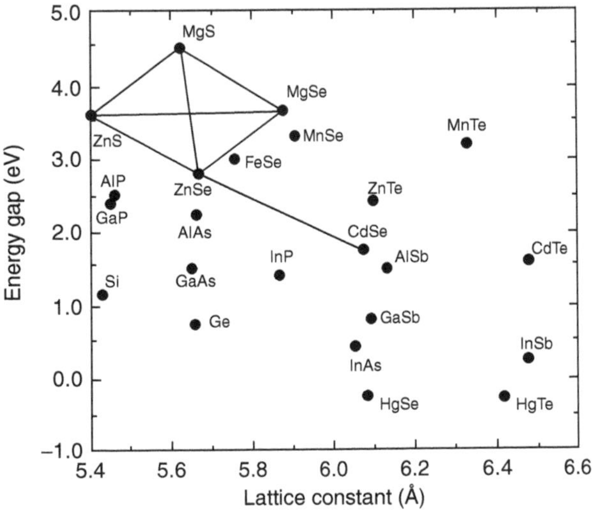

FIGURE 3.1 Energy gaps and lattice parameters for cubic group IV, III–V and II–VI semiconductors.

TABLE 3.2 Tetrahedral Radii r_i for Group II, III, IV, V, and VI Elements (in Å)

II	III	IV	V	VI
Zn	Al	Si	P	S
1.225 Å	1.230 Å	1.173 Å	1.128 Å	1.127 Å
Cd	Ga	Ge	As	Se
1.405 Å	1.225 Å	1.225 Å	1.225 Å	1.225 Å
Hg	In	α-Sn	Sb	Te
1.402 Å	1.405 Å	1.405 Å	1.405 Å	1.405 Å

the II–VIs and either III–V or Group-IV materials. It is also clear that, except for the elements in the first row (i.e., Al, Si, P. and S), elements in the same row have the same tetrahedral radii. This means that II–VI compounds will have approximately the same lattice constant as the III–V counterpart, which involves elements closest in the Periodic Table (Table 3.2) to the II–VI elements. For example, ZnSe has a lattice constant very close to that of GaAs, noting that Ga and As are the Group-III and Group-V elements closest to Zn and Se, respectively.

Although the offset of a given semiconductor heterostructure is often a complicated matter, the type of band alignment is often the same between the counterparts of II–VI and III–VI compounds (defined by the closest elements in the Periodic Table, as was used above). For instance, ZnCdSe/ZnSe and GaInAs/GaAs are type-I systems, and ZnTe/CdSe and GaSb/InAs have type-II band alignments.

Furthermore, there are also analogies in doping II–VI and III–V compounds. Doping II–VIs has been a more difficult task in most cases. For a given II–VI compound, the relative difficulty of n-type and p-type doping is similar to that for its III–V counterpart. For example, in ZnSe n-type doping is much easier than p-type doping, which is the same as for GaAs. All of these similarities can be traced back to the atomic states of these elements that affect the absolute positions of the conduction and the valence bands, which involve complicated arguments. From a practical point of view, however, such similarities can be used as simple guides for exploring new heterostructures and for transferring techniques used in dealing with existing compounds to their lesser known counterparts.

3.3 Quantum Confinement in Wide-Gap II–VI Semiconductor Heterostructures

Research and development of II–VI semiconductor heterostructures has long been driven by the possibility of optoelectronic applications in the blue region. Two systems, namely, ZnSe- and ZnTe-based quantum well structures, have attracted most of the attention and have experienced some degree of success. We will focus on the quantum confinement effects in wide-gap II–VI semiconductor quantum wells and will present a qualitative description of the reason for quantum confinement being strong in some systems and weak in others.

Modification of electronic states of bulk materials using semiconductor heterostructures provides the possibility of tailoring the electronic and optical properties of the structures. Excitonic effects in wide gap II–VI semiconductors are particularly strong, because of the ionic nature of the bonds in these materials (which results in smaller dielectric constants and consequently stronger Coulomb interaction) and of the large effective masses of the carriers that lead to more tightly bound excitons (i.e., smaller Bohr radii). One of the most important properties that can be achieved in quantum well structures is the quantum confinement of carriers. It is responsible for a list of characteristics in type-I quantum wells, which have been utilized for various purposes, such as the increase of optical transition probability, enhancement of the exciton binding energy, and quantum confined stark effect (QCSE), to name a few.

The degree of confinement is determined by the depths of the quantum wells in the conduction and the valence bands, which in turn are derived from the energy gap difference between the two materials comprising the heterostructure and their alignment (or offset). The band alignment of wide gap II–VI

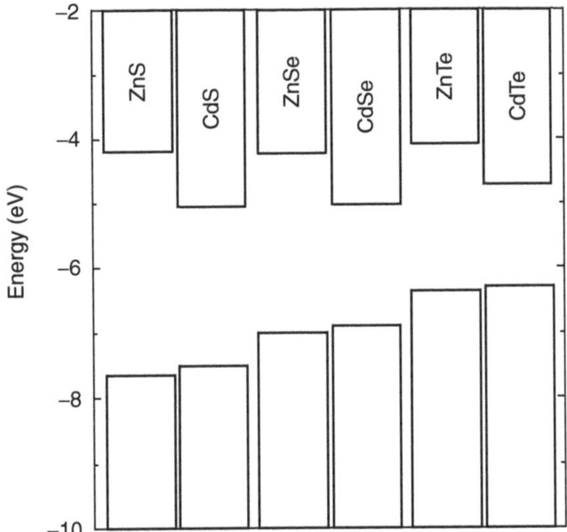

FIGURE 3.2 Band alignment for wide-gap II–VI semiconductors on an absolute scale. Note the relatively good agreement with the common-anion and common-cation rule.

semiconductors relative to the vacuum level is shown in Figure 3.2. Examining the positions of the edges in the conduction and the valence bands for the various compounds, one finds a small conduction band offset when two compounds share the same cation, and a small valence band offset when there is a common anion. In other words, although the "common anion rule" has been shown to be inaccurate, it does provide a very good rule of thumb in II–VI systems.

Most type-I quantum wells consisting of II–VI compounds and alloys that were fabricated before 1990 did not have sufficient well depths to confine either electrons, or holes at room temperature. One of the extensively studied-systems is ZnMnSe/ZnSe (a common anion system) (Kolodziejski et al. 1985), which showed reasonable confinement effects of electrons and holes at low temperatures. The lack of deep wells in the valence band was evidenced by the temperature dependence of exciton absorption in this system. The strong exciton absorption at low temperatures deteriorates rapidly as the temperature is increased and totally disappears at room temperature. Attempts to observe stimulated emission in these structures revealed the same temperature behavior (Bylsma et al. 1985).

The reason for the temperature dependence of excitons originates from the strong electron-LO phonon coupling (i.e., Fröhlich interaction). The LO phonon energies are typically larger than the exciton binding energies. In the case of ZnSe, the LO phonon energy is 32 meV, whereas the exciton binding energy is 18 meV. Thus at high temperatures, all excitons dissociate into electron-hole pairs by LO phonons. The confinement of electrons and holes increases the exciton binding energy E_{ex}. When E_{ex} becomes larger than the LO phonon energy, the exciton dissociation rate will be greatly reduced. The enhancement of the Coulomb interaction, however, strongly depends on the height of the confining potentials in both the conduction and the valence bands. With the band alignments shown in Figure 3.2, it appears difficult to find a ternary system in which wells in both the conduction and the valence bands are sufficiently deep. The first room temperature exciton absorption was observed in ZnCdTe/ZnTe quantum wells. Although it is one of the common anion systems, a strong carrier confinement effect was observed when well width was reduced to values less than that of the exciton Bohr radius. The same effect was soon observed in another common anion system, namely, ZnCdSe/ZnSe quantum wells (Ding et al. 1990).

The effect of carrier confinement in ZnCdSe/ZnSe quantum wells was demonstrated by the study of the exciton absorption as a function of well width and temperature. In Figure 3.3, the absorption coefficient is plotted for two multiquantum well structures, one with a well width of 200 Å and the other 35 Å, both

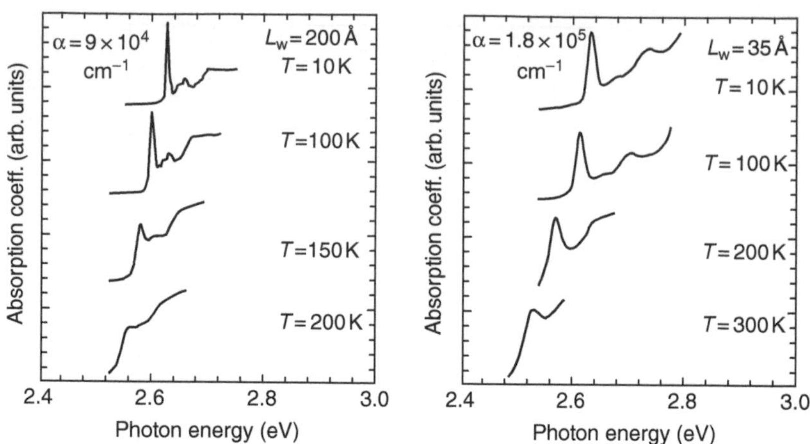

FIGURE 3.3 Absorption coefficient of two ZnCdSe/ZnSe multi-quantum well samples with well thicknesses $L_w = 200$ Å (left panel) and $L_w = 35$ Å (right panel), as a function of temperature.

having a Cd concentration of 24%. The confinement effect can be seen from the blue shift of the exciton peak when the well width is decreased. The temperature dependence of the exciton absorption coefficients α in the two systems is also drastically different. The coefficient α in the 200 Å wells shows a bulk like behavior, disappearing when the temperature approaches 200 K. In the 35 Å wells, however, the exciton absorption survives at much higher temperatures and is observable at temperatures as high as 400 K. This indicates that the excitons in the 35 Å wells have a binding energy greater than the LO phonon energy.

The linewidth broadening of the quasi-2D case is plotted in Figure 3.4 as a function of temperature, together with the 3D case in the one-phonon scattering model (see Chemla et al. 1988)

$$\Gamma(T) = \Gamma_0 + \frac{\Gamma_{LO}}{e^{h\omega_{LO}/kT} - 1}, \tag{3.1}$$

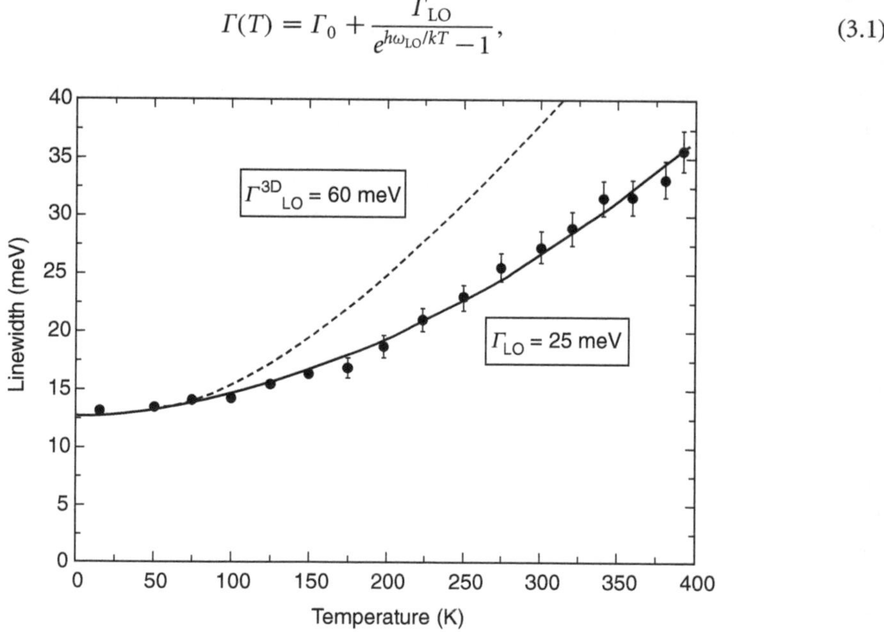

FIGURE 3.4 The linewidth of the $n = 1$ HH excition absorption peak as a function of temperature of a multiquantum well sample with $L_w = 30$ Å. The Fröhlich interaction parameter obtained from a theoretical fit (the solid line) is $\Gamma_{LO} = 28$ meV. For comparison, the behavior of bulk ZnSe is shown by the dashed line.

where Γ_{LO} is the extrapolated linewidth at $T=0$, and Γ_{LO} a phenomenological parameter, which describes the strength of the Frölich interaction. The value of Γ_{LO} is estimated from the data to be 25 meV, which is significantly less than the bulk value of 60 meV. The quasi-2D nature is further studied by magnetoabsorption experiments in fields up to 24 T. Additional information is provided by the simultaneous observation of both is ($n=1$) heavy-hole (HH) exciton add 2s ($n=1$) HH exciton absorption peaks. The low field behavior of the two peaks is very similar. At fields greater than 10 T, the 2s exciton peak shows a much larger field dependence. At $B=23.5$ T, the field-induced shift of the 2s exciton peak is six times larger than that of the 1s state, which is consistent with the results of 2D excitons (Akimoto and Hasegawa 1966). The advantage of observing the 2s exciton peak is that it allows an estimate of the exciton binding energy. The difference between the 1s and the 2s exciton transition energies is 0.75 Rydberg^{3D} in the 3D case, and 0.88 Rydberg^{2D} for 2D excitons. Therefore, the exciton binding energy E_{ex} of the observed quasi-2D exciton can be estimated from $0.75<(E_{2s}-E_{1s})/E_{ex}<0.88$. The observed value of $E_{2s}-E_{1s}$ is 31 meV. Thus, the exciton binding energy in the quasi-2D case should be 40 meV$>E_{ex}>$35 meV, which is larger than the LO phonon energy of 32 meV.

Detailed studies are needed to illustrate why some structures demonstrate large confining potentials for carriers (such as ZnCdSe/ZnSe and ZnCdTe/ZnTe), whereas other structures show a weaker confinement effect (e.g., ZnMnSe/ZnSe and ZnSe/ZnSSe). Most of the observations can be qualitatively understood by considering the effects of strain in these systems. It can be seen in Figure 3.1 that all common-anion or common-cation ternary combinations of wide-gap II–VIs will be lattice-mismatched. Because the offsets can be very small in these two systems, the offset introduced by strain can be significant. When a layer is under compressive strain (in the plane of the layer, as is the case in quantum well structures), the conduction band will move to a higher energy. Meanwhile, the heavy hole and the light hole bands split, with the heavy hole moving up in energy and the light hole to a lower energy. On the other hand, when the layer is under tensile strain, all three bands mentioned above move in opposite directions compared to the case of compressive strain. In all wide-gap II–VI compounds, the heavy-hole to conduction-band transition is stronger than that of the light hole to the conduction band. Thus, the effects of strain on the heavy hole and the conduction band are particularly important.

When the deformation potentials of the two materials involved in a quantum well structure are similar, the offset for any given band (e.g., the heavy-hole band) is determined mainly by the lattice mismatch and is not sensitive to whether the wells or the barriers are strained. We will use an example in which the wells are strained. For common-anion quantum wells (i.e., shallow wells in the valence band), it will be desirable to have the layer of the well under compressive strain, thus, moving the heavy-hole band edge up in the well, increasing the well depth for the heavy-hole quantum well. On the other hand, in dealing with shallow quantum wells in the conduction band in the case of common-cation systems, a tensile strain will lower the conduction band edge in the well.

For common anion structures, such as ZnMnSe/ZnSe and ZnMnTe/ZnTe quantum wells, the lattice constants of the well layers (ZnSe and ZnTe) are smaller than that of the barriers. Thus, the quantum well layers are under tensile strain. As a result, the effect of strain reduces the already small band offset for the heavy holes. This is consistent with the lack of success in achieving sufficient quantum confinement in these quantum wells. For ZnCdSe/ZnSe and ZnCdTe/ZnTe quantum wells, which have common anions, the quantum well layers (ZnCdSe and ZnCdTe) have a larger lattice constant than that in the barriers. Therefore, the heavy hole confining potentials will be larger due to the compressive strain, which is responsible for the observations of room temperature exciton absorption in such quantum wells, as discussed earlier. The same effect is also true for common-cation quantum wells. For example, in the ZnSe/ZnSSe system, the well layers (i.e., ZnSe), have the larger lattice constant of the two constituents. The compressive strain in the ZnSe layers, therefore, lifts its conduction-band edge, reducing the well depth in the conduction band, which is consistent with experimental observations.

In the examples discussed above, we have assumed that the barriers are lattice-matched to the buffer layers and only the wells are strained. The effect, however, remains qualitatively the same when the condition is changed to the opposite limit where the barriers are strained and the wells are unstrained. Furthermore, the same conclusion can be extended to situations where both the wells and the barriers are

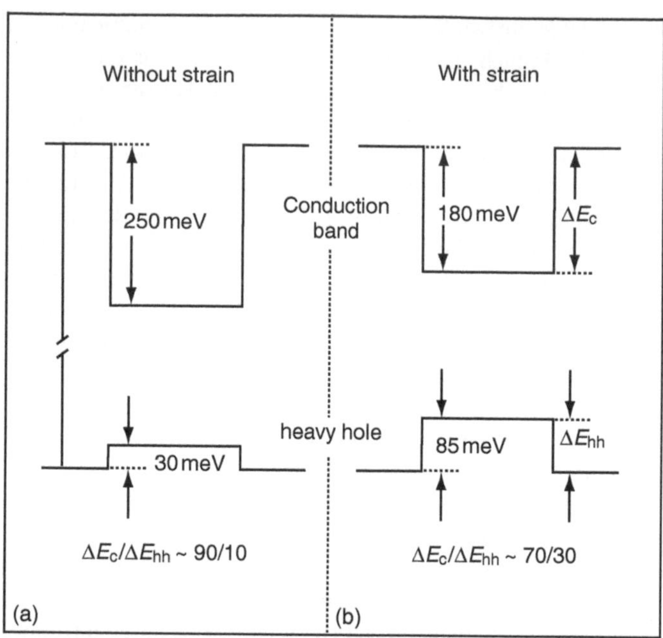

FIGURE 3.5 Band alignment in the $ZnSe/Zn_{0.75}Cd_{0.25}Se$ quantum well system. Note the effect of strain, leading to an increased band offset in the valence band.

strained, which can be caused, for example, by a buffer layer with a lattice constant different from those in either the wells or the barriers.

To illustrate the significance of this effect, we examine the case of a $Zn_{0.75}Cd_{0.25}Se/ZnSe$ quantum well. The lattice constant of CdSe is about 7% larger than that of ZnSe. Detailed studies indicate that the effect of strain, in fact, provides a major portion of the band discontinuity for the heavy holes (more than the initial offset in strain-free structures). This is illustrated in Figure 3.5, which shows band offsets in the conduction and the valence bands of a $Zn_{0.75}Cd_{0.25}Se/ZnSe$ quantum well in two cases, namely, with strain absent (Figure 3.5a) and included (Figure 3.5b). Before considering the effect of strain, the valence band offset is estimated to be 30 meV. With the band gap difference, this would lead to a conduction-band to valence-band offset ratio of 9:1. Such a ratio is not favorable for room temperature lasing because the hole-confining potential is too shallow. With the effect of strain, this ratio changes to 7:3, which is much closer to that in GaAs/GaAlAs quantum wells. Furthermore, the ratio should remain qualitatively the same for different Cd concentrations because the energy gap, the offset, and the effect of strain all have close to linear dependence on the Cd concentration.

3.4 ZnSe-Based Blue–Green Light Emitting Devices

All wide-gap II–VI semiconductors have direct energy gaps and large effective masses (thus, large density of states), which result in intrinsically efficient light absorption and emission. Therefore, they are ideally suited for optical studies of quasi-2D electronic confinement in quantum well systems and possible photonic device applications. The quest for fabricating short wavelength semiconductor lasers and light emitting diodes started long before the growth of wide-gap semiconductor heterostructures became a reality. Among II–VI semiconductors, ZnSe appears to be well-suited for blue light emitting devices not only because of its band gap (2.7 eV at room temperature), but also because its lattice constant is very close to that of GaAs. Although there had been low-temperature observations of lasing action in bulk ZnSe with various pumping schemes, it was very clear that, for a practical device, one would have to employ a suitable quantum well structure. One of the consequences of the large effective mass in ZnSe is that, for a specific quantum well configuration (i.e., for a given well depth and width), the ground state is much better localized in wide gap quantum wells. The disadvantage for the ZnSe system, however, is that there is no common Group-II element which forms a compound with Se in the zincblende structure (as an example, CdSe has the wurtzite structure in the bulk). As a result, it is difficult to form a common-anion heterostructure in the form of ZnXSe/ZnSe (similar to AlGaAs/GaAs), where X is a Group-II element.

Due to the lack of a suitable candidate of Group-II elements, one of the compounds investigated was ZnMnSe, where Mn (being a transition metal) is introduced to substitute for Zn atoms. Because the MnSe has a rock-salt structure in the bulk, a zincblende phase of ZnMnSe can be formed in the bulk only at low Mn concentrations. However, cubic ZnMnSe can be grown by MBE for the whole composition range. The ZnMnSe/ZnSe heterostructures studied first involved compositions higher than 10% Mn, in which case the ZnMnSe and ZnSe layers form barriers and wells, respectively (Kolodziejski et al. 1985). Because of the attractive wavelength of the light emission in these quantum wells, lasing experiments with optical pumping were carried out, and were reasonably successful at low temperatures (Bylsma et al. 1985). The insufficient carrier confinement discussed earlier results in optical characteristics that deteriorate rapidly as the temperature is raised. When ZnMnSe is used as barriers, the internal transitions in Mn^{++} in the yellow region of the spectrum were also detrimental to achieving strong luminescence from the quantum wells in the blue, due to relaxation of the exciton to the localized states in Mn^{++}. In common-cation system, namely, ZnSSe/ZnSe quantum wells, in which ZnSe is the well, the conduction-band offset is small. Despite some early success in achieving lasing action (Cammack et al. 1987; Suemune et al. 1989; Zmudzinski et al. 1990; Sun et al. 1991), it was recognized that such a system was not suitable for room temperature applications. As a result, early experiments in this area were plagued with a number of problems, some of which are intrinsic to wide-gap materials in general, such as doping, whereas others are material-specific, e.g., finding a heterostructure suitable for fabricating the active region of a short-wavelength laser. Therefore, there was no report of lasing action in the blue which showed promise for realistic device applications (CW lasing action at room temperature) until 1990. The interest in blue lasers was revived by a series of breakthroughs since 1990. The demonstration of lasing action in ZnSe/ZnCdSe quantum wells identified a viable active-region configuration for the fabrication of blue lasers (Jeon et al. 1990), and p-type doping of ZnSe by atomic nitrogen was successfully carried out (Park et al. 1990; Ohkawa et al. 1991). These results were immediately followed by the successful fabrication of the first laser diode (Haase et al. 1991). In this section, we will focus on ZnSe-based blue–green laser structures, specific achievements that led to the current status of these devices, and describe obstacles that still exist.

3.4.1 Quantum Well Structures for Blue–Green Lasers

As already mentioned, early choices of ZnSe-based quantum wells, such as ZnSe/ZnMnSe, ZnSe/ZnSeS, and so on, were shown to have confining potentials that are very shallow either for electrons (e.g., ZnSe/ZnSeS) or for holes (e.g., ZnSe/ZnMnSe). Even when these wells are sufficiently deep to provide some confinement at low temperatures, thermal excitation degrades the degree of localization at room temperature below what would be required for efficient radiative recombination. It is important to notice that the condition for exciton absorption was also shown later to apply to lasing action. In other words, quantum well depth and width suitable for exciton absorption at room temperature are also optimum for lasing action, although the number of wells needed for lasing can be more than that needed for observing exciton absorption due to wave guiding (which can be necessary when the structure is only ZnCdSe/ZnSe, where the quantum wells are used for both carrier and light confinement). After exhausting all alloys based on II–VI zincblende binary compounds, it became clear that other alloys involving compounds that do not occur in the zincblende structure in nature need to be explored. One such possibility involves ZnCdSe alloys. The CdSe component has a wurtzite structure in the bulk. The first problem in such an approach is that of structural integrity, namely, whether the compound can be grown in the zinc blende phase with reasonable crystalline quality. It was demonstrated that zincblende CdSe can indeed be grown by MBE on (100) GaAs substrates, although the density of stacking faults turns out to be rather high (Samarth et al. 1989), because CdSe is heavily lattice mismatched to GaAs ($\sim$7% mismatch). Because the natural structure of CdSe is not zinc blende, the crystal quality does not improve significantly at thicknesses beyond the critical thickness, in contrast to other "natural" zinc blende compounds, such as ZnTe or CdTe on GaAs. Significantly, however, it was demonstrated that when cubic CdSe is grown on ZnTe buffer layers (which is nearly lattice-matched to CdSe), its crystal quality improves dramatically (Luo et al. 1991). These initial efforts in perfecting the structure of CdSe led to the subsequent MBE growth of ZnCdSe alloys. It was shown that such alloys—especially at lower Cd concentrations—can be grown with an impressively high degree of perfection (Samarth et al. 1990a). Furthermore, the optical properties of this new family of alloys were favorable for light-emitting devices, including "clean" PL spectra, with relatively narrow band-edge luminescence and with no significant luminescence below the band edge. With the successful growth of ZnCdSe alloys and ZnCdSe/ZnSe quantum well and superlattice structures, the first room temperature lasing action in a ZnSe/ZnCdSe superlattice was reported in 1990 using optical pumping (Jeon et al. 1990).

3.4.2 Doping of ZnSe

The active region of any type of laser is unquestionably the most important component for its operation. At the same time, the formation of a suitable p–n junction configuration is a crucial link in the series of steps leading to the ultimate fabrication of laser diodes. Problems with doping ZnSe, especially with p-type doping, constituted a major and long-standing bottleneck. Part of the difficulty in doping any wide-gap semiconductor arises intrinsically from the size of the energy gap itself. The Fermi level for an undoped semiconductor lies in the middle of the energy gap. Thus, in trying to dope a wide-gap semiconductor, one is essentially shifting the Fermi level by half of the energy gap to the edge of either the conduction band (for n-type doping) or the valence band (for p-type doping). For ZnSe, which has an energy gap of 2.7 eV at room temperature, this corresponds to an energy change of 1.35 eV, which is enough to promote compensation through defect formation.

Although doping of II–VIs is in general difficult, n-type doping of ZnSe has been considerably easier, compared to p-type doping. This can be partly attributed to the relatively low position of the conduction band with respect to the vacuum level. Two promising dopants, Ga and Cl, have been the most exhaustively studied elements in this connection. The doping level of Ga in ZnSe can easily reach 10^{17} cm^{-3}. However, attempts to increase the carrier concentration by introducing higher concentrations of Ga during growth usually result in either saturation or even a decrease of the actual carrier concentration (Venkatesan et al. 1989), which is believed as due to the so-called "self-compensation" effect. When Cl is introduced (usually from a $ZnCl_2$ source in MBE growth) as the n-type dopant, doping levels can be successfully raised to above 10^{19} cm^{-3} (Ohkawa et al. 1987). Deep levels associated with Ga and Cl donors were compared by deep level transient spectroscopy (Karczewski et al. 1994). These results indicate that Cl is far superior to Ga as an n-type dopant, because no deep levels directly associated with Cl atoms can be detected, in contrast with Ga-doping, which results in a significant concentration of defects involving Ga atoms.

As mentioned earlier, p-type doping of ZnSe has been and, to some extent, still remains a major obstacle. The problem encountered in p-type doping of ZnSe stems from the strong lattice relaxation associated with most—and especially the heavier—group-V dopant candidates, which leads to the formation of deep centers and consequently to heavy compensation (Chadi et al. 1989). The discovery of incorporating nitrogen (by using a plasma source in MBE growth) made it possible to dope ZnSe p-type with hole concentrations close to what is needed for a laser diode. Although the use of the plasma source, as well as other means of exciting nitrogen gas, such as by microwaves, still awaits a thorough investigation, it is clear that the breakup of N_2 molecules into single N atoms is responsible for the success of this doping process. The highest net hole concentrations reached by this method are generally around 2×10^{18} cm^{-3}. As additional nitrogen is incorporated in ZnSe beyond the above concentration, compensation becomes a problem that, as yet, has not been overcome. It is widely believed that at the 10^{18} cm^{-3} level, nitrogen begins to form complexes responsible for this self-limiting behavior. Although the doping level is sufficiently high for the operation of laser diodes, in practice, higher doping levels are still very desirable. This is because all metal contacts to p-type ZnSe create high Schottky barriers, since their work functions are without exception much higher than the valence band of ZnSe. If the hole concentration in ZnSe could be further increased, the Schottky barriers would eventually become so narrow that tunneling would be sufficient for injecting the holes into the p-ZnSe. Thus, further improvement of p-type doping of ZnSe remains critical to developing practical blue laser diodes.

Doping of ZnSe-based alloys involved in current laser diodes, namely, ZnSSe, ZnMgSe, and ZnMgSSe, remains a problem, especially in the case of nitrogen doping. The mechanisms responsible are yet to be identified. Because the structures of such diodes consist mostly of these alloys, better doping schemes are needed.

3.4.3 ZnSe-Based Laser Diodes

Although significant progress has been made in recent years, there are still problems in fabricating practical ZnSe-based laser diodes, which are all, in one way or another, related to the very high lasing threshold. A reduced threshold will, undoubtedly, prolong the lifetime of the device. Three key factors are particularly important in reducing the threshold current: (1) quantum efficiency of the active region determined by the confinement of charge carriers in the quantum wells; (2) light confinement for maximum overlap between light in the laser cavity and carriers in the quantum wells; and (3) contact resistance. Below, we will discuss the efforts made in these three areas.

3.4.3.1 Electron and Light Confinement

The degree of quantum confinement is affected by two factors: the height of the confining potential and the well width. Because of the large difference in the ZnSe and CdSe lattice constants, there is a practical limit for the Cd concentration in the ZnCdSe quantum well region if one wishes to avoid excessive dislocation densities. Furthermore, it was demonstrated that, as Cd concentration increases, the overall crystalline quality of ZnCdSe deteriorates. Finally, a high Cd content will also reduce the energy gap of the active region. With all these factors considered, the Cd composition yielding the best results is currently around 25–30%. As mentioned earlier, the ZnSe/ZnCdSe system exhibits strong carrier confinement when the well width is reduced below the exciton Bohr radius. This includes not only an increase in the exciton binding energy, but also an enhancement of the oscillator strength due to high spatial overlap of the electron and hole states. It was demonstrated that the lasing threshold is closely related to such efficient confinement of carriers. The quantum well width in current laser diodes is around 50 Å. With the introduction of S and Mg as alloy constituents, the well depth can be further increased in both the conduction band (by adding Mg) and in the valence band (with S). Thus, in principle, the system ZnMgSeS/ZnCdSe can achieve large confining potentials for both electrons and holes. Other considerations, involving trade-offs between electron confinement and light confinement, and lattice matching (see below), suggest that the optimal configuration of quantum wells currently in use is most commonly in the form of ZnSeS/ZnCdSe (Gaines et al. 1993).

Light confinement in the ZnSe-based lasing structures is also a crucial issue in achieving a low operating threshold, as in the case of II–V semiconductor lasers. Although ZnSe/ZnCdSe quantum wells themselves, have a significant waveguiding effect (ZnSe has a lower index of refraction than ZnCdSe), materials with an even lower refractive index are needed to sandwich the quantum well active region to further improve light confinement. The fact that the index of refraction varies inversely with the energy gap has been used in selecting the II–VI materials to optimize light confinement in the present structures. The first II–VI blue laser diode fabricated at the 3M Company consisted of ZnSe/ZnCdSe quantum wells for the active region, sandwiched between ZnSeS layers for the purpose of optical confinement. A more recent structure demonstrated by North American Philips Laboratories, shown in Figure 3.6, involved $Zn_{1-x}Mg_xSe_{1-y}S_y$ (Gaines et al. 1993), a quaternary alloy first grown by Sony to achieve separate

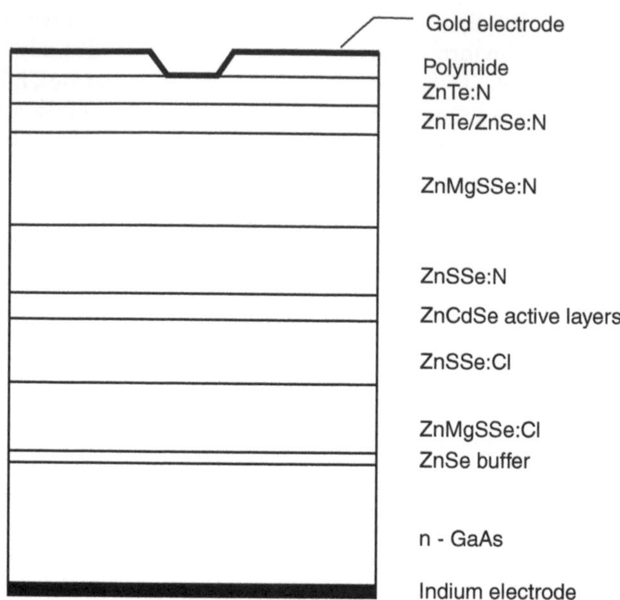

FIGURE 3.6 Schematic diagram of a separate-confinement laser diode structure involving ZnMgSeS.

confinement. This alloy not only has the largest energy gap among the ZnSe-based alloys (thus, ideal for the cladding layer), but also can be tailored to match the lattice constant with ZnSeS. Thus, allows the use of ZnSeS, lattice matched as barriers. We note here that the use of S in the barriers not only improves the structural integrity of the system by reducing stacking faults and dislocations through a better lattice match, but also increases the valence band offset between ZnSeS and ZnCdSe and, thus, the confinement of holes. This is achieved via two independent but additive mechanisms: the presence of S lowers the valence band of the barrier, and the reduction of the barrier lattice parameter due to the presence of S increases compression of the ZnCdSe layer, moving the valence band of the quantum well upward.

3.4.3.2 Ohmic Contacts

As mentioned earlier, ohmic contacts to wide-gap semiconductors constitute a major problem because ideally one would wish to use metals with work functions above the bottom of the conduction band on the n-type side of the p–n junction and below the top of the valence band on the p-type side, to avoid formation of Schottky barriers. As the gap increases, metals with requisite work functions are, thus, harder and harder to find. When such metals do not exist, there are two methods that can be used to overcome the unavoidable Schottky barriers. One of them is to heavily dope the semiconductor layer to which the contact is to be made, such that the depletion layer associated with the Schottky barrier is sufficiently thin to promote carrier injection by tunneling. For n-ZnSe, this approach does indeed provide contacts that appear to be ohmic, owing to the high doping levels achievable with Cl-doping and the relatively small difference between the bottom of the conduction band and the metal work function. The dominant problem is making an ohmic contact to p-ZnSe, because there is no metal that has a work function near the top of the valence band. The problem is further compounded by the inability to attain high doping levels in p-type ZnSe. A solution to this problem is to use a graded alloy, so as to move the top of the valence band close to the metal work function, as was done in the case of n-type GaAs/Au contacts by using GaInAs (Woodall et al. 1981). For p-type ZnSe, the alloy ZnSeTe was used (because of the high valence band position of ZnTe, as discussed earlier; see also Figure 3.2) to raise the valence band edge of the overall structure on its p-doped side close to the work function of Au, as shown in Figure 3.7. It is also significant in this context that ZnTe can be doped p-type to a much higher level than ZnSe.

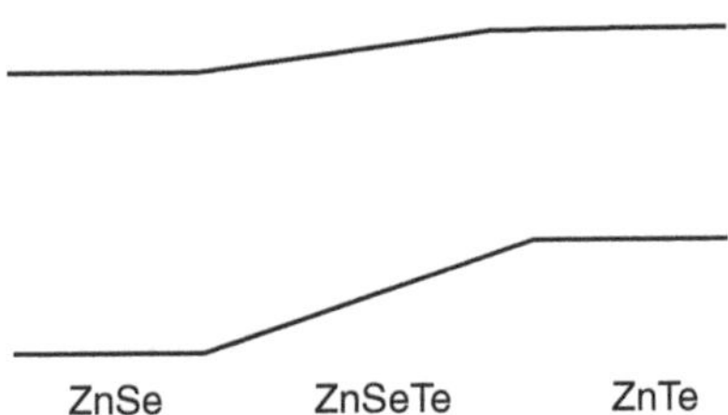

FIGURE 3.7 Band profile of linearly graded ZnSeTe region. The graded region is used to reduce Schottky barriers at p-ZnSe/Au contacts.

Together, these effects result in a dramatically reduced contact resistance (Fan et al. 1992). It was also shown that, by growing HgSe on p-ZnSe, the contact resistance can be reduced in a similar way (Lansari et al. 1992).

A recent theoretical study indicates that the profile of the graded region is also very crucial to the contact resistance (Yang et al. 1995). So far, all of the graded-alloy regions have used a linear profile, as shown in Figure 3.7. By changing to a parabolic grading profile, it was shown that the length of the residual barrier can be further reduced. With the same doping level, the grading length (i.e., the thickness of the graded region) can, therefore, be much smaller (by almost a factor of 5) for the parabolic grading than for the linear case. This is additionally advantageous because ZnTe has a large lattice mismatch with ZnSe, and parabolic grading reduces the thickness of the part in the graded region where a high density of dislocations is expected.

3.4.4 The Lasing Mechanism in ZnSe-Based Laser Diodes

Semiconductor lasers that have been well studied, such as III–V semiconductor lasers, involve strongly interacting states. In such cases, the large Bohr radii of the ground-state (1S) excitons lead to strong overlap, as enough excitons are generated to reach the lasing threshold. Thus, the excitonic nature of the underlying recombination transition is replaced by a plasma of electrons and holes, and stimulated emission arises from recombining such electrons and holes in the plasma. This situation is naturally linked to the exciton population and to the size of the excitons. In principle, if the Bohr radius of the exciton is reduced, the mechanism would revert to the recombination of individual excitons. The large effective masses in ZnSe result in Bohr radii that are much smaller than those in GaAs and are further reduced by quantum confinement. It was shown that in ZnSe/ZnCdSe quantum well lasers exhibiting strong confinement, exciton recombination remains even under lasing condition, especially at low temperatures (Ding et al. 1992). There have been recent studies showing that in certain ZnSe/ZnCdSe quantum well lasers the mechanism involves an electron-hole plasma (see, for example, Cingolani et al. 1994). However, in all the studies reported so far which showed the electron-hole plasma as the active medium, there is substantially less confinement (either wide quantum wells or shallow wells corresponding to small Cd concentrations) compared to the quantum wells used by Ding et al. (1992). Thus to some extent the question still remains unresolved. The results available seem to suggest that when Bohr radii close to those in bulk ZnSe favor the electron-hole plasma situation. On the other hand, a reduced Bohr radius in narrow and deep ZnSe/ZnCdSe quantum wells, with their strong confining effects, can tip the scales from the band-to-band lasing involving an electron-hole plasma to excitonic lasing involving discrete exciton levels. Considerable systematic experimental and theoretical work is needed to identify the mechanisms and the exact conditions to which these mechanisms correspond.

3.4.5 Future Research and Development

Although significant improvements have been achieved in a remarkably short time, current ZnSe-based blue–green lasers have one outstanding problem—their short operating lifetime at room temperature.

There appear to be several areas, not necessarily independent of one another, where improvements are still needed. The short lifetime itself has been shown to be related to defects which propagate (irreversibly) as the temperature is raised during the operation of the laser (Guha et al. 1993a; Hua et al. 1994). One can therefore categorize the problem as consisting of two issues: defect formation during growth and defect expansion due to the heat that is generated because of the high threshold current. Since defect densities in even closely lattice-matched systems are still high ($>10^3$ cm^{-2}), which causes failures of laser diodes, it appears that improvement in growth and other procedures related to material quality of ZnSe-based systems is needed. It has been demonstrated that MBE-grown GaAs buffer layers on GaAs substrates greatly reduce the density of dislocations (Xie et al. 1992; Guha et al. 1993b), which improve the lifetimes of laser diodes. Homoepitaxy is another area that has been studied in this connection (Ohishi et al. 1988; Ohkawa et al. 1992; Harsch et al. 1994). The key to such an approach is the availability of high quality ZnSe substrates. Recent reports by the collaboration between the North Carolina State University and Eagle–Picher have demonstrated the successful growth (by seeded physical vapor transport technique) of undoped ZnSe substrates. X-ray diffraction rocking curves with full width at half maximum as low as 11 arc sec are observed on ZnSe substrates grown in this way, which is comparable to that of GaAs substrates (Harsch et al. 1994). Attempts in doping these substrates are also underway and have shown moderate success with carrier concentrations reaching high 10^{17} cm^{-3} (Harsch et al. 1994). However, such substrates are currently not commercially available, and therefore, the advantages of using ZnSe substrates, especially for laser diodes, have not been fully explored. The second issue, that of high threshold, is a complicated one, involving several possible factors, such as still less than perfect contact resistance, the overall design of the laser structure with regard to both carrier and optical confinement, and device fabrication. Thus, all three major achievements that are emphasized in this article—quantum well structure, doping, and ohmic contact—all have room for improvement.

There is yet another question which is much harder to address at this stage, namely, whether the heat generated by the lowest possible threshold current for room temperature laser operation can itself generate defects in an initially defect-free structure. The answer to this, however, cannot be properly arrived at before the problems discussed above are fully resolved. Finally, as more attention is directed toward GaN-based, wide gap structures, a comparison between GaN-based and ZnSe-based devices seems appropriate. Although there are advantages to GaN-based light emitting diodes (e.g., extremely long lifetimes) (Nakamura et al. 1994), it is much more difficult to compare laser diodes fabricated from the two materials, because the requirements for LEDs and for laser diodes are significantly different and because a practical GaN laser diode remains to be fabricated. The issue of finding a suitable substrate, the thermal mismatch with such a substrate, and the growth of GaN-based heterostructures are all major problems. Although GaN belongs to the well-studied III–V family, the nitrides differ in many ways from the phosphides, arsenides, and antimonides, and it may very well be the case that integration with the standard III–V semiconductors will, in the end, prove much easier for ZnSe-based systems than for the nitrides. It is very clear that both ZnSe-based and GaN-based systems still face significant difficulties. Therefore, it appears much more appropriate to explore both systems to assess their potential.

3.5 Type-II Heterostructures

Because of the large offset between Se and Te compounds, as shown in Figure 3.2, it is possible to fabricate type-II structures with extremely large offsets both in the conduction and the valence bands. One of the type-II heterostructures in the II–VI semiconductor family that is fairly well understood is the ZnSe/ZnTe short period superlattice (Kobayashi et al. 1986). Due to the large lattice mismatch (around 7%), individual layers in this structure have to be very thin (a few monolayers) to keep the density of dislocations low. The valence band offset in this case is close to 1 eV. Optical experiments were carried out on such structures, and demonstrated the type-II band alignment. There has not been significant effort in utilizing this structure itself for optoelectronic applications. The large value of the valence band offset, however, proved to be useful for forming ohmic contacts, as discussed earlier.

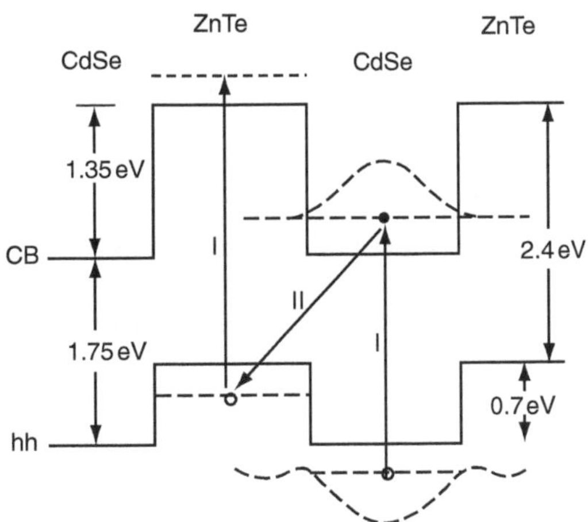

FIGURE 3.8 Schematic diagram of band alignment for a type-II CdSe/ZnTe superlattice. Wave functions of both localized and quasilocalized states are shown by the dashed lines. Arrows indicate optical transitions.

Another type-II structure in the II–VI family is CdSe/ZnTe (Luo et al. 1991; Yu et al. 1991), which is the exact analog of GaSb/InAs. As expected from Table 3.1, it is also a nearly lattice-matched system. Here, we will describe two fundamentally important phenomena. The first is structural, involving a switch of atoms at the interface (Kemner et al. 1992). Either the cations or the anions on the two sides of the interface exchange positions, which results in a few monolayers of CdTe and ZnSe at the interface.

Because of the large offsets in both the conduction and the valence bands, the wide energy gaps, and the absence of the X-point in the energy range of interest, CdSe/ZnTe heterostructures provide an ideal system for fundamental studies of optical properties of type-II alignment. The band alignment of this type-II system is schematically shown in Figure 3.8. It had been a common belief that type-II superlattices are optically less active compared to type-I superlattices and quantum wells, because the type-II transitions are spatially indirect (i.e., the initial and the final states are localized in different layers of the superlattice). However, the studies of quasilocalized states above the barriers indicated that there are direct (type-I) transitions in type-II superlattices that involve both localized and quasilocalized states, as indicated by arrows in Figure 3.8 (Luo and Furdyna 1992). Optical studies of this system revealed indeed that type-I excitons can form in type-II superlattices, which results in large oscillator strengths (Luo et al. 1993a). It was observed that the absorption coefficients corresponding to such type-I transitions can be as large as $5\times10^4\ \mathrm{cm}^{-1}$, comparable to those in type-I quantum wells with similar energy gaps.

Furthermore, there are two types of type-I excitons in a type-II superlattice, respectively, localized in the two layers (i.e., CdSe and ZnTe layers). One type of such excitons involves a confined electron and a localized above-barrier hole, both in the CdSe layers. The other kind consists of a confined hole and a localized above-barrier electron, both in the ZnTe layers. The behavior of both excitons shown resemble type-I excitons in type-I quantum wells, such as the confinement energy that increases with a reduction of layer thickness (Zhang et al. 1993). The dynamic properties of such excitons, however, dramatically differ from their counterparts in type-I systems. Because the localized above-barrier states are not the ground state, an electron (or a hole) in such a state will quickly relax to lower energy states in the same band. Such a relaxation process (typically in the sub-picosecond range) is several orders of magnitude faster than optical recombination processes which are on the order of a few nanoseconds. As a result, the type-I transitions can be observed easily only in measurements involving the excitation process, such as absorption or reflectivity experiments, but not in measurements of optical recombination, e.g., photoluminescence.

It is important to point out that the observed type-I transitions in type-II CdSe/ZnTe superlattices, which originate from the presence of quasilocalized states in the barriers, should be present in all type-II systems, such as InAs/GaSb superlattices. These transitions should be taken into account in analyzing the optical properties of GaAs/AlGaAs superlattices, where the band alignments at the Γ-point and the X-point can be of different types (Lew Yan Voon et al. 1993).

3.6 HgTe/CdTe Type-III Heterostructures

As mentioned in the Introduction, there have been review articles (see, for example, Meyer et al. 1993), in which HgTe/CdTe systems are discussed in detail. Here, we will briefly describe the unique properties of type-III superlattices and their potential for device applications.

The band structure of HgTe is characterized by a symmetry-induced zero energy gap, i.e., both the conduction and the valence bands belong to the Γ_8 bands that are degenerate at the Γ-point, as shown in Figure 3.9a and b. The light-hole band is now the conduction band. The Γ_6 band, which is typically the conduction band in an open-gap semiconductor, is now in the valence band. The band diagram consisting of a zero-gap compound, e.g., HgTe, and an open-gap compound is referred to as the type-III alignment.

It is worth noting that not only the band structures in HgTe/CdTe are qualitatively different, but the band gaps and the effective masses are also drastically different. In addition to the new properties which can be expected from such a novel electronic configuration, type-III superlattices are of interest from the applied point of view, because the effective energy gap (the energy difference between the valence and conduction subbands allowed by the superlattice geometry) and the effective masses can be tuned over a wide range. We will discuss these two aspects—the physical properties and the capabilities for infrared detector applications.

The band structure of a HgTe/CdTe superlattice is illustrated in Figure 3.10 for two combinations of well and barrier dimensions, showing two qualitatively distinct regions of the band structure (Figure 3.10a and b). The difference in band structure is determined primarily by the well width, which, to some extent, is similar to Hg concentration in CdHgTe. For wide wells, the HHl subband (originating from the bulk Γ_8 band) lying above the E_1 subbands (originating from Γ_6) at the center of

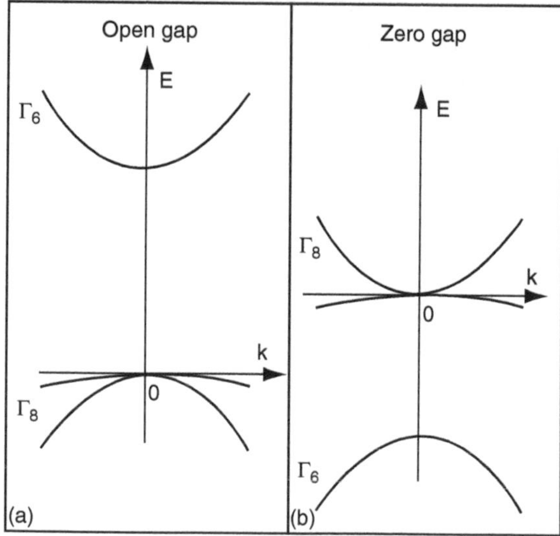

FIGURE 3.9 Schematic diagrams of band structures of an open-gap (left panel) and a zero-gap (right panel) semiconductor. A heterostructure consisting of the two materials is referred to as having type-III band alignment.

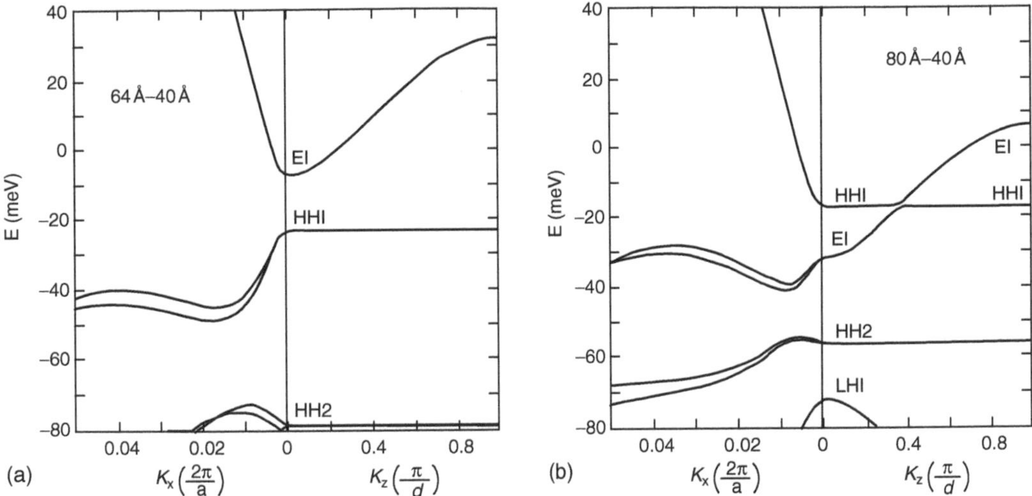

FIGURE 3.10 Energy band structures for two representative cases of HgTe/CdTe superlattices: (a) shows an open-gap semiconductor system, and (b) corresponds to a semimetallic system. Note that there is a true overlap between El and the HH1 bands at the center of the Brillouin zone.

the Brillouin zone. We will refer to superlattices corresponding to the wide-well limit as semimetallic. Note that—unlike bulk HgTe and its alloys HgCdTe with less than 10% Hg, which are zero-gap semiconductors—the superlattices in this range behave as true semimetals, i.e., they have a significant overlap of the conduction and the valence bands. As expected, the widths of the bands displayed in the figures vary inversely as a function of barrier width.

As the well width decreases, the E1 band of the superlattice is gradually pushed upward, eventually ending up above the HH1 band. In this region, the superlattice resembles in many ways, the alloy $Hg_{1-x}Cd_xTe$ for $x>0.15$, and we shall refer to it as semiconducting. The gap between the Γ_6-like E1 conduction band and the Γ_8-like HH1 valence band continues to increase as the well width decreases. One should note that, for any constant barrier width, a decrease of the well width represents a physical increase in the Cd content of the superlattice system as a whole and, in this sense, the behavior of the superlattice is similar to that of the $Hg_{1-x}Cd_xTe$ alloy with increasing x.

This similarity notwithstanding, there are major differences between $Hg_{1-x}Cd_xTe$ alloys and HgTe/CdTe superlattices. Most obvious is the anisotropy (represented by the two directions, the growth direction, z, and the in-plane direction, x) in the band structure diagram. Note that, in the valence band, the E vs. k dispersion in the growth direction is very flat, indicating that the hole mass characterizing motion along z is extremely heavy, whereas the mass in-plane is very light. Second, we see that, for x-motion the effective mass of the holes is nonparabolic. Such nonparabolicity (where a mass can increase, e.g., from $0.01m_0$ to a value two orders of magnitude larger in a span of only a few tens of meV) is not found in any other known material. A closely related feature is the mass broadening effect, where one must take into account a spectrum of masses (for different k-values) to properly describe the behavior of carriers in a given subband. These novel properties and the effects that follow are unique to type-III superlattices and do not occur in any other material.

The ability to tune the energy gap of HgTe/CdTe superlattices by varying the superlattice parameters (especially the well width) makes such structures of interest for infrared and far infrared detector applications, as an alternative to HgCdTe alloys. We note here two advantages of the superlattice in this context. First, it can be argued that the superlattice, grown by deposition of binary compounds, lends itself to better composition control and avoids band-gap broadening effects due to alloy fluctuations.

Second, and probably more important, the large electron and hole mass in the *z*-direction (already discussed in connection with Figure 3.10) guarantees a reduction in the background noise arising from dark current, which is expected to flow in any photovoltaic device. Whether these advantages will, in the end, compensate for the comparatively greater cost of producing HgTe/CdTe superlattices over HgCdTe alloys still remains unresolved.

3.7 Diluted Magnetic Semiconductor Heterostructures

Transition metals, such as Mn, Fe, Co, and so on, can be alloyed into II–VI semiconductors, forming what is referred to as diluted magnetic semiconductors (DMSs) (Furdyna 1988). The exchange interaction between free carriers and localized electrons in these magnetic ions (sp–d exchange interaction) leads to a wide range of unique properties, combining semiconductor physics and magnetism. For example, the sp–d exchange interaction leads to enormous band Zeeman splittings (two orders of magnitude larger than those in nonmagnetic II–V materials). The efforts in utilizing this property for practical applications have been very limited, because in most cases the large splittings are present only when the temperature is low. Nevertheless, such efforts played an important role in the studies not only of DMSs, but semiconductors, in general.

In bulk DMSs, the transition metal concentrations are typically low because of the different crystal structures of the compounds involved, which is the reason for the use of the word *diluted*. Modern crystal growth techniques, e.g., MBE, have greatly extended the composition range of the magnetic component to 100% for the cases of Mn and Fe (Kolodziejski et al. 1986; Durbin et al. 1989; Jonker et al. 1988), for example MnTe. Despite the extended composition range, we will still use the name DMS for historic reasons. The large composition of magnetic ions has revealed new magnetic properties, such as novel forms of long range antiferromagnetic order, which had not been possible in bulk samples because of the limited magnetic ion compositions. Although most recent studies in this area have focused on II–VI DMS heterostructures, efforts have been made to introduce magnetic ions, i.e., Mn^{++}, into III–V semiconductors using the technique of MBE (Munekata et al. 1989). Both structural and magnetic properties have been studied (Krol et al. 1993; von Molnar et al. 1991), which greatly enriched the family of DMSs. The introduction of magnetic ions in heterostructures has exciting implications because of its multicomponent nature—semiconductor physics, magnetism and quantum confinement of charge carriers—and the sp–d exchange interaction. In this section, we will focus our attention on II–VI DMS heterostructures, namely, their optical properties (originating from the sp–d exchange interaction) and examples of applying such properties to the understanding of semiconductor heterostructures in general.

The most interesting DMS heterostructures involve a combination of DMS and non-DMS layers, such as ZnMnSe/ZnSe and ZnFeSe/ZnSe. Because of the drastically different electron and hole Zeeman splittings in DMS and non-DMS compounds, band offsets can be significantly changed by an externally applied magnetic field, to the extent that even the type of the band alignment can be changed (e.g., from type-I to type-II). The advantage of such a choice is twofold. First, one can have a continuous range of offset values in one sample, rather than growing a number of samples, as is often done for non-DMS heterostructures. Therefore, the properties of a given structure can be studied over a wide parameter range without ambiguity concerning alloy compositions and structural dimensions, which is often encountered in dealing with different samples. The second is the accurate mapping of the wave function for the carrier states. For example, in optical studies, one can quantitatively determine the shape of wave functions by measuring the Zeeman splitting of a given transition that reflects the location of the confined states and their penetration into the adjacent layers. These unique effects have led to a series of novel observations, some of which strictly belong to DMS heterostructures (such as long range antiferromagnetic order observed by Giebultowicz et al. 1992, and magnetic polarons by Yakovlev et al. 1990), whereas others can be applied to non-DMS heterostructures (e.g., III–V heterostructures). We will discuss selected examples of general interest in the following.

3.7.1 Magnetic Field Induced Type-I to Type-II Transition

As discussed above, band edges in DMS layers can be tuned over a wide energy range by an applied magnetic field, which in some cases can exceed the zero-field offsets. As a result, it is possible to convert the types of the band alignment of a heterostructure that consists of DMS and non-DMS layers. As an example, consider the conduction band and the heavy-hole valence band of a single non-DMS quantum well with DMS barriers, as shown in Figure 3.11. The structure corresponds to a type-I band alignment at $B=0$, with a deep conduction band well and a shallow well in the valence band (e.g., a common-anion system). Note that the heavy-hole spin splitting is typically much larger than that in the conduction band (Furdyna 1988). Upon applying a magnetic field, one spin orientation ($m_j=+1/2$ electrons, $m_j=+3/2$ holes) will experience a deepening of the wells in both the conduction and the valence bands (enhancing the type-I nature). For the other spin states ($m_j=-1/2$ electrons, $m_j=-3/2$ holes), however, the enormous Zeeman splitting will, at some field, exceed the valence band offset, and the original valence band well will be transformed into a barrier, as shown in Figure 3.11 for $B\neq 0$.

The conversion from type-I to type-II is manifested by a dramatic change in optical properties, because, in a type-II structure, the carriers involved in the strongest optical transitions are now physically separated in different layers. The phenomenon of such a conversion was first observed in ZnFeSe/ZnSe (Liu et al. 1989), and later in CdMnTe/CdTe (Deleporte et al. 1990).

It is important to note that this phenomenon is particularly important to the determination of band offsets. The Zeeman splitting of a given DMS layer can be measured accurately. Thus, the splitting at the field that corresponds to the type-I to type-II transition can provide direct information concerning the zero-field offset. Such a direct measurement of the band offsets, using simple optical techniques, has been a challenge for non-DMS heterostructures (e.g., III–V heterostructures).

3.7.2 Wave Function Mapping

The degree of confinement of charge carriers is of great importance for fundamental properties and device applications, as discussed earlier. The shape of the wave function, characterized by its localization in the quantum well and by its penetration into the adjacent layers, determines the confinement characteristics of a given state. Most of the experimental evidence concerning wave-function shapes (thus the degree of confinement) has been indirect, which includes measurements of the transition energies or deviation of the exciton states from the 3-D hydrogenic model. For low-lying states (such as the ground state), such estimates can provide reasonably accurate distribution of wave functions, but the uncertainty increases dramatically for higher energy states.

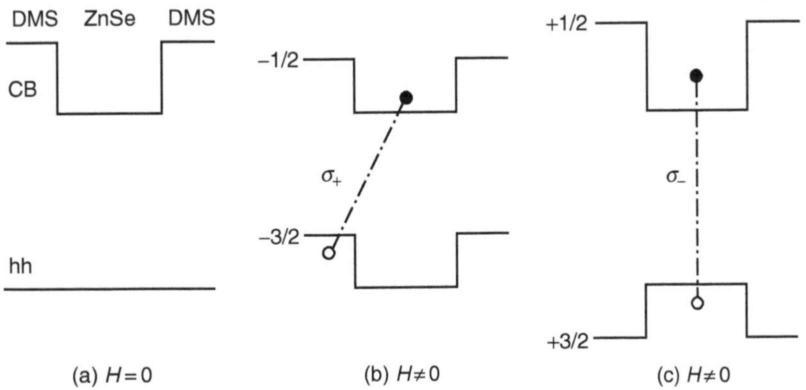

FIGURE 3.11 Schematic diagram of the conduction and the valence band alignment in ZnSe/$Zn_{0.9}Fe_{0.1}Se$ quantum well structures: (a) $B=0$; (b) $B\neq 0$, $m_j=-1/2$ electrons, $m_j=-3/2$ holes; (c) $B\neq 0$, $m_j=+1/2$ electrons, $m_j=+3/2$ holes. The dot-dashed lines indicate the allowed heavy hole exciton transitions in the Faraday geometry.

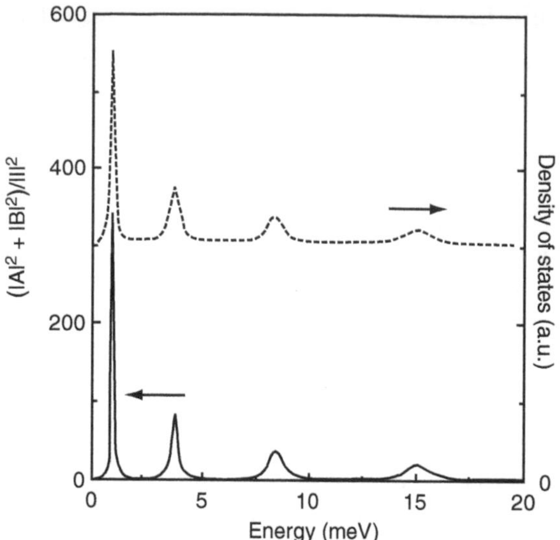

FIGURE 3.12 Quasilocalization and density of states associated with a single quantum barrier.

One can improve the situation by exploiting the fact that, in heterostructures consisting of DMS and non-DMS layers, the Zeeman splitting of a given state will reflect its probability distribution over these two media. In other words, the Zeeman splitting will be determined effectively by how many localized magnetic moments the electron "sees". It can, thus, be used as a tool for mapping the wave function (or probability), providing spatial identification of electronic states. This technique has been essential in observing several novel effects.

One such observation involved type-I excitons in type-II superlattices (Luo et al. 1993a). Type-II heterostructures have been studied for a long time and have been considered optically less active because of the spatial separation of the initial and final states. The strong transitions observed in the type-II system CdSe/ZnTe discussed earlier have been ellusive for several years after the first observation. The final identification of the origins of the optical transitions was made in CdSe/ZnTe structures, in which Mn^{++} was introduced either in the CdSe or ZnTe layers.

Another nonintuitive phenomenon is the quasilocalized states in a single quantum barrier and related structures. A single barrier is characterized by a continuous energy spectrum, which appears to be uninteresting in optical studies, presumably due to lack of discrete states. It was shown, however, quasilocalized states in a single barrier behave similarly to a single quantum well (Luo et al. 1993b). Such a similarity is characterized by quasilocalization of wave functions in the barrier region (measured by $|A|^2+|B|^2$, A and B being the coefficients of the plane waves in the barrier region, normalized to the probability of the incoming wave, $|I|^2$) and peaks in the density of states, which resemble the discrete states in a quantum well, as shown in Figure 3.12. The ability to spatially identify the optical transitions greatly simplified the analysis of such quasilocalized states. The structures used consist of ZaMnSe layers (barriers) sandwiched between ZnSe. The absorption peaks, in a system with a *continuous energy spectrum* between the quasilocalized states in the valence and the conduction bands, exhibited large Zeeman splitting (comparable to that in a ZnMnSe epilayer with the same Mn concentration) (Luo et al. 1993b). Thus, one can unambiguously identify that the transitions occur in the barrier region.

3.7.3 Spin Superlattices

A "spin superlattice" is a superlattice in which carriers with opposite spin states are confined in different layers of the structure. To achieve such spin-modulation, one must look for a structure having a small

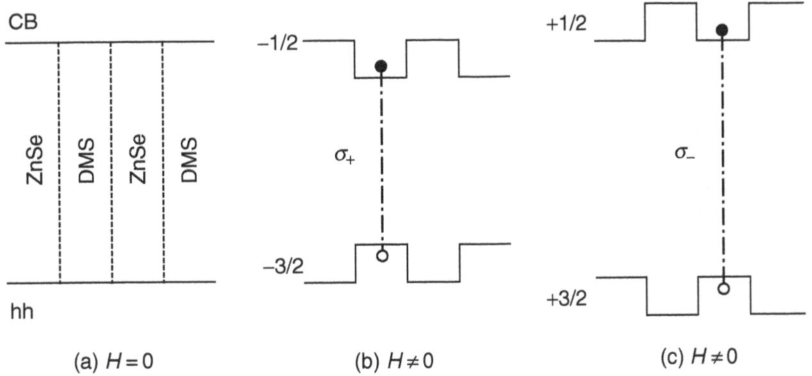

FIGURE 3.13 Schematic diagram of the conduction and the valence band alignment in spin superlattices: (a) $B=0$; (b) $B\neq0$, $m_j=-1/2$ electrons, $m_j=-3/2$ holes; (c) $B\neq0$, $m_j=+1/2$ electrons, $m_j=+3/2$ holes. The vertical dot-dashed lines represent the allowed heavy-hole exciton transitions in the Faraday geometry.

energy gap difference between the constituent layers and small band offsets in the absence of a magnetic field. In other words, an ideal system would be one in which both the conduction and the valence bands are flat without an applied field. When a magnetic field is applied, the large Zeeman splitting of the band edges in the DMS layers results in band offsets so as to produce a spatial separation of the spin-up and the spin-down states, as shown in Figure 3.13. The band offsets are created by the magnetic field. Layers of the superlattice that behave as quantum wells for one spin state will be the barriers for the other spin state in both the conduction and the valence bands. The spatial separation of the spin states can be monitored by optical experiments. Two systems have been used for this purpose, namely, ZnFeSe/ZnSe and ZnMnSe/ZnSe, both of which exhibited spin superlattice behavior (Chou et al. 1991; Dai et al. 1991).

3.8 Concluding Remarks

The future of II–VI semiconductor heterostructures relies strongly on the development of device applications, which are not necessarily limited to currently emphasized ZnSe-based laser diodes. Other optoelectronic applications, e.g., modulators, that have fewer restrictions on operating power (and thus present fewer problems to the lifetime issue), will add to the variety. The efforts in fabricating ZnTe- and ZnSe-based modulators have already cleared the way for more sophisticated optoelectronic devices (Wang et al. 1993; Partovi et al. 1991). Despite the significant advances in the fabrication of blue–green laser diodes, device processing of II–VIs is still in its infancy. Many techniques commonly used for Si and III–V materials, such as ion implantation, have so far not been adequately applied to II–VI semiconductors.

The fundamental studies of II–VI heterostructures are still far behind what has been accomplished in their III–V and Si-based counterparts. This is particularly true for structures of lower dimensions, namely, quantum wires and dots. Effects related to growth processes, such as spontaneous chemical ordering and the formation of quantum dots on lattice mismatched substrates or buffer layers, also present opportunities for future studies. There are numerous problems which have not been properly addressed because of the limited progress in device applications. Such problems include the surface structures of II–VI compounds, surface electronic states, and growth and characterization of oxide layers. The development in these areas, however, will, to a great extent, depend on the progress in device applications.

Theoretical studies have traditionally concentrated on III–V and Si-related problems. Recent activities in the area of blue–green laser diodes have stimulated a great deal of interest in II–VI materials, which has added strength to the overall effort in this area and will continue to be important in the future. For example, a thorough understanding of stabilities of the II–VIs and defect formation related to doping will provide important input to device fabrication. II–VI semiconductor heterostructures have experienced a great deal of development in the last few years, which, in turn, stimulated worldwide interest. The field as a whole, however, faces tremendous technological challenges.

Acknowledgments

The authors would like to acknowledge the support of the Center for Electronic and Electrooptical Materials (CEEM) at SUNY, NSF (DMR 9223054) and ONR/DoD under the MFEL program. We would also like to thank E. H. Lee, H. C. Chang, and M. S. Salib for their help in preparing this manuscript.

References

Akimoto, O. and Hasegawa, H. 1966. *J. Phys. Soc. Jpn.* 22, 181.

Bylsma, R., Becker, W., Bonsett, T., Kolodziejski, L. A., Gunshor, R. L., Yamanishi, M., and Datta, S. 1985. *Appl. Phys. Lett.*, 47, 1039.

Cammack, D. A., Dalby, R., Corneliassen, H., and Khurgin, J. 1987. *J. Appl. Phys.*, 62, 3071.

Chadi, J. and Chang, K. J. 1989. *Appl. Phys. Lett.*, 55, 575.

Chemla, D., Schmitt-Rink, S., and Miller, D. A. B. 1988. In *Optical Nonlinearities and Instabilities in Semiconductors*, H. Haug, ed., p. 83. Academic Press, New York.

Chou, W. C., Petrou, A., Warnock, J., and Jonker, B. T. 1991. *Phys. Rev. Lett.*, 67, 3820.

Cingolani, R., Rinaldi, R., Calcagnile, L., Prete, P., Sciacovelle, P., Tapfer, L., Vanzetti, L., Mula, G., Bassani, F., Sorba, L., and Franciosi, A. 1994. *Phys. Rev. B.*, 49, 16769.

Dai, N., Luo, H., Zhang, F. C., Samarth, N., Dobrowolska, M., and Furdyna, J. K. 1991. *Phys. Rev. Lett.*, 67, 3824.

Deleporte, E., Berroir, J. M., Bastard, G., Delalande, C., Hong, J. M., and Chang, L. L. 1990. *Phys. Rev. B.*, 42, 5891.

Ding, J., Pelekanos, N., Nurmikko, A. V., Luo, H., Samarth, N., and Furdyna, J. K. 1990. *Appl. Phys. Lett.*, 57, 2885.

Ding, J., Jeon, H., Ishihara, T., Hagerott, M., Nurmikko, A. V., Luo, H., Samarth, N., and Furdyna, J. K. 1992. *Phys. Rev. Lett.*, 69, 1707.

Durbin, S. M. 1989. *Appl. Phys. Lett.*, 55, 2087.

Fan, Y., Han, J., He, L., Saraie, J., Gunshor, R. L., Hagerott, M., Jeon, H., Nurmikko, A. V., Hua, G. C., and Otsuka, N. 1992. *Appl. Phys. Lett.*, 61, 3160.

Furdyna, J. K. 1988. *J. Appl. Phys.*, 64, R29.

Gaines, J., Drenten, R., Haberern, K., Marshall, T., Mensz, P., and Petruzzelo, J. 1993. *Appl. Phys. Lett.*, 62, 2462.

Giebultowicz, T. M., Samarth, N., Luo, H., Furdyna, J. K., Klosowski, P., and Rhyne, J. J. 1992. *Phys. Rev. B.*, 46, 12076.

Guha, S., DePuydt, J. M., Haase, M. A., Qiu, J., and Cheng, H. 1993a. *Appl. Phys. Lett.*, 63, 3107.

Guha, S., Munekata, H., and Chang, L. L. 1993b. *J. Appl. Phys.*, 73, 2294.

Haase, M. A., Qiu, J., DePuydt, J. M., and Cheng, H. 1991. *Appl. Phys. Lett.*, 59, 1272.

Harsch, W. C., Cantwell, G., and Schetzina, J. F. 1994. In *Proceedings of the International Workshop on ZnSe-Based Blue–Green Laser Structures.*

Hua, G. C., Otsuka, N., Grillo, D. C., Fan, Y., Han, J., Ringle, M. D., Gunshor, R. L., Hovinen, M., and Nurmikko, A. V. 1994. *Appl. Phys. Lett.*, 65, 1331.

Jeon, H., Ding, J., Nurmikko, A. V., Luo, H., Samarth, N., Furdyna, J. K., Bonner, W. A., and Nahory, R. E. 1990. *Appl. Phys. Lett.*, 57, 2413.

Jonker, B. T., Krebs, J. J., Qadri, S. B., Prinz, G. A., Volkening, F. A., and Koon, N. C. 1988. *J. Appl. Phys.*, 63, 3303.

Karczewski, G., Hu, B., Yin, A., Luo, H., and Furdyna, J. K. 1994. *J. Appl. Phys.*, 75, 7382.

Kemner, K. M., Bunker, B. A., Luo, H., Samarth, N., Furdyna, J. K., Weidmann, M. R., and Newman, K. E. 1992. *Phys. Rev. B.*, 46, 7272.

Kobayashi, M., Mino, N., Katagiri, H., Kimura, R., and Konagai, M. 1986. *J. Appl. Phys.*, 60, 773.

Kolodziejski, L. A., Gunshor, R. L., Bonsett, T. C., Venkatasubramaniam, R., Datta, S., Bylsma, R. B., Becker, W. M., and Otsuka, N. 1985. *Appl. Phys. Lett.*, 47, 169.

Kolodziejski, L. A., Gunshot, R. L., Otsuka, N., Gu, B. P., Hefetz, Y., and Nurmikko, A. V. 1986. *Appl. Phys. Lett.*, 48, 1482.

Krol, A., Soo, Y. L., Huang, S., Ming, Z. H., and Kao, Y. H. 1993. *Phys. Rev. B.*, 47, 7187.

Lansari, Y., Ren, J., Sneed, B., Bowers, K. A., Cook, J. W. Jr., and Schetzina, J. F. 1992. *Appl. Phys. Lett.*, 61, 2554.

Lee, D., Zucker, J., Johnson, A. M., Feldman, R. D., and Austin, R. F. 1990. *Appl. Phys. Lett.*, 57, 1132.

Lew Yan Voon, L. C., Ram-Mohan, L. R., Luo, H., and Furdyna, J. K. 1993a. *Phys. Rev. B.*, 47, 6585.

Liu, X., Petrou, A., Warnock, J., Jonker, B. T., Prinz, G. A., and Krebs, J. J. 1989. *Phys. Rev. Lett.*, 63, 2280.

Luo, H., Samarth, N., Zhang, F. C., Pareek, A., Dobrowolska, M., Furdyna, J. K., and Ostsuka, N. 1991. *Appl. Phys. Lett.*, 58, 1783.

Luo, H. and Furdyna, J. K. 1992. *Bull. Am. Phys. Soc.*, 37, 659.

Luo, H., Chou, W. C., Samarth, N., Petrou, A., and Furdyna, J. K. 1993. *Solid State Commun.*, 85, 691.

Luo, H., Dai, N., Zhang, F. C., Samarth, N., Dobrowolska, M., Furdyna, J. K., Parks, C., and Ramdas, A. K. 1993b. *Phys. Rev. Lett.*, 70, 1307.

Meyer, J. R., Hoffman, C. A., Myers, T. H., and Giles, N. C. 1993. In *Handbook on Semiconductors, 2nd Ed.*, Vol. III, S. Mahajan, ed., p. 1849. Elsevier, Amsterdam.

Munekata, H. et al. 1989. *Phys. Rev. Lett.*, 63, 1849.

Myers, T. H., Meyer, J. R., and Hoffman, C. A. 1993. In *Quantum Wells and Superlattices for Long Wavelength Infrared Detectors*, M.O., Manasreh, ed., p.1697. Artech House, Boston.

Nakamura, S., Mukai, T., and Senoh, M. 1994. *Appl. Phys. Lett.*, 64, 1697.

Park, R. M., Troffer, M. T., Rouleau, C. M., DePuydt, J. M., and Haase, M. A. 1990. *Appl. Phys. Lett.*, 57, 2127.

Ohishi, M., Ohmori, K., Fujii, Y., and Saito, H. 1988. *J. Crystal Growth*, 86, 375.

Ohkawa, O., Mitsuyu, T., and Yamazaki, O. 1987. *J. Appl. Phys.*, 62, 3216.

Ohkawa, K., Karasawa, T., and Mitsuyu, T. 1991. *Jpn J. Appl. Phys.*, 30, L152.

Ohkawa, K., Ueno, A., and Mitsuyu, T. 1992. *J. Crystal Growth*, 117, 375.

Partovi, A., Glass, A. M., Olson, D. H., Feldman, R. D., Austin, R. F., Lee, D., Johnson, A. M., and Miller, D. A. B. 1991. *Appl. Phys. Lett.*, 58, 334.

Pelekanos, N. T., Ding, J., Hagerott, M., Nurmikko, A. V., Luo, H., Samarth, N., and Furdyna, J. K. 1992. *Phys. Rev. B.*, 45, 6037.

Samarth, N., Luo, H., Furdyna, J. K., Qadri, S. B., Lee, Y. R., Ramdas, A. K., and Otsuka, N. 1989. *Appl. Phys. Lett.*, 54, 2680.

Samarth, N., Luo, H., Furdyna, J. K., Qadri, S. B., Lee, Y. R., Ramdas, A. K., and Otsuka, N. 1990a. *J. Electron. Mater.*, 19, 543.

Samarth, N., Luo, H., Furdyna, J. K., Alonso, R. G., Lee, Y. R., Ramdas, A. K., Qadri, S. B., and Otsuka, N. 1990b. *Appl. Phys. Lett.*, 56, 1163.

Suemune, I., Yamada, K., Masato, H., Kan, Y., and Yamanishi, M. 1989. *Appl. Phys. Lett.*, 54, 981.

Sun, G., Shahzad, K., Khurgin, J., and Gaines, J. 1991. Conf. Lasers and Electrooptics, Baltimore.

Venkatesan, S., Pierret, R. F., Qiu, J., Kobayashi, M., Gunshor, R. L., and Kolodziejski, L. A. 1989. *J. Appl. Phys.*, 66, 3656.

von Molnar, S. et al. 1991. *J. Magn. Magn. Mater.*, 93, 356.

Wang, S. Y., Kanakami, Y., Simpson, J., Stewart, H., Prior, K. A., and Cavenett, B. C. 1993. *Appl. Phys. Lett.*, 62, 1715.

Woodall, J. M., Freeout, J. L., Pettit, G. D., Jackson, T., and Kirchner, P. 1981. *J. Vac. Sci. Technol.*, 19, 626.

Xie, W., Grub, D. C., Gunshor, R. L., Kobayashi, M., Jeon, H., Ding, J., Nurmikko, A. V., Hua, G. C., and Otsuka, N. 1992. *Appl. Phys. Lett.*, 60, 1999.

Yakovlev, D. R., Ossau, W., Landwehr, G., Bicknell-Tassius, R. N., Waag, A., and Uraltsev, I. N. 1990. *Solid State Commun.*, 76, 325.

Yang, G. L., Luo, H., Lowandowski, L., and Furdyna, J. K. 1995. *Phys. Stat. Sol. B.*, 187, 435.

Yu, E. T., Phillips, M. C., McCaldin, J. O., and McGill, T. C. 1991. *J. Vac. Sci. Technol. B.*, 9, 2233.

Zmudzinski, C. A., Guan, Y., and Zory, P. S. 1990. *IEEE Photonics Technol. Lett.*, 2, 94.

Zhang, F. C., Luo, H., Dai, N., Samarth, N., Dobrowolska, M., and Furdyna, J. K. 1993. *Phys. Rev. B.*, 47, 3806.

4

III-V Nitrides and Silicon Carbide as Optoelectronic Materials

Seydi Doğan

Ali Teke

Hadis Morkoç

4.1 Introduction

SiC has emerged as the leading candidate for high temperature and high power device applications due in part to the commercial availability of high quality SiC substrates of ever increasing diameter and quality. These accomplishments are due to advances in chemical vapor deposition (CVD) growth of epitaxial structures paving the way for researchers to easily dope both *n*- and *p*-type materials as well as obtaining semi-insulating behavior. The large Si–C bonding energy makes SiC resistant to chemical attack and radiation, and ensures its stability at high temperatures. In addition, SiC has a large bandgap, a large avalanche breakdown field, an excellent thermal conductivity, and a high electron saturation velocity. Due to its above properties, it may replace silicon in high-power, high-voltage switching applications, high-temperature electronics, high-power microwave applications, high-radiation environments, or in some UV optoelectronic devices. Metal semiconductor and metal-oxide-semiconductor transistors (MOSFETs) with outstanding high-temperature performance have already been demonstrated. SiC also forms high-quality native SiO_2 on the surface, which makes it suitable for devices e.g. MOSFETs. SiC substrates are suitable for nitride epitaxy due to their relatively close lattice match and high thermal conductivity. This substrate is also currently used as a template for a good fraction of the world production of green, blue, and ultraviolet light-emitting diodes based on nitride semiconductors. With the recent introduction of a controllable 4H polytype that exhibits large electron mobilities, SiC is certain to attract more attention for high-power electronics applications.

Industries such as aerospace, automobile, petroleum and others have continuously provided the impetus for pushing the development of fringe technologies that are tolerant of increasingly high temperatures and hostile environments. Beyond SiC, other materials such as the III-V nitrides are potentially capable of increased power and higher-temperature operation due to their even larger predicted electron velocities, bandgaps, and lower ohmic contact resistances. Commercialization of bright blue and green LEDs, together with red LEDs made possible by conventional III-V technologies, has paved the way for the development of full-color displays, emitters, low-level lighting, background illuminations, and possibly general lighting. If the three primary-color LEDs, including red, produced by the InGaAlAs system are used in place of incandescent light bulbs in some form of a color-mixing scheme, they would provide not only compactness and longer lifetimes, but also less power consumption for the same luminous flux output. Additional possible applications include use in agriculture as light sources for accelerated photosynthesis, and in health care for diagnosis and treatment. Unlike display and lighting applications, digital information storage and reading require coherent light sources because the diffraction-limited optical storage density increases approximately quadratically with decreasing wavelength. The nitride material system, when adapted to semiconductor lasers in blue and UV wavelengths, offers increased data storage density, possibly as high as 50 Gb per disc with 25 Gb being eminent in the Blu-Ray system. Other equally attractive applications envisioned include printing and surgery.

When used as UV sensors in jet engines, automobiles, and furnaces (boilers), these devices would allow optimal fuel efficiency and control of effluents for a cleaner environment. Moreover, visible-blind and solar-blind nitride-based photodetectors are also ideal candidates for a number of applications, including early missile plume detection, UV astronomy, space-to-space communication, and biological research.

Another area gaining a lot of attention for III-V nitrides is high temperature/high power electronic applications, such as radar, missiles, and satellites, as well as low-cost compact amplifiers for wireless base stations, due to their excellent electron transport properties, including good mobility and high saturation drift velocity. The strongest feature of the III-nitrides compared to other wide-bandgap counterparts is the heterostructure technology that they can support. Quantum wells, modulation-doped heterointerfaces, and heterojunction structures can all be made in this system, giving access to new spectral regions for the optical devices and new operational regimes for electronic devices. Other attractive properties of the nitrides include high mechanical and thermal stability, and large piezoelectric constants.

A suitable semiconductor technology would allow bulky and hazardous aircraft hydraulics to be replaced with actuators controlled with heat-tolerant electronics. In satellites, heat radiators could be reduced in size, resulting in considerable weight reduction. Other bulky power electronics components in power transmitters could also be replaced with more reliable and compact solid-state amplifiers. Both SiC and GaN are characterized not only by large bandgaps, but also by small lattice constants. Figure 4.1 shows the lattice constants and bandgaps of various polytypes of SiC and both hexagonal wurtzite (α) and cubic (β) phases of the nitrides, along with traditional elemental and compound semiconductors.

In this review, we concentrate on optoelectronic device oriented developments and applications of SiC and GaN. The first section covers SiC and describes recent progress in SiC substrates, epitaxy, processing,

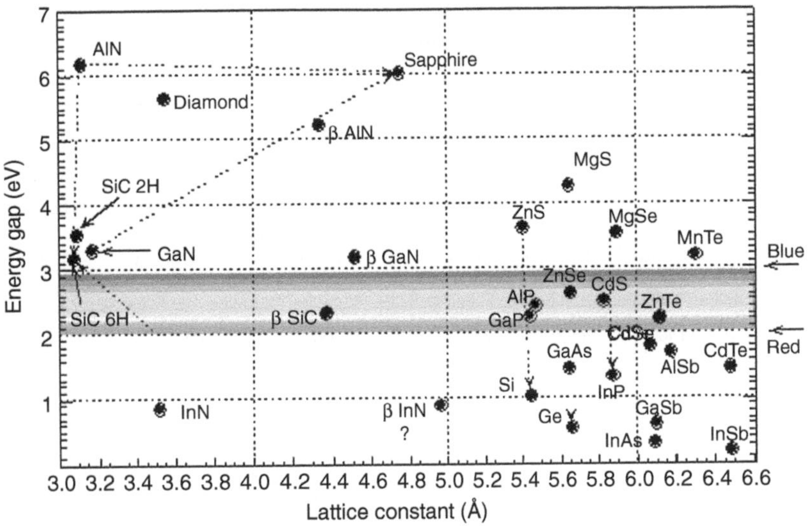

FIGURE 4.1 Zincblende and wurtzite SiC and GaN lattice constants vs. the energy gap.

corresponding devices, and relevant advances. This then segues into an examination of GaN and its alloys with AlN and InN. Among the topics discussed are advances in epitaxial techniques, doping, and device technology.

4.2 Silicon Carbide

SiC, from the IV–IV semiconductor family, was one of the first semiconductors discovered; its large cohesive energy caused some to mistake it for an element. Large Si-C bonding energy with short bond lengths leads to a large energy difference between bonding and antibonding states, resulting in wide bandgaps. Low leakage current in SiC, many orders of magnitude less than that of Si, reduced minority carrier generation rate (emanating from large bandgaps), high-breakdown electric field, good thermal conductivity, and high thermal stability are some of the advantages of SiC. These advantages make SiC highly effective for high-temperature and high-voltage applications (Weitzel et al. 1997; Rottner et al. 1999; Carter et al. 1999; Elasser et al. 2003). An appreciation of the potential of SiC for electronic applications can be gained by examining Table 4.1, which compares relevant material properties of SiC (3C, 4H, and 6H polytypes) and GaN with Si and GaAs, the two most popular semiconductor device technologies, and GaP and diamond, two other contenders for high-temperature applications.

SiC occurs in many different crystal structures called polytypes, (Nishino et al. 1980) owing to different arrangements of Si and C atoms within the SiC crystal lattice (individual bond length and local atomic environment by the stacking order) in the *c*-direction. More than 200 different polytypes of SiC have been reported. In spite of the fact that all SiC polytypes chemically consist of 50% carbon atoms bonded with 50% silicon atoms, each SiC polytype itself exhibits unique fundamental electrical and optical properties, and can be situated in one of three possible positions with respect to the lattice in the *c*-direction while maintaining the tetrahedral bonding scheme of the crystal. These positions each arbitrarily assigned the notation A, B, or C (Morkoç et al. 1994).

Depending on the stacking order, the bonding between Si and C atoms in adjacent bilayer planes is either of a zincblende (cubic) or wurtzite (hexagonal) nature. Zincblende bonds are rotated 60° with respect to nearest neighbors whereas hexagonal bonds are mirror images. Each type of bond provides a slightly altered atomic environment that makes some lattice sites inequivalent in polytypes with mixed bonding schemes and reduces the overall crystal symmetry. These effects are important when considering the substitutional impurity incorporation and electronic transport properties of SiC.

TABLE 4.1 Important Parameters of Semiconductors of Interest for Conventional Electronics, Emerging High-Temperature, High-Power, and High-Frequency Electronics

Property	Si	GaAs	3C–SiC	4H–SiC	6H–SiC	Diamond	GaN
Bandgap (eV) at 300 K	1.1	1.4	2.3	3.2	3.0	5.5	3.4
Relative dielectric constant	11.9	13.1	9.7	9.7	9.7	5.5	9
Melting point (K)	1690	1510	~3100 at 35 atm	~3100 at 35 atm	~3100 at 35 atm	change	2500
Physical stability	Good	Fair	Good	Excellent	Excellent	Very good	Good
Thermal conductivity (W/cm K)	1.5	0.5	3–5	3–5	3–5	20	1.3
Electron mobility at RT (cm^2/V s)	1400	8500	800	900	400	2200	900
Hole mobility at RT (cm^2/V s)	600	400	320	120	90	1600	250–350
Saturated electron velocity ($\times 10^7$ cm/s)	1.0	2	2.5	2	2	2.7	2.7
Breakdown Field ($\times 10^6$ V/cm)	0.3	0.4	1.0	3–5	3–5	10	5

If the stacking sequence is ABCABC…, where A, B, and C represent the three possible positions of the Si-C double layers, a purely cubic zincblende structure is realized, which is commonly abbreviated as 3C SiC (also referred to as β-SiC). The number 3 refers to the three-bilayer periodicity of the stacking and the letter C denotes the overall cubic symmetry of the crystal. 3C SiC is the only cubic polytype. A purely wurtzite ABAB… stacking sequence has also been observed, and is abbreviated as 2H SiC, reflecting its two-bilayer stacking periodicity and hexagonal symmetry. All of the other polytypes are mixtures of the fundamental zincblende and wurtzite bonds. Some common hexagonal polytypes with more complex stacking sequences are 4H and 6H SiC. The schematic models of the four most common SiC polytypes are shown in Figure 4.2. The bilayer sequence ABCB…. for 4H SiC is composed equally of cubic and hexagonal bonds whereas the 6H SiC (ABCACB…) bilayer sequence is two-thirds cubic, despite each having overall hexagonal crystal symmetry. The family of hexagonal polytypes is collectively referred to as α-SiC. Rhombehedral structures such as 15R and 21R have also been documented (Jagodzinski and Arnold 1960).

Different polytypes of SiC have widely ranging physical properties. 3C SiC exhibits a saturation velocity resulting from the increased symmetry of the cubic crystal that reduces phonon scattering. Figure 4.3 shows the change of the exciton energy gap with a variation in hexagonality. The bandgap is dependent upon the polytype or stacking sequence. The bandgaps differ widely among the polytypes, ranging from 2.3 eV for 3C SiC (0% hexagonality) to 3.0 eV in 6H SiC to 3.2 eV for 4H SiC (50% hexagonality), and to 3.3 eV for 2H SiC (Righi et al. 2002). The 2H polytype has the full hexagonal structure with AB stacking sequence. Among the SiC polytypes, 6H is most easily prepared and has been the most extensively studied, while the 4H polytype is steadily attracting more attention. About 95% of all the research deals with three main polytypes: 3C, 4H, and 6H. The polytypism of SiC makes it nontrivial to grow as a single-phase material, but it also offers some potential advantages if crystal growth methods can be sufficiently developed to capitalize on the possibility of polytype (homo/hetero) junctions. Figure 4.4 shows one such possibility—a 6H/3C SiC interface. Such a junction incorporates the advantages of heterojunction band offsets while maintaining a completely charge-free, lattice-matched, and coherent interface.

Powell et al. (1990a, 1990b) have shown that 3C SiC can be grown on well-oriented basal plane (0001) 6H substrates. In this case, the heteroepitaxy is terrace controlled, and the adatoms are free to choose the most energetically favorable stacking sequence in the direction perpendicular to the substrate surface. Proper control of the growth conditions then gives rise to the heteroepitaxy of 3C SiC. When the

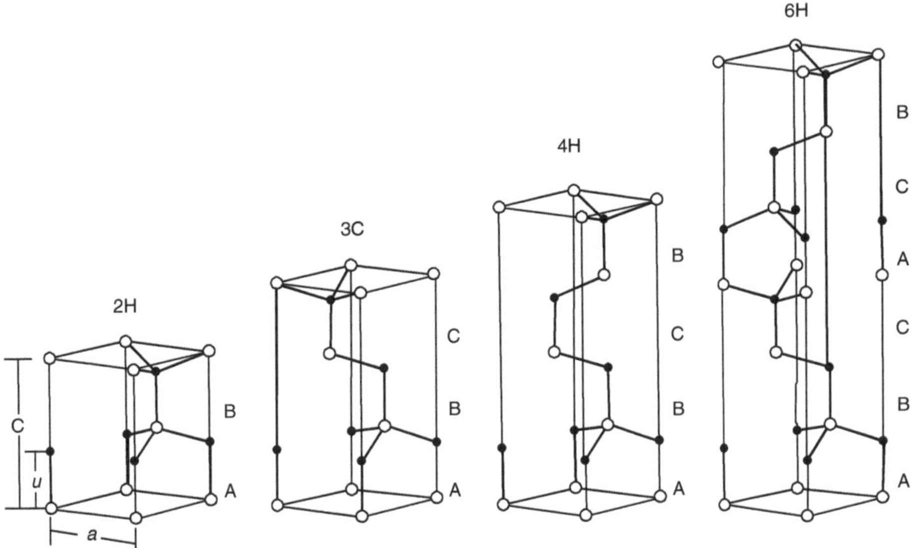

FIGURE 4.2 Hexagonal unit cells of the 2H, 3C, 4H, and 6H polytypes of SiC. The stacking sequence AB (2H), ABC (3C), ABCB (4H), ABCACB (6H), and the lattice parameters *a*, *c*, and *u* are indicated.

substrate is miscut several degrees away from the basal plane towards $(11\bar{2}0)$, steps are prevalent on the growth surface. If the step spacing is less than the surface diffusion length of the adatoms, the resultant growth is step controlled, allowing the underlying 6H material to influence the epitaxy, and homoepitaxial 6H SiC is grown.

4.2.1 SiC Substrate Crystal Growth

One major advantage that SiC enjoys over the other wide-bandgap competitors is a well established commercialized process for the growth of the high-quality substrate materials. 6H SiC substrates have been prepared in the laboratory by three separate processes (Levin et al. 1978; Ikeda et al. 1979;

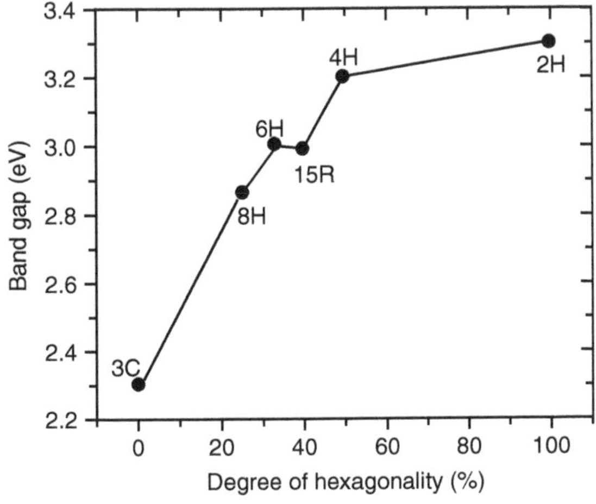

FIGURE 4.3 Exciton energy gap variation of most common SiC polytypes with hexagonality.

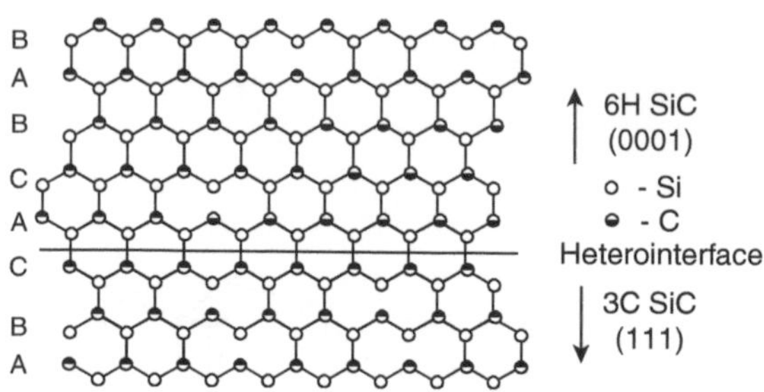

FIGURE 4.4 Polytype (homo/hetero) junctions using a 6H/3C SiC interface.

Tairov and Tsverkov 1981), namely thermal decomposition, growth from a carbon-enriched Si melt, and sublimation. The sublimation growth technique has been used, after years of further development, for the growth of commercial substrates. One-inch diameter 6H SiC wafers were introduced commercially in 1989 by Cree Research, Inc. *n*-type, *p*-type and semi-insulating 6H and 4H SiC wafers with a 3-inch diameter are commercially available at the time of this writing. Although efforts to increase the substrate area are underway, these substrates have not yet reached the marketplace. Substrates are available with both *n*- and *p*-type conductivity over a wide range, as well as high resistivity. High resistivity can be obtained by intrinsic compensation by introducing a metal that produces a deep level for compensation.

In sublimation growth, which paved the way for the large area SiC substrates (Takahashi and Ohtani 1997; Tairov and Tsverkov 1981), SiC is transported in the vapor phase to a SiC seed crystal that is held at a lower temperature compared with a source temperature of 2100°C–2400°C. A thermal gradient of 20°C–35°C cm^{-1} across the growing crystal results in a growth rate of few mm/h. For larger growth rates, higher source and seed crystal temperatures (2300 and 2200°C) in an ambient pressure of roughly less than 20 torr has been successful. In the latter approach, the sublimated SiC clusters must be diffused though porous graphite under carefully controlled thermal and pressure gradients to form high-quality, single-crystal 6H SiC. Reduced defect densities were noted when the source was situated below the seed crystal.

One serious problem faced by crystal growers is the formation of micropipes. Several mechanisms or combinations of mechanisms causing micropipes in SiC have been proposed (Powell and Rowland 2002). Micropipes are voids that run through the wafer and subsequently grown epilayers and are fatal, particularly for the large area devices. This phenomenon is associated with the open core of a micropipe filling with metal contact and shorting the device. Neudeck and Powell (1994) and Powell et al. (1994) have reported *p-n* junction structures with a nominal breakdown voltage of 1000 V. On the contrary, the devices with a junction intersected by a micropipe exhibited a sudden increase in reverse the current at the voltages that are well below the nominal breakdown field. A similar correlation was observed by Koga et al. (1992) in *p-n* junctions containing a micropipe. The diameter of the micropipes can vary, depending on the growth condition and formation mechanism, but are typically 0.5–10 μm in diameter. They are observed to nucleate at the seed crystal due to some local disturbance in the growth process and always have dark hexagonal etch pits or cross-sections. Researchers have had some success eliminating micropipes by preparing seed crystals in which epitaxial layers are grown on the A face prior to sublimation growth of the bulk crystal. Recently, micropipe densities have been steadily reduced by over an order of magnitude. Analysis of a 4-inch diameter, defect-etched 4H SiC wafers has revealed areas up to 25 mm in diameter that are entirely micropipe free, and a total of 18 micropipes have been counted, corresponding to 0.9 cm^{-2}.

Because SiC has an extremely high hardness value and is a very brittle material, the mechanical treatment for SiC such as slicing, lapping and polishing is quite difficult and produces plenty of subsurface damage, defects, scratches, etc. Traditional semiconductor tools such as diamond abrasive are used for slicing the substrates. In one process, the diced substrates underwent mechanical polishing involving a sequence of SiC and diamond pastes (Matsunami, private communication). However, polishing solely by a paste process inherently leaves a residual subsurface damage layer. This damage can be removed by treatment of the surface by reactive ion etching (Palmour, Kong, and Edmond 1990) and chemical mechanical polishing (Zhou et al. 1997). The wafers are then typically characterized by x-ray topography, optical, SEM, and AFM observations. However, H-polishing at temperatures greater than 1650°C results in atomically smooth and damage-free surfaces.

4.2.2 SiC Thin-Film Epitaxy

Although SiC growth technology has made tremendous progress and SiC devices show better performance compared to those fabricated using Si and GaAs in terms of power and voltage, there are still some fundamental and practical problems in the growth technology that prevent one from obtaining outstanding SiC devices. SiC has been successfully grown on many substrates, including sapphire, Si, AlN, TiC, as well as SiC (Khan and Learn 1969; Kong, Glass, and Davis 1989; Pazik, Kelner, and Bottka 1991; Chien et al. 1994; Dmitriev et al. 1994a; Nishino et al. 1994). Only SiC has been utilized as a substrate for the high-performance devices among these substrates.

There are many techniques for forming SiC epitaxial layers. Early SiC epitaxy was achieved using liquid-phase epitaxy (LPE), which has been used for many years (Kimoto et al. 1993; Rendakova, Ivantosov, and Dimitriev 1998). LPE can be performed at lower temperatures (1500°C–1700°C) compared to sublimation epitaxy, which uses a solid SiC as a source material. The early efforts suffered from contamination of C-saturated molten Si by conventional graphite crucibles (Dmitriev et al. 1985; Muench and Kurzinger 1978; Ikeda et al. 1979). This problem was circumvented by introducing a graphite-free technique in which the Si melt was suspended in an electromagnetic field. Recently, LPE has become of interest because of the ability to reduce the micropipe density in modified Lely substrates and capacity to achieve a wide range of doping levels—between 10^{16} and 10^{20} cm^{-3} for both *n*- and *p*-type SiC.

Chemical vapor deposition (CVD) has replaced LPE and sublimation epitaxy as the growth method of choice. Most of the epitaxial growth of the SiC has been carried out by CVD and suitable layers have been achieved for device application (Choyke, Matsunami, and Pensl 1997). Both low and atmospheric-pressure CVD have been successfully applied to SiC epitaxy. Low pressure is preferred when deposition over large-area substrates, such as Si, is attempted. Recently, high-temperature CVD has been developed for epitaxial growth of SiC with higher growth rates, which is desirable for high-power devices, up to hundreds of microns per hour (Kordina et al. 1996; Ellison et al. 1999; Tsushida et al. 2000). Several groups (Fuyuki, Yoshinobu, and Matsunami 1993; Kern et al. 1997a; Kern and Davis 1997b) have also used gas-source molecular-beam epitaxy (GSMBE), using disilane (Si_2H_6) and ethylene (C_2H_4), for SiC homoepitaxial growth, especially for exploring Si and SiC technologies together. In this case, GSMBE allows lower growth temperatures that restrict dopant diffusion in Si. Vapor-phase epitaxy (VPE) is another growth technique for SiC epitaxy and is accomplished by supplying Si- and C-containing gases to a heated SiC single-crystal wafer. In this growth technique, high temperature is required to reduce the high density of defects such as double positioning boundaries (DPBs) (Kong et al. 1988).

Many workers have attempted to grow SiC heteroepitaxially on Si, thereby taking advantage of inexpensive, larger area of these substrates, available in reproducibly high quality, and the potential advantages for Si–SiC device integration (Kitabakate 2000; Chassagne et al. 2002). These advantages are offset by the large thermal and lattice mismatches between the two materials, which have made high-quality materials extremely difficult to achieve. Nevertheless, work continues on improving the quality of SiC/Si, Nishino, Powell, and Will (1983) developed a technique in which the clean Si surface is exposed to a C-containing gas at or above the growth temperature. This recipe has been followed in almost all reported cases and results in the formation of a thin monocrystalline 3C SiC layer that serves as a template for epitaxial growth. The growth of SiC is then performed on the carbonized layer by introducing Si- and C-containing gases.

In addition to threading dislocations, two types of defects dominate SiC/Si epilayers. Planar defects such as microtwins and stacking faults are commonly observed. (Nutt et al. 1987; Lendenmann et al. 2001). This is not surprising given the polytypism of SiC. A stacking fault is simply a temporary disruption of the bulk stacking sequence. During growth on the zincblende (100) surface, a coherent interface such as that shown in Figure 4.4 is no longer possible because a bilayer plane must be terminated. Therefore, in SiC grown on (100) Si, the stacking faults are accompanied by dislocations, and together they relieve the lattice and thermal mismatch strain. In the hexagonal 4H SiC and 6H SiC polytypes, all of the stacking faults act as a quantum well, leading to the localization of electrons. This then results in a forward voltage drop in SiC bipolar devices by interfering with the electron transport along the *c*-axis (Sridhara et al. 2001). When the SiC substrates are stepped and terminated by both types of layers (i.e., normal and twinned), the 3C SiC epilayers are a mixture of both variants. The boundaries between them are referred to as *double positioning* (or equivalently, *inversion* or *antiphase domain*) boundaries (DPB) (Shibahara, Nishino, and Matsunami 1986; Kong et al. 1988). This defect is another prevalent defect and arises from the reduced symmetry of SiC compared to the Si substrate. These defects, with typical densities of about $100\ \text{cm}^{-1}$, make heteroepitaxial structures unsuitable for device application (Skowronski 2003). The best approach to eliminate this defect is to grow homoepitaxy by step-controlled epitaxy growth.

4.2.3 Dopant Considerations

One of the important points for SiC is to achieve *p*- type, *n*-type, and semi-insulating material in situ, both during growth and by ion implantation. Here, particulars dealing with *n*- and *p*-type doping are discussed.

4.2.3.1 *n*-Type Doping

Nitrogen (N) and phosphorous (P), as well as arsenic (As) and antimony (Sb), are the most popular *n*-type dopants and act as shallow donors in SiC (Pensl et al. 1993; Troffer et al. 1996a). N has become the most popular *n*-type dopant because it has a small ionization energy and low atomic mass, and N is the lightest donor of the other possible *n*-type dopants. N-doping and its incorporation mechanisms during epitaxial growth by CVD have been studied extensively by several groups (Rowland, Burk, and Brandt 1998; Zhang et al. 2001; Wang, Bhat, and Chow 2002). For CVD, ammonia has proven to be a suitable dopant source gas for *n*-type doping (Kong et al. 1988). N-doping in SiC is unintentionally introduced during bulk crystal growth and to some extent in epitaxial growth as well. When N-doping is introduced during the growth, carrier concentrations as high as $10^{19}\ \text{cm}^{-3}$ have been realized, despite a solubility limit of $10^{20}\ \text{cm}^{-3}$. It has been observed that if the concentration of N is above $10^{19}\ \text{cm}^{-3}$, electrically inactive dopants are formed, and efforts to further increase the N incorporation resulted in a polycrystalline material. Ion implantation with subsequent Ar annealing has yielded electron concentrations as large as $3\times10^{19}\ \text{cm}^{-3}$ at a N dose of $5\times10^{20}\ \text{cm}^{-3}$ (Cooper and Melloch, 2002). An excellent review of the optical and electrical properties of doped SiC has been published by Pensl and Choyke (1993). The fact that dopants can occupy either hexagonal or cubic sites in the more complex SiC polytypes was pointed out by the authors. It is generally accepted that N occupies a carbon (C) site (Woodbury and Ludwig 1961). Analyzing several samples, Pensl and Choyke (1993) showed that the relative abundance of the various N dopant levels corresponds to the ratio of available binding sites. That is, in 4H SiC, an equal number of donors occupy cubic sites and hexagonal sites. In 6H SiC, the ratio is 2:1, reflecting the fact that two-thirds of the Γ sites possess cubic bonding. In 6H SiC, the measured ionization energy of the hexagonal site was 85.5 meV whereas the energy of the cubic site was 125 meV. For 4H SiC, the hexagonal (h) and cubic (k) binding energies were measured to be 45 meV and 100 meV, respectively. In 3C SiC, a value of 48 meV was determined.

Although there is great interest in investigating N doping, little has been reported on in situ doping using P. In addition, the location of P in the SiC lattice as an isolated impurity, how it participates in complexes, and the corresponding electronic levels are still poorly understood (Devaty and Choyke 1997). P donors can be easily introduced into the SiC lattice after crystal growth because of its diffusion properties (Mokhov et al. 1992). Based on its atomic size, P is expected to substitute preferentially on the Si sites in SiC (Weber et al. 1995; Larkin 1997). Hall measurements (Troffer et al. 1996a; Capano et al. 2000; Wang, Bhat, and Chow 2002) on phosphorous-implanted 6H SiC and 4H SiC epilayers have yielded two ionization energies, ~0.080 and ~0.053 eV for hexagonal sites, and ~0.110 and ~0.093 eV for cubic sites, respectively.

4.2.3.2 *p*-Type Doping

The most common *p*-type dopants for SiC are aluminum (Al), boron (B) and gallium (Ga). Al, however, is most frequently used because of its smaller ionization energy and higher solubility. Al produces acceptor levels at 0.1–0.27 and 0.16–0.23 eV above the valance band maximum for 6H SiC (Raynaud et al. 1995; Schöner et al. 1996) and 4H SiC (Kusnetsov and Zubrilov 1995), respectively. Triethylaluminum (TEA) has been used as a source gas for *p*-type doping (Kong et al. 1988). The ionization energy is 0.30–0.40 eV for 6H SiC (Veinger et al. 1980) and 0.28–0.39 eV for 4H SiC (Jenny et al. 1996a; Troffer et al. 1996b), which is relatively deeper in comparison to Al. Al and B have been found to occupy Si sites in SiC (Petrenko, Teslenko, and Mokhov 1992; Müller et al. 1993; Larkin et al. 1994).

All of the acceptor impurities thus far investigated, namely Al, B, Ga, and Sc, form deep levels in comparison with *n*-type dopants and are difficult to activate. In addition, post implantation requires a high-temperature anneal. The implanted atoms generally occupy interstitial lattice sites where they are usually electrically inactive. The depth of the acceptor levels also leads to strongly varying hole concentrations with temperature, which complicates device design and operation. Al is somewhat difficult to incorporate into the SiC lattice and high carrier concentrations are relatively difficult to achieve. Despite these difficulties, hole concentrations in the range of 10^{19}–10^{20} cm^{-3} have been achieved using TMA in a CVD process on the Si face of 6H SiC. In contrast, growth on the C face permitted only 2×10^{18} cm^{-3} *p*-type doping. The carrier concentration could be easily controlled down to the low 10^{16} cm^{-3} range. On the upper end, particularly above 10^{19} cm^{-3}, the observed hole concentration became a sublinear function of TMA flow. Low-resistivity ($< 0.1\ \Omega$ cm), *p*-type 6H SiC layers of about 500-nm width and targeted acceptor concentrations of 1.5×10^{20} and 5.0×10^{20} cm^{-3} have been produced by the combination of high-dose multienergy Al^+ ion, ion-beam-induced crystallization and subsequent furnace annealing by Heera et al. (2002). The hole concentration at 50°C increased by more than an order of magnitude from 9.0×10^{17} to 1.6×10^{19} cm^{-3} in the case of 1.5×10^{20} cm^{-3} targeted Al concentration, and from 6.1×10^{18} to 8.0×10^{19} cm^{-3} for a targeted Al concentration of 5.0×10^{20} cm^{-3}, as determined by Hall measurements. It has been indicated that the critical concentration for the formation of an impurity band can be achieved if the loss of active Al acceptors by precipitation is reduced in the nanocrystalline layers.

4.2.3.3 Semi-Insulating (SI) Behavior

SI–SiC is required for a variety of applications, including microwave field-effect transistors and other devices for high-power and high-temperature applications to exhibit low dielectric losses and reduce device parasitic losses. In addition, SI–SiC is becoming the dominant substrate material for high-power, GaN-based, modulation-doped, field-effect transistors (MODFETs). One of the methods used to achieve a SI–SiC material is intentional vanadium (V) doping, which introduces a deep-donor or deep-acceptor state and compensates the residual impurities (Hobgood et al. 1995; Jenny et al. 1996b; Bickermann et al. 2001). SI–SiC can also be achieved by intrinsic compensation, i.e., without external doping, but details are not yet available even though SI–SiC prepared by this method is commercially available. The V atoms produce levels in the energy gap at 1.35–1.47 eV and 0.68–0.97 eV below the conduction-band minimum for 6H and 4H SiC, respectively (Dalibor et al. 1997; Lebedev 1999). After Schneider et al. (1990) found deep levels in SiC caused by V impurities, SI behavior was investigated extensively and exploited in SiC (Hobgood et al. 1995; Mitchel et al. 1998). The important point to obtain a high-quality SI–SiC crystal is to control the amount of V such that it should be just enough to compensate the residual dopant. If it is less than a critical value, all the residual dopants are not compensated and the material remains *p*-type or *n*-type. If the solid solubility of V is exceeded, inclusions are formed through segregation of V, which lead to degradation of the crystal quality. The precipitates so-formed are vanadium carbide or vanadium silicide.

4.2.4 Ohmic Contacts to SiC

The properties of the ohmic contact, as well as the Schottky contact, can strongly depend on the surface cleaning procedure used prior to forming the contact. In devices designed for high-temperature and high-power applications, ohmic contacts with low contact resistivity are required because many vital parameters of semiconductor devices, including speed and high-power performance, depend strongly on the ohmic contact resistance (Murakami 1990; Morkoç et al. 1994). As the demand of the high-temperature devices has recently increased for use in a variety of applications, thermally stable and low resistance ohmic contacts are required. This is the one of the key issues and is especially important in wide-bandgap semiconductor systems because of the large Schottky barrier heights involved. This problem can often be alleviated in other semiconductors by choice of a metal with an optimal barrier height. Annealing at temperatures as high as 1000°C is required to yield good, low resistance, ohmic contacts with SiC when conventional contact materials are used. In this vein, TiN, as well as the other Ti-based compounds such as TiC and TiW (Lee, Zetterling, and Östling 2000; Lee et al. 2002), have been proposed as the best candidate materials for ohmic contact on SiC because of high-temperature stability, nonreactivity up to 1100°C, and low resistivity (Glass et al. 1992; Pecz 2001). Ni has been the most widely used metal for ohmic contacts to *n*-type SiC. This process generally requires annealing at high temperatures. Good Ni or Ni-based ohmic contacts (Crofton et al. 1995; Uemoto 1995; Oskam, Searson and Cole 2000) have been reported following various thermal processes that typically center around annealing at 1050°C for up to 10 min. The resultant contacts with specific resistivities on both implanted *n*- and *p*-type materials are as low as 6×10^{-6} $\Omega\,cm^2$ and 1×10^{-4} $\Omega\,cm^2$, respectively (Fursin, Zhoa, and Weiner 2001). Several contact metallization schemes based on Ni, Ti, TiW, Mo, Cr, and W have been reported in the literature as good ohmic contacts to *n*-type SiC (Porter and Davis 1995). Palmour et al. (1991) fabricated ohmic contacts using Ni and subsequent high-temperature annealing. In addition to α-SiC, annealed Ni has also been commonly used for ohmic contact to 3C SiC (β-SiC) (Edmond et al. 1988; Cho et al. 1993; Wan, Capano, and Melloch 2002). Ni ohmic contacts have been obtained after annealing the contacts at 1250°C for 5 min by Edmond et al. (1988). However, Ni ohmic contacts on *p*-type 3C SiC have been reported by Cho et al. (1993) in the as-deposited case. Contacts with good ohmic behavior on 3C SiC using Ti, W, Ni, and TiC (Parsons, Kruaval, and Chaddha 1994; Shor et al. 1994; Jacob et al. 1998; Touati, Takemasa, and Saji 1999) have also been reported. Ohmic contact formation with the aid of Ni/C films on *n*-type 4H SiC has been investigated by Lu et al. (2003) with various thickness of the carbon film, annealing time, and annealing temperatures. The carbon interfacial layer between the Ni film and SiC improved ohmic contact properties and low specific contact resistivities in the range of 10^{-6}–10^{-7} $\Omega\,cm^2$ were achieved on SiC with an acceptor dopant concentration of 3.1×10^{19} cm^{-3}.

Ohmic contacts to the *p*-type materials are more difficult because of the large work function of *p*-type SiC that results from its large bandgap, in spite of the fact that some extensive studies have been carried out (Dmitriev et al. 1994b; Crofton, Porter, and Williams 1997a, Crofton et al. 1997b; Kassamakova et al. 1999; Moscatelli et al. 2003). Ohmic contacts are typically formed by annealed aluminum (Crofton, Porter, and Williams 1997a, Crofton et al 1997b) and this imposes a limitation on the operating temperature of SiC devices with Al-based *p*-type contacts because of the low melting point and oxidation characteristic of Al. The melting point can be increased by using Al alloys such as TiAl. These alloys have been used for many years and the lowest TiAl specific contact resistance, around $1.5\times10^{-4}-1.0\times10^{-6}$ $\Omega\,cm^2$, has been reported in the literature (Crofton et al. 1993; Crofton et al. 1997b; Lin et al. 2000; Crofton et al. 2002; Nakatsuka et al. 2002a) after annealing typically at 1000°C. The effect of the Al concentration, which strongly affects the required annealing temperature, and layer structure on the electrical and microstructural properties of the TiAl ohmic contacts has been investigated by Nakatsuka et al. (2002a). It was found that Al fraction below 75% showed nonohmic behavior and the electrical property of the TiAl contact was quite sensitive to the Al concentration. Crofton et al. (2002) have attempted to determine the best AlTi alloy for use as ohmic contact to *p*-type 6H SiC by exploring four different AlTi compositions. The 70/30 wt% Al–Ti alloy composition produced good results to *p*-type 6H SiC (Johnson and Capano 2003). Lee et al. (2002) have demonstrated low-resistivity Ti-based ohmic contacts (TiC and TiW) to both *n*-and *p*-type 4H SiC for high-power and high-temperature device applications. To reduce the annealing temperature, a transition metal, such as Ni and Co, was proposed as a contact material. Co/Al (Nakatsuka, Koide, and Murakami 2002b) and Ni/Al and Ni/Ti/Al (Konishi et al. 2003) ohmic contacts to *p*-type 4H SiC have been investigated; it was found that these contacts showed ohmic behavior after annealing at 800°C.

4.2.5 Schottky Contacts to SiC

Metal/semiconductor (MS) contact is a major concern in SiC-based devices, and it is imperative to control and understand MS interfaces to obtain reproducible devices. Due to the high breakdown electric field and wide bandgap of the SiC, MS devices on SiC have the potential to replace high-power Si-based devices (Saxena, Su, and Stekl 1999). The barrier height of the MS contact determines the minimum on-state voltage drop in a switching application and in this respect a small barrier height is desired. On the other hand, the reverse current drops exponentially with the barrier height. In this respect, a large barrier height is desired for the device to hold large reverse bias voltages without extensive current flow, which is detrimental. There are many reports on how to lower Schottky barrier heights and how to reduce reverse leakage current in SiC Schottky diode (Schoen et al. 1998; Khemka, Ananthan, and Chow 2000). It has been shown that the Fermi level in SiC Schottky contacts is not pinned (Itoh, Kimoto, and Matsunami 1995a; Itoh and Matsunami 1997; Lee, Zetterling, and Östling 2001a). Consequently, the Schottky barrier height is metal-work-function dependent. The study of the interface properties of metal/SiC contacts has become a popular field of investigation, and some authors have already investigated the properties of the metal/SiC interface to obtain reproducible and stable Schottky devices (Raaghunathan, Alok, and Baliga 1995; Itoh, Kimoto, and Matsunami 1995b; Bozack 1997; Defives et al. 1999). In spite of the fact that the most common metals for Schottky contact to SiC are Ni, Ti, Al, and Au, a number of other metals have been used for electrical contacts in SiC-based devices (Waldrop et al. 1992; Porter and Davis 1995; Saxena and Steckl 1998).

The Schottky barrier height of a number of metals on *n*-type and *p*-type 6H SiC has been measured for different temperature range. It has been observed that the barrier height for *n*-type material did not exhibit temperature dependence. In contrast, the barrier height for *p*-type material showed temperature dependence similar to that of the energy bandgap (Aboelfotoh, Fröjdh, and Petersson et al. 2003). It should be mentioned that the preparation of high-quality Schottky diodes depends on the properties of the substrate. Despite tremendous progress, it is very difficult to obtain SiC with sufficiently low levels of defects. The best substrates have a density of micropipes of approximately 1 cm^{-2}; lowering of this parameter creates many complications (Carter et al. 1999).

4.3 SiC Power Devices

Si-based power devices have been used for some time in power electronics and power system applications (Baliga 1996; Ghandi 1998). However, due to increasing need for size, complexity, and functionality, Si-based devices are not able to keep up with the requirements for high-power, high-voltage, and high-temperature applications. However, SiC and GaN have attracted much attention because of their superior properties (see Table 4.1). The former is well suited for high-voltage switching applications whereas the latter is suited for high-power/high-frequency amplifiers. The SiC technology can be easily developed borrowing concepts developed for the well-established Si technology in terms of Schottky diodes, *p-i-n* diodes, thyristors (gate turn-off thyristors (GTOs) and gate controlled thyristors (GCTs)), and transistors (insulated-gate bipolar transistors (IGBTs), bipolar-junction transistors (BJTs), and metal-oxide semiconductor field-effect transistors (MOSFETs)). Generally, high-voltage power rectifiers are categorized into two classes: the unipolar Schottky rectifier and the bipolar *p-i-n* rectifier. One of these rectifiers is preferred depends on many device parameters, such as reverse blocking voltage, forward current density, maximum allowable reverse current density, operating temperature, and switching frequency. For example, Schottky barrier diodes (SBDs) are not yet sufficient for high-voltage applications because of their high reverse leakage current and low breakdown voltage. On the other hand, *p-i-n* diodes do better in these respects, but the switching speed is limited by the long reverse recovery time and by the large amount of charge stored in the drift region during forward conduction. Hence, an approach combining a SBD and a *p-i-n* diode has been realized by Alexandrov et al. (2003). As a result, the reverse leakage current has been reduced and a fast switching speed has been demonstrated. The *I-V* characteristics of a 4H SiC *p-i-n* diode with high breakdown voltage (19 kV) have been demonstrated by Suguwara et al. (2001). SBDs with 600 V and 1–10 A were released by Cree in 2001 and now these rectifiers are commercially available. Recently, 1.6-kV 4H SiC SBDs having 10-μm-thick blocking layer with 5×10^{15} cm^{-3} *n*-type dopant concentration have been fabricated in SemiSouth Laboratories (Casady et al. 2003). Edge termination techniques, guard rings, field plates, and mesa termination are implemented in SiC *p-i-n* fabrication to achieve blocking voltages close to the theoretically predicted value for an ideal planar junction diode, even at the expense of additional processing complications such as additional implantation and post-implant annealing. Blocking voltages have been demonstrated up to 1.8 kV, and process variants such as gate sidewall implants have been introduced to achieve extremely low resistance (2.77 mΩcm^2 for 1710-V blocking) for vertical-junction FET SiC switch technology (Mitlehner et al. 1999; Zhao et al. 2003). BJT has received considerable attention recently, in both monolithic and hybrid Darlington configurations (Tang et al. 2002; Huijie et al. 2002). The current gain, blocking voltage, and forward current have been reported to be 400, 500 V, and 23 A, respectively, for the hybrid configuration. Recently, 4H SiC GTOs with a 2.6-kV breakover voltage have been successfully demonstrated (Agarwal et al. 2000; Agarwal et al. 2001). In recent years, photoconductive semiconductor switches (PCSSs) have gained much attention and become competitive in high-power switches for certain applications. The PCSS is an important type of electronic device with applications in high-speed photodetectors, high-voltage pulse generators, and electron-beam-pumped lasers. SiC PCSS devices have unique advantages over conventional power switches, such as high breakdown fields, high speeds, long lifetimes, and negligible jitter times. Although Si and GaAs are the most commonly used materials in PCSS (Rosen and Zutavern 1994; Islam, Schumiloglu, and Fleddermann 1998) compared to GaAs and Si, better performance and wider (more) applications are expected from SiC PCSS due to its higher saturation electron drift velocity and higher thermal conductivity. SiC PCSS have been fabricated on SiC, on 6H SiC (Saddow et al. 1994; Cho et al. 1995), 3C SiC (Sheng et al. 1997) and 4H SiC (Doğan et al. 2003). Photoconductivity measurements have carried out on 4H SiC devices at bias voltages up to 1005 V. The average on-state resistance was 20 Ω and the average off-state resistance was 5×10^{12} Ω, for an on/off ratio of 2.5×10^{11}. The prepared PCSS devices were measured with DC bias and showed repeatable behavior over many thousands of pulses (Doğan et al. 2003).

4.4 Remaining Issues and Summary for SiC

SiC device technology has made great strides and has reached the marketplace. The SiC substrates are used for GaN-based light emitters and high-power amplifiers. However, many aspects of the material quality, epitaxial growth, packaging, device processing, and oxide quality need to be improved. Extensive efforts have been expended in exploring SiC for applications in high-temperature, high-frequency, and high-power devices. Availability of SiC substrates has made possible the recent advances in SiC (as well as GaN) material and device quality. One of the most important problems to be solved while manufacturing fully functional SiC devices is the fabrication of low-resistivity ohmic contacts, particularly contacts to the *p*-type material. At the present time, the performance of SiC devices, such as bipolar and unipolar devices, is limited by material quality (increased diameter of the wafer, epitaxial growth, reduced cost, and micropipe density) and the fabrication of thermally stable and low-resistivity ohmic contacts. At the present time, research-grade, 3-inch 4H SiC, 3-inch 6H SiC, 2-inch 6H SiC, and 4H SiC substrates are available from multiple suppliers. The larger the SiC substrate, the lower the cost, and the higher the quality, the more SiC devices will be available for a wider range of applications such as high-power, high-temperature devices. The same is applicable to GaN-based, high-frequency applications, blue and ultraviolet laser diodes, and light-emitting diodes utilizing SiC substrates.

The quality of epitaxial SiC has benefited from SiC substrates and improved CVD growth techniques. Both 6H and 3C SiC can now be deposited on 6H substrates. Although the role of defects caused by substrates or epitaxial layers has long been a subject of active research, a thorough understanding of which defects cause device failure and how they can be eliminated is not yet clear. For this reason, one of the bottleneck problems for the crystal growers is the elimination of defects.

4.5 III-V Nitride Semiconductors

During the last three decades, developments in the field of III-nitrides have been spectacular, with major breakthroughs taking place in the 1990s. Nitride semiconductors are viewed as a highly promising material system for electronic and optoelectronic applications. As members of the III-nitrides family, AlN, GaN, InN, and their alloys are all wide-bandgap materials and can crystallize in both wurtzite and zinc-blend polytypes. The bandgaps of the wurtzite polytypes are direct and range from possibly 0.7–0.8 eV for InN, to 3.4 eV for GaN, and to 6.1 eV for AlN. GaN alloyed with AlN and InN forms a complete series of ternary and quenternary alloys that spans throughout much of the visible spectrum and extends well into the ultraviolet (UV) region. This makes the nitride material system attractive for optoelectronic applications, such as green, blue, or UV light-emitting diodes (LEDs), laser diodes (LDs), and UV detectors. The strongest feature of III-nitrides compared to other wide-bandgap materials such as SiC is the availability of functional heterostructure technology. Quantum wells, modulation-doped heterointerfaces, and other heterojunction structures can all be made in this system, giving access to new spectral regions for optical devices and new regimes for electronic applications.

As of 2003, a report titled "Gallium Nitride 2003—Technology Status, Applications, and Market Forecasts" tabulates that developments and markets for both opto and electronic devices are tremendous. The report tabulates 477 companies and research centers involved in GaN production and development, including 51 government agencies that support GaN research, 184 companies involved in GaN production or development (devices, materials, and equipment), and 293 universities and research centers involved in this field. In addition to dramatic growth in research and development activities, market growth for GaN-based devices has also been robust. Since 1999, the market for devices (primarily blue, green, and white high-brightness LEDs) has grown by 221% to $1.35 billion. Since the first significant commercial shipments of GaN LEDs in 1995, growth has progressed at an average annual rate of 64.5%. Blue–violet laser diodes based on GaN began to ship in 1999, achieving modest levels through 2002. In 2003, the market for blue-violet (405 nm) laser diodes in optical storage began in earnest; several companies are announcing shipments of next-generation high-density DVD recorders. GaN-based DVDs are still under development, with the first significant commercial shipments expected in 2004. Overall, the market for GaN devices of all types is forecast to grow to $4.5 billion by 2007. In short, the evaluation of nitride semiconductors, the most environmentally friendly materials available, will have a great impact in the future. In this review, the most recent developments and achievements in the field of nitride semiconductors and devices will be updated.

4.5.1 Nitride Crystal Growth

Nearly every major crystal-growth technique has been developed, including molecular-beam epitaxy (MBE), hydride vapor-phase epitaxy (HVPE), and metal-organic chemical vapor deposition (MOCVD), in relation to nitride semiconductors. Several modifications to conventional MBE methods have been implemented for III-nitride growth: growth with ammonia or hydrazine (the latter is not attractive due to safety reasons and success of ammonia), plasma-assisted MBE (PAMBE), metal-organic MBE (MOMBE), etc. Among other methods, radio-frequency (RF) plasma sources are the most commonly employed devices to activate the neutral nitrogen species in the MBE environment, with electron-cyclotron resonance (ECR) fading because of its high-energy ion content. Although all of these epitaxial methods contend with problems related to the lack of native GaN substrates and difficulty with nitrogen incorporation, remarkable progress in the growth of high-quality epitaxial layers of III-nitrides by a variety of methods has been achieved over the years (Morkoç 2006).

The major issue in III-V nitride growth by conventional deposition techniques, regardless of the deposition method employed, is the high density of dislocations that result from the lattice mismatch between the buffer layer and the film and/or the buffer layer and the substrate. One approach to reducing the dislocation density used recently for GaN films on SiC and sapphire is the lateral epitaxial overgrowth (LOE), also known as epitaxial lateral overgrowth (ELO), selective area growth (SAG), or epitaxial lateral over growth (ELOG). Selective epitaxial growth of GaN on sapphire was reported as early as 1994 (Kato et al. 1994). This was followed by a transmission electron microscopic study that revealed that these overgrown regions of pyramids contained a much lower density of dislocations (Zheleva et al. 1997). Using the LEO technique, Nakamura reported laser diode lifetimes of about 10,000 h at room temperature (Nakamura et al. 1998), which is substantial in that the longevity of laser diodes hovered around 300 h prior to LEO. Basically, the method relies on growth of GaN on windows opened in a dielectric material such as SiO_2, followed by lateral extension and coalescence. Selective growth, in conjunction with lateral growth, on GaN or AlN nucleation stripes without the dielectric mask has also been developed. The first step in the LEO process is the growth of a few-μm-thick GaN layer on sapphire, SiC, or on SiC/Si followed by deposition of a dielectric (SiO_2 or SiN) mask by CVD or PECVD. A set of parallel stripes separated by window areas is then opened using standard photolithography. During the initial regrowth, either in MOVPE (Zheleva et al. 1997), HVPE (Sakai, Sunakawa, and Usui 1997) or sublimation growth (Kurai, Nishino, and Sakai 1997; Wang et al. 1998), selective area epitaxy is achieved in the windows without any nucleation on the dielectric mask. If the growth parameters are then chosen correctly, and once the GaN growing film reaches the top of the stripes, epitaxial lateral growth takes place over the mask and finally leads to a full coalescence, culminating in a smooth surface. The basic concept and typical corresponding results are shown in Figure 4.5. Above the windows, the microstructure of the underlying GaN template is reproduced with its defect structure to a first extent. In the laterally grown regions (wings—what used to be the area covered by the mask), however, the material contains significantly fewer defects. This is the one-step LEO. With two-step LEO, an additional masking and regrowth is performed in such a way that what used to be unmasked areas are masked. In theory, this would lead to a uniformly extended, defect-free material. It has been demonstrated that several steps lead to ELO GaN with wider usable surfaces, free of emerging threading dislocations. Even better quality can be foreseen with more than two steps at the expense of higher complexity. Nagahama et al. (2000a, 2000b) implemented a three-step LEO process, in combination with very thick GaN buffer layers, in an effort to further reduce the threading dislocation density. The effort led to densities slightly above the mid-10^5 cm^{-2} point, and thereby allowed the fabrication of laser diodes lasting 15,000 h with an output power of 30 mW.

A variant of LEO is termed *pendeo epitaxy*. This method relies on full selective growth of GaN on GaN stripes on SiC or Si substrates (Linthicum et al. 1999; Davis et al. 2001). The method has the advantage of eliminating the SiO_2 mask on which the conventional LEO relies. In this approach, the etched columnar GaN posts are capped with a mask layer; GaN growth then precedes laterally and vertically until it coalesces between and over the mask on top of the columns, thereby creating a continuous layer as schematically shown in Figure 4.6. Also shown is a typical TEM image that revealed a strong reduction of

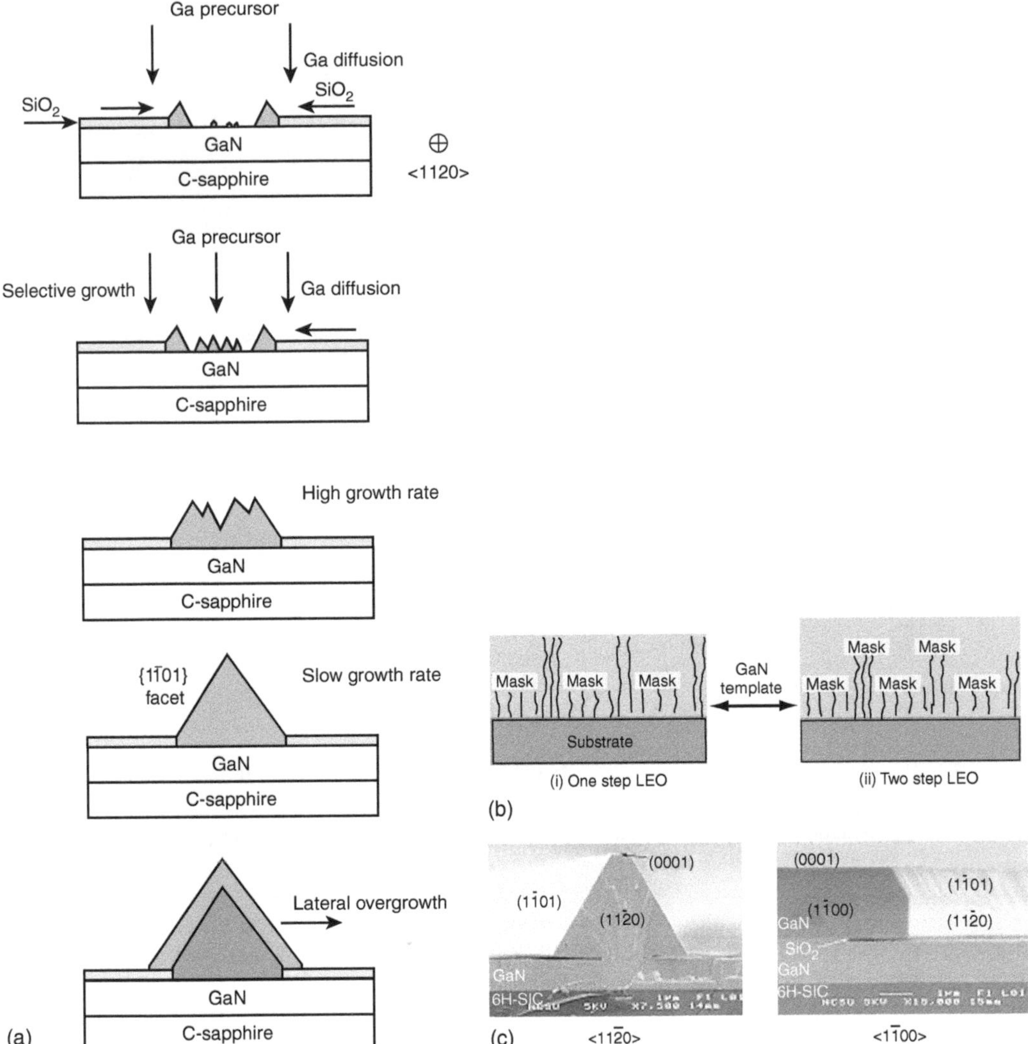

FIGURE 4.5 (a) Evolution of lateral epitaxial growth (LEO). (b) Schematic representation of single step (one step) LEO and two step LEO. The vertical lines represent threading dislocations. (c) Scanning electron micrographs showing the morphologies of GaN layers grown on stripe openings oriented along (i) <11$\bar{2}$0> and (ii) <1$\bar{1}$00> directions. (Courtesy of R. Davis.)

the threading dislocation density, with no dislocations generated in the coalescence region, with some threading dislocations propagating vertically from the GaN seed post. This combined with no tilt having been observed in the wing regions represents a significant improvement by the pendeo-epitaxy method. Compared to conventional LEO, the advantages of PE are: (1) growth initiates from a GaN facet different from (0001), and (2) the substrate can be used as a mask, taking advantage of the selective nature of growth in the case of 6H SiC substrate. In a certain temperature window, no growth occurs on SiC, whereas normal growth takes place over edges of the GaN posts (columns). In short, the growth does not initiate through open windows on the (0001) surface of the GaN seed layer; instead, it is forced to selectively begin on the sidewalls of a tailored microstructure comprised of forms previously etched into this seed layer. Continuation of the pendeo-epitaxial growth of a GaN layer until coalescence over and between these forms results in a layer of lower defect-density GaN. This process has also been extended to

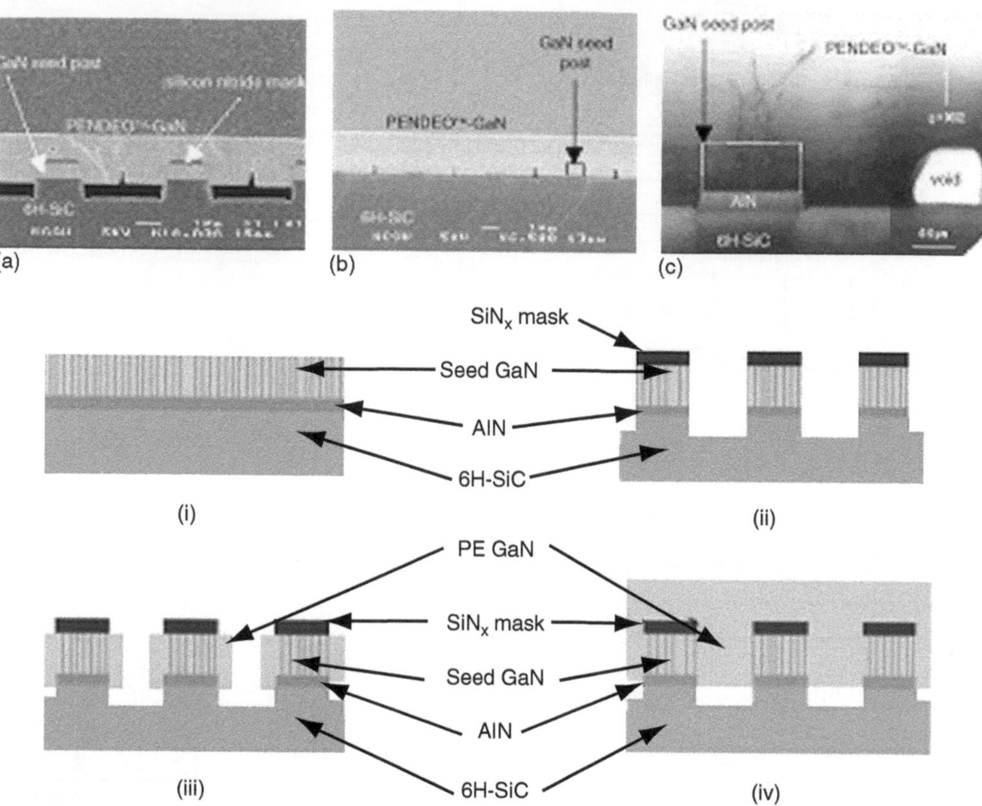

FIGURE 4.6 Schematic representation of different steps for growth in the technique of pendeo-epitaxy (PE): (i) GaN template, (ii) etched GaN columns prior to growth, (iii) partial growth of PE GaN showing growth only from the sidewalls, and (iv) fully coalesced growth of GaN. (a) GaN grown over the masked GaN posts and laterally. (b) GaN grown on the unmasked submicron post and laterally. (c) TEM image of (b). (From Thomson, D., Gehrke, T., Linthicum, K. J., Rajagopal, P., and Davis, R. F. 1999. *J. Nitride Semicond. Res.*, 4S1, G3 37; Gehrke, T., Linthicum, K., Preble, J. E., Rajagopal, P., Ronning, C., Zorman, C., Mehregany, M., and Davis, R. F. 2000. *J. Electron. Mater.*, 29, 306. Courtesy of R. F. Davis.)

include Si (111) substrates, after the conversion of its surface with a 3C SiC layer that is eventually situated between the AlN nucleation layer and the (111) Si substrate. For more comprehensive and detailed information about nitride growth technology, the reader is urged to pursue the book written by Morkoç (2006).

4.5.2 Substrates for Nitride Epitaxy

One of the main difficulties that have hindered III-nitride research is the lack of a lattice-matched and thermally compatible substrate material. Although efforts are ongoing to grow large-size bulk GaN or AlN crystals for substrates, researchers at present have no choice but to grow nitrides heteroepitaxially. A wide variety of materials have been used for nitride epitaxy, including insulating metal oxides, metal nitrides, and other semiconductors. In practice, properties other than the lattice constants and thermal compatibility, including the crystal structure, surface finish, composition, reactivity, chemical and electrical properties, are also important in determining the suitability of a given material as a substrate. To a large extent, the substrate employed determines the crystal orientation, polarity, polytype, surface morphology, strain, and the defect concentration of the epitaxial films. The most promising results thus far have been obtained on more conventional substrates such as sapphire and SiC. Thick, freestanding

GaN templates grown by HVPE that are subsequently separated from the sapphire substrates are emerging as new substrates. As a matter of completeness, III-V nitrides have been grown on Si, ZnO, NaCl, GaP, InP, 3C-SiC, W, $MgAl_2O_4$, TiO_2, and MgO. There are other substrates used for nitride growth, includeing Hf, $LiAlO_2$, and $LiGaO_2$. A summary of crystal structures, lattice constants, and thermal characteristics of a number of candidate substrate materials for nitride growth are given in Table 4.2.

The preference for sapphire substrates for nitride heteroepitaxy can be attributed to their wide availability, hexagonal symmetry, ease of handling, simple pre-growth cleaning requirements, and stability at elevated temperatures. However, as can be seen in Table 4.1, the lattice and thermal mismatch between sapphire and GaN is quite large. Lateral (a-lattice constant) mismatched substrates lead to a substantial density of misfit and threading dislocation in broad-area epitaxially deposited GaN on foreign substrates in the range of 10^8 and 10^{10} cm^{-2}. An appropriate surface preparation such as nitridation, deposition of a low-temperature (LT) AlN, medium-temperature AlN, or GaN buffer layer, selective epitaxy followed by coalescence (LEO) can reduce dislocation densities down to 10^6 cm^{-2}.

However, these dislocation figures are still too high compared to extended defect densities of essentially zero for silicon homoepitaxy, and 10^2 to 10^4 cm^{-2} for GaAs homoepitaxy. Vertical (c-lattice constant) mismatch creates additional crystalline defects besetting the layers, including inversion domain boundaries and stacking faults. In addition, thermal expansion coefficient mismatches between epitaxial films and substrate induce stress that can cause crack formation in the film and substrate for thick films as it is cooled from the deposition temperature. Defects introducing energy states in the forbidden energy bandgap are either directly or indirectly responsible for the creation of nonradiative recombination centers, which reduces the quantum efficiency and minority carrier lifetimes. Other adverse effects of structural and point defects are impurities that diffuse more readily along threading dislocations, and carrier transport that is either impeded, as in laterial transport, or that is aided, as in vertical transport. Because of the high piezoelectric constants of GaN, the local strain surrounding threading dislocations can also lead to inhomogeneities in electric potential and electric field in comparison to the bulk material. A high density of defects, which increase the device threshold voltage and reverse leakage currents, deplete sheet charge carrier density in heterojunction field effect

TABLE 4.2 Lattice Parameters and Thermal Characteristics of a Number of the Prospective Substrate Materials for Nitride Growth, and Their Lattice Mismatch with GaN

Crystal	Structure	a (Å)	b (Å)	c (Å)	κ ($W\ cm^{-1}\ K$)	$\Delta a/a$ $\Delta c/c$ ($\times 10^{-6}\ K^{-1}$)	Mismatch
AlN	Hexagonal	3.111		4.980	3.2	4.2 5.3	−2.7%
GaN	Hexagonal	3.1885		5.185	1.3	5.59 3.17	0%
Sapphire	Hexagonal	4.765		1.2982	0.3–0.5	7.5 8.5	13%
6H–SiC	Hexagonal	3.0806		1.5117	4.9	4.2 4.68	−3.1%
ZnO	Hexagonal	3.2496		5.2065	0.3–0.4	4.75 2.9	2.1%
$LiAlO_2$	Tetragonal	5.169		6.267		7.1 15	1.4%
$LiGaO_2$	Orthorhombic	5.402	6.372	5.007		a:6, b:9 7	0.18%
$MgAl_2O_4$	Cubic	8.083				7.45	
Si	Cubic	5.4301			1.5	3.59	
GaAs	Cubic	5.6533			0.5	6	
3C–SiC	Cubic	4.3596			4.9		
MgO	Rock salt	4.21				10.5	

Source: After Liu, L., and Edgar, J. H. 2002. *Mater. Sci. Eng. R.*, 37, 61; Morkoç, H. 2006. *Nitride Semiconductors and Devices*, Springer Verlag, Heidelburg.

transistor, and reduce the charge carrier mobility and thermal conductivity, which are detrimental for device applications and for achieving optimal performance. Thus, substrates capable of supporting better quality epitaxial layers are always needed for realizing the full potential of nitride-based devices.

4.5.3 Polytypism in the III-V Nitrides

The crystal structures shared by the group-III nitrides are wurtzite, zincblende, and rocksalt. At ambient conditions, the thermodynamically stable phase is wurtzite for bulk AlN, GaN, and InN. The cohesive energy per bond in wurtzite variety is 2.88 eV (63.5 kcal/mol), 2.20 eV (48.5 kcal/mol), and 1.93 eV (42.5 kcal/mol) for AlN, GaN, and InN, respectively (Harris 1980). Although the calculated energy difference $\Delta E_{W\text{-}ZB}$ between wurtzite and zincblende lattice is small (-18.41 meV/atom for AlN, -9.88 meV/atom for GaN, and -11.44 meV/atom for InN) (Yeh et al. 1992), the wurtzite form is energetically preferable for all three nitrides compared to zincblende. The wurtzite structure has a hexagonal unit cell with two lattice parameters, a and c, in the ratio of $c/a=\sqrt{8/3}=1.633$ and belongs to the space group $P6_3mc$. The structure is composed of two interpenetrating hexagonal close-packed (hcp) sublattices, each of which consists of one type of atom displaced with respect to each other along the three-fold c-axis by the amount of $u=3/8=0.375$ in fractional coordinates in an ideal hexagonal lattice. The u parameter represents the ratio of the bond length along the c-direction to the c lattice parameter. Each sublattice includes four atoms per unit cell with a molecular mass of 40.9882, 83.7267, and 128.827 g/mol for AlN, GaN, and InN, respectively. Every atom of one kind (group-III atom) is surrounded by four atoms of the other kind (nitrogen), or vice versa, which are coordinated at the edges of a tetrahedron. For actual nitrides, the wurtzite structure deviates from the ideal arrangement by changing the c/a ratio or the u value (Yeh et al. 1992). It should be pointed out that a strong correlation exists between the c/a ratio and the u parameter in that when the c/a ratio decreases, the u parameter increases such that those four tetrahedral distances remain nearly constant through a distortion of tetrahedral angles due to long-range polar interactions. These two slightly different bond lengths will be equal if the following relation holds:

$$u=\left(\frac{1}{3}\right)\left(\frac{a^2}{c^2}\right)+\frac{1}{4}. \tag{4.1}$$

Because the c/a ratio also correlates with the difference of the electronegativities of the two constituents, components with the greatest differences show largest departure from the ideal c/a ratio (Jeffery et al. 1956). For GaN, the c/a ratio and the value of u are measured as 1.627 and 0.377, respectively, which are close to the ideal values (Schulz and Theimann 1977). AlN deviates significantly from the ideal parameters: $c/a=1.601$ and $u=0.382$. Consequently, the interatomic distance and angles differ by 0.01 Å and 3°, respectively. For InN, though the reliability of the available data is not as good as that for GaN and AlN, the measured u parameter (Ambacher et al. 2002) is 0.379 and $c/a=1.618$.

A phase transition to the rocksalt (NaCl) structure in III nitrides takes place at very high external pressures. This occurs becase the reduction of the lattice dimensions causes the interionic Coulomb interaction to favor the ionicity over the covalent nature. The structural phase transition was experimentally observed at pressures of 22.9 GPa for AlN (Xia, Xia, and Ruoff 1993), 52.2 GPa for GaN (Perlin et al. 1992), and 12.1 GPa for InN (Ueno et al. 1994). The space group symmetry of the rocksalt type structure is $Fm3m$, and the structure is six-fold coordinated. However, rocksalt III-nitrides cannot be stabilized by epitaxial growth.

The zincblende structure is metastable and can be stabilized only by heteroepitaxial growth on cubic substrates, such as cubic SiC (Paisley et al. 1989), Si (Lei et al. 1991), MgO (Powell et al. 1993), and GaAs (Mizita et al. 1986; Strite et al. 1991), reflecting topological compatibility to overcome the intrinsic tendency of forming the wurtzite phase. In the case of highly mismatched substrates, there is usually a certain amount of the zincblende phase of nitrides separated by crystallographic defects from the wurtzite phase. The symmetry of the zincblende structure is given by space group $F\bar{4}3m$ and composed of two interpenetrating face-centered cubic sublattices (fcc) shifted by one-quarter of a body diagonal. There are four atoms per unit cell and every atom of one type (group III nitrides) is tetrahedrally coordinated with four atoms of the other type (nitrogen), and vice versa. The overall equivalent bond length is about 1.623 Å for zincblende structures.

Because of the tetrahedral coordination of wurtzite and zincblende structures, the four nearest neighbors and twelve next-nearest neighbors have the same bond distance in both structures. The main difference between these two structures lies in the stacking sequence of closed-packed diatomic planes. The wurtzite structure consists of triangularly arranged alternating biatomic close-packed (0001) planes, for example Ga and N pairs, thus the stacking sequence of the (0001) plane is AaBbAaBb... in the ⟨0001⟩ direction. In contrast, zincblende structure consists of triangularly arranged atoms in the closed-packed (111) planes along the ⟨111⟩ direction with a 60° rotation that causes a stacking order of AaBbCcAaBbCc... Small and large letters stand for the two different kinds of constituents.

Because none of the three structures described above possess inversion symmetry, the crystal exhibits crystallographic polarity; closed-packed (111) planes in zincblende and rocksalt structures and corresponding (0001) basal planes in the wurtzite structure differ from $(\bar{1}\bar{1}\bar{1})$ and $(000\bar{1})$ planes, respectively. In general, group-III (Al, Ga, or In) terminated planes are denoted as (0001) A plane (referred to as *Ga-polarity*) and group-V (N) terminated planes are designated as $(000\bar{1})$ B plane (referred to as *N-polarity*). Many properties of the material depend on its polarity, for example growth, etching, defect generation and plasticity, spontaneous polarization, and piezoelectricity. In wurtzite nitrides, besides the primary polar plane (0001) and associated direction ⟨0001⟩, which is the most commonly used surface and direction for growth, many other secondary planes and directions exist in the crystal structure.

4.5.4 Fundamental Properties of III-V Nitrides

Over some 30 years, much effort has gone into determining the fundamental physical properties of the III-V nitride semiconductors. Although many applications based on nitride semiconductors have emerged and some of them are commercially available, there is not a convergence on values of the basic physical properties of these materials due to the fact that the quality of the crystals is of primary importance for the precise determination of these properties. In this respect, they are not mature yet. It should be remembered that knowledge of the fundamental properties is crucial not only from the physics point of view but also understanding and optimizing the device structures for better performance. Besides the experimental investigations, many theoretical calculations have been performed on structural and mechanical properties of the group-III nitrides. Below, in Table 4.3, we tabulate the most important properties of GaN, AlN, and InN. The more interested reader is referred to a longer discussion of the fundamental physical properties of group-III nitrides (Edgar et al. 1999; Morkoç 2006).

4.5.5 Optical Properties of III-V Nitrides

The optical properties of a semiconductor are connected with both intrinsic and extrinsic effects. Intrinsic optical transitions take place between the electrons in the conduction band and holes in the valance band, including excitonic effects due to the Coulomb interaction. Excitons are classified into free and bound excitons. Extrinsic properties are related to dopants or defects, which usually create discrete electronic states within the bandgap, and therefore influence both optical absorption and emission processes. In theory, excitons could be bound to neutral or charged donors and acceptors. Each of these fundamental excitons states are expected to have a fine structure due both to exciton polariton longitudinal-transverse splitting and the splitting caused by electron-hole exchange interaction, which are on the order of 1–2 meV (Ivchenko 1982).

The optical spectroscopy of the intrinsic excitons can be measured by employing low temperature PL, absorption, and/or derivative technique like photoreflectance (PR) and calorimetric absorption and reflection techniques (Song and Shan 1998; Monemar, Bergman, and Buyanova 1998). There are several

TABLE 4.3 Some Physical Parameters of Wurtzite AlN, GaN, and InN Obtained by Several Experimental Techniques and Theoretical Calculations

Parameters	AlN	GaN	InN
Room temperature bandgap energy, E_g (Γ) (eV)	6.2	3.42	1.89 or 0.78 (based on latest results)
Index of refraction, n	2.15	2.35	2.80
Dielectric constants	$\varepsilon_r=8.5$ $\varepsilon_\infty=4.77$	$\varepsilon_r=10.4$ $\varepsilon_\infty=5.33$	$\varepsilon_r=15.3$ $\varepsilon_\infty=8.4$
Electron effective mass, m^* (kg)	0.33 m_0	0.20 m_0	0.11 m_0
Saturation drift velocity, V_d (cm/s)	1.4×10^7	2.5×10^7	2.5×10^7
Lattice constants (Å)	$C=3.1106$ $a=4.9795$	$C=3.1892$ $A=5.1850$	$c=3.538$ $a=5.703$
C_{11} (GPa)	345–411	296–396	190–271
C_{12} (GPa)	125–149	120–160	104–124
C_{13} (GPa)	90–127	80–158	92–121
C_{33} (GPa)	373–395	209–405	182–224
C_{44} (GPa)	96–125	81–105	10–48
Poisson's ratio, ν	0.216	0.372	—
Bulk modulus, B (GPa)	160–218	195–245	125–147
dB/dP	5.2–6.3	4–4.5	3.4
Young's modulus, E (GPa)	295–374	150	—
Shear modulus,	117–154	121	43
Yield strength, σ_Y (GPa)	0.3 at 1000°C	0.1−0.2 at 900°C	—
Hardness (GPa)	Micro; 17.7 Nano; 18.0	Micro; 10.2 Nano; 18–20	Nano; 11.2
Electronic deformation potential, E_{ds} (eV)	9.5	8.54	7.10
Thermal expansion (K^{-1})	$\Delta a/a=4.2\times10^{-6}$ $\Delta c/c=5.3\times10^{-6}$	$\Delta a/a=5.59\times10^{-6}$ $\Delta c/c=3.17\times10^{-6}$	$\Delta a/a=2.85\times10^{-6}$ $\Delta c/c=3.75\times10^{-6}$
Thermal conductivity, κ (W/cm K)	3.2	2.3	0.8
Heat of formation, ΔH_{298} (kcal/mol)	−64	−26.4	−4.6
Melting point (°C)	3000	> 1700	1100
Debye temperature, θ_D (K)	950	650	660

Source: Morkoç, H. 2006. *Nitride Semiconductors and Devices*, Springer Verlag, Heidelburg; Edgar, J. H., Strite, S. T., Akasaki, I., Amano, H., and Wetzel, C., 1999. *Properties, Processing and Applications of Gallium Nitride and Related Semiconductors*, INCPEC, London.

reports in literature on reflectance studies for thick GaN epilayers (Gil, Briort, and Aulombard 1995; Smith et al. 1995; Shan et al. 1996) as well as on homoepitaxial layers (Korona et al. 1996; Kornitzer et al. 1999) and bulk GaN (Skromme et al. 2002). Monemar, Bergman, and Buyanova (1998), who have examined a number of thin and thick layers on various substrates, including homoepitaxial layers on GaN templates, concluded that the A, B, and C exciton lines in GaN relaxed to an accuracy of ±2 meV are 3.478, 3.484, and 3.502 eV, respectively, at 2 K.

Photoluminescence (PL) is a very powerful technique for studying exciton structure in semiconductors. Figure 4.7 displays an example of a PL spectrum in the range of fundamental excitons recently taken at 2 K for an approximately strain free thick GaN epilayer grown on sapphire by HVPE. The PL spectrum is typically dominated by strong emission from donor bound exciton related transitions. The recombination of bound excitons typically gives rise to sharp lines, with a photon energy characteristic for each defect. A characteristic phonon coupling, which can involve both lattice modes, and defect-related vibrational modes is also seen for each particular bound exciton spectrum. The photon energy region for DBE spectra is about 3.470–3.4733 eV at 2 K for strain free GaN (Monemar, Bergman, and

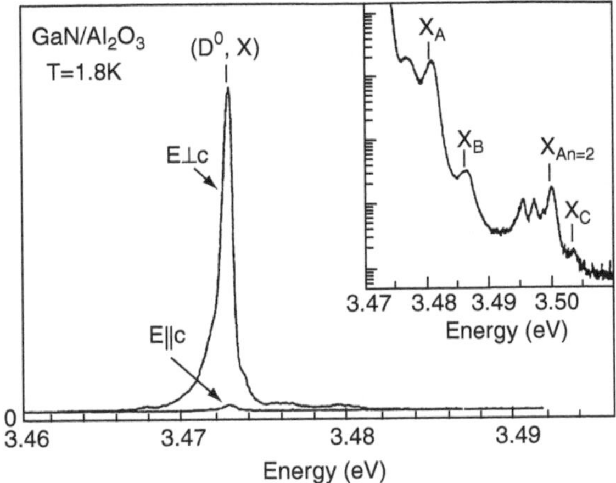

FIGURE 4.7 Photoluminescence spectra at 1.8 K for a 400-µm thick GaN grown on sapphire substrate. The spectrum is dominated by the donor bound exciton, but the intrinsic exciton states are also resolved at higher energies (see inset). (After Eckey, L., Podloswski, L., Goldner, A., Hoffmann, A., Broser, I., Meyer, B. K., Volm, D., Streibl, T., Hiramatsu, K., Detcprohm, T., Amano, H., and Akasaki, I. 1996. *Ins. Phys. Conf. Ser.*, 142, 943. With permission.)

Buyanova 1998; Kornitzer et al. 1999). Thick heteroepitaxial layers grown by HVPE give the best spectroscopic characteristic, with an optical linewidth of about 1 meV or less, as seen in Figure 4.8. It is commonly argued that oxygen and silicon are the two most dominant residual shallow donors in GaN. In homoepitaxial GaN samples grown by MOVPE on a pre-treated GaN single crystal, a better spectral linewidth down to 0.1 meV for the DBE peaked at 3.4709 eV has been reported (Kornitzer et al. 1999). In this case, in addition to unidentified sharp satellite lines on both sides of main DBE peak, shallow DBE is the dominant line and has, in the recent literature, been assigned to substitutional Si donors (Neu et al. 2000). Therefore, the high-energy line would then be O-related, which is known to be a typical contaminant in GaN grown by all techniques.

In the energy region below the principle DBE line, there is usually a rich spectrum of bound excitons assumed to be acceptor-related. However, the situation for acceptor BEs (ABE or I_1) in GaN is somewhat less clear than for the donors. The most prominent neutral ABE is found at about 3.466 eV in strain free

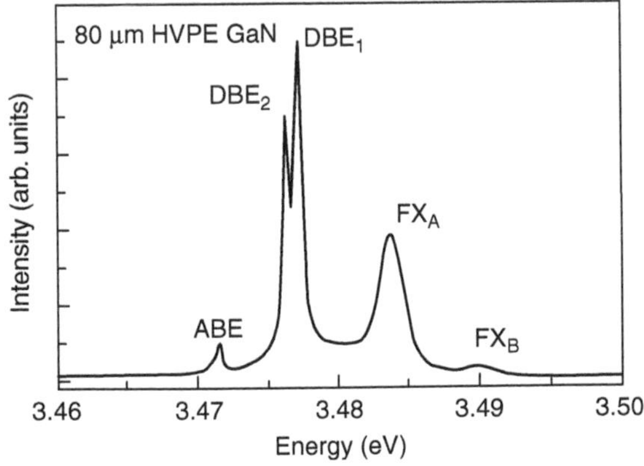

FIGURE 4.8 Photoluminescence spectrum of a thick HVPE GaN layer showing two donor BE lines and an acceptor BE. The spectrum is upshifted by the compressive strain about 6 meV compared with strain free GaN. (After Monemar, B., Paskov, P. P., Paskova, T., Bergman, J. P., Pozina, G., Chen, W. M., Hai, P. N., Buyanova, I. A., Amano, H., and Akasaki, I. 2002. *Mater. Sci. Eng. B.*, 93, 112. With permission.)

GaN and attributed to the exciton bound to shallow acceptor, probably Mg_{Ga}. Further, this peak has been found to be dominant in slightly Mg-doped GaN samples (Leroux et al. 1999; Skromme and Martinez 1999). An alternative interpretation of this BE line, which is related to charged donor bound excitons, has been proposed by several groups (Reynolds et al. 1997; Santic et al. 1997; Mair et al. 1999). The other main peak at about 3.455 eV has been attributed to Zn, but this is not justified so far. The broader peak at 3.44 eV is probably due to a low energy acoustic phonon wing connected to the main peak at 3.455 eV, which is very characteristic for ABEs (Thomas and Hopfield 1962; Saito, Shionoya and Hanamura 1973). Also, phonon replicas are seen in the spectrum, indicating a rather strong LO phonon coupling, much stronger than for the DBE.

A broad PL band peaking at 2.2 ± 0.1 eV, the so-called yellow emission, is almost systematically observed in undoped or *n*-GaN. As studied extensively by Ogino and Aoki (1980) and more recently by others (Neugebauer and Van de Walle 1996; Saarinen et al. 1997; Mattila and Nieminen 1997), there seems to be an agreement that transitions from the conduction band or a shallow donor to a deep acceptor are responsible for this band. Another interpretation of the yellow band as being due to transitions between a deep donor to a shallow acceptor was also proposed to explain the results of magnetic-resonance experiments (Glaser et al. 1995). The exact position of this band and its linewidth slightly differ in numerous publications. The nature of the deep donor has not been established, but several candidates such as the nitrogen vacancy (Perlin et al. 1995), gallium vacancy, or its complexes with shallow donor (Neugebauer and Van de Walle 1996; Mattila and Nieminen 1997) or carbon (Ogino and Aoki 1980; Zhang and Kuech 1998) have been suggested.

The temporal behavior of excitons is of importance for emitters in that it provides a window on the dynamics of recombination processes. For high radiative efficiency, the recombination should ideally be fast to be able to compete efficiently with the nonradiative processes, such as multiphonon emission and defect-related nonradiative centers. PL transient data represent an excellent tool to study both recombination rates of radiative and nonradiative processes in GaN. The temporal behavior at low excitation levels indicates processes, depending on the sample, that are as fast as about 35 ps for free excitons. Typical transients that were obtained for different excitation intensities for a GaN epilayer grown on sapphire are shown in Figure 4.9. Decay times at 2 K were found to vary between 60 ps and 115 ps, dependent on the excitation intensity, indicating strong defect participation under these excitation intensities. The decay in luminescence intensity can more reasonably be described with a combination of a fast decay followed by a slower process with a decay time of about 300 ps at longer times. This slow process may be associated with weak localization of the free excitons, perhaps due to potential fluctuations induced by the inhomogeneous strain field.

In HVPE grown thick layers, the low intensity decay time about 200 ps is observed at 2 K, which is a typical radiative lifetime for DBEs related to shallow neutral donors. In thin strained heteroepitaxial layers with a higher defect density, a shorter lifetime is observed (Pau et al. 1998; Pozina et al. 1999; Pozina et al. 2001), indicating the effect of excitation transfer from the DBE to lower energy states before the radiative recombination takes place. Similarly, homoepitaxial GaN layers also often show a fast DBE decay at low excitation density, due to excitation transfer to point defects (Monemar 1998). A slow component, which is indicative of a slower radiative process overlapping the DBE transition, is also observed in the DBE decay dynamics (Reynolds et al. 1996; Monemar 1998). A recent observation on the DBE decay time for *n*-type GaN layers grown on SiC substrates indicated a clear trend versus the energy position of the DBE line, which in turn correlated with the strain in the layer. The biaxial strain in the GaN layer strongly affects the top valence band states, which in turn is reflected in the DBE wave function (Stepniewski et al. 1999), affecting the oscillator strength and hence the observed radiative lifetime (Skromme et al. 2002).

The observed value of radiative lifetimes is about 0.7 ns for the shallowest ABE peaked at 3.466 eV, as compared to much longer, 3.6 ns, for the deeper acceptor with ABE peak between 3.455–3.46 eV (Pozina et al. 1999). This corresponds to an oscillator strength on the order of 1, very similar to the shallow acceptor BEs in CdS (Henry and Nassau 1970). The transient PL decay times of the YL has been investigated by different research groups. Hoffman et al. (1995) and Korotkov, Reshchikov, and Wessels

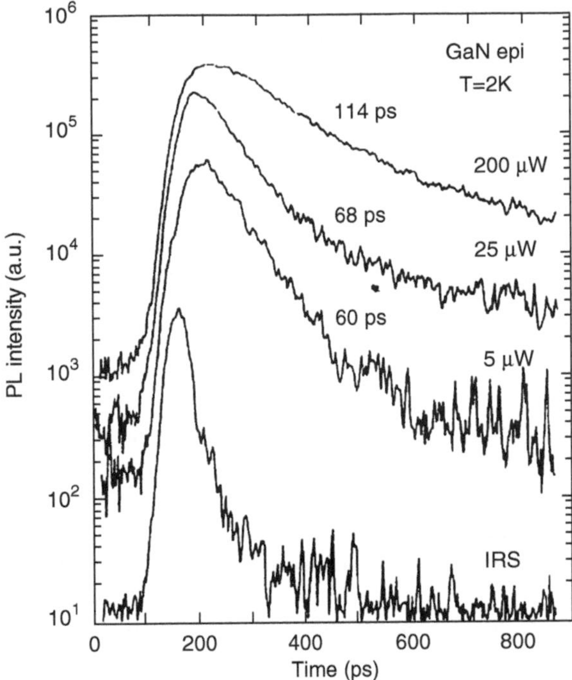

FIGURE 4.9 Photoluminescence transients at 2 K for a GaN epilayer grown on sapphire measured at different excitation power intensity. IRS denotes the instrumental response in the experiment. (After Monemar, B., Bergman, J. P., and Buyanova, I. A. 1998. In *GaN and Related Materials Semiconductor Compounds*, S. J., Pearton, ed., pp. 85–139. Gordon and Breach, New York. With permission.)

(1999) reported rather long and nonexponential decay of the YL in the range 0.1–100 μs at low temperatures, which was quantitatively described in the Thomas–Hopfield model (Thomas, Hopfield, and Augustyniak 1965) of the donor-acceptor pair (DAP) recombination. A very long-lived emission decay time of about 300 ms at 10 K has been observed by Seitz et al. (1997). In contrast, very fast decay times of about 1 ns at 2 K and 20 ps at room temperature have been reported and related to strong contribution of free-to-bound transitions and DAP recombination by Gollewski et al. (1997) and Haag et al. (1999), respectively.

The optical properties of AlN have been investigated in many forms, including powders, sintered ceramics, polycrystals, and single crystal samples. Because an AlN lattice has a very large affinity to oxygen dissolution, oxygen contamination is hard to eliminate in AlN, in which optical properties are influenced by oxygen-related defects. Some oxygen is dissolved in the AlN lattice while the remainder forms an oxide coating on the surface of each powder grain. In recent work, band-edge emission of high-quality AlN grown by MOCVD on sapphire substrate has been investigated by Li et al. (2002a). Band-edge emission lines at 5.960 and 6.033 eV have been observed at room temperature and 10 K, respectively, as shown in Figure 4.10. The peak-integrated emission intensity of the deep-impurity-related emission centered at around 3.2 and 4.2 eV is only about 1% and 3% of that of the band-edge transition at room temperature. The PL emission properties of AlN have been compared with those of GaN and it was shown that the optical quality as well as quantum efficiency of AlN epilayers is as good as that of GaN (Nam et al. 2003). The same group has also studied deep UV picosecond-time-resolved photoluminescence spectroscopy to investigate the observed optical transition in steady-state PL measurements. Two PL emission lines at 6.015 and 6.033 eV were attributed to donor-bound exciton

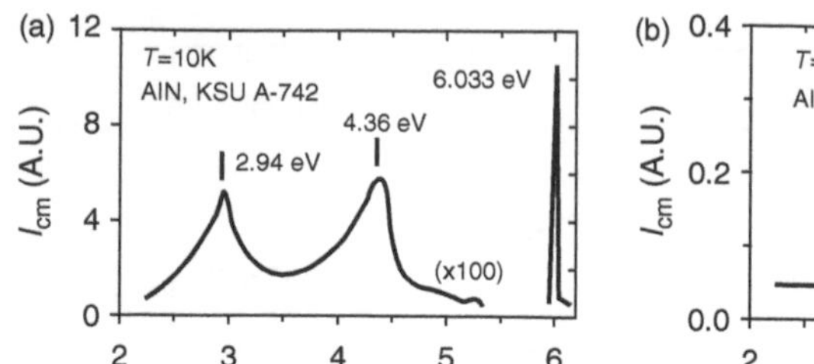

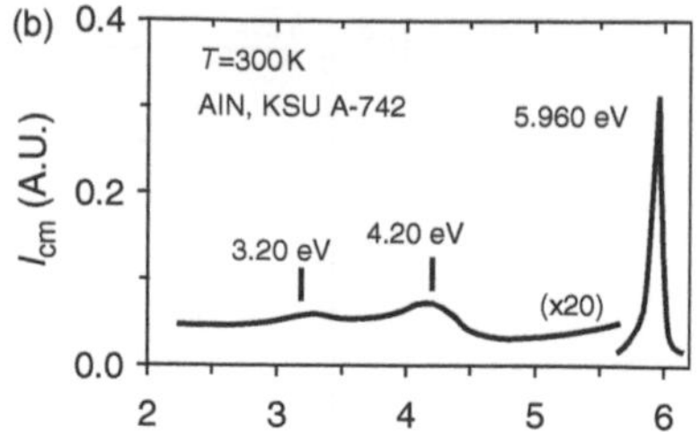

FIGURE 4.10 PL spectra of AlN aepilayers measured (a) at 10 K and (b) at 300 K. The spectra cover a broad spectral range from 2.2 to 6.2 eV. (From Li, J., Nam, K. B., Nakarmi, M. L., Lin, J. Y., Jiang. H. X. 2002a. *Appl. Phys. Lett.*, 81, 3365. With permission.)

and free-exciton transitions, respectively, from which the binding energy of the donor bound excitons in AlN epilayers was determined to be around 16 meV.

Time-resolved PL measurements revealed that the recombination lifetimes were around 80 ps for bound exciton and 50 ps for free exciton (Nam et al. 2003). About 80 meV free exciton binding energy was also deduced from the temperature dependencies of both free exciton radiative decay time and emission intensity.

Among the group-III nitrides, InN is not very well investigated. The bandgap has been frequently measured by transmission experiments. In early studies, due to large background electron concentration of the samples and low crystalline quality, the excitonic features, neither in the absorption spectra nor in the band-edge photoluminescence spectra, had been observed. However, a number of groups have performed optical measurements on InN (Osamura et al. 1975; Kubota, Kobayashi, and Fujimoto 1989; Wakahara, Tsuchiya, and Yoshida 1990; Inushima et al. 1996). Early experimental studies suggested that the band gap of wurtzite InN ranges from 1.7 to 2.07 eV at room temperature. They were estimated from the absorption spectra, which can lead to overestimating the bandgap if the sample quality is poor or if the sample is highly doped. Despite that, the value of 1.89 eV for the InN band gap obtained by Tansley and Foley (1986) is widely cited in the literature. Recent progress in growth techniques methods tailed for InN using MBE and OMVPE has led to improved InN samples, which show photoluminescence as well as a clear absorption edge. Yodo et al. (2002) observed only weak photoluminescence peaks with energies ranging from 1.81 to 2.16 eV on InN grown on Si substrates. Another branch of studies show that in improved InN films strong photoluminescence transitions at energies around 1 eV appear (see Figure 4.11) (Inushima et al. 2001; Davydov et al. 2002; Wu et al. 2002a, 2002b). These new measurements have challenged the previous widely accepted band gap value and suggest that the actual fundamental band gap of InN is much smaller, between 0.7 and 0.8 eV. The substantially large bandgaps reported earlier may be due to O contamination. Even in the early attempts at determining the bandgap of InGaN, the extrapolated bandgap of InN from the compositional dependence of InGaN bandgap and measured of InN were in conflict as pointed out by Matsuoka (1989); Matsuoka et al. (2003). It should be pointed out that, at least in the case of OMVPE, the high quality InN by recent efforts is grown with N-polarity under exceedingly large ammonia flows (Matuoka et al. 2003; Yoshikawa et al. 2003).

4.5.6 Electrical Properties of III-V Nitrides

Control of the III-V nitride electrical properties has traditionally been, and still is, one of the greatest challenges facing nitride researchers. Electron mobility is one of the most important parameters associated with the material with great impact on devices. Experimental investigation of the temperature-dependent carrier mobility and concentration can be used to determine the fundamental material parameters and understand the carrier scattering mechanism along with an accurate

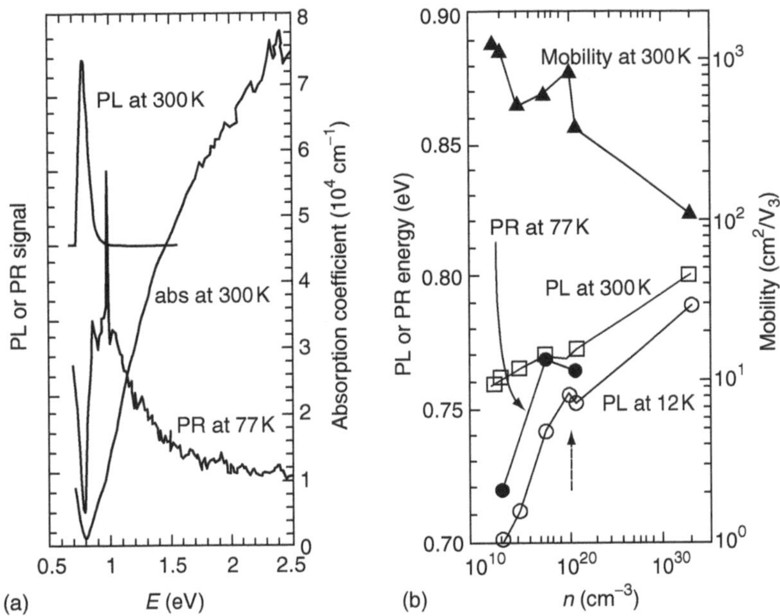

FIGURE 4.11 (a) Optical absorption (300 K), PL (300 K), and PR (77 K) spectra of a typical InN sample. This sample is undoped with room-temperature electron concentration of 5.48×10^{18} cm^{-3}. The spike on the PR spectrum at 0.97 eV is an artifact due to the light source used in the PR measurement. (b) Room-temperature mobility, PL peak energy (300 and 12 K), and the critical energy determined by PR (77 K) as a function of free-electron concentration. (After Wu, J., Walukiewicz, W., Yu, K. M., Ager III, J. W., Haller, E. E., Lu, H., Schaff, W. J., Saito, Y., and Nanishi, Y. 2002C. *Appl. Phys. Lett.*, 80, 3967. With permission.)

comparison with theory (Rode 1975). In GaN films, the earlier transport investigations had to cope with poor crystal quality and low carrier mobility, well below predictions (Chin et al. 1994; Rode and Gaskill 1995). Despite tremendous challenges, early MBE layers exhibited motilities as high as 580 cm^2 V^{-1} s^{-1} at room temperature on SiC substrates, which at that time were not as commonly used substrate as in recent times (Lin et al. 1993). Typically, however, the MBE-grown films produce much lower mobility values of 100–300 cm^2 V^{-1} s^{-1} (Ng et al. 1998). Different models were used to explain the observed low electron mobilities in GaN. Scattering of electrons at charged dislocation lines (Ng et al. 1998; Weimann et al. 1998; Look and Sizelove 1999; Choi, Zhang, and Chua 2001) and scattering through elevated levels of point defects (Wook et al. 1998; Fang et al. 1998), such as nitrogen vacancies (Zhu and Sawaki 2000; Chen et al. 2002) were considered as possible mechanisms responsible for these observations. From the temperature-dependent Hall measurements, it has been argued that the measured mobility is dominated by ionized impurity and dislocation scattering at low temperatures, but polar optical phonon and nitrogen vacancy scattering at high temperatures. It should be pointed out that N vacancy is not considered as relevant as in the past due to early problems achieving sufficient levels of reactive nitrogen. The room-temperature mobility measured by Nakamura, Mukai, and Senoh (1992a) is 600 cm^2 V^{-1} s^{-1} for an electron concentration of $\sim 3\times10^{16}$ cm^{-3}. A higher room temperature mobility of 845 cm^2 V^{-1} s^{-1} at an electron concentration of $\sim$ 6×10^{16} cm^{-3} was achieved by Götz et al. (1998). In later publications, due to the advent of the high-quality samples grown by several growth techniques, there is a significant improvement in reported room temperature mobility values. Nakamura, Mukai, and Senoh (1992b); Binari and Dietrich (1997) were able to obtain 900 cm^2 V^{-1} s^{-1} room-temperature mobility at an electron concentration of 3×10^{16} and 5×10^{16} cm^{-3}, respectively. Recently, even higher electron mobilities of 1100 cm^2 V^{-1} s^{-1} at room temperature and 1425 cm^2 V^{-1} s^{-1} at 273 K were reported by (Yun et al. 2000) and Huang et al. (2001), respectively, for a 200-μm-thick,

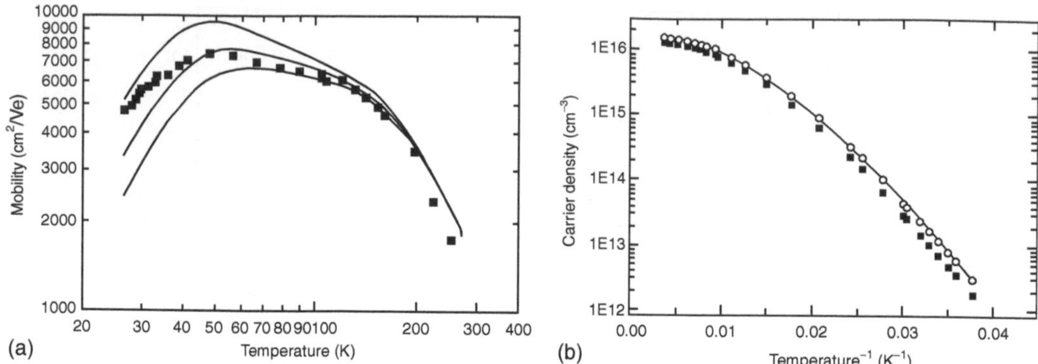

FIGURE 4.12 (a) The measured Hall mobility data (solid squares) from the GaN template grown by HVPE as a function of temperature. The solid line is the calculated result using $N_a = 2.4 \times 10^{15}$ cm^{-3}, representing the best fit to the measured results. The upper and lower dot lines are the calculated results using $N_a = 1.4 \times 10^{15}$ and 3.4×10^{15} cm^{-3}. (b) The measured Hall densities n_H (solid squares) as a function of reciprocal temperature from the GaN template grown by HVPE. The open circles represent the carrier density corrected by the Hall factor, $n = n_H r_H$. The solid line is the fit to the theoretical expression of charge balance with hole and neutral acceptor densities neglected. (After Huang, D., Yun, F., Reshchikov, M. A., Wang, D., Morkoç, H., Rode, D. L., Farina, L. A., Kurdak, Ç., Tsen, K. T., Park, S. S., and Lee, K. Y. 2001. *Solid State Electron.* 45, 711.)

freestanding *n*-type GaN template grown by HVPE at Samsung. This achievement was attributed to the excellent crystalline structure of the GaN sample with low levels of compensation and the defect-related scattering. A quantitative comparison with theoretical calculations demonstrates that the one-layer and one-donor conductance model is sufficient to account for the measured data in the entire temperature range without considering any dislocation scattering and any adjustable parameter other than the acceptor concentration. The measured temperature dependent Hall mobility, carrier concentration and Hall scattering factor are shown in Figure 4.12 along with the best fit theoretical calculation based on Rode's iterative solution of the Boltzmann equation. As shown, quantitative agreement with the measured mobility in the entire temperature range was obtained to within about 30%. Heying et al. (2000) investigated both the morphology and electrical properties of homoepitaxial GaN layers grown by MBE as a function of Ga/N ratio. GaN films grown with higher Ga/N ratios (intermediate regime) showed fewer pits with areas of atomically flat surface, which gives the highest mobility $\sim$1191 cm^2 V^{-1} s^{-1} reported so far at room temperature. Koleske et al. (2000) investigated the effect of the AlN nucleation layer on transport properties of GaN films grown on 6H- and 4H-SiC substrate. Room temperature electron mobilities of 876, 884, and 932 cm^2 V^{-1} s^{-1} were obtained on 6H SiC, 4H SiC, and 3.5° off-axis 6H SiC substrates. They attributed the observed high electron mobilities to the improved AlN morphology and reduction in screw-dislocation density near the AlN/GaN interface.

Due to the low intrinsic carrier concentration, and the deep native defect and impurity energy levels (owing to the wide bandgap of AlN $\sim$6.2 eV at 300 K), the electrical transport properties of AlN have not been studied extensively and have usually been limited to resistivity measurements. Resistivities in the range $\rho = 10^7$–10^{13} Ω cm have been reported for unintentionally doped AlN single crystals (Edwards et al. 1965; Kawabe, Tredgold, and Inyishi 1967), a value consistent with other reports (Cox et al. 1967; Yim et al. 1973; Yoshida et al. 1979). Unintentionally doped *n*AlN films grown by a modified physical transport technique by Rutz (1976) had resistivity as low as 400 Ω cm. Although the source of the electrons has not been determined, Rutz, Harrison, and Cuome (1973) observed an interesting transition in their AIN films in which the resistivity abruptly decreased by two orders of magnitude with an increase in the applied bias. This observation found applications to switchable resistive memory elements that are operated at 20 MHz. It has been concluded by Fara, Bernadini, and Fiorentini (1999), who reported the theoretical evidence, based on ab initio density-functional calculations, for acceptors, donors, and native defects in AlN, for the semi-insulating character of AlN that acceptors are deeper (Be ~ 0.25, Mg ~ 0.45) and less soluble in AlN than in GaN, and both the extrinsic donors, Si_{Al} and C_{Al}, and the native donor, V_N, are found to be deep (about 1 to 3 eV below the conduction). Consequently, doped AlN may in general turn out to be semi-insulating, when attained using Al-rich conditions, in agreement with the known doping difficulties of high Al-content AlGaN alloys. The theoretical calculations have only been made to estimate the mobility in AlN by Chin, Tansley, and Osotchan (1994). Because AlN is an insulator of extremely low carrier concentration, only phonon limited scattering processes were considered in the calculation of the temperature dependence mobility. The mobility was found to decline rapidly at high temperatures, with a value of about 2000 $cm^2\ V^{-1}\ s^{-1}$ at 77 K and dropping to 300 $cm^2\ V^{-1}\ s^{-1}$ at 300 K for optical phonon limited mobility.

Predictions have been made in regard to the mobilities in InN for various temperatures (Tansley and Foley 1985). InN suffers from the lack of a suitable substrate material, causing a high concentration of extended defects in the films. Sapphire substrate is usually used for InN growth, but the difference in lattice constants between InN and sapphire is as large as 25% and even more than 19% with the other nitrides. A large disparity of the atomic radii of In and N is an additional contributing factor to the difficulty of obtaining InN of good quality. Because of all these factors, the electron mobilities obtained from various films have varied very widely. Electrical properties vary also substantially with the choice of growth techniques. A range of film deposition techniques, such as reactive ion sputtering, reactive RF magneton sputtering, metal organic vapor phase epitaxy (MOVPE) and MBE, has been used. However, recent efforts by both MBE and OMVPE have led to consistently high mobilities in excess of 1000 $cm^2\ V^{-1}\ s^{-1}$ with MBE films being slightly better (Morkoç 2006).

In majority, electron mobilities in films up to very recent past often remained relatively poor. This was widely attributed to nitrogen vacancies leading to large background electron concentrations, which is responsible for the observed low electron mobility. An empirical linear relationship between the electron mobility and electron concentration can be deduced from the reported results, although not all films exhibit this type of behavior. A systematic study carried out by Tansley, Egan, and Horrigan (1988) indicates that the electron concentration decreases as the nitrogen density is increased in the growth plasma in reactive ion sputtering. Although the reported time is relatively old, the maximum mobility of 2700 $cm^2\ V^{-1}\ s^{-1}$ at an electron concentration of $5\times10^{16}\ cm^{-3}$ was reported for RF reactive ion sputtered growth of InN, which is higher than the high quality films obtained recently by MBE and OMVPE. In terms of the more recent work, a progressive improvement in electrical properties of InN films grown by vacuum deposition techniques, including MBE and MOVPE is seen. Depending on film quality and background electron concentration electron motilities in the range of 500–2100 $cm^2\ V^{-1}\ s^{-1}$ has been reported in recent works (Yamaguchi et al. 1999; Saito et al. 2002; Lu et al. 2002; Higashivaki and Matsui 2003). Very high unintentional donor concentrations, greater than $10^{18}\ cm^{-3}$, seem to be one of the major problems for further progress in device applications of InN. O_N and Si_{In} and possible interstitial H have been proposed to be the likely dominant defects responsible for high electron concentration for state-of-the-art MBE grown InN, based on their low formation energies (Stampfl et al. 2000; Look et al. 2002). Intractable though is what is recently believed to be the very small bandgap of InN. However, it is possible that the ternary InGaN, which garners some of its properties from InN, may be of some benefit to electronic devices. The opportunity offered by the small bandgap of InN is exploited in stacked solar cells capturing most of the solar radiation for higher conversion efficiency.

The calculated peak mobilities are found to be 25,000, 12,000, and 8000 $cm^2\ V^{-1}\ s^{-1}$ for 10^{16}, 10^{17}, and $10^{18}\ cm^{-3}$, respectively, at different temperatures (between 100 and 200 K) depending on the electron density and compensation ratio. This is due to the interplay of piezoelectric acoustic phonon scattering at low concentrations and ionized impurity scattering at high temperatures. These two mechanisms are the dominant scattering mechanisms below 200 K, while polar mode optical phonon scattering is the most significant process above this temperature. The low-concentration limit for room temperature mobility in uncompensated InN is estimated to be 4400 $cm^2\ V^{-1}\ s^{-1}$ (Chin, Tansley, and Osotchan 1994).

4.5.7 Doping in the III-V Nitrides

As is the case in any semiconductor technology, the control of the electrical properties of GaN and related materials is extremely important in the progress of device development. Doping determines the position of the Fermi level in a semiconductor making the material *n*-type if the Fermi level is close to the conduction band, *p*-type if it is close to the valance band. In general, wide bandgap semiconductors, like III-nitrides, are difficult to dope both *n*- and *p*-type. Unintentionally doped GaN has, in all cases, been observed to be *n*-type with the best samples still showing a compensated electron concentration of about 5×10^{16} cm^3. With the exception of oxygen, no impurity has been found present in a sufficient quantity to account for the carriers, so researchers have attributed the background to "native defects," which are widely believed to be nitrogen vacancies, and/or oxygen.

With a successful reduction of the unintentional *n*-type background, doping GaN with Si has very well been established in both the vapor phase and vacuum deposition techniques. Controllable Si doping of GaN has been demonstrated over a wide range of concentrations (low 10^{17} to mid 10^{19} cm^{-3}). In the dilution limit the donor ionization energy in GaN is about 27 meV and decreases with increasing doping level due to screening effect. PL measurements yielded a value of 22 meV (Götz et al. 1996a). Measurements on high quality GaN layers grown on freestanding GaN wafer indicate the binding energy of Si and O in GaN to be 28.8 and 32.6 meV, respectively (Reshchikov et al. 2001).

Although Si- is widely used for growing *n*-type GaN, Ge can also be used. It has been shown that electron concentrations up to 10^{19} cm^{-3} have been obtained for Ge doping have reported on the properties of Si- and Ge-doped GaN grown by MOVPE (Nakamura, Mukai, and Senoh 1992c). A linear variation of the electron concentration as a function of both the SiH_4 and GeH_4 flow rates was observed across the entire experimental range (see Figure 4.13). Ge incorporation is roughly an order of magnitude less efficient than Si, as judged by the larger GeH_4 flow rates required to obtain similar electron concentrations. Other and much safer sources, such as monomethylsilane, have successfully been employed to obtain *n*-type GaN but their uses are not yet very common. Goldenberg, Zook, and Van Vechten (1993) observed higher *n*-type GaN conductivities as the NH_3 flow was increased during the growth.

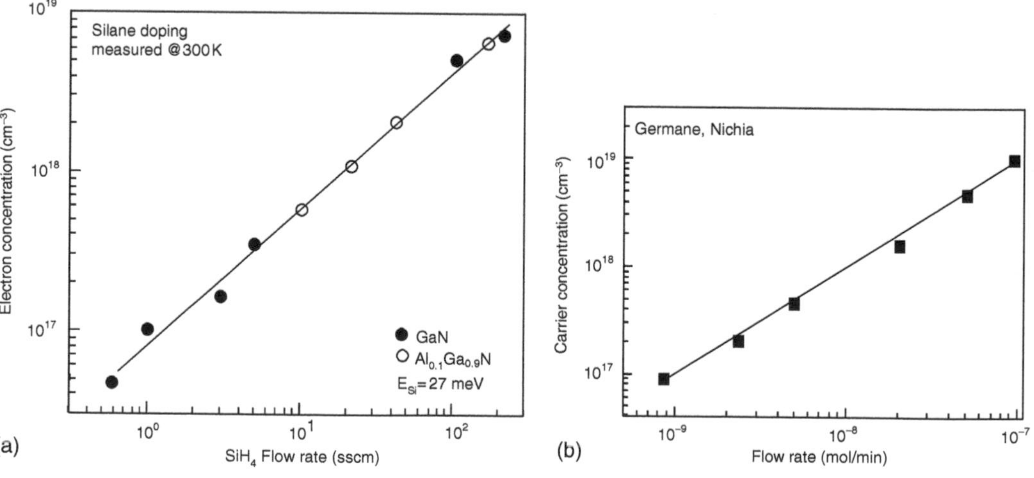

FIGURE 4.13 Incorporation rates as a function of gas flow in MOVPE grown GaN, (a) Si and (b) Ge. Both dopants are well behaved in that the active donor concentrations linearly vary with silane and germane flow rates. However, H4Ge requires a factor of 10 higher flow to obtain the same doping level. (From Nakamura, S., Mukai, T., and Senoh, M., 1992. *Jpn. J. Appl. Phys.*, 31, 195.)

Electron concentrations up to 6×10^{19} cm^3 have been achieved in Se doped GaN MOCVD films (Yi and Wessels 1996). The electron concentration was proportional to the H_2Se flow rate employed as a dopant source. The room-temperature electron mobilities ranged from 10 to 150 cm^2 V^{-1} s^{-1}. The high compensation ratio of $\approx$ 0.4 was nearly constant over the concentration range of $10^{18} - 10^{19}$ cm^3 and became even higher at higher electron concentrations. The compensating acceptor was attributed to the Ga vacancy. There has been much effort aimed at doping GaN and its ternaries *p*-type introducing group-II and group-IV elements. The achievement of a *p*-type doping above the 10^{18} cm^3 level for GaN is a critical step in the development of III-nitrides for device applications such as light emitting diodes and laser diodes. Many potential dopants such as Zn, Cd, Mg, and Be have been incorporated into GaN. However all of these divalent elements form deep acceptors, and some of them has been observed to effectively compensate electrons in GaN, leading to highly resistive material. Neugebauer and Van de Walle (1999) reported the results of a comprehensive first-principles investigation of several possible acceptors in GaN. They found that Mg is still the best acceptor. Optical measurements have revealed that Mg acceptors in GaN and $Al_xGa_{1-x}N$ have activation energy (160–200 meV for GaN) (Tanaka et al. 1994; Götz et al. 1996b; Kim et al. 1997a; Goepfert et al. 1999) much larger than kT above the valance band edge of GaN at room temperature, resulting in low activation and therefore low conductivity in *p*-type GaN and AlGaN. Only after 1990s, *p*-type conductivity has been realized with the work of Amono et al. and Nakamura et al. who demonstrated that post-growth treatments such as low energy electron beam (LEEBI) irradiation and thermal annealing in N_2 atmosphere, respectively, convert Mg compensated GaN into *p*-type material. This is due to dissociation of the neural Mg-H complexes responsible for the semi-insulating nature of as-grown Mg doped GaN. Recently Cheong et al. reported acceptor density and annealing temperature dependent activation energy for Mg in GaN quantitatively (Cheong et al. 2002). For MBE grown Mg-doped GaN, *p*-type conductivity has been reported without post-growth treatment due to the absence of hydrogen and H-N radicals during growth. The Mg incorporation during MBE growth is known to depend on the stoichiometry and polarity of the growing surface (Ptak et al. 2001) as well as on the substrate temperature and growth rate (Smorchkova et al. 2000).

The codoping method, in which both *n*- and *p*-type dopants at the same time under thermal non-equilibrium crystal growth conditions by using (MBE) or (OMVPE) has been used. Yamamoto and Katayama-Yoshida (1998) investigated the role of *n*-type dopants, Si and O, codoping in doping properties of *p*-type GaN doped with Be or Mg using ab initio electronic band structure calculations. They concluded that the *p*-type codoped GaN using Be or Mg as acceptors and Si or O as donors would exhibit an increased incorporation of Be or Mg acceptors compared with *p*-type GaN doped with the acceptors alone. Nowadays, very successful experiments have been achieved by using codoping method to fabricate the low-resistivity *p*-type GaN. Korotkov et al successfully achieved highly conductive *p*-type GaN by using Mg and O codoping using OMVPE growth. The resistivity of codoped layers decreased from 8 to 0.2 Ω upon oxygen codoping. The activation energy of Mg decreased from 170 meV to 135 meV, and the hole concentration is as high as 2×10^{18} cm^{-3} (Korotkov, Gregie, and Wessels 2001). Kipshidze et al. (2002) also reported their successful Mg and O codoping in *p*-type GaN and $Al_xGa_{1-x}N$ ($x < 0.08$) using GSMBE. Kim et al. (2000) used Mg–Zn codoping in GaN by MOCVD. A low resistivity of 0.72 and hole concentration of 8.5×10^{17} were reported.

Another promising method to obtain higher acceptor activation and lower resistivity is using $Al_xGa_{1-x}N$/GaN doped superlattices (Schubert, Grieshaber, and Goepfert 1996; Kozodoy et al. 1999a; Saxler et al. 1999). The use of $Al_xGa_{1-x}N$/GaN superlattice is from the idea of AlGaN/GaN heterojunction for enhancement of *p*-type doping, (Hsu and Walukiewicz 1999; Kozodoy et al. 1999b) which has been proposed as a technique to increase the average hole concentration. Recent work has provided an experimental demonstration of increased hole concentration at room temperature through the use of AlGaN/GaN superlattices (Waldron, Graff, and Schubert 2001).

A few studies have been reported on Mg-doped *p*-type $Al_xGa_{1-x}N$ with low Al content $0 < x < 0.27$ (Tanaka et al. 1994; Suziki et al. 1998; Suigura et al. 1998; Li et al. 2002b). Figure 14 shows activation energies of Mg acceptors in Mg-doped *p*-type $Al_xGa_{1-x}N$ as a function of Al content x with an activation energy as high as 310 meV for $x = 0.27$. Although quite a lot of efforts have been made in *p*-type doping in

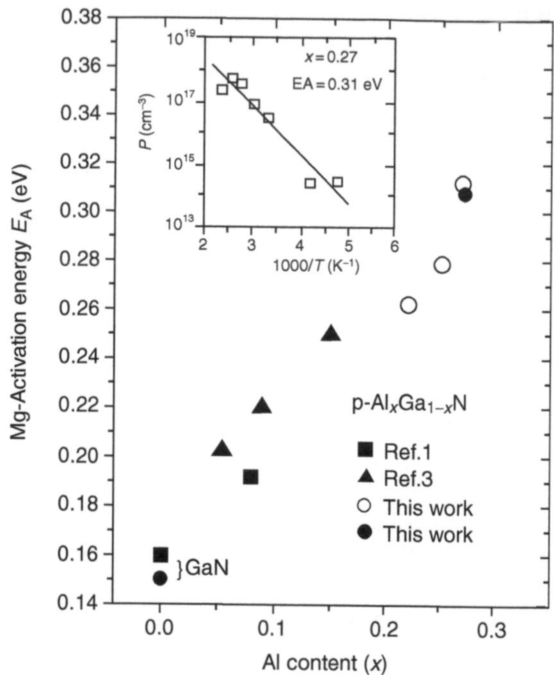

FIGURE 4.14 Activation energies of Mg acceptors in Mg-doped *p*-type $Al_xGa_{1-x}N$ as a function of Al content *x*. Closed squares, closed circles and triangles are data from (From Akasaki, I., Amano. H. 1994. *J. Electrochem. Soc.*, 141, 2266. Akasaki, I., Amano. H. 1997. *Jpn. J. Appl. Phys.*, 36, 5393; Alexandrov, P., Pan, M., Weiner, M., Jiao, L., Zhao, J. H. 2003. *Solid State Electron.*, 47, 263.) respectively, all obtained by Hall measurements. Open circles indicate data obtained by PL measurements from (From Akasaki, I., Amano. H. 1994. *J. Electrochem. Soc.*, 141, 2266.) The inset shows measured temperature dependence of Hall concentration *p* in the Mg-doped *p*-type $Al_{0.27}Ga_{0.73}N$ sample from which $E_A = 0.310$ eV was obtained. (After Li, J., Oder, T. N., Nakarmi, M. L., Lin, J. Y., Jiang. H. X. 2002b. *Appl. Phys. Lett.*, 80, 1210. With permission.)

GaN and AlGaN, however, it is still a limitation for opto-electronic devices, particularly when for the shorter wavelength devices, such as real solar-blind ($\lambda < 290$ nm) detectors, or UV LEDs or LDs, which need *p*-type AlGaN with high Al composition. For example, a UV LED using AlGaN *p-n* junction with emission wavelength shorter than 290 nm can only give submilliwatt power (Adivarahan et al. 2002).

4.6 Properties of Nitride Alloys

Many important GaNbased devices involve heterostructures to achieve an improved device performance. Ternary alloys of wurtzite polytypes of GaN and AlN, and InN have been obtained in the continuous alloy systems whose direct band gap ranges from 0.7 eV for InN to 6.2 eV for AlN. Many of these properties such as the energy bandgap, effective masses of the electrons and holes, and the dielectric constant are dependent on the alloy composition. AlN and GaN are reasonably well latticed matched (2.4%) and for many devices only small amounts of AlN are needed in the GaN lattice to provide sufficient carrier and optical field confinement. The compositional dependence of the lattice constant, the direct energy gap, electrical and cothodoluminescence (CL) properties of the AlGaN alloys was measured (Yoshida, Misawa, and Gonda 1982). In general, the compositional dependence of the optical bandgap of ternary alloys can be predicted by the following empirical expression:

$$E_g(x) = xE_g(\mathrm{AlN}) + (1-x)E_g(\mathrm{GaN}) - x(1-x)b, \tag{4.2}$$

where E_g(AlN) and E_g(GaN) are the optical bandgap of AlN and GaN, respectively, while x and b are the Al N molar fraction and bowing parameter, respectively. In order to determine this relation precise characterization of both bandgap and alloy composition is important. A wide dispersion in the bowing parameters ranging from -0.8 eV (upward bowing) to $+2.6$ eV (downward bowing) has been reported (Shan et al. 1999; Ochalski et al. 1999; Nikishin et al. 2000; Wagner et al. 2000; ÖÖzgür et al. 2001zgür et al. 2001; Jiang et al. 2001). Yun et al. (2002) revisited the bowing parameter using x-ray and analytical techniques, such as secondary ion mass spectroscopy (SIMS) and Rutherford back Scattering (RBS), for composition determination, reflectance and absorption for bandgap determination. The results of this study are depicted in Figure 4.15 in the form of AlGaN bandgap vs. composition along with a least square fit to the data, solid circles, yields a bowing parameter of $b=1.0$ eV for the entire range of alloy compositions. Widening the x-ray diffraction peaks for alloy composition around the midway point has been attributed to be a most likely source of error in determining the bowing parameter. On the other hand, the validity of the characterization techniques used in determination the optical properties of AlGaN alloys is deeply affected by the material crystalline quality and purity.

As far as the electrical and doping issues are concerned, Hall measurement for $n\mathrm{Al}_{0.09}\mathrm{Ga}_{0.91}\mathrm{N}$ demonstrated a carrier concentration of 5×10^{18} cm^3 and a mobility of 35 cm^2 V^{-1}s^{-1} at room temperature (Khan et al. 1991). Other Hall measurements (Tanaka et al. 1994) on Mg-doped $p\mathrm{Al}_{0.08}\mathrm{Ga}_{0.92}\mathrm{N}$ grown by OMVPE addressed the temperature dependence of the mobility. They indicate that the hole mobility decreases with increasing temperature, reaching a value of about 9 cm^2 V^{-1} s^{-1} for a doping density of 1.48×10^{19} cm^3. This low mobility is ascribed to a high carrier concentration and the intergrain scattering present in the samples. While the lattice constant was studied, it was observed to be almost linearly dependent on the Al N mole fraction in AlGaN.

Until recently the resistivity of unintentionally doped AlGaN was believed to increase so rapidly with increasing AlN mole fraction that AlGaN became almost insulating for AIN mole fractions exceeding 20%. As the AlN mole fraction increased from 0 to 30%, the *n*type carrier concentration dropped from 10^{20} to 10^{17} cm^3, and the mobility increased from 10 to 30 cm^2 V^{-1}s^{-1}. An increase in the native-defect ionization energies with increasing AlN may possibly be responsible for this variation. The respond of the dopant atoms such as Si and Mg to the variation of the AlN mole fraction in AlGaN has not been well understood yet. It was suggested that dopant atom moves deeper into the forbidden energy bandgap as the AlN mole fraction increases. For example, Hall effect measurements show that the activation energy of Si donor increases linearly from 0.02 eV in GaN to 0.32 eV in AlN (Stutzmann et al. 1997). However, devices such as lasers, which depend critically on the overall device series resistance, will probably be restricted by the ability to dope high mole fraction AlGaN, especially when low resistivity *p*-type material is required. Fortunately, the emergence of InGaN coupled with the fact that good optical field confinement can be obtained with low AlN mole fraction AlGaN mitigate this problem enormously, and the potential appears great for laser development in this material system.

The ternary InGaN is used mostly for quantum wells, strained to some extent depending on the level of phase segregation, etc., in the active regions of LEDs and lasers, which can emit in the violet or blue wavelength range. Needless to say, high-efficiency blue and green LEDs utilizing InGaN active layers are commercially available. However, added complexities such as the phase separation and other inhomogeneities due to great disparity between Ga and In make the determination of the bandgap of InGaN versus composition a very difficult task. The compositional dependence of InGaN bandgap is a crucial parameter in designs of any heterostructure utilizing it. Similar to the case of AlGaN, the energy bandgap of $\mathrm{In}_x\mathrm{Ga}_{1\text{-}x}\mathrm{N}$ over $0 \leq x \leq 1$ can be expressed by Equation 2 using the bandgap of InN instead of AlN bandgap. When a bandgap of ~ 1.9 eV for InN is assumed as the end point value for InN in regard to InGaN ternary, large and or more than one bowing parameters are required to fit the compositional dependence of the bandgap energy. An earlier investigation of InGaN bowing parameter for alloys with small concentrations of InN (Nakamura and Mukai 1995a) led to a bowing parameter of 1.0 which is in disagreement with 3.2 reported by Amano et al. (1997), who took into consideration strain and piezoelectric fields as well and arrived at a value of 3.2. A bowing parameter of 2.5 eV was obtained from optical absorption measurements and a value of 4.4 eV was obtained from the position of the emission

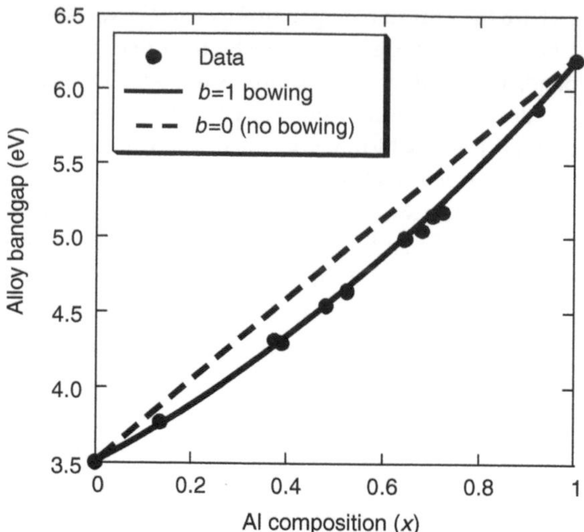

FIGURE 4.15 Experimental data of energy bandgap of AlGaN ($0 \leq x \leq 1$) plotted as a function of Al composition (solid circle), and the least squares fit (solid line) giving a bowing parameter of $b = 1.0$ eV. The dashed line shows the case of zero bowing. As the quality of the near 50:50 alloy layers get better, giving rise to sharper x-ray and PR data, smaller bowing parameters may ensue. Already, bowing parameters as low as 0.7 have been reported. (After Yun, F., Reshchikov, M. A., He, L., King, T., Morkoç, H., Novak, S. W., and Wei, L. 2002. *J. Appl. Phys. Rapid Commun.*, 92, 4837.)

peak (O'Donnell et al. 2001). Nagatomo et al. (1989) noted that the $In_xGa_{1-x}N$ lattice constant varies linearly with the In mole fraction up to at least $x = 0.42$, but it violates Vegard's law for $x > 0.42$, which may be caused by erroneous determination of the composition and speaks well to the difficulty of the problem at hand. Recent observations indicated that these alloys show strong infrared PL signal as expected from an InN bandgap of ~0.8 eV, extending the emission spectrum of the $In_{1-x}Ga_xN$ system to near infrared. Wu et al. (2002a, 2002b, 2002c) revisited the bandgap dependence of InGaN on composition by considering ~0.8 eV for the bandgap of the end binary InN. Shown in Figure 4.16 is the composition dependence of the bandgap of InGaN, determined by using the techniques of photomodulated transmission, (Shan et al. 1998) optical absorption (Pereira et al. 2001) measurements, as a function of GaN fraction. The compositional dependence of the bandgap in the entire composition range can also be fit by a bowing parameter of $b = 1.43$ eV. Shown in the figure with dashed line is the fit to the empirical expression using energy of 1.9 eV for InN and bowing parameter of 2.63 eV to demonstrate that it does represent the Ga rich side of the compositions well. However, the bowing parameter that is good for the entire compositional range is the1.43 eV utilizing 0.77 eV for the bandgap of InN.

There are isolated reports on electrical properties of InGaN alloy. Nagatomo et al. (1989) reported high resistivity InGaN films with In compositions as high as 0.42. Background electron concentration as low as 10^{16} cm^{-3} has been reported for MOVPE grown InGaN films. Conductivity control of both n-type and p-type InGaN was reported by Akasaki and Amano (1997). The growth and mobility of p-InGaN was also discussed by Yamasaki et al. (1995). Yoshimoto, Matsuoka, and Katsui (1991) studied the effect of growth conditions on the carrier concentration and transport properties of $In_xGa_{1-x}N$. They observed that by increasing the deposition temperature of $In_xGa_{1x}N$ grown both on sapphire and ZnO with x ≈ 0.2, results in decrease in carrier concentration and increase in electron mobility. Nakamura and Mukai (1992) discovered that the film quality of $In_xGa_{1-x}N$ grown on highquality GaN films could be significantly improved. Thus, it may be concluded that the major challenge for obtaining high-mobility InGaN is to find a compromise in the growth temperature, since InN is unstable at typical GaN deposition temperatures. This growth temperature would undoubtedly be a function of the dopant atoms, as well as the method (MBE, OMVPE, etc.) used for the growth. This is evident from a study by Nakamura and co-workers, who have since expanded the study of InGaN employing Si (Nakamura, Mukai, and Senoh 1993a) and Cd (Nakamura, Iwasa, and Nagahama 1993b).

$In_{1-x}Al_xN$ is an important compound that can provide a lattice-matched barrier to GaN, low fraction AlGaN and InGaN, and consequently, lattice matched AlInN/AlGaN or AlInN/InGaN heterostructures. The growth and electrical properties of this semiconductor have not yet been extensively studied, as the growth of this ternary is also challenging because of diverse thermal stability, lattice constant, cohesive energy of AlN and InN. Kim et al. (1997b) deposited thin AlInN films and observed an increase of In content in AlInN of up to 8% by lowering the substrate temperature to 600°C. Radio Frequency (RF) sputtering was employed to grow InAlN alloy by Starosta (1981), and later Kubota, Kobayashi, and Fujimoto (1989). Kistenmacher, Ecelberger, and Bryden (1993), on the other hand, used the RF magnetron sputtering (RFMS) from a composite metal target to grow InAlN at 300°C. It was observed that the energy bandgap of this alloy varies between 2.0 and 6.20 eV for x between 0 and 1 (Kubota, Kobayashi, and Fujimoto 1989). Optical properties of 1 μm thick $Al_{1-x}In_xN$ layers for x values up to in the range of 0.19 and 0.44 have been investigated by absorption and photoluminescence (Yamaguchi et al. 1998). Figure 4.17 shows results of the photoluminescence spectra from $Al_{1-x}In_xN$ layers. From the absorption spectra the band gaps are found to shift gradually with varying In content. Tailing of the band gap edge is also found to occur, probably due to fluctuation of In content and due to grain boundaries. The PL peak energy position shifts towards the lower energy region with an increase in In content as expected (Davies 1989). Considering that PL spectral linewidths are much larger than expected from a completely random alloy, mechanisms other than the alloy broadening such as microscopic phase separation might be invoked to account for such broadening. The degree to which the PL peak shifts (the same is true for the absorption edge) with mole fraction, x, is very small might reflect the immiscibility of AlN in InN (or vice versa).

Transport data in InAlN are extremely scarce. The carrier concentration and the mobility of $In_{1x}Al_xN$ for $x=0.04$ were reported 2×10^{20} cm^3 and 35 cm^2 V^{-1}s^{-1}, respectively, and 8×10^{19} cm^3 and 2 cm^2 V^{-1}s^{-1} for $x=0.25$, respectively (Bryden and Kistenmacher 1994). The mobility was found to decrease substantially with an increase in the Al mole fraction and increase with an increase in the In mole fraction in close analogy to the parent compounds AlN and InN.

By alloying InN together with GaN and AlN, the band gap of the resulting alloy(s) can be increased from 1.9 eV (or near 0.8 eV if we use the recently determined InN bandgap) to a value of 6.2 eV, which is critical for making high efficiency visible light sources and detectors. In addition, the bandgap of this quaternary can be changed while keeping the lattice content matched to GaN (Mohammad, Salvador, and Morkoç 1995; Bedair et al. 1997). In quaternary alloys of nitrides the N atoms constitute anion sublattice while three groups elements (In, Ga, Al) constitute the cation sublattice. Use of this quaternary material allows almost independent control of the band offset in AlInGaN-based heterostructures. However, among other difficulties brought about by the 4 component system, the optimal growth temperature is important to optimize and control, as aluminum-based compounds generally require higher growth temperatures, and In based ones require lower temperatures. Higher temperatures are also desirable for reducing the O incorporation in the growing film as oxides of Ga and In desorb from the surface. The growth temperature will therefore govern the limits of In and Al incorporation into the AlGaInN quaternary alloy (Bedair et al. 1997). Quaternary alloys, $(Ga_{1-x}Al_x)In_{1-y}N$, are expected to exist in the entire composition range $0 < x < 1$ and $0 < y < 1$. Unfortunately, as in the case of the InGaN alloy, incorporation of indium in these quaternary alloys is not easy. Ryu et al. (2002) reported on optical emission in this quaternary system and AlInGaN/AlInGaN multiple quantum wells grown by pulsed-metal-organic chemical-vapor deposition. Strong blue shift with excitation intensity was observed in both the quaternary layers and quantum wells, which was attributed to localization. This would imply that the structures are of inhomogeneous nature and or presence of band tail states indicative or early stages of material development and or serious technological problems involved.

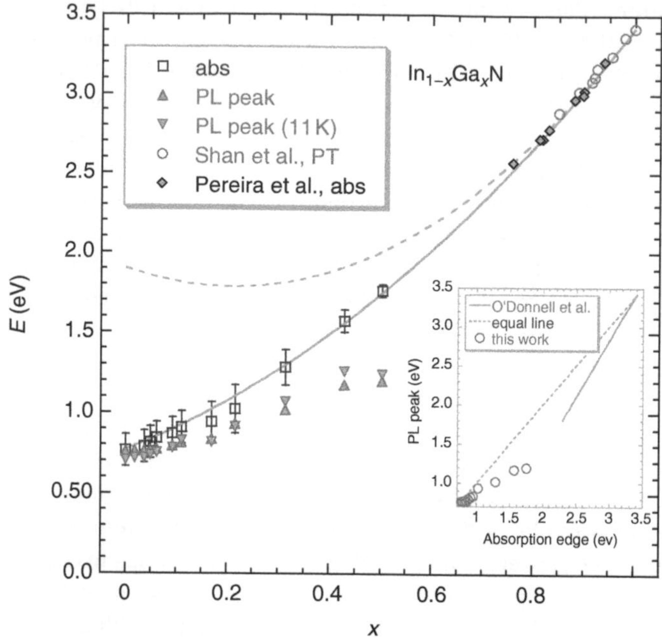

FIGURE 4.16 PL peak energy and bandgap determined by optical absorption as a function of composition inclusive of previously reported data for the Ga-rich end of the ternary which is not as affected by the large InN bandgap pereviously used. The solid curve shows the fit to the bandgap energies (abs and PT) using a bowing parameter $b = 1.43$ eV. The dashed curve is the fit of the bandgap energies on the Ga-rich side assuming a bandgap of 1.9 eV for InN. Inset: PL peak energy plotted against absorption edge energy. The solid line is a least-square fit to experimental data on the Ga-rich side. The dashed straight line shows the relation when the Stokes shift is zero (After Wu, J., Walukiewicz, W., Yu, K. M., Ager III, J. W., Haller, E. E., Lu, H. and Schaff, W. J. 2002a. *Appl. Phys. Lett.*, 80, 4741; Wu, J., Walukiewicz, W., Shan, W., Yu, K. M., Ager III, J. W., Haller, E. E., Lu, H. and Schaff, W.J. 2002b. *Phys. Rev. B.*, 80, 201413 and Wu, J., Walukiewicz, W., Yu, K. M., Ager III, J. W., Haller, E. E., Lu, H., Schaff, W. J., Saito, Y. and Nanishi, Y. 2002c. *Appl. Phys. Lett.*, 80, 3967. With permission.)

4.7 Ohmic Contacts to III-V Nitrides

Low-resistance ohmic contacts are vital for a semiconductor device to be connected to the outside world with no adverse change to its current-voltage characteristics and no additional voltage drop. Ideally, this occurs when the semiconductor and the metal work functions are about the same, and there are no appreciable interface states, which tend to pin the Fermi level. Ohmic contacts to GaN and the related semiconductors have received considerable attention in the last decade. Ohmic contacts to *n*-GaN have been far easier to achieve than that to *p*-GaN due to the non-existence metal with a large-enough work function close to *p*-type GaN. Early studies suggested that specific contact resistances to *n*-type GaN of $10^4 - 10^3\ \Omega\ cm^2$ could be realized using Al and Au metallizations (Foresi and Moustakas 1993). From current-voltage measurements it was determined that the *as-deposited* Al contacts were ohmic. However, the as deposited Au contacts were rectifying and became ohmic after annealing at 575°C, paving the way to the argument that work functions of GaN and Al are very close and that the surface of GaN is defect free. The most frequently reported ohmic contacts to *n*-GaN are variations on the Ti/Al metallization scheme (Lin et al. 1994a; Lester et al. 1996; Luther et al. 1997). Specific contact resistivities below 10^{-7} $\Omega\ cm^2$ have been reported for the Ti/Al/Ni/Au (15 nm/220 nm/40 nm/50 nm) contact when a pre-metallization reactive ion etch was employed at 900°C for 30 s. (Fan et al. 1996). Somewhat higher contact resistivities have been reported when nitride layers, such as TiN (Mohney et al. 1998) and Ti/Al

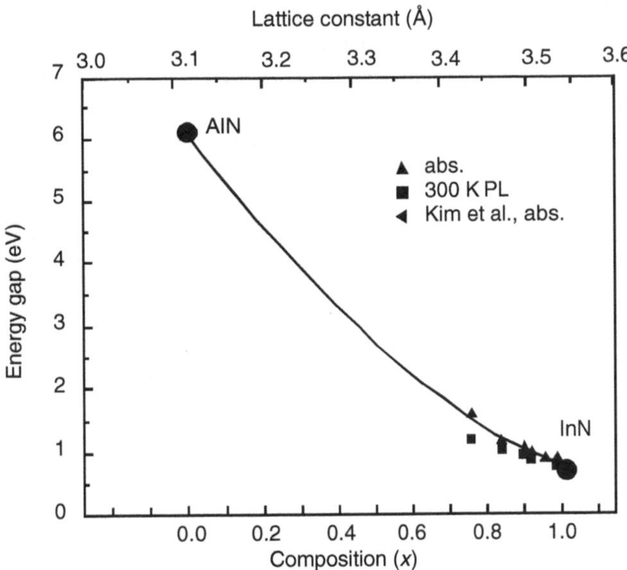

FIGURE 4.17 Bandgap of $In_xAl_{1-x}N$ films for various In concentrations. (Courtesy of Wladek Walukiewicz. With permission.)

on a short period InN/GaN superlattices (Lin, Huang, and Morkoc 1994b), are directly deposited on *n*-type GaN. They presented smooth surface morphologies and good long-term thermal stability. Other approaches, such as evaporation of Ti/Al on GaN implanted by Si, to ohmic contact fabrication have been applied and the contact resistivity of about 4×10^{-8} Ω cm^2 was achieved (Burn et al. 1997).

Many of the low resistance ohmic contacts to *n*-GaN, including those in the Ti/Al family, involve nitride layers in direct contact with the *n*-GaN. Researchers have speculated on the nature of the reactions responsible for these low contact resistances. Mostly accepted one is the formation of a nitride layer through reaction with GaN. One plausible process whereby this may occur involves the solid phase reaction between the Ti and GaN forming TiN. Suppose N is extracted from GaN without decomposing the GaN structure (i.e., N out-diffusion from the GaN lattice), then an accumulation of N vacancies created in the GaN near the junction would provide the desired benefit of increased electron concentration. This region would be heavily doped n-type, which enhances the carrier transport by tunneling processes. We note that only two monolayers of TiN are needed in order to generate a 100-Å layer of GaN with an electron density of 10^{20} cm^{-3}. If there were Al in the metal scheme it would act to passivate the surface and also possibly reacts with Ti forming TiAl (Ruvimov et al. 1996). This very possibility was investigated in contacts formed by Ti/Al, and Ti/Al/Ni/Au before and after annealing by Rutherford backscattering (RBS) by Duxstad (1997) Annealed and unannealed TiAl contacts on GaN, examined by RBS, indicate some Ti and GaN on the surface, which leads to the conclusion that the picture is more complicated. Similar experiments with the multilayered Ti/Al/Ni/Au (500 Å/2000 Å/700 Å/600 Å) contacts revealed the presence of Ti, Al, and Ni on the surface after annealing Duxstad (1997).

The oxygen presented in the annealing ambient has a detrimental effect on the resulting electrical properties of Al/Ti/n-GaN contacts (Pelto et al. 2001). A Ni/Au bilayer directly on the top of Ti/Al that addresses the oxidation issue with much success. In addition, Pt/Au (Cai et al. 1998), Pd/Au (Chor et al. 2001), Ti/Au (Wang et al. 2001), Mo/Au (Kumar et al. 2002) and $TiAl_3$ (Pelto et al. 2002) were also examined as the oxidation resistance layer and showed promise. It should be noted that the annealing of Ti/Al/$TiAl_3$ can be carried out at lower temperature (700°C) and the contact resistance is very insensitive to the amount of oxygen in the annealing ambient. This metallization scheme is a potential candidate for implementation in the devices that require both n- and p-GaN contacts on the same wafer, since it can achieve near optimal performance in an oxidation anneal at roughly the same temperature as that common for Ni/Au contacts to p-GaN annealed in an oxidizing atmosphere. Other techniques such as vacuum annealing (Papanicolaou et al. 2001) minimizing the oxidation of the Al surface, and Cr/Al/Ni/Au requiring considerably lower annealing temperature (550~600°C) have also been explored (Papanicolaou and Zekentes 2002). The achieved specific contact resistances are about $1\sim 3\times10^{-5}$ Ω cm^2.

In contrast to n-GaN, achieving low resistance ohmic contacts to p-GaN are much more difficult and has been challenging due to large work function of p-GaN, residual hydrogen passivation effect, and relatively low hole concentration. The most commonly employed metallization is Ni/Au. Several approaches to processing ohmic contacts have been proposed for the resistivity reduction. These methods include post-metallization anneal to activate the Mg dopant, electrodeposition process for metallization, and annealing in an oxygen-containing environment. Improved Ni/Au ohmic contacts to moderately doped p-GaN ($N_A=10^{17\sim18}$ cm^{-3}) have been demonstrated by annealing in O_2/N_2 and resistivity as low as 4.0×10^{-6} Ω cm^2 was achieved (Ho et al. 1999; Chen et al. 2000). Two mechanisms were proposed for the resistivity reduction, i.e., NiO as a p-type semiconductor forms a thin Schottky barrier with p-GaN by Ho et al. (2000), or the presence of oxygen during annealing helps to remove the residual hydrogen which bonded with Mg or N atoms in the p-GaN epilayer by Koide et al. (1999). Both Hull et al. (2000) and Qiao et al. (2000) found that the presence of oxygen in annealing environment could reduce the resistivity of p-GaN, and their experimental observation supports the argument of Koide et al. (1999). Lee, Lin, and Lee (2001b) routinely obtained a specific contact resistance of 4.5×10^{-6} Ω cm^2 by peroxiding p-GaN at 750°C for 30 min in air and then removing the oxide using $(NH_4)_2S_x$. It is argued that the Ga vacancies acted as acceptors result in more hole concentration due to the removal of the native oxide and GaO_x by $(NH_4)_2S_x$ solution. Moreover, it is proposed that the reversal reactions assisted by oxidation forms a structure of NiO/Au/p-GaN and the presence of Au throughout the entire metallized layer improves the conductivity of the contact further (Chen et al. 1999).

Besides the widely used Ni/Au ohmic contact to p-GaN, other metallization schemes have also been studied recently. On most moderately Mg doped p-GaN, Ni/Pd/Au ohmic contact exhibited low resistivity of 1.0×10^{-4} Ω cm^2 (Chu et al. 2000), Ti/Pt/Au 4.2×10^{-5} Ω cm^2 (Zhou et al. 2000a), Pd/Ni 5.7×10^{-6} Ω cm^2 (Jang et al. 2001), Ru/Ni 4.5×10^{-5} Ω cm^2 (Jang et al. 2002), Pt/Ru 2.2×10^{-6} Ω cm^2 (Jang et al. 2000), Ta/Ti 3×10^{-5} Ω cm^2 (Suzuki et al. 1999), Pd/Ag/Au/Ti/Au 1.0×10^{-6} Ω cm^2 (Adivarahan et al. 2001a). These schemes may need further investigation to be suitable for the fabrication of thermal stable, low resistivity ohmic contact to p-GaN with high reproducibility.

Non-alloyed (as-deposited) ohmic contacts to n^+-InN and n^+-$In_xGa_{1-x}N$ ($x=0.65$) have been demonstrated using W, Ti/Al, and WSi_x metal scheme (Ren et al. 1997). Among these W contacts exhibited the best thermal stability, with a contact resistance of $<10^{-7}$ Ω cm^2 and 1×10^{-7} Ω cm^2 on $In_{0.65}Ga_{0.35}N$ and InN after annealing at 600°C and 300°C, respectively. It is difficult to achieve low contact resistance on AlGaN due to large bandgap of AlN and hence large Schottky barrier height of many metals on AlGaN. Contact resistivities of about 3×10^{-6} Ω cm^2 have been measured for optimized Al/Ti/n-AlGaN/n-GaN contacts for heterostructure devices (Liu et al. 1997). Contact resistivities of 10^{-2} Ω cm^2 have been reported for annealed Ti contacts to n-$Al_{0.5}In_{0.5}N$ by Guohua et al. (1997), and even lower resistance ohmic contact have been achieved on AlInN graded layer followed by deposition of a WSi_x metal (Donovan et al. 1997). Qiao et al. (2001a, 2001b) proposed Ta-based interface ohmic contact, which leads to a resistivity of 5×10^{-7} Ω cm^2, efficient electron emission for e-beam lithographic alignment, and elimination of the capping layer for the ohmic annealing. The specific contact resistances of Ti/Al/Ni/Au ohmic contacts to AlGaN/GaN MODFET structures, with the value of 3.44×10^{-6} Ω cm^2 (Chu et al. 1997) and 7.3×10^{-7} Ω cm^2 (Jacobs et al. 2002), have also been reported. However, as more Al-rich alloys are used, achieving low contact resistivities to these materials may present a considerable challenge.

4.8 GaN-Based LEDs

Light-emitting diodes have undergone a tremendous advancement in performance and are now used in nearly every aspect of life. Among the applications of LEDs are displays, indicator lights, signs, traffic lights, lighting, which requires emission in the visible part of the spectrum, printers, and telecommunications (Mukai et al. 2001). A primary goal of GaN research is to efficiently harness its direct energy bandgap for optical emission. By appropriately alloying GaN with its cousins AlN and InN, the energy bandgap of the resulting Al(In)GaN can be altered for emission in the range of ultraviolet to yellow or even red, as shown in Figure 4.18. The first GaN LED was reported over thirty years ago (Pankove, Miller, and Berkeyheiser 1971) using MIS structures rather than p-n junction due to the inability at the time to dope GaN *p*-type. With the achievement of *p*-type doping, as discussed above, followed by the improvement in the quality of GaN and its alloys In(Al)GaN, the developments in nitride LEDs are tremendous.

The first GaN p-n junction LED consisting of a Mg-doped GaN layer grown sapphire substrate was demonstrated by Amano et al. (1989). The electroluminescence of the devices was dominated by nearbandedge emission at 375 nm, which was attributed to transitions involving injected electrons and Mg-associated centers in the *p*-GaN region. In order to reduce the internal absorption losses and total internal reflections double heterostructures (DHs) design for light-emission devices have been developed (Nakamura, Mukai, and Senoh 1991; Akasaki and Amano 1994). InGaN alloys for emission media are required to achieve other desired colors besides 375 nm emissions. The blue and blue-green LEDs developed by Nichia Chemical initially relied on the transitions to Zn centers, which situated about 500 meV above the valence-band edge of GaN. These LEDs had the undesirable characteristics of wide spectral widths and a saturation of the light output with injection current accompanied by the blue shift. The In mole-fraction content of the active layer was varied and resulted in a shift of the peak wavelengths of the device's electroluminescent spectra from 411 to 420 nm. Impressively, researchers at Nichia Chemical (Nakamura et al. 1995b) were later successfully demonstrated Zn-free blue and green LEDs by reducing the thickness of InGaN emission layers to about 30 Å. This paved the way to use InGaN quantum wells with InN mole fractions up to a maximum of 70% for excellent violet, blue, green, and yellow/amber InGaN LEDs. The commercial LEDs exhibit power levels of 5 and 3 mW at 20 mA injections current for the wavelengths of 450 and 525 nm, respectively. A schematic representation of one such Nichia LED is illustrated in Figure 4.19. Elimination of the Zn centers led to the very important consequence that the FWHM of the emission spectra were reduced to 20, 30, and 80 nm for blue, green, and yellow LEDs, respectively (see Figure 4.20). The In mole fractions used are 15–20, 40–45, and about 60% for the 450, 525, and 590 nm emission, respectively.

As of July 2002, for better device performance in terms of increasing the light extraction, LumiLeds developed large area flip chip mount LEDs operative in green, cyan and blue wavelength (Steranka et al. 2002). It has been demonstrated that the flip chip LEDs can be operated up to 1.0 A without significant power degradation or failure, and supports a thermal resistance of approximately 14°C/W. Moreover, the flip chip LED has higher efficiency, producing 16 lm with about 27 lm/W efficiency at an injection current of 200 mA which corresponds to a current density of about 30 A/cm^2. When the drive current of the flip chip LED is increased to 1 A, a flux value of 48 lm at 445 nm is produced. The forward voltage of the power flip chip LED is 2.95 V at 200 mA compared to 3.15 V for the top-emitting power LED. The flip chip LED external quantum efficiency is also1.6 times greater than the top-emitting power LED from the blue to the green region of the visible spectrum.

The attainment of larger LED chips that can produce large luminous flux, albeit at high current densities will be critical to expanding the range of application for LEDs as well as reducing their cost. In one such effort, LumiLeds produced high brightness LEDs called Luxeon, which is designed to handle high power, and give extremely bright output. Luxeon LEDs mounted on a heat-sink sub-mount deliver average lumen maintenance of 70% through 50,000 h under typical conditions. Shown in Figure 21 are two such output power vs. input drive power for 1×1 mm^2 and 2×2 mm^2 die size Luxeon LEDs, the former producing 259 mW at a drive of 350 mA/3.2V with 22.6% wall-plug efficiency, the latter producing 1.1 W at a drive of 700 mA/7V with 22.4% wall plug efficiency. A Luxeon 5-watt green LED produced a flux per package of 120 lm/LED with a luminous efficiency of 24 lm/W, while the device with 1 W package produced an efficiency of 50 lm/W.

For the amber color LEDs, the InGaN/GaN material system has some advantage over InGaAlP material system in terms of carrier confinement. If high quality InGaN layers with large In concentration could in fact be grown, the temperature stability of these LEDs would be superior to those based on the InGaAlP system. Similar to blue and green LEDs, the amber InGaN LED device structure developed by Nichia consists of an *n*-GaN:Si, an undoped GaN (used as a uniform current-spreading layer), an active layer of undoped InGaN 25 Å thick, a layer of *p*-type $Al_{0.2}Ga_{0.8}N$:Mg and a layer of *p*-type GaN:Mg (Mukai, Narimatsu, and Nakamura 1998a; Mukai et al. 2001). At room temperature, the output powers of amber InGaN based LED produced 1.4 mW at a typical 3.3 V forward voltage and a 20 mA forward current compared to 0.66 mW power output obtained for AlInGaP LEDs. Due to large bandgap discontinuity in the nitride system very weak temperature dependence was observed. When the ambient temperature is increased from RT to 80°C, the output power of amber InGaN LEDs only decreases to 90% of the room temperature value possibly due to additional nonradiative paths becoming available.

For many biological staining/imaging applications, biodetection, and even short range communications, LEDs operating in the UV region of the spectrum are of interest. For this purpose the varieties possessing low mole fractions InGaN (Mukai, Morita, and Nakamura 1998b), GaN, GaN/AlGaN quantum well, AlGaInN quantum well (Adivarahan et al. 2001b), AlGaInN (Hirayama, Kinoshita, and Ayoagi 2001) and AlGaN (Nishida, Saito, and Kobayashi 2001) emission layer has been reported. By reducing the InN mole fraction, the wavelength of operation can be shifted to the UV region of the spectrum. Large area, 1 mm×1 mm GaN and InGaN based UV LEDs with optical power levels of 100 mW at 365 and 380 nm with operating lifetimes of 100,000 have been achieved at Nichia Chemical (Mukai et al. 2001). The forward current to attain this power level is 500 mA and the forward voltage is around 4.6 V. The LEDs grown by LEO technique on patterned sapphire substrates, which results in further reductions in threading dislocation density, and fabricated in the form of flip chip mounted on Si templates exhibited power levels of about 15 mW at 382 nm for an injection current of 20 mA at room temperature. When the injection current increased to 50 mA, a power level of 38 mW is achieved.

To reduce the wavelength further, AlGaN emission layers must be used. Since the effect of dislocations is expected to be more severe, approaches such as incorporation of GaN templates for substrates have been explored. In one such investigation (Nishida, Saito, and Kobayashi 2001), an output power exceeding 3 mW at the peak wavelength of 352 nm for injection current of 100 mA under a bare-chip geometry was reported. The internal quantum efficiency was estimated at more than 80%. The maximum

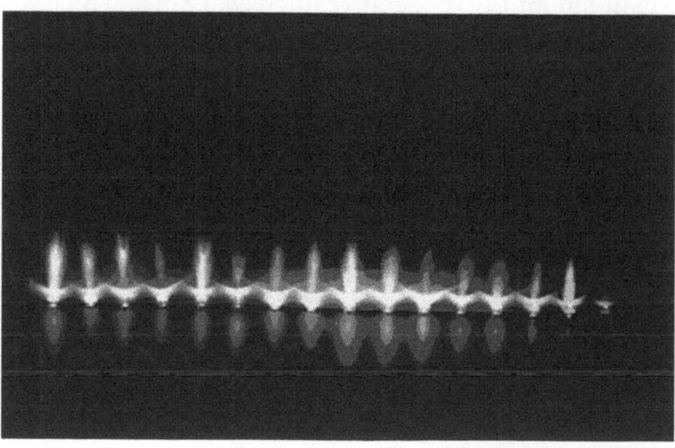

FIGURE 4.18 InGaN LEDs spanning the spectral range from violet to orange. (Courtesy of S. Nakamura, Nichia Chemical Co.)

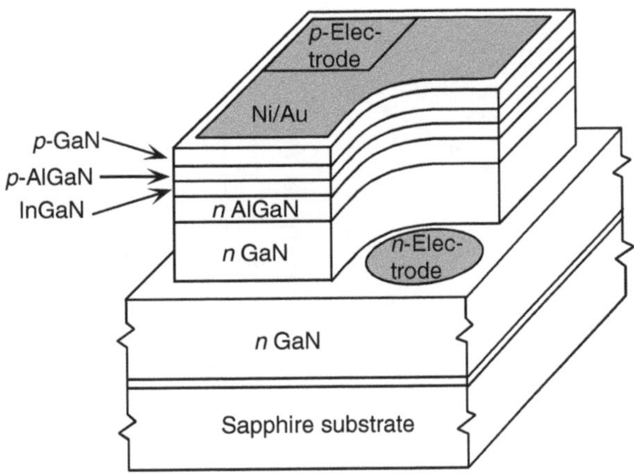

FIGURE 4.19 Artist's view of a Nichia LED with transparent large-area contact to the top *p*-type GaN.

power exceeded 10 mW for a large current injection of 400 mA, with an operation voltage below 6 V. LEDs even with shorter emission wavelengths in the range of 280 to 330 nm are much interest (Adivarahan et al. 2001; Hirayama, Kinoshita, and Ayoagi 2001; Kinoshita et al. 2000) and require bulk AlGaN, AlInGaN, or quantum wells of the two in some combination. However, higher output powers at shorter wavelengths are limited due to the problems associated with both quality and *p*-type doping issues. One approach is to use the quaternary alloy, which allows lattice matching conditions while providing bandgap variability for UV LEDs for its higher quantum efficiency. Room-temperature intense UV emission in the wavelength range of 315–370 nm from quaternary $In_xAl_yGa_{1-x-y}N$ alloys grown by metalorganic vapor-phase epitaxy has been reported (Hirayama et al. 2002). In another approach AlN/AlGaInN superlattices were employed to produce LEDs emitting at 280 nm using reactive MBE with ammonia (Kipshidze et al. 2003). Although the spectrum of the devices consisting of superlattice approach has multiple peaks with the dominant emission occurring at ~330 nm but a

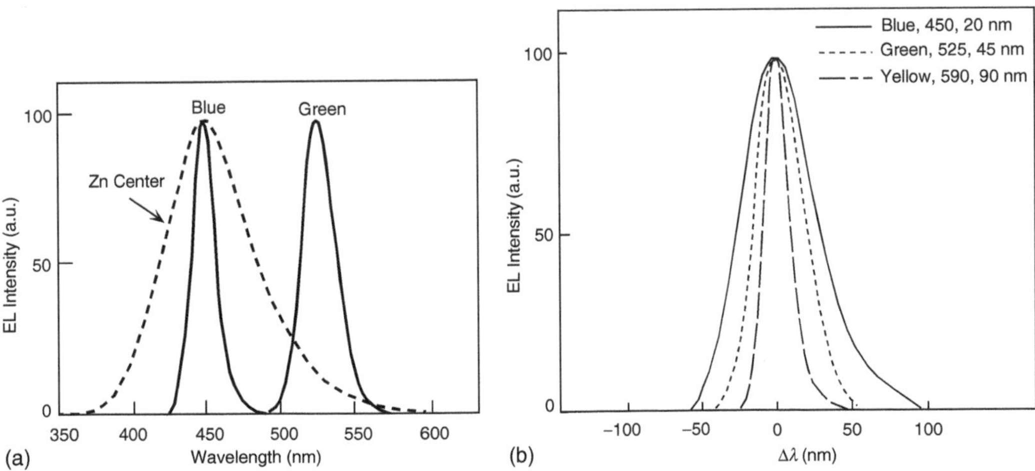

FIGURE 4.20 (a) Electroluminescence spectra of the Nichia blue and green LEDs. By way of comparison to the one with Zn centers, the contrast is drawn to the improved spectrum in devices relying on near band-to-band transitions (b) Spectral linewidth of blue, green and yellow InGaN-based LEDs, the socalled quantum well types. Increasing line broadening due to strain and compositional inhomogeneities with increasing InN mole fraction is noticeable. (Courtesy of S. Nakamura, Nichia Chemical Co.)

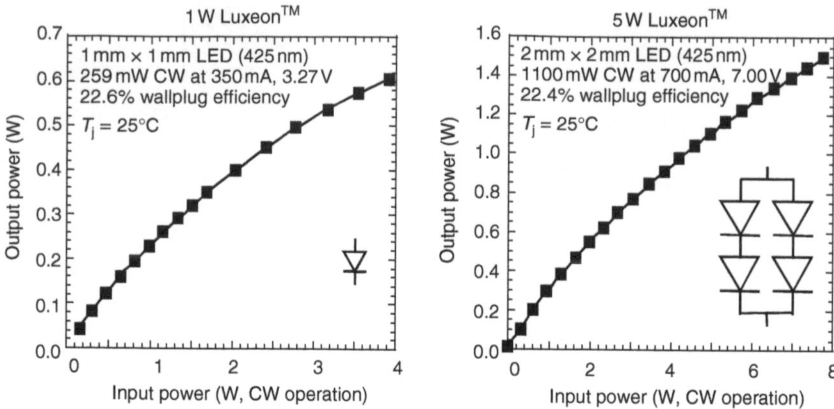

FIGURE 4.21 High power flip chip blue LEDs emitting at 425 nm with packages that can handle electrical input power of 1 W in one class of packages and 5 W in another. The LEDs are produced as part of the Luxeon line of products by LumiLeds. (Courtesy of LumiLeds Lighting.)

second peak at 280 nm is also present. When driven with pulsed current, up to 350 mA, the intensity of the 280 nm begins to dominate. When mesas were etched, the 330 nm emission decreased to a shoulder with the predominant emission at 280 nm.

The resonant cavity enhanced (RCE) LEDs are interesting in that by use of Bragg reflector near monochromatic light (when single mode operation is supported) can be obtained (Salvador et al. 1994). In addition, this device represents the first step toward the fabrication of vertical cavity lasers discussed in the next section. Blue resonant cavity light emitting diodes (RCLEDs), based on InGaN/GaN quantum well (QW) heterostructures have been reported (Song et al. 2000a). Vertical microcavity device with either one or both mirrors forming the cavity are patterned and high reflectivity dielectrics Bragg reflectors have been fabricated for this purpose. The active region in reference (Song et al. 2000a) was grown on 2.0-μm GaN buffer layer and composed of 10 $In_xGa_{1-x}N$ quantum wells ($L_w=30$ Å) with GaN barriers ($L_B=50$ Å), surrounded by $Al_{0.07}Ga_{0.93}N$ upper and lower outer cladding layers (approximately 3600 Å and 0.5 μm, respectively). The In concentration was varied to pan the wavelength range from about 430 to 480 nm. The structures were capped by a 1000 Å thick *p*-GaN contact layer. The schematic diagram and spectral response of an RCLED is shown in Figure 4.22 for which the light emission was collected along the optical axis of the device, within an approximately 5° forward solid angle. The cavity for the device is relatively thick, approximately 16λ. The spectrum of the device encompassing two dielectric DBRs is centered at around $\lambda=430$ nm and underscores the impact of a high quality resonator on the definition of the cavity modes, with the inset showing a particular mode. The modal linewidth is approximately 0.6 nm, implying a cavity Q-factor of approximately 750. This device where the bottom reflector is made of a metallic mirror with an improved reflectivity of $R > 0.9$ is easier to fabricate and could be a good candidate for short cavity blue RCLED.

4.9 GaN-Based Lasers

Although semiconductor lasers have many applications, for example, in communications as carrier and pumping sources, and mundane applications such as pointers, the most salient and imminent application of GaNbased lasers is in digital versatile disks (DVD for short). Optically pumped stimulated emission from GaN was first observed over thirty years ago (Dingle et al. 1971). Since then, researchers have been working toward both vertical cavity surface emitting lasers (VCSEL) and conventional separate confinement heterostructure edge emitting lasers in GaN. Major improvements have been made in the performance and device durability, especially at the Nichia Chemical Company, and first commercially available laser diodes (LDs) appeared in 2000. It is increasingly likely that 405–415 nm InGaN lasers, or preferable shorter wavelength devices utilizing GaN or AlInGaN layers, will form the basis of the next generation of optical disk technology, to replace the current DVD format. For consumer applications, CW-operation lifetimes on the order of 10,000 hours at 60°C are required. The nitridebased lasers with their inherently short wavelengths, when adopted offer much increased data storage capacity possibly in excess of 50 Gb per compact disk. Figure 4.23 presents a photograph of a Nichia InGaN laser emitting near 400 nm, which is intended for such an application.

Although a semiconductor material such as GaN has a large covalent bonding energy and other mechanical properties, high concentration of point or extended defect present in the material prevents the long-term CW operation of laser. During the evolution of GaN laser longevity, many types of devices exhibited varying lifetimes. For example, lasers grown with conventional growth techniques had operating lifetimes in the tens of hours under CW testing at room temperature (Nakamura et al. 1998). A successful reduction of the dislocation density through the incorporation of LEO processes, discussed above, during the growth of the laser structures has paved the way to improve the lifetimes steadily towards truly long-lived blue and UV diode lasers (lifetime $> 10^4$ hours). Early versions of LEO based lasers produced by Nichia operated at room temperature up to about 3000 h under CW testing. The life testing improved devices where the current drive is automatically adjusted to maintain a CW per facet power of 2 mW led to lifetimes of up to about 1000 h without any marked degradation (Nakamura et al. 1999). Researchers at Sony Laboratories have correlated the photoluminescence efficiency, surface profile of cleaved laser facets, and laser performance on LEO-based templates (Tojyo et al. 2001). The optimized Sony devices (Tojyo et al. 2002) utilizing the LEO process improved on this figure tremendously with extrapolated room temperature lifetimes of about 15,000 h. Figure 4.24 displays the evolution of the current through 13 InGaN MQW lasers over 1000 h when the per facet output power is kept at 60 mW at 60°C. The power levels and temperature are acceptable for commercial write and definitely read applications.

By incorporating the LEO process with a very thick HVPE grown layers (Molnar et al. 1997) has shown lower the defect density and thus higher the longevity of laser diodes. As an example adopting this approach, Nagahama et al. (2000a, 2000b) explored a series of cases. One case involves just the LEO followed by laser structure. The second case involves the use of free standing GaN in conjunction with LEO followed by laser structure. The lifetimes at 30-mW average power increased from a reported 700 h in simple LEO to 15,000 h in two LEO substrate removal and HVPE process at a device case temperature of 60°C (Nurmikko 2003). This experiment clearly illustrates the importance of reducing dislocations in order to increase the longevity of violet diode lasers.

Heat removal and strain-induced cracks are of major concerns in laser diodes. A substrate with poor heat conductivity such as sapphire presents a heat removal problems in a laser diode exhibiting a typically power in excess of 500 mW. The layer thickness and mole fractions can also create a problem associated with strain-induced crack in the carrier and light confinement layers. The active region of violet edge emitting lasers diode is based on the InGaN/GaN/AlGaN MQW separate confinement heterostructure (SCH), grown either on nonconducting substrate such as sapphire or conducting substrate such as SiC.

It should be noted that increased InN mole fraction beyond a certain point and below a certain point brings one into a realm where there is not sufficient gain for lasing. The threshold current density increases rapidly for wavelengths beyond about 420 nm as well as below about 400 nm because of, to a large extent, the reduced recombination efficiency of the emitting medium and to some extent weakened electronic confinement. In going from the violet into the ultraviolet lasers, a binary GaN, ternary AlGaN, or a quaternary AlGaInN QW active medium represents the options below about 360 nm (Schmidt et al. 1996). Each presents its own set of material challenges in terms of point defects in the active medium and extended defects emanating from lattice mismatch with the substrate and also within the heterostructure such as the cladding layer of the SCH structure. Nagahama et al. (2001) demonstrated cw edge emitting lasers at room temperature near 370 nm with both GaN QW and AlInGaN QWs. The light output-current and current-voltage (L-I and I-V) performance of a quaternary QW device featuring an

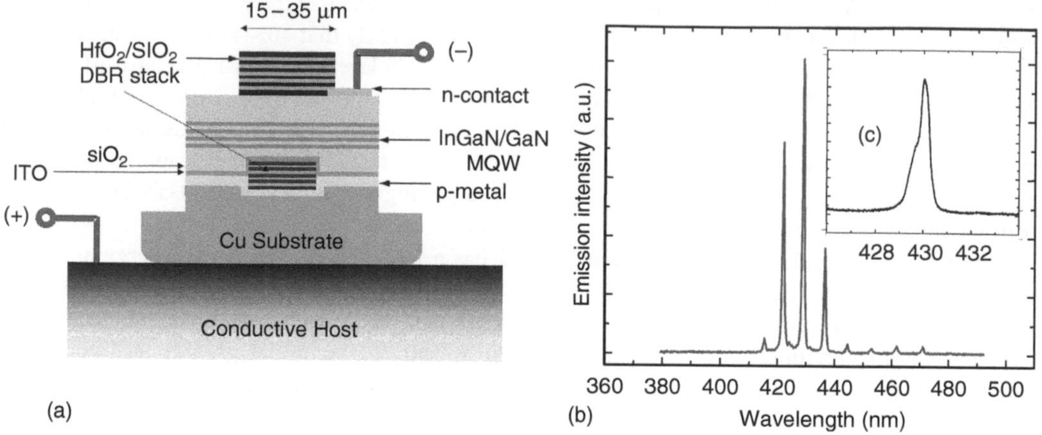

FIGURE 4.22 (a) Schematic representation of a RCLED with two dielectric DBR mirrors. The device also features a low loss intracavity ITO *p*-current spreading layer. Comparison of the emission spectrum of the two types of resonant cavity devices (b) the hybrid structure and (c) a device incorporating two dielectric DBR mirrors. (After Song, Y.-K., Diagne, M., Zhou, H., Nurmikko, A. V., Schneider, R. P., Jr., and Takeuchi, T. 2000. *Appl. Phys. Lett.*, 77, 1744. With permission.)

$Al_{0.03}In_{0.03}Ga_{0.94}N$ QW and the associated emission spectrum centered near 371.7 nm above threshold are shown in Figure 4.25. Note the cw operation at room temperature and the maximum output power reaching several mW. This is red shifted somewhat as the center wavelength under pulsed operation is around 366.4 nm. Although it has been shown the wavelength tunability for different Al and In compositions in the quaternary well, the increasing threshold current density up to about 12 kA/cm^2 for $x_{Al}=0.08$ should be considered. Nagahama et al. (2001) attributed this increase primarily to the worsening the crystal quality of the quaternary $Al_xIn_yGa_{1-x-y}N$, both in terms of general morphology and defects as the Al and also In content are increased. Nevertheless, these attempts expand the basis of knowledge toward developmental efforts in III-nitride materials for light emitters into the deeper ultraviolet.

Vertical-cavity surface-emitting lasers (VCSELs) based on gallium nitride semiconductor heterostructures represent a contemporary focus of research, the aim of which is to develop a new class of planar microdevices in the blue, violet, and near-ultraviolet ranges. In terms of applications, advantages of VCSEL compared to edge emitting lasers include the integration into two-dimensional arrays for massive parallel operation and an increased modulation bandwidth. In an analogue to arsenide- and phosphide-based VCSELs, a natural first choice for the fabrication of high-reflectivity mirrors ($R \geq 99\%$) for a nitride microcavity resonator would focus on in situ epitaxially grown AlGaN/GaN DBRs. Apart from the difficulties of growing AlGaN with high morphological quality (root mean roughness of 2–3 nm across a wafer area of at lest several hundred square micrometers) at aluminum compositions exceeding 25%, the small refractive-index contrast of this alloy demands a large number of layer pairs (> 50) for the required reflectivities. In spite of this the epitaxial growth of such DBRs has been demonstrated (Someya and Arakawa 1998; Langer et al. 1999; Krestnikov et al. 1999; Ng, Moustakas, and Chu 2000). Furthermore, vertical cavity or surface lasing under intense, low-duty-cycle, pulsed optical pumping from GaN or InGaN MQW or thin-film heterostructures that involve in situ grown AlGaN/GaN DBR reflectors has been reported (Redwing et al. 1996; Krestnikov et al. 1999). Someya et al. (1999) have shown stimulated emission under such intense photopumping, employing a hybrid structure comprised of an in situ grown AlGaN/GaN DBR and a dielectric DBR

Optically pumped quasi-cw VCSEL operation has been achieved by 355 nm excitation to secure a predominant electron-hole pair generation in the InGaN QWs (Song et al. 2000b). Several well-defined

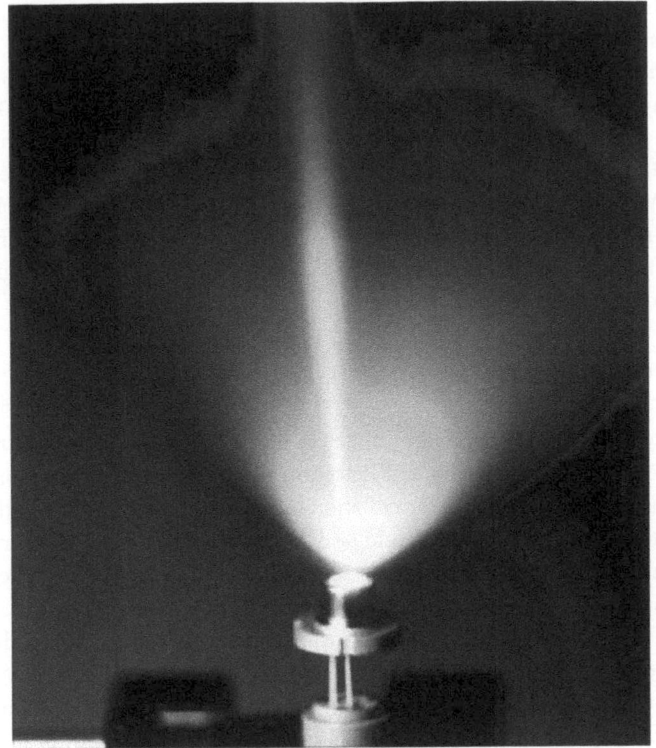

FIGURE 4.23 Photographs of a Nichia InGaN laser in action at about 400 nm. (Courtesy of S. Nakamura, Nichia Chemical Co.)

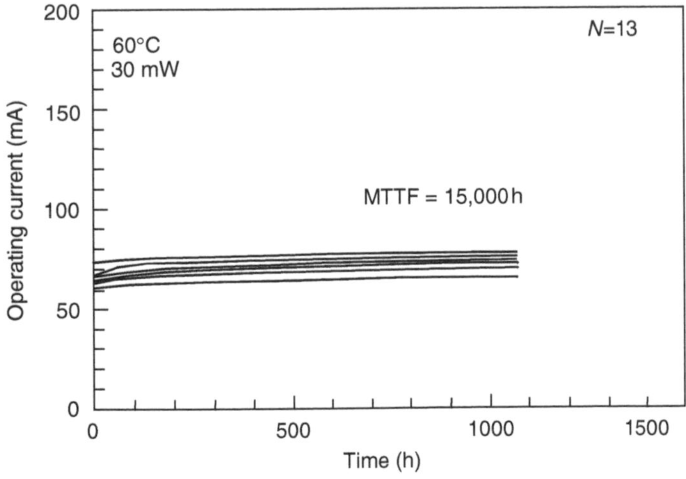

FIGURE 4.24 Evolution of the current through 13 CW Sony InGaN lasers with initial threshold currents in the range of 60–70 mA. The injection current was adjusted to maintain the light output at 60 mW at 60°C. The mean time to failure (MTTF) is estimated at 15,000 h. (After Tojyo, T., Uchida, S., Mizuno, T., Asano, T., Takeya, M., Hino, T., Kijima, S., Goto, S., Yabuki, Y., and Ikeda, M. 2002. *Jpn. J. Appl. Phys.*, 41, 1829.)

cavity modes are observed with a typical modal linewidth of approximately 0.6 nm, which is limited by scattering from residual morphological roughness (see Figure 4.26). The upper trace in the figure presents the spontaneous emission spectrum at $T=258$ K for an average incident power of approximately 17 mW. A hybrid type of DBRs of high quality with a peak reflectivity $R\sim0.991$ have also been used to demonstrate room temperature quasi-continuous wave (cw) operation of an optically pumped $In_xGa_{1-x}N$ ($x\sim0.03$) MQW VCSEL at near $\lambda=383$ nm (Zhou et al. 2000b). An average VCSEL output powers up to 3 mW were reported at pump power of 30 mW. Although possible operation of the electrically injected resonant cavity light emitting diode has been demonstrated in some extent, achieving the stimulated emission in electrically pumped VCSEL structure is really challenging tasks, as carrier injection, particularly for holes, is difficult.

4.10 UV Detectors

A large absorption coefficient and the ability to detect the UV regions of the spectrum make the III-V nitride based devices ideal candidates for UV detection in a number of applications. The ozone layer around the earth absorbs nearly all the solar radiation in the band near 280 nm, so the background seen by a detector operating at that wavelength would be dark. Therefore, they are imperative, particularly for military applications, because they will enable air, sea, and ground equipment to detect and warn against the ground-to-air, air-to-air, air-to-ground, and ground-to-ground missile threats. UV detectors have more applications, including biological and chemical sensors (ozone detection, determination of air pollution levels, biological agent detection), flame sensors (fire alarm systems, combustion engine control), spatial optical communications (intra- and inter-satellite secured communications), emitter calibration, and UV imaging such as solar UV measurements and astronomical studies. In the past decade, there have been several reports on GaN-AlGaN based photoconductors (Kung et al. 1995), *p-n* junction (Monroy et al. 1998), *p-i-n* (Sandvik et al. 2001; Teke et al. 2003a) and *p-π-n* (Osinsky et al. 1997), Schottky barrier (Biyikli et al. 2001; Katz et al. 2001; Teke et al. 2003b), metal-semiconductor-metal (Pau et al. 2002), metal-insulator-semiconductor (Monroy et al. 2001), field-effect transistor or

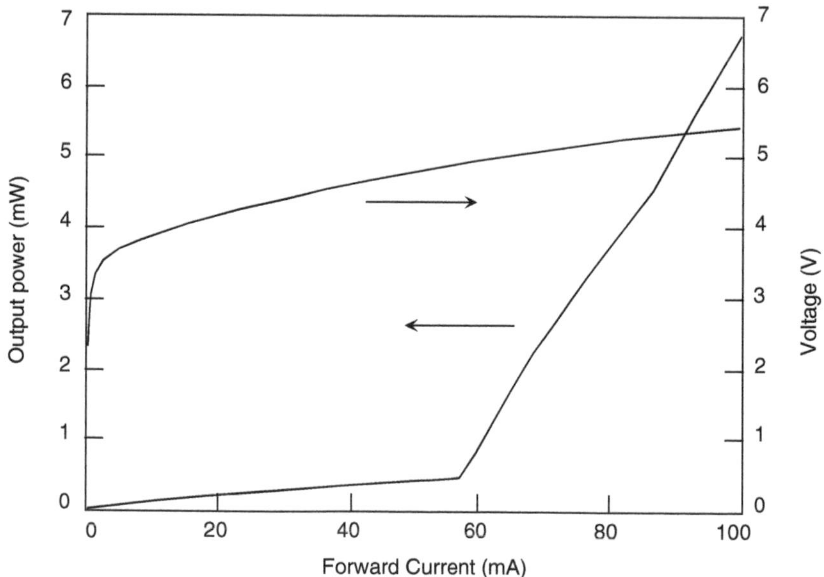

FIGURE 4.25 *L-I* and *I-V* characteristics of $Al_{0.03}In_{0.03}Ga_{0.94}N$ SQW laser under cw operation at 25°C. (After Nagahama, S., Yanamoto, T., Sano, M., and Mukai, T. 2001. *Jpn. J. Appl. Phys.*, 40, L785. With permission.)

bipolar-junction transistor (Moullet et al. 2001), and avalanche photodiode (Carrano et al. 2000). These detector structures were usually grown on sapphire substrates by variations of MBE and MOCVD methods. Improvements in III-nitride materials growth and detector processing have resulted in progressively higher-performance ultraviolet detectors in the last few years. Figure 4.27 illustrates specific detectivity (a figure of merit inversely proportional to the noise level with higher numbers indicating of better performance) of photodetectors covering various parts of the spectrum, namely the ultraviolet (250–400 nm), the visible (400–750 nm), and the near infrared (750 nm and beyond) (Temkin, imprint). For reference, a line denoting the background-limited detectivity expected from an ideal photodetector at 293 K is also shown. Results for solar blind AlGaN detectors are also presented with depictions of a star (Collins et al. 2002; Dupuis 2002;), open circle (Kuryatkov et al. 2003), open square and full square (Kuryatkov et al. 2001), open circle (Brown et al. 1999), and asterisks. AlGaN solar-blind *p-i-n* photodiodes shown with star has 53% quantum efficiency and 3.2×10^{14} cm•$Hz^{1/2}W^{-1}$ specific detectivity at 275 nm. The photomultiplier tube (PMT) based detectors, having traditionally shown the highest specific detectivities, are used as benchmark Design consideration and material quality of $Al_xGa_{1-x}N$ (for 280 nm or shorter wavelengths detection AlN mole fraction of 50% must be employed) based solar blind UV photodetector arrays are the key issues to achieve responsivities, detectivities, gain, speeds, and low noise levels comparable to and preferably better than those of high-gain photo multiplier tube (PMT) detectors (Litton et al. 2002).

A group of researchers from North Carolina State University, Honeywell, Inc., and Night Vision Laboratories has already demonstrated a visible-blind UV camera based on a 32×32 array of back-illuminated GaN/AlGaN *p-i-n* photodiodes (Brown et al. 1999). Each photodiode in the array consisted of an *n*-type base layer of AlGaN ($\sim$20%) followed by an undoped GaN layer topped by a *p*-type GaN layer, all deposited on double side polished sapphire substrate by MOVPE. The photodiode array was connected to a silicon readout integrated circuit using In bump bonds. This visible-blind digital camera is sensitive to radiation from 320 nm to 365 nm in the UV spectral region. Figure 4.28 depicts UV image from an III-V nitride-based photodiode array. Following the preliminary developments of 32×32 pixel imagers, arrays as large as 128×128, consisting of 16,384 individual photodiodes, have been prepared and used in a focal plane array camera.

4.11 Discussion

Wide-bandgap semiconductors have applications in blue displays and recording/reading. In particular SiC boasts a very large thermal conductivity of about 5 W/ $cm^{-1}K^{-1}$ and both SiC and GaN have large cohesive energies. To a large extent the high thermal conductivity of SiC can be utilized for GaN as well by growing the epitaxial layers of GaN on SiC substrates. One must keep in mind that even the relatively thin GaN buffer layers can reduce the thermal conductivity of the overall structure.

Important advantages of GaN over SiC are its direct bandgap and heterostructures with allied AlN and InN binaries and their ternaries. Fortuitously large bandgap disparities between AlN, GaN, and InN render this particular semiconductor system excellent for obtaining heterointerfaces with good carrier confinement. Though heterojunctions can be obtained in SiC as well, it requires exploitation of the many polytypes of SiC which are not necessarily easy to obtain and control.

In the past, many obstacles such as poor quality of material, p-type doping and reliable fabrication procedures related to group III-nitrides have recently been sufficiently overcome, so that electrical and optical devices fabricated from such materials have been demonstrated and partially commercialized. The outstanding physical properties of Group III nitrides enable the fabrication of LEDs, lasers, detectors, transistors, surface acoustic wave devices and offer a valuable combination of electrical, optical and piezoelectrical behavior. With the advent of the LEO technique great improvements in device properties were realized mainly due to a dramatic reduction of extended defects in epitaxial layers and heterostructures caused by the large mismatch of lattice constants and thermal expansion coefficients between group III-nitrides and available substrate materials, such as sapphire and SiC.

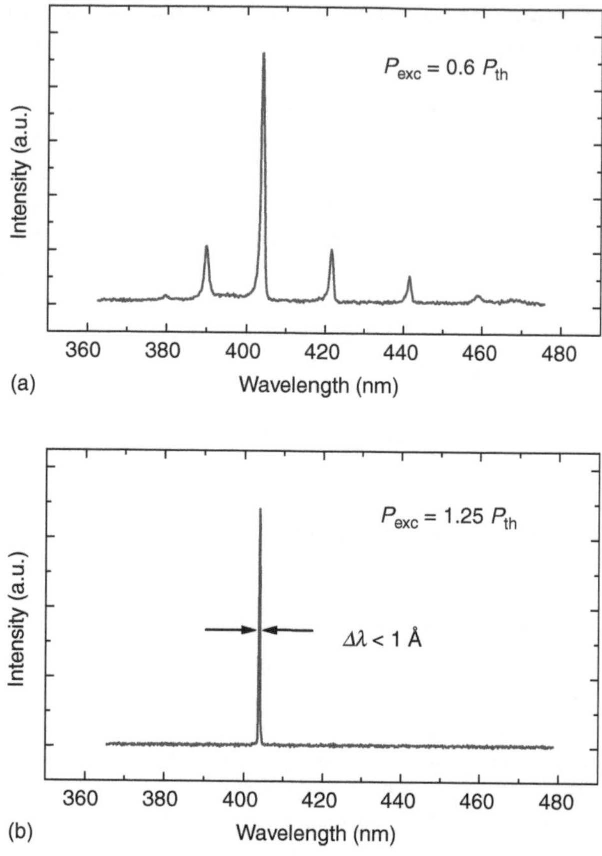

FIGURE 4.26 (a) Upper trace: Spontaneous emission spectra of the optically pumped InGaN MQW VCSEL below threshold. (b) Lower trace: Stimulated emission spectra above threshold under quasi-cw pumping conditions at $T=$ 258 K. (Courtesy of Prof. A. Nurmikko.)

The growth of InGaN, AlInN and AlInGaN layers with high concentrations of indium offer the possibility of bandgap engineering together with the fabrication of lattice-matched heterostructures. The deposition of lattice-matched structures will lead to a further reduction of structural defects and therefore to an improvement of the electrical and optical properties of complex devices. High-luminosity yellow, green, blue, white and UV emitting diodes are now commercialized. For example, Nichia made 1×1 die size LED at 380 nm producing a optical output power of 100 mW at 500 mA driving current has a lifetime of 100,000 h. The prevailing explanation for the excellent lifetime figures in GaN LEDs rests on the resilience of GaN against defect generation during growth and operation. The high threshold for damage generation in GaN, and also in SiC, is responsible for excellent lifetimes. GaN and AlGaN based optical detectors have also demonstrated successful operation of visible-blind and solar-blind ultraviolet detectors. AlGaN solar-blind *p-i-n* photodiodes has shown 53% quantum efficiency and 3.2×10^{14} cm Hz$^{1/2}$ W^{-1} specific detectivity at 275 nm, which is very promising for many applications. While LEDs are enough for display purposes, high-density compact disks require blue and even shorter wavelength coherent sources provided. ELO-grown epilayers on free-standing GaN substrate reduced the threading dislocation density slightly above the mid 10^5 cm^{-2}, and thereby led to the fabrication of laser diodes lasting 15,000 h with an output power of 30 mW. Such a laser diode, operating at room temperature with such lifetime will be instrumental for high-density optical storage capacity. There is still potential for further reduction of the laser wavelength to the near-UV

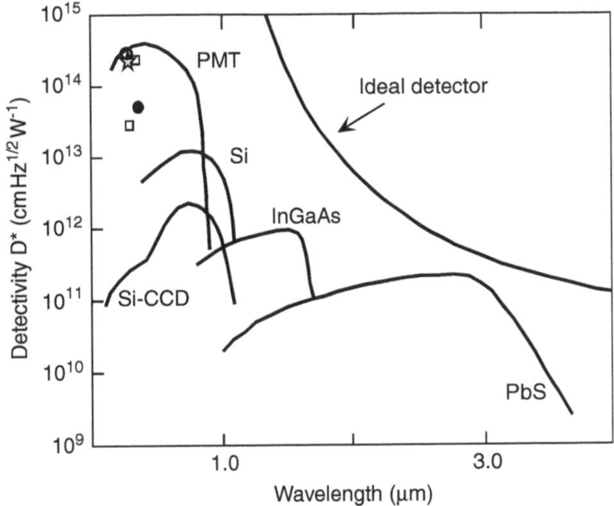

FIGURE 4.27 Specific detectivity D^* plotted as a function of wavelength for a number of photodetectors, from the ultraviolet to the near infrared. Detectivity of an ideal detector, limited by room temperature background radiation, is plotted for reference. (Courtesy of Prof. H. Temkin.)

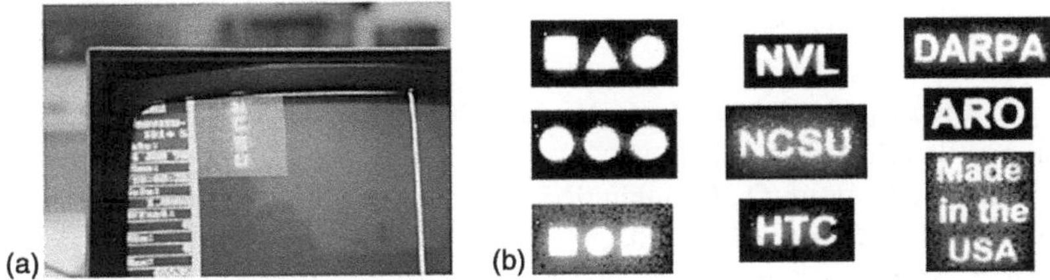

FIGURE 4.28 Demonstration of a visible-blind UV image from a III-V nitride-based photodiode array. The image from the FPA was recorded at NVL on July 28, 1999. (After Brown, J. D., Yu, Z., Matthews, J., Harney, S., Boney, J. F., Schetzina, J. F., Benson, J. D., Dang, K. W., Terril, C., Nohava, T., Yang, W., and Krishnankutty, S. 1999. *MRS Internet J. Nitride Semicond. Res.*, 4, 9; Courtesty of Prof. J. Schetzina. With permission.)

spectral range by using GaN or AlGaN/GaN MQWs as active layers. An additional decrease in wavelength will further enhance the information density of optical storage and perhaps enable the fabrication of optical sensors for molecules with absorbances in the near-UV. Monolithic structures open the door for new combinations of different wide-bandgap semiconductor devices fabricated on the same substrate, which cannot be realized by any other group of materials.

Acknowledgments

The author benefited from many discussions with their colleagues at VCU and elsewhere, and their co-authors whom unfortunately cannot be listed due to space limitation. They have been the real inspiration for H. M. Funding from the Air Force Office of Scientific Research and Office of Naval Research, and National Science Foundation is greatly appreciated. H.M. benefited immensely from participating in the High Temperature Electronics panel organized by AFOSR under the leadership of Dr. Jerry Witt, which helped popularize GaN. H. M. also acknowledges many discussions with and support of Dr. Cole Litton and Max Yoder.

References

Aboelfotoh, M. O., Fröjdh, C., and Petersson, C. S. 2003. *Phys. Rev. B*, 67, 075312.

Adivarahan, V., Lunev, A., Khan, M. A., Yang, J., Simin, G., Shur, M. S., and Gaska, R. 2001a. *Appl. Phys. Lett.*, 78, 2781.

Adivarahan, V., Chitnis, A., Zhang, J. P., Shatalov, M., Yang, J. W., Simin, G., and Khan, M. A. 2001b. *Appl. Phys. Lett.*, 79, 4240.

Adivarahan, V., Zhang, J. P., Chitnis, A., Shuai, W., Sun, J., Pachipulusu, R., Shatalov, M., and Khan, M. A. 2002. *Jpn J. Appl. Phys.*, 41, L435.

Agarwal, A. K., Ryu, S. H., Singh, R., and Palmour, J. W. 2000. *Mater. Sci. Forum.*, 338-342, 1387.

Agarwal, A. K., Ivanov, P. A., Levinshtein, M. E., Palmour, J. W., Rumyantsev, S. L., Ryu, S. H., and Shur, M. S. 2001. *Mater. Sci. Forum.*, 353–356, 743.

Akasaki, I. and Amano, H. 1994. *J. Electrochem. Soc.*, 141, 2266.

Akasaki, I. and Amano, H. 1997. *Jpn J. Appl. Phys.*, 36, 5393.

Alexandrov, P., Pan, W. M., Weiner, M., Jiao, L., and Zhao, J. H. 2003. *Solid State Electron.*, 47, 263.

Amano, H., Kito, M., Hiramatsu, K., and Akasaki, I. 1989. *Jpn. J. Appl. Phys.*, 28, L2112.

Amano, H., Takeuchi, T., Sota, S., Sakai, H., and Akasaki, I. 1997. In *III-V Nitrides*, F. A. Ponce, T. D. Moustakas, I. Akasaki, and B. Menemar, eds., MRS Proc., Vol. 449,.

Ambacher, O., Majewski, J., Miskys, C., Link, A., Hermann, M., Eickhoff, M., Stutzmann, M., Bernardini, F., Fiorentini, V., Tilak, V., Schaff, B., and Eastman, F. 2002. *J. Phys.: Condens. Matter.*, 14, 3399.

Baliga, B. J. 1996. *Physics of Semiconductor Power Devices*, JWS Publishing, New York.

Bedair, S. M., McIntosh, F. G., Roberts, J. C., Piner, E. L., Boutros, K. S., and El-Masry, N. A. 1997. *J. Crystal Growth*, 178, 32.

Bickermann, M., Hofmann, D., Straubinger, T. L., Weingärtner, R., Wellmann, P. J., and Winnacker, A. 2001. *Appl. Surf. Sci.*, 184, 84.

Binari, S. C. and Dietrich, H. C. 1997. In *GaN and Related Materials*, S. J. Pearton, ed., pp. 509–534. Gordon and Breach, New York.

Biyikli, N., Kartaloglu, T., Aytur, O., Kimukin, I., and Ozbay, E. 2001. *Appl. Phys. Lett.*, 70, 2838.

Bozack, M. J. 1997. *Phys. Status Solidi B*, 202, 549.

Brown, J. D., Yu, Z., Matthews, J., Harney, S., Boney, J., Schetzina, J. F., Benson, J. D., Dang, K. W., Terril, C., Nohava, T., Yang, W., and Krishnankutty, S. 1999. MRS Internet. *J. Nitride Semicond. Res.*, 4, 9.

Bryden, W. R. and Kistenmacher, T. J. 1994. In *Properties of Group III Nitrides*, J. H. Edgar, ed., pp. 464–468. INSPEC, London.

Burm, J., Chu, K., Davis, W. A., Schaff, W. J., Eastman, L. F., and Eustis, T. J. 1997. *Appl. Phys. Lett.*, 70, 464.

Cai, S. J., Li, R., Chen, Y. L., Wong, L., Wu, W. G., Thomas, S. G., and Wang, K. L. 1998. *Electron. Lett.*, 34, 2354.

Capano, M. A., Cooper, J. A., Melloch, M. R., Saxler, A., and Mitchel, W. C. 2000. *J. Appl. Phys.*, 87, 8773.

Carrano, J. C., Lambert, D. J. H., Eiting, C. J., Collins, C. J., Li, T., Wang, S., Yang, B., Beck, A. L., Dupuis, R. D., and Campbell, J. C. 2000. *Appl. Phys. Lett.*, 76, 924.

Carter, C. H., Tsvetkov, V. F., Glass, R. C., Henshall, D., Brady, M., Müller, G., Kordina, O., Irvine, K., Edmond, J. A., Kong, H. S., Singh, R., Allen, S. T., and Palmour, J. M. 1999. *Mater. Sci. Eng. B*, 61–62, 1–8.

Casady, J. B., Bonds, J. R., Draper, W. A., Merrett, J. N., Sankin, I., Seale, D., and Mazzola, M. S. 2003. *Mater. Res. Soc. Symp. Proc.*, Vol 764, C 1.3.

Chassagne, T., Pero, G., Chaussende, D., Cauwet, F., Monteil, Y., and Bouix, J. 2002. *Thin Solid Films*, 402, 83.

Chen, L. C., Chen, F. R., Kai, J. J., Chang, L., Ho, J. K., Jong, C. S., Chiu, C. C., Huang, C. N., Chen, C. Y., and Shih, K. K. 1999. *J. Appl. Phys.*, 86, 3826.

Chen, L. C., Ho, J. K., Jong, C. S., Chiu, C. C., Shih, K. K., Chen, F. R., Kai, J. J., and Chang, L. 2000. *Appl. Phys. Lett.*, 76, 3703.

Chen, Z., Yuan, Y., Lu, Da-C., Sun, X., Wan, S., Liu, X., Han, P., Wang, X., Zhu, Q., and Wang, Z. 2002. *Solid State Electron.*, 46, 2069.

Cheong, M. G., Kim, K. S., Kim, C. S., Choi, R. J., Yoon, H. S., Namgung, N. W., Suh, E.-K., and Lee, H. J. 2002. *Appl. Phys. Lett.*, 80, 1001.

Chien, F. R., Nutt, S. R., Carulli, J. M., Buchan, N., Beetz, C. P., and Yoo, W. S. 1994. *J. Mater. Res.*, 9, 2086.

Chin, V. W. L., Tansley, T. L., and Osotchan, T. 1994. *J. Appl. Phys.*, 75, 7365.

Cho, H. J., Hwang, C. S., Bang, W., and Kim, H. J. 1993. *Inst. Phys. Con. Set.*, 37, 663.

Cho, P. S., Goldhar, J., Lee, C. H., Saddow, S. E., and Neudeck, P. 1995. *J. Appl. Phys.*, 77, 1591.

Choi, H. W., Zhang, J., and Chua, S. J. 2001. *Mater. Sci. Semicond. Process.*, 4, 567.

Chor, E. F., Zhang, D., Gong, H., Chen, G. L., and Liew, T. Y. F. 2001. *J. Appl. Phys.*, 90, 1242.

Choyke, W. J. Matsunami, H. Pensl, G. eds. 1997. Phys. Stat. Sol. (B).(Part I), 202.

Chu, K. K., Murphy, M. J., Burm, J., Schaff, W. J., Eastman, L. F. 1997. *IEEE Proceedings, IEEE/Cornell Conference on Advanced Concepts in High Speed Semiconductor Devices and Circuits*, (IEEE, Piscataway, NJ), 399.

Chu, C. F., Yu, C. C., Wnag, Y. K., Tsai, J. Y., Lai, F. I., and Wang, S. C. 2000. *Appl. Phys. Lett.*, 77, 3423.

Collins, C. J., Chowdhury, U., Wong, M. M., Yang, B., Beck, A. L., Dupuis, R. D., and Campbell, J. C. 2002. *Electron. Lett.*, 38, 824.

Cooper, J. A., Melloch, M. Private communication, 2002.

Cox, G. A., Cummins, D. O., Kawabe, K., and Tredgold, R. H. 1967. *J. Phys. Chem. Solilds*, 28, 543.

Crofton, J., Barnes, P. A., Williams, J. R., and Edmond, J. A. 1993. *Appl. Phys. Lett.*, 62, 384.

Crofton, J., McMullin, P. G., Williams, J. R., and Bozack, J. 1995. *J. Appl. Phys.*, 77, 1317.

Crofton, J., Porter, L. M., and Williams, J. R. 1997a. *Phys. Stat. Sol. (B)*, 202, 581.

Crofton, J., Beyer, L., Williams, J. R., Luckowski, E. D., Mohney, S. E., and Delucca, J. M. 1997b. *Solid State Electron.*, 41, 1725.

Crofton, J., Mohney, S. E., Williams, J. R., and Smith, T. I. 2002. *Solid State Electron.*, 46, 109.

Dalibor, T., Pensl, G., Matsunami, H., Kimoto, T., Choyke, W. J., Schoner, A., and Nordell, N. 1997. *Phys. Stat. Sol. A*, 162, 199.

Davies, G. 1989. *Phys. Rep.*, 176, 83.

Davis, R. F., Gehrke, T., Linthicum, K. J., Zheleva, T. S., Preble, E. A., Rajagopal, P., Zorman, C. A., and Mehregany, M. 2001. *J. Cryst. Growth*, 225, 134.

Davydov, Yu, V., Klochikhin, A. A., Seisyan, R. P., Emtsev, V. V., Ivanov, S. V., Bechstedt, F., Furthmuller, J., Harima, H., Mudryi, A. V., Aderhold, J., Semchinova, O., and Graul, J. 2002. *Phys. Stat. Solidi (B)*, 229, R1.

Defives, D., Noblanc, O., Doa, C., Brylisky, C., Barthula, M., and Mayer, F. 1999. *Mater. Sci. Eng. B.*, 61–62, 395.

Devaty, R. P. and Choyke, W. J. 1997. *Phys. Stat. Sol. A*, 162, 5.

Dingle, D., Shaklee, K. L., Leheny, R. F., and Zetterstrom, R. B. 1971. *Appl. Phys. Lett.*, 19, 5.

Dmitriev, V. A., Ivanov, P. A., Korkin, I., Morozenko, Y., Popov, I., Sidorova, T. A., Strslchuk, A. M., and Chelnokov, E. 1985. *Sov. Phys. Tech. Lett.*, 11, 98.

Dmitriev, V. A., Irvine, K. G., Spencer, M. G., and Nikitina, I. P. 1994a. *Silicon Carbide Related Mater*, 137, 67.

Dmitriev, V. A., Irvine, K., Spencer, M., and Kelner, K. 1994b. *Appl. Phys. Lett.*, 64, 318.

Doğan, S., Teke, A., Huang, D., Morkoç, H., Roberts, C. B., Parish, J., Ganguly, B., Smith, M., Myers, R. E., and Saddow, S. E. 2003. *Appl. Phys. Lett.*, 82, 3107.

Donovan, S. M., MacKenzie, J. D., Abernathy, C. R., Pearton, S. J., Ren, F., Jones, K., and Cole, M. 1997. *Appl. Phys. Lett.*, 70, 2592.

Dupuis, R.D. Wocsemmad, Austin, Texas, 2002.

Duxstad, K. J. 1997. Metal Contacts to ZnSe and GaN. PhD diss., Dept. of Mater. Sci., University of California at Berkeley.

Eckey, L., Podloswski, L., Goldner, A., Hoffmann, A., Broser, I., Meyer, B. K., Volm, D., Streibl, T., Hiramatsu, K., Detcprohm, T., Amano, H., and Akasaki, I. 1996. *Ins. Phys. Conf. Ser.*, 142, 943.

Edgar, J. H., Strite, S. T., Akasaki, I., Amano, H., and Wetzel, C. 1999. *Properties, Processing and Applications of Gallium Nitride and Related Semiconductors*. INCPEC, London.

Edmond, J. A., Ryu, J., Glass, J. T., and Davis, R. F. 1988. *J. Electrochem. Soc.*, 135, 359.

Edwards, J., Kawabe, K., Stevens, G., and Tredgold, R. H. 1965. *Solid State Commun.*, 3, 99.

Elasser, A., Kheraluwala, M. H., Ghezzo, M., Steigerwald, R. L., Evers, N. A., Kretchmer, J., and Chow, T. P. 2003. *Transac. Industry Appl.*, 39, 915.

Ellison, A., Zhang, J., Peterson, J., Henry, A., Wahab, Q., Pergamon, J. P., Makarow, Y. N., Vorobev, A., Vehanen, A., and Janzén, E. 1999. *Mater. Sci. Eng. B*, 61–62, 113.

Fan, Z., Noor Mohammad, S., Kim, W., Aktas, O., Botchkarev, E., and Morkoc, H. 1996. *Appl. Phys. Lett.*, 68, 1672.

Fang, Z. Q., Look, D. C., Kim, W., Fan, Z., Botchkarev, A., and Morkoc, H. 1998. *Appl. Phys. Lett.*, 72, 2277.

Fara, A., Bernadini, F., and Fiorentini, V. 1999. *J. Appl. Phys.*, 85, 2001.

Foresi, J. S. and Moustakas, T. D. 1993. *Appl. Phys. Lett.*, 62, 2859.

Fursin, L. G., Zhoa, J. H., and Weiner, M. 2001. *IEEE Electron. Lett.*, 37, 1092.

Fuyuki, T., Yoshinobu, Y., and Matsunami, H. 1993. *Thin Solid Films*, 225, 225.

Gehrke, T., Linthicum, K. J., Preble, E., Rajagopal, P., Ronning, C., Zorman, C., Mehregany, M., and Davis, R. F. 2000. *J. Electron. Mater.*, 29, 306.

Ghandhi, S. K. 1998. *Semiconductor Power Devices*. Wiley, New York.

Gil, B., Briot, O., and Aulombard, R. L. 1995. *Phys. Rev. B*, 52, R17028.

Glaser, E. R., Kennedy, T. A., Doverspike, K., Rowland, L. B., Gaskill, D. K., Freitas, J.A., Jr., Asif Khan, M., Olson, D. T., Kuznia, J. N., and Wickenden, D. K. 1995. *Phys. Rev. B*, 51, 13326.

Glass, R. C., Spellman, L. M., Tanaka, S., and Davis, R. F. 1992. *J. Vac. Sci. Technol. A*, 10, 1625.

Goepfert, I. D., Schubert, E. F., Osinsky, A., and Norris, P. E. 1999. *Electron. Lett.*, 35, 1109.

Goldenberg, B., Zook, J. D., and Van Vechten, J. 1993. *Bull. Am. Phys. Soc.*, 38, 446.

Götz, W., Johnson, N. M., Chen, C., Liu, H., Kuo, C., and Imler, W. 1996a. *Appl. Phys. Lett.*, 68, 3144.

Götz, W., Johnson, N. M., Walker, J., Bour, D. P., and Street, R. A. 1996b. *Appl. Phys. Lett.*, 68, 667.

Götz, W., Romano, L. T., Walker, J., Johnson, N. M., and Molnar, R. J. 1998. *Appl. Phys. Lett.*, 72, 1214.

Guohua, Q., Chen, F., Olowolafe, J. O., Swann, C. P., Unruh, K. M., and Holmes, D. S. 1997. *Mater. Res., Symp. Proc.*, 499, 1073.

Haag, H., Hönerlage, B., Briot, O., and Aulombard, R. L. 1999. *Phys. Rev. B*, 60, 11624.

Harris, W. A. 1980. *Electronic Structure and Properties of Solids*, Dover, New York.

Heera, V., Madhusoodanan, K. N., Mülkich, A., Panknin, D., and Skorupa, W. 2002. *Appl. Phys. Lett.*, 81, 70.

Henry, C. H. and Nassau, K. 1970. *Phys. Rev. B.1*, 1628.

Heying, B., Smorchkova, I., Poblenz, C., Elsass, C., Fini, P., Den Baars, S., Mishra, U., and Speck, J. S. 2000. *Appl. Phys. Lett.*, 77, 2885.

Higashiwaki, M. and Matsui, T. 2003. *J. Cryst. Growth*, 252, 128.

Hirayama, H., Kinoshita, A., and Ayoagi, Y. 2001. *Phys. Stat. Sol. A*, 188, 83.

Hirayama, H., Kinoshita, A., Yamabi, T., Enomoto, Y., Hirata, A., Araki, T., Nanishi, Y., and Aoyagi, Y. 2002. *Appl. Phys. Lett.*, 80, 207.

Ho, J. K., Jong, C. S., Chiu, C. C., Huang, C.-N., and Shih, K. K. 1999. *Appl. Phys. Lett.*, 74, 1275.

Hobgood,H. McD. , Glass, R. C., Augustine, G., Hopkins, R. H., Jenny, J., Skrowonski, M., Mitchel, W. C., and Roth, M. 1995. *Appl. Phys. Lett.*, 66, 1364.

Hofmann, D. M., Kovalev, D., Steude, G., Meyer, B. K., Hofmann, A., Eckey, L., Heitz, R., Detchprom, T., Amano, H., and Akasaki, I. 1995. *Phys. Rev. B*, 52, 16702.

Hsu, L. and Walukiewicz, W. 1999. *Appl. Phys. Lett.*, 74, 2405.

Huang, D., Yun, F., Reshchikov, M. A., Wang, D., Morkoç, H., Rode, D. L., Farina, L. A., Kurdak, Ç., Tsen, K. T., Park, S. S., and Lee, K. Y. 2001. *Solid State Electron.*, 45, 711.

Huijie, Y., Lai, J., Li, X., Luo, Y., Fursin, L., Zhao, J. H., Alexandrov, P., Bright, B., and Weiner, M. 2002. *Ind. Appl. Conf.*, 4, 2609.

Hull, B. A., Mohney, S. E., Venugopalan, H. S., and Ramer, J. C. 2000. *Appl. Phys. Lett.*, 76, 2271.

Ikeda, M., Hayakawa, T., Yamagiva, Z., Matsunami, H., and Tanaka, T. 1979. *J. Appl. Phys.*, 50, 8215.

Inushima, T., Yaguchi, T., Nagase, A., and Shiraishi, T. 1996. *Ins. Phys. Conf. Ser.*, 142, 971.

Inushima, T., Mamutin, V. V., Vekshin, V. A., Ivanov, S. V., Sakon, T., Motokawa, M., and Ohoya, S. 2001. *J. Cryst. Growth*, 227–228, 481.

Islam, N. I., Schumiloglu, E., and Fleddermann, C. B. 1998. *Appl. Phys. Lett.*, 73, 14, 1988.

Itoh, A. and Matsunami, H. 1997. *Phys. Stat. Sol. A*, 162, 389.

Itoh, A., Kimoto, T., Matsunami, H. 1995. Proceedings of the ICSCRM. 685.

Itoh, A., Kimoto, T., and Matsunami, H. 1995b. *IEEE Electron. Devices Lett.*, 16, 280.

Ivchenko, E. L. 1982. Excitons. North Holland Publishing Company, New York.

Jacob, C., Pirouz, P., Kuo, H. I., and Mehregany, M. 1998. *Solid State Electron.*, 42, 2329.

Jacobs, B., Kramer, M. C. J. C. M., Geluk, E. J., and Karouta, F. 2002. *J. Cryst. Growth.*, 241, 15.

Jagodzinski, H. and Arnold, H. 1960. In *Silicon Carbide, A High Temperature Semiconductor*, J. R. O'Connor and J. Smiltens, eds., pp. 136–146. Pergamon, New York.

Jang, J. S., Park, S. J., and Seong, T. Y. 2000. *Appl. Phys. Lett.*, 76, 2898.

Jang, H. W., Kim, K. H., Kim, J. K., Hwang, S.-W., Yang, J. J., Lee, K. J., Son, S. J., and Lee, J.-L. 2001. *Appl. Phys. Lett.*, 79, 1822.

Jang, H. W., Urbanek, W., Yoo, M. C., and Lee, J. L. 2002. *Appl. Phys. Lett.*, 80, 2937.

Jeffery, G. A., Parry, G. S., and Mozzi, R. L. 1956. *J. Chem. Phys.*, 25, 1024.

Jenny, J. R., Skowronski, M., Mitchel, W. C., Hobgood, H. M., Glass, R. C., Augustine, G., and Hopkins, R. H. 1996a. *J. Appl. Phys.*, 79, 2326.

Jenny, J. R., Skowronski, M., Mitchel, W. C., Hobgood, H. M., Glass, R. C., Augustine, G., and Hopkins, R. H. 1996b. *Appl. Phys. Lett.*, 68, 1963.

Jiang, H., Zhao, G. Y., Ishikawa, H., Egawa, T., Jimbo, T., and Umeno, M. 2001. *J. Appl. Phys.*, 89, 1046.

Johnson, B. J. and Capano, M. A. 2003. *Solid State Electron.*, 47, 1437.

Kassamakova, L., Kakanakov, R., Nordell, N., Savage, S., Kakanova-Georgieva, A., and Marinova, T. 1999. *Mater. Sci. Eng. B*, 61–62, 291.

Kato, Y., Kitamura, S., Hiramatsu, K., and Sawaki, S. 1994. *J. Cryst. Growth*, 144, 133.

Katz, O., Garber, V., Meyler, B., Bahir, G., and Salzman, J. 2001. *Appl. Phys. Lett.*, 79, 1417.

Kawabe, K., Tredgold, R. H., and Inyishi, Y. 1967. *Electron. Eng. Jpn*, 87, 62.

Kern, R. S. and Davis, R. F. 1997b. *Appl. Phys. Lett.*, 71, 1356.

Kern, R. S., Jährendahl, K., Tanaka, S., and Davis, R. F. 1997a. *Phys. Stat. Sol. B*, 202, 379.

Khan, I. H. and Learn, A. J. 1969. *Appl. Phys. Lett.*, 15, 410.

Khan, M. A., Van Hove, J. M., Kuznia, J. N., and Olson, D. T. 1991. *Appl. Phys. Lett.*, 58, 2408.

Khemka, V., Ananthan, V., and Chow, T. P. 2000. *IEEE Electron. Devices Lett.*, 21, 286.

Kim, W., Botchkarev, A. E., Salvador, A., Popovici, G., Tang, H., and Morkoç, H. 1997a. *Appl. Phys. Lett.*, 82, 219.

Kim, K. S., Saxler, A., Kung, P., Razeghi, R., and Lim, K. Y. 1997b. *Appl. Phys. Lett.*, 71, 800.

Kim, K. S., Han, M. S., Yang, G. M., Youn, C. J., Lee, H. J., Cho, H. K., and Lee, J. Y. 2000. *Appl. Phys. Lett.*, 77, 1123.

Kimoto, T., Nishino, H., Yoo, W. S., and Matsunami, H. 1993. *J. Appl. Phys.*, 73, 726.

Kinoshita, A., Hirayama, H., Ainoya, M., Aoyagi, Y., and Hirata, A. 2000. *Appl. Phys. Lett.*, 77, 175.

Kipshidze, G., Kuryatkov, V., Borisov, B., Kudryavtsev, Y., Asomoza, R., Nikishin, S., and Temkin, H. 2002. *Appl. Phys. Lett.*, 80, 2910.

Kipshidze, G., Kuryatkov, V., Zhu, K., Borisov, B., Holtz, M., Nikishin, S., and Temkin, H. 2003. *J. Appl. Phys.*, 93, 1363.

Kistenmacher, T. J., Ecelberger, S. A., and Bryden, W. A. 1993. *J. Appl. Phys.*, 74, 1684.

Kitabakate, M. 2000. *Thin Solid Films*, 369, 257.

Koga, K., Fulikawa, Y., Ueda, Y., and Yamaguchi, T. 1992. *Springer Proc. Phys.*, 71, 96.

Koide, Y., Maeda, T., Kawakami, T., Fujita, S., Uemura, T., Shibata, N., and Murakami, M. 1999. *J. Electron. Mater.*, 28, 341.

Koleske, D. D., Henry, R. L., Twigg, M. E., Culbertson, J. C., Binari, S. C., Wickenden, A. E., and Fatemi, M. 2000. *Appl. Phys. Lett.*, 80, 4372.

Kong, H. S., Jiang, B. L., Glass, J. T., Rozgonyi, G. A., and Moore, K. L. 1988. *J. Appl. Phys.*, 63, 2645.

Kong, H. S., Glass, J. T., and Davis, R. F. 1989. *J. Mater. Res.*, 4, 204.

Konishi, R., Yasukochi, R., Nakatsuka, O., Koide, Y., Miroyama, M., and Murakami, M. 2003. *Mater. Sci. Eng. B*, 98, 286.

Kordina, O., Hallin, C., Ellison, A., Bakin, A. S., Ivanov, I. G., Henry, A., Yakimova, R., Touminen, M., Vehanen, A., and Janzen, E. 1996. *Appl. Phys. Lett.*, 69, 1456.

Kornitzer, K., Ebner, T., Grehl, M., Thonke, K., Sauer, R., Kirchner, C., Schwegler, V., Kamp, M., Leszczynski, M., Grzegory, I., and Porowski, S. 1999. *Phys. Stat. Sol. B*, 216, 5.

Korona, K. P., Wysmolek, A., Pakula, K., Stepniewski, R., Baranowski, J. M., Grzegory, I., Lucznik, B., Wroblewski, M., and Porowski, S. 1996. *Appl. Phys. Lett.*, 69, 788.

Korotkov, R. Y., Reshchikov, M. A., and Wessels, B. W. 1999. *Physica B*, 273–274, 80.

Korotkov, R. Y., Gregie, J. M., and Wessels, B. W. 2001. *Appl. Phys. Lett.*, 78, 222.

Kozodoy, P., Smorchkova, Y. P., Hansen, M., Xing, H., DenBaars, S. P., Mishra, U. K., Saxler, A. W., Perrin, R., and Mitchel, W. C. 1999a. *Appl. Phys. Lett.*, 75, 2444.

Kozodoy, P., Hansen, M., DenBaars, S. P., and Mishra, U. K. 1999b. *Appl. Phys. Lett.*, 74, 3681.

Krestnikov, I., Lundin, W., Sakharov, A. V., Semenov, V., Usikov, A., Tsatsulnikov, A. F., Alferov, Zh., Ledentsov, N., Hofmann, A., and Bimberg, D. 1999. *Appl. Phys. Lett.*, 75, 1192.

Kubota, K., Kobayashi, Y., and Fujimoto, K. 1989. *J. Appl. Phys.*, 66, 2984.

Kumar, V., Zhou, L., Selvanathan, D., and Adesida, I. 2002. *J. Appl. Phys.*, 92, 1712.

Kung, P., Zhang, X., Walker, D., Saxler, A., Piotrowsky, J., Rogalski, J., and Razeghi, M. 1995. *Appl. Phys. Lett.*, 67, 3792.

Kurai, S., Nishino, K., and Sakai, S. 1997. *Jpn J. Appl. Phys.*, 36, L184.

Kuryatkov, V. V., Temkin, H., Campbell, J. C., and Dupuis, R. D. 2001. *Appl. Phys. Lett.*, 78, 3340.

Kuryatkov, V., Chandolu, A., Borisov, B., Kipshidze, G., Zhu, K., Nikishin, S., Temkin, H., and Holtz, M. 2003. *Appl. Phys. Lett.*, 82, 1323.

Kusnetsov, N. I. and Zubrilov, A. S. 1995. *Mater. Sci. Eng. B*, 29, 181.

Langer, R., Barski, A., Simon, J., Pelekanos, N., Konovalov, O., Andre, R., and Dang, L. S. 1999. *Appl. Phys. Lett.*, 74, 3610.

Larkin, D. J. 1997. *Phys. Stat. Sol. B.*, 202, 305.

Larkin, D. J., Neudeck, P. G., Powell, J. A., and Matus, L. G. 1994. *Ins. Phys. Conf. Ser.*, 137, 51.

Lebedev, A. A. 1999. *Semiconductors*, 33, 107.

Lee, S. K., Zetterling, C. M., and Östling, M. 2000. *J. Appl. Phys.*, 87, 8039.

Lee, S. K., Zetterling, C. M., and Östling, M. 2001a. *J. Electron. Mater.*, 30, 242.

Lee, C.-S., Lin, Y.-J., and Lee, C.-T. 2001b. *Appl. Phys. Lett.*, 79, 3815.

Lee, S. K., Zetterling, C. M., Östling, M., Palmquist, J. P., and Jansson, U. 2002. *Microelectron. Eng.*, 60, 261.

Lei, T., Fanciulli, M., Molnar, R. J., Moustakas, T. D., Graham, R. J., and Scanlon, J. 1991. *Appl. Phys. Lett.*, 59, 944.

Lendenmann, H., Dahlquist, F., Johansson, N., Soderholm, R., Nilsson, P. A., Bergaman, J., and Skytt, P. 2001. *Mater. Sci. Forum.*, 353–356, 727.

Leroux, M., Grandjean, N., Beaumont, B., Nataf, G., Semond, F., Massies, J., and Gibart, P. 1999. *J. Appl. Phys.*, 86, 3721.

Lester, L. F., Brown, J. M., Ramer, J. C., Zhang, L., and Hersee, S. D. 1996. *Appl. Phys. Lett.*, 69, 2737.

Levin, V. I., Tairov, Y. M., Travazhdyan, M. G., Tsvetkov, F., and Chernov, M. A. 1978. *Sov. Phys. Izvestia*, 14, 830.

Li, J., Nam, K. B., Nakarmi, M. L., Lin, J. Y., and Jiang, H. X. 2002a. *Appl. Phys. Lett.*, 81, 3365.

Li, J., Oder, T. N., Nakarmi, M. L., Lin, J. Y., and Jiang, H. X. 2002b. *Appl. Phys. Lett.*, 80, 1210.

Lin, M. E., Sverdlov, B., Zhou, G. L., and Morkoç, H. 1993. *Appl. Phys. Lett.*, 62, 3479.

Lin, M. E., Ma, Z., Huang, F. Y., Fan, Z. F., Allen, L. H., and Morkoc, H. 1994a. *Appl. Phys. Lett.*, 64, 1003.

Lin, M. E., Huang, F. Y., and Morkoc, H. 1994b. *Appl. Phys. Lett.*, 64, 2557.

Lin, J. Y., Mohney, S. E., Smalley, M., Crofton, J., Williams, J. R., and Smith, T. I. 2000. *Mater. Res. Soc.*, 640, H7.3.

Linthicum, K. J., Gehrke, T., Thomson, D., Carlson, E., Rajagopal, P., Smith, T., and Davis, R. 1999. *Appl. Phys. Lett.*, 75, 196.

Litton, C. W., Schreiber, P. J., Smith, G. A., Dang, T., Morkoç, H. 2001. Design requirements for high-sensitivity UV solar blind imaging detectors based on AlGaN/GaN photodetector arrays: A review. Proc. of 46th SPIE Symposium on Optical Science & Technology, Materials for Infrared and Ultraviolet Detectors, Vol. 4454, 218; Invited Paper Presented at the SPIE Symposium, 29 July–4 August 2001, San Diego, CA.

Liu, L. and Edgar, J. H. 2002. *Mater. Sci. Eng. R.*, 37, 61.

Liu, Q. Z., Yu, L. S., Deng, F., Lau, S. S., Chen, Q., Yang, J. W., and Khan, M. A. 1997. *Appl. Phys. Lett.*, 71, 1658.

Look, D. C. and Sizelove, J. R. 1999. *Phys. Rev. Lett.*, 82, 1237.

Look, D. C., Lu, H., Schaff, W. J., Jasinski, J., and Liliental-Weber, Z. 2002. *Appl. Phys. Lett.*, 80, 258.

Lu, H. W., Schaff, J., Eastman, L. F., Wu, J., Walukiewicz, W., Yu, K. M., Auger III, J. W., Haller, E. E., and Ambacher, O. 2002. *Conference digest the 44*th Electronic Materials Conference, 2.

Lu, W. J., Mitchel, W. C., Landis, G. R., Crenshaw, T. R., and Collins, W. E. 2003. *Solid State Electron.*, 47, 2001.

Luther, B. P., Mohney, S. E., Jackson, T. N., Asif Khan, M., Chen, Q., and Wang, J. W. 1997. *Appl. Phys. Lett.*, 70, 57.

Mair, R. A., Li, J., Duan, S. K., Lin, J. Y., and Jiang, H. X. 1999. *Appl. Phys. Lett.*, 74, 513.

Matsunami, H. 2004. Private communication.

Matsuoka, T. 1989. IOP publications.

Matsuoka, T., Okamoto, H., Nakao, M., Takahata, H., Mitate, T., Mizuno, S., Harima, H., Makimoto, T. 2003. 8th Wide bandgap nitride workshop, September 29–October 2, Richmond, VA.

Mattila, T. and Nieminen, R. M. 1997. *Phys. Rev. B*, 55, 9571.

Mitchel, W. C., Perrin, R., Goldstein, J., Roth, M., Ahoujja, M., Smith, S. R., Evwaraye, A. O., Solomon, J. S., Landis, G., Jenny, J., Hobgood, H. McD. , Augustine, G., and Balakrishna, V. 1998. *Mater. Sci. Forum*, 264–268, 545.

Mitlehner, H., Bartsch, W., Dohnke, K. O., Friedrichs, P., Kaltschmidt, R., Weiss, B., and Stephani, D. 1999. Proceeding of the 11th International Symposium on Power Semiconductor Devices and IC's, 339.

Mizita, M., Fujieda, S., Matsumoto, Y., and Kawamura, T. 1986. *Jpn J. Appl. Phys.*, 25, L945.

Mohammad, S. N., Salvador, A., and Morkoç, H. 1995. *Proc. IEEE*, 83, 1306.

Mohney, S. E., Luther, B. P., Wolter, S. D., Jackson, T. N., Karlicek, R. F. Jr. and Kern, R. S. 1998. In *Proc. 4th Int. Conf. on High Temperature Electronics*, 134.

Mokhov, E. N., Gornushkina, E. D., Didik, V. A., and Kozlovskii, V. V. 1992. *Sov. Phys. Solid State*, 34, 1043.

Molnar, R. J., Götz, W., Romano, L. T., and Johnson, N. M. 1997. *J. Cryst. Growth*, 178, 147.

Monemar, B. 1998. *Semicond. Semimetals*, 50, 305.

Monemar, B., Paskov, P. P., Paskova, T., Bergman, J. P., Pozina, G., Chen, W. M., Hai, P. N., Buyanova, I. A., Amano, H., and Akasaki, I. 2002. *Mat. Sci. Eng. B*, 93, 112.

Monroy, E., Munoz, E., Sanchez, F. J., Calle, F., Calleja, E., Beaumont, B., Gibart, P., Munoz, J. A., and Cusso, F. 1998. *Semicond. Sci. Technol.*, 13, 1042.

Monroy, E., Calle, F., Pau, J. L., Munoz, E., Omnes, F., Beaumont, B., and Gibart, P. 2001. *J. Cryst. Growth*, 270, 541.

Morkoç, H. 2006. *Nitride Semiconductors and Devices,* Springer Verlag, Heidelburg.

Morkoç, H., Strite, S., Gao, G. B., Lin, M. E., Sverdlov, B., and Burns, M. J. 1994. *Appl. Phys.*, 76, 1363.

Moscatelli, F., Scorzoni, A., Poggi, A., Cardinalli, G. C., and Nipoti, R. 2003. *Semicond. Sci. Technol.*, 18, 554.

Moullet, R., Hirano, A., Iwaya, M., Detchprohm, T., Amona, H., and Akasaki, I. 2001. *Jpn. J. Appl. Phys.*, 40, part 2, L498.

Muench, W. and Kurzinger, W. 1978. *Solid State Electron.*, 21, 1129.

Mukai, T., Narimatsu, H., and Nakamura, S. 1998a. *Jpn J. Appl. Phys.*, 37, L479.

Mukai, T., Morita, D., and Nakamura, S. 1998b. *J. Cryst. Growth.*, 189/190, 778.

Mukai, T., Nagahama, S., Iwasa, N., Senoh, M., and Yamada, T. 2001. *J. Phys. : Condens. Matter.*, 13, 7089.

Müller, R., Feege, M., Weber, S. G., and Spaeth, S. M. 1993. *Semicond. Sci. Technol.*, 8, 1377.

Murakami, M. 1990. *Mater. Sci. Rep.*, 5, 273.

Nagahama, S., Iwasa, N., Senoh, M., Matsushita, T., Sugimoto, Y., Kiyoku, H., Kozaki, T., Sano, M., Matsumura, H., Umemoto, H., Chocho, K., and Mukai, T. 2000a. *Jpn J. Appl. Phys.*, 39, L647.

Nagahama, S-I. , Iwasa, N., Senoh, M., Matsushita, T., Sugimoto, Y., Kiyoku, H., Kozaki, T., Sano, M., Matsumura, H., Umemoto, H., Chocho, K., and Mukai, T. 2000b. *Jpn J. Appl. Phys.*, 39, L647.

Nagahama, S., Yanamoto, T., Sano, M., and Mukai, T. 2001. *Jpn J. Appl. Phys.*, 40, L785.

Nagatoma, T., Kuboyama, T., Minamino, H., and Omoto, O. 1989. *Jpn J. Appl. Phys.*, 28, L1334.

Nakamura, S. and Mukai, T. 1992. *Jpn J. Appl. Phys.*, 31, L1457.

Nakamura, S. and Mukai, T. 1995. *J. Vac. Sci. Technol. A.*, 13, 6844.

Nakamura, S., Mukai, T., and Senoh, M. 1991. *Jpn J. Appl. Phys.*, 30, L1998.

Nakamura, S., Mukai, T., and Senoh, M. 1992a. *Jpn J. Appl. Phys.*, 31, part 1, 2883.

Nakamura, S., Mukai, T., and Senoh, M. 1992b. *J. Appl. Phys.*, 71, 5543.

Nakamura, S., Mukai, T., and Senoh, M. 1992c. *Jpn J. Appl. Phys.*, 31, 195.

Nakamura, S., Mukai, T., and Senoh, M. 1993a. *Jpn. J. Appl. Phys.*, 30, L16.

Nakamura, S., Iwasa, N., and Nagahama, S. 1993b. *Jpn J. Appl. Phys.*, 32, L338.

Nakamura, S., Senoh, M., Isawa, N., and Nagahama, S. 1995. *Jpn J. Appl. Phys.*, 34, L797.

Nakamura, S., Senoh, M., Nagahama, S., Iwasa, N., Yamada, T., Matsushita, T., Kiyoku, H., Sugimoto, Y., Kozaki, T., Umemoto, H., Sano, M., and Chocho, K. 1998. *Appl. Phys. Lett.*, 72, 211.

Nakamura, S., Senoh, M., Nagahama, S., Masushita, T., Kiyoku, H., Sugimoto, Y., Kozaki, T., Umemoto, H., Sano, M., and Mukai, T. 1999. *Jpn J. Appl. Phys.*, 38, L226.

Nakatsuka, O., Takei, T., Koide, Y., and Murakami, M. 2002a. *Mater. Transac.*, 43, 1684.

Nakatsuka, O., Koide, Y., and Murakami, M. 2002. Proceedings of the ICSCRM, 885.

Nam, K. B., Li, J., Nakarmi, M. L., Lin, J. Y., and Jiang, H. X. 2003. *Appl. Phys. Lett.*, 82, 1694.

Neu, G., Teisseire, M., Frayssinet, E., Knap, W., Sadowski, M. L., Witowski, A. M., Pakula, K., Leszczynski, M., and Prystawsko, P. 2000. *Appl. Phys. Lett.*, 77, 1348.

Neudeck, P. G. and Powell, J. A. 1994. *IEEE Electron. Device Lett.*, 15, 63.

Neugebauer, J. and Van de Walle, C. G. 1996. *Appl. Phys. Lett.*, 69, 503.

Neugebauer, J. and Van de Walle, C. G. 1999. *J. Appl. Phys.*, 85, 3003.

Ng, H. M., Doppalapudi, D., Moustakas, T. D., Weimann, N. G., and Eastman, L. F. 1998. *Appl. Phys. Lett.*, 73, 821.

Ng, H. M., Moustakas, T. D., and Chu, S. N. G. 2000. *Appl. Phys. Lett.*, 76, 2818.

Nikishin, S. A., Faleev, N. N., Zubrilov, A. S., Antipov, V. G., and Temkin, H. 2000. *Appl. Phys. Lett.*, 76, 3028.

Nishida, T., Saito, H., and Kobayashi, N. 2001. *Appl. Phys. Lett.*, 79, 711.

Nishino, S., Hazuki, Y., Matsunami, H., and Tanaka, T. 1980. *J. Electrochem. Soc.*, 127, 2764.

Nishino, S., Powell, J. A., and Will, H. A. 1983. *Appl. Phys. Lett.*, 42, 460.

Nishino, S., Takahashi, K., Tanaka, H., and Saraie, J. 1994. *Silicon Carbide Relat. Mater.*, 137, 63.

Nurmikko, A. *Short Wavelength III-Nitride Lasers* in "Recent and Evolving Advanced Semiconductor and Organic Nano-technologies" in Vol. 1 "Nanoscale Electronics and Optoelectronics", Academic Press, San Diego, in ISBN 0-12-507061-6 Ed. Hadis Morkoç, 2003.

Nutt, S. R., Smith, D. J., Kim, H. J., and Davis, R. F. 1987. *Appl. Phys. Lett.*, 50, 203.

O'Donnell, K. P., Martin, R. W., Trager-Cowan, C., White, M. E., Esona, K., Deatcher, C., Middleton, P. G., Jacobs, K., van der Stricht, W., Merlet, C., Gil, B., Vantomme, A., and Mosselmans, J. F. W. 2001. *Mater. Sci. Eng. B*, 82, 194.

Ochalski, T. J., Gil, B., Lefebvre, P., Grandjean, M., Leroux, M., Massies, J., Nakamura, S., and Morkoç, H. 1999. *Appl. Phys. Lett.*, 74, 3353.

Ogino, T. and Aoki, M. 1980. *Jpn J. Appl. Phys.*, 19, 2395.

Osamura, K., Nakajima, N., Murakami, Y., Shingu, P. H., and Ohtsuki, A. 1975. *Solid State Commun.*, 46, 3432.

Osinsky, A., Gangopadhyay, S., Gaska, R., Williams, B., Khan, M. A., Kuksenkow, D., and Temkin, H. 1997. *Appl. Phys. Lett.*, 71, 2334.

Oskam, G., Searson, P. C., and Cole, M. W. 2000. *Appl. Phys. Lett.*, 76, 1300.

Özgür, Ü., Webb-Wood, G., Everitt, H. O., Yun, F., and Morkoç, H. 2001. *Appl. Phys. Lett.*, 79, 4103.

Paisley, M. J., Sitar, Z., Posthill, J. B., and Davis, R. F. 1989. *J. Vac. Sci. Technol.*, 7, 701.

Palmour, J. W., Kong, H., and Edmond, J. US Patent 4,946,547, 1990.

Palmour, J. W., Kong, H. S., Waltz, E. D., Edmond, J. A., Carter, C. H. 1991. *1st International High Temperature Electronics Conference.*

Pankove, J. I., Miller, E. I., and Berkeyheiser, J. E. 1971. *RCA Rev.*, 32, 383.

Papanicolaou, N. A. and Zekentes, K. 2002. *Solid State Electron.*, 46, 1975.

Papanicolaou, N. A., Rao, M. V., Mittereder, J., and Anderson, W. T. 2001. *J. Vac. Sci. Technol. B*, 19, 261.

Parsons, J. D., Kruaval, G. B., and Chaddha, A. K. 1994. *Appl. Phys. Lett.*, 65, 444.

Pau, S., Kuhl, J., Khan, M. A., and Sun, C. J. 1998. *Phys. Rev. B*, 58, 12916.

Pau, J. L., Monroy, E., Sanchez-Garcia, M. A., Calleja, E., and Munoz, E. 2002. *Mater. Sci. Eng. B*, 93, 159.

Pazik, J. C., Kelner, G., and Bottka, N. 1991. *Appl. Phys. Lett.*, 58, 1419.

Pecz, B. 2001. *Appl. Surf. Sci.*, 184, 287.

Pelto, C. M., Chang, Y. A., Chen, Y., and Williams, R. S. 2001. *Solid State Electron.*, 45, 1597.

Pelto, C. M., Chang, Y. A., Chen, Y., and Williams, R. S. 2002. *J. Appl. Phys.*, 92, 4283.

Pensl, G. and Choyke, W. J. 1993. *Physica B*, 185, 264.

Pereira, S., Correia, M. R., Monteiro, T., Pereira, E., Alves, E., Sequeira, A. D., and Franco, N. 2001. *Appl. Phys. Lett.*, 78, 2137.

Perlin, P., Jauberthie-Carillon, C., Itie, J. P., San Miguel, A., Grzegory, I., and Polian, A. 1992. *Phys. Rev. B*, 45, 83.

Perlin, P., Suski, T., Teisseyre, H., Leszczynski, M., Grzegory, I., Jun, J., Porowski, S., Boguslawski, P., Berholc, J., Chervin, J. C., Polian, A., and Moustakas, T. D. 1995. *Phys. Rev. Lett.*, 75, 296.

Petrenko, T. L., Teslenko, V. V., and Mokhov, E. N. 1992. *Sov. Phys.-Semicond.*, 26, 874.

Porter, L. M. and Davis, R. F. 1995. *Mater. Sci. Eng. B*, 34, 83.

Powell, A. R. and Rowland, L. B. 2002. *Proc. IEEE*, 90, 942.

Powell, J. A., Larkin, D. J., Matus, L. G., Choyke, W. J., Bradshaw, J. L., Henderson, L., Yoganathan, M., Yang, J., and Pirouz, P. 1990a. *Appl. Phys. Lett.*, 56, 1353.

Powell, J. A., Larkin, D. J., Matus, L. G., Choyke, W. J., Bradshaw, J. L., Henderson, L., Yoganathan, M., Yang, J., and Pirouz, P. 1990b. *Appl. Phys. Lett.*, 56, 1442.

Powell, R. C., Lee, N. E., Kim, Y. W., and Greene, J. E. 1993. *J. Appl. Phys.*, 73, 189.

Powell, J. A., Neudeck, P. G., Larkin, D. J., Yang, J. W., and Pirouz, P. 1994. *Ins. Phys. Conf. Ser.*, 137, 161.

Pozina, G., Bergman, J. P., Paskova, T., and Monemar, B. 1999. *Appl. Phys. Lett.*, 75, 412.

Pozina, G., Edwards, N. V., Bergman, J. P., Paskova, T., Monemar, B., Bremser, M. D., and Davis, R. F. 2001. *Appl. Phys. Lett.*, 78, 1062.

Ptak, A. J., Myers, T. H., Romano, L. T., Van de Walle, C. G., and Northrup, J. E. 2001. *Appl. Phys. Lett.*, 78, 285.

Qiao, D., Jia, L., Yu, L. S., Asbekck, P. M., Lau, S. S., Lim, S. H., Liliental-Weber, Z., Haynes, T. E., and Barner, J. B. 2001a. *J. Appl. Phys.*, 89, 5543.

Qiao, D., Yu, L. S., Lau, S. S., Lin, J. Y., Jiang, H. X., and Haynes, T. E. 2001b. *J. Appl. Phys.*, 88, 4196.

Raaghunathan, R., Alok, D., and Baliga, B. 1995. *IEEE Electron. Devices Lett.*, 16, 226.

Raynaud, C., Richier, C., Brounkov, P. N., Ducroquet, F., Guillot, G., Porter, L. M., Davis, R. F., Jaussaud, C., and Billon, T. 1995. *Mater. Sci. Eng. B*, 29, 122.

Redwing, J. M., Loeber, D. A. S., Anderson, N. G., Tischler, M. A., and Flynn, J. S. 1996. *Appl. Phys. Lett.*, 69, 1.

Ren, F., Vartuli, C. B., Pearton, S. J., Abernathy, C. R., Donovan, S. M., MacKenzie, J. D., Shul, R. J., Zolper, J. C., Lovejoy, M. L., Baca, A. G., Hagerott-Crawford, M., and Jones, K. A. 1997. *J. Vac. Sci. Technol. A*, 15, 802.

Rendakova, S., Ivantsov, V., and Dimitriev, V. 1998. *Mater. Sci. Forum*, 264–268, 163.

Reshchikov, M. A., Huang, D., Yun, F., He, L., Reynolds, D. C., Park, S. S., Lee, K. Y., and Morkoç, H. 2001. *Appl. Phys. Lett.*, 79, 3779.

Reynolds, D. C., Look, D. C., Kim, W., Aktas, O., Botchkarev, A., Salvador, A., Morkoç, H., and Talwar, D. N. 1996. *J. Appl. Phys.*, 80, 594.

Reynolds, D. C., Look, D. C., Jogai, B., Phanse, V. M., and Vaudo, R. P. 1997. *Solid State Commun.*, 103, 533.

Righi, M. C., Pignedoli, C. A., Borghi, G., Di Felice, R., and Bertoni, C. M. 2002. *Phys. Rev.*, 66, 045320.

Rode, D. L. 1975. R. K. Willardson and A. C. Beer, eds., In *Semiconductors and Semimetals, Vol. 10*, pp. 1–90. Academic Press, New York.

Rode, D. L. and Gaskill, D. K. 1995. *Appl. Phys. Lett.*, 66, 1972.

Rosen, A. and Zutavern, F. 1994. *High power optically activated solid-state switches*. Artech House, Boston.

Rottner, K., Frischholz, M., Myrtveit, T., Mou, D., Nordgen, K., Henry, A., Hallin, C., Gustafsson, U., and Schöner, A. 1999. *Mater. Sci. Technol., B*, 61–62, 330.

Rowland, L. B., Burk, A. A., and Brandt, C. D. 1998. *Mater. Sci. Forum*, 264–268, 115.

Rutz, R. F. 1976. *Appl. Phys. Lett.*, 28, 379.

Rutz, R. F., Harrison, E. P., and Cuome, J. J. 1973. *IBMJ. Res. Sev.*, 17, 61.

Ruvimov, S., Liliental Weber, Z., Washburn, J., Duxstad, K. J., Haller, E. E., Mohammad, S. N., Fan, Z., and Morkoç, H. 1996. *Appl. Phys. Lett.*, 69, 1556.

Ryu, M-Y. , Chen, C. Q., Kuokstis, E., Yang, J. W., Simin, G., and Khan, M. A. 2002. *Appl. Phys. Lett.*, 80, 3730.

Saarinen, K., Laine, T., Kuisma, S., Nissilä, J., Hautojärvi, P., Dobrzynski, L., Baranowski, J. M., Pakula, K., Stepniewski, R., Wojdak, M., Wysmolek, A., Suski, T., Leszczynski, M., Grzegory, I., and Porowski, S. 1997. *Phys. Rev. Lett.*, 79, 3030.

Saddow, S. E., Cho, P. S., Goldhar, J., McLean, B. F., Palmour, J. W., and Lee, C. H. 1994. *Inst. Phys. Conf. Ser.*, 137, 573.

Saito, H., Shionoya, S., and Hanamura, E. 1973. *Solid State Commun.*, 12, 227.

Saito, Y., Yamaguchi, T., Kanazawa, H., Kano, K., Araki, T., Nanishi, Y., Teraguchi, N., and Suzuki, A. 2002. *J. Cryst. Growth*, 237–239, 1017.

Sakai, A., Sunakawa, H., and Usui, A. 1997. *Appl. Phys. Lett.*, 71, 2259.

Salvador, A., Sverdlov, B., Lehner, T., Botchkarev, A., Houng, F., and Morkoç, H. 1994. *Appl. Phys. Lett.*, 65, 1880.

Sandvik, P., Mi, K., Shahedipour, F., McClintock, R., Yasan, A., Kung, P., and Razeghi, M. 2001. *J. Cryst. Growth*, 231, 366.

Santic, B., Merz, C., Kaufmann, U., Niebuhr, R., Obloh, H., and Bachem, K. 1997. *Appl. Phys. Lett.*, 71, 1837.

Saxena, V. and Steckl, A. 1998. *SiC Materials and Devices*. Vol. 52 of Semiconductors and Semimetals, 52 pp 77–160. Academic Press, San Diego.

Saxena, V., Su, J. N., and Stekl, A. J. 1999. *IEEE Trans. Electron. Devices*, 46, 456.

Saxler, A., Mitchel, W. C., Kung, P., and Razeghi, R. 1999. *Appl. Phys. Lett.*, 74, 2023.

Schmidt, T. J., Yang, X. H., Shan, W., Song, J. J., Salvador, A., Kim, W., Aktas, Ö., Botchkarev, A., and Morkoç, H. 1996. *Appl. Phys. Lett.*, 68, 1820.

Schneider, J., Müller, H. D., Maier, K., Wilkening, W., Fuchs, F., Dörnen, A., Leibenzeder, S., and Stein, R. 1990. *Appl. Phys. Lett.*, 56, 1184.

Schoen, K. J., Henning, J. P., Woodall, J. M., Copper, J. A., and Melloch, M. R. 1998. *IEEE Electron. Devices Lett.*, 19, 97.

Schöner, A., Nordell, N., Rottner, K., Helbig, R., and Pensl, G. 1996. *Ins. Phys. Conf. Ser.*, 142, 493.

Schubert, E. F., Grieshaber, W., and Goepfert, I. D. 1996. *Appl. Phys. Lett.*, 69, 3737.

Schulz, H. and Theimann, K. H. 1977. *Solid State Commun.*, 23, 815.

Seitz, R., Gaspar, C., Monteiro, T., Pereira, E., Leroux, M., Beaumont, B., and Gibart, P. 1997. *MRS Internet J. Nitride Semicond. Res.*, 2, 36.

Shan, W., Little, B. D., Fischer, A. J., Song, J. J., Goldenberg, B., Perry, W. G., Bremser, M. D., and Davis, R. F. 1996. *Phys. Rev. B*, 54, 16369.

Shan, W., Walukiewicz, W., Haller, E. E., Little, B. D., Song, J. J., McCluskey, M. D., Johnson, N. M., Feng, Z. C., Schurman, M., and Stall, R. A. 1998. *J. Appl. Phys.*, 84, 4452.

Shan, W., Ager, J. W., Yu, K. M., Walukiewicz, W., Haller, E. E., Martin, M. C., McKinney, W. R., and Yang, W. 1999. *J. Appl. Phys.*, 85, 8505.

Sheng, S., Spencer, M. G., Tang, X., Zhou, P., Wongchotigul, K., Taylor, C., and Harris, G. 1997. *Mater. Sci. Eng. B*, 46, 147.

Shibahara, K., Nishino, S., and Matsunami, H. 1986. *J. Cryst. Growth*, 78, 538.

Shor, J. S., Weber, R. A., Provost, L. G., Goldstein, D., and Kurtz, A. D. 1994. *J. Electrochem. Soc.*, 141, 579.

Skowronski, M. In *Wide energy bandgap electronic devices*, F. Ren, and J.C., Zopler, ed. 291, 2003.

Skromme, B. J. and Martinez, G. L. 1999. *Mater. Res. Soc. Symp. Proc.*, 595, W98.

Skromme, B. J., Palle, K., Poweleit, C. D., Yamane, H., Aoki, M., and Disalvo, F. J. 2002. *J. Crys. Growth*, 246, 299.

Smith, M., Chen, G. D., Lin, J. Y., Jiang, H. X., Salvador, A., Kim, W. K., Aktas, O., Botchkarev, A., and Morkoç, H. 1995. *Appl. Phys. Lett.*, 67, 3387.

Smorchkova, I. P., Haus, E., Heying, B., Kozodoy, P., Fini, P., Ibbetson, J. P., Keller, S., Den Baars, S. P., Speck, J. S., and Mishra, U. K. 2000. *Appl. Phys. Lett.*, 76, 718.

Someya, T. and Arakawa, Y. 1998. *Appl. Phys. Lett.*, 73, 3653.

Someya, T., Werner, R., Forchel, A., Catalano, M., Cingolani, R., and Arakawa, Y. 1999. *Science*, 285, 1905.

Song, J. J. and Shan, W. 1998. *Group III Nitride Semiconductor Compounds*, B. Gil, ed., pp. 182–241. Clarendon, Oxford.

Song, Y.-K., Diagne, Zhou, M. H., Nurmikko, A. V., Schneider, R. P., and Takeuchi, T. 2000a. *Appl. Phys. Lett.*, 77, 1744.

Song, Y.-K., Nurmikko, A. V., Schneider, R. P., Kuo, C. P., Krames, M. R., Kern, R. S., Carter-Coman, C., and Kish, F. A. 2000b. *Appl. Phys. Lett.*, 76, 1662.

Sridhara, S. G., Carlsson, F. H. C., Bergaman, J. P., and Janzen, E. 2001. *Appl. Phys. Lett.*, 79, 3944.

Stampfl, C., Van de Walle, C. G., Vogel, D., Kruger, P., and Pollmann, J. 2000. *Phys. Rev. B.*, 61, R7846.

Starosta, K. 1981. *Phys. Status Solidi A*, 68, K55.

Stepniewski, R., Potemski, M., Wysmolek, A., Pakula, K., Baranowski, J. M., Lusakowski, J., Grzegory, I., Porowski, S., Martinez, G., and Wyder, P. 1999. *Phys. Rev. B*, 60, 4438.

Steranka, F. M., Bhat, J., Collins, D., Cook, L., Craford, M. G., Fletcher, R., Gardner, N., Grillot, P., Götz, W., Keuper, M. et al. 2002. High Power LEDs—Technology status and market applications. Presented at International Workshop on Nitrides, Nagoya, Japan.

Strite, S., Ruan, J., Li, Z., Manning, N., Salvador, A., Chen, H., Smith, D. J., Choyke, W. J., and Morkoç, H. 1991. *J. Vac. Sci. Technol. B*, 9, 1924.

Stutzmann, M., Ambacher, O., Cros, A., Brandt, M. S., Angerer, H., Dimitrov, R., Reinacher, N., Metzger, T., Hopler, R., Brunner, D., Freudenberg, F., Handschuh, R., Deger, C. Presented at the E-MRS Straβburg, Symposium L, 1997.

Sugiura, L., Suzuki, M., Nishio, J., Itaya, K., Kokubun, Y., Ishikawa, M. 1998. *Jpn J. Appl. Phys., (Part 17)*, 3878.

Suguwara, Y., Takayama, D., Asano, K., Singh, R., Palmour, J., Hayashi, T. 2001. *Proceedings of International Symp. Power Semiconductor Devices and ICs*, 27–30.

Suzuki, M., Nishio, J., Onomura, M., and Hongo, C. 1998. *J. Cryst. Growth*, 189/190, 511.

Suzuki, M., Kawakami, T., Arai, T., Kobayashi, S., Koide, Y., Uemura, T., Shibata, N., and Murakami, M. 1999. *Appl. Phys. Lett.*, 74, 275.

Tairov, Y. M. and Tsverkov, R. 1981. *J. Cryst. Growth*, 52, 146.

Takahashi, J. and Ohtani, N. 1997. *Phys. Stat. Sol. B*, 202, 163.

Tanaka, T., Watanabe, A., Amano, H., Kobayashi, Y., Akasaki, I., Yamazaki, S., and Koike, M. 1994. *Appl. Phys. Lett.*, 65, 593.

Tang, Y., Banerjee, S. and Chow, T. P. 2002. *Proceeding of the 11th International Symposium on Power Semiconductor Devices and IC's*, 53.

Tansley, T. L., Foley, C. P. *Proc. 3rd Intl Conf. on Semiinsulating III–V Materials* (Warm Springs, OR 1984), ed. by J. S. Blakemore (Shiva, London, pp. 497–500, 1985.

Tansley, T. L. and Foley, C. P. 1986. *J. Appl. Phys.*, 59, 3241.

Tansley, T. L., Egan, R. J., and Horrigan, E. C. 1988. *Thin Solid Films*, 164, 441.

Teke, A., Morkoç, H. 2005. Group III-nitrides. *Handbook of electronic and optoelectronic materials*, S. Kasap and P. Capper, ed., Kluwer Academic.

Teke, A., Dogan, S., Le, H., Huang, D., Yun, F., Mikkelson, M., Morkoç, H., Zhang, S. K., Wang, W. B., and Alfano, R. R. 2003a. *J. Electron. Mater.*, 32, 307.

Teke, A., Dogan, S., Yun, F., Reshchikov, M. A., Le, H., Liu, X. Q., Morkoç, H., Zhang, S. K., Wang, W. B., and Alfano, R. R. 2003b. *Solid-State Electron.*, 47, 1401.

Thomas, D. G. and Hopfield, J. J. 1962. *Phys. Rev.*, 128, 2135.

Thomas, D. G., Hopfield, J. J., and Augustyniak, W. M. 1965. *Phys. Rev.*, 140, A202.

Thomson, D., Gehrke, T., Linthicum, K. J., Rapagopal, P., and Davis, R. F. 1999. *J. Nitride Semicond. Res.*, 4S1, G337.

Tojyo, T., Asano, T., Takeya, M., Hino, T., Kijima, S., Goto, S., Uchida, S., and Ikeda, M. 2001. *Jpn J. Appl. Phys.*, 40, 3206.

Tojyo, T., Uchida, S., Mizuno, T., Asano, T., Takeya, M., Hino, T., Kijima, S., Goto, S., Yabuki, Y., and Ikeda, M. 2002. *Jpn J. Appl. Phys.*, 41, 1829.

Touati, F., Takemasa, K., and Saji, M. 1999. *IEEE Trans. Electron. Dev.*, 46, 444.

Troffer, T., Peppermuller, C., Pensl, G., Rottner, K., and Schöner, A. 1996a. *J. Appl. Phys.*, 80, 3739.

Troffer, T., Hä, C., Pensyl, G., Hölzlein, K., Mitlehner, H., and Völkl, J. 1996b. *Ins. Phys. Conf. Ser.*, 142, 281.

Tsushida, H., Kamata, I., Jikomoto, T., and Izumi, K. 2000. *Mater. Sci. Forum*, 338–342, 145.

Uemoto, T. 1995. *Jpn J. Appl. Phys.*, 34, L7.

Ueno, M., Yoshida, M., Onodera, A., Shimommura, O., and Takemura, K. 1994. *Phys. Rev. B*, 49, 14.

Veinger, A. I., Vodakov, Y. A., Kulev, Y., Lomakina, G. A., Mokhov, V, E. N., Oding, G., and Sokolov, V. I. 1980. *Sov. Techn. Phys. Lett.*, 6, 566.

Wagner, J., Obloh, H., Kunzer, M., Maier, M., Kohler, K., and Johs, B. 2000. *J. Appl. Phys.*, 89, 2779.

Wakahara, A., Tsuchiya, T., and Yoshida, A. 1990. *J. Cryst. Growth*, 99, 385.

Waldron, E. L., Graff, J. W., and Schubert, E. F. 2001. *Appl. Phys. Lett.*, 79, 2737.

Waldrop, J. R., Grant, R. W., Wang, Y. C., and Davis, R. F. 1992. *J. Appl. Phys.*, 72, 557.

Wan, J., Capano, M. A., and Melloch, M. R. 2002. *Solid State Electron.*, 46, 1227.

Wang, J., Tottori, S., Sato, H., Hao, M.-S., Ishikawa, Y., Sugahara, T., Yamashita, K., and Sakai, S. 1998. *Jpn J. Appl. Phys.*, 37, 4475.

Wang, D. F., Shiwei, F., Lu, C., Motayed, A., Jah, M., Mohammad, S. N., Jones, K. A., and Salamanca-Riba, L. 2001. *J. Appl. Phys.*, 89, 6214.

Wang, R., Bhat, I. B., and Chow, T. P. 2002. *J. Appl. Phys.*, 92, 7587.

Weber, S. G., Fegee, M., Spaeth, J. M., Kalabukhova, E. N., Lukin, S. N., and Mokhov, E. 1995. *Solid State Commun.*, 93, 393.

Weimann, N. G., Eastman, L. F., Doppalapudi, D., Ng, H. M., and Moustakas, T. D. 1998. *J. Appl. Phys.*, 83, 3656.

Weitzel, C. S., Palmour, J. W., Carter, C. H., Moore, K., Carter, K. J. *Proc. Int. Symp. Power Semiconductor Devices and ICs*, pp. 25–32, 1997.

Woodbury, H. H. and Ludwig, G. W. 1961. *Phys. Rev.*, 124, 1083.

Wook, K., Botohkarev, A. E., Morkoc, H., Fang, Z. Q., Look, D. C., and Smith, D. J. 1998. *J. Appl. Phys.*, 84, 6680.

Wu, J., Walukiewicz, W., Yu, K. M., Ager, J. W., Haller, E. E., Lu, H., and Schaff, W. J. 2002a. *Appl. Phys. Lett.*, 80, 4741.

Wu, J., Walukiewicz, W., Shan, W., Yu, K. M., Ager, J. W., Haller, E. E., Lu, H., and Schaff, W. J. 2002b. *Phys. Rev. B.*, 80, 201413.

Wu, J., Walukiewicz, W., Yu, K. M., Ager, J. W., Haller, E. E., Lu, H., Schaff, W. J., Saito, Y., and Nanishi, Y. 2002c. *Appl. Phys. Lett.*, 80, 3967.

Xia, Q., Xia, H., and Ruoff, A. L. 1993. *J. Appl. Phys.*, 73, 8198.

Yamaguchi, S., Kariya, M., Nitta, S., Takeuchi, T., Wetzel, C., Amano, H., and Akasaki, I. 1998. *Appl. Phys. Lett.*, 73, 830.

Yamaguchi, S., Kariya, M., Nitta, S., Takeuchi, T., Wetzel, C., Amano, H., and Akasaki, I. 1999. *J. Appl. Phys.*, 85, 7682.

Yamamoto, T. and Katayama-Yoshida, H. 1998. *J. Cryst. Growth*, 190, 532.

Yamasaki, S., Asami, S., Shibata, N., Koike, M., Manabe, K., Tanaka, T., Amano, H., and Akasaki, I. 1995. *Appl. Phys. Lett.*, 66, 1112.

Yeh, C. Y., Lu, Z. W., Froyen, S., and Zunger, A. 1992. *Phys. Rev. B*, 46, 10086.

Yi, C.-C. and Wessels, B. W. 1996. *Appl. Phys. Lett.*, 69, 3026.

Yim, W. M., Stotko, E. J., Zanzucchi, P. J., Pankove, J., Ettenberg, M., and Gilbert, S. L. 1973. *J. Appl. Phys.*, 44, 292.

Yodo, T., Yona, H., Ando, H., Nosei, D., and Harada, Y. 2002. *Appl. Phys. Lett.*, 80, 968.

Yoshida, S., Misawa, S., Fujii, Y., Takada, S., Hayakawa, H., Gonda, S., and Itoh, A. 1979. *J. Vac. Sci. Technol.*, 16, 990.

Yoshida, S., Misawa, S., and Gonda, S. 1982. *J. Appl. Phys.*, 53, 6844.

Yoshikawa, A., Xu, K., Hashimoto, N., Terashima, W., Yoshitani, M., Choe, S., Ishitani, Y. 2003. 8th Wide bandgap nitride workshop, September 29–October 2, 2003, Richmond, VA.

Yoshimoto, N., Matsuoka, T., and Katsui, A. 1991. *Appl. Phys. Lett.*, 59, 2251.

Yun, F., Reshchikov, M. A., Jones, K., Visconti, P., Morkoç, H., Park, S. S., and Lee, K. Y. 2000. *Solid State Electron.*, 44, 2225.

Yun, F., Reshchikov, M. A., He, L., King, T., Morkoç, H., Novak, S. W., and Wei, L. 2002. *J. Appl. Phys. Rapid Commun.*, 92, 4837.

Zhang, R. and Kuech, T. F. 1998. *Appl. Phys. Lett.*, 72, 1611.

Zhang, J., Ellison, A., Henry, A., Linnarsson, M. K., and Janzen, E. 2001. *J. Cryst. Growth*, 226, 267.

Zhao, J., Tone, K., Alexandrov, P., Fursin, L., and Weiner, M. 2003. *IEEE Electron. Device Lett.*, 24, 81.

Zheleva, T. S., Nam, O-H., Bremser, D. M., and Davis, R. F. 1997. *Appl. Phys. Lett.*, 71, 2472.

Zhou, L., Audurier, V., Prouz, P., and Powell, J. A. 1997. *J. Electrochem. Soc.*, 144, L161.

Zhou, L., Lanford, W., Ping, A. T., Adesida, I., Yang, J. W., and Khan, M. A. 2000a. *Appl. Phys. Lett.*, 76, 3451.

Zhou, H., Diagne, M., Makarona, E., Nurmikko, A. V., Han, J., Waldrip, K. E., and Figiel, J. J. 2000b. *Electron. Lett.*, 36, 1777.

Zhu, Q. S. and Sawaki, N. 2000. *Appl. Phys. Lett.*, 76, 1594.

III

Materials—Inorganic

5 Nonlinear Optics

Andrew J. Stentz

Robert W. Boyd

5.1 Introduction

Nonlinear optical phenomena occur when the response of a material system to an applied optical field depends nonlinearly on the strength of the applied field. The subsequent interaction of the optical field with this nonlinear response generates a myriad of interesting physical processes. In some applications, these nonlinear processes are merely limitations to the performance of linear systems, whereas in photonics, nonlinear processes promise to be much more. Applications of nonlinear optics include the frequency doubling of semiconductor lasers, the generation of ultrashort laser pulses, dispersion compensation in communication systems, and all-optical switching. In this chapter, we provide a brief summary of this field, emphasizing important results and qualitative features. A more complete, pedagogical introduction to nonlinear optics is presented in Boyd (1992). The reader is also directed to the textbooks by Shen (1984), Butcher and Cotter (1990), and Zernike and Midwinter (1973).

Two systems of units are commonly used in nonlinear optics. The SI (or mks) system is typically used in applied nonlinear optics and photonics and is the primary system of units used in this handbook. The Gaussian systems of units is often used in fundamental studies of nonlinear optics, and most of the original papers in nonlinear optics are written with these units. To facilitate transformation between these systems of units, every equation quoted in this chapter can be interpreted as an SI equation as written or as a Gaussian-system equation by omitting the prefactors (often ϵ_0 or $4\pi\epsilon_0$) that appear in square brackets. For example, Equation 2.1 is correct as written as an SI equation and becomes $\tilde{P}(t) = \chi^{(1)}\tilde{E}(t)$ in the Gaussian system.

5.2 Nonlinear Optical Susceptibility

In conventional linear optics, it is assumed that the applied optical field is sufficiently weak that the induced polarization $\tilde{P}(t)$ depends linearly on the applied electric field strength $\tilde{E}(t)$ such that

$$\tilde{P}(t) = [\epsilon_0]\chi^{(1)}\tilde{E}(t) \tag{5.1}$$

where the constant of proportionality $\chi^{(1)}$ is the linear susceptibility. When the applied field is sufficiently strong, one can often accurately describe the induced polarization as a power series in the applied electric field strength such that

$$\tilde{P}(t) = [\epsilon_0][\chi^{(1)}\tilde{E}(t) + \chi^{(2)}\tilde{E}^2(t) + \chi^{(3)}\tilde{E}^3(t) + \cdots] \tag{5.2}$$

where $\chi^{(2)}$ and $\chi^{(3)}$ are known, respectively, as the second-and third-order nonlinear optical susceptibilities. The two lowest order nonlinear terms, which are displayed explicitly in Equation 5.2, are responsible for most nonlinear optical interactions. For simplicity, we have neglected the vector nature of the polarization and the electric field and the tensor nature of the optical susceptibility in writing Equation 5.1 and Equation 5.2. As the strength and nature of nonlinear optical interactions are dependent on the magnitude and form of the nonlinear susceptibility tensor, in the following section, we present a proper definition of susceptibility and describe the symmetry properties that are reflected in its form.

5.2.1 Definition of Nonlinear Optical Susceptibility

If we assume that the electric field $\tilde{\mathbf{E}}(\mathbf{r},t)$ can be decomposed into a sum of discrete frequency components $\mathbf{E}(\omega_n)$ as

$$\tilde{\mathbf{E}}(\mathbf{r},t) = \sum_n \mathbf{E}(\omega_n)e^{-i\omega_n t} + c.c., \tag{5.3}$$

then, we can define the components of the second-order nonlinear susceptibility tensor as

$$P_i(\omega_n + \omega_m) = [\epsilon_0]\sum_{jk}\sum_{(nm)}\chi^{(2)}_{ijk}(\omega_n + \omega_m,\omega_n,\omega_m)E_j(\omega_n)E_k(\omega_m) \tag{5.4}$$

where i, j, k refer to the Cartesian coordinates, and the notation (nm) indicates that the sum over n and m is to be made with the sum $\omega_n + \omega_m$ held constant. In general, the summation over n and m may be performed to obtain the result

$$P_i(\omega_n + \omega_m) = [\epsilon_0]D\sum_{jkl}\chi^{(2)}_{ijk}(\omega_n + \omega_m,\omega_n,\omega_m)E_j(\omega_n)E_k(\omega_m) \tag{5.5}$$

where D is known as the degeneracy factor and is equal to the number of distinct permutations of the frequencies ω_n and ω_m. Similarly, we define the components of the third-order nonlinear susceptibility tensor as

$$P_i(\omega_0 + \omega_n + \omega_m) = [\epsilon_0]D\sum_{jkl}\chi^{(3)}_{ijk}(\omega_0 + \omega_n + \omega_m,\omega_0,\omega_n,\omega_m)E_j(\omega_0)E_k(\omega_n)E_l(\omega_m). \tag{5.6}$$

To completely describe all second-order nonlinear interactions with the nonlinear susceptibility described in Equation 5.5, one would need to know 324 complex numbers. This figure accounts for all permutations of the Cartesian coordinates and of the frequency components for both positive and negative frequencies. Fortunately, the symmetries summarized in the following section greatly reduce the number of independent components.

5.2.2 Symmetry Properties of Nonlinear Optical Susceptibility

General symmetry properties of nonlinear susceptibility are reviewed in Table 5.1. Detailed discussions of their derivations are given by Boyd (1992). In addition to the symmetry properties found in Table 5.1, the spatial symmetries of a given material may place additional constraints on the allowed form of the nonlinear susceptibility tensor. See Boyd (1992) for the symmetry properties of various crystal point groups.

The symmetry with the most striking impact on nonlinear susceptibility is inversion symmetry. When a material possesses a center of inversion (i.e., is centrosymmetric), the second-order susceptibility is identically zero. This is the case for all gases, liquids, and amorphous solids as well as for some crystal classes. Inversion symmetry also has a dramatic impact on the number of independent elements in the third-order nonlinear susceptibility, reducing the number of possible independent elements from 81 to 3.

The conditions for the validity of Kleinman's symmetry are so frequently met that a contracted notation for the second-order nonlinear susceptibility under these conditions has been developed. We introduce this notation by defining the tensor d_{ijk} as

$$d_{ijk} = \frac{1}{2}\chi^{(2)}_{ijk}. \tag{5.7}$$

Under the conditions of Kleinman symmetry and, in general, for second-harmonic generation, the tensor d_{ijk} is symmetric in its last two indices. Therefore, we can introduce the contracted matrix d_{il}

TABLE 5.1 Symmetry Properties of Nonlinear Optical Susceptibility

Property	Conditions of Validity	Transformation That Preserves Numerical Value of χ	Example with Second-Order Susceptibility
Reality of fields	Always true	Change sign of all frequency components and take complex conjugate	$\chi^{(2)}_{ijk}(-\omega_n-\omega_m,-\omega_n,-\omega_m)$ $=\chi^{(2)}_{ijk}(\omega_n+\omega_m,\omega_n,\omega_m)^*$
Intrinsic permutation symmetry	Always true	Simultaneously interchange last two frequency components and last two Cartesian coordinates	$\chi^{(2)}_{ijk}(\omega_n+\omega_m,\omega_n,\omega_m)$ $=\chi^{(2)}_{ijk}(\omega_n+\omega_m,\omega_m,\omega_n)$
Reality of nonlinear susceptibility tensor	Lossless medium	Take complex conjugate	$\chi^{(2)}_{ijk}(\omega_n+\omega_m,\omega_n,\omega_m)$ $=\chi^{(2)}_{ijk}(\omega_n+\omega_m,\omega_n,\omega_m)^*$
Full permutation symmetry	Lossless medium	Simultaneously interchange any frequency components and the corresponding Cartesian coordinates	$\chi^{(2)}_{ijk}(\omega_n+\omega_m,\omega_n,\omega_m)$ $=\chi^{(2)}_{ijk}(-\omega_n,\omega_m,-\omega_n-\omega_m)$ and, therefore, with the above results, $\chi^{(2)}_{ijk}(\omega_n+\omega_m,\omega_n,\omega_m)=\chi^{(2)}_{ijk}(\omega_n,-\omega_m,\omega_n+\omega_m)$
Kleinman's symmetry	χ is independent of frequency and medium is lossless (e.g. $\omega_{opt.} << \omega_{reson.}$)	Interchange cartesian coordinates without interchanging frequency components	$\chi^{(2)}_{ijk}(\omega_n+\omega_m,\omega_n,\omega_m)$ $=\chi^{(2)}_{ijk}(\omega_n+\omega_m,\omega_n,\omega_m)$
Inversion symmetry	Medium is centrosymmetric	All values of second-order susceptibility are zero	$\chi^{(2)}_{ijk}(\omega_n+\omega_m,\omega_n,\omega_m)=0$

according to the prescription

$jk:$	11	22	33	23,32	31,13	12,21
$l:$	1	2	3	4	5	6

The nonlinear susceptibility then takes the following form:

$$d_{il} = \begin{bmatrix} d_{11} & d_{12} & d_{13} & d_{14} & d_{15} & d_{16} \\ d_{21} & d_{22} & d_{23} & d_{24} & d_{25} & d_{26} \\ d_{31} & d_{32} & d_{33} & d_{34} & d_{35} & d_{36} \end{bmatrix}. \tag{5.8}$$

When the Kleinman symmetry condition is valid, only 10 of these elements are independent and the matrix takes the following general form:

$$d_{il} = \begin{bmatrix} d_{11} & d_{12} & d_{13} & d_{14} & d_{15} & d_{16} \\ d_{16} & d_{22} & d_{23} & d_{24} & d_{14} & d_{12} \\ d_{15} & d_{24} & d_{33} & d_{23} & d_{13} & d_{14} \end{bmatrix}. \tag{5.9}$$

In addition, for fixed frequencies and geometry, one often finds the notation d_{eff} or simple d used to represent the susceptibility with the appropriate summation over j, k in Equation 5.5.

5.3 Second-Order Nonlinearities

Second-order nonlinearities are commonly employed to generate new frequency components from one or more incident laser beams. For example, frequency-doubling the output of near-infrared semiconductor lasers is a viable option for the generation of compact blue–green laser sources. Optical parametric oscillators (OPOs) based on difference-frequency generation are routinely used to generate frequency-tunable radiation throughout the optical spectrum. In this section, we review the basic equations governing sum-and difference-frequency generation and second-harmonic generation.

5.3.1 Nonlinear Wave Equation

Propagation of light in a nonlinear, nonmagnetic medium with no currents and no free charges is governed by the nonlinear wave equation

$$\nabla \times \nabla \times \tilde{\mathbf{E}} + \frac{1}{[\epsilon_0]c^2}\frac{\partial^2}{\partial t^2}\tilde{\mathbf{D}}^{(1)} = -\frac{4\pi}{[4\pi\epsilon_0]}\frac{1}{c^2}\frac{\partial^2}{\partial t^2}\tilde{\mathbf{P}}^{\mathrm{NL}}, \tag{5.10}$$

where we have split the polarization $\mathbf{P}$ into its linear and nonlinear parts such that

$$\tilde{\mathbf{P}} = \tilde{\mathbf{P}}^{(1)} + \tilde{\mathbf{P}}^{\mathrm{NL}}, \tag{5.11}$$

and where we have introduced the linear displacement field $\mathbf{D}^{(1)}$ as

$$\tilde{\mathbf{D}}^{(1)} = [\epsilon_0]\tilde{\mathbf{E}} + \frac{4\pi}{[4\pi]}\tilde{\mathbf{P}}^{(1)}. \tag{5.12}$$

In a dispersive medium, each frequency component needs to be considered separately, and, therefore, we decompose the fields into their discrete frequency components as

$$\tilde{\mathbf{E}}(\mathbf{r},t) = \sum_n \mathbf{E}_n(\mathbf{r})e^{-i\omega_n t} + c.c., \tag{5.13a}$$

$$\tilde{\mathbf{D}}^{(1)}(\mathbf{r},t) = \sum_n \mathbf{D}_n^{(1)}(\mathbf{r})e^{-i\omega_n t} + c.c., \tag{5.13b}$$

and

$$\tilde{\mathbf{P}}^{\mathrm{NL}}(\mathbf{r},t) = \sum_n \mathbf{P}_n^{\mathrm{NL}}(\mathbf{r})e^{-i\omega_n t} + c.c.. \tag{5.13c}$$

If the medium is also lossless, $\mathbf{D}^{(1)}$ and $\mathbf{E}_n$ are related by a real, frequency-dependent dielectric tensor as

$$\tilde{\mathbf{D}}_n^{(1)}(\mathbf{r},t) = \epsilon^{(1)}(\omega_n)\tilde{\mathbf{E}}_n(\mathbf{r},t). \tag{5.14}$$

Therefore, under most conditions, the nonlinear wave equation reduces to the form

$$\nabla^2\mathbf{E}_n(\mathbf{r}) + \frac{1}{[\epsilon_0]c^2}\omega_n^2\epsilon^{(1)}(\omega_n)\mathbf{E}_n(\mathbf{r}) = -\frac{4\pi}{[4\pi\epsilon_0]}\frac{1}{c^2}\omega_n^2\mathbf{P}_n^{\mathrm{NL}}. \tag{5.15}$$

5.3.2 Sum-Frequency Generation

Sum-frequency generation, depicted schematically in Figure 5.1, is a common nonlinear optical process used to produce light at the sum frequency of the incident optical waves. If the incident beams are of the same frequency, the process is known as second-harmonic generation. The coupled-wave equations that govern this interaction and their solutions in a few illustrative regimes are presented below.

5.3.2.1 Coupled-Wave Equations

If we consider sum-frequency generation in a lossless, nonlinear medium with monochromatic plane waves, we may assume that the electric fields and nonlinear polarizations take the forms

$$\tilde{E}_i(z,t) = A_i e^{i(k_i z - \omega_i t)} + c.c., \tag{5.16}$$

where A_i is known as the slowly varying envelope of the electric field and

$$k_i = \frac{n_i\omega}{c}, \quad n_i = \left(\frac{\epsilon^{(1)}(\omega_i)}{[\epsilon_0]}\right)^{1/2}. \tag{5.17}$$

Similarly, we may assume that the nonlinear polarizations takes the form

$$\tilde{P}_i(z,t) = P_i e^{-i\omega_i t} + c.c. \tag{5.18}$$

By considering which terms of the nonlinear polarization oscillate at the frequency $\omega_3=\omega_1+\omega_2$, we find from Equation 5.6 through Equation 5.9 that the contribution to the nonlinear polarization that leads to sum-frequency generation is of the form

$$P_3 = [\epsilon_0]4\,dA_1A_2e^{i(k_1+k_2)z}. \tag{5.19}$$

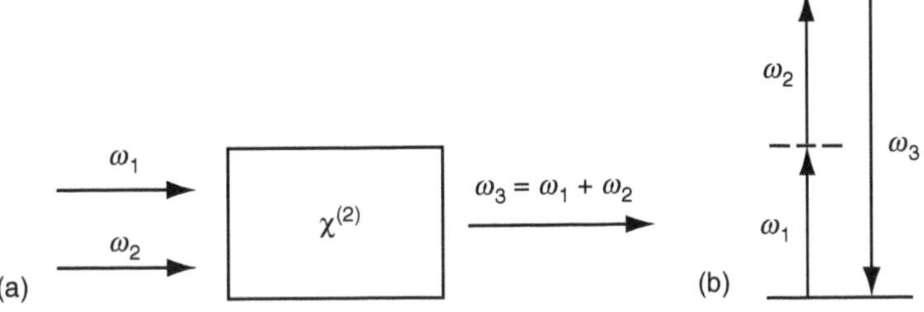

FIGURE 5.1 Sum-frequency generation. (a) Geometry of the interaction, (b) Energy-level description.

Using Equation 5.15 through Equation 5.19, we obtain the nonlinear wave equation

$$\frac{d^2A_3}{dz^2} + 2ik_3\frac{dA_3}{dz} = -\frac{16\pi}{[4\pi]}\frac{d\omega_3^2}{c^2}A_1A_2e^{i(k_1+k_2-k_3)z}. \tag{5.20}$$

In most cases, we can neglect the first term on the left-hand side of the this equation because it is much smaller than the second term. This simplification is known as the slowly varying amplitude approximation. We, thereby, obtain the following equation for the spatial evolution of the sum-frequency field A_3:

$$\frac{dA_3}{dz} = \frac{8\pi}{[4\pi]}\frac{id\omega_3^3}{k_3c^2}A_1A_2e^{i\Delta kz} \tag{5.21}$$

where the quantity

$$\Delta k = k_1 + k_2 - k_3 \tag{5.22}$$

is known as the wave vector mismatch. Similarly the coupled-wave equations for the other two optical fields are given by

$$\frac{dA_1}{dz} = \frac{8\pi}{[4\pi]}\frac{id\omega_1^3}{k_1c^2}A_3A_2^*e^{-i\Delta kz} \tag{5.23a}$$

and

$$\frac{dA_2}{dz} = \frac{8\pi}{[4\pi]}\frac{id\omega_2^3}{k_2c^2}A_3A_1^*e^{-i\Delta kz}. \tag{5.23b}$$

5.3.2.2 The Manley–Rowe Relations

If we now consider the spatial evolution of the intensities of the three interacting waves, it can easily be shown from Equation 5.21 through Equation 5.23 that the following relationships hold for sum-frequency generation:

$$\frac{d}{dz}\left(\frac{I_1}{\omega_1}\right) = \frac{d}{dz}\left(\frac{I_2}{\omega_2}\right) = -\frac{d}{dz}\left(\frac{I_3}{\omega_3}\right) \tag{5.24}$$

where the intensities are given by

$$I_i = \frac{[4\pi\epsilon_0]}{2\pi}n_ic|A_i|^2 \tag{5.25}$$

Equation 5.24 are known as the Manley–Rowe relations. They tell us that the rate at which a photon at ω_1 is destroyed equals the rate at which a photon at ω_2 is destroyed which equals the rate at which a photon at frequency ω_3 is created. The relations can also be useful in deriving expressions for the spatial evolution of the intensities during nonlinear interactions.

5.3.2.3 Undepleted Pump Regime

The general solution of the three coupled equations for sum-frequency generation was derived by Armstrong et al. (1962). Here, we present the solution of the equations in a few specific cases to illustrate the general properties of the nonlinear interaction. When the efficiency of sum-frequency generation is sufficiently small that depletion of the incident waves can be neglected, the spatial evolution of the intensity of the sum-frequency wave is given by

$$I_3 = \left[\frac{1}{64\pi^3\epsilon_0}\right]\frac{512\pi^5 d^2 I_1 I_2}{n_1 n_2 n_3 \lambda_3^2 c} z^2 \sin c^2\left(\frac{\Delta kz}{2}\right). \tag{5.26}$$

Note the dependence on the wave vector mismatch Δk. For distances z greater than $2/\Delta k$, the sum-frequency field becomes out of phase with its driving polarization, and the power of the sum-frequency field flows back into the incident fields. For this reason, the coherence length of the interaction is defined as

$$L_C = \frac{2}{\Delta k}. \tag{5.27}$$

5.3.2.4 Upconversion

We now consider the case where one of the incident beams is weak and other is sufficiently strong that we can treat its amplitude as constant. This case is known as upconversion. Taking A_2 as the strong field amplitude, we find that the coupled-amplitude equations governing the evolution of A_1 and A_3 are

$$\frac{dA_1}{dz} = \kappa_1 A_3 e^{-i\Delta kz} \tag{5.28a}$$

and

$$\frac{dA_3}{dz} = \kappa_3 A_1 e^{i\Delta kz} \tag{5.28b}$$

where

$$\kappa_1 = \frac{8\pi}{[4\pi]}\frac{i\omega_1^2 d}{k_1 c^2} A_2^* \tag{5.29}$$

and

$$\kappa_3 = \frac{8\pi}{[4\pi]}\frac{i\omega_3^2 d}{k_3 c^2} A_2.$$

The general solution to these equations is given by

$$A_1(z) = \left[A_1(0)\cos gz + \left(\frac{\kappa_1}{g}A_3(0) + \frac{i\Delta k}{2g}A_1(0)\right)\sin gz\right] e^{-i\Delta kz/2} \tag{5.30a}$$

and

$$A_3(z) = \left[A_3(0)\cos gz + \left(\frac{\kappa_3}{g}A_1(0) - \frac{i\Delta k}{2g}A_3(0)\right)\sin gz\right] e^{i\Delta kz/2} \tag{5.30b}$$

where g is the positive square root of the expression

$$g = \sqrt{-\kappa_1\kappa_3 + \frac{1}{4}\Delta k^2}. \tag{5.31}$$

Consider the special case in which there is no sum-frequency field incident upon the nonlinear medium. In this case, the evolution of the intensity of sum-frequency field is given by

$$I_3(z) = \frac{n_3}{n_1} I_1(0) \frac{|\kappa_3|^2}{g^2} \sin^2 gz. \tag{5.32}$$

Note that the maximum sum-frequency intensity is generated when $\Delta k = 0$ and that the maximum decreases by the factor $\kappa_1\kappa_3/\left(\kappa_1\kappa_3 - \frac{1}{4}\Delta\kappa^2\right)$ for nonzero wave vector mismatch. The evolution of the intensity of the sum-frequency field is illustrated in Figure 5.2. Note the sinusoidal flow of energy into and then out of the sum-frequency field. The characteristic length of evolution of the interaction is g^{-1} so that the frequency of the oscillations increases with the wave vector mismatch.

5.3.2.5 Second-Harmonic Generation

Second-harmonic generation displays the same basic features as illustrated above. See the work of Armstrong et al. (1962) for the general plane-wave solutions. In practice, however, to achieve the highest conversion efficiency, it is necessary to consider detailed aspects of the problem such as the need for a small positive wave vector mismatch and the spatial walk-off of the focused fundamental and second-harmonic beams. A detailed review of this problem is beyond the scope of this chapter, and readers are referred to the text by Zernike and Midwinter (1973) and to the exhaustive treatment by Boyd and Kleinman (1968).

5.3.3 Difference-Frequency Generation

The process of difference-frequency generation is schematically illustrated in Figure 5.3. Some aspects of the evolution of the difference-frequency wave are qualitatively different from the processes treated above. The following treatment illustrates those differences.

5.3.3.1 Coupled-Wave Equations

An treatment analogous to that outlined in Section 5.3.2.1 for sum-frequency generation leads to the following coupled-wave equations for the spatial evolution of the difference-frequency field

$$\frac{dA_1}{dz} = \frac{8\pi}{[4\pi]} \frac{id\omega_1^2}{k_1 c^2} A_3 A_2^* e^{-i\Delta kz}, \tag{5.33a}$$

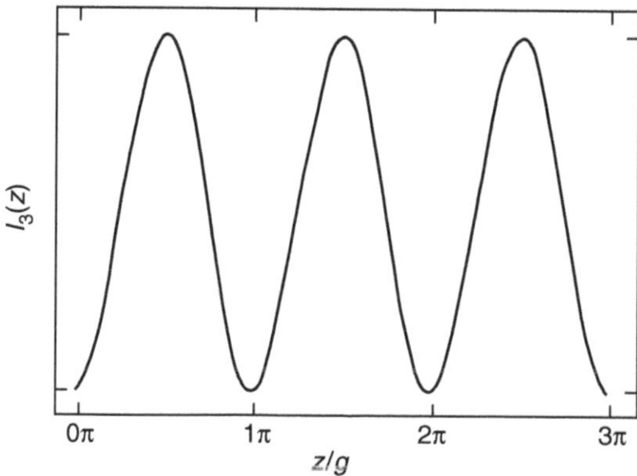

FIGURE 5.2 Spatial evolution of the intensity of the sum-frequency wave in the undepleted pump regime.

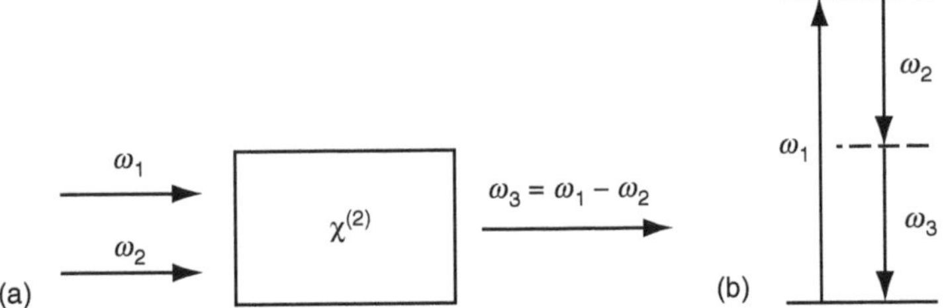

FIGURE 5.3 Difference-frequency generation. (a) Geometry of the interaction, (b) energy-level description.

$$\frac{\mathrm{d}A_2}{\mathrm{d}z} = \frac{8\pi}{[4\pi]}\frac{i\,\mathrm{d}\omega_2^2}{k_2c^2}A_1A_3^*\mathrm{e}^{i\Delta kz}, \tag{5.33b}$$

and

$$\frac{\mathrm{d}A_3}{\mathrm{d}z} = \frac{8\pi}{[4\pi]}\frac{i\,\mathrm{d}\omega_3^2}{k_3c^2}A_1A_2^*\mathrm{e}^{i\Delta kz} \tag{5.33c}$$

where

$$\Delta k = k_1 - k_2 - k_3. \tag{5.34}$$

5.3.3.2 Undepleted Pump Regime

If we assume that the applied pump field A_1 is sufficiently strong that it can be treated as constant, the general solutions to Equations 5.33 are

$$A_2(z) = \left[A_2(0)\left(\cosh gz - \frac{i\Delta k}{2g}\sinh gz\right) + \frac{\kappa_2}{g}A_3^*(0)\sinh gz\right]\mathrm{e}^{i\Delta kz/2} \tag{5.35a}$$

and

$$A_3(z) = \left[A_3(0)\left(\cosh gz - \frac{i\Delta k}{2g}\sinh gz\right) + \frac{\kappa_3}{g}A_2^*(0)\sinh gz\right]\mathrm{e}^{i\Delta kz/2} \tag{5.35b}$$

where

$$g = \left[\kappa_2\kappa_3^* - \left(\frac{\Delta k}{2}\right)^2\right]^{1/2} \tag{5.36}$$

and

$$\kappa_j = \frac{8\pi}{[4\pi]}\frac{i\,\mathrm{d}\omega_j^2A_1}{k_jc^2}. \tag{5.37}$$

The spatial evolution of the intensity of the A_2 and A_3 fields for the boundary condition $A_3(0) = 0$ and zero wave vector mismatch is shown in Figure 5.4. Notice that both fields experience monotonic growth. This is qualitatively different from the oscillatory behavior of sum-frequency generation.

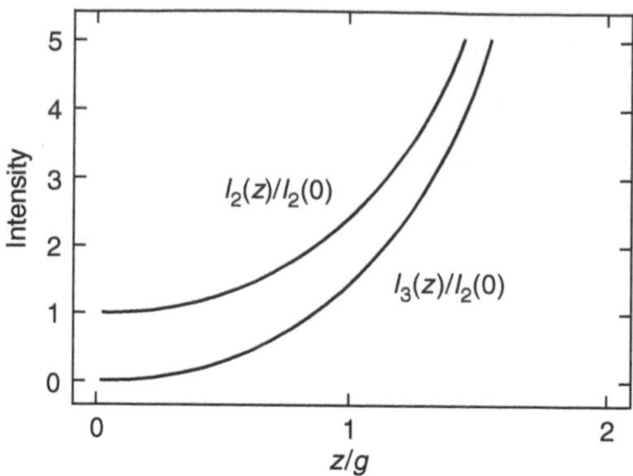

FIGURE 5.4 Spatial evolution of the intensities I_2 and I_3 for difference-frequency generation in the undepleted pump regime.

Processes such as difference-frequency generation, in which the initial and final quantum-mechanical states are identical, are known as parametric processes. For this reason, the amplification of the field A_2 as illustrated in Figure 5.4 is often referred to as parametric amplification. If the purpose of the nonlinear process is to amplify the A_2 field, this field is referred to as the signal wave, and the field A_3 is called the idler wave. If an optical resonator that is highly reflective at frequency ω_2 is placed around the nonlinear medium, oscillation can occur due to the gain provided by parametric amplification. This type of a device, shown schematically in Figure 5.5, is known as an OPO. The efficiency of an OPO can be increased at the expense of simplicity and stability if the device is made resonant at both the signal and idler frequencies. For a review of the current status of OPOs, see the review article by Tang et al. (1992).

5.3.4 Phase-Matching Considerations

As illustrated in the above discussion, the conversion efficiency of nonlinear mixing processes is critically sensitive to the amount of wave vector mismatch. To maximize the conversion efficiency, one usually needs to satisfy the phase-matching condition $\Delta k = 0$. For example, for the case of sum-frequency generation, the phase-matching condition $\Delta k = 0$ implies that

$$n_1\omega_1 + n_2\omega_2 = n_3\omega_3 \tag{5.38}$$

where

$$\omega_1 + \omega_2 = \omega_3. \tag{5.39}$$

Because most optically transparent materials exhibit normal dispersion (i.e., their refractive indices monotonically increase with frequency), this phase-matching requirement is impossible to satisfy without using special techniques of the sort described below.

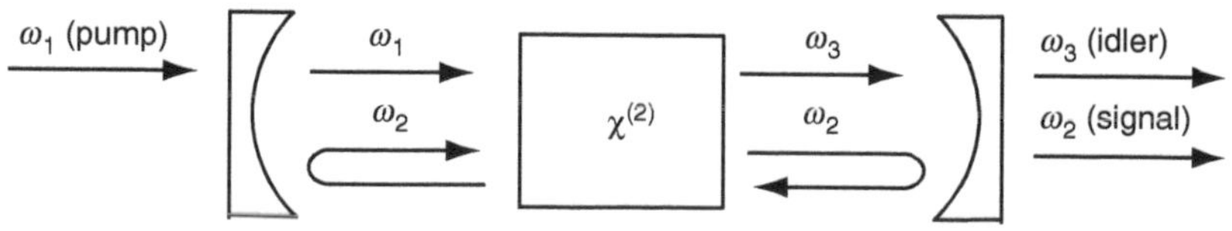

FIGURE 5.5 Schematic illustration of an OPO.

5.3.4.1 Angle Tuning

Angle-tuning is by far the most common method employed to achieve phase matching. With this method, optical fields of different polarizations propagate in a birefringent crystal at the orientation necessary to achieve phase matching. For example, in the case of sum-frequency generation, if the optical fields with the lower two frequencies ω_1 and ω_2 are polarized so that they experience the ordinary refractive index of a uniaxial crystal and the sum-frequency field ω_3 is polarized so that it experiences the extraordinary index, the phase-matching conditions is given by

$$n_1^{\mathrm{o}}\omega_1 + n_2^{\mathrm{o}}\omega_2 = n_3^{\mathrm{e}}\omega_3. \tag{5.40}$$

This condition can be satisfied if the crystal is a negative, uniaxial crystal such that $n_{\mathrm{e}} < n_{\mathrm{o}}$. The case (assumed in Equation 5.40) in which the lower frequency waves have the same polarization is known as type I phase matching, and the case where the lower frequency waves are orthogonally polarized is known as type II phase matching.

Of course, only rarely will the principal axis of a crystal provide the refractive indices necessary to achieve phase matching. In practice, the angular orientation of the crystal must be adjusted. In the case of a uniaxial crystal, the extraordinary polarization experiences the refractive index $n_{\mathrm{e}}(\theta)$ defined by the relationship

$$\frac{1}{n_{\mathrm{e}}(\theta)^2} = \frac{\sin^2\theta}{\bar{n}_{\mathrm{e}}^2} + \frac{\cos^2\theta}{n_{\mathrm{o}}^2} \tag{5.41}$$

where θ is the angle between the propagation vector and the optic axis. By properly adjusting the angle θ, the value of $n_{\mathrm{e}}(\theta)$ can be altered to satisfy the phase-matching condition $\Delta k = 0$.

5.3.4.2 Temperature Tuning

If the phase matching angle θ of Equation 5.41 has a value other than 90°, the Poynting and the propagation vectors of extraordinary waves propagating within the crystal are not parallel. This effect leads to a spatial walk-off of the interacting beams and can sufficiently decrease the efficiency of the nonlinear mixing process as to make angle tuning impractical. In some cases, this problem can be overcome by using temperature to tune the birefringence of the crystal.

5.3.4.3 Quasi Phase Matching

Another technique for achieving efficient sum-frequency generation is known as quasi phase matching. With this technique, parameters of the nonlinear medium are periodically altered along the length of the medium to prevent the flow of energy from the sum-frequency field. Consider the sinusoidal evolution of the sum-frequency field with nonzero wave vector mismatch, as illustrated in Figure 5.2. If the sign of the wave vector mismatch or the sign of the nonlinear susceptibility is periodically changed at multiples of $\pi/2\,g$, energy will monotonically flow from the pump fields into the sum-frequency field. This technique was first proposed by Armstrong et al. (1962). The recent work of Suhara and Nishihara (1990) has shown that the optimum solution is to sinusoidally modify the nonlinear susceptibility. This may be done with lithium niobate, for example. Great strides have been made in recent years in implementing this scheme in nonlinear guided-wave devices. See the work of Lim et al. (1989) as an example.

5.3.4.4 Cherenkov Radiation in Waveguides

Another technique specific to nonlinear waveguides has been very effective in overcoming phase-matching limitations and efficiently generating second-harmonic radiation. With this technique, the fundamental wave is propagated in a nonlinear waveguide with a phase velocity that is less than the phase velocity of the second-harmonic wave freely propagating in the substrate. This situation leads to the efficient generation of second-harmonic radiation in the form of Cherenkov radiation into the substrate. For more information on this technique, see the early work of Tien et al. (1970) and the review by Stegeman (1992).

5.3.5 Materials with Second-Order Nonlinear Susceptibilities

Properties of common second-order nonlinear materials can be found in Table 5.2. The nonlinear susceptibilities are listed in the contracted notation described above. Most of the values for the inorganic materials are taken from Singh (1986). BBO, LBO, and KTP are the most recently developed inorganic materials listed in the table and have become quite popular for use in OPOs. See Eimerl et al. (1987), Chen et al. (1989), and Bierlein et al. (1989) for more information on these materials. Although they are not yet commercially available, several organic materials are included in the table. Single crystals of these materials were grown to characterize their susceptibility tensors. The values for the organic materials are taken from Singh (1986), Kerkoc et al. (1989), Bierlein et al. (1990), and Ledoux et al. (1990). The dispersion coefficients for most of the materials found in Table 5.2 are listed in Table 5.3. These coefficients allow for the calculation of phase-matching angles as described above.

TABLE 5.2 Properties of Common Second-Order Nonlinear Optical Materials

Material	Point Group	d_{il} (pm/V)	$\lambda_{meas.}$ (μm)	Transmission Range (μm)
Inorganics				
Ammonium dihydrogen phosphate (ADP) $NOH_4H_2PO_4$	$\overline{4}2m$	$d_{36}=0.53$	1.064	0.2–1.1
Barium borate (BBO) β-BaB_2O_4	$3m$	$d_{11}=1.8$	1.064	0.2–2.2
Barium sodium niobate $Ba_2NaNb_5O_{15}$	$mm2$	$d_{15}=-13$	1.064	0.25–5
		$d_{24}=-13$		
		$d_{31}=-13$		
		$d_{32}=-13$		
		$d_{33}=-18$		
Cadmium germanium arsenide $CdGeAs_2$	$\overline{4}2m$	$d_{36}=460$	10.6	2.5–18
Cadmium selenide CdSe	$6\,mm$	$d_{15}=31$	10.6	0.75–8
		$d_{31}=-29$		
		$d_{33}=55$		
Cadmium sulfide CdS	$6\,mm$	$d_{15}=14$	1.058	0.55–15
		$d_{31}=-13$		
		$d_{33}=26$		
Gallium arsenide GaAs	$\overline{4}3m$	$d_{14}=230$	1.058	0.9–17
		$d_{36}=210$		
Lithium iodate $LiIO_3$	6	$d_{31}=-5.0$	1.064	0.3–>5.6
		$d_{33}=-5.2$		
Lithium niobate $LiNbO_3$	$3m$	$d_{22}=2.6$	1.058	0.35–4.5
		$d_{31}=-4.9$		
		$d_{33}=-44$		
Lithium *tri*borate (LBO) LiB_3O_5	$mm2$	$d_{31}=-1.2$	1.079	0.16–2.6
		$d_{32}=1.2$		
Potassium dideuterium phosphate (KD*P) KD_2PO_4	$\overline{4}2m$	$d_{14}=0.37$	1.058	0.2–1.6
		$d_{36}=0.38$		
Potassium dihydrogen phosphate (KDP) KH_2PO_4	$\overline{4}2m$	$d_{14}=0.44$	1.064	0.2–1.4
		$d_{36}=0.44$		
Potassium titanyl phosphate (KTP) $KTiOPO_4$	mm	$d_{15}=6.1$	1.064	0.35–4.5
		$d_{24}=7.6$		
		$d_{31}=6.5$		
		$d_{32}=5.0$		
		$d_{33}=14$		
Silver gallium selenide $AgGaSe_2$	$\overline{4}2m$	$d_{36}=50$	10.6	0.75–20

(continued)

TABLE 5.2 (Continued)

Material	Point Group	d_{il} (pm/V)	$\lambda_{meas.}$ (μm)	Transmission Range (μm)
Organics				
DAN 4-(N,N-dimethylamino)-3-acetamidonitrobenzene 4-(CH_3NCH_3)-3-(CH_2CONH_2)$NO_2C_6H_3$	2	$d_{21}=1.5$ $d_{22}=5.2$ $d_{23}=50$ $d_{25}=1.5$	1.064	0.49–2.2
MAP Methyl-(2,4-dinitrophenyl)-amino-propanoate $C_{10}H_{12}N_3O_6$	2	$d_{21}=2.4$ $d_{22}=12$ $d_{23}=11$ $d_{25}=-0.35$	1.064	0.55−2.3
MNA 2-methyl-4-nitroaniline CH_3-NH_2-NO_2-C_6H_4	m	$d_{11}=160$ $d_{24}=24$	1.064	0.48–2
MMONs 3-methy-4-methoxy-4′-nitrostilbene 3-CH_3-4-$CH_3OC_6H_5C=CH(C_6H_4NO_2$-4)	$mm2$	$d_{24}=71$ $d_{32}=41$ $d_{33}=184$	1.064	0.51–1.6
NPP N-4-nitrophenyl-(l)-prolinol 4-$NO_2C_6H_4C_5H_8NO_3$	2	$d_{21}=57$ $d_{22}=19$	1.34	0.52–1.6
POM 3-methyl-4-nitropyridine-1-oxide, $NO_2CH_3NOC_5H_4$	222	$d_{36}=6.4$	1.064	0.49–2.1

See text for references.

5.4 Third-Order Nonlinearities

5.4.1 Third-Harmonic Generation

Third-harmonic generation is the process that leads to the creation of a wave at frequency 3ω in response to an applied wave at frequency ω. The source of the third-harmonic wave is the nonlinear polarization

$$P(3\omega) = [\epsilon_0]\chi^{(3)}(3\omega,\omega,\omega,\omega)E(\omega)^3. \tag{5.42}$$

Third harmonic generation is often used as a means of measuring the value of $\chi^{(3)}$ of optical materials.

5.4.2 Four-Wave Mixing

Four-wave mixing is the name given to a process in which input waves at frequencies ω_1, ω_2, and ω_3 interact in a nonlinear material by means of the nonlinear susceptibility $\chi^{(3)}(\omega_4, \omega_3, \omega_2, \omega_1)$ to create an output wave at frequency $\omega_4 = \omega_1 + \omega_2 + \omega_3$.

5.4.3 Degenerate Four-Wave Mixing and Phase Conjugation

Degenerate four-wave mixing refers to the interaction of four-waves all of the same frequency by means of the nonlinear susceptibility $\chi^{(3)}(\omega, \omega, \omega, -\omega)$. When the four waves interact in the geometry of Figure 5.6, this process leads to optical phase conjugation, that is, the generation of an output wave, (the "conjugate" wave) whose complex amplitude A_4 is proportional to the complex conjugate of that of the signal wave A_3. In particular, a simple model of this interaction shows that (Yariv and Pepper 1977)

TABLE 5.3 Dispersion Parameters of Common Second-Order Nonlinear Optical Materials

Material	Dispersion Eq. #[a]	Index	A	B	C	D	E
Inorganics							
ADP	2	n_o	1.0	1.28196	0.01069		
		n_e	1.0	1.15607	0.00890		
$AgGaSe_2$	3	n_o	3.9362	2.9113	0.38821	1.7954	40
		n_e	3.3132	3.3616	0.38201	1.7677	40
$Ba_2NaNb_5O_{15}$		n_x	1.0	3.94655	0.040179		
		n_y	1.0	3.95233	0.040252		
		n_z	1.0	3.60287	0.032149		
BBO	1	n_o	2.7405	0.0184	0.0179	0.0155	
		n_e	2.3730	0.0128	0.0156	0.0044	
$CdGeAs_2$	3	n_o	10.1064	2.2988	1.0872	1.6247	1370
		n_e	11.8018	1.2152	2.6971	1.6922	1370
CdS	1	n_o	5.235	0.1819	0.1651		
		n_e	5.239	0.2076	0.1651		
CdSe	3	n_o	4.2243	1.7680	0.2270	3.1200	3380
		n_e	4.2009	1.8875	0.2171	3.6461	3629
GaAs	4	n	3.5	4.5×10^9	1.4×10^5	2.45×10^4	269
KDP	2	n_o	1.0	1.24361	0.00959		
		n_e	1.0	1.12854	0.00841		
KD[a]P	3	n_o	1.661824	0.585337	0.016017	0.691221	30
		n_e	1.687522	0.447488	0.017039	0.596216	30
KTP	2	n_x	2.1146	0.89188	0.20861	0.01320	
		n_y	2.1518	0.87862	0.21801	0.01327	
		n_z	2.3136	1.00012	0.23831	0.01679	
LBO	1	n_x	2.4517	0.01177	0.00921	0.00960	
		n_y	2.5279	0.01652	−0.005459	0.01137	
		n_z	2.5818	0.01414	0.01186	0.01457	
$LiIO_3$	2	n_o	1.0	2.40109	0.021865		
		n_e	1.0	1.91359	0.01940		
$LiNbO_3$	1	n_o	4.9048	0.11768	0.04750	0.027169	
		n_e	4.5820	0.099169	0.044432	0.021950	
Organics							
DAN	2	n_x	2.1390	0.1474	0.1355		
		n_y	2.3290	0.3072	0.1547		
		n_z	2.5379	0.7196	0.1759		
MAP	2	n_x	2.7523	0.6079	0.1606	0.05361	
		n_y	2.3100	0.2258	0.17988	0.01886	
		n_z	2.1713	0.10305	0.16951	0.01667	
MMONS	2	n_x	1.987	0.314	0.363		
		n_y	2.184	0.405	0.403		
		n_z	2.507	1.130	0.421		
NPP	2	n_x	2.3532	1.1299	0.1678	−0.0392	
		n_y	2.8137	0.3655	0.2030	0.0816	
		n_z	2.1268	0.0527	0.155	0.0608	
POM	2	n_x	2.5521	0.7962	0.1289	0.0941	
		n_y	2.4315	0.3556	0.1276	0.0579	
		n_z	2.4529	0.1641	0.1280	0.0	

[a] Dispersion Eq. # Dispersion Equation

1	$n^2 = A + \frac{B}{(\lambda^2 - C)} - D\lambda^2 (\lambda \text{ in } \mu\text{m})$
2	$n^2 = A + \frac{B\lambda^2}{(\lambda^2 - C)} - D\lambda^2 (\lambda \text{ in } \mu\text{m})$
3	$n^2 = A + \frac{B\lambda^2}{(\lambda^2 - C)} + \frac{D\lambda^2}{(\lambda^2 - E)} (\lambda \text{ in } \mu\text{m})$
4	$n^2 = A + \frac{B}{D^2 - \gamma^2} + \frac{C}{E^2 - \gamma^2} (\gamma \text{ in cm}^{-1})$

See text for references.

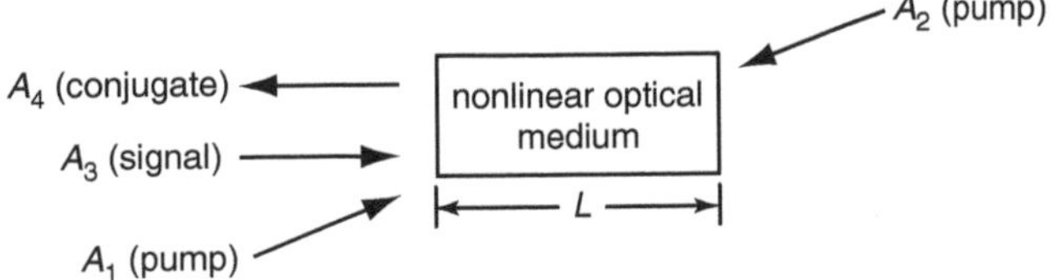

FIGURE 5.6 Geometry of degenerate four-wave mixing that leads to optical phase conjugation.

$$A_4 = \frac{i\kappa}{|\kappa|} A_3^* \tan|\kappa|L \tag{5.43a}$$

where

$$\kappa = \frac{12\pi}{[4\pi]} \frac{\omega}{nc} \chi^{(3)} A_1 A_2. \tag{5.43b}$$

The importance of being able to generate a wave proportional to the complex conjugate of another wave is that this is exactly the transformation required to remove the effects of aberrations from certain optical systems. Optical phase conjugation thus, is, useful in aberration correction and other optical signal processing applications.

5.4.4 Intensity-Dependent Refractive Index

The process of degenerate four-wave mixing can be understood conceptually as a process in which the refractive index of a material becomes modified in proportion to the local intensity $I = [4\pi\epsilon_0](nc/2\pi)|E|^2$ of the light field according to

$$n = n_o + n_2 I \tag{5.44}$$

where

$$n_2 = \left[\frac{1}{16\pi^2\epsilon_0}\right] \frac{12\pi^2}{n_0^2 c} \chi^{(3)}(\omega,\omega,\omega,-\omega). \tag{5.45}$$

This conceptual picture can be used to obtain an understanding of the nature of the degenerate four-wave mixing process illustrated in Figure 5.6. One can imagine the signal wave A_3 interfering with one of the pump waves (A_1) to create a spatially varying intensity distribution which modulates the refractive index in accordance with Equation 5.45 to form a transient grating in the material. The other pump wave, then, scatters from the grating to create the output, conjugate wave. Typical values of n_2 arising from various physical mechanisms are listed in Table 5.4.

TABLE 5.4 Typical Values of the Nonlinear Refractive Index

Physical Mechanism	n_2 (cm^2/W)	Response Time
Thermal effects	10^{-6}	10^{-3}
Electrostriction	10^{-14}	10^{-9}
Molecular orientation	10^{-14}	10^{-12}
Electronic polarization	10^{-16}	10^{-15}

5.4.5 Self-Focusing

Under many circumstances, the nonlinear refractive index coefficient n_2 is positive, and, in this case, the propagation of light through such a material can be unstable to a process known as self-focusing. In this process, rays of light are brought to a focus inside the material as a consequence of the lensing action induced by the tendency of the material to have a larger refractive index near the center of the beam where the intensity is largest. Self-focusing can occur only if the tendency of a beam to form a focus overcomes the tendency of the beam to expand due to diffraction effects, and a simple model of the self-focusing process predicts that this can occur only if the power in a beam exceeds the critical power whose value is given by

$$P_{\rm cr} = \frac{\pi(0.61)^2}{8n_o n_2}\lambda^2. \tag{5.46}$$

For $P << P_{\rm cr}$, the distance from the entrance of the nonlinear optical medium to the point of the self-focus is given by

$$z_{\rm f} = \frac{2n_0}{0.61}\frac{w_0^2}{\lambda}\frac{1}{(P/P_{\rm cr})^{1/2}} \tag{5.47}$$

where w_0 is the beam waist radius at the entrance face of the nonlinear medium.

5.4.6 Tensor Nature of the Third-Order Susceptibility

The third-order susceptibility is a fourth-rank tensor and, thus, possesses 81 components, each of which is independent in the most general case. However, for degenerate four-wave mixing in an isotropic material, $\chi^{(3)}(\omega, \omega, \omega, -\omega)$ possesses only two independent components and the nonlinear polarization can be expressed as (Maker and Terhune 1965)

$$\mathbf{P} = [\epsilon_0]\left[A(\mathbf{E}\cdot\mathbf{E}^*)\mathbf{E} + \frac{1}{2}B(\mathbf{E}\cdot\mathbf{E})\mathbf{E}^*\right] \tag{5.48}$$

where $A = 6\chi^{(3)}_{1122}$ and $B = 6\chi^{(3)}_{1221}$.

5.4.7 Two-Photon Absorption

At sufficiently large laser intensity, the usual law of linear absorption becomes modified to

$$\frac{\mathrm{d}I}{\mathrm{d}z} = -\alpha I - \beta I^2 \tag{5.49}$$

where α is the usual linear absorption coefficient and β is a new constant that describes processes in which two photons are simultaneously removed from the light beam. Two-photon absorption can be understood as arising from the imaginary part of $\chi^{(3)}(\omega, \omega, \omega, -\omega)$.

5.4.8 Enhancement of Nonlinear Interactions in Optical Waveguides

The nonlinear susceptibility of many waveguide materials is orders of magnitude smaller than the susceptibility of common bulk nonlinear materials. Nonetheless, nonlinear effects in many waveguides can often be observed at modest power levels due to the small mode size and low loss of the waveguides. The strength of third-order nonlinear interactions depends on the product of the intensity and the effective interaction length. For a focused Gaussian beam, this product is given by

$$(IL_{\rm eff})_{\rm gaussian} = \left(\frac{P}{\pi w_0^2}\right)\left(\frac{\pi w_0^2}{\lambda}\right) = \frac{P}{\lambda} \tag{5.50}$$

where w_0 is the beam waist radius. Notice that the product is independent of w_0. For a waveguide, this product is given by

$$(IL_{\text{eff}})_{\text{waveguide}} = \left(\frac{P}{\pi w_0^2}\right)\left(\frac{1}{\alpha}\right) = \frac{P}{\pi w_0^2 \alpha} \tag{5.51}$$

where w_0 is the mode radius and L_{eff} is taken as one Beer's length. The ratio of these two product is

$$\frac{(IL_{\text{eff}})_{\text{waveguide}}}{(IL_{\text{eff}})_{\text{Gaussian}}} = \frac{\lambda}{\pi w_0^2 \alpha}. \tag{5.52}$$

This is the enhancement factor for the strength of nonlinear interactions occurring in a waveguide rather than in a bulk geometry. In some cases, this enhancement factor is exceedingly large. For example, near the minimum-loss wavelength of 1.55 μm in optical fibers, the enhancement factor is $\sim 10^9$. This huge enhancement causes nonlinear optical processes, like stimulated Raman scattering, to limit the power levels of fiber communication systems. In the next section, we discuss a situation in which nonlinear optics can be used to enhance the performance of communication systems.

5.4.9 Soliton Propagation

The propagation of pulses of picosecond duration in optical fibers is often governed by the nonlinear Schrodinger equation shown below:

$$\frac{\partial A}{\partial z} = -\frac{i}{2}\beta_2 \frac{\partial^2 A}{\partial \tau^2} + i\gamma |A|^2 A. \tag{5.53}$$

Here, A is the slowly varying envelope of the electric field, z is the longitudinal coordinate, τ is the local time defined by $\tau = t - zv_g$ where v_g is the group velocity, γ is a nonlinear response parameter given by $\gamma = \{[4\pi\epsilon_0]/2\pi\} n_2 n_0 \omega_0$, and β_2 is the group velocity dispersion parameter defined by

$$a\beta_2 = \left(\frac{\partial^2 \beta}{\partial \omega^2}\right)\Big|_{\omega=\omega_0} \tag{5.54}$$

Where β is the wave number of the fundamental mode of the fiber. The first term on the right-hand side of Equation 5.53 treats the effects of group-velocity dispersion. At the minimum-loss wavelength of standard communication fibers, pulses experience anomalous group-velocity dispersion. In this regime, the higher frequency components of the pulse propagate at a faster group velocity than do lower frequency components, and, thus, the pulse broadens and becomes down-chirped. This effect can be balanced against the effects of self-phase modulation which are treated by the second term on the right-hand side of Equation 5.53. Shown in Figure 5.7 is the nonlinear phase shift experienced by a pulse propagating under the influence of self-phase modulation in a medium with a positive nonlinear refractive index. This nonlinear phase modulation induces an instantaneous frequency shift given by

$$\delta\omega = -\frac{\partial \phi_{\text{ne}}}{\partial \tau}. \tag{5.55}$$

The leading edge of the pulse is shifted to lower frequencies whereas the trailing edge is upshifted. This effect can counteract the pulse-broadening effects of group-velocity dispersion for the proper pulse shape and intensity. The result is a pulse known as the fundamental bright soliton which has the form

$$A = \sqrt{\frac{1}{L_D \gamma}} \text{sech}\left(\frac{\tau}{\tau_p}\right) e^{iz/2L_D} \tag{5.56}$$

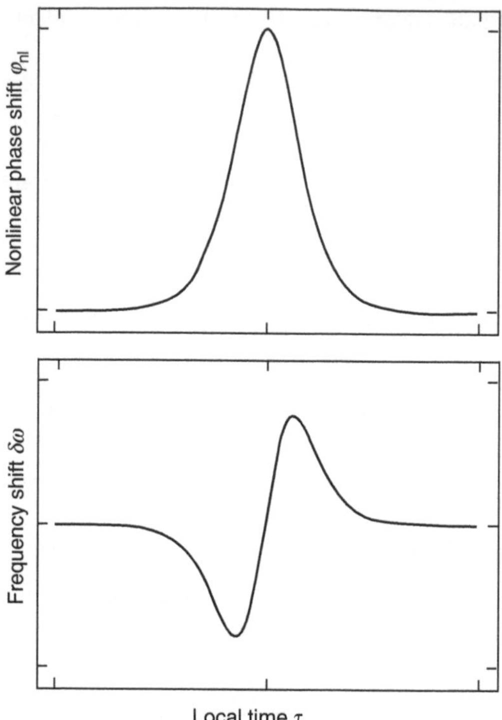

FIGURE 5.7 Nonlinear phase shift and instantaneous frequency shift induced by self-phase modulation.

where

$$L_D = \frac{\tau_p^2}{|\beta_2|}. \tag{5.57}$$

Notice that this pulse propagates without changing shape and without becoming chirped. In addition, the pulse accumulates a uniform nonlinear phase shift across its profile. This feature makes solitons natural bits of information for the all-optical switching schemes discussed in the next section. For a derivation of the above equations and for an extensive discussion of nonlinear fiber optics, see Agrawal (1989). See the work of Mollenauer (1993) as an example of recent soliton-transmission experiments.

5.4.10 Prototypical Nonlinear Optical Switches

Four examples of nonlinear optical switches are depicted in Figure 5.8. For simplicity, the first example is a nonlinear Mach–Zehnder interferometer. One arm of the interferometer is constructed with a highly nonlinear material to produce an intensity-dependent nonlinear phase shift and, thereby, nonlinear switching. Nonlinear common-path interferometers are also being investigated (Blow et al. 1989). In particularly, nonlinear loop mirrors (i.e., nonlinear Sagnac interferometers constructed with all-fiber optic components) are being investigated as all-optical switches in communication systems (Mollenauer et al. 1993) and have been used to mode lock fiber lasers (Duling 1991). The nonlinear directional coupler depicted in Figure 5.8b is another promising waveguide switch. In this device, evanescent coupling occurs between two nonlinear waveguides. Nonlinear directional coupling has been demonstrated in glass fibers (Gusovskii et al. 1985), in polymers (Townsend et al. 1989), and in semiconductors (Jin et al. 1990). The switch depicted in Figure 5.8c is based on nonlinear polarization evolution.

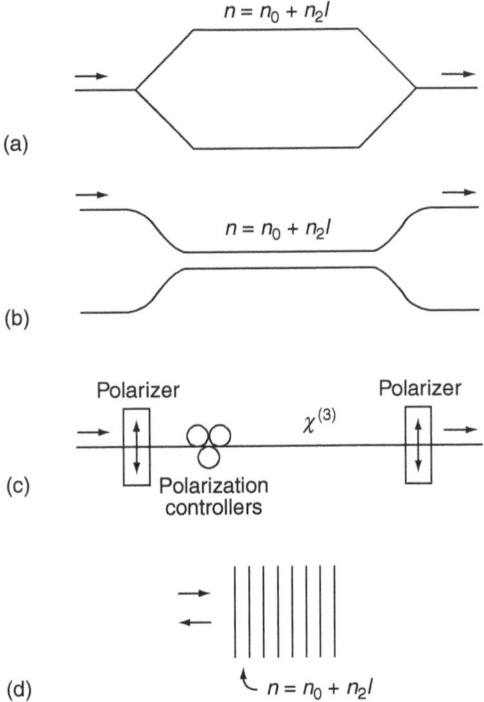

FIGURE 5.8 Four examples of nonlinear optical switches: (a) nonlinear Mach–Zehnder interferometer, (b) nonlinear directional coupler, (c) nonlinear polarization evolution switch, and (d) nonlinear Bragg reflector.

The input light is passed through a linear polarizer and, then, polarization controllers to produce elliptically polarized light. Due to the tensor nature of the third-order susceptibility, the light, then, undergoes nonlinear polarization evolution, creating an intensity-dependent transmission through the output polarizer. This device has recently been used to mode lock fiber lasers (Matsas et al. 1992). Depicted in Figure 5.8d is a nonlinear Bragg reflector, constructed with nonlinear material to make the reflectivity intensity-dependent. For more details, see Ehrlich (1990).

Two-photon absorption and nonlinear index saturation limit the performance of the nonlinear switches described above. To identify materials suited for nonlinear switching applications, the following figures of merit have been developed:

$$W = \frac{\Delta n_{\max}}{\alpha\lambda}, \quad T = \frac{2\beta\lambda}{n_2} \tag{5.58}$$

where α and β are the linear and two-photon absorption coefficients, respectively, as defined in Equation 5.49, and $\Delta n_{\max}$ is the limiting value of the nonlinear index change. Useful optical switching requires $W > 2$ and $T > 1$. See Stegeman (1987), Mizrahi (1989), and Stegeman (1992) for more details.

5.4.11 Materials with Third-Order Nonlinear Susceptibilities

Parameters of some common third-order nonlinear materials are listed in Table 5.5. In each case, the nonlinear refractive index experienced by linearly polarized light is listed. The values of the parameters for the liquids and the large-bandgap crystals are from Smith (1986), the semiconductor parameters are from Sheik-Bahae et al. (1991), and the parameters for the glasses are from Milam and Weber (1976).

TABLE 5.5 Properties of Common Third-Order Nonlinear Materials

Material	n_o	$n_2(10^{-16}$ cm^2/W)	$\lambda_{meas.}$ (μm)	Pulse Duration (pis)
Acetone $(CH_3)_2$ CO	1.35	24	1.064	10
Borosilicate glass (BK-7)	1.52	3.4	1.064	125
Carbon disulfide CS_2	1.63	514	1.064	10
		290	1.064	1000
Carbon tetrachloride CCl_4	1.45	15	1.064	10
Cadmium sulfide CdS	2.34	500	1.064	30
Diamond C	2.42	12.6	0.545	4000
Gallium arsenide GaAs	3.47	−3300	1.064	30
Germanium Ge	4.00	2800	10.6	30
Lithium flouride LiF	1.39	1.05	1.064	125
Nitrobenzene $C_6H_5NO_2$	1.54	685	1.064	10
Potassium titanyl phosphate (KTP) $KTiOPO_4$	1.78	31	1.064	30
Sapphire Al_2O_3	1.76	3.1	1.064	30
Silica, fused SiO_2	1.458	2.7	1.064	125
Yttrium aluminum garnet (YAG) $Y_3Al_5O_{12}$	1.83	7.2	1.064	150

See text for references.

5.5 Stimulated Light Scattering

5.5.1 General Features of Stimulated Light Scattering

Light can be removed from an incident laser beam either by spontaneous or by stimulated light scattering. Spontaneous light scattering is the dominant process at low intensities, whereas stimulated scattering dominates at higher intensities (greater than approximately 100 MW/cm^2). The distinction between these processes is that spontaneous light scattering involves the interaction of light with material excitations created either by thermal excitation or by quantum mechanical zero-point fluctuations, whereas stimulated light scattering involves material excitations created or enhanced by the presence of the light field within the material. The connection between stimulated and spontaneous light scattering was first elucidated by Hellwarth (1963), who showed, theoretically, that the rate at which photons are scattered into some particular mode S (which, for historical reasons, is known as the Stokes mode) can be represented as

$$\frac{dm_S}{dt} = Dm_L(m_S + 1) \tag{5.59}$$

where m_L and m_S are the number of photons per mode for the laser and Stokes fields, respectively, and D is a constant of proportionality. In the factor $m_S + 1$, the m_S and the 1 represent the stimulated and spontaneous contributions to light scattering, respectively. It is generally believed that Equation 5.59 is valid for all light scattering processes and, thus, that a stimulated version of every spontaneous light scattering process must exist. Table 5.6 shows the properties of some typical light scattering processes. Note that stimulated Brillouin scattering (SBS) has highest gain coefficient of these processes, but has a rather slow response time of ~1 ns. Thus, SBS tends to be the dominant scattering process for laser pulses sufficiently long to excite SBS. For shorter pulses, stimulated Raman scattering (SRS) tends to be the dominant process. For these reasons, the most important stimulated light scattering processes are SBS and SRS, and the ensuing discussion will be restricted to these processes.

Boyd (1992) has presented a pedagogical description of stimulated light scattering.

TABLE 5.6 Typical Light Scattering Processes

Process	Material Excitation	Frequency Shift	Response Time	Gain (cm/MW)
Raman	Molecular vibrations (optical phonons)	30 THz	1 ps	5×10^{-3}
Brillouin	Sound waves (acoustic phonons)	3 GHz	1 ns	10^{-2}
Rayleigh	Density variations	0	10 ns	10^{-4}
Rayleigh wing	Molecular orientation	0	1 ps	10^{-3}

5.5.2 Stimulated Raman Scattering

Raman scattering is light scattering by the interaction of light with the vibrational degree of freedom of a molecule. The scattered light is shifted in frequency by a vibrational frequency of the molecule that scatters the light. Raman scattering can be described by the energy level diagram of Figure 5.9, which shows a laser photon of frequency ω_L being scattered into a photon of frequency $\omega_S=\omega_L-\omega_V$ with the molecule being excited from its ground state $v=0$ to its first excited vibrational state $v=1$. These states differ in energy by $\hbar\omega_V$.

Stimulated Raman scattering can be studied either in the amplifier configuration of Figure 5.10 or the generator configuration of Figure 5.10b. In the amplifier configuration, both the laser wave and a (typically weak) Stokes wave are applied to the optical medium and the Stokes wave experiences exponential growth according to

$$|A^S(z)|^2 = |A^S(0)|^2 e^{Gz} \tag{5.60}$$

where the gain coefficient G can be represented as

$$G = \frac{24\pi}{[4\pi]} i \frac{\omega_s}{n_s c} \chi_R(\omega_S)|A_L|^2, \tag{5.61}$$

where $\chi_R(\omega_S) \equiv \chi^{(3)}(\omega_S = \omega_S + \omega_L - \omega_L)$ is the Raman susceptibility. The gain coefficient can often be represented in terms of molecular parameters as

$$G = \frac{2\pi N}{n_S c m \gamma} \frac{\omega_S}{\omega_v} \left(\frac{\partial\alpha}{\partial q}\right)_0^2 |A_L|^2, \tag{5.62}$$

where N is the number density of molecules, m is the reduced mass associated with the vibrational degree of freedom, γ is the width of the Raman resonance, and $(\partial\alpha/\partial q)_0$ is a measure of how rapidly the electronic polarizability α changes with the vibrational coordinate q. Analogous considerations show that a weak wave at the anti-Stokes frequency $\omega_a=\omega_L+\omega_V$ experiences attenuation according to

$$|A^a(z)|^2 = |A^a(0)|^2 e^{-Gz}. \tag{5.63}$$

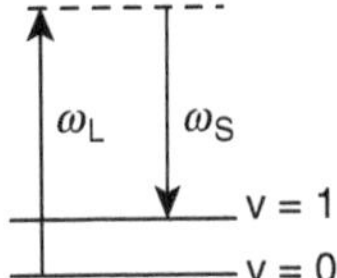

FIGURE 5.9 Energy-level description of stimulated Raman scattering.

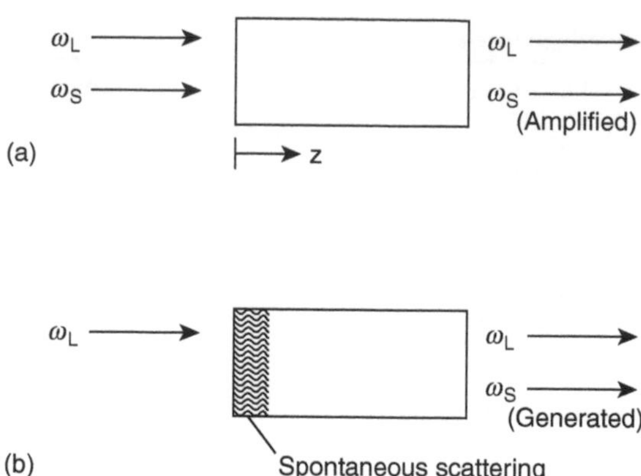

FIGURE 5.10 (a) Amplifier and (b) generator configuration for stimulated Raman scattering.

Stokes amplification (anti-Stokes attenuation) is an example of an induced gain (absorption) process. The presence of gain, that is, does not depend on the fulfillment of a phase-matching condition, and the value of the gain coefficient G does not depend critically on the angle between the laser and Stokes beams (which for simplicity is shown as zero in Figure 5.10).

The properties of some common Raman materials are shown in Table 5.7.

While the Raman amplifier configuration of Figure 5.10a lends itself to a simple theoretical description, most applications of SRS use the generator configuration of Figure 5.10b. In this case, no Stokes beam is applied externally; instead, SRS is initiated by light produced at the Stokes frequency by spontaneous Raman scattering, which occurs as a consequence of quantum fluctuations of the material medium. The theoretical analysis of this situation is quite complicated, although detailed models of the initiation of SRS have been presented by Raymer and Mostowski (1981). As a good approximation, SRS in the generator configuration is a threshold process, and the condition for the occurrence of SRS is given by

$$GL > (GL)_{\text{th}}, \tag{5.64}$$

where L is the length of the interaction region and the value (GL)th of the product GL at the threshold for SRS is approximately 30.

One application of SRS is as a frequency shifter for light fields. As an example, recall that radiation at a wavelength of 1.06 μm can be generated with high efficiency by the Nd:YAG laser. This radiation can be shifted to different wavelengths through SRS. Note that molecular hydrogen has the largest frequency shift of any known material (see Table 5.7) and, thus, is particularly useful for Raman frequency shifting.

TABLE 5.7 Properties of Some Raman Media

Substance	Frequency Shift (cm^{-1})	Gain Factor G/I_L (cm/GW)
Benzene	992	2.8
Nitrobenzene	1345	2.1
SiO_2	467	0.8
H_2 gas (P > 10 atm)	4155	1.5
CS_2	656	24

Source: Kaiser, W. and Maier, M. 1972. *Laser Handbook, Vol. 2*, North-Holland, Amsterdam, part E2. Penzkopfer, A., Laubereau, A., and Kaiser, W. 1979. *Prog. Quantum Electron.*, 6, 55.

5.5.3 Stimulated Brillouin Scattering

Brillouin scattering is the scattering of light from sound waves. Like SRS, SBS is an induced gain process, i.e., the Stokes wave intensity I_S experiences exponential growth of the form

$$I_S(z) = I_S(0)e^{gI_Lz}. \tag{5.65}$$

Nonetheless, unlike SRS, SBS cannot occur in the exact forward direction. The reason is that the acoustic wave must obey the dispersion relation $\Omega = v|\mathbf{q}|$, where Ω is the angular frequency of the acoustic wave and $\mathbf{q}$ is its wavevector, and consequently, in the forward direction, only a zero-frequency photon can couple to the laser and Stokes waves so that $\omega_S = \omega_L - \Omega$ and $\mathbf{k}_S = \mathbf{k}_i - \mathbf{q}$. Consequently, SBS is usually observed only in the backward and near backward directions, in the configuration illustrated in Figure 5.11. In the backward direction, the frequency shift associated with the SBS process is given by

$$\Omega = 2\omega \frac{v}{c/n}. \tag{5.66}$$

where v denotes the velocity of sound in the material.

Detailed consideration (Boyd 1992) of the coupling between the acoustic and optical waves through use of Maxwell's equation and the equation of hydrodynamic shows that the gain factor g of Equation 5.64 is given by

$$g = \frac{\gamma^2\omega^2}{[\epsilon_0^2]nvc^3\rho_0\Gamma_B}, \tag{5.67}$$

where $\gamma \equiv \rho(\partial\epsilon/\partial\rho)$ is the electrostrictive coupling coefficient, ρ_0 is the material density, and Γ_B is the width (full width at half-maximum in angular frequency units) of the Brillouin resonance.

Some of the properties of materials commonly used for SBS are listed in Table 5.8.

Like SRS, SBS can be excited either in an amplifier configuration (in which both laser and Stokes waves are externally applied to the interaction region) or a generator configuration (in which only a laser wave is applied externally). Most applications involve a generator configuration. In a generator configuration, SBS displays a threshold behavior; the threshold for the occurrence of SBS is given by (Boyd et al. 1990)

$$gI_LL \geq (gI_LL)_{th} \tag{5.68}$$

where $(gI_LL)_{th} \approx 25 - 30$.

One application of SBS is in the field of optical phase conjugation. Detailed considerations of the transverse structure of the laser and Stokes fields show (Zel'dovich et al. 1972; Boyd and Grynberg 1992) that, under many circumstances, the generated Stokes wave is a phase-conjugate replica of the incident laser wave, that is $E_S^{out} \propto (E_L^{in})*$. The generation of phase-conjugate wavefronts is important because, under certain situations, phase conjugation can be used to remove the influence of aberrations from optical systems.

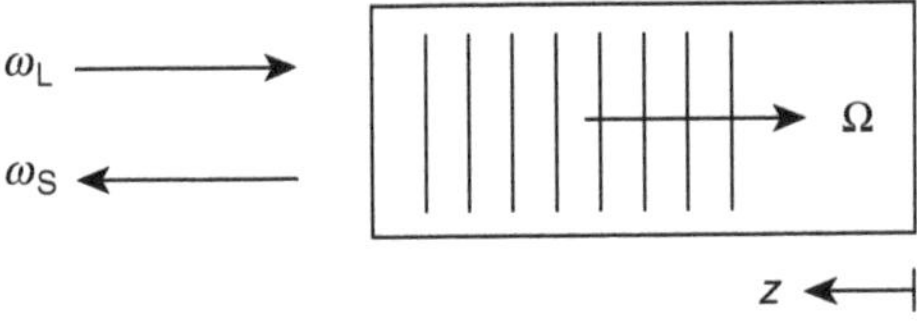

FIGURE 5.11 Stimulated Brillouin scattering involves scattering of the optical wave ω_L from a retreating sound wave to produce a backward Stokes wave.

TABLE 5.8 Properties of Several Materials Used for SBS

Material	Frequency Shift Ω/2x (MHz)	Linewidth Γ/2π (MHz)	Gain Factor g (cm/GW)
CS_2	5,850	52.3	150
CCl_4	4,390	520	6
Methanol	4,250	250	13
Ethanol	4,550	353	12
H_2O	5,690	317	4.8
Optical glasses	11,000–17,000	10–106	4–25
Fused silica	17,000	78	4.5

Values quoted for a wavelength of 0.694 μm.
Source: Kaiser, W. and Maier, M. 1972. *Laser Handbook, Vol. 2*, North-Holland, Amsterdam, part E2.

Another application of SBS is laser pulse compression. Under appropriate conditions, an appreciable fraction of the laser pulse energy can be transferred to an output Stokes pulse that is considerable shorter than the incident laser pulse. Pulse compression by a factor of 10 is possible by means of this effect. The origin of this effect is purely kinematic. The leading edge of the Stokes pulse is more efficiently amplified than the trailing edge because the laser and Stokes pulses are conterpropagating and the trailing edge interacts with a portion of the pump pulse that that has been depleted by the leading edge of the Stokes pulse. One necessary condition for the occurrence of pulse shortening is that the longitudinal extent of the interaction region must exceed the physical length of the laser pulse.

Acknowledgments

The authors would like to acknowledge the assistance of George Fischer in preparing the tables on organic materials.

References

Agrawal, G. P. 1989. *Nonlinear Fiber Optics, 1st Ed.*, Academic Press, San Diego.

Armstrong, J. A., Bloembergen, N., Ducuing, J., and Pershan, P. S. 1962. Interactions between light waves in a nonlinear dielectric. *Phys. Rev.*, 127, 1918.

Bierlein, J. D. and Vanherzeele, H. 1989. Potassium titanyl phosphate: Properties and new applications. *J. Opt. Soc. Am. B.*, 6, 622.

Bierlein, J. D., Cheng, L. K., Wang, Y., and Tam, W. 1990. Linear and nonlinear optical properties of 3-methyl-4-methoxy-4'-nitrostilbene single crystals. *Appl. Phys. Lett.*, 56, 423.

Blow, K. J., Doran, N. J., and Nayar, B. K. 1989. Experimental demonstration of optical soliton switching in an all-fiber nonlinear Sagnac interferometer. *Opt. Lett.*, 14, 754.

Boyd, G. D. and Kleinman, D. A. 1968. Parametric interaction of focused Gaussian light beams. *J. Appl. Phys.*, 39, 3597.

Boyd, R. W. 1992. *Nonlinear Optics*, Academic Press, Boston.

Boyd, R. W. and Grynberg, G. 1992. *Contemporary Nonlinear Optics, 1st Ed.*, Acadamic Press, San Diego. Chap. 3.

Boyd, R. W., Rzazewski, K., and Narum, P. 1990. Noise initiation of stimulated Brillouin scattering. *Phys. Rev. A.*, 42, 5514.

Butcher, P. N. and Cotter, D. 1990. *The Elements of Nonlinear Optics*. Cambridge University Press, Cambridge.

Chen, C., Wu, Y., Jiang, A., Wu, B., You, G., Li, R., and Lin, S. 1989. New nonlinear-optical crystal: LiB_3O_5. *J. Opt. Soc. Am. B.*, 6, 616.

Duling, I. N. 1991. All-fiber ring soliton laser mode locked with a nonlinear mirror. *Opt. Lett.*, 16, 539.

Ehrlich, J. E., Assanto, G., and Stegeman, G. L. 1990. All-optical tuning of waveguide nonlinear distributed feedback gratings. *Appl. Phys. Lett.*, 56, 602.

Eimerl, D., Davis, L., Velsko, S., Graham, E. K., and Zalkin, A. 1987. Optical, mechanical, and thermal properties of barium borate. *J. Appl. Phys.*, 62, 1968.

Gusovskii, D. D., Dianov, E. M., Maier, A. A., Neustruev, V. B., Shklovskii, E. I., and Shcherbakov, I. A. 1985. Nonlinear light transfer in tunnel-coupled optical waveguides. *Sov. J. Quantum Electron.*, 15, 1523.

Hellwarth, R. W. 1963. Theory of stimulated Raman scattering. *Phys. Rev.*, 130, 1850.

Jin, R., Sokoloff, J. P., Harten, P. A., Chuang, C. L., Lee, S. G., Warren, M., and Gibbs, H. M. 1990. Peyghambarian, ultrafast modulation with subpicosecond recovery time in a GaAs/AlGaAs nonlinear directional coupler. *Appl. Phys. Lett.*, 56, 993.

Kaiser, W. and Maier, M. 1972. *Laser Handbook, Vol. 2,* North-Holland, Amsterdom, part E2.

Kerkoc, P., Zgonik, M., Sutter, K., Bosshard, Ch., and Gunter, P. 1989. Optical and nonlinear optical properties of 4-(*N,N*-dimethylamino)-3-acetamidonitrobenzene single crystals. *Appl. Phys. Lett.*, 54, 2062.

Ledoux, I., Lepers, C., Perigaud, A., Bandan, J., and Zyss, J. 1990. Linear and nonlinear optical properties of *N*-4-nitrophenyl-1-prolinol single crystals. *Opt. Commun.*, 80, 149.

Lim, E. J., Fejer, M. M., and Byer, R. L. 1989. Blue light generation by frequency doubling in periodically poled lithium niobate channel waveguide. *Electron. Lett.*, 25, 731.

Maker, P. D. and Terhune, R. W. 1965. Study of optical effects due to an induced polarization third order in the electric field strength. *Phys. Rev.*, 137, A801.

Matsas, V. J., Newson, T. P., Richardson, D. J., and Payne, D. N. 1992. Self-starting passively mode-locked fibre ring soliton laser exploiting nonlinear polarisation rotation. *Electron Lett.*, 28, 1391.

Milam, D. and Weber, M. J. 1976. Measurement of nonlinear refractive-index coefficients using time-resolved interferometry: Application to optical materials for high-power neodymium lasers. *J. Appl. Phys.*, 47, 2497.

Mizrahi, V., DeLong, K. W., Stegeman, G. I., Saifi, M. A., and Andrejco, M. J. 1989. Two-photon absorption as a limitation to all-optical switching. *Opt. Lett.*, 14, 1140.

Mollenauer, L. F., Lichtman, E., Neubelt, M. J., and Harvey, G. T. 1993. Demonstration, using sliding-frequency guiding filters, of error-free soliton transmission over more than 20 Mm at 10 Gbits/s, single channel, and over more than 13 Mm at 20 Gbit/s in a two-channel wdm. *Electron. Lett.*, 29, 910.

Penzkopfer, A., Laubereau, A., and Kaiser, W. 1979. High intensity Raman interactions. *Prog. Quantum Electron.*, 6, 55.

Raymer, M. G. and Mostowski, J. 1981. Stimulated Raman scattering: Unified treatment of spontaneous initiation and spatial propagation. *Phys. Rev. A.*, 24, 1980.

Sheik-Bahae, M., Hutchings, D. C., Hagan, D. J., and Van Stryland, E. W. 1991. Dispersion of bound electronic nonlinear refraction in solids. *IEEE J. Quantum Electron.*, 27, 1296.

Shen, Y. R. 1984. *The Principles of Nonlinear Optics*. Wiley, New York.

Singh, S. 1986. *Handbook of Laser Science and Technology, Vol. III,* CRC Press, Boca Raton, FL, Section 1.1.

Smith, W. L. 1986. *Handbook of Laser Science and Technology, Vol. III,* CRC Press, Boca Raton, FL, Section 1.3.

Stegeman, G. I. 1987. Parameter trade-offs in nonlinear directional couplers: Two level saturable nonlinear media. *Opt. Commun.*, 63, 281.

Stegeman, G. I. 1992. *Contemporary Nonlinear Optics*. Academic Press, *1st Ed.*, San Diego, Chap. 1.

Suhara, T. and Nishihara, H. 1990. Theoretical analysis of waveguide second-harmonic generation phase matched with uniform and chirped gratings. *IEEE J. Quantum Electron.*, 26, 1265.

Tang, C. L., Bosenberg, W. R., Ukachi, T., Lane, R. J., and Cheng, L. K. 1992. Optical parametric oscillators. *Proc. IEEE.*, 80, 365.

Tien, P. K., Ulrich, R., and Martin, R. J. 1970. Optical second-harmonic generation in form of coherent Cerenkov radiation from a thin-film waveguide. *Appl. Phys. Lett.*, 17, 447.

Townsend, P. D., Jackel, J. L., Baker, G. L., Shelburne, J. A., and Etemad, S. 1989. Observation of nonlinear optical transmission and switching phenomena in polydiacetylene-based directional couplers. *Appl. Phys. Lett.*, 55, 1829.

Yariv, A. and Pepper, D. M. 1977. Amplified reflection, phase conjugation, and oscillation in degenerate four-wave mixing. *Opt. Lett.*, 1, 16.

Zel'dovich, B. Y., Popovichev, V. I., Ragulsky, V. V., and Faizullov, F. S. 1972. Connection between the wave fronts of the reflected and excited light in stimulated Mandel'shtam–Brillouin scattering. *JETP Lett.*, 15, 109.

Zernike, F. and Midwinter, J. E. 1973. *Applied Nonlinear Optics, 2nd Ed.*, Wiley, New York.

6

Ferroelectric Materials

Venkatraman Gopalan

Kenneth L. Schepler

Volkmar Dierolf

Ivan Biaggio

6.1 General Characteristics of Ferroelectric Materials

Ferroelectrics are substances in which a built-in spontaneous polarization, P_s (in units of electrical dipole moments per unit volume, C-m/m^3), exists in a certain range of temperatures and isotropic pressures in the absence of an external electric field [1]. This spontaneous polarization has two or more orientational states and can be switched from one state to the other by an external electric field, or in some cases, by a mechanical stress.

As noted by Lines and Glass [1], this definition is not very strict, since the ability to switch the polarization depends on the experimental conditions and the sample properties, such as the crystalline perfection and electrical conductivity. For example, it is difficult to distinguish ferroelectric hysteresis loops from nonlinear dielectric losses in some materials with large dielectric losses. However, ferroelectric switching is an important feature in distinguishing it from pyroelectric crystals. A pyroelectric material shows a flow of charge to and from its surfaces upon a temperature change, which changes its spontaneous polarization. While all ferroelectric materials are also pyroelectric, (see Figure 6.1) the converse is not true [1].

We now describe some general characteristics of ferroelectrics.

6.1.1 Non-Centrosymmetry

In general, all ferroelectric materials in their ferroelectric phase are pyroelectric, piezoelectric, and non-centrosymmetric (e.g., they lack center of inversion symmetry). The lack of inversion symmetry is a particularly important feature of ferroelectrics, since it endows them with possible nonlinear optical

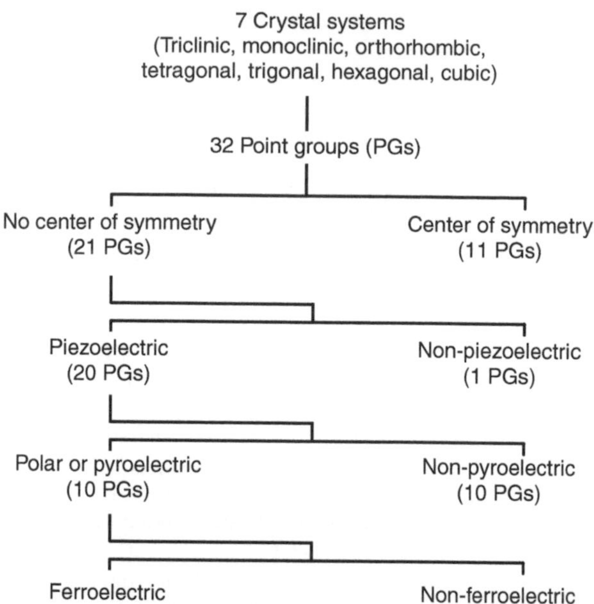

FIGURE 6.1 A schematic tree of the different classes of crystal systems and their property classifications. Piezoelectricity refers to elastic strains in the material in response to an external electric field. Pyroelectricity refers to a presence of spontaneous polarization in the material tehat can be changed in magnitude with temperature, resulting in charge flow to and from the material surfaces. A ferroelectric material therefore is pyroelectric, piezoelectric, and non-centrosymmetric. (From Gopalan, V. Structure-optical property correlation in ferroelectric potassium niobate thin films, PhD diss., Cornell Univ., 1995. With permission.)

responses that are odd-rank polar tensor properties, which would otherwise disappear in a centrosymmetric material. Examples include spontaneous polarization (P_s: 1st rank), electro-optic effect (r_{ijk}: 3rd rank), and sum–difference frequency generation (d_{ijk}: 3rd rank). Indeed, some of the best-known electro-optic and nonlinear optical materials in lasers and optics, such as lithium niobate ($LiNbO_3$), lithium tantalate ($LiTaO_3$), barium titanate ($BaTiO_3$), and potassium niobate ($KNbO_3$), etc., are ferroelectrics.

6.1.2 Polarization, Domains, and Domain Walls

Ferroelectric samples typically exhibit separate regions called domains which differ in their spontaneous polarization direction. The number of distinct spontaneous polarization directions in a ferroelectric depends on its point group symmetry. All polar classes can be potentially ferroelectric because they contain at least one unique axis that does not change with the symmetry operations of the point group. Table 6.1 summarizes these point groups and the possible directions of ferroelectric polarizations in them. The unit cell of the crystal structure is typically described in *crystallographic axes* (*a*,*b*,*c*) which need not be orthogonal. All the physical properties of the crystals are described in tensor form through the *crystal physics axes* (x_1, x_2, x_3), which form an orthogonal set. The relationship between the two sets of coordinates is described in Table 6.1. For all the uniaxial ferroelectric crystals, and the biaxial orthorhombic ferroelectrics, the polarization, P_3, can exist only along their crystallographic *c*-axis, (or x_3 axes in crystal physics coordinates). For the monoclinic ferroelectrics, the polarization can exist either along the 2-fold rotation axis, *b* (as in point group 2), or in any direction within the mirror plane, *a*–*c* (as in point group *m*). There is no restriction on the direction of polarization in a triclinic system.

Regions in a ferroelectric with uniform polarization, *P*, are called ferroelectric *domains*. For a polarization in direction $+P$, there can be a polarization in the direction $-P$, forming at least two

TABLE 6.1 Ten Polar Point Groups That Can Possess Ferroelectricity

Polar Point Groups, International (Schönflies)	Crystal System, Crystallographic Axes,[a] Crystal Physics Axes[b]	Possible Ferroelectric Polarization[c] P	Possible Domain Walls	Optical Class, Second Rank-Tensor,[c,d] T_{ij}
$1\ (C_1)$	*Triclinic* $a \neq b \neq c, \alpha \neq \beta \neq \gamma$	(P_1, P_2, P_3)	48 types [5]	*Biaxial* $\begin{bmatrix} T_{11} & T_{12} & T_{13} \\ T_{21} & T_{22} & T_{23} \\ T_{31} & T_{32} & T_{33} \end{bmatrix}$
$2(C_2)$	*Monoclinic* $a \neq b \neq c, \alpha = \gamma = 90° \neq \beta,\ x_2//b$	$(0, P_2, 0)$	24 types [5]	*Biaxial* $\begin{bmatrix} T_{11} & 0 & T_{13} \\ 0 & T_{22} & 0 \\ T_{31} & 0 & T_{33} \end{bmatrix}$
$m(C_s)$		$(P_1, 0, P_3)$		
$mm2(C_{2v})$	*Orthorhombic* $a \neq b \neq c, \alpha = \beta = \gamma = 90°\ x_1//a,\ x_2//b,\ x_3//c$	$(0, 0, P_3)$	4 types:60°, 120°: {100} 90°, 180°: {110}	*Biaxial* $\begin{bmatrix} T_{11} & 0 & 0 \\ 0 & T_{22} & 0 \\ 0 & 0 & T_{33} \end{bmatrix}$
$4(C_4)$	*Tetragonal* $a = b \neq c, \alpha = \beta = \gamma = 90°\ x_1//a,\ x_2//b,\ x_3//c$	$(0, 0, P_3)$	2 types: 90° {110} 180° {001}	*Uniaxial* $\begin{bmatrix} T_{11} & 0 & 0 \\ 0 & T_{11} & 0 \\ 0 & 0 & T_{33} \end{bmatrix}$
$4mm(C_{4v})$		$(0, 0, 0\ P_3)$		
$3(C_3)$	*Trigonal* $a = b \neq c, \alpha = \beta = 90°, \gamma = 120°\ x_1//a,\ x_3//c$	$(0, 0, P_3)$	1 type: 180° ⊥ to (00.1)	
$3m(C_{3v})$		$(0, 0, P_3)$		
$6(C_6)$	*Hexagonal* $a = b \neq c, \alpha = \beta = 90°, \gamma = 120°\ x_1//a,\ x_3//c$	$(0, 0, P_3)$		
$6mm(C_{6v})$		$(0, 0, P_3)$		

[a] Crystallographic axes are (a, b, c) and angles between (a&b, b&c, c&a) axes are (γ, α, β).
[b] Crystal Physics axes are (x_1, x_2, x_3) and they are orthogonal to each other.
[c] In crystal physics coordinates.
[d] T_{ij} can be any 2nd rank tensor property. A common optical tensor property is the relative impermittivity, $\beta_{ij} = (1/n_{ij})^2$, which appears in expression for the optical index ellipsoid construction. Here, n_{ij} are the refractive index components. Optical class here refers to either *uniaxial* (single unique optical axis), or biaxial (two unique optical axes) crystals.

degenerate polarization states. Additional degeneracies can come from the number of distinct polarization directions possible in a crystal system. When two domains with distinct polarization directions P meet commensurately within a crystal, they form *domain walls* that separate the two domains. The domain walls are also restricted by symmetry to distinct crystallographic planes within the crystal depending on the nature of the walls (see Table 6.1). All ferroelectrics can possess 180° domain walls (or *antiparallel* domain walls) that separate $+P$ and $-P$ polarization domains. For uniaxial trigonal and hexagonal materials, only 180° walls can exist. Controlling and patterning these domain walls in uniaxial ferroelectric materials such as lithium niobate ($LiNbO_3$) and lithium tantalate ($LiTaO_3$) is extremely important for many electro-optic and frequency conversion devices, as described below. For tetragonal systems, 90° walls can also exist, where the latter refers to walls separating two perpendicular polarization directions; this is an equivalent of crystallographic twinning (Figure 6.2). Example materials include ferroelectric barium titanate ($BaTiO_3$), lead titanate ($PbTiO_3$), and strontium barium niobate

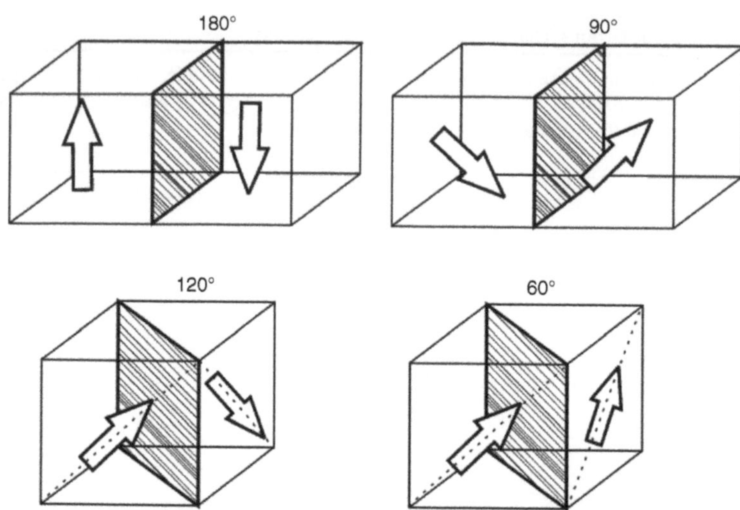

FIGURE 6.2 Schematic examples of some common types of ferroelectric domain walls. Broad arrows indicate the direction of polarizations, P_s, in each domain, and the hatched plane indicates the domain wall.

($Sr_{1-x}Ba_xNb_2O_6$, also called SBN). Orthorhombic crystal systems such as potassium niobate ($KNbO_3$) and potassium titanyl phosphate ($KTiOPO_4$) can have 60, 90, 120, and 180° domain walls. The domain wall orientations become more numerous and complex for monoclinic and triclinic systems [3]. The most common symmetries among materials in nature are the low symmetries, such as monoclinic, triclinic, and orthorhombic systems [4]. However, many useful optical ferroelectrics used today fall outside these classes.

A fictitious, infinite ferroelectric crystal (without surfaces) in equilibrium state would be composed of a single domain, since creation of domain walls costs energy. However, the formation of domains and domain walls in a real finite ferroelectric can be understood by considering a slab of a uniaxial ferroelectric crystal with its polarization direction perpendicular to the slab surfaces. Starting from a single domain state, the surfaces will be charged with charge density $\sigma = +P$ on the surface where the positive end of the polarization terminates, and $\sigma = -P$ on the surface where the negative end of the polarization terminates. An electric field, $E_d = -P/\varepsilon_0$, is created between these uniformly charges surface sheets of opposite polarities, which is called the *depolarizing field*, since it points in the direction opposite to the spontaneous polarization, and hence destabilizes it. In order to stabilize the domain, compensating charges of opposite polarities are required on each face, which would neutralize the charges and minimize the depolarization field. These charges, in practice, can be provided by an external circuit, or by the atmosphere, through crystal conductivity if present, or through surface reconstruction, which then stabilizes the single domain state. However, if insufficient compensating charges are present on the time scale that a spontaneous polarization arises, then an alternate way for the crystal to compensate the depolarization field is to split up into a multi-domain state. Domain walls also possess energy, W_w. The balance between a decrease in the depolarization energy W_d through multi-domain creation, and an increase in wall energy, W_w, through domain wall creation determines the final domain structure in such a ferroelectric. For a simple periodic domain structure composed of 180° walls, the equilibrium domain width in the absence of external compensating charge mechanism was derived originally by Kittel [6] as,

$$d = \left(\frac{\sigma_w t}{\varepsilon P^2}\right)^{1/2} \tag{6.1}$$

where, σ_w is the wall energy (per unit area), ε is a constant dependent on the dielectric constant of the ferroelectric, P is the spontaneous polarization, and t is the slab thickness. In this situation, domains of width greater than this value are suppressed by the depolarization energy, and those smaller than this are suppressed by the domain wall energy. Furthermore, since d is proportional to $t^{1/2}$, decreasing the thickness of the ferroelectric slab until the domain width is of the order of the wall thickness will result in a cessation of ferroelectricity.

6.1.3 Domain Reversal and Polarization Hysteresis Loops

The pyroelectric materials that are ferroelectric are distinguished by the fact that the polarization direction in a ferroelectric can be changed at least partially by an externally applied electric field. Domains with polarization direction closer to the applied field are favored, and they grow by nucleation and growth, or by the movement of existing domain walls.

A ferroelectric hysteresis loop, as shown in Figure 6.3, is a measurement of the extent of the *average* polarization reversal in a material as a function of applied electric fields. Consider a ferroelectric material in a nearly single domain state achieved by orienting all the domains in the direction of a large externally applied electric field. This single domain state orientation can be nearly perfect in a single crystal material, while it is more in an *average* sense in a polycrystalline ferroelectric comprised of multiple single crystal grains oriented randomly within the material. At very small fields of opposite polarity to the domains, the polarization increases linearly with the field just like a linear dielectric. This is the region of reversible domain wall displacements. At higher fields, nucleation of new domains parallel to the field and irreversible motion of domain walls results in a faster than linear response of the polarization with applied field. The field at which the ferroelectric has precisely half the switchable ferroelectric dipoles pointing in the direction of the applied field, and the other half opposite to the field, corresponds to the point where the *net* polarization, P of the material is zero. This field, E, corresponds to the coercive field, E_c. Finally, at saturation polarization, the ferroelectric starts behaving like a linear dielectric again. This corresponds to a complete reversal of the switchable part of the spontaneous polarization. Any further increase in the induced polarization due to applied field is not due to reversal of domains, but rather due to electronic and ionic displacement processes. Upon reducing the field after saturation, the ferroelectric polarization does not return to its original value but to a value of P_r at zero field, which is the *remanent polarization* of the ferroelectric. The spontaneous polarization is the value of the saturation polarization extrapolated to zero field. The values of P_r and P_s can be different due to reverse nucleation before changing the field direction.

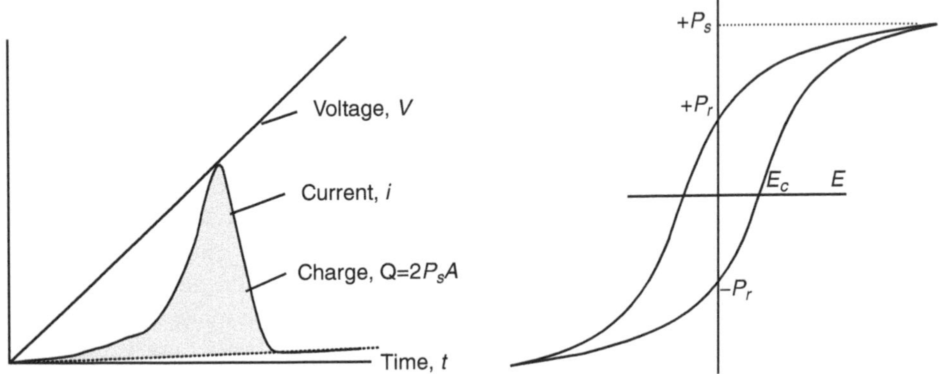

FIGURE 6.3 (Left) Measurement of one arm of a ferroelectric hysteresis loop, P vs. E. A linearly ramped voltage, V across a sample of thickness d, (field $E = V/d$) results in a transient current, i when domains are reversed. The charge Q (t) under this curve (excluding any baseline current from sample conductivity), when integrated over time, t results in one arm of the loop. (Right) Schematic of a ferroelectric hysteresis loops indicating the spontaneous polarization P_s, remnant polarization P_r, and coercive field, E_c.

If the field polarity is now reversed, and a field E greater than the coercive field E_c is applied again, the material can be taken back to its original domain state (in an average sense).

The area inside the curve is twice the energy required to switch the polarization from $+P_s$ to $-P_s$. As pointed out by Lines and Glass [1], some caution must be exercised in interpreting the hysteresis curve. If part of the P_s is clamped (and unswitchable), then the area inside the curve gives only the work done to switch the unclamped part of P_s. Also, in a lossy dielectric material (with large electrical conductivity), application of a periodic electric field can result in a lossy dielectric loop in addition to the response of any ferroelectric switching in the material. Indeed, such losses, when high, may even prevent ferroelectric domain reversal due to effective screening of the applied field by the crystal conductivity.

The measurement of polarization hysteresis loops described above can be performed as follows (see Figure 6.3): a ferroelectric slab of thickness t is electroded on both surfaces and connected to an external circuit across the terminals of a voltage supply. A linearly ramped voltage, $V(t)$, is applied starting from zero voltage. At a voltage, V, corresponding to a field $V/t \sim E_c$, domain reversal begins, resulting in a flow of current in the external circuit to compensate for the depolarization charges being created on the sample surface by domain reversal. This current, called the *transient current*, is measured as a function of time by measuring the voltage across a small resistor in series with the sample. When the domain reversal is complete, the current drops down to a non-switching value corresponding to the response due to any crystal conductivity. A baseline conductivity plot can be obtained by reapplying the field starting from zero to the maximum value in the same direction as the final polarization, thus excluding any domain switching contributions to the baseline current. This baseline conductivity current value is subtracted from the switching current versus time plot, and the resultant current data, i, is integrated with time, t, to yield a charge $Q(t) = \int_0^t i dt$. The plot of $Q(t)$ versus $E(t)$ should qualitatively reflect one arm of the polarization hysteresis loop. (The other arm can be obtained by repeating the domain reversal process in the opposite direction.) This charge value, $Q(t)$, is related quantitatively to the changes in the spontaneous polarization value, $P(t)$. For example, for a 180° domain switching across a ferroelectric slab where the polarization P_s is perpendicular to the sample surface, a change in polarization from $+P_s$ to $-P_s$ results in a total change of polarization of $2P_s$ over an area A, and a charge of $Q_{max} = 2P_sA$. For a polycrystalline sample, the polarization P_s should be replaced by P_n, which is the component of the polarization *normal* to the sample surfaces.

6.1.4 Ferroelectric Phase Transitions

As the temperature is increased, a ferroelectric undergoes a phase transition, which is accompanied by the disappearance of spontaneous polarization and by a change in the symmetry of the crystal lattice. This temperature is called the *Curie temperature*, denoted by T_c. Above T_c the material becomes *paraelectric.* This phase with the highest symmetry is called the *prototype phase.* In most, but not all cases, this phase is non-polar and occurs in the highest temperature phase.

Phase transitions can be classified as first or second order transitions. In a second order transition, the second derivative of the thermodynamic potential shows a discontinuous change, such as dielectric susceptibility, dielectric constants, specific heat, thermal expansion coefficient, elastic modulus, etc. In a first order phase transition, both the first and second derivatives of the thermodynamic potential show discontinuous change. The first derivatives are spontaneous polarization, volume, entropy, etc.

Figure 6.4 depicts how the energy potential, the polarization (P), and the dielectric permittivity (ε) change across a first-order and a second-order transition. These are phenomenologically described by the Landau theory of phase transitions [1]. The dielectric permittivity (ε) above the Curie temperature varies with temperature (T) as follows:

$$\varepsilon = 1 + \frac{C}{T - T_0} \tag{6.2}$$

where C is the Curie constant. The Curie temperature $T_c = T_0$ for a second order transition (Figure 6.4d–f), and $T_c > T_0$ for a first-order transition, as shown in Figure 6.4a–c. As per the

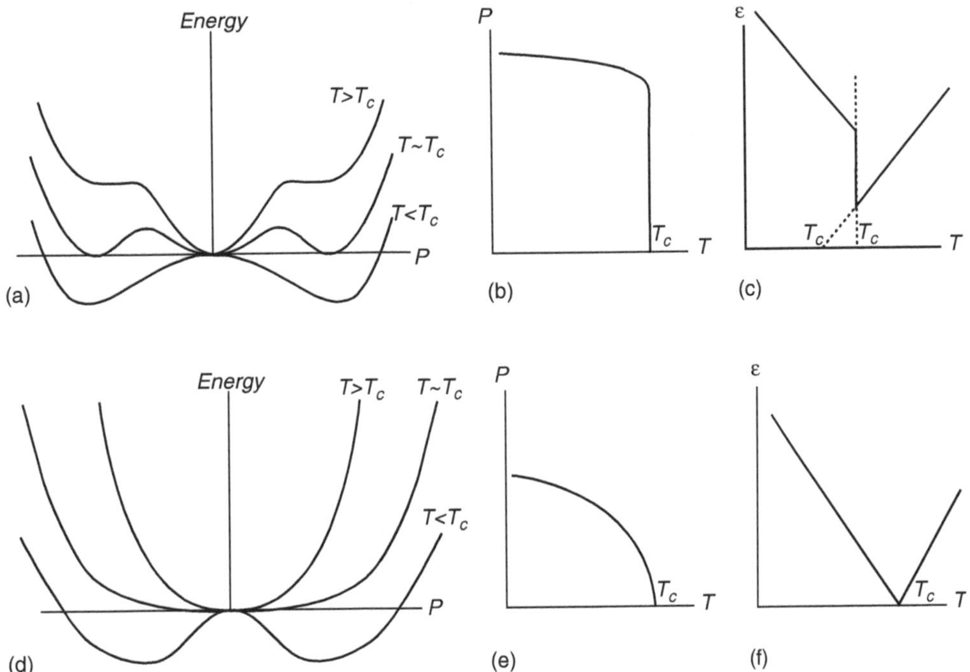

FIGURE 6.4 The Landau–Ginzburg–Devonshire free energy potential as (a) a function of polarization and temperature, and (b) the corresponding dependence of the polarization, P, and (c) the dielectric permittivity, ε as a function of temperature for a first order ferroelectric phase transition. The corresponding plots for a second order phase transition are shown in (d), (e) and (f), respectively.

discussions above, $BaTiO_3$ shows a first order transition, while lithium tantalate shows a second order transition.

Table 6.2 lists the series of phase transitions in some common optical ferroelectrics. Also note that multiple ferroelectric phases can exist in a material. For example, both $BaTiO_3$ and $KNbO_3$ show transition from the high temperature prototype cubic phase to ferroelectric phases of tetragonal, orthorhombic, and rhombohedral structures successively, as the temperature is lowered.

6.2 Materials and Structures

Important *optical* ferroelectric materials can be classified into *oxygen-octahedra* ferroelectrics and *phosphate/arsenate/sulphate/selenate-type* ferroelectrics. These materials are either used as single crystals or *thin films*. These are briefly described below. Table 6.2 lists example materials in these structural classes.

6.2.1 Oxygen Octahedra Ferroelectrics

Classification of oxygen octahedra ferroelectrics based on structure leads to *perovskite-type*, *Aurivillius-type*, *$LiNbO_3$-type*, and *tungsten-bronze type* ferroelectric structures, as shown in Figure 6.5. All these ferroelectrics have a common motif structure of BO_6 composed of six oxygens (O) surrounding a central tertravalent or pentavalent transition metal ion, here labeled generally as B. The oxygen cage has 8 faces; hence, it is called an octahedron. The highest symmetry phase (typically the highest temperature phase) is called the paraelectric phase, where the time-averaged mean position of the cation B occupies the geometric center of the oxygen octahedron. At lower temperature, a Jahn–Teller distortion of the octahedron results in the creation of ferroelectricity. This is accompanied by a displacement of the B ion

TABLE 6.2 Phase Transition and Switching Properties of Some Common Ferroelectric Materials

Material	Point Group (**Bold**: Room Temperature)	Phase Transition Temperature (°C)	Spontaneous Polarization[c] P_s (μC/cm^2)	Coercive Field (kV/cm)
$BaTiO_3$	$m3m \rightarrow \mathbf{4mm} \rightarrow mm2 \rightarrow 3m$	120, 5, −90	26	∼1
$KNbO_3$	$m3m \rightarrow 4mm \rightarrow \mathbf{mm2} \rightarrow 3m$	426, 203, −50	30	55 (1 kHz) [7]
$KTa_{1-x}Nb_xO_3$ (x=0.32)	$m3m \rightarrow 4mm \rightarrow mm2 \rightarrow 3m$	−24, −77, −114	∼18–20 (< −60°C)	∼5, [8] 8.7 [9] (−10, −48°C)
$Pb_{1-x}La_x(Zr_yTi_{1-y})_{1-0.25x}V_{0.25x}O_3$ (PLZT) x=0.08, y=0.65 [10]	$m3m \rightarrow$ mixed $\rightarrow \mathbf{3m}$	125, 75	32 (25°C)	∼9 (25°C)
Lithium Niobate: $(Li_{1-x}Nb_{0.8x}V^{Li}{}_{0.2x})\ NbO_3$ (V^{Li} : Li-vacancy)	$\bar{3}m \rightarrow \mathbf{3m}$	1198 (x=0.05)	∼80	210
		1138 (x=0.01)		<40
Lithium tantalate: $(Li_{1-x}Ta_{0.8x}V^{Li}{}_{0.2x})\ TaO_3$ (V^{Li} is Li-vacancy)	$\bar{3}m \rightarrow \mathbf{3m}$	610 (x=0.05)	55–60	210
		703(x∼0)		<1
$Ba_{0.4}Sr_{0.6}Nb_2O_6$ (SBN)	$(4/m)mm \rightarrow \mathbf{4mm} \rightarrow m$	75, −213	∼28	∼2 [11]
$Ba_2NaNb_5O_{15}$ (BNN)	$(4/m)mm \rightarrow 4mm \rightarrow \mathbf{mm2} \rightarrow mm2 \rightarrow 4mm$	560, 300, 275, −163	40	8.3 [12]
$K_3Li_2Nb_5O_{15}$ (KLN)	$(4/m)mm \rightarrow \mathbf{4mm}$	430	∼22	?
$(K_xNa_{1-x})_{0.4}\ (Sr_yBa_{1-y})_{0.6}\ Nb_2O_6$ (KNSBN) [13]	$(4/m)mm \rightarrow \mathbf{4mm}$	150–250	∼4	5.3
KH_2PO_4 (KDP) [14]	$\bar{4}2m \rightarrow mm2$	−150	4.8	∼1.5–10
RbH_2PO_4 (RDP)	$\bar{4}2m \rightarrow mm2$	−126	5.6	∼8–10
$NH_4H_2PO_4$ (ADP)	$\bar{4}2m \rightarrow 222$	−125	∼5	–
KH_2AsO_4 (KDA)	$\bar{4}2m \rightarrow mm2$	−176	5	∼5
$(NH_2CH_2COOH)_3.H_2SO_4$ (TGS)	$2/m \rightarrow \mathbf{2}$	49	2.8	0.25–1.75 [15]
$KTiOPO_4$ (KTP)	$mmm \rightarrow \mathbf{mm2}$	943	∼17	20
$KTiOAsO_4$ (KTA)	$mmm \rightarrow \mathbf{mm2}$	861	36	90
$RbTiOPO_4$(RTP) [16]	$mmm \rightarrow \mathbf{mm2}$	829	19.8	26.5
$RbTiOAsO_4$(RTA)	$mmm \rightarrow \mathbf{mm2}$	765	40	26 [17]
$CsTiOAsO_4$(CTA) [18, 19]	$m\bar{3}m \rightarrow \mathbf{mm}2$	671	14.4	13
$Bi_4Ti_3O_{12}$ [20]	$(4/m)mm \rightarrow \mathbf{m}$	675	50 (a-axis), 4(c-axis)	50, 3.5

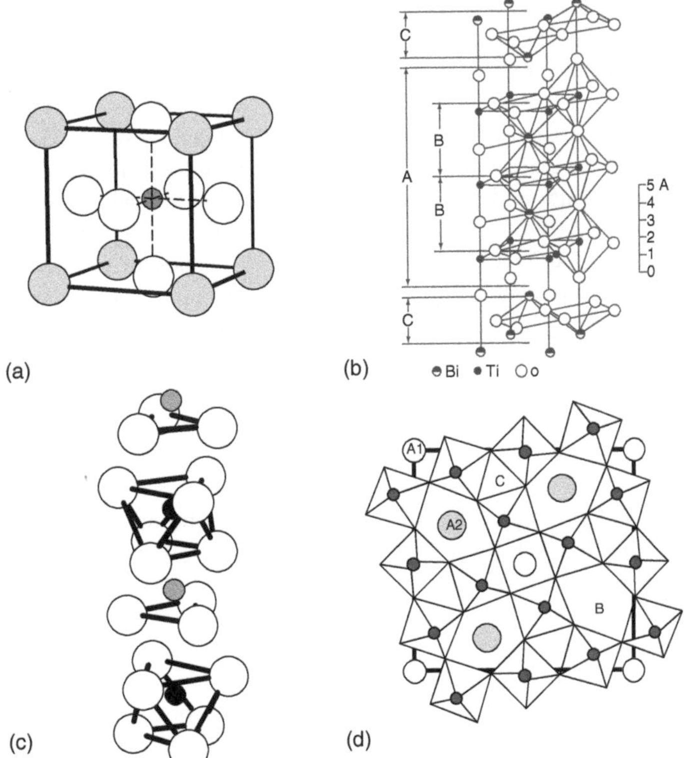

FIGURE 6.5 Schematic diagrams of units cells of (a) *Perovskite* ABO_3: Open circles are oxygen, light grey circles are *A* atoms, and dark gray circle (center) is the transition element *B*. (b) *Aurivillius* phase of the general formulae $(Bi_2O_2)^{2+}(Me_{x-1}R_xO_{3x+1})^{2-}$, where in this example of $Bi_4Ti_3O_{12}$, *Me* is a Bi-atom (half black-half white circles), *R* is a Ti-atom (black circles), and open circles are oxygen. (c) $LiNbO_3$-type structure, where grey circles are Li, black circles are Nb, and open circles are oxygen. (d) *Tungsten Bronze* structure with the general formulae $(A1)_x(A2)_{5-x}Nb_{10}O_{30}$. The black circles are Nb, placed inside oxygen octahedras (squares), open circles are *A*1 atoms, and grey circles are *A*2 atoms. (Figure (b) reprinted with permission from Subbarao, E. C. 1961. *Physical Review*, 122, 804, With permission.)

from the geometric center of the cage, which occurs during a ferroelectric phase transition and contributes in large part to the ferroelectricity in these structures.

Perovskite-type ferroelectrics of the general formulae ABO_3 are the most common type of optical ferroelectrics (see Figure 6.5a), where A are monovalent or divalent metal ions that occupy the corners of the pervoskite cube, the oxygens that form the octahedra occupy the six face centers, and the transition element B occupies the center of the cube. The perovskite structures form a larger lattice by stacking of the cubes, leading to a network of vertices sharing octahedra. The most common optical ferroelectrics of this type are $BaTiO_3$, $KNbO_3$, $KTa_{1-x}Nb_xO_3$ (also called KTN:*x*), $PbTiO_3$, and $Pb_{1-x}La_x(Zr_yTi_{1-y})_{1-0.25x}V^B{}_{0.25x}O_3$, (also called PLZT:*x*/*y*/1−*y*), all of which are primarily useful for electro-optic applications, and $Ba_{1-x}Sr_xTiO_3$, which is used in pyroelectric detectors. Note that multiple ferroelectric phases can exist with a single material in different temperature ranges. For example, $BaTiO_3$ and $KNbO_3$ both show a high temperature cubic paraelectric phase, which on cooling results successively in tetragonal, orthorhombic, and rhombohedral ferroelectric phases. These are respectively accompanied by displacement of the cation B from the geometric center of the perovskite unit cell in the directions of ⟨100⟩, ⟨110⟩ and ⟨111⟩. These displacements are typically small (of the order of ∼0.1 Å). KTN and PLZT are examples where multiple cations occupy the A and the B sites. For example, KTN can be considered to be a solid solution (on atomic scale) of $(1-x)$ $KTaO_3+xKNbO_3$, where $KNbO_3$ is a ferroelectric with a cubic-to-tetragonal T_c of 435°C (708 K) and $KTaO_3$ is paraelectric down to 1.6 K [22]. By varying the relative composition, x, from 0 to 1, a continuous gradation of ferroelectric properties such as Curie temperature, polarization, electro-optic, nonlinear-optical coefficients, and dielectric response can be achieved [23]. $PbTiO_3$ is a tetragonal ferroelectric below a $T_c=493$°C, and $PbZrO_3$ is orthorhombic and antiferroelectric below a Neel temperature, $T_N=230$°C. Their solid solution, $PbZr_xTi_{1-x}O_3$, also called PZT:x, is the most commonly used piezoelectric material today, though its application in optics is limited. However, the addition of La to the Pb-site helps in sintering these materials into highly transparent ceramics, $Pb_{1-x}La_x(Zr_yTi_{1-y})_{1-0.25x}V^B{}_{0.25x}O_3$, ($V^B$ stands for vacancies in the B site) which are useful for electro-optic modulators due to their large electro-optic coefficients [24].

Aurivillius phases (Figure 6.5b) are closely related to the perovskite type structures but contain a layered approach to building each unit cell, which can be very large. They are composed of perovskite-like layers of $(Me_{n-1}R_nO_{3n+1})^{2-}$ slabs regularly interleaved with $(Bi_2O_2)^{2+}$ layers, where Me is a monovalent or divalent element such as Bi^{2+}, Ba^{2+}, Sr^{2+}, K^{1+}, Ca^{2+}, Na^{1+}, Pb^{2+}, etc., and R can be Ti^{4+}, Nb^{5+}, Ta^{5+}, Mo^{6+}, W^{6+}, Ga^{3+}, Fe^{3+}, Cr^{3+}, and n can be 1–8. Their overall general formulae is therefore $Me_{n-1}R_nO_{3n+3}$ [21,25]. Examples of ferroelectrics in this class are $Bi_4Ti_3O_{12}$, $SrBi_2Nb_2O_9$, $SrBi_2Ta_2O_9$, $BaBi_2Nb_2O_9$, $BaBi_2Ta_2O_9$, $PbBi_2Nb_2O_9$, $PbBi_2Ta_2O_9$, and other more complex multi-cation compounds such as $BaBi_3Ti_2NbO_{12}$, etc. Of these, the only one to receive some attention for optical applications is bismuth titanate, $Bi_4Ti_3O_{12}$ with $n=3$ [26–28]. Strontium bismuth niobate and tantalate have received considerable attention for non-volatile ferroelectric memory, but their optical properties have not been investigated.

Lithium Niobate type ferroelectrics, particularly $LiNbO_3$ and $LiTaO_3$ (Figure 6.5c), are widely used optical ferroelectrics today, with applications in high speed electro-optic and acousto-optic modulators, domain engineered devices for optical frequency conversion, beam steering, dynamic focusing and beam shaping, and pyroelectric detectors, as described below. Unlike perovskites, the oxygen octahedra units O_6 are joined *not* by connecting vertices, but by connecting the octahedral faces [29–32]. This results in a stack of distorted oxygen octahedra on top of each other joined by their faces along the polar c-axis. Unlike perovskites, *not all* of these stacked octahedra are filled by B cations. Along the c-axis of the trigonal structure, the octahedra are filled in the sequence "Li, Nb (or Ta), vacancy, Li, Nb(or Ta), vacancy,…." Two formulae units of $LiNbO_3$ form one unit cell because of staggered oxygens that repeat after two consequentive operations of the *c-glide* symmetry. The high temperature paraelectric phase is trigonal $\bar{3}m$, which transforms to the ferroelectric, trigonal $3m$ phase below the T_c of ~1200°C for $LiNbO_3$ and ~620°C for $LiTaO_3$. This transition is accompanied by a small displacement of the Nb(Ta) from the center of its octahedron to an asymmetric position along the c-axis within the same octahedron, and a corresponding motion of the Li from one octahedron to the adjacent vacant octahedron through the intermediate close-packed oxygen plane. The direction of motion of these cations defines the positive end of the spontaneous polarization, P_s. Given the only two degrees of freedom for this motion, $\pm c$, only two possible domain orientations can exist, $\pm P_s$ as can also be deduced from Table 6.1. Recent studies have shown that the properties of both $LiNbO_3$ and $LiTaO_3$ are extremely sensitive to the $C=Li/(Li+M)$ ratio (M=Nb, Ta). As C varies from 0.5 (*stoichiometric*) to 0.485 (*congruent*) composition, it causes large shifts in T_c, two orders of magnitude increase in coercive field, E_c, for domain reversal; creation of large internal fields; increased photorefraction; lower electro-optic coefficients; and unexpected creation of optical birefringence, strain, and local electric fields adjacent to domain walls (see Figure 6.6) [33,34]. Most studies in the literature over the last 40 years, and most present day devices are based on congruent compositions, since they are easy to grow as high quality crystals from the high temperature melt using the Czhochralski technique. Historically, $LiNbO_3$ has been studied far more extensively than $LiTaO_3$. Recent developments over the last decade of double crucible Czhochralski (DCCZ), and vapor transport equilibration (VTE) techniques have enabled synthesis of crystals close to stoichiometric compositions [35,36]. Stoichiometric and near-stoichiometric compositions, particularly of lithium tantalate, are currently under intense investigation for optical frequency conversion devices (Section 6.3.2) because of their high optical damage resistance and very low coercive fields [37].

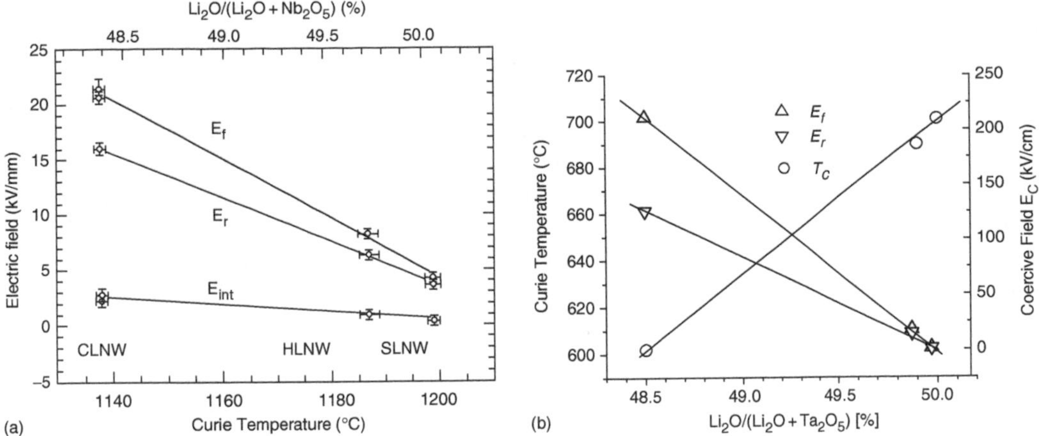

FIGURE 6.6 A change in the coercive field (E_f is forward domain switch, E_r is reverse domain switch) and Curie temperature (T_c), as a function of % ratio of $Li_2O/(Li_2O+Ta_2O_5)$ in (a) $LiNbO_3$ and (b) $LiTaO_3$. The presence of large internal fields, $E_{int}=(E_f\text{-}E_r)/2$, is clearly correlated to the presence of non-stoichiometry. (From Gopalan, V., et al. 1988. *Applied Physics Letters*, 72, 1981; and Tian, L.; and Gopalan, V. 2004. *Applied Physics Letters*, 85, 4445, With permission.)

Tungsten Bronze (TB) *ferroelectric structures* are so called because they are closely related to the tetragonal tungsten bronzes K_xWO_3 and Na_xWO_3. The general TB compositions are considered to be close to one of the following [1]: $(A1)_x(A2)_{5-x}Nb_{10}O_{30}$ (when A1 and A2 are alkaline earth ions), $(A1)_{4+x}(A2)_{2-2x}Nb_{10}O_{30}$ (when A1 is alkaline and A2 is alkali), and $(A1)_{6-x}(A2)_{4+x}Nb_{10}O_{30}$ (when A1 and A2 are both alkali ions). The well-known optical ferroelectrics of this type are strontium barium niobate ($Sr_{5-x}Ba_xNb_{10}O_{30}$, also called SBN), barium sodium niobate ($Ba_{4+x}Na_{2-2x}Nb_{10}O_{30}$, also called BNN), and potassium lithium niobate ($K_{6-x-y}Li_{4+x}Nb_{10+y}O_{30}$, also called KLN), which are useful for electro-optics and holography. They also have a frame of BO_6 octahedron in an ABO_6 lattice. The tetragonal unit cell consists of 10 BO_6 octahedra linked by their corners to form three different types of channels (labeled as A1, A2 and C) running through the structure parallel to the *c*-axis (Figure 6.5d) [1,38,39]. In addition, there are two different B-cation sites, B1 and B2. In SBN, for example, the unit cell contains 5 formulae units (10NbO_6 octahedra) (Figure 6.5d). The A (Sr, Ba) atoms take ~5 interstitial sites out of 6 available sites (2 A1+4 A2) among the octahedron. The A1 site has 4-fold symmetry and contains the smaller Sr ion, while A2 is pentagonal and larger in size, preferred by the Ba, and also the remaining Sr. The interstitial sites (C) are empty in SBN and BNN because the A1 and A2 site ions are too big to enter the C sites. This is not the case in KLN where the small Li atoms fill the C sites, along with some of the excess Nb. The B1 and B2 sites are occupied by Nb in all three materials. The O–Nb–O octahedral axes in the octahedra are not precisely parallel to the *c*-axis. They are tilted through about 8° from the polar axis (*c*-axis) in SBN. The small displacements of Sr/Ba and Nb atoms in SBN along the *c*-axis from their adjacent oxygen planes at room temperature create the permanent dipole inside the crystal and make them ferroelectric at room temperature. The ferroelectric properties reflect the disorder in the structure. For example, a diffuse phase transition occurs, where the transition temperature shifts from ~330 K for $x=1.25$ to ~469 K for $x=3.75$.

6.2.2 Phosphate, Arsenate, Sulphate, Selenate Ferroelectrics

There are two classes of such optical ferroelectrics: *KDP-type* ferroelectrics, and *KTP-type* ferroelectrics.

KDP-type ferroelectrics are potassium dihydrogen phosphate, KH_2PO_4 (KDP), deuterated KD_2PO_4 (KD*P), a related family of isomorphs such as Rb or Ce substitution for K, or As substitution for P (leading to RDP, RDA, KDA, CDA, etc.), and triglycine sulphate, $(NH_2CH_2COOH)_3H_2SO_4$ (also called TGS). Substitution of ammonium (NH_4) for potassium (K) in KDP leads to antiferroelectric properties, $(NH_4)H_2PO_4$ (called ADP), and deuteration makes it AD*P. The hydrogen ion is a common feature in all of these compounds. Single crystals of all these materials with high optical quality can be grown from water solutions, which are useful for electro-optic modulators and pyroelectric infrared detectors. The T_c in KDP is ~123 K, below which it is orthorhombic (point group *mm*2) and ferroelectric, and above which it is tetragonal and paraelectric (point group $\bar{4}2m$), though still non-centrosymmetric. KDP at room temperature, therefore, still has third rank tensor properties such as linear electro-optic effect and second harmonic generation effects. The ferroelectricity primarily arises from the deformation of the KH_2^{3+}–PO_4^{3-} complex along the *c*-axis. The crystal structure of KDP consists of two interpenetrating body-centered lattices of PO_4 tetrahedra, as well as two interpenetrating body-center lattices of K. These two types of lattices are displaced along the *c*-axis. Each PO_4 is linked to four neighbouring PO_4 groups through hydrogen bonds, which are nearly perpendicular to the *c*-axis. The main problem with KDP is its deliquescence, which can be overcome by sealing or applying a protective coating on the surface. KDP shows a dramatic effect on deuteration to KD*P, with a near doubling of the T_c to ~220 K and a 20–25% increase in spontaneous polarization from ~5 to ~6.2 μC/cm^2. This is attributed to a change in the tunneling frequency of the H between two equilibrium positions, on substituting it with a heavier deuterium [1]. The isomorphs KDA, RDA, CDA, RDP (R is Rb, A is As, C is Cs) all show similar structure and behavior with T_c of 96, 110, 146, and 147 K respectively, and increasing trend of first-order nature of phase transition, respectively. Deuteration again increases their T_c's. ADP is isomorphous with KDP above a T_c of ~150 K. Below T_c, it develops antiparallel dipoles along the *a*- or *b*-axis, which then becomes the antiferroelectric axis. TGS, used as a pyroelectric material, shows a second order phase transition at a T_c~322 K, transforming from a centrosymmetric 2/*m* monoclinic phase above T_c to a polar group 2 below T_c. The polarization axis is along the 2-fold axis. TGS does not show a strong deuteration effect. The main problem with TGS is its low T_c, leading to depolarization into a multidomain state. This can be improved by doping it with L-alanine, phosphoric acid, and arsenic acid [40–42]. Replacing sulfur in TGS by Se results in isomorphous TGSe, and replacing SO_4 group in TGS by BeF_4 (fluoroberyllate) results in TGFB. Mixed crystals of TGS–TGSe and TGS–TGFB have better properties.

KTP-type ferroelectrics consist of potassium titanyl phosphate $KTiOPO_4$ (KTP) and its isomorphs obtained by Rb, Cs, Ti, and NH_4 substitution of the K, or As substitution of P, leading to KTA, RTP, RTA, CTA, TTP, ATP, and isomorphs such as KRTP by co-doping of multiple cations (K and Rb in this example). Figure 6.7 shows the crystal structure of KTP. KTP-type ferroelectrics show a phase transition from orthorhombic paraelectric phase of *mmm* to a ferroelectric orthorhombic phase of *mm*2. Ferroelectricity primarily arises from the displacement of Ti in the TiO_6 octahedra and a displacement of K, Rb, Cs, etc., in the *c*-axis. Polarization occurs in the ⟨011⟩ directions. KTP and its isomorphs have a key application in nonlinear quasi-phase-matched and birefringently phase-matched frequency conversion devices and to some extent in electro-optic devices.

6.2.3 Single Crystals and Thin Films

A majority of ferroelectric optical device applications today use single crystals. The primary reasons for this are the high crystalline quality, which leads to low propagation losses, and the possibility of ferroelectric domain control into single domain or controlled patterning into different domain shapes like gratings, prisms, etc. Ferroelectrics such as $LiNbO_3$ and $LiTaO_3$ are available in 3-inch wafers at reasonable cost, which allows mass production and application of wafer level processing using modern nanofabrication facilities. However, this cannot be said of other optical ferroelectrics that remain small, difficult to grow, and expensive. Primary crystal growth techniques are Czhochralski (CZ) and flux growth techniques. CZ growth involves rotating and slowly pulling a seed crystal from the melt. For off-congruent compositions, specialized techniques such as DCCZ [36,43] and VTE [34] techniques have been developed. For multi-cation solid solution systems such as potassium tantalate niobate (KTN) and

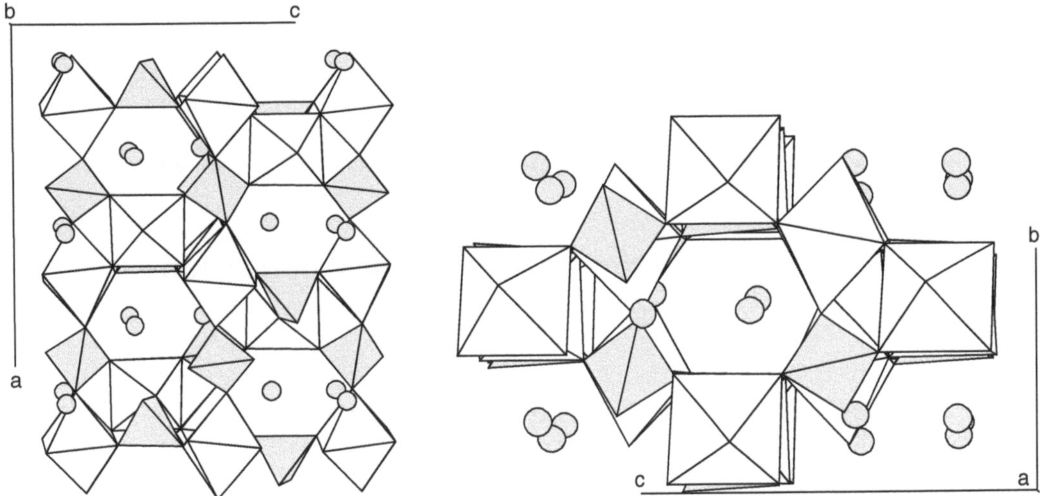

FIGURE 6.7 Projection of the crystal structure of $KTiOPO_4$ ferroelectric looking down two different zone axes: (Left) *b*-axis <010>, (right) *a*-axis <100>. The shaded elements are tetrahedra PO_4, which connect the (unshaded) octahedra TiO_6. Grey circles are K atoms.

strontium barium niobate (SBN), CZ results in compositional striations and optical inhomogeneities in the crystal. In such cases, Stepanov's method [44] of pulling crystals from the melt without rotation is employed.

There are many reasons to desire thick and thin films of ferroelectrics. They substantially reduce the voltage requirements for electro-optic devices. They can provide much more flexibility in the design of step index waveguides, and in choosing different combination of films and substrates. Thin films of even complex multi-cation ferroelectrics are easier to grow than single crystal growth, and potential integration with silicon optoelectronics is possible. However, despite over 15 years of intense activity in this area, this promise has only recently begun to be realized, and much work remains ahead.

Thick film (1–10 μm) growth efforts have seen more substantial progress in growing single crystal quality optical ferroelectric films. Liquid phase epitaxy [45] is one of the techniques used, which involves starting from melt on a homoepitaxial substrate and cooling it down slowly under conditions similar to crystal growth. Crystal ion slicing [46] has recently been demonstrated, emulating the "smart cut" technologies used in silicon and other semiconductors. This involves implanting high energy (MeV) alpha particles in single crystals of ferroelectrics, such that a damage layer is created at a precise depth (1–10 μm) from the surface, which is then thermally separated or etched away, resulting in a surface "slice" of single crystal quality. Often, annealing is performed to heal any lattice damage in this layer due to implantation and to restore the original bulk optical properties of the material slice. Controlling these processes, bonding the slice to a substrate, and processing optical devices in this slice are all topics of current research.

Thin films (1 nm–1 μm) of ferroelectrics have been attempted by many techniques, including radio frequency (RF) magnetron sputtering, sol–gel method, metallorganic deposition, metallorganic chemical vapor deposition, pulsed laser deposition, and molecular beam epitaxy. While numerous studies exist that show promising material properties in thin film form, a common problem among most of them is simply optical propagation losses, which tend to be unacceptably high (10–20 dB/cm). In this regard, the best films (losses order of 1–2 dB/cm) to date have been prepared by chemical vapor deposition (CVD), or molecular beam epitaxy (MBE). In particular, high quality thin films of $BaTiO_3$ and $LiNbO_3$ have been demonstrated with promising electro-optic responses [47,48]. A further key problem is their multi-grain, multi-domain microstructure. This gives rise to optical losses, as well as leads to randomly *averaged* optical properties instead of access to individual tensor coefficients of properties such as the electro-optic effect or the second harmonic generation. This continues to be a problem, though very recent results on controlling the epitaxy of perovskite oxide films through closely lattice matched substrates such as scandates appear to be able to give pure *c*-axis oriented $BaTiO_3$ films, in addition to enhancing ferroelectric properties [49]. This still leaves the issue of achieving a single-domain film (with no or controlled domain walls). One can safely say that this is currently an active area of research with much promise as well as much work ahead!

6.3 Intrinsic Optical Properties and Their Applications

6.3.1 Electro-Optics

The electro-optic effect is a change in the refractive index of a material in response to an external electric field. The fastest optical modulators today are based on the electro-optic effect in ferroelectric $LiNbO_3$ operating at 10–40 GHz. The electro-optic effect is also used for beam steering, dynamic focusing, and beam shaping. Using tensor notation, one can in general write the change in the refractive index. Δn_{ij} due to applied fields of $E_{k,l}$ as,

$$\Delta\left(\frac{1}{n_{ij}}\right)^2 = r_{ijk}E_k + R_{ijkl}E_kE_l \tag{6.3}$$

where r_{ijk} and R_{ijkl} are the linear (Pockel's effect) and quadratic (Kerr effect) electro-optic tensor coefficients, and each of the subscripts *i*, *j*, *k* can be any one of the orthogonal *crystal physics* axes (*x*,*y*,*z*) [also (1,2,3)] defined in the unit cell of each material. The relationship between the *crystallographic* (*a*,*b*,*c*), and crystal physics, (*x*,*y*,*z*) axes for each point group is given in the literature [4]. Upon further simplifying, we get,

$$\Delta n_{ij} = -\frac{1}{2}n_{ij}^3(r_{ijk}E_k + R_{ijkl}E_kE_l) \tag{6.4}$$

This is also sometimes written in terms of polarization P_k, as follows:

$$\Delta n_{ij} = -\frac{1}{2}n_{ij}^3(f_{ijk}P_k + g_{ijkl}P_kP_l) \tag{6.5}$$

If we rewrite the polarization P_k in terms of an electric field E_n as $P_k=\varepsilon_0(\varepsilon_{kk}-1)E_k$, where ε_{kk} is a *principal* dielectric constant, then we have the following relations between these two notations:

$$r_{ijk} = f_{ijk}\varepsilon_o(\varepsilon_{kk}-1) \tag{6.6}$$

$$R_{ijkl} = g_{ijkl}\varepsilon_0^2(\varepsilon_{kk}-1)(\varepsilon_{ll}-1) \tag{6.7}$$

Symmetry dictates that odd-rank tensor properties such as the r_{ijk} are present only in crystal symmetries that *lack* a center of symmetry, while odd rank tensor properties can exist for all crystal symmetries. Since ferroelectrics by definition are polar, non-centrosymmetric, and exhibit a first rank tensor property (polarization, P_i), they therefore also exhibit the linear electro-optic effect. The quadratic electro-optic effect can exist in all materials; however, quadratic electro-optic coefficients are typically large in many ferroelectrics because of the presence of polarization.

Neumann's law [4] dictates that a physical property of a material must at least contain the symmetry of the point group of that material. For example, the refractive index ellipsoid of a cubic material is isotropic (e.g., contains ∞-fold rotation axis) which includes the symmetry of the cubic point group (e.g., contains a 4-fold rotation axis). Some common ferroelectric materials and their symmetry dictated linear electro-optic tensors r_{ijk} (in abbreviated notation) are given below. [Rules for abbreviation of r_{ijk} into r_{nk} are, $ij\rightarrow n$, where $11\rightarrow1$, $22\rightarrow2$, $33\rightarrow3$, $(23, 32)\rightarrow4$, $(31, 13)\rightarrow5$, and $(12, 21)\rightarrow6$].

Point group $\bar{4}2\,m$(e.g., KDP)

$$\begin{pmatrix} 0 & 0 & 0 \\ 0 & 0 & 0 \\ 0 & 0 & 0 \\ r_{41} & 0 & 0 \\ 0 & r_{41} & 0 \\ 0 & 0 & r_{63} \end{pmatrix}$$

Point group $4mm$ (e.g., $BaTiO_3$, SBN)

$$\begin{pmatrix} 0 & 0 & r_{13} \\ 0 & 0 & r_{13} \\ 0 & 0 & r_{33} \\ 0 & r_{51} & 0 \\ r_{51} & 0 & 0 \\ 0 & 0 & 0 \end{pmatrix}$$

Point group $3m$ (e.g., $LiNbO_3$, $LiTaO_3$)

$$\begin{pmatrix} 0 & -r_{22} & r_{13} \\ 0 & r_{22} & r_{13} \\ 0 & 0 & r_{33} \\ 0 & r_{51} & 0 \\ r_{51} & 0 & 0 \\ -r_{22} & 0 & 0 \end{pmatrix}$$

Point group $mm2$ (e.g., KTP)

$$\begin{pmatrix} 0 & 0 & r_{13} \\ 0 & 0 & r_{23} \\ 0 & 0 & r_{33} \\ 0 & r_{42} & 0 \\ r_{51} & 0 & 0 \\ 0 & 0 & 0 \end{pmatrix}$$

Table 6.3 summarizes a number of ferroelectrics and their electro-optic properties. We will now consider two examples of materials from an application viewpoint: $LiNbO_3$ and $BaTiO_3$.

The largest and most useful linear electro-optic coefficient in $LiNbO_3$ (and $LiTaO_3$) is the $r_{33} \equiv r_{333}$ component, which is ~32.5 (and 30.5) pm/V, respectively. An external electric field in the 3 (also called c- or z-) direction gives rise to a change in the $n_{33} \equiv n_3$ index (also called n_e or the extraordinary index), as well as a change in the ordinary refractive index $n_{11} = n_{22} \equiv n_1 \equiv n_2$ (also called the ordinary index, n_o) given by,

$$\Delta n_e = -\frac{1}{2} n_e^3 r_{33} E_3 \quad \text{and} \quad \Delta n_o = -\frac{1}{2} n_o^3 r_{13} E_3 \tag{6.8}$$

The index change Δn_3 is particularly used in today's high-speed Mach–Zhender interferometeric modulators operating at 10–40 GHz as shown in Figure 6.8. In this case, phase modulation is employed through controlling the difference in optical path lengths (product of $n_{\text{eff}}L$ where n_{eff} is the effective index of the extraordinary polarized guided mode, and L is the actual length of the arm) between two arms of the interferometer. For extraordinary guided modes, the n_e in Equation 6.8 above is replaced by n_{eff}. When this phase difference $\Delta\phi = (2\pi/\lambda)\Delta(n_{\text{eff}}L)$ between the two arms approaches $(2m+1)\pi$ (m is an integer), destructive interference is observed, and when it approaches $2m\pi$, constructive interference is observed. The voltage V_π is defined as the voltage required to switch from the constructive (on-state, or 1-state) to the destructive (off-state or 0-state) interference condition. In thin film geometry where voltage could be applied in a parallel plate geometry across the film (top and bottom), this voltage is simply the product of field E times the film thickness, d. For ion-exchanged waveguides formed in single crystal modulators as shown in Figure 6.8 (right), the field distribution is more complex due to surface electrodes and fringing fields. Equation 6.8 also indicates another very useful feature. When the electric field E_3 is positive (defined as being parallel to the spontaneous polarization, $+P_s$), both the extraordinary and ordinary refractive indices decrease. For negative fields ($E_3 // -P_s$), the indices increase by the same amount. This indicates that if a uniform electric field E_3 is applied *across* a 180° domain wall separating $+P_s$ and $-P_s$

TABLE 6.3 Electro-Optic Properties of Ferroelectric Materials

Material	Point Group	r_{ij} (10^{-12} m/V) or R_{ij} (10^{-18} m^2/V^2) g_{ij} (m^4/C^2)	λ(μm)	Index of Refraction	Dielectric Constant (ε_o)	Transparency Range (μm)
$BaTiO_3$	4*mm*	$r_{13}=r_{23}=10.2(S)-8(T)$ $r_{33}=40.6(S)-105(T)$ $r_{51}=r_{42}=730(S)-1300(T)$	0.6328	$n_o=2.412$ $n_e=2.360$	$\varepsilon_1=\varepsilon_2=2200(S)-4400(T)$ $\varepsilon_3=56(S)-129(T)$	~0.41–8
			0.546	$n_o=2.44$ $n_e=2.37$		
$BaTiO_3$ (T_c~120°C)	*m*2*m*	$R_{11}-R_{12}=2290$ $(T>T_c)$ $n_o^3(R_{11}-R_{12})=72{,}000(T\sim T_c)$ $n_o^3R_{44}=44{,}000(T\sim T_c)$	0.6328 0.550 0.550	$n=2.42$ $(T>T_c)$	$\varepsilon_1=\varepsilon_2=3600(T)$ $\varepsilon_3=135(T)$	~0.41–8
$KNbO_3$	*mm*2	$r_{13}=21(S)-33.5(T)$, $r_{23}=7.1(S)-6(T)$ $r_{33}=35(S)-63.3$ (T), $r_{51}=27.8(S)-120(T)$ $r_{42}=360(S)-450(T)$	0.6328	$n_a=2.2801$, $n_b=2.3296$, $n_c=2.1687$	$\varepsilon_{11}=34(S)-154(T)$, $\varepsilon_{22}=780(S)-985(T)$, $\varepsilon_{33}=24(S)-44(T)$	~0.41–10
$LiNbO_3$ (Li/Nb=0.972) (congruent)	3*m*	$r_{13}=9.1(S)-10.1(T)$ $r_{33}=31.2(S)-31.4$ (T) $r_{51}=18.1(S)-34.0(T)$ $r_{22}=3.4(S)-6.6(T)$	0.6328	$n_o=2.2909$ $n_e=2.2003$	$\varepsilon_1=\varepsilon_2=45.5(S)-84.5(T)$ $\varepsilon_3=26.2(S)-27.8(T)$	0.325–5.2
			1.32	$n_o=2.2198$ $n_e=2.2200$		
$LiNbO_3$ (Li/Nb=0.99) (near-stoichiometric)	3*m*	$r_{13}=10.4\pm1.4(T)$	0.6328	$n_o=2.1898$	$\varepsilon_1=\varepsilon_2=87(T)$	[61] 0.305–5.2
		$r_{33}=38.3\pm1.4(T)$ $r_{61}=9.89\pm1.4(T)$		$n_e=2.2865$		
Doped-$LiNbO_3$ (congruent)	3*m*				—	
10 mol % MgO		$r_{13}=11.2(T)$; $r_{33}=36(T)$	0.6328	—		~0.31–4.5
		$r_{13}=10.3(T)$; $r_{33}=34.1(T)$	1.32	—		
3 mol% Ti		$r_{13}=10.7(T)$; $r_{33}=36.6(T)$	0.6328	$\Delta n_e\sim0.02$ $\Delta n_o\sim0.01$		
		$r_{13}=9.3(T)$; $r_{33}=33.6(T)$	1.32	$\Delta n_e\sim0.016$ $\Delta n_o\sim0.006$		
$LiTaO_3$ (Li/Ta=0.946) (congruent)	3*m*	$r_{13}=8.1\pm0.4(T)$	0.6328	$n_o=2.176$	$\varepsilon_3=41$ (1–50 kHz)	0.256–5
		$r_{33}=30\pm0.5(T)$		$n_e=2.180$		
$LiTaO_3$ (Li/Ta=0.992) (near-stoichiometric)	3*m*	$r_{13}=9.2\pm0.2(T)$	0.6328	$n_e=2.1762$	$\varepsilon_3=50$ (1 KHz)	0.260–5
		$r_{33}=34.3\pm0.2(T)$			$\varepsilon_3=46$ (50 kHz)	

$LiTaO_3$ (Li/Ta = 1.00 stoichiometric)		$r_{13}=10.3\pm0.3$(T)	0.6328	$n_e=2.1737$	$\varepsilon_3=48$ (1–50 kHz)	0.275–5
		$r_{33}=35.8\pm0.1$(T)				
$KTa_xNb_{1-x}O_3$ $x=0.65$ ($T_c=280$K)	$4mm$	$\lvert r_{33}-r_{13}\rvert\sim500$	0.633	$n_o=2.318$	~6000–7000 (RT)	0.4–6
		$r_{51}\sim16000$		$n_e=2.270$		
$x=0.52$ ($T_c=87.5$ K)		$r_{33}\sim260$ (300K)	0.546			
$x=0.53$ ($T_c=296$ K)		$r_{42}=5770$	0.5145	$n_o=2.345$		
				$n_e=2.307$		
$KTa_xNb_{1-x}O_3$ $x=0.63$ ($T_c=280$ K)	$m3m$	$R_{11}-R_{12}=2890$ ($T\sim20$°C)	0.633	$n=2.29$	~6000–7000 (RT)	0.4–6
$x=0.62$ ($T_c=283$ K) [62]		$g_{11}=0.136$; $g_{12}=-0.038$				
		$g_{44}=+0.147$				
PLZT(La/Zr/Ti)	∞m				At RT	0.37–14
7/62/38		$R_c''=443$[c]	0.633	$n=2.5$–3.0	~2000	
8/65/35 (10 μm[d])		$R_c'=523$			~3000	
8/65/35 (3 μm)		$R_c'=612$				
12.5/40/60		$n_e^3r_{33}-n_o^3r_{13}=2320$	0.546	$n_o=2.55$		
8/65/35		$N^3(R_{33}-R_{13})=26000$ (63°C)	0.550	$n=2.450$		
9/65/35		$(L_{33}-L_{13})=150$(S) $-$ 911.6(T)	0.633			
$Sr_{0.25}Ba_{0.75}Nb_2O_6$	$4mm$	$r_c'=41$(S)	0.633	$n_o=2.3144$	$\varepsilon_{33}=250$ (10 KHz)	
				$n_e=2.2596$		
$Sr_{0.46}Ba_{0.54}Nb_2O_6$		$r_{13}=35$(T), $r_{33}=180$(T)		—	—	
				$n_o=2.3123$	$\varepsilon_{33}=500$ (10 kHz)	0.4–8
$Sr_{0.50}Ba_{0.50}Nb_2O_6$		$r_c'=90$(S) $-$ 205(T)		$n_e=2.2734$		
				$n_o=2.370$	$\varepsilon_{33}=900$	0.4–5
$Sr_{0.61}Ba_{0.39}Nb_2O_6$		$r_{13}=48.9$(T), $r_{33}=420$(T), $r_{42}=38$(T)		$n_e=2.335$	$\varepsilon_{11}=450$	
		$r_{13}=67$(T), $r_{33}=1293$-1340(T)		$n_o=2.3117$	$\varepsilon_{33}=6500$ (10 kHz)	
$Sr_{0.75}Ba_{0.25}Nb_2O_6$		$_{51}=42$(T), $ar_c'=1066$(S) $-$ 1400(T)		$n_e=2.2987$		0.4–5
$Ba_2NaNb_5O_{15}$ (BNN)	$mm2$	$r_{13}=15$; $r_{23}=13$; $r_{33}=48$	0.633	$n_a=2.3222$	$\varepsilon_{11}=222$(S) $-$ 235(T)	0.4–5
		$r_{42}=92$; $r_{51}=90$		$n_b=2.3205$	$\varepsilon_{22}=227$(S) $-$ 247(T)	
		$r_{33}-(n_x/n_z)^3r_{13}=31.9$		$n_c=2.2177$	$\varepsilon_{33}=32$(S) $-$ 51(T)	
		$r_{33}-(n_x/n_z)^3r_{23}=35.0$				
$K_3Li_2Nb_5O_{15}$ (KLN)	4mm	$n_e^3r_{33}=790$	0.633	$n_o=2.277$	$\varepsilon_{11}=309$	0.4–5
		$n_o^3r_{33}=105$		$n_e=2.163$	$\varepsilon_{33}=100$ at 30°C	
$(K_xNa_{1-x})_{0.4}$ $(Sr_yBa_{1-y})_{0.6}$ Nb_2O_6 ($x=0.5$, $y=0.6$)	4mm	$r_{33}=270$	0.633	$n_o=2.3066$	$\varepsilon_{11}=360$	0.4-5.00
		$r_{51}=400$		$n_e=2.2490$	$\varepsilon_{33}=120$	

(continued)

TABLE 6.3 (Continued)

Material	Point Group	r_{ij} (10^{-12} m/V) or R_{ij} (10^{-18} m^2/V^2) g_{ij} (m^4/C^2)	λ(μm)	Index of Refraction	Dielectric Constant (ε_o)	Transparency Range (μm)
KH_2PO_4 (KDP)	$\bar{4}2m$	$r_{63}=9.7$(S) to -10.5(T) $r_{41}=8.6$ (T)	0.656	$n_o=1.5064$ $n_e=1.4664$	$\varepsilon_1=\varepsilon_2=44(S)-42$(T) $\varepsilon_3=21$(S,T)	0.2–1.55
		$r_{63}=-10.3$(T) $r_{41}=18.77$ (T)	0.546	$n_o=1.5115$ $n_e=1.4698$		
KH_2PO_4 (KDP) (RT)	$\bar{4}2m$	$n_e^3(R_{33}-R_{13})=31$ $n_o^3(R_{31}-R_{11})=13.5$ $n_0^3(R_{12}-R_{11})=8.9$ $n_o^3R_{66}=3.0$	0.540	$n_o=1.5115$ $n_e=1.4698$	$\varepsilon_1=\varepsilon_2=44(S)-42$(T) $\varepsilon_3=21$(S,T)	0.2–1.55
KD_2PO_4 (KD*P)	$\bar{4}2m$	$r_{63}=$-17.2(S) to -26.4(T) $r_{41}=8.8$ (T)	0.546	$n_o=1.5079$ $n_e=1.4683$	$\varepsilon_1=\varepsilon_2=58$(S) $\varepsilon_3=48$(S)-50(T)	0.2–2.15
$NH_4H_2PO_4$ (ADP)	$\bar{4}2m$	$r_{63}=-4.77$(S) to -8.5(T) $r_{41}=24.5$ (T)	0.656	$n_o=1.52098$ $n_e=1.5721$	$\varepsilon_1=\varepsilon_2=58(S)-56$(T) $\varepsilon_3=14$(S)-15(T)	0.19–1.4
$NH_4H_2PO_4$ (ADP)	$\bar{4}2m$	$n_e^3(R_{33}-R_{13})=24$ $n_o^3(R_{13}-R_{11})=16.5$ $n_o^3(R_{12}-R_{11})=5.8$ $n_o^3R_{66}=2$	0.540	$n_o=1.5266$ $n_e=1.4808$	$\varepsilon_1=\varepsilon_2=58(S)-56$(T) $\varepsilon_3=14$(S)-15(T)	0.19–1.4
$ND_4D_2PO_4$ (AD*P)	$\bar{4}2m$	$r_{63}=10$ (T) $r_{41}=40$ (T)	0.633	$n_o=1.516$ $n_e=1.475$	$\varepsilon_1=\varepsilon_2=72$(T) $\varepsilon_3=24$(T)	0.19–1.4
		$r_{63}=11.9$ (T) $r_{41}=55.5$ (T)	0.546			
KH_2AsO_4 (KDA)	$\bar{4}2m$	$r_{63}=-10.9$(T) $r_{41}=12.5$ (T)	0.656	$n_o=1.5632$ $n_e=1.5721$	$\varepsilon_1=\varepsilon_2\sim80$(T) $\varepsilon_3\sim30$(T)	0.246–$\sim$1.5
		$r_{63}=-10.9$(T) $r_{41}=12.5$ (T)	0.546	$n_o=1.5707$ $n_e=1.5206$		
RbH_2AsO_4 (RDA)	$\bar{4}2m$	$r_{63}=13.0$	0.546	$n_o=1.5613$ $n_e=1.5241$	$\varepsilon_3\sim30$(T)	$\sim$0.35–1.6
RbH_2PO_4 (RDP)	$\bar{4}2m$	$r_{63}=-12.5$ $r_{41}=7.7$	0.633	$n_o=1.5053$ $n_e=1.4773$	$\varepsilon_3\sim25$(T) $\varepsilon_3\sim44$(T)	$\sim$0.35–1.4
		$r_{63}=-14$ $r_{41}=9$	0.546			
$KTiOPO_4$(KTP)	$mm2$	$r_{13}=9.5\pm0.5$ $r_{23}=15.7\pm0.8$ $r_{33}=36.3\pm1.8$ $r_{51}=7.3\pm0.7$	0.633	$n_1=1.7634$ $n_2=1.7717$ $n_3=1.8639$	$\varepsilon_3\sim3600$ (100 Hz)	0.35–4.0

		$r_{42}=9.3\pm0.9$				
$KTiOAsO_4$(KTA)	$mm2$	$r_{13}=11.5$, $r_{23}=15.4$ $r_{33}=37.5$	0.633	$n_1=1.8075$ $n_2=1.8138$ $n_3=1.9044$	20 (1 MHz)	0.35–5.3
$RbTiOAsO_4$(RTA)	$mm2$	$r_{13}=13.5$, $r_{23}=17.5$ $r_{33}=40.5$	0.633	$n_1=1.8294$ $n_2=1.8379$ $n_3=1.9184$	19 (10 GHz)	1–5.5
$RbTiOPO_4$(RTP)	$mm2$	$r_{13}=10.9$, $r_{23}=15.0$ $r_{33}=33.0$	0.633	$n_1=1.7884$ $n_2=1.7992$ $n_3=1.8858$	$\varepsilon_1=11$ $\varepsilon_2=15$ $\varepsilon_3=18.54$	0.4–4.3
$CsTiOAsO_4$(CTA)	$mm2$	$r_{13}=14.2$, $r_{23}=18.5$ $r_{33}=38.0$	0.633	$n_1=1.8795$ $n_2=1.8947$ $n_3=1.9596$	$\varepsilon_1\sim110$ $\varepsilon_3\sim50$ ~29 (10 GHz)	0.4–5
$(NH_2CH_2COOH)_3H_2S0_4$ (TGS)	2	$n^3_\gamma r_{12}-n^3_\alpha r_{22}\sim30$	0.535	$n_\alpha\sim1.486$, $n_\gamma\sim1.589$	$\varepsilon_3\sim6$, $\varepsilon_1\sim9.4$ (1 MHz)	0.25–2.5

[a] $r_c=r_{33}-(n_o/n_e)^3r_{13}$
[b] $r'_c=r_{33}(n_e/n_o)^3-r_{13}$
[c] $r''_c=r_{33}(n_c/n_a)^3-r_{13}$
[d] Grain size

Source: From Xu, Y. 1991. *Ferroelectric Materials and their Applications*. Elsevier Science Publishers B.V., Amsterdam, The Netherlands.

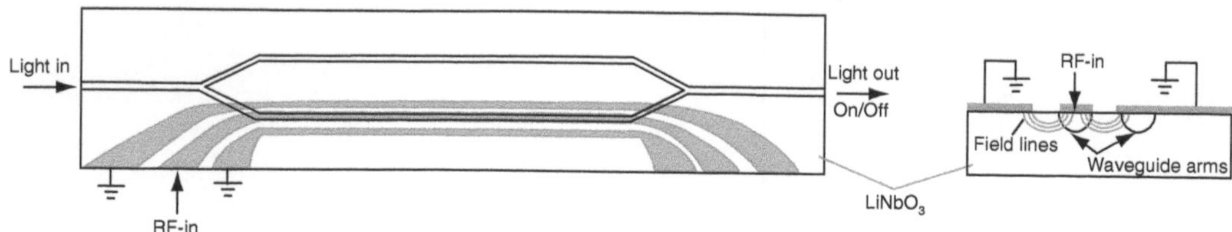

FIGURE 6.8 High-speed Mach-Zender interferometric electro-optic modulator based on surface waveguides in a ferroelectric $LiNbO_3$ modulator.

domains, then a field tunable extraordinary index change of $\Delta n = 2\Delta n_e$, and ordinary index change of $\Delta n = 2\Delta n_o$ is created across the domain wall. By shaping these domain walls into prisms, lenses and other features, electro-optic tunable integrated optical devices can be created, such as electro-optic beam steering, dynamic focusing, and beam shaping (see Figure 6.9). Many of these devices have been demonstrated [50–54]. In ferroelectric crystals with 4*mm* symmetry, similar effects can be observed by exploiting the r_{33} coefficient. For example, in tetragonal 4*mm* $BaTiO_3$ and SBN crystals, three independent electro-optic coefficients exist: r_{13}, r_{33}, and r_{51}. Under a uniform applied field E_3 along the *c*-axis of a crystal and with the light propagating perpendicular to the *c*-axis of the crystal, the refractive index changes are again given according to Equation 6.8, and they can be similarly exploited as in $LiNbO_3$. As seen from Figure 6.1, one can exploit *ferroelectricity*, *electro-optic effect* (index change with applied fields), *photorefractive effects* (index changes with light intensity), *optical nonlinearity* (for frequency conversion), *light amplification and lasing*, and *acousto-optics* (index changes with sound waves) all on the same chip.

In $BaTiO_3$, however, the largest electro-optic coefficient is $r_{51} = r_{42}$, whose reported value ranges from ~ 631(S) to 1261(T) pm/V (at 1550 nm wavelength), which is significantly larger than the r_{33} coefficient (~ 28 pm/V) used in $LiNbO_3$. Since $r_{51} \equiv r_{131} \equiv r_{311} = r_{42} \equiv r_{232} \equiv r_{322}$, an electric field in 1 or 2 directions (E_1 or E_2) rotates the refractives index ellipsoid in the 1–3 (*x–z*) or 2–3 (*y–z*) plane by an angle θ. In addition, it changes the eigenvalues of the indices in that plane from (n_o, n_e) before applying a field, to (n_1, n_2) after applying a field of E_1. For a field of $E_1 < 10^6$ V/m, one can write approximate equations relating these indices as follows:

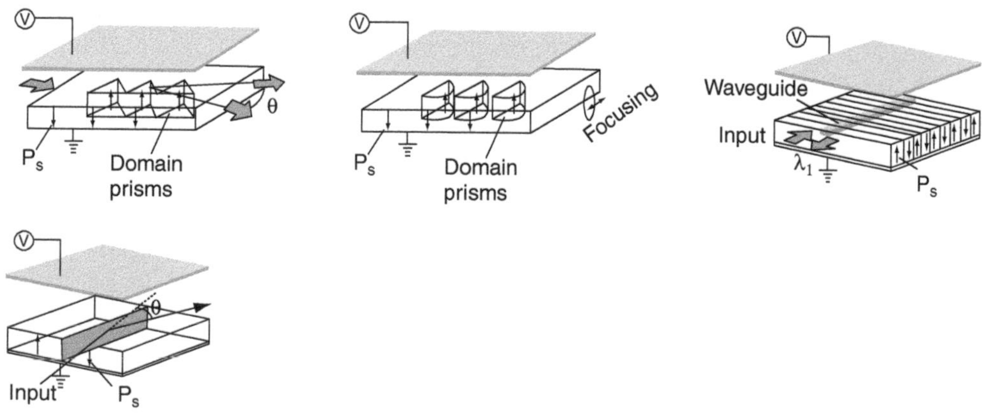

FIGURE 6.9 Examples of electro-optic devices based on ferroelectric domain engineering: (Left to Right): electro-optic scanner, electro-optic lenses, tunable electro-optic grating, total internal reflection mirror.

$$n_1 = n_o + \frac{1}{2}n_o^3 \frac{r_{42}E_1^2}{((1/n_e^2)-(1/n_o^2))}, \qquad n_2 = n_e - \frac{1}{2}n_e^3 \frac{r_{42}E_1^2}{((1/n_e^2)-(1/n_o^2))} \tag{6.9}$$

$$\tan 2\theta = \frac{2r_{42}E_1}{((1/n_e^2)-(1/n_o^2))} \tag{6.10}$$

Note that even though r_{42} is a *linear* electro-optic effect, its off-diagonal nature results in an index change that is proportional to E_1^2, which is *quadratic*. Figure 6.10a shows the birefringence change, n_1-n_2 as a function of field for $BaTiO_3$. As a comparison, the birefringence change, n_o-n_e, in $LiNbO_3$ based on Equation 6.8 is also shown. One way to exploit this advantage in $BaTiO_3$ is through cross-polarizer geometry, as shown in Figure 6.10b. The incident beam is polarized in the *x–z* plane at 45° to each axis, and the output analyzer is cross polarized to achieve a null intensity. Application of an electric field E_1 will change the intensity as follows:

$$I = I_o \sin^2\left(\frac{\Gamma}{2}\right), \tag{6.11}$$

where, $\Gamma=(2\pi/\lambda)(n_1\text{-}n_2)L$, and L is the length of travel through $BaTiO_3$. The voltage required for intensity change from 0 to I_o is called the V_π. Recently, such modulation in $BaTiO_3$ up to 40 GHz has been shown [55], as well as the growth of very high quality single crystal $BaTiO_3$ films [47]. Mach–Zender interferometry has also been shown in $BaTiO_3$ films, using phase modulation [56].

The *quadratic electro-optic* effect can arise in many ways: (1) fourth rank tensor properties, R_{ijkl} that exists in all materials, (2) effective quadratic effect through off-diagonal r_{ijk} terms, and (3) multidomain microstructure switchable with an external field. An example of effect (2) has been described for $BaTiO_3$ above. Effects (1) and (3) are described below.

Some examples of quadratic electro-optic tensors $R_{ijkl}\equiv R_{mn}$ in abbreviated notation are given below for important point groups.

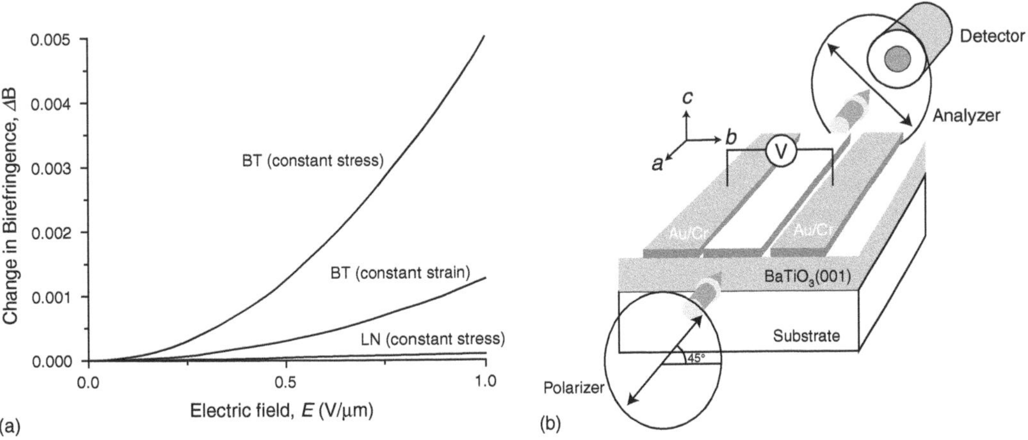

FIGURE 6.10 (a) A comparison of the bulk $BaTiO_3$ (BT) and $LiNbO_3$ (LN) electro-optic responses at 1550-nm wavelength. The LN EO-response due to r_{33} and r_{13} are linear effects, where the change in birefringence, $\Delta B=\Delta(n_o-n_e)\propto(n_e^3r_{33}-n_o^3r_{13})E$. The BT has an effective quadratic effect, given by $\Delta B=\{(n_o^3+n_e^3)/(1/n_e^2-1/n_o^2)\}(r_{42}E_1)^2$. (b) Electro-optic modulation scheme in $BaTiO_3$ (001) single crystal or thin film based on field-dependent birefringence change, ΔB.

Point group $mm2$, 222, mmm

$$\begin{pmatrix} R_{11} & R_{12} & R_{13} & 0 & 0 & 0 \\ R_{21} & R_{22} & R_{23} & 0 & 0 & 0 \\ R_{31} & R_{32} & R_{33} & 0 & 0 & 0 \\ 0 & 0 & 0 & R_{44} & 0 & 0 \\ 0 & 0 & 0 & 0 & R_{55} & 0 \\ 0 & 0 & 0 & 0 & 0 & R_{66} \end{pmatrix}$$

Point group 32, $3m$, $\bar{3}m$

$$\begin{pmatrix} R_{11} & R_{12} & R_{13} & R_{14} & 0 & 0 \\ R_{12} & R_{11} & R_{13} & -R_{14} & 0 & 0 \\ R_{13} & R_{13} & R_{33} & 0 & 0 & 0 \\ R_{41} & -R_{41} & 0 & R_{44} & 0 & 0 \\ 0 & 0 & 0 & 0 & R_{44} & 0 \\ 0 & 0 & 0 & 0 & R_{14} & \frac{1}{2}(R_{11}-R_{12}) \end{pmatrix}$$

Point groups 422, $4mm$, $\bar{4}2m$, $4/mm$

$$\begin{pmatrix} R_{11} & R_{12} & R_{13} & 0 & 0 & 0 \\ R_{12} & R_{11} & R_{13} & 0 & 0 & 0 \\ R_{31} & R_{31} & R_{33} & 0 & 0 & 0 \\ 0 & 0 & 0 & R_{44} & 0 & 0 \\ 0 & 0 & 0 & 0 & R_{44} & 0 \\ 0 & 0 & 0 & 0 & 0 & R_{66} \end{pmatrix}$$

Point group 432, $m3m$, $\bar{4}3m$

$$\begin{pmatrix} R_{11} & R_{12} & R_{12} & 0 & 0 & 0 \\ R_{12} & R_{11} & R_{12} & 0 & 0 & 0 \\ R_{12} & R_{12} & R_{11} & 0 & 0 & 0 \\ 0 & 0 & 0 & R_{44} & 0 & 0 \\ 0 & 0 & 0 & 0 & R_{44} & 0 \\ 0 & 0 & 0 & 0 & 0 & R_{44} \end{pmatrix}$$

Note that the 432 and $m3m$ are cubic point groups which contain inversion symmetry and hence do not possess any linear electro-optic coefficients, r_{nk}, but do possess non-zero quadratic R_{mn}. The phenomenological theory developed by Gunter et al. predicts the dependence of the electro-optic coefficients, r_{ijk} on the spontaneous polarization, P_s as follows [57]:

$$r_{ijk} = 2g_{ijkl}P_l\varepsilon_o(\varepsilon_{kk}-1) = \frac{2R_{ijkl}P_l}{\varepsilon_o(\varepsilon_{ll}-1)} \tag{6.12}$$

where $P_l = P_s$ is the spontaneous polarization, ε_{kk} is the principal *relative* permittivity (in units of ε_o), and g_{ijk} is the *quadratic polarization-optic tensor.*

As an example, bulk $BaTiO_3$ is paraelectric and cubic $m3m$ phase above the phase transition temperature T_c of ~120°C, and ferroelectric $4mm$ between 5 and 120°C, orthorhombic $mm2$ between −90 and 5°C, and trigonal $3m$ below −90°C. In the paraelectric phase above T_c, ferroelectrics exhibit only quadratic electro-optic effect. In the ferroelectric phase, they exhibit both linear and quadratic effects (Equation 6.3). However, typically the linear effect is stronger than the quadratic terms and dominates. In paraelectric $BaTiO_3$ phase, R_{11}, R_{12}, and R_{44} are the only independent R_{mn} coefficients. As $BaTiO_3$ develops ferroelectricity P_s in the z-direction below T_c, the linear r_{ijk} also develops. From the above tensor for $m3m$ and Equation 6.12, one can deduce that in the tetragonal phase, $r_{42} = r_{51} = 2R_{44}/(\varepsilon_0(\varepsilon_{33}-1))$, $r_{13} = r_{23} = 2R_{12}/(\varepsilon_0(\varepsilon_{33}-1))$, and $r_{33} = 2R_{11}/(\varepsilon_0(\varepsilon_{33}-1))$.

Multidomain ferroelectric microstructures can also give rise to a response that can effectively look like a quadratic (or even higher order) electro-optic effect. This is typically seen in ferroelectric ceramics such as lead lanthanum zirconate titanate (PLZT), or thin films of most ferroelectrics with a mixture of switchable domains. The quadratic (and higher order) nature of the electro-optic effect arises because of the simultaneous occurrence of two effects as electric field is applied to the material: an averaged effect of the linear electro-optic effect from a mixture of grains and domains and a simultaneous reorientation of these domains toward the direction of the applied field as the field is increased that gives rise to a more "oriented" ferroelectric domain microstructure. The domain reorientation is highly a nonlinear process with applied field as seen from the hysteresis loops in Figure 6.3, and this contributes to the nonlinearity and hysteresis of the electro-optic response. Reversing the field simply repeats this domain reversal process in the reverse direction, thus giving rise to a symmetric index change (of the same sign) even with reversed field polarity. (This symmetry can be broken by built-in internal fields in the ceramic due to compositional doping or nonstoichiometry.) This is in contrast to a pure linear electro-optic effect (without simultaneous domain reversal) where the index change, Δn, reverses its sign with electric field reversal. A quasi-linear response can be achieved by "poling" a multidomain microstructure in one direction by a large field and observing the response at low fields close to zero, which would not significantly "depole" the domain microstructure.

Typically, such quadratic effects are characterized by an *effective* quadratic electro-optic coefficient, R^*. In the PLZT system, R^* is the largest (8×10^{16} m^2/V^2) for the composition $(Pb_{0.865}La_{0.09})(Zr_{0.65}Ti_{0.35})O_3$ (also called PLZT 9/65/35). Higher values of R^* can be achieved by doping PZT with other materials instead of La alone, such as $Sr(La_{0.5}Nb_{0.5})O_3$ and $Ba(La_{0.5}Nb_{0.5})O_3$ [58,59]. PLZT is used for segmented displays, modulators, protective goggles, shutters, etc. [60]. The advantage of such materials is the ease of ceramic processing in contrast to single crystal growth, and much larger index changes ($\Delta n\sim 0.01$–0.1) in comparison to the linear electro-optic effect in materials such as $LiNbO_3$ ($\Delta n\sim 10^{-4}$–10^{-3}). The disadvantages are high optical losses for waveguide devices requiring propagation lengths of millimeters or more, hysteretic effects, and slower response (in the order of 50 μs) due to domain switching, in contrast to the pure electro-optic effect that can approach 40–100 GHz.

6.3.2 Nonlinear Optics

In conventional (linear) optics, the induced polarization (the dipole moment per unit volume) of a medium is proportional to the applied electric field. But for higher field strengths, like those available with lasers, instantaneous polarization can be generalized as a nonlinear response expressed as a Taylor series expansion:

$$P(t) = \chi^{(1)}E(t) + \chi^{(2)}E^2(t) + \chi^{(3)}E^3(t) + \ldots \tag{6.13}$$

with the first term representing linear effects, the second term representing second-order nonlinear effects, the third term representing third-order nonlinear effects, and so on. Switching to frequency (ω) space, the instantaneous polarization for second-order effects can be expressed in terms of the Fourier components of the electric field $E(\omega)$ by the basic equation

$$\begin{aligned} P_i(\omega_3) &= 2\textstyle\sum_{jk}\chi^{(2)}_{ijk}(\omega_3=\omega_1\pm\omega_2)E_j(\omega_1)E_k(\omega_2) \quad \text{Gaussian units} \\ P_i(\omega_3) &= \textstyle\sum_{jk}\varepsilon_o d^{(2)}_{ijk}(\omega_3=\omega_1\pm\omega_2)E_j(\omega_1)E_k(\omega_2) \quad \text{MKS units} \end{aligned} \tag{6.14}$$

where the third-rank tensor d_{ijk} [or χ_{ijk}] is the second-order nonlinear dielectric susceptibility, also called a nonlinear optical coefficient. The second-order nonlinear optical effects resulting from Equation 6.14 are the optical sum-frequency generation or difference-frequency generation effect for $\omega_1\neq\omega_2$. Second-harmonic generation (+) or the optical rectification effect (−) are nonlinear effects for $\omega_1=\omega_2$. A variety of differences in the definition and notations of second-order nonlinear optical susceptibility can be found in the literature. Detailed discussions can be found in the chapter on Nonlinear Optics or most nonlinear optics textbooks. We will continue with Gaussian units only here.

Similar to the electrooptic coefficients, the third rank tensor d_{ijk} can be contracted to a 3×6 matrix, d_{ij} using simplified subscript indices since j and k can be exchanged based upon intrinsic permutation symmetry arguments [63]. Thus d_{ijk} can be represented by a matrix with $3\times6=18$ components. According to the crystal symmetry, the number of independent components can be further reduced. The contracted d_{ij} tensors for point groups observed in ferroelectric materials are shown as follows.

Point group $\bar{4}2m$

$$\begin{pmatrix} 0 & 0 & 0 & d_{14} & 0 & 0 \\ 0 & 0 & 0 & 0 & d_{14} & 0 \\ 0 & 0 & 0 & 0 & 0 & d_{36} \end{pmatrix}$$

Point group $4mm$

$$\begin{pmatrix} 0 & 0 & 0 & 0 & d_{15} & 0 \\ 0 & 0 & 0 & d_{24} & 0 & 0 \\ d_{31} & d_{31} & d_{33} & 0 & 0 & 0 \end{pmatrix}$$

Point group $3m$

$$\begin{pmatrix} 0 & 0 & 0 & 0 & d_{15} & -d_{22} \\ -d_{22} & d_{22} & 0 & d_{15} & 0 & 0 \\ d_{31} & d_{31} & d_{33} & 0 & 0 & 0 \end{pmatrix}$$

Point group $mm2$

$$\begin{pmatrix} 0 & 0 & 0 & 0 & d_{15} & 0 \\ 0 & 0 & 0 & d_{24} & 0 & 0 \\ d_{31} & d_{32} & d_{33} & 0 & 0 & 0 \end{pmatrix}$$

Point group 3

$$\begin{pmatrix} d_{11} & -d_{11} & 0 & d_{14} & d_{15} & -d_{22} \\ -d_{22} & d_{22} & 0 & d_{15} & -d_{14} & -d_{11} \\ d_{31} & d_{31} & d_{33} & 0 & 0 & 0 \end{pmatrix}$$

Point group $6mm$

$$\begin{pmatrix} 0 & 0 & 0 & 0 & d_{15} & 0 \\ 0 & 0 & 0 & d_{15} & 0 & 0 \\ d_{31} & d_{31} & d_{33} & 0 & 0 & 0 \end{pmatrix}$$

Nonlinear optic coefficients of some ferroelectric crystals are listed in Table 6.4.

As shown in Equation 6.14, oscillating electric fields at two frequencies can generate instantaneous polarizations at new frequencies. Second order effects involve sums or differences of integer multiples of the driving frequencies. The instantaneous polarization in turn is a driving function for the wave equation, i.e.,

$$\nabla^2\mathbf{E} - \frac{n^2}{c^2}\frac{\partial^2\mathbf{E}}{\partial t^2} = \frac{4\pi}{c^2}\frac{\partial^2\mathbf{P}}{\partial t^2} \quad \text{(gaussian units)} \tag{6.15}$$

where n is the refractive index, c is the speed of light in vacuum, and $c/n = \omega/k$ is the phase velocity. Thus, we can use nonlinear materials to generate electromagnetic waves with new frequencies. This frequency conversion property is an important application of nonlinear optical materials.

However, to obtain a significant conversion of the input waves to the new wave, energy and momentum must be conserved. For sum-frequency generation (we will drop generality for simplicity) the conservation laws can be written

$$\begin{aligned} \hbar\omega_3 &= \hbar\omega_1 + \hbar\omega_2 \quad \text{energy conservation} \\ \hbar\mathbf{k}_3 &= \hbar\mathbf{k}_1 + \hbar\mathbf{k}_2 \quad \text{momentum conservation} \end{aligned} \tag{6.16}$$

where $\mathbf{k}$ describes the propagation vector of a wave and

$$|k| = \frac{n\omega}{c}. \tag{6.17}$$

But refractive index n is a function of frequency and thus normal dispersion, where n increases with increasing frequency, will prevent momentum conservation (or phase matching) from taking place as the three interacting waves propagate through the nonlinear material at different speeds. This phase mismatch is an inherent difficulty for nonlinear frequency conversion processes. Fortunately, two techniques, birefringent phase matching (BPM) and quasi-phase matching (QPM), have been developed to achieve efficient frequency conversion.

6.3.2.1 Birefringent Phase Matching

Birefringent phase matching uses the fact that different polarizations of light have different indices of refraction to meet the requirements of Equation 6.16. In general, any propagation direction in a

TABLE 6.4 Second-Order Nonlinear Optical Properties of Ferroelectric Materials (d_{ij} Values Given for 1064-nm SHG When Available)

Material	Point Group	d_{ij} (10^{-12} m/V)	Nonlinear Coefficient Reference	Dispersion Reference
$BaTiO_3$	4*mm*	$d_{15}=+13.6$	[93–95]	[96]
		$d_{31}=-14.4$		
		$d_{33}=-5.5$		
$KNbO_3$	*mm*2	$d_{31}=-(11.8^{a},8.9^{c})$		[100]
		$d_{32}=-(13.7^{a},10.8^{b},12.4^{c})$	^{a}[97], ^{b}[98], ^{c}[99]	
		0$d_{33}=-(20.5^{a},19.6^{b},21.9^{c})$		
		$d_{15}=-(12.4^{a},9.2^{c})$		
		$d_{24}=-(12.8^{a},12.5^{b},13.0^{c})$		
$LiNbO_3$ (Li/Nb=0.946)	3*m*	$d_{31}=-5.95^{d}$	^{d}[101]	[102–104]
(congruent)		$d_{33}=-34.4^{d}$	^{e}[98]	
		$d_{22}=+2.3^{d}$		
		$d_{33}=-25^{e}$		
$LiNbO_3$ (Li/Nb=0.99) (near-stoichiometric)	3*m*	$d_{33}=-29$	[61]	[105]
MgO:$LiNbO_3$ (near-stoichiometric) 5 mol% Mg	3*m*	$d_{33}=-30$	[61]	
$LiTaO_3$ (congruent)	3*m*	$d_{31}=-1.0^{f}$	^{f}[106]	[107]
		$d_{33}=-15.6^{f}$	^{g}[98]	
		$d_{33}=-13.8^{g}$		
$LiTaO_3$ (near-stoichiometric)	3*m*	$d_{33}=-14.5$	[61]	[108]
$Ba_2NaNb_5O_{15}$	*mm*2	$d_{31}=-13.2$	[109]	[109]
		$d_{32}=-13.2$		
		$d_{33}=-18.2$		
		$d_{15}=-13.2$		
		$d_{24}=-12.5$		
KH_2PO_4 (KDP)	$\bar{4}2m$	$d_{36}=0.39$	[95]	[110]
KDPO (KD*P)	$\bar{4}2m$	$d_{36}=0.37$	[111]	[110]
$NH_4H_2PO_4$ (ADP)	$\bar{4}2m$	$d_{36}=0.53$	[112]	[110] [95]
$ND_4D_2PO_4$ (AD*P)	$\bar{4}2m$	$d_{36}=0.52$	[113]	[95]
KH_2AsO_4 (KDA)	$\bar{4}2m$	$d_{36}=0.52$	[114]	[95]
$KTiOAsO_4$(KTA)	*mm*2	$d_{31}=2.3$	[115]	[116]
		$d_{32}=3.7$		
		$d_{33}=15.5$		
		$d_{15}=2.3$		
		$d_{24}=3.6$		
$KTiOPO_4$(KTP)	*mm*2	$d_{31}=.1$	[117]	[118]
		$d_{32}=3.8$	[115]	[119]
		$d_{33}=5.4$		
		$d_{15}=2.0$		
		$d_{24}=3.6$		
RbH_2PO_4 (RDP)	$\bar{4}2m$	$d_{36}=0.40$	[112]	[95]
RbH_2AsO_4 (RDA)	aE_i^{ω}	$d_{36}=0.39$	[120]	[95]
$RbTiOAsO_4$(RTA)	*mm*2	$d_{31}=2.2$	[115]	[116]
		$d_{32}=3.9$	[121]	
		$d_{33}=5.9$		
		$d_{15}=2.2$		
		$d_{24}=3.9$		
$RbTiOPO_4$(RTP)	*mm*2	$d_{31}=2.0$	[115]	[122]
		$d_{32}=3.8$		
		$d_{33}=15.6$		
		$d_{15}=2.0$		

(continued)

TABLE 6.4 (Continued)

Material	Point Group	d_{ij} (10^{-12} m/V)	Nonlinear Coefficient Reference	Dispersion Reference
		$d_{24}=4.0$		
$CsTiOAsO_4$(CTA)	$mm2$	$d_{31}=2.1$	[122]	[116]
		$d_{32}=3.4$		
		$d_{33}=18.1$		
$(NH_2CH_2COOH)_3$_H_2SO_4 (TGS)	2	$d_{23}=0.32$	[123]	

birefringent crystal will have two orthogonal eigenpolarizations that do not mix as the beam propagates. In addition, each eigenpolarization direction has a corresponding eigenindex that is a function of propagation angle relative to the crystal principal axes. Thus, it is sometimes, but not always, possible to find a crystal propagation direction (or phase matching angle) for which the three frequencies in Equation 6.16 can have a solution. As an example, consider the simple case of frequency doubling where the fundamental frequency $\omega=\omega_2=\omega_3$ and the second harmonic frequency $2\omega=\omega_1$. In addition, we will assume that we are using a negative uniaxial crystal (extraordinary index $n_e<n_o$) with normal dispersion (refractive index increases with frequency) and pump photons with the same polarization, i.e., Type I BPM. Assuming collinear propagation of all waves, the momentum conservation equation becomes

$$n_o^{\omega} = n_e^{2\omega}(\theta) \tag{6.18}$$

where θ is the angle between the propagation direction of the waves and the c-axis, n_o^{ω} is the ordinary index at the fundamental frequency and $n_e^{2\omega}$ is the extraordinary refractive index at the second harmonic frequency. The extraordinary eigenindex for a given propagation direction is given by the following function of principal index values and θ:

$$\frac{1}{n_e^2(\theta)} = \frac{\cos^2\theta}{n_o^2} + \frac{\sin^2\theta}{n_e^2}. \tag{6.19}$$

The angle that provides the perfect phase matching solution of these equations is given by

$$\sin^2\theta_m = \frac{(n_o^{\omega})^{-2} - (n_o^{2\omega})^{-2}}{(n_e^{2\omega})^{-2} - (n_o^{2\omega})^{-2}} \tag{6.20}$$

and is shown graphically in Figure 6.11. However, the existence of a real phase matching solution for θ_m depends upon the details of the dispersion for the particular material being used. Therefore, BPM may not always be possible.

In addition, the strength of the energy transfer is proportional to the effective nonlinear coefficient (d_{eff}). The magnitude of d_{eff} depends upon the propagation direction through the crystal and the point group symmetry of the crystal since d_{ijk} is a 3rd rank tensor. For lithium niobate with $3m$ point group, we have

$$d_{\text{eff}} = d_{31}\sin\theta_m - d_{22}\cos\theta_m \sin 3\phi \tag{6.21}$$

where d_{31} and d_{22} are contracted d-tensor elements and ϕ is the propagation angle relative to the crystallographic x-axis in the x–y plane. See Zernike and Midwinter [64] for a complete listing of d_{eff} for various point group symmetries. Also, Roberts [65] provides an excellent overview of BPM nomenclature and formulas for both uniaxial and biaxial materials. Note from Equation 6.21 that BPM cannot use the

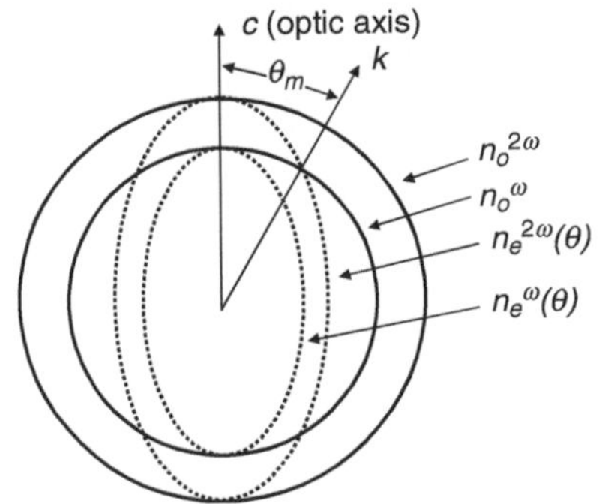

FIGURE 6.11 Negative uniaxial Type I birefringent phasematching.

largest d-tensor element available in most cases. For our lithium niobate example, the d_{33} tensor element is 5 times larger than d_{31} and 10 times larger than d_{22}. Also, BMP has a limited interaction length for actual laser beam interactions. The extraordinary beam and ordinary beam do not propagate in exactly the same direction. This phenomenon is known as spatial walkoff and is caused by the fact that the Poynting vector, which describes the direction in which the beam energy is traveling, is not the same for the ordinary and extraordinary beams. Since nonlinear frequency conversion can only take place where the beams overlap, it limits the interaction length and ultimately conversion efficiency in actual devices.

6.3.2.2 Quasi Phase Matching

In 1962, Armstrong et al. [66] proposed QPM as a means to achieve high nonlinear frequency conversion. To understand the reason for this terminology, let us consider what happens if phase matching is not perfect, i.e.,

$$\Delta k = k_1 + k_2 - k_3 \neq 0 \tag{6.22}$$

for a sum frequency mixing interaction. Then, with the usual assumptions of slowly varying amplitude, lossless medium and full permutation symmetry of $\chi^{(2)}$, a phase term appears in the coupled nonlinear equations as shown below [63]:

$$\begin{aligned}
\frac{\partial A_1}{\partial z} &= -\mathrm{i}\kappa A_3 A_2^* \mathrm{e}^{-\mathrm{i}\Delta kz} \\
\frac{\partial A_2^*}{\partial z} &= \mathrm{i}\kappa A_1 A_3^* \mathrm{e}^{\mathrm{i}\Delta kz} \\
\frac{\partial A_3}{\partial z} &= -\mathrm{i}\kappa A_1 A_2 \mathrm{e}^{\mathrm{i}\Delta kz}
\end{aligned} \tag{6.23}$$

where A_i is the field amplitude of the ith wave, κ is a coupling coefficient proportional to d_{eff} and z is the depth into the nonlinear material along the propagation direction of the interacting waves. Notice that the change in field amplitudes has a phase that changes with propagation distance z of $\exp(\pm\mathrm{i}\Delta kz)$. Let us assume that A_3 is increasing with increasing z (this depends on the initial conditions). So $\mathrm{d}A_3/\mathrm{d}z$ has a positive sign at $z=0$, and A_3 will continue to increase as the wave propagates in the $+z$ direction until $\exp(-\mathrm{i}\Delta kz)$ changes sign, i.e., when $\Delta kz=\pi$ (see Figure 6.12). Since the derivative changes sign at this point, A_3 begins to decrease. The result is that when $\Delta k\neq 0$ the field amplitudes, and thus the photon flux densities oscillate sinusoidally with the distance traveled through the nonlinear crystal. This is shown in the No QPM trace in Figure 6.12. L_c, the coherence length, is the distance over which phase does not change sign, i.e.,

$$L_c = \frac{\pi}{|\Delta k|}, \tag{6.24}$$

and power continues to flow in the same direction between the interacting waves. Remember that BPM works by making Δk small or making the coherence length long compared to the crystal or interaction length.

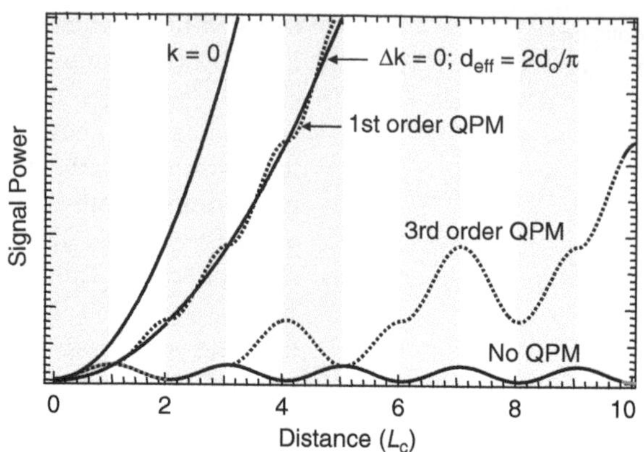

FIGURE 6.12 Nonlinear conversion as a function of propagation distance into the nonlinear crystal for quasi phase matching. Distance is given in terms of coherence length L_c.

The rate at which $|A_3|^2$ (proportional to photon flux density) increases for perfect BPM is shown for the $\Delta k = 0$ line with d-coefficient d_o and the assumption of undepleted pump.

Armstrong et al. [66] proposed a technique to counteract the change in sign of the phase term when $\Delta k \neq 0$. Note that a change of the sign of κ (or of d_{eff}) in Equation 6.23 could counteract the change in the sign of the phase. Armstrong et al. proposed that slabs of nonlinear material with alternating orientations be used to compensate for the changes in the phase factor as shown in Figure 6.13.

As an example, consider a rotation of 180° about the x-axis in a crystal. The third rank d-tensor transforms under rotation in the same way as the product of three vectors. So $d_{33} \to -d_{33}$ since $(z)(z)(z) \to (-z)(-z)(-z) = -(z)(z)(z)$ under a 180° rotation about the x-axis. So, we could slice up our nonlinear crystal into pieces, $L = \pi/\Delta k$ thick, to achieve QPM. We can rotate every other piece by 180° and then fuse them back together to get a structure with periodic changes in the sign of d. The result is shown in the wavy but monotonically increasing curve labeled first-order QPM in Figure 6.12. Because of the wavy nature of QPM, the average gain acts as $d_{\text{eff}} = d_{33} 2/\pi$. But of course, d_{33} cannot be used for BPM. A more valid comparison is to compare the gain possible for the two types of phase matching in lithium niobate as follows:

$$\frac{\text{QPM(gain)}}{\text{BPM(gain)}} = \frac{4d_{33}^2}{d_{31}^2 \pi^2 \sin^2\theta} \cong 13 \quad \text{(ignoring the } \theta \text{ dependence)} \tag{6.25}$$

Thus, over an order of magnitude, more gain is possible with QPM. Another advantage of QPM is that all interacting waves can be the same polarization. That means that walkoff is not the problem it is for BPM, and beams can be tightly focused in nonlinear devices to achieve excellent overlap over long interaction distances.

QPM also offers much more flexibility in engineering the desired nonlinear properties of the material. The phase matching of desired frequencies can be engineered by fabricating the size of the domains that alter their signs. The phasematching equation for QPM with the polarizations of all three interacting waves parallel and with collinear propagation becomes

$$\Delta k = k_1 + k_2 - k_3 = \frac{\pi}{L_c} = \frac{2\pi}{\Lambda} \tag{6.26}$$

where Λ is the period of the domain switching pattern. Thus, the crystal orientation pattern or grating makes up for the Δk phase mismatch as shown in Figure 6.14. Typical periods for a $LiNbO_3$ optical parametric oscillator (OPO) pumped by a 1.064-μm laser to produce mid-infrared signal wavelengths

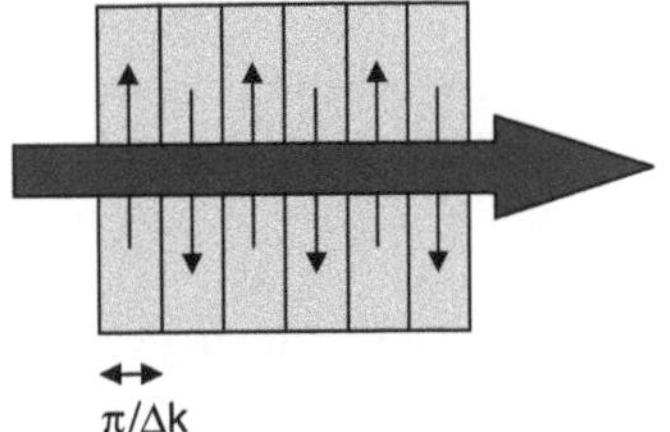

FIGURE 6.13 Alternating sign of d_{eff} in alternating orientations of nonlinear material with the thickness of each slab equal to a coherence length.

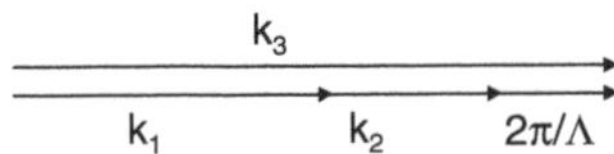

FIGURE 6.14 QPM phase matching diagram.

($\sim$3–4 μm) are 25–30 μm. Periods become very small (just a few microns) for conversion to blue or ultraviolet frequencies. Obviously, if we change the period of our domain grating and assume that we have the same input beam (k_3), then the output beams in an OPO will change frequencies to maintain optimum gain. Simply changing the period of the QPM material provides output frequency tuning.

It is also possible to achieve higher order QPM where the domain size is made to equal an odd integral multiple of the coherence length. The case of third-order QPM is shown in Figure 6.12. The first two coherence lengths of propagation simply convert and then back-convert input power. The last coherence length provides conversion before the domain switch occurs. The next two coherence lengths again cancel each other with the 6th coherence length providing net gain. The net gain for higher order QPM is smaller than first-order QPM but it does provide an advantage where the domain sizes required for first-order QPM are too small to be easily fabricated. As shown in the figure, third-order QPM domain sizes are 3 times as large as first-order QPM domains.

While the QPM technique was proposed in 1962 [66], BPM was the dominant technique used for over 30 years because there was no simple technique to make the many precisely fabricated domain thicknesses (hundreds or even thousands of them) needed for efficient nonlinear frequency interactions. A key breakthrough came with the demonstration of electric field driven periodic poling in ferroelectric materials [67–69]. Initially, chemical indiffusion [70] was used to make periodic structures of good quality; however, the pattern was limited to shallow layers only sufficient for waveguide devices, and the structure lacked the power handling capability required for many applications. Bulk periodically poled lithium niobate (PPLN) had been made by modulation of the crystal growth process, but difficulty in controlling axial variations in domain periodicity significantly degraded efficiency [71,72]. Electron beam writing can produce bulk PPLN with good periodicity, but uniformity is poor and the process does not lend itself to manufacturability. Several groups of researchers [67–69] developed techniques for ferroelectric domain reversal with an external electric field in waveguide structures and produced PPLN for QPM second-harmonic generation. They controlled domain periodicity with a precisely defined lithographic mask and used standard microfabrication techniques. Myers et al. [73] extended this technique to bulk 0.5-mm thick wafers of congruent lithium niobate, making high power operation possible.

As discussed in Section 6.3, the orientation of a ferroelectric domain can be changed with the application of an electric field. The field strength needed to reverse the polarization in a ferroelectric material is called the coercive field and can be quite large (several kV/mm) in materials such as congruent $LiNbO_3$ and $LiTaO_3$. Known coercive fields for ferroelectric nonlinear materials are listed in Table 6.2.

6.3.2.3 Nonlinear Applications of Ferroelectrics

The development of photolithographic techniques to generate any desired pattern of domain structures revolutionized the field of nonlinear frequency conversion. This advance suddenly made QPM available for practical devices. All of the advantages of QPM over BPM, higher gain, no walkoff, and tight focusing combined with the engineerability of the poling technique, resulted in major technological advances.

CW OPOs had been demonstrated before the advent of periodically poled nonlinear devices as early as the 1960s [74], but such devices used resonators that were highly reflective for both the signal beam and the idler beam. While these doubly resonant OPOs (and even triply resonant devices that also resonated the pump wavelength) reduced the threshold of the device to acceptable levels, the devices had severe problems with frequency and amplitude stability. Singly resonant OPO resonators were theoretically possible but had an order of magnitude higher threshold and thus required extremely high pump powers with narrow bandwidth. QPM in ferroelectric materials made singly resonant CW OPO devices practical and efficient. Bosenberg et al. [75] demonstrated 93% pump depletion in a CW ring singly resonant OPO using PPLN as the nonlinear material. They also achieved 3.5 W of CW output with a simple commercial 13.5-W Nd:YAG laser with no special effort to narrow the pump frequency bandwidth. Their pump laser operated with 9 axial modes, 2.2 GHz linewidth. QPM OPO frequency tuning was easily achieved by fabricating several domain periods on the same wafer and then mechanically translating the pump beam to different parts of the patterned wafer [76,77]. Fine tuning of wavelength between fixed domain periods was achieved through temperature tuning. An alternative technique for continuous tuning on one wafer of PPLN was achieved via use of a fan pattern where the domain periodicity changes smoothly from one edge of a wafer to the other [78]. Translating the wafer across the pump beam continuously accesses different periods and changes the OPO output.

Because PPLN has such a high coercive field, ~22 kV/mm, the thickness of practical devices has been limited to 1 mm or less. This limits the power handling capability due to surface damage, particularly where Q-switched pulses are used. Efforts to circumvent this limit in thickness have included use of elliptical pump beams [79], diffusion bonded stacks of PPLN [80], and a conversion of the pump beam into a 2D array of beamlets focused into unbonded stacks of PPLN wafers [81]. The beamlet array approach has achieved a record best 66 mJ of 1.5-μm output.

This observed high coercive field, however, is true only for congruently grown lithium niobate. Techniques to grow stoichiometric lithium niobate have been developed [35] and stoichiometric lithium niobate is found to have a much lower coercive field [33]. This allows poling of much thicker wafers. The key technical problem is maintaining the exact stoichiometric ratio uniformly throughout the boule during growth since the natural incorporation ratio of Li to Nb ions is not 1:1. Any variation in Li and Nb concentration ratio in the single crystal wafer results in large variations in coercive field and thus large variations in poling quality. However, better quality stoichiometric lithium niobate and lithium tantalate continue to be produced and both are now available commercially.

Periodic poling is also possible in other ferroelectric materials. Lithium tantalate and barium nitrate have been periodically poled. These materials have nonlinear properties very similar to lithium niobate. Barium nitrate samples poled with poor uniformity across a sample [82]. Lithium tantalate has an advantage over lithium niobate in that it is transparent down to shorter wavelengths in the ultraviolet region [83,84].

Another major class of ferroelectrics of interest for nonlinear frequency conversion is potassium titanyl phosphate (KTP) and its isomorphs such as rubidium titanyl arsenate (RTA) and potassium titanyl arsenate [85]. These crystals have also been successfully periodically poled. While KTP and its isomorphs have smaller nonlinear coefficients than lithium niobate by a factor of 2, they are more robust with surface damage thresholds about an order of magnitude higher. They also do not suffer from photorefractive effects. PPLN wafers must typically be run at temperatures above 80°C in nonlinear devices pumped by cw near-IR or shorter wavelengths. The heating allows charge generated through short wavelength pumping to be annealed out before it accumulates to levels that would overwhelm nonlinear conversion gain. KTP and its isomorphs can operate at room temperature and do not require any oven. KTP and its isomorphs also have lower coercive fields than congruent lithium niobate, allowing poling at up to 3-mm thicknesses [86].

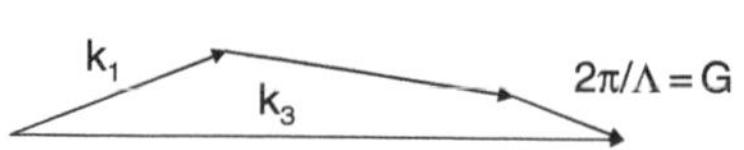

FIGURE 6.15 Noncollinear phasematching. G is a reciprocal lattice vector defined by the periodicity of the domain regions.

The engineerability and high gain of ferroelectric periodically poled materials has made a number of nonlinear innovations possible. Spatially shaped interactions are possible since different areas of the wafer can be poled with different periods. This allows broadband or comb-type frequency output from a single OPO or OPG device [87]. Multiple nonlinear interactions can be achieved on one chip, such as cascade OPO interactions where the signal generated by one nonlinear interaction becomes the pump for a 2nd nonlinear interaction. This has been used to generate idler output with greater than 100% quantum efficiency by designing the 2nd OPO interaction to generate the same idler as the first OPO interaction [88]. Sum frequency mixing processes have also been demonstrated, e.g., summing two Nd laser beams in PPLN to generate red light [89]. Ultrashort pulse frequency conversion has also taken advantage of the engineerability of QPM to compensate for group velocity dispersion (GVD) [90]. The slower frequencies were doubled at the beginning of the wafer, and the fast frequencies were doubled at the end. With the input beam stretched to place the slow frequencies first and the fast frequencies last, the output 2nd harmonic pulse maintained the ultrashort nature of the input pulse in spite of the GVD in the nonlinear material.

The high gain of ferroelectric quasi-phasematching even allows noncollinear interactions to occur with significant efficiency. Researchers have considered varying the ferroelectric domains both aperiodically and periodically in 2D periodic arrays [91]. Broderick et al. [92] observed noncollinear SHG in 2D hexagonally poled lithium niobate. Phase matching was analyzed in terms of reciprocal lattice vectors **G** shown in Figure 6.15. Russell et al. [87] showed that a fan grating can be considered a 2D QPM structure with phase matching solutions no longer points in a 2D lattice structure but continuous regions in reciprocal space. Pumping the full fan grating structure resulted in a continuum of noncollinear OPO signal and idler wavelengths.

The diversity and versatility of poled ferroelectric nonlinear devices has only begun to be developed. The discussion above is only a sample of the results achieved since quasi-phase-matched nonlinear devices became practical with the advent of electric field poling of ferroelectric materials.

6.3.3 Pyroelectricity

6.3.3.1 Pyroelectric Effect

The pyroelectric effect is the generation of a current (or voltage) caused by a temperature change in a material. This effect can be used for detecting radiation via thermal effects once it is absorbed. The pyroelectric effect is only found in polar crystals. The mechanism of pyroelectric infrared detection is explained as follows. Usually, the surface charge (due to spontaneous polarization) of a pyroelectric crystal is neutralized by free carriers from the interior and exterior of the crystal. The average time needed for interior free carriers to neutralize surface charge is $\tau_E = \varepsilon/\sigma$, where ε is the dielectric constant and σ is the conductivity of the crystal. The value of τ_E ranges from 1 to 1000 s for most of the pyroelectric materials. When a light beam with modulated frequency f is irradiated on a pyroelectric material, the light will be absorbed by the material (or by an additional absorbing layer), and the temperature of the material will be modulated at the same frequency. The change in temperature alters the lattice spacing of the pyroelectric material and produces a change in spontaneous polarization. An electric current is generated to balance the polarization change, and thus the surface charge of the material will also be modulated with frequency f. If $f > 1/\tau_E$, free carriers from the interior of the material do not have enough time to compensate for the alternating surface charge. An alternating current therefore will flow in an external circuit connected to the pyroelectric material. The magnitude of this current is given by

$$I = A\left(\frac{\mathrm{d}P_s}{\mathrm{d}t}\right) = A\left(\frac{\mathrm{d}P_s}{\mathrm{d}T}\right)\left(\frac{\mathrm{d}T}{\mathrm{d}t}\right) = Ap\left(\frac{\mathrm{d}T}{\mathrm{d}t}\right) \tag{6.27}$$

where A is the area of electrodes, $\mathrm{d}P_s/\mathrm{d}t$ is the spontaneous polarization change rate, $\mathrm{d}T/\mathrm{d}t$ is temperature change rate, and $\mathrm{d}P_s/\mathrm{d}T$ is, by definition, the pyroelectric coefficient p. When the temperature change ΔT is small, the pyroelectric coefficient can be considered to be a constant. Then, the magnitude of signal current I is proportional to $\mathrm{d}T/\mathrm{d}t$. The responsivity of a pyroelectric detector defined as output signal (voltage) divided by input radiant power is given by Kruse [124] as

$$R_s = \frac{\eta \omega p A R}{G(1+\omega^2\tau_E^2)^{1/2}(1+\omega^2\tau_T^2)^{1/2}} \tag{6.28}$$

where η is the absorption efficiency, ω is the angular frequency, R is the loss resistance, G is the thermal conductance, τ_E is the electrical response time and τ_T is the thermal response time. Thermal response time is defined as the ratio of heat capacity C to thermal conductance G. Note that responsivity has both a thermal and an electrical response time. The equation also shows that pyroelectric detectors have no dc ($\omega=0$) response. They must employ radiation modulators or be continuously panned across a scene.

Ideal materials for pyroelectric applications should have large dT/dt (temperature change rates). A large dT/dt can be achieved with materials that have a large IR absorption coefficient, a small specific heat (c_p), low density, and small physical dimensions. High Curie temperatures (T_c), high pyroelectric coefficients (p), as well as low dielectric constants (ε), and low dielectric loss R, especially for high frequency applications, are desirable material properties for pyroelectric infrared detectors. The quality of a pyroelectric sensor depends not only on the sensitivity, but also on the signal to noise ratio. Researchers use a variety of figures of merit to describe this material goodness quality, and the details depend upon the operational details. One figure of merit defined by Whatmore [125] is given by $F_V=p/(\varepsilon\varepsilon_o c_p)$. Table 6.5 below lists pyroelectric properties of some of the more commonly used ferroelectric materials. Note that F_V increases as T_c decreases. But as T_c approaches sensor operating temperatures, careful temperature control must be maintained to keep it below T_c since p goes to zero above the Curie temperature.

6.3.3.2 Pyroelectric Applications

Thousands of publications cover pyroelectricity, with topics ranging from materials to processes to devices [1,127]. Pyroelectric detectors are common commercially available devices. Laser power measurement is a major application of pyroelectric materials. Pyroelectric detectors are typically made using a pyroelectric material sandwiched between two electrodes and coated with an absorbing material such as graphite. These detectors exhibit a broad spectral response in the ultraviolet, visible, and infrared parts of the spectrum, limited only by the characteristics of the absorbing material and the transmittance of any window. The radiation absorbed by the absorbing layer in a pyroelectric detector is converted into heat which raises the temperature of the pyroelectric material. As shown in Equation 6.27, this leads to a change in voltage (or current) proportional to incident power. Pyroelectric detectors respond only to pulsed, modulated or chopped radiation sources. CW sources can be measured by mechanically chopping the beam. Pyroelectric detectors are also inherently piezoelectric and are

TABLE 6.5 Pyroelectric Properties of Some Ferroelectrics

Material	P (μC m^{-2} K^{-1})	ε	c_p (10^6 J m^{-3} K^{-1})	$F_v=p/(\varepsilon\varepsilon_0 c_p)$	T_c (°C)
$LiTaO_3$, single crystal	230	47	3.2	0.17	620
TGS, single crystal	280	38	2.3	0.36	49
SBN-50, single crystal	550	400	2.3	0.07	121
$(Pb,Ba)_5Ge_3O_{11}$, Single crystal	320	81	2.0	0.22	70
$PbZrTiO_3$	380	290	2.5	0.06	230
$PbTiO_3$, ceramic	180	190	3.0	0.04	490
$Ba_xSr_{1-x}TiO_3$, ceramic					17

Source: From Muralt, P. 2001. *Reports Progress Physical*. 64,1339–1388. With permission.

therefore acoustically sensitive. Domain engineering of the spontaneous polarization of ferroelectric pyroelectric detectors has been employed to reduce acoustic sensitivity [128].

The quality of new materials must be judged with respect to existing ones. An overview of the pyroelectric materials most commonly used in current devices is therefore given here.

$LiTaO_3$ and sometimes $LiNbO_3$ are the pyroelectric detector materials of choice in commercial laser power meters. They are grown in large quantities for other applications and thus are readily available. Although the pyroelectric coefficient of $LiTaO_3$ is smaller than those in many other pyroelectric materials, the Curie temperature is higher. Moreover, $LiTaO_3$ and $LiNbO_3$ have stable physical and chemical properties, which some pyroelectric materials lack, especially the water-soluble crystals. Therefore, both crystals—especially $LiTaO_3$, which has a higher pyroelectric coefficient and a lower dielectric constant—are very attractive as materials for infrared detectors. An infrared detector employing $LiTaO_3$ crystals is able to withstand high-energy infrared radiation and can be built to have a fast response time [129,130]. Picosecond response time has been demonstrated by using $LiTaO_3$ crystals doped with Cu^{2+} impurities in detecting 10-ps pulses from a mode-locked Nd:glass laser [131]. By combining a $LiTaO_3$ plate and a Si-CCD (charge coupling device), an infrared image sensor CCD has been developed [132].

Even though it is hygroscopic, triglycine sulfate (TGS) is a favorite material for high sensitivity applications. It is widely used as the pyroelectric material in infrared detectors for Fourier Transform Infrared (FTIR) spectrometers. TGS has a large pyroelectric coefficient of 280 $\mu C\ m^{-2}\ K^{-1}$ at room temperature and the largest response sensitivity D^* among known pyroelectric materials. In doped TGS, D^* reaches a value of 2.5×10^9 cm $Hz^{1/2}\ W^{-1}$. However, there are certain shortcomings of TGS for pyroelectric applications. The main problem is the depolarization owing to its low Curie point of 49°C. (In fact, even at room temperature, partial depolarization occurs.) To solve the depolarization problem, Fang et al. [41] doped TGS crystals with L-alanine and phosphoric acid. The crystals were then deuterated to form a material labeled ADTGSP. The Curie temperature increased to 57.5°C and the p/ε figure of merit increased from 9 to 13 relative to TGS. In ADTGSP crystals, an “internal bias field” was found to exist and its ferroelectric hysteresis loop shifted to one side. Thus, at temperatures below T_c, ADTGSP maintained its saturation polarization state without any external field.

Strontium barium niobate (SBN) shows strong pyroelectric effects with $p = 550\ \mu C\ m^{-2}\ K^{-1}$. Like $LiTiO_3$, SBN is stable in air and does not need protective windows. The characteristics of SBN detectors such as the Curie temperature can be adjusted by changing the ratio of Sr/Ba. For pyroelectric applications, SBN crystals are frequently modified by doping with Pb and La. For example, doping with 0.01% Pb improves the pyroelectric properties of $Ba_{0.50}Sr_{0.50}Nb_2O_6$ [133] and $Ba_{0.52}Sr_{0.48}Nb_2O_6$ over undoped SBN; laser damages are also reduced. La-doped SBN has larger pyroelectric coefficients and dielectric constants and lower Curie temperatures. However, SBN is not as widely used as lithium tantalate because its advantage over the more readily available lithium tantalate is not always commercially apparent.

Major advances in the development of uncooled infrared imaging arrays have occurred in recent years. Buser and Tompsett [134] present an historical look at this development. Early infrared detector systems were single, cooled, detector elements mechanically scanned to capture an image. Reasonably fast collection rates of images required fast and sensitive detectors. Pyroelectric detectors did not have the required speed. Pyroelectric vidicon cameras [135] were the next advance and allowed the surface of a pyroelectric disk to be scanned at video rates by an electron beam in a cathode tube. Such cameras are useful devices for high temperature applications (e.g., scene viewing by firefighters).

However, commercial vidicon cameras failed to reach below the 1 K level of noise-equivalent temperature difference (NETD) at a target. An alternative approach is to use an array of detectors. But lower NETD requires better control of the thermal properties of the thermal detector. Thermal fluctuation noise defines the ultimate detection limit and it is proportional to the thermal conductance, $\sqrt{G}$. Lowering G requires thermally insulating the detector elements from each other and from the substrates, electrical connections, and housing. In addition, a large responsivity is needed to reduce the effect of background noise. Lowering the heat capacity, H, of the detector element will increase the temperature change and thus the response. Both G and H can be lowered by reducing the thickness of the detector structure. This is the basis for the strong push to develop infrared detector arrays based upon pyroelectric thin films. The ferroelectric-hybrid focal-plane array was developed in the 1980s. Detector elements made of lead strontium titanate (PST) or barium strontium titanate (BST) were commonly used [136]. These materials have Curie temperatures close to room temperature. The devices operated at temperatures above the Curie level (in the so-called paraelectric phase) and an applied electric field induced the pyroelectric effect. An excellent review of the state of the art in thin-film pyroelectric detectors is provided by Muralt [126]. As an example of what can be accomplished, Hanson et al. [137] demonstrated a thin-film pyroelectric device in 1997 with an NETD of 10 mK; arrays operated at ~500 mK NETD although the authors believed much improvement was still possible.

A key question under debate is whether thin film pyroelectric detectors can compete in the effort to further down-scale array pixel size, decrease NETD, and integrate them with silicon technology. Depositing thin films of pyroelectrics requires high temperatures and oxidative atmospheres which are incompatible with the silicon-based structures used. Still, some progress has been made in overcoming this issue with demonstration [138] of a 320×240 pixel thin-film pyroelectric array with an NETD of 210 mK. The resistive bolometric array is an alternative technology receiving much attention. Resistive devices use materials, such as vanadium oxide (not a ferroelectric), which change resistance with a change in temperature. Micromachining of microbolometers on silicon has resulted in impressive performance. Wood et al. [139] reported operation of a 240×336 pixel camera with an NETD of 39 mK in 1993.

6.4 Extrinsic Optical Properties and Their Applications

6.4.1 Extrinsic Optical Properties Via Dopants

Ferroelectrics have intrinsically very attractive optical properties that enable various functionalities in optical applications. This suite of functional groups can be optimized and further be extended by intentionally doping the material with a variety of ions. Areas of application for particular ions generally fall into the following:

- Increase or decrease of photorefractive sensitivities
- Active ions for laser and optical amplifier applications
- Modification of the refractive index for waveguide production

The ability to modify the material selectively by doping the material opens the possibility of realizing integrated optical devices by combining different functional groups on one chip. This capability has made $LiNbO_3$ and $LiTaO_3$ one of the main materials used in integrated optics, as seen in Table 6.6.In this integration, interaction of functional groups with each other and with the intrinsic properties need to be kept in mind. Examples of this are the reduction of the nonlinear coefficient in some proton exchanged H-doped regions [140,141] and the inhomogeneous broadening of optical transitions in the presence of Ti in Ti:$LiNbO_3$ waveguides [142]. Extrinsic dopants in the ferroelectric oxide materials of perovskite structure ($BaTiO_3$ and $KNbO_3$) have mainly been used to optimize the photorefractive properties of these materials.

TABLE 6.6 Functions of Different Ions in $LiNbO_3$ and $LiTaO_3$

Function	Ions
Active ions	Rare earth ions (Er, Pr, Yb, Nd, Ho), Cr^{3+}
Increase photorefractive sensitivity	Fe^{2+}/Fe^{3+}, Cu^{+}
Increase resistively against optical damage	Mg^{2+}, Zn^{2+}, In
Change refractive index for waveguide production	Ti, Zn, H

6.4.2 Dopants and Their Microscopic Structure

6.4.2.1 Lattice Site

In order to understand the function of a particular ion, it is essential to have an understanding of its incorporation into the lattice; therefore, this aspect has been studied for many ions. Most detailed results are available for $LiNbO_3$ and $BaTiO_3$ due to their promise and relevance for applications. Lattice sites have been determined by Rutherford backscattering techniques [143–145] and magnetic resonance [146–148]. While the first is usually only able to give information about the majority defect site, the latter technique requires that the dopant ions have an electronic spin. For ions with no free spins, changing the valence state by light illumination has often been used to determine their site [149].

Many of the dopants carry charges that are different from the ions that they are replacing, and for this reason, charge-compensating defect structures are required. The different ways to achieve charge neutrality leads to many different defect configurations around the dopant ion. Details of the charge compensation are in most cases very complicated and not completely clarified. The multi-site behavior has been studied extensively by site selective optical spectroscopy and magnetic resonance. In optical spectroscopy, the technique of combined excitation-emission spectroscopy, in which a large number of emission spectra are recorded for continuously varying excitation wavelengths, has proven to be very effective in the distinction of defect sites and "fingerprinting" of the energy levels of different defect sites [143,150].

In the study of defects that play a role in the photorefractive effect, the dopants appear in different charge states that can be manipulated by oxidation and reduction treatments as well as by light illumination. The number of different valences makes identification more complex. To tackle this problem, a combined optical/magnetic resonance technique was successfully applied. In this technique, electron spin resonance (EPR) and optical absorption spectra are obtained simultaneously after the sample has been exposed to light of continuously varying wavelengths. The resulting data set of absorption changes as a function of absorption and bleach wavelength can be interpreted by correlation with the EPR results. The current state of knowledge is summarized in the following section.

6.4.2.2 $LiNbO_3$ and $LiTaO_3$

In these ferroelectrics, metal ions are incorporated mostly on Li-sites with a slight displacement from the original Li-positions [151,152]. The requirement of charge compensation favors certain lattice environments for which different dopants are competing with each other. Once these sites are used up, other lattice sites are occupied by the dopants. For this reason, it is observed that some defects also occupy the Nb or the Ta-sites in particular, for higher doping levels and under co-doping with high levels of Zn [153,154] and Mg [155–158].

$LiNbO_3$ and $LiTaO_3$ are most easily grown in a congruent composition that has a significantly lower Li to Nb or Ta ratio than the stoichiometric one. To compensate for the resulting Li deficiency, a large number of intrinsic defects occur. The number and kind of intrinsic defects can be varied by co-doping and by the control of the stoichiometry [156]. The presence of these intrinsic defects has an influence on the incorporation of the dopant ions because they are the prime candidates for local and distant charge compensation. It has been observed that for low concentration of intrinsic defects, trivalent dopants (Fe, Cr, Er) [146,156,159] are incorporated preferentially without a local charge compensation. As the number of intrinsic defects increases, local charge compensation occurs, thus lowering the local C_{3v} symmetry of the dopant ions.

The role of the intrinsic defects in the charge compensation is further reflected in the observation that the number of trace impurities is reduced in the stoichiometric composition. In the absence of intrinsic defects, incorporation of dopants becomes more difficult. When doped intentionally, the requirement for charge compensation increases the number of intrinsic defects as the dopant concentration is increased. This again leads to additional defects that have local charge compensation. The increase of intrinsic defects with dopant level is also apparent in a shift of the band-edge absorption and increased coercive fields for doped crystals with nominally stoichiometric composition.

In most cases, the charge compensation is only characterized in terms of its effect on the local defect symmetry. Details of the charge compensation been obtained by electron-nuclear double resonance (ENDOR) methods for only for a few selected defect configurations [160].

Due to their small size, hydrogen ions defects do not follow the rules described above. In low concentration, at least three different H sites have been identified using decomposition of the IR absorption band [161], and Raman spectrum [162]. All of them exhibit a C_1 type symmetry as determined by polarized Raman spectroscopy and evaluation using the behavior-type method [162]. Based on these results, models have been proposed that assume the H bond is directed toward one of the six nearest O_2-ions located in an oxygen plane that is perpendicular to the *c*-axis. There are three different O–O bond lengths in this plane, giving rise to 6 different sites for the proton. It is assumed that the longest bonds are not occupied, leaving four different sites [163]. In the presence of other impurities, additional H-defects occur. The influence of intrinsic defects on the hydrogen defect is evident by the changes of bandshape and peak position with changes in the Li/Nb(Ta) ratio and with Mg-concentration [164]. The hydrogen stretch mode is further sensitive to the changes occurring during domain inversion, but does not play an active role in determining the required switching or intrinsic fields [165]. Hydrogen diffuses very easily even at fairly low temperatures (200°C). This is used, for instance, in the thermal fixing of photorefractive gratings, but also leads to the observation that crystals surfaces are often contaminated with H-related defects. For higher concentration of H defects as obtained in proton exchange in waveguide production, various different phases of $H_xL_{1-x}NbO_3$ are produced [166].

6.4.2.3 Perovskites

The most comprehensive studies of the various charged states [167–174] of an extrinsic defect in an oxide with cubic perovskite structure have been performed for Fe in $KTaO_3$. It exemplifies the complexity that is generally present for other hosts and defects as well. The Fe-defect in $KTaO_3$ occurs in numerous charge states (1+, 2+, 3+, 4+, 5+) and different charge compensators (oxygen vacancies, interstitial oxygen, local symmetry [cubic, axial, octahedral, rhombic], and lattice sites [Ta, K]). The large number of different defect types for a single dopant which additionally is easily changed by light illumination and chemical treatments (like reduction and oxidation) is a major challenge for the charge transfer processes, electron and hole trapping, and other processes involved in the photorefractive effect. It is no surprise that in most cases, a definite model is virtually impossible to achieve and that usually competing models co-exist.

Due to its importance as a photorefractive material, a large number of extrinsic defects ions have been investigated in $BaTiO_3$. Among the most important ones are the following: $Cr^{2+,3+,4+,5+}$ [147,175–177], $Fe^{2+,3+,4+,5+}$ [147,149,178–180], and $Rd^{2+,3+,4+,5+}$ [147,149]. In most cases, the lattice location is the Ti-site, and charge compensation is obtained by oxygen vacancies which can be nearest neighbors or somewhat farther away. The monovalent Ni, K, Na ions are examples that are located in the Ba-site [181,182] instead of the Ti-site. Hydrogen in $BaTiO_3$ is on an interstitial position within the TiO-plane. It introduces a shallow donor level [183].

6.4.3 Optical Properties of Doped Materials

6.4.3.1 Absorption and Refractive Index Changes

Incorporation of dopants induces, in most cases, additional optical transitions, which are either transitions between levels of the defects that are influenced by the hosts or charge transfer transitions in which the electron is excited out of the defects into levels of the host ions. For many defect ions, photo-ionization can occur. The majority of dopants induce additional absorption changes in the vicinity of the band-gap. For instance, in $LiNb(Ta)O_3$ the UV absorption edge moves to longer wavelengths with increasing concentrations. This effect is due to the reduction of the Li/Nb(Ta) ratio. In contrast to this behavior, the band edge moves to higher energy for Mg-doping. This effect is attributed to the reduction of Nb_{Li}.

Others dopants (e.g., Cr, rare earth ions) show discrete absorption and emission lines in the band gap of the host. In all these aspects, ferroelectrics are quite similar to the behavior of other insulating materials that are described in various textbooks [184]. The ferroelectric axis, however, breaks the inversion symmetry, giving rise to a preferential direction and to optical properties that are unique to the ferroelectrics:

1. The energy states are polarized, leading a linear Stark effect in the optical transitions that has been measured for several ions in $LiNbO_3$ [185,186] and which makes these transitions more sensitive to changes in the internal electric field. This property was exploited to use intentionally doped defects as probes for the intrinsic electric field in integrated optical device structures in $LiNbO_3$ [187,188].
2. In charge transfer transitions and optical transitions that induce photo-ionization, the intrinsic ferroelectric dipole gives a preferential direction for the motion of the free electrons and holes, giving a space charge field that leads in conjunction with the electro-optical to a photorefractive effect.

The introduced dopants have different polarizabilities compared to the host ions and therefore lead to changes in the refractive index. For heavier ions, the refractive index generally increases and the property can be exploited to create optical waveguides (see Section 6.4.7).

Through the requirement of charge compensations and their sensitivity to intrinsic electric fields, many dopants are very sensitive in their optical properties to the presence of intrinsic defects. This sensitivity can be utilized to determine the stoichiometry through the observed spectra for dopants. The hydrogen defect shows a characteristic change in its vibrational energy as function of stoichiometry [164]. Due to the increase in local order, the inhomogeneous linewidths of zero-phonon optical transitions of rare earth ions decrease [189]. Similarly, the resonance linewidth of paramagnetic ions in ENDOR spectra [146] is strongly reduced with the number of intrinsic defects.

6.4.3.2 Localized Excitations

The 4f-shell electrons of rare earth ions are only weakly disturbed by the host ions. For this reason, rather sharp and almost host independent absorption and emission lines are observed. The crystal environment just leads to a host and environment dependent (Stark) splitting of the multiplets, partially lifting their degeneracys according to the local symmetry [190,191]. As a result, small spectral shifts can be observed. On this basis, many site-selective spectroscopic studies have been conducted for these ions, in particular those that find application as active laser ions (Er, Yb, Nd, Pr). Common to all ions and hosts, several different environments for the ions can be distinguished by their differences in excitation and emission transitions (Figure 6.16). In $LiNbO_3$, at least six such environments can be identified [159,192,193]. At least four sites are found at least for $LiTaO_3$ [194]. Only a few, less detailed studies exist for $BaTiO_3$ [195].

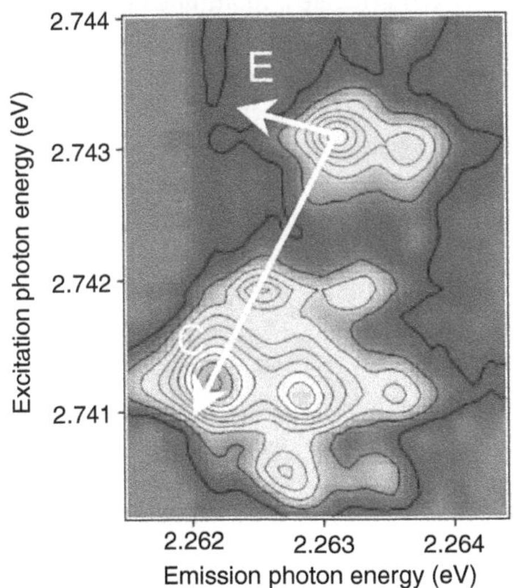

FIGURE 6.16 Combined site selective spectroscopy of Er ions in $LiNbO_3$. Each mountain in the image/contour plot indicates a different incorporation site of the ion. The arrows indicate the spectral shift of excitation and emission under application of an electric field (e) and hydrostatic pressure (c). (From Dierolf, V. et al. 2000. *Journal of Luminescence*, 87–89, With permission.)

Due to its potential application for wide band tunable optical amplifiers, IR lasers, and self-frequency doubled visible laser, Cr^{3+} has been investigated in several ferroelectrics ($LiNbO_3$, $LiTaO_3$, and $BaTiO_3$). Due to fairly efficient temperature quenching of the emission, this dopant has not fulfilled its promise yet. From a spectroscopic point of view, this ion is interesting since it represents an intermediate case between strong and weak crystal field coupling that leads to sharp (R-type) emission lines for some defect configurations (like ruby) and to broad emission (as in Cr: LiSAF) for others. Through application of hydrostatic pressure, the defects can be transferred from the weak to the strong coupling case [196].

6.4.4 Photoconductivity

The presence of defect-related energy levels in the energy gap between valence and conduction band leads to the possibility of light-induced excitation of electrons and holes, and therefore of photoconductivity through sub-bandgap illumination. Impurity and defect states have been used in photoconductivity-related experiments (mostly the photorefractive effect discussed below) in many ferroelectric materials, among them $LiNbO_3$, $LiTaO_3$, $BaTaO_3$, $KNbO_3$, KH_2PO_4, $KTa_{1-x}Nb_xO_3$, $K_3Li_2Nb_5O_{15}$, $Sr_xBa_{1-x}Nb_2O_6$, $Ba_2NaNb_5O_{15}$, $Ba_{0.77}Ca_{0.23}TiO_3$, and $Sn_2P_2S_6$, to cite only a few.

In the following, we use the example of oxygen octahedra perovkites to discuss the main features of photoconductivity in ferroelectric materials. In ferroelectric perovskites like $BaTiO_3$ or $KNbO_3$, the titanium or niobium atom (with a valence of 4+or 5+, respectively) is encaged in an octahedron consisting of 6 oxygen ions (with a valence of 2−). The 2p orbitals of the oxygen ions are completely filled, and they constitute the states in the upper part of the valence band. The 4d orbitals of the titanium ion (the 5d orbitals for niobium) are unoccupied and constitute the bottom edge of the conduction band. The distance between valence band and conduction band at room temperature in $BaTiO_3$ or $KNbO_3$ is larger than 3 eV. These materials are insulators and are transparent in the visible spectral range.

6.4.4.1 Energy Levels Connected to One Extrinsic Impurity

As previously mentioned for the case of Fe in $KTaO_3$ and for other materials and impurity atoms—and in contrast to the well known case of highly covalent semiconductors—in these dielectric oxides a single transition metal impurity can introduce a number of different electronic energy levels in the band gap [197–199]. The position of impurity energy levels in ABO_3 materials has been extensively investigated for different kinds of dopants, with a lot of early theoretical work concentrated on $SrTiO_3$ and on cubic and tetragonal $BaTiO_3$ [200–206], which showed how an impurity can generate a multitude of defect centers with different electron-binding energies. The necessity of charge compensation has been already mentioned. As an example, iron ions can enter the $BaTiO_3$ crystal lattice in place of the Ti ion. While a Fe^{4+} ion can directly substitute the Ti^{4+} ion, an Fe^{3+} ion requires charge compensation. Any difference in ionic charge between the introduced impurity and the Ti ion it replaces is normally easily compensated by the removal of oxygen atoms from the lattice during crystal growth, creating oxygen vacancies V_O^{2+}. Oxygen vacancies can be a next neighbor to the impurity atom (local charge compensation) but can also appear elsewhere in the lattice, and become active as electron traps that are generally relatively close (some fraction of an eV) to the edge of the conduction band.

The fact that one impurity atom leads to several energy levels, exacerbated by the possible presence of local charge-compensation, leads to a complex time-, intensity-, and wavelength-dependence of the photoconductivity. A general picture of the photoexcitation possibilities caused by an impurity atom in a ferroelectric can be drawn following the pioneering work of Wechsler and Klein [207]. It is presented in Figure 6.17. It describes an impurity atom A that is present in different valence states in the same crystal. The distance between the horizontal bars in this figure and the bottom of the conduction band is the minimum energy required to excite an electron localized on that impurity energy level into the conduction band (e.g., $A^{2+} \leftrightarrow A^{3+} + e$). They do not indicate the electron binding energies in the different ground and excited states of the impurity. A gray dot on a horizontal bar means that there is an electron that can be excited in such a way. The absence of a gray dot means that the impurity atom can trap an electron that will then have the given excitation energy. There are two levels labeled A^{2+}/A^{3+} in Figure 6.17 because the center A^{2+} can have two occupied electron states with different excitation energies and photoexcitation of either electron always leads, after thermalization, to an A^{3+} center. When a gray dot is present in the upper A^{2+}/A^{3+} level, we have an A^{2+} center, and there is no horizontal bar at the A^{3+}/A^{4+} energy. An empty bar in the upper A^{2+}/A^{3+} level represents an A^{3+} center, which can be excited according to the energy of the

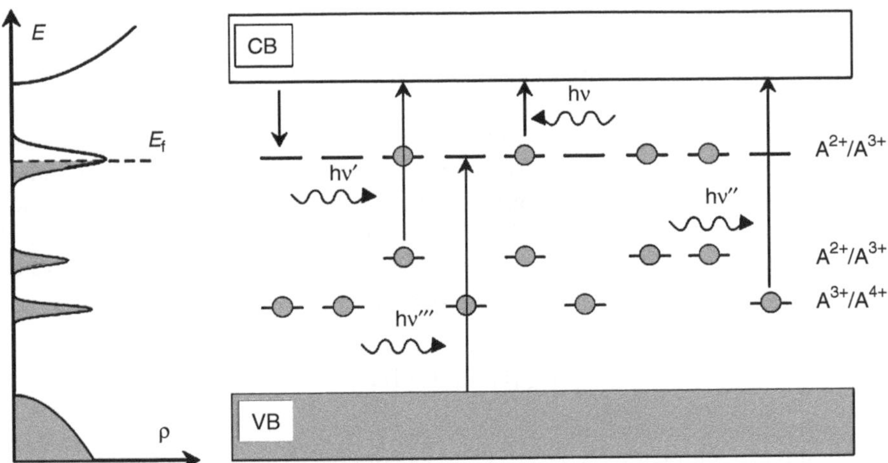

FIGURE 6.17 Qualitative representation of the intra band-gap energy levels introduced by an impurity atom *A*. On the left, a hypothetical electronic energy vs. density of states plot is shown; on the right is the corresponding energy band diagram. Each level represents a photoionization energy. The ability to trap an electron and to obtain a centre with such a photoionization energy is indicated by a line in the diagram on the right. The presence of a photoexcitable electron is indicated with a grey dot. The levels are labeled with the photoionization transition they represent.

A^{3+}/A^{4+}, to create an A^{4+} center and a free electron in the conduction band, and which is able to trap an electron to become the center A^{2+}. The excitation energy from the A^{3+} center corresponds to the A^{3+}/A^{4+} energy level, whose horizontal bars are all occupied by a gray dot in the figure, because we assume a thermodynamic equilibrium situation where the concentration of A^{4+} is zero and the A^{3+}/A^{4+} level is "full". This picture is typical of impurities in ferroelectrics. In wide band gap ionic crystals, impurity states are much more complex than what happens in covalent semiconductors.

The representation of Figure 6.17 allows a more direct understanding of the ionization and recombination processes that are of primary importance for understanding the wavelength and intensity dependence of the photoconductivity and the photorefractive effect. As an example, in Figure 6.17, three possible electron photoexcitation transitions are drawn. Which transition will be favored under illumination depends on the photoionization cross section of the impurities and on the wavelength of the light. The number of active donors is given by the number of filled centers that can be photoexcited using a certain wavelength. The number of active traps is given by the empty levels. In the example shown in the diagram of Figure 6.17, the center A^{3+} (in the upper A^{2+}/A^{3+} level) acts as an electron trap with a small ionization energy. This trap disappears if A^{3+} is photoionized to produce A^{4+}, which is an electron trap with much larger photoionization energy.

In thermal equilibrium, there can be only one partially filled level (the distance between different levels is generally greater than k_BT), and a given impurity center is expected to appear in the host crystal in at most two valence states. However, consider the example of Figure 6.17 under illumination. Then electrons can be photoexcited from A^{3+} centers and recombine on other A^{3+} centers to form the center A^{2+}. When the illumination is switched off, there will be A^{2+} as well as A^{3+} and A^{4+} centers present in the crystal. The crystal starts to approach thermal equilibrium when electrons are thermally excited from the A^{2+} state. The minimum time needed to reach equilibrium is limited by the probability for thermal excitation of the A^{2+} center. If the thermal excitation energy is substantially larger than k_BT, it may take days or months or even years for such a wide band gap crystal to reach thermal equilibrium. This example underlines an additional complexity typical of these wide bandgap dielectrics.

Above we described only the photoexcitation of electrons to the conduction band. The symmetric case of photoexcitation of an electron from the valence band to an empty level like A^{3+} is obviously also possible.

The changes in level occupation induced by this transition are analogous to what was described above in the case of transitions to the conduction band; we will not discuss them here any further.

One should note that the picture of Figure 6.17 only shows levels directly connected to the impurity atom. In reality, the picture is even more complex because of the necessity of charge compensation. As an example, iron in $KNbO_3$ can be present as a substituent for Nb in the valence states Fe^{2+}, Fe^{3+}, and Fe^{4+}, similar to the A atom in Figure 6.17. But the deficit of positive charge introduced by the iron with respect to the Nb^{5+} ion also implies the presence of oxygen vacancies for charge compensation, either isolated (V_O^{2+}), or in close proximity to the iron ion ($Fe^{3+}-V_O^{2+}$). This introduces a series of related defect levels which must fulfill a charge-compensation balance. In this example, the required concentration of isolated oxygen vacancies depends on the concentration of all other uncompensated defects and is

$$[V_O^{2+}] = \frac{1}{2}[V_O^{+}] + \frac{3}{2}[Fe^{2+}] + \frac{1}{2}[Fe^{2+}-V_O^{2+}] + [Fe^{3+}] \tag{6.29}$$

Note that for this example of $KNbO_3$, $Fe^{3+}-V_O^{2+}$ does not need charge compensation and therefore its concentration does not appear (for simplicity, we did not consider Fe^{4+}). In thermal equilibrium, Equation 6.29 can be simplified because several of the concentrations on the right-hand side must vanish. As an example, for a highly reduced crystal in thermal equilibrium $[Fe^{3+}]=0$, while for an oxidized crystal $[V_O]=0$. Anyhow, the requirement of charge compensation creates, for an individual impurity atom, a large family of energy levels that affect the photoconductive properties. Since charge compensation happens through oxygen vacancies, it is possible to shift the position of the chemical potential in the band gap by a reduction or oxidation treatment. At any given wavelength, this will change the density of centers that can be photoexcited to create electrons in the conduction band or holes in the valence band, and will change the type of the dominant charge carriers. Because of the difference between the mobilities of electrons and holes, this will also change the magnitude of the photoinduced conductivity and, for the photorefractive effect, its response speed (see below).

6.4.4.2 Photocurrent

Because of the presence of impurity levels, illumination with light energies lower than the band gap can lead to excitation of both electrons and holes into the conduction and valence band. In the presence of an applied field E the resulting current density

$$j_i = e(\mu_{ij}^{h} n_h + \mu_{ij}^{e} n_e)E_j, \tag{6.30}$$

depends on the densities n_h and n_e of photoexcited electrons and holes, respectively, and on their mobilities. The mobilities μ_{ij} are, in general, a second rank tensor that can be diagonalized to obtain the charge carrier mobilities along the three main axis of a ferroelectric crystal. This mobility can be highly anisotropic. As an example, a hole drifting in a direction perpendicular to the spontaneous polarization in $BaTiO_3$ has a mobility that is 20 times larger than for a hole drifting along it [208]. The anisotropy is less than a factor of 3 for $KNbO_3$, and the reason for its high value in $BaTiO_3$ is not known [208,209]. Depending on the length- and time-scale over which charge transport takes place, charge carriers can undergo many trapping and thermal excitation events from shallow traps while drifting. The densities and mobilities in Equation 6.30 can be either those of free carriers in the conduction or valence band, or they can be defined as the total density of carriers contributing to the charge transport and a trap-limited mobility that takes into account the trapping process [210–213]. The intrinsic electron mobility along the polar axis of $KNbO_3$ was found to be 0.5 $cm^2/(Vs)$ [214], and studies of the Hall mobility of electrons in $BaTiO_3$ have been recently performed between 120 and 220 K [215], but the intrinsic mobility of charge carriers in ferroelectrics and its properties are in general not well known, with values expected to be less than 1 $cm^2/(Vs)$ at room temperature, limited by small-polaron effects.

The densities of photoexcited holes and electrons depend on the structure of the impurity levels in the bandgap. When there is just a single donor level in the band gap the density n of electrons under illumination depends on the equilibrium between photoexcitation and recombination and is proportional to the light intensity,

$$n = \tau_0 \frac{\phi \alpha I}{h\nu} \tag{6.31}$$

where τ_0 is the carrier recombination time, α is the power absorption induced by the impurity level in the bandgap, ϕ is the quantum efficiency that an absorption process results in a free carrier, I is the light intensity, and $h\nu$ is the energy of a photon. In the presence of additional levels, photoinduced transfer from one level to the other is important and the carrier density can increase sublinearly with the light intensity [210,216–219]. Clearly, it is also possible to use short wavelength and interband photoexcitation of electrons and holes, which leads to a photoconductivity near the surface of a sample that increases as the square root of the light intensity increases.

6.4.5 Photogalvanic Effect

The photogalvanic effect leads to a photoinduced current in a ferroelectric crystal that is connected to an external circuit and illuminated by sub-bandgap light [220–224]. From the microscopic point of view, it can be understood as an asymmetric charge transfer process that takes place upon photoexcitation of an electron or hole from an impurity [220,221,225]. The charge displacement upon photoexcitation corresponds to a current. The contribution of a fixed density of impurities to the photogalvanic current should be proportional to the photoexcitation probability and should be linear in the light intensity. It follows that it has to be bilinear in the electric field vector of the light [225]. It can then be written as

$$j_i = \beta_{ijk} E_j^{\omega} E_k^{\omega}, \tag{6.32}$$

where β_{ijk} is a third rank tensor symmetric in its last two indices, j_i is the ith component of a current density vector $\mathbf{j}$, and E_i^{ω} are the components of the optical electric field. The photogalvanic tensor is a property of the dopants, and it is also proportional to the number of dopants and their photoexcitation cross-section. Sometimes the photogalvanic current is expressed in terms of a length L_{ph} (often called photovoltaic length) by $\mathbf{j} = ne\mathbf{L}_{ph}$, where n is the density of carriers excited per unit time, e is the unit charge, and the vector $\mathbf{L}_{ph}$ has modulus L_{ph} and a direction obtained from Equation 6.32. In $LiNbO_3$, β_{333} and β_{311} can be of the order of 10^{-7} V^{-1} [226], while L_{ph} can be of the order of 10–1000 pm.

The photogalvanic current produces a charge accumulation on the end faces of the crystal or in the shadow region outside the illuminated zone, which leads to an electric field in the illuminated regions. In steady state the photogalvanic current will be compensated by the drift current and the potential drop caused by this space-charge field leads to a photoinduced voltage. This is the reason why this effect has been often called the *photovoltaic* effect. However, the size of the photoinduced voltage depends not only on the properties of the impurities as described by Equation 6.32, but it also depends on the conductivity: the average amount of photoexcited free charge carriers and their mobility. The highest photovoltaic effects have been observed in materials with inefficient charge transport (low carrier mobilities and low photoconductivities) and they are not necessarily related to a large photogalvanic effect. In general, it is best to talk about the photogalvanic effect because the microscopic origin of the effect is a current.

6.4.6 Photorefractive Effect

In ferroelectric crystals, any electric field caused by a displacement of charges modulates the refractive index through the linear electro-optic (Pockels) effect. Any inhomogeneous illumination that leads to a spatially modulated charge carrier photoexcitation will create, through charge-carrier diffusion (or drift in an applied field), a redistribution of charge with associated space-charge electric field and refractive index modulation. This is the origin and meaning of photorefractive effect when the term is applied to this class of materials.

There is a vast literature in the form of textbooks [227,228] and collections of review articles [229–234] dedicated to the photorefractive effect for interested readers. The following is a discussion of the most basic principles.

6.4.6.1 Photoinduced Space-Charge Field in the Interference Pattern of Two Plane Waves

One of the most controlled ways to apply an inhomogeneous illumination to a crystal is to use the interference pattern between two intersecting laser beams, which can give a spatially modulated intensity

$$I(z) = I\,\mathrm{Re}[1 + m\exp(i\mathbf{k}\cdot\mathbf{r})] = I\,\mathrm{Re}[1 + m\exp(ikz)], \tag{6.33}$$

where I is the average intensity, m is a modulation index, and the wavevector $\mathbf{k}$ defines the direction (along the z coordinate) and spatial period of the interference pattern. The realization of this experimental configuration with sub-bandgap light leads, through photoinduced rearrangement of charges inside relatively deep impurity levels, to a refractive index modulation that in the simplest case is (like the interference pattern) in the form of a plane wave described by a wavevector $\mathbf{k}$. For small values of the modulation index, m, the photoinduced space charge field is of the form

$$E_{sc}(z,t) = \mathrm{Re}[E_{sc}(t)\exp(ikz)]. \tag{6.34}$$

For continuous wave (cw) illumination, the evolution of the complex amplitude of the photoinduced space-charge field after any small change in the illumination pattern (assumed to have a small modulation) is given by

$$E_{sc}(t) - E_{sc}(0) = E_{sc}[1 - \exp(-t/\tau)], \tag{6.35}$$

where τ is a photorefractive relaxation time that, in general, can become complex when an electric field is applied to the photorefractive crystal, leading to an oscillatory behavior in the build-up dynamics of the space-charge field amplitude. The same response time applies both to the build-up of a photorefractive grating and to its decay, as long as the average intensity and photoconductivity are kept equal in the two cases.

Even though there are potentially a multitude of levels that can be responsible for carrier photoexcitation and trapping, in thermodynamic equilibrium only one of the deep levels is partially occupied, while shallow levels with short thermal excitation lifetimes often only influence the speed of (photoinduced) charge redistribution in the deeper levels. Because of this, the main properties of the space-charge field can be described in terms of only one donor/trap level that in the dark and at zero temperature has N_D occupied centers and N_A empty centers. Under these assumptions of small modulation, cw illumination, and only one level responsible for charge carrier excitation and trapping, and with the additional assumptions that only one type of carrier is photoexcited and that no electric field is applied to the crystal, the steady state amplitude of the space-charge field and the relaxation time are

$$E_{sc} = \pm im\frac{k_B T}{e}k_0\frac{k_0/k}{1 + k_0^2/k^2} = \pm im\frac{E_q E_d}{E_q + E_d}, \tag{6.36}$$

$$\tau = \tau_{\mathrm{die}}\frac{1 + k^2/k_d^2}{1 + k^2/k_0^2}, \tag{6.37}$$

where the modulation index m depends on the contrast in the local conductivity inside the interference pattern. It is the same as in Equation 6.33 when the photoconductivity is much larger than the dark conductivity, but it decreases rapidly when the photoconductivity is not dominant anymore, leading to a small space charge field and the disappearance of the photorefractive effect in crystals with a larger conductivity. In Equation 6.36, the upper sign applies when the photoexcited charge carriers are positive and the lower one when they are negative. The other quantities used in Equation 6.36 and Equation 6.37 are

$$E_q = e\frac{N_{eff}}{k\varepsilon_0\varepsilon_{eff}} = \frac{k_B T}{e}\frac{k_0^2}{k} = E_d\frac{k_0^2}{k^2}, \tag{6.38}$$

$$E_d = \frac{k_B T}{e}k \tag{6.39}$$

$$N_{eff} = \frac{N_D N_A}{N_D + N_A}, \tag{6.40}$$

$$k_0^2 = \frac{e^2 N_{eff}}{\varepsilon_0 \varepsilon_{eff} k_B T}, \tag{6.41}$$

$$k_d^2 = \frac{e}{\mu\tau_0 k_B T}, \tag{6.42}$$

$$\tau_{die} = \frac{\varepsilon_0\varepsilon_{eff}}{\sigma}, \tag{6.43}$$

where i is the imaginary unit, k_B is Boltzmann's constant, T is the absolute temperature, e is the unit charge, ε_0 is the permittivity of vacuum, μ is the carrier mobility for drift in the z-direction, τ_0 is the lifetime of an average photoexcited carrier before being retrapped in the deep level, ε_{eff} is an effective dielectric constant that depends on the grating wavevector $\mathbf{k}$ (see below), and σ is the average conductivity in the interference pattern, which depends on the illumination intensity. In the simplest case $\sigma = n\mu e$, where the average electron density n is given by Equation 6.31. However, it should be noted that Equation 6.43 and Equation 6.37 often apply in general for any given light intensity also when the conductivity depends in a more complex way from the illumination.

The space-charge field amplitude and the photorefractive response time are plotted in Figure 6.18 as a function of the grating spacing. The space-charge amplitude is limited by both the trap-limited field E_q—which takes into account the limited number of energy levels available for charge redistribution, expressed by the effective density N_{eff}—and by the diffusion field E_d—which is the maximum field that can be reached by charge diffusion even when there are an infinite number of charges available for redistribution. The space charge field becomes small whenever the energy level in the band gap is almost full or almost empty and when the grating spacing $2\pi/k$ is either too small—so that the space-charge modulation is limited by the available trap density and the corresponding space charge field decreases with grating spacing—or too large, so that charge diffusion does not matter anymore. The maximum space-charge field is obtained at a grating spacing such that $k = k_0$, where k_0, given by Equation 6.41, is the inverse of the *Debye screening length*. A measurement of the space-charge field amplitude as a function of grating spacing (while controlling the modulation index m, which can be easily kept constant by polarizing the two beams that create the interference pattern perpendicular to the plane of incidence) is a much-used way to determine the effective trap density.

The response time, Equation 6.37, tends to the dielectric relaxation time, Equation 6.43, at large grating spacings, where the relaxation speed of a charge distribution is similar to the RC time constant of an electric circuit consisting of capacitors connected in series by wires with constant resistivity. The deviation from this limit at shorter grating spacings is described by the inverse diffusion length k_d, Equation 6.42, and by the Debye screening length. For $k_d > k_0$ (short diffusion lengths), the photorefractive response becomes faster towards shorter grating spacing. This is generally the case for most ferroelectrics. A measurement of the response time versus grating spacing can be used to determine the effective trap density as well as the mobility-lifetime product $\mu\tau_0$, which determines the diffusion length. A measurement of the response time in the long-grating spacing limit is a useful method to determine conductivity values all-optically for a specific drift direction, and without the need of applying electrical contacts to a sample [208].

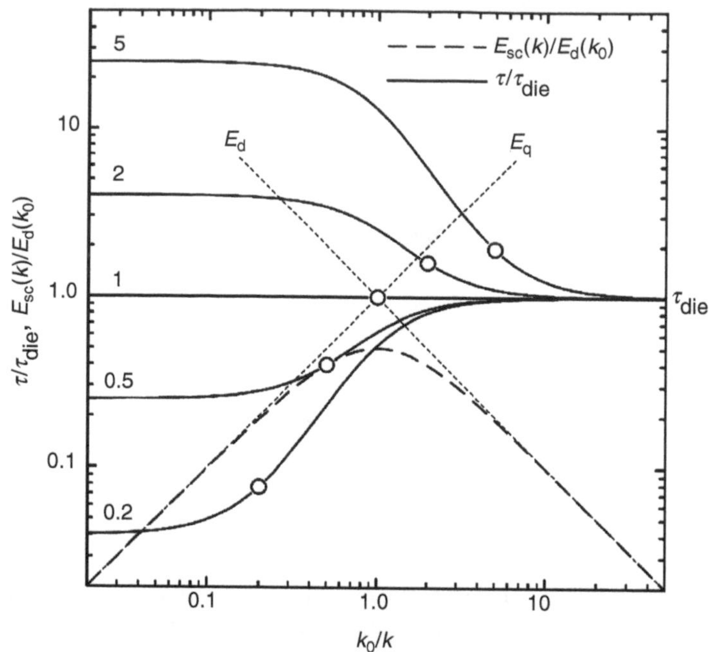

FIGURE 6.18 Photorefractive response time and space-charge field as a function of the normalized grating spacing k_0/k, expressed in units of 2π times the Debye screening length k_0^{-1}. The solid curves give the photorefractive response time, Equation 6.37, normalized to the dielectric relaxation time, Equation 6.43, for different values of the diffusion length k_d^{-1}. The curve parameter is the ratio k_0/k_d between diffusion length and Debye screening length, and the white circles indicate the grating spacings where $k=k_d$ for every curve. The response time at long grating spacings is in all cases given by the dielectric relaxation time. The dashed curve gives the space-charge field amplitude, Equation 6.36, normalized to the value of the diffusion field at $k=k_0$. The thin dashed curves give the two limiting fields E_d and E_q (see Equation 6.38 and Equation 6.39), also in units of the diffusion field at $k=k_0$.

In the presence of an applied electric field directed along the grating wavevector, both the space-charge field amplitude and the response time of the photorefractive effect are modified. The response time can become a complex number, leading to an oscillatory dynamic in the space-charge field. The space-charge field acquires a real part, and its phase-shift with respect to the illumination pattern is modified. At long grating spacings, it is no longer limited by the diffusion field. In this long grating-spacing limit, the photorefractive effect can be understood as a light-induced screening (through the photoexcited mobile charges) of the applied field. We refer to the literature for a deeper discussion of the effect of an applied electric field.

Even though the above description is limited to cw illumination and only considers the simplest situation, it contains all the important general features: a inhomogeneous illumination leads to a space-charge field that is based on a cumulative charge-redistribution, and as such, it can lead to comparatively large refractive index changes and light-light interaction effects even at very small light intensities (as long as they give a significant photoconductivity when compared to the dark conductivity). However, at small light intensities it will take a very long time (given by Equation 6.43) to reach the steady-state situation described by Equation 6.36. In this sense, and even though it leads to light-light interaction effects, this charge-transport mediated photorefractive effect cannot be put in the same class as off-resonant parametric third-order nonlinear optical processes [235] which are described by a direct multi-photon interaction and which have an essentially instantaneous (and intensity-independent) response time.

6.4.6.2 Photoinduced Refractive Index Modulation, Diffraction, and Amplification

The space-charge modulation induces a space-charge field that depends on the dielectric properties of the crystal. But since the space-charge is modulated like a plane wave in space, there are constraints on the strain pattern that it can induce in the crystal, corresponding to a boundary condition that partially clamps the crystal [236,237] and that is especially important in ferroelectrics. The effective dielectric tensor to be used to calculate the space-charge field from the space-charge modulation depends on the direction of the grating wavevector, and it is neither the stress-free nor the strain-free one. The scalar effective dielectric constant [237] used in the expressions above is the projection of a wavevector-dependent effective tensor $\varepsilon_{ij}(\mathbf{k})$ [238] on the grating wavevector, $\varepsilon_{\text{eff}}=\varepsilon_{ij}(\mathbf{k})k_ik_j/k^2$, and it depends on the elastic and piezoelectric properties of the crystal [237].

The same argument applies to the linear electro-optic Pockels tensor that gives the refractive index change as a function of the space-charge field. Because of the partial clamping of the crystal caused by the periodic modulation of the space-charge field, an effective, wavevector-dependent, electro-optic tensor [238] must be used to calculate the individual changes in the refractive index matrix as a function of the electric-field amplitude [237]. The effective electro-optic tensor $r_{ijk}(\mathbf{k})$ depends on the elastic, piezoelectric, and elasto-optic properties of the crystal.

The use of the effective tensors $\varepsilon_{ij}(\mathbf{k})$ and $r_{ijk}(\mathbf{k})$, when quantitatively analyzing the photorefractive effect and its time dynamics, is very important because they can be very different from either the strain-free or the stress-free tensors, and they have important geometry dependences that can lead to dramatic effects, such as the variations in the photorefractive response time observed in $BaTiO_3$ as a function of the direction of the grating wavevector [208]. The full data on the elastic, piezoelectric, electro-optic, dielectric, and elasto-optic properties, and tabulations of the values of the effective electro-optic coefficients, are available for $KNbO_3$ [237], $BaTiO_3$ [239], and $LiNbO_3$ [240].

The photoinduced refractive index grating can lead to Bragg diffraction. The complex amplitude of the space-charge field of Equation 6.36 is purely imaginary, which means that the phase grating induced by the space-charge field is shifted by $\pm\pi/2$ in the direction of the grating wavevector when compared to the position of the intensity pattern. When the two beams that create the interference pattern diffract from the phase grating induced by the space-charge field, they propagate in the same direction as the transmitted beams, and this $\pm\pi/2$ phase-shift adds to an additional $\pi/2$ phase-shift connected with Bragg-diffraction. As a consequence, one of the diffracted beams will be in phase with the transmitted beam, which it will coherently amplify, while the other diffracted beam will be in counterphase, causing an attenuation of the transmitted beam by destructive interference. This energy-transfer effect from one optical beam to the other is known as *two-beam coupling* or two-wave mixing. When a strong "pump" beam amplifies a weaker "signal" beam, the intensity of the latter grows exponentially with propagation length. Two-beam coupling can be used for coherent beam amplification as well as to determine the sign of the photoexcited charge carriers from the direction of energy transfer. If one of the transmitted beams is fed back into the crystal, it is possible to obtain a four-wave mixing configuration that leads to phase-conjugation [241,242] or to a ring-cavity with a mode sustained by photorefractive amplification.

In order to properly calculate diffraction from a phase grating in a ferroelectric material, it is important to take into account the anisotropy of its material parameters. As an example, this must be done in a dichroic material when calculating the diffraction from a refractive index grating [243] or when calculating the photoconductivity [243] from the illumination pattern. These anisotropy effects have a strong influence on photorefractive two-wave mixing [244,245].

There is a great variety of applications where diffraction and coherent amplification from refractive index holograms photoinduced through the photorefractive effect can be used, and they are extensively cited in the literature on the photorefractive effect [227–234]. Among them we mention dynamic holography, self-adapting detection systems [246], optical storage [247–249], multi-mode to single-mode conversion [250], and single-mode operation of lasers [251,252].

6.4.6.3 Generalized and Specialized Models of the Photorefractive Effect

The one-level, one-carrier model of the photorefractive effect discussed above can often be used at one particular illumination wavelength even though the influence of additional levels accessible with higher photon energies can be seen as a wavelength-dependent effective trap density [253]. It can also be used around a particular illumination intensity even though multiple levels would lead to a sublinear intensity dependence [210,216–219]. However, several important generalizations of this model have been introduced. The model has been expanded over the years to take into account the presence of more than one energy level in the band gap [217–219,254,255] or of shallow traps [210,212,213,256–258], the contribution of drifting ions at high temperatures [259–262], simultaneous excitation of both electrons and holes [261,263,264], direct interband photoexcitation of electrons and holes [265], and the use of short pulses [211,266–271].

An additional effect that is of special relevance to ferroelectrics is the possibility of domain inversion or of depolarization induced by the space-charge field [272–274]. The presence of a large effective trap density, Equation 6.40, leads to strong photorefractive effects at small grating spacings, which can be desirable for many applications based on coherent light amplification. However, the space-charge field obtained in this case can be so strong that it becomes larger than the coercitive field in some materials, leading to partial depolarization. As an example, the magnitude of the diffusion field, Equation 6.39, is 3.2 kV/cm for a grating spacing of 0.5 μm, while the critical field for depoling $BaTiO_3$ can be of the order of 0.5 kV/cm [272]. When an electric field is applied to the material, the combination of the space-charge field and of the applied field can lead to depoling at even longer grating spacings.

6.4.6.4 Space-Charge Field Photoinduced by a Single Beam: Photorefractive Self-Focusing and Solitons

A Gaussian beam propagating through a photorefractive crystal causes a inhomogeneous photoexcitation that, through charge transport, also leads to a space-charge field. The effect of the refractive index change induced by this space-charge field on beam propagation can assume different forms, depending on background illumination, the ratio between dark and photoconductivity, and the presence of applied fields or of the photogalvanic effect.

In the absence of photogalvanic effects or applied field, diffusion of photoexcited carriers away from the central part of the beam leads to a space-charge field that points to the center of the beam. The focusing and defocusing effects of the corresponding refractive index modulation will affect the beam profile. In addition, interference between the incoming beam and scattered light can lead to photorefractive gratings that grow from noise and amplify the light scattering. This produces a phenomenon known as photorefractive *beam fanning* that causes strong scattering to one side of a laser beam and effectively destroys the beam profile.

Under specially chosen conditions, however, the space-charge field induced by a single beam can have much more constructive effects. The simplest situation occurs in the presence of an applied field where the effect of the photoconductivity is simply to screen the field in the regions illuminated by the beam. Under the appropriate conditions, the screened region can have a higher refractive index than the surrounding regions, which leads to a photoinduced lens and to self-focusing. For appropriate illumination conditions, the self-focusing can compensate the diffraction effects in a stable way and give rise to spatial solitons. Photorefractive spatial solitons have been studied under many different circumstances and have been ordered into an ever-growing classification of effects. An overview of the field and of the many relevant publications can be obtained from [275,276], which present the topic from slightly different points of views. The interested reader should therefore consider working with both publications in order to get a full picture of the field of photorefractive solitons.

6.4.6.5 Figures of Merit

Over the years, many figures of merit have been introduced in order to give guidelines for the optimization of photorefractive materials for particular applications, or as a tool for comparing different materials. The difficulty in doing this lies in the fact that the main features and limitations of the photorefractive effect depend on how a specific experiment and application is set up. It is therefore not possible to compare different materials independently of the application they should be used for with one single figure of merit. A discussion of different possible figures of merit for the photorefractive effect and a comparison between different materials has been given by [277]. The following is a short review.

In general, one can say that for most applications it is good to have a large refractive index modulation, which is proportional to $n^3\ r_{eff}\ E_{sc}$, which means that a large refractive index and especially a large effective electro-optic coefficient are normally desirable qualities. For some applications, in many materials, these quantities are large enough, and it is not important to optimize the amplitude of the space-charge field by optimizing the grating spacing. Rather, one can work at longer grating spacings where the space-charge field is limited by the diffusion field. For other applications (like holographic storage), it may be more important to work at shorter grating spacings, where the space-charge field is limited by the trap-limited field, Equation 6.38. By looking at these two cases we can get a good idea of the material quantities involved in the photorefractive performance.

At long grating spacings, the space-charge field amplitude is limited by the diffusion field, Equation 6.39, which is a material-independent quantity, and therefore only the product $n^3 r_{eff}$ contributes to a figure of merit. In addition, a fast response is often desired either because it allows one to reach the steady-state more rapidly or also because one may need to produce a given refractive index amplitude with as little optical energy as possible. It is therefore necessary to include the response time in the figure of merit. Since we assumed above that the space-charge field was given by the diffusion field, then we can obtain a figure of merit by multiplying $n^3 r_{eff}$ with the inverse of the dielectric relaxation time, Equation 6.43. By keeping only the material dependent quantities, one finds a figure of merit that must depend on the product between n^3 $r_{eff}/\varepsilon_{eff}$, $\phi\mu\tau_0$, and the dopant-induced absorption α. The latter depends on the density of dopants and it can be tuned, so one can choose to define a figure of merit $\phi\mu\tau_0 n^3 r_{eff}/\varepsilon_{eff}$.

At short grating spacings, the space-charge field is limited by the trap-limited field, while the response time is limited by the "excitation time" $\tau_{ex} = \tau_{die} k_0^2/k_d^2 = h\nu\ N_{eff}/(\phi\alpha I)$. We get a contribution to a figure of merit of $n^3\ r_{eff} N_{eff}/\varepsilon_{eff}$ from the grating amplitude, and of $\phi\alpha/N_{eff}$ from the response time. Removing once again the absorption α, which depends on the dopant density, one ends up for this case with a figure of merit $\phi n^3 r_{eff}/\varepsilon_{eff}$, similar to that for long grating spacings, but without the mobility-lifetime product $\mu\tau_0$. The effective trap density disappears from the figure of merit because, in this limit, both the response time and the space-charge field amplitude become proportional to it.

For the cases where only the final amplitude of the grating matters, one should use the figure of merit $n^3 r_{eff}/\varepsilon_{eff}$ for long grating spacing applications and the figure of merit $n^3 r_{eff} N_{eff}/\varepsilon_{eff}$ for short grating spacing applications. If one also cares about the response time in order to obtain a large photorefractive sensitivity, the figure of merits are $\Phi\mu\tau_0 n^3 r_{eff}/\varepsilon_{eff}$ for long-grating-spacing applications and $\phi n^3 r_{eff}/\varepsilon_{eff}$ for short-grating-spacing applications. The combination $n^3 r_{eff}/\varepsilon_{eff}$ appears in all cases, but the mobility-lifetime product and the quantum efficiency for carrier generation are also important quantities because they determine the sensitivity. Reference [277] gives tables where these figure of merits are compared for various ferroelectrics and other materials.

It is important to note that this discussion of photorefractive sensitivity does not give a complete picture for the case where one is interested in *minimizing* the sensitivity in order to avoid the photorefractive effect. Making the photorefractive trap density smaller is one way to decrease the photorefractive sensitivity, but it may not be possible to reach a small enough value in the case where the light is modulated over a long spatial period (such as at long grating spacings or when the illumination comes from a single beam of several tens of micrometers in diameter). In this case the only way to avoid the photorefractive effect is to increase the conductivity of a material by doping it in such a way that any photoconductivity becomes a small disturbance on top of a large background conductivity, which correspondingly decreases any photorefractive space-charge accumulation caused by the illumination pattern.

6.4.7 Applications

6.4.7.1 Waveguides

The ability to modify the linear refractive index by doping the material allows the creation of waveguides. For this purpose, the doping has to be locally defined and can therefore not be done during crystal growth. Due to the requirement of fairly large samples with good optical quality, waveguides have mainly been studied in the more mature ferroelectric materials such as $LiNbO_3$, $LiTaO_3$, $KNbO_3$, $BaTiO_3$ and crystals of the KTP family. Waveguides find their application in electro-optical Mach–Zehnder type modulators, acousto-optical filters, integrated optical laser and nonlinear frequency conversion devices. For the latter, the increased interaction length and high confinement of the pump laser intensity leads to significantly reduced thresholds.

Several methods have been developed to allow post-growth incorporation of dopants. In-diffusion of an evaporated layer of metal (e.g., Ti) has proven to produce the waveguides with the lowest losses in $LiNbO_3$, $LiTaO_3$ [278,279]. Due to the slow diffusion, waveguides are generally limited to the surface.

Due to the higher diffusion coefficients of hydrogen, so-called proton exchange (PE), in which Li-ions are replaced by an equal number of H-ions when $LiNb(Ta)O_3$ crystals are placed in hot benzoic acid [280], offers larger versatility and allows embedded waveguides (through reverse proton exchange [281]); however, the changes in refractive index are complicated functions of concentration and crystalline direction such that only waveguides for one light polarization at a time can be produced. Before annealing, the electro-optical and nonlinear properties are adversely affected and must be restored by post-annealing [166]. This process is referred to as annealed proton exchange (APE). During the proton exchange, at least seven different phases of $H_xLi_{1-x}Nb(Ta)O_3$ can be produced, which have drastically different linear and nonlinear optical properties, thus making comparison of results from different groups often very difficult. A review of the current state of understanding is found in [166].

Ion-exchange is another method to introduce dopants in the crystal at relatively low temperatures and has been used for a variety of host materials. However, it should be noted that in $LiNbO_3$ and $LiTaO_3$ processes first assigned to ion-exchange of Ag and Tl have been identified as Li-outdiffusion and proton exchange processes [282–284].

The most versatile tool for post-growth doping and waveguide production is ion implantation. It has been used to create waveguides in a wide variety of ferroelectrics, including $LiNb(Ta)O_3$, $BaTiO_3$, KTP, and SBN. The implantation of the light elements ions (He, H, Ar), leads to a dose dependent contraction or expansion of the lattice in the damaged layer and hence to a change of refractive index [285]. As a result of the ion-implantation with light elements, a two-layered structure is formed. In the near surface layer, the crystal properties are only slightly modified by the interaction with an ion beam. This layer works as a light guiding region. At some depth the nuclear collisions become the dominant mechanism of energy transfer, and ion stopping is actively developed. Typically, a drastic index decrease happens at the depths near the projected range R_p, which is the depth of the maximum concentration of the implant. Therefore, the layer with the reduced index can be formed by nuclear processes, achieving the waveguide mode confinement between this buried layer and the surface [286]. In general, the chemical nature of the implanted ion is not critical. Since the waveguiding effect is caused by damage created during the deposition of kinetic energy, the energy lost per unit length is the decisive parameter. By tailoring the impact energy of the implanted ions, the depth of the buried layer can be controlled, and buried waveguides can be realized by producing two reduced index layers at different depths. Details of the mechanism for the reduction of the refractive index are still under discussion. Due to the fact that the implanted ions pass through the guiding zone, defects are introduced in this zone as well that will lead to increased guiding losses. Fortunately, annealing after the implantation reduces this drawback of the method. It is still fair to say that waveguide losses tend to be higher compared to other methods.

Waveguides for the KTP-family of ferroelectrics are produced mainly through ion exchange [287]. For fabricating channel waveguides, the patterned KTP sample is immersed into an ion exchange bath that consists of a molten nitrate salt of Rb, Cs, or TI or various mixtures of these salts and is heated to temperatures around 350°C. Buried waveguides in KTP have been produced by a combination of ion exchange and He ion implantation [288]. Table 6.7 lists waveguide production techniques for several host and dopant combinations.

6.4.7.2 Optical Damage and Countermeasures

The usefulness of $LiNb(Ta)O_3$ for optical applications that require high power in the visible is limited by the photorefractive effect that occurs due to the presence of trace ionic defects (such as Fe^{2+}/Fe^{3+}) even in nominally pure materials as well as due to the intrinsic defects related to any non-stoichiometry in the material. This optically induced change in refractive index, commonly referred to as optical damage, leads to distortions of light beams in the material and is very detrimental for stable device performance. A general strategy to reduce the optical damage effect is to increase the dark conductivity. This has been achieved by improving the stoichiometry [310] and by co-doping the material with high levels of certain metal ions (Mg, Zn, In). The improvement has been correlated with a reduction of $Nb(Ta)_{Li}$ antisites [311]. For the (Mg, Zn) a minimum (threshold) concentration is required to achieve an effective reduction of the optical damage. These values for congruent $LiNbO_3$ are 5.5 mol% MgO and 7 mol% ZnO. Doping with Indium already proves to be effective at low concentrations. It has also been observed that the incorporation of Ti ions for waveguide production increases the optical damage. This increase can be compensated by performing the Ti-indiffusion in a wet atmosphere [279].

6.4.7.3 Photorefractive Gratings and Filters

The photorefractive effect has been exploited to produce distribute Bragg reflectors and filters [312]. To produce such grating for instance in $LiNbO_3$, in which a section of the sample is doped with a photorefractive ion (such as Fe), the sample is illuminated with a spatially modulated light intensity produced by interference. This leads to a modulation in the $Fe^{3+/2+}$ concentration and hence to space charge fields. These fields produce a modulation in the refractive index. Dark conductivity limits the stability of the gratings. In order to produce more stable gratings, a process called thermal fixing [313] is exploited. In the process, the interferrometric illumination is performed at elevated temperatures that allow protons to diffuse in the space charge fields and adjust such that the $Fe^{3+/2+}$ pattern is compensated. At room temperatures, the ions are no longer mobile, and after homogeneous incoherent light illumination (that equalizes the $Fe^{3+/2+}$ ratio), a grating due to the spatial modulation of protons is obtained. Due to dark conductivity, these gratings are not long-term stable but can always be refreshed by light illumination. It has been found that by doping for $Cu^{+/2+}$, this adverse compensation process is less pronounced.

6.4.7.4 Lasers

Active ions can be incorporated and optical gain and laser emission have been achieved for several host materials and doping. Just by itself, ferroelectrics do not offer specific advantages as host materials, and hence simple cw free running bulk lasers are of little practical use and are seen just as proof of principle. The strength of the ferroelectric material is unleashed when the gain is combined with specific electro-optic, nonlinear, acousto-optic, and photorefractive properties; and unique integrated laser systems are realized in the bulk and integrated optical configuration. Combining the gain section with the properties of the ferroelectric material, the following laser types can be realized:

- Waveguide laser: all the following can be combined with the waveguide as well
- Self-frequency converting and doubling laser by combining the active ion with a QPM structure
- Mode locked and Q-switched lasers by combining the active ion with an electro-optical switch
- Narrow-band laser using distributed feedback or Bragg gratings that are produced by holographic exposure and thermal fixing, exploiting the photorefractive effect (see Figure 6.19)
- Tunable lasers, by placing an acousto-optical filter within the laser cavity

All these combinations have been realized in $LiNbO_3$ and some also in $LiTaO_3$. For these hosts, lasers have been demonstrated to exploit the rather narrow optical transitions of rare earth ions (Er^{3+}, Nd^{3+},...) as active ions. For the Nd ion, several of its potential laser transitions have been realized, while for Er so far only lasers for the 1.5-μm region exist. Attempts to realize up-conversion lasers (at 550 nm) and lasers

TABLE 6.7 Waveguide Production Techniques for Several Host Materials

Dopant	Incorporation Method	Δn	Modes	Reference	Comments
			$LiNbO_3$		
Ti	In-diffusion	$\Delta n_e \approx 0.03$, $\Delta n_o \approx 0.01$	TE, TM	[278, 279, 289]	Increases optical damage
Zn	In-diffusion	$\Delta n_o \approx 0.03$	TE, TM	[290]	High damage resistance
Ag, Tl	"Ion exchange"	$\Delta n_o \approx 0.1$	TE	[279, 283, 291, 292]	Due to Li outdiffusion or proton exchange and not due to Ag, Tl
H	Proton exchange	$\Delta n_e \approx 0.1$, $\Delta n_o \approx -0.04$	TE	[166]	Several structural phases with drastic different properties
H	Reversed Proton exchange	$\Delta n_e \approx 0.1$, $\Delta n_o \approx -0.04$	TM	[293–295]	TM-Buried waveguides possible
H, He, B	Ion Implantation	$\Delta n_e \approx -0.1$, $\Delta n_o \approx -0.1$	TE, TM	[296, 297]	Losses higher than in Ti diffused waveguides
			$LiTaO_3$		
Ti	indiffusion	$\Delta n_e \approx 0.024$, $\Delta n_o \approx -0.01$	TE, TM	[298]	Less favorable because diffusion temperature is above Curie temperature
H	Proton exchange	$\Delta n_e \approx 0.14$	TE	[166, 299, 300]	Several structural phases with drastic different properties exist
H	Reverse proton exchange		TM	[295]	TM-Buried waveguides possible
Ag, Tl	Ion exchange	$\Delta n_e \approx 0.02$	TE	[279, 291, 301]	Due to Li outdiffusion or proton exchange and not due to Ag, Tl
H, He	Ion Implantation	$\Delta n_o \approx -0.03$	TE, TM	[296, 302]	
			$BaTiO_3$		
H,He	Ion Implantation	$\Delta n_o \approx -0.1$, $\Delta n_e \approx -0.025$	TE, TM	[303]	
He	Ion Implantation	$n_o \approx -0.1$, $n_e \approx -0.1$	TE, TM	[304]	
			$KNbO_3$		
H, He	Ion Implantation	$\Delta n_x \approx -0.061$	TE, TM	[305, 306]	1 dB/cm loss
			KTP family		
Rb, Cs. Tl	Ion exchange	$\Delta n_{x,y,z} \approx -0.02$	TE, TM	[287]	Losses <0.4 dB/cm
He	Ion Implantation	$\Delta n_{x,y,z} \approx -0.08$	TE, TM	[288, 307]	Buried waveguides possible
			SBN		
He, B	Ion Implantation	$\Delta n \approx -0.07$	TE, TM	[308, 309]	Losses down to 0.2 db/cm

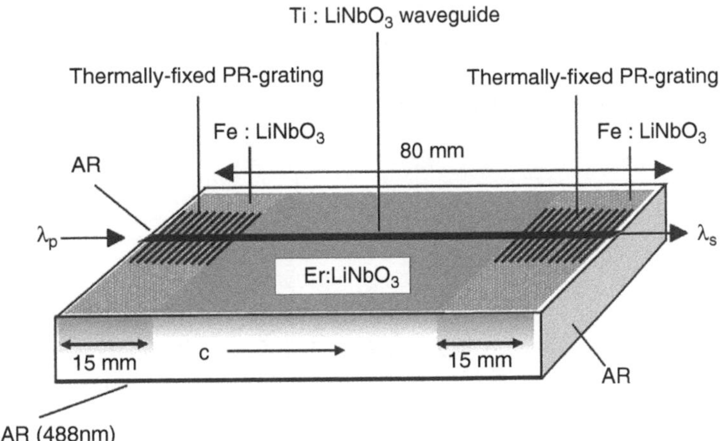

FIGURE 6.19 Schematic structure of an integrated optical DBR laser on $LiNbO_3$. (From Das, B. K., et al. 2001. *Applied Physics B.*, B73.)

utilizing the 3-μm transition have not yet been successful [315]. Optical amplification has been demonstrated for the wide band vibrational transition of Cr^{3+} ions [316]. However, once material with high optical quality is available, other ferroelectric hosts are suitable as well, as has been demonstrated for $BaTiO_3$ [195] by the observation of optical gain (Table 6.8).

6.4.7.5 Laser Beam Cleanup and Self-Adaptive Detection Systems Based on the Photorefractive Effect

The photorefractive effect in ferroelectric materials has been used to demonstrate a large variety of potential applications based on the diffraction properties of photorefractive grating and on the energy exchange and coherent amplification effects that they cause [227–234]. Some of these applications exploit the refractive index modulation that can be "written" by light, and the fact that it can be written inside the bulk of a thick crystal, to create holographic optical storage systems, for example [247–249]. Others rely on the energy exchange effects caused by the phase-shift between refractive index grating and light interference pattern to create "phase conjugating" mirrors that reflect an incoming intensity pattern back onto itself [241,242]. One of the most unique applications of the photorefractive effect exploits the very fact that the amplitude of the photorefractive grating is determined by an equilibrium that is established with a characteristic relaxation time. Depending on materials and light intensity, this time can range all the way between microseconds and seconds. When the wavevector or phase distribution in an illumination pattern changes with time, the photorefractive grating adapts to slow frequency changes in the input fields, but acts like a static hologram when the changes are faster. This can be used to build a detection system that adapts to changing conditions dynamically and that has a high-pass characteristics with respect to the frequency of an input "disturbance." As an example, by an appropriate choice of materials and illumination wavelength, the cut-off frequency could be chosen to be of the order of several kilohertzs, meaning that the holographic detection system dynamically adapts to any noise typical of acoustic and other vibrations. This principle has been at the center of various self-adapting detection systems where slower environmental influences can be filtered out while a fast signal is observed with great sensitivity. An example is detectors of ultrasound waves that adapt to any slow acoustic change [246]. The same self-adapting abilities can be used to convert the multi-mode output of a laser diode into a coherent diffraction limited beam [250] and to stabilize laser resonators [251,252].

TABLE 6.8 Lasers in $LiNbO_3$ and $LiTaO_3$[a]

Dopant	Laser Type	Performance Parameters [b]	Ref.
		$LiNbO_3$bulk	
Nd	cw WL: 1094 nm/547 nm/SH power 8 mW	Self-frequency doubled	[317, 318]
	Cw	WL: 929 nm/max:7 mW/SE: 6%, Zn-doped	[319]
Yb	self-frequency doubled	Low SH Power	[320]
		$LiNbO_3$waveguide lasers	
Er	cw, free-running	WL: 1561 nm/max: 60 mW/SE: 30%	[321, 322]
		WL: 1531 nm/max: 15 mW/SE: 21%	
	cw, acousto-optical tunable	WL:1550 nm/max: 0.3 mW/40nm tuning range	[323]
	Q-switched	PW: 4.5 ns/PP:2.5 kW	[323]
Er	Mode-locked	RR:10 GHz/PW:1565 nm/AP:10 mW	[324]
	DBR-laser	WL: 1560 nm/LW: 6 pm/max:1.5 mW	[314]
	DFB-laser	WL: 1531 nm/max:1 mW (using additional gain section)	[324, 325]
	Self-frequency doubling laser using QPM	WL: 1531 nm/max 3 mW and WL:765.5 nm/3 μW, Th87 mW	
	Cw, free running	WL:1084 nm/max.:100 mW/SE: 55%, Ti-waveguide	[150, 326–330]
		WL:1084 nm/max: 14 mW/SE: 34%, PE waveguides	
		WL:1084 nm/max: 0.2 mW/SE: 20%, Zn-diffused waveguides	
		WL: 1373 nm/max; 0.1mW/SE: 0.5%, Zn-diffused waveguides	
Nd	Cw, tunable, using E-O effect	WL: 1093 $\pm$ 1.2 nm	[331, 332]
	Q-switched	WL:1085 nm, PP: 350 W/PW 300 ps	[333]
	Mode-locked	WL: 1085 nm, PP:250 W/PW: 45 ps/RR: 6.7 GHz	[334]
Yb	Cw, free running	WL: 1008,1030,1060/max: 7 mW/SE 16%	[335]
		$LiTaO_3$bulk	
Nd	cw	WL: 1082 nm and 1092 nm/max; 11 mW/SE: 17%	[336]
		$LiTaO_3$waveguide laser	
Nd	Cw, free running	WL: 1078 nm and 1092nm, max; 3 mW/SE: 33%	[337, 338]

[a] The quoted performance parameters are typical values given in the quoted references and depend strongly on operation conditions.

[b] WL, Wavelength; max, Maximum reported output power; SE, slope efficiency; Th, threshold; PE, pulse width; PP, peak power; AW, average power; RR, repetition rate.

References

1. Lines, M. E. and Glass, A. M. 1977. *Principles and Applications of Ferroelectrics and Related Materials*, Clarendon Press, Oxford.
2. Gopalan, V. 1995. Structure–optical property correlation in ferroelectric potassium niobate thin films. PhD dissertation, Cornell University.
3. Janovec, V. 1972. Group analysis of domains and domain pairs. *Czechoslovak Journal of Physics, Section B*, 22, 974–994.
4. Newnham, R. E. 2005. *Properties of Materials: Anisortopy, Symmetry, Structure*. Oxford University Press, Oxford.
5. Erhart, J. 2004. Domain wall orientations in ferroelastics and ferroelectrics. *Phase Transitions*, 77, 989–1074.
6. Kittell, C. 2005. *Introduction to Solid State Physics, 8th Ed.*, Wiley, chap. 16.
7. Tuttle, B. A. et al. 1989. In *Ferroelectric potassium niobate thin films*. 91st Annual Meeting of the American Ceramic Society, Indianapolis, IN.
8. Wang, S. et al. 2000. Dielectric, ferroelectric properties of $KTa_{0:65}Nb_{0:35}O_3$ thin films prepared by sol–gel process on pt(111)/Ti/MgO(100) substrates. *Journal of Sol–Gel Science and Technology*, 17, 159–162.
9. Yogo, T. et al. 1995. Synthesis of highly oriented $K(Ta,Nb)O_3$ (Ta:Nb=65:35) film using metal alkoxides. *Journal of American Ceramic Society*, 78, 8, 2175–2179.
10. Haertling, G. H. 1971. *Journal of American Ceramic Society*, 54, 303–309.
11. Tian, L. et al. 2005. Real-time study of domain dynamics in ferroelectric $Sr_{0.61}Ba_{0.39}Nb_2O_6$. *Journal of Applied Physics*, 97, 114111(1–7).
12. Yoichiro, M. et al. 1995. Optical properties of $Ba_2NaNb_5O_{15}$ film fabricated by RF magnetron sputtering method. *Japanese Journal of Applied Physics*, 34, 5124–5131.
13. Xia, H. R. et al. 1997. Polarization and dipole moments of manganese-modified potassium sodium strontium barium niobate crystals. *Japanese Journal of Applied Physics*, 36, 1/4A, 2179–2182.
14. Barkla, H. M. and Finlayson, D. M. 1953. *Philosophical Magazine*, 44, 109–130.
15. Fugiel, B. et al. 2002. The transverse field effect in the triglycine sulphate ferroelectric. *Journal of Physics: Condensed Matter*, 14, 11837–11848.
16. Jiang, Q. et al. 2000. Ferroelectricity, conductivity, domain structure and poling conditions of rubidium titanyl phosphate. *Journal of Physics D: Applied Physics*, 33, 2831–2836.
17. Hu, Z. W. et al. 1996. Domain inversion in RTA using electric field poling. *Journal of Physics D: Applied Physics Rapid Communication*, 29, 6, 1681–1684.
18. Loiacono, G. M. et al. 1993. Crystal growth and characterization of ferroelectric $CsTiOAsO_4$. *Journal of Crystal Growth*, 131, 3–4, 323–330.
19. Loiacono, G. M. et al. 1993. *Journal of Crystal Growth*, 131, 323.
20. Dawley, J. T. et al. 2001. Sol–gel derived bismuth titanate thin films with *c*-axis orientation. *Journal of Sol–Gel Science and Technology*, 20, 85–93.
21. Subbarao, E. C. 1961. *Physical Review*, 122, 804–807.
22. Wemple, S. H. 1964. Tech. Rep. No. 425. Mass Inst. Technology, Research Laboratory Electronics.
23. Mitsui, T. et al. 1981. Landolt-Börnstein—Group III Condensed Matter, Ferroelectrics and Related Substances—Oxides, Vol. 36/Subvolume A, Springer-Verlag GmbH.
24. Kirkby, C. J. 1981. *Ferroelectrics*, 37, 567–570.
25. Newnham, R. E. et al. 1971. *Materials Research Bulletin*, 6, 1029–1040.
26. Taylor, G. W. 1970. *Ferroelectrics*, 1, 79–86.
27. Wu, S. Y. et al. 1976. Optical switching characteristics of epitaxial bismuth titanate films for matrix-addressed display. *Ferroelectrics*, 10, 209–213.
28. Barad, Y. et al. 2001. Probing domain microstructure in ferroelectric $Bi_4Ti_3O_{12}$ thin films by optical second harmonic generation. *Journal of Applied Physics*, 89, 1387–1392; Barad, Y. et al. 2001. *Erratum*, 89, 5230.

29. Abrahams, S. C. and Bernstein, L. J. 1967. Ferroelectric lithium fantalate—1. Single crystal x-ray diffraction study at 24°C. *Journal of Physics and Chemistry of Solids*, 28, 1685–1692.
30. Abrahams, S. C. et al. 1966. Ferroelectric lithium niobate. 4. Single crystal neutron diffraction study at 24°C. *Journal of Physics and Chemistry of Solids*, 27, 1013–1018.
31. Abrahams, S. C. et al. 1966. Ferroelectric lithium niobate. 5. Polycrystal x-ray diffraction study between 24° and 1200°C. *Journal of Physics and Chemistry of Solids*, 27, 1019–1026.
32. Abrahams, S. C. et al. 1966. Ferroelectric lithium niobate. 3. Single crystal x-ray diffraction study at 24°C. *Journal of Physics and Chemistry of Solids*, 27, 997–1012.
33. Gopalan, V. et al. 1988. The role of stoichiometry in 180° domain switching in $LiNbO_3$ crystals. *Applied Physics Letters*, 72, 1981–1983.
34. Tian, L. and Gopalan, V. 2004. Domain reversal in stoichiometric lithium tantalate prepared by vapor transport equillibration. *Applied Physics Letters*, 85, 4445–4447.
35. Kitamura, K. et al. 1998. Crystal growth and low coercive field 180° domain switching characteristics of stoichiometric $LiTaO_3$. *Applied Physics Letters*, 73, 21, 3073–3075.
36. Kitamura, K. et al. 1992. *Journal of Crystal Growth*, 116, 327–332 (see also *Ferroelectrics*, 21, 202, 1997).
37. Gopalan, V. et al. 2001. Crystal growth, characterization, and domain studies in ferroelectric lithium niobate and tantalate, In *Ferroelectrics and Dielectrics, Vol. 4*, H. S. Nalwa, ed., pp. 57–114. Academic Press, New York.
38. Jamieson, P. B. et al. 1968. *The Journal of Chemical Physics*, 48, 5048–5057.
39. Francombe, M. H. 1960. *Acta Crystallographica*, 13, 131–140.
40. Fang, C. S. et al. 1989. *Ferroelectrics*, 91, 349–356.
41. Fang, C. S. et al. 1989. ADTGSP single crystal with high pyroelectric figure of merit. *Ferroelectrics*, 91, 373–377.
42. Fang, C. S. et al. 1983. Pyroelectric properties of a new alanine and phosphate-substituted triglycine sulphate (ATGSP) crystal. *Materials Letters*, 2, 134–136.
43. Gopalan, V. et al. 2001. *Crystal Growth, Characterization, and Domain Studies in Ferroelectric Lithium Niobate and Tantalate, Vol. 4*, pp. 57–114. Academic Press, New York.
44. Takekawa, S. et al. 2001. Double crucible Stepanov technique for the growth of striation-free SBN single crystal. *Journal of Crystal Growth*, 229, 238–242.
45. Khachaturyan, O. A. 1987. Liquid-phase electroepitaxy of lithium niobate. *Pis'ma v Zhurnal Tekhnicheskoi Fizika*, 13, 1–2, 55–57.
46. Levy, M. et al. 1998. Fabrication of single-crystal lithium niobate films by crystal ion slicing. *Applied Physics Letters*, 73, 16, 2293–2295.
47. Tang, P. et al. 2004. Low-loss electrooptic $BaTiO_3$ thin film waveguide modulator. *IEEE Photonics Technology Letters*, 16, 8, 1837–1839.
48. Xie, H. and Raj, R. 1993. Epitaxial $LiTaO_3$ thin film by pulsed metalorganic chemical vapor deposition from a single precursor. *Applied Physics Letters*, 63, 23, 3146–3148.
49. Choi, K. J. et al. 2004. Enhancement of ferroelectricity in strained $BaTiO_3$ thin films. *Science*, 306, 1005–1009.
50. Chiu, Y. et al. 1999. Integrated optical device with second-harmonic generator, electrooptic lens, and electrooptic scanner in $LiTaO_3$. *Journal of Lightwave Technology*, 17, 462–465.
51. Gahagan, K. T. et al. 2001. Integrated high power electro-optic lens and large-angle deflector. *Applied Optics*, 31, 5638–5642.
52. Scrymgeour, D. A. et al. 2002. Cascaded electro-optic scanning of laser light over large angles using domain microengineered ferroelectrics. *Applied Physics Letters*, 81, 3140–3142.
53. Chen, X. et al. 2003. Optimal design of broadened flat bandpass electro-optic phase modulator based on aperiodic domain-inverted grating. *Journal of Optics A: Pure and Applied Optics*, 5, 3, 159–162.
54. Boyland, A. J. et al. 2001. Total internal reflection switching in electrooptically addressable domain-engineered LiNbO3. *Electronics Letters*, 37, 9, 585–587.

55. Tang, P. et al. 2004. Electrooptic modulation up to 40 GHz in a barium titanate thin film waveguide modulator. *Optics Express*, 12, 24, 5962–5967.
56. Petraru, A. et al. 2003. Integrated optical Mach–Zehnder modulator based on polycrystalline $BaTiO_3$. *Optics Letters*, 28, 24, 2527–2529.
57. Bernasconi, P. et al. 1995. Temperature dependence and dispersion of electro-optic and elasto-optic effect in perovskite crystals. *Journal of Applied Physics*, 78, 2651–2658.
58. Nagata, K. et al. 1983. Anisotropies of hot-pressed transparent (Pb, Ba, La)Nb_2O_6 ceramics. *Japanese Journal of Applied Physics*, 22, 9, 1353–1356.
59. Nagata, K. et al. 1986. In *Electrical, Optical and Switching Measurements on (Pb,La)(Zn,Nb,Zr,Ti)O_3 electro-optic ceramics*, CFI–Ceramic Forum International—Berichte der Deutschen Keramischen Gesellschaft, pp. 280–284.
60. Haertling, G. H. 1987. PLZT electrooptic materials and applications—a review. *Ferroelectrics*, 75, 1–2, 25–55.
61. Kitamura, K. et al. 2002. Single crystal of lithium niobate or tantalate and its optical element, and process and apparatus for producing an oxide single crystal. 6,464,777 B2, 15 October 2002.
62. Chen, F. S. et al. 1966. *Journal of Applied Physics*, 37, 388–398.
63. Boyd, R. W. 1992. *Nonlinear Optics*, Academic Press, New York.
64. Zernike, F. and Midwinter, J. E. 1973. *Applied Nonlinear Optics*, Wiley, New York.
65. Roberts, D. A. 1992. Simplified characterization of uniaxial and biaxial nonlinear optical crystals: A plea for standardization of nomenclature and conventions. *IEEE Journal of Quantum Electronics*, 28, 10, 2057–2074.
66. Armstrong, J. A. et al. 1962. Interactions between light waves in a nonlinear dielectric. *Physical Review*, 127, 6, 1918–1939.
67. Ito, H. et al. 1991. Fabrication of periodic domain grating in $LiNbO_3$ by electron beam writing for application of nonlinear optical processess. *Electronics Letters*, 27, 1221–1222.
68. Yamada, M. et al. 1993. First-order quasi-phase matched $LiNbO_3$ waveguide periodically poled by applying an external field for efficient blue second-harmonic generation. *Applied Physics Letters*, 62, 5, 435–436.
69. Burns, W. K. et al. 1994. *IEEE Photonics Technology Letters*, 6, 252.
70. Lim, E. J. et al. 1989. Second-harmonic generation of green light in periodically poled planar lithium niobate waveguide. *Electronics Letters*, 25, 174–175.
71. Jundt, D. H. et al. 1991. Periodically poled $LiNbO_3$ for high-efficiently second harmonic generation. *Applied Physics Letters*, 59, 2657–2659.
72. Lu, Y. et al. 1994. Blue-light generation by frequency doubling of an 810-nm cw GaAlAs diode laser in a quasi-phase-matched $LiNbO_3$ crystal. *Optics Letters*, 19, 1037–1039.
73. Myers, L. E. et al. 1995. Quasi-phase-matched 1.064-μm-pumped optical parametric oscillator in bulk periodically poled $LiNbO_3$. *Optics Letters*, 20, 1, 52–54.
74. Smith, R. G. et al. 1968. Continuous optical parametric oscillation in $Ba_2NaNb_5O_{15}$. *Applied Physics Letters*, 12, 9, 308–310.
75. Bosenberg, W. R. et al. 1996. 93% Pump depletion, 3.5-W continuous wave, singly resonant optical parametric oscillator. *Optics Letters*, 21, 17, 1336–1338.
76. Myers, L. E. et al. 1995. Quasi-phase-matched optical parametric oscillators in bulk periodically poled $LiNbO_3$. *Journal of the Optical Society of America B*, 12, 11, 2102–2116.
77. Myers, L. E. et al. 1996. Multigrating quasi-phase-matched optical parametric oscillator in periodically poled $LiNbO_3$. *Optics Letters*, 21, 8, 591–593.
78. Powers, P. E. et al. 1998. Continuous tuning of a continuous-wave periodically poled lithium niobate optical parametric oscillator by use of a fan-out grating design. *Optics Letters*, 23, 3, 159–161.
79. Missey, M. et al. 2000. Elliptical energy scaling with monolithic PPLN OPO's. *Proceedings of the SPIE*, 3928, 16–24.

80. Missey, M. J. et al. 1998. Diffusion-bonded stacks of periodically poled lithium niobate. *Optics Letters*, 23, 9, 664–666.
81. Russell, S. et al. 2001. In *Optical Parametric Generation of Greater Than 30 mJ Signal Energies in PPLN*, Conference on Lasers and Electro–Optics, Baltimore, MD, p. CPD24, Optical Society of America, Baltimore, MD, 2001.
82. Missey, M. et al. 2000. *Air Force Research Lab*, unpublished.
83. Meyn, J.-P. and Fejer, M. M. 1997. Tunable ultraviolet radiation by second-harmonic generation in periodically poled lithium tantalate. *Optics Letters*, 22, 16, 1214–1216.
84. Champert, P. A. et al. 2000. Efficient second-harmonic generation at 384 nm in periodically poled lithium tantalate by use of a visible Yb–Er-seeded fiber source. *Optics Letters*, 25, 17, 1252–1254.
85. Schepler, K. L. and Kokta, M. 1988. Spectroscopic properties of GdScAl garnet doped with Ti^{3+}. *Journal of Applied Physics*, 63, 960–962.
86. Hellstrom, J. et al. 2000. High-power optical parametric oscillation in large-aperture periodically poled $KTiOPO_4$. *Optics Letters*, 25, 3, 174–176.
87. Russell, S. et al. 2001. Broadband mid-infrared generation with two-dimensional quasi-phase-matched structures. *IEEE Journal of Quantum Electronics*, 37, 7, 877–887.
88. Dearborn, M. E. et al. 1998. Greater than 100% photon-conversion efficiency from an optical parametric oscillator with intracavity difference-frequency mixing. *Optics Letters*, 23, 759–761.
89. Kretschmann, H. M. et al. 1997. All-solid-state continuous-wave doubly resonant all-intracavity sum-frequency mixer. *Optics Letters*, 22, 1461–1463.
90. Arbore, M. A. et al. 1997. Engineerable compression of ultrashort pulses by use of second-harmonic generation in chirped-period-poled lithium niobate. *Optics Letters*, 22, 1341–1343.
91. Berger, V. 1998. Nonlinear photonic crystals. *Physical Review Letters*, 81, 19, 4136–4139.
92. Broderick, N. G. R. et al. 2000. HeXLN: A 2-dimensional nonlinear periodic crystal. *Physical Review Letters*, 84, 4345–4348.
93. Miller, R. C. and Nordland, W. A. 1970. Absolute signs of second-harmonic generation coefficients of piezoelectric crystals. *Physical Review B*, 2, 12, 4896–4902.
94. Miller, R. C. et al. 1963. Quantitative studies of optical harmonic generation in CdS, $BaTiO_3$, and KH_2PO_4 type crystals. *Physical Review Letters*, 11, 4, 146–149.
95. Eimerl, D. 1987. Electro-optic, linear, and nonlinear optical properties of KDP and its isomorphs. *Ferroelectrics*, 72, 95.
96. Buse, K. et al. 1993. Refractive indices of single domain $BaTiO_3$ for different wavelengths and temperatures. *Physica Status Solidi A*, 135, 2, K87–K89.
97. Biaggio, I. et al. 1992. Refractive indices of orthorhombic KNbO3. II. Phase-matching configurations for nonlinear-optical interactions. *Journal of the Optical Society of America B*, 9, 4, 507–517.
98. Shoji, I. et al. 1997. Absolute scale of second-order nonlinear-optical coefficients. *Journal of the Optical Society of America B*, 14, 9, 2268–2294.
99. Pack, M. V. et al. 2003. Measurement of the $\chi(2)$ tensor of the potassium niobate crystal. *Journal of the Optical Society of America B*, 20, 10, 2109–2116.
100. Zysset, B. et al. 1992. Refractive indices of orthorhombic $KNbO_3$ I. Dispersion and temperature dependence. *Journal of the Optical Society of America B*, 9, 3, 380–386.
101. Choy, M. M. and Byer, R. L. 1976. *Physical Review B*, 14, 1693–1706.
102. Smith, D. S. et al. 1976. Refractive indices of lithium niobate. *Optics Communications*, 17, 332–335.
103. Smith, D. S. et al. 1977. Errata. *Optics Communications*, 20, 108.
104. Jundt, D. H. 1997. Temperature-dependent Sellmeier equation for the index of refraction, n_e, in congruent lithium niobate. *Optics Letters*, 22, 20, 1553–1555.
105. Jundt, D. H. et al. 1990. Optical properties of lithium-rich lithium niobate pabricated by vapor transmport equilibration. *IEEE Journal of Quantum Electronics*, 26, 1, 135–138.
106. Miller, R. C. and Harris, S. E. 1968. Power and bandwidth of spontaneous parametric emission. *Physical Review*, 168, 1064–1068.

107. Bruner, A. et al. 2002. Refractive index dispersion measurements of congruent and stoichiometric $LiTaO_3$. *SPIE*, 4628, 66–73.
108. Bruner, A. et al. 2003. Temperature-dependent Sellmeier equation for the refractive index of stoichiometric lithium tantalate. *Optics Letters*, 28, 3, 194–196.
109. Singh, S. et al. 1970. Optical and ferroelectric properties of barium sodium niobate. *Physical Review B*, 2, 7, 2709–2724.
110. Ghosh, G. C. and Bhar, G. C. 1982. Temperature dispersion in ADP, KDP, and KD*P for nonlinear devices. *IEEE Journal of Quantum Electronics*, QE-18, 2, 143–145.
111. Eckardt, R. C. et al. 1990. Absolute and relative nonlinear optical coefficients of KDP, KD*P, BaB_2O_4, $LiIO_3$, MgO:$LiNbO_3$, and KTP measured by phase-matched second-harmonic generation. *IEEE Journal of Quantum Electronics*, 26, 5, 922–933.
112. Levine, B. F. and Bethea, C. G. 1972. Nonlinear susceptibility of GaP; relative measurement and use of measured values to determine a better absolute value. *Applied Physics Letters*, 20, 8, 272–275.
113. Suvorov, V. S. et al. 1968. Some nonlinear optical properties of crystals of KDP group. *Soviet Physics JETP*, 26, 33–37.
114. Suvorov, V. S. and Sonin, A. S. 1966. Nonlinear optical materials. *Soviet Physics Crystallography*, 11, 711–723.
115. Pack, M. V. et al. 2004. Measurement of the $\chi(2)$ tensors of $KTiOPO_4$, $KTiOAsO_4$, $RbTiOPO_4$ and $RbTiOAsO_4$ crystals. *Applied Optics*, 43, 16, 3319–3323.
116. Feve, J. P. et al. 2000. Refined Sellmeier equations from phase-matching measurements over the complete transparency range of $KTiOAsO_4$, $RbTiOAsO_4$ and $CsTiOAsO_4$. OSA. *Trends in Optics and Photonics*, 34, 575–577.
117. Vanherzeele, H. and Bierlein, J. D. 1992. Magnitude of the nonlinear coefficients of KTiOPO4. *Optics Letters*, 17, 982–984.
118. Bierlein, J. D. and Vanherzeele, H. 1989. Potassium titanyl phosphate: Properties and new applications. *Journal of the Optical Society of America B*, 6, 622.
119. Fan, T. Y. et al. 1987. Second harmonic generation and accurate index of refraction measurments in flux-grown $KTiOPO_4$. *Applied Optics*, 26, 2390.
120. Kato, K. 1974. High efficient UV generation at 3472 A in RDA. *IEEE Journal of Quantum Electronics*, QE-10, 622–624.
121. Cheng, L. K. et al. 1994. Crystal growth and characterization of $KTiOPO_4$ isomorphs from the self-fluxes. *Journal of Crystal Growth*, 137, 107–115.
122. Cheng, L. T. et al. 1993. Linear and nonlinear optical properties of the arsenate isomorphs of KTP. *Proceedings of SPIE*, 1863, .
123. Sonin, A. S. and Suvorov, V. S. 1967. Nonlinear optical properties of triglycine sulfate single crystals. *Soviet Physics Solid State*, 9, 1437.
124. Kruse, P. 1997. *Principles of Uncooled Infrared Focal Plane Arrays, Vol. 47*, Academic Press, New York.
125. Whatmore, R. W. 1986. *Reports on Progress in Physics*, 49, 1335–1386.
126. Muralt, P. 2001. Micromachined infrared detectors based on pyroelectric thin films. *Reports on Progress in Physics*, 64, 1339–1388.
127. Lang, S. B. 1974. *Sourcebook of Pyroelectricity*, Gordon and Breach Science, London.
128. Lehman, J. H. and Aust, J. A. 1998. *Applied Optics*, 37, 4210–4212.
129. Putley, E. H. 1970. Pyroelectric detectors (review), In *Semiconductors and Semimetals Vol. 5*, R. K. Willardson and Albert C. Beer, eds., pp. 259–285. Academic Press, New York.
130. Stokowski, S. E. 1976. Thermal modulation transfer function analysis of pyroelectric device characteristics. *Applied Optics*, 15, 7, 1767–1774.
131. Auston, D. H. and Glass, A. M. 1972. Optical generation of intense picosecond electrical pulses. *Applied Physics Letters*, 20, 10, 398–399.
132. Okuyama, M. et al. 1989. Pyroelectric infrared-CCD image sensor using $LiTaO_3$. *Sensors and Actuators*, 16, 3, 263–271.

133. Maciolek, R. B. and Liu, S. T. 1973. Preparations and properties of low loss $Sr_{1-x}Ba_xNb_2O_6$ ferroelectric single crystals. *Journal of Electronic Materials*, 2, 2, 191–200.
134. Buser, R. and Tompsett, M. F. 1997. Historical overview in uncooled infrared focal plane arrays. In *Semiconductors and Semimetals, Vol. 47,* Paul W. Kruse and David Skatrud, eds., pp. 1–14. Academic Press, New York.
135. Tompsett, M. F. 1997. Pyroelectric vidicon in uncooled infrared focal plane arrays, In *Semiconductors and Semimetals, Vol. 47*, Paul W. Kruse and David Skatrud, eds., pp. 219–225. Academic Press, New York.
136. Lang, S. B. 2005. Pyroelectricity: From ancient curiosity to modern imaging tool. *Physics Today*, 58, 8, 31–36.
137. Hanson, C. M. et al. 1998. Advances in monolithic ferroelectric uncooled IRFPA technology. *Proceedings of the SPIE—The International Society for Optical Engineering*, 3379, 60–68.
138. Tidrow, M. Z. et al. 1998. Uncooled infrared detectors and focal plane arrays. *Proceedings of the SPIE—The International Society for Optical Engineering*, 3553, 178–187.
139. Wood, R. A. 1997. Monolithic silicon microbolometer arrays in uncooled infrared focal plane arrays, In *Semiconductors and Semimetals*, Paul W. Kruse and David Skatrud, eds., *Vol. 47*, pp. 45–119. Academic Press, New York.
140. Cao, X. 1991. Recovery of second-order optical nonlinearity in annealed proton-exchanged $LiNbO_3$. *IEEE Photonics Technology Letters*, 3.
141. Korkishko, Y. N. 2000. SHG-response of different phases in proton exchanged lithium niobate waveguides. *IEEE Journal on Selected Topics in Quantum Electronics*, 6, 132–142.
142. Dierolf, V. 2001. Spectral line broadening mechanism of Er^{3+} transitions in Er:Ti:$LiNbO_3$ channel waveguides. *Applied Physics B: Lasers and Optics*, 73, 443–448.
143. Kling, A. 1999. Incorporation of hexavalent impurities into $LiNbO_3$. *Radiation Effects and Defects in Solids*, 150, 641–645.
144. Herreros, B. 1998. Photoluminescence and Rutherford backscattering spectrometry study of ion-implanted Er^{3+}-doped $LiNbO_3$ planar waveguides. *Journal of Physics: Condensed Matter*, 10, 3275–3283.
145. Rebouta, L. 1992. Lattice site for transition and rare-earth impurities in $LiNbO_3$ by ion-beam methods. *Journal of Materials Research*, 7, 130–135.
146. Malovichko, G. 1999. Interrelation of intrinsic and extrinsic defects—congruent, stoichiometric, and regularly ordered lithium niobate. *Applied Physics B: Lasers and Optics*, 68, 785–793.
147. Possenriede, E. 1992. Paramagnetic defects in $BaTiO_3$ and their role in light-induced charge transport. I. ESR studies. *Journal of Physics: Condensed Matter*, 4, 4719–4742.
148. Schirmer, O. F. 1991. Defects in $LiNbO_3$. I. Experimental aspects. *Journal of Physics and Chemistry of Solids*, 52, 185–200.
149. Schirmer, O. F. 2001. Combined EPR/optical investigations of light-induced charge transfers in photorefractive materials. *Optical Materials*, 18, 1–6.
150. Gill, D. M. 1992. Method for the local incorporation of Er into $LiNbO_3$ guided wave optic devices by Ti co-diffusion. *Applied Physics Letters*, 60, 1067–1069.
151. Hauer, B. et al. 1994. Lattice site of Ti in $LiNbO_3$. *Journal of Physics: Condensed Matter*, 6, 267–274.
152. Gog, T. et al. 1994. Lattice location of Ti and Er atoms in $LiNbO_3$: an X-ray standing wave study. *Ferroelectrics*, 153, 249–254.
153. Dierolf, V. et al. 2003. Comparative studies of Er^{3+} ions in $LiNbO_3$ waveguides produced by different methods. *Radiation Effects and Defects in Solids*, 158, 263–267.
154. Torchia, G. A. et al. 1998. Compositional effect on Cr^{3+} site distribution in MgO or ZnO codoped $LiNbO_3$:Cr congruent and stoichiometric crystals. *Journal of Physics: Condensed Matter*, 10, 341–345.
155. Kling, A. et al. 1998. Lattice site determination of Cr in low doped lithium niobate single crystals using PIXE/channeling. *Nuclear Instruments and Methods in Physics Research B*, 136/138, 426–430.

156. Malovichko, G. et al. 1999. Axial and low-symmetry centers of trivalent impurities in lithium niobate: Chromium in congruent and stoichiometric crystals. *Physical Review B: Condensed Matter*, 9113–9125.
157. Boker, A. et al. 1990. Two sites of Fe^{3+} in highly Mg-doped $LiNbO_3$. *Journal of Physics: Condensed Matter*, 2, 6865–6868.
158. Basun, S. A. et al. 1999. Novel luminescent center in $LiNbO_3$:Cr:Mg crystals. *Journal of Luminescence*, 83–84, 435–439.
159. Dierolf, V. and Koerdt, M. 2000. Combined excitation–emission spectroscopy of Er^{3+} ions in stoichiometric $LiNbO_3$: the site selectivity of direct and up conversion excitation processes. *Physical Review*, B65, 8043–8052.
160. Malovichko, G. et al. 2002. ENDOR study of Cr^{3+} centers substituting for lithium in lithium niobate. *Physical Review*, B65, 224116–224121.
161. Wohlecke, M. and Kovacs, L. 2001. OH^- ions in oxide crystals. *Critical Reviews in Solid State and Materials Sciences*, 26, 1–86.
162. Klauer, S. and Wohlecke, M. 1994. Incorporation of hydrogen in cubic and uniaxial oxidic crystals deduced from polarized Raman scattering. *Physical Review B: Condensed Matter*, 49, 158–181.
163. Cabrera, J. M. et al. 1996. Hydrogen in lithium niobate. *Advances in Physics*, 45, 349–392.
164. Kovacs, L. et al. 1984. Stoichiometry dependence of the OH^- absorption band in $LiNbO_3$ crystals. *Solid State Communications*, 52, 1029–1031.
165. Gopalan, V. et al. 1998. The role of nonstoichiometry in 180° domain switching of $LiNbO_3$ crystals. *Applied Physics Letters*, 72, 16, 1981–1983.
166. Korkishko, Y. N. et al. 2003. Proton exchanged $LiNbO_3$ and $LiTaO_3$ optical waveguides and integrated optic devices. *Microelectronic Engineering*, 69, 228–236.
167. Rytz, D. et al. 1982. Anomalous temperature effect on the Fe^{3+} cubic crystalline field splitting in $KTaO_3$ [EPR study]. *Journal of Physics C-Solid State Physics*, 15, 3371–3379.
168. Faust, B. et al. 1996. Optically detected magnetic resonance of Fe^{4+} OI in $KTaO_3$. *Solid State Communications*, 98, 445–447.
169. Reyher, H. J. et al. 2000. A magnetic circular dichroism and optically detected magnetic resonance investigation of Fe^{2+} and Fe^{3+} centres in $KTaO_3$. *Journal of Physics: Condensed Matter*, 12, 10599–10610.
170. Reyher, H. J. et al. 1996. ODMR and EPR investigations of Fe centers in $KTaO_3$. *Applied Physics B*, 63, 331–337.
171. Laguta, V. V. et al. 1985. Investigation of the width of ESR lines of axial Fe^{3+} centers in $KTaO_3$. *Fizika Tverdogo Tela*, 27, 162–166.
172. Bykov, I. P. et al. 1983. Electron spin resonance of axial Fe^{3+} centers in $KTaO_3$. *Fizika Tverdogo Tela*, 25, 3586–3590.
173. Abragam, A. and Bleaney, B. 1970. *Electron Paramagnetic Resonance of Transition Ions*. Clarendon, London.
174. Pechenyi, A. P. et al. 1992. Rhombic Fe^{3+} centers in $KTaO_3$. *Physica Status Solidi B*, 174, 325–333.
175. Schirmer, O. F. 1999. Light-induced charge transfer between defects in photorefractive materials. *Radiation Effects and Defects in Solids*, 149, 1–11.
176. Muller, K. A. et al. 1985. Paramagnetic resonance and local position of Cr^{3+} in ferroelectric $BaTiO_3$. *Physical Review*, B32, 5837–5844.
177. Possenriede, E. et al. 1990. Light sensitive paramagnetic defects in $BaTiO_3$. *Ferroelectrics*, 107, 313–318.
178. Mazur, A. et al. 1997. Optical absorption and light-induced charge transport of Fe^{2+} in $BaTiO_3$. *Applied Physics Letters*, 70, 2395–2397.
179. Sakudo, T. and Unoki, H. 1964. *Journal of Physical Society of Japan*, 19, 2109–2112.
180. Possenriede, E. et al. 1989. ESR identification of Fe containing defects in $BaTiO_3$. *Ferroelectrics*, 92, 245–252.
181. Scharfschwerdt, R. et al. 1996. Oxygen vacancies in $BaTiO_3$. *Physical Review*, B54, 15284–15288.

182. Lenjer, S. et al. 2000. An off-center ion near a Ba site in $BaTiO_3$ as studied by EPR under uniaxial stress. *Solid State Communications*, 116, 133–136.
183. Xiong, K. and Robertson, J. 2004. Hydrogen-induced defects and degradation in oxide ferroelectrics. *Applied Physics Letters*, 85, 2577–2579.
184. Henderson, B. and Imbusch, G. F. 1989. *Optical Spectroscopy of Inorganic Solids: Monographs on the Physics and Chemistry of Materials*. Clarendon, Oxford.
185. Skvortsov, A. P. et al. 1997. Stark effect on Cr^{3+} spectra in $LiNbO_3$. *Journal of Luminescence*, 72–74, 716–718.
186. Dierolf, V. et al. 2000. High-resolution site selective optical spectroscopy of rare earth and transition metal defects in insulators. *Journal of Luminescence*, 87–89, 989–991.
187. Dierolf, V. and Sandmann, C. 2003. In *Proceedings of 11th European Conference on Integrated Optics*. pp. 119–121.
188. Dierolf, V. and Ostendorf, A. 2001. Double-resonance excitation-emission spectroscopy on Er:Ti:$LiNbO_3$ waveguides. *Radiation Effects and Defects in Solids*, 155, 211–215.
189. Dierolf, V. and Koerdt, M. 1999. High resolution and site-selective optical spectroscopy of Er^{3+} ions in stoichiometric $LiNbO_3$ crystals and Ti:$LiNbO_3$ waveguides. *Radiation Effects and Defects in Solids*, 149, 323–328.
190. Henderson, B. and Bartram, R. 2000. *Crystal-Field Engineering of Solid-State Laser Materials*. Cambridge University Press, Cambridge.
191. Henderson, B. and Imbusch, G. F. 1989. *Optical Spectroscopy of Inorganic Solids*, Clarendon Press, Oxford.
192. Sole, J. G. et al. 1998. Rare earth and transition metal ion centers in $LiNbO_3$. *Spectrochimica Acta, Part A: Molecular and Biomolecular Spectroscopy*, 54A, 1571–1581.
193. Gill, D. M. et al. 1994. Site characterization of rare-earth-doped $LiNbO_3$ using total site selective spectroscopy. *Applied Physics Letters*, 64, 2483–2485.
194. Sokolska, I. et al. 1998. Optical spectroscopy of doped $LiTaO_3$ crystals. *Spectrochimica Acta, Part A: Molecular and Biomolecular Spectroscopy*, 54A, 1685–1694.
195. Teren, A. R. et al. 2002. Erbium-doped barium titanate thin film waveguides for integrated optical amplifiers, In *Proceedings of Ferroelectric Thin Films X*, pp. 413–415.
196. Basun, S. A. et al. 2001. Dominant Cr^{3+} centers in $LiNbO_3$. *Fizika Tverdogo Tela*, 43, 1010–1017.
197. Michel-Calendini, F. M. 1981. Impurity levels associated to Fe^{3+}–Vo, Fe^{4+}–Vo and Ti^{4+}–Vo centers in cubic perovskite crystals: Xa cluster analysis. *Ferroelectrics*, 37, 499–502.
198. Michel-Calendini, F. M. 1984. Electronic structure and term energy calculations for cubic and axial iron defects in $SrTiO_3$. *Solid State Communications*, 52, 167–172.
199. Moretti, P. and Michel-Calendini, F. M. 1988. Theoretical active impurity levels of iron, chromium, and manganese for photorefractive effects in barium titanate. *Journal of the Optical Society of America B*, 8, 1697–1701.
200. Morin, J. and Oliver, J. R. 1973. Energy levels of iron and aluminium in $SrTiO_3$. *Physical Review B*, 8, 5847–5854.
201. Siegel, E. and Müller, K. A. 1979. Structure of transition-metal-oxygen-vacancy pair centers. *Physics Review B*, 19, 109–120.
202. Müller, K. A. 1981. Paramagnetic point and pair defects in oxide perovskites. *Journal de Physique*, 42, 551–557.
203. Michel-Calendini, F. M. and Chermette, H. 1981. Incidence of ferroelectric phase transition on the electronic structure of potassium niobate. *Journal of Physics C*, 14, 1179–1192.
204. Michel-Calendini, F. M. and Moretti, P. 1983. Electronic Structures of Co(II) and Co(III) Impurities in Cubic Perovskite Hosts. *Physical Review B*, 27, 763–770.
205. Michel-Calendini, F. M. et al. 1985. Cubic to tetragonal phase transition effects on the electronic structures of pure and iron doped barium titanate. *Solid State Commununications*, 54, 951–956.
206. Klein, M. B. and Schwartz, R. N. 1986. The photorefractive effect in $BaTiO_3$: Microscopic origins. *Journal of the Optical Society of America B*, 3, 293–305.

207. Wechsler, B. A. and Klein, M. B. 1988. Thermodynamic point defect model of barium titanate and application to the photorefractive effect. *Journal of the Optical Society of America B*, 5, 1711–1723.
208. Bernasconi, P. et al. 1997. Anisotropy of the electron and hole drift mobility in $KNbO_3$ and $BaTiO_3$. *Physical Review Letters*, 78, 106–109.
209. Mahgerefteh, D. et al. 1996. Anisotropy of the hole drift mobility in barium titanate. *Physical Review B*, 53, 7094–7098.
210. Mort, J. et al. 1976. *Photoconductivity and Related Phenomena*. Elsevier, Amsterdam.
211. Pauliat, G. and Roosen, G. 1990. Photorefractive effect generated in sillenite crystals by picosecond pulses and comparison with the quasi-continuous regime. *Journal of the Optical Society of America B*, 7, 2259–2267.
212. Nouchi, P. et al. 1992. Temperature dependence of the electron mobility in photorefractive $Bi_{12}SiO_20$. *Journal of the Optical Society of America B*, 9, 1428–1431.
213. Biaggio, I. and Roosen, G. 1996. Influence of shallow traps on the enhancement of the photorefractive grating amplitude by a high-frequency alternating electric field: A probabilistic analysis. *Journal of the Optical Society of America B*, 13, 2306–2314.
214. Ewart, M. et al. 1994. Pulsed-photoexcitation studies in photorefractive $KNbO_3$. *Physical Review B*, 49, 5263–5273.
215. Kolodiazhnyi, T. et al. 2003. Thermoelectric power, hall effect, and mobility of n-type $BaTiO_3$. *Physical Review B*, 68, 085205-1–5.
216. Rose, A. 1978. *Concepts in Photoconductivity and Allied Problems*, R.E. Krieger Publishing Co., New York.
217. Holtmann, L. 1989. Model for the nonlinear photoconductivity of barium titanate ($BaTiO_3$). *Physica Status Solidi A*, 113, K89–K93.
218. Mahgerefteh, D. and Feinberg, J. 1990. Explanation of the apparent sublinear photoconductivity of photorefractive barium titanate. *Physical Review Letters*, 64, 2195–2198.
219. Brost, G. A. and Motes, R. A. 1990. Origin of the sublinear photorefractive response time in $BaTiO_3$. *Optics Letters*, 15, 1194–1196.
220. Glass, A. M. et al. 1974. High-voltage bulk photovoltaic effect and the photorefractive process in $LiNbO_3$. *Applied Physics Letters*, 25, 233–235.
221. Glass, A. M. et al. 1975. Excited state polarization, bulk photovoltaic effect and the photorefractive effect in electrically polarized media. *Journal of Electronic Materials*, 4, 915–943.
222. Belinicher, V. I. and Sturman, B. I. 1980. The photogalvanic effect in media lacking a center of symmetry. *Soviet Physics Uspekhi*, 23, 199–223.
223. Sturman, B. 1980. The photogalvanic effect-a new mechanism of nonlinear wave interaction in electrooptic crystals. *Soviet Journal of Quantum Electronics*, 10, 276–278.
224. von Baltz, R. et al. 1983. Nonlocal photovoltaic response function for the interpretation of hologram writing in ferroelectric crystals. *Applied Physics A*, 32, 13–18.
225. Sturman, B. I. and Fridkin, V. M. 1992. *The Photovoltaic and Photorefractive Effects in Noncentrosymmetric Materials*. OPA, Amsterdam.
226. Sommerfeldt, R. et al. 1988. Influence of Mg doping and composition on the light induced charge transport in $LiNbO_3$. *Physica Status Solidi A*, 106, 89–98.
227. Yeh, P. 1993. *Introduction to Photorefractive Nonlinear Optics*, Wiley, New York.
228. Solymar, L. et al. 1996. *The Physics and Applications of Photorefractive Materials*. Clarendon Press, Oxford.
229. Günter, P. et al. 1988. *Photorefractive Materials and their Applications I: Fundamental Phenomena*. Springer, Berlin.
230. Günter, P. et al. 1988. *Photorefractive Materials and their Applications II: Survey of Applications*. Springer, Berlin.
231. Nolte, D. D. ed. 1995. *Photorefractive Effects and Materials*, Kluwer Academic Press, Dordrecht, The Netherlands.

232. Yu, F. et al. 2000. *Photorefractive Optics: Materials, Properties, and Applications*. Academic Press, San Diego, CA.
233. Kuroda, K. ed. 2002. *Progress in Photorefractive Nonlinear Optics*, Taylor and Francis, London.
234. Günter, P. and Huignard, J. P., eds. 2006–2007. *Photorefractive Materials and their Applications, Vol. I–III*, Springer series in Optical Sciences, Springer Verlag, Berlin.
235. Hellwarth, R. W. 1977. Third-order optical susceptibilities of liquids and solids. *Progress in Quantum Electronics*, 5, 1–68.
236. Günter, P. and Zgonik, M. 1991. Clamped–Unclamped electro-optic coefficient dilemma in photorefractive phenomena. *Optics Letters*, 16, 1826–1828.
237. Zgonik, M. et al. 1993. Material constants of $KNbO_3$ relevant for electro- and acousto-optics. *Journal of Applied Physics*, 74, 1287–1297.
238. Biaggio, I. 1999. Nonlocal contributions to degenerate four wave mixing in noncentrosymmetric materials. *Physical Review Letters*, 82, 193–196.
239. Zgonik, M. et al. 1994. Dielectric, elastic, piezoelectric, electro-optic, and elasto-optic tensors of $BaTiO_3$ crystals. *Physical Review B*, 50, 5941–5949.
240. Jazbinsek, M. and Zgonik, M. 2002. Material tensor parameters of $LiNbO_3$ relevant for electro- and elasto-optics. *Applied Physics B*, 74, 407–414.
241. Hellwarth, R. W. 1977. Generation of time-reversed wavefront by nonlinear refraction. *Journal of the Optical Society of America*, 67, 1–3.
242. Zel'dovich, B. Y. et al. 1985. *Principles of Phase Conjugation*, Springer, Berlin.
243. Montemezzani, G. and Zgonik, M. 1997. Light diffraction at mixed phase and absorption gratings in anisotropic media for arbitrary geometries. *Physical Review E*, 55, 1035–1047.
244. Montemezzani, G. 2000. Optimization of photorefractive two-wave mixing by accounting for material anisotropies: $KNbO_3$ and $BaTiO_3$. *Physical Review A*, 62, 053803.
245. Montemezzani, G. et al. 1997. Charge carrier photoexcitation and two-wave mixing in dichroic materials. *Physical Review Letters*, 79, 3403–3406.
246. Dewhurst, R. J. and Shan, Q. 1999. Optical remote measurement of ultrasound. *Measurement Science and Technology*, 10, R139–R168.
247. Psaltis, D. and Burr, G. W. 1998. Holographic data storage. *Computer*, 31, 52–60.
248. Burr, G. W. 2003. Holographic storage, In *Encyclopedia of Optical Engineering*, R. B. Johnson and R. G. Driggers, eds., Marcel Dekker, New York.
249. Hesselink, L. et al. 2004. Holographic data storage systems. *Proceedings of the IEEE*, 92, 1231–1280.
250. MacCormack, S. et al. 1997. Powerful, diffraction-limited semiconductor laser using photorefractive beam coupling. *Optics Letters*, 22, 1–3.
251. Huot, N. et al. 1999. Laser mode manipulation by intracavity dynamic holography: Application to mode selection. *Applied Physics B*, 69, 155–157.
252. Maerten, S. et al. 2002. Laser diode made single-mode by a self-adaptive photorefractive filter. *Optics Communication*, 208, 183–189.
253. Delaye, P. et al. 1997. Wavelength dependent effective trap density in CdTe: Evidence for the presence of two photorefractive species. *Optics Commununication*, 134, 580–590.
254. Brost, G. A. et al. 1988. Intensity-dependent absorption and photorefractive effect in barium titanate. *Journal of the Optical Society of America B*, 5, 1879–1885.
255. Buse, K. and Krätzig, E. 1995. Three-valence charge-transport model for explanation of the photorefractive effect. *Applied Physics B*, 61, 27–32.
256. Strohkendl, F. P. 1989. Light induced dark decays of photorefractive gratings and their observation in $Bi_{12}SO_20$. *Journal Applied Physics*, 65, 3773–3780.
257. Biaggio, I. et al. 1990. Build-up and dark decay of transient photorefractive gratings in reduced $KNbO_3$. *Optics Commununication*, 77, 312–317.
258. Tayebati, P. and Mahgerefteh, D. 1991. Theory of the photorefractive effect for $Bi_{12}SO_20$ and $BaTiO_3$ with shallow traps. *Journal of the Optical Society of America B*, 8, 1053–1064.

259. Amodei, J. J. and Staebler, D. L. 1971. Holographic pattern fixing in electro-optic crystals. *Applied Physics Letters*, 18, 540–542.
260. Carrascosa, M. and Agulló-López, F. 1990. Theoretical modeling of the fixing and developing of holographic gratings in $LiNbO_3$. *Journal of the Optical Society of America B*, 7, 2317–2322.
261. Montemezzani, G. et al. 1993. Photorefractive charge compensation at elevated temperatures and application to $KNbO_3$ crystals. *Journal of the Optical Society of America B*, 10, 171–185.
262. Yariv, A. et al. 1996. Holographic storage dynamics in lithium niobate: Theory and experiment. *Journal of the Optical Society of America B*, 13, 2513–2523.
263. Strohkendl, F. P. et al. 1986. Hole electron competition in photorefractive gratings. *Optics Letters*, 11, 312–314.
264. Valley, G. C. 1986. Simultaneous electron/hole transport in photorefractive materials. *Journal of Applied Physics*, 59, 156–157.
265. Montemezzani, G. et al. 1994. Interband photorefractive effects: Theory and experiments in $KNbO_3$. *Physical Review B*, 49, 2484–2502.
266. Valley, G. C. 1983. Short-pulse grating formation in photorefractive materials. *IEEE Journal of Quantum Electronics*, QE-19, 1637–1645.
267. Partanen, J. P. et al. 1990. Direct determination of electron mobility in photorefractive $Bi_{12}SiO_{20}$ (BSO) by holographic time-of-flight technique. *Applied Physics Letters*, 57, 2404–2406.
268. Partanen, J. P. et al. 1991. Comparison between holographic and transient-photocurrent measurements of electron mobility in photorefractive $Bi_{12}SiO_{20}$. *Physical Review B*, 44, 1487–1491.
269. Biaggio, I. et al. 1992. Photorefractive effects induced by picosecond light pulses in reduced $KNbO_3$. *Journal of the Optical Society of America B*, 9, 1480–1487.
270. Biaggio, I. 2003. Holographic time of flight, In *Photoexcited Processes, Diagnostics and Applications*, A. Peled, ed., Kluwer Academic Publishers, Dordrecht.
271. Biaggio, I. et al. 1997. Band mobility of photoexcited electrons in $Bi_{12}SiO_{20}$. *Physical Reviews Letters*, 78, 891–894.
272. Cudney, R. S. et al. 1993. Photorefractive and domain gratings in barium titanate. *Applied Physics Letters*, 63, 3399–3401.
273. Cudney, R. S. et al. 1994. Enhancement of the amplitude and lifetime of photoinduced space-charge fields in multidomain ferroelectric crystals. *Physical Review Letters*, 72, 3883–3886.
274. Kewitsch, A. S. et al. 1995. Optical and electrical Barkhausen noise induced by recording ferroelectric domain holograms. *Journal of the Optical Society of America B*, 12, 1460–1467.
275. Stegeman, G. I. A. et al. 2000. Optical spatial solitons: Historical perspectives. *IEEE Journal of Selected Topics in Quantum Electronics*, 6, 1419–1427.
276. Królikowski, W. et al. 2003. Photorefractive Solitions. *Journal of Quantum Electronics*, 39, 3–12.
277. Montemezzani, G. et al. 2000. The photorefractive effect in inorganic and organic materials, In *Nonlinear Optical Effects and Materials*, P. Günter, ed., pp. 301–373. Springer, Berlin.
278. Stulz, L. W. 1979. Titanium in-diffused $LiNbO_3$ optical waveguide fabrication. *Applied Optics*, 18, 2041–2044.
279. Jackel, J. L. et al. 1981. Optical damage resistance of monovalent ion diffused $LiNbO_3$ and $LiTaO_3$ waveguides. *Journal of Applied Physics*, 52, 4855–4856.
280. Jackel, J. L. et al. 1982. Proton exchange for high-index waveguides $LiNbO_3$. *Applied Physics Letters*, 41, 607–608.
281. Fedorov, V. A. and Korkishko, Y. N. 1994. Reverse proton exchange in lithium tantalate crystals. *Journal of Optical Communications*, 15, 155–158.
282. Chen, Y.-X. et al. 1982. Characterization of $LiNbO_3$ waveguides exchanged in $TlNO_3$ solution. *Applied Physics Letters*, 40, 10–12.
283. Jackel, J. 1980. High-delta n optical waveguides in $LiNbO_3$: Thallium–lithium ion exchange. *Applied Physics Letters*, 40, 739–741.
284. Jackel, J. L. and Rice, C. E. 1982. Variation in waveguides fabricated by immersion of $LiNbO_3$. *Applied Physics Letters*, 41, 508–510.

285. Zolotoyabko, E. et al. 1996. Strain profiles in He-implanted waveguide layers of $LiNbO_3$ crystals. *Materials Letters*, 27, 17–20.

286. Atuchin, V. V. 2000. Causes of refractive indices changes in He-implanted $LiNbO_3$ and $LiTaO_3$ waveguides. *Nuclear Instruments and Methods in Physics Research, Section B: Beam Interactions with Materials and Atoms*, 168, 498–502.

287. Bierlein, J. D. and Vanherzeele, H. 1989. Potassium titanyl phosphate: Properties and new applications. *Journal of the Optical Society of America B*, 6, 622–633.

288. Schrempel, F. et al. 2004. Properties of buried waveguides produced by He-irradiation in KTP and Rb:KTP. *Nuclear Instruments and Methods in Physics Research B*, 218, 209–216.

289. Schmidt, R. V. and Kaminow, I. P. 1974. Metal-diffused optical waveguides in $LiNbO_3$. *Applied Physics Letters*, 25, 458–460.

290. Herreros, B. and Lifante, G. 1995. $LiNbO_3$ optical waveguides by Zn diffusion from vapor phase. *Applied Physics Letters*, 66, 1449–1452.

291. Jackel, J. L. 1980. In *Integrated and Guided-wave Optics Technical Digest, 28–30 January 1980*, Incline Village, NV, USA.

292. Shah, M. L. 1975. Optical waveguides in $LiNbO_3$ by ion exchange technique. *Applied Physics Letters*, 26, 652–653.

293. Korkishko, Y. N. et al. 1995. *Functional Photonic Integrated Circuits, February 9–10 1995*, San Jose, CA, *Vol. 2401*, pp. 149–150.

294. Jackel, J. I. and Johnson, J. J. 1991. Reverse exchange method for burying proton exchanged waveguides. *Electronics Letters*, 27, 1360–1361.

295. Korkishko, Y. N. et al. 1995. Reverse proton exchange in $LiTaO_3$ and $LiNbO_3$ for buried optical waveguides. In *Proceedings of the SPIE*, 2401, pp. 149–161.

296. Mignotte, C. et al. 1996. Fabrication by H^+ or He^+ ion implantation in lithium niobates and tantalates doped with rare earths and optical characterization. *Journal of Optics*, 27, 259–264.

297. Destefanis, G. L. et al. 1979. The formation of waveguides and modulators in $LiNbO_3$ by ion implantation. *Journal of Applied Physics*, 50, 7898–7905.

298. Tangonan, G. L. et al. 1977. High optical power capabilities of Ti-diffused $LiTaO_3$ waveguide modulator structures. *Applied Physics Letters*, 30, 238–239.

299. Hadi, K. E. et al. 1997. Ordinary and extraordinary waveguides realized by reverse proton exchange on $LiTaO_3$. *Optics Communications*, 140, 23–26.

300. Spillman, W. B. et al. 1983. Optical waveguides in $LiTaO_3$ formed by proton exchange. *Optics Letters*, 8, 497–498.

301. Jackel, J. L. 1980. Optical waveguides in $LiTaO_3$: Silver lithium ion exchange. *Applied Optics*, 19, 1996–1999.

302. Wang, K.-M. et al. 2001. Waveguide formation in $LiTaO_3$ and LiB_3O_5 by keV hydrogen ion implantation. *Optics Communications*, 196, 215–219.

303. Youden, K. E. et al. 1992. Photorefractive planar waveguides in $BaTiO_3$ fabricated by ion-beam implantation. *Optics Letters*, 17, 1509–1511.

304. Moretti, P. et al. 1990. Waveguides in barium titanate by helium implantation. *Physica Status Solidi A*, 117, 85–88.

305. Fluck, D. et al. 1993. Modeling of refractive index profiles of He^+ ion-implanted $KNbO_3$ waveguides based on the irradiation parameters. *Journal of Applied Physics*, 74, 6023–6031.

306. Beckers, L. et al. 1998. Potassium niobate waveguides: He^+ implantation in bulk single crystals and pulsed laser deposition of thin films. *Journal of Applied Physics*, A253, 292–295.

307. Bindner, P. et al. 1998. Refractive index behaviors of helium implanted optical planar waveguides in $LiNbO_3$, $KTiOPO_4$ and $Li_2B_4O_7$. *Nuclear Instruments and Methods in Physics Research*, B142, 329–337.

308. Chen, F. et al. 2002. Planar waveguides in Ce:SBN and Cu:KNSBN crystals by 6.0 MeV B^{3+} ion implantation. *Applied Surface Science*, 202, 86–91.

309. Kip, D. et al. 1995. Low-loss planar optical waveguides in strontium barium niobate crystals formed by ion-beam implantation. *Optics Letters*, 20, 1256–1258.
310. Holtmann, F. et al. 2004. Photorefractive properties of undoped lithium tantalate crystals for various composition. *Journal of Applied Physics*, 96, 7455–7459.
311. Volk, T. et al. 1994. Optical-damage-resistant impurities in lithium niobate. *Journal of the Optical Society of America B: Optical Physics*, 11, 1681–1687.
312. Muller, R. et al. 1993. In *Proceedings of the 16th Congress of the International Commission for Optics*, pp. 558–559.
313. Breer, S. et al. 1997. In *Proceedings of the 1997 Conference on Lasers and Electro-Optics*, 11, 9–10.
314. Das, B. K. et al. 2001. Single-frequency Ti:Er–$LiNbO_3$ distributed Bragg reflector waveguide laser with thermally fixed photorefractive cavity. *Applied Physics B*, B73, 439–442.
315. Dierolf, V. et al. 1999. Towards new lasers in Ti:Er:$LiNbO_3$ waveguides: A study of the excited Er^{3+} states. *Applied Physics*, B68, 767–775.
316. Vossler, G. L. et al. 1996. Chromium indiffused $LiNbO_3$ waveguide amplifier. In *Proceedings of the 1996 Conference on Lasers and Electro-Optics, CLEO'96*, pp. 531–532.
317. Zhang, K. et al. 1996. Laser-diode-pumped cw Nd:MgO:$LiNbO_3$ self-frequency-doubling laser around room temperature. *Applied Optics*, 35, 3200–3202.
318. Fan, T. Y. et al. 1986. Nd:MgO:$LiNbO_3$ spectroscopy and laser devices. *Journal of the Optical Society of America B: Optical Physics*, 3, 140–148.
319. Jaque, D. et al. 2004. Continuous-wave laser oscillation at 929 nm from a Nd^{3+}-doped $LiNbO_3$:ZnO nonlinear laser crystal: A powerful tool for blue laser light generation. *Applied Physics Letters*, 85, 19–21.
320. Montoya, E. et al. 1999. In *Technical Digest CLEO'99. Conference on Lasers and Electro-Optics.* pp. 90–91.
321. Baumann, I. et al. 1996. Ti:Er:$LiNbO_3$ waveguide laser of optimized efficiency. *IEEE Journal of Quantum Electronics*, 32, 1695–1706.
322. Brinkmann, R. et al. 1991. Continuous-wave erbium-diffused $LiNbO_3$ waveguide laser. *Electronics Letters*, 27, 415–417.
323. Baumann, I. et al. 1996. Er-doped integrated optical devices in $LiNbO_3$. *IEEE Journal of Selected Topics in Quantum Electronics*, 2, 355–366.
324. Becker, C. et al. 2000. Advanced Ti:Er:$LiNbO_3$ waveguide lasers. *IEEE Journal of Selected Topics in Quantum Electronics*, 6, 101–113.
325. Das, B. K. et al. 2003. Integrated optical distributed feedback laser with Ti:Fe:Er:$LiNbO_3$ waveguide. *Applied Physics*, 82, 1515–1517.
326. Paolo, R. E. D. et al. 2001. Continuous wave waveguide laser at room temperature in Nd^{3+}-doped Zn:$LiNbO_3$. *Applied Physics Letters*, 79, 4088–4090.
327. Lallier, E. et al. 1991. Nd:MgO:$LiNbO_3$ channel waveguide laser devices. *IEEE Journal of Quantum Electronics*, 27, 618–625.
328. Lallier, E. et al. 1990. Efficient Nd:MgO:$LiNbO_3$ waveguide laser. *Journal of Quantum Electronics*, 26, 927–928.
329. Lallier, E. et al. 1990. Nd:MgO:$LiNbO_3$ waveguide laser and amplifier. *Optics Letters*, 15, 682–684.
330. Hempstead, M. et al. 1992. Waveguide lasers operating at 1084 nm in neodymium-diffused lithium niobate. *IEEE Photonics Technology Letters*, 4, 852–855.
331. Domenech, M. and Lifante, G. 2004. Continuous-wave laser operation at 1.3 μm in Nd^{3+}-doped Zn:$LiNbO_3$ channel waveguides. *Applied Physics Letters*, 84, 3271–3273.
332. Amin, J. et al. 1994. Tunable coupled-cavity waveguide laser at room temperature in Nd-diffused Ti:$LiNbO_3$. *Optics Letters*, 19, 1541–1543.
333. Lallier, E. et al. 1993. Short pulse, high power q-switched Nd:MgO:$LiNbO_3$ waveguide laser. *Electronics Letters*, 29, 175–176.
334. Lallier, E. et al. 1991. Integrated Nd:MgO:$LiNbO_3$ FM mode-locked waveguide laser. *Electronic Letters*, 27, 936–937.

335. Jones, J. K. et al. 1995. Channel waveguide laser at 1 μm in Yb-indiffused $LiNbO_3$. *Optics Letters*, 20, 1477–1479.
336. Abedin, K. S. et al. 1995. Ordinary and extraordinary continuous wave lasing at 1.092 and 1.082 μm in bulk Nd:$LiTaO_3$ crystal. *Journal of Applied Physics*, 78, 691–693.
337. Nouh, S. et al. 1995. Fabrication parameter optimization of a low-threshold high-efficiency proton-exchanged waveguide laser in Nd:$LiTaO_3$. *Optics Letters*, 20, 1468–1470.
338. Sanford, N. A. et al. 1992. Nd:$LiTaO_3$ waveguide laser. *Optics Letters*, 17, 1578–1580.

IV

Materials—Organic

7

Second- and Third-Order Nonlinear Optical Materials

Larry Dalton
Philip Sullivan
Alex K.-Y. Jen

7.1 Introduction

Second- and third-order optical nonlinearity can perhaps be best understood as the coefficients of the second and third terms in the power series expansion of molecular and macroscopic polarization in terms of applied electric fields.

$$P_i = \alpha_{ij}E_j + \beta_{ijk}E_jE_k + \gamma_{ijkl}E_jE_kE_l + \ldots \tag{7.1}$$

$$P_i = \chi^{(1)}{}_{ij}E_j + \chi^{(2)}{}_{ijk}E_jE_k + \chi^{(3)}{}_{ijkl}E_jE_kE_l + \ldots \tag{7.2a}$$

$$P_i = \chi^{(1)}{}_{ij}E_{j\omega}\cos(\omega t - kz) + (1/2)\chi^{(2)}{}_{ijj}E_{jj\omega}[1 + \cos(2\omega t - 2kz)] + \ldots \tag{7.2b}$$

The terms β and γ represent the first and second molecular hyperpolarizabilities, whereas the terms $\chi^{(2)}$ and $\chi^{(3)}$ represent the second and third-order nonlinear material (macroscopic) susceptibilities. Each term arises from the nonlinear interaction of applied electric fields with the quasi-delocalized electron distribution of molecules and materials. Moreover, each of these terms can give rise to a variety of nonlinear responses reflecting different frequency dependences.[1–13] Second-order terms give rise to second harmonic generation (see Equation 7.2b), difference frequency (e.g., terahertz frequency) generation, optical rectification (see Equation 7.2b), and electro-optic modulation (the Pockels effect). Third-order optical nonlinearity gives rise to third harmonic generation, phase conjugation, optical limiting, optical parametric effects, and all-optical modulation (the Kerr effect). As is typical for power series expansions, the second-order coefficients are larger than the third-order coefficients. Although commercial applications have been realized for both second- and third-order inorganic materials, such as lithium niobate electro-optic and titanium/sapphire optical parametric materials, organic nonlinear optical materials are still struggling to obtain a beachhead on the commercial landscape. To the present time, second-order nonlinear optical organic materials appear closer to commercial application and will thus receive greater attention in this chapter.

It is very difficult to define a universal figure-of-merit (FOM) for either second- or third-order nonlinear optical materials, as practical device performance will often depend on the details of device design as well as intrinsic material properties. However, the most simplistic and yet somewhat realistic figure-of-merit can be expressed as $\mathrm{FOM}=\chi^{(2 \text{ or } 3)}/\tau\alpha$, where τ is the response time for the system reacting to an electric field perturbation and α is the optical loss. For π-electron organic materials, the response time, τ, is the phase relaxation time of the conjugated π-electrons, which is typically on the order of tens of femtoseconds. If device bandwidths are determined by the intrinsic material response time, bandwidths of tens of terahertz are possible. The optical loss, α, typically includes both absorption and scattering contributions. Optical loss will, thus, be influenced by material heterogeneity as well as molecular structure. Of course, practical device applications may also require many additional material properties including stability (e.g., thermal and photochemical) and processability (e.g., solubility in spin-casting solvent, appropriate sublimation temperatures for vapor deposition, appropriate glass transition temperatures for nanoimprint lithography). The FOM defined above is most commonly used in ranking third-order nonlinear optical materials. With second-order nonlinear optical materials, other factors such as the resistivity of metal electrodes used in electro-optic modulator devices can limit bandwidths, so the material FOM is commonly simplified to $\chi^{(2)}/\alpha$ or as $\chi^{(2)}/\alpha\varepsilon$, where ε is the material dielectric constant. For example, for electro-optic applications, it is important to match the velocity of propagating optical and radiofrequency waves. Velocity matching is optimized when $n^2=\varepsilon$, where n is the material index of refraction.

Quantum mechanical calculations have proven useful in investigating the relationship of β and γ to molecular (chromophore) structure.[14–24] For simple polyenes, the variation of molecular hyperpolarizability with the length of the conjugated π-electron structure and with bond length alternation is reasonably well predicted by theoretical calculations. Third-order nonlinear optical activity can be observed for both isotropic and anisotropic materials. From the above polarization equations, it can be seen that the symmetry requirement for second-order optical nonlinearity requires chromophores to exhibit either dipolar[23,24] or octupolar[25–27] symmetry. For materials to exhibit second-order optical nonlinearity, macroscopic dipolar or octupolar symmetry must exist; such symmetry is frequently introduced by electric field poling or by sequential synthesis/self assembly from a functionalized surface. Because of the additional symmetry requirement for second-order nonlinear optical materials, statistical mechanical calculations have proven useful in guiding the optimization of desired nano- and mesoscopic order.

In the following discussion, greater attention will be paid to second-order nonlinear optical organic materials than for third-order materials. The reason for this disproportionate focus relates to the fact that more commercial attention has been focused on second-order materials, whose design issues have therefore received more intense research scrutiny. For example, very little attention has been given to the optical loss and photostability of third-order nonlinear optical materials, whereas these properties have been extensively studied for a number of second-order materials.

7.2 Second-Order Nonlinear Optical Materials

The primary applications of second-order nonlinear optical organic materials include electro-optic modulation, second harmonic generation, optical rectification, and terahertz radiation generation and detection. With recent success in the development of visible wavelength lasers and light emitting diodes, second harmonic generation has received less attention. Also, absorption at visible wavelengths of second harmonic light by organic materials is problematic and has inhibited the practical utilization of materials for this application. An excellent review of second harmonic generation, including an extensive discussion of the concept of phase-matching, has been given by Stegeman.[28] This review remains highly relevant.

The most common application of organic second-order nonlinear optical materials is electro-optic modulation. Electro-optic modulation involves the application of a low (relative to optical frequencies) frequency electrical field to a material. This low frequency field is often referred to as a radiofrequency field, although actual frequencies can range from DC to tens of terahertz. The applied field perturbs the π-electron distribution of the material, which in turn alters the velocity of light propagating in the material. Thus, electro-optic activity can be viewed as voltage control of the refraction index of a material. The primary focus of this chapter will be a review of organic electro-optic materials and devices.

An increasingly popular application of second-order nonlinear optical materials is terahertz generation and detection. This phenomenon is relevant to a variety of sensing applications, ranging from homeland security to medical imaging. Organic materials also have potential for promoting the development and increased utilization of terahertz spectroscopy. Terahertz generation is an example of optical rectification or "difference frequency" phenomena. Like second harmonic generation, terahertz generation involves the interaction of two optical fields with the charge density of the material.

7.2.1 Electro-Optic Materials

Organic electro-optic materials include single crystal materials such as 4′-dimethylamino-*N*-methyl-4-stibazolium tosylate (DAST),[13,29] chromophore/polymer composite materials,[30–33] polymeric materials containing covalently incorporated chromophores (including heavily crosslinked materials),[30,31,34–40] single-chromophore-containing dendrimers,[24,41–44] multichromophore-containing dendrimers,[44,45] chromophore-containing dendronized polymers,[44,46–51] doped chromophore materials (including binary chromophore systems),[52] and materials prepared by sequential synthesis/self-assembly of chromophores from a functionalized surface by Langmuir–Blodgett or modified Merrifield techniques.[13,53–57] The vast majority of systems studied involve dipolar chromophores; the reader is referred elsewhere for an introduction to octupolar materials.[25–27] In like manner, most devices are currently prepared by electric field poling of polymeric or dendritic materials. For materials prepared by electric field poling or by sequential synthesis/self-assembly, only two nonzero electro-optic tensor elements (r_{33} and r_{13}) exist. These are given approximately by:

$$r_{33} = 2Nf(0)\beta_{zzz} < \cos^3\theta > /(n_{0e})^4 \quad (7.3a)$$

$$r_{13} = Nf(0)\beta_{zzz} < \sin^2\theta \cos\theta > /(n_{0o})^2(n_{0e})2, \quad (7.3b)$$

where N is the chromophore number density (molecules/cm^3), $f(\omega)$ are local field factors that account for the dielectric nature of the media surrounding chromophores, n_{0o} and n_{0e} are the ordinary and extraordinary linear refractive indices, and the order parameters, $<\cos^3\theta>$ and $<\sin^2\theta \cos\theta>$, define the orientational distribution of chromophores. Equation 7.3 neglects the minor elements of the molecular first hyperpolarizability tensor.

7.2.1.1 Optimizing Electro-Optic Activity

The process of optimizing electro-optic activity is typically a two-step process, in which quantum mechanical calculations are used to guide the improvement of molecular first hyperpolarizability,β_{ijk}, values[14–24] and statistical mechanical calculations are employed to optimize the product of the chromophore number density and the order parameter.[58–64] Quantum mechanical calculations have guided the development of a number of chromophores, leading to dramatic improvements in molecular first hyperpolarizability, i.e., to β_0 values on the order of or greater than 1000×10^{-30} esu. The tricyanovinylfuran (TCF) acceptor moiety shown in Scheme 7.1 has facilitated the development of workhorse chromophores appropriate for prototype device development.[65–72] In recent years, the synthesis of chromophores has been greatly aided by the utilization of microwave-assisted synthesis techniques.[73,74] Of course, for a chromophore to be carried forward for development of device-appropriate materials, it must exhibit appropriate thermal, chemical, and photochemical stability, acceptably low levels of absorption at anticipated device operational wavelengths, and appropriate processability (e.g., solubility in spin casting solvents, etc.). These features will be addressed later in this chapter; note that the comments that apply for electro-optic materials also apply to other second-order materials and, to some extent, to third-order materials.

The theoretically-inspired development of chromophores with improved molecular first hyperpolarizability can be divided into two categories: (1) variations of the fundamental donor, bridge, and acceptor blocks of modular dipolar chromophores; and (2) investigation of novel chromophore architectures such as "X-shaped"[18,19,75,76] and "twisted"[20,21] chromophores. The former strategy has proven to be very effective in the past and significant future improvement may be possible following this strategy. The latter strategy is much newer, but may afford dramatic improvements in molecular first hyperpolarizability, while permitting desirable auxiliary properties such as high transparency at operating wavelengths.

Molecular first hyperpolarizability (β) values are commonly measured by hyper-Rayleigh scattering (HRS).[77–79] Such measurements are complicated by two-photon fluorescence and by molecular aggregation. A variety of modifications have been made to he HRS technique in efforts to circumvent these problems, including the use of femtosecond pulse techniques, measurements at a number of wavelengths (using laser wavelength agility afforded by optical parametric devices), and measurements as a function of concentration with measured β values determined exploiting extrapolations to zero concentration.[77–79] Moreover, HRS measurements are normally carried out to determine relative (to a standard solvent such as chloroform) β values. Absolute values are most frequently defined using an integrating sphere approach. A problem arises in the comparison of β values between theory and

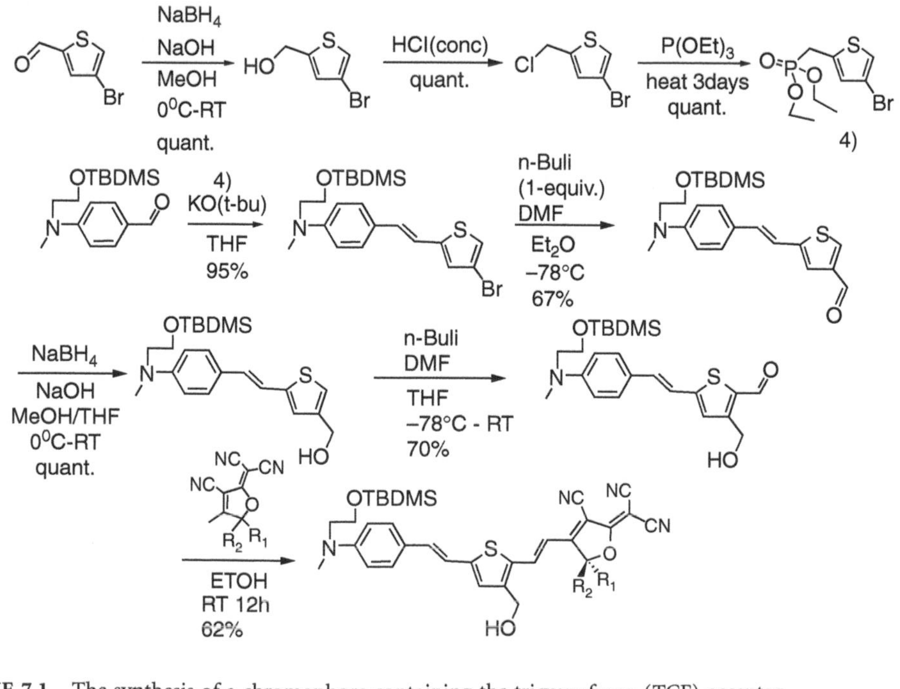

$NaBH_4$, NaOH, MeOH, 0^0C-RT, quant.
HCl(conc), quant.
$P(OEt)_3$, heat 3days, quant.
4)
4), KO(t-bu), THF, 95%
n-Buli (1-equiv.), DMF, Et_2O, −78°C, 67%
$NaBH_4$, NaOH, MeOH/THF, 0^0C-RT, quant.
n-Buli, DMF, THF, −78°C - RT, 70%
ETOH, RT 12h, 62%

SCHEME 7.1 The synthesis of a chromophore containing the tricyanofuran (TCF) acceptor.

experiment. Theoretical β values are calculated for isolated particles in the long wavelength or zero frequency limit. On the other hand, experimental β values are measured in solutions of varying dielectric constants and at finite infrared wavelengths such as 1.9 μm. To avoid comparing data corresponding to different conditions, relative β values are frequently compared, e.g., both theoretical and experimental β values referenced to a standard such as paranitroaniline.[23]

The product of dipole moment, μ, and molecular first hyperpolarizability, β, is also a useful quantity, particularly as the slope of the curve of r_{33} versus N in the low concentration limit is given by $2f(0)[\mu\beta]E_p/5kTn^4$, where for the sake of simplicity the subscripts on β and n have been dropped. The product $\mu\beta$ can be measured by electric field induced second harmonic generation (EFISH) techniques. Like HRS measurements, many factors can complicate the measurements and great care must be exercised to obtain meaningful data. If molecular first hyperpolarizability values relevant to electro-optic response are to be extracted from EFISH data, then care must be exercised to avoid multiphoton resonance contributions. Again, like HRS measurements, EFISH measurements should ideally be made at a number of wavelengths.

The second aspect of optimizing electro-optic activity involves optimizing the product $N\langle\cos^3\theta\rangle$. It is now well-appreciated that intermolecular electrostatic (e.g., dipole–dipole) interactions involving prolate ellipsoid-shaped π-electron chromophores can lead to serious attenuation of electro-optic activity.[58–64] The Monte Carlo calculations in Figure 7.1 illustrate this point. These calculations were carried out with the restriction that the chromophores maintain a uniform lattice distribution. Different results are obtained if a non-uniform chromophore distribution is permitted. This latter treatment permits aggregation effects to be taken into account. The discussion of this latter treatment, which can be

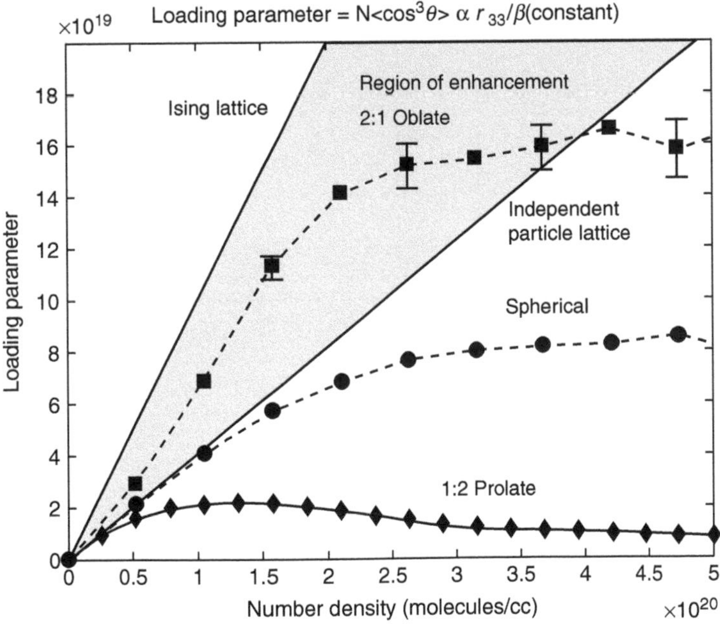

FIGURE 7.1 Calculation (employing pseudo-atomistic Monte Carlo methods) of $N\langle\cos^3\theta\rangle$ vs. N for chromophore shapes ranging from prolate to oblate ellipsoidal. Calculations are based on an "on-lattice" approximation, which means that chromophores are restricted to uniform spacing among chromophores. Different results, particularly at high loading, are obtained for "off-lattice" calculations, in which chromophores are permitted to assume a non-uniform lattice distribution (e.g., to aggregate). For off-lattice calculations, results are much more sensitive to the details of chromophore structure. Pseudo-atomistic calculations mean that π-conjugated segments are treated in the united atom approximation whereas σ-bonded molecular fragments are treated fully atomistically.

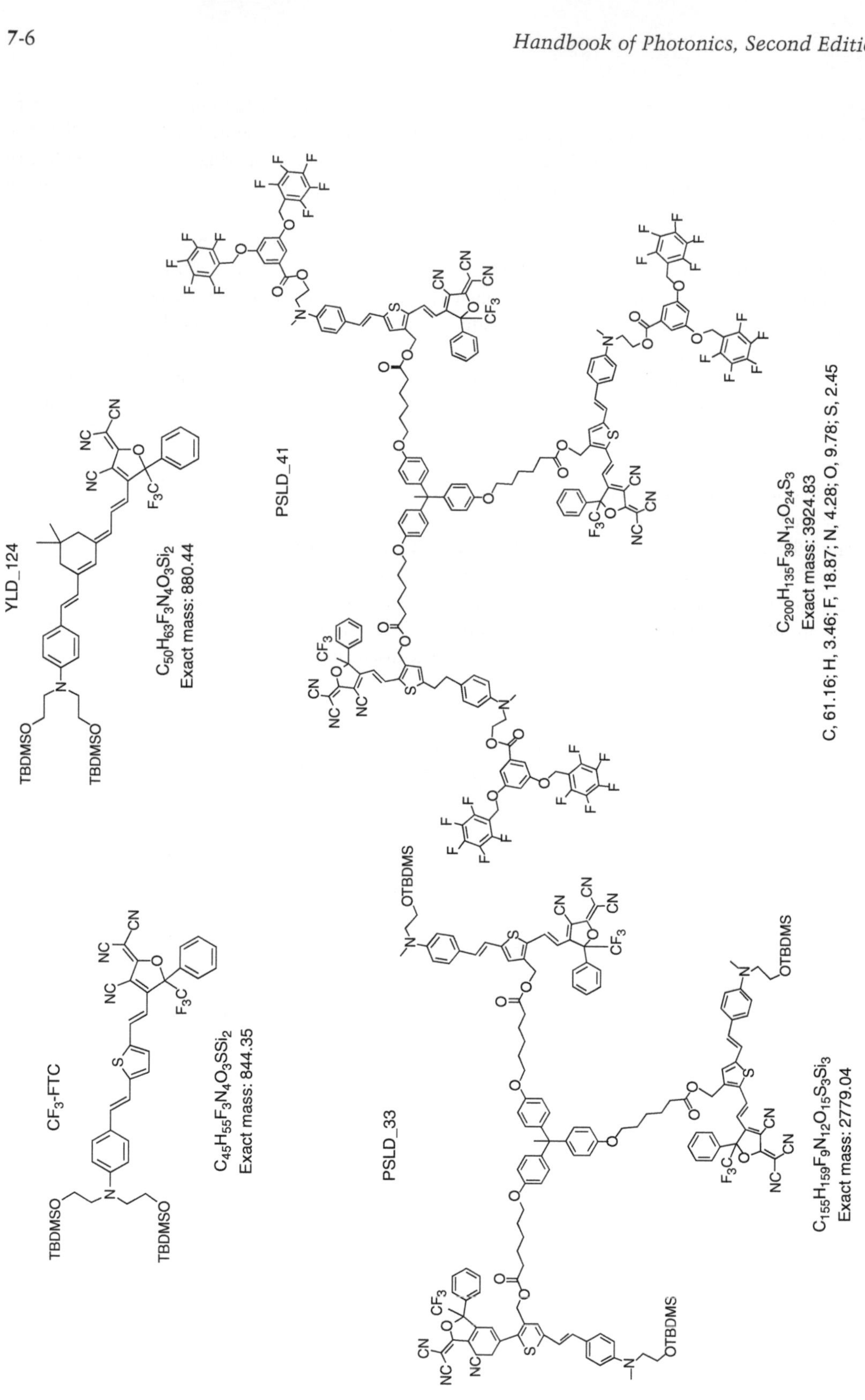

YLD_124

$C_{50}H_{63}F_3N_4O_3Si_2$
Exact mass: 880.44

CF$_3$-FTC

$C_{45}H_{55}F_3N_4O_3SSi_2$
Exact mass: 844.35

PSLD_41

$C_{200}H_{135}F_{39}N_{12}O_{24}S_3$
Exact mass: 3924.83
C, 61.16; H, 3.46; F, 18.87; N, 4.28; O, 9.78; S, 2.45

PSLD_33

$C_{155}H_{159}F_9N_{12}O_{15}S_3Si_3$
Exact mass: 2779.04

FIGURE 7.2 Four chromophore structures related to the data presented in Figure 7.3.

critical for considering very large number densities (high chromophore loading), is sufficiently complicated to be beyond the discussion presented here. In the present discussion, we restrict our consideration to the more simplistic case of a uniform chromophore distribution. As shown in this figure, intermolecular electrostatic interactions can also augment poling-induced noncentrosymmetric order. It has been suggested in at least two studies that chromophore-polymer interactions can also influence order.[80,81] Indeed, in addition to chromophore shape effects illustrated in Figure 7.1, the spatial and dynamical restrictions associated with covalent bonds impact poling-induced order. This is illustrated in studies of multichromophore-containing dendrimers (MCCD), as depicted in Figure 7.2 and Figure 7.3. In Figure 7.3, a significant enhancement in electro-optic activity is observed for a MCCD, relative to the behavior observed for the same chromophore in a polymer composite material. Indeed, behavior for the MCCD lies between that expected for a chromophore in a spherically-symmetric environment and the independent particle limit (see Figure 7.1). Even more intriguing behavior is observed when a second chromophore is doped into the MCCD. The electro-optic activity, as shown in Figure 7.3, increased nearly linearly with a slope more than twice the initial slope of the r_{33} versus N plot for the same chromophore in an amorphous polycarbonate (APC) polymer host.[32,33] A simplistic analysis suggests that such behavior may reflect intermolecular electrostatic interactions, acting to increase $N<\cos^3\theta>$ in a manner analogous to that seen in Figure 7.1. However, unlike the case for chromophore/polymer composite and undoped dendrimer materials, this behavior has not yet been quantitatively reproduced by theoretical calculations. Moreover, all necessary control experiments to rule out other potential contributions to the unusual behavior shown in Figure 7.3 have not yet been completed. Nevertheless, the realization of electro-optic activities greater than 300 pm/V for doped single-chromophore-containing dendrimers, MCCDs, and dendronized polymers is an important milestone and provides the potential (providing further improvement in molecular first hyperpolarizability is forthcoming) of realizing electro-optic activity on the order of 1000 pm/V. Of course, for large electro-optic activity to be meaningful, it must be accompanied by acceptable optical transparency,

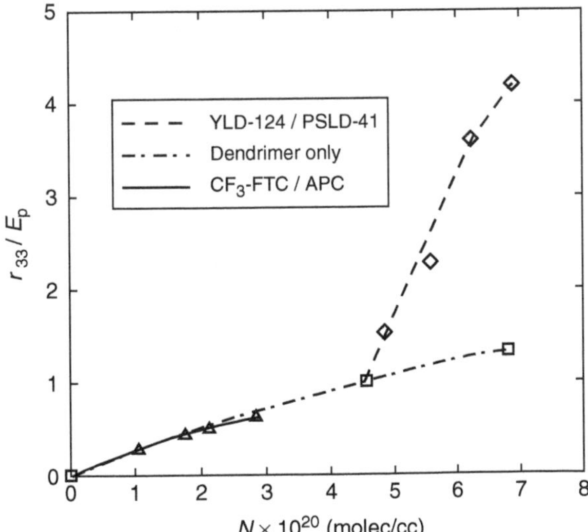

FIGURE 7.3 Data for the four chromophore structures of Figure 7.2. Since a linear dependence of electro-optic activity, r_{33}, on electric poling field strength is observed in all cases, r_{33}/E_p vs. N is plotted to facilitate comparison among data sets. Data for composite materials consisting of the chromophore CF3-FTC in amorphous polycarbonate (APC, Aldrich Chemicals) are indicated by triangles. Data for the two dendrimer materials PSLD 33 and 41 are indicated by squares. Data for samples of YLD 124 doped into PSLD 41 are indicated by diamonds.

thermal stability and photostability. We now turn attention to discussion of these issues, after a few comments about the measurement of electro-optic tensor components.

Electro-optic activity of thin film samples is most commonly measured by the following techniques: Teng-Man simple reflection[82–84] attenuated total reflection (ATR),[85–87] Fabry–Perot interferometry, FPI,[81,88,89] and two-slit interference.[90] Electro-optic activity can also be measured in a variety of waveguide (e.g., Mach Zehnder interferometry, MZI),[91–93] ring microresonator,[66,72] and etalon devices. Each technique has its particular advantages and limitations, and in general it is desirable to obtain electro-optic tensor components from multiple measurements using different techniques. It can also be important to define the dispersion of tensor components. A variety of techniques, such as ATR, FPI, and MZI devices permit both r_{33} and r_{13} to be determined. Such complete tensor determination permits a more definitive characterization of chromophore orientational order. These various characterization techniques can also be adapted by the introduction of temperature control and DC voltage application stages to provide in situ monitoring of the introduction of noncentrosymmetric order by electric field poling and the subsequent relaxation of that order.[45]

7.2.1.2 Minimizing Optical Loss

This discussion will first focus on "material" loss. Throughout the 1990s, most of the chromophores being investigated had a charge transfer absorption maxima, λ_{max}, of less than 600 nm. For such materials, optical absorption loss at 1.3 and 1.55 micron telecommunication operating wavelengths was most frequently dominated by hydrogen vibrational overtone absorptions. For dendrimer materials, optical loss values as low as 0.2 dB/cm have been observed.[94] When optical loss of greater than 2 dB/cm was observed, it was normally indicative of light scattering arising from material heterogeneity associated with various processing conditions.[35] More recently, chromophores with interband absorption maxima approaching 800 nm have come into use. For these materials, and even for some earlier materials,[32,33] absorption loss at telecommunication wavelengths is dominated by electronic charge transfer absorption. Thus, increasing $\chi^{(2)}$ may not lead to an improvement in FOM, because it is accompanied by a corresponding increase in α. The appearance of exciton bands may even lead to a decrease in FOM. In designing new chromophores for improved optical nonlinearity, it is critical to consider optical loss.

Optical loss due to electronic absorptions (both interband charge transfer and excitonic absorptions associated with aggregation) is strongly influenced by solvatochromic and line broadening effects, particularly as these effects influence the long wavelength tails of absorptions. Line broadening is frequently defined by the heterogeneity of the chromophore environment and thus can be influenced by chromophore order and by the mode of attachment of the chromophore to its surrounding matrix. The dielectric properties of the surrounding matrix will, of course, have a profound effect on solvatochromic shifts. Researchers at Lockheed Martin[32,33] appear to be the first to focus on an effort to control absorption contributions to optical loss by a systematic consideration of the roles played by the structure of the chromophore and of the surrounding matrix.

Quite different effects on absorption loss can be observed for different types of materials at telecommunication wavelengths. In this regard, dendritic materials may afford significant advantages relative to chromophore/polymer composite materials due to the fact that they permit control of the local chromophore environment and chromophore solubility in the surrounding matrix. For example, chromophores incorporated in fluorinated dendrimers typically exhibit blue (hypsochromic) shifted absorption maxima relative to the same chromophore in a polymer such as amorphous polycarbonate (APC, Aldrich Chemical). Moreover, the "solubility" of the chromophore in dendrimers is controlled by covalent bond attachment. Of course, comments made regarding dendrimers can also apply to chromophores covalently incorporated into polymers, provided that access of chromophores to each other is inhibited by the covalent incorporation. A high concentration of chromophores does not necessarily imply disastrous optical loss, as illustrated in Figure 7.4, which shows the same chromophore in a covalent-bonding-defined chromophore "bundle" and in APC.[95] The absorption maximum of the chromophore in the bundle is shifted to higher energy and no detectable line broadening is observed. The molecular hyperpolarizability of the chromophore bundle is nearly three times that of the isolated

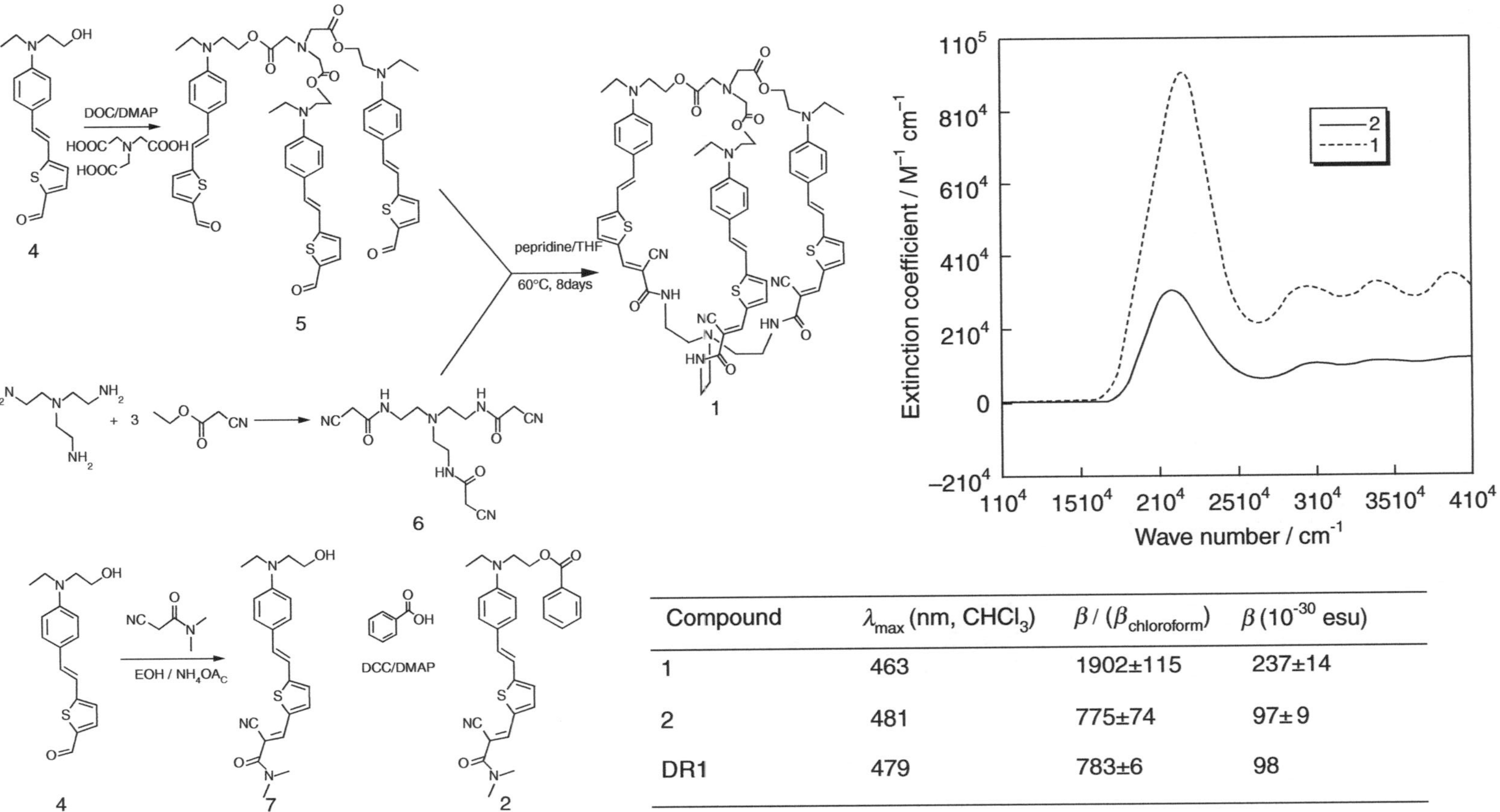

Compound	λ_{max} (nm, $CHCl_3$)	$\beta / (\beta_{chloroform})$	β (10^{-30} esu)
1	463	1902±115	237±14
2	481	775±74	97±9
DR1	479	783±6	98

FIGURE 7.4 The comparison of optical spectra and molecular hyperpolarizability for the same chromophore in APC and in a three-chromophore bundle (right hand side of figure). On the left hand side of the figure, the syntheses of the three-chromophore bundle and the analogous single chromophore are shown.

chromophore. The experimental results shown here are in good agreement with theoretical calculations carried out for the respective structures. Care should be exercised with respect to extrapolating the results of Figure 7.4 to other materials; in general, the exact optical behavior will be determined by the precise translational and orientational positioning of chromophores, which will vary from structure to structure.

Optical loss is also influenced by scattering losses. These frequently arise from material heterogeneities introduced by spin casting, electric-field poling, crosslinking (lattice hardening), and by various device processing steps (e.g., reactive ion etching of waveguides and deposition of cladding layers). The problem is typically worse (ignoring for the moment issues associated with crosslinking or lattice hardening) for chromophore/polymer composites than for chromophores covalently incorporated into dendrimers or polymers. Again, the major issue is control of chromophore "solubility." For chromophore/polymer composites, problems include: the differential solubility of the chromophore and polymer host in spin casting solvents; sublimation of chromophores during baking (to drive off residual spin casting solvents); sublimation of chromophores during electric field poling; and "electrophoretic" phase separation during poling. "Covalent-tailoring" of chromophores and their incorporation into host materials is a very attractive means of controlling chromophore solubility and packing (void volumes) in the final material. As we shall see shortly, crosslinking (or lattice hardening) is typically necessary to achieve adequate thermal stability. Some crosslinking chemistries can lead to phase separation and lattice strain and thus dramatic increases in optical loss due to light scattering. New cycloaddition crosslinking chemistries involving the "soft" free-radical chemistry of the fluorovinyl ether group or the Diels-Alder/retro-Diels-Alder reaction lead not only to lattice hardening without attenuation of poling efficiency, but also to materials with low optical loss.

Optical loss can also arise from the process of fabricating buried channel waveguides. When this is done by techniques such as reactive ion etching, care must be exercise to avoid pitting (waveguide wall roughness) due to reactive ions with excess kinetic energy (i.e., a physical rather than chemical etch). When care is employed, the excess waveguide loss can be 0.01 dB/cm or less.[96] Loss can also be introduced in deposition of cladding layers if the solvent used to deposit the cladding layer attacks the electro-optic material. For both reactive ion etching and deposition of cladding layers, the lowest loss is typically observed when very hard electro-optic materials are used. Loss can also arise from material damage (dielectric breakdown) during electric field poling.

Optical loss can be influenced by the structure of the device, e.g., bending loss for ring microresonators. Of course, one of the greatest contributions to total device insertion loss is coupling loss. With organic electro-optic materials, the dominant contribution to coupling loss arises from a mismatch in mode size and shape between light propagating in silica transmission fibers and that in the organic EO waveguides. The solution to this problem is typically to employ a "mode transformer" structure.[97–101] Mode transformers permit per facet coupling losses to be kept to a few tenths of a dB. The overall objective is typically to achieve a total insertion loss of less than 5 dB (the current standard for lithium niobate devices). Thus, if device lengths of 2 cm or less are to be used, then material loss must be kept to less than 2 dB/cm. Short device lengths, of course, have the advantage of permitting greater operational bandwidths by minimizing electrical signal loss occurring in metal drive electrodes. The calculated performance of a Mach Zehnder device for typical EO, cladding, and electrode material conditions is shown in Figure 7.5.

7.2.1.3 Maximizing Thermal Stability

The stability of electro-optic activity following the cessation of electric field poling is critical for device applications. In the simplest sense, thermal stability relates to the temperature difference between the operating temperature and the material glass transition temperature, T_g. With composite materials, incorporation of chromophores results in plasticization and a corresponding reduction of glass transition temperature with increasing dopant (chromophore) concentration. Thus, to realize sufficient thermal stability to satisfy Telcordia standards, it is necessary to use a host polymer material with an initial glass transition temperature on the order of 150°C or greater. Polycarbonates, polyquinolines, and polyimides have been the most commonly employed polymers for producing composite materials appropriate for

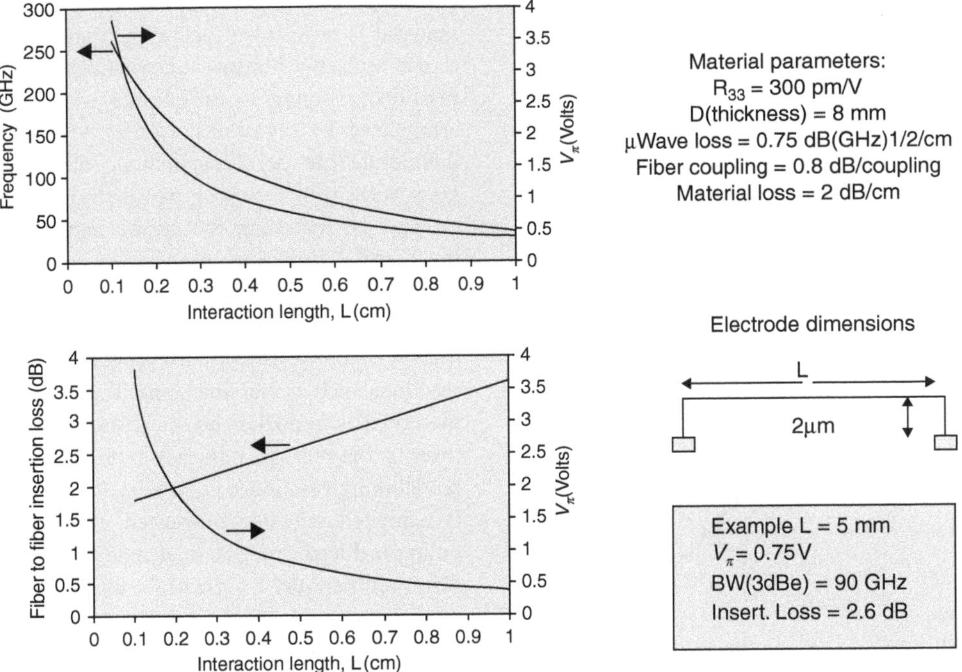

FIGURE 7.5 The variation of critical Mach Zehnder device parameters (bandwidth, drive voltage, fiber-to-fiber insertion loss) with device (electrical/optical field interaction) length. An example of performance for a 5-mm length Mach Zehnder electro-optic modulator expected with current materials is also shown.

prototype device fabrication. One of the problems with very high-glass-transition-temperature composite materials, such as polyimides, is that very high poling temperatures are required. Such temperatures promote chromophore sublimation and the decomposition of chromophores, and can also be incompatible with processing methodologies such as nanoimprint lithography. High glass transition materials also commonly exhibit poor solubility in solvents used for spin casting. A convenient means of avoiding this requirement for high temperature processing at intermediate stages of electro-optic material production is to achieve thermal stability by crosslinking (lattice hardening) in the final stages of electro-optic material production. With crosslinking, thermal stability will be defined by the density of crosslinks and the flexibility of intervening segments. Throughout the 1990s, crosslinking chemistries used to elevate final material glass transitions temperatures frequently involved condensation (e.g., urethane and sol gel) chemistry.[34,35] Such reactions give off water as an elimination product and are influenced by atmospheric moisture. Moreover, the expulsion of gaseous elimination products can result in lattice disruption and increased light scattering. As the condensation reaction proceeds, increased lattice strain (e.g., lattice contraction) can also be a problem, which can be serious with sol-gel glasses. Both thermal and photo-induced crosslinking chemistries have been explored.[30,31] The latter has been plagued by competition for light absorption involving the photo-initiator and the EO chromophore. More recently, cycloaddition chemistries[36–38,44,46–52,102] have become popular for realizing high glass transition temperature (e.g., to 200°C) materials without the attenuation of electro-optic activity or an increase in optical loss associated with earlier crosslinking reactions. These chemistries are pictorially illustrated in Figure 7.6. The fluorovinyl ether crosslinking reaction illustrated in Figure 7.6a is a soft free radical reaction that has the advantage of yielding high glass transition materials that are also characterized by very low optical loss at telecommunication wavelengths due to low hydrogen content in the final material. The Diels-Alder/retro-Diels-Alder reaction of Figure 7.6b has the added advantage

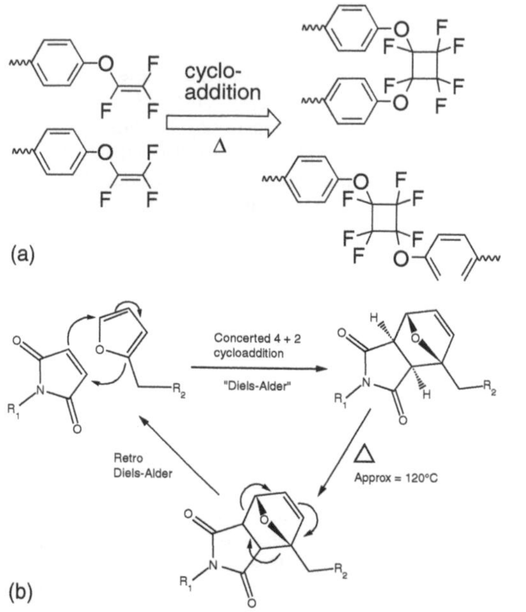

FIGURE 7.6 (a) Fluorovinyl ether and (b) Diels-Alder/retro-Diels-Alder cycloaddition crosslinking reactions.

(in some cases) of reversibility. That is, the material is crosslinked below the glass transition temperature but becomes uncrosslinked at the T_g, permitting poling to be effected without being attenuated by crosslinks. Choice of diene and dienophile can permit systematic tuning of the glass transition temperature, including for the purpose of matching the poling temperature to the thermal initiation temperature of the fluorovinyl ether crosslinking reaction (when both types of cycloaddition crosslinking are used together). Obviously, for irreversible crosslinking reactions such as reaction of the fluorovinyl ether moiety, it is important to match poling temperatures to the thermal initiation temperature of the crosslinking reaction so that effective crosslinking is achieved without unwanted attenuation of poling-induced order. A final material glass transition temperature on the order of 200°C is more than adequate for satisfying Telcordia standards for thermal stability (long-term stability at an operating temperature of 85°C). Cycloaddition crosslinking has been demonstrated to be effective in providing such stability.

7.2.1.4 Maximizing Photostability

Photostability has been shown to be largely a matter of avoiding singlet oxygen chemistry.[103–108] Stegeman and coworkers[103–106] have demonstrated most of the critical features of photodecomposition of organic electro-optic materials, including the absence of contributions from multiphoton absorption. Indeed, they defined a single photostability figure-of-merit, B/σ, where B^{-1} is the probability of photodecay from the LUMO (lowest unoccupied molecular orbital) charge transfer state and σ is the interband (charge transfer) absorption coefficient. This definition has been used by subsequent researchers, although the data analysis of Stegeman and coworkers may have been somewhat overly simplistic, consequently over-estimating photo-instability. For example, more detailed analyses demonstrate that the decay data cannot be fit with a single exponential and that "observer" power in the Stegeman pump–probe experiment may lead to artificially fast decay. Moreover, Stegeman and coworkers failed to carry out measurements at telecommunication wavelengths; this was addressed in subsequent work by researchers at Corning.[107,108] The Corning group demonstrated that the photostability FOM for a given chromophore structure could vary over four orders of magnitude depending on conditions that influence singlet oxygen chemistry. Even larger variation has been observed by other groups and photostability has been shown to improve with the use of small quantities of singlet oxygen quenchers. In addition to pump-probe experiments carried out by Stegeman and coworkers, researchers at Corning, Gunter and coworkers, and Dalton and coworkers, photostability has also been investigated in operating Mach Zehnder devices by Steier and coworkers[69] and by Ashley and coworkers.[109] Again, in these studies, photo-instability could be attributed to singlet oxygen chemistry, with good photostability being observed for materials and devices where this chemistry was partially inhibited.

In summary, it appears that good photostability can be achieved with appropriate materials modification or with appropriate packaging of devices to minimize the presence of oxygen. In this latter regard, the problems faced with organic electro-optic materials are analogous (but not quite so severe) as those faced with organic light emitting device (OLED) materials. It should be noted that dense crystals such as DAST exhibit excellent photostability, again consistent with the role played by oxygen.

7.2.1.5 Radiation Hardness

Very little work has been carried out examining the performance of organic electro-optic materials and devices in the presence of high-energy gamma rays and protons associated with space environments. Part of the problem of evaluating the impact of high-energy radiation on device performance is isolating changes attributable to organic electro-optic materials from changes of performance induced by events unrelated to the organic electro-optic materials (e.g., damage to silica input fibers). Nevertheless, the one published report[110] on this subject appears promising with respect to good stability exhibited by organic electro-optic materials.

7.2.1.6 Fabrication of Prototype Devices

The most common prototype device fabrication involves stripline Mach Zehnder modulators; however, ring microresonator, cascaded prism, and etalon structures have also been produced and evaluated. A critical issue with the fabrication of prototype devices is how to deal with the properties of cladding and electrode materials. Devices are typically multilayer structures consisting of bottom (ground) electrode/bottom cladding/electro-optic waveguide/top cladding/top (drive) electrode. If materials are poled through cladding layers, the poling field can be attenuated due to the resistivity of the cladding layers. A potential solution to this problem would be to identify cladding materials with significant conductivity. Unfortunately, cladding materials with prerequisite conductivity have also exhibited, to the present time, unacceptably high levels of optical loss. The presence of cladding layers thus makes it difficult to realize the same electro-optic activity in device structures that have been achieved in thin films. Another issue in the fabrication of prototype devices is the impact of electrode materials and device structure on performance, including operational bandwidth. For stripline devices, such as Mach Zehnder interferometers, resistive losses in metal electrodes typically define operational bandwidths. Shorter electrode structures (see Figure 7.5) lead to higher bandwidths, but at a price of increased drive voltage (V_π, the voltage required to produce a π phase shift). As already noted, shorter devices afford the advantage of reduced insertion loss and are more appropriate for high-density integration of many modulators on a single chip. For resonated devices, such as ring microresonators and etalon devices, bandwidth is limited by the optical lifetime in the resonated structure. This is defined by the quality factor, Q, of the resonant device. High quality factors have the advantage of affording reduced drive voltage operation but limit the bandwidth of the device. Very high center operational frequencies can still be obtained, but the bandwidth about the center frequency is limited by the Q. Ring microresonators have the added advantage of reduced size, which can facilitate high-density integration. In discussing bandwidth, one needs to distinguish between digital and analog signals. This point will be illustrated in the following simplified discussion of bandwidth and drive voltage requirement for resonant device structures.

For a resonator, the 3-dB electrical modulation bandwidth (the detected voltage is down by 3 dB) is given by

$$\Delta f_{3\text{dBe}} = c/\lambda Q = \Delta f_{\text{FWHM}}, \tag{7.4}$$

where c is the speed of light, λ is the wavelength of light, Δf_{FWHM} and is the full-width at half-maximum of the bandpass of the resonant device. The case of digital modulation is considered first.

For a 10-dB contrast in the digital pulses, the voltage induced frequency shift of the bandpass must be

$$\Delta f = 3\Delta_{\text{FWHM}}/2. \tag{7.5}$$

The frequency shift with voltage is

$$(\Delta f / V) = (n_{\text{eff}})^2 r_{33} c / (2d\lambda), \tag{7.6}$$

where d is the electrode spacing. Combining,

$$V_{10\text{dB}} = (3d\lambda \Delta f_{3\text{dBe}}) / [(n_{\text{eff}})^2 r_{33} c]. \tag{7.7}$$

In digital systems, the required bandwidth depends on the modulation format. Conservatively assuming that $\Delta f_{3\text{dBe}} = B$ (the bit rate), then

$$V_{10\text{dB}} = 3d\lambda B / [(n_{\text{eff}})^2 r_{33} c]. \tag{7.8}$$

For example, if $B = 10$ Gb/s, $n = 1.6$, $r_{33} = 300$ pm/V, $d = 6$ μm, $\lambda = 1.3$ μm, and $Q = 5 \times 10^3$, then $V_{10\text{dB}} = 1$ V (with the result obviously scaling linearly with the bit rate).

Now consider analog signal modulation. The optical wavelength is set to the point of maximum slope of the resonator transmission (bandpass) curve:

$$\Delta\lambda = \Delta\lambda_{\text{FWHM}} / (2\sqrt{3}) \tag{7.9}$$

It is helpful to define a $V\pi_{\text{equiv}}$ so that it corresponds to an equal $V\pi$ of a Mach Zehnder modulator:

$$(V_{\pi\,\text{equiv}} / \Delta f_{3\text{dBe}}) = 4\pi d\lambda / [(3\sqrt{3})(n_{\text{eff}})^2 r_{33} c]. \tag{7.10}$$

For example, if $n = 1.6$, $r_{33} = 300$ pm/V, $d = 6$ μm and $\lambda = 1.3$ μm, then $(V\pi_{\text{equiv}}/\Delta f_{3\text{dBe}}) = 0.08$ V/GHz. It is clear from the above analyses for digital and analog signal processing that the simplest route to improving the performance of resonant devices is to increase the electro-optic coefficient, r, of the material used to fabricate the device. The current performance of organic electro-optic materials moves resonant devices close to practical application.

Device structures are most commonly fabricated using reactive ion etching, although ring microresonator and Mach Zehnder devices structures have also been made by nanoimprint lithography.[68] Nanoimprint lithography has the potential advantage of allowing complex electro-optic circuitry to be mass-produced in a cost-effective manner.

The range of application of organic electro-optic materials has been recently broadened by the incorporation of these materials into silicon photonic device structures.[66] The small dimensions of these structures and the obvious potential for convenient integration with silicon electronics is very attractive. Moreover, the small (nanoscopic) dimensions of silicon photonic circuitry result in an optical field concentration, which has recently been exploited to achieve optical rectification of light at mW optical powers.[66] Thus, the same device structure can be employed for electrical/optical and optical/electrical signal transduction.

7.2.1.7 Applications

Applications of electro-optic materials and devices include electrical-to-optical signal transduction, optical switching, optical beam steering, radiofrequency signal generation, phased array radar (radiofrequency beam steering), optical gyroscopes, analog-digital conversion, frequency conversion (time stretching), and sensing (of both physical and chemical phenomena). The term *radiofrequency* describes frequencies in the range 0–30 THz. Electro-optic technology is at the heart of “RF Photonics,” or the delivery of radiofrequency signals via optical transmission. Many applications of electro-optic materials lie in the arena of defense and homeland security, but interest is growing in the areas of computer chip manufacture, transportation, telecommunications (both fiber and wireless), civil engineering, and medicine. For example, in the development of next-generation computer chips, photonics may be used to route high frequency information among various components of the chip, avoiding the problems of signal loss and heating associated with moving electrons through metal connectors. The field of embedded network sensing combines sensors, computer processors, and communication components on a single chip. Electro-optic devices provide critical signal transduction and routing on embedded network sensing platforms. Embedded network sensing is finding increasing applications in medicine and infrastructure monitoring (civil engineering).

7.2.2 Optical Rectification Including Terahertz Radiation and Detection

Optical rectification is the difference frequency analog of second harmonic generation. Like second harmonic generation, it will occur whenever the electric field component of the optical field is large enough to produce a nonlinear perturbation of the charge distribution of the nonlinear optical material. Normally, intense laser powers are required, but the nanoscopic dimensions of silicon photonic circuitry can concentrate optical fields from milliwatt diode lasers to the point of producing optical rectification.[66] Thus, organic EO/silicon hybrid devices can act as both electrical-to-optical signal transducers and optical-to-electrical signal transducers. In other words, they have the potential to compete with photodiodes as photodetectors, providing that sufficiently large material second-order optical nonlinearity can be obtained.

Another manifestation of optical rectification/difference frequency generation is terahertz signal generation/detection. Recently, Hayden and coworkers,[111,112] Gunter and coworkers,[113] and researchers in Japan[114–116] have pioneered the use of organic electro-optic materials for terahertz applications (imaging and spectroscopy). An advantage of organic materials is that optical and terahertz waves propagate with comparable velocities in organic materials, permitting phase matching of the optical and terahertz radiation. Also, the second-order optical nonlinearity of organic materials is orders of magnitude greater than inorganic crystalline materials such as zinc telluride (ZnTe). A problem encountered with poled organic (polymer) materials is that it is difficult to effectively pole thick (millimeter) films. Such film thickness would be ideal for terahertz applications.

Among the promising sensor applications of terahertz radiation is the ability to image plastic weapons. It is an attractive alternative to magnetic (metal detector) sensing in the arena of homeland security. The possible applications in biomedical imaging are also attractive.

7.3 Third-Order Nonlinear Optical Materials

Third-order organic nonlinear optical materials have very little in common with second-order nonlinear optical materials, other than the fact that both involve significant π-conjugation. Since no symmetry requirement exists for third-order activity, a much broader range of materials can give rise to third-order optical nonlinearity, $\chi^{(3)}$. Indeed, third-order materials range from conjugated polymers such as polyacetylene, to molecules such as C_{60} and C_{70}, to metallomacrocyclic complexes. When charge transfer molecules are investigated, molecules with quadrupolar[117] symmetry (e.g., donor–acceptor–donor or acceptor–donor–acceptor) frequently exhibit larger optical nonlinearity than corresponding dipolar (donor–acceptor) molecules. Dendritic materials containing connected π-electron segments have also been observed to give rise to large third-order optical nonlinearities[118, 119].

Although the third-order optical nonlinearity for organic materials has been increased over the past two decades, values are still too low for practical applications. A possible exception may be for materials incorporated into silicon photonic circuitry. Scherer and coworkers[120] have demonstrated all-optical modulation to greater than 5 THz with mW pump powers derived from a diode laser operating at telecommunication wavelengths. The concentration of optical power in waveguides of nanometer dimensions results in an amplification of optical field intensities by orders of magnitude, permitting much smaller $\chi^{(3)}$ values to be utilized effectively.

Recently, interest in $\chi^{(3)}$ materials has increased because of potential applications of materials with large two-photon absorption coefficients. These applications range from sensor protection to biomedical imaging, photodynamic therapy, and two-photon photolithography.

Because no symmetry requirement exists for nonzero third-order optical nonlinearity, lattice hardness is a less serious problem than for second-order materials, although little research has been conducted on this feature of third-order materials. Optical loss is as important for third-order materials as it is for second-order materials, but very little research attention has been paid to this topic. Third-order materials should afford comparable advantages in processability compared to second-order organic nonlinear optical materials; however, it is unlikely that auxiliary properties will receive much attention unless significant improvement in $\chi^{(3)}$ can be achieved.

7.4 Summary

Extensive π-conjugation of organic materials leads to significant second- and third-order optical nonlinearities. However, until the present decade, values of $\chi^{(2)}$ and $\chi^{(3)}$ have been too small to promote significant commercial application. Currently, electro-optic coefficients in the range 300–400 pm/V are observed for a variety of dendrimer and dendronized polymer materials containing chromophores with large first hyperopolarizabilities. Such values are an order of magnitude greater than values for the commercial standard lithium niobate. New organic electro-optic materials afford the possibility of more compact and lightweight devices operating with bandwidths of 100 GHz or greater and with drive voltages of less than one volt. Realization of gain in RF photonics becomes a possibility for the first time. Many issues remain to be addressed before significant commercialization is likely, including the systematic control of optical loss, thermal stability, and photochemical stability. Moreover, better utilization needs to be made of the processing advantages of organic electro-optic materials, including for the production of integrated electronic/photonic circuitry exploiting a high density of organic electro-optic devices on a single chip. The integration of organic electro-optic materials with silicon photonic circuitry appears to afford some impressive new opportunities not only for high bandwidth electro-optic modulation and optical switching but also for optical rectification. One distinct advantage of organic electro-optic materials is that their processability may permit new device structures and applications to be considered. For example, new sensor technologies appear possible, e.g., by exploiting ring microresonators positioned on side-polished optical fibers.

The prognosis for third-order organic nonlinear optical materials is somewhat less optimistic, as $\chi^{(3)}$ values still need to be increased by one to two orders of magnitude for many applications. Incorporating third-order organic nonlinear optical materials into resonant structures (e.g., ring microresonators) and into silicon photonic waveguides may help reduce the performance demands on $\chi^{(3)}$ materials. At any rate, the development of third-order organic nonlinear optical materials is in a more immature stage than the development of second-order materials, as very little attention is given to auxiliary properties such as optical loss, stability, and processability. The most promising application of third-order organic materials appears to involve materials with large two-photon cross-sections for applications such as biomedical imaging, photodynamic therapy, two-photon photolithography, and ultrafast-responding sensor protection (Table 7.1 through Table 7.3).

Acknowledgments

Support from the National Science Foundation, the Air Force Office of Scientific Research, and the Defense Advanced Research Projects Agency is gratefully acknowledged. The authors thank their colleagues, particularly Professors William Steier and Bruce Robinson, for many helpful discussions.

TABLE 7.1 Experimental Results of Molecular studies. λ_{max}, Absorption Peak Wavelength of Lowest Energy Strong Trainsition; μ, Dipole Moment; β_{μ} (λ), EFISH Determined First Hyperpolarizability at a Fundamental Wavelength Given in Microns; β_0, as Defined in Section 6.4; S Indicates Solvatochromic Result; HR Indicates Hyper-Rayleigh Scattering Results

Structure	Solvent	λ_{max}(nm)	μ (10^{-18} esu)	$\beta_{\mu}(\lambda)^{ref.}$ (10^{-30} esu)
Unconjugated derivatives				
	Chloroform	<240	3.5	0.5(1.9)[156]
		<200	4.6	2.3 (1.06)[37]
K^+	H_2O			7 (1.06HR)[265]
Benzene derivatives				
$X=CH_3$	Neat		0.34	0.18 (1.06)[141]
	Neat	260	0.38	<0.2 (1.9)[46]
X=OH	Neat			0.17 (1.06)[193]
	Neat		1.5	0.17 (1.06)[191]
	p-Dioxane	270	1.5	<0.2 (1.9)[46]
X=SH	Neat		1.4	<0.2 (1.9)[46]
$X=OCH_3$	Neat	275	1.4	<0.2 (1.9)[46]
$X=SCH_3$	Neat		1.3	<0.2 (1.9)[46]
$X=NH_2$	Neat		1.5	0.89 (1.06)[141]
	Neat			1.2 (1.06)[193]
	Neat		1.5	1.1 (1.06)[190]
	Neat		1.5	1.2 (1.06)[191]
	Neat	284	1.5	0.55 (1.9)[46]

(continued)

TABLE 7.1 (Continued)

Structure	Solvent	λ_{max}(nm)	μ (10^{-18} esu)	$\beta_{\mu}(\lambda)^{ref.}$ (10^{-30} esu)
X = $N(CH_3)_2$	Neat			1.8 (1.06)[193]
	Neat		1.8	1.3 (1.06)[191]
	Neat	293	1.6	1.1 (1.9)[46]
X = F	Neat		1.5	0.44 (1.06)[141]
	Neat			0.53 (1.06)[193]
	Neat		1.6	0.70 (1.06)[191]
	Neat	266	0.63	<0.2 (1.9)[46]
X = Cl	Neat		1.5	0.28 (1.06)[141]
	Neat			0.22 (1.06)[193]
	Neat		1.7	0.33 (1.06)[191]
	Neat	272	0.44	<0.2 (1.9)[46]
X = Br	Neat		1.4	0.04 (1.06)[141]
	Neat			0.02 (1.9)[193]
	Neat		1.7	0.2 (1.06)[191]
	Neat	274	0.81	<0.2 (1.9)[46]
X = I	Neat		1.1	0.28 (1.06)[141]
	Neat			0.7 (1.06)[193]
	Neat		1.7	0.46 (1.06)[191]
	Neat	304	0.96	<0.2 (1.9)[46]
X = SO_2CH_3	*p*-Dioxane		1.1	<0.2 (1.9)[46]
X = SO_2F	Neat		1.7	0.3 (1.9)[46]
X = CN	Neat			0.51 (1.06)[193]
	Neat		4.1	0.48 (1.06)[191]
	Neat	230	3.9	0.36 (1.9)[46]
X = COH	Neat	320	2.8	0.8 (1.9)[46]
X = $COCF_3$	*p*-Dioxane	300	3.3	1.3 (1.9)[46]
X = NO	*p*-Dioxane	340/745	3.1	1.7 (1.9)[46]
X = NO_2	Neat			2.0 (1.06)[145]
	Neat			2.3 (1.06)[193]
	Neat		4.0	2.2 (1.06)[190]
	Neat		4.0	2.3 (1.06)[191]
	Hexane		4.2	1.1 (1.06)[232]
	Neat	268	4.0	1.9 (1.9)[46]
X = $C_2H(CN)_2$	*p*-Dioxane	303	4.8	3.1 (1.9)[46]

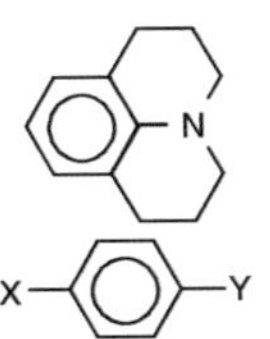		p-Dioxane	310	1.6	1.3 (1.9)[46]
$X=SO_2CH_3$	$Y=OH$	p-Dioxane	290	3.4	1.3 (1.9)[46]
$X=SO_2CH_3$	$Y=N(CH_3)_2$	Chloroform		$\mu\beta=26\times10^{-48}$	(1.9)[264]
$X=CN$	$Y=CH_3$	DMSO	232	4.4	2.9 (1.9)[56]
		Neat		4.4	0.7 (1.9)[46]
$X=CN$	$Y=Cl$	p-Dioxane		2.3	0.8 (1.9)[46]
$X=CN$	$Y=Br$	p-Dioxane		2.4	1.1 (1.9)[46]
		p-Dioxane	240	2.5	2.2 (1.59)[87]
$X=CN$	$Y=OC_6H_5$	p-Dioxane		4.1	1.2 (1.9)[46]
$X=CN$	$Y=OCH_3$	DMSO	247	4.8	4.8 (1.9)[56]
		p-Dioxane	248	4.8	1.9 (1.9)[46]
$X=CN$	$Y=SCH_3$	p-Dioxane		4.4	2.8 (1.9)[46]
$X=CN$	$Y=NH_2$	DMSO	269	5.9	13 (1.9)[56]
		p-Dioxane	270	5.0	3.1 (1.9)[46]
		p-Dioxane	271	6.4	2.4 (1.59)[87]
$X=CN$	$Y=N\ (CH_3)_2$	DMSO	297	6.6	14 (1.9)[56]
		p-Dioxane	290	5.6	5.0 (1.9)[46]
		CCl_4	290	$\mu\beta=44\times10^{-48}$	(1.06)[211]
		Chloroform	292	$\mu\beta=50\times10^{-48}$	(1.06)[211]
		H_2CCl_2	292	$\mu\beta=53\times10^{-48}$	(1.06)[211]
		H_2CCN	291	$\mu\beta=48\times10^{-48}$	(1.06)[211]
		H_3COH	295	$\mu\beta=47\times10^{-48}$	(1.06)[211]
$X=COH$	$Y=CH_3$	Neat		3.0	1.7 (1.9)[46]
$X=COH$	$Y=OC_6H_5$	Neat	269	2.8	1.9 (1.9)[46]
$X=COH$	$Y=OCH_3$	Neat	269	3.5	2.2 (1.9)[46]
$X=COH$	$Y=SCH_3$	Neat	310	3.1	2.6 (1.9)[46]
$X=COH$	$Y=N(CH_3)_2$	p-Dioxane	326	5.1	6.3 (1.9)[46]
$X=SO_2C_3H_7$	$Y=OCH_3$	Chloroform	290	5.4	3.3 (1.9)[46]
$X=SC_6H_5$	$Y=NH_2$	Acetone	320	3.0	14 (1.06)[16]
$X=COCH_3$	$Y=NH_2$	DMSO	310	4.5	2.4 (1.9)[56]
$X=COCF_3$	$Y=OC_6H_5$	p-Dioxane	292	3.5	3.6 (1.9)[46]
$X=COCF_3$	$Y=OCH_3$	p-Dioxane	292	4.0	3.6 (1.9)[46]

(continued)

TABLE 7.1 (Continued)

Structure		Solvent	λ_{max}(nm)	μ (10^{-18} esu)	$\beta_\mu(\lambda)^{ref.}$ (10^{-30} esu)
X=$COCF_3$	Y=N $(CH_3)_2$	*p*-Dioxane	356	5.9	10 (1.9)[46]
X=NO	Y=N $(CH_3)_2$	*p*-dioxane	407	6.2	12 (1.9)[46]
X=NO_2	Y=CH_3	DMSO	280	4.0	8 (1.9)[56]
		p-Dioxane	272	4.2	2.1 (1.9)[46]
X=NO_2	Y=F	Neat		2.7	2.1 (1.06)[141]
X=NO_2	Y=Br	*p*-Dioxane	274	3.0	3.3 (1.9)[46]
X=NO_2	Y=OH	*p*-Dioxane	304	5.0	3.0 (1.9)[46]
X=NO_2	Y=OC_6H_5	*p*-Dioxane	294	4.2	4.0 (1.9)[46]
X=NO_2	Y=OCH_3	DMSO	314	4.5	16 (1.9)[56]
		p-Dioxane		4.9	6 (1.06)[98]
				5.3	4 (1.9S)[195]
		p-Dioxane	302	4.6	5.1 (1.9)[46]
X=NO_2	Y=SCH_3			4.6	6.7 (1.9S)[195]
		p-Dioxane	322	4.4	6.1 (1.9)[46]
		Acetone	335	4.4	20 (1.06)[16]
X=NO_2	Y=SC_6H_5	Acetone	334	4.3	19 (1.06)[16]
X=NO_2	Y=N_2H_3	*p*-Dioxane	366	6.3	7.6 (1.9)[46]
X=NO_2	Y=NH_2	Methanol			36 (1.06)[193]
		Melt		7.2	21 (1.06)[140]
		Methanol		6.2	35 (1.06)[190]
		DMSO	378	6.1	47 (1.9)[56]
		p-Dioxane			9.6 (1.91)[250]
		p-Dioxane			12 (1.37)[250]
		p-Dioxane			17 (1.06)[250]
		p-Dioxane			25 (0.91)[250]
		p-Dioxane			40 (0.83)[250]
		DMSO	384	$\mu\beta=75\times10^{-48}$	(1.58)[231]
				7.1	9.6 (1.9S)[195]
		Chloroform			23 (1.06HR)[48]
		Acetone	365	6.2	9.2 (1.9)[46]
		p-Dioxane	354	7.0	16 (1.06)[237]
		Chloroform	348	6.4	17 (1.06)[237]
		H_3CCN	366	6.2	29 (1.06)[237]
		Methanol	370	6.1	32 (1.06)[237]
		NMP	386	6.8	38 (1.06)[237]
		p-Dioxane		$\mu\beta=69\times10^{-48}$	(1.9)[107]
		Chloroform	347	6.2	20 (1.06)[117]

X=NO_2	Y=NH(NH_2)	DMSO	378	$\mu\beta=10\times10^{-47}$	(1.36)[231]
X=NO_2	Y=N$(CH_3)_2$	DMSO	418	6.8	52 (1.9)[56]
		Chloroform		$\mu\beta=14\times10^{-47}$	(1.9)[264]
		Acetone	388	6.9	26 (1.06)[16]
		DMSO	404	$\mu\beta=14\times10^{-47}$	(1.36)[231]
		Acetone	376	6.4	12 (1.9)[46]
X=NO_2	Y=CN	*p*-Dioxane		0.9	0.6 (1.9)[46]
X=NO_2	Y=COH	*p*-Dioxane	376	2.5	0.2 (1.9)[46]
X=$C_2H(CN)_2$	Y=OCH_3	*p*-Dioxane	345	5.5	9.8 (1.9)[46]
X=$C_2H(CN)_2$	Y=N$(CH_3)_2$	DMSO	440	$\mu\beta=27\times10^{-47}$	(1.36)[231]
		DMSO	440	$\mu\beta=62\times10^{-47}$	(1.06)[101]
				7.1	12 (1.9s)[195]
		Chloroform	420	7.8	32 (1.9)[46]
X=$C_2H(CN)_2$	Y=N$(C_2H_5)_2$	*p*-Dioxane	419	$\mu\beta=30\times10^{-47}$	(1.9)[106]
X=$C_2H(CN)_2$	Y=NC_5H_{10}	Chloroform	440	8.5	17 (β_0)[177]
X=$C_2H(CN)_2$	Y=N(4-$CH_3C_6H_4)_2$	Chloroform	452	7.6	25 (β_0)[177]
X=$C_2(CN)_3$	Y=NH_2	CH_2Cl_2	498	7.8	39 (1.9)[46]
X=$C_2(CN)_3$	Y=N$(CH_3)_2$	CH_2Cl_2	516	8.2	50 (1.9)[46]
		DMSO	528	$\mu\beta=85\times10^{-47}$	(1.36)[231]
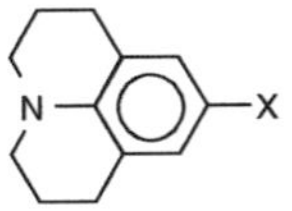 X=COH		Chloroform		5.7	14 (1.9)[43]
X=$C_2H(CN)_2$		CH_2Cl_2	458	8.0	44 (1.9)[46]
		Chloroform		8.3	51 (1.9)[43]
X=$C_2(CN)_3$		CH_2Cl_2	556	1.6	60(1.9)[46]
		DMSO	561	$\mu\beta=73\times10^{-47}$	(1.36)[231]
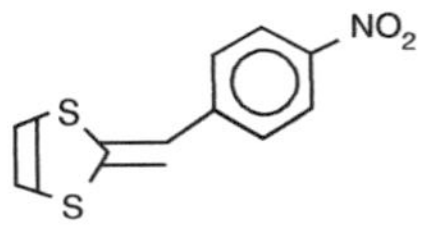		DMSO	458	$\mu\beta=36\times10^{-47}$	(1.36)[231]

(continued)

TABLE 7.1 (Continued)

Structure	Solvent	λ_{max}(nm)	μ (10^{-18} esu)	$\beta_{\mu}(\lambda)^{ref.}$ (10^{-30} esu)
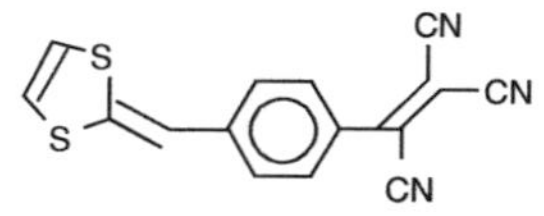	DMSO	602	$\mu\beta = 12 \times 10^{-46}$	(1.58)[231]
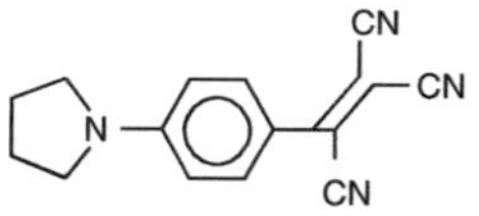	*p*-Dioxane	505	$\mu\beta = 71 \times 10^{-47}$	(1.36)[203]
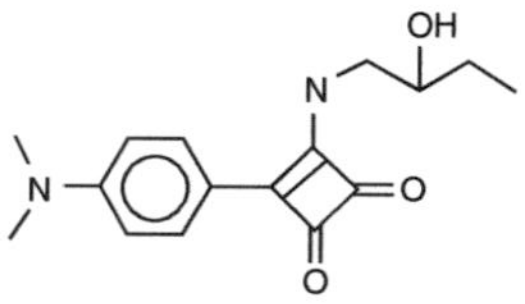		397	$\beta = 171 \times 10^{-30}$	(1.06S)[200]
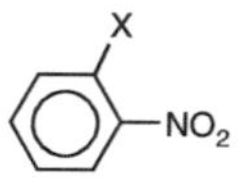				
X=CH_3	Neat		3.9	1.0 (1.9)[46]
X=F	Neat		5.0	1.8 (1.9)[141]
X=Br	*p*-Dioxane		4.0	0.4 (1.9)[46]
X=OH	*p*-Dioxane	348	3.4	1.2 (1.9)[46]
X=OCH_3	Neat	318	3.8	1.4 (1.9)[46]
X=NH_2	Melt		5.0	6.4 (1.32)[140]
	Acetone		4.3	10 (1.06)[190]
	p-Dioxane		4.1	2.5 (1.9)[46]
X=CN	*p*-Dioxane		5.5	1.2 (1.9)[46]
X=COH	*p*-Dioxane		4.0	0.8 (1.9)[46]

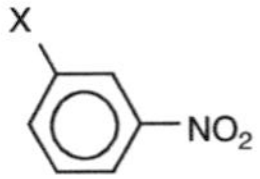				
$X=CH_3$	Neat		3.9	1.5 (1.9)[46]
$X=F$	Neat		3.6	- 1.6 (1.06)[141]
$X=Cl$	Benzene		3.4	1.6 (1.06)[191]
$X=Br$	Benzene		3.4	1.2 (1.06)[191]
	p-Dioxane		3.4	1.0 (1.9)[46]
$X=OH$	*p*-Dioxane		3.6	0.8 (1.9)[46]
$X=OCH_3$	*p*-Dioxane	326	3.9	1.6 (1.9)[46]
$X=NH_2$	Melt		5.5	4.2 (1.32)[140]
	Acetone		4.9	6 (1.06)[190]
	p-Dioxane	396	4.7	1.9 (1.9)[46]
$X=CN$	*p*-Dioxane		3.7	0.8 (1.9)[46]
$X=COH$	*p*-Dioxane		2.8	1.7 (1.9)[46]
$X=NO_2$	Benzene		3.9	1.2 (1.06)[191]
	p-Dioxane			9.5 (1.91)[250]
	p-Dioxane			13 (1.37)[250]
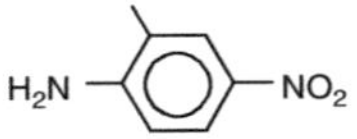				
	p-Dioxane			17 (1.06)[250]
	p-Dioxane			27 (0.91)[250]
	p-Dioxane			45 (0.83)[250]
	p-Dioxane	361	6.2	8.7 (1.9)[46]
	p-Dioxane		7.4	18 (1.06)[25]
	p-Dioxane			9.1 (1.32)[25]
	p-Dioxane			7.6 (1.91)[25]
	p-Dioxane		5.7	13 (1.06)[262]
	p-Dioxane	350	5.9	6.8 (1.9)[46]
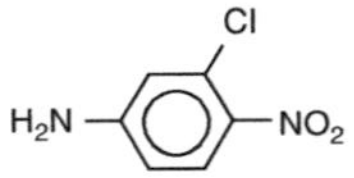				
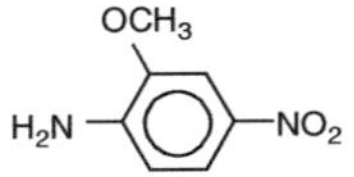	*p*-Dioxane		6.0	8.7 (1.9)[46]

(continued)

TABLE 7.1 (Continued)

Structure	Solvent	λ_{max}(nm)	μ (10^{-18} esu)	$\beta_{\mu}(\lambda)^{ref.}$ (10^{-30} esu)
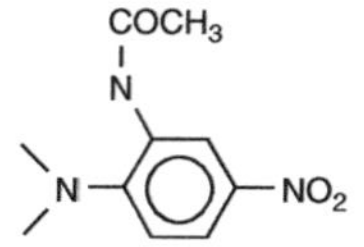	*p*-Dioxane		7.4	18 (1.06)[262]
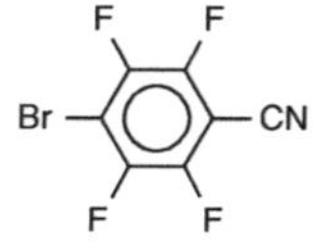			8.1	9.5 (1.06S)[195]
	p-Dioxane	240	3.5	1.4 (1.59)[87]
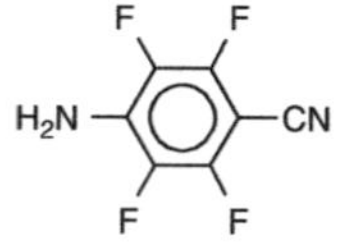	*p*-Dioxane	274	6.5	2.6 (1.59)[87]
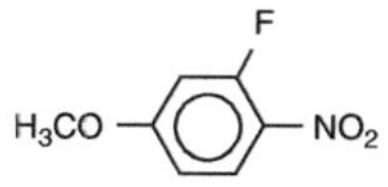	*p*-Dioxane	304	4.4	2.5 (1.9)[46]
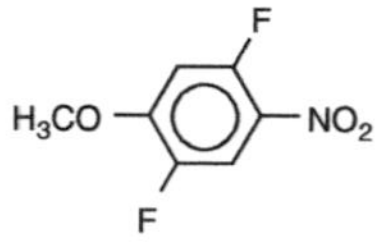	*p*-Dioxane	304	4.9	2.6 (1.9)[46]
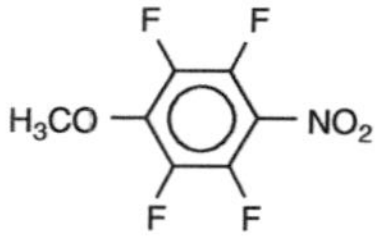	*p*-Dioxane	270	4.2	1.7 (1.9)[46]

	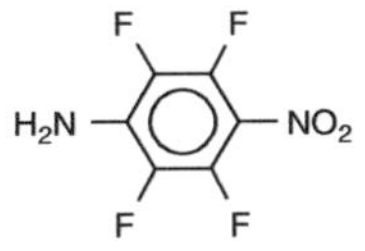	p-Dioxane	319	6.0	5.5 (1.59)[87]
	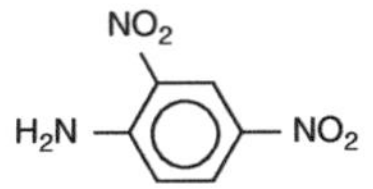	Acetone		5.5	21 (1.06)[190]
	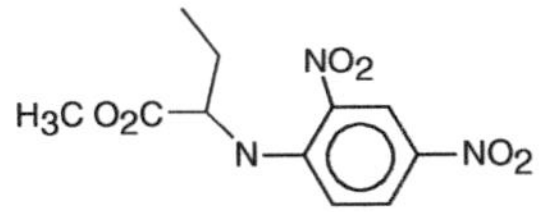	p-Dioxane		$\mu\beta = 49 \times 10^{-48}$	(1.9)[107]
		Acetone		5.6	22 (1.06)[190]
	Styrene derivatives				
	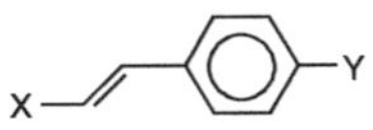				
$X=CN$	$Y=OCH_3$	Chloroform	304	4.2	7.0 (1.9)[45]
$X=CN$	$Y=N(CH_3)_2$	Chloroform	364	6.0	23 (1.9)[45]
$X=COH$	$Y=Br$	Chloroform	298	2.0	6.5 (1.9)[45]
$X=COH$	$Y=OCH_3$	DMSO	320	$\mu\beta = 80 \times 10^{-48}$	(1.9)[55]
		Chloroform	318	4.2	11 (1.9)[45]
$X=COH$	$Y=N(CH_3)_2$	DMSO	387	$\mu\beta = 37 \times 10^{-47}$	(1.9)[55]
		Chloroform	384	$\mu\beta = 32 \times 10^{-47}$	(1.34)[14]
		Chloroform	384	5.6	30 (1.9)[45]
$X=COCH_3$	$Y=OCH_3$	Chloroform	316	4.0	8.9 (1.9)[45]
$X=COCF_3$	$Y=N(CH_3)_2$	Chloroform		6.6	64 (1.9)[43]
$X=NO_2$	$Y=H$	Chloroform	312	3.8	8.0 (1.9)[45]

(continued)

TABLE 7.1 (Continued)

Structure		Solvent	λ_{max}(nm)	μ (10^{-18} esu)	$\beta_{\mu}(\lambda)^{ref.}$ (10^{-30} esu)
$X=NO_2$	$Y=OH$	Chloroform	352	5.1	18 (1.9)[45]
$X=NO_2$	$Y=OCH_3$	p-Dioxane		5.5	12 (1.06)[98]
		Chloroform	352	4.6	17 (1.9)[45]
$X=NO_2$	$Y=N(CH_3)_2$	Chloroform		7.9	220 (1.06)[189]
		Chloroform	438	6.5	50 (1.9)[45]
$X=CHC(CN)_2$	$Y=N(CH_3)_2$	DMSO	500	$\mu\beta=36\times10^{-46}$	(1.06)[102]
		DMSO	500	$Re(\mu\beta)=34\times10^{-46}$	(1.06)[101]
				$Im(\mu\beta)=36\times10^{-48}$	(1.06)[101]
$X=N(CH_3)_2$	$Y=NO_2$	Chloroform	438	5.9	35 (1.9)[45]
	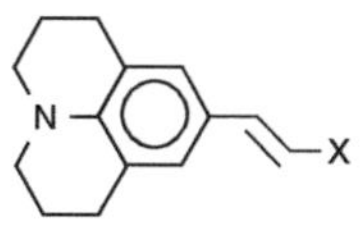				
	$X=COH$	Chloroform		6.3	48 (1.9)[43]
	$X=NO_2$	Chloroform		7.1	90 (1.9)[43]
	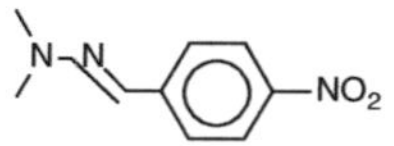	p-Dioxane		$\mu\beta=15\times10^{-47}$	(1.9)[107]
	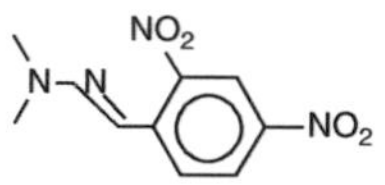	p-Dioxane		$\mu\beta=16\times10^{-47}$	(1.9)[107]
	Biphenyl derivatives				
	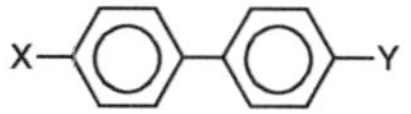				
$X=CN$	$Y=H$	p-Dioxane	272	4.0	1.9 (1.9)[45]

$X=COCH_3$	$Y=H$	p-Dioxane	280	3.1	2.0 (1.9)[45]
$X=NO_2$	$Y=H$	p-Dioxane	304	3.8	4.1 (1.9)[45]
$X=SO_2CH_3$	$Y=N(CH_3)_2$	Chloroform	340	6.0	13 (1.9)[45]
$X=CN$	$Y=OH$	p-Dioxane	292	4.8	6.3 (1.9)[45]
$X=COCH_3$	$Y=OCH_3$	p-Dioxane	304	3.4	4.9 (1.9)[45]
$X=NO_2$	$Y=Br$	p-Dioxane	306	2.7	4.4 (1.9)[45]
$X=NO_2$	$Y=OH$	p-Dioxane	334	4.9	7.7 (1.9)[45]
$X=NO_2$	$Y=OCH_3$	p-Dioxane	332	4.5	9.2 (1.9)[45]
$X=NO_2$	$Y=NH_2$	Chloroform	372	5.0	24 (1.9)[45]
$X=NO_2$	$Y=NH_2$	NMP		7.8	24 (1.9)[45]
$X=NO_2$	$Y=N(CH_3)_2$	Chloroform	390	5.5	50 (1.9)[45]
		Toluene	401	7.5	89 (1.06)[271]
$X=NO_2$	$Y=N(C_6H_{13})_2$	$C_6H_{11}CH_3$	400	7.5	73 (1.06)[271]
	Fluorene derivatives				
	X–(fluorene)–Y				
$X=CN$	$Y=H$	p-Dioxane	284	3.9	3.0 (1.9)[45]
$X=NO_2$	$Y=H$	p-Dioxane	328	4.1	5.1 (1.9)[45]
$X=NO_2$	$Y=Br$	p-Dioxane	330	2.8	6.0 (1.9)[45]
$X=NO_2$	$Y=OCH_3$	p-Dioxane	356	4.7	11 (1.9)[45]
$X=NO_2$	$Y=N(CH_3)_2$	p-Dioxane	410	5.6	40 (1.9)[45]
		Chloroform	417	6.0	55 (1.9)[45]
	Diphenyl ethers and derivatives				
	$CH_3O-C_6H_4-O-C_6H_4-SO_2-$	Chloroform	252	5.2	6.3 (1.9)[43]
	$H_2N-C_6H_4-O-C_6H_4-NO_2$	p-Dioxane		6.2	15 (1.06)[210]
		p-Dioxane	296	4.9	4.5 (1.9)[43]

(continued)

TABLE 7.1 (Continued)

Structure	Solvent	λ_{max}(nm)	μ (10^{-18} esu)	$\beta_{\mu}(\lambda)^{ref.}$ (10^{-30} esu)
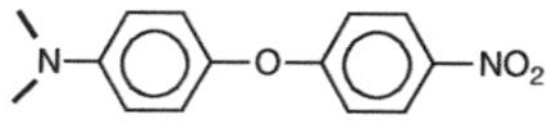 N — O — NO_2	*p*-Dioxane	294	5.2	5.3 (1.9)[43]
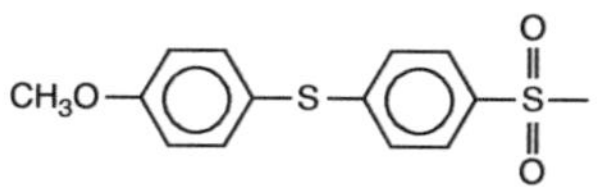 CH_3O — S — SO_2 —	Chloroform	295	5.3	8.9 (1.9)[43]
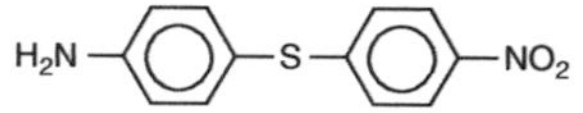 H_2N — S — NO_2	Benzene		6.0	26 (1.06)[210]
	Acetone	341	5.8	28 (1.06)[16]
	p-Dioxane	334	5	8.7 (1.9)[43]
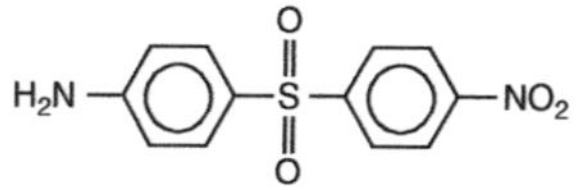 H_2N — SO_2 — NO_2	Acetone	332	8	19 (1.06)[16]
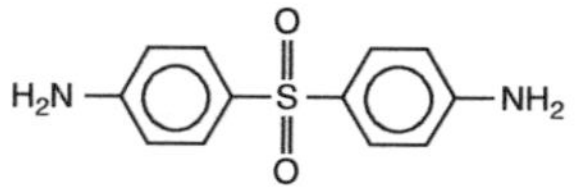 H_2N — SO_2 — NH_2	Acetone	295	7.8	8 (1.06)[16]
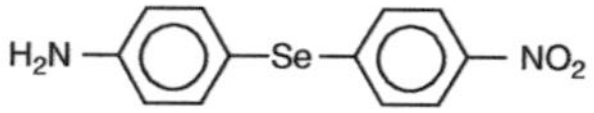 H_2N — Se — NO_2	*p*-Dioxane		5.9	27 (1.06)[210]
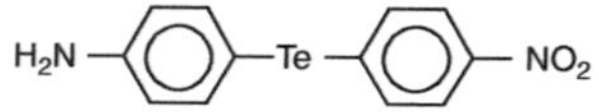 H_2N — Te — NO_2	*p*-Dioxane		5.6	27 (1.06)[210]

	$n=1$	Chloroform	320	6	16 (1.06)[164]
	$n=2$	Chloroform	334	7	22 (1.06)[164]
	$n=3$	Chloroform	385	6.8	38 (1.06)[164]
	Tolane derivatives				
$X=SO_2CH_3$	$Y=H$	*p*-Dioxane	310	5.3	3.8 (1.06)[32]
$X=SO_2CH_3$	$Y=OCH_3$	*p*-Dioxane	310	5.9	11 (1.06)[32]
$X=SO_2CH_3$	$Y=CH_3S$	*p*-Dioxane	320	5.2	16 (1.06)[32]
$X=SO_2CH_3$	$Y=NH_2$	Chloroform	338	6.5	13 (1.9)[45]
$X=SO_2CH_3$	$Y=N(CH_3)_2$	*p*-Dioxane	358	7.5	5.6(1.06)[32]
$X=SO_2CF_3$	$Y=OCH_3$	*p*-Dioxane	327	6.2	21 (1.06)[32]
$X=SO_2CF_3$	$Y=N(CH_3)_2$	*p*-Dioxane	388	8.4	40 (1.06)[32]
$X=CO_2CH_3$	$Y=SCH_3$	Chloroform	328	2.9	8 (1.9)[45]
$X=CO_2CH_3$	$Y=NH_2$	Chloroform	332	3.8	15 (1.9)[45]
$X=COCH_3$	$Y=SCH_3$	Chloroform	336	3.7	9.8(1.9)[45]
$X=COCH_3$	$Y=NH_2$	Chloroform	334	3.3	12 (1.9)[45]
$X=COC_6H_5$	$Y=NH_2$	Chloroform	352	3.7	19 (1.9)[45]
$X=CN$	$Y=SCH_3$	Chloroform	333	4.0	15 (1.9)[45]
$X=CN$	$Y=NH_2$	Chloroform	342	5.2	20 (1.9)[45]
$X=CN$	$Y=NHCH_3$	Chloroform	358	5.7	27 (1.9)[45]
$X=CN$	$Y=N(CH_3)_2$	Chloroform	372	6.1	29 (1.9)[45]
$X=NO_2$	$Y=H$	*p*-Dioxane	326	4.6	16 (1.06)[32]
$X=NO_2$	$Y=Br$	Chloroform	335	3.0	10 (1.9)[45]
$X=NO_2$	$Y=OCH_3$	*p*-Dioxane	356	4.4	14 (1.9)[45]
$X=NO_2$	$Y=SCH_3$	Chloroform	362	4.0	20 (1.9)[45]
$X=NO_2$	$Y=NH_2$	Chloroform	380	5.5	24 (1.9)[45]
		NMP	410	5.5	40 (1.9)[45]
$X=NO_2$	$Y=NHCH_3$	Chloroform	400	5.7	46 (1.9)[45]
$X=NO_2$	$Y=N(CH_3)_2$	Chloroform	415	6.1	46 (1.9)[45]
		p-Dioxane	402	7.1	102 (1.06)[32]
		Chloroform	416	6.6	33 (β_0)[177]

(continued)

TABLE 7.1 (Continued)

Structure		Solvent	λ_{max}(nm)	μ (10^{-18} esu)	$\beta_\mu(\lambda)^{ref.}$ (10^{-30} esu)
X=NO_2	Y=$N(C_6H_5)_2$	Chloroform	418	4.8	28 (β_0)[177]
	Stilbene derivatives				
X=H	Y=OCH_3		320	2.8	6.1 (1.06)[175]
X=H	Y=NH_2	Benzene		2.1	12 (1.06)[189]
		p-Dioxane	332	2.2	7.4 (1.9)[46]
X=H	Y=$N(CH_3)_2$	Benzene		2.4	29 (1.06)[189]
		p-Dioxane	340	2.1	10 (1.9)[46]
X=Cl	Y=H	Benzene		1.5	3.6 (1.06)[189]
X=Cl	Y=$N(CH_3)_2$	Chloroform		4.0	42 (1.06)[189]
X=CF_3	Y=OCH_3	*p*-Dioxane	323	4.3	12 (1.06)[244]
			326	4.2	16 (1.06)[175]
X=CF_3	Y=OH	*p*-Dioxane	327	4.7	12 (1.06)[244]
X=SO_2CH_3	Y=H	*p*-Dioxane	364	4.4	58 (1.06)[32]
X=SO_2CH_3	Y=OCH_3	Chloroform	336	6.5	10 (1.9)[46]
		p-Dioxane	335	6.1	9.1 (1.06)[32]
X=SO_2CH_3	Y=SC_6H_5	*p*-Dioxane	344	4.4	19 (1.06)[32]
X=SO_2CH_3	Y=$N(CH_3)_2$	*p*-Dioxane	376	6.9	66 (1.06)[32]
X=SO_2CH_3	Y=$N(CH_2C_2H_3)_2$	Chloroform	391	$\mu\beta=57\times10^{-47}$	(1.9)[264]
X=SO_2CF_3	Y=OCH_3	*p*-Dioxane	347	7.8	14 (1.9)[46]
		p-Dioxane	350	6.6	34 (1.06)[32]
X=$SO_2C_6F_{13}$	Y=$N(CH_3)_2$	Chloroform		8.0	59 (1.9)[43]
X=$COCF_3$	Y=OCH_3	*p*-Dioxane	368	4.2	16 (1.9)[46]
X=COH	Y=$N(CH_3)_2$	Chloroform	360	3.5	24 (1.9)[254]
X=CN	Y=OH	*p*-Dioxane	344	4.5	13 (1.9)[46]
X=CN	Y=OCH_3	DMSO	342	$\mu\beta=82\times10^{-47}$	(1.9)[55]
		Chloroform	340	3.8	19 (1.9)[46]
X=CN	Y=$N(CH_3)_2$	DMSO	390	$\mu\beta=98\times10^{-48}$	(1.9)[55]
		Chloroform	382	5.7	36 (1.9)[46]
X=NO_2	Y=H	Benzene		4.6	29 (1.06)[189]
		p-Dioxane	345	4.2	11 (1.9)[46]
X=NO_2	Y=CH_3	DMSO	368	$\mu\beta=20\times10^{-48}$	(1.9)[55]
		p-Dioxane	351	4.7	15 (1.9)[46]

$X=NO_2$	$Y=Cl$	Chloroform		3.1	39 (1.06)[189]
$X=NO_2$	$Y=Br$	*p*-Dioxane	344	3.2	14 (1.9)[46]
		Chloroform	356	3.4	18 (1.9)[46]
$X=NO_2$	$Y=OH$	Chloroform			93 (1.06HR)[48]
		p-Dioxane	370	5.5	17 (1.9)[46]
$X=NO_2$	$Y=OC_6H_5$	*p*-Dioxane	350	4.6	18 (1.9)[46]
$X=NO_2$	$Y=OCH_3$	*p*-Dioxane		5.7	81 (1.06)[98]
		Chloroform			105 (1.06HR)[48]
		p-Dioxane	364	4.5	28 (1.9)[46]
		Chloroform	370	4.5	34 (1.9)[46]
		p-Dioxane	364	4.5	60 (1.06)[32]
$X=NO_2$	$Y=SCH_3$	*p*-Dioxane	374	4.3	26 (1.9)[46]
		Chloroform	380	4.3	34 (1.9)[46]
		p-Dioxane	378	5.1	68 (1.06)[32]
$X=NO_2$	$Y=NH_2$	Acetone		7.5	260 (1.06)[189]
		Chloroform	402	5.1	40 (1.9)[46]
$X=NO_2$	$Y=N(CH_3)_2$	Acetone		7.4	450(1.06)[193]
		DMSO	447	$\mu\beta=42\times10^{-46}$	(1.9)[55]
		DMSO	453	$\mu\beta=76\times10^{-47}$	(1.36)[231]
				7.1	323 (1.06S)[195]
		Chloroform	427	6.6	73 (1.9)[46]
		NMP		7.2	70 (1.9)[46]
		p-Dioxane		$\mu\beta=58\times10^{-47}$	(1.9)[107]
		Chloroform	438	6.7	42 (β_0)[177]
$X=NO_2$	$Y=N(CH_2C_2H_3)_2$	Chloroform	452	$\mu\beta=57\times10^{-47}$	(1.9)[264]
$X=NO_2$	$Y=N(C_6H_5)_2$	Chloroform	436	4.8	37 (β_0)[177]
$X=NO_2$	$Y=COOCH_3$	CH_2Cl_2	350	4.0	4 (1.9)[46]
$X=NO_2$	$Y=COH$	*p*-Dioxane	352	4.1	6 (1.9)[46]
$X=CHC(CN)_2$	$Y=N(CH_3)_2$	Chloroform		7.8	210 (1.9)[43]
$X=CHC(CN)_2$	$Y=N(C_2H_5)_2$	CH_2Cl_2	485	8.2	180 (1.58)[54]
		p-Dioxane	468	$\mu\beta=11\times10^{-46}$	(1.9)[106]
$X=Br$	$Y=OCH_3$	*p*-Dioxane	325	4.0	25 (1.9)[46]

(continued)

TABLE 7.1 (Continued)

Structure	Solvent	λ_{max}(nm)	μ (10^{-18} esu)	$\beta_{\mu}(\lambda)^{ref.}$ (10^{-30} esu)
	Chloroform	438	7	96 (1.9)[46]
	Chloroform	360	3.8	4.4 (1.9)[46]
	Chloroform	370	3.7	1.6 (1.9)[46]
	Chloroform	390	3.5	3.8 (1.9)[46]
	Chloroform	320	4.4	5.5 (1.9)[46]
	Chloroform	292	3.9	4.5 (1.9)[46]

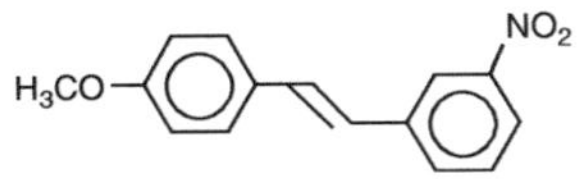 H_3CO, NO_2	*p*-Dioxane	318	3.9	5.3 (1.9)[46]
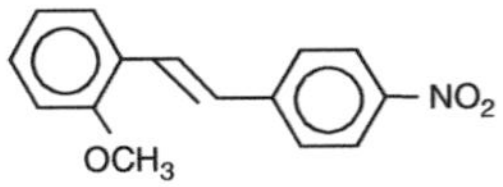 OCH_3, NO_2	Chloroform	362	5.0	22 (1.9)[46]
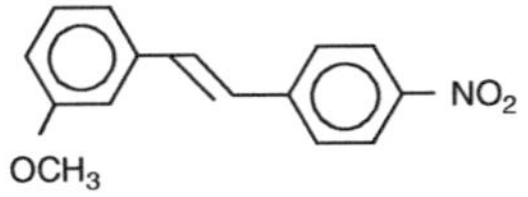 OCH_3, NO_2	Chloroform	352	4.0	21 (1.9)[46]
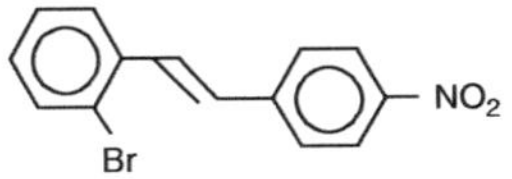 Br, NO_2	Chloroform	346	4.6	12 (1.9)[46]
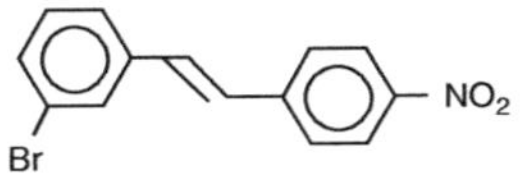 Br, NO_2	Chloroform	346	3.4	14 (1.9)[46]
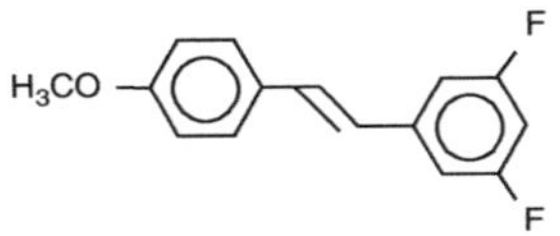 H_3CO, F, F		322	3.1	9.2 (1.06)[175]

(continued)

TABLE 7.1 (Continued)

Structure	Solvent	λ_{max}(nm)	μ (10^{-18} esu)	$\beta_\mu(\lambda)^{ref.}$ (10^{-30} esu)
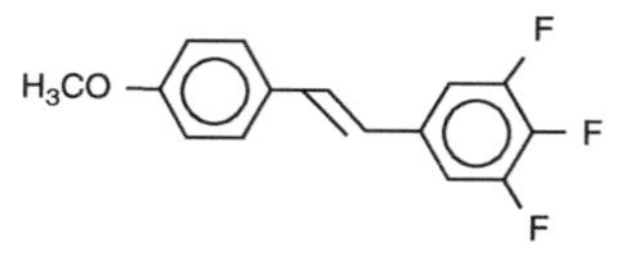		320	4.6	6.8 (1.06)[175]
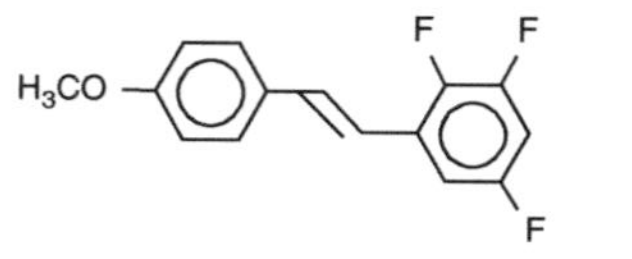		322	3.4	10 (1.06)[175]
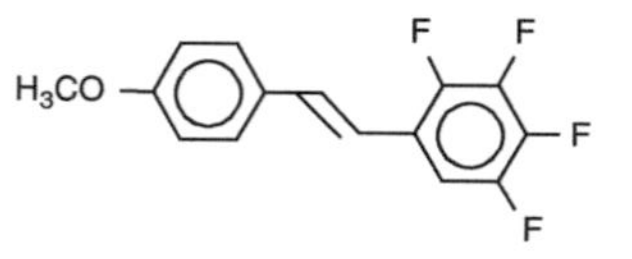		324	3.5	12 (1.06)[175]
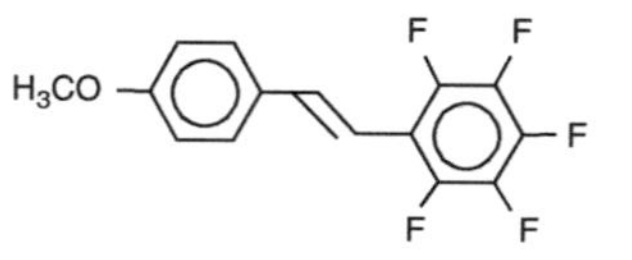		322	3.3	16 (1.06)[175]
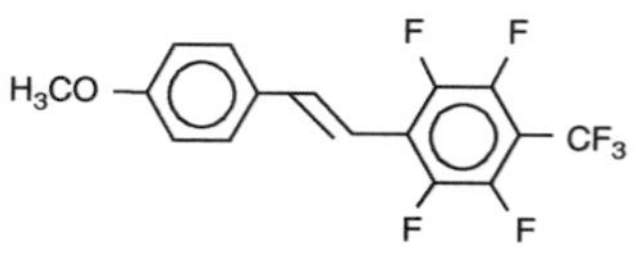		334	4.0	23 (1.06)[175]

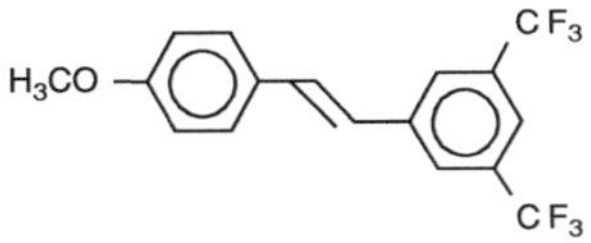		326	4.2	11 (1.06)[175]
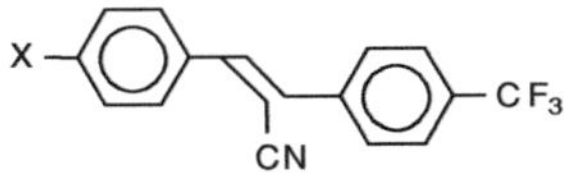				
X=H	p-Dioxane	314	4.7	4.5 (1.06)[244]
X=Cl	p-Dioxane	318	4.2	5.1 (1.06)[244]
X=Br	p-Dioxane	320	4.6	8.1 (1.06)[244]
X=CH_3	p-Dioxane	323	4.8	7.4 (1.06)[244]
X=OCH_3	p-Dioxane	340	5.3	15 (1.06)[244]
X=OH	p-Dioxane	348	5.5	16 (1.06)[244]
X=SCH_3	p-Dioxane	362	5.0	16 (1.06)[244]
X=$N(CH_3)_2$	p-Dioxane	410	6.4	29 (1.06)[244]
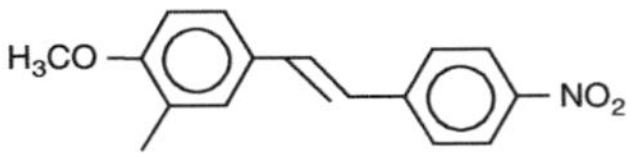	p-Dioxane	366	5.2	26 (1.9)[46]
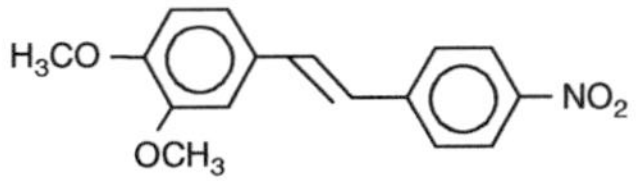	p-Dioxane	380	4.7	23 (1.9)[46]
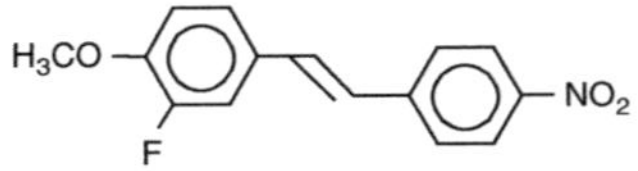	p-Dioxane	363	4.1	18 (1.9)[46]

(continued)

TABLE 7.1 (Continued)

Structure	Solvent	λ_{max}(nm)	μ (10^{-18} esu)	$\beta_{\mu}(\lambda)^{ref.}$ (10^{-30} esu)
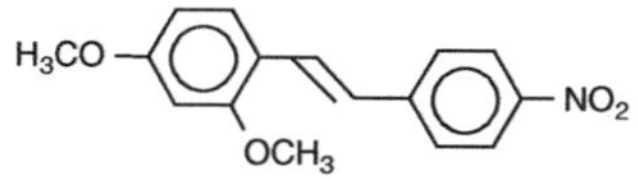 H_3CO, OCH_3, NO_2	*p*-Dioxane	395	4.8	32 (1.9)[46]
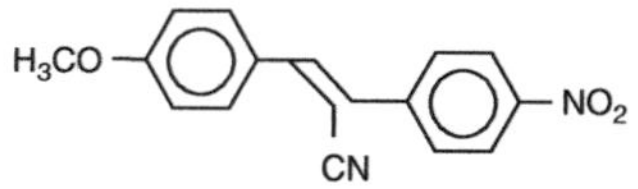 H_3CO, CN, NO_2	*p*-Dioxane	361	5.3	21 (1.9)[46]
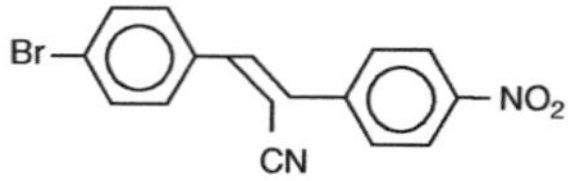 Br, CN, NO_2	*p*-Dioxane	340	4.6	8 (1.9)[46]
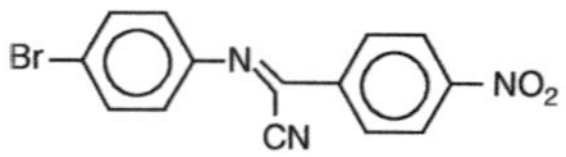 Br, N, CN, NO_2	*p*-Dioxane	382	4.1	2.1 (1.9)[46]
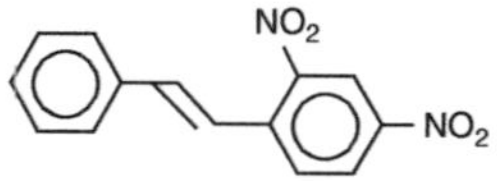 NO_2, NO_2	*p*-Dioxane	355	4	10 (1.9)[46]
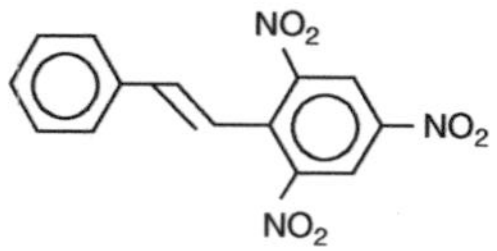 NO_2, NO_2, NO_2	*p*-Dioxane	354	2	5 (1.9)[46]

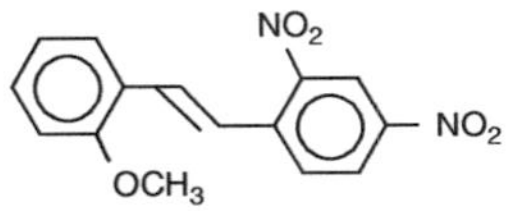	*p*-Dioxane	378	5.0	12 (1.9)[255]
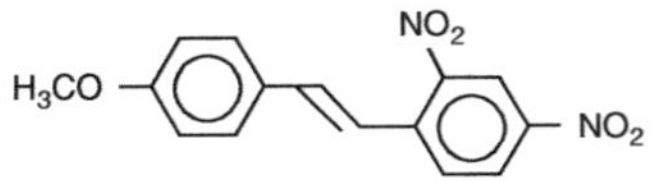	*p*-Dioxane	384	4.7	22 (1.9)[46]
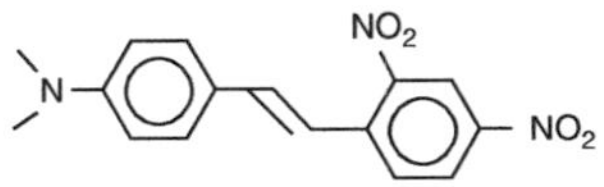	*p*-Dioxane *p*-Dioxane	466	7.0 $\mu\beta = 66 \times 10^{-47}$	57 (1.9)[255] (1.9)[107]
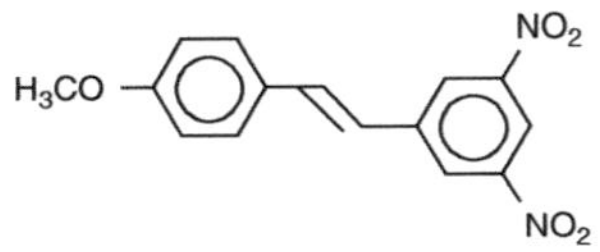	Chloroform		4.1	15 (1.9)[43]
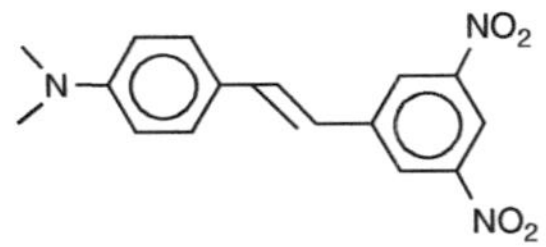	Chloroform		6.2	45 (1.9)[43]
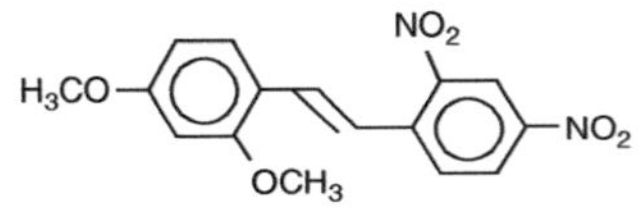	*p*-Dioxane	404	5.6	25 (1.9)[46]

(continued)

TABLE 7.1 (Continued)

Structure	Solvent	λ_{max}(nm)	μ (10^{-18} esu)	$\beta_{\mu}(\lambda)^{ref.}$ (10^{-30} esu)
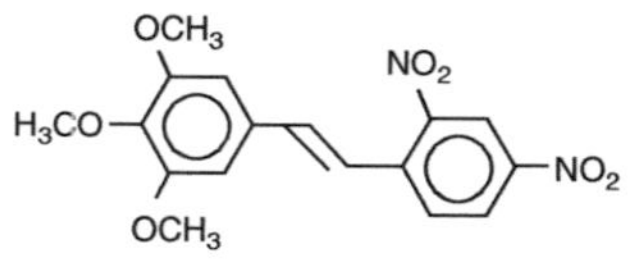 H_3CO, OCH_3, OCH_3, NO_2, NO_2	*p*-Dioxane	390	3.1	11 (1.9)[46]
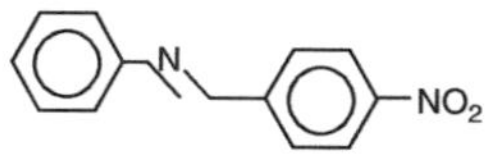 N, NO_2	*p*-Dioxane	346	4.4	4.9 (1.9)[46]
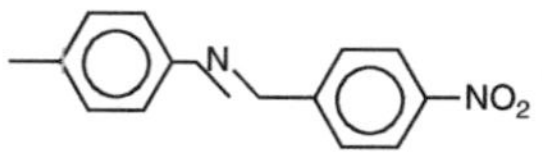 N, NO_2	*p*-Dioxane	351	4.7	15 (1.9)[46]
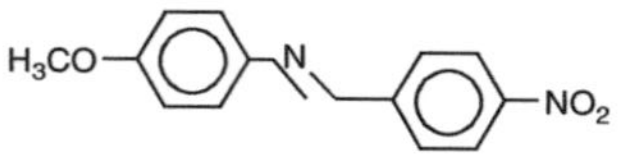 H_3CO, N, NO_2	*p*-Dioxane	376	4.4	14 (1.9)[46]
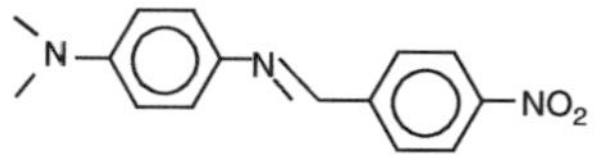 N, N, NO_2	DMSO	458	$\mu\beta = 50 \times 10^{-47}$	(1.36)[231]
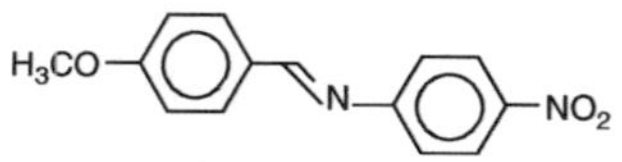 H_3CO, N, NO_2	*p*-Dioxane	349	5.4	6.6 (1.9)[46]
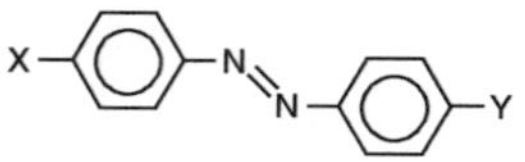 X, N, N, Y				

$X=SO_2CH_3$	$Y=N(C_4H_9)_2$	Chloroform	461	$\mu\beta=51\times10^{-47}$	(1.9)[264]
$X=NO_2$	$Y=NH_2$	*p*-Dioxane	420	5.8	29 (1.9)[46]
		DMSO	470	$\mu\beta=77\times10^{-47}$	(1.36)[231]
$X=NO_2$	$Y=N(CH_3)_2$	Chloroform	498	$\mu\beta=13\times10^{-46}$	(1.9)[264]
		Chloroform	480	7.7	40 (β_0)[177]
$X=NO_2$	$Y=N(C_2H_5)_2$	Chloroform	494	8.0	50 (β_0)[177]
$X=N0_2$	$Y=N(C_2H_5)C_2H_4)OH$	CH_2Cl_2	480	8.9	90 (1.57)[54]
		p-Dioxane	455	7.0	49 (1.9)[46]
		DMSO	508	$\mu\beta=11\times10^{-46}$	(1.36)[231]
$X=NO_2$	$Y=N(C_6H_5)_2$	Chloroform	486	5.9	54 (β_0)[177]
$X=CHC(CN)_2$	$Y=N(CH_3)_2$	DMSO	492	$\mu\beta=27\times10^{-46}$	(1.36)[231]
$X=C_2(CN)_3$	$Y=N(C_2H_5)_2$	DMSO		$\mu\beta=41\times10^{-46}$	(1.58)[264]
$X=C_2(CN)_3$	$Y=N(C_2H_5)(C_2H_4)OH$	CH_2Cl_2	513	10	190 (1.57)[54]
	Pyridine derivatives				
	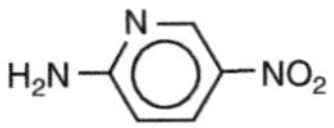	Acetone		6.5	3.7 (1.9)[46]
	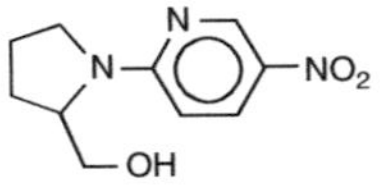	*p*-Dioxane	376	7.2	18 (1.06)[25]
		p-Dioxane			11 (1.32)[25]
		p-Dioxane			11 (1.91)[25]
		p-Dioxane		5.5	17 (1.06)[262]
	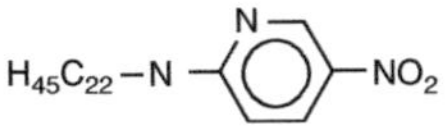	*p*-Dioxane	357	6.7	13 (1.06)[25]
		p-Dioxane			9.3 (1.32)[25]
		p-Dioxane			6.0 (1.91)[25]
	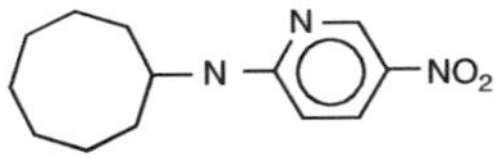	*p*-Dioxane	361	6.8	22 (1.06)[25]
		p-Dioxane			12 (1.32)[25]
		p-Dioxane			11 (1.91)[25]

(continued)

TABLE 7.1 (Continued)

Structure	Solvent	λ_{max}(nm)	μ (10^{-18} esu)	$\beta_{\mu}(\lambda)^{ref.}$ (10^{-30} esu)
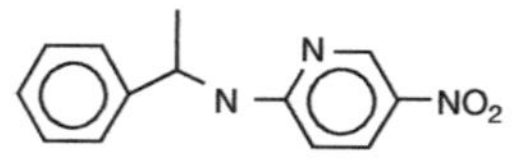 N, N, NO_2	*p*-Dioxane		6.1	15 (1.06)[262]
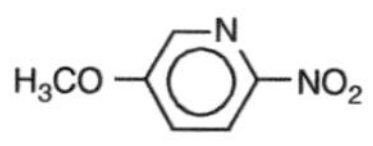 H_3CO, N, NO_2	*p*-Dioxane		3.5	2.2 (1.9)[46]
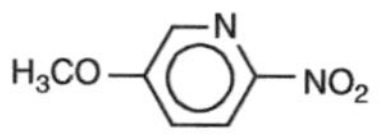 H_3CO, N, NO_2	*p*-Dioxane		3.5	2.2 (1.9)[46]
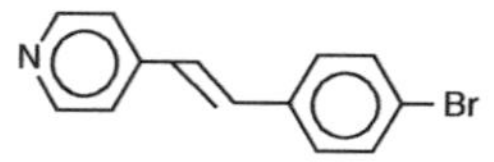 N, Br	Chloroform		0.9	10 (1.9)[46]
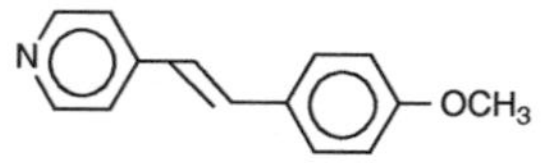 N, OCH_3	Chloroform	335	3.8	16 (1.9)[43]
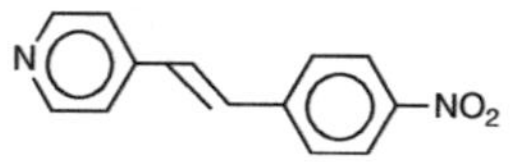 N, NO_2	Chloroform		1.3	8 (1.9)[46]

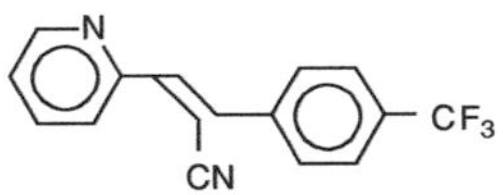	*p*-Dioxane	312	6.2	4.3 (1.06)[244]
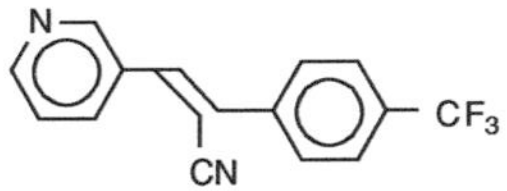	*p*-Dioxane	309	5.5	4.4 (1.06)[244]
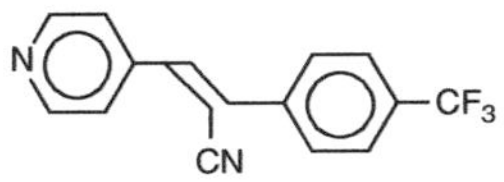	*p*-Dioxane	307	5.1	4.2 (1.06)[244]
Coumarin derivatives				
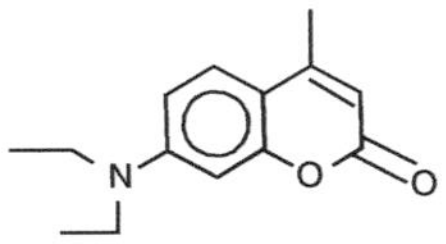	Chloroform		5	15 (1.9)[43]
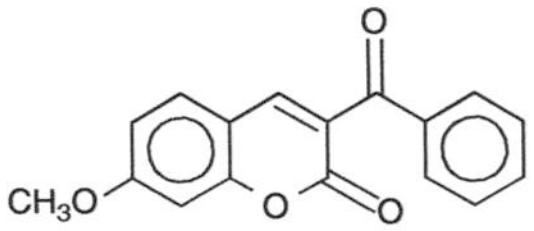	Chloroform		5	8 (1.9)[43]
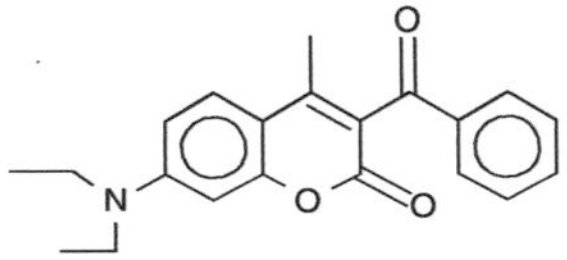	Chloroform		7.3	30 (1.9)[43]

(continued)

TABLE 7.1 (Continued)

Structure	Solvent	λ_{max}(nm)	μ (10^{-18} esu)	$\beta_{\mu}(\lambda)^{ref.}$ (10^{-30} esu)
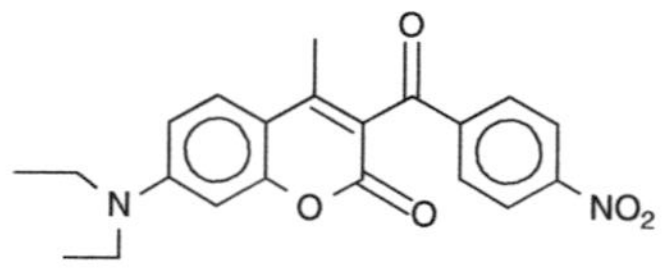	Chloroform		8.8	50 (1.9)[43]
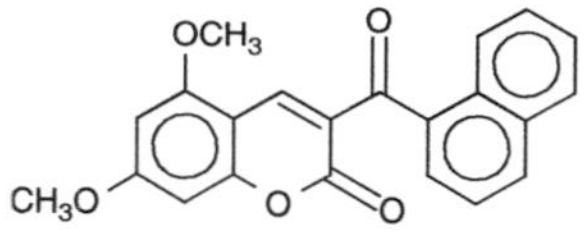	Chloroform		5.8	9.5 (1.9)[43]
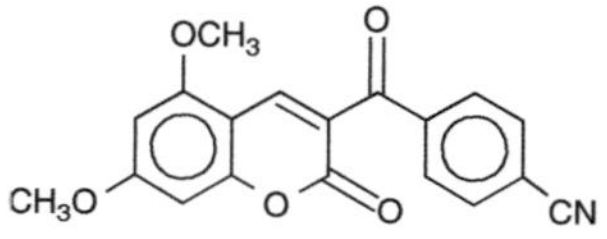	Chloroform		7.3	15 (1.9)[43]
Other polycyclic aromatic derivatives				
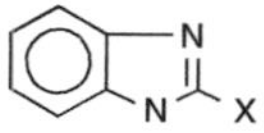 X=H	*p*-Dioxane		3.6	<1 (1.9)[43]
X=SH	*p*-Dioxane		4	<1 (1.9)[43]
X=COOH	*p*-Dioxane		2.6	<1 (1.9)[43]
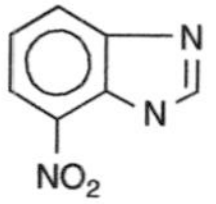	*p*-Dioxane			
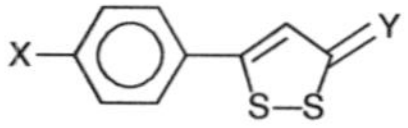 X=H Y=O	*p*-Dioxane	340		4.8 (1.34)[28]
X OCH$_3$ Y=O	*p*-Dioxane	312		13 (1.34)[28]

X=H Y=S	p-Dioxane	438		11 (1.34)[28]
$X=OCH_3$ Y=S	p-Dioxane	438		21 (1.34)[28]
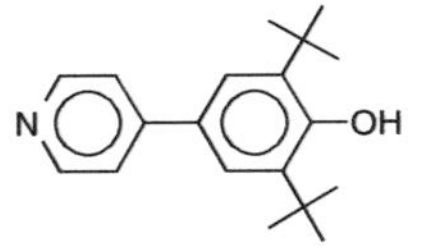		333	4.0	$\beta_0=12(S)$[7]
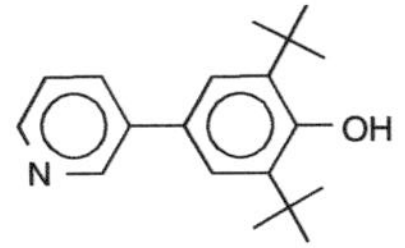		356	6.2	$\beta_0=11(S)$[7]
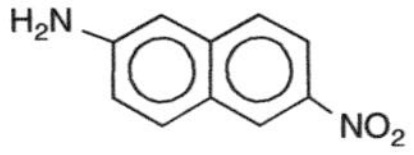	DMSO		$\mu\beta=23\times10^{-47}$	(1.36)[231]
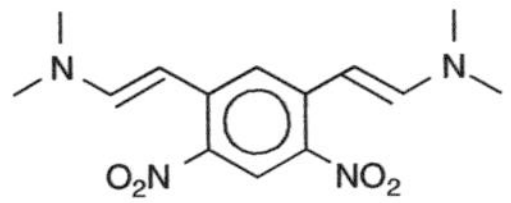	Chloroform		7.6	36 (1.9)[43]
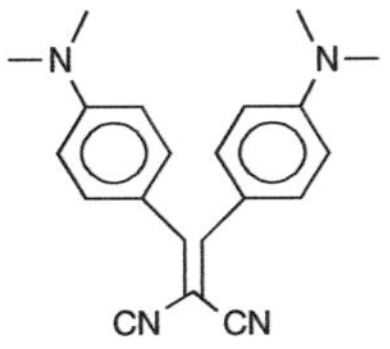	Chloroform		7.3.	20 (1.9)[43]

(continued)

TABLE 7.1 (Continued)

Structure	Solvent	λ_{max}(nm)	μ (10^{-18} esu)	$\beta_{\mu}(\lambda)^{ref.}$ (10^{-30} esu)
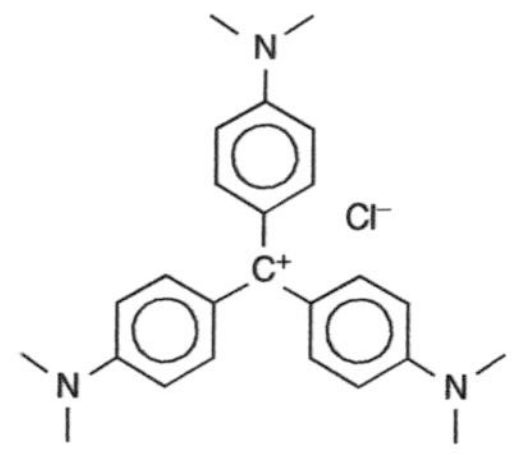	CCl_4	590		580 (1.06HR)[291]
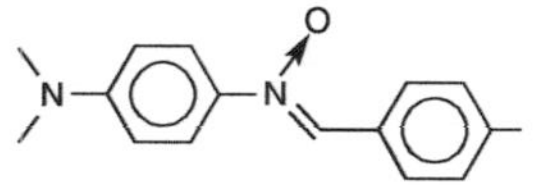	Chloroform		4	10 (1.9)[43]
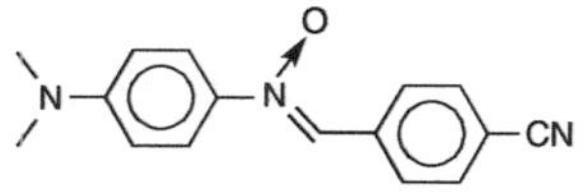	Chloroform	400	6.7	31 (1.9)[43]
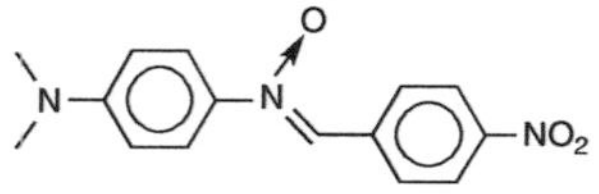	Chloroform	423	7	45 (1.9)[43]
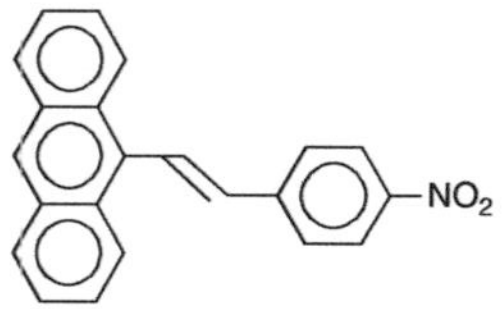	*p*-Dioxane		6	5 (1.9)[43]

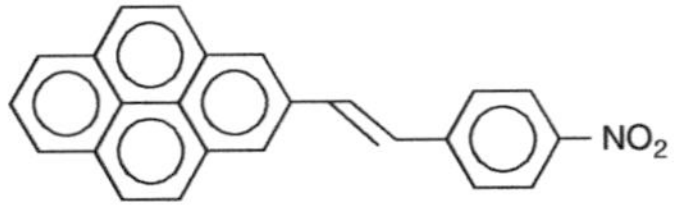	p-Dioxane		8	11 (1.9)[43]
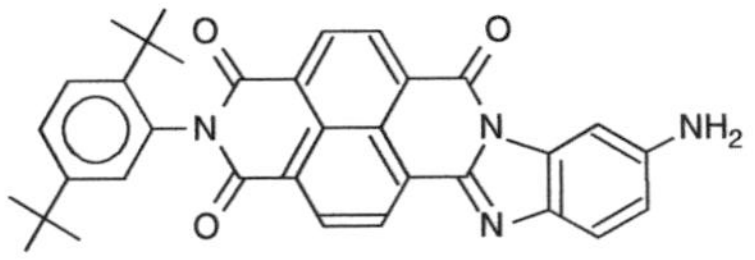	p-Dioxane	575	$\mu\beta = 30 \times 10^{-47}$	($\approx$2)[226]
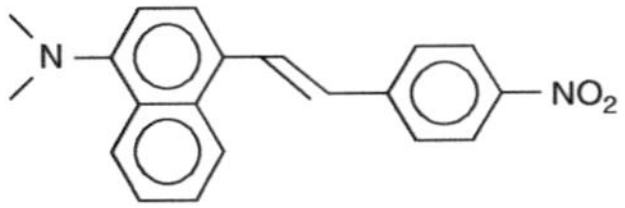	Chloroform		6.5	35 (1.9)[43]
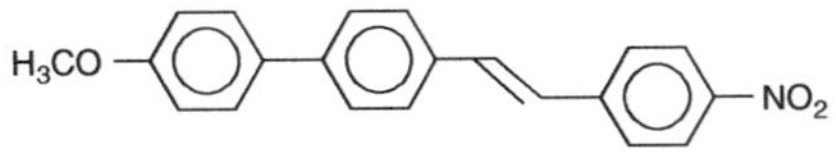	NMP	364	7	18 (1.9)[43]
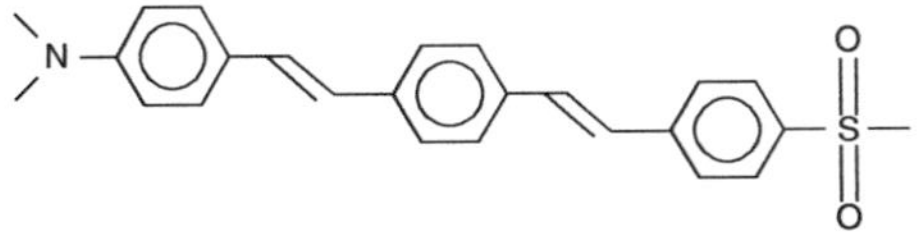	NMP	406	8	30 (1.9)[43]
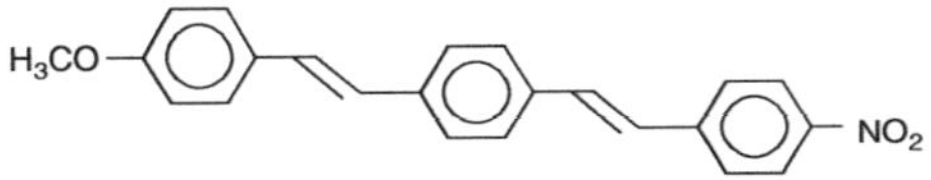	Chloroform	394	8	52 (1.9)[43]
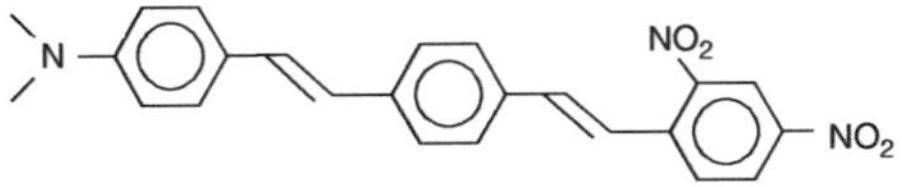	p-Dioxane		$\mu\beta = 13 \times 10^{-46}$	(1.9)[107]

(continued)

TABLE 7.1 (Continued)

	Structure		Solvent	λ_{max}(nm)	μ (10^{-18} esu)	$\beta_{\mu}(\lambda)^{ref.}$ (10^{-30} esu)
	Polyene derivatives					
	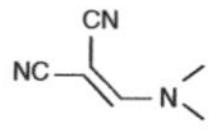		Chloroform	352	7.6	1 (1.9)[156]
	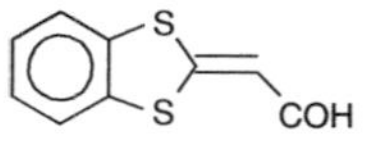		Chloroform	372	$\mu\beta = 30 \times 10^{-48}$	(1.34)[14]
$n=1$	 X=COH	Y=N$(CH_3)_2$	Chloroform	284	6.3	3.3 (1.9)[156]
$n=2$	X=COH	Y=N$(C_2H_5)_2$	Chloroform	363	6.5	20 (1.9)[156]
$n=3$	X=COH	Y=N$(CH_3)_2$	Chloroform	422	6.9	53 (1.9)[156]
$n=1$	X=CHC$(CN)_2$	Y=N$(CH_3)_2$	Chloroform	374	8.9	6.1 (1.9)[156]
$n=2$	X=CHC$(CN)_2$	Y=N$(C_2H_5)_2$	Chloroform	476	10.7	45 (1.9)[156]
$n=3$	X=CHC$(CN)_2$	Y=N$(CH_3)_2$	Chloroform	550	9.9	211 (1.9)[156]
$n=1$	X=NO_2	Y=N$(CH_3)_2$	Chloroform		6.3	4.8 (1.9)[43]
$n=2$	X=NO_2	Y=N$(CH_3)_2$	Chloroform		6.7	21 (1.9)[43]
$n=3$	X=NO_2	Y=N$(CH_3)_2$	Chloroform		8.4	73 (1.9)[43]
$n=3$	X=SO_2CF_3	Y=N$(CH_3)_2$	Chloroform		9.8	40 (1.9)[43]
	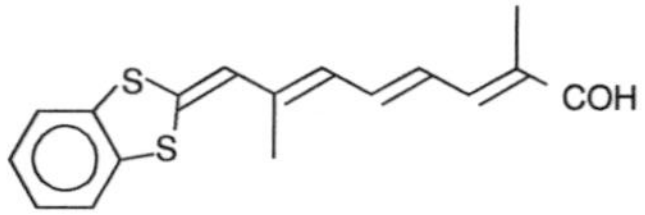		Chloroform	456	$\mu\beta = 12 \times 10^{-46}$	(1.34)[14]
	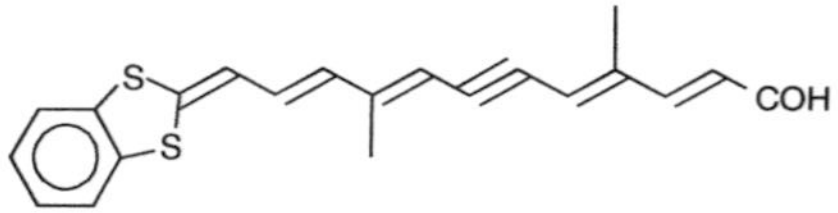		Chloroform	466	$\mu\beta = 22 \times 10^{-46}$	(1.34)[14]

	Chloroform	500	$\mu\beta = 73 \times 10^{-46}$	(1.34)[14]
	DMSO	380	$\mu\beta = 23 \times 10^{-47}$	(1.06)[101]
	DMSO	470	$\mathrm{Re}(\mu\beta) = 13 \times 10^{-46}$ $\mathrm{Im}(\mu\beta) = 15 \times 10^{-46}$	(1.06)[101] (1.06)[101]
	DMSO	480	$\mathrm{Re}(\mu\beta) = 14 \times 10^{-46}$ $\mathrm{Im}(\mu\beta) = 35 \times 10^{-46}$	(1.06)[101] (1.06)[101]
	DMSO	440	$\mathrm{Re}(\mu\beta) = 15 \times 10^{-46}$ $\mathrm{Im}(\mu\beta) = 17 \times 10^{-46}$	(1.06)[101] (1.06)[101]
	$Cl_2CHCHCl_2$	470	$\mathrm{Re}(\mu\beta) = 10 \times 10^{-47}$ $\mathrm{Im}(\mu\beta) = 44 \times 10^{-46}$	(1.06)[101] (1.06)[101]
	Chloroform	476	$\mu\beta = 96 \times 10^{-47}$	(1.9)[53]

(continued)

TABLE 7.1 (Continued)

	Structure		Solvent	λ_{max}(nm)	μ (10^{-18} esu)	$\beta_\mu(\lambda)^{ref.}$ (10^{-30} esu)
			$Cl_2CHCHCl_2$	500	$Re(\mu\beta) = -20 \times 10^{-46}$	$(1.06)^{101}$
					$Im(\mu\beta) = 25 \times 10^{-46}$	$(1.06)^{101}$
			$Cl_2CHCHCl_2$	570	$Re(\mu\beta) = 35 \times 10^{-46}$	$(1.06)^{101}$
					$Im(\mu\beta) = 47 \times 10^{-46}$	$(1.06)^{101}$
			Chloroform	566	$\mu\beta = 44 \times 10^{-46}$	$(1.9)^{53}$
					$Re(\mu\beta) = -12 \times 10^{-45}$	$(1.06)^{101}$
			$Cl_2CHCHCl_2$	510	$Im(\mu\beta) = 73 \times 10^{-46}$	$(1.06)^{101}$
	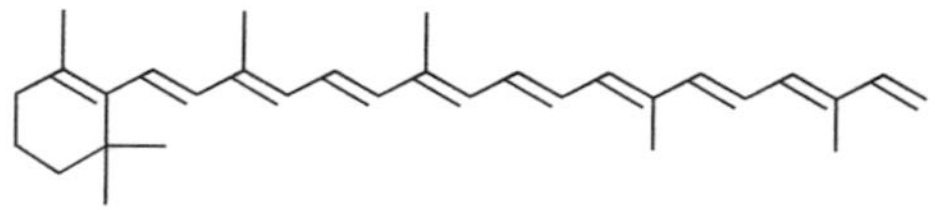		Chloroform	502	$\mu\beta = 15 \times 10^{-46}$	$(1.9)^{53}$
	α-Phenylpolyene Derivatives 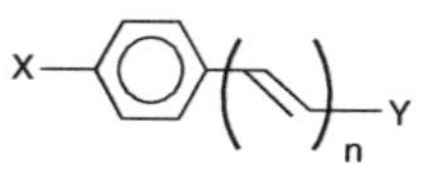					
$n=2$	X=COH	$Y=OCH_3$	Chloroform	350	4.3	28 $(1.9)^{45}$
$n=3$	X=COH	$Y=OCH_3$	Chloroform	376	4.6	42 $(1.9)^{45}$
$n=2$	X=COH	$Y=N(CH_3)_2$	Chloroform	412	6.0	52 $(1.9)^{45}$
$n=3$	X=COH	$Y=N(CH_3)_2$	Chloroform	434	6.3	88 $(1.9)^{45}$
			Chloroform		6.6	105 $(1.9)^{43}$

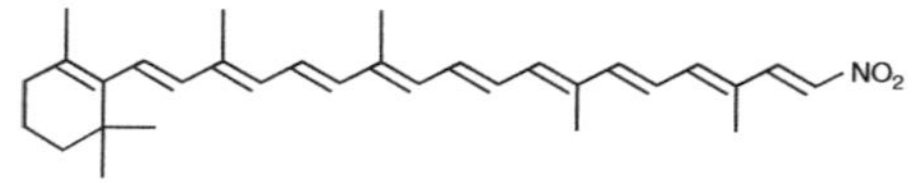

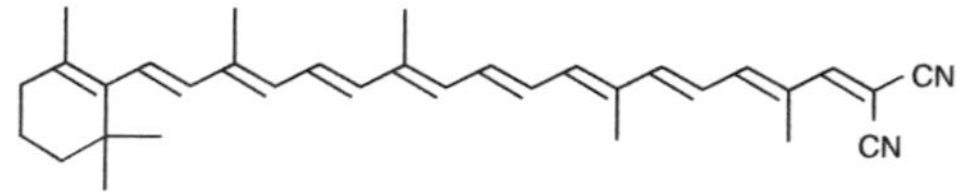

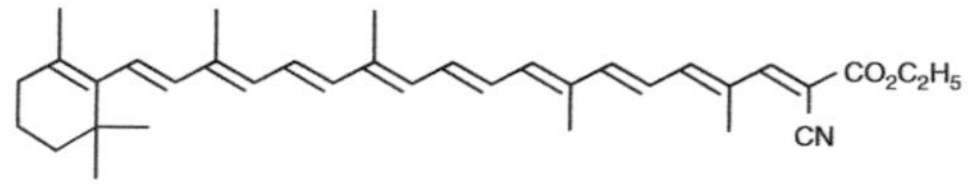

$n=4$	$X=COH$	$Y=N(CH_3)_2$	Chloroform		8.0	138 (1.9)[43]
$n=2$	$X=COCF_3$	$Y=N(CH_3)_2$	Chloroform		6.6	126 (1.9)[43]
$n=2$	$X=NO_2$	$Y=OCH_3$	DMSO	370	$\mu\beta=81\times10^{-48}$	(1.06)[102]
			Chloroform		4.6	42 (1.9)[45]
$n=3$	$X=NO_2$	$Y=OCH_3$	DMSO	400	$\mu\beta=30\times10^{-47}$	(1.06)[102]
$n=2$	$X=NO_2$	$Y=N(CH_3)_2$	Acetone		8.8	630 (1.06)[189]
			DMSO	460	$\mu\beta=17\times10^{-46}$	(1.06)[102]
			Chloroform	466	6.5	140 (1.9)[254]
$n=3$	$X=NO_2$	$Y=N(CH_3)_2$	DMSO	490	$\mu\beta=55\times10^{-46}$	(1.06)[102]
			Chloroform	487	6.6	240 (1.9)[254]
$n=4$	$X=NO_2$	$Y=N(CH_3)_2$	Chloroform	502	7.6	280 (1.9)[254]
$n=2$	$X=CHC(CN)_2$	$Y=N(CH_3)_2$	DMSO	500	$\mu\beta=36\times10^{-46}$	(1.06)[102]
			Chloroform	520	9.0	163 (1.9)[45]
$n=3$	$X=CHC(CN)_2$ 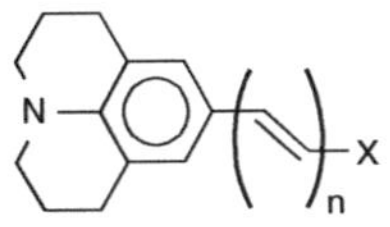	$Y=N(CH_3)_2$	Chloroform		8.8	432 (1.9)[43]
$n=3$		$X=COH$	Chloroform		7.1	162 (1.9)[43]
$n=3$		$X=NO_2$	Chloroform		7.8	287 (1.9)[43]
$n=4$		$X=NO_2$	Chloroform		$\mu\beta=26\times10^{-46}$	(1.9)[43]
$n=3$		$X=CHC(CN)_2$	Chloroform		8.7	485 (1.9)[43]
	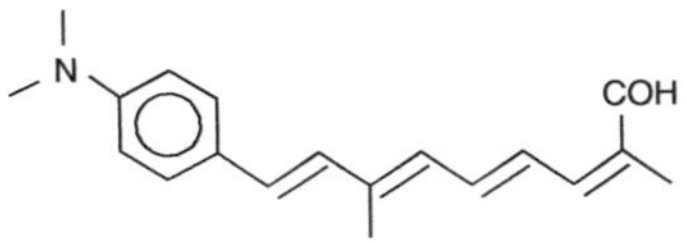		Chloroform	450	$\mu\beta=20\times10^{-46}$	(1.34)[14]
	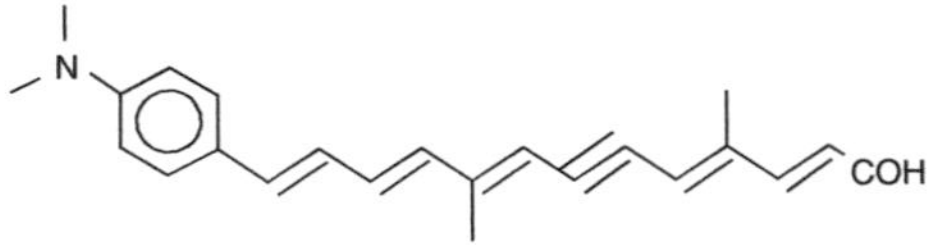		Chloroform	461	$\mu\beta=42\times10^{-46}$	(1.34)[14]

(continued)

TABLE 7.1 (Continued)

Structure			Solvent	λ_{max}(nm)	μ (10^{-18} esu)	$\beta_\mu(\lambda)^{ref.}$ (10^{-30} esu)
	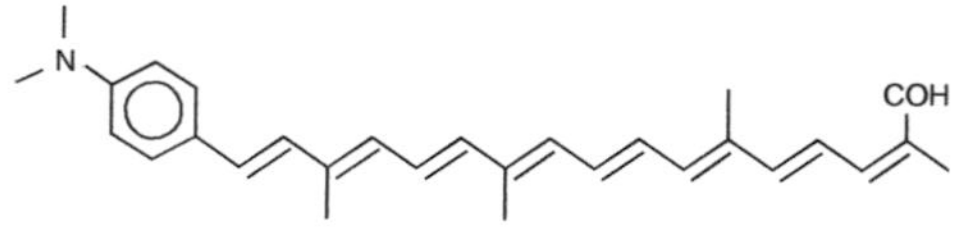		Chloroform	498	$\mu\beta=89\times10^{-46}$	(1.34)[14]
	Diphenylpolyene derivatives					
	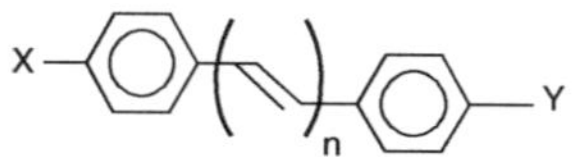					
$n=2$	X=CN	Y=OCH_3	Chloroform	360	4.3	27 (1.9)[45]
$n=3$	X=CN	Y=OCH_3	Chloroform	380	4.6	40 (1.9)[45]
$n=2$	X=NO_2	Y=Br	Chloroform	378	3.5	21 (1.9)[45]
$n=3$	X=NO_2	Y=Br	Chloroform	400	3.8	35 (1.9)[45]
$n=2$	X=NO_2	Y=OCH_3	*p*-Dioxane		6.0	135 (1.06)[98]
			Chloroform	397	4.8	47 (1.9)[45]
$n=3$	X=NO_2	Y=OCH_3	*p*-Dioxane		6.6	274 (1.06)[98]
			Chloroform	414	5.1	76 (1.9)[45]
$n=4$	X=NO_2	Y=OCH_3	*p*-Dioxane		6.7	367 (1.06)[98]
			Chloroform	430	5.8	55 (1.9)[45]
$n=5$	X=NO_2	Y=OCH_3	*p*-Dioxane		7.0	623 (1.06)[98]
$n=2$	X=NO_2	Y=SCH_3	Chloroform	398	4.5	101 (1.9)[45]
$n=2$	X=NO_2	Y=$N(CH_3)_2$	Chloroform	442	7.6	107 (1.9)[45]
			p-Dioxane		$\mu\beta=75\times10^{-47}$	(1.9)[107]
$n=3$	X=NO_2	Y=$N(CH_3)_2$	Chloroform	458	8.2	131 (1.9)[45]
$N=4$	X=NO_2	Y=$N(CH_3)_2$	Chloroform	464	9	190 (1.9)[45]
$N=2$	X=C_2HCN_2	Y=$N(CH_3)_2$	DMSO	481	$\mu\beta=13\times10^{-46}$	(1.36)[231]
	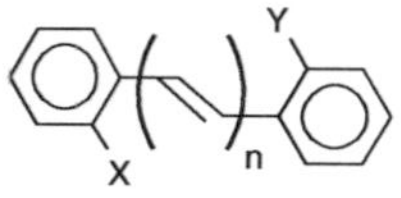					
$n=2$	X=CN	Y=OCH_3	Chloroform	354	3.8	4.5 (1.9)[45]
$n=3$	X=CN	Y=OCH_3	Chloroform	376	3.8	7.1 (1.9)[45]

$n=2$	$X=NO_2$	$Y=OCH_3$	Chloroform	376	3.7	6.4 (1.9)[45]
$n=3$	$X=NO_2$	$Y=OCH_3$	Chloroform	392	4.1	11 (1.9)[45]
	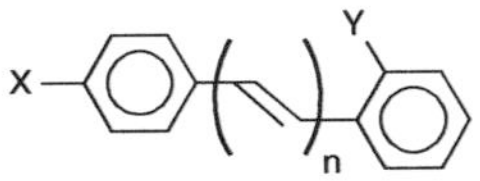					
$n=2$	$X=OCH_3$	$Y=CN$	Chloroform	356	3.9	2.6 (1.9)[45]
$n=3$	$X=OCH_3$	$Y=CN$	Chloroform	378	3.9	4.3 (1.9)[45]
$n=2$	$X=CN$	$Y=OCH_3$	Chloroform	358	4.9	4.3 (1.9)[45]
$n=2$	$X=OCH_3$	$Y=NO_2$	Chloroform	376	3.8	4.9 (1.9)[45]
$n=3$	$X=OCH_3$	$Y=NO_2$	Chloroform	392	3.8	11 (1.9)[45]
$n=2$	$X=NO_2$	$Y=OCH_3$	Chloroform	380	4.3	17 (1.9)[45]
$n=3$	$X=NO_2$	$Y=OCH_3$	Chloroform	412	4.8	56 (1.9)[45]
	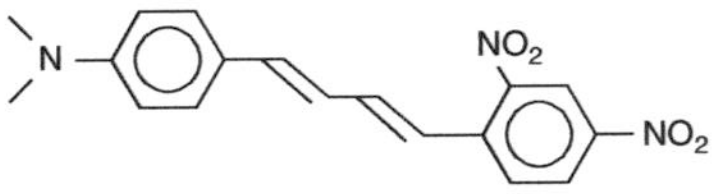		*p*-Dioxane	480	7.1	98 (1.9)[255]
			p-Dioxane		$\mu\beta=13\times10^{-46}$	(1.9)[107]
	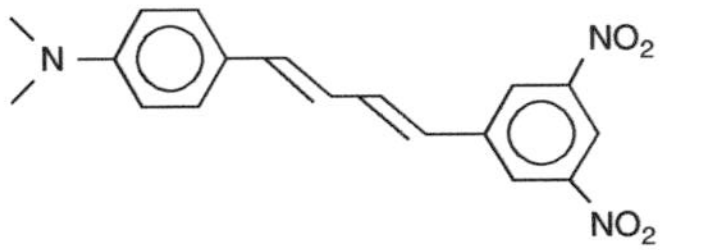		Chloroform		6.2	71 (1.9)[255]
	α,ω-Diphenylpolyene derivatives					
	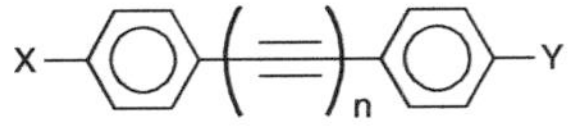					
$n=2$	$X=CN$	$Y=SCH_3$	Chloroform	330	3.7	17 (1.9)[45]
$n=2$	$X=CN$	$Y=NH_2$	NMP	388	$\mu\beta=11\times10^{-47}$	(1.9)[45]

(continued)

TABLE 7.1 (Continued)

	Structure		Solvent	λ_{max}(nm)	μ (10^{-18} esu)	$\beta_{\mu}(\lambda)^{ref.}$ (10^{-30} esu)
$n=2$	$X=NO_2$	$Y=SCH_3$	Chloroform	338	3.9	17 (1.9)[45]
$n=2$	$X=NO_2$	$Y=NH_2$	Chloroform	334	6.3	28 (1.9)[45]
$n=2$	$X=NO_2$	$Y=NH_2$	NMP	416	$\mu\beta=24\times10^{-47}$	(1.9)[45]
$n=3$	$X=NO_2$	$Y=NH_2$	NMP	440	$\mu\beta=41\times10^{-47}$	(1.9)[45]
	Cumulene derivatives					
	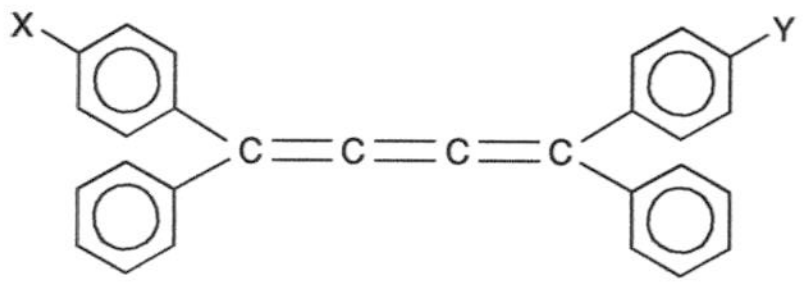 X, Y, C=C=C=C					
	$X=NO_2$	$Y=H$	*p*-Dioxane	442	$\mu\beta=15\times10^{-47}$	(1.9)[61]
	$X=NO_2$	$Y=H$	*p*-Dioxane	442	$\mu\beta=66\times10^{-47}$	(1.06)[61]
	$X=NO_2$	$Y=CH_3$	*p*-Dioxane	448	$\mu\beta=60\times10^{-47}$	(1.06)[61]
	$X=NO_2$	$Y=OCH_3$	*p*-Dioxane	459	$\mu\beta=85\times10^{-47}$	(1.06)[61]
	$X=NO_2$	$Y=O(CH_2)_{11}$ CH_3	*p*-Dioxane	461	$\mu\beta=28\times10^{-47}$	(1.9)[61]
	α,ω-Polyphenyl derivatives					
	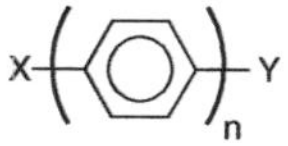 X, Y, n					
$n=3$	$X=NO_2$	$Y=OCH_3$	*p*-Dioxane	340	5.0	11 (1.9)[45]
$n=3$	$X=NO_2$	$Y=NH_2$	NMP	360	7.8	24 (1.9)[45]
$n=4$	$X=NO_2$	$Y=NH_2$	NMP	344	7.6	16 (1.9)[45]
	Pyrrol derivatives					
	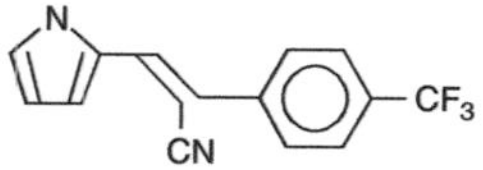 N, CN, CF_3		*p*-dioxane	376	5.8	11 (1.06)[244]
	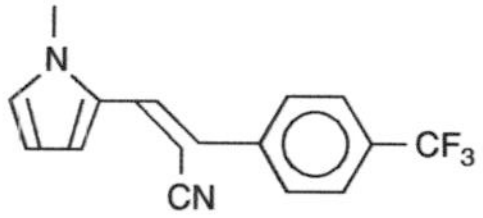 N, CN, CF_3		*p*-dioxane	381	7.2	12 (1.06)[244]

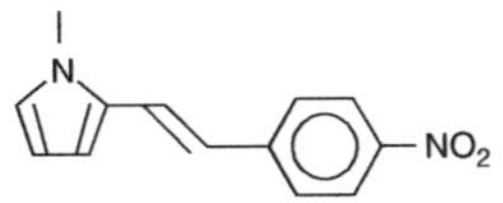		p-dioxane	410	5.5	26 (1.9)[43]
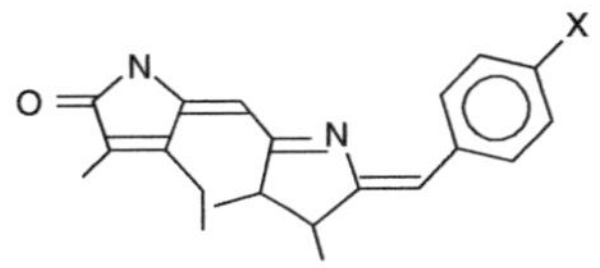 $X=OCH_3$		DMSO	485	1.4	32 (1.9S)[64]
$X=N(CH_3)_2$		DMSO	553	1.2	439 (1.9S)[64]
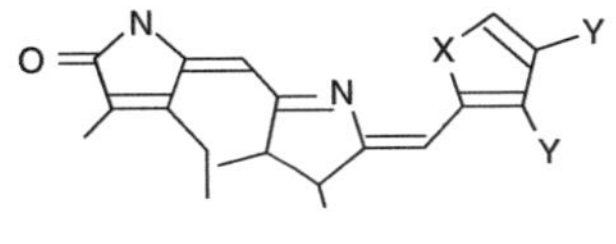 X=NH	$Y=CH_3$	DMSO	539	3.7	20 (1.9S)[64]
X=S	Y=H	DMSO	493	2.5	10 (1.9S)[64]
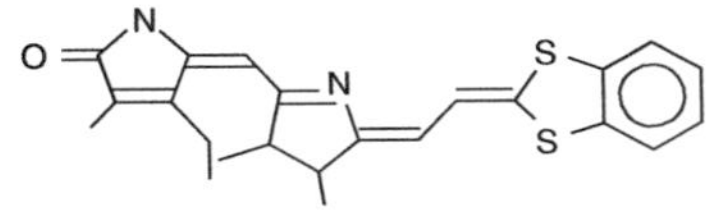		DMSO	553	2.5	578 (1.9S)[64]
Furan derivatives					
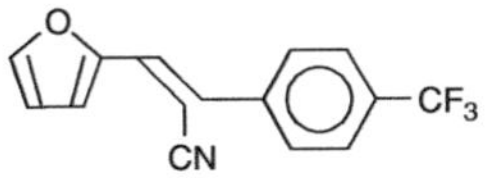		p-Dioxane	345	5.7	7.7 (1.06)[244]
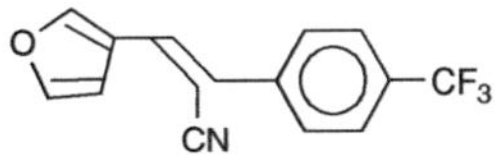		p-Dioxane	314	5.4	7.1 (1.06)[244]

(continued)

TABLE 7.1 (Continued)

Structure	Solvent	λ_{max}(nm)	μ (10^{-18} esu)	$\beta_{\mu}(\lambda)^{ref.}$ (10^{-30} esu)
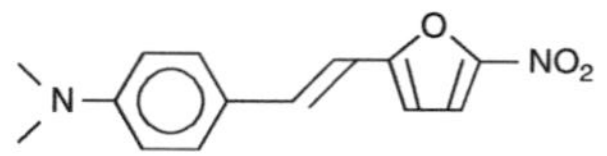 N, NO_2	Chloroform	478	6.9	83 (1.9)[45]
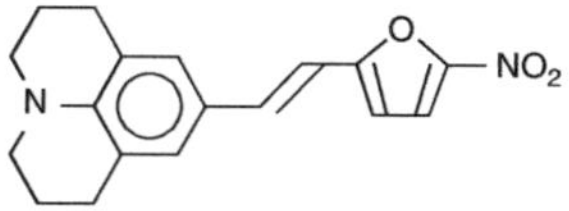 N, NO_2	Chloroform		5.9	173 (1.9)[43]
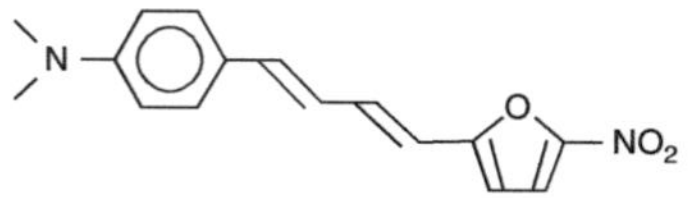 N, NO_2	Chloroform	488	7.2	113 (1.9)[45]
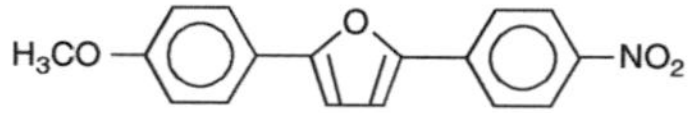 H_3CO, NO_2	Chloroform	400	5.2	40 (1.9)[45]
Thiophene derivatives				
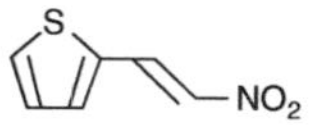 S, NO_2	Chloroform	351	5.2	20 (1.06)[117]
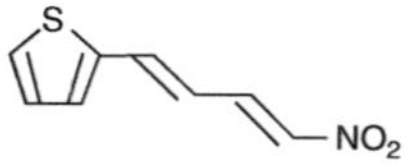 S, NO_2	Chloroform	382	5.4	67 (1.06)[117]

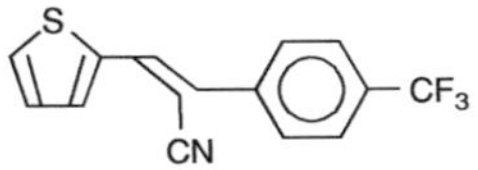	*p*-Dioxane	346	6.8	6.6 (1.06)[244]
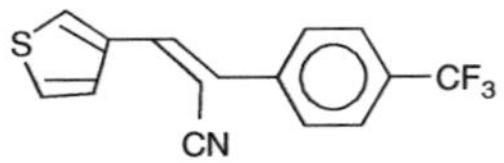	*p*-Dioxane	321	6.7	5.4 (1.06)[244]
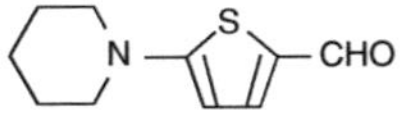	Chloroform		5.8	7.5 (1.9)[43]
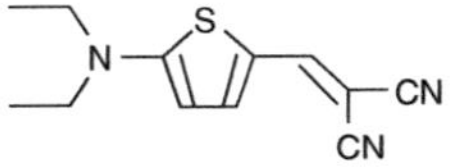	Chloroform		9.0	21 (1.9)[43]
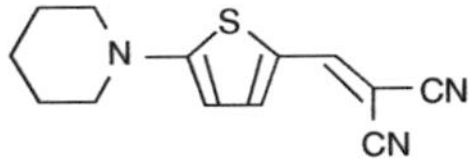	Chloroform		8.8	21 (1.9)[43]
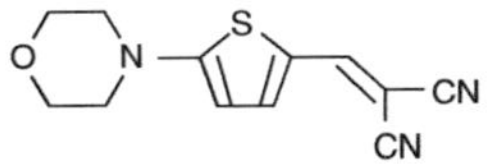	Chloroform		5.9	23 (1.9)[43]

(continued)

TABLE 7.1 (Continued)

Structure	Solvent	λ_{max}(nm)	μ (10^{-18} esu)	$\beta_{\mu}(\lambda)^{ref.}$ (10^{-30} esu)
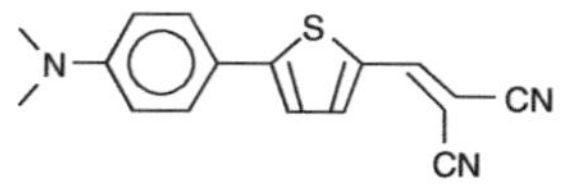	In PMMA	510	$\mu\beta_0 = 60 \times 10^{-47}$ (from EO at 633 nm)[165]	
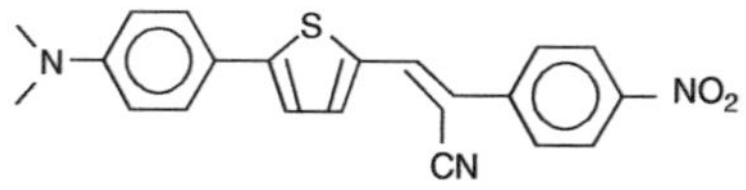	In PMMA	499	$\mu\beta_0 = 26 \times 10^{-47}$ (from EO at 633 nm)[165]	
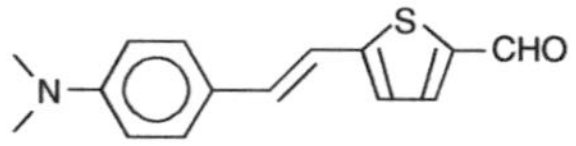	Chloroform		3.7	30 (1.9)[43]
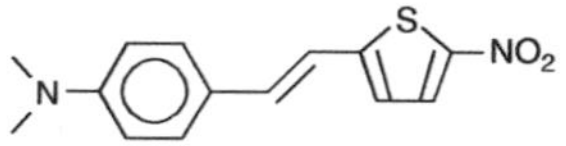	Chloroform	492	7.4	98 (1.9)[45]
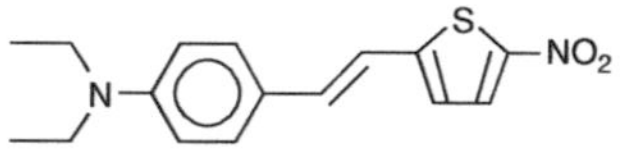	*p*-Dioxane	478	$\mu\beta = 60 \times 10^{-47}$	(1.9)[203]
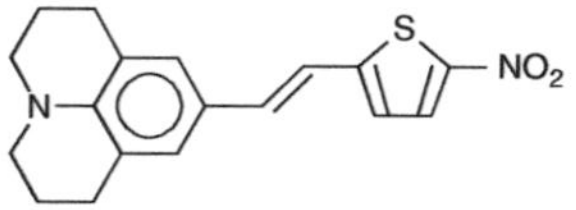	Chloroform		7.0	197 (1.9)[43]

	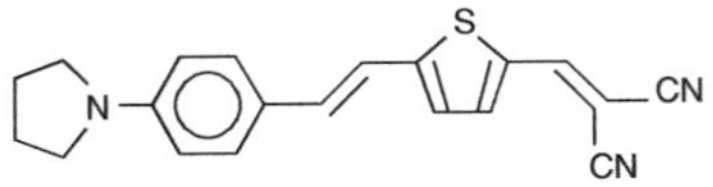	p-Dioxane	584	$\mu\beta=26\times10^{-46}$	(1.9)[106]
	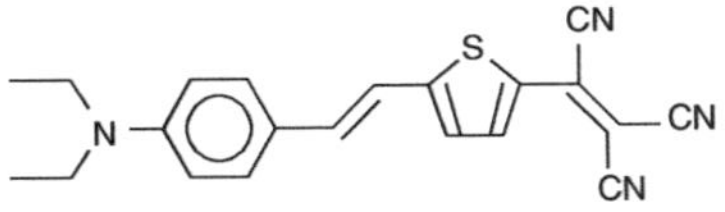	p-Dioxane	640	$\mu\beta=62\times10^{-46}$	(1.9)[203]
	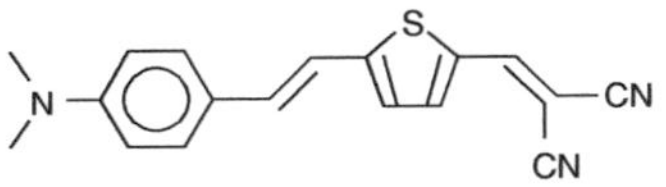	Chloroform		7.5	161 (1.9)[43]
	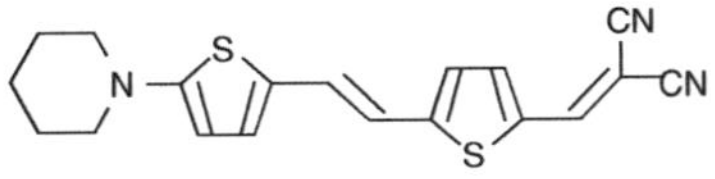	Chloroform		7.6	250 (1.9)[72]
	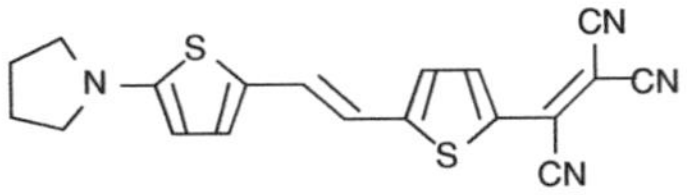	p-Dioxane	718	$\mu\beta=69\times10^{-46}$	(1.9)[203]
	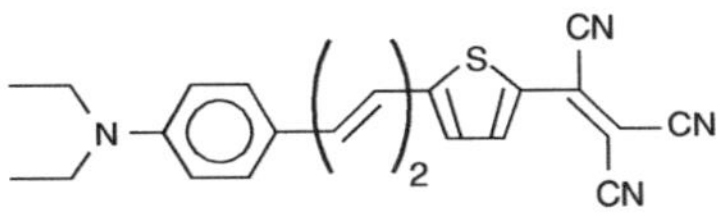	p-Dioxane	662	$\mu\beta=91\times10^{-46}$	(1.9)[203]
	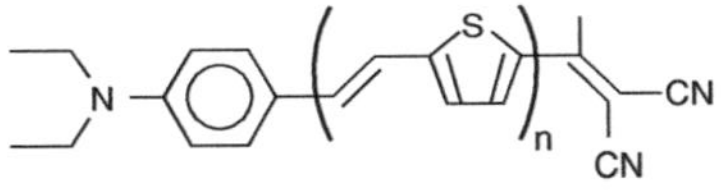				
$n=1$		p-Dioxane	513	$\mu\beta=13\times10^{-46}$	(1.9)[106]
$n=2$		p-Dioxane	547	$\mu\beta=23\times10^{-46}$	(1.9)[106]
$n=3$		p-Dioxane	556	$\mu\beta=38\times10^{-46}$	(1.9)[106]

(continued)

TABLE 7.1 (Continued)

Structure			Solvent	λ_{max}(nm)	μ (10^{-18} esu)	$\beta_\mu(\lambda)^{ref.}$ (10^{-30} esu)
	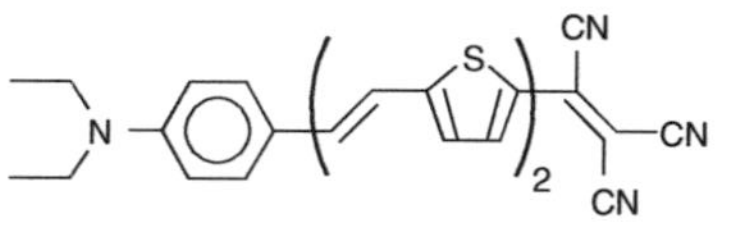		*p*-dioxane	653	$\mu\beta = 74 \times 10^{-46}$	(1.9)[203]
	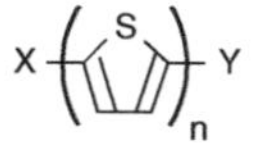					
$n=1$	$X=NO_2$	$Y=OCH_3$	CCl_4	349	5.5	7.6 (1.06)[271]
$n=2$	$X=NO_2$	$Y=H$	$C_6H_{11}CH_3$	378	5.0	19 (1.06)[271]
$n=2$	$X=NO_2$	$Y=OCH_3$	$C_6H_{11}CH_3$	411	6.0	41 (1.06)[271]
$n=2$	$X=NO_2$	$Y=OCH_3$	CCl_4	418	6.0	46 (1.06)[271]
$n=2$	$X=NO_2$	$Y=SCH_3$	$C_6H_{11}CH_3$	394	5.0	49 (1.06)[271]
$n=2$	$X=NO_2$	$Y=N(CH_3)_2$	$C_6H_{11}CH_3$	472	8.0	319 (1.06)[271]
$n=3$	$X=NO_2$	$Y=OCH_3$	CCl_4	452	6.0	162 (1.06)[271]
	Azole derivatives					
	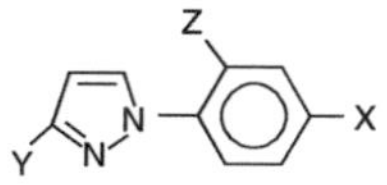					
$X=NO_2$	$Y=CH_3OC_6H_4$	$Z=H$	*p*-Dioxane	354	6.6	31 (1.06)[166]
$X=NO_2$	$Y=C_6H_5CH–CH–$	$Z=H$	*p*-Dioxane	346	6.1	32 (1.06)[166]
$X=NO_2$	$Y=p\text{-}CH_3OC_6H_4CH–CH–$	$Z=H$	*p*-Dioxane	368	5.7	40 (1.06)[166]
$X=SO_2CH_3$	$Y=p\text{-}CH_3OC_6H_4CH–CH–$	$Z=H$	*p*-Dioxane	330	5.6	16 (1.06)[166]
$X=NO_2$	$Y=p\text{-}CH_3SC_6H_4CH–CH–$	$Z=NO_2$	*p*-Dioxane	354	4.1	42 (1.06)[166]
$X=CH_3O$	$Y=p\text{-}O_2NC_6H_4CH–H–$	$Z=H$	*p*-Dioxane	370	7.0	48 (1.06)[166]

$X = NO_2$	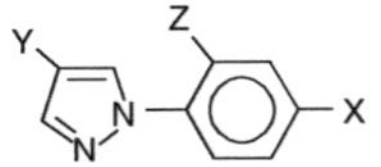 $Y = p\text{-}CH_3OC_6H_4C{\equiv}C\text{–}$	$Z = H$	*p*-dioxane	340	5.6	20 (1.06)[166]
$X = NO_2$	$Y = p\text{-}CH_3OC_6H_4CH\text{–}CH\text{–}$	$Z = H$	*p*-Dioxane	358	5.1	63 (1.06)[166]
$X = NO_2$	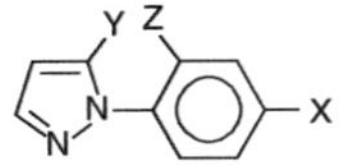 $Y = CH_3OC_6H_4$	$Z = H$	*p*-Dioxane	306	6.3	13 (1.06)[166]
$X = NO_2$	$Y = C_6H_5CH{=}CH\text{–}$	$Z = H$	*p*-Dioxane	298	4.6	7 (1.06)[166]
$X = NO_2$	$Y = p\text{-}CH_3OC_6H_4CH{=}CH\text{–}$	$Z = H$	*p*-Dioxane	302	5.5	14 (1.06)[166]
$X = NO_2$	$Y = p\text{-}CH_3SC_6H_4CH{=}CH\text{–}$	$Z = NO_2$	*p*-Dioxane	312	3.3	12 (1.06)[166]
$X = NO_2$	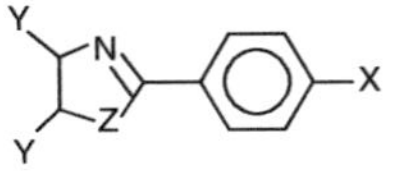 $Y = p\text{-}(CH_2)_6N$	$Z = N$	Chloroform	476	8.3	79 (1.9)[176]
$X = NO_2$	$Y = p\text{-}CH_3OC_6H_4C{\equiv}C\text{–}$	$Z = N$	*p*-Dioxane	344	8.0	53 (1.06)[176]
$X = NO_2$	$Y = p\text{-}(CH_2)_5N$	$Z = N$	Chloroform	438	7.2	46 (1.9)[176]
$X = O_2NC_6H_4C{\equiv}C\text{-}$	$Y = p\text{-}CH_3OC_6H_4$	$Z = N$	*p*-Dioxane	400	8.1	69 (1.06)[176]
$X = NO_2$	$Y = p\text{-}CH_3OC_6H_4S$	$Z = N$	*p*-Dioxane	398	6.6	56 (1.06)[176]
$X = NO_2$	$Y = p\text{-}CH_3OC_6H_4$	$Z = N$	*p*-Dioxane	410	7.0	52 (1.06)[176]
$X = NO_2$	$Y = p\text{-}C_4H_9CH(C_2H_5)CH_2OC_6H_4$	$Z = N$	Chloroform	416	6.3	25 (1.9)[176]
$X = NO_2$	$Y = p\text{-}CH_3OC_6H_4$	$Z = N$	Chloroform	412	6.4	20 (1.9)[176]
$X = SO_2C_4F_9$	$Y = p\text{-}C_4H_9CH(C_2H_5)CH_2OC_6H_4$	$Z = N$	Chloroform	384	6.5	14 (1.9)[176]
$X = SO_2C_6H_5$	$Y = p\text{-}CH_3OC_6H_4$	$Z = N$	*p*-Dioxane	362	8.0	10 (1.06)[176]
$X = O_2NC_6H_4C{\equiv}\text{-}$	$Y = p\text{-}CH_3OC_6H_4$	$Z = O$	*p*-Dioxane	378	7.1	64 (1.06)[176]
$X = SO_2C_6H_4CF_3$	$Y = p\text{-}CH_3OC_6H_4$	$Z = O$	Chloroform	370	8.0	32 (1.06)[176]

(continued)

TABLE 7.1 (Continued)

Structure			Solvent	λ_{max}(nm)	μ (10^{-18} esu)	$\beta_{\mu}(\lambda)^{ref.}$ (10^{-30} esu)
X = NO_2	Y = *p*-$CH_3OC_6H_4$	Z = O	*p*-Dioxane	390	6.3	47 (1.06)[176]
X = $SO_2C_4F_9$	Y = *p*-$CH_3OC_6H_4$	Z = O	*p*-Dioxane	378	7.9	31 (1.06)[176]
X = $SO_2C_6H_5$	Y = *p*-$CH_3OC_6H_4$	Z = O	*p*-Dioxane	358	7.0	17 (1.06)[176]
X = SO_2CH_3	Y = *p*-$CH_3OC_6H_4$	Z = O	*p*-Dioxane	352	7.0	15 (1.06)[176]
X = C_6F_{16}	Y = *p*-$CH_3OC_6H_4$	Z = O	*p*-Dioxane	344	5.7	13 (1.06)[176]
X = CF_3	Y = *p*-$CH_3OC_6H_4$	Z = O	*p*-Dioxane	338	5.3	9 (1.06)[176]
X = NO_2	Y = *p*-$CH_3OC_6H_4$	Z = S		400	8.1	21 (β_0)[176]
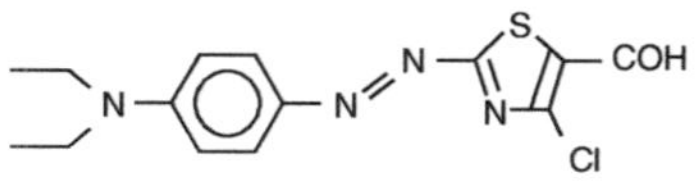			Chloroform	564	6.2	71 (1.9)[43]
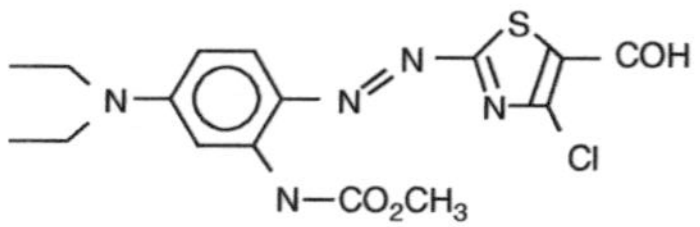			Chloroform	578	6.1	83 (1.9)[43]
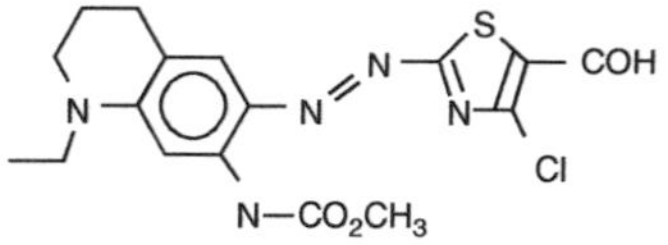			Chloroform	596	6.5	75 (1.9)[43]
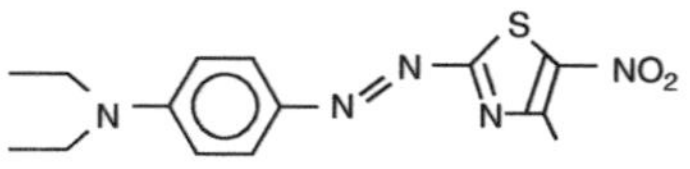			CH_2Cl_2	579	8.5	260 (1.58)[54]

	Chloroform	582	9.0	52 (β_0)[177]
	Chloroform	582	6.9	68 (β_0)[177]
	CH_2Cl_2	645	10	530 (1.58)[54]
	Chloroform	634	6.9	100 (1.9)[43]
	Chloroform	670	6.9	130 (1.9)[43]
Azulene derivatives				
	Chloroform		1.2	<1 (1.9)[43]
	Chloroform		2.5	<1 (1.9)[43]

(continued)

TABLE 7.1 (Continued)

Structure	Solvent	λ_{max}(nm)	μ (10^{-18} esu)	$\beta_{\mu}(\lambda)^{ref.}$ (10^{-30} esu)
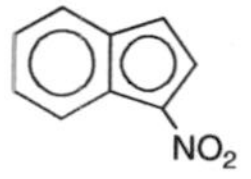 NO_2	Chloroform		4	<1 (1.9)[43]
Pentafulvene derivatives				
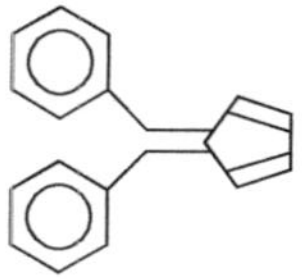	Chloroform		1	7 (1.9)[43]
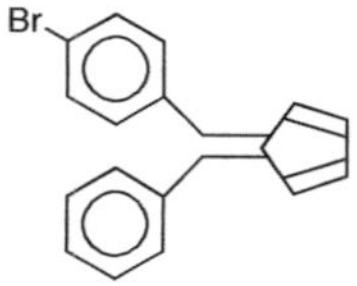 Br	Chloroform		1	5 (1.9)[43]
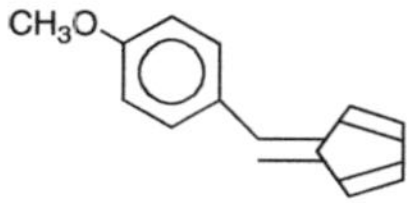 CH_3O	Chloroform		2	10 (1.9)[43]
N	Chloroform		3	30 (1.9)[43]
	DMSO	420	$\mu\beta = 44 \times 10^{-47}$	(1.06)[102]
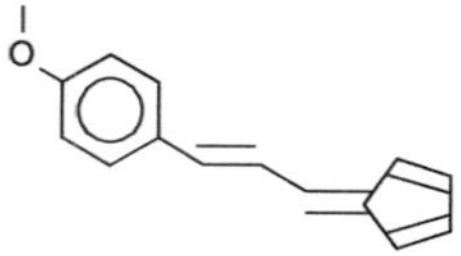 O	DMSO	370	$\mu\beta = 70 \times 10^{-48}$	(1.06)[102]

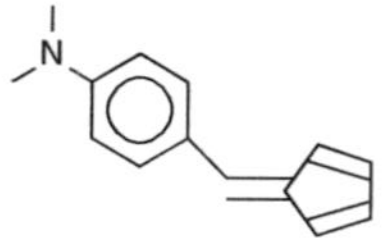

	Chloroform		3.6	74 (1.9)[43]
	DMSO	450	$\mu\beta = 20 \times 10^{-46}$	(1.06)[102]
	Chloroform		$\mu\beta = 73 \times 10^{-47}$	(1.9)[43]
	Chloroform		4.0	0.9 (1.9)[43]
	Chloroform		3.5	4.6 (1.9)[43]
$C_{18}H_{37}-N^+$ … O^-	Methanol	444	26	−100 (1.3)[144]
	Pyridine	606	17	−210 (1.3)[144]
	DMSO	570	8	1000 (1.9)[57]
	Chloroform	590	4.0	190 (1.9)[153]

(continued)

TABLE 7.1 (Continued)

Structure	Solvent	λ_{max}(nm)	μ (10^{-18} esu)	$\beta_{\mu}(\lambda)^{ref.}$ (10^{-30} esu)
	Chloroform	580	4.0	79 (1.9)[153]
	Chloroform	610	4.3	91 (1.9)[153]
X=H	Chloroform	428	1.5	5.9 (1.9)[154]
X=Br	Chloroform	432	1.3	13 (1.9)[154]
X=OCH_3	Chloroform	469	2.4	17 (1.9)[154]
X=OCH_3; *ortho*: OCH_3	Chloroform	498	3.4	19 (1.9)[154]
X=SCH_3	Chloroform	476	2.3	17 (1.9)[154]
X=NH_2	Chloroform	—	3.3	38 (1.9)[154]
X=$N(CH_3)_2$	Chloroform	558	3.9	78 (1.9)[154]
X=CH–CH–C_6H_4–OCH_3	Chloroform	497	2.6	48 (1.9)[154]
X=CH–CH–C_6H_4–$N(CH_3)_2$	Chloroform	512	3.7	116 (1.9)[154]
	DMSO	470	$\mu\beta = 74 \times 10^{-47}$	(1.06)[104]
	DMSO	550	$\mathrm{Re}(\mu\beta) = 48 \times 10^{-45}$	(1.06)[104]

					$Im(\mu\beta)=61\times10^{-45}$	(1.06)[104]
	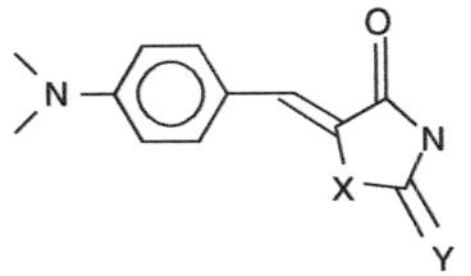		DMSO	550	$Re(\mu\beta)=48\times10^{-45}$	(1.06)[104]
					$Im(\mu\beta)=61\times10^{-45}$	(1.06)[104]
X=N		Y=O	DMSO	390	$\mu\beta=12\times10^{-47}$	(1.06)[103]
X=S		Y=S	DMSO	480	$Re(\mu\beta)=11\times10^{-46}$	(1.06)[103]
					$Im(\mu\beta)=31\times10^{-47}$	(1.06)[103]
	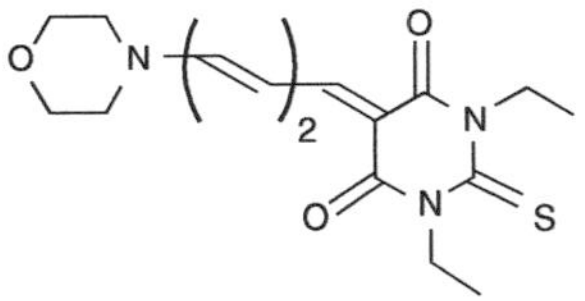		DMSO	530	$Re(\mu\beta)=21\times10^{-45}$	(1.06)[104]
					$Im(\mu\beta)=30\times10^{-45}$	(1.06)[104]
$n=2$	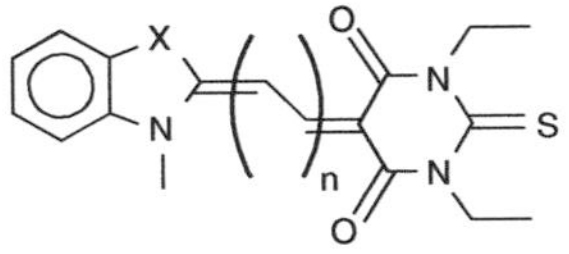	$X=C(CH_3)_2$	DMSO		$Re(\mu\beta)=17\times10^{-45}$	(1.06)[104]
					$Im(\mu\beta)=24\times10^{-45}$	(1.06)[104]
$n=3$		X=S	DMSO		$Re(\mu\beta)=13\times10^{-45}$	(1.06)[104]
					$Im(\mu\beta)=23\times10^{-45}$	(1.06)[104]
$n=3$		$X=C\text{–}CH_2$	DMSO		$Re(\mu\beta)=20\times10^{-45}$	(1.06)[104]
					$Im(\mu\beta)=27\times10^{-45}$	(1.06)[104]
	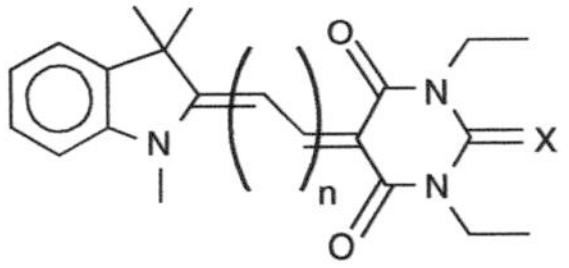					
$n=3$		X=O	Chloroform	6.9		360 (1.9)[43]
$n=1$		X=S	Chloroform	7.4		−46 (1.9)[43]

(continued)

TABLE 7.1 (Continued)

Structure				Solvent	λ_{max}(nm)	μ (10^{-18} esu)	$\beta_{\mu}(\lambda)^{ref.}$ (10^{-30} esu)
$n=2$		$X=S$		Chloroform		8.9	52 (1.9)[43]
$n=3$		$X=S$		Chloroform		9.1	300 (1.9)[43]
	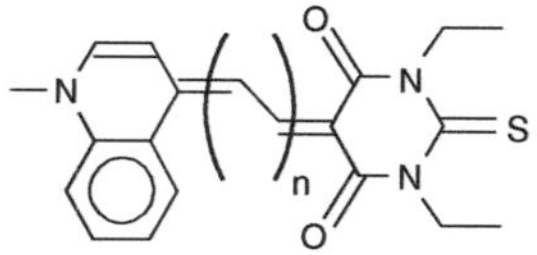 $n=1$			DMSO		$Re(\mu\beta)=13\times10^{-45}$	(1.06)[104]
						$Im(\mu\beta)=15\times10^{-45}$	(1.06)[104]
	$n=3$			DMSO		$Re(\mu\beta)=27\times10^{-45}$	(1.06)[104]
						$Im(\mu\beta)=49\times10^{-45}$	(1.06)[104]
	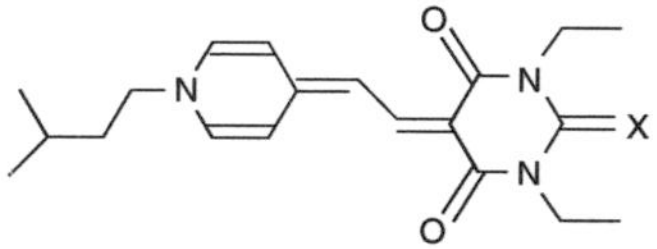 $X=O$			Chloroform	510	$\mu\beta=-37\times10^{-47}$	(1.9)[60]
	$X=S$			Chloroform	526	$\mu\beta=-60\times10^{-47}$	(1.9)[60]
$n=0$	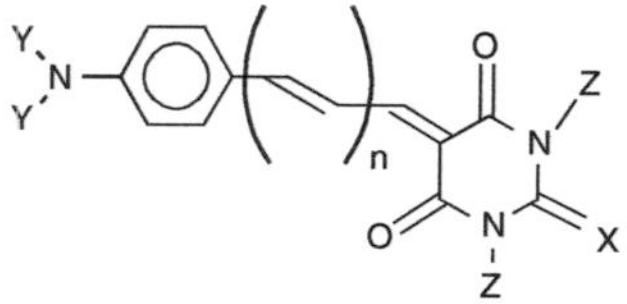 $X=O$	$Y=CH_3$	$Z=H$	DMSO	470	$\mu\beta=63\times10^{-47}$	(1.06)[103]
$n=0$	$X=O$	$Y=C_2H_5$	$Z=CH_3$	Chloroform	470	6.2	20 (β_0)[177]
$n=0$	$X=O$	$Y=4$-$CH_3C_6H_4$	$Z=CH_3$	Chloroform	486	4.5	40 (β_0)[177]
$n=0$	$X=O$	$Y=CH_3$	$Z=C_2H_5$	Chloroform	464	5.1	47 (1.9)[43]
$n=1$	$X=O$	$Y=CH_3$	$Z=CH_3$	DMSO	540	$Re(\mu\beta)=13\times10^{-46}$	(1.06)[103]
						$Im(\mu\beta)=52\times10^{-46}$	(1.06)[103]

$n=1$	X=O	Y=C_2H_5	Z=C_6H_5	Chloroform	558	7.1	91 (β_0)[177]
$n=1$	X=O	Y=4-$CH_3C_6H_4$	Z=C_6H_5	Chloroform	560	4.0	97 (β_0)[177]
$n=1$	X=O	Y=CH_3	Z=C_2H_5	Chloroform	532	5.1	173 (1.9)[43]
$n=2$	X=O	Y=CH_3	Z=C_2H_5	Chloroform	554	5.8	338 (1.9)[43]
$n=3$	X=O	Y=CH_3	Z=C_2H_5	Chloroform	572	5.3	660 (1.9)[43]
$n=0$	X=S	Y=CH_3	Z=H	DMSO	500	$\mathrm{Re}(\mu\beta)=12\times10^{-46}$	(1.06)[103]
						$\mathrm{Im}(\mu\beta)=60\times10^{-47}$	(1.06)[103]
$n=0$	X=S	Y=CH_3	Z=C_2H_5	Chloroform	494	6.4	20 (β_0)[177]
				Chloroform	484	5.4	68 (1.9)[155]
$n=0$	X=S	Y=C_2H_5	Z=C_2H_5	Chloroform	500	7.1	20 (β_0)[177]
$n=0$	X=S	Y=4-$CH_3C_6H_4$	Z=C_2H_5	Chloroform	518	5.2	45 (β_0)[177]
$n=0$	X=S	Y=C_2H_5	Z=C_6H_5	Chloroform	502	7.6	25 (β_0)[177]
$n=0$	X=S	Y=4-$CH_3C_6H_4$	Z=C_6H_5	Chloroform	520	5.7	57 (β_0)[177]
$n=0$	X=S	Y=CH_3	Z=C_6H_5	Chloroform		5.5	58 (1.9)[43]
$n=0$	X=S	Y=CH_3	Z=4-$NO_2C_6H_4$	Chloroform		9.0	80 (1.9)[43]
$n=1$	X=S	Y=C_2H_5	Z=C_6H_5	Chloroform	588	8.3	94 (β_0)[177]
$n=1$	X=S	Y=4-$CH_3C_6H_4$	Z=C_6H_5	Chloroform	594	6.5	119 (β_0)[177]
$n=1$	X=S	Y=CH_3	Z=C_2H_5	Chloroform	572	5.7	256 (1.9)[155]
$n=2$	X=S	Y=CH_3	Z=C_2H_5	Chloroform	604	6.2	636 (1.9)[155]
$n=3$	X=S	Y=CH_3	Z=C_2H_5	Chloroform	624	6.6	1490 (1.9)[155]
				Chloroform		4.2	299 (1.9)[43]

(continued)

TABLE 7.1 (Continued)

Structure		Solvent	λ_{max}(nm)	μ (10^{-18} esu)	$\beta_{\mu}(\lambda)^{ref.}$ (10^{-30} esu)
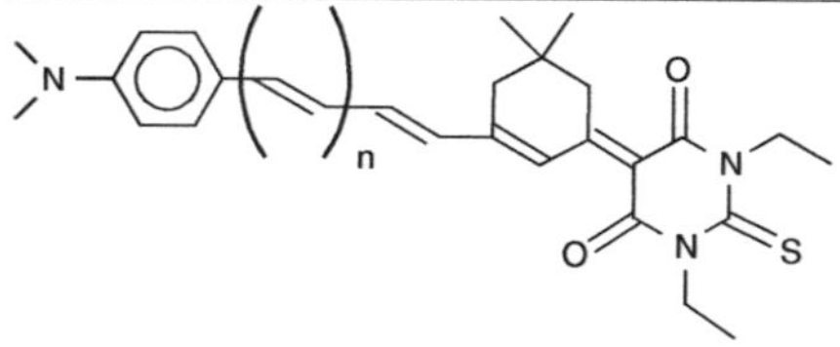					
$n=2$		Chloroform		6.8	630 (1.9)[43]
$n=3$		Chloroform		7.2	771 (1.9)[43]
$n=4$		Chloroform		7.5	893 (1.9)[43]
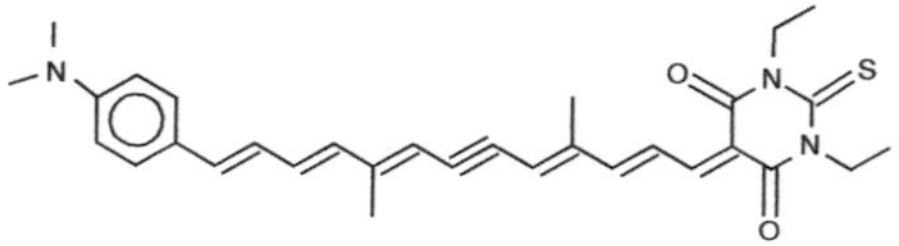		Chloroform		$\mu\beta=71\times10^{-46}$	(1.9)[43]
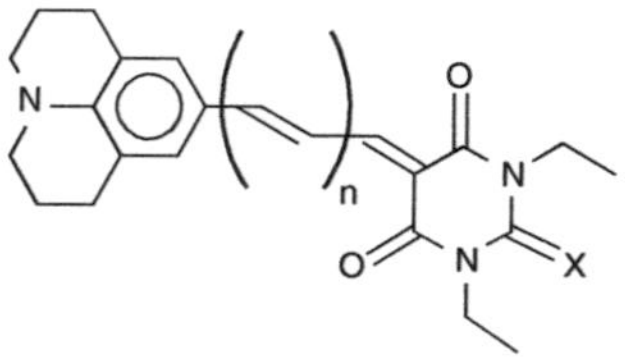					
$n=0$	X=O	Chloroform	480	5.7	64 (1.9)[43]
$n=1$	X=O	Chloroform	574	5.9	213 (1.9)[43]
$n=2$	X=O	Chloroform	616	6.5	504 (1.9)[43]
$n=3$	X=O	Chloroform		7.1	935 (1.9)[43]
$n=0$	X=S	Chloroform	522	7.0	87 (1.9)[155]
$n=1$	X=S	Chloroform	614	6.6	355 (1.9)[155]

$n=2$	X=S	Chloroform	680	6.3	1141 (1.9)[155]
$n=3$	X=S	Chloroform	686	8.8	2169 (1.9)[155]
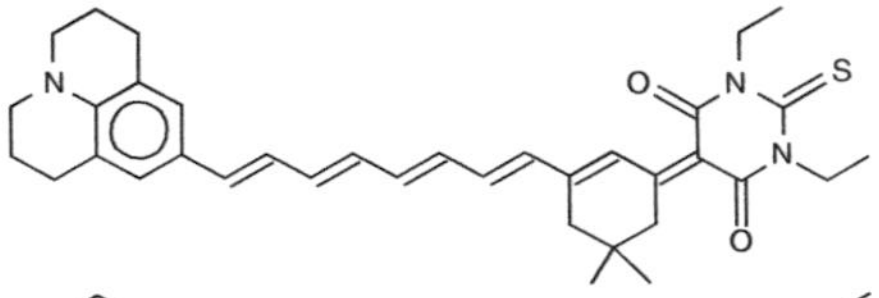		Chloroform		$\mu\beta=19\times10^{-45}$	(1.9)[43]
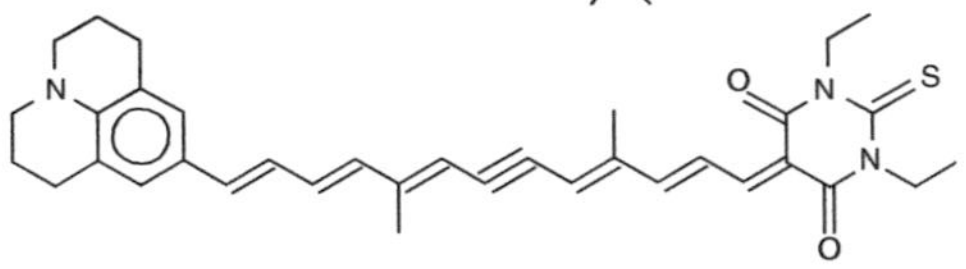		Chloroform		$\mu\beta=12\times10^{-45}$	(1.9)[43]
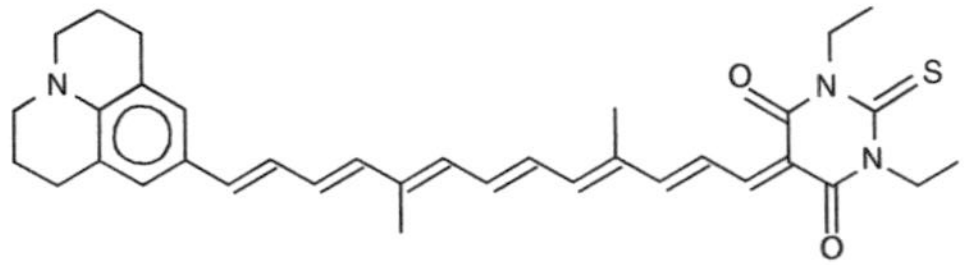		Chloroform	680	$\mu\beta=35\times10^{-46}$	(1.9)[155]
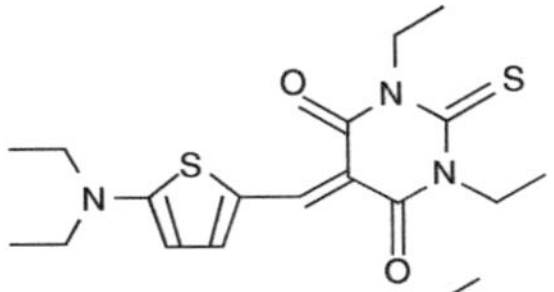		Chloroform		7.9	11 (1.9)[43]
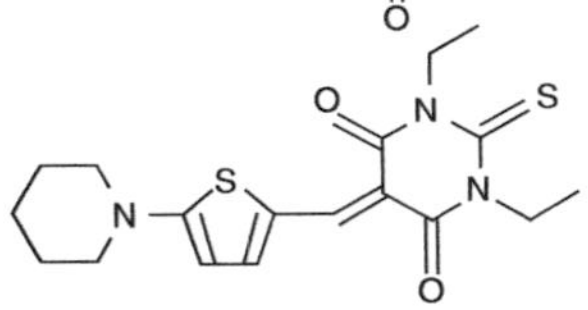		Chloroform		7.5	13 (1.9)[43]
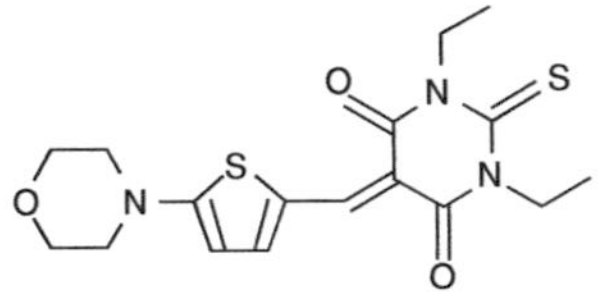		Chloroform		5.5	14 (1.9)[43]

(continued)

TABLE 7.1 (Continued)

Structure	Solvent	λ_{max}(nm)	μ (10^{-18} esu)	$\beta_{\mu}(\lambda)^{ref.}$ (10^{-30} esu)
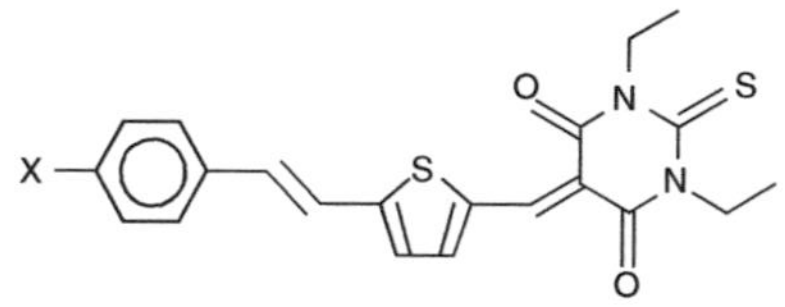				
$X=OCH_3$	Chloroform		2.7	171 (1.9)[43]
$X=N(CH_3)_2$	Chloroform		5.6	339 (1.9)[43]
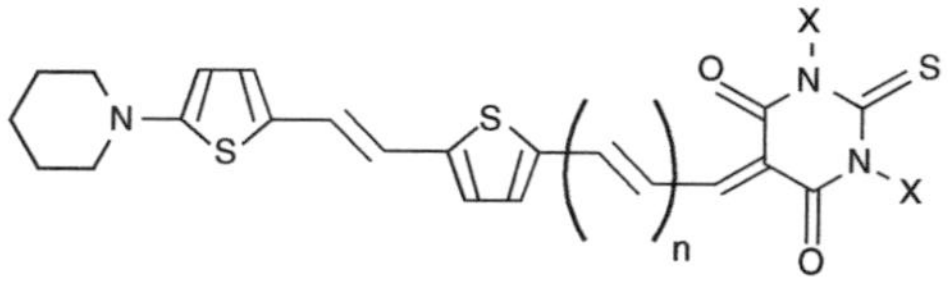				
$n=0$ $X=C_2H_5$	Chloroform		$\mu\beta=37\times10^{-46}$	(1.9)[43]
$n=1$ $X=C_2H_5$	Chloroform		$\mu\beta=71\times10^{-46}$	(1.9)[43]
$n=1$ $X=C_6H_5$	Chloroform		$\mu\beta=86\times10^{-46}$	(1.9)[74]
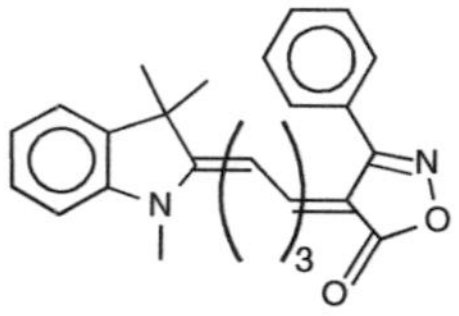	Chloroform		11	280 (1.9)[43]
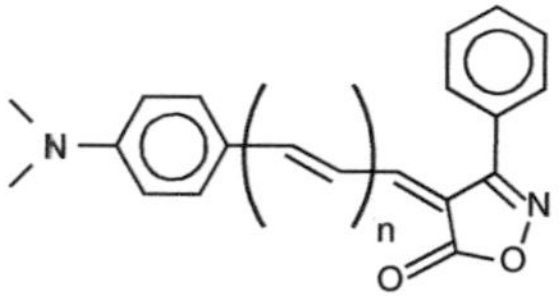				

$n=0$	Chloroform	478	8.3	37 (1.9)[155]
$n=1$	Chloroform	530	8.6	140 (1.9)[155]
$n=2$	Chloroform	562	8.7	362 (1.9)[155]
$n=3$	Chloroform	582	8.9	918 (1.9)[155]
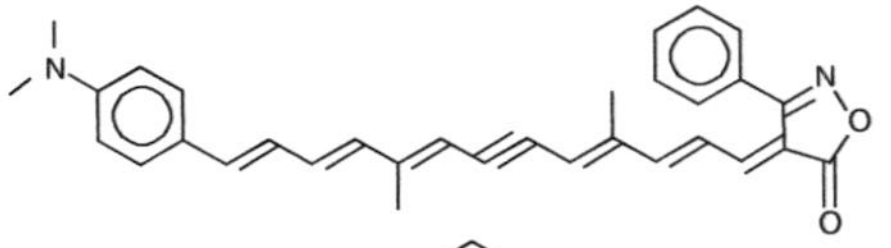	Chloroform		$\mu\beta=67\times10^{-45}$	(1.9)[43]
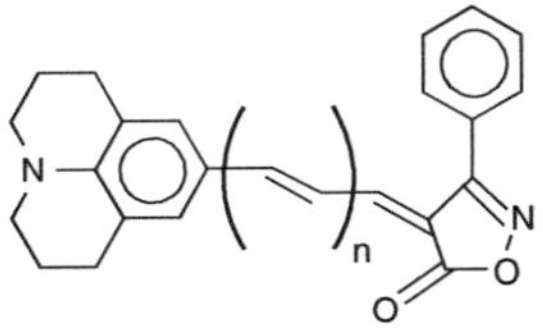 $n=0$	Chloroform	504	9.5	51 (1.9)[155]
$n=1$	Chloroform	586	9.1	180 (1.9)[155]
$n=2$	Chloroform	620	9.0	656 (1.9)[155]
$n=3$	Chloroform	640	9.8	995 (1.9)[155]
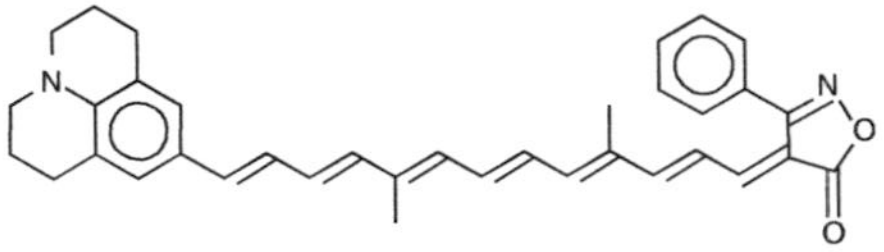	Chloroform	647	$\mu\beta=29\times10^{-45}$	(1.9)[155]
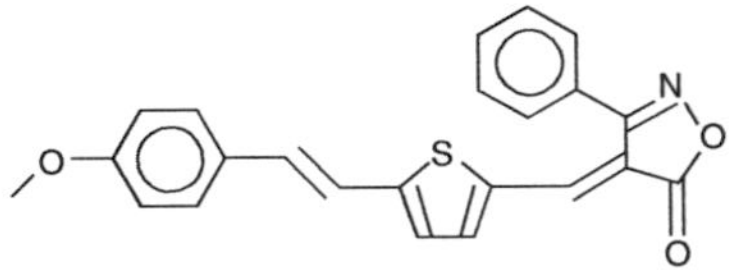	Chloroform		6.5	46 (1.9)[43]

(continued)

TABLE 7.1 (Continued)

Structure	Solvent	λ_{max}(nm)	μ (10^{-18} esu)	$\beta_{\mu}(\lambda)^{ref.}$ (10^{-30} esu)
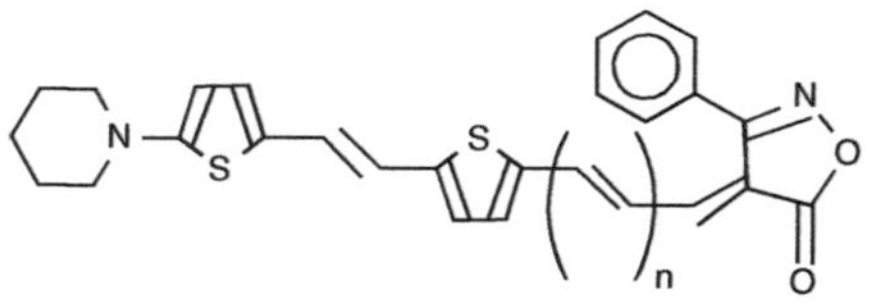				
$n=0$	Chloroform		$\mu\beta=25\times10^{-46}$	(1.9)[72]
$n=1$	Chloroform		$\mu\beta=41\times10^{-46}$	(1.9)[72]
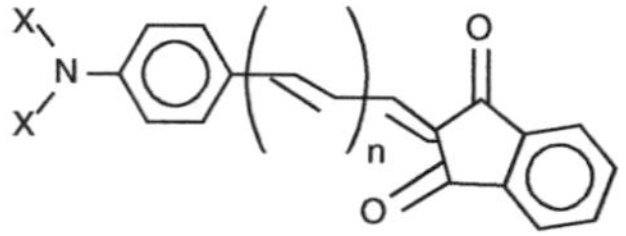	Chloroform		4.4	500 (1.9)[43]
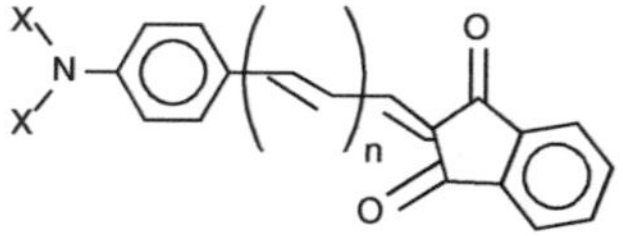				
$n=0$ $X=C_2H_5$	Chloroform	490	3.7	38 (β_0)[177]
$n=0$ X=4-$CH_3C_5H_4$	Chloroform	498	2.0	64 (β_0)[177]
$n=0$ $X=CH_3$	Chloroform		3.7	58 (1.9)[43]
$n=1$ $X=C_2H_5$	Chloroform	552	4.4	92 (β_0)[177]
$n=1$ X=4-$CH_3C_5H_4$	Chloroform	544	2.0	93 (β_0)[177]
$n=1$ $X=CH_3$	Chloroform		3.6	191 (1.9)[43]
$n=2$ $X=CH_3$	Chloroform		4.9	266 (1.9)[43]

$n=3$	$X=CH_3$	Chloroform		5.3	347 (1.9)[43]
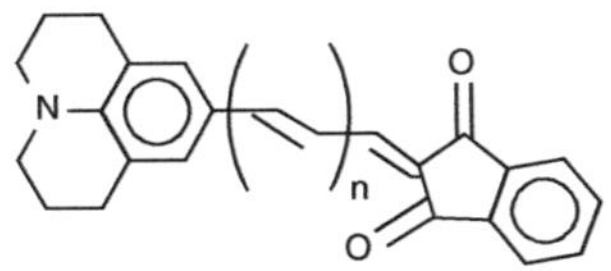					
$n=0$		Chloroform		4.0	104 (1.9)[43]
$n=1$		Chloroform		4.5	215 (1.9)[43]
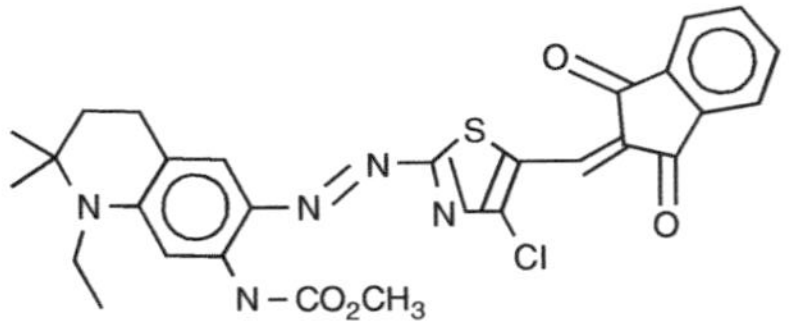		Chloroform	670	6.1	84 (1.9)[43]
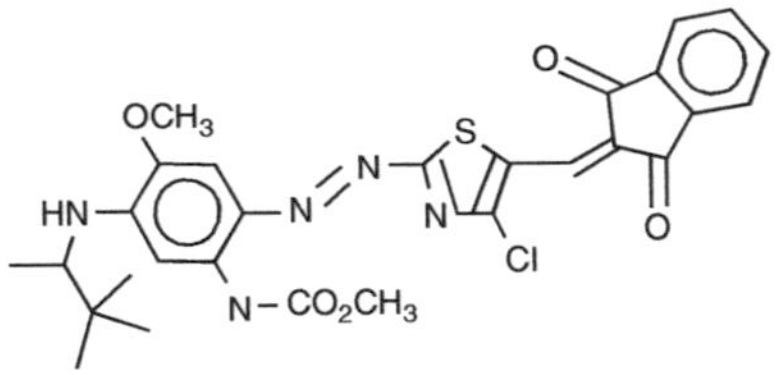		Chloroform	680	6.6	118 (1.9)[43]
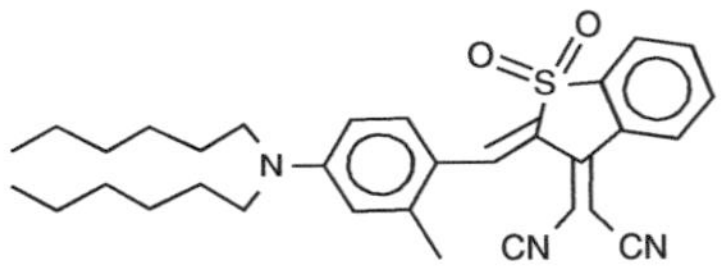		DMSO	620	6.8	163 (1.58)[34] −118 (0.95)[34]

(continued)

TABLE 7.1 (Continued)

Structure	Solvent	λ_{max}(nm)	μ (10^{-18} esu)	$\beta_{\mu}(\lambda)^{ref.}$ (10^{-30} esu)
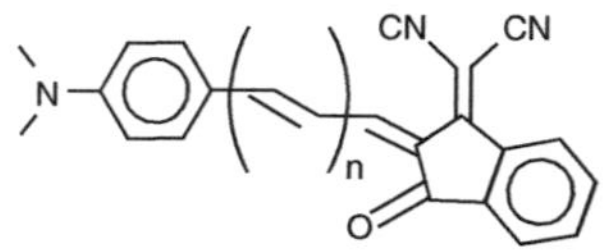				
$n=0$	Chloroform		5.0	98 (1.9)[43]
$n=1$	Chloroform		6.0	264 (1.9)[43]
$n=2$	Chloroform		6.9	466 (1.9)[43]
$n=3$	Chloroform		6.5	750 (1.9)[43]
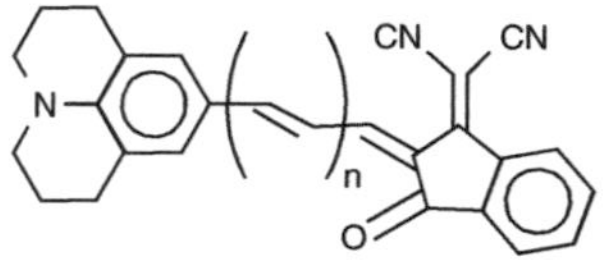				
$n=0$	Chloroform		6.3	124 (1.9)[43]
$n=1$	Chloroform		6.1	268 (1.9)[43]
$n=2$	Chloroform		6.8	534 (1.9)[43]
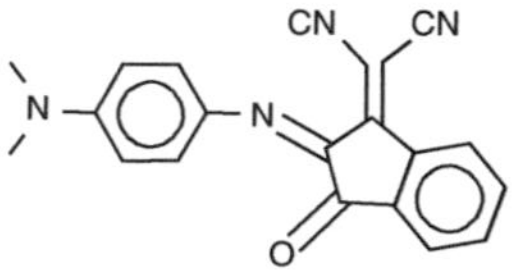	Chloroform		5.0	98 (1.9)[43]
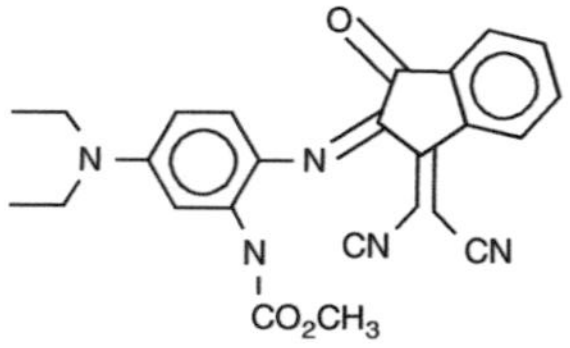	Chloroform		5.6	64 (1.9)[43]

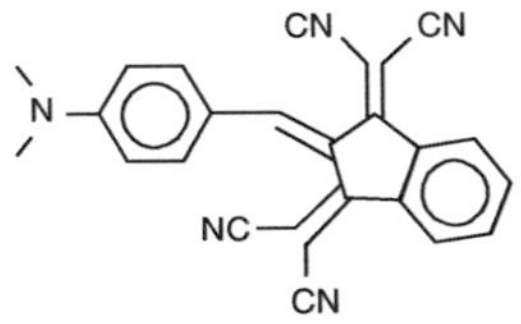	Chloroform	6.7	97 (1.9)[43]
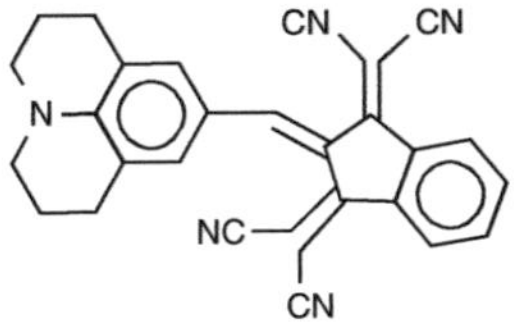	Chloroform	8.0	89 (1.9)[43]
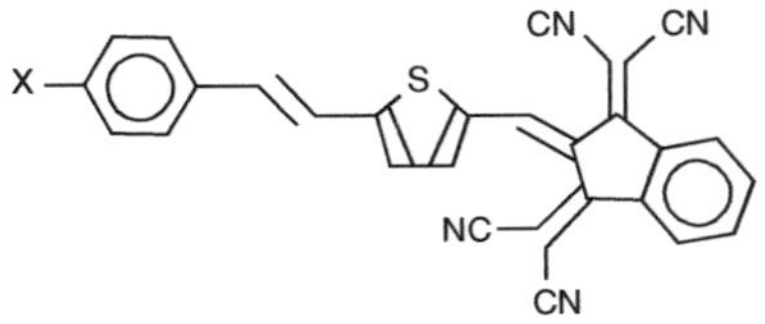 $X = OCH_3$	Chloroform	6.0	165 (1.9)[43]
$X = N(CH_3)_2$	Chloroform	6.0	1024 (1.9)[43]
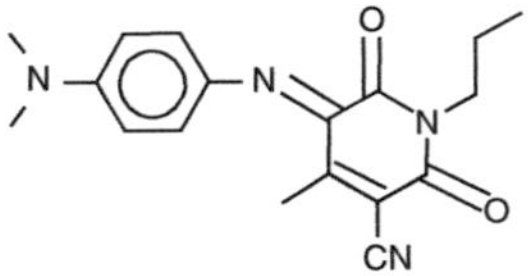	Chloroform	7.8	169 (1.9)[43]
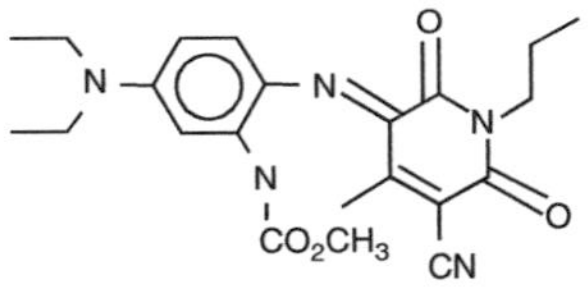	Chloroform	7.7	83 (1.9)[43]

(continued)

TABLE 7.1 (Continued)

Structure	Solvent	λ_{max}(nm)	μ (10^{-18} esu)	$\beta_{\mu}(\lambda)^{ref.}$ (10^{-30} esu)
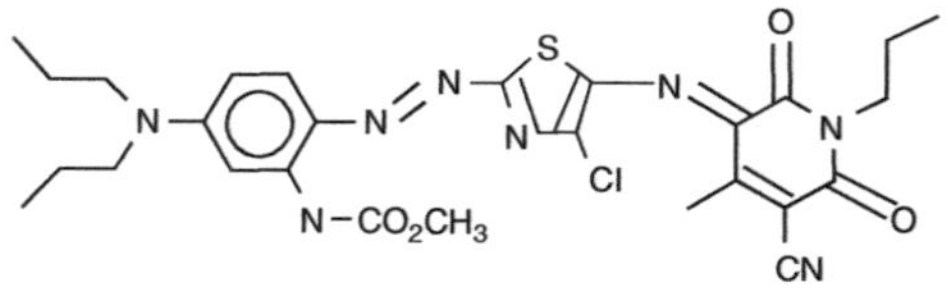	Chloroform	696	8.1	432 (1.9)[43]
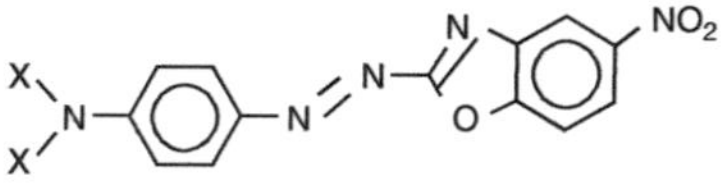 $X=C_2H_5$; $X=HO(CH_2)_6$	Chloroform	548	9.5	60 (β_0)[177]
$X=4\text{-}CH_3C_5H_4$	Chloroform	550	7.2	72 (β_0)[177]
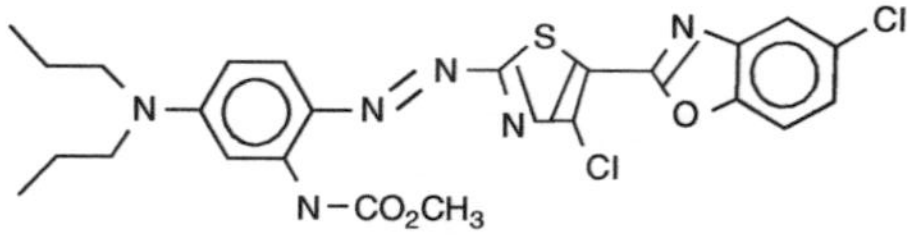	Chloroform		5.5	164 (1.9)[43]
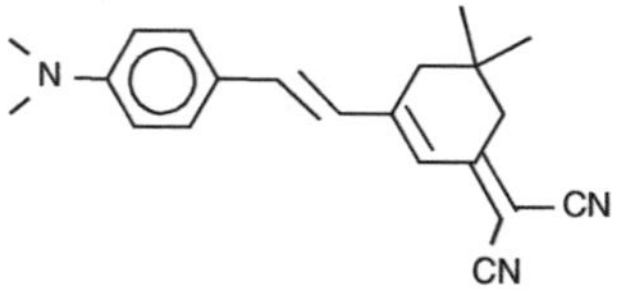	Chloroform		8.7	129 (1.9)[43]
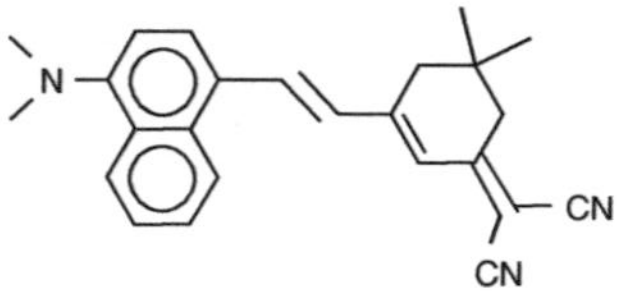	Chloroform		7.0	87 (1.9)[43]

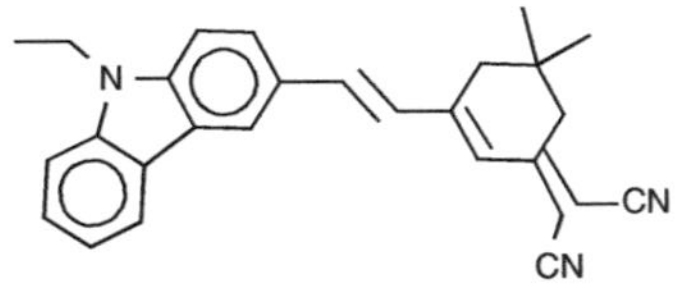	Chloroform	7.5	93 (1.9)[43]
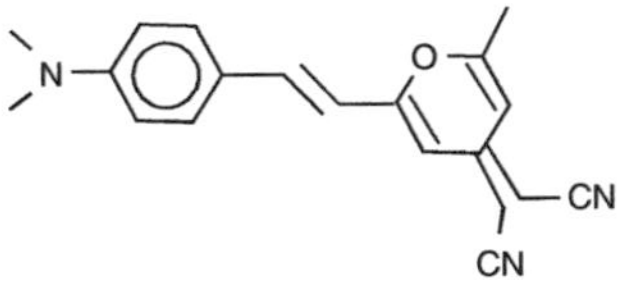	Chloroform	8.6	82 (1.9)[43]
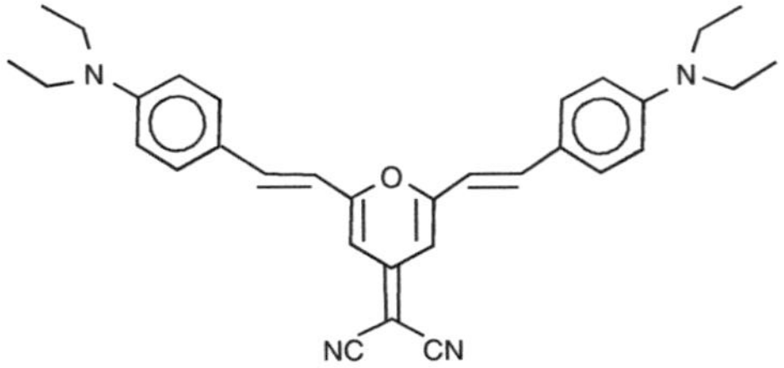	Chloroform	7.9	102 (1.9)[43]
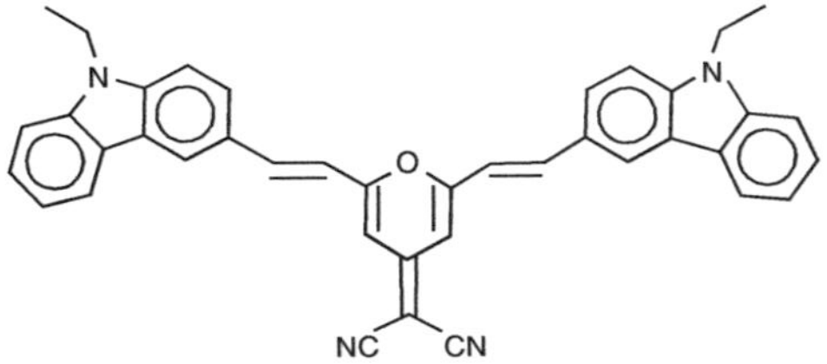	Chloroform	8.5	95 (1.9)[43]

TABLE 7.2 Single Crystal Results on Organic Nonlinear Materials: n_i, Refractive Index at 633 nm; $\lambda_{NC}(\theta)$, θ-Noncritical Phase-Matching Wavelength; $\lambda_{NC}(\lambda)$, λ-Noncritical Phase-Matching Wavelength; and at 1.064 μm: DT(τ), Optical Damage Threshold (Pulse Duration); d_{eff}, Effective Nonlinearity (Phase-Matching Type); $\Delta\theta l$, Angular Acceptance; ΔTl, Temperature Acceptance; $d\theta_{pm}/dT$, Temperature Tuning of Phase-Matching Angle; and ρ, Walk-Off Angle (Phase-Matching Type)

Abstract					
Structure and Nomenclature (acronym)	Point Group	SHG $d_{ij}(\lambda)$ & EO $r_{ij}(\lambda)$ (pm/v) (μm)	Cut-Off λ (nm)	Ref.	Properties
H_2N, H_2N, =O Urea	m	$d_{36}(1.06) \approx 1.3$ $r_{41}(0.63) \approx 1.9$ $r_{63}(0.63) \approx 0.83$	200 & 1800	212 80 18,21, 172	DT(10 ns) = 5 GW/cm^2. $n_o = 1.485$, $n_e = 1.567$.
O, O, S, HO, OH p,p'-Dihydroxydiphenyl sulfone (DHDPS)	$mm2$	$d_{33}(1.06) \approx 7$ $d_{32}(1.06) \approx 0.4$ $D_{11}(1.06) \approx 3$	300 & 1500	275	Relatively high hardness, nonhygroscopic, and photochemically stable. Crystals cleave at input power > 400 MW/cm^2, $n_x = 2.009$, $n_y = 2.000$, $n_z = 1.921$. $\lambda_{NC}(\theta) = 865$ nm.
OCH_3, HO, CHO 3-Methoxy-4-hydroxy-benzaldehyde (MHBA)	2	$d_{13}(1.06) \approx 13$ $d_{11}(1.06) \approx 9.8$ $d_{12}(1.06) \approx 3.9$ $d_{14}(1.06) \approx 3.2$	370	248 286	DT(10 ns) = 2 GW/cm^2. $n_x = 1.545$, $n_y = 1.685$, $n_z = 1.780$. $\Delta\theta l = 0.9$ mrad-cm. $\rho = 6°$
O, O, N, O 8-(4′-Acetylphenyl)-1,4-dioxa-8-azaspiro[4,5]decane(APDA)	$mm2$	$d_{33}(1.06) \approx 50$ $d_{32}(1.06) \approx 7$	384	217	$n_x = 1.56$, $n_y = 1.66$, $n_z = 1.68$. $d_{eff}(1.06) \approx 14.9$ pm/V.
H, N, O, NO_2, N, H, O 5-Nitrouracil (5NU)	222	$d_{14}(1.06) \approx 8.7$	410 & 1550	201	Relatively transparent at near IR, DT(10 ns) = 3 GW/cm^2. $n_x = 1.569$, $n_y = 1.901$, $n_z = 1.707$. $\Delta\theta l = 8$ mrad-cm, $\rho = 7°$, $\lambda_{NC}(\lambda) = 1440$ nm.

m-Aminophenol (*m*AP)	*mm*2	$d_{33}(1.06)\approx 3.3$ $d_{32}(1.06)\approx 2.4$ $d_{31}(1.06)\approx 0.7$	320 & 1700	36	Poor mechanical strength, $n_x=1.659$, $n_y=1.765$, $n_z=1.578$.
m-Chloronitrobenzene (*m*CNB)	*mm*2	$d_{33}(1.06)\approx 7.8$ $d_{32}(1.06)\approx 4$ $d_{31}(1.06)\approx 4.5$	400 & 2000	36	Cleaves easily, not phase-matchable for SHG at 1.064 μm, $n_x=1.676$, $n_y=1.684$, $n_z=1.649$.
m-Bromonitrobenzene (*m*BNB)	*mm*2	$d_{33}(1.06)\approx 8$ $d_{32}(1.06)\approx 4.5$ $d_{31}(1.06)\approx 4$	420 & 2100	36	$\lambda_{NC}(\theta)$ at 1.064 μm for solid solution of $mC_{0.95}Br_{0.05}NB$, $n_x=1.649$, $n_y=1.729$, $n_z=1.678$.
4-Nitrophenol sodium dihydrate (NPNa)	*mm*2	Type I $d_{eff}(1.06)\approx 8$	515	167	Vickers hardness: 34, good thermal conductivity.
m-Nitroaniline (*m*NA)	*mm*2	$d_{33}(1.06)\approx 20$ $d_{32}(1.06)\approx 1.5$ $d_{31}(1.06)\approx 20$ $r_{33}(0.63)\approx 17$ $r_{23}(0.63)\approx 0.1$ $r_{13}(0.63)\approx 7.5$	500 & 1900	256 36 240	Cleaves easily, melt grown into channel waveguide (unfavourable orientation for SHG), $n_x=1.752$, $n_y=1.715$, $n_z=1.665$.
2-Methyl-4-nitroaniline (MNA)	*m*	$d_{11}(1.06)\approx 167$ $d_{12}(1.06)\approx 25$ $r_{11}(0.63)\approx 67$	500 & 1900	143 170 181 150	Melt grown into channel waveguide (orientation controlled by electric or temperature gradients). DT(20 ns) = 0.2 GW/cm^2. $n_x=2.001$, $n_y=1.658$, $n_z=1.435$, near optimum molecular alignment for EO.

(continued)

TABLE 7.2 (Continued)

Structure and Nomenclature (acronym)	Point Group	SHG $d_{ij}(\lambda)$ & EO $r_{ij}(\lambda)$ (pm/v) (μm)	Cut-Off λ (nm)	Ref.	Properties
		Abstract			
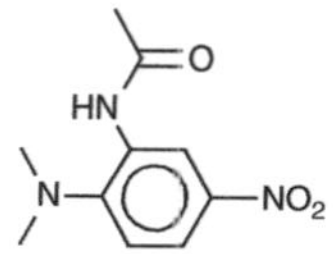 4-(*N*,*N*-dimethylamino)-3-acetamidonitrobenzene (DAN)	2	$d_{21}(1.06)\approx 1.5$ $d_{22}(1.06)\approx 5.2$ $d_{23}(1.06)\approx 50$ $d_{25}(1.06)\approx 1.5$ $r_{11}(0.63)\approx 13$	485 & 2270	121 122 123	Fiber waveguide allows full use of nonzero d_{ij}, DT(15 ns) = 80 MW/cm^2 $d_{eff}\approx 35.5$ (I) & 9 (II) pm/v. $n_x=1.539$, $n_y=1.682$, $n_z=1.949$. $\Delta\theta l\approx 1.5$ mrad-cm.
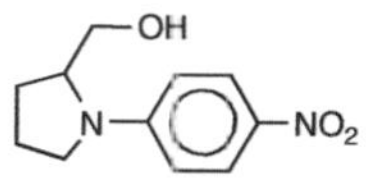 *N*-(4-nitrophenyl)-(*s*)-prolinol (NPP)	2	$d_{22}(1.06)\approx 28$ $d_{21}(1.06)\approx 85$	500 & 2000	289 138	Near optimum molecular alignment for largest d_{21}. $n_x=2.066$, $n_y=1.876$, $n_z=1.478$. $\lambda_{NC}(\theta)=1.15$ μm using d_{21} $\lambda_{NC}(\lambda)=1.5$ μm.
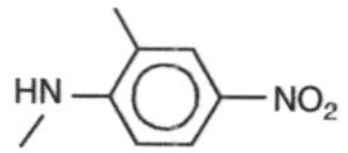 2-Methyl-4-nitro-*N*-methylaniline (MNMA)	*mm*2	$d_{33}(1.06)\approx 2.6$ $d_{31}(1.06)\approx 13$ $d_{15}(1.06)\approx 12$ $r_{13}(0.63)\approx 8$ $r_{33}(0.63)\approx 7.5$	510	245	$n_x=2.148$, $n_x=1.520$.
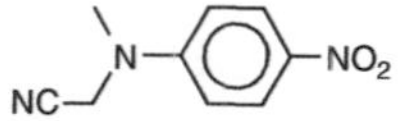 *N*-(4-nitrophenyl)-*N*-aminoacetonitrile (NPAN)	*mm*2	$d_{33}(1.06)\approx 27$ $d_{32}(1.06)\approx 57$ $d_{33}(1.34)\approx 24$ $d_{32}(1.34)\approx 48$	500	171 15	Near optimum molecular alignment for largest d_{32}. $\Delta\theta l\approx 2$ mrad-cm. $\Delta T1\approx 5$°C-cm
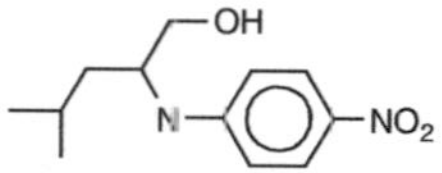 L-*N*-(5-nitro-2-pyridyl)leucinol(NPLO)	2	Type I: $d_{eff}(1.06)\approx 37$ Type II: $d_{eff}(1.06)\approx 3$	480	263	Vickers hardness: 34, nonhygroscopic, DT(8 ns) = 6 GW/cm^2. $n_x=1.457$, $n_y=1.631$, $n_z=1.933$. $\rho=0.22$ (I) & 0.24 (I) mrad.

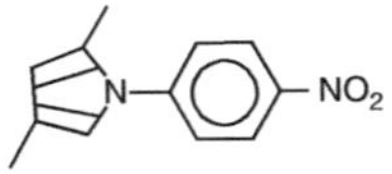 3,5-Dimethyl-1-(4-nitrophenyl) pyrazole (DMNP)	*mm*2	$d_{33}(0.84) \approx 29$ $d_{32}(0.84) \approx 90$	450	83 100	(100) oriented core fibers allow full use of d32, $\lambda_{NC}(\theta) = 944$ nm using d_{32}.
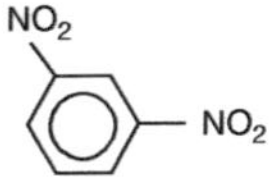 *m*-Dinitrobenzene (*m*DB)	*mm*2	$d_{33}(1.06) \approx 0.7$ $d_{32}(1.06) \approx 2.7$ $d_{31}(1.06) \approx 1.8$	400 & 2200	20	$n_x = 1.738$, $n_y = 1.680$, $n_z = 1.483$.
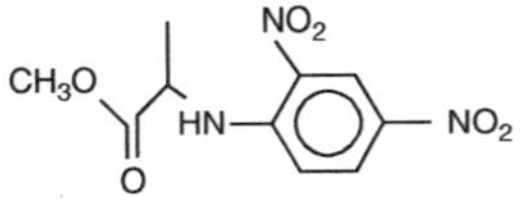 Methyl-(2,4-dinitrophenyl)-aminopropanoate (MAP)	2	$d_{22}(1.06) \approx 18$ $d_{21}(1.06) \approx 17$ $d_{23}(1.06) \approx 3.7$	500 & 2000	192	DT(10 ns) = 3 GW/cm^2. $d_{eff} \approx 3.8$ (I) & 8.8 (II) pm/v. $n_x = 1.531$, $n_y = 1.653$, $n_z = 1.935$. $\Delta\theta l \approx 1.5$ mrad-cm. $\rho = 11.5°$ (I) & 2.4°(I)
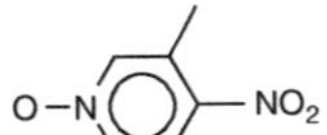 3-Methyl-4-nitropyridine-*N*-oxide (POM)	222	$d_{14}(1.06) \approx 10$ $r_{52}(0.63) \approx 5.2$	450 & 2100	288 230	DT(20 ps) = 2 GW/cm^2. $d_{eff} \approx 7.9$ (I) & 4.0 (II) pm/v. $n_x = 1.663$, $n_y = 1.829$, $n_z = 1.625$. $\rho = 6.3°$ (I) 1.4° (I).
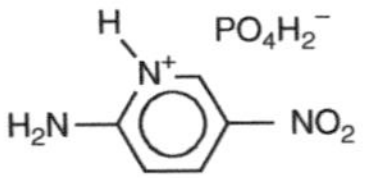 2-Amino-5-nitropyridinium-dihydrogen phosphate (2A5NPDP)	*mm*2	$d_{33}(1.06) \approx 12$ $d_{15}(1.06) \approx 7$ $d_{24}(1.06) \approx 1$	420 & 2000	135	$n_x = 1.752$, $n_y = 1.715$, $n_z = 1.665$. $\lambda_{NC}(\theta) = 1084$ & 1129 nm. $\lambda_{NC}(\lambda) = 1340$ nm. Not phase-matchable for type-II SHG at 1.064 μm. Type-I $d_{eff} \approx 2$ pm/V.. $d\theta_{pm}/dT = 2.4'/°C$ at 1.3 μm.
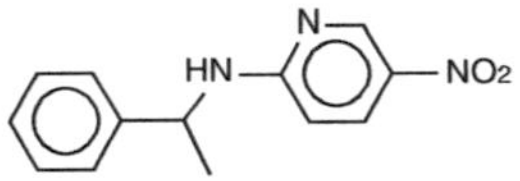 (−)2-(α-Methylbenzylamino)-5-nitropyridine (MBANP)	2	$d_{22}(1.06) \approx 60$[58] $d_{22}(1.06) \approx 35$[57]	430 & 1500	12 11 134	DT(425 ns) = 1 GW/cm^2. $n_y = 1.813$, $n_c = 1.676$. $d_{eff} \approx 1.2 \times LiIO_3 \approx 6$ pm/V.

(continued)

TABLE 7.2 (Continued)

Structure and Nomenclature (acronym)	Point Group	SHG $d_{ij}(\lambda)$ & EO $r_{ij}(\lambda)$ (pm/v) (μm)	Cut-Off λ (nm)	Ref.	Properties
Abstract					
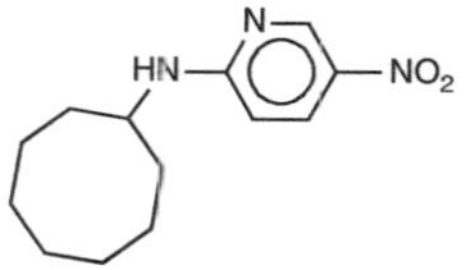 2-*N*-cyclooctylamino-5-nitropyridine (COANP)	*mm2*	$d_{33}(1.06) \approx 14$ $d_{32}(1.06) \approx 32$ $d_{31}(1.06) \approx 15$ $r_{33}(0.63) \approx 15$ $r_{13}(0.63) \approx 3.4$ $r_{23}(0.63) \approx 13$	470	77 26 27	$\lambda_{NC}(\theta) = 1023$ nm using d_{32}. $\lambda_{NC}(\theta) = 1413$ nm using d_{31}. $n_x = 1.68$, $n_y = 1.78$, $n_z = 1.64$. $d_{eff} \approx 24$ pm/V $r_{33}(0.52) \approx 28$, $r_{33}(1.06) \approx 7.7$. $r_{13}(0.52) \approx 6.8$, $r_{13}(1.06) \approx 0.9$. $r_{23}(0.52) \approx 26$, $r_{23}(1.06) \approx 6.3$.
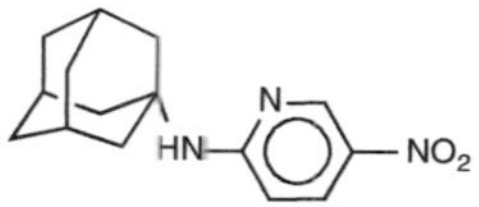 2-Adamantylamino-5-nitropyridine (AANP)	*mm2*	$d_{33}(1.06) \approx 60$ $d_{31}(1.06) \approx 80$	460	257	At 533 nm: $n_x = 1.77$, $n_y = 1.61$, $n_z = 1.86$. At 1.06 μm: $n_x = 1.67$, $n_y = 1.59$, $n_z = 1.71$.
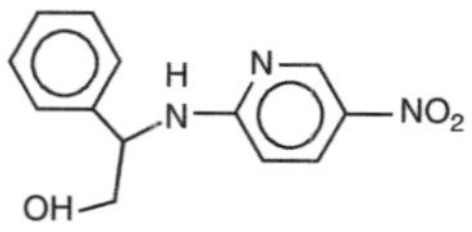 *N*-(4-nitro-2-pyridinyl)-(*s*)-phenylalaninol (NPPA)	2	$d_{22}(1.06) \approx 2.6$ $d_{21}(1.06) \approx 0.4$ $d_{23}(1.06) \approx 31$ $d_{16}(1.06) \approx 0.5$ $d_{34}(1.06) \approx 25$	480	261 247	$\lambda_{NC}(\theta)$ for SHG and SFG within absorption edge, calculated $d_{eff} \approx 31$ pm/V. $n_x = 1.524$, $n_y = 1.694$, $n_z = 1.907$.
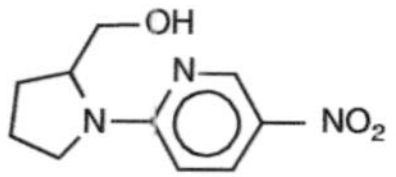 2-(*N*-prolinol)-5-nitropyridine (PNP)	2	$d_{22}(1.06) \approx 17$ $d_{21}(1.06) \approx 48$ $r_{22}(0.63) \approx 13$	490 & 2080	246 27	$\lambda_{NC}(\theta) = 1020$ nm using d_{21}. $d_{eff} \approx 47$ pm/V. $n_x = 1.990$, $n_y = 1.788$, $n_z = 1.467$. $\rho = 7°$ $r_{22}(0.52) \approx 28$, $r_{12}(0.52) \approx 20$. $r_{22}(1.06) \approx 8$, $r_{12}(1.06) \approx 9$.

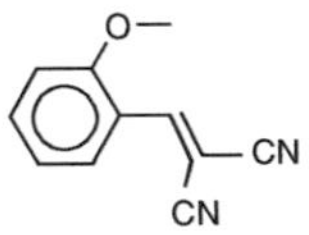 2-Dicyanovinylanisole (DIVA)	2	$d_{22}(1.06) \approx 10$		78	$d_{eff}(1.06) \approx 20$ pm/V. $n = 1.65$.
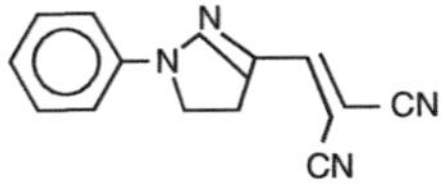 3-(1,1-Dicyanoethenyl)-1-phenyl-4,5-dihydro-1*H*-pyrazole (DCNP)	*m*	$r_{33}(0.63) \approx 87$	700 & 1600	3	High melting point: 190°C, near optimum molecular alignment for EO, $n_x = 1.9$, $n_z = 2.7$.
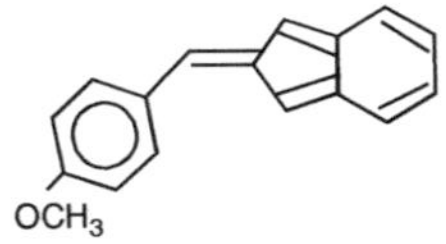 ω-(*p*-Methoxyphenyl) benzofulvene (MPBF)	*mm2*	$d_{eff}(1.06) \approx 7$	>450	133	$d_{eff} \approx 7$ pm/V. $\Delta\theta l = 0.32$ mrad-cm. $\Delta T1 \approx 4.6$°C-cm.
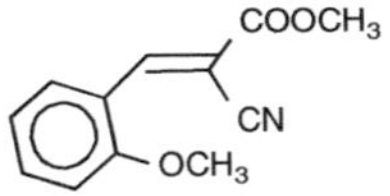 2-Cyano-3-(2-methoxyphenyl)-2-propenoic acid methyl ester (CMP-methyl)	2	$d_{22}(1.06) \approx 29$	410	183	$n_y = 1.85$
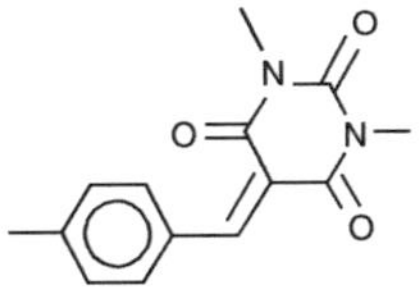 *p*-Methylbenzal-1,3-dimethylbarbituric acid	2	$d_{eff}(1.06) \approx 8$	460	132	Vickers hardness: 25.5.

(continued)

TABLE 7.2 (Continued)

Abstract					
Structure and Nomenclature (acronym)	Point Group	SHG $d_{ij}(\lambda)$ & EO $r_{ij}(\lambda)$ (pm/v) (μm)	Cut-Off λ (nm)	Ref.	Properties
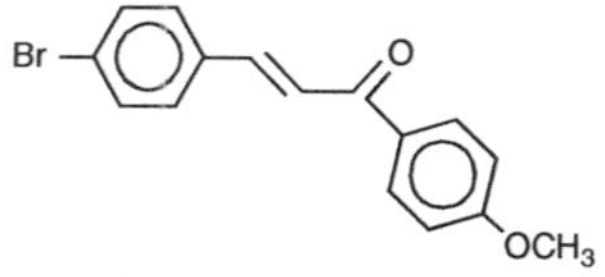 Br, O, OCH_3 4-Br-4′-methoxychalcone	*m*	$d_{33}(1.06) \approx 6$ $d_{13}(1.06) \approx 27$	420	285	$n_x = 1.55$, $n_y = 1.47$, $n_z = 1.90$.
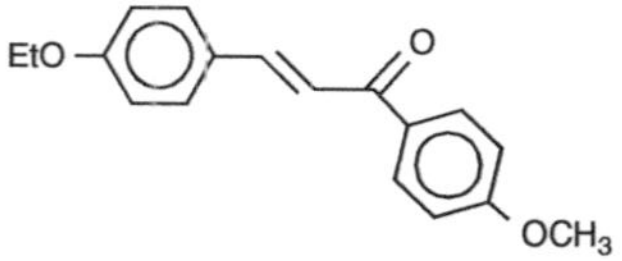 EtO, O, OCH_3 4-Ethoxy-4′-methoxychalcone	*mm2*	$d_{eff}(1.06) \approx 5.7$	430	129	Low hardness, DT(1 ns) > 30 GW/cm^2. At 532 nm: $n_x = 1.493$, $n_y = 1.710$, $n_z = 1.983$. At 1.06 μm: $n_x = 1.477$, $n_y = 1.663$, $n_z = 1.850$. $d_{eff} \approx 3.5$(I) & 5.7 (II) pm/v.
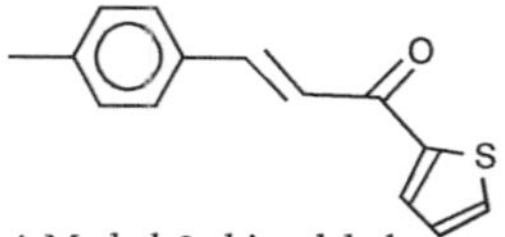 O, S 4-Methyl-2-thienylchalcone	2	$d_{eff}(1.06) \approx 7$	430	128	Low hardness, $n_x = 1.648$, $n_y = 1.696$, $n_z = 1.775$. $d_{eff} \approx 7$ pm/V. $\Delta\theta l = 0.9$ mrad-cm. $\rho = 3.6°$ $\Delta T1 = 2.2°$C-cm
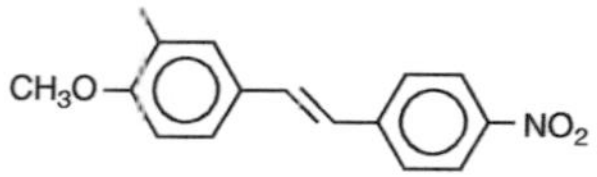 CH_3O, NO_2 3-Methyl-4-methoxy-4′-nitrostilbene (MMONS)	*mm2*	$d_{33}(1.06) \approx 184$ $d_{32}(1.06) \approx 41$ $d_{24}(1.06) \approx 71$ $r_{33}(0.63) \approx 40$	515 & 2000	22 220	$\lambda_{NC}(\theta) = 1028$ nm using d_{24}. $d_{eff} \approx 43$ pm/V.[71] $n_x = 1.569$, $n_y = 1.693$, $n_z = 2.129$. $\rho = 9.6°$ $\Delta T1 = 0.17°$C-cm

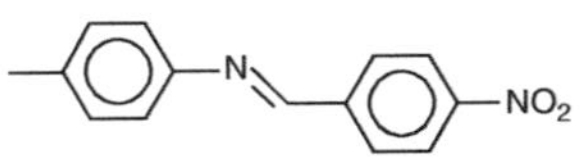 4-Nitro-4′-methylbenzylidiene aniline (NMBA)	*m*	$d_{11}(1.06) \approx 139$ $d_{33}(1.06) \approx 0.6$ $d_{31}(1.06) \approx 41$ $r_{11}(0.63) \approx 25$	480 & 1600	8, 9	Near optimum molecular alignment for EO, $\lambda_{NC}(\theta)$ for SHG within absorption edge, $\lambda_{NC}(\lambda) =$ 1500 nm. $n_x = 1.951$, $n_y = 1.657$, $n_z = 1.510$. $d\Delta n/dT = 15.8 \times 10^{-5}\ K^{-1}$. $d_{eff} \approx 2$ pm/V.
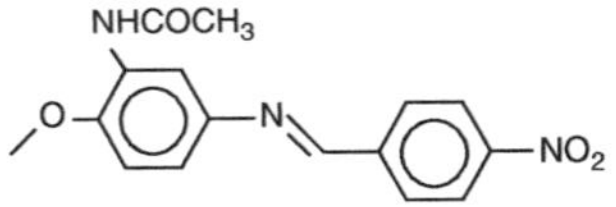 4′-Nitrobenzylidene-3-acetamino-4-methoxylaniline (MNBA)	*m*	$d_{11}(1.06) \approx 175$ $d_{31}(1.06) \approx 2$ $d_{33}(1.06) \approx 2$ $r_{11}(0.63) \approx 29$ $r_{13}(0.63) \approx 0.5$ $r_{33}(0.63) \approx 2.4$	505	131	Near optimum molecular alignment for EO, $n_x = 2.024$, $n_y = 1.648$, $n_z = 1.583$.
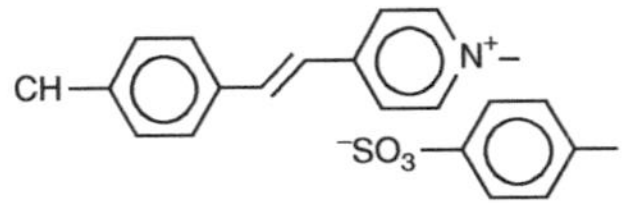 Cyanostilbazolium *p*-toluenesulfonate complex	*mm2*	$d_{33}(1.06) \approx 21$	415	258	Melting point: 279°C, $n_z = 1.775$.
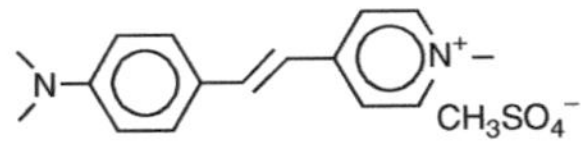 4′-Dimethyamino-*N*-methyl-4-stilbazolium methyl sulfate (SPCD)	*mm2*	$r_{33}(0.63) \approx 430$	600	282	Only thin-film cystal reported, $n_y = 1.31$, $n_z = 1.55$.
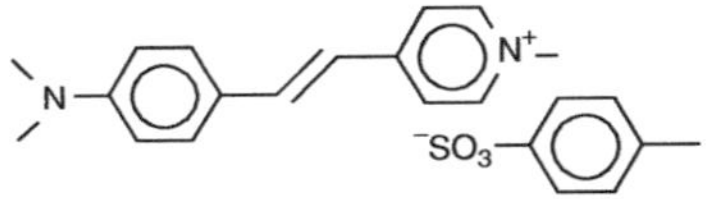 4′-Dimethyamino-*N*-methyl-4-stilbazolium tosulate (DAST)	*m*	$d_{11}(1.91) \approx 600$ $d_{22}(1.91) \approx 100$ $d_{12}(1.91) \approx 30$ $r_{33}(0.82) \approx 400$	700 & 2000	157 198	Near optimum molecular alignment for EO. At 820 nm: $n_x = 2.216$, $n_y = 1.66$, $n_z = 1.65$. Dielectric constant: $\varepsilon_{ab} = 5.1$, $\varepsilon_c = 3.1$.

TABLE 7.3 Experimental Results of Poled Polymer Studies: M_w, Molecular Weight; $\rho\#$, Chromophore Number Density; α(Wavelength), Optical Propagation Loss; τ_1 and τ_2, Relaxation Time in $d_{eff}(t)/d_{eff}(0) = Ae^{-t/\tau 1} + Be^{-t/\tau 2}$

Structure and Nomenclature (Ref)	Properties
Guest-host polymer composites 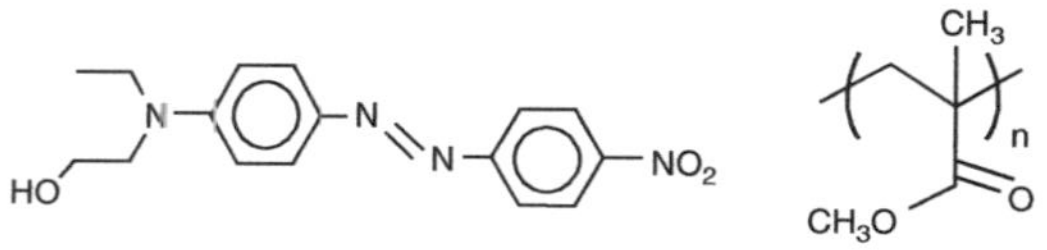 (4[*N*-ehtyl-*N*-(2-hydroxyethyl)]amino-4′-nitroazobenzene) (DR1) doped in poly-(methylmethacrylate) (PMMA) (Singer et al. 1986)	$T_g \approx 100°C$. $\lambda_{max} \approx 470$ nm. $\rho^{\#} = 2.4 \times 10^{20}$ cm^{-3} low dielectric constant: $\varepsilon = 3.6$. contact poled with 62 V/μm at 100°C. d_{33}(1.58 μm) = 2.5 pm/V. Novelty: first study of amorphous guest–host system.
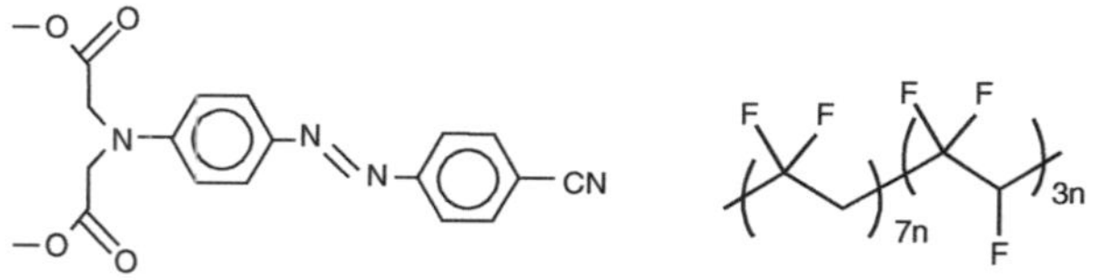 4-(4′-Cyanophenylazo)-*N*,*N*-bis-(methoxycarbonylmethyl)-aniline doped in copolymer of vinylidene fluoride and trifluoroethylene (Foraflon®, 70:30 mol%) (Pantelis et al. 1988; Hill et al. 1987)	$T_g \approx 100°C$. $\lambda_{max} = 400$ nm. UP to 10 wt% doping. α(1 μm) ≈ −1.5 dB/cm. Corona poled at 25°C. d_{33} (1.06 μm) up to 2.6 pm/V. Stable nonlinearity after 300 days at ambient condition. Novelty: ferroelectric host polymer provides a stable internal field of 150 V/μm.
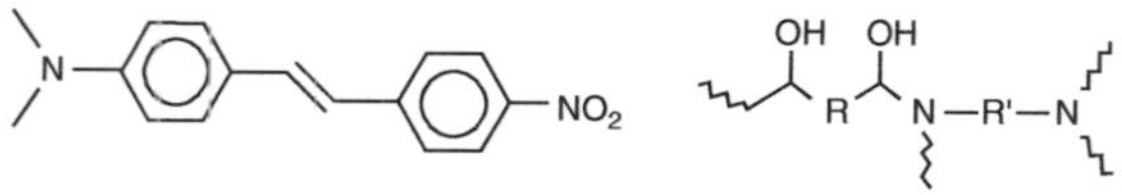 4-*N*,*N*-dimethylamino-4′-nitrostilbene doped in thermosetting epoxy (EPO-TEK® 301-2) (Hubbard et al. 1989)	Precure at 80°C prior to poling. $\lambda_{max} \approx 430$ nm. $\rho^{\#} = 0.2 \times 10^{20}$ cm^{-3} Contact poled at 60 V/μm. Temporal stability: τ_1 (25°C) = 7 days and τ_2 (25°C) = 72 days. Novelty: use of crosslinked polymer as host.
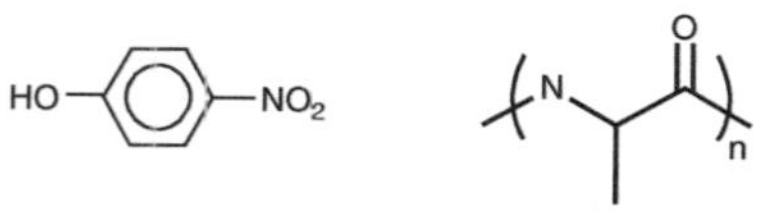 *p*-Nitrophenol doped in type A gelatin (Ho et al. 1992)	$T_g \approx 60$–70°C. $\lambda_{max} < 350$ nm. UP to 35 wt% doping. Spin coating from aqueous solution. Contact poled at 40 V/μm. r_{33} (633 nm) = 10–40 pm/V. 40% activity remains after 5 days. Novelty: use of cross-linked biopolymer as host.

Material	Properties
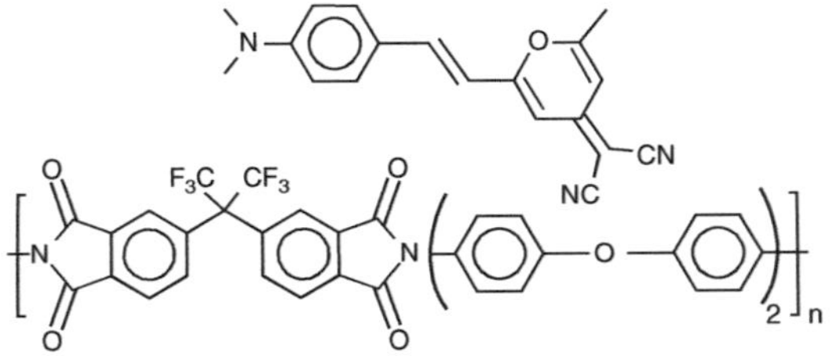 (Dicyanomethylene)-2-methyl-6-(*p*-dimethylaminostyryl-4*H*-pyran (DCM) doped in polymide (Amoco Ultradel®) (Ermer et al. 1992)	$T_g = 220°C$. $\lambda_{max} = 474$ nm. Doping level: 20 wt% n (830 nm) = 1.651 α (830 nm) ≈ −1.5 dB/cm. Multilayers contact poled with 312 $V/\tilde{\mu}$ 190°C. r_{33} (830 nm) = 3.4 pm/V. Low poling field at NLO layer due to high conductivity. Novelty: use of high T_g host polymer.
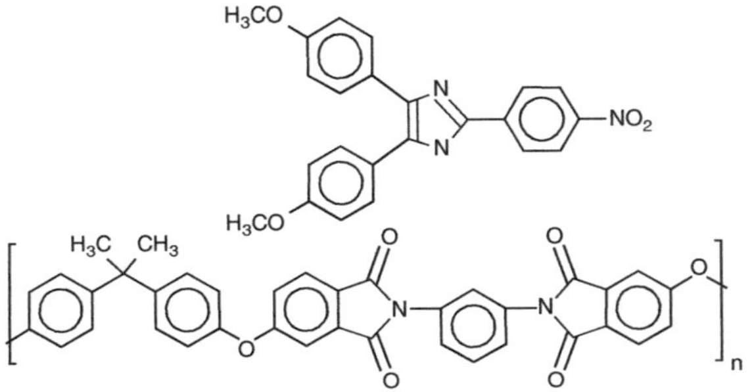 2,4,5-Triarylimidazole derivative (lophine) doped in polyimide (Ultem®) (Stahelin et al. 1992a)	$T_g = 210$–150°C for 0–35 wt%. $\lambda_{max} = 410$ nm. Chromophore: $\mu g = 7 \times 10^{-18}$, $\beta_0 = 18 \times 10^{-30}$ (esu). High loading, up to 35 wt%. For 0–35 wt% doping, corona poling gives d_{33} (1.047 μm) = 6–17 pm/V. For 20 wt% doping, corona poling gives $T_g = 180°C$, d_{33} = 10.5 pm/V., τ (80°C) = 1.5 year. Novelty: use of thermally stable chromophore.
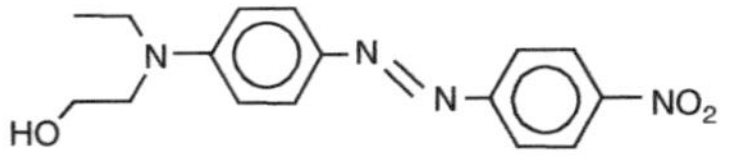 DR1 doped in phenylsiloxane polymer (Allied Signal Accuglass 204®) (Jeng et al. 1992a)	$\lambda_{max} = 493$ nm. n (533 nm) = 1.628. Doping level: 35 mg dye in 2 g of A204. Corona poled at 200°C for 10 min. d_{33} (1.06 μm) = 11.43 pm/V. d_{33} (40 h, 100°C) = 3.8 pm/V. Novelty: Use of sol-gel as host polymer.

(continued)

TABLE 7.3 (Continued)

Structure and Nomenclature (Ref)	Properties
Side-Chain acrylate polymers 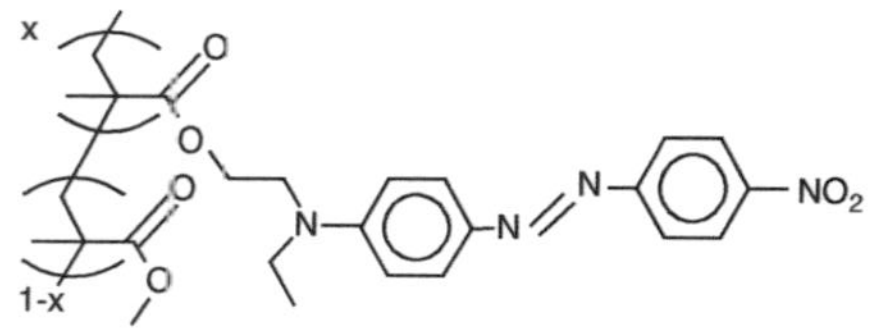 Methylmethacrylate (MMA)/DR1 functionalized methacrylate copolymer (Esselin et al. 1988)	T_g=128–134°C for 1–19 mol%. λ_{max}=470 nm. $\rho^{\#}$: upto 7.5 × 10^{20} cm^{-3}. Contact poled with 90 V/μm at 130°C. d_{33} (1.064 μm)=3–58 pm/V. for 1–19 mol% Novelty: high $\rho^{\#}$.
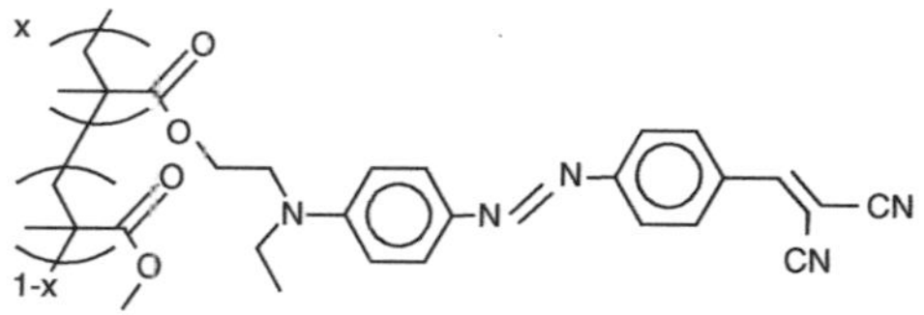 MMA/4-dicyanovinyl-4′-(*N*,*N*-dialkylamino)azobenzene functionalized methacrylate copolymer (Singer et al. 1988)	T_g=127°C. $\rho^{\#}$=8×10^{20} cm^{-3} n (800 nm)=1.58. Corona poled above T_g. d_{33} (1.58 μm)=21 pm/V. r_{33} (799 nm)=15 pm/V. 90% of activity remains stable after 35 days at ambient conditions.
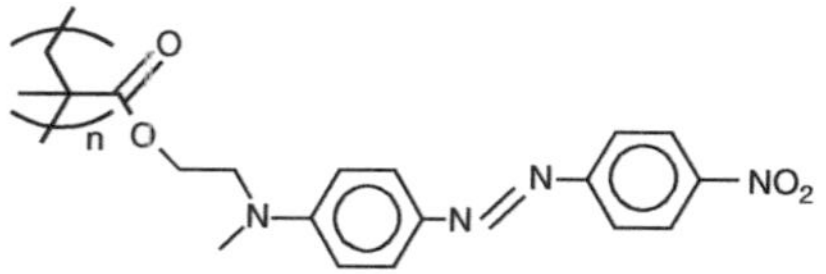 Poly-4-(4′-nitrophenylazo)-*N*-methyl-*N*-(2-acroyloxyethyl)aniline (Hill et al. 1989, 1988)	T_g=105°C. λ_{max}≈470 nm. M_w=4.9×10^3. Contact poling at 190 V/μm gives d_{33} (1.06 μm)=55 pm/V. Contact poling at 20 V/μm gives r_{33} (633 nm)=30 pm/V. Activity remains stable for over 2 years at ambient conditions.

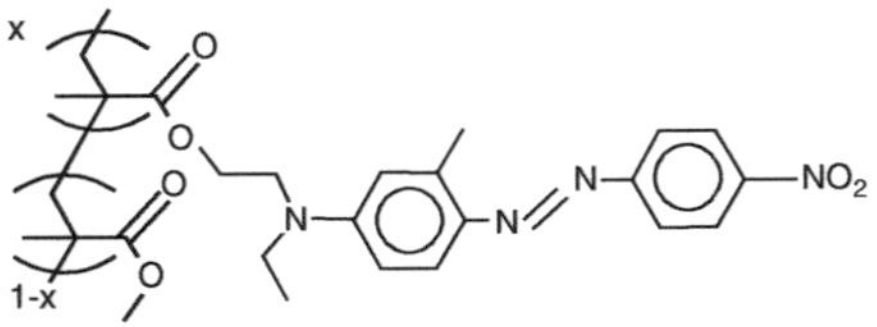 MMA/4-(*N*-ethyl-*N*-2-methacryloxyethoxy)-2-methyl-4′-nitroazobenzene copolymer (Ore et al. 1989)	$\lambda_{max}=486$ nm. $M_w=10\times10^3$. $\rho^{\#}=9.2\times10^{20}$ cm^{-3} Corona poled d_{33} (1.06 μm) = 41 pm/V. Stable nonlinearity after 75 days at ambient conditions.
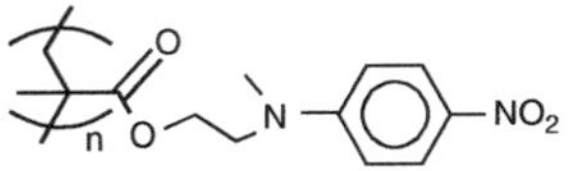 Poly-*N*-(2-methacryloxyethyl)-*N*-methyl-4′-nitroaniline (Hayashi et al. 1992)	$T_g=100$°C. $\lambda_{max}=390$ nm. $M_w=11\text{–}17\times10^3$. n (633 nm) = 1.66. Corona poled d_{33} (1.58 μm) = 30 pm/V. Activity decays to 65% in 40 days at ambient conditions.
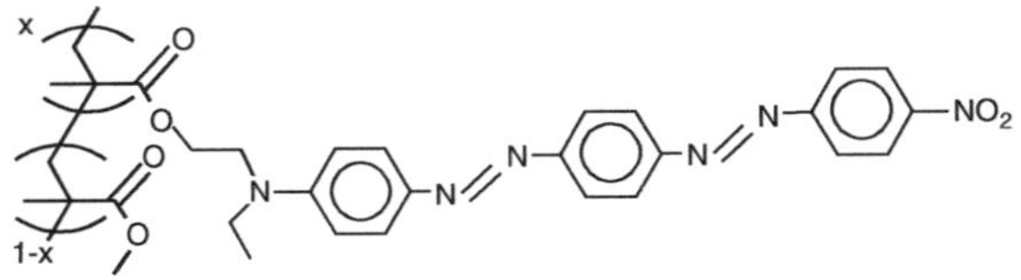 MMA/4-*N*-(2-methacryloxyethyl)-*N*-ethyl-4′-aminophenylazo-4″-nitroazobenzene copolymer (Amano et al. 1990)	$T_g\approx100$°C. $\lambda_{max}=500$ nm. $\rho^{\#}=4.3\times10^{20}$ cm^{-3} Corona poled at 100°C. d_{33} (1.06 μm) = 142 pm/V. d_{33} (1.7 μm) = 70 pm/V.
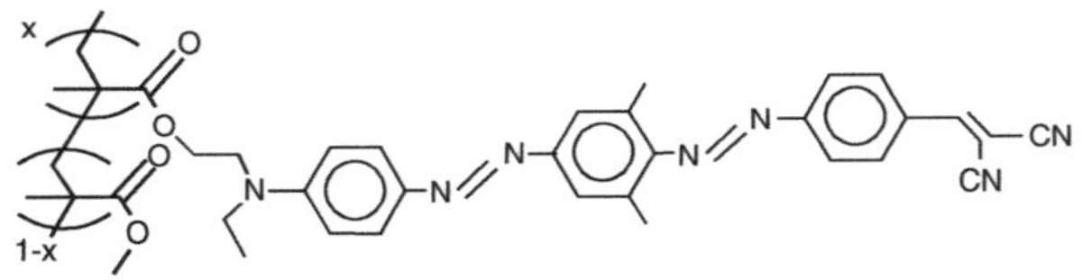 MMA/2,5-dimethyl-4-*N*-(2-methacryloxyethyl)-*N*-ethyl-4′-aminophenylazo-4″-dicyanoazobenzene copolymer (Shuto et al. 1991)	$T_g\approx140$°C. $\lambda_{max}=510$ nm. $\rho^{\#}=4\times10^{20}$ cm^{-3} α (633 nm) = −50 dB/cm due to absorption. Corona poling with 200 V/μm at 140°C gives d_{33} (1.06 μm) = 417 pm/V. Contact poling at 150 V/μm at 140°C gives r_{33} (633 nm) = 40 pm/V. Excellent temporal stability at 80°C.

(continued)

TABLE 7.3 (Continued)

Structure and Nomenclature (Ref)	Properties
MMA/4-(*N*-methyl-*N*-2-methacroyloxyethoxy)-4′-cyanoabenzene homo-and copolymers (S'heeren et al. 1993a)	Corona poled near T_g. d_{33} values are measured 10 days after poling at 1.064 μm.

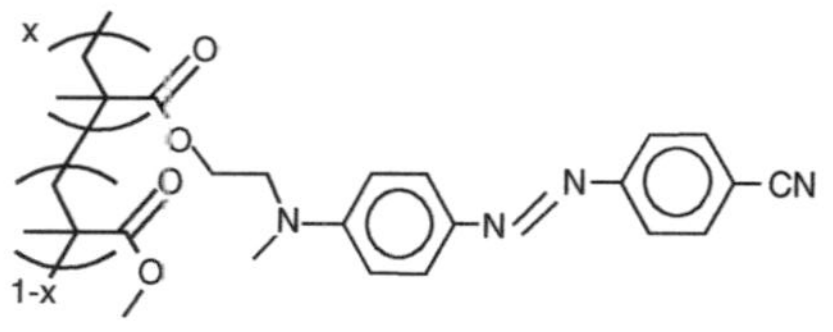

dye (mol%)	T_g (°C)	Mw (×10³)	d_{33} (pm/V)
12	129	59	5.2
16	128	44	21
32	134	59	68
44	128	39	45
49	125	98	32
72	124	36	31
100	181	15	26

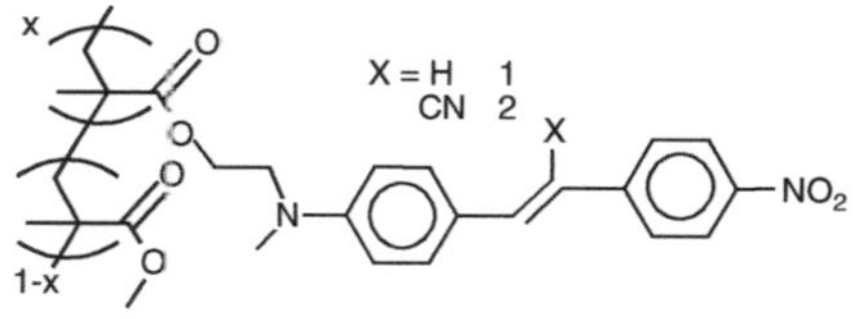

Structure and Nomenclature (Ref)	Properties
X = H 1, CN 2 — MMA/4-(*N*-methyl-*N*-2-methacroyloxyethoxy)-4′-nitrostilbene co-polymers (S'heeren et al. 1993c)	Corona poled near T_g. d_{33} values are measured several days after poling at 1.064 μm.

dye	(mol%)	T_g (°C)	Mw (×10³)	d_{33} (pm/V)
1	11	122	69	18
	19	120	70	31
	31	124	69	44
	39	126	66	15
	52	120	48	66
2	9	124	56	22
	18	131	62	41
	33	126	60	20

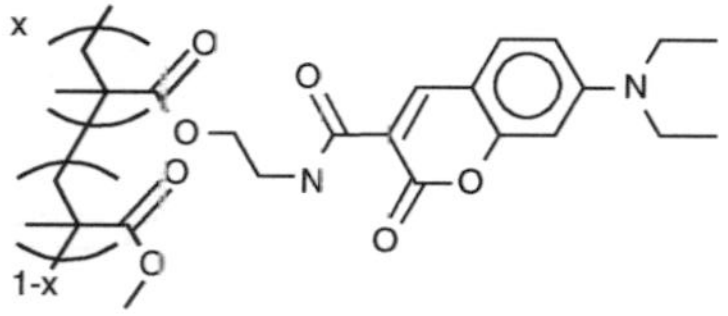

Structure and Nomenclature (Ref)	Properties
MMA/*N*-(3-methacryloxyalky)-7-diethylaminocoumarin-3-carboxamide (76:23 mol%) copolymer (Mortazavi et al. 1991)	$T_g = 135°C$. $\lambda_{max} = 410$ nm. $\rho^{\#} = 9.8 \times 10^{20}$ cm^{-3} $M_w = 89 \times 10^3$. Corona poled for SHG and contact poled with 100 V/μm for EO at 60°C. d_{33} (1.06 μm) = 13 pm/V. r_{33} (477–1100 nm) = 2–12 pm/V. d_{33} decays by 25% at 100°C in 50 h.

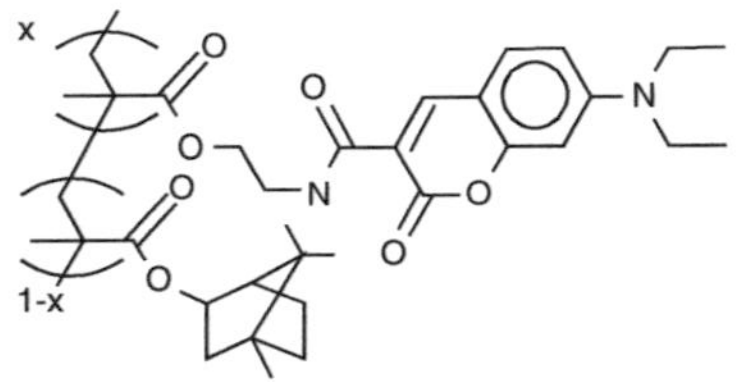

Isobornyl methacrylate/*N*-(3-methacryloxyalkyl)-7-diethlaminocoumarin-3-carboxamide (76:23 mol%) copolymer (Mortazavi et al. 1991)

T_g = 170°C.
λ_{max} = 410 nm.
$M_w = 50 \times 10^3$.
Corona poled at 200°C.
d_{33} (1.06 μm) = 11 pm/V.
d_{33} decays by 10% at 100°C and by 40% at 140°C in 50 h.

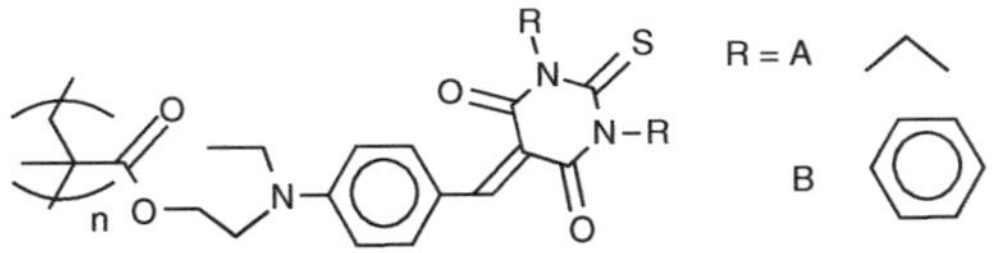

Poly(*p*-*N*-(2-methacroyloxethyl)-*N*-ethylaminobenzall-1-3-diethyl (or diphenyl thiobarbituric acid) (Cheng and Tan 1993)

R = A (ethyl), B (phenyl)

Corona poled near T_g.
d_{33} and r_{33} values are measured at 1.064 μm.
At 100°C d_{33} of A(B) decays by 80% (10%) in 150 min (15 days)

	T_g(°C)	Mw ($\times 10^3$)	d_{33} (pm/V)	r_{33} (pm/V)
A	125	74	33	14
B	194	60	40	18

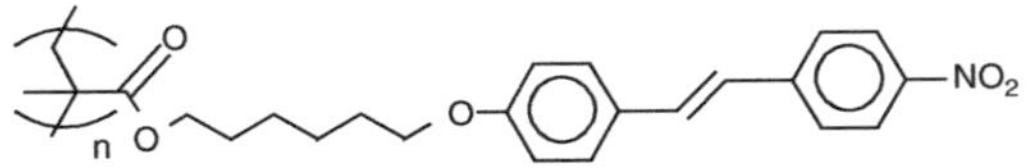

Poly-4-(6-acroyloxyhexyloxy)-4′-nitrostilbene (Huijts et al. 1989b)

T_g = 65°C.
λ_{max} = 380 nm.
$\rho^{\#}$:18×10^{20} cm^{-3}
n (633 nm) = 1.62.
α (1.3 μm) ≈ −1.5 dB/cm.
Contact poled with 23 V/μm at 65°C.
r_{33} (633 nm) = 0.9 pm/V.

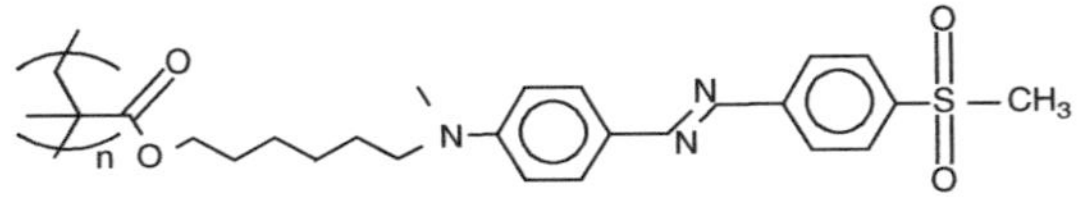

Poly-4′-*N*-(6-methacroyloxyhexyl)-*N*-methyl-amino-4-methylsulfonylazobenzene (Robello et al. 1991a, 1992)

T_g = 99°C.
λ_{max} = 446 nm.
$\rho^{\#} = 17 \times 10^{20}$ cm^{-3}
$M_w = 89 \times 10^3$.
n (633 nm) = 1.76.
α (830 nm) = −1 dB/cm.
Contact poled with 90 V/μm at 100°C.
r_{33} (633 nm) = 39 pm/V.; r_{33} (860 nm) = 13 pm/V.

(continued)

TABLE 7.3 (Continued)

Structure and Nomenclature (Ref)	Properties
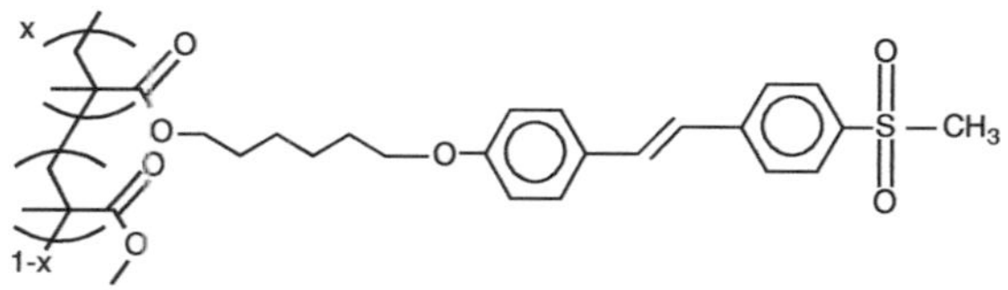 MMA/4′-(6-methacroyloxyhexloxy)-4-methylsulfonylstilbene (55:45 wt%) copolymer (Rikken et al. 1991; Seppen et al. 1991)	$T_g = 117°C$. $\lambda_{max} = 355$ nm. Low absorption at 420 nm. Corona poled with 120 V/μm at 100°C. d_{33} (820 nm) = 9 pm/V. Activity decreases to a quasi-stable 70% value after 120 days at ambient. Waveguide formation by UV photobleaching.
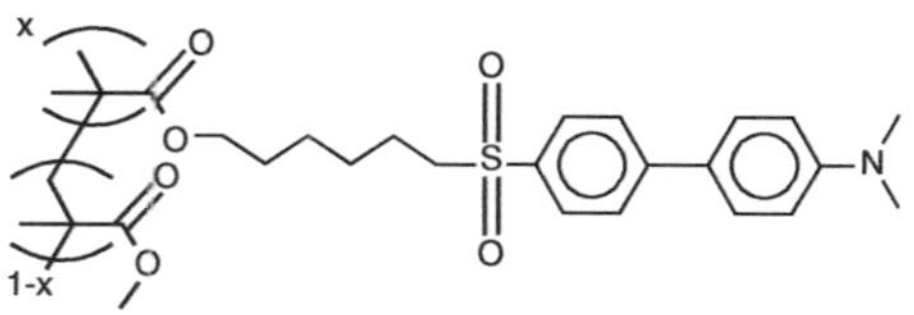 MMA/4′-(6-methacroyloxyhexylsulfony)-4-*N,N* dimethylamino-biphenyl (50:50 wt%) copolymer (Rikken et al. 1992)	$T_g \approx 100°C$. $\lambda_{max} = 340$ nm. Corona poled with 120 V/m at 100°C. d_{33} (820 nm) = 45 pm/V. poor temporal stability at 60°C.
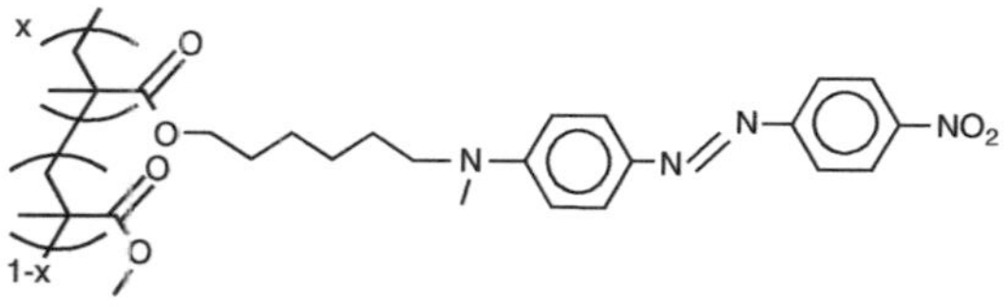 MMA/4-(4′-nitrophenylazo)-*N*-methyl-*N*-(6-methacroyloxyhexyl) aniline (81:19 mol%) copolymer (Muller et al. 1992a)	$T_g = 104°C$. $\lambda_{max} \approx 470$ nm. $M_w = 100 \times 10^3$. Thermally decomposes at 267°C. Poled with 110 V/μm. r_{33}–r_{13} (1.3 μm) = 9 pm/V.

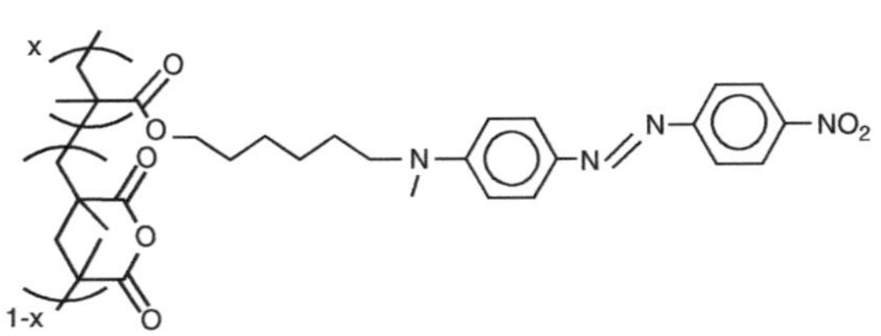 Methacrylic anhydride/4-(4′-nitrophenylazo)-*N*-methyl-*N*-(6-methacroyloxyl)aniline (33:67 mol%) copolymer (Strohriegl 1993; Muller et al. 1992a)	$T_g=90°C$. $\lambda_{max}\approx 470$ nm. Thermally decomposes at 227°C. Poled with 110 V/μm. r_{33}–r_{13} (1.5 μm) = 19 pm/V.
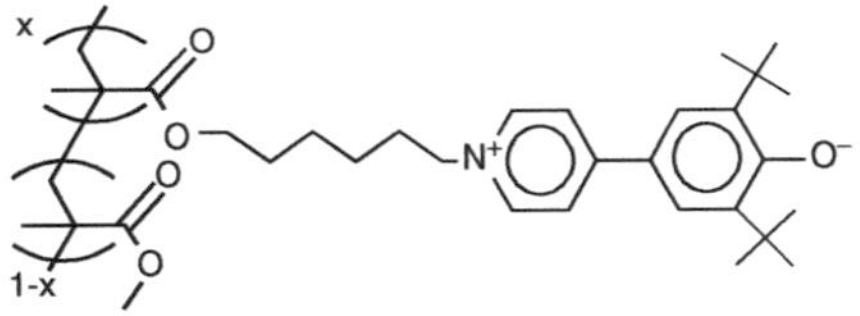 MMA/[2,6-di-*tert*-butyl-4-(1-ω-methacroyloxyalyl)-4-pyridino]phenolates (85:15 mol%) copolymer (Combellas et al. 1992)	$T_g=130°C$. $\lambda_{max}\approx 525$ nm. Contains zwitterionic chromophore with $\beta_0=-30\times 10^{-30}$ esu. Soluble in THF.
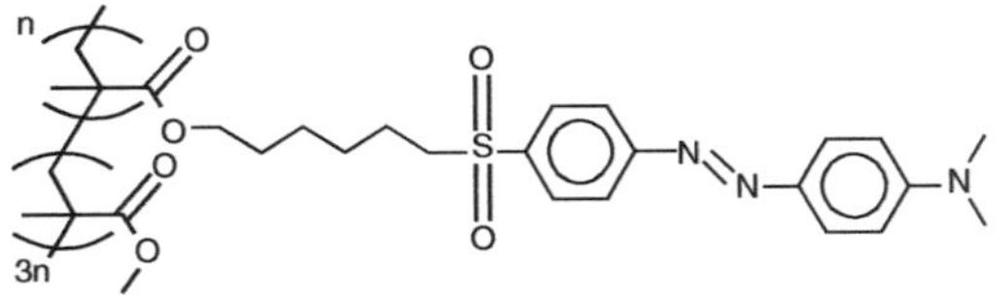 MMA/4′-(6-methacroyloxyhexlsulfonyl)-4-*N*,*N*-dimethylamino-azobenzene (25 mol%) copolymer (Xu et al. 1993)	$T_g=124°C$. $\lambda_{max}=437$ nm. $M_w=17\times 10^3$. Corona poled at 125°C for 45 min. d_{33} (1.06 μm) = 100 pm/V. Activity stabilizes at 95% value after 10 days at ambient conditions.

(continued)

TABLE 7.3 (Continued)

Structure and Nomenclature (Ref)	Properties
Functionalized homo-and copolymers of isocyanatoethyl methacrylate, MMA, and dimethylaminoethyl methacrylate. (Cheng et al. 1992)	Corona poled near T_g. d_{33} values are measured at 1.064 μm. d_{33} of A & C are stable at room temperature. Polymer B is stable at 80°C. $\alpha(1.06\ \mu m) \approx -(1.5\text{–}3)$ dB/cm.
Side-chain Non-acrylate polymers	
Poly-4-[*N*-(4′-nitrophenyl)amino-methyl]ethylene (Eich et al. 1989b)	$T_g = 125$°C. $\lambda_{max} \approx 390$ nm. Thermally decomposes at 260°C. Corona poled at 140°C. d_{33} (1.06 μm) = 31 pm/V. d_{33} relaxes to 19 pm/V in 5 days.
Poly-4-[*N*-methyl-*N*-(4′-nitrophenyl)amino-methyl]styrene (Hayashi 1991)	$T_g = 103$°C. $\lambda_{max} = 393$ nm. $M_w = 12 \times 10^3$. n(633 nm) = 1.73. α (633 nm) = −10 dB/cm. Corona poled at 110°C. d_{33} (1.06 μm) = 28 pm/V which stabilizes to 18 pm/V in 5 months at ambient conditions.
Poly-4-[*N*-(4′-nitrophenyl)amino-methyl]styrene (Hayashi et al. 1992)	$T_g = 123$°C. $\lambda_{max} = 383$ nm. $M_w = 15 \times 10^3$. n(633 nm) = 1.70. Corona poled at T_g. d_{33}(1.06 μm) = 10 pm/V. Low d_{33} due to poling-induced decomposition.

(x, y, z)	T_g(°C)	ρ^a (10^{20}/cm³)	d_{33} (pm/V)
A(l, 1, 0)	120	6.8	19
B(l, 0, 0)	130	13	44
C(l, 0, 0)	140	13	88
C(l, 0, 1)	130	9.6	121
C(l, 0, 2)	135	12	90

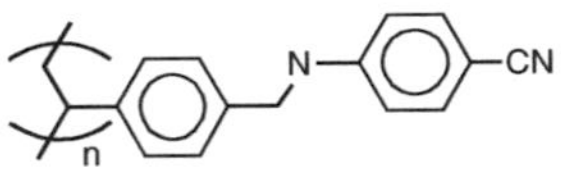 Poly-4-[*N*-(4′-cyanophenyl)amino-methyl]styrene (Hayashi, et al. 1992)	$T_g = 132°C$. $\lambda_{max} = 288$ nm. $M_w = 15 \times 10^3$. $n(633\ nm) = 1.65$. Corona poled at T_g. $d_{33}(1.06\ \mu m) = 1$ pm/V.
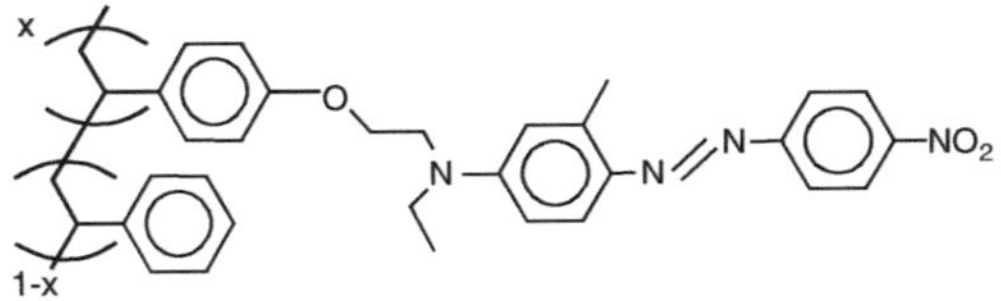 Styrene/*p*-[4-nitro-4′-(*N*-ethyl-*N*-2-oxyethyl)azobenzene] methylstyrene copolymer (88:12 mol%) (Ye et al. 1987)	$T_g = 110°C$. $\lambda_{max} \approx 470$ nm. Contact poled with 30 V/μm at ≈ 110°C. d_{33} (1.06 μm) = 1.1 pm/V. Stable activity at room temperature.
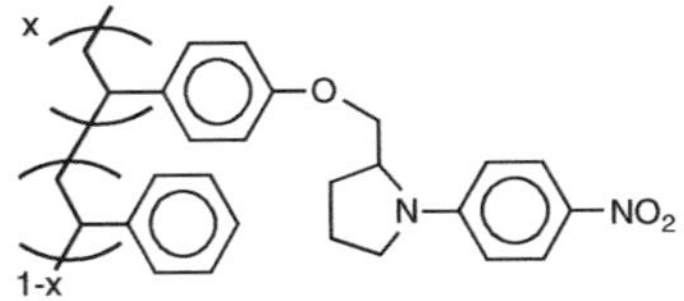 Styrene/*N*-(4-nitrophenyl)-*S*-prolinoxy methylstyrene copolymer (64:36 mol%) (Ye et al. 1989)	$T_g = 110°C$. $\lambda_{max} \approx 380$ nm. Contact poled with 70 V/μm at ≈ 110°C. d_{33} (1.06 μm) = 1.6 pm/V. Low d_{33} due to material impurity.
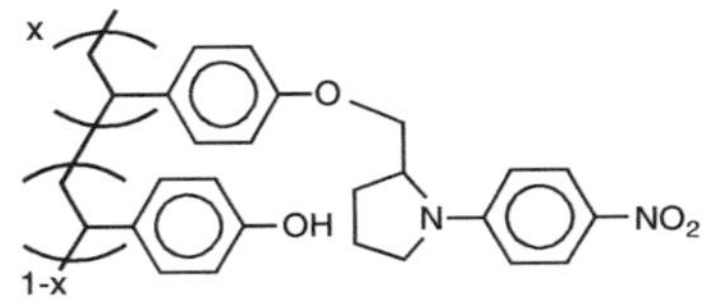 *p*-hydroxystyrene/*N*-(4-nitrophenyl)-*S*-prolinoxy styrene copolymer (10:90 mol%) (Ye et al. 1989)	$T_g = 96°C$. $\lambda_{max} \approx 380$ nm. $\rho^{\#} = 2.3 \times 10^{21}\ cm^{-3}$ Corona poled at T_g. d_{33} (1.06 μm) = 33 pm/V. T_g increase to 146°C at 16 mol% functionalization.

(continued)

TABLE 7.3 (Continued)

Structure and Nomenclature (Ref)	Properties
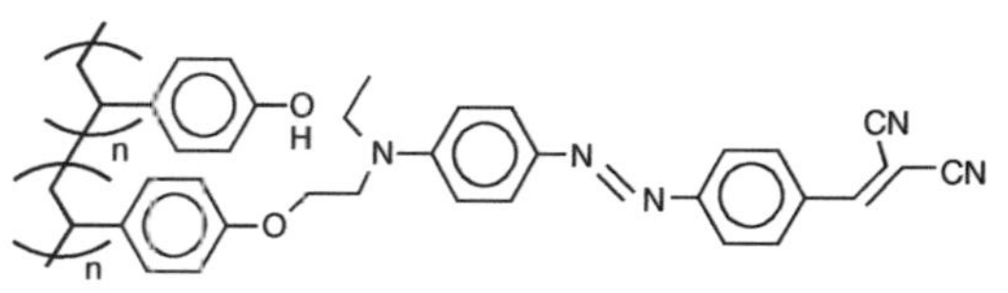 *p*-hydroxystyrene/*p*-[4-(2,2-dicryanovinyl)-4′-(*N*-ethyl-*N*-2-oxyethyl)azobenzene] styrene copolymer (Ye et al. 1992)	Corona poled at 117°C for 15 h. Thermal and electrochemical decomposition observed at 147°C.
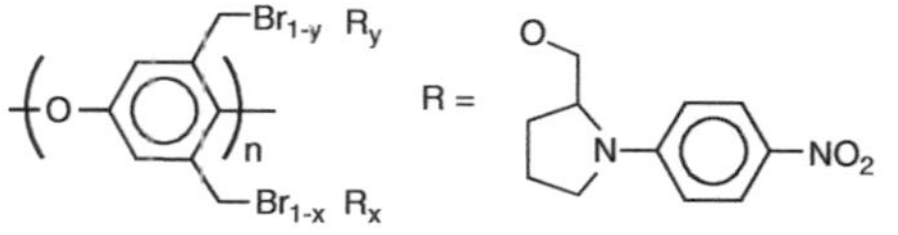 Poly-(2,6-dimethylbromo-1,4-phenylene oxide) partially functionalized with *N*-(4-nitrophenyl)-*S*-prolinol (Dai et al. 1990)	$T_g = 170$°C. $\rho^{\#} = 2.6 \times 10^{21}$ cm^{-3} n (633 nm) = 1.584. Corona poled at 190°C for 30 min. $\alpha = -1$ dB/cm. Temporal stability: τ_1(25°C) = 0.3 days and τ_2(25°C) = 39 days.
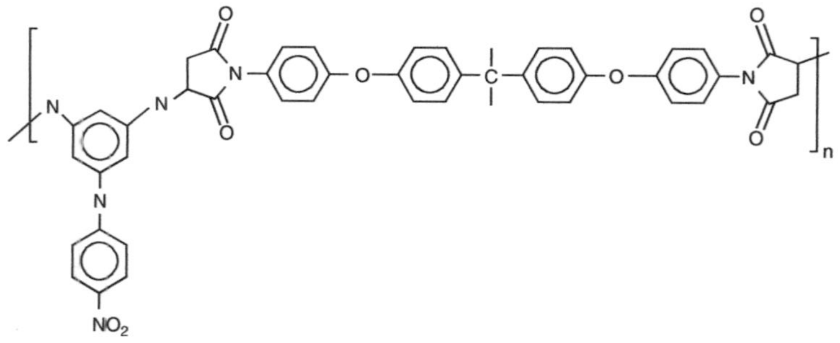 *p*-nitroaniline functionalized polyimide (Lin et al. 1992)	$T_g = 236$°C. $\lambda_{max} = 390$ nm. $\rho^{\#} = 7 \times 10^{20}$ cm^{-3} Films are cured and corona poled at 240°C. d_{33} (1.06 μm) = 5.4 pm/V. d_{33} (24 h, 85°C) ≈ 5 pm/V and stable.

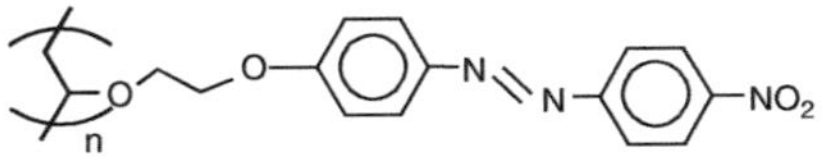 Poly-(4-nitro-4′-(vinylooxyethyloxy)azobenzene, 1, and other poly-(vinyl ethers) (S′heeren et al. 1992)	
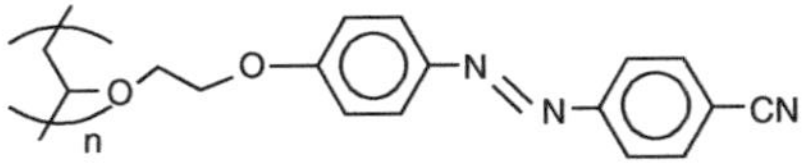 Poly-(4-cyano-4′-(vinylooxyethyloxy) azobenzene 2	
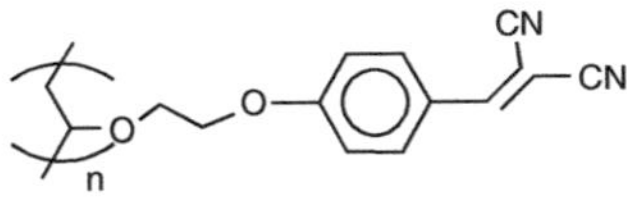 Poly-(4-dicyanovinyl-4′-(vinylooxyethyloxy) benzene 3	
Poly-(4-cyano-4-carbomethoxyvinyl-4′-(vinylooxyethyloxy) benzene 4	
Polyvinylalcohol/*N*-ethyl-*N*-methylamino nitroaniline copolymer(Sasaki 1993)	$T_g = 120°C$. $\lambda_{max} \approx 380$ nm. Corona poled at T_g. d_{33} (1.06 μm) = 10 pm/V. d_{33} relaxes to 7 pm/V in 40 days.

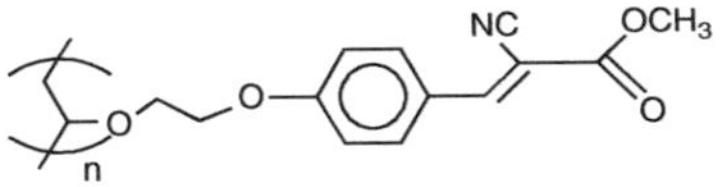

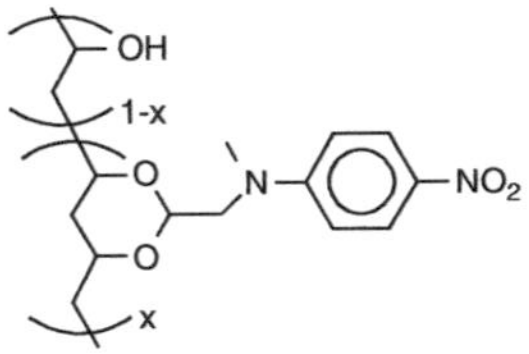

(continued)

TABLE 7.3 (Continued)

Structure and Nomenclature (Ref)	Properties
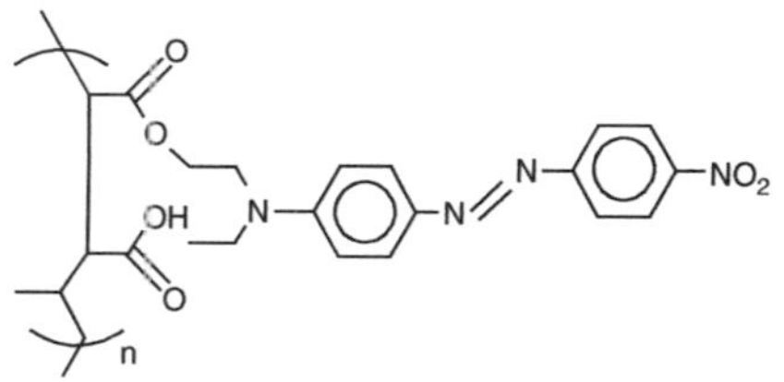 DR1/Poly(maleic anhydride-co-propylene copolymer (16:84 mol%) (Bauer et al. 1993)	T_g=180°C. $\lambda_{max}\approx$470 nm. Corona poled at 185°C. r_{33} (780 nm)=6 pm/V. with 3°C/min heating rate, r_{33} is stable up to 100°C.
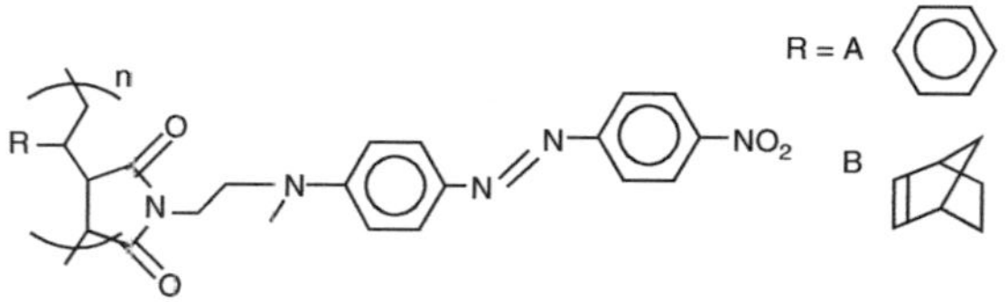 R = A, B Azo dye/Poly (maleic anhydride-*co*-styrene or co-norbornadiene copolymer (Ahlheim and Lehr 1994)	(see table below)
Main-Chain Polymers	
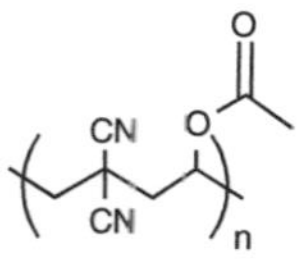 Vinylidene cyanide/vinyl acetate copolymer (50:50 mol%) (Azumai et al. 1990; Eich et al. 1988; Sato et al. 1987)	T_g=180°C. M_w=470×10^3. n(2.94 μm)=1.434. α(2.94 μm)=1.4 mm^{-1}. Corona poled at 180°C for 2 h. d_{33}(1.064 μm)=0.3 pm/V.[Eich]. Note: large discrepancy in SHG *d* coefficients.

	T_g (°C)	Mw $(\times10)^3$	λ_{max} (nm)	dye (mol%)
A	170	54	468	92
B	180			96

Aromatic polyurea (Nagamori et al. 1992; kajikawa et al. 1991)

Thin film prepared by vapor deposition polymerization.
T_g => 150°C.
$\lambda_{cut\text{-}off}$=400 nm.
Corona poled at 180°C for 3 min.
d_{33}(1.06 μm)=1.7 pm/V.
Negligible relaxation of activity over 2 months at ambient conditions.
Activity remains stable after short-term heating to 200°C.

N-phenylated polyurea (Nalwa et al. 1993a; Nalwa et al. 1993b; Azumai et al. 1990; Stato et al. 1987)

T_g=123°C.
λ_{max}=253 nm.
n(633 nm)=1.577.
α(633 nm)=−1.2 dB/cm.
Corona poled at 130°C for 1 h.
d_{33}(1.064 μm)=5.5 pm/V.
90% of activity remains stable after 40 days at ambient.

Epoxy polymers containing 4-amino-4′-nitroazobenzene (Teraoka et al. 1991)

Polymers prepared by melt condensation at 150°C under N_2.
d_{33} values are measured at 1.064 μm during corona poling.

	1	2
T_g(°C)	86	77
λ_{max}(nm)	392	412
Mw(×10³)	7	5
Poling temp. (°C)	100	85
d_{33}(pm/V)	31	12
τ(hour)	24000	770
β	0.32	0.34

(continued)

TABLE 7.3 (Continued)

Structure and Nomenclature (Ref)	Properties
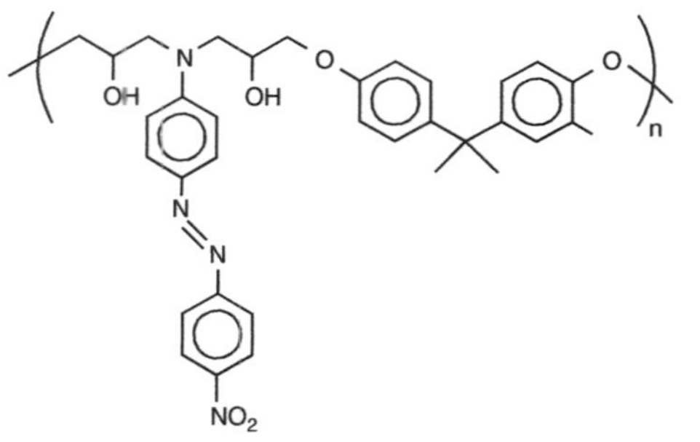 Epoxy polymers containing *p*-aminonitrobenzene (Gadret et al. 1991)	T_g = 77°C. λ_{max} = 480 nm. Corona poled after procuring. d_{33}(1.06 μm) = 25 pm/V. Stable activity at ambient conditions for > 20 days. Activity decreases rapidly to zero when heated above 80°C.
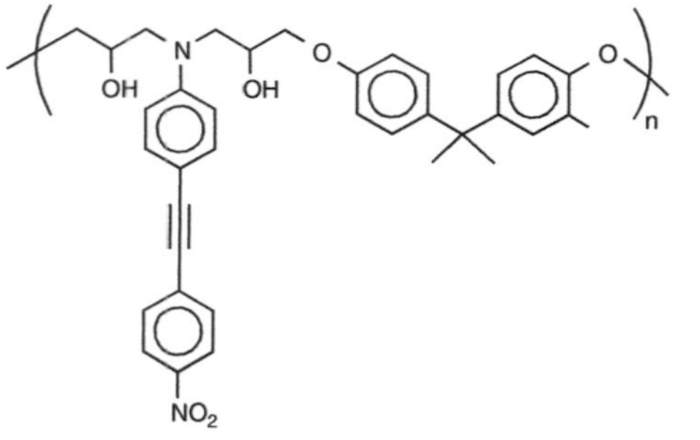 Epoxy polymer containing 4-amino-4′nitrotolane (Jungbauer et al. 1991)	T_g = 125°C. λ_{max} = 418 nm. n (633 nm) = 1.71. Corona poled at 135°C for 1 h. d_{33} (1.064 μm) = 89 pm/V. r_{13} (633 nm) = 8 pm/V. Stable birefringence at ambient conditions. Birfringence decays with τ = 448 h and β = 0.45 at 100°C.

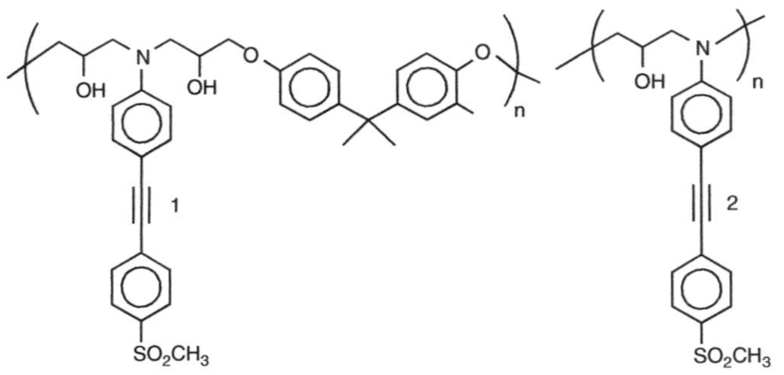

Epoxy polymers containing 4-amino-4′methylsulfonyltolane (Twieg et al. 1992)

Polymers prepared by melt condensation at 150°C.
d_{33} values are measured at 1.064 μm during corona poling.

	1	2	3
T_g (°C)	74	94	116
λ_{max} (nm)	471	476	
Mw (×10³)	11	15	31
n (830 nm)	1.69	1.69	1.70
χ_{333} (pm/V)	14.8	10.3	

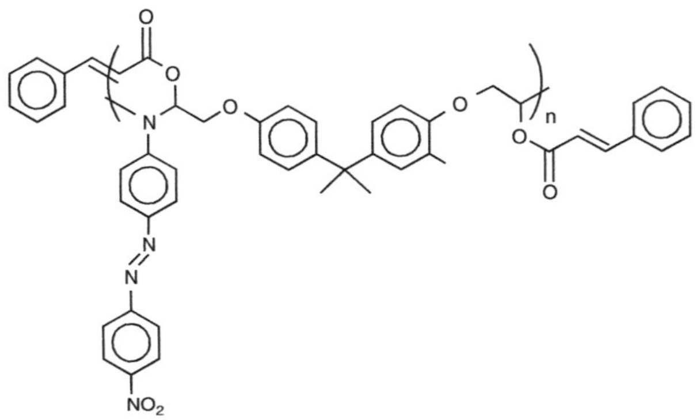

Epoxy polymers containing 4-amino-4′nitroazobenzene (Jeng et al. 1992c)

$T_g = 115°C$.
$\lambda_{max} = 461$ nm.
n (533 nm) = 1.718.
Corona poled at 115°C for 1 h.
d_{33} (1.06 μm) = 34 pm/V.
70% of SHG activity remains stable after 20 days at ambient conditions.

(continued)

TABLE 7.3 (Continued)

Structure and Nomenclature (Ref)	Properties
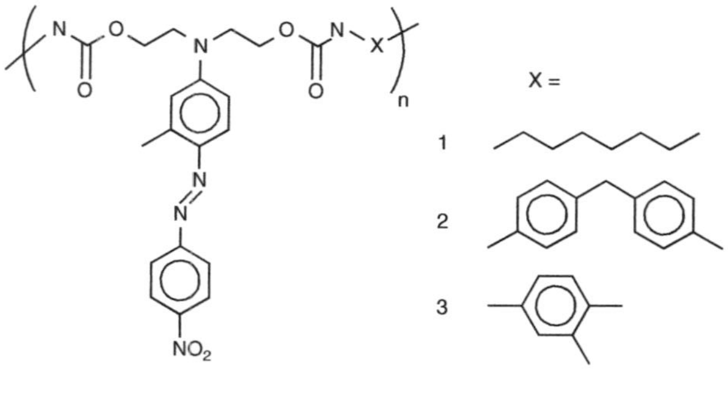 Polyurethanes containing 4-amino-4′nitroazobenzene [Meyrueix et al. 1991a, 1991b]	Polymers prepared by melt condensation at 150°C. χ_{333} (−ω, ω, 0) values are measured at 830 nm after contact poling with 25 V/μm at 100°C.
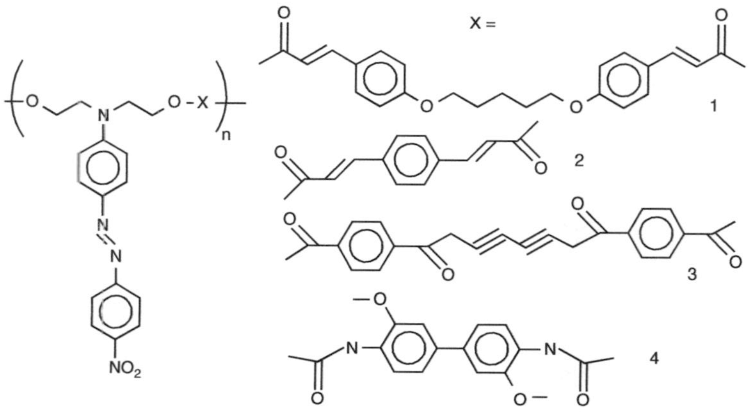 Polyurethanes containing 4-amino-4′nitroazobenzene [Chen et al. 1991]	Polymers prepared by condensation in dioxane solution. d_{33} is stabilized value measured at 1.064 μm after corona poling at about 130°C. Activity is stable at ambient conditions.

	1	2	3
T_g (°C)	74	94	116
λ_{max} (nm)	471	476	
Mw (×10³)	11	15	31
n (830 nm)	1.69	1.69	1.70
χ_{333} (pm/V)	14.8	10.3	

	1	2	3	4
T_g (°C)	92	120	122	140
λ_{max} (nm)	460	470	455	463
Dye (wt%)	48	68	47	52
Thickness (μm)	0.04	0.05	0.22	0.28
d_{33} (pm/V)	108	250	119	223

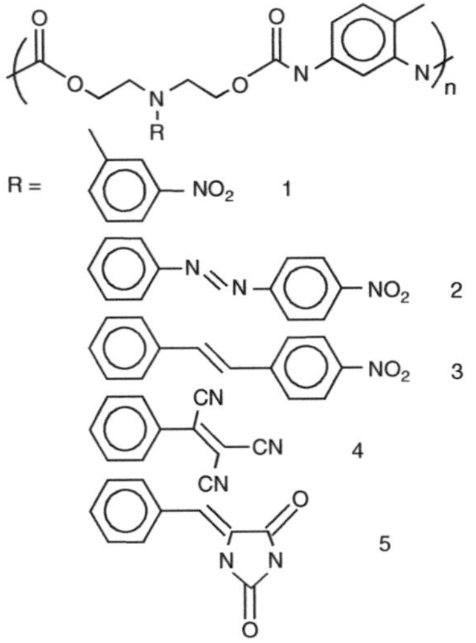

In situ corona poling during melt polymerization at 120°C.
d_{33} values are measured at 1.064 μm; stabilized values after 180 days at 25°C are given in *italics*.

	1	2	3	4	5
T_g (°C)	98	131	121	125	178
λ_{max} (nm)	373	472	431	518	388
Mw (×10^3)	2.3	2.5	2.0	1.7	2.1
d_{33} (pm/V)	9	10	2.6	3.6	2
	0.5	5.5	6	0.9	2.5

Polyurethanes derived from 2,4-toluenediisocyanate/2-methyl-4-nitro-[*N*,*N*-bis(2-hydroxyethyl)]-aniline and other diols [Kitipichai et al.1993]

(continued)

TABLE 7.3 (Continued)

Structure and Nomenclature (Ref)	Properties
1 2 3 Main-chain polymers derived from (1) α-cyano-ester quinodimethanes, (2) *p*-oxy-α-cyanocinnamates, (3) *p*-thio-α-cyanocinnamates [Fuso 1991;Hall et al. 1988; Green 1987a, 1987b]	Polymers prepared by high temperature transesterification. Homopolymers are typically insoluble and high melting. Unit molecular nonlinearity μβ/n values are measured at 1.064 μm.
Main-chain homopolymers derived from (4-*N*-ethyl-*N*-(2-hydroxyethyl)amino-α-cyanocinnamates (1) and the corresponding acid (2) [Stenger-Smith et al. 1991, 1990]	Polymerization: (1) melt transesterification at 160°C; (2) solution polycondensation at 25°C.

	1	2	3
T_g(°C)	None	<25	45
Dye (mol%)	50	67	100
Mw (×10³)		70	
μβ/n (×10⁻⁴⁸ esu)		140	

	1	2
T_g(°C)	100–110	60–90
λ_{max} (nm)	442	442
Mw (×10³)	27	5
Poling temp. (°C)	120	85–100
d_{33} (pm/V) at 1.064 μm	7	7

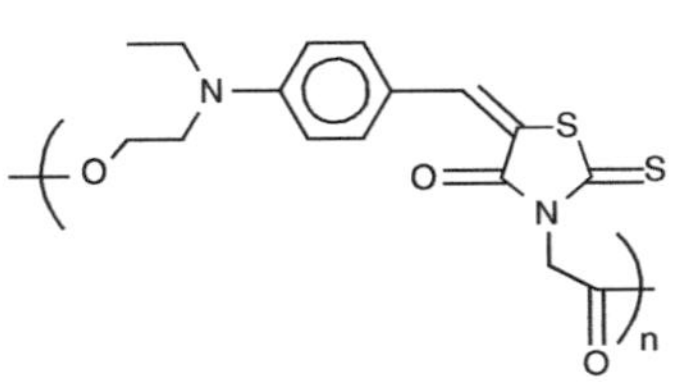

Main-chain polymer derived from 3-[(methyoxycarbonyl)methyl]-5-[4′-[*N*-ethyl-*N*-(2″-hydroxyethyl)amino] benzylidene] rhodanine [Francis et al. 1993b]

Polymers are prepared by transesterication in melt at 140°C for 8 h.
$T_g = 63°C$.
$M_w = 13 \times 10^3$.
$\lambda_{max} = 473$ nm.
n (790 nm) = 1.760.
Contact poled with 43 V/μm at 65°C.
d_{33} (1.58 μm) = 7.3 pm/V.

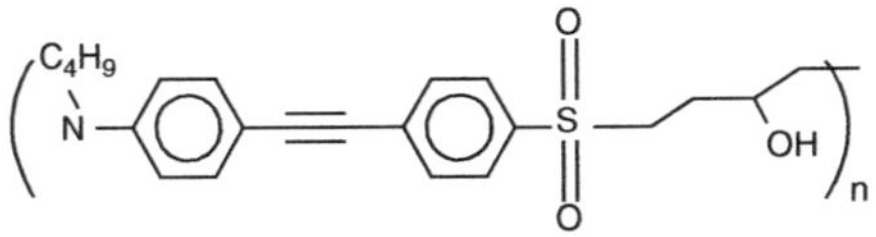

4-amino-4′-alkylsulfonyltolane main-chain polymer [Zentel et al. 1993]

Polymers are corona poled during melt polymerization at 150°C for a few hours.
$T_g = 60°C$.
$\lambda_{max} = 366$ nm.
d_{33}(1.064 μm) = 12.5 pm/V.
Activity decays by 40% in 25 days at ambient conditions.

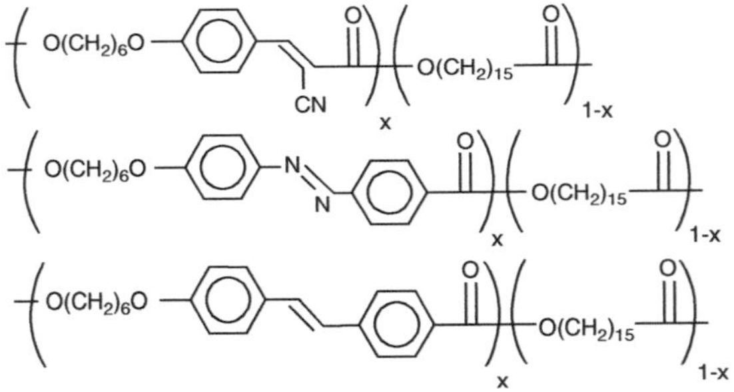

Main-chain polyesters derived from 6-hydroxyhexyloxyphenyl propenoic, azobenzoic, or ethenylbenzoic acids [S'heeren et al. 1993b]

Polymers are prepared by melt polycondensation at 220°C, 1.3 mbar for 4 h.
Films are of pale yellow in color.
d_{33} values are measured at 1.064 μm after corona poling at 10°C above T_g.

	1	2	3	4
T_g (°C)	48	41	51	53
Dye (mol%)	30	43	63	34
Mw (×10³)		35	115	99
d_{33} (pm/V)		0.04	0.5	0.3

(*continued*)

TABLE 7.3 (Continued)

Structure and Nomenclature (Ref)	Properties
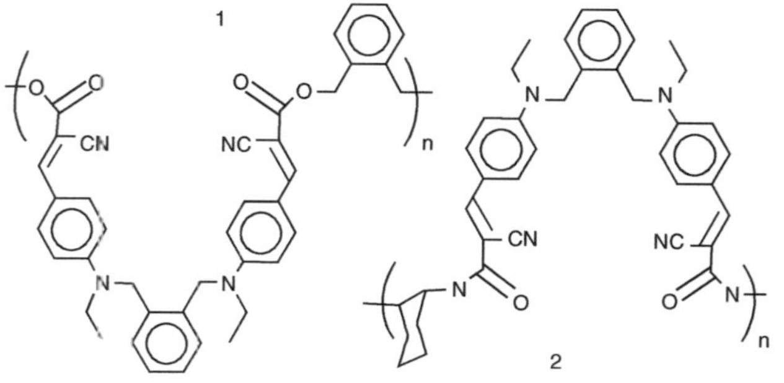 Main-chain accordionpolymers of α-cyano cinnamamides [Lindsay et al. 1992; Wang and Guan 1992]	Polymer 2 was corona poled. Activity is found to be stable at room temperature.
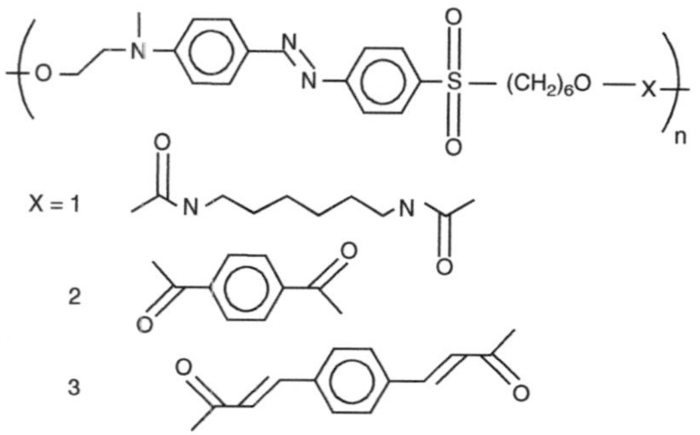 Random main-chain polymers containing 4-*N,N*-dialkylamino-4′-hexylsulfonylazobenzene [Xu et al. 1993, 1992a]	Polymers are prepared by Condensation in dioxane solution. d_{33} values are measured at 1.064 μm after corona poling at about 120°C.

	1	2
T_g(°C)	143	193
λ_{max} (nm)	≈425	
Mw (×10³)	55	
Poling temp. (°C)		215
d_{33} (pm/V) at 1.064 μm		3.5

	1	2	3
T_g(°C)	62	114	108
λ_{max} (nm)	440	436	443
Mw (×10³)	8	7	9
d_{33} (pm/V)	60	125	150
d_{33} (10 days) at ambient	12		140

Cross-linked polymers 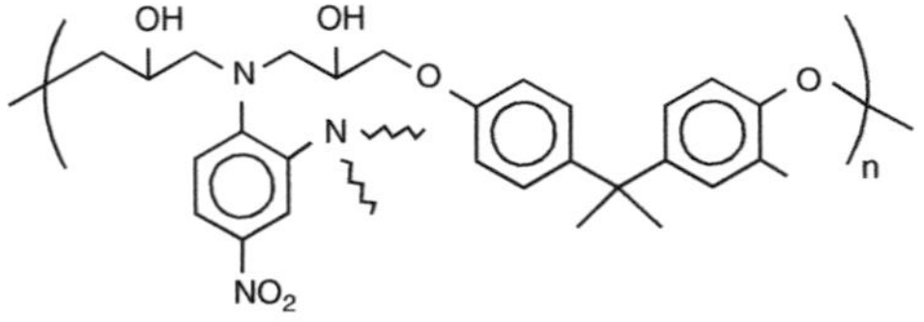 Cross-Linked epoxy polymer from 4-nitro-1, 2-phenylenediamine and bisphenol-A diglycidylether [Eich et al. 1989a]	Thermal cross-linking is achieved by procuring at 100°C for 3 h at 140°C for 16 h under a corona poling field. λ_{max}=410 nm. n(633 nm) =1.629. d_{33} (1.06 μm)=13.5 pm/V. No decay of SHG activity is observed for 36 mins at 80°C.
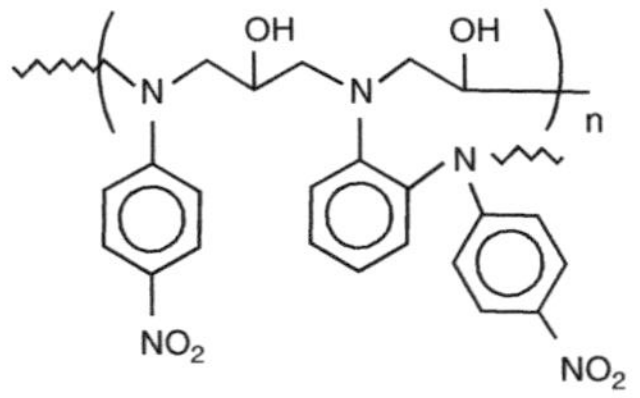 Cross-linked epoxy polymer from *N,N*-(diglycidyl)-4-nitroaniline and *N*-(2-aminophenyl)-4-nitroaniline [Jungbauer et al. 1990]	Thermal cross-linking is achieved by procuring at 130°C for 4 min and at 120°C for 24 h under a corona poling field. λ_{max}=397 nm. High dye content: 63 wt%. n (633 nm)=1.74. d_{33} (1.064 μm)=50 pm/V. Stable activity is observed at 80% of initial value at 80°C.

(continued)

TABLE 7.3 (Continued)

Structure and Nomenclature (Ref)	Properties
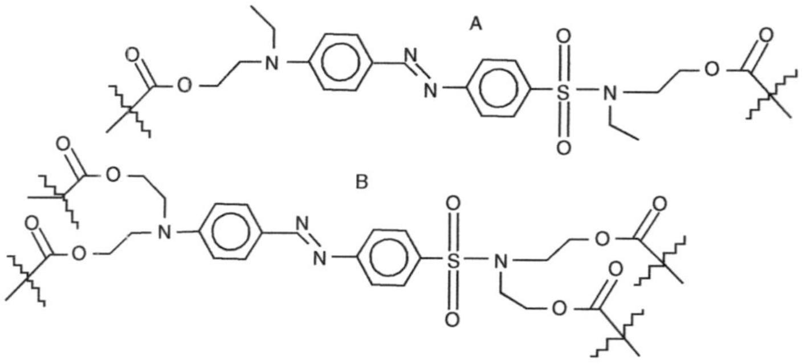 Cross-linked methacrylate polymer containing 4-*N,N*-dialkylamino-4′-*N,N*-dialkylaminosulfonylazobenzene [Wang et al. 1993; Allen et al. 1991]	Cross-linking is initiated with free radicals under a corona poling field. $\lambda_{max}=480$ nm. r_{33} (633 μm)=36 pm/V; r_{33} (1.3 μm)=6.6 pm/V. No decay of activity is observed over 8 months at ambient conditions for A. Higher temporal stability found for B.
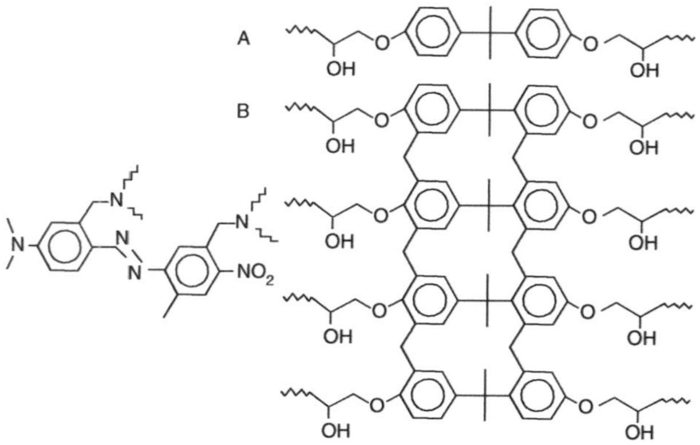 Cross-linked polymer from a reactive diamine derivative of 4-*N*, *N*-dimethylamino-4′-nitroazobenzene and oligomeric derivative of diglycidyl ether of bisphenol A [Hubbard 1992]	Thermal cross-linking is achieved procuring at 100°C for 3 h and at 130°C for 2 hour a corona poling field. $\rho^{\#}=6\times10^{20}$ cm^{-3} for B. d_{33} (1.064 μm)=3(A)-6(B) pm/V. Temporal stability: $\tau_1(25°C)=6(A)–4.1(B)$ days and $\tau_2(25°C)=120(A)-300(B)$ daya; $\tau_1(85°C)=1.6(B)$ days and $\tau_2(85°C)=120(B)$ days.

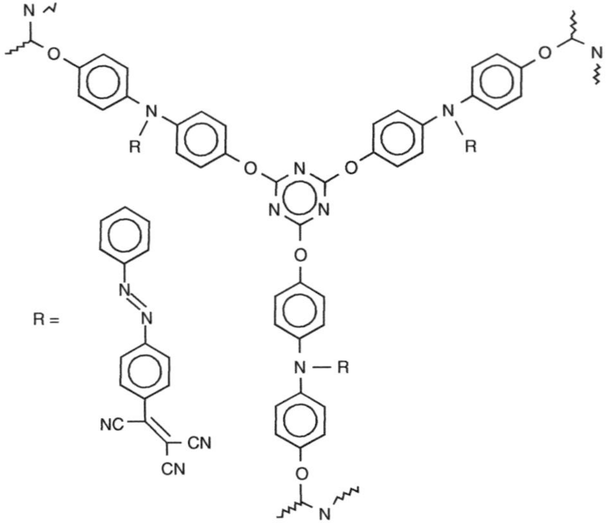

Triazine cross-linked polymer obtained from *p*(*N*,*N*-bis(4′-cyanatobenzyl)amino-p′-(2,2dicyanovinyl)azobenzene (Holland and Fang 1992; Singer et al. 1991)	Polymers are prepared by poly-acyclotrimerization at 150°C under a poling field for several hours. r_{33}(830 nm) = 11 pm/V. High stability is observed at 85°C.

(continued)

TABLE 7.3 (Continued)

Structure and Nomenclature (Ref)	Properties
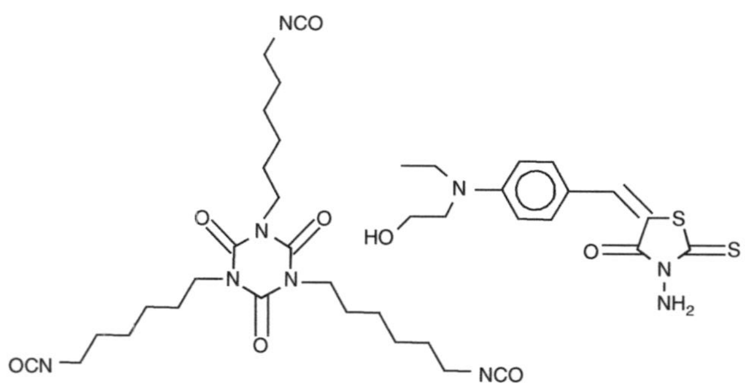 Isocyanate cross-linked polymer derived from tris-1-hexamethyleneisocyanate isocyanurate and 3-amino-5-[4′(*N*-ethyl-*N*-(2″-hydroxyethyl)amino)benzylidene]-rhodanine (Francis et al. 1993a)	Thermally cross-linked at 135°C for 16 h. λ_{max} = 470 nm. n(1.3 μm) = 1.611. Doping level: 60 mol%. Contact poled with 100 V/μm at 135°C. d_{33}(1.58 μm) = 6.9 pm/V. r_{33}(1.3 μm) = 3.6 pm/V. Activity decays by 30% in 150 days at 100°C due to thermal decomposition.
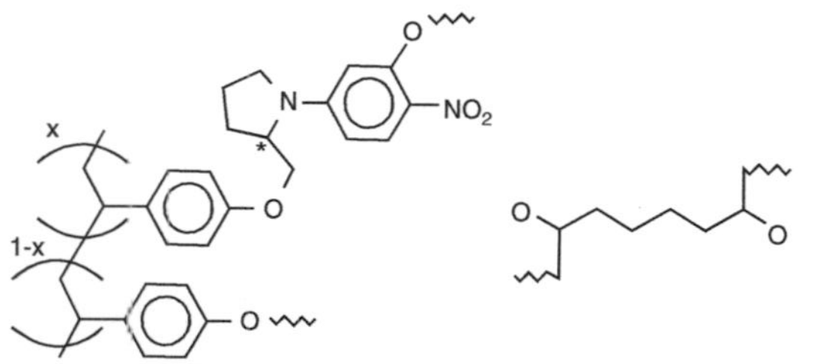 Cross-linked epoxy polymer from 1,2,7,8-diepoxyoctane and *N*-(3-hydroxy-4-nitrophenly)-(S)-prolinoxy functionalized poly(*p*-hydroxystyrene) copolymer (Park et al. 1990)	Thermal cross-linking is achieved by procuring at 100°C for 24 h and at 180°C for 1 hour under a corona poling field. High dye content: 16 wt%. diepoxide/phenol ratio: 0.5. d_{33}(1.06 μm) = 3 pm/V. Temporal stability: τ_1(25°C) = 79 days and τ_2(25°C) = 100 days.

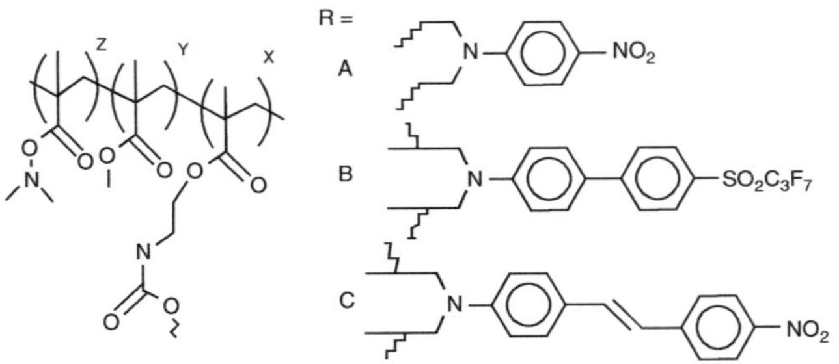

Cross-linked homo-and copolymers of isocyanatoethyl methacrylate, MMA, and dimethylaminoethyl methacrylate. (Cheng et al. 1992)

Corona poled near T_g.
d_{33} values are measured at 1.064 μm.
d_{33} of A are stable at room temperature.
Activity decays by 10% in 100 days at 80°C for polymer B and C.
$\alpha(1.06\ \mu m) \approx -(1.5\text{-}3)$ dB/cm.

(x, y, z)	$\rho^{\#}$ (10/cm³)	d_{33} (pm/V)
A(1, 0, 0)	11	11
B(1, 1, 0)	6	10
C(1, 0, 0)	9	34
C(1, 0, 1)	6	39

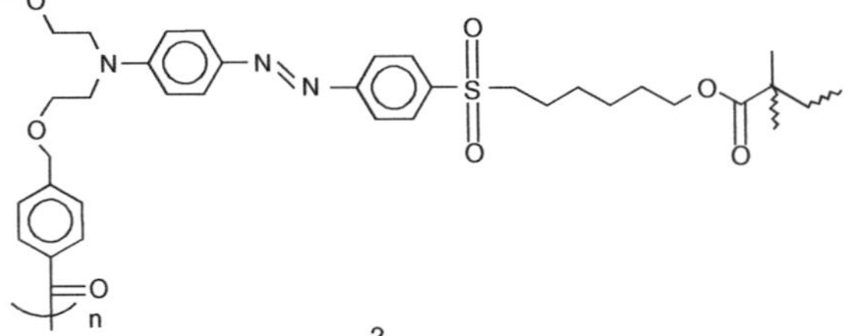

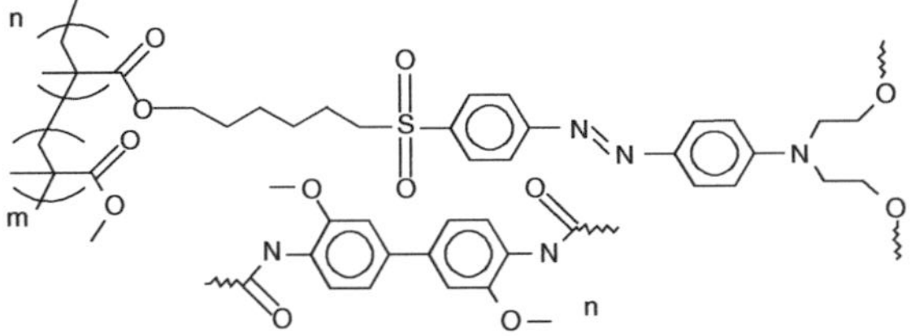

Cross-linked side-chain polymers containing 4-*N*,*N*-dialkylamino-4′-alkylsulfonylazobenzene [Xu et al. 1992b; Shi et al. 1993]

Thermal cross-linking under corona poling field: (1) with initiators; (2) without initiator.

	1	2
λ_{max} (nm)	≈440	≈440
Poling temp. (°C)	115	160–180
d_{33}(pm/V) at 1.064 μm	30	50
d_{33}(100 °C 4 days)		47.5

(continued)

TABLE 7.3 (Continued)

Structure and Nomenclature (Ref)	Properties
0.4; 0.6; with 1,2, or 3; 1; 2; 3 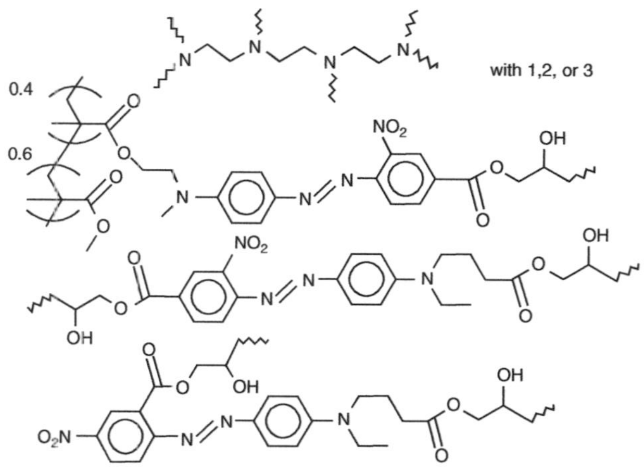 Cross-linked epoxy polymers containing dialkyamino nitroazobenzene derivatives [Muller et al. 1993]	Cross-linked polymers are prepared with 10–20 wt% PMMA as a binder. Films are characterized as having good to excellent optical quality. r_{33} values are measured at 1.32 μm after corona poling at about 140°C for 8 h.
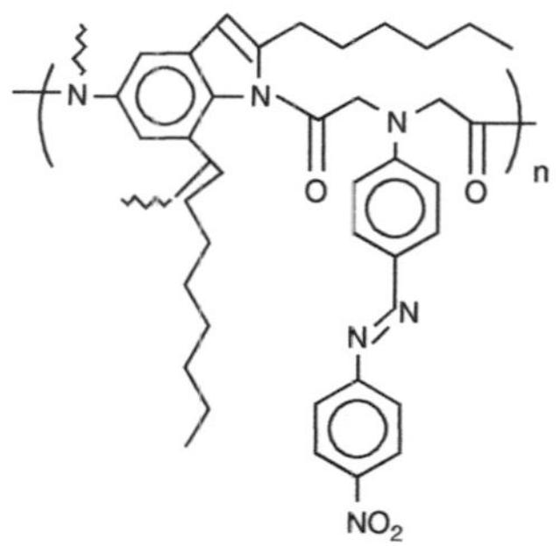 Cross-Linked polymer containing 4-amino-4′-nitroazobenzene (Yu et al. 1992)	Prepolymer containing ethynyl group is thermally cross-linked at 190°C for 2 h under a corona poling field. $\lambda_{max} \approx 470$ nm. $d_{33}(1.064\ \mu m) = 20$ pm/V. Activity decays to 75% of initial value after 30 days at 90°C.

	1	2	3
T_g(°C)	165	150	140
Precure (°C)	none	140	140
Precure (hr)	none	4.5	2.5
r_{33} (pm/V)	1.6	0.9	3.5
r_{33} (80 °C for 30 min)	8.8	7.7	1.5

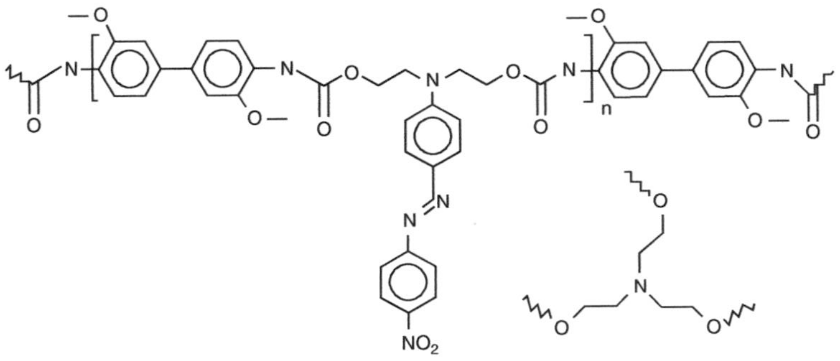 Cross-Linked polyurethane containing 4-*N*,*N*-dialkylamino-4′-nitroazobenzene(Chen et al. 1992; Shi et al. 1992)	Oligomeric prepolymer is thermally cross-linked with triethanolamine at 160°C for an hour under a corona poling field. λ_{max}=475 nm. n(633 nm)=1.753; n(800 nm)=1.692. d_{33}(1.06 μm)=120 pm/V. r_{13}(633 μm)=13 pm/V.; r_{13}(800 nm)=5 pm/V. No decay of activity is observed at ambient for 4 months; activity stablizes to 70% of initial value after 4 months at 90°C.
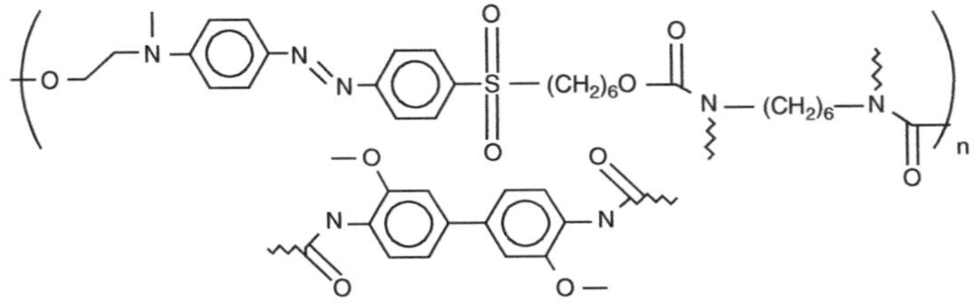 Cross-Linked random main-chain polymer containing 4-*N*,*N*-dialkamino-4′-hexylsulfonylazobenzene and 3,3′-dianisidine diisocyanate (Xu 1992; Ranon et al. 1993; Xu et al. 1993)	Thermally cross-linked under a corona poling field at 125°C for 2 h. Prepolymer $M_w=8\times10^3$. λ_{max}=440 nm. d_{33}(1.064 μm)=40 pm/V. 90% activity remains stable at ambient conditions for >3 months.

(continued)

TABLE 7.3 (Continued)

Structure and Nomenclature (Ref)	Properties
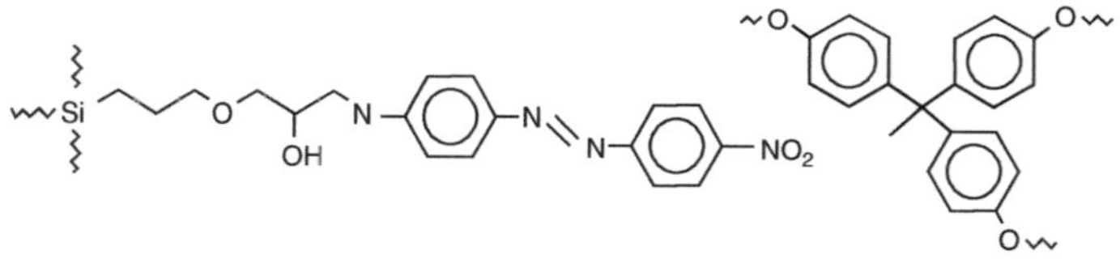 Alkoxysilane derivative of 4(-4′nitrophenylazo)-phenylamine cross-linked with 1,1,1-tris(4-hydroyphenyl)ethane (Jeng et al. 1993)	Thermally cross-linking occurs at 200°C. T_g=110°C. λ_{max}=493 nm. n(533 nm)=1.744. Doping level: 50 mol%. Corona poled at 200°C for 30 min. d_{33}(1.06 μm)=77 pm/V. d_{33}(24 hrs, 105°C)=62 pm/V. No decay of activity is observed for 7 days at ambient conditions.
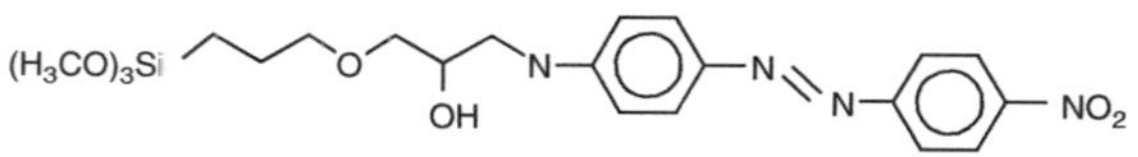 $[(SiO)_{>1}(C_6H_5)_{<0.5}(OC_2H_5)_{<0.5}(OH)_{<0.5}]_n$ Alkoxysilane derivative of 4-(4′nitrophenylazo)-phenylamine cross-linked with phenylsiloxane polymer (Allied Signal Accuglass 204®) (Jeng et al. 1992a)	Thermal cross-linking occurs at 200°C. λ_{max}=493 nm. n(533 nm)=1.537. Doping level: 0.1 g dye in 4 g of A204. Corona poled at 200°C for 10 min. d_{33}(1.06 μm)=5.28 pm/V. d_{33}(40 h, 100°C)=2.9 pm/V.
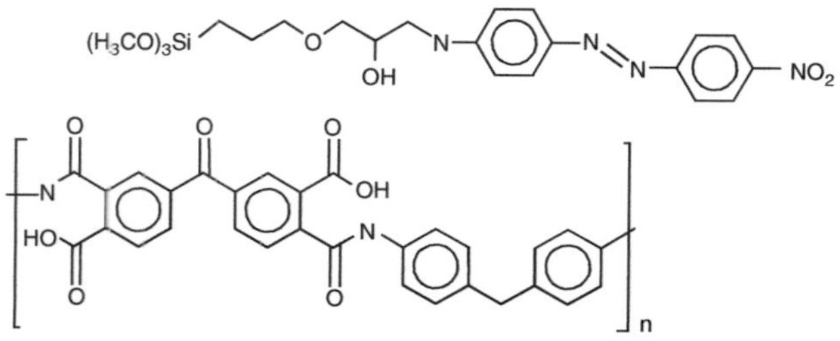 Alkoxysilane derivative of 4-(4′-nitrophenylazo)phenylamine/polyimide(Skybond®) composite (Jeng et al. 1992b)	T_g > 275°C. λ_{max}=466 nm. Doping level:16 wt%. Excellent optical quality. Corona poled at 220°C. d_{33}(1.06 μm)=13.7 pm/V. Stable d_{33} (168 h, 120°C)=10 pm/V. Novelty: composite of polyimide and Si-O-Si network.

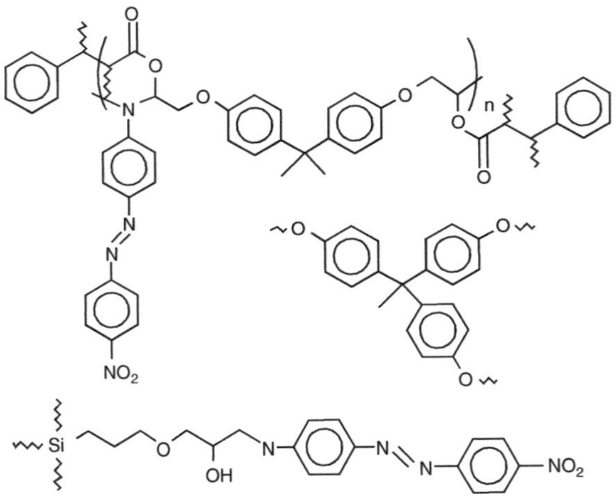 Interpenetrating network of epoxy and silicon-based cross-linked polymers(Marturunkakul et al. 1993)	Equal weight ratio of the two polymers is heated at 200°C for 1 h under a corona poling field. Films are high optical quality. $T_g = 176°C$. $\lambda_{max} = 458$ nm. N(533 nm) = 1.708. $d_{33}(1.064\ \mu m) = 33$ pm/V. No decay of activity is observed at 100°C for 7 days, 50% decay is observed after 15 h at 160°C.
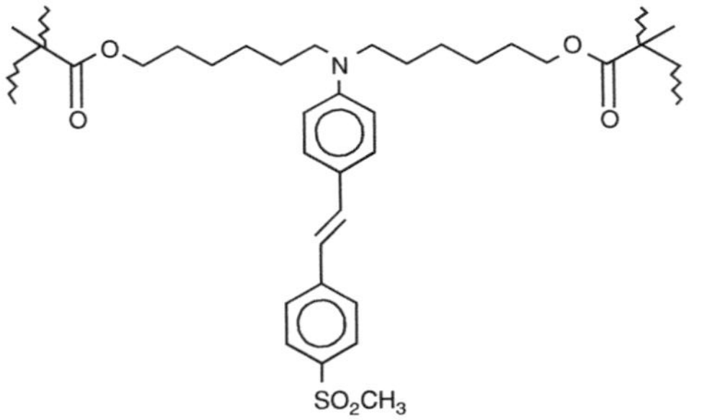 Cross-linked acrylate polymer of 4′-*N*,*N*-bis (6-methacroyloxyhexyl)-amino-4-methylsulfonylstilbene (Robello et al. 1991b)	Liquid monomers are cross-linked by UV irradiation with initiator at room temperature. $T_g < 70°C$. Good optical quality. Contact poled with 6.7 V/μm. $d_{33}(1.06\ \mu m) = 0.7$ pm/V. Activity decays to zero in several months at ambient conditions.

(continued)

TABLE 7.3 (Continued)

Structure and Nomenclature (Ref)	Properties
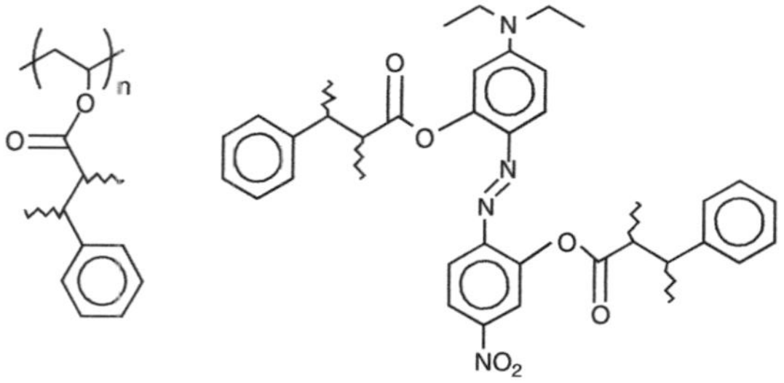 Cross-linked polymer of polyvinylcinnamate doped with 3-cinnamoyloxy-4-[4-(*N*,*N*-diethylamino)-2-cinnamoyloxy phenylazo] nitrobenzene (Mandal et al. 1991a, 1991b)	Cross-linking is induced by UV irradiation (2 mw/cm^2 at 254 nm for 3 to 10 min) at 70°C. T_g=84°C (before cross-linking). λ_{max}=520 nm. Doping level: 10 wt%. n(633 nm)=1.677. Corona poled at 70°C during UV irradiation. d_{33}(1.064 μm)=11.5 pm/V. d_{33}(1.54 μm)=3.7 pm/V. r_{33}(633 nm)=9pm/V. No decay of activity is observed in 22 h at ambient conditions
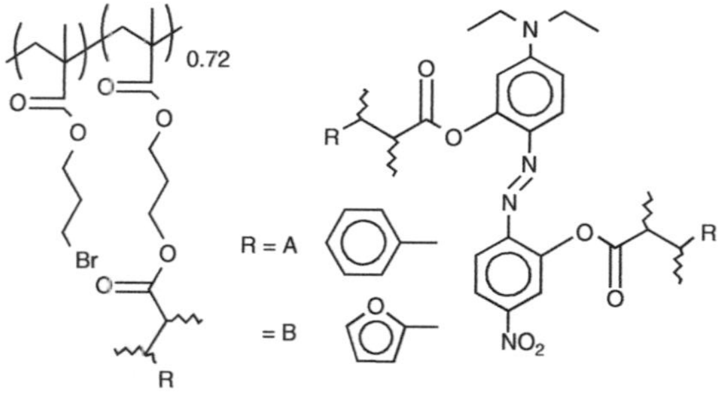 Cross-linked side-chain cinnamate or furylacrylate polymers doped with cinnamoyloxy or furylacryloyloxy functionalized 4-*N*,*N*-diethylamino 4′-azonitrobenzene (Muller et al. 1992b)	Photo-cross-linking with 1.8 mw/cm^2 at 312 nm at 75°C for 4 (A) and 2 (B) h to achieve 60% reaction. B cannot be poled under irradiation and is poled with a cycle of poling (10 min at 25°C) and cross-linking (20 min at −15°C) for a total exposure time of 2 h. r_{33}(1.32 μm)=0.6 pm/V. Activity remains stable after 3 h at 80°C.

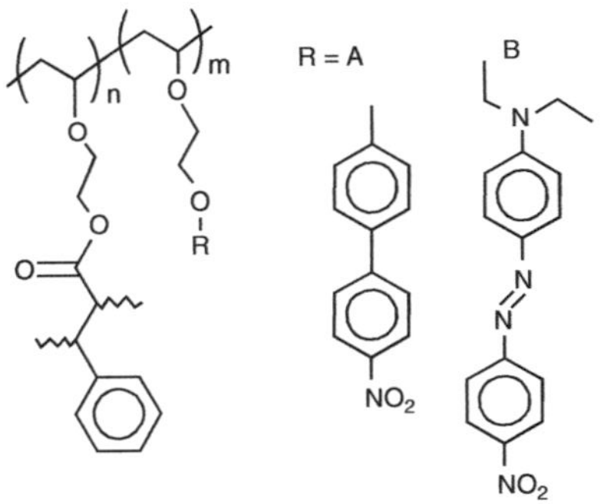

R = A B

Cross-linked side-chain cinnamate copolymers functionalized with 4-alkoxy-4′-biphenyl of 4-*N*,*N*-diethylamino 4′-azonitrobenzene [Kato et al. 1993]

Photo-cross-linking with UV light under corona poling field: (A) 1.5 min at 50°C; (B) 10 min at room temperature. Both polymers were subsequently poled at 150°C for 20 min.
Activities are stable at room temperature.

	mol%	d_{33}(pm/V)
A	10	6
	33	20
	57	30
B	27	150

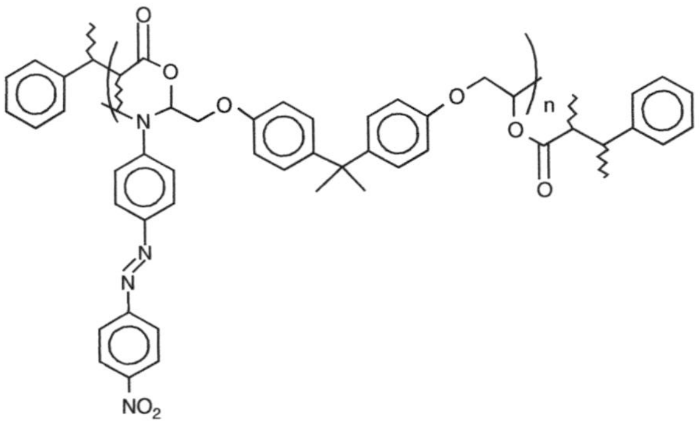

Cross-linked epoxy polymer containing 4-amino-4′-nitroazobenzene (Jeng et al. 1992c)

Prepolymer containing cinnamoyl group is cross-linked by UV irradiation at 3 MW/cm^2 for 10 min.
λ_{max} = 461 nm.
n(533 nm) = 1.718.
Corona poled at 115°C for 1 hour.
d_{33}(1.064 μm) = 22 pm/V.
95% of SHG activity remains stable after 20 days at ambient conditions.

(continued)

TABLE 7.3 (Continued)

Structure and Nomenclature (Ref)	Properties
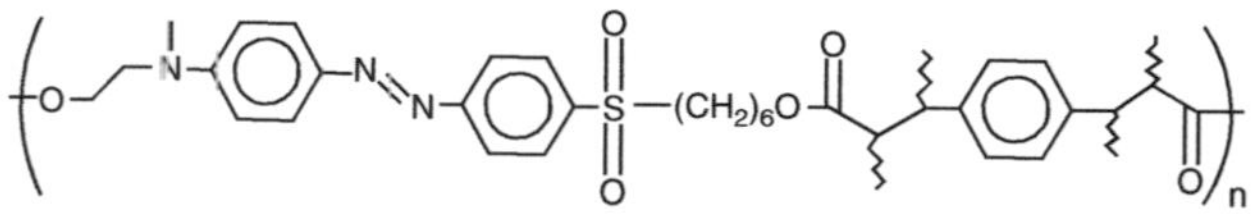 Cross-linked, random, main-chain polymer containing 4-*N*,*N*-dialkyamino-4′-hexylsulfonylazobenzene and cinnamoyl groups (Xu et al. 1993; Chen et al. 1991)	Photo-cross-linked under a corona poling field. Prepolymer $M_w = 8.5 \times 10^3$. Dye content = 53 wt%. $\lambda_{max} = 443$ nm. $d_{33}(1.064\ \mu m) = 150$ pm/V. 90% activity remains stable at ambient conditions for 25 days.

References

1. Shen, Y. R. 1984. *Principles of Nonlinear Optics*. Wiley, New York.
2. Butcher, P. N. and Corter, D. 1990. *The Elements of Nonlinear Optics*. Cambridge University Press, New York.
3. Prasad, P. and Williams, D. J. 1991. *Introduction to Nonlinear Optical Effects in Molecules and Polymers*. Wiley, New York.
4. Hornak, L. A. ed. 1992. *Polymers for Lightwave and Integrated Optics*, pp. 609–661. Marcel Dekker, New York.
5. Zyss, J. ed. 1994. *Molecular Nonlinear Optics*, pp. 609–661. Academic Press, New York.
6. Nalwa, H. S. and Miyata, S. eds. 1997. *Nonlinear Optics of Organic Molecules and Polymers*, pp. 609–661. CRC Press, Boca Raton, FL.
7. Dalton, L. R. 1998. Polymers for electro-optic modulator waveguides, In *Electrical and Optical Polymer Systems: Fundamentals, Methods, and Applications*, D. L. Wise, T. M. Cooper, J. D. Gresser, D. J. Trantolo, and G. W. Wnek, eds., pp. 609–661. World Scientific, Singapore.
8. Kuzyk, M. G. and Dirk, C. W. eds. 1998. *Characterization Techniques and Tabulations for Organic Nonlinear Optical Materials*, pp. 8–33. Marcel Dekker, New York.
9. Dalton, L. R., Harper, A. W., Ren, A., Wang, F., Todorova, G., Chen, J., Zhang, C., and Lee, M. 1999. Polymeric electro-optic modulators: from chromophore design to integration with semiconductor VLSI electronics and silica fiber optics. *Industrial and Engineering Chemistry Research*, 38, 8–33.
10. Dalton, L. R. 2001. Nonlinear optical polymeric materials: from chromophore design to commercial applications. *Advances in Polymer Science*., Vol. 158, pp. 1–86, Heidelberg.
11. Dalton, L. R. 2003. Rational design of organic electro-optic materials. *Journal of Physics: Condensed Matter*, 15, R897–R934.
12. Dalton, L. R. 2004. Nonlinear optics—applications: electro-optics, EO. R.D. Guenther, ed., In *Encyclopedia of Modern Optics*, vol. 3, pp. 121–129. Elsevier, Amsterdam.
13. Bosshard, C., Sutter, K., Pretre, P., Hulliger, J., Florsheimer, M., Kaatz, P., and Gunter, P. 1995. In *Organic nonlinear optical materials*, A. Garito and F. Kajzar, eds., In *Advances in Nonlinear Optics*, vol. 1, pp. 776–789. Gordon Breach, Basel.
14. Agawal, G. P., Cojan, C., and Flytzanis, C. 1978. Nonlinear optical properties of one-dimensional semiconductors and conjugated polymers. *Physical Review B, Condensed Matter*, 17, 776–789.
15. Heflin, J. R., Wong, K. Y., Zamani-Kharmiri, O., and Garito, A. F. 1988. Nonlinear optical properties of linear chains and electron-correlation effects. *Physical Review B, Condensed Matter*, 38, 1573–1576.
16. Kanis, D. R., Ratner, M. A., and Marks, T. J. 1994. Design and construction of molecular assemblies with large second-order optical nonlinearities. Quantum chemical aspects. *Chemical Reviews*, 94, 195–242.
17. Albert, I. D. L., Marks, T. J., and Ratner, M. A. 1998. Practical computational approaches to molecular properties, In *Characterization Techniques and Tabulations for Organic Nonlinear Optical Materials*, M. G. Kuzyk and C.W. Dirk, eds., pp. 37–109. Marcel Dekker, New York.
18. Qin, A., Bai, F., and Ye, C. 2003. Design of novel X-type second-order nonlinear optical chromophores with low ground state dipole moment and large first hyperpolarizabilities. *Journal of Molecular Structure: Theochem*, 631, 79–85.
19. Kang, H., Zhu, P., Yang, Y., Facchetti, A., and Marks, T. J. 2004. Self-Assembled electrooptic thin films with remarkably blue-shifted optical absorption based on an X-shaped chromophore. *Journal of the American Chemical Society*, 126, 15974–15975.
20. Keinan, S., Zojer, E., Bredas, J., Ratner, M., and Marks, T. 2003. Twisted π-system electro-optic chromophores: a CIS vs. MRD-CI theoretical investigation. *Journal of Molecular Structure: Theochem*, 633, 227–235.

21. Kang, H., Facchetti, A., Zhu, P., Jiang, H., Yang, Y., Cariati, E., Righetto, S., Ugo, R., Liu, Z., Ho, S., and Marks, T. 2005. Exceptional molecular hyperpolarizabilities in twisted π-system chromophores. *Angewandte Chemie & Angewandte Chemie, International Edition in English*, 44, 7922–7925.
22. Breitung, E. M., Shu, C. F., and McMahon, R. J. 2000. Thiazole and thiophene analogues of donor–acceptor stilbenes: Molecular hyperpolarizabilities and structure-property relationships. *Journal of the American Chemical Society*, 122, 1154–1160.
23. Liao, Y., Eichinger, B., Firestone, K., Haller, M., Luo, J., Kaminsky, W., Benedict, J., Reid, P., Jen, A., Dalton, L., and Robinson, B. 2005. Systematic study of the structure-property relationship of a series of ferrocenyl nonlinear optical chromophores. *Journal of the American Chemical Society*, 127, 2758–2766.
24. Dalton, L. R., Robinson, B. H., Jen, A. K. Y., Reid, P., Eichinger, B., Sullivan, P., Akelaitis, A., Bale, D., Haller, M., Luo, J., Liu, S., Liao, Y., Firestone, K., Bhatambrekar, N., Bhattacharjee, S., Sinness, J., Hammond, S., Buker, N., Snoeberger, R., Lingwood, M., Rommel, H., Amend, J., Jang, S. H., Chen, A., and Steier, W. 2005. Acentric lattice electro-optic materials by rational design. *Proceedings of SPIE*, 5912, 43–54.
25. Minhaeng, C., Sun, Y. A., Hochan, L., Ledoux, I., and Zyss, J. 2002. Nonlinear optical properties of tetrahedral donor–acceptor octupolar molecules. *Journal of Chemical Physics*, 116, 9165–9173.
26. Le Bozec, H., Renouard, T., Bourgault, M., Dhenaut, C., Brasselet, S., Ledoux, I., and Zyss, J. 2001. Molecular engineering of octupolar tris(bipyridyl) metal complexes. *Synthetic Metals*, 124, 185–189.
27. Brunel, J., Juland, A., Ledoux, I., Zyss, J., and Blanchard-Desce, M. 2001. Boomerang-shaped octupolar molecules derived from triphenylbenzene. *Synthetic Metals*, 124, 195–199.
28. Stegeman, G. I. and Torreuellas, D. 1994. Issues in organics for nonlinear optics, In *Electrical, Optical, and Magnetic Properties of Prganic Solid State Materials*, A. F. Garito, A. K. Y. Jen, C. Y. C. Lee, and L. R. Dalton, eds., pp. 397–412. Materials Research Society, Pittsburgh.
29. Perry, J. W., Marder, S. R., Perry, K. J., Sleva, E. T., Yakymyshyn, C., Stewart, K. R., and Boden, E. P. 1991. Organic salts with large electro-optic coefficient. *Proceedings of SPIE*, 1560, 302–309.
30. Burland, D. M., Miller, R. D., and Walsh, C. A. 1994. Second-order nonlinearity in poled-polymer systems. *Chemical Reviews*, 94, 31–75.
31. Dalton, L. R., Harper, A. W., Ghosn, R., Steier, W. H., Ziari, M., Fetterman, H., Shi, Y., Mustacich, R., Jen, A. K. Y., and Shea, K. J. 1995. Synthesis and processing of improved second order nonlinear optical materials for applications in photonics. *Chemistry of Materials*, 7, 1060–1081.
32. Barto, R., Bedworth, P., Frank, C., Ermer, S., and Taylor, R. 2005. Near-infrared optical-absorption behavior in high-beta nonlinear optical chromophore-polymer guest-host materials. II. Dye spacer length effects in an amorphous polycarbonate copolymer host. *Journal of Chemical Physics*, 122, 23, 1–14. 234907.
33. Barto, R., Frank, C., Bedworth, P., Ermer, S., and Taylor, R. 2004. Near-infrared optical absorption behavior in high-β nonlinear optical chromophore-polymer guest-host materials. 1. Continuum dielectric effects in polycarbonate hosts. *Journal of Physical Chemistry B: Materials, Surfaces, Interfaces & Biophysical*, 108, 8702–8715.
34. Oviatt, H. W., Shea, K. J., Kalluri, S., Shi, Y., Steier, W. H., and Dalton, L. R. 1995. Applications of organic bridged polysilsesquioxane xerogels to nonlinear optical materials by the sol-gel method. *Chemistry of Materials*, 7, 493–498.
35. Mao, S. S. H., Ra, Y., Guo, L., Zhang, C., Dalton, L. R., Chen, A., Garner, S., and Steier, W. H. 1998. Progress towards device-quality second-order nonlinear optical materials: 1influence of composition and processing conditions on nonlinearity, temporal stability and optical loss. *Chemistry of Materials*, 10, 146–155.
36. Suresh, S., Chen, S., Topping, C., Ballato, J., and Smith, D. 2003. Novel perfluorocyclo-butyl (PFCB) polymers containing isophorone derived chromophore for electro-optic [EO] applications. *Proceedings of SPIE*, 4991, 530–536.

37. Jin, D., Londergan, T., Huang, D., Wolf, N., Condon, S., Tolstedt, D., Guan, H., Cong, S., Johnson, E., and Dinu, R. 2004. Achieving large electro-optic response: DH-type chromophores in both crosslinked systems and linear high Tg systems. *Proceedings of SPIE*, 5351, 44–56.
38. Zhang, C., Zhang, H., Oh, M., Dalton, L., and Steier, W. 2003. What the ultimate polymeric electro-optic materials will be: guest-host, crosslinked, or side-chain. *Proceedings of SPIE*, 4991, 537–551.
39. Bai, Y., Song, N., Gao, J. P., Yu, G., Sun, X., Wang, X., and Wang, Z. Y. 2005. A new approach to highly electro-optically active materials using cross-linkable hyperbranched chromophore-containing oligomers as a macromolecular dopant. *Journal of the American Chemical Society*, 127, 2060–2061.
40. Song, N., Men, L., Gao, J. P., Bai, Y., Beaudin, A. M. R., Yu, G., and Wang, Z. Y. 2004. Cross-linkable zwitterionic polyimides with high electrooptic ceofficients at telecommunication wavelengths. *Chemistry of Materials*, 16, 3708–3713.
41. Ma, H., Chen, B., Takafumi, S., Dalton, L. R., and Jen, A. K. Y. 2001. Highly efficient and thermally stable nonlinear optical dendrimer for electro-optics. *Journal of the American Chemical Society*, 123, 986–987.
42. Jen, A. K. Y., Ma, J., Sassa, T., Liu, S., Suresh, S., Dalton, L. R., and Haller, M. 2001. Highly efficient and thermally stable organic/polymeric electro-optic materials by dendritic approach. *Proceedings of SPIE*, 4461, 172–179.
43. Ma, H., Liu, S., Luo, J., Suresh, S., Liu, L., Kang, S. H., Haller, M., Sassa, T., Jen, A. K. Y., and Dalton, L. R. 2002. Highly efficient and thermally stable electro-optical dendrimers for photonics. *Advanced Functional Materials*, 12, 565–574.
44. Luo, J., Haller, M., Ma, H., Liu, S., Kim, T. D., Tian, Y., Chen, B., Jang, S. H., Dalton, L. R., and Jen, A. K. Y. 2004. Nanoscale architectural control and macromolecular engineering of nonlinear optical dendrimers and polymers for electro-optics. *Journal of Physical Chemistry B*, 108, 8523–8530.
45. Sullivan, P. A., Akelaitis, A. J. P., Lee, S. K., McGrew, G., Lee, S. K., Choi, D. H., and Dalton, L. R. 2006. Novel dendritic chromophores for electro-optics: influence of binding mode and attachment flexibility on EO behavior. *Chemistry of Materials*, 18, 344–351.
46. Luo, J., Liu, S., Haller, M., Li, H., Kim, T. D., Kim, K. S., Tang, H. Z., Kang, S. H., Jang, S. H., Ma, H., Dalton, L. R., and Jen, A. K. Y. 2003. Recent progress in developing highly efficient nonlinear optical chromophores and side-chain dendronized polymers for electro-optics. *Proceedings of SPIE*, 4991, 520–529.
47. Luo, J., Kim, T. D., Ma, H., Liu, S., Kang, S. H., Wong, S., Haller, M. A., Jang, S. H., Li, H., Barto, R. R., Frank, C. W., Dalton, L. R., and Jen, A. K. Y. 2003. Nanoscale architectural control of organic functional materials for photonics. *Proceedings of SPIE*, 5224, 104–112.
48. Luo, J., Liu, S., Haller, M., Kang, J. W., Kim, T. D., Jiang, S. H., Chen, B., Tucker, N. H., Li, H., Tang, H. Z., Dalton, L. R., Liao, Y., Robinson, B. H., and Jen, A. K. Y. 2004. Recent progress in developing highly efficient and thermally stable nonlinear optical polymers for electrooptics. *Proceedings of SPIE*, 5351, 36–43.
49. Haller, M., Luo, J., Li, H., Kim, T. D., Liao, Y., Robinson, B., Dalton, L. R., and Jen, A. K. Y. 2004. A novel lattice-hardening process to achieve highly efficient and thermally stable nonlinear optical polymers. *Macromolecules*, 37, 688–690.
50. Tucker, N., Li, H., Tang, H., Dalton, L. R., Liao, Y., Robinson, B. H., Jen, A. K. J., Luo, J., Liu, S., Haller, M., Kang, J., Kim, T., Jang, S., and Chen, B. 2004. Recent progress in developinghighly efficient and thermally stable nonlinear optical polymers for electro-optics. *Proceedings of SPIE*, 5351, 36–43.
51. Jen, A., Luo, J., Kim, T. D., Chen, B., Jang, S. H., Kang, J. W., Tucker, N. M., Hau, S., Tian, Y., Ka, J. W., Haller, M., Liao, Y., Robinson, B., Dalton, L., and Herman, W. 2005. Exceptional electro-optic properties through molecular design and controlled self-assembly. *Proceedings of SPIE*, 5935, 5935061–5935113.

52. Dalton, L., Robinson, B., Jen, A., Ried, P., Eichinger, B., Sullivan, P., Akelaitis, A., Bale, D., Haller, M., Luo, J., Liu, S., Liao, Y., Firestone, K., Sago, A., Bhatambrekar, N., Bhattacharjee, S., Sinness, J., Hammond, S., Buker, N., Snoeberger, R., and Lingwood, M. 2005. Optimizing electro-optic activity in chromophore/polymer composites and in organic chromophore glasses. *Proceedings of SPIE*, 5990, 59000C-1-10.
53. Yitzchaik, S. and Marks, T. J. 1996. Chromophore self-assembled superlattices. *Accounts of Chemical Research*, 29, 197–202.
54. Zhao, Y. G., Wu, A., Lu, H. L., Chang, S., Lu, W. K., Ho, S. T., Van der Boom, M., and Marks, T. J. 2001. Traveling wave electro-optic phase modulators based on intrinsically polar self assembled chromophoric superlattices. *Applied Physics Letters*, 79, 587–589.
55. Zhu, P., van der Boom, M. E., Kang, H., Evmenenko, G., Dutta, P., and Marks, T. J. 2002. Realization of expeditious layer-by-layer siloxane-based self-assembly as an efficient route to structurally regular acentric superlattices with large electro-optic responses. *Chemistry of Materials*, 14, 4982–4989.
56. Marks, T. J., Ho, S. T., Liu, Z., Zhu, P., Sun, D. G., Ma, J., Ziao, Y., and Kang, H. 2003. Electro optic waveguide modulators by the integration of self-assembled superlattices with polymeric and semiconductor materials. *Proceedings of SPIE*, 4991, 133–143.
57. Li, J., Neyman, P. J., Vercellino, M., Heflin, J. R., Duncan, R., and Evoy, S. 2004. Active photonic crystal devices in self-assembled electro-optic polymeric materials. *Materials Research Society Symposia Proceedings*, 817, 133–138.
58. Dalton, L. R., Harper, A. W., and Robinson, B. H. 1997. The role of London forces in defining noncentrosymmetric order of high dipole moment-high hyperpolarizability chromophores in electrically poled polymeric thin films. *Proceedings of the National Academy of Sciences of the United States of America*, 94, 4842–4847.
59. Dalton, L. R., Harper, A. W., Chen, J., Sun, S., Mao, S. S., Garner, S., Chen, A., and Steier, W. H. 1997. The role of intermolecular interactions in fabricating hardened electro-optic materials. *Proceedings of SPIE*, CR68, 313–321.
60. Harper, A. W., Sun, S., Dalton, L. R., Garner, S. M., Chen, A., Kalluri, S., Steier, W. H., and Robinson, B. H. 1998. Translating microscopic optical nonlinearity to macroscopic optical nonlinearity: The role of chromophore-chromophore electrostatic interactions. *Journal of the Optical Society of America B, Optical Physics*, 15, 329–337.
61. Robinson, B. H. and Dalton, L. R. 2000. Monte Carlo statistical mechanical simulations of the competition of intermolecular electrostatic and poling field interactions in defining macroscopic electro-optic activity for organic chromophore/polymer materials. *Journal of Physical Chemistry*, 104, 4785–4795.
62. Dalton, L. R., Robinson, B. H., Jen, A. K. Y., Steier, W. H., and Nielsen, R. 2003. Systematic development of high bandwidth, low drive voltage organic electro-optic devices and their applications. *Optical Materials*, 21, 19–28.
63. Nielsen, R. D., Rommel, H. L., and Robinson, B. H. 2004. Simulation of the loading parameter in organic nonlinear optical materials. *Journal of Physical Chemistry B: Materials, Surfaces, Interfaces and Biophysical*, 108, 8659–8667.
64. Pereverzev, Y. V., Prezhdo, O. V., and Dalton, L. R. 2004. Macroscopic order and electro-optic response of dipolar chromophore-polymer materials. *Chemical Physics, Chemistry*, 5, 1–11.
65. Shi, Y., Zhang, C., Zhang, H., Bechtel, J. H., Dalton, L. R., Robinson, B. H., and Steier, W. H. 2000. Low (sub-1 volt) halfwave voltage polymeric electrooptic modulators achieved by control of chromophore shape. *Science*, 288, 119–122.
66. Baehr-Jones, T., Hochberg, M., Wang, G., Lawson, R., Liao, Y., Sullivan, P. A., Dalton, L. R., Jen, A. K. Y., and Scherer, A. 2005. Optical modulation and detection in slotted silicon waveguides. *Optics Express*, 13, 5216–5226.
67. Song, H., Oh, M., Ahn, S., and Steier, W. 2003. Flexible low-voltage electro-optic polymer modulators. *Applied Physics Letters*, 82, 4432–4434.

68. Paloczi, G., Huang, Y., Yariv, A., Luo, J., and Jen, A. 2004. Replica-molded electro-optic polymer Mach-Zehnder modulator. *Applied Physics Letters*, 85, 1662–1664.
69. Zhang, C., Dalton, L. R., Oh, M.-C., Zhang, H., and Steier, W. H. 2001. Low V_{π} electrooptic modulators from CLD-1: chromophore design and synthesis, materials processing, and characterization. *Chemistry of Materials*, 13, 3043–3050.
70. Sun, L., Kim, J., Jang, C., An, D., Lu, X., Zhou, Q., Taboada, J. M., Chen, R. T., Maki, J. J., Tang, S., Zhang, H., Steier, W. H., Zhang, C., and Dalton, L. R. 2001. Polymeric waveguide prism based electro-optic beam deflector. *Optical Engineering*, 40, 1217–1222.
71. Zhang, H., Oh, M.-C., Szep, A., Steier, W. H., Zhang, C., Dalton, L. R., Erlig, H., Chang, Y., Chang, D. H., and Fetterman, H. R. 2001. Push pull electro-optic polymer modulators with low half-wave voltage and low loss at both 1310 and 1550 μm. *Applied Physics Letters*, 78, 3136–3138.
72. Rabiei, P., Steier, W. H., Zhang, C., and Dalton, L. R. 2002. Polymer micro-ring filters and modulators. *Journal of Lightwave Technology*, 20, 1968–1975.
73. Liu, S., Haller, M. A., Luo, J., Jang, S. H., Ma, H., Dalton, L. R., and Jen, A. K. Y. 2003. Focused microwave-assisted synthesis of 2,5-dihydrofuran derivatives as electron acceptors for highly efficient nonlinear optical chromophores. *Materials Research Society Symposia Proceedings*, 771, 375–380.
74. Liu, S., Haller, M. A., Ma, H., Dalton, L. R., Jang, S. H., and Jen, A. K. Y. 2003. Focused microwave-assisted synthesis of 2,5-dihydrofuran derivatives as electron acceptors for highly efficient nonlinear optical chromophores. *Advanced Materials*, 15, 603–607.
75. Zhou, Z., Shaojun, L., and Cheng, Y. 2003. Poling properties of guest-host polymer films of X-type nonlinear optical chromophores. *Synthetic Metals*, 147, 1519–1520.
76. Sullivan, P. A., Bhattacharjee, S., Eichinger, B. E., Firestone, K., Robinson, B. H., and Dalton, L. R. 2004. Exploration of series type multifuctionalized nonlinear optical chromophore concept. *Proceedings of SPIE*, 5351, 253–259.
77. Firestone, K., Bale, D., Liao, Y., Casmier, D., Clot, O., Dalton, L., and Reid, P. 2005. Frequency-agile Hyper-Rayleigh Scattering studies of electro-optic chromophores. *Proceedings of SPIE*, 5935, 59350P1–59350P9.
78. Firestone, K. A., Reid, P., Lawson, R., Jang, S. H., and Dalton, L. R. 2004. Advances in organic electro-optic materials and processing. *Inorganic Chemistry Acta*, 357, 3957–3966.
79. Jang, S. H., Luo, J., Tucker, N. M., Leclercq, A., Zojer, E., Haller, M. A., Kim, T. D., Kang, J. W., Firestone, K., Bale, D., Lao, D., Benedict, J. B., Cohen, D., Kaminsky, W., Kahr, B., Bredas, J. L., Reid, P., Dalton, L. R., and Jen, A. K. Y. 2006. Pyrroline chromophores for electro-optics. *Chemistry of Materials*, 18, 13, 2982–2988.
80. Kim, W. K. and Hayden, L. M. 1999. Fully atomistic modeling of an electric field poled guest-host nonlinear optical polymer. *Journal of Chemical Physics*, 111, 5212–5222.
81. Nagtegaele, P., Brasselet, E., and Zyss, J. E. 2003. Anisotropy and dispersion of a Pockels tensor: a benchmark for electro-optic organic thin-film assessment. *Journal of the Optical Society of America B: Optical Physics*, 20, 1932–1936.
82. Teng, C. and Man, H. 1990. Simple reflection technique for measuring the electro-optic coefficient of poled polymers. *Applied Physics Letters*, 56, 1734–1736.
83. Schildkraut, J. 1990. Determination of the electrooptic coefficient of a poled polymer film. *Applied Optics*, 29, 2839–2841.
84. Park, D., Kang, J., Luo, J., Lim, T., Jen, A., Lee, C., and Herman, W. 2005. Nonlinear ellipsometric analysis of poled organic glasses having very large electro-optic coefficients. *Proceedings of SPIE*, 5935, 59350O-1-12.
85. Chen, A., Chuyanov, V., Garner, S., Steier, W., and Dalton, L. 1997. Modified attenuated total reflection for the fast and routine electrooptic measurements of nonlinear optical polymer thin films. In *Organic Thin Films for Photonic Applications, Vol. 14*, OSA Technical Digest Series, pp. 158–159. Optical Society of America, Washington DC.

86. Dumont, M. and Levy, Y. 1989. Measurement of electrooptic properties of organic thin films by attenuated total reflection, In *Nonlinear Optics of Organics and Semiconductors*, T. Kobayashi, ed., pp. 256–266. Springer, Berlin.
87. Levy, Y., Dumont, M., Shastaing, E., Robin, P., Chollet, P. A., Gadret, G., and Kajzar, F. 1993. Reflection method for electro-optic coefficient determination in stratified thin film structures. *Molecular Crystals and Liquid Crystals Science and Technology, Section B*, 4, 1–6.
88. Meyrueix, R., Lecomte, J., and Tapolsky, G. 1991. A Fabry-Perot interferometric technique for the electro-optic characterization of nonlinear optical polymers. *Nonlinear Optics*, 1, 201–211.
89. Meyrueix, R., Dickens, M., Lemonnier, O., Lecomte, J., and Tapolsky, G. 1994. Fabry-Perot interferometry applied to the study of piezoelectric and electro-optic properties of a poled NLO polyurethane. *Optics Communications*, 110, 445–455.
90. Kalluri, S., Garner, S., Ziari, M., Steier, W. H., and Dalton, L. R. 1996. Simple two-slit interference electrooptic coefficients measurement technique and efficient coplanar electrode poling of polymer thin films. *Applied Physics Letters*, 69, 275–277.
91. Donval, A., Toussaere, E., Hierle, R., and Zyss, J. 2000. Polarization insensitive electro-optic polymer modulator. *Journal of Applied Physics*, 87, 3258–3262.
92. Shin, M., Cho, H., Han, S., and Wu, J. 1998. Analysis of a Mach-Zehnder interferometry measurement of the Pockels coefficients in a poled polymer film with a reflection configuration. *Journal of Applied Physics*, 83, 1848–1853.
93. Chen, A., Chuyanov, V., Zhang, H., Garner, S., Steier, W. H., Chen, J., Zhu, J., He, M., Mao, S. S. H., and Dalton, L. R. 1998. "Demonstration of the full potential of second order nonlinear optic polymers for electrooptic modulation using a high $\mu\beta$ chromophore and a constant bias field. *Optics Letters*, 23, 478–480.
94. Gopalan, P. and Campbell, V. 2005. Low-loss electro-optic polymers for active wave guide components, *Pacifichem*, Honolulu, Hawaii.
95. Liao, Y., Firestone, K. A., Bhattacharjee, S., Luo, J., Haller, M., Hau, S., Anderson, C. A., Lao, D., Eichinger, B. E., Robinson, B. H., Reid, P., Jen, A. K. Y., and Dalton, L. R. 2006. Linear and nonlinear optical properties of a macrocyclic trichromophore bundle with parallel-aligned dipole moments. *Journal of Physical Chemistry B: Materials, Surfaces, Interfaces & Biophysical*, 110, 5434–5438.
96. Steier, W. H., Kalluri, S., Chen, A., Garner, S., Chuyanov, V., Ziari, M., Shi, Y., Fetterman, H., Jalali, B., Wang, W., Chen, D., and Dalton, L. R. 1996. Applications of electro-optic polymers in photonics. In *Materials Research Society Symposium Proceedings, Electrical, Optical and Magnetic Properties of Organic Solid State Materials, Vol. 413*, pp. 147–158, Materials Research Society, Pittsburgh.
97. Chen, A., Chuyanov, V., Marti-Carrera, F. I., Garner, S., Steier, W. H., Chen, J., Sun, S., and Dalton, L. R. 1997. Integrated polymer waveguide mode size transformer with a vertical taper for improved fiber coupling. *Proceedings of SPIE*, 3005, 65–76.
98. Chen, A., Chuyanov, V., Marti-Carrera, F. I., Garner, S. M., Steier, W. H., Chen, J., Sun, S. S., and Dalton, L. R. 2000. Vertically tapered polymer waveguide mode size transformer for improved fiber coupling. *Optical Engineering*, 39, 1507–1516.
99. Oh, M. C., Zhang, C., Lee, H. J., Steier, W. H., and Fetterman, H. R. 2002. Low-loss interconnection between electrooptic and passive polymer waveguides with a vertical taper. *IEEE Photonics Technology Letters*, 14, 1121–1123.
100. Zang, D. Y., Shu, G., Downing, T., Lin, W., Oh, C., and Bechtel, J. 2003. Insertion loss reduction in high speed polymer electrooptic modulators using tapered waveguide, fiber tip lenses and modification of waveguide structures. *Proceedings of SPIE*, 4991, 601–609.
101. Chang, D. H., Azfar, T., Kim, S. K., Fetterman, H. R., Zhang, C., and Steier, W. H. 2003. Vertical adiabatic transition between silica planar waveguide and electro-optic polymer fabricated using grayscale lithography. *Optics Letters*, 28, 869–871.

102. Ma, H., Wu, J., Herguth, P., Chen, B., and Jen, A. K. Y. 2000. A novel class of high performance perfluorocyclobutane-containing polymers for second-order nonlinear optics. *Chemistry of Materials*, 12, 1187–1189.
103. Zhang, Q., Canva, M., and Stegeman, G. 1998. Wavelength dependence of 4-dimethylamino-4'-nitrostilbene polymer thin film photodegradation. *Applied Physics Letters*, 73, 912–914.
104. Galvan-Gonzalez, A., Canva, M., Stegeman, G. I., Twieg, R., Kowalczyk, T. C., and Lackritz, H. S. 1999. Effect of temperature and atmospheric environment on the photodegradation of some Disperse Red 1 type polymers. *Optics Letters*, 24, 1741–1743.
105. Galvan-Gonzalez, A., Canva, M., Stegeman, G. I., Twieg, R. J., Chang, K. P., Kowalczyk, T. C., Zhang, X. Q., Lackritz, H. S., Marder, S., and Thayumanavan, S. 2000. Systematic behavior of electro-optic chromophore photostability. *Optics Letters*, 25, 332–334.
106. Galvan-Gonzalez, A., Canva, M., Stegeman, G. I., Sukhomilinova, L., Twieg, R. J., Chang, K. P., Kowalczyk, T. C., and Lackritz, H. S. 2000. Photodegradation of azobenzene nonlinear optical chromophores: The influence of structure and environment. *Journal of the Optical Society of America B: Optical Physics*, 17, 1992–2000.
107. DeRosa, M. E., He, M., Cites, J. S., Garner, S. M., and Tang, Y. R. 2004. Photostability of high $\mu\beta$ electro-optic chromophores at 1550 nm. *Journal of Physical Chemistry B: Materials, Surfaces, Interfaces & Biophysical*, 108, 8725–8730.
108. He, M., Leslie, T., Garner, S., DeRosa, M. E., and Cites, J. 2004. Synthesis of new electrooptic chromophores and their structure-property relationship. *Journal of Physical Chemistry B: Materials, Surfaces, Interfaces & Biophysical*, 108, 8731–8736.
109. Ashley, P., Sangnadasa, M., and Lindsay, G. 2005. Investigation of long term thermal stability of electrically poled CLD and FTC chromophores in amorphous polycarbonates, *Pacifichem*, Honolulu, Hawaii.
110. Taylor, E. W., Nichter, J. E., Nash, F. D., Haas, F., Szep, A. A., Michalak, R. J., Flusche, B. M., Cook, P. R., McEwen, T. A., McKeon, B. F., Payson, P. M., Brost, G. A., Pirich, A. R., Castaneda, C., Tsap, B., and Fetterman, H. R. 2005. Radiation resistance of electro-optic polymer-based modulators. *Applied Physics Letters*, 86, . 201122-1-3.
111. Sinyukov, A. M. and Hayden, L. M. 2004. Efficient electrooptic polymers for THz applications. *Journal of Physical Chemistry B*, 108, 8515–8522.
112. Sinyukov, A. M., Leahy, M. R., Hayden, L. M., Haller, M., Luo, J., Jen, A. K. Y., and Dalton, L. R. 2004. Resonance enhanced THz generation in electro-optic polymers near the absorption maximum. *Applied Physics Letters*, 85, 5827–5829.
113. Schneider, A. and Gunter, P. 2005. Spectrum of terahertz pulses form organic DAST crystals. *Ferroelectrics*, 318, 83–88.
114. Kawase, K., Hatanaka, T., Takahashi, H., Nakamura, K., Taniuchi, T., and Ito, H. 2000. Tunable terahertz-wave generation from DAST crystal by dual signal-wave parametric oscillation of periodically poled lithium niobate. *Optics Letters*, 25, 1714–1716.
115. Kawase, K., Mizuno, M., Sohma, S., Takahashi, H., Taniuchi, T., Urata, Y., Wada, S., Tashiro, H., and Ito, H. 1999. Difference-frequency terahertz-wave generation from 4-dimethylamino-N-methyl-4-stilbazolium-tosylate by use of an electronically tuned Ti:sapphire laser. *Optics Letters*, 24, 1065–1067.
116. Adachi, H., Tamiuchi, T., Yoshimura, M., Brahadeeswaran, S., Higo, T., Takagi, M., Mori, Y., Sasaki, T., and Nakanishi, H. 2004. High-quality organic 4-dimentylamino-N-methyl-4-stilbazolium tosylate (DAST) crystals for THz wave generation. *Japanese Journal of Applied Physics*, 43, L1121–L1123.
117. Zojer, E., Beljonne, D., Kogij, T., Vogel, H., Marder, S. R., Perry, J. W., and Bredas, J. L. 2002. Tuning the two-photon absorption response of quadrupolar organic molecules. *Journal of Chemical Physics*, 116, 3646–3658.

118. Brousmiche, D. W., Senn, J. M., Frechet, J. M. J., He, G. S., Lin, T.-C., Chung, S.-J., Prasad, P. N., Kannan, R., and Tan, L.-S. 2004. Fluorescence resonance energy transfer in novel multiphoton absorbing dendritic structures. *Journal of Physical Chemistry B*, 108, 8592–8600.
119. Drobizhev, M., Karotki, A., Rebane, A., and Spangler, C. 2001. Dendrimer molecules with record large two-photon absorption cross section. *Optics Letters*, 26, 1081–1083.
120. Hochberg, M., Baehr-Jones, T., Wang, G., Parker, J., Harvard, K., Liu, J., Chen, B., Shi, Z., Lawson, R., Sullivan, P., Jen, A. K. J., Dalton, L., and Scherer, A. Submitted. All optical modulator in silicon with terahertz bandwidth, *Nature Materials*.

V

Devices

8

Photovoltaic Devices

Bryce S. Richards
Avi Shalav

8.1 Introduction

Photovoltaic (PV) cells directly convert light to electricity. The most common source of light on earth is the sun. The photosphere, or the external region of the sun, emits radiation closely approaching that of a thermodynamic "black body," or perfect radiator, at a temperature of 6000 K, with a spectral distribution governed by Planck's radiation law, as shown in Figure 8.1.[1] The radiant power per unit area perpendicular to the direction of the sun just outside the earth's atmosphere is essentially constant and is known as the *solar constant*. This is also known as the *air-mass zero (AM0) spectrum*, which has the value of 1.367 kW/m^2. The AM0 spectrum, plotted in Figure 8.1, is used for determining the performance of PV cells in space. Sunlight is attenuated by at least 30% while passing through the earth's atmosphere due to scattering of light by molecules, aerosols, and dust particles, as well as absorption by gases such as oxygen, ozone, water vapor, and carbon dioxide. In addition, sunlight passes through more or less of the atmosphere depending on the time of day, with the most direct path being when the sun is directly overhead, defined as air-mass one (AM1). Therefore, terrestrial sunlight varies in both intensity and spectral composition. To allow a meaningful comparison between the performance of solar cells at different locations, a terrestrial standard has been defined that corresponds to the sun being 48° off overhead, known as AM1.5. This spectrum is also plotted in Figure 8.1, and the total power density is normalized to $P_{in} = 1$ kW/m^2, which is close to the actual value received at the earth's surface.

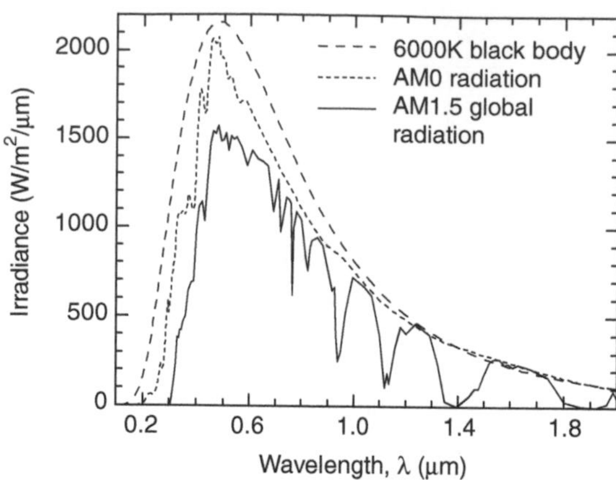

FIGURE 8.1 Reference solar spectra used for determining the performance of space (air-mass zero, AM0) and terrestrial (AM1.5) photovoltaic devices. The emission from the sun closely follows that of a black-body radiator at a temperature of 6000 K, except for absorption peaks due to gases in the earth's atmosphere.

At current consumption rates, all of the world's readily-exploitable fossil-fuel reserves are expected to be depleted within the next 50 (gas) to 200 (coal) years. In 2002, electricity generated from renewable energy sources such as hydro, wind, biomass, geothermal, solar thermal, and photovoltaics amounts to 14% of total electricity generated.[2] However, several groups, including Shell Petroleum, have predicted that more than half of the world's energy supply will come from renewable resources by 2050.[3] To put into perspective the amount of energy that is constantly being radiated by the sun, the amount of solar energy that strikes earth in a period of a few days is greater than the amount of fuel burnt over the entire course of human history.[4] With regard to the required land area, it has been estimated that PV cells with a sunlight-to-electricity conversion efficiency of $\eta = 15\%$ would need to cover just 0.25% of the global pastoral area to meet all of the world's primary energy requirements.[5] Although these simplistic statements may sound like a gross oversimplification of the impending energy crisis facing humankind, it encourages one to think about ways in which this energy could be effectively harnessed to satisfy our ever-increasing demand for energy. The major barrier that prevents electricity generated by PV from competing with fossil-fuel-fired power stations is cost, and cost reductions of about an order of magnitude are required before photovoltaics are expected to be able to compete head-to-head with conventional forms of electricity generation. The cost of solar cells is decreasing markedly with increased production, as shown in Figure 8.2, and this can be attributed to both economies of scale and new, lower-cost photovoltaic technologies,[4] with today's photovoltaic devices costing about US$3.50 per peak-watt ($W_P$). The peak-watt rating is for a cell or module that produces 1 W electrical output at standard test conditions (1000 W/m^2 of the global AM1.5 spectrum at 25°C). If current trends continue, an order of magnitude reduction in price over year-2000 levels in photovoltaic cells (all measured in 1998 US$) could be expected to occur by the time the accumulated shipments reach 10^9 MW_P, in about the year 2020. This would result in installed system prices that can compete with the price of energy generated by today's fossil-fuel-fired power stations of about US$0.03–0.04/kWh.

For the last 9 years (1997–2005), the photovoltaic industry has experienced a period of unprecedented growth, at an average annual rate of nearly 35%, and this growth is expected to increase at least until 2010. This result is plotted in Figure 8.3, where the data is broken down into the leading production regions. The top ten manufacturers accounted for 85% of global production in 2003, with Sharp (Japan) producing 26% of all photovoltaic cells and nearly three times the production capacity of its next largest rival, Shell (USA and The Netherlands), followed by Kyocera (Japan) and BP Solar (world-wide).[6]

The discovery of the photovoltaic effect is often attributed to Becquerel in 1839, however the photocurrent that Becquerel observed by illuminating silver chloride- or silver bromide-coated platinum

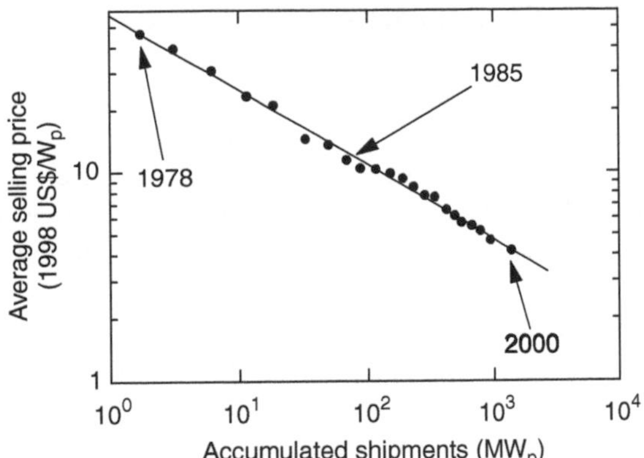

FIGURE 8.2 The price of photovoltaic modules closely follows a trend of halving for each order of magnitude increase in production. (Adapted from Green, M. A. and Hansen, J. 2003. *Catalog of Solar Cell Drawings*, University of New South Wales.)

electrodes in an aqueous solution is more accurately described as a photoelectrochemical effect.[5] The observation of photoconductivity in solid selenium in 1873 led to the discovery of the photovoltaic effect in that material in 1877, and the realization of the first practical photovoltaic cell, consisting of a thin selenium layer on an iron baseplate with a semi-transparent top gold contact, was achieved by Fritts in 1883.[1,5] Selenium photodetectors were commercialized in Germany in the 1930s and are still in use today as a photographic light meters.

The conversion efficiency of these metal–semiconductor devices was well below 1% due to the large dark currents. A semiconductor–semiconductor device offered better properties in this regard, and, in 1941, Russell Ohl of Bell Telephone Laboratories formed the first *p–n* junction in a silicon crystal. Further work over the years by Bell researchers lead to the first diffused *p–n* junction photovoltaic cell in 1954 that had an efficiency of about 4.5%.[5] Within 18 months, conversion efficiencies of 10% had been demonstrated, and by 1958 solar cells had established their first market—aboard the Vanguard I spacecraft. Following the oil crises of the early 1970s, the first terrestrial PV systems were installed in 1973,[7] and since this time, silicon has formed the backbone of the photovoltaic industry.

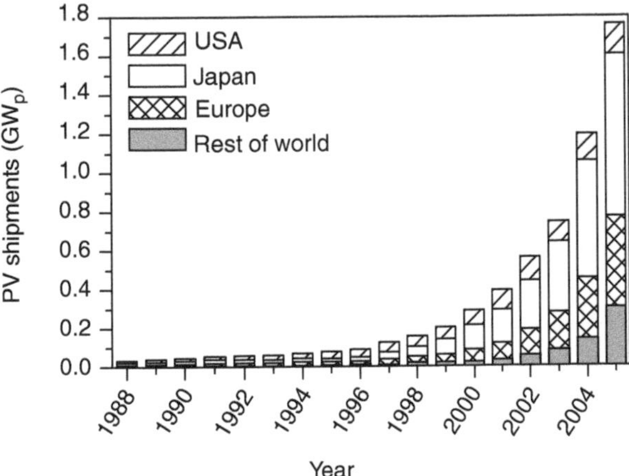

FIGURE 8.3 Shipments of photovoltaic cells have experienced an annual average growth rate of about 35% over the last 7 years. (From Maycock, P. 2004. *Renewable Energy World*, July–August, 86. With permission.)

8.2 Generations of Photovoltaic Devices

In general, photovoltaic cells can be classified into three "generations." The first-generation technology of cells are based on silicon wafers that are thick enough to be self-supporting, typically about 300 μm thick. Production of these single-crystal and multicrystalline wafer technologies comprise nearly 91% of the current PV market.[6] These photovoltaic cells typically have an efficiency of 12–17%. The primary limitation of first-generation technologies is their cost, which is dominated by the cost of the high-purity silicon wafer.

The second generation represents thin-film technologies that aim for lower cost at the expense of efficiency by using thin layers of semiconducting material.[8] Because the active layer is only a few microns thick, a rigid substrate or transparent superstrate is required for mechanical support. The thin films are deposited rapidly over large areas of the sub/superstrate and are subsequently patterned to form multiple cells. Therefore, no soldering is required to connect the individual cells and the basic production unit is the module rather than the cell. The four most important thin-film technologies are amorphous silicon (a-Si), polycrystalline silicon, cadmium telluride (CdTe), and copper indium diselenide (CIS), with current production efficiencies ranging from 4% to 10%. While thin-film cells have been expected to overtake the first-generation cells' dominance for decades, this has not yet occurred for a variety of reasons. More recently, organic thin-film devices have emerged as a possible path for achieving truly low-cost PV devices, albeit with a lower conversion efficiency. These devices, which only exist in laboratories at the moment, will be discussed, along with the reasons for their typically low efficiencies of 0.1–5%.

Third-generation approaches are eventually intended to achieve both high-efficiency and low cost.[9] One example of a third-generation photovoltaic cell that is already commercially produced is the tandem cell. Two types of tandem cells exist. The first, a thin-film technology based on layers of a-Si and microcrystalline silicon (μc-Si) or a-Si and amorphous germanium (a-Ge) is a low-cost and low-efficiency device. The second, made from elements of groups III and V of the periodic table (known simply as III–V cells), achieves ultra-high efficiency, but at a high cost premium.

Figure 8.4 summarizes the efficiency and price (in US$/m^2 and US$/W$_p$) targets of the three generations of PV technology. In general, several thin-film technologies (region II) are meeting the targets for second-generation devices, whereas the first-generation PV modules still lie in the upper price range defined by region I. The thermodynamic limits will be discussed in more detail in the section on third-generation photovoltaics, region III.

The final device presented here is the dye-sensitized solar cell (DSSC).[10] This solar cell does not rely on the photovoltaic effect to convert sunlight into electricity, but rather the same photoelectrochemical effect as observed by Becquerel in 1839. Consequently, this device is not included in the generations of PV technology outlined above and is discussed in a separate section.

8.3 Photovoltaic Cell Operation

Solar cells can be regarded as a large-area, illuminated semiconductor diode. The semiconductor's properties are exploited in two ways. First, light is absorbed and generates free charge carriers within the material. Second, a junction in the semiconductor separates the negative and positive charge carriers, producing a unidirectional electrical current through the two contacts that have a voltage difference between them. As shown in Figure 8.5, the separation of charge is usually performed by a *p–n* junction, where regions of material that have been "doped" with different impurities to create an excess of free electrons in *n*-type regions and a shortage of free electrons in *p*-type regions. This establishes a region at the interface that has a built-in electrostatic field, sweeping electrons one way and holes the other. In a bulk silicon device, the junction is typically located at about 0.5 μm from the front surface. The remaining 300 μm of wafer is required to absorb the majority of above-bandgap light and to provide mechanical strength.

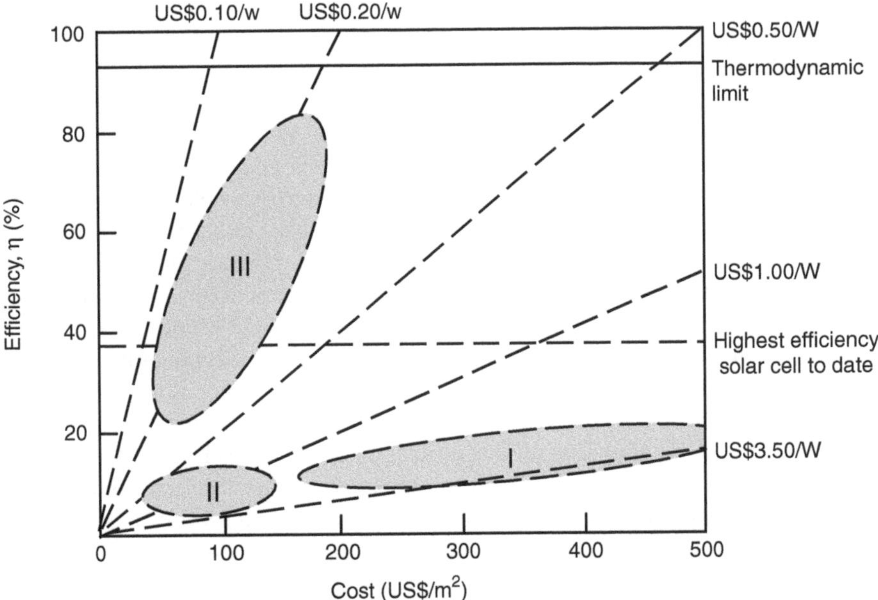

FIGURE 8.4 Efficiency and cost targets for first-, second- and third-generation photovoltaic technology, along with the present limits of today's PV technology and the upper thermodynamic limits. (Adapted from Green, M. A. and Hansen, J. 2003. *Catalog of Solar Cell Drawings*, University of New South Wales.)

The main processes of solar-cell operation are shown in Figure 8.5. The sun's photons strike the front surface of the solar cell. High-energy (ultraviolet or blue) photons are absorbed strongly near the front surface of the device, but as the photon energy decreases towards the infrared, these photons are more weakly absorbed and penetrate deeper into the device. The absorption of a photon results in an electron in the valence band of the semiconductor being excited up to the conduction band, leaving behind a positively charged electron vacancy or "hole." In some materials, known as *direct bandgap semiconductors*, the absorption involves just the photon and the created carrier pair; however in indirect bandgap semiconductors, the process also requires interaction with one or more phonons. Indirect bandgap semiconductors, such as silicon (Si), exhibit weaker absorption due to the required presence of this third

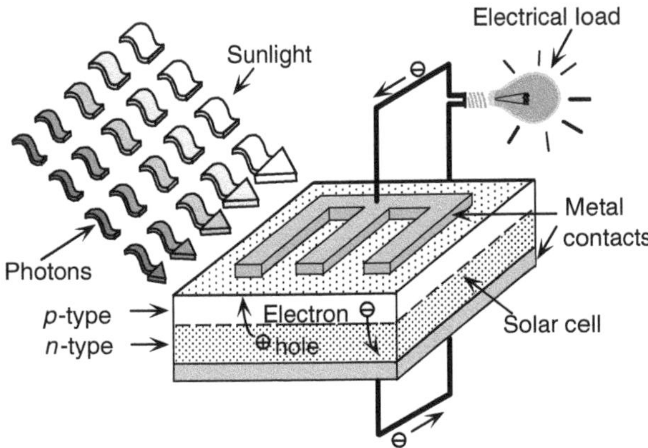

FIGURE 8.5 Operation of a *p–n* junction solar cell. (Adapted from Green, M. A. and Hansen, J. 2003. *Catalog of Solar Cell Drawings*, University of New South Wales.)

"particle" for carrier generation, while gallium arsenide (GaAs) is a good example of a direct bandgap semiconductor.

Once created, the free electrons and holes are mobile and they will move due to two mechanisms: electric field ("drift"), or diffusion, to regions of lower carrier concentration. The flow of minority carriers (electrons in *p*-type material and holes in an *n*-type material) principally determines the performance of the majority of photovoltaic cells (a typical *p–n*-junction band diagram can be seen in Figure 8.18 in the Section 8.6. The minority carriers from electron–hole pairs generated near the junction are swept across it by the strong electric field. As each crosses the junction to become a majority carrier, a real contribution to the cell's output current is made. Minority carriers that are generated far away from the junction can be transported to the junction by diffusion, as a lower minority carrier concentration exists near the junction due to the field action there. However, until a minority carrier is collected across the junction, there is always the possibility of it recombining with a surrounding majority carrier. Recombination can be either radiative, which is the inverse of the optical generation process that produced the electron–hole pair and energy is lost in the production and emission of a new photon, or nonradiative, where energy is dissipated as heat in the cell. Nonradiative recombination mechanisms dominate in silicon solar cells, with crystal defects, unwanted impurities, and the abrupt termination of the crystal structure at the surfaces all serving as strong nonradiative recombination sites. Metal contacts at the front and rear of the device allow physical connection of the photovoltaic cell to an electrical load and the photogenerated current to flow. The front contact is normally in the form of a fine metallic grid to minimize the fraction of the front surface area that is shaded, preventing light from passing through into the semiconductor. The rear side of the device is typically covered with the second metal contact.

The light-generated current flows in what is normally regarded as the reverse direction in diode theory, with electrons flowing out of the cell into the circuit from the *n*-type contact and back into the cell through the *p*-type contact, even though the voltage across the cell is in the forward bias direction (i.e., positive at the *p*-type contact). An ideal solar cell can be considered as being a current source connected in parallel with a rectifying diode, as illustrated by the circuit with the solid lines in Figure 8.6.

The current–voltage (I–V) relationship of an ideal device can be described by

$$I = I_L - I_0[\exp(qV/kT) - 1], \tag{8.1}$$

where k is the Boltzmann constant (1.381×10^{-23} J/K), q is the electron charge (1.602×10^{-19} C), T is absolute temperature, V is the voltage at the terminals of the device, I_L is the photogenerated current, and I_0 is the dark saturation current. The light-generated current can be regarded as being roughly independent of the voltage across the device. The dark current can be thought of as simultaneously flowing in the opposite direction to the light-generated current. As indicated by the solid line in Figure 8.7, the dark current increases strongly with the voltage; at the open circuit voltage (V_{oc}), it completely cancels the light-generated current. Thus, a typical I–V characteristic for a photovoltaic cell appears as a dashed line in Figure 8.7. The (forward) dark current is added to the (reverse) light-generated current (I_L) to make the resultant current negative for a range of positive voltages. In this

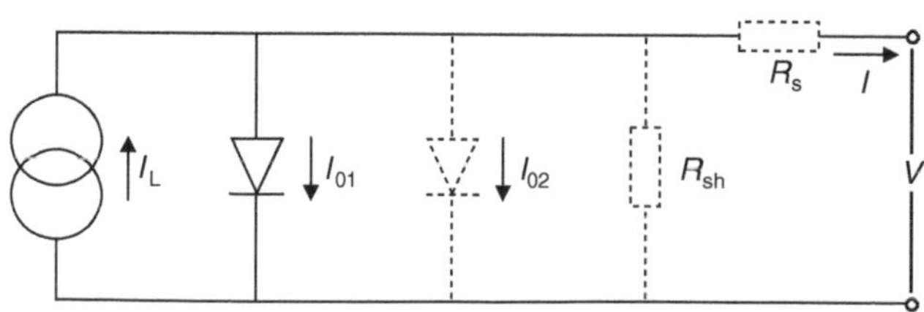

FIGURE 8.6 The equivalent circuit of a solar cell, showing the configuration for both an ideal (solid lines) and nonideal device (dotted lines).

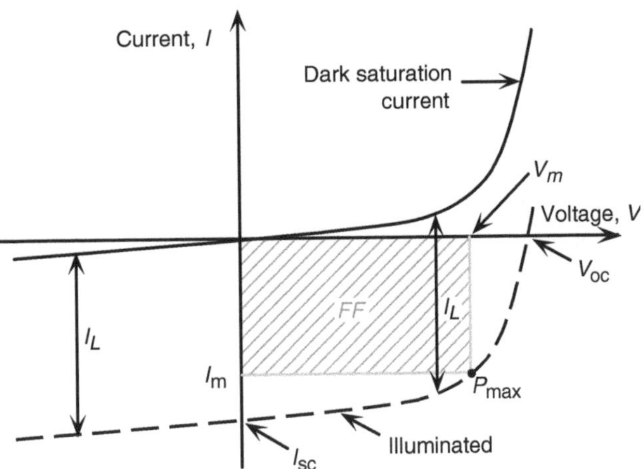

FIGURE 8.7 Typical *I–V* characteristic for a photovoltaic cell. (Adapted from Green, M. A. and Hansen, J. 2003. *Catalog of Solar Cell Drawings*, University of New South Wales.)

quadrant, the cell can generate power. The power curve (not shown) is zero at both zero volts and V_{oc}, with maximum power (P_{max}) being produced near the "knee" of the *I–V* curve. The voltage (V_m) and current (I_m) at this point are used to define the fill factor (FF) of the solar cell:

$$\mathrm{FF} = \frac{I_m V_m}{I_{sc} V_{oc}} = \frac{P_{max}}{I_{sc} V_{oc}}, \tag{8.2}$$

where the short-circuit current (I_{sc}) is equal to the light-generated current I_L in the ideal case. Ideally, electrical loads connected to the solar cell should be chosen to keep the operating point close to the optimum knee-point during normal operation. The conversion efficiency (η) of a solar cell is defined as the ratio of generated electrical power to the incident solar radiation, P_{in}:

$$\eta = \frac{P_{max}}{P_{in}}. \tag{8.3}$$

In reality, the performance of practical PV devices departs from the ideal case described above, and additional factors are required to model the behavior of the device. These include a second diode with a diode-ideality factor of 2, series resistance (R_s), and shunt resistance (R_{sh}), as given in the following equation:

$$I = I_L - I_0\left[\exp\left(\frac{q(V + IR_s)}{kT}\right) - 1\right] - I_{02}\left[\exp\left(\frac{q(V + IR_s)}{2kT}\right) - 1\right] - \frac{V + IR_s}{R_{sh}}. \tag{8.4}$$

Series resistance can arise from the bulk resistance of the semiconductor, the front and rear metal contacts, and the contact resistance between the metal contacts and the semiconductor. The shunt resistance is typically caused by a nonideal *p–n* junction or shorting of the junction. The effect of both resistances is noticed first in the reduction of the FF, although highly shunted devices will result in a reduction in V_{oc}, whereas a highly increased R_s will cause a decrease in I_{sc}.

Thus far, the electrical operation of a solar cell has been described. However, because these are optoelectronic devices, techniques for optimizing the optical performance of the PV cell will now be described. As a polished silicon wafer reflects at least 30% of photons that are incident upon it, the front surface reflection must first be reduced to allow efficient absorption; there are two common techniques

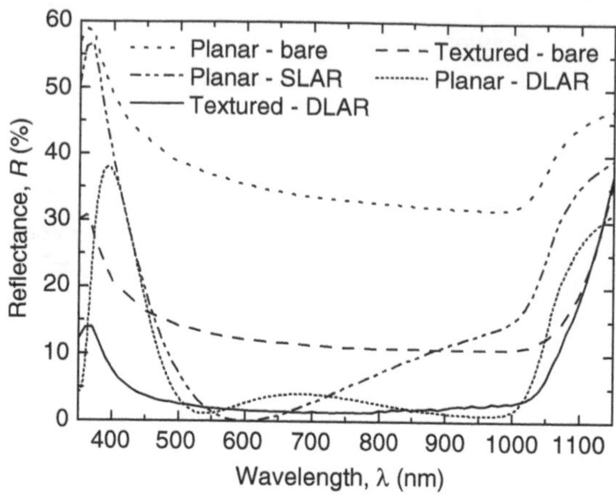

FIGURE 8.8 Measured reflectance of silicon wafers (polished versus chemically textured) both with and without AR coatings. Note that the increased reflectance beyond 1000 nm is due to reflectance from the back surface of the silicon wafer, which is becoming transparent in the NIR.

for achieving this. First, a monocrystalline wafer that has the (100) planes of the silicon atoms at the surface can be textured in an alkaline etch to expose the pyramids of the (111) planes. This roughened or "textured" surface of pyramids results in the majority of the light having two chances to be coupled into the silicon wafer. Second, an antireflection (AR) coating made of transparent dielectric materials such as amorphous silicon nitride (a-SiN_x) and titanium dioxide (TiO_2) is applied to the surface.[11,12] In industry, a single antireflection (DLAR) coating is normally applied, whereas to achieve the highest conversion efficiencies, laboratory cells usually have a double-layer antireflection (DLAR) coating. The effect of both of these techniques on the measured front-surface reflection is shown in Figure 8.8 for planar and textured silicon wafers.

The final measure of PV device performance relies on a measurement of the spectral response of the solar cell to light, ranging from UV to NIR. The internal quantum efficiency (IQE) of a solar cell is defined as the fraction of photons absorbed to the number of electron–hole pairs created. This is a measure of the electronic performance of the device. However, due to other reflectance and parasitic absorption losses, a second concept is commonly used to compare the optoelectronic performance of a solar cell—the external quantum efficiency (EQE). The EQE is defined as the fraction of photons incident on the front surface of the solar cell to the number of electron–hole pairs created. Figure 8.9[15,16] compares the IQE of several different types of solar cells, including high-efficiency silicon and two types of multijunction solar cells (one III–V device based on gallium arsenide technology and one amorphous silicon).

8.4 First-Generation PV

The first-generation photovoltaic technology is based upon thick silicon wafers. Silicon enjoys several advantages as a photovoltaic material. First, it is nontoxic and the most abundant element in the earth's crust. Second, its native oxide (silicon dioxide) can electronically passivate the surfaces of a wafer to reduce the level of recombination at the terminated crystal planes. Third, silicon solar-cell research has benefited greatly from the vast amount of research carried out by the microelectronics industry and, until recently, off-specification silicon rejected by the electronics industry supplied most of the feedstock for the photovoltaic industry. Finally, useful wet-chemical etches are available that enable monocrystalline silicon to be textured, exposing pyramids on the surface. This roughened surface gives photons two chances to strike the semiconductor surface and be absorbed. Texturing also redirects photons into oblique paths within the cell, giving longer paths for the infrared photons that are weakly absorbed

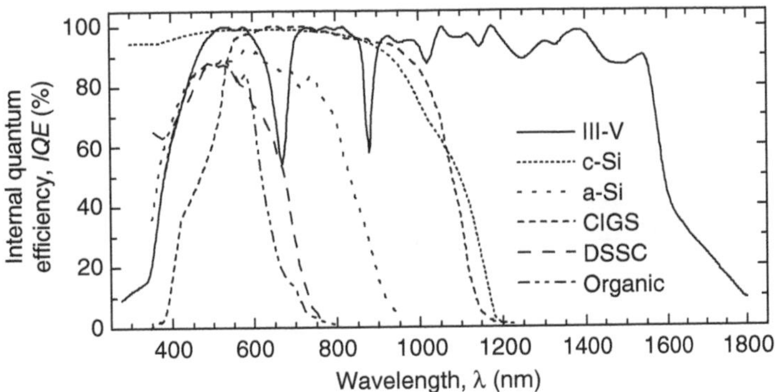

FIGURE 8.9 Comparison of the IQE of several different types of solar cells, including a III–V triple junction [Ref.], a high-efficiency crystalline silicon (c-Si), an amorphous silicon (a-Si) triple junction, a copper indium gallium selenide (CIGS) thin film on glass, a TiO_2 dye-sensitized solar cell (DSSC), and an organic thin-film device. (Adapted from Zhao, J. et al. 1998. *Applied Physics Letters.* 73, 1991–1993; Yang, J., Banerjee, A., and Guha, S. 1997. *Applied Physics Letters.* 70(22), 2975–2977; Contreras, M. A. et al. 1999. *Progress in Photovoltaics.* 7, 311–316; Grätzel, M. 2001. *Nature.* 414, 338–344; Brabec, C. J. 2004. *Solar Energy Materials and Solar Cells.* 83, 273.)

into silicon and providing more opportunity for absorption via light trapping. Silicon-wafer-based technologies have dominated the photovoltaic industry since 1954 and continue to do so today. The efficiency increase of silicon solar cells over time is shown in Figure 8.10, up to the present record of nearly 25%,[13] whereas the efficiency of PV modules are up to 16.8%.[18] Four common types of silicon are monocrystalline and multicrystalline wafers, as well as amorphous and polycrystalline thin films. Wafer-based technologies comprised nearly 91% of all solar cells manufactured in 2003, while amorphous silicon had the greatest production capacity for thin-film devices.[6]

8.4.1 Monocrystalline Silicon

Large (up to 300 mm diameter), cylindrical ingots of extremely pure, single-crystal silicon are grown from molten silicon by the Czochralski technique.[1] The entire ingot is doped, usually *p*-type with boron,

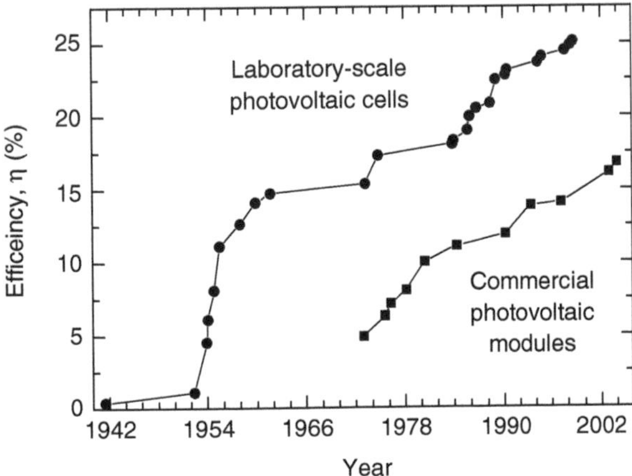

FIGURE 8.10 Increase in the efficiency of laboratory-scale silicon photovoltaic cells and commercial photovoltaic modules over time. Laboratory-scale devices are seen to be approaching the theoretical upper limit for a single-junction device (31%), whereas commercial modules lag further behind. (Adapted from Green, M. A. and Hansen, J. 2003. *Catalog of Solar Cell Drawings*, University of New South Wales.)

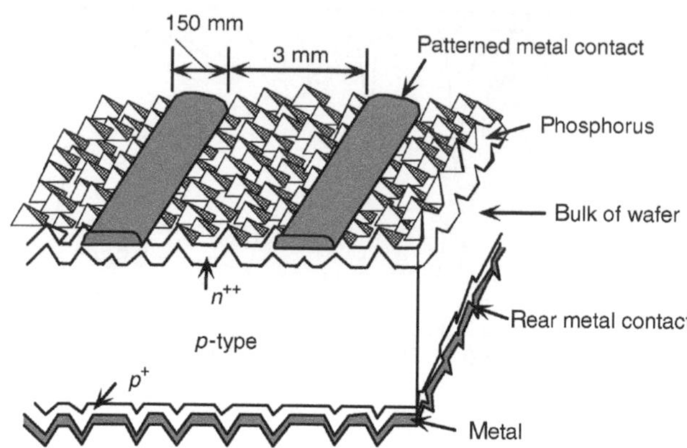

FIGURE 8.11 Schematic diagram of a screen-printed solar cell fabricated on a textured monocrystalline silicon wafer. (Adapted from Green, M. A. and Hansen, J. 2003. *Catalog of Solar Cell Drawings*, University of New South Wales.)

during the melt phase and is sawn into circular wafers of 300 μm thickness, from which solar cells are made. The most common photovoltaic technology produced on both monocrystalline and multicrystalline silicon is, by far, the screen-printed solar cell (as shown in Figure 8.11) that is produced by many different manufacturers. The basic cell fabrication process is as follows:[19] (1) surfaces are textured to reduce reflection losses and increase the light trapping within the wafer; (2) a heavily doped (n^{++}-type) emitter is diffused into the surface of the *p*-type wafer using a phosphorus source to form the *p–n* junction; (3) an antireflection coating is deposited on the front surface; (4) metal contacts on the front and rear are formed by the screen printing of a metallic paste; (5) and finally, the contacts are fired briefly to ensure good electrical contact to the silicon. The paste and firing regimes are designed such that they can be "fired-through" the thin antireflection-coating layer.

Screen-printed photovoltaic cells are typically 125×125 to $150 \times 150\ \text{mm}^2$ in size and exhibit conversion efficiencies of 13–15% under standard testing conditions (AM1.5, 25°C), producing just over 2 W_P at a voltage of about 0.6 V. To produce useful voltages and currents, 36 or 72 of these cells are soldered together and encapsulated under a sheet of glass to form weatherproof solar modules that produce about 75 or 150 W, respectively. In addition, the module protects the fragile silicon wafers from environmental and physical damage. A photovoltaic module typically operates at about 60°C, and at these temperatures the conversion efficiency is reduced by a temperature coefficient in the range -0.4 to -0.5%/°C.[1] Once encapsulated, typical screen-printed PV module efficiencies are 12–14%.

8.4.2 Buried-Contact and Point-Contact Cells

The buried-contact solar cell, shown in Figure 8.12, was developed in the mid-1980s at the University of New South Wales and offered several advantages over the screen-printed technology.[20] The primary difference is that the contacts on the front surface are plated into heavily doped (n^{++}-type) grooves that are formed by laser or mechanical scribing. This permits lower metallic shading of the surface, and the lower number of impurities required in the emitter results in a better response to short-wavelength light. This technology is manufactured by BP Solar in Spain, and is more popular for applications that require a higher power density.

SunPower Corporation (USA) produces point-contact silicon cells that have both the *n*- and *p*-type contacts located on the rear, leaving the front surface totally unshaded. This technology was originally developed for operating under highly concentrated sunlight with a conversion efficiency of 26.8%. However, a one-sun version of this solar cell, the A-300, has an efficiency of 21.5%.[21] SunPower has

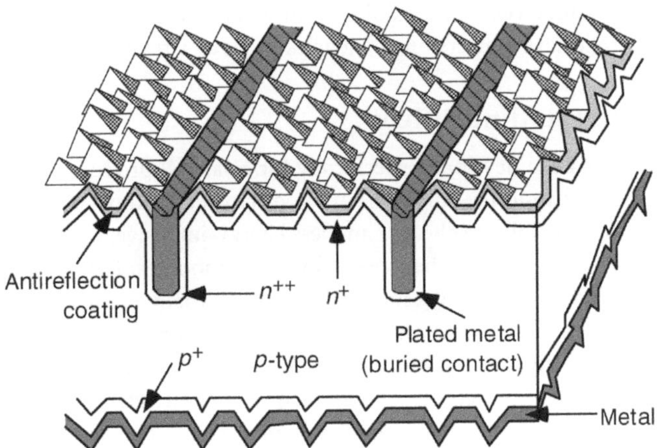

FIGURE 8.12 Schematic diagram of a buried-contact solar cell, indicating the reduced metal shading area and lightly-doped emitter that lead to reduced losses and enhanced conversion efficiencies. (Adapted from Green, M. A. and Hansen, J. 2003. *Catalog of Solar Cell Drawings*, University of New South Wales.)

recently begun shipping 210 W_p modules using the A-300 cells produced at their manufacturing plant in the Philippines, with a module efficiency of 16.8%.[18]

8.4.3 HIT

Sanyo's novel HIT (heterojunction with intrinsic thin-layer) cell,[22] shown schematically in Figure 8.13, combines both wafer-based and amorphous-silicon technologies (amorphous silicon is discussed in more detail below). Thin films of intrinsic (*i*-type) and doped amorphous silicon are deposited onto both sides of *n*-type monocrystalline silicon wafers, and an electric field is established across the *p–i–n* region, forming a heterojunction. The role of the *i*-type amorphous layers is to reduce recombination at the surfaces. Electrical contact to the highly-doped *n*- and *p*-type films is achieved via metal electrodes and a transparent conducting oxide (TCO) layer. Although the TCO and doped amorphous layers exhibit some parasitic absorption that does not result in a photogenerated current, this is largely compensated by the excellent passivation of the crystalline silicon by the amorphous. A further advantage of the HIT cell is that its negative temperature coefficient (−0.33%/°C) is smaller than most other silicon solar cells.

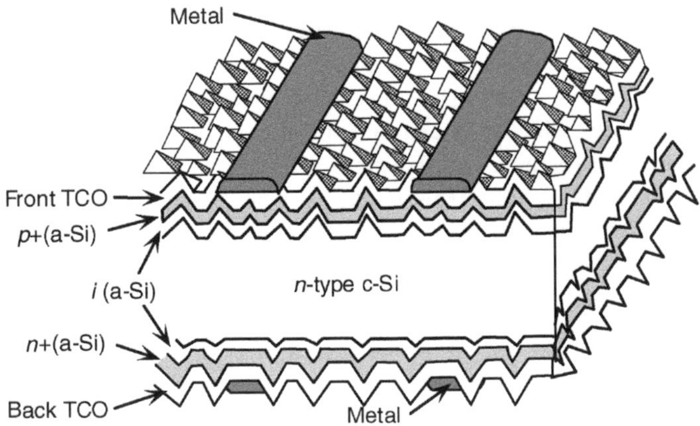

FIGURE 8.13 Structure of the Sanyo HIT cell. (Adapted from Green, M. A. and Hansen, J. 2003. *Catalog of Solar Cell Drawings*, University of New South Wales.)

Although large-area laboratory cells have exhibited an efficiency of 21%, Sanyo have recently begun shipping HIT modules that have conversion efficiencies of greater than 16%.

8.4.4 Cast Multicrystalline Silicon

Multicrystalline silicon has been developed specifically for solar cell production—a departure from the reliance of the photovoltaic industry upon the microelectronics industry. Cast wafers produced by cooling molten silicon in a crucible are cheaper and less energy-intensive, with the added advantage that they are square and can be packed together more closely in modules, leaving less unproductive space between. However, as the name implies, the material consists of many randomly oriented silicon crystals (called *grains*). This results in lower-quality material due to the poor electronic properties of boundaries between the grains, and a greatly reduced effectiveness of the standard wet-chemical etch used to texture silicon. For this reason, an antireflection coating such as silicon nitride that is also able to afford passivation to surface and bulk recombination sites is important. The screen-printed technology has adapted well to multicrystalline silicon wafers, achieving cell conversion efficiencies up to 15% and, in 2003, this product comprised nearly 62% of global photovoltaic cell production.[6]

8.4.5 Ribbon and Sheet Silicon

Several different techniques are used to avoid the wasteful sawing process and to directly produce silicon sheets of thickness suitable for photovoltaic cell production. Edge-defined film growth (EFG) cells and modules are produced by RWE Schott Solar (Germany). A large, hollow nonagon of silicon is drawn through a graphite dye, and wafers are cut from each side. The material is multicrystalline with large, elongated grains and the resulting cells have efficiencies of 13–15%. Evergreen Solar (USA) grows cell material by a string ribbon process, whereby molten silicon forms a film between two carbon strings as they are withdrawn from the crucible, achieving similar efficiencies. It is estimated that ribbon or sheet crystalline silicon cells accounted for about 4% of world production in 2003, continuing a decreasing trend over the previous 2 years.[23]

8.4.6 Other Notable Silicon Technologies

Several approaches have been developed to grow crystalline silicon films on substrates from which they are detached and the substrates reused. As the silicon wafer comprises more than half the cost of a completed module, these methods are aimed at dramatically reducing silicon usage. While none of these technologies are in production, the Sliver-cell concept, developed at the Australian National University in conjunction with Origin Energy, is entering the pilot-production stage.[24] In this technology, about 1000 "slivers" of silicon are formed by micromachining narrow grooves through each 1 mm thick silicon wafer, resulting in long, thin (less than 70 μm) strips that can be fabricated into efficient solar cells (up to 20%) and modules (currently at about 18% with selected material).[25]

Spheral cells are currently being developed by Spheral Solar Power (Canada) in partnership with Photowatt (France) and by Kyosemi Corp. (Japan). Small silicon spheres, each a tiny cell of 0.5–2 mm diameter, are interconnected in sheets. This results in a flexible and lightweight cell structure that uses significantly less silicon than conventional wafer cells, although the efficiency of these devices is significantly lower than other silicon-wafer PV technologies. The first flexible modules using this technology have recently been released by Spheral Solar Power and exhibit an efficiency of about 7%.

8.5 Second-Generation PV

The focus of thin-film (second-generation) technologies is to achieve a much lower cost per unit area than silicon wafer technologies, albeit at a lower efficiency. The four dominant technologies are based on the following inorganic thin-film materials: amorphous silicon, microcrystalline and polycrystalline silicon, cadmium telluride, and copper indium diselenide. Most of these films exhibit a direct bandgap and, therefore, a higher absorption coefficient than silicon; they therefore require a much thinner layer of active material. Organic materials are also being investigated for photovoltaics due to their potential high-throughput manufacture and low cost, estimated at about $1/$W_p$.[26] This section on second-generation PV devices therefore summarizes the latest developments of both inorganic and organic solar cells.

8.5.1 Inorganic Solar Cells

8.5.1.1 Amorphous and Microcrystalline Silicon Cells

There is no periodic arrangement of the silicon atoms in amorphous silicon (a-Si) and many atoms have bonds that are unsatisfied by neighboring silicon atoms. For this reason, hydrogen is incorporated into the material to passivate these "dangling" bonds and reduce recombination, forming hydrogenated amorphous silicon (a-Sigh). However, a light-induced degradation effect has severely restricted the early promise of this technology. Degradation has been minimized by making the cells as thin as possible—a typical film thickness is approximately 0.4 μm—and textured TCO surfaces have been regaining some of the absorption lost by thinning. As the bandgap of a-Sigh is higher than optimal (1.7 eV), another group-IV element, germanium, is often alloyed with the silicon.

To achieve a high hydrogen concentration, a-Si films are usually deposited by plasma-enhanced chemical vapor deposition (PECVD) at temperatures of about 200°C. Amorphous cells have a *p–i–n* structure, using a thicker intrinsic layer between the *p*- and *n*-doped layers due to its better electronic properties. The electric field extends across the intrinsic region, creating a thicker absorbing region for the generation and collection of carriers. Up to three junctions can be stacked in series for higher output, and triple-junction tandem modules are commercially available. The tandem approach provides the opportunity to have a different bandgap in each junction, thus giving a better match to the solar spectrum; this approach is discussed in greater detail in Section 8.6.

Amorphous modules have only a small negative temperature coefficient of efficiency (−0.1%/°C), making them less subject to output reduction due to heat, and they are more tolerant of reduced intensity and diffuse light. In addition, the low required deposition temperatures enable plastics and other substrates to be used to make flexible modules. They also lend themselves to use in semi-transparent building products and car sunroofs. The share of the photovoltaic market for a-Sigh has been declining over the years, from over 12% in 1999 to 3.4% in 2003.[6] Due to the light-induced degradation in these devices, the stabilized efficiency that is quoted takes these losses into account. The best reported stabilized efficiency for a triple-junction alloy device is 12.1%,[14] whereas commercial triple-junction modules are around 6% efficient.

PECVD is also used to deposit films of microcrystalline silicon (μc-Si), which results in the growth of columns of crystallites separated by amorphous regions. Again, hydrogen is necessary to passivate defects, this time at the grain boundaries. Hydrogenated microcrystalline silicon (μc-Sigh) exhibits similar optical and electronic properties to bulk silicon.

Kaneka (Japan) is using μc-Sigh films as an alternative to silicon/germanium alloys in hybrid structures with a-Sigh and has achieved a stabilized efficiency of 10% for a 39 W module.[27] The hybrid modules are also reported to be more stable than traditional a-Sigh modules. Particular measures are necessary to allow the amorphous layers to be kept thin enough to avoid light-induced degradation while producing similar current to the microcrystalline cell in series. The University of Neuchâtel (Switzerland) has developed a μc-Si:H/a-Si:H tandem (double-junction) design with around 11% efficiency on a laboratory scale.

8.5.1.2 Polycrystalline Thin-Film Silicon Cells

Various methods are being investigated to develop thin-film silicon cells deposited on foreign (nonsilicon) substrates. Most involve deposition and subsequent processing at high temperatures,[28] however CSG Solar (Germany), formerly Pacific Solar (Australia), is approaching commercial production with a process in which a thin-film silicon cell is formed on a textured glass superstrate at relatively low temperatures. Craters are formed through the active material by laser or ink-jet printing to contact the *n*-type layer closest to the glass. The concentration of active material is low as-deposited, but is then improved by subsequent thermal steps. The best efficiency achieved at the time of writing is 8.9% on a 100 cm^2 submodule, and work is in progress to improve the efficiency and scale-up the process for large-scale production.[29]

8.5.1.3 Cadmium-Telluride Cells

A cadmium telluride (CdTe) thin-film module is formed by taking a glass superstrate, depositing a TCO layer as front contact, an *n*-type cadmium sulfide (CdS) window layer, *p*-type CdTe, followed by a metallic rear contact. The contact films are scribed to divide the module into cells that are series-connected to produce the required voltage. Record CdTe thin-film cell and module efficiencies of 16.5% and 10.6% have been achieved,[30] respectively; however, commercial modules are generally less than 7% efficient. CdTe comprised 1% of global cell production in 2003, mostly for indoor use in consumer products. Potential restrictions on market growth are the toxicity of cadmium, limited global supply of tellurium, and the cell's sensitivity to moisture. The toxicity issue is controversial and affects public acceptance of the technology. Relatively small-scale production is underway in the United States and Germany, but production by major players in Japan and the United States ceased around 2002.

8.5.1.4 Copper-Indium-Diselenide Cells

Copper indium gallium diselenide ($Cu(In,Ga)Se_2$, abbreviated to CIGS (or to CIS if the gallium is absent) is another promising thin-film PV material. A molybdenum rear contact is deposited on a soda-lime glass substrate and patterned by laser to divide into cells. The elements of the CIGS absorber layer are evaporated simultaneously before a heterojunction is formed with a CdS layer. A TCO front electrode, mechanical scribing, contact formation, and encapsulation complete the module. Large grains of this material can be produced on foreign substrates, grain boundaries can be passivated, and the material is tolerant of deviations in concentrations of the constituent elements. Excellent efficiencies on laboratory-scale devices have proven difficult to transfer to reliable mass production. Several problems remain to be overcome, including the need for custom-built deposition equipment, the complexity of co-depositing many elements, demanding process control, difficulty in achieving homogeneous films over large areas, and moisture sensitivity of unencapsulated cells. Two other potential problems include the limited global supplies of indium, which may be a restriction on widespread use in the long term, and the toxicity of the cadmium, although this may possibly be avoided by use of alternative zinc-based compounds. Confirmed best cell and module efficiencies are currently 19.2% and 13.4%, respectively,[30] although commercial production in 2003 amounted to less than 1% of the world solar cell total. Flexible modules for portable, personal power provision is one high-value market being targeted by CIGS manufacturers.

8.5.2 Organic Solar Cells

Inorganic thin-film technology requires high-temperature processing in high-vacuum environments. Manufacturing costs are still quite high due to energy-intensive processes and the use of relatively large, expensive substrates that can withstand the high temperatures during processing. Organic materials are being investigated for PV due to their potential high-throughput, low-temperature manufacture and their subsequently lower cost, estimated at about $\$1/W_p$.[17] Organic solar cells are not restricted to batch processing onto rigid substrates and could benefit from existing roll-to-roll technology from an already-developed polymer/plastic industry. Flexible organic PV devices could find their way into textiles, labels and consumer packaging.

Organic materials, including conjugated polymers, dyes, and organic glasses, can show *p*-type or *n*-type semiconducting properties. These materials have very high absorption coefficients that make them ideal for thin-film PV technology. Film thicknesses on the order of only a few hundred nanometers would be required. The development of organic solar cells could benefit from already-available commercial organic photoconductive materials currently used in laser printing and organic light-emitting diode (OLED) technology.[31] Unfortunately, organic PV devices still have very low power conversion efficiencies and stabilities (compared with inorganic PV devices) and, as a result, are not currently commercially viable. Their performance is limited by poor absorption of longer-wavelength light, poor charge transport, and low stability.

Organic PV technology has improved over recent years with the maturation of innovative organic materials and device structures. The development of new materials, however, will be required to produce efficiencies close to inorganic thin-film PV technology (~ 10%). It is believed that with the optimization of already-available materials, conversion efficiencies greater than 5% could be achieved.[32]

8.5.2.1 History of Organic PV Development

In the 1950s and 1960s, it was discovered that many common organic dyes had semiconducting properties; these dyes were consequently were among the first organic materials to exhibit the PV effect. By utilizing the first step of the biological process of photosynthesis (involving the absorption of light), many biological molecules were also used to demonstrate the PV effect. These early devices were rather simple in construction, utilizing only a single organic homojunction layer sandwiched between two electrodes. The electrodes have different work functions and provide an electric field that can separate the photogenerated exciton; the fields also aid the drift of these carriers towards the appropriate contacts. Conversion efficiencies between 0.01% and 0.1% are typical for such devices. In 1978, an impressive 0.7% power conversion efficiency was obtained from a single-layer device using merocyanine dye sandwiched between a metal–metal oxide and a metal electrode, enhancing the Schottky-barrier effect within the device.[33] In the early 1980s, the first conjugated polymers were investigated in PV single-layer cells; however, these exhibited conversion efficiencies below 0.1%. A major breakthrough occurred in 1986, when Tang published the concept of a bilayer heterojunction device that incorporated a donor–acceptor interface.[34] Tang demonstrated conversion efficiencies of around 1% for simulated AM2 illumination.. In the mid 1990s, the next development occurred with the introduction of a bulk heterojunction (electron donor and acceptor are blended together). Conversion efficiencies between 3% and 4% have been obtained from at least three different device structures based on donor–acceptor systems,[35] and efficiencies close to 5% are currently obtainable using bulk polymer–fullerene heterojunction devices.[17]

8.5.2.2 Commonly Used Organic Materials for PV Applications

Early single-layer devices utilized organic dyes and molecules—for example, methylene blue, anthracene, chlorophylls, and carotenes. Polymers, including polyacetylene and poly(sulfur nitride), were found to be highly conductive in the presence of certain dopants and were used in the 1980s in single-layer devices, but efficiencies did not noticeably improve. The use of polymers for PV applications has significant advantages (for example, long shelf/operating lifetimes and larger dielectric constants) compared with organic materials.[36] Tang used a phtalocyanine derivative as a *p*-type semiconductor and a perylene derivative as an *n*-type for his record-breaking bilayer heterojunction device.[34] His concept of using a heterojunction has since been investigated in a large number of donor–acceptor cells, including dye–dye, polymer–dye, polymer–polymer and polymer–fullerene. The organic fullerene C_{60} molecule has a high electron affinity (required for an *n*-type material), is relatively transparent, and has a good electron conductance. Polymer–fullerene devices currently dominate bilayer and bulk heterojunction device research.

8.5.2.3 Organic Solar-Cell Design

Both PV and LED structures are similar and operate using reverse principles: a PV device "uses" light and "creates" electricity and an LED "uses" electricity and "creates" light. Organic solar cells and OLED's have a similar basic diode structure, as do inorganic LEDs and conventional PV structures. The basic organic diode structure consists of four primary layers: a transparent superstrate (typically glass or plastic); a semi-transparent electrode that can be the same TCO as used for inorganic thin-film solar cells, such as indium–tin–oxide (ITO), or a thin metal layer; an organic material; and a rear metal electrode (usually Al, Mg, Ca, or Au), as illustrated in Figure 8.14. The organic material sandwiched between the electrodes includes either a single layer of a homogeneous material (homojunction), a double layer using two different materials (bilayer heterojunction), or a mixture of two materials (bulk/blend or dispersed heterojunction). A fourth type of layer based on both the bilayer and bulk heterojunction is called the *diffuse-layer heterojunction*.

8.5.2.4 Homojunction Device Operation

The operation of a single-layer organic PV device can generally be divided into four stages, as shown in Figure 8.15:

Light absorption (①). Photons are absorbed within an organic molecule and an electron is excited to the highest occupied molecular orbital (HOMO) from the lowest unoccupied molecular orbital (LUMO), thereby forming an exciton (a bound electron–hole pair). In crystalline inorganic semiconductors, the individual LUMOs and HOMOs form a conduction and valence band, respectively, throughout the material. This is not possible for most organic semiconductors because the intermolecular forces are too weak to form structured lattices. Instead, charge transport occurs via hopping between localized states. Charge-carrier mobilities are therefore low because no "bands" exist for the transport of the charge carriers through the device. Furthermore, the high degree of disorder within organic solids also contributes to the low mobilities. In an inorganic device, absorption of a photon leads to free electrons and holes that are not bound to each other. At room temperature, excited organic semiconductor electrons are bound to the hole due to the low dielectric constant of the organic material.

Exciton diffusion (②). Excitons cannot directly contribute to a photocurrent and must be separated into electrons and holes that then need to make their way to the appropriate electrode. Diffusion of the exciton occurs until it reaches a contact; however, excitons can only diffuse a few nanometers before they decay and the charges are lost. Consequently, only the layer of organic material that lies within an exciton diffusion length (1–10 nm) of a contact may contribute to the photocurrent.

Charge separation/dissociation (③). The rear metal contact (anode) has a high work function, ϕ_a, than that of the TCO (cathode), ϕ_b. The difference in the work function creates an electric field through the organic material. If the electrodes are connected to the cell and the device is short-circuited, the Fermi levels of the electrodes will align, thereby skewering/distorting the HOMO and LUMO within the molecule. The shape of the distorted HOMO and LUMO is dependent upon the conductance of the organic material. Figure 8.16b shows the skewering of the HOMO and LUMO for the case when the organic material acts as an insulator. It can be seen that the field profile changes linearly throughout the organic material. In this case, the electric field is not usually strong enough to break up the photogenerated exciton within an organic material and only aids the drift of the separated carriers. In general, organic materials (especially most polymers) are hole-conducting (p-type). Generated holes

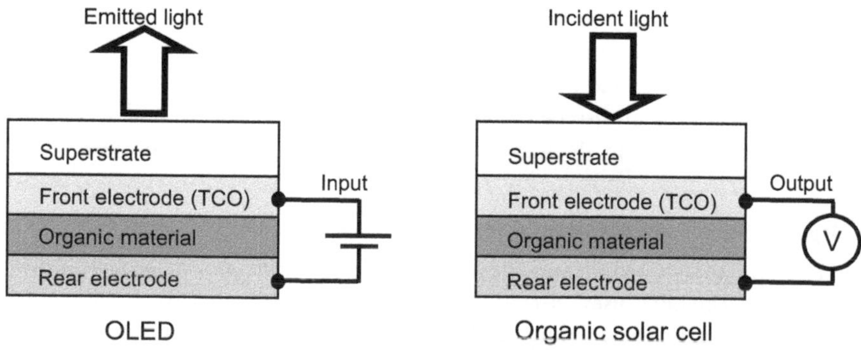

FIGURE 8.14 Basic OLED and organic solar-cell operational structures. The transparent substrate is usually glass with the organic material usually sandwiched between transparent TCO (front) and a metal (rear) electrodes.

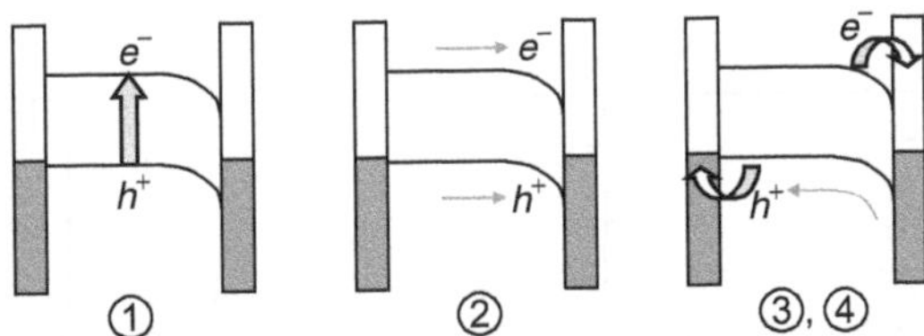

FIGURE 8.15 ① Light absorption, ② exciton diffusion, ③ charge separation/dissociation, and ④ charge transport and collection, as described in the text.

redistribute through the hole-conducting material, "flattening" the HOMO and LUMO near the higher work-function electrode and forming a Schottky junction (see Figure 8.16c). Orbital/band bending occurs close to the lower work function, creating a depletion region, *W*. The electric field is strong enough in this region to separate the charges of the exciton. Therefore, it is desirable that the exciton should diffuse toward the lower work-function electrode.

Charge transport and collection (④). The separated carriers drift to the appropriate contacts due to the presence of the internal electric field. If they reach the electrode and do not recombine, they can be collected and used to supply a direct current to a load.

8.5.2.5 Heterojunction Device Operation

Figure 8.17a shows the basic device structure for a bilayer heterojunction. Two different materials are used, each with different electron affinities and ionization potentials. A donor material is able to "donate" electrons (hole-condcucting *p*-type), whereas an "acceptor" is able to accept electrons (electron-conducting *n*-type). A localized field exists (denoted by E_2) in Figure 8.17a at the heterojunction interface (similar to a *p–n* junction in conventional inorganic solar cell). This localized electric field dissociates excitons much more efficiently than at the electrode interface. Similar to the standard bilayer heterojunction, the diffuse bilayer heterojunction also provides a spatially uninterrupted pathway for

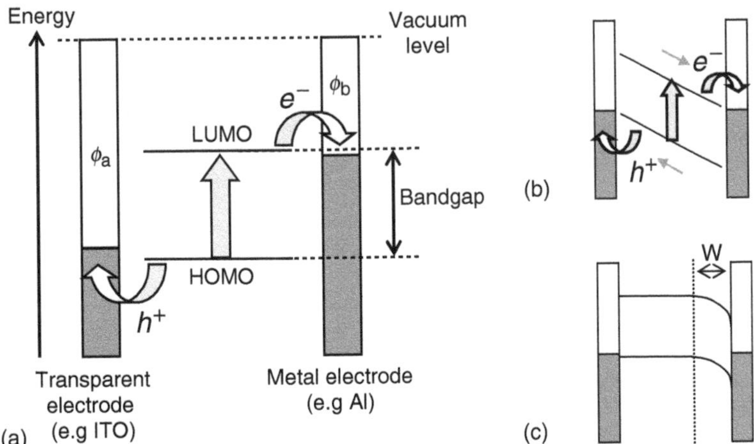

FIGURE 8.16 (a) Energy band diagram within an isolated organic PV structure. Absorption of a photon (up-arrow) excites an electron from the HOMO to the LUMO. The excited electron and the associated hole must then be collected at the cathode and anode, respectively. The electrodes in this diagram are not in contact with organic material and the Fermi levels are not aligned. (b) The electrodes are now attached to the solar cell, aligning the Fermi levels and skewering the HOMO and LUMO levels. The organic layer is presumed to be an insulator. (c) A more realistic Schottky organic device, indicating the formation of a depletion region, *W*, near the cathode. In this case, the organic material is hole-conducting (*p*-type).

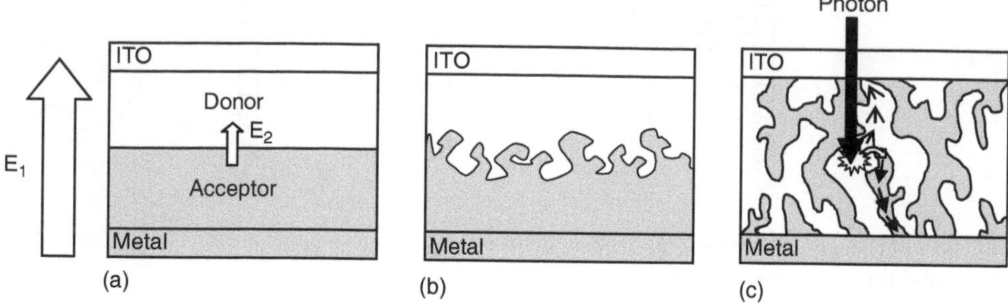

FIGURE 8.17 (a) Device structure for a bilayer heterojunction device. The electric potential created due to the difference in workfunction, E_1, is used for separated charge transport. The localized electric field is used for exciton dissociation. (b) Device structure for a diffuse heterojunction device. These can be created by lamination, spin coating, or annealing techniques. (c) Device structure for a bulk heterojunction device. A photon is absorbed and an exciton is formed. This exciton is dissociated at the donor–acceptor interface. Holes are then transported (up arrows) to the top ITO electrode through the *p*-type donor material and electrons are transported (down arrows) to the bottom metal electrode through the *n*-type acceptor material.

the separated charge carriers, as seen in Figure 8.17b. One advantage of the diffuse bilayer structure compared to the standard bilayer is that it has a larger donor–acceptor interface where excitons can be dissociated. The bulk heterojunction structure takes this idea one step further (refer to Figure 8.17c). In a bulk heterojunction, the donor and acceptor materials are mixed with the aim of creating a donor–acceptor interface less than the exciton diffusion length from a photon-absorbing site. The diffuse-layer heterojunction is often used for polymer–polymer devices because it is difficult to control the mixing of polymers in a bulk device.

8.6 Third-Generation PV

In the first- and second-generation technologies described above, the choice of bandgap energy of the semiconductor from which a photovoltaic cell is made establishes a fundamental upper limit for its conversion efficiency. One major loss mechanism in single-junction solar cells is lattice thermalization, whereby the photoexcited pair quickly loses energy in excess of the bandgap, illustrated by process ① on a typical *p*–*n* junction band diagram in Figure 8.18. This energy is lost as heat within the device

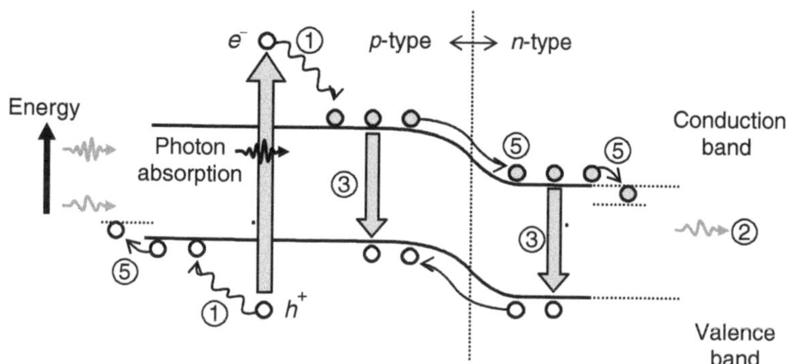

FIGURE 8.18 Loss processes in a standard solar cell: ① lattice thermalization loss; ② transparency; ③ recombination loss, ④ junction loss, and ⑤ contact-voltage loss. (Adapted from Green, M. A. and Hansen, J. 2003. *Catalog of Solar Cell Drawings*, University of New South Wales.)

for low-bandgap materials. This allows a larger fraction of photons in sunlight to be absorbed, thereby allowing the generation of a potentially higher current. A solar cell can only extract an amount of energy from each photon that is about the same as the bandgap. This means that a low-energy red photon is just as effective as a much higher-energy blue photon in the creation of an electron–hole pair, and the extra unutilized energy is dissipated as heat. The other major loss mechanism is the transparency of photons with energy less than the bandgap of the semiconductor (see ② in Figure 8.18). These two losses alone limit the conversion efficiency of a solar cell to about 44%.[9] Therefore, the best single-junction solar cell materials have a bandgap that gives a compromise between these two effects.

Another important loss mechanism is the recombination of photoexcited electron–hole pairs (see ③ in Figure 8.18). The impact of this parameter can be kept to a minimum by using semiconductor material with appropriate properties, including high lifetimes for the photogenerated carriers. The performance of photovoltaic cells is also governed by the laws of thermodynamics. Shockley and Queisser demonstrated in 1961, by using the principle of detailed balance between incident and escaping photons and extracted electrons, that the efficiency limit for a single-material cell is around 31% for an optimal bandgap of around 1.3 eV for a 1-sun spectrum. This assumes that the only unavoidable losses from the device are the emission of photons produced by radiative recombination, and includes voltage drops across the contacts and junction (loss mechanisms ④ and ⑤ in Figure 8.18).

In determining the upper efficiency limits of a photovoltaic device, the sun is assumed to behave as a black body at 6000 K, whereas the cell is usually assumed to be another perfect radiator at 300 K (27°C). The Carnot limit for solar conversion under these temperatures is 95% (assuming an infinitesimal electrical output) because the cell would need to emit light at the same spectrum and intensity as the sun to avoid entropy production in the energy transfer. The Landsberg efficiency for direct sunlight conversion (93.3%) is the ultimate solar conversion efficiency and results from the assumption that the cell generates no entropy during absorption and emission of light; this efficiency, however, is lower than the Carnot limit because the light emitted by a cell is considered lost, not returned to the sun for later use. Within this framework, a range of different solar cell concepts are theoretically able to achieve efficiency limits ranging from 31% to 68.2% for one and an infinite number of bandgaps, respectively, operating under unconcentrated sunlight. Under maximum solar concentration of around 46,000 times, these values increase to 40.8% and 86.8%, respectively.[9]

8.6.1 Tandem and Multijunction III–V Cells

Tandem cells are one approach to exceeding the efficiency limits of single-material cells. They reduce the two main losses mentioned earlier: the thermalization of the excess energy of high-energy photons and transparency to low-energy photons. Tandem cells are stacks of *p–n* junctions, each of which is formed from a semiconductor of different bandgap energy. Each responds to a different section of the solar spectrum, yielding higher overall efficiency. The cells are stacked in order of decreasing bandgap, such that the light is automatically filtered as each cell extracts photons that exceed its bandgap. This technique is commercially used to fabricate triple-junction amorphous cells.

Tandem cells are also fabricated from compounds of elements from groups III and V of the periodic table, such as gallium arsenide (GaAs). GaAs is used extensively in the optoelectronics and electronics industries and may be used to make excellent, but expensive, solar cells. High-efficiency photovoltaic cells based on GaAs are produced almost exclusively for space applications. Certain elements used in these devices, such as gallium and indium, are not abundant, implying high cost and a limited ability to supply a large fraction of the global energy needs in terrestrial applications. There have long been plans for extensive terrestrial use in solar concentrators, where the concentration of the light from a large collection area onto a cell may be able to justify such expense, but this has not yet occurred to a significant extent. The best terrestrial cell efficiency is currently 27.6% for a single-junction GaAs cell under 255 times concentration.

The use of crystalline III–V compounds to fabricate high-efficiency tandem devices presents several challenges. The materials used in the various junctions must be compatible with each other over a wide

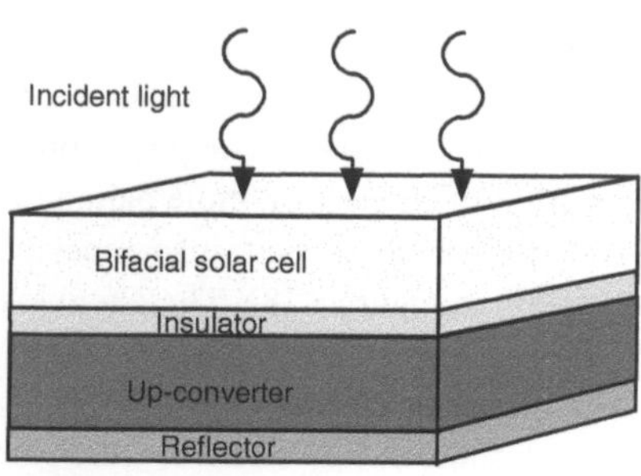

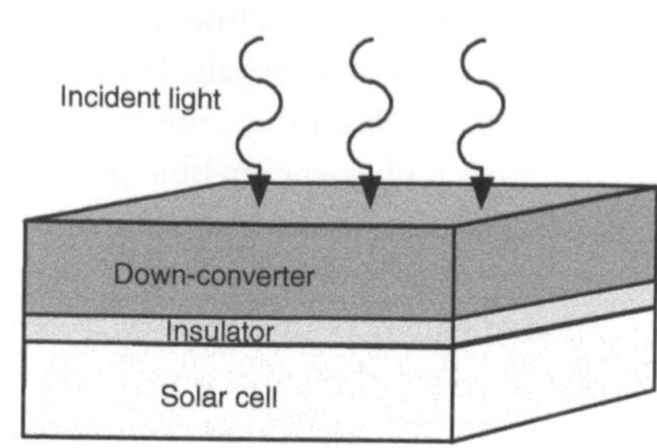

FIGURE 8.19 Schematic diagram of (left) a bifacial solar cell with an up-converter placed at the rear to capture sub-bandgap light, and (right) a down-converter placed at the front of a solar cell to convert each high-energy photon into multiple low-energy photons. (Adapted from Trupke T., Green, M. A., and Würfel, P. 2002. *Journal of Applied Physics*, 92(7), 4117–4122; Trupke T., Green, M. A., and Würfel, P. 2002. *Journal of Applied Physics*, 92(3), 1668–1674.)

temperature range. In particular, the thermal expansion coefficients should be similar (lattice matched) and the bandgaps should be a good match to the spectrum. The overall record cell efficiency of 36.9% is for a triple stacked GaInP/GaAs/Ge tandem cell operating under 309-times concentration.[30,37] while researchers at the Fraunhofer Institute for Solar Energy (Germany) developed the first five-junction tandem cell.

8.6.2 Luminescence Up- and Down-Conversion

The transparency and thermalization losses of single-junction solar cells can also be reduced by using passive optical devices called up- and down-converters, respectively. Down-conversion (DC) results in the generation of more than one photon with energy just greater than the semiconductor bandgap (E_g) being generated per incident high-energy photon, while up-conversion (UC) generates one photon with energy greater than E_g for every two or more sub-bandgap photons absorbed. Diagrams of a solar cell in combination with an UC and DC are presented in Figure 8.19.[38,39]

The authors have recently reported application of UC layers based on rare-earth (RE) luminescence centers contained within polymer sheets.[40,41] Erbium-doped (20%) sodium yttrium fluoride ($NaYF_4:Er^{3+}$) phosphors were mixed into an acrylic-based binder at a concentration of 30 wt% and placed on the rear of bifacial silicon solar cells with a front and rear side efficiency of 15% and 12%, respectively. Reflective white paint was applied to the rear side of the of UC layer, as shown in Figure 8.20. Results indicate that the peak EQE of the solar cell/UC system in the range $\lambda = 1480$–1570 nm is 2.5%,[41] as shown in Figure 8.21. The IQE of the luminescence process, however, is

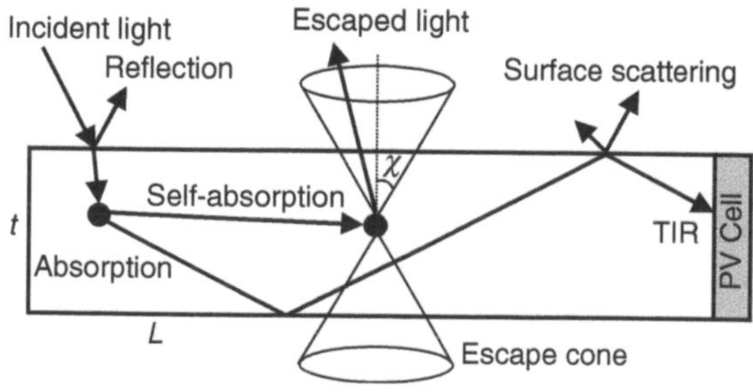

FIGURE 8.20 Cross-sectional diagram of a traditional LSC with luminescence centres (●) and PV cells attached to the edge. Light is transported to the edge via TIR, incurring some losses along the way, e.g., self-absorption, scattering, and the escape of light emitted within the critical angle (χ).

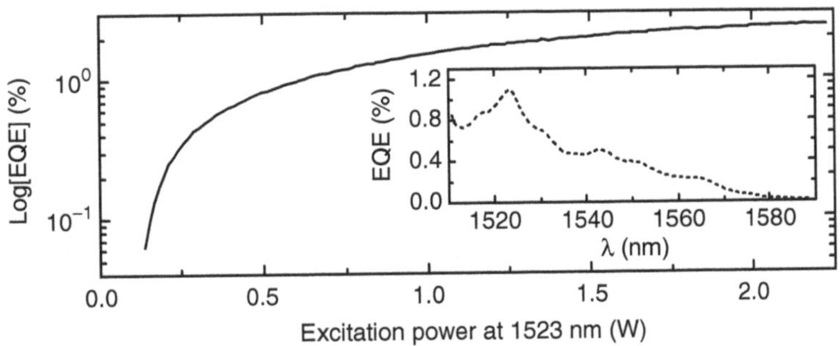

FIGURE 8.21 EQE of the silicon solar cell and rear UC layer at 1523 nm, and (inset) the spectral response over the $\lambda = 1510–1590$ nm range.

estimated to be closer to 4% due to various optical losses in the system and the Si solar cell response. Further improvements are expected to be made by using: (1) polymer hosts with a higher refractive index and lower absorption coefficient in the NIR; (2) higher efficiency bifacial solar cells; and, (3) a higher RE ion concentration. An examination of possible experimental mechanisms for the DC of the highest-energy photons in terrestrial sunlight (UV to blue) into two NIR photons ($\lambda < 1200$ nm) did not yield very promising results.[42]

A final luminescence process, down-shifting (DS), is similar to DC except that the EQE of the DS process is less than unity. The most widely researched application of DS to silicon PV has been for luminescent solar concentrators (LSC)[43] because they can potentially generate 20% more electrical energy per day than a PV device of the same peak watt (W_p) rating at a lower cost per watt-peak ($\$/W_p$). While the DS process exhibits a sub-unity EQE, the geometric ratio of a large thin sheet ($G = A_{\text{front surface}}/A_{\text{perimeter}}$) results in the concentration of the majority of the isotropically-emitted luminescence at the edges via total internal reflection (TIR), as shown in Figure 8.21. The lower cost potential arises due to a much smaller number of silicon solar cells required to cover the perimeter, as opposed to the incident area, of the sheet. In addition, the LSC is a static device and is able to concentrate diffuse light without tracking the sun.

8.7 Other Technologies

8.7.1 Dye-Sensitized Cells

In 1991, O'Regan and Grätzel published an article in *Nature* describing the development of a new dye-sensitized nanocrystalline photovoltaic cell.[10] These cells are now commonly referred to as *Grätzel cells* and are fundamentally different from the others discussed here in that they do not rely on semiconductor *p–n* junctions. Instead, they are electrochemical devices in which the optical absorption and carrier-collection processes are separate. As shown in Figure 8.22, a porous film of a wide-bandgap semiconductor, usually titanium dioxide (TiO_2), is coated with a redox charge-transfer dye. Figure 8.23 shows the basic operating and chemical principles of a Grätzel cell using TiO_2, an organic dye, and an iodide solution as an electrolyte. The dye is excited by the absorption of a photon (a), an electron is injected into an excited energy level of the oxide (b), and the dye is regenerated by a liquid electrolyte (c). The excited electron in the oxide reaches an electrode and is then able to migrate through an external circuit, perform useful work in the load, then pass to the other electrode where it regenerates the electrolyte (d). The thick TiO_2 film is deposited onto TCO-coated glass that forms the front contact. They work despite the high defect densities at the interface because of the very fast (picosecond) spatial separation of electrons from ions at the dye–TiO_2 interface; this makes them effectively majority-carrier devices.

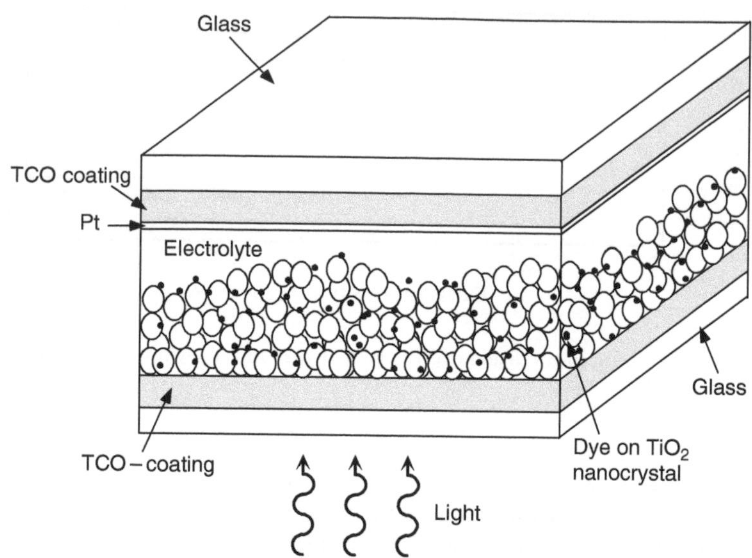

FIGURE 8.22 Structure of the nanocrystalline dye-sensitized solar cell. (Adapted from Green, M. A. and Hansen, J. 2003. *Catalog of Solar Cell Drawings*, University of New South Wales.)

The underlying operating principles of Grätzel cells are not considered difficult. Simple devices using natural dyes from berries can quite easily be made and used to demonstrate chemical and physical topics that are taught at a senior-high-school level.[44]

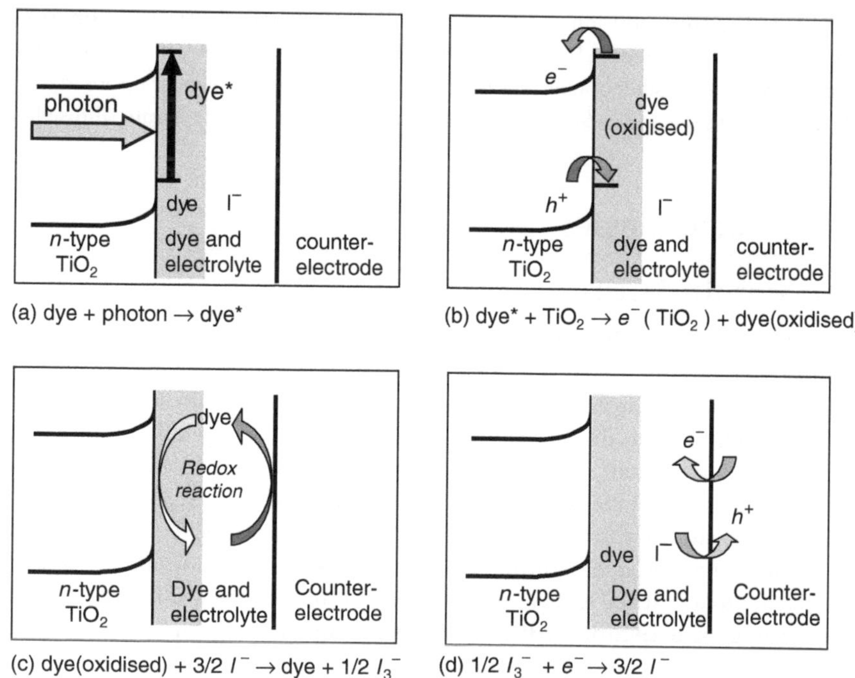

FIGURE 8.23 Operating and chemical principles of a Grätzel cell. (a) Photons excite the dye. (b) Electrons are transferred to the *n*-type matieral from the dye where they can be collected by an electrode. (c) The dye is regenerated (redox reaction) by the electrolyte. (d) An electron from the counter-electrode regenerates the electrolyte.

The best-demonstrated efficiency for a laboratory device is 11%,[45] and reasonable stability under illumination has been demonstrated. Module efficiencies using dye-sensitized cells are similar to those of commercially available thin-film amorphous silicon modules. Dye-sensitized cells are coming into small-scale commercial production in Australia and the U.S., and the installation of the first wall panel has been performed in Australia. Dye-sensitized cells exhibit promise as potentially low-cost cells, in terms of both finance and energy, with the added advantage that their temperature coefficient can be positive, increasing the output power at elevated operating temperatures. The main challenges are to improve high-temperature stability, scale-up production, and, preferably, replace the liquid electrolyte with a solid. Tandem dye-sensitized cells are particularly aimed at the direct splitting of water for hydrogen production rather than at electricity production. In one experimental system, yielding 4.5% conversion efficiency from sunlight to chemical energy, two semiconductors, TiO_2 and WO_3 or Fe_2O_3, are used, with water being the electrolyte.

8.8 Photovoltaic System Applications

Photovoltaic modules are used in many cost-effective applications today, supplying both direct current (DC) and alternating current (AC) electrical loads. For supplying DC electricity, a photovoltaic system usually consists of the solar panel itself, an appropriately sized battery to supply a continuous load, and the necessary control electronics to keep the photovoltaic cell operating at its maximum power point and maintaining the battery state of charge. In AC systems, an inverter is required to transform the DC output of the solar panel to 120 or 240 V AC. The final possible system component is a solar tracker that mechanically moves the solar panels through one-axis or two-axes to keep them normal to the incident sunlight, thus maximizing the output power. The increase in system complexity and the decrease in reliability means that these trackers are not often used. The primary applications of photovoltaic cells is described below.

Many of the highest-efficiency photovoltaic cells produced over the years have been destined for space applications, especially for powering satellites. In this market, high specific power (kW/kg) and hardness to cosmic radiation are more important than price. High efficiency is important because if fewer devices are required to power a satellite, then more mass can be used for scientific instruments. A major advantage of space operation is the constant presence of the sun, however the cells are not protected by the Earth's atmosphere from high-energy particles and radiation; they consequently have a life expectancy of only about 7 years. Two recent applications of photovoltaic cells in such applications are the Mars rover and an unmanned high-altitude solar plane that is intended to replace satellites.

The first terrestrial markets to use solar cells were typically located in remote areas and capitalized on the weatherproof nature of photovoltaic modules. Applications included powering navigation buoys, lighthouses, repeater stations for telecommunications, and weather-monitoring stations. Powering such applications from photovoltaics was economical due to the high cost of replacing batteries in these systems, the low power consumption (10–100 W) of the load, and the high system availability, or low loss-of-load probability (LOLP) required. Where a low LOLP was critical, up to 15 days of battery storage was included in the system. Another traditional market for solar cells include powering cathodic protection systems to provide an electrical current that counteracts the natural electrochemical currents that cause corrosion, such as oil and gas pipes.

Screen-printed monocrystalline silicon cells were used in the majority of the applications mentioned above. For many years, amorphous silicon solar cells have been used in consumer products such as calculators and watches. Using the photovoltaic technology obviates the need for batteries, as these devices can operate under extremely low (indoor) light levels.

Photovoltaic cells have played a role in bringing power to some of the estimated 2 billion people in the world that do not have access to electricity. Solar home systems (SHS) are specifically designed for application in developing countries, and typically include one 50 W solar panel, a battery, a charge controller, and some light emitting diode (LED)-based lamps. In addition, owners might typically add

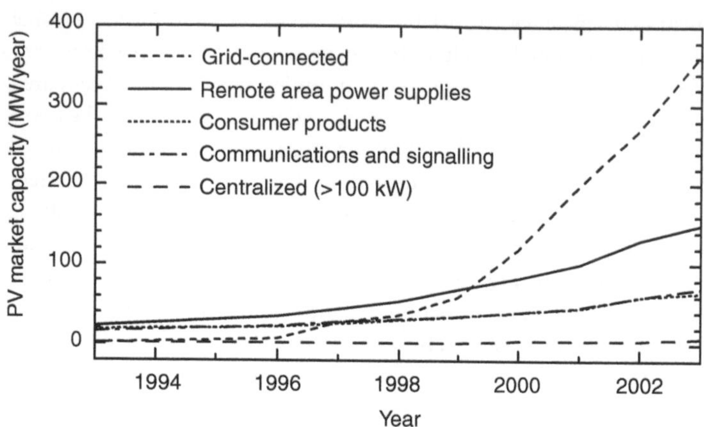

FIGURE 8.24 Breakdown of the PV market by application for the 10 years 1993–2003. (Data from Maycock, P. 2004. *Renewable Energy World*, July–August, 86.)

a radio or black and white television to the system. By providing light in the evenings, children are able to be better educated, adult community members can work to earn additional income for their family, and air pollution from kerosene lamps is diminished. Street lights that include an integrated solar panel are becoming more common in both developing and developed countries. In the latter case, a photovoltaic power light becomes more economical than a grid-connected light as soon as the necessary trenches for cables are more than a few meters in length.

Photovoltaic systems are used to power water pumping systems in a remote community, to bring water from the local river or well. This water is either used for human consumption or irrigation. Where the water quality is too poor to drink, photovoltaics have been used to power the pumps in desalination systems that force the water through membranes to filter out the unwanted particulates, bacteria, viruses, and salts.[7] The source water for such systems is typically brackish, but the desalination of seawater is commonplace in the Middle East.

The biggest growth market for photovoltaics over the past few years are for on-grid residential rooftop systems, as illustrated in Figure 8.24. On-grid systems do not require batteries as the electricity grid is used as "storage," with excess power being exported to the grid during the day, and bought back at night as required. The installation of such systems is being driven by astute policy-making decisions in places such as Japan, Germany, and California, with rebates for offsetting installation costs or attractive buyback rates for generated electricity being offered. Although they have yet to become a common sight in our everyday lives, photovoltaics have been used to power many vehicles, including solar boats and racing cars.

The final measure of the performance of a PV module or system is its energy yield ratio (EYR)—the ratio of the electrical energy generated over the 25-year lifetime of a PV module to the energy used in the fabrication of a PV module from raw materials. The EYR of PV modules is high; today's PV modules generate between 5.6 to 12.4 times more energy than went into their fabrication.[7] The EYR is not significantly reduced for a rooftop grid-connected PV system because there is minimal framing and the embodied energy of the inverter is low. However, for stand-alone PV systems that rely on battery storage, the EYR is significantly lower, although the value is still nearly three times greater—even for a region with a poor solar-radiation resource.[7]

8.9 Conclusions

Since the mid-1950s, wafer-based silicon solar cells have dominated the photovoltaic market. Although the production of this first-generation technology is increasing each year, silicon feedstock and photovoltaic devices have also matured, with a move away from monocrystalline wafers used by the semiconductor to more economic and less energy-intensive multicrystalline silicon wafers. This has presented to industry the second generation of thin-film technologies with a moving target. Although the latter technologies that are based on thin films of amorphous, microcrystalline, or polycrystalline semiconductors deposited onto cheap substrates or superstrates offer a lower cost per area, the efficiency of these devices is also lower and therefore the cost per peak-watt generated is similar. Several scientific, materials engineering, production, and environmental issues have hampered the market penetration of the thin-film cells. These challenges must be overcome before second-generation technologies pose a serious threat to the huge scale of production that is being constantly increased with first-generation technologies. The most prominent second-generation technology is amorphous silicon, with an established, but decreasing, market share. Polycrystalline silicon, which has very similar properties to bulk silicon wafers, may eventually come to dominate the second generation of solar-cell production. Finally, the third generation of photovoltaic cells, currently represented by the tandem solar-cell structure, seeks to make more efficient use of the solar spectrum by reducing the thermalization and sub-bandgap losses inherent in first- and second-generation devices.

Although their development has been slow, organic semiconducting materials offer significant advantages compared with traditional inorganic semiconducting materials in terms of cost and versatility. Organic PV devices may not initially be able to compete with large-scale inorganic PV modules that offer higher efficiencies and longer operating lifetimes. Initial organic PV devices may initially find their own niche market in smaller consumer-based products. Organic electronic devices have only recently become commercially available (for example, OLEDs) but have shown that they can be a cost-effective alternative to the dominant silicon semiconductor industry. With increasing scientific and commercial interest, the potential for organic PV has become increasingly realizable.

For many years, a large traditional market for solar cells was for powering systems in remote areas where no electricity grid was present. Such systems include health clinics, refrigeration units, weather and other monitoring stations, navigation aids, and telecommunication repeater stations. Photovoltaic cells are also an economic substitution for powering many small items of equipment that traditionally require batteries. The photovoltaic industry has experienced an average growth of 35% over the past 7 years and, buoyed by current strong policy in Germany and Japan and many new emerging markets, it appears that solar cells face a bright future.

Acknowledgments

The Centre of Excellence for Advanced Silicon Photovoltaics and Photonics is supported by the Australian Research Council's Centres of Excellence Scheme.

References

1. Green, M. A. 1986. *Solar Cells: Operating Principles, Technology and System Applications*. University of New South Wales, Sydney, Australia.
2. International Energy Agency. 2003. *Key World Energy Statistics*. International Energy Agency, Paris.
3. Van der Veer, J. 1997. *Shell International Renewables*. Press Release from Shell International, London, October.
4. Green, M. A. 2000. *Power to the People: Sunlight to Electricity using Solar Cells*. University of New South Wales Press, Sydney, Australia.
5. Archer, M. D. and Hill, R., eds. 2001. *Clean Electricity from Photovoltaics.*: Imperial College Press, London.
6. Maycock, P. 2004. PV market update. *Renewable Energy World*, July–August, 86.
7. Richards, B. S. and Watt, M. E. 2007. Permanently dispelling a myth of photovoltaics via the adoption of a new net energy indicator. *Renewable and Sustainable Energy Reviews*, 11, 162–172.
8. Green, M. A., 2004. Thin-film photovoltaics, In *Advances in Solar Energy*, K. W. Böer, ed., pp. 187–214. Boulder, American Solar Energy Society, CO.

9. Green, M. A. 2004. *Third Generation Photovoltaics: Ultra-high Efficiency at Low Cost.* Springer, Berlin.
10. O'Regan, B. and Grätzel, M. 1991. A low-cost, high efficiency solar cell based on dye-sensitized colloidal TiO_2 films. *Nature*, 353, 737–740.
11. Richards, B. S. 2004. Comparison of dielectric coatings for buried-contact solar cells: A review. *Progress in Photovoltaics*, 12, 253–281.
12. Richards, B. S. 2003. Single-material TiO_2 double-layer antireflection coatings. *Solar Energy Materials and Solar Cells*, 79, 3, 369–390.
13. Zhao, J. et al. 1998. Novel 19.8% efficient 'honeycomb' textured multicrystalline and 24.4% monocrystalline silicon solar cell. *Applied Physics Letters*, 73, 1991–1993.
14. Yang, J., Banerjee, A., and Guha, S. 1997. Triple-junction amorphous silicon alloy solar cell with 14.6% initial and 13.0% stable conversion efficiencies. *Applied Physics Letters*, 70, 22, 2975–2977.
15. Contreras, M. A. et al. 1999. Progress toward 20% efficiency in Cu(In,Ga)Se polycrystalline thin-film solar cell. *Progress in Photovoltaics*, 7, 311–316.
16. Grätzel, M. 2001. Photoelectrochemical cells. *Nature*, 414, 338–344.
17. Brabec, C. J. 2004. Organic photovoltaics: Technology and market. *Solar Energy Materials and Solar Cells*, 83, 273–292.
18. Aschenbrenner, P. 2004. SunPower Corporation, Sunnyvale, California. Private Communication.
19. Szlufcik, J. et al. 1997. Low-Cost industrial technologies of crystalline silicon solar cells. *Proceedings of the IEEE*, 85, 5, 711–730.
20. Green, M. A. et al. 1985. Improvements in silicon solar cell efficiency. In *Proceedings 18th IEEE Photovoltaics Specialists Conference*, p. 39.
21. McIntosh, K. R. et al. In press. The choice of silicon wafer for the production of low-cost rear-contact solar cells. In *Proceedings 3rd World Conference Photovoltaic Energy Conversion*, pp. 971–974.
22. Tanaka, M. et al. 2003. Development of HIT solar cells with more than 21% conversion efficiency and commercialization of the highest performance HIT modules. In *Proceedings. 3rd World Conference. Photovoltaic Energy Conversion.*
23. Photon International. 2003. *Market Survey on PV Modules.* Solar Verlag GmbH, Aachen, Germany.
24. Stocks, M. J. et al. In press. 65-micron thin monocrystalline silicon solar cell technology allowing 12-fold reduction in silicon usage. In *Proceedings 3rd World Conference Photovoltaic Energy Conversion*, pp. 184–187.
25. Blakers, A. et al. 2004. SLIVER modules—a crystalline silicon technology of the future. In *Proceedings 19th European Photovoltaic Solar Energy Conference.*
26. Nelson, J. 2002. Organic photovoltaic films. *Current Opinion in Solid State & Materials Science*, 6, 1, 87–95.
27. Tawada, Y., Yamagishi, H., and Yamamoto, Y. 2003. Mass production of thin film silicon PV modules. *Solar Energy Materials and Solar Cells*, 78, 647–662.
28. Brendel, R. 2003. *Thin-film Crystalline Silicon Solar Cells: Physics and Technology.* Wiley–VCH, Berlin.
29. Basore, P. A. 2003. Large-area deposition for crystalline silicon on glass modules. In *Proceedings 3rd World Conference Photovoltaic Energy Conversion.*
30. Green, M. A. et al. 2004. Solar cell efficiency tables (version 23). *Progress in Photovoltaics*, 12, 55.
31. Halls, J. J. M. and Friend, R. H. 2001. Organic PV devices, *In Clean Electricity from PVs—Series on Photoconduction of Solar Energy*, M. D. Archer and R. Hill, eds., Vol. 1, Imperial College Press, London, U.K..
32. Waldauf, C. et al. 2004. Material and device concepts for organic photovltaics: Towards competitive efficiencies. *Thin Solid Films*, 451–452, 503–507.
33. Morel, D. L. et al. 1978. High efficiency organic solar cells. *Applied Physics Letters*, 32, 495.
34. Tang, C. W. 1986. Two-layer organic PV cell. *Applied Physics Letters*, 48, 183–185.

35. Nelson, J. 2003. Organic and plastic solar cells, In *Practical Handbook of PVs—Fundamentals and Applications*, T. Markvart and L. Castañer, eds., pp. 483–511. Elsevier, Oxford, U.K.
36. Spanggaard, H. and Krebs, F. C. 2004. A brief history of the development of organic and polymeric PVs. *Solar Energy Materials & Solar Cells*, 83, 125.
37. King, R. R. et al. In press. Lattice-matched and metamorphic GaInP/GaInAs/Ge concentrator solar cells. In *Proceedings 3rd World Conference Photovoltaic Energy Conversion*, pp. 622–625.
38. Trupke, T., Green, M. A., and Würfel, P. 2002. *Journal of Applied Physics*, 92, 7, 4117–4122.
39. Trupke, T., Green, M. A., and Würfel, P. 2002. *Journal of Applied Physics*, 92, 3, 1668–1674.
40. Shalav, A. et al. In press. The application of up-converting phosphors for increased solar cell conversion efficiencies. In *Proceedings 3rd World Conference Photovoltaic Energy Conversion*, pp. 114–117.
41. Shalav, A. et al. 2005. The application of $NaYF_4$:Er^{3+} up-converting phosphors for enhanced near infrared silicon solar cell response. *Applied Physics Letters*, 86, 013505.
42. Richards, B. S. 2006. Luminescent layers for enhanced silicon solar cell performance: Down-conversion. *Solar Energy Materials and Solar Cells*, 90, 1129–1207.
43. Richards, B. S., Shalav, A., and Corkish, R. P. In press. A low escape-cone-loss luminescent solar concentrator. In *Proceedings 19th European Photovoltaic and Solar Energy Conversion Conference*, 113–116.
44. Smestad, G. P. and Grätzel, M. 1998. Demonstrating electron transfer and nanotechnology: A dye-sensitized nanocrystalline energy converter. *Journal of Chemical Education*, 75, 6, 752.
45. Grätzel, M. 2000. Perspectives for dye-sensitised nanocrystalline solar cells. *Progress in Photovoltaics*, 8, 171–185.
46. Richards, B. S. and Schäfer, A. I. 2003. Photovoltaic-powered desalination system for remote Australian communities. *Renewable Energy*, 28, 2013–2022.

9 Optoelectronic Devices

Ilesanmi Adesida
James J. Coleman

9.1 Introduction

The advent of low-loss fibers has been a key to the recent emergence of optical methods as the backbone of long distance communications. The sheer amount of information that can be moved via optical communications is destined to have a huge impact on every aspect of human endeavor. The enormous scientific and economic potential of the foregoing fact is largely responsible for the dynamic pace of research and development activities in the areas of optoelectronic devices and optoelectronic integrated circuits (OEICs). These devices and circuits along with optical fibers constitute the basic components for the physical network needed to transmit information. This network has been dubbed the information superhighway because of the dense amount of traffic it is expected to support. Other areas is which optoelectronic devices have found applications are optical storage systems and entertainment systems, such as compact disks.

The basic optoelectronic devices which are usually made with semiconductor materials are heterojunction lasers, light emitting diodes (LEDs), and photodetectors. The integration of lasers or photodetectors with electronic devices, such as field effect transistors and/or bipolar transistors constitute OEICs. This integration can be of the hybrid or the monolithic form. Although silicon is the predominant material for microelectronics, it is not suitable for optoelectronic devices in which energy conversion from electrical to optical and vice versa is the fundamental mechanism of operation. Optoelectronic devices are made from compound semiconductors which have direct energy gaps. Energy conversion is more efficient in these materials. III–V compound semiconductors have been the traditional materials used for optoelectronic devices. However, the II–VI compounds have recently demonstrated the potential for optoelectronic devices operating at short wavelengths. The wide bandgap nitrides, which also have direct bandgaps, have demonstrated impressive results for lasers and photodetectors at short wavelengths. The properties of all these materials are discussed in other chapters of this handbook.

The intent of this chapter is to discuss the general principles and characteristics of semiconductor lasers and photodetectors. Although the types of lasers and photodetectors available are diverse, there is an underlying unity in the basic principles and characteristics so that figures of merit can be used to compare the various devices. Tranverse structure, spatial optical profiles, and modulation are some of the important laser characteristics that are discussed. Quantum efficiency, responsivity, bandwidth, and noise equivalent power are the important parameters used to characterize photodetectors. These are discussed and the relevant basic equations are provided. The different types of photodetectors—photoconductors, *p–i–n* photodiodes, and avalanche photodiodes, commonly used along with the relatively new metal–semiconductors–metal photodiodes, are included in this chapter.

9.2 Semiconductors Diode Lasers

9.2.1 Principles

The semiconductors diode laser is a two terminal optoelectronic device that can very efficiently convert electrical energy in the form of a current to optical energy in the form of a coherent, single-frequency beam of light. Thus, characterizing a laser diode requires understanding the electrical properties of the device, which are similar to those of a conventional semiconductors junction diode, and the optical properties of the device, which include the spatial and spectral properties of the emitted light. Injected current results in a large number of electron–hole pairs. These can recombine to yield photons with nearly unitary quantum efficiency. The photons propagate in a waveguide resonant cavity, establish optical gain, and, eventually, laser oscillation. A useful portion of the light is emitted from the laser cavity.

In concise terms, the basic processes are

1. Injection of electron–hole pairs by means of an injected current
2. Efficient recombination of electron–hole pairs to form photons and
3. Generation of gain and oscillation from photons propagating in a resonant cavity

9.2.2 Transverse Structure

The physical structure of a semiconductor of a semiconductors laser generally consists of a number of epitaxial layers, typically five or more, engineered in thickness, composition, and doping to provide an electrically and optically efficient device. The schematic cross-section of a typical $Al_xGa_{1-x}As$–GaAs semiconductor laser is shown in Figure 9.1. Similar structures are used with other material systems. The *p–n* junction structure consists of two relatively thick confining layers, two thinner inner barrier

FIGURE 9.1 Schematic cross-section of a typical $Al_xGa_{1-x}As$–GaAs semiconductor laser diode.

layers, and an undoped active layer. The cap layer and substrate are heavily doped to facilitate low-resistance ohmic contacts. The energy band structure, consisting of the conduction band edge E_c and the valence band E_v, for this structure, called the separate confinement heterostructure (SCH), are shown in Figure 9.2a, where x is measured into the wafer from the surface. The confining layers are chosen to have the widest bandgap energy, the active layer has the smallest bandgap energy, and the inner barrier layers have an intermediate bandgap energy. This structure provides for efficient confinement of injected electrons and holes in the active layer.

The index of refraction profile of the SCH structure, shown in Figure 9.2b, provides effective optical confinement and determines the optical field profile of the laser shown in Figure 9.2c. A key parameter in describing the transverse waveguide is the optical confinement factor Γ, defined as the fraction of the total optical field that overlaps the active layer. Another typical laser structure is the graded-index, separate confinement heterostructure (GRIN-SCH) laser, shown in Figure 9.3. In this structure, the inner barrier layers are replaced by a parabolically graded layer, similar to the graded-index optical fiber. There are many other laser structures, all of which provide electrical and optical confinement in a similar manner, but differ slightly in detail.

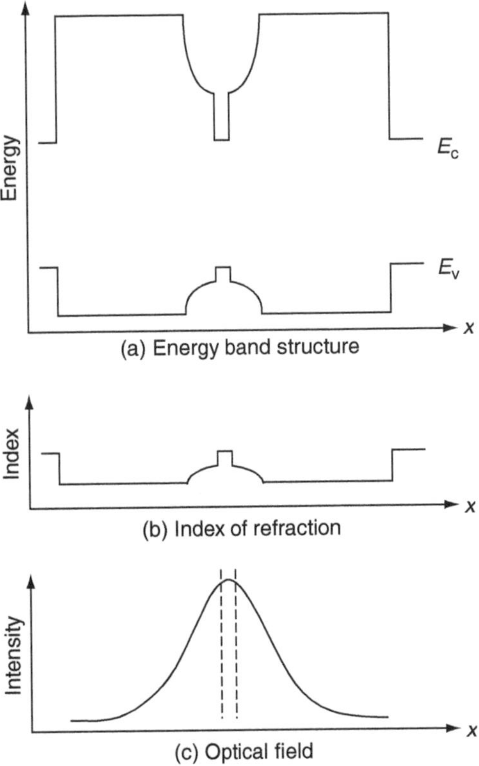

FIGURE 9.3 (a) Energy band structure, (b) index of refraction profile, and (c) optical field profile for the graded index separate confinement, heterostructural (GRIN-SCH) laser.

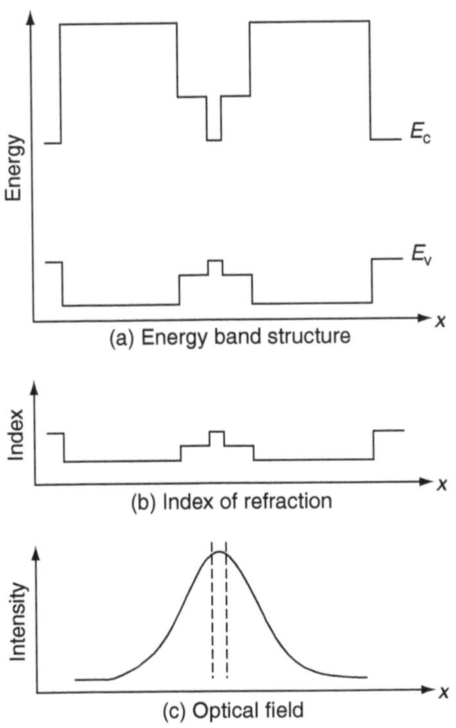

FIGURE 9.2 (a) Energy band structure, (b) index of refraction profile, and (c) optical field profile for the separate confinement, heterostructural (SCH) laser.

The quantum well laser differs only slightly from the structures described above. The chief difference is that the active layer is thin enough, usually less than 200 Å, so that quantum size effects are important. For these structures, the optical confinement factor Γ is considerably smaller, but the optical gain is larger, requiring much less current. In some cases, a number of closely coupled quantum wells are utilized in place of a single active layer.

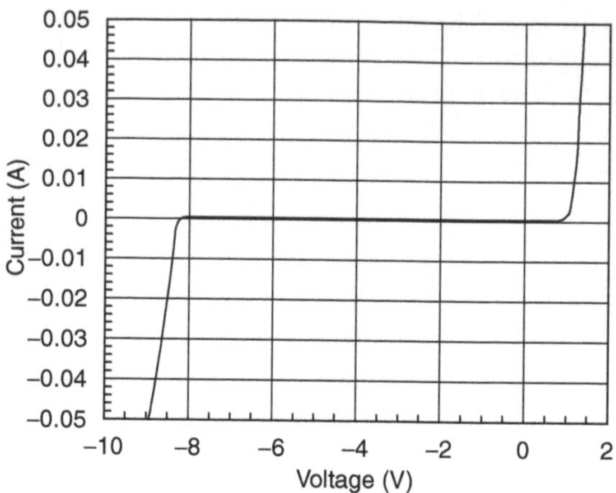

FIGURE 9.4 Typical *i–V* characteristic of a semiconductor laser.

9.2.3 *i–V* Characteristic

The *i–V* characteristic of a semiconductor laser is identical to that of any *p–n* junction diode. A typical *i–V* characteristic is shown in Figure 9.4. The electrical *p–n* junction located in the active layer so the turn-on voltage, or contact potential V_0 for the diode is given approximately by

$$V_0 = qE_g, \tag{9.1}$$

where E_g is the active layer bandgap energy and q is the fundamental charge. Similarly, the emission wavelength λ of the laser is related to the active layer bandgap energy by

$$\lambda = \frac{hc}{E_g}, \tag{9.2}$$

where h is Planck's constant and c is the speed of light. For commonly used units of wavelength and energy, this gives

$$\lambda(\mu\text{m}) = \frac{1.239852}{E_g(\text{eV})}. \tag{9.3}$$

The series resistance of the laser diode R_s is given by

$$\frac{1}{R_s} = \frac{dI}{dV}, \tag{9.4}$$

where the slope is taken above the turn-on voltage. The largest contribution to series resistance is usually the contact resistance from the metal contacts. Typical and acceptable values for the series resistance are $< 1\ \Omega$. Because of the increasing depletion width with reverse bias, the reverse breakdown voltage is determined by the doping, composition, and thickness of the active layer and the surrounding layers. Typical values are in the range of 8–10 V.

9.2.4 Light–Current Characteristic

The light–current characteristic for a typical layer diode is shown in Figure 9.5. At low drive current levels, the power output is a relatively low-level, spontaneous emission. At laser threshold, indicated by a knee in the light–current characteristic, the gain in the resonant cavity equals the losses and laser action begins. Shown as a dashed line in Figure 9.5 is a linear fit to the characteristic above the threshold. The intercept along the x axis is considered to be the laser threshold current whereas the slope is a measurement of the external, or differential, quantum efficiency η_{ext}. The threshold current is given by

$$I_{th} = wLJ_{th} \tag{9.5}$$

where w is the width, L is the cavity length, and J_{th} is the threshold current density given by

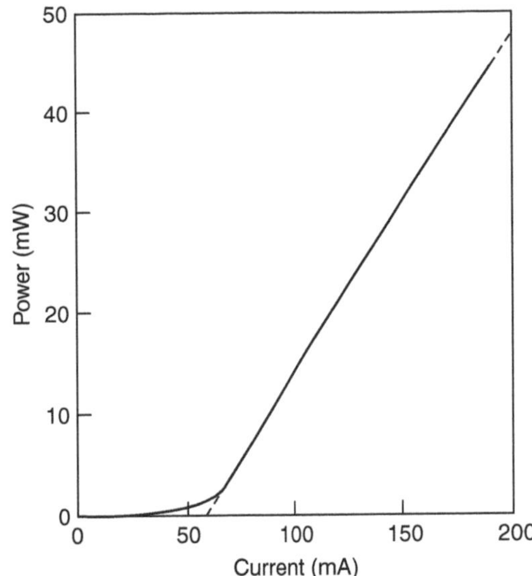

FIGURE 9.5 Light–current characteristic for a typical laser diode.

$$J_{th}(\text{A cm}^{-2}) = \frac{J_0}{\eta} + \frac{1}{\eta_{\beta \Gamma}}\left[\alpha_i + \frac{1}{2L}\ln\frac{1}{R_1R_2}\right], \tag{9.6}$$

where η is the internal quantum efficiency (typically greater than 0.90), r is the confinement factor, J_0 is the transparency current density, and β is the linear gain coefficient, which is defined as the differential of gain with respect to current density ($\gamma_i = \beta[J - J_0]$). R_1 and R_2 are the facet reflectivities. In conventional double heterostructure lasers, the transparency current density overwhelmingly dominates the threshold current density. For quantum well heterostructure lasers, a nonlinear gain coefficient, taken from $\gamma_i = J_0\beta \ln(J/J_0)$, is more appropriate, and results in a threshold current density given by

$$J_{th}(\text{A cm}^{-2}) = \frac{J_0}{\eta}\exp\left(\frac{1}{J_0\beta\Gamma}\left[\alpha_i + \frac{1}{2L}\ln\frac{1}{R_1R_2}\right]\right). \tag{9.7}$$

The very small confinement factor in thin, single quantum well, heterostructure lasers can be increased while maintaining the key elements of the quantum size effect by utilizing a multiple quantum well structure. For m wells, the total confinement factor, and, hence, model gain, increases by approximately m times, with the penalty that the transparency current density also increases by factor of m.

The threshold for the equations above is defined by setting the gain equal to the loss in the laser cavity. This loss consists of undesirable distributed loss α_i, such as residual optical absorption, and useful mirror loss α_m arising from light from light transmitted from the ends of the laser. The distributed loss is usually small in high-performance lasers ($\alpha_i = 3$–$15\ \text{cm}^{-1}$) whereas the mirror loss, a function of cavity length and facet reflectivity, is given by

$$\alpha_m = \frac{1}{2L}\ln\frac{1}{R_1R_2}. \tag{9.8}$$

The slope efficiency of the laser is given in terms of either a percent where

$$\eta_{ext} = \frac{1}{V}\frac{dP}{dI} \tag{9.9}$$

or simply in the units of W/A where

$$\eta_{\text{ext}} = \frac{dP}{dI}. \tag{9.10}$$

The external and internal quantum efficiencies are related by

$$\eta_{\text{ext}} = \eta_{\text{i}} \frac{\frac{1}{2L}\ln\left(\frac{1}{R_1R_2}\right)}{\alpha_{\text{i}} + \frac{1}{2L}\ln\left(\frac{1}{R_1R_2}\right)}. \tag{9.11}$$

The wall plug efficiency, at any value of drive current, is given by

$$\frac{P(I)}{I(V_0 + IR_s)}. \tag{9.12}$$

The uncoated facets of a semiconductor laser have a natural facet reflectivity of $R_1=R_2=0.30$. In most cases, it is desirable to have significant light emission from only a single facet, which can be obtained by providing dielectric optical coatings on the facets. High reflectivity (HR) facet coatings have a typical facet reflectivity of $R_1=0.90-0.95$, and antireflection (AR) facet coatings typically have a reflectivity of $R_2=0.10-0.05$. It is common to design the AR–HR coatings so that the product of R_1 and R_2 is ~0.09 as it is for uncoated facets.

9.2.5 Lateral Structure

Threshold current densities of optimized double heterostructure lasers are typically in the range of 500–750 A/cm^2 and, for quantum well heterostructure lasers, this value can be as small as 150 A/cm^2. The typical cavity length of a semiconductor laser is in the range of 300–1000 μm and cannot be significantly reduced without increasing the threshold current density. Reasonable drive circuitry and heat sinking require $I_{\text{th}}<30$ mA, so that the width of the laser must be adjusted accordingly. Ideally, the width is as small as possible. In addition to confining the current to a narrow stripe, it is also important to consider the beam shape. The transverse waveguide described above very strongly guides the optical wave and results in a transverse beam width of 1 μm or less at the facet. To minimize beam astigmatism, a similar strong narrow waveguide in the lateral direction is desirable.

An important example of a common, index-guided laser structure is the ridge waveguide laser shown in Figure 9.6. After epitaxial growth of the structure, narrow mesa stripes are patterned by conventional

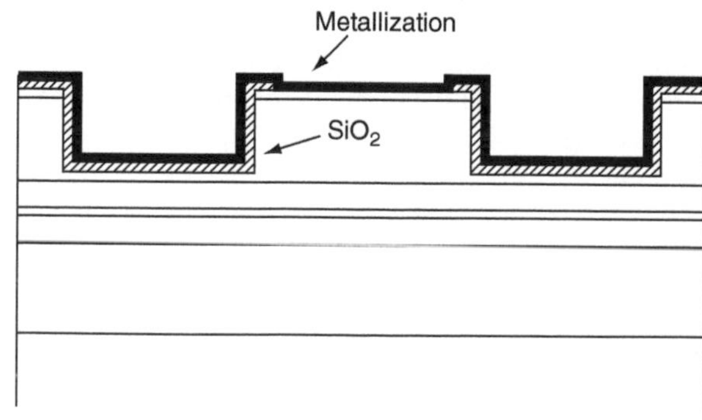

FIGURE 9.6 Cross-section of a ridge waveguide, index-guided laser structure.

lithography and etched with wet chemicals or by one of the dry etching methods near (<0.25 μm), but not through, the active region. The optical field outside the stripe is distorted by the proximity of the etched surface, oxide, and metallization resulting in a relatively large step in the lateral effective index. Because the index step is structural, the lateral waveguide formed is independent of the drive current and, if the stripe width and index step are appropriate with respect to the wavelength, fundamental lateral mode operation can be maintained over a wide range of drive current. In addition, the stripe width necessary for fundamental mode operation is usually only 1–3 μm, resulting in threshold of less than 20 mA for typical double heterostructure, ridge waveguide lasers.

9.2.6 Spatial Optical Profiles

The spatial near-field emission patterns of a semiconductors laser are defined by the transverse and lateral index of refraction profiles. As described above, the transverse index profile is generally characterized by a large index step and a thin active layer. In contrast, the lateral index step is usually much smaller with a wider guiding layer. Shown in Figure 9.7 is the astigmatic near-field emission pattern at the facet for a typical laser diode. Shown for reference are the cross-sectional intensity profiles in the lateral and transverse directions with the index of refraction profiles as dashed lines. The near-field beam width in each direction is characterized by a full width at half maximum (FWHM), typically in units of microns.

The far-field emission pattern of a semiconductor diode laser is, of course, the Fourier transform of the near-field emission pattern. The relationship is shown schematically in Figure 9.8. The far-field beam width in each direction is characterized by a FWHM beam divergence, typically in units of degrees.

9.2.7 Optical Spectrum

The optical spectrum of a semiconductor laser diode is a convolution of the material gain spectrum of the active layer of the laser and the spectral response of the resonant cavity. The gain spectrum, shown in Figure 9.9, results from the approximately parabolic band structure of semiconductor materials multiplied by a Fermi–Dirac occupancy probability function. This is a relatively broad function, usually many hundreds of angstroms wide. The cavity mode spectrum shown in Figure 9.9 is for a typical Fabry–Perot resonator formed by cleaving and perhaps, later coating plane, parallel, reflecting facets oriented along a fundamental crystal plane of the laser material. These cleaved facets yield a large number of strong resonances with a spacing given by

$$\Delta\lambda = \frac{\lambda^2}{2\pi L[n - \lambda\partial n/\partial\lambda]} \tag{9.13}$$

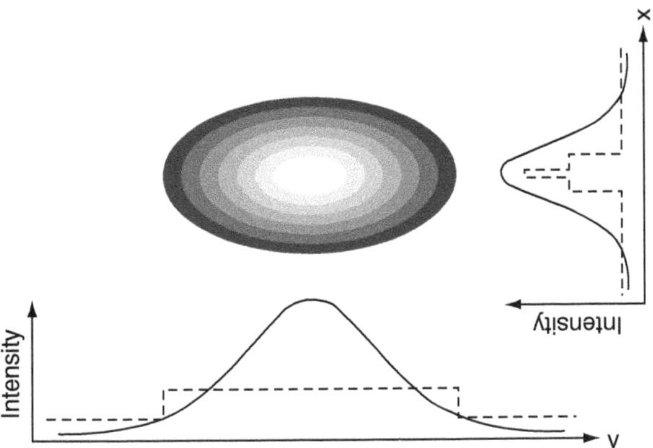

FIGURE 9.7 The astigmatic near-field emission pattern at the facet for a typical laser diode.

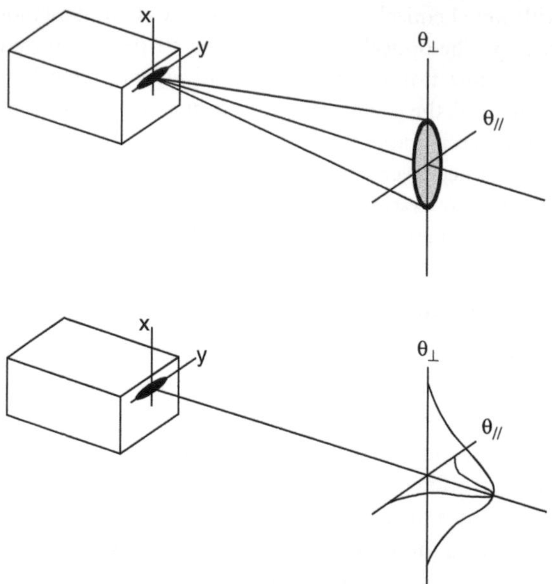

FIGURE 9.8 Schematic description of the Fourier transform relationship between the far-field emission pattern and the near-field emission pattern of a semiconductor diode laser.

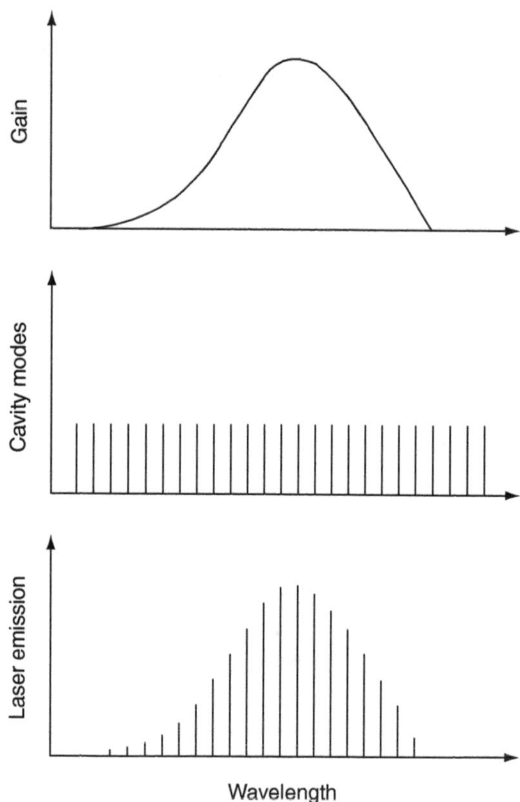

FIGURE 9.9 The gain spectrum, Fabry–Perot cavity mode spectrum, and the emission spectrum of a semiconductor laser.

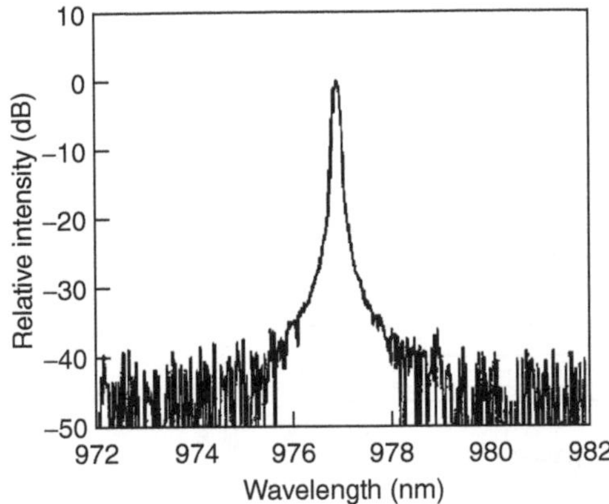

FIGURE 9.10 Longitudinal mode spectrum of a ridge waveguide, distributed Bragg reflector (DBR) laser.

where L is the cavity length and $[n - \lambda\partial n/\partial\lambda]$ is the effective index of refraction, including dispersion. Typical values of $\Delta\lambda$ for normal cavity lengths are in the range of 3 Å. The product of material gain and cavity resonance is also shown in Figure 9.9. At higher drive currents, a smaller number of longitudinal modes, or a single mode, often dominates the laser emission. With heating or additional drive current, however, mode hopping to nearby adjacent cavity modes can easily occur.

When the longitudinal mode stability of the Fabry–Perot resonator is insufficient for an application, greater stability against wavelength shift with temperature and mode hopping can be obtained by using a distributed feedback (DFB) or a distributed Bragg reflector (DBR) laser structure. These structures, processed with a wavelength-sensitive Bragg grating either distributed along the gain path (DFB) or at the end of the gain path (DBR), are characterized by a very narrow linewidth, single, resonant frequency. Shown in Figure 9.10 is the longitudinal mode spectrum of a ridge waveguide DBR laser. The side mode suppression ratio for this family of structures is usually greater than 35 dB.

9.2.8 Modulation

The simplest and most direct method for modulating a semiconductor laser diode is modulation of the laser drive current. Direct current modulation is best described by rate equations for both carriers and photons. Solution of the equations gives the modulation frequency response,

$$\frac{\partial P}{\partial J} = \frac{-(1/qd)\Gamma A P_0}{\omega^2 - i\omega/\tau_s - i\omega A P_0 - A P_0/\tau_p}, \tag{9.14}$$

where P is the photon density, J is the current density, A is a constant related to the gain coefficient, d is the active region thickness, τ_s is the spontaneous recombination lifetime, τ_p is the photon lifetime, and ω is the frequency. The modulation frequency response is shown for two output power levels in Figure 9.11. The frequency response is flat to frequencies approaching 1 GHz, rises to a peak value at some characteristic frequency, and, then, quickly rolls off. The peak in the frequency response is given approximately by

$$\omega_{\max} = \left[\frac{A P_0}{\tau_p}\right]^{1/2}. \tag{9.15}$$

Because A is related to the gain of the laser, higher frequency operation requires greater gain, greater photon density, and decreased photon lifetime.

Modulation of the laser drive current modulates the injected carrier density which, in turn, results in a modulation of the quasi-Fermi levels. Because the emission wavelength and, hence, frequency is related to the difference between the quasi-Fermi levels, modulation of the carrier density results in modulation of the emission frequency. Thus, amplitude modulation (AM) of the laser power output results in a corresponding frequency modulation (FM). The change and broadening of the emission frequency is called "chirp."

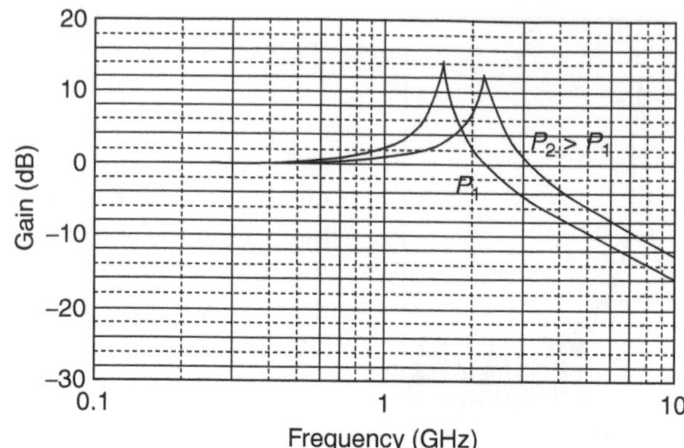

FIGURE 9.11 Direct current modulation frequency response of a semiconductor laser diode for two output power levels.

9.2.9 Temperature Effects

There are two key design parameters associated with variation of the operating junction temperature of a semiconductor laser diode. The first is the variation of the laser threshold current with temperature. This is usually quantified as a characteristic temperature T_0 where the threshold current is given by

$$I_{th}(T) = I_0 \exp(T/T_0) \tag{9.16}$$

where I_0 and T_0 are only valid around a specific temperature, usually room temperature. Values of $T_0 > 125$ K are observed for AlGaAs lasers whereas, for InGaAsP lasers, values in the range of 50–60 K are typical. The highest values are most desirable.

The second parameter is the change in wavelength with operating temperature $d\lambda/dT$. The wavelength for peak gain in most semiconductor materials varies with temperature in the range of 3–5 Å/°C. This is also the temperature variation of wavelength observed in Fabry–Perot lasers where there are a large number of closely spaced, cavity modes. Actually, of course, there is also mode hopping taking place. In DFB and DBR lasers, the mode spacing is much larger, and the inherent temperature dependence of the Bragg wavelength is much smaller, typically less than 0.8 Å/°C.

9.2.10 Reliability

The failure mechanisms for semiconductor lasers can be separated into two relatively broad categories. The first category is catastrophic failure associated with exceeding the maximum safe value of some operating parameter. Perhaps the most important example is catastrophic facet damage. When the optical power density at the laser facet reaches a certain value, catastrophic optical damage (COD) occurs. COD is a function of the shape of the near-field pattern, the drive current amplitude and pulse length, and the presence or absence of passivation on the facets. COD occurs rapidly and irreversibly at the laser's upper limit for power output.

The second general category of semiconductor laser diode failure is gradual degradation resulting from long-term effects related to materials, such as defects or contaminants, or processing, such as handling damage. An estimate of the long-term reliability and mean time between failure (MTBF) of a particular laser design is an important part of commercial laser development, especially for remote or space-based laser systems. The measurements utilized to establish reliability parameters include current constant power measurements, where device failure is defined as the laser output power falling to half its initial value, or constant output power current adjustment, where device failure is defined as the laser drive current rising to twice its initial value.

As semiconductor laser lifetimes have approached those of other solid state electronic devices, measured in tens of thousands of hours, accelerated life testing methods have become important. By measuring the failure rate or median lifetime as a function of temperature above room temperature (typically 30°–70°C), an activation energy E_a can be determined from the slope of the best fit line. This activation energy allows a preliminary analysis of the dominant failure mechanism, because certain failure mechanisms have characteristic activation energies, and extrapolation of a room temperature lifetime.

9.3 Photodetectors

9.3.1 Principles

A photodetector is a device that converts optical energy to electrical energy. The principal mechanism responsible for this transformation is photoconductivity. This property is exhibited by all semiconductors and it is the increase in conductivity brought about by the absorption of photons. The absorption of a photon results in the generation of an electron–hole pair. The electrons and holes separate to become mobile carriers which are transported through the semiconductor under the influence of an externally applied electric field. The transport of these carriers enhances the conductivity of the semiconductor.

In concise terms, the three basic processes involved in this conversion are

1. The absorption of photons and the resulting generation of carriers.
2. The transport of the generated carriers across the absorption or drift region under the influence of an applied field. Internal amplification of carriers can occur at this stage via various mechanisms. An example of a mechanism is impact ionization which occurs with the application of large electric fields.
3. Collection of carriers constituting a photocurrent which flows through an external circuitry.

Perhaps, the simplest type of photodetector is the photoconductor which is simply a slab of an intrinsic semiconductor with two contacts. The electron–hole pairs, generated by absorption of photons in the material, are collected by oppositely biased contacts to constitute a photocurrent. Other types of photodetectors are photodiodes which are based on either the *p–n* junction or the metal–semiconductor junction (also called a Schottky-barrier). A *p–i–n* (or PIN) photodiode is a reverse-biased *p–n* junction with an intrinsic layer interposed between the p and n layers. Because the depletion area is the only region supporting an electric field in a *p–n* junction, the intrinsic layer serves to increase the depletion layer width and therefore the photon absorption region of the device. The PIN photodiode is normally operated in a bias mode in which the device does not exhibit gain. Another type of photodetector based on a *p–n* junction is the avalanche photodiode (APD). APDs are operated at electric fields which are high enough to cause impact ionization and, thereby, generate more carriers, leading to the avalanche effect. The net effect of the avalanche process is a multiplication of carriers, resulting in gain for the output photocurrent of the device. As mentioned above, another important class of photodetectors, which have recently gained prominence, are the metal–semiconductor photodiodes which are made from metal–semiconductor junctions. The relative ease with which they can be fabricated has made them and the metal–semiconductor–metal (MSM) photodiodes attractive for some applications requiring monolithic integration.

As described above, there are some basic properties exhibited by all semiconductor photodetectors. These general properties can be quantified, to a certain extent, and have become figures of merit used in comparing photodetectors. The properties can be quantified in terms of quantum efficiency, responsivity, bandwidth, and noise equivalent power (NEP).

9.3.2 Quantum Efficiency

The quantum efficiency η of a photodetector is the number of electron–hole pairs generated per incident photon collected at the contacts. Quantum efficiency is determined by many factors. These include the

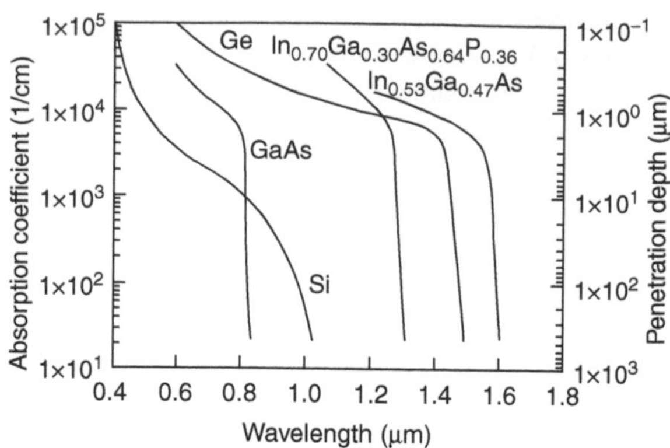

FIGURE 9.12 Wavelength dependence of the optical absorption coeffiecients of several semiconductor materials.

fact that not all photons incident on the semiconductor will produce electron–hole pairs and that some of the photons may be reflected at the surface of the semiconductor. All these factors combine to reduce η. Quantum efficiency can therefore be given by

$$\eta = (1-r)\zeta[1-\exp(-\alpha L)] \tag{9.17}$$

where r is the optical power reflectance at the surface of the detector, ζ is the fraction of electron–hole pairs that actually contribute to the photocurrent, α is the absorption coefficient of the detector material per centimeter, and L is the width of the detector's absorption region. By applying an antireflection coating on the detector's surface for the wavelength of operation, reflection can be reduced and, thereby, the factor $(1-r)$ can be maximized. It is difficult to estimate the factor ζ because it depends on the quality of the materials. Carriers can be lost through recombination at the surface or in the bulk of the material which reduces the photocurrent. Modern epitaxial growth methods are now capable of producing high quality materials, and therefore, for a practical estimation of quantum efficiency, ζ can be assumed as unity. The last factor $[1-\exp(-\alpha L)]$ denotes the fraction of the incident optical power absorbed in bulk of the detector.

In terms of the quantities easily measured in the laboratory, quantum efficiency is given by

$$\eta = \frac{I_p/q}{p_i/h\nu} = \frac{h\nu}{q}\cdot\frac{I_p}{p_i} \tag{9.18}$$

where I_p is the detector photocurrent, P_i is the incident optical power, and $h\nu$ is the photon energy. The quantity η given above is known as the external quantum efficiency η_{ext}. The quantum efficiency is dependent on the absorption coefficient α which is a function of wavelength λ; therefore, η is dependent on λ. Figure 9.12 shows the dependence of α on λ for some detector materials. Because only photons with energy greater than or equal to the bandgap energy E_g can be absorbed (i.e., $hc/\lambda \geq E_g$), the long-wavelength limit for a practical detector is the bandgap wavelength. Bandgap energies at 300 K for representative photodetector materials are displayed in Table 9.1. There is also a short-wavelength limit because α is very large at short wavelengths for most semiconductors, and, consequently, all of the incident photons are absorbed near the surface of the detector.

9.3.3 Responsivity

The responsivity $\mathcal{R}$ of a detector is the photocurrent in the device divided by the input optical power and is given by

$$\Re = \frac{I_p}{P_i} = \frac{\eta q}{h\nu} = \eta\frac{\lambda(\mu\text{m})}{1.24}. \tag{9.19}$$

TABLE 9.1 Bandgap Energies (in eV) at 300 K for Some Photodiode Materials

Material	Bandgap Energy (eV)
GaAs	1.42
GaSb	0.73
$GaAs_{0.88}Sb_{0.12}$	1.15
Ge	0.67
InAs	0.35
$In_{0.53}Ga_{0.47}As$	0.75
InP	1.35
Si	1.14

The unit of responsivity in A/W. It is seen from this expression that, for a constant η, $\mathcal{R}$ should increase with λ. This is illustrated in Figure 9.13. However, because α depends on λ, there is a region between the short- and long-wavelength limits over which $\mathcal{R}$ increases. For photodetectors which exhibit gain, the gain factor G can be accommodated in a more general equation for responsivity given by

$$\Re = G\eta \frac{\lambda(\mu\text{m})}{1.24}. \tag{9.20}$$

It is possible to degrade the responsivity of a detector by applying excessive incident optical power. The detector becomes saturated, thus, limiting its linear dynamic range, which is the range over which the relationship between the detector's output and the incident optical power in linear.

9.3.4 Bandwidth

The bandwidth B of a photodetector measures the shortest response time of the device. This property becomes very important when a photodetector is used in a data transmission circuit. The faster a detector can respond to a stream of optical pulses, the higher the density of the transmitted data can be. The response time of a photodetector is determined by three factors—transit time, diffusion time, and the device RC time constant.

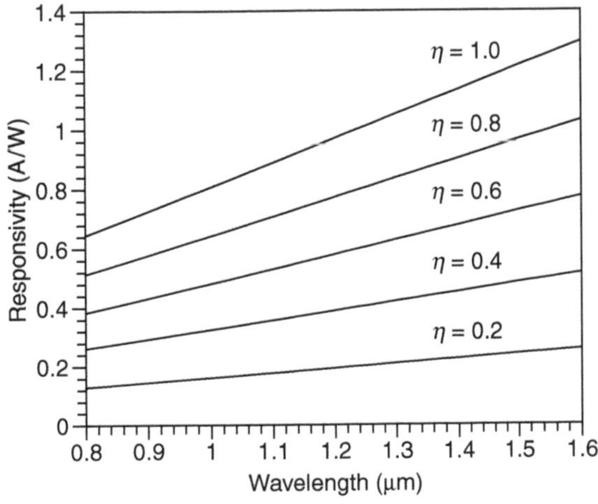

FIGURE 9.13 Responsivity versus wavelength for various external quantum efficiencies.

Electron–hole pairs created by photons in the active region of a photodetector move in directions opposite to the contacts for collection under an applied electric field. The carriers move by drift and diffusion. If the electric field is sufficiently large, most of the carriers travel by drift, and they reach their scattering-limited or saturation velocity in the material. The velocity of holes is usually smaller than that of electrons, therefore, the time (i.e., transit time) it takes holes to drift across the active region of the detector limits the response time of the device. If electron–hole pairs are generated uniformly throughout the material, then, a severe transit time spread between electrons and holes can occur. Diffusion time limitations can occur only at low bias where the drift field is low. Because the diffusion process is slow, it can be a severe problem even though only a small number of carriers may be involved. A judicious design of the active area of the detectors and the application of an appropriate bias can make this limitation insignificant. The last factor is the resistance R and the capacitance C of the device and its associated circuitry. This composite RC network integrates the output current of the detector and, therefore, increases the response time. Different types of photodetectors are influenced by different combinations of these limitations which set their bandwidths. However, photodetectors of a given design and material do exhibit a constant gain-bandwidth product.

9.3.5 Noise Equivalent Power

Photodetectors are subject to several sources of noise that degrades their performance. The inherent randomness in the arrival of photons and the absorption of photons in the device serve as sources of noise. Various sources of current generation exist in all photodetectors. Some of these include current due to the incoming optical signal, current due to background radiation, and the dark current that is due to surface leakage, tunneling, and thermal generation of electron–hole pairs in or around the active region. All of these currents are generated randomly and contribute to shot noise. The amplification process that produces gain in some detectors is the avalanche mechanism. This is a random effect, and, therefore, there is a gain noise associated with such detectors. Another source of noise involves the random motion of carriers in resistive electrical materials at finite temperatures. There are parasitic resistances intrinsic to photodetectors and also resistances in circuits in which photodetectors are utilized. An example is a receiver circuit in which a detector serves as a source of input current to a preamplifier. The noise generated by these resistive elements is called thermal, or Johnson, or Nyquist noise. This noise is given by

$$\langle i_j^2 \rangle = \frac{4kT_{\text{eff}}B}{R_{\text{eff}}} \tag{9.21}$$

where R_{eff} is the parallel combination of the detector and the preamplifier input resistances, B is the bandwidth, and T_{eff} is the effective temperature which is related to the noise figure NF of the amplifier:

$$T_{\text{eff}} = T(10^{\text{NF}/10} - 1) \tag{9.22}$$

where T is the ambient temperature. It is, therefore, evident that, in the operation of a detector, the output signal must be above the noise level. The signal-to-noise ratio (SNR) is, therefore, an important characteristic in photodetectors, and it is related to sensitivity. The sensitivity of a photodetector is the minimum optical input power needed to achieve a given value of SNR. A measure of sensitivity is called noise equivalent power (NEP). NEP is the optical power (or photocurrent) required for the SNR to be unity over a 1-Hz bandwidth. Essentially, this measures when the photocurrent is exactly equal to the noise current. Thus, NEP measures the minimum detectable power in a photodetector. NEP depends on bandwidth, and, to find the optical power required to produce a SNR of unity for an entire measurement bandwidth, we have

$$P_{\text{i}} = \text{NEP}\sqrt{B}. \tag{9.23}$$

Another figure of merit also useful for determining the ultimate detection limit is detectivity D^* given by

$$D^* = \frac{\sqrt{AB}}{\text{NEP}}(\text{cm. Hz}^{1/2}.\ \text{W}^{-1}) \tag{9.24}$$

where A is the area of the photodetector on which light is incident. As with NEP, the reference bandwidth is taken as 1 Hz. D^* is usually expressed as $D^*(\lambda, f, 1)$ where λ is the wavelength and f is the modulation frequency of the input optical signal. It must be noted that NEP and D^* are not equal to system sensitivity in actual applications because other noise sources, such as preamplifier noise, may dominate, especially in high speed (GHz) systems.

9.3.6 Photoconductors

Essentially, a photoconductor is a semiconductor material with two alloyed ohmic contacts. Its operation is based on the increase in conductivity of the semiconductor material due to the absorption of photons. The photogenerated electrons and holes are transported to the contacts under the influence of the electric field applied to the material by an external voltage source The conductivity of the device is increased, and the signal is detected as an increased current flow under a constant-voltage bias. The conductivity of the material increases as a function of the photon flux. A practical form of photoconductors is shown in Figure 9.14 consisting of interdigitated contacts on a semiconductor material. The interdigitated contacts are designed to minimize transit time and to maximize light transmission into the semiconductor material. To further improve quantum efficiency, an antireflection coating can be deposited on the device surface or light can be coupled in from the bottom if the substrate material has a wider bandgap than the absorption region. This is the case in Figure 9.14 where InP is the substrate and, therefore, transparent to wavelengths longer than 0.92 μm at 300 K.

The quantum efficiency of a photoconductor is given by Equation 9.17 and the responsivity is given by Equation 9.20, where G is the internal photocurrent gain of the device. The gain is brought about by the fact that the photogenerated carriers contribute to current until they recombine. This gain is given by

$$G = \frac{\tau}{t_{tr}} \tag{9.25}$$

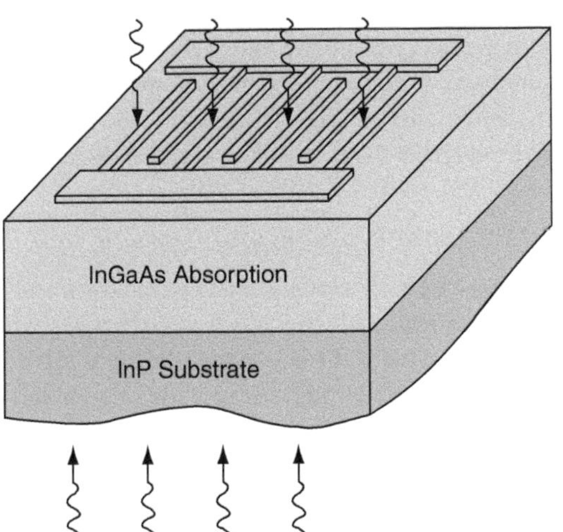

FIGURE 9.14 Schematic of an InGaAs photoconductor which is illuminated on the front and back sides.

where τ is the excess-carrier recombination lifetime and t_{tr} is the transit time of the majority carrier. The transit time is given by

$$t_{tr} = \frac{L}{v} \tag{9.26}$$

where L is the channel length (distance between contacts) and v is the carrier velocity. It is seen from the these equations that, if the recombination lifetime is greater than the majority-carrier transit time, then many carriers will travel between the contacts before recombination takes place. This is photoconductor gain, and it can be below unity or well above unity depending on various factors including semiconductor material, size of the device, and the magnitude of the applied voltage.

The bandwidth of a photoconductor is given by

$$B = \frac{1}{2\pi\tau}. \tag{9.27}$$

It is seen that, whereas a long recombination time makes for a high gain, it also reduces bandwidth. So, a trade-off between gain and bandwidth exists for photoconductors. The gain-bandwidth product of a photoconductive detector is given by

$$\mathrm{GB} = \frac{1}{2\pi t_{\mathrm{tr}}} \tag{9.28}$$

where t_{tr} is a constant for a given material and detector configuration.

The primary contributions to noise in photoconductive detectors are made by thermal or Johnson noise and the generation–recombination noise. The thermal noise given by Equation 9.21 results from the random motion of carriers with average energy of kT contributing to the dark current of the device. The generation–recombination noise is due to fluctuations in the generation and recombination of carriers which, in turn, leads to fluctuations in the conductivity of the device. The generation–recombination noise is given by

$$\left\langle i_{\mathrm{G-R}}^2 \right\rangle = \frac{4qI_0\mathrm{GB}}{1+\omega^2\tau^2} \tag{9.29}$$

where I_0 is steady-state output photocurrent and ω is the angular modulation frequency of the input optical signal.

To describe the overall noise performance of a photoconductor, the NEP is given by

$$\mathrm{NEP} = \frac{8h\nu}{\eta}\left[1 + \frac{kT}{qG}(1+\omega^2\tau^2)\frac{G_c}{I_0}\right] \tag{9.30}$$

where G_c is the conductance of the channel. The dominant noise mechanism in a photoconductive detector is the thermal noise of the conducting channel. An increase in channel resistance is necessary to reduce thermal noise. If the thickness of the channel is reduced to increase the resistance, then quantum efficiency is reduced. To obtain the highest resistance and, hence, the lowest thermal noise achievable while maintaining high gain and quantum efficiency, it is necessary to utilize materials with the lowest carrier concentration. Thermal noise, generation–recombination noise, and dark current are high in semiconductors with smaller bandgaps. Although, photoconductors can have large gains, the gain may not be enough to surmount the inherent noise limitations to make them useful in many applications.

9.3.7 PIN Photodiodes

A *p–i–n* (PIN) photodetector is a *p–n* junction with an intrinsic (i) layer sandwiched between the *p* and *n* layers. In practice, the *i*-layer is either a p^- or n^- layer (i.e., lightly doped) which is inserted between the p^+ and n^+ layers. This structure is illustrated in Figure 9.15 along with the energy band diagram under an applied bias. The PIN photodiode is operated in the reverse-bias mode, and, because the *i*-region has a low concentration of free carriers, it can be depleted with a minimum amount of voltage. Therefore, the depletion region extends through the entire *i*-region. When photons with energy greater than or equal to the bandgap energy are incident on the photodiode, electron–hole pairs are created. Carriers generated within a diffusion length of the depletion region diffuse into the *i*-region. These carriers along with all the carriers generated in the depletion region are transported by drift due to the applied reverse bias and are collected. The electric field in the depletion region is high and sufficiently uniform so that the carriers travel at saturation velocities. If all of the photogenerated carriers are collected, the quantum efficiency is

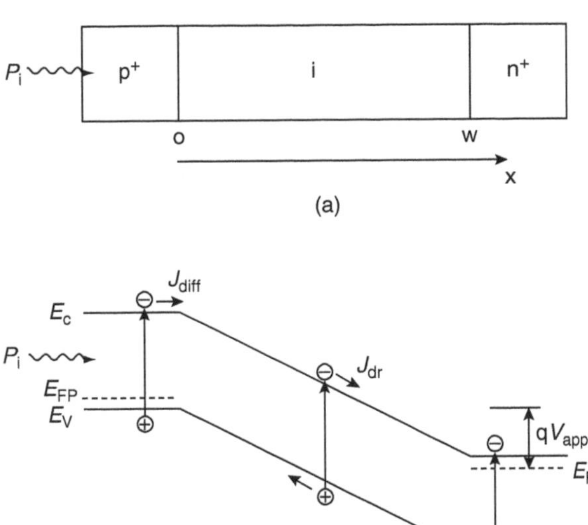

FIGURE 9.15 The *p–i–n* photodiode: (a) the layer structure and (b) the associated energy band diagram under an applied bias.

given by Equation 9.17 with L being the thickness of the region where the light is absorbed and $\zeta = 1$. L is usually assumed to be thickness of the i-region. As mentioned earlier, r can be made negligible by utilizing an antireflection coating to obtain good quantum efficiency. The internal quantum efficiencies $(1 - \exp(-\alpha L))$ for InGaAs and GaAs at various operating wavelengths are shown in Figure 9.16. Because there is no internal optical gain associated with the PIN diode, the maximum internal quantum efficiency that can be expected is 100%. The external quantum efficiency is given by Equation 9.17 with $\zeta = 1$, and the responsivity is given by Equation 9.20 with unity gain.

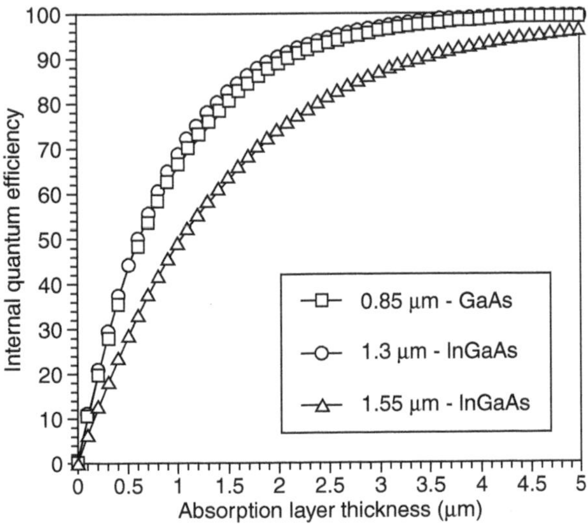

FIGURE 9.16 Internal quantum efficiency for GaAs and InGaAs at certain wavelengths and various absorption layer thickness.

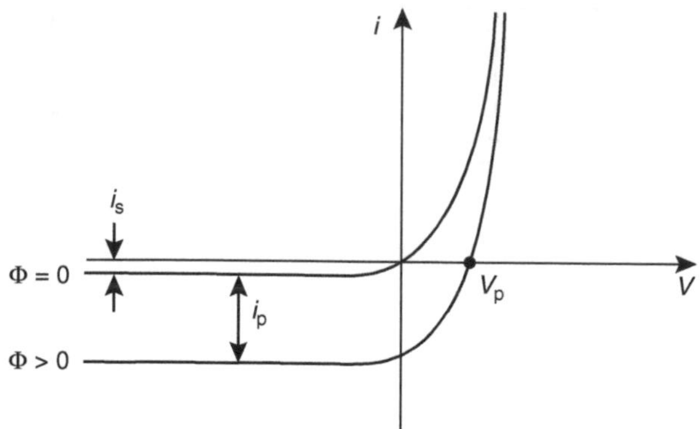

FIGURE 9.17 An *i–V* relationship of a typical photodiode. The two different curves represent the *i–V* relationship when the incident photon flux is equal to zero and greater than zero.

A schematic of a typical *i–V* characteristic of a PIN photodiode is shown in Figure 9.17. This is the usual *i–V* characteristic of a *p–n* junction but with an added current $-i_p$ which is proportional to the incoming photon flux. The *i–V* relationship is given by

$$i = i_s[e^{eV/kT}-1]-i_p. \tag{9.31}$$

The applied reverse bias (V_A) needed to deplete the i-region of thickness L is given by

$$L = \frac{\sqrt{2\varepsilon(V_0 + V_A)}}{qN_c} \tag{9.32}$$

where ε is the dielectric constant, V_0 is the built-in voltage, and N_C is the reduced carrier concentration, $N_C = N_A N_D / N_A + N_D$.

For high quantum efficiency or responsivity, it is necessary that $\alpha L \gg 1$. Of course, if L is large, the transit time of carriers across the i-region becomes large and device speed suffers. It is desirable to make α large, but it is a material property which cannot be changed. Therefore, there are many factors that must be considered to realize high-performance PIN photodiodes.

The response speed of PIN photodiodes can be limited by (a) the transit time of the photogenerated carriers across the depletion region, (b) the diffusion time of carriers generated outside the depletion region, and (c) the *RC* time constant with *C* consisting of junction capacitance and any other parasitic capacitances. The transit time is limited by the slower carriers, usually the holes. Another factor which can limit speed is charge trapping at heterojunctions. For the *RC* time constant, the junction capacitance of the photodiode is given by

$$C = \frac{\varepsilon A}{L} \tag{9.33}$$

where A is the cross-sectional area of the device. As seen here, the device capacitance can be minimized by increasing the thickness of the depletion region and by reducing the diameter of the device. Other parasitic capacitances associated with packaging can also be reduced. As mentioned previously, transit-time considerations suggest the use of a very thin i-layer to obtain high speed response. However, that increases parasitic junction capacitance which, in turn, reduces bandwidth. So, for a given detector area, an optimum i-layer thickness exists to obtain the highest speed possible. The transit-time limited

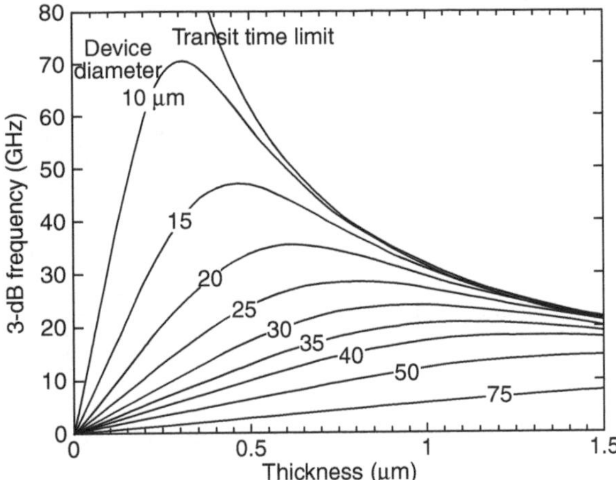

FIGURE 9.18 The theoretical frequency response of a *p–i–n* photodiode for various active layer thicknesses and dimensions.

frequency response of a PIN detector is given by

$$\frac{i(\omega)}{i(0)} = \frac{1}{(1-e^{-\alpha L})}\left[\frac{e^{j\omega\tau_n-\alpha L}-1}{j\omega\tau_n-\alpha L} - e^{-\alpha L}\frac{(e^{j\omega\tau_n}-1)}{j\omega\tau_n} + \frac{e^{j\omega\tau_p}-1}{j\omega\tau_p} - e^{-\alpha L}\frac{(e^{j\omega\tau_p+\alpha L}-1)}{j\omega\tau_p+\alpha L}\right] \tag{9.34}$$

where $i(\omega)$ is the detected current at an angular modulation frequency of ω, $i(0)$ is the dc current, L is the thickness of the i-layer, and $\tau_n = L/v_n$ and $\tau_p = L/v_p$ are the electron and hole transit times, and v_n and v_p are the electron and hole saturation velocities, respectively. For InGaAs, the saturation electron and hole velocities are 6.5×10^6 and 4.8×10^6 cm/s, respectively. Using this equation and the frequency response of a parallel *RC* network, the theoretical 3 dB bandwidth of InGaAs/InP PIN detectors of several diameters are plotted against i-layer thickness in Figure 9.18. The results were calculated for 1.3 μm operation (i.e., $\alpha = 1.16\ \mu m^{-1}$) and for a load resistance of 50 Ω. In practice, there are parasitic resistances (the shunt or junction resistance and the series resistance) and a stray capacitance that should be taken into account for higher accuracy. The transit-time limited response is also shown in the figure. The figure dearly demonstrates the compromises that are needed to design PIN photodiodes and also shows that devices with bandwidths over 60 GHz can be realized if the small diameter needed is not limiting for the particular application in hand.

An approximate expression for bandwidth for a detector with a very thin i-layer is

$$B = \frac{0.45v}{L} \tag{9.35}$$

for which $v_n = v_p = v$ has been assumed. Therefore, for a high speed detector where $\alpha L \ll 1$, the quantum efficiency from Equation 9.17 is $(1-r)\alpha L$, and consequently, the bandwidth-efficiency product is

$$B \cdot \eta = 0.45\alpha v(1-r). \tag{9.36}$$

Once again, the critical importance of a large α is clear. The large α for InGaAs at long wavelengths (1.3 and 1.55 μm) gives it an advantage over Si and Ge as a material of choice for PIN photodetectors for optical communication applications.

The sources of noise in a PIN photodiode are from (a) the current due to the photocurrent I_p, (b) the dark current due to generation in the depletion region I_D, and (c) the current due to background

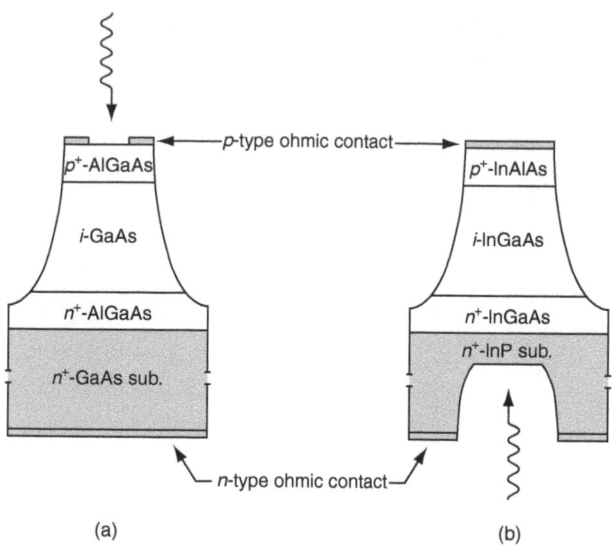

FIGURE 9.19 Schematics of *p–i–n* photodiodes utilizing (a) front side illumination and (b) back side illumination.

radiation I_B. Due to the random generation of these currents, they contribute to shot noise which can be expressed as

$$\langle i_s^2 \rangle = 2q(I_P + I_D + I_B)B \tag{9.37}$$

where B is the bandwidth. In addition, there is thermal or Johnson noise contributed with the shunt resistance of the diode and the input resistance of the following preamplifier stage. The Johnson noise for the PIN photodiode is given by Equation 9.21. From these considerations, the NEP in units of watts for a PIN photodiode is given by

$$\text{NEP} = \frac{h\nu}{q\eta}\left[2q(I_P + I_D + I_B) + \frac{4kT}{R_{\text{eff}}}\right]^{1/2}. \tag{9.38}$$

It is seen that the key to increasing the sensitivity of the photodiode is to make R_{eff} and η as large as possible and make the I_B and I_D as small as possible. I_B, which is due to background radiation is usually very small, and I_D is also very small in a PIN photodiode because the device operates in the reverse-bias mode. The shot noise contributed by I_D is very small compared to that in photoconductors. Johnson noise is usually dominant in PIN detectors, but it can be minimized by optimizing the device and the circuit parameters. Therefore, PIN detectors are very useful for high speed, low noise applications such as encountered in optical communications.

In these applications which require operation at long wavelengths, photodiodes are made of InP-based materials, such as $In_{0.53}Ga_{0.47}As$ and InGaAsP. Heterojunctions of InP/InGaAs and InAlAs/InGaAs are needed for photodiodes to obtain higher breakdown and lower reverse leakage current than can be obtained using homojunctions. Practical heterostructures for PIN devices operating at infrared and long wavelengths are shown in Figure 9.19a and Figure 9.19b.

9.3.8 Avalanche Photodiodes

An avalanche photodiode (APD) is essentially a p–n junction operated at high reverse-bias voltages close to the breakdown voltage. At such high voltages, photogenerated carriers in the depletion region gain kinetic energy from the induced electric field and travel at their saturation velocities in the host material.

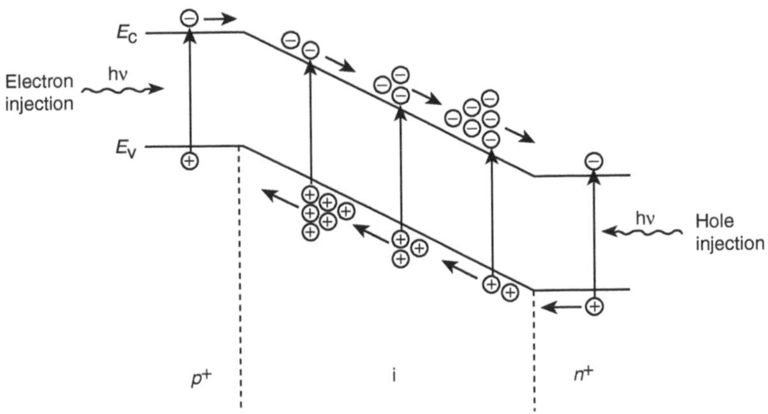

FIGURE 9.20 The energy band diagram of an APD illustrating the avalanche gain process.

The carriers can acquire sufficient kinetic energies to undergo inelastic collisions with the lattice to create secondary electron–hole pairs. These secondary electron–hole pairs along with the primary carriers continue to drift and produce tertiary electron–hole pairs. To have an ionizing collision, a carrier must gain a threshold energy greater than the bandgap energy. Therefore, the critical field required to create ionizing collisions is material dependent and ranges from 10^4 to 10^5 V/cm. The process of creating carriers through ionizing collisions is called impact ionization, and the subsequent increase in the number of generated carriers is termed avalanche multiplication. This process is illustrated in the band diagram of Figure 9.20 showing the direction of electron and hole injection via optical illumination. The avalanche multiplication process is an internal gain mechanism. This gain is typically much higher than that associated with a photoconductor. The quantum efficiency of an APD is given by Equation 9.17 with L being the thickness of the absorption region.

As described above, the avalanche process is characterized by impact ionization. The electron–hole pairs create drift in opposite directions, and, with sufficient energies, each carrier type can undergo impact ionization. The number of ionization collisions per unit length is designated α_n and α_p, the ionization coefficients for holes and electrons, respectively. Among other things, α_n and α_p depend on materials and their band structures. Ionization coefficients are also dependent on the applied electric field, and an approximate relationship is given by

$$\alpha_n, \alpha_p = a \exp\left[-\frac{b}{E}\right] \tag{9.39}$$

where a and b are constants dependent on the type of material and doping density and E is the applied electric field. The field-dependent impact ionization coefficients for various semiconductor materials have been determined and reported in the literature. In general, the two rates approach each other at fields higher than 3×10^5 V/cm for many semiconductors. Table 9.2 shows the values of a and b of Equation 9.39 for various semiconductors.

The avalanche process produces a carrier multiplication factor M or an avalanche gain which is dependent on α_n and α_p. The responsivity of an APD is expressed by Equation 9.20 where G is the gain and it is equivalent to the multiplication factor M. The low-frequency gain for electrons is given by

$$M = \left\{1 - \int_0^W \alpha_n \exp\left[-\int_0^x (\alpha_n - \alpha_p)\mathrm{d}x'\right]\mathrm{d}x\right\}^{-1} \tag{9.40}$$

TABLE 9.2 The Impact Ionization Coefficients for Various Semiconductors

	a (cm^{-1})		b ($V cm^{-1}$)		
Semiconductor Material	Electrons	Holes	Electrons	Holes	Reference
GaAs	1.1×10^7	5.5×10^6	2.2×10^6	2.2×10^6	—[a]
GaAs	3.82×10^4	4.50×10^4	3.80×10^5	3.10×10^5	—[b]
$In_{0.15}Ga_{0.63}Al_{0.22}As$	3.19×10^4	4.03×10^4	4.03×10^5	3.25×10^5	—[b]
$In_{0.2}Ga_{0.8}As$	5.9×10^4	6.8×10^4	3.0×10^5	3.02×10^5	—[b]
Si	9.2×10^5	2.4×10^5	1.45×10^6	1.64×10^6	—[c]
InP	5.5×10^6	1.98×10^6	3.10×10^6	2.29×10^6	—[d]
$In_{0.53}Ga_{0.47}As$	2.27×10^6	3.95×10^6	1.13×10^6	1.45×10^6	—[d]
$In_{0.67}Ga_{0.33}As_{0.70}P_{0.30}$	3.37×10^6	2.94×10^6	2.29×10^6	2.40×10^6	—[d]

[a] Ando, H. and Kanbe, H., Ionization coefficient measurement in GaAs by using multiplication noise characteristics 19.81. *Solid State Elec.*, 24, 629–634.

[b] Chen, Y. C. and Bhattacharya, P.K., Impact ionization coefficients for electrons and holes in strained $In_{0.2}Ga_{0.8}As$ and $In_{0.15}Ga_{0.63}Al_{0.22}As$ channels embedded in $Al_{0.3}Ga_{0.7}As$. 1993. *J. Appl. Phys.*, 73(1), 465–467.

[c] Woods, M. H., Johnson, W. C., and Lampert, M. A., 1973. Use of a Schottky barrier to measure impact ionization coefficients in semiconductors, *Solid State Elec.*, 16, 381–394.

[d] Osaka, F., Mikawa, T., and Kaneda, T., 1985. Impact ionization coefficients of electrons and holes in (100)-Oriented $Ga_{1-x}In_xAs_yP_{1-y}$. *IEEE J. Quantum Elec.*, 21(9), 1326–1338.

where W is the width of the depletion region. When $\alpha_n \neq \alpha_p$ and the values of both ionization rates are independent of position as *p–i–n* diodes, then,

$$M = \frac{(1-\alpha_p/\alpha_n)\exp[\alpha_n W(1-\alpha_p/\alpha_n)]}{1-(\alpha_p/\alpha_n)\exp[\alpha_n W(1-\alpha_p/\alpha_n)]}. \tag{9.41}$$

For $\alpha_n = \alpha_p = \alpha$, the multiplication factor M takes the form

$$M = \frac{1}{(1-\alpha W)}. \tag{9.42}$$

It is observed that M goes to ∞ when $\alpha W = 1$ which signifies the condition for junction breakdown. Therefore, high values of M can be obtained for photodiodes biased near the breakdown voltage. A schematic of the avalanche gain in an APD is illustrated in Figure 9.21 showing the sharp increase in gain as the reverse bias approaches the breakdown voltage. The critical field at which breakdown is initiated depends on impurity concentration in semiconductors and be calculated from the one-sided abrupt *p–n* junction approximation.

The steady-state photocurrent in an APD in the presence of an avalanche gain is given by

$$I_p = \frac{q\eta P_i}{h\nu} M \tag{9.43}$$

where P_i is the incident optical power. For the case of an intensity-modulated incident optical signal, the root mean square (rms) signal with a modulation index m is expressed as

$$i_p = \frac{q\eta P_i m}{\sqrt{2}h\nu} M. \tag{9.44}$$

The avalanche process is a regenerative process and it can be time-consuming. It takes time to build up which implies that the higher the avalanche gain, the longer the avalanche process persists. This results in the presence of a large number of secondary carriers in the depletion region after the primary carriers

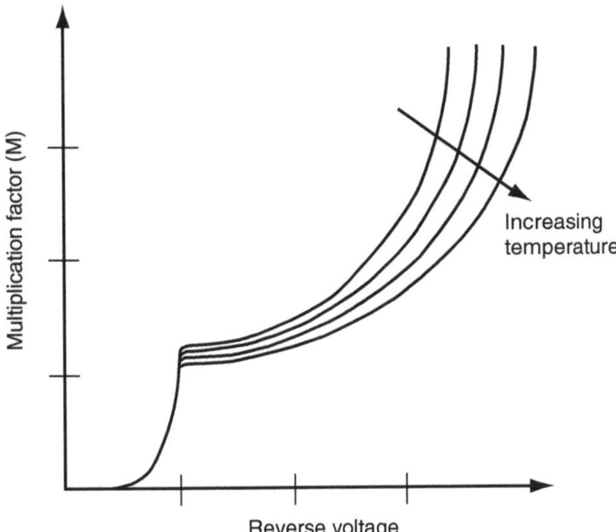

FIGURE 9.21 A representation of the multiplication factor as a function of reverse bias voltage and temperature for an APD.

have been collected. Therefore, the gain and bandwidth of an APD are inextricably linked together. This results in a finite gain-bandwidth product fixed for a given material and device structure. The gain-bandwidth product is given by

$$B \cdot M = \frac{1}{2\pi\tau_1} \tag{9.45}$$

where $\tau_1 = (\tau_n + \tau_p)/2$, τ_n is the electron transit time, and τ_p is the hole transmit time. τ_n and τ_p are given by W/v_n and W/v_p where v_n and v_p are electron and hole saturation velocities, respectively. The equation above has only considered the transit-time effects on the response speed of a photodiode. Other usual effects are diffusion and *RC* limitations. Another effect particular to APDs is the time it takes carriers to complete the avalanche process, which is called the avalanche build-up time. Due to the randomness of the multiplication process, the build-up time is itself random. The dependence of the multiplication factor on frequency for $M_0 > \alpha_n/\alpha_p$ is of the form

$$M(\omega) \approx \frac{M_0}{\sqrt{1 + \omega^2 M_0^2 \tau_t^2}} \tag{9.46}$$

where M_0 is the dc value of the multiplication factor and t_t is an effective transit time through the avalanche region. A functional form that describes the effective transit time is

$$\tau_t = N\tau_1(\alpha_n/\alpha_p) \tag{9.47}$$

where τ_1 is the real transit time through the avalanche region and N is a number slowly varying from $N = 1/3$ to 2 as α_p/α_n varies from 1 to 0.001.

Apart from the usual shot noise limitations in photodiodes, the randomness associated with the avalanche or multiplication process makes it a principal source of noise in APDs. These random fluctuations create a distribution of gain which causes excess noise in the device. For the condition where

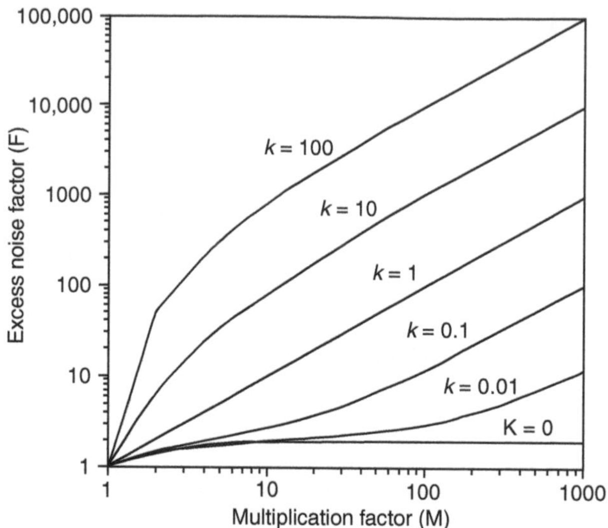

FIGURE 9.22 Excess noise factor as a function of multiplication gain and the ratio of ionization coefficients.

the avalanche is initiated by electrons (i.e., $\alpha_n > \alpha_p$), the excess noise factor is given by

$$F = M\left[1-(1-k)\left(\frac{M-1}{M}\right)^2\right] \tag{9.48}$$

where k is the ratio of the ionization factor, i.e., α_p/α_n. An equivalent expression also exists for a hole-induced avalanche gain in which k is replaced by $k' = \alpha_n/\alpha_p$. There are two special cases, $k=0$ and $k=1$. The first case results in

$$F = 2-\frac{1}{M} \tag{9.49}$$

whereas the second, denoting where both ionization coefficients are equal, results in

$$F = M. \tag{9.50}$$

The noise factor for various multiplication gains and ratios of ionization coefficient is shown in Figure 9.22.

In an APD, the dark current and the current due to background radiation are all enhanced by the multiplication gain similar to that of the signal current in Equation 9.43. Therefore the root mean shot noise is given by

$$\left\langle i_s^2 \right\rangle = 2q(I_P + I_D + I_B)FBM^2. \tag{9.51}$$

The Johnson or thermal noise for an APD is identical to that of a PIN photodiode. Therefore, the resulting NEP for an APD is

$$\text{NEP} = \frac{h\nu}{q\eta M}\left[2q(I_P + I_D + I_B)FM^2 + \frac{4kT}{R_{\text{eff}}}\right]^{1/2}. \tag{9.52}$$

It is observed that the NEP of an APD is almost identical to that of a PIN photodiode except for an extra gain factor M which appears in the denominator. It is evident, therefore, that the gain factor does reduce

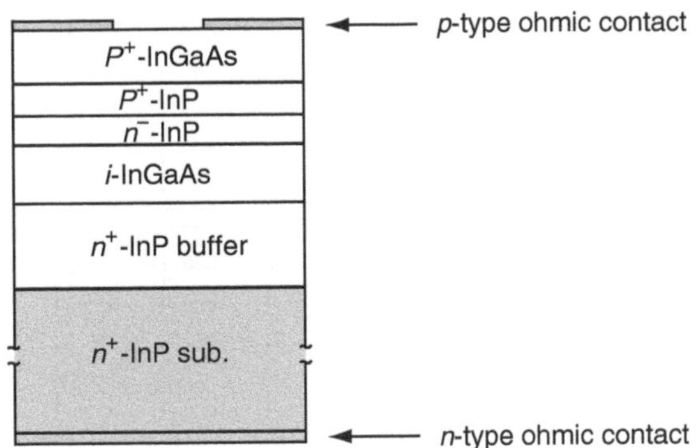

FIGURE 9.23 Schematic diagram of a separate avalanche and multiplication avalanche photodiode (SAM-APD).

the NEP of an APD, making the device more sensitive than PIN photodiodes. Indeed, APDs have shown superior sensitivity compared to other photodiodes.

Important issues for consideration in designing and fabricating APDs concern dark current and high-speed performance. The contributions to the dark current of an APD include generation–recombination in the depletion region, tunneling of carriers across the bandgap, and leakage across junctions. Various methods have been developed to alleviate these problems. For example, guard rings and dielectric film deposition are used in Si APDs to minimize various forms of leakage currents. Modern APDs utilized for fiber-optic communications depend on InP related compounds. Lightwave operation at 1.55 μm relies on $In_{0.47}Ga_{0.53}As$ which has an energy bandgap of 0.75 eV. Large tunneling current is a problem if a material with such a low bandgap is used in homojunction devices. Therefore, a separate absorption and multiplication (SAM) structure, in which a low-field InGaAs region is utilized as the absorption region and a high-field Inp region is the avalanching area, is now used for APDs. Figure 9.23 shows a typical InP/InGaAs SAM-APD heterostructure where the p^+–n^- junction is located in the high-bandgap (1.35 eV) InP material. In practice, a graded bandgap material (InGaAsP) is placed at the InP/InGaAs heterojunction to prevent a sharp energy bandgap discontinuity which can trap carriers and result in slow device response. This modification results in a structure called separate absorption-graded-multiplication APD (or SAGM-APD). These devices combine low dark current and high gain-bandwidth operations.

Other types of APDs have been designed to ensure preferential multiplication of one carrier type (i.e., k or k' approaches zero) over the other. These are the superlattice and the multiquantum-well APDs.

9.3.9 Metal–Semiconductor–Metal Photodiode

A metal–semiconductor–metal (MSM) photodiode consists of two interdigitated electrodes which form back-to-back Schottky diodes on a semiconductor absorbing layer. The MSM is a planar device which can be fabricated easily. This ease in fabrication along with the ease in integrating the MSM with conventional field-effect-transistor (FET) technology have provided the underlying impetus to gradually improve the MSM's efficiency, dark current, and bandwidth to the point where it can now compete with the PIN photodiode. The schematic of the MSM is shown in Figure 9.24. The interdigitated electrodes are similar to those of the photoconductors of Figure 9.14 but differ in the type of contacts made to the semiconductor. The contacts of photoconductors are ohmic. Photogenerated electrons and holes in MSMs are transported along the electric field lines to the oppositely biased contacts as shown in Figure 9.24b. The electric field distribution is nonuniform due to the electrode geometry, with fields being strongest near the semiconductor's surface. Therefore, the applied bias must be sufficiently large to ensure the depletion of the entire absorbing region for the carriers to travel at their saturation velocities. The energy band diagram in Figure 9.24c shows the flow of the generated electrons and holes and also

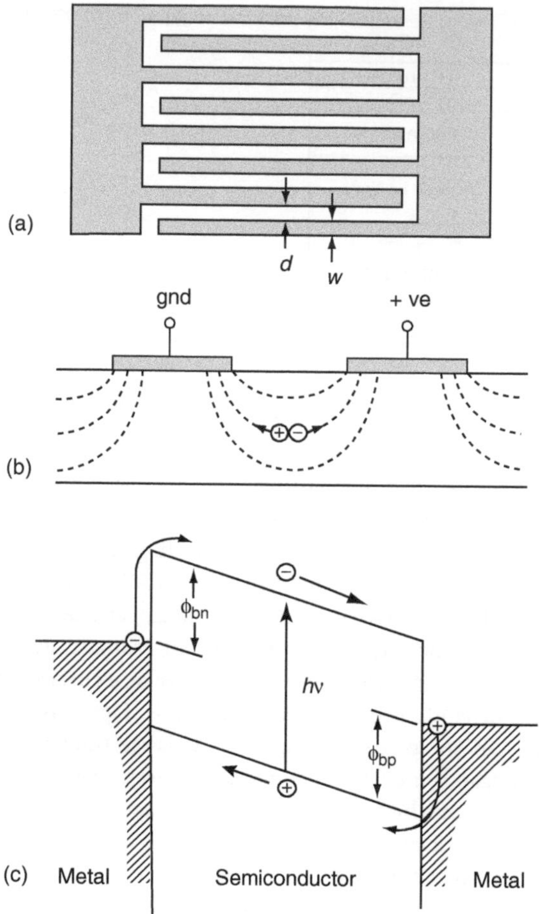

FIGURE 9.24 (a) Schematic of a MSM photodiode, (b) cross-sectional view of the device displaying the direction carriers travel is response to an electric field, and (c) the associated energy band diagram.

shows how the Schottky barriers limit the injection of carriers from the metal contacts to the semiconductor.

The external quantum efficiency is basically given by Equation 9.17, but taking the opaque interdigitated electrodes into account, we have

$$\eta = (1-r)\frac{d}{d+w}\zeta[1-\exp(-\alpha L)] \tag{9.53}$$

where w is the electrode finger width and d is the finger spacing. It is seen that η is reduced by the opacity of the electrodes. The use of transparent Schottky contacts, such as indium tin oxide and cadmium tin oxide, to avoid the opacity of metal electrodes has been demonstrated. The shortcoming of this method is that carries generated in the low-field region under the contacts degrade the bandwidth of the device. The responsivity of MSMs is given by Equation 9.20. Ideally, the gain G should be unity. However, two possible gain mechanisms have been identified in MSMs. These are (i) the gain due to avalanche multiplication in high-field region of the device, for example, around the sharp edges of the contacts and (ii) the gain due to trapping of carriers. Trapping of carriers can occur at trapping centers on semiconductor surface (between fingers), can have a detrimental influence on the dark current and high-speed response. The use of high quality epitaxial materials (low trap density), fabrication of high

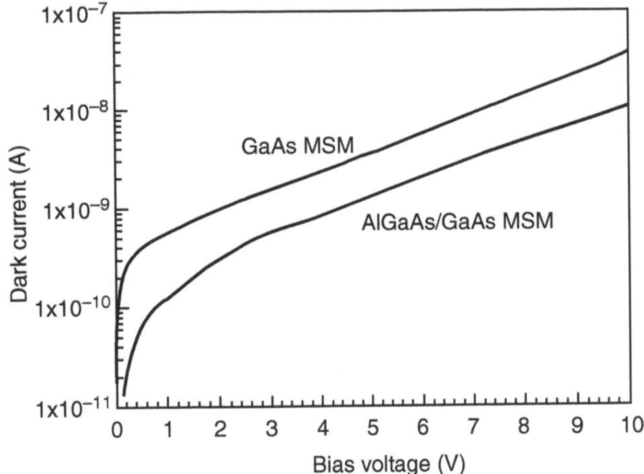

FIGURE 9.25 Comparison of the dark current of a 1 μm by 1 μm GaAsMSM photodiade with and without an AlGaAs cap layer. (From Seo, J. W., 1993. Metal–Semiconductor–Metal Photodetectors for Optoelectronic Receiver Applications, Ph.D. Thesis, University of Illinois at Urbana-Champaign, Urbana, IL. With permission.)

quality Schottky diodes (no interfacial oxides), and appropriate biasing of devices (avoid high fields at contacts) are some means of minimizing these effects.

The dark current of an epitaxial GaAs MSM is shown in Figure 9.25. The photodiode has an area of 50×50 μm, a finger width of 1 μm a finger spacing of 1 μm, and an absorbing layer thickness of 1 μm. A symmetrical trace of the above result should be obtained for negative voltages due to the back-to-back Schottky diode configuration. The dark current of an MSM is dominated by thermionic emission over the barrier as illustrated in Figure 9.24. A simple one-dimensional model gives the flat band voltage V_{fb} at the forward-biased contract as

$$V_{\mathrm{fb}} = \frac{qN_{\mathrm{D}}d^2}{2\varepsilon} \tag{9.54}$$

where q is the electron charge, ε is the dielectric constant of the semiconductor, N_{D} is the residual doping in the semiconductor, and d is the electrode spacing. The maximum V_{fb} is limited by the avalanche breakdown voltage near the reverse-biased contact. At low bias voltage compared to V_{fb}, electron injection at the reverse-biased contact is the dominant factor in carrier transport. At higher bias, hole injection at the forward-biased contact become dominant. At higher voltages, there is a reach-through condition, when the edges of the depletion regions of the two Schottky diodes merge. The total current density for a voltage higher than the reach-through voltage is given by

$$J_{\mathrm{tot}} = A_{\mathrm{n}}^{*}T^2\exp\left[\frac{-q(\phi_{\mathrm{bn}} - \Delta\phi_{\mathrm{bn}})}{kT}\right] + A_{\mathrm{p}}^{*}T^2\exp\left[\frac{-q(\phi_{\mathrm{bp}} - \Delta\phi_{\mathrm{bp}})}{kT}\right] \tag{9.55}$$

where A_{n}^{*} and A_{p}^{*} are Richardson constants for electrons and holes, ϕ_{bn} and ϕ_{bp} are the barrier heights for electrons and holes, and, finally, $\Delta\phi_{\mathrm{bn}}$ and $\Delta\phi_{\mathrm{bp}}$ are the respective image force barrier lowerings. A_{n}^{*} and A_{p}^{*} are 8.15 and 74.4 A/cm^2/K^2 for GaAs, respectively. The dark current in Figure 9.25 compares to that of a PIN photodiode. Improvements in MSM dark current have been shown to be possible by choosing an appropriate metal system for Schottky contacts and also by using surface epitaxial layers with higher bandgap than the absorbing layer. The latter example is demonstrated for an AlGaAs/GaAs MSM in Figure 9.25 where the Schottky contacts were made to an $Al_{0.3}Ga_{0.7}As$ layer grown on the GaAs absorbing layer.

The bandwidth of a MSM photodiode is determined by the carrier transit time and the *RC* time constant, and both depend on the physical dimensions of the device. The carrier transit time is proportional to the distance between the electrodes if the effects of velocity overshoot and fringing electric field are neglected. The time is limited by the slowest carriers (holes). Using a dc model, the resistance of a MSM photodiode is given by

$$R_{\text{tot}} = \frac{2R_0}{N} \tag{9.56}$$

where R_0 is the resistance of a single electrode and N is the number of electrodes on each side of the two sets of interdigitated electrodes. The resistance of the electrodes is usually less than the 50 Ω load resistance. A major advantage of an interdigitated MSM is its low capacitance. For an undoped and infinitely thick semiconductor, the capacitance is given by

$$C = \frac{\kappa(k)}{\kappa(k')}\varepsilon_0(1+\varepsilon_r)\frac{A}{w+d} \tag{9.57}$$

where ε_r is the relative dielectric constant of the semiconductor, A is the active area of the MSM, w and d are the finger width and spacing, respectively, and $\kappa(k)$ is the complete elliptic integral of the first kind given by

$$\kappa(k) = \int_0^{\pi/2} \frac{d\Phi}{\sqrt{1-k^2\sin^2\Phi}}, \tag{9.58}$$

$$k = \tan^2\left[\frac{\pi w}{4(w+d)}\right], \tag{9.59}$$

and

$$k' = \sqrt{1-k^2}. \tag{9.60}$$

Using these expressions, it has been shown that the capacitance of a MSM is significantly lower than that of a PIN photodiode of corresponding size.

The transit-time limited response of the a surface-illuminated PIN photodiode can be adapted to the treatment of MSMs. The upper bound of the transit-time limited bandwidth of a MSM is given by Equation 9.35 where the interdigitated spacing of the MSM is equal to the intrinsic layer thickness of the PIN. Because the effects of the carrier transit time and the *RC* time constant on the bandwidth are cumulative, then, the bandwidth of a MSM is given by

$$B = \left(\frac{d}{0.45\nu} + R_L C\right)^{-1} \tag{9.61}$$

where R_L is the MSM load resistance and $\nu_n = \nu_p = \nu$. Although, the finger with to finger spacing ratio is a design parameter for MSM, if a ratio of unity is assumed for practical reasons, then, Equation 9.61 becomes

$$B = \left(\frac{d}{0.45\nu} + \frac{\pi K R_L}{d}\right)^{-1} \tag{9.62}$$

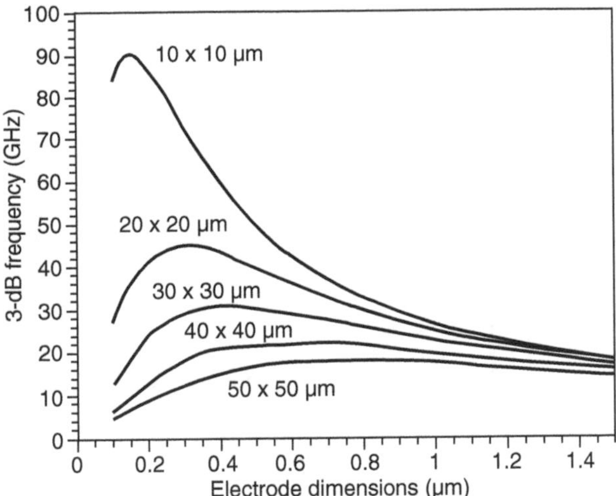

FIGURE 9.26 The theoretical frequency response of an InGaAs MSM photodiode as a function of electrode and active area dimensions for a device with equal electrode width and spacing.

where K is a constant given by

$$K = \frac{K(k_0)}{K(k_0')}\varepsilon_0(1+\varepsilon_r)A \tag{9.63}$$

with

$$k_0 = \tan^2\left(\frac{\pi}{8}\right). \tag{9.64}$$

The elliptic integrals in Equation 9.63 are $K(k_0)=1.58$ and $K(k_0')=3.17$ for a device with $d=w$. Figure 9.26 shows a plot of bandwidth as a function of finger spacing for an InGaAs MSM with the active area as a variable parameter. As shown in the curve, for small values of d the response in RC-limited for larger values of d the response is transit-time limited.

It is possible to obtain the finger spacing d_{max}, which gives a maximum available bandwidth, by equating the derivative of Equation 9.62 with respect to d to zero, yielding

$$d_{max} = \sqrt{0.45\pi vKR_L}. \tag{9.65}$$

Therefore, for a device with a $50\times50\ \mu m^2$ active area, an $In_{0.47}Ga_{0.53}As$ absorbing layer, and a 50 Ω load resistance, we obtain a d_{max} and B_{max} of 1 μm and 21.4 GHz, respectively. Figure 9.27 shows the pulse response waveforms of InA1As/InGaAs MSMs with Ti/Au Schottky contacts at a bias of 10 V for various device finger widths/finger spacings and a $50\times50\ \mu m^2$ active area. The full-width-at-half-maximums (FWHMs) of the response pulses for the devices are 42, 47, and 52 ps, respectively. Corresponding estimated bandwidths are 20, 14, and 11 GHz, respectively.

Various schemes have been adopted to improve the high speed response of MSMs. The use of thin absorbing layers and small finger spacing to finger width ratio will ensure very fast transit time but low quantum efficiency. The smaller ratio also increases the capacitance. Ion implantation in the absorbing region reduces transit time but also reduces sensitivity. This may contribute to excess noise and gain. In GaAs MSM design, the use of an AlGaAs barrier layer in the buffer region has been shown to improve response with a slight reduction in the dc responsivity. The barrier impedes slow photogenerated carriers

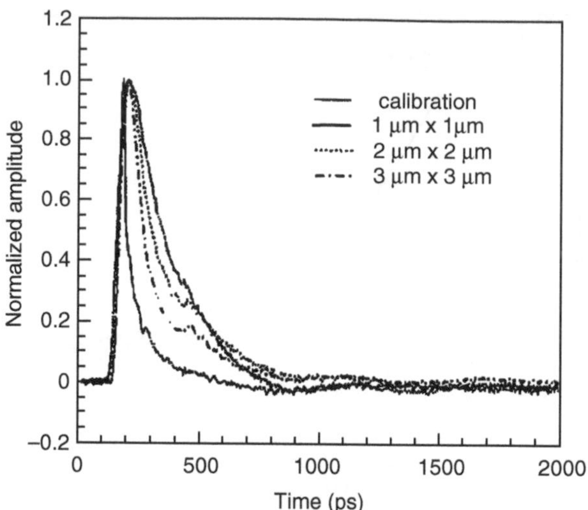

FIGURE 9.27 A typical time-domain pulse response of an InGaAs MSM for devices with different electrode dimensions. (From Seo, J. W. 1993. Metal–Semiconductor–Metal Photodetectors for Optoelectronic Receiver Applications, Ph.D. Thesis, University of Illinois at Urbana-Champaign, Urbana, IL. With permission.)

from deep in the substrate, where the electric field is week, from being collected by the electrodes. This leads to reduced transit time and a suppressed tail in the pulse response. To maximize charge collection and also reduce transit time, a graded layer is usually placed at the heterostructure interface to prevent a sharp, energy-band discontinuity.

Much like the PIN photodiode, the source of shot noise in MSM are the photocurrent, the dark current, and the current due to background radiation. There is also thermal noise component. The MSM photodiode has a lateral current flow with strong interactions with semiconductor interfaces like a FET device. Therefore, it is expected that MSM will exhibit a low-frequency, excess noise component with a $1/f^{\alpha}$-like dependence. This excess $1/f^{\alpha}$ noise has been observed at low frequencies and high biases close to breakdown. Evidence shows that this excess noise is very dependent on the presence of traps. A significant excess noise was obtained at frequencies below 35 MHz for an InGaAs MSM which had a strained GaAs schottky barrier enhancement layer. However, for a high quality InGaAs MSM with an InA1As enhancement layer, the device was shot-noise dominated at frequencies as low as 1 MHz. Presumably, the excess noise for the GaAs/InGaAs MSMs is due to the large number of traps present at the strained interface between GaAs and InGaAs. Therefore, for MSMs fabricated on high quality epitaxial layers, the dominant noise properties will be similar to those of PIN photodiodes.

Acknowledgments

The authors gratefully acknowledgement the superb assistance of Walter Wohlmuth in preparing editing the photodetectors aspects of this chapter. The work was supported by funding from the National Science Foundation (ECD 89-43166 and DMR 89-20538), the Advanced Research Projects Agency (MDA972-94-1-0004), and the Joint Services Electronics Program (N0014-90-J-1270).

References

Adachi, S. 1992. *Physical Properties of III–V Semiconductor Compounds, InP, InAs, GaAs, GaP, InGaAs, and InGaAsP.* Wiley, New York.

Agarwal, G. P. and Dutta, N. K. 1993. *Semiconductor Lasers, 2nd Ed.*, Van Nostrand Reinhold, New York.

Bhattacharya, P. 1993. *Properties of Lattice-Matched and Strained Indium Gallium Arsenide.* INSPEC, Institution of Electrical Engineers, London.

Bhattacharya, P. 1994. *Semiconductors OptoElectronic Devices.* Prentice-Hall, Englewood Cliffs, NJ.

Bowers, J. E. and Burrus, C. A. 1987. Ultrawide-band long-wavelength p–i–n photodetectors. *J. Lightwave Tech.*, 5 (10), 1339.

Bowers, J. W., Burrus, C. A., and McCoy, R. J. 1985. InGaAs PIN photodetectors with modulation response to millimeter wavelengths. *Elec. Lett.*, 21 (18), 812.

Brennan, K. F. 1998. Field and spatial geometry dependencies of the electron and hole ionization rates in GaAs/A1GaAs multiquantum well APD's. *IEEE Trans. Elec. Dev.*, 35 (5), 634.

Butler, J. K. 1979. *Semiconductors Injection Lasers.* IEEE Press, New York.

Capasso, F. 1982. New ultra-low noise avalanche photodiode with separated electron and hole avalanche regions. *Elec. Lett.*, 18 (1), 12.

Capasso, F., Mohammed, K., and Cho, A. Y. 1985. Tunable barrier heights and band discontinuities via doping interface dipoles: an interface engineering technique and its device applications. *J. Vac. Sci. Tech., B*, 3 (4), 1245.

Cheo, P. K. 1989. *Handbook of Solid State Lasers.* Marcel Dekker, New York.

Chou, S. Y., Liu, Y., and Fischer, P. B. 1992a. Terahertz GaAs metal–semiconuctor–metal photodetectors with 25 nm finger spacing and finger width. *Appl. Phys. Lett.*, 61 (4), 477.

Chou, S. Y., Liu, Y., Khalil, W., Hsiang, T. Y., and Alexandrou, S. 1992b. Ultrafast nanoscale metal–semiconductor–metal photodetectors on bulk and low-temperature grown GaAs. *Appl. Phys. Lett.*, 61 (7), 819.

Coleman, J. J. 1992. *Selected Papers on Semiconductors Diode Lasers.* SPIE Optical Engineering Press, Bellingham, WA.

Dentan, M. and de Cremoux, B. 1990. Numerical simulation of the nonlinear response of a p–i–n photodiode under high illumination. *J. Lightwave Tech.*, 8 (8), 1137.

Dutta, N. K., Lopata, J., Berger, P. R., Wang, S. J., Smith, P. R., Sivco, D. L., and Cho, A. Y. 1993. 10 GHz bandwidth monolithic p–i–n modulation-doped field effect transistor photoreceiver. *Appl. Phys. Lett.*, 63 (15), 2115.

Ferry, D. K. 1985. *Gallium Arsenide Technology.* H.W. Sams, Indianpolis, IN.

Gooch, G. H. 1973. *Injection Electroluminescent Devices.* Wiley, New York.

Hanatani, S., Nakamura, H., Tanaka, S., and Ido, T. 1994. Flip-Chip InA1As/InGaAs superlattice avalanche photodiodes with back-illuminated structures. *Microw. Opt. Tech. Lett.*, 7 (3), 103.

Harder, Ch. S., van Zeghbroeck, B. J., Kesler, M. P., Meier, H. P., Vettiger, P., Webb, D. J., and Wolf, P. 1990. High-Speed GaAs/A1GaAs optoelectronic devices for computer applications. *IBM J. Res. Dev.*, 34 (4), 568.

He, L., Lin, Y., van Rheenen, A. D., van der Ziel, A., van der Ziel, J. P., and young, A. 1990. Low-Frequency noise in small InGaAs/InP p–i–n diodes under different bias and illumination conditions. *J. Appl. Phys.*, 68 (10), 5200.

Hunsberger, R. G. 1984. *Integrated Optics: Theory and Technology, 2nd Ed.*, Springer, Berlin.

Jhee, Y. K., Campbell, J. C., Holden, W. S., Dentai, A. G., and Plourde, J. K. 1985. The effect of nonuniform gain on the multiplication noise of InP/InGaAsP/InGaAs avalanche photodiodes. *IEEE J. Quantum Elec.*, 21 (12), 1858.

Kahraman, G., Saleh, B. E. A., Sargeant, W. L., and Teich, M. C. 1992. Time and frequency response of avalanche photodiodes with arbitrary structure. *IEEE Trans. Elec. Dev.*, 39 (3), 553.

Kawaga, T., Kawamura, Y., and Iwamura, H. 1993. A wide-bandwidth low-noise InGaAsP–InA1As superlattice avalanche photodiode with a flip-chip structure for wavelengths of 1.3 and 1.55 μm. *IEEE J. Quantum Elec.*, 29 (5), 1387.

Kim, J. H., Griem, H. T., Friedman, R. A., Chan, E. Y., and Ray, S. 1992. High-Performance back-illuminated InGaAs/InA1As MSM photodetector with a record responsivity of 0.96 A/W. *IEEE Photonics Tech. Lett.*, 4 (11), 1241.

Klingenstein, M., Kuhl, J., Rosenzweig, J., Moglestue, C., and Axmann, A. 1991. Transit time limited response of GaAs metal–semiconductor–metal photodiodes. *Appl. Phys. Lett.*, 58 (22), 2503.

Kressel, H. 1982. *Topics in Applied Physics. Semiconductor Devices for Optical Communication, Vol. 39*, Springer, Berlin.

Kressel, H. and Butler, J. K. 1977. *Semiconductor Lasers and Heterojunction LEDs*. Academic Press, New York.

Kuhl, D., Hieronymi, F., Böttcher, E. H., Wolf, T., Krost, A., and Bimberg, D. 1990. Very high-speed metal–semiconductor–metal InGaAs:Fe photodetectors with InP:Fe barrier enhancement layer grown by low pressure metalloorganic chemical vapour deposition. *Elec. Lett.*, 26 (25), 2107.

Kuhl, D., Hieronymi, F., Böttcher, E. H., Wolf, T., Bimberg, D., Kuhl, J., and Klingenstein, M. 1992. Influence of space charges on the impulse response of InGaAs metal–semiconductor–metal photodetectots. *J. Lightwave Tech.*, 10 (6), 753.

Lim, Y. C. and Moore, R. A. 1968. Properties of alternately charged coplanar parallel strips by conformal mappings. *IEEE Trans. Elec. Dev.*, 15 (3), 173.

Lu, J., Surridge, R., Pakulski, G., van Driel, H., and Xu, J. M. 1993. Studies of high-speed metal–semiconductor–metal photodetector with a GaAs/A1GaAs/GaAs heterostructure. *IEEE Trans. Elec. Dev.*, 40 (6), 1087.

Lucovsky, G., Schwarz, R. F., and Emmons, R. B. 1963. Transit-time considerations in p–i–n diodes. *J. Appl. phys.*, 35 (3), 622.

Miller, S. E. and Kaminow, I. P. 1988. *Optical Fiber Telecommunications II*. Academic Press, New York.

Newman, D. H. and Ritchie, S. 1986. Sources and detectors for optical fiber communications applications: the first 20 years. *IEEE Proc. Pt. J.*, 133 (3), 213.

Panish, M. B. and Casey, H. C. 1978a. *Heterostructure Lasers, Part A: Fundamental Principles*. Academic Press, New York.

Panish, M. B. and Casey, H. C. 1978b. *Heterostructure Lasers, Part B: Materials and Operating Characteristics*. Academic Press, New York.

Pankove, J. I. 1971. *Optical Processes in Semiconductors*. Dover, New York.

Parker, D. G. 1988. The theory, fabrication and assessment of ultra high speed photodiodes. *GEC J. Res.*, 6 (2), 106.

Pearsall, T. P. 1982. *GaInAs Alloy Semiconductors*. Wiley, New York.

Peredo, E., Decoster, D., Gouy, J. P., Vilcot, J. P., and Constant, M. 1994. Comparison of InGaAs/InP photodetectors for microwave applications. *Microw. Opt. Tech. Lett.*, 7 (7), 332.

Peterson, R. L. 1987. Numerical study of currents and fields in a photoconductive device. *IEEE J. Quantum Elec.*, 23 (7), 1185.

Powers, J. P. 1993. *An Introduction to Fiber Optical Systems*. Richard D. Irwin and Aksen Associates, Homewood, IL.

Ridley, B. K. 1985. Factors affecting impact ionisation in multilayer avalanche photodiodes. *IEE Proc. Pt. J.*, 132 (3), 177.

Sabella, R. and Merli, S. 1993. Analysis of InGaAs P–I–N photodiode frequency response. *IEEE J. Quantum Elec.*, 29 (3), 906.

Saleh, B. E. A. and Teich, M. C. 1991. *Fundamentals of Photonics*. Wiley, New York.

Schichijo, H. and Hess, K. 1981. Band-structure dependent transport and impact ionization in GaAs. *Phys. Rev. B*, 23 (8), 4197.

Schumacher, H., LeBlanc, H. P., Soole, J. B. D., and Bhat, R. 1988. An investigation of the optoelectronic response of GaAs/InGaAs MSM photodetectors. *IEEE Elec. Dev. Lett.*, 9 (11), 607.

Soole, J. B. D. 1992. InGaAs MSM photodetectors for long wavelength fiber communications. *SPIE, High Speed Elec. OptoElectronics*, 1680, 153.

Soole, J. B. D. and Schumacher, H. 1991. InGaAs metal–semiconductor–metal photodetecrors for long wavelength optical communications. *IEEE J. Quantum Elec.*, 27 (3), 737.

Stillman, G. E. and Wolfe, C. M. 1977. Avalanche photodiodes. In *Semiconductors and Semimetals*, R. K. Willardson and A. C. Beer, eds., pp. 291–393. Academic Press, New York.

Stillman, G. E., Robbins, V. M., and Tabatabaie, N. 1984. III–V Compound semiconductor devices: optical detectors. *IEEE Trans. Elec. Dev.*, 31 (11), 1643.

Streetman, B. G. 1990. *Solid State Electronics Devices, 3rd Ed.*, Prentice-Hall, Englewood Cliffs, NJ.

Thompson, G. H. B. 1980. *Physics of Semiconductor Laser Devices*. Wiley, New York.

Tiwari, S. and Tischler, M. A. 1992. On the role of mobility and saturated velocity in the dynamic operation of p–i–n and metal–semiconductor–metal photodetectors. *Appl. Phys. Lett.*, 60 (9), 1135.

Tsang, W. T. 1985a. *Semiconductors and Semimetals, Lightwave Communications Technology, Part A Material Growth Technologies., Vol. 22*, Academic Press, New York.

Tsang, W. T. 1985b. *Semiconductors and Semimetals, Lightwave Communications Technology, Part B Semiconductor Injection Lasers I., Vol. 22*, Academic Press, New York.

Tsang, W. T. 1985c. *Semiconductors and Semimetals, Lightwave Communications Technology, Part C Semiconductor Injection Lasers II., Vol. 22*, Academic Press, New York, NY.

Tucker, R. S., Taylor, A. J., Burrus, C. A., Eisenstein, G., and Wersenfeld, J. M. 1986. Coaxially mounted 67 GHz bandwidth InGaAs photodiode. *Elec. Lett.*, 22 (17), 917.

van Zeghbroeck, B. J., Patrick, W., Halbout, J.-M., and Vettiger, P. 1988. 105-GHz bandwidth metal–semiconductor–metal photodiode. *IEEE Elec. Dev. Lett.*, 9 (10), 527.

Verdeyen, J. T. 1989. *Laser Electronics, 2nd Ed.*, Prentice-Hall, Englewood Cliffs, NJ.

Wada, O., Nobuhara, H., Hamaguchi, H., Mikawa, T., Tackeuchi, A., and Fuji, T. 1989. Very high speed GaInAs metal–semiconductor–metal photodiode incorporating an A1InAs/GaInAs graded superlattice. *Appl. Phys. Lett.*, 54 (1), 16.

Waynant, R. W. and Ediger, M. N. 1994. *Electro-Optics Handbook*. McGraw-Hill, New York.

Wilson, J. and Hawkes, J. F. B. 1989. *Optoelectronics: An Introduction, 2nd Ed.*, Prentice-Hall, Englewood Cliffs, NJ.

Wojtczuk, S. J., Ballantyne, J. M., Wanaga, S., and Chen, Y. K. 1987. Comparative study of easily integratable photodetectors. *J. Lightwave Tech.*, 5 (10), 1365.

Wright, D. R., Keir, A. M., Pryce, A. M., Birbeck, J. C. H., Heaton, J. M., Norcross, R. J., and Wright, P. J. 1988. Limits of electroabsorption in high purity GaAs and the optimisation of waveguide devices. *IEE Proc. Pt. J.*, 135 (1), 39.

Yariv, A. 1989. *Quantum Electronics, 3rd Ed.*, Wiley, New York.

Yi, M. B., Paslaski, J., Lu, L.-T., Margalit, S., Yariv, A., Blauvelt, H., and Lau, K. 1985. InGaAsP p–i–n photodiodes for optical communication at the 1.3 μm wavelength. *J. Appl. Phys.*, 58 (12), 4730.

Zory, P. S. 1993. *Quantum Well Lasers*. Academic Press, New York.

10

Miniature Solid-State Lasers

John J. Zayhowski*

David Welford

J. Harrison

10.1 Introduction

This chapter presents fundamental concepts and formulas for designing and using miniature solid-state lasers. Diode-pumped, miniature, monolithic, solid-state lasers offer an efficient, compact, and robust means of generating diffraction-limited, single-frequency radiation. In addition, their diminutive size results in high-speed tuning capabilities and short-pulsed operation unmatched by larger devices. Application areas are as diverse as communications, spectroscopy, remote sensing, nonlinear optics, projection displays, and micromachining.

* The Lincoln Laboratory portion of this work was sponsored by the Department of the Air Force under Air Force Contract F19628-00-C-0002. Opinions, interpretations, conclusions, and recommendations are those of the authors and are not necessarily endorsed by the United States Government.

The text is structured to provide adequate background for those readers who are new to the area. Section 10.2 through Section 10.4 present a broad introduction to laser parameters and modeling, gain media, and resonators. Section 10.5 discusses the interactions in the gain media that determine spectral purity and presents methods for controlling the frequency of a laser. Section 10.6 contains a thorough review of issues specific to various modes of pulsed operation. There is a detailed discussion of Q switching, which presents analytic results of detailed temporal modeling. Passive Q switching is discussed in considerable detail, reflecting the relative importance that passively Q-switched miniature lasers have obtained in recent years. The final technical section, Section 10.7, offers a few comments on some important adjuncts to miniature lasers, including amplifiers, frequency converters, and modulators.

Although the presentation in this chapter is tailored specifically toward miniature solid-state lasers, the concepts are generally applicable to other classes of lasers. A section entitled Further Reading has been included at the end of the chapter to help direct the interested reader to references on other laser types.

10.2 Fundamental Concepts

10.2.1 Laser Components

A laser generally consists of three components: an active medium with energy levels that can be selectively populated, a pump to produce population inversion between some of these levels, and a resonant electromagnetic cavity that contains the active medium and provides feedback to maintain the coherence of the electromagnetic field. In a continuously operating laser, coherent radiation will build up in the cavity to the level required to balance stimulated emission and cavity losses. The system is then said to be lasing, and radiation is emitted in a direction defined by the cavity.

10.2.2 Stimulated Emission

10.2.2.1 Absorption

Quantum theory shows us that matter exists only in certain allowed energy levels or states. In thermal equilibrium, lower energy states are preferentially occupied, with an occupation probability proportional to $\exp(-E_s/k_BT)$, where E_s is the energy of the state, T is absolute temperature, and k_B is Boltzmann's constant. When light interacts with matter, it is possible for one quantum of optical energy, a photon, to be absorbed while simultaneously exciting the material into a higher-energy state. In this process the energy difference between the states is equal to the energy of the absorbed photon, and energy is conserved.

Consider a simple material system with only two energy states, and assume that essentially all optically active sites within the material system are in the lowest-energy state in thermal equilibrium. The probability that a randomly chosen photon in an optical field of cross-section A will be absorbed by a given absorption site with a radiative cross-section σ_r as it passes through a material is σ_r/A. If there are g_u identical (degenerate) high-energy states, the probability becomes $g_u\sigma_r/A$. The product $g_u\sigma_r$ is known as the absorption cross-section σ_a. When all of the photons in an optical field of intensity I and all of the absorption sites in a material of length dl are accounted for, the intensity of an optical field passing through the material changes by

$$dI = -I\rho_l\sigma_a dl, \tag{10.1}$$

where ρ_l is the density of absorption sites in the material. This equation has the solution

$$I(l) = I(0)\exp(-\alpha l), \tag{10.2}$$

where $I(0)$ is the intensity of the optical field as it enters the material at position $z=0$ and $\alpha=\rho_l\sigma_a$ is the absorption coefficient of the material.

10.2.2.2 Population Inversion, Stimulated Emission, and Gain

In the material system discussed above, there is the possibility that some sites will be in an upper energy state. Such sites are referred to as inverted. In the presence of an optical field, a transition from an upper state to a lower state can be induced by the radiation, with the simultaneous emission of a photon in phase (coherent) with the stimulating radiation. This stimulated-emission process is the inverse of the absorption process.

The probability that a randomly chosen photon in an optical field of cross-section A will stimulate a given inverted site with a radiative cross-section σ_r as it passes through a material is σ_r/A. The radiative cross-section of the transition is proportional to the dipole strength of the transition and is the same for absorption and emission. If the low-energy state is degenerate, with a degeneracy of g_l, the probability of stimulated emission becomes $g_l\sigma_r/A$. The product $g_l\sigma_r$ is known as the emission or gain cross-section σ_g. When all of the photons in the optical field are accounted for and absorption and stimulated emission are included, Equation 10.2 becomes

$$I(l) = I(0)\exp[(\sigma_g\rho_u - \sigma_a\rho_l)l], \tag{10.3}$$

where ρ_u is the density of inverted sites. If the material is forced out of thermal equilibrium (pumped) to a sufficient degree, so that $\sigma_g\rho_u > \sigma_a\rho_l$, stimulated emission occurs at a higher rate than absorption. This leads to the coherent growth or amplification of the optical field. The material is now said to have gain, with a gain coefficient

$$g = \left[\rho_u - \left(\frac{g_u}{g_l}\right)\rho_l\right]\sigma_g. \tag{10.4}$$

The term $[\rho_u - (g_u/g_l)\rho_l]$ is referred to as the effective inversion density ρ_{eff}, and $g = \rho_{\text{eff}}\sigma_g$. When $\sigma_g\rho_u = \sigma_a\rho_l$ there is no change in the intensity of an optical field as it passes through the material and the material is said to be in a state of transparency.

10.2.2.3 Spontaneous Emission and Lifetime

In the absence of an optical field, materials with an inverted population will evolve toward thermal equilibrium. Transitions from an upper state to a lower state may be facilitated through interactions with the lattice and accompanied by the emission of a phonon (quantum of lattice vibration) or the generation of heat. Alternatively, they can take place through the spontaneous emission of a photon.

The time associated with the spontaneous decay of the upper state is known as the spontaneous lifetime τ and is dominated by spontaneous optical transitions in many laser gain media. For dipole transitions, the radiative lifetime τ_r is related to the gain cross-section such that (Siegman 1986)

$$\int \sigma_g(\nu)\mathrm{d}\nu = \frac{3^*\lambda_0^2}{8\pi n_g^2\tau_r}, \tag{10.5}$$

where ν is frequency, λ_0 is the free-space wavelength of the gain peak, n_g is the refractive index of the gain medium, and 3^* is a number between 0 and 3, depending on the alignment of the electronic wave functions with the polarization of the optical field. The combination of large gain cross-section and large bandwidth implies short lifetime. In general, the spontaneous lifetime is related to the radiative lifetime τ_r and nonradiative lifetime τ_{nr} according to

$$\tau^{-1} = \tau_r^{-1} + \tau_{nr}^{-1}. \tag{10.6}$$

10.2.2.4 Bandwidth

In any material system, the energy levels have a finite spectral (energy) width. This results in a bandwidth for the optical transitions of the system. The bandwidth may be determined by effects that are common to all sites within the system, resulting in a homogeneously broadened transition, or may be determined by local variations in material properties, leading to an inhomogeneously broadened transition. From the above discussion of Spontaneous Emission and Lifetime, the gain cross-section σ_g of an optical transition is related to the bandwidth of the transition and the radiative lifetime τ_r. A broad transition generally implies a small gain cross-section or a short spontaneous lifetime.

10.2.3 Models of Gain Media

10.2.3.1 Four-Level Gain Media

The energy-level structure of a laser plays an important role in obtaining inversion. Let us try to understand pumping and relaxation in an "ideal" four-level laser with the aid of Figure 10.1. The pumping process, indicated by the upward arrow, is assumed to excite the system from the lowest energy level, denoted by g for ground level, to the highest level, denoted by e for excited level. Pumping might occur in a variety of ways, one of which could be through radiative excitation using light whose frequency coincides with the transition frequency between g and e. Level e is assumed to relax to the upper laser level u. The population of the upper laser level is radiatively transferred, either through spontaneous or stimulated emission, to the lower laser level l. Finally, the lower laser level can either relax to the ground level or absorb the laser radiation and repopulate the upper laser level.

Several conclusions concerning optimal operation can be made from this model. First, the relaxation rates from e to u and from l to g should be as rapid as possible to maintain the maximum population inversion between u and l. Second, the pumping rate between g and e should be sufficiently rapid to overcome the spontaneous emission from u to l. Third, the thermal-equilibrium population of l should be as small as possible. Fourth, decay of e to any level other than u should be as slow as possible (for optical pumping, e can decay radiatively to g) and the nonradiative decay of u should be slow. For radiative pumping, it is advantageous to have e be distinct from u and not to have rapid radiative decay of u to g.

10.2.3.2 Three-Level Gain Media

In a three-level laser, such as ruby, the lower laser level and the ground level are the same (Figure 10.2). Therefore, a large fraction of the ground level must be depopulated to obtain population inversion. For this reason, pumping of a three-level laser requires an extremely high-intensity source.

10.2.3.3 Quasi-Three-Level Gain Media

Energy levels in ionic gain media are grouped into manifolds, as discussed in Section 10.3. Most diode-pumped systems that are referred to as "three-level" lasers have a lower laser level that is slightly above the ground level, typically within the same ground-state manifold (see Section 10.3). The term quasi-three-level is sometimes used to distinguish these systems from true three-level lasers (e.g., ruby). In quasi-three-level lasers, the lower laser level is partially occupied in thermal equilibrium, and such lasers have properties that are intermediate between true three-level systems and four-level lasers.

Even though the populations of the different energy-level manifolds may be out of equilibrium, there is often rapid thermalization within each manifold. When thermalization within the manifolds occurs on a timescale that is short compared to changes in the optical field, it can be modeled as instantaneous. With this approximation, the population of a level within a manifold is given by $g_s f_s N_m$, where g_s is the degeneracy of states in the level, N_m is the total occupation of the manifold, and the Boltzmann coefficients

$$f_s = \frac{\exp(-E_s/k_B T)}{\sum_m g_m \exp(-E_m/k_B T)}, \tag{10.7}$$

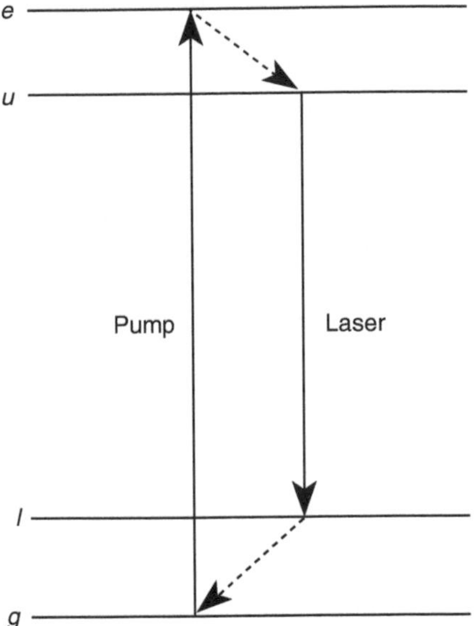

FIGURE 10.1 Schematic representation of a four-level system. Population is pumped from g to e and laser operation occurs on the transition between u and l.

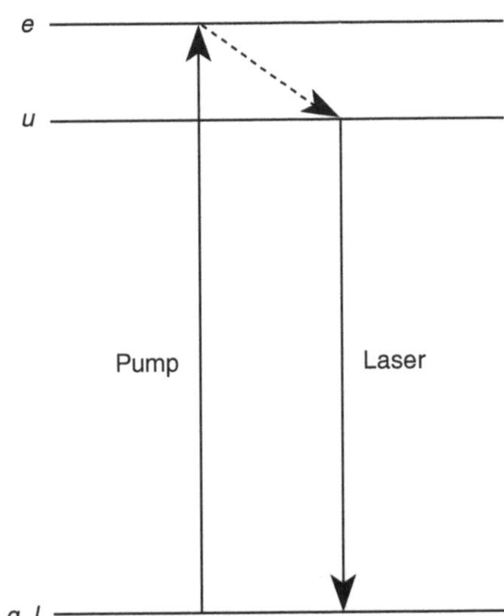

FIGURE 10.2 Schematic representation of a three-level system. Note that g and l are now the same level.

with the sum taken over all levels in the manifold. The population of the lower laser level N_l is, therefore, $g_l f_l N_{lm}$, where the subscript l denotes the lower laser level and lm denotes the lower manifold. Similarly, the population of the ground level N_g is $g_g f_g N_{lm}$. To make the model as general as possible, the upper laser level is assumed to be in an upper manifold with a population $N_u = g_u f_u N_{um}$.

The quasi-three-level model reduces to the three-level model when $g_g f_g = g_l f_l = 1$ and $g_u f_u = 1$. The four-level model is obtained when $g_g f_g = 1$, $g_l f_l = 0$, and $g_u f_u = 1$. In the discussion of Rate Equations given below (Section 10.2.4.1), it is shown that the four-level model is also obtained without the restriction $g_u f_u = 1$ if the rate equations are written using an effective emission cross-section $\sigma_{g,eff} = g_u f_u \sigma_g$ instead of the spectroscopic gain cross-section σ_g. For parallel reasons, $g_u(f_u + f_l)\sigma_g$ is often referred to as the effective cross-section for quasi-three-level systems (see Section 10.2.4.1). When multiple closely spaced transitions overlap at a given frequency, the effective cross-section at that frequency is the sum of $g_u(f_u + f_l)\sigma_g$ over all overlapping transitions. One important example of this is the ~1.064-μm line in room-temperature Nd:YAG, where a pair of transitions centered at 1.06415 and 1.0644 μm both contribute to the effective cross-section.

10.2.4 Rate-Equation Model of Lasers

10.2.4.1 Rate Equations

Many of the properties of a laser can be determined from a rate-equation model for the populations of the laser levels and the number of photons in the laser cavity. The rate equations provide a simple and intuitive, yet accurate, picture of the behavior of lasers. In the most simplified form, the increase in photon number within the laser cavity is balanced by the decrease in the population difference between the upper and lower laser levels. In addition, the population difference increases on account of pumping, whereas the photon number decreases because of intracavity losses (including absorption and scattering), diffraction of the beam out of the cavity, and transmission through the mirrors.

The rate-equation model can be derived as an approximation to the fundamental equations relating the electromagnetic field, the material polarization, and the quantum-state populations. The validity of the rate equations requires that the material polarization can be accurately approximated by assuming that it instantaneously follows the field; this is a situation that applies to most lasers. To describe the problem in terms of total population and total photon number within a laser cavity, it is necessary that the gain of the laser be small during one pass through the cavity, that the laser operate in a single longitudinal mode, and that the population and optical field be uniform within the cavity. (Many of these assumptions are reexamined in the discussion of Single-Frequency Operation in Section 10.5.)

As stated above, the number of photons q within the laser cavity is affected by two types of events, the emission of a photon by the gain medium ($\dot{q}_e$) and the escape of a photon from the cavity or absorption by unpumped transitions ($\dot{q}_l$). Photon emission can be either stimulated ($\dot{q}_{st}$) or spontaneous ($\dot{q}_{sp}$). Once a laser is above threshold the stimulated-emission rate is much greater than the spontaneous rate into the oscillating mode and, to first order, spontaneous emission can be ignored. We will return to the issue of spontaneous emission in the next section when we discuss the build-up of a laser from noise.

The stimulated-emission rate is proportional to the number of photons within the cavity q; the effective population inversion $N_{eff}=\rho_{eff}A_g l_g=N_u-(g_u/g_l)N_l$, where A_g is the cross-sectional area of the laser mode in the gain medium and l_g is the length of the gain medium; and the probability per unit time B that a given photon will interact with a given inverted site. The interaction probability B is the product of the probability that a photon will pass within the gain cross-section σ_g of a given inverted site as it traverses the gain medium and the number of times the gain medium is traversed during a round trip within the laser cavity, divided by the round-trip transit time. Mathematically, this reduces to $B=\sigma_g/A_g\times 2^\ddagger/t_{rt}$, where $2^\ddagger$ is the number of times the gain medium is traversed (the notation $2^\ddagger$ is used as a reminder that for the most common type of laser, a standing-wave laser, $2^\ddagger=2$), $t_{rt}=l_{op,rt}/c$ is the round-trip time of light in the laser cavity, $l_{op,rt}=\oint n dl$ is the round-trip optical length of the cavity, c is the speed of light in vacuum, n is the local refractive index, and the contour integral defining $l_{op,rt}$ is performed along the optical path. Therefore, the stimulated-emission rate $\dot{q}_{st}=2^\ddagger q N_{eff}\sigma_g/t_{rt}A_g$.

The escape of photons from the laser cavity and their absorption within the cavity are characterized by the cavity lifetime in the absence of any inversion, $\tau_c=t_{rt}/\gamma_{rt}$, where $\gamma_{rt}=-\ln(1-\Gamma_{rt})$ is the round-trip loss coefficient and Γ_{rt} is the round-trip loss including transmission through the output coupler. The corresponding decrease in photon number $\dot{q}_l=q\gamma_{rt}/t_{rt}$. Thus, the total rate of change of the number of photons within the laser cavity is

$$\dot{q}=\dot{q}_{st}-\dot{q}_l=\frac{q}{t_{rt}}\left(\frac{2^\ddagger N_{eff}\sigma_g}{A_g}-\gamma_{rt}\right). \tag{10.8}$$

We will now derive the rate equations for the population inversion of both a four-level and quasi-three-level laser (see the above discussion of Models of Gain Media in Section 10.2.3). In both cases, the pump excites the active medium from the ground level to the excited level. It is assumed that the excited level quickly decays to the upper laser level (or manifold), so that the population of the excited level is nearly zero. Lasing occurs between the upper laser level and the lower laser level.

To derive the rate equation for the effective population inversion N_{eff}, we start by considering the population of the upper laser level N_u. The population of the upper level is affected by pumping $\dot{N}_p$, stimulated emission $\dot{N}_{u,st}$, and spontaneous decay $\dot{N}_{u,sp}$. For most pumping schemes, the pump rate is proportional to the number of ions in the ground level and can be written as $\dot{N}_p=W_pN_g$. The stimulated-emission process decreases the population of the upper laser level by one for every photon created, so that $\dot{N}_{u,st}=-\dot{q}_{st}$. Spontaneous decay is characterized by the spontaneous lifetime τ, corresponding to $\dot{N}_{u,sp}=-N_u/\tau$. Thus, the rate equation for the upper-level population is

$$\dot{N}_u=\dot{N}_p+\dot{N}_{u,st}+\dot{N}_{u,sp}=W_pN_g-\frac{2^\ddagger q N_{eff}\sigma_g}{t_{rt}A_g}-\frac{N_u}{\tau}. \tag{10.9}$$

In an ideal four-level laser there is a very rapid decay of the lower laser level to the ground level, so that N_l and $\dot{N}_l$ are approximately equal to zero and $N_{eff}\approx N_u$. Because the total number of active ions N_t is constant, $N_t \approx N_g + N_u$. Therefore, the rate equation for the effective population inversion of a four-level laser is

$$\dot{N}_{eff} = W_p(N_t - N_{eff}) - \frac{2^{\ddagger} q N_{eff} \sigma_g}{t_{rt} A_g} - \frac{N_{eff}}{\tau}. \tag{10.10}$$

In a quasi-three-level laser, the population is distributed in the upper and lower manifolds, and $N_t = N_{um} + N_{lm}$. When thermalization of the manifolds is fast relative to changes in the optical field, the population of the upper laser level $N_u = g_u f_u N_{um}$, the population of the lower laser level $N_l = g_l f_l N_{lm}$, and the population of the ground level $N_g = g_g f_g N_{lm}$. As a result, the rate equation for the effective population inversion $N_{eff} = N_u - (g_u/g_l) N_l$ of a quasi-three-level laser reduces to

$$\dot{N}_{eff} = g_g f_g W_p(g_u f_u N_t - N_{eff}) - \frac{2^{\ddagger} g_u (f_u + f_l) q N_{eff} \sigma_g}{t_{rt} A_g} - \frac{g_u f_l N_t + N_{eff}}{\tau}. \tag{10.11}$$

Equation 10.11 assumes instantaneous thermalization of the populations within the laser manifolds. Within this approximation, all of the levels in the upper manifold rapidly replenish the upper laser level (on a timescale quicker than changes in the optical field), and the laser can extract energy from the entire upper-manifold population. Likewise, the population of the lower laser level is rapidly depopulated through thermalization with the other levels in the lower manifold.

One of the features of miniature pulsed lasers is that they can produce very short pulses and can challenge the instantaneous thermalization approximation. When the thermalization times for the manifolds (they are generally different for the upper and lower manifold) are long compared to the pulse width, repopulation of the upper laser level and depopulation of the lower level through thermalization cannot occur during the pulse. As a result, the effective values of f_u and f_l approach unity; every laser transition reduces the upper-lever population by one and increases the population of the lower laser level by the same amount. The Boltzmann distribution (Equation 10.7) determines the initial population of the upper and lower levels before the onset of lasing, but thermalization does not occur during the pulse. In this case, the laser can only extract energy from the upper-laser-level population; it does not have access to the rest of the population in the upper manifold.

When the thermalization times for the manifolds are comparable to the pulse width, the redistribution of the populations within the manifolds will affect the dynamics of pulse formation and more sophisticated analysis is required (Degnan, Coyle, and Kay 1998).

In many important laser systems, the lower laser level is essentially empty, and all of the ions are either in the ground level or the upper laser manifold. For such systems, when instantaneous thermalization is a valid approximation, Equation 10.8 and Equation 10.11 can by written as

$$\dot{q} = \frac{q}{t_{rt}} \left(\frac{2^{\ddagger} N_{um} \sigma_{g,\,eff}}{A_g} - \gamma_{rt} \right) \tag{10.12}$$

and

$$\dot{N}_{um} = W_p(N_t - N_{um}) - \frac{2^{\ddagger} q N_{um} \sigma_{g,eff}}{t_{rt} A_g} - \frac{N_{um}}{\tau}, \tag{10.13}$$

where $\sigma_{g,eff} = g_u f_u \sigma_g$. These equations are the same as the rate equations for an ideal four-level laser except that the emission cross-section has been replaced by an effective emission cross-section and the population of the upper laser level has been replaced by the population of the upper-level manifold. The concept of an effective emission cross-section has been extended to quasi-three-level lasers with significant population in the lower laser level; $\sigma_{g,eff} = g_u (f_u + f_l) \sigma_g$.

10.2.4.2 Build-Up from Noise

In the photon rate equations derived above, the term corresponding to spontaneous emission was left out. Laser action is initiated by spontaneous emission, or noise. As a result, these rate equations cannot account for the onset of lasing, as is seen by setting $q=0$ at time $t=0$. When spontaneous emission is properly taken into account, the photon rate equation becomes

$$\dot{q} = \frac{2^{\ddagger}(qN_{\text{eff}} + 2^{\dagger}N_{\text{u}}/2)\sigma_{\text{g}}}{A_{\text{g}}t_{\text{rt}}} - \frac{q\gamma_{\text{rt}}}{t_{\text{rt}}}, \tag{10.14}$$

where

$$2^{\dagger} = \begin{cases} 1 & \text{for a traveling-wave laser} \\ 2 & \text{for a standing-wave laser} \end{cases}.$$

The difference between a standing-wave laser and a traveling-wave laser can be understood by realizing that the intracavity optical field for a standing-wave laser is the sum of two counterpropagating traveling waves, each with its own one-half photon of noise. This one-half photon of noise stimulates optical transitions and initiates lasing.

10.2.4.3 Threshold

The threshold inversion for lasing is derived by requiring that the photon rate equation have a nontrivial solution in steady state, where $\dot{q}=0$. This results in the condition $2^{\ddagger}N_{\text{eff}}\sigma_{\text{g}}/A_{\text{g}}-\gamma_{\text{rt}}=0$. Physically, the number of photons leaving the cavity must be balanced by the number of photons generated through stimulated emission. The threshold inversion, therefore, is given by

$$N_{\text{th}} = \frac{\gamma_{\text{rt}}A_{\text{g}}}{2^{\ddagger}\sigma_{\text{g}}}. \tag{10.15}$$

The pump rate W_{p} required to reach threshold is obtained by setting $\dot{N}_{\text{eff}}=0$, $\dot{q}=0$, and $N_{\text{eff}}=N_{\text{th}}$. For a four-level laser,

$$W_{\text{p,th}} = \frac{N_{\text{th}}}{(N_{\text{t}}-N_{\text{th}})\tau} \approx \frac{\gamma_{\text{rt}}A_{\text{g}}}{2^{\ddagger}N_{\text{t}}\sigma_{\text{g}}\tau}, \tag{10.16}$$

where we have assumed that $N_{\text{th}} \ll N_{\text{t}}$. For a quasi-three-level laser,

$$W_{\text{p,th}} = \frac{g_{\text{u}}f_{\text{l}}N_{\text{t}} + N_{\text{th}}}{g_{\text{g}}f_{\text{g}}(g_{\text{u}}f_{\text{u}}N_{\text{t}} - N_{\text{th}})\tau}. \tag{10.17}$$

For the same value of τ, the threshold pump rate for a four-level laser is usually much smaller than the threshold pump rate for a quasi-three-level system. This is the basis of the superior performance of a four-level system over a quasi-three-level system in cw (continuous wave) operation.

10.2.4.4 Gain Saturation

The gain coefficient g of an active medium is defined as the fractional change in optical intensity per unit length as a light beam passes through. From the above discussions of Stimulated Emission and Rate Equations (Section 10.2.2 and Section 10.2.4.1), it follows that $g=N_{\text{eff}}\sigma_{\text{g}}/V_{\text{g}}$, where V_{g} is the volume of active medium that interacts with the optical field. In the presence of a strong optical field, the population inversion is reduced and the gain is saturated. The rate equation for the population inversion (of a three- or four-level gain medium) in steady state can be rewritten in the form

$$g = \frac{g_0}{1 + I/I_{sat}}, \tag{10.18}$$

where g_0 is the unsaturated gain coefficient (the gain coefficient in the absence of an optical field), $I = qh\nu_o/t_{rt}A_g$ is the circulating optical intensity ($h\nu_o$ is the energy of one photon), and I_{sat} is the circulating saturation intensity (saturation intensity divided by $2^{\ddagger}$). For a four-level laser,

$$g_0 = \frac{N_t \sigma_g W_p \tau}{V_g (W_p \tau + 1)} \tag{10.19}$$

and

$$I_{sat} = \frac{h\nu_o}{2^{\ddagger} \sigma_g \tau}(W_p \tau + 1). \tag{10.20}$$

It is often true that $W_p\tau \ll 1$, resulting in the simplification $g_0 = \dot{N}_p \sigma_g \tau / V_g$ and $I_{sat} = h\nu_o/2^{\ddagger}\sigma_g\tau$. The additional terms in Equation 10.19 and Equation 10.20 correspond to bleaching (or saturating) of the pump transition by the pump. For a quasi-three-level laser,

$$g_0 = \frac{N_t \sigma_g (g_u f_u g_g f_g W_p \tau - g_u f_l)}{V_g (g_g f_g W_p \tau + 1)} \tag{10.21}$$

and

$$I_{sat} = \frac{h\nu_o (g_g f_g W_p \tau + 1)}{2^{\ddagger} g_u (f_u + f_l) \sigma_g \tau}. \tag{10.22}$$

10.2.4.5 Laser Efficiency

The steady-state photon rate equation ($\dot{q} = 0$) predicts that, above threshold ($q > 0$), the inversion density and the gain of a laser are clamped at their threshold values; the round-trip gain of the cavity is equal to the round-trip loss. With increased pumping, the gain remains fixed while the photon number and the output of the laser increase. For a four-level laser,

$$q = \frac{t_{rt}}{\gamma_{rt}} \left[W_p (N_t - N_{th}) - \frac{N_{th}}{\tau} \right]; \tag{10.23}$$

for a quasi-three-level laser,

$$q = \frac{t_{rt}}{g_u (f_u + f_l) \gamma_{rt}} \left[g_g f_g W_p (g_u f_u N_t - N_{th}) - \frac{g_u f_l N_t + N_{th}}{\tau} \right]. \tag{10.24}$$

The output power of a laser

$$P_o = q h\nu_o \gamma_o / t_{rt}, \tag{10.25}$$

where $\gamma_o = -\ln(1 - T_o)$ is the output-coupling coefficient and T_o is the transmission of the output coupler. The use of γ_o in this expression for output power treats the output coupler as if it were distributed throughout the cavity, consistent with the use of a single photon number q to describe the optical intensity in the cavity. (For small T_o, $\gamma_o \approx T_o$).

Because the photon number and the output power of the laser are linear functions of the pump rate, the efficiency of a laser is often discussed in terms of the slope efficiency η_s. Slope efficiency is defined as the ratio of the change in output power to the change in pump power of a laser once it has reached threshold, and is determined by five factors: the pump efficiency η_p, the area efficiency η_a, the inversion efficiency η_i, the quantum efficiency η_q, and the output-coupling efficiency η_o. Mathematically,

$$\eta_s = \eta_p \eta_a \eta_i \eta_q \eta_o. \tag{10.26}$$

The pump efficiency η_p is the ratio of the energy absorbed by the gain medium to the energy of the pump source. In an optically pumped laser, part of the incident optical energy may be reflected by the gain medium and part may be transmitted. Both of these effects decrease the pump efficiency. The area efficiency η_a is a measure of how well the pumped volume is used by the oscillating mode. If the cross-section of the pumped volume is much larger than the cross-section of the lasing mode, only a small portion of the pumped volume contributes to the gain of the system, and the area efficiency will be low. For a longitudinally pumped laser, the area efficiency can often be approximated by $\eta_a = 1/(r_p^2/r_m^2 + 1)$, where r_p is the radius of the pump beam and r_m is the radius of the oscillating mode. The inversion efficiency η_i is the ratio of the energy of a photon created during lasing ($h\nu_o$) to the energy required to invert one active site. In an optically pumped system, the inversion efficiency is the ratio of the energy of a photon at the oscillating frequency to the energy of an absorbed pump photon. The difference between the energies of the absorbed and emitted photons is referred to as the quantum defect. The quantum efficiency η_q is the fraction of the inverted sites that emit a photon into the oscillating mode. For an optically pumped system, 100% quantum efficiency implies that each absorbed pump photon results in one photon at the oscillating frequency. A laser operating with 100% quantum efficiency is said to have quantum-limited performance. Finally, the output-coupling efficiency η_o is the ratio of the output coupling to the total round-trip loss of the laser cavity.

The total efficiency η of a laser (power out divided by power in) is dependent on the slope efficiency and the laser threshold. For a constant slope efficiency,

$$\eta = \eta_s \left(1 - \frac{P_{th}}{P_p}\right), \tag{10.27}$$

where P_p is the total pump power and P_{th} is the pump power required to reach threshold.

10.3 Solid-State Gain Media

10.3.1 Rare-Earth Dopants

Crystals and glasses doped with rare earths are the most important subclass of solid-state gain media both from an applications perspective and in terms of the development of miniature lasers. Neodymium-doped yttrium aluminum garnet (Nd^{3+}:$Y_3Al_5O_{12}$ or Nd:YAG) has been the most widely used gain medium because of its unusual combination of favorable optical, thermal, and mechanical properties. Other rare-earth ions of particular interest are ytterbium (Yb^{3+}), thulium (Tm^{3+}), holmium (Ho^{3+}), and erbium (Er^{3+}). All of these have been operated as conventional lasers and as miniature lasers, including diode-pumped monolithic devices. The cerium ion (Ce^{3+}) offers the possibility of broadly tunable laser operation in the UV. The ability to pump several of the Ce-doped fluorides with the fourth harmonic of a Nd laser gives these material systems the potential to be operated as miniature laser systems.

The optical transitions of interest for Nd^{3+}, Yb^{3+}, Tm^{3+}, Ho^{3+}, and Er^{3+} are similar in that they correspond to transitions among an incomplete set of 4f electrons that are, generally, well shielded from the host crystal field by a complete xenon shell ($5s^2 5p^6$). (The physics of the Ce^{3+} transition is different than that of the other rare-earth ions discussed here, and is discussed separately in the Section 10.3.1.6).

The dopant ion is triply ionized, having given up a pair of 6s electrons and a single 4f electron. The transitions are characterized by sharp lines that vary little in energy from host to host. Transitions for the entire rare-earth series have been charted by Dieke and Crosswhite (1963) in a form that is convenient for reference. Figure 10.3 shows the positions of the energy manifolds of interest for diode pumping of Nd^{3+}, Yb^{3+}, Tm^{3+}, Ho^{3+}, and Er^{3+}. According to convention, the levels are designated by the spectroscopic notation ${}^{2S+1}L_J$, where the quantum numbers S (total spin) and J (total angular momentum) are either integers or half-integers, and L (total orbital angular momentum) is assigned a letter value (S, P, D, F, G, H, I, K, L, M,... corresponding to 0, 1, 2, 3, 4, 5, 6, 7, 8, 9,...). Each manifold consists of $2J+1$ states. In manifolds with half-integer values of J, each state is at least doubly degenerate (there are at most $J+1/2$ energetically distinct levels). The symmetry of the host crystal field determines any further degeneracy. Energy levels of atomic systems strongly affected by the host crystal field, such as the transition-metal dopants (discussed below), are designated by a different notation which reflects the symmetry of the modified electronic waveforms and the degeneracy of the levels.

Note that, although the interaction between the optically active electrons of the dopant ions and the host medium is weak, it is sufficient to provide electric-dipole strength to what are otherwise parity-forbidden dipole transitions. The resulting transitions are characterized by narrow linewidths and long lifetimes (~0.1–10 ms). The former property leads to the generalization that rare-earth gain media are line tunable (the tuning range is a very small fraction of the oscillating frequency) as opposed to broadly tunable. The long lifetimes are particularly useful for pump-energy storage and the generation of high peak power via Q switching of the laser cavity.

10.3.1.1 Neodymium

The three most commonly used Nd lines correspond to transitions originating in the ${}^4F_{3/2}$ manifold. The highest gain occurs near 1.06 μm, terminating in the ${}^4I_{11/2}$ manifold. The 1.32-μm line (${}^4F_{3/2} \rightarrow {}^4I_{13/2}$) has important applications in fiber networks (where the wavelength coincides with the propagation-loss minimum in silica fibers), although the effective cross-section $\sigma_{g,\mathrm{eff}}$ is roughly 1/5 of that at 1.06 μm. The 0.95-μm line terminates in the ${}^4I_{9/2}$ ground-state manifold and, typically, operates as a quasi-three-level system near room temperature.

Historically, Nd:YAG was developed for lamp-pumped systems in which pump absorption is achieved in a handful of discrete spectral bands located throughout the visible and near-infrared, where the broad lamp emission overlaps relatively narrow Nd transitions. In this process, the upper laser level is populated through the rapid, nonradiative relaxation of excited ions. One means of improving the net absorption is to co-dope the material with a broadband absorber that better matches the lamp spectrum while efficiently transferring the pump energy to the Nd system. The efficiency of the transfer process is sensitive to the crystal field and, although it is not effective in Nd:YAG, Cr^{3+} has been used as a

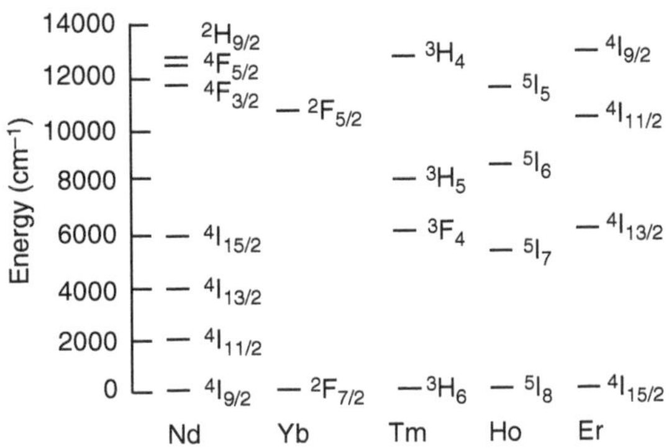

FIGURE 10.3 Partial energy-level diagrams for Nd, Yb, Tm, Ho, and Er.

sensitizing ion in low-field garnet hosts. Primary among these are gadolinium scandium aluminum garnet ($Gd_3Sc_2Al_3O_{12}$ or GSGG) and yttrium scandium aluminum garnet ($Y_3Sc_2Al_3O_{12}$ or YSGG).

The optimal host material for Nd, in a given application, is determined by the detailed source requirements. The garnets have been extensively developed because of their excellent mechanical and thermal properties (e.g., hardness and thermal conductivity), and because garnet materials can be grown as relatively large boules. Among noncubic crystalline hosts, two uniaxial materials, yttrium lithium fluoride ($LiYF_4$ or YLF) and yttrium orthovanadate (YVO_4), are the most commonly used in applications involving miniature lasers.

YLF was originally developed as an alternative to YAG for systems in which thermo-optic distortions in Nd:YAG limited the performance at high average power. It is characterized by a negative bulk-lensing coefficient ($dn/dT<0$) which offsets pump-induced changes in the optical path length due to thermal expansion. As a result, thermal lensing is greatly reduced relative to that in other Nd-doped materials, and YLF is often referred to as an athermal material. Furthermore, the large natural birefringence provides immunity to pump-induced birefringence. Another favorable property is the relatively low refractive index of YLF which increases parasitic thresholds in high-gain systems. Although Nd:YLF can now be grown in large boules of high optical quality, its relatively low fracture threshold continues to be a limitation both in high-average-power lamp-pumped systems and in high-intensity diode-pumped applications.

The combination of very high gain cross-section and short lifetime make Nd:YVO_4 a favorable gain medium in high-repetition-rate *Q*-switched systems. In addition, it has received attention for diode-pumped systems required to operate over wide temperature ranges because the continuous range of useful absorption near 0.81 μm is quite broad (>10 nm). Historically, poor crystal quality has limited the use of YVO_4. However, with renewed interest in the material for diode-pumped lasers have come significant improvements in crystal growth, and good material is now generally available.

Apart from the basic optical, thermal, and mechanical considerations, the practical viability of a given crystalline host is critically dependent on the ability to synthesize large boules of high optical quality. A feature of diode-laser pumping is that the relatively high absorption coefficients achieved near 0.8 μm permit the use of smaller crystals. Figure 10.4 shows the absorption coefficient measured for Nd:YAG nominally doped at 1 at.% Nd, in the spectral region typically used for diode pumping. The peak absorption scales linearly with the Nd concentration, which is usually limited to ~1% to avoid a reduction in lifetime due to concentration quenching:

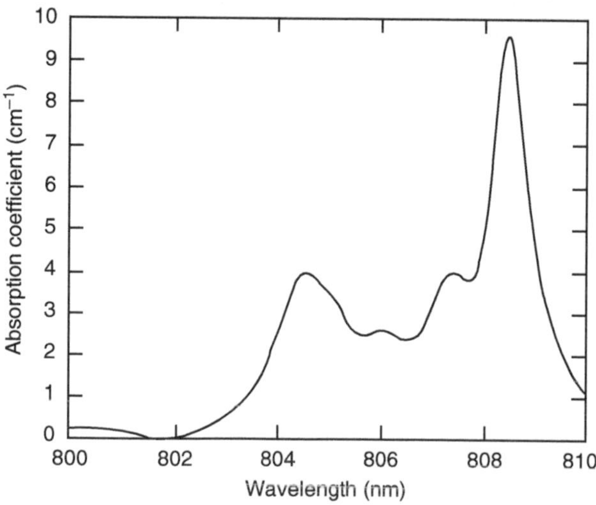

FIGURE 10.4 Absorption spectrum of Nd:YAG (1 at.%) in the region commonly used for diode pumping.

$$\tau \approx \frac{\tau_1}{1 + (\rho_d/\rho_{cq})^2}, \quad (10.28)$$

where τ_1 is the spontaneous lifetime of the upper-state population in the limit of low doping, ρ_d is the doping concentration, and ρ_{cq} is a parameter used to characterize the effects of concentration quenching. For Nd:YAG, $\rho_{cq}=2.63$ at.%; similar values are obtained for other host crystals.

Concentration quenching stems from the presence of the $^4I_{15/2}$ manifold midway between the upper laser level and the ground state. With a sufficient density of Nd ions, energy transfer among ions enables a cross-relaxation process in which a single upper-level ion combines with a single ground-state ion to yield two additional ions in the $^4I_{15/2}$ manifold. One facet of the effort to develop miniature, diode-pumped Nd lasers has been the development of stoichiometric Nd materials in which concentration quenching is effectively inhibited even at very high Nd concentration. The inhibiting mechanism is believed to be the deviation from resonance of the intermediate $^4I_{15/2}$ levels. NdP_5O_{14} (NPP) and $LiNdP_4O_{12}$ (LNP) were early examples of stoichiometric Nd materials (Danielmeyer 1975; Weber 1975).

In applications where extremely high peak powers are generated, such as laser fusion, very large gain elements are required to spread the incident beam over sufficient area to prevent irreversible damage to the optic. In such cases, a glass host is often used, in spite of the fact that the thermal characteristics are generally inferior to those of crystals. Also, unlike the case with crystalline hosts, Nd:glass is inhomogeneously broadened near room temperature due to local field variations inherent in disordered hosts.

Tables of material characteristics and laser parameters for Nd:YAG, Nd:YLF, and $Nd:YVO_4$ are included below.

The miniature solid-state lasers in use today derive from early efforts to pump with GaAs light-emitting diodes and laser diodes (Newman 1963; Keyes and Quist 1964). With the advent of high-brightness GaAlAs lasers, it has become possible to operate miniature Nd lasers in end-pumped configurations with slope efficiencies (optical-to-optical) in excess of 50%.

Much of the work on monolithic cw Nd lasers has been devoted to the development of single-frequency sources for coherent applications in communications and remote sensing. One geometry that has proven successful is the nonplanar ring, illustrated in Figure 10.5. In this device, the conventional approach toward eliminating spatial hole burning (see the discussion of Spatial Hole Burning in Section 10.5) has been implemented in a compact, multifaceted Nd gain element commonly referred to as a MISER (Monolithic Isolated Single-mode End-pumped Ring) (Kane and Byer 1985; Kane, Nilsson, and Byer 1987). Unidirectional operation is achieved by a combination of the Faraday effect (induced by an applied magnetic field), the nonplanar geometry of the resonator, and the polarization dependence of the reflectivity of the coated output facet (Nilsson, Gustafson, and Byer 1989). To date, this approach has been limited to optically isotropic host materials [YAG, GGG (Day et al. 1989)]. The generation of nearly 1 W in a single-frequency, diffraction-limited beam has been reported from a diode-pumped, Nd:YAG MISER laser operating at 1.06 μm with a slope efficiency (output power vs. incident pump power) of 47% (Cheng and Kane 1991). The same paper reported over 300 mW of output power with a slope efficiency of 60% for the same device pumped by a lower-power, higher-brightness diode laser.

A second monolithic geometry that has proven successful is the microchip laser, illustrated in Figure 10.6. In a microchip laser, two faces of the gain element are polished flat and parallel and are dielectrically coated to form the cavity mirrors. The resonator is stabilized by pump-induced thermal guiding due to a combination of bulk lensing and end-face curvature (Zayhowski 1991a; MacKinnon and Sinclair 1992). In cubic crystals, unidirectional, transverse stress can be used to break the polarization degeneracy of the crystal and obtain linearly polarized output (Zayhowski and Mooradian 1989b). Single-frequency operation well above threshold can be achieved if the longitudinal-mode spacing of the monolith is large compared to the gain bandwidth (Zayhowski and Mooradian 1989a). For Nd:YAG at 1.06 μm, this dictates microchips with submillimeter crystal lengths. Due to the short absorption path, design optimization can include unusually high Nd concentrations in a trade-off between pump utilization (absorption) and radiation efficiency (concentration quenching).

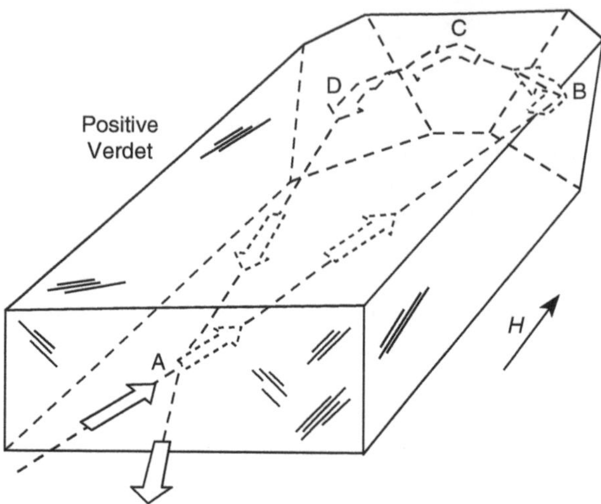

FIGURE 10.5 Illustration of a monolithic isolated single-mode end-pumped ring (MISER) laser. Polarization selection takes place at the curved, partially transmitting face (point A). At points B, C, and D, total internal reflection occurs. A magnetic field *H* is applied to establish unidirectional oscillation. Magnetic rotation takes place along segments *AB* and *DA*. The focused pump laser beam enters the crystal at point *A*, and the output beam emerges at the same point. (From Kane, T. J. and Byer, R. L. 1985. Monolithic, unidirectional single-mode Nd:YAG ring laser. *Opt. Lett.*, 10, 65.)

The earliest work on microchip lasers concentrated on Nd-based stoichiometric gain media characterized by large absorption coefficients, and used an argon-ion laser as a pump source (Winzer et al. 1976; Winzer, Möckel, and Krühler 1978; Krühler et al. 1979). Renewed interest in microchip lasers followed the first demonstration of a diode-laser-pumped device in Nd:YAG (Zayhowski and Mooradian 1989a). Early diode-pumped, Nd microchip lasers oscillating at ~1.06 and 1.32 μm were reported in many host materials, including YAG (Zayhowski and Mooradian 1989a), NPP (Zayhowski and Mooradian 1989a), LNP (Zayhowski and Mooradian 1989c; Dixon, Lingvay, and Jarman 1989), GSGG (Zayhowski and Mooradian 1989a), YVO_4 (Taira et al. 1991), $LaMgAl_{11}O_{19}$ (Mermilliod, François, and Wyon 1991), YCeAG (Gavrilovic et al. 1992), YLF (Leilabady, Anthon, and Gullicksen 1992), $Nd_xY_{1-x}Al_3(BO_3)_4$ (Amano 1992), La_2O_2S (Zarrabi et al. 1993), and $MgO{:}LiNbO_3$ (MacKinnon, Norrie, and Sinclair 1994). $NdAl_3(BO_3)_4$ microchip lasers were also operated in a "self-frequency-doubled" mode, with output at 531 nm (Amano 1992).

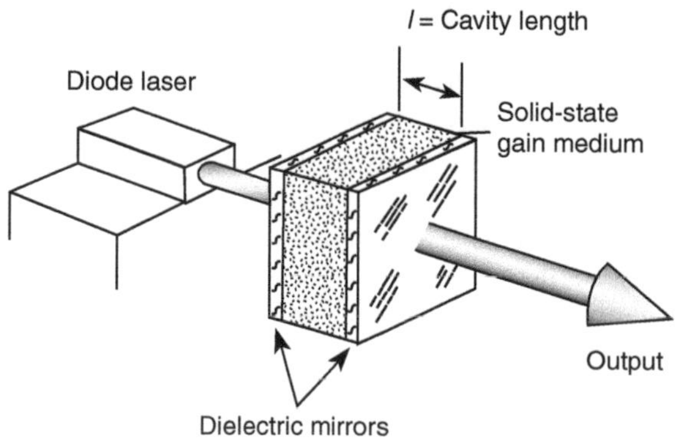

FIGURE 10.6 Illustration of a microchip laser.

10.3.1.2 Ytterbium

The relevant energy-level structure of Yb^{3+} is unique among the rare-earth ions of interest in that it consists entirely of two manifolds. By pumping into the upper manifold with a source near 0.94 μm, it is possible to obtain quasi-three-level laser operation at 1.03 μm near room temperature. Owing to the absence of higher-lying manifolds, the efficiency of lamp-pumped systems is very poor. Renewed interest in Yb in recent years is the result of the development of strained-layer InGaAs diode lasers, and a large number of host crystals have been evaluated for use in diode-pumped Yb lasers (DeLoach et al. 1993). A room-temperature, diode-pumped microchip laser has been demonstrated in Yb:YAG (Fan 1994), where the width of the ground-state manifold is close to 800 cm^{-1} ($\Delta\nu/c = 800\ cm^{-1}$). The key laser parameters for Yb:YAG are included in Table 10.8.

Yb:YAG provides an excellent example of how high-brightness diode lasers have enabled useful applications of otherwise unfriendly laser media. Efficient operation of Yb:YAG requires a spectrally narrow pump source. Furthermore, that source must have adequate beam quality to allow efficient energy transfer to the laser mode while uniformly saturating the ground-state absorption throughout the gain region. Given the high efficiency of diode lasers and the low quantum defect inherent in the Yb^{3+} system, diode-pumped Yb:YAG is being pursued as an alternative to Nd:YAG in high-power systems. The small quantum defect provides an additional benefit in that the heat load imparted to the host is substantially lower at a given pump level than that of a comparable Nd laser (i.e., pumped near 0.8 μm for operation at 1.06 μm). Also, the long upper-state lifetime (~1 ms, compared to ~240 μs in Nd:YAG) reduces the pump requirements in low-repetition-rate *Q*-switched systems.

10.3.1.3 Thulium

The primary Tm laser transition is from the $^3F_4$ manifold to the ground state ($^3H_6$), corresponding to laser action at ~2.0 μm. This wavelength is near the peak absorption of liquid water at 1.94 μm (especially important to medical applications) and is useful for eye-safe systems where Nd lasers are generally unsuitable. The key laser parameters for Tm:YAG at room temperature are listed in Table 10.9. In addition to YAG and other garnet hosts, YLF and YVO_4 have received considerable attention for Tm lasers.

The long lifetime of Tm is well suited to energy storage. Cr^{3+} sensitization has been employed to improve the efficiency of lamp-pumped Tm lasers in several garnet hosts including YAG (where a fortuitous overlap between the Cr ^2E lines and the Tm $^3H_4$ manifold enables efficient energy transfer). Because of the low gain cross-section, however, damage often limits *Q*-switched operation.

For diode-pumped operation, GaAlAs lasers are employed at wavelengths just below 0.8 μm to pump the $^3H_4$ Tm manifold. Although the quantum defect is large, it is typically compensated for by a remarkable cross-relaxation process in which a single excited $^3H_4$ ion combines with a single ground-state ion to produce two ions in the $^3F_4$ upper laser level. The pump quantum efficiency of this process can approach 200% with adequate Tm concentration (~4 at.% in YAG). Note that this cross-relaxation process, sometimes referred to as self-quenching, is equivalent to concentration quenching in Nd. Early reports of diode-pumped, monolithic Tm lasers include a Tm:YAG MISER laser (Kane and Kubo 1991), and Tm:YAG (Storm 1991) and Tm:YVO_4 (Zayhowski et al. 1995a) microchip lasers.

Selection of the Tm concentration involves a fundamental trade-off in the design of room-temperature, diode-pumped Tm lasers. High concentrations are required to efficiently absorb the pump power and to provide for efficient self-quenching. However, ground-state absorption increases in proportion to the Tm concentration. In practice, this problem has been overcome by employing Tm as a sensitizer for Ho, as described below.

10.3.1.4 Holmium

Table 10.10 includes the laser parameters for Ho:YAG. As a laser-active ion, Ho is similar to Tm. The relevant transition ($^5I_7$ to the $^5I_8$ ground state) is close in wavelength and lifetime to the $^3F_4 \rightarrow {}^3H_6$ Tm transition, although the effective gain cross-section is roughly five times larger. The key to efficient, low-threshold operation of Ho has been to use Tm as the pump-absorbing ion in co-doped crystals (e.g., Tm,Ho:YAG and Tm,Ho:YLF). This works because the upper laser level of Ho ($^5I_7$) lies just below the $^3F_4$ Tm level in energy. As a result, in thermal equilibrium the excited-ion population is shared according to the Boltzmann distribution between the Tm and Ho systems. Figure 10.7 illustrates the pump process. As long as the emission rate of the Ho laser is low compared to the equilibration rate, the Tm system acts as a reservoir for the Ho upper laser level. In *Q*-switched operation, the emission rate can be sufficiently high that only the energy stored in the Ho system is extracted in a single pulse. Typically, the concentration of Ho is an order of magnitude less than that of Tm and roughly half of the excited ions are in each system at room temperature. Doubly sensitized material, Cr,Tm,Ho:YAG (CTH:YAG), is commonly used in lamp-pumped systems.

In diode-pumped operation, upconversion can severely limit performance in Tm,Ho:YAG lasers (Fan et al. 1988). The upconversion rate constant has been shown to be five times lower in Tm,Ho:YLF (Hansson, Callens, and Nelsson 1993), making it the preferred material in many cases. The list of early diode-pumped monolithic Ho^{3+} lasers includes a Tm,Ho:YAG MISER laser (Kane and Kubo 1991), and Tm,Ho:YAG (Storm and Rohrbach 1989) and Tm,Ho:YLF (Harrison and Martinsen 1994a) microchip lasers.

It is worth noting that the advent of room-temperature, mid-infrared diode lasers raises the prospect of directly pumping the $^3F_4$ Tm level in co-doped materials or resonantly pumping either Tm or Ho. An example of the latter, Ho:YAG pumped by a GaInAsSb/AlGaAsSb diode laser, was first reported by Nabors et al. (1994). Ho lasers can also be resonantly pumped by Tm lasers. An interesting example of this uses of a Tm-doped fiber laser as the pump source (Abdolvand et al. 2003).

10.3.1.5 Erbium

Er has transitions of interest near 2.9 and 1.5 μm. The longer wavelength has unique medical potential because it is close to a liquid-water absorption peak that is two orders of magnitude stronger than the peak near 2 μm. The proximity of the shorter wavelength to the dispersion minimum in silica fiber has led to considerable interest in Er lasers as sources for fiber-based communications systems. In addition, it is suitable for eye-safe remote-sensing applications. Diode-pumped microchip lasers have been reported at 2.8 μm in Er:YSGG (Harrison and Martinsen 1994b) and 1.5 μm in Yb,Er:glass (Laporta et al. 1993).

For lamp-pumped operation, low-field garnets can be sensitized with Cr^{3+} for efficient energy transfer to the $^4I_{9/2}$ Er manifold. From there, the excited ions rapidly decay ($\tau < 1$ μs) to the $^4I_{11/2}$ manifold, the source of the 2.9-μm transition ($^4I_{11/2} \rightarrow {}^4I_{13/2}$). For diode-pumped operation, the $^4I_{9/2}$ level can be pumped directly near 0.97 μm. Based on the spontaneous lifetimes, the $^4I_{11/2}$ ($\tau = 1.5$ ms) to $^4I_{13/2}$ ($\tau = 5.1$ ms) transition appears to be a poor candidate for cw operation due to the apparent lower-level

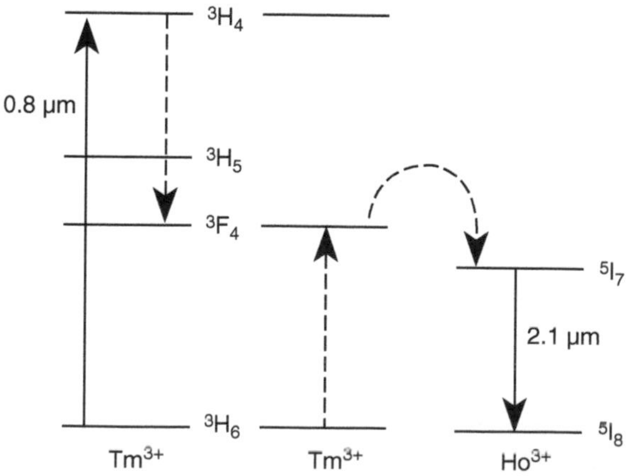

FIGURE 10.7 Energy transfer in diode-pumped materials co-doped with Tm and Ho.

bottleneck. However, near-quantum-limited slope efficiency has been achieved in diode-pumped monolithic lasers (Dinerman, Harrison, and Moulton 1994) with several different host crystals. This surprising result is due to a fortuitous upconversion process ($^4I_{13/2}+^4I_{13/2}\rightarrow^4I_{15/2}+^4I_{9/2}$) that depletes the lower level while partially repumping the upper level ($^4I_{9/2}\rightarrow^4I_{11/2}$). In fact, by pumping directly into the $^4I_{11/2}$ manifold (resonant pumping), a slope efficiency consistent with a quantum efficiency in excess of 100% has been demonstrated in Er:GSGG (Stoneman and Esterowitz 1992).

Yb^{3+} sensitization is used to enable direct pumping of the ground-state transition near 1.5 μm. CW operation is generally restricted to glass media for two reasons. The first is that the efficiency of the upconversion process is significantly less in glasses than in crystalline hosts (for uncertain reasons). The second is that the lifetime of the $^4I_{11/2}$ manifold is generally much shorter in glasses (because nonradiative coupling is facilitated by the relatively low phonon frequencies), so that the effect of any upconversion is reduced.

10.3.1.6 Cerium

Cerium-doped fluorides offer the possibility of broadly tunable lasers in the 280- to 330-nm spectral region. Unlike the other rare-earth-doped crystals discussed in this section, laser action in Ce^{3+}-doped crystals is based on a 5d→4f transition. This transition is electric-dipole allowed, giving rise to large emission cross-sections (up to 10^{-17} cm^2) and radiative lifetimes on the order of a few tens of nanoseconds. This is in strong contrast to the 4f→4f transitions used in Nd:YAG lasers, which have lifetimes of a couple of hundred microseconds and emission cross-sections on the order of 10^{-19} cm^2.

To achieve efficient lasing in the UV, a wide-bandgap crystal with good UV/VUV transmission is required to ensure minimal excited-state absorption into the conduction band, a problem that has historically plagued lasers based on the Ce^{3+} ion. Many fluorides have the necessary wide bandgaps and excellent UV and VUV transmission. As a result, lasing has been demonstrated in several Ce-doped fluoride crystals (Marshall et al. 1994).

Depending on the crystal host, strong crystal-field splitting of the 5d levels leads to four or five broad absorption bands. The lowest band forms the upper laser level. The lower laser level is the 4f level. This level is split by the spin-orbit interaction into $^2F_{5/2}$ and $^2F_{7/2}$ levels, which weakly interact with the crystal field to form two bands. As a result, Ce lasers are broadly tunable in the UV, with the host material determining the tuning range.

Three important Ce-doped fluorides are lithium calcium aluminum fluoride (Ce^{3+}:$LiCaAlF_6$ or Ce:LiCAF), lithium strontium aluminum fluoride (Ce^{3+}:$LiSrAlF_6$ or Ce:LiSAF), and lithium lutetium fluoride (Ce^{3+}:$LiLuF_4$ orCe:LiLuF or Ce:LLF). Typically, Ce:LiCAF operates with higher efficiency (up to 46% slope efficiency) and broader tunability than Ce:LiSAF (Pinto, Esterowitz, and Quarles 1995), although mixed composition Ce:$LiSr_xCa_{1-x}AlF_6$ crystals are being developed and show promise. Ce:LiCAF has a strong absorption band at 266 nm, and is readily pumped by the fourth harmonic of Nd:YAG. Therein lies its greatest potential to be operated in a miniature laser system (Zayhowski, Buchter, and Wilson 2001). Several researchers have demonstrated tunability of Ce:LiCAF from 280 to 316 nm, and average output powers in excess of 600 mW have been obtained at 1 kHz from a commercial system (Govorkov et al. 1998).

Cerium-doped LiLuF offers the opportunity to extend the range of Ce-doped fluorides to longer UV wavelengths, near 330 nm. However, Ce:LiLuF does not have any absorption at 266 nm and cannot be pumped by the fourth harmonic of Nd lasers. While the fifth harmonic of Nd:YAG does overlap an absorption band of LiLuF, pumping at 213 nm is much less efficient (Sarkura et al. 1998).

Solid-state Ce-doped fluoride lasers promise a simple and efficient option to nonlinear frequency conversion for a variety of spectroscopic applications, in fields such as atmospheric sensing and combustion diagnostics, which require tunable UV light sources.

10.3.2 Transition-Metal Dopants

The second major subclass of dopant materials is the transition-metal ions. Primary among these are chromium (Cr^{2+}, Cr^{3+}, Cr^{4+}), titanium (Ti^{3+}), and cobalt (Co^{2+}). These are especially interesting in cases where the ion-host interaction results in a broadly tunable four-level system. Such lasers are powerful tools in applications, such as spectroscopy and remote sensing, where it is desirable to have a single source to investigate multiple subjects. In addition, the broad fluorescence spectra are well suited to the generation of subpicosecond pulses.

The transition metals are characterized by a number of 3d valence electrons in addition to a closed inner shell ($3s^23p^6$). In contrast to the rare-earth ions, the energy-level structure of the transition metal ions is strongly influenced by the crystal field of the host. As a result, not all electronic levels have the same spatial atomic distribution. Broad tunability is possible when the optically active transition involves a pair of differently configured levels (Moulton 1992). In such cases, the electronic transitions require phonon emission to reach the minimum vibrational energy within an electronic level. The atomic motion occurs slowly compared to photon emission, and optical transitions can occur corresponding to a broad range of vibrational states. Coupling between the electronic levels of the ions and the vibrational states of the host allows numerous single-photon/multiple-phonon transitions that satisfy the energy requirements of the system, resulting in vibronic broadening of the absorption and emission spectra.

The transition-metal lasers in use today represent a small portion of the systems that have been investigated, dating back to early work at Bell Laboratories (Johnson, Dietz, and Guggenheim 1963; Johnson, Dietz, and Guggenheim 1964; Johnson, Guggenheim, and Thomas 1966; Johnson, Guggenheim, and Thomas 1967). Excited-state absorption and fluorescence quantum efficiency are often important factors in determining the viability of transition-metal-doped materials. In some cases, cryogenic operation is required to limit the nonradiative transition rate. The laser performance of potential vibronic materials is difficult to predict accurately; each one requires careful experimental characterization. However, the promise of broadly tunable lasers, especially where there are gaps in the available spectrum, continues to spur active development of new materials.

10.3.2.1 Chromium

Cr^{3+} is the active ion in the ruby laser ($Cr{:}Al_2O_3$), the first laser system to be demonstrated (Maiman 1960). It operates as a true three-level laser at 0.69 μm on a transition between the ^2E state and the $^4A_2$ ground state. The transition energy is relatively insensitive to the crystal field and the two levels have the same spatial distribution at equilibrium. Lamp pumping occurs via the vibronically broadened $^4A_2 \rightarrow {}^4T_2$ and $^4A_2 \rightarrow {}^4T_1$ transitions, with rapid, efficient decay to the upper laser level.

The position of the $^4T_2$ state is extremely sensitive to the host (its separation from the ground state is proportional to the strength of the crystal field). In excited low-field materials, it can have significant thermal population at room temperature due to the thermal distribution of excitations among the $^4T_2$, ^2E, and intermediate $^2T_1$ levels, and may be the basis of a broadly tunable, four-level laser. Such is the case for alexandrite (Cr^{3+}-doped chrysoberyl or $Cr{:}BeAl_2O_4$) (Walling et al. 1985) which, though initially developed as a substitute for ruby, was the first vibronic laser based on Cr^{3+}. In $Cr{:}BeAl_2O_4$, the $^4T_2$ level ($\tau = 6.6$ μs) lies approximately 800 cm^{-1} above the ^2E level ($\tau = 1.5$ ms). If the emission rate remains low compared to the thermalization rate, the ^2E level can serve as an excitation reservoir for the near-infrared vibronic transition ($\sim$0.7–0.8 μm). This leads to interesting thermal effects. For example, with increasing temperature the effective gain cross-section increases, the effective upper-state lifetime decreases, and the gain peak shifts toward the red (partially due to thermal population of the ground-state manifold). In many cases, optimal laser performance on the vibronic transition is achieved well above room temperature.

Figure 10.8 shows a comparison of the relevant energy levels in generic high-field and low-field Cr^{3+}-doped materials. Although both might operate as three-level lasers on the $^2E \rightarrow {}^4A_2$ transition, only the low-field material promises operation as a vibronic laser. Note that while it can be operated efficiently on the vibronic transition, $Cr{:}BeAl_2O_4$ is actually an intermediate-field material in that the ^2E state remains between the $^4T_2$ and $^4A_2$ states. Although vibronic operation has been demonstrated in many Cr^{3+}-doped materials, few have developed as practical sources in the near-infrared.

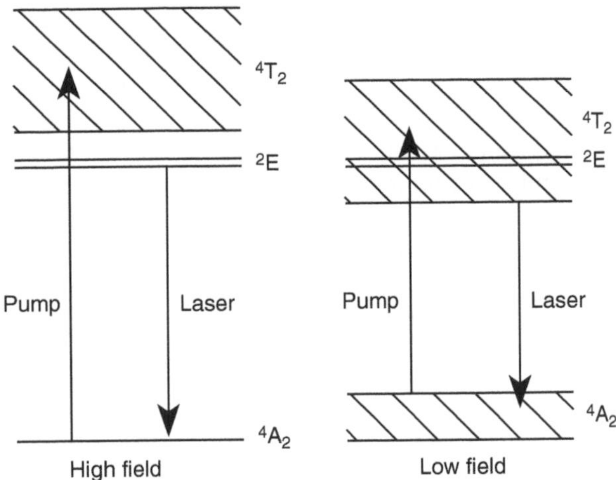

FIGURE 10.8 Partial energy-level diagrams for Cr^{3+} in high- and low-field hosts.

A number of related hexafluoride crystals have shown unusual promise for tunable laser operation in the near-infrared (Payne et al. 1988; Payne et al. 1989; Smith et al. 1992; Chai et al. 1992; Smith et al. 1993), particularly $Cr:LiCaAlF_6$ (Cr:LiCAF) and $Cr:LiSrAlF_6$ (Cr:LiSAF). Both are true Cr^{3+}-doped, low-field materials as depicted in Figure 10.8, although of the two, Cr:LiCAF has a considerably higher field. As a result, Cr:LiSAF has a broader tuning range (~0.8–1.0 μm) but poorer mechanical and thermal properties.

Direct diode-laser pumping of the ^2E state is a promising means of achieving an efficient, broadly tunable, near-infrared, miniature laser. Materials that can be doped to high levels, such as Cr:LiSAF, are particularly attractive. As a first step, diode-pumped operation was demonstrated in $Cr:BeAl_2O_4$ (Scheps et al. 1990), Cr:LiCAF (Scheps 1991), Cr:LiSAF (Scheps et al. 1991), and $Cr:LiSrGaF_6$ (Scheps 1992) using AlGaInP lasers operating near 0.67 μm. As a second step, a tuning range of 80 nm was demonstrated with a diode-pumped Cr:LiSAF laser system consisting of a 0.5-mm-long microchip gain medium coupled to a dispersive external resonator (Zhang et al. 1992). More recently, diode-pumped Cr:LiSAF microchip lasers have been demonstrated (Sutherland et al. 1995), and low thresholds with good output beam quality have been obtained over a wide range of wavelengths. Diode-pumped, mode-locked Cr:LiSAF lasers have also been reported (Balembois, Georges, and Brun 1993; French et al. 1993), and have generated femtosecond pulses from very compact systems (Agate et al. 2001).

When vibronic laser operation of Cr-doped fosterite ($Cr:Mg_2SiO_4$) was first reported in 1988 (Petricevic, Gayen, and Alfano 1988a) the active ion was believed to be Cr^{3+}. This material is of particular interest because it operates within the otherwise inaccessible ~200-nm-wide tuning range centered near 1.25 μm. The fluorescence spectrum is now attributed to a vibronic transition of Cr^{4+} (Petricevic, Gayen, and Alfano 1988b; Verdun et al. 1988; Moncorge et al. 1991). The broad absorption band is suitable for lamp pumping. It also allows pumping with a Nd:YAG laser at 1.06 μm, although the much stronger peak absorption near 0.75 μm suggests a means for diode-pumped operation. Currently, material quality and poor thermal properties remain important issues limiting the performance of $Cr:Mg_2SiO_4$ lasers. In addition, nonradiative decay impedes efficient cw operation at room temperature. However, at 77 K, pumped with a 12-W, 1.06-μm Nd:YAG laser, 2.8 W of cw output was obtained (Carrig and Pollock 1993). Going from room temperature to 77 K, τ increased from 3.6 to 27 μs.

Cr^{4+}:YAG is broadly tunable between 1.3 and 1.6 μm, and has been used as a gain medium for miniature femtosecond mode-locked lasers (Collings et al. 1996; Tomaru and Petek 2000; Tomaru 2001;

Leburn et al. 2003) and miniature gain-switched lasers (Zayhowski, Buchter, and Wilson 2001). This material is typically pumped by the ~1.06-μm output of a Nd laser. Although the laser efficiency is limited by excited-state absorption (Borodin et al. 1990; Kück, Peterman, and Huber 1991; Kück et al. 1996), the YAG host has excellent mechanical and thermal properties. The short nonradiative decay time, 4.1 μs, impedes efficient cw operation at room temperature.

In addition to being a useful laser gain medium, Cr^{4+}:YAG has become one of the most commonly used saturable absorbers for miniature solid-state lasers (see Section 10.6.2.3).

More recently, Cr^{2+}-doped II–VI semiconductor chalcogenides have emerged as broadly tunable gain media in the 2.1–3.4 μm region (DeLoach et al. 1996; Page et al. 1997; McKay, Roh, and Schepler 2002). These materials have broad pump bands between 1.5 and 2 μm, and have been pumped with the output of diode lasers, Er fiber lasers, and other solid-state lasers. CW Cr^{2+}:ZnS and Cr^{2+}:ZnSe microchip lasers have demonstrated output powers of 63 and 100 mW at 2.32 and 2.52 μm, with slope efficiencies of 53 and 20%, respectively (Mirov et al. 2002).

10.3.2.2 Titanium

Since its discovery as a vibronic laser material (Moulton 1982), Ti^{3+}-doped sapphire (Ti:Al_2O_3) has proven to be a nearly ideal laser material (Moulton 1986). Laser pumped near 0.5 μm (by an argon-ion laser, a frequency-doubled Nd laser, a copper-vapor laser, or a dye laser), it can be operated with nearly unitary quantum efficiency at room temperature either gain switched, cw, or mode locked. The peak gain near 0.8 μm has a radiative cross-section similar to that of Nd:YAG at 1.06 μm, and Ti:Al_2O_3 has been tuned (in both pulsed and cw operation) from below 0.7 μm to beyond 1.1 μm. The laser operates on a single-electron ($3d^1$) transition ($^2E \rightarrow {}^2T_2$); there are no higher-lying levels to support excited-state absorption. Nonradiative decay only becomes significant above room temperature. The lifetime at room temperature ($\tau = 3.2$ μs at 300 K) is only slightly below that at cryogenic temperatures ($\tau = 3.8$ μs at 77 K). Furthermore, Al_2O_3 has excellent mechanical and thermal properties. Laser-pumped Ti:Al_2O_3 lasers have generated up to 2.0 J in pulsed operation (Muller et al. 1988) and 43 W cw (Erbert et al. 1991). Additionally, its large gain bandwidth makes Ti:Al_2O_3 an ideal gain medium for mode-locked lasers, and Ti:Al_2O_3 lasers have produced the shortest pulses yet achieved, with a pulse width of less than 5 fs (less than 2 oscillations of light) (Sutter et al. 1999; Ell et al. 2001).

Early work on the development of Ti:Al_2O_3 lasers was plagued by a broad, featureless parasitic absorption that extended throughout the fluorescence band. Material was characterized in terms of the figure of merit, defined as the ratio of the peak absorption near 0.5 μm to the absorption at 0.8 μm. Figures of merit of a few tens were typical of early laser material. Later, the parasitic absorption was attributed to the presence of Ti^{4+}, which can be reduced by annealing (Aggarwal et al. 1988). Currently, the figure of merit of commercial material is commonly above 200, leading to negligible parasitic loss in most cases.

One drawback of Ti:Al_2O_3 is the difficulty of direct flashlamp pumping due to a short upper-state lifetime and limited overlap of the pump spectrum with typical lamp spectra. However, this has not prevented the introduction of commercial, lamp-pumped systems. Flashlamp-pumped pulse energies as high as 6.5 J have been reported (Brown and Fisher 1993). The prospects for miniature, diode-pumped Ti:Al_2O_3 lasers are less clear. Direct pumping is likely to be achieved in the long term, assuming continued success in the development of green diode lasers. Currently, diode-pumped systems use diode-pumped, frequency-doubled Nd lasers to pump Ti:Al_2O_3. Miniature lasers of this type have been demonstrated in both pulsed (Maker and Ferguson 1990; Zayhowski, Buchter, and Wilson 2001; Zayhowski and Wilson 2002) and cw operation (Harrison et al. 1991).

10.3.2.3 Cobalt

Cryogenic laser operation of Co^{2+}-doped magnesium fluoride (Co:MgF_2) was first achieved in 1964 (Johnson, Dietz, and Guggenheim 1964), but it was not until 1988 that room temperature, pulsed operation was reported (Welford and Moulton 1988). The laser transition ($^4T_2 \rightarrow {}^4T_1$) is relatively weak (Harrison, Welford, and Moulton 1989) but extremely broad (Lovold et al. 1985). Tuning over the ranges 1.51–2.28 μm and 1.75–2.50 μm has been demonstrated at 77 K and 299 K, respectively (Moulton 1992). The absorption peak near 1.3 μm is well suited to Nd-laser pumping. Although excited-state absorption is insignificant in $Co:MgF_2$, nonradiative decay is a major challenge in noncryogenic operation. The lifetime increases from 36 μs at room temperature to 1.3 ms at 77 K (Moulton 1985). The material has good mechanical and thermal properties (in particular, low thermal lensing due to high thermal conductivity and low dn/dt), and boules can be grown with high optical quality.

Near-quantum-limited slope efficiency was demonstrated at room temperature by employing a low-loss resonator and pump pulses with sufficiently short rise times to achieve quasi-cw emission in a time short compared to τ (Welford and Moulton 1988). Cryogenic temperatures are still required for cw operation. Currently, the prospects for direct diode pumping of $Co:MgF_2$ lasers are poor because there are few applications to drive the development of high-power diode lasers at 1.3 μm. A more realistic goal may be to utilize a diode-pumped Nd laser operating at 1.3 μm as a pump source in an all-solid-state system (Zhang et al. 2002).

10.3.3 Developments in Diode-Pumped Laser Materials

In recent years, a large number of Nd- and Yb-doped materials have been developed to take advantage of diode-laser pumping. The focus of the work has been multifold. Host materials that allow high dopant concentrations were sought to make efficient use of the diode pump light, particularly in longitudinally pumped laser systems. Materials with broad absorption were sought to relax the spectral requirements on the pump diodes, and to minimize the related need to accurately control their temperature (which consumes a considerable amount of power in most diode-pumped solid-state lasers). Improved mechanical and thermal properties were also desired, with particular emphasis on increased thermal conductivity to reduce thermo-optic effects and thermo-mechanical stresses that can lead to device fracture. For applications in short-pulse mode-locked lasers, or broadly tunable devices, broad emission spectra were desired. Finally, there has been a considerable amount of effort to develop materials that oscillate at particular wavelengths (see the discussion on Apatites, Fluorophosphates, and Fluorovanadates in Section 10.3.3.1) or serve multiple functions within the laser cavity, such as gain medium and frequency converter (see the discussions on Borates and Oxyborates in Section 10.3.3.2, and Tungstates in Section 10.3.3.3).

Historically, most laser materials were developed for use with flashlamp pump sources, and are generally limited to garnets, fluorides, and, to a lesser extent, glasses. The most commonly used materials in the 1-μm region are Nd:YAG, Nd:YLF, $Nd:YVO_4$ and Yb:YAG, with Nd:YAG serving as the standard, against which all other materials are judged. Some of the earliest diode-laser-pumped miniature lasers used pentaphosphate hosts (NPP and NLP, described in Section 10.3.1.1 on Neodymium), which are related to the more recently developed orthophosphates that incorporate lutetium, thereby enabling higher lanthanide concentrations (Rapaport et al. 1999). Similar developments in fluoride hosts include Nd-doped lutetium lithium fluoride ($Nd:LuLiF_4$ or Nd:LuLiF) (Barnes et al. 1998).

Other recent additions to the list of "traditional" materials include Nd-doped calcium niobium gallium garnet (Nd:CNGG) and Nd-doped calcium lithium niobium gallium garnet (Nd:CLNGG) (Agnesi et al. 2001; Lupei et al. 2000). Nd:CNGG and Nd:CLNGG are disordered materials that provide high dopant concentration, broad absorption bandwidth, and improved thermal conductivity, but suffer from undesirable energy transfer processes as a result of their disordered structure.

Yb-doped media offer a reduced quantum defect, increased absorption and gain bandwidths, and increased energy storage capability compared to Nd-doped media. These attributes are desirable for use in high-power, high-energy, and ultrashort pulsed lasers, and have motivated a search for new Yb hosts.

In particular, the apatites, oxyborates, and tungstates described below are predominantly used to host Yb (Krupke 2000).

10.3.3.1 Apatites, Fluorophosphates, and Fluorovanadates

There is a relatively large class of host media characterized by either the fluorophosphate, $P_3O_{12}F$, or fluorovanadate, $V_3O_{12}F$, groups that are collectively referred to as apatites. The emission and absorption cross-sections of Y^{3+} in the apatites are substantially larger than those of other Yb-doped media. The most common apatites are Yb-doped calcium fluorapatite (Yb:$Ca_5P_3O_{12}F$ or Yb:FAP) (Payne et al. 1994; Scheps et al. 1993) and Yb-doped strontium fluorapatite (Yb:$Sr_5P_3O_{12}F$ or Yb:SFAP) (Marshall et al. 1996; Bayramian et al. 1999).

Nd-doped apatites include FAP, SFAP, calcium fluorovanadate [$Ca_5(VO_4)_3F$], strontium fluorovanadate [$Sr_5(VO_4)_3F$ or SVAP] (Matthews et al. 1996), gadolinium lanthanum orthovanadate ($GdLaVO_4$) (Ostroumov et al. 1996), and barium chlorapatite ($Ba_5P_3O_{12}Cl$) (Noginov et al. 2000). In addition, the Nd-doped mixed-composition hosts $(Sr_{1-x}Ba_x)_5P_3O_{12}F$, $Sr_5(P_{1-x}V_x)_3O_{12}F$ and $Ba_5(P_{1-x}V_x)_3O_{12}F$ have been investigated as a means to accurately tune the $^4F_{3/2}\rightarrow{}^4I_{9/2}$ Nd-transition (Noginov et al. 2001). A similar compositional tuning scheme was demonstrated for Nd:YAG_xYSAG_{1-x} operating at 944.11 nm for an atmospheric water sensing application (Walsh et al. 2001).

10.3.3.2 Borates and Oxyborates

Yb-doped yttrium aluminum borate [$YAl_3(BO_3)_4$ or YAB] and yttrium strontium borate [$YSr_3(BO_3)_4$ or YSB] show significant promise as materials for the generation of femtosecond laser pulses, with gain bandwidths on the order of 50 nm. Additionally, YAB possesses a large nonlinear coefficient for the second-harmonic process that enables efficient self-frequency doubling. Self-frequency doubling of Nd:YAB to 531 nm (green) (Amano 1992; Jaque et al. 1998) and tunable operation from 510 to 545 nm (green) and 560 to 570 nm (yellow) with self-frequency doubled Yb:YAB have been demonstrated (Burns et al. 2002a; Dekker et al. 2003). Lanthanum scandium borate [$LaSc_3(BO_3)_4$ or LSB] has a much smaller emission bandwidth (4 nm) but has the largest saturation intensity of all known Nd-doped media, 1.2×10^4 W cm^{-2} (Meyn, Jensen, and Huber 1994; Beier et al. 1994).

Yttrium calcium oxyborate [$Ca_4YO(BO_3)_3$ or YCOB] (Chai et al. 1998; Mougel et al. 1998; Zhang et al. 2001) and gadolinium calcium oxyborate [$Ca_4GdO(BO_3)_3$ or GdCOB] are also capable of self-frequency doubling, as demonstrated in a Nd:YCOB passively *Q*-switched device that clearly illustrates the potential of these materials for the efficient generation of visible radiation in miniature diode-pumped lasers (Lucas-Leclin et al. 2000a; Lucas-Leclin et al. 2000b). Yb:GdCOB has significantly reduced reabsorption losses compared to other Yb-doped media as a result of its large ground-state energy-level splitting and an extremely large (90 nm) emission bandwidth (Auge et al. 2000). Yb:YCOB (210 fs, Valentine et al. 2000) and Yb:GdCOB (90 fs, Hönninger et al. 2000) lasers have generated femtosecond pulses.

10.3.3.3 Tungstates

The Yb-doped tungstates potassium gadolinium tungstate [$KG(WO_4)_2$ or KGW] (150 fs, Courjaud, Deguil, and Salin 2002) and potassium yttrium tungstate [$KY(WO_4)_2$ or KYW] (240 fs, Brunner et al. 2002) also operate with large emission bandwidths suitable for femtosecond pulse generation and demonstrate high slope efficiency. In addition, the tungstates show self-frequency shifting due to the Raman effect (Grabtchikov et al. 1999; Grabtchikov et al. 2000; Grabtchikov et al. 2002), which led to the development of Nd-doped lead tungstate ($PbWO_4$), sodium yttrium tungstate [$NaY(WO_4)_2$], sodium gadolinium tungstate [$NaGd(WO_4)_2$], and calcium tungstate ($CaWO_4$) as laser-active Raman scattering media (Faure et al. 1996; Kaminskii et al. 1999). These same host media, and other tungstates, may also be used as Raman scattering media in their undoped state. Other notable achievements in the tungstates are the demonstation of a green Nd:KGW microchip laser (Demidovich et al. 1997; Demidovich et al. 1998), passive *Q* switching of Yb:KGW (Lagatsky, Abdolvand, and Kuleshov 2000a), and high-power thin-disk lasers (Erhard et al. 2001).

10.3.3.4 Longer-Wavelength Materials

Some, if not all, of the Yb-doped media just discussed can be co-doped with Er^{3+} to produce eye-safe laser output in the 1.5-μm spectral region. To date, Er,Yb:YCOB has demonstrated efficient cw operation at 1.5 μm in microlaser form (Burns et al. 2002b). Similarly, the use of Tm^{3+} in Tm:$GdVO_4$ (Zagumennyi et al. 1996), Tm:YLF (Yokozawa, Izawa, and Hara 1998), and Tm,Yb:KGW (Demidovich et al. 2002) has led to the demonstration of efficient diode-pumped laser operation at 1.97 μm. Systems in the 2-μm region also include a flashlamp-pumped Ho-doped apatite (Ho:FAP) laser at 2.06 μm (Henningsen, Feitchner, and Melamed 1971), and diode-pumped Tm,Ho:$GdVO_4$ (Zagumennyi et al. 1996) and Ho:KGW (Pujol et al. 2002) lasers. Eu:GdCOB and Tm:GdCOB have been investigated for longer-wavelength operation (Dominiak-Dzik et al. 2000).

10.3.3.5 Ceramics

Another recent development in laser materials is the processing of ceramic laser gain media (Ikesue et al. 1995; Lu et al. 2002). Ceramics may make otherwise impractical high-growth-temperature oxides such as Y_2O_3 (Lu et al. 2001) more commercially viable as laser media, may allow the fabrication of very large gain media, and may offer advantages in the mass production of small systems.

10.3.4 Properties of Select Host Crystals

Table 10.1 through Table 10.4 list several important properties of four common host crystals: yttrium aluminum garnet (YAG), yttrium lithium fluoride (YLF), yttrium orthovanadate (YVO_4), and sapphire (Al_2O_3). The values listed are typical of those reported in the literature, although, in many cases, there is a significant spread in the reported values. All values listed are for room temperature (300 K).

TABLE 10.1 Properties of Yttrium Aluminum Garnet ($Y_3Al_5O_{12}$)

Property	Symbol	Value	Units	Comments
Crystal symmetry				Cubic
Lattice constant	a_0	12.01	Å	
Refractive index	n	1.818		1.064 μm
Temperature index variation	dn/dT	7.8×10^{-6}	K^{-1}	1.064 μm
Thermal expansion coefficient	a_e	6.1×10^{-6}	K^{-1}	
Thermal conductivity	k_c	0.11	$W\ cm^{-1}\ K^{-1}$	
Specific heat	c_{sh}	0.6	$J\ g^{-1}\ K^{-1}$	
Thermal diffusivity	a_d	0.044	$cm^2\ s^{-1}$	
Mass density	ρ_m	4.56	$g\ cm^{-3}$	
Elastic constant	C_{11}	3.2×10^9	$g\ cm^{-2}$	
Poisson's ratio		0.25		
Hardness		1215	$kg\ mm^{-2}$	Knoop
		8–8.5		Moh's scale
Tensile strength		2.0×10^6	$g\ cm^{-2}$	
Melting point		1950	°C	

TABLE 10.2 Properties of Yttrium Lithium Fluoride ($LiYF_4$)

Property	Symbol	Value	Units	Comments
Crystal symmetry				Tetragonal Positive uniaxial
Lattice constant	a_0	5.242	Å	
	c_0	11.37	Å	
Refractive index	n_π	1.4704		E‖c, 1.053 μm
	n_σ	1.4482		E⊥c, 1.053 μm
Temperature index variation	dn/dT	-6.6×10^{-6}	K^{-1}	E‖c, 1.053 μm
		-4.6×10^{-6}	K^{-1}	E⊥c, 1.053 μm
Thermal expansion coefficient	α_e	10×10^{-6}	K^{-1}	‖c
		14×10^{-6}	K^{-1}	⊥c
Thermal conductivity	k_c	0.072	$W\ cm^{-1}\ K^{-1}$	‖c
		0.053	$W\ cm^{-1}\ K^{-1}$	⊥c
Specific heat	c_{sh}	0.79	$J\ g^{-1}\ K^{-1}$	
Thermal diffusivity	α_d	0.023	$cm^2\ s^{-1}$	‖c
		0.017	$cm^2\ s^{-1}$	⊥c
Mass density	ρ_m	3.98	$g\ cm^{-3}$	
Elastic constant	C_{11}	7.6×10^8	$g\ cm^{-2}$	
Poisson's ratio		0.33		
Hardness		260–325	$kg\ mm^{-2}$	Knoop
		4–5		Moh's scale
Tensile strength		3.4×10^5	$g\ cm^{-2}$	
Melting point		825	°C	

TABLE 10.3 Properties of Yttrium Orthovanadate (YVO_4)

Property	Symbol	Value	Units	Comments
Crystal symmetry				Tetragonal Positive uniaxial
Lattice constant	a_0	7.119	Å	
	c_0	6.290	Å	
Refractive index	n_π	2.168		E‖c, 1.064 μm
	n_σ	1.958		E⊥c, 1.064 μm
Temperature index variation	dn/dT	2.9×10^{-6}	K^{-1}	E‖c, 632.8 nm
		8.5×10^{-6}	K^{-1}	E⊥c, 632.8 nm
Thermal expansion coefficient	α_e	11×10^{-6}	K^{-1}	‖c
		4.4×10^{-6}	K^{-1}	⊥c
Thermal conductivity	k_c	0.052	$W\ cm^{-1}\ K^{-1}$	‖c
		0.051	$W\ cm^{-1}\ K^{-1}$	⊥c
Specific heat	c_{sh}	0.51	$J\ g^{-1}\ K^{-1}$	
Thermal diffusivity	α_d	0.23	$cm^2\ s^{-1}$	
Mass density	ρ_m	4.23	$g\ cm^{-3}$	
Elastic constant	C_{11}		$g\ cm^{-2}$	
Poisson's ratio				
Hardness		480	$kg\ mm^{-2}$	Knoop
		5.5		Moh's scale
Tensile strength			$g\ cm^{-2}$	
Melting point		1750–1940	°C	Dependent on oxygen pressure

TABLE 10.4 Properties of Sapphire (Al_2O_3)

Property	Symbol	Value	Units	Comments
Crystal symmetry				Rhombohedral Positive uniaxial
Lattice constant	a_0	4.759	Å	
	c_0	12.99	Å	
Refractive index	n_π	1.755		E‖c, 694.3 nm
	n_σ	1.763		E⊥c, 694.3 nm
Temperature index variation	dn/dT	17×10^{-6}	K^{-1}	E‖c, 546.1 nm
		13×10^{-6}	K^{-1}	E⊥c, 546.1 nm
Thermal expansion coefficient	α_e	5.3×10^{-6}	K^{-1}	‖c
		4.5×10^{-6}	K^{-1}	⊥c
Thermal conductivity	k_c	0.40	$W\ cm^{-1}\ K^{-1}$	‖c
		0.33	$W\ cm^{-1}\ K^{-1}$	⊥c
Specific heat	c_{sh}	0.75	$J\ g^{-1}\ K^{-1}$	
Thermal diffusivity	α_d	0.14	$cm^2\ s^{-1}$	
Mass density	ρ_m	3.99	$g\ cm^{-3}$	
Elastic constant	C_{11}	3.5×10^9	$g\ cm^{-2}$	
Poisson's ratio		0.27		
Hardness		2100	$kg\ mm^{-2}$	Knoop
		9		Moh's scale
Tensile strength		5.5×10^6	$g\ cm^{-2}$	
Melting point		2040	°C	

10.3.5 Laser Parameters for Select Gain Media

Table 10.5 through Table 10.14 list important laser parameters for several gain media: Nd:YAG, Nd:YLF, Nd:YVO_4, Yb:YAG, Tm:YAG, Ho:YAG, Tm,Ho:YAG, Tm,Ho:YLF, Er:YAG, and Ti:sapphire. The values listed are typical of those reported in the literature, although, in many cases, there is a significant spread in the reported values. All values listed are for room temperature (300 K).

TABLE 10.5 Laser Parameters for Neodymium-Doped YAG (Nd^{3+}:$Y_3Al_5O_{12}$)

Property	Symbol	Value	Units	Comments
Nd^{3+} concentration (1.0 at.%)		1.39×10^{20}	cm^{-3}	
Quantum efficiency	η_q	1.0		
Wavelength at gain peak	λ_0	946	nm	
		1.064	μm	
		1.319	μm	
Linewidth	$\Delta\lambda_g$	0.8	nm	946 nm
		0.6	nm	1.064 μm
		0.6	nm	1.319 μm
Spontaneous lifetime	τ	230	μs	1.0 at.%
Concentration-quenching parameter	ρ_{cq}	2.63	at.%	
Emission cross-section	σ_g	0.4×10^{-19}	cm^2	946 nm
		6.5×10^{-19}	cm^2	1.06415 μm
		1.2×10^{-19}	cm^2	1.0644 μm
		1.7×10^{-19}	cm^2	1.319 μm
Occupation probability	f_u	0.60		946 nm
	f_l	0.008		946 nm
	f_u	0.40		1.06415 μm
	f_l	0		1.06415 μm
	f_u	0.60		1.0644 μm
	f_l	0		1.0644 μm

(continued)

Table 10.5 (Continued)

Property	Symbol	Value	Units	Comments
	f_u	0.40		1.319 μm
	f_l	0		1.319 μm
Energy of lower level	E_l	857	cm^{-1}	946 nm
Effective emission cross-section	$\sigma_{g,eff}$	0.3×10^{-19}	cm^2	946 nm
		3.3×10^{-19}	cm^2	Sum of 1.06415 and 1.0644 μm
		0.7×10^{-19}	cm^2	1.319 μm
Pump absorption coefficient	α_p	9.5	cm^{-1}	808.5 nm

TABLE 10.6 Laser parameters for Neodymium-doped YLF (Nd^{3+}:$LiYF_4$)

Property	Symbol	Value	Units	Comments
Nd^{3+} concentration (1.0 at.%)		1.4×10^{20}	cm^{-3}	
Quantum efficiency	η_q	1.0		
Wavelength at gain peak	λ_0	1.047	μm	E‖c
		1.053	μm	E⊥c
Linewidth	$\Delta\lambda_g$	1.2	nm	E‖c, 1.047 μm
		1.4	nm	E⊥c, 1.053 μm
Spontaneous lifetime	τ	460	μs	1.0 at.%
Concentration-quenching parameter	ρ_{cq}	2.77	at.%	
Emission cross-section	σ_g	4.4×10^{-19}	cm^2	E‖c, 1.047 μm
		2.2×10^{-19}	cm^2	E⊥c, 1.053 μm
Occupation probability	f_u	0.43		E‖c, 1.047 μm
	f_l	0		E‖c, 1.047 μm
	f_u	0.57		E⊥c, 1.053 μm
	f_l	0		E⊥c, 1.053 μm
Effective emission cross-section	$\sigma_{g,eff}$	1.9×10^{-19}	cm^2	E‖c, 1.047 μm
		1.2×10^{-19}	cm^2	E⊥c, 1.053 μm
Pump absorption coefficient	α_p	2.4	cm^{-1}	E‖c, 806 nm
		3.6	cm^{-1}	E‖c, 797 nm
		8.0	cm^{-1}	E⊥c, 797 nm
		11	cm^{-1}	E⊥c, 792 nm

TABLE 10.7 Laser Parameters for Neodymium-Doped YVO_4 (Nd^{3+}:YVO_4)

Property	Symbol	Value	Units	Comments
Nd^{3+} concentration (1.0 at.%)		1.252×10^{20}	cm^{-3}	
Quantum efficiency	η_q	1.0		
Wavelength at gain peak	λ_0	1.064		
Linewidth	$\Delta\lambda_g$	0.96	nm	
Spontaneous lifetime	τ	90	μs	1.0 at.%
Concentration-quenching parameter	ρ_{cq}	2.76	at.%	
Emission cross-section	σ_g	25.0×10^{-19}	cm^2	E‖c
		7.0×10^{-19}	cm^2	E⊥c
Occupation probability	f_u	0.52		
	f_l	0		
Effective emission cross-section	$\sigma_{g,eff}$	13×10^{-19}	cm^2	E‖c
		3.6×10^{-19}	cm^2	E⊥c
Pump absorption coefficient	α_p	41	cm^{-1}	E‖c, 808.5 nm
		11	cm^1	E⊥c, 808.5 nm

TABLE 10.8 Laser Parameters for Ytterbium-Doped YAG (Yb^{3+}:$Y_3Al_5O_{12}$)

Property	Symbol	Value	Units	Comments
Yb^{3+} concentration (5.5 at.%)		7.6×10^{20}		
Quantum efficiency	η_q	1.0		
Wavelength at gain peak	λ_0	1.030	μm	
Linewidth	$\Delta\lambda_g$	9	nm	
Spontaneous lifetime	τ	0.95	ms	
Emission cross-section	σ_g	3.1×10^{-20}	cm^2	
Occupation probability	f_u	0.7		
	f_l	0.047		
Energy of lower level	E_l	612	cm^{-1}	
Effective emission cross-section	$\sigma_{g,eff}$	2.3×10^{-20}	cm^2	
Pump absorption coefficient	α_p	4.9	cm^{-1}	968 nm
		5.1	cm^{-1}	941 nm

TABLE 10.9 Laser Parameters for Thulium-Doped YAG (Tm^{3+}:$Y_3Al_5O_{12}$)

Property	Symbol	Value	Units	Comments
Tm^{3+} concentration (6.0 at.%)		8.3×10^{20}		
Quantum efficiency	η_q	1.8		
Wavelength at gain peak	λ_0	2.013	μm	
Spontaneous lifetime	τ	10	ms	
Emission cross-section	σ_g	3.6×10^{-21}	cm^2	
Occupation probability	f_u	0.46		
	f_l	0.017		
Energy of lower level	E_l	588	cm^{-1}	
Effective emission cross-section	$\sigma_{g,eff}$	1.7×10^{-21}	cm^2	
Pump absorption coefficient	α_p	6.2	cm^{-1}	786 nm
		5.4	cm^{-1}	781 nm

TABLE 10.10 Laser Parameters for Holmium-Doped YAG (Ho^{3+}:$Y_3Al_5O_{12}$)

Property	Symbol	Value	Units	Comments
Ho^{3+} concentration (4.0 at.%)		5.6×10^{20}	cm^{-3}	
Quantum efficiency	η_q	1.0		
Wavelength at gain peak	λ_0	2.092	μm	
Spontaneous lifetime	τ	7	ms	
Emission cross-section	σ_g	1×10^{-19}	cm^2	
Occupation probability	f_u	0.10		
	f_l	0.016		
Energy of lower level	E_l	458	cm^{-1}	
Effective emission cross-section	$\sigma_{g,eff}$	1×10^{-20}	cm^2	
Pump absorption coefficient	α_p	5.8	cm^{-1}	1.907 μm

TABLE 10.11 Laser Parameters for Thulium–Holmium YAG ($Tm^{3+},Ho^{3+}:Y_3Al_5O_{12}$)

Property	Symbol	Value	Units	Comments
Tm^{3+} concentration (6.0 at.%)		8.3×10^{20}	cm^{-3}	
Ho^{3+} concentration (0.4 at.%)		5.6×10^{19}	cm^{-3}	
Quantum efficiency	η_q	1.8		
Wavelength at gain peak	λ_0	2.092	μm	
Spontaneous lifetime	τ	8	ms	
Effective lifetime		1–3	ms	Due to upconversion
Emission cross-section	σ_g	1×10^{-19}	cm^2	
Occupation probability	f_u	0.10		
	f_l	0.016		
Energy of lower level	E_l	458	cm^{-1}	
Effective emission cross-section	$\sigma_{g,eff}$	1×10^{-20}	cm^2	
Fraction of excitation in Ho^{3+}		0.58		
Pump absorption coefficient	α_p	6.2	cm^{-1}	786 nm
		5.4	cm^{-1}	781 nm

TABLE 10.12 Laser Parameters for Thulium–Holmium YLF ($Tm^{3+},Ho^{3+}:LiYF_4$)

Property	Symbol	Value	Units	Comments
Tm^{3+} concentration (6.0 at.%)		8.3×10^{20}		
Ho^{3+} concentration (0.4 at.%)		5.6×10^{19}	cm^{-3}	
Quantum efficiency	η_q	1.8		
Wavelength at gain peak	λ_0	2.067	μm	E‖c
Spontaneous lifetime	τ	15	ms	
Effective lifetime		~5	ms	Due to upconversion
Emission cross-section	σ_g	1.4×10^{-19}	cm^2	
Occupation probability	f_u	0.13		
	f_l	0.032		
Energy of lower level	E_l	315	cm^{-1}	
Effective emission cross-section	$\sigma_{g,eff}$	2.3×10^{-20}	cm^2	
Fraction of excitation in Ho^{3+}		0.56		
Pump absorption coefficient	α_p	2.7	cm^{-1}	E⊥c, 795 nm
		4.3	cm^{-1}	E‖c, 792 nm
		6.1	cm^{-1}	E‖c, 781 nm

TABLE 10.13 Laser Parameters for Erbium-Doped YAG ($Er^{3+}:Y_3Al_5O_{12}$)

Property	Symbol	Value	Units	Comments
Er^{3+} concentration (30 at.%)		4.2×10^{21}	cm^{-3}	
Quantum efficiency	η_q	1.6		
Wavelength at gain peak	λ_0	2.937	μm	
Linewidth	$\Delta\lambda_g$	8	nm	
Spontaneous lifetime	τ	100	μs	
Emission cross-section	σ_g	2.7×10^{-20}	cm^2	
Occupation probability	f_u	0.21		
	f_l	0		
Effective emission cross-section	$\sigma_{g,eff}$	5.6×10^{-21}	cm^2	
Pump absorption coefficient	α_p	6.2	cm^{-1}	974 nm
		13	cm^{-1}	965 nm
		11	cm^{-1}	960 nm

TABLE 10.14 Laser Parameters for Titanium-Doped Sapphire (Ti^{3+}:Al_2O_3)

Property	Symbol	Value	Units	Comments
Ti^{3+} concentration (0.09 wt.%)		3×10^{19}	cm^{-3}	
Quantum efficiency	η_q	0.10		
Wavelength at gain peak	λ_0	790	nm	
Linewidth	$\Delta\lambda_g$	300	nm	
Spontaneous lifetime	τ	3.15	μs	
Emission cross-section	σ_g	3.5×10^{-19}	cm^2	E‖c, 790 nm
Pump absorption coefficient	α_p	2.2	cm^{-1}	E‖c, 532 nm
		2.9	cm^{-1}	E‖c, 490 nm

10.4 Cavity Design

10.4.1 Issues in Laser Design

The design of a laser is dictated by many interdependent factors, including the requirements placed on the output beam (wavelength, spectral purity, tunability, beam quality, polarization, power, power stability), the operating environment (temperature, humidity, vibration, acceleration, externally applied forces), and practical considerations (size, cost, available power, pump-source characteristics). There is an increasingly large number of gain media, cavity designs, and pump configurations that have been employed in lasers, and several texts have been written on the subject of laser design. No one design is well suited to all applications; every laser is optimized for operation at one point in the multidimensional parameter space outlined above.

A very important issue in the design of many lasers is the extraction of heat from the gain medium. In the process of pumping the gain medium, heat is generated. As the temperature of the gain medium changes, so too do its physical length and refractive index. Each of these contributes to changes in the optical length and resonant frequencies of the laser cavity. (These changes are especially significant in miniature lasers where the active volume constitutes much of the resonator.) Nonuniform heating results in a nonuniform refractive index and internal stress. Index gradients lead to thermal lensing, which changes the confocal parameters of the laser cavity and can destabilize an otherwise stable cavity, or vice versa. Internal stress leads to stress birefringence and, eventually, stress-induced fracture.

Other issues that must be considered in high-power lasers are nonlinear optical effects and optical damage. The electrical field within the optical beam of a high-power laser can be large enough to damage optical components. This is particularly important in high-peak-power pulsed lasers. At optical intensities below the optical damage level, deleterious nonlinear optical interactions can still degrade the performance of the laser and even destroy the device. One example is stimulated Brillouin (acoustic wave) scattering in fiber lasers. In this case, nonlinear interactions create acoustic waves that can shatter the ends of the fiber.

10.4.2 Fundamental-Transverse-Mode Operation

10.4.2.1 Conventional Cavity Designs

One common feature of many applications for miniature lasers is that mode quality is at least as important as total power. As a result, miniature lasers are usually designed to operate in the fundamental transverse mode (see the discussion of Pump Considerations in Section 10.4.3). Most of the results and formulas presented in this chapter are derived for fundamental-mode operation and may require some modification for lasers operating in multiple transverse modes.

The transverse modes of a laser are determined by the cavity design and the pump energy deposition profile. The cavities for many miniature solid-state lasers are small versions of larger devices, designed according to the same principles (Hall and Jackson 1989). However, the use of longitudinal pumping allows additional possibilities for obtaining stable cavity modes.

10.4.2.2 Fabry-Pérot Cavities

Thermal Guiding. Efficient lasers can be produced using small, longitudinally pumped standing-wave laser cavities defined by two plane mirrors. The planar uniformity of such a cavity is broken by the pump beam, which deposits heat as it pumps the crystal. The heat diffuses away from the pump beam, generally resulting in a radially symmetric temperature distribution. In materials with a positive change in refractive index n with temperature T ($dn/dT>0$), such as Nd:YAG, this results in a thermal waveguide. In addition, when the cavity mirrors are deposited on the gain medium, there is some thermally induced curvature of the mirrors as the warmer sections of the gain medium expand or contract. In materials with a positive thermal expansion coefficient α_e this effect also contributes to the stabilization of the transverse mode. In some materials, such as Nd:YLF, this term dominates and can lead to stable transverse mode operation in an otherwise flat-flat cavity, despite a negative dn/dT. Another effect is strain-induced variation of the refractive index caused by nonuniform heating and expansion of the gain medium. This effect tends to be less important than the others in determining the transverse mode characteristics of the cavity, although it can cause significant local birefringence. It will be ignored in this section.

When both index change and thermal expansion are considered, the variation in the optical length of a material nl (where l is the physical length of the material) as a function of temperature is given by

$$\frac{\delta(nl)}{nl} = \left(C\alpha_e + \frac{1}{n}\frac{dn}{dT}\right)\delta T, \tag{10.29}$$

where C is a number between 0 and 1, depending on whether the material is constrained ($C=0$) or free to expand ($C=1$). For the case of nonuniform heating, the thermal expansion of the warmer sections of the material will be constrained by the cooler regions, and C may be a function of the thermal gradients. If the cavity length is short compared to the confocal parameter, as in the case of microchip lasers, the total change in optical length as a function of transverse cavity position can be modeled as a simple lens between the two flat mirrors or as an axially uniform waveguide with a radially varying index.

To simplify the analysis, we will assume that C is independent of position, so that $\delta(nl)=nl\Delta_{nl,T}\delta T$, where $\Delta_{nl,T}=C\alpha_e+dn/ndT$ is a constant. For a monolithic, longitudinally pumped, short cavity with radial heat flow, the radius of the oscillating mode r_m is given by (Zayhowski 1991a)

$$r_m^2 = \lambda_o\left[\frac{(r_0^2+r_p^2)lk_c}{\pi\kappa P_a n\Delta_{nl,T}}\right]^{1/2}, \tag{10.30}$$

where λ_o is the free-space wavelength of the oscillating mode, l is the cavity length, k_c is the thermal conductivity of the gain medium, κ is the heat-generating efficiency of the gain medium, P_a is the absorbed pump power, r_p is the average radius of the pump beam within the laser cavity, and r_0 determines the waist size of the oscillating mode for extremely small pump radii. Once $\Delta_{nl,T}$ and r_0 are determined, this equation does an excellent job of describing the pump-power dependence of the oscillating-mode radius.

For a thermally guided Fabry-Pérot laser cavity to create a symmetric fundamental transverse mode, parallelism between the cavity mirrors is critical. The maximum angle ψ_{max} that can be tolerated between the mirrors (for symmetric fundamental-mode operation) is given by (Zayhowski 1991a)

$$\psi_{max} = 1.8\frac{\kappa P_a\Delta_{nl,T}}{4\pi k_c r_p}. \tag{10.31}$$

A linear thermal gradient across the oscillating mode will contribute to the effective wedge between the two mirrors, through both the thermal expansion and the temperature-induced change in index of the material.

Aperture Guiding. In three-level or quasi-three-level lasers, there can be significant absorption of the oscillating radiation in unpumped regions of the gain medium. For longitudinally pumped devices, this creates a radially dependent loss (aperture) that can restrict the transverse dimensions of the lasing mode, resulting in smaller mode radii than predicted by Equation 10.30. Aperture guiding may be important in quasi-three-level systems such as Yb:YAG microchip lasers (Fan 1994).

Gain Guiding. The absence of gain in the unpumped regions of the gain medium is sufficient to define a stable transverse mode (Kogelnik 1965). This effect is similar to aperture guiding. It is usually insignificant compared to thermal guiding, but can be important in low-duty-cycle lasers where thermal effects are minimized. It can also be important in lasers operating near threshold. Near threshold there is little saturation of the population inversion, the round-trip amplification of the mode is highest near the center of the mode, and gain guiding will tend to reduce the mode cross-section. Well above threshold the population inversion at the center of the mode is saturated and residual gain in the wings (if there is any) can lead to a slight increase in the mode radius (Kemp et al. 1999).

Gain-Related Index Guiding. Optical gain provides dispersion. Laser modes that are spectrally detuned from the center of the gain profile will see a refractive index that is a function of their detuning. Modes that fall on the long-wavelength side of the gain profile will see an increased refractive index; modes on the short-wavelength side will see a decreased refractive index.

If we assume a Lorenzian gain profile, the change in refractive index seen by a mode as it moves away from the gain peak is

$$\Delta n(r, z, \lambda) = \frac{\lambda_0(\lambda - \lambda_0)g(r, z, \lambda)}{2\pi\Delta\lambda}, \tag{10.32}$$

where $g(r, z, \lambda)$ is the saturated, spatially and spectrally dependent gain in the presence of laser oscillation. Hence, $\Delta n(r, z, \lambda)$ depends on the mode profile and must be calculated self-consistently. (This can be done iteratively starting with the mode profile due to thermal considerations alone.) Near threshold the gain will have a maximum near the center of the mode; well above threshold the gain will be strongly saturated near the mode center and will have a local minimum on axis. In either case, the deformation of the optical path near the center of the mode due to gain-related index guiding can usually be modeled as a spherical (positive or negative) intracavity lens.

As a result of gain-related index guiding, each of the longitudinal modes of a laser has a slightly different spatial profile, a different amount of overlap with the pump, and a different far-field divergence. Gain-related index guiding has been shown to play an important role in Nd:YVO_4 microchip lasers (Kemp et al. 1999), and can lead to interesting effects such as self-Q switching (Conroy et al. 1998).

Self-Focusing. The refractive index of a material can be affected by the presence of a high-intensity optical field through the optical Kerr effect:

$$n = n_0 + n_2 I, \tag{10.33}$$

where n_0 is the refractive index in the absence of an optical field and n_2 is the nonlinear Kerr index. Light traveling through a Kerr medium will experience self-focusing. When the Kerr medium is inside a laser cavity, this will cause a reduction in the diameter of the oscillating mode. Self-focusing is usually insignificant for miniature cw lasers, but is the driving mechanism in Kerr-lens mode-locked lasers (see Section 10.6.5.2) and can play a role in determining the mode diameter in some miniature *Q*-switched and gain-switched lasers. For the miniature *Q*-switched and gain-switched lasers demonstrated to date, self-focusing introduces only a small perturbation to the mode diameter that would result in its absence.

10.4.3 Pump Considerations

To ensure oscillation in the fundamental transverse mode, it is necessary for the fundamental mode to use most of the gain available to the laser. If the radius of the fundamental mode is much less than the radius of the pumped region of the gain medium, higher-order transverse modes will oscillate.

For longitudinally pumped lasers the length of the gain medium l_g and the desired round-trip gain $g_{rt} = 2^{\dagger} g l_g$ together determine the required brightness of the pump source B_p, where brightness is defined as intensity per unit solid angle. For an incoherent pump source, the required brightness is given by (Fan and Sanchez 1990)

$$B_{\mathrm{p}} > \frac{4}{P_{\mathrm{p}}}\left(\frac{h\nu_{\mathrm{p}}g_{\mathrm{rt}}l_{\mathrm{g}}}{2^{\ddagger}n_{\mathrm{g}}\sigma_{\mathrm{g,eff}}\tau\eta_{\mathrm{p}}}\right)^{2}, \tag{10.34}$$

where P_{p} is the total pump power, $h\nu_{\mathrm{p}}$ is the energy of a pump photon, and η_{p} is the fraction of the pump power absorbed. For a diffraction-limited pump source, the brightness requirement is reduced by a factor of four (Fan and Sanchez 1990):

$$B_{\mathrm{p}} > \frac{1}{P_{\mathrm{p}}}\left(\frac{h\nu_{\mathrm{p}}g_{\mathrm{rt}}l_{\mathrm{g}}}{2^{\ddagger}n_{\mathrm{g}}\sigma_{\mathrm{g,eff}}\tau\eta_{\mathrm{p}}}\right)^{2}. \tag{10.35}$$

For single-transverse-mode operation the pump light should be focused within the volume of the fundamental oscillating mode.

10.4.4 Polarization Control

It is often desirable for a laser to oscillate in a single linear polarization. For lasers with isotropic gain media, this can be ensured by including a polarizing element, such as a Brewster plate, within the cavity. In single-frequency devices, the presence of any tilted optical element may be sufficient to polarize a laser because the reflection coefficient of surfaces is often different for *s*- and *p*-polarized waves; only a small amount of modal discrimination is needed to select one of two frequency-degenerate polarizations (see the discussion of Single-Frequency Operation in Section 10.5.1). If the design of a cavity does not favor a given polarization, the polarization degeneracy of an isotropic gain medium can often be removed by applying uniaxial transverse stress.

If there is very little polarization selectivity within a laser cavity, feedback from external surfaces can determine the polarization of the oscillating mode. Although this effect is usually undesirable, it has been used to controllably switch the polarization of microchip lasers at rates up to 100 kHz (Zayhowski 1991b).

Finally, in the absence of any strong polarizing mechanism, the polarization of the pump light may determine the polarization of the laser.

10.4.5 Additional Thermal Effects

Temperature affects many of the properties of the materials in a miniature laser. In addition to the refractive index, it affects the population distributions within the laser manifolds of the gain medium, the thermalization times, and the magnitude and spectral position of the gain. Thermal effects are most important in lasers that require a high average inversion density.

The rate-equation analysis presented in Section 10.6.2.3 shows that the pulse parameters of a passively *Q*-switched microchip laser are fixed by the material parameters and cavity design. To first order, the pulse width is independent of pulse repetition rate (Zayhowski and Dill 1994; Shimony et al. 1996). However, short-pulse lasers require aggressive pumping, and the temperature of the active region in the gain medium can increase significantly at high repetition rates. As it does, the population distributions within the laser manifolds and the thermalization times change. In short-pulse passively *Q*-switched Nd:YAG-based microchip lasers this often results in shorter pulses at higher repetition rates.

Some of the thermal effects in miniature lasers are dependent on changes in temperature across the active area of the laser. Thermal guiding, discussed in Section 10.4.2.2, is a very important example. Another such effect, which can limit the performance of these devices, is the temperature dependence of the spectral position of the gain peak. (The 1.064-μm transition in Nd:YAG shifts to longer wavelengths by about 0.46 nm per 100 °C variation in temperature.) As the temperature gradients across the active area of the gain medium increase, the spectral position of the gain peak at the center of the lasing mode will begin to deviate from the spectral position of the gain in the wings. The average gain coefficient for the mode therefore decreases, and a larger inversion is required to reach threshold. A larger inversion requires additional pumping, which further increases the temperature gradients, and the process can run away. For passively Q-switched Nd:YAG/Cr^{4+}:YAG microchip lasers, this effect contributes to making it difficult to obtain very short pulses at high repetition rates (Zayhowski and Wilson 2003a).

There are numerous additional subtleties that can only be captured by a full three-dimensional model of the laser, including the temperature dependences of the relative material parameters at all points in the active volume of the device. Such an analysis is outside of the scope of this chapter. The concepts and analysis presented here, supplemented with a small amount of experimentation, are adequate for the design of miniatures lasers and for understanding most of their characteristics. They are an essential part of any more complete model.

10.5 Spectral Control

10.5.1 Single-Frequency Operation

10.5.1.1 Multimode Operation

The frequency spacing between adjacent longitudinal modes of a cavity (also known as the free spectral range) is given by

$$\Delta\nu_{\text{fsr}} = \frac{c}{l_{\text{op,rt}}} = \frac{1}{t_{\text{rt}}}. \tag{10.36}$$

For most solid-state lasers, this is much less than the gain bandwidth. For example, the free spectral range of a 10-cm-long empty cavity is 1.5 GHz; the gain bandwidth of commonly used solid-state laser media is greater than 100 GHz. As a result, lasers tend to oscillate at several frequencies simultaneously.

Although the above statement is true, the reasons are more subtle than they may initially seem. In the early days of lasers, it was believed that lasers with homogeneously broadened gain spectra should operate in a single longitudinal mode. The reasoning behind this can be understood from the rate equations. If the optical intensity within the laser cavity is assumed to be uniform, the steady-state solution to the photon rate equation fixes the inversion density at its threshold value. The first cavity mode to lase (the one with the highest net gain) clamps the inversion density and no other mode can reach threshold. The flaw in this reasoning lies in the assumption of uniform optical intensity. Lasers with both homogeneously and inhomogeneously broadened gain media tend to oscillate in several longitudinal modes as a result of spatial and spectral hole burning.

10.5.1.2 Spatial Hole Burning and Energy Diffusion

Spatial hole burning (Tang, Statz, and DeMars 1963) and energy diffusion are responsible for shaping the population inversion in standing-wave laser cavities. The spatial distribution of the population inversion determines the gain of the different cavity modes and, therefore, which mode or modes will oscillate.

Spatial Hole Burning. In standing-wave laser cavities, the coherent superposition of the optical fields traveling in opposite directions within the cavity results in a sinusoidal intensity distribution. At the maxima of the intensity distribution, there is strong gain saturation, and the population inversion is depleted. However, at nulls in the optical field, the oscillating mode is unable to deplete the inversion. As a result, the inversion density is no longer uniform, but has "holes" at the positions corresponding to the peaks in the optical intensity. This phenomenon is known as spatial hole burning. The gain at the nulls in the optical field will continue to increase as the gain medium is pumped harder. Because other cavity modes have a different spatial profile than the first mode and can use the population inversion at these positions, this can lead to multimode operation. These ideas are illustrated in Figure 10.9.

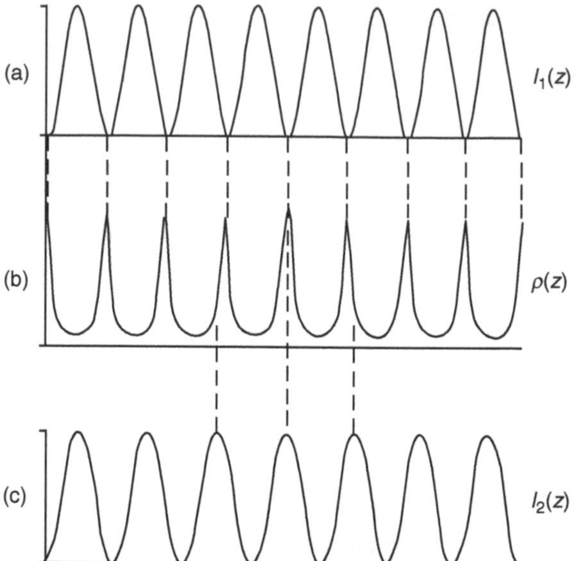

FIGURE 10.9 Illustration of spatial hole burning showing (a) the intensity profile of the first longitudinal mode to lase $I_1(z)$, (b) the population inversion in the presence of the first oscillating mode $\rho(z)$, and (c) the intensity profile of the second longitudinal mode to lase $I_2(z)$.

If we make the assumption that the round-trip loss of the cavity is relatively low, so that all of the maxima are of nearly equal intensity, the intensity of mode m within the gain medium can be written as

$$I_m(z) = 4I_m \sin^2(k_m z + \phi_m), \tag{10.37}$$

where I_m is the circulating intensity of mode m, k_m and ϕ_m are determined by the oscillating frequency and the geometry of the cavity, and z is the dimension along the cavity axis. For a cavity with a homogeneous four-level gain medium located between $z=0$ and $z=l_g$, and assuming that energy diffusion occurs slowly compared to the stimulated relaxation of the inversion density (the effects of energy diffusion will be discussed shortly), we can write the cw rate equations for each of the oscillating modes as

$$\sigma_m \int_0^{l_g} \rho(z) I_m(z) \mathrm{d}z = \gamma_{\mathrm{rt},m} I_m, \tag{10.38}$$

and the rate equation for the upper-manifold population density $\rho(z)$ as

$$\dot{\rho}_\mathrm{p}(z) = \sum_m \frac{\sigma_\mathrm{m} \rho(z) I_m(z)}{h\nu_m} + \frac{\rho(z)}{\tau}, \tag{10.39}$$

where σ_m is the value of the effective emission cross-section at the frequency ν_m corresponding to mode m, $\gamma_{\mathrm{rt},m}$ is the round-trip loss coefficient for mode m, and bleaching of the pump transition is neglected under the assumption that the pump rate $\dot{\rho}_\mathrm{p}(z)$ is independent of the inversion density. (For a spatially uniform optical intensity and inversion density, Equation 10.38 and Equation 10.39 reduce to Equation 10.12 and Equation 10.13.)

For single-mode operation (oscillating in mode $m=1$), we can solve Equation 10.38 and Equation 10.39 for the intensity I_1 (Casperson 1980):

$$\frac{4I_1}{I_{\text{sat},1}} = \frac{2\langle\rho_0\rangle}{\rho_{\text{th}}} - \left(\frac{2\langle\rho_0\rangle}{\rho_{\text{th}}} + \frac{1}{4}\right)^{1/2} - \frac{1}{2}, \tag{10.40}$$

where

$$\langle\rho_0\rangle \equiv \frac{1}{l_g}\int_0^{l_g} \rho_0(z)\mathrm{d}z \tag{10.41}$$

is the average unsaturated inversion density within the gain medium (which is proportional to the absorbed pump power), $\rho_{\text{th}} = \gamma_{\text{rt},1}/2\sigma_1 l_g$ is the threshold inversion density, $I_{\text{sat},1} = h\nu_1/\sigma_1\tau$, and it has been assumed that the active portion of the gain medium is much longer than the wavelength of oscillation. Because the output intensity of the laser is $T_o I_1$, where T_o is the transmission of the output coupler, Equation 10.40 shows that, in contrast to the results predicted in the discussion of the rate-equation Model of Lasers in Section 10.2.4, for a single-mode standing-wave laser, the output power is not a linear function of the pump power. The neglect of spatial hole burning (the use of Equation 10.12 and Equation 10.13) leads to an overestimate of the laser intensity; for a single-frequency laser, the laser intensity is overestimated by a factor of 1.5 near threshold.

The condition for single-mode operation reduces to

$$\int_0^{l_g} \rho(z)\left[\frac{I_1(z)}{I_1} - \beta(1,2)\frac{I_2(z)}{I_2}\right]\mathrm{d}z < 0, \tag{10.42}$$

where the discrimination factor $\beta(1,2)$ is given by

$$\beta(1,2) \equiv \frac{\sigma_1\gamma_{\text{rt},2}}{\sigma_2\gamma_{\text{rt},1}}, \tag{10.43}$$

and the subscript 2 denotes the second mode to oscillate. For a Lorentzian gain profile, if we assume that one cavity mode falls exactly at line center and that the cavity losses are the same for all potential oscillating modes,

$$\beta(1,2) = 1 + \left[\frac{2\Delta\lambda(1,2)}{\Delta\lambda_g}\right]^2, \tag{10.44}$$

where $\Delta\lambda_g$ is the full width at half-maximum of the gain peak, $\Delta\lambda(1,2) = \Delta m\lambda_0^2/ct_{\text{rt}}$ is the difference in the free-space wavelength of the first mode to oscillate and a potential second mode, Δm is the difference in the longitudinal mode numbers of the first and second modes (the difference in the number of intensity peaks along the cavity axis), and λ_0 is the free-space wavelength of the gain peak. (The Lorentzian-lineshape assumption must be used with care in solid-state lasers. Even if a given transition has a Lorentzian spectrum, the second mode to oscillate may correspond to a different transition.)

The solution to Equation 10.42 is (Zayhowski 1990a)

$$\langle\rho_0\rangle < \zeta_{\text{SH}}(1,2)\rho_{\text{th}}, \tag{10.45}$$

where $\zeta_{\text{SH}}(1,2)$ is the ratio of the maximum single-mode inversion density to the threshold inversion density and is a function of the cavity parameters, the gain cross-sections at the frequencies of modes 1 and 2, and the phase relationship between modes 1 and 2:

$$\zeta_{SH}(1,2) \equiv \left(\frac{\beta(1,2)-1}{1-\langle\psi(1,2)\rangle}+1\right)\left(\frac{2[\beta(1,2)-1]}{1-\langle\psi(1,2)\rangle}+1\right). \tag{10.46}$$

Here, the mode correlation factor $\langle\psi(1,2)\rangle$ is the weighted spatial average of the cosine of the phase difference between the standing-wave intensity patterns generated by modes 1 and 2:

$$\langle\psi(1,2)\rangle \equiv \frac{1}{l_g\langle\rho_0\rangle}\int_0^{l_g}\rho_0(z)cos[2(k_1-k_2)z+2(\phi_1-\phi_2)]dz. \tag{10.47}$$

The correlation factor $\langle\psi(1,2)\rangle$ can be solved in closed form for many common laser configurations, including lasers with uniformly pumped gain media and longitudinally pumped lasers. For a longitudinally pumped gain medium positioned adjacent to the end mirror of the cavity through which it is pumped,

$$\langle\psi(1,2)\rangle = \frac{1-\exp(-\alpha_p l_g)cos[2(k_1-k_2)l_g]-[2(k_1-k_2)/\alpha_p]\sin[2(k_1-k_2)l_g]}{[1-\exp(-\alpha_p l_g)][1+4(k_1-k_2)^2/\alpha_p^2]}, \tag{10.48}$$

where α_p is the absorption coefficient of the gain medium at the pump wavelength. When the active region fills the entire cavity (as is common in microchip lasers),

$$\langle\psi(1,2)\rangle = \frac{1}{1+[2\pi\Delta m/(\alpha_p l_g)]^2}. \tag{10.49}$$

For a uniformly pumped gain medium, Equation 10.48 and Equation 10.49 reduce to

$$\langle\psi(1,2)\rangle = \frac{\sin(2\pi\Delta m F_g)}{2\pi\Delta m F_g} \tag{10.50}$$

and

$$\langle\psi(1,2)\rangle = 0, \tag{10.51}$$

respectively, where F_g is the fraction of the cavity's optical length occupied by the gain medium.

$\zeta_{SH}(1,2)$ is plotted as a function of $\beta(1,2)$ for $\langle\psi(1,2)\rangle=\{1,0,-1\}$ in Figure 10.10. $\langle\psi(1,2)\rangle=1$ corresponds to perfect correlation between modes 1 and 2. In this case, the first mode to lase efficiently depletes the gain for the second mode, and the second mode will never reach threshold. Two orthogonally polarized modes of a two-mirror cavity with the same wave vector k are one example of perfect correlation (if we assume that each inverted site contributes equally to gain in both polarizations). $\langle\psi(1,2)\rangle=-1$ corresponds to the case of anticorrelated modes, where the maxima of one mode line up with the minima (nulls) of the other. This is the most favorable situation for the onset of a second lasing mode and occurs when a thin gain medium is located in the center of a standing-wave cavity.

Energy Diffusion. In the presence of spatial hole burning, energy diffusion moves some of the inverted population away from the nulls in the optical field toward the maxima, where it can be effectively depleted. As a result, energy diffusion allows single-mode operation of lasers at power levels well above the multimode threshold determined by Expression 45. In the presence of energy diffusion,

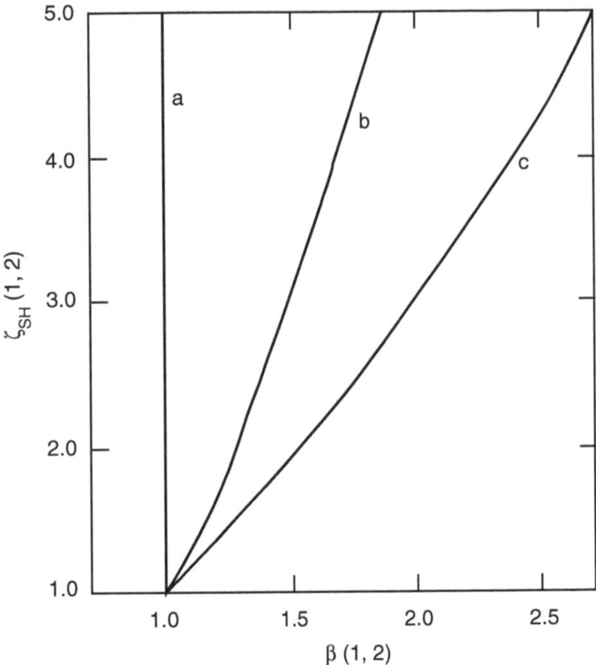

FIGURE 10.10 $\zeta_{SH}(1,2)$ as a function of $\beta(1,2)$ for (a) $\langle\psi(1,2)\rangle=1$, (b) $\langle\psi(1,2)\rangle=0$, and (c) $\langle\psi(1,2)\rangle=-1$.

Equation 10.39 takes the form

$$\dot{\rho}_{\mathrm{p}}(z) = \sum_{m} \frac{\sigma_m \rho(z) I_m(z)}{h\nu_m} + \frac{\rho(z)}{\tau} - D_{\mathrm{e}} \frac{\mathrm{d}^2\rho(z)}{\mathrm{d}z^2}, \tag{10.52}$$

where D_{e} is the energy diffusion constant. From this equation, a conservative estimate for the onset of lasing in a second mode can be obtained. In the presence of energy diffusion, single-mode operation is maintained as long as (Zayhowski 1990c)

$$\langle\rho_0\rangle < \zeta_{\mathrm{D}}(1,2)\rho_{\mathrm{th}}, \tag{10.53}$$

where

$$\zeta_{\mathrm{D}}(1,2) \equiv (4k_1^2 D_{\mathrm{e}}\tau - 1)\left(\frac{\beta(1,2)-1}{1-\langle\psi(1,2)\rangle}\right) + 4k_1^2 D_{\mathrm{e}}\tau\left(\frac{[\beta(1,2)-1]}{1-\langle\psi(1,2)\rangle}\right)^2. \tag{10.54}$$

Expressions 10.45 and 10.53 are complementary. Expression 10.45 accounts for spatial hole burning and neglects the effects of energy diffusion. Expression 10.53 accounts for energy diffusion in the presence of strong spatial hole burning. Whichever predicts a larger single-mode inversion ratio will provide a better estimate of the maximum single-mode value of $\langle\rho_0\rangle/\rho_{\mathrm{th}}$. It is, therefore, reasonable to define a function $\zeta(1,2)$ which is the larger of $\zeta_{\mathrm{SH}}(1,2)$ and $\zeta_{\mathrm{D}}(1,2)$. The inversion ratio $\zeta(1,2)$ is shown as a function of $[\beta(1,2)-1]/[1-\langle\psi(1,2)\rangle]$ for several values of $4k_1^2 D_{\mathrm{e}}\tau$ in Figure 10.11.

10.5.1.3 Spectral Hole Burning and Cross-Relaxation

Homogeneous gain broadening occurs when each optically active site sees exactly the same environment and, therefore, the excited states of each site have the same energy distribution. This is often the case in

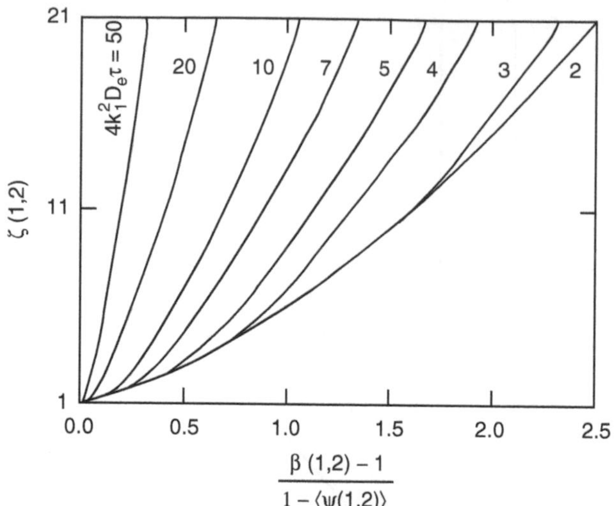

FIGURE 10.11 $\zeta(1,2)$ as a function of $[\beta(1,2)-1]/[1-\langle\psi(1,2)\rangle]$ for several values of $4k_1^2D_e\tau$.

crystalline solid-state gain media at room temperature. In other materials (e.g., glasses), each site sees a slightly different environment. The result is a collection of homogenous lines with slightly different center frequencies. The ensemble effect is called inhomogeneous gain broadening.

Spectral Hole Burning. In an inhomogeneous system, only those excitations with gain at the lasing frequency are able to participate efficiently in the stimulated-emission process. As a result, only those excitations become depleted, producing a gain spectrum that has a dip at the lasing frequency. This is known as spectral hole burning and is illustrated in Figure 10.12.

If the homogeneous linewidth is small compared to the inhomogeneous width then the hole is easily seen experimentally. For many of the laser glasses used, the room-temperature homogeneous and inhomogeneous linewidths are comparable and it is difficult to observe the spectral hole directly; spectral hole burning is more readily seen in glasses at low temperature, or in Doppler-broadened gas systems. Nevertheless, spectral hole burning does play a role in the performance of glass lasers, even at room temperature, and the general concept is discussed below.

For weak hole burning in a system with an inhomogeneous linewidth that is much greater than the homogeneous linewidth, the width of the spectral hole is typically between one and two times the homogeneous linewidth, depending on the homogeneous line shape. As the intracavity optical field increases, it can draw energy from an increasingly large inversion density because many of the homogeneous lines associated with different center frequencies will have spectral tails extending to the

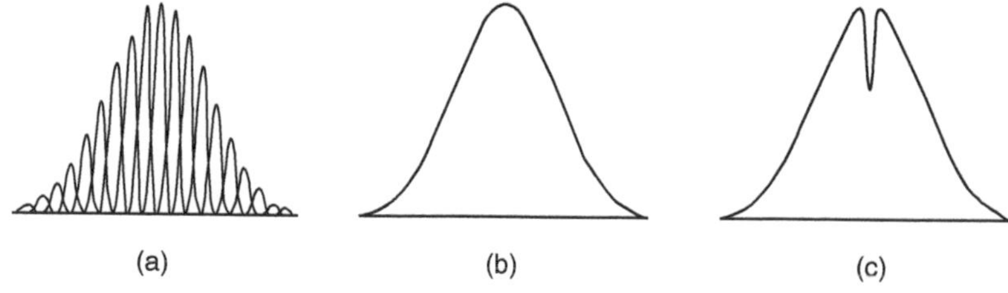

FIGURE 10.12 Illustration of spectral hole burning showing (a) several closely spaced homogeneously broadened spectra, (b) the inhomogeneously broadened spectrum resulting from the sum of several homogeneously broadened spectra, and (c) the inhomogeneously broadened spectrum with a spectral hole burned in the center due to the saturation of one of the homogeneously broadened components.

oscillating frequency. As a result, the width of the spectral hole increases with increasing optical intensity, and the inhomogeneous gain saturates more slowly than any of the individual homogeneous lines. For an inhomogeneous Gaussian ensemble of homogeneous Lorentzian lines (Siegman 1986), the width of the spectral hole is twice the Lorentzian linewidth at low optical intensities and increases as $\sqrt{I/I_{sat}}$; the gain at the oscillating frequency saturates as $1/\sqrt{1+I/I_{sat}}$ rather than as $1/(1+I/I_{sat})$. As in the case of spatial hole burning, the output power of the laser is a superlinear function of pump power near threshold.

Cross-Relaxation. Excitations that do not contribute to the lasing process for the first oscillating mode may contribute to the onset of lasing for other modes. If the separation between the potential oscillating modes is greater than the width of the spectral holes, the inversion densities seen by the modes are nearly independent. The condition for single-mode operation is, therefore,

$$\langle \rho_0 \rangle < \beta(1,2)\rho_{th}. \tag{10.55}$$

At higher inversion densities (pump powers), multimode behavior results.

In many material systems there is some cross-relaxation (or spectral diffusion) of the individual homogenous lines in an inhomogeneous gain spectrum. This is usually a small but significant effect and is directly analogous to energy diffusion in the case of spatial hole burning. In the limit of very fast cross-relaxation, the inhomogeneous line becomes effectively homogenous for all practical purposes. The mechanisms responsible for cross-relaxation in solids include dipole interactions between neighboring sites, radiation trapping, and phonon coupling.

10.5.1.4 Single-Frequency Lasers

There are many techniques for obtaining single-frequency operation of a laser. Several of these involve introducing a dispersive element into the cavity to provide a frequency-dependent loss, thereby increasing the modal discrimination factor $\beta(1,2)$ enough to select an individual longitudinal cavity mode. Examples of such elements are prisms, gratings, Fabry-Pérot etalons, and combinations of birefringent filters and polarizing elements. In some cases, a cavity may require multiple dispersive elements to obtain enough frequency selectivity to ensure single-frequency operation.

One variation on the use of Fabry-Pérot etalons that has gained momentum in recent years is the use of coupled cavities to control the mode structure of the laser system. When the Fabry-Pérot etalon is oriented with its reflecting surfaces normal to the optic axis of the laser the mode structure becomes more complicated, but is nonetheless easily calculated. This approach can be extended to an arbitrary number of coupled cavities (Penderson and Skettrup 1996) and is compatible with the mass fabrication of microchip lasers. Proper choice of mirror reflectivities will ensure single-frequency operation in microchip lasers with very broad gain media, such as Cr:LiSAF (Sutherland et al. 1995), Tm:YAG (Heine and Huber 1998), Tm:YLF (Yokozawa, Izawa, and Hara 1998), Tm:Ho:YLF (Izawa et al. 2000), and Yb:YAB (Burns et al. 2002a). While most of the reports of single-frequency coupled-cavity microchip lasers used an air-gap Fabry-Pérot for laboratory proof-of-concept demonstrations, the Cr:LiSAF devices used solid etalons to produce robust devices suitable for field use.

Alternatively, single-mode operation can be obtained in homogeneously broadened gain media by designing the laser cavity to reduce or eliminate the effects of spatial hole burning, resulting in a large inversion ratio $\zeta(1,2)$ (see the discussion of Spatial Hole Burning and Energy Diffusion in Section 10.5.1.2). This approach usually requires interferometric control of the cavity length because it is desirable to have the oscillating mode located at the center of the gain peak so that the discrimination factor $\beta(1,2)$ is as large as possible [if two modes straddle the gain peak $\beta(1,2)\approx 1$ and $\zeta(1,2)\approx 1$]. Nevertheless, there are several practical devices available that rely on this concept.

Spatial hole burning is almost completely eliminated in traveling-wave lasers, such as unidirectional ring lasers (Clobes and Brienza 1972; Kane and Byer 1985). (A small amount of hole burning is still present if the gain medium is located near a mirror, where fields propagating in different directions overlap.) In such lasers, there is a uniform intracavity optical intensity and the first mode to oscillate clamps the gain so that no other mode can reach threshold. A variation of the same idea, the "twisted-mode" laser, involves placing a quarter-wave plate on either side of the gain medium in a two-mirror cavity (Evtuhov and Siegman 1965; Wallmeroth 1990). In such a laser, the optical fields traveling in opposite directions in the gain medium are orthogonally polarized and do not interact coherently. The optical intensity within the gain medium is uniform, and there is no spatial hole burning.

Spatial hole burning is also reduced in standing-wave cavities with a large output coupling. In this case, the standing-wave field of the cavity is modified by the asymmetry between oppositely traveling waves. The wave traveling toward the output coupler is more intense than the wave traveling away from the output coupler. The more the ratio of the intensities of these two traveling waves deviates from unity, the smaller the contrast between the maximum and minimum axial intensity, and the smaller the effect of spatial hole burning. There is, therefore, more uniform gain saturation.

Standing-wave lasers will also operate in a single longitudinal mode at higher powers when the active region of the gain medium is strongly localized near one end mirror of the cavity than when it is located near the center. This is due to the pinning of the nodes of all cavity modes at the mirror, resulting in a strong correlation between modes ($\langle\psi(1,2)\rangle \approx 1$). Localization of the active region may be the result of positioning a thin gain element at one of the mirrors. Alternatively, it may result from longitudinally pumping a gain medium that strongly absorbs the pump radiation, even though the gain medium itself may be large (Kintz and Baer 1990). Examples of appropriate gain media include Nd:YVO_4 and stoichiometric materials. The best results are obtained by using a short gain element with a large absorption coefficient placed adjacent to one of the cavity end mirrors.

Single-mode operation has also been obtained by reducing the length of the cavity so that the longitudinal mode spacing is comparable to, or greater than, the gain bandwidth (Zayhowski and Mooradian 1989a). For very short cavity lengths, $\zeta(1,2)$ is a strong function of cavity length. Nd:YAG microchip lasers have operated in a single longitudinal mode at pump powers in excess of 22 times threshold; Nd:YVO_4 microchip lasers have produced 240 mW of single-frequency output.

A saturable absorber in a standing-wave laser cavity can also help to discriminate against multi-frequency oscillation. As the oscillating mode burns holes in the gain, it also burns holes in the loss. In cw lasers this approach can severely reduce the efficiency of the device. In passively *Q*-switched lasers, which already contain a saturable absorber, locating the saturable absorber near the center of the cavity will provide the highest discrimination against multimode behavior.

In some gain media there is a large amount of energy diffusion. As discussed above, energy diffusion mitigates the effects of spatial hole burning and can result in single-mode oscillation. Energy diffusion is expected to be most important in gain media with a high density of excited states, such as stoichiometric materials, but still plays a significant role in more dilute materials. In semiconductor materials, where the excited states (electrons and holes) themselves diffuse, energy diffusion can completely eliminate the effects of spatial hole burning.

10.5.2 Linewidth

One contribution to the finite spectral width of all lasers is the coupling of spontaneous emission (or noise) to the oscillating mode (Schawlow and Townes 1958; Lax 1966), which results in a Lorentzian power spectrum. For many ultrastable lasers, this contribution alone determines the fundamental linewidth, and it is common practice to obtain the fundamental linewidth of such a laser by fitting the tails of the measured power spectrum to a Lorentzian curve. In very small lasers, there is a second important contribution to the fundamental linewidth — the thermal fluctuations of the cavity length at a constant temperature (Jaseja, Javan, and Townes 1963). This contribution is expected to result in a Gaussian power spectrum and, unlike the Lorentzian contribution discussed above, is independent of the output power of the device. These two fundamental noise sources introduce fluctuations on different time scales, corresponding to the relaxation frequency of the laser (see Section 10.6.1.1) and the acoustic resonances of the laser (or its components), if they occur below the relaxation frequency. In lasers with a very small mode volume, such as microchip lasers, the contribution due to thermal fluctuations is much larger than that due to spontaneous emission (Zayhowski 1990b). However, because a Gaussian curve decays more quickly than a Lorentzian curve, the tails of the power spectrum still correspond to the Lorentzian contribution.

The output of a laser contains both stimulated and spontaneous emission. Because the phase of the spontaneous emission is not correlated with the phase of the stimulated emission, this leads to a random variation in the net phase of the optical field. As a result, the fundamental linewidth of a laser has a full width at half-maximum of

$$\delta\nu_{\mathrm{L}} = \frac{1}{2\pi}\frac{\dot{q}_{\mathrm{sp}}}{2q}, \tag{10.56}$$

where the 2 in the denominator of $\dot{q}_{\mathrm{sp}}/2q$ results from averaging over all possible phases of the spontaneous emission relative to the stimulated emission and the prefactor $1/2\pi$ converts phase uncertainty to frequency width. If we approximate the optical field within the laser cavity as uniform (i.e., neglect spatial hole burning), the spontaneous-emission rate is clamped at the threshold emission rate (see the discussion of the Rate-Equation Model of Lasers in Section 10.2.4):

$$\dot{q}_{\mathrm{sp}} = \frac{2^{\dagger}N_{\mathrm{u,th}}}{2N_{\mathrm{th}}}\frac{\gamma_{\mathrm{rt}}}{t_{\mathrm{rt}}}, \tag{10.57}$$

where $N_{\mathrm{u,th}}$ is the population of the upper laser level at threshold. The resulting Lorentzian linewidth is given by

$$\delta\nu_{\mathrm{L}} = \frac{2^{\dagger}N_{\mathrm{u,th}}}{2N_{\mathrm{th}}}\frac{h\nu_0}{4\pi P_{\mathrm{o}}}\frac{\gamma_{\mathrm{rt}}\gamma_{\mathrm{o}}}{t_{\mathrm{rt}}^2}, \tag{10.58}$$

where h is Planck's constant, ν_0 is the center frequency of the laser, P_{o} is the output power, $\gamma_{\mathrm{rt}}=\gamma_{\mathrm{rt,p}}+\gamma_{\mathrm{o}}$ is the round-trip loss coefficient, $\gamma_{\mathrm{rt,p}}$ is the parasitic round-trip cavity loss coefficient (not including transmission through the output coupler), $\gamma_{\mathrm{o}}=-\ln(1-T_{\mathrm{o}})$ is the output-coupling loss coefficient, and T_{o} is the transmission of the output coupler.

The spectral-broadening effects of thermal fluctuations in cavity length are easily calculated using the principle of equipartition of energy. This principle, derived from classical mechanics, states that whenever the energy of a system can be written as a sum of independent terms, each of which is quadratic in the variable representing the associated degree of freedom, then in equilibrium at absolute temperature T, each of the terms (that is, each degree of freedom) contributes $k_{\mathrm{B}}T/2$, where k_{B} is Boltzmann's constant, to the energy of the system. In a monolithic standing-wave cavity this leads to the expression (Zayhowski 1990b)

$$C_{11}\left\langle\left(\frac{\delta l}{l}\right)^2\right\rangle V_{\mathrm{g}} = k_{\mathrm{B}}T, \tag{10.59}$$

where C_{11} is the longitudinal elastic constant of the gain medium, δl is the change in the cavity length l, V_{g} is the active volume of the gain medium, and the brackets indicate averaging over time. This results in a Gaussian contribution to the fundamental linewidth, with a full width at half-maximum given by

$$\delta\nu_{\mathrm{G}} = \nu_0\left[\frac{8k_{\mathrm{B}}T\ln(2)}{C_{11}V_{\mathrm{g}}}\right]^{1/2}. \tag{10.60}$$

For microchip lasers, the value of $\delta\nu_{\mathrm{G}}$ varies with pump power and pump-beam diameter, but is typically between 5 and 7 kHz, compared to a Lorentzian contribution to the linewidth, $\delta\nu_{\mathrm{L}}$, of only a few hertz (Zayhowski 1990b).

Other factors that contribute to the linewidth of a laser include fluctuations of the pump power, optical feedback into the laser cavity, mechanical vibrations, and temperature variations. These contributions are less fundamental, however, and can be controlled. In addition, they tend to occur on a longer time scale than spontaneous emission and fundamental thermal fluctuations. Although they are often not important factors in attempts to measure the fundamental linewidth of a laser, they are important contributions to the frequency fluctuations of lasers on time scales of practical interest.

All fluctuations in the frequency of a laser, including those resulting from spontaneous emission and thermal fluctuations, can be actively compensated for, and a laser can be locked to an external reference with extreme precision (Hough et al. 1984; Man and Brillet 1984; Day, Gustafson, and Byer 1990). Note that reducing the frequency fluctuations below the fundamental limits may entail an increase in the intensity noise (Ohtsu and Kotajima 1985).

10.5.3 Frequency Tuning

10.5.3.1 Mode Selection

Frequency tuning of a laser can occur in one of two ways. If the longitudinal mode spacing of the laser cavity is much less than the gain bandwidth, the cavity is capable of supporting several modes, each at a different frequency. A single frequency can be selected through the insertion of one or more frequency-selective dispersive elements (see the discussion of Single-Frequency Lasers in Section 10.5.1.4). In all of the examples listed there, a small repositioning of the frequency-selective element results in the selection of a new longitudinal mode (and, hence, a new operating frequency). The frequency-selective element is used to select one of the several cavity modes, and discrete tuning is obtained.

Rapid tuning can be obtained through the use of electro-optic components. However, if the frequency selectivity of the tuning element is weak, cavity dynamics rather than the response time of the electro-optic element determine how quickly the laser will switch frequencies. This is often the case in single-frequency unidirectional ring lasers. In such lasers, the additional loss introduced at the original oscillating frequency to induce a mode switch may be small, and the original mode will continue to oscillate, depleting the population inversion. Therefore, the net gain at the newly selected frequency is initially small, and radiation in the new mode builds up slowly. Eventually the old mode decays, releasing the gain to the new mode, which then rapidly reaches its cw intensity. Under these conditions, the switching time is approximately given by the expression (Schulz 1990)

$$t_{sw} = \frac{\tau_c}{\delta\gamma_{rt}} = \frac{t_{rt}}{\gamma_{rt}\delta\gamma_{rt}}, \tag{10.61}$$

where $\delta\gamma_{rt}$ is the change in the round-trip cavity loss coefficient induced by the frequency-selective element for the original mode. In some cases it requires less time to turn a laser off and restart it in the new mode than to switch modes while it is oscillating.

10.5.3.2 Mode Shifting

The other way a laser can be tuned is to change the frequency of a given cavity mode by changing the optical length of the cavity:

$$\frac{\delta\nu_o}{\nu_o} = -\frac{\delta(l_{op,rt})}{l_{op,rt}} = -\frac{\delta(t_{rt})}{t_{rt}}. \tag{10.62}$$

Because the cavity's optical length can be changed continuously, this leads to continuous tuning. This type of tuning is limited by the free spectral range of the cavity (except under transient conditions). Once the cavity modes are shifted by a full free spectral range, an adjacent cavity mode is positioned at the frequency where the initial mode started. For the same reasons that the initial mode was originally favored, the adjacent mode is now favored, and the laser will have a tendency to mode hop (the longitudinal mode number will change by one), returning to the original frequency.

The optical length of the elements in a laser cavity can be changed by a variety of techniques including thermal tuning, stress tuning, and electro-optic tuning. Pump-power modulation represents a special case of thermal tuning. Each of these techniques allows continuous frequency modulation of a single longitudinal cavity mode and is discussed below. To calculate the total change in the optical length of the laser cavity, one simply sums the changes in the optical lengths of each of the intracavity elements, keeping in mind that the total change in the physical length of the cavity may be constrained by the mechanical support structure of the device. For example, the expansion of the gain medium may be partially offset by the shortening of an adjacent air gap.

Thermal Tuning. Changing the temperature of an element in a laser cavity is often the simplest way to tune the device. The response time of the cavity, however, is limited by the thermal diffusion time, which is usually relatively long (typically several milliseconds or more). The temperature-induced change in the optical length of an unconstrained piece of material is given by

$$\frac{\delta(nl)}{nl} = \left(\alpha_e + \frac{1}{n}\frac{\mathrm{d}n}{\mathrm{d}T}\right)\delta T, \tag{10.63}$$

where α_e is the thermal expansion coefficient, n is the refractive index, l is the length of the material, and δT is the change in temperature.

Stress Tuning. The cavity modes of a resonator will also tune as elements within the cavity are squeezed. For squeezing transverse to the optic axis, the main effect is usually an elongation of the material along the optical axis of the resonator (Owyoung and Esherick 1987). Superimposed on this effect is the stress-optic effect. In crystals with cubic symmetry, the stress-optic effect can split the frequency degeneracy of orthogonally polarized optical modes because the stress-optic coefficients are different for light polarized parallel to and perpendicular to the applied stress. For squeezing along the optical axis of the cavity, there is a compression of the element and the frequency degeneracy of orthogonally polarized modes remains unchanged. By using a piezoelectric transducer to squeeze a monolithic microchip laser, tuning has been obtained at modulation frequencies up to 20 MHz, although linear (nonresonant) response was obtained only at modulation frequencies up to 80 kHz (Zayhowski and Mooradian 1989b). The nonresonant tuning response in that experiment was 300 kHz V^{-1}. Much larger tuning responses can be obtained by piezoelectrically driving a separate cavity mirror.

Electro-Optic Tuning. For applications including frequency-modulated optical communications and chirped coherent laser radar, extremely high rates of tuning are required. These rates can only be achieved electro-optically.

The frequency response of a laser whose optical length is varied is well understood. When a linear voltage ramp is applied to an intracavity electro-optic crystal, the frequency of the laser undergoes a series of steps whose spacing in time is the cavity round-trip time t_{rt} (Genack and Brewer 1978). When the rise time T_r of the voltage is long compared to the cavity round-trip time, if we assume that the change in the optical length of the cavity tracks the applied voltage, the frequency has an approximately linear chirp with a fractional deviation from linearity of $t_{rt}/2T_r$. (This deviation from linearity can be corrected by using a voltage waveform that is quadratic during the first cavity round-trip time and linear thereafter.) Also, because two steps are required to define a modulation frequency (one step up and one step down), the maximum response frequency of the cavity is $1/2t_{rt}$.

The sensitivity of electro-optic voltage-to-frequency conversion increases linearly with the percentage of the cavity length occupied by the electro-optic element. If we ignore piezoelectric and electrostrictive effects,

$$\frac{\delta(\nu_o)}{\nu_o} = -F_{eo}n^2 r_{eo}E, \tag{10.64}$$

where F_{eo} is the fraction of the cavity's round-trip optical length occupied by the electro-optic material, r_{eo} is the appropriate electro-optic coefficient, and E is the magnitude of the applied electric field.

For high-sensitivity tuning, it is desirable to fill the cavity with as large a fraction of electro-optic material as possible. However, it is often still important to keep the total cavity length as small as possible, for the reasons discussed above. In addition, as the length of the electro-optic crystal increases, the capacitance between the electrodes on the crystal rises (for fields applied transverse to the crystal length), resulting in higher energy requirements and slower electrical response.

Most electro-optic crystals are piezoelectrically active. As a result, when a voltage is applied to the crystal, both stress (and the concomitant strain) and refractive index are modulated. Although the effects of stress may normally be small compared to the index modulation, a free-standing crystal can act as a high-Q acoustic cavity. At the resonant frequencies, the piezoelectric effect can cause a greatly enhanced frequency-modulation response of the laser. Depending on the dimensions of the crystal, these resonances can fall between a few kilohertz and several megahertz. To eliminate these resonances, the electro-optic material can be bonded to materials with a similar acoustic impedance to transmit the electrically excited acoustic waves out of the crystal (Schulz and Henion 1991a).

For the reasons discussed above, the high-speed tuning of a laser is most linear when the shortest possible laser cavity is used. Short cavity length offers an advantage in tuning range as well, because the amount of frequency tuning that can be obtained without mode hopping is limited by the free spectral range of the laser (Equation 36). As a result, the short length of composite-cavity electro-optically tuned microchip lasers makes them particularly attractive as solid-state lasers capable of high sensitivity, a large continuous tuning range, and a high tuning rate (Zayhowski et al. 1993). Such devices have been continuously tuned over a 30-GHz range with a tuning sensitivity of $\sim$14 MHz V^{-1}. The tuning response was relatively flat for tuning rates from dc to 1.3 GHz. Even higher rates of modulation should be possible.

Pump-Power Modulation. Changes in pump power typically induce frequency changes in the output of solid-state lasers. As the pump power increases, more thermal energy is deposited in the gain medium, raising its temperature and changing both the refractive index and length. Because frequency tuning via pump-power modulation relies on thermal effects, it is often thought to be too slow for many applications. In addition, modulating the pump power has the undesirable effect of changing the amplitude of the laser output. However, for very small lasers, significant frequency modulation can be obtained at relatively high modulation rates with little associated amplitude modulation. For example, pump-power modulation has been used with 1.32-μm microchip lasers to obtain 10-MHz frequency modulation at a 1-kHz rate and 1-MHz frequency modulation at a 10-kHz rate with an associated amplitude modulation of less than 5% (Zayhowski and Keszenheimer 1992). This technique has been used to phase lock two lasers, and introduced less than 0.1% amplitude modulation on the slave laser (Keszenheimer, Balboni, and Zayhowski 1992). When it can be used, pump-power modulation has advantages over other frequency modulation techniques since it requires very little power, no high-voltage electronics, no special mechanical fixturing, and no additional intracavity elements. Another important reason to understand the frequency response of a laser to changes in pump power is that pump-power fluctuations are often responsible for a large portion of the frequency fluctuations of the laser output.

The following results are obtained for a small, longitudinally pumped, monolithic solid-state laser. Both the pump beam and oscillating mode are assumed to be Gaussian, with constant amplitude and diameter as they propagate along the cavity axis. Because the thermal conductivity of most solid-state gain media is much greater than that of either dielectric coatings or air, it is assumed that all heat flow is directed in the plane normal to the cavity axis. To further simplify the analysis, the transverse dimensions of the gain medium are initially modeled as infinite. A discussion of the effect of heat sinking the gain medium at a finite distance from the cavity axis is included later in this section.

The variation in the optical length of the laser cavity as a function of temperature is given by Equation 10.29. For the case of nonuniform heating, the thermal expansion of the warmer sections of the crystal will be constrained by the cooler regions, and Equation 10.29 will not be completely accurate (see the discussion of Fabry-Pérot Cavities in Section 10.4.2.2). In addition, there will be internal stress that could modify the refractive index through the stress-optic coefficient. These effects usually represent only small corrections to Equation 10.29 and are neglected in the following analysis.

With the approximations stated above, the crystal temperature averaged over the volume of a laser mode with waist radius r_{m} in a cavity of length l is given by

$$T_{\mathrm{ave}} = \frac{2}{\pi r_{\mathrm{m}}^2 l} \int_0^l \int_0^{2\pi} \int_0^{\infty} T\exp\left(-\frac{2r^2}{r_{\mathrm{m}}^2}\right) r\mathrm{d}r\mathrm{d}\theta\mathrm{d}z. \quad (10.65)$$

The temperature distribution T within the gain medium satisfies the equation

$$\frac{\mathrm{d}T}{\mathrm{d}t} = \alpha_{\mathrm{d}}\nabla^2 T + \frac{Q_{\mathrm{h}}}{\rho_{\mathrm{m}} c_{\mathrm{sh}}}, \quad (10.66)$$

where c_{sh} is the specific heat, ρ_{m} is the mass density, $\alpha_{\mathrm{d}}=k_{\mathrm{c}}/\rho_{\mathrm{m}} c_{\mathrm{sh}}$ is the thermal diffusivity, k_{c} is the thermal conductivity, and Q_{h} is the rate of heat deposition per unit volume. The spectral components of the rate of heat deposition $Q_{\mathrm{h}}(\omega,r)$ are proportional to the pump power $P_{\mathrm{p}}(\omega)$ at frequency ω:

$$Q_{\mathrm{h}}(\omega,r)\sin(\omega t) = \frac{2\kappa P_{\mathrm{p}}(\omega)}{\pi r_{\mathrm{m}}^2 l}\exp\left(-\frac{2r^2}{r_{\mathrm{p}}^2}\right)\sin(\omega t), \quad (10.67)$$

where κ is the heat-generating efficiency of the pump and r_{p} is the radius of the pump beam.

The frequency response of the laser to pump-power modulation, obtained from the above equations, is given by (Zayhowski and Keszenheimer 1992)

$$\left|\frac{\mathrm{d}\nu_{\mathrm{o}}(\omega)}{\mathrm{d}P_{\mathrm{p}}(\omega)}\right| \sin[\omega t + \phi(\omega)] = \frac{\nu_{\mathrm{o}}\kappa}{4\pi k_{\mathrm{c}} l}\left(C\alpha_{\mathrm{e}} + \frac{1}{n}\frac{\mathrm{d}n}{\mathrm{d}T}\right)$$
$$\times\left\{\left[\frac{\pi}{2} - \mathrm{Si}\left(\frac{\omega}{\omega_0}\right)\right]\cos\left[\omega\left(t+\frac{1}{\omega_0}\right)\right] + \mathrm{Ci}\left(\frac{\omega}{\omega_0}\right)\sin\left[\omega\left(t+\frac{1}{\omega_0}\right)\right]\right\}, \quad (10.68)$$

where

$$\omega_0 = \frac{8\alpha_{\mathrm{d}}}{r_{\mathrm{m}}^2 + r_{\mathrm{p}}^2}, \quad (10.69)$$

and Si and Ci are the sine and cosine integrals. The magnitude and phase of this frequency response function are shown in Figure 10.13. Asymptotically,

$$\lim_{\omega\gg\omega_0}\left|\frac{\mathrm{d}\nu_{\mathrm{o}}(\omega)}{\mathrm{d}P_{\mathrm{p}}(\omega)}\right| = \frac{\nu_{\mathrm{o}}\kappa}{4\pi k_{\mathrm{c}} l}\left(C\alpha_{\mathrm{e}} + \frac{1}{n}\frac{\mathrm{d}n}{\mathrm{d}T}\right)\frac{\omega_0}{\omega}, \quad (10.70)$$

$$\lim_{\omega\gg\omega_0}\phi(\omega) = \frac{\pi}{2}, \quad (10.71)$$

$$\lim_{\omega\ll\omega_0}\left|\frac{\mathrm{d}\nu_{\mathrm{o}}(\omega)}{\mathrm{d}P_{\mathrm{p}}(\omega)}\right| = \frac{\nu_{\mathrm{o}}\kappa}{4\pi k_{\mathrm{c}} l}\left(C\alpha_{\mathrm{e}} + \frac{1}{n}\frac{\mathrm{d}n}{\mathrm{d}T}\right)\left|\gamma + \ln\left(\frac{\omega}{\omega_0}\right)\right|, \quad (10.72)$$

$$\lim_{\omega\ll\omega_0}\phi(\omega) = \pi, \quad (10.73)$$

where $\gamma\approx 0.577$ is Euler's constant.

The logarithmic divergence of the response function for small ω is a result of the approximation that the gain medium has infinite transverse dimensions. In the quasi-static limit ($\omega \lesssim \omega_{\mathrm{qs}}$, with ω_{qs} to be

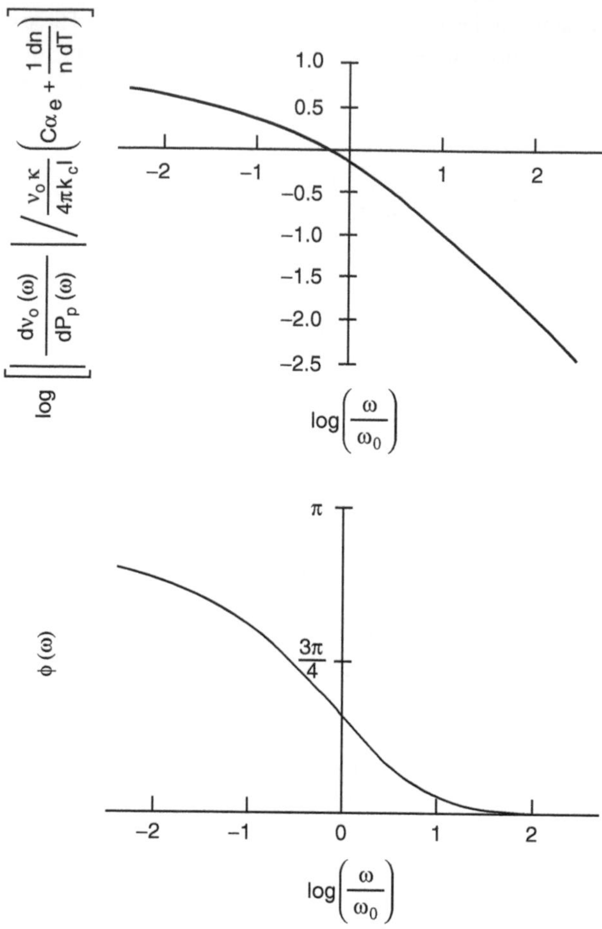

FIGURE 10.13 Magnitude and phase of the frequency response of a miniature solid-state laser to pump-power modulation.

defined below), with a radially symmetric heat sink of infinite capacity at a distance r_s from the cavity axis, the change in laser frequency is expressed by (Zayhowski and Keszenheimer 1992)

$$\frac{d\nu_o(\omega_{qs})}{dP_p(\omega_{qs})} = -\frac{\nu_o\kappa}{4\pi k_c l}\left(C\alpha_e + \frac{1}{n}\frac{dn}{dT}\right)$$
$$\times\left\{\gamma + \ln\left(\frac{2r_s^2}{r_m^2 + r_p^2}\right) + \mathrm{Ei}\left[-\frac{2r_s^2(r_m^2 + r_p^2)}{r_m^2 r_p^2}\right] - \mathrm{Ei}\left(-\frac{2r_s^2}{r_m^2}\right) - \mathrm{Ei}\left(-\frac{2r_s^2}{r_p^2}\right)\right\}, \quad (10.74)$$

where Ei is the exponential integral. This function is illustrated in Figure 10.14. Asymptotically,

$$\lim_{r_s \gg r_m, r_p} \frac{d\nu_o(\omega_{qs})}{dP_p(\omega_{qs})} = -\frac{\nu_o\kappa}{4\pi k_c l}\left(C\alpha_e + \frac{1}{n}\frac{dn}{dT}\right)\left\{\gamma + \ln\left(\frac{2r_s^2}{r_m^2 + r_p^2}\right)\right\}. \quad (10.75)$$

In this limit and in the limit $\omega_{qs} \ll \omega_0$, Equation 10.68 is equivalent to Equation 10.74 if

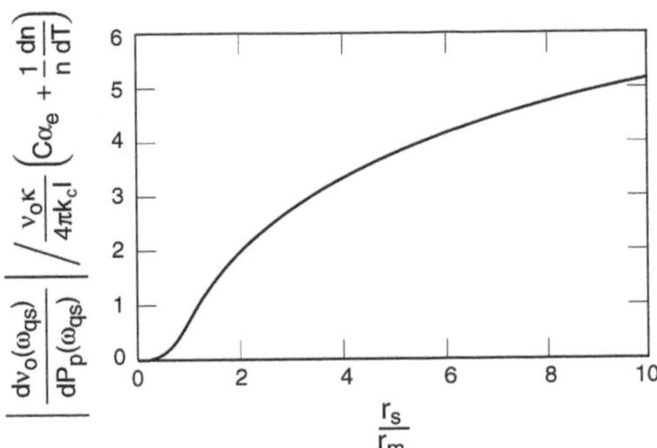

FIGURE 10.14 Magnitude of the quasi-static frequency response of a miniature solid-state laser to pump-power modulation as a function of r_s for $r_m = r_p$.

$$\omega_{qs} = \frac{4\alpha_d}{r_s^2}\exp(-2\gamma). \tag{10.76}$$

Equation 10.76 can be interpreted as defining the quasi-static frequency limit. For modulation frequencies below ω_{qs}, the response function is approximately given by Equation 10.74; at higher modulation frequencies Equation 10.68 becomes more accurate.

A very important aspect of the frequency response of a laser to pump-power modulation is its dependence on the dimensions of the laser. The characteristic frequencies of the system ω_0 and ω_{qs} increase as $1/(r_m^2 + r_p^2)$ and $1/r_s^2$, respectively. For the highest frequency response, all of the transverse dimensions should be kept as small as possible. The magnitude of the response function is inversely proportional to the cavity length. This also means that, for a laser with a given slope efficiency, the ratio of the frequency modulation to amplitude modulation [FM(ω)/AM(ω)], for a given pump modulation, increases as the inverse of the cavity length. For large FM(ω)/AM(ω), the cavity length should be kept as short as possible. With their short cavity lengths (typically $l < 1$ mm) and small transverse dimensions (r_m, $r_p \approx 0.1$ mm, $r_s \approx 0.5$ mm), microchip lasers are well suited for frequency tuning via pump-power modulation.

10.6 Pulsed Operation

10.6.1 Quasi-CW Operation

Quasi-cw operation (also called long-pulse or normal-mode operation) refers to a pulsed laser with a pulse duration long enough for all relevant parameters within the system to approach their steady-state values. Although the behavior of the system is cw-like at the end of the pulse, it is, in general, quite different at the beginning of the pulse.

Consider a laser with a step-function pump source. The pump may quickly create a population inversion. It will take some time, however, for an oscillating mode to build up from spontaneous emission. During this time, the inversion density may greatly exceed the cw threshold value. The large inversion density can result in an optical intensity well in excess of the cw value. This, in turn, will drive the inversion density below threshold, substantially reducing the laser intensity. The entire process may then start again. For a single-mode laser, this often leads to regular spiking at the beginning of the pulse. The process is damped, however, and with time the intensity of the spiking decreases. Eventually, spiking gives way to damped oscillations (known as relaxation oscillations) in the optical intensity and, finally, cw-like behavior.

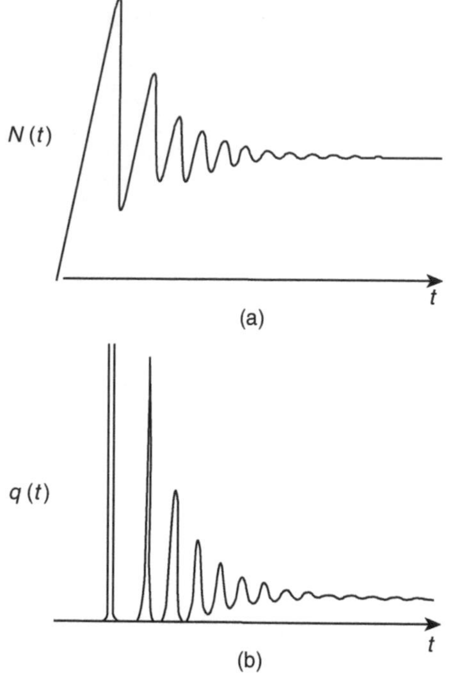

FIGURE 10.15 Computer solutions to the rate equations for a laser with a step-function pump source showing (a) the population inversion $N(t)$ and (b) the photon number $q(t)$.

In high-power lasers, the initial spike can be sufficiently intense to damage the laser. One technique used to reduce the intensity of this spike and damp the relaxation oscillations is to include a nonlinear crystal within the cavity (Statz, DeMars, and Wilson 1965; Kennedy and Barry 1974; Jeys 1991). Nonlinear frequency generation in the crystal limits the initial spike by serving as an intensity-dependent loss.

Figure 10.15 shows computer solutions to the rate equations for a quasi-cw laser. In these simulations, the spiking was heavily damped and cw-like behavior was quickly obtained. In a multimode laser, the interaction between modes can lead to mode hopping, mode beating, and prolonged, highly irregular spiking.

10.6.1.1 Relaxation Oscillations

Relaxation oscillations occur whenever the population inversion of a laser is disturbed from its steady-state value. They are a result of the coupling between the population inversion and the photon density within the laser cavity, as described above. For most solid-state lasers, the relaxation oscillations are underdamped and many oscillations occur before steady-state is reached.

The relaxation-oscillation frequency ω_r and damping time τ_0 can be derived by linearizing the rate equations (Equation 10.8, Equation 10.10, and Equation 10.11) about the steady-state operating point. If we assume $\tau_0 > 1/\omega_r$, the results for a single-mode four-level laser are

$$\tau_0 = \frac{2\tau(1 - F_{\mathrm{N,th}})}{R_{\mathrm{W}}} \tag{10.77}$$

and

$$\omega_r = \left[\frac{(R_{\mathrm{W}} - 1)\gamma_{\mathrm{rt}}}{\tau t_{\mathrm{rt}}} - \frac{1}{\tau_0^2}\right]^{1/2}, \tag{10.78}$$

where $R_{\mathrm{W}} = W_{\mathrm{p}}/W_{\mathrm{p,th}}$ is the number of times the pump rate factor exceeds its threshold value and $F_{\mathrm{N,th}} = N_{\mathrm{th}}/N_{\mathrm{t}}$ is the ratio of the effective population inversion at threshold to the total population of active sites in the active region of the gain medium. Because $F_{\mathrm{N,th}} \ll 1$ for most four-level lasers, this term is often dropped from the numerator of Equation 10.77. The oscillatory behavior of a quasi-three-level laser is similar to that of a four-level laser, with

$$\tau_0 = \frac{2\tau F_{\mathrm{N,th}}(g_u f_u - F_{\mathrm{N,th}})}{g_u f_u R_{\mathrm{W}}(g_u f_l + F_{\mathrm{N,th}}) - g_u f_l (g_u f_u - F_{\mathrm{N,th}})} \tag{10.79}$$

and

$$\omega_r = \left[\frac{(R_W - 1)(g_u f_l + F_{N,th})\gamma_{rt}}{\tau t_{rt} F_{N,th}} - \frac{1}{\tau_0^2}\right]^{1/2}. \tag{10.80}$$

Note that if $\tau_0 < 1/\omega_r$ the oscillations are overdamped and spiking will not occur. Although this condition is not satisfied in solid-state lasers, it is common in gas lasers because of the relatively short upper-state lifetimes of laser gasses.

Because oscillations in the intensity of a laser generate spectral sidebands, the frequency spectrum of a free-running laser typically shows a pronounced feature at the relaxation frequency. This resonance can be reduced through the use of electronic feedback to the pump (Kane 1990).

10.6.1.2 Amplitude Modulation

The output power of a laser can be controlled by changing the pump power, the output coupling, or the intracavity loss. This type of amplitude modulation is usually limited to frequencies below the frequency of the relaxation oscillations. The relaxation frequency characterizes the response time of the cavity. There is resonant enhancement of the modulation response near the relaxation frequency; above the relaxation frequency, the response rolls off.

Methods used for direct amplitude modulation of a laser may have the side effect of introducing frequency modulation as well. For example, changing the pump power affects the thermal load on the gain medium and, therefore, the temperature. This, in turn, affects the refractive index, changing the optical length of the cavity and the oscillating frequency. For applications where frequency stability is critical, it is often better to modulate the laser power external to the cavity.

10.6.2 Q-Switched Operation

Q switching is a means of generating short, high-peak-power pulses with relatively low-power pump sources. In a *Q*-switched laser, the loss of the cavity is maintained at a high level until a large population inversion is achieved. At such time, the loss is rapidly decreased so that the inversion is well above its new threshold value, resulting in a short, high-power output pulse. *Q* switching relies on the fact that the lifetime of the population inversion is much longer than the output pulse width. The gain medium is therefore able to store energy, which can be quickly released in the form of an optical pulse. The cavity loss is used to control the performance of the laser.

The first report of *Q*-switched output pulses, called giant pulses at the time, was from an actively *Q*-switched ruby laser (McClung and Hellwarth 1962). Shortly thereafter, passive *Q* switching, using an organic dye as a saturable absorber, was reported (Kafalas, Masters, and Murray 1964; Soffer 1964; Sorokin et al. 1964). Since 1985, the advent of diode-laser-pumped solid-state lasers and solid-state saturable absorbers has generated renewed interest in *Q*-switched lasers. The majority of the smaller solid-state systems are monolithic microchip lasers that are typically less than 1-cm long. This implementation was first demonstrated in 1994 (Zayhowski and Dill 1994) using Nd:YAG as the gain medium and Cr^{4+}:YAG as the saturable absorber. The first passively *Q*-switched microchip laser reported generated 11-μJ, 337-ps pulses at 1.064 μm, at a pulse rate of 6 kHz.

In this section, we analyze the performance of *Q*-switched lasers under the ideal assumption of instantaneous *Q* switching. Instantaneous *Q* switching is a reasonable approximation for many practical systems, and the resulting equations are a useful starting point for the design of practical systems. We then discuss actively *Q*-switched lasers before going into a detailed discussion of passively *Q*-switched devices.

10.6.2.1 Rate-Equation Model — Instantaneous Q Switching

The standard tools for analyzing the performance of a *Q*-switched laser are the rate equations (Wagner and Lengyel 1963). To model the *Q*-switched operation of a laser, we will start with the rate equations derived in Section 10.2 and assume uniform cw pumping of the gain medium. Because the *Q*-switched output pulses from a laser are typically much shorter than both the spontaneous lifetime and the pump period (time between output pulses for a cw-pumped, repetitively Q-switched laser), spontaneous relaxation and pumping can be safely neglected during the development of the output pulse. (As shown in the discussions of Build-Up from Noise in Section 10.2.4.2 and Build-Up Time in Section 10.6.2.1, spontaneous emission is still important in initiating the lasing process.) This reduces the rate equations for a single-mode laser to

$$\frac{\dot{q}t_{\mathrm{rt}}}{\gamma_{\mathrm{rt}}} = q\left(\frac{N_{\mathrm{eff}}}{N_{\mathrm{th}}} - 1\right) \tag{10.81}$$

and

$$\frac{\dot{N}_{\mathrm{eff}}t_{\mathrm{rt}}}{\gamma_{\mathrm{rt}}} = -\frac{2^{*}qN_{\mathrm{eff}}}{N_{\mathrm{th}}}, \tag{10.82}$$

where

$$2^{*} = \begin{cases} 1 & \text{for a four-level laser} \\ g_{\mathrm{u}}(f_{\mathrm{u,eff}} + f_{\mathrm{l,eff}}) & \text{for a quasi-three-level laser} \\ 2 & \text{for a nondegenerate three-level laser} \end{cases},$$

and $f_{\mathrm{u,eff}}$ and $f_{\mathrm{l,eff}}$ are the effective values of f_{u} and f_{l}, as discussed in Section 10.2.4.1.

Through $f_{\mathrm{u,eff}}$ and $f_{\mathrm{l,eff}}$, the value of 2^{*} is dependent on the duration of the output pulse from the laser. If the thermalization time for the upper laser manifold is short compared to the pulse width $f_{\mathrm{u,eff}}=f_{\mathrm{u}}$; if it is long compared to the pulse width $f_{\mathrm{u,eff}}\approx 1$. Likewise, if the thermalization time for the lower laser manifold is short compared to the pulse width $f_{\mathrm{l,eff}}=f_{\mathrm{l}}$; if it is long compared to the pulse width $f_{\mathrm{l,eff}}\approx 1$. Similarly, if the lower level of a four-level system has a decay time longer than the output pulse, the system will behave like a three-level system under Q-switched operation (Fan 1988). The thermalization times and lower-level decay times for even the most commonly used laser gain media are not well known, but there is experimental reason to believe that they may be comparable to or longer than the pulse widths of short-pulse miniature Q-switched lasers (several hundred picoseconds) in many material systems, including Nd:YAG (Bibeau, Payen, and Powell 1993; Palombo et al. 1993; Buzelis et al. 1995; Degnan, Coyle, and Kay 1998). Fortunately, there are many useful results that can be obtained from the rate equations without an accurate knowledge of the thermalization times.

These rate equations (Equation 10.81 and Equation 10.82) will tend to overestimate the efficiency of standing-wave Q-switched lasers because they do not include the effects of spatial and spectral hole burning (see the discussions of Spatial Hole Burning and Energy Diffusion in Section 10.5.1.2, and Spectral Hole Burning and Cross-Relaxation in Section 10.5.1.3). In addition, many Q-switched lasers operate with a large round-trip gain, so that the optical field within the cavity is not uniform. Although this mitigates the effects of spatial hole burning, it violates one of the assumptions used to reduce the rate equations to such a simple form. The net result is to overestimate the peak power and energy efficiency obtained from the device, and to distort the pulse shape (Stone 1992). These effects are often small, however, and Equation 10.81 and Equation 10.82 are extremely powerful tools.

For most of this section, we will assume that the loss of the laser cavity is rapidly switched at time $t=0$, such that for $t<0$ the effective inversion N_{eff} within the cavity is below its threshold value and for $t>0$ it is above the new threshold inversion N_{th}. With these assumptions, we can derive expressions for the maximum peak power, maximum pulse energy, and minimum pulse width of a single Q-switched output pulse, along with the corresponding cavity output couplings. The maximum power efficiency of a repetitively Q-switched laser and the corresponding output coupling are also obtained. The instantaneous switching approximation will be examined in the discussion of Pulse Build-Up Time below, followed by a discussion of Single-Frequency Operation, Afterpulsing, and Pulse Bifurcation.

Maximum Peak Power. To determine the maximum peak power that can be achieved from a Q-switched laser, we start by dividing Equation 10.81 by Equation 10.82 and integrating with respect to population inversion, to obtain

$$2^*(q-q_0) = N_0 - N_{\text{eff}} - N_{\text{th}}\ln\left(\frac{N_0}{N_{\text{eff}}}\right), \tag{10.83}$$

where q_0 and N_0 are the number of photons in the cavity and the total population inversion at the time the output pulse begins to develop. Using the fact that the inversion density is equal to its threshold value at the peak of the pulse ($\dot{q}=0$) and neglecting the initial number of photons in the cavity, we obtain a peak output power

$$P_{\text{po}} = \frac{h\nu_o N_{\text{th}}}{2^* t_{\text{rt}}}\left[\frac{N_0}{N_{\text{th}}} - 1 - \ln\left(\frac{N_0}{N_{\text{th}}}\right)\right]\gamma_o, \tag{10.84}$$

where $h\nu_o$ is the energy of a photon at the oscillating wavelength, $\gamma_o = -\ln(1-T_o)$ is the output-coupling coefficient, and T_o is the transmission of the output coupler. The peak power is maximized when

$$\frac{\gamma_{\text{rt}} + \gamma_o}{\gamma_{\text{rt}}}\ln\left(\frac{N_0}{N_{\text{th}}}\right) = \frac{N_0}{N_{\text{th}}} - 1. \tag{10.85}$$

P_{po} is plotted as a function of $1/\gamma_o$ in Figure 10.16 for the case of negligible parasitic loss ($\gamma_{\text{rt}} = \gamma_o$). In this case, the maximum peak power (Zayhowski and Kelley 1991)

$$P_{\text{po,max}} = \frac{0.102 N_0^2 h\nu_o \gamma_{\text{rt}}}{2^* N_{\text{th}} t_{\text{rt}}}, \tag{10.86}$$

and is obtained for

$$\gamma_o = 0.28 g_{\text{rt}}, \tag{10.87}$$

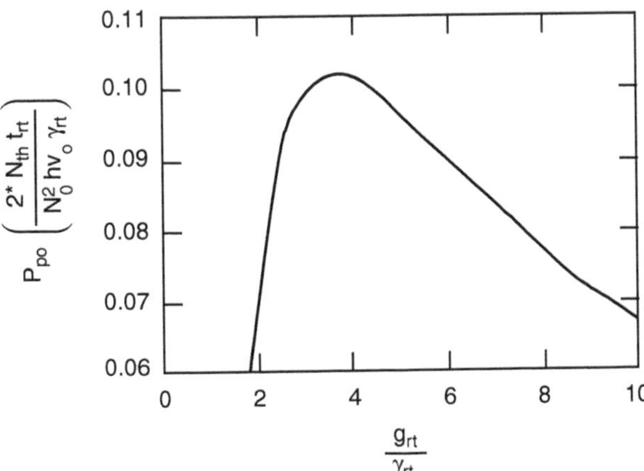

FIGURE 10.16 Peak output power as a function of the inverse cavity output coupling for a cavity with negligible parasitic loss.

where $g_{rt}=2^{\ddagger}N_0\sigma_g l_g/V_g$ is the round-trip gain coefficient at the time the pulse begins to develop. The presence of parasitic loss increases the optimal value of γ_o.

From Equation 10.84 and Equation 10.86 it is clear that the peak power is dependent on the value of 2* and, therefore, the thermalization times of the laser manifolds. When the output pulse width is longer than the thermalization times of the laser manifolds, thermalization of the upper manifold repopulates the upper laser level as it gets depleted by the optical pulse and thermalization of the lower manifold keeps the population of the lower laser level as small as possible. This allows the maximum amount of energy extraction from the system and relatively high peak powers. When the pulse width is short compared to the thermalization times only the energy in the upper laser level (not the upper laser manifold) can be extracted by the laser pulse, and the laser will act like a three-level system.

When the pulse width of the laser is comparable to the thermalization times of the laser manifolds it is not possible to accurately calculate the peak output power of the laser pulses without accurately knowing the thermalization times. However, the output coupling that maximizes the peak power is independent of the thermalization times.

Maximum Pulse Energy. The quantum extraction efficiency η_e of a pulse is determined by the change in the population inversion density during the pulse:

$$\eta_e = \frac{N_0 - N_f}{N_0}, \tag{10.88}$$

where N_f is the population inversion well after the peak of the output pulse. By combining Equation 10.88 and Equation 10.83 we obtain the implicit relationship

$$\frac{N_0}{N_{th}} = -\ln\left(\frac{1-\eta_e}{\eta_e}\right). \tag{10.89}$$

The energy in the output pulse E_o is proportional to the extraction efficiency and is given by

$$E_o = \frac{\eta_e N_0 h\nu_o \gamma_o}{2^* \gamma_{rt}}. \tag{10.90}$$

The maximum value of E_o (Degnan 1989)

$$E_{o,max} = \frac{N_0 h\nu_o \gamma_{rt,p}}{2^* g_{rt}}\left[\frac{g_{rt}}{\gamma_{rt,p}} - 1 - \ln\left(\frac{g_{rt}}{\gamma_{rt,p}}\right)\right], \tag{10.91}$$

and is obtained for

$$\gamma_o = -\gamma_{rt,p}\left[\frac{(g_{rt}/\gamma_{rt,p}) - 1 - \ln(g_{rt}/\gamma_{rt,p})}{\ln(g_{rt}/\gamma_{rt,p})}\right]. \tag{10.92}$$

Like the peak power, and for the same reasons, the pulse energy is dependent on 2*, but the optimal value of output coupling is not. In a repetitively Q-switched short-pulse laser, some of the energy not extracted from the upper laser manifold by one pulse will be available for a subsequent pulse and the efficiency of the system will improve at high repetition rates.

Minimum Pulse Width. The width of a pulse t_w (full width at half-maximum) can be determined from the energy in the pulse E_o, the peak power of the pulse P_{po}, and the shape of the pulse:

$$t_w = \frac{S_p E_o}{P_{po}} = \frac{S_p \eta_e N_0 t_{rt}}{N_{th}[N_0/N_{th} - 1 - \ln(N_0/N_{th})]\gamma_{rt}}, \tag{10.93}$$

where S_p is a number that characterizes the pulse shape. Single-mode Q-switched pulses have a pulse shape factor $S_p \approx 0.86$ (Zayhowski and Kelley 1991), which is midway between the pulse shape factor for sech- and sech^2-shaped pulses ($S_p = 0.84$ and 0.88, respectively). The rising edge of the pulse is exponential with a rise time determined by the gain of the cavity; the exponential decay of the trailing edge is determined by the cavity lifetime.

The exact solution to the rate equations (Equation 10.81 and Equation 10.82) for the pulse width is shown as a function of the inverse cavity loss $1/\gamma_{rt}$ in Figure 10.17. The minimum pulse width (Zayhowski and Kelley 1991)

$$t_{w,min} = \frac{8.1 t_{rt}}{g_{rt}}, \tag{10.94}$$

is obtained for

$$\gamma_o = 0.32 g_{rt} - \gamma_{rt,p}. \tag{10.95}$$

The term 2^* does not appear in Equation 10.94, indicating that the pulse width is independent of the thermalization times of the laser manifolds and whether the gain medium is a three- or four-level system. A more careful analysis shows that the laser will have the same pulse width if the thermalization times are much longer than or much shorter than the pulse width. However, when the pulse width and thermalization times are comparable, thermalization of the manifolds will affect the dynamics of pulse formation, changing the pulse shape and its duration (Degnan, Coyle, and Kay 1998). In a system optimized for minimum pulse width, the rising edge of the laser pulse is sharper than the falling edge. As the pulse starts to decay, thermalization of the laser manifolds will pump the laser, slowing its decay and broadening the pulse. When the thermalization times and the pulse width are nearly equal, this can increase the pulse width of an optimized system by ~50%.

Maximum Power Efficiency of a Repetitively Q-Switched Laser. For a repetitive train of identical pulses, the inversion not used by one pulse may contribute to the next pulse, improving the net efficiency of the system. With the assumptions of continuous pumping and instantaneous thermalization of the laser manifolds, the extraction efficiency of a repetitive system becomes

$$\eta_{e,rep} = \frac{\eta_e[1 - \exp(-\tau_p/\tau)]\tau}{[1 - (1 - \eta_e)\exp(-\tau_p/\tau)]\tau_p}, \tag{10.96}$$

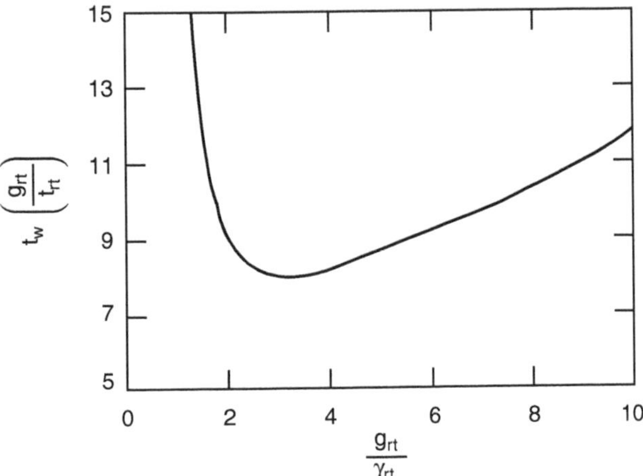

FIGURE 10.17 Pulse width as a function of the inverse cavity loss.

where τ_p is the time between Q-switched pulses. (This equation needs to be modified when the pulse duration is comparable to or shorter than the thermalization times of the manifolds.) The total efficiency of the laser is

$$\eta = \eta_p \eta_a \eta_i \eta_{e,rep} \eta_o, \tag{10.97}$$

where $\eta_o = \gamma_o / \gamma_{rt}$, and is maximized when (Zayhowski and Kelley 1991)

$$\gamma_o \left[\frac{1 - \exp(-\tau_p/\tau)}{1 - (1 - \eta_e)\exp(-\tau_p/\tau)} \right] = -\gamma_{rt,p} \left[\frac{\eta_e}{(1 - \eta_e)\ln(1 - \eta_e)} + 1 \right]. \tag{10.98}$$

In the absence of parasitic loss, the maximum energy efficiency is obtained by making the output coupling as small as possible to maximize the single-pulse extraction efficiency and minimize the effects of spontaneous emission. On the other hand, if the effect of spontaneous emission is unimportant (e.g., when the interpulse spacing is small compared to the spontaneous lifetime), the output coupling should be made as large as possible to minimize the parasitic loss. In miniature Q-switched lasers, efficiency is often sacrificed to obtain the desired pulse width or peak power.

Pulse Build-Up Time. All of the above results for the Q-switched operation of lasers were obtained on the assumption that the Q of the laser cavity is changed instantaneously. In practice, actively changing the Q often requires the switching of high voltages or mechanical motion of components, and is difficult to do rapidly. Passive Q switching also requires time. As a result, the instantaneous Q-switching approximation may not be valid. Fortunately, the magnitude and shape of the Q-switched output pulse are determined by the cavity Q at the time of the output pulse (within t_w of the peak) and are insensitive to the details of the way in which the Q is switched. The results obtained above are, therefore, valid as long as the cavity Q does not change significantly during the output pulse. Because the pulse build-up time is usually much greater than the pulse width, this is often the case. To illustrate this, it is helpful to calculate the pulse build-up times for the cases of instantaneous Q switching and gradual Q switching.

Because the optical intensity is small during most of the pulse build-up time, the inversion can be approximated as constant during the pulse development. This allows us to solve Equation 10.81 analytically to obtain an approximate expression for the pulse build-up time. If the cavity Q is changed instantaneously,

$$t_b = \ln\left(\frac{q_p}{q_0}\right) \frac{t_{rt}}{\gamma_{rt}} \frac{N_{th}}{N_{eff} - N_{th}}, \tag{10.99}$$

where q_p is the number of photons in the laser cavity at the peak of the output pulse (typically in the range 10^{11} to 10^{18}). For a laser cavity that is sufficiently lossy between output pulses, the pulse must build up from spontaneous emission (unless intentionally seeded). The pulse build-up time for a minimum-width pulse is given by (Zayhowski and Kelley 1991)

$$t_b \approx \frac{\ln(0.63 N_0) t_{rt}}{2.1 \gamma_{rt}}, \tag{10.100}$$

which is typically a factor of 5–8 greater than the pulse width. If the value of the cavity loss changes linearly with time, lowering the threshold inversion N_{th} from an initial value of slightly greater than N_{eff} to a value of $0.32N_{eff}$ just before the peak of the output pulse, the pulse build-up time can be increased by a factor of ~3.3. The switching time is further increased if the initial value of N_{th} is much greater than N_{eff} or if N_{th} does not reach $0.32N_{eff}$ by the time the pulse forms. Also, the change in cavity Q can continue beyond the time when the output pulse is generated, as long as it is reversed early enough to prevent the generation of a second pulse.

Single-Frequency Operation. When Q-switched lasers operate in more than one longitudinal mode, the output pulse is intensity modulated as a result of mode beating; single-longitudinal-mode lasers produce pulses with a smoothly varying temporal profile. Smooth temporal profiles are desirable for many applications in high-resolution ranging (Abshire et al. 2000; Afzal et al. 1977; Degnan 1993; Degnan 2000), altimetry (Degnan et al. 2001), and light detection and ranging (LIDAR). Traditionally, the methods used to obtain single-frequency operation in a Q-switched laser involve the addition of intracavity frequency-selective optical elements or the injection of an optical signal from a cw single-frequency oscillator (optical seeding).

To derive a criterion for single-frequency operation of Q-switched lasers, we will require that the number of photons in the primary mode (mode 1) is at least 100 times greater than the number in any other mode (mode 2) by the time the output pulse forms (t_b). [Other authors have suggested that a factor of 10 is sufficient (Sooy 1965); the number is somewhat arbitrary.] In this case, the primary mode will extract most of the stored energy in the cavity, leaving very little energy for the second mode. From Equation 10.81 it follows that, for the case of instantaneous Q switching, our criterion reduces to

$$\frac{N_{eff}(N_{th,2}-N_{th,1})}{N_{th,2}(N_{eff}-N_{th,1})} > \frac{\ln(100)}{\ln(q_p)} = \frac{4.6}{\ln(q_p)}, \tag{10.101}$$

where $N_{th,1}$ and $N_{th,2}$ are the cw thresholds for modes 1 and 2 when the cavity is in the low-loss state, and we have assumed that both modes build up from noise.

An alternative criterion for single-frequency operation is that the build-up time for two competing modes differ by an amount comparable to or greater than the pulse duration (Isyanova and Welford 1999). The first mode to lase (mode 1) extracts most of the stored energy in the cavity, slowing the build-up of the second mode (mode 2) and greatly decreasing its intensity. For the case of instantaneous Q switching, this criterion reduces to

$$\frac{N_{eff}(N_{th,2}-N_{th,1})}{(N_{eff}-N_{th,2})(N_{eff}-N_{th,1})} > \frac{2.6t_w}{\ln(q_p)t_{w,min}}, \tag{10.102}$$

where $t_{w,min}$ is the minimum obtainable pulse width as given by Equation 10.94. Equation 10.101 and Equation 10.102 give similar results for lasers designed to produce short pulses; both ways of treating the problem are equally valid.

For an instantaneously Q-switched laser designed to produce pulses with the minimum possible duration, with typical values of q_p in the range of 10^{11} to 10^{18}, the thresholds for modes 1 and 2 (as calculated with Equation 10.101 or Equation 10.102) must differ by more then 7% to 12% in order to ensure single-frequency operation. This requirement can be relaxed if the cavity Q is switched more slowly, since the resulting longer pulse build-up times lead to more optical passes through any mode-selective elements in the resonator (Sooy 1965). Nevertheless, for typical gain linewidths in solid-state media, the gain differential between longitudinal modes is not enough to ensure single-frequency operation for a cavity length greater than ~1 cm and, if single-frequency operation is desired, intracavity frequency-selective loss elements may be required. For shorter cavity lengths the gain differential may be sufficient, if care is taken to keep one of the longitudinal modes near the peak of the gain profile.

Afterpulsing. The first output pulse from a laser will not extract all of the energy stored in the laser cavity. The extraction efficiency is never 100%, even for traveling-wave cavities, and is lower for standing-wave cavities due to spatial hole burning. The residual gain left by the first output pulse can contribute to the continued development of a pulse in a competing longitudinal mode, resulting in a second pulse, delayed in time with respect to the first pulse and much smaller in magnitude.

There is an additional mechanism for afterpulsing when the thermalization times of the laser manifolds are long compared to the Q-switched output pulse. Themalization of the manifolds can rapidly pump the laser transition and lead to gain-switched pulses (see Section 10.6.3) following the primary Q-switched pulse. These gain-switched pulses can be in the same longitudinal mode as the primary pulse and can be very stable in amplitude and time.

In actively Q-switched lasers, the second pulse, or afterpulse, can be prevented by rapidly decreasing the cavity Q after the first output pulse exits the system.

Pulse Bifurcation. At high pulse repetition rates, when the pulse-to-pulse spacing is comparable to or shorter than the upper-state lifetime of the gain medium, some of the energy not extracted by one pulse will still be present in the laser cavity when the next pulse forms. As a result of spatial hole burning, the residual gain may be greater at positions where it favors the development of a pulse in a different longitudinal mode than the first pulse. This is more likely to be the case when the difference between the thresholds of the two modes is very small, and in standing-wave cavities with relatively high Q. The second pulse will then preferentially deplete the gain most available to it, creating a situation where the first mode is favored in the subsequent pulse. As a result, it is not uncommon to see pulse bifurcation (where pulses alternate between two longitudinal modes) in Q-switched lasers operating at high repetition rates. At still higher pulse repetition rates, more modes can come into play, creating trains of three or more modes occurring in a regular sequence or, depending on the stability of the system, chaotically. Because of gain-related index-guiding effects (see Section 10.4.2.2), each of the modes will have slightly different transverse dimensions, energies, and divergences.

10.6.2.2 Active Q Switching

Intracavity Q Switches. Conventional Q switches are located within the laser cavity and control the amount of cavity loss. The first devices used to Q switch a laser were mechanical switches, such as rotating mirrors. Their relatively slow speed has led to their replacement by acousto-optic and electro-optic Q switches, except in cases where optical damage limits the use of the alternative technologies.

There are two contributions to the switching time of an acousto-optic Q switch. First, the transducer is usually part of a resonator with a finite response time. The second component is the time required for the acoustic excitation to propagate across the diameter of the optical beam. Acousto-optic TeO_2 Q switches have been reported that operate at a frequency of 108 MHz, with a switching speed of 18 ns, limited by the transit time of the acoustic signal across the oscillating mode ($\sim$100-μm diameter). Using these switches, Q-switched pulses as short as 600 ps have been generated (Plaessmann et al. 1994).

Electro-optic Q switches are capable of faster switching than acousto-optic devices because the speed of light is faster than the speed of sound. However, they usually require high-voltage electronics. When used in a coupled-cavity configuration, electro-optic Q switches have produced the shortest Q-switched pulses yet obtained from an actively Q-switched solid-state laser (Zayhowski and Dill 1992). (Shorter Q-switched pulses have been obtained from passively Q-switched devices, as discussed below.)

Coupled-Cavity Q Switches. Coupled-cavity electro-optically Q-switched microchip lasers (illustrated in Figure 10.18) have, to date, demonstrated the shortest Q-switched pulses obtained from an actively Q-switched solid-state laser. From a Nd:YAG device, pulse widths of 270 ps have been obtained, and repetition rates of 500 kHz were demonstrated (Zayhowski and Dill 1992). From a Nd:YVO_4 laser, pulse widths of 115 ps and repetition rates of 2.25 MHz were obtained (Zayhowski and Dill 1995). Coupled-cavity electro-optically Q-switched microchip lasers have also produced the highest peak power ($>$90 kW) of any laser pumped by a single (1 W) laser diode (Zayhowski and Dill 1995).

The principle behind the operation of the coupled-cavity Q-switched laser is that an etalon formed by an electro-optic element serves as a variable-reflectivity output coupler for a gain cavity (defined by two mirrors adjacent to the gain medium). The potential lasing modes of the device are determined primarily by the gain cavity. In the low-Q state, the variable etalon must have a high transmission for all potential lasing modes so that none can reach threshold. In the high-Q state, the reflectivity of the etalon is high for the desired mode, and a Q-switched output pulse develops. To ensure that all potential modes of the gain cavity can be simultaneously suppressed, the optical length of the variable etalon must be nearly an integral multiple of the optical length of the gain cavity. The higher the Q of the etalon, the tighter the tolerance on length. For an isotropic gain medium such as Nd:YAG, the length tolerance imposes a

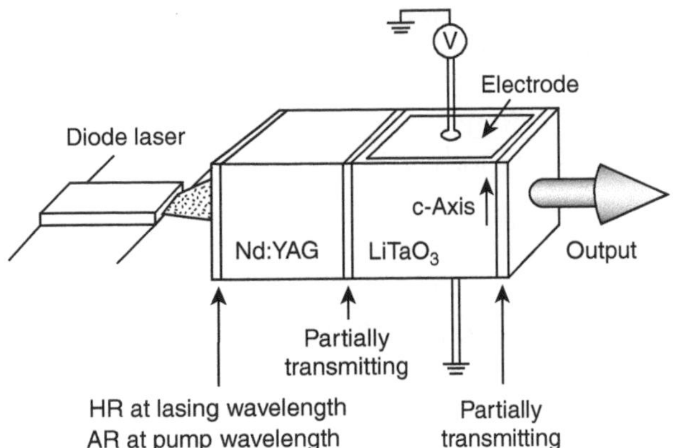

FIGURE 10.18 Illustration of one embodiment of a coupled-cavity electro-optically *Q*-switched microchip laser (HR, highly reflecting; AR, antireflecting).

restriction on the birefringence of the electro-optic material, because oscillation must be suppressed for modes of both polarizations.

Before Equation 10.95 can be used to optimize the output coupling of a coupled-cavity microchip laser, we must have an expression for the effective output coupling. Similarly, before Equation 10.94 can be used to determine the minimum pulse width, we must have an expression for the effective round-trip time of the laser cavity.

The transmission of the electro-optic etalon is dynamically dependent on the rate of formation and decay of the pulse. For example, if the gain cavity and the etalon have the same optical length, the light within the gain cavity is amplified by the round-trip gain of the cavity before it is recombined with light that is reflected from the far mirror of the etalon. As a result, the effective transmission of a lossless etalon, as seen at the interface between the gain medium and the etalon during the early formation of the pulse, is given by

$$T_{\mathrm{et},t=0} = \frac{[1-R_1][1-R_2\exp(-m_1 g_{\mathrm{rt}})]}{1+R_1R_2\exp(-m_1 g_{\mathrm{rt}})-2\cos(\phi_{\mathrm{et}})\sqrt{R_1R_2}\exp(-m_1 g_{\mathrm{rt}}/2)}, \tag{10.103}$$

where R_1 is the reflectivity of the interface between the gain medium and the electro-optic material, R_2 is the reflectivity of the output mirror on the etalon, $m_1 = t_{\mathrm{rt,et}}/t_{\mathrm{rt,g}}$ is the ratio of the round-trip time of light in the etalon $t_{\mathrm{rt,et}}$ to the round-trip time of light in the gain cavity $t_{\mathrm{rt,g}}$, and $\phi_{\mathrm{et}} = 2\pi(t_{\mathrm{rt,et}} - t_{\mathrm{rt,g}})\nu_{\mathrm{o}}$. To hold off lasing, we must be able to satisfy the relationship

$$T_{\mathrm{et},t=0} \geq 1-\exp(\gamma_{\mathrm{rt,p}} - g_{\mathrm{rt}}). \tag{10.104}$$

Near the peak of the output pulse the gain of the laser is saturated and the effective transmission of the etalon approaches its cw value

$$T_{\mathrm{et,cw}} = \frac{(1-R_1)(1-R_2)}{1+R_1R_2-2\cos(\phi_{\mathrm{et}})\sqrt{R_1R_2}}. \tag{10.105}$$

Because the pulse width of the laser is determined by the transmission of the etalon near the peak of the pulse, it is this value of transmission that should be used as the output coupling T_{o} $[\gamma_{\mathrm{o}} = -\ln(1-T_{\mathrm{o}})]$ in Equation 10.95.

The transit time of light in the etalon affects the behavior of the coupled-cavity *Q*-switched laser in a second way. After leaving the gain cavity, light must escape the etalon, which has an associated decay time. To minimize this effect and reduce the pulse width, it is desirable to keep R_2 as small as possible. The minimum pulse width is therefore obtained by selecting values of R_1 and R_2 that satisfy the equations

$$\frac{[1-R_1][1-R_2\exp(-m_l g_{rt})]}{1+R_1R_2\exp(-m_l g_{rt})-2\sin(2\pi\nu_o\delta_1)\sqrt{R_1R_2}\exp(-m_l g_{rt}/2)}=1-\exp(\gamma_{rt,p}-g_{rt}) \quad (10.106)$$

and

$$\frac{(1-R_1)(1-R_2)}{1+R_1R_2+2\sin(2\pi\nu_o\delta_2)\sqrt{R_1R_2}}=1-\exp(\gamma_{rt,p}-0.32g_{rt}), \quad (10.107)$$

where $\delta_1+\delta_2=\delta(t_{rt,et})\leq 1/2\nu_o$ is the total change in the round-trip time of the electro-optic etalon induced during Q switching and the amount of change associated with δ_1 and δ_2 is determined by the exact length of the etalon before Q switching.

For low-gain operation ($g_{rt}\ll 1$), the solution to Equation 10.106 and Equation 10.107 yields a high value for R_1 and a low value for R_2, and the effective round-trip time of a properly optimized coupled-cavity laser is approximately the round-trip time of the gain cavity. In very high-gain systems, R_2 becomes larger than R_1, and the effective round-trip time of the coupled cavity approaches the sum of the round-trip times of the gain cavity and the etalon. A simple expression that satisfies these asymptotic constraints and has a reasonable fit to numerical simulations is

$$t_{rt}=t_{rt,g}\left[1+\frac{m_l(\gamma_{rt,p}-0.32g_{rt})}{\ln(R_2)}\right] \quad (10.108)$$

Figure 10.19 shows the resulting normalized effective round-trip time of an optimized coupled-cavity laser with $\delta_1=\delta_2=1/4\nu_o$ as a function of gain for $m_l=1$ and 2, for $\gamma_{rt,p}=0$. Over a large part of the potential operating range of the coupled-cavity Q-switched laser, the effective cavity length is the length of the gain cavity; the length of the Q switch does not contribute significantly.

10.6.2.3 Passive Q Switching

A passively Q-switched laser contains a gain medium and a saturable absorber. As the gain medium is pumped it accumulates stored energy and emits photons. Over many round trips in the resonator the photon flux sees gain, fixed loss, and saturable loss. If the gain medium saturates before the saturable absorber, the laser will tend to oscillate cw. On the other hand, if the photon flux builds up to a level that saturates, or bleaches, the saturable absorber first, the resonator will see a dramatic reduction in intracavity loss and the laser will Q switch, generating a short, intense pulse of light.

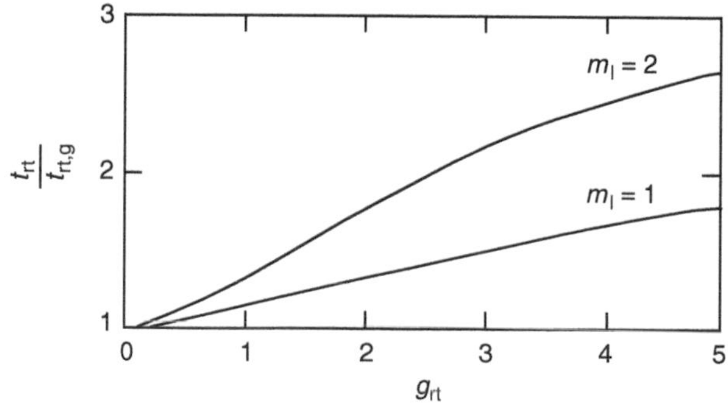

FIGURE 10.19 Effective cavity round-trip time versus gain for an optimized coupled-cavity Q-switched laser with $m_l=1$ and $m_l=2$.

In addition to the simplicity of implementation, the major advantages of a passively *Q*-switched laser include the generation of pulses with a well defined energy and duration that are insensitive to pumping conditions. Additionally, it is much easier to obtain single-frequency operation in a passively *Q*-switched laser than in an actively *Q*-switched device. These benefits are achieved at the expense of increased pulse timing jitter compared to actively *Q*-switched lasers.

Types of Saturable Absorbers. Most of the commonly used solid-state saturable absorbers are vibrationally broadened transition-metal-doped single-crystal materials or, to a lesser degree, glasses. The most commonly used material for passive *Q* switching of miniature lasers is Cr^{4+}:YAG. The use of Cr^{4+}:YAG dates back to 1991 (Andrauskas and Kennedy 1991). It has been used to *Q* switch Nd:YAG microchip lasers operating at 1.064 μm (Zayhowski and Dill 1994), 946 nm (Zayhowski et al. 1996), and 1.074 μm (Zayhowski et al. 2000); and Nd:YVO_4 and Nd:$GdVO_4$ microchip lasers operating at 1.064 and 1.062 μm, respectively (Jaspan et al. 2000; and Liu et al. 2003; respectively). This material has been very successful despite the presence of significant excited-state absorption and polarization anisotropy (Eilers et al. 1992; Il'ichev et al. 1997). The latter property does not impose any significant limitations on passively *Q*-switched laser performance, and can be used to help fix the polarization of passively *Q*-switched lasers with isotropic gain media, such as Nd:YAG. Some of the relevant material characteristics of Cr^{4+}:YAG are listed in Table 10.15. Table 10.16 lists several gain medium/bulk saturable absorber combinations that have been used at a variety of wavelengths.

Semiconductor saturable-absorber mirrors (SESAMs) (Kärtner et al. 1995; Keller et al. 1996; Spühler et al. 1999) are used to generate *Q*-switched pulses with durations $<$ 200 ps, and have been used with a variety of gain media operating at a wide range of wavelengths. They are, however, typically limited to pulse energies in the nanojoule regime by the onset of optical damage (Braun et al. 1996; Braun et al. 1997).

Bulk semiconductor saturable absorbers (Gu et al. 2000; Kajave and Gaeta 1996; Tsou et al. 1993) and semiconductor-doped glasses (Bilinsky et al. 1998), although demonstrated, are rarely used for passive *Q* switching. Although organic dyes may be used as saturable absorbers, they are generally not used in solid-state devices operating in the near infrared because photothermal degradation leads to impractically short operational lifetimcs.

Model of Saturable Absorber. Many saturable absorbers can be understood with the four-band model illustrated in Figure 10.20. [This model is particularly applicable to Cr^{4+}:YAG (Eilers et al. 1992; Il'ichev et al. 1997; Sennaroglu and Yilmaz 1997; Burshtein et al. 1998).] If the saturable absorber is initially in the ground state, absorption of a photon will excite the system from its lowest energy level *g* to the excited band *e*. Band *e* is assumed to relax rapidly to upper level *u*. From level *u* the system can relax back to the ground state *g* or absorb another photon and become excited to the higher-energy band *h*. Band *h* is assumed to relax rapidly back to upper level *u*.

Rate Equations for Saturable Absorber. The photon rate equation for a laser containing a saturable absorber must be modified to include the absorption of a photon by a saturable-absorber ion in either the ground state or the upper level. Following the logic of Section 10.2.4.1, ground-state absorption leads to the additional term $\dot{q}_{s,g,ab} = 2_s^{\ddagger} q N_{s,g} \sigma_{s,g} / t_{rt} A_s$ and upper-level absorption leads to the term $\dot{q}_{s,u,ab} = 2_s^{\ddagger} q N_{s,u} \sigma_{s,u} / t_{rt} A_s$, where $N_{s,g}$ is the population of saturable-absorber ions in the ground state, $\sigma_{s,g}$ is the ground-state absorption cross-section at the oscillating wavelength, $N_{s,u}$ is the population of saturable-absorber ions in the upper level, $\sigma_{s,u}$ is the upper-level absorption cross-section at the oscillating wavelength, A_s is the cross-sectional area of the laser mode in the saturable absorber, and

TABLE 10.15 Properties of Cr^{4+}:YAG Saturable Absorber

Property	Symbol	Value	Units	Comments
Ground-state absorption cross-section	$\sigma_{s,g}$	7×10^{-18}	cm^2	1.064 μm
		2×10^{-18}	cm^2	808 nm
Upper-level absorption cross-section	$\sigma_{s,u}$	2×10^{-18}	cm^2	1.064 μm
Upper-level lifetime	τ_s	4.1×10^{-6}	s	Room temperature

TABLE 10.16 Passively *Q*-Switched Laser Gain Media and Bulk Saturable Absorber Combinations

Wavelength (μm)	Gain Medium	Saturable Absorber	Reference
0.694	Ruby	Cr^{4+}:GSGG	Chen et al. (1993)
0.694	Ruby	$Cr^{4+}:CaGd_4(SiO_4)_3$	Yumashev et al. (1998)
0.781–0.806	$Cr^{3+}:LiCaAlF_6$	$Cr^{4+}:Y_2SiO_5$	Kuo, Huang, and Birnbaum (1995)
0.946*	Nd:YAG	Cr^{4+}:YAG	Zayhowski et al. (1996)
1.03*	Yb:YAG	Cr^{4+}:YAG	Zhou et al. (2003)
1.04*	Yb:KGW	V^{3+}:YAG	Lagatsky, Abdolvand, and Kuleshov (2000b)
1.062	$Nd:GdVO_4$	Cr^{4+}:YAG	Liu et al. (2003)
1.064*	Nd:YAG	Cr^{4+}:YAG	Andrauskas and Kennedy (1991) and Zayhowski and Dill (1994)
1.064*	$Nd:YVO_4$	Cr^{4+}:YAG	Bai et al. (1997); Jaspan et al. (2000)
1.064	Nd:YAG	$Cr:Mg_2SiO_4$	Demchuk et al. (1992)
1.064	Nd:YAG	F^{2-}:LiF	Isyanova and Welford (1993)
1.064	Nd:YAG	Cu_2Se:glass	Yumashev et al. (2001)
1.067	Nd:KGW	$Cu_{2+x}Se$:glass	Yumashev et al. (2001)
1.074*	Nd:YAG	Cr^{4+}:YAG	Zayhowski et al. (2000)
1.08	Nd:YALO	$Cr^{4+}:SrGd_4(SiO_4)_3$	Yumashev et al. (1998)
1.32	Nd:YAG	$Cu_{2+x}Se$:glass	Yumashev et al. (2001)
1.34*	Nd:YAG	V^{3+}:YAG	Jabczynski et al. (2001)
1.34*	Nd:YAP	V^{3+}:YAG	Jabczynski et al. (2001)
1.34*	$Nd:YVO_4$	V^{3+}:YAG	Jabczynski et al. (2001)
1.34	Nd:YAP	Cu_2Se:glass	Yumashev et al. (2001)
1.34	Nd:YAG	$Co^{2+}:MgAl_2O_4$	Wu et al. (2000)
1.34	Nd:YALO	$Co^{2+}:LaMgAl_{11}O_{19}$	Yumashev et al. (1998)
1.44	Nd:KGW	$Co^{2+}:MgAl_2O_4$	Wu et al. (2000)
1.44*	Nd:YAG	V^{3+}:YAG	Kuleshov et al. (2000)
1.5*	Yb,Tm:YLF	$Co^{2+}:LaMgAl_{11}O_{19}$	Braud et al. (2000)
1.53	Er:glass	$U^{2+}:CaF_2$	Stultz, Camargo, and Birnbaum (1995)
1.54*	Er:glass	$Co^{2+}:LaMgAl_{11}O_{19}$	Thony et al. (1999); Yumashev et al. (1999)
1.54*	Er:glass	$Co^{2+}:MgAl_2O_4$	Wu et al. (2000)
1.54	Er:glass	Co^{2+}:glass	Malyarevich et al. (2001a)
1.54	Er:glass	$Cu_{2+x}Se$:glass	Yumashev et al. (2001)
2.0	Tm,Cr:YAG	Ho:YLF	Kuo, Birnbaum, and Chen (1994)
2.1	Ho:YAG	PbSe:glass	Malyarevich et al. (2001b)

*Denotes demonstration in a miniature or microchip laser.

$2_s^{\ddagger}$ is the number of times the same cross-sectional area of the saturable absorber is traversed by light during one round trip within the laser cavity. If we assume that the decay of the excited and higher-energy bands is very rapid, all of the saturable-absorber ions (total population N_s) are either in the ground state or upper level, and the new photon rate equation becomes

$$\dot{q} = \frac{q}{t_{rt}}\left[\frac{2^{\ddagger}N_{eff}\sigma_g}{A_g} - \frac{2_s^{\ddagger}N_{s,g}(\sigma_{s,g}-\sigma_{s,u})}{A_s} - \frac{2_s^{\ddagger}N_s\sigma_{s,u}}{A_s} - \gamma_{rt}\right]. \quad (10.109)$$

From this equation, it is clear that the presence of upper-level absorption (excited-state absorption or ESA) decreases the saturable component of loss in the laser and increases the parasitic loss, both undesirable effects.

The rate equation for the gain medium is unaffected by the presence of the saturable absorber. To complete the modeling of a laser containing a saturable absorber, we need a rate equation for the saturable-absorber ground-state population $N_{s,g}$. The population of the saturable-absorber ground state decreases by one for every photon absorbed, resulting in a term $\dot{N}_{s,g,ab} = -\dot{q}_{s,g,ab}$. Decay from the upper level is characterized by the lifetime τ_s, corresponding to $\dot{N}_{s,g,d} = N_{s,u}/\tau_s = (N_s - N_{s,g})/\tau_s$. Many important saturable absorbers have broad absorption bands, and for optically pumped systems where some of the pump light is incident on the saturable absorber we must include a term in the rate

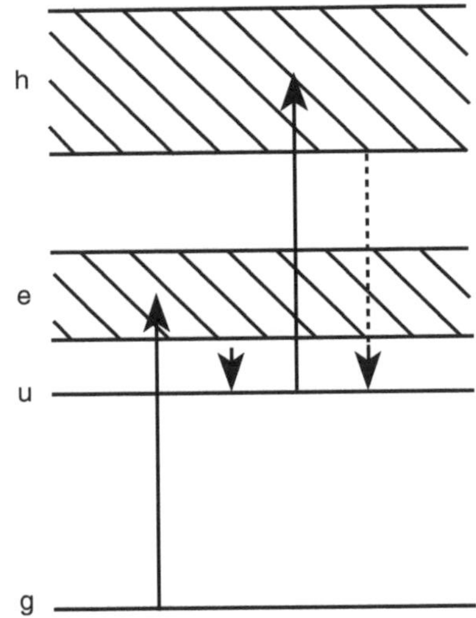

FIGURE 10.20 Schematic representation of a four-band saturable absorber.

equation to account for absorption of the pump light. Consistent with the form of the pump term used throughout this chapter, $\dot{N}_{s,u,p} = W_{p,s}N_{s,g}$. Thus, the rate equation for the ground state of the saturable absorber is

$$\dot{N}_{s,g} = \dot{N}_{s,g,ab} + \dot{N}_{s,g,d} + \dot{N}_{s,g,p} = -\frac{2_s^{\ddagger} q N_{s,g}\sigma_{s,g}}{t_{rt}A_s} + \frac{(N_s - N_{s,g})}{\tau_s} - W_{p,s}N_{s,g}. \tag{10.110}$$

Initial Conditions. For many systems of practical interest, the upper-state lifetime τ_s is much longer than the Q-switched output pulse, but much shorter than the pump period preceding it. This allows us to solve Equation 10.110 piecewise. Prior to the onset of lasing ($q = \dot{q} = 0$) we are in the quasi-steady state and

$$N_{s,g,0} = \frac{N_s}{1 + W_{p,s}/\tau_s}. \tag{10.111}$$

For optically pumped systems where some of the pump light is incident on the saturable absorber

$$N_{s,g,0} = \int \frac{\rho_s}{1 + I_p/I_{p,sat}} dV, \tag{10.112}$$

where ρ_s is the density of saturable-absorber ions, I_p is the pump intensity in the saturable absorber, $I_{p,sat} = h\nu_p/\sigma_{s,g,p}\tau_s$ is the saturation intensity of the saturable absorber at the pump wavelength, $\sigma_{s,g,p}$ is the ground-state absorption cross-section of the saturable absorber at the pump wavelength, and the integration is over the oscillating-mode volume in the saturable absorber.

For longitudinally pumped, passively Q-switched microchip lasers in the geometry illustrated in Figure 10.21, the pump light reaching the saturable absorber can approach the saturation intensity $I_{p,sat}$, greatly reducing the number of saturable-absorber ions that can participate in the passive Q-switching process (Welford 2001; Zayhowski and Wilson 2003a; Jaspan et al. 2004). This increases the minimum-possible Q-switched pulse width and decreases the pulse energy. If the saturable absorber is separate from the gain medium (as opposed to a single crystal that acts as both the gain medium and the saturable absorber), the amount of pump light reaching the saturable absorber can be reduced by any of the following: using a high-concentration gain medium with a large absorption coefficient at the pump wavelength, tuning the pump laser to fall exactly on the absorption peak of the gain medium, narrowing the linewidth of the pump source so that it better overlaps the gain-medium absorption, or putting a dichroic interface between the gain medium and the saturable absorber that reflects the pump light. Alternatively, the saturation intensity could be increased if we could quench the saturable-absorber excited-state lifetime. Finally, in some cases, we can focus the pump light in such a way that it diverges significantly by the time it reaches the saturable absorber, decreasing its intensity.

Rate-Equation Model — Passive Q Switching. The output pulses from a passively Q-switched laser are usually much shorter than the spontaneous lifetime of the gain medium, the upper-state lifetime of the saturable absorber, and the pump period prior to the formation of the pulse. This allows us to neglect spontaneous relaxation and pumping during the development of the output pulse, reducing the rate

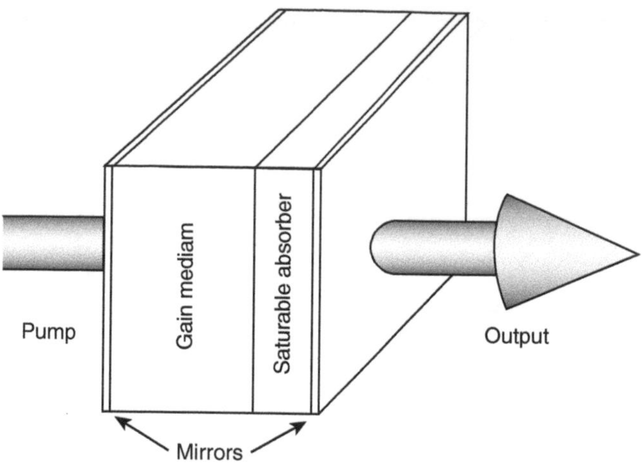

FIGURE 10.21 Simple variation of a passively Q-switched microchip laser.

equations for a single-mode passively Q-switched laser to

$$\dot{q} = \frac{q}{t_{\mathrm{rt}}}\left[\frac{2^{\ddagger}N_{\mathrm{eff}}\sigma_{\mathrm{g}}}{A_{\mathrm{g}}} - \frac{2_{\mathrm{s}}^{\ddagger}N_{\mathrm{s,g}}(\sigma_{\mathrm{s,g}} - \sigma_{\mathrm{s,u}})}{A_{\mathrm{s}}} - \frac{2_{\mathrm{s}}^{\ddagger}N_{\mathrm{s}}\sigma_{\mathrm{s,u}}}{A_{\mathrm{s}}} - \gamma_{\mathrm{rt}}\right], \tag{10.113}$$

$$\dot{N}_{\mathrm{eff}} = -\frac{2^{*}2^{\ddagger}qN_{\mathrm{eff}}\sigma_{\mathrm{g}}}{t_{\mathrm{rt}}A_{\mathrm{g}}}, \tag{10.114}$$

and

$$\dot{N}_{\mathrm{s,g}} = -\frac{2_{\mathrm{s}}^{\ddagger}qN_{\mathrm{s,g}}\sigma_{\mathrm{s,g}}}{t_{\mathrm{rt}}A_{\mathrm{s}}}. \tag{10.115}$$

It is worth noting that these equations are analogous to the set used by Degnan (1995) modified to include the effects of excited-state absorption in the saturable absorber (Xiao and Bass 1997; Xiao and Bass 1998) and thermalization of the energy levels in the gain medium (Degnan, Coyle, and Kay 1998). There have been several other notable publications on the rate-equation analysis of passive Q switching (Zhang et al. 1997; Peterson and Gavrielides 1999; Zhang et al. 2000; Chen et al. 2001; Liu et al. 2001; Patel and Beach 2001; Chen et al. 2002; Zheng et al. 2002); most cover the whole topic, while two focus on estimating the temporal profile of the output pulses (Peterson and Gavrielides 1999; Liu et al. 2001), one presents analysis of the quasi-three-level gain medium Yb:YAG (Patel and Beach 2001), and another presents analysis of passively mode-locked, Q-switched operation (Chen et al. 2002).

In order to obtain passively Q-switched operation the system must, at some time $t=0$, go from being absorbing to having net gain. At that transition time the net gain experienced by a photon in the laser cavity is zero, $\dot{q}=0$, and must be increasing, $\ddot{q}>0$. This leads to the requirement

$$\alpha_{\mathrm{r}} \equiv \frac{2_{\mathrm{s}}^{\ddagger}\sigma_{\mathrm{s,g}}A_{\mathrm{g}}}{2^{*}2^{\ddagger}\sigma_{\mathrm{g}}A_{\mathrm{s}}} > \frac{2_{\mathrm{s}}^{\ddagger}N_{\mathrm{s,g,0}}(\sigma_{\mathrm{s,g}} - \sigma_{\mathrm{s,u}}) + 2_{\mathrm{s}}^{\ddagger}N_{\mathrm{s}}\sigma_{\mathrm{s,u}} + A_{\mathrm{s}}\gamma_{\mathrm{tr}}}{2_{\mathrm{s}}^{\ddagger}N_{\mathrm{s,g,0}}(\sigma_{\mathrm{s,g}} - \sigma_{\mathrm{s,u}})}, \tag{10.116}$$

where the right-hand side of the expression is the ratio of the coefficient of total round-trip loss ($\gamma_{\mathrm{rt,t}}$) to the coefficient of saturable round-trip loss ($\gamma_{\mathrm{rt,s}}$) at $t=0$. Expression 10.116 is often referred to as the second threshold condition. In systems dominated by saturable loss ($\gamma_{\mathrm{rt,t}}/\gamma_{\mathrm{rt,s}} \approx 1$), Expression 10.116 reduces to a statement that the loss of the saturable absorber must saturate more quickly than the gain of the gain medium. When Expression 10.116 is satisfied by a large factor the saturable absorber acts as an ideal, instantaneous Q switch — when it is satisfied by a factor of greater than 10, the instantaneous-Q-switch model is an excellent representation of the system; for a factor of 3, the instantaneous-Q-switch model gives results that are within a factor of ~2 of being correct (Szabo and Stein 1965). When Expression 10.116 is not satisfied, the laser will tend to oscillate cw.

The initial population inversion for a passively Q-switched laser is determined by setting the right-hand side of the photon rate equation (Equation 10.113) equal to 0 at the time the pulse formation begins:

$$N_0 = \frac{A_g}{2^\ddagger \sigma_g}\left[\frac{2_s^\ddagger N_{s,g,0}(\sigma_{s,g}-\sigma_{s,u})}{A_s} + \frac{2_s^\ddagger N_s \sigma_{s,u}}{A_s} + \gamma_{rt}\right]. \tag{10.117}$$

In contrast to actively Q-switched lasers, the initial population inversion of passively Q-switched lasers is fixed by material parameters and the design of the laser, and, in the absence of pulse bifurcation (see below), is identical from pulse to pulse. This can result in extremely stable pulse energies and pulse widths, with measured stabilities of better than 1 part in 10^4. The stability of the pulse energy and pulse width is achieved at the expense of pulse timing stability. Pulse formation will not occur until the population inversion has reached the proper value, and fluctuations in the pump source are the primary cause of pulse-to-pulse timing jitter in many passively Q-switched lasers.

Peak Power. To determine the maximum peak power that can be achieved from a passively Q-switched laser, we start by dividing Equation 10.115 by Equation 10.114 and integrating, to obtain

$$N_{s,g} = N_{s,g,0}\left(\frac{N_{eff}}{N_0}\right)^{\alpha_r}. \tag{10.118}$$

We then divide Equation 10.113 by Equation 10.114 and substitute Equation 10.118 into the result, which we integrate with respect to population inversion to obtain

$$2^*(q-q_0) = N_0 - N_{eff} - \frac{A_g}{2^\ddagger \sigma_g}\left(\frac{2_s^\ddagger N_s \sigma_{s,u}}{A_s} + \gamma_{rt}\right)\ln\left(\frac{N_0}{N_{eff}}\right)$$
$$- N_{s,g,0}\frac{2^*(\sigma_{s,g}-\sigma_{s,u})}{\sigma_{s,g}}\left[1-\left(\frac{N_{eff}}{N_0}\right)^{\alpha_r}\right], \tag{10.119}$$

where q_0 is the number of photons in the cavity at the time the output pulse begins to develop.

The maximum number of photons in the cavity occurs when $\dot{q}=0$. Thus, from Equation 10.113 we obtain an implicit expression for the population inversion N_p at the time of peak output power,

$$N_p = \frac{A_g}{2^\ddagger \sigma_g}\left[\frac{2_s^\ddagger N_s \sigma_{s,u}}{A_s} + \gamma_{rt} + \frac{2_s^\ddagger N_{s,g,0}(\sigma_{s,g}-\sigma_{s,u})}{A_s}\left(\frac{N_p}{N_0}\right)^{\alpha_r}\right]. \tag{10.120}$$

Since α_r is much greater than unity for systems of interest, N_p can be obtained by iterative application of this equation using its value for $\alpha_r = \infty$ as a starting point. Neglecting the initial number of photons in the cavity, the peak output power is given by

$$P_{po} = \frac{h\nu_o \gamma_o}{2^* t_{rt}}\left\{N_0 - N_p - \frac{A_g}{2^\ddagger \sigma_g}\left(\frac{2_s^\ddagger N_s \sigma_{s,u}}{A_s} + \gamma_{rt}\right)\ln\left(\frac{N_0}{N_p}\right) - N_{s,g,0}\frac{2^*(\sigma_{s,g}-\sigma_{s,u})}{\sigma_{s,g}}\left[1-\left(\frac{N_p}{N_0}\right)^{\alpha_r}\right]\right\} \tag{10.121}$$

where $h\nu_o$ is the energy of a photon at the oscillating wavelength, $\gamma_o = -\ln(1 - T_o)$ is the output-coupling coefficient, and T_o is the transmission of the output coupler.

In the limit of large α_r and negligible absorption from the upper level of the saturable absorber, Equation 10.121 asymptotically approaches the result obtained for instantaneous Q switching (Equation 10.84). Optimization of the peak power is easily performed numerically, once the material parameters of the system are known.

Pulse Energy. The total pulse energy E_o of a passively Q-switched laser is obtained by integrating the output power P_o using the following technique (Degnan 1989):

$$E_o = \int_0^\infty P_o \mathrm{d}t = \frac{h\nu_o \gamma_o}{t_{rt}} \int_0^\infty q \mathrm{d}t = \frac{h\nu_o A_g \gamma_o}{2^* 2^\ddagger \sigma_g} \int_{N_0}^{N_f} \frac{\mathrm{d}N_{eff}}{N_{eff}}, \tag{10.122}$$

where the last step used Equation 10.114 and N_f is the population inversion well after the peak of the output pulse. The final integration leads to

$$E_o = \frac{h\nu_o A_g \gamma_o}{2^* 2^\ddagger \sigma_g} \ln\left(\frac{N_0}{N_f}\right), \tag{10.123}$$

and the final population inversion is obtained by setting the right-hand side of Equation 10.119 equal to 0, yielding the implicit expression

$$N_0 - N_f - \frac{A_g}{2^\ddagger \sigma_g}\left(\frac{2_s^\ddagger N_s \sigma_{s,u}}{A_s} + \gamma_{rt}\right) \ln\left(\frac{N_0}{N_f}\right) - N_{s,g,0} \frac{2^*(\sigma_{s,g} - \sigma_{s,u})}{\sigma_{s,g}} \left[1 - \left(\frac{N_f}{N_0}\right)^{\alpha_r}\right] = 0. \tag{10.124}$$

Again, in the limit of large α_r and negligible absorption from the upper level of the saturable absorber, Equation 10.123 asymptotically approaches the result obtained for instantaneous Q switching (Equation 10.90) and optimization is easily performed numerically once the material parameters of the system are known.

Bleaching of the saturable absorber consumes stored energy, which can result in lower pump-to-output efficiencies for passively Q-switched lasers than for comparable actively Q-switched lasers. The intrinsic losses of the saturable absorber, including excited-state absorption, also limit the efficiency of a passively Q-switched device. Quite often, these inefficiencies result in a small amount of energy consumption compared to the energy required to drive an active Q switch.

Pulse Width. The duration t_w of a passively Q-switched output pulse (full width at half-maximum) is obtained from the energy in the pulse E_o, the peak power of the pulse P_{po}, and the shape of the pulse S_p:

$$t_w = \frac{S_p t_{rt}}{\gamma_{rt} + \dfrac{2_s^\ddagger N_s \sigma_{s,u}}{A_s}} \times \left\{ \frac{N_0 - N_f - N_{s,g,0} \dfrac{2^*(\sigma_{s,g} - \sigma_{s,u})}{\sigma_{s,g}} \left[1 - \left(\dfrac{N_f}{N_0}\right)^{\alpha_r}\right]}{N_0 - N_p - \dfrac{A_g}{2^\ddagger \sigma_g}\left(\dfrac{2_s^\ddagger N_s \sigma_{s,u}}{A_s} + \gamma_{rt}\right) \ln\left(\dfrac{N_0}{N_p}\right) - N_{s,g,0} \dfrac{2^*(\sigma_{s,g} - \sigma_{s,u})}{\sigma_{s,g}} \left[1 - \left(\dfrac{N_p}{N_0}\right)^{\alpha_r}\right]} \right\}. \tag{10.125}$$

For systems with a large α_r the pulse duration is determined primarily by the ratio of the resonator length to some combination of the fixed and the saturable loss. The pulse energy depends on the same parameters, but is also is directly proportional to the laser mode volume, which provides the means to scale the pulse energy independent of the pulse duration.

CW-Pumped Passively Q-Switched Lasers. For a cw-pumped passively Q-switched laser, the repetition rate of the pulses increases as the pump power is increased above threshold, while the pulse energies and durations remain essentially independent of the pump power (Zayhowski and Dill 1994; Shimony et al. 1996). The average interpulse period is given by (Liu et al. 2003)

$$\tau_\text{p} = \tau \ln\left[\frac{P_\text{a} - P_\text{i} - (P_\text{th} - P_\text{i})N_\text{f}/N_0}{P_\text{a} - P_\text{th}}\right], \tag{10.126}$$

where P_a is the total pump power absorbed within the lasing mode volume, P_i is the absorbed pump power required to reach inversion ($N_\text{eff}=0$), and P_th is the absorbed pump power necessary to obtain $N_\text{eff}=N_0$. This expression does not include the effects of intramanifold thermalization, which could decrease the interpulse period of short-pulse systems at high pulse repetition rates. For low-power devices the upper limit to the pulse repetition rate (the inverse of the interpulse period) is set by the saturable absorber recovery time; at higher pulse rates the laser will deteriorate into unstable pulse generation and ultimately cw operation. (In high-power devices the maximum pulse repetition rate is limited by thermal effects.) Pulse rates as high as 300 kHz have been demonstrated with a Cr^{4+}:YAG saturable absorber (Jaspan et al. 2000; Zheng et al. 2002). The use of semiconductor saturable structures, with short upper-level lifetimes (see below), allows picosecond Q-switched pulses to be generated at megahertz rates (Keller et al. 1996).

Single-Frequency Operation. In an actively Q-switched laser, potential lasing modes start to build up when the cavity Q is switched. All of the modes start from spontaneous emission and the amount of gain seen by the mode determines how quickly it builds up (see Section 10.6.2.1). In a passively Q-switched laser, the cavity Q is not switched until after a mode starts to build up; it is the light in the first mode that causes Q switching to occur. As a result, when the cavity Q switches, that initial mode already contains a significant photon population and has a headstart in the race to form a pulse. Since the initial mode has the highest net gain, the number of photons in the first mode at any time during the pulse build-up will exceed the number in any other mode by a factor greater than the number of photons in the first mode at the time the second mode reaches threshold. Since a factor of 100 is more than enough to ensure single-frequency operation, single-frequency pulses are easily achieved in passively Q-switched systems.

Once one mode (mode 1) in a passively Q-switched laser begins to oscillate, the ground-state population of saturable-absorber ions $N_\text{s,g}$ and inversion N_eff begin to change rapidly. By the time a second mode (mode 2) reaches threshold, they have changed by

$$\Delta N_\text{s,g}(1,2) \approx \frac{2^\ddagger N_0 A_\text{s}(\sigma_\text{g,1} - \sigma_\text{g,2})}{2_\text{s}^\ddagger A_\text{g}(\sigma_\text{s,g} - \sigma_\text{s,u})}\left[1 - \frac{N_0 \sigma_\text{s,g}}{2^* \alpha_\text{r}^2 N_\text{s,g,0}(\sigma_\text{s,g} - \sigma_\text{s,u})}\right]^{-1} \tag{10.127}$$

and

$$\Delta N_\text{eff}(1,2) \approx \frac{2^\ddagger N_0^2 A_\text{s}(\sigma_\text{g,1} - \sigma_\text{g,2})}{2_\text{s}^\ddagger \alpha_\text{r} N_\text{s,g,0} A_\text{g}(\sigma_\text{s,g} - \sigma_\text{s,u})}\left[1 - \frac{N_0 \sigma_\text{s,g}}{2^* \alpha_\text{r}^2 N_\text{s,g,0}(\sigma_\text{s,g} - \sigma_\text{s,u})}\right]^{-1}, \tag{10.128}$$

where we have assumed that the difference in thresholds between modes 1 and 2 is the result of a small difference in the effective gain cross-sections $\sigma_\text{s,1}$ and $\sigma_\text{s,2}$. (A similar result can be derived if we assume a small difference in cavity loss.) The bracketed term on the right-hand side of these equations approaches unity for large α_r and can often be ignored.

To calculate the photon population of the first mode at the onset of the second mode we use the approximation

$$q(1,2) = q_0 + \Delta N_{\mathrm{eff}}(1,2)q'(0) + \frac{[\Delta N_{\mathrm{eff}}(1,2)]^2 q''(0)}{2!}, \quad (10.129)$$

where the primes indicate differentiation with respect to N_{eff}, q_0 is assumed to be negligible (1/2 photon per mode), and the last term is included because $q'(0)=0$. From Equation 10.119,

$$q''(0) = -\frac{\alpha_r^2 N_{s,g,0}(\sigma_{s,g} - \sigma_{s,u})}{N_0^2 \sigma_{s,g}} \left[1 - \frac{\gamma_{rt,t}}{\alpha_r \gamma_{rt,s}}\right], \quad (10.130)$$

where $\gamma_{rt,t}/\gamma_{rt,s}$ is the fraction that appears on the right-hand side of Expression 10.116, and is of order unity. Thus,

$$q(1,2) = \frac{N_0}{2^* 2\alpha_r} \left(\frac{\sigma_{g,1} - \sigma_{g,2}}{\sigma_{g,1}}\right)^2 \left[1 - \frac{\gamma_{rt,t}}{\alpha_r \gamma_{rt,s}}\right] \left[1 - \frac{N_0 \sigma_{s,g}}{2^* \alpha_r^2 N_{s,g,0}(\sigma_{s,g} - \sigma_{s,u})}\right]^{-2}, \quad (10.131)$$

where the two bracketed terms approach 1 for large α_r and can usually be ignored.

To gain an appreciation for how easy it is to obtain single-frequency operation with a passively Q-switched laser, consider a 1.064-μm Nd:YAG/Cr^{4+}:YAG microchip laser, with $A_g = A_s$ and $\alpha_r = 21$. If we assume an output pulse energy of 100 μJ and an extraction efficiency of 80% ($N_0 = 6.7 \times 10^{14}$), and consider only one possible polarization, $q(1,2) > 100$ for $(\sigma_{g,1} - \sigma_{g,2})/\sigma_{g,1}$ as small as 2.5×10^{-6}. If we make no attempt to control the position of the cavity modes relative to the gain peak, the probability of multimode operation is

$$P_{\mathrm{mm}} = \left(\frac{\delta\nu_L}{2\Delta\nu}\right)^2 \left(\frac{\sigma_{g,1} - \sigma_{g,2}}{\sigma_{g,1}}\right)_{q(1,2)=100}, \quad (10.132)$$

where $\delta\nu_L$ is the (Lorentzian) linewidth of the gain medium (160 GHz for Nd:YAG), $\Delta\nu$ is the mode spacing of the cavity, and the value of $(\sigma_{g,1} - \sigma_{g,2})/\sigma_{g,1}$ is determined by setting $q(1,2) = 100$ in Equation 10.131. With a cavity length of 1 cm, the probability of the laser producing a multimode pulse is less than 2.5×10^{-4}. Note that this is a conservative estimate of the maximum likelihood of multimode operation, since it does not include the fact that the first mode to oscillate will continue to develop more quickly than the other modes after the second mode reaches threshold (see the discussion of Single-Frequency Operation in Section 10.6.2.1), and we are insisting that the energy in the primary mode be at least 100 times greater than the energy in any other mode.

Afterpulsing. The cavity Q of passively Q-switched lasers cannot be rapidly decreased once the Q-switched output pulse exits the system. The residual gain left by the first output pulse will contribute to the continued development of pulses in competing longitudinal modes. In addition, when the thermalization times of the laser manifolds are long compared to the Q-switched output pulse, themalization of the manifolds can rapidly pump the laser transition and lead to gain-switched pulses (see Section 10.6.3) following the primary Q-switched pulse. These gain-switched pulses can be in the same longitudinal mode as the primary pulse. In either case, afterpulsing is difficult to prevent.

When the afterpulse is in a different longitudinal mode than the primary pulse, as is often the case, it is minimized when there is a large difference in the gain of the primary and competing modes. This is most easily accomplished in cavities with a large mode spacing (short cavity length) and when the primary mode is positioned at the peak of the gain profile. The latter can often be accomplished by controlling the pump power to the laser or its temperature, or by applying a controlled amount of stress (see Section 10.5.3). As the primary pulse gets closer to the peak of the gain profile the time delay between the primary pulse and the afterpulse increases, and the magnitude of the afterpulse decreases.

Afterpulsing that results from competing longitudinal modes is also minimized by positioning the saturable absorber near the center of a standing-wave passively *Q*-switched laser cavity. Because of spatial hole burning (see Section 10.5.1.2), the first mode to lase does not completely bleach the saturable absorber for the second mode. The differential loss for adjacent modes is greatest when the saturable absorber is located in the center of the cavity.

For applications where afterpulsing cannot be tolerated, and the afterpulse is in a different longitudinal mode than the primary pulse, it may be possible to eliminate the afterpulse with spectral filtering (an appropriately tuned etalon) outside of the laser cavity, if the laser is bifurcation free (see below). In uncontrolled environments, it may be necessary for the filter to actively track the laser, or visa versa.

Pulse Bifurcation, Pulse-to-Pulse Amplitude Stability. At high pulse repetition rates some of the energy not extracted by one pulse will still be present in the laser cavity when the next pulse forms. As a result of spatial hole burning in the gain medium, the residual gain will be greater at positions where it favors the development of a pulse in a different longitudinal mode than the first pulse. When the gain differential between modes is extremely small, even small amounts of spatial hole burning can cause pulse bifurcation, and the effects can be significant even when the pulse-to-pulse spacing is several times the spontaneous lifetime of the gain medium. The second pulse will preferentially deplete the gain most available to it, creating a situation where the first mode is favored in the subsequent pulse. At higher pulse repetition rates more modes can come into play, creating trains of three or more modes occurring in a regular sequence or, depending on the stability of the system, chaotically. In systems with small mode-gain differentials operating at high repetition rates, time-averaged spectra will show the presence of multiple oscillating modes — this should not be interpreted as proof of multimode pulses; each individual pulse may still be single frequency with a smooth temporal profile, free of mode beating.

Pulse bifurcation in a cw-pumped passively *Q*-switched laser usually results in alternating strong and weak pulses. The timing interval between the pulses typically varies in accordance with the amplitudes; the period preceding a weak pulse is shorter than the period preceding a strong pulse. At higher pulse repetition rates, as the pulse train subdivides into pulses of more longitudinal modes, there will be a greater variation in pulse amplitudes and pulse-to-pulse timing. Because of gain-related index-guiding effects (see Section 10.4.2), each of the modes will have slightly different transverse dimensions and divergences.

For some applications, pulse-to-pulse amplitude stability is critical and pulse bifurcation cannot be tolerated. To maximize the pulse repetition rate at which bifurcation-free operation is achieved, the gain differential between pulses must be maximized. Techniques that reduce afterpulsing also encourage bifurcation-free operation. Thus, observations of the afterpulse can be useful in tuning a passively *Q*-switched laser for maximum pulse-to-pulse stability.

Since the amplitude and pulse width of a passively *Q*-switched laser are determined by material characteristics and cavity design, they can be extremely stable. Passively *Q*-switched microchip lasers operating on a single longitudinal mode and producing pulse energies of 10 μJ have demonstrated pulse-to-pulse amplitude stabilities better than one part in 10^4 at pulse repetition rates of 7 kHz.

Pulse-to-Pulse Timing Stability. The pulse-to-pulse timing stability of passively *Q*-switched lasers is generally worse than that of actively *Q*-switched systems (Mandeville et al. 1996) and, in bifurcation-free systems, is often limited by noise in the pump source to ~1% of the pump duration. The timing stability is improved by using a pulsed pump source, operating at as high a peak power as possible for just long enough to produce an output pulse. When operated in this fashion, the pump pulses are turned on by an external clock and turned off when a *Q*-switched output pulse is detected. This approach has the added benefit of minimizing thermal effects in passively *Q*-switched lasers since it minimizes efficiency losses due to spontaneous decay of the gain medium, leading to additional benefits in performance.

In a system that uses a pulsed pump source, the pulse-to-pulse timing jitter of the passively *Q*-switched laser gets smaller as the pump period decreases. The minimum possible pump duration is limited by the power (or cost) of the available pump diodes. One approach to minimize this limitation is to cw pump the laser just below threshold and then quickly raise the pump power to its maximum value just before an output pulse is desired. This technique results in the greatest pulse-to-pulse timing stability.

Other techniques used to reduce the pulse timing jitter include hybrid active/passive loss modulation (Arvidsson et al. 1998) and external optical synchronization (Dascalu et al. 1996). These approaches have resulted in timing jitter as low as 65 ps (Hansson and Arvidsson 2000), at the expense of added system complexity and some sacrifice in the amplitude and pulse-width stability of a true passively Q-switched system.

Transverse Mode Definition. The transverse mode of miniature passively Q-switched lasers, especially microchip lasers, is nearly diffraction limited. In addition to the mode-defining mechanisms discussed in Section 10.4.2, the saturable absorber acts as a soft aperture and spatially filters the resonator modes, which helps to suppress higher-order transverse modes. However, since the saturable absorber bleaches at the center of the beam first, where the optical intensity is greatest, this soft aperture evolves as the pulse forms. As a result, there can be a measurable change in the output beam profile during the pulse (Yao et al. 1994; Arvidsson 2001).

Semiconductor Saturable-Absorber Mirrors. All of the formalism developed in this section was based on bulk saturable absorbers. Semiconductor saturable-absorber mirrors (SESAMs) are a second class of devices used as passive Q switches in solid-state lasers. SESAMs consist of an antiresonant semiconductor Fabry-Pérot etalon formed by a semiconductor layer grown on top of a highly reflecting semiconductor Bragg mirror and covered by a dielectric reflector. The semiconductor layer typically consists of absorptive quantum-well layers in an otherwise transparent medium. The bandgap of the quantum wells can be engineered to provide saturable absorption at a wide variety of wavelengths.

By their design, SESAMs are high reflectors with a saturable loss γ_s. In addition, there is an undesirable parasitic loss $\gamma_{s,p}$, which typically accounts for ~20% of the unsaturated loss. The effective absorption cross-section of the saturable loss, determined by the absorption cross-section of the semiconductor quantum wells and the reflectivity of the dielectric reflector, is usually much larger than the gain cross-section of the gain medium, leading to a very large α_r in most applications. As a result, SESAMs are, in some ways, ideal saturable absorbers and can often be modeled as instantaneous Q switches. This is almost always the case when they are used in microchip and other miniature lasers, where typical values of α_r are several thousand.

In a cavity containing a SESAM, the SESAM is used as one of the cavity mirrors, with the oscillating light incident on the dielectric reflector. The total round-trip cavity loss $\gamma_{rt,t}$ consists of three components: an output-coupling loss γ_o, a parasitic loss $\gamma_{rt,p}$ (which now includes the unsaturable loss of the SESAM), and the saturable loss of the SESAM γ_s,

$$\gamma_{rt,t,t<0} = \gamma_o + \gamma_{rt,p} + \gamma_s. \tag{10.133}$$

The intracavity loss is given by Equation 10.133 until the laser reaches threshold, at which time it rapidly decreases to

$$\gamma_{rt,t,t>0} = \gamma_o + \gamma_{rt,p}. \tag{10.134}$$

The change in cavity Q, for the very large values of α_r typically associated with SESAMs, is rapid enough to treat the transition as instantaneous for most purposes. The inversion within the cavity at the time Q switching occurs has the value

$$N_0 = \frac{(\gamma_0 + \gamma_{rt,p} + \gamma_s)A_g}{2^{\ddagger}\sigma_g}, \tag{10.135}$$

and the resulting pulse width is

$$t_w = \frac{S_p t_{rt}}{\gamma_s}\left[\frac{R_s(1+R_s)\eta_e}{R_s - \ln(1+R_s)}\right], \tag{10.136}$$

where $R_s = \gamma_s/(\gamma_{rt,p} + \gamma_o)$ is the ratio of saturable to unsaturable cavity losses. For a given amount of saturable loss, the pulse width asymptotically approaches its minimum value of $t_w = 4S_p t_{rt}/\gamma_s$ in the limit of large unsaturable losses ($\gamma_{rt,p} + \gamma_o \gg \gamma_s$). However, this minimum pulse width is obtained at the expense of high threshold and low efficiency. In the opposite limit ($\gamma_{rt,p} + \gamma_o \ll \gamma_s$) the pulse width asymptotically approaches $t_w = S_p t_{rt}/\gamma_{rt,p} + \gamma_o$, the threshold of the laser is reduced, and the extraction efficiency can be high. The best compromise between pulse width, threshold, and efficiency will depend on the application of the laser.

While SESAMs act like instantaneous *Q* switches in many ways, they still benefit from being saturable absorbers in their single-frequency performance; it takes extremely large values of α_r before this benefit disappears.

SESAMs offer both advantages and limitations when compared to bulk saturable absorbers. The advantages include the fact that they can be engineered to operate at a wide variety of wavelengths, making SESAMs particularly interesting where good gain medium/bulk saturable-absorber combinations have not been identified. The effective absorption cross-sections are very large, leading to large values of α_r and nearly instantaneous *Q* switching. The physical length of the saturable-absorber region is small and its contribution to the round-trip time of light in the laser cavity is negligible, resulting in the shortest possible pulses. The dielectric top reflector on the SESAM can be (and often is) designed to be highly reflecting at the wavelength of the pump light so that there is no pump-induced bleaching of the saturable absorber.

One of the main limitations of SESAMs is their relatively low damage threshold. Typical values for the damage fluence of the semiconductor materials are $\sim$10 mJ cm^{-2}. The damage fluence of the SESAMs can be increased beyond this value by using highly reflective dielectric layers. This, however, reduces the saturable loss γ_s, which generally reduces the pulse energy and increases the pulse duration. Typical values for the saturable loss are in the range of 10% or less. The largest pulse energy reported for a microchip laser using a SESAM is 4 μJ, with fractions of a microjoule being more typical. Other limitations of SESAMs are that they cannot be used as input or output couplers, and that their thermal expansion coefficients are not matched to the gain media, precluding thc possibility of bonding the gain medium to the saturable absorber in a quasi-monolithic fashion.

Finally, the carrier recombination time in semiconductors (the upper-state lifetime of the SESAM) can be quite short, ranging from $\sim$10 ps to $\sim$1 ns, depending on the material and the growth conditions. When the upper-state lifetime is comparable to the output pulse width, it can introduce a significant parasitic loss. On the other hand, it effectively closes the *Q* switch and can prevent the formation of afterpulses. It also allows SESAMs to be operated at extremely high pulse repetition rates, and 7-MHz operation has been reported (Braun et al. 1997).

As a result of their advantages and limitation, SESAMs are most attractive in applications requiring short, low-energy pulses, and where requirements on the system's robustness can be relaxed. In this regime, they can be operated at very high repetition rates and can be engineered to work with gain media at many different wavelengths.

Demonstrated Passively Q-Switched Microlasers. Monolithic, diode-pumped, passively *Q*-switched lasers using Cr^{4+},Nd^{3+}:YAG as both the gain medium and saturable absorber have demonstrated output pulses with durations between 3.0 ns (Zhou et al. 1993) and 290 ps (Wang et al. 1995). Composite-cavity microchip lasers containing a Nd^{3+}:YAG gain medium diffusion bonded to a Cr^{4+}:YAG saturable absorber have demonstrated pulses as short as 200 ps with peak powers in excess of 30 kW (Zayhowski and Dill 1994; Zayhowski, Ochoa, and Dill 1995b) when pumped with a 1-W fiber-coupled diode laser. By using a 15-W fiber-coupled diode array as the pump, 565-kW peak powers have been obtained in 380-ps pulses (Zayhowski 1998), and pulses as short as 150 ps have been demonstrated (Zayhowski and Wilson 2003a). Pulses as short as 180 ps have been obtained by *Q*-switching $Nd{:}LaSc_3(BO_3)_4$ microchip lasers with semiconductor antiresonant Fabry-Pérot saturable absorbers (Braun et al. 1996); pulses as short as 37 ps were obtained using the same method in a $Nd{:}YVO_4$ laser (Spühler et al. 1999). Although the devices based on semiconductor saturable absorbers operate at lower pulse energies and peak powers than Cr^{4+}:YAG devices, they have been pulsed at repetition rates up to 7 MHz. Another route to short, energetic pulses is simultaneous passive mode locking and *Q* switching with the same saturable absorber (Chen and Tsai 2001) (see the discussion of Saturable-Absorber Mode Locking in Section 10.2).

All of the passively *Q*-switched microlasers mentioned above operate on the ~1.06-μm Nd^{3+} transition. Passively *Q*-switched microchip lasers operating on the 1.32-μm (Keller 1996), 0.946-μm (Zayhowski et al. 1996), and 1.074-μm (Zayhowski et al. 2000) Nd^{3+}:YAG transitions have also been demonstrated. Passively *Q*-switched microchip lasers operating at 1.03 μm in Cr,Yb:YAG (Zhou et al. 2003), 1.34 μm in Nd:YVO_4 (Fluck et al. 1997), and 1.55 μm in Er:glass (Thony et al. 1999) have also been demonstrated. A slightly larger miniature passively *Q*-switched 1.5-μm Yb:Tm:YLF laser (Braud et al. 2000) has also been demonstrated.

Passively *Q*-switched microlasers excel in generating subnanosecond, single-frequency pulses, but often at the expense of pulse energy, since the shrinking volume of the gain medium reduces the stored energy capacity. Zayhowski has demonstrated passively *Q*-switched microchip lasers with a range of pulse energies from 4.0 μJ to 250 μJ, and pulse durations of 148 ps to 2.2 ns (Zayhowski 1998; Zayhowski et al. 1999; Zayhowski and Wilson 2003a). Using an unstable resonator with a much larger mode, pulse energies as high as 2.15 mJ have been obtained at low repetition rates (Liu, Zhou, and Chen 1998).

The subnanosecond pulse durations of passively *Q*-switched microlasers are extremely useful for high-resolution ranging, and these lasers have been directly used for short-range imaging of buildings and industrial structures. However, larger pulse energies are required for some applications. The solution is to let the microlaser define the pulse width and pulse rate and use an amplifier to increase the pulse energy. For a 1.064-μm Nd:YAG/Cr^{4+}:YAG passively *Q*-switched microlaser either Nd:YAG (Druon et al. 1999; Zayhowski and Wilson 2002; Zayhowski and Wilson 2003b) or Nd:YVO_4 (Isyanova et al. 2001; Zayhowski and Wilson 2004) can be used for the amplification stage. For microlasers with low pulse energies, the need to drive the amplifier into saturation for efficient extraction of energy makes Nd:YVO_4 the preferred gain medium because of its increased emission cross-section and lower saturation energy. Since the strong, relatively broad pump bands of Nd:YVO_4 overlap those of Nd:YAG, Nd:YVO_4 amplifiers can be pumped with the unabsorbed pump light transmitted by a Nd:YAG microlaser, significantly increasing the system efficiency (Zayhowski and Wilson 2004). For higher-energy microlasers, where reaching saturation in the amplifier is not an issue, the mechanical and thermal properties of Nd:YAG make it the preferred gain medium. Passively *Q*-switched microchip oscillator/amplifier systems producing 500-μJ pulses with peak powers in excess of 1 MW at a 500-Hz repetition rate (Zayhowski and Wilson 2003b), or 200-μJ pulses with peak powers in excess of 400 MW at 10 kHz (Zayhowski and Wilson 2002), have been packaged in optical heads occupying a volume of less than 0.25 liters.

The high peak powers that can be obtained from passively *Q*-switched microlasers enable a variety of miniature nonlinear optical devices, including harmonic generators (Zayhowski 1996a; Zayhowski 1996b; Druon et al. 1999; Isyanova et al. 2001; Zayhowski and Wilson 2003b), parametric amplifiers (Zayhowski 1997a; Zayhowski and Wilson 2000), parametric oscillators (Zayhowski 1997b; Nettleton et al. 2000; Capmany et al. 2001), combined harmonic and parametric generators (Bäder et al. 1999), Raman frequency converters (Grabtchikov et al. 1999; Demidovich et al. 2003), broad-band fiber-based Raman amplifiers (Zayhowski 2000; Agrawal 1995), self-frequency doubling lasers (Amano 1992; Lucas-Leclin et al. 2000b; Zhang et al. 2001), and miniature gain-switched devices (Zayhowski, Buchter, and Wilson 2001; Zayhowski and Wilson 2002; see below). Such devices have been used to frequency convert the output of passively *Q*-switched Nd-based microlasers to wavelengths covering the entire spectrum from 0.2 μm to 5 μm in extremely compact optical systems.

10.6.3 Gain-Switched Operation

Gain switching is another way to obtain short, high-peak-power pulses from a laser. The idea is to rapidly increase the pump power so that the population inversion of the laser is well in excess of its threshold value by the time the first output spike develops. The optical pulse then drives the population inversion below its threshold value, as described in the above discussion of Quasi-CW Operation (Section 10.6.1). The pump power is subsequently reduced so that the inversion remains below threshold and a single output pulse is obtained. Gain-switched lasers are an attractive alternative to *Q*-switched devices when the gain medium has a short upper-state lifetime, giving it a limited capacity to accumulate the energy from low-power cw or quasi-cw pump sources, or when energetic, pulsed pump sources are more readily available. They are also an attractive option for passive systems when good saturable absorbers have not been identified for the gain medium.

The equations that describe a *Q*-switched pulse also describe a gain-switched pulse. The difference is that the operation of the laser is controlled by gain rather than loss. Once we know the inversion density at the beginning of the output pulse, all of the formalism developed in the above discussion of *Q*-Switched Operation (Section 10.6.2) applies. Essentially, given the temporal pump profile, we must calculate the pulse build-up time and integrate the pump rate up to that time to obtain N_0.

Gain-switched Nd:YAG microchip lasers, pumped by 36-ns-long pulses from a Ti:Al_2O_3 laser, have produced pulses as short as 760 ps (Zayhowski, Ochoa, and Mooradian 1989). Diode-pumped, 1.3-μm Nd:YAG microchip lasers, pumped with a peak power of 1 W, have been gain switched to produce a 100-kHz train of pulses with durations of 170 ns and peak powers of 1.8 W (Zayhowski 1990b). In this case, the microchip lasers were resonantly pumped at their relaxation frequency to minimize both the pulse width and the fluctuations in pulse amplitude.

A more interesting variation of a gain-switched laser system uses a passively *Q*-switched microchip lasers as the pump. This approach is particularly attractive when the absorption of the gain medium is well matched to the fundamental or one of the harmonics of Nd:YAG. It can result in efficient, compact, and robust sources of short-pulsed laser radiation at diverse wavelengths. Optimal performance is obtained when the inversion density in the gain-switched laser cavity at the time the pulse forms is about three times above threshold. If the goal is to have a single output pulse extract the maximum amount of energy from each pump pulse, the duration of the pump pulse should typically be less than 3–10 times the output pulse width. Longer pump pulses can lead to multiple output pulses.

Miniature Cr^{4+}:YAG gain-switched lasers, pumped by the 80-μJ fundamental of a passively *Q*-switched Nd:YAG microchip laser, have produced 4-ns output pulses at a wavelength of 1.44 μm, with a pulse energy of 2 μJ (Zayhowski, Buchter, and Wilson 2001). This system should be tunable between 1.3 and 1.6 μm.

Ti:Al_2O_3 is a nearly ideal gain medium for gain-switched laser oscillation in the near infrared. Using a frequency-doubled passively *Q*-switched microchip laser as the pump source, gain-switched Ti:Al_2O_3 lasers have been demonstrated with output pulses as short as 350 ps at wavelengths near 800 nm, and linewidths (in a different experimental setup) as narrow as 0.05 nm (Zayhowski, Buchter, and Wilson 2001). The same authors demonstrated tuning of miniature gain-switched Ti:Al_2O_3 lasers over 65 nm, limited by the free spectral range of the tuning element. The energy conversion efficiency, from input 532-nm pump energy to output 800-nm energy, approaches 50%. More recently, miniature gain-switched Ti:Al_2O_3 lasers have been packaged in optical heads with a volume of less then 0.2 litres (Zayhowski and Wilson 2002). With the appropriate optics, it should be possible to build miniature gain-switched Ti:Al_2O_3 lasers operating at any wavelength within the extremely broad gain bandwidth of the Ti:Al_2O_3 system, ~650–1100 nm.

Miniature gain-switched Ce^{3+}-doped-fluoride lasers, pumped by the fourth harmonic of passively *Q*-switched Nd:YAG microchip lasers, should be able to operate throughout broad regions of the near ultraviolet, from ~280 to 330 nm (Marshall et al. 1994). There is a large number of other gain media that could be used in miniature gain-switched laser systems to generate output over a large range of wavelengths from the infrared through the ultraviolet.

Gain-switched lasers with spectrally broad gain bandwidths offer an alternative to optical parametric devices as tunable short-pulse sources. As pulse durations get shorter, the thresholds of optical parametric oscillators (OPOs) increase because of the cavity build-up time, and there is a large penalty (in threshold) for including passive elements within the OPO cavity (Byer 1975; Zayhowski 1997b). This is a result of the fact that the gain of the system is only present during the pump pulse. Gain-switched lasers do not suffer the same penalty, since energy (hence gain) is stored in the gain medium until it is extracted. In tunable systems, particularly when narrow linewidths are desired, it is often necessary to use intracavity frequency-selective elements. For short-pulsed applications where narrow linewidth is required, gain-switched lasers can have a lower threshold than OPOs and provide more robust operation.

10.6.4 Cavity-Dumped Operation

In principle, cavity dumping allows the energy in a laser cavity to be output in a time comparable to the cavity round-trip time. The concept is to rapidly introduce (within a cavity round-trip time) a large output coupling (~ 100%) into a cavity that previously had no output coupling. This works well for large laser cavities. As the length of a laser cavity becomes small, however, the duration of a cavity-dumped pulse becomes limited by the switching time of the output-coupling mechanism (see the discussion of Active Q Switching in Section 10.6.2.2). For small laser cavities, Q switching is more efficient and produces shorter pulses. Methods of cavity dumping include the use of an electro-optic Pockels cell and polarizing beamsplitter, and acousto-optic devices.

10.6.5 Mode-Locked Operation

Mode locking refers to the situation where the phases of several longitudinal cavity modes are fixed (or locked) with respect to each other such that the coherent addition of the electric fields adds constructively for a short period of time. This allows the generation of a periodic train of high-peak-power, ultrashort pulses. To understand this, consider the case of $2n+1$ equally spaced longitudinal modes oscillating with the same amplitude E_0. Assume that the phases ϕ_m of the modes are locked according to $\phi_m-\phi_{m-1}=\Delta\phi$, where $\Delta\phi$ is a constant. The total electric field is the sum of all of these modes:

$$E(t) = \sum_{m=-n}^{n} E_0\exp\{2\pi i[(\nu_0 - m\Delta\nu)t + m\Delta\phi]\} = E'(t')\exp(2\pi i\nu_0 t), \tag{10.137}$$

where ν_0 is the frequency of the center mode, $\Delta\nu$ is the frequency difference between two adjacent modes, $t'=t+\Delta\phi/\Delta\nu$, and

$$E'(t') = E_0\frac{\sin[\pi(2n+1)\Delta\nu t']}{\sin[\pi\Delta\nu t']} \tag{10.138}$$

is the electric field envelope. Equation 10.137 shows that $E(t)$ can be represented in terms of a wave with a carrier frequency ν_0 whose amplitude is time dependent. The power of this wave is proportional to $E'^2(t')$; the waveform consists of a train of pulses whose peak power is proportional to $E'^2(0)=(2n+1)^2E_0^2$, pulse width $t_w=1/(2n+1)\Delta\nu$, and separation between pulses $\tau_p=1/\Delta\nu$. Since the total oscillating bandwidth $\Delta\nu_{osc}$ is given by $(2n+1)\Delta\nu$ and the frequency separation $\Delta\nu$ between modes is $1/t_{rt}$, the mode-locked laser just described produces a train of output pulses with a pulse width given by the inverse of the oscillating bandwidth and a separation between pulses equal to the cavity round-trip time. Figure 10.22 shows $E'^2(t')$ for $n=5$.

In the above example, the phases of all of the modes were locked so that the output pulse had the minimum possible duration. Such a pulse is referred to as transform limited, because the temporal profile is the Fourier transform of the spectral profile. This need not be the case — it is possible to obtain longer pulses (but not shorter pulses). The above example also had only one pulse circulating in the optical cavity. This is known as fundamental mode locking, and will be assumed for the rest of this section. It is possible, but less common, to increase the pulse repetition rate of a cavity by having multiple pulses circulating, a situation referred to as harmonic mode locking. Finally, it is worth noting that, unlike the other pulsed schemes described in this section, a mode-locked laser is a cw device and there is phase coherence between pulses.

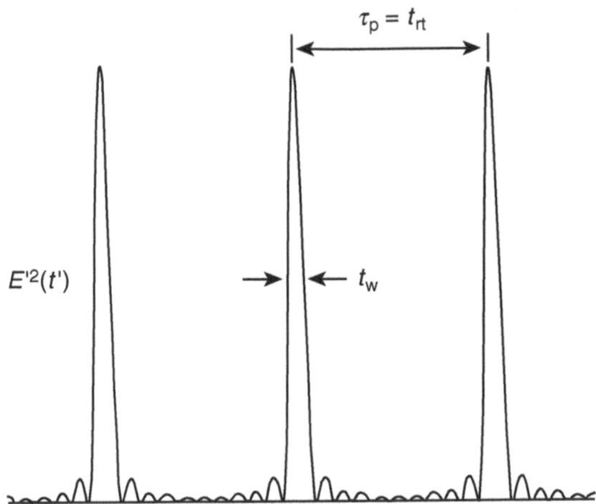

FIGURE 10.22 Train of mode-locked pulses made up of 11 modes of equal intensity.

The discussion above tells us what mode locking is, but sheds little light on how or why it occurs. Mode locking can be broken into two categories, active mode locking and passive mode locking. Active mode locking can be further subdivided into AM mode locking (produced with an amplitude modulator within the laser cavity) and FM mode locking (produced with a frequency or phase modulator within the laser cavity). Each of these is discussed below. In general, mode locking will occur if the net gain for a mode-locked train of pulses is greater than the net gain for any other combination of cavity modes.

10.6.5.1 Active Mode Locking

AM Mode Locking. In AM mode locking the loss of some element in the laser cavity is modulated at the round-trip cavity frequency. This modulates the amplitude of the optical field and generates sidebands that are resonant with cavity modes. With the proper phase relationship between these modes, most of the light is incident on the modulator during its minimum loss. This combination of modes sees a lower loss than any other combination of modes and is, therefore, favored. The same result occurs if the gain of the cavity is modulated. Gain modulation through modulation of the pump source is known as synchronous pumping.

In an ideal mode-locked system, the relative phases of the cavity modes remain constant after a complete round trip in the laser cavity. In such a system, once the phase relationship between the modes is established, it will persist even in the absence of amplitude modulation. One effect that can destroy the phase relationship and, therefore, disrupt mode locking, is dispersion. Others are self-phase modulation (a nonlinear effect) and spontaneous emission (noise), since the phase of the noise is uncorrelated with the phase of the oscillating modes. Although dispersion is normally disruptive to the mode-locking process, the combination of anomalous dispersion and self-phase modulation can lead to pulse shortening and soliton formation. This effect is extremely important in fibers (Agrawal 1995) and can lead to pulse shortening even in miniature solid-state lasers. Mode locking will occur when there is sufficient loss (or gain) modulation to overcome the disruptive effects of net round-trip dispersion.

The pulse width of a mode-locked laser is ultimately determined by a balance between the pulse shortening and broadening mechanisms. The duration of the mode-locked pulse and the net intracavity gain are strongly dependent on the net intracavity dispersion. Excess dispersion will increase the duration of the pulses. Typically, pulses formed in dispersive cavities are frequency chirped, with a different optical frequency at the beginning of the pulse than at the end. To minimize the round-trip dispersion of the cavity, prisms or chirped mirrors (which add a spectrally dependent phase delay to the reflected light) are often used to compensate for the dispersion of other intracavity elements, such as the gain medium.

The most commonly used tool to analyze mode-locked lasers is the Master Equation (Haus 1975a; Haus 1975b). The Master Equation is written in terms of the electric field envelope $E'(t)$. It describes the propagation of the pulse through the cavity and is solved by requiring that the pulse reproduce itself after one round trip. The Master Equation for an actively mode-locked laser with negligible dispersion includes a term for the frequency-dependent gain, a term for the time-dependent loss, and a term for the passive loss. With reasonable approximations, the equation yields an analytic solution (Ippen 1994):

$$E'(t) = E'(0)\exp\left(-\frac{t^2}{t_w^2}\right) \tag{10.139}$$

where

$$t_w = \left(\frac{g_{rt}}{2\gamma_a \pi^4 \delta\nu_g^2 \delta\nu_m^2}\right)^{1/4}, \tag{10.140}$$

γ_a is the depth of active loss modulation, $\delta\nu_g = c\Delta\lambda_g/\lambda_0^2$ is the gain bandwidth, and $\delta\nu_m$ is the modulation bandwidth. The pulse width will be shorter for large gain bandwidths, high modulation bandwidths, and large modulation strengths. The minimum pulse width obtainable by active mode locking is often limited by the bandwidth of the active modulation.

FM Mode Locking. In FM mode locking, the optical length of the laser cavity (physical length or refractive index) is modulated at the round-trip cavity frequency. As in the case of AM mode locking, the modulation generates sidebands that are resonant with cavity modes. For simplicity, consider the case where one of the cavity mirrors is moving sinusoidally along the direction of the cavity axis. Light incident on a moving mirror will be Doppler shifted. As a result, it will not reproduce itself after one round trip and will not produce a coherent oscillating mode. Light incident on the mirror at its turning points (maximum or minimum cavity length) is reflected by a stationary mirror and does not experience a Doppler shift. Modulation of the refractive index at some point in the cavity has the same effect. Mode locking will occur when there is sufficient modulation to overcome the effects of dispersion and noise. FM mode locking typically produces chirped pulses. The duration of the mode-locked pulse and the net intracavity gain are (as in the case of AM mode locking) strongly dependent on intracavity dispersion.

Examples of miniature FM mode-locked laser systems include an acousto-optically mode-locked 1.05-μm Nd:YLF laser that produced a train of 13-ps pulses at a 5-GHz repetition rate (3-cm cavity length) (Schulz and Henion 1991b) and electro-optically mode-locked 1.085-μm Nd:$LiNbO_3$ microchip lasers with pulse widths as short as 18.6 ps and repetition rates up to 20-GHz (3.48-mm cavity length) (Vieira, Herczfeld, and Contarino 1997; Vieira et al. 2001).

10.6.5.2 Passive Mode locking

Passive mode locking can occur when a laser cavity contains a nonlinear optical element, such as a saturable absorber. In this case, the more intense the light incident on the saturable absorber, the less the total absorption. The total loss of the cavity is, therefore, minimized by putting all of the energy into short pulses. This is, essentially, self-induced AM mode locking. A similar effect is obtained with the combination of a Kerr lens (often formed within the gain medium itself) and an aperture within the cavity (sometimes a soft aperture defined by the pump profile in the gain medium). Other techniques include the use of interferometric elements containing nonlinear media. One example is additive pulse mode locking (Haus, Fujimoto, and Ippen 1991; Haus, Fujimoto, and Ippen 1992).

Passive mode locking must be initiated by the presence of a pulse within the cavity. If the optical intensity within the cavity is uniform in time, there is no loss or gain element that is modulated at the round-trip cavity frequency to induce mode locking. Noise is capable of introducing a small amplitude modulation on the optical field. In some lasers, this small modulation is sufficient to start the mode-locking process. Such lasers are referred to as self-starting. In other systems a pulse (or AM modulation) must be intentionally introduced into the cavity to start the mode-locking process. Once started, however, mode locking may persist indefinitely.

Saturable-Absorber Mode Locking. Saturable absorbers, including SESAMs, can be used to mode lock a laser in much the same way that they are used to Q switch a laser (see Section 10.6.2.3). To work most effectively, the recovery time of the saturable absorber should be shorter than the round-trip time of light in the cavity. As in the case of passive Q switching, the absorption cross-section of the saturable absorber must be greater than the gain cross-section of the gain medium; Expression 10.116 must be obeyed. Also, the round-trip intracavity dispersion must be small enough so that the superposition of the modes after one round trip produces a pulse of sufficient peak power to maintain the required nonlinear steady state. Finally, in the case of cw mode locking, the pump must replace the amount of population inversion extracted by the pulse during each round-trip in the cavity.

The amount of loss and gain saturation that occurs during cw passive mode locking is usually quite small compared to what occurs in Q switching. The pulse width is often limited by the recovery time of the saturable absorber, but can reach the limits imposed by the bandwidth of the gain medium. When the saturable absorber recovers quickly enough, the minimum pulse width can be estimated by (Paschotta and Keller 2001)

$$t_w = \frac{1.07}{\delta\nu_g}\left(\frac{g_{rt}}{\gamma_{rt,sat}}\right)^{1/2}, \tag{10.141}$$

where the influence of spatial hole burning has been neglected [spatial hole burning can increase the effective gain bandwidth when the gain medium is close to an end mirror of the cavity (Braun et al. 1995)]. The minimum pulse width is obtained when the intracavity pulse energy is 3–5 times the saturation energy of the saturable absorber. Equation 10.141 is valid when the recovery time of the saturable absorber is shorter than ~28 times the pulse width (Paschotta and Keller 2001), provided that it is also short enough for nearly complete recovery of the saturable absorber between pulses. Longer saturable absorber recovery times can lead to unstable behavior.

When the saturable absorber does not recover within the round-trip cavity time, Q-switched mode locking often occurs. This was commonly observed in the early development of passive solid-state mode-locked lasers with long upper-state lifetimes, due to the limited parameter range of available saturable absorbers. More recently, the Q-switching instability has been eliminated in many solid-state mode-locked lasers through a better understanding of the interaction between the saturation of the saturable absorber and of the gain medium (Haus 1976; Kärtner et al. 1995; Hönninger et al. 1999) and the engineering of semiconductor saturable absorbers, such as SESAMs (Keller et al. 1992; Keller et al. 1996; Keller 1998).

The relevant absorber parameters of semiconductor saturable absorbers can be engineered over several orders of magnitude. For optimal performance, the lifetime of the saturable absorber should be greater than the pulse width and smaller than the cavity round-trip time. When this is true and the saturation of the gain medium and saturable absorber are the dominant pulse-forming mechanisms, a rate-equation analysis shows that cw mode locking is stable if the intracavity pulse energy $E_i = E_o/T_o$ is greater than a critical energy (Hönninger et al. 1999)

$$E_{i,c} \equiv \left(E_{g,sat}E_{s,sat}\gamma_{rt,sat}\right)^{1/2}, \tag{10.142}$$

where $\gamma_{rt,sat}$ is the round-trip saturable loss coefficient, $E_{g,sat} = h\nu_0 A_g/2^{\ddagger}\sigma_g$ is the saturation energy of the gain medium, and $E_{s,sat} = h\nu_0 A_s/2_s^{\ddagger}\sigma_{s,eff}$ is the saturation energy of the saturable absorber (for the bulk saturable absorbers discussed in Section 10.6.2.3, $\sigma_{s,eff} = \sigma_{s,g} - \sigma_{s,u}$; for SESAMs it is defined such that $E_{s,sat}$ is the pulse energy required to bleach the saturable absorption to $1/e$ of its initial value). When the intracavity pulse energy is less than than $E_{i,c}$, Q-switched mode locking often occurs. This stability criterion is in excellent agreement with experimental results obtained with picosecond mode-locked lasers. The required minimum intracavity pulse energy can be demanding on the pump source and/or the saturable absorber.

For ultrashort pulses, where soliton effects are important, an increase in pulse energy results in self-phase modulation and/or self-amplitude modulation that broadens the pulse spectrum. A broader spectrum reduces the gain because of the finite gain bandwidth of the system, providing negative feedback that decreases the critical pulse energy required to maintain stable cw mode locking, and a more relaxed stability criteria is obtained (Hönninger et al. 1999). *Q* switching can also be suppressed through nonlinear intracavity loss such as two-photon absorption (Jiang et al. 1999; Schibli et al. 2000; Thoen et al. 2000). Other approaches include the use of active feedback to prevent *Q* switching (Schibli, Morgner, and Kärtner 2001a; Schibli et al. 2002; Valentine et al. 2002).

SESAMs have been used to produce cw mode locking in miniature 1.064-μm $Nd:YVO_4$ lasers with cavity lengths as short as ~0.45 mm, resulting in pulse repetition rates up to 157 GHz, with pulse durations of less than 3 ps and average output powers of 10–100 mW (Krainer et al. 1999; Krainer et al. 2000; Krainer et al. 2002a). Picosecond SESAM mode-locked Er:Yb:glass lasers have operated at pulse rates up to 40 GHz (Zeller et al. 2003), and have demonstrated broad tunability across the telecommunications C-band (1.529–1.569 μm) (Krainer et al. 2002b; Spühler et al. 2003). Similar devices have also demonstrated less than 350-fs timing jitter (Spühler et al. 2002). Cr:LiSAF lasers, operating at a center wavelength of 857 nm, mode locked with a SESAM, have produced 146-fs pulses at 1 GHz (Kemp et al. 2001).

For some applications in nonlinear frequency conversion, precise micromachining, and surgery, *Q*-switched mode locking may be attractive because of the large pulse energies that can be obtained in the ultrashort mode-locked pulses. For good definition of the mode-locked pulses under a *Q*-switched envelope, the mode-locked pulses must build up quickly. In general, mode-locked pulses build up most quickly in actively mode-locked systems and most slowly in passively mode-locked systems that rely on fast nonlinear optical interactions. Passive *Q*-switched mode locking has been realized in 1.5-μm Cr^{4+}:YAG microchip lasers using semiconductor saturable absorbers and has produced 10-GHz mode-locked trains of 500-fs pulses at a *Q*-switched repetition rate of 1 MHz, with 42-ns *Q*-switched pulse envelopes and an average power of 100 mW (Schibli et al. 2001b).

Kerr-Lens Mode Locking. The shortest mode-locked laser pulses are obtained using fast nonlinear optical effects, such as the Kerr nonlinearity, to modulate the transmission of some element (or combination of optical elements) within the laser cavity (Haus, Fujimoto, and Ippen 1991; Haus, Fujimoto, and Ippen 1992). Fast optical nonlinearities follow the pulse envelope instantaneously (or nearly so). When the cavity is properly dispersion compensated and the loss coefficient of the nonlinear element can be written

$$\gamma_{\mathrm{nl}}(t) = \gamma_0 - \Gamma_{\mathrm{nl}} P(t), \tag{10.143}$$

where $P(t)$ is the instantaneous power in the nonlinear loss element, the Master Equation for the cavity yields the analytic solution (Haus, Fujimoto, and Ippen 1991)

$$E'(t) = E'(0)\,\mathrm{sech}\left(\frac{t}{t_{\mathrm{w}}}\right), \tag{10.144}$$

where

$$t_{\mathrm{w}} = \frac{g_{\mathrm{rt}}}{\Gamma_{\mathrm{nl}} E_{\mathrm{i}} \pi^2 \delta\nu_{\mathrm{g}}^2}, \tag{10.145}$$

E_{i} is the intracavity pulse energy, and $\delta\nu_{\mathrm{g}}$ is the gain bandwidth.

Although passively mode-locked systems based on fast nonlinear optics can produce extremely short pulses, quite often they are not self-starting and require another mechanism, such as an intracavity saturable absorber, to initiate the pulse formation. Once the pulse reaches a high enough peak power, the fast nonlinear process takes over.

Kerr-lens mode-locked Ti:sapphire lasers have produced pulses as short as 5 fs (less than 2 oscillations of light) (Sutter et al. 1999; Ell et al. 2001). Miniature passive Kerr-lens mode-locked femtosecond Cr^{4+}:YAG lasers have demonstrated repetition rates up to 4 GHz (3.75-cm cavity length) with pulse durations as short as 68 fs (Collings et al. 1996; Tomaru and Petek 2000; Tomaru 2001; Leburn et al. 2003; Tomaru 2003).

10.7 Amplifiers, Frequency Converters, and Modulators

There are several important adjuncts to miniature lasers, including amplifiers, frequency converters, and modulators. Although several of these have been mentioned throughout the chapter, a detailed discussion of any is beyond the chapter's scope. Nevertheless, a few comments are offered below. A more detailed discussion of miniature amplified and frequency-converted passively *Q*-switched microchip lasers can be found elsewhere (Zayhowski 2004; Zayhowski, to be published).

10.7.1 Amplifiers

As the required output power of a laser system increases, the need for amplification stages becomes apparent. Master oscillators (lasers) that operate at low powers can be easily controlled to produce a desired output format. Successive amplification stages can be designed to provide increasingly higher output powers by increasing the pumping power and the amplifier aperture. Limiting the gain of each stage can eliminate parasitic oscillation. Isolators are used to prevent feedback from subsequent amplification stages. The design of each stage can also be optimized for heat removal and low optical distortion. Beam cleanup between stages can be used to suppress unwanted spatial frequencies and, thereby, maintain single-transverse-mode operation. Amplifiers can also be paralleled by splitting the master oscillator beam; if a coherent output is desired, the outputs of the amplifiers must be phased.

There are two basic types of amplifier, regenerative and traveling wave. A regenerative amplifier provides a feedback loop and can oscillate without the laser (master oscillator) input. Control of the regenerative amplifier is achieved by injection seeding it with the master oscillator signal. In a traveling-wave configuration, the amplifier simply boosts the signal injected by the master oscillator. Regenerative amplifiers are, by nature, multi-pass. Traveling-wave amplifiers can be designed so that the radiation travels through the amplifying medium more than once without retracing its path.

The gain media in solid-state amplifiers can be in the form of thin discs, rods, slabs, or fibers. Each of these has been realized in compact formats, with the size of the unit ultimately limited by pumping and thermal considerations.

10.7.2 Frequency Converters

Nonlinear optical techniques are used to extend the frequency coverage of lasers. Frequency conversion is a very important adjunct, converting the output of practical lasers to regions where primary laser sources may not exist or may not be very practical. Harmonic generation, frequency mixing, optical parametric oscillation, and stimulated Raman scattering have been used for frequency conversion.

For high-peak-power lasers, efficient nonlinear frequency conversion can be achieved during a single pass of the output beam through an appropriate nonlinear crystal. For lower-power applications the nonlinear crystal is often contained in a resonant cavity, whose resonance must be carefully locked to the output frequency of the laser. In all cases, the nonlinear crystal must be properly aligned with the excitation so that phase matching of the exciting and generated optical waves is achieved.

The most frequently used frequency-conversion techniques are second-harmonic and sum-frequency generation. Typically, the output of infrared lasers in the 1-μm region, such as Nd:YAG, is frequency doubled to the green. Shorter wavelengths may be obtained, for example, by summing the green radiation with the infrared or frequency doubling the green. There are a variety of uses for short-wavelength output, including pumping short-wavelength lasers and optical parametric oscillators, and fluorescence spectroscopy. For second-harmonic generation to the green, conversion efficiencies as high as 92% have been reported (Matveets et al. 1978). Materials such as KH_2PO_4 (KDP), KD_2PO_4 (KD*P), $KTiOPO_4$ (KTP), periodically poled $LiNbO_3$ (PPLN), LiB_3O_5 (LBO), and β-BaB_2O_4 (BBO) are used.

Difference-frequency generation and optical parametric conversion are used to provide tunable sources of radiation for various spectroscopic applications, primarily in the infrared. Continuous-output difference-frequency generation, while generally producing very low average power with low efficiency, is used for ultrahigh-resolution molecular spectroscopy in the mid-infrared. Optical parametric converters are useful sources of tunable pulsed output with high peak power and high efficiency.

Stimulated Raman scattering has been used to frequency shift the output of pulsed lasers with nearly unity quantum efficiency, at a variety of wavelengths. Raman combined with other nonlinear effects in fibers can be used to create extremely broadband, spatially coherent optical continua.

10.7.3 Modulators

Acousto-optic or electro-optics modulators are often used after a cw laser to impose amplitude, frequency, or phase modulations on the output beam, with applications in optical communications, spectroscopy, and lidar. Phase modulation is also a critical technology for the coherent combining of multiple individual emitters.

List of Symbols

$$2^* = \begin{cases} 1 & \text{for a four-level laser} \\ g_u(f_u + f_l) & \text{for a quasi-three-level laser} \\ 2 & \text{for a nondegenerate three-level laser} \end{cases}$$

$$2^\dagger = \begin{cases} 1 & \text{for a traveling-wave laser} \\ 2 & \text{for a standing-wave laser} \end{cases}$$

$2^\ddagger$ number of times the same cross-sectional area of the gain medium is traversed by light during one round trip within the laser cavity

$2^\ddagger_s$ number of times the same cross-sectional area of the saturable absorber is traversed by light during one round trip within the laser cavity

3^* number between 0 and 3, depending on the alignment of the electronic wave functions with the polarization of the optical field

A cross-sectional area of the optical field (cm^2)

A_g cross-sectional area of the oscillating mode in the gain medium (cm^2)

A_s cross-sectional area of the oscillating mode in the saturable absorber (cm^2)

B probability per unit time that a photon will interact with a given inverted site (s^{-1})

B_p pump brightness (W cm^{-2} st^{-1})

c speed of light in vacuum (cm s^{-1})

c_{sh} specific heat (J g^{-1} K^{-1})

C number between 0 and 1 that characterizes mechanical boundary conditions

C_{11} longitudinal elastic constant (Pa)

D_e energy-diffusion constant (cm^2 s^{-1})

E electric-field amplitude (V cm^{-1})

$E(t)$ electric-field amplitude function (V cm^{-1})

$E'(t)$ electric-field envelope function (V cm^{-1})

E_0 electric-field amplitude of modes in a mode-locked laser (V cm^{-1})

$E_{g,sat}$ saturation energy of the gain medium (J)
E_i intracavity pulse energy of a mode-locked laser (J)
$E_{i,c}$ critical intracavity pulse energy for stable cw mode locking of a laser (J)
E_l energy of the lower laser level relative to the ground level (cm^{-1})
E_o output pulse energy (J)
$E_{o,max}$ maximum obtainable output pulse energy (J)
E_s energy of a state (J)
$E_{s,sat}$ saturation energy of the saturable absorber (J)
f_g thermal occupation probability of states in the ground level
f_l thermal occupation probability of states in the lower laser level
$f_{l,eff}$ effective value of f_l used in simplified rate equations
f_s thermal occupation probability of a state within a manifold
f_u thermal occupation probability of states in the upper laser level
$f_{u,eff}$ effective value of f_u used in simplified rate equations
F_{eo} fraction of the optical length of the laser cavity filled with electro-optic material
F_g fraction of the optical length of the laser cavity filled with gain medium
$F_{N,th}$ fraction of the total population of active ions in the gain medium inverted at threshold
g gain coefficient (cm^{-1})
$g(r,z,\lambda)$ saturated, spatially and spectrally dependent gain coefficient (cm^{-1})
g_0 unsaturated gain coefficient (cm^{-1})
g_g degeneracy of states in the ground level
g_l degeneracy of states in the lower laser level
g_{rt} round-trip gain coefficient at the time a Q-switched or mode-locked pulse begins to develop
g_s degeneracy of a state
g_u degeneracy of states in the upper laser level
h Planck's constant (J s)
I circulating optical intensity ($W\ cm^{-2}$)
I_g circulating optical intensity in the gain medium ($W\ cm^{-2}$)
I_m circulating optical intensity of mode m ($W\ cm^{-2}$)
$I_m(z)$ total optical intensity of mode m at position z ($W\ cm^{-2}$)
I_p pump intensity in the saturable absorber ($W\ cm^{-2}$)
$I_{p,sat}$ saturation intensity of the saturable absorber at the pump wavelength ($W\ cm^{-2}$)
I_s circulating optical intensity in the saturable absorber ($W\ cm^{-2}$)
I_{sat} circulating optical saturation intensity ($W\ cm^{-2}$)
k_B Boltzmann's constant ($J\ K^{-1}$)
k_c thermal conductivity ($W\ cm^{-1}\ K^{-1}$)
k_m wave vector of mode m (cm^{-1})
l length of the material (cm)
l_g length of the gain medium (cm)
$l_{op,rt}$ round-trip optical length of the laser cavity (cm)
l_s length of the saturable absorber (cm)
m mode identifier
m_l ratio of the optical length of the etalon cavity to the optical length of the gain cavity in a coupled-cavity Q-switched laser
Δm difference in longitudinal mode numbers
n refractive index of the material
n_0 refractive index in the absence of an optical field
n_2 nonlinear Kerr index ($cm^2\ W^{-1}$)
n_g refractive index of the gain medium
$\Delta n(r,z,\lambda)$ spatially and spectrally dependent change in the refractive index seen by an oscillating mode as it moves away from the gain peak
N_0 effective population inversion immediately before Q switching
N_{eff} effective population inversion
N_f effective population inversion well after the peak of a Q-switched output pulse
N_g population of the ground level
N_l population of the lower laser level
N_{lm} population of the lower laser manifold
N_m population of a manifold
N_p effective population inversion at the time of peak output power

N_s population of saturable absorber ions
$N_{s,g}$ population of saturable absorber ions in the ground state
$N_{s,g,0}$ population of saturable absorber ions in the ground state at the time an optical pulse begins to form
$N_{s,u}$ population of saturable absorber ions in the upper level
N_t population of active ions in the gain medium
N_{th} effective population inversion at threshold
$N_{th,1}$ effective population inversion at threshold for the primary mode of a Q-switched laser in the low-loss state
$N_{th,2}$ effective population inversion at threshold for the second mode of a Q-switched laser in the low-loss state
N_u population of the upper laser level
$N_{u,th}$ population of the upper laser level at threshold
N_{um} population of the upper laser manifold
$\Delta N_{eff}(1,2)$ change in the effective population inversion between the times when the first and second modes of a pulsed laser begin to oscillate
$\Delta N_{s,g}(1,2)$ change in the population of saturable absorber ions in the ground state between the times when the first and second modes of a pulsed laser begin to oscillate
$\dot{N}_p$ pump rate (s^{-1})
$\dot{N}_{u,sp}$ spontaneous decay rate of the upper-level population (s^{-1})
$\dot{N}_{u,st}$ stimulated decay rate of the upper-level population (s^{-1})
$\dot{N}_{s,g,ab}$ rate of change of the saturable-absorber ground-state population due to absorption of photons at the oscillating wavelength (s^{-1})
$\dot{N}_{s,g,d}$ rate of change of the saturable-absorber ground-state population due to spontaneous decay of the upper level (s^{-1})
$\dot{N}_{s,g,p}$ rate of change of the saturable-absorber ground-state population due to absorption of photons at the pump wavelength (s^{-1})
$\dot{N}_{s,u,p}$ rate of change of the saturable-absorber upper-level population due to absorption of photons at the pump wavelength (s^{-1})
$P(t)$ instantaneous power (W)
P_a absorbed pump power (W)
P_i absorbed pump power required to obtain inversion (W)
P_{mm} probability of multimode operation in a passively Q-switched laser
P_o output power (W)
P_p pump power (W)
P_{po} peak output power (W)
$P_{po,max}$ maximum obtainable peak output power (W)
P_{th} pump power at threshold (W)
q number of photons in the laser cavity
$q(1,2)$ number of photons in the first mode at the time the second mode reaches threshold
q_0 number of photons in the laser cavity at the time a Q-switched pulse begins to form
q_p number of photons in the laser cavity at the peak of the output pulse
$\dot{q}_e$ photon emission rate (s^{-1})
$\dot{q}_l$ photon loss rate (s^{-1})
$\dot{q}_{s,g,ab}$ absorption rate of photons by saturable-absorber ions in the ground state (s^{-1})
$\dot{q}_{s,u,ab}$ absorption rate of photons by saturable-absorber ions in the upper level (s^{-1})
$\dot{q}_{sp}$ spontaneous-emission rate of photons (s^{-1})
$\dot{q}_{st}$ stimulated-emission rate of photons (s^{-1})
Q "quality" of the optical cavity
Q_h rate of heat deposition ($W\ s^{-1}$)
r radial coordinate (cm)
r_0 characteristic, thermally defined radius of the oscillating mode (cm)
r_{eo} electro-optic coefficient ($cm\ V^{-1}$)
r_m radius of the oscillating mode (cm)
r_p radius of the pump beam (cm)
r_s radial distance to the heat sink (cm)
R_1 reflectivity of the interface between the gain cavity and the etalon cavity of a coupled-cavity Q-switched laser
R_2 reflectivity of the output coupler of the etalon cavity of a coupled-cavity Q-switched laser

R_s ratio of saturable to unsaturable cavity losses
R_W ratio of the pump rate factor W_p to its threshold value
S_p pulse shape factor
t time (s)
t_b pulse build-up time (s)
t_{rt} cavity round-trip time (s)
$t_{rt,et}$ round-trip time of the etalon cavity of a coupled-cavity Q-switched laser (s)
$t_{rt,g}$ round-trip time of the gain cavity of a coupled-cavity Q-switched laser (s)
t_{sw} switching time (s)
t_w pulse width (full width at half-maximum) (s)
$t_{w,min}$ minimum obtainable pulse width (full width at half-maximum) (s)
T temperature (K or °C)
$T_{et,cw}$ cw transmission of an etalon
$T_{et,t=0}$ effective transmission of an etalon during the early formation of a Q-switched pulse
T_o transmission of the output coupler of a laser
T_r voltage rise time (s)
V_g active volume of the gain medium (cm^3)
W_p pump rate factor (s^{-1})
$W_{p,s}$ pump rate factor for the saturable absorber (s^{-1})
$W_{p,th}$ pump rate factor at threshold (s^{-1})
z coordinate along the cavity axis (cm)
α absorption coefficient of a material (cm^{-1})
α_d thermal diffusivity ($cm^2\ s^{-1}$)
α_e thermal expansion coefficient (K^{-1})
α_p pump absorption coefficient (cm^{-1})
α_r ratio of the optical power required to saturate the gain medium to the optical power required to saturate the saturable absorber
$\beta(1,2)$ mode-discrimination factor
γ Euler's constant
γ_0 time-independent portion of the round-trip loss coefficient
γ_a depth of active loss modulation
$\gamma_{nl}(t)$ time-dependent loss coefficient for a nonlinear element
γ_o output-coupling loss coefficient
γ_{rt} round-trip loss coefficient, excluding loss due to a saturable absorber
$\gamma_{rt,m}$ round-trip loss coefficient for mode m
$\gamma_{rt,p}$ round-trip parasitic loss coefficient
$\gamma_{rt,s}$ round-trip saturable loss coefficient
$\gamma_{rt,t}$ total round-trip loss coefficient, including loss due to the saturable absorber
$\gamma_{rt,t,t<0}$ total round-trip loss coefficient prior to the onset of lasing
$\gamma_{rt,t,t>0}$ total round-trip loss coefficient after the laser has Q switched
γ_s saturable loss coefficient of a SESAM
$\gamma_{s,p}$ unsaturable, parasitic loss coefficient of a SESAM
Γ_{nl} coefficient of nonlinear loss (W^{-1})
Γ_{rt} round-trip loss
$\Delta_{nl,T}$ fractional thermally induced change in optical length (K^{-1})
ζ single-mode inversion ratio
ζ_D single-mode inversion ratio resulting from diffusion
ζ_{SH} single-mode inversion ratio resulting from spatial hole burning
η efficiency
η_a area efficiency
η_e pulse extraction efficiency
$\eta_{e,rep}$ extraction efficiency of a repetitively pulsed system
η_i inversion efficiency
η_o output-coupling efficiency
η_p pump efficiency
η_q quantum efficiency
η_s slope efficiency
θ angular coordinate
κ heat-generating efficiency of the pump
λ free-space wavelength (measured in vacuum) $= c/\nu$ (μm or nm)
λ_0 free-space wavelength of the gain peak (μm or nm)
λ_o free-space wavelength of the oscillating mode (μm or nm)
$\Delta\lambda(1,2)$ difference in the free-space wavelength of the first mode to oscillate and a potential second mode (nm)
$\Delta\lambda_g$ full width at half-maximum of the gain peak (nm)
ν optical frequency (s^{-1})
ν_0 optical center frequency (s^{-1})
ν_m optical frequency of mode m (s^{-1})

ν_o optical frequency of the oscillating mode (s^{-1})
ν_p optical frequency of the pump (s^{-1})
$\delta\nu_g$ gain bandwidth (full width at half-maximum) (s^{-1})
$\delta\nu_G$ Gaussian linewidth (full width at half-maximum) (s^{-1})
$\delta\nu_L$ Lorentzian linewidth (full width at half-maximum) (s^{-1})
$\delta\nu_m$ modulation bandwidth (s^{-1})
$\Delta\nu$ frequency difference between adjacent longitudinal modes of a cavity (s^{-1})
$\Delta\nu_{fsr}$ free spectral range (s^{-1})
$\Delta\nu_{osc}$ oscillating bandwidth (s^{-1})
$\rho(z)$ population density of the upper laser manifold at position z (cm^{-3})
ρ_0 unsaturated inversion density (cm^{-3})
ρ_{cq} parameter used to characterize concentration quenching (at.%)
ρ_d doping concentration (at.%)
ρ_{eff} effective inversion density (cm^{-3})
ρ_l density of sites in the lower laser level (cm^{-3})
ρ_m mass density (g cm^{-3})
ρ_s density of saturable-absorber ions (cm^{-3})
ρ_{th} threshold population inversion density (cm^{-3})
ρ_u density of sites in the upper laser level (cm^{-3})
$\dot{\rho}_p(z)$ pump rate at position z ($cm^{-3}\ s^{-1}$)
σ_a absorption cross-section (cm^2)
σ_g stimulated-emission cross-section (cm^2)
$\sigma_{g,1}$ effective stimulated-emission cross-section of the first mode to lase in a Q-switched laser (cm^2)
$\sigma_{g,2}$ effective stimulated-emission cross-section of the second mode to lase in a Q-switched laser (cm^2)
$\sigma_{g,eff}$ effective stimulated-emission cross-section of the gain medium (cm^2)
σ_m effective stimulated-emission cross-section for mode m (cm^2)
σ_r radiative cross-section of a transition (cm^2)
$\sigma_{s,eff}$ effective absorption cross-section of the saturable absorber (cm^2)
$\sigma_{s,g}$ ground-state absorption cross-section of the saturable absorber at the oscillating wavelength (cm^2)
$\sigma_{s,g,p}$ ground-state absorption cross-section of the saturable absorber at the pump wavelength (cm^2)
$\sigma_{s,u}$ upper-level absorption cross-section of the saturable absorber at the oscillating wavelength (cm^2)
τ spontaneous lifetime of the upper-state population in the gain medium (s)
τ_0 damping constant (s)
τ_c cavity lifetime (s)
τ_l spontaneous lifetime of the upper-state population in the limit of low doping (s)
τ_{nr} nonradiative lifetime of the upper-state population (s)
τ_p temporal separation between pulses (s)
τ_r radiative lifetime of the upper-state population in the gain medium (s)
τ_s radiative lifetime of the upper-state population in the saturable absorber (s)
$\phi(\omega)$ phase
ϕ_{et} phase length of an etalon
ϕ_m phase of mode m
$\Delta\phi$ phase difference between modes m and $m-1$
ψ_{max} maximum angle that can be tolerated between the mirrors of a Fabry–Pérot cavity
$\langle\psi(1,2)\rangle$ mode-correlation factor
ω frequency (s^{-1})
ω_0 characteristic frequency (s^{-1})
ω_{qs} quasi-static frequency (s^{-1})
ω_r relaxation frequency (s^{-1})

References

Abdolvand, A., Shen, D. Y., Cooper, L. J., Williams, R. B., and Clarkson, W. A. 2003. Ultra-efficient Ho:YAG laser end-pumped by a cladding-pumped Tm-doped silica fiber laser, In *OSA TOPS 83, Advanced Solid-State Photonics*, J. J. Zayhowski, ed., p. 7. Optical Society of America, Washington, DC.

Abshire, J., Ketchum, E., Afzal, R., Millar, P., and Sun, X. 2000. The geoscience laser altimeter system (GLAS) for the ICEsat mission. *Conf. Lasers Electro-Optics, Tech. Dig.*, 602.

Afzal, R. S., Yu, A. W., Zayhowski, J. J., and Fan, T. Y. 1997. Single-mode, high-peak-power passively *Q*-switched diode-pumped Nd:YAG laser. *Opt. Lett.*, 22, 1314.

Agate, B., Kemp, A. J., Hopkins, J. M., Sibbett, W., Keller, U., and Burns, D. 2001. Highly compact femtosecond Cr:LiSAF lasers. *Conf. Lasers Electro-Optics, Tech. Dig.*, 27.

Aggarwal, R. L., Sanchez, A., Stuppi, M. M., Fahey, R. E., Strauss, A. J., Rapoport, W. R., and Khattak, C. P. 1988. Residual infrared absorption in as-grown and annealed crystals of Ti:Al_2O_3. *IEEE J. Quantum Electron.*, 24, 1003.

Agnesi, A., Dell'Acqua, S., Guandalini, A., Reali, G., Cornacchia, F., Toncelli, A., Tonelli, M., Shimamura, K., and Fukuda, T. 2001. Optical spectroscopy and diode-pumped laser performance of Nd^{3+} in the CNGG crystal. *IEEE J. Quantum Electron.*, 37, 304.

Agrawal, G. P. 1995. *Nonlinear Fiber Optics, 2nd Ed.*, Academic, San Diego, CA, chap. 8.

Amano, S. 1992. Microchip NYAB green laser. *Rev. Laser Eng.*, 20, 723 (in Japanese).

Andrauskas, D. M. and Kennedy, C. 1991. Tetravalent chromium solid-state passive *Q* switch for Nd:YAG laser systems, In *OSA TOPS 10, Advanced Solid-State Lasers*, G. Dubé and L. Chase, eds., p. 393. Optical Society of America, Washington, DC.

Auge, F., Druon, F., Balembois, F., Georges, P., Brun, A., Mougel, F., Aka, G., and Vivien, D. 2000. Theoretical and experimental investigations of a diode-pumped quasi-three-level laser: the Yb^{3+}-doped $Ca_4GdO(BO_3)_3$ (Yb:GdCOB) laser. *IEEE J. Quantum Electron.*, 36, 598.

Arvidsson, M. 2001. Far-field timing effects with passively *Q*-switched lasers. *Opt. Lett.*, 26, 196.

Arvidsson, M., Hansson, B., Holmgren, M., and Lindstrom, C. 1998. A combined actively and passively *Q*-switched microchip laser. *SPIE*, 3265, 106.

Bäder, U., Meyn, J. P., Bartschke, J., Weber, T., Borsutzky, A., Wallenstein, R., Batchko, R. G., Fejer, M. M., and Byer, R. L. 1999. Nanosecond periodically poled lithium niobate optical parametric generator pumped at 532 nm by a single-frequency passively *Q*-switched Nd:YAG laser. *Opt. Lett.*, 24, 1608.

Bai, Y., Wu, N., Zhang, J., Li, J., Li, S., Xu, J., and Deng, P. 1997. Passively *Q*-switched Nd:YVO_4 laser with a Cr^{4+}:YAG crystal saturable absorber. *Appl. Opt.*, 36, 2468.

Balembois, F., Georges, P., and Brun, A. 1993. Quasi-continuous-wave and actively mode-locked diode-pumped Cr^{3+}:$LiSrAlF_6$ laser. *Opt. Lett.*, 18, 1730.

Barnes, N. P., Walsh, B. M., Murray, K. E., Quarles, G. J., and Castillo, V. K. 1998. Nd:LuLF operating on the $^4F_{3/2}\rightarrow{}^4I_{11/2}$ and $^4F_{3/2}\rightarrow{}^4I_{13/2}$ transitions. *J. Opt. Soc. Am. B*, 15, 2788.

Bayramian, A. J., Marshall, C. D., Schaffers, K. I., and Payne, S. A. 1999. Characterization of Yb^{3+}:$Sr_{5-x}Ba_x(PO_4)_3F$ crystals for diode-pumped lasers. *IEEE J. Quantum Electron.*, 35, 665.

Beier, B., Meyn, J.-P., Knappe, R., Boller, K.-J., Huber, G., and Wallenstein, R. 1994. A 180 mW Nd:$LaSc_3(BO_3)_4$ single-frequency TEM_{00} microchip laser pumped by an injection-locked diode-laser array. *Appl. Phys. B*, 58, 381.

Bibeau, C., Payen, S. A., and Powell, H. T. 1993. Evaluation of the terminal level lifetime in sixteen Neodymium-doped crystals and glasses, *OSA Proceedings on Advanced Solid-State Lasers 15*, A. A. Pinto and T. Y. Fan, eds., p. 74. Optical Society of America, Washington, DC.

Bilinsky, I. P., Fujimoto, J. G., Walpole, J. N., and Missaggia, L. J. 1998. Semiconductor-doped-silica saturable-absorber films for solid-state laser mode locking. *Opt. Lett.*, 23, 1766.

Borodin, N., Zhitnyuk, V., Okhrimchuk, A., and Shestakov, A. 1990. Oscillation of a $Y_3Al_5O_{12}$:Cr^{4+} laser in wave length region of 1.34–1.6 μm. *Bull. Acad. Sci. U.S.S.R. Physical Series*, 54, 1500.

Braud, A., Girard, S., Doualan, J. L., and Moncorge, R. 2000. Wavelength tunability and passive Q-switching of a (Yb,Tm):YLF laser operating around 1.5 μm. *Conf. Lasers Electro-Optics, Tech. Dig.*, 463.

Braun, B., Weingarten, K. J., Kärtner, F. X., and Keller, U. 1995. Continuous-wave mode-locked solid-state lasers with enhanced spatial hole-burning, Part I: Experiments. *Appl. Phys. B*, 61, 429.

Braun, B., Kärtner, F. X., Keller, U., Meyn, J.-P., and Huber, G. 1996. Passively Q-switched 180-ps Nd:$LaSc_3(BO_3)_4$ microchip laser. *Opt. Lett.*, 24, 405.

Braun, B., Kärtner, F. X., Zhang, G., Moser, M., and Keller, U. 1997. 56-ps passively Q-switched diode-pumped microchip laser. *Opt. Lett.*, 22, 381.

Brown, A. J. W. and Fisher, C. H. 1993. A 6.5-J flashlamp-pumped Ti:Al_2O_3 laser. *IEEE J. Quantum Electron.*, 29, 2513.

Brunner, F., Sudmeyer, T., Innerhofer, E., Paschotta, R., Morier-Genoud, F., Keller, U., Gao, J. et al. 2002. 240-fs pulses with 22-W average power from a passively mode-locked thin-disk Yb:$KY(WO_4)_2$. *Conf. Lasers Electro-Optics, Tech. Dig.*, 24.

Burns, P. A., Dawes, J. M., Dekker, P., Piper, J. A., Li, J., and Wang, J. 2002a. Coupled-cavity, single-frequency, tunable cw Yb:YAB yellow microchip laser. *Opt. Commun.*, 207, 315.

Burns, P., Dawes, J. M., Dekker, P., Piper, J. A., Zhang, H., and Wang, J. 2002b. CW diode-pumped microlaser operation at 1.5–1.6 μm in Er,Yb:YCOB. *IEEE Photon. Technol. Lett.*, 14, 1677.

Burshtein, Z., Blau, P., Kalisky, Y., Shimony, Y., and Kokta, M. R. 1998. Excited-state absorption studies of Cr^{4+} ions in several garnet host crystals. *IEEE J. Quantum Electron.*, 34, 292.

Buzelis, R., Dement'ev, A. S., Kosenko, E. K., and Murauskas, E. 1995. Stimulated-Brillouin-scattering compression of pulses from an Nd:YAG laser with a short cavity and measurement of the nonradiative relaxation time of the lower active level. *Quantum Electron.*, 25, 540.

Byer, R. L. 1975. Optical parametric oscillators, In *Quantum Electronics: A Treatise*, H. Rabin and C. L. Tang, eds., p. 587. Academic, New York.

Capmany, J., Bermudez, V., Callejo, D., and Dieguez, E. June, 2001. Microchip OPOs operate in the infrared. *Laser Focus World*, 143.

Carrig, T. J. and Pollock, C. R. 1993. Performance of a continuous-wave fosterite laser with krypton ion, Ti:sapphire, and Nd:YAG pump lasers. *IEEE J. Quantum Electron.*, 29, 2835.

Casperson, L. W. 1980. Laser power calculations: Sources of error. *Appl. Opt.*, 19, 422.

Chai, B. H. T., Lefaucheur, J.-L., Stalder, M., and Bass, M. 1992. Cr:$LiSr_{0.8}Ca_{0.2}AlF_6$ tunable laser. *Opt. Lett.*, 17, 1584.

Chai, B. H. T., Eichenholz, J. M., Qing, Y., Jang, W. K., and Richardson, M. 1998. Visible light generation by self-frequency doubling Nd:YCOB. *Conf. Lasers Electro-Optics, Tech. Dig.*, 325.

Chen, Y. F. and Tsai, S. W. 2001. Simultaneous Q-switching and mode-locking in a diode-pumped Nd:YVO_4–Cr^{4+}:YAG laser. *IEEE J. Quantum Electron.*, 37, 580.

Chen, W., Spariosu, K., Stultz, R., Kuo, Y. K., Birnbaum, M., and Shestakov, A. V. 1993. Cr^{4+}:GSGG saturable absorber Q-switch for the ruby laser. *Opt. Commun.*, 104, 71.

Chen, Y. F., Lan, Y. P., and Chang, H. L. 2001. Analytical model for design criteria of passively Q-switched lasers. *IEEE J. Quantum Electron.*, 37, 462.

Chen, Y. F., Lee, J. L., Hsieh, H. D., and Tsai, S. W. 2002. Analysis of passively Q-switched lasers with simultaneous modelocking. *IEEE J. Quantum Electron.*, 38, 312.

Cheng, E. A. P. and Kane, T. J. 1991. High-power single-mode diode-pumped Nd:YAG laser using a monolithic nonplanar ring resonator. *Opt. Lett.*, 16, 478.

Clobes, A. R. and Brienza, M. J. 1972. Single-frequency traveling-wave Nd:YAG laser. *Appl. Phys. Lett.*, 21, 265.

Collings, B. C., Stark, J. B., Tsuda, S., Knox, W. H., Cunningham, J. E., Jan, W. Y., Pathak, R., and Bergman, K. 1996. Saturable Bragg reflector self-starting passive mode locking of a Cr^{4+}:YAG laser pumped with a diode-pumped Nd:YVO_4 laser. *Opt. Lett.*, 21, 1171.

Conroy, R. S., Lake, T., Friel, G. J., Kemp, A. J., and Sinclair, B. D. 1998. Self-Q-switched Nd:YVO_4 microchip laser. *Opt. Lett.*, 23, 457.

Courjaud, A., Deguil, N., and Salin, F. 2002. High-power diode-pumped Yb:KGW ultrafast laser. *Conf. Lasers Electro-Optics, Tech. Dig.*, 501.

Danielmeyer, H. G. 1975. Stoichiometric laser materials, In *Festkörperprobleme (Advances in Solid State Physics) XV*, H.-J. Queisser, ed., p. 253. Pergamon Vieweg, Braunschweig.

Dascalu, T., Pavel, N., Lupei, V., Philipps, G., Beck, T., and Weber, H. 1996. Investigation of a passive *Q*-switched, externally controlled, quasicontinuous or continuous pumped Nd:YAG laser. *Opt. Eng.*, 35, 1247.

Day, T., Nilsson, A. C., Fejer, M. M., Farinas, A. D., Gustafson, E. K., Nabors, C. D., and Byer, R. L. 1989. 30 Hz-linewidth, diode-laser-pumped, Nd:GGG nonplanar ring oscillators by active frequency stabilization. *Electron. Lett.*, 25, 810.

Day, T., Gustafson, E. K., and Byer, R. L. 1990. Active frequency stabilization of a 1.062-μm Nd:GGG, diode-laser-pumped nonlinear ring oscillator to less than 3 Hz of relative linewidth. *Opt. Lett.*, 15, 221.

Degnan, J. J. 1989. Theory of the optimally coupled *Q*-switched laser. *IEEE J. Quantum Electron.*, 25, 214.

Degnan, J. J. 1993. Millimeter accuracy satellite laser ranging: a review. *Contributions of Space Geodesy to Geodynamics: Technology, AGU Geodynamics Series*, 25, 133.

Degnan, J. J. 1995. Optimization of passively *Q*-switched lasers. *IEEE J. Quantum Electron.*, 31, 1890.

Degnan, J. J. 1999. Engineering progress on the fully automated photon-counting SLR2000 satellite laser ranging station. *SPIE*, 3865, 76.

Degnan, J. J., Coyle, D. B., and Kay, R. B. 1998. Effects of thermalization on *Q*-switched laser properties. *IEEE J. Quantum Electron.*, 34, 887.

Degnan, J. J., McGarry, J., Zagwodzki, T., Dabney, P., Geiger, J., Chabot, R., Steggerda, C., Marzuok, J., and Chu, A. 2001. Design and performance of an airborne multikilohertz, photon-counting microlaser altimeter. *Int. Arch. Photogrammetry Remote Sensing*, XXXIV-3/W4, 9.

Dekker, P., Burns, P. A., Dawes, J. M., Piper, J. A., Li, J., Hu, X., and Wang, J. 2003. Widely tunable yellow-green lasers based on the self-frequency-doubling material Yb:YAB. *J. Opt. Soc. Am. B*, 20, 706.

DeLoach, L. D., Payne, S. A., Chase, L. L., Smith, L. K., Kway, W. L., and Krupke, W. F. 1993. Evaluation of absorption and emission properties of Yb^{3+} doped crystals for laser application. *IEEE J. Quantum Electron.*, 29, 1179.

DeLoach, L. D., Page, R. H., Wilke, G. D., Payne, S. A., and Krupke, W. F. 1996. Transition metal-doped zinc chalcogenides: spectroscopy and laser demonstration of a new class of gain media. *IEEE J. Quantum Electron.*, 32, 885.

Demchuk, M. I., Mikhailov, V. P., Zhavoronkov, N. I., Kuleshov, N. V., Prokoshin, P. V., Yumashev, K. V., Livshits, M. G., and Minkov, B. I. 1992. Chromium-doped fosterite as a solid-state saturable absorber. *Opt. Lett.*, 17, 929.

Demidovich, A. A., Kuzmin, A. N., Rybtsev, G. I., Batay, L. E., Titov, A. N., and Yakobson, V. 1997. Low power Nd:KGW microchip green laser. *Conf. Lasers Electro-Optics, Tech. Dig.*, 522.

Demidovich, A. A., Shkadarevich, A. P., Danailov, M. B., Apai, P., Gasmi, T., Gribkovskii, V. P., Kuzmin, A. N., Ryabtsev, G. I., and Batay, L. E. 1998. Comparison of cw laser performance of Nd:KGW, Nd:YAG, Nd:BEL, and Nd:YVO_4 under laser diode pumping. *Appl. Phys. B*, 67, 11.

Demidovich, A. A., Kuzmin, A. N., Nikeenko, N. K., Titov, A. N., Mond, M., and Kueck, S. 2002. Diode-pumped laser operation of Yb,Tm:$KY(WO_4)_2$ around 1.97 μm. *Conf. Lasers Electro-Optics, Tech. Dig.*, 500.

Demidovich, A. A., Apanasevich, P. A., Batay, L. E., Grabtchikov, A. S., Kuzmin, A. N., Lisinetskii, V. A., Orlovich, V. A. et al. 2003. Sub-nanosecond microchip laser with intracavity Raman conversion. *Appl. Phys. B*, 76, 509.

Dieke, G. H. and Crosswhite, H. M. 1963. The spectra of the doubly and triply ionized rare earths. *Appl. Opt.*, 2, 675.

Dinerman, B. J., Harrison, J., and Moulton, P. F. 1994. CW and pulsed operation at 3 μm in Er^{3+}-doped crystals, In *OSA Proceedings on Advanced Solid-State Lasers 20*, T. Y. Fan and B. H. T. Chai, eds., p. 168. Optical Society of America, Washington, DC.

Dixon, G. J., Lingvay, L. S., and Jarman, R. H. 1989. Properties of close-coupled, monolithic lithium neodymium tetraphosphate laser. *SPIE*, 1104, 107.

Dominiak-Dzik, G., Ryba-Romanowski, W., Golab, S., and Pajaczkowska, A. 2000. Optical investigation of Eu^{3+} and Tm^{3+} in the $Ca_4GdO(BO_3)_3$ crystal, *Lasers and Electro-Optics Europe, Conf. Dig.*, Paper CWF19.

Druon, F., Balembois, F., Georges, P., and Brun, A. 1999. High-repetition-rate 300 ps pulsed ultraviolet source with a passively *Q*-switched microchip laser and a multipass amplifier. *Opt. Lett.*, 24, 499.

Eilers, H., Hoffman, K. R., Dennis, W. M., Jacobsen, S. M., and Yen, W. M. 1992. Saturation of 1.064 μm absorption in Cr,Ca:$Y_3Al_5O_{12}$ crystals. *Appl. Phys. Lett.*, 61, 2958.

Ell, R., Morgner, U., Kärtner, F. X., Fujimoto, J. G., Ippen, E. P., Scheuer, V., Angelow, G. et al. 2001. Generation of 5-fs pulses and octive-spanning spectra directly from a Ti:sapphire laser. *Opt. Lett.*, 26, 373.

Erbert, G., Bass, I., Hackel, R., Jemkins, S., Kanz, K., and Paisner, J. 1991. 43-W, cw Ti:sapphire laser. *Conf. Lasers Electro-Optics, Tech. Dig.*, 390.

Erhard, S., Gao, J., Giesen, A., Contag, K., Lagatsky, A. A., Abdolvand, A., Kuleshov, N. V. et al. 2001. High power Yb:KGW and Yb:KYW thin disk laser operation. *Conf. Lasers Electro-Optics, Tech. Dig.*, 333.

Evtuhov, V. and Siegman, A. E. 1965. A 'twisted-mode' technique for obtaining axially uniform energy density in a laser cavity. *Appl. Opt.*, 4, 142.

Fan, T. Y. 1988. Effect of finite lower level lifetime on *Q*-switched lasers. *IEEE. J. Quantum Electron.*, 24, 2345.

Fan, T. Y. 1994. Aperture guiding in quasi-three-level lasers. *Opt. Lett.*, 19, 554.

Fan, T. Y. and Sanchez, A. 1990. Pump source requirements for end-pumped lasers. *IEEE J. Quantum Electron.*, 26, 311.

Fan, T. Y., Huber, G., Byer, R. L., and Mitzscherlich, P. 1988. Spectroscopy and diode laser-pumped operation of Tm,Ho:YAG. *IEEE J. Quantum Electron.*, 24, 924.

Faure, N., Bore, C., Couchaud, M., Basset, G., Templier, R., and Wyon, C. 1996. Optical properties and laser performance of neodymium doped scheelites $CaWO_4$ and $NaGd(WO_4)_2$. *Appl. Phys. B*, 63, 593.

Fluck, R., Braun, B., Gini, E., Melchoir, H., and Keller, U. 1997. Passively *Q*-switched 1.34 μm Nd:YVO_4 microchip laser with semiconductor saturable-absorber mirrors. *Opt. Lett.*, 22, 991.

French, P. M. W., Mellish, R., Taylor, J. R., Delfyett, P. J., and Florez, L. T. 1993. Mode-locked all-solid-state diode-pumped Cr:LiSAF laser. *Opt. Lett.*, 18, 1934.

Gavrilovic, P., O'Neill, M. S., Meehan, K., Zarrabi, J. H., and Singh, S. 1992. Temperature-tunable, single frequency microcavity lasers fabricated from flux-grown YCeAG:Nd. *Appl. Phys. Lett.*, 60, 1652.

Genack, A. Z. and Brewer, R. G. 1978. Optical coherent transitions by laser frequency switching. *Phys. Rev. A*, 17, 1463.

Govorkov, S., Wiebner, A., Genter, P., Stamm, U., Zschocke, W., Schroeder, T., and Basting, D. 1998. High-average-power Ce:LiCaF laser at 1-kHz repetition rate tunable from 280 nm to 315 nm. *SPIE*, 3247, 60.

Grabtchikov, A. S., Kuzmin, A. N., Lisinetskii, V. A., Orlovich, V. A., Ryabtsev, G. I., and Demidovich, A. A. 1999. All solid-state diode-pumped Raman laser with self-frequency conversion. *Appl. Phys. Lett.*, 75, 3742.

Grabtchikov, A. S., Kuzmin, A. N., Lisinetskii, V. A., Orlovich, V. A., Ryabtsev, G. I., and Demidovich, A. A. 2000. All solid-state diode-pumped Raman laser with self-frequency conversion. *Conf. Lasers Electro-Optics, Tech. Dig.*, 228.

Grabtchikov, A. S., Kuzmin, A. N., Lisinetskii, V. A., Orlovich, V. A., Demidovich, A. A., Danailov, M. B., Eichler, H. J., Bednarkiewicz, A., Strek, W., and Titov, A. N. 2002. Laser operation and Raman self-frequency conversion in Yb:KYW microchip laser. *Appl. Phys. B*, 75, 795.

Gu, J., Tam, S. C., Lam, Y. L., Chen, Y., Kam, C. H., Tan, W., Xie, W. J., Zhao, G., and Yang, H. 2000. Novel use of GaAs as a passive *Q*-switch as well as an output coupler for diode-infrared solid state lasers. *SPIE*, 3929, 222.

Hall, D. R. and Jackson, P. E., eds. 1989. *The Physics and Technology of Laser Resonators*, Adam Hilger, Bristol, U.K.

Hansson, B. and Arvidsson, M. 2000. *Q*-switched microchip laser with 65 ps timing jitter. *Electron. Lett.*, 36, 1123.

Hansson, G., Callenås, A., and Nelsson, C. 1993. Upconversion studies in laser pumped Tm,Ho:$YLiF_4$, *OSA Proceedings on Advanced Solid-State Lasers 15*, A. Pinto and T. Y. Fan, eds., p. 446. Optical Society of America, Washington, DC.

Harrison, J. and Martinsen, R. J. 1994a. Thermal modeling for mode-size estimation in microlasers with application to linear arrays in Nd:YAG and Tm,Ho:YLF. *IEEE J. Quantum Electron.*, 30, 2628.

Harrison, J. and Martinsen, R. J. 1994b. Operation of linear microlaser arrays near 1 μm, 2 μm, and 3 μm, In *OSA Proceedings on Advanced Solid-State Lasers 20*, T. Y. Fan and B. H. T. Chai, eds., p. 272. Optical Society of America, Washington, DC.

Harrison, J., Welford, D., and Moulton, P. F. 1989. Threshold analysis of pulsed lasers with application to a room-temperature Co:MgF_2 laser. *IEEE J. Quantum Electron.*, 25, 1708.

Harrison, J., Finch, A., Rines, D. M., Rines, G. A., and Moulton, P. F. 1991. Low-threshold, cw, all-solid-state Ti:Al_2O_3 laser. *Opt. Lett.*, 16, 581.

Haus, H. A. 1975a. Theory of mode locking with a fast saturable absorber. *J. Appl. Phys.*, 46, 3049.

Haus, H. A. 1975b. Theory of modelocking with a slow saturable absorber. *IEEE J. Quantum Electron.*, QE-11, 736.

Haus, H. A. 1976. Parameter ranges for cw passive modelocking. *IEEE J. Quantum Electron.*, QE-12, 169.

Haus, H. A., Fujimoto, J. G., and Ippen, E. P. 1991. Structures for additive pulse mode locking. *J. Opt. Soc. Am. B*, 8, 2068.

Haus, H. A., Fujimoto, J. G., and Ippen, E. P. 1992. Structures for additive pulse and Kerr lens mode locking. *IEEE J. Quantum Electron.*, 28, 2086.

Heine, F. F. and Huber, G. 1998. Tunable single frequency thulium-YAG microchip laser with external feedback. *Appl. Opt.*, 37, 3268.

Henningsen, T., Feitchner, J., and Melamed, N. 1971. Frequency doubling of 2.06-μm Ho-doped apatite laser output in Ag_3AsS_3. *IEEE J. Quantum Electron.*, 7, 248.

Hönninger, C., Paschotta, R., Morier-Genoud, F., Moser, M., and Keller, U. 1999. *Q*-switching stability limits of continuous-wave passive mode locking. *J. Opt. Soc. Am. B*, 16, 46.

Hönninger, C., Courjaud, A., Salin, F., Druon, F., Balembois, F., Georges, P., Brun, A. et al. 2000. 90-fs pulses from diode-pumped Yb:GdCOB laser. *Conf. Lasers and Electro-Optics, Tech. Dig.*, 119.

Hough, J., Hils, D., Rayman, M. D., Ma, L.-S., Hollberg, L., and Hall, J. L. 1984. Dye-laser frequency stabilization using optical resonators. *Appl. Phys. B*, 33, 179.

Ikesue, A., Kinoshita, T., Kamata, K., and Yoshida, K. 1995. Fabrication and optical properties of high-performance polycrystalline Nd:YAG ceramics for solid-state lasers. *J. Am. Ceram. Soc.*, 78, 1033.

Il'ichev, N. N., Kir'yanov, A. V., Gulyamova, E. S., and Pashinin, P. P. 1997. An influence of passive shutter Cr^{4+}:YAG latent anisotropy on output energy and polarisation characteristics of neodymium laser at passive *Q*-switching, In *OSA TOPS 10, Advanced Solid-State Lasers*, C. R. Pollock and W.R Bosenberg, eds., p. 137. Optical Society of America, Washington DC.

Ippen, E. P. 1994. Principles of passive mode locking. *Appl. Phys. B*, 58, 159.

Isyanova, Y. and Welford, D. 1993. 2.4-ns Pulse generation in a solid-state, passively *Q*-switched, laser-diode-pumped Nd:YAG laser, In *OSA Proceedings on Advanced Solid-State Lasers 15*, A. A. Pinto and T. Y. Fan, eds., p. 20. Optical Society of America, Washington, DC.

Isyanova, Y. and Welford, D. 1999. Temporal criterion for single-frequency operation of passively *Q*-switched lasers. *Opt. Lett.*, 24, 1035.

Isyanova, Y., Manni, J. G., Welford, D., Jaspers, M., and Russell, J. A. 2001. High-power, passively *Q*-switched microlaser-power amplifier system, In *OSA TOPS 50, Advanced Solid-State Lasers*, C. Marshall, ed., p. 186. Optical Society of America, Washington, DC.

Izawa, J., Nakijima, H., Hara, H., and Arimoto, Y. 2000. A tunable and longitudinal mode oscillation of a Tm,Ho,YLF microchip laser using an external etalon. *Opt. Commun.*, 180, 137.

Jabczynski, J. K., Kopczynski, K., Mierczyk, Z., Agnesi, A., Guandalini, A., and Reali, G. C. 2001. Application of V^{3+}:YAG crystals for *Q*-switching and mode-locking of 1.3-μm diode-pumped neodymium lasers. *Opt. Eng.*, 40, 2802.

Jaque, D., Capmany, J., Garca-Sole, J., Luo, Z. D., and Jiang, A. D. 1998. Continuous-wave laser properties of the self-frequency-doubling $YAl_3(BO_3)_4$:Nd crystal. *J. Opt. Soc. Am. B*, 15, 1656.

Jaseja, T. S., Javan, A., and Townes, C. H. 1963. Frequency stability of He–Ne masers and measurements of length. *Phys. Rev. Lett.*, 10, 165.

Jaspan, M. A., Welford, D., Xiao, G., and Bass, M. 2000. Atypical behavior of Cr:YAG passively *Q*-switched Nd:YVO_4 microlasers at high-pumping rates. *Conf. Lasers Electro-Optics, Tech. Dig.*, 454.

Jaspan, M. A., Welford, D., and Russell, J. A. 2004. Passively *Q*-switched microlaser performance in the presence of pump-induced bleaching of the saturable absorber. *Appl. Opt.*, 43, 2555.

Jeys, T. H. 1991. Suppression of laser spiking by intracavity harmonic generation. *Appl. Opt.*, 30, 1011.

Jiang, M., Sucha, G., Fermann, M. E., Jimenez, J., Harter, D., Degenais, M., Fox, S., and Hu, Y. 1999. Nonlinearly limited saturable-absorber mode locking of an erbium fiber laser. *Opt. Lett.*, 24, 1074.

Johnson, L. F., Dietz, R. E., and Guggenheim, H. J. 1963. Optical maser oscillation from Ni^{2+} in MgF_2 involving simultaneous emission of phonons. *Phys. Rev. Lett.*, 11, 318.

Johnson, L. F., Dietz, R. E., and Guggenheim, H. J. 1964. Spontaneous and stimulated emission from Co^{2+} ions in MgF_2 and ZnF_2. *Appl. Phys. Lett.*, 5, 21.

Johnson, L. F., Guggenheim, H. J., and Thomas, R. A. 1966. Phonon-terminated optical masers. *Phys. Rev.*, 149, 179.

Johnson, L. F., Guggenheim, H. J., and Thomas, R. A. 1967. Phonon-terminated coherent emission from V^{2+} ions in MgF_2. *J. Appl. Phys.*, 38, 4837.

Kafalas, P., Masters, J. I., and Murray, E. M. E. 1964. Photosensitive liquid used as a nondestructive passive *Q*-switch in a ruby laser. *J. Appl. Phys.*, 35, 2349.

Kajave, T. T. and Gaeta, A. L. 1996. *Q* switching of a diode-pumped Nd:YAG laser with GaAs. *Opt. Lett.*, 21, 1244.

Kaminskii, A. A., Eichler, H. J., Ueda, K., Klassen, N. V., Redkin, B. S., Li, L. E., Findeisen, J. et al. 1999. Properties of Nd^{3+}-doped and undoped tetragonal $PbWO_4$, $NaY(WO_4)_2$, $CaWO_4$ and undoped monoclinic $ZnWO_4$ and $CdWO_4$ as laser-active and stimulated Raman scattering-active crystals. *Appl. Opt.*, 38, 4533.

Kane, T. J. 1990. Intensity noise in diode-pumped single-frequency Nd:YAG lasers and its control by electronic feedback. *IEEE Photon. Technol. Lett.*, 2, 244.

Kane, T. J. and Byer, R. L. 1985. Monolithic, unidirectional single-mode Nd:YAG ring laser. *Opt. Lett.*, 10, 65.

Kane, T. J. and Kubo, T. S. 1991. Diode-pumped single-frequency lasers and *Q*-switched lasers using Tm:YAG and Tm,Ho:YAG, In *OSA Proceedings on Advanced Solid-State Lasers 6*, H. P. Jenssen and G. Dubé., eds., p. 136. Optical Society of America, Washington, DC.

Kane, T. J., Nilsson, A. C., and Byer, R. L. 1987. Frequency stability and offset locking of a laser-diode-pumped Nd:YAG monolithic nonplanar ring oscillator. *Opt. Lett.*, 12, 175.

Kärtner, F. X., Brovelli, L. R., Kopf, D., Kamp, M., Calasso, I., and Keller, U. 1995. Control of solid state laser dynamics by semiconductor devices. *Opt. Eng.*, 34, 2024.

Keller, U. 1996. Modelocked and Q-switched solid-state lasers using semiconductor saturable absorbers. *Conf. Proceedings of 9th Annual Meeting IEEE Lasers and Electro-Optics Society*, 1, 50.

Keller, U. 1998. Semiconductor nonlinearities for solid-state laser modelocking and Q-switching, In *Nonlinear Optics in Semiconductors 59*, A. Kost and E. Garmire, eds., p. 211. Academic, Boston, MA.

Keller, U., Miller, D. A. B., Boyd, G. D., Chiu, T. H., Ferguson, J. F., and Asom, M. T. 1992. Solid-state low-loss intracavity saturable absorber for Nd:YLF lasers: an antiresonant semiconductor Fabry-Pérot saturable absorber. *Opt. Lett.*, 17, 505.

Keller, U., Weingarten, K. J., Kärtner, F. X., Kopf, D., Braun, B., Jung, I. D., Fluck, R., Hönninger, C., Matuschek, N., and Aus der Au, J. 1996. Semiconductor saturable absorber mirrors (SESAMs) for femtosecond to nanosecond pulse generation in solid-state lasers. *IEEE J. Sel. Topics Quantum Electron.*, 2, 435.

Kemp, A. J., Conroy, R. S., Friel, G. J., and Sinclair, B. D. 1999. Guiding effects in Nd:YVO_4 microchip lasers operating well above threshold. *IEEE J. Quantum Electron.*, 35, 675.

Kemp, A. J., Stormont, B., Agate, B., Brown, C. T. A., Keller, U., and Sibbett, W. 2001. Gigahertz repetition-rate from directly diode-pumped femtosecond Cr:LiSAF laser. *Electron. Lett.*, 37, 1457.

Kennedy, C. J. and Barry, J. D. 1974. Stability of an intracavity frequency-doubled Nd:YAG laser. *IEEE J. Quantum Electron.*, QE-10, 596.

Keszenheimer, J. A., Balboni, E. J., and Zayhowski, J. J. 1992. Phase locking of 1.32-μm microchip lasers through the use of pump-diode modulation. *Opt. Lett.*, 17, 649.

Keyes, R. J. and Quist, T. M. 1964. Injection luminescent pumping of CaF_2:U^{3+} with GaAs diode lasers. *Appl. Phys. Lett.*, 4, 50.

Kintz, G. J. and Baer, T. 1990. Single-frequency operation in solid state laser materials with short absorption depths. *IEEE J. Quantum Electron.*, 26, 1457.

Kogelnik, H. 1965. On the propagation of Gaussian beams of light through lenslike media including those with a loss or gain variation. *Appl. Opt.*, 4, 1562.

Krainer, L., Paschotta, R., Spühler, G. J., Moser, M., and Keller, U. 1999. 29 GHz modelocked miniature Nd:YVO_4 laser. *Electron. Lett.*, 35, 1160.

Krainer, L., Paschotta, R., Moser, M., and Keller, U. 2000. 77 GHz soliton modelocked Nd:YVO_4 laser. *Electron. Lett.*, 36, 1846.

Krainer, L., Paschotta, R., Lecomte, S., Moser, M., Weingarten, K. J., and Keller, U. 2002a. Compact Nd:YVO_4 lasers with pulse repetition rates up to 160 GHz. *IEEE J. Quantum Electron.*, 38, 1331.

Krainer, L., Paschotta, R., Spühler, G. J., Klimov, I., Teisset, C. Y., Weingarten, K. J., and Keller, U. 2002b. Tunable picosecond pulse-generating laser with repetition rate exceeding 10 GHz. *Electron. Lett.*, 38, 225.

Krühler, W. W., Plättner, R. D., and Stetter, W. 1979. CW oscillation at 1.05 and 1.32 μm of LiNd$(PO_3)_4$ lasers in external resonators and in resonators with directly applied mirrors. *IEEE J. Quantum Electron.*, QE-14, 840.

Krupke, W. F. 2000. Ytterbium solid-state lasers. The first decade. *IEEE Sel. Topics Quantum Electron.*, 6, 1287.

Kück, S., Peterman, K., and Huber, G. 1991. Spectroscopic investigation of the Cr^{4+}-center in YAG, In *OSA Proceedings on Advanced Solid State Lasers 10*, G. Dúbe and L. Chase, eds., p. 92. Optical Society of America, Washington, DC.

Kück, S., Schepler, K. L., Peterman, K., and Huber, G. 1996. Excited state absorption and stimulated emission measurements of Cr^{4+}-doped $Y_3Al_5O_{12}$, $Y_3Sc_{0.9}Al_{4.1}O_{12}$, and $CaY_2Mg_2Ge_3O_{12}$, In *OSA TOPS 1, Advanced Solid State Lasers*, S. A. Payne and C. Pollock, eds., p. 94. Optical Society of America, Washington, DC.

Kuleshov, N. V., Podlipensky, A. V., Yumashev, K. V., Kretschmann, H. M., and Huber, G. 2000. V:YAG saturable absorber as a Q-switch for diode-pumped Nd:YAG-lasers at 1.44 μm and 1.34 μm. *Conf. Lasers Electro-Optics, Tech. Dig.*, 228.

Kuo, Y. K., Birnbaum, M., and Chen, W. 1994. Ho:YLiF$_4$ saturable absorber *Q*-switch for the 2 μm Tm,Cr:Y$_3$Al$_5$O$_2$ laser. *Appl. Phys. Lett.*, 65, 3060.

Kuo, Y. K., Huang, M. F., and Birnbaum, M. 1995. Tunable Cr^{4+}:YSO *Q*-switched Cr:LiCAF laser. *Appl. Phys. Lett.*, 65, 3060.

Lagatsky, A. A., Abdolvand, A., and Kuleshov, N.V. 2000a. Diode-pumped passively *Q*-switched Yb:KGW laser with V:YAG saturable absorber. *Lasers and Electro-Optics Europe. Conf. Dig.*, Paper CTuK36.

Lagatsky, A. A., Abdolvand, A., and Kuleshov, N. V. 2000b. Passive *Q* switching and self frequency Raman conversion in a diode-pumped Yb:KGd(WO$_4$)$_2$ laser. *Opt. Lett.*, 25, 616.

Laporta, P., Taccheo, S., Longhi, S., and Svelto, O. 1993. Diode-pumped microchip Er-Yb laser. *Opt. Lett.*, 18, 1232.

Lax, M. 1966. Quantum noise V: Phase noise in a homogeneously broadened maser, In *Physics of Quantum Electronics*, P. K. Kelley, B. Lax, and P. E. Tannenwald, eds., p. 735. McGraw-Hill, New York.

Leburn, C. G., Lagatsky, A. A., Brown, C. T. A., and Sibbett, W. 2003. Three-element mode-locked femtosecond Cr^{4+}:YAG laser operating up to 3.6-GHz repetition rate, *Conf. Lasers and Electro-Optics, Tech. Dig.*, Paper CThF6.

Leilabady, P. A., Anthon, D. W., and Gullicksen, P. O. 1992. Single-frequency Nd:YLF cube lasers pumped by laser diode arrays. *Conf. Lasers and Electro-Optics, Tech. Dig.*, 54.

Liu, H., Zhou, S. H., and Chen, Y. C. 1998. High-power monolithic unstable-resonator solid-state laser. *Opt. Lett.*, 23, 451.

Liu, J., Shen, D., Tam, S. C., and Lam, T. L. 2001. Modeling pulse shape of *Q*-switched lasers. *IEEE J. Quantum Electron.*, 37, 888.

Liu, J., Ozygus, B., Yang, S., Erhard, J., Seelig, U., Ding, A., Weber, H. et al. 2003. Efficient passive *Q*-switching operation of a diode-pumped Nd:GdVO$_4$ laser with a Cr^{4+}:YAG saturable absorber. *J. Opt. Soc. Am. B*, 20, 652.

Lovold, S., Moulton, P. F., Killinger, D. K., and Menyuk, N. 1985. Frequency tuning characteristics of a *Q*-switched Co:MgF$_2$ laser. *IEEE J. Quantum Electron.*, QE-21, 202.

Lu, J., Lu, J., Murai, T., Takaichi, K., Uematsu, T., Ueda, K., Yagi, H., Yanagitani, T., and Kaminskii, A. A. 2001. Nd^{3+}:Y$_2$O$_3$ ceramic laser. *Jpn. J. Appl. Phys.*, 40, L1277.

Lu, J., Ueda, K., Yagi, H., Yanagitani, T., Akiyama, Y., and Kaminskii, A. A. 2002. Neodymium doped yttrium aluminum garnet (Y$_3$Al$_5$O$_{12}$) nanocrystalline ceramics—a new generation of solid state and optical materials. *J. Alloys Compounds*, 341, 220.

Lucas-Leclin, G., Balembois, F., Georges, P., Brun, A., Mougel, F., Aka, G., Vivien, D. et al. 2000a. Microchip lasers based on Nd:Ca$_4$GdO(BO$_3$)$_3$. *Lasers and Electro-Optics Europe. Conf. Dig.*, Paper CTuK55.

Lucas-Leclin, G., Augé, F., Auzanneau, S. C., Balembois, F., Georges, P., Brun, A., Mougel, F., Aka, G., and Vivien, D. 2000b. Diode-pumped self-frequency-doubling Nd:GdCa$_4$O(BO$_3$)$_3$ lasers: towards green microchip lasers. *J. Opt. Soc. Am. B*, 17, 1526.

Lupei, A., Lupei, V., Rogobete, L., Osiac, E., Gheorghe, L., and Petraru, A. 2000. Inhomogeneous broadening effects in optical spectra of Nd^{3+} in CNGG and CLNGG, *Lasers and Electro-Optics Europe, Conf. Dig.*, Paper CFH2.

MacKinnon, N. and Sinclair, B. D. 1992. Pump power induced cavity stability in lithium neodymium tetraphosphate (LNP) microchip lasers. *Opt. Commun.*, 94, 281.

MacKinnon, N., Norrie, C. J., and Sinclair, B. D. 1994. Laser-diode-pumped, electro-optically tunable Nd:MgO:LiNbO$_3$ microchip laser. *J. Opt. Soc. Am. B*, 11, 519.

Maiman, T. H. 1960. Stimulated optical radiation in ruby masers. *Nature*, 187, 493.

Maker, G. T. and Ferguson, A. I. 1990. Ti:sapphire laser pumped by a frequency-doubled diode-pumped Nd:YLF laser. *Opt. Lett.*, 15, 375.

Malyarevich, A. M., Denisov, I. A., Yumashev, K. V., Dymshits, O. S., and Zhilin, A. A. 2001a. Co^{2+}-doped glass ceramic as saturable absorber *Q* switch for 1.54 μm Er-glass laser, *OSA TOPS 50, Advanced Solid-State Lasers*, C. Marshall, ed., p. 241. Optical Society of America, Washington, DC.

Malyarevich, A. M., Savitski, V. G., Prokoshin, P. V., Yumashev, K. V., and Lipovskii, A. A. 2001b. Passive Q-switch operation of PbSe-doped glass at 2.1 μm. *SPIE*, 4350, 32.

Man, C. N. and Brillet, A. 1984. Injection locking of argon-ion lasers. *Opt. Lett.*, 9, 333.

Mandeville, W., Dindorf, K. M., and Champigny, N. E. 1996. Characterization of passively Q-switched microchip lasers for laser radar. *SPIE*, 2748, 358.

Marshall, C. D., Speth, J. A., Payne, S. A., Krupke, W. F., Quarles, G. J., Castillo, V., and Chai, B. H. T. 1994. Ultraviolet laser emission properties of Ce^{3+}-doped $LiSrAlF_6$ and $LiCaAlF_6$. *J. Opt. Soc. Am. B*, 11, 2054.

Marshall, C. D., Smith, L. K., Beach, R. J., Emanuel, M. A., Schaffers, K. I., Skidmore, J., Payne, S. A., and Chai, B. H. T. 1996. Diode-pumped ytterbium-doped $Sr_5(PO_4)_3F$ laser performance. *IEEE J. Quantum Electron.*, 32, 650.

Matthews, D. G., Boon, J. R., Conroy, R. S., and Sinclair, B. D. 1996. A comparative study of diode pumped microchip laser materials: Nd-doped YVO_4, YOS, SFAP, and SVAP. *J. Mod. Opt.*, 43, 1079.

Matveets, Y. A., Nikogosyan, D. N., Kabelka, V., and Piskarskas, A. 1978. Efficient second harmonic generation in a KDP crystal pumped with picosecond YAG:Nd^{3+} laser pulses of 0.5 Hz repetition frequency. *Sov. J. Quantum Electron.*, 8, 386.

McClung, F. J. and Hellwarth, R. W. 1962. Giant optical pulsations from ruby. *J. Appl. Phys.*, 33, 828.

McKay, J. B., Roh, W. B., and Schepler, K. L. 2002. Extended mid-IR tuning of a Cr^{2+}:CdSe laser, In *OSA TOPS 68, Advanced Solid-State Lasers*, M. E. Fermann and L. R. Marshall, eds., p. 371. Optical Society of America, Washington, DC.

Mermilliod, N., François, B., and Wyon, Ch. 1991. $LaMgAl_{11}O_{19}$:Nd microchip laser. *Appl. Phys. Lett.*, 59, 3520.

Meyn, J. P., Jensen, T., and Huber, G. 1994. Spectroscopic properties and efficient diode-laser-pumped laser operation of neodymium-doped lanthanum scandium borate. *IEEE J. Quantum Electron.*, 30, 913.

Mirov, S. B., Fedorov, V. V., Graham, K., Moskalev, I. S., Badikov, V. V., and Panyutin, V. 2002. Erbium fiber laser-pumped continuous-wave microchip Cr^{2+}:ZnS and Cr^{2+}:ZnSe lasers. *Opt. Lett.*, 27, 909.

Moncorge, R., Cormier, G., Simkin, D. J., and Capobianco, J. A. 1991. Fluorescence analysis of chromium-doped fosterite (Mg_2SiO_4). *IEEE J. Quantum Electron.*, 27, 114.

Mougel, F., Aka, G., Benitez, J. M., Kahn-Harari, A., Salin, F., and Pelenc, D. 1998. Crystal growth and nonlinear optical characterization of $Ca_4YO(BO_3)_3$. *Eur. Quantum Electron. Conf., Conf. Dig.*, 19.

Moulton, P. F. 1982. Ti:Al_2O_3—a new tunable solid state laser. *M.I.T. Lincoln Lab., Solid State Res. Rep.*, DTIC AD-A124305/4, 15.

Moulton, P. F. 1985. An investigation of the Co:MgF_2 laser system. *IEEE J. Quantum Electron.*, QE-21, 1582.

Moulton, P. F. 1986. Spectroscopic and laser characteristics of Ti:Al_2O_3. *J. Opt. Soc. Am. B*, 3, 125.

Moulton, P. F. 1992. Tunable solid-state lasers. *Proc. IEEE*, 80, 348.

Muller III, C. H., Lowenthal, D. D., Kangas, K. W., Hamil, R. A., and Tisone, G. C. 1988. 2.0-J Ti:sapphire laser oscillator. *Opt. Lett.*, 13, 380.

Nabors, C. D., Ochoa, J., Fan, T. Y., Sanchez, A., Choi, H., and Turner, G. 1994. 1.9-μm-diode-laser-pumped 2.1-μm Ho:YAG laser. *Conf. Lasers Electro-Optics, Tech. Dig.*, 172.

Nettleton, J. E., Schilling, B. W., Barr, D. N., and Lei, J. S. 2000. Monoblock laser for a low-cost, eyesafe, microlaser range finder. *Appl. Opt.*, 39, 2428.

Newman, R. 1963. Excitation of Nd^{3+} fluorescence in $CaWO_4$ by recombination radiation in GaAs. *J. Appl. Phys.*, 34, 437.

Nilsson, A., Gustafson, E. K., and Byer, R. L. 1989. Eigenpolarization theory of monolithic nonplanar ring oscillators. *IEEE J. Quantum Electron.*, 25, 767.

Noginov, M. A., Loutts, G. B., Bonner, C. E., Taylor, S., Stefanos, S., Wynne, R. M., and Lalsley, B. A. 2000. Crystal growth and characterization of a new laser material, Nd:$Ba_5(PO_4)_3Cl$. *J. Opt. Soc. Am. B*, 17, 1329.

Noginov, M. A., Loutts, G. B., Lucas, B. D., Fifer, D., Higgins, P. T., Truong, P. T., Nogin, N., Barnes, N. P., and Kuck, S. 2001. Development of Nd-doped solid-state laser materials for 944-nm operation. *IEEE J. Quantum Electron.*, 37, 469.

Ohtsu, M. and Kotajima, S. 1985. Linewidth reduction of a semiconductor laser by electrical feedback. *IEEE J. Quantum Electron.*, QE-21, 1905.

Ostroumov, V. G., Huber, G., Zagumennyi, A. I., Zavartsev, Y. D., Studenikin, P. A., and Shcherbakov, I. A. 1996. Spectroscopic properties and lasing of $Gd_{0.5}La_{0.5}VO_4$ crystals. *Opt. Commun.*, 124, 63.

Owyoung, A. and Esherick, P. 1987. Stress-induced tuning of a diode-laser-excited monolithic Nd:YAG laser. *Opt. Lett.*, 12, 999.

Page, R. H., Schaffers, K. I., DeLoach, L. D., Wilke, G. D., Patel, F. D., Tassano, J.B. Jr., Payne, S. A., Krupke, W. F., Chen, K.-T., and Burger, A. 1997. Cr^{2+}-doped zinc chalcogenides as efficient widely tunable mid-infrared lasers. *IEEE J. Quantum Electron.*, 33, 609.

Palombo, F. K., Matthews, S., Sheldrake, S., and Kapps, D. 1993. Determination of the effective lower level lifetime for Nd:YLF and Nd:YAG through experimental measurements and computer modeling, *OSA Proceedings on Advanced Solid-State Lasers 15*, A. A. Pinto and T. Y. Fan, eds., p. 78. Optical Society of America, Washington, DC.

Paschotta, R. and Keller, U. 2001. Passive mode locking with slow saturable absorbers. *Appl. Phys. B*, 73, 653.

Patel, F. D. and Beach, R. J. 2001. New formalism for the analysis of passively *Q*-switched laser systems. *IEEE J. Quantum Electron.*, 37, 707.

Payne, S. A., Chase, L. L., Newkirk, H. W., Smith, L. K., and Krupke, W. F. 1988. $LiCaAlF_6:Cr^{3+}$: A promising new soild-state laser material. *IEEE J. Quantum Electron*, 24, 2243.

Payne, S. A., Chase, L. L., Smith, L. K., Kway, W. L., and Newkrik, H. W. 1989. Laser performance of $LiSrAlF_6:Cr^{3+}$. *J. Appl. Phys.*, 66, 1051.

Payne, S. A., Smith, L. K., DeLoach, L. D., Kway, W. L., Tassano, J. B., and Krupke, W. F. 1994. Laser, optical and thermomechanical properties of doped fluorapatite. *IEEE J. Quantum Electron.*, 30, 170.

Penderson, C. and Skettrup, T. 1996. Laser modes and threshold conditions in N-mirror resonators. *J. Opt. Soc. Am. B*, 13, 926.

Peterson, P. and Gavrielides, A. 1999. Pulse train characteristics of a passively *Q*-switched microchip laser. *Opt. Express*, 5, 149.

Petricevic, V., Gayen, S. K., and Alfano, R. R. 1988a. Laser action in chromium-activated fosterite. *Appl. Phys. Lett.*, 52, 1040.

Petricevic, V., Gayen, S. K., and Alfano, R. R. 1988b. Laser action in chromium-activated fosterite for near-infrared excitation: Is Cr^{4+} the lasing ion? *Appl. Phys. Lett.*, 53, 2590.

Pinto, J. F., Esterowitz, L., and Quarles, G. J. 1995. High performance $Ce^{3+}:LiSrAlF_6/LiCaAlF_6$ UV lasers with extended tunability. *Electron. lett.*, 31, 2009.

Plaessmann, H., Yamada, K. S., Rich, C. E., and Grossman, W. M. 1994. Subnanosecond pulse generation from diode-pumped acousto-optically *Q*-switched solid-state lasers. *Appl. Opt.*, 32, 6616.

Pujol, M. C., Massons, J., Aguilo, M., Diaz, F., Rico, M., and Zaldo, C. 2002. Emission cross sections and spectroscopy of Ho^{3+} laser channels in $KGd(WO_4)_2$ single crystal. *IEEE J. Quantum Electron.*, 38, 93.

Rapaport, A., Moteau, O., Bass, M., Boatner, L. A., and Deka, C. 1999. Optical spectroscopy and lasing properties of neodymium-doped lutetium orthophosphate. *J. Opt. Soc. Am. B*, 16, 911.

Sarkura, N., Liu, Z., Izumida, S., Dubinskii, M. A., Abdulsabirov, R. Y., and Korableva, S. L. 1998. All-solid-state tunable ultraviolet subnanosecond laser with direct pumping by the fifth harmonic of a Nd:YAG laser. *Appl. Opt.*, 37, 6446.

Schawlow, A. L. and Townes, C. H. 1958. Infrared and optical masers. *Phys. Rev.*, 12, 1940.

Scheps, R. 1991. $Cr:LiCaAlF_6$ laser pumped by visible laser diodes. *IEEE J. Quantum Electron.*, 27, 1968.

Scheps, R. 1992. Laser-diode-pumped $Cr:LiSrGaF_6$ laser. *IEEE Photon. Technol. Lett.*, 4, 548.

Scheps, R., Gately, B. M., Myers, J. F., Krasinski, J. S., and Heller, D. F. 1990. Alexandrite laser pumped by semiconductor lasers. *Appl. Phys. Lett.*, 56, 2288.

Scheps, R., Myers, J. F., Serreze, H. B., Rosenberg, A., Morris, R. C., and Long, M. 1991. Diode-pumped Cr:LiSrAlF$_6$ laser. *Opt. Lett.*, 16, 820.

Scheps, R., Myers, J. F., and Payne, S. A. 1993. End-pumped Yb-doped fluorapatite laser. *IEEE Photon. Technol. Lett.*, 5, 1285.

Schibli, T. R., Thoen, E. R., Kärtner, F. X., and Ippen, E. P. 2000. Suppression of Q-switched mode-locking and break-up into multiple pulses by inverse saturable absorption. *Appl. Phys. B*, 70 (suppl.), S41.

Schibli, T. R., Morgner, U., and Kärtner, F. X. 2001a. Control of Q-switched mode-locking by active feedback. *Opt. Lett.*, 26, 148.

Schibli, T. R., Kremp, T., Morgner, U., Kärtner, F. X., Butendeich, R., Schwarz, J., Schweizer, H., Scholz, F., Hetzler, J., and Wegener, M. 2001b. Continuous-wave and Q-switched mode locking of Cr^{4+}:YAG microchip lasers. *Opt. Lett.*, 26, 941.

Schibli, T. R., Robinson, K. E., Morgner, U., Mohr, S., Kopf, D., and Kärtner, F. X. 2002. Control of Q-switching instabilities in passively mode-locked lasers, In *OSA TOPS 68, Advanced Solid-State Lasers*, M. E. Fermann and L. R. Marshall, eds., p. 498. Optical Society of America, Washington, DC.

Schulz, P. A. 1990. Fast electro-optic wavelength selection and frequency modulation in solid state lasers. *Lincoln Lab. J.*, 3, 463.

Schulz, P. A. and Henion, S. R. 1991a. Frequency-modulated Nd:YAG laser. *Opt. Lett.*, 16, 578.

Schulz, P. A. and Henion, S. R. 1991b. 5-GHz mode locking of a Nd:YLF laser. *Opt. Lett.*, 16, 1502.

Sennaroglu, A. and Yilmaz, M. B. 1997. Experimental and theoretical study of thermal loading in chromium-doped YAG saturable absorbers, In *OSA TOPS 10, Advanced Solid-State Lasers*, C. R. Pollock and W. R. Bosenberg, eds., p. 132. Optical Society of America, Washington DC.

Shimony, Y., Burshtein, Z., Baranga, B. A., Kalisky, Y., and Strauss, M. 1996. Repetitive Q-switching of a cw Nd:YAG laser using Cr^{4+}:YAG saturable absorbers. *IEEE J. Quantum Electron.*, 32, 305.

Siegman, A. J. 1986. *Lasers*. University Science Books. Mill Valley. CA.

Smith, L. K., Payne, S. A., Kway, W. L., Chase, L. L., and Chai, B. H. T. 1992. Investigation of the laser properties of Cr^{3+}:LiSrGaF$_6$. *IEEE J. Quantum Electron.*, 28, 2612.

Smith, L. K., Payne, S. A., Krupke, W. F., DeLoach, L. D., Morris, R., O'Dell, E. W., and Nelson, D. J. 1993. Laser emission from the transition-metal compound LiSrCrF$_6$. *Opt. Lett.*, 18, 200.

Soffer, B. H. 1964. Giant pulse laser operation by a passive, reversibly bleachable absorber. *J. Appl. Phys.*, 35, 2251.

Sooy, W. R. 1965. The natural selection of modes in a passive Q-switched laser. *Appl. Phys. Lett.*, 7, 66.

Sorokin, P. P., Luzzi, J. J., Lankard, J. R., and Petit, G. D. 1964. Ruby laser Q-switching elements using phthalocyanine molecules in solution. *IBM J. Res. Div.*, 8, 182.

Spühler, G. J., Paschotta, R., Fluck, R., Braun, B., Moser, M., Zhang, G., Gini, E., and Keller, U. 1999. Experimentally confirmed design guidelines for passively Q-switched microchip lasers using semiconductor saturable absorbers. *J. Opt. Soc. Am. B*, 16, 376.

Spühler, G. J., Dymott, M., Klimov, I., Luntz, G., Baraldi, L., Kilburn, I., Crosby, P. et al. 2002. 40 GHz pulse generating source with less than 350 fs timing jitter. *Electron. Lett.*, 38, 1031.

Spühler, G. J., Golding, P. S., Krainer, L., Kilburn, I. J., Crosby, P. A., Brownell, M., Weingarten, K. J. et al. 2003. Multi-wavelength source with 25 GHz channel spacing tunable over C-band. *Electron. Lett.*, 39, 778.

Statz, H., DeMars, G. A., and Wilson, D. T. 1965. Problem of spike elimination in lasers. *J. Appl. Phys.*, 36, 1510.

Stone, D. H. 1992. Effects of axial nonuniformity in modeling Q-switched lasers. *IEEE J. Quantum Electron.*, 28, 1970.

Stoneman, R. C. and Esterowitz, L. 1992. Efficient resonantly pumped 2.8-μm Er^{3+}:GSGG laser. *Opt. Lett.*, 17, 816.

Storm, M. E. 1991. Spectral performance of monolithic holmium and thulium lasers, In *OSA Proceedings on Advanced Solid-State Lasers 5*, M. L. Shand and H. P. Jenssen, eds., p. 186. Optical Society of America, Washington, DC.

Storm, M. E. and Rohrbach, W. W. 1989. Single-longitudinal-mode lasing of Ho:Tm:YAG at 2.091 μm. *Appl. Opt.*, 28, 4965.

Stultz, R. D., Camargo, M. B., and Birnbaum, M. 1995. Passive *Q*-switching at 1.53 μm using divalent uranium ions in calcium fluoride. *J. Appl. Phys.*, 78, 2959.

Sutherland, J. M., Ruan, S., Mellish, R., French, P. M. W., and Taylor, J. R. 1995. Diode-pumped single-frequency Cr:LiSAF microlaser. *Opt. Commun.*, 113, 458.

Sutter, D. H., Steinmeyer, G., Gallman, L., Matuschel, N., Morier-Genoud, F., Keller, U., Scheuer, V., Angelow, G., and Tschudi, T. 1999. Semiconductor saturable-absorber mirror-assisted Kerr-lens mode-locked Ti:sapphire laser producing pulses in the two-cycle regime. *Opt. Lett.*, 24, 631.

Szabo, A. and Stein, R. A. 1965. Theory of laser giant pulsing by saturable absorber. *J. Appl. Phys.*, 36, 1562.

Taira, T., Mukai, A., Nozawa, Y., and Kobayashi, T. 1991. Single-mode oscillation of laser-diode-pumped Nd:YVO_4 microchip lasers. *Opt. Lett.*, 16, 1955.

Tang, C. L., Statz, H., and DeMars, G. 1963. Spectral output and spiking behavior of solid-state lasers. *J. Appl. Phys.*, 34, 2289.

Thoen, E. R., Grein, M. E., Koontz, E. M., Ippen, E. P., Haus, H. A., and Kolodziejski, L. A. 2000. Stabilization of an active harmonically mode-locked fiber laser using two-photon absorption. *Opt. Lett.*, 25, 948.

Thony, P., Fulbert, L., Besesty, P., and Ferrand, B. 1999. Laser radar using a 1.55-μm passively *Q*-switched microchip laser. *SPIE*, 3707, 616.

Tomaru, T. 2001. Two-element-cavity femtosecond Cr^{4+}:YAG laser operating at a 2.6-GHz repetition rate. *Opt. Lett.*, 26, 1439.

Tomaru, T. 2003. Mode-locking operating points of a three-element-cavity femtosecond Cr^{4+}:YAG laser. *Opt. Commun.*, 225, 163.

Tomaru, T. and Petek, H. 2000. Femtosecond Cr^{4+}:YAG laser with an L-fold cavity operating at a 1.2-GHz repetition rate. *Opt. Lett.*, 25, 584.

Tsou, Y., Garmire, E., Chen, W., Birnbaum, M., and Asthana, R. 1993. Passive *Q* switching of Nd:YAG lasers by use of bulk semiconductors. *Opt. Lett.*, 18, 1514.

Valentine, G. J., Kemp, A. J., Birkin, D. J. L., Burns, D., Balembois, F., Georges, P., Bernas, H. et al. 2000. Femtosecond Yb:YCOB laser pumped by narrow-stripe laser diode and passively mode-locked using ion-implanted saturable-absorber mirror. *Electron. Lett.*, 36, 1621.

Valentine, G. J., Burns, D., Ferguson, A. I., and Bente, E. 2002. High power, passively mode-locked quasi-cw Nd:YLF laser with feedback for *Q*-switch and spike supression. *Conf. Lasers Electro-Optics, Tech. Dig.*, 150.

Verdun, H. R., Thomas, L. M., Andrauskas, D. M., and McCollum, T. 1988. Chromium-doped fosterite laser pumped with 1.06 μm radiation. *Appl. Phys. Lett.*, 53, 2593.

Vieira, A. J. C., Herczfeld, P. R., and Contarino, V. M. 1997. 20 GHz mode-locked Nd:$LiNbO_3$ microchip laser. *Conf. Lasers Electro-Optics, Tech. Dig.*, 141.

Vieira, A. J. C., Herczfeld, P. R., Rosen, A., Ermold, M., Funk, E. E., Jemison, W. D., and Williams, K. J. 2001. A mode-locked microchip laser optical transmitter for fiber radio. *IEEE Trans. Microwave Theory Tech.*, 49, 1882.

Wagner, W. G. and Lengyel, B. A. 1963. Evolution of the giant pulse in a laser. *J. Appl. Phys.*, 34, 2040.

Walling, J. C., Heller, D. F., Samelson, H., Harter, D. J., Pete, J. A., and Morris, R. C. 1985. Tunable alexandrite lasers: Development and performance. *IEEE J. Quantum Electron.*, QE-21, 1568.

Wallmeroth, K. 1990. Monolithic integrated Nd:YAG laser. *Opt. Lett.*, 15, 903.

Walsh, B. M., Barnes, N. P., Hutcheson, R. L., and Equall, R. W. 2001. Compositionally tuned 0.94-μm laser: a comparative laser material study and demonstration of 100-mJ *Q*-switched lasing at 0.946 and 0.9441 μm. *IEEE J. Quantum Electron.*, 37, 1203.

Wang, P., Zhou, S.-H., Lee, K. K., and Chen, Y. C. 1995. Picosecond laser pulse generation in a monolithic self-*Q*-switched solid-state laser. *Opt. Commun.*, 114, 439.

Weber, H. P. 1975. Review of Nd-pentaphosphate lasers. *Opt. Quantum Electron.*, 7, 431.

Welford, D. 2001. Passively *Q*-switched lasers: short pulse duration, single frequency sources. LEOS Annual Meeting, San Diego, CA, Paper MP1.

Welford, D. and Moulton, P. F. 1988. Room temperature operation of a Co:MgF_2 laser. *Opt. Lett.*, 13, 975.

Winzer, G., Möckel, P. G., Oberbacher, R., and Vité, L. 1976. Laser emission from polished NdP_5O_{14} crystals with directly applied mirrors. *Appl. Phys.*, 11, 121.

Winzer, G., Möckel, P. G., and Krühler, W. W. 1978. Laser emission from miniature $NdAl_3(BO_3)_4$ crystals with directly applied mirrors. *IEEE J. Quantum Electron.*, QE-14, 840.

Wu, R., Myers, J. D., Myers, M. J., Denker, B. I., Galagan, B. I., Sverchkov, S. E., Hutchinson, J. A., and Trussel, W. 2000. Co^{2+}:$MgAl_2O_4$ crystal passive *Q*-switch performance at 1.34, 1.44, and 1.54 micron. *SPIE*, 3929, 42.

Xiao, G. and Bass, M. 1997. A generalized model for passively *Q*-switched lasers including excited state absorption in the saturable absorber. *IEEE J. Quantum Electron.*, 33, 41.

Xiao, G. and Bass, M. 1998. Additional experimental confirmation of the predictions of a model to optimize passively *Q*-switched lasers. *IEEE J. Quantum Electron.*, 34, 1142.

Yao, G., Lee, K. K., Chen, Y. C., and Zhou, S. 1994. Characteristics of transverse mode of diode-pumped self-*Q*-switched microchip laser, In *OSA Proc. Advanced Solid-State Lasers 20*, T. Y. Fan and B. H. T. Chai, eds., p. 28. Optical Society of America, Washington, DC.

Yokozawa, T., Izawa, J., and Hara, H. 1998. Mode control of a Tm:YLF microchip laser by a multiple resonator. *Opt. Commun.*, 145, 98.

Yumashev, K. V., Psonov, N. N., Denisov, I. A., Mikhailov, V. P., and Moncorge, R. 1998. Nonlinear spectroscopy and passive *Q*-switching operation of Cr^{4+}-doped $SrGd_4(SiO_4)_3O$ and $CaGd_4(SiO_4)_3O$ crystals. *J. Opt. Soc. Am. B*, 15, 1707.

Yumashev, K. V., Denisov, I. A., Posnov, N. N., Mikhailov, V. P., Moncorge, R., Vivien, D., Ferrand, B., and Guyot, Y. 1999. Nonlinear spectroscopy and passive *Q*-switching operation of a Co^{2+}:$LaMgAl_{11}O_{19}$ crystal. *J. Opt. Soc. Am. B*, 16, 2189.

Yumashev, K. V., Prokoshin, P. V., Zolotovskaya, S. A., Gurin, V. S., Prokopenko, V. B., and Alexeenko, A. A. 2001. Copper selenide-doped glass saturable absorbers for solid-state lasers of 1.0–1.5-μm region, In *OSA TOPS 50, Advanced Solid-State Lasers*, C. Marshall, ed., p. 77. Optical Society of America, Washington, DC.

Zagumennyi, A. I., Zavartsev, T. D., Studenikin, P. A., Shcherbakov, I. A., Umyskov, F., Popov, P. A., and Ufimtsev, V. B. 1996. $GdVO_4$ crystals with Nd^{3+}, Tm^{3+}, Ho^{3+}, Er^{3+} ions for diode pumped microchip laser. *SPIE*, 2698, 182.

Zarrabi, J. H., Gavrilovic, P., Williams, J. E., O'Neill, M. S., and Singh, S. 1993. Single-frequency, diode-pumped, neodymium-doped lanthanum oxysulfide microchip laser. *Conf. Lasers Electro-Optics, Tech. Dig.*, 588.

Zayhowski, J. J. 1990a. Limits imposed by spatial hole burning on the single-mode operation of standing-wave laser cavities. *Opt. Lett.*, 15, 431.

Zayhowski, J. J. 1990b. Microchip lasers. *Lincoln Lab. J.*, 3, 427.

Zayhowski, J. J. 1990c. The effects of spatial hole burning and energy diffusion on the single-mode operation of standing-wave lasers. *IEEE J. Quantum Electron.*, 26, 2052.

Zayhowski, J. J. 1991a. Thermal guiding in microchip lasers, In *OSA Proc. Advanced Solid-State Lasers 6*, H. P. Jenssen and G. Dubé, eds., p. 9. Optical Society of America, Washington, DC.

Zayhowski, J. J. 1991b. Polarization-switchable microchip lasers. *Appl. Phys. Lett.*, 58, 2746.

Zayhowski, J. J. April, 1996a. Microchip lasers create light in small places. *Laser Focus World*, 73.

Zayhowski, J. J. 1996b. Ultraviolet generation with passively *Q*-switched microchip lasers. *Opt. Lett.*, 21, 588 (Ultraviolet generation with passively *Q*-switched microchip lasers: errata. *Opt. Lett.*, 21, 1618).

Zayhowski, J. J. 1997a. Periodically poled lithium niobate optical parametric amplifiers pumped by high-power passively *Q*-switched microchip lasers. *Opt. Lett.*, 22, 169.

Zayhowski, J. J. 1997b. Microchip optical parametric oscillators. *IEEE Photon. Technol. Lett.*, 9, 925.

Zayhowski, J. J. 1998. Passively *Q*-switched microchip lasers and applications. *Rev. Laser Eng.*, 26, 841.

Zayhowski, J. J. 2000. Passively *Q*-switched Nd:YAG microchip lasers and applications. *J. Alloys Compounds*, 303–304, 393.

Zayhowski, J. J. 2004. Compact solid-state sources and their applications. *SPIE*, 5620, 155.

Zayhowski, J. J., to be published. Passively *Q*-switched microchip lasers. *Solid-State Lasers and Applications*, chap. 1. A. Sennaroglu, ed., CRC, Boca Raton, FL.

Zayhowski, J. J. and Dill III, C. 1992. Diode-pumped microchip lasers electro-optically *Q* switched at high pulse repetition rates. *Opt. Lett.*, 17, 1201.

Zayhowski, J. J. and Dill III, C. 1994. Diode-pumped passively *Q*-switched picosecond microchip lasers. *Opt. Lett.*, 19, 1427.

Zayhowski, J. J. and Dill III, C. 1995. Coupled-cavity electro-optically *Q*-switched Nd:YVO_4 microchip lasers. *Opt. Lett.*, 20, 716.

Zayhowski, J. J. and Kelley, P. L. 1991. Optimization of *Q*-switched lasers. *IEEE J. Quantum Electron.*, 27, 2220 (Corrections to optimization of *Q*-switched lasers. *IEEE J. Quantum Electron.*, 29, 1239).

Zayhowski, J. J. and Keszenheimer, J. A. 1992. Frequency tuning of microchip lasers using pump-power modulation. *IEEE J. Quantum Electron.*, 28, 1118.

Zayhowski, J. J. and Mooradian, A. 1989a. Single-frequency microchip Nd:YAG lasers. *Opt. Lett.*, 14, 24.

Zayhowski, J. J. and Mooradian, A. 1989b. Frequency-modulated Nd:YAG microchip lasers. *Opt. Lett.*, 14, 618.

Zayhowski, J. J. and Mooradian, A. 1989c. Microchip lasers, In *OSA Proceedings on Advanced Solid State Lasers 5*, M. L. Shand and H. P. Jenssen, eds., p. 288. Optical Society of America, Washington, DC.

Zayhowski, J. J. and Wilson, A. L. 2000. Miniature sources of subnanosecond 1.4–4.0-μm pulses with high peak power, In *OSA TOPS 34, Advanced Solid-State Lasers*, U. Keller, H. Injeyan, and C. Marshall, eds., p. 308. Optical Society of America, Washington, DC.

Zayhowski, J. J. and Wilson, A. L. Jr. 2002. Miniature, pulsed Ti:sapphire laser system. *IEEE J. Quantum Electron.*, 38, 1449.

Zayhowski, J. J. and Wilson, A. L. Jr. 2003a. Pump-induced bleaching of the saturable absorber in short-pulse Nd:YAG/Cr^{4+}:YAG passively *Q*-switched microchip lasers. *IEEE J. Quantum Electron.*, 39, 1588.

Zayhowski, J. J. and Wilson, A. L. Jr. 2003b. Miniature, high-power 355-nm laser system, In *OSA TOPS 83, Advanced Solid-State Photonics*, J. J. Zayhowski, ed., p. 357. Optical Society of America, Washington, DC.

Zayhowski, J. J. and Wilson, A. L. Jr. 2004. Energy-scavenging amplifiers for miniature solid-state lasers. *Opt. Lett.*, 29, 1218.

Zayhowski, J. J., Ochoa, J., and Mooradian, A. 1989. Gain-switched pulsed operation of microchip lasers. *Opt. Lett.*, 14, 1318.

Zayhowski, J. J., Schulz, P. A., Dill, III, C., and Henion, S. R. 1993. Diode-pumped composite-cavity electrooptically tuned microchip laser. *IEEE Photon. Technol. Lett.*, 5, 1153.

Zayhowski, J. J., Harrison, J., Dill, III, C., and Ochoa, J. 1995a. Diode-pumped Tm:YVO_4 microchip laser. *Appl. Opt.*, 34, 435.

Zayhowski, J. J., Ochoa, J., and Dill, III, C. 1995b. UV generation with passively *Q*-switched picosecond microchip lasers. *Conf. Lasers Electro-Optics, Tech. Dig.*, 15, 139.

Zayhowski, J. J., Fan, T. Y., Cook, C., and Daneu, J. L. 1996. 946-nm passively *Q*-switched microlasers. *M.I.T. Lincoln Lab., Solid State Res., Quarterly Tech. Rep., 1996:3*, ESC-TR-96-096, 5.

Zayhowski, J. J., Dill, III, C., Cook, C., and Daneu, J. L. 1999. Mid- and high-power passively *Q*-switched microchip lasers, In *OSA TOPS XXVI, Advanced Solid-State Lasers*, M. M. Fejer, H. Injeyan, and U. Keller, eds., p. 178. Optical Society of America, Washington DC.

Zayhowski, J. J., Cook, C. C., Wormhoudt, J., and Shorter, J. H. 2000. Passively *Q*-switched 214.8-nm Nd:YAG/Cr^{4+}:YAG microchip-laser system for the detection of NO, In *OSA TOPS 34, Advanced Solid-State Lasers*, U. Keller, H. Injeyan, and C. Marshall, eds., p. 409. Optical Society of America, Washington, DC.

Zayhowski, J. J., Buchter, S. C., and Wilson, A. L. 2001. Miniature gain-switched lasers, In *OSA TOPS 50, Advanced Solid-State Lasers*, C. Marshall, ed., p. 462. Optical Society of America, Washington, DC.

Zeller, S. C., Krainer, L., Spühler, G. J., Weingarten, K. J., Paschotta, R., and Keller, U. 2003. Passively mode-locked 40-GHz Er:Yb:glass laser. *Appl. Phys. B*, 76, 787.

Zhang, Q., Dixon, G. J., Chai, B. H. T., and Kean, P. N. 1992. Electronically tuned diode-laser-pumped Cr:$LiSrAlF_6$ laser. *Opt. Lett.*, 17, 43.

Zhang, X., Zhao, S., Wang, Q., Zhang, Q., Sun, L., and Zhang, S. 1997. Optimization of Cr^{4+}-doped saturable absorber *Q*-switched lasers. *IEEE J. Quantum Electron.*, 33, 2286.

Zhang, X., Zhao, S., Wang, Q., Ozygus, B., and Weber, H. 2000. Modeling of passively *Q*-switched lasers. *J. Opt. Soc. Am. B*, 17, 1166.

Zhang, X., Zhao, S., Wang, Q., Zhang, S., Sun, L., Liu, X., Zhang, S., and Chen, H. 2001. Passively *Q*-switched self-frequency-doubled Nd^{3+}:$GdCa_4O(BO_3)_3$ laser. *J. Opt. Soc. Am. B*, 18, 770.

Zhang, Z. M., Cui, Y. B., Li, F. L., Zhang, G. B., Pu, Q. R., and Xu, G. J. 2002. Polarization characteristic of a room-temperature Co:MgF_2 laser. *Appl. Opt.*, 41, 1071.

Zheng, J., Zhao, S., and Chen, L. 2002. Laser-diode end-pumped passively *Q*-switched Nd:YVO_4 laser with Cr^{4+}:YAG saturable absorber. *Opt. Eng.*, 41, 2271.

Zhou, S., Lee, K. K., Chen, Y. C., and Li, S. 1993. Monolithic self-*Q*-switched Cr,Nd:YAG laser. *Opt. Lett.*, 18, 511.

Zhou, Y., Thai, Q., Chen, Y. C., and Zhou, S. 2003. Monolithic *Q*-switched Cr,Yb:YAG laser. *Opt. Commun.*, 219, 365.

Further Reading

Popular discussions of lasers and their applications and history are given in:

Hecht, J. 1992. *Laser Pioneers*, rev. ed., Academic, Boston, MA.

Hecht, J., and Teresi, D. 1998. *Laser: Light of a Million Uses*, Dover, Mineola, NY.

A general discussion of laser science and technology, accessible to anyone with a background of electromagnetic theory, basic quantum mechanics, and calculus, can be found in any of the following:

Eastham, D. A. 1986. *Atomic Physics of Lasers*, Taylor & Francis, London.

Kaminow, I. P., Siegman, A. E. 1973. *Laser Devices and Applications*, IEEE, New York.

Shimoda, K. 1991. *Introduction to Laser Physics, 2nd Ed.*, Springer, New York.

Siegman, A. E. 1986. *Lasers*, University Science, Mill Valley, CA.

Silfvast, W. T. 2004. *Laser Fundamentals, 2nd Ed.*, Cambridge, Cambridge.

Svelto, O. 1998. *Principles of Lasers, 4th Ed.*, Plenum, New York.

Verdeyen, J. T. 1995. *Laser Electronics, 3rd Ed.*, Prentice Hall, Englewood Cliffs, NJ.

Yariv, A. 1989. *Quantum Electronics, 3rd Ed.*, Wiley, New York.

Detailed theoretical discussion of lasers is contained in:

Haken, H. 1986. *Laser Theory, reprint Ed.*, Springer, New York.

Sargent III, M., Scully, M. O., and Lamb, W. E., Jr. 1974. *Laser Physics*, Addison-Wesley, Reading, MA.

The underlying physics of laser resonators and its application to the design of many types of lasers in use today is the topic of:

Hall, D. R., and Jackson, P. E., eds. 1989. *Physics and Technology of Laser Resonators*, Adam Hilgar, Bristol, U.K.

Specific types of lasers are discussed in:

Brown, D. C. 1981. *High-Peak-Power Nd:Glass Laser Systems*, Springer, New York.

Casey, H. C., Jr. and Panish, M. B. 1978. *Heterostructure Lasers*, Parts A and B, Academic, New York.

Coldren, L., and Corzine, S. W. 1995. *Diode Lasers and Photonic Integrated Circuits*, Wiley, New York.

Duarte, F. J., ed. 1995. *Tunable Lasers Handbook*, Academic, San Diego, CA.

Duarte, F. J. and Hillman, L. W., eds. 1990. *Dye Laser Principles*, Academic, Boston, MA.

Garrett, C. G. B. 1967. *Gas Lasers*, McGraw-Hill, New York.

Kaminskii, A. A. 1990. *Laser Crystals, 2nd Ed.*, Springer, New York.

Kapon, E., ed. 1999. *Semiconductor Lasers I: Fundamentals*, Academic, San Diego, CA.

Koechner, W. 1999. *Solid-State Laser Engineering, 5th Ed.*, Springer, New York.

Koechner, W., Bass, M., and Roth, H. 2003. *Solid State Lasers: A Graduate Text*, Springer, New York.

Mollenauer, L. F., White, J. C., and Pollock, C. R., eds. 1992. *Tunable Lasers, 2nd Ed.*, Springer, New York.

Schäfer, F. P., ed. 1990. *Dye Lasers, 3rd Ed.*, Springer, New York.

Willett, C. S. 1974. *Introduction to Gas Lasers: Population Inversion Mechanisms*, Pergamon, Oxford.

A fairly comprehensive review of early work in diode-pumped solid-state lasers is provided by the two articles:

Hughes, D. W. and Barr, J. R. M. 1992. Laser diode pumped solid-state lasers, *J. Phys. D: Appl. Phys.* 25, 563.

Fan, T. Y. and Byer, R. L. 1988. Diode laser pumped solid-state lasers, *IEEE J. Quantum Electron.* 24, 895.

A concise source of data in tabular and graphical form is provided for workers in the areas of laser research and development in the series:

Weber, M. J., ed. 1982. *Handbook of Laser Science and Technology*, CRC, Boca Raton, FL, Vols. I and II; 1986, *ibid.*, Vols. III and IV; 1987, *ibid.*, Vol. V; 1991, *ibid.*, suppl. 1; 1995, *ibid.*, suppl. 2.

Weber, M. J., ed. 2003. *Handbook of Lasers*, CRC, Boca Raton, FL.

11

Optical Modulators

John N. Lee

11.1 Introduction

Optical modulators are devices that alter the temporal and spatial character of a light beam. Such devices are required in virtually all areas of optical technology, particularly optical communication, storage, signal and image processing, and optical sensing. Other chapters in this Handbook provide in-depth descriptions of application areas, such as optical communications (Chapter 12 and Chapter 20), optical data storage (Chapter 12 and Chapter 16), and optical signal processing (Chapter 19). Many of the technologies and processes used in the construction of optical modulators are important aspects of other chapters, especially the chapters on waveguide optics (Chapter 13 and Chapter 14).

One can clearly appreciate that there is generally no single "best" modulator; the choice of modulator is often highly dependent on the specific application. However, there are general performance factors and device limitations that the user should consider. Six important device considerations are:

1. Speed of operation. While higher speed is invariably thought more desirable, often speed is not the most important factor.

2. Linearity of output signal versus input drive signal level. This factor will affect the operating dynamic range of a modulator, since nonlinearities will result in spurious output signals. This is an especially important consideration for manipulation of analog signals.
3. Efficiency, or modulated output power per unit of drive power.
4. Power handling capability of the device. It is desirable that modulation performance be invariant with optical power when modulating optical beams with high power density.
5. Optical characteristics of the modulated beam. Any changes in optical beam characteristics, such as amplitude, phase, and polarization, can represent either a desired modulation, or undesired by-products to be minimized.
6. Wavelength dependence in any of the above modulation characteristics.

This chapter aims to provide a systematic description and categorization of modulators, describing their basic operating principles, current device-performance levels, fundamental limitations, and status of specific devices. Attention will often focus on the six factors listed above, so the user will gain an understanding of the tradeoffs in selecting one device over another. Also, the impact of recent developments in nanotechnologies and in display technology on many of the topics of this chapter will be discussed.

The first broad categorization provided here is according to modulation mechanism, with a second according to whether the optical modulator is electrically or optically addressed. We feel that identification of modulation mechanism generally identifies and bounds a number of parameters important to performance and implementation (e.g., whether the requisite speed, frame rate and resolution are achievable, and whether the device size, power consumption, and mechanical and electrical circuit configuration are compatible with an application). The major categories of modulation mechanisms used here are:

- Electro-optic
- Acousto-optic
- Magneto-optic
- Mechanical and micromechanical
- Thermal

We also briefly cover some specialized and emerging approaches, such as use of nanostructures (Krauss and DeLaRue 1999; Prather et al. 2002), photonic bandgap structures, directly modulated light sources, and devices based on biological materials.

11.1.1 General Modulator Considerations

The input light going into a modulator can be described very generally as the spatio-temporal amplitude and phase function, with the amplitude function $A(x,y,t)$ and the phase information represented by the expression in brackets as seen in Equation 11.1.

$$A(x,y,t)\exp\{-j[\Omega t + k_1 x + k_2 y + \phi(x,y,t)]\} \tag{11.1}$$

Optical modulators then generally modify the phase, polarization, and amplitude or intensity of the input light. An optical switch would be the simplest example of amplitude modulation. Optical phase modulation is often desired, but phase modulation can also be converted into an amplitude modulation by techniques discussed below. Large modulations in phase can also be interpreted as optical-frequency modulation. Finally, modulation of optical wavelengths, i.e., extremely large changes in frequency Ω ($\sim$terahertz), can also be desired. Usually, it is desirable to modulate only one optical parameter at a time; simultaneous changes in the other parameters must then be avoided in an acceptable modulator.

The simplest type of modulator is the temporal-only or point modulator, where the modulation signal is identical within any cross-section perpendicular to the propagation of the light beam. A more general modulation device performs both spatial and temporal modulation; such a device is usually called a spatial light modulator (SLM). With spatial light modulators there is the additional objective of simultaneously maximizing the frame rate and, within a given frame, the spatial resolution (or equivalently, the number of pixels or the space-bandwidth product).

The linear operating range of optical modulators is often limited. For example, there is often a fundamental sine-squared transfer function, such as might be obtained with the transmission of polarized light through crossed polarization-selective components. Quasi-linear operation with a sine-squared device is obtained either by operating in the small-signal regime, or by operating around the quadrature ($\pi/4$) point. Pre-distortion of the input signals might be performed to give a linear output relative to the original input, but this is often difficult in practice, especially for spatial light modulators. Deviations from linearity will lead to undesired harmonic responses and third-order mixing terms in the output if the modulator is operated at high drive level, over large bandwidth, and with a multi-component drive signal. These undesirable outputs reduce the usable dynamic range of the modulator, and can be quantitatively modeled from Taylor-series expansions of the modulator response (Hecht 1976; Johnson and Rousell 1988; Korotsky and DeRidder 1990). Dynamic range is also affected by the efficiency of the modulator. High efficiency usually results in a larger quasi-linear response range.

Fundamental noise limits on dynamic range should always be kept in mind in the application of optical modulators. These can be either thermal noise or quantum noise. The thermal noise limit arises in all electrical components of the modulator, and is given by kTB, where k is the Boltzmann constant, T is the absolute temperature, and B is the temporal bandwidth of the drive signal. In the optical domain, photon shot-noise is the fundamental limit, and often dominates over thermal noise. The discrete-photon nature of light give a quantum noise limit of hf, where h is Planck's constant and f is the frequency of the light. As an example, at $T=300$ K thermal noise, kT is 4.14×10^{-21} J. For an optical beam at 850 nm wavelength, the photon noise limit is 2.34×10^{-19} J. Thus, for equal amounts of power and the same detection bandwidth, the noise in the optical beam is 18 dB higher.

Spatial light modulation requires additional considerations. Such modulation may be either one-dimensional (1-D) or two-dimensional (2-D). The SLM may be either optically or electrically addressed, as will be discussed below. Figure 11.1 illustrates an optically addressed 2-D SLM. In an electrically addressed SLM, the write beam is replaced by electrical signals, such as in matrix-addressed displays and electron-beam-addressed cathode ray tubes. The modulation may be either spatially continuous or within discrete pixels. The performance characteristics of an SLM are strongly related to the spatial

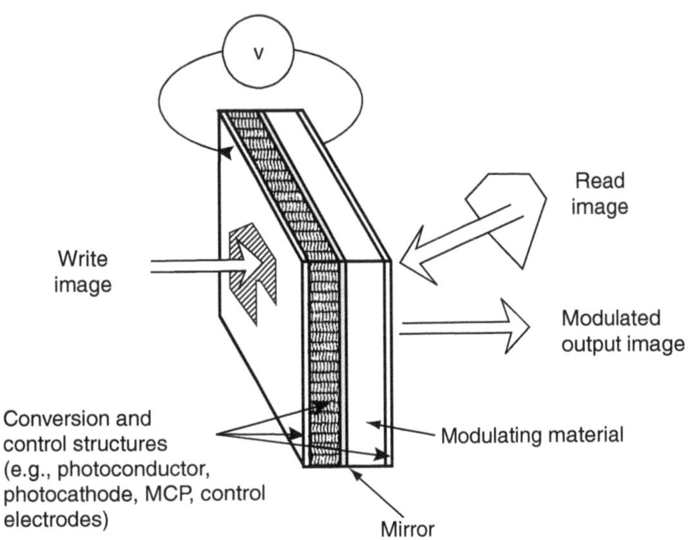

FIGURE 11.1 Generic sandwich construction of a two-dimensional spatial light modulator with optical addressing.

dimensions of the data display (e.g., pixel pitch, size, and shape (Turner, Johnson, and Serati 1995)). If the pixel pitch is d mm, then the maximum spatial frequency (line pairs/mm) is,

$$f_n = 1/(2d) \tag{11.2}$$

The information capacity of an SLM of area A, also known as the space-bandwidth product, is given by,

$$N^2 = 4A(f_{max})^2 \tag{11.3}$$

Note that large flat-panel displays usually also have a large value for d, thus limiting the value of N^2 and correspondingly increasing the size of the associated imaging optics. Therefore, such displays occupy a small, specialized niche in satisfying modulator applications.

For applications where the spatial spectrum of the 2-D data is important, a pixilated modulator introduces an envelope function, $\mathrm{sinc}(f_x a)\mathrm{sinc}(f_y b)$, onto the spectrum, where a and b are the dimensions of a rectangular pixel, and f_x and f_y are the spatial frequencies in the x and y directions.

For optical processing applications, the purpose of an SLM is not only to perform light modulation, but also to provide three-terminal device operation somewhat analogous to that of a transistor, but employing spatio-temporal optical beams instead of temporal electrical signals. Essentially, the three ports of the three-terminal SLM are the input optical beam (for input signal), the readout optical beam (for second signal input port and power supply for the output), and the final output beam (output signal). A three-terminal device isolates the input from the output and provides signal gain and storage. In addition, such devices can add, subtract, and multiply arrays of data in parallel, and perform linear and nonlinear transformations. If the SLM is optically addressed, it is also possible to perform photodetection and optical conversions, such as from incoherent to coherent light and from one wavelength to another.

The definition of adequate SLM performance can vary greatly. The major parameters are resolution, frame rate, and dynamic range. At one extreme is a goal of very high resolution (>million pixels); at another extreme is a goal for extremely high speed (10 ns or less), but usually at much lower resolution. It is generally not necessary to simultaneously achieve the two extreme figures. It can be expected that a variety of SLMs should be developed, each optimal for particular needs. A measure of comparison among the various SLM devices is to look at the product of frame resolution and frame rate, which produces a figure of merit in units of pixels/second. A figure as high as 10^{12} pixels/s can arise in demanding applications. The noise and dynamic range performance for an SLM is affected by a wide range of factors, in addition to the general ones given above. These include spatial uniformity of response, crosstalk mechanisms, such as electrical crosstalk and optical scatter (Brown, Craig, and Lee 1989), and coherent effects, such as speckle. Some analyses have been done for specific SLMs by modeling them as a component in a numerical linear system (Taylor and Casasent 1986; Batsell et al. 1990).

We now consider optical and electrical methods for addressing 1-D and 2-D SLMs.

11.1.2 Optically Addressed Modulators

Much of the research motivation for optically addressed spatial light modulators is the desire to process 2-D data arrays, especially imagery, completely within the optical domain, avoiding the inefficiencies and latencies associated with present conversion from optical domain into the electrical for processing. Optically addressed SLMs would be crucial to such "all-optical" concepts, even if the SLM involves electronic processes at the microscopic level. Many optically addressed SLMs employ the basic sandwich structure of Figure 11.1. A bias voltage applied to the sandwich is shunted within the illuminated regions of the photosensitive layer. In the reflective configuration illustrated, there is a mirror and light-blocking layer at the center of the sandwich, which allows the written input information to be read out by reflection from the modulating-material side of the SLM. An optically addressed SLM generally requires a photosensor integrated with the optical material in order to maximize speed and to reduce drive power and voltages. In some cases, the modulating material can also function as the photosensor (as will be discussed below), but in most cases, one deposits a photosensitive layer onto the modulating material, using exactly the geometry shown in Figure 11.1. Photoconductors, such as CdS, PVK:TNF, amorphous Se, CdS, ZnS, and ZnCdS, have been deposited onto various modulating materials. The performance of these photoconductors can limit the device performance. For example, photoconductors such as amorphous Se exhibit carrier-trapping effects that limit frame rates to $<$1000/s (Armitage, Anderson, Karr 1985). Some SLM devices have therefore tended to use amorphous silicon (a-Si) or polysilicon thin film transistors (Ashley, Davis, and Oh 1988) and photodiodes (Armitage, Anderson, and Karr 1985).

11.1.3 Electrically Addressed Modulators

Electrical addressing of 2-D devices often leverages the tremendous amount of development from the commercial display industry. The principle electrical addressing mechanisms are the following:

- Electron-beam. An electron beam, such as in a cathode ray tube, may be scanned across the modulating material or control structures, resulting in an activating electric field. It is also possible for a light beam to be scanned across a photosensitive material to produce electrons that impinge onto the modulating material; this case should be considered electrical addressing, since the scanning action must be electrically controlled.
- Electrode Matrix. The electrode matrix can be either one- or two-dimensional. For 2-D SLMs, a rectangular grid of electrodes is placed on the modulating material, and circuits placed at the ends of row and column electrodes. The orthogonal lines are usually in the same plane, but it is possible to place them on opposite faces of the material. Pixels are addressed at the intersections of the grid lines. A simple "passive matrix" of electrodes suffers from the need to distinguish between a pixel where both the row and the column electrode are activated (true signal) and a pixel where only one of the electrodes is activated (half-select); this generally requires the material or device to have a nonlinear response.

Circuitry can also be placed on the grid intersections. Use of arrays of thin-film transistors has been developed in the display industry (e.g., "active matrix" displays). Representative of the 2-D active matrix-addressed devices are the various commercial flat panel liquid crystal televisions (LCTVs)(Blechman 1986), cell phone displays, and Digital Light Processing (DLP©) micromechanical displays (Brooks 1985; Hornbeck 1989). Active matrix addressing eliminates the half-select problem. The transistors are generally fabricated using a-Si:H deposited on a glass substrate; polycrystalline silicon, and, less commonly, CdSe are also used. The transistor then connects to an electrode over the modulating material. The electrode must be optically transparent for materials such as liquid crystal.

In principle, it is possible to place more circuitry at the grid intersections using microchip fabrication techniques. This would increase the functionality of an SLM, especially to perform nonlinear operations on the data, such as logarithm, square root, inversion, hardclipping, switching, and general arithmetic functions, all of which are difficult to perform using only the physical properties of the modulating materials.

- Traveling Wave Addressing. Spatial information can be introduced into an optical modulator in the form of a traveling wave. The prime example of this type of addressing is the acousto-optic device to be discussed in detail in a later section. The spatial information is transitory, the duration depending on the velocity of the wave in the medium.

It is also possible to use a traveling wave to perform the addressing functions in an SLM. Surface acoustic waves in a piezoelectric medium have been primarily employed, since there is a strong acousto-electric field accompanying the wave at the surface of the medium that can be used for activation purposes. Thus, an acoustic impulse can be used to probe or activate a thin-film optical medium (which might also be the piezoelectric acoustic medium).

In the following sections, the various modulators will be grouped according to their modulation mechanism. For each group, we shall first discuss the modulation mechanism and available materials. Devices are then tabulated and described. Within each group, devices are identified as either electrically or optically addressed. Performance numbers can be expected to change with time if a technology is not mature. Hence, descriptions will indicate whether technological limits have been reached, and what those limits might be.

11.2 Electro-Optic Devices

11.2.1 Electro-Optic Effect and Materials Considerations

The classic electro-optic effect is an alteration of the optical refractive index due to application of an electric field E. This effect is employed in materials without a center of symmetry, as can be identified by crystal class (Koster et al. 1963), and as in an electrically poled semi-amorphous material. The application of an external electric field to such non-centrosymmetric materials change in the inverse of the refractive index (n, a real quantity for now) according to,

$$\Delta(1/n^2)_i = \sum_{j=1}^{3} r_{ij}E_j \quad i = 1,2\ldots,6 = x,y,z,yz,xz,xy \tag{11.4}$$

where r_{ij} are called the electro-optic coefficients, and the index i corresponds to the coordinates of the general "index ellipsoid" that specifies light velocity and polarization with direction when an electric field is applied (Yariv 1989).

Figure 11.2 shows how a change in one of the refractive indices of a birefringent material can be used to produce optical intensity modulation. The birefringent electro-optic material is placed between crossed polarization-selective elements, as illustrated in Figure 11.2. A voltage (V) applied to electrodes deposited on the birefringent material produces an electric field within the material. In Figure 11.2, this field is along the direction of light propagation, although the field may also be transverse to the light as described by Equation 11.4. The field changes the extraordinary and ordinary indices of refraction (n_e and n_o, respectively) by different amounts. Hence, linearly polarized incident light will exit as elliptically-polarized light. Crossed polarizers then transmit a fraction of the light intensity, according to a sine-squared function of applied voltage,

$$I_{out}/I_{in} = \sin^2\{(\pi/2)(V/V_\pi)\} \tag{11.5}$$

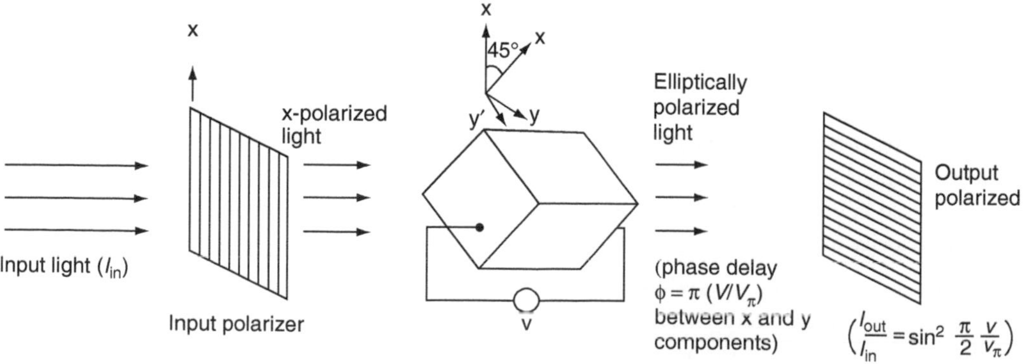

FIGURE 11.2 Intensity modulation with a birefringent electro-optic material between crossed polarizers.

where the half-wave voltage V_π is as defined following Equation 11.11 below. Note that Equation 11.5 represents a non-linear transfer. The sine-squared function occurs often in optical modulation, where the modulated and unmodulated light beams comprise a two-mode coupled system (Yariv 1989). Further note that the configuration of Figure 11.2 is intrinsically wavelength dependent.

The origin of the classic electro-optic effect is the macroscopic polarizability of a material due to an external field E. The polarizability of any medium is obtained by summing over all molecular sites k, and is given by

$$P = \Sigma P_k = P_o + \chi^{(1)} E_{opt} + \chi^{(2)} E_{opt} E + R E_{opt} EE \tag{11.6}$$

where E_{opt} is the optical field. The quantity $\chi^{(2)}$ in Equation 11.6 is a measure of the *linear electro-optic* effect, often referred to as the Pockels effect. $\chi^{(2)}$ is related to r_{ij} by

$$(\varepsilon/\varepsilon_0)^2 r_{ij} = \chi^{(2)}.$$

The fourth term in Equation 11.6, with coefficient R, represents a quadratic dependence on the applied field, often referred to as the Kerr effect. Devices using the quadratic effect will have the advantage of no dependence on optical polarization. Other triple-field terms in Equation 11.6 are possible if E_{opt} replaces either or both E factors, giving rise to two-beam mixing (or second-harmonic optical generation) and three-beam mixing (Yariv 1989). Multi-beam mixing has been explored for all-optical switching operations, but those operaitons are not a topic for this chapter.

In the above discussion, the optical index of refraction, n, has been considered a real quantity. If the index is defined as a complex quantity, the optical field in Equation 11.6 can be rewritten as

$$E = A \exp\{j[\omega t - \omega n_o (1 - j\kappa) z/c]\} \tag{11.7}$$

if the refractive index is more broadly defined as

$$n' = n_o (1 - j\kappa) \tag{11.8}$$

The complex component of Equation 11.8 corresponds to absorption in the medium. Hence, an absorption change induced by an applied field should also be considered an electro-optic effect under this generalized definition. Further, it is important to remember that a change in the real part of the refractive index accompanies the change in absorption coefficient. The real component of n is related to the imaginary component by the well-known Kramers–Kronig relationships. Hence, an almost pure phase modulator is also possible by operating off the peak absorption point, but where significant phase change is still possible.

Table 11.1 lists the most widely used inorganics. The most familiar and available electro-optic materials are generally crystalline inorganics. Lithium niobate ($LiNbO_3$) and lanthanum-modified lead zirconate titanate ($Pb_{1-x}\ La_x\ (Zr_y\ Ti_{1-y})_{1-x/4}O_3$, or PLZT) exemplify materials with real linear and quadratic refractive-index changes with field, respectively. The speed of response for crystalline inorganics is very high, sub-picosecond, since the microscopic contributions to the external polarization in Equation 11.6 are at the ionic level. However, intrinsic speed can be compromised if the material is in polycrystalline form, such as for PLZT.

Organic electro-optic materials have recently been developed which try to exploit intrinsic capabilities of certain organic molecules and long-chain polymers to provide orders of magnitude larger effect at lower drive powers than traditional inorganics (Singer and Garito 1981; Garito, Teng, and Wong 1984; Singer et al. 1988) (see also Chapter 7). Although charge transport processes that give rise to refractive index changes are at the molecular level, the intrinsic response speed is still high. However, as will be discussed below, demonstrated device speeds are not yet limited by the intrinsic response.

TABLE 11.1 Selected Inorganic Electro-Optic Materials and Properties at $T=300$ K

Material	Electro-Optic Coefficient	Index of Refraction	$n_o^3 r$ (10^{-12} m/V)	$\varepsilon/\varepsilon_o$
Linear Electro-optic Materials				
$\lambda=550$ nm	Linear coeff. (10^{-12} m/V)			
KD*P	r63=23.6	$n_o=1.51$	80	‖ c-50
(KD_2PO_4)		$n_e=1.47$		
KDP	r41=8.6	$n_o=1.51$	(r41) –29	⊥ c-20
(KH_2PO_4)	r63=10.6	$n_e=1.47$	(r63)–34	‖ c-45
$LiNbO_3$	r33=30.8	$n_o=2.29$	(r33)–328	⊥ c-98
	r13=8.6			
	r22=3.4	$n_e=2.20$	(r22)–37	‖ c-50
	r42=28			
$LiTaO_3$	r33=30.3	$n_o=2.175$		
	r13=5.7	$n_e=2.18$	(r33)–314	‖ c-43
$BaTiO_3$	r33=23	$n_o=2.437$		⊥ c-4300
	r13=8		(r33)–334	
	r42=1640	$n_e=2.365$		‖ c-106
$Sr_xBa_{1-x}NbO_3$	r33=450	$n_o=2.346$	(r33)–5800	⊥ c-3000
(SBN)	For $x=0.75$	$n_e=2.310$		
$KTa_x\ Nb_{1-x}\ O_3$	r33=1400	2.25	(r33)–16000	⊥ c-3000
(KTN)	r42=3000		(r42)–34000	
Quartz (SiO_2)	r41=0.2	$n_o=1.54$	(r41)–0.7	⊥ c-4.3
	r63=0.93	$n_e=1.55$	(r63)–3.4	‖ c-4.3
GaP	r41=0.97	$n_o=3.31$	(r41)–29	
ZnTe	r41=3.9	$n_o=2.79$	(r41)–77	7.3
CdTe	r41=6.8	$n_o=2.60$	(r41)–120	
GaAs	r41=1.6	$n_o=3.34$	(r41)–59	11.5
(10600 nm)				
Quadratic Electro-optic Materials				
$\lambda=550$ nm	Quad. coeff. (10^{-16} m²/V²)	Index of refraction	$n_o^3 R$ (10^{-16} m²/V²)	$\varepsilon/\varepsilon_o$
$KTa_x\ Nb_{1-x}\ O_3$ (KTN)	$R>0.8$	2.25	9	3000
$Pb_{1-x}\ La_x\ (Zry\ Ti_{1-y})1-x/4\ O_3$ (PLZT)	3.8, for $x=0.07$, $y=0.65$	2.55	63	4500

Source: From Kaminow, I. P. and Turner, E. 1971. *Handbook of Lasers*, Chemical Rubber Co., Cleveland, OH; Yariv, A. 1989. *Quantum Electronics, 3rd Ed.*, Wiley, New York.

Nevertheless, a variety of other effects can be included under the electro-optic category. These include:

- The photorefractive effect, which is a manifestation of the classic electro-optic effect on a microscopic scale (Pepper, Feinberg, and Kukhtarev 1990). The speed at which significant electro-optic change can be obtained in photorefractive materials depends on the scale of charge transport, which in turn depends on specific material. Response speed can range from the nanosecond to millisecond regimes.
- Alteration of optical properties of liquid-crystal materials, both twisted nematic and ferroelectric (DeJeu 1980; Blinov 1983; Goodby and Patel 1986). Since the electro-optic effect requires the movement of large molecular structures, the intrinsic response speed vary from milliseconds for twisted nematic material to sub-microsecond for ferroelectric liquid crystals.
- Alteration of material absorptivity (the imaginary part of the complex index of refraction). Materials engineered with III–V compounds, such as very thin alternating layers of InGaAs and InGaAlAs "multiple quantum well (MQW)" structures (Miller et al. 1984), are the major material class exhibiting strong electroabsorption. Since the basic modulation mechanism involves a wavelength shift of an exciton absorption in MQW material, the intrinsic speed of response is again very high. Introduction of free carriers into silicon can also produce rapid changes in absorptivity (Soref and Bennett 1987).

In actual devices, speed will often be determined more by device construction than by intrinsic material speed, as will be discussed in detail throughout this chapter.

11.2.1.1 Inorganic Materials

11.2.1.1.1 Linear Electro-Optic Materials

The basics of classic linear electro-optic modulation in inorganic crystals can be illustrated by considering a uniaxial crystal with the E-field along the z-axis ($i=j=3$). Differentiating Equation 11.4 gives

$$\Delta n_3 = -(n^3/2)r_{33}E_3 \tag{11.9}$$

Inserting Equation 11.9 into the phase term of the optical wave

$$E = A\exp\{j(\omega t - \omega n_o z/c)\} \tag{11.10}$$

results in a phase change of

$$\Delta\phi = -(\omega/c)(n^3/2)r_{33}V \tag{11.11}$$

where $V=E\,Z$ is the applied voltage across a sample of thickness Z. If $\Delta f=\pi$, then $V=V_\pi$ is known as the half-wave voltage. For a two-dimensional SLM, minimum thickness Z is desired in order to maximize number of pixels (or space-bandwidth), since the minimum pixel dimension is roughly equal to Z due to the fringing electric fields in small spatial features. Thus, a 100-mm-thick sample requires $\Delta n=0.005$ to produce an optical path difference $\Delta nZ=\lambda/2$, i.e., $\Delta f=\pi$, at $\lambda=1000$ nm.

Regardless of whether an electro-optic material is used in an optically or electrically addressed device (see Section 11.1 and electro-optic modulators, below, for a discussion of such SLMs), an RC time constant can be the limit on the device speed. We can model the electro-optic crystal as a capacitor of area A, thickness Z, and capacitance C. The half-wave charge for this capacitor, Q_π, is related to the half-wave voltage V_π by

$$Q_\pi = CV_\pi = (\varepsilon AV_\pi)/(4\pi\varepsilon_o Z) \tag{11.12}$$

where ε is the DC dielectric constant, and ε_0 is the permittivity of free space. It is instructive to consider the response time for an electrically addressed device where R corresponds to an external load resistor R_L. The optically addressed case can be described by equivalent circuit models. Since $R_L=V/I$, we obtain the dependence of the device response, on ε, V_π, and I, the current in the drive circuit.

$$t = (\varepsilon AV_\pi)/(4\pi\varepsilon_o ZI) \tag{11.13}$$

In particular, note that the speed of response scales inversely with ε.

The drive power requirements for high-frequency modulation in the electrically addressed case can be derived using Equation 11.11. The above electro-optic crystal capacitor operating into a load resistor R_L, requires a power to modulate over the bandwidth $\Delta f=1/(R_LC)$ of

$$P = \left(\frac{V^2}{2R_L}\right) = \frac{(\Delta\phi)^2\lambda^2 A\varepsilon(\Delta f)}{4\pi Zn^6r^2} \tag{11.14}$$

Note, in particular, the importance of material properties n and ε. Dependence on λ is both explicit in Equation 11.14 and implicit in the λ dependence of n. The important objective of minimizing P requires maximizing the quantity

$$F = n^3r/\varepsilon \tag{11.15}$$

F, therefore, represents an important figure of merit for material evaluation. Using a material having small ε will not only maximize F, but will also shorten the time response according to Equation 11.13. The quantities n, ε, and n^3r are listed in Table 11.1.

In Table 11.1, it is worth noting that one of the strongest crystalline electro-optic materials is $LiNbO_3$ ($r_{33}=30.8\times10^{-12}$ m/V). However, this material, with $n=2.29$ and $\varepsilon=98$, requires relatively high drive powers. Using Equation 11.8, the half-wave index change $\Delta n=0.005$, in a 100-μm-thick sample of $LiNbO_3$, requires $E=2.5\times10^7$ V/m, or a voltage of 2500 V across the sample. SLMs developed using classic electro-optic materials, such as $LiNbO_3$ and potassium dihydrogen phosphate (KDP), or its deuterated version KDP, show limitations on frame rate from the considerations of Equation 11.13 and Equation 11.14, especially when high spatial resolution is desired (e.g., use of very thin samples which have large capacitance). If ε can be decreased, and if electro-optic materials can be produced in thin films, both resolution and speed should be increased. Thin-film deposition of electro-optic materials on a variety of substrates would also allow easier integration with electronic circuits. Efforts to deposit $LiNbO_3$ have continued, and progress has been made in major problem areas, such as the tendency for the films to be polycrystalline and for the electro-optic coefficients to be lower than for the bulk materials (Griffel, Ruschin, and Croitrou 1989). However, significant work remains to be done at this time.

In the past decade, new inorganics have been developed with high electro-optic coefficients. Most notable among these are tungsten–bronze compounds like strontium barium niobate, $Sr_xBa_{1-x}NbO_3$ or SBN (Neurgaonkar and Cory 1986), and $KTa_x\,Nb_{1-x}\,O_3$ (KTN), which can be grown in thin film form on silicon and GaAs.

11.2.1.1.2 Quadratic Electro-Optic Materials.

A major class of solid-state inorganics that exhibits strong quadratic effect is lanthanum-modified lead zirconate titanate ($Pb_{1-x}\,La_x\,(Zr_y\,Ti_{1-y})_{1-x/4}O_3$ or PLZT) material. Commercially-available samples are ceramics composed of hot-pressed microcrystallites. As an example of the strength of the quadratic electro-optic effect, commercial PLZT with $x=0.07$ and $y=0.65$ has $R=3.8\times10^{-16}$ m^2/V^2. Comparing PLZT to a typical linear material, about 100 V are required for the quadratic material to produce the same effect as 1 V in the linear material; but if the linear material requires 1000 V, such as across a platelet of $LiNbO_3$, the voltage requirement for the quadratic material is comparable (3000 V). Since solid-state quadratic materials generally consist of hot-pressed material, the speed at which one can change the state of a device is typically limited to about 50 ms. However, quadratic materials such as KTN can still be used in high-speed applications, by employing latent pre-stored holograms that can be accessed and processed in nanoseconds with the application of an external electric field signal (Agranat et al. 2002).

As with linear electro-optic materials, it is desirable to produce thin-film single-crystal quadratic materials. There has been significant progress in thin-film deposition of PLZT and other quadratic electro-optic materials. Single-crystal, micrometer-thick PZT and PLZT films have been produced on crystalline and glass substrates with quadratic behavior consistent with expected bulk behavior (Preston and Haertling 1992; Tunaboylu, Harvey, and Esener 1998). $KTa_x\,Nb_{1-x}\,O_3$ (KTN) also exhibits a quadratic effect, and as noted above, has been grown in thin films. Rf- sputtered films of KTN have been grown on substrates of silicon and GaAs with quadratic coefficients $R>0.8\times10^{-16}$m^2/V^2.

11.2.1.2 Organic Materials

A major motivation for continuing research in organic materials is the generally lower ε as compared to inorganic materials, and thus the potential to lower modulator drive power, or provide higher bandwidths. Optical organic materials are covered in detail in Chapter 7. For optical modulation, these materials possess large polarizabilities, arising from large changes between ground and excited states in the electron wave functions of the molecular π bonds, and resulting in large, second-order molecular susceptibilities. Organic crystals have been investigated extensively (Chemla and Zyss 1987). The most developed of these is 2-methyl-4-nitroaniline (MNA). Values of $\chi^{(2)}$ of 67 pm/V have been measured for MNA, which compares favorably with $LiNbO_3$. However, this falls considerably short of the potential performance derived from intrinsic hyperpolarizabilities for organic dye molecules, which predict 50X larger electro-optic coefficient than for MNA (Marowsky et al. 1988).

Major issues with organic crystals have historically included fabrication of large samples with good optical quality, fabrication of thin-film samples (as discussed previously for the inorganics), and material lifetime. Hence, continuing efforts have been directed towards polymer systems. One of the advantages of polymers is their ability to be deposited as thin films onto a variety of substrates using spinning techniques. Active species, or chromophores, such as dye molecules, are introduced into a polymer host. Early research was performed with guest–host systems, such as the red dye molecule, hemicyanine, in PMMA (Singer et al. 1988). Material is poled with a strong electric field, such as with a corona discharge that avoids problems with electric-field breakdown in the material. Poling is performed at a temperature above the glass-transition temperature of the polymer to orient all the electric dipoles, and the electric field is kept on as the sample is slowly cooled to below the glass-transition temperature. Retention of the poled character is an intrinsic problem; the poled state is not one of thermodynamic equilibrium, as the natural tendency is for neighboring dipoles to orient anti-parallel. Stability in the guest–host systems has been found to be especially problematic, with less than half of the poling remaining after a few days. More recently, the chromophores have been attached as side groups along the length of a stiff polymer chain, such as those in polystyrene, to produce a so-called *copolymer*. Further progress in lifetime was achieved through cross-linking of the polymer chains, and use of polymers with very high glass transition temperatures compared to room temperature (Mandal et al. 1991; Stahelin et al. 1992). Retention of about 75% of the poling strength eight years after fabrication has been demonstrated (Man and Yoon 1998). However, additional work remains to be done at this time to insure further increase in lifetime under conditions of repeated use, elevated temperature, and environmental stress. Most present electro-optic organic polymers have electro-optic performance somewhat less than that of $LiNbO_3$.

11.2.1.3 Liquid Crystals

For a given applied voltage, liquid crystals produce the largest electro-optic effects of any present material. Liquid crystals are long-chain polymers that possess phases that are uniaxial. These phases include the uniaxial smectic and nematic phases, including the technologically important twisted-nematic phase. These uniaxial phases exhibit very strong birefringence. The change in birefringence of twisted nematic material can be exploited using the crossed-polarizer configuration, shown in Figure 11.2, to convert polarization changes in the transmitted light into optical intensity modulation. Twisted nematic materials are widely employed for commercial displays, such as computer screens and hand-held televisions. We will discuss various liquid crystal modulators in the next subsection. Liquid crystal materials can also be made very sensitive to temperature changes; devices that employ these changes are discussed in Section 11.6 on thermal devices.

Typically, $\Delta n = n_e - n_o \sim 0.2$–$0.4$ are possible in nematic liquid crystals. Δn generally increases with decreasing wavelength and temperature (DeJeu 1980; Blinov 1983). Application of relatively small electric fields, corresponding to a few volts, can change the direction of the axis for the liquid crystal chains, with correspondingly large refractive-index changes. An alternative interpretation is that the field rotates the index ellipsoid. Values of $\chi^{(2)}$ for nematic liquid crystals can be very large, on the order of 10^{-6} m/V; by comparison, values for inorganics are in units of 10^{-12} m/V. The large field-induced refractive-index changes in liquid crystals are due to physical movement of the molecular chains. Larger electric fields will speed molecular reorientation to an extent. The electric field induces electric dipoles in the twisted-nematic material. The dipolar field interaction then causes the molecular chains to reorient along the field direction. Larger fields increase the dipolar fields, and hence the speed of response, up to certain limits due to viscosity. But reversion to the original state must occur naturally at a characteristic relaxation rate that is typically milliseconds. The fastest nematic response observed is still $\sim$1-ms, employing a voltage undershoot effect in the nematic material (Sayyah et al. 1992). Due to this, there has been great interest in ferroelectric liquid crystal (FLC) materials (Clark and Lagerwall 1980) for faster response times.

Unlike the twisted-nematic material where the dipolar field is induced, FLC molecules possess permanent electric dipoles, allowing rapid switching between different optical polarization states. Intrinsic speeds at a few volts are typically on the order of a few microseconds, again limited by material viscosity. Primarily, two types of material have been employed—smectic C and smectic A. The surface-stabilized smectic C FLC material constrains the molecular director (the average direction of the elongated liquid-crystal molecules) to only one particular value of tilt angle from the axis of symmetry. Material with a tilt angle of 22.5 degrees has been extensively employed in actual devices. This tilt angle defines a polarization axis. The material is used as a switchable half-wave plate between crossed polarizers. An applied field switches the optic axis of the half-wave plate (or equivalently the molecular director of the FLC) through 45 degrees—twice the tilt angle—so light polarized along the original axis has its polarization rotated by 90 degrees. When placed between crossed polarizers, optical transmission is switched between on and off states. Contrast ratio depends on how perfectly the FLC behaves as a half-wave plate; this tends to be a more complex manufacturing problem for the FLC material than for inorganic electro-optic materials. For example, the thickness and composition of the FLC film must be carefully controlled, or else plane-polarized input light results in elliptically-polarized output light.

The smectic C material operated as a half-wave plate provides only two- or three-state (binary or ternary) intensity modulation, although devices can be designed to use either time- or space-multiplexing to provide the equivalent of gray scale (see Electro-optic Modulators below). The smectic A material allows a continuous variation of the optic-axis tilt angle with applied voltage (Anderson et al. 1988) so that devices with analog intensity modulation are possible when using this material in conjunction with a polarizer. Increased dynamic range (number of gray levels) and temperature sensitivity are present areas of development for smectic A devices. To maximize the dynamic range, a tilt angle of 22.5 degrees must be achieved, since this angle allows in a single film full on-off intensity swing when placed between crossed polarizers. The induced tilt angle is a strong function of temperature, so that device temperature needs to be controlled in order to have reproducible grey levels. Finally, while a large number of analog gray-scale levels are desirable, it is also desirable to keep activation voltages below 5 V; such voltages may limit the number of gray levels. Smectic C devices are presently much more available.

A promising avenue for increasing the strength of electro-optic effect and the modulation speed in FLCs has been the recent development of bent-core FLC material. Effective values of $r \sim 10$ pm/V, compared to $r \sim 3$pm/V for conventional FLC materials, and response speeds up to 40 MHz have been shown feasible (Rickard et al. 2005).

11.2.1.4 Photorefractive Materials

The discussion so far on inorganic, organic, and liquid crystal electro-optic materials relates to optical changes induced in bulk samples, such as via electrodes attached to the sample. It is also possible to consider the classic linear electro-optic effect on a microscopic scale. Charge (electrons or holes) is produced at illuminated regions of certain photoconductive materials. Electric fields then cause the charge to be transported to non-illuminated regions where they become trapped. The resultant charge pattern produces a corresponding spatial electric-field pattern that alters the refractive index of the material (Pepper, Feinberg, and Kukhtarev 1990), and these refractive-index changes are optically read out; this results in a very simple SLM structure. Use of such *photorefractive* materials in optically addressed SLMs is particularly attractive, since the material is both the photosensor and the modulator. Optical control of other optical beams occurs via interaction at the atomic and molecular level.

The first materials in which the photorefractive effect was observed were $LiNbO_3$ and $LiTaO_3$ (Chen 1969). The effect was observed in both the pure and doped materials. The charge transport mechanism in the pure material is a result of electron charge cloud distortions in the vicinity of the ions. For material doped with iron, it was found that photoconductivity causes charge transfer between Fe^{2+} and Fe^{3+} impurity sites. Since these early findings, other ferroelectrics exhibiting photorefraction have been extensively studied; the most popular of these are $BaTiO_3$ (White and Yariv 1984) and the tungsten–bronze compounds like strontium barium niobate, $Sr_xBa_{1-x}NbO_3$ (Neurgaonkar and Cory 1986; Rakuljic et al. 1988).

Very efficient photorefractive materials have been developed from photoconductors with high charge mobility. The photoconductor that has undergone the most study is bismuth silicon oxide ($Bi_{10}SiO_{12}$ or BSO)(White and Yariv 1984). The drift velocity of the charge can be increased by application of large electric fields, so that charge separation is maximized and response times of <1 ms are possible.

An important consideration for photorefractives is wavelength of activation. The band-gap for BSO is relatively large, thus requiring the use of visible wavelengths to induced photoconduction. It is desirable to have materials responsive in the near-IR region corresponding to the semiconductor-laser wavelengths, which are becoming more common in optical processing systems (Lee and Fisher 1987). Doped semiconductors have therefore been investigated as photorefractives. CdTe:V has been demonstrated at 1500 nm (Partovi et al. 1990), and GaAs:Cr (Albanese, Kumar, and Steier 1986) and InP:Fe (Valley, McCahon, and Klein 1988) have been demonstrated at 1060 nm.

Readout of photorefractives must be done at a longer wavelength than for activation to avoid further charge generation that would erase the stored information.

11.2.1.5 III–V Compounds and Multiple Quantum Well (MQW) Materials

Application of electric fields to semiconductors in the III–V class can strongly alter the phase and amplitude of transmitted light. In particular, it is possible to obtain strong electroabsorption (change in absorption coefficient κ with applied field) using quantum-well-structures. III–V semiconductors are of great importance both in electronics and photonics; they include not only basic compounds formed from Group III and V elements of the Periodic Table, such as GaAs, GaP, and InP, but also the ternary compounds like GaAlAs and quaternary compounds like InGaAsP. The percentage of the various Group III and V constituents can be adjusted to vary the band-gap and refractive index of the material, and the varied compounds layered to produce LEDs and lasers. Because the band-gap of these materials can be altered, optical devices can be made for various wavelengths. For example, devices with GaAlAs/GaAs layers generally operate at 850 nm, while InGaAsP/InP devices can be made for operation at either the 1300 or 1500 nm region to match wavelengths of interest for optical fiber systems.

It has been difficult to obtain large changes in absorption in bulk GaAs through mechanisms such as the Franz–Keldysh effect—a red shift of the band edge to longer wavelengths with an applied electric field. However, with multiple-quantum-well (MQW) structures, orders of magnitude improvements have been achieved (Miller et al. 1984).

Quantum wells are very thin ($\sim$10-nm) layers of various III–V materials, grown by methods such as molecular or chemical beam epitaxy. For MQW structures there are two or more types of doped or undoped layers that are stacked in some alternating scheme. Excitons in the lower-band-gap layers are quantum-confined, so that the binding energy between electrons and holes is increased. As a result, when electric fields are applied, the excitons do not become ionized, and the exciton absorption line remains sharp. Band shifting still occurs with application of the field, and when large changes in absorption are possible. The absorption change is maximal at an optical wavelength that depends on the material composition. As an example, for a multiple quantum-well stack of alternating GaAlAs and GaAs layers, optimized at 850 nm wavelength, the absorptivity changes from 6×10^3 to 8×10^3 cm^{-1} (Aull et al. 1988). Thus, even with the larger effects with quantum wells, these absorptivity values still require relatively thin layers to minimize the residual absorption. As a result, contrast enhancement is often needed. One enhancement technique is to fabricate the MQW material into an asymmetric Fabry-Perot device, although this can limit spatial resolution and narrows the wavelength band of operation. For optical addressing at high resolution and frame rate, it is possible to use a two-beam coupling configuration between reference and signal beams to create transient spatial gratings within the material (Bowman et al. 1994). These gratings then diffract a third readout beam.

A potential attraction of modulators fabricated with III–V materials is the hope that light sources, modulators, and drive circuits can all be fabricated from these materials and integrated on a single chip. This integration has not been totally achieved at this time because the fabrication processes for the three types of devices are either not entirely compatible, or the entire fabrication process for an integrated chip is complex compared to current capabilities.

11.2.1.6 Optically Absorptive Materials: Silicon and Other Materials

Recent device research has shown promise in the use of silicon as an optical modulating material. The motivations are (i) potential for monolithic integration of photonic and electronic devices using silicon as a common material, and (ii) leveraging of the well-developed and very large silicon integration technology and manufacturing base for silicon photonic devices. However, silicon does not possess significant r_{ij} values for the classic electro-optic effect which involves changes in the real index of refraction. Nevertheless, according to the more general definition of a complex, index of refraction, Equation 11.7 and Equation 11.8, it is possible for an electrical signal to produce an electro-optic effect based on absorption changes in silicon and other materials. A large absorption of light in an optical waveguide of silicon can be obtained with the introduction of free carriers into the waveguide (Soref and Bennett 1987).

11.2.2 Electro-Optic Modulators

11.2.2.1 Temporal Modulators

Major objectives in temporal electro-optic modulators are ultrahigh modulation bandwidth, linearity, and high sensitivity (or low drive power). The primary technique for achieving high bandwidth with good depth of modulation without inordinately high drive power is to use integrated- or guided-optical-wave devices. Integrated optic techniques and devices are described in detail in Chapter 13 and Chapter 14. Bulk modulators, consisting of discrete samples of electro-optic material in the configuration of Figure 11.2 (to obtain amplitude modulation), are possible, but only with high drive power and voltages, and generally at lower wavelengths <1600 nm, as determined by Equation 11.14. Various commercial devices operate from DC to 10 GHz with 1–2% modulation bandwidth. Bulk electro-optic modulators compete with acousto-optic temporal modulators (see Acousto-optic Modulators below). However, the low latency time and high speed of modulation of the electro-optic device can make them attractive in spite of higher power and voltage requirements. Finally, the modulation scheme of Figure 11.2 is inherently wavelength and polarization sensitive.

Efficient phase modulation in integrated-optic waveguide devices is a result of confinement of the optical wave, and from matching of optical velocities with electrical-signal velocities in transmission-line electrodes along a long section of guide ($\sim$cm). Construction of two-arm waveguide interferometers, such as a Mach-Zehnder interferometer with one modulated arm, converts the phase change, produced by the electro-optic effect (see Equation 11.11), into an amplitude-modulated output. Mach-Zehnder device performance in excess of 100 GHz has been shown possible in $LiNbO_3$ devices (Noguchi, Mitomi, and Miyazawa 1998) at the 1330 and 1550-nm bands used in fiber communications. More recent work has concentrated on producing commercial devices with reduced V_π, e.g., ~ 1 V at 28-GHz bandwidth (Nakazawa 2002). Use of organic electro-optic materials, which have comparatively lower values of ε, has resulted in the demonstration of potential devices with <1 V halfwave voltage (Shi et al. 2000), that perform at high bandwidths, >100 GHz (Chen et al. 1997), and that have relatively good thermal stability (Oh 2001).

Temporal modulation has been obtained with guided-wave devices in III–V materials employing either phase modulation (Wakita, Kotaka, and Asai 1992), or electroabsorption in multiple quantum wells (Suzuki, Tanaka, and Matsushima 1992; Devaux et al. 1994). Direct intensity modulation bandwidth as high as 40 GHz has been demonstrated with electroabsorption in InGaAs/InAlAs MQW devices in packaged modules (Yoshino et al. 1999) for fiber-optic system use.

Non-guided-wave temporal modulation has been demonstrated using III–V materials in MQW corner-retroreflector devices for application to free-space optical communication (Rabinovich et al. 2005). Such devices have been demonstrated at both 980 and 1550 nm, wavelengths better able to penetrate many atmospheric conditions. Although device speed is intrinsically high, the length of the free-space link determines the feasible bandwidth.

Use of either a waveguide Mach–Zehnder or a bulk phase modulator in the configuration of Figure 11.2 produces an optical intensity output that varies as sine-squared with drive signal, as in Equation 11.5. Quasi-linear operation is obtained either by operating in the small-signal regime, or by operating around the quadrature ($\pi/4$) point. Pre-distortion of the input signals might be performed to give a linear output relative to the original input (Childs and O'Byrne 1990), but this is often difficult in practice. Operating within the quasi-linear regions has additional negative consequences. For a broadband, multiple-component input signal, the sine-squared transfer function produces spurious in-band intermodulation-product signals. The usable dynamic range is therefore reduced. The production of spurious third-order signals can be seen from a Taylor-series expansion of the modulated quadrature-point signal $1 \pm \sin\{(\pi V(t)/V_\pi)\}$ as $1 \pm (\pi V(t)/V_\pi) \pm \{(1/6)(\pi V(t)/V_\pi)^3 \pm$ higher-order terms$\}$. The nonlinear third-power term will cause a triple-product mixing of the frequencies of two different signal components. Linearized modulators that greatly reduce third-order distortion have been demonstrated using multiple Mach–Zehnder devices, either in parallel or in series (Korotsky and DeRidder 1990; Johnson and Rousell 1988). The third-order term in the Taylor series can be canceled to more than 20 dB by carefully controlling the splitting of the optical power and the rf drive signal to each of the Mach-Zehnder devices (Skeie and Johnson 1991).

11.2.2.2 Spatial Light Modulator Devices

Most spatial light modulator (SLM) concepts and devices have been developed over the course of many decades. Recent activities aim to use newer technologies and materials to improve performance. The development of SLMs is closely tied to developments in the display industry, and leverages many of the same technologies. But, for many SLM applications, as has already been mentioned in the Introduction, differing parameters need to be addressed than those for displays.

In this section, we will tabulate the most important of the electro-optic spatial light modulators (SLMs), and provide some descriptions of optically and electrically addressed SLM devices to guide the reader through Table 11.2. In Table 11.2, the performance parameters are defined as follows:

- Space-bandwidth/resolution is that at 50% MTF.
- Frame rate is for the space-bandwidth/resolution indicated.

The organization of Table 11.2 is as follows:

- The SLM is identified first by modulating material.
- If there is a common name for the modulator it is then given.
- The SLM is then identified as either electrically or optically addressed.
- If more than one method exists for addressing, then performance figures are provided for each.
- Unique aspects of particular SLMs are provided under "Remarks." Availability and development status are also given.

11.2.2.3 Optically Addressed Devices

We will now discuss the various optically addressed devices that employ electro-optic modulating materials.

Electro-optic inorganic crystals are employed in various devices, such as KD*P in the Phototitus (Donjon et al. 1973; Casasent 1978a; Armitage, Anderson, and Karr 1985) and $LiNbO_3$ in the microchannel-plate spatial light modulator (MSLM) (Schwartz, Wang, and Warde 1985). These devices use photoconductors and photocathodes, respectively, as the light-sensitive element. The microchannel plate in the MSLM multiplies the photoelectrons from a photocathode, allowing an alternate means of signal gain (other than a stronger readout beam).

Twisted-nematic liquid-crystal devices are currently the most available of all SLMs. The twisted-nematic liquid crystal light valve (LCLV) SLM was first developed more than two decades ago (Bleha et al. 1978; Augborg et al. 1982); this is an optically addressed device using a layer of CdS or CdSe as the photosensitive layer. Resolution and sensitivity depend on film quality. Alternatively, an array of silicon Schottky diodes has been used for optically addressed LCLV (Sayyah et al. 1991); resolution is limited by the Schottky diode periodicity.

Binary-only ferroelectric devices have been optically addressed with a photosensitive layer of hydrogenated amorphous silicon (a-Si:H) over the ferroelectric liquid crystal (Jared and Johnson 1991; Roe and Schehrer 1993).

TABLE 11.2 Selected Electro-Optic Spatial Light Modulators and Performance

Device/Name	Address Method	Frame Rate (Hz)	Space or Time Bandwidth/Resolution (pixels/mm)	Remarks
KD*P (KD_2PO_4)/Titus	Electrical: e-beam	30	20/mm resolution	Limited commercial availability
KD*P (KD_2PO_4)/Phototitus	Optical:			
	Se film	500	15/mm	
	Si diodes	500	10/mm	
Bismuth Silicon Oxide (BSO) or Bismuth Germaniun Oxide (BGO)/PROM, PRIZ	Optical: BSO, BGO	500	10/mm	Limited commercial availability
$LiNbO_3$/MSLM	Electrical: e-beam (Micro- channel plate)	100	10/mm	Long storage times due to large DC dielectric constant (days to months) Commercially available
$LiNbO_3$/MSLM	Optical: Photo-cathode + Micro-channel plate	100	10/mm	Very high optical sensitivity: 10 pJ/cm^2 Long storage times: days to months Limited Commercial availability
$LiNbO_3$/TIR	Electrical: Si Chip	10,000	20/mm	Commercially available 1D modulator
PLZT	Electrical/Optical:Si transistor/phototransistor	10,000	10 pixels/mm	Laboratory prototype Analog modulation
Twisted Nematic Liquid Crystal/LCTV	Electrical: Thin-film transistor arrays	30	$>1000\times1000$	Widely available
Twisted Nematic Liquid Crystal/LCLV	Optical: CdS/CdSe BSO Si diode array amorph. Si	30	30/mm	Intrinsic decay time limits frame rate Commercially available
Ferroelectric Liquid Crystal	Electrical:	10,000	1024×1024	Commercially available. Binary Modulation/Analog with space/time multiplexing
Ferroelectric Liquid Crystal	Optical:	10,000	32/mm	Commercially available. Binary Modulation
GaAs, Multiple Quantum Well	Optical	6×10^5	160/mm	Laboratory Prototype
GaAs, Multiple Quantum Well	Electrical: Si chip	6×10^5	128×128	Laboratory Prototype
GaAs, Multiple Quantum Well/SEED array	Optical: Si chip	10^8	256×256	Limited availability. High power required 100 MHz capability
GaAs, Multiple Quantum Well	Electrical: Surface acoustic wave (SAW)	10^5	1000 spots time-bandwidth	1D device Laboratory Prototype
Photorefractive $LiNbO_3$	Electrical readout: SAW Optical writing	10^5	1000 spots time-bandwidth	1D device Laboratory Prototype

Ferroelectric ceramics, such as in the PLZT devices (Land 1978), can be optically addressed by overlaying the material with an array of phototransistors. Islands of recrystallized polysilicon are deposited on a PLZT substrate; each Si island contains a phototransistor, which controls the potential across the surface of an adjacent region of the PLZT (Lee et al. 1986). A read beam transmitted through this PLZT region, and on through the substrate, is thus modulated by the transverse electro-optic effect. Alternatively, reflective readout can be obtained by the addition of a mirror and use of a two-pass readout beam configuration. PLZT islands can be deposited onto a silicon chip (Jin et al. 1994).

The electro-optic modulating material itself can also act as the photosensor. The Pockels readout optical modulator (PROM) (Horowitz and Corbett 1978; Owechko and Tanguay 1982) and the PRIZ (Petrov 1981) device use photoconductors, either BSO ($Bi_{12}SiO_{20}$) or BGO ($Bi_{12}GeO_{20}$). In the PROM, the modulating material is sandwiched between two transparent electrodes to which an external field bias voltage is applied. In the PRIZ, the fields are transverse, across the writing face; this configuration was shown to be useful for edge enhancement of images and to have better resolution than the PROM configuration. The bias voltage causes drifting of the photocharge produced by the optical writing beam with a wavelength shorter than the material band-gap. Thus, a spatial refractive-index pattern corresponding to the input pattern is produced by the electro-optic effect (e.g., photorefraction). To avoid producing more photocharge during readout of the device, a beam at a wavelength longer than that corresponding to the band-gap is used. Hence, it is also possible to use such devices as wavelength converters. Other photorefractive materials that have been employed include $BaTiO_3$, KTN, SBN and iron-doped $LiNbO_3$.

Optical activation of electroabsorption changes in III–V MQW, where structures can be obtained via the photovoltages from a photosensor array. Optical readout contrast is enhanced by the MQW material within an asymmetric Fabry-Perot cavity, which maximizes the electroabsorption. However, this results in very narrow operating optical bandwidth, and the spatial extent of the individual Fabry–Perot cavities limits device resolution.

Another means of optical addressing and readout is to use two-beam coupling between a reference and signal beam to create transient spatial gratings in the MQW material that then diffract a third laser beam (Bowman et al. 1994). High spatial resolution (<6 μm) and diffraction efficiency ($>1\%$) is obtained by proton implantation in buffering surface layers to prevent transverse movement of charge, while speed is unaffected since the non-implanted optically active layer retains high charge mobility across the layer. Writing/erasure at rates $>600{,}000$ frames/sec have been shown possible with a 1-μm thick sample.

III–V quantum-well devices can provide nonlinearities needed for optical logic or threshold functions In particular, a transfer function for incident intensity versus output intensity has a bistable character, similar to that of an electrical diode. A MQW structure is sandwiched between layers of bulk GaAs; the interfaces between the MQW section and the bulk material form p-n junctions. This Self Electro-optic Device (SEED) thus acts as a photodetector of the incident light. However, the absorption of light produces a change in the bias voltage on the MQW, altering the optical transmission of the device. The major attraction is that a large 2-D array of SEED devices can be used to implement a very high frame rate SLM for handling 2-D digital data. While the band-shifting mechanisms occur on a picosecond time frame, the speed of actual devices is dependent on incident optical power, and ultimately, the speeds are limited by the capacity for the device to dissipate energy. For example, if a SEED array can dissipate a maximum of $1 W/cm^2$, a maximum speed of 100 MHz results (Yu and Forrest 1993).

11.2.2.4 Electrically Addressed Devices

Electrically addressed SLMs can employ the same electro-optic materials as the optically addressed SLMs. Any material-dictated performance properties will be common to the two.

- *Electron-beam devices.* A scanning electron beam, as in a cathode ray tube, is used with electro-optic crystals, such as KD*P in the Titus (Groth and Marie 1970; Casasent 1978b) and $LiNbO_3$ in the MSLM (Schwartz, Wang, and Warde, 1985). In the latter, faster writing is achieved with a microchannel plate that amplifies the electron beam (rather than photoelectrons as in the optically addressed MSLM).

- *Electrode matrix devices*. An electro-optic modulator with a 1-D array of electrodes is the total internal reflection (TIR) modulator. A VLSI silicon ship, with as many as 5000 electrodes and driver circuitry, is overlaid onto an electro-optic plate of $LiNbO_3$ (Johnson et al. 1983).

Representative of the most mature 2-D active and passive matrix-addressed devices are the various commercial flat panel liquid crystal televisions (LCTVs) (Blechman 1986) and digital light processing (DLP©) micromechanical displays (Brooks 1985; Hornbeck 1989). LCTVs have been modified to perform as inexpensive SLMs (McEwan et al. 1985; Liu, Davis, and Lilly 1985), even though the large pixel size is a disadvantage (see discussion in the Introduction above). However, the small pixel spacing available in liquid-crystal cell phone displays has been used to implement a compact optical-correlation recognition and tracking system (Hartman and Farr, 2003). Customized binary-intensity-only ferroelectric liquid crystal (FLC) modulators have been used in a prototype page-oriented holographic digital data systems with 10 Gbits/s data-transfer rate (Orlov et al. 2004).

It is important to realize that commercial display devices are not specifically built for use with coherent light; their flatness and cosmetic qualities may not be adequate for maintaining optical amplitude and phase. For example, a radius of curvature of less than 10 m will reduce the peak of a correlation function (relative to its sidelobes) by more than 10% (Turner, Johnson, and Serati 1995). Further, large-size pixels in many liquid crystal displays can result in poor diffraction efficiency, as discussed in the Introduction. These considerations are especially critical for applications requiring analog intensity modulation. However, it is possible to perform a limited degree of analog modulation with good accuracy using mature binary-intensity-only modulators. For devices with a large number of pixels, it is possible to use spatial multiplexing to obtain effective grey-scale display from materials that provide only binary response. For example, a 4×4 "super pixel" would provide 16 grey levels. If high-speed response is available from both the electrode matrix and the material (e.g., FLC (compared to twisted nematic) material), then time multiplexing is also possible to provide effective grey scale for applications that temporally integrate light signals (light is switched on for varying times up to the integration period). The maximum number of grey levels is the ratio of the integration period to the device response time.

In the 2-D active matrix devices, placing more circuitry at the grid intersections, as described in the Introduction, makes possible an SLM with "smart pixels." Such SLMs are optically addressable if the circuitry includes a photodetector with drive amplifier for modulation, such as with the PLZT device mentioned above (Lee et al. 1986).

- *Traveling wave addressing*. Acoustic-wave devices have been used for fast serial addressing of electrical signal-storage devices (Cafarella 1978), and of a photorefractive optical memory device (Lee 1983). More recently, addressing of a multiple quantum well optical modulator has been demonstrated using surface acoustic waves (SAWs) (Jain and Bhattacharjee 1989). The device material is essentially the p-i-n GaAs/GaAlAs structure of the SEED device previously described in Section 11.2. Since GaAs is a material that is both piezoelectric and electro-optic, it can be addressed with the electric fields that accompany the SAW (e.g., the acoustoelectric field). A SAW is propagated in one of the bulk GaAs layers, resulting in electroabsorption modulation of the MQW layer. Limitations of such acoustic-wave devices are (1) they are intrinsically 1-D, and (2) the information continuously flows through the device at the sound velocity. These limitations are the same as those discussed below in Acousto-Optic Devices.

11.3 Acousto-Optic Devices

11.3.1 Acousto-Optic Effect and Materials Considerations

In acousto-optic materials, the ultrasonic wave induces refractive-index changes (Δn) via the photoelastic effect. These index changes have the periodicity, amplitude and phase modulation of the acoustic wave, and they act as a phase grating to diffract the incident optical beam. As in any grating, there are generally multiple diffraction orders, but the generation of multiple orders is usually not technologically desirable; this Raman–Nath regime (Lee and VanderLugt 1989) arises when the grating is weak (Δn small) and the interaction distance short. Hence, the interaction length is increased. With a sufficiently thick diffraction grating (thickness Z) at a particular incident angle, only one diffraction order persists with destructive interference occurring at all other diffraction angles. In an isotropic medium, maximum diffraction efficiency is obtained when the incident light direction is at an angle θ_B to the acoustic wave-front. The diffracted light is also at θ_B (deflected by an angle $2\theta_B$). The case of anisotropic media will be discussed in Acousto-optic Modulators. This mode of diffraction is called the Bragg regime, in analogy with the Bragg effect occurring with x-ray diffraction in crystal lattices. θ_B is called the Bragg angle and, for the isotropic case, is given by (Berg and Lee 1983)

$$\sin\theta_{\mathrm{B}} = \lambda/(2\Lambda) = \lambda f/2v \tag{11.16}$$

where λ is the optical wavelength, Λ is the acoustic carrier wavelength, v is the acoustic velocity and f is the acoustic frequency. Bragg diffraction is therefore the basis for all acousto-optic light devices.

The basic construction of an acousto-optic device is shown in Figure 11.3. The device is electrically addressed by a temporal signal. A piezoelectric material is used to construct an electrical-to-acoustic transducer that is attached to the acousto-optic medium. The transducer is driven by signal r(t), giving the acoustic space/time signal $f(t-x/v)$. The thinner the transducer is, the higher the operating frequency (Young and Yao 1981). Mechanical thinning techniques allow devices that operate at rf frequencies up to 4 GHz; however, acoustic attenuation (discussed below) may limit frequency before this point.

The medium is illuminated by a light beam of wavelength λ that may be modulated in both space and time. The thickness of the acoustic signal Z is usually made sufficiently large so that one operates in the Bragg regime, where there is only one diffraction order of significant amplitude, as described by coupled-mode analysis involving the diffracted and undiffracted light (Phariseau 1956). Quantitatively, we characterize the transition from multiple-order diffraction (Raman–Nath regime) to the Bragg regime with the expression (Klein and Cook 1967)

$$Q = (2\pi\lambda Z)/n_o\Lambda^2 \tag{11.17}$$

For $Q>7$, the intensity of the first diffraction order exceeds 90% of the maximum.

Acousto-optic modulators can also be fabricated using thin-film optical waveguide structures (Berg and Lee 1983). Then, surface acoustic waves (SAWs) are employed. A SAW interdigital transducer is deposited on top of the optical waveguide. Materials that are both piezoelectric and photoelastic are favored, since this results in the simplest construction. Hence, materials such as $LiNbO_3$ and $LiTaO_3$ are

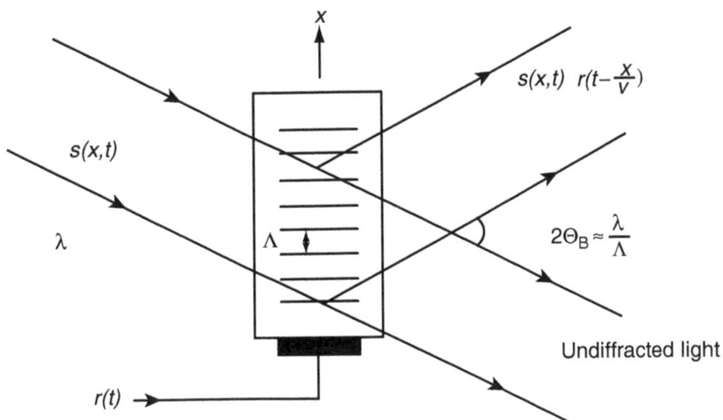

FIGURE 11.3 Multiplication operation with an acousto-optic cell.

popular for acousto-optic-SAW devices. Acoustic attenuation at higher frequencies (see Equation 11.22) and line width limitations in the construction of practical SAW transducers constrain these devices to about 1 GHz operating frequency.

In addition to a deflection of the beam in the acousto-optic interaction, there is also a multiplication of the data on the light carrier by that on the acoustic wave. The multiplicative operation of an acousto-optic cell is illustrated in Figure 11.3, where an electrical signal $r(t)$ is transducted into an acoustic signal, which then diffracts an spatio-temporal optical signal $s(x,t)$ at wavelength λ. The amplitude and phase of the diffracted light are the products of those for the incident light and the acoustic wave. Further, the frequency of the light is shifted by the frequency of the acoustic wave in a direction so as to conserve momentum in the phonon–photon interaction, or equivalently, to obey the Doppler-shift relationship. The diffracted light therefore represents single-sideband modulated information. It is possible to have either Doppler-frequency upshift or downshift in the diffracted light; the upshift case is shown in Figure 11.3, while the downshift case occurs for the $-\theta_B$ angle of incidence. For a given drive signal, the data on the diffracted light in these two cases are complex conjugates of each other. The drive signal $r(t)$ must be a real signal, and it must be on a sinusoidal carrier, since acousto-optic cells cannot support low temporal or spatial frequencies (Lee and VanderLugt 1989). However, $r(t)$ can represent complex quantities (Berg and Lee 1983); for example, $r(t)$ can be made the sum of two real components whose carriers are in quadrature.

Aside from the traveling-wave character of the index changes, the optical modulation is entirely analogous to the method employed on many SLMs using data encoded on a spatial carrier. The product of the time aperture (T) and the bandwidth (B) of the acousto-optic cell is equal to the number of deflection spots (or pixels) possible in an acousto-optic spatial light modulator. A diffraction-limited spot has an angular extent given by

$$\delta\theta = \lambda/D = \lambda/(vT) \tag{11.18}$$

but the total angular capability of the device can be very closely approximated as, from Equation 11.16,

$$\lambda\Delta f/v = \lambda B/v. \tag{11.19}$$

The ratio of Equation 3.4 to Equation 3.3 is therefore the number of resolvable spots, or TB.

Material parameters strongly affect achievable device diffraction efficiency, bandwidth, and/or aperture size. Most good acousto-optic materials have been identified by noting the strong dependence of various figures of merit on the index of refraction, n, and the acoustic velocity, v. The popular figure of merit, M_2, is obtained from the expression for diffraction efficiency (Dixon 1967a, 1967b),

$$I_o/I_{in} = C_o\sin^2\{\sqrt{M_2}(\pi^2/2\lambda^2)(L/H)C_{rf}P_{rf}\} \tag{11.20}$$

where C_o and C_{rf} are constants relating to gain/loss in the optical and rf systems, L and H are the acoustic transducer dimensions along and perpendicular to the light, respectively. The sine-squared relationship again arises from a coupled-mode formalism (Yariv 1989), as described in Electro-optic Devices. The constant, M_2, is related to material properties by (Uchida and Niizeki 1973)

$$M_2 = (n^6p^2)/(\rho v^3) \tag{11.21}$$

where p is the photoelastic constant and ρ is the material density.

Table 11.3 summarizes properties for the most important existing and emerging acousto-optic materials. It is worth noting that it is not always desirable to focus on the optimum for any one single parameter. For example, $LiNbO_3$ and TeO_2 (Uchida and Ohmachi 1969) are popular in large measure due to their availability, in spite of being non-optimum in figure of merit and attenuation, respectively. Availability of material means that good, homogeneous optical transmission, low defect levels (to minimize optical loss), and low optical scatter are routinely obtained (Brown, Craig, and Lee 1989).

TABLE 11.3 Selected Acousto-Optic Materials and Properties at Room Temperature

Material	Index of Refraction for Optical Mode	Acoustic Mode and Velocity (10^5 cm/s)	Figure of Merit, M_2 (10^{-18} s^3/g)	Acoustic Attenuation at 500 MHz (dB/ s)
λ=633 nm				
Fused silica (SiO2)	1.46–⊥	Long. 5.96	1.56	1.8
Dense flint glass (SF-59)	1.95–‖	Long. 3.26	19	
LiNbO3	2.20–‖	Long.[100] 6.57	7	0.03
		Shear[100] 3.60	13	
TeO2	2.27–⊥	Long.[001] 4.26	34.5	1.6
	2.27–‖	Shear[110] 0.617	793	73.0
		Long[a].[110] 4.65	4.4	0.6
		Shear[a] 14.6 deg. from [110] 2.31	5.7	1.0
GaP	3.31–‖	Long.[110] 6.32	44.6	0.6
		Shear[100] 4.13		
	3.31–arb.		24.1	
TiO_2	2.58–[010]	Long.[100] 8.03	3.9	0.11
PbMoO4	2.39–⊥	Long.[001] 3.66	36.1	1.4
	2.26–‖		36.3	
H_2O	1.33–arb	Long. 1.5	126	90.
λ=1150 nm				
As_2S_3 glass	2.46–‖	Long. 2.6	347	1.7
Tl_3AsS_4	2.63–‖	Long.[001] 2.15	510	0.3
λ=10,600 nm				
Ge	4.00–‖	Long.[111] 5.50	840	1.4
	4.00 –arb.	Shear[100] 4.13	290	0.7
Te	4.80–‖	Long.[100] 2.2	4400	6.0

Note: M_2 varies with wavelength and temperature, due to dependency of the parameters in M_2 on such ‖ and ⊥ indicate optical polarization directions with respect to acoustic-wave propagation direction

[a] (Pape 1998).

Note that in Table 11.3 values of M_2 are specified at particular wavelengths at room temperature. Equation 3.5 and Equation 3.6, as well as Table 11.3, show significant dependencies on wavelength, polarization, and temperature, both explicitly and implicitly in quantities such as n.

To find the maximal values of M_2 for a given material requires detailed calculations. The stiffness tensor of a material is used in solving the Christoffel matrix (Dieulesaint and Royer 1980), giving acoustic-mode velocities and polarizations. The strain tensor associated with these modes is used to calculate a strain-perturbed index ellipsoid, which in turn is related to an optical polarization direction to give an effective photoelastic constant, and hence the resultant M_2. Even though there are only a few good acousto-optic materials, calculations are needed for each to identify optimum acoustic modes and directions for particular applications. Hence, Table 11.3 provides data on different acoustic modes for a given material.

Also listed in Table 11.3 is acoustic attenuation, another material parameter of importance to acousto-optic device design. Attenuation of the acoustic wave limits the center frequency, bandwidth of operation and the aperture size. As noted above, the number of spots in a deflector is a product of the time aperture and the bandwidth of the deflector. Further, high attenuation can result in heating of the acoustic material, which in turn can lead to effects such as change in the acoustic velocity (and hence change in M_2) and defocusing of the optical beam. Attenuation is shown in Table 11.3 in units of dB per microsecond of acoustic propagation time. While the basic mechanisms of acoustic attenuation are known (Woodruff and Ehrenreich 1961; Uchida and Niizeki 1973), observed values depend on the material quality. Theoretically, the dependence of attenuation on frequency (f) is expected to be f^2 (Woodruff and Ehrenreich 1961)

$$a = (f^2 T)/(v^4) \tag{11.22}$$

where T is the absolute temperature. Therefore, the acoustic attenuation coefficients in Table 11.3 are normalized to the value at 500 MHz. Deviations from quadratic are often observed in practice (Chang and Lee 1983). This is usually ascribed to variation in material quality.

Materials such as $LiNbO_3$ and rutile (TiO_2) have the lowest attenuation coefficients (tenths of a dB per cm at 1 GHz acoustic frequency). $LiNbO_3$ has benefited from intense materials-growth efforts to improve its quality, but the M_2 of $LiNbO_3$ is still not particularly high compared to other materials such as TeO_2 (7 vs. 793). However, acoustic attenuation limits the usable bandwidth of TeO_2 Bragg cells to about 50 MHz. The throughput limit for an acousto-optic system is determined by the *product* of the bandwidth and the time-bandwidth (Lee and VanderLugt 1989). The extraordinarily slow acoustic velocity of TeO_2 thus allows for large time-bandwidth product devices (>5000). It is therefore of interest to search for acousto-optic materials with even slower velocities than TeO_2 for larger time-bandwidth capability. There has been promising crystal-growth work with the mercurous halide compounds, Hg_2Cl_2 and Hg_2Br_2 (Goutzoulis and Gottlieb 1988), which have about half the acoustic velocity of TeO_2.

The number of possible acoustic modes covered by Table 11.3 is somewhat limited due to space considerations. Extensive compilations of the acoustic and acousto-optic properties of materials can be found elsewhere for those with unique requirements (Uchida and Niizeki 1973; Blistanov et al. 1982).

11.3.2 Acousto-Optic Modulators

Acousto-optic devices have been very effective as temporal point modulators and one-dimensional (1-D) spatial light modulators with unique advantages for use as a tool in many optical systems, such as for optical processing systems (Berg and Lee 1983). The 1-D acousto-optic SLMs (or Bragg cells) have, perhaps, undergone the most successful development of any SLM. Primary issues in such devices are bandwidth and diffraction efficiency, and linearity. Diffraction efficiency is given by Equation 11.20, and has been discussed in the previous section, primarily from a materials perspective. However, the nonlinear sine-squared response of acousto-optic devices in Equation 11.20 places limitations on wide-bandwidth, or multiple-frequency performance, due to generation of harmonic distortion and in-band frequency components. As in the case of the electro-optic modulators described above, one can consider quasi-linear operation about the quadrature point, but this does not totally eliminate the production of such intermodulation products. Quadrature point operation has the additional difficulty of requiring much more electrical drive power for the modulator; the consequent requirement for high-power, distortion-free amplifiers is difficult to satisfy for high bandwidth operation. For the multi-frequency, small signal case, the usable dynamic range is the ratio of the intensity of intermodulation products to the signal intensity at the true frequency. To keep the intermodulation-free dynamic range under control, it is necessary to limit the diffraction efficiency per frequency to about 1% (Hecht 1976).

11.3.2.1 Temporal Modulators

For temporal-only modulation, we can visualize the acousto-optic cell in Figure 11.3 to have zero length, so that the space/time signal $f(t-x/v)$ degenerates to $f(t)$, giving a "point" modulator. The modulation bandwidth is inversely proportional to the transit time of the acoustic wave through the incident light beam, and diffraction efficiency is highest near the transducer for the acousto-optic cell. Hence, the cell is illuminated with a focused cone of light near the transducer. The diffracted light is a similar cone modulated by $f(t)$. With optical beam focusing, typical bandwidths of commercial devices are $>$ 100 MHz, with >30 dB extinction ratio and 10% diffraction efficiency. However, tradeoffs exist. If diffraction efficiency is maximized by increasing the value of Q (Equation 11.14), then the output cone becomes elliptical due to the optical-acoustic wave mismatching (Young and Yao 1981). For high diffraction efficiencies, $\sim$90%, a longer acousto-optic interaction length (Z) and a wider optical beam are required, and modulation bandwidth is greatly reduced, to $\sim$5%. Finally, it is not possible to place a strongly focused cone of light infinitesimally close to the acoustic transducer, so there is an inevitable latency in the application of the drive signal and optical response.

The acousto-optic modulator has several uniquely attractive features.

- While such modulators are generally used in conjunction with laser beams, it is also possible to modulate several wavelengths simultaneously, or even white light.
- The modulator also shifts the frequency of the optical beam, which is desired in some applications.
- Although an alternative is to directly modulate the current in a light source, such as a laser diode, sometimes the acousto-optic modulator provides better coherence in the modulated beam than in directly modulated lasers or LEDs.

11.3.2.2 Spatial Modulators

11.3.2.2.1 One-Dimensional Devices.

1-D Bragg cells perform optical deflection, and they fill particular, but important roles in optical processing. However, since they are 1-D, a single acousto-optic cell exploits only a fraction of the capability of optics for processing. While optical capability can be increased by using a multiplicity of cells, operating in parallel, such devices still only handle 1-D data—temporal data. Further, there is no information-storage capability as the information is carried on an acoustic wave that flows continuously through the cell. Nonetheless, development of 1-D optical processors has been pursued for real-time high-bandwidth time-series data processing, and such processors are generally more advanced than true 2-D processors. Reasons for this include achievement of very high throughput with 1-D light modulators, and slower development of 2-D SLMs. For example, with acousto-optic devices it is possible to input data at several gigasamples per second. Further, it is possible to use acousto-optic devices to perform some 2-D processing operations (e.g., ambiguity functions and folded spectra [Turpin 1981]); these are operations that can be factored into multiple 1-D terms and entered into orthogonally-situated acousto-optic cells.

For some advanced Bragg cells constructed of birefringent material, low values of M_2 (low efficiency), have been overcome by employing anisotropic Bragg diffraction. With the isotropic case described by Equation 11.16, there is a strong tradeoff between the cell bandwidth and the diffraction efficiency due to phase mismatch (Dixon 1967a) (i.e., momentum is not conserved over the entire interaction bandwidth), since, except for at one acoustic frequency, the optical momentum (K) vectors are not perfectly connected by the acoustic K-vector ($K_{acoustic}$). Phase matching is achieved over a wide bandwidth, but with reduced diffraction efficiency, achieved by using acoustic transducers with very small lateral dimensions; acoustic diffraction (due to the finite extent of the transducer) distributes the acoustic energy over a wide range of angles. In birefringent materials and over specific frequency ranges, the loss of diffraction efficiency can be minimized and large bandwidths obtained, with the technique of tangential phase matching (Dixon 1967a), which acousto-optically couples oppositely polarized diffracted and undiffracted modes. This technique is illustrated in Figure 11.4. The light vectors K_{in} and K_{out} each terminate on one of two optical-velocity (or refractive-index) surfaces (for ordinary or extraordinary polarized light). The incident and diffracted light directions are generally not symmetrically situated with respect to $K_{acoustic}$ as in the isotropic case, since the optical-velocity surfaces can be ellipsoidal rather than spheroidal in the isotropic case. $K_{acoustic}$ connects optical K vectors, as desired, except that it is almost tangential to one optical-velocity surface. Only small changes in the angular direction of the acoustic wave-vector with frequency are required to maintain near-conservation of momentum. Hence, the need for acoustic diffraction is minimal, and it is possible to increase the length of the transducer along the direction of the optical beam, thereby recovering diffraction efficiency. As the acoustic frequency increases, for a fixed direction for $K_{acoustic,}$ exact phase matching occurs at two points, f_2 and f_1, giving the wideband double-peaked band frequency the response seen in Figure 11.4. The degree to which this can be achieved depends on the shape of the optical velocity surfaces of the birefringent material, and hence on the crystal "cut" or the acoustic-wave direction. From the definition of M_2, we want a crystal cut where the photoelastic constant, p, and the optical index, n, are maximized, and the acoustic velocity, v, is minimized. Using such an optimization procedure, shear mode $LiNbO_3$ cells have been demonstrated with diffraction efficiencies of 44%/rf watt over the 2–3 GHz band and 12%/rf watt over the 2–4 GHz band at 633 nm wavelength (Chang and Lee 1983; Anderson et al. 1991).

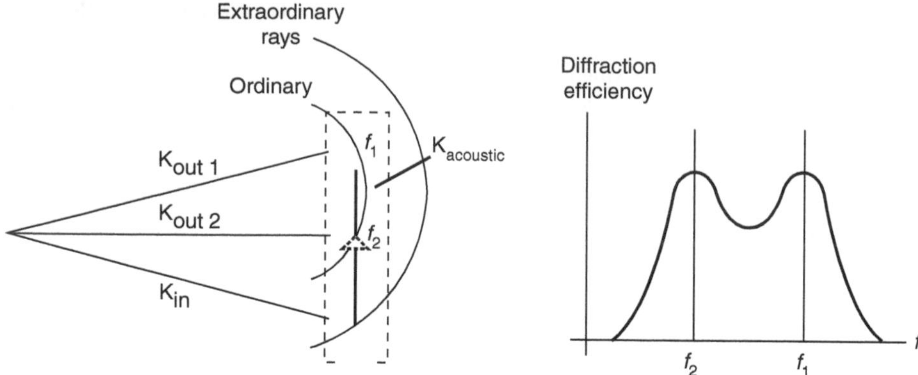

FIGURE 11.4 Anisotropic acousto-optic Bragg diffraction.

11.3.2.2.2 Two-Dimensional Devices.

Acousto-optic devices can be extended to two-dimensional architectures by either, (i) constructing an array of 1-D devices in a single crystal, or by (ii) constructing a device with acoustic transducers on orthogonal edges of a crystal. In either of these approaches a large, high quality crystal is required. For multi-channel, acousto-optic devices, a large number of channels is desirable, but in the absence of electrical crosstalk, the major fundamental limitation is acoustic diffraction. Acoustic diffraction depends not only on the dimensions of the transducer, but also on material. Ideally, the near-field acoustic wavefront is planar over the aperture of the transducer, and diffraction effects are observed in the far-field. Hence, it is desirable to maintain a near-field condition over as long a propagation distance as possible. In an isotropic medium, the transition from near to far field occurs at a distance from the transducer of approximately

$$D = H^2/8\Lambda \tag{11.23}$$

where H is the vertical dimension of the transducer and Λ is the acoustic wavelength. However, for an anisotropic medium, D can be increased (Hecht and Petrie 1980). Because of the severe acoustic spreading in shear mode TeO_2, multi-channel devices have been constructed only of longitudinal mode TeO_2. A 32-channel longitudinal mode TeO_2 device at a center frequency of 250 MHz has been demonstrated (VanderLugt, Moore, and Mathe 1983; Amano and Roos 1987).

The second 2-D approach uses orthogonally propagating acoustic waves. Using two separate 1-D cells, orthogonally situated, and employing anamorphic optics between the cells to pass light from one to the other, one can perform 2-D beam steering and 2-D processing (Turpin 1981). A much more compact alternative to individual 1-D devices is to employ a single large square crystal with transducers along two edge facets and light propagating perpendicular to the square aperture. This approach is feasible only in cubic materials. TeO_2 satisfies this criterion, and shear-mode devices have been demonstrated. However, if the two acoustic waves propagate within the same volume of crystal, nonlinear mixing of the signals will occur at lower power levels; this is particularly true with TeO_2, which exhibits acoustic nonlinearities at relatively low power levels. Thus, devices are made with the acoustic transducers offset from each other along the optical-beam direction. The transducer dimension along the optical path and the offset distance must be minimized to prevent loss of resolution, or to avoid use of optics with large depth of focus.

11.3.2.3 Optical Wavelength Control and Filtering

Many optical modulators are sensitive to the wavelength of operation. However, acousto-optic modulators are unique in that, without a large degree of modification, they can be used as wavelength-selective elements. Specific wavelengths can be selected from a broadband input spectrum, by driving an acousto-optic device with an appropriate set of rf frequencies. This filtering action is notable in that a multiplicity of wavelengths may be selected simultaneously, unlike optical glass and interference filters, and resonant tunable structures such as Fabry–Perot cavities. This multiple-wavelength capability allows applications such as:

- Color correction in displays (Young and Belfatto 1993).
- Wavelength and spectral modulation as an added dimension in information processing, such as in wavelength division multiplexed fiber optic systems (Smith et al. 1990).
- Use in laser light shows.

Acousto-optic tunable filters generally operate in the anisotropic diffraction mode (see Figure 11.4 and the earlier description) in order to maximize the distance over which optical and acoustic waves are phase matched (Chang 1983), since the longer the resonance interaction length the higher the spectral resolution. Therefore, the rf frequencies required for specific wavelengths is governed by phase matching conditions for anisotropic diffraction (Cohen 1967; Dixon 1967a). Both bulk and waveguide devices are possible. The proper color is extracted by selecting the polarization of the diffracted light. Waveguide devices are able to be integrated with optical fiber, and utilize the polarization anisotropy in thin-film optical waveguides, in addition to material polarization anisotropy (Smith et al. 1990) for wavelength selection.

11.4 Magneto-Optic Devices

If plane-polarized light passes through an optically transparent magneto-optic material, the Faraday effect rotates the plane of polarization of the light as a function of magnetic field strength, H, and sample thickness, d, along the light-propagation direction according to

$$\vartheta = VHd \tag{11.24}$$

where V is the Verdet constant, defined as the rotation per unit field and path length. The polarization change can be employed for either phase or amplitude optical modulation using polarizers such as described for electro-optic modulation (Figure 11.2). ϑ depends only on the direction of H and the sign of V, not on the light propagation direction, making the effect useful for isolation against back-reflected optical beams. The polarization rotation arises from a difference in the refractive indices for right-handed and left-handed circularly polarized light; it can be shown that the normal modes of a light wave in a magnetized medium are circularly polarized (Jenkins and White 1976). Faraday rotation can occur in materials either with or without a permanent internal magnetization. An example of the former case is Terbium Gallium Garnet, with $V = .002$deg/(cm-Oersted) at a wavelength of 1064 nm. The latter case involves materials with paramagnetic or diamagnetic atomic species such as doped glasses. A representative paramagnetic glass is terbium-doped borosilicate glass with $V = 0.0058$ deg/(cm-Oersted) at a wavelength of 531 nm (Butler and Venturini 1987).

For materials with internal magnetization, an externally-applied field is necessary only to saturate the magnetic state. The polarization rotation in the magnetically saturated state is described by the Faraday rotation ϕ_F with units of degrees/cm. Ferrite garnets are perhaps the best materials with strong Faraday rotation, and include yttrium iron garnet (YIG) with $\phi_F = 280$ deg/cm at a wavelength of 1000 nm (Dillon 1978). Bismuth-doped YIG films exhibit $\sim 10 \times$ larger values of ϕ_F. Values of ϕ_F are strongly dependent on wavelength; f_F usually increases steeply at shorter wavelengths, but optical absorption also tends to increase commensurately. Hence, a figure of merit is often the ratio of ϕ_F to absorption.

A second method of employing magneto-optics is to use magnetostatic waves for light modulation. 1-D spatial light modulators have been constructed using magnetostatic, traveling waves in ways analogous to 1-D acousto-optic devices (Fisher et al. 1982; Tsai et al. 1985). Magnetostatic waves (MSWs) are slow-velocity electromagnetic waves that propagate in magnetic ferrites (Damon and Eshbach 1961), such as thin-film YIG, under application of a bias magnetic field. MSW-optical devices are electrically addressed in a similar way to surface-acoustic-wave acousto-optic devices (see Section 11.3); the electrical-to-MSW transducer is usually a simple microwave stripline antenna on the surface of the film. The major differences with acousto-optics are that modulation can be performed at high microwave frequencies, up to 30 GHz, the frequency of operation and wave velocity are adjustable with the bias field, and the devices must be thin-film (see Magnetostatic Wave Interactions and Devices).

Faraday-rotation and MSW spatial light modulator devices are presently either specialty commercial items, or still in the research phase, as discussed in the following paragraphs.

11.4.1 Faraday-Rotation Materials and Devices

Both temporal and spatial light modulators employing Faraday rotation have been constructed. Faraday rotation is used in a crossed-polarizer configuration, like that of Figure 11.2, to produce intensity modulation, with the magneto-optic device replacing the electro-optic device in the Figure. A temporal modulator is generally constructed of garnets such as TGG, YIG, and bismuth-doped YIG. An external magnetic field allows control of the amount of polarization rotation. 2-D spatial light modulators have been constructed by fabricating an array of Faraday-rotation elements from films of garnet material such as bismuth-doped YIG (Ross, Psaltis, and Anderson 1983). To date, only electrical matrix-addressed SLM devices have been constructed; the addressing issues are as discussed in the Introduction.

In the electrically addressed 2-D SLMs, thin-film material is pixelated into mesas, with the surrounding channels containing the addressing electrode. Application of current to the surrounding electrodes transforms a mesa into a single magnetic domain in one of two opposite magnetic orientations. The two orientations produce equal but opposite Faraday rotation in the polarization of a transmitted readout beam, while an unaddressed mesa is observed to maintain the input polarization. If the magneto-optic material is placed between crossed polarizers, as in Figure 11.2, either two or three discrete amplitude modulation levels can be obtained. In the former, the rotated polarization state due to one orientation, say the clockwise-rotated state, is aligned with the input polarization, and every pixel is addressed; this mode maximizes contrast. For the latter, or ternary-state case, the unrotated state is also used, and its amplitude transmittance sensed. Using an optical arrangement that senses the direction of polarization rotation, polarization- or phase-encoding is possible, since the optical phases for the two orientations are 180-degrees to each other (Ross, Psaltis, and Anderson 1983).

Operation of the device requires that one corner of each mesa be modified to allow a pixel to be activated, but without activating any of the other three neighboring pixels. In one type of device, one corner of a pixel is ion-implanted to lower the threshold for domain switching. The conjunction of currents in both crossing matrix electrodes at the ion-implanted corner produces high magnetic fields that induce reversal of the magnetic domain of the pixel. In another type of device, a small resistive element at a corner thermally induces domain flipping. Framing rates of 1000 frames/s, and display sizes of 512×512 have been demonstrated in commercial devices.

One of the advantages of the magneto-optic SLM is its intrinsic, long-term storage capability, since domain orientation remains constant in the absence of an addressing signal. Therefore, the magneto-optic SLM does not require periodic frame updating, and random pixel-by-pixel addressing is possible. Random pixel addressing can also reduce average power consumption. The required currents for switching can lead to the perception that magnetic devices have high power consumption. To reduce switching current requirement, a thinner mesa with higher Faraday rotation can be used. An added benefit is that larger arrays at a given power level are possible. Hence, bismuth-doped films have been employed in more recent devices. Faraday rotation increases with bismuth content. However, lattice constants increase with doping. Therefore, to maintain lattice matching between the grown film and the substrate requires so-called large-lattice-constant substrates of doped gallium gadolinium garnet (GGG).

11.4.2 Magnetostatic Wave Interactions and Devices

Optical modulation with MSWs is very similar to that shown for acousto-optics in Figure 11.3. The MSWs induce refractive index changes in the optical medium. Conservation of momentum results in deflection of the incident light and a frequency shift by the MSW frequency. But, there are major differences between MSW-optical and acousto-optic device parameters.

- MSW velocities are roughly 10 × higher than typical acoustic velocities. Thus, MSW delay lines are several hundreds of nanoseconds in length, compared to microseconds for acoustic-optic delay lines. Consequently, frame time for MSW 1-D modulators can be as fast as 100 ns, but time-bandwidth product is presently limited to about 100.
- MSW center frequencies are in the microwave region, from 1–30 GHz, and can be tuned with an applied external magnetic field. By contrast, acousto-optic devices operate at no more than 4 GHz presently, due to acoustic attenuation. While modulation at these high microwave frequencies is also possible with laser diodes and with traveling-wave electro-optic modulators, these alternatives provide only temporal, not spatial, light modulation. MSWs, therefore, allow direct light modulation at microwave frequencies with potentially large time-bandwidth product.
- MSWs must generally be propagated in thin-film geometries. In bulk samples of ferrite, shape-factor demagnetizing fields within the sample are generally not spatially constant, leading to spatial inhomogeneities in MSW energy density and non-constant wave propagation. However, a thin, infinite sheet has constant shape-factor demagnetization. Thus, these optical modulators are best employed in a guided-wave configuration. For example, a YIG film is grown on a lattice-matched substrate of gallium gadolinium garnet (GGG), and serves as both magneto-optic medium and optical waveguide. Highest quality growth is usually with liquid-phase epitaxy techniques.
- MSWs are excited with a stripline antenna, as opposed to either a bulk piezoelectric plate or suface-acoustic-wave transducer for acousto-optic devices.

Optical interactions with MSWs were first demonstrated in epitaxial YIG films (Fisher et al. 1982). Subsequently, devices for spectrum analysis and correlation (Fisher 1985; Tsai et al. 1985) have been demonstrated. Diffraction efficiencies of about 4%/watt were observed at an optical wavelength of 1300 nm, and bandwidths of 30 MHz in a homogeneous bias field, and 350 MHz in gradient fields (Wey et al. 1986) have been achieved. The gradient field approach, which biases different spatial regions of the sample to different center frequencies, has potential for bandwidths as high as 1 GHz, but with lower diffraction efficiency than the homogeneous case.

Higher diffraction efficiencies are desired, and bismuth-doped iron-garnet films such as Bi:YIG have been investigated for this purpose (Tamada et al. 1988; Butler et al. 1992). The theory for the optical-MSW interaction shows that the efficiency scales as the square of ϕ_F (Fisher et al. 1982; Fisher 1985). Bismuth-doped films have ~10 × larger Faraday rotation constants than pure YIG, giving the potential to provide 100 × higher diffraction efficiencies. Experimentally, a diffraction efficiency of 12% over a tuning range of 3.7–12 GHz, and a bandwidth of 150 MHz using Bi:YIG with $\phi_F = -2313$ deg/cm has been reported (Young and Tsai 1989). Since optical transparency, MSW attenuation, and Faraday rotation must all be simultaneously optimized, to date there has been limited use of these doped materials in optical-MSW devices.

11.5 Mechanical and Micromechanical Devices

Mechanical devices are perhaps the oldest method for light modulation; these include high-speed shutters, such as found on cameras, rotating slotted-disk optical choppers, and vibrating resonant structures, such as tuning forks (Woodruff 1991). Such mechanical devices are generally limited to temporal modulation only. Micromechanical mechanisms, employed in micro-electromechanical structures (MEMS), have also been employed to temporally and spatially modulate light via the physical movement of very thin (<0.1 mm) membranes of polymer materials or silicon structures (e.g., cantilevered beams, torsion bars, optical waveguides, etc.)(Somers 1972; Brooks 1985; Pape 1985; Fisher et al. 1986; Hornbeck 1989; Ohkawa, Izutsu, and Sueta 1989; Monteverde 2004). These modulators are summarized below and in Table 11.4.

11.5.1 Mechanical Devices

Temporal modulation is achieved with mechanical shutters and choppers. A shutter allows both the frequency and the duty cycle to be varied, while in a chopper these parameters are fixed. For example, a tuning-fork chopper provides a sinusoidal intensity modulation, and a rotary slotted-plate chopper gives a triangular or trapezoidal modulation depending on slot size relative to the beam. The maximum frequencies for shutters and choppers are generally in the 1–10 kHz regime, limited by inertia. Advantages of these mechanical systems include independence of modulation on wavelength, and the ability to handle high optical power levels. In practice, the opaque portions of shutters and choppers must completely absorb or redirect the optical beam without scattering light into the modulated beam direction. A mechanical modulation device that combines most of the desired features is the magnetic flexure shutter, consisting of a single flexing, optically reflective, blade at a near-grazing angle to the incident optical beam direction. It offers high speed (>10 kHz), reliability, and high optical power-handling capability (up to 100 MW) with little optical backscatter (10^{-6} of incident) (Woodruff 1991).

11.5.2 Silicon Mechanical Devices

For temporal modulation only, guided-wave Mach-Zehnder interferometer devices have been demonstrated with either glass waveguides (Ohkawa, Izutsu, and Sueta 1989), or with a silicon dioxide/silicon nitride stack (Vadekar, Nathan, and Huang 1994) deposited onto a silicon substrate. These devices have been used for optical sensing of acoustic signals. One arm of the interferometer is subjected to an applied pressure, resulting in a change in optical path length and thus, an optical phase shift with respect to the reference arm. Mode attenuation due to bending has been shown to be negligible, and efficiency was increased by acoustic impedance matching to the applied source (Vadekar, Nathan, and Huang 1994).

For spatial modulation, the most popular approach is to use an array of optically reflective micromechanical silicon structures to reflect a light beam in or out of a readout optical system. A microscopic array of structures, such as thin, cantilevered beams (Brooks 1985) and torsion beams supported by posts above a corresponding array of addressing circuits on a silicon ship, are fabricated entirely of silicon by an etching process (Hornbeck 1989). These structures are the basis for the widely available commercial Digital Light Processing (DLP©) flat-panel display that is presently widely available, although the literature on its principles originates from more than two decades ago. A related device, the Grating Light Valve (GLV), employs an array of parallel silicon microribbons to diffract light through the Schlieren readout system (Monteverde 2004). The silicon circuit below these micromechanical structures provides addressing and activation voltages that deflect the structures and hence, the incident light at each pixel position. The deflection changes the angular position of the reflected light beam, producing intensity modulation in the readout system. Recent efforts have been focused on large 2-D arrays (e.g., 1920×1080 pixels for digital display, HDTV, printing, and scene simulation applications). Addressing of such arrays at a high frame rate is a problem. Electrical addressing can require a number of parallel high-speed lines. Optical addressing requires the integration of photodetectors as part of a large optoelectronic integrated circuit. Other optical and electrical addressing issues, as discussed in the Introduction, generally involve VLSI fabrication issues rather than materials issues.

TABLE 11.4 Selected Micromechanical Spatial Light Modulators and Performance

Device/Name	Address Method	Frame Rate (Hz)	Space-Bandwidth/Resolution (pixels/mm)	Remarks
Deformable membrane/Membrane light modulator—MLM	Optical:			Limited availability
	Si photodiodes	100	10/mm	
	Electrical:	100	100×100	
Deformable elastomer/RUTICON	Optical:			Limited availability
	Amorphous Se	100	100/mm	
	Electrical: e-beam	25	15/mm	
Deformable membrane device—DMD	Optical:			Limited availability:
	Si photo—transistors	500	128×128	Superceded by cantilevered and torsion beam devices
	Electrical: Si CCDs/transistors	500	128×128	
Cantilevered micro-mechanical beams—SiO_2 or metal film —DMD	Electrical: Silicon circuits	500	256×256	Commercially available
		30	640×480	
Thermoplastic	Optical: PVK:TNF	10	1000/mm	Commercially available, limited lifetime
Photoemitter membrane light modulator	Optical: Photo-cathode + Microchannel plate	1000	80/mm	Laboratory prototype, very high optical sensitivity: 10 pJ/cm^2. Storage time: days
Membrane light modulator	Electrical: e-beam + charge transfer plate	30	256×256	Limited Commercial Availability

11.5.3 Membrane Devices

Membrane devices are closely related to the silicon mechanical devices. Instead of a moveable silicon structure deflecting a light beam, a membrane is deflected to modulate the optical phase as the light reflects off either the membrane or a deposited reflective metallic layer. Membrane devices can be either optically or electrically addressed. Addressing issues are as discussed in the Introduction.

A variety of membrane-based devices have been developed over the past three decades. Most of these devices are either specialty commercial items or are R&D laboratory items. The discussion in the following paragraphs, however, provides insights from past research activity on achievable performance levels with MEMS technology

Membranes must be deposited onto some substrate that performs the conversion and control function indicated in Figure 11.1. Membrane devices that have undergone the most development are the membrane light modulator (MLM) (Preston 1969; Reizman 1969), where a thin conducting membrane is stretched over an array of insulating holes, the deformable membrane device, or DMD (Pape 1985), where an insulating membrane is laid onto silicon circuits, and the photoemitter membrane light modulator, or the PEMLM (Somers 1972; Fisher et al. 1986), where an insulating membrane is laid onto a microchannel plate.

Membrane devices exhibit the maximum framing speed for those SLMs based on mechanical movements because the required membrane motion is only a fraction of an optical wavelength and the membrane element is thin and can be very small. Membranes a few micrometers in diameter have been demonstrated in the PEMLM device (Fisher et al. 1986), where the membrane diameter is equal to the pore diameter of the microchannel plate. The ultimate speed of a membrane device is determined by the resonance frequency, which for a circular geometry of the PEMLM membrane has been calculated as (Fisher, Ling, and Lee 1984)

$$f_o = 0.38R\sqrt{\tau/M} \tag{11.25}$$

where τ is the membrane tension, R the radius, and M the mass density of the membrane. The membrane material used in the initial PEMLM and DMD devices was nitrocellulose. For this material, a 40-nm-thick circular membrane with a 30-nm reflective indium coating and $R=12$ mm, $M=2.85\times 10^{-4}$ kg/m^2, and $\tau=0.8$ N-m, results in a 1.66-MHz frequency response. However, the framing rate of a membrane device is invariably much less than this speed. Limitations are usually first encountered in the drive current required to charge the membrane for a given membrane deflection Z. For the DMD, the limit is the current for the silicon circuit, while for the PEMLM it is the amplified current from the microchannel plate. Membrane deflection amplitude is given by (Somers 1972)

$$Z = (E^2R^2\varepsilon_o)/(8\tau) \tag{11.26}$$

where E is the electrostatic field at the surface of the membrane (V/m), and ε_o is the dielectric constant of free space. For example, for a PEMLM device with a nitrocellulose membrane to produce a deflection of $Z=\lambda/4$ (full-contrast reversal), the drive-current requirements limit the frame rate to about 1.5 kHz (Fisher et al. 1986). However, Equation 11.26 shows that speed can be optimized by reducing the required value for E, which, for a given Z, requires lowering the membrane tension. τ can be lowered by using a polymer material that has lower surface energy than nitrocellulose (Rolsma and Lee 1990). Results such as those in Equation 11.25 and Equation 11.26 have analogs in other mechanical systems, such as the silicon mechanical devices.

The contrast ratio is limited by device construction. The fundamental thermal noise contribution to the deflection Z is much less than the $\lambda/4$ required for full contrast reversal.

11.6 Thermal Devices

Modulation employing thermo-optic effects involves the alteration of optical properties by a change in temperature. Many of the modulating materials discussed above can be considered candidates for thermo-optic materials to the extent that their optical properties vary with temperature. For most materials, thermal addressing speeds are generally slow compared to those for inorganic electro-optic materials. However, thermal modulation is attractive for its simplicity, and can be either applied in situations not requiring high speed, or applied to microstructures where thermal time constants and power consumption can be minimized. As an example of the latter, thermo-optic modulation has been examined in silicon optical waveguides, to allow 2.2 Mb/s digital transmission (Iodice, Della Corte, Rendina 2003). However, there are certain materials that exhibit especially large optical changes with temperature, which can be classified as follows.

- Plastics and oil films that deform when heated.
- Materials that undergo a phase transition at a specific temperature. The most well-known of these is the phase transition of liquid crystals from the smectic to the nematic phase. Another material that has been used is vanadium dioxide (VO_2), which exhibits a metal-to-insulator phase transition.
- More recently, engineered materials, such as very compact photonic crystal waveguide structures, have shown high thermal response speed.

Both electrically and optically addressed devices have been demonstrated, and addressing issues are as discussed in the Introduction. A tabulation of the thermal devices described below is given in Table 11.5.

11.6.1 Thermoplastic and Oil-Film Devices

Plastics and oil films can be deformed with heat. Local heating can be employed for spatial light modulation. In both types of devices, heating has been performed electrically with a scanning electron beam (Turpin 1974; Schneeberger et al. 1979). Optical addressing has been achieved in the oil-film device by a light-absorbing plastic substrate (Schneeberger et al. 1979). Deformations on the order of one

TABLE 11.5 Selected Thermal Spatial Light Modulators and Performance

Device/Name	Address Method	Frame Rate (Hz)	Space-Bandwidth/Resolution (pixels/mm)	Remarks
Deformable elastomer/RUTICON	Optical: Amorph.Se	100	100/mm	Limited availability
	Electrical: e-beam	25	15/mm	
Thermoplastic	Optical: PVK:TNF	10	1000/mm	Commercially available. Limited lifetime
Vanadium dioxide—VO_2 metal-to-insulator phase transition	Optical:		150/mm	Limited availability. High writing power needed
	Electrical:		20/mm	
Oil film—Talaria/Eidophor	Electrical: e-beam	30	100/mm (1000×1000)	Limited commercial availability
Oil film	Optical: Heat absorp. in substrate	5	4/mm	Limited commercial availability
Smectic-to-nematic liquid crystal phase transition	Optical IR laser heating	0.1	16/mm (2200×3400)	Commercial display product

optical wavelength phase-modulate the readout beam, as in the membrane devices discussed in Mechanical and Micromechanical Devices.

Thermoplastics have been employed in optically addressed devices in a sandwiched geometry, almost exactly that shown in Figure 11.1. The thermoplastic material is sandwiched between a metallic electrode layer and a photosensor layer (Colburn and Chang 1978). Materials used for the photosensor layer have included PVK:TNF, amorphous Se, CdS, ZnS, and ZnCdS. Electrically addressed devices using thermoplastics and other elastomers, such as those for the Ruticon devices (Ralston and McDaniel 1979), have also been constructed using an ionized gas or conducting liquid rather than a metallic electrode to apply voltage. The various electrode configurations have overcome some, but not all the problems with limited lifetime, especially in thermoplastics. Thus, there has been relatively little recent work (Colburn and Chang 1978; Ralston and McDaniel 1979).

11.6.2 Phase-Transition Devices

A highly ordered smectic-phase of liquid crystals is possible at room temperature. Heating of the material, with an electron/optical beam or a pixel heating element, changes it into the nematic phase. Upon cooling, the material transforms into a disordered smectic phase that strongly scatters light. Schlieren readout optics produces the intensity-modulated output. Grey scale is obtained by partial writing/erasure. The nematic-to-smectic mechanism has been employed for very large high definition displays (2200×3400), using near-IR semiconductor lasers (Kahn et al. 1987). However, the response speed is slow—on the order of a second, and the spatial temperature uniformity of the device must usually be closely controlled.

VO_2 exhibits a phase transition between a highly reflective metallic phase and a low-reflectance insulating phase (Eden 1979). Generally, the VO_2 phase-transition devices provide only binary intensity modulation, with typically a 10:1 contrast ratio. However, the temperature of the material must be held close to the transition temperature in order to minimize the required switching energy and time. The response time of the material can be reasonably fast, a few hundred nanoseconds (Eden 1979), for temporal switching applications. For spatial light modulation, it is usually the addressing mechanisms that are the limit on frame rate.

11.6.3 Thermo-Optic Microstructure Devices

Optical structures can be fabricated very compactly in photonic crystals, as described in Specialized and Emerging Devices. The degree of compaction can be two orders of magnitude compared to conventional fabrication (e.g., ~ 10 μm for a passive optical coupler in a photonic crystals versus ~ 1000 μm for conventional construction) (Sharkawy et al. 2002). Because of these small dimensions, active devices can be made with reasonably high-speed response (~ 10 MHz) using thermal addressing. Specific devices are described in Specialized and Emerging Devices.

11.7 Specialized and Emerging Devices

11.7.1 Photonic Bandgap Materials

Recent years have seen large growth in the field of nanotechnology, which involves the research and development of structures with dimensions on the order of 1–1000 nm. Use of nanotechnology to engineer optical materials has resulted in "photonic crystals," a periodic dielectric structure possessing a "band gap" in the photon density of states (Prather et al. 2002; Krauss and DeLaRue 1999). The periodic structure is generally a lattice of composite materials of high and low dielectric constant (e.g., silica spheres with air in the interstitial volume, a semiconductor with an etched lattice of holes, or silicon pillars in a 2-D lattice). The key features in such structures arise because (i) the periodicity in the photonic crystal is much less than the wavelength of operation; passive waveguide devices can be made with much reduced dimensions, shorter with much larger bend radii compared to conventional waveguides, and (ii) one can produce a band gap response at optical frequencies. The cut-on and cut-off frequencies of the band gap can be used as the basis for optical modulation.

Because of its simplicity, thermo-optic addressing has been used for many initial devices, rather than direct electro-optic addressing.

- The earliest approaches have been towards the implementation of very compact versions of standard optical devices, especially the Mach-Zehnder interferometer for temporal modulation. A thermo-optically controlled AlGaAs/GaAs epitaxial waveguide Mach-Zehnder device has been demonstrated at 1550 nm wavelength, requiring only 42 mW for a π-phase shift (Camargo, Chong, and De La Rue 2004).
- The cut-off frequency of a silicon-based photonic crystal waveguide device in the vicinity of 1550 nm has been shifted by up to 60 nm by locally increasing the temperature of the device (Tinker and Lee 2005), resulting in 25–50 dB extinction ratio and $\sim$10 μs response time.

A wide variety of other opportunities exist for developing optical modulators with photonic band gap materials.

- Devices have been explored for electrically tuning or modulating the band gap, in order to produce an electro-optic shutter in a three-dimensional structure. By filling the interstices of a photonic crystal with liquid crystal material, an electrically induced phase separation can result in switchable Bragg diffraction (Sutherland et al. 2002).
- Modulation of the active photonic band gap, by a short optical pulse of an InGaAs/GaAs quantum well structure, results in a switchable mirror that either reflects or transmits, potentially operating at terahertz rates (Prineas et al. 2002).
- Photonic waveguide switches can be made with dimensions less than the wavelength of operation. Devices have been explored that can electrically or optically drive a 100% exchange of the energy between two coupled photonic crystal waveguides over a very short propagation distance (Sharkawy et al. 2002).
- Tunable optical filtering for telecommunication applications has been explored. Multiple photonic band gap quarter-wave stacks, consisting of alternating layers of Si and SiO2 separated by a tunable air gap, can be electrically tuned for wavelength transmission (Yi et al. 2002).
- All-Optical switching, where one optical beam controls a second, higher-power beam, is a long-standing goal for ultra-high-speed telecommunication and processing systems. Such systems might be enabled using ultra-small high-Q photonic crystal nanocavities that can switch in less than 100 ps, and use energies of less than 100 femtojoules (Tanabe et al. 2005).

While nanostructures can reduce the size of optical devices, further device-size reduction has been achieved by employing surface plasmons within the nanostructure (Barnes, Drereux, and Ebbesen 2003). Light incident on a metal-dielectric nanostructure (e.g., a waveguide sandwich of metal films and dielectrics such as polymer films) produces confined surface plasmons with a wavelength much smaller than that of the light; the plasmons are converted back into the optical domain at the output. With metal–polymer waveguide films in a Mach-Zehnder configuration, thermal effects from an electrical drive signal can modulate the output optical power at millisecond speeds (Nikolajsen et al. 2004).

Beyond the few examples cited above, it appears that a large number of modulation approaches with unique characteristics appear possible.

11.7.2 Direct Modulation of Light Sources

Modulation can be achieved by directly controlling the electrical power to an optical source. The primary, directly-modulated light sources are semiconductor laser diodes and light emitting diodes (LEDs) because of their small size (μm-sized emission facet), high electrical-to-optical conversion efficiency, and high optical output power. Laser diodes and LEDs are available at wavelengths from the blue (~400 nm) to the mid-wave IR (~3000 nm). Modulation is achieved by controlling the drive current in these devices. A laser diode is often advantageous because its light is highly monochromatic with well-defined polarization. An LED with appropriate optics can produce a Lambertian output illumination of >100 lumens per watt of electrical drive power. But, the light emission is over a fairly wide wavelength band or even white, driven by development for display applications.

Since the most popular laser diodes and LEDs are fabricated with III–V compounds, such as GaAlAs and InGaAsP, there is the hope that such light sources can be integrated on a single chip together with high-speed drive circuitry fabricated with these same III–V materials.

11.7.2.1 Temporal Modulation

Temporal modulation of laser diodes and LEDs is quite common and capable of very high bandwidth (tens of GHz), but is presently used primarily in single-channel systems. Achievable optical power, dynamic range and linearity of modulation are major issues. Good linearity of modulation is possible with LEDs, but the optical power output is low, typically <10 mW. Power from presently available, single-element single-mode laser diodes is >100 mW, but modulation over large dynamic range is still a challenge. In addition, the laser drive current must exceed a threshold before lasing with a linear output vs. drive current. This threshold current has been successfully reduced using quantum well structures in the active gain region (Lau, Derry, and Yariv 1988), although an LED-like threshold will always be present.

High-bandwidth modulation of a laser diode can cause modal instability, with attendant intensity fluctuations (e.g., even if a laser is single-mode, the intensity can abruptly jump if the modulation amplitude puts the laser near a point where mode competition and hopping occurs). In practice, one must also be careful to avoid back reflection into the laser cavity, since this will contribute to mode instability. Another serious problem is shifting or chirping of the optical frequency with modulation. Chirping arises from changes in the carrier density (N) during modulation such that the later portions of a driving signal see a different refractive index, n, (which is proportional to N) than the earlier portions. The linearity and optical frequency stability of laser diodes has been improved with the use of distributed feedback (DFB). These lasers have a periodic grating structure within the cavity that continuously Bragg-scatters a portion of the light backwards; this limits the number of longitudinal modes which can propagate, leading to better mode stability, higher spectral purity, and suppression of chirping during modulation (Nakamura et al. 1973). Present laser diodes can be modulated at bandwidths greater than 20 GHz. Quantum-well laser structures can theoretically reach the 30-GHz regime (Uomi, Mishima, and Chinone 1985), based on limits due to relaxation oscillations of charge carrier density in the laser cavity. By using an external-cavity configuration, where a well-behaved low-power laser is used to control the behavior of a high-power laser, the dynamic range at high modulation bandwidths can be increased.

11.7.2.2 Spatial Modulation

Arrays of either incoherent light sources or laser diodes with individual element addressing can be employed for spatial light modulation. The addressing issues are as discussed in the Introduction.

One approach is to use incoherent light by adapting existing luminous displays (Apt 1985; Blechman 1986). This approach, like the case described earlier for SLMs based on transmissive liquid-crystal televisions, takes advantage of the vast amount of technological development put into realizing large, high-resolution displays. A variety of emissive display technologies are available, including cathode-ray tubes, gas plasma displays, electroluminescent panels, and LED displays. However, the major problems with this approach are (i) luminous intensity is often insufficient for optical processing, (ii) with incoherent light, phase cannot be processed without using some encoding scheme that sacrifices a parameter such as resolution or dynamic range, (iii) only electrical addressing is available for these devices, so that a single device cannot perform image multiplication, and iv) the cosmetic quality and flatness are often not sufficient for SLM applications.

A coherent-light approach, using an array of semiconductor laser diodes, can be implemented with either electrical or optical addressing. For example, a laser amplifies an input optical at the cavity wavelength. An approach that has excited much interest is use of a large number of individually addressable semiconductor lasers on a single wafer. In principle, such an array can be either optically or electrically addressed, but the latter is presently the rule. Conventional edge-emitting (in-plane) lasers can be sliced from a wafer to produce a line array of sources. Stacking such line arrays to produce a large 2-D array, has been achieved, but mainly as a high-power optical source; it is difficult and costly to provide addressing. A 2-D array of surface-emitting lasers integrated on a single wafer can be achieved in either of two ways. The first is to combine with each laser in an edge-emitting array a 45-degree mirror structure to redirect the beam upwards (Liau, Walpole, and Tsang 1984). The second approach is to construct the array using vertical-cavity surface emitting lasers (VCSELs) (Chang-Hasnain et al. 1991). Construction of such integrated laser arrays involves the III–V compounds discussed above, but with the more-complex structures involved in laser fabrication. Several materials-related issues arise. High electrical-to-optical efficiency is required for the lasers in order to keep within power dissipation capabilities of a wafer. Use of a quantum-well structure in the construction of the semiconductor lasers is imperative in order to minimize the lasing threshold, and to maximize the electrical-to-optical efficiency (Lau, Derry, and Yariv 1988). For the VCSEL approach, devices also tend to exhibit high resistance (Tai et al. 1990), primarily due to the mirror layers for the laser cavity; this further increases the heating of the wafer.

11.7.3 Biological Materials

Biological compounds, often embedded in membranes, constitute a large subset of organics, and remain largely unexplored for optical modulation. The compound of most historic interest for optical modulation is bacteriorhodopsin (Oesterhelt and Stoeckenius 1971; Hampp et al. 1992). Bacteriorhodopsin consists of a chain of amino acids to which a chromophoric group is bound. To make an SLM, the material is placed in a 2-D matrix embedded in a cell membrane. Light activation produces a gradient in the hydrogen ion (or proton) concentration across the membrane. This gradient is normally needed for synthesis of other biological compounds. Light at $\sim$570 nm excites the bacteriorhodopsin from the so-called B state to the long-lived intermediate M state, which has a lifetime of about 10 ms. The M state would naturally relax back to the B state, but 412 nm light can also initiate the transition. Hence, optical beams at 570 and 412 nm form the writing and read beams, respectively, of the SLM. Although bacteriorhodopsin-based devices are presently still in the developmental stage, they indicate the potential for biological materials in SLM devices.

11.8 Summary

Many materials and devices have been explored for optical modulation. Some are quite mature, such as mechanical devices and many inorganic electro-optic and acousto-optic materials. In other cases, the materials are well-developed, but significant device development still remains. Finally, in some case, both materials and devices are immature. In general, temporal and 1-D modulators, such as acousto-optic modulators, are generally much more mature than 2-D spatial light modulators. New approaches and implementations are fostered by research and development in areas such as nanotechnology and display technology.

References

Agranat, A. J. 2002. Electroholographic switching devices and applications, Paper CMH1, 2002 CLEO Conference Technical Digest, Optical Society of America, Washington, 37.

Albanese, G., Kumar, J., and Steier, W. H. 1986. Investigation of the photorefractive behavior of chrome-doped GaAs by using two-beam coupling. *Opt. Lett.*, 11, 650.

Amano, M. and Roos, E. 1987. 32-channel acousto-optic Bragg cell for optical computing. Acoustooptic, electrooptic, and magnetooptic devices and applications. *Proc. SPIE*, 753, 37.

Anderson, G., Dahl, I., Kuczynski, W., Lagerwall, W. T., Sharp, K., and Stebler, B. 1988. The soft-mode ferroelectric effect. *Ferroelectrics*, 84, 285.

Anderson, G. W., Webb, D., Spezio, A. E., and Lee, J. N. 1991. Advanced channelization devices for rf, microwave, and millimeterwave applications. *Proc. IEEE*, 79, 355.

Apt, C. M. 1985. Perfecting the Picture. *IEEE Spectr.*, 22, 7, 60.

Armitage, D., Anderson, W. W., and Karr, T. J. 1985. High-speed spatial light modulator. *IEEE J. Quantum Electron.*, QE-21, 1241.

Ashley, P. R., Davis, J. H., and Oh, T. K. 1988. Liquid crystal spatial light modulator with a transmissive amorphous silicon photoconductor. *Appl. Opt.*, 27, 1797.

Augborg, P., Huignard, J. P., Hareng, M., and Mullen, R. A. 1982. Liquid crystal light valve using bulk monocrystalline $Bi_{12}SiO_{20}$ as the photoconductive material. *Appl. Opt.*, 21, 3706.

Aull, B. F., Kirby, B. N., Goodhue, W. D., and Burke, B. E. 1988. *Multiple-Quantum-Well CCD Spatial Light Modulators. Vol. 8*, Spatial Light Modulators and Applications. OSA 1988 Technical Digest Series., Optical Society of America, Washington, DC.

Barnes, W. L., Drereux, A., and Ebbesen, T. W. 2003. Surface plasmon subwavelength optics. *Nature*, 422, 824.

Batsell, S. G., Jong, T. L., Walkup, J. F., and Krile, T. F. 1990. Noise limitations in optical linear algebra processors. *Appl. Opt.*, 29, 2084.

Berg, N. J. and Lee, J. N. eds. 1983. *Acousto-Optic Signal Processing: Theory and Implementation*, pp. 371–1376. Marcel-Dekker, New York.

Blechman, F. 1986. Pocket Television Receivers, Radio Electronics 57, Pt. 1, July, 39 and Pt. 2, Aug., 47.

Bleha, W. P., Lipton, L. T., Weiner-Avnear, E., Grinberg, J., Rief, P. G., Casasent, D., Brown, H. B., and Markevitch, B. V. 1978. Application of the liquid crystal light valve to real-time optical data processing. *Opt. Eng.*, 17, 371.

Blinov, L. M. 1983. *Electro-Optical and Magneto-Optical Properties of Liquid Crystals*. Wiley, New York.

Blistanov, A. A., Bondarenko, V. S., Perelomova, N. V., Strizhevskaya, F. N., Chkalova, V. V., and Shaskol'skaya, M. P. 1982. *Acoustic Crystals*. Publishing House "Nauka," Moscow.

Bowman, S. R., Rabinovich, W. S., Kyono, C. S., Katzer, D. S., and Ikossi-Anastasiou, K. 1994. High resolution spatial light modulators using GaAs/AlGaAs multiple quantum wells. *Appl. Phys. Lett.*, 65, 956.

Brooks, R. E. 1985. Micromechanical light modulator on silicon. *Opt. Eng.*, 24, 101.

Brown, R. B., Craig, A. E., and Lee, J. N. 1989. Predictions of stray light modeling on the ultimate performance of acousto-optic processors. *Opt. Eng.*, 28, 1299.

Butler, M. A. and Venturini, E. L. 1987. High frequency Faraday rotation in FR-5 glass. *Appl. Opt.*, 26, 1581.

Butler, J. C., Kramer, J. J., Lee, J. N., Ings, J. B., and Belt, R. F. 1992. Optical and electrical characterization of magnesium-doped bismuth-substituted lutetium iron garnet thin films. *J. Appl. Phys.*, 71, 924.

Cafarella, J. H. 1978. Acoustoelectrical signal-processing devices with charge storage. *Proc. 1978 IEEE Ultrasonics Symp., IEEE Cat. No. 78CH1344-1SU*, 767.

Camargo, E., Chong, H., and De La Rue, R. 2004. 2D photonic crystal thermo-optic switch based on AlGaAs/GaAs epitaxial structure. *Opt. Express.*, 12, 588.

Casasent, D. 1978a. Photo DKDP light valve: A review. *Opt. Eng.*, 17, 365.

Casasent, D. 1978b. E-beam DKDP light valves. *Opt. Eng.*, 17, 344.

Chang, I. C. 1983. Acousto-optic tunable filters. *Acousto-optic Signal Processing: Theory and Implementation*. Marcel-Dekker, New York.

Chang, I. C. and Lee, S. 1983. Efficient wideband acousto-optic cells, *Proc. 1983 IEEE Ultrasonics Symp., IEEE No. 83CH 1947-1*, 427.

Chang-Hasnain, C. J., Maeda, M. W., Harbison, J. P., Florez, L. T., and Lin, C. 1991. Monolithic multiple wavelength surface emitting laser arrays. *J. Lightwave Technol.*, 9, 1665.

Chemla, D. S. and Zyss, J. eds. 1987. *Nonlinear Optical Properties of Organic Molecules and Crystals*, pp. 3389–1376. Academic Press, London.

Chen, F. S. 1969. Optically induced changes of refractive indices in $LiNbO_3$ and $LiTaO_3$. *J. Appl. Phys.*, 40, 3389.

Chen, D., Fetterman, H. R., Chen, A., Steier, W. H., and Dalton, L. R. 1997. Demonstration of 110-GHz electro-optic polymer modulators. *Appl. Phys. Lett.*, 7025, 3335.

Childs, R. B. and O'Byrne, V. A. 1990. Multichannel AM video transmission using a high-power Nd:YAG laser and linearized external modulator. *IEEE J. Select. Areas Commun.*, 8, 1369–1376.

Clark, N. A. and Lagerwall, S. 1980. Submicrosecond bistable electro-optic switching in liquid crystal. *Appl. Phys. Lett.*, 36, 899–901.

Cohen, M. G. 1967. Optical study of ultrasonic diffraction and focusing in anisotropic media. *J. Appl. Phys.*, 38, 3821.

Colburn, W. S. and Chang, B. J. 1978. Photoconductor-thermoplastic image transducer. *Opt. Eng.*, 17, 334.

Damon, R. W. and Eshbach, J. R. 1961. Magnetostatic modes of a ferromagnet slab. *J. Phys. Chem. Solids*, 19, 308.

DeJeu, W. H. 1980. *Physical Properties of Liquid Crystalline Materials*. Gordon and Breach, New York.

Devaux, F., Muller, S., Ougazzaden, A., Mircea, A., Ramdane, A., Krauz, P., Semo, J., Huet, F., Carre, M., and Carenco, A. 1994. Zero-loss multiple-quantum-well electroabsorption modulator with very low chirp. *Appl. Phys. Lett.*, 64, 954.

Dieulesaint, E. and Royer, D. 1980. *Elastic Waves in Solids*. Wiley, Chichester.

Dillon, J. F. 1978. Magneto-Optical Properties of Magnetic Garnets, *Physics of Magnetic Garnets*, New York, North-Holland.

Dixon, R. W. 1967a. Acoustic diffraction of light in anisotropic media. *IEEE J. Quantum Electron.*, QA-3, 85.

Dixon, R. W. 1967b. Photoelastic properties of selected materials and their relevance for applications to acoustic light modulators and scanners. *J. Appl. Phys.*, 38, 5149.

Donjon, J., Dumont, F., Grenot, M., Hazan, J. P., Marie, G., and Pergrale, J. 1973. A Pockels-effect light valve: Phototitus. Applications to optical image processing. *IEEE Trans. Elec. Dev.*, ED-20, 1037.

Eden, D. D. 1979. Some applications involving the semiconductor-to-metal phase transition in VO_2. *Opt. Process. Syst. Proc. SPIE*, 185, 97.

Fisher, A. D. 1985. *Optical signal Processing with Magnetostatic Waves, Circuits, Systems, and Signal Processing*, 4, 265.

Fisher, A. D., Lee, J. N., Gaynor, E. S., and Tveten, A. B. 1982. Optical guided-wave interactions with magnetostatic waves at microwave frequencies. *Appl. Phys. Lett.*, 41, 779.

Fisher, A. D., Ling, L. C., and Lee, J. N. 1984. A high performance photo-emitter membrane spatial light modulator, in spatial light modulators and applications. *Proc. SPIE*, 465, 36.

Fisher, A. D., Ling, L. C., Lee, J. N., and Fukuda, R. C. 1986. The photo-emitter membrane light modulator. *Opt. Eng.*, 25, 271.

Garito, A. F., Teng, C. C., Wong, K. Y., and Zammani'Khamiri, O. 1984. Molecular optics: Nonlinear optical processes in organic and polymer crystals. *Mol. Cryst. Liq. Cryst.*, 106, 219.

Goodby, J. W. and Patel, J. S. 1986. Properties of ferroelectric crystals. Liquid crystals and spatial light modulators. *Proc. SPIE*, 52.

Goutzoulis, A. P. and Gottlieb, M. S. 1988. Design and performance of optical activity based Hg_2Cl_2 Bragg cells. *Adv. Opt. Inf. Process. III, Proc. SPIE*, 936, 119.

Griffel, G., Ruschin, S., and Croitrou, N. 1989. Linear electro-optic effect in sputtered polycrystalline $LiNbO_3$ films. *Appl. Phys. Lett.*, 54, 1385.

Groth, G. and Marie, G. 1970. Information input in an optical pattern recognition system using a relay tube based on the Pockels effect. *Opt. Commun.*, 2, 133.

Hampp, N., Thoma, R., Oesterhelt, D., and Brauchle, C. 1992. Biological photochrome bacteriorhodopsin and its genetic variant ASP96(Asn as media for optical pattern recognition. *Appl. Opt.*, 31, 1834.

Hartman, R. L. and Farr, K. B. 2003. Demonstration of the ULTOR target recognition and tracking system. *Opt. Pattern Recognit. XIV, Proc. SPIE*, 5106, 30.

Hecht, D. L. 1976. Spectrum analysis using acoustooptic devices. *Acoustooptics, Proc. SPIE.*, 90, 148.

Hecht, D. L. and Petrie, G. W. 1980. Acousto-optic diffraction from acoustic anisotropic shear modes in gallium phosphide, *Proc. 1980 IEEE Ultrasonics Symp., IEEE No. 80CH1602-2*, 474.

Hornbeck, L. J. 1989. Deformable-mirror spatial light modulators. *Spat. Light Modulators and Appl. III, Proc. SPIE*, 1150, 86.

Horowitz, B. A. and Corbett, F. J. 1978. The PROM—theory and application of the pockels readout optical modulator. *Opt. Eng.*, 17, 353.

Iodice, M., Della Corte, F. G., and Rendina, I. 2003. Transient analysis of a high-speed thermo-optic modulator integrated in an all-silicon waveguide. *Opt. Eng.*, 42, 169.

Jain, F. C. and Bhattacharjee, K. K. 1989. Multiple quantum well optical modulator structures using surface acoustic wave induced stark effect. *IEEE Photon. Technol. Lett.*, 1, 307.

Jared, D. A. and Johnson, K. M. 1991. Optically addressed thresholding very-large-scale-integration liquid crystal spatial light modulator. *Opt. Lett.*, 16, 967.

Jenkins, F. A. and White, H. E. 1976. *Fundamentals of Optics, 4th Ed.*, McGraw-Hill, New York.

Jin, M. S., Wang, J. H., Ozguz, V., and Lee, S. H. 1994. Monolithic integration of a silicon driver circuit onto a lead lanthanum zirconate titanate substrate for smart spatial light modulator fabrication. *Appl. Opt.*, 33, 2842.

Johnson, L. M. and Rousell, H. V. 1988. Reduction of intermodulation distortion in interferometric optical modulator. *Opt. Lett.*, 13, 928.

Johnson, R. V., Hecht, D. L., Sprague, R. A., Flores, L. N., Steinmetz, D. L., and Turner, W. D. 1983. Characteristics of the linear array total internal reflection TIR electrooptic spatial light modulator for optical information processing. *Opt. Eng.*, 22, 665.

Kahn, F. J., Kendrick, P. N., Leff, J., Livoni, L. J., Loucks, B. E., and Stepner, D. 1987. A paperless plotter display system using a laser smectic liquid-crystal light valve. *1987 SID International Symposium, Digest of Technical Papers*. Paper 14.1, 254.

Kaminow, I. P. and Turner, E. 1971. *Linear Electrooptic Materials. Handbook of Lasers*. Chemical Rubber Co., Cleveland, OH.

Klein, W. R. and Cook, B. D. 1967. Unified approach to ultrasonic light diffractio. *IEEE Trans. Sonics Ultrason.*, SU-14, 123.

Korotsky, S. K. and DeRidder, R. M. 1990. Dual parallel modulation schemes for low-distortion analog optical transmission. *IEEE J. Select. Areas Commun.*, 8, 1377.

Koster, G. F., Dimmock, J. O., Wheeler, R. G., and Statz, H. 1963. *Properties of the Thirty-Two Point Groups*. MIT Press, Cambridge, MA.

Krauss, T. F. and DeLaRue, R. M. 1999. *Photonic Crystals in the Optical Regime—Past, Present, and Future. Progress in Quantum Electronics*. Pergamon Press, New York.

Land, C. E. 1978. Optical information storage and spatial light modulation in PLZT ceramics. *Opt. Eng.*, 17, 317.

Lau, K. Y., Derry, P. L., and Yariv, A. 1988. Ultimate limit in low threshold quantum well GaAlAs semiconductor lasers. *Appl. Phys. Lett.*, 52, 88.

Lee, J. N. 1983. Signal processing using an acousto-optic memory device. *Acousto-Optic Signal Processing: Theory and Implementation*, Marcel-Dekker, New York.

Lee, J. N. and Fisher, A. D. 1987. Device developments for optical information processing. *Adv. Electron. Elec. Phys.*, 69, 115.

Lee, J. N. and VanderLugt, A. 1989. Acousto-optic signal processing and computing. *Proc. IEEE*, 7, 1528.

Lee, S. H., Esner, S. C., Title, M. A., and Drabik, T. J. 1986. Two-dimensional silicon/PLZT spatial light modulators: design considerations and technology. *Opt. Eng.*, 25, 250.

Liau, Z. L., Walpole, J. N., and Tsang, D. Z. 1984. Mass-transported GaInAsP/InP buried-heterostructure lasers and integrated mirrors. *Technical Digest of the 7th Topical Meeting on Integrated and Guided-Wave Optics*. Optical Society of America, Washington, DC.

Liu, H. K., Davis, J. A., and Lilly, R. A. 1985. Optical data processing properties of a liquid-crystal television light modulator. *Opt. Lett.*, 10, 635.

Man, H-T. and Yoon, H. N. 1998. Long term stability of a poled side-chain nonlinear optical polymer. *Appl. Phys. Lett.*, 72, 540.

Mandal, B. K., Chen, Y. M., Lee, J. Y., Kumar, J., and Tripathy, S. 1991. Cross-linked stable second-order nonlinear optical polymer by photochemical reaction. *Appl. Phys. Lett.*, 58, 2459.

Marowsky, G., Chi, L. F., Mobius, D., Steinhoff, R., Shen, Y. R., Dorsch, D., and Rieger, B. 1988. Non-linear optical properties of hemicyanine monolayers and the protonation effect. *Chem. Phys. Lett.*, 147, 420.

McEwan, J. A., Fisher, A. D., Rolsma, P. B., and Lee, J. N. 1985. Optical processing characteristics of a low cost liquid crystal display. *J. Opt. Soc. Am. A*, 2, 13, 8.

Miller, D. A. B., Chemla, D. S., Damen, T. C., Gossard, A. C., Wiegmann, W., Wood, T. H., and Burrus, C. A. 1984. Novel hydrid optically bistable switch: The quantum well self-electro-optic effect device. *Appl. Phys. Lett.*, 45, 13.

Monteverde, R. J. 2004. Spatial light modulators illuminate a wide variety of application spaces. *Laser Focus World* 93.

Nakamura, M., Yen, H. W., Yariv, A., Garmire, E., and Somekh, S. 1973. Laser oscillation in epitaxial GaAs waveguides with corrugation feedback. *Appl. Phys. Lett.*, 23, 224.

Nakazawa, T. 2002. Microwave photonics 2002. *Japan* 5.

Neurgaonkar, R. R. and Cory, W. K. 1986. Progress in photorefractive tungsten bronze crystals. *J. Opt. Soc. Am*, B, 3, 274.

Nikolajsen, T., Leosson, K., and Bozhevolnyi, S. I. 2004. Surface plasmon polariton based modulators and switches operating at telecom wavelengths. *Appl. Phys. Lett.*, 85, 5833.

Noguchi, K., Mitomi, O., and Miyazawa, H. 1998. Millimeter-Wave Ti: $LiNbO_3$ Optical modulators. *J. Lightwave Technol.*, 16, 615.

Oesterhelt, D. and Stoeckenius, W. 1971. Rhodopsin-like protein from the purple membrane of *Halobacterium halobium*. *Nature*, 233, 149.

Oh, M.-C. 2001. Recent advances in electrooptic polymer modulators incorporating highly nonlinear chromophore. *IEEE J. Select. Top. Quantum Electron.*, 705, 826.

Ohkawa, M., Izutsu, M., and Sueta, T. 1989. Integrated optic pressure sensor on silicon substrate. *Appl. Opt.*, 28, 5153.

Orlov, S. S., Phillips, W., Bjornson, E., Takashima, Y., Sundaram, P., Hesselink, L., Okas, R., Kwan, D., and Snyder, R. 2004. High-transfer-rate high-capacity holographic disk data-storage system. *Appl. Opt.*, 43, 4902.

Owechko, Y. and Tanguay, A. R. 1982. Effects of operating mode on electrooptic spatial light modulator resolution and sensitivity. *Opt. Lett.*, 7, 587.

Pape, D. R. 1985. Optically addressed membrane spatial light modulator. *Opt. Eng.*, 24, 107.

Pape, D. R. 1998. New self collimating mode TeO_2 multichannel Bragg cells. 1998 *IEEE Ultrasonics Symposium*, 651.

Partovi, A., Millerd, J., Garmire, E. M., Ziari, M., Steier, W. H., Trivedi, S. B., and Klein, M. B. 1990. Photorefractivity at 1.5 mm in CdTe:V. *Appl. Phys. Lett.*, 57, 846.

Pepper, D. M., Feinberg, J., and Kukhtarev, N. V. 1990. The photorefractive effect. *Sci. Am.*, 263, 34.

Petrov, M. P. 1981. Electrooptic photosensitive media for image recording and processing. *Current Trends in Optics*, Taylor Francis, London.

Phariseau, P. 1956. On the diffraction of light by progressive supersonic waves. Oblique incidence: Intensities in the neighborhood of the Bragg angle. *Proc. Indian Acad. Sci.*, 44A, 165.

Prather, D., Sharkawy, A., and Shouyuan, S. 2002. Photonic crystals design and applications. *Handbook of Nanoscience, Engineering, and Technology, Electrical Engineering Handbook*, CRC Press, Boca Raton, FL.

Preston, K. P. 1969. The membrane light modulator and its application to optical computers. *Opt. Acta*, 16, 579.

Preston, K. D. and Haertling, G. H. 1992. Comparison of electro-optic lead-lanthanum zirconate titanate films on crystalline and glass substrates. *Appl. Phys. Lett.*, 60, 2831.

Prineas, J. P., Zhou, J. Y., Kuhl, J., Gibbs, H. M., Khitrova, G., Koch, S. W., and Knorr, A. 2002. Ultrafast ac Stark effect switching of the active photonic bandgap from bragg-periodic semiconductor quantum wells. *Appl. Phys. Lett.*, 81, 4332.

Rabinovich, W., Mahon, R., Burris, H. R., Gilbreath, G. C., Goetz, P., Moore, C. I., Stell, M. F., Vilcheck, M. J., Witkowsky, J. L., Swingen, L., Suite, M. R., Oh, E., and Koplow, J. 2005. Free-space optical communications link at 1550 nm using multiple-quantum-well modulating retroreflectors in a marine environment. *Opt. Eng.*, 44, 5, 056001.

Rakuljic, G. A., Sayano, K., Agranat, A., Yariv, A., and Neurgaonkar, R. R. 1988. Photorefractive properties of Ce- and Ca-doped $Sr_{0.6}$ $Ba_{0.4}$ NbO_3. *Appl. Phys. Lett.*, 53, 1465.

Ralston, L. M. and McDaniel, R. V. 1979. Experimental evaluation of an elastomer storage device. *Opt. Process. Syst. Proc. SPIE*, 185, 86.

Reizman, F. 1969. *Proc. of the 1969 Electrooptic. Syst. Des. Conf.*, 225.

Rickard, M., Nakata, M., Takezoe, H., Watanabe, J., and Clark, N. 2005. Electronic electrooptic phase modulation using bentcore liquid crystals. *Appl. Phys. Lett.*, 87, 261115.

Roe, M. G. and Schehrer, K. L. 1993. High-speed and high-contrast operation of ferroelectric liquid crystal optically addressed spatial light modulators. *Opt. Eng.*, 327, 1662.

Rolsma, P. B. and Lee, J. N. 1990. Polymer membrane properties and structures for membrane light modulator applications. *Opt. Lett.*, 15, 712.

Ross, W. E., Psaltis, D., and Anderson, R. H. 1983. Two-dimensional magneto-optic spatial light modulator for signal processing. *Opt. Eng.*, 22, 485.

Sayyah, K., Efron, U., Forber, R. A., Goodwin, N. W., and Reif, P. G. 1991. Schottky diode silicon liquid crystal light valve. *Liq. Cryst. Dev. Mat., Proc. SPIE*, 1455, 249.

Sayyah, K., Wu, C. S., Wu, S. T., and Efron, U. 1992. Anomalous liquid crystal undershoot effect resulting in a nematic liquid crystal-based spatial light modulator with one millisecond response time. *Appl. Phys. Lett.*, 61, 883.

Schneeberger, B., Laeri, F., Tschudi, T., and Mast, F. 1979. Realtime spatial light modulator. *Opt. Commun.*, 31, 13.

Schwartz, A., Wang, X.-Y., and Warde, C. 1985. Electron-beam-addressed microchannel spatial light modulator. *Opt. Eng.*, 24, 119.

Sharkawy, A., Shi, S., Prather, D. W., and Soref, R. A. 2002. Electro-optical switching using coupled photonic crystal waveguides. *Opt. Express*, 10, 1048.

Shi, Y., Zhang, C., Zhang, H., Bechtel, J. H., Dalton, L. R., Robinson, B. H., and Steir, W. H. 2000. *Science*, 288, 119.

Singer, K. D. and Garito, A. F. 1981. Measurements of molecular second order optical susceptibilities using DC induced second harmonic generation. *J. Chem. Phys.*, 757, 3572.

Singer, K. D., Kuzyk, M. G., Holland, W. R., Sohn, J. E., Lalama, S. J., Comizzoli, R. B., Katz, H. E., and Schilling, M. L. 1988. Electro-optic phase modulation and optical second-harmonic generation in corona-poled polymer films. *Appl. Phys. Lett.*, 53, 1800.

Skeie, H. and Johnson, R. V. 1991. Linearization of electro-optic modulators by a cascade coupling of phase modulating electrodes. *Integr. Opt. Circuits, Proc. SPIE*, 1583, 153.

Smith, D. A., Baran, J. E., Cheung, K. W., and Johnson, J. J. 1990. Polarization-independent acoustically tunable filter. *Appl. Phys. Lett.*, 60, 1538.

Somers, L. E. 1972. The photoemitter-membrane light modulator image transducer. *Adv. Elec. Electron. Phys.*, 33A, 493.

Soref, R. A. and Bennett, B. 1987. Electrooptical effects in silicon. *IEEE J. Quantum Electron.*, 23, 123.

Stahelin, M., Burland, D. M., Ebert, M., Miller, R. D., Smith, B. A., Tweig, R. J., Volksen, W., and Walsh, C. A. 1992. Re-evaluation of the thermal stability of optically nonlinear polymeric guest-host systems. *Appl. Phys. Lett.*, 61, 1626.

Sutherland, R. L., Tondiglia, V., Natarajan, S., Chandra, S., Tomlin, D., and Bunning, T. 2002. Switchable orthorhombic F photonic crystals formed by holographic polymerization-induced phase separation of liquid crystal. *Opt. Express*, 10, 1074.

Suzuki, M., Tanaka, H., and Matsushima, Y. 1992. InGaAsP electroabsorption modulator for high-bit-rate EDFA system. *Photon. Technol. Lett.*, 4, 586.

Tai, K., Yang, L., Wang, Y. H., Wynn, J. D., and Cho, A. Y. 1990. Drastic reduction of series resistence in doped semiconductor distributed Bragg reflectors for surface emittting lasers. *Appl. Phys. Lett.*, 56, 1942.

Tamada, H., Kaneko, M., and Okamoto, T. 1988. TM–TE optical-mode conversion induced by a transversely propagating magnetostatic wave in $BiLu_3Fe_5O_{12}$ film. *J. Appl. Phys.*, 64, 554.

Tanabe, T., Notomi, M., Shinya, A., Mitsugi, S., and Kuramochi, E. 2005. All-optical switches and memories fabricated on a silicon chip using photonic crystal nanocavities. *Appl. Phys. Lett.*, 87, 151112.

Taylor, B. K. and Casasent, D. P. 1986. Error-source effects in a high accuracy optical finite-element processor. *Appl. Opt.*, 25, 966.

Tinker, M. T. and Lee, J. B. 2005. Thermo-optic photonic crystal light modulator. *Appl. Phys. Lett.*, 86, 221111.

Tsai, C. S., Young, D., Chen, W., Adkins, L., Lee, C. C., and Glass, H. 1985. Noncollinear coplanar magneto-optic interaction of guided optical waves and magnetostatic surface waves in yttrium iron garnet-gadolinium gallium garnet waveguide. *Appl. Phys. Lett.*, 47, 651.

Tunaboylu, B., Harvey, P., and Esener, S. C. 1998. Characterization of dielectric and electro-optic properties of PLZT 9/65/35 films on sapphire for electro-optic application. *IEEE Trans. Ultrason. Ferroelec. Freq. Control.*, 45, 1105.

Turner, R. M., Johnson, K. M., and Serati, S. 1995. High-speed compact optical correlator design and implementation, In *Design Issues in Optical Processing,* Cambridge Press, Cambridge.

Turpin, T. 1974. Real time input transducer for coherent optical processing. *Digest of the 1974 International Optical Computing Conference*, p. 34.

Turpin, T. M. 1981. Spectrum analysis using optical processing. *Proc. IEEE*, 69, 79.

Uchida, N. and Niizeki, N. 1973. Acousto-optic deflection materials and techniques. *Proc. IEEE*, 61, 1073.

Uchida, N. and Ohmachi, Y. 1969. Elastic and photoeleastic properties of of TeO_2 single crystal. *J. Appl. Phys.*, 42, 4692.

Uomi, K., Mishima, T., and Chinone, N. 1985. Ultra-high relaxation oscillation frequency up to 30 GHz of highly p-doped GaAlAs multiquantum well lasers. *Appl. Phys. Lett.*, 51, 78.

Vadekar, A., Nathan, A., and Huang, W. P. 1994. Analysis and design of an integrated silicon ARROW Mach-Zehnder micromechanical interferometer. *J. Lightwave Technol.*, 12, 157.

Valley, G. C., McCahon, S. W., and Klein, M. B. 1988. Photorefractive measurement of photoionization and recombination cross sections in InP:Fe. *J. Appl. Phys.*, 64, 6684.

VanderLugt, A., Moore, G. S., and Mathe, S. S. 1983. Multichannel Bragg cell compensation for acoustic spreading. *Appl. Opt.*, 22, 3906.

Wakita, K., Kotaka, I., and Asai, H. 1992. High-speed InGaAlAs/InAlAs multiple quantum well electrooptic phase modulators with bandwidths in excess of 20 GHz. *Photon. Technol. Lett.*, 4, 29.

Wey, A. C. T., Tuan, H. S., Parekh, J. P., Craig, A. E., and Lee, J. N. 1986. Inhomogeneous field MSFVW-optical interaction. *Integr. Opt. Circuit Eng. IV, Proc. SPIE*, 704, 51.

White, J. O. and Yariv, A. 1984. Photorefractive crystals as optical devices, elements, and processors. *Solid State Opt. Control Dev. Proc. SPIE*, 464, 7.

Woodruff, D. C. 1991. New shutter designs meet challenge of laser systems. *Laser Focus World*, 27, 129.

Woodruff, T. O. and Ehrenreich, H. 1961. Absorption of sound in insulators. *Phys. Rev.*, 123, 1553.

Yariv, A. 1989. *Quantum Electronics, 3rd Ed.*, Wiley, New York.

Yi, Y., Bermel, P., Wada, K., Duan, X., Joannopoulos, J. D., and Kimerling, L. C. 2002. Tunable multichannel optical filter based on silicon photonic band gap materials actuation. *Appl. Phys. Lett.*, 81, 4112.

Yoshino, K., Takeshita, T., Kotaka, I., Kondo, S., Noguchi, Y., Iga, R., and Wakita, K. 1999. Compact and stable electroabsorption optical modulator modules. *J. Lightwave Technol.*, 17, 1700.

Young, E. H. and Belfatto, R. V. 1993. Polychromatic acousto-optic modulators let users tailor output wavelengths. *Laser Focus World*, November, 179.

Young, D. and Tsai, C. S. 1989. X-band magneto-optic Bragg cells using bismuth-doped yttrium iron garnet wavegguides. *Appl. Phys. Lett.*, 55, 2242.

Young, E. H. and Yao, S. K. 1981. Design considerations for acousto-optic devices. *Proc. IEEE*, 69, 54.

Yu, S. and Forrest, S. 1993. Implementations of smart pixels for optoelectronic processors and interconnection systems: I. Optoelectronic gate technology, and II. SEED-based technology and comparison with optoelectronic gates. *J. Lightwave Technol.*, 11, 1659.

12

Optical Fibers

Daniel A. Nolan

12.1 Introduction

Optical communication technology continues to expand after many years of rapid growth. It was over twenty years ago that Corning Incorporated (Kapron et al. 1970) demonstrated that the attenuation in doped silica fiber was low enough that long-distance communication through optical fibers would be advantageous over copper. Soon thereafter, fiber-optic systems proved feasible. In the 70's, the first commercial systems used multimode fibers to better capture the available light and because the state of the art of the associate components was better suited for a large-core fiber. In the 80's, long-haul transmission proved, beyond a doubt, that fiber-optic transmission was a commercial reality. The use of single-mode fiber with laser diodes enabled transmission rates to reach the gigabit-per-second rate over tens of kilometers. Almost immediately, single-mode fiber replaced multimode fiber in trunk line applications. Multimode fiber, however, found important applications in local area networks.

At this point, research activities in the area of new optical fibers and applications of such fibers have continued to expand at a rapid pace. Areas of research, for example, have included new index profiles for transmission in the 1550 nm window, rare earth-doped fibers for optical amplifiers and fiber lasers, the demonstration of soliton transmission in fiber, the demonstration of a number of important nonlinear effects in fibers, and the development of numerical techniques to simulate optical propagation. Meanwhile, process research has led to rapid reduction in the price of fiber. Commercially, the deployment of fiber has expanded beyond the trunk lines to feeder lines, to major business users and, in a number of cases, to the home. All of these activities have progressed in parallel with the development of new active and passive components and advances in communication systems and networks.

In this chapter, we will review the fundamental properties of optical fibers. This will include the material and fabrication properties and optical propagation fundamentals. Propagation in the optical fibers used in today's communication systems is best described using the concept of guided modes. In Section 12.2, the concept of modes of propagation will be introduced. Multimode (Section 12.3) and single-mode fiber will be discussed with particular attention being paid to single-mode fiber. This will include single-mode fiber for light transmission at 1.31 μms and 1.55 μms. Index profiles suitable for transmission in the 1.55 μm window, known as dispersion-shifted fibers, are receiving much attention today because the intrinsic loss of the fiber is low in this window and, more recently, because erbium-doped, fiber-optic amplifiers are expected to change the transmission architecture. These amplifiers are effective only for light transmission in the 1.55 μm window. In Section 12.4, we will discuss the causes of attenuation in optical fiber. Attenuation of the optical signal in today's fiber approaches the theoretical minimum due to Rayleigh scattering. In Section 12.5, the important single-mode parameters will be introduced and in Section 12.6, the dispersion-shifted profiles are shown capable of low-dispersion propagation at 1.55 μm.

In Section 12.7, we will discuss the optical fiber fabrication process. This includes techniques for fabricating the fiber blank and drawing the blank into fiber. Today's blank manufacturing processes include the outside vapor deposition process, in which soot particles are collected on a bait rod and then sintered to glass at elevated temperatures, the chemical deposition process, in which glass particles are formed inside a glass tube and deposited onto the wall, and the axial deposition process where the soot stream of the outside deposition process is deposited axially. Important to the reliability is the organic coating put on the fiber during the fiber-drawing process. The finished fiber is cabled after manufacturing to be ready for deployment.

In Section 12.8, we will discuss more advanced topics related to the propagation of electromagnetic waves in a fiber. This will include the subject of modes of polarization. Polarization-mode dispersion (PMD) has become an important issue for ultralong-distance communication and for analog transmission for video transport. For these systems, polarization-mode dispersion must be kept to a minimum value. A significant amount of research has been recently published addressing the concept requirements and measurement of PMD. On the other hand, polarization-maintaining fiber enables one to retain a given polarization through the fiber. These fibers are used in sensors, such as the fiber-optic gyroscope.

The fiber-optic amplifier, Section 12.9, represents a major advance in fiber-optic technology. Commercialization of this device is in the initial stages. Amplification of the signal enables one to communicate over ultralong distances without the need for expensive electronic regeneration. This is expected, in part, to enable two important technologies: wavelength division multiplexing and new applications in nonlinear optics, such as soliton pulse propagation. The use of amplifiers in optical communication systems is not straightforward from a system's point of view because the optical nonlinearities that develop at these higher powers can significantly affect signal quality, Section 12.10. Research into new applications of fiber optics via amplifiers will, no doubt, continue to expand. The promise of optical communication using many wavelengths and exploiting nonlinearities, such as soliton propagation, Section 12.11, is indeed exciting.

12.2 Optical Propagation

The phenomenon of total internal reflection first reported by Tyndall (1884), enables us to understand how one can transmit light over many kilometers through a fiber. Geometrical optics can be used to describe the repeated reflection of light through the guide. The description is useful for analyzing optical propagation through multimode fiber and can be used, to a limited extent, to understand single-mode propagation, Figure 12.1 shows the variation of the index squared with radius for gradient-index multimode and step-index, single-mode fibers. For a step-index fiber, light undergoes total internal reflection if the angle made by the ray is less than the critical angle. The critical angle θ_c is given

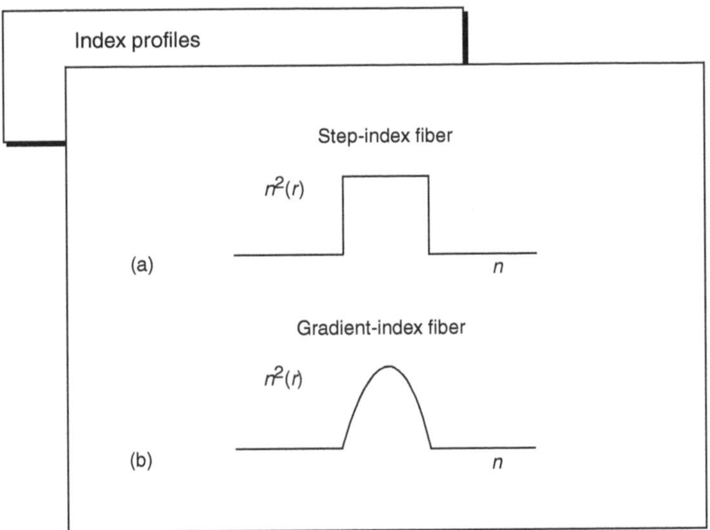

FIGURE 12.1 Index profiles (a) step index; (b) parabolic index.

(assuming launch from air into fiber) by

$$\sin(\theta_c) = \left(n_{co}^2 - n_{cl}^2\right)^{1/2}. \qquad (12.1)$$

Therefore, light propagating at angles less than the critical angle (θ_c) is guided and is described with an acceptance angle called the numerical aperture (NA), defined by

$$\text{NA} = \sin(\theta_c). \qquad (12.2)$$

Light is, of course, refracted when input to the guide as described by Snell's law,

$$\sin(\varphi_i) = n_{co} \sin(\varphi_r) \qquad (12.3)$$

The index of the core n_{co} is slightly higher than that of the cladding n_{cl} which, for silica, is 1.457 in the visible. Rays incident with different angles will propagate along the fiber with different speeds. This, of course, results in a distortion of the signal or pulse and, therefore, a decrease in the possible signal rate or temporal bandwidth. To circumvent this problem, two approaches have been considered. The first is the use of graded-index multimode fibers. In a step-index, multimode fiber, the rays of higher propagation angle are required to traverse along a greater optical path. The optical path can be affected by altering the index with radius. In a graded-index multimode fiber, the use of a parabolic index profile (Gloge and Marcateli 1973) effectively balances the optical path of the higher angle rays with the optical path of the rays making a smaller traverse angle with the axis of the fiber. Later, Olshansky and Keck (1975) showed that a more optimum profile is of the form

$$n^2(r) = n_{co}^2(0) - 2\Delta(r/a)^{\alpha} \qquad (12.4a)$$

where a is the core diameter, the exponent α depends on the wavelength of operation and Δ is determined from the core and clad indices by

$$\Delta = (n_{co}^2(0) - n_{cl}^2)/n_{co}^2(0). \tag{12.4b}$$

The exponent α is a function of the index change with wavelength or index dispersion.

The use of single-mode fiber eliminates the need to balance the optical paths of different rays because only one or mode propagates through the fiber. Analysis of pulse transmission through fibers is most effective using the concept of modal propagation rather than a ray-tracing analysis. Here, the scalar wave equation is used to determine the modes of propagation within the fiber. The scalar wave equation is derived from Maxwell's equations.

$$\overline{\nabla} \times \overline{E} = -\partial\overline{B}/\partial t, \tag{12.5}$$

$$\overline{\nabla} \times \overline{H} = \partial\overline{D}/\partial t, \tag{12.6}$$

$$\overline{\nabla} \cdot \overline{D} = 0, \tag{12.7}$$

and

$$\overline{\nabla} \cdot \overline{B} = 0. \tag{12.8}$$

Here, $\bar{E}$ and $\bar{H}$ are the field vectors. $\overline{D}$ and $\overline{B}$ are the respective flux densities and can be expressed as

$$\overline{D} = \varepsilon_0\overline{E} + \overline{P} \tag{12.9}$$

and

$$\overline{B} = \mu_0\overline{H} + \overline{M}. \tag{12.10}$$

In Equation 12.9 and Equation 12.10, P and M are the induced electric and magnetic polarizations, and ε_0 and μ_0 are the vacuum permitivity and permeability. The scalar wave equation is obtained by taking the curl of Equation 12.5

$$\overline{\nabla} \times \overline{\nabla} \times \overline{E} = -\partial\overline{\nabla} \times \overline{B}/\partial t \tag{12.11}$$

using

$$\overline{\nabla} \times \overline{B} = \overline{\nabla} \times [\mu_0\overline{H} + \overline{M}]. \tag{12.12}$$

Because glass fibers are an insulator, we set

$$\overline{\nabla} \times \overline{M} = 0 \tag{12.13}$$

and, then,

$$\overline{\nabla} \times \overline{\nabla} \times \overline{E} = -\partial\overline{\nabla} \times \mu_0\overline{H}/\partial t = \mu\partial/\partial t\partial\overline{D}/\partial t. \tag{12.14}$$

The vector wave equation can now be written

$$\overline{\nabla} \times \overline{\nabla} \times \overline{E} = -\mu_0 \varepsilon_0 \partial^2 \overline{E} / \partial t^2 - \mu_0 \partial^2 \overline{P} / \partial t^2. \tag{12.15}$$

Using the expression

$$\overline{\nabla} \times \overline{\nabla} \times \overline{E} = \overline{\nabla}(\overline{\nabla} \cdot \overline{E}) - \nabla^2 \overline{E} \tag{12.16}$$

and because glass fibers are an insulating material,

$$\overline{\nabla} \cdot \overline{E} = 0, \tag{12.17}$$

and the vector wave equation becomes

$$\overline{\nabla}(\overline{\nabla} \cdot \overline{E}) = -\mu_0 \varepsilon_0 \partial^2 \overline{E} / \partial t^2 - \mu_0 \partial^2 \overline{P} / \partial t^2. \tag{12.18}$$

The susceptibility relates the electric field with the polarization according to

$$\varepsilon \overline{E} = \varepsilon_0 \overline{E} + \overline{P} \tag{12.19}$$

and because the speed of light in a vacuum is related to the permitivity and permeability

$$c = \frac{1}{\sqrt{\varepsilon_0 \mu_0}}, \tag{12.20}$$

The vector wave equation becomes

$$\nabla^2 \overline{E} = -\mu_0 \varepsilon \frac{\partial^2 \overline{E}}{\partial t^2}. \tag{12.21}$$

In the case of optical fibers, the dielectric constant is the index squared times ε_0 so that the vector wave equation is

$$\nabla^2 \overline{E} = -\mu_0 \varepsilon_0 n^2 \frac{\partial^2 \overline{E}}{\partial t^2} = -\frac{n^2}{c^2} \frac{\partial^2 \overline{E}}{\partial t^2}. \tag{12.22}$$

We can separate the spatial (r) and temporal (t) variables by assuming

$$\overline{E}(\overline{r}, t) = \overline{E}(\overline{r}) e^{iwt} \tag{12.23}$$

so that

$$\nabla^2 \overline{E}(r) - n^2 \frac{\omega^2}{c^2} \overline{E}(r) = -n^2 k_0^2 \overline{E}(r) \tag{12.24}$$

where the free-space wave number is given by

$$k_0 = \frac{2\pi}{\lambda}. \tag{12.25}$$

In the scalar wave approximation, the radial and azimuth components of the electric and magnetic fields can be determined from the axial components. For this reason, we need solve only Equation 12.23 for the axial component.

$$\frac{\partial^2 E_z}{\partial r^2} + \frac{1}{r}\frac{\partial E_z}{\partial \tau} + \frac{1}{r^2}\frac{\partial^2 E_z}{\partial z^2} + n^2 k_0^2 E_z = 0. \tag{12.26}$$

We assume that

$$E_z(r,\theta,z) = \Psi(r)e^{i\beta z} \tag{12.27a}$$

and

$$\frac{\partial^2 \Psi(r)}{\partial r^2} + \frac{1}{r}\frac{\partial \Psi(r)}{\partial r} + \left(n^2(r)k_0^2 - \frac{l^2}{r^2} - \beta^2\right)\Psi(r) = 0. \tag{12.27b}$$

The term β is referred to the propagation constant, and its value is bounded by

$$\frac{2\pi}{\lambda} n_{co} > \beta > \frac{2\pi}{\lambda} n_{cl} \tag{12.28}$$

where n_{co} is the maximum index of the core and n_{cl} is the index of the cladding.

At this point, it is useful to analyze Equation 12.27 for different fiber-index distributions. These solutions enable one to determine the modes of propagation for a given index profile. Profiles usually considered (Snyder and Love 1983) include step, parabolic and gaussian distributions of the core index.

The step-index profile can be described using Equation 12.4 with $\alpha = \infty$. The number of bound modes depends on the numerical aperture. Equation 12.2, or Δ, Equation 12.4b. The solutions to Equation 12.27 for the step-index profile are the well-known Bessel functions. So that the fields are finite at $r=0$ and ∞,

$$\Psi_{\mathrm{m}} = N^{1/2} J_l(u)_{r \le a} \tag{12.29a}$$

$$= N^{1/2} K_l(w)_{r \ge a}. \tag{12.29b}$$

Here, N is a normalization factor,

$$u^2 = (n_{co}^2 k_0^2 - \beta^2)a^2, \tag{12.30}$$

$$w^2 = (\beta^2 - n_{cl}^2 k_0^2)a^2, \tag{12.31}$$

and

$$v^2 = u^2 + w^2. \tag{12.32}$$

A set of eigenvalue equations determines the propagation constant β. The eigenvalue equations are determined by requiring that both the tangential electric and magnetic fields are continuous at the core-clad boundary. Both the tangential electric and magnetic fields can be derived from the axial components (Agrawal 1992). However, Gloge (1971) has shown that the longitudinal electric and magnetic field components are essentially zero for small Δ. In this case, one characteristic equation can be used for all modes:

$$u[J_{l-1}(u)/J_l(u)] = -w\{[k_{l-1}(w)/k_l(w)]\}. \tag{12.33}$$

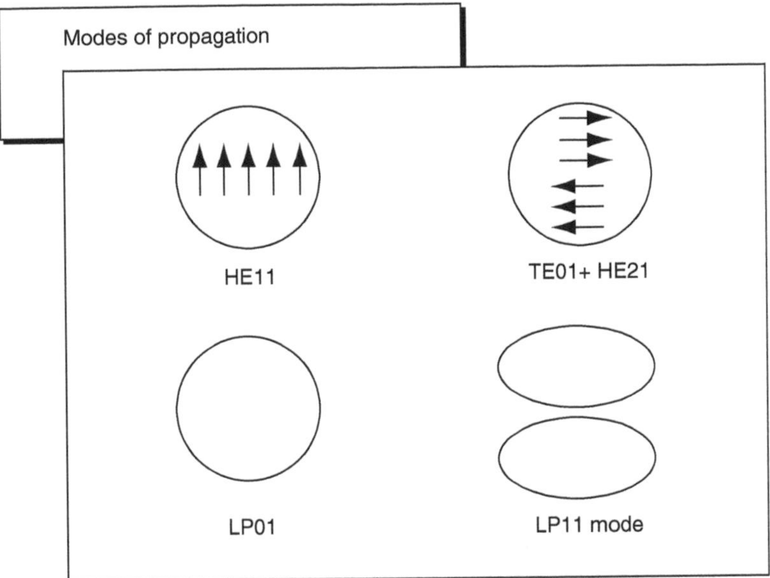

FIGURE 12.2 Schematic of the LP modes.

This expression is termed the linearly polarized (LP) mode approximation (Figure 12.2). For $l \geq 1$, each solution comprises four modes: the two polarizations and the orientation of the lobes associated with l. The fields are of the form

$$\Psi_1 = \frac{1}{2}[J_1(ur/a)/J_1(u)][\cos(l+1)0 + \cos(l-1)\phi]r \leq a$$

and

$$\frac{1}{2}[k_1(wr/a)/k_1(w)][\cos(l+1)\,0 + \cos(l-1)\phi]r \geq a. \tag{12.34}$$

The linearly polarized mode approximation is very useful for analyzing the key attributes of dispersion in both multimode and single-mode fibers. It is also useful, to a limited extent, in analyzing the propagation of polarized modes within fibers.

12.3 Multimode Fibers

As mentioned above, gradient-index cores are used to minimize mode propagation differences within fibers. The scalar wave equation can be used to determine the mode-delay differences as a function of the index profile. For gradient-index profiles the parameters of Equation 12.4 are useful for determining the number of modes within the fiber and the mode-delay differences. This parametric description of the index is used in Equation 12.27 to determine the number of guided modes. The solutions to Equation 12.27 with $\alpha = 2$ are the Laguerre–Gauss functions:

$$n^2(r) = n_{\text{co}}^2(1 - g^2r^2), \tag{12.35}$$

with (Vassallo 1993)

$$g^2 \equiv 2\Delta/a^2.$$

The solutions are

$$\varnothing_{nr}(r) = \mathrm{e}^{-kn_{co}^2 gr^2/2}\left(\sqrt{kn_{co}r}\right)^{\eta} L_{\mu}^{\eta}(kn_{co}gr^2). \tag{12.36}$$

The propagation constants $\beta_{n,m}$ are determined from the eigenvalues

$$\beta_{m,n}^2 = k^2 n_{co}^2 - 2(\mu + 2\eta + 1)kn_{co}g. \tag{12.37}$$

The associated propagation delay time is given by

$$\tau_{\mu,\eta} = \frac{L}{c}\frac{\partial \beta}{\partial k}. \tag{12.38}$$

Solving Equation 12.36 for $\tau_{\mu,\eta}$, it has been shown that $\tau_{\mu,\eta}$ is independent of μ and η to the first order in Δ and, therefore, the parabolic profile optimizes the index for a temporal bandwidth. However, Olshansky and Keck (1975) have shown that, because the index of glass is a function of wavelength, the parabolic profile is, in fact, not optimized for bandwidth. They used the exponent α to characterize the radial dependence of the index and, using the WKB approximation, showed that, for minimum pulse distortion,

$$\alpha = 2 + y - \Delta\frac{(4+y)(3+y)}{5+2y}. \tag{12.39a}$$

where

$$y = \frac{-2n_1}{n_1 - \frac{\mathrm{d}n_{co}}{\mathrm{d}\lambda}}\frac{\lambda}{\Delta}\frac{\mathrm{d}\Delta}{\mathrm{d}\lambda}.$$

As can be seen from Equation 12.38, the temporal bandwidth depends on the value of Δ. In practice, bandwidths as high as a Ghz/km have been achieved with Δs of 2%.

The WKB method is used to solve equations of the form

$$\frac{\mathrm{d}^2 f}{\mathrm{d}x^2} = P(x)f(x) = 0. \tag{12.40}$$

For circular fibers, the transformation

$$r = \mathrm{e}^{\xi} \tag{12.41}$$

is used to express the radial wave equation in the form of Equation 12.27, after which one obtains an equation for the propagation constant

$$\int_{r_1}^{r_2}\left[k^2 n^2(r) - \beta_{m,n}^2 - \frac{m^2}{r^2}\right]^{1/2} \mathrm{d}r = \frac{\pi}{2} + N\pi. \tag{12.42}$$

Here, r_1 and r_2 are the turning points (where the bracket expression is zero) and, for $m=0$, r_1 is chosen as zero. The WKB is most useful for fibers with a large number of modes, i.e., core radius much longer than the wavelength of the propagating light. For more information on the application of the WKB method to circular guides, see Morse and Feschbach 1953.

12.4 Attenuation

In addition to the requirement of minimum pulse broadening during pulse propagation, minimum loss in optical power is also required. The optical fiber fabrication process minimize the introduction of common transition-metal impurities, such as iron, copper, cobalt, etc., into the glass. The important source of attenuation in fiber can be classified as due to molecular vibrational resonance, vibrational resonances of the OH ion, and Rayleigh scattering. The molecular vibrational resonance in optical fiber glasses occurs in the infrared region beyond 5 μms, but these bands have tails that extend into the region at which optical signals propagate (0.8–1.6 μms). Although small, these bands contribute a loss to optical propagation on the order of 0.5 dB/km and less. Vibrational resonances in fibers can also result from the inclusion of OH ions in the silica matrix. An important vibrational resonance of the OH ion occurs at 2.71 μms. There are overtones and combination bands at 0.95, 1.25, and 1.38 μms due to the 2.71-μm resonance (Agrawal 1992). Optical fiber fabrication processes limit the amount of OH ions through proper doping procedures, and it is possible to fabricate fibers with OH ion inclusions on the order of parts per billion. The windows of transmission of 0.8, 1.3 and 1.55 μm are chosen so that losses due to OH ion vibrational resonances are limited.

Rayleigh scattering of light is strongly wavelength-dependent and varies inversely with the fourth power of the wavelength. This source of optical loss results from concentration fluctuations at high temperatures that are frozen in place as the glass cools through the transformation region. One can calculate an effective loss coefficient due to Rayleigh scattering by the expression

$$\alpha_{\mathrm{R}} = \frac{8\pi^3}{3\lambda^4}(n^2-1)KTB \tag{12.43}$$

Here, B represents the isothermal compressibility of the glass, T is the fictive temperature and K is Boltzman's constant. At 1550 nms, losses resulting from Rayleigh scattering are on the order of 0.13 dB/km. This window is minimum with respect to optical attenuation as it is optimized with respect to the losses resulting from Rayleigh scattering (which increase with lower wavelength) and those from molecular vibrational resonances (which increase at higher wavelengths). Losses in this window are as low as 0.2 dB/km, whereas in the 1310 nm window, they are on the order of 0.35 dB/km (Figure 12.3).

12.5 Single-Mode Propagation

The gradient-index profile is especially useful for minimizing temporal pulse broadening in multimode fibers. In single-mode fibers, this balance is, of course, not required because the information is carried by one optical mode. Pulse broadening does result, however, from both intramode dispersion and polarization-mode dispersion. Polarization-mode dispersion will be discussed in Section 12.8. Intramode dispersion results, primarily, from the fact that the index varies with wavelength (material dispersion), and the confinement of the light within the guide varies with wavelength. Light of high wavelength occupies more of the cladding and, as such, experiences a slight shift in time delay (profile dispersion). The material dispersion results from the fact that the index varies with wavelength and in germania-doped silica is a minimum near 1.3 μms. The wavelength at which the minimum occurs is a function of the dopant level. The wavelength dependence of the index is generally described using the Sellmeier coefficients

$$n^2-1 = \sum_{j=1}^{N} \frac{A_j\lambda^2}{\lambda^2-l_j^2}. \tag{12.44}$$

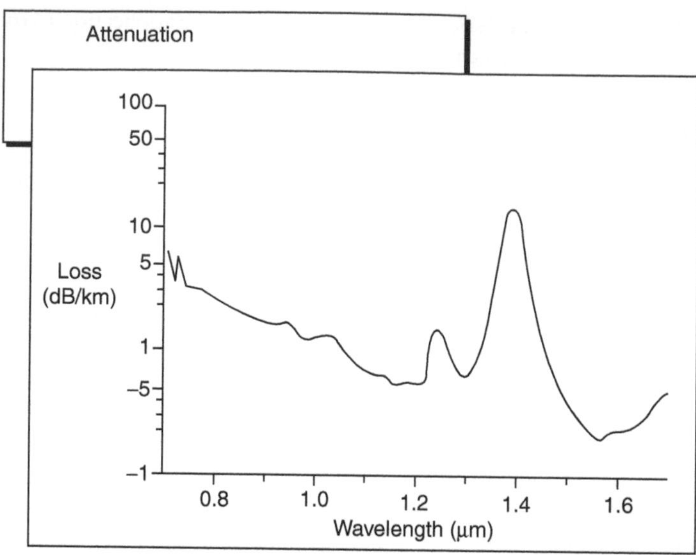

FIGURE 12.3 Absorption, including OH, with wavelength.

The parameters A_j and I_j are obtained by fitting the index data and depend on the glass-dopant levels. These parameters represent the oscillator strength and resonance wavelength respectively.

Pulse broadening in single-mode fibers is characterized by the term dispersion. Dispersion represents the change in time delay with wavelength and is measured in psecs/(nm. km)

$$D = \frac{\partial}{\partial \lambda}\left[\frac{\partial \beta}{\partial k}\right]\frac{L}{c}. \quad (12.45)$$

The dispersion can be further expressed as

$$D = \frac{\partial}{\partial \lambda}\left[\frac{\partial \lambda}{\partial k}\cdot\frac{\partial \beta}{\partial \lambda}\right]\frac{L}{c} = -\frac{\lambda}{\pi}\left[\frac{\partial \beta}{\partial \lambda} + \frac{\lambda}{2}\frac{\partial^2 \beta}{\partial \lambda^2}\right]\frac{L}{c}. \quad (12.46)$$

In step-index, single-mode fibers, the dispersion varies nearly linearly from approximately -1 ps/(nm km) at 1280 nm to 15 ps/(nm km) at 1550 nm. The propagation constant β is determined from the eigenvalue equation, Equation 12.32, for a step-index guide.

Another important parameter for single-mode guides is the spot size. The field distribution calculated using the eigenfunction can be approximated with the expression

$$\Psi(r) = \exp\left(-\frac{1}{2}\left(\frac{r}{r_0}\right)^2\right). \quad (12.47)$$

The spot size for standard single-mode fibers or mode-field diameter is on the order of 9 μms. For a given index profile, the spot size increases with wavelength on the order of 1 μm per 200 nms. Larger spot size enables simpler fiber-to-fiber connection and splicing, but bending losses, both from macro and micro bending, increase with increasing spot size.

12.6 Dispersion-Shifted Fiber

As mentioned above, the attenuation of silica-based fibers is a minimum near 1.55 μms. For this reason, there is considerable interest in transmitting information at this wavelength. Also, with the recent availability of erbium-doped fiber amplifiers at 1.55 μms, this window of operation is most advantageous for long distance communication. Step-index, single-mode fibers, however, exhibit considerable pulse broadening in this window, approximately 18 ps/(nm. km). Figure 12.4 shows the segmented core profile of a dispersion-shifted fiber (Bhagavatula et al. 1984) and that of the platform profile. The segmented core profile was the first dispersion-shifted profile capable of shifting the λ_0 while maintaining other important parameters.

Solution of the radial equation, Equation 12.27, using the index profile shown, shows that the dispersion minimum can be easily shifted to the 1.55 μm window, while at the same time maintaining other important properties of the fiber (Bhagavatula et al. 1984). Other important parameters include both bend loss, spot size (r_0), and cutoff wavelength (λ_c). The cutoff wavelength refers to the wavelength below which multimode propagation exists. These dispersion-shifted profiles are now finding application in transoceanic communication and the trunk lines. The use of dispersion-shifted fibers with erbium-doped amplifiers requires a minimum dispersion, on the order of 0.2 ps.(nm. km) because optical nonlinearities distort the signal at the zero dispersion wavelength λ_0. Of particular concern is four-wave mixing. Later, we will summarize the important nonlinearities and how they affect signal distortion.

12.7 Fiber Fabrication

Modern fiber-optic processes enable the fabrication of low-cost fibers of excellent quality. In all glass-forming processes, silica tetrachloride and the dopants, such as germanium tetrachloride, are delivered to the reaction region as vapors, where silica and germania are formed. The byproduct is chlorine gas, Cl_2. In addition to glass blank fabrication, high-speed draws (greater than one meter per second) are used to draw and coat the fiber with an organic material that protects the fiber from handling and from the environment.

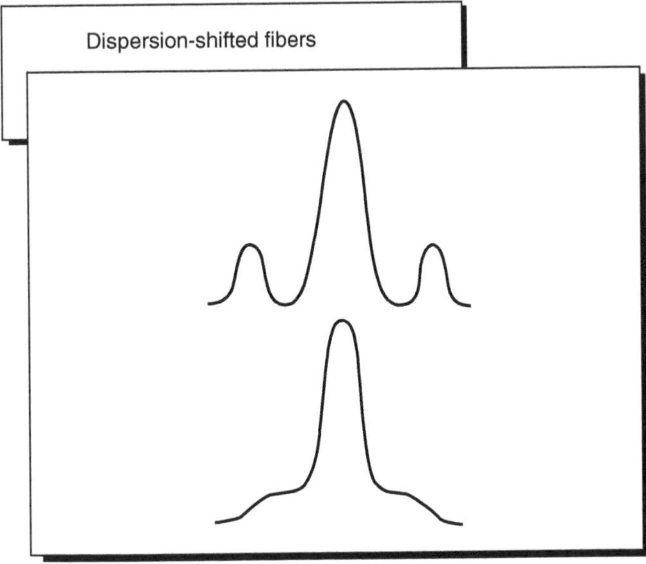

FIGURE 12.4 Dispersion-shifted profiles.

There are basically three methods used to form the glass blank. The modified chemical vapor deposition process (MCVD) (Nagel et al. 1985), the outside vapor deposition process (OVD) (Morrow et al. 1985) and the vapor-axial deposition process (VAD) (Niizeki et al. 1985). In the MCVD process, successive layers of SiO_2 and dopants, which include germania, phosphorous, and fluorine, are deposited on the inside of a fused silica tube by mixing the chloride vapors and oxygen at a temperature on the order of 1800°C. The temperature is maintained using a burner which traverses the outside of the tube. After the dopants are deposited, the temperature of the tube is raised, via the burner, to collapse the tube. In the OVD process, the core and cladding layers are deposited on a rotating mandrel by the flame hydrolysis process. Upon completing the deposition process, the mandrel is removed from the preform, and the preform is sintered in a furnace to form the fiber blank. The VAD process is also a flame hydrolysis process, but, in this technique, the soot is deposited axially. Figure 12.5 is a schematic of the three blank fabrication processes.

In the drawing process, the blank is fed from above into the drawing portion of the furnace while being drawn from the bottom using tractors. The fiber is then wound onto a drum while being monitored for tensile strength. The temperature during draw is on the order of 2000°C. After exiting the furnace, the fiber is coated with a UV-curable coating before winding on the drum.

12.8 Polarization

Single-mode fibers capable of maintaining an input linear polarization are known as polarization-preserving fibers. These fibers employ either an elliptical core or stress rods placed 180° from each other and outside the core. The elliptical core and/or stress rods introduce a birefringence that removes the degeneracy of the orthogonally polarized modes. The birefringence is defined as the difference between the effective indices of the two orthogonal modes and is on the order of 10^{-4}. Figure 12.6 shows the index profile for stress rod and elliptical core polarization-maintaining fibers. Polarization-preserving fibers are not used at present in telecommunication systems but rather find applications in the area of fiber-optic sensors. As an example, fiber-optic gyroscopes use polarization-preserving fiber to prevent coupling from one polarization to the other in the gyroscope coil. Such coupling leads to inaccurate sensing and is eliminated by using only one polarization. An important parameter for these fibers is the beat length, defined as the wavelength divided by the birefringence (δn). This parameter describes the beating which results from the fact that the propagation speed along the fast axis is considerably larger than that along the slow axis. Highly polarized fiber exhibits beat lengths on the order of less than a centimeter. Standard fibers have very small birefringence and exhibit beat lengths greater than ten meters.

The slight differences in propagation speeds for the two polarization modes in standard single-mode fiber lead to PMD (Rashleigh 1983). Polarization-mode dispersion for standard fiber is, typically, significantly less than 1 ps/sqrt(km). The inverse square root length dependence results from the fact that the polarization randomly couples as the light propagates through the fiber. Transoceanic systems using optical amplifiers eliminate the need for expensive electronic regenerators. Without signal regeneration, polarization-mode dispersion can become significant over hundreds of kilometers. Also, CATV analog transmission systems can be sensitive to PMD when component polarization-dependent loss and significant laser chirping occur. In this situation, composite second-order distortion (CSO) (Peterman 1981) can significantly degrade the signal.

As mentioned, small-core eccentricities, on the order of a few percent, can be the source of PMD. Significant literature exists on the effect of core ellipticity on polarization dispersion for step-index fibers. However, these perturbation models predict that the polarization dispersion increases linearly with length. The square root length dependence that is experimentally observed results from random coupling of the polarization modes. In step-index fibers, polarization dispersion (Kapron et al. 1972) results from both form birefringence (Snyder and Love 1983) and stress birefringence (Rashleigh 1983), both of which depend on core ellipticity and are approximated in the low ellipticity limit as

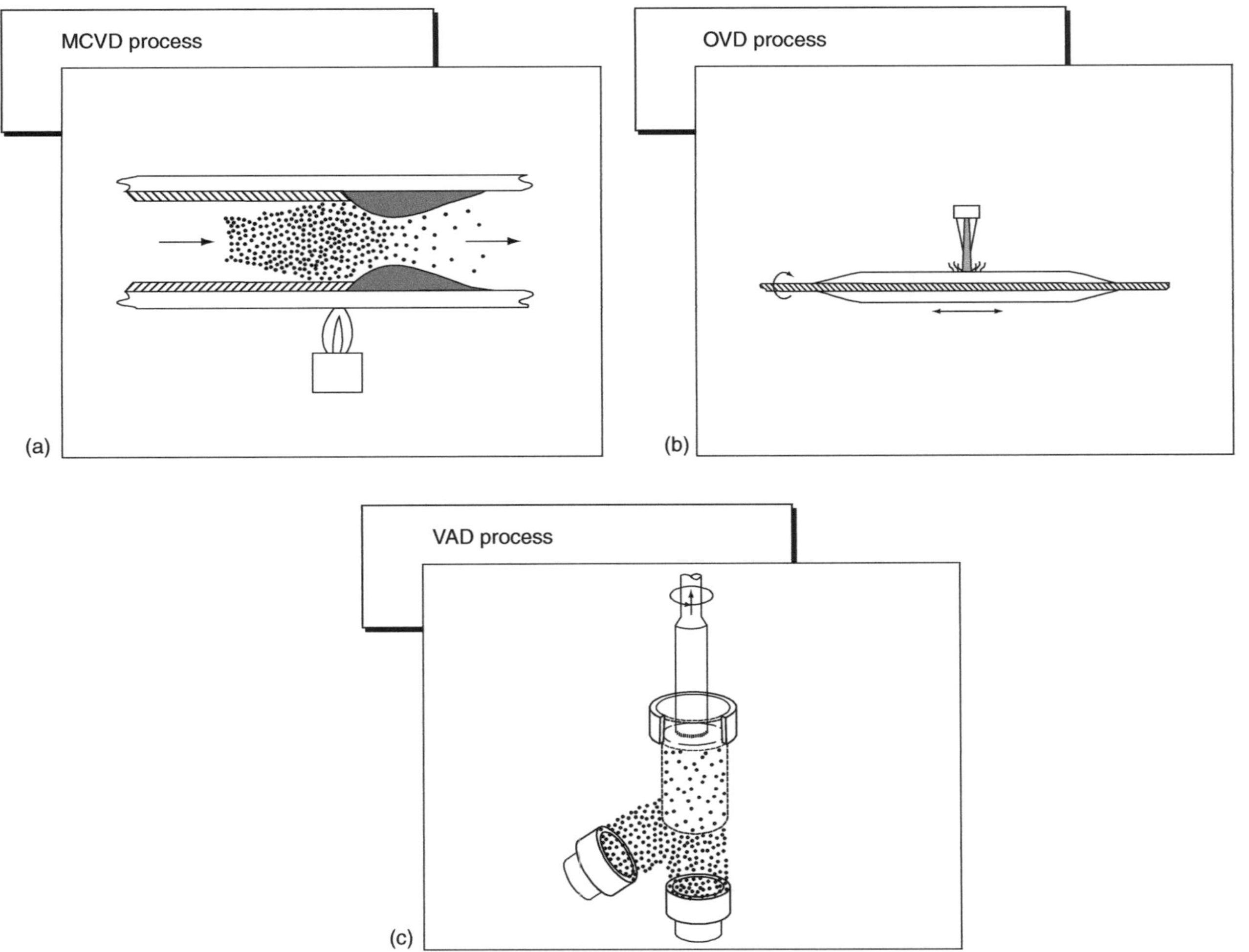

FIGURE 12.5 Three blank fabrication processes: (a) MCVD; (b) OVD; (c) VAD.

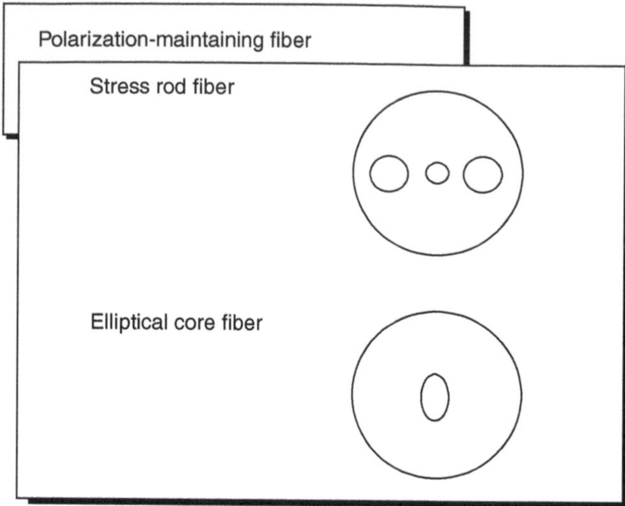

FIGURE 12.6 Polarization-maintaining fibers, index profiles.

$$\Delta\beta(\text{FORM}) = \frac{E(2\Delta)^{3/2}}{\rho}\frac{4}{V^3}\frac{(\ln \nu)^3}{1+\ln \nu} \tag{12.48}$$

and

$$\Delta\beta(\text{Stress}) = \left(1-\frac{u^2}{\nu^2}\right)\frac{C_s}{1-\eta_\rho}\Delta\alpha\Delta T. \tag{12.49}$$

In the above equations, u and v are defined in Equation 12.30 through Equation 12.32, and e is the eccentricity which is defined as

$$e = \frac{A-B}{A+B} \tag{12.50}$$

and

$$E = 1-\frac{B^2}{A^2},$$

where A and B are the major and minor diameters. The stress birefringence depends on the expansion differences of the core and cladding, $\Delta\alpha$, which, in turn, depend on the dopant concentrations. The stress birefringence also depends on the fictive temperature T which is used to approximate a point below which the glass structure cannot change on the same time scale as the cooling rate, a stress optic coefficient C_s, and Poisson's ratio h_r.

The measured value of PMD can be sensitive to fiber deployment. This is because the amount of mode coupling varies with deployment and with the environment. The effect of mode coupling on PMD is modeled using the statistics of random occurrences (Matera and Someda 1992). This analysis explains the square root length dependence of PMD and the distribution of PMD values with changes in the environment and the fact that repeated measurements of PMD show that the observed values obey a

Maxwellian process:

$$P(\Delta\tau) = \frac{2\tau}{\sqrt{2\pi q^3}} e^{-(\Delta\tau^2/2q^2)} \tag{12.51}$$

In Equation 12.51, q^2 is the variance. Another aspect of the random coupling of the polarization modes is a significant wavelength dependence of PMD (Poole and Wagner 1986; Matera and Someda 1992).

One final and important aspect of polarization modes in fibers is the evolution of the polarization states as light propagates through a fiber. Light linearly polarized as it enters the fiber not only evolves into circularly or elliptically polarized light, as it propagates, but significant mode coupling occurs. This complicates the analysis and measurement (Hoffner 1993), of PMD. Recently, techniques to minimize the effects of mode coupling on the stability of PMD have been described by Judy (1994).

12.9 Optical Fiber Amplification

An important aspect of optical fiber research is rare-earth doping for amplification and lasing. Amplification in optical fiber has not only renewed interest in the materials and propagative aspects of fiber research, but it has also significantly affected the systems aspects of optical communication. Doping of fibers for optical amplification has been under study since the 1960s (Koester and Snitzer 1964) when neodymium was used as a dopant. Interest in rare-earth doping was renewed in the 1980s when research scientists at the University of Southampton (Mears et al. 1987) showed amplification in the 1.55 μm region, which coincidentally is the low-loss transmission window. The fact that erbium-doping enables amplification in the transmission window of lowest loss has attracted much interest in this aspect of optical fiber research.

In the erbium-doped fiber amplifier (in its simplest form), Figure 12.7, an erbium-doped fiber with lengths on the order of meters and dopant levels on the order of 2 ppm, is spliced to a wavelength-dependent, fiber-optic coupler. The coupler enables one to continuously pump the erbium-doped fiber with light emitted from a high-power semiconductor laser diode at 980 or 1480 nm. Filters and optical isolators are often included to minimize spontaneous emission noise and reflections. The pump light is used to excite ions from the ground state to an excited state. Signal light entering the fiber initiates stimulated emission and is coherently amplified. Years of research at many laboratories has led to

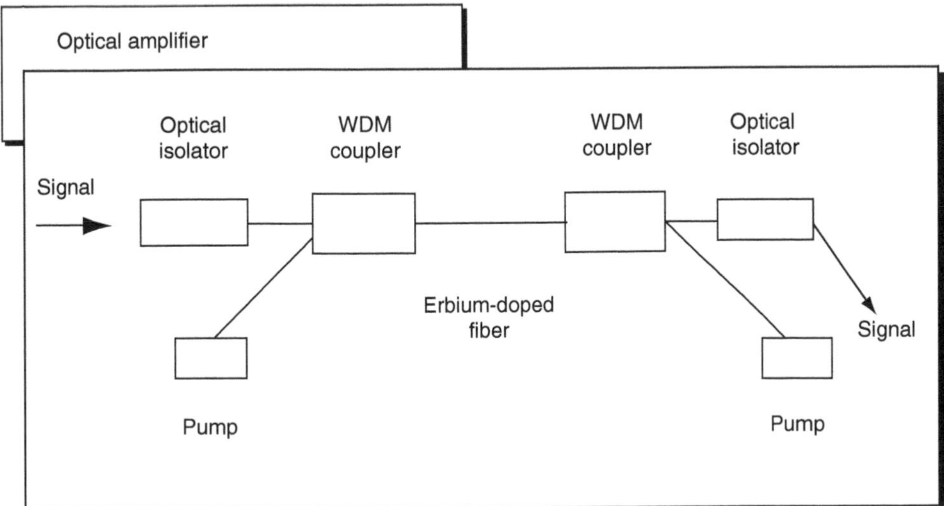

FIGURE 12.7 Schematic (simple) of the optical amplifier.

the development of erbium amplifiers. Such technical issues as wavelength dependence of gain, gain saturation, polarization dependence and spontaneous emission, among others, have been carefully studied. Spontaneous emission occurs when ions in the excited state spontaneously relax to the ground state contributing to noise. This phenomenon in itself significantly affects the signal-to-noise ratio of an amplifier-based communication system. Another important parameter of the optical amplifier is the concentration of erbium ions. An optimum concentration of erbium ions avoids ion clustering which alters the excited states and results in elevating one ion to a higher state and emission to the ground of neighboring ions.

Research in the areas of rare-earth doped optical fibers is far from complete. Issues, such as multiple wavelength amplification, need to be further addressed. Research into the possibilities of using erbium-doped fibers in a lasing configuration is useful for picosecond pulse sources. Another aspect to rare-earth doping is the interest in amplification at 1.3 μms. Amplification in the low-loss transmission windows requires dopants other than erbium (praseodymium. co-doped with neodymium, for example), and also requires the use of a fluoride glass host (Ohishi et al. 1991). In a silica glass, the phonon vibrational spectrum affects the amplification process. In a fluoride-based glass, however, the phonon edge is shifted to higher frequencies, thereby, enabling reasonable amplification in the 1.3 μm window. Other issues remain, however, including the amount of gain and the wavelength dependence of the gain.

Fluoride fibers are melted at temperatures far below that of silica and, therefore, cannot be fused to silica fibers. The index is not well matched to silica and, therefore, leads to strong back reflections. More importantly, the fabrication technology of fluoride-based fibers is less advanced than that of silica. Nonetheless, interest in amplification at 1.3 μms remains because there is a huge installed base of optical fiber that is optimized for minimum pulse distortion at 1.3μms rather than 1.55 μms.

12.10 Optical Nonlinearities in Fibers

Nonlinear effects in silica fibers have been considered unimportant. However, the use of optical amplifiers and the promise of dense wavelength-division multiplexing imposes a number of limitations (Chraplyvy 1990) on the ultimate bandwidth of optical communication. Optical nonlinearities will be discussed in detail in other chapters, but it is important to mention the limitation that these nonlinearities impose on communication bit rates. These nonlinearities are stimulated Brillouin scattering, four-wave mixing, cross-phase modulation, and stimulated Raman scattering.

Multiple wavelength mixing was first observed in optical fibers by Hill et al. (1978). The output power (Inoue 1992) of light generated at a fourth wavelength can be written as

$$P_F(L) = \frac{1024\pi^6}{n^4\lambda^2c^2}(D_x)^2\frac{P_i(0)P_j(0)P_k(0)}{A_{\text{eff}}^2}\mathrm{e}^{-\alpha L}\frac{(1-\mathrm{e}^{-\alpha L})^2}{\alpha^2}\eta \tag{12.52}$$

In Equation 12.52, i, j and k represent the three input wavelengths at $Z=0$, P_F is the power at wavelength F and at $Z=L$. A_{eff} is the effective area of the guide, α is the loss, and D_x is a degeneracy factor. The efficiency factor η is given by

$$\eta = \frac{\alpha^2}{\alpha^2(\Delta\beta)^2}\left[1+\frac{4\mathrm{e}^{-\alpha L}\sin^2(\Delta\beta L/2)}{(1-\mathrm{e}^{-\alpha L})^2}\right] \tag{12.53}$$

where

$$\Delta\beta = \beta(f_i)+\beta(f_j)-\beta(f_k)-\beta(f_F). \tag{12.54}$$

In Equation 12.54, $\Delta\beta$ is the difference in propagation constants, which depends on the dispersion of the fiber. Numerical and systems studies have shown the importance of using a finite amount of dispersion to avoid generating new wavelengths while depleting the signal wavelength. Finite dispersion is also important at single channel operation over very long lengths because multiple wavelength mixing occurs through the wavelengths generated with amplifiers by amplified spontaneous emission (Marcuse 1991). A major impact of four-wave mixing is that it forces designers to operate their signal wavelength away from a dispersion zero. Over long distances, this require chromatic or dispersion compensation.

Brillouin scattering is the interaction of light with acoustical vibrations in fiber. The signal or carrier wave is shifted to longer wavelengths (Stokes shifted) with the simultaneous emission of an acoustical phonon. The amount of power generated can be characterized with an exponential gain coefficient, g_B (Stolen 1979):

$$g_B = 4 \times 10^{-9}\ \text{cm/W}. \tag{12.55}$$

The gain coefficient enables one to calculate the amount of stimulated Brillouin scattering with length as a function of incident light power. The wavelengths generated are separated from the carrier by less than one thousandth of a nanometer. This slight shift in wavelength would not be expected to cause problems, except for the fact that the generated light is scattered backwards, depleting the carrier signal and, at times, affecting the transmission laser. It is estimated that Brillouin scattering becomes an issue when the power in an optical fiber is on the order of a milliwatt (Chraplyvy 1990).

Raman scattering results from an interaction of the incident light with molecular vibrations in the fiber. As with Brillouin scattering, Raman scattering can be characterized with a gain coefficient (Stolen and Ippen 1973)

$$g_R = 7 \times 16^{-12}\ \text{cm/W}. \tag{12.56}$$

Comparison of Equation 12.55 and Equation 12.56 shows that the threshold for Raman scattering occurs at power levels three orders of magnitude higher than that for Brillouin scattering. Raman scattering can be a problem, however, because the wavelength or bandwidth of interaction is greater than a hundred nanometers. Raman scattering can cause serious cross-talk for multiple channel systems of significant power and will ultimately limit the information transmission capacity of optical fiber systems. It is estimated that Raman scattering becomes an issue (Chraplyvy 1990) when the total power in an optical fiber is on the order of 1w.

Self- and cross-phase modulation refers to the fact that the index of glass is intensity-dependent (Islam 1992):

$$n = n_0 + n_2 I \tag{12.57}$$

where n_0 represents the linear index and the second terms includes the intensity-dependent refractive index n_2 and the optical intensity I.

In silica, n_2 is on the order of $3 \times 10^{-16}\ \text{cm}^2/\text{W}$ and $6 \times 10^{-16}\ \text{cm}^2/\text{W}$ for self- and cross-phase modulation, respectively. Both cross- and self-modulation affect the phase and, hence, the arrival time of a pulse. Changes in power and the modulation of power with other carriers limit the amount of power in fibers. It is estimated (Chraplyvy 1990) that self- and phase-modulation become an issue when the power in a fiber is on the order of 10mw.

12.11 Solitons in Optical Fiber

In this section, we will briefly introduce the concept of soliton pulses in optical fiber. This is a fascinating research topic, and it is expected that the use of soliton pulses in optical fibers win enable transmission over transoceanic distances at the highest of possible bit rates.

Soliton propagation in fiber is a nonlinear phenomenon, and such pulses are sensitive to the amount of optical power. Attenuation severely limits the distances over which a soliton pulse can travel without significant distortion. The development of the erbium-doped optical amplifier has spurred research activity toward commercializing transoceanic soliton systems. The coincidence of minimum fiber attenuation in the 1550 nm telecommunication window and the strong gain in this same window with the erbium amplifier is, indeed, encouraging. In fact, by amplifying the signal every 30 kms, soliton pulses can travel thousands of kilometers without distortion. One can also use multiple wavelengths wavelength division multiplexing (WDM) without the imposition of a number of the nonlinearities mentioned in the previous section. For these reasons, it is expected that, ultimately, soliton transmission will be the technique of choice for transoceanic communication. However, more research is required before the potential of soliton communication can be fully realized.

Solitons occur in nature in many different ways, optical pulses being one of them. Scott Russel is credited with first observing and recording solitary waves in a barge canal in Great Britain in 1938. He followed the "large solitary elevation" on horseback and noted that it did not change in form or speed for miles. He derived equations describing the velocity of the waves and reported on his work at the Liverpool meeting of the British Association for the Advancement of Science. In 1967, a group of mathematical physicists (Gandner et al. 1967) from Princeton University solved the so-called Korteweg-de Vries equations describing the nonlinear hydrodynamic wave. The Russian pair, Zahkharov and Shabat (1971) considered the nonlinear propagation of optical waves in a two-dimensional medium. They showed that the nonlinear Schroedinger equation could be solved using the inverse scattering theorem and pointed out the relationship between spatial dispersion and optical intensity.

Hasegawa and Tappert (1973) showed, theoretically, that temporal solitons should exist in fiber and pointed out the possibility of using them for optical communication. Mollenauer et al. (1980) were the first to observe solitons experimentally in fiber. They built a "color center" laser that enabled them to generate and launch narrow temporal pulses of power levels significant enough to develop into soliton pulses. This experimental observation can be considered to mark the beginning of a new technology aimed at enabling dispersionless transmission over transoceanic distances. There are, however, many obstacles yet to be overcome before this technology can be fully utilized.

Nonlinear pulse propagation in fiber is described with the so-called nonlinear Schroedinger equation

$$-i\frac{\partial u}{\partial z} = \frac{\lambda^2 D}{4\pi c}\frac{\partial^2 u}{\partial t^2} + \frac{2\pi n_2}{\lambda A_{\text{eff}}}|u|^2 u. \tag{12.58}$$

Here, D is the dispersion and u is the pulse amplitude which varies both spatially and temporally. The soliton pulse has both a temporal and spectral width, both of which are described with hyperbolic secant functions (Islam 1992):

$$u(t) = \text{sech}[t]\text{e}^{\text{i}z/2} \tag{12.59}$$

and

$$\bar{u}(w) = \frac{1}{2}\text{sech}\left[(w - w_0)\frac{\pi t_c}{2}\right]. \tag{12.60}$$

In Equation 12.58, t_c is the temporal width and w_0 the center frequency. An important parameter is the power P_c at which the nonlinearity and dispersion balance (Islam 1992):

$$P_c = \frac{\lambda A_{\text{eff}}}{2\pi n_2 Z_c} \tag{12.61}$$

where

$$Z_c = \frac{2\pi c}{\lambda^2 D} t_c^2. \tag{12.62}$$

The parameter Z_c characterizes the distance at which the pulse begins to spread. Another important parameter is the soliton period, characterized by Z_c times $\pi/2$.

At this point, we will terminate our introduction of soliton pulses in optical fibers. We should mention, however, that current research interest is in the areas of minimizing pulse jitter (Gordon–Haus jitter) using frequency filters (Mollenauer et al. 1992), soliton–soliton interactions, and the use of WDM with solitons.

The possibility of switching light at ultrafast speeds using solitons is also a topic for research. Soliton pulses have properties that make them attractive in this regard.

Defining Terms

Attenuation: Propagation loss of an optical input or signal. Quantified with an exponential coefficient and measured in dB/km.

Birefringence: Effective index difference of the two polarization modes in single-mode fiber.

Brillouin scattering: Scattering of light by an acoustical phonon. In an optical fiber, the light predominantly scatters in the background direction.

Critical angle: The angle below which total internal reflection occurs, dependent on the index difference between the propagation and the surrounding medium.

Dispersion-Shifted fiber: Optical fiber in which the wavelength of minimum dispersion is shifted further to the infrared by an index profile design.

Fiber blank: Core and clad glass composite which is drawn into fiber.

Fiber draw: Equipment used to draw fiber from the blank to 125 μms and less. Simultaneously, the fiber is coated with an organic coating to protect the glass from the environment.

Fiber-Optic amplifier: Device composed of rare-earth-doped fiber and components, such as couplers and isolators. The device is capable of amplifying an input signal by over three orders of magnitude.

Gradient-Index profile: Fiber-optic index profile where the core index is graded with core radius. Typically, it is parabolic with radius.

Maxwell's equations: Differential equations describing the relationships of the electric and magnetic field vectors of light.

MCVD process: Optical blank making process, where gases are input inside a rotating tube, react and, under a thermophoresis process, are deposited onto the wall as glass.

Multimode fibers: Optical fiber capable of guiding more than one mode. Typically hundreds of modes are guided.

Nonlinear index: The index component dependent on optical power. In silica based fiber, this parameter becomes above about one milliwatt.

Outside vapor deposition process: Optical blank making process, where glass soot particles are deposited onto a rotating mandrel and, then, consolidated to glass.

Polarization: In optical fiber, the vector components of the propagating light. Includes both linear and circular components.

Raman scattering: Scattering of light at high optical powers, resulting from an interaction of light with the molecular vibrations of the glass.

Rayleigh scattering: Scattering of light from particles with differences in index. Affects the fundamental limit of attenuation in an optical fiber.

Single-Mode fiber: Fiber capable of propagating only one mode at the wavelength of interest, but includes modes of both polarizations.

Self-Phase modulation: Results from the fact that the index of light is intensity-dependent. Power-dependent changes in index affect the phase delay of the signal.

Step-Index profile: The fiber-optic index profile where the core and clad index are constant with radius, the clad glass being less in index than the core.

References

Agrawal, G. P. 1992. *Fiber-Optic Communication Systems*, Wiley, New York.

Bhagavatula, V. A., Spotz, M. S., and Love, W. F. 1984. Dispersion shifted segmented core single-mode fibers. *Opt. Lett.*, 9, 186.

Born, M. and Wolf, E. 1980. *Principles of Optics*. Pergamon Press, Oxford.

Chraplyvy, A. R. 1984. Optical power limits in multichannel WDM systems due to stimulated Brillouin scattering. *Electron. Lett.*, 20, 58.

Chraplyvy, A. R. 1990. Limitations on lightwave communications imposed by optical-fiber nonlinearities. *J. Lightwave Technol.*, 8, 1548.

Fleming, J. W. 1978. Material dispersion in light guide glasses. *Electron. Lett.*, 14, 326.

Gardner, C. S., Greene, J. M., Krusal, M. D., and Miura, R. M. 1967. Method for solving the Korteweg–de Vries equation. *Phys. Rev. Lett.*, 19, 1095.

Gloge, D. 1971. Weakly guiding fibers. *Appl. Opt.*, 10, 2252.

Gloge, D. and Marcateli, E. A. J. 1973. Multimode theory of graded-core fibers. *Bell Syst. Tech. J.*, 52, #9, 1563.

Hasegawa, A. and Tappert, F. D. 1973. Theory of stationary nonlinear optical pulses in dispersive dielectric fiber. 1. Anomalous dispersion. *Appl. Phys. Lett.*, 23, 142.

Heffner, B. L. 1993. Accurate automated measurement of differential group delay dispersion and principal state variation using Jones matrix eigenanalysis. *IEEE Photon Tech. Lett.*, 5, 814.

Hill, K. O., Johnson, D. C., Kawasaki, B. S., and MacDonald, R. I. 1978. CW three-wave mixing in single-mode optical fibers. *J. Appl. Phys.*, 49, 5098.

Inoue, K. 1992. Four-wave mixing in an optical fiber in the zero-dispersion wavelength region. *J. Lightwave Technol.*, 10, 1553.

Islam, M. N. 1992. *Ultrafast Fiber Switching Devices*. Cambridge University Press, Cambridge, UK.

Judy, A. F. 1994. Improved PMD stability in optical fibers and cables. *International Wire and Cable Symp. Proc.*

Kapron, K. P., Keck, D. B., and Maurer, R. D. 1970. Radiation losses in optical waveguides. *Appl. Phys. Lett.*, 17, 423.

Kapron, F. P., Borrelli, N. F., and Keck, D. B. 1972. Birefringence in dielectric optical waveguides. *IEEE J. Quantum Electron.*, 8, 222.

Keck, D. B. 1981. Optical fiber waveguides. In *Fundamentals of Optical Fiber Communications*, M. K. Barnoski, ed.,

Koester, C. J. and Snitzer, E. A. 1964. Amplification in a fiber laser. *Appl. Opt.*, 3, 1182.

Marcuse, D. 1991. Single channel operation in very long nonlinear fibers with optical amplifiers at zero dispersion. *J. Lightwave Technol.*, 9, 356.

Matera, F. and Someda, C. G. 1992. In *Random Birefringence and Polarization Dispersion in Anisotropic and Nonlinear Optical Waveguides*, C. G. Someda and G. Stegeman, eds., p. 1026. Elsevier, Amsterdam.

Mears, R. J., Reekie, L., Jauncey, I. M., and Payne, D. N. 1987. *Electron. Lett.*, 23, 1026.

Mollenauer, L. F., Stolen, R. H., and Gordon, J. P. 1980. Experimental observation of picosecond pulse narrowing and solitons in optical fibers. *Phys. Rev. Lett.*, 45, 1095.

Mollenauer, L. F., Gordon, J. P., and Islam, M. N. 1986. Soliton propagation in long fiber with periodically compensated loss. *IEEE J. Quantum Electron.*, QE-22, 157.

Mollenauer, L. F., Gordon, J. P., and Evangelides, S. G. 1992. The sliding-frequency guiding filter: an improved form of soliton jitter control. *Opt. Lett.*, 17, 22, 1575.

Morrow, A. J., Sarkar, A., and Schultz, P. C. 1985. In *Optical Fiber Communications*, T. Li, ed., Vol. 1, p.59. *Academic Press*, San Diego, CA, Chap. 1.

Morse, P. M. and Feschbach, H. 1953. *Methods of Theoretical Physics*. McGraw-Hill, New York.

Nagel, S. R., MacChesney, J. B., and Waller, K. L. 1985. T. Li, ed., In *Optical Fiber Communications*, Vol. 1, p. 59. Academic Press, San Diego, CA, chap. 1.

Niizeki, N., Inagaki, N., and Edahiro, T. 1985. T. Li, ed., In *Optical Fiber Communications*, Vol. 1, p. 59. Academic Press, San Diego, CA, chap. 3.

Ohishi, Y., Kanamori, T., Kitagawa, T., Tagahashi, S., Snitzer, E., and Sigel, G. H., Jr. 1991. Pr^{3+}-Doped fluoride fiber amplifier operating at 1.31 μm, Optical Fiber Conf., San Diego, CA (PD-2).

Olshansky, R. and Keck, D. B. 1975. Material effects on minimizing pulse broadening, Optical Fiber Transmission Conf., OSA, TUC5.

Petermann, K. 1981. Nonlinear transmission behavior of a single-mode fiber transmission line with polarization coupling. *J. Opt. Commun.*, 2, 59.

Poole, C. D. and Wagner, R. E. 1986. Phenomenological approach to polarization dispersion in long single-mode fibers. *Electron. Lett.*, 22, 1029.

Rashleigh, S. C. 1983. Origins and control of polarization effects in single-mode fibers. *J. Lightwave Technol.*, 1, 312.

Snyder, A. W. and Love, J. P. 1983. *Optical Waveguide Theory.* Chapman and Hall, London.

Stolen, R. H. 1979. Nonlinear properties of optical fibers In *Optical Fiber Telecommunications*, S. E Miller and A. G. Chynoweth, eds., p. 1062. Academic Press, New York.

Stolen, R. H. and Bjorkholm, J. E. 1982. Parametric amplification and frequency conversion in optical fibers. *IEEE J. Quantum Electron.*, 18, 1062.

Stolen, R. H. and Ippen, E. P. 1973. Raman gain in glass optical waveguides. *Appl. Phys. Lett.*, 22, 276.

Tyndall, J. 1884. *Proc. Roy. Inst.*, 1, 466.

Vassallo, C. 1993. *Optical Waveguide Concepts*. Elsevier, Amsterdam.

Zahkharov, V. E. and Shabat, A. B. 1971. *Zh. Eksp. i Teor. Fiz.*, 61, 118.

13

Ion-Exchanged Glass Waveguides

Brian West

13.1 Introduction

Optical materials may be broadly classified as step-index or gradient-index (GRIN). The former (which includes bulk materials) consists of discrete regions of constant refractive index, and is typical of fabrication processes that consist of repeated deposition and etching. The latter exhibits an index profile that varies smoothly over one or more spatial coordinates. Several technologies have been employed to produce index gradients in optical materials. In semiconductors, ion implantation and post-diffusion, or quantum-well intermixing produce GRIN profiles. In ferroelectric materials, such as lithium niobate, titanium in-diffusion or proton exchange—well-established technologies for electro-optic device fabrication—produce an inherent index gradient. The most common fabrication process, however, is ion exchange in glass (Doremus 1964; Ramaswamy and Srivastava 1988; Ross 1989; Nikonorov and Petrovskii 1999). A comprehensive discussion of this technology is the focus of this chapter. We begin by describing the physical mechanism of ion exchange, the types of substrates in common use, and the various processing configurations by which the exchange has been accomplished (Section 13.2). Section 13.3 contains a discussion of the modeling of these processes, which in general will require numerical methods to solve for the index distributions. Several important optical properties of ion-exchanged waveguides are investigated in Section 13.4. Finally, in Section 13.5, novel functionality in ion-exchanged specialty glasses is discussed, and examples of devices that utilize the unique properties of this technology are given.

Ion exchange is a diffusive process that can occur in multicomponent oxide glasses. These glasses consist of three general types of oxides, classified according to the relative strength of the bonds between the cations and their neighboring oxygen atoms. Oxides with strong bonds (e.g., SiO_2, GeO_2, B_2O_3, P_2O_5) are called *network formers*, and largely determine physical properties such as viscosity and glass transition temperature. *Network intermediates* such as PbO and ZnO are not able to form glasses by themselves, but still contribute to the strength of the glass network. Oxides with the weakest bonds, such as Na_2O, CaO, and K_2O, are referred to as *network modifiers*, and aid in preventing crystallization during cooling of the melt. The cations of network modifiers (usually Na^+) may be replaced by those of a different species, by methods detailed in Section 13.2. Because of the different size and polarizability of the exchanged ions, this process locally alters the electrical properties, optical properties, and state of stress of the glass. It is this property that makes ion exchange a viable technology for the production of optical components.

Ion exchange as an engineering process was originally used to improve the surface mechanical-properties of structural glass (Schulze 1914; Kistler 1962). As glass fails in tension, the introduction of a compressive stress at the surface will increase the modulus of rupture. This can be accomplished in a process often termed "ion stuffing," by exchanging sodium ions in the glass for ions of greater size, such as silver. This concept has also been applied to improving the thermal shock resistance of laser glasses (Jiang et al. 1995). The fabrication by ion exchange of optical waveguides in glass was first achieved in 1972 using a melt containing thallium ions (Izawa and Nakagome 1972). Subsequently, Giallorenzi et al. (1973) produced waveguides using a melt containing silver ions, which today is by far the most common process. Other dopant ions include K^+, Cs^+, Rb^+, and Li^+ (Ramaswamy and Srivastava 1988).

Ion-exchanged glass waveguides have several favorable properties that will be developed in detail later in this chapter:

- The process is relatively simple and economical. No complicated material growth steps are required following the photolithography (which is common to all integrated optics technologies), and batch processing is possible.
- There is much flexibility with respect to the fabrication parameters, allowing waveguides to be optimized for a wide range of applications.
- The diffusive nature of ion exchange acts to "smooth out" any roughness in the photolithography, thus minimizing scattering losses.
- Through proper design, ion-exchanged waveguides can have near-circular cross-sections that are almost perfectly mode-matched to optical fiber.
- The glass substrates have very low optical loss in the technologically important near-infrared region of the spectrum. Additionally, the amorphous nature of glass results in no intrinsic birefringence (although form birefringence may still be present; this will be discussed in Section 13.4).
- A variety of active or nonlinear dopant species may be introduced into the glass substrates, allowing for a wide range of functionality (Section 13.5).
- As the ion-exchanged substrates remain planar, it is relatively simple to hybridize the waveguides with other material systems such as polymers and semiconductors (Section 13.5).

13.2 Fabrication of Ion-Exchanged Glass Waveguides

As they are only loosely bound to the glass network, network modifier cations have a finite mobility, and can migrate to different sites within the network. Thermal exchange occurs when a substrate containing ions of type A^+ is immersed in a molten salt that contains chemically similar ions of type B^+, which are essentially interchangeable in the glass. At the interface between the glass and the melt, the abrupt discontinuity in both ionic concentrations represents a nonequilibrium condition. Random collisions at the surface cause an exchange of ions to restore equilibrium; the exchange is one-to-one in order to preserve charge neutrality. The A^+ ions released into the melt move away from the surface rapidly by convection, so the concentration of both types of ions in the melt is generally considered to be a constant. Meanwhile, the B^+ ions diffuse further into the glass in response to their concentration gradient. The ionic mobility has a strong temperature dependence, which greatly accelerates this process at higher temperatures. When the substrate returns to room temperature, the mobility becomes negligible, and the ionic concentrations are effectively frozen in place. A thorough description of ionic kinetics at the melt/substrate interface can be found in Garfinkel (1968).

Field-assisted ion exchange is accomplished by setting up a potential difference between two salt melts on either side of the substrate. This produces an ionic current in the substrate, with B^+ ions entering through one surface and A^+ ions exiting through the other. The field-assisted process cannot be accomplished without a thermal component; elevated temperatures are required to provide sufficient ionic mobility (and hence conductivity).

13.2.1 Substrates for Ion Exchange

Glasses used for ion exchange must possess several properties. Foremost among these is a sizeable concentration of mobile network modifiers. If this concentration is too low, the maximum possible index change will be too small to create practical waveguides. The temperature at which ion exchange is performed must be above the melting temperature of the salts involved, and must also be high enough that the mobility of the exchanging ions in the glass is significant. On the other hand, the temperature cannot approach the annealing temperature of the substrate; otherwise local changes in the optical properties of the glass may occur. For this reason, substrates with high annealing temperatures are preferred. Chemical and physical resistance to the harsh alkali salt melts is of utmost importance, as is the ability to withstand the polishing process required to produce low-loss end facets.

Early research into ion exchange used ordinary microscope slides as substrates. These consist of soda-lime glass, which has the high sodium content required for efficient exchange (sodium ions are highly mobile in silicate glasses). While waveguiding was observed in these slides, compositional variations among batches made reproducible results impossible. Additionally, metallic impurities may be at an unacceptably high level in microscope slides, and while these may not cause significant absorption in the bulk glass, they can donate an electron to an exchanged ion, reducing it to a metallic state. While optical glasses (for example, Corning 0211 and Schott BK7) have been used to great success, researchers have developed glasses exhibiting all of the above properties (for example, IOT BGG31 and Schott IOG-10) specifically for ion-exchange applications. Compositions of several glasses that have been used for ion exchange are given in Table 13.1 and Table 13.2. Ion exchange substrates with active or nonlinear constituents will be discussed in Section 13.5.

13.2.2 Melts for Ion Exchange

The choice of salt melt for ion exchange is determined to a large extent by the desired maximum index change and exchange temperature. Nitrate salts are generally used due to their low melting points (Table 13.3), and using eutectic mixtures of two or more salts can lower the required temperature even further (Jackel 1988). The maximum index change is a property of the substrate as well as the type of alkali ion, but as it is roughly proportional to the polarizability and size of the exchanging ions, it is considerably larger for silver than for potassium. Thallium, as used in the pioneering work of Izawa and Nakagome (1972), produces a larger index change than potassium, but is highly toxic and is seldom used today. The polarizabilities and radii of several typical network modifier cations are listed in Table 13.4. Much flexibility in the surface index change can be achieved by using dilute mixtures of a few percent

TABLE 13.1 Composition of Common Ion-Exchange Substrates, in Weight % Unless Indicated Otherwise

	Typical Soda–Lime[a]	Schott IOG-1[b] (mol%)	Schott IOG-10[c]	Fisher Premium Slides[d]	Corning 0211[d]	Schott BK7[a]	Pyrex Corning 7740[a]	Bausch and Lomb 2046[a]
SiO_2	71–74	—	63.6	72.2	65.0	69.6	81	67
P_2O_5	—	60	—	—	—	—	—	—
Na_2O	13–15	24	9.2	14.3	7.0	8.4	4	25.6
K_2O	0.1–2	—	6.8	1.2	7.0	8.4	—	—
Al_2O_3	0.1–1.2	13	3.0	1.2	2.0	—	2	7.4
CaO	6–11	—	—	6.4	—	—	—	—
MgO	3–5	—	2.8	1.2	—	—	—	—
Fe_2O_3	—	—	—	0.03	—	—	—	—
SO_3	—	—	—	0.30	—	—	—	—
B_2O_3	—	—	1.0	—	9.0	9.9	13	—
BaO	—	—	—	—	—	2.5	—	—
ZnO	—	—	12.5	—	7.0	—	—	—
TiO_2	—	—	0.6	—	3.0	—	—	—
R_2O_3 (R= rare earths)	—	3	—	—	—	—	—	—
Other	1–2	—	0.5	3.17	—	1.2	—	—

[a] From Albert, J. 1992. In *Introduction to Glass Integrated Optics*, S. I. Najafi, ed., pp. 7–38. Artech House, Norwood, MA.
[b] From Veasey, D. L., Funk, D. S., Peters, P. M., Sanford, N. A., Obarski, G. E., Fontaine, N., Young, M., Peskin, W. C., Houde-Walter, S. N., and Hayden, J. S. 2000. *J. Non-Cryst. Solids*, 263–369, 381.
[c] From Li, G., Winick, K. A., Griffin, H. C., and Hayden, J. S. 2006. *Appl. Opt.*, 45, 1743–1755.
[d] From Najafi, S. I. 1988. *Appl. Opt.*, 27, 3728–3731.

$AgNO_3$ in $NaNO_3$. The relationship between melt dilution and maximum index change is determined by equilibrium considerations involving the thermodynamic activities of both types of ions in the melt and glass phases, and is highly nonlinear (Laity 1957; Stewart and Laybourn 1978). As such, there is a strong dependence of maximum index change on silver mole fraction in the melt, for highly diluted melts; saturating to a nearly constant value for silver fractions above 10% (Stewart and Laybourn 1978; Chludzinski, Ramaswamy and Anderson 1987). This can be used to achieve the small index changes

TABLE 13.2 Composition of IOT BGG31 (in at.%)

O	Si	Al	Na	B	F	As	K
58.3	17.6	7.1	6.8	6.0	4.2	0.06	<0.01

Source: Ross, L. 1989. *Glass Sci. Technol.*, 62, 285–297.

TABLE 13.3 Common Salts Used in Ion Exchange, and their Melting Points

Salt	Melting Point (°C)
$AgNO_3$	212
AgCl	455
$NaNO_3$	307
KNO_3	334
KNO_3–$AgNO_3$ (37:63 mol%)	132
KNO_3–$NaNO_3$ (50:50 mol%)	220
$TlNO_3$	206
$CsNO_3$	414

Source: Albert, J. 1992. In *Introduction to Glass Integrated Optics*, S. I. Najafi, ed., pp. 7–38. Artech House, Norwood, MA.

TABLE 13.4 Typical Monovalent Network Modifier Cations and their Polarizability and Radius

Ion	Polarizability (Å^3)	Radius (Å)
Na^+	0.43	0.95
K^+	1.33	1.33
Ag^+	2.40	1.26
Li^+	0.03	0.65
Tl^+	5.20	1.49
Rb^+	1.98	1.49
Cs^+	3.34	1.65

Source: Ramaswamy, R. V. and Srivastava, R. 1988. *J. Lightwave Technol.*, 6, 984–1002.

required for single-mode waveguides, and the reduced amount of silver in the waveguide lowers the concentration of absorbing metallic colloids. Additionally, it has been shown that the use of diluted melts in the case of Ag^+–Na^+ exchange relaxes the tolerances on fabrication conditions such as temperature stability (Millar and Hutchins 1978), and from a more practical standpoint, using diluted melts is much more economical due to the relatively high cost of silver salts. In practice, calculating the ion exchange parameters for a dilute melt is impractical, and most researchers use empirical parameter extraction methods, as described in Section 13.3.

13.2.3 Processing Configurations

Researchers have identified no fewer than seven different processing configurations for ion exchange (Tervonen 1990); these are shown in Figure 13.1. For purposes of illustration, Ag^+–Na^+ exchange is considered here and throughout this chapter, unless otherwise noted; however, the configurations described here are more or less applicable to all cation pairs. Figure 13.1a illustrates thermal exchange from a molten salt. Ag^+ ions are driven into the substrate from the melt by random thermal collisions, as explained previously. Once in the glass phase, Ag^+ ions are redistributed by thermal diffusion. Note that a slab waveguide is formed on the back of the substrate, which is unmasked. This can be utilized in subsequent characterization of the process, as will be described in Section 13.3.

Thermal exchange may also be assisted by an applied electric field. A voltage V_a is applied across the substrate, with the anode plane at the masked surface, as shown in Figure 13.1b. Under the influence of the resulting electric field, the Ag^+ ions migrate deeper into the glass than they would in a purely diffusive process, and approach a step-like profile under the mask opening. This process is preferred when a waveguide requires strong optical confinement. There is no need for the electrodes to make physical contact with the substrate, as the melt is conductive. Unlike the configuration for thermal exchange from a molten salt (Figure 13.1a), the melts contacting each side of the substrate must be electrically isolated from each other, increasing the complexity of this process.

The field-assisted burial process step is shown in Figure 13.1c. The substrate containing a surface waveguide is placed in a melt containing Na^+ ions, as opposed to Ag^+. A voltage V_a is applied across the substrate, as with the field-assisted thermal exchange. As the Ag^+ ions migrate deeper into the glass, they are replaced by Na^+ ions near the surface. Burial of the waveguide is beneficial in several ways. Reduction of the optical mode's overlap with the substrate surface decreases both propagation loss (which is caused by surface scattering), and birefringence (which arises due to the asymmetric boundary condition of the dielectric constant). Furthermore, while surface guides are roughly semicircular in shape (Zhou and Liu 2005), buried guides can be made to have nearly circular symmetry, which improves coupling efficiency to optical fiber. For active devices, circular waveguide symmetry is often required to maximize the overlap of pump and signal modes. It is also possible to utilize a mask during this step to alter the profile of the buried waveguide. This will be discussed in detail in Section 13.4.

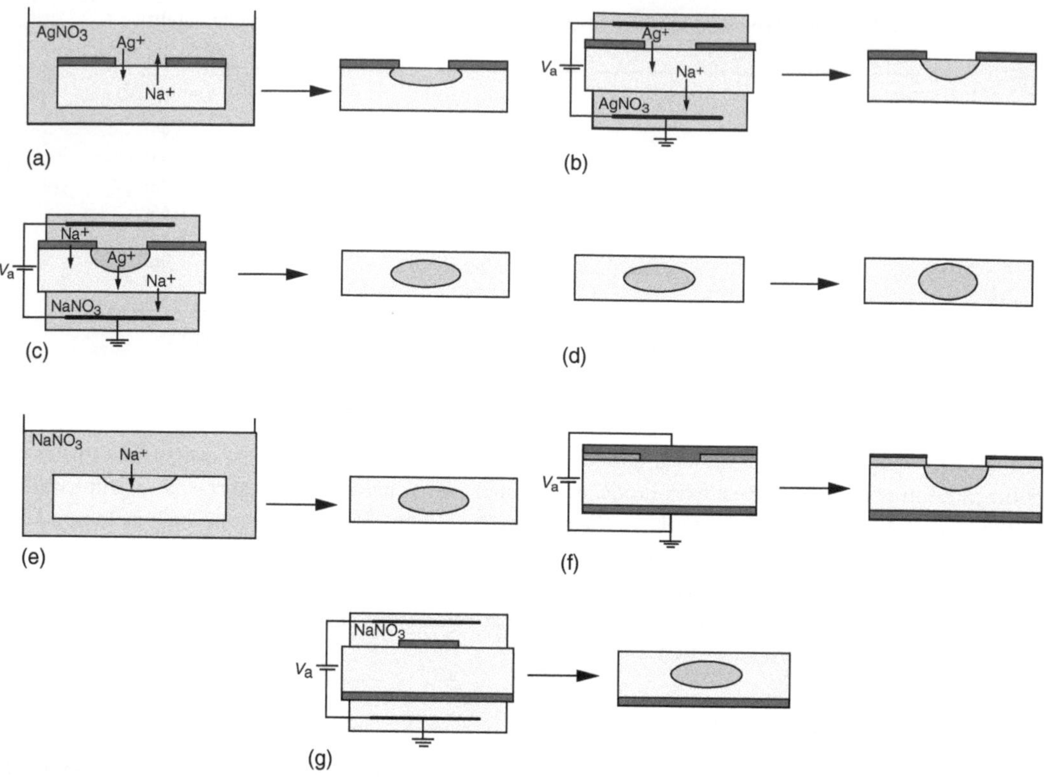

FIGURE 13.1 Processing configurations for ion exchange: (a) thermal exchange from a molten salt, (b) field-assisted thermal exchange from a molten salt, (c) field-assisted burial, (d) thermal annealing, (e) burial by reverse ion exchange, (f) field-assisted exchange from a solid silver film, and (g) one-step field-assisted exchange and burial.

Figure 13.1d illustrates the process step of thermal annealing. No external ion source or applied field is present, so the existing Ag^+ ions are redistributed entirely through thermal diffusion at an elevated temperature. This process provides for fine adjustment of the mode shape and birefringence.

A seldom-used thermal burial process is shown in Figure 13.1e. The sample is placed in a melt containing Na^+ ions, which replace Ag^+ ions near the surface of the glass. The Ag^+ ion gradient causes the waveguide to migrate away from the surface through thermal diffusion. This process requires no external field and is thus technologically simpler, but results in shallower and less confined waveguides.

The final two configurations utilize a thin solid film of silver as the ion source. Figure 13.1f illustrates a field-assisted ion exchange (Chartier et al. 1978; Viljanen and Leppihalme 1980; Honkanen 1992). The waveguide geometry is determined by the pattern of the silver film anode. Upon application of an electric field at elevated temperatures, silver ions are released into the glass through an electrochemical reaction in which the metallic silver is oxidized: $Ag \rightarrow Ag^+ + e^-$. This process is carried out at a lower temperature than the thermal exchange, allowing for greater confinement of the resulting waveguide due to reduction of thermal diffusion. The silver film technique has some additional benefits. First, by accurately controlling the film thickness, precise control can be achieved over the amount of silver introduced into the substrate; once the film is exhausted, the ion exchange ceases (Tervonen, Honkanen, and Leppihalme 1987). Alternatively, the total amount of exchanged ions can be monitored—and abruptly cut off if required—by integrating the ionic current over time. Additionally, as the glass surface is not exposed to a hot salt melt, surface damage is reduced. This process is especially important in the fabrication of active devices, which often use phosphate rather than silicate glasses in order to increase the solubility of rare earth ions. As phosphate glasses are much more susceptible to damage from the alkali salts, this dry process is preferred.

A one-step process for fabrication of buried waveguides from a silver film is shown in Figure 13.1g (Pantchev 1986). The anode consists of a melt containing Na^+ ions. When the silver film is consumed, the Na^+ ions exchange into the glass, while the Ag^+ ions migrate away from the surface. Moreover, the presence of the melt provides a more homogeneous electric field in the glass, which aids in lateral confinement of the waveguide.

13.2.4 Masking

Figure 13.2 illustrates the procedure by which a metal-oxide mask is processed photolithographically on the substrate. This mask is used to define the device geometry for the exchanges from a molten salt. A representative masking procedure is as follows. Following a series of ultrasonic cleaning in acetone, ethanol, and isopropyl alcohol, a titanium layer of 100–150 nm in thickness is deposited onto the substrate. A thin layer of positive photoresist (PR) is spin-coated onto the titanium. The PR is exposed to ultraviolet light through a mask and subsequently developed, removing the PR from the regions where the researcher wishes ion exchange to occur. The exposed regions of titanium are etched away using an acid solution (ethylene diamine tetraacetic acid (EDTA)/H_2O_2/deionized water), and then all remaining PR is removed with acetone. Finally, the wafer is placed into an oven at 250–300 K for roughly 16 h to fully oxidize the metal film. This oxidation step is performed in order to inhibit the formation of silver metal colloids at the surface during the exchange process; such colloids can produce unacceptably high levels of optical loss. In addition, the metal oxide film can act as an electrical insulator for certain field-assisted processes, as will be discussed in Section 13.3. Following the ion exchange, the mask is removed using the same etchant as before.

The masking process for silver film ion exchange is detailed in Figure 13.3. For this exchange configuration, it is critical that no silver remains in contact with the substrate in regions outside of the waveguides, and for this reason, a PR mask is used (Pöyhönen et al. 1991). The substrate is polished and

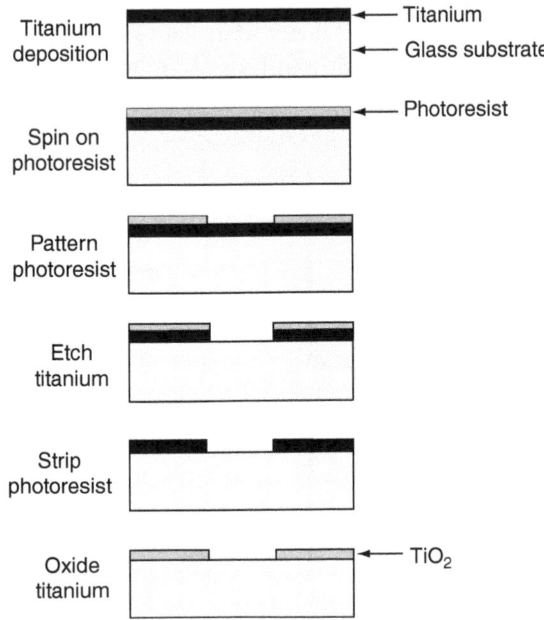

FIGURE 13.2 Photolithographic processing of an oxidized titanium mask.

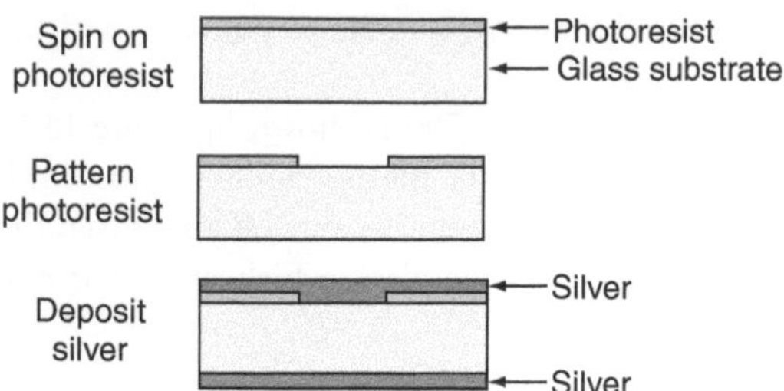

FIGURE 13.3 Photolithographic processing for ion exchange from a solid silver film.

cleaned as detailed above, then coated with PR. The PR is exposed, developed, and cured, exposing the substrate surface in the regions where one wishes to form waveguides. Finally, both surfaces are coated with 100–150 nm of silver. The metal film on the bottom surface of the substrate will be used as a cathode, as no ionic melt is present in this type of exchange.

13.2.5 Ionic Masking

An additional procedure for the fabrication of channel waveguides was reported by Honkanen et al. (1992b) and is depicted in Figure 13.4. Here, the waveguides are defined using metal strips (usually aluminum) with the glass surface exposed outside of the waveguide. Immersing the substrate in molten KNO_3 forms a shallow region of high K^+ concentration beneath the surface, outside of the masked region. No aluminum enters the glass because of the absence of any applied electric field. The mask is then removed, and the substrate is immersed in molten $AgNO_3$ at a lower temperature. Here, the potassium ions have very low mobility in comparison to the silver ions, and thus the Ag^+–Na^+ exchange occurs almost entirely in the previously masked area. This process is beneficial in that the initial step of ionic masking effectively reduces lateral diffusion during the Ag^+–Na^+ exchange, resulting in a more highly confined waveguide. The index increase due to potassium ions is relatively small in comparison to that of silver ions, so the waveguiding in the thin K^+ layer is negligible. In addition, the absence of a metallic mask during the silver ion exchange yields lower waveguide losses, as explained in Section 13.4.

13.2.6 Edge Polishing

Following the ion exchange, the glass wafer is diced into individual samples and cleaned using the same procedure as that detailed above. Several samples are stacked together (with the exchanged surfaces facing each other) in an edge-polishing jig. It is critical that the samples are in intimate contact during this step; observation of widely spaced Newton's fringes under white light serves as confirmation of this. Lapping the edges with a 3-μm Al_2O_3 slurry on a cast-iron plate serves to even out the edges of the samples. Finally, the edges are polished on 1-μm and 0.3-μm polyurethane polishing pads.

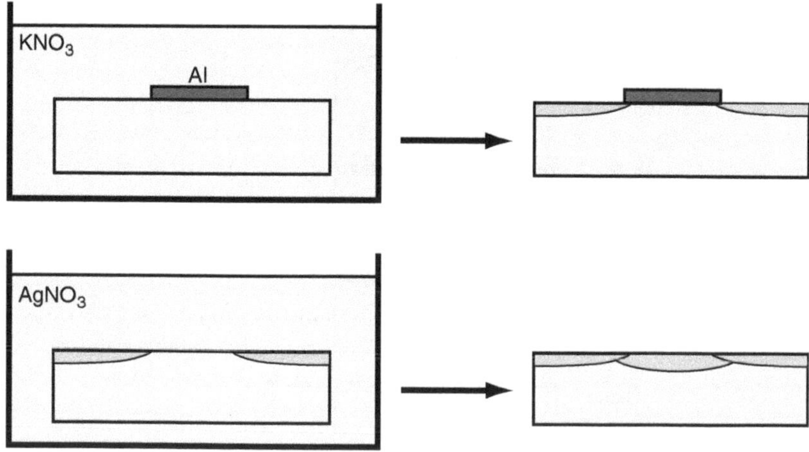

FIGURE 13.4 Ion exchange using ionic masking.

13.3 Theoretical Analysis

Determining the refractive index profile that results from the processes described in Section 13.2 is of paramount importance in device design. Calculating the index requires modeling of the binary ion-exchange process to determine the concentration distribution of the exchanged ions, and subsequent correlation of the concentration to a change in the refractive index. The term "binary" refers to the fact that ions of two species, possibly with different mobilities, are transported between the glass and melt, and within the glass. Other mobile ionic species may be present in the glass, but their presence can be neglected if their concentration is low, or if their mobility is appreciably smaller than that of the two species under consideration. The nonlinear diffusion equation for binary ion exchange is (Albert and Lit 1990):

$$\frac{\partial C_{\mathrm{A}}}{\partial t} = \frac{D_{\mathrm{A}}}{1-(1-M)C_{\mathrm{A}}}\left[\nabla^2 C_{\mathrm{A}} + \frac{(1-M)(\nabla C_{\mathrm{A}})^2}{1-(1-M)C_{\mathrm{A}}} - \frac{q\mathbf{E}_{\mathrm{ext}}}{kT}\nabla C_{\mathrm{A}}\right] \tag{13.1}$$

In this equation, C_{A} is the concentration of incoming ions, D_{A} is the self-diffusion coefficient of incoming ions, and M is the ratio of self-diffusion coefficients between the incoming and outgoing ions. Some authors refer instead to the *Stewart coefficient*, $\alpha = 1-M$ (Stewart et al. 1977). $\mathbf{E}_{\mathrm{ext}}$ is the externally applied electric field, if applicable. T, q, and k are absolute temperature, electron charge, and Boltzmann's constant, respectively. It is important to note that D_{A} and M can be strongly dependent upon the processing conditions, and establishing their values requires some effort. This is discussed later in this section.

In practice, one can define C_{A} as the concentration of incoming ions normalized with respect to the saturated concentration, eliminating the dependence on absolute ion concentration. The saturated concentration is dependent on the stoichiometry of the substrate and melt. Its exact value is generally unknown, but this problem is overcome by setting $C_{\mathrm{A}} = 1$ at the surface of the substrate that is in contact with the melt, and relating C_{A} to an index change that is determined experimentally.

The model described above is valid in any number of spatial dimensions. For slab waveguides, only one dimension is required by symmetry. One-dimensional modeling will be applied later in this section to determine the ion exchange parameters. For channel waveguides, two-dimensional cross-sections are required. This is by far the most common application of the model. The need to apply the model in three dimensions is rare, because of the slow variation of waveguide geometry in the propagation direction for these weakly-guided waveguides.

13.3.1 Analytical Solutions

In a number of idealized situations, Equation 13.1 can be solved analytically. In one dimension, with no applied field and equal ion mobilities ($M = 1$), Equation 13.1 reduces to

$$\frac{\partial C_{\mathrm{A}}}{\partial t} = D_{\mathrm{A}} \frac{\partial^2 C_{\mathrm{A}}}{\partial y^2}, \tag{13.2}$$

which has the solution

$$C_{\mathrm{A}}(y,t) = C_{\mathrm{A0}}\,\mathrm{erfc}\left(\frac{y}{2\sqrt{D_{\mathrm{A}}t}}\right), \tag{13.3}$$

where C_{A0} is a constant and the complementary error function is defined as:

$$\mathrm{erfc}(\alpha) = \frac{2}{\sqrt{\pi}}\int_{\alpha}^{\infty} \mathrm{e}^{-\gamma^2}\,\mathrm{d}\gamma. \tag{13.4}$$

Other functional forms that have been assumed for planar ion-exchanged waveguides include a modified Fermi distribution for the case of a dominant drift component (Yip, Noutsios, and Kishioka 1990), and a Rayleigh distribution for a K^+–Na^+ guide formed by thermal exchange and subsequent thermal back-diffusion (Noutsios and Yip 1991), which becomes a shifted Rayleigh distribution in the case of a diluted burial melt (Mounier et al. 1993). Notwithstanding these efforts, with the powerful numerical modeling tools available today, the assumption of analytical concentration profiles has become obsolete.

13.3.2 Numerical Solutions

For slab waveguides in the case where $M \neq 1$, or for channel waveguides, no analytical solutions exist for Equation 13.1, and numerical techniques must be employed. The simplest algorithm for solving Equation 13.1 is Euler's method (Gerald and Wheatley 1999). It is explicit, in that the unknown concentration at the advanced time is a function of known concentrations at the current time. As such, no matrix inversion is required. This method is only conditionally stable, however, and the short time steps necessary to achieve stability on the fine spatial grid that is required for precise optical simulations render it inefficient.

The explicit Dufort–Frankel algorithm (Tervonen 1992), which is widely quoted in the literature as possessing unconditional stability, is the method most often used in the field. It must be noted however that this stability property exists only with respect to the linear diffusion equation. In the case of Equation 13.1, stability is lost for values of M that are less than ~ 0.1, with only minor dependence on node spacing or time step. The value of M in the Ag^+–Na^+ system depends on the glass composition. For typical silicate glasses it is less than unity (Doremus 1964), and can be very close to the stability boundary. Many other systems of interest fall well beyond the range of stability.

Stability is approached by turning to implicit methods, in which the unknown concentration at a node at a given time is expressed as a function of unknown concentrations at adjacent nodes at the same time. The Crank–Nicolson (CN) method (Gerald and Wheatley 1999) is an example of an implicit method. For a two-dimensional problem, the simplest implementation of the CN method requires the solution of a pentadiagonal matrix at each time step, that matrix containing $(n_x n_y)^2$ elements for an n_x by n_y domain. This is computationally quite intensive for simulations containing a large number of nodes. The CN method is well suited to the one-dimensional problem, however.

The alternating direction implicit method of Peaceman and Rachford (PR-ADI) (Peaceman and Rachford 1955; Masalkar 1994) reduces computational effort by requiring solution of a matrix for only one dimension at a time. Each time step is divided into two half-steps. In the first, partial derivatives are calculated explicitly with respect to one dimension, and implicitly with respect to the other. The process is then reversed for the second half-step. The problem is thus reduced to solving n_x matrices of n_y^2 elements, followed by n_y matrices of n_x^2 elements, at each time step. More importantly, the matrices are tridiagonal, and very efficient algorithms exist for the solution of such systems (Ferziger 1981). As with the Dufort–Frankel algorithm, stability is a concern for very small values of M. However, the speed of the PR-ADI algorithm allows for more freedom in choosing a suitable time step. Methods incorporating adaptive time steps and node spacings have been used to maximize efficiency within the stability regime (Saarikoski et al. 1997). A comprehensive derivation of the PR-ADI solution to Equation 13.1 has been presented in West (2004); see also West et al. (2004).

In actuality, the diffusion coefficients are implicitly dependent on the local concentration of silver ions (Poling and Houde-Walter 2000). Provided that the concentration dependence of the parameters D_A and M is known and varies slowly over the area of the waveguide, it can be incorporated into the PR-ADI algorithm simply by utilizing the appropriate value at each point in the calculation domain. For the implicit half-time step, the concentration profile obtained in the previous time step must be used, a justifiable approximation for moderate values of ion flux, or alternatively, for small time steps.

A numerical solution of Equation 13.1 requires known conditions of the Ag^+ concentration on all boundaries of the computational domain, shown in Figure 13.5. When the sides ($|x| = w/2$) and bottom

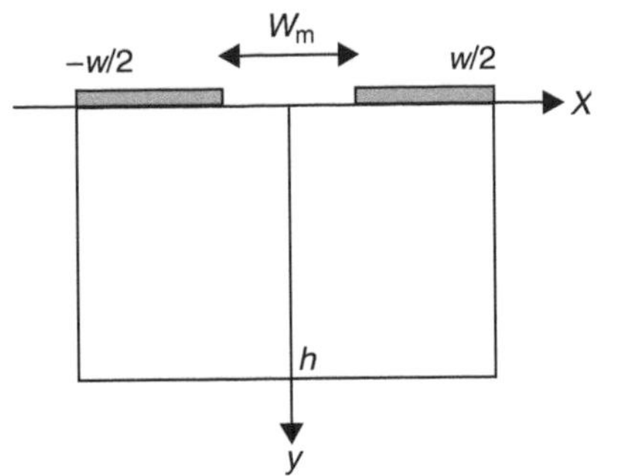

FIGURE 13.5 Computational domain used in ion exchange modeling.

($y=h$) are sufficiently far from the waveguide, the Dirichlet condition $C_A=0$ is used. For the top of the computational domain, the boundary conditions are determined by the mask geometry. Where the mask is open, the normalized silver concentration $C_A=1$ (for exchange) or $C_A=0$ (for burial) is assumed. Where the mask is closed, the metal oxide effectively acts as a barrier to ion exchange between the glass and melt. Here, the Neumann condition $\partial C_A/\partial y=0$ is applied, indicating zero ion flux normal to the surface. Transverse ion flux is present however, so there can be no definite condition on C_A directly under the mask. In cases where the waveguide exhibits symmetry in the x-direction, modeling only one half of the waveguide can reduce computational effort. In this case, an appropriate boundary condition is $\partial C_A/\partial x|_{x=0}=0$ (zero net ion flux across the axis of symmetry). Boundary conditions for common ion exchange processes are summarized in Table 13.5.

13.3.3 Electrostatic Modeling

The electric field $\mathbf{E}_{ext}$ in Equation 13.1 exhibits some subtleties that require considerable effort for accurate modeling. To a first approximation, the electric field in the vicinity of the waveguide for unmasked processes is simply the voltage drop across the substrate divided by the substrate thickness. However, the conductivity of the glass is nonhomogeneous in the region of the waveguide, owing to the differing conductivities of the two mobile ionic species. The result is a perturbation of both the magnitude and direction of the electric field lines in the vicinity of the waveguide (Cheng et al. 1997; Hazart and Minier 2001). It has been observed that the depth (Madasamy et al. 2003b) and vertical symmetry of buried single-mode waveguides vary when nonhomogeneous conductivity is accounted for in the Ag^+–Na^+ system. In addition to the variation in conductivity, some processes (in particular, the field-assisted exchange from a silver film) utilize anodes that contain structural features near the waveguide. If the exchanging ions have equal mobility, the field between a strip of width W and an infinite ground plane a distance d from the top surface is given by the well-known triplate transmission line equation (Lilienhof et al. 1982),

TABLE 13.5 Boundary Conditions on Ag^+ Concentration Used in Ion-Exchange Modeling

Process Step	Surface ($y=0$)	Bottom of Domain ($y=h$)	Sides of Domain ($\|x\|=w/2$)	Center of Domain ($\|x\|=0$)
Thermal exchange from molten salt	$C_{Ag}=1$ (mask open) $\partial C_{Ag}/\partial y=0$ (mask closed)	$C_{Ag}=0$	$C_{Ag}=0$	$\partial C_{Ag}/\partial x=0$[a]
Field-assisted burial: full	$C_{Ag}=0$	$C_{Ag}=0$	$C_{Ag}=0$	$\partial C_{Ag}/\partial x=0$[a]
Field-assisted burial: selective	$C_{Ag}=0$ (mask open) $\partial C_{Ag}/\partial y=0$ (mask closed)	$C_{Ag}=0$	$C_{Ag}=0$	N/A
Annealing	$\partial C_{Ag}/\partial y=0$	$C_{Ag}=0$	$C_{Ag}=0$	$\partial C_{Ag}/\partial x=0$[a]

[a] This condition assumes that the concentration profile is x-symmetric prior to the process step in question.

$$\mathbf{E}_{\text{ext}} = E_x + iE_y = i\frac{\pi}{2}\frac{V_{\text{a}}}{dK(m)}\left(\frac{\tanh^2(u)-1}{\tanh^2(u)-m^2}\right)^{1/2}, \tag{13.5}$$

where $m=\tanh(\pi W/4d)$, $u=(\pi/2)(x-iy)/d$, and $K(m)$ is the Jacobian elliptic function:

$$K(m) = \int_0^{\pi/2} \frac{d\theta}{[m^2 + (1-m^2)\sin^2\theta]^{1/2}}. \tag{13.6}$$

For numerical solution of the field, note that the electrical potential $\varphi(x,y)$ everywhere within the simulation domain satisfies the nonstandard Laplace equation

$$\sigma(x,y,C_{\text{A}})\nabla^2\phi(x,y) + \nabla\sigma(x,y,C_{\text{A}}) \cdot \nabla\phi(x,y) = 0, \tag{13.7}$$

where $\sigma(x,y,C_{\text{A}})$ is the concentration-dependent ionic conductivity. The spatial dependence of σ is shown explicitly for emphasis. This is calculated by assuming a linear relationship between concentration and conductivity

$$\sigma(x,y,C_{\text{A}}) = [1-C_{\text{A}}(x,y)]\sigma_{\text{B}} + C_{\text{A}}(x,y)\sigma_{\text{A}}, \tag{13.8}$$

with the conductivity of species A (B) given by the Nernst–Einstein relationship

$$\sigma_{\text{A(B)}} = \frac{D_{\text{A(B)}}c_{\text{A(B)}}q^2}{kT}, \tag{13.9}$$

where $c_{\text{A(B)}}$ is the absolute (nonnormalized) concentration of the particular ionic species. Combining Equation 13.8 and Equation 13.9 results in

$$\sigma(x,y,C_{\text{A}}) = \frac{D_{\text{A}}c_0q^2}{kT}\left\{\frac{1}{M}[1-C_{\text{A}}(x,y)] + C_{\text{A}}(x,y)\right\}. \tag{13.10}$$

Knowledge of the constants outside the braces in Equation 13.10 is not necessary for the solution of Equation 13.7 due to its (mathematical) homogeneity. Note also that when $M=1$, the concentration dependence of σ is eliminated, and the conductivity is said to be homogenous throughout the waveguide.

The solution of Equation 13.7, subject to the condition of Equation 13.10, is discussed in detail in West (2004). Note that the conductivity profile changes slowly over time as the ion concentration profile evolves. As such, it is necessary to re-calculate the electric field at each time step during evaluation of the diffusion equation. Appropriate boundary conditions for the solution of the electrical potential are listed in Table 13.6. An additional consideration in the electrostatic modeling of masked field-assisted processes (Figure 13.1) is the presence of a thin negative space-charge layer underneath the mask (Pantchev, Danesh, and Nikolov 1993). This occurs because the mask serves as a blocking anode; it prevents cations from replacing those that have drifted toward the cathode. Mrozek, Mrozek and Lukaszewicz (2006) have suggested utilization of the Poisson equation rather than the Laplace equation to account for this space-charge layer.

An example of the modeling procedure described above is shown in Figure 13.6. The first step is a 75-min thermal exchange, through a mask opening of 3 μm. Diffusion parameters are $D_{\text{A}}=1\times10^{-15}$ m²/s, $M=0.2$, and $T=280$°C. A subsequent 40-min field-assisted burial (with no burial mask) is modeled using $D_{\text{A}}=3.5\times10^{-16}$ m²/s ($T=250$°C), with 320 V applied across a 2-mm substrate. The figures indicate that the buried waveguide has a considerably lower peak ion concentration.

TABLE 13.6 Boundary Conditions on Potential Used in Electric Field Modeling

Process Step	Surface ($y=0$)	Bottom of Domain ($y=h$)	Sides of Domain ($\|x\|=w/2$)
Full burial	$\varphi=U$	$\varphi=0$	$\partial\varphi/\partial x=0$
Selective burial or field-assisted thermal exchange	$\varphi=U$ (mask open) $\partial\varphi/\partial y=0$ (mask closed)	$\varphi=0$	$\partial\varphi/\partial x=0$

13.3.4 Converting Ion Concentration to Refractive Index Profile

To model the optical properties of an ion-exchanged waveguide, one must first convert the ion-concentration profile to a refractive-index distribution. The presence of exchanged ions locally perturbs the index via three physical mechanisms; ionic size, ionic polarizability, and induced stress (Fantone 1983). It has been suggested that the contribution of the stress-optic effect is small for Ag^+–Na^+ ion-exchanged guides (Brandenburg 1986; Gonella 1992), although for K^+–Na^+ ion exchange, the index change is dominated by the stress component (Albert and Yip 1987). From the viewpoint of waveguide birefringence, the presence of the glass surface produces boundary conditions on dielectric constant and stress, such that stress-induced birefringence has an opposite sign to that of form birefringence, making their sum very small in magnitude. This will be elaborated upon in Section 13.4. We will therefore concentrate on the other two contributions. A basic model (Tervonen 1992) predicts that both produce a local change in refractive index that is linearly proportional to the ion concentration for small absolute concentrations, and thus it is not necessary to explicitly separate the two effects. This can be shown by manipulating the frequency-independent Lorentz–Lorenz equation (Rao 2002):

$$R_M = \left(\frac{n^2-1}{n^2+2}\right)\frac{M_w}{\rho} = \frac{4\pi}{3}\sum_i N_i\alpha_i, \tag{13.11}$$

where R_M is molar polarization, n is the index of refraction, N_i is the molar number density of particles of substance i in the glass, α_i is the atomic polarizability of substance i, M_w is the molecular weight of the glass, and ρ is the glass density. If the refractive index prior to ion exchange is denoted as n_{sub}, then Equation 13.11 takes the form

$$\left(\frac{n_{sub}^2-1}{n_{sub}^2+2}\right)\frac{M_w}{\rho} = \frac{4\pi}{3}\left(N_A\alpha_A + N_B\alpha_B + \sum_j N_j\alpha_j\right) \tag{13.12}$$

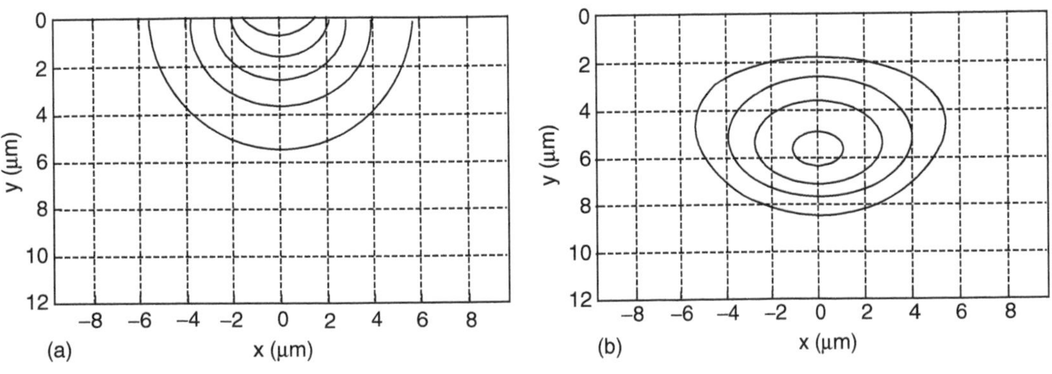

FIGURE 13.6 Concentration profile of exchanged ions ($C_{Ag}=0.1, 0.3, \ldots 0.9$): (a) following thermal exchange, and (b) following field-assisted burial. Ion exchange parameters are given in the text.

for binary ion exchange, where the summation is over all ionic species not involved in the exchange. After the exchange, a number density ΔN of species A are substituted by an equal number density (to preserve charge neutrality) of species B. Thus

$$\left(\frac{n^2-1}{n^2+2}\right)\frac{M_w}{\rho}=\frac{4\pi}{3}\left[(N_A-\Delta N)\alpha_A+(N_B+\Delta N)\alpha_B+\sum_j N_j\alpha_j\right]. \tag{13.13}$$

Subtracting Equation 13.12 from Equation 13.13 gives

$$\left(\frac{n^2-1}{n^2+2}-\frac{n_{sub}^2-1}{n_{sub}^2+2}\right)\frac{M_w}{\rho}=\frac{4\pi}{3}\Delta N(\alpha_B-\alpha_A). \tag{13.14}$$

For $\Delta n = n - n_{sub} \ll 1$, this can be simplified as:

$$\Delta n \approx \left(\frac{n_{sub}^2+2}{2n_{sub}}\right)\frac{4\pi\rho(\alpha_B-\alpha_A)}{3M_w}\Delta N. \tag{13.15}$$

This shows the linear relationship between density of exchanged ions and increase of refractive index. The refractive index profile of the waveguide cross-section can then be expressed as

$$n(x,y,\lambda) = n_{sub}(\lambda) + \Delta n_0(\lambda)C_A(x,y), \tag{13.16}$$

where $n_{sub}(\lambda)$ is the substrate index prior to ion exchange and $\Delta n_0(\lambda)$ is the increase in refractive index resulting from $C_A = 1$, which is determined experimentally.

13.3.5 Determination of Ion Exchange Parameters

To accurately determine the index profile using Equations 13.1 and Equation 13.16, it is critical that the diffusion parameters D_A and M, and the maximum index difference $\Delta n_0(\lambda)$, are precisely known. These values are not routinely provided by the substrate manufacturers, and are strongly dependent on processing conditions such as temperature and melt composition. As such, they must be determined experimentally. This procedure generally consists of the following steps: (1) processing of a slab waveguide with no applied electric field; (2) measurement of the resulting index profile, or more commonly, measurement of the effective indices of the guided modes, with subsequent reconstruction of the index profile; and (3) determination of the parameters that produced this index profile. Early efforts to this end generally assumed a functional form for the index distribution (Yip and Albert 1985), allowing D_A and Δn_0 to be calculated rapidly using the Wentzel–Kramer–Brillouin (WKB) dispersion relation (Chiang 1991). However, assuming a particular functional form of the refractive index is equivalent to specifying M. As the index profile can resemble a complementary error function (for $M = 1$) (Gortych and Hall 1985), parabola ($M \approx 0.5$) (Tsoneb and Savatinova 1988), Gaussian ($M \approx 0.1$) (Yip and Albert 1985), or step-like profile ($M \ll 0.1$) (Albert 1992), it is inadvisable to make such an assumption when there is no *a priori* information on the ion exchange system in question. A related semi-analytical method assumes that the index profile is a linear combination of the extreme cases ($M = 1$ and $M < 0.1$), with the relative contributions determined by a numerical variational method (Sebastiani et al. 2005).

Direct simultaneous determination of D_A, M, and Δn_0 by comparison of measured and modeled effective mode indices has been achieved (Opilski et al. 1997; Choo et al. 2003). In this procedure, the diffusion equation (Equation 13.1) is solved numerically for assumed values of D_A and M, and the resulting concentration profile is converted to an index profile for an assumed value of Δn_0. The effective mode indices are solved for (Chiang 1991), and compared in a squared-error sense with those measured from the fabricated waveguide using a prism coupler technique (Tien and Ulrich 1970). The parameters are subsequently altered until the squared error is acceptably minimized. Given the nonlinearity of Equation 13.1, this can be a very labor-intensive process. Recently, a method has been developed in which the modeled and measured modes can be matched using a genetic algorithm to seek out the optimal values of the ion exchange parameters (West and Honkanen 2005).

As stated previously, the self-diffusion coefficients are strongly temperature-dependent. Given that ionic mobility depends on a microscopic level on the probability that an ion's kinetic energy exceeds the activation energy required to escape from its site in the glass matrix, D_A exhibits Arrhenius behavior (Albert 1992)

$$D_A(T) = D_{A0} \exp\left(\frac{-E_a}{kT}\right), \tag{13.17}$$

where E_a is the activation energy and D_{A0} is a constant. This can be used to verify the calculated self-diffusion coefficients. Plotting $\ln(D_A)$ vs. $(1/kT)$ should produce a straight line of slope $-E_a$. Any data that deviates considerably from this line is likely erroneous. A similar procedure may be used to verify the calculated values of M, although the temperature variation of this parameter is much smaller than that of D_A. In addition to improving the accuracy of the parameters, knowledge of the temperature dependence allows for parameter values to be extrapolated to the lower temperatures commonly encountered in field-assisted or annealing processes. At these temperatures, D_A may be sufficiently low that fabricating a slab waveguide by thermal exchange from a salt melt would take a prohibitive length of time. In fact, it may be impossible if the temperature of interest falls below the salt's melting temperature.

The ion exchange model used in this simulation describes the exchange of two species of mobile ions. Although this is sufficient in most common substrates in current use, the need may arise for modeling of ternary (or higher-order) ion exchange, for substrates that contain two or more species of network modifiers with substantial concentration and mobility. In this case, Equation 13.1 is no longer strictly true. Fortunately, this issue is avoided through the combined efforts of parameter extraction and process modeling. As the parameter extraction algorithm uses experimentally obtained effective-index data as input, the calculated parameters can be thought of as "effective" physical constants, which may have been perturbed by any number of additional physical effects. These same parameters are subsequently used in the process modeling, which also utilizes Equation 13.1 as described earlier in this section, making the full modeling process self-consistent from an engineering standpoint.

13.4 Properties of Ion-Exchanged Waveguides

13.4.1 Mode Profiles

For all optical modeling in this section, the electric field E_n of each scalar mode supported by the waveguide is found by solving the Helmholtz equation

$$(\nabla^2 + k^2)E_n = \beta_n^2 E_n, \tag{13.18}$$

where $k = k_0 n(x,y) = 2\pi n(x,y)/\lambda$ is the wavenumber and β_n is the propagation constant of the nth mode. While the scalar modes are useful in determining such waveguide characteristics as burial depth and fiber coupling efficiency, they are insufficient for many applications such as birefringence modeling. In this case, quasi-vectorial modes are used, which requires modifications to Equation 13.18 to account for the discontinuity in normal field components (Stern 1988). Equation 13.18 is solved by using a five-point finite difference method developed by Kim and Ramaswamy (1989). The use of variable node spacing is recommended in order to efficiently model the optical mode in the vicinity of the substrate surface.

Figure 13.7 shows TE modes at a wavelength of 1.55 μm, for the waveguides modeled in Figure 13.6, using a substrate index of 1.4525 and a maximum index increase $\Delta n_0 = 0.03$. The surface waveguide supports a tightly confined mode that exhibits vertical asymmetry, while the mode of the buried waveguide is less confined and clearly more circular in profile.

13.4.2 Birefringence

For polarization-independent device operation, waveguides must exhibit very low birefringence. As nonfiber-based devices generally lack any type of rotational symmetry, form birefringence is prevalent in technologies such as silica-on-silicon, requiring post-production techniques such as etching of strain-relief grooves or overlayer growth (Takahashi et al. 1993) to compensate. Even so, compensation is optimized for only one waveguide width, which is problematic in devices containing both very wide and very narrow guides, such as multimode interference or arrayed waveguide grating devices (Buchold and Voges 1996). In contrast, ion-exchanged glass waveguides, particularly those which have been buried beneath the surface, have exhibited very low birefringence on the order of 10^{-5} or lower, over a range of mask openings from 2 to 10 μm, with a linear dependence on mask opening width (Yliniemi, West, and Honkanen 2005).

Waveguide birefringence has both form and stress-induced components. In all ion-exchange processes (with the possible exception of ionic masking), the exchanged ion concentration profile is wider in the lateral direction than in the depth direction. The resulting index distribution produces a positive form birefringence, defined as $B = n_{TE} - n_{TM}$.

The stress component of birefringence occurs due to the difference in size between the two species of ions (Table 13.4). In Ag^+–Na^+ exchange, the larger Ag^+ ion causes a compressive stress. The resulting refractive indices can be expressed as (Brandenburg 1986)

$$\begin{aligned} n_{TE} &= n_0 + C_1 \varsigma_x + C_2(\varsigma_y + \varsigma_z) \\ n_{TM} &= n_0 + C_1 \varsigma_y + C_2(\varsigma_x + \varsigma_z), \end{aligned} \tag{13.19}$$

where n_0 is the substrate index before exchange, ς_i ($i = x, y, z$) is the stress component in direction i, and C_1 and C_2 are the elasto-optic coefficients ($C_2 > C_1$). These terms are generally unknown—even values tabulated for a particular substrate are unreliable in the exchanged region—but Equation 13.19 can be analyzed qualitatively to determine the sign of birefringence. As the waveguide can expand in the direction normal to the surface but not in the transverse direction (which is essentially infinite), it is clear that $\varsigma_y < \varsigma_x$. With $\varsigma_z = 0$ for straight (z-independent) waveguides, it can be concluded that the presence of compressive stress results in a negative contribution to birefringence.

The effect of stress was not included in the modeling described in Section 13.3. However, it is possible to estimate the stress component by tracking the birefringence of a buried waveguide as it is annealed.

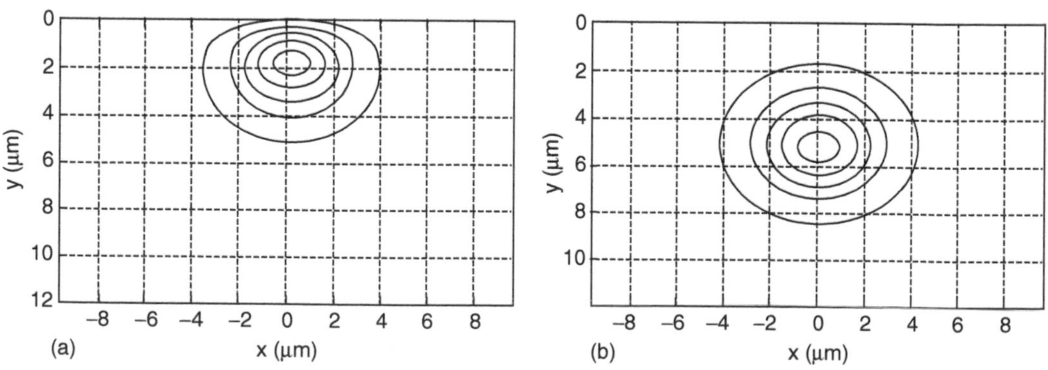

FIGURE 13.7 Intensity profile of the TE mode for the waveguides shown in Figure 13.6: (a) following thermal exchange, and (b) following field-assisted burial. Contours represent 0.1, 0.3,…0.9 of peak power.

Here, the form birefringence asymptotically vanishes, while the stress birefringence essentially remains (stress is relieved negligibly at waveguide annealing temperatures, which are considerably lower than the stress relaxation temperature of the glass). In Figure 13.8, measured and modeled birefringence for simultaneously-processed buried waveguides of various widths are compared, showing that the stress in the waveguide does indeed give a negative contribution to B (Yliniemi, West, and Honkanen 2005). Figure 13.9 shows measured values of B as these waveguides are annealed, and it is evident that after a certain annealing duration, the only residual birefringence is due to stress. A point exists during the annealing step where form- and stress-induced birefringence are in perfect balance, resulting in a waveguide with zero overall birefringence. An identical phenomenon has been observed in ring resonators by Carriere et al. (2004).

While the use of burial and annealing processes offers a simple and flexible procedure for birefringence control, it results in weakly confined waveguides. For strongly-confined surface waveguides with low birefringence, a double ion exchange process is possible (Äryäs et al. 1998). An initial K^+–Na^+ thermal exchange produces waveguides with negative birefringence, as the stress component outweighs that due to the small index change. Subsequent thermal Ag^+–Na^+ exchange reverses this effect, resulting in waveguides with arbitrarily low birefringence.

13.4.3 Losses in Ion-Exchanged Waveguides

Waveguide losses occur both in propagation and in coupling to optical fiber. Through an appropriate sequence of exchange, burial, and annealing, the guided modes can essentially match that of single-mode fiber, lowering coupling loss to negligible values. Propagation loss occurs because of two phenomena; scattering at the substrate surface and volume absorption within the waveguide. As mentioned previously, buried waveguides support modes that do not significantly interact with the surface. However, for some applications, surface guides are necessary, and in this case, it is advisable to use the dry silver-film process in order to preserve the surface quality of the substrate if it is susceptible to chemical attack; this is a problem mostly for phosphate glasses.

Absorption in ion-exchanged waveguides, particularly in the visible range, can increase over that of the bulk substrate glass, if the exchange process promotes the formation of metal colloids. A fundamental requirement of all ion exchange substrates is the elimination of impurities that can reduce Ag^+ to metallic silver. Additionally, the use of metallic masks will cause silver particles to form by electrolytic deposition (Walker, Wilkinson, and Wilkinson 1983); these remain in the glass even after mask removal

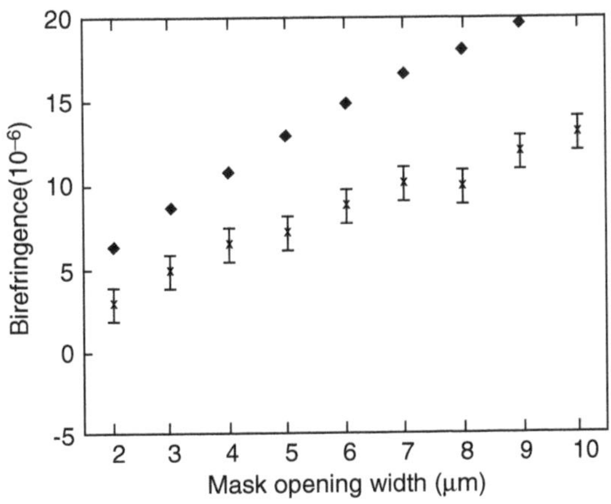

FIGURE 13.8 Measured (crosses) and modeled (diamonds) birefringence for waveguides from various mask opening widths. (From Yliniemi, S., West, B. R., and Honkanen, S. 2005. *Appl. Opt.*, 44, 3358–3363. With permission.)

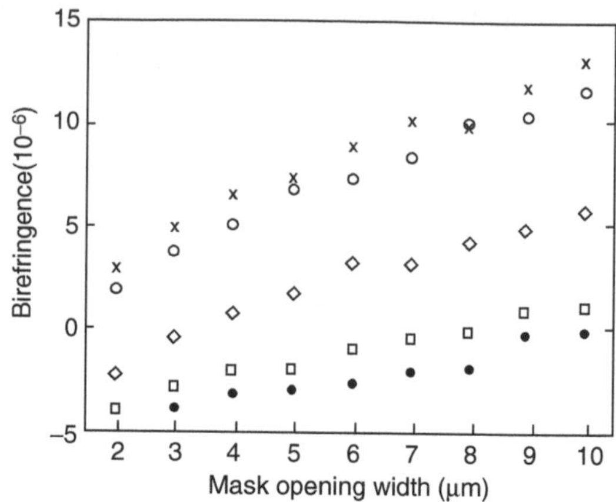

FIGURE 13.9 Measured birefringence before annealing (crosses), and after annealing at 250°C for 15 min (open circles), 45 min (diamonds), 75 min (squares), and 105 min (filled circles). (From Yliniemi, S., West, B. R., and Honkanen, S. 2005. *Appl. Opt.*, 44, 3358–3363. With permission.)

and result in a yellow tint in the exchanged layer of the glass. For this reason, dielectric masks are used for Ag^+–Na^+ exchange. Using potassium rather than silver as the exchanging species reduces this effect considerably (Honkanen et al. 1993; Wang et al. 1993), but the small index changes obtained using K^+–Na^+ exchange are often insufficient to produce a well-confined waveguide. A double ion exchange process similar to that used for birefringence control has been described by Li et al. (1991), in which an initial K^+–Na^+ exchange is followed by an Ag^+–Na^+ exchange. After the first exchange, there is a high K^+ concentration (and lowered Na^+ concentration) near the mask edge. During the second exchange, few Ag^+ ions can diffuse into this region, as the mobility of K^+ is much lower than that of Na^+. Effectively, the potassium ions shield the silver from the mask edge, reducing the occurrence of colloid formation. Propagation losses for surface guides of 0.2–0.3 dB/cm were obtained using this process (Wang et al. 1993). An extension to this process has recently been proposed by Zou and Chen (2006), in which losses are reduced by etching a 0.25-μm layer off the waveguide surface. As the majority of the metallic silver resides in this thin layer, the absorption losses are lowered. In recent years, the use of dielectric masks has rendered the double ion exchange process obsolete for absorption minimization, and most researchers using well-established processing methods with optimized substrates report propagation losses of 0.1 dB/cm or below (McCourt 1993).

13.4.4 Proximity Effect

When closely spaced waveguides such as those found in directional couplers are fabricated using ion exchange, the presence of each guide perturbs both the local Ag^+ concentration and the local electric field in the vicinity of the other guide (Honkanen et al. 1987; West et al. 2003). Due to the presence of fringing fields, this perturbation occurs even in the case of $M=1$ if the field is applied through openings in a mask. As a result, the index profile of the coupled system is not identical to that of each individual guide superimposed side-by-side, a type of proximity effect. In particular, there will be a noticeable difference in the index between the guides, a region that is critical to the coupling characteristics. This effect is more pronounced when the initial thermal exchange is field-assisted. It has been shown that the optical coupling between these perturbed waveguides can either increase or decrease, depending on the relative effects of drift and diffusion in the coupling region (West 2004); thus, it is advisable to perform process modeling and optical modeling on the coupled system. A plot of concentration contours and field lines for closely spaced waveguides is shown in Figure 13.10.

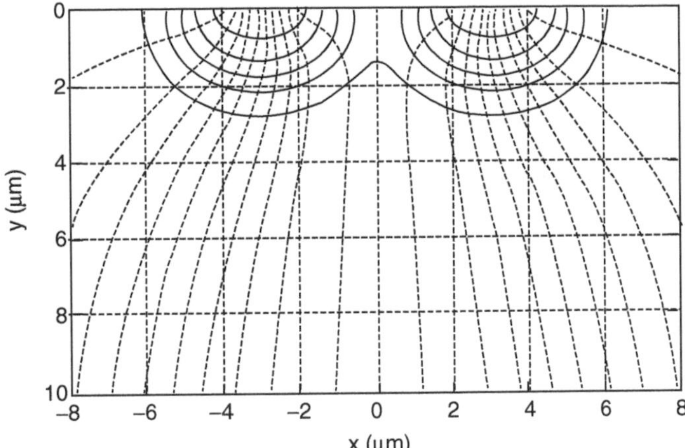

FIGURE 13.10 An example of the proximity effect between closely spaced waveguides. Solid lines represent Ag^+ concentration contours ($C_{Ag}=0.1, 0.3,\ldots 0.9$). Dashed lines represent the electric field. (From West, B. R., Madasamy, P., Peyghambarian, N., and Honkanen, S. 2003. Paper ThGG4, Presented at OSA Annual Meeting, Tucson, AZ. With permission.)

13.4.5 Tapers: Vertical and Radial

As previously discussed, buried ion-exchanged waveguides have desirable properties with respect to propagation loss, birefringence, and mode shape. Later in this chapter, various applications of ion-exchanged waveguides that require the guide to exist at the substrate surface will be discussed. To transfer optical energy between these two types of waveguides, an adiabatic vertical transition is required. This is a rather complex feature in other planar waveguide technologies, generally accomplished using grayscale lithography, diffusion-limited etching, dip-etching, or an adiabatic coupler with two vertically stacked waveguides (see Moerman, Daele, and Demeester 1997 for a review). In ion exchange technology, the vertical taper is easily accomplished using a procedure known as *selective burial* (Rehouma, Persegol, and Kevorkian 1994; Rehouma et al. 1995; Frantz et al. 2003). Here, the field-assisted burial is performed using a metal-oxide burial mask whose edge intersects the waveguide at a very shallow angle, as shown in Figure 13.11. The end of the waveguide that is far from the mask experiences a full burial. However, the end of the waveguide that lies entirely beneath the mask is insulated from the burial field, and therefore, remains at the surface (although annealed to a small extent, which lowers the effective index, thus shortening the required adiabatic transition length). The intermediate regions see a smooth transition between these two extremes. Evolution of the index profile and fundamental mode are depicted in Figure 13.12 and Figure 13.13 (West et al. 2004). Evidently, the vertical transition is accompanied by a horizontal shift. This is due to the fringing fields in the region of the mask edge, and is not detrimental to the mode propagation as long as the intersection angle between the initial waveguide and the burial mask edge is sufficiently small.

It often is the case that an ion-exchanged waveguide must exhibit strong confinement (for example, to facilitate tighter bends), but must be mode-matched to single-mode optical fiber as well. Merely incorporating a transverse taper into the mask design may not solve this problem, as the increased number of exchanged ions can result in a multimode input end. Clearly, the increase in waveguide dimension must be accompanied by a decrease in Δn. Mahapatra and Connors (1988) have described a simple process by which such a taper can be formed. Following the initial exchange, the substrate was placed on a microscope slide. A heating coil was wrapped around one end of the slide, while the other end was clamped to a metal block, which acted as a heat sink. In this fashion, a thermal gradient as high as

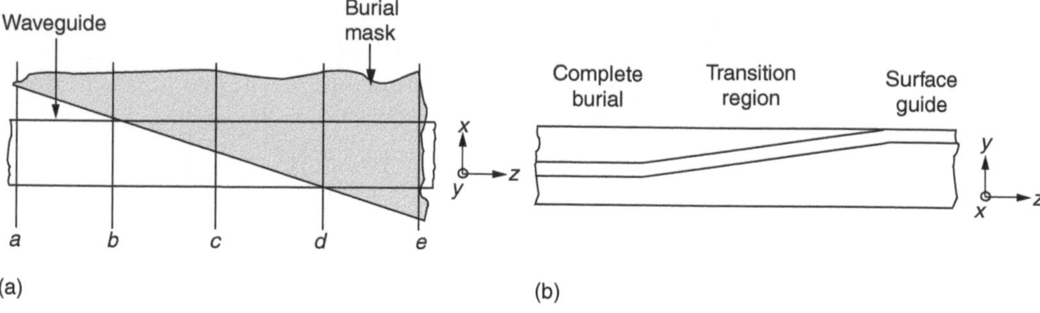

FIGURE 13.11 Mask orientation for selective burial. The unmasked section of the waveguide is buried, while the masked section remains at the surface: (a) top view, (b) side view after burial. Cross-section labels refer to Figure 13.12 and Figure 13.13. (Reprinted from West, B. R., Madasamy, P., Peyghambarian, N., and Honkanen, S. 2004. *J. Non-Cryst. Solids*, 347, 18–26. With permission.)

30°C/mm was formed along the axis of the waveguide to a maximum temperature of 325°C. This caused an axially-varying annealing of the waveguide, resulting in an almost three-fold increase of the mode size at a wavelength of 800 nm.

13.4.6 Photosensitivity

Many waveguide devices achieve wavelength selectivity through the use of Bragg gratings. Originally, these were relief-etched directly into the substrate surface (Li et al. 1991), but these structures suffer from severe polarization dependence in coupling strength (Li and Najafi 1993). Although this is suitable (even preferred) for polarization selectivity in waveguide lasers, most applications require polarization-independent volume gratings. As is the case with glass fibers, the substrates used for ion exchange may have their optical properties locally altered by exposure to ultraviolet irradiation. Early studies in UV writing of gratings in ion-exchanged substrates involved the use of experimental glasses with γ-irradiation pretreatment (Roman and Winick 1993). Subsequently, the fabrication of gratings in both bulk (Provenzano et al. 1999) and ion-exchanged (Geraghty et al. 1999, 2000) BGG31 was reported, with index changes of $\sim 3.9 \times 10^{-5}$ observed. A further study of gratings written in BGG31 show stability even after more than five hours of annealing at 230°C (Yliniemi et al. 2006b). Such thermal stability is a major factor in qualifying grating-based devices for telecom deployment.

A study of the photosensitivity of Ag^+–Na^+ ion-exchanged Er^{3+}-doped phosphate glass (Schott IOG-1) to 248 nm radiation from a KrF laser has been performed by Pissadakis et al. (2004). Whereas the maximum index change in the bulk glass is $\sim 10^{-5}$, an index-change of $\sim 2 \times 10^{-3}$ was observed in the exchanged regions, even with $AgNO_3$ concentrations as low as 0.2 wt% in the melt. This was attributed to thermal migration of the silver ions and photoionization of Ag^0 and Ag^+ species. Yliniemi et al. (2006a) used 193 nm radiation from an ArF laser to form gratings in both undoped and Er^{3+}-doped IOG-1, to utilize that stronger absorption band. They formed the grating prior to the ion exchange; no significant degradation of the grating was observed following the silver film exchange process.

13.5 Devices and Applications

13.5.1 Passive Devices

Many passive devices have been successfully demonstrated in ion exchange. The earliest of these were wavelength filters (Najafi et al. 1992; Lefebvre et al. 1993) and multiplexers (Najafi 1992) based on Mach-Zehnder interferometers, as well as y-branch splitters (Ross 1989; Pöyhönen et al. 1991). $1 \times N$

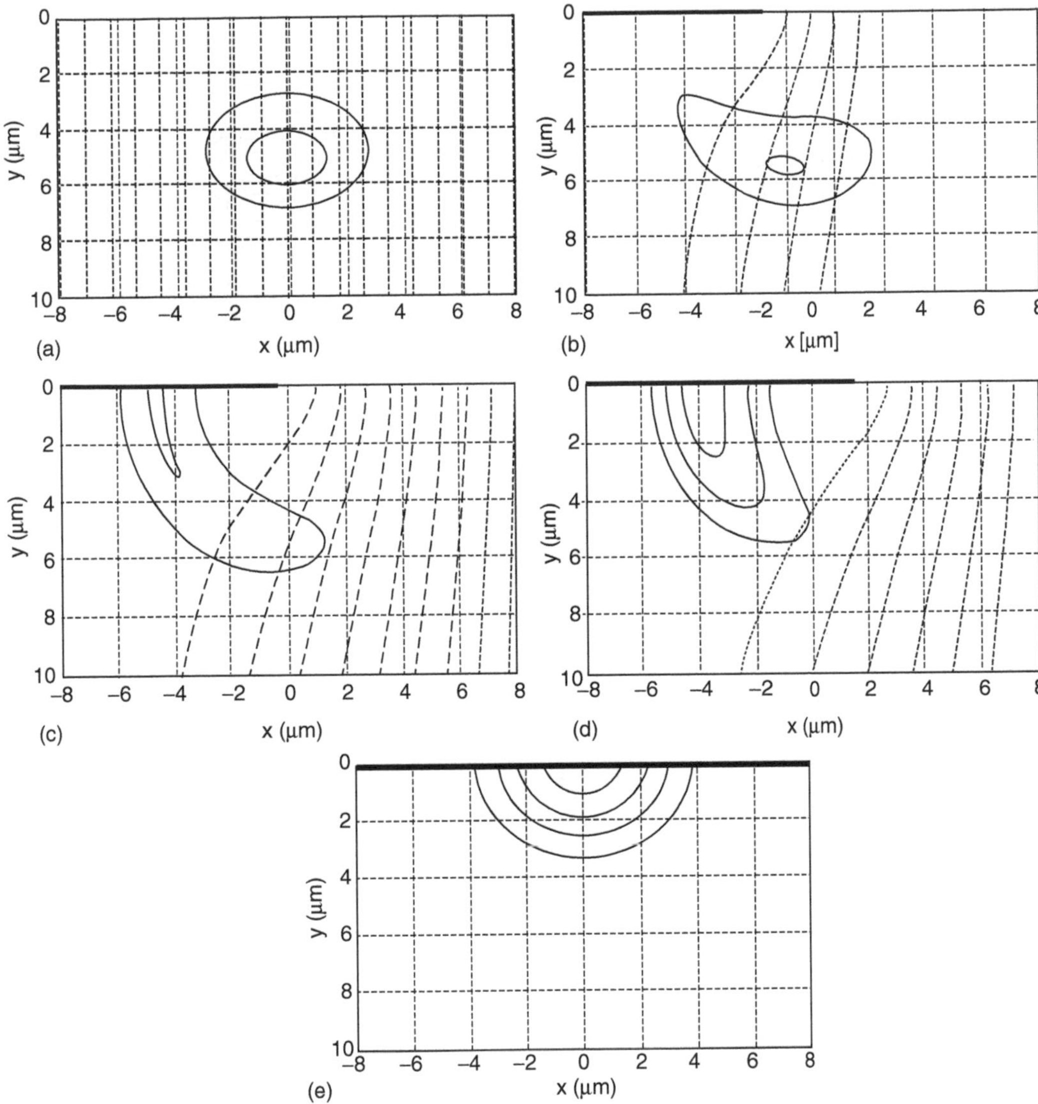

FIGURE 13.12 Solid lines represent simulated Ag^+ concentration contours at various transverse cross-sections of a selectively buried waveguide ($C_{Ag}=0.1, 0.2 \ldots$). Dashed lines represent electric field lines at the end of the burial process. Locations of the cross-sections are shown in Figure 13.11a. (Reprinted from West, B. R., Madasamy, P., Peyghambarian, N., and Honkanen, S. 2004. *J. Non-Cryst. Solids*, 347, 18–26. With permission.)

polarization-maintaining splitters are commercially available (TEEM)[1]. Add/drop multiplexers based on tilted gratings in a dual-mode waveguide (Geraghty et al. 2001; Castro et al. 2004), or four-port Mach–Zehnder interferometers with amorphous silicon overlay gratings (Kim, Li, and Winick 2004) have been demonstrated and show great promise in conjunction with the low-birefringence concepts developed in the previous section. Ring resonator filters (Wang et al. 1992) and gyroscope rotation

[1]TEEM photonics (France), http://www.teemphotonics.com, accessed on September 13, 2006.

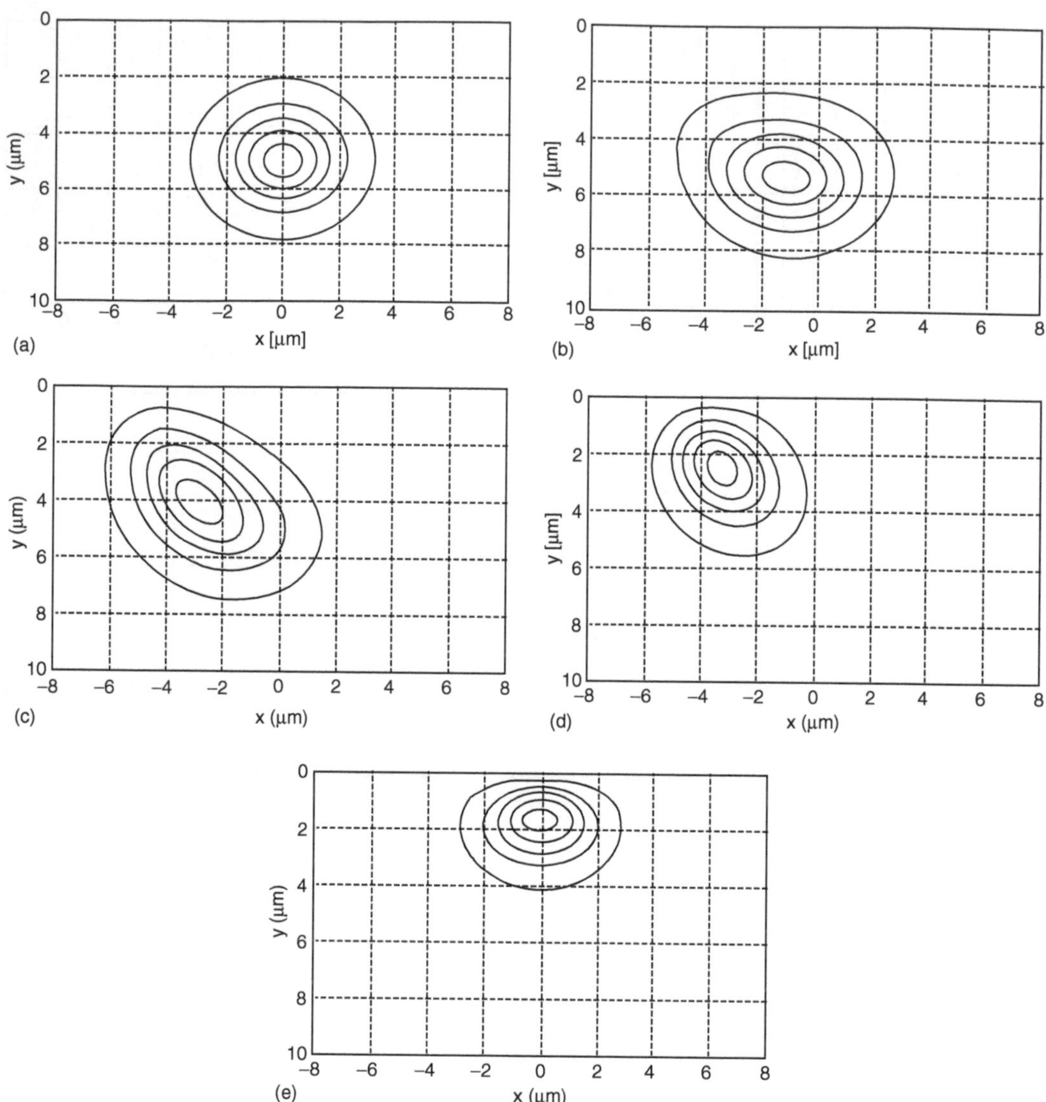

FIGURE 13.13 Normalized power contours (0.1, 0.3,...0.9) of the fundamental guided mode at various transverse cross-sections of a selectively buried waveguide. Locations of the cross-sections are shown in Figure 13.11. (Reprinted from West, B. R., Madasamy, P., Peyghambarian, N., and Honkanen, S. 2004. *J. Non-Cryst. Solids*, 347, 18–26. With permission.)

sensors (Monovoukas, Swiecki, and Maseeh 2000; Carriere 2004) have been demonstrated, with the latter author reporting sensitivity on the order of 1 degree/$h^{1/2}$.

13.5.2 Nonlinear and Active Substrates

By introducing dopant species with nonlinear or active properties into the glass melt, waveguide devices with many interesting properties can be fabricated. Aitchison et al. (1991) reported the observation of spatial solitons in ion-exchanged slab waveguides in two types of glass. Their initial experiments utilized a K^+–Na^+ exchange and subsequent burial in Schott B270 glass (nonlinear coefficient $n_2 \approx 3.4 \times 10^{-16}$ cm^2/W). Subsequently, an experimental silicate glass containing the transition elements Ti and Nb was produced with $n_2 \approx 4 \times 10^{-15}$ cm^2/W. A waveguide was formed by K^+–Tl^+ exchange and burial. Stable spatial solitons were observed in both waveguides, although the presence of enhanced two-photon absorption (TPA) led to soliton breakup at high input powers.

It is possible to produce glass of high optical quality that is doped with semiconductor quantum dots (QDs) with narrow size distribution (Borrelli and Smith 1994). Using glass as a host material is less expensive than growing quantum-confined structures and offers more semiconductor material flexibility. PbS QDs with radii of 2–5 nm exhibit strong quantum confinement, resulting in large optical nonlinear effects including bleaching and gain. Additionally, PbS offers QD-resonance tunability across the near-IR (wavelength increases with QD size) due to its narrow bulk bandgap. While the optical nonlinearities have been measured in bulk QD-doped glasses (Wundke, Auxier, and Schültzgen 1999; Wundke et al. 2000), these effects can be enhanced through the use of waveguiding. Drawing these QD-doped glasses into fiber is difficult due to the high drawing temperature, which destroys the QDs. So, instead of drawing fiber, Auxier et al. (2004, 2006a) have studied the fabrication of waveguides using both thermal K^+–Na^+ exchange from a melt and for smaller mode size (hence more enhanced nonlinear effects), silver film ion exchange (Auxier et al. 2006b). As the host glass is not optimized for ion exchange, the purely thermal process is very slow; K^+–Na^+ exchanges of 200–260 h in duration at 370°C were required. The authors reported the production of very low-loss guides (propagation loss of 0.2–0.7 dB/cm). Most importantly, the photoluminescence spectra of the substrates were not noticeably changed by either ion exchange process, indicating that the QDs are not physically altered during the process.

Perhaps the most dynamic area of ion exchange research today is the design of active devices, waveguide amplifiers and lasers. Ion exchange has been used to produce active waveguides since shortly after the first ion-exchanged waveguides were reported (Saruwatari and Izawa 1974). Most early work in this area was concentrated on neodymium-doped devices amplifying at a wavelength of 1.06 μm (Aoki, Marayama and Asahara 1990), but with the advent of wavelength division multiplexing (WDM) technology and its associated components in the mid 1990s, research efforts turned to erbium and erbium–ytterbium codoped glasses. Unlike fiber-based active components, which utilize a doped inner core in an otherwise passive structure, active ion-exchanged components are fabricated by forming a waveguide in a homogeneously doped substrate. The choice of host glass for such components must account not only for ion exchange capability, but also for the solubility of rare earth ions. Phosphate glasses are recommended for these applications, as silicate glasses exhibit relatively low solubility of rare earth ions, leading to clustering and a reduction in conversion efficiency (Giles and Desurvire 1991). However, phosphate glasses exhibit poor chemical durability when exposed to the molten salts used for ion exchange (Najafi 1992). For this reason, the dry silver film exchange process is preferred (Honkanen et al. 1991).

The earliest example of an ion-exchanged waveguide amplifier in Er/Yb-codoped phosphate glass was produced by Barbier et al. (1995). This device, operating in a double-pass configuration, exhibited 34 dB of internal gain and 7 dB of total gain over a 41 mm length, and was highly reproducible. Fiber-to-guide coupling losses were minimized by utilizing a burial process.

A lossless 1×2 splitter at 1.55 μm has been proposed by Camy et al. (1996). The device consists of a y-branch exchanged into Er/Yb-codoped borosilicate glass and buried, with pump/signal multiplexing performed externally. Thermal tapers were used to mode-match the input and output waveguides to optical fiber. Lossless splitting was achieved over a range of 1534–1548 nm with a noise figure of 8 dB. The performance of this device was limited by the splitting of the pump mode, which reduced the achievable gain in each output arm. A subsequent design by Jaouën et al. (1999) separates the amplification and splitting functions, with net gain observed in each of 8 WDM channels following 1×8 splitting.

The first Er-doped ion-exchanged laser emitting at 1540 nm was reported by Feuchter et al. (1992). The substrate used was BK7 glass with 0.5 wt% Er_2O_3, and a thermal exchange with KNO_3 was employed to produce the waveguides. Dielectric mirrors were bonded to each end of the chip to form a Fabry–Perot cavity. A pump threshold of 150 mW and slope efficiency of 0.55% for this nonoptimized structure was achieved.

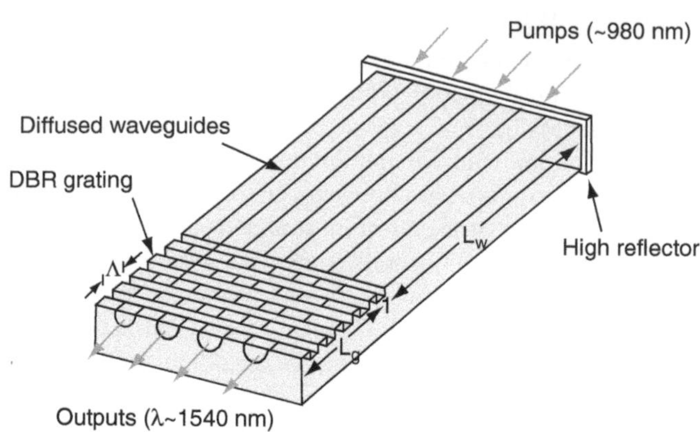

FIGURE 13.14 Er–Yb codoped DBR waveguide laser array. (Reprinted from Veasey, D. L., Funk, D. S., Peters, P. M., Sanford, N. A., Obarski, G. E., Fontaine, N., Young, M., Peskin, W. C., Houde-Walter, S. N., and Hayden, J. S. 2000. *J. Non-Cryst. Solids*, 263, 369–381. With permission.)

Veasey et al. (2000) described arrays of Er^{3+}–Yb^{3+} co-doped distributed Bragg reflection (DBR) lasers in a newly-developed phosphate glass containing Al_2O_3 to improve chemical durability. Surface waveguides of 3–8 μm width were formed using K^+–Na^+ thermal exchange, and a 6-mm long DBR grating with a period of 508 nm was formed by Ar ion etching (Figure 13.14) through a Cr-enhanced holographic photoresist mask. The same grating was used for all lasers in the array; wavelength selectivity was achieved by the difference in effective index among the waveguides. A high-reflectivity mirror completed the laser cavity. The waveguides with widths of 5 μm and greater exhibited lasing action over a range of 0.3 nm; for narrower waveguides, the mode volume provided insufficient gain to overcome transmission losses through the grating. For a representative laser within this array, a slope efficiency of 26% was estimated, and single longitudinal mode lasing with full-width half-maximum (FWHM) linewidth of ~500 kHz was observed for coupled pump powers below 300 mW. The use of surface guides and an etched grating promoted robustly single-polarization (TE) operation. A similar DBR laser array was later described by Madasamy et al. (2002). Here, the tuning range was increased to 2.1 nm by using the silver film ion exchange process to produce waveguides of larger index change in the same substrate.

An array of 15 distributed feedback (DFB) Er^{3+}–Yb^{3+} codoped waveguide lasers spanning a wavelength range of nearly 10 nm has been reported by Blaize et al. (2003). The waveguides were formed by Ag^+–Na^+ thermal exchange with mask openings between 0.5 and 6 μm in width, followed by field-assisted burial to about 5 μm below the surface (Figure 13.15). A 17 mm long grating was formed using reactive ion etching through a holographic photoresist mask to a depth of 150 nm; the phase shift required to allow a lasing mode in the Bragg stop-band was created by erasing a short section of the mask prior to the grating formation. After etching, a thick dielectric overlayer was deposited on the surface to reduce grating diffraction losses at the pump wavelength. For the waveguide with 2.5 μm mask opening, slope efficiencies of 5.7% (back facet) and 3.4% (front facet) were obtained with a threshold pump power of 14 mW; the higher power from the back facet was attributed to longitudinal variation of gain due to the strong pump absorption. All lasers in the array lased in a single longitudinal mode with FWHM linewidth below 130 MHz.

One important issue with rare-earth-doped substrates is that they absorb strongly when unpumped. This precludes such substrates from being used for the passive components required for amplifiers and lasers (pump/signal multiplexers, for instance). To address this problem, Conzone et al. (2001) have developed a low-temperature method for joining active and passive phosphate glasses to create hybrid

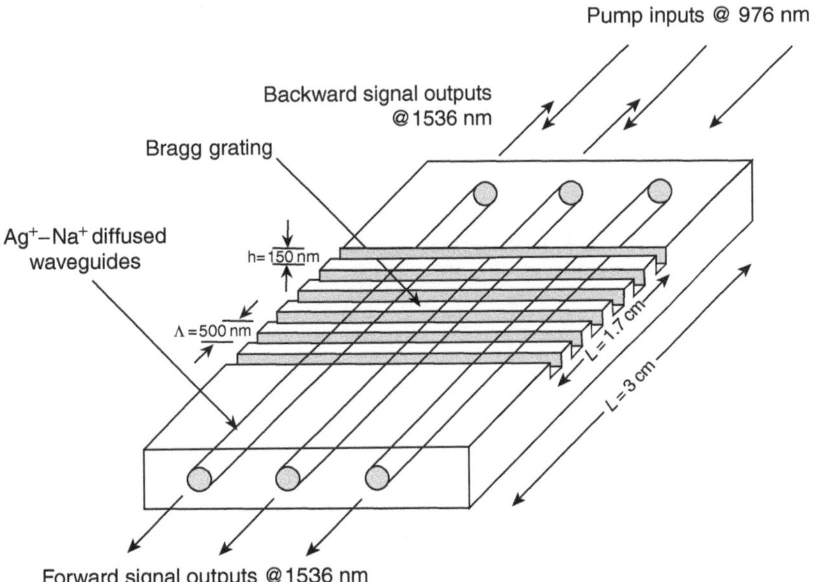

FIGURE 13.15 Er–Yb codoped DFB waveguide laser array. (From Blaize, S., Bastard, L., Cassagnètes, C., and Broquin, J. E. 2003. *IEEE Photonics Technol. Lett.*, 15, 516–518. With permission.)

preforms, which are subsequently diced into hybrid substrates. In the passive section, the active rare earth elements are replaced with La^{3+}, a nonlasing species, in order to equalize the passive optical and physical properties between the two glasses. An alternate approach described by Gardillou, Bastard, and Broquin (2004) involves vertical hybridization of passive and active glasses. A surface waveguide is exchanged into a low-index, passive substrate. An active substrate of slightly higher index is wafer-bonded to this surface and subsequently thinned to maintain single transverse mode operation. The optical mode is confined mostly to the high-index active region, with lateral confinement provided by the ion-exchanged strip. The authors used a Tl^{+}–K^{+} exchange in Schott NFK5 silicate glass; the similar ionic radius of the exchanging cations resulted in a flatter surface, facilitating the bonding. An earlier design in which the active glass had a refractive index between the surface- and substrate-values in the passive glass, was suggested by Honkanen, Najafi, and Wang (1992a). Using the vertical transition described in Section 13.4, passive components can be formed in buried waveguides, where the guided modes do not interact with the active region. This suggests the possibility of reducing the component footprint in comparison to Conzone's method, as the waveguides do not need to wind back and forth in plane between passive and active regions.

Madasamy et al. (2003a) utilized the hybrid substrates described by Conzone to design a single-mode tapered waveguide laser that can be pumped by an inexpensive high-power broad-area multimode laser. As shown in Figure 13.16, the laser waveguide consists of a 100-μm wide region, tapered down to a single-mode region. A surface-relief photoresist grating reflecting at 1538 nm is formed in the single-mode region; this, together with a broadband dielectric mirror (reflective in the 1550 nm band but transparent at the pump wavelength of 965 nm), forms the laser cavity. The grating ensures single longitudinal mode operation, while higher-order transverse modes are filtered out in the taper. The joint between active and passive regions is positioned in the middle of the taper; this allows for more efficient pumping in the tapered region, as the intensity of lower-order pump modes increases to compensate for the radiation of higher-order modes. To form the waveguide, the authors used a silver film process in order to reduce propagation losses due to surface damage.

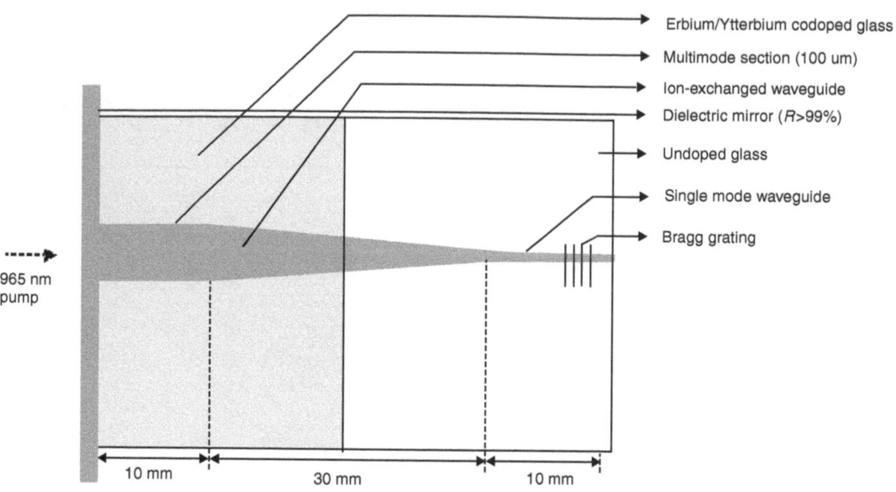

FIGURE 13.16 Schematic diagram of the Er-Yb tapered waveguide laser pumped by a broad area diode laser. (Reprinted from Madasamy, P., Honkanen, S., Geraghty, D. F., and Peyghambarian, N. 2003. *Appl. Phys. Lett.*, 82, 1332–1334. With permission.)

A novel application of hybrid substrates, in which ion exchange actually occurs during the substrate bonding, has been described by Gawith et al (2002). The authors used two substrates. One was similar to BK7 but with a higher potassium content and correspondingly lower sodium content. The second substrate was an experimental neodymium-doped photosensitive germano-borosilicate glass (SGBN) with substantial sodium content, and a thermal expansion coefficient matched to the BK7. When the substrates were bonded at 350°C, intersubstrate K^+–Na^+ ion exchange formed a slab waveguide on the SGBN side of the interface (Figure 13.17). The Nd^{3+}-doped side was polished down to a thickness of 200 μm. By focusing a frequency-doubled argon laser at 244 nm onto the sample from above (such that the beam waist coincided with the ion-exchanged layer) and translating the sample, a channel waveguide was formed. Propagation losses less than 0.3 dB/cm were reported for these waveguides. Enclosed by plane mirrors to form a Fabry–Perot laser cavity, this nonoptimized device exhibited a slope efficiency of 9% when pumped at 808 nm, with threshold absorbed pump power of 4.5 mW.

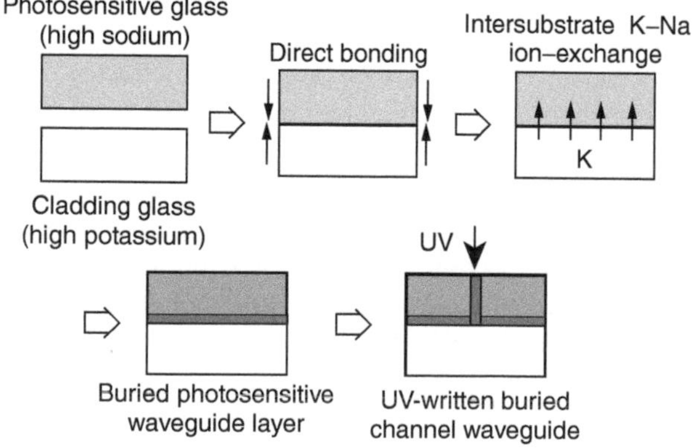

FIGURE 13.17 Processing stages for a buried channel waveguide laser by a combination of direct-bonding and direct-UV-writing techniques. (Reprinted from Gawith, C. B. E., Fu, A., Bhutta, T., Hua, P., Shepherd, D. P., Taylor, E. R., and Smith, P. G. R. 2002. *Appl. Phys. Lett.*, 81, 3522–3524. With permission.)

13.5.3 Hybrid Devices

The applications of ion-exchanged waveguides can be increased dramatically through hybridization. This term is used to describe structures that utilize more than one material system, with physical contact between the systems achieved by a bonding or coating process. Ion-exchanged waveguides are particularly suitable for hybridization for two reasons. First, the waveguides exist at the surface of the substrate, where the guided modes can easily interact with an overlayer. Second, as the glass substrates do not undergo etching during waveguide fabrication, they remain planar, which is beneficial for most types of hybridization. Many authors have described hybrid ion-exchanged waveguide devices; a few of these are described here.

By utilizing a nonlinear polymer as an overlayer, second harmonic generation (Alshikh Khalil et al. 1999) and optical parametric amplification (Alshikh Khalil et al. 2000) have been achieved. In both cases, a planar waveguide was formed by first performing a Na^+–Tl^+ ion exchange in Schott KF3 glass, then spinning on an overlayer of *para*-nitroaniline/poly(methyl methacrylate) (PNA–PMMA). The polymer was baked to eliminate the host solvent, and then poled by corona discharge to achieve a net dipole orientation. In this composite waveguide, modal dispersion was used to phase-match the required frequency components; this was facilitated by the large difference in dispersion between the two materials. As the polymer index was higher than the surface index of the glass, modal overlap was primarily in the nonlinear material, increasing frequency conversion efficiency.

Optical control, a prerequisite for optical switching, has been demonstrated in a hybrid polymer/glass waveguide device (Rodríguez et al. 2001). The configuration is shown in Figure 13.18. The waveguides are fabricated by Na^+–Tl^+ thermal exchange, and an active region is defined by an overlayer of PMMA, sensitized with the azo dye Disperse Red 1. An input signal is split at a y-branch, and interacts with the polymer. Both arms pass through the same active region to equalize absorption, but only one arm of the device is irradiated by an argon laser control beam at 514.5 nm, to effect a phase shift between the two waveguides. The arms are collimated with tapered transitions and then interfered in a planar waveguide. The spatial interference pattern thus created is altered by the control beam.

Ion-exchanged waveguides have been successfully hybridized with III–V semiconductor membranes. Yi-Yan et al. (1989) achieved the first integration of a GaAs metal–semiconductor–metal (MSM) photodetector with an ion-exchanged surface waveguide. A 250-nm thick GaAs membrane was grown by molecular beam epitaxy (MBE) over a sacrificial AlAs layer on a GaAs substrate. A buffered HF solution preferentially etched the AlAs layer, causing liftoff of the membrane. This was subsequently grafted onto the surface of a glass substrate containing an ion-exchanged surface waveguide; a strong bond between the semiconductor and glass was achieved by van der Waals forces. Schottky metal was then evaporated and patterned to complete the MSM photodetector. This technique was later used by the same authors to integrate GaInAs/InP *pin* photodetectors (Yi-Yan et al. 1991) and InGaAsP light emitting diodes (Yi-Yan et al. 1992) with ion-exchanged waveguides. Nannini et al. (2005) describe a process by which InP is bonded to the glass substrate using plasma activation; the low temperatures used in this process are necessary so as not to alter the profile of the ion-exchanged guide. An O_2 plasma is used to attach oxygen atoms to the surfaces. Subsequently, a dip into deionized water links hydrogen atoms with the oxygen layer. A drop of water is placed onto this hydrophilic layer, then, the surfaces are placed into physical contact while under low vacuum, allowing for strong covalent bonding between the materials. Finally, the InP is thinned, first mechanically, and then by wet etching up to an InGaAsP etch stop layer. Because of the high refractive index of III–V materials in comparison to glass, these structures act as inverted strip-loaded waveguides. The mode is confined primarily in the semiconductor, with weak lateral guiding provided by the ion-exchanged region, as shown in Figure 13.19. Hybrid structures such as this show great potential for nonlinear or active devices.

Ion-exchanged waveguides have long been used as the basis for sensing elements, in both absorptive and interferometric configurations. In both cases, surface waveguides are used in order to allow the guided mode to interact with an overlayer or adsorbed surface layer. Interferometric chemical- or bio-sensors generally assume a Mach–Zehnder configuration (Gauglitz and Ingenhoff 1993), where one

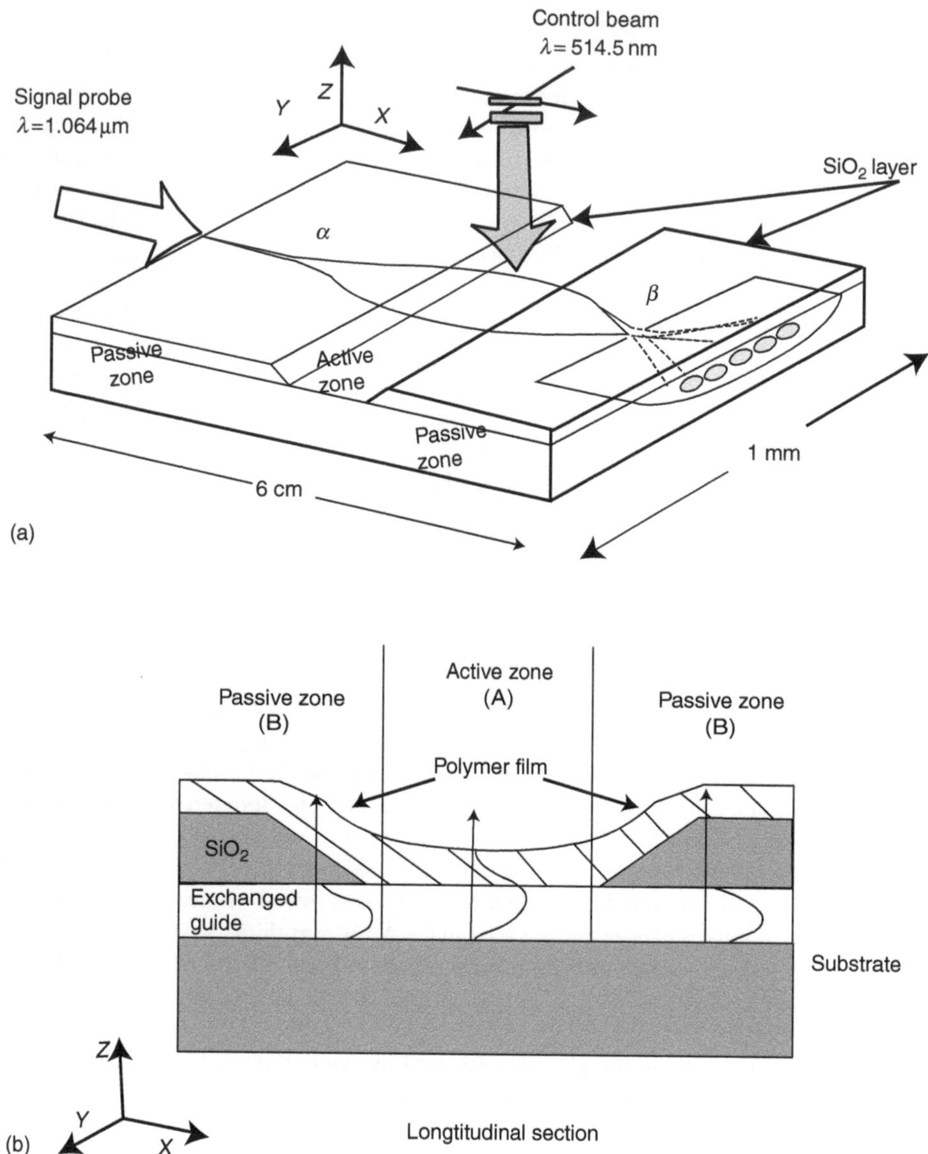

FIGURE 13.18 (a) Schematic view of a hybrid polymer/glass interferometer for optical switching. (b) Longitudinal section of the device in the active region. (Reprinted from Rodríguez, A., Vitrant, G., Chollet, P. A., and Kajzar, F. 2001. *Appl. Phys. Lett.*, 79, 461–463. With permission.)

arm of the device is exposed to the analyte. Displacement sensors have been demonstrated using a Michelson configuration, facilitated by y-branches and an integrated loop mirror (Poffo et al. 2005). An example of an absorption-based integrated biosensor has been described by Araci et al. (2005). In this configuration (Figure 13.20), a straight surface waveguide is fabricated by ion exchange, and a sol–gel overlayer is spin-coated and patterned to expose a small region of the surface. Grayscale lithography is used to taper the overlayer in order to reduce scattering losses. When horse-heart cytochrome *c* protein, dissolved in an aqueous phosphate buffer solution, was introduced to the sensing region, a spectroscopically selective attenuation by the adsorbed monolayer of a probe signal was measured. The authors estimated a sensitivity of 5% of a full monolayer.

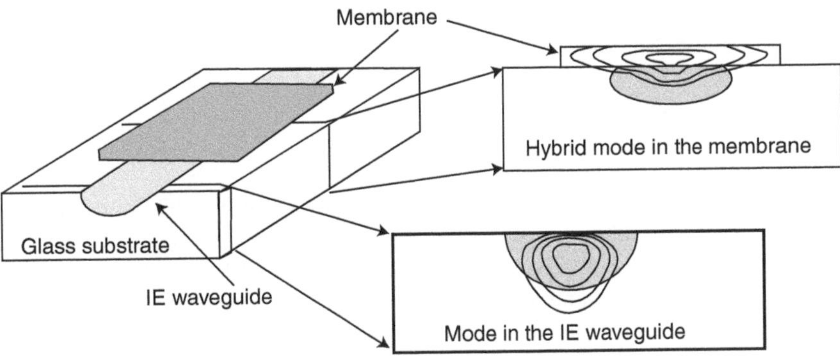

FIGURE 13.19 Hybridization of III-V semiconductor membrane with ion-exchanged waveguide. (From Nannini, M., Grondin, E., Gorin, A., Aimez, V. and Broquin, J. E. 2005. *IEEE J. Sel. Topics Quantum Electron.*, 11, 547–554. With permission.)

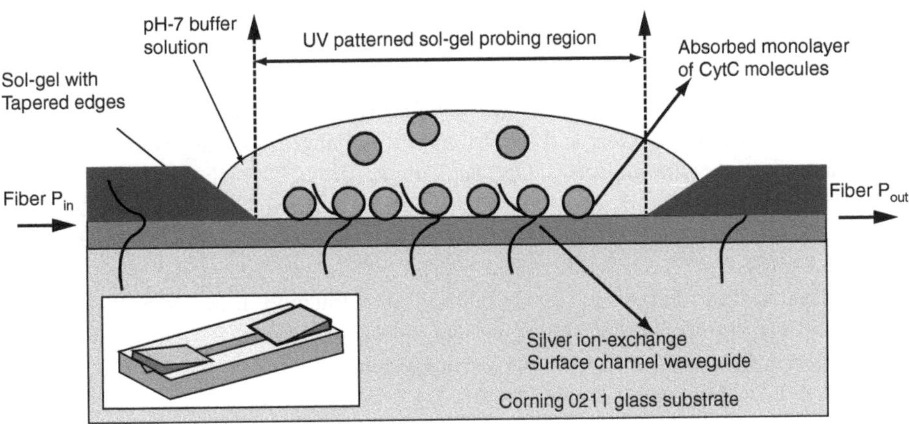

FIGURE 13.20 Hybrid ion-exchanged glass/sol-gel waveguide structure for biosensing. (From Araci, I., Yurt, N., Honkanen, S., Mendez, S., and Peyghambarian, N. 2005. *Proc. SPIE*, 6004, 149–155. With permission.)

One factor that considerably reduces the sensitivity of these devices is the large index difference between the surface waveguide ($n \approx 1.5$) and the analyte ($n \approx 1$ or 1.33 for gaseous/aqueous overlayers). This limits the penetration of the guided mode into the analyte, lowering the achievable modification of the effective index. A composite optical waveguide (COWG) described by Qi et al. (2002) can be used to significantly increase this field penetration. The COWG is produced by sputtering a thin TiO_2 film onto the surface waveguide. As the film has a higher index than that of the glass, the optical power will be guided laterally in the film; the ion-exchanged waveguide below provides transverse waveguiding. However, as the film is only on the order of 25 nm thick, the lateral confinement is poor, and thus penetration of the guided mode into the overlayer is significantly enhanced.

References

Aitchison, J. S., Silberberg, Y., Weiner, A. M., Leaird, D. E., Oliver, M. K., Jackel, J. L., Vogel, E. M., and Smith, P. W. E. 1991. Spatial optical solitons in planar glass waveguides. *J. Opt. Soc. Am. B.*, 8, 1290–1297.

Albert, J. 1992. Ion exchange from salt melts. In *Introduction to Glass Integrated Optics*, S. I. Najafi, ed., pp. 7–38. Artech House, Norwood, MA.

Albert, J. and Lit, J. W. Y. 1990. Full modeling of field-assisted ion exchange for graded index buried channel optical waveguides. *Appl. Opt.*, 29, 2798–2804.

Albert, J. and Yip, G. L. 1987. Stress-induced index change for K^+–Na^+ ion exchange in glass. *Electron. Lett.*, 23, 737–738.

Alshikh Khalil, M., Vitrant, G., Raimond, P., Chollet, P. A., and Kajzar, F. 1999. Modal dispersion phase-matched frequency doubling in composite planar waveguide using ion-exchanged glass and optically nonlinear poled polymer. *Opt. Commun.*, 170, 281–284.

Alshikh Khalil, M., Vitrant, G., Raimond, P., Chollet, P. A., and Kajzar, F. 2000. Optical parametric amplification in composite polymer/ion exchanged planar waveguide. *Appl. Phys. Lett.*, 77, 3713–3715.

Aoki, H., Marayama, O., and Asahara, Y. 1990. Glass waveguide laser. *IEEE Photonics Technol. Lett.*, 2, 459–460.

Araci, I., Yurt, N., Honkanen, S., Mendez, S., and Peyghambarian, N. 2005. Hybrid glass and sol-gel integrated optical waveguides for absorption-based biosensing. *Proc. SPIE*, 6004, 149–155.

Äryäs, P., Nunzi Conti, G., Honkanen, S., and Peyghambarian, N. 1998. Birefringence control for ion-exchanged channel glass waveguides. *Appl. Opt.*, 37, 8400–8405.

Auxier, J. M., Morrell, M. M., West, B. R., Honkanen, S., Schülzgen, A., Sen, S., Borrelli, N. F., and Peyghambarian, N. 2004. Ion-exchanged waveguides in glass doped with PbS quantum dots. *Appl. Phys. Lett.*, 85, 6098–6100.

Auxier, J. M., Honkanen, S., Schützgen, A., Morrell, M. M., Leigh, M., Sen, S., Borrelli, N. F., and Peyghambarian, N. 2006a. Silver and potassium ion-exchanged waveguides in glasses doped with PbS semiconductor quantum dots. *J. Opt. Soc. Am. B.*, 23, 1037–1045.

Auxier, J. M., Honkanen, S., Morrell, M. M., Leigh, M., Sen, S., Borrelli, N. F., and Schültzgen, A. 2006b. Small mode-size waveguides in quantum-dot-doped glasses by Ag-film ion exchange. *J. Appl. Phys.*, 99, 12610.

Barbier, D., Delavaux, J. M., Kevorkian, A., Gastaldo, P., and Jouanno, J. M. 1995. Postdeadline paper PD-3, Yb/Er integrated optics amplifiers on phosphate glass in single and double pass configurations, *Optical Fiber Communications Conference.*

Blaize, S., Bastard, L., Cassagnètes, C., and Broquin, J. E. 2003. Multiwavelengths DFB waveguide laser arrays in Yb–Er codoped phosphate glass substrate. *IEEE Photonics Technol. Lett.*, 15, 516–518.

Borrelli, N. F. and Smith, D. W. 1994. Quantum confinement of PbS microcrystals in glass. *J. Non-Cryst. Solids*, 180, 25–31.

Brandenburg, A. 1986. Stress in ion-exchanged glass waveguides. *J. Lightwave Technol.*, 4, 1580–1593.

Buchold, B. and Voges, E. 1996. Polarisation insensitive arrayed-waveguide grating multiplexers with ion-exchanged waveguides in glass. *Electron. Lett.*, 32, 2248–2250.

Camy, P., Román, J. E., Willems, F. W., Hempstead, M., van der Plaats, J. C., Prel, C., Béguin, A., Koonen, A. M. J., Wilkinson, J. S., and Lerminiaux, C. 1996. Ion-exchanged planar lossless splitter at 1.5 μm. *Electron. Lett.*, 32, 321–323.

Carriere, J. T. A. 2004. Ion-Exchanged Ring Resonator Integrated Optic Devices. PhD dissertation. The University of Arizona.

Carriere, J. T. A., Frantz, J. A., Youmans, B. R., Honkanen, S., and Kostuk, R. K. 2004. Measurement of waveguide birefringence using a ring resonator. *IEEE Photonics Technol. Lett.*, 16, 1134–1136.

Castro, J. M., Geraghty, D. F., West, B. R., and Honkanen, S. 2004. Fabrication and comprehensive modeling of ion-exchanged Bragg optical add-drop multiplexers. *Appl. Opt.*, 43, 6166–6173.

Chartier, G. H., Jaussaud, P., de Oliveira, A. D., and Parriaux, O. 1978. Optical waveguides fabricated by electric-field controlled ion exchange in glass. *Electron. Lett.*, 14, 132–134.

Cheng, D., Saarienen, J., Saarikoski, H., and Tervonen, A. 1997. Simulation of field-assisted ion exchange for glass channel waveguide fabrication: effect of nonhomogeneous time-dependent electric conductivity. *Opt. Commun.*, 137, 233–238.

Chiang, K. S. 1991. Effective-index function method for the analysis and design of inhomogeneous planar waveguides based on the WKB equation. *Opt. Commun.*, 84, 258–263.

Chludzinski, P., Ramaswamy, R. V., and Anderson, T. J. 1987. Ion exchange between soda-lime–silica glass and sodium nitrate–silver nitrate molten salts. *Phys. Chem. Glasses*, 28, 169–173.

Choo, P. Y., Frantz, J. A., Carriere, J. T. A., Mathine, D. L., Kostuk, R. K., and Peyghambarian, N. 2003. Measurement and modeling of ion-exchange parameters for IOG-10 glass. *Opt. Eng.*, 42, 2812–2816.

Conzone, S. D., Hayden, J. S., Funk, D. S., Roshko, A., and Veasey, D. L. 2001. Hybrid glass substrates for waveguide device manufacture. *Opt. Lett.*, 26, 509–511.

Doremus, R. H. 1964. Exchange and diffusion of ions in glass. *J. Phys. Chem.*, 68, 2212–2218.

Fantone, S. D. 1983. Refractive index and spectral models for gradient-index materials. *Appl. Opt.*, 22, 432–440.

Ferziger, J. H. 1981. *Numerical Methods for Engineering Application*, pp. 253–254. Wiley, New York.

Feuchter, T., Mwarania, E. K., Wang, J., Reekie, L., and Wilkinson, J. S. 1992. Erbium-doped ion-exchanged waveguide lasers in BK-7 glass. *IEEE Photonics Technol. Lett.*, 4, 542–544.

Frantz, J., Carriere, J., Enami, Y., Kostuk, R., Honkanen, S., Mathine, D., Peyghambarian, N., Jen, A. and Zheng, L. 2003. Selectively buried ion-exchanged waveguides as an interface to electro-optic polymers. *Optical Fiber Communications Conference, Vol. 1*, p. 318.

Gardillou, F., Bastard, L., and Broquin, J. E. 2004. 4.24 dB Gain in a hybrid silicate/phosphate glasses optical amplifier made by wafer bonding and ion-exchange techniques. *Appl. Phys. Lett.*, 85, 5176–5178.

Garfinkel, H. M. 1968. Ion-exchange equilibrium between glass and molten salts. *J. Phys. Chem.*, 72, 4175–4181.

Gauglitz, G. and Ingenhoff, J. 1993. Integrated optical sensors for halogenated and non-halogenated hydrocarbons. *Sens. Act. B.*, 11, 207–212.

Gawith, C. B. E., Fu, A., Bhutta, T., Hua, P., Shepherd, D. P., Taylor, E. R., and Smith, P. G. R. 2002. Direct UV-written buried channel waveguide lasers in direct-bonded intersubstrate ion-exchanged neodymium-doped germano-borosilicate glass. *Appl. Phys. Lett.*, 81, 3522–3524.

Geraghty, D. F., Provenzano, D., Marshall, W. K., Honkanen, S., Yariv, A., and Peyghambarian, N. 1999. Gratings written in ion-exchanged glass channel waveguides. *Electron. Lett.*, 35, 585–586.

Geraghty, D. F., Provenzano, D., Morrell, M. M., Ingenhoff, J., Drapp, B., Honkanen, S., Yariv, A., and Peyghambarian, N. 2000. Polarisation-independent Bragg gratings in ion-exchanged glass channel waveguides. *Electron. Lett.*, 36, 531–532.

Geraghty, D. F., Provenzano, D., Morrell, M. M., Honkanen, S., Yariv, A., and Peyghambarian, N. 2001. Ion-exchanged waveguide add/drop filter. *Electron. Lett.*, 37, 829–831.

Gerald, C. F. and Wheatley, P. O. 1999. *Applied Numerical Analysis 6th Ed.*, pp. 455–459, 610–611. Addison-Wesley, Reading, MA.

Giallorenzi, T. G., West, E. J., Kirk, R., Ginther, R., and Andrews, R. A. 1973. Optical waveguides formed by thermal migration of ions in glass. *Appl. Opt.*, 12, 1240–1245.

Giles, C. R. and Desurvire, E. 1991. Modeling erbium-doped fiber amplifiers. *J. Lightwave Technol.*, 9, 271–283.

Gonella, F. 1992. Stress-induced optical effects in Ag^+–Na^+ ion-exchanged glass waveguides. *Opt. Lett.*, 17, 1667–1669.

Gortych, J. E. and Hall, G. 1985. Fabrication of optical waveguides by K^+ ion exchange in BK7 glass. *Opt. Lett.*, 11, 100–102.

Hazart, J. and Minier, V. 2001. Concentration profile calculation for buried ion-exchanged channel waveguides in glass using explicit space-charge analysis. *IEEE J. Quantum Electron.*, 37, 606–612.

Honkanen, S. 1992. Silver-film ion-exchange technique. In *Introduction to Glass Integrated Optics*, S. I. Najafi, ed., pp. 39–71. Artech House, Norwood, MA.

Honkanen, S., Tervonen, A., von Bagh, H., Salin, A., and Leppihalme, M. 1987. Fabrication of ion-exchanged channel waveguides directly into integrated circuit mask plates. *Appl. Phys. Lett.*, 51, 296–298.

Honkanen, S., Najafi, S. I., Pöyhönen, P., Orcel, G., Wang, W. J., and Chrostowski, J. 1991. Silver-film ion-exchanged single-mode waveguides in Er-doped phosphate glass. *Electron. Lett.*, 27, 2167–2168.

Honkanen, S., Najafi, S. I., and Wang, W. J. 1992a. Composite rare-earth-doped glass waveguides. *Electron. Lett.*, 28, 746–747.

Honkanen, S., Najafi, S., Wang, I., Lefebvre, J., and Li, M. J. 1992b. Single-mode glass channel waveguides by ion exchange with ionic masking. *Opt. Commun.*, 94, 54–58.

Honkanen, S., Pöyhönen, P., Tervonen, A., and Najafi, S. I. 1993. Waveguide coupler for potassium- and silver-ion-exchanged waveguides in glass. *Appl. Opt.*, 32, 2109–2111.

Izawa, T. and Nakagome, H. 1972. Optical waveguide formed by electrically induced migration of ions in glass plates. *Appl. Phys. Lett.*, 21, 584–586.

Jackel, J. L. 1988. Glass waveguides made using low melting point nitrate mixtures. *Appl. Opt.*, 27, 472–475.

Jaouën, Y., du Mouza, L., Barbier, D., Delavaux, J. M., and Bruno, P. 1999. Eight-wavelength Er–Yb doped amplifier: combiner/splitter planar integrated module. *IEEE Photonics Technol. Lett.*, 11, 1105–1107.

Jiang, S., Myers, J. D., Wu, R., Bishop, G. M., Rhonehouse, D. L., Myers, M. J., and Hamlin, S. J. 1995. Chemically strengthened Er^{3+}, Nd^{3+} doped phosphate laser glasses. *Proc. SPIE*, 2379, 17–25.

Kim, C. M. and Ramaswamy, R. V. 1989. Modeling of graded-index channel waveguides using nonuniform finite difference method. *J. Lightwave Technol.*, 7, 1581–1589.

Kim, J., Li, G., and Winick, K. 2004. Design and fabrication of a glass waveguide optical add-drop multiplexer by use of an amorphous-silicon overlay distributed Bragg reflector. *Appl. Opt.*, 43, 671–677.

Kistler, S. S. 1962. Stresses in glass produced by nonuniform exchange of monovalent ions. *J. Am. Ceram. Soc.*, 45, 59–68.

Laity, R. W. 1957. Fused salt concentration cells with transference—Activity coefficients in the system silver nitrate–sodium nitrate. *J. Am. Chem. Soc.*, 79, 1849–1851.

Lefebvre, P., Honkanen, S., Najafi, S. I., and Tervonen, A. 1993. Nonsymmetrical potassium ion-exchanged Mach–Zehnder interferometers in glass. *Opt. Commun.*, 96, 36–40.

Li, M. J., Honkanen, S., Wang, W. J., Leonelli, R., Albert, J., and Najafi, S. I. 1991. Potassium and silver ion-exchanged dual-core glass waveguides with gratings. *Appl. Phys. Lett.*, 58, 2607–2609.

Li, G., Winick, K. A., Griffin, H. C., and Hayden, J. S. 2006. Systematic modeling study of channel waveguide fabrication by thermal silver ion exchange. *Appl. Opt.*, 45, 1743–1755.

Li, M. J. and Najafi, S. I. 1993. Polarization dependence of grating-assisted waveguide Bragg reflectors. *Appl. Opt.*, 32, 4517–4521.

Lilienhof, H. J., Voges, E., Ritter, D., and Pantschew, B. 1982. Field-induced index profiles of multimode ion-exchanged strip waveguides. *IEEE J. Quantum Electron.*, QE-18, 1877–1883.

Madasamy, P., Nunzi Conti, G., Pöyhönen, P., Hu, Y., Morrell, M. M., Geraghty, D. F., Honkanen, S., and Peyghambarian., N. 2002. Waveguide distributed Bragg reflector laser arrays in erbium doped glass made by Ag film ion exchange. *Opt. Eng.*, 41, 1084–1086.

Madasamy, P., Honkanen, S., Geraghty, D. F., and Peyghambarian, N. 2003a. Single-mode tapered waveguide laser in Er-doped glass with multimode-diode pumping. *Appl. Phys. Lett.*, 82, 1332–1334.

Madasamy, P., West, B. R., Morrell, M. M., Geraghty, D. F., Honkanen, S., and Peyghambarian, N. 2003b. Buried ion-exchanged glass waveguides: burial depth dependence on the waveguide width. *Opt. Lett.*, 28, 1132–1134.

Mahapatra, A. and Connors, J. M. 1988. Thermal tapering of ion-exchanged channel guides in glass. *Opt. Lett.*, 13, 169–171.

Masalkar, P. J. 1994. Calculation of concentration profile in ion-exchange waveguides by finite difference ADI method. *Optik*, 95, 168–172.

McCourt, M. 1993. Status of glass and silicon-based technologies for passive components. *Eur. Trans. Telecomm.*, 4, 685–689.

Millar, C. A. and Hutchins, R. H. 1978. Manufacturing tolerances for silver-sodium ion-exchanged planar optical waveguides. *J. Phys. D.*, 11, 1567–1576.

Moerman, I., Daele, P. V., and Demeester, P. 1997. A review on fabrication technologies for the monolithic integration of taper with III–V semiconductor devices. *IEEE J. Sel. Topics Quantum Electron.*, 3, 1308–1320.

Monovoukas, C., Swiecki, A., and Maseeh, F. 2000. Integrated optical gyroscopes offering low cost, small size and vibration immunity. *Proc. SPIE*, 936, 293–300.

Mounier, E., Yip, G. L., Noutsios, P. C., Chartier, G., Benech, P., and Rimet, R. 1993. Improved loss characteristics of buried waveguides. *Opt. Lett.*, 18, 1624–1626.

Mrozek, P., Mrozek, E., and Lukaszewicz, T. 2006. Side diffusion modeling by the explicit consideration of a space-charge buildup under the mask during strip waveguide formation in the Ag^+–Na^+ field-assisted ion-exchange process. *Appl. Opt.*, 45, 619–625.

Najafi, S. I. 1988. Optical behavior of potassium ion-exchanged glass waveguides. *Appl. Opt.*, 27, 3728–3731.

Najafi, S. I. 1992. Waveguides and devices. In *Introduction to Glass Integrated Optics*, S. I. Najafi, ed., pp. 137–167. Artech House, Norwood, MA.

Najafi, S. I., Lefebvre, P., Albert, J., Honkanen, S., Vahid-Shahidi, A., and Wang, W.J 1992. Ion-exchanged Mach–Zehnder interferometers in glass. *Appl. Opt.*, 31, 3381–3383.

Nannini, M., Grondin, E., Gorin, A., Aimez, V., and Broquin, J. E. 2005. Hybridization of III–V semiconductor membranes onto ion-exchanged waveguides. *IEEE J. Sel. Topics Quantum Electron.*, 11, 547–554.

Nikonorov, N. V. and Petrovskii, G. T. 1999. Ion-exchanged glasses in integrated optics: the current state of research and prospects (a review). *Glass Phys. Chem.*, 25, 16–55.

Noutsios, P. C. and Yip, G. L. 1991. Diffusion and propagation characteristics of buried single-mode waveguides in glass. *IEEE J. Quantum Electron.*, 27, 549–555.

Opilski, A., Rogoziński, R., Błahut, M., Karasiński, P., Gut, K., and Opilski, Z. 1997. Technology of ion exchange in glass and its application in waveguide planar sensors. *Opt. Eng.*, 36, 1625–1638.

Pantchev, B. 1986. Multimode strip waveguide formed by ion-electro-diffusion from solid state silver: side diffusion reduction. *Opt. Commun.*, 60, 373–375.

Pantchev, B., Danesh, P., and Nikolov, Z. 1993. Field-assisted ion exchange in glass: the effect of masking films. *Appl. Phys. Lett.*, 62, 1212–1214.

Peaceman, D. W. and Rachford, H. H. Jr. 1955. The numerical solution of parabolic and elliptic differential equations. *J. Soc. Ind. Appl. Math.*, 3, 28–41.

Pissadakis, S., Ikiades, A., Hua, P., Sheridan, A. K., and Wilkinson, J. S. 2004. Photosensitivity of ion-exchanged Er-doped phosphate glass using 248 nm excimer laser radiation. *Opt. Express*, 12, 3131–3136.

Poling, S. A. and Houde-Walter, S. N. 2000. Relaxation behavior of silver-sodium interdiffusion in a boroaluminosilicate glass. *J. Non-Cryst. Solids*, 272, 39–45.

Poffo, L., Lemaître-Auger, P., Benech, P., and Benech, P. 2005. Improvement of a glass integrated optical displacement sensor interferometer using acousto-optic effect. *Meas. Sci. Technol.*, 16, 1341–1348.

Pöyhönen, P., Honkanen, S., Tervonen, A., Tahkokorpi, M., and Albert, J. 1991. Planar 1/8 splitter in glass by photoresist masked silver film ion exchange. *Electron. Lett.*, 27, 1319–1320.

Provenzano, D., Marshall, W. K., Yariv, A., Geraghty, D. F., Honkanen, S., and Peyghambarian, N. 1999. Grating formation in BGG31 glass by UV exposure. *Electron. Lett.*, 35, 1332–1334.

Qi, Z., Matsuda, N., Itoh, K., Murabayashi, M., and Lavers, C. R. 2002. A design for improving the sensitivity of a Mach–Zehnder interferometer to chemical and biological measurands. *Sens. Act. B*, 81, 254–258.

Ramaswamy, R. V. and Srivastava, R. 1988. Ion-exchanged glass waveguides: a review. *J. Lightwave Technol.*, 6, 984–1002.

Rao, R. J. 2002. *Structural Chemistry of Glasses.* Elsevier, New York.

Rehouma, F., Persegol, D., and Kevorkian, A. 1994. Optical waveguides for evanescent field sensing. *Appl. Phys. Lett.*, 65, 1477–1479.

Rehouma, F., Elflein, W., Persegol, D., Kevorkian, A., Clauss, G., Benech, P., and Rimet, R. 1995. Improved structures for evanescent wave sensors. *Appl. Phys. Lett.*, 66, 1461–1462.

Rodríguez, A., Vitrant, G., Chollet, P. A., and Kajzar, F. 2001. Optical control of an integrated interferometer using a photochromic polymer. *Appl. Phys. Lett.*, 79, 461–463.

Roman, J. E. and Winick, K. A. 1993. Photowritten gratings in ion-exchanged glass waveguides. *Opt. Lett.*, 18, 808–810.

Ross, L. 1989. Integrated optical components in substrate glasses. *Glass Sci. Technol.*, 62, 285–297.

Saarikoski, H., Salmio, R. P., Saarinen, J., Eirola, T., and Tervonen, A. 1997. Fast numerical solution of nonlinear diffusion Equation 13. For the simulation of ion-exchanged micro-optics components in glass. *Opt. Commun.*, 134, 362–370.

Saruwatari, M. and Izawa, T. 1974. Nd-glass laser with three-dimensional optical waveguide. *Appl. Phys. Lett.*, 24, 603–605.

Schulze, G. 1914. Versuche über die diffusion von silber in glas. *Angew Physik.*, 40, 335–367.

Sebastiani, S., Berneschi, S., Brenci, M., Nunzi Conti, G., Pelli, S., and Righini, G. C. 2005. Simple approach to calculate the refractive index profile of ion-exchanged waveguides. *Opt. Eng.*, 44, 054602.

Stern, M. S. 1988. Semivectorial polarized finite difference method for optical waveguides with arbitrary index profiles. *IEE Proc. J.*, 135, 56–63.

Stewart, G., Millar, C. A., Laybourn, P. J. R., Wilkinson, C. D. W., and DeLaRue, R. M. 1977. Planar optical waveguides formed by silver-ion migration in glass. *IEEE J. Quantum Electron.*, 13, 192–200.

Stewart, G. and Laybourn, P. J. R. 1978. Fabrication of ion-exchanged optical waveguides from dilute silver nitrate melts. *IEEE J. Quantum Electron.*, 14, 930–934.

Takahashi, H., Hibino, Y., Ohmori, Y., and Kawachi, M. 1993. Polarization-insensitive arrayed-waveguide wavelength multiplexer with birefringence compensating film. *IEEE Photonics Technol. Lett.*, 5, 707–709.

Tervonen, A. 1990. A general model for fabrication processes of channel waveguides by ion exchange. *J. Appl. Phys.*, 67, 2746–2752.

Tervonen, A. 1992. Theoretical analysis of ion-exchanged glass waveguides. In *Introduction to Glass Integrated Optics*, S. I. Najafi, ed., Artech House, Norwood, MA.

Tervonen, A., Honkanen, S., and Leppihalme, M. 1987. Control of ion-exchanged waveguide profiles with Ag thin-film sources. *J. Appl. Phys.*, 62, 759–763.

Tien, P. K. and Ulrich, R. 1970. Theory of prism-film coupler and thin-film light guides. *J. Opt. Soc. Am.*, 60, 1325–1327.

Tsoneb, L. and Savatinova, I. 1988. Investigation of refractive-index profiles of ion-exchanged planar waveguides. *J. Mod. Opt.*, 35, 919–924.

Veasey, D. L., Funk, D. S., Peters, P. M., Sanford, N. A., Obarski, G. E., Fontaine, N., Young, M., Peskin, W. C., Houde-Walter, S. N., and Hayden, J. S. 2000. Yb/Er-codoped and Yb-doped waveguide lasers in phosphate glass. *J. Non-Cryst. Solids.*, 263, 369–381.

Viljanen, J. and Leppihalme, M. 1980. Fabrication of optical strip waveguides with nearly circular cross section by silver ion migration technique. *J. Appl. Phys.*, 51, 3562–3565.

Wang, W. J., Honkanen, S., Najafi, S. I., and Tervonen, A. 1992. New integrated optical ring resonator in glass. *Electron. Lett.*, 28, 1967–1968.

Wang, W. J., Honkanen, S., Najafi, S. I., and Tervonen, A. 1993. Loss characteristics of potassium and silver double-ion-exchanged glass waveguides. *J. Appl. Phys.*, 51, 1529–1533.

Walker, R. G., Wilkinson, C. D. W., and Wilkinson, J. A. H. 1983. Integrated optical waveguiding structures made by silver ion-exchange in glass 1: the propagation characteristics of stripe ion-exchanged waveguides; a theoretical and experimental investigation. *Appl. Opt.*, 22, 1923–1928.

West, B. R., Modeling and Analysis of Ion-Exchanged Photonic Devices, PhD dissertation. The University of Arizona, 2004.

West, B. R. and Honkanen, S. 2005. Determination of ion exchange parameters by a genetic algorithm. *Opt. Eng.*, 44, 094602.

West, B. R., Madasamy, P., Peyghambarian, N., and Honkanen, S., 2003. Proximity Effects in Ion-exchanged Glass Waveguides: Numerical Modeling, Paper ThGG4, Presented at OSA Annual Meeting, Tucson AZ.

West, B. R., Madasamy, P., Peyghambarian, N., and Honkanen, S. 2004. Modeling of ion-exchanged glass waveguide structures. *J. Non-Cryst. Solids.*, 347, 18–26.

Wundke, K., Auxier, J.M., Schültzgen, A., Peyghambarian, N.F., and Borrelli, N.F. 1999. Room-temperature gain at 1.3 μm in PbS-doped glasses. *Appl. Phys. Lett.*, 75, 3060–3062.

Wundke, K., Pötting, S., Auxier, J. M., Schültzgen, A., Peyghambarian, N. F., and Borrelli, N. F. 2000. PbS quantum-dot-doped glasses for ultrashort-pulse generation. *Appl. Phys. Lett.*, 76, 10–12.

Yip, G. L. and Albert, J. 1985. Characterization of planar optical waveguides by K^+-ion exchange in glass. *Opt. Lett.*, 10, 151–153.

Yip, G. L., Noutsios, P. C., and Kishioka, K. 1990. Characteristics of optical waveguides made by electric-field-assisted K^+-ion exchange. *Opt. Lett.*, 15, 789–791.

Yi-Yan, A., Chan, W. K., Gmitter, T. J., Florez, L. T., Jackel, J. L., Yablonovich, E., Bhat, R., and Harbison, J. P. 1989. Grafted GaAs detectors on lithium niobate and glass optical waveguides. *IEEE Photonics Technol. Lett.*, 15, 379–380.

Yi-Yan, A., Chan, W. K., Nguyen, C. K., Gmitter, E., Bhat, R., and Jackel, J. L. 1991. GaInAs/InP pin photodetectors integrated with glass waveguides. *Electron. Lett.*, 27, 87–89.

Yi-Yan, A., Chan, W. K., Ravi, T. S., Gmitter, E., Bhat, R., and Yoo, K. H. 1992. Grafted InGaAsP light emitting diodes on glass channel waveguides. *Electron. Lett.*, 28, 341–342.

Yliniemi, S., West, B. R., and Honkanen, S. 2005. Ion-exchanged glass waveguides with low birefringence for a broad range of waveguide widths. *Appl. Opt.*, 44, 3358–3363.

Yliniemi, S., Albert, J., Wang, Q., and Honkanen, S. 2006a. UV-exposed Bragg gratings for laser applications in silver-sodium ion-exchanged phosphate glass waveguides. *Opt. Express.*, 14, 2898–2903.

Yliniemi, S., Albert, J., Laronche, A., Castro, J. M., Geraghty, D., and Honkanen, S. 2006b. Negligible birefringence in dual-mode ion-exchanged glass waveguide gratings. *Appl. Opt.*, 45, 6602–6606.

Zhou, Z. G. and Liu, D. 2005. Semicircular glass waveguide dependence on exchange-time and window width. *Opt. Commun.*, 247, 313–317.

Zou, J. and Chen, R. T. 2006. Improvement of two-step K^+–Na^+ and Ag^+–Na^+ ion-exchanged glass waveguides by surface removal. *Appl. Phys. Lett.*, 88, 081102.

14

Design Methodology for Guided-Wave Photonic Devices

G.L. Yip

14.1 Introduction

Since the early 1970s, there has been an explosion of research activities in realizing planar integrated waveguide devices fabricated by the technologies of ion-exchange in glass (Izawa and Nakagome 1972; Giallorenzi et al. 1973), Ti diffusion into $LiNbO_3$ crystals (Schmidt and Kaminow 1974) and proton exchange in $LiNbO_3$ substrates (Jackel et al. 1982). Early efforts involving diffusion processes were directed toward conceptualization and experimental realization of such devices, which have great signal processing capabilities in optical fiber communication and sensor applications. Some of these capabilities include power division, wavelength division multiplexing/demultiplexing, switching, modulation, polarization splitting, and so on. However, to obtain high performance in these devices, it is important, first, to develop accurate designs. For this purpose, detailed and accurate information on the characteristics of slab and channel guides in respective substrates must be known in relation to their fabrication conditions because slabs and channels often form the basic units in the more complicated structures of waveguide devices. Therefore, "design methodology" does not merely involve a mathematical or computational process. It is more important, first, to establish an accurate refractive-index model for the waveguide, fabricated from the results of experimental characterization, before mathematical and computer-aided design procedures can become effective. It goes without saying, however, that clever mathematics and efficient, accurate numerical algorithms would help provide accurate designs.

In this chapter, we will first explain in Section 14.2 the popularly used effective-index method (EIM) for analyzing waveguide properties and then start with the characterizations of slab waveguides in Section 14.3 and of channel waveguides in Section 14.4. The characterization of channel guides is tackled via that of the slab guides. Section 14.5 presents the beam propagation method (BPM) as a very useful design tool. All of the materials presented in Section 14.2 through Section 14.5 are, then, used in the illustrative examples discussed in Section 14.6, where both the design and measured device performance parameters are compared and discussed. Finally, we conclude our chapter in Section 14.7. It is hoped that the methodology presented here will help researchers and engineers in university, industry, and government laboratories to achieve design-optimized guided-wave devices.

14.2 The Effective Index Method

14.2.1 Two-Dimensional, Step-Index Waveguides

A simple technique, which can be used to analyze and provide good predictions for the propagation characteristics of a channel guide, is the well-known "Effective Index Method" (EIM) (Krox and Toulios 1970). Assuming a uniform index in the direction of propagation along the z axis, the method reduces a 3-D problem to a 2-D one in the transverse plane. In the step-index case, the solution of the 2-D problem starts by taking the equivalence of the rectangular waveguide to the problem of two slab waveguides, one in the x direction (the depth problem) and the other in the y direction (the lateral problem) as illustrated in Figure 14.1.

Let us consider the asymmetric waveguide structure in Figure 14.1b and define the effective guide index $N = \beta_d/k_0$ for the depth mode, where β_d is the related propagation constant and k_0 the wave number in free space. The effective index N should satisfy the following dispersion relationship (Tamir 1988)

$$Kh = m\pi + \Phi_s\Phi_c, \quad m = 0,1,2,\ldots \tag{14.1}$$

where m is the mode order and Φ_s and Φ_c are the phase shifts related to the total internal reflection of light from the film-substrate and film-cover interfaces, respectively. The propagation constant K and the decay constants γ_s and γ_c in the substrate and cover, respectively, along the x direction are given by

$$K = k_0\sqrt{n_f^2 - N^2},$$
$$\gamma_s = k_0\sqrt{N^2 - n_b^2},$$

and

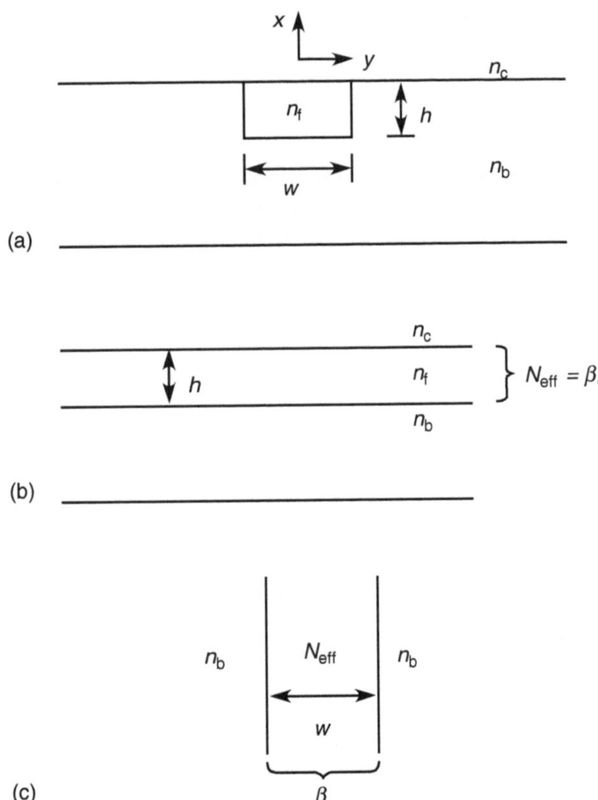

FIGURE 14.1 The Effective Index Method (EIM).

$$\gamma_c = k_0\sqrt{N^2 - n_c^2}. \tag{14.2}$$

Usually, $n_f > N > n_c$.

For TE modes with field components E_y, H_x, and H_z, it is convenient to define a normalized guide index b_E, an asymmetric measure a_E, and a normalized frequency V as follows:

$$\begin{aligned} b_E &= (N^2 - n_b^2)/(n_f^2 - n_b^2), \\ a_E &= (n_b^2 - n_c^2)/(n_f^2 - n_b^2) \end{aligned} \tag{14.3}$$

and

$$V = k_0 h\sqrt{n_f^2 - n_b^2},$$

and the expression for the phase shifts Φ_s and Φ_c are

$$\tan \Phi_s = \gamma_s/K, \tag{14.4}$$

and

$$\tan \Phi_c = \gamma_c/K$$

Using Equation 14.2 through Equation 14.4, the dispersion relationship, Equation 14.1 can be written in the normalized form

$$V\sqrt{(1-b_E)} = m\pi + \tan^{-1}\sqrt{b_E/(1-b_E)} + \tan^{-1}\sqrt{(b_E + a_E)/(1-b_E)}. \quad (14.5)$$

Given the parameters V and a_E, Equation 14.5 can be solved via a root-search technique to yield b_E. For the TM modes with field components E_x, H_y, and E_z, it is necessary to define separately the normalized guide index b_M and the asymmetry measure a_M as (Nishihara et al. 1989)

$$b_M = [(N^2 - n_b^2)/(n_f^2 - n_b^2)][n_f^2/(n_b^2 q_s)], \quad (14.6)$$

where

$$q_s = \frac{N^2}{n_f^2} + \frac{N^2}{n_b^2} - 1$$

and

$$a_M = \left(\frac{n_f}{n_c}\right)^4 \frac{(n_b^2 - n_c^2)}{(n_f^2 - n_b^2)} \quad (14.7)$$

Using Equation 14.6 and Equation 14.7, the corresponding dispersion relationship for the TM modes in the normalized form is

$$V\left[\sqrt{q_s}\frac{n_f}{n_b}\right]\sqrt{1-b_M} = m\pi + \tan^{-1}\sqrt{\frac{b_M}{1-b_M}} + \tan^{-1}\sqrt{\frac{b_M + a_M(1-b_M d_f)}{1-b_M}}, \quad (14.8)$$

$$m = 0,1,2,$$

where

$$d_f = \left(1 - \frac{n_b^2}{n_f^2}\right)\left(1 - \frac{n_c^2}{n_f^2}\right)$$

Now, let us consider the lateral problem as depicted in Figure 14.1. If we have examined the TE mode (E_y, H_x, and H_z) in the above depth problem, the guided mode of interest in the lateral problem should be regarded as the TM mode because E_y is polarized along the y direction and, hence, perpendicular to the waveguide boundaries. The EIM method uses the propagation characteristics of the equivalent slab guide in Figure 14.1c to yield those of the original channel guide in the y direction. This can be achieved by simply substituting the effective index $N_{eff}=N$ and the film and substrate indices, n_f and n_b in Equation 14.8, the dispersion equation for the TM mode.

14.2.2 Two-Dimensional, Graded-Index Waveguides

The method outlined above can be extended to analyze 3-D, graded-index channel guides, whose index variations are confined to the transverse directions x and y. The index in the direction of the propagation z is taken as uniform.

The procedure of analysis is as indicated in Figure 14.2a–e. Because the calculation of the effective index distribution and the propagation constant of the channel guide necessitates the solutions of some wave equations, the pertinent wave equations are given below, Applying Maxwell's equations to an inhomogeneous medium with a refractive index given by $n(x,y)$, we can get the following vector wave

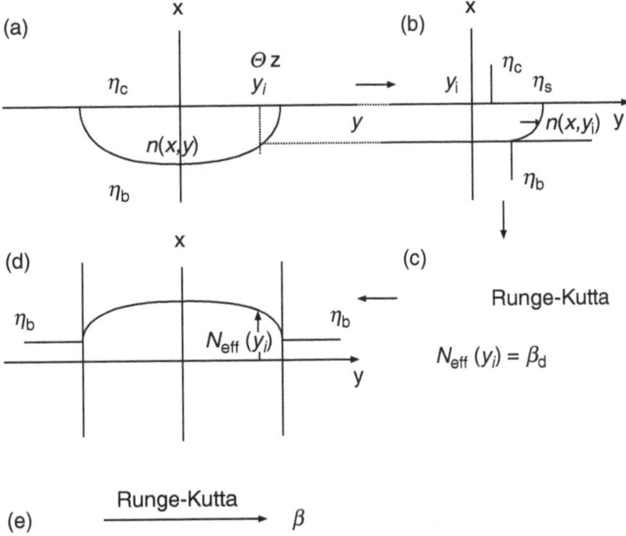

FIGURE 14.2 Procedure analysis.

equations for the electric field **E** and magnetic field **H**:

$$\nabla^2\mathbf{E} + \left[\frac{\mathbf{E}\cdot\nabla n^2}{n^2}\right] + k_0^2 n^2\mathbf{E} = 0 \tag{14.9}$$

and

$$\nabla^2\mathbf{H} + \left[\frac{\nabla n^2}{n^2}\times(\nabla\times\mathbf{H})\right] + k_0^2 n^2\mathbf{H} = 0. \tag{14.10}$$

With a uniform index in the z direction, vector wave equations just for the transverse field components can be separated from Equation 14.9 and Equation 14.10 as

$$\nabla_t^2\mathbf{E}_t + \nabla_t\left\{\mathbf{E}_t\cdot\left[\frac{\nabla_t n^2(x,y)}{n^2(x,y)}\right]\right\} + \left[k_0^2 n^2(x,y) - \beta^2\right]\mathbf{E}_t = 0 \tag{14.11}$$

and

$$\nabla_t^2\mathbf{H}_t + \frac{\nabla_t n^2(x,y)}{n^2(x,y)}\times(\nabla_t\times H_t) + \left[k_0^2 n^2(x,y) - \beta^2\right]H_t = 0, \tag{14.12}$$

where it is assumed that the traveling wave term is, of the form exp $j(\omega t-\beta z)$ and

$$\nabla_t^2 = \frac{\partial^2}{\partial x^2} + \frac{\partial^2}{\partial y^2}.$$

For diffused waveguides, whose index variations in the transverse directions (x, y) are slow, the term $\nabla_t n^2(x,y)$ in Equation 14.11 and Equation 14.12 can sometimes be neglected. Under this so-called scalar wave approximation, scalar wave equations for the transverse field components are

$$\nabla_t^2 \Psi + [k_0^2 n^2(x,y) - \beta^2]\Psi = 0, \tag{14.13}$$

where $\Psi = E_y(x,y)$ for the TE modes and $\Psi = H_y(x,y)$ for the TM modes. Using the separation of variables, we look for the solutions of the form $\Psi(x, y) = F(x,y) \cdot G(y)$. Substituting this in Equation 14.13 and assuming that the variations of $F(x,y)$ with y can be neglected (which has been numerically verified with practical problems), $F(x, y)$ and $G(y)$ can be shown to satisfy the following equations respectively:

$$\frac{\partial^2 F}{\partial x^2} + k_0^2[n^2(x,y) - N_{\text{eff}}^2(y)]F = 0 \tag{14.14}$$

for the depth problem, and

$$\frac{\partial^2 G}{\partial y^2} + [k_0^2 N_{\text{eff}}^2(y) - \beta^2]G = 0 \tag{14.15}$$

for the lateral problem.

Starting from the channel guide with a given index distribution $n(x,y)$ as shown in Figure 14.2a, we fix a specific position at y_i along the y-direction and calculate the local effective index $N_{\text{eff}}(x,y_i)$ ($=\beta_d$) for the depth mode by solving Equation 14.14. This is equivalent to solving the wave equation for a one-dimensional slab guide in the y direction. If we carry out this process for a sufficient number of positions y_i's, we will be able to construct a lateral effective index profile $N_{\text{eff}}(x,y)$ as shown in Figure 14.2d for the lateral problem in order to get β, the propagation constant for the original channel guide. This second step is again equivalent to solving the wave equation for a slab guide in the x direction.

For inhomogeneous stab guides occurring in practical situations, exact analytical field solutions are rare. We have to resort to numerical methods. Among an almost innumerable number of numerical methods, our laboratory has had experience with two very accurate and reliable techniques, namely the Raleigh–Ritz variational procedure and the Runge–Kutta method. We will encounter them in Section 14.4.

14.3 Experimental Characterization and Theoretical Modeling of Slab Waveguides

14.3.1 Waveguides by Ion Exchange in Glass

Due to space limitations, the characterization of slab waveguides by K^+-ion exchange in soda-lime glass substrates is used for illustration. For $AgNO_3$ melt, Stewart et al. (1977) found a linear relationship between d and $t^{1/2}$ given by

$$d = \sqrt{D_e t} \tag{14.16}$$

where d is the diffusion depth, t the diffusion time in seconds and D_e has been defined as the effective diffusion constant. Further, they also found that the dependence of D_e on the inverse temperature $1/T$ was given by the exponential relationship

$$D_e = C_1 \exp(-C_2/T)\ \text{m}^2\ \text{s}^{-1}. \tag{14.17}$$

Hence, from Equation 14.16 and Equation 14.17

$$d = (60C_1)^{1/2} \times 10^6 t^{1/2} \exp(-C_2/2T)\ \mu\text{m} \tag{14.18}$$

where t is in minutes, T in degrees Kelvin, and C_1 and C_2 are constants. It was intuitively thought (Yip and Albert 1985) that, for ion-exchanged waveguides employing soda-lime glass substrates and a potassium nitrate (KNO_3) melt, relationships similar to Equation 14.16 through Equation 14.18 should hold. Our task was, therefore, to determine C_1 and C_2 experimentally for KNO_3. The importance of establishing such relationships for KNO_3 is that, given (T,t) of such a waveguide, one can establish the waveguide's important parameters quickly without the need for further measurement.

To fabricate the diffused planar waveguides, the soda-lime glass microscope slides were immersed in molten KNO_3 at different temperatures and for different durations. The melt temperatures ranged from 360 to 440°C, and the diffusion times ranged from 30 min to 24 h. The effective indices of the guided modes in each sample were obtained by measuring the synchronous angles with a prism coupler (flint glass, $n_p = 1.785$) (Tien and Ulrich 1970). An accuracy of $\pm 1'$ was obtained in the measured synchronous angle θ_i, corresponding to an accuracy of $\pm 1 \times 10^{-4}$ in the measured effective index n_e. The measurements were initially performed with a He–Ne laser at 0.63 μm.

Assuming a gaussian index distribution in the diffused waveguide

$$n(x) = n_b + \Delta n_s \exp(-x^2), \tag{14.19}$$

where $\Delta n_s = n_s - n_b$, $x = x/d$, n_b is the substrate index n_s is the surface index and d the effective guide depth such that $n(d) = n_b + \Delta n_s/e$, the well-known WKB dispersion relationship for guided modes [Equation 2.1 of Stewart et al. 1977] was used. This relationship can be expressed as

$$d = \frac{(m + 3/4)\pi}{k_0 \int_0^{x_t} [n^2(x) - N_e^2]^{1/2} dx}, \quad m = 0,1,2,\ldots \tag{14.20}$$

where x_t is the normalized turning point and N_e is the effective mode index. The phase shift on reflection at the glass–air interface has been taken to be $-\pi$, here, for both TE and TM modes. Equation 14.20 contains two unknowns, n_s and d. Hence, given any pair of measured mode indices for a waveguide, one can eliminate d by using Equation 14.20 and determine n_e by a root-search technique in the resulting equation. Then, n_s can be substituted in Equation 14.20 to get d. This procedure can be repeatedly applied to all the possible pairs of the mode indices for a specific waveguide sample, and the average values of its n_s and d can be determined.

Our measurement data has confirmed the previous observation that the surface index value n_s is affected only by the diffusion temperature and is independent of the diffusion time (Chartier et al. 1978). In Figure 14.3, the theoretical dispersion curves computed from Equation 14.20 for the TE modes in the samples prepared at 400°C are presented together with experimentally measured data. The agreement seems good. The difference between the measured mode index and the theoretical index was calculated

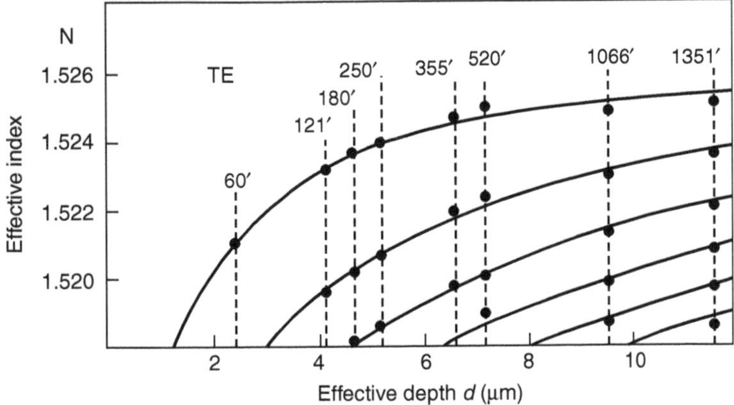

FIGURE 14.3 Theoretical dispersion curves. *SPIE* (*Optics Letters*). (From Yip, G. L. and Albert, J. 1985. Characterization of planar optical waveguides by K^+-ion exchange in glass. *Opt. Lett.*, 10, 151–153. With permission.)

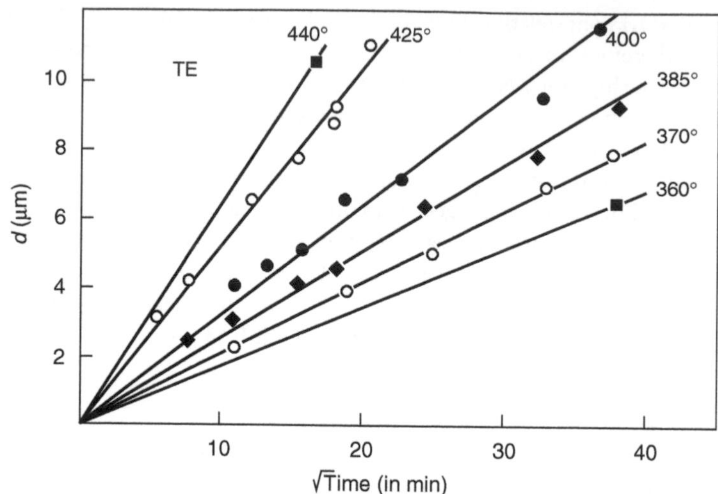

FIGURE 14.4 Effective guide depth. *SPIE* (*Optics Letters*). (From Yip, G. L. and Albert, J. 1985. Characterization of planar optical waveguides by K^+-ion exchange in glass. *Opt. Lett.*, 10, 151–153. With permission.)

for each of the 167 measured mode indices, and the average of this difference was found to be $|n_e$ (meas.) $- n_e$(theo.)$|$av. $= (1.6 \pm 1.2) \times 10^{-4}$ with the largest single deviation being 6×10^{-4}. Similar, though not identical, results for the TM modes are not presented here for brevity.

The measured data of d versus $t^{1/2}$ are presented in Figure 14.4 for the TE modes also showing a linear relationship between d and $t^{1/2}$ for KNO_3 as for $AgNO_3$ (Stewart et al. 1977). Table 14.1 summarizes the results of our measurements and values of D_e obtained from Equation 14.16 and Figure 14.4. The results in Table 14.1 were, then, used to plot $\log_e D_e$ versus $(1/T)$ in Figure 14.5, where, with the help of a linear-regression method, $C_1^{TE} = 7.82 \times 10^{-6}$ m^2/s, $C_2^{TE} = 1.489 \times 10^4$ K and $C_2^{TM} = 1.604 \times 10^{-5}$ m^2/s, $C_2^{TM} = C_2^{TM} = 1.604 \times 10^{-5}$ m^2/s, $C_2^{TM} = 1.54 \times 10^4$ K respectively, were determined. To characterize a slab waveguide, with (T,t) given, its effective guide depth can be obtained by using Equation 14.18). The value of n_s can be obtained by referring to Table 14.1 using interpolation, if necessary. Anisotropy in ion-exchanged wave guides, as suggested by the differences in Δn_s^{TE} and Δn_s^{TM} was also observed previously (Izawa and Nakagome 1972).

Because the practical significance of the ion-exchanged waveguide devices now becomes more and more evident in single-mode wideband fiber communication systems operating at 1.31 and 1.55 µm, the diffused slab waveguides made by the K^+-ion exchange were also characterized experimentally at these waveguides by the same prism-coupler technique. Figure 14.6 shows the measurement setup, where a He–Ne laser beam at 0.63 µm was used to help with alignment because the He–Ne laser beam is invisible at 1.152 and 1.523 µm.

TABLE 14.1 Measured Surface Index Change

Temp (°C)	Δn_s^{TE} (×10^{-3})	Δn_s^{TM} (×10^{-3})	D_e^{TE} (m^2/s) (×10^{-16})	D_e^{TM} (m^2/s) (×10^{-16})
360	9.2±0.3	11.3±0.3	4.91	4.53
370	9.1±0.5	10.7±0.4	7.18	6.50
385	8.8±0.2	10.8±0.4	10.82	10.61
400	8.7±0.3	10.5±0.5	17.14	18.02
425	8.7±0.3	10.4±0.4	45.21	42.16
440	8.4±0.5	10.0±0.6	66.78	68.81

[a] All of the readings must be multiplied by the factor indicated to yield the actual values.
SPIE (*Optics Letters*).
Source: From Yip, G. L. and Albert, J. 1985. Characterization of planar optical waveguides by K^+-ion exchange in glass. *Opt. Lett.*, 10,151–153. With permission.

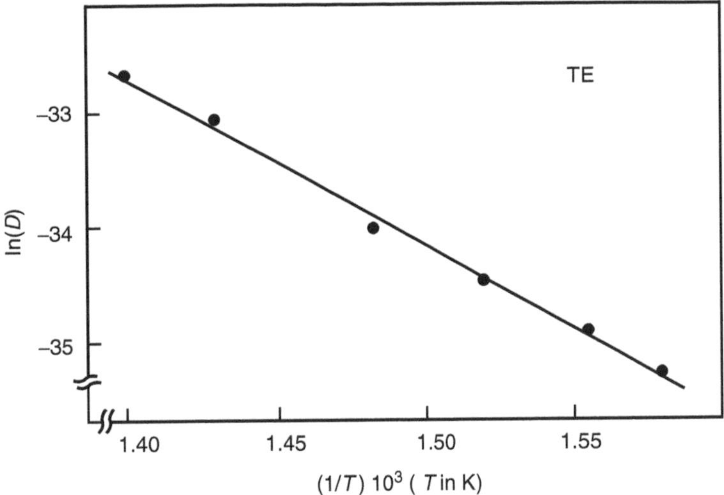

FIGURE 14.5 The relation between $\log_e$. *SPIE* (*Optics Letters*). (From Yip, G. L. and Albert, J. 1985. Characterization of planar optical waveguides by K^+-ion exchange in glass. *Opt. Lett.*, 10, 151–153. With permission.)

The refractive indices of the prism coupler (flint glass), at 1.152 and 1.523 μm, were found by interpolating, from data in Optics Guide 4 (1988), to be $n_p = 1.7523$ and 1.74485, respectively The corresponding indices for the substrate glass $n_b = 1.5030 \pm 1 \times 10^{-4}$ and $1.4984 \pm 1 \times 10^{-4}$, respectively, were determined by employing the multilayer Brewster angle measurement method (Xiang and Yip 1992). Because a He–Ne laser at 1.31 μm is unavailable, the data at this wavelength cannot be obtained experimentally but can be calculated by interpolation of the data at 1.152 and 1.523 μm.

Some of our measured results are summarized in Table 14.2 and Table 14.3, and results for other diffusion times show that the gaussian profile yields the best fit to the measured mode indices, whereas the exponential profile fits worst.

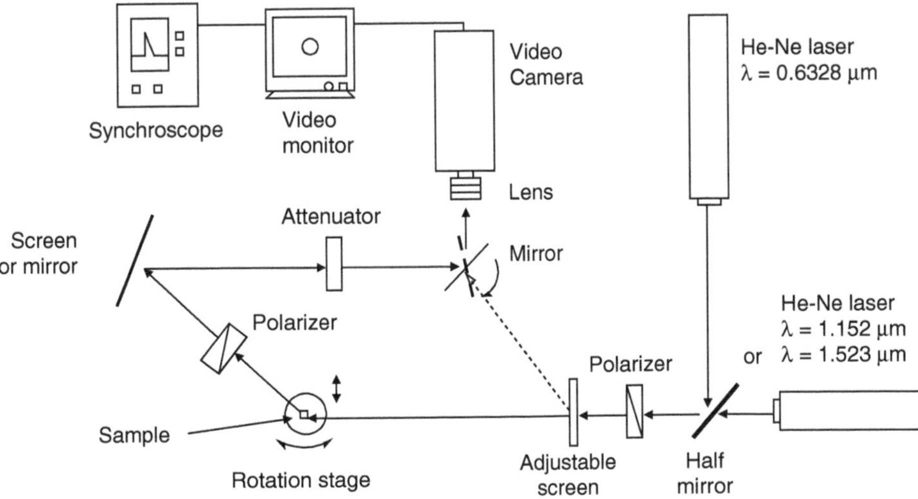

FIGURE 14.6 The measurement set-up. (From Yip, G. L. et al., 1991. Characterization of planar optical waveguides by K^+-ion exchange in glass at 1.152 and 1.523 μm. *SPIE Proceeedings of the Integrated Optical Circuits*, K. K. Wong, ed., *Vol. 1583*, 3–4, September, pp. 14–18, Boston, MA. With permission.)

TABLE 14.2 Measured Surface Index Change

Δ (μm)	Δn_s^{TE} (×10^{-3})	Δn_s^{TM} (×10^{-3})	D_e^{TE} (m^2/s) (×10^{-16})	D_e^{TM} (m^2/s) (×10^{-16})
1.152	8.56±0. 11	10.05±0.36	10.33	10.04
1.523	8.57±0.26	10.34±0.24	11.53	11.27

Source: From Yip, G. L. et al., 1991. Characterization of planar optical waveguides by K^+-ion exchange in glass at 1.152 and 1.523 μm. *SPIE Proc. Integrated Optical Circuits*, K. K. Wong, ed., *Vol. 1583*, 3–4, September, pp. 14–18, Boston, MA. With permission.

Characterization of ion-exchanged waveguides with diluted KNO_3 was also reported by Kishioka (1991).

14.3.2 Waveguides by Field-Assisted Ion Exchange in Glass

Field-assisted ion exchange has the advantage of being a much faster process capable of producing various index profiles (Chartier et al. 1978), including the step-index profile (Izawa and Nagakome 1972; Abou-el-Liel and Leonberger 1988). It is also a simple and attractive technique in producing deep buried waveguides through a two-step process (Chartier et al. 1980), important for index-profile tailoring to achieve efficient coupling with optical fibers. Figure 14.7 gives the schematic of a typical experimental setup (Yip et al. 1990).

For Ag^+ ions, it was found (Ramaswamy and Najafi 1986) that a linear combination of the diffusive and electromigrative terms was more appropriate for expressing the guide depth. For K^+ ions, a similar approach can be adopted. Hence,

$$d = \sqrt{D_e t} + F_e t + K_0 \tag{14.21}$$

where F_e is defined as the effective field-dependent coefficient and K_0 is a constant. The electromigrative coefficient F_e is linearly proportional to the total applied field E_a^t so that

$$F_e = \mu_k E_a^t \tag{14.22}$$

where μ_k is a constant defined as the K^+ ion mobility in glass. Relationships in Equation 14.21 and Equation 14.22 again yield the guide depth, given the fabrication conditions (E_a^t,t).

For fabrication details of planar waveguides by field-assisted ion exchange in our scheme, the readers are referred to (Yip et al. 1990). To avoid cracking the glass substrates immediately after the ion-exchange process, the samples were kept in the furnace for an additional ten minutes before being exposed for

TABLE 14.3 Comparisons of Measured TM-Mode Indices

		N (Theor.)			Error (×10^{-4})[a]		
m	*N* (Meas.)	Gaussian[b]	exp.[c]	erfc[d]	Δ[a]	Δ[b]	Δ[c]
0	1.510577	1.510490	1.512143	1.511102	0.87	15.66	5.25
1	1.507414	1.50746	1.507863	1.507611	0.43	4.49	1.97
2	1.505037	1.505152	1.505395	1.505230	0.15	3.58	1.93
3	1.503524	1.503417	1.503949	1.503675	1.07	4.25	1.51

[a] $\Delta = |N\ (\text{Meas.}) - N\ (\text{Theor.})\ |$.
[b] $n_s = 1.5116$, $d = 11.4$ μm.
[c] $n_s = 1.5198$, $d = 7.27$ μm.
[d] $n_s = 1.5161$, $d = 14.85$ μm.

Source: From Yip, G. L. et al., 1991. Characterization of planar optical waveguides by K^+-ion exchange in glass at 1.152 and 1.523 [mu]m. *SPIE Proc. Integrated Optical Circuits*, K. K. Wong, ed. *Vol. 1583*, 3–4, Sept., pp. 14–18, Boston, MA. With permission.)

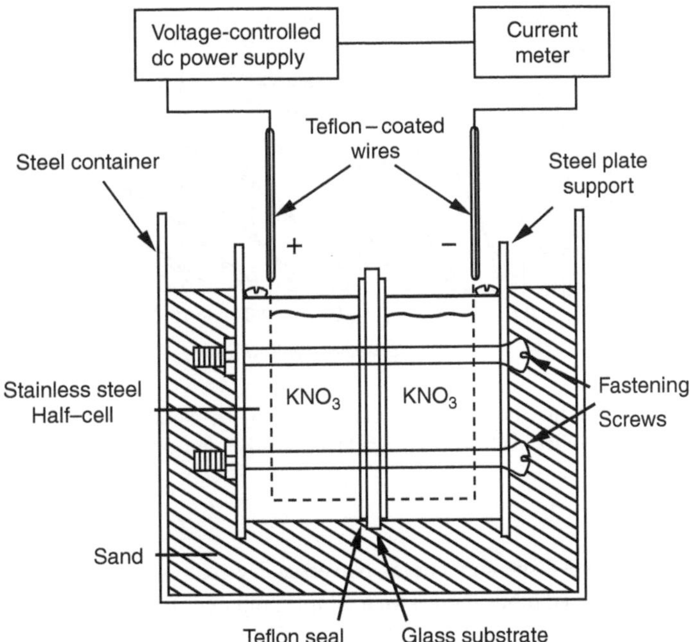

FIGURE 14.7 Schematic of the experimental apparatus. (From Yip, G. L., Noutsios, P. C., and Kishioka, K. 1990. Characteristics of optical waveguides made by electric-field-assisted K^+-ion exchange. *Opt. Lett.*, 15, 789–791. With permission.)

cooling in air. This extra time contributes to the K_0 value in Equation 14.21. Another practical point to note is the observed existence, by several research groups, of a battery effect manifested as a potential drop across the electrodes even after the removal of the applied voltage. This potential drop (1.3 V in our case) can be represented by a constant field E_0 between the two electrodes so that $E_a^t = E_a - E_0$, where E_a refers to the applied field.

Like the case of purely thermal diffusion, the surface index n_s, here, again depends only on the temperature and is independent of the diffusion time. The refractive index (concentration) profile of constant, applied-current (Abou-el-Liel and Leonberger 1988) or electric-field-assisted (Miliou et al. 1989) K^+-ion exchanged planar waveguides has been found to be step like, and, hence, theoretical dispersion curves can be plotted with this step model. Although this step profile may be suitable for the case of either high current or electric field, it may not yield accurate results in our experiments using lower fields. For the more general case, we have proposed a modified Fermi (MF) index profile (Chen and Yang 1985), which was previously used to model the Tl^+-ion-exchanged glass waveguides, namely,

$$n(x) = n_b + \Delta n_s \left\{ 1 - \exp\left(-\frac{d}{a}\right) + \exp\left(\frac{x-d}{a}\right) \right\}^{-1} \tag{14.23}$$

where a is a fitting parameter associated with the profile shape. The effective guide depth is defined as $n(d) = n_b + \Delta n_s[2 - \exp(-d/a)]$. The value of d/a in our experiments varies from 4 to 23 so that the effective guide depth occurs at almost the half-point of the profile.

With a set of fabrication conditions $a(T, E_a^t, t)$ and using Equation 14.20 and Equation 14.23, given any pair of measured mode indices of a waveguide sample, one can determine the set of parameters $(\Delta n_s, d, a)$ by a root-search technique similar to the way just discussed in the previous section. This procedure is normally repeated for all the possible pairs of the mode indices for a specific waveguide sample, and a is chosen so as to yield the minimum deviations of Δn_s and d from their average values.

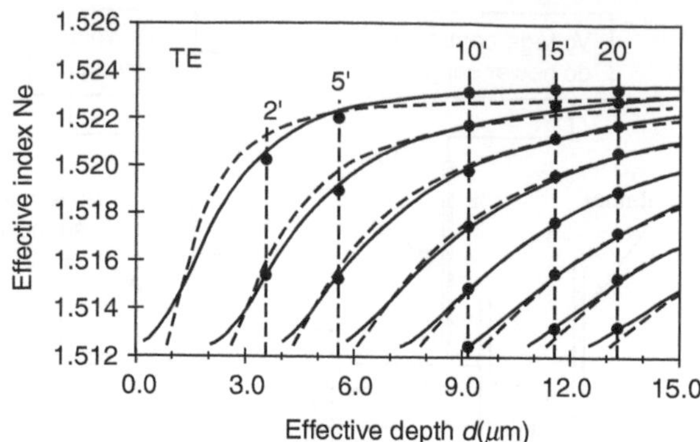

FIGURE 14.8 Comparison of the theoretical TE. (From Yip, G. L., Noutsios, P. C., and Kishioka, K. 1990. Characteristics of optical waveguides made by electric-field-assisted K^+-ion exchange. *Opt. Lett.*, 15, 789–791. With permission.)

In Figure 14.8, the TE mode dispersion curves, N_e versus d, were plotted, using a step-index profile, in dashed curves with the measured mode indices for a sample prepared at $T=385°C$ and $E_a=21.1$ V/mm. In comparison, the solid dispersion curves, using the MF profile and the WKB dispersion relationship yield better agreement with the measured data. In Figure 14.9, the plots of d versus t clearly established a linear relationship between d and t for all of the applied fields. By using a least-squares fit, values of F_e and K_0 can be obtained for each E_a. In Figure 14.10, values of F_e are plotted against E_a^t, where the use of a linear regression yields the constant $\mu_k(\text{TE})=21.33\ \mu\text{m}^2/\text{V}$ min and $\mu_k(\text{TM})=20.52\ \mu\text{m}^2/\text{V}$ min. The relevant coefficients and index changes determined at different applied fields are summarized in Table 14.4 and Table 14.5. These results are believed to cover most of the practical conditions under which passive devices can be fabricated at $T=385°C$. In Table 14.4, each value of a entered is the average over all the values of a obtained through optimization for all the samples prepared under each specified field.

14.3.3 $LiNbO_3$ Waveguides by Ti Diffusion and Proton Exchange

Similarly, although more complex in detail, $LiNbO_3$ waveguides by Ti diffusion and proton exchange (P.E.) can also be characterized. For brevity, only Ti-diffused waveguides are discussed here.

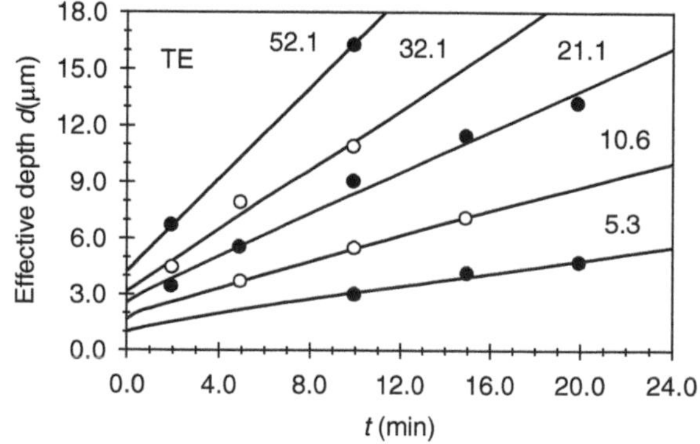

FIGURE 14.9 Effective guide depths versus. (From Yip, G. L., Noutsios, P. C., and Kishioka, K. 1990. Characteristics of optical waveguides made by electric-field-assisted K^+-ion exchange. *Opt. Lett.*, 15, 789–791. With permission.)

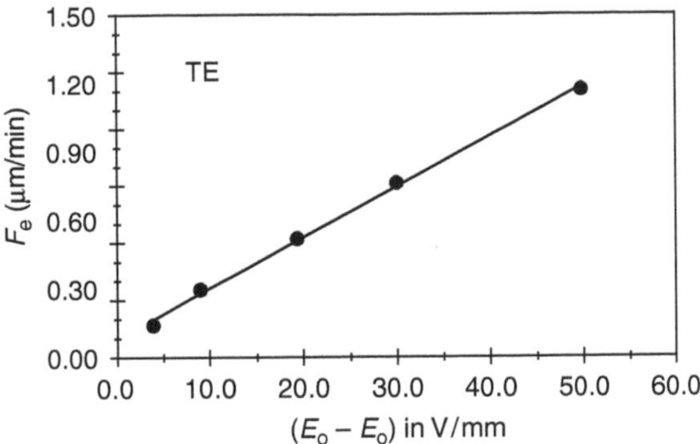

FIGURE 14.10 Variation of coefficient F. (From Yip, G. L., Noutsios, P. C., and Kishioka, K. 1990. Characteristics of optical waveguides made by electric-field-assisted K^+-ion exchange. *Opt. Lett.*, 15, 789–791. With permission.)

For lengthy diffusion times compared to the time required for the titanium film to completely enter the crystal, an appropriate analytical concentration profile for the diffused Ti ions is given by Schmidt and Kaminow (1974); Burns et al. (1979):

$$c(x,t) = \left(\frac{2}{\sqrt{\pi}}\right)\left(\frac{\alpha\tau}{d_x}\right)\exp\left(-\frac{x^2}{d_x^2}\right), \tag{14.24}$$

where α is the atomic density of the Ti film and τ is its thickness. d_x is the diffusion depth, given by

$$d_x = 2\sqrt{D_x t} \tag{14.25}$$

and D_x, the diffusion coefficient, is given by

$$D_x = D_0 \exp\left(-\frac{\in_a}{k_B T}\right) \tag{14.26}$$

where D_0 is a constant, $\in_a$ is the thermal activation energy, k_B is the Boltzmann constant, and T is the temperature.

As before, if the index change is assumed to be proportional to the concentration profile,

$$\Delta n(x,t) = Kc(x,t) = \Delta n_s, \exp\left\{-\left(\frac{x}{d}\right)^2\right\}. \tag{14.27}$$

TABLE 14.4 Measured Surface Index at 385°C

E_a (V/mm)	Δn_s(TE) ($\times 10^{-3}$)	Δn_s(TM) ($\times 10^{-3}$)	a(TE) (μm)	a(TM) (μm)
5.3	10.8 ± 0.0	13.7 ± 0.3	0.53	0.64
10.6	11.2 ± 0.2	13.3 ± 0.1	0.69	0.52
21.1	11.1 ± 0.1	13.4 ± 0.2	0.65	0.57
32.1	11.2 ± 0.1	13.6 ± 0.2	0.61	0.60
52.1	11.1 ± 0.0	13.5 ± 0.2	0.78	0.79

Source: From Yip, G. L., Noutsios, P. C., and Kishioka, K. 1990. Characteristics of optical waveguides made by electric-field-assisted K^+-ion exchange. *Opt. Lett.*, 15,789–791. With permission.

TABLE 14.5 Diffusion Coefficients and Depth Parameters for Various Applied Fields at 385°C

E_a (V/mm)	F_e(TE) (μm/min)	F_e(TM) (μm/min)	K_0(TE) (μm)	K_0(TM) (μm)
5.3	0.14	0.14	0.96	0.81
10.6	0.29	0.29	1.74	1.76
21.1	0.51	0.50	2.56	2.72
32.1	0.74	0.76	3.09	3.01
52.1	1.15	1.11	4.08	4.30

Source: From Yip, G. L., Noutsios, P. C., and Kishioka, K. 1990. Characteristics of optical waveguides made by electric-field-assisted K^+-ion exchange. *Opt. Lett.*, 15, 789–791. With permission.

Hence,

$$\Delta n_s = \left\{ \frac{2}{\sqrt{\pi}} \alpha \frac{dn}{dc} \right\} \frac{\tau}{d_x} \tag{14.28a}$$

where the surface index change $\Delta n_s = n_s - n_b$, n_s is the surface index, n_b the bulk index, and $dn_e/dc = 1.6 \times 10^{-23}$ cm^3 for Ti.

The functional dependence of $c(x,t)$, d_x, D, and $\Delta n(x, t)$ on the various parameters as expressed in Equation 14.24 through Equation 14.27 have been verified experimentally by several research groups employing different measurement techniques, such as mode index measurements (Naitoh et al. 1977), the x-ray microanalyzer method (XMA) (Minakata et al. 1978), and the secondary-ion mass spectrometry (SIMS) techniques (Burns et al. 1979). Our interest will be focused on the experimental data obtained via the prism-coupler method as discussed in the previous section. Naitoh et al. (1977) have shown that the experimental linear relationships between the various parameters as plotted in Figure 14.11 through Figure 14.13 qualitatively agree with their relationships expressed in Equation 14.24 through Equation 14.27 for the TE modes propagating in a y-cut Ti:LiNbO$_3$ waveguide. Ctyroky et al.

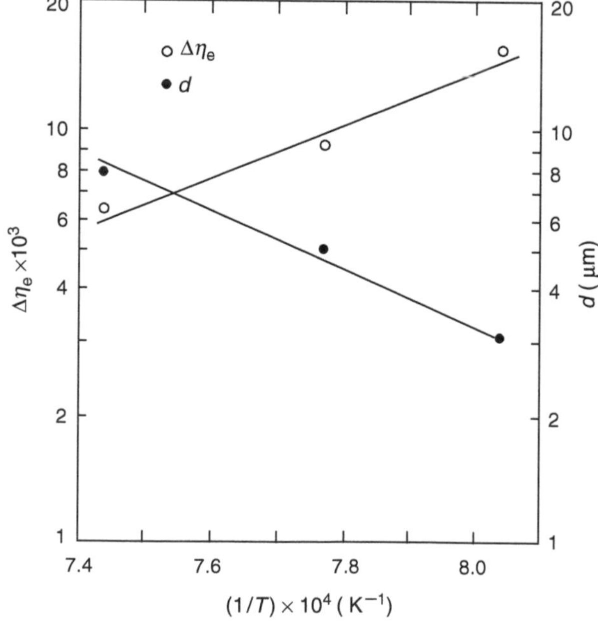

FIGURE 14.11 Variations of the index change. (From Naitoh, H., Nunoshita, M., and Nakayama, T. 1977. Mode control of Ti-diffused LiNbO$_3$ slab optical waveguides. *Appl. Pot.* 16(9), L2546–2549. With permission.)

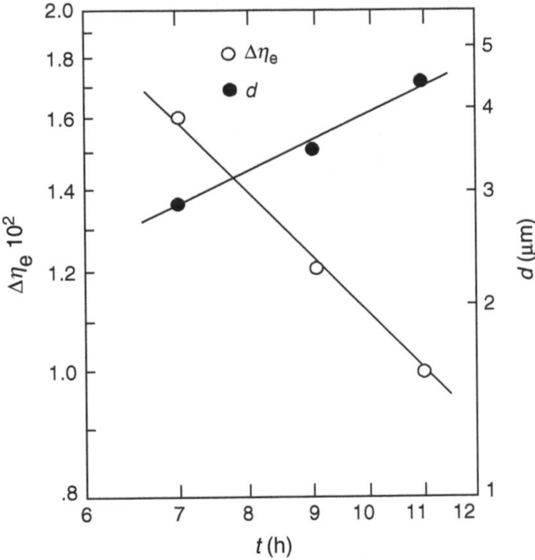

FIGURE 14.12 Variations of the index. (From Naitoh, H., Nunoshita, M., and Nakayama, T. 1977. Mode control of Ti-diffused $LiNbO_3$ slab optical waveguides. *Appl. Pot.* 16(9), L2546–2549. With permission.)

1984 observed, through their measurements of mode indices in *y*-cut and *z*-cut $LiNbO_3$ substrates that no significant differences have been found between these two kinds of substrates.

14.4 Characterizing and Modeling Channel Guides

14.4.1 Ion-Exchanged Channel Waveguides

For a channel guide the index distribution in the depth direction is still modeled by a gaussian distribution. In the lateral direction, a simple way is to use the step-index model based on the EIM, where the channel width is defined by the photomask aperture for the channel guide. This model offers

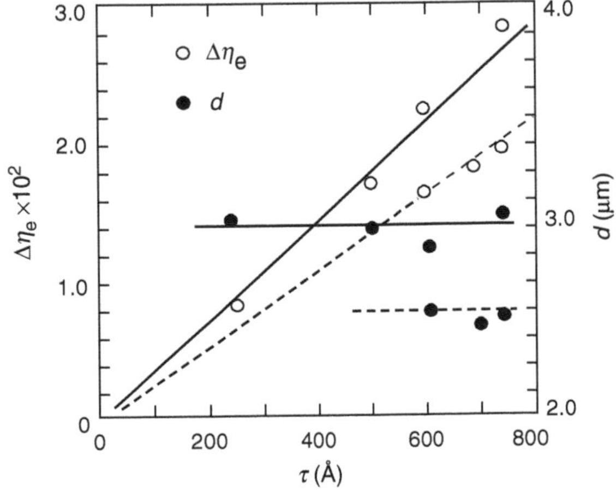

FIGURE 14.13 Variations of the index. (From Naitoh, H., Nunoshita, M., and Nakayama, T. 1977. Mode control of Ti-diffused $LiNbO_3$ slab optical waveguides. *Appl. Pot.* 16(9), L2546–2549. With permission.)

reasonable accuracy as illustrated in the problem of a three-branch power divider to be presented in Section 6.1.

A more accurate model can be constructed as outlined in Section 2.2. To do this, the 2-D refractive index distribution $n(x,y)$ must first be established. The index $n(x,y)$ is assumed to be proportional to the normalized concentration $C(x,y)$ of incoming ions diffused into the glass as indicated in Equation 14.28a and Equation 14.28b, and C satisfies the diffusion Equation 14.29 (Crank 1975):

$$n(x,y) = n_b + \Delta n_s C(x,y), \tag{14.28b}$$

$$\frac{\partial c}{\partial t} = \frac{\partial}{\partial x}\left[\frac{D_1}{1-\alpha c}\frac{\partial C}{\partial x}\right] + \frac{\partial}{\partial y}\left[\frac{D_1}{1-\alpha c}\frac{\partial c}{\partial y}\right], \tag{14.29}$$

with $C = C_1/C_0$ and $\alpha = 1 - D_1/D_0$, where C_1 and D_1 are the concentration and self-diffusion coefficients of the incoming ions and C_0 and D_0 are those of the outgoing ions in the glass prior to the exchange. The values of D_1 for K^+-ions at 0.63 μm, for example, can be obtained from our previous characterizations of the planar K^+-ion exchange in soda-lime glass (Yip and Albert 1985) and at 1.33 and 1.55 μm from Yip et al. (1991). α has been determined to be 0.998 (Doremus 1969).

To increase accuracy and save computer time, a nonuniform grid, finer in the region near (0,0) and coarser away from the origin should be used as shown in Figure 14.14.

As an illustrative example, an explicit three-level, finite-difference scheme was used (Gerald 1970) to solve Equation 14.29 for a two-step ion-exchange process (Yip and Finak 1984) (developed in our laboratory to produce a wide, weakly guiding, single-mode channel 10-μm wide to be compatible for butt-coupling with a single-mode fiber). This process consists of an initial exchange (time t_1) through a finite opening in an aluminum mask, to define a guiding channel, and a second exchange (time t_2), after the removal of the Al mask, over the whole plane of the substrate to adjust the lateral waveguiding properties of the channels by modifying the effective index of the surrounding areas. Figure 14.15 shows typical concentration contours for $T = 385°C$, $t_1 = 17$ min, and $t_1 + t_2 = 1$ h.

The method used to solve Equation 14.14 for this example, was the Raleigh–Ritz variational procedure with Hermite–Gaussian basis functions $\phi_i(x)$ (Taylor 1976; Albert and Yip 1988). The eigenvalues $N_{\text{eff}}(\gamma)$ of Equation 14.14 can be obtained by finding the field $\Psi(x)$, where

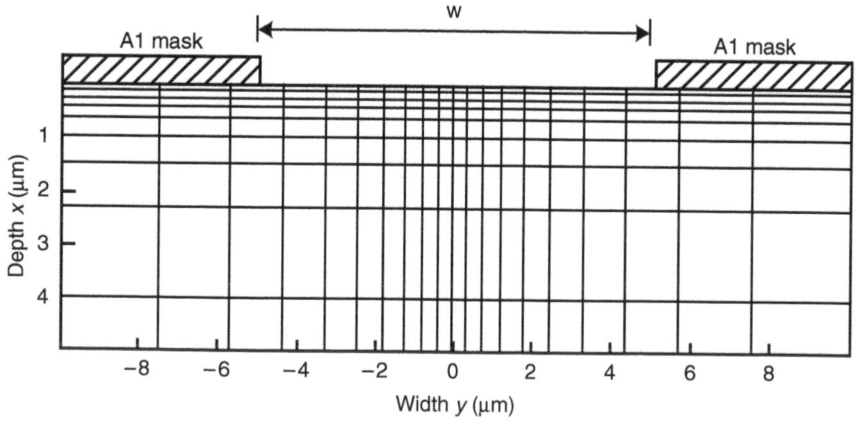

FIGURE 14.14 A nonuniform computation. (From Albert, J. and Yip, G. L. 1988. Wide single-mode channels and directional coupler by two-step ion-exchange in glass. *IEEE J. Lightwave Technol.*, 6, 552–563. With permission.)

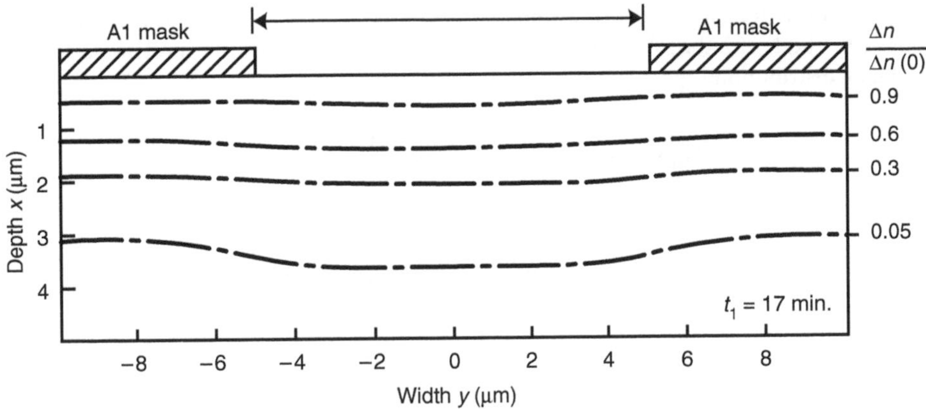

FIGURE 14.15 An example of concentration. (From Albert, J. and Yip, G. L. 1988. Wide single-mode channels and directional coupler by two-step ion-exchange in glass. *IEEE J. Lightwave Technol.*, 6, 552–563. With permission.)

$$\Psi(x) = \sum_{i=1}^{M} C_i \phi_i(x) \tag{14.30}$$

which minimizes the functional,

$$N_{\text{eff}}^2(y) = \max_{\psi} \int_{-\infty}^{\infty} \frac{\mathrm{d}x[n^2(x,y)\Psi^2(\nabla\Psi)^2/k_0^2]}{\int_{-\infty}^{\infty}\mathrm{d}x\Psi^2}. \tag{14.31}$$

It was found that $N_{\text{eff}}(\gamma)$ is correct to one part in 10^5 for $M=21$. Figure 14.16 shows some simple plots for $N_{\text{eff}}(\gamma)$ for $t_1=9$ min and 17 min, t_1+t_2 being kept at 1 h.

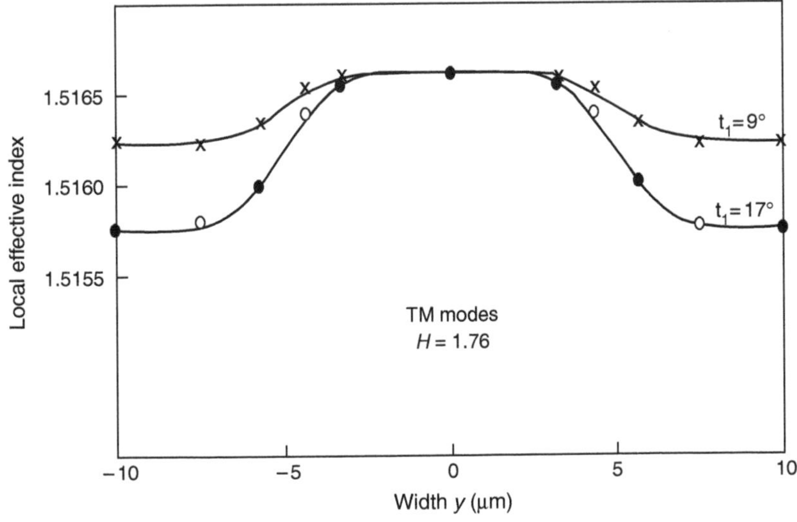

FIGURE 14.16 Local effective indices. (From Albert, J. and Yip, G. L. 1988. Wide single-mode channels and directional coupler by two-step ion-exchange in glass. *IEEE J. Lightwave Technol.*, 6, 552–563. With permission.)

The sets of values of $N_{eff}(\gamma)$ numerically obtained for each sample with a fixed t_1 can be fitted, via a least-squares method, with the following two error functions (erf) to provide a good model for the lateral effective index distribution:

$$N_{eff}(y) = N(\infty) + \frac{[N(0) - N(\infty)]}{2} \times \left[\text{erf} \frac{(y + w/2)}{H} - \text{erf} \frac{(y - w/2)}{H} \right] \quad (14.32)$$

where H is a fitting parameter $N(0) = N_{eff}(y=0)$, $N(\infty) = N_{eff}(y=\infty)$ and w is the aperture width in the photomask. Here, H can be approximated by $H = (D_y t_2)^{1/2}$ with $D_y = D_e$. The good agreement between the fitting function given by Equation 14.32 and the numerical results can be readily seen in Figure 14.16. *The channel guide formed by a single ion-exchange process can be treated as a special case of the two-step process, where t_2, becomes zero.* Thus the combination of Equation 14.32 and $H = (D_e t_2)^{1/2}$ yields a model for the lateral effective index profile of two-step, ion-exchanged waveguides, which depends exclusively on data derived from planar waveguide characterizations (i.e., $N(0)$, $N(\infty)$, and D_e). Of course, for the more general cases, $t_1 + t_2$ will not be kept at one hour, but would vary.

Having established the effective index model $N_{eff}(y)$, the propagation characteristics of the channel guide (i.e., the lateral problem) can be studied by solving Equation 14.15 for β and $G(y)$ by either the WKB method or a single-function, variational method (Sharma et al. 1980) with good agreement between the two. For the latter, we use a single function for the first two modes, namely, a Gaussian for the fundamental mode

$$G_0(y) = e^{-\sigma_0 y^2} \quad (14.33)$$

and the Hermite–Gaussian of order 1 for the second mode,

$$G_1(y) = \sqrt{\sigma_1} y e^{-\sigma_1 y^2}. \quad (14.34)$$

Some dispersion curves of the channel guides are presented in Figure 14.17, where

$$b = \frac{(\beta/k_0)^2 - N^2(\infty)}{N^2(0) - N^2(\infty)},$$

and

$$V = k_0 w \sqrt{N^2(0) - N^2(\infty)} \quad (14.35)$$

In Figure 14.2 in Section 2.2, it was indicated that the numerical Runge-Kutta (R-K) method can be used to calculate the effective index in the depth problem and, hence, to construct the lateral effective-index distribution for a channel guide as shown in Figure 14.2b–d. Finally, it can also be used to calculate the channel guide's propagation constant as shown in Figure 14.2d–e. The R-K method will not be outlined (Ralston and Wilf 1960; Gill 1951) but its use will be illustrated through two specific examples.

The constant concentration contours of two single-channel guides, with a width $w = 6$ μm and $w = 12$ μm, respectively, and produced by a single K^+-ion exchange, are shown in Figure 14.18a and b with their fabrication conditions specified. Using Equation 14.29, the 2-D index distribution $n(x, y)$ can be calculated. Following the procedures indicated in Figure 14.2a and b, $N_{eff}(y_i)$ can be computed at discrete points, as shown in Figure 14.19a and b, by the R-K method for the TM modes at 1.523 μm.

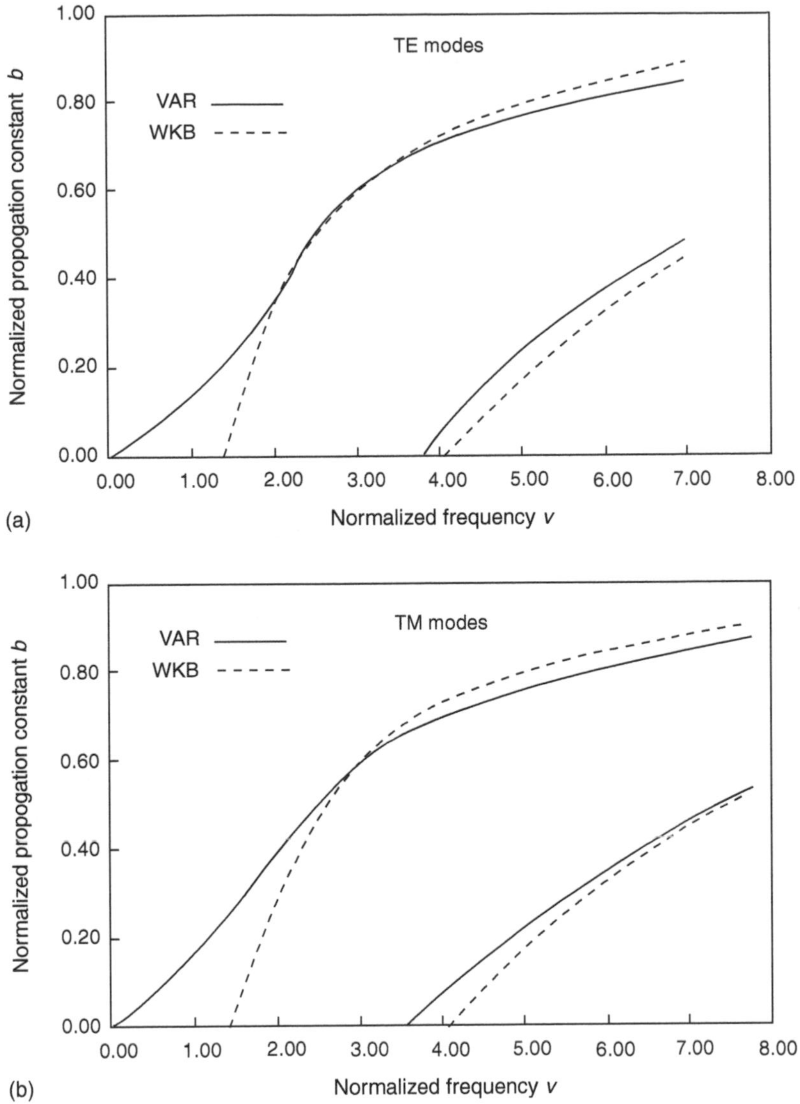

FIGURE 14.17 (a), (b) Normalized dispersion curves. (From Albert, J. and Yip, G. L. 1988. Wide single-mode channels and directional coupler by two-step ion-exchange in glass. *IEEE J. Lightwave Technol.*, 6, 552–563. With permission.)

Under the same fabrication conditions, the diffusion problem of two parallel channel guides, for example, used in a directional coupler, can also be simulated for different waveguide spacings s (separation distance) and p (distance between the centers of the guides and $s=p-w$) as shown in Figure 14.20a and b. The corresponding effective-index distributions $N_{eff}(y_i)$ are presented in Figure 14.21a and b for the TM modes at 1.523 μm. It can be seen that, in the special use of zero separation distance ($s=0$), the two guides merge into one with a width of $w=12$ μm. The effective index $N_{eff}(y_i)$ of this composite guide is, then, identical to that shown in Figure 14.19b. In the case of very large separation distances ($s>12$ μm, say), the two guides have less and less interaction between them, and the effective index $N_{eff}(y_i)$ looks more and more like that for each individual guide, centered at $y=p/2$ and $-p/2$ as shown in Figure 14.21b.

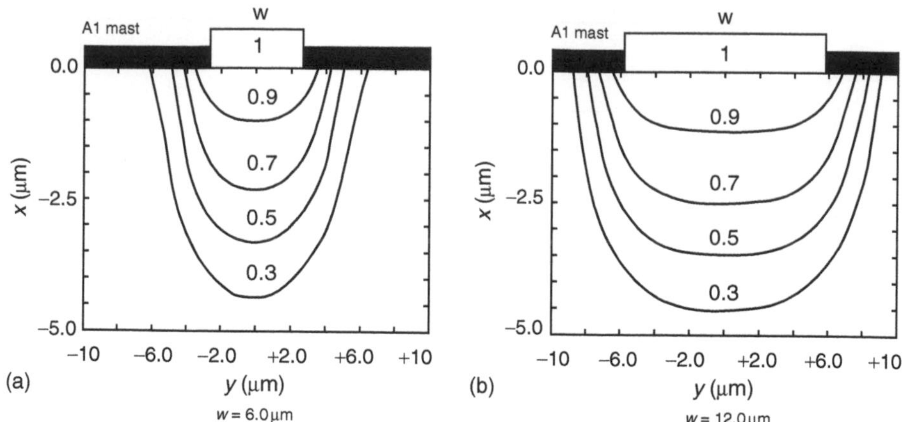

FIGURE 14.18 Constant concentration contours. (a,b) L. Babin, Thesis. (From Babin, L. 1993. Optimization and fabrication of a widened x-branch optical demultiplexer in glass. M. Eng. thesis. Department of Electrical Engineering, McGill University, Montreal, Quebec, Canada. With permission.)

In Figure 14.19 and Figure 14.21, the values of N_{eff} resulting from the exact numerical R-K solution, the erf fit, and the step-index approximation are compared for the TM modes. The results under the step-index approximation have been obtained by neglecting side-diffusion ($\partial/\partial y=0$) in Equation 14.29 and by assuming $N_{eff}(y)=N_{eff}(0)$ within the width of the channel guide and $N_{eff}(y)=n_b$ outside. The erf fit is very good for the middle of the guide, but the agreement deteriorates near the guide's two edges, where, however, the contribution of the index values to the guide's propagation constant becomes less important. The step-index model appears to be a poorer one, especially for narrower guides seen in Figure 14.19a, where its $N_{eff}(0)$ is significantly higher than that obtained from the exact solution. For a two-guide structure with $0<s<12$ μm, the erf fit for $N_{eff}(y)$ is typically as presented in Figure 14.21a, where it is seen to deviate somewhat from that by the R-K method. This should not be surprising because the erf fit attempted, here, is based on the single channel fit represented by Equation 14.32. However, in spite of this deviation, the resulting propagation constant β calculated will still be of sufficient accuracy as discussed in the following paragraph.

In Table 14.6, the effective index $N_{eff}=\beta/k_0$, of the first-order and second-order TM modes for the two different channel guides, calculated by the EIM, employing the lateral effective index $N_{eff}(y)$ established

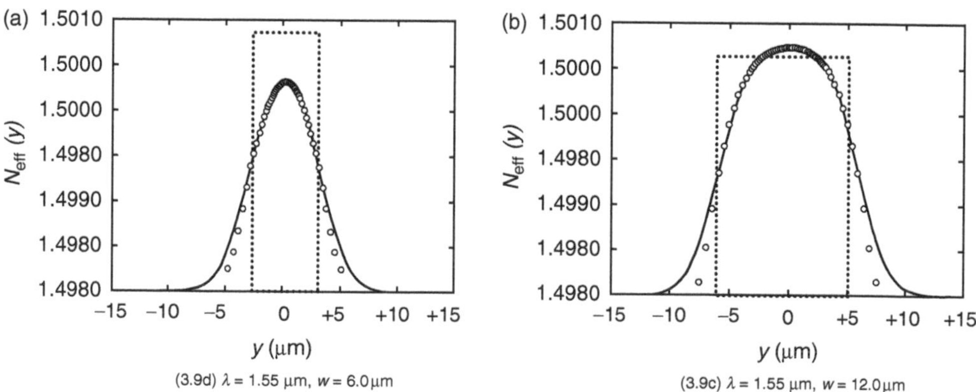

FIGURE 14.19 Discrete values of the local effective. (a,b) Babin, Thesis. (From Babin, L. 1993. Optimization and fabrication of a widened x-branch optical demultiplexer in glass. M. Eng. thesis. Department of Electrical Engineering, McGill University, Montreal, Quebec, Canada. With permission.)

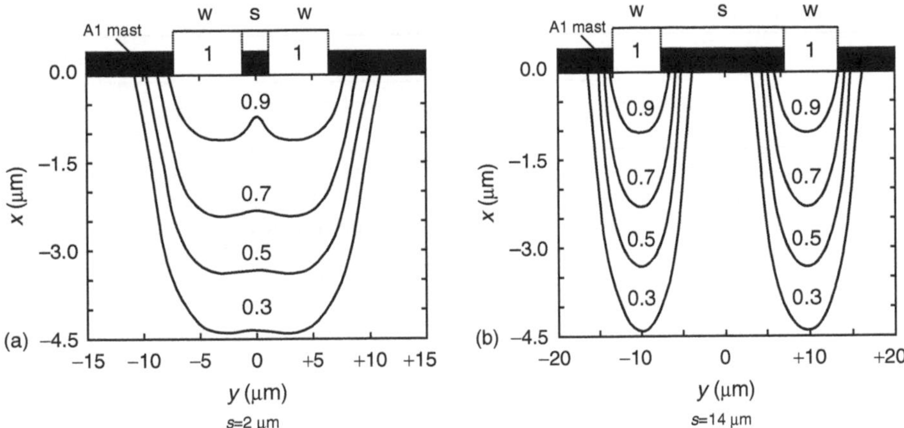

FIGURE 14.20 Constant concentration contours resulting. (a, b) Babin, Thesis. (From Babin, L. 1993. Optimization and fabrication of a widened x-branch optical demultiplexer in glass. M. Eng. thesis. Deptartment of Electrical Engineering, McGill University, Montreal, Quebec, Canada. With permission.)

via the R-K method, the erf and the step-index model, respectively, are presented and compared. The errors are relative to the exact numerical results. The erf model is seen to yield closer results. The errors in β are of the order of $(5–9)\times 10^{-5}$.

Using the EIM and step-index or erf fit model for $N_{\mathrm{eff}}(y)$, extensive, normalized, universal dispersion curves can be plotted as shown in Figure 14.22 for channel guides built by a single K^+–Na^+ ion exchange. For this, four normalized waveguide parameters should be defined (Hocker and Burns 1977) as follows:

$$\begin{aligned} b_0 &= (N_p^2 - n_b^2)/(n_s^2 - n_b^2), \\ V_0 &= k_0 d\sqrt{(n_s^2 - n_b^2)}, \\ b' &= \left[\left(\frac{\beta}{k_0}\right)^2 - n_b^2\right]\Bigg/(N_p^2 - n_b^2), \end{aligned} \qquad (14.36)$$

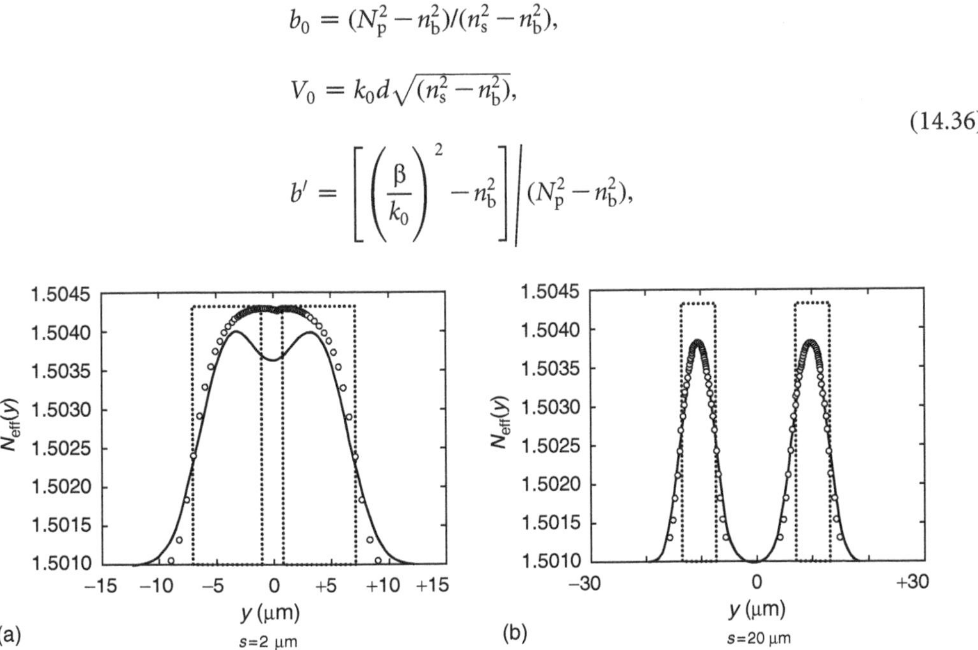

FIGURE 14.21 Discrete values of the local effective index. (a, b) Babin, Thesis. (From Babin, L. 1993. Optimization and fabrication of a widened x-branch optical demultiplexer in glass. M. Eng. thesis. Dept. of Electrical Engineering, McGill University, Montreal, Quebec, Canada. With permission.)

TABLE 14.6 Calculated Effective Index

	R-K Numerical Solution	"erf fit" N_{eff}	"erf fit" Error ($\times 10^{-5}$)	Step-Index Model N_{eff}	Step-Index Model Error ($\times 10^{-5}$)
1st order mode	1.503608	1.503666	+5.8	1.503800	+19.2
2nd order mode	1.502075	1.502148	+7.3	1.502309	+23.4
w=12.0 μm, t_d=270 min, λ=1.31 μm, TM modes					
1st order mode	1.499946	1.499955	/+4.9	1.500110	+16.4
2nd order mode	1.498351	1.498441	+9.0	1.498486	+13.5
w=12.0 μm, t_d=270 min, λ=1.55 μm, TM modes					
1st order mode	1.502705	1.502611	−9.4	1.503043	+43.2
w=6.0 μm, t_d=270 min, λ=1.31 μm, TM modes					
1st order mode	1.499008	1.499107	+9.9	1.499361	+35.2
w=6.0 μm, t_d=270 min, λ=1.55 μm, TM modes					

Source: From Babin, L. 1993. Optimization and fabrication of a widened x-branch optical demultiplexer in glass. M. Eng. thesis. Department of Electrical Engineering, McGill University, Montreal, Quebec, Canada. With permission.

and

$$V' = k_0 w \sqrt{(N_p^2 - n_b^2)} = v_0 b_0^{1/2} \frac{w}{d}$$

where N_p is the effective index of a slab waveguide ion exchanged for a time t. All of the other parameters have been defined previously.

For weakly guiding K^+–Na^+ ion-exchanged guides, $n_s \simeq n_b$ and, hence, the universal curves $b' - V'$ are roughly the same for both polarizations. However, channel guides are still polarization-dependent because values of N_p differ for the TE and TM modes. Within the step-index model, the $b' - V'$ curves are b_0-independent, but are b_0-dependent within the erf model, which includes side-diffusion effects. The parameters N_p and b_0 strongly depend on the diffusion time t, increasing with t. For short diffusion times, side-diffusion effects are limited and do not contribute much difference in the propagation constants of the channel guides. However, for 200 min < t < 400 min (used in a design example in Section 6.3) side-

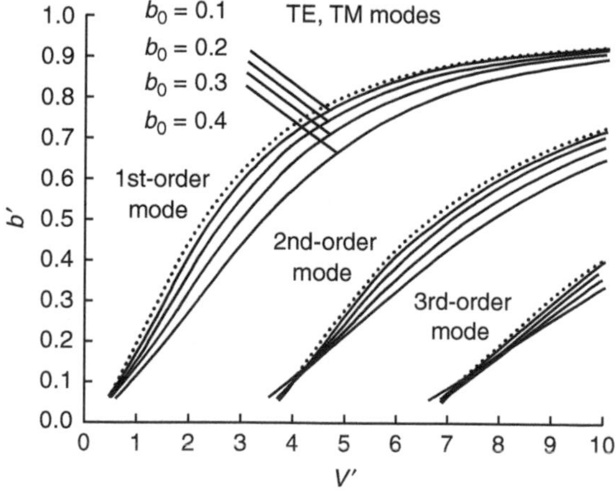

FIGURE 14.22 Normalized dispersion curve. Babin, Thesis. (From Babin, L. 1993. Optimization and fabrication of a widened x-branch optical, demultiplexer in glass. M. Eng. thesis. Department of Electrical Engineering, McGill University, Montreal, Quebec, Canada. With permission.)

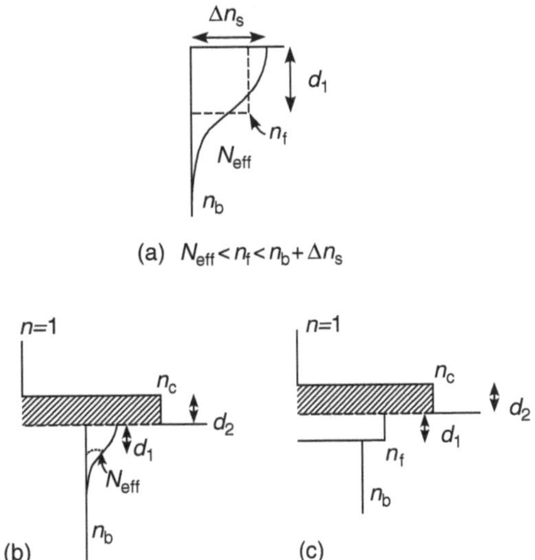

FIGURE 14.23 Modeling of a slab. (a, b, c) M. Belanger, Doctoral Theses. (From Belanger, M. 1986. Theoretical and experimental studies on active and passive 3-branch waveguides; and their derivatives. Ph.D. thesis. Department of Electrical Engineering, McGill University, Montreal, Quebec, Canada. With permission.)

diffusion effects could contribute significant differences to the propagation constants of the channel guide modes (Figure 14.23).

It is important to point out, here, that this section has shown how the channel guides can be modeled and their propagation characteristics analyzed with the EIM, *using experimental data from slab guide characterizations only.* The accuracies in the lateral-index profile models will be tested in the design and implementation of some channel guide devices to be presented in Section 14.6.

14.4.2 Effect of a Dielectric Cladding

In many situations, a dielectric cladding is deposited on a slab or channel guide to enhance the guide's effective index for light beam processing. In $LiNbO_3$ electro-optical waveguide devices, a dielectric cladding on a channel guide is usually deposited to serve as a buffer layer between the guide and an electrode to reduce device losses. Hence, it is important to know the change in the guide's effective index due to dielectric cladding in the device design.

Assuming a Gaussian index distribution in the diffused slab or channel guide, an equivalent homogeneous slab model is chosen with the criteria that the effective indices N_{eff} in both waveguides are equal and that the thickness of the homogeneous model is given by the penetration depth $d_1 = x_t$ (also called the turning point). The value of N_{eff} for an inhomogeneous slab guide can be calculated from the WKB dispersion relationship, Equation 14.20, which can be rewritten in terms of the normalized waveguide parameters ν_1 and b_1 as

$$V_1 \int_0^{\bar{x}_t} [f(\bar{x}) - b_1]^{1/2}\, \overline{\mathrm{d}x} = (m + 3/4)\pi; \quad m = 0,1,2,\ldots \tag{14.37}$$

where $f(\bar{x}) = \exp(-\bar{x}^2)$ and $\nu_1 = \nu_0$, $\mathrm{b}_1 = \mathrm{b}_0$ as defined in Equation 14.36.

Then, N_{eff} can be substituted in the dispersion Equation 14.5 for the TE modes or Equation 14.8 for TM modes in an equivalent homogeneous slab waveguide to obtain its equivalent n_f, where $n_c = 1$ for free space.

A dispersion relationship for this equivalent homogeneous slab waveguide of thickness d_1, covered on top by a uniform cladding layer of thickness d_2 and index n_c, and then free space, and bounded below by an infinite medium of index n_b, can, then, be derived, in the case of the TM modes, as given in Equation 14.38:

$$k_2\left[1-\frac{k_2}{\gamma_3}\tan k_2d_2\right]\left[\tan k_1d_1+\frac{k_1}{\gamma_4}\right]-k_1\left[1-\frac{k_1}{\gamma_4}\tan k_1d_1\right]\left[\tan k_2d_2+\frac{k_2}{\gamma_3}\right]=0 \tag{14.38}$$

where

$$\begin{aligned} k_1 &= (k_0^2n_f^2-\beta^2)^{1/2},\\ k_2 &= (k_0^2n_c^2-\beta^2)^{1/2},\\ \gamma_3 &= (\beta^2-k_0^2)^{1/2},\\ \gamma_4 &= (\beta^2-k_0^2n_b^2)^{1/2}, \end{aligned} \tag{14.39}$$

and β is the propagation constant of the fundamental mode in the equivalent homogeneous slab guide with a film index n_f. Equation 14.38 can then be used to compute the new values of $N_{eff}=\beta/k_0$, and, hence, the index change ΔN caused by the cladding (n_c).

14.4.3 $LiNbO_3$ Waveguides by Ti Diffusion and Proton Exchange

Like the ion-exchanged channel guides in glass, the index distribution along the depth direction can be modeled by a gaussian function. In the lateral direction, the simpler step-index model, which neglects side diffusion, can yield reasonably accurate design predictions. A more accurate model, which takes into consideration the effects of side diffusion, is, again, given by the combination of two error functions. Hence, the general index distribution for a two-dimensional channel guide can be represented by (Fukuma et al. 1978; Burns et al. 1979; Ctyroky et al. 1984)

$$\Delta n(x,y)=\Delta n_s\exp\left(-\left(\frac{x}{d}\right)^2\right)\left\{\mathrm{erf}\left(\frac{W/2-y}{d_y}\right)+\mathrm{erf}\left(\frac{W/2+y}{d_y}\right)\right\} \tag{13.40a}$$

where Δn_s is the surface index change at the center of the channel guide, and

$$\begin{aligned} d_x &= 2\sqrt{D_xt},\\ d_y &= 2\sqrt{D_yt}, \end{aligned} \tag{14.40b}$$

where D_x and D_y are the diffusion coefficients along the depth and lateral direction and d_x and d_y are the corresponding diffusion lengths.

14.4.4 Electro-Optical Index Change in Ti:$LiNbO_3$ Waveguides

Due to the excellent electro-optical properties and low propagation loss of $LiNbO_3$ Crystals, electro-optical devices, such as modulators switches, polarization splitters, etc. have been designed and implemented in $LiNbO_3$ substrates (Alferness 1988). Fundamentals of crystal optics and applications to electro-optical devices are discussed in Kaminow (1974). A very brief account of the index change due to voltage-induced electro-optical effects through an electrode system on the crystal surface will be given here.

$LiNbO_3$ is a negative, uniaxial crystal with a dielectric tensor given by (Alferness 1988)

$$\overline{\in} = \begin{bmatrix} \in_x & 0 & 0 \\ 0 & \in_y & 0 \\ 0 & 0 & \in_z \end{bmatrix} = \begin{bmatrix} n_x^2 & 0 & 0 \\ 0 & n_y^2 & 0 \\ 0 & 0 & n_z^2 \end{bmatrix} \tag{14.41}$$

where $\in_x = \in_y \neq \in_z$, $n_x = n_y = n_0$ and $n_z = n_e$ ($< n_0$, hence the name negative). For waves with an electric field polarized along either the x or y axes, the propagating waves "see" only the indices n_x or n_y. Such waves are called ordinary waves and x, y correspond to the ordinary axis. For waves with an electric field polarized along the z axis, they "see" only the index n_z and propagate with a different phase velocity. They are, hence, called extraordinary waves and the z axis in this case corresponds to the optical or c axis.

Electromagnetic wave propagation in a crystal is characterized by the so-called indicatrix or index ellipsoid. The orientation of this ellipsoid model is related to the crystal axes, which correspond to the ellipsoid's principal axes. The semiaxes of the ellipsoid along the x, y and z directions give the principal refractive indices n_x, n_y, and n_z, respectively. Index changes via electro-optic effects can, therefore, be described in terms of changes in the semiaxes of the index ellipsoid. The index ellipsoid is given by (Kaminow 1974; Nishihara et al. 1989)

$$\frac{x^2}{n_x^2} + \frac{y^2}{n_y^2} + \frac{z^2}{n_z^2} = 1 \tag{14.42}$$

where $i = D_i/\sqrt{2\in_0 \omega_e}$, $\quad i = x,y,z$. D_i is the electric displacement vector, $\in_0$ the free space permittivity and ω_e the stored electric energy density in the crystal.

Active integrated optical devices utilize the electro-optic index change $\delta n_{e.o.} = \delta n$ induced in certain optical crystals by an applied electric field. The induced index change causes the index ellipsoid to deform. This induced index change $\Delta(n^{-2})$ is given by (Kaminow 1974).

$$\Delta(n^{-2}) = rE + R^2E^2. \tag{14.43}$$

The first term varies linearly with the applied electric field and is known as the Pockels effect. The second term varies with the square of the applied field and is known as the Kerr effect. Our subsequent discussions will be confined to the linear Pockels effect. The induced electro-optic index change alters the shape, size, and orientation of the ellipsoid. The general expression for the deformed index ellipsoid is given by

$$a_{11}x^2 + a_{22}y^2 + a_{33}z^2 + 2a_{23}yz + 2a_{31}zx + 2a_{12}xy = 1, \tag{14.44}$$

where x, y, z, generally, no longer form the principal axes as in Equation 14.42. The constants a_i, in Equation 14.44 are related to the three components of the applied electric field E through the following 6×3 electro-optic tensor:

$$\begin{bmatrix} a_{11} - n_x^{-2} \\ a_{22} - n_y^{-2} \\ a_{33} - n_z^{-2} \\ a_{23} \\ a_{31} \\ a_{12} \end{bmatrix} = \begin{bmatrix} r_{11} & r_{12} & r_{13} \\ r_{21} & r_{22} & r_{23} \\ r_{31} & r_{32} & r_{33} \\ r_{41} & r_{42} & r_{43} \\ r_{51} & r_{52} & r_{53} \\ r_{61} & r_{62} & r_{63} \end{bmatrix} \cdot \begin{bmatrix} E_x \\ E_y \\ E_z \end{bmatrix} \tag{14.45}$$

where r_{ij} are called electro-optic or Pockels coefficients, and n_x, n_y, n_z are the refractive indices in the respective directions. When the applied E-field is zero, Equation 14.44 reduces to Equation 14.42.

For $LiNbO_3$, the $[r_{ij}]$ tensor is given by (Alferness 1988)

$$\begin{bmatrix} 0 & -r_{22} & r_{13} \\ 0 & r_{22} & 0 \\ 0 & 0 & r_{33} \\ 0 & r_{51} & 0 \\ r_{51} & 0 & 0 \\ -r_{22} & 0 & 0 \end{bmatrix} \tag{14.46}$$

where $r_{13}=8.6\times$, $r_{33}=30.8\times$, $r_{22}=3.4\times$ and $r_{51}=28\times10^{-12}$ m/V respectively.

When $n_x=n_y=n_0$, $n_z=n_e$, and $E_x=E_y=0$, through Equation 14.45 and Equation 14.46, Equation 14.44 becomes

$$(n_0^{-2}+r_{13}E_z)x^2+(n_0^{-2}+r_{13}E_z)y^2+(n_e^{-2}+r_{33}E_z)z^2=1. \tag{14.47}$$

Because the electro-optically induced index changes are usually very small, i.e., $|r_{13}n_0^2E_z|\ll 1$ and $|r_{33}n_e^2E_z|\ll 1$, Equation 14.47 can be rewritten as

$$\frac{x^2}{\left(n_0-\frac{n_0^3}{2}r_{13}E_z\right)^2}+\frac{y^2}{\left(n_0-\frac{n_0^3}{2}r_{13}E_z\right)^2}+\frac{z^2}{\left(n_e-\frac{n_e^3}{2}r_{33}E_z\right)^2}=1 \tag{14.48}$$

The refractive index changes due to an applied field E_z are, hence, respectively,

$$\Delta n_x=\Delta n_y=-\frac{1}{2}r_{13}n_0^3E_z$$

and

$$\Delta n_z=-\frac{1}{2}r_{33}n_e^3E_z. \tag{14.49}$$

For $LiNbO_3$, $n_0=2.286$ and $n_e=2.2$ at 0.633 μm.

14.4.5 Electrode Analysis and Design

For active electro-optic devices, an electrode system is always needed on a crystal surface to effect electro-optic control of guided waves, for example, in switching and modulation and also for fine tuning refractive indices in slab and channel guide devices, such as directional couplers and Mach-Zender interferometers, etc. A low-voltage and low-power loss to achieve the desired device functions are important design considerations.

The electric field distribution in the substrate can often be obtained analytically by conformal mapping for electrodes deposited directly on the $LiNbO_3$ surface (Vandenbulcke and Lagasse 1974; Ramer 1982). However, metal electrodes would attenuate optical waves, and, in practice, they are normally separated from the crystal surface by a thin buffer layer with a lower index to reduce the attenuation of optical waves. The presence of a buffer layer, nevertheless, causes a weakening of the applied electric field strength for the intended electro-optic effects and also makes the conformal mapping solution of the field distribution difficult, if not impossible. Fortunately, a semianalytical and numerical solution of the potential problem with a buffer layer is now available (Thylén and Gravestrand 1986) and will be employed here.

A three-electrode system is used to illustrate this method, as shown in Figure 14.24, where the space is divided into three regions marked by $-1(x<0)$, $0(0<x<d_0)$, and $1(x>d_0)$, respectively. The potential (due to the electrodes) to be solved must satisfy Laplace's equation:

$$\in_x \frac{\partial^2 \phi}{\partial x^2} + \varepsilon_y \frac{\partial^2 \phi}{\partial y^2} = 0 \tag{14.50}$$

with the following boundary conditions:

$$\begin{cases} \phi(x \to \infty, y) = 0, \\ \text{at the electrodes, } \phi(0,y) = V, \text{ the applied voltage, and} \\ \text{between the electrodes, the normal derivative } \dfrac{\partial \phi}{\partial n} \text{ is zero.} \end{cases} \tag{14.51a}$$

At the buffer-substrate interface, there should be

$$\begin{cases} \text{continuity of potential, } \phi_1(d_0,y) = \phi_0(d_0,y), \text{ and} \\ \text{continuity of } D_x, \in_x \dfrac{\partial \phi_1}{\partial x}(x,y) \rfloor_{x=d_0} = \in_i \dfrac{\partial \phi_0}{\partial x}(x,y) \rfloor_{x=d_0}. \end{cases} \tag{14.51b}$$

The semianalytical method of solution uses the Fourier transform technique. Let $\phi(x,y)$ be the original field and $\Phi(x,v)$ its Fourier transform in the y direction, where v is the Fourier transform variable. Fourier transforming Equation 14.50), one gets (Thylén and Gravestrand 1986)

$$\phi_1(x,y) = \int_{-\infty}^{\infty} dv \Phi_1(d_0,v) e^{j2\pi vy} e^{-2\pi|v|(x-d_0)\sqrt{\varepsilon_y/\varepsilon_z}} \tag{14.52}$$

throughout the substrate, provided $\Phi_1(d_0,\text{v})$ is known. $\Phi_1(d_0,\text{v})$ is related to $\Phi_0(0,\text{v})$ by

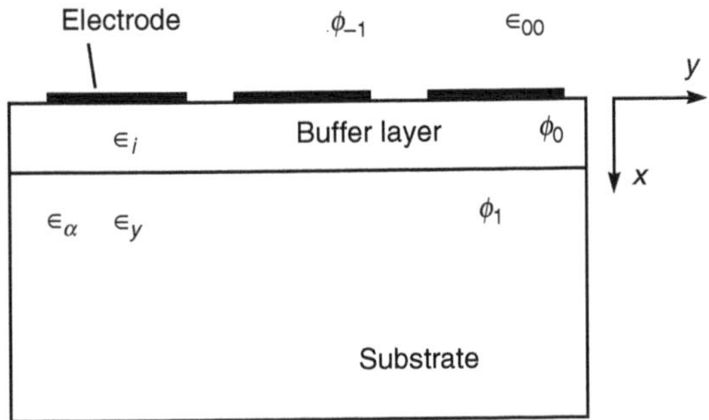

FIGURE 14.24 Electrode configuration. (From Sekerka-Bajbus, M. A. 1989. The BPM design optimization and implementation of 3-branch waveguide devices. M. Eng. thesis. Department of Electrical Engineering, McGill University, Montreal, Quebec, Canada. With permission.)

$$\Phi_1(d_0,v) = \Phi_0(0,v)/[\cosh(2\pi|v|d_0) + \sinh(2\pi|v|d_0\alpha_1)] \tag{14.53}$$

where $\alpha_1 = (\in_x \in_y)^{1/2}/\in_i$

Hence, $\Phi_0(0, v)$ must be known, but $\Phi_0(0, v)$ (Thylén and Gravestrand 1986) must satisfy

$$\int_{-\infty}^{\infty} dv|v|\Phi_0(0,v)e^{j2\pi vy}\left[\in_i \frac{\sinh(2\pi|v|d_0) + \alpha_1\cosh(2\pi|v|d_0)}{\cosh(2\pi|v|d_0) + \alpha_1\sinh(2\pi|v|d_0)} + \in_\infty\right] = 0, \tag{14.54}$$

where $\Phi_0(0,v)$ can be solved iteratively.

The numerical procedure in the electrode analysis and design can be summarized as follows:

1. Initially, guess of the surface potential $\phi_0(0,y)$.
2. FFT this potential to get $\Phi_0(0,v)$.
3. Multiply by $-2\pi|v|$, and use an inverse FFT to obtain the corresponding electric displacement D_x at the surface $x=0$. $D_x = \in_0(\partial\phi/\partial x)(0,y)$.
4. Modify D_x by setting all values *between* the electrodes to zero, i.e., $D_x=0$ at $x=0$.
5. FFT the resulting potential and divide by $-2\pi|v|$.
6. Inverse FFT to obtain the modified surface potential $\phi(0,y)$. Set the potential on the electrodes to the applied voltage, and use an interpolation scheme to make the potential continuous between the electrodes.
7. Repeat the iteration from step 2 until the potential and the electric displacement D_x satisfy their boundary conditions.

The above procedure yields numerical results which converge very fast even for very rough initial guesses. Once the surface potential for a specific electrode configuration has been determined, the potential and field distributions can be calculated, using the analytical expressions given in Equation 14.52 through Equation 14.54. The corresponding index change, induced electro-optically, at any point, can also be determined. Consequently, the initial gaussian index profile is perturbed by the electro-optic effect:

$$n(x) = n_s + \Delta n_s \exp(-x^2/d_x^2) + \delta n(x)_{\text{e.o.}}. \tag{14.55}$$

For the three-electrode configuration shown in Figure 14.24, the resulting effective index change is shown in Figure 14.25 for a voltage V applied to the central electrode while the potentials at the two outer electrodes are maintained at zero. The normalized electric field E_x for various electrode parameters is

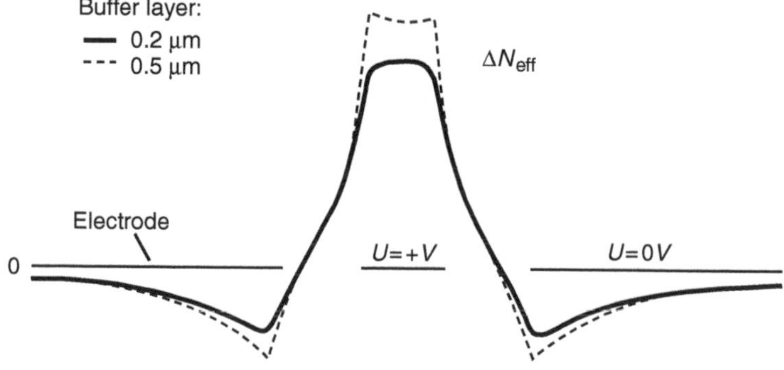

FIGURE 14.25 Electro-optically induced change. (From Sekerka-Bajbus, M. A. 1989. The BPM design optimization and implementation of 3-branch waveguide devices. M. Eng. thesis. Dept. of Electrical Engineering, McGill University, Montreal, Quebec, Canada. With permission.)

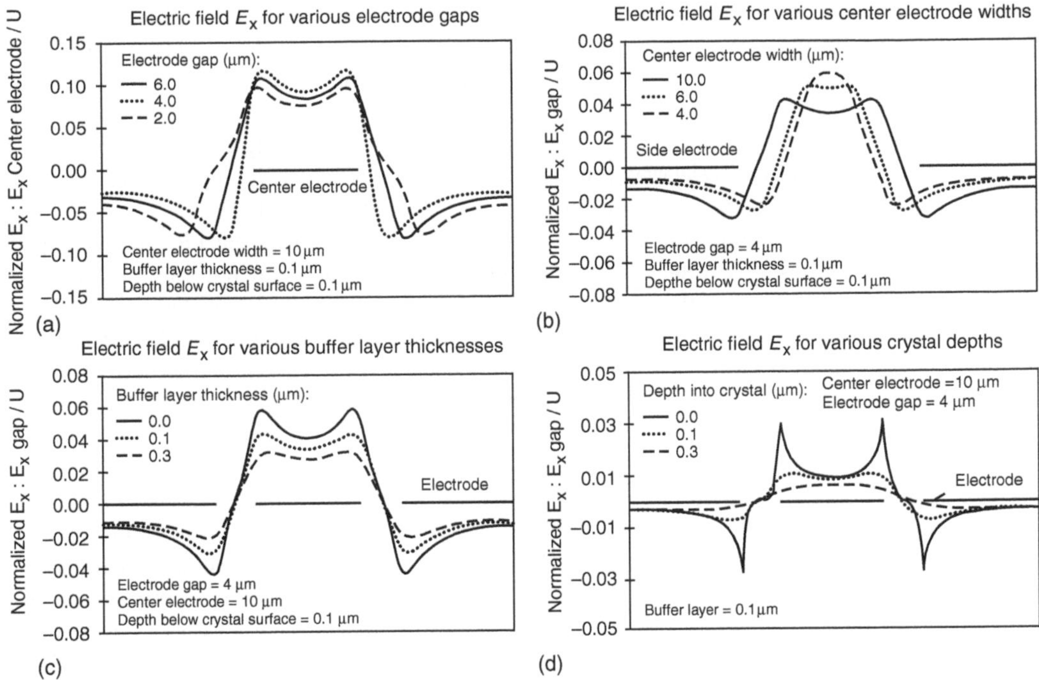

FIGURE 14.26 Normalized electric field E_x. (From Sekerka–Bajbus, M. A. 1989. The BPM design optimization and implementation of 3-branch waveguide devices. M. Eng. thesis. Dept. of Electrical Engineering, McGill University, Montreal, Quebec, Canada. With permission.)

shown in Figure 14.26a through d. Reducing the electrode gap, the central electrode width, and the buffer layer thickness increases the electric field strength.

14.5 The Beam Propagation Method

Fleck et al. (1976) proposed a new numerical method for solving the scalar Helmholtz equation. The usefulness of this technique, which was originally developed to trace laser beam propagation in the atmosphere, has been successfully demonstrated in the problems involving beam propagation in optical fibers later (Feit and Fleck 1978; and 1980). It has since been known as the beam propagation method (BPM) and permits a combined treatment of both the guided and radiation modes in weakly guiding waveguides. More recently, this method has also been used in analyzing and designing integrated optical waveguide devices (Feit et al. 1983).

There are, two popular formulations of the BPM. Both can be implemented with the numerical Fast Fourier Transform (FFT). The more commonly used BPM formulation is based on the parabolic or Fresnel approximation of the scalar Helmholtz equation and is applicable to problems where the optical fields vary slowly along the propagation direction over distances of the order of a wavelength. When the Fresnel approximation is not valid, another more accurate form of the BPM based on the full scalar Helmholtz equation can be derived. For most problems, however, both formulations would lead to about the same numerical results. These schemes are referred to as FFT–BPM.

One drawback in the BPM technique is that, if the electric field propagates to the edge of the computational window in the transverse (x,y) plane, it will be folded back to the opposite side of the window and superposed on the propagation field during succeeding steps, thus, causing high-frequency numerical stability of the solution. This was pointed out by Saijonmaa and Yevick (1983) in their study of losses in bent optical waveguide fibers. They suggested that this can be avoided by absorbing the field at the edge of the grid, either by setting the field to zero at a few grid points close to the edge of the window or by inserting a large imaginary component in the refractive index at these points, thus, simulating the effect of a lossy cladding. The absorber function must be set up to ensure that the electric field is absorbed gradually near the window boundaries.

Besides the conventional FFT–BPM, the beam propagation method using finite differences, called FD–BPM, to solve the scalar paraxial wave equation has also been developed (Chung and Dagli 1990) and compared to FFT–BPM (Chung and Dagli 1990; Scarmozzino and Osgood 1991).

The aforementioned schemes, however, can solve only the scalar wave equation under paraxial approximation. This may be adequate for some weakly guiding devices but not for those which are polarization-sensitive or those with abrupt index changes (Liu and Li 1991): In the latter cases, one cannot neglect the vectorial nature of the guided waves in the analysis and simulation. Recently, a vector beam propagation method based on a finite difference scheme, called FD–VBPM, has been developed, accompanied by a detailed analysis and assessment performed for two-dimensional (2-D) waveguide structures (Huang et al. 1992).

Instead of the conventional absorbing boundary condition, a numerical, *transparent* boundary condition (TBC) developed more recently (Hadley 1991), which allows the passage of the traveling waves through the edges of the computation window, can be recommended for the FD–VBPM scheme.

Unlike the absorber boundary condition, the algorithm for the TBC does not contain a parameter to be adjusted for each specific problem and is, therefore, problem-independent. It adapts naturally to a finite-difference scheme and has been shown to be accurate and robust for both 2-D and 3-D problems.

The reference index n_0 should be chosen so that the Fresnel or paraxial approximation is satisfied. If the variation of the refractive index over the transverse cross-section of the waveguide structure is small, the refractive index of the cladding can be chosen to be n_0. Otherwise, the reference index n_0 has to be chosen so as to minimize the variation of the transverse field components along the z direction. For single-mode waveguides, the propagation constant β of the fundamental mode can be chosen to be the reference index i.e., $n_0=\beta/k_0$. For dual-mode waveguides, such as directional couplers and two-mode interference (TMI) cross-channel waveguides, n_0 can be chosen to be given by $n_0=(\beta_e+\beta_0)/2$ where β_e and β_0 are the propagation constants of the even and odd mode of the structure, respectively.

Due to the page limitation, mathematical details of the various BPM schemes are not given here but have been fully documented in the references cited above.

14.6 Illustrative Examples: Design and Fabrication of Some Guided-Wave Photonic Devices

The characterization and modeling procedures described in the previous sections have been employed in the theoretical design and experimental realization of several guided-wave photonic devices. In this section, we will present some of these devices as illustrative examples. Wherever possible, design and experimental results are both presented and comparisons made.

14.6.1 A Passive Three-Branch Power Divider by K^+-Ion Exchange in Glass

This device consists of an input channel guide followed by a widening taper which divides the incoming optical power into three output branches. The geometry and dimension of this device are given in Figure 14.27. A layer of dielectric cladding can be deposited onto the relevant regions to increase the index in the taper and branch regions of arm 1 and 3, to channel more power into the outer arms 1 and 3 for equal power division. The design must be pursued with predetermined fabrication conditions, which can be optimized for the design outcome in an iterative manner. The channel guides were fabricated by K^+-ion exchange in a soda-lime glass substrate ($n_b=1.512$) through an Al mask immersed in molten KNO_3 at 370°C for one hour. This yielded single-mode channel guides with $d=1.53$ μm and $\Delta n_s=0.010$.

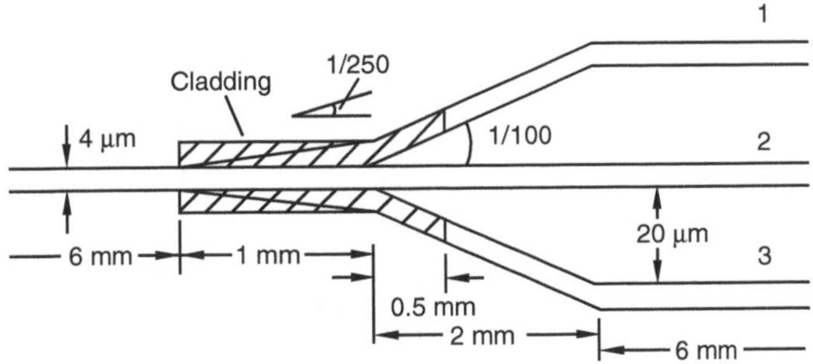

FIGURE 14.27 Configuration of a three-branch power divider. (From Yip, G. L. and Serkerka-Bajbus, M. A. 1988. Design of symmetric and asymmetric passive 3-branch power dividers by beam propagation method. *Electron. Lett.*, 25, 1584–1586. With permission.)

Also, a layer of cladding, using Corning glass 7059 with $n_c \approx 1.544$ was Rf-sputter deposited. With these fabrication data, the effective index of the diffused channel guides can be calculated by solving the dispersion relationship for an equivalent homogeneous slab waveguide. The increase in the effective index in the waveguide regions under the cladding is obtained by solving the dispersion relationship for the guided modes in the homogeneous slab guide, covered on top by a uniform cladding of thickness $d_{2,i}$ $(i = 1,3)$ and index n_c, then, free space, and bounded below by an infinite medium of index n_b (substrate) as discussed in Section 4.2. In the lateral direction, the simplest step-index model was chosen for the channel guide. The optical field distribution in the device was calculated by the FFT beam propagation method (FFT–BPM) as presented in Section 5.1 and Yip and Sekerka-Bajbus (1988) and shown in Figure 14.28 for the case of equal power division. The power distribution in the individual branches, as a function of the cladding thickness, is given in Figure 14.29. The designed device has been fabricated and measured (Haruna et al. 1985). Figure 14.30 shows the measured output light intensity profile by a TV camera for the case of equal power division. The theoretical cladding thickness required for equal power division can be estimated from Figure 14.29 to be 0.215 μm. This compares favorably with the measured result of 0.2 μm (Haruna et al. 1985).

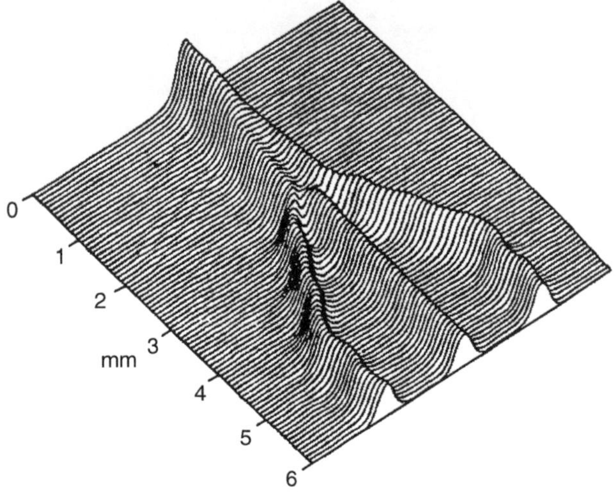

FIGURE 14.28 Optical field distribution. (From Yip, G. L. and Serkerka–Bajbus, M. A. 1988. Design of symmetric and asymmetric passive 3-branch power dividers by beam propagation method. *Electron. Lett.*, 25, 1584–1586. With permission.)

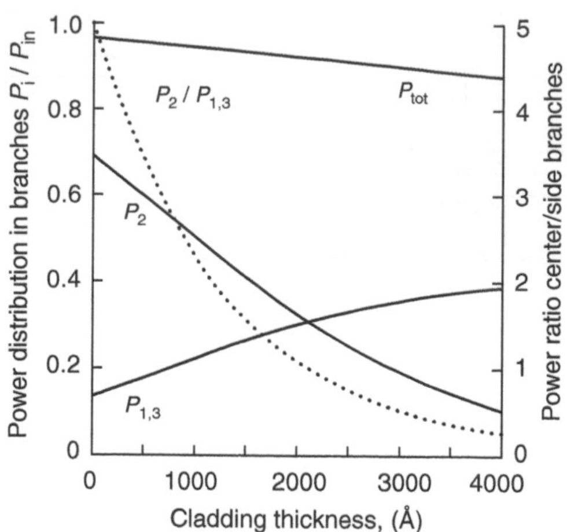

FIGURE 14.29 Power distribution in individual branches, *Electron. Lett.* (From Yip, G. L. and Serkerka-Bejbus, M. A. 1988. Design of symmetric and asymmetric passive 3 -branch power dividers by beam propagation method. *Electron. Lett.*, 25, 1584–1586. With permission.)

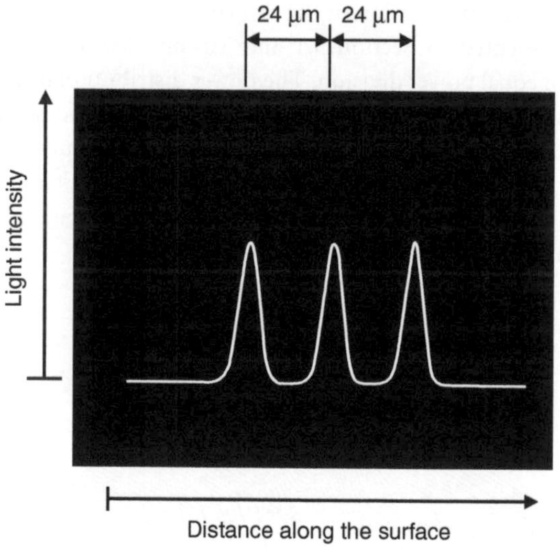

FIGURE 14.30 The output light intensity profile. (From Haruna, M., Belanger, M. N. and Yip, G. L. 1985. Passive 3-branch optical power dividers by K^+-ion exchange in glass. *Electron. Lett.*, 21, 535–536. With permission.)

14.6.2 A Directional Coupler Power Divider by Two-Step K^+-Ion Exchange in Glass

The characterization and modeling of a channel guide by a two-step, ion-exchange process as discussed in Section 4.1 were used in realizing a directional coupler power divider (Albert and Yip 1988) as shown in Figure 14.31.

A two-step ion-exchange process (Yip and Finak 1984) was conceived and adopted for realizing a single-mode channel guide because it has the advantage of relaxing the channel width to 10 μm instead of

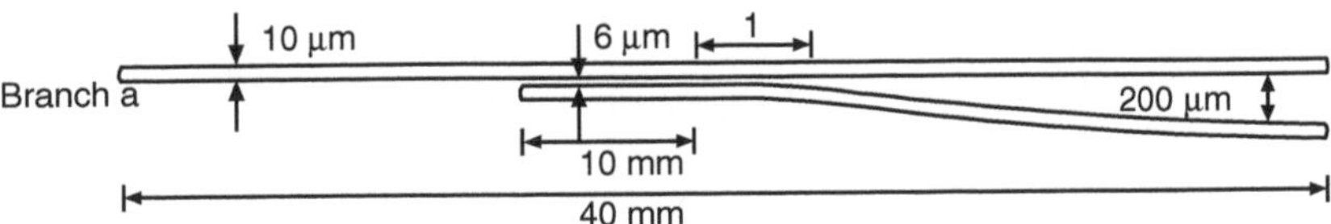

FIGURE 14.31 Design of a directional coupler. (From Albert, J. and Uip, G. L. 1988. Wide single-mode channels and directional coupler by two-step ion-exchange in glass. *IEEE J. Lightwave Technol.*, 6, 552–563. With permission.)

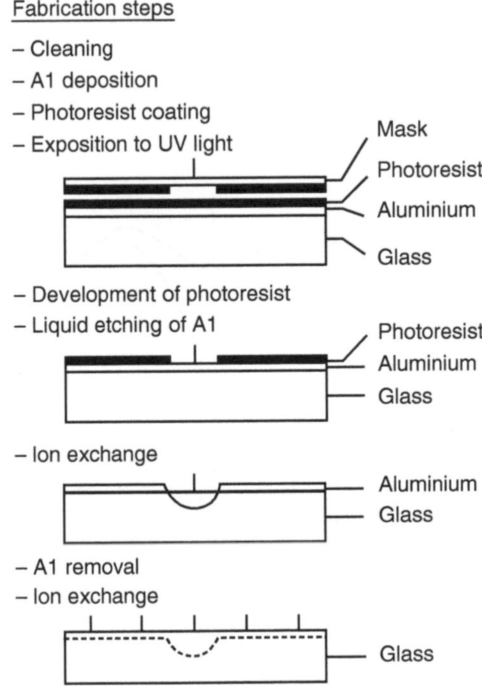

FIGURE 14.32 Diagram of fabrication procedure. (From Albert, J. and Uip, G. L. 1988. Wide single-mode channels and directional coupler by two-step ion-exchange in glass. *IEEE J. Lightwave Technol.*, 6, 552–563. With permission.)

4–6 μm to render the channel guide more compatible for coupling to a single-mode fiber, whose core diameter is typically 8–10 μm. This fabrication procedure is outlined in Figure 14.32.

The approach adopted to analyze, and design the directional coupler is to calculate the normal modes of the structure consisting of two parallel channels. Then, by projecting the optical field incident in one branch onto the symmetric and antisymmetric modes of the coupler, the output characteristics can be obtained by propagating the two modes, thus, launched with their respective propagation constants β_e and β_0, respectively, down the length L of the coupling region. The interference between the two normal modes causes the power to oscillate between the two channels until their separation increases to a value where coupling ceases. The normal modes can be found by solving Equation 14.14 with the following index profile $N_{eff}(y)$, using Equation 14.32:

$$N_{eff}(y) = N(y + w/2) + N(y - w/2) - N(\infty) \tag{14.56}$$

and the trial function, presented in Equation 14.33

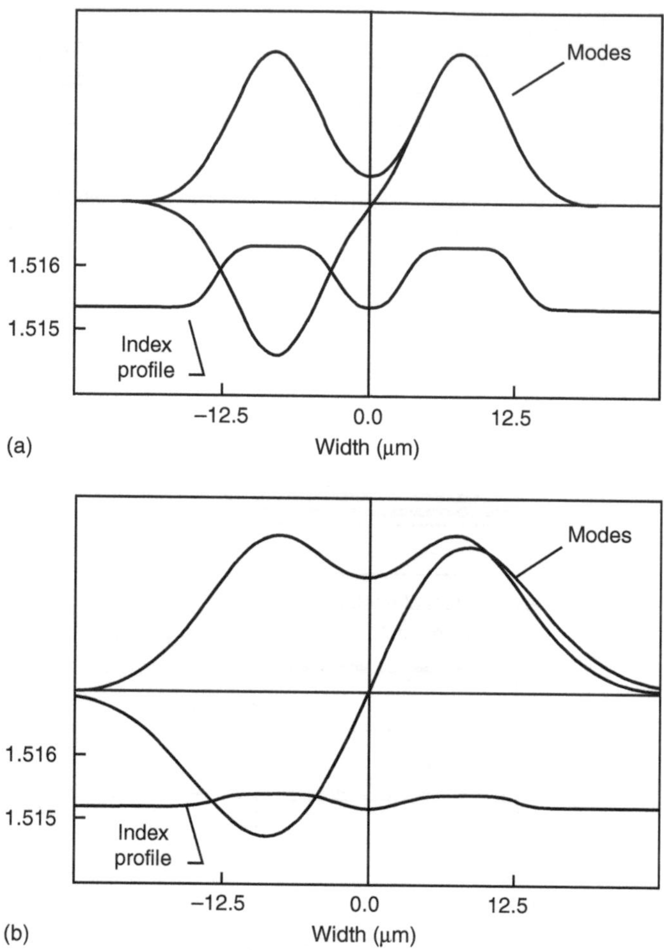

FIGURE 14.33 Two examples of the normal lateral. (From Albert, J. and Uip, G. L. 1988. Wide single-mode channels and directional coupler by two-step ion-exchange in glass. *IEEE J. Lightwave Technol.*, 6, 552–563. With permission.)

$$G_0^e(y) = \exp\left[-\alpha_0^e(y + w/2)^2\right] + s\exp\left[-\alpha_0^e(y - w/2)^2\right] \quad (14.57)$$

where $s=1$ for the even mode e and -1 for the odd mode 0. Figure 14.33 shows some index profile and corresponding normal mode plots. The power transfer efficiency η is defined as

$$\eta = P_b/(P_a + P_b) \quad (14.58)$$

where P_a and P_b are the respective powers at the outputs of channels a and b. P_a and P_b can be calculated theoretically and also measured. The design optimization in our context means calculating the performance parameters of this device, e.g., η, for a set of waveguide design parameters, in an iterative manner, until the best desired performance parameters are achieved. The device is, then, ready for fabrication.

The device shown in Figure 14.31 was fabricated at 385°C by the two-step, ion-exchange process depicted in Figure 14.32, keeping $t_1+t_2=1$ h. to maintain a single mode in the depth direction. The outputs of the two channel guides were measured either with a scanning optical power meter or a CCD

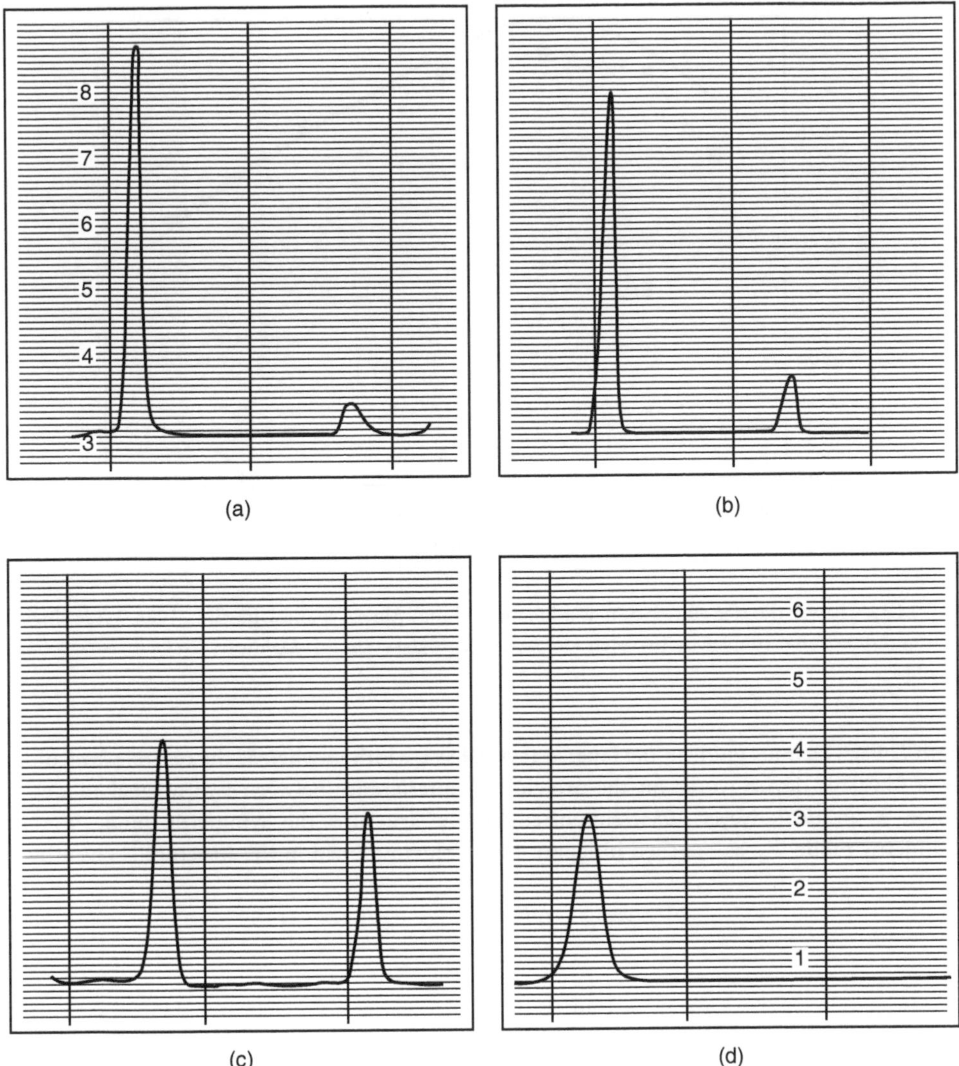

FIGURE 14.34 Typical lateral output profiles. *IEEE J. Lightwave Technol.* (From Albert, J. and Uip, G. L. 1988. Wide single-mode channels and directional coupler by two-step ion-exchange in glass. *IEEE J. Lightwave Technol.*, 6, 552–563. With permission.)

camera at 0.63 μm. Some typical lateral output power profiles, as observed on a chart recorder, are presented in Figure 14.34 for device samples fabricated with different t_1. The mode in each channel seems to be well confined. For (c) and (d), it is evident that the same sample could yield a different power transfer efficiency, depending upon the critical launching condition at the input. In Figure 14.35, both the theoretically calculated and measured values of η for device samples fabricated with different t_1 are presented, and the overall agreement appears good.

14.6.3 A Widened X-Branch Wavelength Demultiplexer at 1.3 μm and 1.55 μm by K^+-Ion Exchange in Glass

The device configuration, shown in Figure 14.36, consists of input and output waveguides in the form of two identical, tapered, directional couplers of length l_t connected adiabatically to a two-mode waveguide

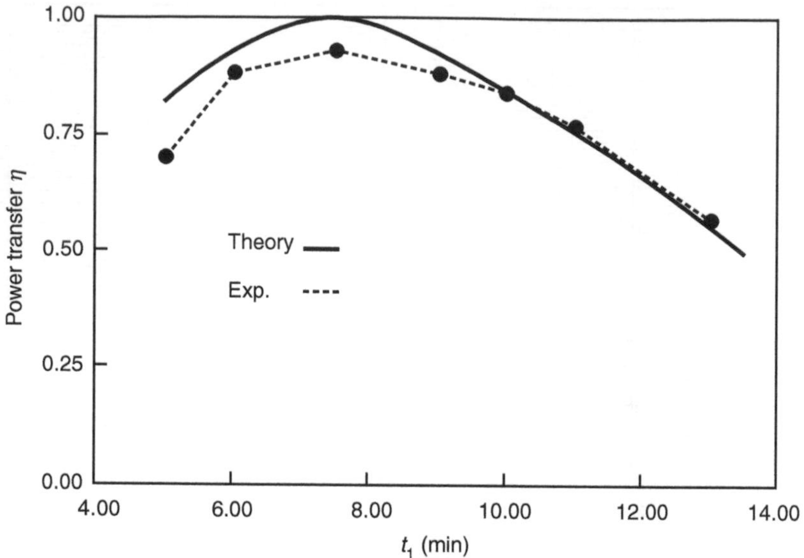

FIGURE 14.35 Power transfer efficiency η as a function. (From Albert, J. and Uip, G. L. 1988. Wide single-mode channels and directional coupler by two-step ion-exchange in glass. *IEEE J. Lightwave Technol.*, 6, 552–563. With permission.)

of length L. Each channel guide in the tapered region has a width w, whereas the central waveguide has a width of $2w$. This device works on the principle of interference between two normal modes (TMI) Papuchon et al. 1977; Neyer 1983, supported by the structure, the symmetric mode Ψ_s and the antisymmetric mode Ψ_a, as sketched in Figure 14.37. The total field in the structure is given by

$$\Psi(x,y,z) = \alpha_s \Psi_s(x,y,z) + \alpha_a \Psi_a(x,y,z) \tag{14.59}$$

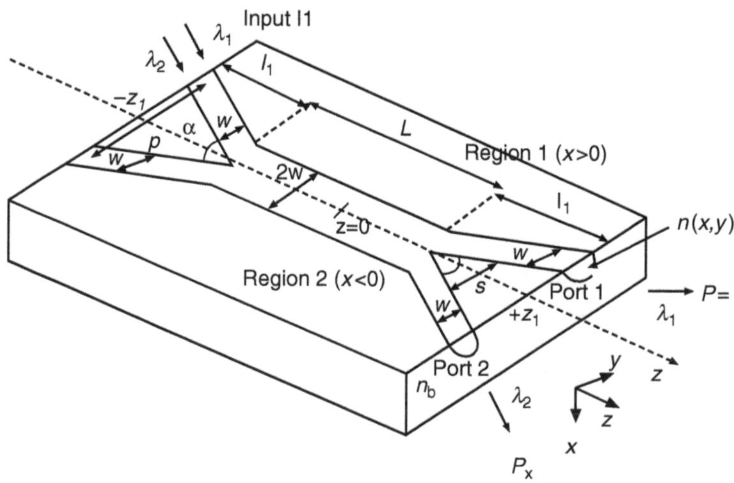

FIGURE 14.36 Widened X-branch. (From Yip, G. L. and Babin, L. 1995. Design optimization and fabrication of a widened x-branch demultiplexer by ion-exchanger in glass. In *Proc. Guided-Wave Optoelectronics Device Characterization, Analysis, and Design*. T., Tamer, G., Griffel, and H. L., Bertoni, eds., pp. 221–229. Plenum Press, New York. With permission.)

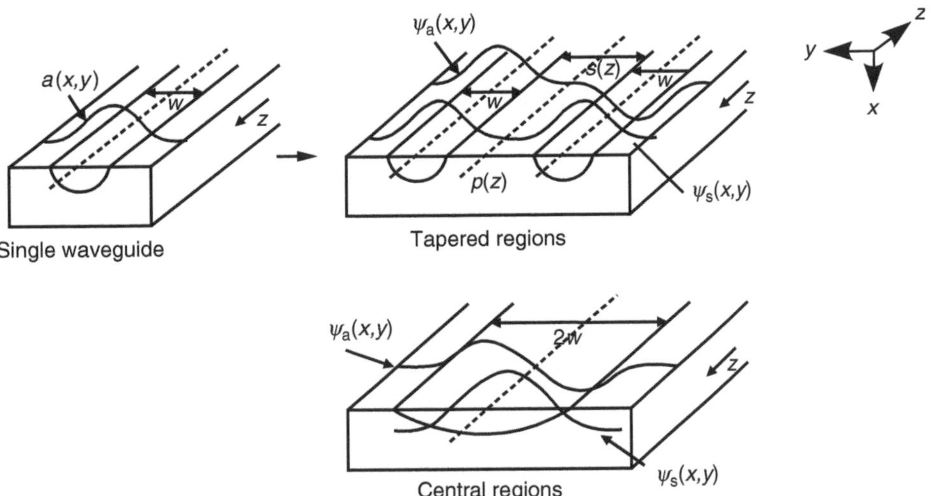

FIGURE 14.37 The symmetric and antisymmetric modes. (From Babin, L. 1993. Optimization and fabrication of a widened x-branch optical demultiplexer in glass. M. Eng. thesis. Department of Electrical Engineering, McGill University, Montreal, Quebec, Canada. With permission.)

where α_s and α_a are the relative excitation amplitudes of the symmetric (even) mode and antisymmetric (odd) mode, respectively, so that $\alpha_s^2 + \alpha_a^2 = 1$.

In the tapered regions ($z < -L/2$ or $z > L/2$), the symmetric and anti-symmetric modes can be considered as two different combinations of the two fundamental modes of the two input channel guides taken separately (Marom and Ruschin 1984; Forber and Marom 1986). Hence,

$$\Psi_s(x,y,z) = [a_1(x,y) + a_2(x,y)]e^{-j\beta_s^z} \tag{14.60a}$$

and

$$\Psi_a(x,y,z) = [a_1(x,y) - a_2(x,y)]e^{-j\beta_a^z} \tag{14.60b}$$

where $a_1\ (x, y) = a(x, y - p/2)$ and $a_2(x, y) = a(x, y + p/2)$ and $a(x, y)$ is the normal $(\int\int |a(x,y)|^2 \mathrm{d}x\mathrm{d}y = 1)$ fundamental mode of a single waveguide of width w, centered $y = 0$. The two propagation constants β_s and β_a depend on the spacing between the input or output waveguides and, hence, on z.

In the central waveguide of width $2w$, β_s, and β_a are the propagation constants of the fundamental mode Ψ_s (even) and the second-order mode Ψ_a (odd), respectively.

Due to mode orthogonality, Ψ_s and Ψ_a do not exchange energy along the entire device. However, because $\beta_s \neq \beta_a$, the two modes accumulate a relative phase difference $\phi = \int(\beta_s - \beta_s)\ dz$ along the device length. Combining Equation 14.59 and Equation 14.60a and Equation 14.60b, the power output at ports 1 and 2 can be evaluated to yield (Forber and Marom 1986)

$$\text{at output port } 1; (y < 0): P = P_{in} \cos^2(\phi/2), \tag{14.61a}$$

and

$$\text{at output port } 2; (y < 0): P_x = P_{in} \sin^2(\phi/2), \tag{14.61b}$$

where P_{in} is the input power at port 1 for either λ_1 (1.31 μm) or λ_2 (1.55 μm), as shown in Figure 14.36, and ϕ is the accumulated phase difference over the two tapered and the central regions. P_{in} can be set to the normalized value of unity. The operating conditions, which must be satisfied, for demultiplexing are (Chung et al. 1989)

$$\phi(\lambda_1) = \Delta\beta_c(\lambda_1)L + 2\phi_t(\lambda_1) = m\pi \tag{14.62a}$$

and

$$\phi(\lambda_2) = \Delta\beta_e(\lambda_2)L + 2\phi_t\lambda_2 = (m-1)\pi \tag{14.62b}$$

where $\Delta\beta_c = \beta_s - \beta_a$ over the central region and is constant, whereas $\phi_t = \int_{z1}^{z2} \Delta\beta_t(z)dz$ over either tapered region, z_1 and z_2 being either $-z_t$ and $-L/2$ or $L/2$ and z_t and $\Delta\beta_t(z) = \beta(z) - \beta_a(z)$.

If $m(=2n)$ is an even integer, using Equation 14.61a and Equation 14.61b, we can show that $P=(\lambda_1)=1$, $P_x(\lambda_1)=0$, and $P=(\lambda_2)=0$, and $P_x(\lambda_2)=1$. Hence, the light at λ_1 is received at the output port 1, and λ_2 is received at the output port 2. If $m(=2n-1)$ is an odd integer, the situation is reversed so that the light at λ_1 is received at the output port 2, and λ_2 at the output port 1. The extinction ratios (ER) at both wavelengths are defined and calculated as

$$\mathrm{ER}(\lambda) = 10\log\left|\frac{P_x(\lambda)}{P=(\lambda)}\right| = 10\log\left|\tan^2\left(\frac{\phi}{2}\right)\right| \tag{14.63}$$

Theoretically, $\mathrm{ER}(\lambda_1)=\mathrm{ER}(\lambda_2)=-\infty$ because $\tan[\phi(\lambda_{1,2})/2]=0$ when $\phi=2n\pi$ or $(2n+1)\pi$. Any deviation of ϕ from these values, denoted by $\Delta\phi$, due to waveguide parameter errors, caused by computational errors in design, for example, an error in L, or fabrication errors, for example, an error Δw in the guide width w in the tapered regions or $\Delta(2w)$ in the central guide width $2w$, could result in a seriously degraded extinction ratio.

To see this, we can get, from Equation 14.63, near $\mathrm{ER}(\lambda)=-\infty$ and assuming $\Delta\phi$ to be small,

$$\begin{aligned}\mathrm{ER}(\lambda) &= 10\log|\tan^2(\Delta\phi(\lambda)/2| \\ &\approx 20\log|\Delta\phi(\lambda)/2|.\end{aligned} \tag{14.64}$$

For example, a small error of $\Delta\phi/2\approx 0.1$ rad. (6°) would lead to a degradation of $\mathrm{ER}\approx 20\log 0.1\approx -20$ dB.

Using a reduced parameter R, similar to that defined by Cheng and Ramaswamy (1991), we can start our design problem by solving Equation 14.62a and Equation 14.62b for L.

$$\begin{aligned}R = \frac{m\pi - 2\phi_t(\lambda_1)}{(m-1)\pi - 2\phi_t(\lambda_2)} &= \frac{\Delta\beta_c(\lambda_1)}{\Delta\beta_c(\lambda_2)} \\ = R_1^m \quad &= R_2.\end{aligned} \tag{14.65}$$

It is important to point out that $\Delta\beta_c$ depends on the wavelength λ, the waveguide width $2w$, hence, w, the waveguide diffusion depth d, hence, t_d (recalling $d=(D_e\, t_d)^{1/2}$ in Section 14.3). The phase ϕ_t also depends on these parameters, but, in addition, significantly, on the branching angle α. Hence, $\Delta\beta_c$ can be expressed as $\Delta\beta_c\,(\lambda, w, t_d)$ and ϕ_t as $\phi_t\,(\lambda, w, t_d, \alpha)$. If the parameters m, w and α are all fixed and λ_1 and λ_2 are set to 1.31 μm and 1.55 μm, respectively, t_d can be found from Equation 14.65 by a root-search technique, namely, by seeking the solution of

$$R_1^m(t_d) - R_2(t_d) = 0 \tag{14.66}$$

After t_d has been found, L can be determined from either Equation 14.62a and Equation 14.62b, namely,

$$L = \frac{m\pi - 2\phi_t(\lambda_1 = 1.3\ \mu\text{m},\alpha,w,t_d)}{\Delta\beta_c(\lambda_1 = 1.31\ \mu\text{m},w,t_d)} = \frac{(m-1)\pi - 2\phi_t(\lambda_2 = 1.55\ \mu\text{m},\alpha,w,t_d)}{\Delta\beta_c(\lambda_2 = 1.55\ \mu\text{m},w,t_d)}. \tag{14.67}$$

Most of the phase shift needed to effect demultiplexing is achieved over the central waveguide region. The contribution from ϕ_t over either of the tapered regions is very small. The design procedure can be represented by the flow chart in Figure 14.38.

Using the procedure outlined in Figure 14.38, one can calculate the lateral effective index profile of a single channel and a directional coupler as presented in Section 4.1. Hence, the lateral index profile of the X-branch could be similarly established as shown in Figure 14.39a and b. Applying the 2-D FD–VBPM presented in Section 5 to the side-diffused model in Figure 14.39b, the BPM simulations of the X-branch demultiplexer were performed. In Equation 14.65), $\Delta\beta_c$ ahd $\Delta\beta_t$ can be computed, employing a procedure described in Feit and Fleck (1980). Using Equation 14.66 and Equation 14.67, t_d and L are plotted versus w for different values for m in Figure 14.40 as design curves for both TM and TE modes. The optical fields, propagating through the device at 1.31 μm and 1.55 μm, are depicted in Figure 14.41 for the TM modes, where all the relevant parameters are also specified. From our numerous computational data, we have observed that, although the lengths L for the TM and TE modes could be about the same, the corresponding diffusion times t_d required are quite different. Hence, a device optimized for the TM mode propagation is *not* optimized for the TE mode operation and vice versa.

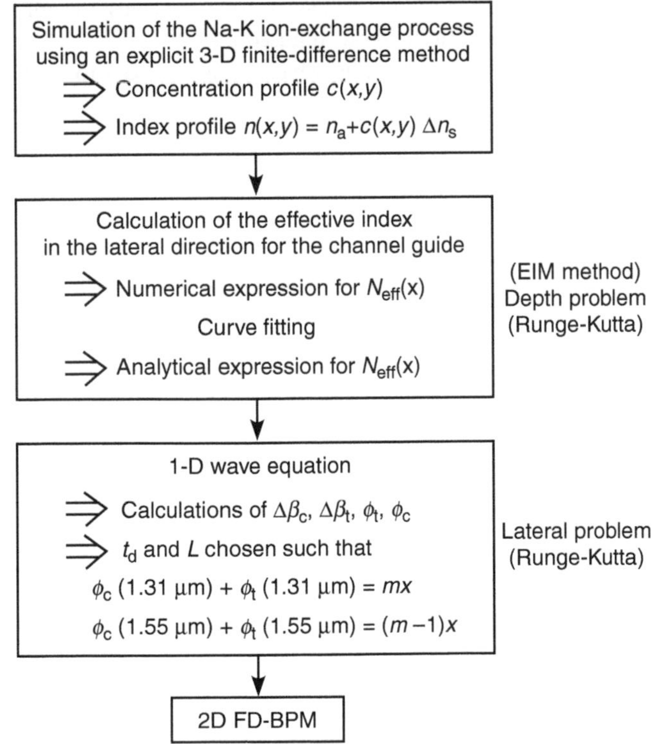

FIGURE 14.38 Design procedure flow chart. (From Babin, L. 1993. Optimization and fabrication of a widened x-branch optical demultiplexer in glass. M. Eng. thesis. Dept. of Electrical Engineering, McGill University, Montreal, Quebec, Canada. With permission.)

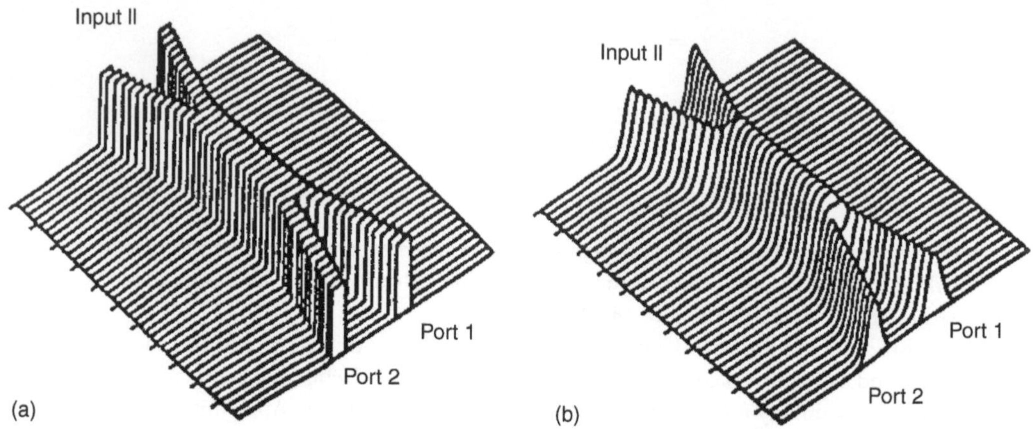

FIGURE 14.39 Lateral effective index. (From Yip, G. L. and Babin, L. 1995. Design optimization and fabrication of a widened x-branch demultiplexer by ion-exchanger in glass. In *Proc. Guided-Wave Optoelectronics: Device Characterization, Analysis, and Design*. T., Tamer, G., Griffel, and H. L., Bertoni, eds. pp., 221–229. Plenum Press, New York. With permission.)

Some caution should be exercised in choosing the excitation field at Input I1. In general, there is no exact analytical expression for the fundamental mode of a diffused channel guide. In our BPM simulations, the fundamental mode of an input branch with a width w and a lateral step-index profile, taken as N_{eff} $(y=0)$, was used as an approximation due to the availability of its analytical expression. After causing small radiation in the first few propagation steps, the approximate input mode will eventually excite the fundamental mode in our diffused input branch.

To calculate the extinction ratios ER and the radiation loss, it is necessary to evaluate the power in each output branch. The output field can be written in each branch as

$$\psi_{\mathrm{out}} = C_{\mathrm{g}}\phi_{\mathrm{g}} + \sum_i C_i\phi_i \tag{14.68}$$

where ϕ_{g} is the normalized fundamental guided mode of either output branch, and $\phi_i s$ are the corresponding radiative modes. The output power normalized to the input power in the rth branch

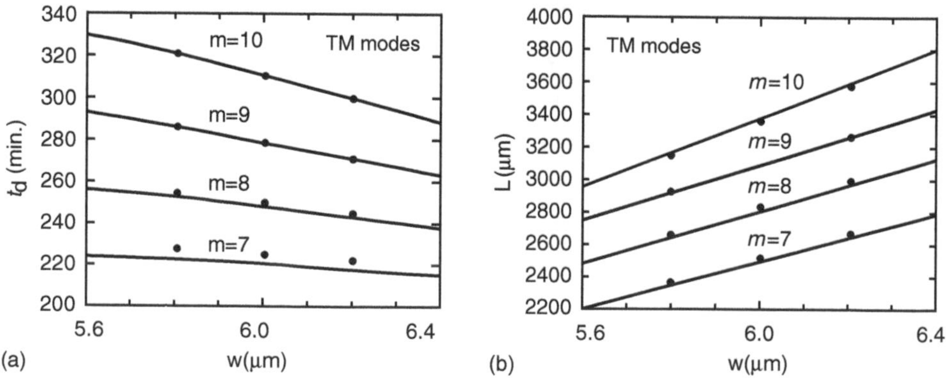

FIGURE 14.40 (a) t_{d} and (b) L versus w calculated by the Runge-Kutta method. (From Babin, L. 1993. Optimization and fabrication of a widened x-branch optical demultiplexer in glass. M. Eng. thesis. Dept. of Electrical Engineering, McGill University, Montreal, Quebec, Canada. With permission.)

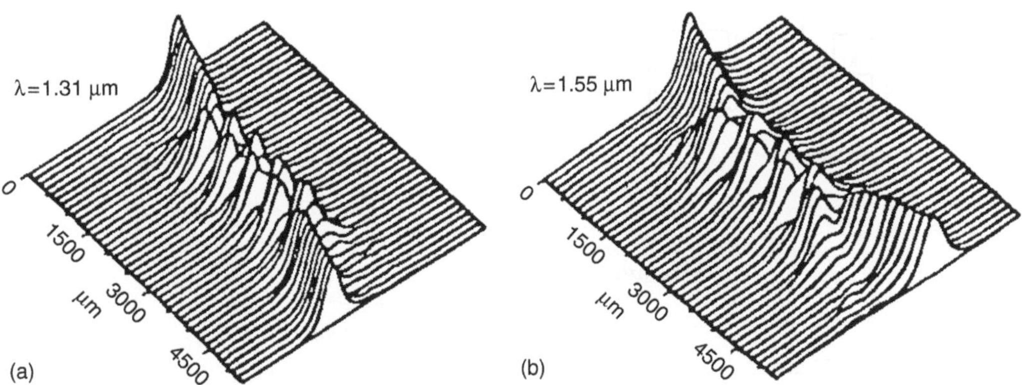

FIGURE 14.41 (a,b) BPM simulations for TM modes. (From Yip, G. L. and Babin, L. 1995. Design optimization and fabrication of a widened x-branch demultiplexer by ion-exchanger in glass. In *Proceeding of the Guided-Wave Optoelectronics. Device Characterization, Analysis, and Design.* T., Tamer, G., Griffel, and H. L., Bertoni, eds. pp. 221–229. Plenum Press, New York. With permission.)

($r=1$ or 2) is given by $|C_{g,r}|^2$. Using the orthogonality between the guided and radiative modes, it is possible to derive, from Equation 14.68, that

$$P_{g,r} = \left(\int_r \psi_{out} \cdot \phi^*_{g,r} dy \right)^2 \tag{14.69}$$

where $P_{g,1}=P_{=}$and $P_{g,2}=P_x \cdot \phi^*_{g,r}$ represents the BPM field calculated for each output branch, which is identical to that for the input branch. Because the BPM output field Ψ_{out} eventually approaches $\phi_{g,r}$, we can replace $\phi^*_{g,r}$ by Ψ^*_{out} in Equation 14.69). The radiation loss can, therefore, be expressed as

$$P_{rad} = 1-(P_{=} + P_x). \tag{14.70}$$

The radiation loss is represented in dB as follows:

$$L_R = 10 \log(1-P_{rad}), \tag{14.71}$$

so that $L_R=0$ dB when $P_{rad}=0$ and $L_R=-\infty$ dB when $P_{rad}=1$.

With the results of the BPM design simulation, a photomask containing several devices with different dimensions can be designed and used to fabricate several devices simultaneously on the same glass substrate. Several samples, each containing several devices, were prepared as depicted in Figure 14.32, and the device performance, such as the extinction ratios (ER) were measured as indicated in the schematic setup in Figure 14.42. Figure 14.43a,b show the near-field light spots at the two output arms for 1.31 and 1.55 μm, respectively, for the TM modes in one particular design. The corresponding measured extinction ratio (ER) is 20 dB at 1.31 μm and 15 dB at 1.523 μm, compared with the theoretical values of 35 dB at both wavelengths shown in Figure 14.41.

Figure 14.40a, b are the results obtained assuming no errors in photolithography, namely, $w_t=w$, $2w_c=2w$. Assuming an error of -0.4 μm in both $2w_c$ and w_t due to imperfect photolithography, namely, $2w_c=11.6$ μm, $w_t=5.6$ μm, and $2w_c \neq 2w_t$, a BPM simulation was carried out separately (because the results in Figure 14.40 could not be used for this case) with $m=9$, $\alpha=0.5°$, $t_d=270$ min, and $L=$ 2930 μm for the TM modes. The corresponding theoretical ER's for this design are 35 dB at both 1.31 μm and 1.523 μm, whereas the measured ER's are 20 dB at both wavelengths.

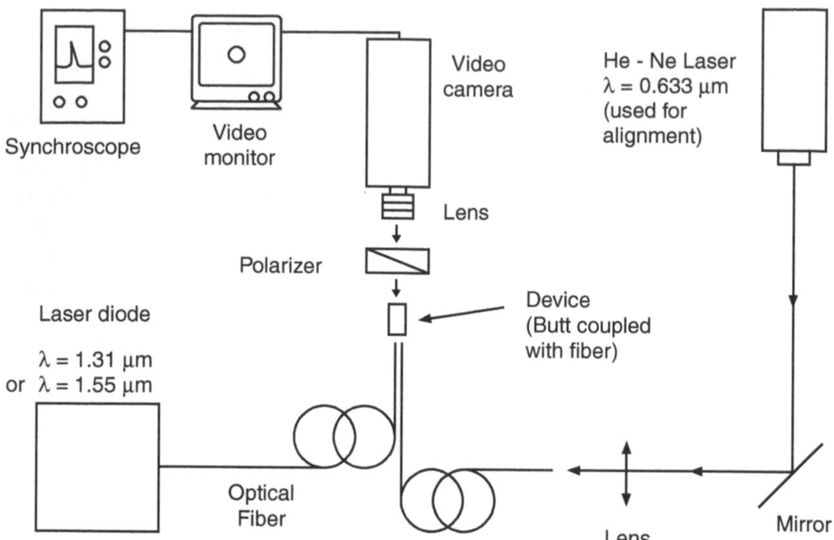

FIGURE 14.42 The measurement setup. (From Yip, G. L. and Babin, L. 1995. Design optimization and fabrication of a widened x-branch demultiplexer by ion-exchanger in glass. In *Proceedings of the Guided-Wave Optoelectronics: Device Characterization, Analysis, and Design*. T., Tamer, G., Griffel, and H. L., Bertoni, eds, pp. 221–229. Plenum Press, New York. With permission.)

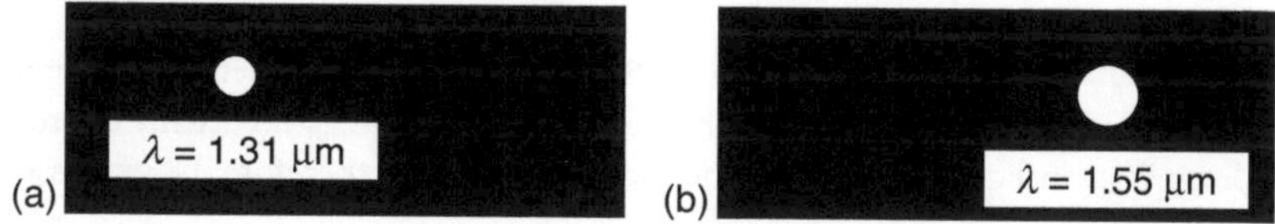

FIGURE 14.43 (a,b) Near-field output light spots. (From Yip, G. L. and Babin, L. 1995. Design optimization and fabrication of a widened x-branch demultiplexer by ion-exchanger in glass. In *Proceedings of the Guided-Wave Optoelectronics, Device Characterization, Analysis, and Design*. T., Tamer, G., Griffel, and H. L., Bertoni, eds, pp. 221–229. Plenum Press, New York. With permission.)

To date, the findings reported represent the best theoretical and experimental results obtained by us. Although, there are still significant discrepancies between the design and measurement ERs, the general agreement appears reasonable. Improvements are still possible, especially, in terms of fabrication and measurement.

14.6.4 An Electro-Optic Ti-diffused $LiNbO_3$, Ridge Waveguide, Mode-Confinement Modulator

The device configuration of the line modulator is shown in Figure 14.44. The device consists of two channel ridge waveguides in regions 1 and 3 connected by a slab waveguide in region 2. The guided modal field in the input ridge guide, on entering the slab region, begins to diverge, so that, at the mouth of the output guide, the overlap between the transmitted field and the guided modal field of the output guide is reduced, resulting in a fraction of the input power being transmitted into the output guide. An increase in the lateral confinement of the modes in the slab waveguide can be achieved by electro-optically inducing a channel waveguide in region 2, producing an increased modal field overlap and, hence, increased transmission between the input and output channel guides. If the voltage polarity between the

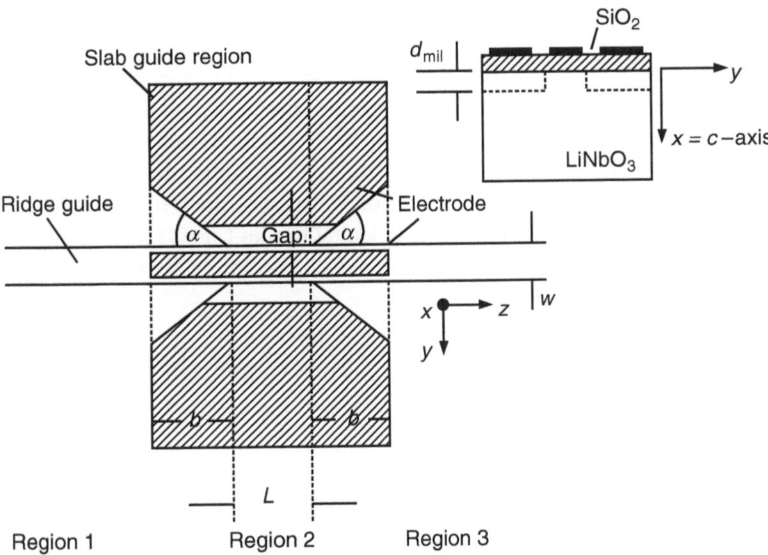

FIGURE 14.44 Device configuration. (From Sekerka-Bajbus, M. K., Yip, G. L., and Goto, N. 1990. BPM design-optimization and experimental improvements of a Ti:LiNbO$_3$ ridge waveguide linear-mode confinement modulator. *IEEE J. Lightwave Technol.*, 8, 1742–1749.)

electrodes is reversed, the optical waves in region 2 will be scattered away from the output guide, yielding a transmission reduction. Hence, with the electrodes, the device performs as an intensity modulator for a voltage signal applied across the electrodes. The branch structure at the end of the slab region helps to guide away the scattered fields at the output junction.

Because the BPM can treat both the guided and radiation (scattered) modes, it is well suited for calculating the scattered modes in region 2 of the modulator. A theoretical analysis and design of this device requires the calculation of the transmitted power across the slab region into the output guide as a function of the applied electrode voltage. This can be accomplished with a combination of the EIM and FFT–BPM as discussed in Section 5 (Sekerka-Bajbus et al. 1990).

As discussed in Section 14.2 and Section 14.3, the lateral effective-index distribution for the device can be found by replacing each lateral point, along the y direction of the crystal surface with a homogeneous slab guide of the same effective index $n_{\text{eff}}(y, z)$ as that of the diffused guide and solving the following WKB relationship:

$$2k_0 \int_{x_0}^{x_t} \left[n(x,y,z)^2 - n_{\text{eff}}^2(y,z)^{1/2} \right] \text{d}x = \left(2m + \frac{3}{2} \right) \pi, \quad m = 0,1,\ldots \tag{14.72}$$

where $m=0$ for a single mode waveguide and $n(x_t, y, z) = n_{\text{eff}}(y, z)$. The lower integration limit x_0 is zero for the ridge channel guide but equal to the milling depth d_{mil} in the etched regions (by ion milling). The index distribution $n(x, y, z)$ is assumed to be Gaussian along the x axis and, below the electrodes, is perturbed by the electro-optical effect. Hence,

$$n(x,y,z) = n_b + \Delta n_s \exp\left(-x^2/d^2\right) + \left[\frac{1}{2} \gamma_{33} n_b^3 E_x(x,y,z) \right] \tag{14.73}$$

where γ_{33} is the electro-optical coefficient along the x axis (C axis of LiNbO$_3$), E_x is the applied electric field in the x direction, and all other parameters have been defined previously. Given any voltage applied across the electrodes, E_x can be calculated by using a method as described in Section 4.5 (Thylén and

Gravestrand 1985), taking into consideration the attenuating influence of the SiO_2 buffer layer. The effective-index distribution in the slab region (region 2) for the on- and off-state voltage polarities are shown in Figure 14.45. The optical-field distribution in the device can, then, be calculated by propagating an eigenmode along the two-dimensional effective-index models in regions 1, 2, and 3, using the FFT–BPM. The output power in a particular mode can be determined from the BPM data by overlapping the output optical field in region 3 with the specific normalized guide mode.

The effective index of the ridge structure can be controlled accurately through control of the milling depth d_{mil}, which helps to determine the number of lateral modes $(m+1)$ a ridge guide of width w can support. Using the well known relationship [Equation 2.1.24 in Tamir (1988)],

$$m = \mathrm{INT}\left[2\frac{W}{\lambda}\sqrt{\left(n_{\mathrm{eff}}^{\mathrm{ridge}}\right)^2 - \left(n_{\mathrm{eff}}^{\mathrm{milled}}\right)^2}\right], \tag{14.74}$$

$\Delta N = \left[\left(n_{\mathrm{eff}}^{\mathrm{ridge}}\right)^2 - \left(n_{\mathrm{eff}}^{\mathrm{milled}}\right)^2\right]^{1/2}$ is plotted versus λ/w in Figure 14.46 with m as a parameter. The $m=1$ can be used to estimate the maximum permissible value of ΔN for single-mode operation. $N_{\mathrm{eff}}^{\mathrm{ridge}}$ is fixed by the diffusion conditions and independent of the subsequent milling. It can be determined by using Equation 14.72 and Equation 14.73. The desired $N_{\mathrm{eff}}^{\mathrm{milled}}$, determined from Figure 14.46a, can be substituted in Equation 14.72 to obtain the required d_{mil}, iteratively. The results are plotted in Figure 14.46b for a gaussian diffusion profile with a surface-index increase of $\Delta n_s=0.01$ and 0.05, respectively, and a diffusion depth of $d=2.0$ μm at $\lambda=0.6328$ μm. The curves in Figure 14.46a and b are very useful in designing ridge waveguides supporting $(m+1)$ lateral modes.

The BPM simulations of the optical fields for a TM mode, propagating through a multimode modulator at an applied voltage of +5 and −5 V (side-electrodes grounded), respectively, are shown in Figure 14.47. The device's design parameters are given in Table 14.7 and are confirmed by the parameters of an actual fabricated device (sample 2 to be discussed later). The optical mode confinement in the induced channel guide in region 2 is clearly observed in Figure 14.47a. The scattered power along the branch structure, away from the mouth of the output ridge guide, is also clearly visible, especially for the off state. The presence of the second and third-order modes in the output ridge guides can also be seen for both voltage polarities. About 10–15% of the transmitted power propagates in these higher order modes. This amount of power can be calculated by overlapping the output modal field in region 3 with its normalized guided mode of the appropriate order m.

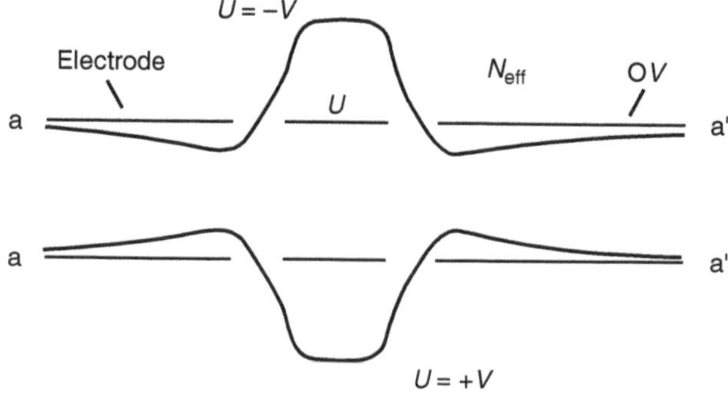

FIGURE 14.45 Effective-index distribution. *IEEE J. Lightwave Technol.* (From Sekerka-Bajbus, M. K., Yip, G. L., and Goto, N. 1990. BPM design-optimization and experimental improvements of a Ti:$LiNbO_3$ ridge waveguide linear-mode confinement modulator. *IEEE J. Lightwave Technol.*, 8, 1742–1749.)

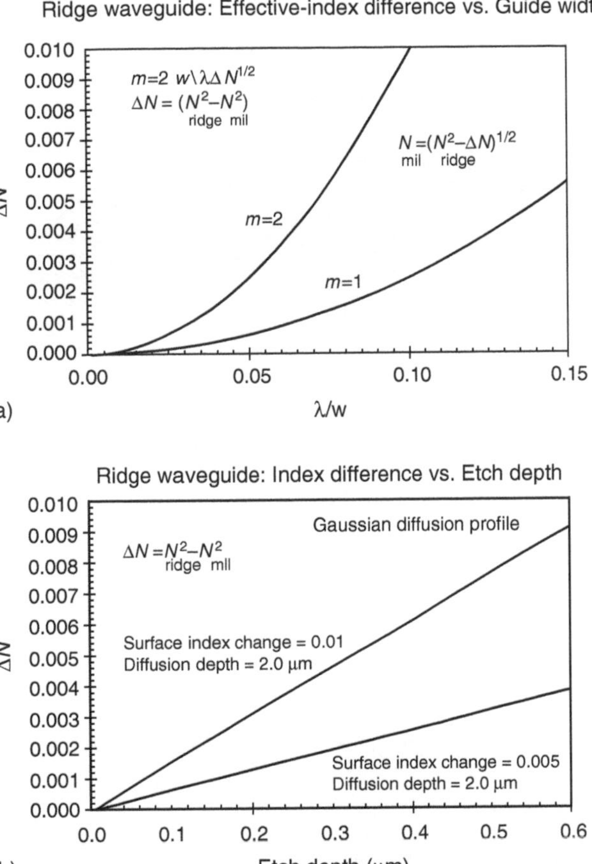

FIGURE 14.46 (a,b) Ridge waveguide design curves. (From Sekerka-Bajbus, M. K., Yip, G. L., and Goto, N. 1990. BPM design-optimization and experimental improvements of a Ti:LiNbO$_3$ ridge waveguide linear-mode confinement modulator. *IEEE J. Lightwave Technol.*, 8, 1742–1749.)

The theoretical model and analysis, just presented, makes it possible to study the effects of varying certain design parameters on the modulation characteristics. Such design calculations are very informative in obtaining a modulator design with some desired performance figures before proceeding on to the actual experimental work. Many BPM simulations have been performed to improve the modulator design for single-mode operations and to achieve important device characteristics such as (1) good modulation depth ($>90\%$) and linearity of the output intensity versus the applied voltage, (2) a small drive voltage ($<\pm 10$ V, say), and (3) small device dimensions, etc. Fabrication tolerances, however, usually impose practical limitations on an achievable device design. In general, increasing the modulation length reduces the drive voltage required to switch off the power in the output guide in the off state. Reducing the branch angle α has the advantage of better separating the guided and scattered modes along the branch structure, so that large radiation intensity will not appear near the output field, especially in the device's off state. It is found that device performance is critically dependent on milling depth, but its proper value can be determined within the range permitted for single-mode operation for a particular device design.

Our design methodology is now used to improve the performance of a similar device fabricated previously with a modulation depth $[(I_{\max}-I_{\min})/I_{\max}]$ of 67% and a drive voltage of ± 20 V (Bélanger

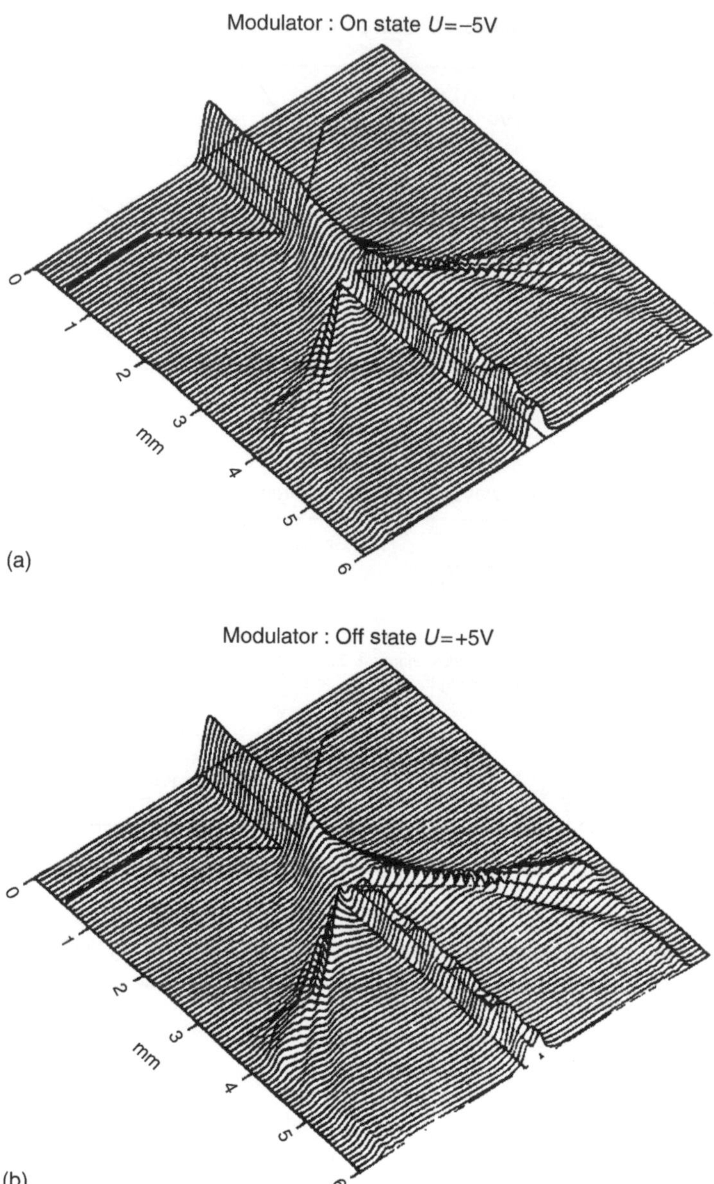

FIGURE 14.47 (a,b) Optical field distribution in the modulator. (From Sekerka-Bajbus, M. K., Yip, G. L., and Goto, N. 1990. BPM design-optimization and experimental improvements of a Ti:$LiNbO_3$ ridge waveguide linear-mode confinement modulator. *IEEE J. Lightwave Technol.*, 8, 1742–1749.)

and Yip 1986), by optimizing its photomask-independent parameters, while using the same photomask as before. With a buffer layer thickness of 0.1 μm, Δn_s, = 0.01 and d = 2 μm, the influence of the milling depth d_{mil} is shown in Figure 14.48 for a device supporting up to three lateral modes. The device performance is, ideally, best under single-mode operation. However, the required milling depth of 0.05 μm is difficult to achieve experimentally, and coupling into such a shallow ridge guide using an input prism is also not easy. To avoid these difficulties, a milling depth of 0.3 μm is chosen. The resulting design supports three lateral modes and provides good modulation depth and linearity with a relatively

TABLE 14.7 Device Parameters

Sample Number	1		2
Surface index change Δn_s	0.01		0.01
Diffusion depth D	2.0 μm		2.0 μm
Ridge height d^{mil}	0.3 μm		0.3 μm
Buffer layer thickness d	0.116 μm		0.08 μm
Modulation length L		1.0 mm	
Branch angle a		2.58°	
Branch length b		1.0 mm	
Ridge width w		10.0 μm	
Center electrode width		8.0 μm	
Inter electrode gap		4.0 μm	
Number of modes supported in-/output guide	3		3
NB: electrode axial misalignment	<3μm		Negligible

Source: From Yip, G. L. and Babin, L. 1995. Design optimization and fabrication of a widened x-branch demultiplexer by ion-exchanger in glass. In *Proc. Guided-Wave Optoelectronics: Device Characterization, Analysis, and Design*. T., Tamer, G., Griffel, and H. L. Bertoni, eds. pp. 221–229. Plenum Press, New York. With permission.

small drive voltage of ± 10 V. These results are supported by experimental measurements to be presented.

Two line modulators with a ridge width of 10 μm were fabricated from the photomask designed previously without using the BPM (Sekerka-Bajbus et al. 1990). A 115-Å thick Ti film was thermally evaporated onto a cleaned $LiNbO_3$ *z*-cut substrate. The titanium was diffused into the crystal at 975°C for 4 h in a flowing argon atmosphere and a further 2 h in flowing oxygen to prevent lithium from diffusing out. This process generated a single-mode, stab waveguide with an estimated depth d and surface-index charge Δn_s of 2.0 μm and 0.01, respectively (using data as presented in Section 14.3). By photolithography, the appropriate ridge structure of height d_{mil} was formed out of the slab waveguide by ion-beam milling. After milling, the sample was annealed at 500°C for 2 h to repair the surface damage caused by the milling process. The output end was, then, polished to permit end fire coupling. Next, a thin SiO_2 buffer layer was RF sputtered onto the sample, followed by further annealing in oxygen for 2 h. Finally,

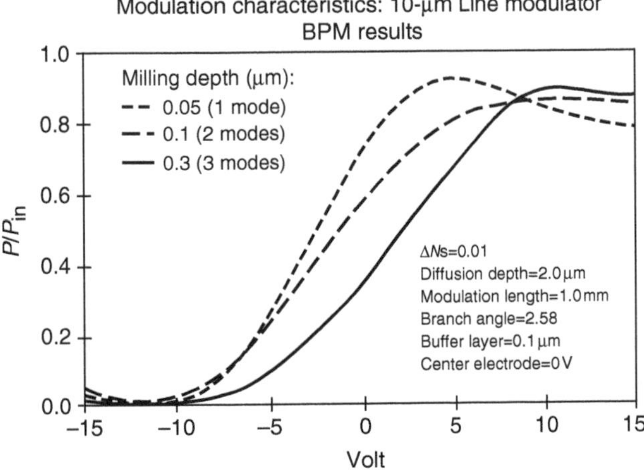

FIGURE 14.48 Modulation characteristics of a 10-μm modulator. (From Babin, L. 1993. Optimization and fabrication of a widened x-branch optical demultiplexer in glass. M. Eng. thesis. Department of Electrical Engineering, McGill University, Montreal, Quebec, Canada. With permission.)

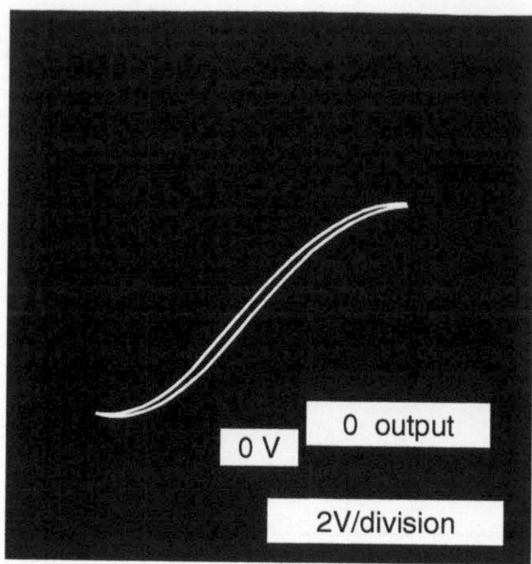

FIGURE 14.49 Measured output intensity. (From Babin, L. 1993. Optimization and fabrication of a widened x-branch optical demultiplexer in glass. M. Eng. thesis. Department of Electrical Engineering, McGill University, Montreal, Quebec, Canada. With permission.)

aluminum electrodes were deposited on top of the buffer layer by thermal evaporation, again through photolithography. The fabricated devices were, then, ready for measurements.

A He–Ne laser beam at 0.6328 μm was TM polarized and focused through a microscope objective lens onto a prism which coupled the light into the input ridge guide. At the output end, light leaving the ridge guide was focused through a second microscope lens onto a photodetector connected to an oscilloscope. The drive voltage applied to the electrodes was a slowly varying ramp from a function generator. Figure 14.49 shows the measured intensity versus the drive voltage for sample 2. Table 14.8 summarizes the measured along with the theoretically predicted performance. The higher drive voltage of ± 15 V required for sample 1 is partly due to the slightly thicker buffer layer compared to sample 2 and an electrode misalignment. The much reduced drive voltage to ± 8 V and the modulation depth of 97% for sample 2 constitute considerable improvements over previously published results (Bélanger and Yip 1986). This is primarily due to a thinner buffer layer and a reduced ridge height, hence, reducing the number of lateral modes. Figure 14.50 compares the measured modulation characteristics with the theoretical design calculations, using the BPM. The good agreement confirms the usefulness of the BPM simulations in device design optimization.

14.6.5 Design and Fabrication of a Y-Branch TE–TM Mode Splitter in $LiNbO_3$ by Proton Exchange and Ti Diffusion

The geometry of this device on a *z*-cut $LiNbO_3$ substrate (Goto et al. 1989) is shown in Figure 14.51. The input waveguide and arm 1 are fabricated by Ti diffusion (TI) whereas arm 2 is by proton exchange (PE).

TABLE 14.8 Comparison between Experimental and Theoretical Results

	Drive Voltage Range (V)		Modulation Depth	
	Theor.	Exp.	Theor.	Exp.
Sample 1	−15 V, +20 V	−15 V, +15 V	95%	98%
Sample 2	−10 V, +10V	−8 V, +8 V	99%	97%

Source: From Babin, L. 1993. Optimization and fabrication of a widened x-branch optical demultiplexer in glass. M. Eng. thesis. Department of Electrical Engineering, McGill University, Montreal, Quebec, Canada. With permission.

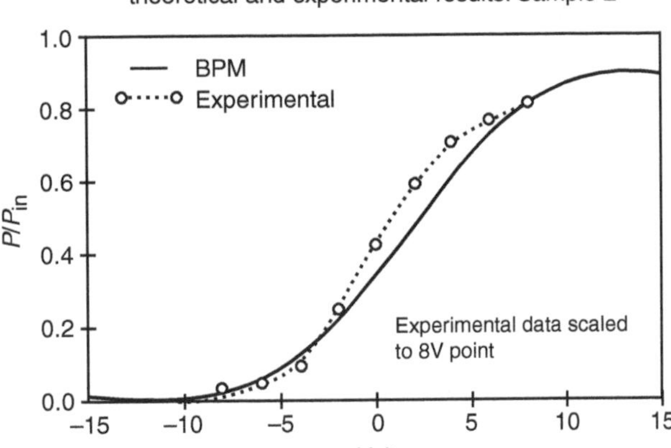

FIGURE 14.50 Comparison of the theoretical modulation characteristics. (From Yip, G. L. and Babin, L. 1995. Design optimization and fabrication of a widened x-branch demultiplexer by ion-exchanger in glass. In *Proc. Guided-Wave Optoelectronics: Device Characterization, Analysis, and Design*. T., Tamer, G., Griffel, and H. L., Bertoni, eds, pp. 221–229. Plenum Press, New York. With permission.)

Because PE increases only the extraordinary index while having very little effect on the ordinary index, arm 2 is "seen" only by the E^x (TM) modes. If the effective index in arm 2 is higher than in arm 1, the fundamental E^x_{11} mode will propagate into arm 2. On the other hand, the existence of arm 2 has negligible effect on the E^y (TE) modes, and the fundamental E^x_{11} mode propagates along arm 1, hence, the TE–TM mode splitting. The taper angle θ_1 and branch angle θ_2 should be sufficiently small to avoid an appreciable mode conversion. They were chosen to be 0.0032 and 0.01 rad., respectively.

Again, the design must start with predetermined fabrication conditions, which are to be optimized with the design outcome in an iterative manner. Hence, the fabrication conditions are briefly described here. After Ti was diffused followed by one hour in flowing oxygen, a Ta mask pattern for PE was formed. The PE arm was formed by immersing the sample in phosphoric acid at 200°C for about 20 min, and its

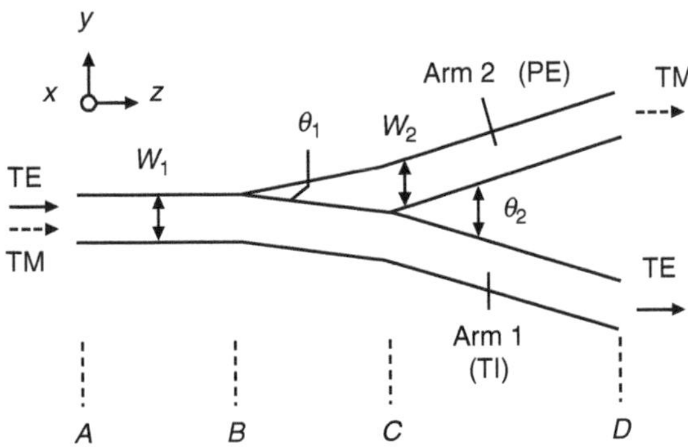

FIGURE 14.51 Comparison of the theoretical modulation characteristics. (From Yip, G. L. and Babin, L. 1995. Design optimization and fabrication of a widened x-branch demultiplexer by ion-exchanger in glass. In *Proceeding of the Guided-Wave Optoelectronics: Device Characterization, Analysis, and Design*. T., Tamer, G., Griffel, and H. L., Bertoni, eds, pp. 221–229. Plenum Press, New York. With permission.)

TABLE 14.9 Experimental Results of TE-TM Mode Splitter

Sample No.	Fabrication Conditions	Effective Index	Extinction Ratio
1	Ti-diffusion: Ti 107 Å Proton exchange: 20 min at 200°C Annealing: 15 min at 300°C	Arm 1 (TI): $2.2889 + 0.0005(E^y_{11})$ $2.2008 + 0.0005(E^x_{11})$ Arm 2 (PE): $2.2024 + 0.0005(E^x_{11})$	TE: >20 dB TM: $=20$ dB
2	Ti-diffusion: Ti: 115 Å Proton exchange: 18 min. at 200°C Annealing: 60 min at 300°C	Arm 1 (TI): $2.2950 + 0.0005(E^y_{11})$ $2.2015 + 0.0005(E^x_{11})$ Arm 2 (PE): $2.2015 + 0.0005(E^x_{11})$	TE: >20 dB TM: $=-1.6$ dB

Source: From Yip, G. L. and Babin, L. 1995. Design optimization and fabrication of a widened x-branch demultiplexer by ion-exchanger in glass. In *Proceedings of the Guided-Wave Optoelectronics: Device Characterization, Analysis, and Design*. T. Tamer, G. Griffel, and H. L. Bertoni, eds., pp. 221–229. Plenum Press, New York. With permission.

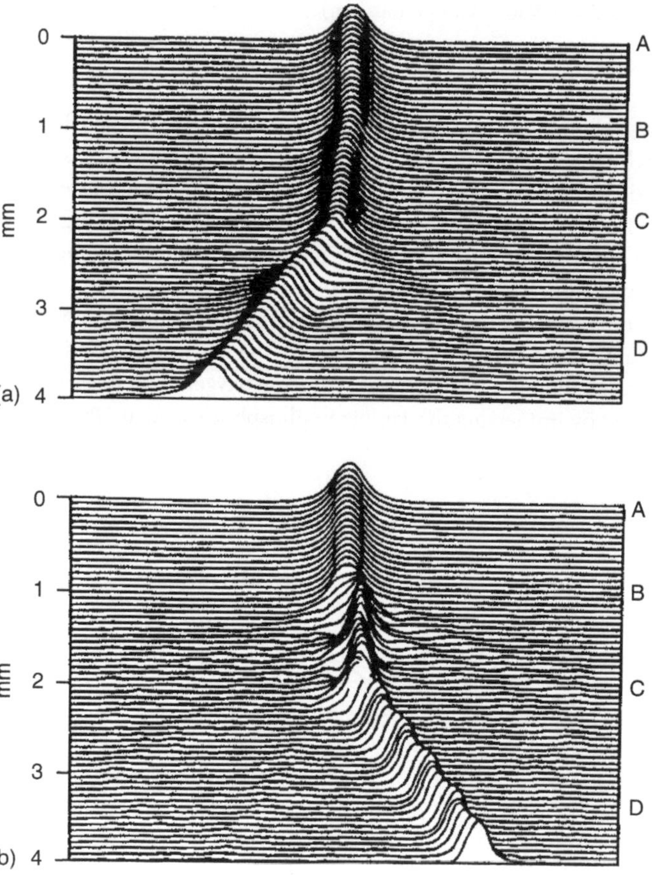

FIGURE 14.52 Intensity distributions of fields propagating. (From Yip, G. L. and Babin, L. 1995. Design optimization and fabrication of a widened x-branch demultiplexer by ion-exchanger in glass. In *Proceeding of the Guided-Wave Optoelectronics: Device Characterization, Analysis, and Design*. T., Tamer, G., Griffel, and H. L., Bertoni, eds, pp. 221–229. Plenum Press, New York. With permission.)

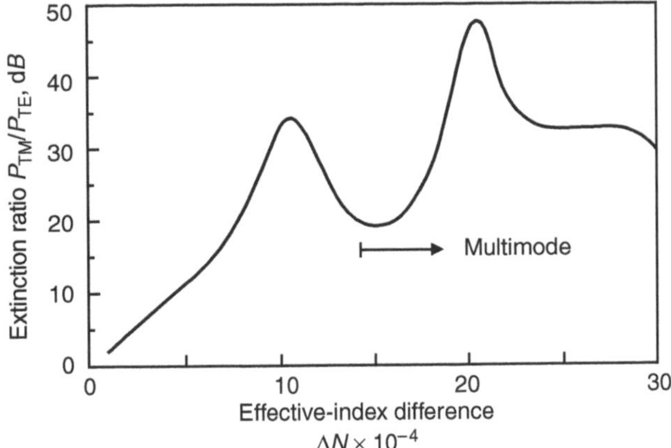

FIGURE 14.53 Extinction ratio of E^z wave. (From Yip, G. L. and Babin, L. 1995, Design optimization and fabrication of a widened X-branch demultiplexer by ion-exchanger in glass. In *Proceedings of the Guided-Wave Optoelectronics. Device Characterization, Analysis, and Design*. T., Tamer, Griffel, G., and Bertoni, H. L., eds, pp. 221–229. Plenum Press, New York. With permission.)

effective index adjusted by postannealing at 300°C for 20–60 min, After the PE process, the Ta film was removed, and the end facets of the device were polished for experimental characterization. The experimental results are presented in Table 14.9.

The effective index of diffused channel guides can be calculated by solving the dispersion equation for an equivalent homogeneous slab waveguide. The index distribution in the depth direction is assumed to

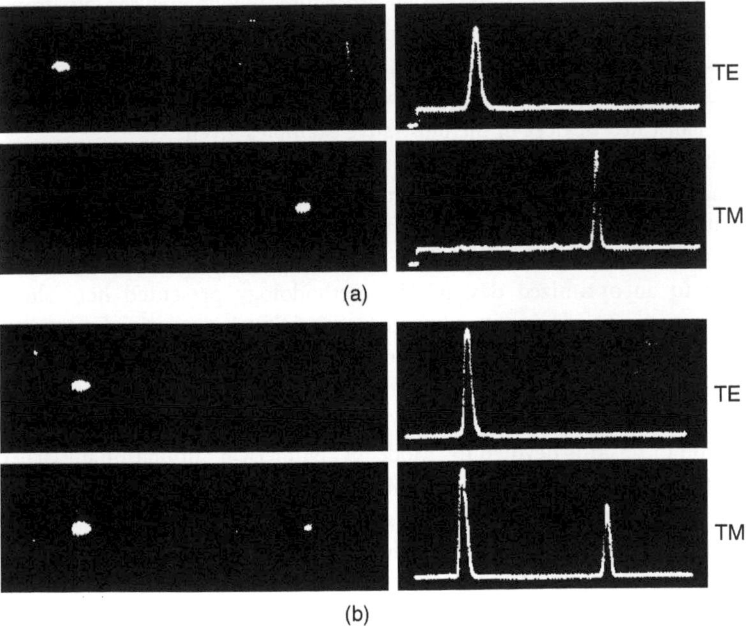

FIGURE 14.54 Near-field patterns and their intensity profiles. (From Yip, G. L. and Babin, L. 1995. Design optimization and fabrication of a widened x-branch demultiplexer by ion-exchanger in glass. In *Proceedings of the Guided-Wave Optoelectronics. Device Characterization, Analysis, and Design*. T., Tamer, G., Griffel, and H. L., Bertoni, eds, pp. 221–229. Plenum Press, New York. With permission.)

be Gaussian for both the TI and APE (PE followed by annealing) waveguides. In the lateral direction, the simple step-index model can be used. The optical field distribution in the device can, then, be calculated by propagating an eigenmode along this effective-index model, using the FFT–BPM. The power in each branch can be calculated by overlapping the optical field in the branch with a normalized eigenmode. The optical field distributions in the device are presented in Figure 14.52a and b for the TE and TM modes, respectively, using the data in Table 13.9. In Figure 14.52b the coupling of the TM modes into the PE waveguide along the taper is clearly visible. Most of the TM mode energy is transferred into the PE arm, whereas a small amount of energy is radiated into the substrate.

The amount of TM mode energy propagation into the PE arm is critically dependent on the difference in the effective indices between the two arms, ΔN $(=N_{\mathrm{Pe}}-N_{\mathrm{Ti}})$. Figure 14.53 is presented the extinction ratio for the E^z mode as a function of ΔN. Clearly, an extinction ratio of more than 20 dB for a single-mode device can be achieved only for a very narrow range of ΔN. This requires strict control of fabrication conditions and is consistent with experimental results obtained.

The near-field patterns of the outputs in arm 1 and arm 2 and the Intensity profiles are shown in Figure 14.54. The extinction ratio of the outputs between the two arms in given in Table 14.9. Since the effective indices of the two arms of sample 2 are almost the same, the E^z (TM) wave is seen to split into both arms. However, sample 1 shows good TE–TM splitting because, in this case, ΔN at 16×10^{-4} has a proper value. These experimental results are in good agreement with theoretical predictions in Figure 14.53.

14.7 Conclusions

In this chapter, we have presented the theoretical and experimental characterizations of slab and channel waveguides by K^+-ion exchange in glass and Ti diffusion into $LiNbO_3$ and the modeling of their index profiles. These serve as illustrations for waveguides fabricated by diffusion-related technologies. The information obtained is important to the analysis, design, and implementation of planar waveguide devices, made by such technologies and employing slab and channel guides as basic structures. Once the index profiles are accurately modeled, the design procedure, involving a combination of the effective-index method and beam propagation method as presented in Sections 6.1, 6.3–6.5, or a combination of the effective-index method and variational method as presented in Section 6.2, then leads to design calculations of device performance parameters, which are well supported by measured performance parameters of the fabricated devices in the illustrative examples given.

We conclude that the key to accurate index modeling of a particular waveguide structure lies in a detailed and clear understanding of the physical processes going on in the fabrication stage with prescribed fabrication conditions, for example, diffusion and annealing temperature and time. Only after establishing an accurate index model, would clever mathematics play a significant role in the design stage, eventually leading to an optimized design. The methodology presented here should be generally applicable to all planar waveguide devices fabricated by the diffusion process, for example, ion exchange in glass involving other ion sources and substrates, Ti diffusion into $LiNbO_3$ and $LiTa_2O_3$ and proton exchange in the two latter substrates. In a wider sense, the design philosophy should be equally applicable to waveguides made from III–V compounds and polymers. Regrettably, we cannot elaborate here.

With the inclusion of sessions devoted to "Modeling, Numerical Simulation and Theory" since 1989, in such an important technology-related topical meeting (IEEE) as "Integrated Photonics Research" and the numerous workshops and symposia on "Design and Simulation of Guided-Wave Optoelectronic Devices", time seems to have arrived to achieve devices, which do not simply work on clever ideas but which have also been design optimized. Design predictions can be matched with the experimental performance of fabricated devices. In the process, much experimental effort can be saved.

Acknowledgments

The author would sincerely like to thank Dr. M. Gupta for his kind invitation to contribute this chapter. Former research associates and graduate students, J. Albert, M. Bélanger, L. Babin, J. Y. Chen, J. Finak,

N. Goto, M. Haruna, K. Kishioka, R C. Noutsios, M. K. Sekerka-Bajbus, and F. Xiang have contributed to the materials presented in this chapter. Our research was supported by NSERC (Natural Sciences and Engineering Research Council of Canada) operating and strategic grants. P. Jorgensen, L. Chen, and A. Yip typed the manuscript. Last but not least, the author would like to express his appreciation to his wife, Alice, and children, Gwendolyn, Alexandra, and Santa, for their understanding and support.

List of Symbols

Symbol	Definition
a_E	asymmetric measure for TE modes
a_M	asymmetric measure for TM modes
b_E	normalized guide index for TE modes
b_m	normalized guide index for TM modes
$c(x,t)$	ion concentration density
C_1, C_2	arbitrary constants associated with D_e
d	diffusion depth
D_0	a constant associated with D_x or D_y
D_e	effective diffusion coefficient
D_i, $i=x, y, z$	electric displacement vector along the i direction
D_x, D_y	diffusion coefficient along the x or y direction
E	electric field
E_a	applied field
E_a^t	total applied field
E_o	a constant field
F_e	effective field-dependent coefficient
H	magnetic field
He–Ne	helium–neon
k_1, k_2	transverse propagation constants
k_B	Boltzmann constant
k_o	wave number in free space
K_0	a constant associated with field-assisted ion-exchange
K^+	potassium ion
m	mode order
n_0	ordinary index
n_b	substrate or bulk index
n_c	cover index
n_e	extraordinary index
n_f	film index
n_p	prism index
n_s	surface index
Δn_s	change in the surface index over the substrate index
$n(x,y)$	local refractive index of an inhomogeneous medium
N	effective guide index
Na^+	sodium ion
N_{eff}	same as N
r_{ij}	electro-optic or Pockels coefficients
t	diffusion time
T	temperature
Ti	titanium
Tl^+	thallium ion
V	normalized frequency of an optical waveguide
w	channel guide width
x	normalized waveguide depth
x_t	normalized turning point
x	depth in the waveguide
β	propagation constant in the z direction
β_p	propagation constant for the depth mode
γ_3, γ_4	transverse decay constants
γ_c	decay constant in the cover
γ_s	decay constant in the substrate
κ	transverse propagation constant in the slab waveguide
λ	wavelength
μ_k	mobility of potassium ion in glass
ν	Fourier transform variable
ρ	atomic density
σ_0	constant associated with a Gaussian function
σ_1	constant associated with the Hermite–Gaussian function
τ	thickness of Ti film
ϕ	electric potential
	Other symbols used are as defined in the text.
Φ_c	phase shift from the film-cover interface
Φ_s	phase shift from the film-substrate interface
Ψ	transverse field component
ω	angular frequency
$\in_a$	thermal activation energy
$\in_i$, $i=x, y, z$	relative permittivity along the i direction

References

Abou-el-Liel, M. and Leonberger, F. 1988. Model for ion-exchanged waveguides in glass. *J. Am. Cer. Soc.*, 71, 497.

Albert, J. and Yip, G. L. 1988. Wide single-mode channels and directional coupler by two-step ion-exchange in glass. *IEEE J. Lightwave Technol.*, 6, 552.

Alferness, R. C. 1988. Titanium-Diffused Lithium Niobate Waveguide Devices, In *Guided-Wave Optoelectronics*, T. Tamir, ed., pp. 145–210. Springer, chap 4

Babin, L. 1993. Optimization and fabrication of a widened x-branch optical demultiplexer in glass. M. Eng. thesis, McGill University.

Babin, L. J. M., Yip, G. L. 1994. Design and realization of a widened x-branch demultiplexer at 1.31 and 1.55 (m by K^+–Na^+ ion exchange in a glass substrate, OSA Topical Meeting on Integrated Photonics Research, February. San Francisco, Calif., Tech. Dig., pp. 241–243.

Bélanger, M. and Yip, G. L. 1986. A novel Ti:$LiNbO_3$ ridge waveguide linear mode confinement modulator fabricated by reactive ion-beam etching. *IEEE J. Lightwave Technol.*, 22, 252.

Burns, W. K., Klein, P. H., West, E. J., and Plew, L. E. 1979. Ti diffusion in Ti:$LiNbO_3$ planar and channel optical waveguides. *J. Appl. Phys.*, 50, 6175.

Chartier, G. H., Jaussaud, P., de Oliveira, A. D., and Parriaux, O. 1978. Optical waveguides fabricated by electric-field controlled ion exchange in glass. *Electron. Lett.*, 14, 132.

Chartier, G., Collier, P., Guez, A., Jaussaud, P., and Won, Y. 1980. Guided-Index surface or buried waveguides by ion-exchange in glass. *Appl. Opt.*, 19, 1092.

Chen, T. R. and Yang, Z. L. 1985. Modes of a planar waveguide with Fermi index profile. *Appl. Opt.*, 24, 2809.

Cheng, H. C. and Ramaswamy, R. V. 1991. Symmetrical directional coupler as a wavelength multiplexer-demultiplexer: Theory and experiment. *IEEE J. Quantum Electron.*, QE-27, 567.

Crank, J. 1975. *Mathematics of Diffusion. 2nd Ed.* Clarendon Press, London.

Čtyroký, J., Hofman, M., Janta, J., and Schröfel, J. 1984. 3-D analysis of Ti:$LiNbO_3$ channel waveguides and directional couplers. *IEEE J. of Quantum Electron.*, QE-20, 400.

Chung, Y. 1989. Analysis of a tunable multi-channel two-mode-interference wavelength division multiplexer/demultiplexer. *IEEE J. Lightwave Technol.*, 7, 766.

Chung, Y. and Dagli, N. 1990. Explicit finite difference beam propagation method: application to semiconductor rib waveguide Y-junction analysis. *Electron. Lett.*, 26, 711.

Doremus, R. H. 1969. Ion exchange in glass. J.A. Marinsky, ed., In *Ion-Exchange, Vol. 2*, pp. 3990–4118. Marcel Dekker, New York.

Feit, M. D. and Fleck, J. A. 1978. Light propagation in graded-index optical fiber. *Appl. Opt.*, 17, 3990.

Feit, M. D. and Fleck, J. A. Jr. 1980. Computation of mode eigenfunctions in graded-index optical fibers by the propagation beam method. *Appl. Opt.*, 19, 2240.

Feit, M. D., Fleck, J. A., and McCaughen, L. 1983. Comparison of calculated and measured performance of diffused channel-waveguide couplers. *J. Opt. Soc. Am.*, 73, 1296.

Fleck, J. A., Morris, J. R., and Feit, M. D. 1976. Time-dependent propagation of higher energy laser beams through the atmosphere. *Appl. Phys.*, 10, 129.

Forber, R. A. and Marom, E. 1986. Symmetric directional coupler switches. *IEEE J. Quantum Electron.*, 2E-22, 911.

Fukuma, M., Noda, J., and Iwasaki, H. 1978. Optical properties in titanium-diffused $LiNbO_3$ strip waveguides. *J. Appl. Phys.*, 49, 3693.

Gerald, G. F. 1970. *Applied Numerical Analysis*. Addison-Wesley.

Giallorenzi, T. G., West, E. J., Kirk, R., Ginther, R., and Andrews, R. A. 1973. Optical waveguides formed by the thermal migration of ions in glass. *Appl. Opt.*, 12, 1240.

Gill, S. 1951. A process for the step by step integration of differential equations in an automatic digital computing machine. *Proc. Camb. Philos. Soc.*, 47, 96.

Goto, N., Sekerka-Bajbus, M. R., and Yip, G. L. 1989. BPM analysis of Y-branch TE-TM mode splitter in $LiNbO_3$ by proton exchange and Ti diffusion. *Electron. Lett.*, 25, 1732.

Hadley, G. R. 1991. Transparent boundary conditions for beam propagation. *Opt. Lett.*, 16, 624.

Haruna, M., Bélanger, M., and Yip, G. L. 1985. Passive 3-branch optical power dividers by K^+-ion exchange in glass. *Electron. Lett.*, 21, 535.

Hocker, G. B. and Burns, W. K. 1977. Mode dispersion in diffused channel waveguide by the effective index method. *Appl. Opt.*, 16, 113–118.

Huang, W. P., Xu, C. L., Chu, S. T., and Chaudhuri, S. K. 1992. The finite difference vector beam propagation method: analysis and assessment. *IEEE J. Lightwave Technol.*, 10, 295.

Izawa, T. and Nakagome, H. 1972. Optical waveguide formed by electrically induced migration of ions in glass plates. *Appl. Phys. Lett.*, 21, 584.

Jackel, J. L., Rice, C. E., and Veselka, J. J. 1982. Proton-exchange for high index waveguides in $LiNbO_3$. *Appl. Phys. Lett.*, 41, 607.

Kaminow, I. P. 1974. *An Introduction to Electro-Optic Devices*. Springer, Berlin.

Kishioka, K. 1991. Characterization of ion exchange waveguide made with diluted KNO_3. *SPIE Proc.*, 1583, 19.

Knox, P. M. and Toulios, P. P. March, 1990. Integrated circuits for the millimeter through optical frequency range, In *Proc. for Symp. on Submillimeter Waves*, J. Fox, ed., Poly. Inst. of Brooklyn,.

Liu, P. L. and Li, B. J. 1991. Study of form birefringence in the waveguide devices using the semi-vector beam propagation method. *IEEE Photon. Techno. Lett.*, 3, 913.

Marom, E. and Ruschin, S. 1984. Relation between normal mode and coupled mode-analysis of parallel waveguides. *IEEE J. Quantum Electron.*, QE-20, 1311.

Miliou, A., Zhenguang, H., Cheng, H. C., Srivastava, R., and Ramaswamy, R.. 1989. 1889. Fiber-compatible K^+–Na^+ ion-exchanged channel waveguides: fabrication and characterization. *IEEE J. Quantum Electron.* 25.

Minakata, M., Saito, S., Shibata, M., and Miyazama, S. 1978. Precise determination of refractive-index changes in Ti-diffused $LiNbO_3$ optical waveguides. *J. Appl. Phys.*, 49, 4677.

Naitoh, H., Nuroshita, M., and Nakayama, T. 1977. Mode control of Ti-diffused $LiNbO_3$ slab optical waveguldes. *Appl. Opt.*, 16, 2546.

Neyer, A. 1983. Electro optic x-switch using single mode $LiNbO_3$ channel waveguides. *Electron. Lett.*, 19, 553.

Nishihara, H., Haruna, M., and Suhara, T. 1989. *Optical Integrated Circuits*. McGraw-Hill,, Chap. 2.

Optical Materials, Optics Guide 4, Melles Griot Corporation, Irvine, Calif, pp. 3–10 1988.

Popuchon, M., Roy, A., and Ostrowsky, D. B. 1977. Electrically active bifurcation: BOA. *Appl. Phys. Lett.*, 31, 266.

Ralston, A. and Wilf, H. S. 1960. *Mathematical Methods for Digital Computers* Vol. 1. 1 Wiley, pp. 111–120.

Ramaswamy, R. V. and Najafi, S. 1. 1986. Planar, buried, ion-exchanged glass waveguides: diffusion characteristics. *IEEE J. Quantum Electron.*, 22, 883.

Ramer, O. G. 1982. Integrated optic electrooptic modulator electrode analysis. *IEEE J. Quantum Electron.*, QE-18, 386.

Saijonmaa, J. and Yevick, D. 1983. Beam propagation analysis of loss in bent optical waveguides and fibers. *J. Opt. Soc. Am.*, 73, 1785.

Scarmozzino, R. and Osgood, R. M. 1991. Comparison of finite difference and Fourier transform solutions of the parabolic wave equation with emphasis on integrated optics applications. *J. Opt. Soc. Am.*, 8, 724.

Schmidt, R. V. and Kaminow, I. 1974. P Metal diffused optical waveguides in $LiNbO_3$. *Appl. Phys. Lett.*, 25, 458.

Sekerka–Bajbus, M. K., Yip, G. L., and Goto, N. 1990. BPM design-optimization and experimental improvements of a Ti:$LiNbO_3$ ridge waveguide linear-mode confinement modulator. *IEEE J. Lightwave Technol.*, 8, 1742.

Sharma, A., Sharma, E., Goyal, I. C., and Ghatak, A. K. 1980. Variational analysis of directional couplers with graded index profile. *Opt. Commun.*, 34, 39.

Stewart, G., Miller, C. A., Laybourn, P. J. R., Wilkinson, C. D. W., and De La Rue, R. M. 1977. Planar optical waveguides formed by silver ion migration in glass. *IEEE J. Quantum Electron.*, QE-13, 192.

Tamir, T., ed. 1988. *Guided-Wave Optoelectronics*, p. 748. Springer. Chapter 2.

Taylor, H. F. 1976. Dispersion characteristics of diffused channel waveguides. *IEEE J. Quantum Electron.*, QE-12, 748.

Thylén, L. and Gravestrand, P. 1986. Integrated optic electro-optic device electrode analysis: The influence of buffer layers. *J. Opt. Commun.*, 7, 11.

Tien, P. K. and Ulrich, R. 1970. Theory of prism-film coupler and thin-film light guide. *J. Opt. Soc. Am.*, 60, 1325.

Vandenbulcke, P. and Lagasse, P. E. 1974. Static field analysis of thin film electro-optic modulators and switches. *Wave Electron.*, 1, 295.

Xiang, F. and Yip, G. L. 1992. Simple technique for determining substrate indices of isotropic materials by a multisheet Brewster angle measurement. *Appl. Opt.*, 31, 7570.

Yip, G. L. and Finak, J. 1984. Directional-coupler power divider by two-step K^+-ion exchange. *Opt. Lett.*, 9, 423.

Yip, G. L. and Albert, J. 1985. Characterization of planar optical waveguides by K^+-ion exchange in glass. *Opt. Lett.*, 10, 151.

Yip, G. L. and Sekerka-Bajbus, M. A. 1988. Design of symmetric and asymmetric passive 3-branch power dividers by beam propagation method. *Electron. Lett.*, 25, 1584.

Yip, G. L., Noutsios, P. C., and Kishioka, K. 1990. Characteristics of optical waveguides made by electric-field-assisted K^+-ion exchange. *Opt. Lett.*, 15, 789.

Yip, G. L., Kishioka, K., Xiang, F., and Chen, J. Y. Sept. 1991. Characterization of planar optical waveguides by K^+-ion exchange in glass at 1.152 and 1.523 μm. SPIE Proc. on Integrated Optical Circuits, Vol. 1583, pp. 14–18.

Yip. G. L., and Babin, L. J. M. BPM design of a widened x-branch dmultiplexer in a glass substrate by K^+ ion exchange for $\lambda = 1.31$ μm and $\lambda = 1.55$ μm, OSA Topical Meet. on Integrated Photonics Research, March 1993, Palm Springs, Calif., Tech. Dig. pp. 72–754.

15

Optical Data Storage

Charles F. Brucker

Terry W. McDaniel

Mool C. Gupta

15.1 Introduction

Optical disk storage technology for data storage typically finds application as a computer, video, or audio peripheral device that competes with other storage devices over a full spectrum of attributes including entry cost, media cost, media removability, form factor, power consumption, data rate, data accessing, reliability, ruggedness, interface support, and standardization. Optical tape drives and optical card readers would have a similar list of attributes. Different subsets of these attributes claim priority for different applications. By general consensus, the strongest attributes of optical disk storage technology are:

- Unlimited storage capacity because of media removability; cost per bit can become negligible as capacity per optical disk drive rises;
- Reliability—the media is very long-lived and the head-disk interface has extreme robustness because the separation is several orders of magnitude greater than that in magnetic recording;
- Random data access, providing near-line access to vast data capacity;
- Extremely robust servo systems and data error correction capability.

On the other hand, present-day optical disk drives suffer some disadvantages relative to other random-access data storage devices, specifically, hard disk drives that utilize magnetic recording:

- Optical disk drives are slower in data access because of larger head mass and lower disk rotation speed, and data throughput is less, due to unavailability or awkwardness of direct overwrite, limited writing laser power, and constraints on linear data density;
- Optical disk drive entry cost is somewhat higher because of drive cost, driven mainly by the cost of fabricating and precisely aligning bulky optical components, and also by the relatively low production volumes for optical disk drives.

The first of these disadvantages is being addressed by development work on integrated optics and micromechanical actuation devices, higher power solid-state lasers or semiconductor diode lasers at shorter wavelengths, and improved schemes for direct overwrite. Advances in these areas will benefit bit-by-bit serial recording approaches, if they arrive in time, or, in the longer term, they will be a boon for implementing compact parallel recording devices (for example, holographic systems). The cost issue is essentially a "chicken and egg" situation. Nothing drives cost down more effectively than volume production. However, to penetrate user applications on a massive scale requires clear performance advantages coupled with competitive costs, or else a clear cost advantage. To date, it has proven difficult to achieve costs/prices for read/write optical disk drives competitive enough to enable them to penetrate mass applications, such as personal computer drive slots. It may require either a production cost breakthrough or manufacturer risk-taking coupled with undeniably advantageous features or applications to push optical disk drives over the threshold to mass acceptance.

The noncontacting nature of optical recording is increasingly perceived as the key differentiating feature compared to magnetic storage systems, where the gradient of the dipole magnetic field requires that the recording transducer be in intimate proximity, or virtual contact, with the surface of the storage medium. The exceptional reliability of optical storage technology is inherent in the use of an optical stylus, which completely avoids any mechanical contact during the write/read process. The distance between the focusing lens and disk surface is on the order of a millimeter, which also enables removability and enhanced archivability. New multifunction optical disk drive systems, offering the choice of read-only (ROM), write-once-read-many (WORM), and rewritable disk memories are facilitating new applications not otherwise possible with strictly (rewritable) magnetic storage systems. The "borrowing" of components already designed, tested, and proven for magnetic hard drives is an effective strategy for closing the price gap and improving performance versus magnetic hard disk drives. Already, optical disk media are competitive with magnetic disk media on a cost per bit basis.

The mechanism of writing on an optical disk or tape involves focusing a semiconductor diode laser beam, whose light is absorbed and converted into thermal energy. The typical laser-beam power at the recording layer is about 10 mW, the focused spot size is about one micron, and the beam duration during writing is on the order of 50 ns. This provides a power density of 1 MW/cm^2 or an energy density of 50 mJ/cm^{2}at the recording layer. If all of the power is absorbed in a typical recording medium thickness of 100 nm, a temperature rise of several hundred °C can be generated in a few nsec. This rise in temperature induces a physical, chemical, or magnetic change in the film and produces an optical contrast between recorded and nonrecorded regions. This is the basic principle of optical information storage, whether in the form of disk, tape, card, etc. The physical change can be ablation of material, substrate deformation, bleaching, crystalline to amorphous phase change, or reversal of magnetic polarity, to name a few. Figure 15.1 schematically illustrates some important recording media configurations. In the case of ablation, if there is a metal reflector under the ablated film, the ablated region will have higher reflectivity (due to lower light absorption) than the adjacent area, and this difference in reflectivity indicates the location of marks. For ablative systems, organic dyes and some metals have been used for recording. In the case of phase change, if the coated film is crystalline, then the laser beam can form amorphous regions, which have lower reflectivity, again providing optical contrast for locating recorded marks. The phase change medium can be erasable if the reverse transition from amorphous to crystalline can be achieved in practical time scales of less than a microsecond. In magneto-optical (MO) recording, the writing process involves heating the irradiated region sufficiently close to a magnetic ordering temperature, e.g., T_{Curie} or T_{Neel}, so that the magnetic polarity can be switched by a small, reverse-applied magnetic field. This is the functional equivalent of thermally assisted magnetic recording and, like magnetic recording, has been shown to exhibit virtually unlimited rewrite cyclability. Readout contrast is achieved by exploiting the magneto-optical Kerr effect using polarization-sensitive detection optics. For all media types, the readout-beam power is chosen to be safely below the writing threshold, typically on the order of a milliwatt.

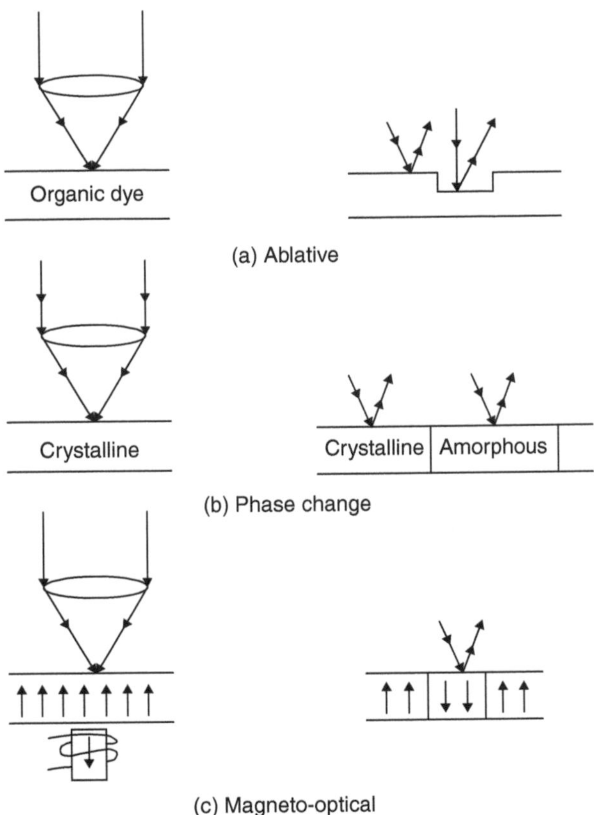

FIGURE 15.1 Schematic illustrations of (a) ablative write-once recording using an organic dye, (b) write-once recording using a crystalline-to-amorphous, phase-change medium, and (c) rewritable recording using magneto-optic media.

Organic dye-based and phase change WORM, magneto-optical rewritable, and other optical disk storage systems have been commercialized (Table 15.1) and are available in one or more of several formats designated by disk diameter: 64 mm (21/2′), 86 mm (31/2′), 120 mm (CD), 130 mm (51/4′), 304 mm (12′), and 356 mm (14′). A "low-end" storage system might consist of a CD-ROM drive and single CD-ROM disk. A typical CD has a track pitch of 1.6 μm, a recorded mark length range of about 1 to 3 μm, and a capacity (one side) of about 650 MB. At the "high-end," complete library systems based on the 356-mm format are available with a disk capacity of about 15 GB and a total system capacity exceeding 1 TB. Many choices are available to satisfy requirements for on-line capacity, total capacity, data rate, access performance, unalterability or rewritability, and cost. Among the smaller diameters ($\leq$130 mm), it is well recognized that so many choices create confusion and raise costs due to lack of standardization. It will be interesting to see whether the need to be more interchangeable and manufacturing cost effective will force convergence toward just one or two of these formats, e.g., the 120-mm form factor of the ubiquitous CD.

Historically, progress in the areal density achievable by optical recording has been closely related to laser development. A perspective on the technology migration path specific to 130-mm magneto-optic media in Figure 15.2 shows previous and anticipated improvement of areal density over time. In addition to decreasing laser wavelength, improvements in the substrate, format, recording media, and data channel can be expected to significantly impact user areal density. This chapter reviews both the current state of the art and newly proposed technologies for advancing optical storage solutions.

TABLE 15.1 Mass Storage Matrix Comparing Basic System Attributes and Prices to the User for Several Optical and Magnetic Storage Formats (Prices as of Late 1995/Early 1996)

Type of Storage	Form Factor	Capacity	Access Time (ms)	Data Transfer Rate[a] (MB/s)	Drive Price ($US)	Recording Media Price ($US)	Media Price per MB ($US)
Rewritable	3.5′	128 MB	33–136	0.4–0.7	350	10	0.078
Optical	5.25′ (1×)	650 MB	25–77	0.6–2.1	1,200	43	0.066
disk (MO)	5.25′ (2×)	1.3 GB	36	1.6	1,800	100	0.077
	5.25′ (4×)	2.6 GB			2,200	130	0.050
Write-Once	5.25′	650 MB	60–125	0.5–1.4	1,250	100	0.15
Optical disk	12′	6.5 GB	100–750	0.3–1.6	19,000	800	0.12
	14′	14.8 GB	170	1	40,000	925	0.063
CD-ROM (4× drive)	120 mm (4.72′)	640 MB	120–250	0.6	100	3	0.0047
CD-R (4× drive)	120 mm	640 MB	500	0.6	1,400	10	0.016
Minidisc	2.5″	140 MB	300	0.15	400	20	0.14
PD	120 mm	650 MB	165	0.8	500	40	0.062
Optical tape reel	880 m (12″ dia × 35 mm)	1 TB	65,000	3	250,000	10,000	0.010
Hard disk	3.5″	420 MB	15	1.5	100	NA	0.24
Drive	5.25″	1–2 GB	12–16	2–4	600	NA	0.30
8-mm magnetic tape helical scan	5.25″	2.3–7 GB	30,000	0.25–0.5	2,000	15	0.0021

Useful sources for this type of information are the Freeman Reports (Santa Barbara, CA), the Disk/Trend Reports (Mountain View, CA), and the Santa Clara Consulting Group Reports (Santa Clara, CA).

[a] Sustained read.

15.2 Design and Modeling

In this section, we give an introductory treatment of some areas of physical design in optical disk storage systems in which computer modeling has proven particularly useful. These topics can be associated with two of the primary components of the recording system-the storage medium and the recording/readout head. Because the currently predominant optical storage medium takes the form of a rotating disk, this section will deal exclusively with disk media formats. Nevertheless, many of the physical design issues related to fundamental optics, thin film optical behavior, and medium thermal performance can be easily translated to the case of nonrotating media, such as optical tape or cards. Examples of commercially successful optical storage media solutions made according to these design principles are discussed in Section 15.3. Our discussion of optical heads is virtually independent of the physical configuration of the recording medium. Practical aspects of commercial head design and performance are discussed in Section 15.4.

The primary design issues associated with multilayer thin film stacks for optical data storage media relate to

a. optical response, primarily for readout, but secondarily for writing and erasing;
b. thermal response, primarily for writing and erasing, but secondarily for readout;
c. material chemical and physical stability suitable for archival purposes; and
d. cost and manufacturability.

Considerations (a) and (b) are particularly amenable to analysis and optimization with computer models, and we focus on them here, leaving discussion of items (c) and (d) to Section 15.3 on media.

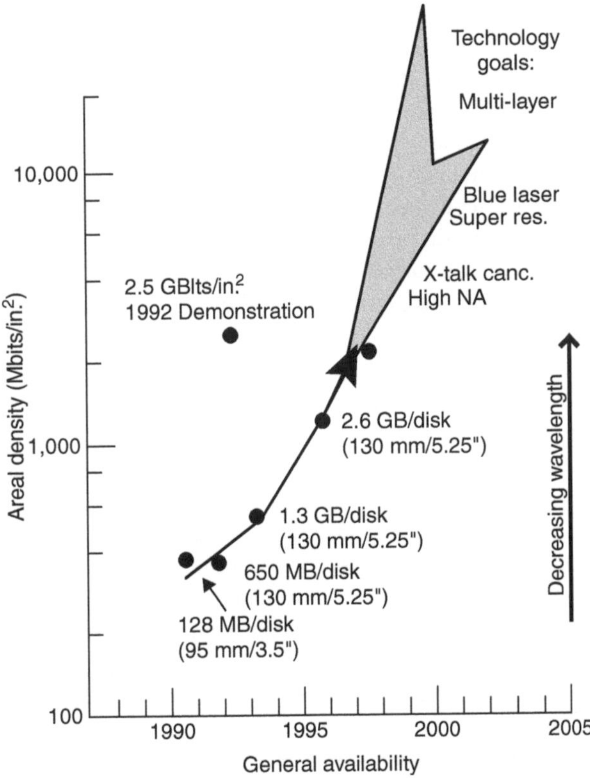

FIGURE 15.2 Technology migration path for magneto-optical recording, showing technology approaches for future density increases. (After Lenth, W. 1994. *Laser Focus World*, p. 87.)

Although we touch on most topics of importance, due to space constraints, we use references to more detailed treatments as an important part of our presentation for the reader who requires greater depth.

We would like to recommend a few general sources to supplement the abbreviated treatments of optical and thermal design. For optical response, the reader is referred to Chapter 5 through Chapter 8 in Mansuripur (1995); McDaniel and Bartholomeusz (1992); Marchant (1990), and Chapter 13 in Reference 4. For thermal design issues, see Chapter 11 and Chapter 17 in Mansuripur (1995), McDaniel and Bartholomeusz (1996) and Marchant (1990), Chapter 5 and Chapter 13 in Hurst and McDaniel (1996), and Connell and Bloomberg (1988).

15.2.1 Optical Thin-Film Design

Beam-addressable, sequential bit-by-bit optical recording utilizes a focused laser directed onto a moving storage medium. The primary optical parameters of importance to the interaction of the beam with the medium for storage or readout are reflectance and absorption. A storage process encompasses writing and erasing, which, by definition, perform a physical alteration of the medium that may or may not be reversible. (Neither of these processes is performed by a user for read-only systems such as CD-ROM, a data format very similar to CD Audio. These media have information impressed on them at the time of manufacture.) A readout or reading process is intended to detect information nondestructively on the medium, and no irreversible physical change of the medium should occur. Thus, the processes of storage and readout are distinguished by different regimes of interaction strength between the laser and the medium.

Today, all optical disk writing and erasing processes are thermally driven, meaning that a temperature rise resulting from absorption of optical energy is responsible for the physical change induced in the medium. The only optical parameter adjusted by the user between storage and readout is the level of optical power directed onto the medium. It follows from this line of reasoning that the thermal effect induced in the medium during readout must be held below some threshold value. In terms of optical design, one must choose a medium absorption that allows the user to conveniently span the necessary thermal regimes with available laser power. Our discussion suggests that thermal effects can be ignored in analyzing readout (once we assure that reading power does not lift us over the threshold of medium thermal "damage"). Conversely, for storage processes, thermal analysis is paramount, and the only pertinent optical issue is whether enough of the available laser power is absorbed to induce the required heating for writing or erasure.

Turning to the principle optical issues in readout, we note that virtually all optical storage systems detect written or imprinted information as a modulation of reflected or recollected light power from the medium. Writing typically creates a "mark" (physical alteration) against a uniform background. The mark may have simple reflective contrast to the background, or it may be a topographical feature that is a diffractive element. In the case of MO media, the mark is a reversed magnetic domain which produces a modulated rotation of the direction of the incident linear polarization (MO Kerr effect). The modulated polarization state is converted to a detected power modulation by a suitable configuration of polarization-sensitive optics in the head (see Section 15.4).

The analysis of readout optical response can be greatly simplified from the actual configuration used by considering only normally incident rays. Optical storage systems typically combine available low-cost laser diode radiation ($\lambda = 650$–850 nm) with intermediate numerical aperture optics (NA $= 0.4$–0.6) to achieve focused spot sizes in the submicron range (full width at half-maximum intensity FWHM $=$ 0.65–1.0 μm). This implies that peripheral rays in a converging cone of a focused spot may approach 20° from the normal direction to the films deposited on the disk substrate (polycarbonate or glass, the incident medium adjacent to the films). However, this beam is formed from a Gaussian weighting of intensity across the focusing lens aperture, with the population of rays in the full beam centered about 0° (normal) incidence. Numerous studies (see Chapter 7 of Mansuripur 1995 and Sarid 1988) have established that, for this class of optical systems, the error in integrated beam effects associated with treating only normally incident rays and neglecting the effects of peripheral rays is very minor. Consequently, much of the detailed analysis of Chapter 5 and Chapter 7 of Mansuripur (1995) can be bypassed. The consideration of the interaction of a normally incident ray with a multilayer film stack is considerably simplified, and one can instead invest modeling effort in studying a wide range of film structures in search of optimally performing ones (McDaniel et al. 1991; Atkinson et al. 1992; McDaniel et al. 1994).

Reflectance R and absorption A are primary design parameters for an optical disk. They are not independent, however, because conservation of energy requires that $R+T+A=1$, with T being transmittance. For single storage layer optical disks, T represents wasted light, so usually T will be made close to zero. T is practically zero for a quadrilayer MO disk, for example, and then $A=1-R$, leaving just one of the three parameters independent. Common constraints of the film stack design include the following:

- At the operating wavelength, R and A must be adjusted to allow writing/erasing with available write laser power and enable tracking with limited laser read power;
- The desired modulation of R between the background film and the written marks must be accommodated;
- One or more additional optical (or magneto-optical) parameters must also lie in a desired range for acceptable performance (e.g., Kerr rotation and ellipticity).

The problem of calculating the classical optical response of a thin film stack for normally incident rays is well-documented (Mansuripur 1995, Chapter 5; Heavens 1955). The physical issues are

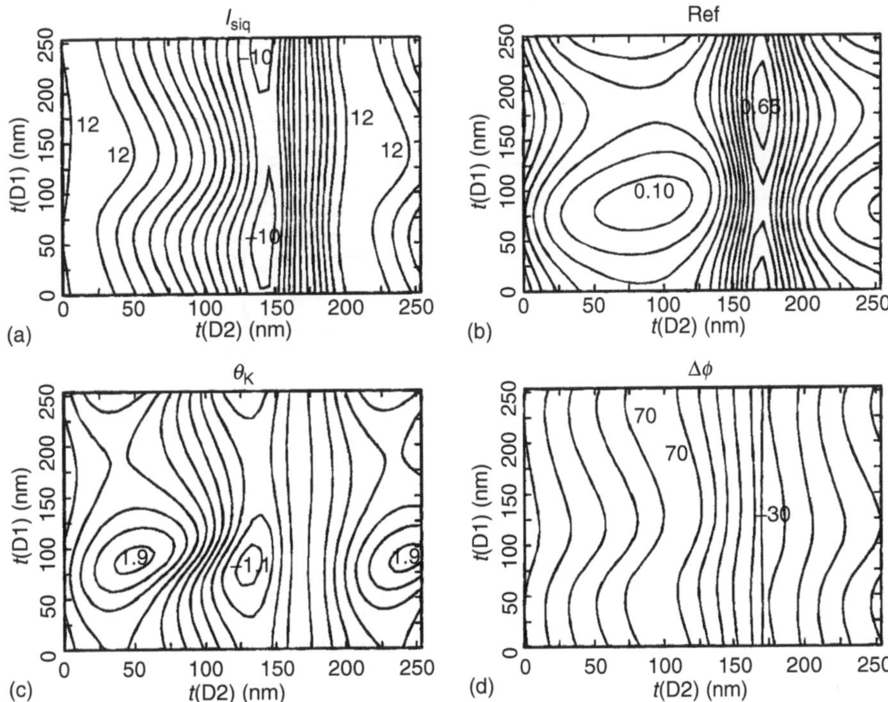

FIGURE 15.3 Optical and MO responses of a quadrilayer structure versus the variation of thicknesses of the encapsulating dielectrics, D1 and D2 (1.3X optical period); (a) MO signal current (arbitrary units); contours range from–10 to 12 in steps of 2; (b) reflectance in steps of 0.05 from 0.10 to 0.65; (c) Kerr rotation Θ_K in steps of 0.3° from $-1.1°$ to 1.9°; (d) MO phase shift $\Delta\varphi$ in steps of 20° from $-30°$ to 70°.

straightforward—given a set of film materials and their complex refractive indices, one needs to account for the cumulative effects of partial reflection and transmission at all film interfaces plus the effect of propagative phase delay through film thicknesses. Either matrix or iterative algebraic formulas are compatible with computer evaluation. Figure 15.3 illustrates a variety of typical responses from a four-layer (quadrilayer) film stack containing an MO storage layer. The MO signal from a differential detection head (see Section 15.4) is expressed by $S_{MO}=cR \sin 2\theta_{K,max} \cos \Delta\varphi$, where c is a constant, $\theta_{K,max}$ is the maximum Kerr rotation (occurring at 0° phase shift), and $\Delta\varphi$ is the phase shift of the Kerr polarization component light relative to the unrotated, reflected Fresnel component. Alternatively, one can write $S_{MO} = cR \sin 2\theta_k \cos 2\varepsilon_K$, where ε_K is the Kerr ellipticity. The MO Kerr effects and the phase shift $\Delta\varphi$ are thus interrelated by $\sin 2\theta_{K,max} \cos \Delta\varphi = \sin 2\theta_K \cos 2\varepsilon_K$, an identity among the parameters of polarized light.

A quadrilayer film stack with an MO layer, as shown in Figure 15.4, is a good example for discussion of optimization of an optical storage medium for readout. First, light is usually incident on the storage films from an optically thick substrate; 0.6–1.2 mm is a common thickness range. Given that the depth of focus λ/NA^2 of intermediate NA optical heads is, at most, a few micrometers, the substrate serves as a protective cover keeping dust, debris, and handling damage, such as scratches away from the focal plane. The thickness of the first dielectric material deposited on the substrate is chosen to adjust the stack reflectance to an optimal value for good readout performance. The metallic storage film is chosen thick enough to produce adequate MO response (Kerr effect), but yet thin enough to have moderate transmittance (5%–40%). This transmissive property is helpful in allowing some light to propagate onto the reflector film that caps the deposited stack, thereby creating a potentially resonant cavity in the stack. A useful resonance phenomenon in this application arises from multiple transmissions through the MO film to

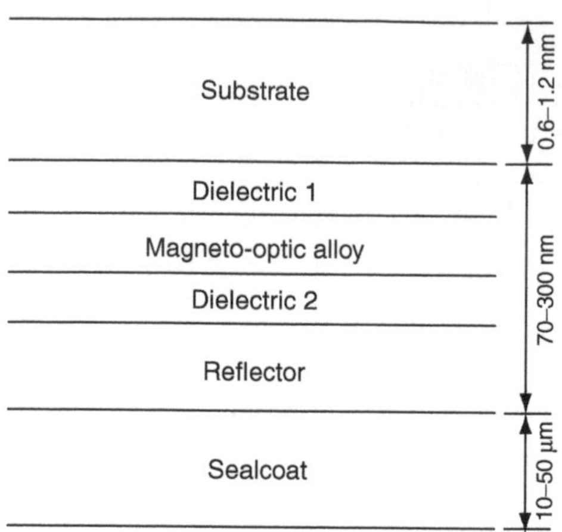

FIGURE 15.4 Generic quadrilayer structure with MO film.

produce additional MO (Faraday) effects supplementing the reflective Kerr effect. (The reflector is thick enough-usually 50 nm or more-to insure that the stack transmittance $T \approx 0$.) Design analysis and experiment confirm that 1–5 dB signal enhancement relative to a single reflective MO layer can be realized by optimally tuning a "resonant" quadrilayer stack. A key part of the tuning is adjusting the dielectric thicknesses for optimal phasing for reflectance and the Kerr effect, which can result in an optimal value of signal or signal-to-noise ratio (SNR) (McDaniel and Bartholomeusz 1996; Atkinson et al. 1993).

Finally, we note that the structure in Figure 15.4 is well-suited for two additional purposes: (a) encapsulation of a highly chemically reactive MO material, often a rare earth-transition metal (RE-TM), to provide a stable archival storage medium, and (b) provision of a structure with considerable thermal design flexibility (next section). Most of the thin-film optical concepts introduced here, e.g., the use of antireflective overcoats and/or reflective and phase-adjusting backcoats, are applicable and useful for other optical storage thin-film designs (write once, phase change rewritable, etc.).

15.2.2 Thermal Design and Modeling

Laser-beam heating of a moving film medium is well described by classical heat conduction and the Fourier law of heat transfer $Q = -k\nabla T$, where Q is heat flux, k is thermal conductivity, and T is temperature. The thermal performance of thin-film optical storage media must satisfy several requirements:

a. adequate thermal efficiency to allow temperature elevation for writing and erasure at available laser powers;
b. limited lateral heat spreading to minimize thermal crosstalk between adjacent marks or tracks; thermal crosstalk can contribute to unwanted intersymbol interference in written data;
c. avoid excessive peak temperature rise to prevent medium damage or even excessive irreversible property changes;
d. effectively dissipate heat imparted to the medium by the readout beam to prevent irreversible medium damage or partial erasure of written data.

Requirement (a) conflicts with (b) and (d), in general. Consequently, trade-offs and compromise are necessary. Item (b) is usually achieved by designing a low thermal resistance path from the MO film to the reflector, as in Figure 15.4, and this tends to diminish the "efficiency" called for in (a).

The heating of thin-film structures with laser irradiation has been studied for applications ranging from laser annealing and machining to optical data storage (Madison and McDaniel 1989; Holtslag 1989). Analytical methods are numerous (Carslaw and Jaeger 1988), but their utility is usually limited to simple geometries and approximate estimates. Nevertheless, analytical results are very helpful for understanding the fundamental physics of heat flow and certain limiting conditions. One interesting benchmark is the adiabatic limit wherein one supposes that all of the energy absorbed from incident irradiation is converted into local thermal energy without diffusion loss. The assumption of no diffusion loss of heat from the incident radiation spot volume is very unrealistic, but this limit establishes an upper bound on estimated temperature rise for a given irradiation. The expression for temperature rise in this limit is $\Delta T = Q/C = P_{inc}(1-R)\Delta t/cm = P_{inc}(1-R)\Delta t/c\rho V$, where Q is the heat deposited, C is the heat capacity of the media layer(s), P_{inc} is the laser power incident on the disk, R is the reflectance, Δt is the laser pulse duration, c is the specific heat, m is the mass of the volume heated (without diffusion), ρ is the medium mass density, and V is the medium volume heated. Two other limits of interest in laser marking are discussed by Marchant (1990). The *static marking limit* applies when a laser pulse of finite duration is completed before the medium displacement under the beam is a significant fraction of the beam diameter. This is often expressed as $v\tau \ll s$, where v is the medium speed, τ is the pulse duration, and s is the measure of the beam diameter (often FWHM or $1/e^2$ width). The opposite limit is the *scanned marking limit* for which $v\tau \geqq s$. Note that these limits can be approached by varying v or τ or both to achieve the desired product. Neither of these limits that Marchant reviews ignore diffusion effects. Simple analytical relations for the power requirement to achieve a marking condition are available for each limit. One particularly interesting relationship for a case of $v = 0$ and $\tau = \infty$, but including heat diffusion, is $\Delta T_{max} = P(1-R)/2(2p)^{1/2}k\sigma_{eff}$, where ΔT_{max} is the peak temperature rise under the stationary beam, P is the beam power, R is reflectance, k is the thermal conductivity, and σ_{eff} is the (Gaussian) beam size parameter (McDaniel and Bartholomeusz 1996).

It is interesting to compare the temperature rise ΔT predicted by the above formulas with that from a full computer simulation. We take common values $v = 0$ m/s, $P = P_{inc} = 20$ mW, $R = 0.20$, and $\Delta t = 10$ ns with material parameters k, c, and ρ as consistent average values for a quadrilayer MO film stack. In addition, we choose V and σ_{eff} consistent with a heated volume diameter, that is, beam diameter ≈ 1 µm and an assumed heating depth of 50 nm. The adiabatic limit ΔT_{ave} is 1,330 K, the stationary CW (continuous wave) beam $\Delta T_{max} = 1{,}140$ K, and the computer simulation predicts $\Delta T_{max} = 160$ K and $\Delta T_{ave} \approx 80$ K. The striking observation is how effective heat diffusion is in removing thermal energy. Notice that in the adiabatic case energy input is a finite 0.2 nJ, whereas CW heating has a continuous energy input of 20 mW = 20 mJ/s = 20 nJ/µs that results in a peak temperature achieved of nearly 200 K less. When the 10 ns pulse heats a realistically diffusive film structure, the temperature rise is nearly 20 times less than that for the adiabatic limit!

Because heat flow in a three-dimensional film structure, such as Figure 15.4 in which the layer materials have a wide range of thermal properties (Table 15.2) is complicated, computer simulations are usually the only way to achieve accurate estimates of thermal profiles for arbitrary power pulsing of a moving laser beam (McDaniel and Mansuripur 1987; Shih, to be published). As an example of computer simulations that provide some insight into the behavior of temperature distribution in a moving optical

TABLE 15.2 Thermal Conductivities and Specific Heats of Common MO Disk Materials

Film Material	Thermal Conductivity (W/m-K)	Specific Heat (J/kg-K)
Polycarbonate	0.15–0.2	200–300
Glass	0.5–1.0	860
SiN_x	1.5–3.5	600–900
TbFeCo	6–9	320–400
Al or AlX	25–50	≈900

disk during writing/erasing, consider the "thermal characteristic curves" shown in Figure 15.5. Here, we use the simplifying approximation that medium marking is a threshold phenomenon that occurs instantaneously when a temperature threshold is passed. These plots show the laser-pulse power required to write a mark of a specified width (here equal to 0.666 μm) as a function of disk velocity and laser-pulse duration. An additional parameter that varies for the five panels is the baseline, continuous, laser power in which the writing pulse is embedded. These curves illustrate a wide, continuous range of behavior between the two limits discussed earlier ($v\tau$ ranges from 0 to ∞). An interesting variation of the characteristic curves is obtained if one employs "power cutting" before and after the pulsing event. Figure 15.6 gives an example of power cutting and shows that the effect raises and spreads the curves in power, because the medium experiences brief cooling intervals before and after the writing heat deposition. "Cut pulses" in writing serve as a convenient means of increasing the magnitude of temperature gradients that define the formation of mark boundaries, albeit at a cost of somewhat higher pulse power. High-temperature gradients are generally observed to improve the accuracy and quality (i.e., lower readout jitter) of written patterns in optical storage.

Thermal modeling is extremely valuable in designing the thermal response of the medium and in developing a writing or erasing strategy. Medium design for thermal performance clearly must be done in concert with optical design of the film stack and material selection (McDaniel and Sequeda, 1992). Fortunately, this design sequence can often occur serially. In a quadrilayer MO structure, for example, after the MO and top dielectric layer (D1) thicknesses are chosen primarily for optical and MO performance, the reflector and lower dielectric (D2) thicknesses, which play a secondary role optically, can be chosen to optimize the heat flow and overall power efficiency of the disk.

Hurst and McDaniel (1996) have an extensive discussion of the use of thermal models to assist in designing writing strategies for high-performance optical recording. The flexibility of a good computer simulation supplements experimental analysis by aiding in the understanding and optimization of pulsing strategies for write compensation. Without careful control of applied laser power, inaccurate or irregular mark shapes are written, which diminishes performance margins and, ultimately, limits achievable data density, particularly for pulsewidth modulation (PWM) recording.

15.2.3 Optical Design and Modeling

The literature on designing and modeling optical components and subsystems for optical data storage devices is vast. Our intent in this section is to raise the reader's awareness of this material and to provide a brief overview of how computer analytical tools have been instrumental in bringing insight into topics involving the propagation of electromagnetic radiation through the optical path of the recording head from the source to the detectors, including the interaction of the light with the storage medium.

A particularly versatile tool for optical path and component analysis is the program DIFFRACT developed at the University of Arizona by Professor Mansuripur. A recently published book (Mansuripur 1995, particularly, Chapter 8) by the program's developer contains a very complete review of these issues. Undoubtedly, other analytical programs exist, many not in the public domain, with similar capabilities. However, given the availability of the results from this program, and now the program itself (For information about DIFFRACT availability, contact MM Technology, Attention: Dr. Masud Mansuripur, Optical Sciences Center, University of Arizona, Tucson, AZ 85721), discussing DIFFRACT as a generic example is probably most useful.

DIFFRACT allows a user to specify the characteristics of a propagating beam and the optical components (lenses, apertures, mirrors, beam splitters, detectors, etc.) in the system path. Indeed, virtually any optical element of interest in an optical storage system can be defined and positioned in a beam for analysis. The program is based on scalar diffraction theory, treating each vector component of the electromagnetic field as a scalar field (Mansuripur, 1989). This analysis accounts for many vector field effects, even though it bypasses exact solutions of Maxwell's equations with boundary conditions. In spite of the absence of a rigorous vector diffraction treatment of light interaction with diffracting features, experience has shown that DIFFRACT accounts for all of the important phenomena observed

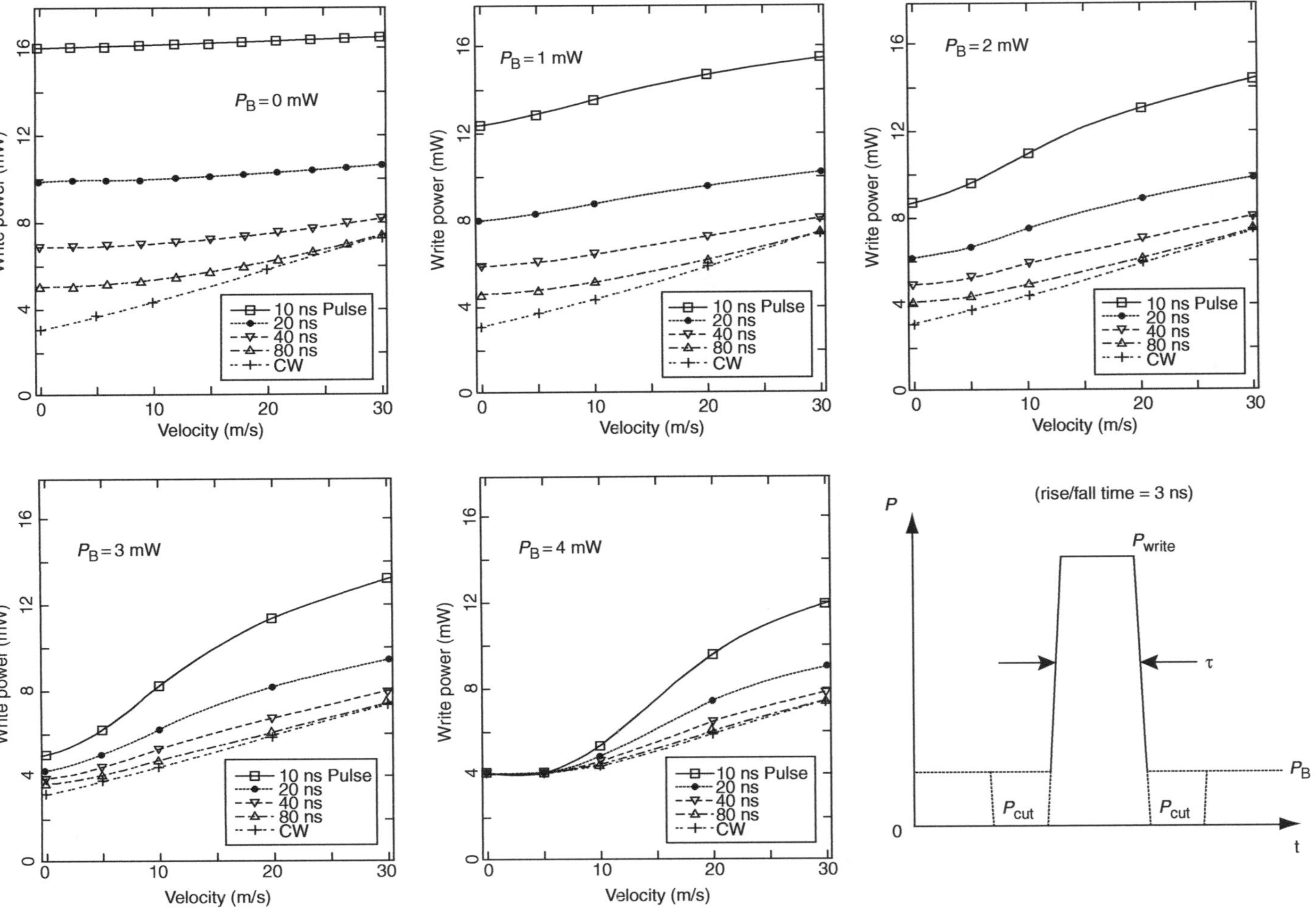

FIGURE 15.5 Thermal characteristic curves for MO media; the writing criterion was to determine the power which would expand the 175 K temperature rise contour to a width of 0.666 μm for constant mark width writing. Independent parameters are disk speed, write pulse duration, and baseline CW power. The 0 mW "cut powers" apply to the characteristic curves in the next figure.

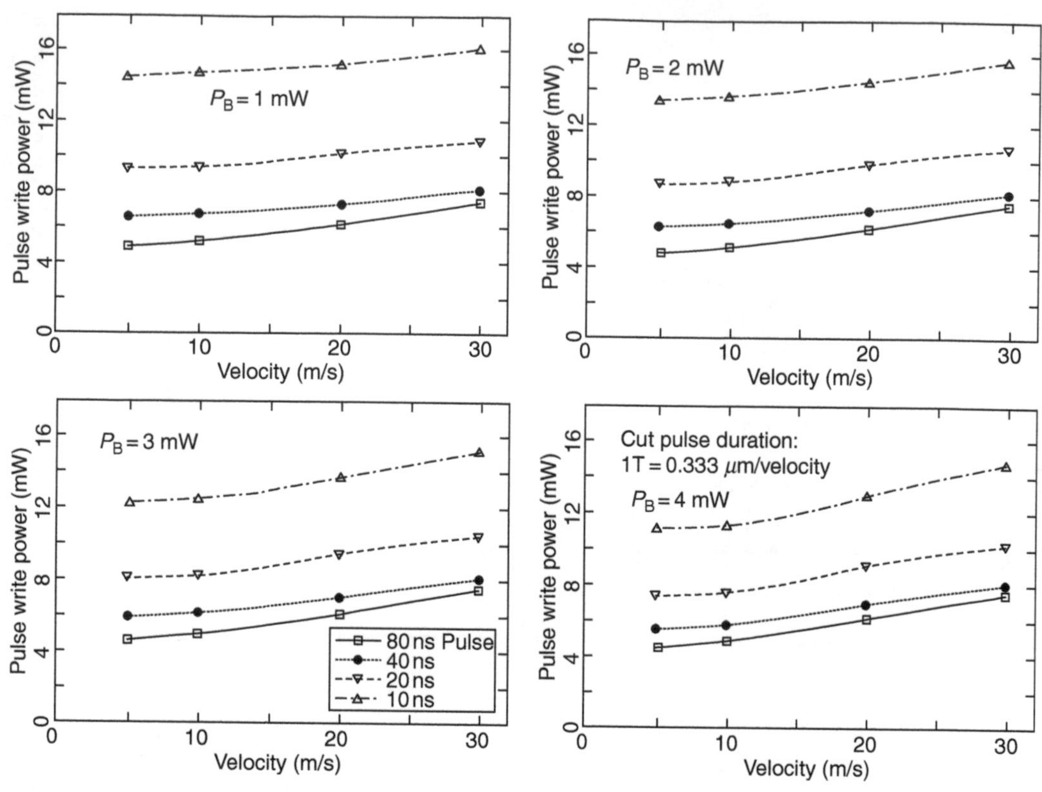

FIGURE 15.6 Thermal characteristic curves for MO media corresponding to the previous figure with the writing pulse modified to have a leading and trailing 0 mW cut power of one clock duration (see power-time profile in previous figure). When using cut powers, a zero baseline power is not allowed.

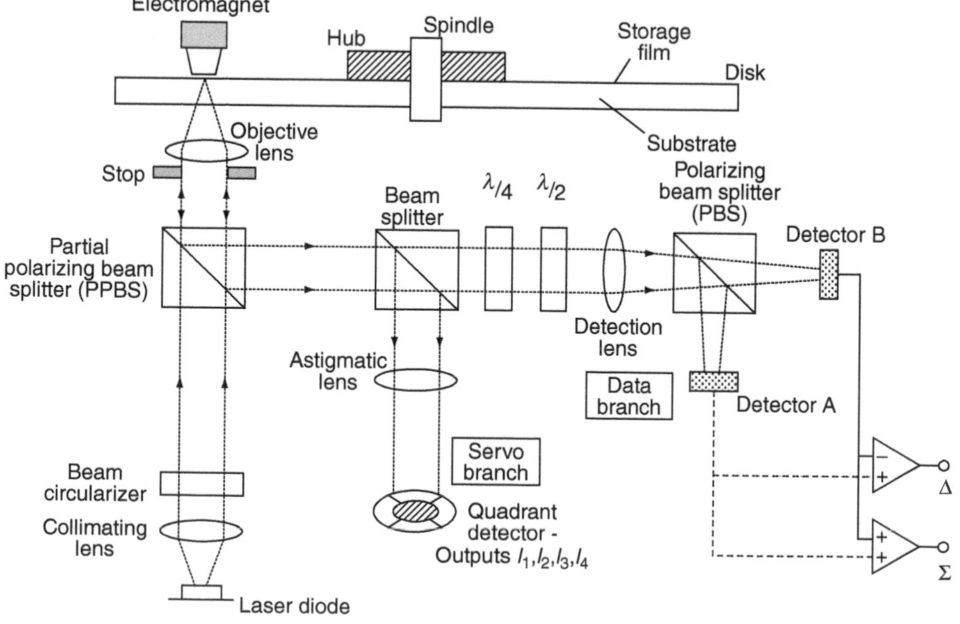

FIGURE 15.7 Generic schematic of an optical head for rewritable MO recording.

experimentally in intermediate NA optical recording systems. The use of scalar theory provides great benefits in computational throughput because it is compatible with well-developed auxiliary analytical techniques, such as plane-wave decomposition and two-dimensional, fast Fourier transformation.

We next illustrate a few examples of the capability of DIFFRACT. A schematic of a generic optical head for an MO data storage device, shown in Figure 15.7, will be helpful in representing the elements of a system which DIFFRACT can simulate. In the first example, Figure 15.8 shows contour plots of the *x*, *y*, and *z* polarization components of the intensity of light in the focal plane of an objective lens. The light incident on the circularly symmetric objective lens was a plane wave with a Gaussian intensity profile over the *x*-*y* plane and linearly polarized along *x*. Notice that the dominant focused power is for *x* polarization as expected, but, interestingly,

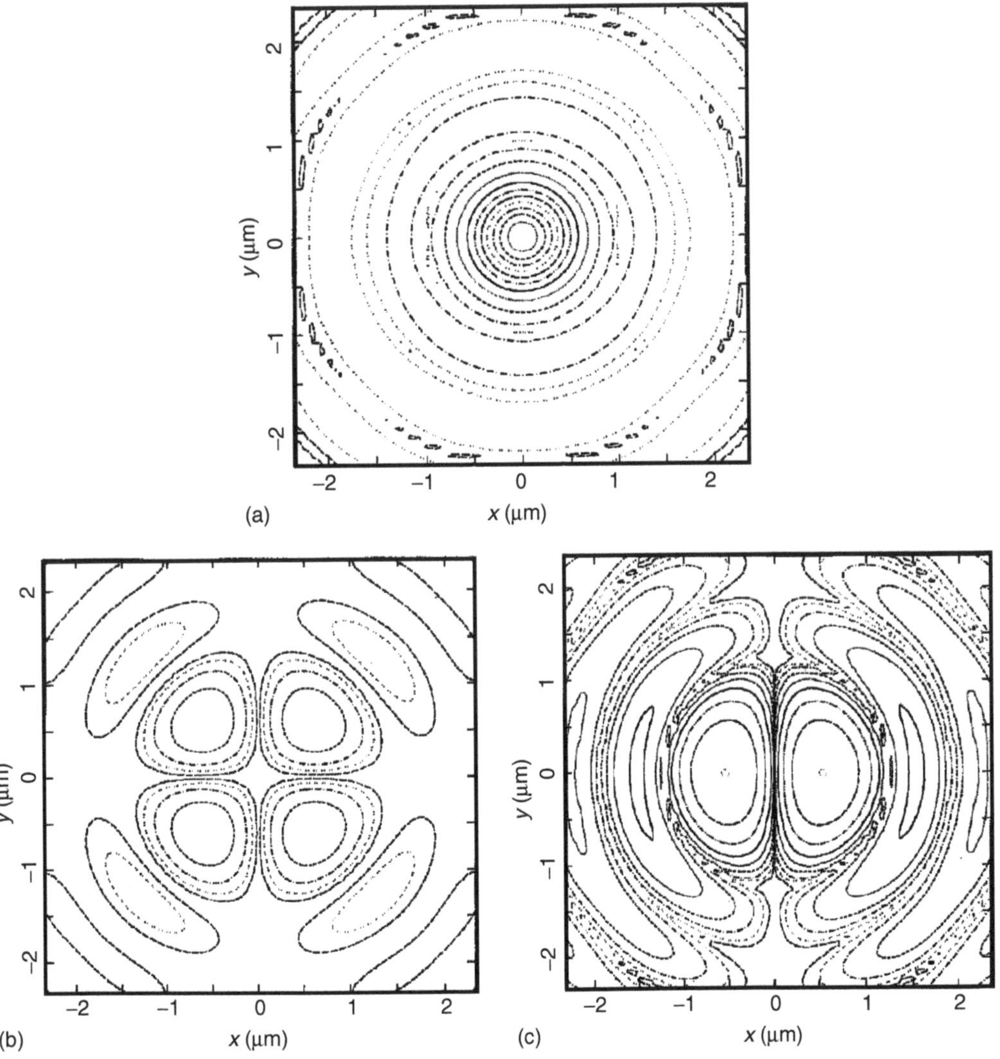

FIGURE 15.8 Contours of $I_i \propto |E_i|^2$, $i = x,y,z$ in (a),(b),(c) from DIFFRACT; in (a) the levels range from 0.9 (center) to 0.00003; in (b) and (c) the range is 10^{-2}, $10^{-2.5}$,..., 10^{-6}; a Gaussian incident beam overfilled a circular lens aperture with a ratio of $d_{1/e2}/d_{aper} = 1.09$, with polarization along *x*; $\lambda = 780$ nm and NA $= 0.55$.

Fresnel refraction in passing through the objective lens results in noticeable *y* and *z* components with four- and twofold symmetry in the focal plane, respectively. This illustrates that vector component effects are available from this quasi-vector analysis. All parameters of the incident beam, the objective lens, and the detector are under the control of the modeler for design studies.

Figure 15.9 shows an example of predicted readout signals as a focused beam traverses both embossed pits on the surface of the substrate and written MO reverse domains placed sequentially down the centerline of an optical disk track. These signals are detected by the sum and difference channels of the data detection arm, respectively, illustrated in Figure 15.7. Figure 15.10 shows track crossing signals seen in a servo tracking error signal (TES) channel utilizing a split detector difference signal which is sensitive to diffractive imbalance on the detector as the focused beam crosses embossed tracking pregrooves. Again, in both of these examples, the parameters of the incident beam, the diffractive features on the disk

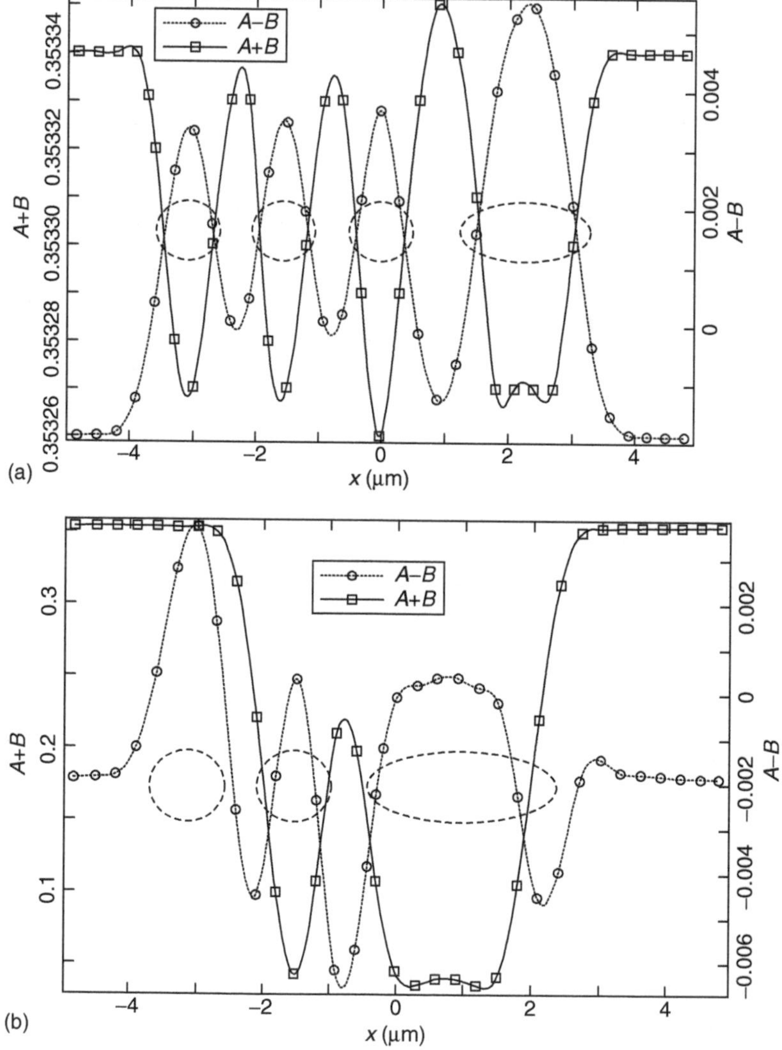

FIGURE 15.9 Modeled readout signals from DIFFRACT for reading down the centerline of embossed pits and/or reversed MO domains; mark and spot sizes, data density, and track width are representative of second generation MO drive products. (a) four MO reversed domains; (b) one MO reversed domain, plus one small and one elongated embossed pit.

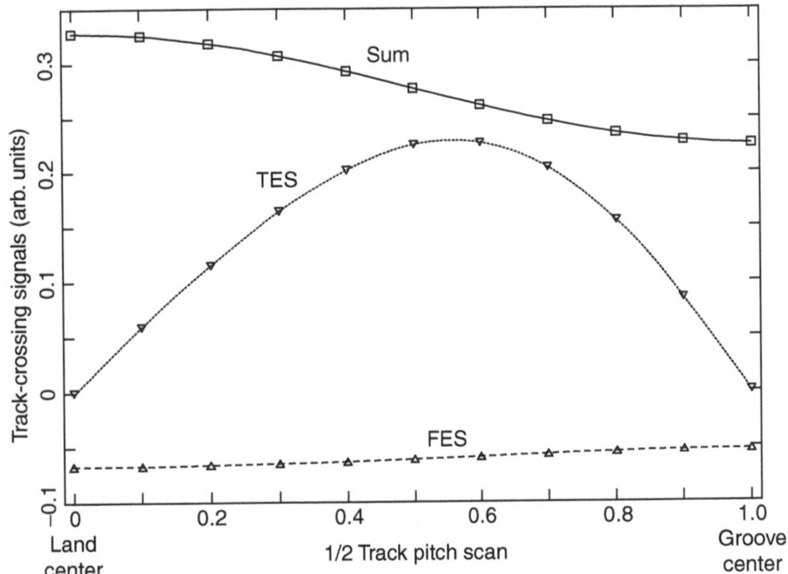

FIGURE 15.10 A tracking error (track crossing) signal, focus error signal, and quad detector sum signal for radial displacement of a readout beam over one-half period of modeled pregrooves in DIFFRACT. Note the diffraction loss in the sum signal over the tracking groove.

surface, the transmitting optics, and the signal-producing detectors are defined by the user. The modeling tool is a very powerful design aid and a device for understanding the physics governing the operation of the optical storage system.

15.3 Optical Storage Media

In this section, we present several examples of actual optical recording media products, which, for the most part, can be considered commercially successful. The theoretical underpinnings of optical media design have been reviewed in Section 15.2; here we emphasize practical material, fabrication, and performance issues which have arisen during the development of real optical media solutions.

15.3.1 Compact Disk

Strictly speaking, read-only products, such as audio CD and CD-ROM, do not qualify as recording systems, but they are important to our discussion because they are forging the consumer optical disk marketplace and because they provide the technical base for more advanced optical recording systems. CD systems also set challenging *de facto* standards for WORM and rewritable systems regarding (i) *optical contrast*, so that readout functions are as easy to deliver as in a read-only system, (ii) *mark permanence*, so that the medium can be used for archival storage and that repeated readout at reduced laser power will not alter the medium, and (iii) *fidelity*, so that mark sizes and shapes repeatably correspond to the areas of laser exposure. Information is "written" by replication of small physical pits on the polycarbonate substrate during the injection molding process. The depth of replicated pits is about 100 nm, carefully chosen to be about one-quarter of the wavelength of the laser light in polycarbonate, and the readout mechanism exploits the optical phase difference between light reflected from the mirror-like land surface and a molded pit.

The original audio CD was one of the most successful new audio electronic products ever introduced. CD disk and drive technology advanced the state of audio art in every important respect—fidelity, convenience, and ruggedness. In 1990, eight years after its introduction, the CD had achieved a household penetration of 28 percent, with annual sales of 9.2 million players and 288 million disks in the United States (Pohlman 1992). Annual worldwide demand for CDs passed the 1 billion mark in 1991. CD-ROM quickly started its own acceleration into the market of mass storage, and CD-I's future looks promising as an interactive audio-video medium. In 1988, the CD-R write-once standard was introduced, and, in 1990, the photo CD became available. In 1994, prototype rewritable CDs began appearing at industry trade shows.

Here, we describe one of several fabrication processes for read-only CDs, such as audio CDs and CD-ROMs. Substrate manufacturing methods for read-only disks are used to make substrates for write-once and rewritable disks. Most CD pressing facilities utilize photoresist mastering and injection molding, as illustrated in Figure 15.11. Disk mastering begins with a highly polished glass plate to which an adhesive chrome film or silane coupling agent is applied, followed by spin coating a layer of photoresist. The master glass plate coated with photoresist is placed on a "lathe" and exposed with a "cutting" laser to form a spiral track, with the laser intensity modulated according to a channel bit stream representing the original audio or data information. After exposure in the master cutter, the exposed areas are etched away using a developing fluid to create pits in the resist surface. Following development, a metal coating (usually silver or nickel) is evaporated onto the photoresist layer in preparation for electroforming. At this point, the disk master, which has a positive impression of the CD pit track, can be played on a master disk player to ascertain the accuracy of the disk information and pit geometry. After verification, the silvered master disk, which is now electrically conductive, is placed in a galvanic nickel electrolyte bath and a thick conformal Ni coating is gradually built up. The electroformed Ni plate is then separated from the glass master and rinsed. Because this usually damages the master's photoresist layer, masters may be used only once. The resulting nickel copy, called the "father," is a negative impression which, in cases of limited production, can be used to replicate CDs. More generally, the father can be used to galvanically generate four or five "mothers." Each mother, if acceptable, can, in turn, be used to generate four or five negative impression "sons" or "stampers" by the same process. When mounted to a die, stampers are used in replicating injection molding machines to produce CD disks.

The injection molding of CD disks presents considerable challenges. To achieve satisfactory mechanical flatness, optical homogeneity, and pit replication, consideration must be given to the quality and dryness of the polycarbonate resin, environmental cleanliness, nozzle and hopper temperature regulation, mold temperature regulation, adjustment and stability of injection volume and time, control of edge flashing due to seams between mold cavity moving parts, and stamper quality. Typically, the resin is heated to about 350°C for good fluidity and injected at high velocity and pressure into the mold, maintained at about 110°C. Injection molded polycarbonate disks tend to have large birefringence, or a polarization-dependent index of refraction, which can affect phase relationships during reading and lead to errors. Birefringence arises due to fluid shear during the injection process, which tends to align the polycarbonate chains in the plane of the disk. Because the polarizability, hence refractive index n, is higher when the E-field of the light is parallel to the polymer chains, n_φ and n_r tend to be much larger than n_z, where, φ, r, and z correspond to E along the in-plane circumferential, in-plane radial, and normal-to-plane directions, respectively. Birefringence can be exacerbated by stress resulting from shrinkage, flow lines, and inclusions. Experience, based on a wide body of empirical knowledge, has shown that birefringence is usually worst near the outer disk radius and can be minimized, but not eliminated, with high melt temperature, rapid injection into the mold, low mold pressure, and short mold time. More dramatic improvements in this regard are being sought, primarily, through the development of new plastics.

The final steps in the CD fabrication process involve reflector metallization, lacquer protection, and label printing. The through-substrate reflectance coefficient in mirror areas is specified to be between 70 and 90%, which allows the use of Al, Cu, and Ag metals and Au as an expensive alternative. Typically, a 50–100-nm thick layer of Al or corrosion-stabilized Al alloy is magnetron sputter deposited. The reflector layer is, then, protected from physical abuse and oxidation by spin coating and curing a 6–7-µm thick plastic layer, e.g., UV curable acrylic with good adhesion to aluminum and high scratch resistance.

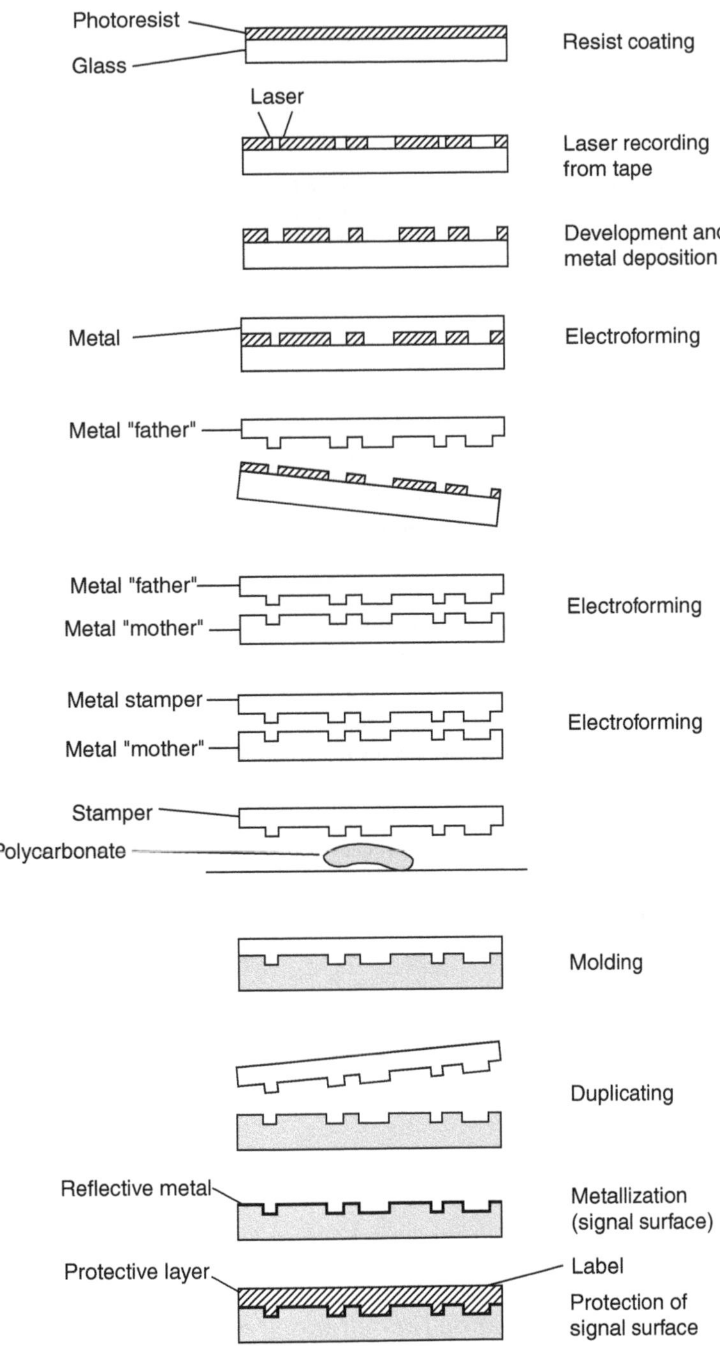

FIGURE 15.11 Schematic CD fabrication process utilizing photoresist mastering, injection molding, reflector metallization, and lacquer protection. (After Pohlman, K. C. 1992. *The Compact Disc Handbook, 2nd Ed.*, Madison, Wisc. A-R Editions.)

Finally, a multicolor label is printed on the lacquer layer using sophisticated silk screening or pad printing machines. This label does not interfere with disk performance because light is incident from the polycarbonate side and what little light returns from the lacquer/label interface is thoroughly defocused. Basic CD physical parameters are summarized in Table 15.3.

TABLE 15.3 Basic Compact Disk Physical Parameters

Disk	Recording
120-mm Diameter	Continuous spiral from center
1.2-mm Thick	15,900 Tracks per inch (1.6-μm track pitch)
4 Layers:	
Polycarbonate substrate	0.6-μm Pit width
Aluminum reflector	0.9–3.3-μm Pit length
Protective lacquer	6×10^8 bits/in.^{2} density
Ink label	Constant linear velocity

15.3.2 Write-Once Ablative Media

As mentioned previously, the total power output of a diode laser in an optical recording system is not very great, but the power density of the optical stylus is impressively high. Light absorbed by a surface at such a high-power density can cause rapid alteration or damage by a variety of mechanisms. Actually, the fact that the energy arrives in the form of light is of secondary importance, i.e., the important mechanisms are thermal. In ablative recording using thin metal films, surface energy differences between the film and substrate can lead to a thermodynamic tendency to bead up or form voids (Kivits et al. 1982). Because the continuous film is in a local energy minimum, a hole of some minimum size must be created before surface tension effects begin to dominate the fluid flow and create a stable void. For high-speed recording, localized melting may not be adequate to open a hole; usually, vaporization of the film surface dominates the initial hole-opening dynamics, hence the term *ablative* recording. Marks formed in light absorbing organic films also look like holes, but, in this case, the mark formation has little to do with surface tension. In fact, the surface energy may actually be increased by the recording process. Because the thermal conductivity of organic materials is much lower than that of metals, extremely high temperatures can be reached, and material ablation is the dominant process. Figure 15.12 shows a typical layer structure used for an organic recording medium (Gupta 1988). The recording layer thickness is chosen for high absorption to provide high recording sensitivity and high read-out optical contrast. Numerical modeling of the reflectivity of the layer structure using optical constants of the various layers is extremely helpful in determining the optimum recording layer thickness. It may be desirable to contain the ablative debris using an encapsulating layer. Figure 15.13 shows such a structure and a typical plot of reflectivity versus layer thickness for an organic dye recording medium (Gupta 1984).

15.3.3 Phase-Change Media

Some materials can exist stably in two or more structural phases which display different optical properties. Some phase-change materials of interest are group IV, V, and VI and low melting point elements, such as Se, Te, Ge, In, Sn, Sb, and their alloys. In *burn-bright* media, the heat of recording anneals the recording material from a metastable amorphous phase to a more favorable polycrystalline state, which is usually more reflective. Phase-change media can also exhibit the opposite transition, i.e., from crystalline to amorphous. The SbInSn phase-change, write-once medium developed by Eastman Kodak Company has demonstrated some of the highest performance in commercial optical disk and tape systems (Tyan et al. 1992). Figure 15.14a shows the reflectivity variation with film thickness for amorphous and crystalline SbInSn phases, and some actual recorded marks are shown in Figure 15.14b. Because no net material motion is required for phase-change recording, i.e., the motion involves atomic reordering as opposed to macroscopic material displacement, in principle, it is the "cleanest" of the write-once mechanisms. It can also be difficult to control, because post recording phase change can occur in response to high storage temperature, contaminants, or autocatalytic reactions. Small volume changes may also occur upon phase reordering, leading to localized mechanical stresses in the recorded thin film.

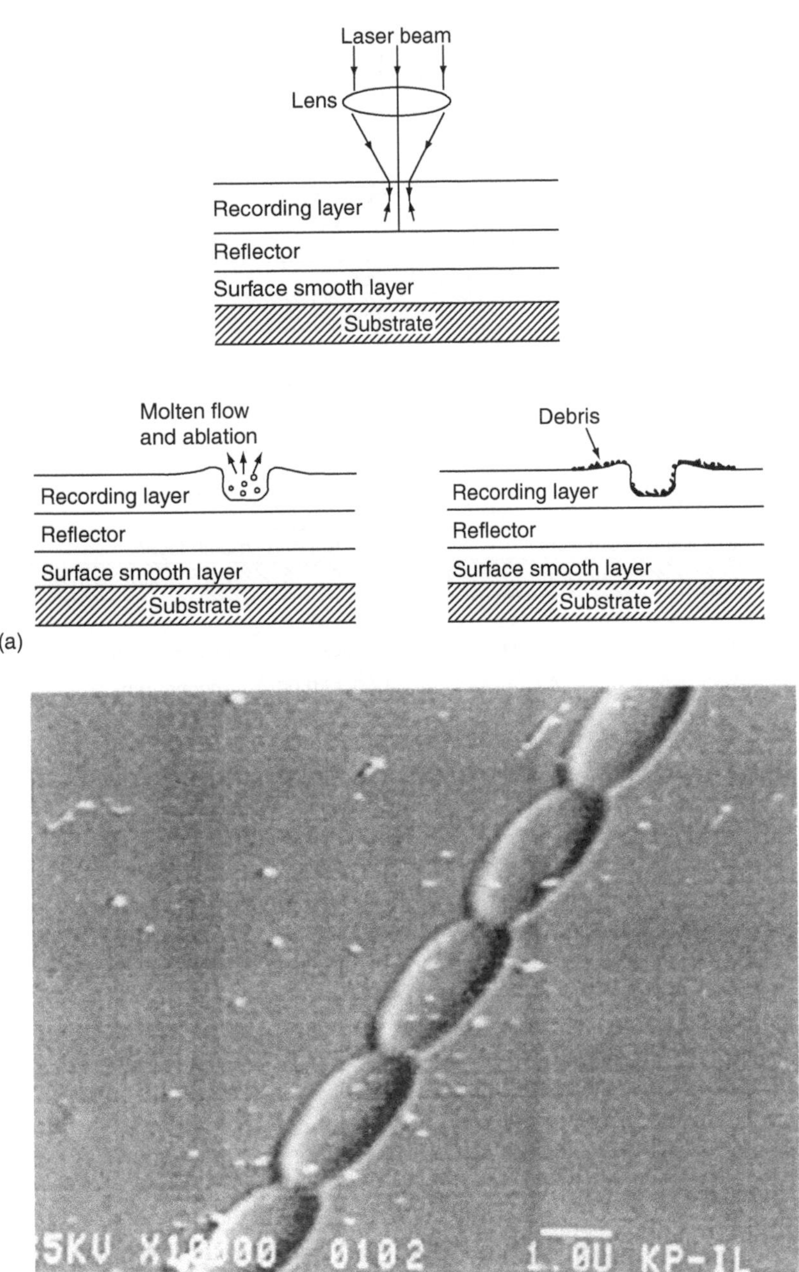

FIGURE 15.12 (a) Debris generation during laser recording process. (b) Debris particles on the organic dye surface due to a single recorded track. (After Gupta, M. C. 1984. *Appl. Opt.*, 23, 3950; Gupta, M. C. 1988. *J. Mater. Res.*, 3, 1187.)

Some phase-change materials can be reversibly converted between phases on time scales sufficiently short to permit practical rewritable media. Figure 15.15 shows a rewritable, phase-change, optical disk structure with a quadrilayer structure consisting of a bottom dielectric layer, active phase-change layer, upper dielectric layer, and reflection layer (Ohta et al. 1995). The bottom dielectric layer protects the polycarbonate substrate against thermal damage and is also used to control the recording sensitivity and

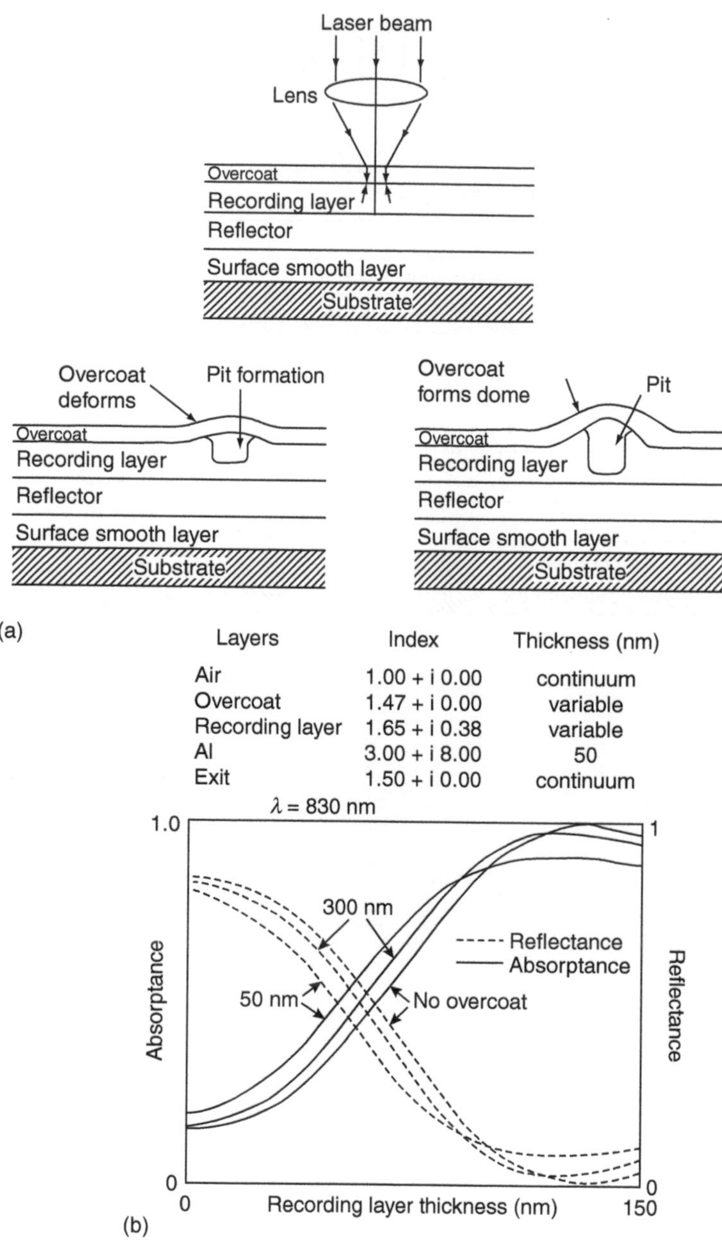

FIGURE 15.13 (a) Suppression of debris by a thin film coated over the laser recording layer (b) Absorptance variation with organic dye thickness for different thicknesses of overcoat on the organic dye. Note that, in the table of indices, the sign of the imaginary component is a matter of convention. (After Gupta, M. C. 1984. *Appl. Opt.*, 23, 3950.)

cooling rate of the active layer. The use of amorphous-to-crystalline, phase-change materials for optical memory applications was first demonstrated by Ovshinsky et al. 1968 using a TeGeSSb chalcogenide alloy system. This material undergoes a crystalline-to-amorphous transition on the order of 100 ns, but the reverse (erase) recrystallization transition requires on the order of microseconds. Recent developments in materials have reduced the recrystallization time to the nanosecond time scale as shown in Figure 15.16. Table 15.4 and Table 15.5 show refractive indices of materials typically used in rewritable, phase-change,

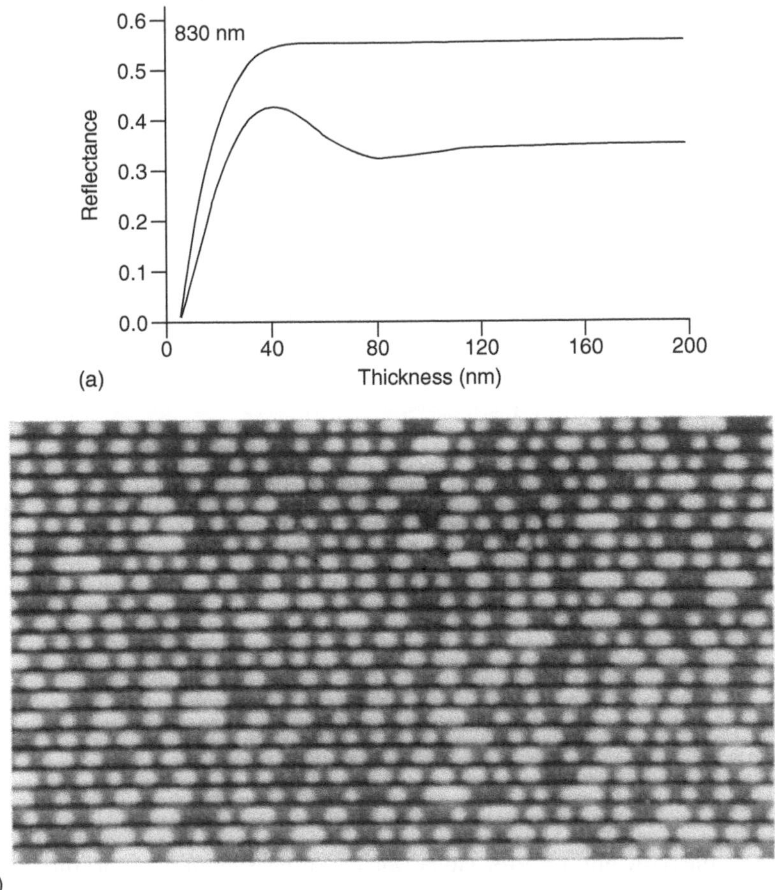

FIGURE 15.14 (a) Calculated reflectivity of an SbInSn thin film for amorphous and crystalline states as a function of film thickness. The refractive indices of the amorphous phase at 830 nm are about $5.07+i2.62$ and those of the crystalline phase $3.64+i5.52$. The reflectivity of the alloy, therefore, increases as a result of crystallization. (From Tyan, Y.-S., Preuss, D. R. Olin, G. R. Vazan, F. Pan, K.-C. Raychaudhuri, P. K. 1992. *Proc. Goddard Conf, Mass Storage Systems and Technologies II*, Kobler B. Hariharan, P. C. eds. NASA Goddard Space Flight Center, pp. 499–511.) (b) Photomicrograph or some laser-recorded marks in an SbInSn thin-film. The white features are the recorded marks, the track pitch is 1.6 μm, and the modulation code is EFM. (After Tyan, Y.-S., Preuss, D. R., Olin, G. R., Vazan, F., Pan, K.-C., Raychaudhuri P. K. 1992. *Proc. Goddard Conf, Mass Storage Systems and Technologies II*, Kobler B., and Hariharan, P. C. eds., NASA Goddard Space Flight Center, 499–511.)

disk structures and optical characteristics calculated from these indices. Using such materials and structures, single-pass overwriting has been demonstrated. The thicknesses of the layers are chosen to obtain the optimum combination of large reflectivity difference and large optical absorption. As shown in Table 15.5, the reflectivity difference between the amorphous and crystalline states is of the order of 20%–30% at practical wavelengths (Ohta et al. 1995).

Rewritable, phase-change media in a sectored CD format are already commercially available, called phase-change dual (PD). Although rewritable magneto-optical media are perhaps a more mature media technology (see below), rewritable phase-change media offer easier compatibility with the large installed base of CD drives which detect data based on reflectance modulation (versus polarization modulation in the case of magneto-optic detection). PD drives can read current generation CD-ROM disks, but PD media cannot be read on current CD-ROM drives due to differing reflectance levels.

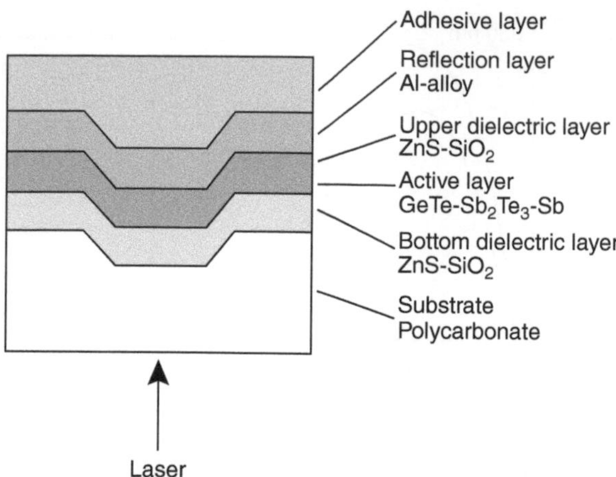

FIGURE 15.15 Cross-sectional view of a phase-change optical disk with rapid cooling sturcture. (After Ohta, T., Akahira, N., and Satoh. I. 1995. *Optoelectronics Devices Technol.*, 10, 361.)

15.3.4 Magneto-Optical Media

Magneto-Optical (MO) media exploit the polar Kerr effect to achieve optical contrast between magnetic domains of "up" versus "down" polarity. Current generation media are based exclusively on rare earth-transition metal (RE-TM) amorphous alloys for the active MO layer. Generally, the RE-TM material is chemically unstable due to the high affinity of the rare-earth element for oxygen and is structurally metastable. As a consequence, the issue of long-term chemical and physical stability has been foremost in MO media design. Of necessity, the RE-TM layer is typically protected by sandwiching it between chemically stable dielectric and reflector layers which also provide optical and thermal enhancement. Properly designed and fabricated, this optical layer stack provides virtually unlimited rewritability and an operational lifetime of at least a decade. On the other hand, the manufacturing complexity introduced by

TABLE 15.4 Refractive Indices of Materials

Material	Refractive Index 830 nm	680 nm 4.39 –*i*1.53
TeGeSb amorphous	4.61–*i*1.05	
Crystal	5.67–*i*3.01	4.84–*i*3.53
$ZnS–SiO_2$	2.0	2.1
Al alloy	2.2– ì7.5	1.8–*i*6.1
Polycarbonate	1.58	1.58

TABLE 15.5 Calculated Optical Characteristics of the Disk

λ	ΔR	A_w	A_d
830 nm	27.7%	82.9%	67.4%
680 m	19.8%	75.6%	62.0%

λ, wavelength of laser diode; ΔR, reflectivity difference between amorphous state and crystalline state; A_w, absorption of amorphous state; A_d, absorption of crystalline state.

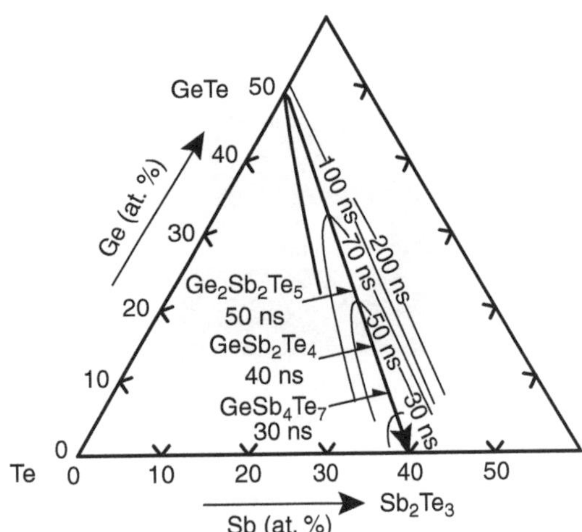

FIGURE 15.16 Threshold crystallizing laser pulse duration for Ge–Sb–Te ternary alloy films. The active layer films are sandwiched by protective layers. (After Ohta, T., Akahira, N., and Satoh, I. 1995. *Optoelectronics Devices Technol.*, 10, 361.)

the need for a multilayer stack, in particular, thick dielectric layers, contributes significantly to cost due to high cycle time and fabricating yield loss. Rewritable phase-change media exhibit similar manufacturing complexity. (Figure 15.16)

An example of an MO disk structure embodying the conventional "quadrilayer" structure is shown in Figure 15.17. The theoretical basis for the optical and thermal response of the basic quadrilayer structure has already been provided in section A. The quadrilayer forms an optical resonant cavity which provides near-optimum readback performance. The dielectric layers sandwiching the RE-TM layer are, typically, very stable, oxygen free compounds, such as nitrides of Si or Al and are deposited with utmost care to eliminate permeation pathways for oxidation and corrosion agents. The metal reflector layer, typically, a corrosion stabilized alloy of aluminum, must also provide environmental protection in addition to a measure of optical and thermal tuning. Ideally, all light which is not otherwise absorbed in the RE-TM layer is returned to the detector. From an optical point of view, the practical design objective is to efficiently maximize θ_k while providing the level of reflectance required for adequate tracking and focusing of signals during readback. Thermally, the chief objectives are to provide adequate recording sensitivity, given the available power density in the writing beam, and to achieve appropriate temperature gradients at the leading and trailing edges of the written mark to freeze the domain wall position repeatably from mark to mark. A more in-depth discussion of practical design, fabrication, and performance issues for real MO recording media is provided in Brucker (1996).

It is rather remarkable that reliable MO disk systems with performance approaching that of magnetic hard disk drives have been achieved using alloys such as Tb–Fe–Co with an intrinsic Kerr rotation of only 0.2–0.3 degrees. Given the installed manufacturing capacity for RE-TM MO media, it is quite conceivable that RE-TM media will remain the mainstay in MO recording systems for some time to come, perhaps with additives to enhance performance at blue-green wavelengths. At the same time, it is clear that materials with superior magneto-optical properties and/or reduced cost could enable new applications in this media sector. A larger θ_k, for example, could increase the signal-to-noise-ratio, allowing greater information-carrying capacity in the data channel of which the storage medium is a part or could make practical the use of thin, semitransparent MO layers required for multilevel and near-field optical recording (see Section 15.5). Materials with larger θ_k are known, but few satisfy the practical prerequisites of appropriate magnetic ordering temperature, sufficient perpendicular magnetic anisotropy and coercivity, high reflectance, low medium noise, environmental

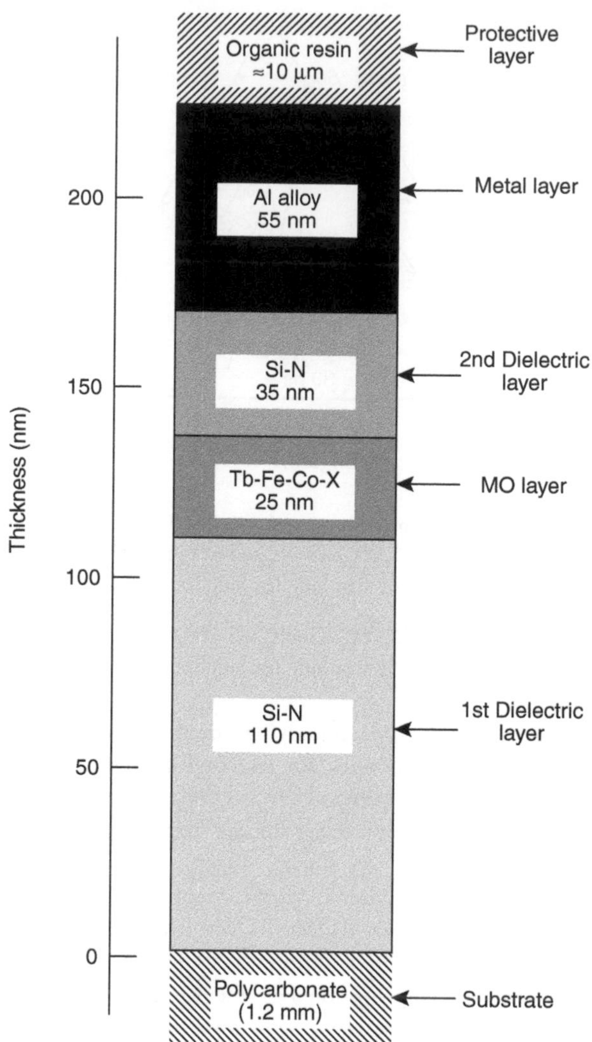

FIGURE 15.17 Commercial quadrilayer film stack based on RE-TM as the functional MO layer. (Adapted from Ikeda, E., Tanaka, T., Chiba, T., and Yoshimura, H. 1993. *J. Magn. Soc. Jpn.*, 26, 335.)

stability, and process compatibility with other materials used in the disk structure. A listing of materials with θ_k comparable to or greater than TbFeCo is shown in Table 15.6. Those materials considered "leading candidates" for practical MO recording media are compared in Table 15.7. Garnets and Mn-based materials have strong Kerr effects, but have some fundamental shortcomings which must be overcome before they can be considered commercially viable. The Co/Pt multilayers possess all the requisite properties for high-density recording, are process compatible with temperature-sensitive polycarbonate substrate material, and have sufficient chemical stability to allow simpler stack structures and alternative dielectric materials, e.g., oxides. The measured carrier-to-noise ratio for Co/Pt disk media is comparable to that of RE-TM. (Zeper et al. 1992). Largely due to its excellent environmental stability, Co/Pt was the material of choice for the recent demonstration of 45 Gbits/in.^{2} data density using near-field optical scanning microscopy (see Section 15.5).

TABLE 15.6 Selected Materials with High Kerr Rotation

Material	θ_k(deg)	Corrosion Stability	Comments	References
Co (film, hexagonal)	0.35		For reference; T_c=1,125°C; $n+ik$=2.5+i4.0 at λ=633 nm	Buschow et al. (1983), Hansen and Andenko Ward (1991)
Fe	0.41		For reference; T_c =770°C(, p 669); $n+ik$ =3.0+i3.2 at λ=633 nm.	Lynch and Hunter (1991)
$Tb_{22}Fe_{67}Co_{11}$	0.27	Poor	R=0.6, $n+ik$ =3.2+i3.5 at λ=780 nm; T_{prep}= room temp; K=7 W/m-K	
Co–Pt alloy	0.40	Good	λ=633 nm; T_{prep}=300°C	Weller et al. (1993)
Fe–Pt alloy	0.5–0.6		λ=633 nm; T_{prep}=600°C	Lairson et al. (1993)
Co-Pt multiplayer	0.25	Good	R=0.7, $n+ik$ =2.6+i5.2 at λ=780 nm; T_{prep}= room temp; K=25W/m-K	
Fe-Ag bilayer	0.50		λ=633 nm	Zhou et al. (1991) Buschow et al. (1983)
PtMnSb	0.93	Good	λ=633nm	
MnBi	0.6–0.7	Poor	λ=633 nm; T_{prep}= room temp to 300°C	
Garnet	High	Excellent	T_{prep} ≧ 400°C	
CeSb	14		λ=2480 nm; $T_c(T_N)$ = −258°C(−270°C)	Reim et al. (1986)
EuS	10		λ=520 nm; T_N=−268°C	Gambino et al. (1993)
USbTe	9		R=0.5 at λ=1770 nm; T_c=−69°C	Reim et al. (1984)

θ_k is the intrinsic (bulk or thick film) Kerr rotation.
R is the reflectance.
$n+ik$ is the complex index of refraction.
λ is the measurement wavelength.
$T_c(T_N)$ is the Curie (Neel) magnetic ordering temperature.
T_{prep} is the highest temperature required to fabricate the material.
K is the coefficient if thermal conductivity.

TABLE 15.7 Comparison of "Leading Candidate" MO Thin-Film Materials with Potential Application for Blue-Green Recording

Material	FOM × 10^{3a} (400–550 nm)	Comments
TbFeCoTa	9–11	"Industry standard," i.e., current generation MO media is based exclusively on RE-TM amorphous alloys for the active MO layer. Constrained by chemical instability. Sputtering targets difficult to fabricate.
Co/Pt	12–14	Recording performance comparable to RE- TM. Offers the distinct advantage of good chemical stability, permitting simpler optical stack structures and reduced media cost.
MnBi	13–15	High θ_k and strong perpendicular anisotropy. Problems include high grain boundary noise, structural phase instability upon write-erase cycling, and high film preparation temperature.
PtMnSb	16–25	Very high θ_K. Problems include narrow compositional tolerance and lack of perpendicular anisotropy.
BiFe Garnet	23–32	Outstanding structural and chemical stability, very high θ_k. Recording performance limited by media noise which appears to be process dependent. High film preparation temperature.

[a] The intrinsic magneto-optical figure-of-merit is defined by FOM ≡ $|\varepsilon_{xy}|/(2Im\varepsilon_{xx})$, where ε_{xx} and ε_{xy} are the diagonal and off-diagonal elements of the dielectric tensor of the MO material. The FOM is equal to the upper bound of the shot-noise-limited, signal-to-noise ratio in MO readout and can be approached in a properly designed quadrilayer structure. θ_k is related to the FOM through the (approximate) expression $\theta_K+i\varepsilon_K=i\varepsilon_{xy}/[\epsilon_{xx}^{1/2}(1-\epsilon_{xx})]\ (1-\varepsilon_{xx})]$, where ε_K is the Kerr ellipticity. After H. Fu and co-workers (1995).

15.3.5 Substrates

The optical disk substrate is the rigid mechanical structure which supports and protects the storage films. It serves several specific functions:

a. It provides a rigid, stable platform to carry the delicate thin-film data storage structure, keeping it reasonably close to the laser focal point as the disk spins rapidly.
b. It serves as a protective dust cover to keep dirt, dust, and debris far from the laser focal plane and to protect against handling damage; to some degree, it may also act as a barrier to external material contaminants.
c. Considering function (b), the substrate is an important part of the system optical path, being the final element from the laser to the storage film; it follows the objective lens, whose function is to form a well-focused, unaberrated spot on the disk.
d. It carries embossed servo and/or pre-format information to aid the servo and data channels in following, accessing, organizing, and interpreting data stored on the disk.

Function (a) places stringent constraints on the shape and stability of substrate materials, which are, most commonly, injection molded polymers or tempered glass. Disk flatness and warp are controlled by specified limits on axial displacement, acceleration, and tilt angle (local slope in the radial and tangential directions) relative to a reference disk plane. Axial deflections under $\pm$200–400 μm and tilt magnitudes under 5–10 milliradians are commonly required for data storage applications. Function (d) is often addressed through embossing concentric or spiral track pregrooves from a precision manufactured master disk, and the total radial runout of these tracks relative to the disk center is usually limited to 50–100 μm, peak-to-peak.

The quality of a disk substrate as a material diffusion barrier in function (b) depends on the material choice. In general, dense glasses are superior protective materials, whereas porous polymers tend to allow slow diffusion of some contaminant species, including water vapor. Water uptake in plastic substrates is known to contribute to warpage, and special steps to guard against this may be necessary. Fortunately, many optical disk designs consist of two laminated substrates with the deposited films positioned near the center of the sandwich. These balanced mechanical packages, are, to a first approximation, immune from curvature and warpage due to similar stressing on the two halves. However, inexpensive single-sided optical disks (e.g., CD formats) are ubiquitous, and the performance demands being placed on them are increasing rapidly, so that the material absorption issue is likely to become more important in the future. This is particularly so for the newly emerging, all-purpose DVD format (see Section 15.5) which will, some day, supplant CD.

Function (c) is extremely critical, as already alluded to in Section 15.2. Therefore, aberrations and distortions on the propagating optical wavefront, caused by the substrate, must be strictly limited to preserve the optical quality in the beam focused on the storage films. For this reason, the refractive index and thickness uniformity of the substrate must be carefully controlled to assure that a corrected objective lens can provide a low-aberration, focused beam. Indeed, the objective lens must compensate for the additional optical path represented by the substrate material, which, otherwise, would introduce excessive spherical aberration into the focused beam for an uncorrected objective. Even more troublesome is a variable and unpredictable aberration introduced by optical disk substrate birefringence. Birefringence is anisotropy in the refractive index of a transmitting optical material, and optical disk substrate materials, especially molded polymers, are prone to two anisotropies that manifest themselves in spot distortion and, thus, data signal and servo system degradation. Notice that this problem will be especially troubling for optical recording techniques utilizing polarized light, because different polarization states of propagating beams sense refractive indices along different directions. Lateral (in-plane) and vertical (axial) birefringence each cause signature degradations of signal or focused beam quality. These issues have been quite thoroughly analyzed (Mansuripur 1995; McDaniel and Bartholomeusz 1996; Marchant 1990), and most of the burden for handling this problem falls on the manufacturer of disk substrates who must carefully control the material selection and injection molding processes to limit the two types of birefringence. Although the variation of refractive index with direction is still quite small in absolute terms (even in the most birefringent plastics, $\Delta n \approx 10^{-5}$ for the in-plane component, whereas $\Delta n \approx 10^{-4}$ to 10^{-3} for the vertical component), birefringence gives rise to an integrated effect on a propagating beam. Different polarization components experience relative phase delays (retardation), depending on the propagative direction of the light ray. The dephasing effects are equivalent to induced aberration and the wavefront quality of the focused beam is degraded.

15.3.6 Optical Tape

The need to archive the ever increasing amount of data being generated is challenging conventional methods of storage, such as paper, microfilm, magnetic tape, and magnetic and optical disks. Optical tape offers the potential of being a high-capacity, low-cost, and low-maintenance technology not only for storing but also for archiving data. Though magnetic tape is also economical, using it for long-term archiving is hardly ideal. Over time, the magnetic tape medium is degraded by creep, track deformation, and magnetic print-through in the tightly wound cartridge. Every few years, therefore, data should be transferred to fresh tape, an expensive inconvenience which does not arise with optical tape. The introduction of a commercial optical tape recorder (Gelbar 1990) has demonstrated the technical feasibility of optical tape systems, but advances in hardware and media costs and performance are needed to make this technology widely accepted in the future.

Compared to disk data storage, tape storage enjoys exceptional volumetric density, because the support is so thin, at the price of greatly increased access time, because so much area must be linearly searched. Tape makes more effective use of the available surface area than disks, with the recorded area exceeding 80% of the total area available for half-inch tape versus about 65% for a 130-mm disk and 40% for a CD (single-sided). For the same areal density, optical tape has about 25 times the volumetric density of CD, and there is no fundamental reason why optical tape substrates cannot be reduced in thickness from the current 0.076 mm (0.003″) to 0.013 mm (0.0005″) or less. Just as optical storage brings higher areal density than magnetics to disks, so also with tape. Through the use of parallel read/write channels, optical tape can also support high data rates, on the order of 3 MB/s in current systems. Several companies worldwide are investigating optical tape not only for its storage and performance advantages, but also because the necessary recording materials, lasers, optics, channels, and tape handling methods are fairly straightforward extensions of existing optical disk and magnetic tape storage technologies.

The most serious drawbacks of optical tape have to do with reduced access time and engineering difficulties such as tape guiding, debris generation, contamination, and wear. For a cartridge holding several tens of meters of tape which can be searched at a rate of 10 m/s, the average access time is several seconds. For tape, access time is roughly proportional to cartridge capacity and inversely proportional to tape handling speed.

Just as with magnetic tape, optical tape in mass production would be an extremely low-cost medium because continuous web manufacture costs much less than the batch processing used to make disks (see Table 15.1). The cartridging, packaging, and testing would dominate the manufacturing cost, not the tape media itself. Optical tape is uniquely suited for archiving large quantities of data requiring relatively infrequent access, an application which takes good advantage of the economical nature of tape while not exacerbating its handling difficulties.

15.4 Optical Storage Systems

15.4.1 Optical Heads

In general, an optical head contains the following elements: (i) a semiconductor diode laser, (ii) focus and tracking actuators, (iii) a focusing/collector lens, (iv) optical components, and (v) detectors. Current diode lasers are made of GaAlAs/GaAs-based semiconductors emitting around 800-nm wavelength. The write power incident on the media is on the order of 10–50 mW and the read power is a few milliwatts.

The laser has a single spatial mode and can be pulsed with a rise time of a few nanoseconds. Two actuators are used, one for maintaining focus on the rotating (and wobbling) disk surface and one for following the recorded track. While no part of an optical head has to "fly" next to the surface of the medium like a magnetic hard disk drive, the height of the objective lens must be maintained with just as much accuracy and precision given a typical depth of focus of ± 0.5 μm. The tracking tolerance is also very stringent, typically ± 0.1 μm, due to the extremely small track pitch of optical recording systems. The submicron focus and tracking tolerances are maintained by active servo systems composed of optical position sensors feedback coupled to high-bandwidth actuators. Maintaining focus and tracking on a rapidly spinning disk which invariably exhibits wobble and eccentricity is no mean task and represents challenges both to the design of servo systems capable of high radial and vertical acceleration and to the fabrication of disk media with high mechanical precision.

Practical focus sensors generally detect deviations from collimation of the reflected beam. Examples include half aperture and astigmatic focus sensors. Figure 15.18 shows the working principle of an astigmatic focus sensor. The astigmatic field lens has two different focal lengths along orthogonal axes and focuses the beam to two perpendicular line foci. A four-element detector positioned between these foci senses the beam profile as circular when in focus $A+D=B+C$ or elliptical when out of focus $A+D\neq B+C$. An important strength of this scheme is its relative insensitivity to lateral misalignment of the detector, because it reacts, primarily, to the shape of the spot, not its position. Tracking sensors rely on grooves, data, or Preformatted servo fields to develop a corrective error signal. Examples include push-pull, outrigger, and sampled servo tracking sensors. Push-pull tracking requires continuous grooves or quasi-continuous tracks of pits to produce $\pm$1st order diffraction beams which are collected along with the 0th order reflected beam and sensed by a split detector, divided parallel to the groves. When centered exactly over a groove, the $\pm$1st orders interfere equally with the 0th order, but, when off-center, an error signal is generated by relative aperture illumination effects indicating which side of the aperture is receiving more light. The optimum optical depth of pits on read-only disks for maximum push-pull tracking signal, in general, differs from the optimum phase depth of pits for maximum data contrast, requiring a design compromise. Also, whereas push-pull tracking is sensitive and convenient, false error signals can be created by out-of-focus defects or dirt and other effects which obscure the light in one area of the reflected beam.

Several alternative tracking techniques have been developed to avoid the difficulties of push-pull tracking, in which the tracking and focus systems are decoupled, by eliminating the continuous grooves. In the three-spot, or outrigger, tracking sensor, the laser beam is passed through a weak large-period phase grating before entering the objective lens. The $\pm$1st order diffraction beams form low-power secondary spots on either side of the main spot. The grating is aligned so that the outrigger spots are displaced by a fraction of the track pitch in the cross-track direction relative to the main beam. The difference between the two outrigger signals gives a direct indication of the main beam's tracking position. The outrigger tracking scheme is rugged, reliable, and relatively inexpensive, and has been

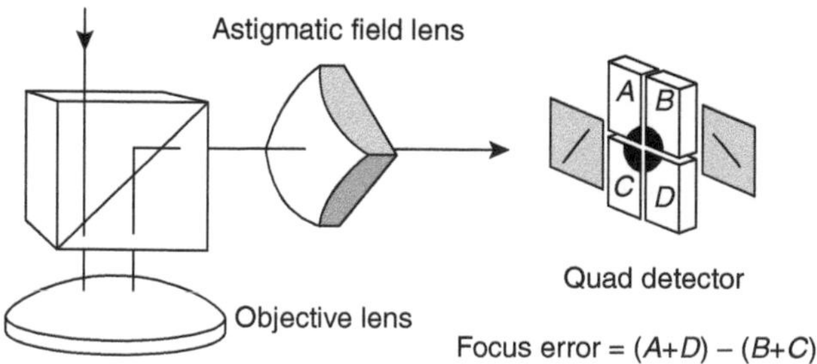

FIGURE 15.18 An astigmatic focus sensor showing conversion of defocus to a change in spot shape. (From Marchant, A. B. 1990. *Optical Recording—A Technical Overview*. Addison-Wesley.)

implemented by Philips in a practical head for CD readout. Of course, the scheme requires the constant presence of a data signal and is, therefore, not useful in WORM or rewritable systems where blank, unrecorded tracks must be followed. This deficiency can be overcome by the sampled tracking method, analogous to the tracking scheme employed in high-end magnetic hard disk drives. Many small servo fields are molded into the substrate or servowritten after the disk is manufactured, guaranteeing that some "data" marks are always present. This scheme is simple to implement and, in effect, transfers the difficult job to the mastering station or servowriter where the servo fields are created. Sampled tracking has been implemented by Kodak in its 14-in. WORM media. The overhead requirement of many (several thousand) servo fields on each track, of course, impacts the storage capacity available to the user, especially for small disks.

Figure 15.7 shows a typical configuration for a differential detection MO head. The head serves these key functions:

- Prepare, propagate, and focus the laser beam in its path from the laser to the disk; this applies for write, erase, and read functions;
- Collect, propagate, and direct reflected (diffracted) light from the disk to the data detection branch;
- Collect, propagate, and direct reflected (diffracted) light from the disk to the servo data detection branch;
- Access any radial track position on the disk;
- Maintain servo control whereby the laser beam retains focus on the storage feature plane while accessing tracks or track following.

A magneto-optical head utilizes polarized light (usually linear; directed parallel or perpendicular to the track direction, depending on the design). Because the writing and erasing functions are required simply to deliver thermal energy to the medium, the polarization state is irrelevant for those functions. MO data readout is fundamentally dependent on the Kerr effect, an interaction between polarized light and material magnetization, so that the readout processing of polarized light reflected (diffracted) from the MO disk is critical. Servo information detection normally does not use polarization information, only the reflection (diffraction) of the unrotated Fresnel light component.

Because the "light-source arm" of the head is shared among the write, erase, and read functions, and the redirection of readout light into the data detection branch involves polarization, careful design consideration must be given to properties of the beam splitters, partially polarizing or otherwise. The issue is preserving efficiency in the power splitting for light passage in each direction for each function. Notice that the partially polarizing beam splitter (PPBS) in the light-source arm will waste a fraction of the light from the laser because the reflection of the PPBS for unrotated light must be nonzero on the return pass from the disk to send usable light on for detection. A typical unrotated transmission is 70% and its reflection is 30%. For rotated (Kerr) light returning from the disk, the reflectance is usually designed to be near 100%. It turns out that this intentional attenuation of the unrotated component effectively increases the Kerr rotation θ_k, although the overall effect on SNR must be considered in designing the splitting fractions of the PPBS (McDaniel et al. 1994).

Optical recording requires a very high-quality focused spot at the disk. Therefore, the beam aberration and wavefront distortion must be held to a very small fraction of a wavelength to have a small, symmetric focused beam at the disk. Excessive coma can degrade intersymbol interference, resolution, and adjacent track cross talk. Excessive astigmatism gives rise to focus offset errors and undesired trade-offs between data and servo signal quality (both focus and tracking) (Sugaya and Mansuripur 1994).

Access (focus acquire, track seek) and servo (focusing, tracking) performance are critical head attributes. The mass of the moving components (objective lens and its mount and actuator only for "split optics" head designs) is a key determinant for the access speed and bandwidth of the servo actuation systems. Reduced mass is also important for better shock resistance and ruggedness. The capability of an optical disk servo system is seen to be very impressive when one considers that the reading or writing beam focal point ($\approx 1\ \mu m^3$ volume) must be held in the storage layer while it translates radially and axially by as much as 50–100 μm in times of the order of milliseconds due to disk runout at several thousand rpm disk rotation rates.

The light-source arm has optics to condition a highly elliptical beam from the laser diode to a nearly circular gaussian beam. Adjustment of overfill at the objective lens aperture is an important factor in determining focused spot size and shape. In the servo branch, Figure 15.7 shows an astigmatic lens coupled with a quadrant detector. As previously discussed, the astigmatic servo derives a focus error signal (FES) as FES α $(I_1 + I_3) - (I_2 + I_4)$. Diffraction from tracking grooves, as the beam moves radially on the disk, provides a tracking error signal (TES) as TES α $(I_1 + I_2) - (I_3 + I_4)$. Often, these signals are normalized by the sum signal from the four quadrants. In the data branch, a polarized beam splitter (PBS) is oriented with its fast axis at 45° to the unrotated polarization direction. The output beams to detectors A and B have amplitude $u+r$ and $u-r$, where u is the unrotated (Fresnel) and r is the rotated (Kerr) light amplitude at each detector. The difference signal is $\Delta\ \alpha\ 4ur$ and the sum signal is $\Sigma\ \alpha\ 2(u^2 + r^2)$. Remember that, due to the weakness of the MO Kerr effect, $r \ll u$. We see that the differential readout signal scales linearly with the unrotated light amplitude u, illustrating that the design of the PPBS polarization splitting is a factor in the data SNR.

15.4.2 Data Encoding and Decoding

The Institute of Electrical and Electronic Engineers defines a code as "a plan for representing each of a finite number of values or symbols as a particular arrangement or sequence of discrete conditions or events." Thus, coding is the process of transforming messages or signals in accordance with a definite set of rules. Binary *source* data is usually unconstrained, i.e., a randomly selected bit has an equal probability of being 1 or 0, and arbitrarily long sequences of all 1's or 0's may occur. Data emerging from the modulation encoder, on the other hand, is usually *d, k* constrained, where d and k specify the minimum and maximum number of 0's between any two 1's. Such *d, k* constrained *channel* data is said to be run-length-limited (RLL) and can be represented, strictly, by segments of binary data starting with a 1 and followed by at least d but not more than k contiguous 0's. The constraint d is used to control pulse crowding effects, whereas k is used to assure self-clocking ability. For a constant channel data rate to correspond to a constant source data rate, practical RLL codes are usually further constrained to map some m bits of source data onto n bits of channel data, where $n > m$. The ratio m/n (code rate) provides a first order measure of code efficiency. The need to be dc-free, essential for magnetic recording systems which sense domains inductively, is somewhat relaxed for optical recording systems. However, most codes still have problems maintaining an accurate decoding threshold in the presence of low-frequency content due to, e.g., dc instability. The code must also limit the maximum run length to maintain clock synchronization during readback and to facilitate decoding.

Foremost in the choice of modern modulation codes is efficiency, the ability to maximize the net linear density while maintaining satisfactory data reliability. Some simple examples illustrate the trade-offs the channel designer faces in achieving an efficient, yet, reliable system. The simplest way to represent binary information is to pulse the signal for each 1 and leave the signal low for each 0. This is referred to as return-to-zero (RZ) coding because the signal drops back to zero at the end of each bit interval. If the pulse length is equal to the bit cell, the signal remains high during consecutive 1's, known as non-return-to-zero (NRZ). In a related scheme, the signal level is switched whenever a 1 is encountered, but left in the prior high or low state for any 0 (NRZI). Although very efficient, these schemes are seldom used for recording. Their most serious deficiency is the lack of clock stability during readout. With arbitrarily long intervals between transitions permitted ($k = \infty$), a change in media velocity can desynchronize the clock. A simple solution to the timing problem is the biphase (or Manchester) code, for which the signal is high during one-half of each bit cell and low during the other half, with the phase determined by whether the data bit is 0 or 1. Biphase modulation is not only self-clocking but also dc-free, however, at the price of low data density. In essence, the source data stream is converted into a channel bit stream by defining two channel bits for each source data bit; a datum 1 becomes a 10, and 0 becomes a 01. A biphase encoded

TABLE 15.8 Comparison of Selected Codes

Code	*d*	*K*	*m*	*n*	*m*/*n*	Max dc
NRZ	0	.	1	1	1	100%
bi-Δ	0	1	1	2	0.5	0
MFM	1	3	1	2	0.5	33
FOOF 4/15	0	16	8	15	0.533	47, uniform
RCA 3-Δ	1	7	2	3	0.667	56
IBM 2,7	2	7	1	2	0.5	40
EFM 8/14	2	10	8	17	0.47	0

track (m/n = 1/2) can store only half as much information as NRZ or NRZI ($m/n = 1$). These and other selected codes are summarized in Table 15.8.

Simultaneously satisfying the objectives of efficiency, clock synchronization, and dc stability is a challenging undertaking, and a host of codes have been developed, each touted as superior to the others, at least, for its specific application. Two different modulation codes are specified for the first-generation, 325 MB/side, 130-mm MO disk storage system (International Standard ISO/IEC 10089). In the first of these, the so-called "four-out-of-fifteen" (FOOF) 4/15 code (Steenbergen et al. 1985), each byte of data is mapped into 15 channel bits, exactly 4 of which are 1s, using a look-up table. In comparison with biphase, of which it is an elaborate variant, 4/15 has slightly improved data density and high, but uniform, dc content. FOOF code is designed with a spectral null at half the channel bit frequency to permit the use of a substrate with a buried clock in the form of tracking grooves with slight periodic width modulations. In this way, a separate clock signal, useful for determining the location of the optical stylus on the disk, can be filtered from the readout channel. Codes, such as FOOF 4/15, are known as block codes because they are based on fixed-length data blocks.

The second modulation code specified for 130-mm MO is the (2,7) RLL code developed by IBM to enhance the data rate and capacity of magnetic disk drives. Although the *m*/*n* efficiency is slightly less for (2,7) versus FOOF coding (Table 15.8), for various reasons it has become the format of choice for first-generation systems and, in fact, is the only code specified for second-generation, 654 MB/side, 130-mm MO (2× MO ISO standard). The (2,7) RLL code is a variable length code in which source data segments, 2, 3, or 4 bits long, are mapped to channel data segments, 4, 6, or 8 bits long, such that the catenated channel segments obey (2,7) constraints. To satisfy further constraints on allowed source data patterns, padding bits may be added which result in variable length, channel data blocks.

The last code listed in Table 15.8 is eight-to-fourteen modulation (EFM), a block code designed for CD. In EFM, each 8-bit source byte is translated to a 14-bit channel word using a look-up table obeying (2,10) constraints. By design, pattern "uniqueness" is greater in the 14-bit words than in the original 8-bit symbols, providing a measure of error prevention. Blocks of 14 bits are subsequently linked by three merging bits, two of which (0s) are required to prevent the occurrence of successive is between serial words, a violation of the coding scheme. A major design objective for EFM was the elimination of problems with low-frequency content. The remaining merging bit is, thus, chosen to maintain the signal's average digital sum value at zero (at a cost of 6% efficiency). During demodulation, the three merging bits are discarded.

There are several useful references which the interested reader may consult to supplement this brief discussion of coding/decoding in the digital channel, as well as other key system attributes, such as track and sector format, measures of signal quality, error detection and correction strategies, and certification and verification to place guaranteed limits on the error distribution. Many of the cornerstones of the optical data storage channel were first developed for magnetic recording systems, as described by Patel (1988) and Jorgensen (1988). A comprehensive overview specific to optical recording systems is provided by Marchant (1990). In-depth and up-to-date discussions specific to CD and MO can be found in Pohlman (1992) and Howe (1996), respectively.

15.4.3 ISO Standards

A uniform industry standard is an important requirement for the success of many products. The floppy disk and the compact disk owe much of their acceptance to a single, industrywide standard. Historically, the optical drive industry has, on more than one occasion, simultaneously introduced incompatible disk formats and media types, which, undoubtedly, has confused customers, resulted in higher than necessary initial costs, and inhibited growth. On the one hand, this is hardly surprising given the great variety of materials systems that lend themselves to write-once or rewritable optical media (although the selection can be sharply narrowed when a constraint, such as backward compatibility with an existing format, is imposed). On the other hand, an awareness has developed that the entire industry can be stunted by zealous promotion of a company's unique technology without regard for interchangability and open systems.

The optical disk industry is learning its lesson in the area of standards and is making good progress in negotiating worldwide standards. Work on standards is organized primarily by disk diameter, including 2.5 in., 3.5 in., 120 mm, 130 mm, 12 in. and 14 in. (Table 15.9). Development and formal approval of a standard is a lengthy process, usually taking at least a year including legal and procedural reviews and a period for public comment. The resultant standards document is a fairly exhaustive specification as to the dimensional and mechanical characteristics of the disk and case (if any), the interface between the disk and the drive, characteristics of the substrate and recording layer, common test procedures to ensure interchangability within a range of environmental conditions, and, more recently, the structure for organizing and labeling files on optical disk media. Normally, formal standards do not specify a necessary lifetime within certain operating or storage environments; rather, such numbers are treated as quality

TABLE 15.9A Survey of "Basic" Standards for CD and MD Media

Basic CD system	Color/ Designator	Year	Description	Standard "Owners"
CD-digital audio (CD-DA)	Red Book	1982	Establishes physical format standards for CD-DA digital audio disks so they can be physically readable on CD-DA drives made by different manufacturers.	Philips and Sony
CD-read only memory (CD-ROM)	Yellow Book	1985	Differs from Red Book in redefinition of the data into groups of 2352 bytes and an extra error detection and correction level.	Philips and Sony
CD-ROM volume and file structure	"High Sierra Group"	1988	Enables data placed on CD-ROM disks by different developers and manufacturers to be readable on different computers using different operating systems and configured with different CD-ROM drives. With enhancements, became "ECMA-119"/"ISO 9660" standard.	"High Sierra Group" ECMA-119 ISO 9660
CD rewritable (CD-MO)	Orange Book Part I	1990	CD-MO format gives the possibility for both audio and data recording (Obsolete).	Philips and Sony
CD write once (CD-WO or CD-R)	Orange Book Part II	1990	CD-DA backward compatible, used in Photo CD.	Philip and Sony
Mini disk (MD)	Rainbow Book	1991	Portable "CD" based on CD-DA with data compression, additional interleave on data, and use of semiconductor memory. If ATRAC Digital Audio Compression technology of the MD-format was applied to the standard CD disk, the playing time would be multiplied by a factor of 5 for a play time of 6 h.	Sony
CD-R volume and file structure	"Frankfurt Group"	1992	Revision of CD-ROM High Sierra format for interchanging files for CD-ROM and CD-R media. Became "ECMA-168" standard.	"Frankfurt Group" ECMA-168

TABLE 15.9B Survey of "Application" Standards Based on CD Media Standards

Basic CD system	Color/ Designator	Year	Description	Standard "Owners"
CD-interactive (CD-I)	Green Book	1989	Interactive real-time multimedia system (text, picture, sound, video) using dedicated real-time operating system (RTOS). Initially conceived and prototyped by Philips in 1983–1984. Final specification released to Philip's and Sony's 150 licensees in 1987.	Philips and Sony
CD-TV	Amiga CD	1990	Enhanced CD-ROM format for interactive multimedia or 60 min of full motion video.	Commodore
DVI		1990	60 min of full motion video.	Intel (RCA)
CD-ROM XA	Yellow Book Extension	1991	Extension of CD-ROM with ADPCM audio and specific computer video modes as specified in CD-I. Additional hardware needed to play XA disks.	Philips and Sony
Photo CD	CD-ROM XA Bridge Disk	1991	Blend of CD-DA, CD-ROM XA, CD-I, and CD-R. The CD bridge disk is a special type of CD-ROM XA disk with a CD-I application program.	Kodak and Philips
3DO multimedia format		1993	New CD-based concept for TV-oriented multimedia applications. Direct competition for CD-I. Based on RISC processor for optimum graphical performance.	3DO
Video CD, logical development of Karaoke CD launched in Japanese market	White Book	1993	CD bridge disk format for 74 min of full-motion video at VHS quality. The Video CD can be played on a CD-I player extended with a video decompression board ($\approx$\$200 in 1994).	Philips, JVC, Sony and Matsushita
Kaleida's PC-based multimedia applications		1994	Blend of CD-DA, CD-ROM, CD-ROM XA, CD-I, and CD-R formats for interactive multimedia applications based on PC's. Under development.	Kaleida: Apple, IBM, Microsoft,...

TABLE 15.9C Developing Standards Based on CD

Basic CD system	Color/ Designator	Year	Description	Standard "Owners"
PD phase-change disk			New CD-WORM standard using "Burn Bright" recording media and sectored data format. Not Red Book compatible. Areal density about 7× CD-ROM.	
DVD-ROM			Laminated construction, single-or double-sided, single or double layer. About 135 min of full-motion video per surface.	

TABLE 15.9D Survey of Standards for 300 mm (12-in.) and 356 mm (14-in.) Optical Disks

International Standard No.	Description	Standard "Owner"
IS 13614	Nominal 300-mm diameter, double-sided, write-orice, optical disk cartridge using sampled servo format (SSF) method of information interchange.	—
IS 13403	Nominal 300-mm diameter, double-sided, write-once optical disk cartridge using continuous composite servo (CCS) method of information interchange.	—
IS 10085	Nominal 356-mm diameter, double-sided, write-once, optical disk cartridge using sampled servo format (SSF) method of information interchange.	Eastman Kodak Co.

features specific to individual products. Life testing is often performed by the medium manufacturer, which can lead to difficulties if the results are not published in a scientific, journal or if tests are performed on laboratory samples, rather than commercially available media. Nonvendor life testing by organizations such as the National Institute of Standards and Technology (NIST) and the Optical Storage Testing and Preservation within the Library of Congress is being encouraged to improve the standardization and objectivity of lifetime estimation.

Especially encouraging is the gathering consensus on a single format for high density DVD. (The initials originally stood for Digital Video Disk, but that term has not been adopted because "video" is an incomplete description in view of audio and computer applications). The two contending formats, "MMCD" and "SD," are being successfully merged into a common format with the physical structure of the double-sided SD and the signal modulation technology originally developed for MMCD. An optical pickup with two switchable lenses can provide backward compatibility with existing CD-ROMs. Hopefully, a common, rewritable CD format can also be arrived at, although, so far, progress in negotiation between the rival CD-E and PD camps has not been very visible. Success (or lack thereof) in this effort could impact the eventual introduction of rewritability to DVD drives.

15.5 Future Optical Storage

Dramatic improvements in optical storage technology can be expected in storage capacity, access time, and cost. Much of the improvement will come through the use of shorter wavelength light for increased storage capacity and enhanced data transfer rate and through the use of miniaturized optical heads. Innovations in substrate, format, recording media, and data channels will also impact future optical storage systems.

15.5.1 Heads and Systems

15.5.1.1 Integrated Heads

Current optical heads are bulky and heavy and require careful alignment and stabilization of the discrete optical components. As a consequence, optical recording, in general, has poorer access time and higher cost than competitive magnetic technology. The use of integrated optics is one approach to achieving the long sought goal of a compact and integrated optical head, but the technology is not yet practical. Ura et al. (1986) showed an integrated optical head whose schematic is shown in Figure 15.19. The laser beam is coupled to a waveguide device and outcoupled and focused on the disk surface using a focus grating coupler. The reflected light beam returning from the disk surface is coupled back into the waveguide. Focus, tracking, and RF signals are attained with an integrated optical head design. The difficulties in practical applications are loss of laser power in optical components, diode laser coupling, sensitivity to laser wavelength fluctuation, and beam quality. Improved integrated head designs have been discussed (Strasser and Gupta 1993), but so far they have not reached a practical stage.

15.5.1.2 Short Wavelength Lasers

For optimum optical recording, the laser beam is focused to a diffraction-limited spot of size approximately 1.2 λ/NA (full beam width), where NA is the numerical aperture of the lens and λ is the wavelength of the light beam. For a given NA, the shorter the wavelength, the smaller the spot size. If the currently used 800 nm IR light is replaced by 400 nm blue light, then, the spot size will be halved, allowing an increase in the in-track storage density and data transfer rate ($\propto$ NA/λ) by a factor of two. With smaller recorded marks, the tracks can also be closer by about a factor of two, giving a net increase in storage capacity by a factor of about four. The optimum recording power is proportional to $(\lambda/\text{NA})^2$, so that recording at shorter wavelengths will require less power. The depth of focus, which varies as $\lambda/(\text{NA})^2$, will decrease for shorter wavelengths, requiring tighter control on the focus actuator. Largely

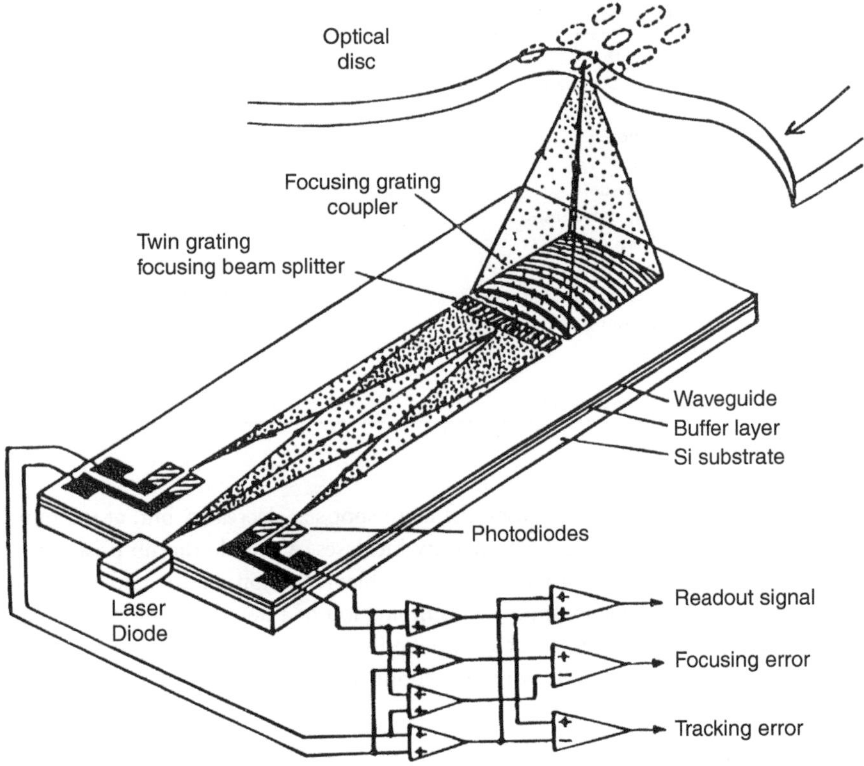

FIGURE 15.19 An integrated optical head. (After Ura, S., Suhara, T., Nishihara, H., and Koyama, J. 1986. *IEEE J. Lightwave Technol.*, LT-4, 913.)

due to the inverse squared dependence on NA for depth of focus, it is far more desirable to reduce wavelength than increase NA to achieve a smaller spot.

Short wavelength sources will be very important in optical recording, and significant research and development has been directed toward their development (Risk, 1990). Short wavelength sources will also have many other applications, such as red, green and blue (RGB) sources for display and printing. Two basic approaches have been pursued for developing short wavelength laser sources, one based on direct conversion using II-VI or III-N materials and the other based on indirect conversion using existing GaAlAs/GaAs lasers. In indirect conversion, using second-harmonic generation in bulk $KNbO_3$ crystals, 54 mW of 428 nm output power has been demonstrated (Hurst and Kozlovsky 1993) with an electrical to optical conversion efficiency of 12%. The frequency conversion approach using bulk crystals or waveguide approaches has been shown (Gupta et al. 1994; Yamada et al. 1993) to produce tens of mW of blue wavelength power, but these systems will be expensive and, hence, may find applications only in high cost products. Direct conversion lasers have not reached a commercial stage because of lifetime issues and will likely require several years of further development. Recently, direct conversion lasing has been demonstrated using II-VI materials (Haase et al. 1991) and III-N materials (Nakamura et al. to be published). The performance requirements for short wavelength laser sources are similar to the requirements for GaAs based IR lasers as shown in Table 15.10.

15.5.1.3 Optical Super Resolution

High-density magneto-optical disk recording has been demonstrated using superresolving optics. It is well known that, by introducing a shading band at the center of the collimated beam, the focused main spot can be reduced to less than the diffraction-limited size, accompanied by the appearance of

TABLE 15.10 Laser Requirements for Optical Recording

Wavelength	400–800 nm
Wavelength variation	$<$0.5 nm/°C
	0.1 nm/mW
Pulsed power	20 mW
CW power	$>$2 mW
Spatial mode	Single
Div. ang. aspect ratio	$<$4:1
Wavefront quality	$<$0.02 RMS
Polarization ratio	$>$30:1
RIN (relative intensity noise)	< -120 dB/Hz
Modulation speed	$>$1 MHz
Rise/fall time	$<$4 ns
Lifetime	10,000 h
Operating temperature	10°–55 °C

sidelobes. As the shading bandwidth increases, the main spot size decreases, but, at the same time, the main spot power decreases and the sidelobe intensity increases. A double rhomb prism in place of the shading band has been shown to provide a comparable reduction in focused spot size without appreciable light power loss. Using an optimized combination of a double rhomb prism and a readout slit to provide spatial rejection of undesirable sidelobe intensity, an improvement in linear recording density of about 1.2 times that possible with conventional optics has been demonstrated (Yamanaka et al. 1990).

15.5.1.4 Near-Field Methods

Another means of circumventing the classical diffraction limit to resolution in conventional optics is near-field optical techniques. Although the imaging process is not understood as well as in conventional far-field optical systems, near-field optical techniques have been applied to a wide range of disciplines including microscopy (Betzig and Trautman 1992) and lithography (Froelich et al. 1992). More important for this discussion, the possibility of extremely high-density data storage has been demonstrated using near-field methods. One such approach utilizes the recently introduced solid immersion lens (SIL) (Mansfield and King 1990), which exploits the reduced wavelength of light in a high index glass ($n = 1.83$) to achieve a spot size below the minimum achievable in air. By combining SIL optics with the air-bearing "slider" technology used in magnetic recording to fly magnetic heads above spinning disks, an effective spot size of about 360 nm has been demonstrated using 830-nm light bsed on read-back resolution from a conventional MO TbFeCo storage medium (Terris et al. 1995). For this demonstration, the beam was incident from the air side instead of the substrate side, and a Si wafer substrate was used to promote rapid cooling and avoid erasure during read. The SIL approach (and the NSOM approach described below) requires a small head-to-medium spacing to avoid evanescent decay of the light beam, which can be achieved by adapting the SIL optics to a flying slider. Based on time interval analysis of (2,7) coded pseudorandom data, a minimum practical mark length of 0.57 μm has been demonstrated. Data rate was limited only by the low linear velocity of 1.25 m/s needed to maintain a low flying height; a modified air bearing design would permit more competitive data rates.

An alternative near-field optical technique utilizes a small aperture between the sample and the light source. If the aperture-to-sample distance is kept smaller than a wavelength, resolution will be determined by the aperture size rather than by the diffraction limit. In particular, the use of a metallized tapered optical fiber, in the end of which is an aperture, has been adapted to a near-field scanning optical microscopy (NSOM) technique to image and record domains. In this technique, a subwavelength-sized source or detector of light is placed in close proximity to a sample and raster-scanned to generate an image. NSOM can provide imagery with polarization contrast. Using a thin, semitransparent MO film as a storage medium, resolution of 30–50 nm has been consistently obtained using NSOM in the imaging mode, whereas domains on the order of 60 nm have been reproducibly written in the recording mode (Betzig et al. 1992). Such small domain sizes correspond to data densities of ≈45 GB/in.^{2}, well in excess of current magneto-optic or magnetic technologies. These densities require a probe-disk separation of ≈10 nm, which, for a practical system, necessitates improvements in the surface cleanliness and flying height of the head. It has been shown that MO materials exist, e.g., Co/Pt multilayer films (see Section 15.3, Magneto-Optical Media) which can, in principle, support 100 Gbit/in.^{2} with a shot-noise-limited signal-to-noise-ratio adequate to achieve data rates in excess of 100 MHz (Kryder 1995). It remains to be seen if a commercially viable combination of parameters can be found, but near-field optical disk storage is certainly well positioned to benefit from advances in both hard disk drive and optical media technologies. Some proposed recording methods using near-field optics mounted on flying air bearing sliders, similar to those used in magnetic hard disk drives are shown in Figure 15.20.

15.5.2 Media

15.5.2.1 Double-Sided and Multiple Data Layer CDs

Existing CD products are restricted to two-dimensional storage on a single recording surface. Driven by the desire to store high-quality digital moving images on a small optical disk, double-sided and

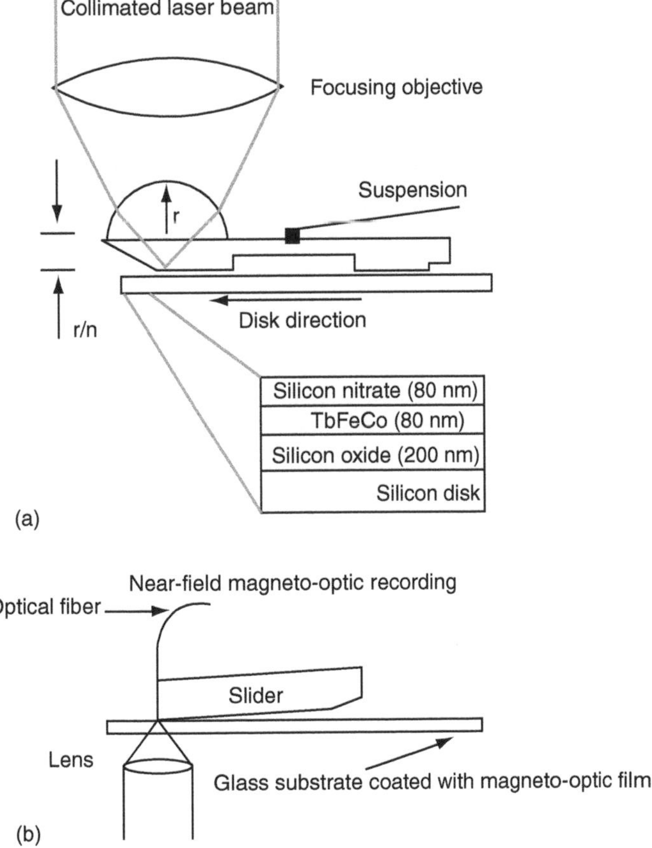

FIGURE 15.20 Schematic diagrams illustrating proposed near-field optical techniques and methods of coupling light to the disk surface using air bearing sliders, (a) Flying solid immersion lens and disk structure. (After Terris, B. D., Mamin, H. J., and Rugar. D. 1995. *Appl. Phys. Lett.*, 68, 141.) (b) Flying near-field aperture and collection lens on opposite sides of a semitransparent disk structure. (After Kryder, M. H. 1995. *Optoelectronics*, 10, 297.)

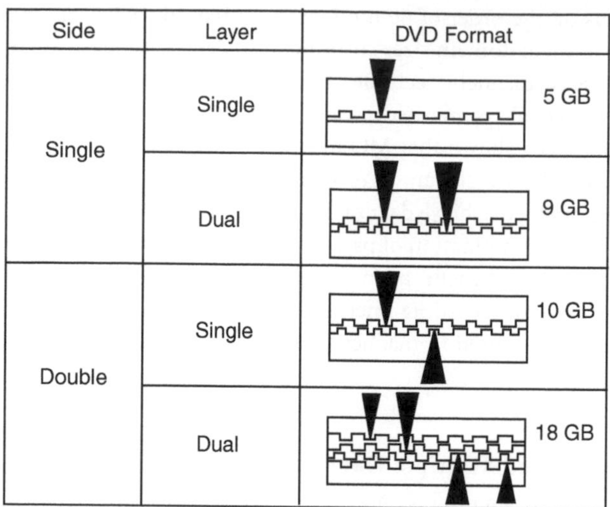

FIGURE 15.21 Schematic illustration of proposed DVD disk structure, with approximate read-only capacities indicated. Write-once and rewritable capacities will be somewhat less.

multiple data layer approaches, which extend optical storage into the third dimension and achieve dramatic increases in disk capacity, are being developed. The double-sided DVD format disk (Figure 15.21) is formed by face-to-face bonding of two half-thickness, 0.6-mm disks. The resulting DVD package is physically compatible with the current generation 1.2-mm CD for backward compatibility. The DVD concept in its single layer embodiment achieves up to 5 GB capacity per side, about 7.5 times the 650-MB capacity of a standard CD, or up to 10-GB total capacity. This capacity increase is anticipated using the combination of reduced wavelength to 650 nm, reduced track pitch to 0.73 μm, increased numerical aperture to 0.6, and more efficient signal modulation and error correction protocols. Use of digital image compression technology according to the MPEG-2 standard, and encoding at a high-speed variable transfer rate allows each side of a DVD disk to carry up to 142 min of high quality moving images and sound, enough for most full-length feature movies. The laminated DVD structure offers improved mechanical stability against environmental changes and improved margin against tilt-induced aberration due to the shorter optical path through polycarbonate. Disadvantages include slightly greater fabrication complexity, increased vulnerability to dirt and scratch-induced defocusing due to the thinner substrate, and the inability to access both data surfaces from one side, i.e., the disk must be flipped or two laser scanners must be used, Note that, in principle, single-side access could be achieved (see below), but at considerable increase in disk fabrication complexity. Of course the DVD format also lends itself to computer-related applications, as a very large capacity ROM, and to very long play, high-fidelity audio disks. Double-sided CD optical storage technology can expand bit capacity with minimal technical risk. Negotiation of industry wide interchange standards remains a key milestone.

The aluminized surface of an audio CD returns much of the incident light back to the read head, an important feature when laser power must be conserved. Stimulated by the realization that an unmetallized polycarbonate CD can provide adequate readback signal if enough laser power is available, an intriguing approach using multiple semitransparent data layers, all accessible from the same side, is also being actively investigated (Rubin et al. 1994) (Figure 15.22). Selective access to each of the data surfaces can be achieved simply by moving the laser focusing lens up and down. Continuous play between layers can be maintained by reversing the sense of the spiral groove on successive surfaces and by the use of a modest send-ahead data buffer during the laser refocus jump. Conversely, successive layers could be formatted differently, e.g., DVD on one layer and CD on the other. Given the short depth of

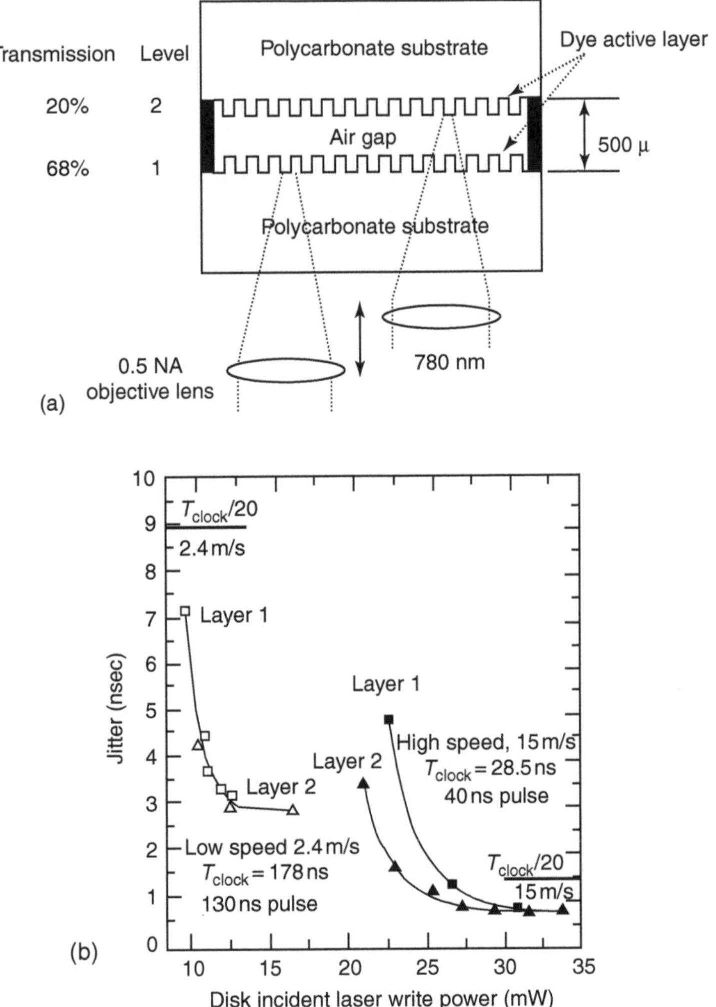

FIGURE 15.22 (a) Two-level multilevel optical storage disk structure built using an air gap between the levels. Level one was optimized to have high transmission of 68% and reasonable reflectivity of 18%. Level two was designed to have a high reflectivity of 30% to compensate for attenuation through level one during readback. (b) Read and write performance of two-level structure at low (2.4 m/s) and high (15 m/s) media speeds. The vertical axis is the jitter from the closely spaced 3T marks. The horizontal axis is the laser power incident on level one necessary to write on level one or two. (After Rubin, K., Rosen, H., Strand, T., Imaino, W., and Tang. W. 1994. *Tech. Dig. Series*, 10, 104.)

focus of a 0.5 NA lens, a layer separation of less than 100 μm is sufficient for effective suppression of interlayer crosstalk during reading and writing. The number of usable recording surfaces is, ultimately, limited only by the available laser power. In the case of read-only applications, there appear to be no fundamental hurdles for disks with 10 or more surfaces. In the case of write-once applications, each recording surface must provide sufficient absorption for writing data. Four-level, write-once recording with product-level signal-to-noise-ratio has been demonstrated (Rubin et al. 1994). Challenges for multiple data layer media include the demonstration of an economical fabrication approach, particularly, regarding formatting (e.g., pregrooves) of each data surface and the commercial production of low reflectance coatings with adequate uniformity.

15.5.2.2 Magneto-Optic Direct Overwrite and Magnetically Induced Super-Resolution

Opportunities for future generation MO media arise not only from superior MO materials, as discussed previously, but also from the clever use of existing materials to enhance performance. One example of this is the functional separation of layers based on exchange-coupled multilayers (ECML), an approach for which conventional RE-TM alloys are particularly well-suited due to their ferrimagnetic ordering and subsequent compensation temperature behavior. The ability to utilize ECML sandwich structures of two or more layers with complementary properties has opened up new and exciting possibilities which would be difficult or impossible with a single MO layer. The most spectacular examples are the achievement of single-pass, direct overwrite (DOW) with laser modulation and magnetically induced superresolution (MSR). Figure 15.23(a) illustrates one such DOW scheme utilizing an ECML medium structure and an initialization magnet (Saito et al. 1987). Several ECML DOW structures have been described which meet or exceed the ISO CNR specification of ≧45 dB at 0.75-μm mark length. Some of these structures do not require the extra initialization magnet, which impedes downward compatibility and occupies precious drive space.

Figure 15.23(b) shows an ECML scheme for realizing MSR (Aratani et al. 1991). The principle of superresolution in microscopy indicates that resolving power exceeding the classical diffraction limit can be obtained by putting an aperture directly against the object, which intercepts all light outside a certain region. In MO MSR, a dynamic aperture is induced magnetically in an exchange-coupled readout layer. Dramatically improved mark length performance using 780-nm light has been demonstrated compared with conventional MO disk structures. Even though these ECML structures increase media complexity and cost, their great versatility will likely maintain their appeal for future product developments.

15.5.2.3 Land and Groove Recording

In most current generation optical disk media, a guardband is provided between successive data tracks to reduce crosstalk interference. This guardband represents space unused for data storage and limits the track density. Land and groove recording achieves virtually zero guardband width, i.e., roughly a doubling of the track density, using the optical phase difference between land and groove regions to cancel crosstalk from neighboring tracks. It has been demonstrated that a 0.6 μm track pitch, half that of current CDs, is feasible using this approach (Honma et al. 1994). The effectiveness of crosstalk cancellation is very sensitive, however, to small deviations in parameters, such as groove depth, domain length, and Kerr ellipticity.

15.5.3 Holographic Storage

Holographic storage provides a three-dimensional optical storage system where terabytes of information can be stored with high data rates in the Gb/s range (Psaltis and Mok, 1995; Hesselink and Bashaw, 1993). The other novel features of holographic storage are fast access time (microseconds), parallel processing of pages as one unit during writing and reading, and no moving parts. In addition to data rate and capacity advantage, holographic storage can be readily adaptable to optical signal processing techniques, such as image recognition, etc.

Figure 15.24 compares recording with a volume holographic system and a current optical recording system (Asthana and Finkelstein 1995). When two laser beams interfere with each other in a light-sensitive material, its optical properties are changed by the grating formed by the intersecting beams. A hologram of the page of data is created when the signal beam meets the reference beam in the photosensitive material. After data has been recorded, the page can be holographically reconstructed by shining the reference beam into the crystal. The reference beam is diffracted, so that it creates the image of the original page. The diffracted beam is then projected onto a charge-coupled device (CCD) thereby reading the stored information. The major components in the holographic system are a spatial light modulator for data input, a laser diode source or two-dimensional laser array, a holographic recording medium, and a charge-coupled device (CCD) as a data readout device. Further improvements in

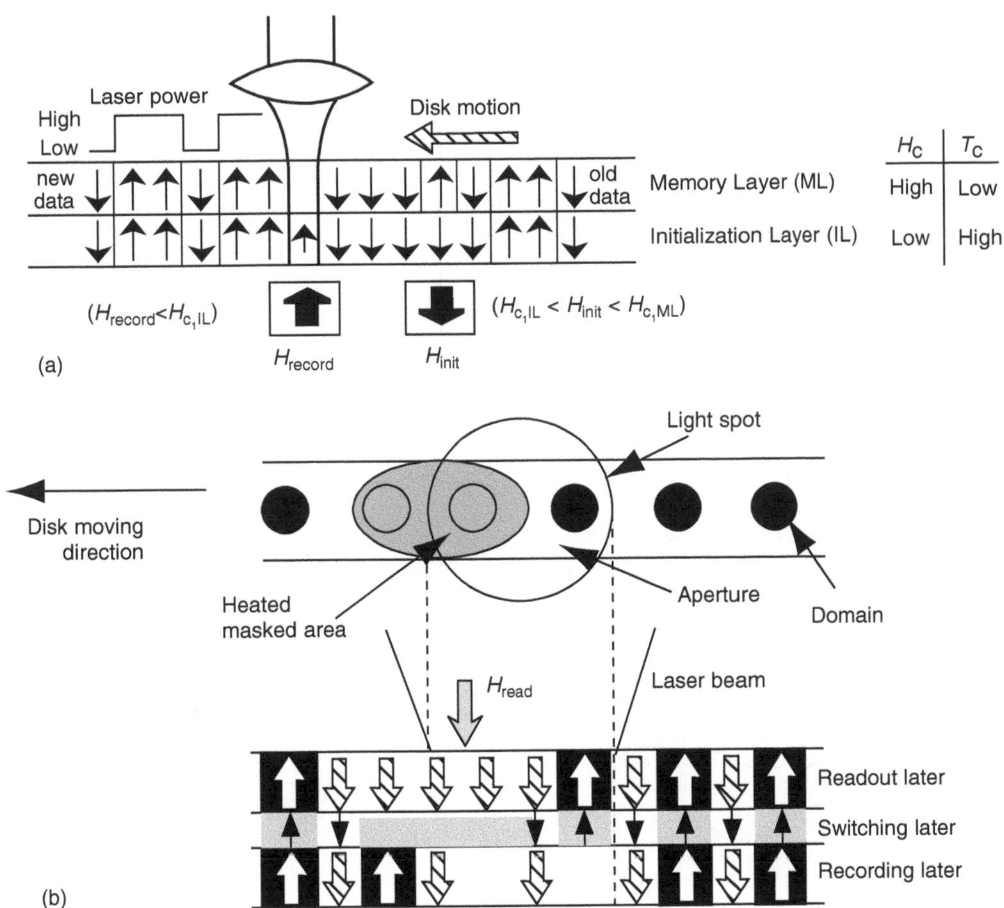

FIGURE 15.23 (a) Principle of direct overwrite in exchange-coupled bilayers. Moving from right to left, the initialization layer is dc-erased downward by the initialization magnet, with no effect on previously written domains in the memory layer. Under high laser power, both layers are heated near their T_c, and upward domains are first frozen into the initialization layer by the recording magnet and, subsequently, copied to the memory layer by exchange coupling upon further cooling. Under low laser power, only the memory layer is heated near its T_c, and downward domains are copied from the unchanged initialization layer to the memory layer upon cooling. (After Saito, J., Sato, M., Matsumoto, H., and Akasawa. H. 1987. *Jpn. J. Appl. Phys.*, 26, 155.) (b) Principle of magnetically induced superresolution (MSR) in exchange-coupled multilayers using front aperture detection (FAD). In MSR-FAD, an intermediate switching layer with low Curie temperature $T_{c,switch}$ is used to control the effective exchange field $H_{exchange}$ between recording and readout layers by conventional light condition $H_{exchange} > H_{record} + H_{c,read}$ is reached. During readout, motion of the disk to the left readout layer magnetization is aligned with H_{read} (because $H_{exchange} = 0$ and $H_{c,read}$). This effectively dc-erases the heated region, which, thus, does not contribute to signal modulation and acts as a magnetic mask. The signal is modulated by unswitched domains in the remaining chevron-shaped aperture under the irradiated spot. After readout, recorded domains are restored in the readout by $H_{exchange}$ just as in the original recording process. (After Aratani, K., Fukumoto, A., Ohta, M., Kaneko, M., and Watanabe, K. 1991. *SPIE*, 1499, 209.)

components are required, such as (i) high-speed scanning devices, (ii) high bandwidth and large size spatial light modulators, (iii) high-speed parallel readout CCD devices, and (iv) high-power visible or near-infrared diode lasers. Currently, one of the major hindrances to commercialization of holographic storage has been appropriate recording media. Presently, there are two types of recording media used for holographic storage: photopolymers and iron-doped $LiNbO_3$ and $Sr_xBa_{1-x}Nb_2O_6$ (SBN). The issues with photopolymers are (i) data loss due to shrinkage of the photopolymer, (ii) limited thickness ($\leq$40 mm)

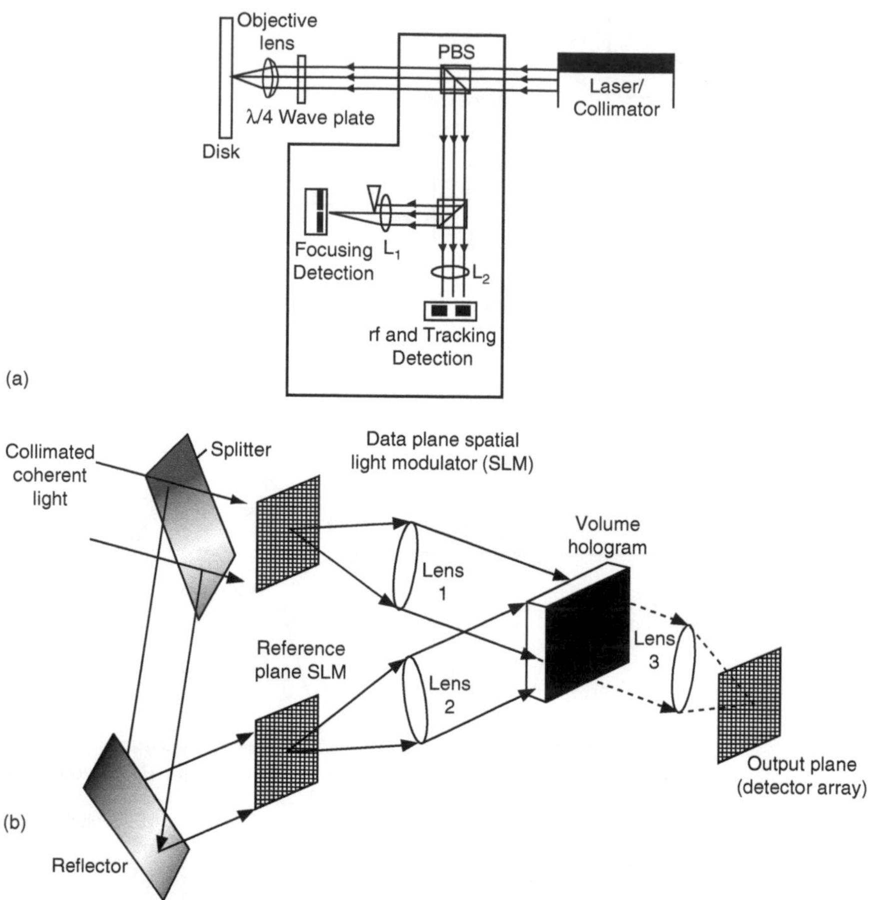

FIGURE 15.24 (a) Layout of a typical bulk optical head that is read-write capable. (b) A volume holographic system. During recording, a collimated laser beam is split into two parts, one for reference and the other for carrying data. The latter passes through a spatial light modulator, which reformats an input bit stream into a page form suitable for holographic storage and modulates it onto the beam. The modulated beam, then, penetrates the holographic material, interacting with the reference beam to form a fringe pattern which the material records. If a second spatial light modulator is used to change the reference beam's incident angle, more than one set of data can be recorded in the same spot. The recording medium can be a sandwich of thin holographic films and transparent buffer layers or a bulk single crystal. (After Asthana, p. and Finkelstein, B. 1995. *IEEE Spectrum*, 25–37.)

which restricts the number of angularly multiplexed holograms that can be stored, (iii) no erasability of the media, and (iv) low signal-to-noise-ratio and high bit-error rates due to low photorefractive response. The difficulties with iron-doped $LiNbO_3$ are slower recording response time and the need to "fix" the recorded data by a thermal process.

Intersecting signal and reference beams generate an interference pattern with regions of high intensity of the form $I=I_0[1+m\cos(Kx)]$, in which I is the intensity, m is the modulation factor and K is the grating vector given by $K=2\pi/\Lambda$, where Λ is grating period and x is transverse to the direction of propagation. The photons in the high intensity regions excite electrons into the conduction band, and electrons then diffuse into darker regions of the grating. The separation of electrons from positive ions establishes an electric field in the crystal. Through electro-optic effects, this electric field produces a change in refractive index. The refractive index variation grating, formed inside the crystal, causes the reference beam to diffract for image retrieval. In a holographic medium that does not display nonlinear beam coupling, the diffraction efficiency is given as $\eta = \sin^2(\pi mn^3 rE/2\lambda \cos\theta)$, where E is the space charge field, n is the index of refraction of the crystal, r is the electro-optic coefficient, λ is the wavelength of light, and 8 is the angle of incidence.

The information capacity of 3-D data storage is approximately the holographic volume divided by the volume of the smallest feature. If we consider that a hologram is an information channel with W independent levels, and a resolving power of at least $1\backslash\lambda$ cycles per dimension, the storage capacity $\rho = \log_2 W\backslash\lambda^3$ (bits/mm^3). For $W = 1{,}024$ and $\lambda = 500$ nm, $\rho = 8 \times 10^{10}$ bits/mm^3. This calculation does not assume any particular multiplexing scheme. The theoretical limits are not realized because noise tends to reduce the usable capacity before geometric or wavelength limits are approached. The storage density can be also written as $\rho_1 = N_p^2 N_o/\omega^2$, where N_p is the number of pixels, N_o is the number of holograms, and ω is the beam size. The number of holograms by angular multiplexing is given by $h_1 = \Delta\theta_r/\Delta\theta$, where $\Delta\theta = \lambda\cos\theta_s/[L \sin(\theta_s + \theta_r)]$, where θ_r and θ_s are reference and signal beam angles and L is the thickness of the sample. For $\lambda = 500$ nm, $\theta_r = \theta_s = 20°$, $\Delta\theta_r = 4.2°$ and $L = 1$ cm, $h_1 = 1000$ holograms. In a demonstration, 10,000 pages of information have been stored at a single location in an iron-doped $LiNbO_3$ crystal (Psaltis et al. 1995).

Holographic storage is most efficient when large data blocks or pages are being recorded. With data blocks as large as 1 Mbit, the recording of digital images is a natural application. If we couple this with very high-speed access to thousands of pages (using, e.g., acousto-optic deflectors with 10–50 (μs access) and data rates in excess of 1 Gbit/s (e.g., a 1 Mbit page is parallel detected in 1 ms), this form of storage is particularly well suited for multiple user servicing. Because of this, a host of applications can be envisioned including video-on-demand, on-line catalogues and manuals, distribution of satellite images and maps, geophysical data, medical and radiological imagery, digital libraries, etc. Commercialization of holographic memories is still several years away. However, entry-level versions of this storage technology may be available by the turn of the century.

References

Aratani, K., Fukumoto, A., Ohta, M., Kaneko, M., and Watanabe, K. 1991. Magnetically induced super-resolution in novel magneto-optical disk. *SPIE*, 1499, 209.

Asthana, P. and Finkelstein, B. 1995. Superdense optical storage. *IEEE Spectrum*, pp. 25–37. August.

Atkinson, R., Salter, I. W., and Xu, J. 1992. Design, fabrication, and performance of enhanced magneto-optic quadrilayers with controllable ellipticity. *Appl. Opt.*, 31, 4847.

Atkinson, R., Salter, I. W., and Xu, J. 1993. Angular performance of phase-optimized magneto-optic quadrilayers. *Optical. Eng.*, 32, 3288.

Betzig, E. and Trautman, J. K. 1992. Near-field optics: Microscopy, spectroscopy, and surface modification beyond the diffraction limit. *Science*, 257, 189.

Betzig, E., Trautman, J. K., Wolfe, R., Gyorgy, E. M., Finn, P. L., Kryder, M. H., and Chang, C.-H. 1992. Near-field magneto-optics and high density data storage. *Appl. Phys. Lett.*, 61, 142.

Brucker, C. F. 1996. Magneto-optical thin film recording materials in practice, In *Handbook of Magneto-Optical Data Recording—Materials, Subsystems, Techniques*, T. W. McDaniel and R. H. Victora, eds., pp. 1–S661. Noyes, New Jersey.

Buschow, K. H. J., van Engen, P. G., and Jongebreur, R. 1983. *J. Magn. Magn. Mater.*, 38, 1.

Carslaw, H. S. and Jaeger, J. C. 1988. *Conduction of Heat in Solids, 2nd Ed.*, Oxford University.

Connell, G. A. N. and Bloomberg, D. S. 1998. Magnetooptical recording. In *Magnetic Recording*, Vol. II, C. D. Mee and E. D. Daniel, eds., pp. 173–S661. McGraw-Hill.

Deeter, M. N. and Sarid, D. 1988. Effects of incident angle on readout in magnetooptic storage media. *Appl. Opt.*, 27, 713.

Froelich, E., Milster, T., and Uber, R. 1992. High-Resolution optical lithography with a near-field scanning subwavelength aperture. *Proc. SPIE*, 1751, 312.

Fu, H., Yan, Z., Lee, S. K., and Mansuripur, M. 1995. Dielectric tensor characterization and evaluation of several magneto-optical recording media. *J. Appl. Phys.*, 78, 4076.

Gambino, R. J., Fumagali, P., and Ruf, R. R. 1993. Metastable films of giant magneto-optic rotators. *J. Magn. Soc. Jpn*, 17, 276.

Gelbar, D. 1990. Optical data storage. *SPIE*, 1316, 65.

Gupta, M. C. 1984. Laser recording on an overcoated organic dye binder medium. *Appl. Opt.*, 23, 3950.

Gupta, M. C. 1988. A study of laser marking of thin films. *J. Mater. Res.*, 3, 1187.

Gupta, M. C., Kozlovsky, W., and Nutt, A. C. G. 1994. Second harmonic generation in bulk and waveguide $LiTaO_3$ with domain inversion induced by electron beam scanning. *Appl. Phys. Lett.*, 64, 3210.

Haase, M. A., Qui, J., DePuydt, J. M., and Cheng, H. 1991. Blue green laser diodes. *Appl. Phys. Lett.*, 59, 1272.

Hansen, M. and Anderko, K. 1958. *Constitution of Binary Alloys*. McGraw-Hill, New York.

Heavens, O. 1995. *Optical and Properties of Thin Solid Films*. Academic.

Hesselink, L. and Bashaw, M. C. 1993. Optical memories implemented with photorefractive media. *Optical Quantum Electron.*, 25, S611–S661.

Holtslag, A. H. M. 1989. Calculations on temperature profiles in optical recording. *J. Appl. Phys.*, 66, 1530.

Honma, H. Iwanga, T., Kayamura, K., Nakada, M., Katayama, R., Kobayashi, K., Itoi, S., and Inada. H. 1994. High density land/groove recording using PRML technology. Optical Data Storage Topical Meeting, WD1.

Howe, D. 1996. Data reliability and errors. In *Handbook of Magneto-Optical Data Recording—Materials, Subsystems, Techniques*, T. W. McDaniel and R. H. Victora, eds., pp. 5301–76. Noyes, New Jersey.

Hurst, J. E. and Kozlovsky, W. 1993. Optical recording at 2.5 $Gbit/in.^2$ using a frequency doubled diode laser. *Jpn. J. Appl. Phys.*, 32, 5301.

Hurst, J. E. and McDaniel, T. W. 1996. Writing and erasing in magneto-optical recording, In *Handbook of Magneto-Optical Data Recording—Materials, Subsystems, Techniques*, T. W. McDaniel and R. H. Victora, eds., pp. 335–76. Noyes, New Jersey.

Ikeda, E., Tanaka, T., Chiba, T., and Yoshimura, H. 1993. The properties of Sony recordable mini disc. *J. Magn. Soc. Jpn*, 26, 335.

International Standard (ISO/IEC 10089: Information Technology—130 mm Rewritable Optical Disk Cartridge for Information Exchange).

Jorgensen, F. 1988. *The Complete Book of Magnetic Recording, 3rd Ed.*, TAB Books, Blue Ridge Summit, PA.

Kivits, P., de Bont, R., Jacobs, B., and Zalm, P. 1982. The hole formation process in tellurium layers for optical data storage. *Thin Solid Films*, 87, 215.

Kryder, M. H. 1995. Near-field optical recording: An approach to 100 $Gbit/in.^2$ recording. *Opto-electronics*, 10, 297.

Lairson, B. M., Visokay, M. R., Sinclair, R., and Clemens, B. M. 1993. *J. Magn. Soc. Jpn.*, 17, 40.

Lenth, W. Dec.1994. Optical storage: A growing mass market for lasers. *Laser Focus World*, p. 87.

Lynch, D. W. and Hunter, W. R. 1991. In *Handbook of Optical Constants of Solids II*, E. D. Palik, ed., pp. 5738–76. Boston, Academic.

Madison, M. R. and McDaniel, T. W. 1989. Temperature distributions produced in an N-Layer film structure by a static or scanning laser or electron beam with applications to magneto-optical media. *J. Appl. Phys.*, 66, 5738.

Mansfield, S. M. and Kino, G. S. 1990. *Appl. Phys. Lett.*, 57, 2615.

Mansuripur, M. 1989. Distribution of light at and near the focus of high numerical aperture objectives. *J. Optical Soc. Am.*, A3, 2086.

Mansuripur, M. 1989. Certain computational aspects of vector diffraction problems. *J. Optical Soc. Am.*, A6, 786.

Mansuripur, M. 1995. *The Physical Principles of Magneto-Optical Recording*. Cambridge University Press, United Kingdom.

Marchant, A. B. 1990. *Optical Recording—A Technical Overview*. Addison-Wesley.

McDaniel, T. W. and Bartholomeusz, B. 1996. Modeling magneto-optical recording processes, In *Handbook of Magneto-Optical Data Recording—Materials, Subsystems, Techniques*, T. W. McDaniel and R. H. Victora, eds., pp. 2943–76. Noyes, New Jersey.

McDaniel, T. W. and Mansuripur, M. 1987. Numerical simulation of thermomagnetic writing in RE-TM films. *IEEE Trans. Magn.*, 23, 2943.

McDaniel, T. W., Rubin, K. A., and Finkelstein, B. I. 1994. Optimum design of optical storage media for drive compatibility. *IEEE Trans. Magn.*, 30, 4413.

McDaniel, T. W. and Sequeda, F. O. 1992. Design and material selection for a thin film magneto-optic disk. *Appl. Phys. Commun.*, 11, 427.

McDaniel, T. W., Sequeda, F. O., McGahan, W., and Woollam, J. A. 1991. Optical and magneto-optical performance of optimized disk structures, Proc. MORIS '91. *J. Magn. Soc. Jpn.*, 15, S1, 361.

Nakamura, S., Senoh, M., Nagahama, S., Iwasa, N., Yamada, T., Matsushita, T., Kiyoku, H., and Sugimoto, Y. 1996. InGaN-based multi-quantum-well structure laser diodes. *Appl. Phys. Lett.*, 68, 3269.

Ohta, T., Akahira, N., and Satoh, I. 1995. High density phase change optical recording. *Optoelectronics Devices Technol.*, 10, 361.

Ovshinsky, S. R. 1968. *Phys. Rev. Lett.*, 21, 1450.

Patel A.M. 1988. In *Magnetic Recording Vol. III*, C. D. Mee and E. D. Daniel, eds., pp. 70–76. McGraw-Hill, New York.

Pohlman, K. C. 1992. *The Compact Disc Handbook, 2nd Ed.*, A-R Editions, Madison, WI.

Psaltis, D. and Mok, F. 1995. Holographic memories. *Sci. Am.*, 70–76. May.

Reim, W., Schoenes, J., and Wachter, P. 1984. New high efficiency Kerr rotators. *IEEE Trans. Magn.*, 20, 1045.

Reim, W., Schoenes, J., Hulliger, F., and Vogt, O. 1986. *J. Magn. Mater.*, 54, 1401.

Risk, W. P. 1990. Compact Blue Laser Devices. *Optics Photonics News*, pp. 10–15. May.

Rubin, K., Rosen, H., Strand, T., Imaino, W., and Tang, W. 1994. Optical Data Storage Topical Meeting. *Tech. Dig. Series*, 10, 104.

Saito, J., Sato, M., Matsumoto, H., and Akasawa, H. 1987. Direct overwrite by light power modulation on magneto-optical multilayered media. *Jpn. J. Appl. Phys.*, 26, 155.

Shih, O. W., A steady-state heat conduction solution for optical disk recording using constant or periodically modulated laser radiation, to be published in *J. Appl. Phys.*

Steenbergen, C., Lou, D., and Verboom, H. Working document for development of a standard for a modulation code to be used with an optical media unit for digital information interchange using a 130 mm nominal diameter disk. ANSI submission 1985 X3B11/85-X3B11135.

Strasser, T. A. and Gupta, M. C. 1993. Integrated optic grating-coupler based optical head. *Appl. Optics.*, 32, 7454.

Sugaya, S. and Mansuripur, M. 1994. Effects of substrate birefringence on focusing and tracking servo signals in magneto-optical disk data storage. *Appl. Opt.*, 33, 5073.

Terris, B. D., Mamin, H. J., and Rugar, D. 1995. Near-field optical data storage. *Appl. Phys. Lett.*, 68, 141.

Tyan, Y.–S., Preuss, D. R., Olin, G. R., Vazan, F., Pan, K.–C., and Raychaudhuri, P. K. 1992. Kodak phase-change media for optical tape application, In *Proc. Goddard Conf, Mass Storage Systems and Technologies II*, B. Kobler and P. C. Hariharan, eds., pp. 499–511. NASA Goddard Space Flight Center.

Ura, S., Suhara, T., Nishihara, H., and Koyama, J. 1986. An integrated-optic disk pickup device. *IEEE J. Lightwave Technol.*, LT-4, 913.

Ward, L. 1991. In *Handbook of Optical Constants of Solids II*, E. D. Palìk, ed., p. 72. Academic, Boston.

Weller, D., Hurst, J., Notarys, H., Brandle, H., Farrow, R. F. C., Marks, R., and Harp, G. 1993. *J. Magn. Soc. Jpn.*, 17, 72.

Yamada, M., Nada, N., Saitoh, M., and Watanabe, K. 1993. *Appl. Phys. Lett.*, 62, 435.
Yamanaka, Y., Hirose, Y., Fuji, H., and Kubota, K. 1990. High density recording by superresolution in an optical disk memory system. *Appl. Opt.*, 29, 3046.
Zeper, W. B., Jongenelis, A. P. J., Jacobs, B. A. J., van Kesteren, H. W., and Carcia, P. F. 1992. *IEEE Trans. Magn.*, 28, 2503.
Zhou, S. M., Lu, M., and Shai, H. R. 1991. *Phys. Status Solidi (b)*, 168, 651.

16

Holographic Data Storage

Sergei S. Orlov

16.1 Introduction

Both magnetic and conventional optical data storage technologies store individual bits of information as distinct magnetic or optical changes of the surface of the recording medium. Currently, both of these techniques approach physical limits of storage density beyond which the bits are either too small to detect or the tolerances imposed on the storage media become too stringent. In magnetic storage, the superparamagnetic limit seems to be not far away, and recording head scaling becomes a growing problem. Optical storage technology, such as blue DVD, is reaching its practical limits of capacity (25 gigabytes per layer in a CD-size disk). To further improve capacity and transfer rates, several options are available, including increasing the NA beyond 0.85, reducing the wavelength below 400 nm, or adding more layers. All these options present significant challenges.

The alternatives to conventional storage techniques include near-field recording, which employs near-field coupling of light in and out of the medium and requires close proximity of the optical stylus and the media; optically assisted magnetic recording; probe tip storage, using a large number of submicron scale probes and direct interaction with the recording medium surface; and volumetric recording, in which the data is recorded throughout the volume of the storage medium rather than on the surface. Holographic data storage is a volumetric approach which, although conceived decades ago, has made recent progress toward practicality largely through advances in the enabling technologies, such as spatial light modulators, low-cost visible light lasers, and CCD and CMOS detectors.

16.2 Holographic Storage Principles

16.2.1 Information Storage via Holography

Holographic storage relies on the principle of holography, i.e., the ability to reconstruct the complete object's wavefront from the imprinted interference fringes, which were created by interfering two optical beams in a photosensitive material (Gabor 1948; Leith and Upatnieks 1962; Denisyuk 1963). In holographic data storage, the information is presented in the form of pixelated pages, which can be rather large (as large as 1 million pixels in each), and stored in the form of volumetric gratings. The writing is done by intersecting two mutually-coherent laser beams within the photosensitive storage material (Figure 16.1). The object beam contains the information to be stored, and the reference beam is designed to be simple to reproduce—for example, a simple collimated beam with a planar wavefront, or a spherical wave. The resulting optical interference pattern causes chemical and/or physical changes in the photosensitive medium: A replica of the interference pattern is stored as a change in the refractive index, absorption, or thickness of the photosensitive medium. When the stored interference grating is illuminated with one of the two waves used during recording (Figure 16.2[a]), some of the incident light is diffracted by the stored grating reconstructing the other wave. Illuminating the stored grating with the reference wave reconstructs the object wave, and vice versa (Figure 16.2[c]). Interestingly, a backward-propagating or phase-conjugate reference wave, illuminating the stored grating from the back side, reconstructs an object wave that also propagates backward toward its original source (Figure 16.2[b]).

Due to the Bragg selectivity property of the volume holograms, a large number of data pages can be superimposed in the same location of the thick piece of medium and accessed independently by

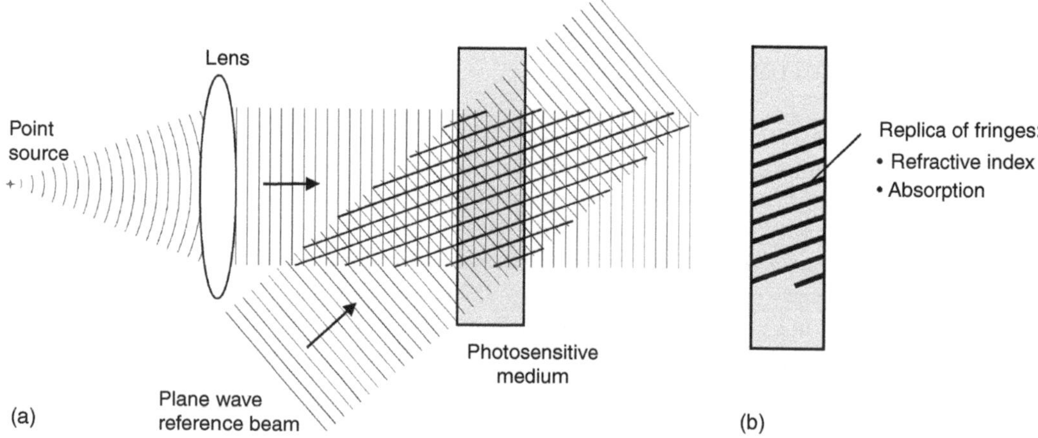

FIGURE 16.1 Storage of a single bit of information as a hologram by (a) interfering the light from a point source with the coherent reference planewave inside photosensitive medium, (b) recording the interference fringes in the form of a volumetric index or absorption grating.

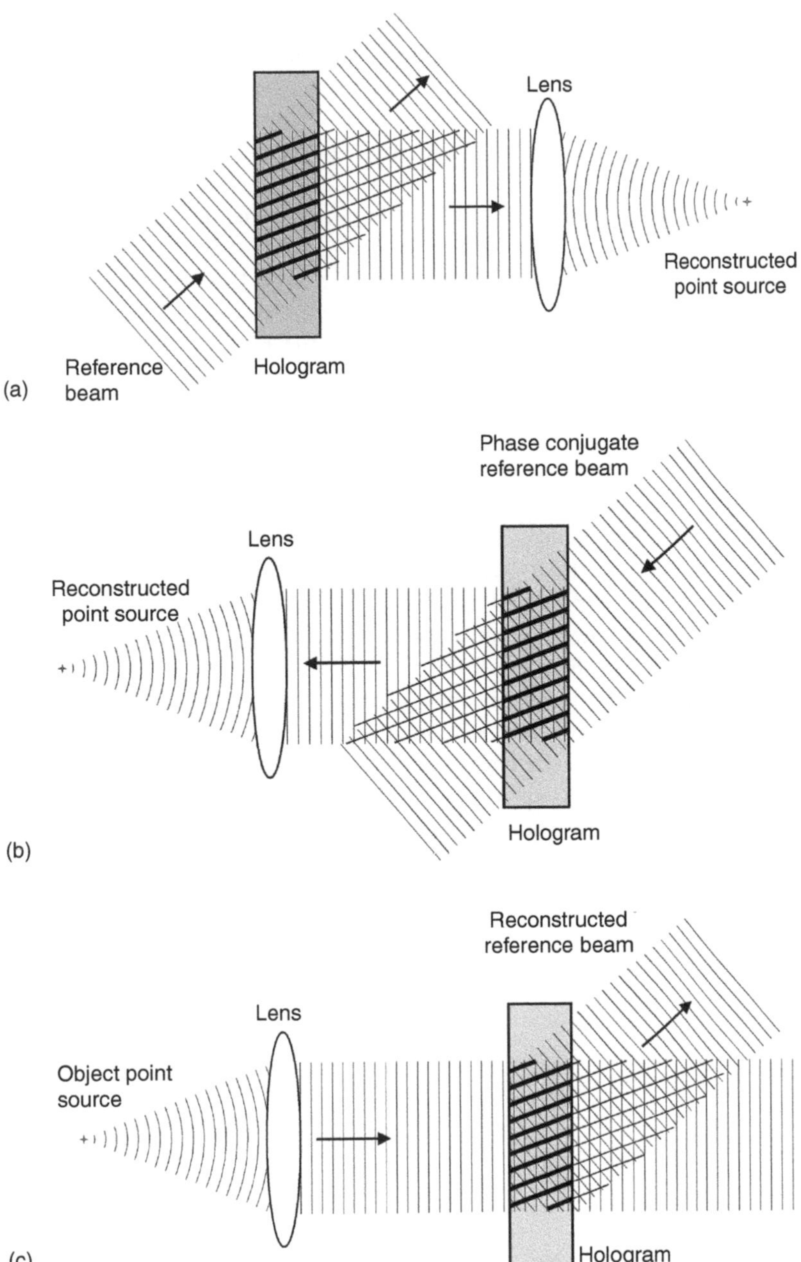

FIGURE 16.2 Retrieving the stored information by: (a) Illumination by the reference beam and reconstructing the stored information bit. (b) Illumination by a phase-conjugate (or counter-propagating in case of a planewave) reference beam reconstructing a phase-conjugate object beam which back-traces the original object beam path. (c) Associative retrieval: illumination by the original object beam reconstructs the corresponding reference.

presenting an appropriate reference beam as long as the gratings belonging to the different pages are distinguishable. The process of superimposing multiple holograms is referred to as multiplexing. Basic multiplexing techniques include angular (d'Auria, Huignard, and Spitz 1973), wherein the reference beams differ by angle of their incidence; wavelength (Rakuljic, Leyva, and Yariv 1992), wherein the references differ in optical wavelength; and shift (Psaltis et al. 1995), in which the same reference beam of curved wavefront is used, but the recording is performed at a shifted position of the medium. In addition to Bragg techniques, there exists another class of multiplexing in which a complex wavefront shape (e.g., speckled) of reference beams are used and addressing of different pages is performed exploiting the spatial correlation properties (Darskii and Markov 1988) of the reference beam combined with shift of the recording medium or the beam itself.

Because of the use of the entire volume of the medium rather than just its surface, holography offers high volumetric storage density with a theoretical limit of $\sim V/\lambda^3$ (van Heerden 1963), which translates into tens of terabits per cubic centimeter. Data pages can contain large numbers of data bits or pixels; practically up to 1 million bits/page have been demonstrated. Because a page is stored and recalled as a whole, data transfer rates can be extremely high, exceeding 10 Gb/s. Among other unique properties of holographic data storage is the possibility of realizing extremely fast access time, as short as 50 μs or less, because the optical reference beam can be moved very rapidly (e.g., with the use of acoustooptic deflectors) without any mechanical motion and inertia.

In a basic holographic storage system, the laser light is split in to two arms: the signal and the reference. The data to be stored are imprinted onto the object beam using a pixelated input device called spatial light modulator (SLM) (Figure 16.3). The SLM is typically either a liquid crystal panel, or a MEMS-based device (like Texas Instruments' deformable mirror array) similar to those used in screen projectors. During the writing stage, the object beam and the corresponding reference are both projected on the storage medium, creating a complex interference fringe pattern, which is replicated by the storage medium in the form of, for instance, an index grating. This set of index fringes distributed throughout the volume now contains the relative phase and amplitude of the object's wavefront, and, thus, contains the information imprinted on the object beam. During the readout stage, the corresponding reference beam is projected on the medium and the diffracted wavefront represents a reconstructed replica of the original page, which is directed onto the pixelated output device—a charge couple detector (CCD) array or a CMOS imager. The pixels in

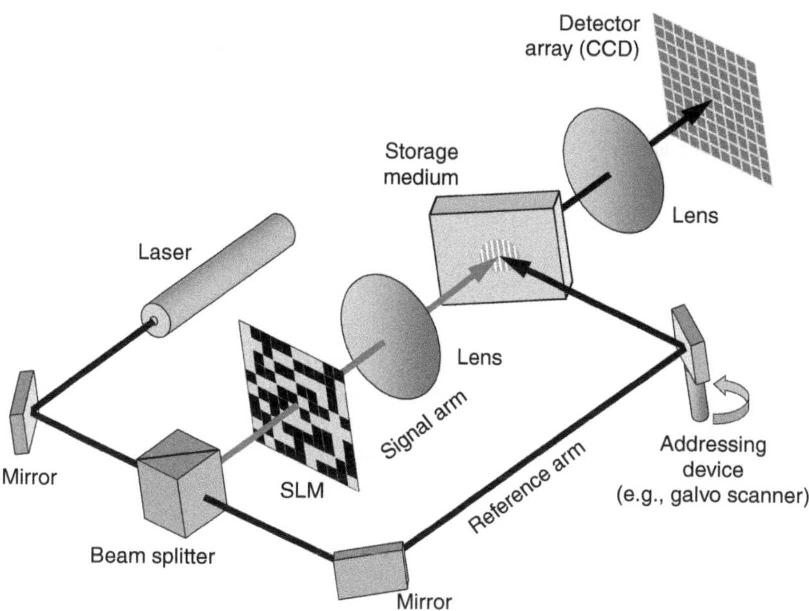

FIGURE 16.3 Basic holographic data storage system comprising a coherent laser source, beam splitter, page composer (SLM), imaging optics, CCD or CMOS detector array, reference beam control unit, and storage medium.

the image captured by the camera must match the pixels of the CCD array (within ± 0.1 pixel) to assure a sufficiently low raw error rate during readout. Although the information pages usually include modulation channel coding, data interleaving, and error correction in order to make the data more robust with respect to optical imperfections and errors, the image of the data page needs to be as close as possible to the original object. Any optical aberrations in the imaging system or defocus of the detector array would spread energy from one pixel to its neighbors. Optical distortions (where pixels on a square grid at the SLM are not imaged to a square grid) or errors in magnification will move the image pixel off its intended receiver, and either of these problems (blur or shift) will introduce errors in the retrieved data. An alternative to high-quality imaging optics is the phase conjugate approach, where the stored page is retrieved using a phase conjugate (or pseudo-phase conjugate) version of the corresponding writing reference beam. In this case, an exact-phase conjugate of the image is reconstructed and, after propagating back through the same optics, the data is recreated in the same optical plane as the original SLM image. Due to phase conjugation, most of the shortcomings of the optics and the storage media are compensated. Both approaches, however, still require a very precise alignment of the SLM and the CDD devices.

An intriguing feature of holographic data storage is associative retrieval: by placing a partial search data pattern on the object beam and illuminating the storage media, all of the reference beams which were used for recording will be reconstructed (Figure 16.2[c]). The relative intensity of the references is proportional to the similarity between the content of the data pages and the search pattern. Thus, by finding the reconstructed reference beam, which has the highest intensity, the address of the closest match in the database can be found. Thousands of holograms multiplexed within the same volume of the media can be searched simultaneously, leading to extremely high search rates of 100 Gbit/s—substantially higher than the current standard found in relational database searches based on conventional storage and search technologies.

16.2.2 Basic System Architectures

Holography is a uniquely different approach to data storage, when compared to conventional optical data recording, in that the different bits of information share the very same space of the storage medium. This provides certain special features, such as fast associative retrieval capability and partial immunity to small-scale point defects in the storage material, but also makes the architectural design rather different from the surface storage, or the volumetric bit-by-bit recording. A variety of holographic architectures exist, including transmission, reflection, and 90-degree, which are differentiated by the way the object and reference beams are arranged with the respect to each other and the medium (Figure 16.4 and Figure 16.5). The phase conjugate retrieval is possible in each of these cases via reading out the stored pages with a phase conjugate replica of the corresponding reference beams. Apart from different holographic arrangements, there are several optical architectures that determine how the SLM image is projected onto the CCD array (Figure 16.6). These options include the image plane storage, wherein the SLM is first imaged into the medium and the reconstruction is re-imaged onto the CCD array; Fourier (or partial Fourier) plane storage, in which the double Fourier transform imaging is used and the storage medium is located at or near the Fourier plane; Fresnel zone storage, wherein a single lens system is used to project the SLM onto the CCD array and the medium is located in the intermediate plane (usually close to the focusing point); and van der Lugt holographic arrangement, where the SLM is illuminated by a converging beam.

Fourier geometry usually provides the highest storage density and lowest level of imaging distortions, but may lead to recording nonlinearities and distortions in the material due to high-intensity Fourier spiking, which can be mitigated either by shifting the medium outside the exact Fourier plane (partial Fourier plane storage) or by using a phase mask on the SLM. Image plane recording does not suffer from the recording nonlinearity problem, but, in general, has a higher degree of optical distortions for short-focal-length lens systems and produces larger spotsize. The advantage of the Fresnel imaging is that it does not require lens optics upon readout, which can be beneficial for ROM applications.

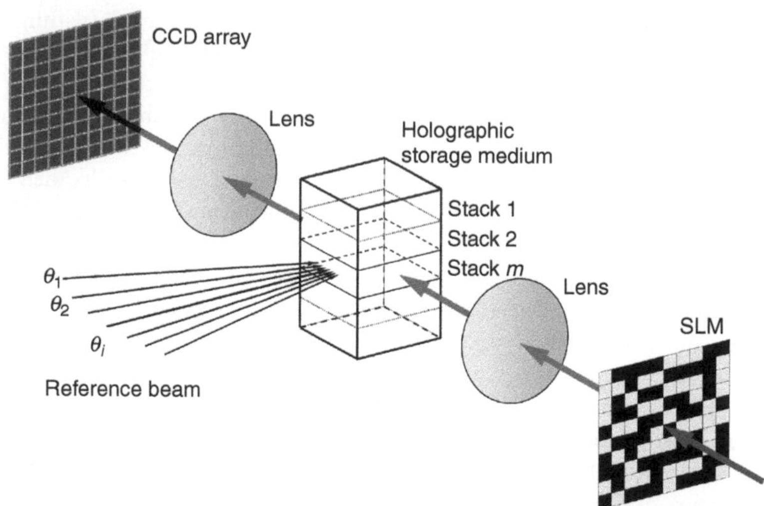

FIGURE 16.4 Schematic of 90-degree geometry holographic architecture. The data are arranged in stacks, individual holograms within a stack are addressed by different incidence angles (or wavelength) of the reference beam.

The choice of optical and holographic architecture is influenced by both the system requirements and the recording media properties. The high-density systems tend to employ double Fourier transform optics, because it allows the smallest hologram spot size for the same effective focal length. 90-Degree geometry (Figure 16.4) is not suitable for thin photopolymer media, but it is the architecture of choice for thickinorganic crystals (several millimeters to a few centimeters). The scattering noise level in the most commonly-used photorefractive iron-doped lithium niobate is extremely small, while the optical quality is high—close to standard optical-grade glasses—allowing storage of thousands of high-resolution pages in the same storage location. The 90-degree systems can have very short access time

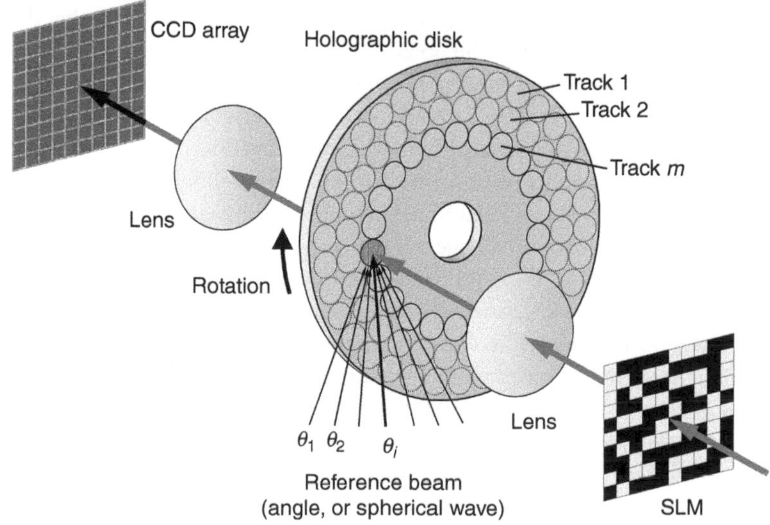

FIGURE 16.5 Holographic disk (transmission) architecture. The data is arranged in tracks; different tracks are accessed by lateral medium positioning; holograms within location are addressed using one of the multiplexing methods (angular, shift, or peristrophic).

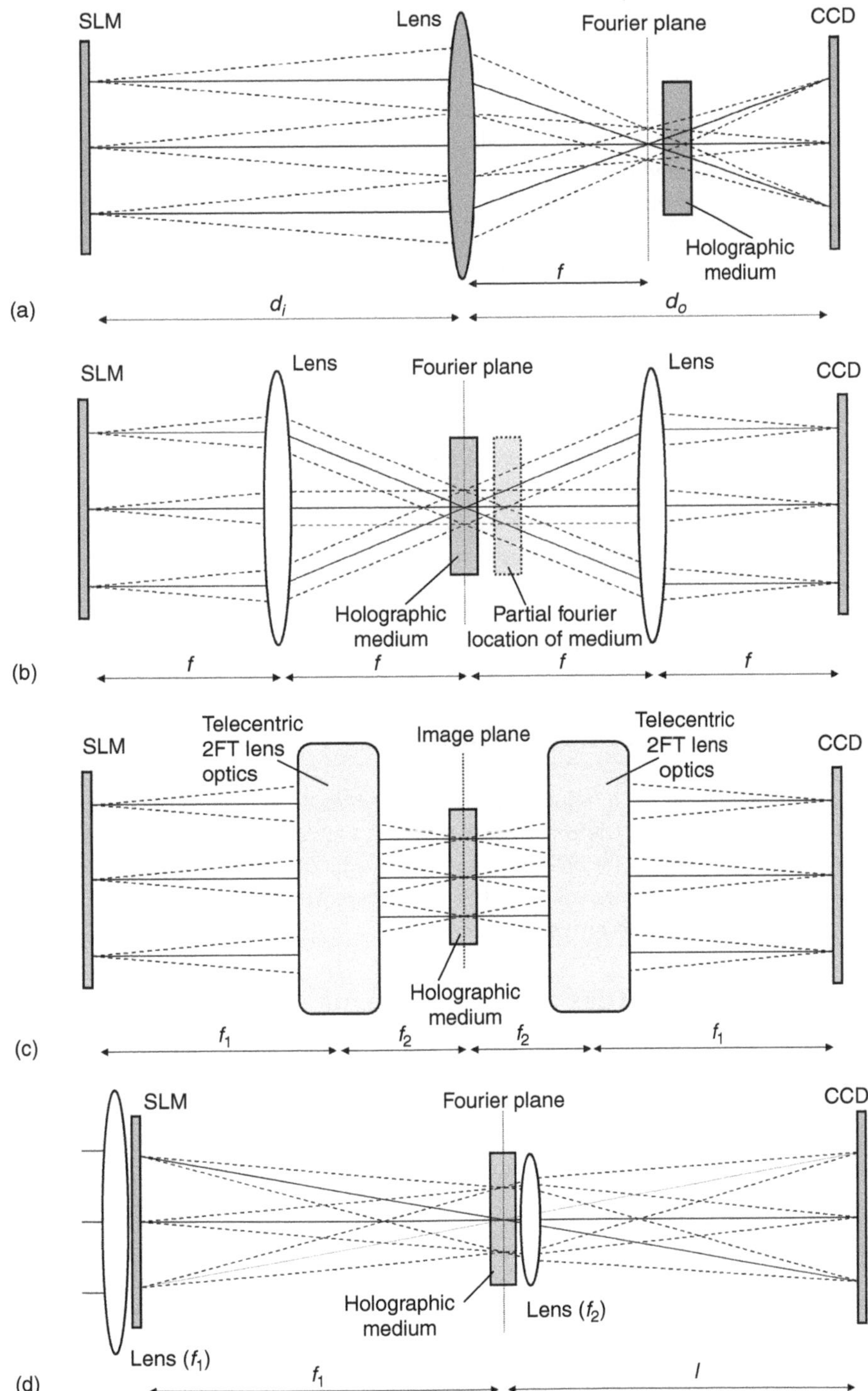

FIGURE 16.6 Optical imaging architectures for holographic storage: (a) Fresnel imaging (the imaging condition is given by the lens formula: $1/f = 1/d_i + 1/d_o$); (b) double Fourier transform; (c) image plane using telecentric double Fourier transform optics (SLM image is de-magnified by f_2/f_1, details of imaging optics are similar to case (b)); (d) van der Lugt system (the imaging condition is $1/f_1 + 1/l = 1/f_2$, magnification is f_1/l).

(e.g., 50 μs or less within the same stack), because the reference beam steering can be accomplished by fast acousto-optic or electro-optic deflectors (Hong et al. 1995). The scaling of the areal density into total system capacity is difficult in 90-degree architecture due to mechanical constraints: since the two beams enter the medium from two different faces, increasing the area of the medium to enhance the capacity can be easily done only in one dimension (e.g., height), while scaling up of the other two dimensions can be accomplished only to a limited degree.

The holographic disk architecture (Figure 16.5) is most suitable for relatively-thin media (e.g., photopolymer). A typical photopolymer disk consists of 200 μm to 1 mm of photosensitive material sandwiched between two rigid, optically-transparent glass or polymer substrates. The hologram recording is performed via optically-initiated chemical reaction in the illuminated area of the disk. The holograms are typically arranged in tracks, which are accessed via lateral motion of the disk, while the data inside each track can be accessed by disk rotation. Disk system can employ either continuous rotation or stop-and-go motion, in which the pages within a single location can be selectively accessed by changing, for example, the incidence angle of the reference beam. Shift multiplexing with spherical or speckle reference beam, however, is the most natural solution for the disk system, because it does not require any action other than rotation of the disk to address the holograms. In the shift multiplexing case, the holograms are recorded at slightly shifted positions along the track and multiple tracks are recorded in radial direction. Because the signal and the reference beams are incident from the same side of the medium, the entire medium area can be accessed by radial translation and/or rotation, which provides true scalability of the recording density and medium size into the total data storage capacity.

16.3 Recording Media and Materials

Performance and architecture of a holographic storage system depend on the recording material and medium properties, and complex tradeoffs arise in adapting a given material to yield the best results. The properties of foremost impotence for holographic storage media can be broadly characterized in terms of optical quality, recording properties, and stability. These parameters directly affect the storage density, obtainable data transfer rates, and bit-error-rate (BER).

16.3.1 Performance Metrics and Requirements

Optical quality of the recording media depends on both the intrinsic properties of the photosensitive material (scattering, index of refraction homogeneity), and on optical and mechanical properties of the entire medium itself (substrate quality, flatness, wedge). Medium scattering is defined as a fraction of the incident light scattered into the direction of the detector (or the direction of the CCD array used for data capture) normalized to the total solid angle occupied by the detector (Figure 16.7). Optical homogeneity is best characterized by the quality of high-resolution test images projected through the medium. The material characterization is typically done using well-calibrated test systems built with nearly ideal optics, so that the distortion floor of the straight-through images without the medium is extremely low, thus allowing direct measurement of the fidelity degradation due to the medium itself. An example of such system is the IBM Holographic Optical System Tester (HOST) described by Bernal et al. (1996). Because the BER will invariably degrade after actual holographic recording and reconstruction, initial starting BER of the imaged-through data should be sufficiently low (10^{-6} or less). Media flatness is another important parameter: if the medium is to be moved to access different areas, such motion must not lead to pixel misregistration and image quality degradation.

Recording properties of the medium include dynamic range, sensitivity, and thickness. In the majority of the current digital holographic storage systems, the recording mechanism of the material is a change in the index of refraction, rather than absorption, or thickness change. In this case, the diffraction efficiency (i.e., the ratio of the diffracted and incident signal intensities) of a volume phase hologram read out at the

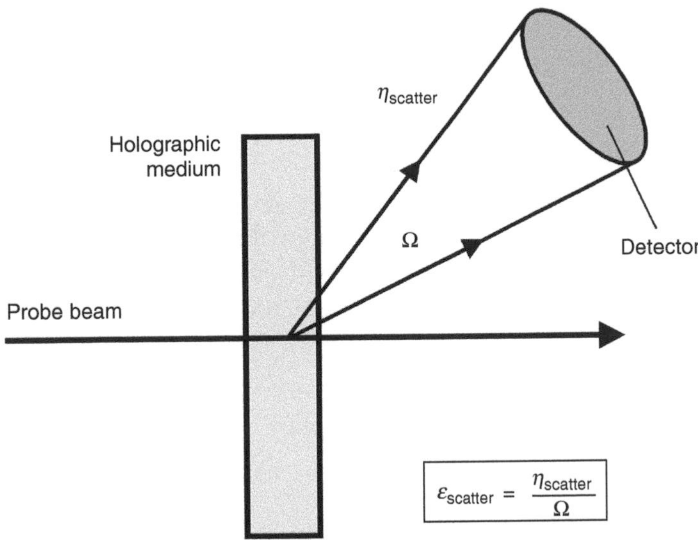

FIGURE 16.7 The scattering amount is defined as a ratio of the scattered light into the unit solid angle in the direction of the detector or CCD array used for the holographic data capture.

Bragg peak maximum is given by Kogelnik (1969),

$$\eta = \sin^2\left(\frac{\pi \delta n L}{\lambda \cos \theta}\right), \tag{16.1}$$

where δn is the amplitude of the spatially sinusoidal index modulation, λ is the optical wavelength, L is the thickness of the material, and θ is the angle between the propagation direction of the read beam and the vector normal to the surface of the medium (i.e., $L/\cos\theta$ is the effective interaction length). Thus, the higher the index change or the thickness of the medium, the larger the resulting diffracted signal strength.

The term *dynamic range* refers to the total response of the medium when the total index change is divided up between multiple holograms sharing the same volume. Dynamic range is characterized quantitatively by a parameter called *M#* ("M-number"), defined as follows (Mok, Burr, and Psaltis 1996):

$$M\# = \sum_{i=1}^{N} \sqrt{\eta_i}. \tag{16.2}$$

That is, it represents a sum of the square roots of diffraction efficiencies of the holograms multiplexed in the volume. Dynamic range is proportional to the maximum index change of the medium and to its thickness ($M\# \propto \delta n_{max} L$). *M#* describes the scaling of the diffraction efficiency with increasing number of holograms, *N*:

$$\eta = \left(\frac{M\#}{N}\right)^2 \tag{16.3}$$

Equation 16.3 assumes that recording is performed in such a fashion that the diffraction efficiencies of all of the multiplexed holograms are the same. The procedure of designing a recording process that results in equal diffraction efficiencies is usually referred to as *scheduling*. Although dynamic range is an intrinsic property of the material, the usable dynamic range depends on the nature of holographic recording. Namely, the *M#* measured when multiple holograms are created by two planewaves is typically several times higher than what is obtained when high-resolution data pages are recorded. This is because the digital holograms have less than unity interference fringe modulation depth, and also because of optical distortions (due to, for instance, optical damage or material shrinkage upon recording) that prevent the usage of the entire index change range for holograms recording.

The sensitivity of the medium is characterized by the rate of grating formation upon holographic exposure:

$$S = \frac{\partial\sqrt{\eta}}{\partial t}\frac{1}{I_w L}, \tag{16.4}$$

where η is the diffraction efficiency, L is the thickness of the medium, I_w is the writing intensity.

Performance of an HDS system greatly depends upon the quality and physical properties of the recording medium. Typical minimum material parameters for optimized holographic performance are listed below (Orlov 1999; Shelby 2000).

Material:

- Thickness: $L > 0.5$ mm.
- Sensitivity: $S > 500$ cm/J (or $-SL > 20$ cm^2/J).
- Dynamic range: $(M\#)_{planewave} > 5.0$; $(M\#)_{image} > 2.0$
- Shrinkage: 0.05% (for 500 μm).
- Scatter: $< 10^{-5}$ s rad^{-1}.
- Wavelength response range: visible (532 nm or blue).

Medium:

- Disk wedge: < 0.5 m rad.
- Format: 120-mm to 6.5-in. diameter disk.
- Total wavefront error over ~ 2 mm spot: $\lambda/10$.
- Scatter (including substrate): $< 10^{-5}$ s rad^{-1}.

The two major groups of the recording materials, which are most widely used in digital holographic storage, are photorefractive crystals and photopolymerizable organic compounds (photopolymers). Photorefractive crystals (e.g., iron-doped lithium niobate) have excellent optical properties, low scatter, and can be produced to have a substantial thickness (~ 10 mm), although the cost of the material and sample preparation (polishing, etc.) is relatively high. The sensitivity of photorefractive media is usually low and not sufficient for realistic applications unless the memory is used as a ROM. In order to retain the information during continuous readout, a fixing procedure needs to be performed. Thermal fixing (in, for example, doped $LiNbO_3$) provides data lifetime of several years, but requires thermal cycling of the material to at least 100 to 120°C after the recording. The dynamic range of iron-doped lithium niobate can be rather high ($M/\# > 10$) due to strong photovoltaic effect, but the usable portion of it is limited by the phenomenon called "optical damage" (Ashkin et al. 1966), which, basically, is an accumulative build-up of optical distortion when a medium with local (e.g., photovoltaic) response is exposed to localized optical illumination.

Photopolymer materials usually have a much higher sensitivity, due to chemical amplification effects and a high dynamic range, but are rather limited in their thickness (1 mm or less). Other shortcomings of the photopolymers include a significantly increased scatter level compared to single-crystal photorefractive materials and volumetric shrinkage upon photopolymerization. The latter effect can be somewhat mitigated by use of high-molecular weight monomers and proper choice of the material chemistry. Shrinkage and scattering limit the available usable thickness of the photopolymer medium. On the other hand, compared to inorganic photorefractive crystals, photopolymers offer more flexibility in material engineering, which includes the choice of specific polymerization chemistry, monomers, and other chemical components, as well as the components' relative concentrations and molecular weights. The cost of the photopolymer material is usually fairly low, but the requirement for high optical flatness tends to increase the overall cost of the holographic disk medium.

Table 16.1 and Table 16.2 present the summary of current holographic storage materials' recording properties, as well as of their optical quality (data obtained by IBM HOST testers Bernal et al. 1996; Shelby 2000). Evidently, the crystalline materials exhibit significantly lower level of intrinsic scatter as

TABLE 16.1 Holographic Storage Materials Properties (Data Collected at the International Workshop on Holographic Data Storage, Nice, France, March 8–11, 1999)

Material	Sensitivity (cm/J)	Thickness (mm)	*M/#*	δn, max	Scatter, s rad^{-1}	Sensitivity Range (nm)	Shrinkage	Fixing Mechanism	Comments
Photorefractive crystals									
Fe:$LiNbO_3$	0.01 –0.1	1–20	10–20	3×10^{-3}	10^{-7} (90 degree)	400–600	–	Thermal $T>$ 100°C	R/W + ionic fixing
$LiNbO_3$ two-photon undoped	0.03 (trans.)	1–10	10	5×10^{-4}	10^{-7} (90 degree)	650–900	–	Optical gating	R/W gating: $\lambda<$ 500 nm
$LiNbO_3$ two-photon Fe-doped	0.04 (trans.)	1–10	10	5×10^{-4}	10^{-7} (90 degree)	650–900	–	Optical gating	R/W gating: $\lambda<$ 500 nm
$KNbO_3$ Fe-doped	600	1–10	1–2	10^{-4}	Higher than $LiNbO_3$	400–600	–	Thermal $T>$ 90°C	Short dark storage time (days)
Organic recording media									
Photo-refractive polymers	2×10^4	0.01–0.05	–	10^{-2}	?	480–830	?	None	Ext. field 10–100 V/μm
Hetero-photoaddressable polymers	0.1–100	0.01–0.2	1	0.1	?	400–600	0	None	R/W high thermal stability
Polaroid CROP	500–5000	0.2–0.5	7–15	5×10^{-3}	10^{-3}	480–540	0.1%	Complete poly-merization	WORM, thermal or opt. preexp.
Lucent acrylate	<500	1.0	40?	9×10^{-3}	?	500–532	0.35%	Complete poly-merization	WORM, two-chemistry
DuPont HRF-150	~4000	0.038	0.5–>1	6×10^{3}–2.8×10^{-3}	?	400–600	2.5–3%	Complete poly-merization	WORM, UV post-cure

TABLE 16.2 Optical Quality of Selected Holographic Storage Materials Characterized as Bit-Error-Rate of Straight-Through Images and Holograms of Different Resolutions (Data Obtained Using IBM Holographic Storage Tester Shelby 2000)

Material	Scatter, $\eta_{effective}$	Transmitted Image			Single Hologram		
		256×256	512×512	1024×1024	256×256	512×512	1024×1024
Photorefractives:							
$Fe:LiNbO_3$	$10^{-7.5}$	–	10^{-17}	10^{-6}	–	10^{-11}	$10^{-5.5}$
$Ce:Sr_{0.61}Ba_{0.39}Na_2O_6$	10^{-7}	–	10^{-11}	10^{-3}	–	$10^{-10.5}$	10^{-2}
$Rh,Ce:BaTiO_3$	10^{-8}	10^{-13}	$10^{-5.5}$	–	10^{-11}	10^{-5}	–
Organic materials:							
Polaroid CROP, 0.2 mm	$10^{-5.5}$	10^{-16}	10^{-11}	–	10^{-14}	10^{-3}	–
Photorefractive polymer	$10^{-4.5}$	10^{-12}	10^{-5}	–	10^{-9}	–	–
Photochromic glass, 0.1 mm	10^{-6}	10^{-12}	$10^{-4.5}$	–	10^{-8}	10^{-2}	–
Photochromic glass, 1.0 mm	10^{-4}	10^{-11}	–	–	$10^{-2.5}$	–	–

compared to organic media, with photorefractive iron-doped lithium niobate being of the highest optical quality. In addition, the larger is the page size, the more stringent are the requirements to the optical quality and to the refractive index homogeneity of the material, because optical imaging of larger size data pages requires lower amounts of optical wavefront aberrations.

16.3.2 Photorefractive Crystals

Photorefractive effect is a change of the index of refraction of the medium upon light exposure. Most photorefractive materials (Günter and Huignard 1988) are electrooptic, non-centrosymmetric, wide-bandgap crystalline media, e.g., niobates, silicates, or germinates. The recording is usually performed with light of photon energy below the bandgap of the material, while the absorption is provided by extrinsic donor dopants or charged intrinsic point defects (Figure 16.8). Grating recording is produced via optical photoexcitation of mobile carriers from extrinsic donor or acceptor impurities, or intrinsic trap defects, carrier diffusion and drift due to internal or externally applied electric field and photovoltaic effect; subsequent retrapping on ionized deep traps, and, finally, the build-up of internal space charge. In the steady state, the carrier transport due to the internal space charge field balances the diffusion and photovoltaic current. The refractive index perturbation, which represents the recorded volume hologram, is produced by liner (Pockels) or biased quadratic electrooptic effect (Figure 16.9). A comprehensive band-transport model of photorefraction in electrooptic crystals was first proposed by Kukhtarev (1976), then Kukhtarev et al. (1979), and, later, was expanded to incorporate both hole and electron photoconductivity (Valley 1986), shallow traps (Mahgerefteh and Feinberg 1990), spontaneous polarization variation (Orlov, Psaltis, and Neurgaonkar 1993), and ionic transport (Carrascosa and Agullo-Lopez 1990; Yariv, Orlov, and Rakuljic 1996).

Charge gratings in photorefractive crystal decay upon continuous readout (or under uniform illumination) due to the same mechanism that is responsible for their creation, namely, optical carrier excitation and transport. Thus, a mechanism for permanent storage (fixing) is necessary to assure the data longevity. Hologram fixing techniques established so far involve reproduction of the original electronic pattern into the form in which the species responsible for the formation of the hologram are not optically active. The two mechanisms that have been proven to be most feasible and effective include thermal ionic and ferroelectric polarization fixing. In thermal ionic fixing, the space charge pattern of optically active deep trap centers is transferred into the space charge of conducting ions present in the crystalline medium (Figure 16.10), while the ferroelectric fixing is a process of converting the space

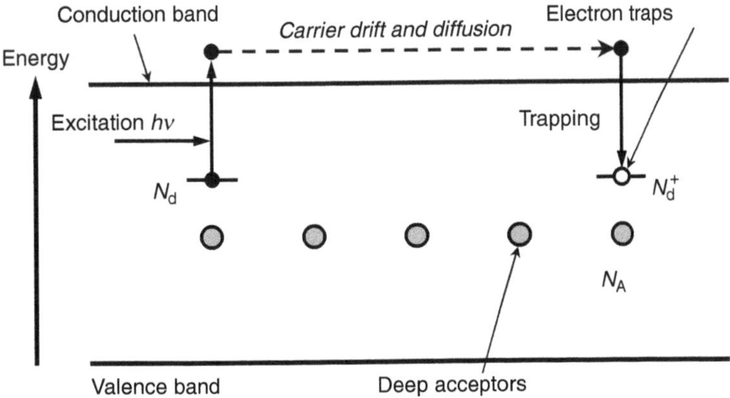

FIGURE 16.8 Photorefractive band transport model. The charge neutrality is maintained by deep acceptors (e.g., intrinsic point defects) which are created during crystal growth (i.e., $N_d^+ = N_A$ in the dark). In some materials (e.g., $LiNbO_3$) the effective acceptor density in the material can be changed by post-growth oxidation or reduction treatments.

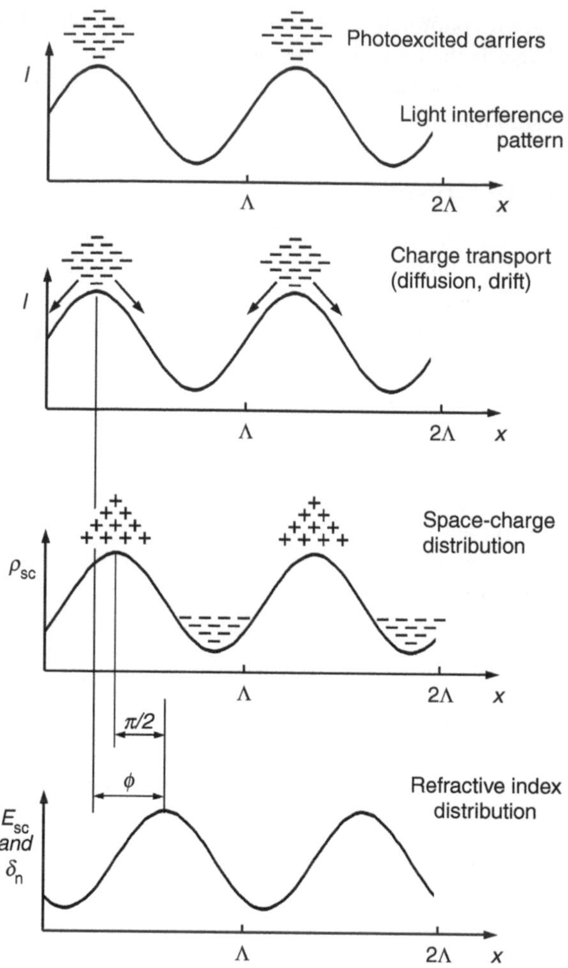

FIGURE 16.9 Photorefractive grating formation mechanism. A sinusoidal light distribution due to signal and reference optical interference leads to spatially nonuniform density of excited carriers and photocurrent. The carrier diffuse and drift due to electric field and are retrapped on deep traps, leading to non-uniform trapped charge density which mimics the original optical field (and may be shifted in phase). The internal space charge field modulates the index of refraction via electrooptic effect; the spatially sinusoidal index (or, in some cases absorption) perturbation represents a volume hologram.

charge distribution of the hologram into optically inactive distribution of ferroelectric microdomains (Micheron and Bismuth 1972). The domain walls produce the space charge electric fields due to discontinuity at the head-to-head domain interfaces, and thus, modulate the index of refraction vie electro-optic effect. Thermal ionic fixing has been demonstrated in a variety of photorefractive materials, including $LiNbO_3$ (Amodei and Staebler 1971), $Bi_{12}SiO_{20}$ (Arizmendi 1989), $KNbO_3$ (Montemezzani and Günter 1990), $BaTiO_3$ (Kirillov and Feinberg 1991), and KLTN (Leyva et al. 1995).

Since its discovery in the early 1970s, ionic fixing of holograms has been extensively studied and characterized, particularly in lithium niobate. The ionic species that become mobile at elevated temperature and which are responsible for ionic hologram formation were determined to be impurity protons (Vormann et al. 1981) in as-grown and hydrated crystals, while in strongly dehydrated samples which possess a lower ionic conductivity but substantially higher fixed grating longevity, the mobile species which dominate the ionic conductivity at elevated temperature are lithium interstitial ions (Yariv, Orlov, and Rakuljic 1996).

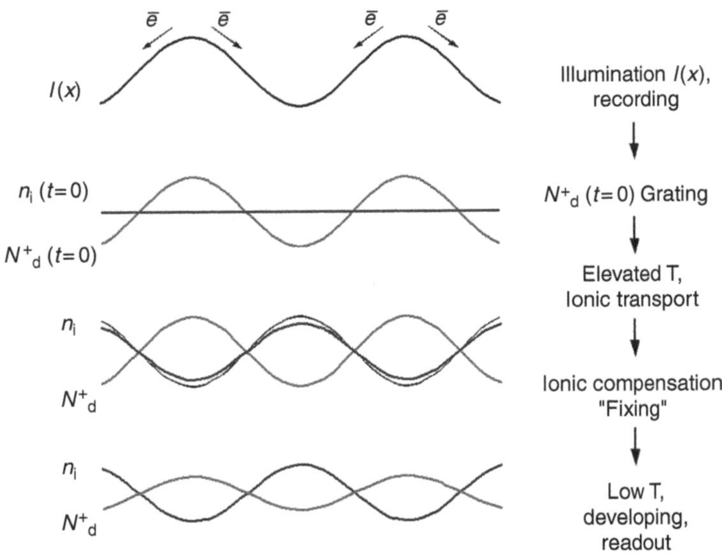

FIGURE 16.10 Mechanism of hologram fixing via thermally assisted ionic drift.

Ionic and electronic conductivity each obey the Arrhenius-type dependence on the temperature, T:

$$\sigma_{\rm i} = en_{\rm i}\mu_0 \exp\left(-\frac{E_{\rm a}}{k_{\rm B}T}\right), \tag{16.5}$$

where E_a is activation energy (approximately 1.2 eV for hydrogen impurity and 1.4 eV for interstitial lithium migration), n_i is the density of conducting ions, e is the electron charge, and μ_0 is a characteristic constant in units of carrier mobility. Efficient ionic fixing is based on the great disparity between the dark electronic conductivity at elevated temperatures and the ionic conductivity. At elevated temperature ($T> \sim 70°C$ in lithium niobate), ionic conductivity is dominant, and ions readily compensate the holographic electric field pattern created by photoexcited electrons by mimicking their spatial distribution. At low temperature, ionic conductivity is very low, enabling a quasi-permanent storage of the ionic replica of the initial electronic hologram. The residual ionic conductivity at low temperature thus determines the lifetime of the fixed hologram.

At room temperature, the ionic hologram is quasi-stabilized due to partial compensation by the electronic grating. This means that the space charge fields of electronic and ionic components have opposite signs (and may be shifted in phase with respect to each other), which leads to weaker total internal field (thus, slower charge grating decay), and reduced diffraction efficiency as compared to the originally recorded, unfixed electronic grating. The shelf lifetime of the fixed grating depends largely on ionic impurity concentration, n_i, and the degree of electronic compensation. The lifetimes of uncompensated ionic gratings can range from a few months (in as-grown $LiNbO_3$) to 2–3 years in dehydrated lithium niobate (Figure 16.11) (Yariv, Orlov, and Rakuljic 1996). Strongly compensated gratings with low reconstruction efficiency may possess lifetime of 10 years or more.

In digital hologram fixing, data is typically first recorded at room temperature and, later, the medium undergoes the thermal cycle. Fixing of thousands of holograms has been successfully demonstrated using this technique (Heanue et al. 1996; An, Psaltis, and Burr 1999). Much stronger signals and much higher effective $M\#$ can be obtained, however, if the data itself is recorded at elevated temperature when ionic conductivity is larger or comparable to photoconductivity. High temperature recording can be employed for recording a single strong grating or a few high diffraction efficiency gratings, which can be useful in,

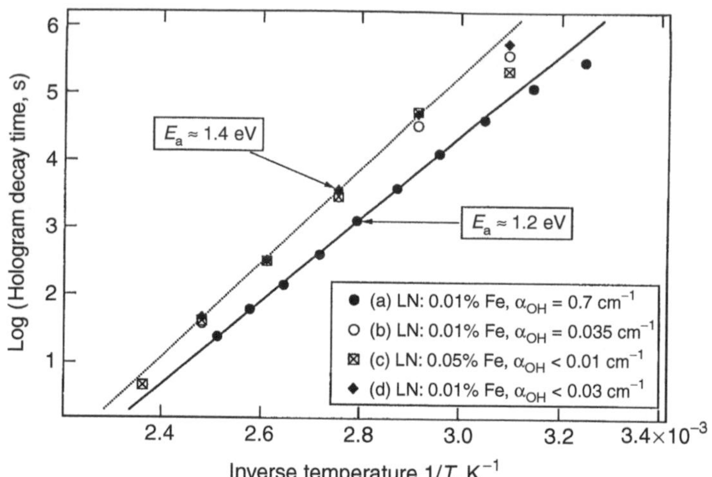

FIGURE 16.11 Arrhenius plot of the ionic hologram lifetime in (a) as-grown congruent ($[Li_2O]=48.6$ mol%) crystal ($E_a=1.2$ eV), (b, c) samples with low hydrogen content and different Fe-doping ($E_a=1.4$ eV), (d) near-stoichiometric (vapor transport equilibrated) crystal ($[Li_2O]=49.5$ mol%) with low hydrogen content ($E_a=1.4$ eV). α_{OH} denotes the peak absorption at 2.87 μm of OH^- stretching vibration. The roll-off of estimated lifetime below ~60°C is due to the small polaron hopping conductivity which becomes dominant in this temperature range due to its low activation energy (~0.4 eV); real room-temperature lifetime of ionic grating is extrapolated from $T>60$°C data.

for instance, WDM telecommunications applications (Leyva, Rakuljic, and O'Conner 1994). Prolonged recording at elevated temperature (in heavily-doped lithium niobate with $[Fe]\geq 0.1$ mol%) usually results in nearly fully-compensated gratings with little or no electro-optic contribution, but with a sufficiently strong index perturbation arising from the photochromic effect due to substantial modulation of the Fe^{2+} ion concentrations. Such gratings possess prolonged lifetimes (over 20 years or more) and do not require visible light for development. Unfortunately, high-temperature recording is not suitable for high-resolution digital data pages, due to thermal Bragg mismatch and distortions when the data is read back at low operating (i.e., room) temperature.

Another technique of permanent or semi-permanent rewritable storage is two-photon gated recording (von der Linde, Glass, and Rodgers 1974). This technique is based on an all-optical control of absorption and sensitivity at writing wavelength (which is usually in the near-infrared range of ~800 nm) using a second, uniform-intensity gating light beam (which is usually near ultraviolet or green). During the recording, the gating light excites the electrons from the deep traps, which are not accessible to the longer wavelength writing light. Electrons are retrapped on the shallow levels and are subsequently re-exited into the conduction band by the near-infrared recording light (Figure 16.12). The photorefractive hologram is formed by the transport mechanisms, similar to the single-photon recording, and is stored as spatial distribution of electrons in the deep traps. When the gating light is switched off and the thermal equilibrium is re-established, the shallow traps become depopulated (i.e., populated only through the thermal excitations, which are negligible in large-bandgap materials), which renders the medium insensitive to the near-infrared wavelength recording light, allowing for non-destructive hologram readout (Figure 16.13).

The two-photon gated holographic recording process is nonlinear with writing light intensity because of the finite lifetime of the carriers trapped in the shallow levels. Highly-efficient recording is possible if the material defect structure is modified to provide sufficiently long lifetimes (> 100 ms) for the excited electrons in the shallow levels. This was achieved by using near-stoichiometric undoped or lightly iron-doped lightly reduced lithium niobate (Figure 16.14) with low concentration of deep traps (Hesselink

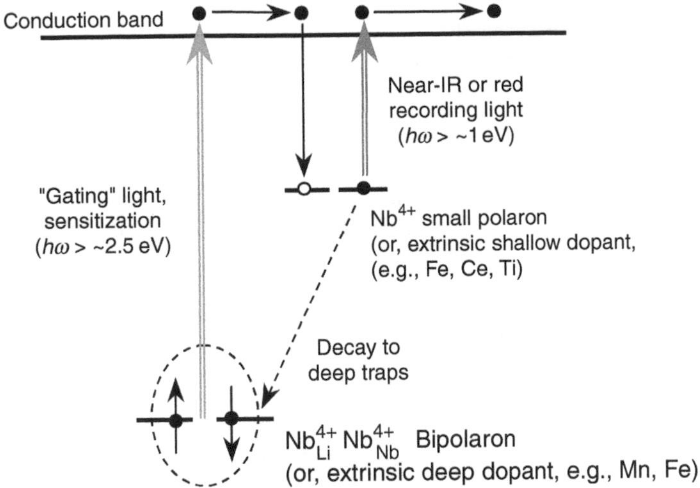

FIGURE 16.12 Mechanism of optically gated recording in lithium niobate.

et al. 1998), or using extrinsic dopants (e.g., iron and manganese) of optimum concentration to provide extrinsic deep and shallow levels (Staebler and Phillips 1974; Buse, Adibi, and Psaltis 1998).

Digital hologram storage has been demonstrated in two-photon gated crystals of near-stoichiometric lithium niobate (Lande et al. 1997), but, so far, with a rather limited number of superimposed holograms and storage density, which is mostly due to high writing intensity requirement and lower optical quality of the near-stoichiometric crystals as compared to standard iron-doped lithium niobate.

16.3.3 Photopolymer and Organic Materials

In general, photopolymer storage materials are complex systems of organic molecules, which can undergo photoinitiated polymerization to record volume phase holograms. They are comprised of several components and typically include one or more monomers, a photoinitiation system, and a chemically-inactive component (binder). The photoinitiation subsystem may include a photosensitizer

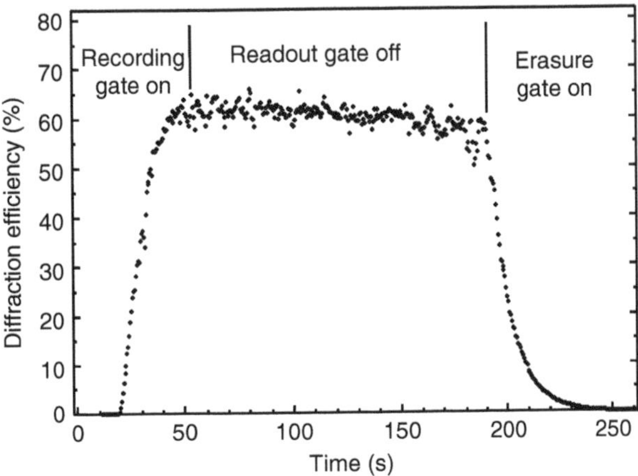

FIGURE 16.13 Optically gated recording, nondestructive readout, and erasure of hologram in near stoichiometric undoped lightly reduced lithium niobate crystal.

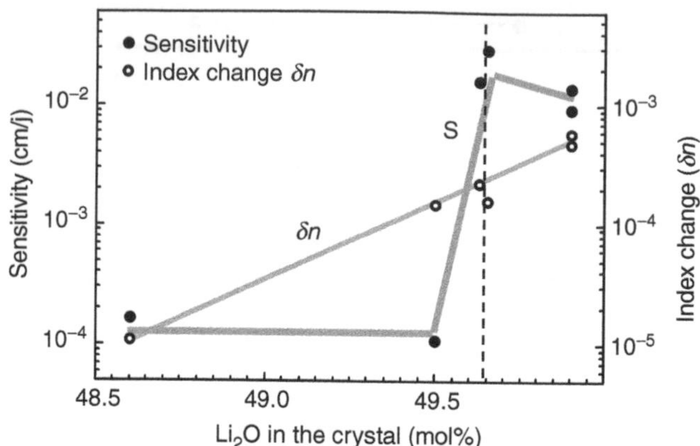

FIGURE 16.14 Two-photon sensitivity and index change as a function of stoichiometry in undoped lightly reduced $LiNbO_3$.

that is specially chosen to sensitize the medium to a specific light wavelength range. Other components may be added to the formulation to control various properties, such as shelf life and viscosity of the initial fluid. An optically-inert binder with an index of refraction different from the polymer system is used to improve the media dynamic range. The recording medium is prepared in one of two ways: either by spin coating on an optically-transparent substrate, or by containing the material between two transparent glass or polymer substrates (e.g., disks). In its initial state, the photopolymer is typically a viscous fluid or a soft solid with a low glass-transition temperature.

The photopolymerization mechanism consists of photoionization that creates a reactive center, which, when combined with a monomer, initiates growth of the polymer chain, which proceeds by addition of monomers to the reactive chain end. The reaction eventually terminates due to several mechanisms, including a reaction with a contaminant species or self-entanglement, when the reactive chain becomes trapped in a location into where the monomers cannot diffuse. The photopolymerization is accompanied by monomer diffusion, due to the change of their local densities, and component segregation (e.g., binder outdiffusion from the areas of higher degree of polymerization). After the reaction and components diffusion reach the steady state, a volumetric phase hologram is formed due to the induced difference in refractive indices in the dark and bright regions of the interference pattern (Figure 16.15). An important parameter for holograms recording in photopolymerizable systems is the ratio of the polymerization rate to the monomer diffusion rate (Zhao and Mouroulis 1994). If the diffusion rate is sufficiently fast, the index profile of the recorded hologram closely matches the light intensity distribution, resulting in linear recording and best image fidelity. If the photopolymerization is much faster than the diffusion, the excess polymer collects at the boundary between the bright and dark regions of the interference pattern, resulting in poor correspondence between the original interference pattern and the hologram, and hence, in poor image fidelity. Thus, the parameters that affect the diffusion rates and index contrast, such as initial formulation viscosity, monomer concentration and molecular weight, binder and plastisizer concentrations, need to be carefully optimized (through theoretical modeling means such as molecular dynamics simulation, as well as carefully designed chemical experiments) for the best material performance.

A large variety of monomers has been studied and used in photopolymer systems. Vinyl monomers, such as acrylate and methacrylate esters, polymerize through free radical mechanisms (Oster and Yang 1968), while the more recently developed materials use cationic ring opening mechanism of polymerization (Figure 16.16) (Crivello and Lee 1989; Waldman, Li, and Horner 1997). Frequently, multifunctional monomers (i.e., monomers with more than one functional group) are added into the formulation to promote cross-linking between growing polymer chains. The cross-linking produces

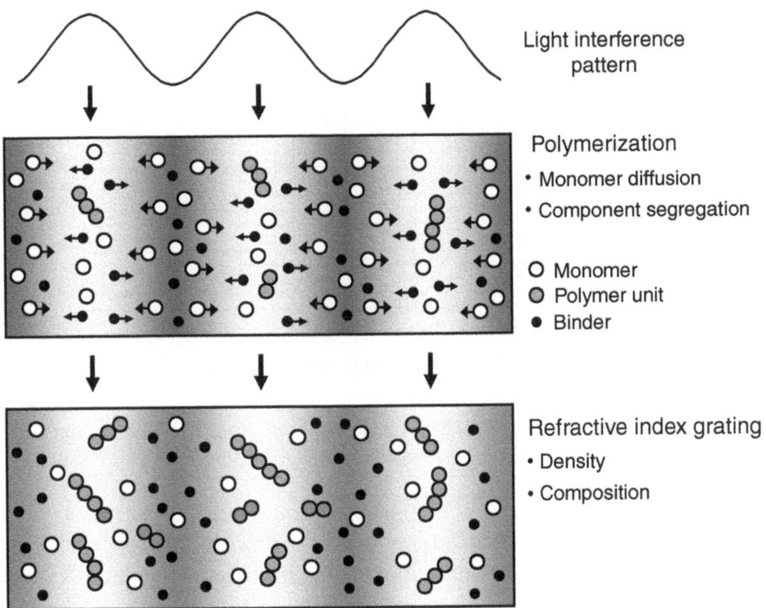

FIGURE 16.15 Mechanism of hologram formation in photopolymer medium. The optical interference pattern establishes a spatial pattern of polymerization of active monomers. The unreacted monomers diffuse to re-establish spatially uniform distribution, while binder molecules drift out of more polymerized regions due to free volume reduction, creating components segregation.

a molecular architecture comprising a stable cross-linked polymeric network, thus improving the dimensional stability and final image quality of the recorded holograms.

The modern photopolymer media do not require any post-exposure processing steps, due to the fact that the material becomes light-insensitive, allowing nondestructive readout after the recording is complete and the medium is completely polymerized. Because the chemistry of most of the photopolymerizable compounds is irreversible, the photopolymers are particularly suited for the write-once-read-many (WORM) data-storage applications.

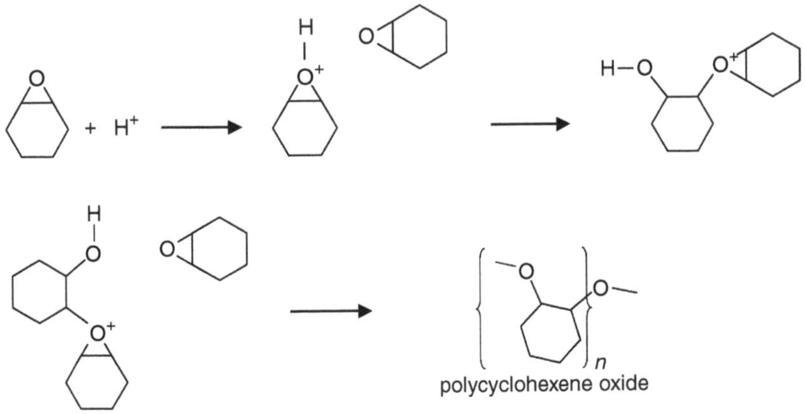

FIGURE 16.16 Generic mechanism of polymerization and reactive chain propagation in cationic ring-opening photopolymer.

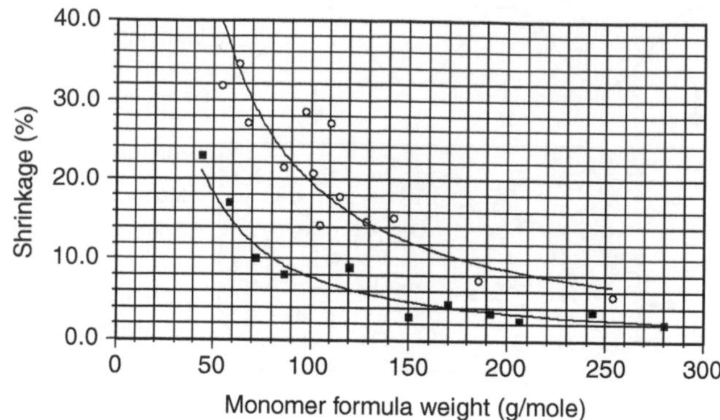

FIGURE 16.17 Comparison of the volume shrinkage on polymerization of vinyl (open circles) and cationic ring-opening (filled squares) monomers as a function of molecular weight. Courtesy of D.A. Waldman (Aprilis, Inc.).

The choice of the specific monomer chemistry influences the recording properties of the medium: the sensitivity, dynamic range, shrinkage amount, and the environmental stability. A comparison of the properties of different photopolymer systems is given in Table 16.1. Shrinkage upon hologram recording occurs practically in all photopolymer systems. Addition of a monomer to a growing polymer chain decreases the volume of the system because, upon the reaction, a covalent bond replaces a near-van der Waals contact. The key advantage of the cationic ring opening mechanism compared with acrylate-based systems is the reduced amount of volumetric change per reaction (Figure 16.17), which leads to a smaller amount of hologram distortion at the same amount of holographic index change. In case of CROP, the volume reduction is partially compensated by increase of interatomic distances due to the ring opening (see Figure 16.16). In some polymer systems, particularly when the degree of polymerization and cross-linking is high and the material represents itself as a very viscous liquid, the shrinking occurs rather slowly as compared to the polymerization itself. The volumetric relaxation can occur over a period of varying from minutes to days. Such slow relaxation may lead to hologram quality degradation with time, particularly in thick media (500 μm or more).

During the recording of multiple holograms, the sensitivity of the photopolymer medium gradually changes (Figure 16.18), because as the degree of the polymerization increases, the medium becomes more

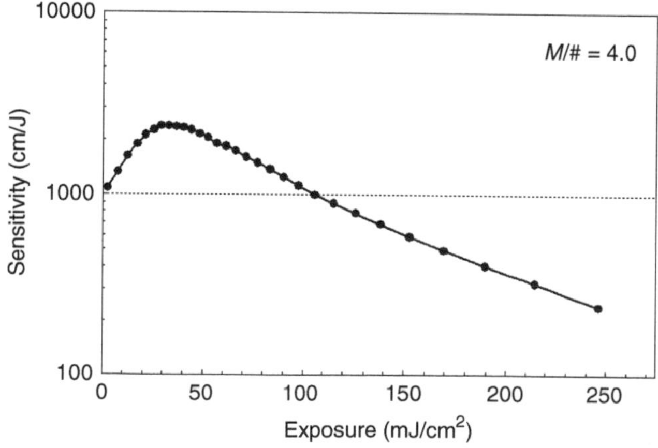

FIGURE 16.18 Sensitivity versus cumulative optical exposure in Aprilis CROP photopolymer medium. Material was heat-treated at 65°C to reduce the residual shrinkage amount.

"solid," leading to slower monomer diffusion rate (and shorter chain growth length). At the same time, the optical absorption is also reducing due to the depletion of the dye and photosensitizer, making the medium even less sensitive. The sensitivity change requires a specially-calibrated scheduling method to achieve the uniform holographic diffraction efficiencies between holograms recorded in the beginning and in the end of the exposure. Generally, for a photopolymer media, the recording times increase within a given multiplexing run to maintain the same signal strength.

In general, the use of higher-molecular-weight monomers tends to reduce the shrinkage amount (Figure 16.17) at the same level of *M#*. Volume shrinkage and hologram distortions can also be reduced to acceptable amount by partial polymerization before holographic recording, which can be accomplished by optical preexposure with coherent or incoherent light or thermal pretreatment. Partial polymerization converts the photopolymer film into a soft-solid physical state, which produces a mechanically-stable recording medium that can be used in, for instance, rotating disk applications. Another approach is the two-chemistry system (Dhar et al. 1998), in which the initial formulation contains two chemically different types of monomers. After the material is cast into the desired geometrical shape, the first group of monomers is polymerized by non-optical means (e.g., at elevated temperature), producing a rigid, cross-linked polymer matrix (which prevents mechanical volume change), while the second polymer group is used for optical recording of the holograms.

Alternatively, the shrinkage can be reduced to almost zero levels if the photopolymerizable medium is encapsulated in certain materials, for instance, inorganic nanoporous glass matrix (Schnoes et al. 1999). Such materials, however, exhibit rather low sensitivity (typically ~10 cm/J,) due to inhibited monomer diffusion, and require very accurate matching between the refractive index of the matrix and the photopolymer to prevent significant scattering.

Along with photopolymerizable materials, several other classes of organic compounds were studied as holographic recording media. These include photorefractive polymers (Kippelen 2000), azo-dye doped "photoaddressable" polymers (Bieringer 2000), and chromophoric dye doped poly(methyl methacrylate), or PMMA (Veniaminov, Goncharov, and Popov 1991).

The mechanism of hologram recording in photorefractive polymers is different from inorganic photorefractive crystals. Polymers are amorphous media and, therefore, an externally-applied electric field (~100 V/μm) is necessary in order to break the central symmetry. Furthermore, the charge transport mechanism in polymers is mainly due to hopping of photoexcited carriers between adjacent molecules, as opposed to conduction band transport. Under external voltage bias, the index of refraction modulation includes two components: one due to electrooptic effect and the other due to spatially nonuniform orientational birefringence (Moerner et al. 1994; Kippelen 2000). The sensitivity of these materials can be quite high due to enhanced carrier transport in a strong electric field—comparable to that of photopolymerizable systems. The dynamic range of the optimized photorefractive polymers (e.g., DMNPAA:PVK:ECZ:TNF) is usually modest (*M#*~1), but the low thickness (100mm or less), as well as the requirement of the externally-applied voltage, make these type of media not very practical for holographic storage systems.

The azo-dye doped photoaddressable polymers (such as those developed by Bayer (Bieringer 2000)) can record holographic gratings via the cis-trans transformation mechanism of the dye chromophore molecule. Upon exposure to polarized optical radiation, the azo dyes (which are attached to the main polymer chain backbone) undergo isotropic alignment, which induces birefringence. The effect of dye orientation on index of refraction can be enhanced by addition of mesogenic groups, which do not interact with light directly, but rather stabilize and amplify the reorientation of the chromophores. The photoaddressible polymers by Bayer exhibit relatively high sensitivity (~400 cm/J), moderate *M#*~1, and excellent optical quality at low thickness (~40 microns). The scalability of the thickness to few hundred microns and techniques for permanent non-destructive storage (i.e., hologram fixing) have yet to be demonstrated.

Phenanthrenequinone (PQ)-doped PMMA (Veniaminov, Goncharov, and Popov 1991; Steckman et al. 1998) has excellent optical imaging quality. The holographic recording mechanism is based on a photoreaction-induced chemical bonding between the dopant and the host polymer matrix, which has to be followed by diffusion of unreacted free chromophore at elevated temperature in order to reveal the hologram. The thermal procedure requires a long (about 48 h at 55°C) thermal post-exposure treatment, which significantly increases the diffraction efficiency, but is a disadvantage from a system perspective. After the heat treatment, a final uniform optical exposure causes bonding of the remaining PQ molecules, thereby preventing any further holographic recording upon readout. The *M#* of optimized material can be as high as five in three-millimeter-thick samples, while the sensitivity is fairly low compared to other organic media, within the range of 1–10 cm/J (calculated from the diffraction efficiency after the post-exposure bake).

At present, the photopolymerizable polymers—due to their high recording sensitivity, dynamic range, and relatively low cost—remain the most promising media candidates for commercial digital holographic storage systems. On the other hand, due to their high optical quality, large available thickness, and re-writability, the photorefractive crystals—and, most notably—iron-doped lithium niobate, represent the unsurpassed test-bed recording materials for development of fundamental aspects of the technology, such as holographic optics, channel coding, and multiplexing techniques.

16.4 Multiplexing Methods

The high density achievable by holographic memories comes from the use of the third dimension and the ability to record and selectively retrieve multiple holograms recorded within the same volume of the storage medium. The process of superimposing multiple holograms is referred to as multiplexing. This is possible due to the Bragg effect; that is, individual holograms are accessed selectively in a fashion similar to selective detection of different periodicities in crystal lattice using Bragg diffraction. An alternative to Bragg-based methods is correlation multiplexing, in which the reference wave has a complex wavefront either generated by the passive random or pseudo-random diffuser, or created using a phase-spatial light modulator. In such cases, different holograms can be superimposed and addressed selectively based on the correlation properties of the reference beam.

As an example, consider a grating recorded by two planewaves and reconstructed by the third planewave propagating at a different angle (Figure 16.19a). Due to volumetric interaction of the readout light with the recorded grating, as the angle of incidence deviates from the original, the strength of reconstruction will diminish. This is because the partial reconstructions from different portions of the volume will no longer be aligned in phase, and, consequently, will no longer add up coherently in amplitude. The strength of the reconstructed signal can be calculated using coupled waves theory and, for planewaves in angular multiplexing scheme is given by Kogelnik (1969):

$$\eta(\Delta\theta) = \eta_0 \sin c^2\left(\frac{2L(\Delta\theta)\sin\theta}{\lambda}\right), \tag{16.6}$$

where η_0 is the on-Bragg diffraction efficiency given by Equation 32.1, θ is the incidence angle, $\Delta\theta$ is the angular deviation from exact on-Bragg condition, λ is the optical wavelength, L is the medium thickness, and $\text{sinc}(x) \equiv \sin(\pi x)/\pi x$.

In general, in relatively thin media, the strength of reconstruction is determined by the medium thickness and the Bragg-mismatch parameter dK_z, which is the z-axis projection of the vectorial sum of the incident wave vector $\boldsymbol{k}_\mathbf{r}$ and grating vector $\mathbf{K}_\mathrm{g}$ ($\mathbf{K}_\mathrm{g} = \mathbf{k}_\mathrm{sig} - \mathbf{k}_\mathrm{w}$), minus the diffracted wave vector $\mathbf{k}_\mathrm{d}$ (Figure 16.20):

$$\eta(\mathrm{d}K_z) = \eta_0 \sin c^2\left(\frac{\mathrm{d}K_z L}{2\pi}\right) \tag{16.7}$$

The diffracted signal is a propagating wave vector, which is found by projecting the vector sum of reading beam and grating vector $\mathbf{K}_\mathrm{g} + \boldsymbol{k}_\mathrm{r}$ (which is the spatial frequency of the polarization wave inside the media arising from the interaction of the incident beam with the index grating) onto the k-sphere

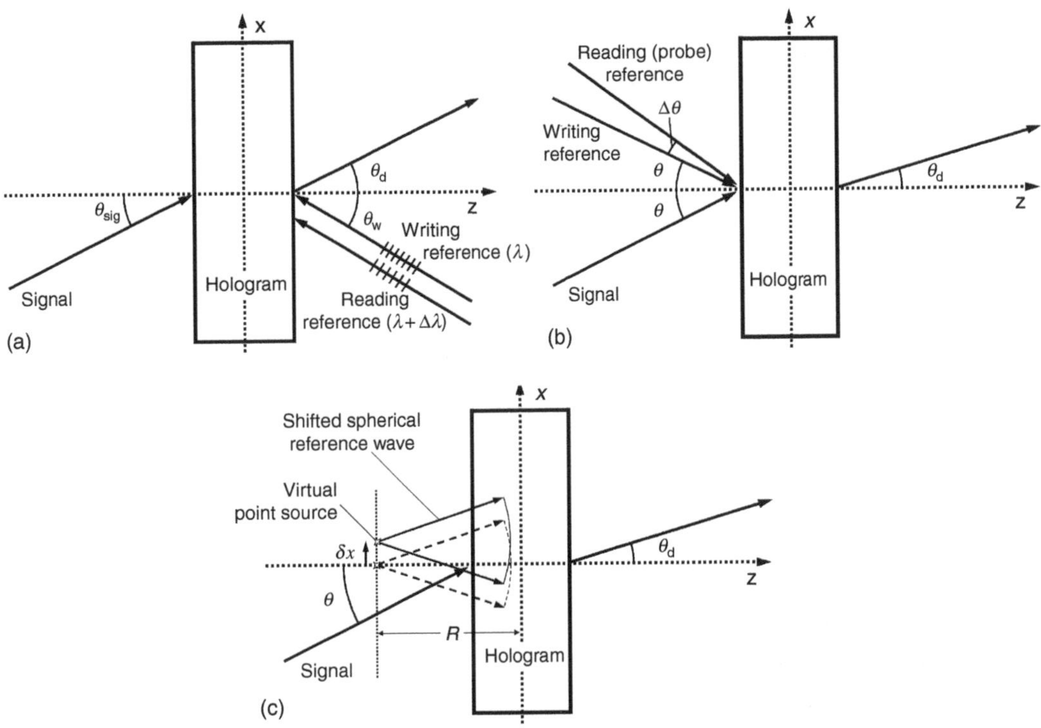

FIGURE 16.19 Schematic of grating reconstruction by (a) probe of different wavelength in reflection geometry; (b) angular rotated probe in transmission geometry; (c) shifted spherical reference beam (or similar reconstruction from a shifted medium).

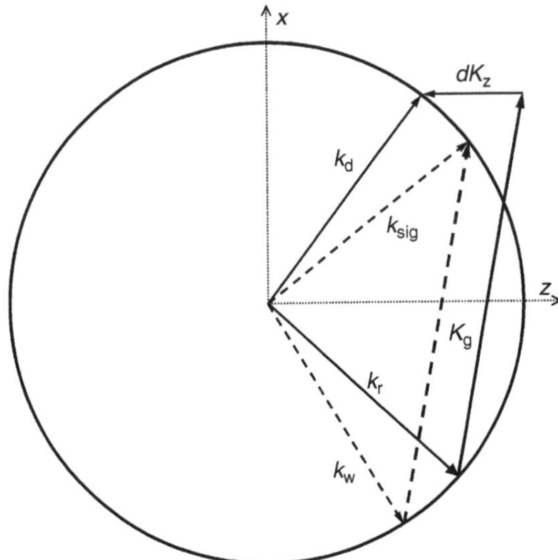

FIGURE 16.20 Bragg diffraction on the Ewald's k-sphere. An index grating with wavevector $\mathbf{K}_g$ originally created by interfering the signal beam $\mathbf{k}_{sig}$ with the writing beam $\mathbf{k}_w$ is reconstructed by a third wave, $\mathbf{k}_r$. The propagation direction of the diffracted wave is given by vector $\mathbf{k}_d$, while its relative strength is determined by the Bragg mismatch parameter dK_z (Equation 16.7).

along the z-axis. The Bragg mismatch parameter characterizes the efficiency of coupling of the electromagnetic polarization wave into the propagating electromagnetic wave.

For a more general case, it can be shown that in relatively thin holographic media, the mismatch parameter (and the Bragg condition) can be expressed in terms of the local derivative of the phase functions of the interacting waves (Orlov and Hesselink 2003):

$$dK_z(x,y) = \frac{2\pi n}{\lambda}((f_x(x,y)-\varphi_x(x,y))(\phi_x(x,y)-\varphi_x(x,y)) + (f_y(x,y)-\varphi_y(x,y))(\phi_y(x,y) - \varphi_y(x,y))) \tag{16.8}$$

where f(x,y), ϕ(x,y), ϕ(x,y) are the optical phase functions (eikonals) of the signal, original reference, and readout wavefront, respectively, evaluated in the center of the holographic medium ($z=0$).

The Bragg-based multiplexing methods can be generally classified into the following groups (after Barbastathis and Psaltis 2000; Figure 16.21):

- Angular multiplexing (Figure 16.21a), in which the reference beam is a planewave and the addressing mechanism is the change of angle of incidence of the reference (in-plane and out-of-plane). The Bragg selectivity curve is given by Equation 16.6 (Figure 16.22).
- Wavelength multiplexing (Figure 16.21b), in which addressing is performed via changing the reference optical wavelength. The wavelength selectivity in reflection geometry (Figure 16.19b) is given by (Figure 16.23):

$$\eta(\Delta\theta) = \eta_0 \sin c^2\left(\frac{2L(\Delta\lambda)\cos^2(\theta_w + \theta_{sig})}{\lambda^2 \cos\theta_{sig}}\right), \tag{16.9}$$

where $\Delta\lambda$ is the wavelength change, and the corresponding angles are defined in Figure 16.19b. The selectivity is maximized when $\theta_w = -\theta_{sig}$, i.e., when the signal and the reference beams are exactly counter-propagating.

- Shift multiplexing with a reference spherical wave (Figure 16.21c), in which addressing is performed by positioning of the medium without its rotation. Because the reference is a spherical wave, the medium shift effectively changes the local angle of incidence of the reference on the medium (Figure 16.19c), leading to local Bragg-mismatch much like in angular multiplexing. The shift selectivity curve (in paraxial approximation) is given by Figure 16.24:

$$\eta(\delta) = \eta_0 \sin c^2\left(\frac{\delta x u L}{\lambda R}\right), \tag{16.10}$$

where R is the radius of curvature of the reference beam inside the medium, $u=\tan(\theta)$ is the paraxial angle between the reference and the signal in the center of the hologram, and δx is the shift amount in Bragg-selective direction.

- Phase-code multiplexing (Figure 16.21d), in which different holograms are recorded using a different phase-modulated reference beams, created by (for example) a phase SLM device.
- Peristrophic multiplexing (Figure 16.21e), in which the addressing of different holograms is performed via rotating the medium around the axis, which is usually normal to the medium surface. In each of these peristrophic angular positions, holograms can be multiplexed by one of the methods described above.
- Spatial multiplexing, in which the holograms are recorded in spatially separated regions (or stacks), in each of which the holograms are multiplexed using one of the multiplexing methods above (e.g., spatio-angular multiplexing).

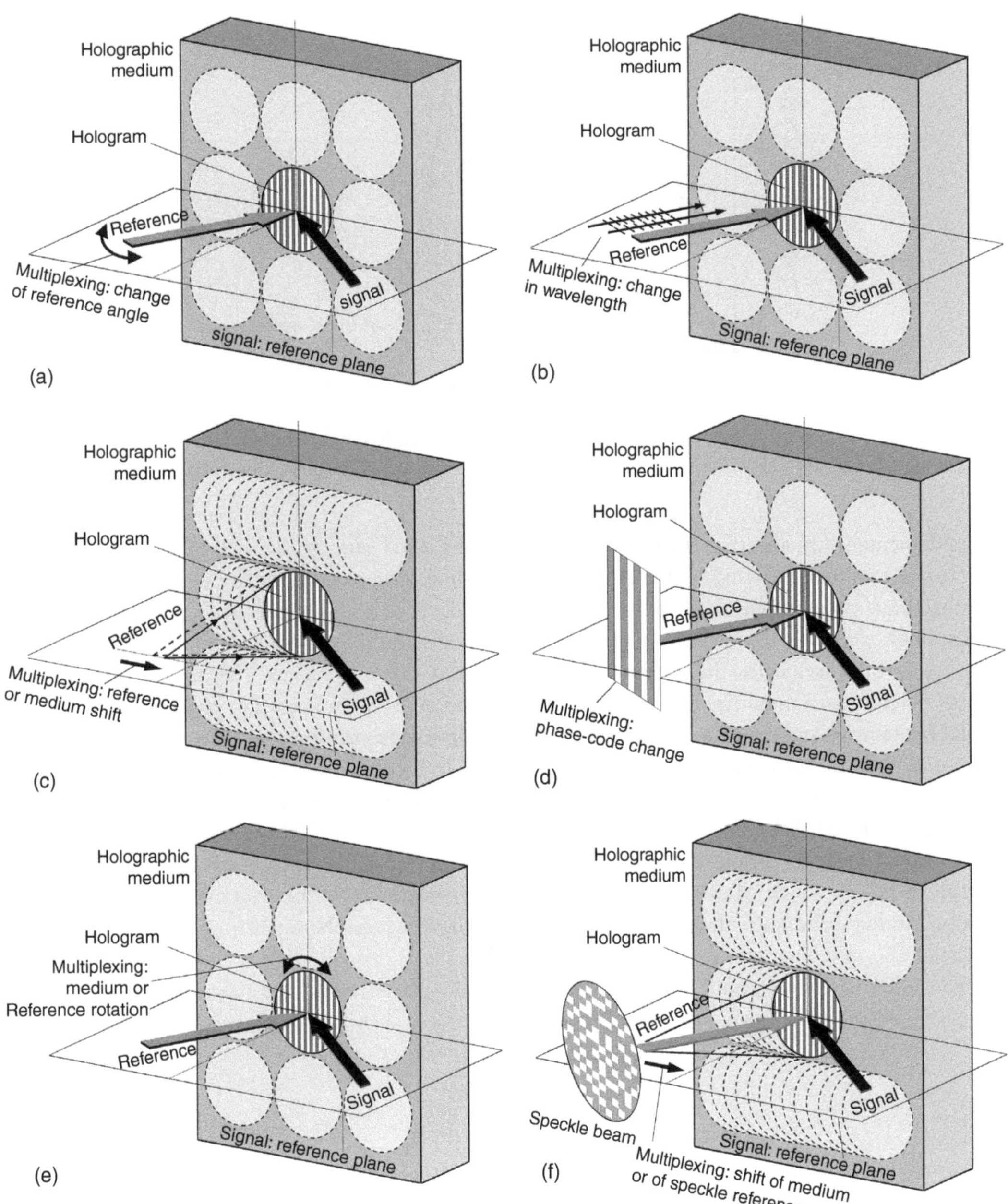

FIGURE 16.21 Holographic multiplexing methods: (a) angular multiplexing; (b) wavelength multiplexing; (c) shift multiplexing with a spherical reference beam; (d) phase-code multiplexing; (e) peristrophic multiplexing; (f) correlation shift multiplexing with a speckle beam.

The multiplexing methods can be combined (with certain restrictions to avoid Bragg-degeneracy) in order to increase overall storage density. In the current state-of-the-art systems, a combination of angular or shift and peristrophic multiplexing is commonly used. Spatial multiplexing can be combined with any of the multiplexing methods in holographic disk configuration, leading to direct scalability of the storage density and medium area into the total system capacity.

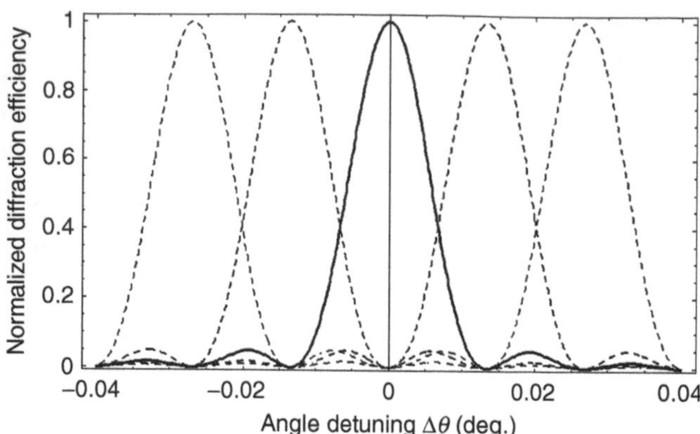

FIGURE 16.22 Normalized diffraction efficiency versus angular deviation $\Delta\theta$ for several multiplexed holograms ($L = 500$ μm, $\theta = 30°$, $n = 1$). Reconstruction of adjacent holograms at on-Bragg condition represents cross-talk.

Fundamentally, if a perfect medium is assumed, the total storage density of three-dimensional memory is eventually limited by the crosstalk noise, which originates from the partial reconstruction of other data pages when a selected page is addressed. This is illustrated in Figure 16.22 through Figure 16.24, where the reconstruction of adjacent holograms and the main hologram is shown. Typically, for angular or wavelength multiplexing, the adjacent holograms are placed at the Bragg-nulls of its nearest neighbors to minimize the crosstalk. However, because the information page contains multiple spatial frequencies and because of other effects, such as optical absorption which distorts the ideal Bragg curve and leads to nulls uplift, adjacent holograms have to be placed in the second or third null of the Bragg curve. This leads to additional reduction of storage density, particularly for thin media such as holographic optical disks.

An alternative to Bragg-detuning multiplexing methods is the correlation multiplexing (Figure 16.21[f]), which relies on the spatial correlation properties of the reference beam. Similarly to Bragg-detuning techniques, the number of the holograms, which can be superimposed, is limited by the crosstalk. The shift correlation approaches employ either statistically-random speckle fields (Darskii and Markov 1988), or pseudorandom (deterministic) spatially phase-modulated beams (Dhar et al. 1998).

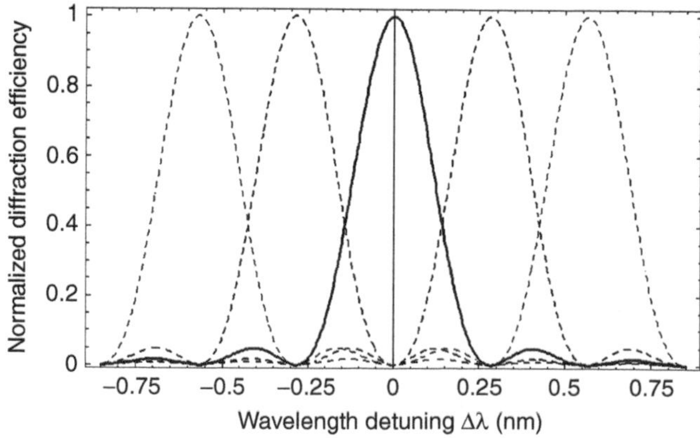

FIGURE 16.23 Normalized diffraction efficiency versus wavelength change ($L = 500$ μm, $n = 1$, $\theta_w \equiv \theta_r = \theta_{sig} = 0$, i.e., the signal and the reference are counter-propagating). Similar to Figure 16.22, reconstruction of adjacent holograms represents crosstalk.

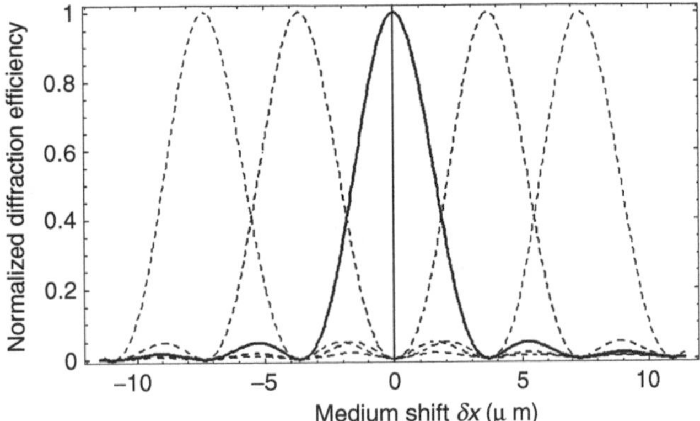

FIGURE 16.24 Normalized diffraction efficiency for several holograms shift-multiplexed with spherical reference beam versus medium shift δx ($L=500$ μm, $n=1$, $R=2$ mm, $\theta=30°$ [i.e., $u=\tan(\theta)=0.58$]).

For the random Gaussian speckle reference beam, the hologram shift selectivity is determined by the autocorrelation function of the optical field and is given by Darskii and Markov (1988):

$$\eta(\delta) = \eta(0)|2J_1(kNA\delta)/(kNA\delta)|^2 \tag{16.11}$$

where δ is the shift amount and $k=2\pi/\lambda$. The speckle selectivity does not depend on the direction of the media shift (Bragg-selective or Bragg-degenerate) and is independent of the media thickness.

A complete analytic model of crosstalk in correlation multiplexing has not been fully developed so far. According to experimental data and qualitative analysis, the crosstalk builds up proportionally with number of holograms and depends on the media thickness, relative angle between the beams, and, notably, with the total angular extent of both the reference and the signal beams.

Experimentally, the random speckle technique allowed a multiplex of as many as 100 low-*NA* planewave holograms (Figure 16.25) in 200-micron-thick media with little crosstalk. The multiplexing run of 30 holograms (with a 0.2 *NA* signal beam) and crosstalk buildup is shown in Figure 16.26. The crosstalk buildup is usually much faster when the medium is shifted in Bragg-degenerate direction (and in thinner media). The buildup of the crosstalk with increasing number of holograms is due to the random nature of the speckle field used in the experiments; i.e., the previously recorded holograms still are reconstructed but with random, spatially-dependent phases, which produce a largely reduced but

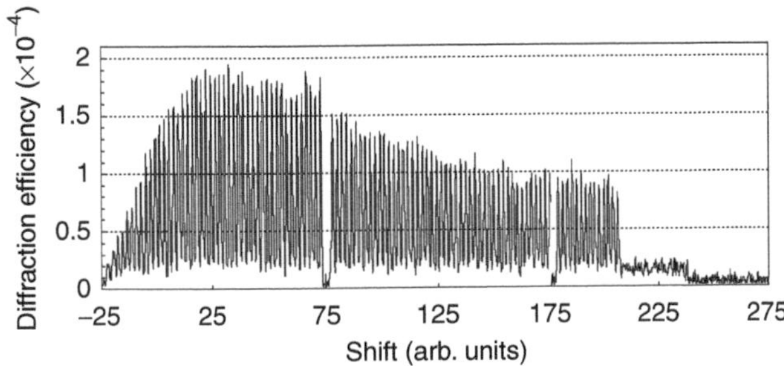

FIGURE 16.25 Readout of 100 holograms superimposed by speckle-shift multiplexing in 200-μm-thick photopolymer CROP medium (hologram separation, approximately 5 μm; $M\# \sim 1.0$). The exposure per hologram was fixed at 0.5 mJ/cm^2 (no scheduling); thus, the change in diffraction efficiency reflects the change in material sensitivity with cumulative exposure (see Figure 16.18).

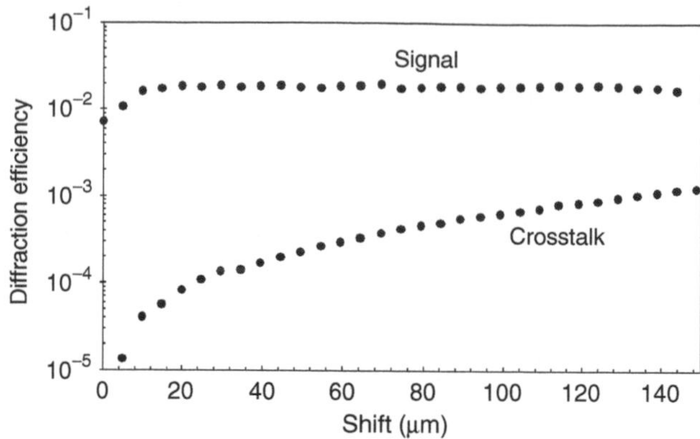

FIGURE 16.26 Signal and cross-talk build-up in random Gaussian speckle-shift multiplexing. Manual scheduling was used to produce a uniform diffraction efficiency distribution among 30 multiplexed holograms. The photopolymer medium's thickness was 200 μm, hologram separation approximately 5 μm, numerical aperture of both beams was 0.2. Crosstalk and signal were measured after each hologram recording. Crosstalk is independent of the spatial location and depends only on the number of previously written holograms (which increases from left to right).

finite amount of diffracted signal in the form of an optical speckle field, which proportionally increases in intensity with the number of multiplexed holograms. However, despite the crosstalk, the experimental results indicate that a few dozen of large-angular bandwidth signal beams can be multiplexed using this technique even in fairly thin (200 microns) media.

16.5 Holographic System Design Tradeoffs

Recording density and the data transfer rate of a holographic memory depend both on the key medium parameters—such as dynamic range, thickness, and scatter—and on the optical and holographic design of the system and the available laser power. This section describes the complex trade-offs and system design considerations, all of which need to be taken into account in order to achieve optimum performance, and, particularly, which arise when both high storage density and high data transfer rates are simultaneously desired (Orlov et al. 2004).

16.5.1 Storage Density of a Single Hologram

The areal density of holographic disk is proportional to the number of superimposed holograms times the density per individual hologram. The optical parameters, which impact the single hologram's storage density, have been investigated extensively (Mikaelian et al. 1970; van der Lugt 1973). In the case of relatively-thin media and transmission recording (like in a holographic disk format), the density can be estimated as the ratio of the number of recorded data pixels ($P \times P$) divided by the storage location area, which includes the Fourier transform size ($\xi\lambda P/2NA$) and the additional light beam expansion within the media away from the Fourier stop (approximated as $2L\ NA$). The resulting density per hologram D_s depends primarily on the numerical aperture NA ($NA = \sin(\arctan[d/2F])$ of the imaging optics:

$$D_s = \left(\frac{2NA}{\xi\lambda}\right)^2 \left(1 + \frac{4NA^2L}{n\xi\lambda P}\right)^{-2} \approx \left(\frac{2NA}{\xi\lambda}\right)^2, \tag{16.12}$$

where d is the image field, F is the focal length, λ is optical wavelength, n is the media index of refraction, ξ is a parameter relating the Fourier stop size to the Nyquist limit (typically, this value is between 1.0 and

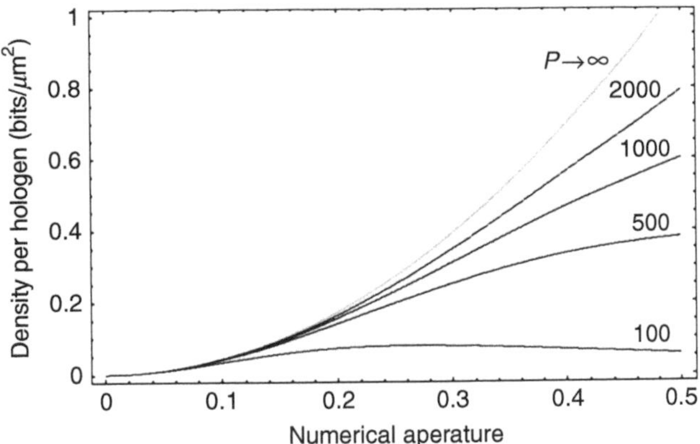

FIGURE 16.27 Individual page storage density vs. numerical aperture of imaging optics for different page sizes P ($L = 500$ μm, $\lambda = 0.532$ μm, $\xi = 1.8$, $n = 1.5$).

1.8 (Bernal et al. 1998)), and L is the medium thickness. In order to maximize the storage capacity, the recording density per hologram needs to be high, which is achieved using high-NA imaging optics (Figure 16.27).

The density per hologram depends primarily on the NA of the holographic image optics, but eventually saturates, as the number of pixels in the page increases (Figure 16.28). Essentially, the diffraction spot size of an individual pixel should be larger than the expansion of the zero order over the thickness of the media. The letter criterion can be expressed in terms of minimum number of pixels (P^2) per page:

$$P >> \frac{4NA^2L}{n\xi\lambda}. \tag{16.13}$$

For example, for $L = 200$ μm, $NA = 0.4$, $\xi = 1.6$, $n = 1.5$, and $\lambda = 0.532$ μm, the page size should be at least 500 × 500 to avoid appreciable penalties in density. Further increase in number of page pixels should be avoided, because that increases the necessary space-bandwidth product of the imaging optics (making it more complex and costly) while not contributing substantially to the achievable density.

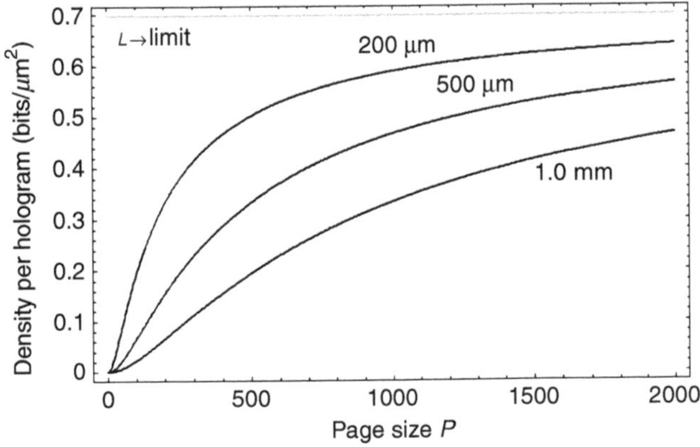

FIGURE 16.28 Individual page storage density vs. page size P for media of different thickness ($NA = 0.4$, $\lambda = 0.532$ μm, $\xi = 1.8$, $n = 1.5$).

16.5.2 Media Dynamic Range and Data Rate

As the number of superimposed holograms (N) increases, the signal strength of the reconstructed holograms drops as $\sim(M\#)/N^2$ (Equation 16.3), where $(M\#)$ is the dynamic range for high-resolution digital pages recording. This total diffraction efficiency of the page is split between P^2 individual image pixels (hence, the effective diffraction efficiency per pixel becomes $(M\#)/N^2/P^2$), and for each of them, the signal collected by its corresponding CCD sensor pixel has to be sufficiently large to overcome the thermal and amplifier noise in order to achieve adequate SNR. Furthermore, the readout data rate R and page size determine the maximum detector integration time per page $\tau \approx P^2/R$, as well as the frame rate $f = R/P^2$.

Quantitatively, when the dominant noise source is the detector noise (thermal and amplifier noise), the maximum number of collocational holograms (N) is limited by the available laser power, the CCD detector noise, and the readout data rate in channel bits per second (Hesselink 2000):

$$N = (M\#)\sqrt{\frac{\eta_{\text{opt}} P_o}{p E_{\text{min}} R}}, \tag{16.14}$$

where P_0 is the average laser power during the read-out (CW or pulsed), η_{opt} is optical efficiency (i.e., the ratio of the reference beam power illuminating the hologram area to the totally emitted laser power), R is the total sustained optical raw data rate (i.e., frame rate in frame-per-second times the number of pixels per page), p is the fraction of ON pixels in the page, and E_{min} is the minimum optical energy per ON pixel to achieve acceptable SNR:

$$E_{\text{min}} \approx \frac{h\nu\, \text{SNR} e_{\text{rms}}}{\text{QE}}, \tag{16.15}$$

where $h\nu$ is the photon energy, QE is detector quantum efficiency (including the fill factor), e_{rms} is the detector noise per pixel (in electrons), and SNR is the signal-to-noise ratio for acceptable BER (for additive white Gaussian noise model). Because the signal strength drops as the number of superimposed holograms increases (Equation 16.3), the achievable density is reduced at higher transfer rates (Figure 16.29):

$$D = (M\#)\sqrt{\frac{\eta_{opt} P_o}{p E_{\text{min}} R}}\left(\frac{2NA}{\xi\lambda}\right)^2\left(1 + \frac{4NA^2 L}{n\xi\lambda P}\right)^{-2}. \tag{16.16}$$

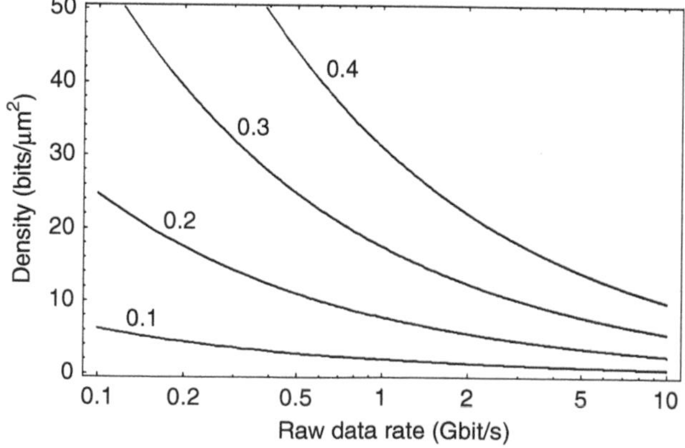

FIGURE 16.29 Achievable storage density vs. raw data rate for different numerical apertures of the imaging optics in the limit of thin media ($L \to 0$) or very large page size ($P \to \infty$) ($P_0 = 200$ mW, $\eta_{\text{opt}} = 0.2$, $p = 0.5$, $e_{\text{min}} = 40 \times 10^{-15}$ J, $M\# = 1.0$, $\lambda = 0.532$ µm, $\xi = 1.8$, $n = 1.5$).

The density limitation due to detector noise is typical for digital hologram storage in thick photorefractive crystals (e.g., 90-degree storage in lithium niobate), while in thin photopolymer media, the multiplexing capacity or medium scattering noise are the major factors limiting the maximum storage density.

16.5.3 Medium Scattering

In the 90-degree geometry in inorganic materials, scattering noise is extremely weak ($\sim 10^{-7}$ s rad^{-1} for Fe:LiNbO$_3$) and usually does not represent a limiting factor for the storage density or the data rate. In most organic (e.g., photopolymer) media (and in transmission geometry), the medium scatter is substantially higher (usually between 10^{-3} to 10^{-4} s rad^{-1}) and, thus, may become a limiting factor for achievable density even at transfer rates in excess of one Gpixel/second. Media scatter noise should be distinguished from the crosstalk noise originating from holographic multiplexing. The latter is related to the holographic properties of the media and the multiplexing scheme, while the media scatter is a characteristic of the properties of the recording material itself. Because scatter light is coherent with the signal, the holographic diffraction efficiency of the signal needs to be at least 30 to 50 times higher than the corresponding scatter power (which is equal to $4\varepsilon_{\text{scatter}}(NA/n)^2$ for a square-shaped detector, Figure 16.7) in order to achieve the acceptable BER. The maximum number of holograms that can be superimposed can be approximated as:

$$N = \frac{n\,(M\#)}{2NA\sqrt{\varepsilon_{\text{scatter}}SScR}}, \tag{16.17}$$

where $\varepsilon_{\text{scatter}}$ is the scatter per steradian into the direction of the CCD array and *SScR* is the required signal-to-scatter intensity ratio. The number of holograms is inversely proportional to the *NA*, because the larger the *NA*, the larger fraction of the scattered light captured by the detector.

The achievable raw density, *D*, is then given in Figure 16.30:

$$D = \frac{2NAn(M\#)}{(\lambda\xi)^2\sqrt{\varepsilon_{scatter}SScR}}\left(1 + \frac{4NA^2L}{n\xi\lambda P}\right)^{-2} \tag{16.18}$$

Higher imaging *NA* still yields higher density, because single hologram density grows as NA^2, while the number of holograms drop only as $1/NA$. A smaller number of holograms means higher diffraction efficiency and stronger signal (Equation 16.3), allowing higher data transfer rate (or reduced laser power for the same data rate) compared to lower *NA* case.

16.5.4 Multiplexing

The ease of accessing large areas of the media at high speed, realizable in rotating disk architecture, is very important for achieving sufficiently high sustainable data rates of one gigabit per second and higher. For holographic disk geometry, shift holographic multiplexing is the most natural solution, as it does not require any beam steering or wavelength change, and is achieved by simply moving the media with respect to the reference and object beams. The standard approaches for shift multiplexing (without moving parts other than disk rotation and radial motion between tracks) include spherical multiplexing (Psaltis et al. 1995) and correlation multiplexing using random (Darskii and Markov 1988) or pseudorandom speckle reference beams (Dhar et al. 1998).

The density of the shift-multiplexed recording has been analyzed in great detail (Psaltis et al. 1995; Steckman, Pu, and Psaltis 2001; Orlov and Hesselink 2003). Here, an estimation of media density based

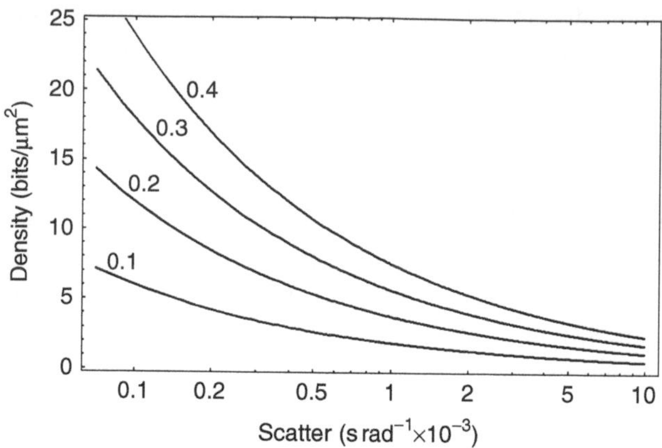

FIGURE 16.30 Storage density vs. media scattering noise $\varepsilon_{scatter}$ for different numerical apertures of the imaging optics ($SScR=30$, $M\#=1.0$, $\lambda=0.532$ μm, $\xi=1.8$, $n=1.5$, thin media ($L\rightarrow 0$) or very large page size ($P\rightarrow\infty$) limit).

upon spherical multiplexing is provided. The maximum number of superimposed holograms can be estimated as a ratio of the reference beam size inside the medium ($2NA_{ref}R/n$) and the minimum spatial separation of the shift multiplexed holograms ($m\lambda R/nLu_s$):

$$N = \frac{2u_s NA_{ref} L}{\lambda m}, \tag{16.19}$$

where u_s is the paraxial angle between the signal and reference in the center of the hologram, NA_{ref} is the numerical aperture of the reference spherical beam (in air), R is the reference beam radius of curvature inside the medium, and m is the order of the Bragg curve null in which multiplexing of adjacent holograms is performed. The resulting storage density for shift multiplexing case is given by Figure 16.31:

$$D = \frac{2u_s NA_{ref} L}{\lambda m}\left(\frac{2NA}{\xi\lambda}\right)^2\left(1+\frac{4NA^2 L}{n\xi\lambda P}\right)^{-2}. \tag{16.20}$$

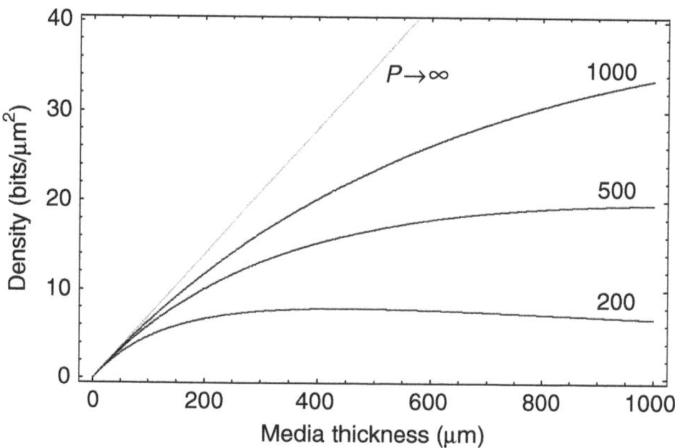

FIGURE 16.31 Achievable density per peristrophic location vs. media thickness L for pages of different sizes P for spherical or angular multiplexing ($NA_{ref}=0.2$, $m=3$, $u_s=0.4$, $NA=0.4$, $\lambda=0.532$ μm, $\xi=1.8$, $n=1.5$). Note the reduction of density with increased thickness for small pages (e.g., 200×200, or less).

It is worth pointing out that the number of holograms which can be multiplexed per location (i.e., the storage density) is the same as in the case of holograms angularly multiplexed using planewave reference beams, which span the angular range from $-NA_{ref}$ to $+NA_{ref}$ around the mean interbeam angle, u_s.

A more detailed, three-dimensional analysis (Orlov and Hesselink 2003) shows that for high-*NA* cases, the estimation given by Equation 16.20 applies only if the parameters of the system are optimized (namely, when the radii of curvature of the signal and the reference are nearly equal inside the medium); and the density can be significantly smaller in non-optimum case due to spatial distribution of crosstalk. Thus, Equation 16.20 represents the theoretical upper bound for shift or angular multiplexing.

The thickness of the holographic disk medium is an important parameter, which should, ideally, lie in the range between 400 and 1000 μm (Figure 16.31). Further increase in thickness does not contribute to the total density, and can even reduce it in the case of small-size data pages.

Substantial increase in storage density is possible, however, if the conventional angular or shift multiplexing schemes are combined with peristrophic multiplexing (Curtis, Pu, and Psaltis 1994). Peristrophic multiplexing (Figure 16.21e) relies on rotation of the medium around the axis normal to the medium plane (or, similarly, rotation of the reference beam around the same axis) and allows a more complete usage of the *k*-space. Combination of this technique with traditional angular or shift multiplexing multiplies the density by the number of possible peristrophic locations. The number of peristrophic locations is limited by the holographic crosstalk and depends on the numerical apertures of the signal and the reference beams. For example, in the case of angular multiplexing with a planewave reference, the maximum number of independent, crosstalk-free peristrophic locations is approximately equal to $\sim\pi\sin(\theta)/NA$, where θ is the mean angle between the object and the reference, and the image beam is normal to the surface (Curtis, Pu, and Psaltis 1994).

An alternative to Bragg-detuning multiplexing methods is correlation multiplexing, which relies on the spatial correlation of the speckled reference beam or employs a phase-spatial light modulator (Figures 16.21d and 16.21f). The number of the holograms, which can be superimposed using correlation technique, is also limited by the holographic crosstalk (Figure 16.26), and is numerically close to the angular or spherical multiplexing case for the same total angular bandwidth of the reference beams (NA_{ref}).

16.5.5 Design Choices for Holographic Systems

In order to achieve optimum system performance, different trade-offs need to be taken into account. Large data density favors reduction in the individual hologram spot size (achieved using high *NA*), while the transfer rate requirement requires having adequately large pages (one megapixel or slightly less), because high frame rates (above several thousand fps) are difficult to implement (even in CMOS cameras, the "size-bandwidth" product is not preserved). Imaging with adequate resolution and large *NA* can be achieved, but only using rather complex optics. Additionally, multiplexing requires that the reference beam's light have access to substantial angular space to provide high data recording density, irrespective of which specific multiplexing technique is used (except for wavelength multiplexing, which is limited by availability of widely tunable lasers). Provision for high angular content of the reference, combined with high *NA* imaging, represents a substantial design challenge due to the mechanical constraints. For a high-data rate system which requires fast media transport, a continuously rotating disk with a pulsed laser recording and readout (Orlov et al. 2004) can be preferred over a stop-and-go approach, due to mechanical reliability considerations. Finally, even in a WORM system, the recording media sensitivity necessarily needs to be reasonably high to maintain meaningful recording time without substantial difference between writing and readout rates. Thus, due to low recording sensitivity, the inorganic photorefractives may be ruled out for use in realistic WORM applications.

16.6 Channel Coding and Signal Processing

In a data-storage system, the goals of error correction coding, data modulation, and signal processing are to reduce the BER to a sufficiently low level, while still achieving such important figures of merit as high density and high data rate. The physical parameters of the system, such as storage density and data rate, are stressed to the limit when the channel is no longer error-free, but introduction of the coding and signal processing still allows user data recovery at acceptable user error rates (e.g., 10^{-9}). The distinct feature of digital holographic data storage is that the data is organized in the form of two-dimensional data pages of substantial size (up to one megapixel or even larger) which are captured by the readout sensor simultaneously, as opposed to the conventional optical or magnetic storage, where the channel bits are usually read out and sampled one by one, forming a temporal sequence of readout signals. The basic characteristics of the data fidelity, such SNR and BER, are still applicable to the holographic channel, but the evaluation of channel in terms of timing jitter and eye-pattern diagrams is not directly relevant. Unlike conventional display holography, where analog images are stored, digital holographic data storage treats each individual pixel (or deterministic blocks of pixels) of the hologram as a data channel, where the pseudo-analog digitized value of the CCD pixel represents a single channel signal strength. This allows the employment of modulation coding and signal-processing techniques to improve the system performance in terms of total capacity at an acceptable error rate.

The definition of the BER is very basic: it is simply the fraction of the bits (zeroes or ones) which are detected incorrectly after passing through the data channel when the size of the bitstream approaches the infinite. This definition is rather independent from the physical implementation of the channel, and includes the errors introduced by the entire channel components chain (medium, optics, electronics). The value of the BER also depends on the channel coding used, and, in the case of block coding, the term *block error rate* is more appropriate, because even if just a few pixels in a block of data are corrupted, the entire block may be decoded incorrectly, and, thus, all the bits which belong to this block may be detected incorrectly (more precisely, they will have statistical error probability of ~50 percent for sufficiently large block size). In practical holographic data storage, the raw BER of coded or uncoded data can be evaluated statistically by using, for instance, the histogram of the received signal amplitudes constructed from a sufficiently large data set. Namely, the CCD pixels are divided into two populations, those corresponding to 0 (OFF pixels) and 1 (ON pixels) of the original image (Figure 16.32). The measured pixel intensities can be used to compute the normalized probability density functions $W_0(x)$ and $W_1(x)$ of the received signal strength for zeroes and ones respectively. Then, by choosing a specific (global) threshold value x_{thr},

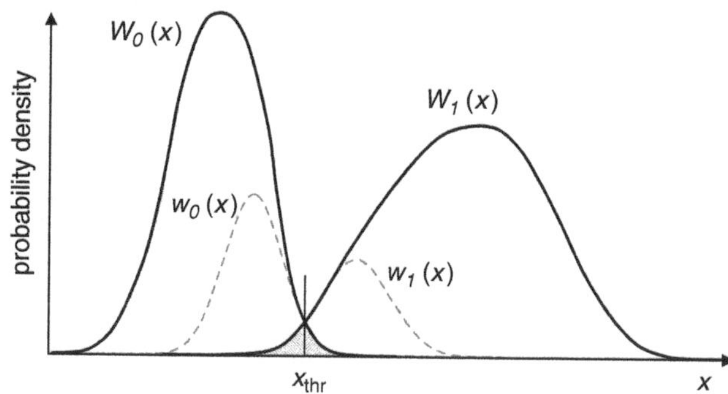

FIGURE 16.32 The distribution functions of received ON and OFF pixels can be estimated from the histogram of a sufficiently large experimental data set. The optimum threshold value, x_{thr}, corresponds to the amplitude where the two distributions cross. The area of the shaded region on the graph is equal to the bit error rate. Shoulders of the distributions can be approximated by analytical functions $w_0(x)$ and $w_1(x)$ (for example, by Gaussian distributions), allowing quantitative evaluation of BER even when extremely low and there were no or few actual errors detected in a given experimental data set (Hoffnagle and Jefferson 2000).

the raw BER of the hologram can be computed as:

$$BER = \frac{1}{2}\left[\int_{x_{thr}}^{\infty} W_0(x)\mathrm{d}x + \int_{0}^{x_{thr}} W_1(x)\mathrm{d}x\right], \tag{16.21}$$

where the pixels whose analog value is below x_{thr} will be decoded as zeroes, and the pixels whose analog values are larger than x_{thr} will be decoded as ones.

An important parameter that affects the BER and is a characteristic of the data fidelity is the SNR of the image, which is usually defined as:

$$SNR = \frac{\mu_1 - \mu_0}{\sqrt{\sigma_1^2 + \sigma_0^2}}, \tag{16.22}$$

where μ_1 and μ_0 are the average values of detected ones and zeroes respectively, and σ_1 and σ_0 are the corresponding variances. For a simple case of added white Gaussian noise (AWGN) channel, the BER can be directly calculated from the measured distribution functions, assuming the simple global threshold detection (see, for example, Saleh and Teich 1991):

$$BER \approx \frac{1}{2} erfc\left(\frac{1}{\sqrt{2}}\frac{\mu_1 - \mu_0}{\sigma_1 - \sigma_0}\right) \tag{16.23}$$

where the optimum threshold is equal to the signal strength at which the 0 and 1 distributions cross each other.

In realistic holographic data-storage systems, the signal distributions rarely obey a pure Gaussian (i.e., incoherent additive noise) or pure Rician (pure coherent noise case) model. Some of the noise sources, such as intersymbol interference and misalignment, are not random in nature, thus affecting the statistic. Furthermore, nonuniformity of light intensity across the information page affects the overall distributions, which now becomes a sum of local distributions with different mean and standard deviation parameters. Therefore, the realistic histogram becomes a complex convolution of local distributions which, by themselves, can also be convolutions of Gaussian and Rician distributions and can have additional contributions from non-random noise sources. In such cases, the SNR of the image can be a measure of the BER, but merely in a qualitative sense, while the quantitative BER needs to be evaluated directly from the data. However, because most errors occur for the signal values where the 0 and 1 distribution functions are near equal, the BER can be estimated rather accurately by locally approximating the portion of the histogram which carries the most errors with simple Gaussian distributions (Hoffnagle and Jefferson 2000). Then, the parameters of these Gaussian distributions can be used to evaluated the final BER by using, for instance, Equation 16.23. This statistical approach is most useful for images of low error rate: i.e., when the direct error counting does not provide adequate BER evaluation accuracy.

The coding signal processing in holographic storage needs to take into account the various different noise sources arising in digital hologram storage, including the nonuniformity of the page intensity (i.e., average signal variations across the page), variability in average diffraction efficiencies of holograms, and local BER variations within a single hologram itself, as well as the intersymbol interference. Intersymbol interference (ISI) occurs when part of the intensity of the pixels spills over to its neighbors, which may happen for a variety of reasons, including light diffraction, pixel misregistration due to optical imaging system distortions, and optical misalignment. Coding and signal processing can involve several qualitatively distinct elements, including error correction coding (ECC), interleaving, holographic modulation encoding, signal predistortion, data storage in the holographic system, hologram retrieval,

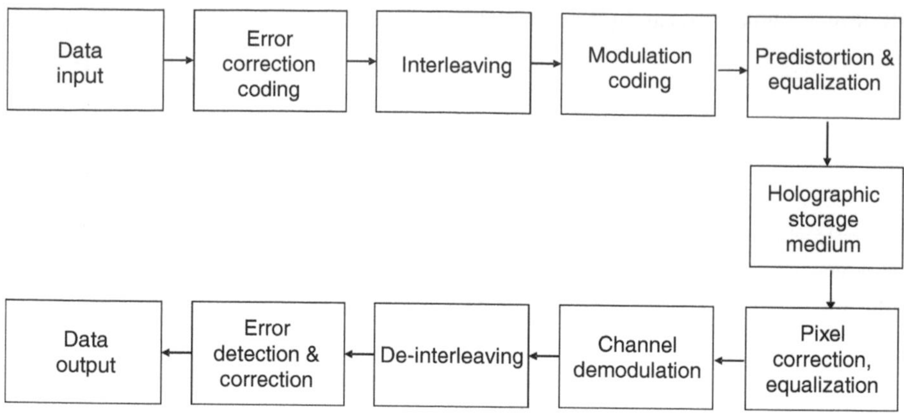

FIGURE 16.33 Signal processing in digital holographic data storage.

analog signal detection and binary digitization (using CCD or CMOS detector array), equalization, holographic channel decoding, and ECC decoding of deinterleaved binary data (Figure 16.33).

ECC processing adds redundancy to the user data in order to provide protection against channel errors. The typical ECC algorithms are Reed-Solomon or turbo product codes. ECC is needed in order to bring the user error rate to extremely low values (e.g., 10^{-9} or less) when the received data is not error-free (BER $\sim 10^{-2}$ to 10^{-3}). The ECC codes are characterized by the code rate (i.e., the ration between the user data and the encoded data block sizes) and coding gain, which is the ratio of BERs of the coded and uncoded data (a function of source data SNR and BER). The coding gain needs to be sufficiently large (e.g., 10^6 or more) and depends on the code rate (the smaller the code rate, the larger redundancy added atop the code, and, hence, the larger the correction capability). The purpose of data interleaving is in reduction of the corrected (after ECC) BER and optimization of ECC performance. The SNR of the source images is usually not uniform across the different image sections (Figure 16.34). However, by distributing data in different sections of the image (a process called interleaving) and de-interleaving the decoded data, one can achieve uniform BERs in all the ECC blocks before error correction operation, thus reducing the error rate of the user data. Because the dependence of corrected BER of the input data error rate is usually highly nonlinear, the uniform error rate distribution among ECC blocks is optimal.

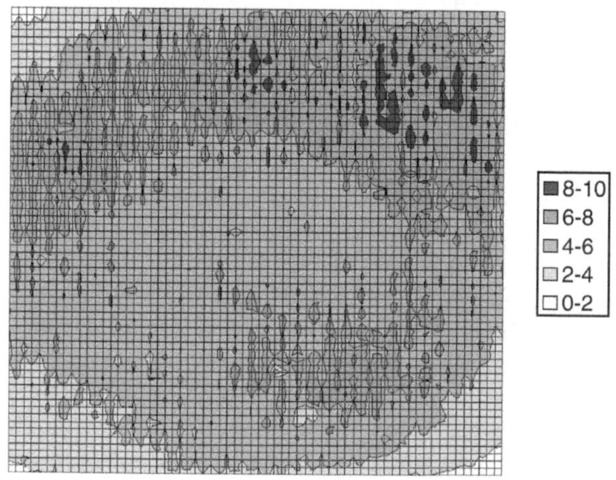

FIGURE 16.34 Example of SNR distribution within a 1-Mpixel holographic page.

Areas of high BER (with SNR $< \sim 2.0$) are usually excluded (masked) from the page because they can actually reduce the system density and capacity.

The purpose of modulation coding is to simplify the detection of the analog signal values from the detector array. The channel coding converts the incoming stream of data bits into a two-dimensional pattern, which is stored in the form of a hologram. Ideally, the outcome of the decoding should not depend on the absolute values of the pixels in a channel block, but only on their relative signal strengths upon readout. This restriction is typically satisfied by using constraints on the types of pixel patterns that are used to code the data bits.

An uncoded data set in which "1" bits are represented by ON pixels and "0" bits are represented by OFF pixel can be decoded using global threshold detection, where a specific threshold value is chosen: any CCD pixel with an intensity above the threshold is declared a one, while the rest are declared zeroes. If all holograms result in equal diffraction efficiencies, a threshold intensity can be chosen to distinguish ON pixels from OFF pixels. The BER, assuming a model of additive white noise independent from pixel to pixel, is than given by:

$$BER = \int_{x_{\text{thr}}}^{\infty} \frac{1}{\sqrt{2\pi}\sigma} \exp\left(-\frac{(x-a_0)^2}{2\sigma^2}\right) \text{d}x, \tag{16.24}$$

where a_0 is the average amplitude of OFF pixels at the CCD array, σ is the standard deviation of the noise, x_{thr} is the threshold value set to be the mean of the ON and OFF values, or, in a more general case of different statistics for ones and zeroes, set to the point where the two distributions cross (which minimizes the BER).

Global threshold coding has a code rate of unity (one bit is encoded as one pixel of the hologram), but in the presence of variation in average signal strength across a data page, does not perform satisfactorily because the optimal value of the threshold becomes dependent on the position within the page, and can be different for different pages. When the threshold value is set to a value different from its optimal value, the BER increases extremely rapidly. One of the ways to mitigate the effect of the intensity variation on the BER is the implementation of a balanced constraint coding. Namely, out of each N pixels in the block, M are always made ON, while the N–M are left OFF. Upon detection, the strongest M pixels are then declared as ON, while the other N–M pixels are classified as OFF. The simplest example is the differential encoding in which each zero and one of the data is encoded in form of a group of two adjacent pixels ***a*** and ***b***, one ON, one OFF (Heanue, Bashaw, and Hesselink 1994). Upon detection and demodulation, if the intensity of pixel ***a*** is lowest, the data bit is declared a value of zero, and if the intensity of pixel ***b*** is lowest, the data bit is declared a value of one. Viewed a different way, an OFF–ON pixel block represents a data bit value of zero and a ON–OFF block represents a data bit value of one. The probability of an error in a differential detection system with additive white noise is given by a following integral:

$$\text{BER} = \int_0^{\infty} \frac{1}{\sqrt{2\pi}\sigma} \exp\left(-\frac{(y-a_{\text{OFF}})^2}{2\sigma^2}\right) \int_0^{y} \frac{1}{\sqrt{2\pi}\sigma} \exp\left(-\frac{(x-a_{\text{ON}})^2}{2\sigma^2}\right) \text{d}x\text{d}y, \tag{16.25}$$

where a_{OFF} and a_{ON} represent the average amplitudes of the OFF and ON pixels, respectively. Most notably, the detection scheme does not rely on the choice of the threshold value, thus making differential encoding very robust and easy-to-implement. This technique largely solves the problems associated with nonuniformity of holographic data pages, but has low code rate of 0.5, due to the constraint to use two pixels to encode a single data bit.

The code rate can be improved by increasing the array size and employing mathematically more sophisticated coding and decoding strategies (Heanue, Bashaw, and Hesselink 1995; Burr et al. 1997a). More advanced balanced modulation codes provide local thresholding capability and relatively high code rates (75 percent and higher). The number of data bits, K, which can be coded into an array of N pixels in which M pixels are ON and (N–M) pixels are OFF is given by:

$$K = \left\lfloor \log_2 \left(\frac{N!}{M!(N-M)!} \right) \right\rfloor \tag{16.26}$$

An example is a 6:8 ($N=8$, $M=4$) balanced block code, in which a data page is divided into 2×4 arrays, each containing exactly four ON pixels and four OFF pixels. The number of permutations for this constraint is 70, which can support up to $K=6$ data bits, i.e., the code rate is 0.75 (Figure 16.35). It should be mentioned that the increased code rate comes at the expense of increased block error rate. For example, the 6:8 code has a block error rate approximately 16 times higher at the same image SNR when compared to simple differential encoding scheme.

In the case of block codes, the detection is based on a sorting or correlation technique, in which the detected block of pixels is compared to the set of valid code words and is declared to be a specific code word based on maximum likelihood criteria. The channel codes with large coded array size (i.e., large N) can also provide intrinsic error correction capability due to highly redundant coding, which allows correction of a limited number of pixel errors directly during the channel decoding step. An example of such code is the 8:12 strip-balanced array code (Burr et al. 1997a), which is capable of correcting a single error per block (i.e., one out of six nominally OFF pixels upon detection can have value larger than one of the nominally ON pixels).

Other modulation codes include parity threshold codes, in which the number of ON and OFF pixels in a large uncoded array is stored in a small coded (with highly redundant code) subarray (Burr et al. 1998); strip codes (Marcus 2000); and soft-decision iterative turbocodes (concatenated with constraint block codes for bias removal).

Predistortion and equalization are techniques which attempt to improve the image SNR and reduce the BER while working in the analog domain. Predistortion (Burr et al. 1997b) manipulates the recording exposure of each individual pixel, either by changing the exposure time or by introducing the grayscale on the SLM during the recording step. Typically, for high-density holograms, different parts of the image have differing amounts of intensity due to overall illumination non-uniformity, as well as differing amounts of optical distortion present as fixed pattern noise. For example, the corners of the image usually experience more distortions and lower illumination, resulting in poorer local SNR. Predistortion calibrates the illumination non-uniformity and adjusts the recording accordingly to provide a nearly uniform SNR distribution across the page (for example, the image corners may receive more exposure or higher brightness on the SLM to boost the signal). Raw BER improvements from 10^{-4} to 10^{-10} have been experimentally demonstrated using this technique. Another approach which can be regarded as

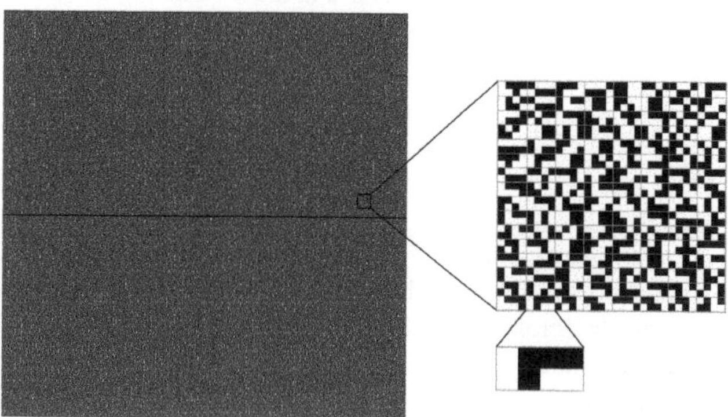

FIGURE 16.35 One-megapixel holographic page encoded with 6:8 modulation code. The page consists of 2×4 pixels subarrays (right); each of these blocks represents a balanced codeword (4 pixels are always ON, 4 pixels are always OFF) and encodes 6 bits of data.

pre-distortion in broader sense is the uniform BER scheduling (Burr et al. 1998). During the multiplexing run in, for instance, iron-doped $LiNbO_3$, the holograms which are recorded in the beginning of the exposure not only become partially erased by optical illumination during continued multiplexing, but are also more distorted due to accumulative optical damage. As a result, even if the holograms have the same diffraction efficiency after the multiplexing (which is achieved by applying uniform efficiency scheduling), the BER of the first images is usually higher than the BER of the images recorded last. By recording the first holograms to even higher initial diffraction efficiency, one can achieve a uniform BER distribution between all holograms after the multiplexing. Experimentally, the uniform BER recording produced a gain of $\sim$20 percent in the total storage capacity over the uniform diffraction efficiency recording (Burr et al. 1998).

Equalization (Vadde and Vijaya Kumar 1999; Chugg, Chen, and Neifeld 1999) manipulates the raw received pixel values before channel decoding in order to compensate for the intersymbol interference. If the channel is linear, the ISI can be represented as a convolution with the original encoded data pattern, and can be reduced in the equalization step by deconvolving the input pattern with a pre-calibrated kernel. Because ISI frequently depends on the position within the page, several different kernels are typically needed for satisfactory performance. The deconvolution kernel can be optimized either by measuring the ISI on the known patterns and creating an inverse, or via iterative optimization, with target function being the best SNR. The difficulty with linear deconvolution approach is that, physically, the ISI occurs at the light amplitude level because the light is coherent, making the channel rather nonlinear, while complete ISI compensation using linear deconvolution is only possible for a linear channel. An alternative to deconvolution is a nonlinear serial equalization process (Burr and Weiss 2001), designed specifically to compensate for the pixel shifts due to optical distortions and misalignment. The serial algorithm reallocates the appropriate portion of the signal detected by each pixel to its neighbors, accounting for both linear and quadratic (i.e., nonlinear) contributions to ISI introduced by coherent illumination. This procedure, although rather computationally expensive, allows compensation of quite large amount of the ISI; as large, in fact, as that which corresponds to a pixel misregistration (due to, for example, misalignment or distortion) amount of nearly 40 percent of the pixel size.

Most holographic data storage systems developed so far perform signal processing and decoding using computer software. This approach is fairly versatile, because a computer can be easily reprogrammed to implement different signal processing operations; however, it lacks speed and synchronization capability. The first (and until now the only) hardware implementation of the holographic channel decoding electronics (Figure 16.36) was developed by Siros Technologies, Inc. (San Jose, CA) and Stanford University as a part of the digital holographic disk system (Orlov et al. 2004; Hesselink, Orlov, and Bashaw 2004), which will be described in more detail in Section 16.8. The channel electronics are based on field-programmable gate array (FPGA) technology and are capable of sustainable one-gigabit-per-second data rate operation. The channel-decoding electronic hardware consists of three main processing blocks: holographic channel decoders, a demultiplexer, and a Reed-Solomon (RS) ECC board. The signal processing steps of the electronics are shown in Figure 16.37.

Equalization and channel decoding are performed by 16 different boards in parallel to gain the necessary computing power because the incoming data rate is very high: eight gigabits per second (i.e., one thousand of one-megapixel images per second; each pixel being represented by eight bits). Equalization is accomplished via 3×3 pixel deconvolution, which is performed on the raw 8-bit pixel values. The 3×3 kernels are computed by software and provide pixel corrections to reduce the effect of optical point-spread function and distortions of the imaging system. Decoder boards provide 6:8 holographic channel decoding. A simple 6:8 code (i.e., 8 pixels corresponding to 6 bits of data, including ECC bits) was chosen, because it provides a fairly high (75 percent) code rate and has sufficiently low complexity to be implementable in hardware. More advanced codes (e.g., 8:12 strip code) may provide better performance (Burr et al. 1997a) with respect to BER and storage capacity, but are significantly more complex for hardware implementation.

Deinterleaving and ECC decoding are performed by two separate boards. After the channel decoding, the average data rate drops to approximately 780 megabits per second, so a single-board computational

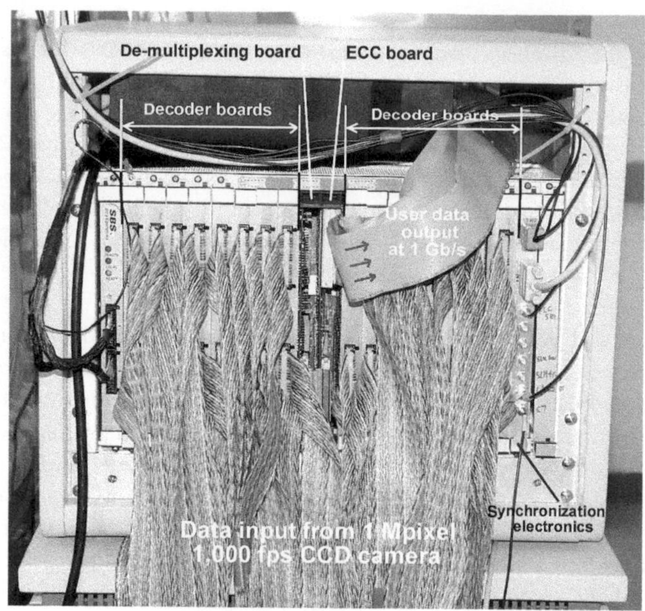

FIGURE 16.36 A photograph of the Siros/Stanford channel electronics.

power is sufficient for each of the operations. The deinterleave hardware reassembles the bits encoded in different parts of the original images into three ECC-encoded byte streams, which are ECC-decoded using commercially-available Reed-Solomon decoding ASICs (produced by Advanced Hardware Architectures). The 8-bit-per-symbol RS code can be configured to accept encoding formats from 235:255 to 135:155. Finally, the user data is output in the form of 12 parallel byte streams with total compound burst data rate of 960 megabits per second.

The channel decoding system described above represents one of the first attempts of full-hardware implementations of holographic channel decoding. The electronics architecture proved to be rather complex and costly, but has successfully demonstrated the capability of fully electronic holographic

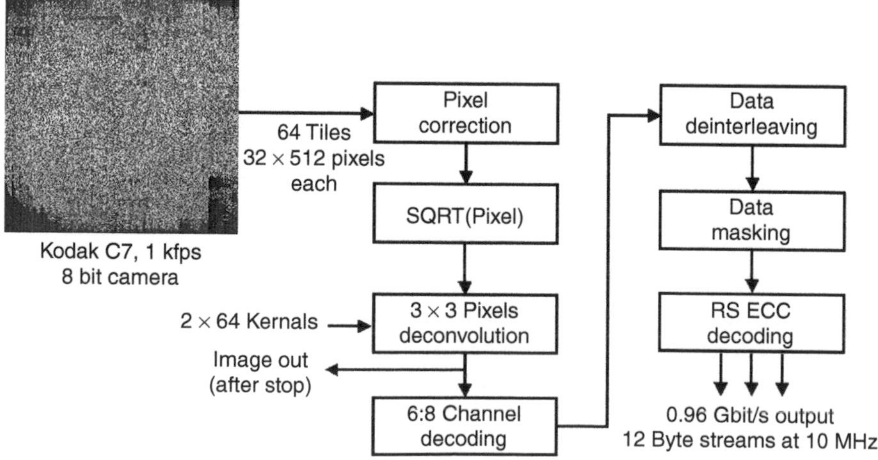

FIGURE 16.37 Operation of Siros/Stanford holographic channel electronics: SQRT (Pixel) denotes a square root operation on the raw 8-bit grayscale pixel values (performed with the use of a lookup table).

channel signal processing at a one-gigabit-per-second raw data rate. The application of ASICs rather than FPGAs and overall higher degree of integration can, however, substantially reduce both the cost and the size of future generation holographic channel decoding devices, which is important for future commercial devices and data storage products based on digital holographic technology.

16.7 Servo Technologies

In any data storage technology, the purpose of the servo system is to provide active feedback on the relative head-to-media position in order to keep the information bits being read or recorded in registration with the reading or writing head. Typically, conventional optical disk drive technology employs pregrooved media and push-pull servo with a quad-detector servo signal detection. Magnetic storage relies on burst servo with pre-recorded servo marks, which are offset from the main track position. Several servo techniques have been proposed and demonstrated for holographic digital data storage; however, most of these approaches are not without certain deficiencies and limitations.

One of the servo approaches (Zhou et al. 1999; Snyder and Daiber 2000) uses alignment marks embedded in the information page (Figure 16.38). After the image of the page is acquired by the camera, the position of the alignment marks is detected by software, and the camera (or one of the imaging lenses) is shifted to the position which provides best pixel-to-pixel alignment. For relatively thin media, the optical shift can also be produced by slightly changing the angle on the reading beam, which, in turn, shifts the position of the image on the CCD. The shortcoming of the alignment-mark approach is that it requires the page to be acquired twice (first to calculate the adjustment amounts, and second to read the actual information). This lowers the data rate, and, in the case of rotating disk systems, does not compensate for the unrepeatable run-out.

Another approach introduced by Horimai, Hesselink, and Inoue (2002) uses conventional pre-grooved medium push-pull servo combined with a holographic engine (Figure 16.39). This approach requires a specially manufactured medium containing a reflective, pre-grooved substrate and thick photopolymer material sandwiched between the grooved substrate and the top cover glass. The system uses two lasers: a near-IR semiconductor laser is used for servoing, while a few-hundred-milliwatt green laser is used for holographic recording and readout. The tracking and focusing is performed just as in a conventional CD-type disk drive, while the data is recorded and read back holographically using so-called "collinear holography" approach. A clear advantage of this technique is that it leverages mature conventional optical disk servo technology. However, exact track following and addressing achieved by this approach is not sufficient to obtain exact pixel matching between the retrieved holograms and the

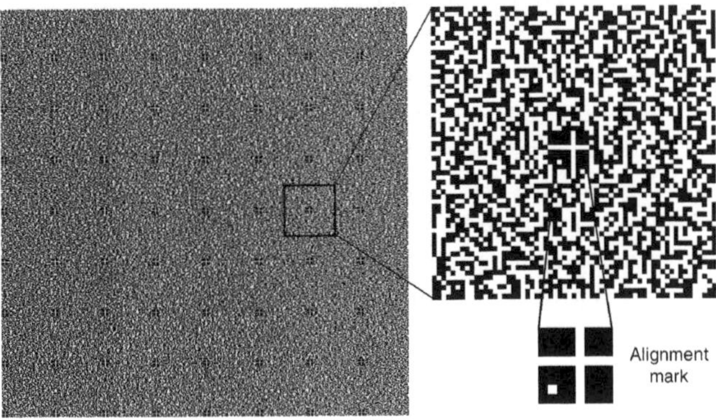

FIGURE 16.38 512×512 pixels holographic page with embedded alignment marks (right). By detecting positions of the marks in the retrieved page, the amount of necessary adjustment to achieve best pixel matching can be calculated.

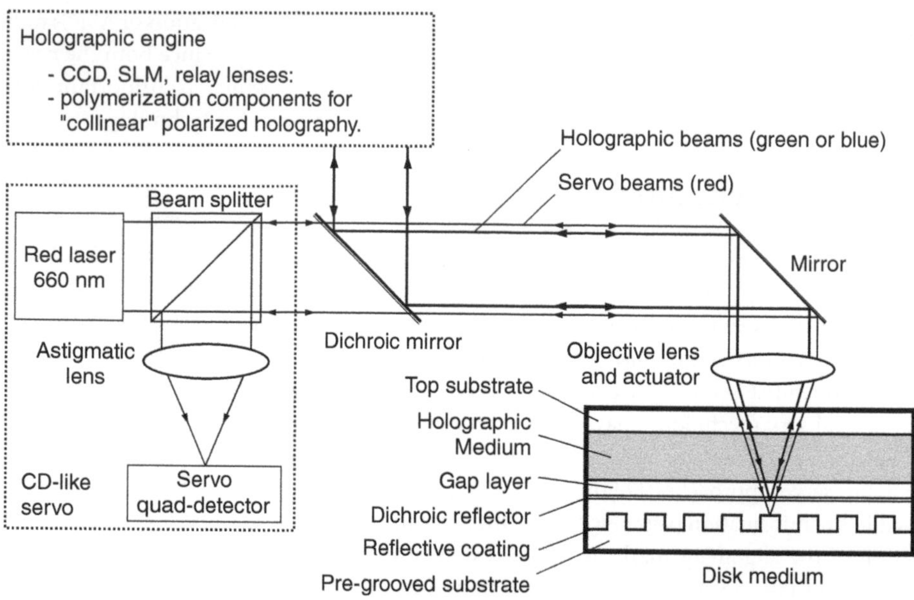

FIGURE 16.39 An optical schematic of Optware servo concept. Objective lens servoing is done similarly to the conventional CD disk drive tracking and focusing push-pull servo using a pre-grooved medium, quad-detector, and astigmatic lens. A dichroic mirror optically decouples the holographic and servo subsystems.

detector array. For example, in the scheme shown in Figure 16.39, the disk tilt or wedge will deflect the reflected light and shift the entire image on the CCD, thus producing the pixel misregistration and errors (while CD beam will still being on the track on the pre-grooved medium). To avoid pixel misregistration, the medium has to be either extremely flat (within less than ~1 mrad or better) and wedge-free, or a second-stage servo loop is required to compensate for the image shift, which increases the overall servo system complexity.

Overall, the servo technology for holographic storage has been, perhaps, one of the least-investigated aspects of the holographic memory, and still may require further development; in addition, more approaches have yet to be devised which would allow servoing with reliability and speed comparable with the performance of established optical and magnetic storage servo technologies.

16.8 Holographic Storage Systems

In this section, the history of the holographic systems development will be briefly reviewed, as well as presenting two case studies of the state-of-the art systems developed within the recent DARPA HDSS research program (1995–2000). These systems demonstrate how different optical and electronic components and technical approaches described in the previous sections are integrated into a complete holographic engine. The first system is an IBM demonstration platform (DEMON II) (Burr et al. 2001), which uses the 90-degree holographic geometry and photorefractive iron-doped lithium niobate as storage material. The second system is based on rotating holographic disk approach, and employs low-shrinkage CROP photopolymer medium (Orlov et al. 2004; Hesselink, Orlov, and Bashaw 2004).

16.8.1 Historical Overview

An excellent overview of the historical developments in holographic data storage can be found in Sincerbox (2000), while the most recently developed systems are described in detail in Coufal, Psaltis, and Sincerbox (2000) and Hesselink, Orlov, and Bashaw (2004). Here, the highlights of the major developments in the field are presented, with a particular emphasis on the complete system implementations rather than theoretical work or demonstrations of specific techniques.

The development of data storage systems based on holography started in the early 1960s, sparked by the invention of the laser, which made the technology practical for storing and retrieving images, as well as a pioneering theoretical work by Polaroid researcher Peter van Heerden, who predicted that storing an interference pattern in a three-dimensional medium can be used to store and retrieve digital information. The early efforts were somewhat hampered by the lack of commercially available components, such as SLMs and CCDs, as well as by deficiencies of the available holographic media. The first systems built (Table 16.3) used photographic films or plates in which the recording layer was rather thin. Thus, massive multiplexing of multiple holograms in the same storage location was not possible, and the storage density was the same as it would have been for a two-dimensional surface storage. Due to unavailability of large dynamic SLMs and integrated CCD cameras (not yet invented at the time), these systems used large sets of static illumination masks (for example, 35-mm slides), which were mechanically transported for new data page imprinting, and arrays of silicon detectors for data capture. These systems, however, still managed to store amounts of data comparable to or higher than, for instance, twister memory storage technologies of those days (few megabits).

It was not until the discovery of the photorefractive effect and emergence of bulk ferroelectric crystals as a new photosensitive medium for holography (Chen, LaMacchia, and Fraser 1968) that the three-dimensional nature of volume hologram storage started to be fully explored. One of the first angular multiplexing systems (d'Auria, Huignard, and Spitz 1973) stored ten holograms per location (10^4 bits per page). Futher developments included the implementation of fast dynamic SLMs, as well as investigation of photorefractive medium re-writability, with as many as 200 write-erase cycles demonstrated (Mikaelian 1976). A substantial effort on improving the properties of the lithium niobate for holographic storage was undertaken at RCA Laboratories, leading to development and optimization of high sensitivity and high dynamic range iron-doped crystals (Phillips, Staebler, and Amodei 1972), and discovery and investigation of ionic hologram fixing (Amodei and Staebler 1971).

In the 1990s, a new wave of research and development activity in the field of holographic data storage systems was triggered by the rapid progress in enabling technologies: compact SLMs and CCD cameras emerged on the display and video consumer applications market, as well as compact all-solid-state lasers being developed for printing and storage applications. These developments allowed the design and construction (at relatively low cost) of holographic storage systems with as many as one million pixels per page, in which the data could be read at video or higher rates. One of the important milestones was the demonstration by researchers from Northrop Grumman (Mok 1993) of multiplexing of as many as 5000 holograms in the same location of iron-doped lithium niobate crystal, which proved the technical feasibility of mass multiplexing in volume holographic storage. Another significant contribution was the application of channel coding and ECC correction (Heanue, Bashaw, and Hesselink 1994), and demonstration that digital data can be stored and retrieved holographically at the user BER levels which are comparable to the conventional commercial storage devices. Further advancements in optics, signal processing, and channel coding ultimately allowed the achievement of ~ 250 Gbit/in.^{2} density in Fe-doped $LiNbO_3$ crystal (Burr et al. 2001), which surpassed the superparamagnetic limit of the conventional magnetic storage (~ 100 Gbit/in.^{2}), thus proving the potential of the holographic technology for high-density storage.

In parallel with the development of systems based on photorefractive crystals, significant progress has been made in the development of the organic recording media. One of the first commercially available photopolymer media specially developed for holographic applications was DuPont HRF film (introduced in the 1980s) which, despite relatively small thickness (~ 40 μm), allowed multiple hologram multiplexing (Curtis and Psaltis, 1992). The key advantage over the previously-used dichromatic gelatin medium was that DuPont film did not require complex wet processing after the holographic exposure. Further improvements in photopolymer medium were made by employing different monomer chemistries (Crivello and Lee 1989; Waldman, Li, and Horner 1997; Dhar et al. 1998); for example, CROP or high-molecular-weight vinyl monomers, which led to improved characteristics, including

TABLE 16.3 Holographic System Demonstrations

Demonstration/Source	Material and Architecture	Key Features	Reference
Historical development			
Bell Labs Digital Page (1968)	Photographic emulsion, WORM, spatial multiplexing only	4 Mbit total system capacity	Anderson 1968
IBM HOSP (1970)	Photographic emulsion, WORM, spatial multiplexing only	35 mm film slides used as SLM 0.5 Mbytes per photographic plate	Lipp and Reynolds 1971
RCA holographic memory (1973)	Photo-thermoplastic, erasable, spatial multiplexing only	1 M pixel total storage capacity, nematic liquid crystal SLM, differential encoding	Steward et al. 1973
Hitachi holographic video disk (1976)	Photoresist for master replication, thin holograms, spatial mux only	300 mm disk containing 30 min of color video, analog image storage	Tsunoda et al. 1976
Thompson/CSF (1974)	Fe-doped $LiNbO_3$, angular multiplexing	First implementation of angular mux using thick volumetric medium	d'Auria, Huignard, and Spitz 1973
A. S. Popov Society of Radio Engineering and Telecommunications (USSR 1973)	Fe-doped LiNbO3, transmission, angular multiplexing	Novel magnetic (Faraday effect) SLM, 40× 40 pixels, <100 μs pixel switching time	Mikaelian 1976
Systems Developed in 1990s			
Tamarack Multistore (1992)	Photopolymer (e.g., DuPont HRF), transmission, angular multiplexing	Holographic disk jukebox, novel optical head design, LCD SLM 256×256	Redfield 2000
Northrop-Grumman (1993)	Fe:LiNbO3, transmission geometry, angular multiplexing	First demonstration of massive angular multiplexing: 5000 analog holograms	Mok 1993
Stanford University (1994)	Fe:$LiNbO_3$, 90 degree geometry, angular multiplexing	Differential holographic channel encoding, error correction coding; first complete HDS	Heanue, Bashaw, and Hesselink 1994
Holoplex fingerprint recognition system (1995)	Photopolymer (DuPont HRF) transmission, angular and peristrophic multiplexing	Fingerprint recognition using holographic correlator	Zhou et al. 1996
Rockwell International (1995)	Fe:$LiNbO_3$, 90 degree geometry, angular multiplexing	<100 μs access time, no moving parts, acousto-optic reference beam steering	Hong et al. 1995; Ma et al. 2000
Optitek video demonstration (1996)	Fe:$LiNbO_3$, 90 degree geometry, angular multiplexing	Fully electronic channel decoding at video rates, several minutes of MPEG compressed video stored and retrieved	Daiber et al. 1997
Lucent Technologies, Inc (1998)	Lucent free radical polymer, transmission, correlation multiplexing	Novel correlation multiplexing technique, new thick (500 μm) photopolymer medium	Dhar et al. 1998

Latest Demonstrations, Including Those Developed by HDSS Consortium Members (1994–2000)			
IBM HOST tester (1996)	Polymer or photorefractive materials, angular multiplexing	System used for material performance evaluation, custom low-distortion optics	Bernal et al. 1996
HDSS Stanford/Siros holographic disk system (2000)	Polaroid CROP, transmission, correlation shift multiplexing	~10 Gbit/s optical transfer rate, full electronic implementation of 1 Gbit/s holographic channel decoding, single-head high NA holographic optics	Orlov et al. 2004
IBM Demon II (2001)	Fe:$LiNbO_3$, 90 degree geometry, angular mlultiplexing	>250 Gpixel/in.^{2} (400 bits/μm^2) density, high NA custom imaging optics	Burr et al. 2001
Optware, Inc (2002)	Photopolymer, “collinear holography” architecture	Integration of holographic optics and conventional CD/DVD type servo	Horimai, Hesselink, and Inoue 2002
Aprilis, Inc. (2002)	Polaroid/Aprilis CROP, transmission, angular + peristrophic multiplexing	>100 bits/μm^2 storage density in novel 400 μm thick ultra-low shrinkage photopolymer	Waldman, Butler, and Raguin 2003

increased thickness (500 microns to one millimeter) and reduced shrinkage (<0.05 percent in CROP media). As an outcome of the progress made in material science, several high-density holographic data storage platforms were developed based on photopolymer disk architecture (Table 16.3).

The necessity of achieving high-density storage in relatively thin media called for novel approaches in both holographic multiplexing methods and holographic optics. Novel techniques such as fractal (Psaltis et al. 1990), peristrophic (Curtis, Pu, and Psaltis 1994), spherical (Psaltis et al. 1995), polytopic (Anderson and Curtis 2004), and correlation (Darskii and Markov 1988) multiplexing were introduced in order to increase the utilization of the available k-space, while the introduction of high-NA optics (Orlov, Snyder, and Bashaw 2000b) specially designed for holographic recording allowed to store the individual data pages at higher density (>0.5 bit/μm^2/page) than had been possible in earlier demonstrations. Using custom high-NA optics and low shrinkage CROP photopolymer, researchers at Aprilis, Inc. demonstrated recently (Waldman, Butler, and Raguin 2003) that storage densities in excess of 100 bits/μm^2 are indeed achievable in relatively thin photopolymer medium (400 μm), if the standard angular or spherical beam multiplexing is combined with peristrophic technique. As of today, most of the companies that pursue commercialization of holographic data storage consider photopolymer disk as the most promising architecture, due to both high recording sensitivity and relatively low medium cost, as well as the relative ease of manufacturing of large size media compared to inorganic crystals.

In order to support research and development efforts (with the final aim at commercialization of the technology), the Defense Advanced Research Projects Agency (DARPA) in 1994 and 1995 initiated two consortia: Photorefractive Information Storage Materials (PRISM) and Holographic Data Storage Systems (HDSS). The PRISM activities were focused on advanced research in new holographic recording materials (both inorganic photorefractives and photopolymers), whereas HDSS focused on development of new optical and electronic components, digital signal processing, holographic channel codes, and digital holographic storage system demonstrators. The two systems described in the following sections were developed as part of these consortia efforts.

16.8.2 IBM DEMON II Platform

The DEMON II holographic storage platform (Burr et al. 2001), shown in Figure 16.40 and Figure 16.41, was designed to achieve high-density holographic data storage using 90-degree geometry with angular multiplexing and iron-doped lithium niobate crystal as the recording medium. This platform was built as a continuation of several holographic storage systems developed at IBM Almaden Research Center, and incorporates large (one megapixel) pages and short-focal-length double Fourier transform custom-built optics.

DEMON II employs a dynamic SLM, which is a reflective device fabricated by IBM Yorktown (Sanford et al., 1998) containing 1024×1024 pixels and illuminated via a polarizing beamsplitter cube. A novel optical element (apodizer) provides uniform illumination over the entire data page without sacrificing input power. The magnification from the 12.8-μm pitch of the SLM pixels to the 12-μm pitch of the 40-Hz CCD camera (Dalsa CA-D4-1024, 1024×1024 pixels, 40 frames per second) is built into the Fourier optics (effective focal length 30 mm). A pair of F-theta (scan) lenses provides a relay of the reference beam from the galvanometrically-actuated mirror to the $LiNbO_3$ crystal with diffraction-limited performance over an angular scan range of ±15 degrees. The experimental platform employs a sophisticated computer control, allowing both direct operator control of the system and unsupervised execution of complex scripted experiments.

The laser light is provided by a diode-pumped solid-state laser (532 nm, doubled Nd–YAG); waveplates and polarizing beamsplitters provide control over the power in the reference beam and object beam. The use of two separate elements in the back Fourier lens (between the storage material and the detector array) allows the magnification of the optical system to be varied over a range of ±0.5 percent. Linear stages provide two axes of motion for the storage material and three axes of motion for the detector array. The entire system, including the laser, occupies two feet×two feet.

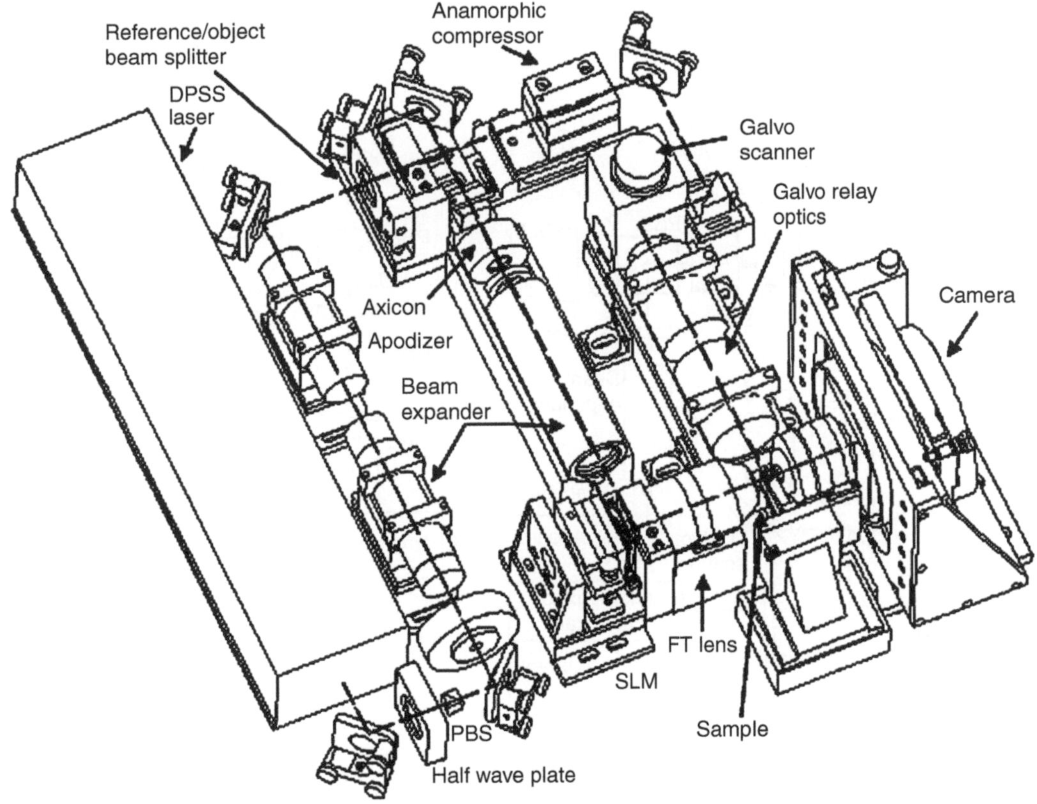

FIGURE 16.40 The layout of the DEMON II holographic digital data storage engine. The system utilizes 90-degree geometry with short focal length double Fourier transform imaging optics. Courtesy of C.M. Jefferson and G.W. Burr (IBM Almaden Research Center).

The short focal length of the DEMON II optics allows the system to demonstrate high areal storage densities (the storage capacity of each stack of holograms, divided by the area of the limiting aperture in the object beam). Custom optics image the SLM pattern through the Fe:LiNbO$_3$ storage material (15 × 15 × 8 mm, *c* axis at 45 ± in the horizontal plane, 0.02 percent iron-doped, absorption coefficient at 532 nm $\alpha \approx 0.8\ \text{cm}^{-1}$) and onto the detector array. The detector sensing elements have a fill factor of 90 percent, and the detected data page was pixel-matched 1:1 over the entire 1024 × 1024 pixels. A square aperture of 1.7 × 1.7 mm, placed directly at the Fourier-transform plane, was found to introduce a tolerable amount of interpixel crosstalk (the Nyquist aperture for this system is 1.25 × 1.25 mm). The object beam entered the 15 mm-×-15 mm face of the crystal 1.5 mm downstream from the aperture and centered 5 mm from the reference beam's orthogonal entrance face.

The reference beam was compressed vertically and apertured to a size of 5 mm wide by 1.9 mm tall. In order to implement angular multiplexing, the collimated reference beam was directed by a galvanometric mirror deflector (Cambridge Instruments 602HC-6450) through a pair of scan lenses in a 4-F configuration and onto the 8 mm × 15 mm face of the crystal.

In order to mitigate the effect of the optical damage build-up during the recording process, the crystal was displaced both horizontally and vertically after each hologram exposure. This effectively blurs the photovoltaic space-charge accumulation at the edges of the illuminated area, thus preventing large index gradients, which cause BER degradation. In effect, the crystal motion allows an increase of the usable *M*#, and hence, the density.

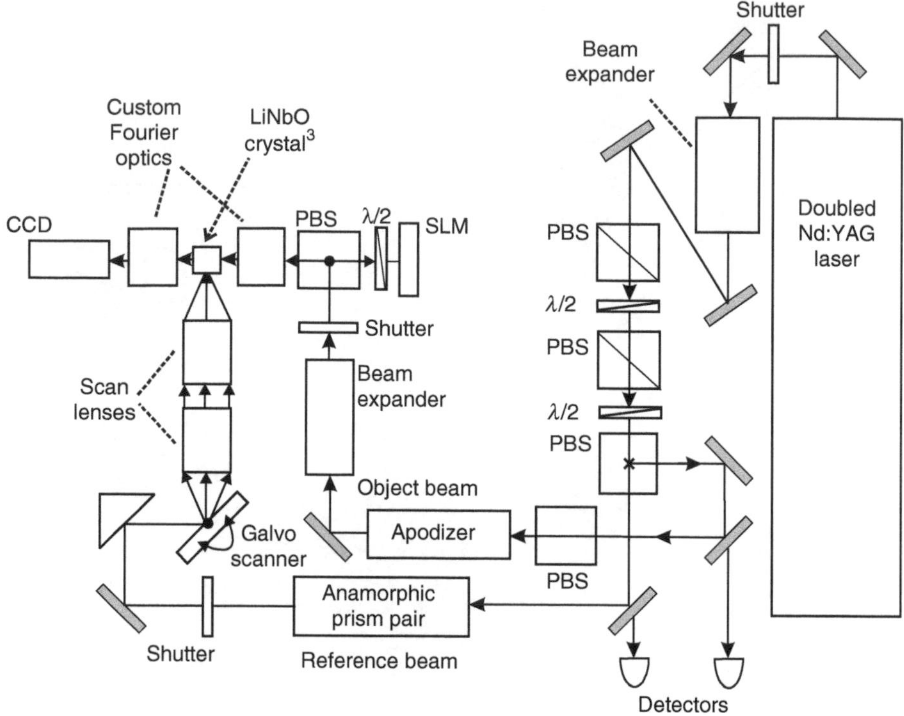

FIGURE 16.41 Optical schematic of DEMON II holographic storage platform. An apodized flat-top beam uniformly illuminates a reflective SLM, which is then imaged to a matched CCD array through a small aperture by a short-focal-length optics. PBS's polarizing beam splitters; $\lambda/2$ half-wave plates. Courtesy of G.W. Burr (IBM Almaden Research Center).

In the high-density recording experiments (Burr et al. 2001), 9 stacks in a 3×3 grid were angular-multiplexed using the same recording schedule, with average recording exposure time of $\sim$0.34 s. Most holograms were spaced in angle by 0.016°, with a total of 1000 holograms in each stack. The average diffraction efficiency was 1.1×10^{-6}, yielding the dynamic range of $M\#\approx1.1$. With each hologram containing 10^6 raw bits (pixels) and effective area defined by stack spacing of 1.6 mm, the resulting recording density was $\sim$400 bits/μm^2 or $\sim$250 Gbits/in.^{2} The data was channel-encoded using 8:12 strip code (Burr et al. 1997a), resulting in 670-Mbit-per-page capacity and $\sim$240 bits/μm^2 user density (assuming a 0.92 code rate Reed-Solomon ECC is used).

To verify the BER of the stored data after the recording, 100 randomly-picked holograms from the central track were sampled at reduced readout power of $\sim$345 mW. For each page, the crystal had to be translated to the position used to write that hologram, thus limiting the data rate. Before channel decoding, the received data were processed using a novel nonlinear equalization algorithm (Burr and Weiss 2001). After equalization, the worst raw bit error rate was found to be 1.1×10^{-3} (average 3.2×10^{-4}), which is sufficiently low and correctable by, for example, Reed-Solomon ECC to yield the desired 10^{-12} user BER target. Because no fixing procedures were implemented in these experiments, the data eventually decays under continuous readout conditions.

The high volumetric density of $\sim$0.072 bit/μm^3 achieved in these experiments still represents only 1.1 percent of the theoretical limit of $1/\lambda^3$ of holographic data storage (van Heerden 1963). In order to utilize the available volumetric storage density more fully, a more-complete coverage of the *k*-space occupied by the grating *k*-vectors of the recording hologram is required. This implies using even higher numerical aperture of the image beam and expanding the range covered by the reference beam optics. Indeed, in a later demonstration, by using imaging optics with $NA \sim 0.2$ and combining angular and peristrophic multiplexing, Waldman, Butler, and Raguin (2003) achieved volumetric density of 0.25 bit/μm^3 (i.e., 100 bit/μm^2 areal density in a 400-μm-thick polymer film).

DEMONII platform was the first holographic storage engine which demonstrated the storage density exceeding the 100 bits/μm^2 benchmark. Importantly, this experimental system has shown that holographic data storage can provide data densities which are higher than the superparamagnetic limit of the conventional (non-patterned medium) magnetic storage media ($\sim$30 Gbit/in.^{2}), and which are about an order of magnitude higher than the storage density envisaged for the future optical blue-DVD disk technology ($\sim$30 bits/μm^2).

16.8.3 Stanford/Siros Holographic Disk System

The Stanford/Siros digital holographic disk storage demonstrator (Orlov et al. 2004) was built as part of the HDSS Consortia effort, with the goal to achieve a readout data rate of at least one Gbit/sec and high recording density using a rotating photopolymer disk architecture. This holographic storage platform represents a complex experimental system that incorporates custom high-NA double Fourier transform optical imaging lens, fully implemented one-gigabit-per-second electronic holographic channel decoding electronics, precision electro-mechanical components, dedicated synchronization electronics, and sophisticated computer control software, and which employs high-performance photopolymer disk medium (Figure 16.42 and Figure 16.43).

Many system components used by the demonstrator were designed and built by the participants of the HDSS and PRISM consortia. These include a 1024×1024 pixel-matched 1000-fps digital Kodak C7 CCD camera and IBM ferroelectric liquid crystal SLM (Sanford et al. 1998). The pixel pitch is 12.8 μm for both devices. The Kodak C7 camera is divided into 64 tiles arranged in a 32×2 array (32×512 pixels in each) and has a frame rate of up to 1000 frames/s with external exposure trigger control. The ferroelectric liquid crystal SLM has up to 2000 frames-per-second switching capability with high-frequency intensity contrast from seven to ten. The SLM is driven by a custom electronic driver designed and built by Displaytech, Inc. (Boulder, Colorado) with a maximum frame rate of 650 fps, limited by the data transfer rate from the control PC to the SLM chip.

Recording and readout are performed with a pulsed doubled Nd:YAG laser (532 nm, 500 μJ/pulse and 20 nsec pulse length at 1 kHz). The holographic photopolymer disk medium is mounted on a precision air-bearing spindle. In most experiments, the spindle rotation rate was 300 revolutions per minute. The angular addressing is performed using a precision optical shaft encoder, which provides 16,384 individual locations along the track of the disk. Synchronization electronics allows addressing each of

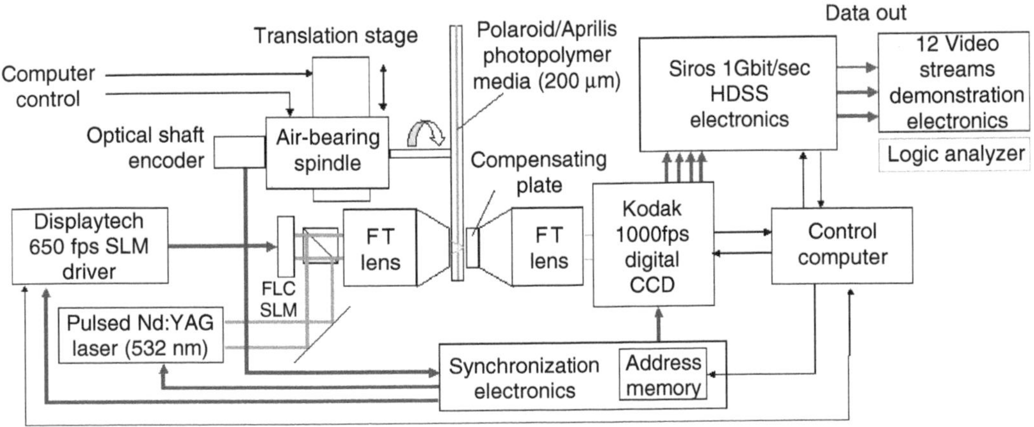

FIGURE 16.42 Architecture of the Stanford/Siros digital holographic disk system.

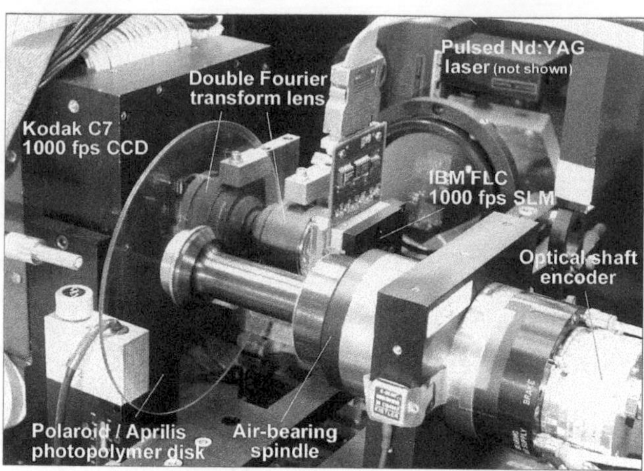

FIGURE 16.43 Photograph of the holographic disk system.

these angular positions on the disk with accuracy and repeatability of better than ±0.1 μm, and provides simultaneous control of the CCD and SLM. High addressing accuracy allows the recording of holograms incrementally, with multiple pulses fired on successive disk rotations. For access to different radial positions, the spindle was horizontally translated on high-precision stage (Newport PM600) with positional repeatability of ± 25 nm.

Real-time signal processing is performed by custom high-speed reprogrammable FPGA-based electronics developed at Siros Technologies and Stanford. The HDSS electronics take input from the CCD camera (8-bit resolution, 1-megapixel images at 1000 fps) and provides channel decoding, ECC decoding, and deinterleaving on-the-fly, producing the user data in the form of 12 parallel byte streams at 10 MHz each (Figure 16.37). The data is encoded using 6:8 block modulation channel code and 135:155 8-bit-per-symbol Reed-Solomon ECC. The details of the holographic channel electronics operation are given in Section 16.6.

The SLM is optically imaged onto the CCD array using short focal length low-distortion (less than ±1.5 μm over the entire CCD field) custom-built optical double Fourier transform lens pair. The total numerical aperture of 0.75 is divided between the central, high-resolution, low distortion portion (*NA* = 0.36), used by the SLM, and the outer area, used by the reference light (Figure 16.44) (Orlov, Snyder, and Bashaw 2000b). The *NA* allocated for each of the reference beam locations is approximately 0.2. The signal and reference beams pass through the same set of optics, minimizing the physical size of the optical system and eliminating mechanical constraints typical for high-*NA* optical systems. The optical design is fully symmetrical in order to minimize the imaging distortion, because a 2.9-mm-thick compensator plate is incorporated into the second objective lens to compensate for the holographic disk thickness. The lens system also incorporates the magnification adjustments and optical distortions correction

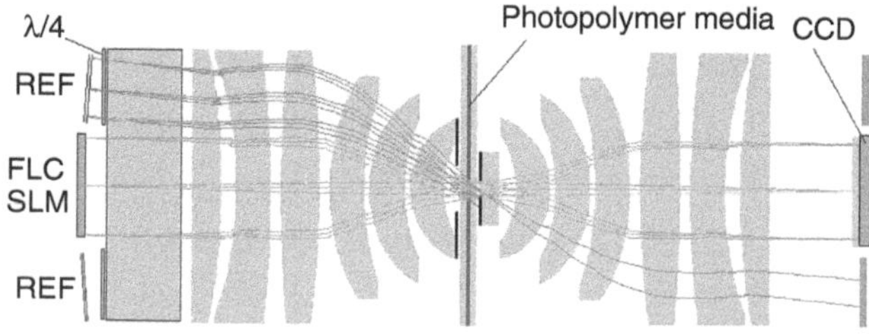

FIGURE 16.44 Optical layout of the recording head.

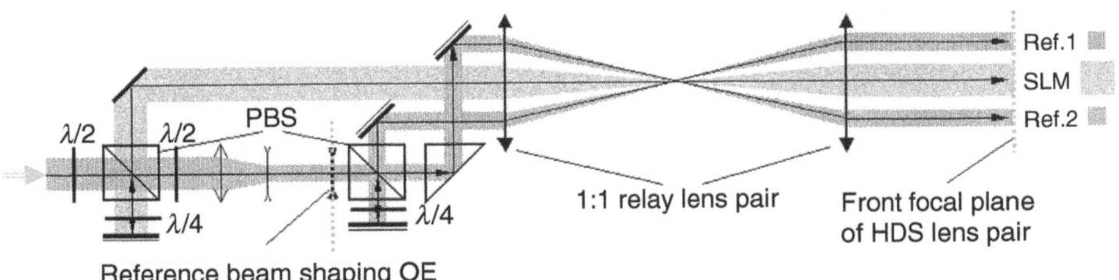

FIGURE 16.45 Beam-splitting and -shaping optics used to provide signal and reference arms for the optical head.

implemented as mechanical movement of the last optical element in the second group (closest to the CCD). The recording media is located in front of the focal plane to avoid recording nonlinearity due to the Fourier spot.

Typical hologram spot size on the photopolymer is approximately 2 mm^2, which provides raw areal density per hologram of 0.5 bits/μm^2. The two separate references can be considered as two separate peristrophic locations that can be used one at a time. The beam shaping optics (shown in Figure 16.45) are designed to provide nominal zero path length difference (in the center of the media) between the image and the reference and to control the intensity ratio of the reference and image beams. Typical intensity beam ratio used in the experiments was approximately 1:10. For speckle correlation multiplexing, the reference beam-generating optical element represents a random Gaussian holographic diffuser, with typical appropriate divergence angle of four to five degrees FWHM. Alternatively, for spherical shift multiplexing, a negative achromatic lens is used with focal length of -100 mm (for 1024×1024 page size). The system employed low-shrinkage 200-μm-thick CROP photopolymer medium (Section 16.3) developed by Polaroid and, later, Aprilis (Waldman, Li, and Horner 1997).

A series of performance demonstrations was carried out using this system. In the first demonstration, source user data were ECC- and channel-encoded, stored in the holographic disk ($\sim$20 holograms), retrieved at 1 Gbit/s, and decoded on-the-fly (Orlov 2000a). The decoded user data was captured by a logic analyzer and converted back into its original file format (JPEG color images). A sample hologram, read out and decoded at 1 Gbit/s, is shown in Figure 16.46. The storage density in the first experiment was rather low (approximately 1 bit/μm^2), and individual holographic page capacity was 65.3 Kbytes (user). In the later demonstration (Phillips et al. 2000), twelve uncompressed digital video streams were holographic channel-encoded and stored in the holographic disk, and later retrieved at 0.65 Gbit/sec and simultaneously displayed during readout on three computer screens (four video streams per monitor). The total amount of data stored and retrieved was roughly 1 gigabyte.

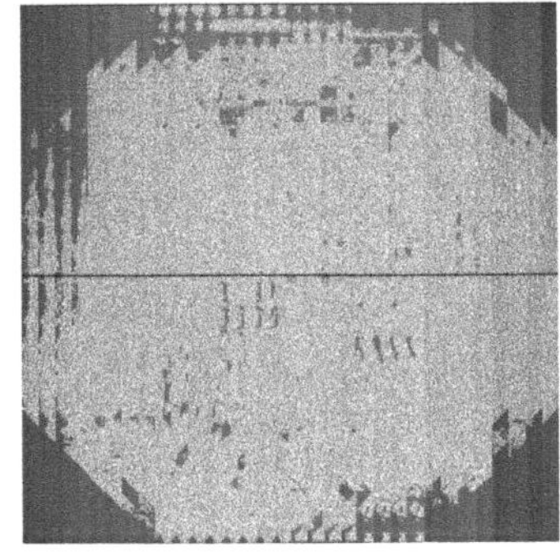

FIGURE 16.46 Sample digital hologram (left) and a retrieved JPEG-encoded (color) image (right); raw byte error rate was 0.3×10^{-3}, $\eta \sim 0.5\%$.

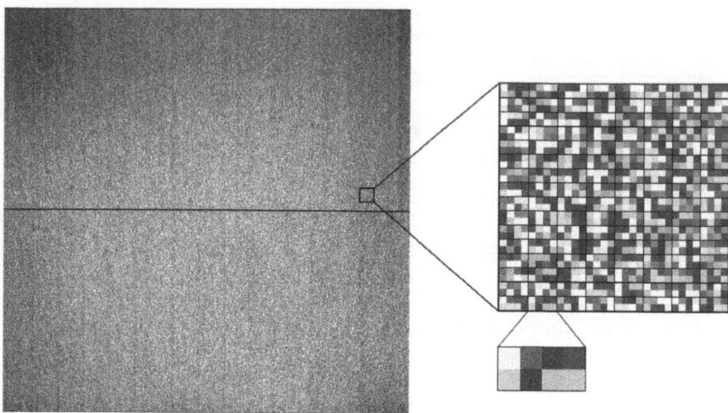

FIGURE 16.47 Sample hologram reconstructed at 10 Gbits/s (byte error rate ~1%, with 10% of the area masked). Channel blocks of 6:8 modulation code are shown at the right (compare with original encoded image shown Figure 16.35).

In high-data-rate experiments (Orlov et al. 2002), sustained optical readout data rates as high as ten Gbit/sec were obtained using higher efficiency optical delivery systems. The latter improvement was essential, due to substantial reduction in the pulse energy with increasing repetition rate (25 μJ/pulse at 10 kHz, compared to 500 μJ/pulse at 1 kHz). The holograms arranged in a single track (2048 pages) were recorded with storage density of approximately 10 bits/μm^2 and reconstructed at 10,240 fps speed (sampled selectively by the CCD at 1 Gbit/s). A sample hologram readout at 10 Gbit/s is shown in Figure 16.47. The raw byte error rate in these experiments was roughly one percent, which still can be corrected by modern ECC codes (e.g., turbocodes).

The storage density in this experimental holographic disk system was limited to approximately ten bits/μm^2 (for each reference beam arm) in 200-μm photopolymer film, due to multiplexing limit (Figure 16.31); while at 1 Gbit/s, the hologram signal strength largely exceeded the camera noise. The signal strength allowed much-higher optical data rates, provided the holograms could be physically transported under the optical readout head and addressed at sufficient speed, which was achieved by employing a continuously rotating disk architecture, pulsed laser, and shift multiplexing.

As of today, the demonstrated ten-Gbit/s optical transfer rate represents the highest data rate ever produced by an optical storage device. This, along with other unique attributes of the system, such as hardware-implemented one-Gbit/s holographic channel electronics and novel holographic head design, makes the Stanford/Siros holographic disk system an important milestone in the development of digital holographic data storage technology.

16.9 Summary

This chapter has described the fundamental principles of holographic data storage, recording materials, multiplexing schemes, signal processing used in digital hologram storage, and system design tradeoffs and limitations, as well as presenting a detailed description of two state-of-the-art systems.

Through the use of the entire volume of the medium, holographic data storage offers enormous theoretical storage capacity (Tbits/cm^3) and, due to highly parallel access to the data, extremely high data rates. Some of these performance figures have been proven realizable, at least when using specially designed sophisticated demonstration platforms: areal densities of over 400 Gbit/μm^2 and data rates as high as 10 Gbit/s have been demonstrated. Until now, deficiencies of holographic materials were considered the main obstacle for the development of commercial holographic storage devices. However, inasmuch as the materials may remain the main obstacle, new advancements in the polymer chemistry and crystal growth have solved the many problems of the past approaches. As a technology, digital holographic storage has progressed tremendously since the time it was conceived in the 1960s—but so have the data storage technologies with which it needs to compete in order to find its way to commercial markets. As of now, it still remains to be seen whether the performance advantages of digital holographic memories outweigh the system cost and engineering complexity associated with this elegant technology.

References

Amodei, J. J. and Staebler, D. L. 1971. Holographic pattern fixing in electro-optic crystals. *Applied Physics Letters*, 18, 540–542.

An, X., Psaltis, D., and Burr, G. W. 1999. Thermal fixing of 10,000 holograms in $LiNbO_3$. *Applied Optics*, 38, 386–393.

Anderson, L. K. 1968. Holographic optical memory for bulk data storage. *Bell Laboratories Record*, 45, 319–326.

Anderson, K. and Curtis, K. 2004. Polytopic multiplexing. *Optics Letters*, 29, 1402–1404.

Arizmendi, L. 1989. Thermal fixing of holographic gratings in $Bi_{12}SiO_{20}$. *Journal of Applied Physics*, 65, 423–427.

Ashkin, A., Boyd, G. D., Dziedzic, J. M., Smith, R. G., Ballmann, A. A., and Nassau, K. 1966. Optically induced refractive index inhomogeneities in $LiNbO_3$ and $LiTaO_3$. *Applied Physics Letters*, 9, 72.

Barbastathis G., and Psaltis, D. 2000. Volume holographic multiplexing methods, In *Holographic Data Storage*, H. J. Coufal, D. Psaltis, and G. T. Sincerbox, eds., *Vol. 76*, Springer Series in Optical Sciences, pp. 21–62. Springer, Berlin.

Bernal, M. -P., Coufal, H., Grygier, R. K., Hoffnagle, J. A., Jefferson, C. M., Macfarlane, R. M., Shelby, R. M., Sincerbox, G. T., Wimmer, P., and Wittmann, G. 1996. A precision tester for studies of holographic optical storage materials and recording physics. *Applied Optics*, 35, 2360–2374.

Bernal, M. -P., Burr, G. W., Coufal, H., and Quintanilla, M. 1998. Balancing interpixel cross talk and detector noise to optimize areal density in holographic storage systems. *Applied Optics*, 37, 5377–5385.

Bieringer, T. 2000. Photoaddressable polymers, In *Holographic Data Storage*, H. J. Coufal, D. Psaltis, and G. T. Sincerbox, eds., *Vol. 76*, Springer Series in Optical Sciences, pp. 209–262. Springer, Berlin.

Burr, G. W. and Weiss, T. 2001. Compensation for pixel misregistration in volume holographic data storage. *Optics Letters*, 26, 542–544.

Burr, G. W., Ashley, J., Coufal, H., Grygier, R. K., Hoffnagle, J., Jefferson, C. M., and Marcus, B. 1997a. Modulation coding for pixel-matched holographic data storage. *Optics Letters*, 22, 639–641.

Burr, G. W., Ashley, J., Coufal, H., Grygier, R. K., Hoffnagle, J., Jefferson, C. M., and Marcus, B. 1997b. Noise reduction of page-oriented data storage by inverse filtering during recording. *Optics Letters*, 23, 289–291.

Burr, G. W., Chou, W. -C., Neifeld, M. A., Coufal, H., Hoffnagle, J. A., and Jefferson, C. M. 1998. Experimental evaluation of user capacity in holographic data storage systems. *Applied Optics*, 37, 5431–5443.

Burr, G. W., Jefferson, C. M., Coufal, H., Jurich, M., Hoffnagle, J., Macfarlane, R. M., and Shelby, R. M. 2001. Volume holographic data storage at an areal density of 250 gigapixels/in.^{2}. *Optics Letters*, 26, 444–446.

Buse, K., Adibi, A., and Psaltis, D. 1998. Non-volatile holographic storage in doubly doped lithium niobate crystals. *Nature*, 393, 665–668.

Carrascosa, M. and Agullo-Lopez, F. 1990. Theoretical modeling of the fixing and developing of holographic gratings in $LiNbO_3$. *Journal of the Optical Society of America B*, 7, 2317–2322.

Chen, F. S., LaMacchia, J. T., and Fraser, D. B. 1968. Holographic storage in lithium niobate. *Applied Physics Letters*, 13, 223–225.

Chugg, K. M., Chen, X., and Neifeld, M. A. 1999. Two-dimensional equalization in coherent and incoherent page oriented optical memory. *Journal of the Optical Society of America A*, 16, 549–562.

Crivello, J. V. and Lee, J. L. 1989. Alkoxy-substituted diaryliodonium salt cationic photoinitiators. *Journal of Polymer Science, Part A: Polymer Chemistry*, 27, 3951–3968.

Coufal, H. J., Psaltis, D., and Sincerbox, G. T. eds. 2000. *Holographic Data Storage*, Springer, New York.

Curtis, K. and Psaltis, D. 1992. Recording of multiple holograms in photopolymer films. *Applied Optics*, 31, 7425–7428.

Curtis, K., Pu, A., and Psaltis, D. 1994. Method for holographic storage using peristrophic multiplexing. *Optics Letters*, 19, 993–995.

Darskii, A. M. and Markov, V. B. 1988. Shift selectivity of holograms with a reference speckle wave. *Optics and Spectroscopy*, 65, 392–395.

d'Auria, L., Huignard, J. P., and Spitz, E. 1973. Holographic read-write memory and capacity enhancement by 3-D storage. *IEEE Transactions on Magnetics*, 9, 83–94.

Daiber, A. J., Snyder, R., Colvin, J., Okas, R., Hesselink, L. 1997. Fully functional digital video holographic storage system. Optical Society of America Annual Meeting, Long Beach, CA, Paper ThR3.

Denisyuk, Yu. N. 1963. On the reproduction of the optical properties of an object by the wave field of its scattered radiation. *Optics and Spectroscopy*, 15, 279–283.

Dhar, L., Curtis, K., Tackitt, M., Schilling, M., Campbell, S., Wilson, W., Hill, A., Boyd, C., Levinos, N., and Harris, A. 1998. Holographic storage of multiple high-capacity digital pages in thick photopolymer systems. *Optics Letters*, 23, 1710–1712.

Gabor, D. 1948. A new microscopic principle. *Nature*, 161, 777–778.

Günter, P. and Huignard, J. -P. eds. 1988. *Photorefractive Materials and Their Applications I: Fundamental Phenomena, Vol. 61*, Springer, Berlin.

Heanue, J. F., Bashaw, M. C., and Hesselink, L. 1994. Volume holographic storage and retrieval of digital data. *Science*, 265, 749–752.

Heanue, J. F., Bashaw, M. C., and Hesselink, L. 1995. Channel coding for digital holography. *Journal of the Optical Society of America A*, 12, 2432–2439.

Heanue, J. F., Bashaw, M. C., Daiber, A. J., Snyder, R., and Hesselink, L. 1996. Thermal fixing for digital holographic data storage. *Optics Letters*, 21, 1615–1617.

Hesselink, L. 2000. Digital Holographic Demonstration Systems by Stanford University and Siros Technologies, In *Holographic Data Storage*, H. J. Coufal, D. Psaltis, G. T. Sincerbox, eds., *Vol. 76*, Springer Series in Optical Sciences, pp. 383–397, Springer, Berlin.

Hesselink, L., Orlov, S. S., Liu, A., Akella, A., Lande, D., and Neurgaonkar, R. R. 1998. Photorefractive materials for nonvolatile volume holographic data storage. *Science*, 282, 1089–1094.

Hesselink, L., Orlov, S. S., and Bashaw, M. C. 2004. Holographic data storage systems. *Proceedings of the IEEE*, 92, 1231–1280.

Hoffnagle, J. A. and Jefferson, C. M. 2000. Bit error rate for holographic data storage, In *Holographic Data Storage*, H. J. Coufal, D. Psaltis, G. T. Sincerbox, eds., *Vol. 76*, Springer Series in Optical Sciences, pp. 91–100, Springer, Berlin.

Hong, J. H., McMichael, I., Chang, T. Y., Christian, W., and Paek, E. G. 1995. Volume holographic memory systems—techniques and architectures. *Optical Engineering*, 34, 2193–2203.

Horimai, H., Hesselink, L., and Inoue M., 2002. Volumetric optical disk storage with collinear polarized holography. Optical Data Storage Topical Meeting, Joint International Symposium on Optical Memory and Optical Data Storage 2002, Waikoloa, HI, USA. In Technical Digest, 43–44.

Kippelen, B. 2000. Overview of photorefractive polymers for holographic data storage, In *Holographic Data Storage*, H. J. Coufal, D. Psaltis, G. T. Sincerbox, eds., *Vol. 76*, Springer Series in Optical Sciences, pp. 159–170, Springer, Berlin.

Kirillov, D. and Feinberg, J. 1991. Fixable complementary gratings in photorefractive $BaTiO_3$. *Optics Letters*, 16, 1520–1522.

Kogelnik, H. 1969. Coupled wave theory for thick hologram gratings. *Bell System Technical Journal*, 48, 2909–2947.

Kukhtarev, N. V. 1976. Kinetics of hologram recording and erasure in electrooptic crystals. *Soviet Technical Physics Letters*, 2, 438–443.

Kukhtarev, N. V., Markov, V. B., Odoulov, S. G., Soskin, M. S., and Vinetskii, V. L. 1979. Holographic storage in electrooptic crystals, I: Steady state. *Ferroelectrics*, 22, 949–960.

Lande, D., Orlov, S. S., Akella, A., Hesselink, L., and Neurgaonkar, R. R. 1997. Digital holographic storage system incorporating optical fixing. *Optics Letters*, 22, 1722–1724.

Leith, E. and Upatnieks, J. 1962. Wavefront reconstruction and communications theory. *Journal of the Optical Society of America*, 52, 1123–1130.

Leyva, V., Rakuljic, G. A., and O'Conner, B. 1994. Narrow bandwidth volume holographic optical filter operating at the Kr transition at 1547.82 nm. *Applied Physics Letters*, 65, 1079–1081.

Leyva, V., Engin, D., Tong, X. -L., Zhang, M., Yariv, A., and Agranat, A. 1995. Fixing of photorefractive volume holograms in $K_{1-y}Li_yTa_{1-x}O_3$. *Optics Letters*, 20, 1319–1321.

Lipp, J. and Reynolds, J. 1971. A high capacity holographic storage system, In *Applications of Holography*, E. S. Barrakette, W. E. Kock, T. Ose, J. Tsujiuchi, and G. W. Stroke, eds., pp. 377–388. Plenum Press, New York.

Ma, J., Chang, T., Choi, S., and Hong, J. 2000. Ruggedized digital holographic data storage with fast access. *Optical and Quantum Electronics*, 32, 383–392.

Mahgerefteh, D. and Feinberg, J. 1990. Explanation of the apparent sublinear photoconductivity of photorefractive barium titanate. *Physical Review Letters*, 64, 2195–2198.

Marcus, B. 2000. Modulation codes for holographic recording, In *Holographic Data Storage*, H. J. Coufal, D. Psaltis, G. T. Sincerbox, eds., *Vol. 76*, Springer Series in Optical Sciences, pp. 283–291, Springer, Berlin.

Markov, V. B., Millerd, J., Trolinger, J., and Norrie, M. 1999. Multilayer volume holographic optical memory. *Optics Letters*, 24, 265–267.

Micheron, F. and Bismuth, G. 1972. Electrical control of fixation and erasure of holographic patterns in ferroelectric materials. *Applied Physics Letters*, 20, 79–81.

Mikaelian, A.L. 1976. Holographic bulk memories using lithium niobate crystals for data recording, In *Optical Information Processing*, E. S. Barrekette, Yu. E. Nesterikhin, G. W. Stroke, and W. E. Kock, eds., *Vol. 2*, Springer Series in Optical Sciences, pp. 217–233, Plenum Press, New York.

Mikaelian, A. L., Bobrinev, V. L., Naumova, S. M., and Sokolova, L. Z. 1970. Design principles of holographic memory devices. *IEEE Journal of Quantum Electronics*, QE-6, 4, 193–198.

Moerner, W. E., Silence, S. M., Hache, F., and Bjorklund, G. C. 1994. Orientationally enhanced photorefractive effect in polymers. *Journal of the Optical Society of America B*, 11, 320–330.

Mok, F. H. 1993. Angle-multiplexed storage of 5000 holograms in lithium niobate. *Optics Letters*, 18, 915–917.

Mok, F. H., Tackitt, M. C., and Stoll, H. M. 1991. Storage of 500 high-resolution holograms in a $LiNbO_3$ crystal. *Optics Letters*, 16, 605–607.

Mok, F. H., Burr, G. W., and Psaltis, D. 1996. System metric for holographic memory systems. *Optics Letters*, 21, 896–898.

Montemezzani, G. and Günter, P. 1990. Thermal hologram fixing in pure and doped $KNbO_3$ crystals. *Journal of the Optical Society of America B*, 7, 2323–2328.

Orlov, S. S. 1999. Overview of holographic recording materials for major system architectures in holographic data storage applications. Paper presented at the National Storage Industry Consortium International Workshop on Holographic Data Storage, Nice, France.

Orlov, S. S. and Hesselink, L. 2003. Holographic shift multiplexing in thin volumetric media. *Journal of the Optical Society of America B*, 20, 1912–1921.

Orlov, S. S., Psaltis, D., and Neurgaonkar, R. R. 1993. Dynamic electronic compensation of fixed gratings in photorefractive media. *Applied Physics Letters*, 63, 2466–2469.

Orlov, S. S., Bjornson, E., Phillips, W., Takashima, Y., Lee, X., Hesselink, L., Okas, R., Snyder, R. 2000. High transfer rate (1 Gbit/s) high capacity holographic disk digital data storage system. In *Conference on Lasers and Electro-optics (CLEO 2000)*, Vol. 39 of OSA Trends in Optics and Photonics Series, Optical Society of America, Washington, DC, 2000. Paper CTuC4 (invited): 190–191.

Orlov, S. S., Snyder R., Bashaw M. C. 2000. Optical relay for pixel-based holographic storage and retrieval. U.S. patent #6,108,110.

Orlov, S. S., Phillips, W., Bjornson, E., Hesselink, L., Okas, R. 2002. High data rate (10 Gbit/s) demonstration in holographic disk digital data storage system. *Conference on lasers and electro-optics (CLEO 2002)*, Vol. 39 of OSA Trends in Optics and Photonics Series, Optical Society of America, Washington, DC, 2002. Paper CMO2: 70–71.

Orlov, S. S., Phillips, W., Bjornson, E., Takashima, Y., Sundaram, P., Hesselink, L., Okas, R., Snyder, R., and Kwan, D. 2004. High-transfer-rate high-capacity holographic disk data-storage system. *Applied Optics*, 43, 4902–4914.

Oster, G. and Yang, N. -L. 1968. Photopolymerization of vinyl monomers. *Chemical Reviews*, 68, 125–151.

Phillips, W., Staebler, D. L., and Amodei, J. J. 1972. Optical and holographic storage properties of transition metal doped lithium niobate. *RCA Review*, 33, 94–109.

Phillips, W., Orlov, S. S., Bjornson, E., Hesselink, L., and Okas, R. 2000. Video demonstration of high data rate holographic disk data storage system. Presented at the 2000 Optical Society of America Annual Meeting, Providence, RI, 22–25 October 2000.

Psaltis, D., Brady, D., Gu, X. G., and Lin, S. 1990. Holography in artificial neural networks. *Nature*, 343, 325–330.

Psaltis, D., Levene, M., Pu, A., Barbastathis, G., and Curtis, K. 1995. Holographic storage using shift multiplexing. *Optics Letters*, 20, 782–784.

Rakuljic, G. A., Leyva, V., and Yariv, A. 1992. Optical data storage by using orthogonal wavelength-multiplexed volume holograms. *Optics Letters*, 17, 1471–1473.

Redfield, S. 2000. Tamarack Optical Head Holographic Storage, In *Holographic Data Storage*, H. J. Coufal, D. Psaltis, G. T. Sincerbox, eds., *Vol. 76*, Springer Series in Optical Sciences, pp. 343–357, Springer, Berlin.

Saleh, B. E. and Teich, M. C. 1991. *Fundamentals of Photonics*. Wiley, New York.

Sanford, J. L., Greier, P. F., Yang, K. H., Lu, M., Olyha, R. S., Narayan, C., Hoffnagle, J. A., Alt, P. M., and Melcher, R. L. 1998. A one-megapixel reflective spatial light modulator system for holographic storage. *IBM Journal of Research and Development*, 42, 411–426.

Schnoes, M. G., Dhar, L., Schilling, M. L., Patel, S. S., and Wiltzius, P. 1999. Photopolymer-filled nanoporous glass as a dimensionally stable holographic recording medium. *Optics Letters*, 24, 658–660.

Shelby, R. M. 2000. Media requirements for digital holographic data storage, In *Holographic Data Storage*, H. J. Coufal, D. Psaltis, G. T. Sincerbox, eds., *Vol. 76*, Springer Series in Optical Sciences, pp. 101–111, Springer, Berlin.

Sincerbox, G. T. 2000. History and Physical Principles, In *Holographic Data Storage*, H. J. Coufal, D. Psaltis, G. T. Sincerbox, eds., *Vol. 76*, Springer Series in Optical Sciences, pp. 3–20, Springer, Berlin.

Snyder, R., and Daiber, A. 2000. Method for holographic storage and retrieval. U.S. Patent #6, 064, 586.

Staebler, D. L. and Phillips, W. 1974. Hologram storage in photochromic $LiNbO_3$. *Applied Physics Letters*, 78, 268–270.

Steckman, G. J., Solomatin, I., Zhou, G., and Psaltis, D. 1998. Characterization of phenanthrenequinone-doped poly(methyl methacrylate) for holographic memory. *Optics Letters*, 23, 1310–1312.

Steckman, G. J., Pu, A., and Psaltis, D. 2001. Storage density of shift-multiplexed holographic memory. *Applied Optics*, 40, 3387–3394.

Steward, W. C., Mezrich, R. S., Cosentino, L. S., Nagle, E. M., Wendt, F. S., and Lohman, R. D. 1973. An experimental read-write holographic memory. *RCA Review*, 34, 3–44.

Tsunoda, Y., Tatsuno, K., Kataoka, K., and Takeda, Y. 1976. Holographic video disk: an alternative approach to optical video disks. *Applied Optics*, 15, 1398–1403.

Vadde, V. and Vijaya Kumar, B. V. K. 1999. Channel modeling & estimation for intrapage equalization in pixel-matched volume holographic data storage. *Applied Optics*, 38, 4374–4386.

Valley, G. C. 1986. Simultaneous electron/hole transport in photorefractive materials. *Journal of Applied Physics*, 59, 3363–3366.

van der Lugt, A. 1973. Design relationships for holographic memories. *Applied Optics*, 12, 1675–1685.

van Heerden, P. 1963. Theory of optical information storage in solids. *Applied Optics*, 2, 393–400.

Veniaminov, A. V., Goncharov, V. F., and Popov, A. P. 1991. Hologram amplification by diffusion destruction of out-of-phase periodic structures. *Optika i Spektroskopiya*, 70, 864–869.

von der Linde, D., Glass, A. A., and Rodgers, K. F. 1974. Multiphoton photorefractive processes for optical storage in $LiNbO_3$. *Applied Physics Letters*, 25, 155–157.

Vormann, H., Weber, G., Kapphan, S., and Wöhlecke, M. 1981. Hydrogen as origin of thermal fixing in $LiNbO_3$:Fe. *Solid State Communications*, 57, 543–545.

Waldman, D. A., Li, H. -Y. S., and Horner, M. G. 1997. Volume shrinkage in slant fringe gratings of a cationic ring-opening volume hologram material. *Journal of Imaging Science and Technology*, 41, 497–514.

Waldman, D. A., Butler, C., Raguin, D. 2003. Advances in Aprilis CROP photopolymer material for holographic data storage at greater than 100 bits/sq micron. In *Organic Holographic Materials and Applications*, Proceedings of SPIE 5216, K. Meerholz. ed.

Yariv, A., Orlov, S. S., and Rakuljic, G. A. 1996. Holographic storage dynamics in lithium niobate: theory and experiment. *Journal of the Optical Society of America B*, 13, 2513–2523.

Zhao, G. and Mouroulis, P. 1994. Diffusion model of hologram formation in dry photopolymer materials. *Journal of Modern Optics*, 41, 1929–1939.

Zhou, G., Qiao, Y., Mok, F., and Psaltis, D. 1996. A holographic memory product for fingerprint identification. *Optics and Photonics News*, 7, 43.

Zhou, G., Psaltis, D., Mok, F., and Pu, A. 1999. Method and system to align holographic images. U.S. Patent # 5,982,513.

17 Binary Optics

Michael W. Farn

17.1 Introduction

17.1.1 A Working Definition

What is binary optics? There is no strict definition because, as a result of the historical development of the field, binary optics means different things to different people. The first element, which was referred to by the term "binary optics," was formally presented in 1980 (Veldkamp 1980). In this laser radar application, the binary optic device shown in Figure 17.1 split a single laser beam into multiple beams, each of which was used as a local oscillator for photomixing. The design of the element was based on a combination of communications theory and scalar diffraction theory, the fabrication was achieved using computer-generated data and VLSI fabrication techniques, and the resulting element was truly a binary surface-relief structure. Since this original element, however, the field has expanded in many directions. Design methods currently range from classical lens design to the numerical solution of Maxwell's equations, application areas run the gamut from the entertainment industry to medical products to military missions, and fabrication technologies can be as diverse as diamond turning and ion milling. In addition, modern-day binary optic devices are usually multilevel or continuous surface reliefs, rather than strictly binary, as in the

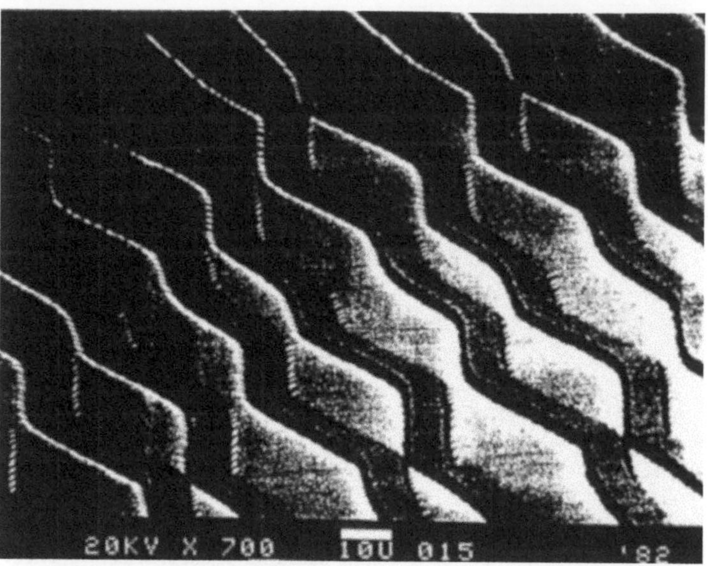

FIGURE 17.1 The original binary optics element. (From Veldkamp, W. B. 1980. *Proc. Soc. Photo-Opt. Instrum. Eng.*, 255, 136.

original application. The "binary" in the name, however, is still relevant because a binary coding technique is used to extend the original fabrication method to create the current multilevel profiles.

As a result of this diversification, it is difficult to come up with an accurate, concise definition for the field. However, many binary optic elements are characterized by the following:

- VLSI-based fabrication—Binary optics are usually manufactured using semiconductor fabrication technology in particular photolithography and micromachining, and the entire fabrication goal is to micromachine a desired surface profile to optical tolerances. Historically, the fabrication techniques are what have distinguished binary optics from other technologies.
- Surface relief—The optical properties of binary optics are solely the result of the surface element. Volume effects, such as those in thick holograms, graded-index optics or photorefractives, are nonexistent. Because the surface profile contains the entire functionality of the device and the profiles are typically thin (on the order of several microns deep), binary optics maybe inexpensively replicated by method, such as embossing.
- Free-space, passive optics—Binary optic components are passive optical elements to be used in free-space applications. Passivity is a result of the surface-relief characteristic (i.e., no transistors, electro-optic effects, etc.), while the exclusion of integrated optics is more for historical reasons than technical ones.

Taken together, these characteristics provide a working definition of binary optics. The definition is "working" in the sense that, even though it is not entirely accurate today and will certainly change tomorrow, it still embodies the crux of the field.

17.1.2 Application Areas

Why or when might one use binary optics? In the majority of successful applications, binary optic elements are used because they offer one of the following two advantages:

- Novel functionality—Most binary optic devices rely on diffraction for the manipulation of light, whereas conventional optical elements rely on refraction or reflection. The fundamental difference between diffraction, refraction, and reflection allows binary optic components to perform functions which could not be achieved otherwise, one example being lenses with simultaneous multiple focal lengths. In addition, the unique fabrication process also allows the manufacture of components which could not be made by other technologies. For example, the ability to make arrays of microoptics, each of which can be of a different shape and perform a different function, is unique to binary optics.

- Increased performance—For existing systems, the use of binary optics may result in increased performance. Typically, this higher performance is a wider field of view, higher numerical aperture, lower mass, lower volume, elimination of exotic materials, or lower cost.

Table 17.1 lists some current applications for binary optics. There are two points of interest in Table 17.1. First, the applications represent a diverse spectrum, ranging from commercial to military, from simple to complex, and from commercially available to still in development. Second, in the majority of applications, binary optic elements are not the end-product. Rather, the binary optic component is used to increase the performance of a larger system. In general, it is true that binary optics is an enabling technology.

17.1.3 Chapter Overview

As shown in Figure 17.2, there are four basic tasks in a typical binary optic project cycle. The first is the optical design of the element, the end result being an optical prescription. The second is the translation of this optical prescription into a fabrication prescription. The third task is the actual fabrication of the element, and the final task is testing of the completed element. This chapter will focus on the two dominant tasks—optical design and fabrication, with more emphasis placed on optical design.

17.1.3.1 Optical Design

Optical design can be further categorized according to the optical theory used to model the element: geometrical optics, scalar diffraction theory, or vector diffraction theory. Each of these methods will be discussed in detail in Section 17.2 through Section 17.4, respectively. The following is a comparison of the three theories.

One way to characterize the three optical design methods is to consider how the binary optic element is modeled and which optical effects are neglected. In the geometrical optics approach, the binary optic model is based on the grating, and rays are traced through the element using the grating equation. This model is the simplest but neglects diffraction effects except at the exit pupil, reflections at material interfaces, polarization effects, and the effect of the finite thickness of the element. In the scalar diffraction approach, binary optic devices are modeled as infinitely thin phase plates and wavefronts are propagated from one surface to another by the scalar diffraction integrals. In this model, most of the previously

TABLE 17.1 Examples of Binary Optics Applications

Application Area	Example Application
Medical	Bifocal intraocular lens
Consumer	Optical heads for compact disks
	Increased sensitivity of CCD cameras
	Stereovision goggles
Industrial	Beam shaping for lasers
	Structured light for machine vision
Optics	Null testing of aspherics
	Hartman-Shack wavefront sensors
	Grating beam splitters
	Microoptics for semiconductor lasers
Military	Aberration correction of wideband imaging systems
	Hardening of detector arrays

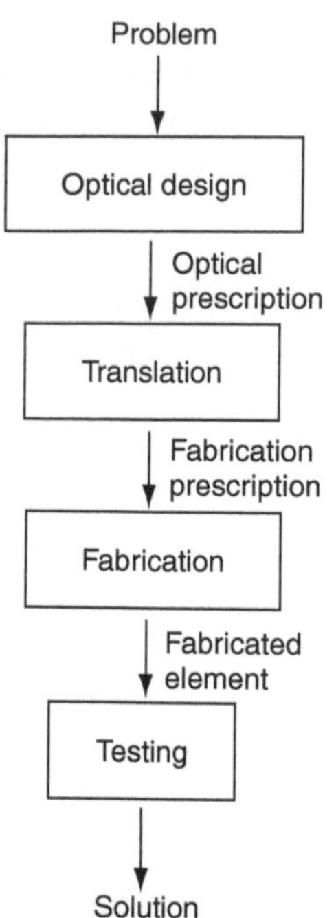

FIGURE 17.2 A typical binary optics project cycle.

neglected diffraction effects are accounted for, but secondary reflections, polarization effects, and the thickness of the element are still neglected. Finally, the vector approach is the most complete, as it is a numerical solution to Maxwell's equations, and all effects predicted by Maxwell's equations, including those mentioned above, are included.

The underlying assumptions determine which applications can be adequately modeled by a given theory. Typical geometrical optic applications include the design of almost all imaging systems, which run the gamut from purely diffractive systems to diffractive modifications of conventional wideband systems, laser systems, and microoptics which are large enough that diffraction from them can be neglected. In short, if an application would normally be designed using geometrical optics, then ray tracing will probably still be adequate to model a version which includes binary optics. The major difference between geometrical optics and scalar theory is that the scalar diffraction integrals include diffraction effects not easily handled by geometrical optics. Therefore, application which rely specifically on these effects must be modeled by scalar theory. Examples include the design of laser cavities, microoptics which are small enough to introduce their own diffraction effect, diffusers, beam shaping optics for lasers, and grating beam splitters (i.e., gratings which split one incoming beam into many outgoing diffraction orders). Finally, vector theory is required when the assumptions implied by scalar theory are no longer valid. In many applications in this regime, the binary optic element is a sort of artificial crystal and the device operates by using the resonances or lack of resonances in the crystal. Examples are antireflection skins, artificially birefringent materials (i.e., form birefringence), artificial index materials, and wavelength filters.

Because each of these theories is quite different from the others, it is also instructive to examine the available tools and users in each case. As one might expect, the geometrical optic model is the most mature, and much of the commercially available lens design software can ray trace diffractive optical elements. However, although codes can accurately predict the direction of different diffraction orders, the calculation of diffraction efficiencies is often inadequate and must be supplemented by other means. Because many of the geometrical optic applications for diffractive optics can be considered classical lens design problems, a lens design background is most appropriate for this type of application. In the scalar and vector regimes, there is not much, if any, general purpose software. As a result, in the scalar diffraction case, the user will code the well-accepted scalar diffraction integrals in the language and on the machine of his choice. To do this, a background in Fourier optics and numerical methods would be most useful. In the vector diffraction case, the designer will also have to code his own tools, but the case differs from the scalar situation in the following respects. First, the vector case is more numerically intense so that the choice of machines is limited and there is a stiffer penalty for inefficient programming. Second, the numerical implementations of the theories are not as mature or as well accepted as in the scalar case. As a result, the designer may have to do some research in addition to any programming. For this type of work, a background in differential equations and numerical methods is useful.

Finally, the three theories are sometimes correlated with the size of the features on the binary optic elements. In cases where the feature is much larger than the wavelength, geometrical optics is used. If the features are on the order of a wavelength or smaller, the vector theory is used, and, in the intermediate case, the scalar theory is appropriate. Although this generalization is not entirely accurate, it does give one a sense of the differences among the approaches.

17.1.3.2 Fabrication

Fabrication can be categorized into the traditional binary optic approach, which uses multiple photomasks, and other fabrication processes. These are discussed in Section 17.5 and Section 17.6, respectively.

In the traditional binary optic process, a continuous surface relief is approximated by a staircase profile, and the staircase profile is created by a process utilizing a set of photolithographic masks. The first mask results in a surface profile with two steps, as in the example of Figure 17.1, and each succeeding mask then doubles the number of steps. Thus, the use of N masks results in a staircase surface profile with 2^N steps. In more detail, the process begins with a translation of the optical prescription to the fabrication prescription, which is the required set of lithographic masks. This conversion is theoretically straightforward but can often result in large amounts of data, and so the major issue is not how to do the conversion but how to do it efficiently, in terms of computation time and data manipulation. The masks are usually fabricated by an electron beam, and the mask set is then used to fabricate the binary optic element. The fabrication of the binary optic device can be broken down into two major steps: photolithography and micromachining. In the photolithography step, the mask set is used to pattern a photoresist, which then acts as a protective layer for part of the underlying optical substrate. The unprotected part of the substrate is subsequently removed by the micromachining step, typically by reactive ion etching or ion milling.

The alternative fabrication processes attempt to generate continuous profiles analogous to numerically controlled milling, except that they occur on a microscopic scale. Examples include diamond turning, laser ablation, and the analog exposure of resist by laser scanning.

Finally, once a master is manufactured, copies may be mass produced by processes such as embossing or casting, as described in Section 17.6.2.

17.2 Designs Based on Ray Tracing

In applications for which ray tracing is adequate, binary optic elements are modeled as a generalization of the linear grating. This section first presents the theory underlying this model, including the generalized grating model and the two common methods for ray tracing binary optics: the grating equation and the Sweatt model. Next, tabulates the phase profiles for some common elements are tabulated.

Finally, this section consider the classes of applications which can be designed using geometrical optics: the correction of monochromatic systems, the correction of wideband systems, and miscellaneous applications. In the monochromatic case, the binary optic element can have significant optical power and is often a direct substitute for a refractive counterpart. In the broadband case, the binary optic structure usually has only enough optical power to balance the chromatic aberrations and/or it may also perform some residual correction of the monochromatic aberrations, similar in function to a Schmidt corrector. In the miscellaneous category, the three most interesting applications are the correction of chromatic aberrations in purely diffractive systems, the thermal correction of systems, and the use of binary optics as microoptics. In geometrical optic designs, binary optics typically offers the following advantages:

- Increased design freedom in correcting aberrations, resulting in better system performance
- Reduction in overall volume, mass, number of elements and/or number of materials
- The generation of arbitrary lens shapes and phase profiles, thereby allowing functionality which could not be realized otherwise

The final section under the geometrical optics heading considers the optical performance of binary optics, which can be broken down into wavefront quality, which usually is not an area of concern, and diffraction efficiency, which almost always is an area of major concern.

17.2.1 Fundamentals

17.2.1.1 The Generalized Grating Model

The geometrical optic model of binary optics is based on the premise that, in a sufficiently local area, a binary optic structure behaves like a linear grating. In particular, a binary optic component with phase profile $\phi(x, y)$, where z is taken as the optical axis and the element is assumed to lie substrate, may be locally modelled as a grating of spatial frequency

$$\mathbf{f}(x, y) = \frac{1}{2\pi}\nabla\phi \tag{17.1}$$

where ∇ is the gradient operator. The direction of the vector **f** indicates the orientation of the local grating, and the magnitude determines the period, which is explicitly given by

$$T(x, y) = 2\pi/|\nabla\phi| \tag{17.2}$$

where $|\,|$ is the vector magnitude. Using this concept of a local grating, result which are derived for linear gratings can be generalized to the case of a binary optic element of arbitrary phase profile. Accordingly much of what follows is derived for the case of linear gratings with the intent that it will be generalized in such a manner. For example, efficiency models for the linear grating are presented later. By using the local grating concept, these linear grating estimates can be used to predict the local efficiency of a binary optic element, which, in turn, can then be combined to derive an average efficiency for the entire element.

To define a binary optic component, one must specify the phase function $\phi(x, y)$, the design wavelength λ_0, and the surface on which the component lies (i.e., the substrate shape). If the phase function is expressed in units of radians, then the design wavelength is not required for ray tracing, but it is still required to determine the blaze depth of the grating. For convenience, this section assumes that the phase function is expressed in radians and that the binary optic element lies on a planar substrate, with (x, y) being the coordinates in the plane and z the coordinate of the optical axis. In the case where the substrate is not planar (e.g., a binary optic structure on one surface of a non-plano lens), the following results are still applicable if z is taken as the coordinate normal to the surface and (x, y) are taken as orthogonal coordinates tangent to the surface, that is, the curved surface is locally approximated as a plane.

The phase function can be defined in any manner as long as it is defined at each point on the surface. Commonly used methods are explicit analytical expression for simple elements, decomposition into polynomials in (x, y), decomposition into polynomials in the radial coordinate $r = \sqrt{x^2 + y^2}$ for circularly symmetric cases, decompositions using other basis functions (e.g., Zernicke polynomials), and interpolation of look-up tables. Of these, the polynomial representations are the most commonly used in ray-tracing software. In addition, many ray-tracing codes also include a spherical term to allow for the convenient representation of optically generated diffractive optics. For binary optics, the spherical term is usually set to zero, and the polynomial decomposition alone is used to describe the profile. The binary optic element is then optimized by optimizing the polynomial coefficients.

17.2.1.2 The Grating Equation

A binary optics element with phase $\phi(x, y)$ can be ray traced by using the local grating concept and applying the grating equation. In the general case, the vector form of the grating equation is

$$n_2(\mathbf{n} \times \mathbf{r}_2) = n_1(\mathbf{n} \times \mathbf{r}_1) + \frac{m\lambda}{2\pi}(\mathbf{n} \times \nabla\phi) \tag{17.3}$$

where $\mathbf{r}_1$ is the unit vector describing the direction of the incident ray, n_1 is the index of the medium on the incident side, $\mathbf{r}_2$ and n_2 are the corresponding quantities for the diffracted ray, $\mathbf{n}$ is the direction of the surface normal, m and λ are the diffraction order and wavelength of interest, and $\times$ is the vector cross-product (Welford 1986). If the substrate surface is planar, then the grating equation reduces to

$$n_2 L_2 = n_1 L_1 + \frac{m\lambda}{2\pi}\frac{\partial \phi}{\partial x} \tag{17.4}$$

and

$$n_2 M_2 = n_1 M_1 + \frac{m\lambda}{2\pi}\frac{\partial \phi}{\partial y} \tag{17.5}$$

where the ray directions are defined by the direction cosines (L, M, N) corresponding to the coordinates (x, y, z). In geometrical designs, the element is usually used in the first order $(m=1)$.

17.2.1.3 The Sweatt Model

In the Sweatt model, the binary optic element is replaced by an equivalent refractive element (Sweatt 1979). The Sweatt model is important because it allows results derived for refractive optics to be directly applied to binary optics and because it allows the modeling of binary optics on software which does not support the grating equation. The refractive equivalent of a binary optic element with phase $\phi(x, y)$ is defined by a thickness profile and refractive index of

$$t(x, y) = \frac{\lambda_0}{n_0 - 1}\frac{\phi(x, y)}{2\pi} + t_0 \tag{17.6}$$

and

$$n(m, \lambda) - 1 = \frac{m\lambda}{\lambda_0}(n_0 - 1), \tag{17.7}$$

where t_0 is a constant chosen to make the refractive element as thin as possible, but still having a positive thickness, and n_0 is the index of the material at the design wavelength λ_0. As n_0 approaches infinity, the Sweatt model approaches the grating equation, and, in practice, values of $n_0=10{,}000$ are sufficiently high for accurate results (Farn 1992). Note that the effective index of refraction is a function of both the wavelength and the diffraction order of the ray being traced.

In the special case of a binary optic lens described by Equation 17.10, the more accurate Sweatt lens can be used, as illustrated in Figure 17.3 (Sweatt 1977). In this special case, the first surface collimates the incoming rays and the second surface, then, focuses them to the image point, thus perfectly imaging the object point (x_1, y_1, z_1) to the image point (x_2, y_2, z_2) just as the diffractive element does. The shapes of these surfaces are hyperbolic with curvatures of

$$c_1 = 1/[(1 - n_0/n_1)z_1] \tag{17.8}$$

and

$$c_2 = 1/[(1 - n_0/n_2)z_2] \tag{17.9}$$

and eccentricities of (n_0/n_1) and (n_0/n_2), with the axis of each surface passing through the respective point source. The refractive index of Equation 17.7 is still valid.

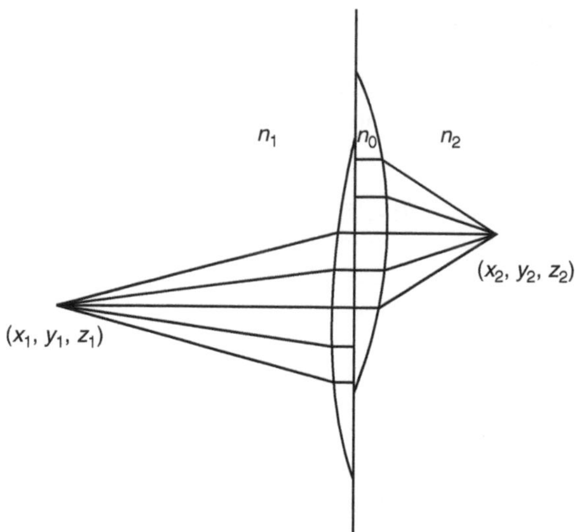

FIGURE 17.3 The Sweatt lens.

17.2.2 Phase Profiles of Common Elements

A lens used to image point (x_1, y_1, z_1) to point (x_2, y_2, z_2) at wavelength λ_0 in the m_0th diffraction order has a phase profile of

$$\phi(x, y) = \frac{k_0}{m_0}[n_1 z_1(\sqrt{(x-x_1)^2/z_1^2 + (y-y_1)^2/z_1^2 + 1} - 1)$$
$$- n_2 z_2(\sqrt{(x-x_2)^2/z_2^2 + (y-y_2)^2/z_2^2 + 1} - 1)] \quad (17.10)$$

where $k_0 = 2\pi/\lambda_0$, n_1 and n_2 are the indices on the object and image side, and the binary optic lens is planar and located at $z = 0$. The paraxial focal length in air of the m_0th order at wavelength λ_0 is given by the lens law,

$$1/f_0 = n_2/z_2 - n_1/z_1. \quad (17.11)$$

The paraxial phase profile of this same lens is given by

$$\phi(r) = \frac{-k_0 r^2}{2m_0 f_0}, \quad (17.12)$$

and the local period, based upon the paraxial phase, is expressed as

$$T(r) = m_0 \lambda_0 f_0 / r. \quad (17.13)$$

The exact local period based on Equation 17.10 usually is not used because the paraxial estimate is accurate enough for most purposes and is much simpler to evaluate. If the lens is used in order m at wavelength λ, then the paraxial focal length in air will be shifted to

$$f = \frac{m_0 \lambda_0}{m\lambda} f_0. \quad (17.14)$$

If the m_0th order of a grating deflects a plane wave of wavelength λ_0 and with direction cosines (L_1, M_1, N_1) to the direction (L_2, M_2, N_2), then, the grating is described by

$$\phi(x,y) = \frac{k_0}{m_0}[x(n_2L_2 - n_1L_1) + y(n_2M_2 - n_1M_1)], \tag{17.15}$$

and the local period is constant at

$$T = m_0\lambda_0/\sqrt{(n_2L_2 - n_1L_1)^2 + (n_2M_2 - n_1M_1)^2}. \tag{17.16}$$

If the grating is used in order m at wavelength λ, then the diffracted direction cosines will be

$$L = \left[n_1L_1 + \frac{m\lambda}{m_0\lambda_0}(n_2L_2 - n_1L_1)\right]/n_2 \tag{17.17}$$

and

$$M = \left[n_1M_1 + \frac{m\lambda}{m_0\lambda_0}(n_2M_2 - n_1M_1)\right]/n_2. \tag{17.18}$$

An axicon, the m_0th order of which diffracts a normally incident plane wave of wavelength λ_0 into the radial direction with direction cosine L_2, is described by

$$\phi(r) = \frac{k_0}{m_0}[r(n_2L_2)] \tag{17.19}$$

with the radial period given by

$$T = m_0\lambda_0/|n_2L_2|. \tag{17.20}$$

The radial direction cosine at order m and wavelegth λ is given by

$$L = \frac{m\lambda}{m_0\lambda_0}L_2. \tag{17.21}$$

A grating is obviously periodic. However, paraxial lenses are also periodic, but in r^2, and axicons are periodic in r. Nonparaxial lenses are also periodic, but with respect to a variable which does not have a simple form. This periodicity allows results derived for gratings to be applied to lenses and axicons.

17.2.3 Design of Monochromatic Systems

The designs presented in this section will generally not be useful for wideband systems due to the inherent chromatic sensitivity of diffractive optics. One significant exception is systems where the wavelengths of interest have a harmonic relationship (e.g., a system containing light and its frequency-doubled harmonic). This special case takes advantage of the fact that the ray trace of the first order of λ_0 is identical to that of the mth order of λ_0/m and, in general, all diffractive ray traces for which $m\lambda$ is constant are identical.

The following is based on Buralli and Morris (1991) and Buralli and Morris (1989). Consider the third-order design of a diffractive singlet which is to be used monochromatically at wavelength λ_0. Let the rotationally symmetric system consist of a diffractive singlet on a substrate of curvature c and a stop located at l, as shown in Figure 17.4. Let the object and image be located in air at l_1 and l_2 with u_1 and u_2

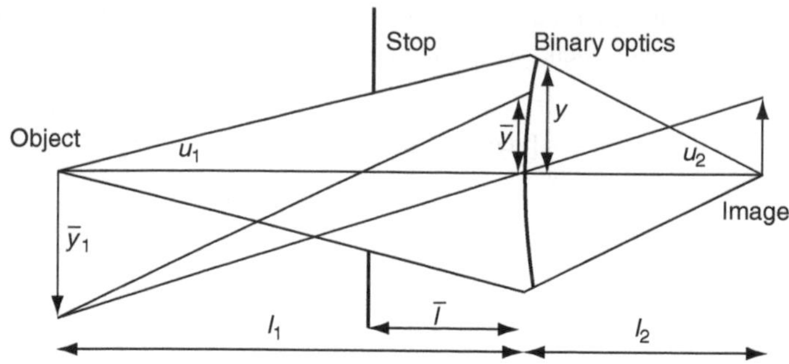

FIGURE 17.4 Paraxial quantities for the diffractive singlet problem.

as the angles of the paraxial marginal rays and y as the paraxial marginal ray height. In addition, let the object have height $\bar{y}_1$ and the paraxial chief ray height $\bar{y}$. Given these constraints, the binary optic singlet will have the form

$$\phi(r) = \frac{-k_0 r^2}{2m_0 f_0} + 2\pi G r^4 \tag{17.22}$$

where f_0 is the focal length defined in Equation 17.11, m_0 and k_0 are the diffraction order and wave number of interest and G is the fourth-order phase coefficient. Higher orders are not of interest in a third-order analysis, and odd orders would result in a cusp at the origin. Now introduce the following intermediate quantities: a bending parameter $B = 2cf_0$, a conjugate parameter $T = (u_1 + u_2)/(u_1 - u_2)$, a stop shift parameter $\epsilon = \bar{y}/y$, and the Lagrange invariant $H = u_1\bar{y}_1$.

As a function of the normalized object coordinate h and normalized pupil coordinates ρ and θ, the aberrations of this system are then given by

$$W(h,\rho,\theta) = \frac{1}{8}S_1\rho^4 + \frac{1}{2}S_2 h\rho^3\cos\theta + \frac{1}{2}S_3 h^2\rho^2\cos^2\theta + \frac{1}{4}(S_3 + S_4)h^2\rho^2 + \frac{1}{2}S_5 h^3\rho\cos\theta, \tag{17.23}$$

where

$$S_1 = \frac{y^4}{4f_0^3}(1 + B^2 + 4BT + 3T^2) - 8\lambda_0 m_0 G y^4, \tag{17.24}$$

$$S_2 = \frac{-y^2 H}{2f_0^2}(B + 2T) + \frac{\epsilon y^4}{4f_0^3}(1 + B^2 + 4BT + 3T^2) - 8\epsilon\lambda_0 m_0 G y^4, \tag{17.25}$$

$$S_3 = \frac{H^2}{f_0} - \frac{2\epsilon y^2 H}{2f_0^2}(B + 2T) + \frac{\epsilon^2 y^4}{4f_0^3}(1 + B^2 + 4BT + 3T^2) - 8\epsilon^2\lambda_0 m_0 G y^4, \tag{17.26}$$

$$S_4 = 0, \tag{17.27}$$

and

$$S_5 = \frac{3\epsilon H^2}{f_0} - \frac{3\epsilon^2 y^2 H}{2f_0^2}(B + 2T) + \frac{\epsilon^3 y^4}{4f_0^3}(1 + B^2 + 4BT + 3T^2) - 8\epsilon^3\lambda_0 m_0 G y^4. \tag{17.28}$$

The quantities S_1 to S_5 represent spherical aberration, coma, astigmatism, Petzval field curvature and distortion. The field curvature is identically zero and the remaining four aberrations may be controlled by the three degrees of freedom; the stop position, the substrate curvature, and the fourth-order phase coefficient. By choosing the phase coefficient

$$G = (1 - T^2)/(32m_0\lambda_0 f_0^3) \tag{17.29}$$

and the stop position so that

$$\epsilon = 2Hf_0/[y^2(B + 2T)], \tag{17.30}$$

the coma and astigmatism terms can be reduced to zero. Designs based on this concept are discussed in further detail in Buralli and Morris (1991) and Buralli and Morris (1989), whereas Buralli and Morris (1992a) extend this approach to the design of two- and three-element telescopes.

Other problems of similar complexity are considered in Welford (1986), including the aberrations introduced when a perfectly corrected system is used at a different wavelength or at different conjugates and a discussion of special imaging conditions (e.g., aplanatic imaging).

Another class of monochromatic applications is the use of binary optics with lasers. In these applications, a binary optic lens or a diffractive/refractive combination is used as a direct replacement of a refractive asphere. Because the diffractive and refractive optical designs perform similar optical functions, the choice of element depends on other factors such as cost, durability, or ease of manufacture (Gruhlke 1992).

17.2.4 Design of Wideband Systems

In wideband systems, binary optic elements are used to correct chromatic or residual monochromatic aberrations. Because their strong wavelength dependence precludes their use with any significant optical power, they are usually combined with conventional elements in so-called hybrid systems and the design of these systems can be described by conventional lens design principles by taking advantage of the Sweatt model.

17.2.4.1 Chromatic Correction

One of the most mature applications of binary optics is the chromatic correction of wideband imaging systems. (Chen 1992; Anderson 1993) describe some recent examples and the following is based on Swanson (1989) and Stone and George (1988), and primarily taken from Optical Society of America (1994), with permission. The chromatic behavior of binary optics can be understood by using the Sweatt model, which states that a binary optic lens behaves like an ultrahigh index refractive lens with an index which varies linearly with wavelength (let $n_0 \rightarrow \infty$ in Equation 17.7). Accordingly, it can be used to correct the primary chromatic aberration of a conventional refractive lens but cannot correct the secondary spectrum. For the design of achromats and apochromats, an effective Abbe number and partial dispersion can be calculated. For example, using the C, d, and F lines, the Abbe number is defined as $V = [n(\lambda_d) - 1]/[n(\lambda_F) - n(\lambda_C)]$. Substituting Equation 17.7 and lettin $n_0 \rightarrow \infty$, then, yields

$$V = \lambda_d/(\lambda_F - \lambda_C) = -3.45. \tag{17.31}$$

Similarly the effective partial dispersion, using the g and F lines, is

$$P = (\lambda_g - \lambda_F)/(\lambda_F - \lambda_C) = .296. \tag{17.32}$$

By using these effective values, the conventional procedure for designing achromats and apochromats can be extended to designs in which one element is a binary optic lens.

Unlike all other materials, a binary optic lens has a negative Abbe number. Thus, an achromatic doublet can be formed by combining a refractive lens and a binary optic lens, both with positive power, thus significantly reducing the required lens curvatures and allowing for larger apertures. Similarly the unusual Abbe number and partial dispersion also allow the design of apochromats with reduced lens curvatures and larger apertures.

17.2.4.2 Correction of Monochromatic Aberrations

The correction of aberrations in monochromatic systems has been discussed previously. In wideband systems, binary optic components are limited to the correction of residual aberrations, as in a Schmidt corrector. As a simple example (Swanson 1989), taken from the Optical Society of America (1994) with permission, consider a refractive system which suffers from third-order spherical aberration given by

$$W(\rho) = \frac{1}{8} S_1 \rho^4, \tag{17.33}$$

where the variables are as defined for Equation 17.23. This is equivalent to a phase aberration of $kW(\rho)$, where $k=2\pi/\lambda$, and a binary optic corrector with phase

$$\phi_C(\rho) = -k_0 \frac{1}{8} S_1 \rho^4 \tag{17.34}$$

will completely correct the aberration at wavelength λ_0. However, at other wavelengths, there will still be a residual spherochromatism of

$$kW(\rho) + \phi_C(\rho) = k\frac{1}{8}[(1-\lambda/\lambda_0)S_1]\rho^4 \tag{17.35}$$

because the change in wavelength affects the phase of the original aberration, but it does not affect the phase produced by the binary optic corrector.

17.2.5 Other Applications

Other applications include the chromatic correction of purely diffractive systems, the use of binary optics for thermal correction, and the use of binary optics as microoptics.

It is often pointed out that an entirely diffractive imaging system with positive power cannot be corrected at two wavelengths (Buralli and Rogers 1989). This statement, while true, implicitly assumes that the diffractive elements are lenslike, the system is imaging in the conventional sense, and only a single diffraction order is utilized. If one or more of these assumptions is relaxed, then the original statement no longer holds, as shown by the following designs. In Figure 17.5a, the chromatic behavior of one diffractive element compensates for the other, thus, bringing two wavelengths to the same focus (Farn and Goodman 1991). A degenerate case of this design occurs when the two elements are linear gratings of the same period, but blazed for opposite orders. For example, the first grating may be blazed for the $+1$ order and the second for the -1 order. In this case, all rays which are parallel upon entering the system are also parallel when they exit the system, regardless of wavelength. Figure 17.5b shows a design based on the one-glass achromat (Levi 1980). The intermediate focus of the longer wavelength insures that the two wavelengths, while focusing at the same point, will have magnifications which are opposite in sign. Finally, as mentioned previously, wavelength/diffraction order combinations which have the same $m\lambda$ product will also ray trace the same, as depicted in Figure 17.5c.

The approach to thermal correction is analogous to that of chromatic correction in that a thermal Abbe number for refractive and diffractive elements can be derived and the thermal variations then balanced (Londono 1993). Physically, a change in temperature results in an index change and a dimensional

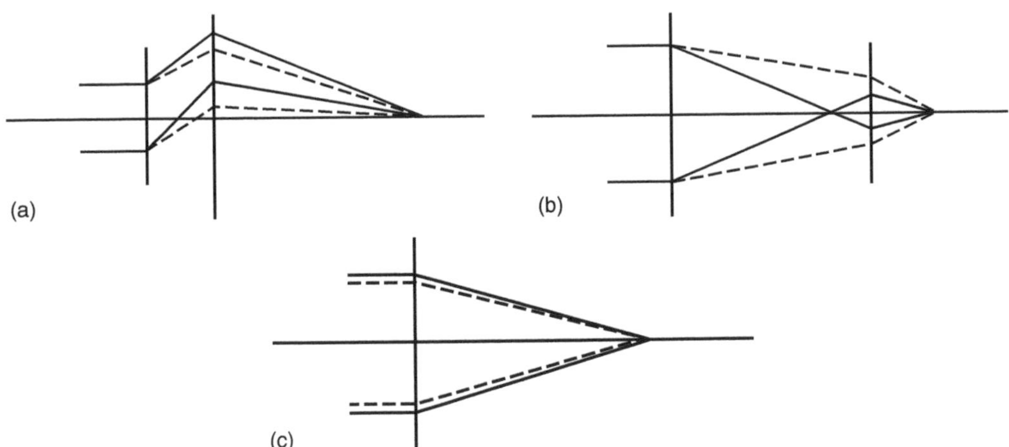

FIGURE 17.5 Unconventional diffractive designs which are corrected at two wavelengths. The shorter wavelength is shown by the dashed line. (a) Use of nonlenses; (b) unconventional imaging; (c) use of multiple orders.

change. For refractive elements, both of these effects will affect a ray trace, whereas, for diffractive elements, only the dimensional change (i.e., change in grating periodicity) changes the ray trace.

As a result of the inherent accuracy of the fabrication process, binary optic technology is well suited for the fabrication of microoptics and microoptic arrays. As with larger binary optics components, the phase profile and aperture of each microoptic are not limited to any specific shapes and the arraying of these microoptics is also not limited to any specific grid patterns. One extreme possibility would be arrays of arbitrarily shaped microoptics on an irregular grid, with each microscopic optimized for a different function. The other extreme would be a regular array of identical microoptics fully covering a large area. As an additional benefit, binary optic technology allows variations across the entire array to be held to optical tolerances, whereas, in most other microoptics technologies, even if each microoptic can meet such tolerances, variations across the array will not.

Although binary optics is usually synonymous with diffractive optics, microoptics is one area where this is not the case. Binary optic structures are constrained by the fact that the overall height of the surface profile is limited, typically, to no more than 3 μm in normal cases, but up to 20 μm in extreme cases. For large optics, this height limitation precludes the fabrication of refractive optics, but this is not so for microoptics. Consider a spherical microlens of curvature c and aperture half-width of r_{max}, where r_{max} is the radius of the circle which would circumscribe the aperture. The height of this spherical cap would be given by

$$h_{max} = c^2_{max}/[1 + \sqrt{1-(cr_{max})^2}] \tag{17.36}$$

which can be less than the height limitation in cases.

Microoptics are often produced as arrays, and Table 17.2 gives the radii of several circles used to characterize arrays. In the table, min{x, y} is the lesser of x and y whereas max{x, y} is the greater of the

TABLE 17.2 Radii of Equivalent Circles for Common Grid Shapes

Tile Shape	Inscribed Circle	Equal Area Circle	Circumscribed Circle
Rectangle: 2a × 2b	$\min\{a, b\}$	$\sqrt{4ab/\pi}$	$\sqrt{a^2+b^2}$
Hexagon: b, 2a, b; c, c	$\min\{c, c(a+b)/\sqrt{b^2+c^2}\}$	$\sqrt{2(2a+b)c/\pi}$	$\max\{(a+b), \sqrt{a^2+c^2}\}$

two quantities. The inscribed circle is the maximum size round lens which could be placed on a given grid pattern, the equal area circle is used to define an equivalent F/# for lenses which fill 100% of the grid pattern, as is typically the case for binary optics devices, and the radius of the circumscribed circle is required to calculate many fabrication parameters.

In their application, microoptics are often combined with sources, commonly laser diodes or detectors. Accordingly, two significant applications of microoptics use them to shape the beam from laser diodes or to concentrate light onto a detector.

17.2.6 Optical Performance

The optical performance of a binary optic element can be described by its wave-front quality and diffraction efficiency. Wave front quality is usually not an issue because the diffractive structure usually introduces an insignificant amount of wave front error. The diffraction efficiency, however, is usually a major issue because the efficiency of even a perfectly fabricated device can be significantly less than 100%. In particular, one must not only be concerned with how much light is diffracted into the order of interest but also with how the remaining light, which is primarily diffracted into other orders, will affect the system performance (Buralli and Morris 1992b). In the following treatment, these issues will be discussed in the context of the linear grating, with the ideas being extended to the general case via the local grating concept introduced previously.

17.2.6.1 Diffraction Efficiency

The diffraction efficiency of a linear grating is a function of the wavelength, polarization and incident angle of the light, the surface profile and period of the grating, and of the indices of the materials involved. The rigorous solution of Maxwell's equations can provide accurate solutions to this problem (Moharam and Gaylord 1982) and should be used when the scalar theory is not adequate, but it is computationally intensive and the results can be nonintuitive. Therefore, it will not be discussed further. Instead, this section focuses on the scalar theory model, which is a simplified solution to this problem, but still quite useful as long as its limitations are kept in mind. In addition, this section assumes that the grating is to be blazed for the $+1$ order, which is taken as diffracting rays in the $+x$ direction, because most geometrical designs are designed for use in this order.

17.2.6.1.1 Scalar Theory Predictions.

The following is based on (Swanson 1991). Consider the problem of Figure 17.6a. The grating has period T in the x direction and height h, light of wavelength λ is incident with direction cosines (L_1, M_1, N_1), and the two media have indices n_1 and n_2. In the scalar theory approach, the grating is modeled as the infinitely thin phase screen of Figure 17.6c. The phase delay is assumed to be linear in x and the only unknown is the optical path difference P between rays which traverse each end of one period. As the period approaches infinity, this path difference can be estimated by considering two rays which traverse a step of height h, as shown in Figure 17.6b. The optical path difference is given by $P = s_1 n_1 - s_2 n_2$. Solving Snel's law and substituting then yields

$$P = h(n_1 N_1 - n_2 N_2), \quad (17.37)$$

where

$$N_2 = \sqrt{1 - (L_2^2 + M_2^2)},$$

$$L_2 = n_1 L_1 / n_2, \quad (17.38)$$

$$M_2 = n_1 M_1 / n_2, \quad (17.39)$$

and where (L_2, M_2, N_2) is the direction of the ray refracted by the step of Figure 17.6b but not the direction of the diffracted ray in Figure 17.6a.

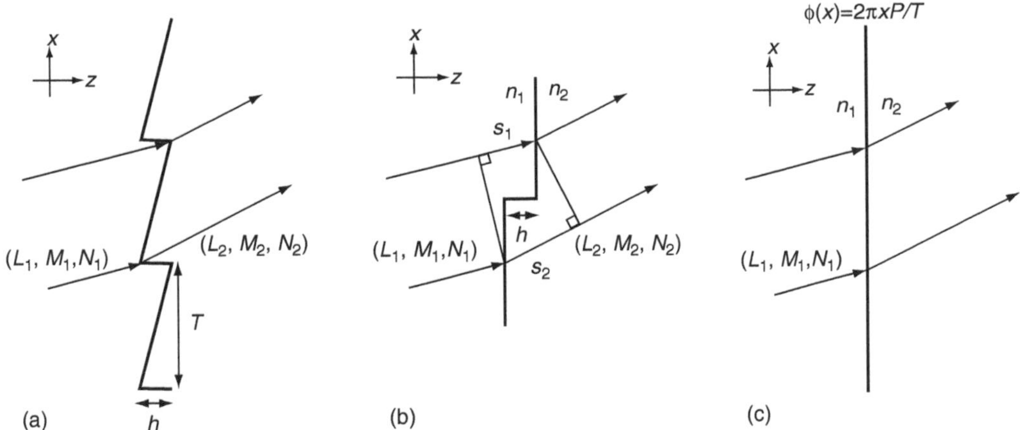

FIGURE 17.6 Diffraction from a grating. (a) The physical problem, *h* positive as shown; (b) optical path difference introduced by a step of height *h*; (c) the resulting scalar phase screen model.

Based on this path difference, the grating can be modeled as phase screen with the phase of one period given by

$$\phi(x) = kPx/T, \tag{17.40}$$

where $k = 2\pi/\lambda$. Based on the approach of Equation 17.62 through Equation 17.64, the Fourier series decomposition of this function predicts a diffraction efficiency of

$$\eta(m) = \text{sinc}^2(P/\lambda - m) \tag{17.41}$$

for the *m*th order, where $\text{sinc}(x) = \sin(\pi x)/(\pi x)$. This equation predicts a first order diffraction efficiency of 100% if $P = \lambda$. For normally incident light at the design wavelength λ_0, combining this condition with Equation 17.37 yields a grating height of

$$h_0 = \lambda_0/(n_1 - n_2). \tag{17.42}$$

Although this blaze height has been derived assuming normally incident light and large grating periods, it is accurate enough to be used as the actual blaze height in most cases and can be used, at least, as an approximate estimate in all cases. The following section presents some ad hoc modifications which will further increase the accuracy of the blaze height estimate.

Similar remarks also apply to the efficiency estimate of Equation 17.41. In particular, in cases where Equation 17.42 is inaccurate, the efficiency equation, as a result of its dependence on Equation 17.37, will also be inaccurate. However, even in these cases, the efficiency equation is still useful because it still shows the dependence of efficiency on incidence angle and wavelength. In general, the effect of incidence angle is negligible except for large angles (e.g., greater than 30°). Of greater concern is the significant efficiency loss which occurs with changing wavelength. As shown in Figure 17.7, light that is not diffracted into the $+1$ order is primarily diffracted into the 0 order at longer wavelengths and into the $+2$ order at shorter wavelengths. For wideband systems, this efficiency loss can be significant and will affect the choice of the design wavelength. For example, if a system has a bandwidth of $\lambda_{min} < \lambda < \lambda_{max}$, then choosing the design wavelength such that $1/\lambda_0$ is the mean of $1/\lambda_{min}$ and $1/\lambda_{max}$ will insure that the efficiencies at the extreme wavelengths are equal. This choice of design wavelength, which results in the efficiency curves of Figure 17.8, maximizes the minimum efficiency but does not maximize the average efficiency.

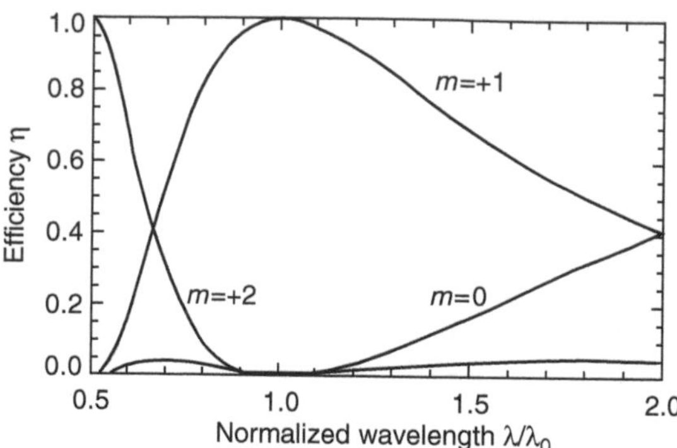

FIGURE 17.7 Scalar diffraction efficiency of a grating blazed for the $m=+1$ order, meterial dispersion neglected. Parasitic orders $m=0$ and $m=+2$ are also shown.

The binary optic fabrication process usually approximates a true blaze by an N-level staircase. In this case, if each step has height h/N, then each step will introduce an optical path difference of P/N, where P is given by Equation 17.37. Note that h is the physical height of the corresponding blazed grating and is not the physical height of the staircase approximation, which is given by $(N-1)h/N$. Following the same approach as previously results in an efficiency (Dammann 1979) of

$$\eta(m) = \text{sinc}^2(m/N)\text{sinc}^2(P/\lambda - m)/\text{sinc}^2[(P/\lambda - m)/N]. \tag{17.43}$$

The quotient in the second and third terms is always less than one, and so, the maximum efficiency is given by the first term and occurs under the same conditions as in the blazed case (i.e., a height of h_0 or step height of h_0/N for normal illumination). For binary optic elements used in the first order, Equation 17.43 yields the often-quoted efficiency estimate of $\text{sinc}^2(1/N)$ for an N-level grating, as given in the second column of Table 17.3.

The overall efficiency response of a staircase grating may be thought of as follows. With respect to wavelength and incidence angle, the first-order efficiency is similar in shape to the blazed case, but reduced by approximately $\text{sinc}^2(1/N)$, with the lost energy being diffracted into orders which are a multiple of N orders away from the first order, as shown in the third and fourth columns 5%, with that 5% going primarily into the -7th, $+9$th and higher orders.

In the reflective case of Figure 17.9, the optical path difference introduced by a step of height h is given by

$$P = h(2n_1N_1), \tag{17.44}$$

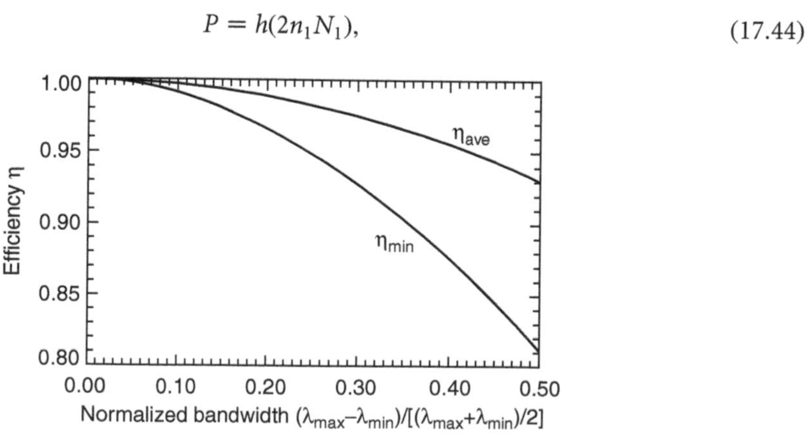

FIGURE 17.8 Average and minimum efficiencies of a blazed grating used in broadband illumination.

TABLE 17.3 Diffraction Efficiencies of an *N*-Level Device at the Design Wavelength

Number of Levels N	$\eta(1)$	$\eta(1-N)$	$\eta(1+N)$
∞	1.000	0.000	0.000
16	0.987	0.004	0.003
12	0.977	0.008	0.006
8	0.950	0.019	0.012
6	0.912	0.036	0.019
4	0.811	0.090	0.032
3	0.684	0.171	0.043
2	0.405	0.405	0.045

where n_1 is the index of the surrounding medium and (L_1, M_1, N_1) are the direction cosines of the incident ray. The corresponding optimal height for normal incidence is given by

$$h_0 = \lambda_0/(2n_1). \tag{17.45}$$

The previous results of this section are still valid for the reflective case except that Equation 17.44 and Equation 17.45 should be used in place of Equation 17.37 and Equation 17.42.

The scalar theory is more accurate for gratings which are thin and have large periods. This concept can be quantified by using the dimensionless parameter

$$Q = 2\pi\lambda h/(\min\{n_1 N_1, n_2 N_2\}T^2), \tag{17.46}$$

where min{x, y} is the lesser of x and y. If $Q \ll 1$, then the scalar theory is applicable (Magnusson and Gaylord 1978). As a rule of thumb, the scalar theory begins to lose validity when the grating period falls below ten wavelengths (i.e., a grating with a period less than $10\lambda_0$ or a lens faster than $F/5$), and lower efficiencies can be expected in these cases (Swanson 1991; Gremaux and Gallagher 1993). In addition to neglecting the influence of grating period and thickness, the scalar theory also neglects polarization effects and Fresnel reflections. Accordingly, if any of these effects are important, then the scalar theory model will be inadequate.

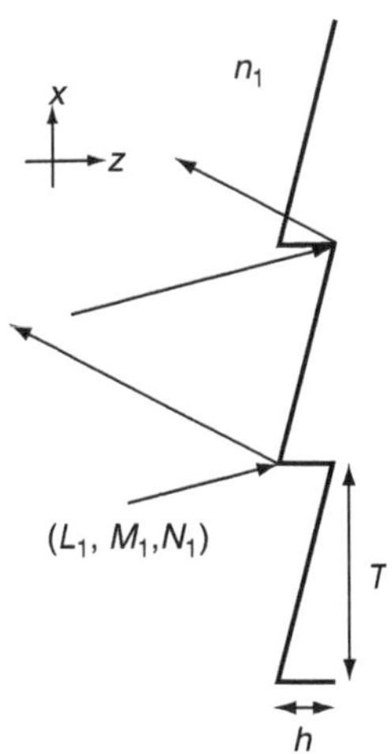

FIGURE 17.9 Diffraction from a reflective grating, h positive as shown.

17.2.6.1.2 Scalar Theory Modifcations

The most significant shortcoming of the scalar theory is that it does not predict the efficiency loss due to small grating periods and finite grating heights. This efficiency loss can be estimated by multiplying the scalar efficiency prediction by the factor (Swanson 1991)

$$F = [1 - h\lambda/(T^2\sqrt{\min\{n_1, n_2\}^2 - (\lambda/T^2)}]^2. \tag{17.47}$$

This multiplier accounts for the degradation in an ad hoc manner and should only be used to predict general behavior. Inmost cases, it is not accurate enough to replace a rigorous solution to Maxwell's equations.

A second shortcoming of the scalar theory is that, for small periods and low index materials, the optimal blaze heights predicted by Equation 17.42 and Equation 17.45 can be off by a significant amount. However, conventional blaze height arguments can be used to modify these estimates in an ad hoc manner (Swanson 1991). For example, consider the grating of Figure 17.6a. The conventional argument states that rays which are diffracted by the grating and those which are refracted off the grating facet should have the same direction cosines. Applying both the grating equation and Shell's law to this situation and solving, then, yields a blaze height of

$$h_0 = \lambda_0/(n_1N_1 - n_2N_2), \tag{17.48}$$

where

$$N_2 = \sqrt{1-(L_2 + M_2^2)}, \tag{17.49}$$

$$L_2 = (n_1L_1 + \lambda_0/T)/n_2, \tag{17.50}$$

$$M_2 = n_1M_1/n_2, \tag{17.51}$$

and (L, M, N) are the direction cosines of the incident and diffracted rays shown in Figure 17.6a. Note that this equation reduces to Equation 17.42 for the case of normal illumination as T approaches infinity. Similarly, the blaze depth for the reflecting case is given by

$$h_0 = \lambda_0/[n_1(N_1 + N_2)], \tag{17.52}$$

where

$$N_2 = \sqrt{1-(L_2^2 + M_2^2)}, \tag{17.53}$$

$$L_2 = L_1 + \lambda_0/(n_1T), \tag{17.54}$$

and

$$M_2 = M_1. \tag{17.55}$$

In addition to the ad hoc modifications given above, there have been some initial efforts to numerically optimize the grating profiles using vector theory, but no simple algorithms have yet been devised.

17.2.6.2 Manufacturing Tolerances

Manufacturing errors can affect both the wave front quality and the diffraction efficiency of a binary optic element. For example, if the periods of a grating are slightly aperiodic, but still ideally blazed, then the aperiodicity will introduce a wavefront error but will not significantly degrade the diffraction efficiency. On the other hand, if the grating is perfectly periodic, but the blaze shape is slightly in error, then the first order will be perfect plane wave, but of lower efficiency. Of these effects, the efficiency issue usually overshadows the wavefront issue. This section presents simple models for these effects, based on the scalar theory.

Errors in the blaze profile can be classified as either height or lateral errors, and both errors will result in lower diffraction efficiency (Emerton 1985; Cox 1991; Ferstl 1992). In the case of height errors, if the actual grating height is in error by δ, as shown in Figure 17.10a, then by Equation 17.41, the efficiency

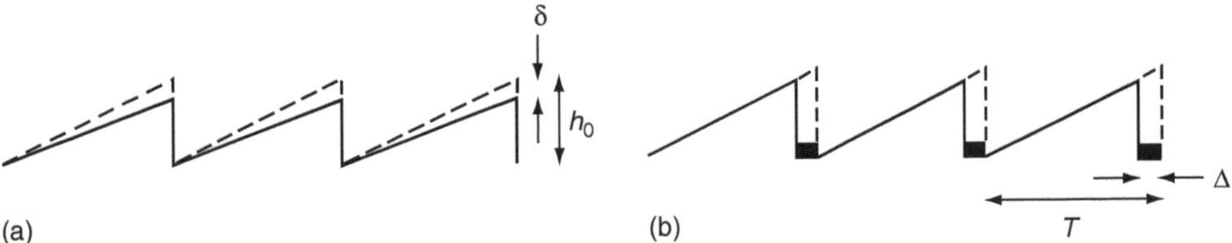

FIGURE 17.10 Models used to estimate the loss of diffraction efficiency due to (a) height errors and (b) lateral errors. The dashed line is the ideal profile.

will be reduced by a multiplicative factor of

$$F = \text{sinc}^2(\delta/h_0), \tag{17.56}$$

where h_0 is the ideal blaze height. In practice, this error is not a major concern because most process can be controlled to $\delta/h_0 < 0.05$ resulting in $F > 0.99$. Lateral errors, however, are of great concern. Lateral errors create portions of the grating which do not contribute to the overall efficiency, and these portions can be modeled as opaque regions, indicated by the black regions in Figure 17.10b. For example, if the grating is diamond-turned, then the finite size of the tool tip will result in incorrect blazing over part of the grating. Similarly, for binary optic components fabricated by multiple mask techniques, misalignments between mask layers will result in areas which deviate substantially from the desired profile, and these areas will not contribute significantly to the first-order efficiency of the grating. If the total length of such areas is Δ, then the efficiency will be reduced by the fill factor squared:

$$F = (1 - \Delta/T)^2. \tag{17.57}$$

Wavefront errors can be calculated by comparing an ideal wavefront with that produced by a grating with fabrication errors. If the error in grating height varies uniformly from $-\delta$ to $+\delta$ as shown in Figure 17.11a, then, the RMS wavefront error in waves is approximately

$$\text{RMS} = \frac{1}{3}(\delta/h_0). \tag{17.58}$$

If the uncertainty in the position of each grating transition is uniformly distributed from $-\Delta$ to $+\Delta$ as shown in Figure 17.11b, then, the RMS wavefront error will be approximately

$$\text{RMS} = \frac{1}{2}(\Delta/T). \tag{17.59}$$

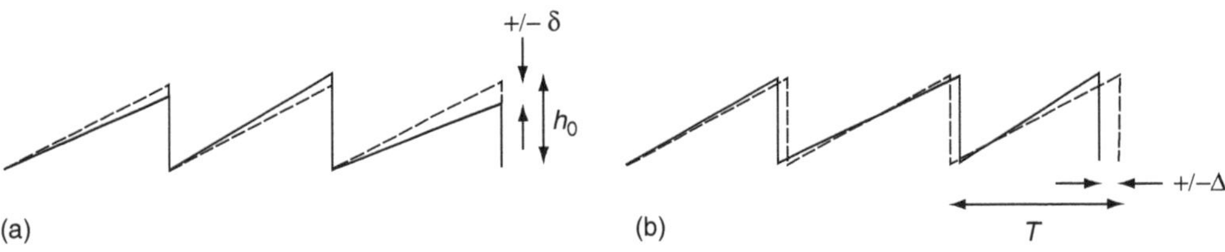

FIGURE 17.11 Models used to estimate wavefront errors due to (a) height errors and (b) lateral errors. The dashed line is the ideal profile.

Choosing worst case numbers of $\delta/h_0 = 0.05$, $T = 4$ μm, and $\Delta = 0.2$ μm yields worst case RMS wavefront errors of 1/60 and 1/40 wave, respectively. Accordingly, wavefront quality is usually insignificantly affected by fabrication errors, but is limited by the quality of the substrate.

17.3 Designs Based on Scalar Diffraction Theory

Designs based on scalar diffraction theory take advantage of the wave nature of light. In particular, the binary optic device adjusts the phase of an incoming wave so that the wave then propagates in some desired manner. For example, the adjusted wave may produce a desired intensity distribution at a point downstream of the binary optic element. The difference between scalar and geometrical designs is that scalar designs rely specifically on diffraction effects which are largely neglected by the geometrical theory.

This section first introduces the foundations for the numerical implementation of the scalar theory, followed by some common applications.

17.3.1 Fundamentals

In the scalar regime, a binary optic component with height profile $h(x, y)$ is modeled as a phase screen introducing a phase delay of

$$\phi(x, y) = k(n_1 - n_2)h(x, y), \tag{17.60}$$

where n_1, and n_2 are the indices of the two media involved. This equation is based on the same assumptions as Equation 17.42. However, there is also a significant difference between the phase profile, as used in the scalar theory, and that used in the geometrical optics case. In particular, note that the phase profile given above is a function of the wavelength, whereas, in the geometrical optic case, the phase profile is independent of wavelength (e.g., see the example of Section 2.4.2). The reason is that in the geometrical optic case, the phase profile is that of only the first-order diffracted wave, whereas, in the scalar case, the phase profile is that resulting from the combination of all orders. In general, as the wavelength changes, the phase profile of each order will remain the same, but the strengths of the diffraction orders will change so that the overall phase profile will also change. In addition, in the geometrical optic case, the phase profile will generally be continuous, spanning many multiples of 2π, whereas the phase profile in the scalar case spans only approximately 2π radians.

In other words, if the binary optics introduces a phase $\phi_0(x, y)$ at the design wavelength λ_0 and this phase accounts for all diffraction orders, then, the phase introduced at wavelength λ, including all diffraction orders, will be given by

$$\phi(x, y) = \frac{\lambda_0}{\lambda}[\phi_0(x, y)\text{mod}(2\pi)], \tag{17.61}$$

where x mod y is the remainder of x divided by y. Contrast this with the geometrical optic case in which $\phi(x, y) = \phi_0(x, y)$ for all wavelengths.

The propagation of wavefronts in the scalar case is usually modeled by the Raleigh–Sommerfeld, Fresnel, or angular spectrum approach, whichever is appropriate. Because these theories are described in length elsewhere, this section concentrates on special cases which occur for binary optics and numerical methods for implementing the integrals.

17.3.1.1 Gratings

As mentioned previously, the case of the linear grating is important because other problems can be solved by generalization or analogy. In the scalar theory, a diffraction grating of period T and phase $\phi(x)$ is decomposed into its Fourier coefficients $C(m)$, with

$$C(m) = \frac{1}{T}\int_0^T \exp[j\phi(x)]\exp(-j2\pi mx/T)\mathrm{d}x, \tag{17.62}$$

$$\exp[j\phi(x)] = \sum_{m=-\infty}^{\infty} C(m)\exp(j2\pi mx/T), \tag{17.63}$$

and the relative intensity or efficiency of the *m*th diffracted order of the grating given by

$$\eta(m) = |C(m)|^2. \tag{17.64}$$

For propagation, each order is typically propagated separately, using the angular spectrum approach, and the resulting orders are then recombined using Equation 17.63 to give the field at the plane of interest. In the angular spectrum approach, a plane wave with direction cosines (L, M, N), propagating over a distance z in air, experiences a transfer function of

$$H(L, M, N) = \exp(jkzN). \tag{17.65}$$

If the incident field is not a plane wave, it can be decomposed into plane waves by the fast Fourier transform (FFT). If the FFT is used for this and the field is sampled at increments of Δx, then the resulting N-point FFT will be sampled in increments of

$$\Delta L = \lambda(N\Delta x). \tag{17.66}$$

If the medium is not air, then the previous two equations are still valid, provided that the wavelength in the medium is used (i.e., replace λ by λ/n).

Several common grating shapes are shown in Figure 17.12 and their Fourier coefficients are given below. The binary optics fabrication process results in piecewise flat elements and, in the general N-step case of Figure 17.12a, the Fourier coefficients take the form

$$C(m) = \sum_{n=0}^{N-1} \exp(j\phi_n)\gamma\delta_n\exp(-j2\pi m\Delta_n)\mathrm{sinc}(m\delta_n) \tag{17.67}$$

where

$$\delta_n = (x_{n+1} - x_n)/T, \tag{17.68}$$

$$\Delta_n = (x_{n+1} - x_n)/(2T), \tag{17.69}$$

ϕ_n are the phases of each step, and x_n are the transition points between steps. If each step of the piecewise flat profile has width Δx, as shown in Figure 17.12b, then, Equation 17.67 reduces to

$$C(m) = \exp(-j\pi m/N)\mathrm{sinc}(m/N)\mathrm{FFT}[\exp(j\phi_n)] \tag{17.70}$$

where

$$\mathrm{FFT}[x_n] = \frac{1}{N}\sum_n x_n\exp(-j2\pi mn/N). \tag{17.71}$$

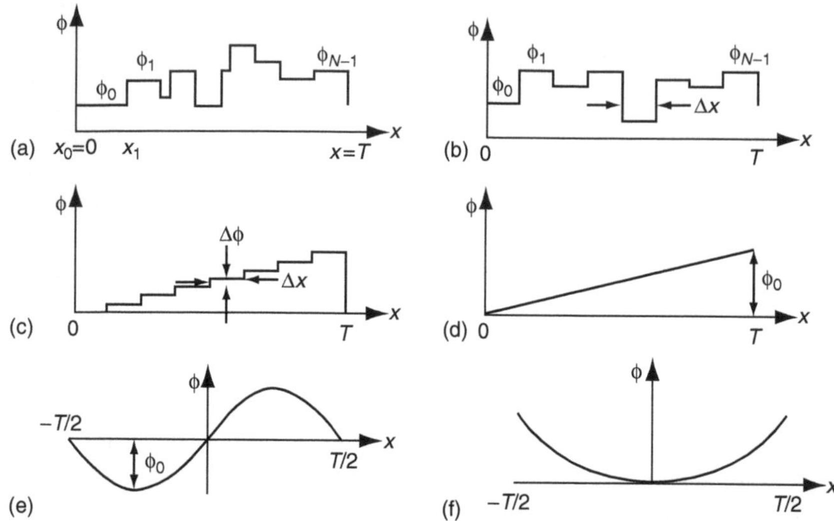

FIGURE 17.12 Common grating profiles: (a) piecewise flat; (b) equal width steps; (c) staircase; (d) linear ramp; (e) paraxial lens array (i.e., parabolic); (f) sinusoidal.

If the phase is also incremented in steps of $\Delta\phi$, as in Figure 17.12c, then, Equation 17.70 further reduces to (Dammann 1979)

$$C(m) = \exp\{j\pi[(N-1)\alpha - m/N]\}\text{sinc}(m/N) \times \text{sinc}(N\alpha)/\text{sinc}\ \alpha, \tag{17.72}$$

where

$$\alpha = \Delta\phi/(2\pi) - m/N. \tag{17.73}$$

The remaining three profiles in Figure 17.12 are common continuous profiles. Figure 17.12d is a linear ramp, with coefficients

$$C(m) = \exp\{j\pi[\phi_0/(2\pi) - m]\}\text{sinc}[\phi_0/(2\pi) - m], \tag{17.74}$$

whereas Figure 17.12e is a sinusoidal phase grating with

$$C(m) = J_m(\phi_0), \tag{17.75}$$

where J_m is the mth order Bessel function of the first kind. Finally, a microlens array is also a grating of parabolic phase profile, as shown in Figure 17.12f. In particular, an array consisting of paraxial lenses of first-order focal length f_0 at wavelength λ_0 will have diffraction orders described by

$$C(m) = \exp[j(\pi/2)(m/\alpha)^2]\text{conj}[\text{Fr}(m/\alpha + \alpha) - \text{Fr}(m/\alpha - \alpha)] \tag{17.76}$$

where

$$\alpha = T/\sqrt{2\lambda_0 f_0}, \tag{17.77}$$

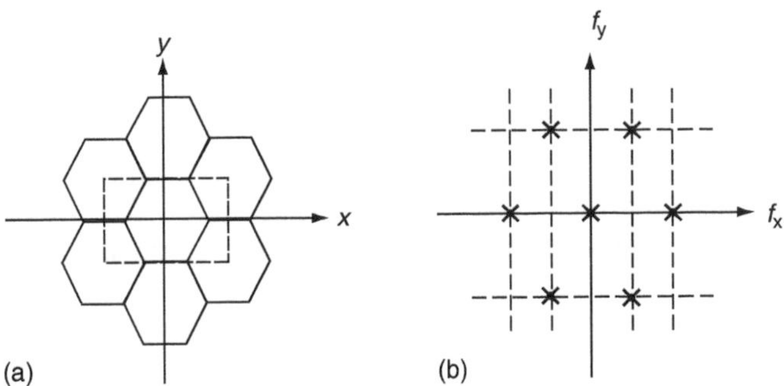

FIGURE 17.13 Hexagonal arrays: (a) the hexagonal array is also a rectangular array, with a basic cell defined by the dashed rectangle; (b) the corresponding diffraction orders.

$$\mathrm{Fr}(x) = \int_0^x \exp(j\pi t^2/2)\mathrm{d}t, \tag{17.78}$$

and conj [ ]denotes the complex conjugate.

The case of hexagonal arrays can be evaluated directly (Roberts 1992) or by taking advantage of the hexagonal pattern's rectangular periodicity, as indicated by the dashed line in Figure 17.13a. The resulting rectangular diffraction pattern is shown in Figure 17.13b by the dashed lines, with the corresponding hexagonal orders marked by *x*'s. The nonhexagonal orders are identically zero due to the symmetry in the original rectangular period.

17.3.1.2 Scalar Diffraction Integrals

A detailed discussion of the numerical methods used to evaluate the scalar diffraction integrals (Barakat 1980) is beyond the scope of this section, which concentrates, instead, on the simplest numerical implementations.

Consider the case in which an incident field $U_1(x, y)$ propagates over a distance z and an FFT implementation of the Fresnel diffraction integral is used to calculate the resulting field $U_2(x, y)$. First, the integrand is sampled in increments of Δx_1 and Δy_1 to yield the series

$$f(m_1, n_1) = \exp\{jk[(m_1\Delta x_1)^2 + (n_1\Delta y_1)^2]/(2z)\}U_1(m_1\Delta x_1, n_1\Delta y_1), \tag{17.79}$$

which consists of $M \times N$ samples, where the number of samples includes any zero padding. The integral can then be approximated by

$$g(m_2, n_2) = \mathrm{FFT}[f(m_1, n_1)], \tag{17.80}$$

where the FFT is the two-dimensional version of Equation 17.71, and m_1, m_2, n_1, and n_2 are index variables. Samples of the desired field are, then, approximated by

$$U_2(m_2\Delta x_2, n_2\Delta y_2) \doteq \frac{\exp(jkz)}{j\lambda z}\exp\{jk[(m_2\Delta x_2)^2 + (n_2\Delta y_2)^2]/(2z)\} \times M\Delta x_1 N\Delta y_1 g(m_2, n_2) \tag{17.81}$$

where

$$\Delta x_2 = \lambda z/(M\Delta x_1) \tag{17.82}$$

and

$$\Delta y_2 = \lambda z/(N\Delta y_1). \tag{17.83}$$

If the medium is not air, then the wavelength in the medium should be used. In an analytical approach, the Fraunhofer approximation is easier to evaluate because the quadratic phase factors are eliminated. However, in a numerical implementation, this advantage is lost, so the Fraunhofer approximation is rarely used.

In the circularly symmetric case, the quasi-fast Hankel transform may be used (Siegman 1977). In this case, the integrand is first represented by the discrete function

$$f(n) = \begin{cases} r_1(n)\exp[jkr_1^2(n)/(2z)]U_1[r_1(n)] & n \in [0, N-1] \\ 0 & n \in [N, 2N-1] \end{cases} \tag{17.84}$$

where

$$r_1(n) = r_{01}\exp(\alpha n) \tag{17.85}$$

which consists of $2N$ samples, the last N of which are zero. The variables r_{01} and α determine the sampling of the incident field, and the continuous form, which is integrated from zero to infinity, will be approximated by a summation over the range $[r_{01}, r_{01}\exp(\alpha N-1)]$. Next, the transform kernel is represented by the $2N$-point sequence

$$h(n) = 2\pi\alpha\frac{r_{01}r_{02}\exp(\alpha n)}{\lambda z}J_0\left[2\pi\frac{r_{01}r_{02}\exp(\alpha n)}{\lambda z}\right], \tag{17.86}$$

where J_0 is the zero-order Bessel function and r_{02} determines the sampling of field U_2. The diffraction integral is approximated by the correlation of these two sequences, which may be implemented by $2N$-point FFT's yielding

$$g(n) = 2N\text{FFT}\{\text{FFT}[f(n)]\cdot\text{IFFT}[h(n)]\} \tag{17.87}$$

where the FFT is as defined in Equation 17.71 and the corresponding inverse IFFT uses the conjugate kernel and drops the $1/N$ factor (actually $1/(2N)$ using the current notation). The resulting field is then given by

$$U_2[r_2(n)] \doteq \frac{\exp(jkz)}{j\lambda z}\exp[jkr_2^2(n)/(2z)] \times \lambda z g(n)/r_2(n) \quad \text{for } n \in [0, N-1] \tag{17.88}$$

where

$$r_2(n) = r_{02}\exp(\alpha n) \tag{17.89}$$

and the last N points in $g(n)$ are discarded.

The use of the FFT in both of these methods increases the speed of the computation. However, it sometimes unnecessarily constrains the sampling of both U_1 and U_2. For example, perhaps U_2 need only be sampled at one point, as is the case for a Strehl calculation, or perhaps the sample points are required on a differently shaped grid. In these cases, other means may be more efficient to evaluate the diffraction integrals (Roose 1993).

17.3.2 Designs Utilizing Multiple Diffraction Orders

In most geometrical designs, the binary optic component is designed to operate in a single order, and the grating profile is chosen to be blazed for that order. In many scalar designs, it is desirable to use more than one order, and the grating profile must be redesigned to achieve this purpose.

17.3.2.1 Grating Beam Splitters

Grating beam splitters are gratings in which the diffraction efficiencies are designed to follow a given weighting. A common example is the 1:*K* fanout problem in which an incident plane wave is divided into *K* diffraction orders of equal intensity. Grating beam splitters can be classified as either continuous phase or binary phase and their performance is evaluated by how closely the actual diffraction efficiencies match the desired weighting and by the total efficiency of the orders of interest. Fanouts of 1:100 with efficiencies over 80% and nonuniformities of less than 10% are typical figures. The continuous designs usually have higher total efficiencies, whereas the binary designs usually are easier to fabricate. Table 17.4 shows upper bounds for the total efficiency in the 1:*K* fanout case (Krackhardt 1992). It should be noted that these bounds are not the tightest possible and that splitting into an odd number of diffraction orders is usually more efficient than splitting into an even number of orders. Table 17.5 gives some common simple designs.

The design of continuous phase gratings is analogous to the phase retrieval problem, and the same methods are used to solve both problems (Wong and Swanson 1993). The resulting phase profiles are "wiggly" in nature, suggesting the superposition of weighted sinusoids. Due to the large number of degrees of freedom, there are no constrains as to what types of efficiency weightings may be attained.

In contrast, the efficiency weightings of a binary or Dammann grating, which is constrained to take on one of two possible phases (typically 0 and π), are constrained to have the following properties. First, the efficiencies of orders m and $-m$ must always b equal. Second, the grating is fully defined by the two phases and the locations of the transitions between the phases. Furthermore, the transition points alone define the relative weightings of all nonzero orders, that is, the ratio $\eta(m)/\eta(n)$, for all m, $n \neq 0$ is determined solely by the transition points and does not change no matter which two phase values are chosen. Third, the two phases determine only zeroth order compared to the nonzero orders. As a result, it is common first to determine the transition points such that the nonzero orders have the correct weightings and then use the choice of phase to balance the zero order. The design of binary gratings is different from that of continuous gratings, and the most common approaches are based on Dammann's methods or search methods (Vasara 1992). In addition, tables of binary designs have been compiled (Killat 1982; Krackhardt 1989).

17.3.2.2 Multifocal Lenses

The concepts used to design gratings with multiple orders can be directly extended to lenses and axicons by taking advantage of the periodicities mentioned previously, and the multiple diffraction orders will appear as multiple foci (King 1985).

TABLE 17.4 Upper Bounds on the Diffraction Efficiency of 1:*K* Grating Beam Splitters

K	Continuous Designs	Binary Designs
2	0.81	0.81
3	0.94	0.94
5	0.96	0.87
7	0.98	0.90
9	0.99	0.88
11	0.98	0.89
13	0.99	0.87
>13	0.97–0.99	0.87–0.88

TABLE 17.5 Grating Multiplexers of Period T, $0<x<T$

Phase Expression	$\eta(-1)$	$\eta(0)$	$\eta(1)$	Remarks
$\phi(x)=\begin{cases}0 & 0<x<T/2\\ \pi & T/2<x<T\end{cases}$	0.41	0.00	0.41	Binary 1:2 splitter
$\phi(x)=\begin{cases}0 & 0<x<T/2\\ 2.01 & T/2<x<T\end{cases}$	0.29	0.29	0.29	Binary 1:3 splitter
$\phi(x)=\pi x/T$	0.05	0.41	0.41	Continuous 1:2 splitter
$\phi(x)=\arctan[2.657\cos(2\pi x/T)]$	0.31	0.31	0.31	Continuous 1:3 splitter

Source: Optical Society of America, *Handbook of Optics*, McGraw-Hill, New York, 1994. With permission.

As an example, taken from the Optical Society of America (1994) with permission, consider the paraxial design of a bifocal lens, as is used in intraocular implants. Half the light should see a lens of focal length f_0, whereas the other half should see no lens. A diffractive lens of focal length f_0, but with the light split evenly between the 0 and +1 orders, will achieve this function. The phase profile of a single focus lens is given by $\phi(r)=-k_0r^2/(2f_0)$. This phase, with the 2π ambiguity removed, is plotted Figure 17.14a as a function of r and in Figure 17.14b as a function of r^2, where the periodicity in r^2 is evident. To split the light between the 0 and +1 orders, the blaze of Figure 17.14b is replaced by the 1:2 continuous splitter of Table 17.5, resulting in Figure 17.14c. This is the final design, and the phase profile is displayed in Figure 17.14d as a function of r.

17.3.3 Beam Shapers and Homogenizers

Another major application area is the reshaping and/or homogenization of laser beams, and these designs can be categorized as either deterministic or statistical, depending on whether or not the characteristics of the incident laser beam are known and stable.

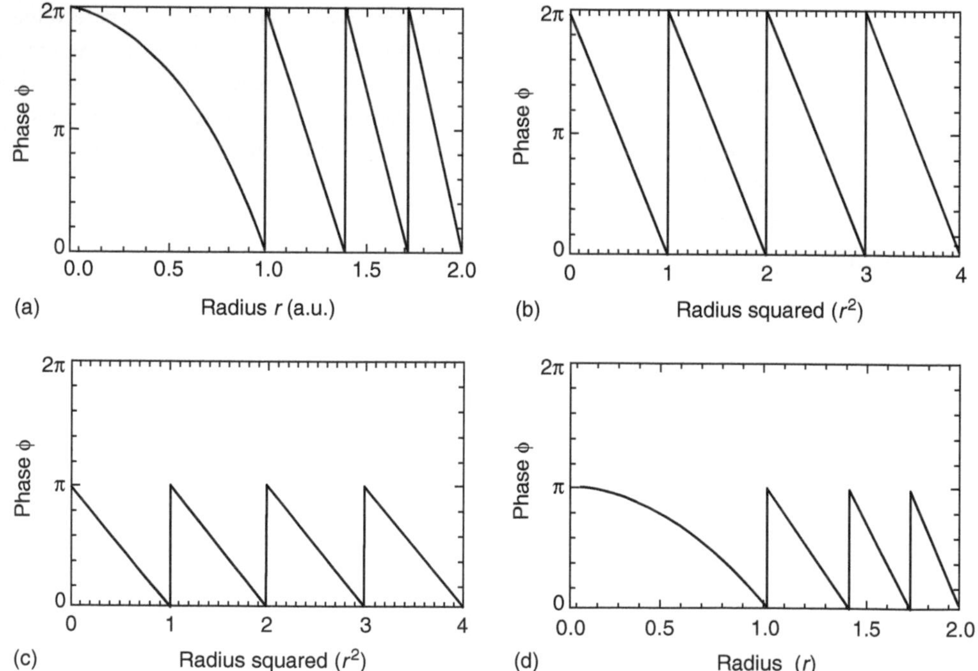

FIGURE 17.14 Designing a bifocal lens: (a) lens with a single focus; (b) same as (a), but showing periodicity in r^2; (c) substitution of a beam-splitting design; (d) Same as (c), but as a function of r.

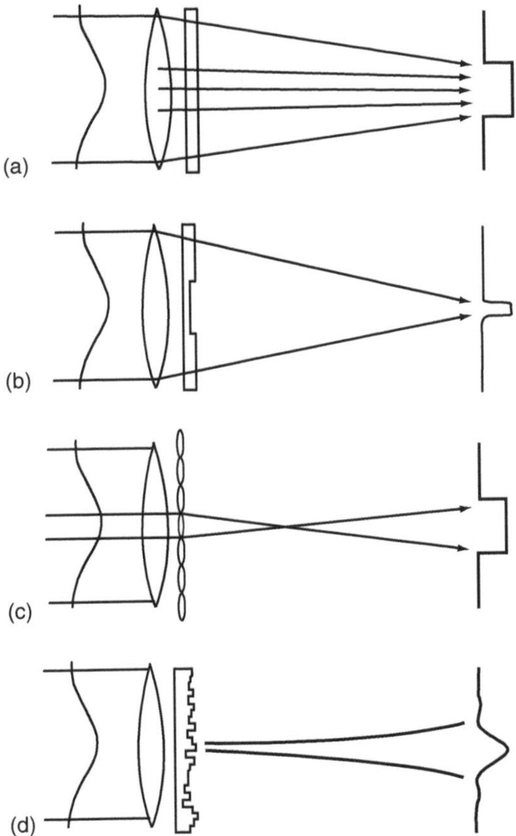

FIGURE 17.15 Methods for reshaping a laser beam. (a) Ray tracing allows the redistribution of energy. (b) In this case, diffraction must be accounted for. (c) Fly's eye approach. (d) Random phase plate.

If the beam is well behaved, then the design problem is greatly simplified and beam shaping can typically be achieved with efficiencies of 90% or greater. If the required output beam is large enough so that diffraction effects are negligible, as shown in Figure 17.15a, then the reshaper can be designed by ray tracing. The crux of this approach is that there is a coordinate transform which maps the power of the laser beam from its original distribution to the desired one, and the binary optic element is then used to implement this transform (Bryngdahl 1974; Hossfeld 1991). The ray-tracing approach is a simple and effective one, but it can sometimes result in unpredicted interference effects, similar in nature to the spot of Arago. An extreme example of this is shown in Figure 17.15b. Here, the desired flat-top distribution is nearly diffraction-limited and the ray-tracing approach is inadequate due to diffraction effects. Accordingly, the design, which is simply a flat plate with a phase-shifted central portion, is accomplished by using an approach similar to the ones used in phase retrieval and apodization problems (Cordingley 1993).

If the beam shape cannot be predicted a priori, then a design which will be robust to fluctuations in the beam is required. Most of these approaches depend on dividing the full aperture of the beam into subapertures, each of which is imaged into the desired output shape. For example, in Figure 17.15c, each fly's eye lens independently images its subaperture onto the full output beam, thus averaging out any local nonuniformities in the incident beam (Deng 1986). In these approaches, it is important to consider whether the beamlets from each subaperture add coherently or incoherently. Continuing the example of Figure 17.15c, if the incident beam is spatially coherent, then the fly's eye array will behave like a grating and the beamlets will interfere to produce sharp diffraction orders rather than the smooth, flat top produced by a spatially incoherent beam. In the case of coherent beams, such a design will still work if the statistical fluctuations in the beam are enough to average out any interference patterns. In an extreme case of this approach, the subapertures are reduced in size until diffraction alone is sufficient to map the subaperture onto the full output beam, as shown in Figure 17.15d. Furthermore, each subaperture is randomly phased with respect to the others to prevent any regular interference patterns, and, hence, the resulting profile is the diffraction pattern of the subaperture modulated by a speckle pattern (Kurtz 1972; Kato 1984).

17.3.4 Other Applications

Other applications include components for optical interconnects (Walker and Jahns 1992), diffractive switches for laser beam steering (Goltsos and Holz 1990), the coherent combination of lasers (Leger 1988), and components for optical disks (Ono 1989), among others.

17.4 Designs Based on Maxwell's Equations

There are two common situations which lead to designs in the vector regime. In the first, the design is based on a more approximate theory or, perhaps, just an intuitive idea, but the resulting surface relief has such fine features that a vector approach is required. In the second, the device is specifically designed to take advantage of a phenomenon, such as form birefringence, which is neglected by more approximate approaches.

17.4.1 Fundamentals

There are a limited number of approaches for attacking problems in the vector regime. The brute force numerical solution of Maxwell's equations is technically possible but limited in usefulness by practical issues. For example, the finite-difference, time-domain approach requires a grid sampling of approximately ten samples per wavelength. In the visible, a cube just 10 μm on a side would require over a million nodes.

Of more practical interest is the solution of Maxwell's equations for a plane wave incident on a periodic structure. The two major approaches to solving this problem are the space harmonic approach and the modal approach (Moharam and Gaylord 1982). In the modal approach, the field in the grating region is represented as a superposition of the eigenmodes of the grating. This approach is more numerically stable and converges more quickly than the space harmonic approach. However, it also requires that the eigenmodes of the grating profile be known and, unfortunately, eigenmodes have only been determined for the simplest cases, such as the square wave grating (Sheng 1982). Accordingly, the space harmonic approach is more widely used. In this approach, the field in the grating region is represented as a set of plane waves which couple energy to each other as they propagate through the grating region. No matter which approach is used, the vector nature of the problem makes the optical design of gratings both nonintuitive and computationally intensive (Sheridan and Sheppard 1990).

If the grating period is much smaller than the wavelength, then it can be argued that the light will not respond to the individual index variations of the grating, but will, instead, respond to some average index (i.e., an effective index). Accordingly, this approach is known as effective medium theory. For the square wave grating shown in Figure 17.16, the effective indices are the solutions to the equations

$$\sqrt{n_1^2-n_O^2}\tan[\pi T(1-f)\sqrt{n_1^2-n_O^2}/\lambda]+\sqrt{n_2^2-n_O^2}\tan[\pi Tf\sqrt{n_2^2-n_O^2}/\lambda]=0 \tag{17.90}$$

and

$$(\sqrt{n_1^2-n_E^2}/n_1^2)\tan[\pi T(1-f)\sqrt{n_1^2-n_E^2}/\lambda]+(\sqrt{n_2^2-n_E^2}/n_2^2)\tan[\pi Tf\sqrt{n_2^2-n_E^2}/\lambda]=0, \tag{17.91}$$

where n_O is the case where the electric field is parallel to the grating grooves, n_E is the case where the electric field cuts across the grating grooves, and f is the fraction of the grating with index n_2 (Rytov 1956). An approximate solution to these equations is given (Raguin and Morris 1993) by

$$n_{O2}^2=n_{O0}^2+\frac{1}{3}[\pi Tf(1-f)/\lambda]^2(n_2^2-n_1^2)^2 \tag{17.92}$$

and

$$n_{E2}^2=n_{E0}^2+\frac{1}{3}[\pi Tf(1-f)/\lambda]^2(1/n_2^2-1/n_1^2)^2n_{E0}^6n_{O0}^2 \tag{17.93}$$

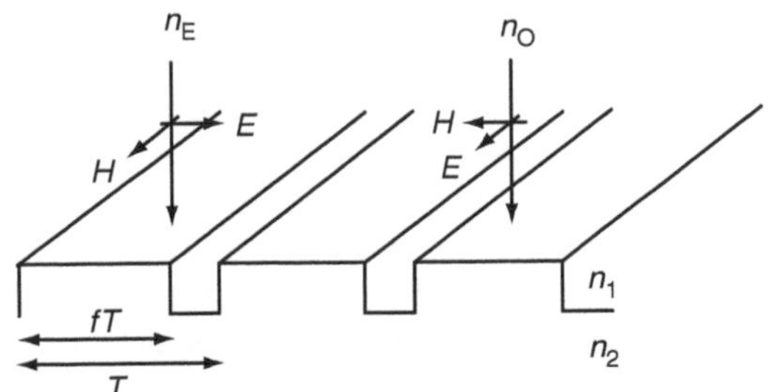

FIGURE 17.16 A grating of small period behaves like a medium of index n_E or n_O depending on polarization.

where n_{O0} and n_{E0} are the zero order solutions

$$n_{O0}^2 = (1-f)n_1^2 + fn_2^2 \tag{17.94}$$

and

$$1/n_{E0}^2 = (1-f)/n_1^2 + f/n_2^2. \tag{17.95}$$

17.4.2 Applications

The effective index, which ranges from n_1 to n_2, is determined by the fill factor f, and a number of applications are based on this concept. In antireflection applications, the fill factor is gradually increased in the direction of light propagation, thus, producing a smooth transition from one index to the other (Raguin and Morris 1993). The resultant structures are similar to the spikes seen in anechoic chambers. If the fill factor is modulated in the transverse direction, the resulting binary structure will be the equivalent of a graded-index structure, similar to the production of gray scale by halftoning (Haidner 1992). As shown in the above equations, the effective index depends on the polarization of the incoming light. Therefore, the structure shown in Figure 17.16 is strongly birefringent because, in essence, it is an artificial crystal (Cescato 1990). Loosely extending this crystal concept, it is possible to develop spectral filters (passband or stopband filters) by engineering the device to have resonances at certain wavelengths. Depending on the exact nature of the resonance, the filter can be either fairly broad (Rhoads 1982) or extremely narrow (Wang and Magnusson 1993).

17.5 Fabrication Using Multiple Masks

The traditional binary optics fabrication process is shown in Figure 17.17. In this process, a set of photolithographic masks is used to produce a staircase approximation to the desired thickness profile. The process begins with an optical substrate, which is coated with photoresist. The resist is then exposed using the first mask of the mask set, as shown in Figure 17.17a, and subsequent development removes resist either in the exposed areas for positive resists or the unexposed areas for negative resists. The substrate areas, which are no longer protected by resist, are then micromachined, as shown in Figure 17.17b, producing the profile depicted by the dashed lines. Usually material is removed, but, in some processes, material is added. After micromachining, the optical substrate will have a surface profile with two levels, and repetition of this process, as shown for the second mask in parts c and d of Figure 17.17, yields a staircase profile with 2^N levels if N masks are used. This process consists of three major steps: generating the mask set, transferring the pattern from the mask to the substrate via photolithography, and micromachining the patterned substrate. The mask generation step is described in some detail because optical designers commonly participate in this step. However, photolithography and micromachining are best handled by personnel specifically trained for these tasks, rather than an optical designer, so the descriptions given here are superficial.

17.5.1 Mask Generation

In the mask generation stage, the optical prescription of the binary optic element, which consists of a phase profile $\phi(x, y)$ and a design wavelength λ_0, is converted to a set of N photolithographic masks. This begins by converting the phase profile to a desired height profile, usually by the equation

$$h(x, y) = \frac{\lambda_0}{2\pi(n_1 - n_2)}(\phi \bmod 2\pi) \tag{17.96}$$

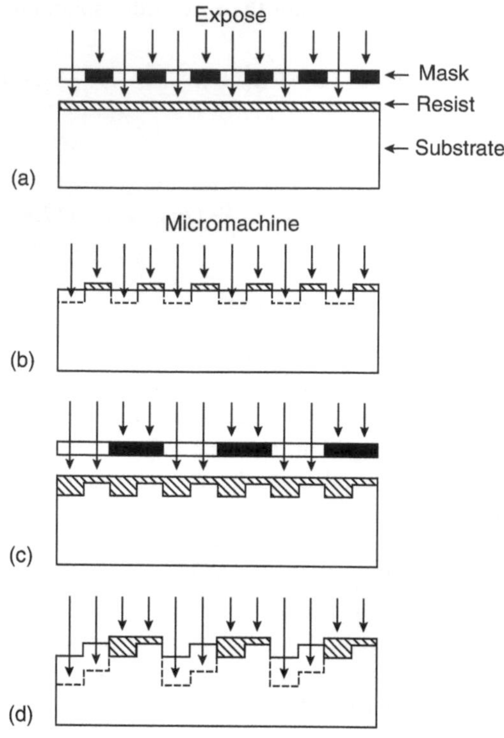

FIGURE 17.17 Binary optics fabrication based on multiple masks: (a) exposure of resist using the first mask; (b) micromachining the exposed areas of the substrate; (c) and (d) repetition of (a) and (b) using the second mask.

which is based on the same assumptions as Equation 17.42. The desired profile, which varies from 0 to $h_0 = \lambda_0/(n_1 - n_2)$, will be approximated by a staircase profile, with the transition points of the staircase occuring along constant thickness contours spaced in increments of h_0/N. If Equation 17.96 is used, these contours will also be contours of constant phase. Table 17.6 defines the masks required to achieve this staircase approximation. The layer one mask is required for a one-mask process, layers one and two for a two-mask process, etc.

The contours can be calculated in several ways, depending on the nature of the thickness profile (Logue and Chisholm 1989). If the thickness profile is described by a simple analytical expression, as is the case for most of the elements described in this chapter, then the contours may be determined analytically. If the phase profile is piecewise fat (e.g., the element consists of a checkerboard pattern with each square having a different phase), then the contours may also be determined analytically, but the resulting contour will consist of a large number of disjoint curves, in contrast to the previous case (e.g., a Fresnel lens) where the contour is usually one simple curve. Also note that, in these two cases, the resulting contours will be exact, which is not true in the following two cases. If the binary optic component is small in size or the phase contours are nearly straight, then, the contours may be estimated by calculating the thickness at every point on a regular grid and interpolating the location of the contours. The limitation to this approach is that the grid must be dense enough to reduce the interpolation error to the required level, but it must also be large enough to cover the entire element. For many elements, this approach is not feasible because of limited computer speed and/or storage. Also note that this approach is not the same as sampling the thickness profile on a regular grid and then pretending that the element is piecewise flat with the thickness of each grid cell given by the sampled value. In particular, the piecewise flat assumption should not be used unless the thickness profile truly is piecewise flat. Finally, if the phase profile is continuous, then the contours can be traced one at a time, using a method similar to those used for tracing fringes on an interferogram (Krishnaswamy 1991). In this approach, the contour may be locally modeled as a circle and the normal and tangent directions to the contour, as well as the radius of the osculating circle, are usually required.

TABLE 17.6 Processing Steps for Binary Optics

Mask Layer	Etch Region, Defined by $h(x, y)$	Etch Region, Defined by $\phi(x, y)$	Etch Depth
1	$0 < h \bmod (h_0) < h_0/2$	$0 < \phi \bmod 2\pi < \pi$	$h_0/2$
2	$0 < h \bmod (h_0/2) < h_0/4$	$0 < \phi \bmod \pi < \pi/2$	$h_0/4$
3	$0 < h \bmod (h_0/4) < h_0/8$	$0 < \phi \bmod \pi/2 < \pi/4$	$h_0/8$
4	$0 < h \bmod (h_0/8) < h_0/16$	$0 < \phi \bmod \pi/4 < \pi/8$	$h_0/16$

Source: Optical Society of America, *Handbook of Optics*, McGraw-Hill, New York, 1994. With permission.

These are depicted in Figure 17.18 given by

$$\hat{n} = \nabla\phi/|\nabla\phi|, \tag{17.97}$$

$$\hat{t} = \hat{n} \times \hat{z}, \tag{17.98}$$

and

$$1/R = [(\phi_y\phi_{yx} - \phi_{yy}\phi_x)^2 + (\phi_{xy}\phi_x - \phi_{xx}\phi_y)^2]^{1/2}/|\nabla\phi|^2 \tag{17.99}$$

where $\hat{n}$ and $\hat{t}$ are unit vectors normal and tangential to the contour but in the plane of the osculating circle, $\hat{z}$ is the unit vector out of the plane, and the notation ϕ_x denotes the partial derivative of ϕ with respect to x, etc.

In addition to mathematically determining the contours, the contours must also be represented in a manner compatible with the mask-making equipment. Common formats include GDSII and CIF (Mead 1980; Rubin 1987). Besides syntax rules, all formats also have rules regarding which shapes can be represented. Of particular note for binary optics, curved contours almost always must be approximated by a series of line segments with the end points of those line segments lying on a predefined grid. Because the grid spacing can, typically, range from 0.05 to 0.5 μm, snapping of coordinates to the grid usually has an insignificant effect if the grid spacing is chosen correctly. However, the approximation of curves by line segments cannot be neglected, and line segments are usually chosen so that the deviation between the actual curve and the line segment matches the grid resolution. In particular, if a circle of radius R is to be approximated with a maximum deviation of δ, then, the line segment may be defined by

$$\theta_1 = 2\arccos(1 - \delta/R_1) \doteq \sqrt{8\delta/R_1}, \tag{17.100}$$

$$R_1 = R, \tag{17.101}$$

or

$$\theta_2 = 2\arccos(1 - 2\delta/R_2)$$

$$\doteq \sqrt{16\delta/R_2}, \tag{17.102}$$

$$R_2 = R + \delta, \tag{17.103}$$

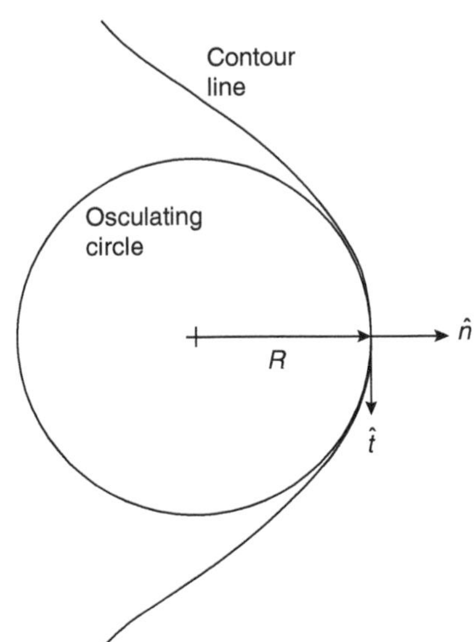

FIGURE 17.18 The contour line may be locally modelled as a circle of radius R.

as shown in Figure 17.19. The aggressive approach results in less data. However, it is more difficult to implement, especially for the case of arbitrarily shaped contours.

If the features on the photomask are coarse (typically greater than 10 μm), then it is possible to produce the photomask optically by photoreduction. However, in most cases, the photomasks are written by electron beams, most of which use the MEBES format as input. Strictly speaking, the formats mentioned previously cannot be used to run a MEBES machine. However, they can be translated or fractured to MEBES with an insignificant loss in fidelity, so that the distinction is transparent to the binary optic designer.

Most commercial mask vendors can routinely write photomasks of sizes up to 7 inches square with a minimum feature size, or critical dimension (CD), of 0.8 μm in the case of general patterns. If the pattern is an array of equal width lines and spaces, then the achievable CD may be as small as 0.3 μm. The CD of a binary optic element is given by the minimum local period, as defined in Equation 17.2, divided by the number of levels in the staircase profile.

In writing the photomask, the MEBES machine effectively represents the entire mask as an array of pixels, which typically range in size from 0.05 to 0.5 μm, and the electron beam is then scanned across the field in a raster fashion, exposing the desired pixels. As a result of this pixelization, curves and gently sloping lines will have noticeably jagged edges when viewed under the microscope. Fortunately, these artifacts usually do not significantly affect the optical performance of the element. In addition, the electron beam is usually scanned by electrostatic deflection in one direction while the mask is mechanically translated in the other direction. This difference in physical mechanisms results in a preferred direction for the machine, and masks for asymetric patterns (e.g., linear gratings) should take advantage of this.

Although most of this section has dealt with feature generation for the binary optic element, more time is usually spent generating the remainder of the mask, which typically includes alignment marks, control features, and labels. Alignment marks are required for aligning the mask to the substrate, aligning the mask or substrate to any jigs, and for aligning the finished piece in the optical system. Control features are diagnostic areas which are used to monitor the fabrication process. For example, the depth of a micromachining step can be checked by witness etch areas, and the fidelity of the photolithographic process can be checked by line and space patterns. Finally, labels allow both the fabricator and the optical designer to easily determine the device name, orientation, mask layer number, etc.

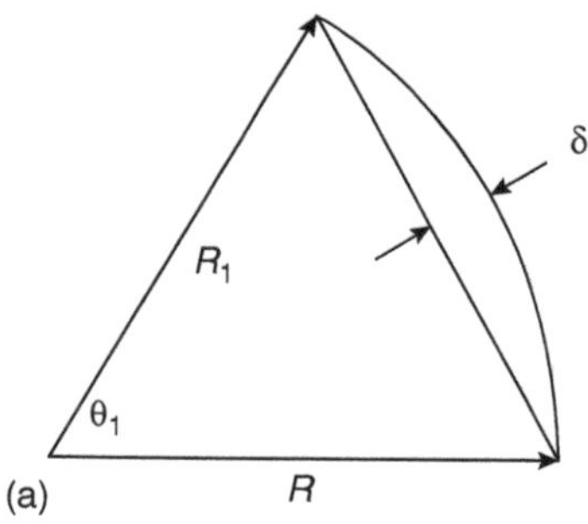

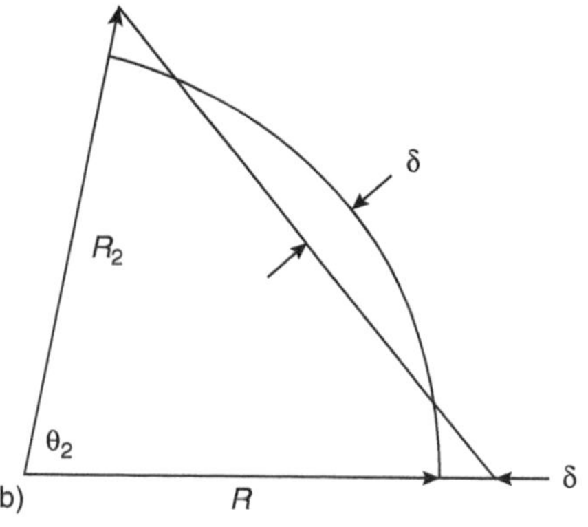

FIGURE 17.19 Approximating a circle of radius R by a line segment: (a) conservative approach; (b) aggressive approach.

17.5.2 Photolithography

The photolithography step transfers the pattern from the mask to the photoresist covering the substrate. The most common methods are contact, proximity, and projection lithography. In contact and proximity printing, the mask is either in direct contact with or in close proximity to the resist, and the pattern is produced by shadow casting. Contact photolithography results in higher resolution and better linewidth fidelity because there are no diffraction effects, whereas proximity printing reduces the chance of physically damaging the resist layer because the mask and resist are separated by 5–50 μm. Both methods require 1:1 masks and, in comparison to projection printing, they offer lower equipment costs and more flexibility in substrate sizes and materials. In projection photolithography, the mask and substrate are separated, and the mask is then imaged onto the resist-covered substrate, with a demagnification ranging from 1 to 20 times. Projection printers are good for high volume manufacturing, such as the production of microprocessors, but, for binary optics, they have limited usefulness because they are expensive, the required alignment accuracies are difficult to achieve, and they have limited fields of view, typically on the order of 2 cm^2.

After exposure, the photoresist is developed. For positive photoresists, the exposed areas are removed by the developer, whereas, for negative resists, the unexposed areas are removed. The underlying substrate in these areas can then be micromachined while the remainder of the substrate is protected by the remaining resist.

Because photolithography is easiest on flat surfaces, the masks are usually processed in order of shallowest etch depth to deepest (i.e., from layer 4 to layer 1 of Table 17.6). Occasionally, a project requires a surface relief which is deeper than can be handled by normal resists, which are typically 0.5–2.0 μm thick. In these cases, more complex processes are required (Stern and Medeiros 1992).

17.5.3 Micromachining Techniques

In the micromachining step, material is removed from the unprotected areas of the substrate. The two most common micromachining methods are reactive ion etching (RIE) and ion milling. In RIE, a plasma is formed at the substrate surface and particles are directed at the surface (Stern 1991a). Substrate material is removed by chemical interactions between the plasma and the substrate and also by mechanical abrasion. Because of the chemical component, the etch attacks different materials at different rates (i.e., the etch is selective), and a different process must be developed for each material. In addition, although fused silica may be etched successfully in this manner, glasses with impurities usually cannot because the impurities react adversely with the chemicals. In contrast, ion milling is similar to sandblasting. The substrate surface is bombarded by inert ions (usually Ar) and material is removed solely by mechanical means (Stern 1991b). Because the process does not involve any chemistry, the process can be used for any material but it is usually less selective and also slower than RIE.

Finally, several other micromachining techniques are worth mentioning. In wet chemical etching, the substrate is immersed in a solution which chemically removes material (Lee 1979). This method can be fast, but it also has lower fidelity because the chemical reaction proceeds in all directions (i.e., the etch is isotropic) and the photoresist provides protection only in one direction. Crystallographic etching is fast but not of general utility, because the etching must follow crystallographic planes. Binary optic structures have also been fabricated by depositing material, rather than removing it, but this approach is more difficult to control (Jahns and Walker 1990).

17.6 Alternate Fabrication Methods

Fabrication based on multiple masks is not the only method for manufacturing binary optics. This section describes some of the other promising approaches as well as the most common replication methods.

17.6.1 Other Micromachining Approaches

Most other fabrication approaches rely on directly micromachining the required surface relief, either in the final substrate material or in an intermediate material, such as a photoresist. Almost all of these processes can be described as a tool traversing a surface, removing material in a specific manner due to the tool shape and tool travel path. The problem of calculating the correct tool path, given the tool shape and desired final profile, is essentially a deconvolution problem which may be solved by an iterative method, such as the van Cittert deconvolution algorithm (Crilly 1991). The feature which distinguishes one fabrication method from another is the tool type, which is discussed below.

Compared to the traditional binary optic fabrication, these micromachining approaches have the advantage of fewer processing steps, and, in particular, layer to layer alignments are no longer required. However, the disadvantage is that the required processes are more difficult to control (e.g., compare the on–off exposure of photoresist used in the traditional approach to the analog exposure required in some of the approaches below) and also take much longer to complete because the tool must traverse the entire surface sequentially.

Of the approaches which directly figure the final surface, diamond turning is the most commercially mature, especially for infrared optics (Riedl and McCann 1991). In this technology, the substrate is mounted on a rotating spindle and then cut by a diamond tool which moves radially. The process is similar to a lathe and a basic trade-off occurs in selecting the tool size. Larger tools remove more material, thus reducing cutting time and tool wear, but smaller tools resolve finer features.

Ion milling, described previously, can also be used to directly figure a surface. In this method, a stream of ions is focused to a spot, which can be as small as 50 nm, and this is then used as the tool (DeFreez 1989). Compared to most other tools, which are on the order of 1 μm or larger, ion milling can resolve much finer features, but the small tool size also results in longer processing times and limited element sizes.

Two other promising techniques for directly figuring substrates are the direct ablation of materials by lasers (Veiko 1993), primarily excimer lasers, and laser-assisted chemical etching of materials (Gratrix and Zarowin 1991).

An alternative to micromachining the surface in a single step is, to first produce a latent image in the material via an exposure step and to then remove the material in a subsequent development step. For example, if a photoresist is used, the resist may be exposed in an analog fashion by scanning a laser across the surface (Gale 1992) or by using a gray-scale mask (Andersson 1990). In the development step, differing amounts of resist are removed due to the variable exposure, and the remaining resist has the desired surface profile. This approach can also be used with electron-beam resists rather than photoresists (Fujita 1982), with the advantage that electron-beams, like ion beams, can be configured to write with much smaller spot sizes. A general disadvantage to this approach is that the original material must be some sort of resist and a second, nontrivial step is required to transfer the surface profile from the resist to the actual substrate material.

17.6.2 Replication Methods

In many cases, the most economical way to produce large quantities of a binary optic element is to fabricate the first element by one of the previously mentioned techniques and then to replicate it by a method, such as embossing, casting, or molding. The first step, in almost all cases, is to produce a metal master which is a negative of the desired surface profile. For examples, this may be achieved by coating the original with a thin conductive layer, immersing the piece in a nickel bath, and then building up a thicker nickel layer via electroplating. The nickel layer can then be separated from the original and used as a metal master. If required, the process may be repeated to produce multiple masters or to reverse the profile.

In embossing, the metal master is pressed into plastic which has been temporarily softened (Cowan and Slafer 1985). Replicas can be formed very inexpensively by wrapping the metal master around a drum and then feeding in the plastic on long sheets, stamping a replica on each revolution of the drum. This method is economical but is limited to shallow, uncomplicated surface reliefs with low aspect ratios. In an alternate embossing process, the metal master is pressed into a photopolymer, the embossing is fixed by exposure to ultraviolet light, and then the master is removed (Shvartsman 1993). This process has been used to reproduce complicated binary optics elements with high fidelity and good dimensional stability.

The remaining replication processes require that the metal master be combined with a second piece, commonly a flat, to form a mold. In conventional injection molding, plastic is heated and then injected into the mold. After the assembly cools, the mold is separated and the replica released (Goto 1989). Alternatively, the injected resin may be cured by exposure to ultraviolet light, thus avoiding the long cooling stage (Tanigami 1989). Sol–gels are processed in a similar fashion, but the resulting material is a glass rather than a polymer (Nogues and Howell 1992). The potential drawbacks to these processes are that only one piece can be formed at a time and shrinkage during cooling or curing can be significant.

Defining Terms

Achromat: Lens corrected at two wavelengths. First-order design is based on the Abbe number.

Angular spectrum approach: Scalar diffraction approach used to propagate fields in the spatial frequency domain.

Apochromat: Lens corrected at three wavelengths. First-order design is based on the Abbe number and partial dispersion.

Blaze height: The grating height required to maximize the efficiency in a desired diffraction order for a given wavelength.

Dammann grating: Grating beam splitter which is binary (i.e., constrained to two phases).

Diffraction efficiency: Energy in a specific diffraction order of a binary optic element, relative to the total incident energy.

Effective medium theory: Model which replaces a fine-featured structure with a medium possessing the "average" properties of the structure.

Fly's eye lens: Type of microlens array.

Form birefringence: Birefringence which is the result of a man-made periodic structure which has a periodicity much less than a wavelength.

Fresnel approximation: Scalar diffraction integral used to propagate fields in the space domain.

Generalized grating model: An approach which locally models a diffractive optic by a linear grating.

Grating beam splitter: Grating with diffraction efficiencies which are designed to follow a specific weighting.

Grating equation: Equation which governs the ray tracing of binary optics.

Hybrid system: System which combines refractive/reflective elements with binary optic elements.

Micromachining: VLSI-based fabrication process which produces a surface relief on a microscopic scale. Common processes include reactive ion etching, ion milling, and wet chemical etching.

Phase screen: Approach which models binary optic elements as instantaneously introducing a phase delay.

Photolithography: Process for transferring a pattern from a photomask to a photoresist. Common types of photolithography include contact, proximity, and projection photolithography.

Photomask: Chrome on glass mask, which is usually written by an electron beam and used to pattern photoresist.

Rigorous coupled wave analysis: Method for solving Maxwell's equations for plane wave diffraction from a grating.

Sweatt lens: Modification of the Sweatt model, in which the binary optic component is a lens.

Sweatt model: Model for binary optics which replaces the binary optic element by an equivalent refractive element.

References

Anderson, J. W. 1993. Thermal weapon sight tws an/pas-13 diffractive optics designed for producibility. *Conf. Binary Optics*, p. 303. NASA Conference Publication 3227.

Andersson, H. 1990. Single photomask, multilevel kinoform in quartz and photoresist: Manufacture and evaluation. *Appl. Opt.*, 29, 4259.

Barakat, R. 1980. Calculation of integrals encountered in optical diffraction theory. In *The Computer in Optical Research Methods and Applications*, B. R. Frieden, ed., pp. 35-80. Springer, New York.

Bryngdahl, O. 1974. Geometrical transforms in optics. *J. Opt. Soc. Am.*, 64, 1092.

Buralli, D. A. and Morris, G. M. 1989. Design of a wide field diffractive landscape lens. *Appl. Opt.*, 28, 3950.

Buralli, D. A. and Morris, G. M. 1991. Design of diffractive singlets for monochromatic imaging. *Appl. Opt.*, 30, 2151.

Buralli, D. A. and Morris, G. M. 1992a. Design of two- and three-element diffractive Keplerian telescopes. *Appl. Opt.*, 31, 38.

Buralli, D. A. and Morris, G. M. 1992b. Effects of diffraction efficiency on the modulation transfer function of diffractive lenses. *Appl. Opt.*, 31, 4389.

Buralli, D. A. and Rogers, J. R. 1989. Some fundamental limitations of achromatic holographic systems. *J. Opt. Soc. Am.*, A6, 1863.

Cescato, L. H. 1990. Holographic quarterwave plates. *Appl. Opt.*, 29, 3286.

Chen, C. W. 1992. Application of diffractive optical elements in visible and infrared optical systems. *Proc. Soc. Photo-Opt. Instrum. Eng.*, CR41, 158.

Cordingley, J. 1993. Application of a binary diffractive optic for beam shaping in semiconductor processing by lasers. *Appl. Opt.*, 32, 2538.

Cowan, J. J. and Slafer, W. D. 1985. Holographic embossing at polaroid: the polaform process. *Proc. Soc. Photo-Opt. Instrum. Eng.*, 600, 49.

Cox, J. A. 1991. Process error limitations on binary optics performance. *Proc. Soc. Photo-Opt. Instrum. Eng.*, 1555, 80.

Crilly, P. B. 1991. A quantitative evaluation of various iterative deconvolution algorithms. *IEEE Trans. Instrum. Meas.*, 40, 558.

Dammann, H. 1979. Spectral characteristics of stepped-phase gratings. *Optik*, 53, 409.

DeFreez, R. K. 1989. Focused ion-beam micromachined diode laser mirrors. *Proc. Soc. Photo-Opt. Instrum. Eng.*, 1043, 25.

Deng, X. 1986. Uniform illumination of large targets using a lens array. *Appl. Opt.*, 25, 377.

Emerton, N. 1985. Manufacturing tolerances for blazed diffractive optical elements. *Acta Polytech. Scand.: Appl. Phys.*, Ph-149, 308.

Farn, M. W. 1992. Quantitative comparison of the general Sweatt model and the grating equation. *Appl. Opt.*, 31, 5312.

Farn, M. W. and Goodman, J. W. 1991. Diffractive doublets corrected at two wavelengths. *J. Opt. Soc. Am.*, A8, 860.

Ferstl, M. 1992. Blazed Fresnel zone lenses approximated by discrete step profiles: effects of fabrication errors. *Proc. Soc. Photo-Opt. Instrum. Eng.*, 1732, 89.

Fujita, T. 1982. Blazed gratings and Fresnel lenses fabricated by electron-beam lithography. *Opt. Lett.*, 7, 578.

Gale, M. T. 1992. Fabrication of kinoform structures for optical computing. *Appl. Opt.*, 31, 5712.

Goltsos, W. and Holz, M. 1990. Agile beam steering using binary optics microlens arrays. *Opt. Eng.*, 29, 1392.

Goto, K. 1989. Plastic grating collimating lens. *Proc. Soc. Photo-Opt. Instrum. Eng.*, 1139, 169.

Gratrix, E. J. and Zarowin, C. B. 1991. Fabrication of microlenses by laser assisted chemical etching. *Proc. Soc. Photo-Opt. Instrum. Eng.*, 1544, 238.

Gremaux, D. A. and Gallagher, N. C. 1993. Limits of scalar diffraction theory for conducting gratings. *Appl. Opt.*, 32, 1948.

Gruhlke, R. 1992. Laser energy distributions produced by the use of diffractive optical elements. *Proc. Soc. Photo-Opt. Instrum. Eng.*, 1834, 152.

Haidner, H. 1992. Zero-order gratings used as an artificial distributed index medium. *Optik*, 89, 107.

Hossfeld, J. 1991. Rectangular focus spots with uniform intensity profile formed by computer generated holograms. *Proc. Soc. Photo-Opt. Instrum. Eng.*, 1574, 159.

Jahns, J. and Walker, S. 1990. Two-dimensional array of diffractive microlenses fabricated by thin film deposition. *Appl. Opt.*, 29, 931.

Kato, Y. 1984. Random phasing of high-power lasers for uniform target acceleration and plasma-instability suppression. *Phys. Rev. Lett.*, 53, 1057.

Killat, U. 1982. Binary phase gratings for star couplers with high splitting ratios. *Fiber Integr. Opt.*, 4, 159.

King, P. R. 1985. Design of diffractive surface relief lenses with more than one focus. *Acta Polytech. Scand.: Appl. Phys.*, Ph-149, 312.

Krackhardt, U. 1989. Binaere Phasengitter als Vielfach-Strahlteiler. Diplomarbeit, Universitaet Erlangen-Nuernberg, Erlangen, Germany.

Krackhardt, U. 1992. Upper bound on the diffraction efficiency of phase-only fanout elements. *Appl. Opt.*, 31, 27.

Krishnaswamy, S. 1991. Algorithm for computer tracing of interference fringes. *Appl. Opt.*, 30, 1624.

Kurtz, C. N. 1972. Transmittance characteristics of surface diffusers and the design of nearly band-limited binary diffusers. *J. Opt. Soc. Am.*, 62, 982.

Lee, W. H. 1979. High efficiency multiple beam gratings. *Appl. Opt.*, 18, 2152.

Leger, J. R. 1988. Coherent laser beam addition: an application of binary-optics technology. *The Lincoln Lab. J.*, 1, 225.

Levi, L. 1980. A one-glass achromatic doublet. *Appl. Opt.*, 27, 1491.

Logue, J. and Chisholm, M. L. 1989. General approaches to mask design for binary optics. *Proc. Soc. Photo-Opt. Instrum. Eng.*, 1052, 19.

Londono, C. 1993. Athermalization of a single-component lens with diffractive optics. *Appl. Opt.*, 32, 2295.

Magnusson, R. and Gaylord, T. K. 1978. Diffraction regimes of transmission gratings. *J. Opt. Soc. Am.*, 68, 809.

Mead, C. A. 1980. *Introduction to VLSI Systems*. Addison-Wesley, Reading, MA.

Moharam, M. G. and Gaylord, T. K. 1982. Diffraction analysis of dielectric surface-relief gratings. *J. Opt. Soc. Am.*, 72, 1385.

Nogues, J. L. R. and Howell, R. L. 1992. Fabrication of pure silica micro-optics by sol–gel processing. *Proc. Soc. Photo-Opt. Instrum. Eng.*, 1751, 214.

Ono, Y. 1989. Computer generated holographic optical elements for optical disk memory read/write heads. *Proc. Soc. Photo-Opt. Instrum. Eng.*, 1052, 150.

Optical Society of America. 1994. *Handbook of Optics*. McGraw-Hill, New York.

Raguin, D. H. and Morris, G. M. 1993. Antireflection structured surfaces for the infrared spectral region. *Appl. Opt.*, 32, 1154.

Rhoads, C. M. 1982. Mid-infrared filters using conducting elements. *Appl. Opt.*, 21, 2814.

Riedl, M. J. and McCann, J. T. 1991. Analysis and performance limits of diamond turned diffractive lenses for the 3–5 and 8–12 micrometer regions. *Proc. Soc. Photo-Opt. Instrum. Eng.*, CR38, 153.

Roberts, N. C. 1992. Binary phase gratings for hexagonal array generation. *Opt. Commun.*, 94, 501.

Roose, S. 1993. An efficient interpolation algorithm for Fourier and diffractive optics. *Opt. Commun.*, 97, 312.

Rubin, S. M. 1987. *Computer Aids for VLSI Design*. Addison-Wesley, Reading, MA.

Rytov, S. M. 1956. Electromagnetic properties of a finely stratified medium. *Sov. Phys. JETP*, 2, 466.

Sheng, P. 1982. Exact eigenfunctions for square-wave gratings: application to diffraction and surface-plasmon calculations. *Phys. Rev. B*, 26, 2907.

Sheridan, J. T. and Sheppard, C. J. R. 1990. An examination of the theories for the calculation of diffraction by square wave gratings: 3. Approximate theories. *Optik*, 85, 135.

Shvartsman, F. P. 1993. Replication of diffractive optics. *Proc. Soc. Photo-Opt. Instrum. Eng.*, CR49, in press.

Siegman, A. E. 1977. Quasi Fast Hankel transform. *Opt. Lett.*, 1, 13.

Stern, M. B. 1991a. Fabricating binary optics: process variables critical to optical efficiency. *J. Vac. Sci. Technol. B*, 9, 3117.

Stern, M. B. 1991b. Binary optics microlens arrays in CdTe. *Mater. Res. Soc. Symp. Proc.*, 216, 107.

Stern, M. B. and Medeiros, S. S. 1992. Deep three-dimensional microstructure fabrication for infrared binary optics. *J. Vac. Sci. Technol. B*, 10, 2520.

Stone, T. and George, N. 1988. Hybrid diffractive-refractive lenses and achromats. *Appl. Opt.*, 27, 2960.

Swanson, G. J. 1989. Binary optics technology: the theory and design of multilevel diffractive optical elements, M.I.T. Lincoln Laboratory Technical Report 854, NTIS Publ. AD-A213-404.

Swanson, G. J. 1991. Binary Optics Technology: theoretical Limits on the Diffraction Efficiency of Multilevel Diffractive Optical Elements, M.I.T. Lincoln Laboratory Technical Report 914.

Sweatt, W. C. 1977. Describing holographic optical elements as lenses. *J. Opt. Soc. Am.*, 67, 803.

Sweatt, W. C. 1979. Mathematical equivalence between a holographic optical element and an ultrahigh index lens. *J. Opt. Soc. Am.*, 69, 486.

Tanigami, M. 1989. Low-wavefront aberration and high-temperature stability molded micro Fresnel lens. *IEEE Photonics Technol. Lett.*, 1, 384.

Vasara, A. 1992. Binary surface-relief gratings for array illumination in digital optics. *Appl. Opt.*, 31, 3320.

Veiko, V. P. 1993. Laser technologies for micro-optics fabrication. *Proc. Soc. Photo-Opt. Instrum. Eng.*, 93, 1874.

Veldkamp, W. B. 1980. Developments in laser-beam control with holographic diffraction gratings. *Proc. Soc. Photo-Opt. Instrum. Eng.*, 255, 136.

Walker, S. J. and Jahns, J. 1992. Optical clock distribution using integrated free-space optics. *Opt. Commun.*, 90, 359.

Wang, S. S. and Magnusson, R. 1993. Theory and applications of guided-mode resonance filters. *Appl. Opt.*, 32, 2606.

Welford, W. T. 1986. *Aberrations of Optical Systems.* Adam Hilger, Boston.

Wong, V. V. and Swanson, G. J. 1993. Design and fabrication of a Gaussian fan-out optical interconnect. *Appl. Opt.*, 32, 2502.

For Further Information

The conferences and special journal issues sponsored by the many professional societies give an accurate picture of current applications and research in binary optics. Volume 9 of the 1992 OSA Technical Digest series is a topical meeting on diffractive optics, and many of these papers also occur in the May 10, 1993 feature issue of *Applied Optics*. The SPIE sponsors two conference series which are of interest. Volume 883 (1988), 1052 (1989), 1211 (1990), 1555 (1991) and 2152 (1994) are part of the *Computer and Optically Generated Holographic Optics* series. Volumes 1544 (1991), 1751 (1992) and 1992 (1993) make up the *Miniature and Microoptics* series.

In the area of optics, Koronkevich's "Computer synthesis of diffraction optical elements" in *Optical Processing and Computing*, Arsenault et al., eds., is an early description of many of the principles currently used in binary optics. In addition, *Computer-Generated Holograms and Diffractive Optics* and *Holographic and Diffractive Lenses and Mirrors*, volumes 33 and 34 of SPIE's. Milestone Series, contain many of the original papers which are pertinent to binary optics today.

For fabrication, Sze's *VLSI Technology* is a general introduction to VLSI fabrication techniques, whereas *Lithography for VLSI*, Einspruch and Watts, eds., and *Plasma Processing for VLSI*, Einspruch and Brown, eds., concentrate specifically on the two areas most pertinent to binary optics. These two books are volumes 16 and 8 of Academic Press' VLSI Electronics Series. Finally, *Holography, Market Place*, Kluepfel and Ross, eds., describes some of the currently available mass replication techniques.

Occasionally, a less in-depth treatment is desirable. Veldkamp and McHugh's "Binary optics" in the May 1992 issue of *Sci. Am.* and Lee's "Recent advances in computer generated hologram applications" in the July 1990 issue of *Opt. and Phot. News* are both short articles that discuss the technical issues and applications of the field. Farn and Veldkamp's "Binary optics: Trends and limitations" in *Conference on Binary Optics*, NASA Conference Publication 3227 (1993), discusses nontechnical issues as well.

18

Electronic Displays

Carlo Infante

18.1 Introduction

This chapter addresses the essential elements of display technology. Displays are the primary communication link between a variety of electronic devices and the ultimate consumer of information, the human being. An enormous variety of display devices have been developed over the years to fill many different needs. These range from the small, low-information content devices found in Personal Information Managers (PIMs), hand-held calculators, and games to the exotic and expensive mammoth displays used in command and control situations or in outdoor stadiums. Important devices are those optimized for television viewing in the home and those capable of reproducing one or more pages of text and graphics, such as used in personal computers, terminals and workstations. Specialized or "niche markets" include helmet-mounted, head-up displays (HUDs) and very high-resolution displays for viewing radiographic images.

It would take several books, not just a single chapter, to fully discuss the principal elements of display technology. The work presented here provides only a brief overview of the subject. Hopefully, the reader will be spurred to look deeper into the topic by consulting some of the excellent texts available.

18.1.1 General Requirements of Electronic Displays

Because displays can be used in a wide variety of applications, it is not surprising that no single technology is applicable to every situation. However, it is useful to pause for a minute and consider some general guidelines as to what a display should be able to do:

- Conform to the needs of the human visual system
- Meet application requirements for
  - Color
  - Gray scale
  - Size
  - Resolution
  - Electrical and physical interface
  - Size, weight, ambient
- Cost

Displays are often used for extended periods of time, thus, viewing should be comfortable and adapted to the needs of the user. This an obvious requirement but not always followed in practice.

18.1.2 Elements of Photometry

What follows is a short discussion on the fundamentals of photometry (Sherr 1993; Infante 1994). It is well known that the human eye does not respond equally to all wavelengths of luminous energy. Although there are differences between individuals, there is something to be said for adopting a standard or prototype eye. This was done some years ago and the result is shown in Figure 18.1.

Given a source of electromagnetic radiation, the luminous flux is defined as

$$F = 683 \int_{0}^{\infty} V(\lambda)P(\lambda)\mathrm{d}\lambda \text{ lumens,}$$

where $V(\lambda)$ is the photopic function above and $P(\lambda)$ is the power spectrum of the source (watts at the wavelength λ). *Illumination* or *illuminance* is the luminous flux per unit area incident on a surface. Illumination is measured in lux or lumens/m^2. *Luminance* is the luminous flux emitted from a surface per unit solid angle. Luminance correlates somewhat with the perceptual sensation of brightness and is measured in nits or candela/m^2 (cd/m^2). Table 18.1 summarizes these terms.

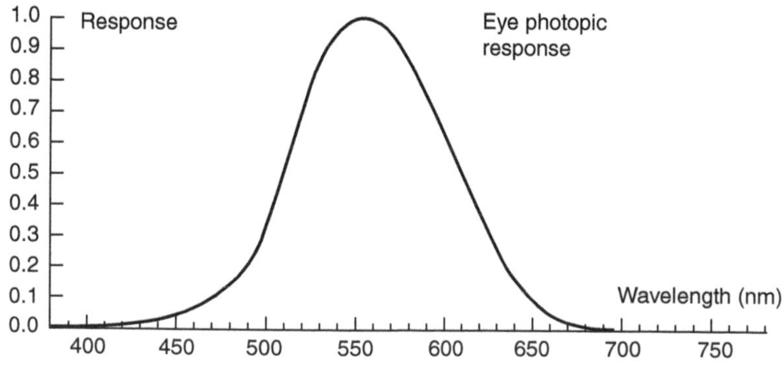

FIGURE 18.1 The response of the human visual system with wavelength.

TABLE 18.1 The Principal Photometric Units

Quantity	Defining Equation	SI Unit	US Unit	Conversion Factor[a]
Luminous flux	$F = K_m \int_0^x V(\lambda)P(\lambda)d\lambda$	Lumens	Lumens	1
Intensity	$I = dF/d\Omega$	Candela (lumens/steradian)	Candela	1
Luminance	$L = dI/dA_e$	Candela/m^2 (A_e is emitting area)	Ft-lamberts (1 cd/π ft^2)	0.2919
Illuminance	$E = dF/dA_i$	Lux (lum/m^2) (A_i is illumin'd area)	Foot-candle (1 lum/ft^2)	0.09294

[a] From metric into US units.

An important fact, often neglected, is that, although we can measure time, distance, and frequency with enormous precision, we can generally measure luminance to no better than a few percent (ref). Thus, care is needed in writing specifications or drawing conclusions from measurements.

The luminance of an image per se, is less important than its contrast with respect to unwanted information. Terms in general use are

$$\text{Contrast} = \frac{L_S - L_B}{L_S + L_S}$$

and

$$\text{Contrast Ratio} = \frac{L_S}{L_B},$$

where L_s is the luminance of the "signal" and L_B is the luminance of the background.

18.1.3 System and Ergonomic Considerations

18.1.3.1 Overview

A description, however condensed, of the key aspects of display technology would not be complete without discussing those factors that make a display system useful. Generally, it is not obvious a priori what characteristics a display should have to provide useful information and comfortable viewing. This problem is the domain of the specialized science of ergonomics or human factors.

Naturally, the most important requirement is that the image be clearly visible: the luminance and contrast should fall within acceptable limits. This, in turn, depends in great part on the ambient illumination in which the display will operate, because of the limited accommodating power of the human eye and the finite reflectance of display materials. As an example, displays intended for use in the cockpit of an aircraft operate from unimpeded sunlight to near-total darkness, an enormous range, approximately 10^5–1 lux. Displays for the cockpit require special design techniques. Conversely, displays designed for the office operate in a much more restricted illumination range of the order of 500–2,000 lux.

This has led to the promulgation and adoption of a number of standards issued by agencies, such as UL, VDE, FCC, ISO, and ANSI. It is beyond the scope of this work to examine each in detail, but an overview and a discussion of some important issues may prove useful.

In the United States, the ANSI/HFS 100 Standard (ANSI 1988) published in 1988 has served as a foundation for a series of European requirements and contains a number of useful recommendations. The most important are summarized in Table 18.2.

TABLE 18.2 Summary of the ANSI/HFS 100-1988 Recommendations[a]

Characteristic	Specification
Ambient illuminance	500 lux nominal
Display luminance	At least 35 cd/m^2
Luminance adjustment	Mandatory
Viewing distance	500 mm nominal
Character height	2.3–6.5 mm @ nom view distance 3.1 mm preferred
Character width	92–93% of height
Image polarity	Both acceptable (black-on-white and white-on-black)
Character format	7×9 minimum
Character modulation	(LM − Lm)/(LM + Lm) > 0.75
Luminance uniformity	Better than 50%
Jitter	Less than 0.1 mm @ nom view distance
Linearity (integral)	Better than 2%
Linearity (differential)	Better than 10%
Flicker	Flicker-free for 90% of viewers

[a] The standard is currently undergoing revision.

18.1.3.2 Flicker

Flicker is the sensation we receive when looking at an image, such as that generated by a CRT, whose repetition frequency is insufficient. The critical fusion frequency (CFF) is defined as the frequency at which the flickering sensation just disappears. Although some of the fundamental principles governing CFF have been known for some time, a complete description of the phenomenon is still lacking. What is known is that CFF depends on a number of factors such as image luminance, ambient illumination, the angle the image forms with the eye, and the age and, possibly, even the gender of the observer. Confounding the problem is the fact that not all individuals are equally sensitive to flicker, so that any recommendation relating to flicker must be, of necessity, statistical. The most comprehensive work to date on the subject is due to Farrell et al. (1987) who published analytical equations that allow predicting flicker performance under a variety of conditions. These equations are at the basis of recommendations as to minimum refresh rates for CRT displays used in the office. Television displays are, generally, viewed at greater distances, thus, viewing angles are much smaller and flicker requirements are not nearly as severe. Thus, in television displays, the practice has been to choose the refresh rate equal to the power line frequency (60 Hz in the United States and 50 Hz in Europe) so as to minimize interference. To minimize the bandwidth over which the signal is transmitted, television systems generally employ interlace. This means that the picture is not scanned progressively from top to bottom. Rather, half the picture is scanned first (e.g., with 250 scanning lines) in 1/30 of a second. During the next "pass" or field, the remaining half of the picture is scanned, with the scanning lines halfway between the old ones.

18.1.3.3 Resolution, Image Quality

One of the most complex and controversial subjects in the ergonomics of displays in general and of CRT displays in particular, is that of image quality. Although this may be surprising, especially in view of the many excellent displays currently on the market, there are good reasons. In the first place, the shape of the scanning aperture is not square, as was shown. Thus, it is not at all obvious, a priori, just what is meant by "resolution." Furthermore, in an optimum display system, we expect the human eye to be the limiting factor, or, at least, a significant contributor to the resolution of the system. Thus even a first-order resolution theory must start with a description that encompasses both the behavior of Gaussian spot profiles and that of the human eye.

This is exactly what has been done by a number of researchers. The description of the hardware aspects of the system are incorporated in the modulation transfer function (MTF) whereas the resolution properties of the human visual system are described by the contrast sensitivity function (CSF). Mathematical expressions have been developed by a number of researchers that allow prediction of image quality once the parameters of the system, such as luminance, viewing distance, and bandwidth are known. Currently, an algorithm known as SQRI (square root integral) has shown the best correlation between predicted values and subjective judgments. SQRI is being considered for inclusion in the revised version of the ANSI-HFS 100 Standard.

18.1.3.4 Radiation: X-rays

A form of radiation that has been of some concern is x-rays. X-rays are emitted, of course, when energetic electrons strike a material, such as the phosphor screen. X-rays penetrate quite deeply into matter including human tissue, especially at shorter wavelengths. In excessive doses, x-rays pose a health hazard. The available energy of the electron beam limits the shortest wavelengths that can be produced. Reasonable design techniques, that limit the beam current together with the maximum beam energy even under failure conditions, insure that x-ray emissions produced by a CRT display will remain below limits imposed by appropriate regulatory agencies.

18.1.3.5 Radiation: Low Frequency Electric and Magnetic Fields

Recently, a number of articles in the popular media have expressed concerns over the potential health effects of the weak electric and magnetic fields produced by CRT monitors. At this time, no conclusive evidence exists that these weak fields do, in fact, cause adverse health effects. To the contrary, a major experiment conducted by the National Institute for Safety and Health (NIOSH) found no increase in miscarriages as a result of exposure to the weak electromagnetic fields emanating from CRT displays. Naturally, a variety of techniques exist that allow reducing of these fields. The techniques include coating the front glass surface of the CRT with a conductive layer, so as to eliminate the electrostatic charging of the front surface. Techniques to reduce dynamic electric and magnetic fields include canceling coils and electrodes, driven in synchronism with the horizontal sweep and arranged so as to cancel the stray flux. Irrespective of the physics of the situation, concerns have developed in the user community. As these market concerns coalesce, some form of regulation, similar in intent to that in effect for x-rays, will undoubtedly emerge.

18.1.4 The Major Display Technologies

A number of different display technologies have achieved a level of technological and market success over the years. These can be classified in a number of different ways. A traditional one has been that of separating the technologies according to whether they generate light (emissive) or simply reflect and modulate the light from separate sources (passive).

We prefer to classify the technologies, as shown in Table 18.3, according to the physical medium on which each depends.

Cathode-ray tubes (CRTs) and field emission devices (FEDs) use electron beams operating in a vacuum. Plasma display panels (PDPs) utilize the controlled electrical breakdown of gas molecules, whereas liquid crystal devices (LCDs) take advantage of the electro-optical properties of a class of materials known as liquid crystals that combine the properties of liquids together with the ordered structure inherent in crystals. Finally, electroluminescent devices (EL) generate light as a result of

TABLE 18.3 The Major Display Technologies

Diplay Medium	Display Technology
Vacuum	CRT, FED
Gas	PDP
Liquid	LCD
Solid	EL, LED

applying an electric field to a solid. Light emitting diodes (LEDs) are described fully in other parts of this volume. They are, of course, semiconductor devices which generate light by forward biasing a p–n junction with the appropriate material and doping it. Each of these technologies will now be discussed in some detail.

18.2 The Cathode-Ray Tube

18.2.1 Overview

Although newer technologies have received a lot of attention, the cathode-ray tube (CRT) remains dominant both in terms of monetary value and the actual number of units shipped (Castellano 1994). The reasons are many and include an established manufacturing base that routinely produces hundreds of millions of the devices at low cost and high yields and the superior ergonomics of CRTs. This situation has made CRTS pervasive in all sorts of applications including personal computers and television, to name just a few.

The block diagram of a CRT display is shown in Figure 18.2. The word "monitor" is generally employed to describe this collection of waveform generators, power supplies together with the display device itself and a mechanical structure or housing. A television set consists of a monitor with the addition of a tuner, and audio and decoding functions.

The heart of the monitor is the CRT device itself shown in Figure 18.3. It consists of an electron source (a thermionic cathode), beam forming and focusing electrodes, a deflection system, and a screen. Naturally, a glass envelope is required to sustain the required vacuum.

In short, electrons are emitted by the cathode, focused into a small bundle by some lenses and accelerated to several kilovolts before hitting the phosphor screen. There, the beam is converted into a luminous spot. The electron beam is modulated in *space* by a deflection field. In addition, the beam current can also be modulated in *intensity* by applying a modulating voltage between the CRT grid and cathode. As a result, a picture appears on the screen and is viewed by the observer. By far the most popular way of addressing a CRT is to repetitively cover the CRT screen with a "striped pattern" or raster as shown in Figure 18.4. Linear deflection waveforms are applied along each axis at frequencies that are

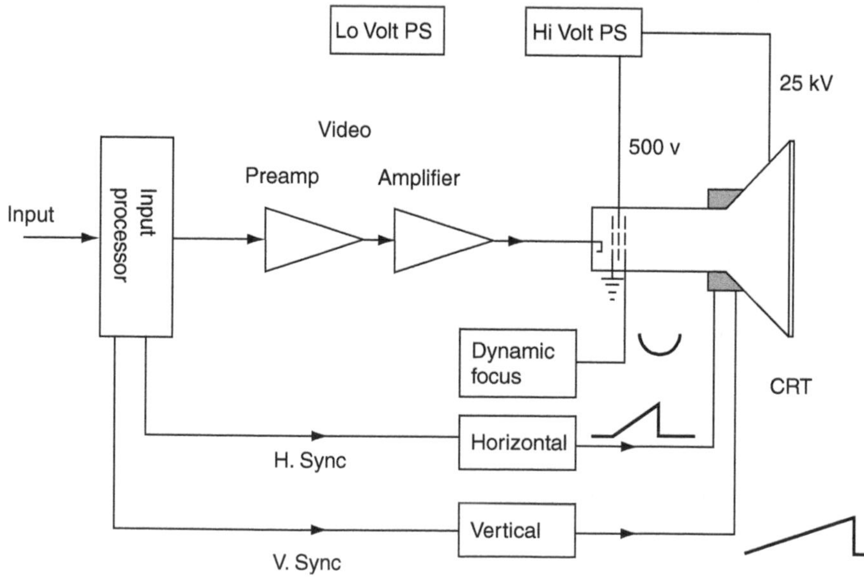

FIGURE 18.2 Monitor block diagram.

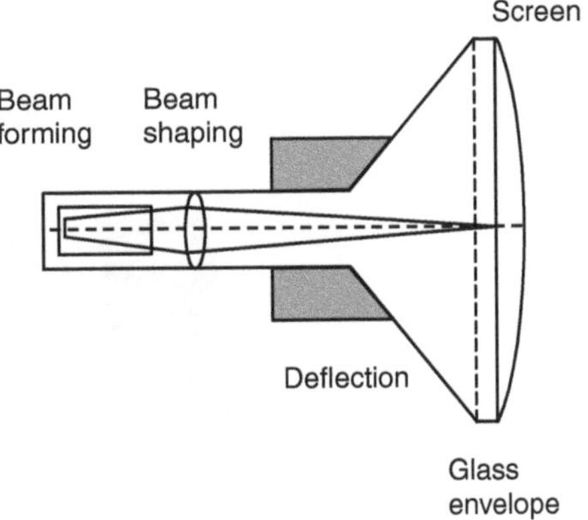

FIGURE 18.3 Monochrome CRT.

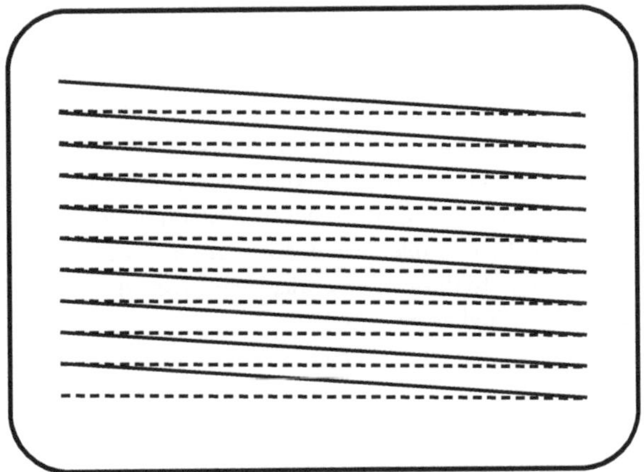

FIGURE 18.4 Raster scanning in a CRT.

fixed and multiples of one another. The electron beam is turned on or off at the right time so that a picture is formed.

18.2.2 Monochrome CRTs

The simple monochrome CRT shown in Figure 18.3 is capable of displaying only images of a single color. Although once dominant because of cost and simplicity, monochrome CRTs have been largely supplanted in the marketplace by their more advanced offspring. Nonetheless, applications exist in which color is not a requirement. These include the display of complex mechanical drawings, simple alpha-numeric terminals, word processing, and x-ray imaging. For these, the superior resolution and ergonomics of the device, together with its inherently lower cost both at the device and at the system level, continue to be attractive. It is also a convenient way to begin the discussion of the technical aspects of CRTs.

18.2.2.1 Elements of Electron Optics

The region in which the beam is formed and focused in a CRT is called the electron gun. In the beam-forming region a filament and cathode, shown in Figure 18.5a and b, generate an electron cloud.

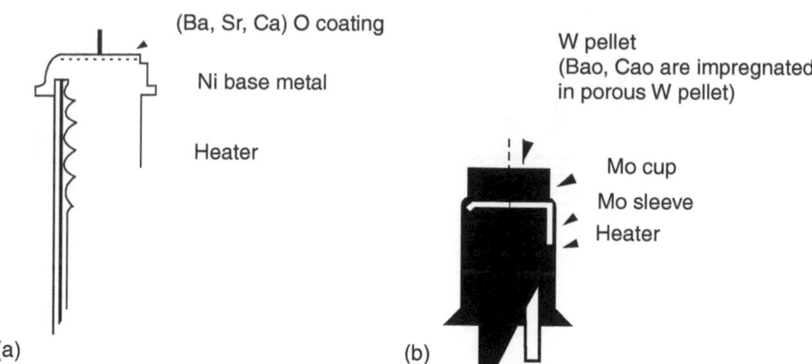

FIGURE 18.5 (a) Conventional cathode, (b) impregnated cathode. (From Yamazaki, E. 1993. *SID Seminar Notes*, 2, F-5, 1–47. With permission.)

These electrons are formed into a bundle called the crossover (ideally a point source) and one then accelerated and focused in the beam-shaping region. The electrons finally impinge on the phosphor screen where their energy is converted into light. This is shown diagrammatically in Figure 18.6. A system of lenses is responsible for forming the beam into a focused spot. The details of the underlying physics are beyond the scope of this chapter. Suffice it to say that the electron paths are described by the same sort of equations that describe light rays. This means that concepts such as index of refraction, focal length, aberration and magnification all have an exact analogies in electron optics (El-Kareh and El-Kareh 1970). More importantly, one can show that simple structures, such as a set of coaxial cylinders at different voltages, have properties very similar to those of an optical lens. Consider Figure 18.7 in which we assume that the right-hand cylinder is at a higher potential.

The electron beam, entering from the left, is subject to the action of the electric field, proportional to the gradient of the potential. The gradient is, of course, perpendicular to the equipotentials at every point in space. Thus, on the *left* side, the beam sees a force deflecting it *toward* the axis. This has the effect of a converging lens. On the *right*-hand side, the gradient is now directed *away* from the axis, thus, the effect is of a diverging lens. The net effect of the whole structure is that of a converging lens. Lenses with differing characteristics and focal lengths are obtained by changing the mechanical structure and the operating potentials. Real lenses, both optical and electron-optical, suffer from various imperfections commonly referred to as "aberrations." Spherical aberrations are particularly important in electron-optics. These increase rapidly as the lens diameter becomes smaller.

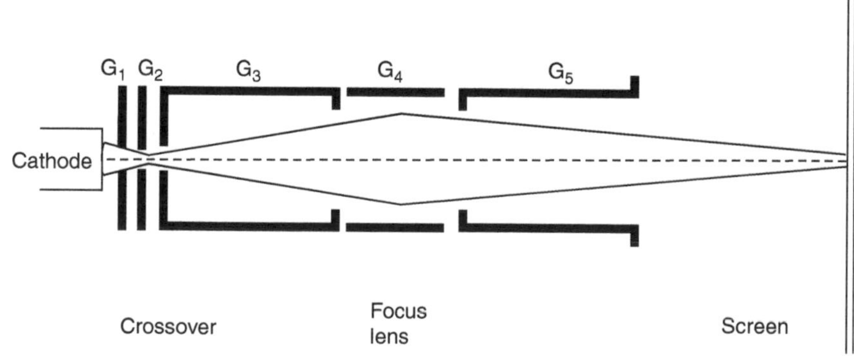

FIGURE 18.6 Operation of an electron gun. The electron cloud is formed into a crossover which is subsequently focused onto the screen.

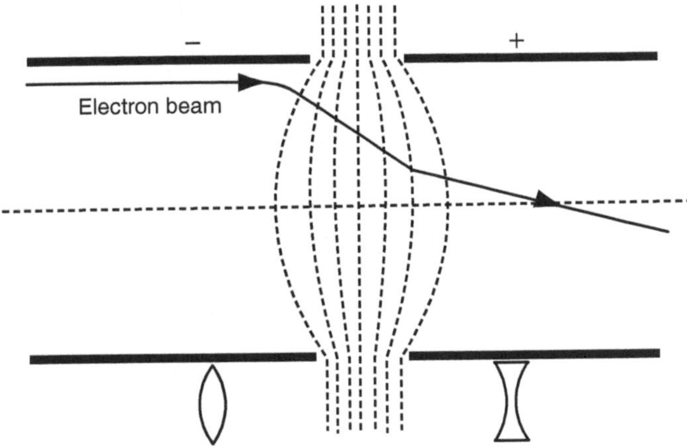

FIGURE 18.7 Operation of an electron-optical lens.

18.2.2.2 Deflection Systems

To generate a picture on the screen, the beam must be deflected in a controlled fashion. There are two known methods to deflect a beam: electrostatic and magnetic. In electrostatic deflection, the beam is made to pass between two sets of conductive deflection plates, orthogonal to one another. The electric field resulting from voltages applied to the plates allows positioning the beam anywhere on the screen. This technique is generally used in applications, such as oscilloscopes, in which deflection speed is a major requirement.

In applications in which a large deflection angle and high luminance are important requirements, magnetic deflection is preferred. Instead of deflection plates, one uses deflection coils, wound orthogonally to the CRT axis, as shown in Figure 18.8. Current is made to flow in the coils and the resultant magnetic field deflects the beam.

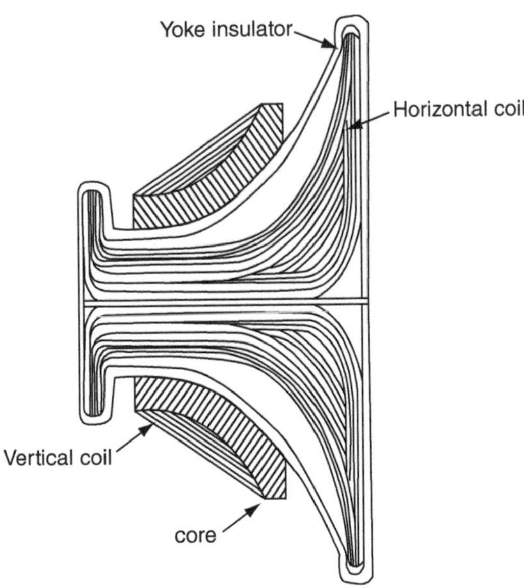

FIGURE 18.8 A CRT deflection yoke. (From Dasgupta, B. B. 1991. *SID Seminar Lecture Notes*, 1, 1–43).

It is important to note that, although the deflection voltage for electrostatic deflection increases *linearly* with the screen voltage E_b, in magnetic deflection, the required field B, hence, the deflection current, increases with the *square root* of E_b. Because display luminance is proportional to E_b, magnetically deflected systems are preferred in most applications.

One of the important characteristics of a deflection yoke is its effect on spot focus as the beam is deflected. If the magnetic field within the yoke is not uniform, different electrons within the beam will be deflected differently and an astigmatic spot results. The distribution of optimum turns required to generate a uniform field is a cosine function of the angle between the CRT axis and the coil loop. A good monochrome yoke is wound with the proper distribution so as to guarantee stability and consistency. This is achieved by using magnetic materials in which slots or teeth are provided and by carefully controlling manufacturing techniques. Any real deflection yoke has another effect on the spot focus. As the beam is deflected away from the axis, it encounters a fringing field and must travel a greater distance before reaching the screen. Both of these effects cause the beam to come to a focus before the screen. So-called dynamic focus techniques to compensate to this effect are often employed and involve changing the focus voltage as a function of the beam position.

18.2.2.3 Phosphors

After the deflection process, the electron beam reaches the screen. The screen consists of a glass faceplate, with the appropriate optical properties, on which a suitable phosphor material has been deposited. Phosphor materials are generally inorganic crystals, with particle sizes in the 5–15 μm range. Due to defects in the crystalline structure, the phosphor materials are able to absorb incident energy and convert it into light. This process as known as cathodoluminescence. Energetic electrons from the electron beam penetrate deep into the phosphor. As the beam is absorbed, it raises some of the phosphor electrons to an excited state. As sketched in Figure 18.9, the excited electrons, then, return to the ground state, releasing the excess energy in the form of light with a wavelength and a persistence dependent on the particular phosphor and its preparation. Phosphors operated above a few kV are usually coated with a thin (150–300 nm) aluminum film that improves luminous efficiency by stabilizing the operating voltage and by reflecting more of the emitted light toward the observer. A wide variety of phosphors are available commercially with differing characteristics (Infante 1986). For satisfactory operation, the right choice of phosphor material and deposition techniques must be made. The characteristics one usually looks for in a phosphor are efficiency, resolution, and good life, together with the appropriate color and persistence.

18.2.3 Color CRTs

Color is a major technological and marketing challenge for displays. Color is important both in computer displays, as it represents an additional information channel, and in the entertainment area, as color displays are much more natural and appealing than monochrome (Morrell et al. 1974).

To display color, one takes advantage of Helmholtz's trichromaticity theory. Thus, each color can be represented by the addition of three independent primaries. As a result, much more information must be processed to display color. Practical devices achieve color by some form of sharing or separation

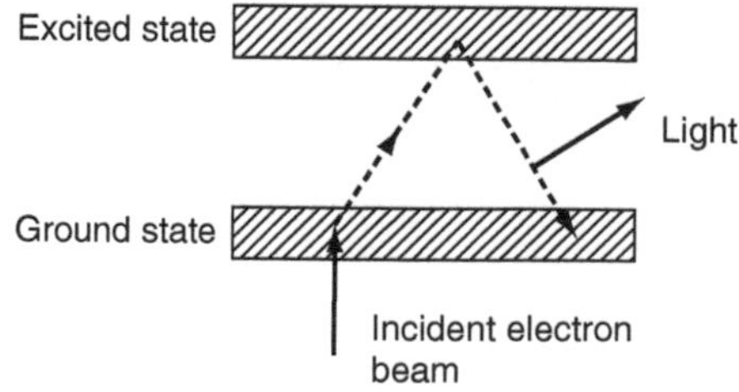

FIGURE 18.9 Phosphor operation.

mechanism in either space or in time. Picture elements in color are obtained by synthesizing either subpixels that are spatially separated or individual monochrome picture frames that are temporally distinct. Each subpixel or subframe is composed of a single primary color. The subpixel technique implies an inherent loss in spatial resolution, whereas the frame sequential technique requires a faster refresh rate to avoid flicker.

A number of technological solutions have been proposed for color CRTs including the CBS spinning disk, penetration CRTs, and the liquid crystal color shutter. None has achieved the market and technological success approaching that of the shadow mask CRT and its successors. The development of this device is linked to the name of David Sarnoff, the founder of RCA, and to the laboratory that bears his name.

18.2.3.1 Ergonomics of Color

Before exploring the hardware aspects of color, we pause for a moment to consider the ergonomic aspects of viewing color (Silverstein 1994; Travis 1991). While seeing color is a natural experience, the science of color is quite complex and as yet not totally developed. As an example of the complexity of the subject, the color of an object is dependent not only on the spectral content of the emitted light, but also on the characteristics of the surround and of the ambient illumination. Furthermore, a number of differing spectral representations exist for the same (perceived) color.

An essential problem is that, of a metric or space with which to represent color in a mathematical sense. There have been a number of proposals, none of them totally satisfactory. Currently, the CIE XYZ system is the most widespread. It is based on a set of color matching functions that define any color by three values.

Two classes of applications are often encountered. The first involves natural images, in which the fidelity of the color representation is essential. A second type of image has no correspondence in the real world, but uses color to distinguish or categorize information. Examples of the latter are color charts representing data or word processing documents in which color is used to highlight information.

18.2.3.2 Shadow-Mask CRTs

A number of alternative architectures for shadow-mask CRTs have evolved over the years. These are shown in Figure 18.10. Although there are substantial variations among the devices, operating principles are similar. Three electron guns are employed and a thin mask or grille is interposed between the guns and the screen. The mask is made of stainless steel and "extrudes" the beams into beamlets insuring that each beamlet lands on the correct phosphor dot element. Naturally, a full frame of a single primary (e.g., red) should appear of uniform color to the eye (purity). For satisfactory performance, the registration between beamlet and phosphor element must be held to no more than a few microns. This is achieved by manufacturing each mask and screen as a matched pair (lighthouse exposure). A further requirement of a color display device is that small images (graphic lines or text characters) appear without fringing colors (convergence). This, in turn, requires that the centroid of each beam coincides with the other two beams over the entire screen to within 100 microns or so.

18.2.3.2.1 Delta Gun

The so-called delta gun is the oldest architecture and the simplest conceptually. Three individual guns are arranged in a triangle or delta inside the tube neck and angled (1° or so) with respect to the axis. The shadow mask consists of a thin (0.1 mm) sheet of steel with an aperture corresponding to each phosphor dot triad. The delta architecture makes good use of the available neck space, allowing relatively large electron-optical lenses. However, maintaining accurate positioning of the electron guns over time, temperature, and manufacturing process variables is very difficult. Delta guns require complex circuitry to compensate for these tolerances and have become practically obsolete.

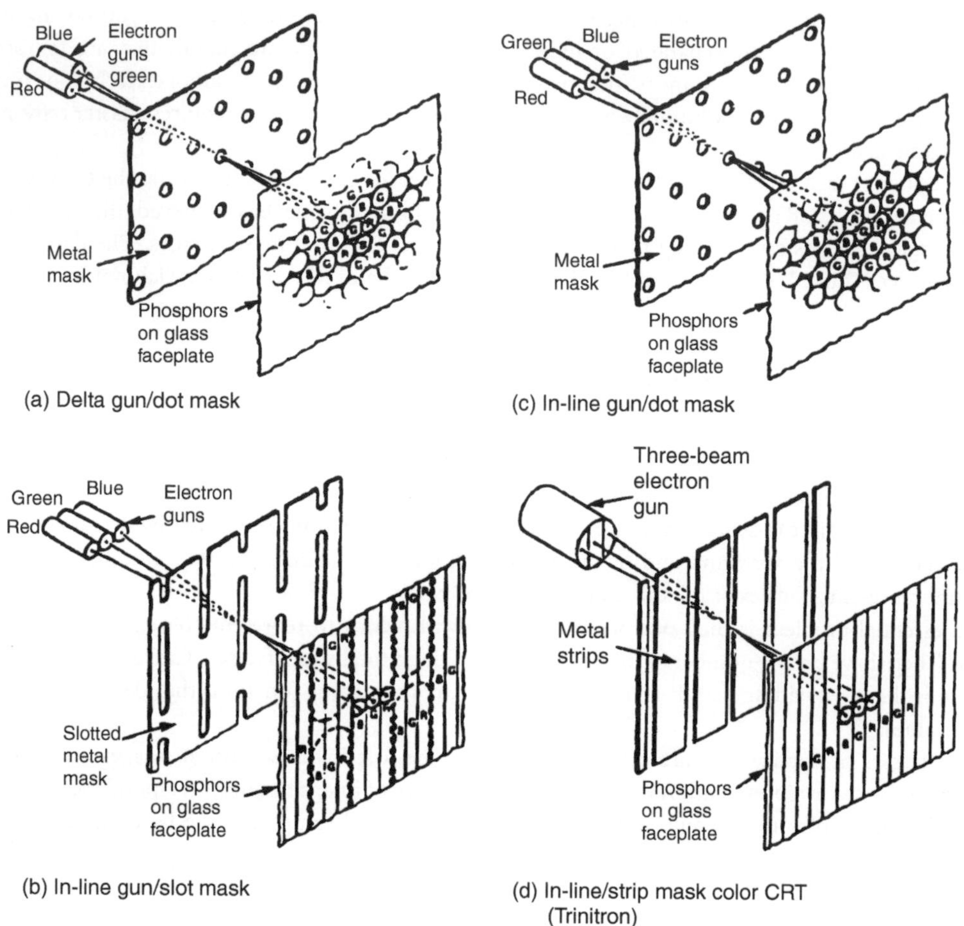

FIGURE 18.10 Shadow-mask architectures. (From Silverstein, L. D. and Merrifield, R. M. 1985. The development and evaluation of color systems for airborne applications. DOT/FAA July 1985 PM-85-19.)

18.2.3.2.2 In-Line Gun

In the in-line structure, the guns are arranged horizontally with the center (green) gun parallel to the axis. The outer guns are tilted inward by about 0.5°. In high resolution applications, a dot screen, similar to the delta dot screen is used, whereas, for lower resolution applications, a slotted mask is preferred for increased brightness. Eliminating the vertical tilt angle between guns and adopting a unitized mechanical structure greatly reduce purity and convergence difficulties to the point where these corrections are routinely built into the deflection yoke. A detailed description of self-converging yokes is beyond the scope of this work. The interested reader should consult the excellent literature on the subject (Dasgupta 1991).

18.2.3.2.3 The Trinitron

A different approach to the problem was developed by Sony with the Trinitron. The guns are, again, arranged horizontally, but the beams go through a common focus lens. This achieves a space utilization factor even better than delta guns, by having a large common focus lens as shown in Figure 18.10. For this to work, Sony had to use a more complex gun structure by adding a prefocus lens, so that most of the electron rays would go through the center of the lens as opposed to the periphery where aberrations are high. The faceplate is cylindrical and the mask is replaced by a grilled aperture. The phosphor is deposited on the screen in stripes. This eliminates the loss in resolution in the vertical direction. In moderate resolution applications, the grille allows higher transmission of the electron beam, thus, increasing luminance. This comes at the added expense and weight of a steel frame required to hold the grille under tension.

18.2.3.2.4 The Flat-Tension Mask

Yet another shadow-mask architecture is the flat-tension mask (FTM) that was developed by Zenith (Figure 18.11). Instead of a domed shadow mask, the device uses a mask held under tension in both vertical and horizontal directions. This requires the use of a flat glass faceplate. Because the mask is under tension, as it absorbs power from the electron beam, the mask itself does not dilate, thereby, affecting purity, but remains flat. As a result, greater luminance levels can be achieved, together with other advantages, such as enabling interchangeable masks and screens.

18.2.4 Important Relationships

There are a number of mathematical relationships that apply to CRT systems. The most important are given below.

The luminance produced by a raster-scanned CRT system is given by

$$\mathrm{Lum} = \frac{\eta T_{\mathrm{G}}}{\pi}\frac{i_{\mathrm{b}}E_{\mathrm{b}}}{A}\delta,$$

where η is the phosphor efficiency in lumens/watt, T_{G} is the glass transmission, i_{b} and E_{b} are the beam current and voltage, and A is the raster area (m^2). δ is the raster duty cycle and Lum is the resulting luminance in nits (cd/m^2).

The deflection angle θ is given by

$$\sin\theta = C_{\mathrm{F}}N_{\mathrm{i}}\sqrt{\frac{q}{2mV_{\mathrm{b}}}}$$

where C_{F} is a constant that depends on the geometry, N_{i} is the number of ampere-turns in the yoke, q and m are the charge and mass of the electron, respectively, and V_{b} is the accelerating voltage.

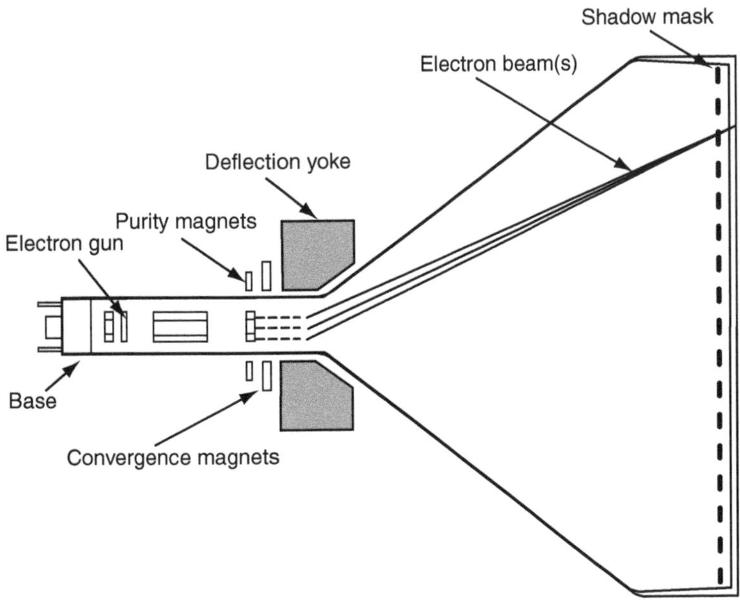

FIGURE 18.11 Structure of the flat-tension-mask CRT.

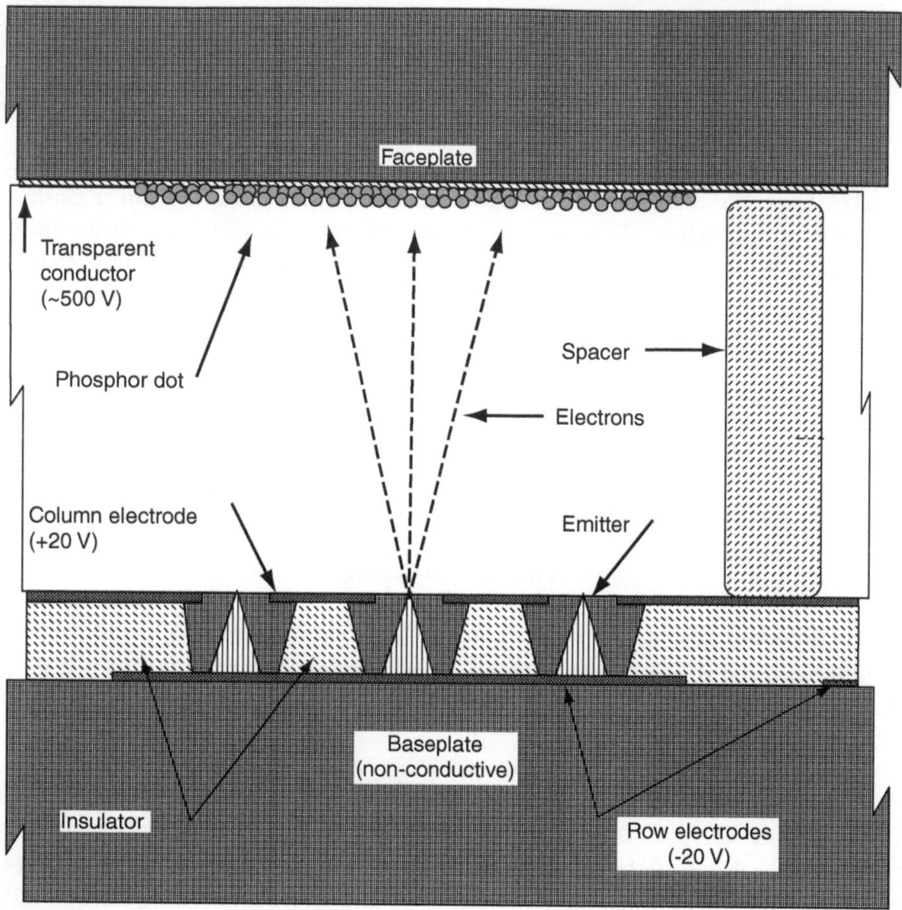

FIGURE 18.12 Structure of field emitter display. (Courtesy, silicon video.)

The size of the spot on the CRT screen is a complex function of the gun parameters and of the CRT geometry. In general, it decreases as V_b is made larger and the gun is made longer. For moderate values of beam current, the shape of the spot is Gaussian.

18.2.5 Other CRT Devices

There have been a number of attempts at devising so-called flat CRTs, i.e., devices with all the advantages of the CRT but with greatly reduced depth. Currently, the so-called field emitting device (FED) (Spindt et al. 1976) is receiving considerable attention from a number of laboratories around the world. It is shown schematically in Figure 18.12.

These devices do not employ thermionic cathodes like CRTs, but depend on cold cathode emission from small microtips. Because of the close proximity to the other electrodes and the shape of the tip, very high fields are present at the tip, which allows electron emission to take place. A large number of these tips can be built simultaneously, many per pixel. The tips are then addressed in a matrix form. Because of the small distance to the screen, there is no need for focusing structures. The electrons strike the screen, coated with phosphor, and light is emitted in much the same way as a CRT.

Clearly, there are many similarities between CRTs and FEDs. Both rely on vacuum, electron optics, and cathodoluminescence. The principal difference is that FEDs operate at much lower voltages, 300–1,000 V, as opposed to the 15–30 kV normally employed with CRTs. This not only makes very thin (a few millimeters) devices possible, but also enables low-voltage matrix addressing of the cathodes and eliminates x-rays. It also means that many of the more common CRT phosphors are unusable and new phosphors such as those used with EL devices, must be chosen. There are number of technological and fabricating issues yet to be resolved. Chief among these are the viability of large area, low-cost devices and the demonstration of long-lived, low-voltage color phosphors with satisfactory chromaticity and luminous efficiency.

18.3 Flat Panel Devices

One of the principal limitations of the cathode-ray tube is its sheer bulk. This is due in large part to the difficulty of deflecting the electron beam by large angles. Thus, in applications in which size and weight are paramount, such as personal information managers, laptop computers, and pocket calculators, alternative display technologies are needed. Concurrently, it seems that something can be gained by abandoning the analog-addressing capability of the CRT, in favor of a digital approach.

To display just a limited set of alphanumeric characters, one can use economical displays featuring just seven segments, such as the one shown in Figure 18.13.

FIGURE 18.13 Schematic diagram of a seven-segment display. (Each segment has an independent connection).

Complex images, however, require a more general arrangement such as the one shown in Figure 18.14.

To display an image, the row electrodes are addressed sequentially, beginning at the top. The appropriate voltages are applied to the column electrodes, thus, producing the image. Figure 18.15 shows a simple matrix display consisting of 3×3 pixels, with the center pixel addressed (Scheffer 1993).

It is apparent that, if simple addressing schemes are used, nonselected pixels are subjected to half the voltage of selected pixels. Unless the optical transfer characteristic of the display is very nonlinear, this can lead to poor contrast. Improved addressing schemes, such as the one shown in Figure 18.15b, are possible. In the general case, one finds (Alt and Pleshko 1974) that optimum performance is achieved when the voltages applied to the nonselect row electrodes and nonselect column electrodes are V_0/b and $2V_0/b$ respectively, where $b=\sqrt{N}+1$ and N is the number of addressed lines. From this, it is clear that, as the resolution of the display increases, the ratio between select and nonselect voltage approaches unity. Thus, for high resolution images, display elements with sharp nonlinearities are required.

18.3.1 Liquid-Crystal Devices (LCDs)

Liquid-crystal displays (LCDs) have been the subject of very intense development efforts in the last decades. Because of this and their inherent properties of low power requirement, low cost, and the ability to display color, LCDs have become a major market presence. LCDs have been known for many years and were first applied to displays in the 1960s, at what was, then, called RCA Laboratories. Liquid crystals are materials that combine the properties typical of ordered structures, such as crystals, while retaining the liquid form which makes them amenable to pouring, etc. There are a number of different material classes that have been developed over the years, such as nematic, cholesteric, and smectic (Chandrasekhar 1992; Steemers 1994). Currently the twisted nematic and its variants, such as supertwist nematic, have become preeminent.

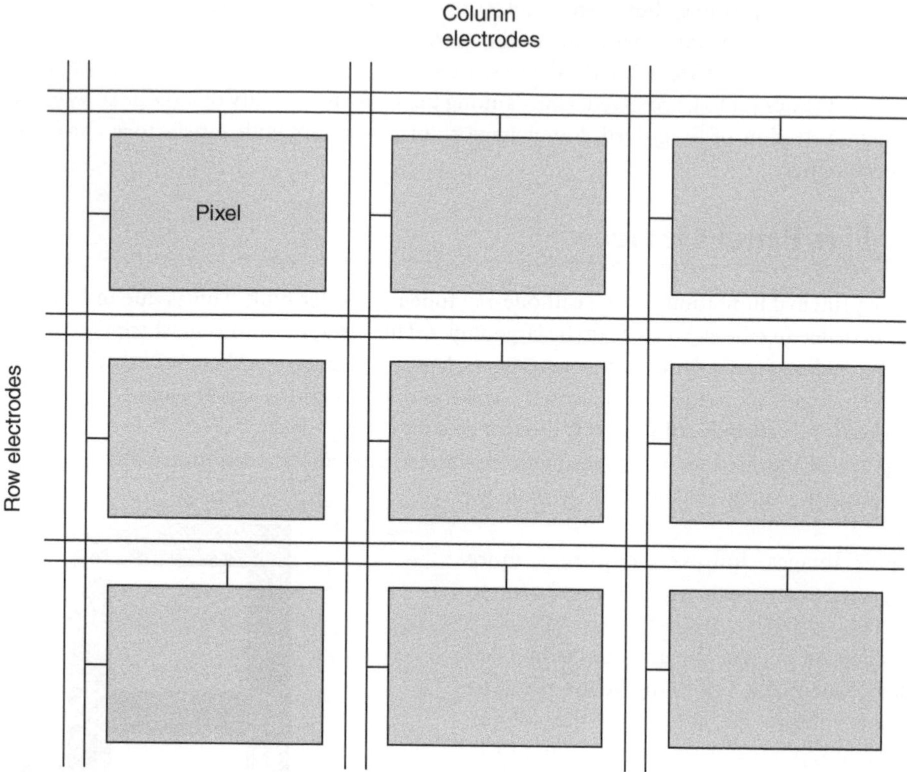

FIGURE 18.14 Generalized matrix display.

The fundamental property of LCs that make them useful as displays is that they are sensitive to an external electric field. This is shown in Figure 18.16.

LC molecules rotate as a result of the electric field. The use of polarizers allows control of the optical transmission. As a result, an undriven cell will appear light, whereas the cell will appear dark when voltage is applied. The optical response is a function of the material and of the applied voltage, with a typical dependency shown in Figure 18.17.

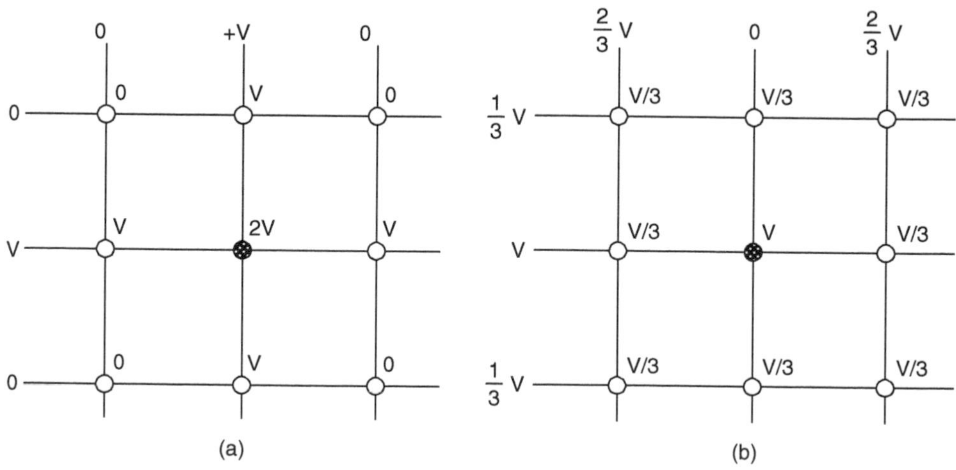

FIGURE 18.15 Matrix display addressing.

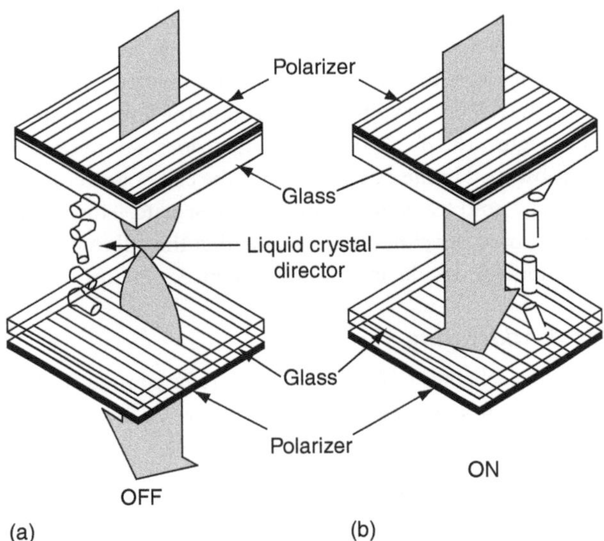

FIGURE 18.16 Operation of liquid-crystal displays. (From Scheffer, T. 1993. *SID Seminar Lecture Notes*, 1, M-7, 1–63.)

Because only a few volts are required for the cell to change states and the current drawn is negligible, the technology has found easy application in hand-held devices and laptop computers. The steepness of the response curve is finite, however. This is important in some applications, as will be seen.

18.3.1.1 Directly Addressed LCDs

18.3.1.1.1 Twisted Nematic Displays.

To make a working display, the LC material is assembled in a cell, shown in Figure 18.16. The cell consists of two glass plates, each coated with a transparent conductor. The cell is assembled so that the liquid-crystal molecules undergo a 90° twist from the top plate to the bottom plate. Polarizers are applied to the outside of each glass plate, with their polarization directions, usually, parallel to the LC direction of each plate.

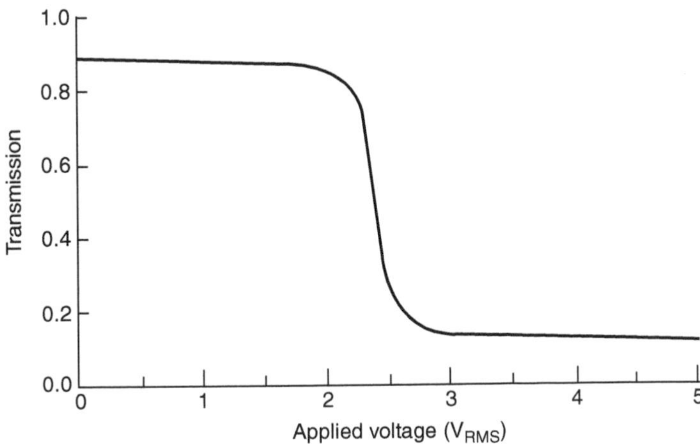

FIGURE 18.17 Transfer characteristic of a typical liquid-crystal cell [18].

The incident light on the cell is polarized linearly, as indicated. With an appropriate choice of LC material and cell gap, the light undergoes a 90° rotation in polarization as it passes through the liquid crystal, before exiting the bottom polarizer. In this mode of operation frequently used in LCDs, the cell is transmissive in its undriven state. When a voltage is applied to the liquid crystal, the molecules will align with the field, as shown in Figure 18.16b. The incident light does *not* undergo rotation in polarization direction and is, therefore, absorbed in the exit polarizer. If the exit polarizer is rotated by 90°, the TN cell is opaque in the undriven state, and transmissive in the driven state. This is called the "normally dark" mode, which is less frequently used. It is important to note that the cell responds to the RMS value of the driving voltage. Any dc component results in drastically reduced cell life.

18.3.1.1.2 Multiplexed Addressing

The straightforward addressing schemes of Figure 18.15 are referred to as multiplexed addressing. With these schemes, the limited steepness of the electro-optic curve of available TN materials resulted in displays with satisfactory contrast and viewing angles only for very small number of elements.

18.3.1.1.3 Supertwisted Nematic Displays

Improving the electro-optic curve by brute force proved to be quite difficult. However, increasing the twist from the top plate to the bottom plate from 90° to 180° or even 270°, resulted in materials with much steeper curves. LC materials made in this way are called supertwist nematic (STN). Although the STN LCD made high-resolution displays possible, it has a number of limitations. Due to the very limited duty cycle of the addressing pulse for high-resolution displays, materials with response times of 150–300 ms are required. This, in turn, limits the display's ability to follow rapid changes in the image, such as mouse or cursor movements.

Another limitation is that the STN cell operates in a black and blue mode, as opposed to the black and white operation of a TN cell. This effect can be compensated for by adding second, undriven, STN cell with opposite liquid-crystal molecule rotation (called the double STN or DSTN cell) at some cost penalty.

18.3.1.1.4 Active Addressing

A method, known as active addressing of STN displays, has been described in the literature (Scheffer 1993). It involves optimizing the driving waveforms and appears to be a significant improvement in addressing technology. An STN display with a 60:1 contrast ratio and a 35-ms response time has been demonstrated.

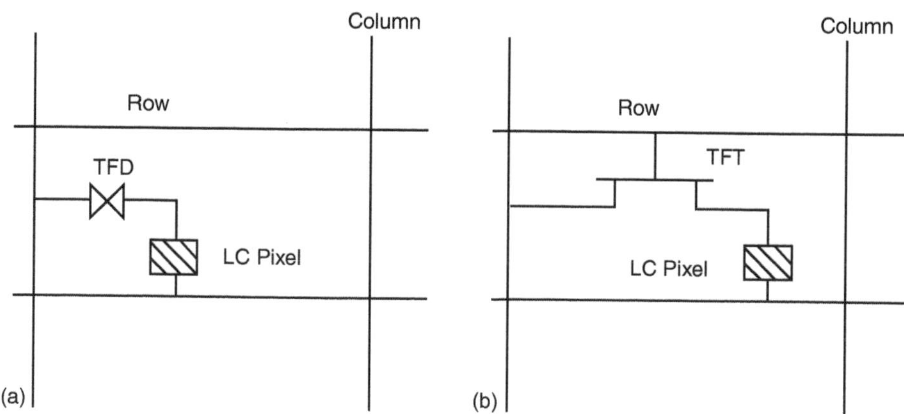

FIGURE 18.18 Active matrix liquid crystals.

18.3.1.2 Active-Matrix LCDs

Two- and Three-Terminal Devices

For a liquid crystal to perform satisfactorily in directly addressed, high-resolution applications, it must demonstrate a steep electro-optic curve, as we have seen, in addition to offering wide viewing angles, good contrast, low-voltage operation, and a wide temperature operating range. A possible solution was proposed in the 1970s consisting of inserting a solid-state circuit device with strong nonlinearities (Morozumi 1993). As Figure 18.18 shows, the nonlinear device can be either a two-terminal device, such as a diode, or a three-terminal device, such as a transistor. The manufacturing processes to make thin-film devices using amorphous silicon have been proven compatible with large area, low-temperature standard glass. Furthermore, the technology can provide satisfactory low-current performance. Thin-film *diodes* (TFDs) have the obvious advantage of being simpler devices that lead to higher yields in manufacturing. Thin-film *transistors* (TFTs), on the other hand, provide increased isolation between the control and signal lines and are less susceptible to manufacturing tolerances of the device. Currently, TFTs enjoy much greater popularity.

In summary, with AMLCDs, there is no inherent limit to the number of elements that can be addressed and, thus, to the resolution of the display. Furthermore, TN materials with good characteristics, such as contrast ratio and switching speed, can be used.

Manufacturing Issues (O'Mara 1994)

The validity of a technology is verified on the manufacturing floor. Although this is somewhat of a truism, the manufacture of LCD panels, both active and passive, poses a unique set of challenges and opportunities. The processes involved are quite complex, ranging from sputter deposition of transparent electrode material to amorphous silicon deposition by plasma-assisted chemical vapor deposition (PECVD) and including, of course, the most stringent clean room and process control techniques. Because a single short circuit or open conductor makes an entire row or column unusable, thus rejecting the entire panel, it is no surprise that manufacturing yields were very poor initially, of the order of 10% for a 10-in. color AMLCD. A series of technological and process advances, including adoption of class 10 clean rooms for lithography greater attention in combating electrostatic discharge, and the incorporation of second-generation manufacturing equipment resulted in the improvements summarized in Table 18.4.

Although the results are very impressive, yield and cost figures are still quite inferior to those achieved by older technologies, such as CRTs (approximately 90% and $60, respectively, for a 19-in. consumer-grade CRT). Only time will tell, of course, whether the continuing cost reductions demanded by the marketplace will be achieved. The continuing flow of massive investments bodes well for this aspect of the technology.

18.3.1.3 Color LCDs

An additive color approach is usually used in direct view displays, similar to the approach used in color CRTs. Pixels are grouped into triads and each aligned with a red, blue, or green color filter, processed on the cover sheet, as shown in Figure 18.11. By controlling the transmission through each of these

TABLE 18.4 Historical Process Yield and Manufacturing Costs for 10-in. AMLCDs

Year	Yield (approx) (%)	Cost ($/module)
1991	10	2,500
1992	32	1,850
1993	53	1,050

Source: O'Mara, W.C. 1994. *SID Seminar Lecture Notes*, M-3, 1–40.

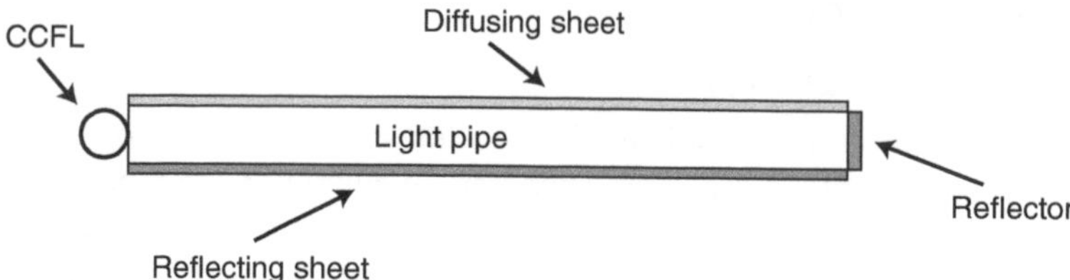

FIGURE 18.19 Schematic drawing of backlight for monochromatic LCDs. (From Lewin, L. 1993. *SID Seminar Lecture Notes*,1, M-9, 1–40.)

subpixels, a large number of colors can be rendered. A number of different color filter mosaics are used, with the "stripe" mosaic commonly used for higher quality "office automation" displays and the "delta" mosaic often seen in 'audiovisual' displays.

To obtain good color performance, in addition to choosing filters with saturated primary colors, one also has to carefully select their transmissions to obtain a balanced white point. This requires careful optimization and good quality control practices in the filter materials.

18.3.1.4 Principles of LCD Lighting

As we seen previously, LDCs do not *generate* light but *modulate* it. Thus for satisfactory operation in a variety of ambients some form of supplementary lighting is necessary. Among the factors contributing to the great success of AMLCDs has been the development of backlighting technology (Lu et al. 1993; Lewin 1993). Illuminating a liquid-crystal panel may seem simple, but the problem is far from trivial. The requirements of uniform illumination, low power consumption, long lamp life, compact physical dimensions, and, of course, low cost are difficult to achieve simultaneously. Initially, many LCDs were used in low-power applications and operated in the reflective mode, but the image quality was unsatisfactory. Currently, the vast majority of LCDs are operated in the transmissive mode and are backlit.

Early light sources for LCD backlighting were the powder EL lamps, which were very thin, but not very efficient, and short-lived. Currently, the most popular backlight technology uses cold cathode fluorescent lamps (CCFLs). These consist of a glass tube coated with phosphor and filled with low pressure gas. Applying a high voltage ionizes the gas and current flows, causing the emission of light, predominantly in the ultraviolet. The UV, in turn, excites the phosphor on the inside walls of the glass.

A typical backlight for a monochromatic STN LCD consists of a single CCFL illuminating a light pipe from the side as shown in Figure 18.19.

The diffusing film at the top of the drawing insures, uniform resulting light, whereas the reflectors at the side and bottom increase the overall efficiency.

Because the optical transmission of black and white STN LCDs is about 20%, a backlight luminance of about 300 cd/m^2 will result in a display luminance of 60 cd/m^2, which is quite acceptable in an office.[1] This is routinely achieved with good uniformity and excellent life.

In the past, color LCDs had much lower optical transmission (about 3%). Thus, backlight designs for color LCDs used a number of CCFLs in a light box placed directly behind the display, with a consequent penalty in cost and power. Recent progress in color filter technology and improvements in TFT LCD design increased the optical transmission of color LCDs to about 4–5%. For a display luminance of 100 cd/m^2, one needs a backlight luminance of 2,000–2,500 cd/m^2. This is achieved with two modern, small diameter CCFLs (one on each end of the light pipe in Figure 18.19) at a power consumption of about 6 watts.

[1]See Table 18.2.

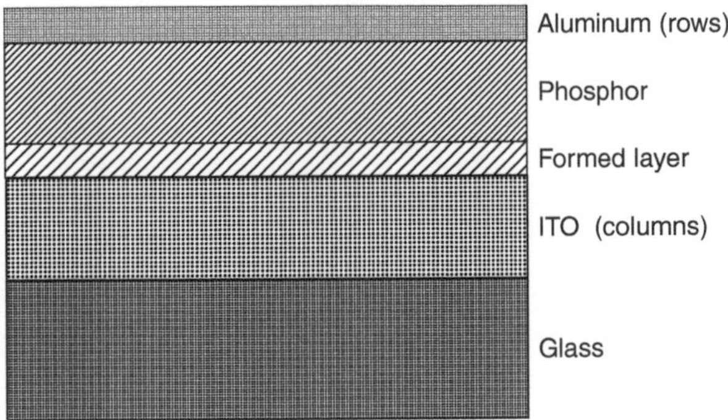

FIGURE 18.20 Structure of a dc flat panel.

18.3.2 Electroluminescent Devices (EL)

18.3.2.1 Introduction

Electroluminescence is defined as nonthermal conversion of electrical energy into luminous energy. This would include the familiar light-emitting diode (LED) covered elsewhere. In *field electroluminescence devices*, commonly referred to as EL, light is generated by impact excitation of light-emitting centers in phosphor materials by high energy ($\approx$200 V) electrons. The electrons gain their high energy from an external electric field, thus, the name (King 1994).

18.3.2.2 DC EL

Phosphorescent materials, such as ZnS, are known as efficient, low-voltage emitters, with a strong nonlinear relationship between the applied voltage and the resultant light. One can build simple flat panels in which the phosphor is directly addressed by conductors. The DCEL structure is shown in Figure 18.20.

Although commercial devices have been made with this technology, it has gained a reputation for unreliability and limited life.

18.3.2.3 Thin Film EL (TFEL)

A more successful approach has been thin-film EL (TFEL) in which the phosphor is not directly in contact with the electrodes, but sandwiched between insulating layers as shown in Figure 18.21.

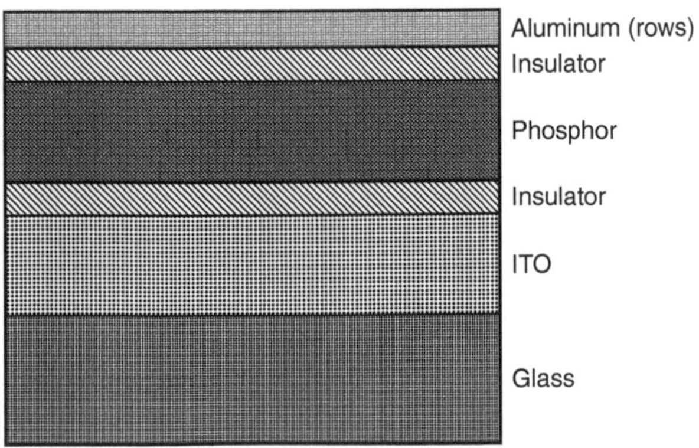

FIGURE 18.21 Structure of a TFEL flat panel.

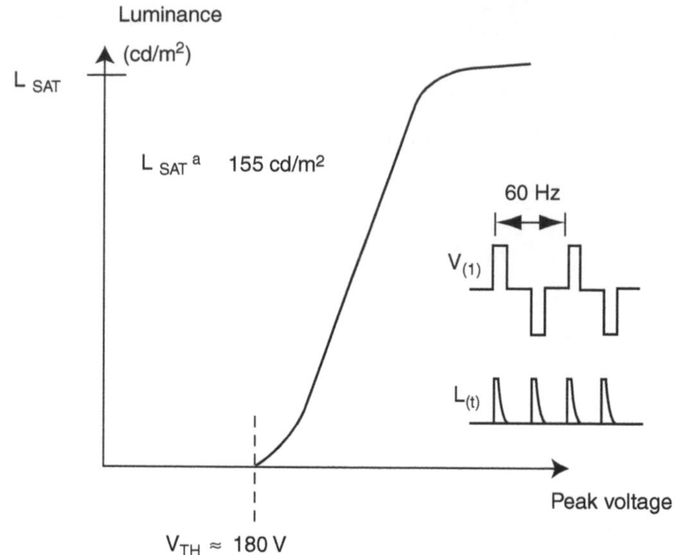

FIGURE 18.22 EO curve of EL.

Because of the capacitive nature of the coupling, excitation must be in AC form. Typical luminance-voltage characteristics are shown in Figure 18.22. The highly nonlinear characteristic curves allow the device to be addressed electrically at a very high multiplexing ratio, while maintaining good contrast. This is what is required for the matrix addressing of flat panel displays with high information content, as was seen previously. Typical performance levels achieved by 640×480 TFEL units on the market are a luminance of 100 cd/m^2 and a contrast ratio of 15:1 in ambients typical of offices. These displays also have wide viewing angles ($\approx 160°$).

18.3.2.4 EL Materials

Much of the performance of an EL display depends on the choice of phosphor material. In general, phosphors, whether used for EL or CRTs, consist of a host material doped with an activator which is the light emission center. The classical EL phosphor consists of a ZnS host lattice doped with Mn atoms for the light emission centers. To be a phosphor host lattice, a material must have a band gap large enough to emit visible light without absorption. This limits the class of possible materials to large-bandgap semiconductors (e.g., >2.5 eV) and insulators. The classical CRT phosphor host materials are the II–VI compounds and the rare-earth oxides and oxysulphides. To date, the oxide phosphors have been unsatisfactory, and only the II–VI materials, such as ZnS, SrS, and CaS, have demonstrated reasonable luminous efficiencies. This restricts the choice of materials for achieving color, as will be seen shortly.

18.3.2.5 Color EL

Any display technology must be capable of full color for long-term viability. This has posed a number of challenges to EL. One approach has been to develop a set of phosphors, similar to what was done for CRTs, with emission spectra in the red, green, and blue, respectively. For satisfactory operation on a system level, this requires matching the *L–V* curves and achieving reasonable efficiencies so that the appropriate white is realized. This has caused considerable difficulty, particularly, with the blue phosphor.

An innovative solution has been reported recently and is shown in Figure 18.23. The front (top) substrate consists of patterned red and green phosphors in an arrangement very similar to a conventional (monochromatic) EL layer structure. The rear substrate is a monochromatic blue-emitting EL panel with transparent top and bottom electrodes.

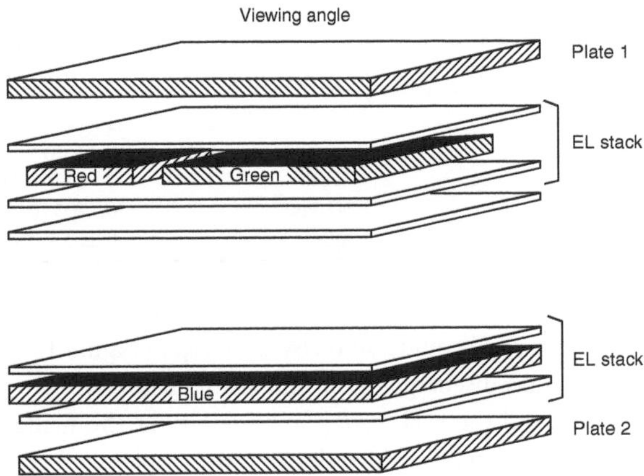

FIGURE 18.23 Construction of dual-substrate color EL. (From King, C. N. 1994. *SID Seminar Lecture Notes*, 1, M-9, 1–38.)

The dual panel structure also allows optimizing the drive electronics independently for each substrate. Furthermore, the threshold voltage of the blue phosphor does not need to match the threshold voltages for the red and green phosphors.

The performance reported for a complete panel based on the concept is summarized in Table 18.5.

A radically different approach is to use a single, broad-band phosphor that can then be filtered to produce an RGB display. This device structure is shown in Figure 18.24. It has the advantage of maintaining the simple manufacturing process of a monochromatic TFEL display. The disadvantage is that it requires a very highly efficient broad-band phosphor.

18.3.3 Plasma Displays (PDPs)

18.3.3.1 Introduction

Plasma displays are one of the few display technologies that can readily be made in large sizes. Monochromatic versions, easily recognizable by their distinctive orange-reddish color, are often used in command and control applications in which their large size and shallow depth make them attractive alternatives to CRTs. PDPs can also achieve color as will be seen.

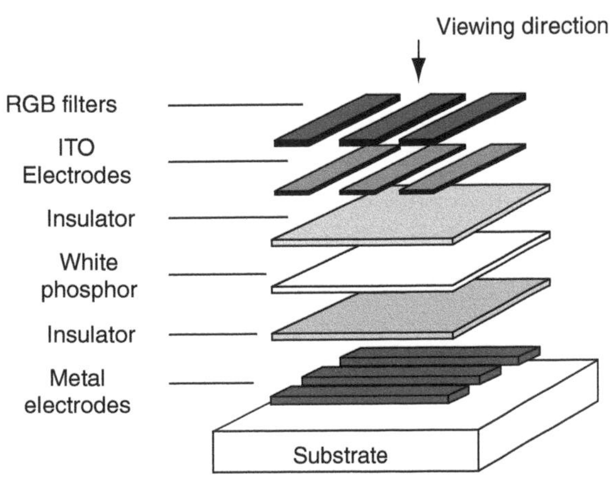

FIGURE 18.24 Construction of color EL employing fitters.

TABLE 18.5 Performance of Color TFEL Panel

Color Pixels		640×480
Displayed colors		16
Pixel pitch		0.31 mm
Subpixel size	Red	0.060×0.210 mm
	Green	0.170×0.210 mm
	Blue	0.210×0.210 mm

The operation of the device can best be understood by referring to Figure 18.25 that shows the voltage–current relationship of a typical gas discharge found in plasma displays.

Note the large current range and the very strong nonlinearity at the firing voltage. This is a major attribute of gas discharges that allows matrix addressing. Plasma displays are generally operated at or near the junction of the normal and the abnormal glow regions.

18.3.3.2 PDP Addressing

Conceptually, there are a number of ways in which a plasma display can be addressed, DC, AC or a mixture of the two. A PDP can also be designed to support only segments or in a dot matrix form. Finally, PDPs may be designed to have inherent memory (i.e., the glow persists for some time after the excitation

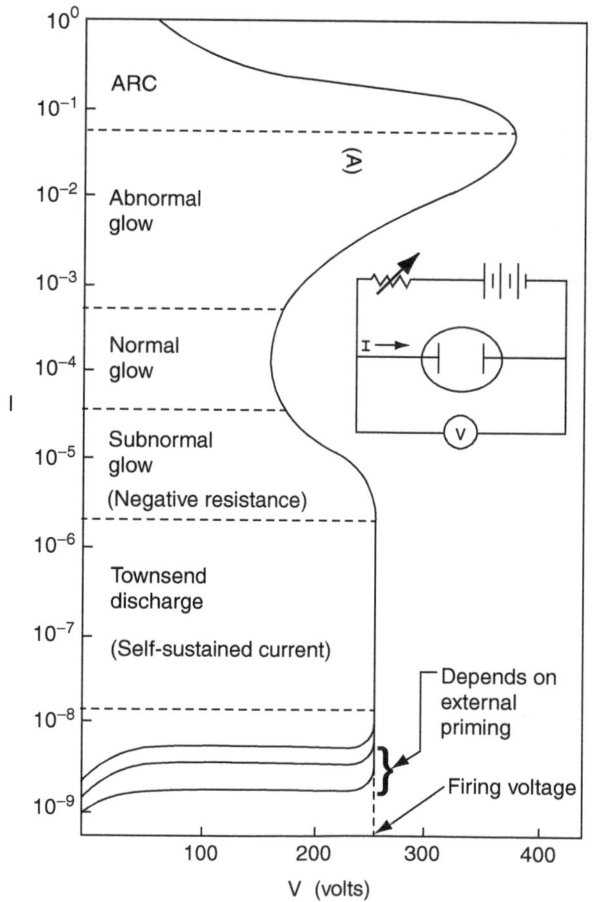

FIGURE 18.25 EO curve for plasma.

is removed). This has led to a variety of product and technological variations by a number of manufacturers who have produced PDPs with individual characteristics and enhancements. Because it is beyond the scope of this chapter to review each in detail, only the fundamental aspects will be covered in what follows. The interested reader is encouraged to pursue some of the excellent references on the subject (Weber 1994).

18.3.3.3 DC PDPs

The simplest plasma product is a DC Plasma that is addressed along the lines of Figure 18.4. A number of manufacturers make segmented displays that have achieved good reputations for cost and reliability under moderate ambient lighting conditions.

It is important to note that the current in a gas discharge must somehow be limited to a predictable value. DC plasma displays use an external resistor or a semiconductor current source. Furthermore, most dc plasma displays have a small amount of mercury added to the neon gas to extend the lifetime of the display by inhibiting the sputtering action of the gas discharge on the cathode.

18.3.3.4 AC PDPS

One undesirable characteristic of most DC displays is a background glow that reduces contrast. This has led to the development of AC plasma displays in which current limiting is achieved by a capacitor placed in series with each pixel. The driving waveforms are coupled capacitively to the gas discharge in much the same way as done in TFEL.

AC displays can achieve memory. When a voltage pulse is applied to an AC panel, the discharge deposits a charge on the wall that reduces the voltage across the gas. After a short time, the discharge will extinguish, and the light output will end until the applied voltage reverses polarity and a new discharge pulse occurs. The net effect of this memory mode is a great increase in the effective duty cycle of the display, thus, an increased brightness, especially, in large displays.

18.3.3.5 Hybrid AC–DC

A very successful implementation of plasma displays is the hybrid AC–DC that was developed by SONY (Amano et al. 1982). This is, basically, a DC plasma panel with the addition of an ac trigger electrode buried under the cathodes. This trigger electrode is not used for display purposes but is used to create priming particles for the DC discharge.

The priming particles have two beneficial effects: they allow use of lower voltages to drive the normal DC address electrodes, which in turn require lower cost circuit drivers. Another advantage of this priming action is the shorter time delay between the start of the address pulse and the ignition of the discharge. Which, in turn, achieves a great increase in luminance uniformity.

18.3.3.6 Color Plasma Displays

After a long period of gestation, full color plasma displays have appeared on the market. These panels generate color by inserting RGB phosphors in the panel and exciting those phosphors with the ultraviolet light of the gas discharge. This creates a new set of problems that have been overcome in a variety ways, although intensive research is still occurring. One problem is, of course, that of luminous efficiency, due to the double energy conversion. A separate problem is the need for eliminating the cross talk between subpixels to insure purity.

Figure 18.26 (Shinoda et al. 1993) shows a 21 in.-diagonal color PDP with 640 × 480 resolution. Note the barrier rib separators between each subpixel. The barrier ribs do not transmit the 147-nm radiation generated by the xenon gas used in most color plasma displays. A number of other manufacturers are working on similar approaches.

18.3.3.7 Conclusion

Some consider PDP a leading candidate for the large color displays required for the next generation of consumer televisions, in particular, for high-definition TV. There are a number of difficulties that must be resolved however. One is the high power requirements. These could be overcome by an increase in

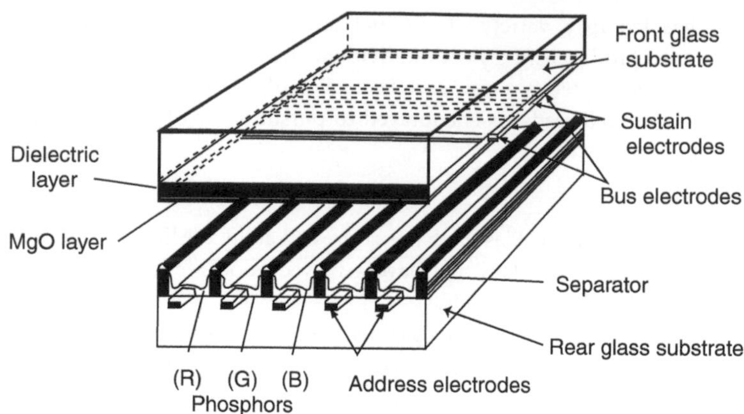

FIGURE 18.26 Example of color PDP.

luminous efficiency. Another major obstacle is the high cost of the rather complex structures involved in PDPs.

18.3.4 Light Emitting Diodes (LEDs)

The physics of light-emitting diodes (LEDs) (Scherr 1993) are addressed elsewhere in this volume. Some of the application considerations will be covered here. LEDs appear to be ideal candidates for matrix-driven displays. They are solid-state, low-voltage-driven, light-emitters with an intrinsic sharp, nonlinear threshold and a fast response. LEDs have been successfully fabricated to emit various colors including red, green, yellow, orange, amber, and blue. The luminous efficiencies that have been achieved depend on a number of factors, including the wavelength of the emitted light, and are of the order of a few lumens/watt or less.

Unfortunately, the fast response characteristic of LEDs has made them unsuitable for the high-resolution displays needed for personal computers and television applications. These displays, typically, require on the order of 0.1–1 million pixels. The resultant duty cycle, or on time, of the driving waveform for each pixel reduces the luminance and contrast to unacceptable levels.

Fortunately this does not apply to applications requiring 10–16 digits for which structures such as the seven-segment display are appropriate. This has made LEDs popular in applications, such as pocket calculators. Single-digit LED displays can also be made in moderately large sizes (2–6 in.) which can be assembled in message boards containing six lines or so of 80 characters.

18.3.5 Other Flat Panel Technologies

There are a number of flat panel technologies under development that deserve mention. *Electrophoretic display* is an intriguing concept that never quite fulfilled its promises. A cell, consisting of front and back glass plates, is filled with liquid containing electrophoretic particles in a suspension.

The glass plates have conducting plates in contact with the liquid. The front plate is transparent and allows viewing of the image. The suspended particles, often made of TiO_2, become charged and are attracted to one of the electrodes by the action of an external electric field, thus forming the image. Prototype displays with excellent contrast have been demonstrated, but keeping the particles from forming large aggregates or migrating to the bottom or the top of the cell over time due to imperfect gravity matching has proven quite difficult.

Another technique with promise is a derivative of liquid-crystal displays, the so-called *ferroelectric displays* (FLCD). Laboratory FLCDs have been made with high switching speeds, high contrast ratios and wide viewing angles. FLCDs displays are bistable: each pixel is driven into a white or black state. The pixel will remain in that state until it is switched again with the opposite polarity. This feature makes it feasible to address many lines at very low duty cycles and without loss of contrast. FLCDs require a very small cell gap (about 1.5 μm) for high performance, which makes mechanical ruggedness difficult. These displays have the potential of low cost because an active matrix is not needed. Large, high-resolution displays with FLCDs have been demonstrated, but commercial production is some years away.

18.4 Display Measurement and Performance Evaluation

The techniques that are the basis for setting, verifying, and monitoring specifications for displays form a subtechnology of their own. Some general principles and guidelines follow.

The most important characteristic of a display is its luminance, determined in great part, by the ambient conditions in which the product will be used. It is good to keep in mind that the human visual system responds logarithmically to luminance, so that an excessive precision in specifying or measuring luminance is inappropriate. Furthermore, the precision of most photometers is no better than a few percent (Infante 1994).

Another characteristic of displays is the reflectance of the viewing surface that can play a large part in determining the usability of the display, particularly, in bright ambients. Reflectance can be specular or diffuse, and different measurement techniques are appropriate for each one. An example of a typical measurement process is shown in Figure 18.27.

A very important characteristic of displays is the ability to present fine detail. This is often referred to as resolution, although the term is often misused and not distinguished from addressability, which is the characteristic of a display that measures how many pixels per unit length are presented. Resolution is a measure of the ability to resolve or distinguish fine detail. In many flat panels, the two are synonymous, whereas for CRTs, the two are quite distinct. A CRT display cannot resolve detail much smaller than the width of its own electron beam, although the beam itself might be stepped into much smaller steps. Measurement of resolution in a CRT display is no simple matter because the profile of the beam itself is generally Gaussian and its determination may be made complex by the shadow mask and screen structure in color CRTs. The effects of the latter will also contribute to the sharpness of the resulting image. Thus, it is not unusual for CRT displays to call out addressability, the size of the electron beam on the screen (at, say, 50% of the Gaussian distribution), and the pitch of the shadow mask.

Another important parameter is a measure of the quality of the color presented. Although this may seem a trivial problem, measuring color is not quite as straightforward as it may seem, because of the influence of the ambient light, the surround, and the vision characteristics of the individual observer. Colorimeters with good accuracy and repeatability are available from a number of vendors.

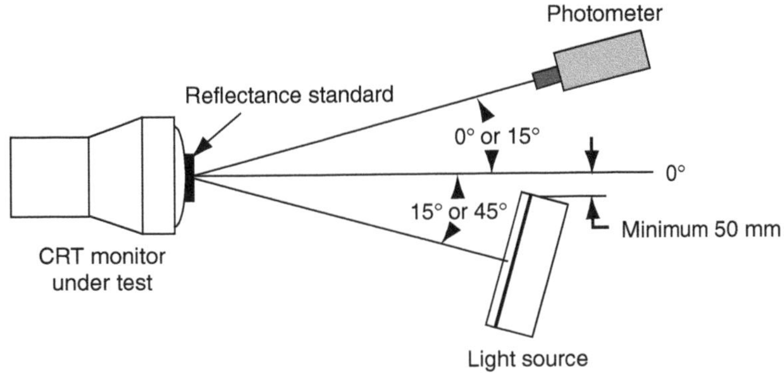

FIGURE 18.27 Reflectance measurement.

Finally, the so-called geometric characteristics of a display can contribute a great deal to the perceived quality. These are factors, such as the apparent straightness of lines, the absence of distortions that transform squares into trapezoids, etc. These again are the province of special-purpose instrumentation.

18.5 Conclusions

The preceding discussion showed the variety of display technologies, all offering differing solutions to the common problem of transferring information to a human being. It should also be apparent that displays are supported by an enormous number of subtechnologies, ranging from detailed investigations of the properties of exotic materials to the very exact and demanding aspects of electron optics. The physical and psychophysical aspects of the human visual system are an underlying thread of all displays. What may not have been apparent is the scope of the continuing effort extant in the field. Older technologies, such as the CRT, continue to be perfected. Newer technologies, such as active matrix liquid crystals, achieve new performance levels on an almost daily basis. Finally, brand-new technologies are being created. If an affordable solution to the age-old problem of the "picture on the wall" still seems elusive, the continued intellectual and market growth of display technology seems assured for the foreseeable future.

References

Alt, P. M. and Pleshko, P. 1974. Scanning limitations of liquid crystal displays. *IEEE Transactions Electron Devices*, ED-21, 7, 146–154.

Amano, Y., Yoshida, K., and Shionoya, T. 1982. High-resolution DC plasma display panel. *SID Symposium Digest* 160–161.

ANSI/HFS-100. 1988. American National Standard for Human Factors Engineering of Visual Display Terminal Workstation. Human Factors Soc.

Castellano, J. A. 1994. The CRT display market: Strategies and trends in the 1990s. *SID Symposium Digest*, 25, 215–218.

Chandrasekhar, S. 1992. *Liquid Crystals, 2nd ed.*, p. 460, Cambridge University, Cambridge, England.

Dasgupta, B. B. 1991. Deflection and convergence technology. *SID Seminar Lecture Notes*, 1, 1–43.

El-Kareh, A. B. and El-Kareh, J. C. 1970. *Electron Beams, Lenses and Optics, Vols. 1 and 2*, Academic Press, New York.

Farrell, J. E., Benson, B. L., and Haynie, C. R. 1987. Predicting flicker thresholds for video display terminals. *Proceedings of the SID*, 28, 4, 449–453.

Infante, C. 1986. Introduction to CRT systems. *SID Seminar Lecture Notes*, 1, 1–106.

Infante, C. 1994. CRT display measurements and quality. *SID Seminar Lecture Notes*, M-7, 1–54.

King, C. N. 1994. Electroluminescent displays. *SID Seminar Lecture Notes*, 1, M-9, 1–38.

Lewin, I. 1993. Principles of LCD lighting. *SID Seminar Lecture Notes*, 1, M-9, 1–40.

Lu, S., Kuo, T., Huang, J., Lin, F. 1993. Bright and thin LCD backlights for monochrome and color LCDs. *SID International Symposium Digest of Technical Papers.*

Morozumi, S. 1993. Active-matrix LCDs. *SID Seminar Lecture Notes*, 1, M-8, 1–31.

Morrell, A. M., Law, H. B., Ramberg, E. G., and Herold, E. W. 1974. *Color Television Picture Tubes*, p. 226, Academic Press, New York.

O'Mara, W. C. 1994. AMLCD manufacturing. *SID Seminar Lecture Notes*, M-3, 1–40.

Scheffer, T. 1993. Supertwisted-Nematic (STN) LCDs. *SID Seminar Lecture Notes*, 1, M-7, 1–63.

Sherr, S. 1993. *Electronic Displays*, 2nd ed., p. 624, Wiley, New York.

Shinoda, T., Wakitani, M., Nanto, T., Yoshikawa, K., et al., 1993. Development of technologies for large-area color AC plasma displays. *SID Symposium Digest*, 161–164.

Silverstein, L. D. 1994. Color in CRT and LC displays. *SID Seminar Lecture Notes*, F-3, 1–70.

Silverstein, L. D. and Merrifield, R. M. 1985. The development and evaluation of color systems for airborne applications. DOT/FAA July 1985 PM-85-19.

Spindt, C. A., Brodie, I., Humphrey, L., and Westerbers, E. R. 1976. Physical properties of thin film field emission cathodes. *Journal of Applied Physics*, 47, 5248.

Steemers, H. 1994. Fundamentals of liquid-crystal displays. *SIS Short Courses*, S-3, 1–66.

Travis, D. 1991. *Effective Color Displays: Theory and Practice*. Academic Press, New York.

Weber, L. F. 1994. Plasma displays. *SID Seminar Lecture Notes*, 1, M-8, 1–35.

Yamazaki, E. 1993. CRT displays. *SID Seminar Lecture Notes*, 2, F-5, 1–47.

VI

Systems

19

Digital Optical Computing and Interconnection

Jürgen Jahns

19.1 Introduction

For some time now, there has been a strong interest in exploring the capabilities of optics for computing. The field of optical computing, in general, started in the early sixties with the advent of the laser. At that time, optical computing was mainly synonymous with *analog* optical signal processing. Analog optical signal processing is based on the ability of a lens to perform a two-dimensional Fourier transformation (Goodman, 1996). This basic operation can be extended to implement optical convolutions and correlations that can be useful for the processing of radar signals or pattern recognition (van der Lugt, 1964). This field has been brought near perfection during the past three decades by an improvement of the algorithms and the hardware, such as temporal and spatial light modulators. Overviews of the area of analog optical computing were given, for example, by Lee (1981); Casasent (1997). Recently, analog computing has seen renewed interest in the context of information security (Javidi, 2005).

Digital computing also started in the early 1960s when people used the newly invented laser to implement nonlinear logic operations. A collection of articles can be found in conference proceedings edited by J. T. Tippett et al. (1965) and in Basov et al. (1972). However, these early trials of optical digital computing were not too successful, and predictions of an enduring inferiority of optics to electronics (Keyes and Armstrong, 1969) discouraged people from pursuing the subject.

A second effort started in the second half of the 1970s. Two developments appear to be of importance: first, the work by Alan Huang (1975), (1984), who pointed out that the limitations of electronic computers might be due to the strength of the associated optics. Huang also introduced the residue number system to the optical computing community (Huang, 1975; Huang et al., 1979); this system was, for a while, the subject of very active research for signal processing applications. The second development of importance was the discovery of optical bistability in gases (Gibbs, McCall, and Venkatesan, 1976). The development of nanoscale fabrication techniques such as molecular-beam epitaxy later led to the investigation of multiple quantum well structures in solid-state materials, gallium arsenide in particular (Miller et al., 1984).

A very important period in the history of optical digital computing was the time from 1985 to 1995, when many activities in research groups around the world led to a groundwork in device research, interconnections, and systems and architectures for optical computers. Significant progress was made in understanding many details of how optics can be used to perform digital logic—and where limitations still exist. The fact that there are still limitations to optics has become evident during recent years, when issues such as speed limitations, complexity of algorithms, and difficulties with the packaging of free-space optics became apparent. Simultaneously, advances in electronics have been impressive, raising doubts whether optics will be able to improve upon what has already been achieved in electronics.

Today's computers have extraordinary processing powers and capabilities. At this writing, electronic processors consist of approximately 5×10^9 transistors running at speeds of several Gigahertz. Further improvements can be expected in microelectronic fabrication technology, so that one can assume that the capabilities of individual processors will still be increased (Semiconductor Industry Association, 2005). Furthermore, there still exists a demand for even faster machines, for example, for complex numerical simulations, real time pattern recognition, and weather forecasting. However, the speed of today's high-performance electronic computers is increasingly limited by communication problems, the number and bandwidth of interconnections, and by data storage and retrieval rates, rather than by processing power. Several limitations exist for the performance of all-electronic computers:

- Architectural limitations such as the often-cited "von Neumann bottleneck" (Backus, 1982; Huang, 1980) that limits the amount of data exchanged between processing unit and memory
- Topological limitations due to the two-dimensional layout of the wiring inherent to electronics (Goodman, 1984)
- Physical limitations such as a limited time-bandwidth product of electric wires that slows down the interconnection speed and requires more energy (Miller, 1989)
- Technological limitations to the lifetime of aluminum wires caused by such effects as electromigration (Keyes, 1987)

In particular, the system bus in an electronic computer has emerged as a bottleneck (Boxer, 1995). Precise timing of the clock signals with minimum skew as well as latency reduction are important issues (Miller, 2000; Lukowicz et al., 2005). Optics, with its large bandwidth and interconnection capabilities, can offer interesting solutions to help alleviate these limitations. Waveguide- and free-space-optical approaches are being considered. Polymer waveguides embedded in a printed circuit board, for example, offer a low cost approach for an optical backplane (Chen et al., 2000; Mederer et al., 2002). Free-space optical propagation allows one to use the third spatial dimension to move signals to and from a chip with a large number of channels, resulting in the ability to use the whole chip area for input/output devices (Lohmann, 1986).

The physical bandwidth limitations of electrical wires are eliminated with optical communications. In addition, it has been argued that optical signal transmission can be more energy efficient, due to the fact that it is not necessary to charge a wire or a cable to transmit a signal (Miller, 1989). This is the basis of fiber optic communications for long and medium distances. However, optical communications can be of advantage energy-wise even for very short distances, as compared to conventional electronic interconnections. Finally, optics offers various degrees of freedom that can be used for signal multiplexing to enhance the throughput of a communication channel, such as wavelength or polarization.

While the area of digital optical processing and optical logic has slowed down during recent years (with a few exceptions), there is now strong interest in the use of optical interconnection for electronic computers. The goal of this chapter is to review the work on optical digital computing in these four areas: optical communications, switching, and processing, and computing architectures. The coverage here cannot be exhaustive; our main purpose is to provide a lexicon of some of the key terminology of optical digital computing. For more information, the reader is referred to the recent literature (Hinton and Midwinter, 1990; Midwinter, 1993; Jahns and Lee, 1993; Jahns, 1998).

19.2 Communicating with Light Signals

The most stringent limits of current electronic computers exist in interconnection technology (Goodman, 1984; Miller, 2000). The task of an interconnection system is to distribute clock and other signals to and among the various circuits or system functions in a computer. This takes place at various levels of a computer system, for example, between chips (interchip), but also on a single chip (intrachip). While integration densities on chips and die sizes continue to increase, the number and capacity of connections cannot be raised at the same rate, due to basic scaling rules (Reisman, 1983). This puts considerable stress on the communication channels at all levels, from intra-chip to board-to-board. These stresses become more and more evident as the processing speeds increase because capacitances and inductivities increase and can cause problems like crosstalk, signal skew, and increasingly power-hungry line drivers. Material effects like electromigration, which results from the formation of grain boundaries in the Al wires at high current densities (Keyes, 1987), have already been mentioned. The latter problem was solved in 1998 by replacing the aluminum wiring in computers with copper wires with silicon dioxide insulators.

Photon- and electron-transmission of signals are similar in some aspects and fundamentally different in others. At high frequencies, an electrical wire acts as a waveguide. Signals are transmitted as electromagnetic pulses, as in optics. The difference is that in the electrical case, the propagation medium is a conductor, whereas in the optical case, it is a dielectric, i.e., an insulator. In both cases, issues such as confinement, crosstalk, and reflection losses play a role. However, in the optical domain, much better confinement is achieved, resulting in lower absorption and crosstalk. In addition, attenuation due to material absorption is usually negligible, and independent of the signal frequency. Several optical waves can be transmitted through a single medium without interaction, thus yielding the potential for multiplexing schemes.

A fundamental aspect of signal transmission relates to the topology of the interconnection. While electrical wires, optical fibers, and waveguides are limited to two dimensions, free-space optical communication offers the possibility of interconnecting chips through the third spatial dimension (Figure 19.1). This allows the entry or extraction of data anywhere on a chip.

Optical waveguides are based on the principle of total internal reflection. Single-mode waveguides offer low temporal dispersion; however, they are more difficult to handle with respect to coupling into the waveguide. For computer applications, waveguide technologies of interest allow for transmission lengths of typically 10–100 cm. This can be achieved by using either fiber bundles or polymer waveguides. The interconnection densities are typically on the order of 10 channels/mm; the standard pitch is 250 μm.

Free-space optics is based on the imaging properties of a lens (Figure 19.2). The theoretical limit of free-space optical interconnection densities is given as the ratio A/λ^2, where A is area of the cross-section and λ is the wavelength). For example, for a wavelength of 1 μm and an area of 1 mm^2, one would obtain a theoretical limit of 10^6 channels/mm^2. Practical values, however, are usually smaller due to the resolution of the imaging optics. The focal spot size formed by an aberration-free optical system with a numerical aperture $NA = n \sin a$, where n is the refractive index and a is the half angle, is $\mu(\lambda/NA)^2$. For a typical numerical aperture of $NA = 0.2$ and the same values for A and λ as above, one would obtain 4×10^4 channels/mm^2, still an impressively large number. Another further limitation to interconnection density might be thermal dissipation on the chip, which imposes a lower limit on the spatial separation of optical input/output devices.

Because of the broadband nature of optical communication channels, the transmission capacity of optical systems is essentially limited by the performance of the transmission and detection components. Fiber optic communications has contributed significantly to the development of useful components and

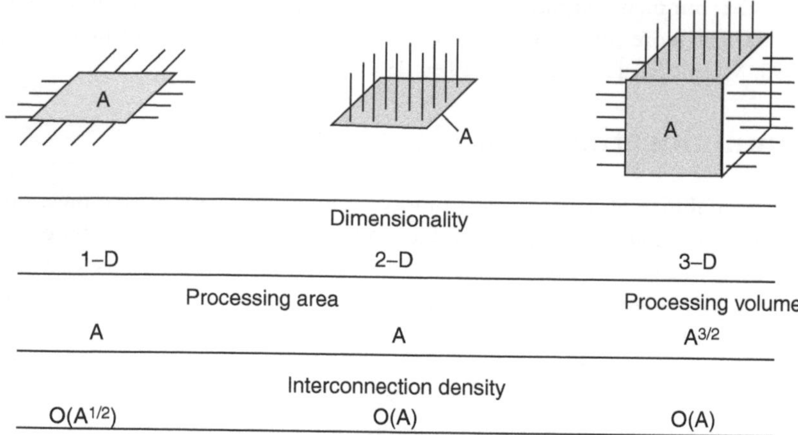

FIGURE 19.1 Scaling of communications topologies. "A" denotes the area or cross-section of a processor. The two-dimensional case provides the optimum situation for the interconnection density as compared to the chip area.

technologies. Laser diodes used as optical transmitters can be modulated at extreme speeds because of their small size; they are also highly efficient in converting electrical to optical energy. Depending on the semiconductor material system that is used (GaAs or InP), operating wavelengths range in the near infrared (above and below 850 nm) or in the mid-infrared around 1.3 μm and 1.55 μm. Modulation frequencies in the range of 10 GHz or more can be achieved. The optical output powers are in the mW range, which is sufficient for communication purposes.

On the detector side, fast and sensitive photodiodes can be built. For near-infrared applications, silicon, an indirect semiconductor, can be used. Silicon is also the most important material for electronics today, although high-speed circuits are more and more often realized in GaAs technology, despite technological difficulties with GaAs electronics. For example, incompatibilities in fabrication technologies for electronic and optoelectronic devices remain to be overcome, but approaches to fully integrated GaAs optoelectronic circuits (OEICs) are promising. For silicon-based systems in the near infrared, hybrid integration techniques are a way to combine detectors, drivers, and logic circuits with GaAs light sources and modulators.

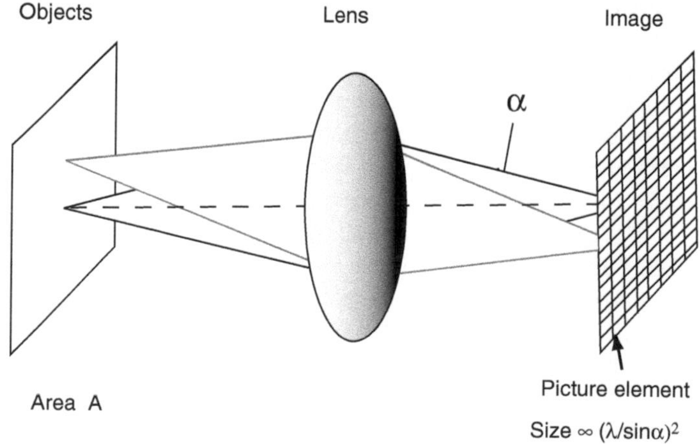

FIGURE 19.2 Optical imaging system. Due to the limited numerical aperture, the resolution in the image plane is finite. In the absence of aberrations, the size of a focal spot is uniform across the image plane.

19.3 Optoelectronic Devices for Processing and Interconnection

In the remainder of this chapter, we will concentrate on free-space optical interconnection and computing technology. For this purpose, the devices of interest can be integrated on one chip as 1D or 2D arrays. Various modulator-based and emitter-based device types have been investigated during the past couple of years. All of them are GaAs devices and operate in the near infrared, mostly at a wavelength around 850 nm. On the side of the modulators, the self-electro-optic effect (SEED) devices (Miller et al., 1984; Lentine et al., 1988; Lentine and Miller, 1993) have been used in the optical computing research projects at AT&T Bell Laboratories. Another device, promoted by NEC, is the VSTEP, which was demonstrated to operate as a modulator and as a laser device (Kasahara, 1993). Vertical cavity surface-emitting laser diodes (Iga et al., 1987) have become very important since they were demonstrated to be feasible at room temperature (Jewell et al., 1989; Morgan et al., 1991). For detailed information about the physics of the devices, the interested reader is referred to the literature and to those sections in this volume that deal with device technology. Here, we want to consider the functional and system aspects of switching and interconnection devices. A more detailed overview of the concepts and devices can be found, for example, in (Jahns, 1998). Here, the discussion is limited to a few typical devices.

SEEDs belong to the class of electroabsorption modulators (EAM). Integrated waveguide versions are now extensively used for high-speed, low-chirp modulation in fiber-optical transmission systems. In an EAM, one makes use of the changes in the optical absorption that can be caused by changes in an electric field normal to the quantum well layers of a semiconductor material (Miller et al., 1984). If the quantum well structure is placed in the intrinsic region of a reverse-bias diode, one can change its absorption by variation of the electric field. A light beam sent onto the device can therefore be modulated by applying an electrical signal to the device (Lentine, 1993).

The SEED can also be used as a detector, in which an incident light beam can generate a photocurrent. By placing two SEEDs in series as shown in Figure 19.3, one can build a structure with optical inputs that control optical outputs. The specific device shown is the symmetric SEED (S-SEED), consisting of two pin-diodes with MQW structures in the intrinsic region. Each serves as the load for the other. There are two optical inputs and two optical outputs. The switching of the device depends on the ratio of the two optical inputs, not on their absolute intensities (within practical limits). This feature is very desirable from a systems point of view because it makes the operation of the device relatively insensitive to power variations. An S-SEED can be operated as a logic gate and as a latch, i.e., a memory device. The contrast for one of the output beams between the high and the low state is typically 10:1.

Another type of device that is interesting for optical interconnections, in particular, is the vertical cavity surface emitting laser diode (VCSEL) (Iga, 1987; Jewell and Olbright, 1993). It consists essentially of a Fabry-Perot resonator between two integrated mirrors. The whole structure operation is fabricated by molecular beam epitaxy, which allows one to achieve extremely high reflectivities for the mirrors, and therefore very large gains. A typical structure is shown in Figure 19.4. VCSELs have been demonstrated with current thresholds of a few volts and single-mode optical output of about 1 mW. Modulation speeds are in the range of several gigahertz.

From a system point of view, there exists an important difference between passive (modulator-based) devices and active (emitter-based) devices. This has to do with the fact that optical power has to be supplied to modulator arrays, while this is not necessary for active device arrays. This requires suitable schemes for splitting up the light from a single laser source, and mechanisms for beam combination using polarization or other optical degrees of freedom (Figure 19.5). The array illumination can be achieved using various techniques summarized by Streibl (1989, 1993). The beam combination can also be done using various techniques discussed in McCormick (1993); Cloonan (1993). One of the most common techniques is to use polarization beam splitters as visualized in Figure 19.5b. We assume that a light beam is s-polarized as indicated in the figure. It is deflected by a polarizing beam splitter and sent onto a device array assumed to

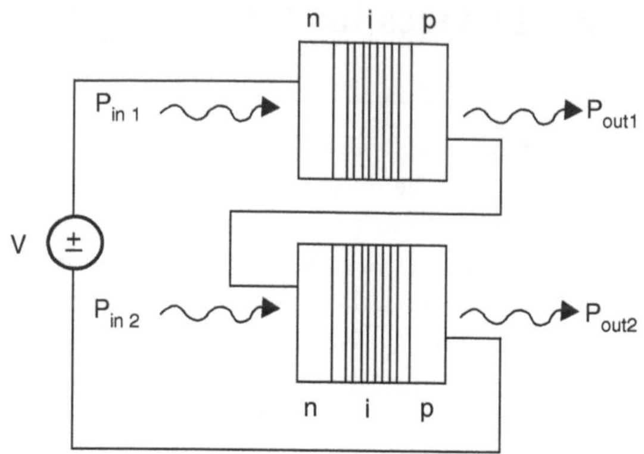

FIGURE 19.3 Symmetric SEED device according to Lentine (1993). Two individual multiple quantum well modulators are connected in series to provide a switching characteristic that is independent of absolute light levels (within practical limits). (From Lentine, A. L. 1993. *Optical computing hardware,* Jahns, J. and Lee, S. H., eds., Academic Press, Boston.)

be operating in reflection (as is usually the case for the SEED devices). By means of a quarter-wave plate, the reflected beam is now p-polarized, and will be transmitted by the beam splitter cube.

Although the modulator concept requires more optical hardware, it offers some interesting advantages, such as potentially longer lifetime and higher modulation speed. The very high current densities in semiconductor lasers may lead to a higher probability of device failure than in the case of a field-driven modulator device. Another aspect relates to the achievable bandwidth: while VCSELs have the prospect of running up to approximately 10 GHz, modulators can possibly be driven faster, provided that a suitable pulsed light source is available. During recent years, practical fs-lasers have become available, also as fiber lasers, which provide good optical quality. The debate continues.

A device concept that has gained significant importance is the concept of the "smart pixel" (Hinton, 1988; Cheng et al., 1993; D'Asaro et al., 1993; Kasahara, 1993b; Wilmsen et al., 1993). In the early work on digital optical computing systems in the mid-1980s, it was assumed that devices like the SEED devices would operate as simple logic gates (AND, NOR, etc.) and that the optical interconnections would be used to directly connect gates in different device arrays. This will be discussed in a later section. However, this concept of low-level interconnections can have problems related to the gate count, for example,

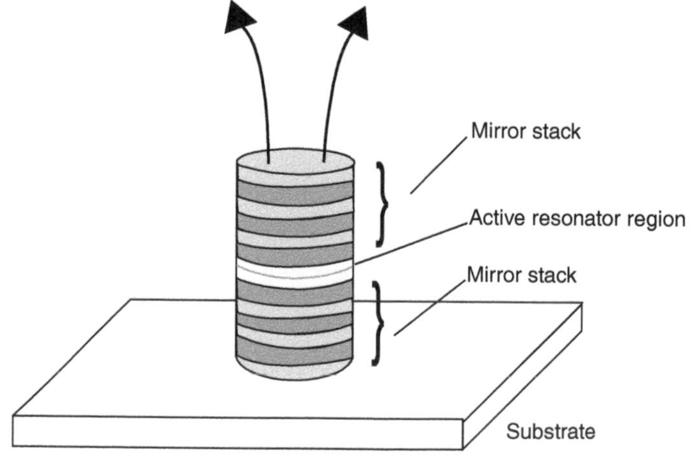

FIGURE 19.4 Vertical cavity surface-emitting laser diode consisting of a quantum well between two mirror stacks.

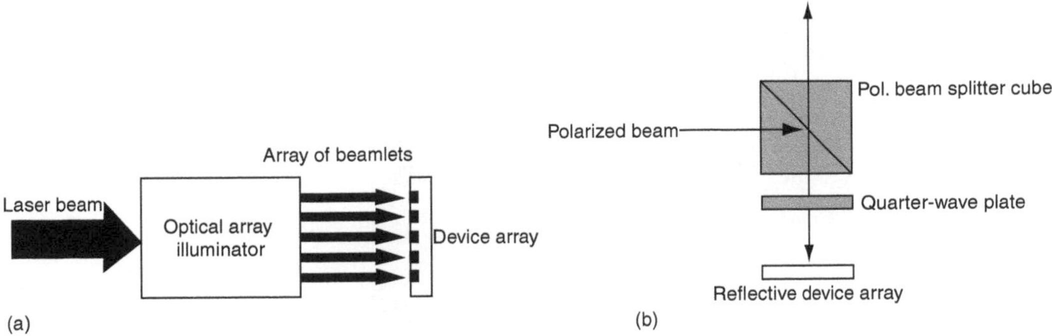

FIGURE 19.5 (a) Principle of optical array illumination. A single laser beam is split up into a 1D or 2D array of equal intensity beams to illuminate an array of modulator devices. (b) Beam combination using a beam-splitter cube.

the number of logic gates required to perform a certain function may be high. For this reason and also for reasons related to the physics of the devices, a current focus of interest is the smart pixel structure. Here, an optical switching device merely serves as an optical input and output device. It is surrounded by electronics which may be used to complement the function of the optical device (amplification, thresholding, etc.) or its function may be on a higher level (digital logic, signal routing, signal processing, etc.)

The use of the smart pixel scheme has consequences with regard to both device technology and the interconnection optics. In the conventional scheme, optical devices and electronic circuitry are separate (Figure 19.6). For the optical interconnections, this has the advantage that the optics is concentrated in a small area (typically 1 mm^2) and that individual devices are spaced very densely. This matches well with the use of conventional imaging optics that provides high resolution over relatively small areas. The optical devices can be combined with electronic circuitry by using hybrid integration mechanisms. The tight spacing of the devices, however, causes problems with the geometry of the electric wires that are used to provide electric power and voltage to the individual devices. The density of the devices is ultimately limited by practical constraints related to the density of the electrical wiring.

In the smart pixel model, the optical device area is split up into many little pieces that are distributed over a much larger chip area (Figure 19.7). The pitch of the devices may be 100 μm or more. The whole chip diameter may be several millimeters. For the optics, this means that one has to use imaging systems

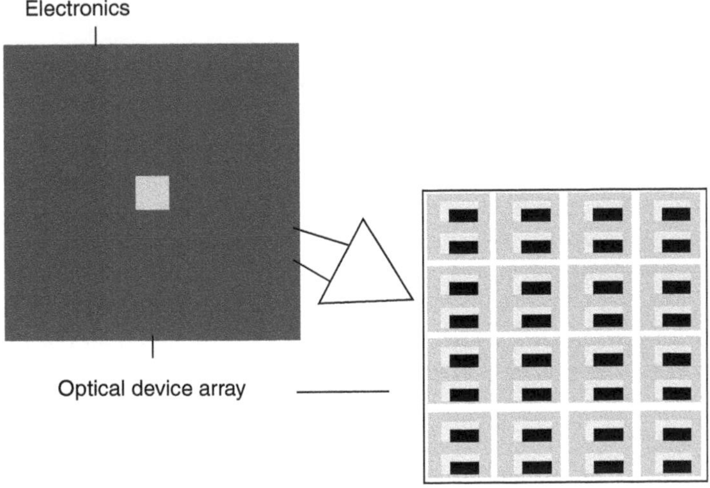

FIGURE 19.6 Conventional array of optical switching devices with supporting electronics. A close-up view is shown on the right of the figure.

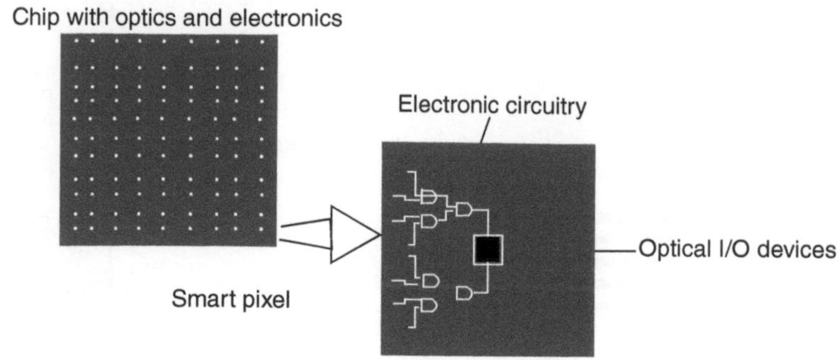

FIGURE 19.7 Array of smart pixels. An individual smart pixel is represented on the right.

that can handle large optical fields at high resolution. Promising approaches based on the use of imaging with lenslet arrays will be discussed in a later section. The problem of the actual implementation of smart pixels may lie in the difficulty of their solution because they require the integration of optoelectronic and electronic functions on the same chip. For an all-GaAs implementation, this brings with it the problems of process compatibility and yield issues. The integration of GaAs devices on Si is a difficult problem that has not yet been solved at a practical level, despite promising approaches (Goossen et al., 1989; Fang et al., 1990). The practical alternative to monolithic integration is hybrid integration by flip-chip bonding, which can provide high-density interconnections between two chips (Krishnamoorthy et al., 1999).

During recent years, there has been a significant increase in the use of organic materials for optical and optoelectronic devices. Examples are organic light-emitting diodes as light sources and the already mentioned polymer waveguides. At the device level, one can observe the development of logic gates based on bacteriorhodopsin (bR). bR is a photochromic retinal protein. Its biological function is the conversion of light energy into chemical energy that is made available to the cell. bR is the simplest natural material that is photon-driven and offers very large nonlinear optical response. All-optical switching has been reported in Mikaelian and Salakhudtinov (1994) and more recently, for example, in Roy et al. (2004). Switching speeds of bR have been demonstrated in the range of tens of milliseconds; however, Mikaelian and Salakhudtinov's (1994) tests and calculations show the possibility of arranging 10^4 switching channels with speeds of about 1 gigabit/s.

19.4 Optical Interconnection Networks

The subject of free-space optical interconnects for digital optical computing systems is closely related to the field of switching networks, in particular space-switching networks (Benes, 1965). A space-switching network consists of arrays of input and output devices and switching nodes with spatially separated inputs and outputs. They are connected to each other through physical channels such as wires, waveguides, or free space. The specific function of the nodes is not assigned at this point. It may vary with the specific application for which the network is used. For a sorting network, for example, the nodes may be compare-exchange units. Each node may, in fact, be a small network of its own.

An interconnect can be represented as a bipartite graph as shown in Figure 19.8a. It consists of arrays of inputs and outputs shown on the left and right, respectively, and lines indicating the interconnections. The specific network shown is a crossbar, where every input is connected to every output. Another representation of the crossbar is shown in Figure 19.8b. In that figure, it becomes obvious that the number of switching nodes (the dark squares) is NM for N input devices and M output devices. The figure shows the special case where $N = M$. Two quantities to describe an interconnection network are the fanout of the switching nodes and the diameter of the network. The fanout F is the number of input positions and

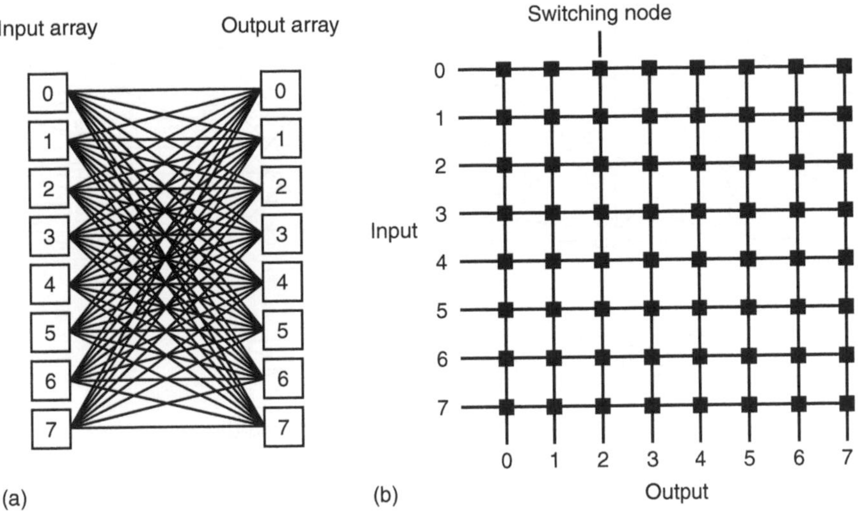

FIGURE 19.8 (a) Crossbar switching network; (b) represents the crossbar as a bipartite graph.

output positions of the nodes. The diameter D is the number of switching stages that are required to link any input to any output device. The crossbar represents an extreme because it consists only of one switching stage, while, on the other hand, the devices have a fanout of $F=N$. The crossbar allows one to route each of the N inputs to any of the M outputs or, in general, arbitrary combinations of outputs. The total number of interconnection patterns supported by a crossbar is N^M (or N^N in the special case discussed here). The crossbar is a nonblocking network, i.e., it allows one to connect an input device to an output device—or to any combination of output devices—independent of already existing connections.

There are two disadvantages to the crossbar. One is the high cost in terms of the required number of switching nodes. In the area of telephone switching networks, this has led to the invention of the Clos network (Clos, 1953). A second disadvantage is the large fanout that the switching nodes have to support. This feature is not desirable for optical devices, as pointed out by Prise et al. (1988). Therefore, other types of interconnection networks have been considered for optical computing and switching systems. Of particular interest are multistage interconnection networks (MINs), which are usually constructed of 2×2 crossbar switches (i.e., $F=2$). Permutation networks based on MINs typically consist of $\log_2 N$ stages (i.e., $D=\log_2 N$), where each stage has N switching nodes. MINs support all $N!$ permutations and some but not all broadcasting connections (Smith, Murdocca, and Stone, 1993).

A well-known MIN is the Perfect Shuffle (Stone, 1971) (Figure 19.9a). The name is derived from a technique for shuffling a deck of N cards by interleaving the upper and the lower halves with one another. After repeating the same operation $\log_2$N times, the initial order of the cards is reestablished. Other types of MINs are the Banyan (Goke and Lipovski, 1973) and the Crossover network (Wise, 1981; Jahns and Murdocca, 1989). These are shown in Figure 19.9b and 19.9c. The Perfect Shuffle, the Banyan, and the Crossover are topologically equivalent. This means that one can be transformed into the other by reordering and renaming the nodes. A formal condition for the equivalence of networks is the "buddy condition" (Agrawal, 1983), which was shown to apply for the Crossover and the Banyan by Jahns and Murdocca (1989).

For the optical implementation of the different types of interconnection networks, three fundamental approaches can be distinguished (Figure 19.10):

- Matrix-vector multiplier setup (Goodman, Dias, and Woody, 1978) for 1D input and output arrays (Figure 19.10a). Light from one input device is spread out over a whole row of spatial light modulators which can be switched to be transmissive or absorptive. A specific output device receives all signals from a whole column in the spatial light modulator array. Thereby, arbitrary interconnections can be set up.

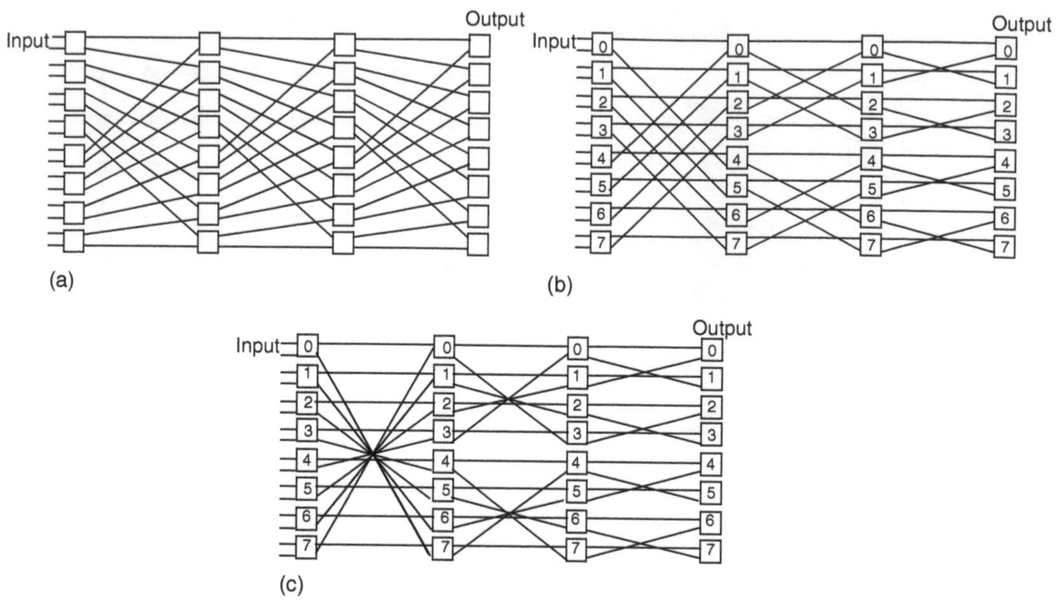

FIGURE 19.9 Three multistage interconnection networks: (a) perfect shuffle, (b) Banyan, (c) crossover network.

- For the implementation of MINs like the perfect shuffle, the use of split-and-combine setups has been used most often. A specific implementation of the perfect shuffle proposed by Brenner and Huang (1988) is shown in Figure 19.10b. A beam-splitter cube is used to form two separate optical paths for the light beams emerging from the input array and to combine the arrays in the output plane. Tilted mirrors with suitable tilts introduce a shift between the two output arrays. In the center of the output plane, one obtains the data in the right sequence as indicated in the figure. Other optical implementations of the perfect shuffle network have been described, for example, by Lohmann et al. (1986), Eichmann and Li (1987), and Sawchuk and Glaser (1988). Optical setups for the Banyan network were described by Jahns (1990); Cloonan and Herron (1989). Optical crossover network implementations were demonstrated by Jahns and Murdocca (1988).
- Multifacet-optics can be used for the implementation of arbitrary interconnects. This can be achieved with microchannel optics as shown in Figure 19.10c. Each channel consists of a separate miniaturized imaging system. An optical element is used to collimate and deflect the light emerging from the input node. A similar element is used to couple the light to the corresponding output node. A microoptic implementation of such an element using "binary optics"-type diffractive elements (Veldkamp, 1997) was demonstrated by Jahns and Däschner (1990) and with diffractive-refractive microoptics by Sauer et al. (1994). Volume holographic elements were used for this purpose by Robertson et al. (1991). It should be noted that in the case of an implementation with multi-faceted optics, the interconnection density is limited by the size of the individual aperture. Typical values for dx might be on the order of 100 μm. Furthermore, for a given aperture size dx, the interconnection length Δz, i.e., the separation between the input and output coupler is determined because of diffraction effects (Leggatt, 1991; Sauer et al., 1994): $\Delta z \leq (\pi/\lambda)dx^2$. Here, λ denotes the wavelength of the light.

Irregular optical interconnections have been discussed in connection with optical interconnections for electronic computers. A scheme that is shown in Figure 19.11 was suggested by Goodman et al. (1984) and has been considered by various groups (Kostuk, Goodman, and Hesselink, 1985; Feldman et al., 1987). It is also the basis of the POEM (programmable optoelectronic multiprocessor) approach (Kiamilev et al.,

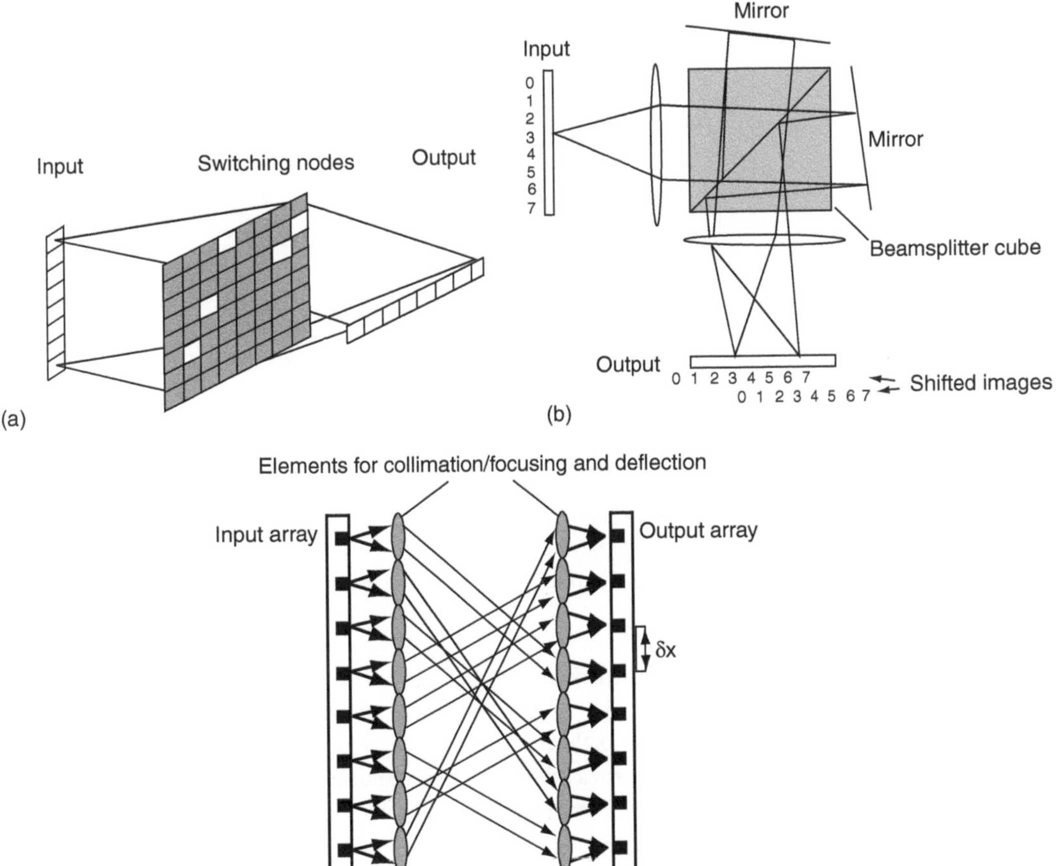

FIGURE 19.10 Optical implementation of switching networks: (a) implementation of a crossbar switch using a matrix-vector multiplier arrangement; (b) implementation of a perfect shuffle according to Brenner and Huang (1988); (c) microchannel optics for the implementation of arbitrary interconnections. This setup can also be used to implement the crossbar or the stages of a MIN such as the perfect shuffle. The channel density is determined by the aperture of the optical elements in the array. (From Brenner, K.-H. 1991. *Proc. Optical Computing Conf.*, 25–28. Optical Society of America, Washington, DC.)

1989). Several processors are interconnected on a wafer-scale through free-space by means of holograms. Intraprocessor communication is realized by electronics. Volume holographic elements can be used to multiplex many routing elements in a single component. Emitters and detectors are integrated with the processing elements on the same chip. A specific configuration of the POEM architecture based on the use of photorefractive reconfigurable holograms is discussed by Ford, Lee, and Fainman (1990).

19.5 Optical Computing Systems and Architectures

Optical interconnections are being considered for various levels in computing and communications systems. It has been argued by Miller (1989); Feldman et al. (1989) that optical communication should be used for all but the shortest distances. Table 19.1 shows which optical media are being used or investigated.

A model of an optical digital computer developed at the former AT&T Bell Laboratories was based on the use of chip-to-chip interconnections (Smith, Murdocca, and Stone, 1993). Optical connections were

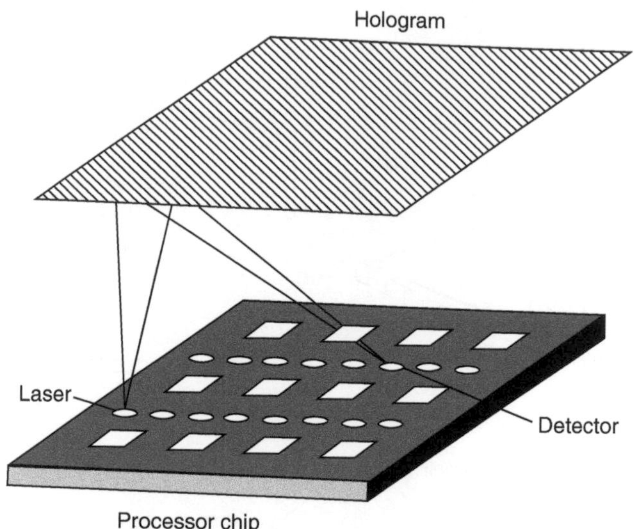

FIGURE 19.11 Optical interconnections for intrachip communications. Emitter and detector elements are distributed across the chip area.

provided between individual logic gates (for example, OR gates) that were implemented as arrays of SEED devices (Figure 19.12). Masks block light at selected positions to customize the interconnects so as to perform specific tasks like addition, subtraction, or sorting. 2 $\log_2 N$ stages are used to process columns and rows sequentially. Feedback may be provided to read the output data back into the system.

For the design of the circuits of the processor, programmable array logic is used (Mead and Conway, 1980; Murdocca et al., 1988). The positions of the logic gates in the array are fixed. All gates have a fanin and fanout of 2. All gates on the same chip perform the same logic operation like an OR or AND. The degrees of freedom available to the designer of the circuit are the positions of the inputs and the outputs and the selection of which mask positions should be open or closed. The general idea is first to generate all minterms of the input variables, and then to select and combine the minterms required to implement specific functions. The generation of minterms of two variables, x and y, is shown in Figure 19.13 (Murdocca et al., 1988). The squares in this figure represent AND gates. The black lines show connections that are used, while dashed lines indicate connections that are blocked by means of masks in the device planes. The design and implemenation of complex circuits were described by Murdocca (1990) and Tanida et al. (1990).

One difficulty of the optical PLA approach is the large number of gates that are required for the design of a specific circuit. In the free-space optical approach, the expense for the interconnection optics between the stages is quite significant, at least if conventional optomechanics is used. In general, the gate count for performing logic operations is relatively high, an issue which has led to doubts about the

TABLE 19.1 Hierarchy of Optical Interconnections

Application	Distance (m)	Medium
Satellite communication	10^6	Free-space
Telecommunication	10^5–10^6	Fiber
Local area networks	10^2–10^3	Fiber
System-to-system	10^1–10^2	Fiber
Board-to-board (intrasystem)	10^{-1}–10^0	Fiber waveguide free-space
Chip-to-chip	10^{-3}–10^{-2}	Waveguide free-space

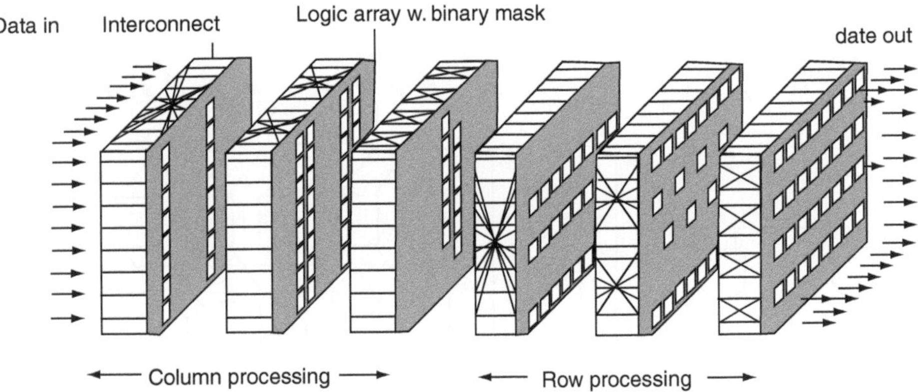

FIGURE 19.12 Architecture of a free-space optical computer according to Smith, Murdocca, and Stone (1993). The individual stages are separated in the figure for clarity. Optical device arrays are interconnected through free space. Masks in the device planes are used to customize the interconnections. (From Smith, D. E., Murdocca, M. J., and Stone, T. W. 1993. Optical computing hardware, Jahns, J. and Lee, S. H., eds., Academic Press, Boston.)

suitability of free-space interconnects at the gate level. Several improvements for reducing the gate count have been discussed (Smith, Murdocca, and Stone, 1993).

A second approach to the implementation of optical logic is computational origami (Huang, 1992). This concept uses highly regular interconnections, as they can be implemented with conventional imaging optics. Data processing is done by transforming the task into a sequence of five basic operations:

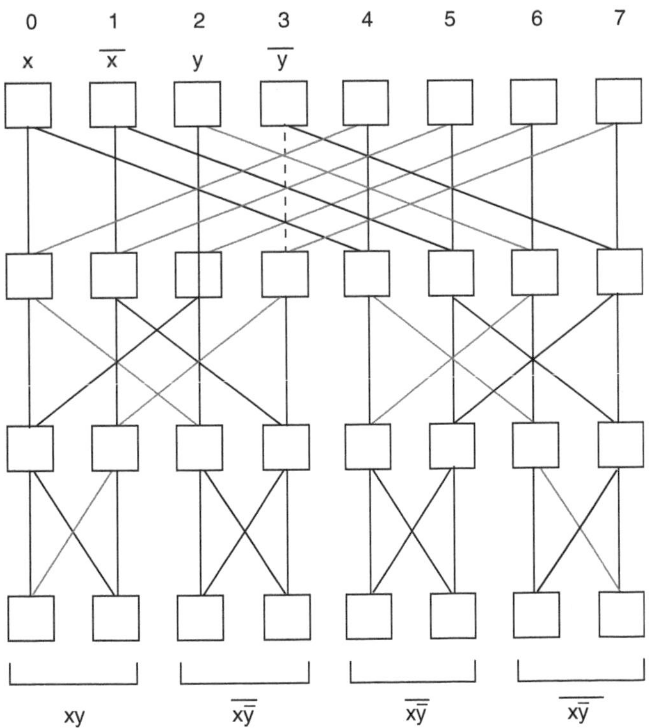

FIGURE 19.13 Interconnection network to generate minterms of two logic variables according to Murdocca et al. (1988). (From Murdocca, M. J., Huang, A., Jahns, J., and Streibl. N. 1988. *Applied Optics*. 27, 1651–1660.)

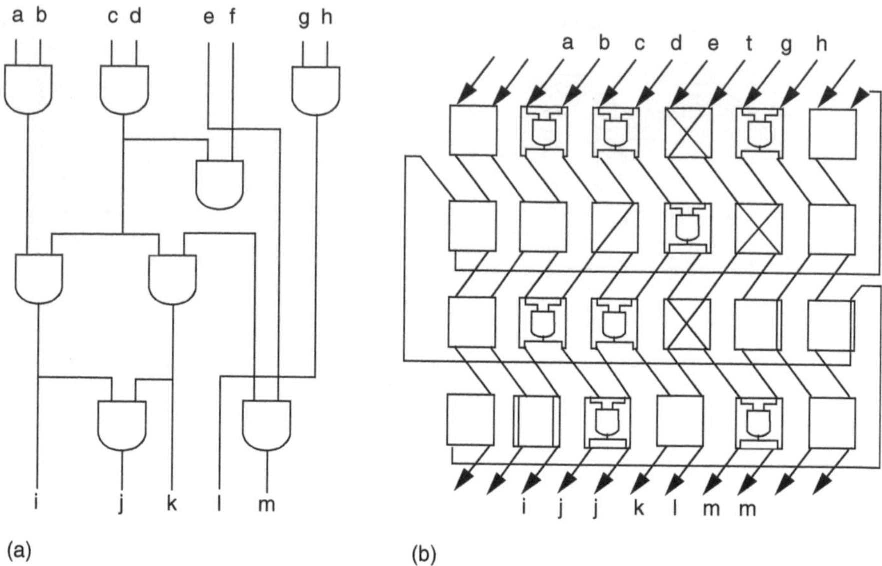

FIGURE 19.14 Computational origami: (a) conventional electronic circuit and (b) regularized version in an origami processor according to Huang (1992). (From Huang, A., 1992. *Applied Optics.* 31, 5419–5422.)

NOR, crossover, bypass, broadcast, and don't care. Figure 19.14a shows a simple conventional logic circuit consisting of eight AND gates. It is redrawn in a folded regular way in Figure 19.14b. A array of logic gates is used with a width of six and a depth of four.

Computational origami is a way of shaping a computational task into a form such that it becomes suitable to the hardware and communications requirements of a processor. The relevance to optics is that computational origami can be used to match the computational task to a highly regular interconnection scheme. It also illustrates how to perform computations with delay lines rather than with random-access memory. The function of memory in a computation is implemented by delaying the data stream until it is needed. Computational origami accomplishes this by using fixed amounts of time for the delays. This avoids the need for random access memory, and therefore reduces the hardware complexity of a computer.

In the heyday of digital optical computing, several groups demonstrated systems based on binary switching devices like SEEDs, using a combination of classical opto-mechanical hardware and microoptical lenslet arrays and beam-splitters. Prototypical demonstrations were shown, for example, by Prise et al. (1991); Craig et al. (1991), and Huang et al. (1993). Limitations of these early devices were speed and system size; nonetheless, important insights were gained into the practical aspects of the systems, which became important for later work on optical interconnection.

Recently, the aspect of latency reduction in symmetric multiprocessor systems was addressed by different research groups (Boxer, 1995; Louri and Kodi, 2003; Lukowicz et al., 2003). In a symmetric multiprocessor, it is essential to maintain cache-coherence, so that every processor knows the contents of every other processor's cache. This is obviously a communication-intensive task that is limited by the interconnection capabilities of the shared bus. As the number of processors in a system grows, the probability of bus contention increases. The increase in processor speed aggravates the problem because the bus cannot run at comparable speeds. Here, optical solutions are being investigated, combining different optics technologies.

19.6 Systems Integration

A specific problem of optical interconnects is the aspect of system packaging. Conventional optical systems rely on the use of mechanical mounting, which results in bulky and expensive setups. Microoptic

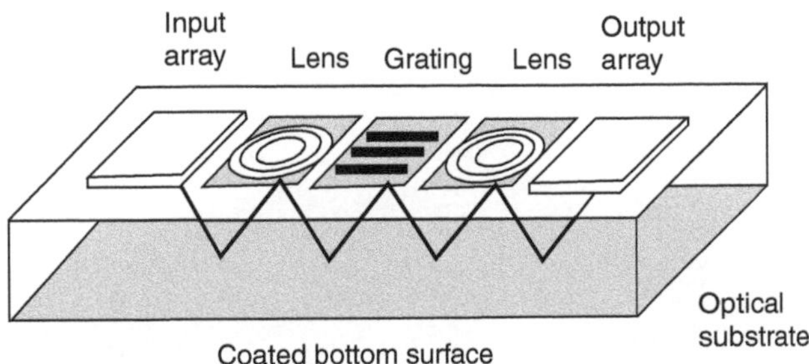

FIGURE 19.15 Principle of planar optical packaging. Free-space interconnects are provided by substrate mode propagation of the light signals. The optics is fabricated by lithographic means.

integration schemes that make use of lithographic fabrication of the optics are an alternative. Three basic approaches were suggested in recent years: "stacked microoptics" (Iga et al., 1982; Brenner, 1991), planar integrated free-space optics (PIFSO) (Jahns and Huang, 1989), and a MEMS (micro-electromechanical systems) approach (Wu, 1997). In the PIFSO approach, microoptic elements are integrated on one or both sides of an optical flat (Figure 19.15). The light signals propagate inside the substrate. Optoelectronic chips are mounted onto the substrate by means of solder bump bonding (Jahns et al., 1992). Parallel interconnections with high interconnection densities (Jahns and Acklin, 1993), optical clock distribution (Walker and Jahns, 1992) and the implementation of space-variant interconnection networks (Jahns and Däschner, 1990; Song et al., 1992) have been demonstrated. An experimental setup including a 2D SEED-modulator-chip, a 1D VCSEL-chip, and electrical wiring integrated onto a functional PIFSO interconnect is shown in Figure 19.16 (Gruber 2004). The interconnect shown can be used for the realization of a vector-matrix multiplier, and also for an optical crossbar switch, as schematically shown in Figure 19.8a (Fey et al., 2000).

An alternative approach to the realization of a crossbar switch has been demonstrated during recent years in context with optical crossconnects; see, for example, Zheng et al. (2003). Here, tiltable micromirror devices with variable angles are used. This allows for an interesting implemention of a nonblocking switch using just two micromirror arrays with $N \times N$ individual mirror devices each (Figure 19.17).

Various aspects of systems integration and packaging have been addressed by research groups over the past years. Issues involve, for example, the microfabrication and packaging of the optics (Thienpont et al., 2000) and aspects of the optical systems design and tolerancing (Christensen et al., 2003; Kirk et al., 2003).

19.7 Conclusion and Outlook

Today's electronic computers are becoming more and more limited by communications problems. The application of optics to computing may offer significant advantages over conventional computer technology and architecture. It is not yet clear how the merging of electronic and optical technology will be achieved, although the interconnection area is the most likely area where optics can be considered a serious alternative.

A critical issue remains, namely the packaging issue. Although many excellent results have been demonstrated at the research level, the problem of integrating optics and electronics cannot be considered as solved in a practical sense. The difficulty lies in the fact that by including optical devices and components, the standard VLSI process is modified because additional steps for bonding, cleaning, etc. have to be added. Some of these have to be carried out at high temperatures. Hybrid integration schemes appear more promising at the moment than monolithic integration schemes. However, they still hold formidable challenges, with regard to thermal management, for example.

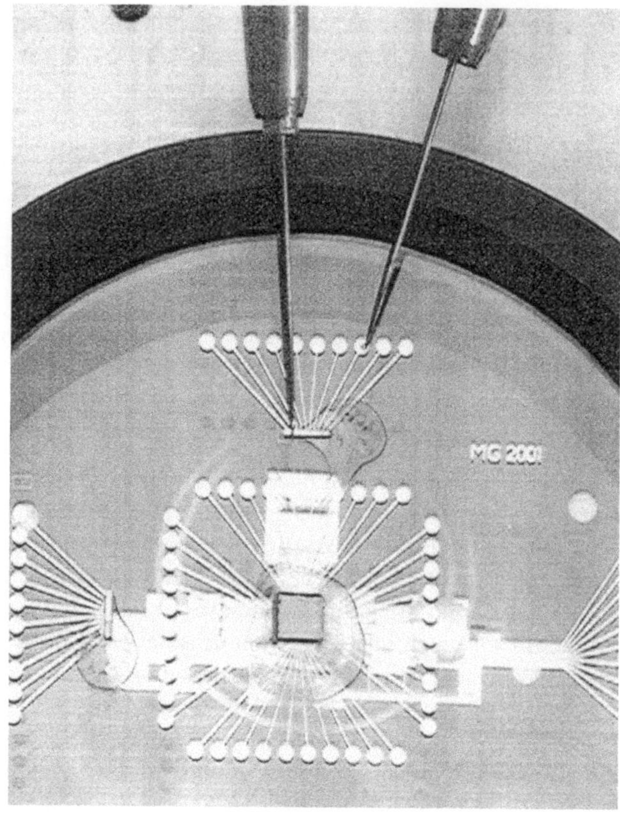

FIGURE 19.16 Optoelectronic multichip module using planar-integrated free-space optics. The optics implements a crossbar interconnect. In the center, a SEED-modulator chip is shown; to the left, signals are entered from a VCSEL-chip.

Optics has become a dominant factor in long-distance communications. In computing and processing, it appears that the dominance of silicon processing will continue. Optical interconnection, however, is an area which allows an optimistic outlook for the use of optics in digital computing.

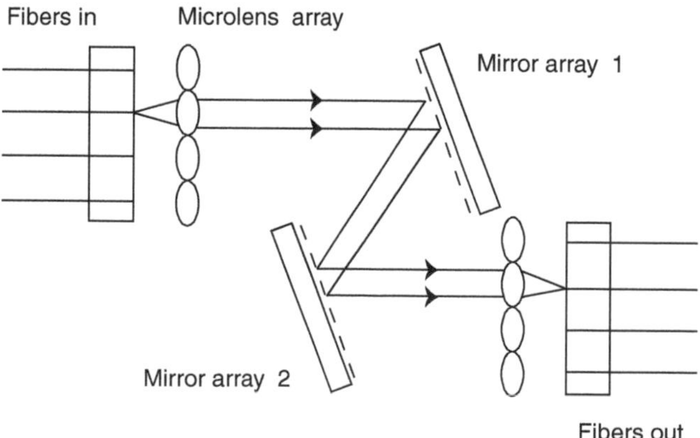

FIGURE 19.17 Architecture of a 3D crossbar switch using MEMS-mirror arrays. Light entering from a single-fiber channel is collimated and deflected by a pair of mirrors to a specific output fiber.

A number of critical issues remain, including multimode versus single mode, modulator devices versus emitters, multiplexing (like WDM or TDM) or no multiplexing, polymer waveguides versus free-space connections, and finally, packaging and thermal management.

For data rates of 10 Gb/s and above (the speed at which optical chip-to-chip solutions begin to be interesting), multimode solutions are challenging because of modal dispersion. VCSEL interconnections operating at more than these data rates have not yet been demonstrated. The lifetime and reliability of emitter devices may be a problem. WDM solutions require a minimal number of input/output paths. Manipulation of WDM is typically space-consuming, however, and requires either multiple-wavelength, precise lasers or a single costly mode-locked laser. Existing WDM devices like arrayed waveguide gratings are made to serve the needs of long-distance communications systems and are not directly applicable to low-cost, short-distance systems.

The quality of polymer materials or suitable realization of a free-space architecture is particularly important when considering reasonable distances between the communicating chips—and more significantly so for single-mode than multimode solutions. With respect to packaging and thermal management, the need to use hybrid integration approaches implies the requirement to bond parts with macro dimensions (on the order of millimeters or even tens of millimeters) with micro precision. On-chip sources will be difficult to manage, especially thermally, for high-I/O-port integrated circuits such as microprocessors. Off-chip sources (like mode-locked lasers) have the advantage of being independently controlled and monitored, yet coupling the off-chip light source to the optical chip or modulator adds complexity.

At the moment, both optical and electronic suppliers are evaluating these paths. The ultimate implementation will lie in a joint solution that fits well into the economic model of all parties.

Lastly, a large amount of literature exists that could not be reviewed in this article. The author would like to refer the reader to the following special issues that provide an excellent overview of earlier work and recent developments: a special issue on "Optical Computing," in the *Proceedings of IEEE* (July 1984); a feature issue in *Applied Optics* edited by Athale and Caulfield (1986); a special issue on "Optical Computing Systems," in *Proceedings of IEEE* (November 1994); a special issue on "Optical Interconnection for Digital Systems," in *Proceedings of IEEE* (June 2000); and a special issue on "Optical Interconnects" in the *IEEE Journal on Selected Topics in Quantum Electronics* (March/April 2003).

References

Agrawal, D. P. 1983. Graph theoretical analysis and design of multistage interconnection networks. *IEEE Transactions on Computers*, C-32, 637–648.

Athale, R. and Caulfield, H. J. 1986. Optical computing. *Applied Optics*, 25, 1541–1542.

Backus, J. 1982. Function-level computing. *Spectrum*, 19, 8, 22–27.

Basov, N. G., Culver, W. H., and Shah, B. 1972. Applications of lasers to computers, ch. F5 In *Laser Handbook Vol.* 2, F. T. Arecchi and E. O. Schulz-Dubois, eds., pp. 1650–1693. North-Holland, Amsterdam.

Benes, V. E. 1965. *Mathematical Theory of Connecting Networks and Telephone Traffic*, Academic Press, New York.

Boxer, A. 1995. Where buses cannot go. *IEEE Spectrum*, 32, 41–45.

Brenner, K.-H., 1991. 3D-Integration of digital optical systems. In *Proc. Topical Meeting on Optical Computing*, pp. 25–28. Optical Society of America, Washington, DC.

Brenner, K.-H. and Huang, A. 1988. Optical implementation of the perfect shuffle interconnections. *Applied Optics*, 27, 135–137.

Casasent, D. 1997. Analog optical data processing, In *CRC Handbook on Photonics*, M. Gupta, ed., pp. 591–604, CRC Press, Boca Raton, FL.

Chen, R. T., Lin, L., Choi, C., Liu, Y., Bihari, B., Wu, L., Tag, S., Wickman, R., Picor, B., Hibbs-Brenner, M. K., and Bristow, J. 2000. Fully embedded board-level guided-wave optoelectronic interconnects. *Proceedings of the IEEE*, 88, 780–793.

Cheng, J., Zhou, P., Sun, S. Z., Hersee, S., Myers, D. R., Zolper, J., and Vawter, G. A. 1993. Surface-emitting laser-based smart pixels for two-dimensional optical logic and reconfigurable interconnections. *IEEE Journal of Quantum Electronics*, 29, 741–756.

Christensen, M. P., Milojkovic, P., McFadden, M. J., and Haney, M. W. 2003. Multiscale optical design for global chip-to-chip optical interconnections and misalignment tolerant packaging. *IEEE Journal of Selected Topics in Quantum Electronics*, 9, 548–556.

Cloonan, T. J. 1993. Architectural considerations for optical computing and photonic switching. In *Optical Computing Hardware*, J. Jahns and S. H. Lee, eds., Academic Press, Boston.

Cloonan, T. J. and Herron, M. J. 1989. Optical implementation and performance of one-dimensional trimmed inverse augmented data manipulator networks for multiprocessor computer systems. *Optical Engineering*, 28, 305–314.

Clos, C. 1953. A study of non-blocking switching networks. *Bell Systems Technical Journal*, 32, 406.

Craig, R. G. A., Wherrett, B. S., Walker, A. C., Tooley, F. A. P., and Smith, S. D. 1991. Optical cellular logic image processor: implementation and programming of a single channel digital optical circuit. *Applied Optics*, 30, 2297–2308.

D'Asaro, L. A., Chirovski, L. M. F., Laskowski, E. J., Pei, S. S., Woodward, T. K., Lentine, A. L., Leibenguth, R. E. et al. 1993. Batch fabrication and operation of GaAs–$Al_xGa_{1-x}As$ field-effect transistor-self-electrooptic effect device (FET-SEED) smart pixel arrays. *IEEE Journal of Quantum Electronics*, 29.

Eichmann, G. and Li, Y. 1987. Compact optical generalized perfect shuffle. *Applied Optics*, 26, 1167–1169.

Fang, S. F., Adomi, K., Iyer, S., Morkoç, H., Zabel, H., Choi, C., and Otsuka, N. 1990. Gallium arsenide and other compound semiconductors on silicon. *Journal of Applied Physics*, 68, R31–R58.

Feldman, M. R., Esener, S. C., Guest, C. C., and Lee, S. H. 1987. Comparison between optical and electrical interconnects based on power and speed considerations. *Applied Optics*, 27, 1742–1751.

Feldman, M. R., Guest, C. C., Drabik, T. J., and Esener, S. C. 1989. Comparison between electrical and free space optical interconnects for fine grain processor arrays based on interconnect density capabilities. *Applied Optics*, 28, 3820–3829.

Fey, D., Erhard, W., Gruber, M., Jahns, J., Bartelt, H., Grimm, G., Hoppe, L., and Sinzinger, S. 2000. Optical interconnects for neural and reconfigurable VLSI architectures. *Proceedings of the IEEE*, 88, 838–848.

Ford, J. E., Lee, S. H., and Fainman, Y. 1990. Application of photorefractive crystals to optical interconnection. *Digital Optical Computing II, Proceedings of SPIE*, 1215, 155–165.

Gibbs, H. M., McCall, S. L., and Venkatesan, T. N. C. 1976. Differential gain and bistability using a sodium-filled Fabry-Perot interferometer. *Physical Review Letters*, 36, 1135–1138.

Goke, L., and Lipovski, G. 1973. Banyan networks for partitioning multiprocessor systems. In *First Annual Symposium on Computer Architecture*, pp. 21–28. ACM/IEEE, New York.

Goodman, J. W. 1996. *Introduction to Fourier Optics. 2nd Ed.* McGraw-Hill, New York.

Goodman, J. W., Dias, A. R., and Woody, L. M. 1978. Fully parallel, high speed incoherent optical method for performing discrete Fourier transforms. *Optics Letters*, 2, 1–3.

Goodman, J. W., Leonberger, F. J., Kung, S., and Athale, R. A. 1984. Optical interconnections for VLSI systems. *Proceedings of the IEEE*, 72, 850–865.

Goossen, K. W., Boyd, G. D., Cunningham, J. E., Jan, W. Y., Miller, D. A. B., Chemla, D. S., and Lum, R. M. 1989. GaAs–AlGaAs multiquantum well reflection modulators grown on GaAs and silicon substrates. *IEEE Photonics Technology Letters*, 1, 304–306.

Gruber, M. 2004. Multichip module with planar-integrated free-space optical vector-matrix-type interconnects. *Applied Optics*, 43, 463–470.

Hinton, H. S. 1988. Architectural considerations for photonic switching networks. *IEEE Journal on Selected Areas in Communications*, 6, 1209–1226.

Hinton, H. S. and Midwinter, J. E. eds. 1990. *Photonics in Switching*, IEEE Press, New York.

Huang, A., 1975. The implementation of a residue arithmetic unit via optical and other physical phenomena. *Proc. International Optical Computing Conference*, p. 14. Washington, DC.

Huang, A., 1980. Design for an optical general purpose computer. In *Proceedings of the 1980 International Optical Computing Conference*, book 2, W. T. Rhodes, ed., SPIE Vol. 232, pp. 119-123, Washington, DC.

Huang, A. 1992. Computational origami: the folding of circuits and systems. *Applied Optics*, 31, 5419–5422.

Huang, A., Tsunoda, Y., Goodman, J. W., and Ishihara, S. 1979. Optical computation using residue arithmetic. *Applied Optics*, 18, 149–162.

Huang, K.-S., Sawchuk, A. A., Jenkins, B. K., Chavel, P., Wang, J.-M., Weber, A. G., Wang, C.-H., and Glaser, I. 1993. Digital optical cellular image processor (DOCIP): experimental implementation. *Applied Optics*, 32, 166–173.

Iga, K., Oikawa, M., Misawa, S., Banno, J., and Kokubun, Y. 1982. Stacked planar optics: an application of the planar microlens. *Applied Optics*, 21, 3456–3460.

Iga, K., Kinoshita, S., and Koyame, F. 1987. Microcavity GaAlAs/GaAs surface emitting lasers with I_{th} = 6 mA. *Electronics Letters*, 23, 134–136.

Jahns, J. 1990. Optical implementation of the Banyan network. *Optics Communications*, 76, 321–324.

Jahns, J. 1998. Free-space optical digital computing and interconnection, In *Progress in Optics, Vol. 28*, E. Wolf, ed., pp. 419–513, North-Holland, Amsterdam.

Jahns, J. and Acklin, B. 1993. Integrated planar optical imaging system with high interconnection density. *Optics Letters*, 18, 1594–1596.

Jahns, J. and Däschner, W. 1990. Optical cyclic shifter using diffractive lenslet arrays. *Optics Communications*, 79, 407–410.

Jahns, J. and Huang, A. 1989. Planar integration of free-space optical components. *Applied Optics*, 28, 1602–1605.

Jahns, J. and Lee, S. H. eds. 1993. *Optical Computing Hardware*, Academic Press, Boston.

Jahns, J. and Murdocca, M. J. 1988. Crossover networks and their optical implementation. *Applied Optics*, 27, 3155–3160.

Jahns, J., Morgan, R. A., Nguyen, H. N., Walker, J. A., Walker, S. J., and Wong, Y. M. 1992. Hybrid integration of surface-emitting microlaser chip and planar optics substrate for interconnection applications. *IEEE Photonics Technology Letters*, 4, 1369–1372.

Javidi, B. ed. 2005. *Optical and Digital Techniques for Information Security*, Springer, New York.

Jewell J.L. Olbright G.R. 1993. Microlaser devices for optical computing. In *Optical Computing Hardware* J. Jahns and S. H. Lee, eds., pp. 99–112, Academic Press Boston.

Jewell, J. L., Scherer, A., McCall, S. L., Lee, Y. H., Walker, S., Harbison, J. P., and Florez, L. T. 1989. Low-threshold electrically pumped vertical-cavity surface-emitting microlasers. *Electronics Letters*, 25, 1123–1124.

Kasahara, K. 1993a. Vertical-to-surface transmission electrophotonic devices, In *Optical Computing Hardware*, J. Jahns and S. H. Lee, eds., Academic Press, Boston.

Kasahara, K. 1993b. VSTEP-based smart pixels. *IEEE Journal of Quantum Electronics*, 29, 757–768.

Keyes, R. W. 1987. *The Physics of VLSI Systems*. Addison-Wesley, Wokingham, England.

Keyes, R. W. and Armstrong, J. A. 1969. Thermal limitations in optical logic. *Applied Optics.*, 8, 2549.

Kiamilev, F., Esener, S., Paturi, R., Fainman, Y., Mercier, P., Guest, C. C., and Lee, S. H. 1989. Programmable optoelectronic multiprocessors and their comparison with symbolic substitution for digital optical computing. *Optical Engineering*, 28, 396–409.

Kirk, A. G., Plant, D. V., Ayliffe, M. H., Châteauneuf, M., and Lacroix, F. 2003. Design rules for highly parallel free-space optical interconnects. *IEEE Journal on Selected Topics in Quantum Electronics*, 9, 531–547.

Kostuk, R. K., Goodman, J. W., and Hesselink, L. 1985. Optical imaging applied to microelectronic chip-to-chip interconnections. *Applied Optics*, 24, 2851–2858.

Krishnamoorthy, A. V., Chirovsky, L. M. F., Hobson, W. S., Leibenguth, R. E., Hui, S. P., Zydzik, C. J., Goossen, K. W. et al. 1999. Vertical cavity surface-emitting lasers flip-chip bonded to gigabit-per-second CMOS circuits. *IEEE Photonics Technology Letters*, 11, 128–130.

Lee, S. H. ed. 1981. *Optical Information Processing: Fundamentals*, Springer, Berlin.

Leggatt, J. S., Hutley, M. C., 1991, Microlens arrays for interconnection of single mode fibre arrays, *Electronics Letters* 27, 238–240.

Lentine, A. L. 1993. Self-electrooptic effect devices for optical information processing, In *ptical Computing Hardware*, J. Jahns and S. H. Lee, eds., pp. 45–72. Academic Press, Boston.

Lentine, A. L. and Miller, D. A. B. 1993. Evolution of the SEED technology: bistable logic gates to optoelectronic smart pixels. *IEEE Journal of Quantum Electronics*, 29, 655–669.

Lentine, A. L., Hinton, H. S., Miller, D. A. B., Henry, J. E., Cunningham, J. E., and Chirovsky, L. M. F. 1988. Symmetric self-electrooptic effect device: optical set-reset latch. *Applied Physics Letters*, 52, 1419–1421.

Lohmann, A. W. 1986. What classical optics can do for the optical computer. *Applied Optics*, 25, 1543–1549.

Lohmann, A. W., Stork, W., and Stucke, G. 1986. Optical perfect shuffle. *Applied Optics*, 25, 1530–1531.

Louri, A. and Kodi, A. K. 2003. Parallel optical interconnection network for address transactions in large-scale cache coherent symmetric multiprocessors. *IEEE Journal on Selected Topics in Quantum Electronics*, 9, 667–676.

Lukowicz, P., Jahns, J., Barbieri, R., Benabes, P., Bierhoff, T., Gauthier, A., Jarczynski, M., Russell, G. A., Schrage, J., Süllau, W., Snowdon, J. F., Wirz, M., and Tröster, G. 2003. Optoelectronic interconnection technology in the HOLMS system. *Journal on Selected Topics in Quantum Electronics*, 9, 624–635.

McCormick F.B. 1993. Free-space interconnection techniques, In *Photonics in Switching* Midwinter, J. E. ed., pp. 169–250. Academic Press, Boston.

Mead, C. and Conway, L. 1980. *Introduction to VLSI Systems*. Addison-Wesley, Reading, MA.

Mederer, F., Michalzik, R., Guttmann, J., Huber, H.-P., Lunitz, B., Moisel, J., and Wiedenmann, D. 2002. 10 Gbit/s data transmission with TO-packaged multimode GaAs VCSELs over 1 m long polymer waveguides for optical backplane applications. *Optics Communications*, 206, 309–312.

Midwinter, J. E. ed. 1993. *Photonics in Switching*, Academic Press, Boston.

Mikaelian, A. L. and Salakhudtinov, V. K. 1994. High-speed multichannel optical switching. *Proceedings of SPIE*, 2144, 84–90.

Miller, D. A. B. 1989. Optics for low-energy communication inside digital processors: quantum detectors, sources, and modulators as efficient impedance converters. *Optics Letters*, 14, 146–148.

Miller, D. A. B. 2000. Rationale and challenges for optical interconnects to electronic chips. *Proceedings of IEEE*, 88, 728–749.

Miller, D. A. B., Chemla, D. S., Damen, T. C., Gossard, A. C., Wiegmann, W., Wood, T. H., and Burrus, C. A. 1984. Novel hybrid optically bistable switch: the quantum well self-electrooptic effect device. *Applied Physics Letters*, 45, 13–15.

Morgan, R. A., Robinson, K. C., Chirovsky, L. M. F., Focht, M. F., Guth, G. D., Leibenguth, R. E., Glogovsky, K. G., Przybylek, G. J., and Smith, L. E. 1991. Uniform 64x1 arrays of individually-addressed vertical cavity top surface emitting lasers. *Electronics Letters*, 16, 1400–1402.

Murdocca, M. J. 1990. *A Digital Design Methodology for Optical Computing*. MIT Press, Cambridge, MA.

Murdocca, M. J., Huang, A., Jahns, J., and Streibl, N. 1988. Optical design of programmable logic arrays. *Applied Optics*, 27, 1651–1660.

Prise, M. E., Streibl, N., and Downs, M. M. 1988. Optical considerations in the design of digital optical computers. *Optical and Quantum Electronics*, 20, 49–77.

Prise, M. E., Craft, N. C., Downs, M. M., LaMarche, R. E., D'Asaro, L. A., Chirovski, L. M. F., and Murdocca, M. J. 1991. Optical digital processor using arrays of symmetric self-electrooptic effect devices. *Applied Optics*, 30, 2287–2296.

Reisman, A. 1983. Device, circuit, and technology scaling to micron and submicron dimensions. *Proceedings of IEEE*, 71, 550–565.

Robertson, B., Restall, E. J., Taghizadeh, M. R., and Walker, A. C. 1991. Space-variant holographic optical elements in dichromated gelatin. *Applied Optics*, 30, 2368–2375.

Roy, S., Sharma, P., Dharmadhikari, A. K., and Mathur, D. 2004. All-optical switching with bacteriorhodopsin. *Optical Communications*, 237, 251–256.

Sauer, F., Jahns, J., Nijander, C. R., Feldblum, A. Y., and Townsend, W. P. 1994. Refractive-diffractive microoptics for permutation interconnects. *Optical Engineering*, 33, 1550–1560.

Sawchuck, A. A. and Glaser, I. 1988. Geometries for optical implementations of the perfect shuffle. *Proceedings of SPIE*, 88, 270.

Semiconductor Industry Association 2001. The national technology roadmap for semiconductors—Interconnect. International Technology Road map for Semiconductors 2005 edition, http://www.itrs.net/Links/2005ITRS/Home2005.htm

Smith, D. E., Murdocca, M. J., and Stone, T. W. 1993. Parallel optical interconnections, In *Optical Computing Hardware*, J. Jahns and S. H. Lee, eds., pp. 193–226. Academic Press, Boston.

Song, S. H., Lee, E. H., Carey, C. D., Selviah, D. R., and Midwinter, J. E. 1992. Planar optical implementation of crossover interconnects. *Optics Letters*, 17, 1253–1255.

Stone, H. S. 1971. Parallel processing with the perfect shuffle. *IEEE Transactions on Computers*, C-20, 153–161.

Streibl, N. 1989. Beam shaping with array generators. *Journal of Modern Optics*, 12, 1559–1573.

Streibl, N. 1993. Multiple beamsplitters, In *Optical Computing Hardware*, J. Jahns and S. H. Lee, eds., pp. 227–248. Academic Press, Boston.

Tanida, J., Nakagawa, J., Yagyu, E., Fukui, M., and Ichioka, Y. 1990. Experimental verification of parallel processing on a hybrid optical parallel array logic system. *Applied Optics*, 29, 2510–2521.

Thienpont, H., Debaes, C., Baukens, V., Ottevaere, H., Vynck, P., Tuteleers, P., Verschaffelt, G., Volckaerts, B., Hermanne, A., and Hanney, M. 2000. Plastic microoptical interconnection modules for parallel free-space inter- and intra-MCM data communications. *Proceedings of IEEE*, 88, 769–779.

Tippett, J. T., Berkowitz, D. A., Clapp, L. C., Koester, C. J., and Vanderburgh, A., eds. 1965. *Optical and Electro-Optical Information Processing*, MIT Press, Cambridge, MA.

van der Lugt, A. B. 1964. Signal detection by complex spatial filtering. *IEEE Transactions on Information Theory*, IT-10, 2.

Veldkamp, W. 1997. Binary optics. In *CRC Handbook on Photonics*, M. Gupta, ed., CRC Press, Boca Raton, FL.

Walker, S. J. and Jahns, J. 1992. Optical clock distribution using integrated free-space optics. *Optics Communications*, 90, 359–371.

Wilmsen, C. W., Beyette, J. F. R., An, X., Feld, S. A., and Geib, K. M. 1993. Smart pixels using the light amplifying optical switch (LAOS). *IEEE Journal of Quantum Electronics*, 29, 769–774.

Wise, D. S. 1981. Compact layout of Banyan/FFT networks. In *Proceedings CMU Conference on VLSI Systems and Computations*, H.T. Kung, B. Sproull, and G. Steele, eds., pp. 186–195. Computer Science Press, Rockville, MD.

Wu, M. C. 1997. Micromachining for optical and optoelectronic systems. *Proceedings of IEEE*, 85, 1833–1856.

Zheng, X., Kaman, V., Yuan, S., Xu, Y., Jerphagnon, O., Keating, A., Anderson, R. C., Poulsen, H. N., Liu, B., Sechrist, J. R., Pusarla, C., Helkey, R., Blumenthal, D. J., and Bowers, J. E. 2003. Three-dimensional MEMS photonic cross-connect switch design and performance. *IEEE Journal on Selected Topics in Quantum Electronics*, 9, 571–578.

20 Optical Communications

Alan E. Willner
Zhongqi Pan

20.1 Introduction

Optical communications is a very old form of data transfer. Line-of-sight primitive digital systems have included lighting bon fires on mountain tops to send a simple one-bit message, smoke signals to send a multiple-bit message, and ship-to-ship broad-incoherent-beam transmission of Morse-code messages. The inventions of the low-loss optical fiber[1] and the high-speed semiconductor laser[2] have caused an explosion in the transmission capacity of optical systems.[3] We will discuss in this chapter the technologies behind sophisticated optical systems, some of which can transmit >10 trillion (10×10^{12}) digital bits of information per second across transpacific distances.[4] In addition to higher capacity, tethered systems have lower loss and higher security than untethered systems. We will therefore focus on the discussion of fiber-based transmission systems.

One of the most basic and undisputed applications of optical rather than electrical communication is for data transmission on long-distance links. Optics is an obvious choice because of the ultra-wide bandwidth (>25 THz) of low-loss (<0.2 dB/km) transmission properties of an optical fiber, while the impedance of the electrical cable increases substantially at GHz speeds to >100 dB/mile. In addition to the ultra-wide bandwidth transmissions, a wavelength-division multiplexing (WDM) optical network also enables straightforward routing and switching of optical data paths, in which the data travels on wavelength-specific "light-paths" from source to destination that can be arranged by a network controller to optimize throughput.

The progress in optical communications over the past 25 years has been astounding. In 1980, AT&T could transmit 672 two-way conversations along a pair of optical fibers.[5] In 1994, an AT&T network connecting Florida with the Virgin Islands was able to carry 320,000 two-way conversations along two pairs of optical fibers. By 2003, Tyco Telecommunications' transoceanic dense WDM (DWDM) system transmitted 128 wavelengths per fiber pair at 10 Gbit/s/wavelength with the total capacity 10 Tbit (8 fiber pairs)—a capability of transmitting more than 100 million simultaneous voice circuits on an eight-fiber pair cable.

The advances of fiber optical systems have progressed in four generations of technology[6] in the last 25 years. The first generation systems used 0.8 μm GaAs semiconductor lasers and multimode fibers. These systems operated at 50–100 Mbit/s and the fiber links were only 10 km in length.[7] The second and third generation systems employed single-mode fibers and 1.3 μm InGaAs and 1.55 μm InGaAsP lasers. The fourth generation systems employed optical amplifiers and WDM technologies. Today, the high-capacity amplified WDM systems have hundreds of channels at 10 Gbit/s with channel spacing as low as 50 GHz and distances to a few thousand kilometers. Systems that operate at 40 Gbit/s/channel rate are commercialized and ready for deployment. The next generation systems will be beyond fiber transmission links to optical networks. There are major efforts to promote the concept of Fiber-to-the-X (FTTX: X=Curb, Home, Desktop, etc.) to support the demand for voice, data, and Internet application across metropolitan areas.

In the first edition, we discussed the technology which has enabled such progress plus the technology which may increase the capacity by an order of magnitude within the next decade. Since that publication, the optical communications has been growing rapidly with new research and business activities. The major explosion came after the maturity of fiber amplifiers and WDM technologies. Due to the wide gain bandwidth of the Erbium-Doped Fiber Amplifier (EDFA) and distributed Raman Fiber Amplifier (DRFA), new fibers and fiber bands are being opened up, unprecedented bit-rate/wavelength is being carried over all-optical distance, and over 100s WDM channels are being simultaneously transmitted over ultra-long distances.[8–11] It is difficult to overstate the impact fiber amplifiers and WDM technologies have had in both generating and supporting the telecommunications revolution during last 10 years.

This second edition is intended to bring these state-of-art technologies into this chapter and reflect the tremendous progress in the past 10 years. Since WDM-related technologies have clearly dominated the research activity and the telecommunications market, we will be exploiting its technical details in a whole new section. In addition, we also add some new materials in this edition, including (1) wideband EDFAs, Raman amplifiers, fiber Bragg gratings, and arrayed waveguide gratings in the section of various components; (2) amplifiers' gain equalization/flattening, power transients control, fiber dispersion and nonlinearity management, polarization mode dispersion and its control, crosstalk, and network monitoring in the section of system technologies; and (3) optical switch and optical networks in a new section.

The demand for network bandwidth is outpacing even the astounding advances of recent years. Figure 20.1 shows the estimated load with the exponential network growth. New applications are driving the need for ultrahigh bandwidth services. The proliferation of online services and network access providers coupled with low cost computers results in exponentially increasing numbers of customers,

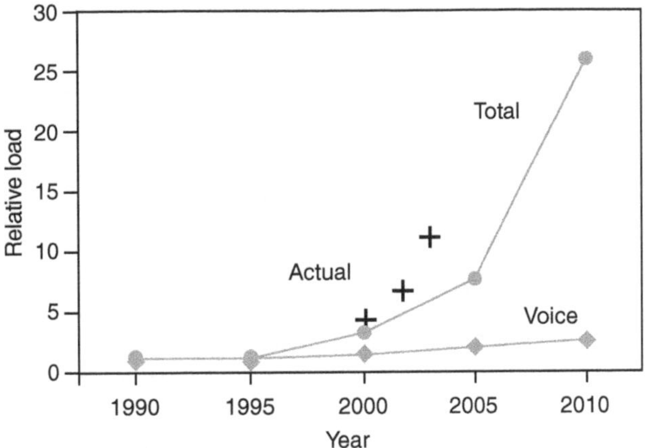

FIGURE 20.1 Estimated network load. (From White, P. E. 1993. *Bellcore*; adapted from Li, Tingye, AT&T.)

with increasing bandwidth demands to support multimedia and other revolutionary applications. Faster processors fuel this demand, as today's computers are outdated tomorrow. More and more people spend more and more time online to perform more and more everyday tasks. Optical transmission is the only viable solution for this ever-increasing bandwidth requirement. Due to its high capacity and performance, optical fiber communications have already replaced many conventional communication systems in point-to-point transmission and networks. The ever-increasing fiber optic base and the acceptance of WDM as an established technology are waiting to fulfill the enormous future potential of next-generation Internet services.

In summary, optical communications is a fairly large and still rapidly-advancing scientific field. In order to treat many ideas in just one chapter, we will cover each topic fairly broadly without too much depth. Only the most basic material will be included, and many references are included to facilitate further study in the reader. Furthermore, since it is not clear which specific optical technology will be most important in future optical systems and networks, we will attempt to include a treatment of each possibly-crucial technology. Finally, optical systems are different from electrical systems in several aspects, but chief of these is the physical implementation. Consequently, the vast majority of this chapter deals with the physical optical technology and its systems implementations.

Read on, and find out more about the physical aspects which will enable future communications systems to connect a multitude of high-speed users throughout the globe. The "information superhighway" may be at hand. Allow your imagination to envision what will travel on this highway of the future!

20.2 Review of Essential Components

As mentioned in the introduction, some photonic technologies that form the basis for optical communication systems are covered in other chapters in this Handbook. They include fibers, sources, detectors, filters, optical amplifiers, and modulators. We will only briefly discuss the essence of these topics as they relate to optical communications. The reader should review these earlier chapters for more in-depth information about these areas.

20.2.1 Optical Fibers

The optical fiber falls into a subset of dielectric optical waveguides. To understand its operation principles, we have to know how the optical beams propagate through the mediums. An optical beam can

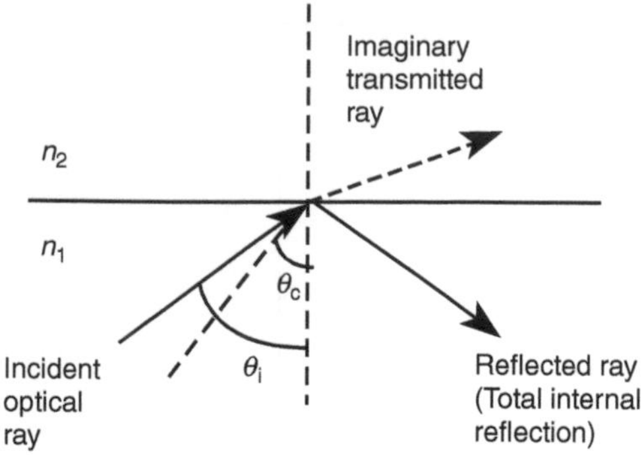

FIGURE 20.2 Total internal reflection at a dielectric boundary when the incident ray's angle exceeds the critical angle.

be thought equivalent to an electromagnetic wave or a large collection of photon particles.[12] Either can be used for instructional purposes, with the choice depending on the specific topic. For the sake of initially explaining optical fibers, we will use a ray-optics approach.

Reflection and refraction will occur when an optical ray, which is propagating in a dielectric medium of index of refraction n_1, impinges on a discontinuous boundary between this medium n_1 and another medium of index of refraction n_2. Snell's Law describes this phenomenon of reflection and refraction, where the transmitted refracted beam follows the relation[13]

$$n_1 \sin \theta_1 = n_2 \sin \theta_2 \tag{20.1}$$

where θ is the angle at which the optical ray hits the boundary with respect to the boundary normal within each medium. The reflected beam "bounces" off the interface back into the original medium 1 at the same original angle with respect to the normal. However, the reflected beam now propagates above and away from the interface normal if the incident beam was below the normal (and vice-versa). The power in the transmitted and reflected beams must equal the total incident power if the process is lossless. Let's assume that the index of the medium in which the ray is propagating is larger than the second medium ($n_1 > n_2$). If the ray hits the boundary at a very grazing angle for which θ approaches 90°, then there is an angle for which the inverse-sin of a number >1 is imaginary, thus resulting in no real solution. If the angle of incidence is greater than this critical angle (θ_c), then (1) the refracted beam itself is imaginary, (2) no power penetrates the dielectric boundary, and (3) total-internal-reflection occurs.[14] (see Figure 20.2) No power has been lost upon reflection at a dielectric boundary.

If we surround the high-index medium of propagation with another medium of lower index, we can achieve lossless waveguiding of an optical beam as long as the beam strikes the boundary at an angle greater than the critical angle. However, there are a discrete set of angles greater than θ_c which will support waveguiding. These discrete angles are eigenvalue solutions to the wave propagation equations which define the waveguide. Figure 20.3 illustrates several possible propagating waves, each of which is called a spatial mode of the waveguide. The number of spatial modes, M, which can be supported in a circular waveguide, or optical fiber, is[15]

$$M \approx \frac{1}{2}\left(\frac{2\pi a}{\lambda}\right)^2 \left(n_1^2 - n_2^2\right) \tag{20.2}$$

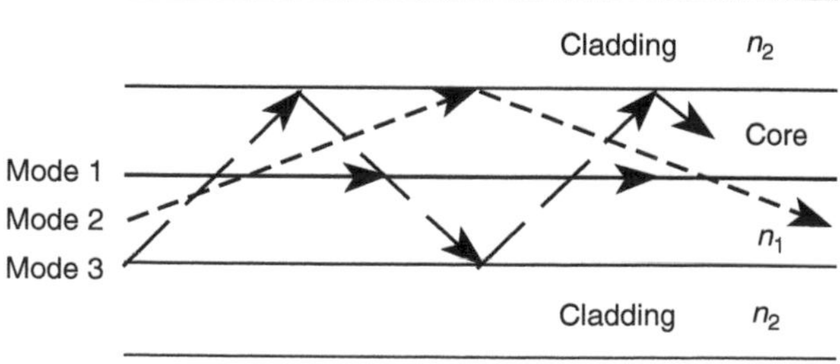

FIGURE 20.3 Ray model illustration of the propagation of many modes in a dielectric waveguide.

where a is the core radius, n_1 is the core index, and n_2 is the cladding index; the key feature of Equation 20.2 is that the number of modes increases when the index difference or the fiber radius is large. An optical fiber is thus "a high-index core surrounded by a lower-index cladding." Guided optical propagation provides a reliable, low loss, and secure medium of transmission.

Figure 20.3 illustrates the ray-model approach of many modes propagating through a multi-mode fiber whose core diameter is fairly large and usually ≥ 50 μm. The lowest order modes, i.e., the modes most easily generated, propagate nearly straight down the fiber at shallow angles to the core-cladding interface and the higher order modes reflect off the core-cladding interface at steeper angles. The most important disadvantage of a multi-mode fiber is intermodal dispersion which arises because a rectangular light pulse of temporal width T input into the multi-mode fiber will generate several modes, with the pulse's power distributed among the lower and higher order modes (see Figure 20.4). However, since the different modes all propagate at the speed of light but at different angles down the fiber, they also will travel different absolute path-length distances even though they all propagate axially down the fiber for the same fixed fiber length. Therefore, each mode will arrive at the fiber end at different times, causing the initial light pulse to spread and disperse temporally, ΔT, by the time it reaches the fiber end. This type of dispersion limits the distance and bit rate which can be transmitted, because some finite time must exist until the subsequent light pulse can be transmitted without mutual interference. Intermodal dispersion can be minimized (but not eliminated) by using a graded-index (GRIN) multi-mode fiber,[16] which was described in the chapter on optical fibers. In general, multi-mode fibers should be avoided for distances larger than several hundred meters and speeds >1 Gbit/s.

The only way to avoid intermodal dispersion is to reduce the core size and allow only one mode to propagate in a fiber. The typical single-mode fiber core diameter is ~8 μm, and the single mode

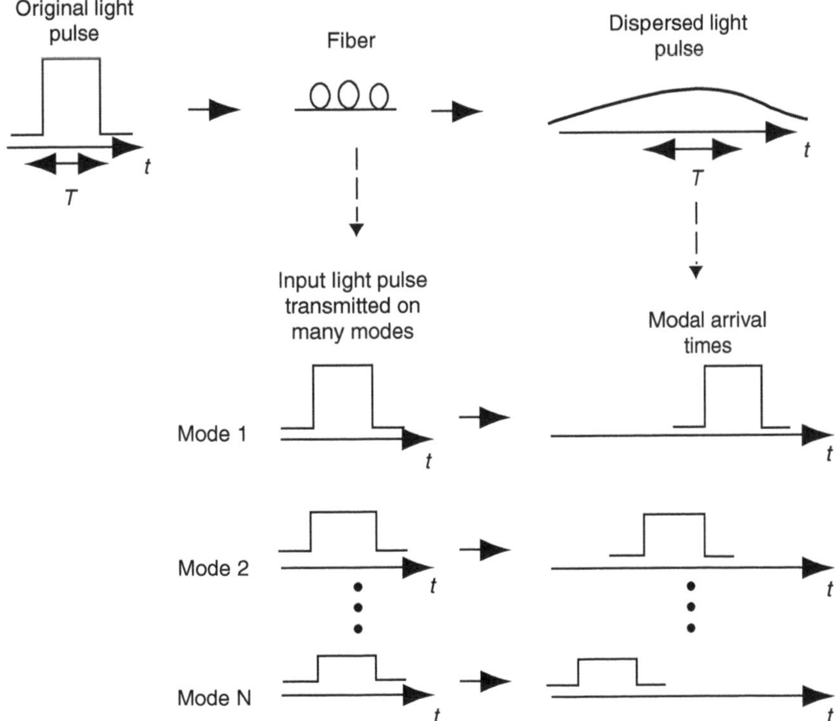

FIGURE 20.4 Illustration of intermodal dispersion in which an input optical pulse has its power transmitted on several different modes, each arriving at different times.

propagates axially straight down the fiber; if we consider this fundamental mode as a real Gaussian beam, the single mode will have most of its energy residing in the core with an evanescent field tail propagating alongside the core within the cladding region. In a single-mode fiber, another type of dispersion will limit the optical transmission, and is known as intramodal dispersion (also called chromatic dispersion).[17] This dispersion is a combination of material and waveguide dispersion in which the fiber material and fiber waveguide are slightly wavelength dependent:[18]

$$\text{Intramodal Dispersion} = \text{Material Dispersion} + \text{Waveguide Dispersion}. \tag{20.3}$$

In fact, in any medium other than vacuum and in any waveguide structure (other than ideal infinite free space), different electromagnetic frequencies travel at different speeds. This is the essence of chromatic dispersion. As the real fiber-optic world is rather distant from the ideal concepts of both vacuum and infinite free space, dispersion will always be a concern when one is dealing with the propagation of electromagnetic radiation through fiber. The velocity in fiber of a single monochromatic wavelength is constant. However, data modulation causes a broadening of the spectrum of even the most monochromatic laser pulse. Thus, all modulated data have a non-zero spectral width which spans several wavelengths, and the different spectral components of modulated data travel at different speeds. In particular, for digital data intensity modulated on an optical carrier, chromatic dispersion leads to pulse broadening, which in turn leads to chromatic dispersion limiting the maximum data rate that can be transmitted through optical fiber (see Figure 20.5).

Considering that the chromatic dispersion in optical fibers is due to the frequency dependent nature of the propagation characteristics for both the material and the waveguide structure, the speed of light of a particular wavelength λ can be expressed as follows using a Taylor series expansion of the value of the refractive index as a function of the wavelength:

$$v(\lambda) = \frac{c_0}{n(\lambda)} = \frac{c_0}{n_0(\lambda_0) + \dfrac{\partial n}{\partial \lambda}\delta\lambda + \dfrac{\partial^2 n}{\partial \lambda^2}(\delta\lambda)^2} \tag{20.4}$$

Here, c_0 is the speed of light in vacuum, λ_0 is a reference wavelength, and the terms in $\partial n/\partial\lambda$ and $\partial^2 n/\partial\lambda^2$ are associated with the chromatic dispersion and the dispersion slope (i.e., the variation of the chromatic dispersion with wavelength), respectively. Transmission fiber has positive dispersion, i.e., longer wavelengths see longer propagation delays. The units of chromatic dispersion are

(a) Photon velocity $(f) = \dfrac{\text{Speed of light in a vacuum}}{\text{Index of refraction } (f)}$

(b) Information bandwidth of data

(c) Temporal pulse spreading → F[distance, (bit rate)^2] → $\dfrac{\text{ps/nm}}{\text{km}}$

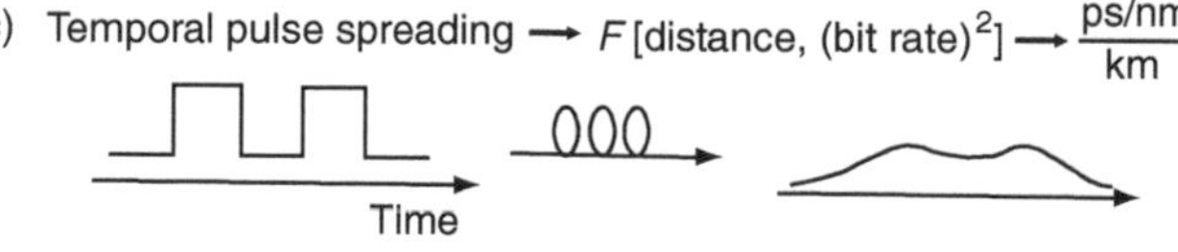

FIGURE 20.5 The origin of chromatic dispersion in data transmission. (a) Chromatic dispersion is caused by the frequency-dependent refractive index in fiber. (b) The non-zero spectral width due to data modulation. (c) Dispersion leads to pulse broadening, proportional to the transmission distance and the data rate.

picoseconds per nanometer per kilometer, meaning that shorter time pulses, wider frequency spread due to data modulation, and longer fiber lengths will each contribute linearly to temporal dispersion. Figure 20.6 shows the dispersion coefficient, D [ps/(nm km)], of a conventional single-mode fiber with the material and waveguide contributions plotted separately.[7] For a given system, a pulse will disperse more over time for a wider frequency distribution of the light and for a longer length of fiber. Higher data rates inherently have both shorter pulses and wider frequency spreads. Therefore, as network speed increases, the impact of chromatic dispersion rises precipitously as the square of the increase in data rate. The quadratic increase with the data rate is a result of two effects, each with a linear contribution. On one hand, a doubling of the data rate makes the spectrum twice as wide, doubling the effect of dispersion. On the other hand the same doubling of the data rate makes the data pulses only half as long (hence twice as sensitive to dispersion). The combination of a wider signal spectrum and a shorter pulse width is what leads to the overall quadratic impact—when the bit rate increases by a factor of four, the effects of chromatic dispersion increase by a whopping factor of 16![19] The reader can find more in-depth information about the chromatic dispersion effects in Section 20.4.

Note that the fiber has an inherent material dispersion coefficient which changes sign and goes through a zero point, resulting in a total intramodal dispersion zero point near 1.3 μm for a conventional fiber.

Another key characteristic of a single-mode fiber is the power attenuation per kilometer, α, and is shown in Figure 20.7 as a function of signal wavelength.[20] The fundamental physical limits imposed on the fiber attenuation are due to scattering off the silica atoms at shorter wavelengths and the material absorption at longer wavelengths. There are two minima in the loss curve, one near 1.3 μm and an even lower one near 1.55 μm. The 1.55 μm loss minimum of ~0.2 dB/km (<1% light lost in 1 km!) is quite close to the theoretical limit and has a bandwidth of ~25 THz. Low loss is extremely important since a light pulse must contain a certain minimum amount of power in order to be detected such that "0" or "1" data bit can be unambiguously detected. If not for dispersion, we would clearly prefer to operate with 1.55 μm light due to its lower loss for long-distance systems.

We now have a trade-off situation for the widely-installed conventional single-mode fiber in which the dispersion minimum is at 1.3 μm but the loss minimum is at 1.55 μm. The choice as to which wavelength to use depends on the practical limits of a particular system. However, there is another type of fiber, known as dispersion-shifted fiber, which has both the dispersion zero and the loss minimum located at 1.55 μm. Since the waveguide dispersion is dependent on the geometry of the

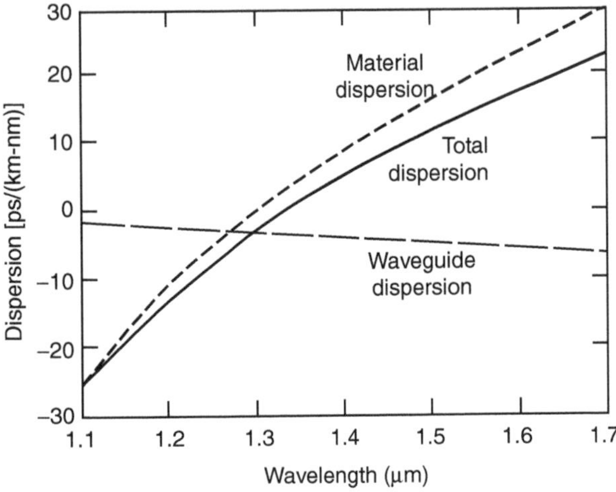

FIGURE 20.6 Dispersion coefficient, D, as a function of wavelength in the conventional silica single-mode fiber. (From Agrawal, G. P. 2002. *Fiber-Optic Communication Systems*. Wiley, New York. With permission.)

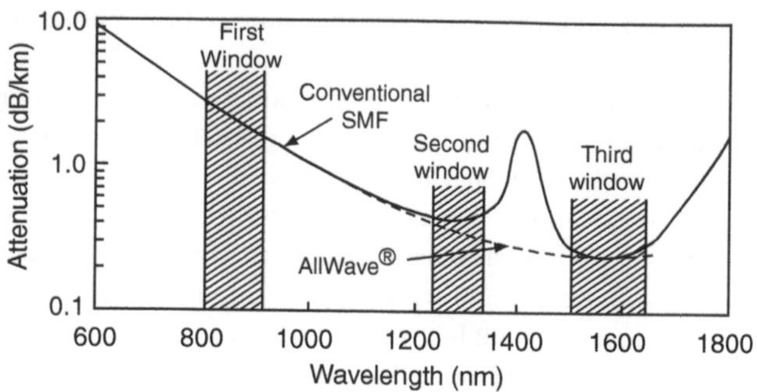

FIGURE 20.7 Attenuation in a single-mode silica fiber. (Adapted from Kaiser, P. and Keck, D. B. 1988. *Optical Fiber Telecommunications II*, S. P. Miller and I. P. Kaminow, eds., Academic Press, New York. With permission.)

waveguide, we can change the fiber doping and core (which effectively changes the waveguiding geometry) to change the material dispersion and thus move the dispersion zero point to 1.55 μm. Figure 20.8 shows the dispersion curve.[20] DSF can be used to attempt the best design for single-channel transmissions.

On the other hand, chromatic dispersion is not always bad for fiber transmission. It is, in fact, a necessary evil for the deployment of WDM systems. When the fiber dispersion is near zero in a WDM system, different channels travel at almost the same speed. Any nonlinear mixing effects that require phase matching between the different wavelength channels will accumulate at a higher rate than if wavelengths travel at widely different speeds (the case of higher dispersion fiber). The deleterious nonlinear effects that tend to destroy the signal integrity are self-phase modulation (SPM), cross-phase modulation (XPM) and four-wave mixing (FWM).[21] Both FWM and XPM are strengthened by interactions between wavelengths over long propagation distances. A dispersion value as small as a few ps/(nm km) is sufficient to make XPM and FWM negligible since the different wavelength channels are not phase-matched and "walk-off" from each other quickly, thus ensuring that they interact with each other only over relatively short distances. We will revisit the fiber dispersion and nonlinearities management in Section 20.4.

To mitigate the effects of nonlinearities, the next generation of fibers introduced relatively modest amounts of chromatic dispersion. The intent was to avoid distorting the signal with too much dispersion, but still introduce enough dispersion to counteract the nonlinear effects. Two of the best-known non-zero dispersion shifted fiber (NZDSF) introduced in the mid-1990s are Corning's large effective area fiber (LEAF)[22] and Lucent's TrueWave fiber. The dispersion of NZDSF is roughly 4–6 ps/(nm km), low enough to allow transmission over longer distances than SMF but

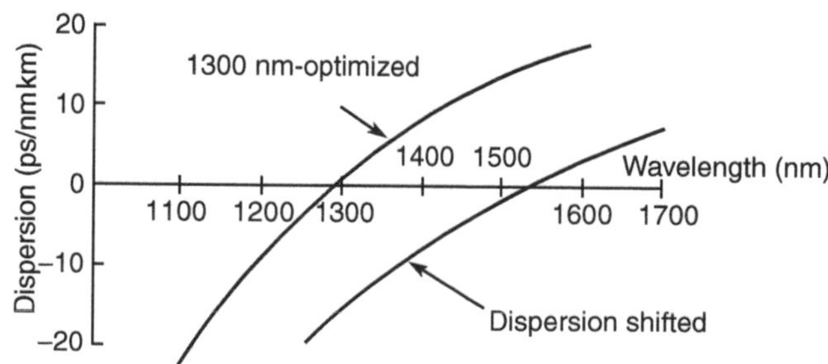

FIGURE 20.8 Dispersion coefficient in a dispersion-shifted fiber. (Adapted from Kaiser, P. and Keck, D. B. 1988. *Optical Fiber Telecommunications II*, S. P. Miller and I. P. Kaminow, eds., Academic Press, New York.)

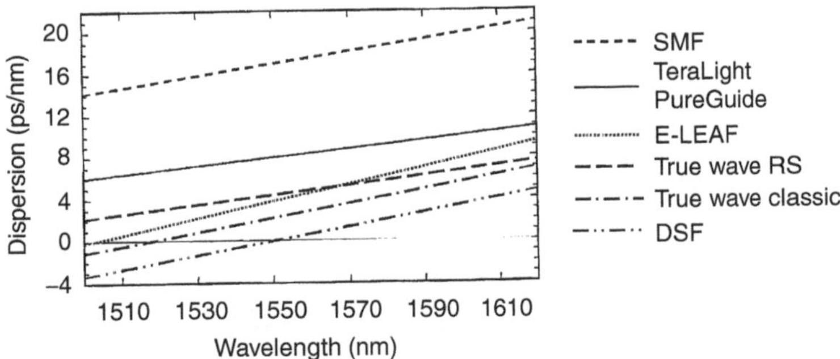

FIGURE 20.9 Chromatic dispersion values for several commercially available types of transmission fiber.

with a dispersion value large enough to reduce FWM and XPM that occurred in DSF. Recent studies show that for very-dense high-channel-count high-speed WDM systems, even the dispersion of NZDSF may be too low.[23] Fiber manufacturers are returning to fibers with larger dispersion. For example, Alcatel has introduced a higher dispersion fiber (TeraLight, 8 ps/(nm km)). Figure 20.9 shows the dispersion curve for several commercialized transmission fibers.

Single-mode fibers actually support two perpendicular polarizations of the original transmitted signal (fundamental mode). In an ideal fiber (perfect) these two modes are indistinguishable, and have the same propagation constants owing to the cylindrical symmetry of the waveguide. However, the core of an optical fiber may not be perfectly circular, and the resultant ellipse has two orthogonal axes. The index-of-refraction of a waveguide, which determines the speed of light, depends on the shape of the waveguide as well as the glass material itself. Therefore, light polarized along one fiber axis travels at a different speed as does light polarized along the orthogonal fiber axis (see Figure 20.10). This phenomenon is called PMD. Fiber asymmetry may be inherent in the fiber from the manufacturing process, or it may be a result of mechanical stress on the deployed fiber. The inherent asymmetries of the fiber are fairly constant over time, while the mechanical stress due to movement of the fiber can vary, resulting in a dynamic aspect to PMD.

Because of random variations in the perturbations along a fiber span, PMD in long fiber spans accumulates in a random-walk-like process that leads to a square root of transmission-length dependence.[24] Furthermore, PMD does not have a single value for a given span of fiber. Rather, it is described in terms of average DGD, and a fiber has a distribution of DGD values over time. The probability of the DGD of a fiber section being a certain value at any particular time follows a Maxwellian distribution (see Figure 20.11). PMD is usually expressed in ps/km$^{1/2}$ in long fiber spans, and the typical PMD parameter (D_p) is 0.1–10 ps/km$^{1/2}$.[25,26]

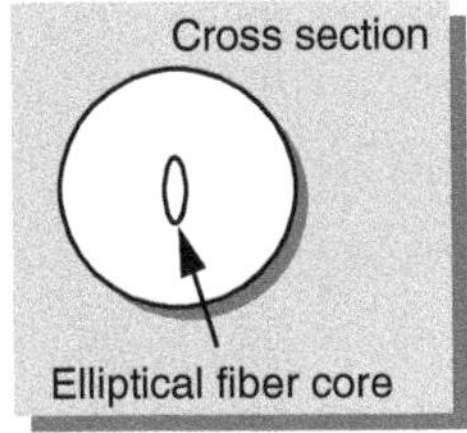

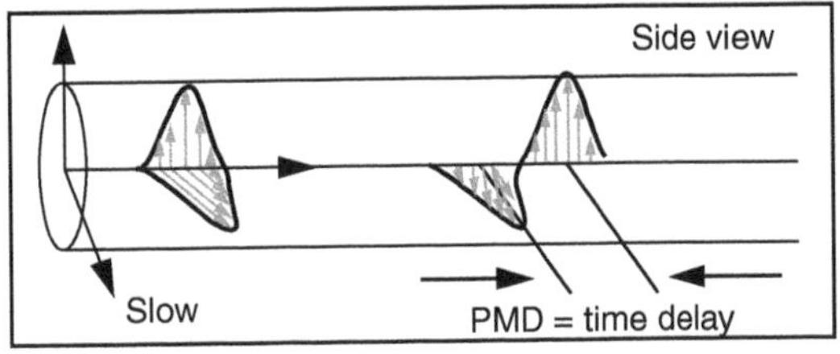

FIGURE 20.10 Illustration of polarization mode dispersion caused by imperfect round fiber core. An input optical pulse has its power transmitted on two orthogonal polarization modes, each arriving at different times.

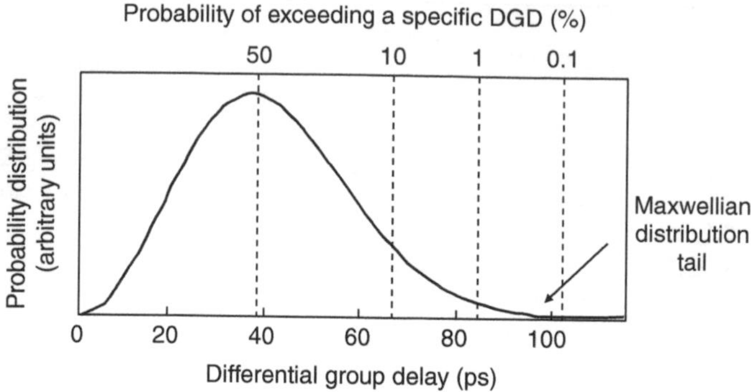

FIGURE 20.11 Probability distribution of DGD in a typical fiber.

Another issue is fiber connection and coupling. Coupling of light into and out of a small-core single-mode fiber is much more difficult to achieve than coupling electrical signals in copper wires since (1) photons are weakly confined to the waveguide, whereas electrons are tightly bound to the wire; and (2) the core of a fiber is typically much smaller than the core of an electrical wire. Light must be coupled into the fiber from a diverging laser beam, and two fibers must be connected to each other. Let's first discuss coupling light into the fiber. Focusing the laser light into the fiber end must be performed with great care since the angle of acceptance cannot be greater than the critical angle for total internal reflection. We must excite only the lowest-order fundamental mode in the 8-μm core, and this mode has a grazing angle with the core-cladding boundary. Therefore, the focusing must be gradual to ensure a small angle and must provide a small focal spot size similar in dimension to the single-mode core. The maximum limit of

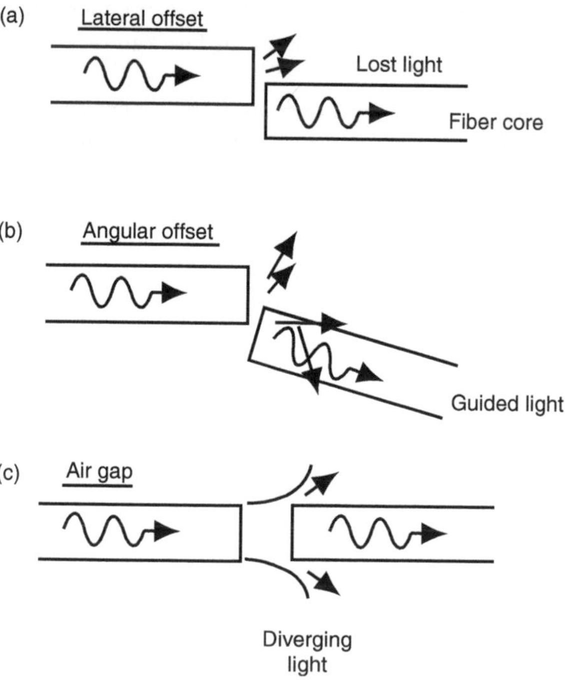

FIGURE 20.12 Sources of loss in a fiber-to-fiber connection.

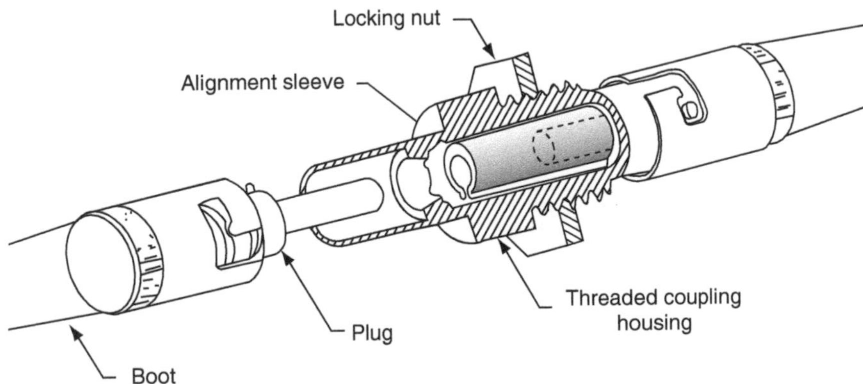

FIGURE 20.13 Diagram of a typical mechanical fiber splice. (Adapted from Carlisle, A. W. 1985. *Proceedings at Conference on Optical Fiber Communications*, San Diego, CA. With permission.)

the acceptance angle for a single- or multi-mode fiber is known as the numerical aperture, NA, and is given by[27]

$$\mathrm{NA} = \sin\theta_c = \sqrt{n_1^2 - n_2^2} \tag{20.5}$$

assuming light is coupled into the fiber from air. The second issue of connecting two different fibers in a system must be performed with great care due to the small size of the cores. We wish to achieve connections exhibiting (1) low loss, (2) low back reflection, (3) repeatability, and (4) reliability. As shown in Figure 20.12, the following will cause losses:

1. An axial offset since the core areas don't fully overlap, causing light to couple into the second-fiber's cladding
2. An angular offset since not all the light will be coupled at an angle above the critical angle
3. An air gap in which both an index mismatch and beam divergence will limit coupling into the second fiber's core

Two popular methods are the permanent splice and the mechanical connector. The permanent "fusion" splice can be accomplished by placing two fiber ends near each other, generating a high-voltage electric arc which melts the fiber ends, and "fusing" the fibers together. Losses and back reflection are extremely low, <0.1 and <-60 dB, respectively. A disadvantage is that the splice is delicate and must be protected. Another is that the splice is permanent. Alternatively, there are several types of mechanical connectors, such as ST and FC/PC.[28] For brevity, we will only discuss the general concept. Figure 20.12 shows that the polished fiber end is permanently placed at the center of a mechanical receptacle. Two such receptacles are mechanically screwed together and a (near) physical contact is made between the fiber cores (Figure 20.13).[29] Losses and back reflection are still fairly good, typically <0.3 and <-45 dB, respectively.

20.2.2 Sources

Photons have an energy which is dependent on the wavelength of the light, λ[30]

$$E = hc/\lambda = h\nu \tag{20.6}$$

where c is the speed of light. Furthermore, semiconductors have an energy bandgap between the electron-rich valence band and the hole-rich conduction bands. If a photon is incident on a semiconductor, the

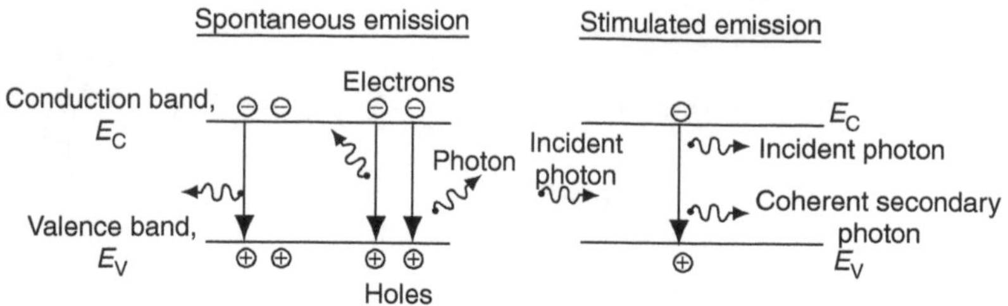

FIGURE 20.14 Spontaneous and stimulated emission in a semiconductor.

photon can be absorbed if its energy is larger than the energy bandgap. In such a case, the photon's energy can be transferred to a valence electron pushing it up into the conduction band and freeing it to move through the semiconductor. This is known as stimulated absorption.[31]

Alternatively, photons can be emitted from a semiconductor if an electron in the conduction band drops down in energy, ΔE, into the valence band, thereby combining with a hole in the valence band and emitting a photon of the same energy as ΔE. This process of photon emission can occur due to two different processes as illustrated in Figure 20.14.[31] The first process is called spontaneous emission in which a finite-lifetime electron in the conduction band randomly combines with a hole to emit a photon. These electrons exist in the conduction band due to their prior pumping into the higher energy level, typically by electrical biasing (i.e., current injection). Since the electrons fill an energy well in the conduction band with a distribution in energy states (see Figure 20.14), the energy drop in the electron upon spontaneous recombination will produce uncorrelated incoherent photons at many different wavelengths (i.e., energies), producing a wide spectral bandwidth for which photon emission can occur. These photons that are random in polarization, phase, frequency, and direction can be considered as noise in the optical system (Figure 20.15).

The second process, called stimulated emission, is when a single photon of a given energy is incident on a semiconductor and causes electron-hole recombination. This stimulated recombination results in the emission of a photon of the same energy as the original incident photon, thus producing two photons

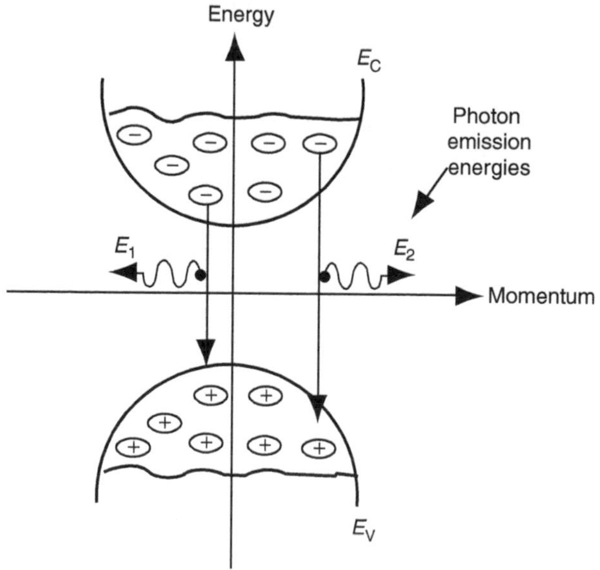

FIGURE 20.15 Energy-band diagram versus momentum in a direct-bandgap semiconductor. E_1 and E_2 represent possible energies of emitted photons.

from an initial one. If the electron population in the conduction band is high enough to sustain continued stimulated emission, then an incident photon at wavelength λ_i will produce two, four, etc., photons that are all coherent with each other and at the same wavelength as the original photon. This process produces gain and the medium is considered active. As the wave traverses through an active medium, it will be amplified as shown in Figure 20.16. The gain, G, is dependent on the gain coefficient per unit length, g, and the length of the medium, L:[32]

$$G(\lambda) = \frac{\text{Signal Output Power}}{\text{Signal Input Power}} = \exp(g(\lambda)) \tag{20.7}$$

The relative rates for stimulated and spontaneous emission are determined by the electron populations in the various energy bands and the external pumping.

Before discussing lasers, it is useful to first introduce a less sophisticated optical source known as the light-emitting diode (LED). In general, light emission requires that carriers be easily pumped into the conduction band and then localized so that they will recombine to emit light. This can be accomplished by forming a p–n junction, hence the word "diode" in the LED name. By selectively tailoring the layers surrounding the p–n junction and creating a potential well in the n region, the electrons and holes can be generated and then localized near the n-semiconductor's depletion region. Carriers are generated through an electrical contact and forward-biasing of the p–n junction. These carriers recombine and spontaneously emit photons in the depletion region. These randomly generated photons will also experience some limited amplification (i.e., they will themselves cause stimulated emission). The generated light is broadband ($\sim$100 nm), incoherent, and is not directionally well-confined. Continued electrical pumping will sustain a light beam emanating from the p–n junction. Since the area of a typical LED is quite large ($\sim$250 μm), the capacitance time constant is also quite large limiting the ON–OFF modulation speed to be limited to less than a few hundred MHz. In addition to this speed limitation, an LED has limited usefulness in high-speed long-distance communications due to the high material dispersion caused by the extremely broadband light and the high intermodal dispersion since a large-core multi-mode fiber is typically used to maximize the capture of the ill-confined light.

A laser is a more complicated system since it involves coherent light generation and feedback over a small bandwidth. Let us first address the issue of feedback for laser oscillation. If we surround a medium in which a light beam is propagating with two near-100% mirrors, then we have formed a frequency-selective resonant cavity which can be explained with the help of Figure 20.17. An optical wave, $L_1(t,z)$, having a given wavelength and phase, is propagating to the left inside this cavity. Upon reflection at the left mirror, a right-propagating wave will be generated, $R_1(t,z)$. Given a long coherence length for the optical wave,[12] the left and right propagating waves will be coherent with each other and have the same frequency, ω, but they each have their own phase, ϕ. This reflection process continues with a reflection of $R_1(t,z)$ at the right mirror which generates a second left-propagating wave of $L_2(t,z)$. If $L_1(t,z)$ and $L_2(t,z)$ are coherent with each other, as would be the case in stimulated emission, then they would interfere with

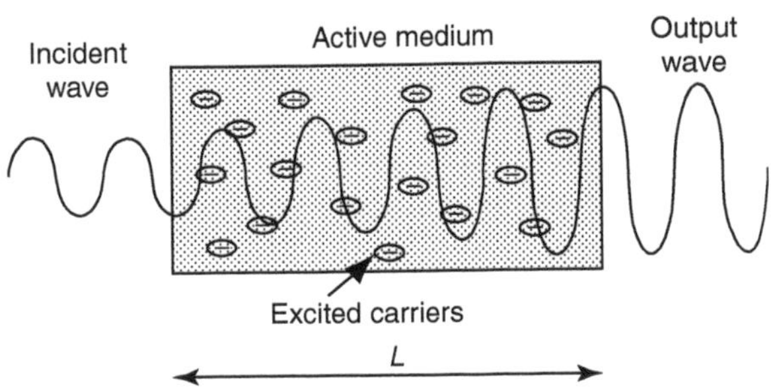

FIGURE 20.16 Amplification of a wave as it propagates through an active medium.

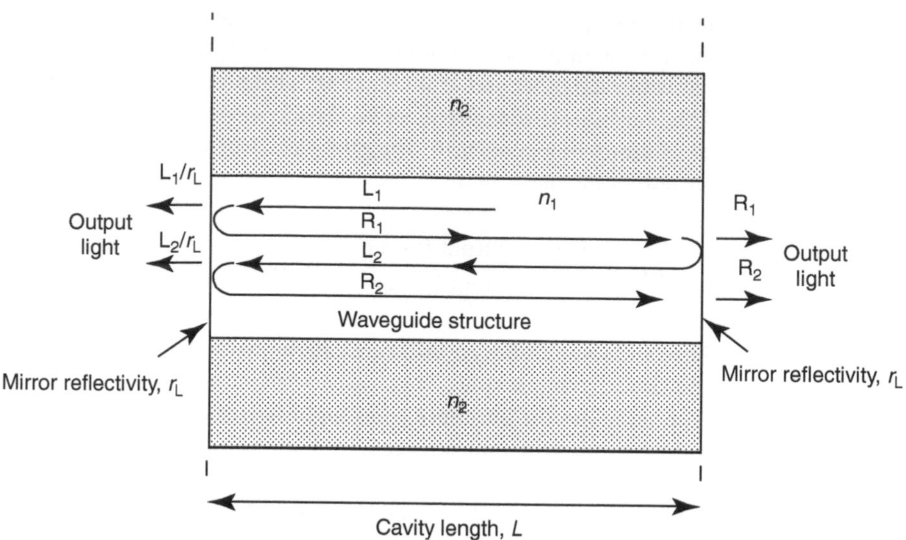

FIGURE 20.17 Multiple coherent reflections in a Fabry-Perot cavity.

each other at the left mirror. If they are in phase with each other (i.e., phase difference equals 0, 2π, 4π...), then they will add constructively and both waves will be mostly reflected and partially transmitted at the left boundary. If they are completely out of phase (i.e., phase difference equals π, 3π, 5π...), then they will add destructively and no wave will be reflected. In reality, there are an infinite number of reflections and all of the reflections must add constructively in phase at the two mirror boundaries. Therefore, the round-trip distance of wave propagation between the two mirrors must be an integral number of 2π phase shifts or an integral number of wavelengths:[32]

$$2L = m\frac{\lambda}{n_1}, \quad m = 1,2,... \tag{20.8}$$

This equation assumes that the waves are confined to the plane of propagation—which can occur if we confine the light to a thin waveguide by using low-index surrounding cladding layers—and that there is no angular dependence. There are an infinite number of integers which will satisfy the in-phase requirement. Therefore, the passband of this resonant cavity is periodic in wavelength, and is known as a Fabry-Perot cavity.[33] Note that the original wave can continually generate by pumping-induced stimulated emission inside an active semiconductor. Additionally, the mirrors in a semiconductor wavelength-selective cavity are formed by the perfectly-parallel smooth cleaved facets of the semiconductor ends. The light would then emanate from the cavity at these $<100\%$ mirrors and be useful as laser light.

The infinite number of reflections in the resonant cavity can be sustained by gain compensating for existing losses. There are two main sources of losses in a semiconductor cavity: (1) $<100\%$ reflectivity of the end mirrors (r_L and r_R being typically $\sim 30\%$ with semiconductor cleaved facets), and (2) non-radiative absorption of light within the semiconductor medium, α_a. Both of these losses are, to first order, wavelength independent. The semiconductor must provide gain within the active p–n junction to overcome these losses and sustain light oscillation in the resonant cavity.[34]

Every laser must therefore fulfill three basic criteria:

1. Gain to provide stimulated emission
2. Gain $>$ loss, to sustain reflections
3. Resonant cavity

These three conditions are displayed graphically in Figure 20.18, which shows that lasing will occur where all the conditions are fulfilled, i.e., within the gain bandwidth, above the loss line, and at the wavelength

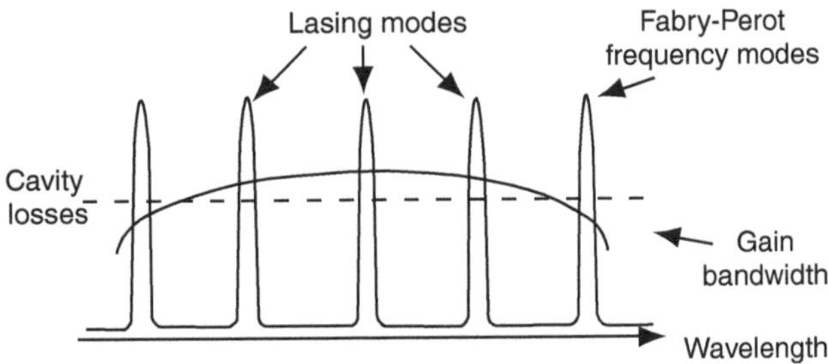

FIGURE 20.18 Illustration of the three lasing conditions of (1) a gain medium, (2) a frequency-selective cavity filter, and (3) gain > loss.

resonances. In the Fabry-Perot diode laser, there are several frequency modes which satisfy all the lasing conditions, and so this laser is called a multi-mode laser with the all modes typically occupying several nanometers in bandwidth.[35] As we discussed in the previous section on fibers, we wish to have a very small frequency bandwidth to minimize intra-spatialmodal dispersion in the single-mode fiber. Therefore, multi-frequency-mode lasers are only useful for shorter-distance (< 50 km) or lower-speed (< 1 GHz) links, but they are much more useful than the simple LED.

Because of intermodal dispersion, we are quite interested in single-frequency-mode lasers. This can be accomplished by referring back to Figure 20.18 and providing a scenario in which only one frequency mode will satisfy all three lasing criteria. The gain bandwidth of the active medium is fixed, and we certainly do not wish to increase the losses in the cavity. The best method is to find a wavelength-selective cavity whose periodic resonance peaks are so far apart that only one passband resides within the active gain bandwidth and above the loss line. This requires an efficient frequency-selective cavity with a high Q-factor, or high Finesse. One such cavity would integrate Bragg gratings to provide the feedback at the ends of the active region instead of using simple mirrors or cleaved facets.[36] Figure 20.19a illustrates the

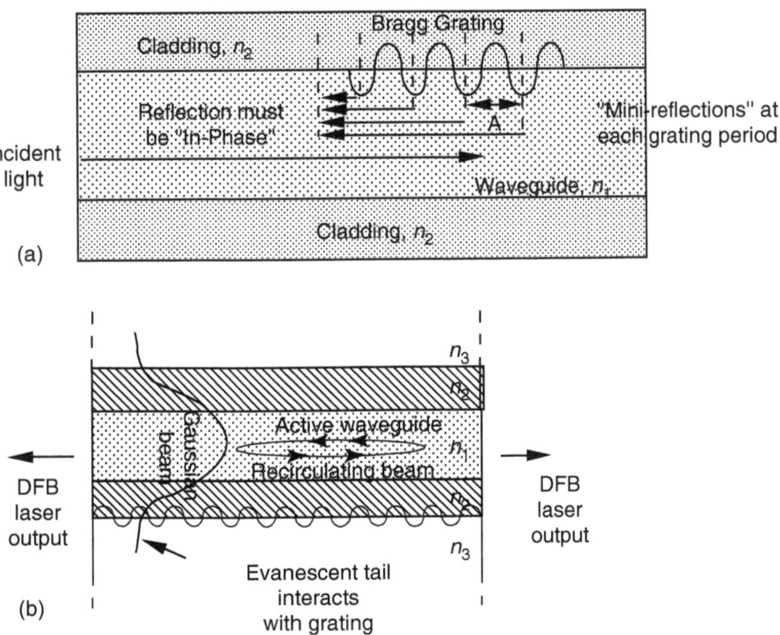

FIGURE 20.19 (a) Small reflections at each period in a Bragg grating can cause frequency-selective reflections, (b) Simple diagram of a distributed feedback laser.

basic concept of the wavelength-selective reflectivity of these gratings. The gratings are formed by a modulation in either the refractive index or the gain; this modulation can exist outside of the waveguiding region and in the cladding region where it will affect the tail, or evanescent field, of the propagating wave. At each period in the grating, a small portion of the propagating wave is reflected. If all the small reflections are in phase, then they will add constructively when they propagate in the reverse direction; of course, wavelengths which do not satisfy the in-phase requirement will not be reflected and will not eventually resonate and lase. This in-phase requirement for the gratings is called the Bragg condition:

$$\Lambda = \frac{\lambda}{2n} \tag{20.9}$$

where Λ is the spacing of the Bragg corrugation. This equation assumes that there is no angular dependence if the wave is confined to the horizontal waveguide.[36]

Not only can the Bragg reflecting regions be placed at the end of the active region, but they can actually be placed within (i.e., above or below) the active region itself. Such intra-cavity distributed Bragg reflection satisfies the eigenvalues of coupled mode theory to provide efficient frequency-selective feedback. This laser is simplistically shown in Figure 20.19b and is called a distributed feedback (DFB) laser.[37]

The light output, L, of a single or multi-mode laser is a function of the input current, I. As the current increases, the gain increases until it is equal to the losses. At this point, known as the laser threshold current (I_{th}), lasing will occur since all three lasing conditions will be satisfied.[30] A typical L-I curve is shown in Figure 20.20. The laser light output can be modulated by directly modulating the bias current injected into the semiconductor active medium. This modulation can replicate the digital bits of some data stream giving rise to light "ON" and "OFF" corresponding to "1" and "0" data bits, respectively.

One final note is that the laser light is generated from stimulated emission in which all the photons are coherent and at the same wavelength. However, this is not strictly true since there are some quantum fluctuations in the energy of the photons emitted by stimulated emission in comparison to the original photon. These energy fluctuations correspond to fluctuations in the wavelength, giving rise to a laser linewidth, which, for a DFB laser, is typically 10 MHz of uncertainty. For high-speed Gbit/s direct-detection systems, this linewidth is negligible, but may cause difficulties for coherent systems (see Section 20.3).

20.2.3 Detectors

Transmitted optical data must be unambiguously recovered and, therefore, photodetection is an extremely important process. Photodetection is quite similar to photogeneration, except in reverse. We have already established that an incident photon will be absorbed by a semiconductor if this photon

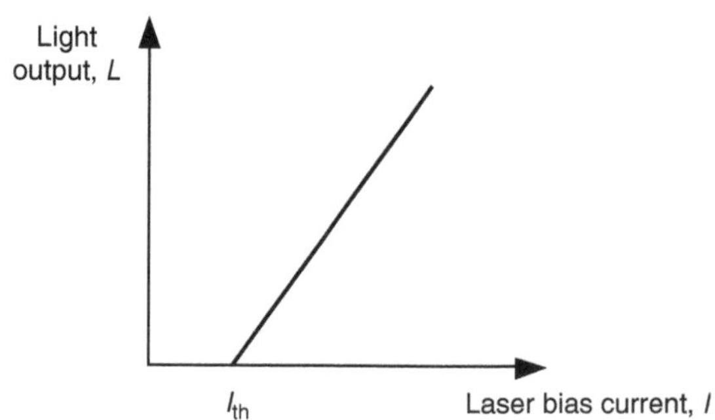

FIGURE 20.20 A typical curve of the laser light output versus input bias current.

has enough energy for an electron to absorb the energy, overcome the energy bandgap, and become a free carrier in the conduction band. We must be able to measure the number of electrons generated by the incident photons, i.e., the photon-generated current. This requires separating and measuring the generated electron-hole pairs before they can recombine. A simple reverse-biased p–n junction can provide the necessary conditions as shown in Figure 20.21. The depletion region of a p–n junction has a built-in electric field due to bound separated ionized cores. If photons generate free electron-hole pairs in the depletion region, then the carriers will be separated by the built-in electric field and can then be measured as a current before recombination occurs. The amount of generated photocurrent, I_{ph}, is:[38]

$$I_{ph} = \frac{\eta q P}{h\nu} \tag{20.10}$$

where η is the quantum efficiency of a given photon producing one electron, P is the incident optical power, $h\nu$ is the photon energy, and q is the electronic charge.

Even if the incident optical power is constant, two main noise sources generated in the detector are thermal noise, σ_{th}^2, and shot noise, σ_{sh}^2.[39] Three important characteristics about these noises are the following:

1. They have a statistical variance.
2. They cover all possible frequencies (i.e., white noise) which are supported by the system's electrical detection bandwidth.
3. They can be approximated as having a Gaussian amplitude distribution centered around the intended photocurrent mean. The Gaussian distribution is centered at a high current level for a "1" bit and a low current level for a "0" bit.

Thermal noise is caused by thermal energy in the detector. This thermal energy is randomly absorbed by electrons which are pushed up into the conduction band and will be mistakenly detected as photocurrent. This thermal noise power is independent of incident optical power and has a statistical variance

$$\sigma_{th}^2 = \frac{4kTB_e}{\Omega} \tag{20.11}$$

where B_e is the low-pass-filter electrical bandwidth of the detector, k is Boltzman's constant, T is the detector temperature, and Ω is the detector resistance.[39] The shot noise is caused by the quantum

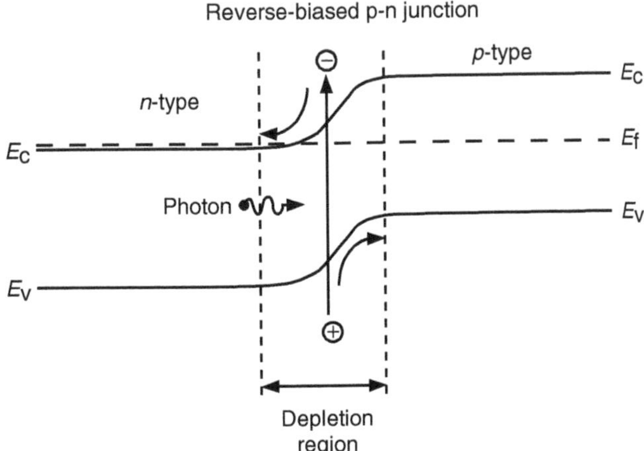

FIGURE 20.21 An energy-band diagram of a p–n junction in which photogenerated carriers are separated and measured as a photogenerated current.

randomness of generating carriers in a detector at random times. It is the most fundamental quantum limit of photodetection since one can never eliminate this noise term. The shot noise variance power is

$$\sigma_{sh}^2 = 2q\left(\frac{q}{h\nu}\right)PB_e \tag{20.12}$$

and is proportional to the absorbed optical power.[39] In direct-detection systems, thermal noise usually dominates shot noise. We will discuss later the effects these noise sources have on the signal-to-noise ratio (SNR) of the recovered data.

Although we have described the detector as a simple p–n junction, there are in reality two different types of detectors which dominate commercial use: the p–i–n detector and the avalanche photodiode (APD). These are shown in Figure 20.22.[40] As its name implies, the p–i–n detector is a p–n junction with a long intrinsic undoped region between the two ionized halves of the depletion region.[39] The electric field remains high throughout the long intrinsic region giving it the ability to efficiently separate photogenerated carriers and produce photocurrent. The effect of the p–i–n detector is to have a much larger region in which photons can be absorbed; in a simple p–n junction, the depletion region by itself is too thin to absorb much of the incoming light beam since absorption in a semiconductor has a characteristic and exponentially-decaying absorption depth. Similarly, the purpose of the APD is to increase the photocurrent generating efficiency of the detector. This is accomplished by adding a region of very highly doped material within the p–i–n detector, thereby creating a small region of extremely high electric field.[40] Photogenerated electrons within this high-electric-field region will be accelerated in velocity and, consequently, gain much energy. The moving electron can collide with a valence-band bound electron and transfer its new excess energy to it, thereby ionizing and freeing it to move; note that this excess energy must be greater than the ionization energy of an electron. The newly freed electron resides within an intense electric field and will itself begin to move in the same direction as the original electron. These two electrons are now moving as photocurrent where only one existed before. This process can continue as long as the electrons are in the highly doped region, achieving a multiplication factor of as large as ~10.

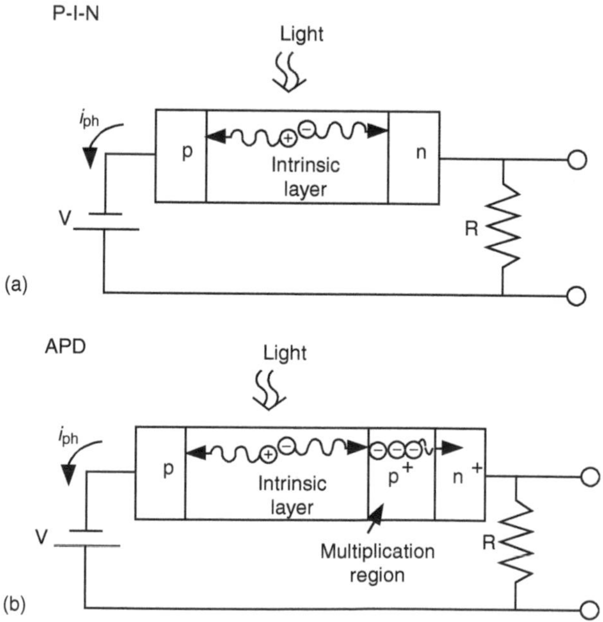

FIGURE 20.22 Diagram of a (a) p–i–n and (b) APD photodetectors.

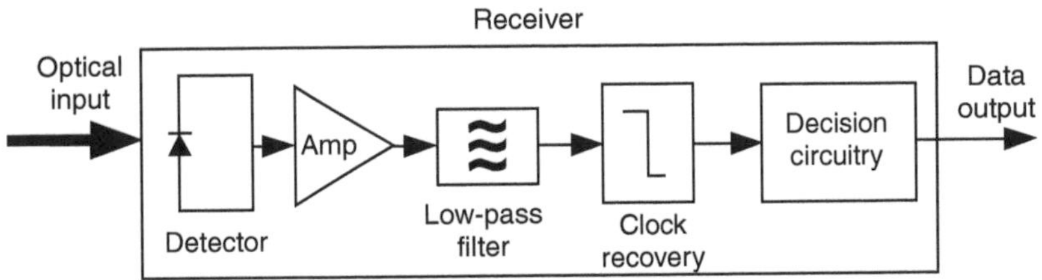

FIGURE 20.23 Block diagram of functions performed in advanced receiver packages.

Although the p–i–n and the APD detectors allow the measurement of a photocurrent, more functions must usually be performed to ensure adequate data recovery. These functions may include[41] (a) electrical amplification by using a transimpedance amplifier, (b) clock recovery to synchronize bit recovery of an arriving data packet, (c) decision circuitry to judge if a "1" or "0" bit has arrived, and (d) electrical filtering to limit the high frequency noise.

All of these functions would be performed in an optical receiver package (see Figure 20.23). At present, high-speed GHz receivers can be commercially purchased. The low-pass filter in the receiver is for the noise reduction (see Section 20.3.5).

20.2.4 Optical Amplifiers

The optical amplifier is ideally a transparent box which provides gain and is also insensitive to the bit-rate, modulation-format, power, and wavelengths of the signal(s) passing through it. The signals remain in optical form during amplification. Optical amplifiers have stolen center stage in the optical telecommunications world due to rapid device progress and revolutionary systems results[42] These devices enable new and exciting optical systems to be conceived and demonstrated during last 15 years. In fact, much of the most relevant recent advances in optical communications (i.e., long-distance NRZ and soliton systems, and wide-area and broadcast multi-channel systems) can be traced to the incorporation of optical amplifiers.

As a simple introduction, optical amplifiers can be thought of as a laser (gain medium) with a low feedback mechanism and whose excited carriers amplify an incident signal but do not generate their own coherent signal.[43] Similar to electronic amplifiers, optical amplifiers can be used to compensate for signal attenuation resulting from distribution, transmission, or component-insertion losses.[7,44] As shown in Figure 20.24, an amplifier provides signal gain, G, but also introduces additive noise (variance $=\sigma^2$) into the system. Each amplifier requires some form of external power to provide the energy for amplification. A voltage source is required for the electrical amplifier and a current or optical source is required for the

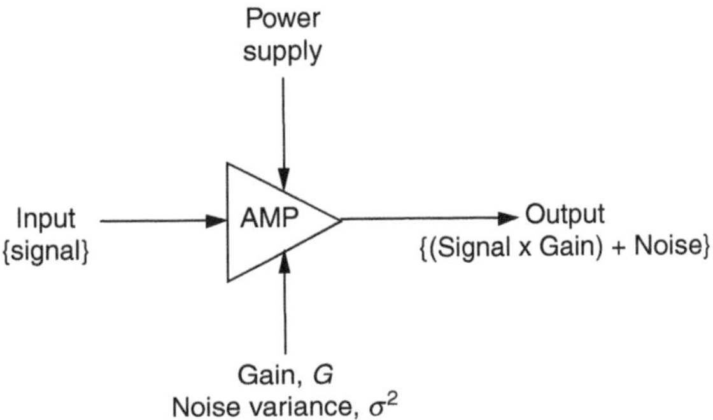

FIGURE 20.24 Basic amplifier characteristics.

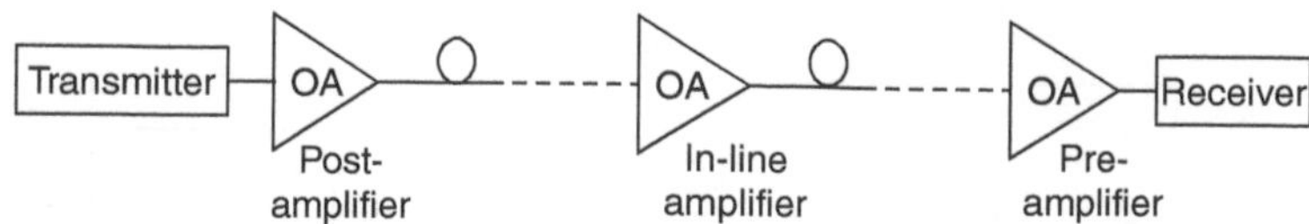

FIGURE 20.25 Three generic configurations for incorporating optical amplifiers into transmission or distribution systems.

optical amplifier. The current or optical source for the optical amplifier is used to pump carriers into a higher energy level which can then decay and emit a photon at the input signal wavelength. One figure of merit for both amplifiers is a low noise figure (NF) to be defined later. Additionally, the amplifier's design, input parameters, and position along a channel must all be optimized for a given system.

The three basic system configurations envisioned for the incorporation of optical amplifiers are shown in Figure 20.25.[45] The first is to place the amplifier immediately following the laser transmitter to act as a "power," or "post," amplifier. This boosts the signal power so that the signal is still above the thermal noise level of the receiver even after attenuation. Furthermore, any noise introduced by the power amplifier will be similarly attenuated together with the signal as they are transmitted through the lossy system. Since the signal power input to the power amplifier is typically large ($0.1 \sim 1$ mW), the key parameter for the power amplifier will be to maximize the saturation output power and not necessarily the absolute gain; if the amplifier can supply no more power than the original laser transmitter, then this configuration produces little advantage. The second is to place the amplifier in-line and may be incorporated at one or more places along the transmission path. The "in-line" amplifier corrects for periodic signal attenuation due either to fiber absorption or network distribution splitting losses.[46] The in-line amplifier may exist in a cascade form, with a given amplifier's output signal and noise feeding into a subsequent amplifier with some attenuation between the amplifiers. Issues such as optical filtering and isolation must be considered. The third possibility is to place the amplifier directly before the receiver and thus functions as a pre-amplifier. In this case the signal has already been significantly attenuated along the transmission path. The main figures of merit are high gain and low amplifier noise since the entire amplifier output is immediately detected. As we shall see later in this chapter, the receiver will be limited by the amplifier noise and not by the receiver's thermal noise.

We will introduce three different types of amplifiers: EDFAs, Semiconductor Optical Amplifiers (SOA), and Raman Amplifiers.

20.2.4.1 Erbium-Doped Fiber Amplifiers

As shown in Figure 20.26, the EDFA is a length of glass fiber which has been doped with the rare-earth metal Erbium ions. These ions act as an active medium with the potential to experience inversion of carriers and emit spontaneous and stimulated emission light near a desirable signal wavelength. The pump is typically another light source whose wavelength is preferentially absorbed by the ions, 0.98 or 1.48 μm for EDF. The pump and signal (1.55 μm) must be combined, typically by a wavelength-selective

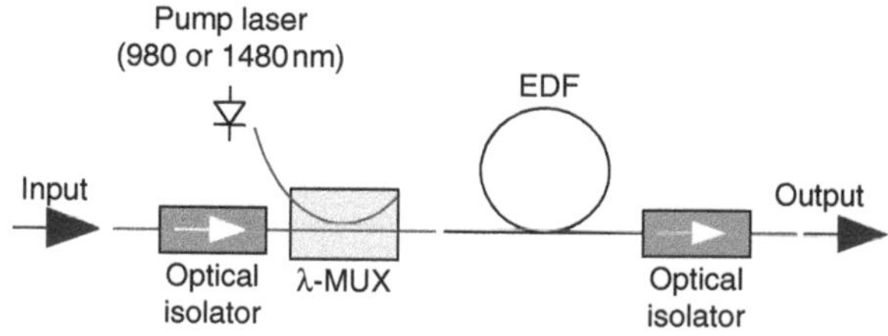

FIGURE 20.26 Basic configuration of Erbium-doped fiber amplifiers.

coupler (WDM), and may co- or counter-propagate with respect to each other inside the doped length of fiber. Therefore, light absorbed by the doped fiber at the pump wavelength will produce gain for a signal at a different wavelength. Since the transmission and the active medium are both fiber-based, the insertion losses are minimal.

Two other components which are not, strictly speaking, essential to an EDFA but which may be required to prevent system degradation are an optical filter and an optical isolator. The broadband amplified spontaneous emission (ASE) emanating from the amplifier will produce spontaneous–spontaneous beat noise. This must be limited in some way to achieve a reasonable SNR at the receiver. A bandpass filter of $\sim$1–2 nm or narrower may be placed at the output of any EDFA. Alternatively for WDM systems incorporating a cascade of amplifiers and many channels located across the gain bandwidth, the wavelength filter can be placed only at the receiver which will minimize noise as well as demultiplex the many channels. Additionally, we wish to prevent reflections back into the amplifier which may cause the NF of the EDFA to increase and may even cause the EDFA to lase if the gain is high enough (i.e., gain greater than loss in a cavity with reflections); note that the NF may increase in the presence of reflections due to the inverted carriers amplifying an unwanted reflected field and not the desired signal. An isolator would prevent these deleterious properties. Furthermore, an isolator may be necessary on the input side to prevent backward-traveling ASE from propagating back into reflection-sensitive components, such as a laser transmitter.

To produce the amplifier gain medium, the silica fiber core of a standard single-mode fiber is doped with Erbium ions. The EDF must be inverted by a pump source.[47] A signal initiates stimulated emission resulting in gain, and spontaneous emission occurs naturally which results in noise. Because of the many different energy levels in Erbium, several wavelengths will be absorbed by the ions. Figure 20.27 shows the energy levels and some of the key wavelengths which can be absorbed.[48] In general, absorption corresponds to a photon being absorbed and causing a carrier (ion) to jump to a higher energy level of energy difference $\Delta E = h\nu$ which roughly matches the energy of the photon. Different wavelengths can cause either ground-state absorption (GSA) or excited-state absorption (ESA).[44] GSA corresponds to a photon exciting a carrier from the ground state to a higher state, and ESA corresponds to a photon exciting a carrier from one of the non-ground-state energy levels to an even higher level. Since the population is greatest in the ground state, then the probability of GSA occurring is much greater than ESA. Once a photon is absorbed and a carrier is excited to a higher energy level, the carrier decays very rapidly to the first excited level. Once the carrier is in the first excited state, it has a very long lifetime of $\sim$10 ms,[68] thereby enabling us to consider the first excited level to be meta-stable. Depending on the external optical excitation signal, this carrier will decay in a stimulated or spontaneous fashion to the ground state and emit a photon. The emission in Erbium is fortuitously near the 1.55 μm loss-minimum of standard silica optical fiber.

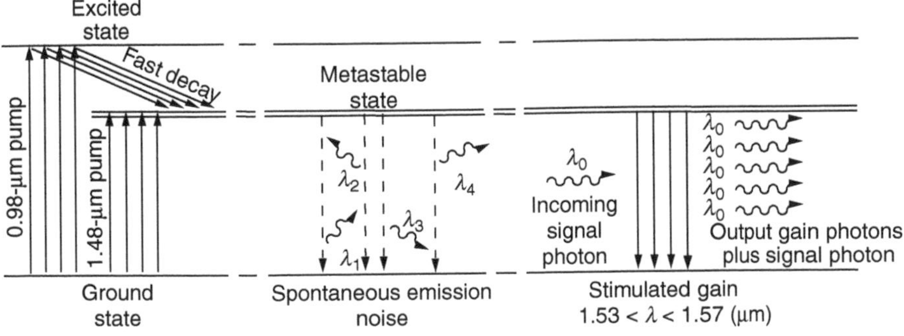

FIGURE 20.27 The energy levels of Erbium.

The absorption is not as strong for all the possible wavelengths, governed critically by the tendency of a pump photon to be absorbed as determined by the cross-section of the Erbium ion with that photon. The wavelength having the strongest absorption coefficient is 0.98 μm, the second strongest is at 1.48 μm, and weaker absorptions occur at 0.53 and 0.8 μm. Fortunately, fabricating high power multi-mode laser-diodes for the 0.98 and 1.48 μm wavelengths can be done by using strained-layer quantum-well material, with output powers >100 mW achievable and commercially available.[49] Laser diode pumps are attractive sources since they are compact, reliable, and potentially inexpensive.

Both the absorption and the emission spectra have an associated bandwidth. These bandwidths depend on the spread in wavelengths which can be absorbed or emitted from a given energy level, allowing multi-moded multiwavelength diode laser light to be absorbed as stated in the previous paragraph. Such a spread in wavelengths is caused by Stark-splitting of the energy levels, allowing a deviation from an exact wavelength.[50] This is highly desirable since (a) the exact wavelength of the pump laser may not be controllable and is impossible for a multi-mode laser, and (b) the signal may be at one of several wavelengths, especially in a WDM system. This makes implementation of the amplifier very flexible. Figure 20.28 shows the bandwidth in the 1.48 μm absorption and 1.55 μm fluorescence spectrum of a typical Erbium-doped fiber (EDF).[51]

Gain Dependencies. The gain in the EDFA depends on the amplifier doping, length, pump power, pump wavelength, and the emission and absorption cross-sections for the signal and pump.[49] The gain and loss coefficient spectra at different inversion levels for EDF with Al and Ge codoping are shown in Figure 20.29. The inversion level is expressed as the fractional percentage $N_1/(N_1+N_0)$, where N_0 and N_1 are the populations of Er ions in ground and excited states.

A typical experimental gain dependence on pump power is shown in Figure 20.30.[48] A minimum pump power is required for the gain to overcome the losses, i.e., achieve transparency. The gain then increases rapidly, followed by a plateau region which represents the fact that all the available carriers have already been inverted throughout the gain medium. The amplifier should be operated near this fully inverted (plateau) region to minimize n_{sp} and obtain a better noise property that we will be discussing in the following section.

Noise Figure. In addition to gain, understanding the effect noise has on the implementation of optical amplifiers in optical communication systems is of paramount importance. We will now discuss the generation of the amplifier noise and its effect on receiver sensitivity.

The noise in an amplifier is inherently due to the random incoherent spontaneous emission events of excited carriers. Each spontaneously decaying carrier can radiate in any solid angle. The fraction of the spontaneous emission that is emitted within the critical angle and allowing it to be coupled into the

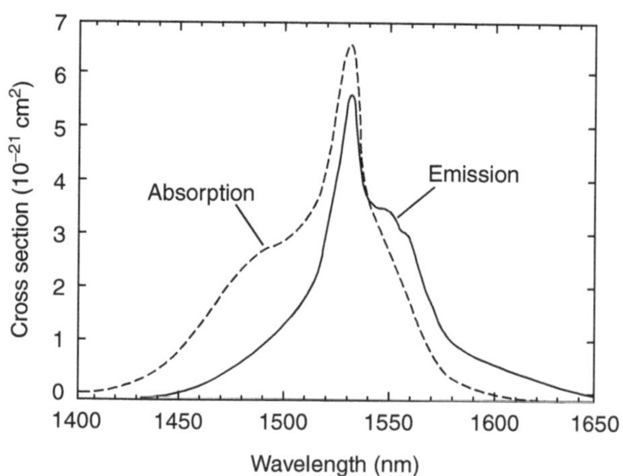

FIGURE 20.28 The absorption and fluorescence spectra for Erbium near 1.5 μm.

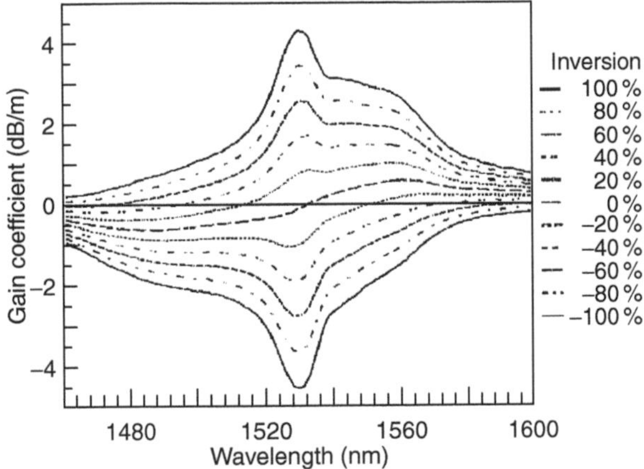

FIGURE 20.29 The Al and Ge codoped EDFs gain/loss coefficients spectra at different inversion levels.

optically-guiding region will itself cause further stimulated emission. Therefore, the spontaneously emitted light that gets coupled into the beam propagation path is subsequently amplified and in total is called the ASE. This ASE is quite broadband, occurring over the entire gain bandwidth. Additionally, since there is only a finite number of excited carriers, the more carriers being utilized for the ASE, the less are available to provide signal gain. This is an additional reason why we wish to suppress forward or backward reflections into the active medium since any reflected non-desired wave depletes the available gain and increases the ASE noise component.

A fairly typical amplified channel is shown in block diagram form in Figure 20.31. The signal, P_{sig}, may initially pass through some lossy components, with a lumped insertion loss of L_{in}. The amplifier then provides gain, G, and adds noise, having variance σ^2. Since the modulated laser-generated signal is typically single-mode (and narrowband) and the ASE spectrum of the amplifier is quite broad (10 s of nm), there will probably be an optical filter present, of bandwidth B_0, to pass the signal and block much of the noise. We can also lump any insertion losses at the output of the amplifier in one term, L_{out}. (We will not include L_{in} and L_{out} in the following analysis; the reader should note that the effect as seen from the receiver of L_{in} is to attenuate the optical signal power, P_{sig}, and L_{out} attenuates both P_{sig} and the

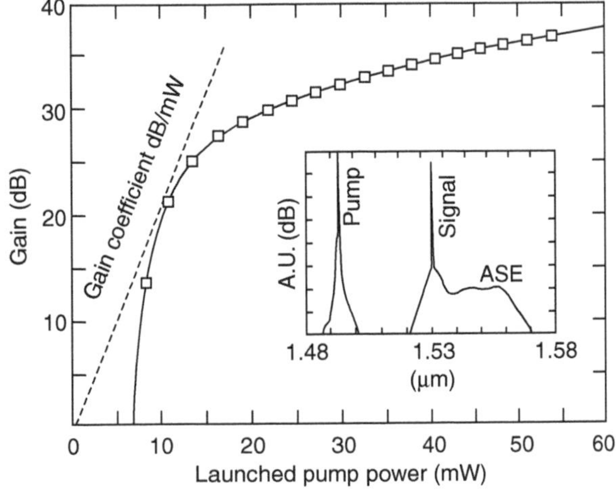

FIGURE 20.30 EDFA gain as a function of 1.48 μm pump power.

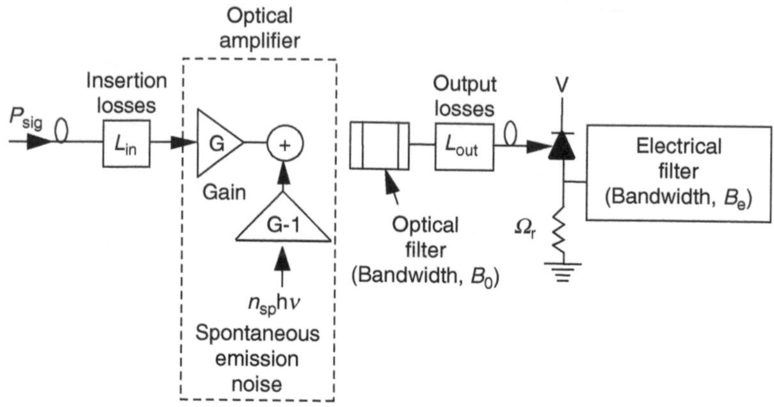

FIGURE 20.31 Block diagram of a signal passing through a typical optical amplifier and then being detected.

spontaneous emission power, P_{sp}.) The optical detector has a characteristic resistance, Ω_r, and the entire receiver has an electrical bandwidth, B_e, to pass the modulated baseband signal and again block higher-frequency noise terms. Note that the signal is in one polarization but the ASE occurs in both polarizations.

If we assume Poisson statistics for the ASE, we can derive the NF and spontaneous emission factor for the Erbium-doped amplifier.[52–54] The NF is by definition and by approximation:

$$\mathrm{NF}(\lambda) = \frac{\mathrm{SNR_{in}}}{\mathrm{SNR_{out}}} \approx \frac{2P_{sp}(\lambda)+1}{G(\lambda)} = \frac{2n_{sp}(\lambda)[G(\lambda)-1]}{G(\lambda)} + \frac{1}{G(\lambda)} \tag{20.13}$$

where

$$n_{sp} = \frac{P_{sp}}{h\nu\Delta\nu(G-1)}. \tag{20.14}$$

The NF of the EDFA is approximately equal to $2n_{sp}$ if $G >> 1$.[48] When factoring in the specific properties of the EDFA, then n_{sp} can be defined as

$$n_{sp} = \frac{N_1}{N_1 - \left(\dfrac{\sigma_{sa}(\lambda)}{\sigma_{se}(\lambda)}\right) N_0} \tag{20.15}$$

and when proper substitutions are made, becomes:[48,55]

$$n_{sp} = \frac{1}{1 - \left(\dfrac{\sigma_{sa}(\lambda)\sigma_{pe}(\lambda)}{\sigma_{se}(\lambda)\sigma_{pa}(\lambda)}\right) - \left(\dfrac{\sigma_{sa}}{\sigma_{se}}\dfrac{P_p^{th}}{P_p}\right)} \tag{20.16}$$

where σ_{se} and σ_{sa} are signal's emission and absorption cross-sections of the Erbium ions. The third term in the denominator of Equation 20.16 represents the contribution to noise due to inadequate pumping. In the high-pumping regime in which all the carriers are inverted along the length of the amplifier, this third term can be neglected. The contribution to noise of this third term is that n_{sp} increases when either pumping is too low or when the amplifier is in deep saturation.

The individual spectra for the absorption and emission (fluorescence) of light near 1.55 μm in an EDFA was shown to be quite non-uniform but also to overlap with each other. That means that if a signal is incident on the amplifier at a wavelength containing significant absorption and emission cross-sections, then the signal will experience both gain and absorption, thereby contributing to some additional ASE. This overlap occurs on the short wavelength end of the gain spectra, and so n_{sp} and the NF will be higher at this end of the spectrum than in the case where the wavelength is higher and the overlap is much less. Figure 20.32 shows the NF as a function of wavelength and how the NF is clearly lower at longer wavelengths.[56] Thus, it is advantageous to operate the signal wavelength not at the gain peak of the amplifier, but at a higher wavelength with a lower NF. Note that the NF is very close to 3 dB, which is the quantum-limited value.

The noise issue of whether to co- or counter-propagate the signal with the pump is more subtle. Certainly, if the two beams are counter-propagating, then the pump does not appear at the detector but must be blocked in the reverse direction by an isolator so as not to propagate back into the signal source. However, the pump should still be blocked from the detector even in the co-propagating case by an in-line optical filter. The subtle noise issue revolves around the accumulation of spontaneous emission noise. The two pump-propagation cases differ when the medium is not fully inverted. This occurs when the pump does not contain sufficient power to fully invert the entire length of the amplifier. An incomplete inversion will then occur at the fiber end in the co-propagating case and at the fiber beginning in the counter-propagating case. The emission dynamics suggest that the better scenario is for co-propagation since in any amplifier system the NF is determined by the first amplifier and not by the last one. However, practical counter-propagation systems will not be affected too adversely since most of these systems will have a high medium inversion along the entire amplifier due to the availability of high-power pump diodes. The slight difference in NF will probably not be significant. Moreover, some systems will employ a combination of both types of pumping schemes to provide high pump powers for high gain and low NF.

Output Power and Saturation. The EDFA gain can be compressed to a small value[57] if the amplifier is saturated by an intense input signal.[58] An amplifier figure-of-merit is to have a high output saturation power, which is especially desirable for power-amplifier applications in which we wish to boost the output of a laser diode to a value higher than what a semiconductor laser or amplifier can provide. The saturation output power is a function of the Erbium concentration, fiber length, and pump power since the more carriers that are inverted in the medium will result in a higher output power. Of course, the limit is reached when all the Erbium ions in the fiber length have been inverted. We plot in Figure 20.33 the signal gain as a function of signal output power, showing the saturation output power of the amplifier.[48] This example is for a given length of fiber and pump power. Saturation output powers as high as ~20 dBm have been measured with extremely high pumping and an extremely large number of Erbium ions.

Broadband EDFAs. In addition to conventional band or C-band (1530–1565 nm), EDFA can also operate in the longer wavelength region or L-band (1565–1625 nm).[59] As shown in Figure 20.29, the EDF has nearly 0.2 dB/m gain coefficient for the inversion level between 20% and 30% (light pumping).

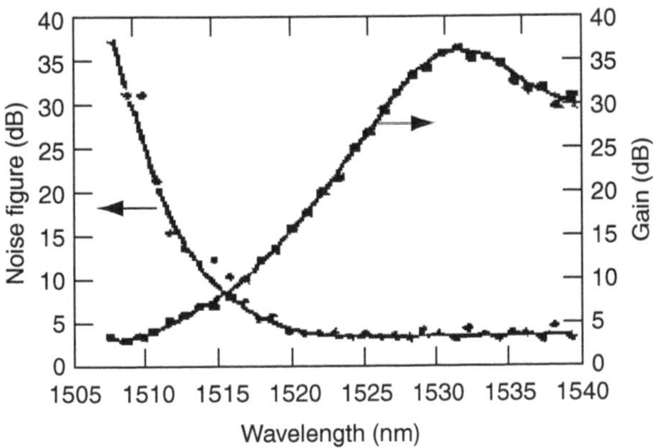

FIGURE 20.32 Noise figure as a function of wavelength in an EDFA pumped with 1.48 μm light.

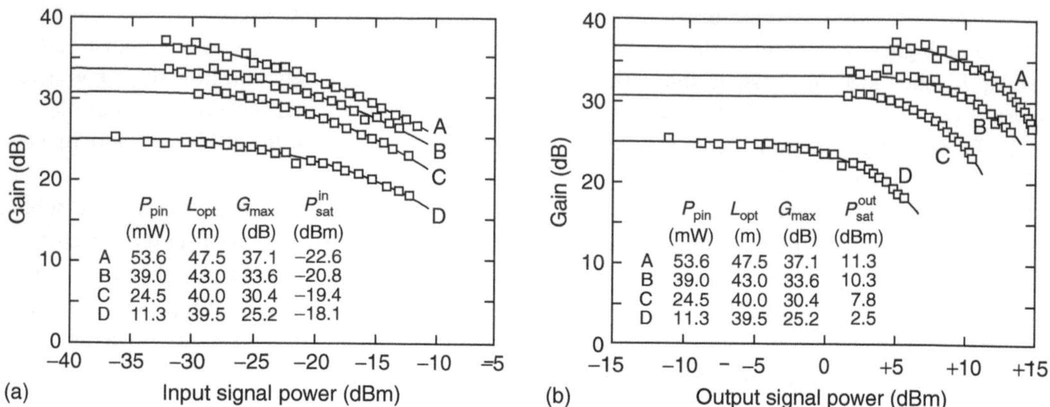

FIGURE 20.33 EDFA gain saturation. (a) signal gain versus the input signal power, (b) signal gain versus the output signal power. (Adapted from Fabry, C. and Perot, A. 1899. *Ann. Chim. Phys.*, 16, 115. With permission.)

The realization of L-band EDFAs has doubled the usable transmission bandwidth. Even though L-band EDFAs operate on the same principle as C-band EDFAs, there are significant differences between C-band and L-band EDFAs. First of all, the gain coefficient in the L-band is much smaller than in the C-band. This will require much longer doped fiber lengths and higher pump powers. Secondly, the gain shape in the L-band is much more uniform as compared to C-band. This means that much less gain flattening filtering is required in L-band EDFAs.

In order to realize ultra wideband amplification, a split-band architecture that combines C and L-band EDFA has been demonstrated, as shown in Figure 20.34. The C-band and L-band signals are first separated by a WDM, then amplified by separate amplifiers, and recombined together afterwards.

The primary applications that have driven EDFAs to commercial development are long-haul, terrestrial, and undersea transport systems. EDFA's have been successfully used as pre-amplifiers in receivers, in-line amplifiers, and power boosting at the output of a laser transmitter.[9–11,56,60–63] Since EDFA's are commercially available now, implementing them anywhere in an optical system has become advantageous and quite easy to achieve. Note that EDFA is the key technology that enables the deployment of multichannel WDM systems. It has been proven to be particularly suitable in the simultaneous amplification of multiple optical channels. However, a few factors make WDM multi-channel systems more difficult to implement than single-channel systems. The uniform spectral

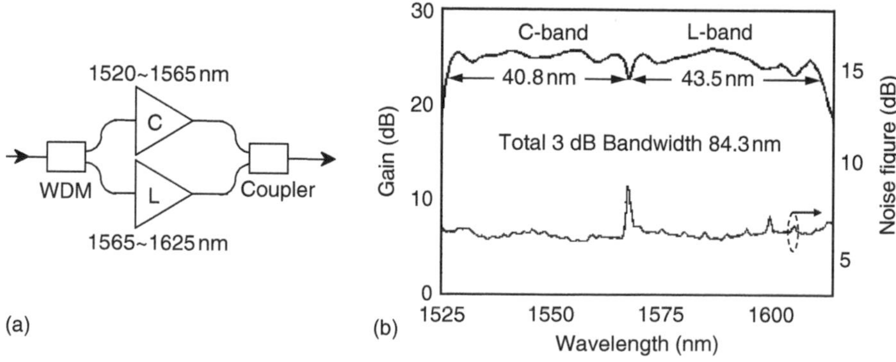

FIGURE 20.34 (a) Basic structure of split band (C+L) EDFA, (b) gain and noise figure spectra. (Adapted from Sun, Y., et al. 1998. *24th Europ. Conf. Opt. Commun.*, 1, 53–54.)

gain and dynamic behavior of the EDFAs should be considered in the WDM systems. We will discuss these issues in-depth in the section of WDM technology.

20.2.4.2 Semiconductor Optical Amplifiers

As shown in Figure 20.35, Semiconductor Optical Amplifiers (SOA)[64–68] is nothing more than a semiconductor laser, with or without facet reflections (the anti-reflection coating reduces the reflections). An electrical current inverts the medium, e.g., electrons are transferred from the valence to the conduction band, which produces spontaneous emission (fluorescence) and the potential for stimulated emission if an external optical field is present. The stimulated emission yields the signal gain. On the other hand, the spontaneous emission is itself amplified and is considered the ASE noise. If we are dealing with a circular-waveguide fiber-based communication system,[69] an external signal must be coupled into and out of the amplifier's rectangular active region producing a mode field mismatch and, consequently, insertion losses.

Gain. The semiconductor active medium is a rectangular waveguide which provides gain to an optical signal that is propagating through the amplifier.[30] Signal gain occurs when (a) carriers are excited from the valence to the conduction band in this quasi-two-level energy system, and (b) an external signal initiates stimulated emission of these conduction band carriers back down to the valence band in this two-level system. We can approximate the unsaturated (i.e., enough excited carriers exist to provide gain to a small signal) spectral gain-coefficient of the active medium to be homogeneously-broadened and to have a Lorentzian lineshape. The gain coefficient represents the likelihood of the occurrence of a spontaneous emission event. Such a spontaneous event means that a carrier makes a transition from the conduction band down to the valence band thus causing a photon to be emitted over a broad range of wavelengths. It should be emphasized that the semiconductor is not, strictly-speaking, a two-level system. In fact, the conduction and valence bands both have a distribution of carriers over a wide energy range.

The general expression for the unsaturated single-pass gain, G_0, through the amplifier medium can be given by

$$G_0 = \exp\left[\frac{L\Gamma C\tau_s}{\nu}\left(\frac{J}{q}-\frac{n_0}{\tau_s}\right)\right] = \exp\left[\frac{Lh\nu}{P_{sat}}\left(\frac{J}{q}-\frac{n_0}{\tau_s}\right)\right] \tag{20.17}$$

where L is the length of the active medium, Γ is the overlap fill factor (i.e., confinement factor) between the active area (containing the electrical carriers) and the optical field accounting for the fact that not all the photons interact with the active gain material, C is a constant of the semiconductor (typically $\sim 10^{-6}$ cm^3/s for 1.3 or 1.55 μm SOA's), τ_s is the carrier (spontaneous emission) lifetime, J is the current density in the active area, q is the charge of an electron, n_0 is the carrier population in the excited state to achieve transparency (i.e., gain equals medium losses), and $h\nu$ is the energy of a photon at frequency ν.[35] P_{sat} is the maximum amount of optical power or saturated optical intensity, which can be output from the active medium:

$$P_{sat} = \frac{h\nu(c/n)}{\tau_s C\Gamma}\left(\frac{W}{m^2}\right) \tag{20.18}$$

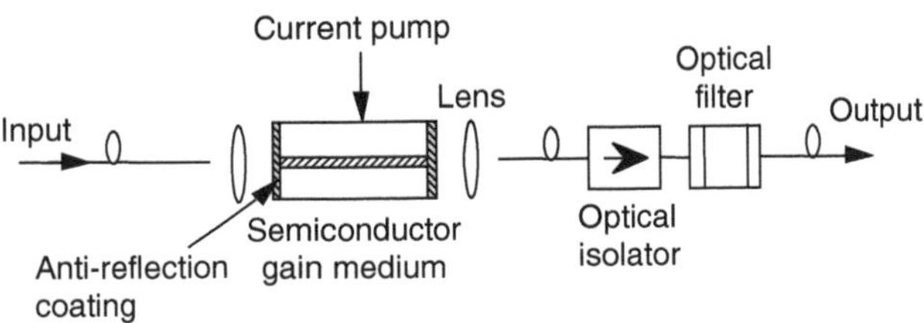

FIGURE 20.35 Block diagram of a semiconductor. The optical isolator and optical filter are included although they may not be required under all circumstances.

Note that the unsaturated small-signal gain, G_0, represents the gain that an amplifier can theoretically provide to a small signal which doesn't consume all the excited carriers available in the amplifier. However, the actual gain, G, provided by the SOA depends on the input-signal intensity. This gain can be much lower if the inverted gain-producing carriers become too depleted to amplify the entire input signal resulting in amplifier saturation. For example, if an incoming signal is so large that there simply are an insufficient number of carriers in the inverted state to allow stimulated emission to occur for all the incoming photons, then the total gain for this intense signal will be less than in the small-signal case. For intense signals, the amplifier gain is diminished and the amplifier itself is considered saturated. Another way of thinking about saturation is that the pump can only provide a maximum number of excited carriers, and therefore an incoming large signal cannot be amplified to the point where more power is output from the amplifier than was initially provided by the pump source. Furthermore, the saturation input (or output) power is usually considered that input (or output) power which will produce a reduction in the small-signal gain by 3 dB. This is an important amplifier figure-of-merit since we desire (a) the amplifier to operate uniformly over a wide dynamic range, and (b) to provide as large a "boost" to the signal if the configuration is as a power amplifier. Figure 20.36 shows how the gain is reduced for an increase in the input signal power for a given initial gain and P_{sat}.

Equation 20.17 provides the basic gain expression but ignores the fact that there may be reflections at the right and left boundaries of the amplifier due to either an index-of-refraction mismatch or a reflective grating. These reflections have a profound impact on the gain achievable from the amplifier. We can determine the effect reflections have on the amplifier gain by considering the wave equations of the optical fields inside and outside the amplifier. Assuming that the reflection coefficients are real, the power gain, G_r, resulting from traversing a gain medium containing facet reflectivities is

$$G_r = \frac{(1-R_1)(1-R_2)G_0}{1+R_1R_2G_0^2-2G_0\sqrt{R_1R_2}\cos(2\beta L)} \tag{20.19}$$

where R_1 and R_2 are the intensity reflectivity coefficients at the input and output boundaries, respectively.[64] The denominator in this equation is periodic, producing periodic Fabry-Perot resonances in the gain spectrum. This equation clearly illustrates the difference between two reflection-dependent types of amplifiers. If the reflections are suppressed and $R_1 = R_2 = 0$, then this is a wideband traveling-wave (TW) amplifier, also known as a "one-pass" amplifier. If there are reflections and R_1, $R_2 > 0$, then this is a narrowband Fabry-Perot amplifier with narrow FP ripples in the wide gain spectrum. The signal travels multiple times between the two facets in the FP amplifier.

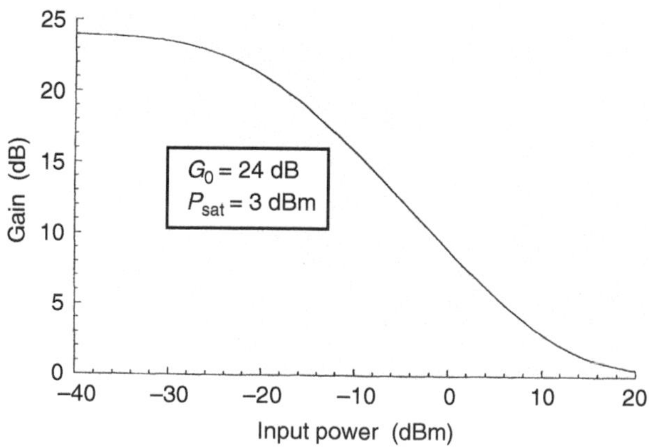

FIGURE 20.36 TW-amplifier gain versus input signal power demonstrating the effects of gain saturation.

Using an AR coating can substantially reduce the reflections at the boundaries back into the FP cavity to a value $\sim 10^{-4}$. However, efficient AR coatings are difficult to fabricate and never reach a value of $R=0$. There are two other methods which are commonly employed to reduce reflections (usually in conjunction with the AR coating): (1) Simply angle the guiding active region in relation to the facet normal.[70] Under this scenario, the optical reflection occurring as the wave hits the semiconductor/air interface upon exiting the cavity will reflect back into the waveguide at the designed angle. If this angle is large enough, i.e., greater than the critical angle necessary for achieving total internal reflection upon reflection back into the guiding cavity, then this reflection will not be coupled into the amplifier waveguide and thus not affect the gain characteristics. (2) Incorporate a buried passive non-guiding window region between the end of the central active layer and each of the facets.[71] As the wave exits the guiding active region, it reaches a non-guiding semiconductor region of the same index-of-refraction. Therefore, no reflection occurs at this interface and the beam diverges. This divergence continues until the beam reaches a facet, at which point a reflection occurs. Since there is still no guiding, the reflected wave continues to diverge, eventually coupling very little light back into the guiding active region.

Unlike fiber amplifiers, the semiconductor gain medium, being rectangular (not square) and having different crystal planes, has a polarization dependent gain inherent in the medium. Furthermore, the confinement fill factor (Γ) of the two polarizations is different, further producing a difference in gain. Operating an SOA with significant polarization dependence in the gain may require using unwanted polarization controllers to control the input signal polarization from the input fiber. Additionally, the AR coatings on the facets must be designed to have a similar response for both polarizations.[72] A difference of <1 dB has been produced by combining these methods plus multiple-quantum-well and strained-layer material.[73] Using semiconductor amplifiers in optical fiber systems requires attention to their polarization dependence since the polarization of any wave along the fiber is typically unknown and can vary. However, if the semiconductor amplifier is integrated on the same chip following a fixed linear-polarization laser transmitter, then only a single polarization passes through the amplifier and the polarization dependence is not a problem. Note that, since the fiber-based amplifiers are circular and the wave in a fiber is typically circularly polarized, the gain medium and signal gain are negligibly dependent on the polarization of the incoming signal.

Noise Figure. The noise considerations of a SOA are quite similar to those for an EDFA. Both amplifiers have ASE noise located in two polarizations, and this noise generates beat-noise terms in a detector. Three differences between the two types of amplifiers are (1) the less wavelength dependence on the NF, of the SOA; (2) the relatively-subtle effect of co- versus counter-propagating the pump light with the signal in an EDFA; and (3) the higher input signal coupling loss from fiber to SOA. When we consider the input coupling efficiency η_i, which approximately equal to 1 for EDFAs, the NF is denoted as:

$$\mathrm{NF} = 2n_{\mathrm{sp}}/\eta_i \tag{20.20}$$

Even though the inversion factor of a SOA can be as high as that of an EDFA, the NF of the SOA is typically 6–8 dB,[7,74] which is not close to the quantum-limitation (3 dB).

Finally, it is important to note that SOAs had been troublesome in WDM multichannel systems due to their speedy gain dynamics, which caused intermodulation distortion (IMD) and crosstalk between WDM channels. The IMD and crosstalk effects are negligible for fiber amplifiers since their carrier lifetime is approximately 10 ms, far too long to produce any IMD for any reasonable-spaced channels and far too long to produce saturation-induced gain fluctuations on the time scale of an individual high-speed bit. Recently, the interest in the SOA is picked up again because the fast gain dynamics. The SOA, on the other hand, can provide various ways for all-optical signal processing, such as wavelength conversion, optical regeneration, and optical TDM. Intermodulation (cross-gain, XPM, and FWM) have been demonstrated for the wavelength conversion, in which the data modulated on one wavelength can be transferred to another signal wavelength.[75–77] The nonlinear response of the SOA also enables regeneration of the data to take place.[78] The fast gain dynamics allow for applications in optical TDM.[79] Undoubtedly, the most attractive advantages of using the SOA as opposed to fiber amplifiers include its small size, potential low cost, and integratability on a chip containing many other optoelectronic components (i.e., lasers and detectors). Beyond providing simple gain, a semiconductor amplifier can be also used as a high-speed switching element in an optical network since the semiconductor will (a) amplify if pumped, and (b) absorb if unpumped. The operation is simply to provide a current pump when an optical data packet is to be passed and discontinue the pump when a data packet is to be blocked. Thus, the passed packet is amplified by the population inversion and the blocked packet is absorbed by the semiconductor since it now lacks a population inversion. When SOAs can deliver on these promises and optical data processing becomes crucial in future optical networks, they will be potentially very useful in the optical communications area.

20.2.4.3 Other Types of Fiber Amplifiers

It is important to mention two other types of fiber amplifiers due to their historic significance in preceding Erbium by many years. These two amplifiers are Brillouin and Raman amplifiers. The fundamental mechanisms involve molecular interactions of light with the fiber, causing the molecules to vibrate. Energy can be transferred from a propagating pump beam to a propagating signal beam. Relative to the amplifiers examined in this section, these amplifiers are much less efficient, requiring high pump power and extremely long lengths of fiber (km). The gain mechanism for these two amplifiers is fundamentally different than the semiconductor and Erbium amplifiers. Therefore, the reader is encouraged to seek further information pertaining to these amplifiers in texts dealing with quantum electronics.

The fundamental principles are as follows.[31] In the Brillouin amplifier, an incident photon will interact with the material forming an acoustic wave (phonon) and a reradiated photon at a lower frequency (Stokes frequency). This photon is shifted ~10 GHz down in frequency, or about 1 Å, since the acoustic wave is uniform in a fiber. The odd part of the Brillouin amplifier is that the pump and signal must be counter-propagating, with no co-propagating interaction possible. In the Raman amplifier, the pump light photon is absorbed and sets the fiber molecules into mechanical vibrations. A photon is again radiated at the Stokes frequency, but since mechanical vibrations are not uniform in a fiber, the Stokes frequency is not a set number. Furthermore, the pump and signal may co- or counter-propagate in the fiber. It is worth mentioning that practical, efficient, and high-power pump sources have diminished the disadvantage of the relatively poor efficiency of the Raman process over the last few years. A tremendous rebirth of the interest in Raman amplification in optical fibers has taken place. Interest in Raman amplification has steadily increased.[80] Raman amplifiers do offer several very attractive advantages over EDFAs and SOAs and find some applications in fiber transmission systems. We will introduce Raman amplifiers briefly in this section.

The most important feature of Raman amplifiers is their capability to provide gain at any signal wavelength, as opposed to EDFAs based on the doped ions in the fibers. The position of the gain bandwidth within the wavelength domain can be adjusted simply by tuning the pump wavelength. Thus, Raman amplification potentially can be achieved in every region of the transmission window of the optical transmission fiber. It only depends on the availability of powerful pump sources at the required wavelengths. Figure 20.37 illustrates the Raman gain coefficient in a few different fibers. The disadvantage of Raman amplification is the need for high pump powers to provide a reasonable gain. However, the Raman effect can be used for signal amplification in transmission windows that cannot be covered properly by EDFAs. The upgrade of already existing systems by opening another transmission window where Raman amplification is applied could be an attractive application. Another application of the Raman effect is given with hybrid EDFA/Raman amplifiers characterized by a flat gain over especially large bandwidths. Repeaters can be built to compensate the nonflatness of the EDFA gain with a more flexible Raman gain. Multiwavelength pumping could be used to shape the Raman gain such that it equalizes for the EDFA gain shaping.

Figure 20.38 shows a typical Raman amplifier that is backward pumped and the gain is distributed over the long transmission fibers.[80,81] The spectral flexibility of Raman amplification allows the gain spectrum

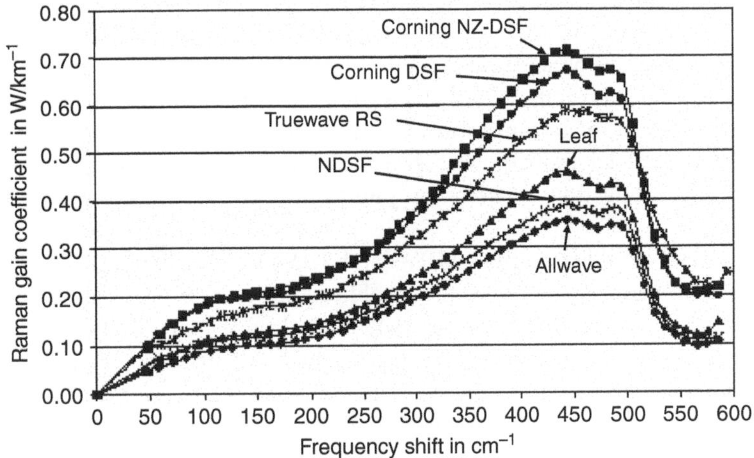

FIGURE 20.37 Raman gain spectra for different commercial fibers (gain peak is shifted 13 THz from the pump wavelength toward longer wavelength). (Adapted from Kogelnik, H. and Shank, C. V. 1972. *J. Appl. Phys.*, 43, 2328. With permission.)

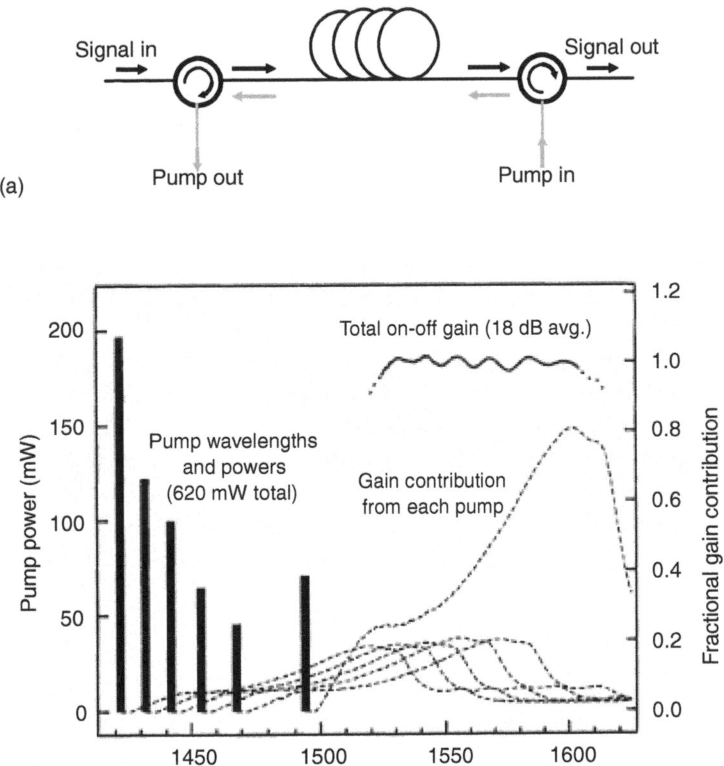

FIGURE 20.38 (a) Basic setup of backward pumped Raman amplifiers, (b) a numerical example of broad-band Raman gain obtained using a broad band spectrum to pump a NZDSF. Bars show the counter-pump wavelengths and its power. Solid line shows the total small-signal on-off gain. Dashed lines show the fractional gain contribution from each pump wavelength. (Adapted from Campbell, J. C. 1989. *Optoelectronic Technology and Lightwave Communication Systems*, C. Lin, ed., Van Nostrand Reinhold. With permission.)

to be shaped by combining multiple pump wavelengths to make a polychromatic pump spectrum. There have been many studies searching for optimization approaches that give the flattest gain with the fewest number of pumps. Using this broad-band pumping approach, amplifiers with gain bandwidths greater than 100 nm have been demonstrated.[82] When designing such broad-band Raman amplifiers, one must consider the strong Raman interaction between the pumps. The short wavelength pumps amplify the longer wavelengths, and so more power is typically needed at the shortest wavelengths (see Figure 20.38b). This interaction between the pumps also affects the noise properties of broad-band amplifiers.

Opposed to lumped amplifiers such as EDFAs and SOAs, the distributed Raman amplifier (DRA) offers improved noise performance. Since the amplification occurs over a significant segment of the transmission fiber itself, typical 20 km, the signal does not drop as much as it would in a conventional lumped amplification. For a large lumped-gain amplifier, the amplification is typically placed near the receiver (a) to provide enough gain to the signal to overcome the receiver thermal noise, and (b) to provide the highest gain without being saturated by an intense signal. The high ASE noise generated will be detected in totality. From the SNR_e point of view, the system using lumped amplifiers allows the signal power to drop to the lowest level and will produce the worst noise performance. The lost SNR_e of a signal can never be recovered (shot-noise limited). While in the DRA, the gain is relocated in a fairly long fiber link and the signal power will never drop to the lowest, as shown in Figure 20.39. This will prevent SNR_e from degrading as quickly as it would have without Raman amplification. Note that the Raman effect can also be used to make the lumped amplifier. A discrete Raman amplifier with 160 nm bandwidth has been demonstrated using tellurite-based fiber.[83]

The noise of a Raman amplifier stems from spontaneous Raman scattering events[84] as well as multi-pass interference (MPI) noise[85] (from unintentional reflections or from Rayleigh scattering). An additional noise source stems from the amplifier pump.[86] While present in every amplifier type, pump noise is mainly significant in Raman amplifiers where it transfers noise to the signals through crosstalk effect. Pump noise is significant in both the linear and nonlinear regimes as it translates directly to gain fluctuations and hence to uncertainties in the amplified signal. This is enhanced by the ultrafast response of the Raman and parametric amplification. In Raman amplifiers, the usual way to reduce such deleterious coupling is to make the pump and signal counter-propagating or use low noise pumps in co-propagation schemes.

Recently, another type of amplifier, so called optical parametric amplifier (OPA) has been intensively studied. The basic operation principle is the four wave mixing in the fiber that may offer wide gain[87] and may be tailored to operate at any wavelength. An OPA is pumped with one or two intense pump waves and may provide gain over more than 200 nm. The main limitation of parametric amplifiers relates to their efficiencies which in turn means that they require high power pump signals. Improvements are possible once larger nonlinearities become available and indeed, highly nonlinear fibers are an active research topic wherever specialized-fiber technology is available. Some special fibers are already available commercially.[88]

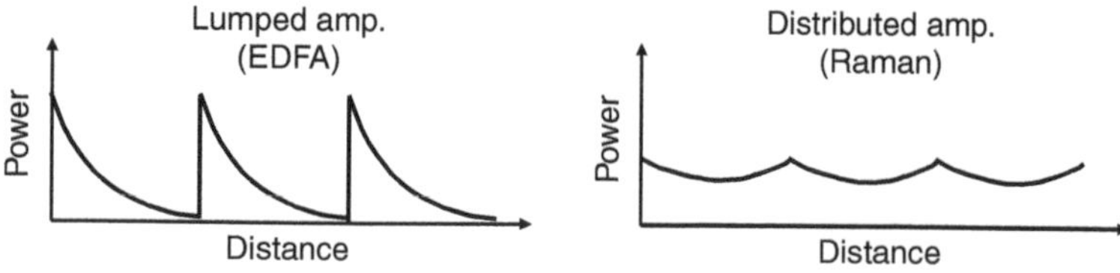

FIGURE 20.39 Power levels in the fiber transmission line: lumped amplification (EDFA) vs. distributed amplification (Raman).

20.2.5 Various Components

20.2.5.1 Filters

Optical filters, like electrical filters, can play a major role in an optical communication system by transmitting a desired wavelength but blocking all others. Here are three common filter applications:

1. The filter is used as a demultiplexer when many different signals are transmitted along a single optical fiber with each signal being on a different wavelength.[89]
2. Some filters can be used as a multiplexer to combine many different signal wavelengths into a single optical fiber.
3. The filter is used to limit the transmitted bandwidth, as would be required at the output of an optical amplifier which generates broadband ASE noise.

There are many different types of optical filters, such as gratings, Fabry-Perot filters, thin film filters, Mach-Zehnder filters, and ring resonators. We will briefly discuss a few widely used filters in this section.

Gratings. A conventional grating is a glass or polished metal surface on which a large number of very fine, parallel grooves or slits are cut. These parallel elements are used to diffract light, generally both separating and spreading different wavelengths or colors. The gratings are used in spectrographic equipment as the principal optical element to separate light into its component wavelengths (frequencies) by diffracting them at various angles, as shown in Figure 20.40.

The most fundamental grating equation is given by

$$a\left[\sin(\theta) + \sin(\theta_d)\right] = m\lambda \tag{20.21}$$

where a is the grating constant, m is an integer $(0, \pm 1, \pm 2, \pm 3, \ldots)$. When the input signal incidents to the grating at the angle θ_i, the constructive interference at a wavelength λ occurs at the direction of angle θ_d if the grating equation is satisfied. Therefore, the grating can be used as a filter or demultiplexer to separate different wavelengths.

Note that the energy at a single wavelength is distributed over all the discrete angles that satisfy this grating equation. However, most of the energy will be concentrated in the zeroth-order $(m=0)$ interference maximum, which occurs at $\theta_i = \theta_d$ for all wavelengths. The light energy in this zeroth-order is then wasted since the wavelengths are not separated. The blazing technology was developed to solve this problem. The readers can learn more about blazing gratings from the reference.[12]

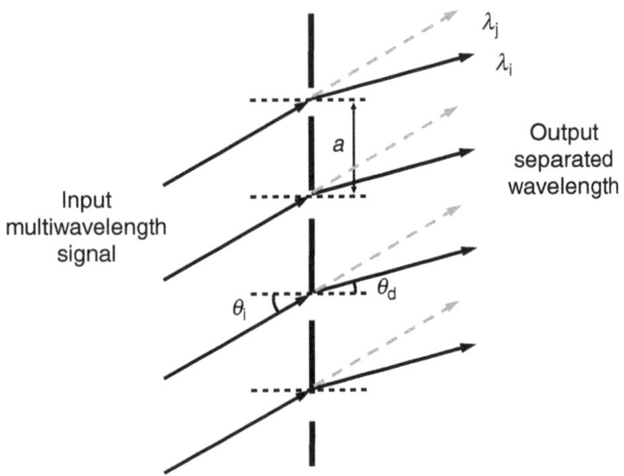

FIGURE 20.40 Principle of operation of a transmission grating.

Fiber Bragg gratings (FBG). In general, any periodic perturbation in the medium serves as Bragg grating. This perturbation is usually the medium's refraction index variation. FBG is produced in an optical fiber by producing a periodic variation of the index of refraction along a short section of a single mode fiber core. There are several different methods for producing this periodic index variation, but most depend on the observed photo-sensitivity of germania doped optical fibers to short wavelength light. In the most general case, the index perturbation $\delta n(z)$ takes the form of a phase and amplitude-modulated waveform

$$\delta n(z) = \delta n_0(z)\left\{1 + \upsilon \cos\left[\frac{2\pi}{\Lambda(z)}z + \phi(z)\right]\right\} \tag{20.22}$$

where z is the position along a grating, $\delta n_0(z)$ is the maximum index modulation spatially averaged over a grating period, υ is the fringe visibility of the index change, $\Lambda(z)$ is the local grating period including grating chirp, and $\phi(z)$ describes phase changes in the grating. The grating filter characteristics can be analyzed by the coupled-mode theory[90] that describes the coupling between the forward- and backward-propagating waves at a given frequency ω.

For uniform gratings, the index change takes the form

$$\delta n(z) = \delta n_0\left\{1 + \upsilon \cos\left[\frac{2\pi}{\Lambda}z\right]\right\}. \tag{20.23}$$

The Bragg resonance wavelength, λ_b, is proportional to the period of the grating, Λ,

$$\lambda_b = 2n_{eff}\Lambda, \tag{20.24}$$

where n_{eff} is the effective core index of refraction and it is a spatial average of $n(z)$. The periodic nature of index variations couples the forward- and backward-propagating waves at wavelengths close to the Bragg wavelength, and as a result, provides frequency-selective reflectivity to the incident signal over a bandwidth determined by the grating strength (see Figure 20.41).

FBGs have been widely used in optical communication systems. Currently, three main applications drive the commercial production of FBGs for telecom applications: wavelength stabilizers for pump lasers, narrowband WDM add/drop filters, and gain-flattening filters. Pump stabilizers are low-reflection gratings inserted in the fiber pigtail of the diode lasers to stabilize the output power and reduce the high-frequency noise of the laser.[91] A second application is when add/drop filters are used to extract or insert channels in WDM systems.[92] Another application is dispersion compensation that uses chirped FBG in which period increases linearly along the length of the component.[93] We will revisit this issue in the section of dispersion management.

It is worth mentioning that another type of fiber gratings so called long-period fiber gratings are very important for optical amplifiers' gain flattening.[94] Most EDAF provide consistent gain across the amplification bandwidth. Adding an element with a spectral transmission that mirrors the amplifier gain yields a net flat gain over the whole bandwidth.

Arrayed Waveguide Gratings (AWG). Over the last few years, arrayed waveguide gratings[95] have become increasingly popular as wavelength multiplexers/demultiplexers for WDM applications. They

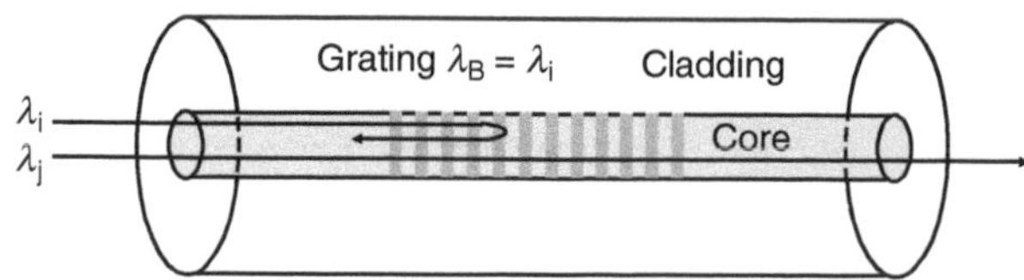

FIGURE 20.41 Illustration of fiber Bragg grating that reflects the signal at the Bragg wavelength.

have proven to be capable of precise demultiplexing of a large number of channels with relatively low losses.

Figure 20.42 shows a schematic representation of the AWG and its spectra. The device consists of two slab waveguide star couplers, connected by a dispersive waveguide array (phase grating). Light propagating in the input waveguide will be coupled into the array via the first star coupler. The array has been designed such that (for the central wavelength of the demultiplexer) the optical path length difference between adjacent array arms equals an integer multiple of the central wavelength of the demultiplexer. As a consequence, the field distribution at the input aperture will be reproduced at the output aperture. Therefore, at this wavelength, the light will focus in the center of the image plane (provided that the input waveguide is centered in the input plane). If the input wavelength is detuned from this central wavelength, phase changes will occur in the array branches. Due to the constant path length difference between adjacent waveguides, this phase change will increase linearly from the inner to outer array waveguides, which will cause the wavefront to be tilted at the output aperture. Consequently, the focal point in the image plane will be shifted away from the center. The positioning of the output waveguides in the image plane allows the spatial separation of the different wavelengths. This is the same principle underlying the operation of conventional gratings.

There are several nice features of AWGs that make them quite a useful item for future optical networks. First of all, AWGs process wavelengths in parallel then can support a large number of channel count. The

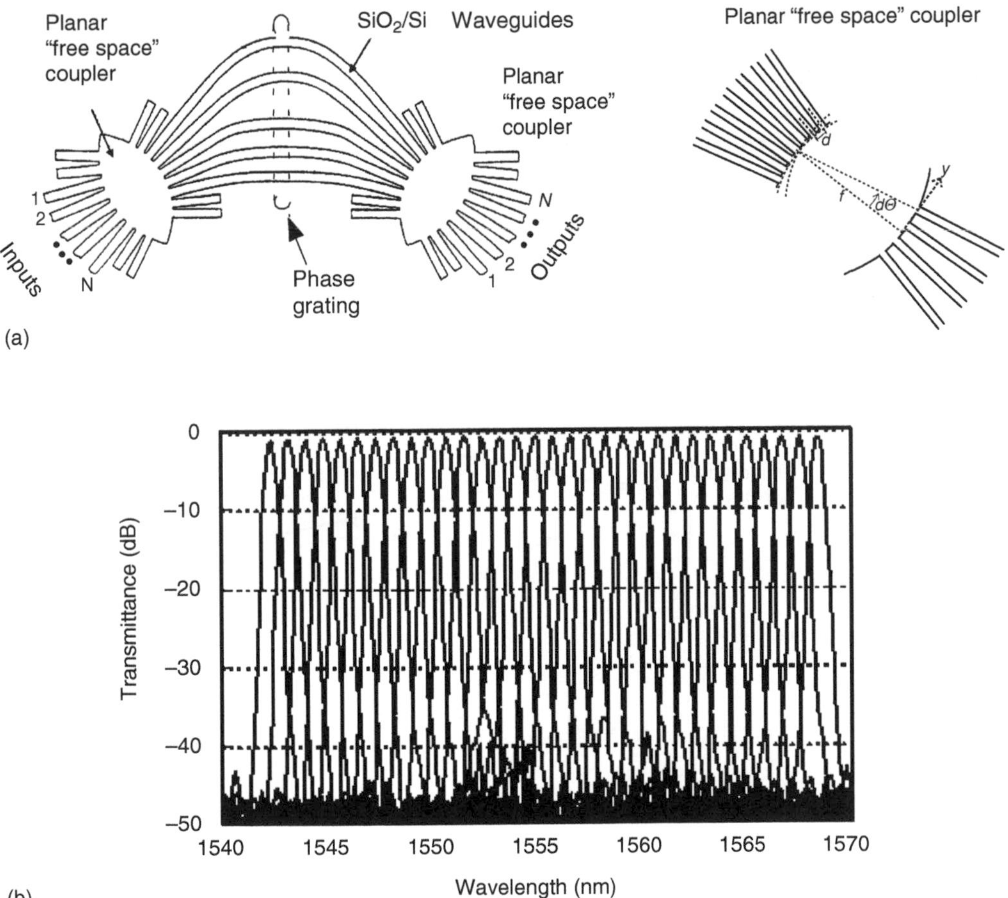

FIGURE 20.42 (a) Schematic representation of the AWG with the magnification of the output star coupler and (b) typical demultiplexing spectra.

capacity of AWG has increased dramatically, to the point that as many as 400 channels can now be incorporated into a single wafer. A capacity of >1000 channels with 25 GHz channel spacing has been reported in tandem AWGs with a multichip configuration.[96] Secondly, they are based in so-called "optical integrated circuits," which means that it is easy to integrate them with other functions like amplification, rather than making them a separate box that makes things more complicated. Note that they may require careful temperature control.

Fabry-Perot Filter. Fabry-Perot operates in much the same manner as does the Fabry-Perot resonant laser cavity. Figure 20.43a shows three media separated by two partially-reflecting mirrors which form a wavelength-selective Fabry-Perot optical filter cavity. An incident right-propagating optical wave at angle θ to the normal of the left mirror is partially transmitted. After this transmitted right-propagating wave traverses the cavity interior, it experiences partial transmission, T_1, and reflection when it encounters the right mirror. The new left-propagating reflected wave will experience reflection and transmission at the left mirror, and subsequently the new right-propagating reflected wave will encounter the right mirror and experience partial transmission, T_2, and reflection. This process continues such that there are an infinite number of transmitted waves at the right mirror, with each higher-number transmitted wave having a smaller intensity due to any losses. Assuming that the cavity length is much smaller than the coherence length of the laser light, all these transmitted waves will be coherent with each other since they originate from the same coherent incident source and will interfere with each other as they are transmitted through the right mirror boundary. If these transmitted waves are in-phase with each other (i.e., phase difference equals 0, 2π, 4π,...) when they encounter the right mirror boundary, then they will all add constructively and the original incident wave will be transmitted thorough the FP optical filter cavity. If the transmitted waves are completely out-of-phase (i.e., phase difference equals π, 3π, 5π,...), then the waves will destroy each other, thereby blocking the original wave from passing through the cavity and reflecting it back away from the filter. If the transmitted waves are slightly

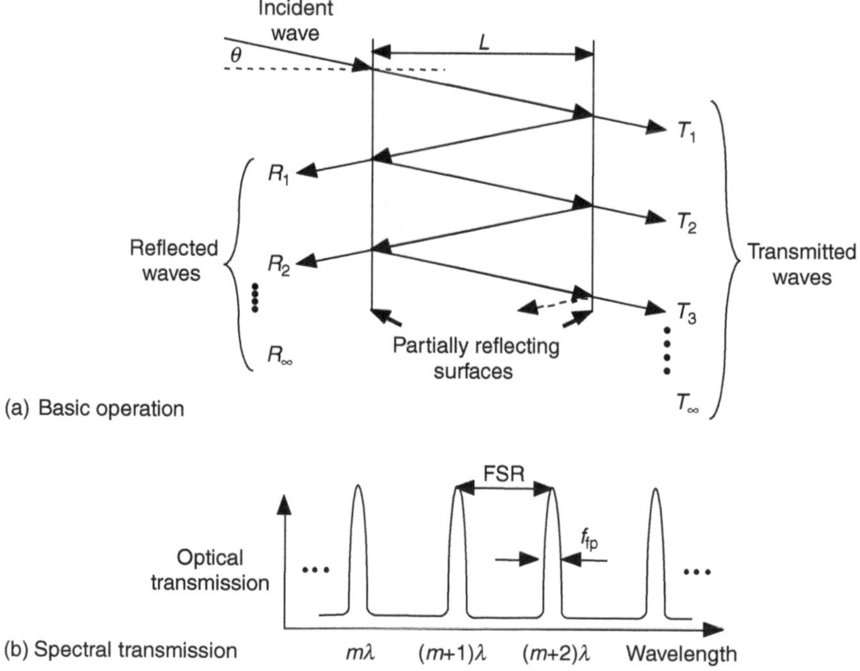

FIGURE 20.43 (a) Basic operation of a Fabry-Perot filter, (b) wavelength-selective optical transmission of a Fabry-Perot filter.

out-of-phase, then they will not add constructively and mostly destructively, with little optical power of the original wave passing through the filter. Since all the waves must be in phase for filter transmission, the round-trip distance of wave propagation inside the cavity must be an integral number of wavelengths:

$$L = \frac{m\lambda}{2n \cos \theta}, \quad m = 1,2,\cdots\cdots \tag{20.25}$$

where n is the refractive index inside the cavity, L is the distance between the two mirrors, and λ is the wavelength of light to be transmitted through the filter.[32] There are an infinite number of integers which will satisfy the in-phase requirement, and the passband of the filter is periodic with wavelength (Figure 20.43b). The optical filter has a transmission resonance bandwidth, f_{FP}, a free-spectral-range between any two resonances, FSR, and a filter Finesse, F, in which

$$F = \frac{\text{FSR}}{f_{\text{FP}}} = \frac{c}{2nLf_{\text{FP}}} = \frac{\pi\sqrt{R}}{1-R} \tag{20.26}$$

where R is the power reflectivity at each of the two mirrors. The Finesse is a figure-of-merit for the optical filter, with a high Finesse corresponding to a more efficient filter. As is evident from Equation 20.38 wavelength selectivity depends on the angle of incidence of the incoming wave since this affects the round-trip propagation distance inside the cavity which is required for phase matching of all the transmitted waves. Therefore, this simple filter can be wavelength-tuned by tilting the angle of the filter in relation to the incoming wave. To first order, angle tuning effectively shifts all the periodic resonances in unison to different transmitting wavelengths.

Multilayer Dielectric Thin-Film Filters. Multilayer thin-film devices can be designed to function as an edge filter or bandpass filter. In either case, wavelength selection is based on the property of destructive light interference. This is the same principle underlying the operation of a Fabry-Perot interferometer.

A thin-film resonant multilayer filter consists of two or more cavities separated by reflective dielectric thin-film layers, as shown in Figure 20.44. On either side of the gap (cavity) are the two reflecting layers. The reflecting layers actually consist of several film layers, each a quarterwave thick. This sandwich of quarterwave layers is made up of an alternating pattern of high and low index material respectively. The quarterwave coatings forming the reflective layer are called a stack. The combination of two stacks and the gap comprise a one cavity bandpass filter. The number of layers in the stack is adjusted to tailor the width of the bandpass. In practice, a single cavity bandpass filter does not exhibit a sharp transition between the passband and out-of-passband wavelengths. To sharpen this cutoff, it is common practice that several cavities are layered sequentially into a multicavity filter design. As more cavities are added, the top of the passband becomes flatter and the skirt becomes sharper, both very desirable filter features.

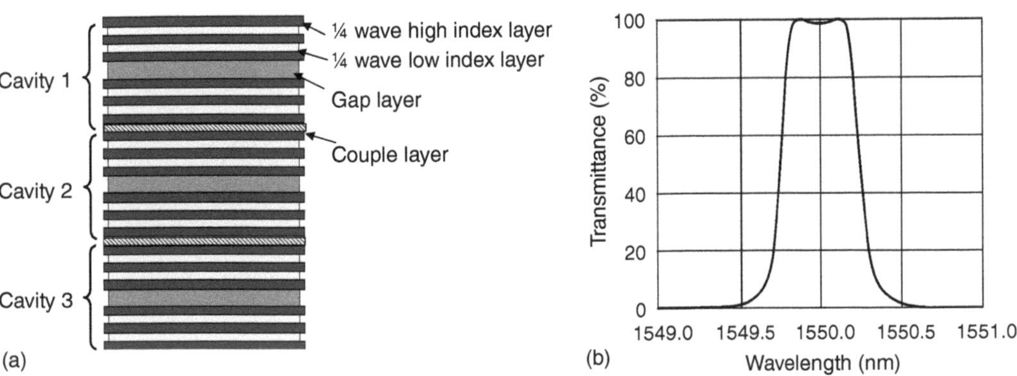

FIGURE 20.44 (a) Illustration of a thin film filter with three cavities and (b) typical filter passband plot.

Thin-film filters are definitely the incumbent technology with the largest revenue share at present. They represent a well understood and established technology, with the attractive characteristic that their cost scales with channel count, because they're modular in design.

Mach-Zehnder Interferometers (MZI). In a MZI (see Figure 20.45a), a device called a beamsplitter splits an input beam into two parts, which pass along different routes, then are combined in a second beamsplitter. The relative phase of the light emerging from the interferometer arms determines its distribution between the two outputs of the output coupler. This phase depends on wavelength as well as the length of the arms. The distribution of light between the two arms changes as the wavelength changes. The transfer function of the MZI is

$$I_1 = I_0 \cos^2\left(\frac{2\pi}{\lambda}\Delta L\right) \tag{20.27}$$

where I_0 is the input optical intensity, ΔL is the light path difference between the two arms. Let's consider the first input and first output as the signal input port and output port. If the distance of propagation is the same in the two arms, then the two waves are in-phase and will add constructively resulting in light appearing at the output. Alternatively, if the distance in the two arms is different by 1/2 a wavelength, then the two waves will be π radians out-of-phase, the waves will add destructively, and no light will appear at the output. Therefore, MZI can be used as both filters and (de)multiplexers as a wavelength selective device.

The MZ is clearly wavelength selective, with the transmission being a raised cosine-squared function of input wavelength and the transmission bandwidth being critically dependent on the path-length difference between the two interferometer arms. If a voltage is applied to an electrode over one arm or if the temperature of the waveguide arms is changed, then the effective path lengths are changed and the filter can be tuned. Such a filter is not very wavelength selective (i.e., it has a low Finesse) due to its simple raised cosine-squared function. However, if several MZ interferometers of decreasing bandwidths can be cascaded, then the overlapping transmission function of several passbands will effectively allow transmission through only the narrowest passband resulting in a high Finesse filter (see Figure 20.45b).

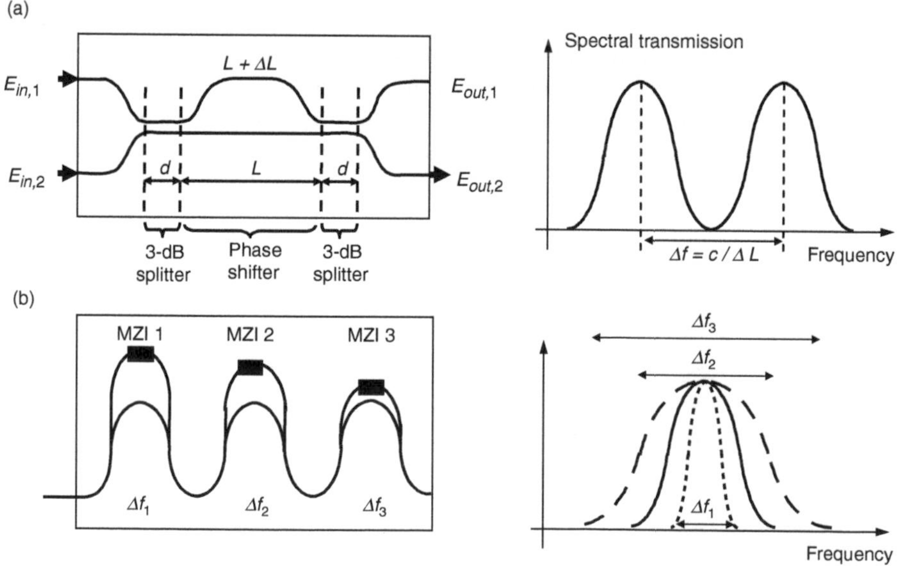

FIGURE 20.45 (a) Mach-Zehnder Interferometer and its transfer function: ΔL denotes the path difference between the two arms, (b) a diagram of 3-stage MZI that uses different path length differences in each stage to achieve high Finesse.

Today's MZI are typically constructed with planar-waveguide technology thus has the potential for integration and fast electronic tuning using several electrodes.

Most of the optical filters we have discussed here can be wavelength tunable. For instance, wavelength selectivity of a FP filter depends on the angle of incidence of the incoming wave since this defines the round-trip distance required for phase matching of all the wave reflections. Therefore, this filter can be wavelength tuned by simply tilting the angle of the filter in relation to the incoming wave or by changing the cavity length. Angle or cavity length tuning effectively shifts the periodic resonances in unison to different transmitting wavelengths. For the MZI, if a voltage is applied to an electrode over one arm or if the temperature of the waveguide arms is changed, then the effective path lengths are changed and the filter can be tuned. Note that tunable filters are one of the key enablers of WDM technology. Such a filter allows one wavelength to pass and blocks all others, thus enabling unambiguous data recovery. The readers can find more in-depth information in the section of WDM technology.

Note that there are many other types of optical filters, such as acousto-optic filter, ring resonators, liquid crystal FP filters, electro-optic tunable filters, and micromachined FP filters that may also have some potential applications in the fiber communication systems. Readers can find the useful information in other chapters on optical devices.

20.2.5.2 Directional Couplers

An optical signal must frequently be split from one input to two output ports for the sake of signal distribution or monitoring. Note that the reverse function of combining is also desirable. This can be accomplished passively by using a directional coupler[97] as shown in Figure 20.46. Two separate single-mode waveguides are brought close together for some interaction length and are then separated. As was mentioned previously, the Gaussian-shaped single-mode propagating in a waveguide will have most of its energy residing in the core with an evanescent field tail propagating alongside the core within the cladding region. A wave propagating in waveguide 1 will have an evanescent tail which partially falls within waveguide 2 while the wave is propagating along the interaction length. Due to coupled-mode theory, this evanescent tail will excite an optical wave within waveguide 2. Power will gradually be transferred from waveguide 1 into waveguide 2. This coupling is resonant and periodic, such that light will oscillate back and forth between these two waveguides if the interaction length is sufficiently long. A 3-dB 50% power splitter can be fabricated by tailoring the interaction length to provide only coupling of half the power from waveguide 1 into waveguide 2. After the waveguides are separated such that no more coupling occurs, light can be independently transmitted from two output ports.

20.2.5.3 Switches

Directional couplers are passive devices in which the ratio of coupling is determined by the interaction length. However, by placing an electrode over the interaction region, we can induce a voltage-controlled electric field. This electric field changes the physical properties and refractive index of the waveguide material. Such a change in the index will change the speed and phase retardation of the optical wave

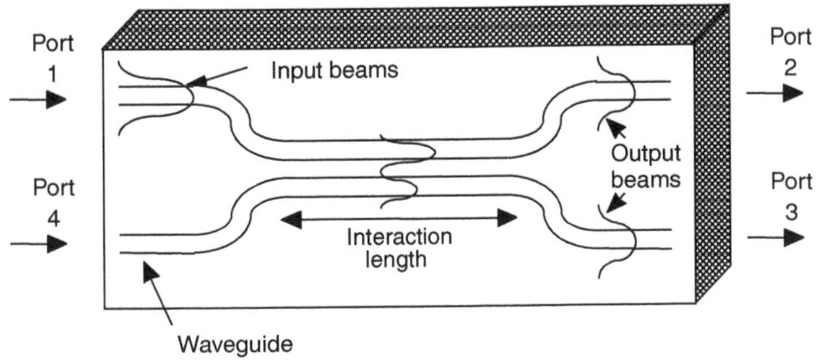

FIGURE 20.46 Optical waveguide interaction in a 4-port directional coupler.

which, consequently, changes the effective interaction length and the coupling ratio.[98] By tuning the voltage, we can tune the optical output ratio between ports 1 and 2. Such a tunable device which can switch an incoming signal between two different output ports is known as a crossbar switch. Note that the MZIs can also be used as a crossbar switch.

There are various different optical switches. For instance, the SOAs can be used as an on-off switch by varying the bias voltage to the device. Mechanical switches move a mirror in and out of the optical path to switch the signal. Micro-electro-mechanical system (MEMS) switches are miniature mechanical switches typically fabricated in silicon substrates. It is worth emphasizing that a large optical switch with port counts ranging from tens to a few hundred even thousand can be realized by cascading many basic 2×2 switches.

20.2.5.4 Modulators

We have mentioned one method of modulating an optical wave with digital data. That method is by directly modulating the bias current of the laser which will modulate the carrier concentration within the active laser semiconductor medium. However, modulating the carrier concentration has a deleterious side-effect since the refractive index of the active medium is slightly dependent on the carrier concentration. Changing the index will change the effective optical length of the cavity between the feedback (or DFB) mirrors. The wavelength selectivity of the cavity depends critically on the phase-matching conditions that the round-trip distance between the cavity mirrors being an integral number of wavelengths. Therefore, a change in the index will change the spectral location of the cavity resonances. As the carrier concentration is changed upon direct bias-current modulation, the cavity will slightly shift and laser light will be produced over a spread of slightly different wavelengths. This wavelength-spreading is known as modulation chirp, and is shown in Figure 20.47 for amplitude-shift-keyed (ASK) (i.e., "ON"–"OFF") modulation of the light.[99] This chirp can have a 3-dB bandwidth as wide as 10 GHz and will limit high-speed and/or long-distance communications due to intermodal dispersion. Furthermore, this wide signal bandwidth will also limit the total number of the multiplexed wavelengths in the WDM systems due to the wide channel spacing.

One method to avoid chirp is to have the laser emit a constant (continuous-wave (CW)) light beam and then externally modulate the light without producing any additive chirp. One type of external modulator is a lithium-niobate MZI[100] and is shown in Figure 20.48. Let us begin with the absolute distance in the two waveguide arms of the interferometer being equal. By placing an electrode over only

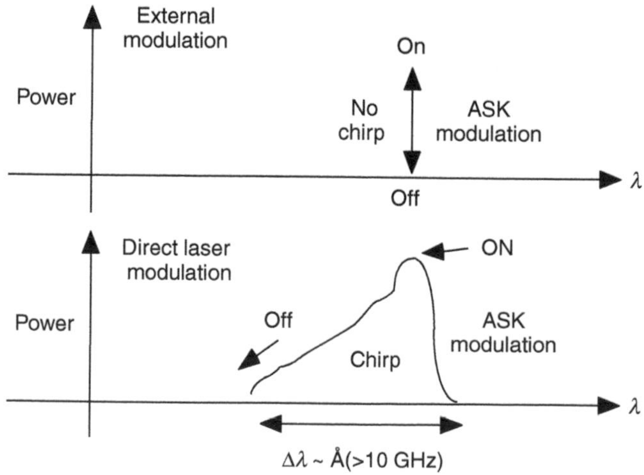

FIGURE 20.47 Optical spectrum showing the chirping of the output wavelength of a directly modulated semiconductor laser.

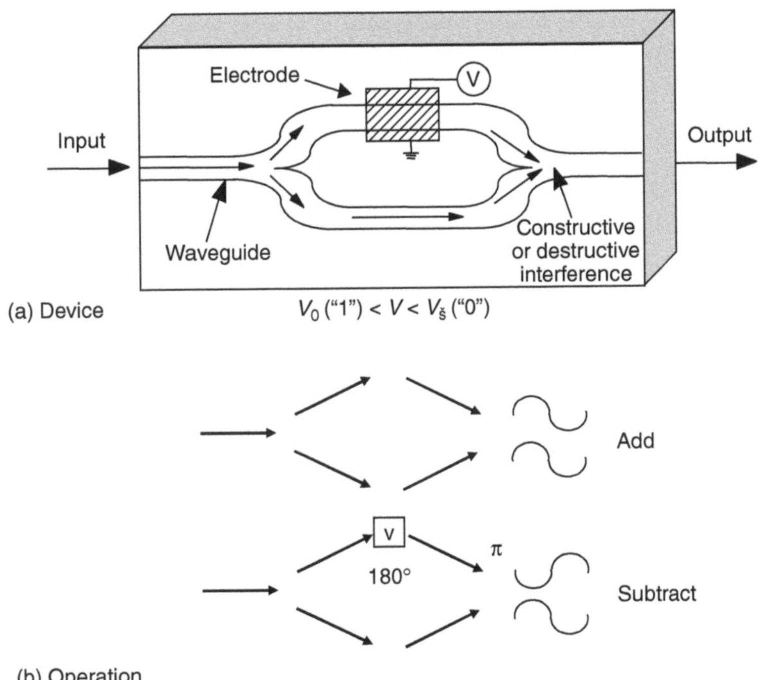

FIGURE 20.48 Mach-Zehnder interferometer acting as an optical modulator: the device itself and its operation.

one of the arms, we can induce a voltage-controlled electric field in that arm which will change the refractive index, retard that wave, and change the phase delay of that arm in relation to the other arm. For a sufficiently high applied voltage, we can induce a π phase shift in one arm, causing the two separate waves to be π out-of-phase when they recombine at the output waveguide. This will result in destructive interference and will cut off the optical transmission at a specific signal wavelength. The ON/OFF transmission characteristics as a function of voltage follow a raised-cosine-squared curve. These modulators can provide a contrast ratio >20 dB, are fairly expensive, and reduce the chirp to a negligible value.

20.2.5.5 Isolators and Circulators

Although an optical fiber itself can accommodate bi-directional optical propagation, many portions of an optical system will require that the light signal only propagate in one direction. This uni-directionality is usually governed by the functions which must be performed by various optical components, i.e., signal generation, detection, filtering, etc. In addition to these basic operations, uni-directional signal propagation is also desired when we wish to severely limit reflections (stimulated Brillouin scattering) which can appear throughout the system and are typically due to slight discontinuities in refractive index at physical interfaces. Most systems require some form of optical isolation which will allow light to pass in one direction but will not allow any reflected light to pass in the other direction. Reflections are detrimental to proper system operation for several reasons:

1. Reflections will slightly change the lasing frequency and threshold current of single-frequency lasers.
2. Reflections impair the ability of the local oscillator to track the incoming signal in coherent detection.
3. If two reflections are close enough so that they are within the coherence length of the signal laser light, then these two reflecting surfaces will establish a self-contained Fabry-Perot cavity with periodic resonances which will modulate the original signal.

4. Reflections will increase the NF of an optical amplifier since the gain-producing carriers are now being used to amplify an undesired reflected wave. Moreover, two strong reflections surrounding the active region of a high-gain optical amplifier may create a cavity and initiate lasing inside the amplifier.

The operation of an optical isolator[44] can be explained by following the light propagation shown in Figure 20.49. Any light beam has a given polarization (i.e., direction of oscillation of the wave amplitude within the plane of propagation), whether it be linearly polarized (such as at the output to a semiconductor laser) or circularly polarized (such as within a fiber). Even circularly polarized light can be decomposed into its two linearly polarized component parts. Let's consider for this simple explanation the case of linear polarization. The isolator consists of 45° Faraday rotator which is surrounded on each side by a polarizer. Each polarizer passes only the light that matches its specific polarization direction. These polarizers are arranged such that the second (i.e., right) polarizer's transmission direction is shifted by +45° in relation to the first (i.e., left) polarizer's transmission direction. The Faraday rotator takes any input optical beam and shifts its polarization by +45° independent of the direction of propagation through the rotator. We can explain the isolator operation by piecing together the above conditions. The left polarizer is aligned so that its transmission polarization direction matches the polarization of a right-propagating optical signal beam, for example 0° to the horizontal plane. Subsequently, the transmitted optical beam passes through the Faraday rotator which shifts the beam's polarization by +45°. The right polarizer is aligned so that its transmission polarization direction matches the polarization of the light emanating from the rotator, i.e., +45°. This original incident light thus passes without attenuation through the isolator. However, if a reflection should occur after the isolator, then this left-propagating reflected wave must pass the isolator in the reverse direction. This is extremely unlikely to happen. Let's assume in the worst case that the reflected wave has a polarization which matches the output right polarizer, thereby passing the reflected wave in the reverse direction. When the reflected wave propagates through the rotator, its polarization is again rotated by

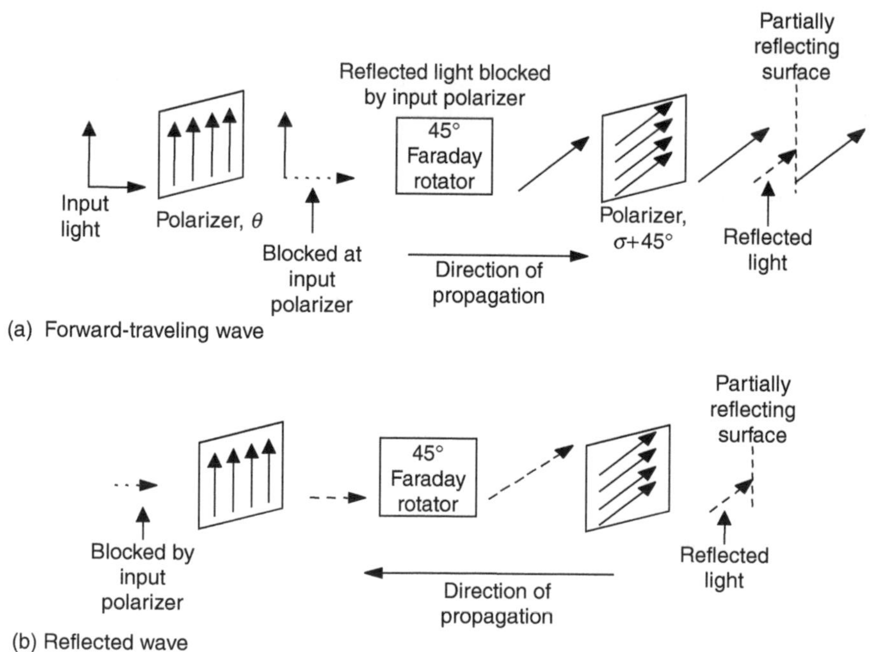

FIGURE 20.49 Illustration of the operation of an optical isolator. The isolator includes polarizers and a 45° Faraday rotator.

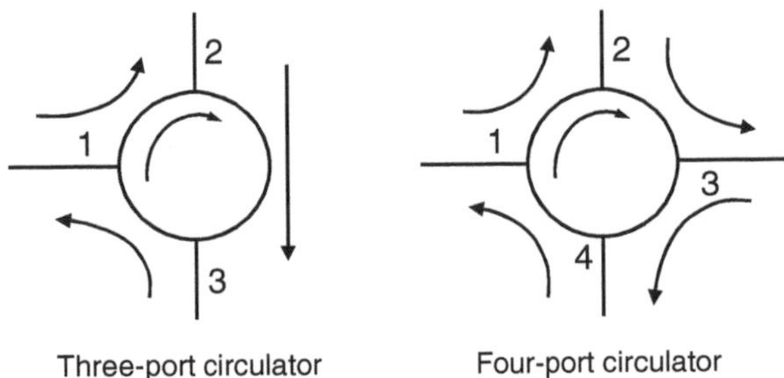

FIGURE 20.50 Functional representation of three-port and four-port optical circulators.

$+45°$, now at $+90°$ to the horizontal. Since the input left polarizer was set to pass only light at 0°, then the reflected wave will be blocked by this polarizer, effectively optically isolating the input side from any reflected wave. In a single isolator, reflections can typically be < -40 dB down from the original signal.

This simple isolator can also be achieved in a polarization-independent fiber-compatible in-line form even though we don't necessarily know the exact polarization of the light inside a fiber. Such isolation is accomplished by separating the incoming wave into its two component polarization parts by using a polarization beam splitter. We then pass each polarization through a separate linear-polarization isolator which function in parallel to isolate the two different polarizations. Then the two light polarizations are combined and coupled into a single output.[44] Generically, separating polarizations and operating on each one independently is known as polarization diversity.

A circulator is similar to an isolator, except that it has multiple ports, typically three or four, as shown in Figure 20.50. In a three-port circulator, input signals on port 1 is sent out on port 2, input signals on port 2 sent out on port 3, and input signals on port 3 is sent out on port 1. Circulators enable separation of signals traveling in opposite directions and are used with the components that operate in reflections, such as Bragg grating filters, thin film filters, and add-drop multiplexers.

20.3 Basic Optical Systems

20.3.1 System Model

A generic optical system can be described in the simple block diagram of Figure 20.51. Since almost invariable the signal origin and destination are in electronic form, the optical medium is simply used to connect the two electrical end points with low loss and high speed. An optoelectronic transmitter and receiver are required to interface the electronic and optical domains. Furthermore, an optical system can

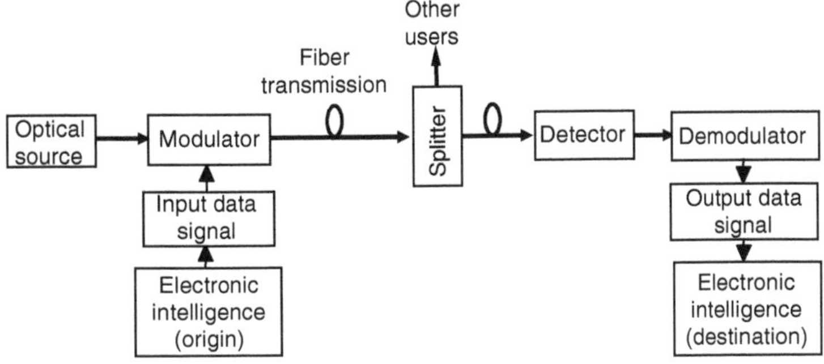

FIGURE 20.51 Block diagram of a generic optical communication system.

be used for transmission or for switching purposes. Unless otherwise stated, fiber will be the medium for optical transmission, although free-space systems will be another option.

In any optical system, the ultimate measure of performance is the SNR of a recovered signal.[101] The higher the SNR when recovering data, the lower will be the probability of error, or bit-error-rate (BER) in a digital system or the loss of fidelity in an analog system. Of course, various types of optical systems will differ in the SNR required for satisfactory system performance. We will devote much discussion throughout this section to find the signal and noise powers in many different optical communication systems in order to analyze the communications channel.

Analog and digital systems are two basic types of communications. Analog can reproduce exactly a given waveform and is compatible with present cable-television systems, but the carrier-to-noise ratio (CNR) required for near-error-free transmission must be extremely high.[102] On the other hand, digital systems are compatible with the way in which computers as well as the modern-telephone-network communicate, and need only distinguish between a “0” and “1” bit.[103] The SNR required is much smaller than in analog, and therefore digital systems are easier to implement and maintain. We devote most of this chapter to digital systems.

20.3.2 Modulation Formats

Information can be classified in two general forms, analog and digital (see Figure 20.52). Analog signals represent quantities which can take on any value within an infinite continuum of amplitude and time. This kind of signal is typical of many measured quantities, such as temperature, volume, frequency, dimensions, etc. Analog signals are, in theory, ultimately accurate. However, to achieve a high-degree of accuracy in signal recovery, the noise must be kept extremely low compared to the signal power (~ -50 dB). Alternatively, digital signals can represent many quantities which may only require measuring of a limited number of discrete levels at specific bit-time intervals. It is only necessary to transmit and detect the difference between discrete levels (i.e., between “0” and “1” levels). Consequently, the noise in relation to the signal that can be tolerated for error-free transmission is much higher than in analog transmission and is typically ~ -20 dB.

We would prefer to transmit digital signals due to the less stringent requirements on the SNR. Even if a signal is originally in analog form, it is fairly straightforward to convert any analog signal into a digital signal and then transmit it. Inherently, we will lose some accuracy since we are taking a continuous-time signal of infinite amplitude levels and forcing it to be represented by a finite number of discrete levels measured at discrete time intervals. The analog signal is periodically sampled at a high-enough sampling rate (i.e., above the Nyquist rate)[104] to adequately recover the highest frequency components in the signal; the highest frequency components in a signal are those portions that experience the fastest transitions in amplitude as a function of time. Each digital amplitude measurement of the analog signal is that value which corresponds most closely to one of the discrete digital levels. Each discrete level can be

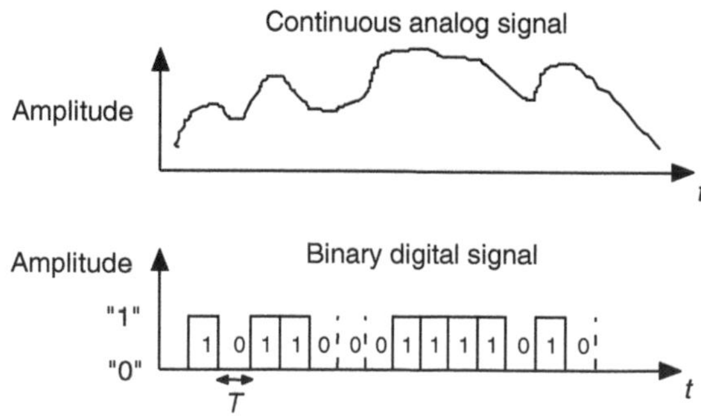

FIGURE 20.52 Analog and digital waveforms.

represented by a string of numbers, typically a binary set of "0"s and "1"s. Note that multi-level digital representation can also be used, but binary is the overwhelmingly popular choice. This is called pulse-code modulation[104] and is represented in Figure 20.53. The more levels and smaller the time interval in sampling the more data bits which must be transmitted per unit time. This means that a higher-bandwidth channel is required to transmit the data. In the limit, infinite bandwidth digital signals correspond to an analog signal. Of course, most quantities do not require such accuracy in representation. Because of the less stringent requirements on digital transmission and reception, we will almost exclusively discuss digital transmission except in Section 20.5, which deals with subcarrier analog transmission.

Although there are several types of modulation coding formats for transmitting binary digital signals, we will discuss only the two most common ones. Figure 20.54 shows non-return-to-zero (NRZ), and return-to-zero (RZ).[103] NRZ is the simplest format in which the amplitude level is high during the bit-time if a "1" is transmitted and is low if a "0" is transmitted. If two or several "1"s are transmitted in sequence, then the level remains high having the result of producing high-frequency components for isolated "1"s and low-frequency components for a string of "1"s. RZ format requires that a "1" always return to the low state during the bit-time even when two "1"s are transmitted in sequence, whereas the "0" bit remains at a low level. This eliminates the possibility of a long string of "1"s producing a constant high level but does not eliminate a long string of "0"s from producing a constant low level. Since NRZ is by far the simplest and most popular format, our discussion will be mainly limited to NRZ format requiring that the system accommodate the full digital signal bandwidth. However, note that RZ format is used for ultra-high-speed pulse transmission and for soliton systems.

We have mentioned the notion of frequency components in a digital signal and the bandwidth which this signal occupies. Figure 20.55a–c show graphically what we mean by frequency components of a digital signal. As is well known, a square pulse in time can be represented in frequency space by taking the Fourier transform. The Fourier transform of the square pulse is a since function and shows that the vast majority of the energy in the pulse resides in the center lobe.[103] However, since an ideal square pulse has

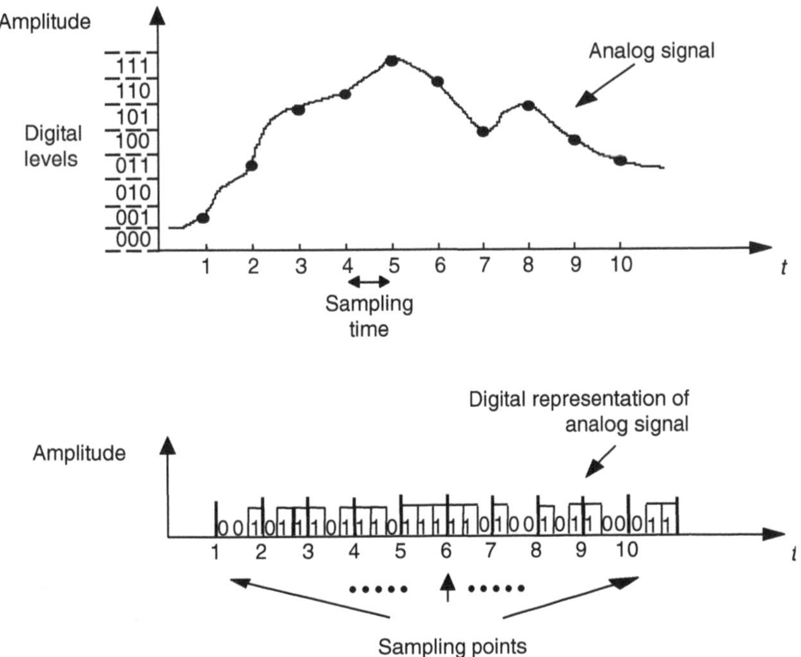

FIGURE 20.53 Pulse-code modulation in which an analog signal can be represented by a string of digital bits.

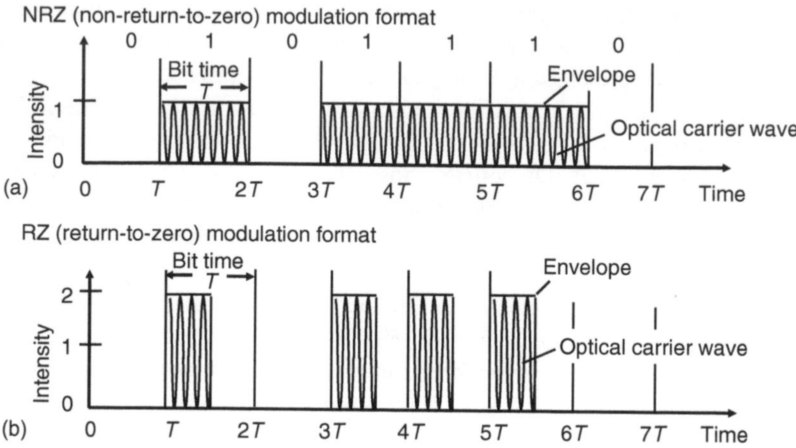

FIGURE 20.54 Modulation formats: NRZ, and RZ. RZ format has higher intensity than NRZ format for a given signal's power due to the narrower pulse width.

an infinitely fast rise and fall transition, some energy of the pulse exists at infinite frequencies; recall that there are infinite frequency components in a delta function. The center of the top flat portion of a pulse represents the lowest frequency components of the pulse since this portion has the slowest transitions. We can retain most of the energy in a pulse and lose the energy in the highest frequencies if we pass a pulse through an electrical low-pass-filter of cutoff frequency which is half the bit rate. In such a case, we have retained only the first term(s) of the infinite Fourier series which is composed of cosines at the discrete harmonics of the signal. The shape of a pulse when passing through such a low-pass-filter becomes rounded at the edges but the center of the pulse still reaches its maximum amplitude level. Since the digital decision of a receiver circuit exists at the center of the bit time, there is little power penalty of decision ambiguity due to this low-pass-filtering. There are two main reasons why we discuss the

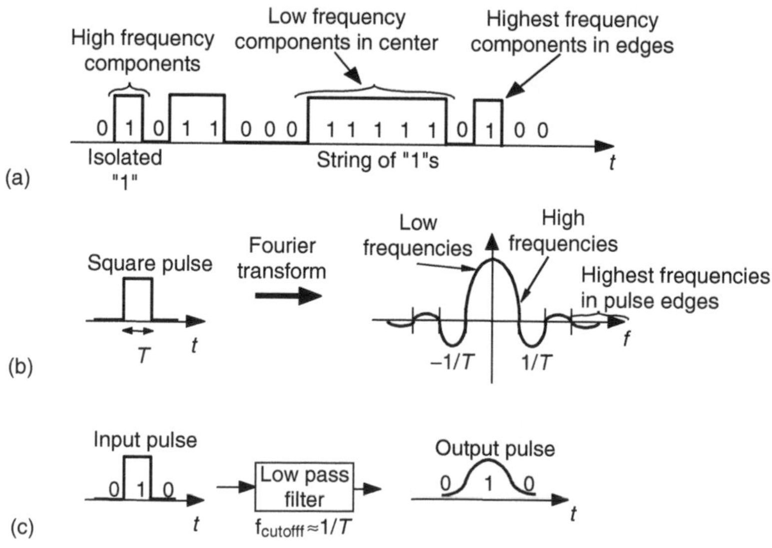

FIGURE 20.55 (a) An illustrative example of the frequency components in a time-based string of bits, (b) a square pulse and its Fourier transform, (c) a square pulse after passing through a low-pass-filter.

low-pass-filtering of a digital signal: (1) the unintentional limiting of the system bandwidth by limited component bandwidth, and (2) the intentional limiting of the bandwidth to limit the excess noise at high frequencies. (We will discuss noise later.)

If we return to consider a random NRZ signal which can have isolated "1"s and strings of "1"s, we see that the highest indispensable frequency components of the data stream, f_H, are in the isolated "1"s and the first and last "1"s in a long sequence. The lowest indispensable frequency components, f_L, are in the center of the longest string of "1"s where the transition rate is the slowest:

$$f_H \approx \frac{1}{T}, f_L \approx \frac{1}{(\text{maximum}\quad \#\text{ of consecutive"1"s}) \times T} \tag{20.28}$$

We will see later that the highest indispensable frequency is approximately $f_H/2$, which corresponds to the Nyquist sampling rate. In order to have good transmission performances, the system must have a uniform response covering all the relevant frequency components in the digital bit stream. Note that the transmitted optical spectrum of an intensity-modulated random NRZ signal has a certain information full-width half-maximum (FWHM) frequency bandwidth which is equal to roughly the bit-rate, R_b.[103]

The digital signal, which may be modulated at approximately Gbit/s rates, is being transmitted on an optical carrier wave whose frequency is in the multi-THz regime. The reason for using the optical carrier is that the optical wave has very low losses in the fiber whereas lower microwave frequencies have extremely high transmission losses in an optical fiber (or in a coaxial cable at Gbit/s speeds). This optical carrier wave, $A(t)$, has an intensity amplitude, A_0, an angular frequency, ω_c, and a phase ϕ:[105]

$$A(t) = A_0 \cos(\omega_c + \phi). \tag{20.29}$$

A binary digital signal implies transmitting two different quantities of anything which can subsequently be detected as representing a "1" and "0", i.e., we can transmit blue and red, and this can represent "1" and "0" in the receiver electronics if blue and red can be distinguished. We can therefore modulate the amplitude, frequency, or phase of the optical carrier between two different values to represent either a "1" or "0", known respectively as amplitude-, frequency-, and phase-shift keying (ASK, FSK, and PSK), with the other two variables remaining constant:

$$A_{ASK}(t) = (A_0 + m(\Delta A)) \cos(\omega_c t + \phi), \quad m = \begin{cases} +1, \text{"1"} \\ -1, \text{"0"} \end{cases} \tag{20.30a}$$

$$A_{FSK}(t) = A_0 \cos([\omega_c + m(\Delta\omega)]t + \phi) \tag{20.30b}$$

$$A_{PSK} = A_0 \cos(\omega_c t + [\phi + m\pi]) \tag{20.30c}$$

where ΔA is the amplitude modulation and is less than A_0, and $\Delta\omega$ is the FSK frequency deviation. ASK has two different light amplitude levels; FSK has two different optical carrier wavelengths; and PSK has two different phases which can be detected as an amplitude change in the center of the bit time for which a "1" or "0" bit can be determined. It is important to emphasize that the differential-phase-shift-keying (DPSK) format, in which the phase of the preceding bit is used as a relative phase reference, has been reemerged in the last few years due to its less OSNR requirement and robustness to fiber nonlinearities.[106–108] The DPSK modulation signal is not the binary code itself, but a code that records changes in the binary stream. The PSK signal can be converted to a DPSK signal by the following rules: a "1" in the PSK signal is denoted by no change in the DPSK, and a "0" in the PSK signal is denoted by a change in the DPSK signal. For a DPSK signal, optical power appears in each bit slot, and can occupy the entire bit slot (NRZ–DPSK) or can appear as an optical pulse (RZ–DPSK).

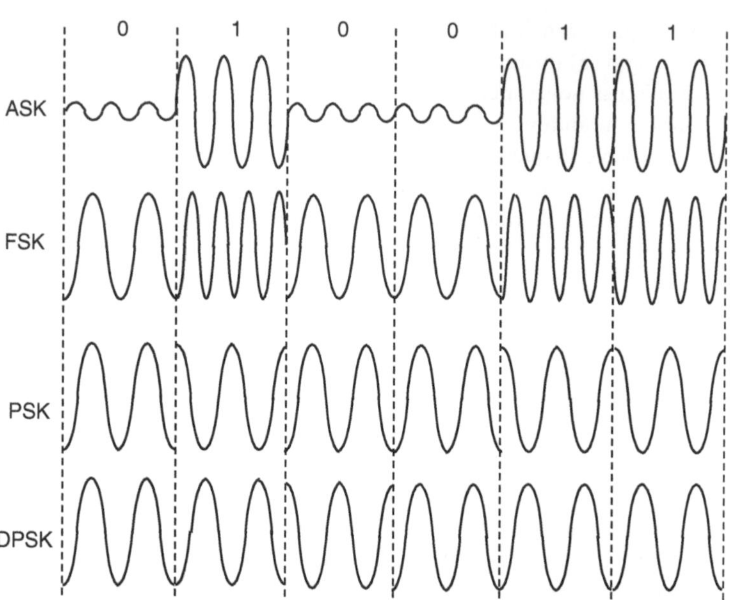

FIGURE 20.56 ASK, FSK, and PSK (DPSK) time modulation while employing an optical carrier wave.

Figure 20.56 shows the impression of a simple digital signal on the optical carrier. These three formats can be implemented by appropriately changing the optical source, whether by modulating the light amplitude, laser output wavelength, or using an external phase shifter. ASK is important since it is the simplest to implement, FSK is important because a smaller chirp is incurred when direct modulation of a laser is used, and PSK is important because, in theory, it requires the least amount of optical power to enjoy error-free data recovery.[6] It should be mentioned that ASK, which is by far the most common form, is called On–Off-Keying (OOK) if the "0" level is really at zero amplitude. Note that the frequency of the optical carrier is so high that the optical detector, whose electronic bandwidth is usually $<<100$ GHz, will not detect it and only the envelope of the Gbit/s data will be electrically recovered.

20.3.3 Detection Schemes

Once a signal has been modulated and transmitted, it must be accurately detected. Two detection schemes include direct detection and heterodyne (i.e., coherent) detection.[109,110] Direct detection can be used with ASK and FSK modulation, and ASK, FSK, and PSK can be used with heterodyne detection. Note that homodyne coherent detection is also possible, but is not considered at present to be the detection scheme of choice due to its relative difficulty of implementation.[111,112] Due to the limited space in this chapter, we will leave homodyne detection to the reader to examine the reference material.

Direct detection, which is by far the simpler of the two schemes, involves detecting the amount of optical power incident on the optical detector. Direct detection of an ASK (or OOK) signal, i.e., light "ON" or light "OFF," is extremely simple to accomplish using a detector and a high-bandwidth power meter (see Figure 20.57a). This ASK signal is recovered by a detector of a certain electrical low-pass-filtering bandwidth. The electrical spectrum of the recovered ASK signal, which is sent in random NRZ format and can be measured on an RF spectrum analyzer, is shown in Figure 20.57b. As mentioned in the previous section, only the first lobe is necessary for recovering the data since only the transition edges would be affected by cutting off the higher lobes. The first lobe is considered the baseband signal representing the data stream.[103]

Directly detecting an FSK signal is a little more complicated since direct detection can only measure optical power but cannot resolve the incoming optical wavelengths which are at carrier frequencies that

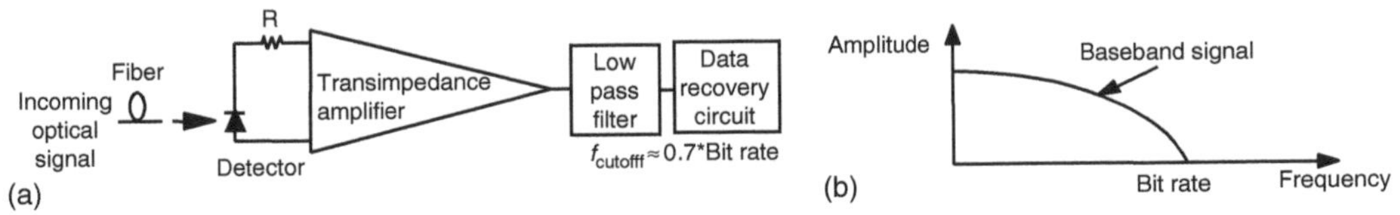

FIGURE 20.57 (a) Direct detection optical system, (b) baseband signal of a directly detected NRZ signal.

are much higher than the detector bandwidth. Direct detection of an FSK signal can be facilitated by using an optical filter to convert the FSK signal into ASK format before detection.[89] To see how this is achieved, Figure 20.58 shows the optical spectrum of an FSK signal. An FSK spectrum is composed of two separate wavelengths, each one being "ON" when the other one is "OFF." Given FSK with a wide deviation between the two wavelengths, each spectrum appears like an independent ASK signal, i.e., instead of blue-red-blue it can be thought of as two separate signals of blue-"OFF"-blue and "OFF"-red-"OFF." If we tune an optical filter to pass only the wavelength representing the "1" bits and block the wavelength for the "0" bits, the output signal appears to be a single ASK signal. Therefore, a narrow optical filter placed before the detector effectively converts an FSK signal into ASK format which is then straightforward to detect. Note that the optical filter has a Lorentzian-lineshaped passband, blocking most of the optical power from the rejected wavelength but allowing some of its optical power to pass. This unwanted optical power will raise the "0" level from zero amplitude to some finite value, thereby reducing the "1"-to-"0" contrast ratio and increasing the system power penalty.

Since photo-detector is inherently insensitive to the optical phase, a detector only converts the optical signal power into an electrical signal. Directly detecting PSK signal is impractical due to the lack of an absolute phase reference. However, we can detect the DPSK signal using a one-bit delayed MZI followed by a balanced direct-detection system. A DPSK signal records the phase changes in the binary stream. Thus the demodulator only needs to determine these changes in the coming signal phase. A typical balanced DPSK receiver is shown in Figure 20.59. The optical DPSK signal is first sent to a MZI with the one-bit period differential delay between the two arms. The MZI lets two adjacent bits interfere with each other at its output ports. This interference leads to the presence (absence) of power at a MZI output port

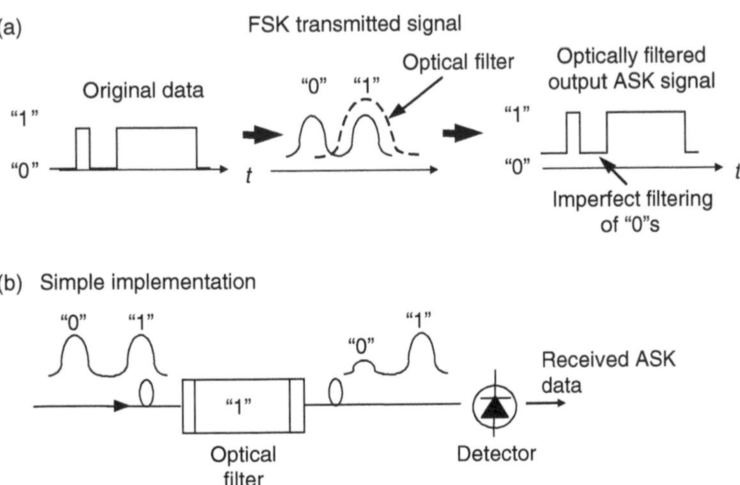

FIGURE 20.58 (a) FSK-to-ASK conversion. The illustration shows the optical spectrum of a transmitted FSK signal and the resultant ASK signal after passing through an optical filter, (b) the hardware implementation showing an optical filter and direct detection.

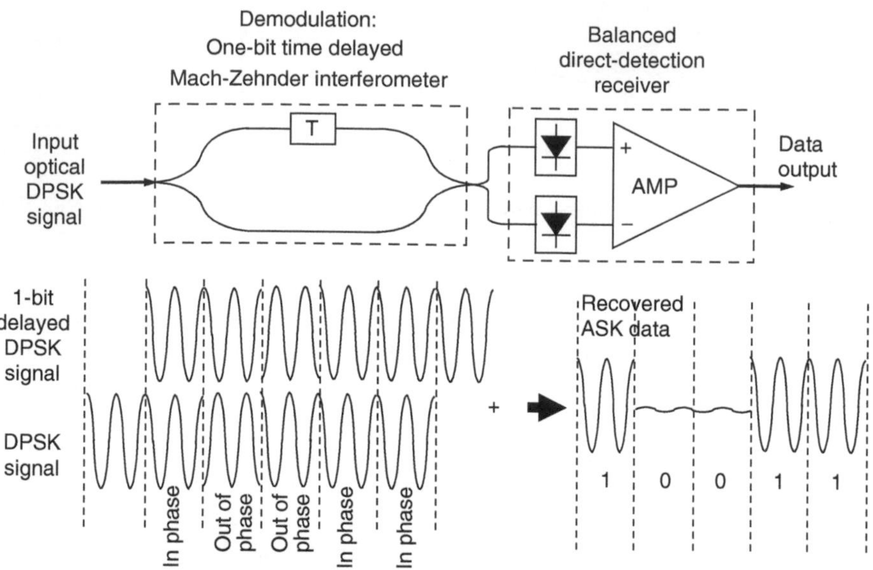

FIGURE 20.59 A typical DPSK receiver.

if two adjacent bits interfere constructively (destructively) with each other. Thus, the preceding bit in a DPSK-encoded bit stream acts as the phase reference for the current bit. No phase change between these two bits will generate a "1" (or "0") at the constructive port (or destructive port). Recall the generation of DPSK from PSK signals, this process is exactly the reverse process. Ideally, one of the MZI output ports is adjusted for destructive interference in the absence of phase modulation ("destructive port"), while the other output port then automatically exhibits constructive interference due to energy conservation ("constructive port"). For the same reason, the two MZI output ports will carry identical, but logically inverted data streams under DPSK modulation.[106]

Using the balanced direct-detection scheme, the DPSK system has the advantage of requiring a lower optical signal-to-noise ratio (OSNR) than OOK to reach a given bit-error ratio (BER). Intuitively, this can be understood by comparing the signal constellations for DPSK and OOK.[103] To achieve the same symbol distance, the average optical power in DPSK is only half as compared to OOK. At 40 Gbit/s, a sensitivity of about 38 photons/bit has been reported using RZ–DPSK.[113] This is approximately 3 dB better than the best OOK results of 78 photons/bit.[114] The lower OSNR requirement of DPSK can be used to extend transmission distance, reduce optical power requirements, or relax component specifications.

It is important to mention that the sensitivity of a direct detection receiver can be improved significantly by using a low noise optical preamplifier that is just before the photodetector. The output of the optical amplifier needs to be sufficiently high so that at the photodetector the noise is dominated by the signal-spontaneous beat noise of the optical preamplifier. This will be discussed further in the next section.

Heterodyne detection is more difficult to accomplish but is considered a more sensitive detection scheme.[111] Heterodyne detection relies on the ability of two waves to "mix" with each other when they are both of the same polarization and incident on a detector. Since an optoelectronic detector is a p–n junction which follows a "square law" between input power and generated current, the detector itself mixes two independent optical waves. The input optical field is squared by the detector, generating square terms and cross-term products. Figure 20.60 shows the combining of two optical waves into a single detector by using a partially-reflecting beam splitter; this function can also be performed by a simple fiber directional coupler. The two optical waves are (1) the modulated transmitted signal, $E_s(t)$; and (2) a CW local oscillator, E_{LO}. By invoking trigonometric relationships, the generated output photocurrent from

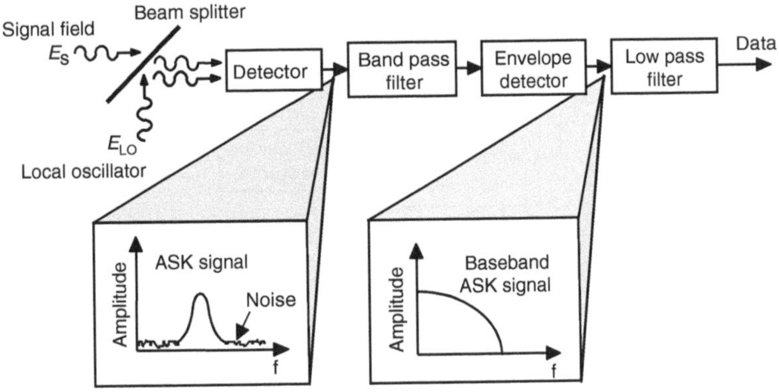

FIGURE 20.60 Heterodyne detection of an incoming signal.

the detector is[115]

$$\begin{aligned} I_{ph} &= (E_s(t) + E_{LO})(E_s(t) + E_{LO})^* \\ &= E_s^2(t) + E_{LO}^2 + 2|E_s||E_{LO}|\cos(\omega_s t)\cos(\omega_{LO} t) \\ &= P_s(t) + P_{LO} + \sqrt{P_s P_{LO}}[\cos([\omega_s + \omega_{LO}]t) + \cos([\omega_s - \omega_{LO}]t)] \end{aligned} \quad (20.31)$$

where the superscript * denotes the complex conjugate of the waves, E_s is the signal electric field and equals $(|E_s|\cos(\omega_s+\phi))$, E_{LO} is the signal electric field and equals $(|E_{LO}|\cos(\omega_{LO}+\phi))$, P represents optical power (or the square of the electromagnetic field), P_{LO} is the DC (or CW) local oscillator power, and P_s is the modulated signal power. We have neglected to write the phases of the signal and local oscillator since we consider them to not change significantly relative to each other over a short time. This requirement means that we must have a narrow linewidth source and narrow linewidth local oscillator to maintain coherence. We can simplify this equation when considering that (1) the local oscillator power is DC and can be filtered out, (2) the local oscillator power can and will be made much greater than the incoming weak signal which has been transmitted over some distance allowing us to eliminate the P_s^2 term, and (3) the term which includes the sum of the two frequencies can be eliminated since the sum frequency is so high as to be unrecoverable by a detector of limited bandwidth or can be easily filtered out in relation to the difference-frequency term. The simplified equation is then:

$$I_{ph} = \sqrt{P_s P_{LO}}\cos([\omega_s - \omega_{LO}]t). \quad (20.32)$$

Instead of direct detection which can achieve a recovered signal power of P_s, we can adjust the local oscillator power in heterodyne detection to be as high as possible, with the effective signal power recovered being increased to the square-root of $P_s P_{LO}$. Our sensitivity has therefore been increased significantly. Although we can increase the local oscillator power to increase the effective signal power, there is a limit to its effectiveness since we will be limited at some point by the high-power-induced shot noise; as the signal power increases, the noise power will increase at the same rate thereby saturating the SNR. Shot noise will be discussed further in the next section.

The center of the recovered "beat" signal's electrical spectrum will appear at the frequency difference between the signal and local oscillator. Therefore, the local oscillator must be wavelength tunable so that the recovered electrical signal can be placed at an intermediate frequency which is within the bandwidth of the detector. Note that even a 1 nm separation corresponds to a beat frequency of $>$ 100 GHz and will not be recovered by the detector. Figure 20.60 shows the full heterodyne receiver which includes (1) an electrical band-pass-filter following the photodetector to allow only the recovered electrical signal to pass and block other channels and other noise-containing frequencies, (2) an electrical envelope detector to

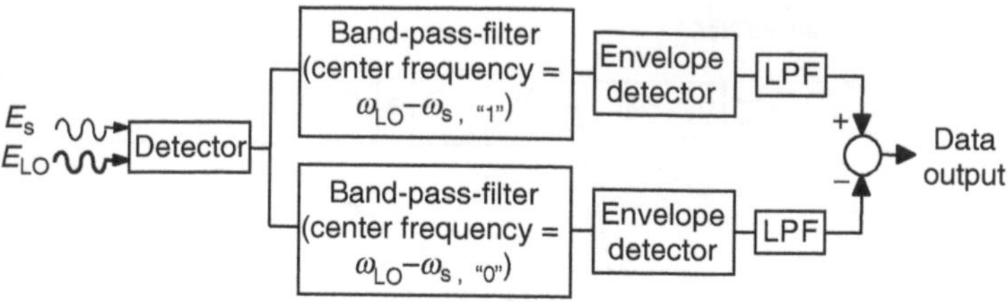

FIGURE 20.61 Heterodyne recovery of an FSK signal.

recover only the data modulation which will then appear as a baseband electrical spectrum, and (3) a low-pass-filter to limit the high frequency noise and recover only the data.[112]

Heterodyne detection can also be used to detect FSK and PSK signals.[110] Figure 20.61 shows the recovery of an FSK signal by heterodyning. The two wavelengths representing the "1" and "0" bits are detected by the photodetector and appear at two different locations in the electrical spectrum corresponding to the frequency difference between that bit's wavelength and the local oscillator wavelength. The electrical band-pass-filter will then discriminate between the two beat frequency signals, passing the "1" bit difference frequency and rejecting the "0" bit difference frequency, and then the envelope detector can easily recover the resultant ASK digital signal.

For PSK signal, what is necessary is the ability to detect the phase of the wave, 0 or π, by mixing the signal with a local oscillator. The difference in these two phases corresponds to a wave whose electric field is either positive or negative in the center of the bit where the digital decision occurs. We can repeat the formalism for PSK heterodyne detection as was done with ASK heterodyne detection. Since we have a generated photocurrent for the "1" and "0" bits which can both be used for data recovery, the "effective" detected photocurrent, I_{PSK}, is:[18]

$$I_{PSK} = 4\sqrt{P_s P_{LO}}\cos(\phi_s - \phi_{LO}). \tag{20.33}$$

A detection circuit must be used to distinguish the two amplitude levels. Note that data recovery is extremely sensitive to the phase of the signal and local oscillator, and to these two phases remaining constant relative to each other. As was mentioned in Section 20.2, the laser has a finite linewidth due to quantum fluctuations in the stimulated emission. This linewidth will cause the phase to wander and make it impossible to recover a signal. Therefore, PSK requires extremely narrow linewidth lasers, much narrower than are needed for even FSK heterodyne detection.

It is important to briefly mention an important issue with heterodyne reception. The polarization of the signal and local oscillator must match. This is difficult since the polarization of light in a fiber is difficult to control and will wander and drift. Therefore, a polarization diversity receiver can be employed in which the two signal polarizations are separated at the receiver input and then separately detected in two parallel heterodyne detectors whose local oscillator polarizations are 90° out of phase with respect to each other (see Figure 20.62).[116]

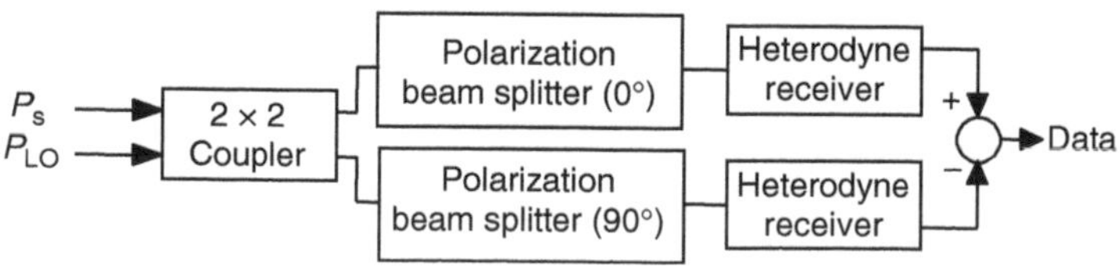

FIGURE 20.62 A polarization-diversity heterodyne receiver.

Although different systems exist, the majority are NRZ–ASK using direct detection. Unless otherwise stated, we will be discussing this generic type of system. Furthermore, due to the brevity of this single chapter, we will not be able to discuss the other types of systems in much more detail except where particularly necessary.

20.3.4 Signal and Noises

The signal power is defined as the mean power in the modulated signal, and the noise power is the statistical variance in the modulated signal.[101] In order to detect a digital bit and accurately decide if it is a "1" or "0" bit, the signal power must be much larger than the noise power. If the noise is too large, then a false "0" or false "1" may be detected. The SNR is considered an indispensable quantity when attempting to evaluate the performance of a system.

The electrically measured optical signal power mean is directly related to the square of the generated photocurrent in a photodetector:[39]

$$S = I_{\text{ph}}^2 = \left(\frac{\eta q P_s}{h\nu}\right)^2 \tag{20.34}$$

The signal power is proportional to the amount of power available for a decision circuit to decide if a "1" or "0" bit was transmitted. Therefore, the effective signal power is the high level in a "1" bit in relation to the low level in a "0" bit; this relationship of ($P_{\text{"1"}}/P_{\text{"0"}}$) is known as the contrast ratio and optimally equal to infinity (see Figure 20.63a). If OOK modulation is used in which the "0" bit is transmitted with 0 power, then Equation 20.48 describes the signal power level. In a case in which the light was not completely turned "OFF" during a "0" bit, then there is a reduction in the contrast ratio between the "1" and "0" bits. Such a reduction affects the effective signal power as a subtraction:

$$S = (P_{\text{"1"}} - P_{\text{"0"}})^2 \tag{20.35}$$

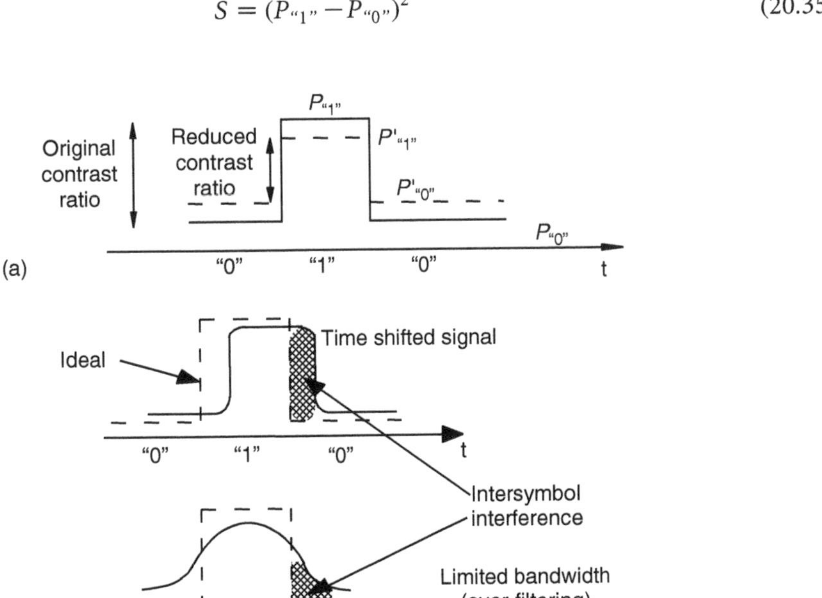

FIGURE 20.63 (a) Contrast ratio of a digital signal, (b) two generic and simple examples of intersymbol interference in a digital stream of bits.

A reduction in the contrast ratio from an increase in the "0" level can arise from several sources, with three of the most common being (1) a non-zero power level when transmitting a "0" bit; (2) optical power from another source incident on the detector, as is the case when another channel at another wavelength is not completely blocked by an optical filter; and (3) rejected wavelength in an FSK signal leaks through the optical filter when using direct detection to perform FSK-to-ASK conversion.[112]

The effective signal power can also be reduced by simple attenuation of a "1" bit due to any number of types of losses (i.e., fiber attenuation, off-peak transmission through an optical filter, insufficient bandwidth in the signal source or detector, etc.).

Another fairly common source for reducing the contrast ratio is called intersymbol interference and is shown graphically and in simple form in Figure 20.63b. Intersymbol interference essentially means that a bit transmitted in one time slot will have some of its energy appear in an adjacent time slot, potentially creating a higher power level for a "0" bit and a lower power level for a "1" bit; this will reduce the signal contrast ratio. This interference can be caused by several mechanisms, but some common causes include (1) insufficient system bandwidth which introduces phase shifts for the bits whose frequency components are outside of the system 3-dB bandwidth, (2) system dispersion (especially due to fiber) which spreads the pulse width and causes a pulse's power to appear in the two adjacent bit times, and (3) system nonlinearities which cause a power-dependent bit transmission velocity.[6]

A reduction in the effective signal power, either by a lowering of the "1" bit level or a raising of the "0" bit level, causes a system power penalty, PP_{CR} (dB), due to a reduced contrast ratio. Note that a system power penalty is the increase in the amount of signal power necessary to unambiguously transmit a data bit given an otherwise well performing system. PP_{CR} is[89]

$$PP_{CR} = 10 \log \left[\frac{P_{“1”} - P_{“0”}}{P'_{“1”} - P'_{“0”}} \right]^2 \tag{20.36}$$

where $P'_{“1”}$ represents the lowered value detected in the "1" bit due to losses and $P'_{“0”}$ is the raised value detected in the "0" bits due to incomplete suppression of the "0" bit or because of intersymbol interference. Unless otherwise stated, we will consider P_S as the effective signal power and that there $P'_{“0”}$ is ~ 0.

The noises that are generated in a typical system are different for direct detection and for heterodyne detection. We will discuss each detection method separately. Furthermore, noise in a p–i–n detector is different than an APD detector. For brevity, we will only discuss the p–i–n detector with the comment that, although current is multiplied in an APD, the noise is also increased.

We consider all the noise terms to be the power in the signal variance. Since these noise terms represent independent incoherent statistical variations in the detector current, we can decouple each term and simply add them together in a total noise term, σ_{tot}^2. The three noise terms in simple direct detection are the shot noise, thermal noise, and circuit noise. Circuit noise is simply variances of electron generation in the receiver circuit. However, due to advances in circuit design, we typically assume that the thermal noise dominates the circuit noise. Retrieving the equations for the thermal and shot noises from Section 20.2 on photodetection, σ_{tot}^2 is[38]

$$\sigma_{tot}^2 = \left[\frac{4kT}{\Omega} + 2q\left(\frac{q}{h\nu}\right) P_s \right] B_e. \tag{20.37}$$

Typically, thermal noise dominates over the shot noise in direct detection, giving rise to the term thermal-noise-limited system reception. These noise powers increase with bandwidth and cover all possible frequencies, and the noise powers are limited only by limiting the frequencies which are supported by the system. The receiver bandwidth must be at least 50–70% of the bit rate in order to adequately recover the transmitted bits without incurring a power penalty. Recall that a receiver bandwidth of approximately half the bit rate will smooth the bit transitions but the center of each bit, where the digital decision is performed, is relatively unaffected. In a well-operated system, any higher bandwidth would not increase the recovered signal much but would allow more noise to be recovered. The receiver will almost always have a low-pass-filter to limit the unwanted high-frequency noise. For example, a 10 Gbit/s signal would require twice the electrical bandwidth as a 5 Gbit/s signal for detection. Therefore, twice the noise is produced and twice the signal power is necessary to achieve the same SNR, thereby incurring a 3-dB system decrease in sensitivity. When combining all the above, the SNR (in decibels (dB)) for direct detection is:

$$\mathrm{SNR} = 10\log\left[\frac{(P'_{\text{“1”}} - P'_{\text{“0”}})}{\left[\frac{4kT}{\Omega} + 2q\left(\frac{q}{h\nu}\right)P_{\mathrm{s}}\right]B_{\mathrm{e}}}\right]. \tag{20.38}$$

If an optical preamplifier is used in the direct detection receiver, we have to consider other noise sources in addition to the shot noise, σ^2_{sh}, and the thermal noise, σ^2_{th}. Recall in Section 20.2, we discussed the NF of an optical amplifier (see Figure 20.31). Since the detector is inherently a square-law device which responds to the intensity (i.e., square) of the incoming optical field, a "beat" term is produced if two different optical waves are incident, $A(t)$ and $B(t)$. Squaring of $(A+B)$ produces A^2 plus B^2 plus the beat term of $2AB$; based on trigonometric identities, the beat term includes the cosine of the sum and difference frequencies between $A(t)$ and $B(t)$. Since the frequencies in the above equation are ultra-high optical frequencies (THz) unable to be detected by the photodetector, the sum frequency term will not be detected and only the difference frequency will appear at the electrical output, as shown in Figure 20.64. The two waves in our system impinging on the detector are the signal and the ASE noise. Therefore, A^2 represents the signal power, B^2 represents the ASE noise power, and $2AB$ represents the signal-spontaneous electrical beat noise. However, the situation is more complicated since the ASE does not exist at a specific wavelength but is broadband consisting of an infinite number of incoherent waves each at a different frequency within the gain spectrum. Therefore, we must integrate over the entire ASE noise passing through the optical and electrical filters, and then beat (i.e., multiply) each thin bandwidth "slice" of ASE with the approximately single-frequency signal term $\vec{P}_{\mathrm{sig}}\vec{N}_1 + \cdots + \vec{P}_{\mathrm{sig}}\vec{N}_{\mathrm{sig}} + \cdots + \vec{P}_{\mathrm{sig}}\vec{N}_\infty$. The resulting signal-spontaneous beat noise, $\sigma^2_{\mathrm{sig-sp}}$, which falls within the optical filter and electrical detector bandwidths (low frequency term: $\vec{P}_{\mathrm{sig}}\vec{N}_{\mathrm{sig}}$), is then[118]

$$\sigma^2_{\mathrm{sig-sp}} = 4q\left(\frac{q}{h\nu}\right)P_{\mathrm{sig}}G(G-1)n_{\mathrm{sp}}B_e. \tag{20.39}$$

The B^2, or ASE noise power, term also must be evaluated since it is not at a single frequency or phase and therefore produces beat terms between one part of the ASE spectrum and another: $\vec{N}_1\vec{N}_2 + \vec{N}_1\vec{N}_3\cdots + \vec{N}_1\vec{N}_\infty + \vec{N}_2\vec{N}_3\cdots + \vec{N}_2\vec{N}_\infty + \cdots$. After integration and convolution, the spontaneous–spontaneous

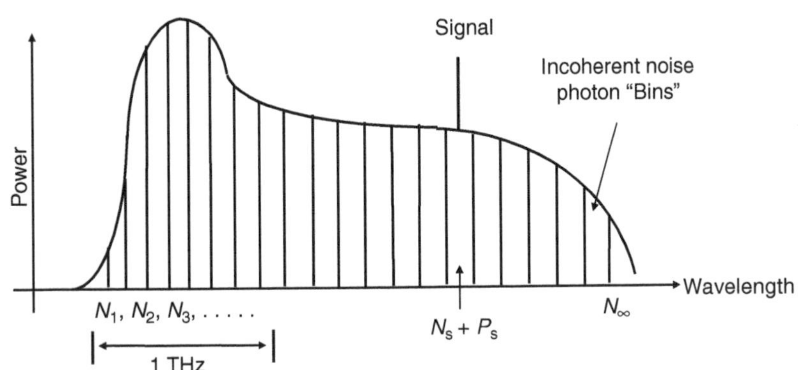

FIGURE 20.64 The noises calculation for a direct detection receiver with an optical preamplifier.

electrical beat noise, $\sigma^2_{\text{sp–sp}}$, is[118]

$$\sigma^2_{\text{sp–sp}} = 4q^2(G-1)^2 n_{\text{sp}}^2 B_e B_o. \tag{20.40}$$

Note that $\sigma^2_{\text{sig–sp}}$ and $\sigma^2_{\text{sp–sp}}$ can be reduced by small optical and electrical filter bandwidths but cannot be eliminated since some ASE must pass the through the filters within the same bandwidth as the signal (see Figure 20.65). Additionally, the majority of the signal-spontaneous beat noise is generated from the small frequency portion (within a few GHz) of the ASE immediately surrounding the signal and so cannot be reduced significantly.

The SNR for the direct-detection receiver with an optical preamplifier is:

$$\text{SNR} = \frac{(GP_{\text{sig}})^2}{\sigma^2_{\text{sh}} + \sigma^2_{\text{th}} + \sigma^2_{\text{sig–sp}} + \sigma^2_{\text{sp–sp}}}. \tag{20.41}$$

We simplify this expression by making three generally valid assumptions: (1) the shot noise is small in comparison to all other terms, (2) the receiver noise will be dominated by the ASE noise and not by the thermal noise (typically true for systems employing a pre-amplifier before the detector), and (3) $G \gg 1$. After making the appropriate substitutions, the SNR can be approximated as

$$\text{SNR} = \frac{P^2_{\text{sig}}}{[2P_{\text{sig}} n_{\text{sp}} h\nu + 2n^2_{\text{sp}}(h\nu)^2 B_o]2B_e}. \tag{20.42}$$

Given all our assumptions, the SNR doesn't change much with an increase in gain since the thermal noise has been overwhelmed by the ASE-generated terms and any higher gain will increase the signal and noise at nearly the same rate as shown in Figure 20.66.

The SNR for heterodyne coherent detection is different than direct detection because now a local oscillator is impinging on the detector at the same time as the signal. As outlined in the previous section, the effective signal power has been enhanced by the local oscillator power:[115]

$$S = P_s P_{\text{LO}}. \tag{20.43}$$

Note that a factor of 2 would be inserted into the right side of Equation 20.43 if a balanced heterodyne receiver was used. The thermal noise is independent of optical power and therefore is not changed due to a local oscillator in heterodyne detection. However, the shot noise is proportional to the total absorbed optical power and is now much larger due to the local oscillator:[115]

$$\sigma^2_{\text{SL}} = 2q\left(\frac{q}{h\nu}\right)\left(\sqrt{P_s P_{\text{LO}}} + P_s + P_{\text{LO}}\right)B_e. \tag{20.44}$$

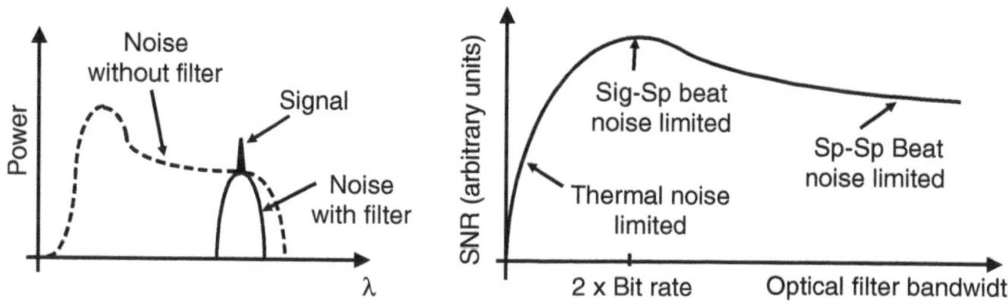

FIGURE 20.65 SNR dependence on optical filtering: for best SNR, must reduce ASE-ASE beat noise to be below SIG-ASE beat noise at the receiver.

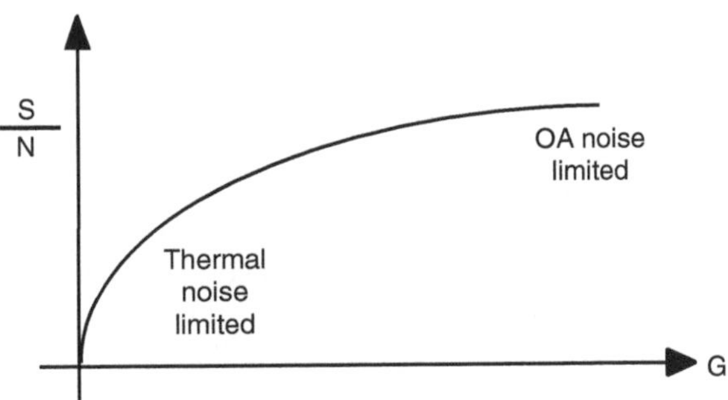

FIGURE 20.66 SNR dependence on amplifier's gain.

Therefore, given that P_{LO} is much greater than P_s, the SNR for a coherent system can be approximated as:[7]

$$\mathrm{SNR} = \frac{2P_s P_{LO}}{\left[2qP_{LO} + \frac{4kT}{\Omega}\right] B_e}. \tag{20.45}$$

We can make the effective signal power as large as we want by increasing the local-oscillator power. However, the shot noise power also increases when the local oscillator power is increased. When the total noise is dominated by the shot noise, the SNR does not continue increasing with increasing P_{LO}. The receiver is then shot-noise-limited, which is the optimal region for achieving the highest SNR in a heterodyne detection system.

20.3.5 Performance Criteria

The undisputed criterion for system performance is determining whether a data bit was correctly recovered as a "1" or "0". This is reflected in determining (and minimizing) the BER of a system. Figure 20.67 shows the voltage or current levels generated from the recovered signal for a "1" and a "0" bit. The received noise has the affect of adding or subtracting electrons when either an optical pulse was transmitted or when no pulse was transmitted. The noise has an approximate Gaussian amplitude

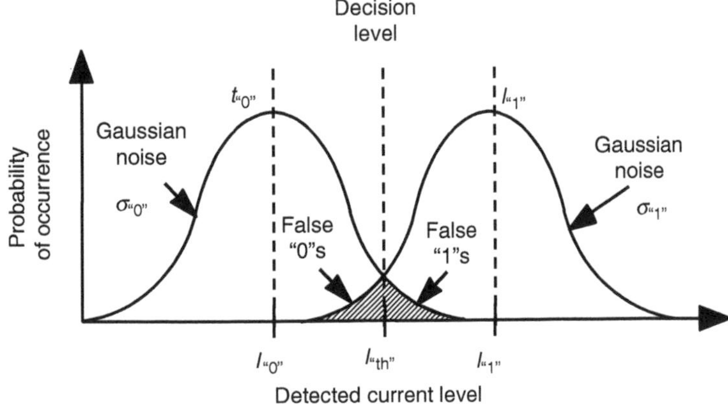

FIGURE 20.67 The probability of current detection when considering the recovered signal level and the Gaussian noise.

distribution around the expected signal value. This represents the variance in the expected number of generated electrons. A decision circuit must set a threshold current level, I_{th}, above which is considered a "1" bit and below which is considered a "0" bit. If we assume for simplicity that the system is thermal-noise-limited and the noise is roughly equal on the "1" and "0" bits, then the decision level will be roughly halfway between the two levels. Note that the noises using an APD receiver are not equal for a "1" and "0" since the APD noises are dependent on the input signal power. Due to the Gaussian noise distribution around a signal mean, a recovered "0" bit will periodically appear above the threshold level with a finite probability and a recovered "1" bit will appear below the threshold level also with a finite probability. These two scenarios represent a false "1" and false "0," respectively, and are considered errors in the system. The probability of error in the system, $P(e)$, is a sum of these two possible errors,[104]

$$P(e) = P(\text{“1”})P(e|\text{“1”}) + P(\text{“0”})P(e|\text{“0”}), \tag{20.46}$$

where $P(e|\text{“1”})$ and $P(e|\text{“0”})$ are the probabilities of error given a "1" and a "0" are transmitted, respectively, and $P(\text{“1”})$ and $P(\text{“0”})$ are the probabilities that a "1" and a "0" had originally been transmitted, respectively.

The key to deriving the BER, or probability of error, is to find the probability of a "0" being above I_{th} and a "1" being below I_{th}. All we need to do is to integrate the probability distribution of the noise from I_{th} to ∞ and from $-\infty$ to I_{th} for the "0" and "1" bits, respectively:[104]

$$P(e) = \left[\frac{1}{\sqrt{2\pi\sigma^2_{\text{“1”}}}}\int_{-\infty}^{I_{th}} e^{-(I-I_{\text{“1”}})^2/2\sigma^2_{\text{“1”}}}\,dI\right] + \left[\frac{1}{\sqrt{2\pi\sigma^2_{\text{“0”}}}}\int_{I_{th}}^{\infty} e^{-(I-I_{\text{“0”}})^2/2\sigma^2_{\text{“0”}}}\,dI\right]. \tag{20.47}$$

Such an integration involves integrating the Gaussian noise distribution. By using some mathematical substitutions, integration of a Gaussian function can be performed by computing the error function of the threshold level in relation to the expected level.[117] The error function, erf(x), where $x=(I_{\text{“1”}}-I_{th})$, finds the probability of error in a transmitted "1," and the complementary error function, erfc(x)$=$ $(1-\text{erf}(x))$, finds the probability of error in a transmitted "0." However, we can make appropriate substitutions and find the total probability of error in terms of erfc(x):[103]

$$\text{BER} = \frac{1}{4}\left[\text{erfc}\left(\frac{I_{\text{“1”}}-I_{th}}{\sigma_{\text{“1”}}\sqrt{2}}\right) + \text{erfc}\left(\frac{I_{th}-I_{\text{“0”}}}{\sigma_{\text{“0”}}\sqrt{2}}\right)\right]. \tag{20.48}$$

The BER is directly related to the SNR, with a higher signal or lower noise contributing to a lower probability of error. In a direct-detection system, the BER is a function of the SNR:[103]

$$\text{BER} = \frac{1}{2}\text{erfc}\left(\frac{Q}{\sqrt{2}}\right). \tag{20.49}$$

Here, Q is related to the SNR and defined as:

$$Q = \frac{I_{\text{“1”}}-I_{\text{“0”}}}{\sigma_{\text{“1”}}+\sigma_{\text{“0”}}}. \tag{20.50}$$

The optimum threshold level, i.e., decision level, I_{th}^{opt}, is the average of the "0" and "1" currents weighted by the noises:[118]

$$I_{th}^{opt} = \frac{\sigma_{\text{“0”}}I_{\text{“1”}}+\sigma_{\text{“1”}}I_{\text{“0”}}}{\sigma_{\text{“0”}}+\sigma_{\text{“1”}}}. \tag{20.51}$$

Figure 20.68 shows the BER as a function of the received optical power.[18] One notices that it follows the form of an error function. This figure shows the sensitivity of a system, i.e., the amount of detected optical power necessary to achieve a certain BER. For this system, an SNR of $\sim$16 dB is required for a BER$=10^{-9}$. We wish to achieve an optical system exhibiting a very low BER, perhaps even 10^{-14}. The following are reasons for requiring such a low BER:

1. We wish to limit the retransmission of data upon error detection. Such retransmission incurs a significant time delay, which negates the benefits of a high-speed optical transmission system.
2. We wish to limit the amount of overhead in a transmitted stream of data which is reserved for error-correction coding. Such overhead bits are a waste of bandwidth given an "error-free" system.

It may seem odd that the noise must be 16 dB below the signal power, roughly a factor of 50, in order to simply distinguish between a "1" and a "0" bit. However, the noise is a statistical quantity which may have a small variance but which will periodically achieve a very high value. In other words, although on average the noise may be small, sometimes a sharp noise spike causing a severe increase or decrease in the electron count will cause an error in a bit. This could be during the bit time of only one out of a billion transmitted bits, corresponding to the probability of error in the Gaussian noise distribution. Therefore the variance and noise powers must be kept extremely low.

In the optically amplified systems, the main source of SNR degradation is the accumulated ASE of the optical amplifiers. Then it is useful to calculate the OSNR, one of the most important system design parameters for optically amplified systems. The OSNR is the ratio of signal power to ASE power in a fixed bandwidth (optically and electrically). Assuming the signal-spontaneous beat noise, $\sigma^2_{\text{sig-sp}}$, is the dominant noise source, the required OSNR to achieve a given BER then can be expressed as:[118]

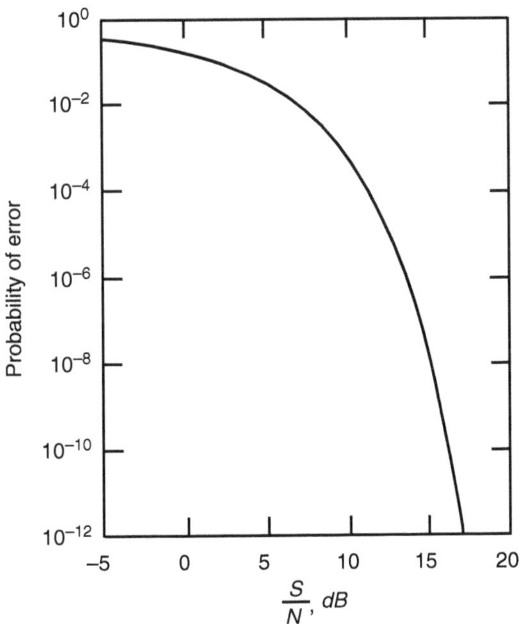

FIGURE 20.68 BER curve as a function of the signal power for a direct detection system. (Adapted from Jones, Jr., W. B. 1988. *Introduction to Optical Fiber Communication Systems*, Holt, Rinehart, and Winston, New York.)

Quantum and (practical) limits

Receiver / Modulation	Direct detection	Heterodyne (asynchronous)	Homodyne
OOK	10 (60) ref. [120]	40 (80) ref. [6]	18 (36) ref. [6]
FSK	---------	40 (80) ref. [6]	--------
PSK	20 (30) ref. [120] (DPSK)	20 (40) ref. [6]	9 (18) ref. [6]

FIGURE 20.69 Comparison of theoretical and practical sensitivity limits in the number of photons-per-bit which must be recovered to achieve a BER of 10–9. The theoretical limit is based on the quantum limit due to shot noise. The practical limits are in parentheses.

$$\mathrm{OSNR_R} = \frac{Q^2 B_e}{B_o} \frac{1 + (P_{\text{“0”}}/P_{\text{“1”}})}{\left(1 - \sqrt{P_{\text{“0”}}/P_{\text{“1”}}}\right)^2}. \tag{20.52}$$

In practical systems, heterodyne detection is more sensitive than direct detection that does not have an optical preamplifier. However, the advent of the EDFA has improved the sensitivity of a direct-detection receiver significantly. If a low NF optical amplifier (preamplifier) is used just before the photo-detector and regenerator, the receiver sensitivity can be improved by 10–20 dB, comparable to that of coherent detection. Since the coherent techniques are much more difficult to deploy than that of the direct-detection, the coherent detection is most likely to be implemented only on the systems that direct-detection is impractical, for example, the multichannel system requiring a small channel spacing (<10 GHz) and the high spectral efficient system using multi-level modulation formats.

Figure 20.69 gives a comparison of the theoretical limits of the various modulation and detection schemes. Actual systems rarely come close to these theoretical limits, and the challenges to reach the lower limits are considerable. Note that the limits are given in photons per bit, which can easily be calculated for any bit rate by noting that the energy in a pulse is $h\nu$, the photon energy, multiplied by the number of photons.

Although the ultimate quantitative measure of performance is the BER (or Q factor), another key performance criteria which is easy to measure on an oscilloscope and gives a qualitative “feel” for system performance is the eye diagram.[104] Figure 20.70 describes how an eye diagram is generated on an oscilloscope from a given pseudo-random NRZ data stream. To begin with, an oscilloscope trace of a series of distinct bits can be seen on an oscilloscope because the bit stream is periodic, repeating continuously so that a nanosecond bit (which can't be seen as a single event by the human observer) is regenerated countless times and thus visible on the oscilloscope screen. This continuous repetition of the bit stream is because the entire bit sequence is repeated and each bit appears during each repetition in the same spatial (i.e., temporal) position on the oscilloscope screen. Note that a truly random bit stream is not periodic and so only a predetermined pseudo-random bit sequence can be displayed. This repetition of the bit stream in which the same bit during each sequence appears in the same location on the screen is accomplished by the bit stream generator sending out a trigger pulse in addition to the clock pulses which synchronize the bits in a time slot. This trigger pulse forces the oscilloscope to begin its trace at the left-most part of the screen for each repetition of the entire bit stream generator sending out a trigger pulse in addition to the clock pulses which synchronize the bits in a time slot. This trigger pulse forces the oscilloscope to begin its trace at the left-most part of the screen for each repetition of the entire bit stream.

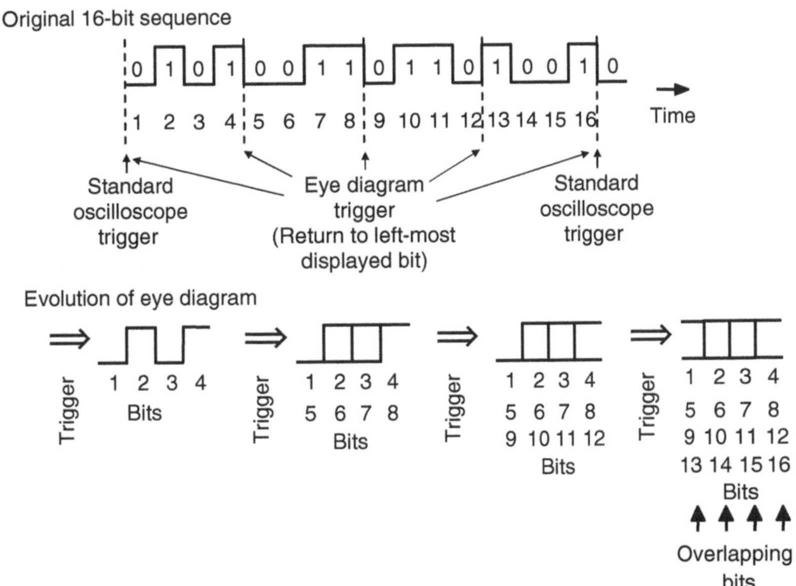

FIGURE 20.70 Generation of a sharp eye diagram.

To generate an eye diagram, the bit-pattern generator triggers the oscilloscope to begin its trace much more often than at the end of the entire bit stream. As an illustrative example, Figure 20.70 shows a 16-bit repetitive sequence which normally is triggered every 16 bits but is now triggered every 4 bits, with the trace beginning at the left part of the screen for bits 1, 5, 9, and 13. The oscilloscope traces and displays bits 1 through 4, returns to the left of the screen, and then traces bits 5 through 9. There are now only four displayed bit time slots into which all the 16 bits must fit. Bits 1, 5, 9, and 13 are displayed in time slot #1, bits 2, 6, 10, and 14 are displayed in time slot #2, bits 3, 7, 11, and 15 are displayed in time slot #3, and bits 4, 8, 12, and 16 are displayed in time slot #4. In our example, each time slot will sometimes have a "1" bit and sometimes have a "0" bit displayed. The total display will eventually look like a series of boxes. As in all optical transmission systems, a low-pass-filter is used to limit the high-frequency noise which exists above the necessary signal bandwidth. This filter will round-off the sharp edges of transmitted rectangular bits. What remains is the "eye" diagram as shown in Figure 20.71a, which literally resembles an "eye." The upper and lower rails have transitions when a "1" bit is followed by a "0" bit, and vice-versa.

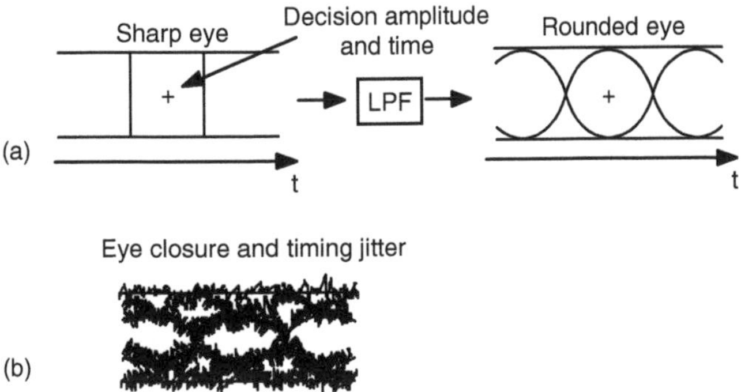

FIGURE 20.71 (a) A low-pass-filtered eye diagram. (b) A closed eye diagram due to contrast ratio degradation, amplitude noise, and timing jitter.

These rails also show no transitions between time slots if a string of "1"s or "0"s is transmitted. These rails also have some amplitude width corresponding to system noise in the vertical direction and timing jitter in the horizontal direction. Figure 20.71a also shows an open eye diagram in which it is fairly easy to distinguish the difference between a "1" and "0" in any given time slot. The decision circuit will decide the identity of the bit at the center of the open eye, usually in the middle of the time slot and at the halfway point between the upper and lower rails. Figure 20.71b shows how the eye closes if either the contrast ratio is too low or if too much noise exists. Also shown in Figure 20.71b is the timing jitter which is an uncertainty in the exact start of a bit time (bit recovery time) and reduces the sensitivity of the receiver. The timing jitter can be caused by faulty circuitry or by various noise terms which result in phase variations. Although the eye diagram cannot represent a 10^{-9} BER since one bit out of a billion could never be seen on a screen, the eye diagram is still an easy and good qualitative indication of proper system performance.

20.3.6 System Design

A system must be designed to function properly and ensure near-error-free data recovery. This implies that enough signal power be detected (i.e., sufficiently high SNR) to meet or exceed the sensitivity of the receiver. From a system design point-of-view, this is accomplished by adding or subtracting the various factors which would contribute to an increase or decrease in the signal power required for proper system performance.

We have discussed many quantities in an optical system in units of decibels (dB) for the reason of ease of system design. Decibels have the property that they compare two quantities, either an input divided by an output or one quantity in relation to a standard value, in terms of a logarithmic relationship:[117]

$$\text{dB} = 10\log\left[\frac{\text{output}}{\text{input}}\right](\text{power}) \tag{20.53a}$$

$$\text{dB} = 20\log\left[\frac{\text{output}}{\text{input}}\right](\text{voltage}) \tag{20.53b}$$

$$\text{dB} = 10\log\left[\frac{x\text{ mW}}{1\text{ mW}}\right](\text{power}) \tag{20.53c}$$

where dBm compares the given value to a milliwatt of power. Although both power and voltage decibels were outlined in Equation 20.61, we will restrict ourselves to discussing decibels in terms of power for the remainder of the chapter. An important point is that decibels are a logarithmic process. Logarithms have the unique characteristic that the logarithm of one number minus the logarithm of another number is equal to the logarithm of the division of the two numbers:[117]

$$\underbrace{10\log\left[\frac{\text{output}}{\text{input}}\right] = 10\log\left[\frac{\text{output}}{1\text{ mW}}\right] - 10\log\left[\frac{\text{input}}{1\text{ mW}}\right]}_{\text{Component Loss (dB)} = \text{Output(dBm)} - \text{Input(dBm)}} \tag{20.54}$$

Division and multiplication are transformed into subtraction and addition, respectively, which are much easier calculations to perform especially when concatenating many system elements for evaluation. If a device has either loss or gain, this can be described by a subtraction or addition of the signal power in decibels. A system power penalty for the losses in a given component (i.e., output/input) would just be a subtraction of some number of decibels. Another elegant quality is that dB by itself is not an absolute number but a description of a ratio which is simply a characteristic of a device. It is therefore independent of the absolute input or output power level of a device.

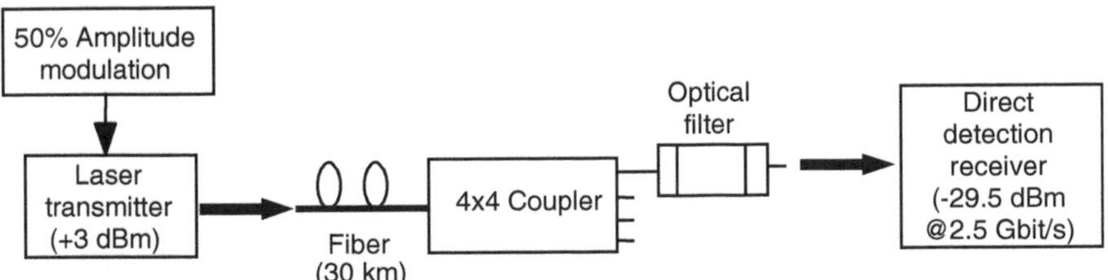

FIGURE 20.72 Example of designing a direct detection optical system.

We will now design a simple optical system using the addition and subtraction of decibels.[109] This simple ASK NRZ direct-detection system is shown in Figure 20.72. The laser puts out a certain amount of optical power in dBm. Any aspect can provide a system addition or subtraction. The end receiver has a certain sensitivity in dBm, which is the optical power necessary for near-error-free data recovery by the receiver. We must ensure that more power than this minimum sensitivity reaches the receiver. In fact, we usually speak of a system power margin, which represents an extra power buffer above the minimum power level required for unambiguous reception. This power margin, which is typically a few dB, allows for slight occasional degradations in system performance without affecting proper system performance. The designed system would be evaluated as follows:

Laser output	+3 dBm
Power penalty due to a 50% contrast ratio	−3 dB
Coupling loss into fiber	−3 dB
Fiber attenuation in 30 km link	−15 dB
Splitting loss in a 4×4 coupler	−6 dB
Insertion loss in an optical filter	−1.5 dB
Requested power margin	−4 dB
Direct-detection receiver sensitivity (BER$=10^{-9}$ at 2.5 Gbit/s)	−29.5 dBm
Total balanced system	0 dBm

Note that any system with different components or characteristics would have different design requirements, i.e., designs would change if there were multi-mode fiber, wavelength dependencies, heterodyne reception, laser chirp, insertion losses, etc.

If the losses were too large for the receiver sensitivity, then optical amplification would be required somewhere in the system. In addition to the power budget, the system has to meet the OSNR requirement too. One can estimate the OSNR (in dB) requirement if the receiver sensitivity P_{rec} is known using

$$\mathrm{OSNR_R} = 58 + P_{rec} - \mathrm{NF} \tag{20.55}$$

where NF is the noise figure of the optically preamplified receiver.

For a long distance transmission link with a chain of N identical amplifiers (see Figure 20.73), the OSNR at the output can be estimated by[119]

$$\mathrm{OSNR} = 58 + P_{in} - \mathrm{NF} - L_\alpha - 10 \log N \tag{20.56}$$

where P_{in} is the launch average power at the given wavelength (at the input of the transmission link) in dBm; NF is the noise figure of the amplifiers, and L_α is the span loss (fiber span loss between two amplifiers), both in dB.

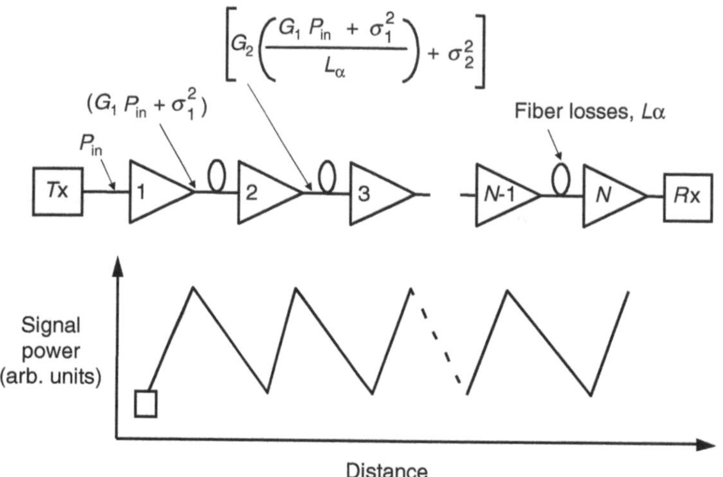

FIGURE 20.73 Cascade of EDFAs with periodic addition of gain, noise, and fiber losses.

Optical amplifiers solve the loss problem but do not correct the dispersion problem. The dispersive effects accumulate over multiple amplified links and become the limit factor for achieving the required BER. We will discuss the techniques of dispersion compensation in the section of dispersion management.

20.3.7 Point-to-Point and Point-to-Multipoint Systems

We have been discussing the basic concepts of simple data transmission and detection. Figure 20.74 shows two generic types of optical systems: point-to-point and point-to-multipoint systems. Point-to-point links can be thought of as transmission of a data stream from point A to point B in which the design rules discussed in the previous section apply. No routing or switching occurs in the optical data path. The most obvious example of a point-to-point system is an undersea long-distance fiber-optic cable connecting two points which are very far apart. By the simplest of calculations, a 1 Gbit/s data stream, which is launched at 0 dBm into a fiber of attenuation 0.2 dB/km and using a receiver of sensitivity −30 dBm, can propagate ~150 km if only attenuation losses and a thermal-noise-limited receiver are considered. In reality, the system would be limited to ~50 km by issues such as power margin, dispersion, and insertion losses. Therefore, an ultra-long-distance (>1000 km) point-to-point optical system would require the signal to be periodically detected and then retransmitted by an optical regenerator.[120] This regenerator is composed of a receiver coupled with a transmitter. The regenerator must have electronics for clock recovery and electronic amplification, both of which are bit-rate specific to the incoming data stream and can only accommodate one signal wavelength at a given time. Regenerators are fairly expensive and not very flexible. As shown in Figure 20.73, the EDFA is an adequate substitute for expensive regenerators. Thus the point-to-point systems for long-distance terrestrial and trans-oceanic communications implemented today are the amplified Wavelength-Division Multiplexed (WDM) systems. WDM technology is the subject of next Section, at which point we will have much more discussion on this topic.

Point-to-multipoint systems can be much more complex than point-to-point systems. A user is connected to many other users through some passive or active optical switching medium. (Note that point-to-multipoint systems in which all the switching is done electrically after an optical signal is detected would be considered a point-to-point optical system.) The inter-communication can be uni-directional, as in a broadcast system, or bi-directional, as in a local-area network (LAN). It is the goal of future systems to have simultaneous transmission of many different signals through the network, thereby

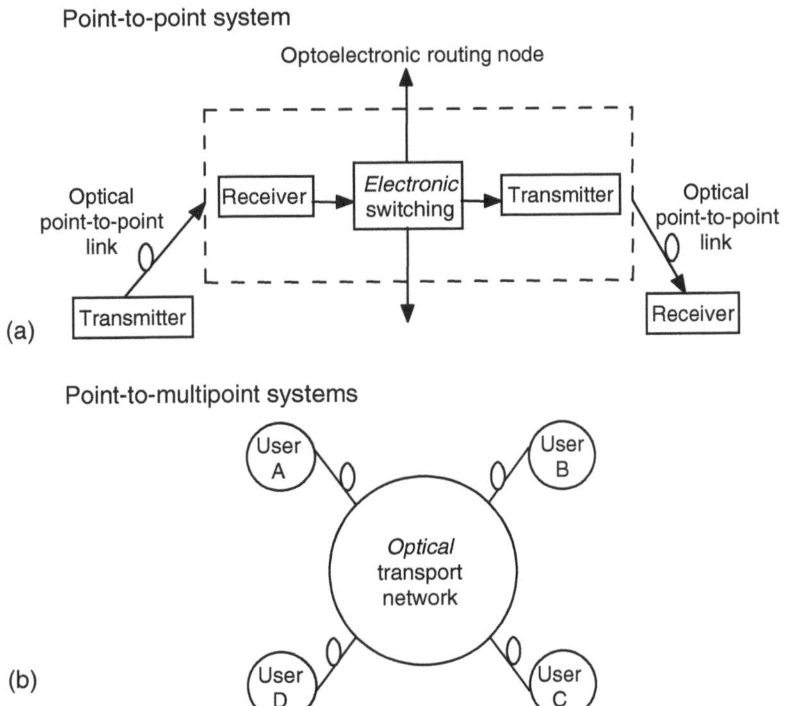

FIGURE 20.74 Point-to-point and point-to-multipoint optical systems.

having an increase in the aggregate system capacity. Signal recovery and design rules still apply, but may be much more complex. In addition, efficient routing, low inter-channel crosstalk, and output-port contention resolution are just some of the intricate issues involved. There will be much more discussion on point-to-multipoint system in the section on optical networks.

Because of the high losses in electrical co-axial systems at speeds exceeding 100s of MHz, it is a foregone conclusion that low-loss high-bandwidth optical fiber transmission is the mode of choice for long-distance, high-capacity point-to-point systems ($\sim$1000 km).[121] Advances in optical network have also followed components and transmission technology innovations. It has dominated metropolitan networks ($\sim$100 km) today. At the same time, practical network implementations uncover the need for added devices, functionality, and very low cost points. However, although point-to-multipoint systems supported by the optical access network do hold the promise of extremely high-bandwidth communications among many users, the technology, network, and cost challenges have still not been definitively solved. The fiber hasn't gotten much closer to residences. The hope is that the last mile problem will be solved and FTTX will become practical within the next few years.

20.4 Wavelength-Division Multiplexing (WDM) Technology

Until the late 1980s, optical fiber communications was mainly confined to transmitting a single optical channel. Due to fiber attenuation, this channel required periodic regeneration which included detection, electronic processing, and optical retransmission. Such regeneration causes a high-speed optoelectronic bottleneck, is bit-rate specific, and can only handle a single wavelength. The need for these single-channel regenerators (i.e., repeaters) was replaced when the EDFAs were developed, enabling high-speed repeaterless single channel transmission. We can think of this single $\sim$Gbit/s channel as a single high-speed "lane" in a highway in which the cars represent packets of optical data and the highway represents the optical fiber. However, recall the single-mode silica fiber described in Section 20.2. its low-loss window near 1.55 μm is approximately 25,000 GHz wide. The high-bandwidth characteristic of the optical fiber implies that a single optical carrier at 1.55 μm can be baseband modulated at ~25,000 Gbit/s, occupying this 25,000 GHz bandwidth. Obviously, this bit rate is impossible for present-day optical devices to achieve, given that heroic lasers, external-modulators, switches, or detectors have bandwidths < 100 GHz; note that practical data links today would be significantly slower, perhaps no more than 40 Gbit/s per channel. As such, a single high-speed channel takes advantage of an extremely small portion of the available fiber bandwidth. It seems natural to dramatically increase the system capacity by transmitting several different independent wavelengths simultaneously down a fiber in order to more fully utilize this enormous fiber bandwidth.[122,123] Therefore, the intent was to develop a multiple-lane highway, with each lane representing data traveling on a different wavelength. This highway cartoon scenario is illustrated in Figure 20.75.

In addition to high-capacity transmissions, WDM technology also enables wavelength routing and switching of data paths in an optical network. By utilizing wavelength-selective component technologies, each data channel's wavelength can be used to determine the routing through the network. Therefore, data can be thought of as traveling not on optical fiber but on wavelength-specific "light-paths" from source to destination that can be arranged by a network controller to optimize throughput.

20.4.1 Basic Operation

The practicality of EDFA which can provide gain to many channels simultaneously over a ~THz wavelength range opened the door to multiplexing signals at many wavelengths onto the same optical fiber. This technique, known as WDM, wonderfully enhances an optical system's capacity. Along with the EDFA, WDM is the other technology that is the key to terabit-per-second optical systems. Conceptually, WDM is the same as the frequency-division multiplexing (FDM) used to place many radio channels on carrier waves of different frequencies. The carrier wave of each optical WDM channel, however, is a million times higher in frequency (terahertz versus megahertz).

In the most basic WDM arrangement as shown in Figure 20.76, the desired number of lasers, each emitting a different wavelength, are multiplexed together by a wavelength multiplexer (or a combiner)

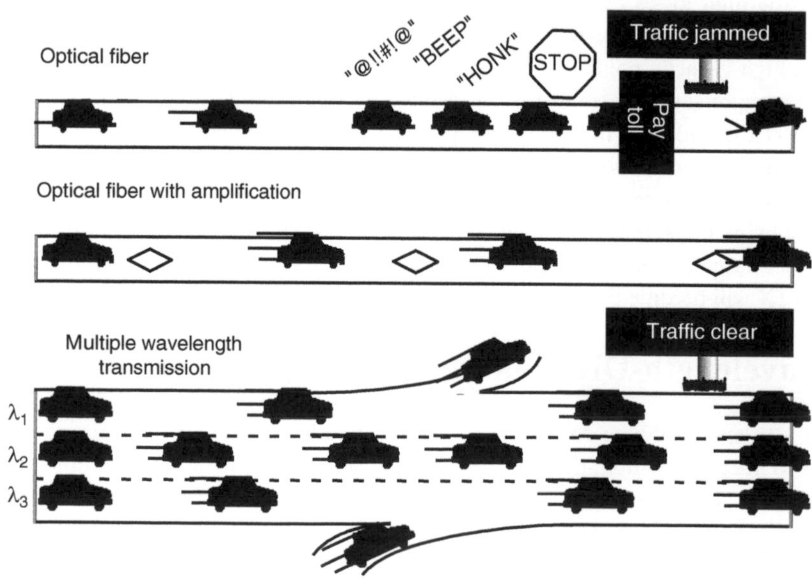

FIGURE 20.75 Multiwavelength optical transmission as represented by a multiple-lane highway.

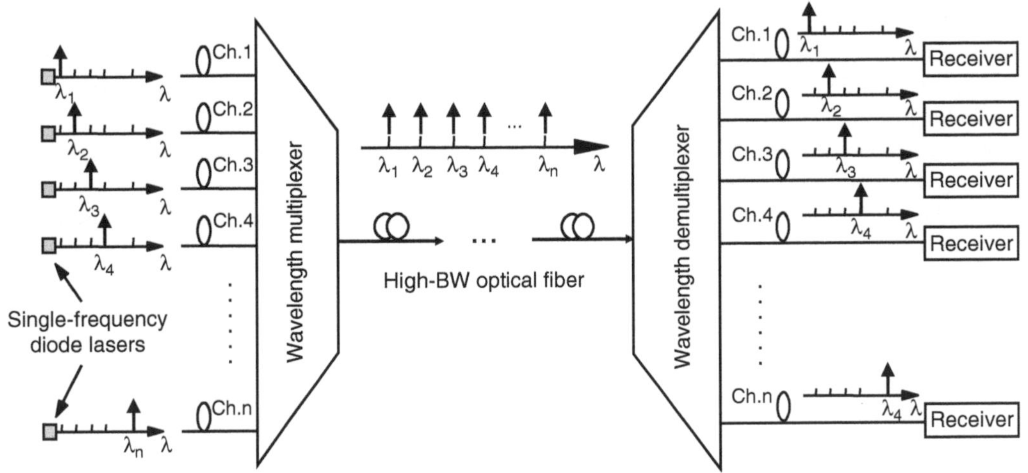

FIGURE 20.76 Diagram of a simple WDM system.

into the same high-bandwidth fiber. After being transmitted through a high-bandwidth optical fiber, the combined optical signals must be demultiplexed by a WDM (or a splitter) at the receiving end by distributing the total optical power to each output port and then requiring that each receiver selectively recover only one wavelength. At the receiver, a narrow-band optical filter is used to select just one of the incoming wavelengths, so that only one signal is allowed to pass and establish a connection between source and destination.

It is important to space channel wavelengths an adequate distance apart. The goal is to minimize transmission of unwanted channels through the filter and to accommodate the drift in the wavelength characteristics of optoelectronic components over time—a change in temperature is just one possible cause. Typical channel spacings range from 0.4 nm to 4 nm (50–500 GHz).

Figure 20.77 illustrates the concept of wavelength demultiplexing using an optical filter. In this example, four channels are input to an optical filter which has a nonideal transmission filtering function. The filter transmission peak is centered over the desired channel, in this case λ_3, thereby transmitting that channel and blocking all other channels. Due to the nonideal filter transmission function, some optical energy of the neighboring channels leaks through the filter causing inter-channel inter-wavelength crosstalk. This crosstalk has the effect of reducing the selected signal's contrast ratio and can be minimized by increasing the spectral separation between channels. Although there is no set definition, a nonstandardized convention exists for defining optical WDM, dense-WDM, and FDM as encompassing a system for which the channel spacing is approximately 10, 1, and 0.1 nm, respectively. However, we will not make any distinction among these system labels in this chapter.

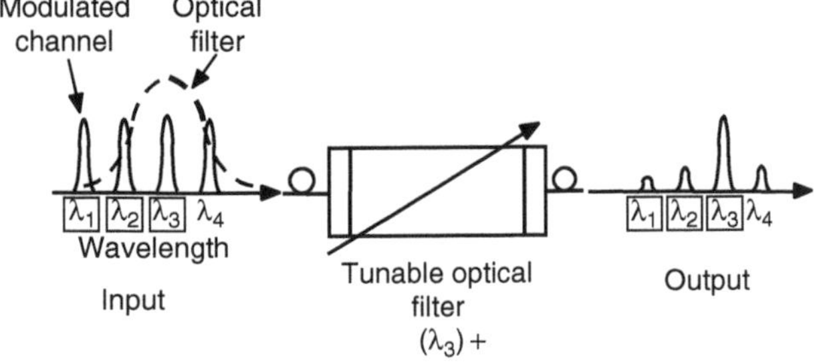

FIGURE 20.77 Optical WDM channels being demultiplexed by an optical filter.

20.4.2 Topologies and Architectures

WDM technology enables the utilization of a significant portion of the available fiber bandwidth by allowing many independent signals to be transmitted simultaneously in one fiber. Additionally, high-bandwidth routing can be also facilitated through a multi-user network. The WDM channels can be routed and detected independently, with the wavelength determining the communication path by acting as the signature address of the origin, destination, or routing. Therefore, the basic system architecture that can take the full advantage of WDM technology is an important issue, and will be discussed in this section.

20.4.2.1 Point-to-Point WDM Links

As shown in Figure 20.74 and Figure 20.76, in a simple point-to-point WDM system, several channels are multiplexed at one node, the combined signals are then transmitted across some distance of fiber, and the channels are demultiplexed at a destination node. This point-to-point WDM link facilitates the high-bandwidth fiber transmission without routing or switching in the optical data path.

20.4.2.2 Wavelength-Routed Networks

Figure 20.78 shows a more complex multi-user WDM network structure, where the wavelength is used as the signature address for either the transmitters or the receivers, and determines the routing path through an optical network. In order for each node to be able to communicate with any other node and facilitate proper link setup, either the transmitters or the receivers must be wavelength tunable; we have arbitrarily chosen the transmitters to be tunable in this network example. Note that the wavelengths are routed passively in wavelength-routed networks.

20.4.2.3 WDM Stars, Rings, and Meshes

Three common WDM network topologies are star, ring, and mesh networks.[124–126] In the star topology, each node has a transmitter and receiver, with the transmitter connected to one of the passive central star's inputs and the receiver connected to one of the star's outputs, as is shown in Figure 20.79a. Rings, as shown in Figure 20.79b, are also popular because (1) many electrical networks use this topology, and (2) rings are easy to implement for any geographical network configuration. In this example, each node in the unidirectional ring can transmit on a specific signature wavelength, and each node can recover any other node's wavelength signal by means of a wavelength-tunable receiver. Although not depicted in the figure, each node must recover a specific channel. This can be performed (1) where a small portion of the

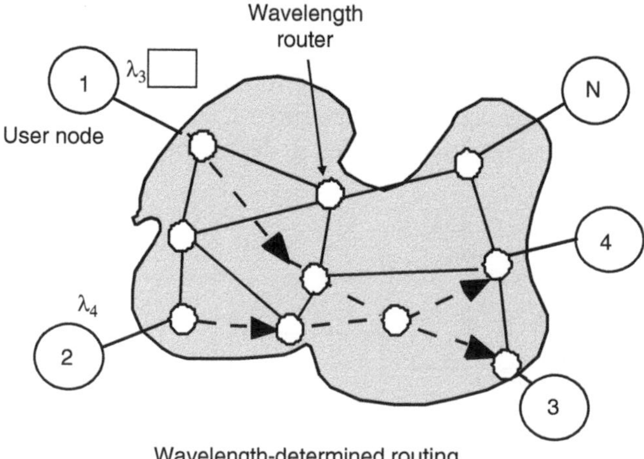

FIGURE 20.78 A generic multi-user network in which the communications links and routing paths are determined by the wavelengths used within the optical switching fabric.

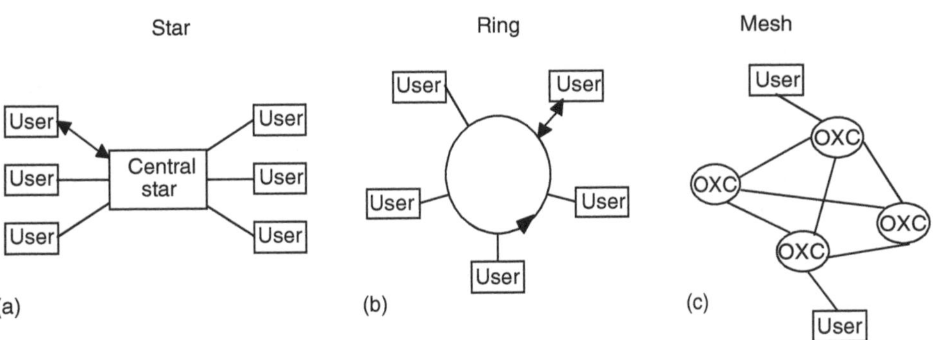

FIGURE 20.79 WDM stars, rings, and meshes.

combined traffic is tapped off by a passive optical coupler, thereby allowing a tunable filter to recover a specific channel; or (2) in which a channel-dropping filter completely removes only the desired signal and allows all other channels to continue propagating around the ring. Furthermore, a synchronous optical network (SONET) dual-ring architecture, with one ring providing service and the other protection, can provide automatic fault detection and protection switching.[127]

In both the star and ring topologies, each node has a signature wavelength, and any two nodes can communicate with each other by transmitting and recovering that wavelength. This implies that *N* wavelengths are required to connect *N* nodes. The obvious advantage of this configuration, known as a single-hop network, is that data transfer occurs with an uninterrupted optical path between the origin and destination; the optical data starts at the originating node and reaches the destination node without stopping at any other intermediate node. A disadvantage of this single-hop WDM network is that the network and all its components must accommodate *N* wavelengths, which may be difficult (or impossible) to achieve in a large network, i.e., present fabrication technology cannot provide and transmission capability cannot accommodate 1000 distinct wavelengths for a 1000-user network!

It is important to mention that reliability is a problem in fiber ring. If a station is disabled or if a fiber breaks, the whole network goes down. To address this problem, a double-ring optical network, also called a "self-healing" ring, is used to bypass the defective stations and loops back around a fiber break, as shown in Figure 20.80. Each station has two inputs and two outputs connected to two rings that operate in opposite directions.

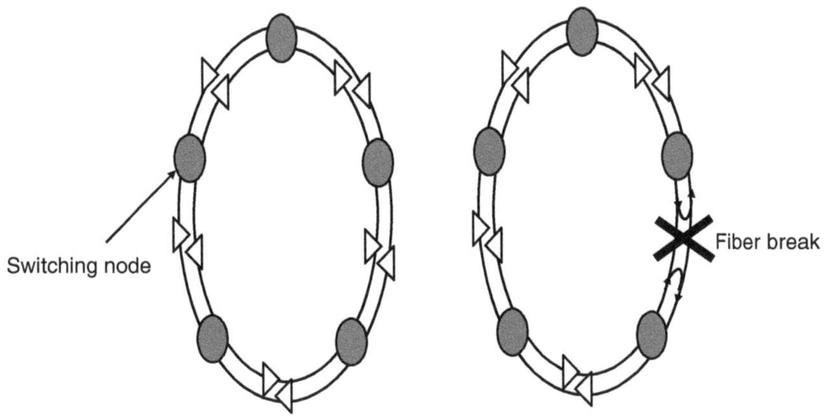

FIGURE 20.80 A self-healing ring network.

An alternative to requiring N wavelengths to accommodate N nodes is to have a multi-hop network (mesh network) in which two nodes can communicate with each other by sending data through a third node, with many such intermediate hops possible, shown in Figure 20.79c. In the mesh network, the nodes are connected by reconfigurable optical crossconnects (OXCs).[128] The wavelength can be dynamically switched and routed by controlling the OXCs. Therefore, the required number of wavelengths and the wavelength tunable range of the components can be reduced in this topology. Moreover, the mesh topology can also provide multiple-paths between two nodes to make network protection and restoration easier to realize. If a failure occurs in one of the paths, the system can automatically find another path and restore communications between any two nodes. However, OXCs with large numbers of ports are extremely difficult to obtain, which limits the scalability of the mesh network.

In addition, there exist several other network topologies, such as a tree network, which is a favorite of broadcast, or distribution, systems. At the "base" of the tree is the source transmitter from which emanates the signal to be broadcast throughout the network. From this base, the tree splits many times into different "branches," with each branch either having nodes connected to it or further dividing into sub-branches. This continues until all the nodes in the network can access the base transmitter. Whereas the other topologies are intended to support bi-directional communication among the nodes, this topology is useful for distributing information uni-directionally from a central point to a multitude of users. This is a very straightforward topology and is in use in many electrical systems, most notably cable television (CATV). We will revisit CATV when discussing SCM.

By introducing Figure 20.81 in which a larger network is composed of smaller ones, we have also introduced the subject of the architecture of the network which depends on the network's geographical extent. The three main architectural types are the local-, metropolitan-, and wide-area networks, denoted by LAN, MAN, and WAN, respectively.[129] Although no rule exists, the generally accepted understanding is that a LAN interconnects a small number of users covering a few km (i.e., intra- and inter-building), a MAN interconnects users inside a city and its outlying regions, and a WAN interconnects significant portions of a country (100s of km). Based on Figure 20.81, the smaller networks represent LANs, the larger ones MANs, and the entire figure would represent a WAN. In other words, a WAN is composed of smaller MANs, and a MAN is composed of smaller LANs. Hybrid systems exist, and typically a wide-area network will consist of smaller LANs, with mixing and

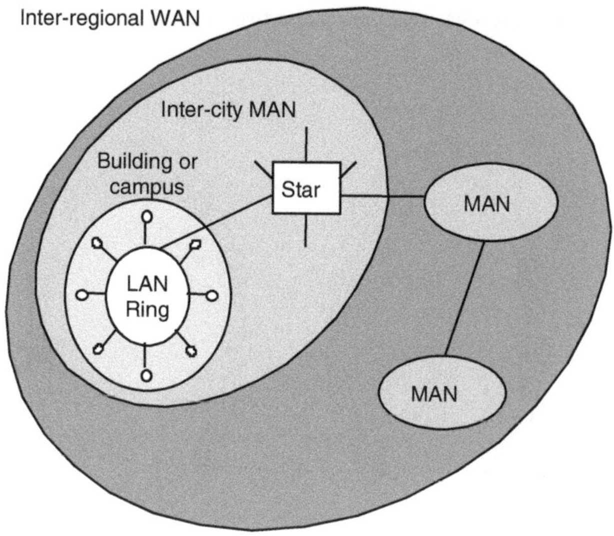

FIGURE 20.81 Hybrid network topologies and architectures woven together to form a large network.

matching between the most practical topologies for a given system. For example, stars and rings may be desirable for LANs whereas buses may be the only practical solution for WANs. It is, at present, unclear which network topology and architecture will ultimately and most effectively take advantage of high-capacity optical systems.

20.4.2.4 Network Reconfigurability

Figure 20.82 shows an example of fixed and reconfigurable optical add/drop nodes that can add and drop wavelengths at intermediate nodes in a communications network. A fixed add/drop multiplexing node can only process the signal(s) at a given wavelength or a group of wavelengths. While in the dynamic reconfigurable node, operators could add or drop any number of wavelengths. This added flexibility would save operating and maintenance costs and would improve network efficiency. In general, a network is reconfigurable if it can provide the following functionality for multi-channel operations: (1) channel add/drop and (2) path reconfiguration for bandwidth allocation or restoration. It appears that a reconfigurable network is highly desirable to meet the requirements of high bandwidth and bursty traffic in future networks.

A reconfigurable network allows dynamic network optimization to accommodate changing traffic patterns, which provides more efficient use of network resources. Figure 20.83 shows blocking probability as a function of call arrival rate in a WDM ring network with 20 nodes.[130] A configurable topology can support 6 times the traffic of a fixed WDM topology for the same blocking probability.

The key component technologies enabling network reconfigurability include wavelength tunable lasers and laser arrays, wavelength routers, optical switches, OXCs, OADMs, and tunable optical filters, etc.

Although huge benefits are possible with a reconfigurable topology, the path to reconfigurability is paved with various degrading effects. As shown in Figure 20.78, the signal may pass through different lengths of fiber links due to the dynamic routing, causing some degrading effects in reconfigurable networks to be more critical than in static networks, such as nonstatic dispersion and nonlinearity accumulation due to reconfigurable paths, EDFA gain transients, channel power nonuniformity, crosstalk in optical switching and crossconnects, and wavelength drift of components. We will discuss some of these effects in the later part of this section.

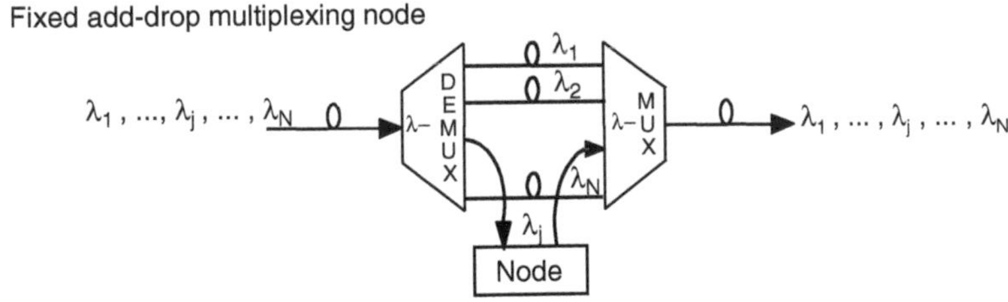

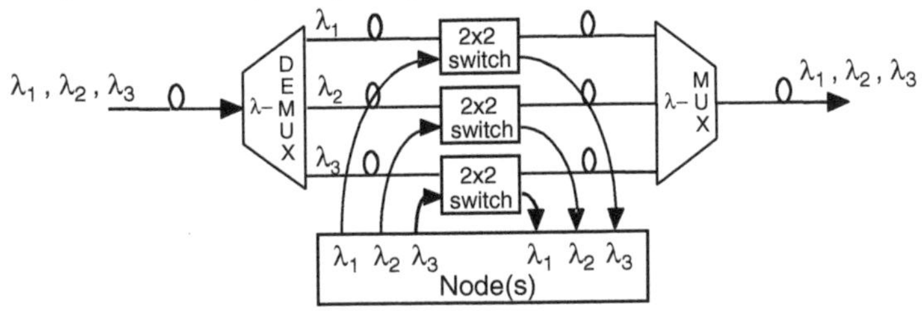

FIGURE 20.82 Optical add-drop multiplexing (OADM) systems: fixed vs. reconfigurable.

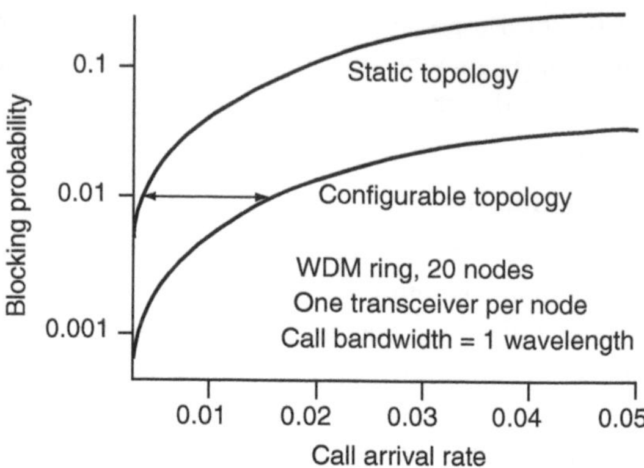

FIGURE 20.83 Blocking probability as a function of call arrival rate in a WDM ring. (Adapted from Chan, V. W. S., Hall, K. L., Modiano, E., and Rauschenbach, K. A. 1998. *IEEE J. Lightwave Technol.*, 16, 2146–2168. With permission.)

20.4.3 WDM Enabling Technologies

Many interesting challenges face the implementation of WDM systems and networks. Several of these challenges involve the control and management of the data through this novel high-speed network. However, as was clear at the outset when research groups were envisioning WDM systems, the required devices were simply not available commercially and were hardly more than prospective research projects. Therefore, the most fundamental problems initially were in the area of providing the unique wavelength-selective devices necessary for good system performance. After several years of research, much progress has been made in the laboratory, with many WDM-related components becoming commercially available already. With all the recent progress, devices still place limits on the design of a WDM network. Moreover, cost will always be a factor in determining which type of network will be implemented, and the cost of these WDM components is still quite high; future work will center on making these devices more affordable and practical for an individual user. This section will deal exclusively with the critical components necessary for a WDM system. For the sake of brevity, we will only include a limited discussion of each technology. It should be emphasized that much of what distinguishes an optical network from an electrical one is in the physical layer implementation, which mainly includes the device and transmission technologies.

Figure 20.84 shows a small subset of critical component technologies typically required for a WDM system, including multiple-wavelength transmitters, wavelength multiplexer and demultiplexer, multi-port star couplers, passive and active wavelength routers, fiber amplifiers, and tunable optical filters. It is important to familiarize the reader with some generic laudable goals which a WDM-device technologist aims to achieve. These include the following:

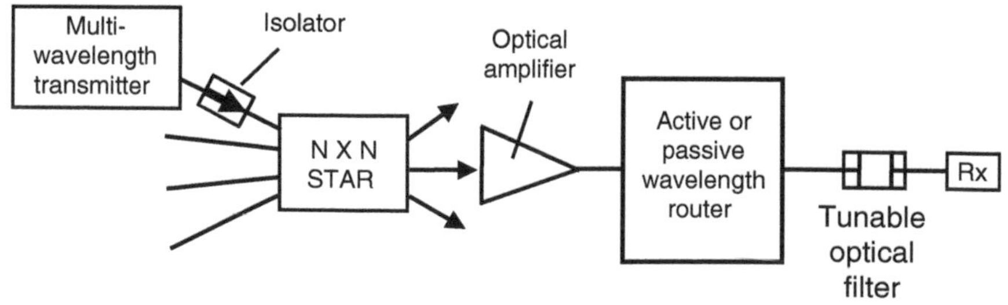

FIGURE 20.84 Schematic of a small subset of enabling device technologies for a WDM system.

1. Large wavelength tuning range
2. Multi-user capability
3. Wavelength stability and repeatability
4. Low crosstalk
5. High extinction ratio
6. Minimum excess losses
7. Fast wavelength tunability (especially for packet switching)
8. High-speed modulation bandwidth
9. Low residual chirp
10. High finesse
11. Low noise
12. Robustness
13. High yield
14. Potential low cost

These device goals enable a system which can accommodate a large number of users and achieve high-speed and packet-switching operation. Of course, this list is not exhaustive, and not all goals apply to each specific component.

20.4.3.1 Wavelength-Tunable Lasers

Multi-wavelength transmitters are required if the wavelength of the transmitter must be tuned to different values corresponding to different destinations. Note that it is quite possible for a fixed-transmitter WDM system to use ordinary nontunable single-frequency DFB[131] lasers that have been hand-selected to be at the desired wavelengths with the appropriate wavelength spacings.[132] However, it is impractical to design a robust system around such discrete-wavelength lasers due to the unpredictability in the exact wavelength of a processed laser, the limited thermal tuning range of DFBs ($\sim$0.1 nm/°C), and the inherent long-term laser wavelength drift. In general, the figures of merit for these devices are (a) large wavelength-tuning range, (b) high-speed wavelength tunability, (c) high data modulation speed for ASK or FSK transmission, (d) rigid wavelength stability and repeatability, and (e) low potential cost.

The two main physical processes underlying wavelength tuning are (1) Bragg reflection[36] and (2) the carrier-dependent refractive index. Bragg gratings can be formed by a periodic longitudinal variation in the real or imaginary part of the index of refraction of a medium. These gratings are quite efficient and highly wavelength selective in their reflectivity, for which the laser light and the gratings must match the Bragg condition

$$\Lambda = \frac{m\lambda}{2n(\cos\theta)} \tag{20.57}$$

where Λ is the periodic distance between adjacent grating peaks, λ is the wavelength of the emitted light, n is the refractive index of the medium, m is the order of the grating, and Θ is the angle between the incident light and the grating normal (i.e., $\Theta = 90°$ for light incident parallel to the grating). We wish to limit our discussion to enabling only the fundamental Bragg condition ($m = 1$), in which incident light is almost totally reflected back towards the direction of the incident wave. This Bragg wavelength condition should also correspond to the wavelength peak of the laser gain medium in order to simultaneously have efficient gratings and efficient gain for lasing. These Bragg gratings can replace end mirrors to provide feedback in a laser cavity, in which the gain-medium waveguide is followed by a waveguiding region with a periodic index grating. The Bragg cavity is more selective than a Fabry-Perot cavity formed by two parallel dielectric mirrors, with the periodic wavelength-selective resonances spaced further apart in frequency (i.e., high Finesse). These widely spaced resonances insure that only one reflectance resonance peak will fall within the gain $>$ loss bandwidth, thus providing single-mode operation at one cavity resonance.

Tuning of the laser wavelength output of a laser given a Bragg reflective grating for feedback can be accomplished by tuning the parameters which govern the condition for Bragg reflection. One method is to change the index of refraction, which is a function of the carrier concentration.[12] If the Bragg region is simply a passive (distributed Bragg reflector [DBR], i.e., undoped and no gain) waveguiding semiconductor region, then we can inject current into that grating region, thereby changing (a) the refractive index, (b) the frequency-selective Bragg reflection function, and ultimately (c) the laser output wavelength. This tuning can be accomplished relatively fast since it depends on carrier dynamics. The second tuning mechanism relates to the refractive index inside the cavity in which the light propagates. As stated before, the refractive index increases as the carrier concentration increases. In a simple sense, injecting more current into a semiconductor will increase its refractive index thereby changing the effective length of and the speed-of-light within a semiconductor cavity. If the effective cavity length is changed, then the wavelength-selective cavity resonances will shift and the lasing mode will appear at a different wavelength or even a different cavity mode will lase. Therefore, changing the carrier concentration in a waveguiding "phase" region will change the wavelength of the emitted light. If the laser light is propagating in a passive section, then the carrier concentration can be controlled by controlling an injected current. However, the situation is more complicated if the laser light is traversing an active gain section of the cavity. In an ideal active section of a laser, the carrier concentration increases until the lasing threshold has been reached (i.e., gain greater than loss for one of the cavity resonances). Once lasing starts, any additional generated carriers should contribute to stimulated emission of photons and not to increasing the carrier concentration. Consequently, it is very difficult to change the refractive index and the lasing wavelength by varying the injected current of any active section of a laser, such as in a standard DFB laser.

Figure 20.85 shows an example of a sampled-grating DBR (SGDBR) tunable laser. The wider tuning range filter is provides by the product of the two differently spaced and independently tuned reflection combs of the SGDBRs at each end of the cavity. High speed switching (<10 ms) is possible since temperature tuning is not required.[133] Good side-mode suppression (>40 dB) has been demonstrated, tuning of over 40 nm is easily accomplished, and fiber-coupled powers of up to 20 mW is reported by the incorporation of another gain section (SOA) on the output side of the output mirror.

For the sake of introducing a mature commercially available source technology, we will also discuss the external-cavity laser which cannot be modulated or tuned rapidly. This laser is composed of a typical Fabry-Perot semiconductor laser with one facet anti-reflection coated. The light output from this low-reflectivity facet propagates until it hits an angle-tuned Bragg grating which forms the second mirror for the frequency-selective cavity. The light output from the uncoated facet is the laser light output. By angle-tuning the Bragg grating, the Bragg condition for the cavity changes and the output wavelength is

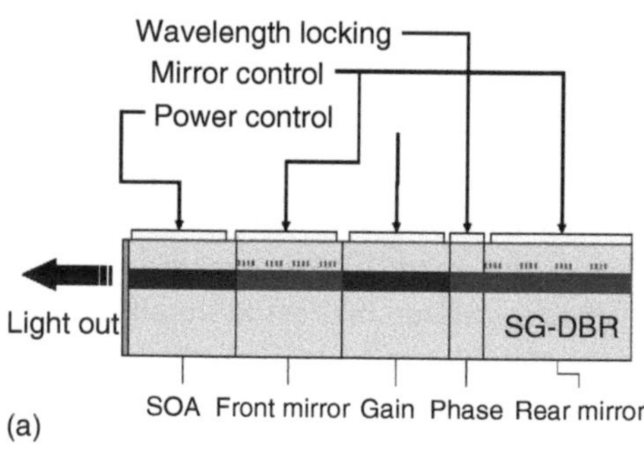

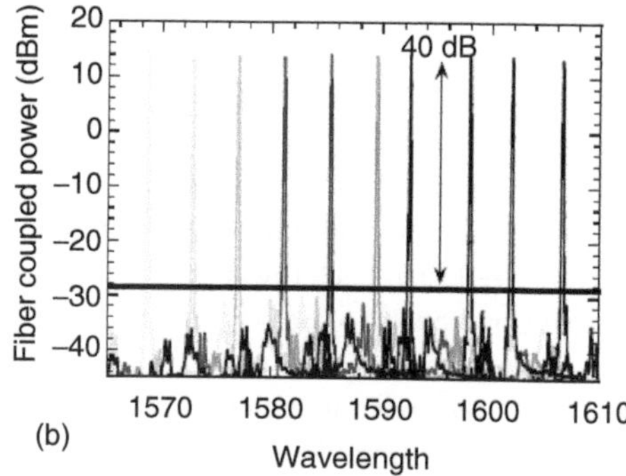

FIGURE 20.85 (a) Schematic of a sampled-grating DBR (SGDBR) with integrated SOA, (b) the output spectra of the SGDBR tunable laser. (Adapted from Coldren, L. A., Fish, G. A., Akulova, Y., Barton, J. S., Johansson, L., and Coldren, C. W. 2004. *IEEE/OSA J. Lightwave Technol.*, 22(1), 193–202. With permission.)

changed. This cavity can be quite long, perhaps even several centimeters. Since this cavity is quite long, small changes in angle can produce large changes in wavelength. Tunability on the order of ∼50 nm is common. However, since the round-trip propagation time of the cavity is so long, it is not possible to modulate or tune the output wavelength at more than 10s of MHz speeds; note that such a long laser cavity also produces a very small laser linewidth, useful in coherent systems. Therefore, this laser is only suitable as a highly tunable source in a fixed-transmitter WDM system.

Recent advances have been also made in developing the vertical-cavity surface-emitting laser (VCSEL).[134,135] The VCSEL cavity is very short, 100–1000 times shorter than that of a typical edge-emitting laser. There is typically one FP wavelength within the gain spectrum; hence the FP wavelength determines the lasing wavelength. The optical thickness variation of the layers in a VCSEL then will change the lasing wavelength.

20.4.3.2 Wavelength-Division (De)Multiplexers

Wavelength-selective multi-spatial routing of signals may be required for signal routing or signal recovery in a WDM network. If passive wavelength routing is required in a network, then $M \times N$ port devices are needed facilitate the selective routing of a signal from a given input port to a given output port according to the input signal's wavelength. These devices are most-commonly fabricated from some type of wavelength-selective grating which spatially deflects an optical beam depending on its wavelength. As examples, we will discuss a few wavelength (de)multiplexers below.

Figure 20.86 shows an $N \times 1$ wavelength multiplexer.[136] N input fibers are coupled to N input planar waveguides, for which a different wavelength would be input to each different fiber. The light from each waveguide would reflect off an integrated Bragg grating. We have already discussed Bragg reflection, and it was shown how the angular reflection off a grating is dependent on the incident wavelength. The multiplexing can occur since (1) each input waveguide is at a different spatial location, and (2) each wavelength has a different angular deflection. When these two mechanisms are combined, the different waveguides with different wavelengths can be arranged so that all light will be focused to the same focal position. Each input waveguide has a specific wavelength associated with it which would satisfy this condition. An output waveguide can be placed at the common focal position and will collect all the light. This multiplexer can be operated in reverse, with 1 input fiber containing N wavelengths demultiplexed into N output fibers.

AWGs (see Figure 20.42) have emerged as a very popular wavelength multiplexers/demultiplexer in WDM systems. It is an integrated frequency router that has a different functionality than either a simple wavelength multiplexer or a passive star coupler.[137] The device has N input ports and N output ports, and the same set of N different wavelengths can be input into each input port. The passbands can be < 0.5 nm and the number of input and output wavelengths/ports can be easily > 20.

Note that wavelength (de)multiplexing is more advanced than is the case for the optical filters for which only one wavelength is passed through the main output port and all the remaining wavelengths can be accessed through the second remaining output port. However, many filters we have discussed in the early section can be designed to support this functionality. For example, a fiber Bragg grating essentially can function as a wavelength-selective directional coupler. It can be used by a local node to

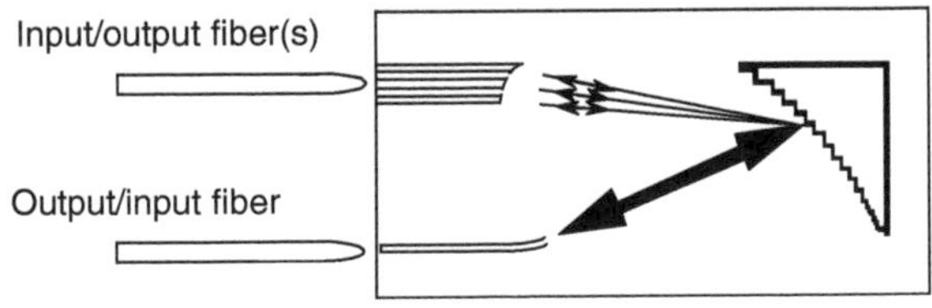

FIGURE 20.86 Integrated grating-based wavelength (de)multiplexer. (Adapted from Soole, J. B. D., et al. 1994. *Technical Digest*, San Jose, CA.)

inject a given wavelength onto a fiber (i.e., add new data) and remove that same wavelength from the fiber (i.e., remove old data). This functionality can be very useful in a ring or bus network configuration.

It is important to mention that even though the use of optical filters and WDMs allow a receiver to recover data on a single wavelength from a multitude of input wavelengths, a receiver array will not only allow single-wavelength demultiplexing but also the simultaneous recovery of several wavelengths at a given receiver. Each receiver in the array can recover a single transmitted wavelength, with the entire receiver array detecting all the possible wavelengths.[138]

20.4.3.3 WDM/SDM Cross-Connects

We have been discussing how a grating or WDM acts to separate wavelengths spatially and can add a new dimension of flexibility by introducing space-division multiplexing (SDM). In order to take advantage of hybridizing a system using two types of multiplexing, we can combine WDM and SDM in a cross-connect configuration as shown in Figure 20.87.[139] In so doing, we can reduce the complexity of each type of multiplexing scheme since, typically, a low-N multiplexing system is much simpler to implement than a high-N system. In this hybrid example, there are four input ports, four output ports, and 4 possible wavelengths at each input port. Each input port has a WDM which routes each wavelength to a different space-division-multiplexed crossbar switch. Each space crossbar switch corresponds to a specific input wavelength and has four inputs derived from each of the four major input fibers. Subsequently, each space switch independently switches a single input wavelength from any of the four original input ports to the appropriate destination output port. The switching is achieved by using some local (electronic) control signals which determine the complete optical path. If there is any output-port contention, the electronic control will reroute one of the signals to a different output port, known as deflection routing. This is considered active routing since there is some control necessary within each of the space switches. The outputs from the individual space switches are distributed to four output-port wavelength multiplexers. In this active WDM/SDM routing, N wavelengths can potentially form N^2 connections, but control signals are necessary for high throughput and effective routing. Subsequent to introducing this WDM/SDM example, the readers should review Section 20.2 on implementing an optical space switch. One advantage of this WDM/SDM approach is that small crossbar switches are

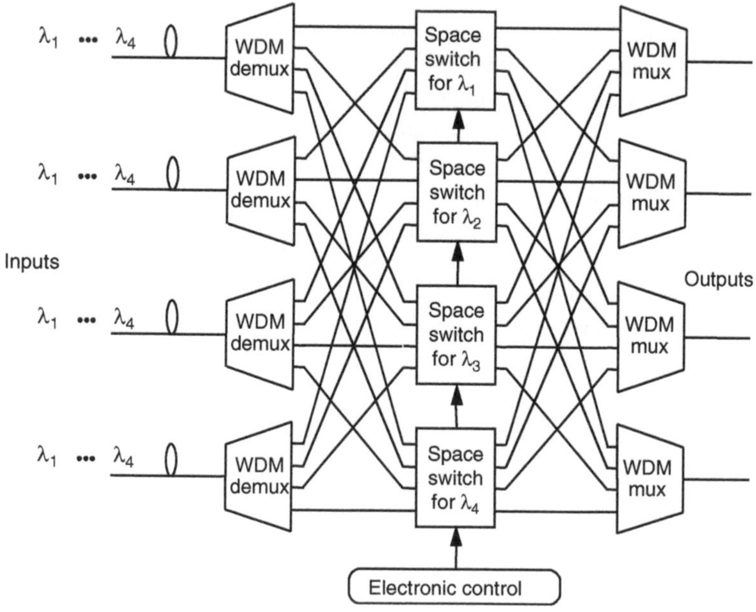

FIGURE 20.87 An optical crossconnect system in reconfigurable optical networks.

commercially available and a fairly mature technology, and can therefore be implemented in the near future. We will revisit in next section the concept of SDM and its generic advantages and disadvantages.

Because of a wealth of information and limited space in this chapter, we have discussed only a few examples of each technology, with many other types left untreated. The reader should review these earlier chapters for more in-depth information about these areas.

20.4.4 System Guidelines

We will now begin to discuss actual WDM system design criteria. It must be emphasized that different WDM systems will require different technologies and operating parameters for implementation.

20.4.4.1 Demultiplexing, Filtering, and Crosstalk

The first issue involves the most basic function of demultiplexing many WDM channels by using an optical filter.[140] We have already discussed the demultiplexing by an optical filter (see Figure 20.77). As an example, we assume that the optical filter is a Fabry-Perot etalon with the FWHM of the filter transmission f_{FP}. The slowly diminishing lineshape with long "tails" of the FP filter will allow some inter-wavelength crosstalk since not all of the optical power from the other channels will be blocked. The total power from many other rejected channels, P_{cr}, will be the crosstalk. We are always concerned about the SNR of the recovered data in the system, and this inter-channel crosstalk will act to reduce the power contrast ratio of the selected recovered signal, $S = P_s - P_{cr}$, where S is the effective recovered signal power, and P_S is the power in the selected wavelength which passes through the filter. The crosstalk due to power leakage from the other channels is deterministic and affects the signal mean (i.e., signal power) and not the signal variance (i.e., noise power). Although the exact tolerable amount of crosstalk depends on the WDM system, crosstalk should not exceed a few percent of the selected channel power in order to maintain good system performance. Note that a narrow filter is desirable since it will limit the amount of power leaking through the filter from neighboring channels. Furthermore, devices must also have a wide wavelength-tuning range in order for the channels to be spaced far apart thereby reducing crosstalk. If the filters have a periodic passband, two different wavelength channels could be transmitted through the filter at different passbands causing ambiguous channel recovery. Therefore, all the channels must be located within one free-spectral-range of the filters.

The previous discussion assumed that the signals had zero bandwidth and acted as if they were delta functions in comparison to the transmission function of the optical filter. In reality, the individual WDM channels have a broadened bandwidth in wavelength which will significantly affect the filtering and demultiplexing in a WDM system. A broadened channel will require (a) a wider optical filter to recover the entire signal and (b) a larger inter-channel spacing. Therefore, fewer channels would be accommodated if the entire system is of limited overall bandwidth. Here are three of the major mechanisms that cause channel broadening:

1. *Information bandwidth*—the channel broadening which is the most fundamental and almost impossible to reduce is the information bandwidth of the transmitted signal. Any modulated signal is spectrally broadened due to the inherent information capacity of that channel.[141] For intensity-modulated channels, the 3-dB FWHM bandwidth is roughly equal to the modulation bit rate, with a 1 Gbit/s signal having a 1 GHz bandwidth. For very fast modulated signals at 20 Gbit/s, the information bandwidth is ~1.5 Å and will affect the required width of the optical filter passband which must be used to recover the entire transmitted signal.
2. *Laser linewidth*—due to phase fluctuations in a laser cavity and the quantum-mechanical energy-level fluctuations of the stimulated emission events, the spectral linewidth of the laser optical output has a nonzero value.[30] The fluctuations produce a broadening of the output spectra even when the light is not being modulated. Typical DFB lasers have output-power-dependent laser linewidths of 10–100 MHz, with most lasers being closer to 10 MHz. Such a value for the laser linewidth would have a severe impact on coherent systems since the input signal and local oscillator must be mixed together and fluctuations inherently reduce the coherence time/length of

an optical signal. For direct-detection systems, the laser linewidth broadens the signal spectrum, requiring a wider optical filter bandwidth to pass the selected channel. However, since most direct-detection optical transmission systems operate at > 100s Mbit/s speeds, the laser linewidth is usually considered negligible compared to the signal information bandwidth.

3. *Laser chirp*—one of the most troubling broadening mechanisms is the chirp (i.e., wavelength broadening) of the laser output when the injected current into the laser is directly modulated to produce a modulation in the output light.[142] Unfortunately, directly modulating the laser current is also considered the simplest way to modulate the light.

In an ideal semiconductor laser, the number of carriers will increase with increasing currents only until the threshold level is reached at which point lasing occurs when there is more gain than loss. After lasing is initiated, any additional injected carriers would only contribute to additional photon generation since the total number of carriers in the cavity will be fixed and clamped at the gain = loss value. However, since nonradiative transitions and undesirable current leakage will occur in a nonideal laser, the carrier density will vary with a change in the injected current even after lasing has begun. A change in the current density causes (1) a slight change in the refractive index, (2) a change in the effective length of the cavity, (3) a change in the wavelength which will satisfy the cavity lasing conditions (as determined by the facet mirrors or the distributed Bragg reflections), and, ultimately, (4) a change in the final output lasing wavelength. Additionally, nonradiative recombinations and a current-density variation will both cause a temperature change in the cavity, thereby changing the length of the cavity and the wavelength which will be output. Direct current injection can cause as much as a ~1 Å (~15 GHz) spectral broadening.

One straightforward method to practically eliminate the effects of chirp is to avoid directly modulating the current of the laser. In Section 20.2 we discussed the external optical modulators. The MZI in which the phase difference between the two arms can be electrically changed from 0 to π radians results in, respectively, constructive and destructive interference at the modulator output. An external modulator is effective but can be more expensive than the laser itself. There has been some recent work to integrate an external modulator on the same chip as the laser itself in order to reduce the cost and complexity of having a separate bulk device. The modulator is external to the lasing cavity but is still located on the same chip. The modulating region controls the optical output by changing the transmissivity and electroabsorption within a quantum-well waveguiding region.

In passing, it should be emphasized that the issue of channel spectral broadening not only affects the channel spacing and filter width in a WDM system, but also critically affects the capacity in a long-distance transmission system. Since different frequencies travel down an optical fiber at different speeds due to a wavelength-dependent refractive index, an optical pulse which is spectrally broad will suffer severe pulse spreading and dispersion as it propagates (i.e., intramodal material dispersion). Therefore, external modulators are exclusively used for transmitting high-speed optical signals over long-distance optical links.

20.4.4.2 Wavelength Stability

Although experimental demonstrations of WDM systems can be performed over the course of a few minutes, real systems must be stable over the course of years. A critical issue in the long-term physical-layer performance of a WDM system is the stability, repeatability, and accuracy of the wavelength characteristics of the various components. The wavelength characteristics of lasers, filters, and multiplexers will change slightly with time due to many factors, chief of which are temperature changes, device aging, and hysteresis. An uncontrolled drift in the wavelength characteristics of the network would wreak havoc and bring this ultra-high-capacity multi-channel high-speed system to a grinding halt.

Wavelength stability can be achieved by various "locking" or control methods. Four basic representative examples include (a) lock channel to an atomic transition,[143] (b) lock filter to a selected channel,[144] (c) lock each channel to a different resonance of a FP cavity,[145] and (d) reference individual wavelength to a central source.[146] Of course, the employment of any type of external control techniques in a WDM system will add to its complexity and cost. Furthermore, these methods can themselves be unreliable over a long period of time. Therefore, it is more desirable to design a system which is itself robust to any slight changes in wavelength characteristics. The most straightforward way to have a robust WDM system is to have the wavelengths spaced very far apart, to have optical filters with large transmission bandwidths, etc. If the optical filter passband is large compared to the channel spacing, then the system will be fairly tolerant to any slight wavelength drift. However, there is still the requirement that the lasers or filters will not drift significantly over the course of years, especially for the dense WDM systems. Currently, many of the commercialized 1.55 μm DFB lasers have very good wavelength stability (~5 pm over 24 h). In one report, by introducing the wavelength-locking feedback loop, the emission wavelength of the laser module drifted by only 0.4 pm from beginning-of-life to end-of-life. The result of 500 cycles of temperature cycling test shows that the wavelength drift was less than 1 pm. The worst overall wavelength drift of these laser modules after 25 years of usage was anticipated to be less than 5 pm.[147] It must be emphasized that a real system may have many wavelength-dependent components cascaded over any given communications path, thereby exacerbating any wavelength-drift problems which may exist (i.e., the effective wavelength passband for the entire link is much narrower for a cascade than for a single element).

20.4.5 Optical Amplifiers in WDM Networks

EDFAs are discussed in great detail in Section 20.2. In this section, we will consider some important issues about EDFAs with regards to their implementation in WDM systems. EDFAs have been used in multi-channel WDM systems to compensate for (1) fiber attenuation losses in transmission, (2) component excess losses, and (3) optical network splitting losses. These optical splitting losses can occur in a passive star, in which the optical power is divided by the number of users (N), or in a ring/bus in which there may possibly be optical tapping losses at each node.

Figure 20.88a shows WDM transmission in a conventional electrically regenerated system. Regenerators can correct for fiber attenuation and chromatic dispersion by detecting an optical signal and then retransmitting it as a new signal using an internal laser. However, regenerators (being a hybrid of optics and electronics) are expensive, bit-rate and modulation-format specific, and waste much power and time in converting from photons to electrons and back again to photons. In contrast, as shown in Figure 20.88b, the EDFA is ideally a transparent box which is insensitive to the bit-rate, modulation-format, power, and wavelengths of the signal(s) passing through it, and most importantly, provides gain for all the WDM channels simultaneously. Since all the channels remain in optical form during

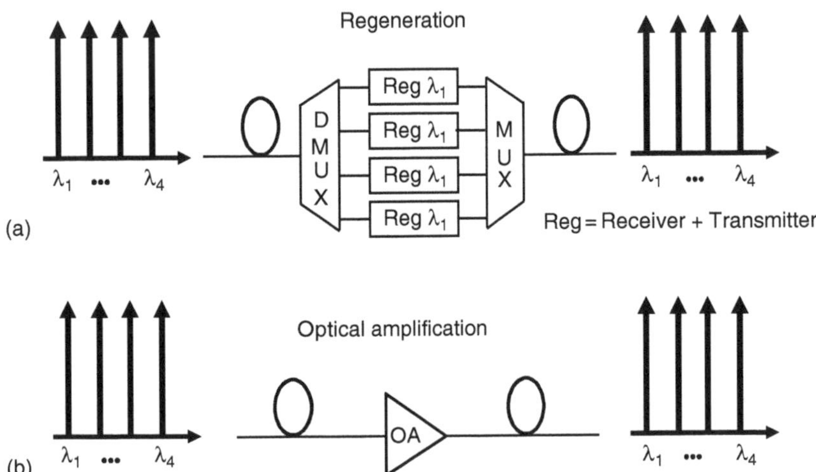

FIGURE 20.88 Wideband amplifiers enable WDM.

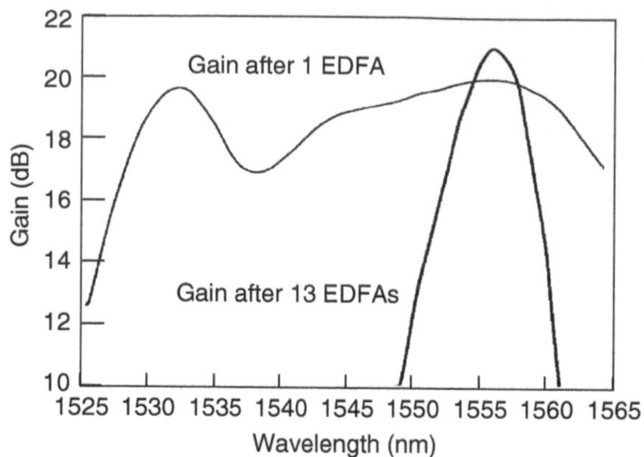

FIGURE 20.89 EDFA gain non-uniformity accumulation.

amplification, optically amplified WDM systems are potentially cheaper and more reliable than electrically regenerated systems.

20.4.5.1 Gain Peaking in EDFA Cascades

The EDFA is an almost ideal optical amplifier for WDM systems except for one major flaw: the gain is not uniform with wavelength, whereas the inter-amplifier losses are nearly wavelength independent.[148–151] For a single amplifier, as shown in Figure 20.89, the gain exhibits a peak at 1530 nm and a relatively flat region near 1555 nm. Moreover, the gain shape of an EDFA is dependent on the inversion of Er^{3+} in the EDF.[119] When the inversion is low, which can be achieved by operating the amplifier in deep saturation, the gain peak at 1530 nm can be suppressed and the gain flatness around 1555 nm would become quite flat.

If several channels are located on the relatively flat shoulder region of the gain spectrum, then the gain differential after a single amplifier will be within a few dB. However, when a cascade of EDFAs is used to periodically compensate for losses, the differential in gain and resultant SNR can become quite severe. A large differential in SNR among many channels can be deleterious for proper system performance. Figure 20.89 shows the gain spectrum after a single amplifier and after 13 cascaded amplifiers. The gain does not accumulate linearly from stage-to-stage, and the resultant wavelength-dependent gain shape dramatically changes in a cascade. Along the cascade, gain is gradually "pulled" away from the shorter wavelengths and made available at the longer wavelengths, resulting in a usable bandwidth of only several nanometers.

20.4.5.2 EDFA Gain Flattening

We have shown the bandwidth reduction due to nonuniform gain in a cascade of EDFAs. It is clear that gain flattening is an important issue in optically amplified networks. Several methods have been reported for equalizing nonuniform EDFA gain:

1. *Long period grating filters.* A long period grating (LPG) with an index-varying period of ~100 μm provides coupling between the core modes and the cladding modes, creating a wavelength dependent loss to equalize the EDFA gain shape,[152–154] as shown in Figure 20.90. Note that in a recent report, similar bandwidth is obtained after more than 9000 km transmission.
2. *Mach-Zehnder filters.* The wavelength dependent transmission characteristics of cascaded Mach-Zehnder filters can be tailored to compensate the gain nonuniformity of EDFAs.[155]
3. *Fluoride-based EDFAs.* A fluoride-based EDFA[156] can provide an intrinsically flat gain over a wide wavelength range from 1530 nm to 1560 nm, which is a much wider bandwidth than for a silica-based EDFA. However, the fluoride-based fiber is extremely difficult to splice to normal fibers, and mechanical connections result in high connection loss and instability.

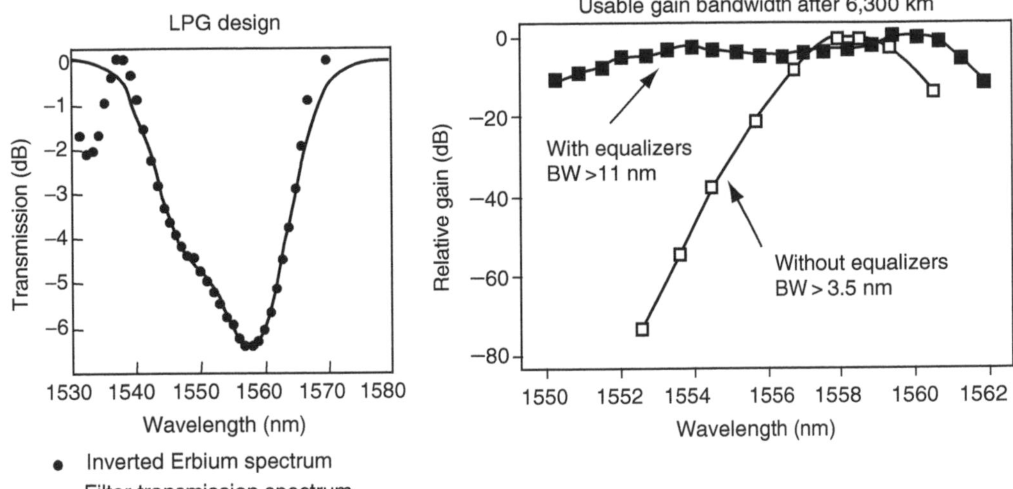

FIGURE 20.90 LPG design and gain equalization results. (Adapted from Vengsarkar, A. M., Lemaire, P. J., Judkins, J. B., Bhatia, V., Erdogan, T., and Sipe, J. E. 1996. *IEEE J. Lightwave Technol.*, 14, 58–65. With permission.)

20.4.5.3 Fast Power Transients

The lifetime of a stimulated Erbium ion is generally ~10 ms, which seems to be long enough to be transparent to signals modulated by data at the rates of several Gbit/s or higher. However, the EDFAs could be critically affected by the adding/dropping of WDM channels, network reconfiguration, or link failures, as illustrated in Figure 20.91. To achieve optimal channel SNRs, the EDFAs are typically operated in the gain-saturation regime where all channels must share the available gain.[157] Therefore, when channels are added or dropped, the power of the remaining channels will increase resulting transient effects.

The transients can be very fast in EDFA cascades.[158] As shown in Figure 20.92, with an increase in the number of cascaded EDFAs, the transients can occur in ~2 μs. These fast power transients in chain-amplifier systems should be controlled dynamically, and the response time required scales as the size of the network. For large-scale networks, response times shorter than 100 ns may be necessary.

From a system point-of-view, fiber nonlinearity may become a problem when too much channel power exists, and a small SNR at the receiver may arise when too little power remains.[159] The corresponding fiber transmission penalty of the surviving channel is shown in Figure 20.93 in terms of the *Q* factor, for varying numbers of cascaded EDFAs. When 15 channels are dropped or added, the penalties are quite severe. Note that this degradation increases with the number of channels *N* simply because of enhanced SPM due to a large power excursion as a result of dropping $N-1$ channels.

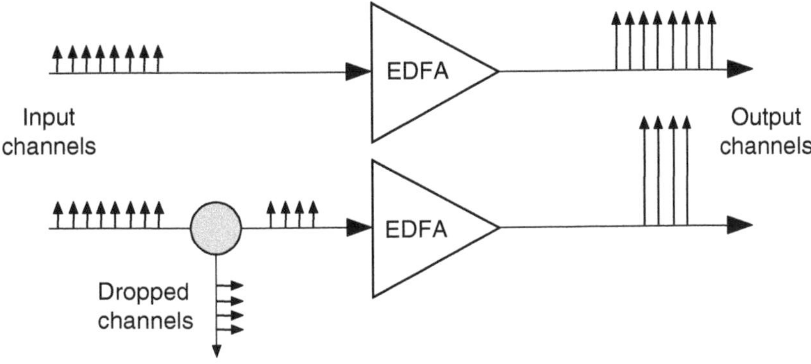

FIGURE 20.91 EDFA gain transients.

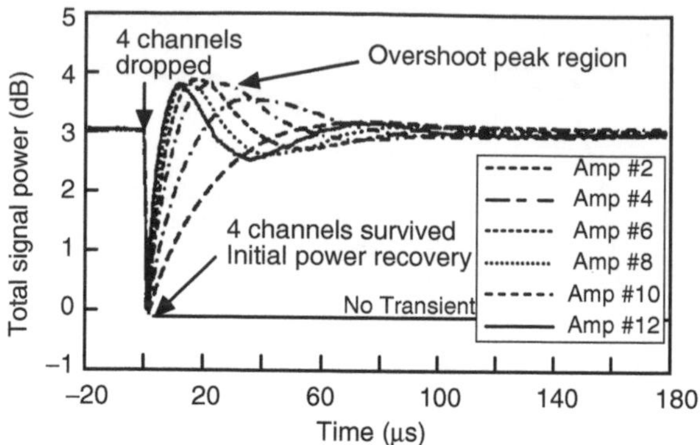

FIGURE 20.92 Fast power transients in EDFA cascades. (Adapted from Zyskind, J., Sun, Y., Srivastava, A., Sulhoff, J., Lucero, A., Wolf, C., and Tkach, R. 1996. *Conference on Optical Fiber Communications*. With permission.)

In order to maintain the quality of service, the surviving channels must be protected when channel add/drop or network reconfiguration occurs. The techniques include (a) optical attenuation, by adjusting optical attenuators between the gain stages in the amplifier to control the amplifier gain;[160] (b) pump power control, by adjusting the drive current of the pump lasers to control the amplifier gain;[161] (c) link control, using a power-variable control channel propagating with the signal channels to balance the amplifier gain;[162] and (d) EDFA gain clamping, by an automatic optical feedback control scheme to achieve all-optical gain clamping.[163]

20.4.5.4 Static Gain Dynamic and Channel Power Equalization

We just discussed EDFA gain flattening, which is a passive channel power equalization scheme effective only for a static link. However, in the nonstatic optical networks, the power in each channel suffers from dynamic network changes, including wavelength drift of components, changes in span loss, and channel add/drop. As an example, Figure 20.94 shows how the gain shape of a cascaded EDFA chain varies significantly with link loss changes due to environmental problems. This is because the EDFA gain spectra are dependent on the saturation level of the amplifiers. The results in Figure 20.94 are for a cascade of 10 gain-flattened EDFAs, each with 20-dB gain, saturated by 16 input-channels with −18 dBm per channel.

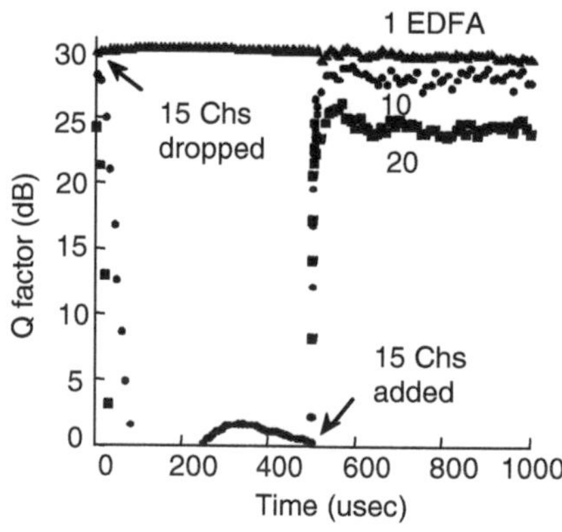

FIGURE 20.93 Q factor versus time for adding and dropping 15 channels of a 16-channel system at a bit rate of 10 Gbit/s. (Adapted from Hayee, M. and Willner, A. 1999. *IEEE Photon. Technol. Lett.*, 11, 889–891. With permission.)

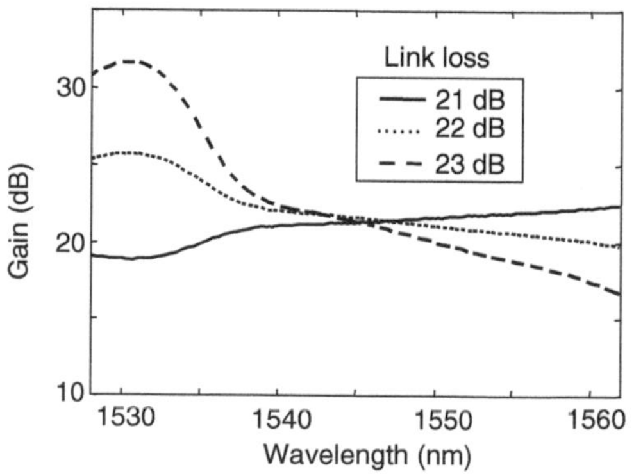

FIGURE 20.94 Gain spectra variation due to link loss changes for a cascade of 10 EDFAs.

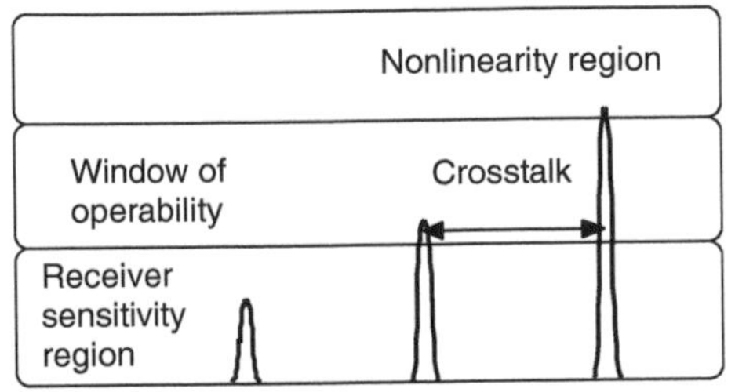

FIGURE 20.95 Degradation due to unequalized WDM channel power.

As illustrated in Figure 20.95, system performance can be degraded due to unequalized WDM channel power. These degrading effects include SNR differential (reduced system dynamic range), widely varying channel crosstalk, nonlinear effects, and low signal power at the receiver. Therefore, channel power needs to be equalized dynamically in WDM networks to ensure stable system performance. To obtain feedback for control purposes, a channel power monitoring scheme is very important. A simple way to accomplish this is to demultiplex all the channels and detect the power in each channel using different photodetectors or detector arrays. To avoid the high cost of many discrete components in WDM systems with large numbers of channels, other monitoring techniques that take advantage of wavelength-to-time mapping have also been proposed including the use of concatenated FBGs or swept acousto-optic tunable filters.

Various techniques have been proposed for dynamic channel power equalization, including parallel loss elements,[164] individual bulk devices (e.g., AOTFs),[165] serial filters,[166] micro-opto-mechanics (MEMS),[167] and integrated devices.[168] As an example, Figure 20.96 shows the parallel loss element scheme, where the channels are demultiplexed and attenuated by separate loss elements. An additional advantage of this scheme is that ASE noise is reduced by the WDM multiplexer and demultiplexer. Possible candidates for the loss elements in this scheme include opto-mechanical attenuators, acousto-optic modulators, or FBGs.

20.4.5.5 Intermodulation and Saturation-Induced Crosstalk in SOAs

We have already discussed the SOAs in the early section. Here, we will be focusing on the IMD and saturation-induced crosstalk in SOAs. These two effects are both nonlinear effects which exhibit significantly different effects for the semiconductor as opposed to the fiber amplifiers due to the vastly different carrier lifetimes involved. Note that even though these two effects have been troublesome in WDM multichannel systems, they are very useful for providing various ways for all-optical signal processing, such as wavelength conversion, optical regeneration, and optical TDM.

When two channels are incident into a closed amplifier system and their combined powers are near the amplifier saturation power, nonlinear effects occur which generate beat frequencies at the cross product of the two optical carrier waves. The carrier density (i.e., gain) will be modulated by the interference between any two optical signals, and this modulation occurs at the various beat (i.e., sum and difference)

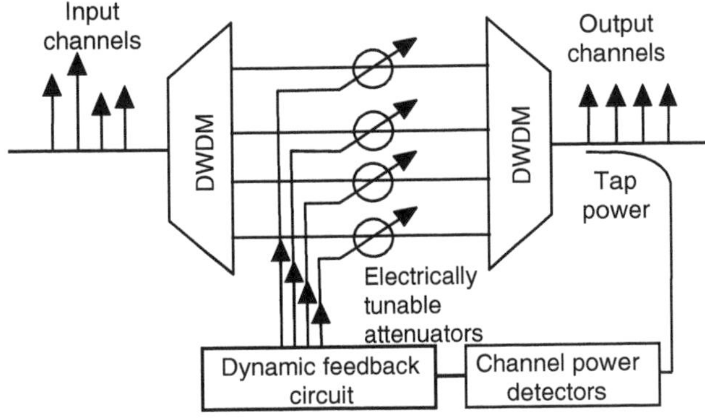

FIGURE 20.96 Parallel loss element scheme for dynamic channel power equalization.

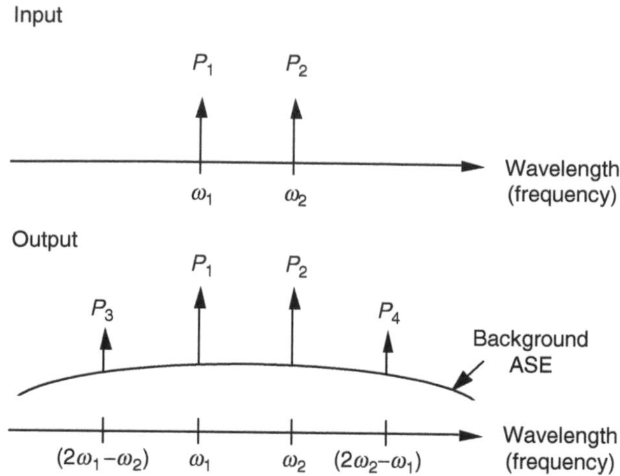

FIGURE 20.97 Example of four-wave mixing with two original signals and two newly produced signals.

frequencies which are generated by the possible combinations of input channels. This carrier density modulation at the beat frequencies produces additional signals which can interfere with the original desired signals. Figure 20.97 illustrates this scenario for 2 input signals which produce 4 output waves. Therefore, this nonlinear effect is called FWM, or, alternatively, IMD.[169–171] The modulated gain will modulate both the amplitude and phase of all signals and can be described by the third-order optical susceptibility of the semiconductor crystal.

Intermodulation distortion will include two-wave products (as in Figure 20.97) and three-wave products (see Figure 20.98). The two-wave products appear at frequencies of $(2\omega_i - \omega_j)$ and $(2\omega_j - \omega_i)$ and the three-wave products appear at $(\omega_i + \omega_j - \omega_k)$, $(\omega_i - \omega_j + \omega_k)$, and $(-\omega_i + \omega_j + \omega_k)$. The number of two-wave and three-wave products is proportional to the square and cube, respectively, of the number of total signals.

Although the IMD produced signals may individually be quite weak when compared to the original signals, many products may appear within the optical bandwidth of an original signal. These products will pass through any optical filter at a receiver and give rise to amplitude fluctuations in the recovered selected signal. Therefore, these products must be summed for their total effect on any multi-channel system.

Crosstalk can also occur in a gain-saturated amplifier. As the intensity of the signals in an amplifier increases beyond the saturation input power, then the gain will decrease. When the input signal intensity

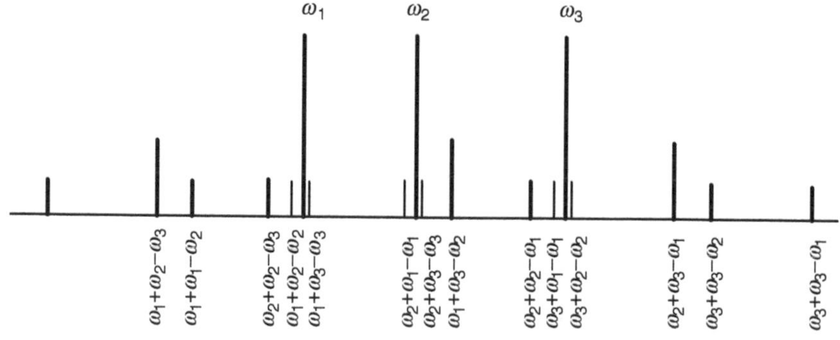

FIGURE 20.98 Example of three-wave products generated due to intermodulation distortion.

drops, the gain increases to its unsaturated value. Thus, the gain and input signal power are inverse functions of each other when the amplifier is saturated. This gain fluctuation occurs as rapidly as determined by the carrier lifetime, τ_s, of the amplifier, again being $\sim$1 ns in a typical semiconductor amplifier and is comparable to the bit-time in a Gbit/s data stream. If we assume a homogeneously broadened amplifier which becomes equally saturated across the entire gain bandwidth independent of the wavelength of the saturating input signal, then an increase in the intensity beyond P_{sat}^{in} of one channel in a two-channel system will necessitate a decrease in the gain of both channels causing crosstalk in the second channel. If the gain can respond on the same time-scale as a bit time in a Gbit/s transmission system, then as one channel is being ASK modulated, the second channel will have its gain also modulated within a bit time, producing signal distortion and power penalties. Therefore, the saturated semiconductor amplifier is bit-rate dependent. This scenario is depicted in Figure 20.99.

To correct both these problems we can increase the saturation output power of the amplifier so that saturation does not occur easily. Furthermore, FWM can be significantly reduced by reducing τ_s. Both these corrective features can be accomplished to some degree by fabricating a multiple-quantum-well (MQW) semiconductor amplifier, which has exhibited a $P_{sat}^{out}(3-\text{dB})$ of $\sim$40 mW and a $\tau_s\sim$0.2 ns.[172,173] However, as we already mentioned in the early section of amplifiers, the nonlinear effects in SOAs can be used for all optical data processing such as wavelength conversion and optical regeneration.

20.4.6 Fiber Dispersion and Nonlinearities Management

As mentioned in Section 20.2, the dispersive effect in an optical fiber has several ingredients including intermodal dispersion in a multimode fiber, waveguide dispersion, material dispersion, chromatic dispersion and polarization mode dispersion. Dispersion is one of the critical roadblocks to increasing the transmission capacity of optical fiber. In particular, chromatic for data rates$\geq$10 Gbit/s, transmission can be quite tricky and the chromatic dispersion-induced degrading effects must be dealt with in some fashion, perhaps by compensation. Furthermore, the effects of chromatic dispersion rise quite rapidly as the bit rate increases—when the bit rate increases by a factor of four, the effects of chromatic dispersion increase by a whopping factor of 16! Even though it is possible to manufacture fiber with zero dispersion, it is not practical to use such fiber for WDM transmission due to large penalties induced by fiber nonlinearities. With zero dispersion, all channels in a WDM system travel at the same speed, in-phase, and a number of deleterious nonlinear effects such as XPM and FWM result. Thus, in WDM systems,

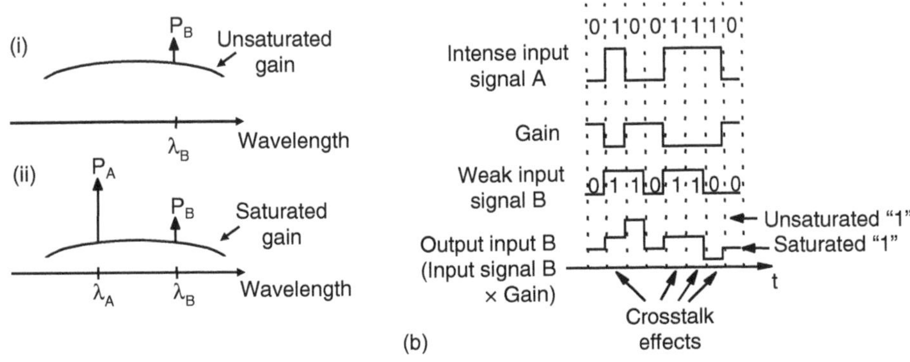

FIGURE 20.99 (a) Signal and gain spectra given a weak signal B "ON" or given intense signal A and weak signal B "ON" simultaneously producing gain saturation. (b) Bit stream sequences for two signals propagating through a semiconductor amplifier. All pulse transitions are sharp since we assume the response time of the amplifier gain to be much greater than the bit rate. If they are comparable, then pulse-rounding effects will occur.

some amount of chromatic dispersion is necessary to keep channels out-of-phase, and as such chromatic dispersion compensation is required.

This section will address the concepts of chromatic dispersion and fiber nonlinearities management followed by some examples highlighting the need for tunability to enable robust optical WDM systems in dynamic environments. PMD effects will be also discussed briefly.

20.4.6.1 Chromatic Dispersion Effects on Fiber Communication Systems

The data rate and the data modulation format can significantly affect the sensitivity of a system to chromatic dispersion. For example, the common NRZ data format, in which the optical power stays high throughout the entire time slot of a “1” bit, is more robust to chromatic dispersion than is the RZ format, in which the optical power stays high in only part of the time slot of a “1” bit. This difference is due to the fact that RZ data has a much wider channel frequency spectrum compared to NRZ data, thus incurring more chromatic dispersion. However, in a real WDM system, the RZ format increases the maximum allowable transmission distance by virtue of its reduced duty cycle (compared to the NRZ format) making it less susceptible to fiber nonlinearities. We will discuss some robust modulation formats in the later section too.

A rule for the maximum distance over which data can be transmitted is to consider a broadening of the pulse equal to the bit period. For a bit period B, a dispersion value D and a spectral width, $\Delta\lambda$, the dispersion-limited distance is given by

$$L_D = \frac{1}{D \cdot B \cdot \Delta\lambda} = \frac{1}{D \cdot B \cdot (cB)} \propto \frac{1}{B^2} \tag{20.58}$$

(see Figure 20.100). For example, for single mode fiber, $D=17$ ps/nm/km, so for 10 Gbit/s data the distance is $L_D=52$ km. In fact, a more exact calculation shows that for 60 km, the dispersion induced power penalty is less than 1 dB.[174] The power penalty for uncompensated dispersion rises exponentially with transmission distance, and thus to maintain good signal quality, dispersion compensation is required.

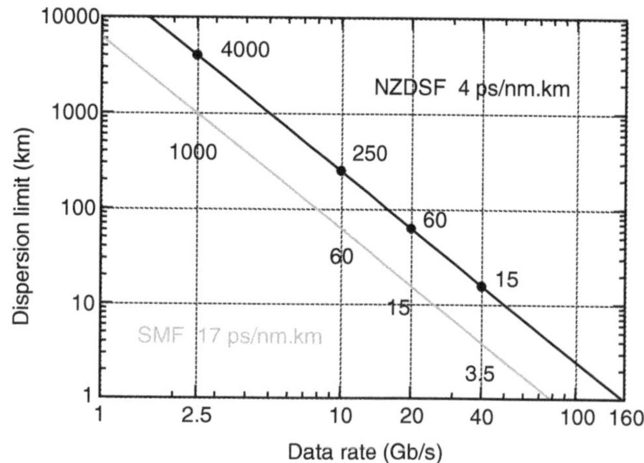

FIGURE 20.100 Transmission distance limitations due to uncompensated dispersion in SMF as a function of data rate for intensity modulated optical signals. (Adapted from Garrett, L. D. 2001. Invited short course at Optical Fiber Communication Conference. With permission.)

20.4.6.2 Optical Nonlinearities as Factors to Be Considered in WDM Systems

Most nonlinear effects originate from the nonlinear refractive index of fiber, which is not only dependent on the frequency of light but also on the intensity (optical power), and is related to the optical power as

$$\bar{n}(f,p) = n(f) + n_2 \frac{P}{A_{\text{eff}}} \tag{20.59}$$

where $n(f)$ is the linear part of the refractive index, P is the optical power inside the fiber, and n_2 is the nonlinear-index coefficient for silica fibers. The typical value of n_2 is 2.6×10^{-20} m^2/W. This number takes into account the averaging of the polarization states of the light as it travels in the fiber. The intensity dependence of the refractive index gives rise to three major nonlinear effects.

SPM. A million photons "see" a different glass than does a single photon, and a photon traveling along with many other photons will slow down. SPM occurs because of the varying intensity profile of an optical pulse on a single WDM channel (see Figure 20.101a). This intensity profile causes a refractive index profile and, thus, a photon speed differential. The resulting phase change for light propagating in an optical fiber is expressed as

$$\Phi_{\text{NL}} = \gamma P L_{\text{eff}} \tag{20.60}$$

where the quantities γ and L_{eff} are defined as

$$\gamma = \frac{2\pi n_2}{\lambda A_{\text{eff}}} \quad \text{and} \quad L_{\text{eff}} = \frac{1 - e^{-\alpha L}}{\alpha} \tag{20.61}$$

where A_{eff} is the effective mode area of the fiber and α is the fiber attenuation loss. L_{eff} is the effective nonlinear length of the fiber that accounts for fiber loss, and γ is the nonlinear coefficient measured in rad/km/W. A typical range of values for γ is between 10^{-30} rad/km/W. Although the nonlinear coefficient is small, the long transmission lengths and high optical powers that have been made possible by the use of optical amplifiers can cause a large enough nonlinear phase change to play a significant role in state-of-the-art lightwave systems.

Cross-Phase Modulation (XPM). When considering many WDM channels co-propagating in a fiber, photons from channels 2 through N can distort the index profile that is experienced by channel 1.

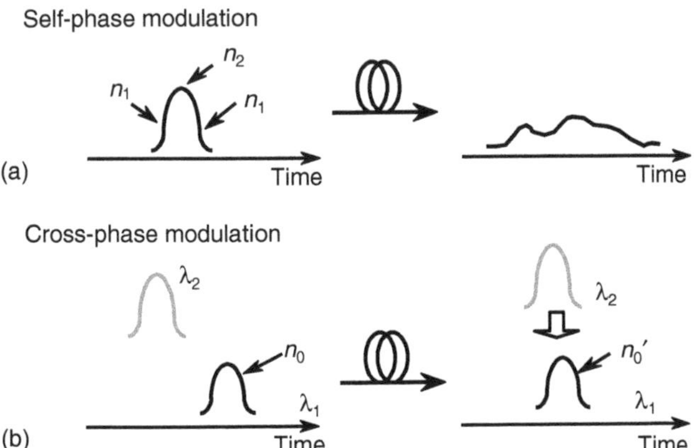

FIGURE 20.101 (a) Self-phase modulation: the photons in the pulse "see" different refraction index, and (b) cross-phase modulation: the glass that a photon in the λ_2 pulse "sees" changes as other channels (with potentially varying power) move to coincide with the λ_2 pulse.

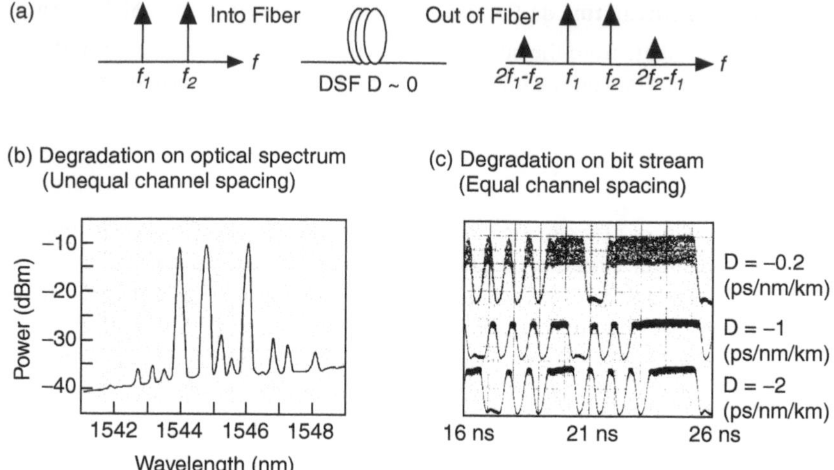

FIGURE 20.102 (a) and (b) FWM induces new spectral components via nonlinear mixing of two wavelength signals, (c) the signal degradation due to FWM products falling on a third data channel can be reduced by even small amounts of dispersion. (After Tkach, R. W., Chraplyvy, A. R., Forghieri, F., Gnauck, A. H., and Derosier, R. M. 1995. *IEEE/OSA J. Lightwave Technol.*, 13(5), 841–849. With permission.).

The photons from the other channels "chirp" the signal frequencies on channel 1, which will interact with fiber chromatic dispersion and cause temporal distortion (see Figure 20.101b). This effect is called XPM. In a two-channel system, the frequency chirp in channel 1 due to power fluctuation within both channels is given by

$$\Delta B = \frac{\mathrm{d}\Phi_{\mathrm{NL}}}{\mathrm{d}t} = \gamma L_{\mathrm{eff}}\frac{\mathrm{d}P_1}{\mathrm{d}t} + 2\gamma L_{\mathrm{eff}}\frac{\mathrm{d}P_2}{\mathrm{d}t} \tag{20.62}$$

where $\mathrm{d}P_1/\mathrm{d}t$ and $\mathrm{d}P_2/\mathrm{d}t$ are the time derivatives of the pulse powers of channels 1 and 2, respectively. The first term on right hand side of the above equation is due to SPM, and the second term is due to XPM. Note that the XPM-induced chirp term is double that of the SPM-induced chirp term. As such, XPM can impose a much greater limitation on WDM systems than can SPM, especially in systems with many WDM channels.

Four-Wave-Mixing (FWM). The optical intensity propagating through the fiber is related to the electric field intensity squared. In a WDM system, the total electric field is the sum of the electric fields of each individual channel. When squaring the sum of different fields, products emerge that are beat terms at various sum and difference frequencies to the original signals. Figure 20.102 shows that if a WDM channel exists at one of the four-wave-mixing beat-term frequencies, the beat term will interfere coherently with this other WDM channel and potentially destroy the data.

20.4.6.3 Dispersion Maps

While zero-dispersion fiber is not a good idea, a large value of the accumulated dispersion at the end of a fiber link is also undesirable. An ideal solution is to have a "dispersion map," alternating sections of positive and negative dispersion as can be seen in Figure 20.103. This is a very powerful concept: at each point along the fiber the dispersion has some nonzero value, eliminating FWM and XPM, but the total dispersion at the end of the fiber link is zero, so that no pulse broadening is induced. The most advanced systems require periodic dispersion compensation, as well as pre- and post-compensation (before and after the transmission fiber).

The addition of negative dispersion to a standard fiber link has been traditionally known as "dispersion compensation"; however, the term "dispersion management" is more appropriate. SMF has positive

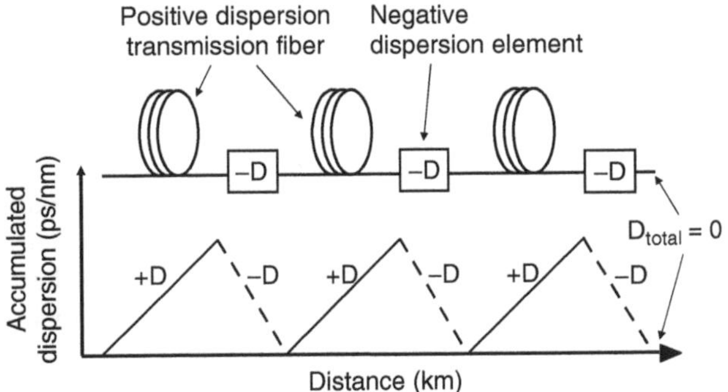

FIGURE 20.103 Dispersion map of a basic dispersion managed system. Positive dispersion transmission fiber alternates with negative dispersion compensation elements such that the total dispersion is zero end-to-end.

dispersion, but some new varieties of nonzero dispersion-shifted fiber (NZDSF) come in both positive and negative dispersion varieties, as shown in Figure 20.104. Reverse dispersion fiber is also now available, with a large dispersion comparable to that of SMF, but with the opposite sign. When such flexibility is available in choosing both the magnitude and sign of the dispersion of the fiber in a link, dispersion-managed systems can fully be optimized to the desired dispersion map using a combination of fiber and dispersion compensation devices (see Figure 20.104).

Dispersion is a linear process, so to first order dispersion maps can be understood as linear systems. However, the effects of nonlinearities cannot be ignored, especially in WDM systems with many tens of channels where the launch power may be very high. In particular, in systems deploying dispersion compensating fiber (DCF), the large nonlinear coefficient of the DCF can dramatically affect the dispersion map.

Chromatic dispersion is a necessity in WDM systems to minimize the effects of fiber nonlinearities. A chromatic dispersion value as small as a few ps/nm/km is usually sufficient to make XPM and FWM negligible. To mitigate the effects of nonlinearities but maintain small amounts of chromatic dispersion, NZDSF is commercially available. Due to these nonlinear effects, chromatic dispersion must be managed, rather than eliminated.

If a dispersion-management system was perfectly linear, it would be irrelevant whether the dispersion along a path is small or large, as long as the overall dispersion is compensated to zero (end to end). Thus,

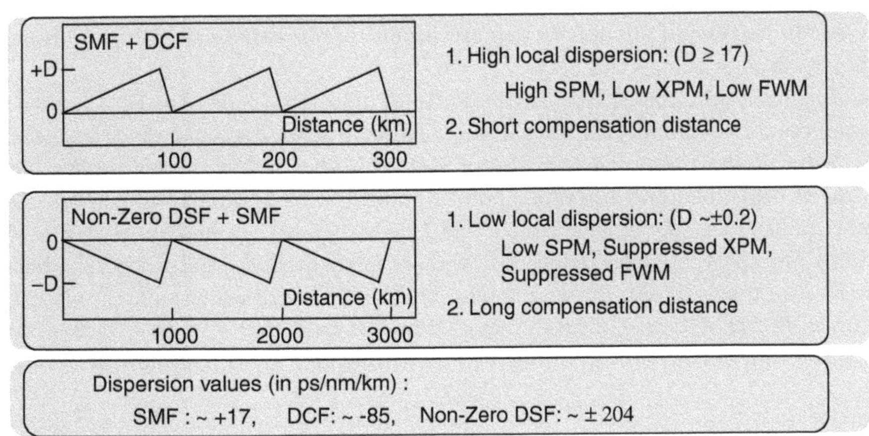

FIGURE 20.104 Various dispersion maps for SMF-DCF and NZDSF-SMF.

in a linear system the performance should be similar, regardless of whether the transmission fiber is SMF, and dispersion compensation modules are deployed every 60 km, or the transmission fiber is NZDSF (with approximately ¼ the dispersion value of SMF) and dispersion compensation modules are deployed every 240 km. In real life, optical nonlinearities are very important, and recent results seem to favor the use of large, SMF-like, dispersion values in the transmission path and correspondingly high dispersion compensation devices. A recent study of performance versus channel spacing showed that the capacity of SMF could be more than four times that of NZDSF. This is because the nonlinear coefficients are much higher in NZDSF than in SMF, and for dense WDM the channel interactions become a limiting factor. A critical conclusion is that not all dispersion compensation maps are created equal: a simple calculation of the dispersion compensation to cancel the overall dispersion value does not lead to optimal dispersion map designs.

Additionally, several solutions have been shown to be either resistant to dispersion, or have been shown to rely on dispersion itself for transmission. Such solutions include chirped pulses (where prechirping emphasizes the spectrum of the pulses so that dispersion does not broaden them too much), dispersion assisted transmission (where an initial phase modulation tailored to the transmission distance leads to full-scale amplitude modulation at the receiver end due to the dispersion), and various modulation formats robust to chromatic dispersion and nonlinearities. We will discuss these issues in the later section.

20.4.6.4 Dispersion Management Solutions

Fixed Dispersion Compensation. From a systems point of view, there are several requirements for a dispersion compensating module: low loss, low optical nonlinearity, broadband (or multichannel) operation, small footprint, low weight, low power consumption, and clearly low cost. It is unfortunate that the first dispersion compensation modules, based on DCF only, met two of these requirements: broadband operation and low power consumption. On the other hand, several solutions have emerged that can complement or even replace these first-generation compensators.

Dispersion Compensating Fiber (DCF). One of the first dispersion compensation techniques was to deploy specially designed sections of fiber with negative chromatic dispersion. The technology for DCF emerged in the 1980s and has developed dramatically since the advent of optical amplifiers in 1990. DCF is the most widely deployed dispersion compensator, providing broadband operation and stable dispersion characteristics, and the lack of a dynamic, tunable DCF solution has not reduced its popularity.[175]

As can be seen in Figure 20.105, the core of the average DCF is much smaller than that of standard SMF, and beams with longer wavelengths experience relatively large changes in mode size (due to the waveguide structure) leading to greater propagation through the cladding of the fiber, where the speed of light is greater than that of the core. This leads to a large negative dispersion value. Additional cladding layers can lead to improved DCF designs that can include negative dispersion slope to counteract the positive dispersion slope of standard SMF.

In spite of its many advantages, DCF has a number of drawbacks. First of all, it is limited to a fixed compensation value. In addition, DCF has a weakly guiding structure and has a much smaller core cross-section, $\sim 19\ \mu m^2$, compared to the $\sim 85\ \mu m^2$ of SMF. This leads to higher nonlinearity, higher splice losses, as well as higher bending losses. Secondly, the length of DCF required to compensate for SMF dispersion is rather long, about one-fifth of the length of the transmission fiber for which it is compensating. Thus DCF modules induce loss, and are relatively bulky and heavy. The bulk is partly due to the mass of fiber, but also due to the resin used to hold the fiber securely in place. One other contribution to the size of the module is the higher bend loss associated with the refractive index profile of DCF; this limits the radius of the DCF loop to 6–8 in., compared to the minimum bend radius of 2 in. for SMF.

Traditionally, DCF-based dispersion compensation modules are usually located at amplifier sites. This serves several purposes. First, amplifier sites offer relatively easy access to the fiber, without requiring any digging or unbraiding of the cable. Second, DCF has high loss (usually at least double that of standard

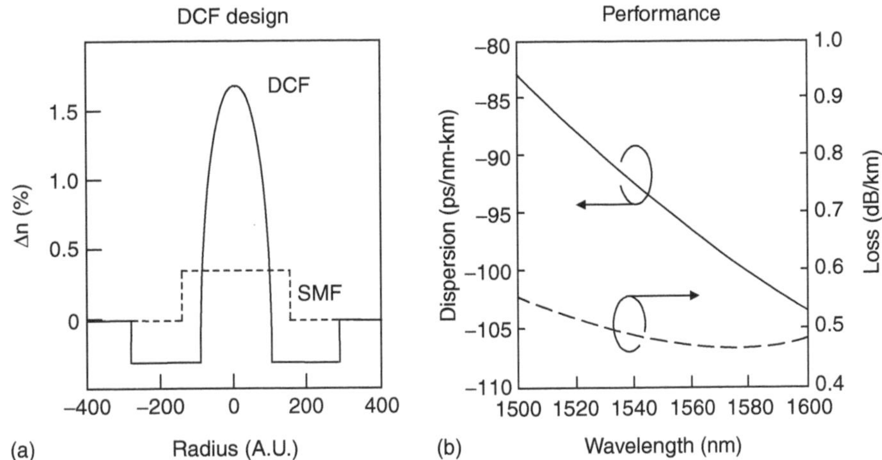

FIGURE 20.105 Typical DCF (a) refractive index profile and (b) dispersion and loss as a function of wavelength. Δn is defined as refractive index variation relative to the cladding.

SMF), so a gain stage is required before the DCF module to avoid excessively low signal levels. DCF has a cross section four times smaller then SMF, hence a higher nonlinearity, which limits the maximum launch power into a DCF module. The compromise is to place the DCF in the midsection of a two-section EDFA. This way, the first stage provides pre-DCF gain, but not to a power level that would generate excessive nonlinear effects in the DCF. The second stage amplifies the dispersion compensated signal to a power level suitable for transmission though the fiber link. This launch power level is typically much higher than could be transmitted through DCF without generating large nonlinear effects. Many newer dispersion compensation devices have better performance than DCF, in particular lower loss and lower nonlinearities. For this reason, they may not have to be deployed at the midsection of an amplifier.

Chirped Fiber Bragg Gratings. FBGs have emerged as major components for dispersion compensation because of their low loss, small footprint and low optical nonlinearity.[176] When the periodicity of the grating is varied along its length, the result is a chirped grating which can be used to compensate for chromatic dispersion. The chirp is understood as the rate of change of the spatial frequency as a function of position along the grating. In chirped gratings the Bragg matching condition for different wavelengths occurs at different positions along the grating length. Thus, the roundtrip delay of each wavelength can be tailored by designing the chirp profile appropriately. Figure 20.106 compares the chirped FBG with uniform FBG. In a data pulse that has been distorted by dispersion, different frequency components arrive with different amounts of relative delay. By tailoring the chirp profile such that the frequency components see a relative delay which is the inverse of the delay of the transmission fiber, the pulse can be compressed back. The dispersion of the grating is the slope of the time delay as a function of wavelength, which is related to the chirp. An optical circulator is traditionally used to separate the reflected output beam from the input beam.

The main drawback of Bragg gratings is that the amplitude profile and the phase profile as a function of wavelength have some amount of ripple. Ideally, the amplitude profile of the grating should have a flat (or rounded) top in the passband, and the phase profile should be linear (for linearly chirped gratings) or polynomial (for nonlinearly chirped gratings). The grating ripple is the deviation from the ideal profile shape. Considerable effort has been expended on reducing the ripple. While early gratings were plagued by more than 100 ps of ripple, published results have shown vast improvement to values close to ± 3 ps.

Ultimately, dispersion compensators should accommodate multi-channel operation. Several WDM channels can be accommodated by a single chirped FBG in one of two ways: (1) fabricating a much

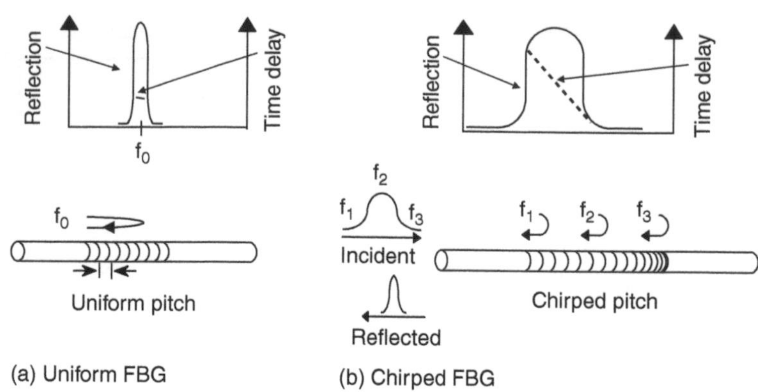

FIGURE 20.106 Uniform and chirped FBGs. (a) A grating with uniform pitch has a narrow reflection spectrum and a flat time delay as a function of wavelength, and (b) A chirped FBG has a wider bandwidth, a varying time delay and a longer grating length. Chirped gratings reflect different frequency components at different locations within the grating.

longer (i.e., meters-length) grating; or (2) using a sampling function when writing the grating, thereby creating many replicas of transfer function of the FBG in the wavelength domain.[177] (see Figure 20.107).

Tunable Dispersion Compensation. In a perfect world, all fiber links would have a known, discrete, and unchanging value of chromatic dispersion. Network operators would then deploy fixed dispersion compensators periodically along every fiber link to exactly match the fiber dispersion. Unfortunately, several vexing issues may necessitate that dispersion compensators have tunability, that they have the ability to adjust the amount of dispersion to match system requirements.

First, there is the most basic business issue of inventory management. Network operators typically do not know the exact length of a deployed fiber link nor its chromatic dispersion value. Moreover, fiber plants periodically undergo upgrades and maintenance, leaving new and nonexact lengths of fiber behind. Therefore, operators would need to keep in stock a large number of different compensator models, and even then the compensation would only be approximate. Second, we must consider the sheer difficulty of 40 Gbit/s signals. The tolerable threshold for accumulated dispersion for a 40 Gbit/s data channel is 16 times smaller than at 10 Gbit/s. If the compensation value does not exactly match the fiber to within a few percent of the required dispersion value, then the communication link will not work. Tunability is considered a key enabler for this bit rate (see Figure 20.108). Third, the accumulated dispersion changes slightly with temperature, which begins to be an issue for 40 Gbit/s systems and 10 Gbit/s ultra long haul systems. In fiber, the zero-dispersion wavelength changes with temperature at a typical rate of 0.03 nm/°C. It has been shown that a not-uncommon 50°C variation along a 1000-km 40 Gbit/s link can produce significant degradation (see Figure 20.109). Fourth, we are experiencing the dawn of reconfigurable optical networking. In such systems, the network path, and therefore the accumulated fiber dispersion, can change. It is important to note that even if the fiber spans are compensated span-by-span, the pervasive use of compensation at the transmitter and receiver suggests that optimization and tunability based on path will still be needed.

Other issues that increase the need for tunability include (1) laser and (de)mux wavelength drifts for which a data channel no longer resides on the flat-top portion of a filter, thereby producing a chirp on the signal that interacts with the fiber's chromatic dispersion; (2) changes in signal power that change both the link's nonlinearity and the optimal system dispersion map; and (3) small differences that exist in transmitter-induced signal chirp.

Approaches to Tunable Dispersion Compensation. A host of techniques for tunable dispersion compensation have been proposed in recent years. Some of these are just interesting research ideas, but several have strong potential to become viable technologies.

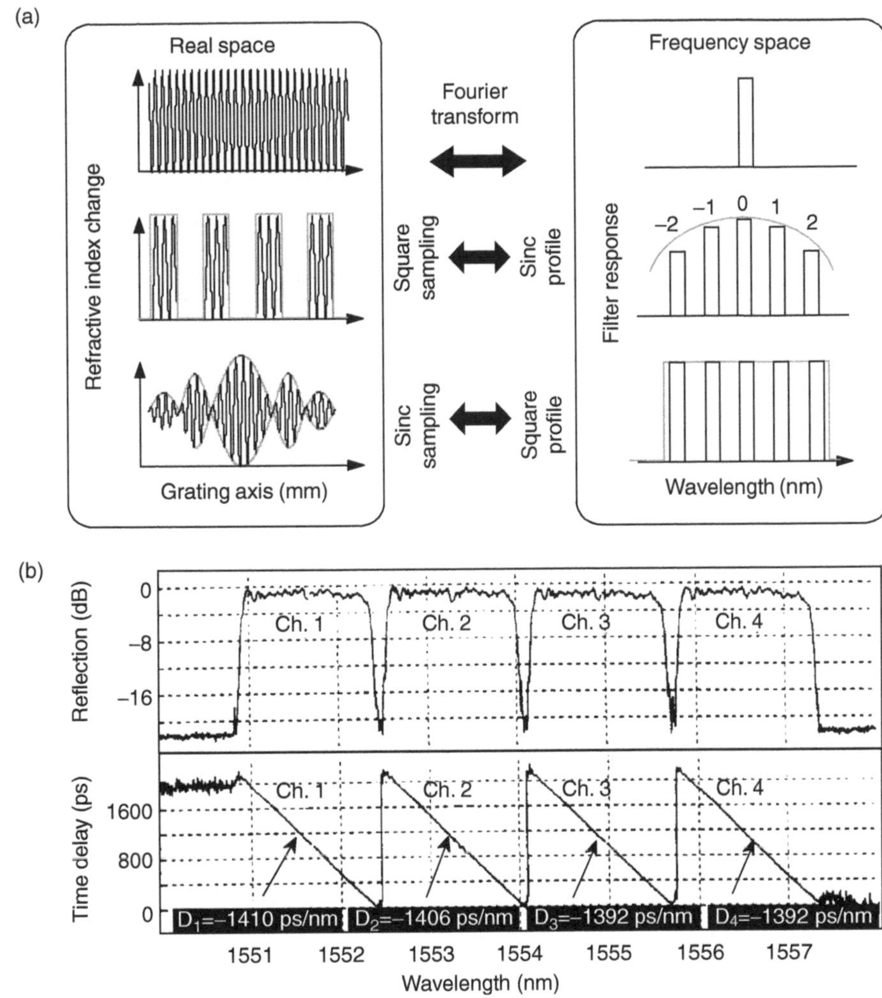

FIGURE 20.107 (a) The concept of "sampled" FBGs, where a superstructure is written on top of the grating that produces a Fourier transform in the frequency domain, leading to multiple grating passbands, (b) amplitude spectra and time delay spectra for a four-channel sinc-sampled dispersion compensating FBG. (Adapted from Ibsen, M., Durkin, M. K., Cole, M. J., and Laming, R. I. 1998. *IEEE Photon. Technol. Lett.*, 10(6), 842–845, With permission.)

If a FBG has a refractive-index periodicity that varies nonlinearly along the length of the fiber, it will produce a time delay that also varies nonlinearly with wavelength (see Figure 20.110). Herein lies the key to tunability. When a linearly chirped grating is stretched uniformly by a single mechanical element, the time delay curve is shifted towards longer wavelengths, but the slope of the ps-vs.-nm curve remains constant at all wavelengths within the passband. When a nonlinearly chirped grating is stretched, the time delay curve is shifted toward longer wavelengths, but the slope of the ps-vs.-nm curve at a specific channel wavelength changes continuously.[93]

Another solution was also reported, which is based on differential heating of the substrate. The thermal gradient induced a chirp gradient, which could be altered electrically[178] that has a major advantage: no moving parts. However, this is countered by the disadvantage of slow tuning, limited to seconds or minutes. Additionally, the technology requires accurate deposition of a thin film of tapered thickness. The process of deposition of the tapered film seems to have some yield issues, making it rather difficult to manufacture.

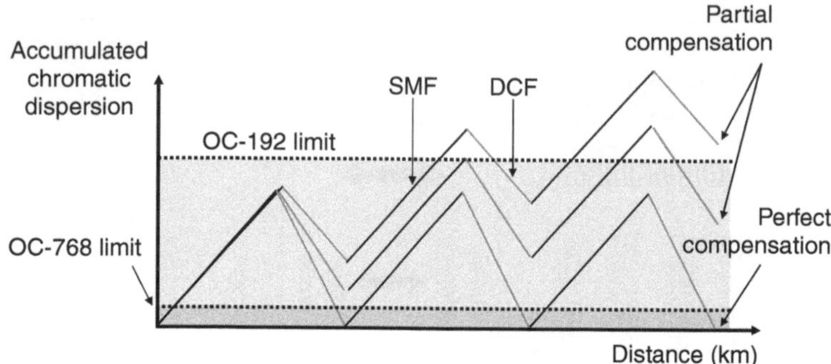

FIGURE 20.108 The need for tunability. The tolerance of OC-768 systems to chromatic dispersion is 16 times lower than that of OC-192 systems. Approximate compensation by fixed in-line dispersion compensators for a single channel may lead to rapid accumulation of unacceptable levels of residual chromatic dispersion.

One free-space-based tunable dispersion compensation device is the virtually imaged phased array (VIPA), based on the dispersion of a Fabry-Perot interferometer.[179] The design requires several lenses, a movable mirror (for tunability), and a glass plate with a thin film layer of tapered reflectivity for good mode matching. Light incident on the glass plate undergoes several reflections inside the plate. As a result, the beam is imaged at several virtual locations, with a spatial distribution that is wavelength-dependent.

Several devices used for dispersion compensation can be integrated on a chip, using either an optical chip media (semiconductor-based laser or amplifier medium) or an electronic chip. One such technology is the micro-ring resonator, a device that when used in a structure similar to that of an all-pass filter (see Figure 20.111) can be used for dispersion compensation on a chip-scale. Although these technologies are not fully mature and not yet ready for deployment as dispersion compensators, they have been used in other applications and have the potential to offer very high performance at low cost.

As the ultimate optical dispersion compensation devices, photonic bandgap fibers (holey fibers) are an interesting class in themselves. These are fibers with a hollow structure, with holes engineered to achieve

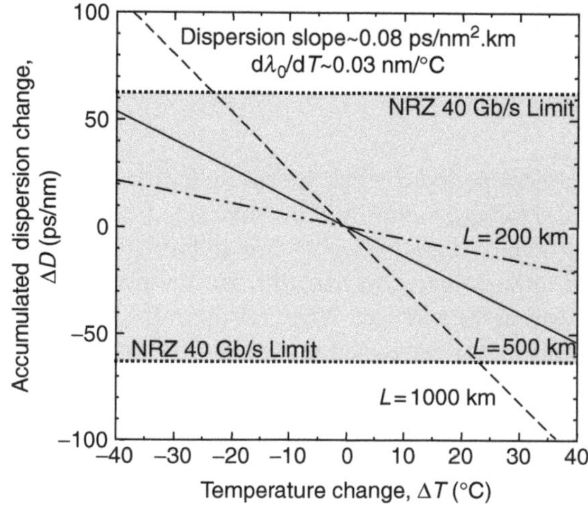

FIGURE 20.109 Accumulated dispersion changes as a function of the link length and temperature fluctuation along the fiber link.

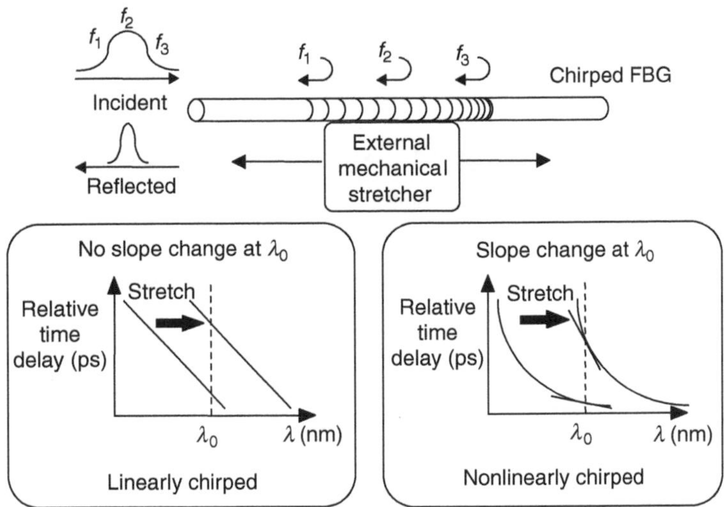

FIGURE 20.110 Tuning results for both linearly- and nonlinearly chirped FBGs using uniform stretching elements. The slope of the dispersion curve at a given wavelength λ_0 is constant when the linearly chirped grating is stretched, but changes as the nonlinear-chirped grating is stretched.

a particular functionality. Instead of being drawn from a solid preform, holey fibers are drawn from a group of capillary tubes fused together. This way, the dispersion, dispersion slope, and the nonlinear coefficients could in principle all be precisely designed and controlled, up to very small or very large values, well outside the range of those of the solid fiber.

Although several technologies have emerged that meet some or all of the above requirements, no technology is a clear winner. The trend is towards tunable devices, or even actively self-tunable compensators, and such devices will allow system designers to cope with the shrinking system margins and with the emerging rapidly reconfigurable optical networks.

20.4.6.5 Dispersion Slope Mismatch

Transmission fiber link, especially fiber with dispersion compensation built in, may suffer from a dispersion slope in which a slightly different dispersion value is produced for each WDM channel (see Figure 20.112). Even though the compensator would be able to cancel the dispersion of the fiber at the design wavelength, there will be residual dispersion left at the other wavelength channels unless the

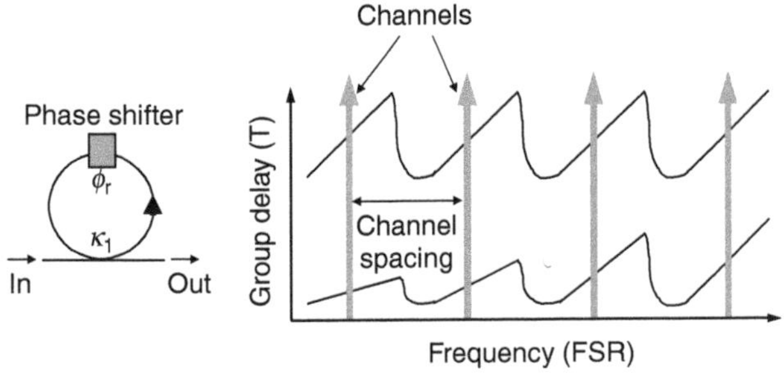

FIGURE 20.111 Architecture of an all-pass filter structure for chromatic dispersion and slope compensation. (After Madsen, C. K., Lenz, G., Bruce, A. J., Cappuzzo, M. A., Gomez, L. T., and Scotti, R. E. 1999. *IEEE Photon. Technol. Lett.*, 11(12), 623–1625. With permission.)

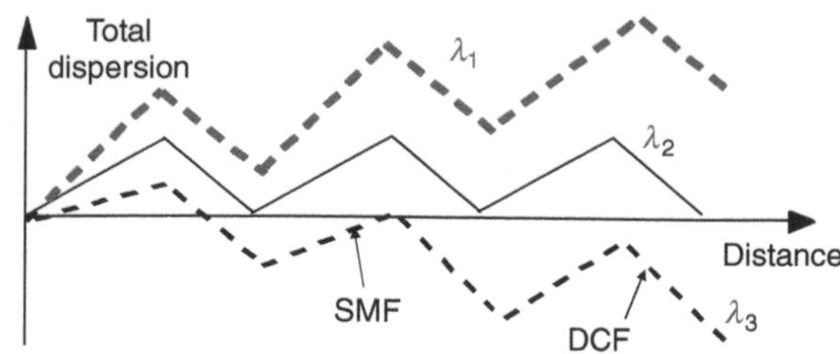

FIGURE 20.112 The chromatic dispersion slope mismatch caused by the different slopes of transmission fiber (SMF or NZDSF) and DCF.

compensator can match the slope of the dispersion curve of the fiber as well. Some solutions for dispersion slope compensation are described in this section.

DCF with negative dispersion slope is a prime candidate for deployment as a dispersion slope compensator even though it cannot easily be made tunable. By designing the DCF with the same ratio of dispersion to dispersion slope as that of a real fiber link, new types of DCF can be used to compensate for both dispersion and dispersion slope much like DCF is used for dispersion compensation today. A third-order nonlinearly chirped FBG can act as a tunable dispersion slope compensator. A simple modification of the nonlinearly chirped FBG allows tuning of the compensated dispersion slope value via stretching the grating. Using an FBG with many spaced thin-film heater sections can also enable tunable dispersion slope. Each heater can be individually electrically controlled, allowing the time delay profile of the grating to be dynamically tuned via changing the temperature along the length of the grating. Combining the VIPA with either a 3D mirror or diffraction grating can also provide tunable free-space dispersion slope compensation. Slope tuning is achieved by dynamically controlling the MEMS-based 3D mirror or the diffraction grating.

20.4.6.6 Solitons

There exists a class of pulses which retains its shape indefinitely, allowing for the potential of ultra-high-speed optical transmission over multi-megameter distances. These pulses, known as solitons,[180,181] take advantage of fiber nonlinear effects, specifically SPM, to overcome the pulse-broadening effects of chromatic dispersion. They have few fundamental limits, and hold the exciting promise that the need for ever-higher data rates anywhere in the world will be met by this technology. Some advanced limitation issues, such as the Gordon–Haus effect, will not be discussed in this section.[182]

An optical pulse is nothing more than a wave which propagates along a fiber. This propagation in a single-mode fiber can be described by a nonlinear Schrödinger equation:[183]

$$-i\frac{\partial u}{\partial z} = \frac{1}{2}\frac{\partial^2 u}{\partial t^2} + |u|^2 u - i\frac{\alpha}{2}u \tag{20.63}$$

where u represents the pulse envelope function, z is the propagation distance along the fiber, t is the retarded time, and α is the energy loss or gain coefficient. The first term on the right side of the equation represents the fiber chromatic dispersion. The second, or nonlinear, term represents the dependence of the fiber index of refraction on the light intensity inside the fiber. The final term represents the fiber loss or gain. Dispersion broadens the optical pulse in time whereas the nonlinear term broadens the pulse frequency spectrum. This nonlinear frequency broadening is known as SPM. If we employ optical amplifiers so that the fiber loss is compensated by the EDFA gain, then we can initially ignore the third term in Equation 20.61. One solution to the above equation is the soliton:[183]

$$u(z,t) = \mathrm{sech}(t)e^{iz/2} \tag{20.64}$$

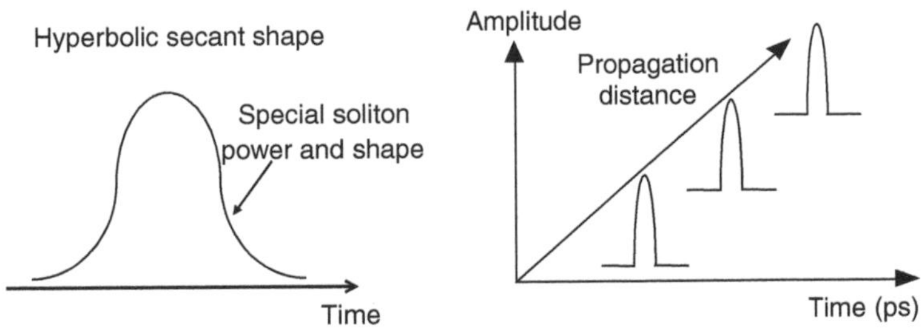

FIGURE 20.113 The fundamental soliton shape (hyperbolic secant) and its propagation along an optical fiber.

The exponential term is simply a phase relationship which does not affect the shape of the pulse. Equation 20.64 shows that the fundamental (zeroth-order) soliton pulse has a hyperbolic secant shape which does not change with distance, as shown in Figure 20.113. This is due to a unique combination of factors for which the dispersion exactly cancels the SPM (i.e., these two effects have the same magnitude but are opposite in sign), making for a nondispersive pulse; note that dispersion can be positive or negative, and so the soliton only can exist in the wavelength range for which the dispersion is opposite to the SPM. An intuitive way of understanding the soliton is that the pulse creates an index "well" which travels along with the pulse. The leading edge of the pulse, which would normally travel faster, is retarded by the intense pulse center changing the index. The trailing edge, which would normally fall behind, speeds up due to the pulse center. This idea is shown in Figure 20.114 in a cartoon illustration of three joggers running on a mattress, a fast leading runner, a large center one, and a slow trailing one.[184] Because the large middle jogger depresses the mattress, the speed of the fast jogger is reduced whereas the speed of the slow jogger is increased. The pulse and the relative positions of the joggers stay the same!

The soliton period has a characteristic length over which the various effects interact. This soliton length could be hundreds of kilometers for short pulses generated near the fiber dispersion minimum. The soliton also has a required peak power, P_c, which will allow exact cancellation between dispersion and SPM.[183] Note that higher-order solitons also exist in which the pulse shapes in an oscillatory fashion over the soliton length, returning to the original fundamental soliton shape after every period.

One of the truly amazing qualities of the soliton is that it is fairly robust to perturbations away from the ideal values. The soliton is the only stable solution to the nonlinear Schrödinger equation, and so the pulse will evolve into a soliton as it propagates along the fiber so long as its shape and peak power are relatively close to the exact solution. It is therefore reasonable to generate a nonideal pulse from a laser which will evolve into a soliton over a short distance. This time evolution into a soliton is accompanied by the generation of low-intensity dispersive waves which are viewed as loss in our soliton system.

FIGURE 20.114 Cartoon illustration of a pulse conceptually composed of three runners representing the pulse's leading edge, center, and trailing edge. (Adapted from Evangelides, S. G., private communication, 1993. *Conference on Optical Fiber Communications '94, Technical Digest*, L. F. Mollenauer, ed., Tutorial TuF, San Jose, CA. With permission.)

We had assumed in our explanation of the basic soliton that the loss term was zero in the nonlinear Schrödinger equation. The robust soliton behavior is manifest in its ability to accommodate lumped amplification and in not requiring that the loss is absolutely zero at every point along the fiber. Lumped amplification can be employed as long as (1) the average path gain compensates for the average path loss and (2) the pulse power remains within the range of the soliton solution. This is also true of the dispersion parameter of the fiber, which is allowed to have even discontinuous changes over small lengths of fiber. Note that the changes in gain and dispersion must occur over distances much smaller than the soliton characteristic length, z_0.

Transmission of solitons requires the generation of short pulses. This is typically accomplished with mode-locked lasers which can easily achieve <50-ps pulse generation at high repetition rates.[185] A train of pulses is produced, and a secondary data bit-stream controls an external modulator which either passes a pulse if a "1" bit is present or blocks the pulse if a "0" bit is present. Since solitons are inherently short pulses, the modulation format is RZ, not NRZ.

Solitons could not exist if optical amplification was not used to compensate for the fiber loss. To this end, EDFAs have been the key enabling technology for long-distance soliton technology. EDFAs can be placed every 10–30 km along an ultra-long-distance link and still provide the appropriate soliton power conditions mentioned above. As with a nonsoliton system, the ASE noise accumulates along the length of the link in a typical soliton system. However, an extremely clever technique has been demonstrated which reduces the accumulation of ASE. This method uses a cascade of sliding-frequency filters in which a transmission filter is placed after each EDFA, with each filter center frequency slightly offset from one filter to the next.[186,187] The nonlinear soliton "slides" in frequency with the changing filter centers, but the linear ASE does not frequency-translate with the changing filter centers and therefore does not accumulate (i.e., experience as much gain) as rapidly as the soliton since part of the ASE under the soliton is always blocked by the next filter in the cascade.

Just as with standard nonsoliton NRZ transmission, we wish to increase the system capacity and take advantage of the broadband fiber to transmit many soliton signals simultaneously down the fiber on different wavelengths, i.e., WDM. Since soliton pulses on different wavelengths will propagate along the dispersive fiber at different speeds, two such WDM pulses will eventually collide with and pass through each other. This collision is characterized by an attraction between two such pulses. It is quite fortuitous that the solitons' attraction is symmetrical, causing one soliton to speed up and then slow down, with the opposite happening to the soliton on the other wavelength.[188] Almost "magically," the two solitons survive without much change after experiencing a collision, even though somewhat-violent interactions take place during the collision. It should be noted that it is undesirable to have a large fraction of a collision occurring inside an amplifier since the two solitons' mutual interaction will be affected by the amplification and will no longer be symmetrical. Therefore, the length of the collision should be 2–3 times longer than the amplifier spacing to ensure that WDM soliton collisions are benign.

Soliton systems evolved dramatically during last 10 years. As shown by Carter et al.,[189] it is advantageous to combine dispersion management with soliton system. The nonlinearity of the system is effectively lowered by the spreading of the solitons, leading to a substantial reduction of the timing jitter. These dispersion-managed solitons (DMS) are no longer stationary, only periodically stationary in that they return to the same pulse shape after every period. On the other hand, one can apply the phase modulation to a raised-cosine RZ pulse train, so called chirped-RZ (CRZ) format. This modulation scheme is essentially the same as DMS. This DMS/CRZ format is remarkably robust and effective, and suitable for long-distance and high-capacity transmission.

20.4.6.7 Modulation Formats

It is important to note that the type of data modulation formats have substantial impacts on the fiber dispersive and nonlinear effects. Some of these formats carry information through on-off keying, but also modulate the optical phase in a noninformation-bearing way in order to enhance the signals' robustness to chromatic dispersion, optical filtering, and/or nonlinearities. This group includes formats such as optical duobinary, chirped return-to-zero (CRZ), and alternating-phase OOK formats such as carrier-suppressed return-to-zero (CSRZ). Single-sideband (SSB) modulations have narrower spectra width that can also reduce the dispersion effect. We will highlight a few examples here and present their advantages and disadvantages based on fiber system performance characterization.

Differential Phase Shift Keying (DPSK). As we mentioned in the early section, PSK formats carry the information in the optical phase itself and DPSK formats carry the information in optical phase changes between bits. Both numerical simulations and experiments have shown DPSK to be more robust to some nonlinear effects than OOK[108] due to the following facts: (1) the optical power is evenly distributed (power is present in every bit slot for DPSK, which reduces bit-pattern-dependent nonlinear effects); and (2) the optical peak power is 3 dB lower for DPSK than for OOK for the same average optical power. Furthermore, an extension to differential quadrature phase-shift keying (DQPSK) and other multilevel formats should enable higher spectral efficiency and greater tolerance to chromatic- and polarization-mode dispersion. Note that experimental results have consistently shown better performance for DPSK than OOK in 40 Gbit/s single-channel and WDM systems.

Carrier-Suppressed Return-to-Zero (CSRZ). CSRZ optical signals have the feature of presenting bits that are π phase-shifted relative to neighboring bits such that on the average all the phases cancel each other out for a net phase of zero. A CSRZ optical data stream may consist of a plurality of pulses where half of the pulses have an alternating phase relationship with the other half of the pulses, which leads to carrier suppression, as shown in Figure 20.115a. Because of this phase inversion between adjacent bit periods that reduces inter-bit interference, CSRZ signals show increased tolerance to dispersion and to nonlinear penalties (see Figure 20.115b).[190]

The CSRZ format can be generated by using a sinusoidal signal to drive the Mach-Zehnder modulator; the drive frequency is one half of the bit rate ($f_r/2$) and the amplitude is $2V_{\pi}$. As shown in, the generated pulse has the repetition rate of f_r, and the phase of the pulses alternate between 0 and π (Figure 20.116).

Optical Duobinary. Optical duobinary has been proven more resilient to dispersion for >10 Gbit/s data and is reasonably simple to implement. As shown in Figure 20.117, optical duobinary signal is

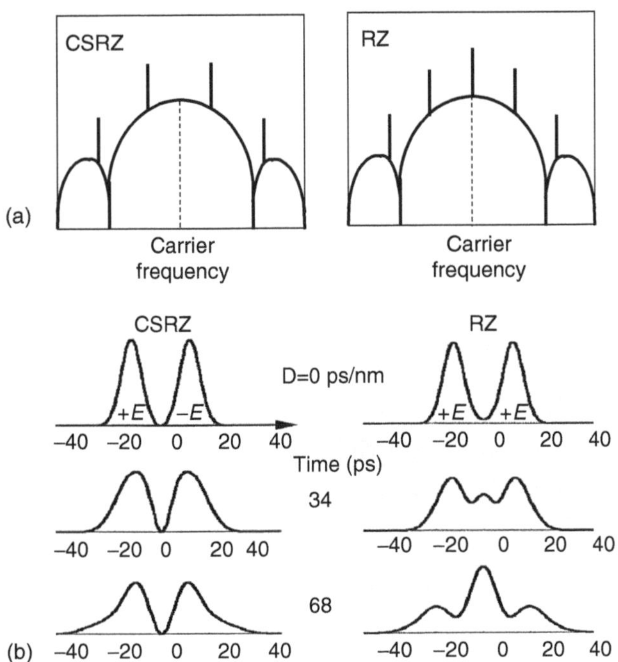

FIGURE 20.115 (a) Optical spectra of CSRZ and RZ formats, (b) 40 Gbit/s CSRZ and RZ pulses under different dispersion values. (Adapted from Sato, K., Kuwahara, S., Miyamoto, Y., Murata, K., and Miyazawa, H. 2002. *IEICE Trans. Commun.*, E85-B, 2, 410–415. With permission.)

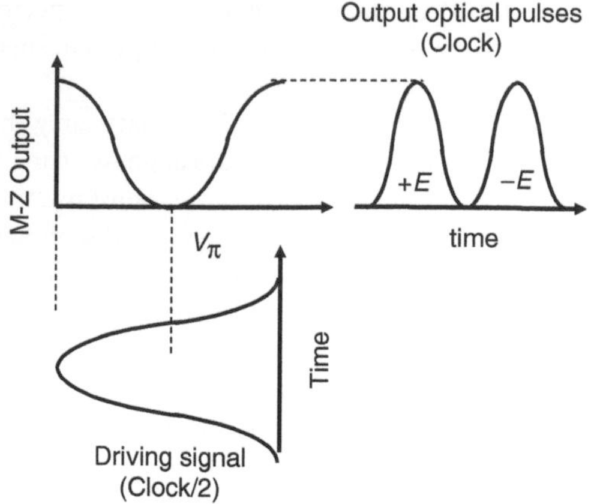

FIGURE 20.116 Generation of CSRZ format using Mach-Zehnder modulator.

commonly generated by applying a baseband duobinary (three-level) electrical signal to the RF input of a lithium niobate ($L_iN_bO_3$) Mach-Zehnder modulator biased at maximum distinction (at V_π). The result is a binary, intensity-modulated optical signal, with a duobinary modulated optical electric field caused by π-radian shift. With a zero ($c_k=1$) input, no light is transmitted, but the $+1$ ($c_k=2$) and -1 ($c_k=0$) inputs are transmitted as $+E$ and $-E$ electrical fields. While this is a three-level signal in terms of the electric field, it is a two-level signal in terms of optical power. The same receiver that is used for a NRZ modulation scheme can be used for duobinary modulation. The power detector squares the electric field to detect power and hence the $+E$ and $-E$ outputs of the fiber get mapped to the same power level and are detected as logical 1s. Note that the received data is the invert of the original binary input. This choice significantly reduces the complexity of the receiver (the first optical duobinary system used a mapping that requires three levels of optical power). One of the key components is a driver that can produce a voltage swing of $2V_\pi$ volts at high data rates such as >10 Gbit/s in order to drive the MZ modulator.

The combination of the duobinary encoder and the above mapping of electric fields help reduce the effects of dispersion in the fiber. The pulses spread out as they travel down the fiber. In an NRZ scheme, a

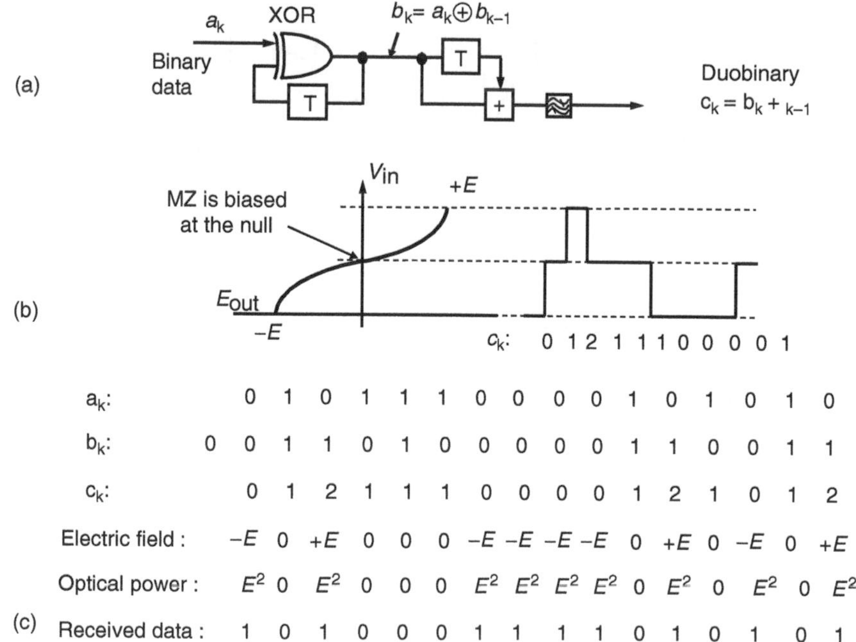

FIGURE 20.117 Optical duobinary modulation format: (a) duobinary encoder, (b) biasing of Mach-Zehnder modulator, and (c) an example of data transformation.

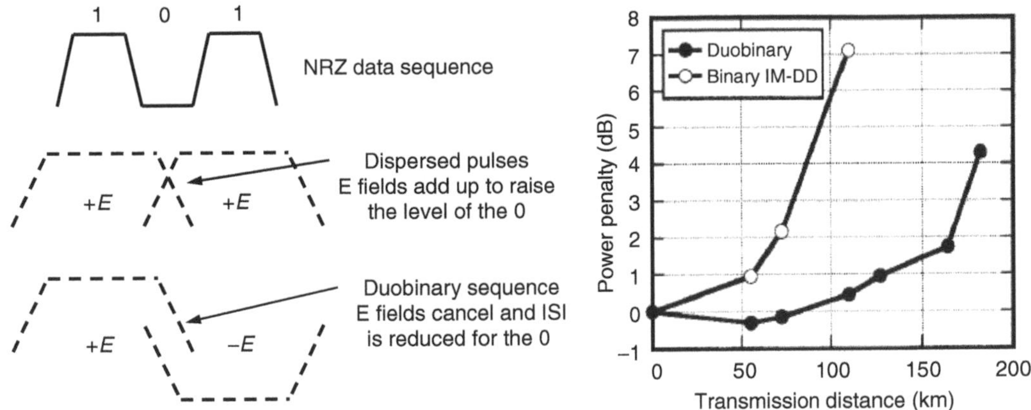

FIGURE 20.118 Dispersion effects on NRZ format and duobinary format. The power penalty due to fiber chromatic dispersion was measured at the BER of 10–9. (Adapted from Yonenaga, K. and Kuwano, S. 1997. *IEEE/OSA J. Lightwave Technol.*, 15(8), 1530–1537. With permission.)

data sequence of 1 0 1 is mapped onto the optical domain as $+E\,0\,+E$. In the encoded duobinary sequence, a 1 0 1 sequence cannot occur, but a 1 0 -1 does occur, which is mapped to $+E\,0\,-E$ in the optical domain. The effect of dispersion in the two cases is shown in Figure 20.118, which depicts why the resulting dispersion is less in the case of duobinary modulation. Figure 20.118 also shows an experimental result that optical duobinary technique expands the transmission distance to >150 km SMF for 10 Gbit/s data.[191]

The optical duobinary technique has been proven in experiment to expand the usable bandwidth and transmission distance in four-channel multiplexed 40 Gbit/s based WDM system without individual channel dispersion slope compensation. That 160 Gbit/s WDM transmission using four 40 Gbit/s optical duobinary channels over a 100 km DSF is successful by virtue of the high dispersion tolerance of the optical duobinary signal.[192] Furthermore, the narrow optical spectrum of optical duobinary signals reduces the coherent crosstalks in the ultradense WDM system.[193]

Single Sideband (SSB) Modulation. The transmitted spectrum of NRZ and RZ data is double sided, with redundant information carried in the upper and lower sideband on either side of the carrier. The advantages of single sideband over double sideband include its optical signal spectral bandwidth is reduced by a factor of two. That is, it achieves the dispersion compensation by (1) reducing the spectral occupancy of the optical signal; and (2) by preserving the phase distortion introduced by the chromatic dispersion such that upon detection, the electrical signal can be equalized. Superior to the SSB scheme is the proposed vestigial-sideband (VSB) formats. Since complete SSB filtering is very difficult and the amplitude characteristic of the VSB can be achieved by using existing optical filters.

With many experimental evidences, bandwidth limitation has great effect on dispersion limit. New modulation and coding schemes can push the dispersion and nonlinearity limit further. However, the modulation scheme alone will not be sufficient to beat dispersion and nonlinearities.

We emphasize that current fiber systems use binary modulation (OOK or DPSK) with error-control coding schemes. The spectral efficiency cannot exceed 1 b/s/Hz per polarization regardless of detection technique, while as shown in Figure 20.119, the spectral efficiency limits of the WDM systems are several b/s/Hz, even considering nonlinear and dispersion effects. At low input power densities, the spectral efficiency increases logarithmically with input power. As the input power density exceeds a critical value, the spectral efficiency decreases exponentially with input power. The critical input power density and thus the spectral efficiency limit, increases with chromatic dispersion and with channel spacing.[194] To achieve spectral efficiencies above 1 b/s/Hz and increase overall capacity of a WDM transmission system, nonbinary modulation, such as quaternary DPSK and PSK, 8-PSK, 8- and 16-QAM, will be needed. Due

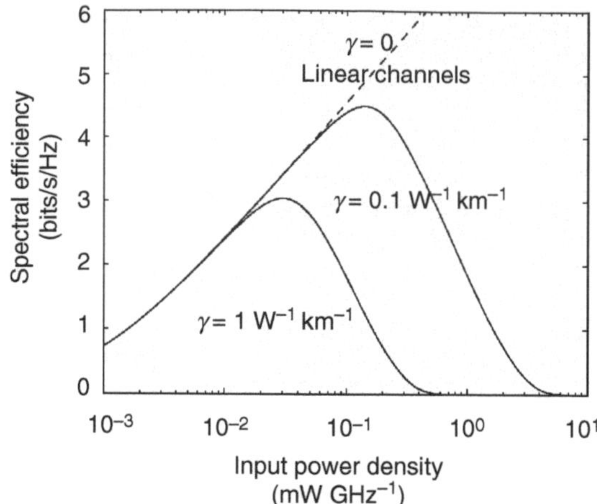

FIGURE 20.119 The spectral efficiency limits of the WDM systems with nonlinear and dispersion effects. (Adapted from Mitra, P. P. and Stark, J. B. 2001. *Nature*, 411, 1027–1030. With permission.)

to the fact that optical signals propagating in fibers offer several degrees of freedom, including amplitude, frequency, phase, polarization, and time, intense research efforts have been made towards the combination coding over these degrees of freedom as a means to increase fiber transmission capacity, especially as a way to combat or benefit from fiber nonlinearity and dispersion.

20.4.6.8 Polarization-Mode-Dispersion and its Control

We have briefly discussed fiber PMD in Section 20.2. Here we will focus on the PMD effects on WDM transmission performances and its control.

Today's fiber has a very low PMD value and is well characterized. But there is still a small residual asymmetry in the fiber core. Moreover, slight polarization dependencies exist in discrete inline components such as isolators, couplers, filters, EDF, modulators, and multiplexers. Therefore, even under the best of circumstances, PMD still significantly limit the deployment of 40 Gbit/s systems (see Figure 20.120).[195]

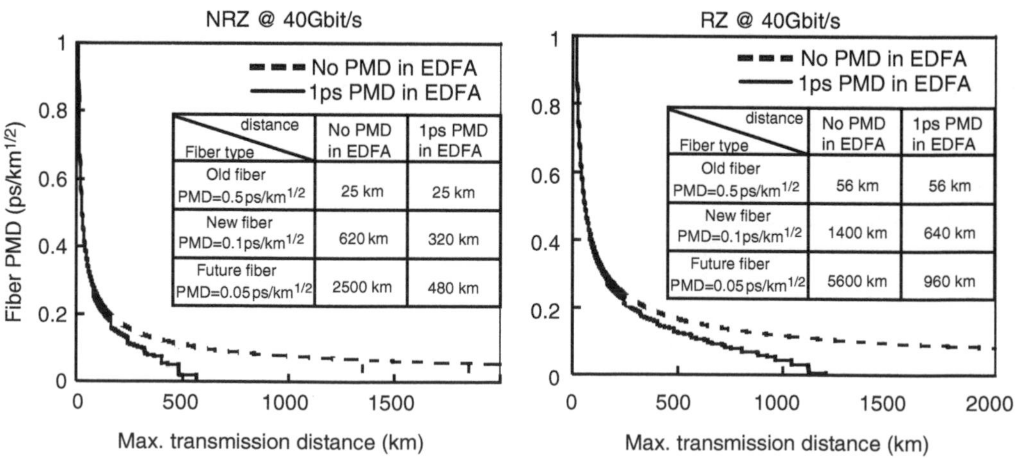

NRZ @ 40Gbit/s

Fiber type \ distance	No PMD in EDFA	1ps PMD in EDFA
Old fiber PMD=0.5 ps/km$^{1/2}$	25 km	25 km
New fiber PMD=0.1 ps/km$^{1/2}$	620 km	320 km
Future fiber PMD=0.05 ps/km$^{1/2}$	2500 km	480 km

RZ @ 40Gbit/s

Fiber type \ distance	No PMD in EDFA	1ps PMD in EDFA
Old fiber PMD=0.5 ps/km$^{1/2}$	56 km	56 km
New fiber PMD=0.1 ps/km$^{1/2}$	1400 km	640 km
Future fiber PMD=0.05 ps/km$^{1/2}$	5600 km	960 km

FIGURE 20.120 Limitations of transmission distances caused by fiber PMD.

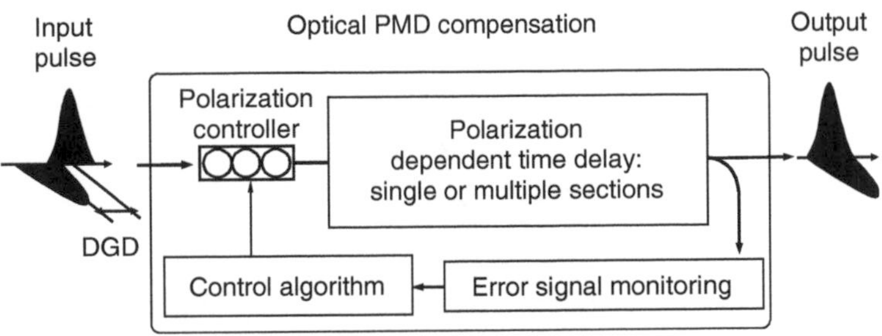

FIGURE 20.121 A generic diagram of an optical PMD compensator.

PMD is a stochastic, random process that changes with time, and roughly follows a Maxwellian probability density function. At one extreme, a Maxwellian distribution has a probability of occurrence for a very small DGD value. At the other extreme, a Maxwellian distribution always has some finite probability of occurrence for values of DGD that extend out to infinity, thereby causing a network outage. It is therefore important to know the average PMD value for a link, since that determines the extent of the deleterious distribution tail of the Maxwellian distribution. It would be quite advantageous to implement a module that would compensate for PMD effects in a fiber link. The generic diagram of a PMD compensator is shown in Figure 20.121. A polarization controller is used at the input to rotate the incoming signal and optimally align it to a DGD element that will correct for some amount of the accumulated PMD. After the DGD element, a feedback loop is necessary to tune the polarization controller and thereby dynamically track changes in the state of polarization of the incoming signal. The feedback loop consists of a PMD monitor and should enable tracking within a few tens of milliseconds. Note that a feed-forward solution can also be implemented. There are several methods for monitoring the PMD, including measuring the signal eye opening, the BER, the radio-frequency power level, and the degree of polarization.

Perhaps the most difficult part of a PMD compensator is the tracking software required in the feedback loop. Such software requires a complicated control algorithm for finding the optimal position to minimize the system power penalty caused by PMD. The space explored by the control software consists of many "pockets" of local PMD minima, whereas there is only one true global minimum.

The DGD element in the compensator can be (1) a single piece of polarization-maintaining (PM) fiber that produces a fixed amount of time delay; (2) a tunable element that produces a varying amount of time delay in which wide variations in accumulated PMD can be accommodated; and (3) multiple sections of independently controlled PM fiber, thereby enabling compensation of higher-order PMD. Each of these options not only progressively increases the compensator's performance, but also its complexity and cost. There are clear tradeoffs to be made in this decision, but the ultimate performance and eventual cost of commercially deployed modules remain unclear.

Another distasteful aspect of PMD is that the effects are different and uncorrelated for each independent channel in a WDM system. Theory and experiments have demonstrated that the bandwidth of the autocorrelation function (ACF) for PMD vector is inversely proportional to the mean DGD of the given fiber link,[196] as shown in Figure 20.122. Therefore, in WDM systems over real fiber, the PMD of different channels is statistically independent. Consequently, PMD compensation modules would seem to require a separate PMD compensator for each channel in a WDM system. If each wavelength in a WDM system requires its own PMD compensation module, this make the economics extremely important in any decision regarding the implementation of PMD compensators.

It should be noted that many of the high-performance 40 Gbit/s systems incorporate forward error correction (FEC) coding. Such coding adds some redundancy into the bits of a data stream to more easily find and correct errors. FEC is implemented using electronic chips, and it adds a system power margin that can ease the deleterious problems associated with fiber nonlinearities, chromatic dispersion, SNR,

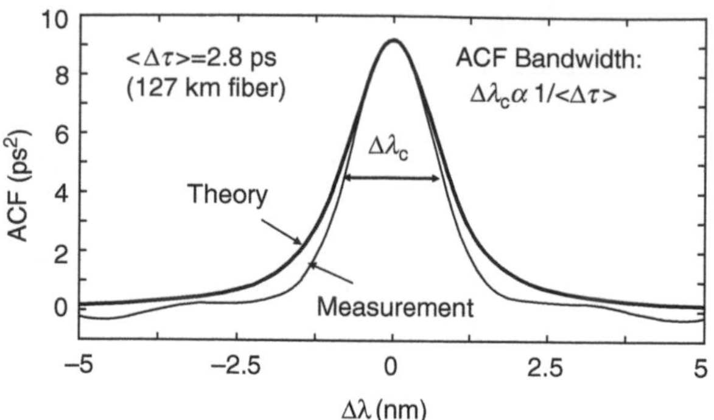

FIGURE 20.122 Autocorrelation function (ACF) in WDM systems over real fiber: the ACF bandwidth is inversely proportional to the mean DGD of the given fiber link. (Adapted from Karlsson, M. and Brentel, J. *Opt. Lett.*, 24(14), 939–941. With permission.)

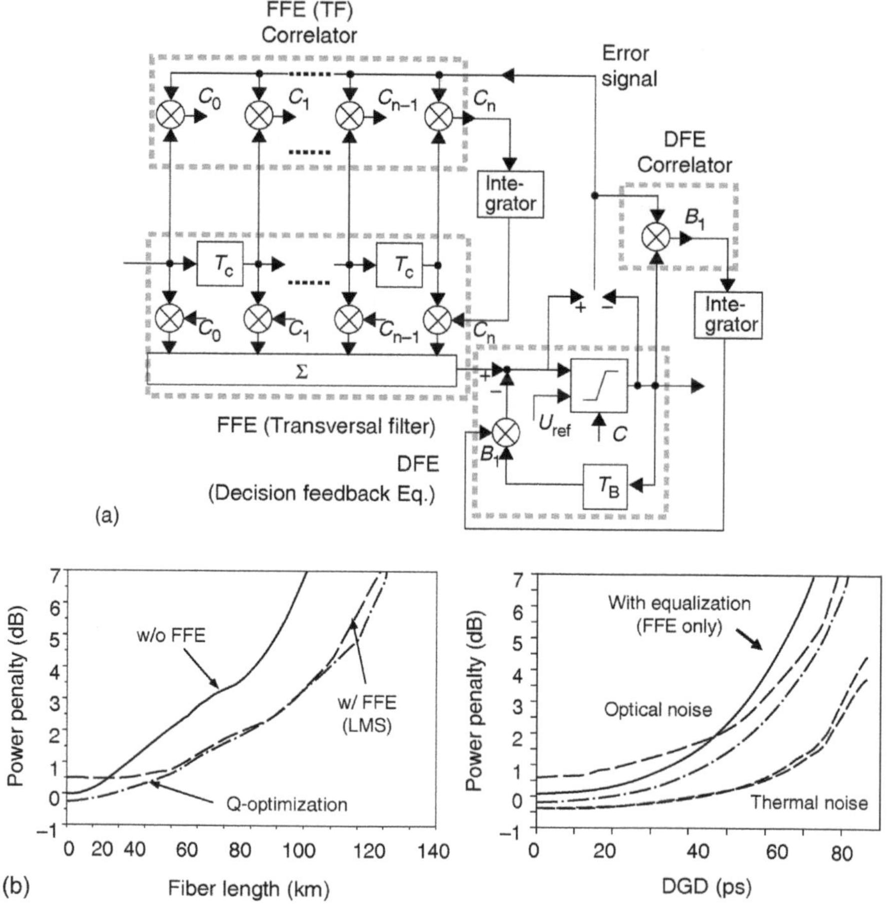

FIGURE 20.123 (a) Configuration of the feed forward + decision feedback equalizer (FFE + DFE), (b) performances of the equalizer for compensating chromatic dispersion and polarization-mode-dispersion. (After Bülow, H. 2002. Optical Fiber Communication Conference and Exhibit, 24–25, Invited paper, TuE4, With permission.)

and, of course, PMD. However, this power margin must be allocated for mitigation of all these problems, and the margin allocated for PMD is only a fraction of the total. It should also be noted that network outages that last for time scales very long as compared to the bit time cannot be mitigated by FEC techniques.

Note that the type of data modulation format will also have an impact on the degree to which PMD will adversely affect the system performance. For example, it seems that chirped RZ may be more robust to PMD that other types of formats.[197]

It is worth mentioning that some of the most promising solutions for fiber chromatic dispersion and PMD compensation are electronic equalizers.[198] These electronic integrated circuits rely on post-detection signal processing including filtering and adaptive signal processing to sharpen up distorted data pulses. As an example, a configuration of a feed forward + decision feedback equalizer (FFE + DFE) and its performances on compensating dispersion effects are shown in Figure 20.123.[201] Because the detection itself is nonlinear, the job of compensating for the linear distortions of chromatic dispersion is quite a bit more complicated. On the other hand, electronic processing is potentially very cheap, and much easier scalable to large volume production, at least at data <10 Gbit/s.

20.4.7 Crosstalk

Incoherent and coherent crosstalks between adjacent channels are major problems that limit the density of WDM channels. In particular, coherent crosstalk in WDM Add/Drop multiplexers (ADMs) and switches places severe requirements on the acceptable crosstalk suppression levels provided by these components. We will introduce these two types of crosstalk and discuss their impact on system performance.

20.4.7.1 Incoherent Crosstalk

WDM channel crosstalk involves the most basic function of demultiplexing many WDM channels using an optical filter.[199] Figure 20.124 shows the demultiplexing of WDM channels by an optical filter so that only one channel is received and the other channel is blocked. However, the slowly diminishing spectral shape of the filter with long "tails" will induce crosstalk to the demultiplexed channel when the signals are detected at the photodetector, as we discussed in the early section of WDM system guidelines.

In the case of incoherent crosstalk, the crosstalk level is determined by the channel spacing and channel powers. This inter-channel crosstalk power P_{cr} will act to reduce the power extinction ratio of the selected recovered signal. The crosstalk due to power leakage from the adjacent channels may raise the level of a "zero," increasing the probability of errors. Although the tolerable amount of crosstalk depends on the specific WDM system, crosstalk should not exceed a few percent of the selected channel power in order to maintain good system performance.

20.4.7.2 Coherent Crosstalk

ADMs are key subsystems in WDM networks.[200] They are typically composed of a WDM followed by a wavelength multiplexer. Ideally, they allow a network node to have access to a single wavelength while allowing all other wavelengths to pass through the node unaffected (see Figure 20.125). Unfortunately,

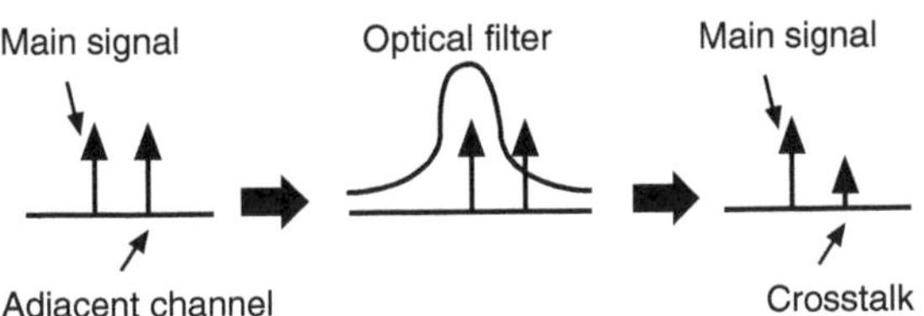

FIGURE 20.124 Incoherent crosstalk from adjacent channels.

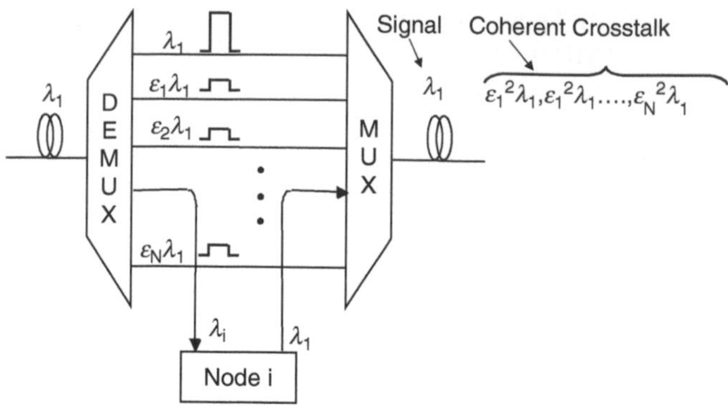

FIGURE 20.125 Coherent crosstalk in an ADM.

wavelength (de)multiplexers are not ideal and allow unwanted crosstalk of any input wavelength onto unintended output ports. For example, a small replica of the signal at wavelength λ_1 will appear at all the demultiplexer output ports, as shown in Figure 20.125. If a nonideal demultiplexer and nonideal multiplexer are placed back-to-back, small replicas of the signal at λ_1 will recombine with the main signal at λ_1 at the multiplexer output.

This ADM structure with crosstalk represents an interferometer since it consists of input and output ports connected by optical paths. Due to coherent interaction of the fields, the amplitude of the output signal will fluctuate as the signal's relative phases and polarization states change.[201]

Coherent crosstalk has been shown to be a serious limiting factor in scaleable WDM networks,[202] since the crosstalk signal has the same frequency as the main signal and the crosstalk is added to the main signal in electric field. Let's consider the case shown in Figure 20.126, where two signals with identical wavelength enter an OXC and then are routed to different paths. Assuming the two channels have the same input power P and the power leakage of the OXC is ε, then the optical power at one of the output ports can be approximated by

$$P_0 \propto (\sqrt{P} + \sqrt{\varepsilon P})^2 = P \pm 2\sqrt{\varepsilon}P + \varepsilon P. \tag{20.65}$$

Therefore, the crosstalk power can be as high as 20% even when the power leakage ratio of the OXC is only 1%.

Figure 20.127 shows scaling limitations due to coherent crosstalk.[202] For the case of a single crosstalk channel, a coherent crosstalk of −25 dB will cause a power penalty of 1 dB, which is much more severe

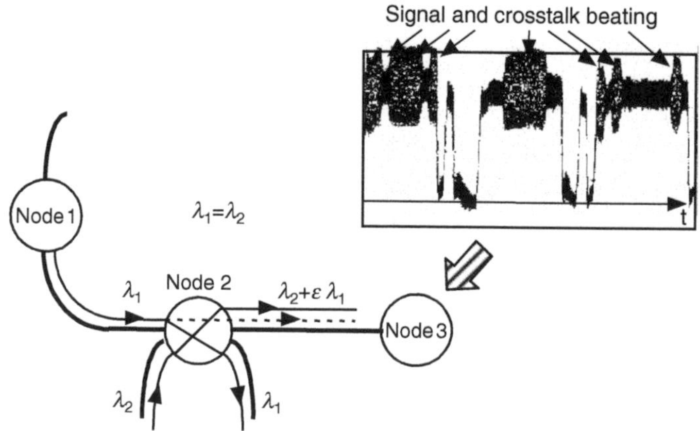

FIGURE 20.126 Power fluctuation due to coherent crosstalk. (Adapted from Goldstein, E. L. and Eskildsen. L. 1995. *IEEE Photon. Technol. Lett.*, 7, 93–94. With permission.)

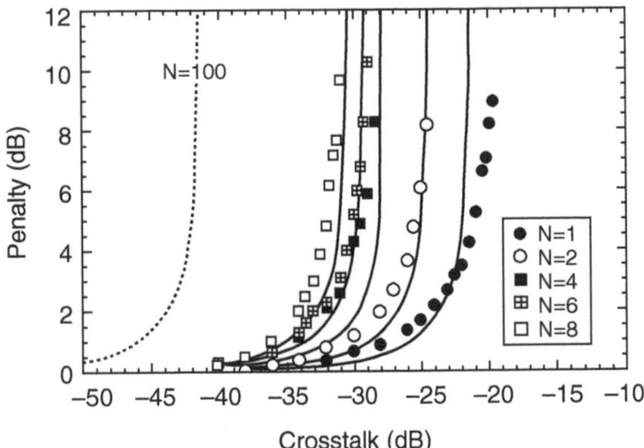

FIGURE 20.127 Scaling limitations due to coherent crosstalk. (Adapted from Goldstein, E. L. and Eskildsen. L. 1995. *IEEE Photon. Technol. Lett.*, 7, 93–94. With permission.)

than the effects of incoherent crosstalk (1 dB power penalty at −8 dB crosstalk[203]). Because of the severity of this penalty, coherent crosstalk critically limits the scalability of an optical network employing ADMs and OXCs. If 100 channels are required, the crosstalk should be lower than −45 dB. However, the typical crosstalk of current components (WDM arrayed-waveguide gratings, switches, and modulators) is only −25 dB.

To mitigate the effects of coherent crosstalk in WDM networks, the following techniques have been proposed: (1) architecture dilation, purposely introducing fiber delays (on the order a bit time) between different crosstalk paths decorrelates the bit pattern of the main signal from the interfering signals and places some of the crosstalk power in the "0" bits of the main signal instead of the "1" bits; (2) low-coherence light sources: using LEDs or chirped distributed feedback lasers (DFBs) with coherence times less than a bit time can reduce coherent crosstalk, but the impact of dispersion would become more critical; (3) polarization scrambling or modulation: polarization scrambling reduces the effect of coherent crosstalk, since interference between the crosstalk and signal decreases when their polarization states are not matched; and (4) phase modulation: modulating the phase of the signal at a rate greater than the bit rate averages out the crosstalk and improves performance, since signal degradation depends on the relative phase between the signal and crosstalk.

20.4.8 System Monitoring

Another important issue related to dispersion management and crosstalk control is the monitoring techniques. For example, in a reconfigurable system, it is necessary to reconfigure any tunable chromatic dispersion compensation modules on the fly as the network changes. An in-line chromatic dispersion monitor should quickly measure the required dispersion compensation value while data is still being transmitted through the optical link. This is very different from the more traditional chromatic dispersion measurement techniques where dark fiber is used and the measurement is done off-line over many hours (or days!).

In fact, optical performance monitoring (OPM) is essential for managing the more complicated high-capacity optical transmission and switching systems. Examples of functions that require OPM include amplifier control, channel identification and signal quality assessment. OPM can be broken down into three layers.[204] First of all, WDM channel management layer monitoring involves a determination of the optical domain characteristics essential for transport and channel management at the WDM layer, such as real time measurements of channel presence, power levels, and the OSNR. Secondly, optical signal quality monitoring involves a single wavelength and performs signal transition sensitive measurements such as the eye statistics, Q-factor, the electronic SNR, and distortion that occur within the eye due to dispersion and nonlinear effects. Finally, the data protocol monitoring involves protocol performance information, such as the BER. In this section, we will briefly discuss the physical layer monitoring of the signal quality, i.e., for the purpose of determining the health of the signal in the optical domain (Figure 20.128).

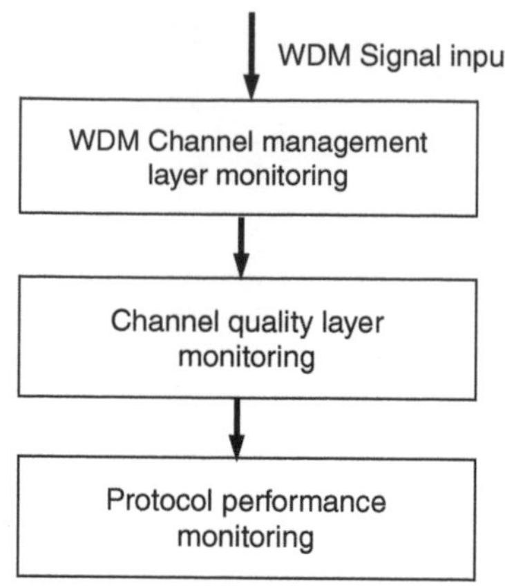

FIGURE 20.128 Three layers of optical performance monitoring: transport monitoring, signal quality monitoring, and protocol monitoring. (Adapted from Kilper, D. C., Bach, R., Blumenthal, D. J., Einstein, D., Landolsi, T., Ostar, L., Preiss, M., and Willner. A. E. 2004. *IEEE/OSA J. Lightwave Technol.*, 22(1), 294–304. With permission.)

Optical impairments can be classified into catastrophic and non catastrophic problems. Catastrophic problem such as component faults include individual or multiple component malfunctions, improperly installed or configured equipment, and damage or intrusion to the network. Impairments due to such faults are as diverse as the components and network designs deployed in the field. In addition to faulty operation, there are many other well-known effects that are always present and must be minimized or controlled. For a reconfigurable network, all of these effects should be controlled dynamically through the network design. The most common effects and the measurement parameters are given in the following lists.[204]

Notable transmission impairments:

1. Amplifier noise (ASE)
2. Amplifier distortion and power transients
3. Chromatic dispersion
4. Polarization-mode dispersion
5. Polarization effects
6. Fiber nonlinearity induced distortion and crosstalk
7. Timing jitter
8. Interference effects (MPI)
9. Pump laser relative-intensity-noise transfer
10. Optical filter distortion
11. Crosstalk

Physical layer measurements parameters:

1. Average power (per wavelength or aggregate)
2. Peak power
3. Pulse/bit shape
4. Eye diagram
5. Intensity/field autocorrelation (including higher order)
6. Amplitude power spectrum (RF spectrum)
7. Polarization state
8. Optical spectrum (wavelength)
9. Amplitude histogram (synchronous and asynchronous)
10. Q-factor/BER (V-curve measurement)
11. Polarization-mode dispersion (DGD, including higher order)
12. Chromatic dispersion
13. Phase/optical carrier characteristics

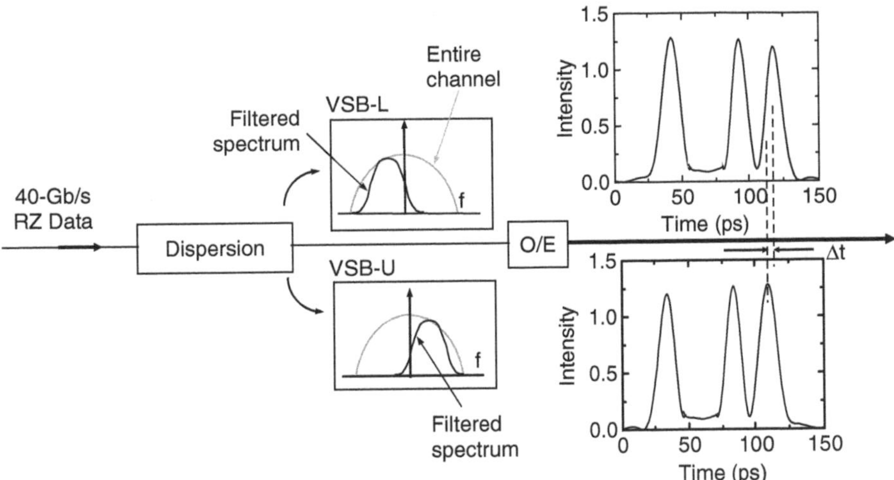

FIGURE 20.129 Conceptual diagram for monitoring chromatic dispersion using optical vestigial-sideband (VSB) filtering: the recovered bits from either part of the spectrum arrive at slightly different times depending on the chromatic dispersion. (Adapted from Kilper, D. C., Bach, R., Blumenthal, D. J., Einstein, D., Landolsi, T., Ostar, L., Preiss, M., and Willner. A. E. 2004. *IEEE/OSA J. Lightwave Technol.*, 22(1), 294–304. With permission.)

Note that monitoring is most often done at the receiving end. As an example, fiber chromatic dispersion can be monitored by using an optical filter to select first the upper and then the lower VSB signals in transmitted optical data at the receiver to determine the relative group delay caused by chromatic dispersion.[205] The group delay can be measured by detecting the relative phase of the clock component in the photodetected data[206] since the time delay between two VSB signals is a function of chromatic dispersion, as shown in Figure 20.129.

The general requirements of OPM include in-line, fast, sensitive, and with relatively low cost. High-performance OPM should be ubiquitous around the network to enable proper diagnosis, such that the monitor should isolate the specific cause and location of the problem, not just simply sound an alarm. It should be noted that high-data-rate optical networks are quite susceptible to "non-catastrophic" problems, in which there is sufficient optical signal power but the data bits themselves are unrecoverable due to various linear and nonlinear dispersive effects. Exhaustive monitoring might be possible with an unlimited budget. Although the future of optical networks is difficult to predict, the value of monitoring increases with the increasing transparency. It might be used to realize new methods of managing traffic. For instant, routing decisions based upon performance monitoring is one possibility. By monitoring the channel quality and link security and updating the routing look-up tables continually, high capacity and priority traffic can be dynamically tuned to high-performance optical channels,[207] thus ensuring that the data channels achieve acceptable BER and the whole network achieves sufficient transmission and protection capacity.

20.5 Other Multichannel Systems

In a large network, it is quite possible that a combination of data formats will be used in the local, regional, or global environments. This may occur if some links use other signaling approaches than WDM more efficiently. We will discuss the time-division-multiplexing (TDM), subcarrier multiplexing (SCM), SDM, and code-division multiplexing (CDM) in this section.

20.5.1 Time-Division Multiplexing (TDM)

At its most basic, optical networks can imitate electrical networks in which TDM is overwhelmingly used for digital data transmission. A fiber can carry many time-multiplexed channels, in which each channel can transmit its data in an assigned time slot.[208,209] That time slot also identifies either the sender's or the receiver's address, but we will assume it is the sender's address for simplicity. Furthermore, it should be emphasized that since the network will allow only one user to transmit during a given time slot, no output-port contention problems exist with TDM.

A typical TDM link is shown in Figure 20.130, in which N transmitters are sequentially "polled" by a fast multiplexer to transmit their data. This time-multiplexed data is transmitted along a fiber link and is then sequentially and rapidly demultiplexed at the receiving node. If a time slot represents the sender's address, then the demultiplexing is accomplished so that a given receiver node can access only that specific time slot which is associated with the desired sender. The multiplexing can be of the bit- or packet-interleaved form, although only bit-interleaving is shown in the figure. In bit interleaving, the multiplexer polls each node for a single bit and then polls the next node. If there are N users who wish to share the same high-bandwidth optical medium, then we must divide each transmitter's bit time, T, into N slots for (de)multiplexing. In packet interleaving, the multiplexer polls a node for an entire packet time and then polls the next node. Although either method is valid, we will limit ourselves to the bit-interleaving case which has more stringent requirements and is the more traditional method of TDM.

Depending on the bit rate of each channel, BR_{ch}, and the number of users, N, it is possible that the multiplexer and demultiplexer may not be sufficiently fast. Specifically, switching (multiplexing and demultiplexing) must be performed at the time-multiplexed speed. The bit rate of the TDM transmission (BR_{TDM}), and, consequently, the speed requirement of the (de)multiplexer is: $BR_{TDM} = N \cdot BR_{ch}$. It is obvious that optical components do not exist that accommodate very large N if each channel operates at moderate speeds, with even the fastest and most complicated TDM switches limited to <100 GHz.

The crucial time (de)multiplexing and demultiplexing functions can be performed in the electrical or optical domain is illustrated in Figure 20.131. Electrical multiplexing essentially time multiplexes several lower-speed channels into one high-speed channel which is then used to modulate a single optical signal; the electrical demultiplexing follows the same argument, in which one optical signal is detected and then electrically time demultiplexed. The only optical procedure is the transmission along the fiber. Electrical TDM is efficient and will almost certainly form the backbone for many high-speed optical signals since few users will individually generate Gbit/sec signals. However, the forefront of research in TDM is in performing the (de)multiplexing in the optical domain. Several lower-speed optical signals are time (de)multiplexed by a fast optical switch. Note that this optical switch, in which the data path remains in

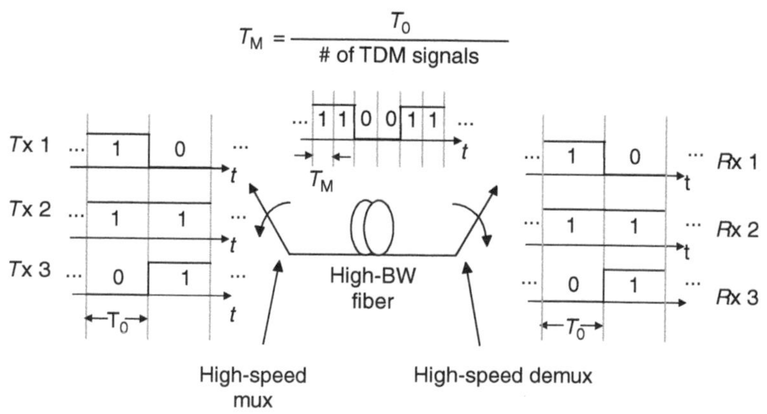

FIGURE 20.130 Concept of bit-interleaving time-division multiplexing.

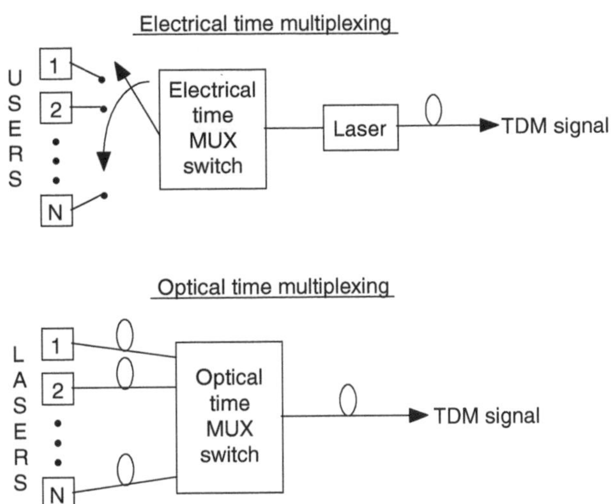

FIGURE 20.131 Time multiplexing and demultiplexing functions performed electrically or optically.

optical form, can be controlled by an external electrical input signal. The importance of this optical TDM technology is that the many lower-speed channels can originate and terminate at locations which are far away from the TDM switch. With electrical TDM, the lower-speed channels must be in close proximity to the TDM switch since even the "lower-speed" signals can still be quite fast and thus incur high transmission losses across an electrical coaxial transmission line. In networks, optical TDM incorporates the ability to perform routing and switching functions based on the specific time-slot of a particular optical bit.

Another important issue in TDM is the critical requirement of synchronizing the incoming bits. Since bits may be arriving at the multiplexer from sources located at some distance away, it is crucial to have the bits synchronized so that the multiplexer is not polling an individual input during a bit transition but only in the middle of a bit. Additionally, the demultiplexer must also know precisely the time slots of the high-speed bits. Therefore, for almost any section of a high-speed optical TDM system, it is critical to recover the transmitted signal's clock within a few clock cycles.

Advantages of TDM include: (1) there is no output-port contention problem since each data bit occupies its own time slot and there is only a single high-speed signal present at any given instant, and (2) the implementation for low-speed photonic networks is quite straightforward and similar to electronic networks. One major disadvantage of TDM is that this scheme requires ultra-high-speed switching components if the individual signals are themselves high-speed and if there are many users. For example, if there are 10 users each transmitting at 2.5 Gbit/s, the photonic switches must have a 25 GHz bandwidth. It is obvious that this method will experience a capacity limitation since the bandwidth of photonic switches will not exceed 10 s of GHz in the near future. Another difficulty of TDM is that network control, stability, and electronic processing become quite difficult (and expensive) to perform efficiently at very high speeds.

Some of the device technologies that may be critical to performing high-speed TDM include: (a) high-speed switches - these could be high-speed lithium-niobate switches in which many 2×2 switches are cascaded to form any $N\times N$ combination,[210] and (b) generation of high-speed pulses—a mode-locked laser could produce picosecond pulses at GHz repetition rates. These pulses could be modulated to produce a high-speed TDM signal, and such a signal would be in RZ format. Additionally, a single high-power short pulse can be passed through several parallel delay lines (see Figure 20.132) that can produce a series of lower-power short pulses each delayed by one bit time.[211]

Much of the recent work concerning optical TDM has centered on reducing the speed bottleneck imposed by the multiplexer and demultiplexer. By using a high-speed lithium-niobate switch and a

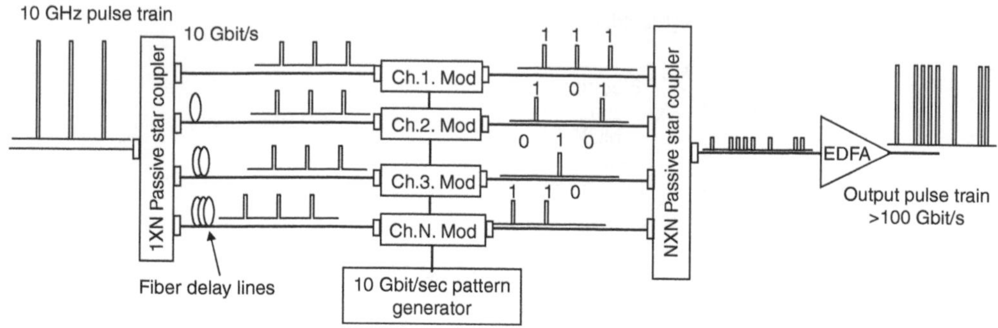

FIGURE 20.132 Generation of a series of short pulses from a single short pulse.

mode-locked laser to generate short optical pulses, time multiplexing was performed at 72 GHz, allowing transmission of two 36 Gbit/sec channels.[212] Slower experiments using conventional NRZ-modulated lasers and lithium-niobate switches have showed 16 Gbit/sec system capacity for either four 4 Gbit/sec channels or sixteen 1 Gbit/sec channels.[210] Another very interesting area of research is the use of an interference-type nonlinear optical loop mirror (NOLM) to enable ultra-high-speed multiplexing and demultiplexing.[213] Results have been demonstrated using the NOLM in which individual bits have been switched at speeds >100 GHz—an astounding achievement. This ultra-high-speed device is still in the research phase and has yet to be embraced as a practical technology.

Although TDM is conceptually straightforward with no output port contention problems, it is obvious that high-speed implementation is difficult and by no means will the ~THz fiber bandwidth be adequately utilized. However, future systems will most probably use some form of TDM (optical or electrical) in conjunction with another type of multiplexing scheme since it is unlikely that individual users will require much greater than Gbit/sec transmission. Therefore, many users would share a single laser or a single optical switch input port.

It is important to mention that TDM and WDM both provide ways to increase the transmission capacity and complement each other. Therefore optical systems today use a combination of TDM and WDM. The whole system may include many WDM channels with each wavelength channel time-division-multiplexed. As shown in Figure 20.133, using a combination of WDM and TDM, systems with transmission of around 1 Tbit/s over a single fiber are commercially available today, and no doubt that systems with higher capacities operating over longer distance will emerge in the future.

20.5.2 Subcarrier Multiplexing (SCM)

This chapter has dealt entirely with digital optical communications due to the potential for very high-speed communication per channel. However, analog optical communication has some very appealing characteristics which make it a possible solution for some near-term (and perhaps long-term) general communications problems. These include Cable Television, video distribution, $\ll 1$ Gbit/sec/channel LANs, and personal communication system interfaces. In this section, we will describe some of the basic elements of SCM and its application to multi-channel and analog systems. The reader should note that WDM is similar to SCM in that each channel is located at a different frequency.

20.5.2.1 Modulation and Demodulation

The essence of SCM is to take all of the (de)modulating, (de)multiplexing, and routing functions, which could be performed optically, and instead perform them electrically.[214] The only optical functions that remain are (1) optical generation using a laser, (2) optical transmission over an optical fiber, (3) optical detection using a photodetector, and (4) some passive optical coupling and splitting. The advantages of performing these functions electrically are that, under present circumstances, (a) electrical components

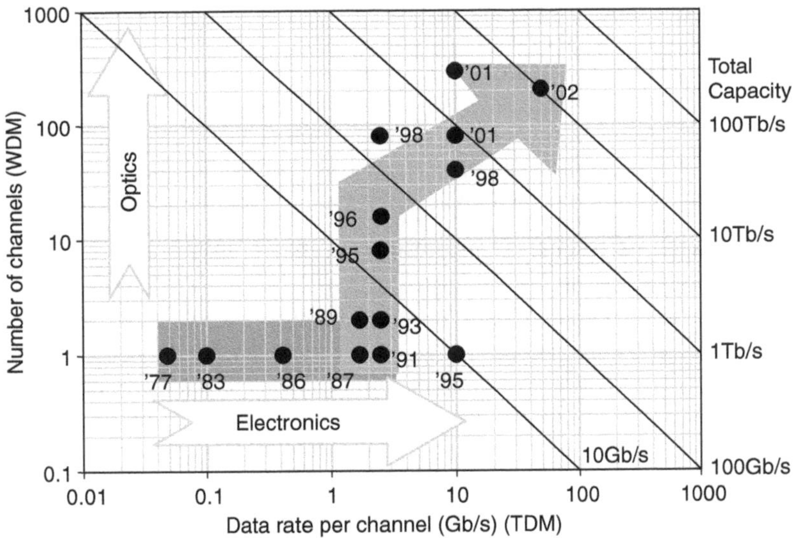

FIGURE 20.133 Capacity of commercial lightwave systems. (Adapted from Kogelnik, H. and Li, T.)

are cheaper and more reliable than optical components, and (b) electrical filters can be of an efficient and near-ideal multi-pole design whereas optical filters are only single pole.

Figure 20.134 shows a single-channel SCM system. In digital transmission, the optical beam is a carrier wave which is directly modulated by the data. We now remove the data one step further from the carrier wave by using a ~GHz subcarrier electrical wave which is modulated by the data and which, in turn, modulates the ~THz optical carrier wave. The ~MHz data is electrically mixed with the electrical subcarrier, producing sum and difference frequencies as results in standard heterodyning. Note that an electrical bandpass filter is used to allow only one product, typically the sum frequency to pass and be transmitted. The resultant modulation directly modulates either the current on a laser or the optical transmission function of an external modulator. The electrical subcarrier is a simple sinusoidal wave which has amplitude, frequency (typically 100 s of MHz or GHz), and phase, in which the optical wave, P, can be described by[215]

$$P = P_c(1 + Am_{\text{data}} \cos(2\pi ft + \varphi_{sc})) \tag{20.66}$$

where f_{sc} is the frequency of the subcarrier, A is the amplitude, m_{data} is the data modulation index, P_c is the unmodulated optical output, and f_{data} is the data modulation frequency. The subcarrier will appear as a modulated tone in the electrical frequency spectrum before optical transmission. After transmission, the optical wave is detected at the receiving end. The optical carrier wave is far above the detector

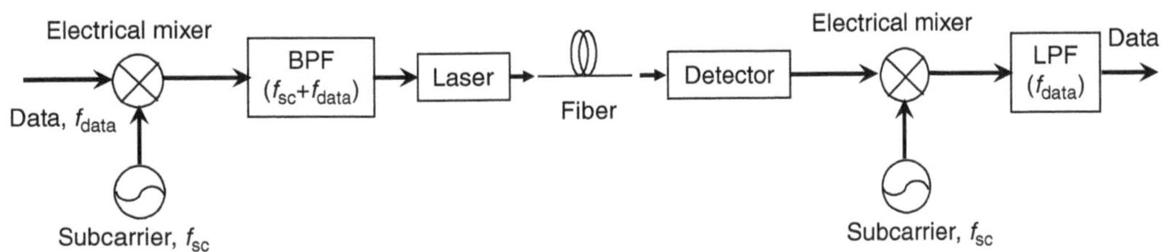

FIGURE 20.134 Basic diagram of subcarrier multiplexing.

bandwidth, but the electrical subcarrier will fall within the detector bandwidth and be detected. We employ the same principle as in the transmitter to demultiplex the data from the subcarrier. At the receiver, the same subcarrier wave frequency of f_{sc} is mixed with the recovered transmitted signal, again producing sum and difference frequencies. The sum frequency is $(2f_{sc}+f_{data})$ and the difference frequency is simply f_{data}. A low-pass filter is then used to allow only the difference frequency to pass, allowing the data to be unambiguously recovered. This method allows all the multiplexing and demultiplexing to occur in the electrical domain, and efficient electrical filters can be incorporated.

20.5.2.2 Performance Criteria and Limitations

One of the key reasons for using analog subcarrier optical transmission is that it is compatible with much of the analog transmission used today in video transmission of CATV signals.[216] The modulation format in CATV is AM–VSB, in which the subcarrier is amplitude modulated and the AM sidebands created by the modulation are manipulated so as to occupy a minimum bandwidth without sacrificing transmission quality.[217] Due to the compatibility issue, we will only discuss AM subcarrier multiplexing, although FM is also quite possible and requires a smaller signal power and larger bandwidth.[218]

The key performance criterion is to achieve a high CNR, which is in comparison to the SNR in digital communications. In a typical system, the receiver still contains random shot noise and thermal noise which are generated in the photodetector. In addition, the laser itself produces what is called relative intensity noise (RIN) over a wide bandwidth.[219] RIN is due to the random phase fluctuations in the laser light (i.e., stimulated emission) which are converted into amplitude fluctuations by reflections back into the laser or by multiple reflections along the transmission path; note that it is critically important to keep reflections low (< -65 dB) in an analog system by using high-quality connectors, splices, and isolators. Even fiber dispersion will have the effect of increasing the RIN. The RIN can be described by the statistical fluctuations in the photogenerated current:[219]

$$\text{RIN} = \frac{\langle i_{\text{ph}}^2 \rangle}{\langle i_{\text{ph}} \rangle^2} \tag{20.67}$$

and the RIN-generated noise power, σ_{RIN}, is

$$\sigma_{\text{RIN}}^2 = (\text{RIN})\bar{P}B_e \tag{20.68}$$

where $\bar{P}$ is the average received optical power, and B_e is the electrical bandwidth of the receiver. RIN is typically a very small value, but AM communications requires an extremely high CNR for good signal fidelity, requiring that the noise be kept to a minimum and that the optical power be kept high. Since analog transmission requires a very high average optical power and since the RIN noise power is proportional to $\bar{P}^4$, RIN usually becomes the dominating noise factor. The entire equation for the CNR for AM optical transmission is:[220]

$$\text{CNR} = \frac{(m\bar{P})^2/2}{\sigma_{\text{sh}}^2 + \sigma_{\text{th}}^2 + \sigma_{\text{RIN}}^2}. \tag{20.69}$$

To achieve high fidelity and "error-free" signal recovery, the CNR typically must be >50 dB, much higher than the ~ 20 dB required in direct-detection ASK digital systems!

When an external modulator is not used, the laser acts to directly convert the electrical signal into an optical signal, with a modulation of the laser bias current producing a direct modulation of the laser light output. Any deviation in the linearity of the laser light as a function of the bias current will produce a decrease in the CNR since the electrical modulation is not exactly replicated by the optical signal output.[221] Since the CNR must be extremely high in an analog system, we therefore require that the laser light output be extremely linear with bias current.

There are several limitations in SCM systems, and we will mention just the issue of "clipping" and "power fading." In general for a communications channel, it is desirable to produce as high a bias swing (i.e., light output swing) in the laser as possible to produce a high modulation index in the carrier power. Moreover, a large bias swing is needed to support many channels since each channel must individually contribute to a minimum modulation swing. However, there is a limit to the bias swing which can be supported by the laser since (1) the laser will produce no light when the bias current is below the laser threshold current, and (2) the laser linearity will degrade above a certain bias current.[222] If the electrical modulation falls below the threshold current, then the SCM signal is "clipped," and the CNR will not be adequate for signal recovery. If the modulation is above a certain current, then nonlinearities will destroy the signal fidelity and produce intermodulation products and distortions in the presence of other channels.

Another issue is the subcarrier power fading. Transmitting traditional double-sideband SCM signals experiences serious radio frequency (RF) power loss due to fiber dispersion. For example, chromatic dispersion produces a deleterious time delay of the two transmitted modulation sidebands relative to the optical carrier, causing dispersion-induced RF power fading that is a function of subcarrier frequency, fiber distance, and accumulated dispersion.[223] Several approaches have been proposed to compensate for dispersion-induced power fading in conventional DSB systems, including using adjustable modulator chirp and linearly chirped FBG's. Single sideband transmission of SCM signal is inherently immune to the problem of chromatic dispersion-induced power fading.[224]

Transmission of SCM signals over fiber will also be severely affected by fiber's PMD.[225] The DGD between the fast and slow polarization states causes a phase difference. Superposition of these two polarization states at the receiver may also lead to serious power fading of the recovered subcarrier signal due to destructive interference that is a function of subcarrier frequency, accumulated DGD, and optical power splitting ratio between the two polarization states. The PMD-induced fading penalty in a microwave fiber link is essentially dependent on the subcarrier frequency (GHz), which can be similar to the bit rates of high-speed optical baseband transmission systems, and not the data rate of the SCM signal itself. Such an optical link may accumulate PMD values of 10 ps over a 100-km distance leading to a significant fading penalty for the transmitted millimeter-wave signals. Furthermore, higher order PMD, such as polarization-dependent chromatic dispersion and depolarization effects, can cause additional distortion and degradation of the transmitted signal. PMD-induced power fading is similar to the fading that occurs in double-sideband systems due to chromatic dispersion. Although it has been shown that SSB transmission is relatively immune to chromatic dispersion, it is clear that SSB intensity modulation is not beneficial in reducing PMD-induced fading.[226] Therefore, the PMD compensation still requires the SCM transmission.

20.5.2.3 Multichannel Frequency-Division Multiplexing

We have explained SCM in terms of single-channel transmission. However, the powerful advantage of SCM is to transmit many channels simultaneously on one laser, with each channel transmitted on its own subcarrier frequency.[227] This is known as FDM and is, conceptually, very analogous to WDM systems. Figure 20.135a shows a schematic of a multi-channel SCM system, with each mixed with a different subcarrier frequency and collectively sharing the same laser transmitter. Each transmitted channel has its own electrical subcarrier generator, and the receiver can have a fixed or tunable oscillator which is used to recover one of the subcarrier frequencies, i.e., to selectively demultiplex the many channels. Shown in Figure 20.135b is a rough representation of the electrical frequency spectrum of the multi-channel transmission. One advantage of this SCM scheme is to have many channels share the cost of the same expensive optical hardware. One potential disadvantage is that the channels sharing the laser must all be in close proximity.

Another major disadvantage of SCM is that there is a significant limitation on the transmission speed of the individual channels and on the number of channels due to the finite bandwidth of the laser transmitter and receiver. Reasonably priced optoelectronic devices do not, at present, exceed several GHz in bandwidth. If we assume that each channel occupies roughly its data rate and we separate the channels

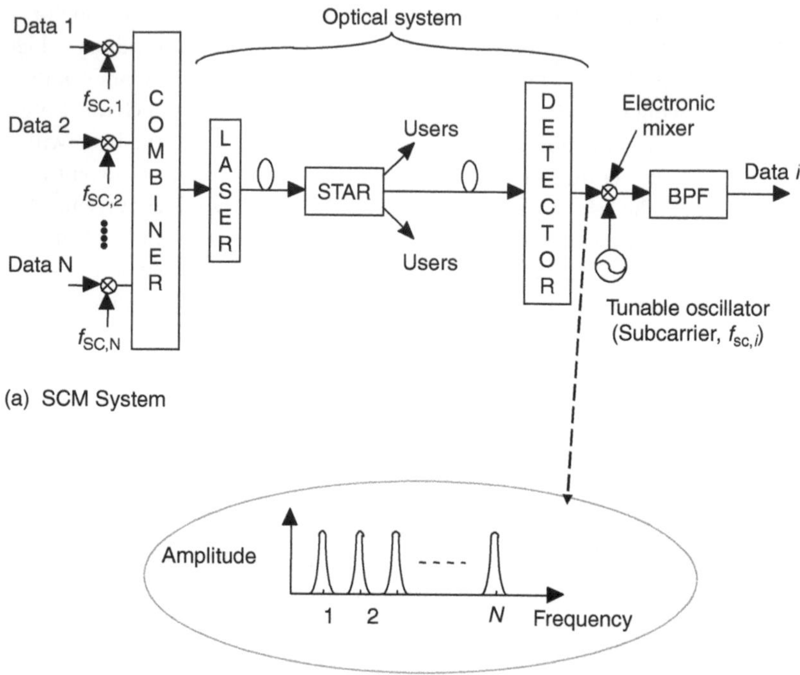

FIGURE 20.135 (a) Frequency-division multiplexed optical SCM system. (b) The electrical spectrum of the transmitted FDM system.

by twice the data rate to allow for filtering (and neglecting the intricacies of nonlinear distortions), then we have the following rough approximation for the total number of channels, N:

$$N \approx \frac{B_e}{2B_{ch}} \tag{20.70}$$

where B_e is the electrical bandwidth of the system, and B_{ch} is the information bandwidth of each channel. It is very difficult to have many 100 MHz channels sharing one laser, and GHz/channel transmission is not considered practical. However, transmitting many video channels (of a few MHz bandwidth each) is quite reasonable.

Since a real system is not perfectly linear in converting the electrical signal into an optical signal, another major limitation in the number of total channels is due to two and three channels mixing together nonlinearly, producing sum- and difference-frequency IMD products.[228] If the optical output as a function of bias current is not exactly linear, then these IMD products will be produced. If we assume three channels for simplicity, then the second-order nonlinear products appear at ($f_1 \pm f_2$ and $f_1 \pm f_3$ and $f_2 \pm f_3$) and the third-order products appear at ($f_1 \pm f_2 \pm f_3$), where the frequencies represent the subcarrier frequencies of the individual channels. Given several original channels, these products may appear at the same frequency as an original channel, decreasing the CNR:[7]

$$CNR = \frac{(m\bar{P})^2/2}{\sigma_{sh}^2 + \sigma_{th}^2 + \sigma_{RIN}^2 + \sigma_{IMD}^2}. \tag{20.71}$$

Here σ_{IMD}^2 is the sum of all the IMD products contributing to system noise within the bandwidth of a single channel. The sums of all the second-order products and the third-order products are called, respectively, the composite second order (CSO) and the composite triple beat (CTB). These are defined as:[229]

$$\text{CSO} = \frac{\text{peak carrier power}}{\text{peak power in composite 2nd-order IM tone}}$$
$$\text{CTB} = \frac{\text{peak carrier power}}{\text{peak power in composite 3rd-order IM tone}} \tag{20.72}$$

The effect of CSO and CTB is to reduce the number of channels which can be accommodated by an analog system. Figure 20.136 shows the predicted relative CSO and CTB performances for 60 CATV channels in the frequency range 50–450 MHz. The effect of CSO is most significant at the passband edges, whereas CTB contributions are most critical at the center of the band.[230]

It is quite likely that analog systems will incorporate EDFAs to compensate for losses. As was discussed in Section 20.2, the EDFA gain spectrum is not uniform with wavelength. If a directly modulated laser produces a chirp of several GHz of the output optical spectrum, then the non-uniform EDFA gain will result in a nonlinear gain (i.e., transfer function) for the different frequency portions of the signal.[231] Higher CSO and CTB will result. There are several methods to reduce this problem, but the most straightforward solution is to use a low-chirp external modulator; using an external modulator also helps to reduce the adverse affects on CNR due to simple fiber dispersion.

Probably the most obvious application of SCM and FDM is for distribution of CATV channels. Several channels can be transmitted on a single laser since the signals all are co-located at the CATV head-end. Furthermore, analog optical transmission of AM-VSB modulation format is completely compatible with the present-day CATV format. Figure 20.137 shows a CATV system in which fiber is used as a backbone, transmitting the channels to a user's local area by optical means and then transmitting the channels for the final short distance along a conventional coaxial-cable electrical system. The optical fiber is high bandwidth and low loss, significantly reducing the number of unreliable cascaded electrical amplifiers which must be traversed between head-end and user. Today's CATV systems use optical transmission. Future ultra-high capacity CATV systems may use more than one laser and use EDFAs. One experiment demonstrated the potential distribution of 16 different wavelengths to 40 million users using EDFAs to compensate for the optical splitting losses.[232]

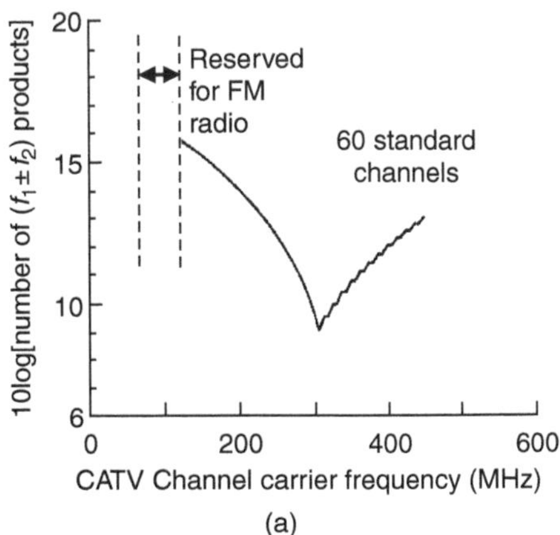

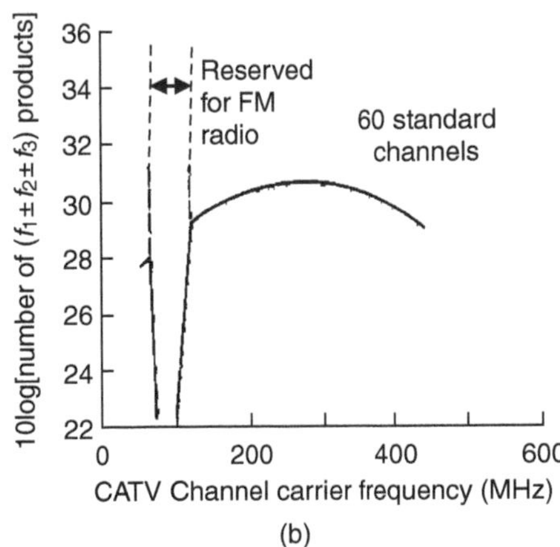

FIGURE 20.136 (a) Predicted relative CSO performance for 60 AM CATV channels, and (b) predicted relative CTB performance for 60 AM CATV channels. (Adapted from Darcie, T. E., Lipson, J. Roxlo, C. B., and McGrath, C. J. 1990. *IEEE LCS*, 1(1), 46–52. With permission.)

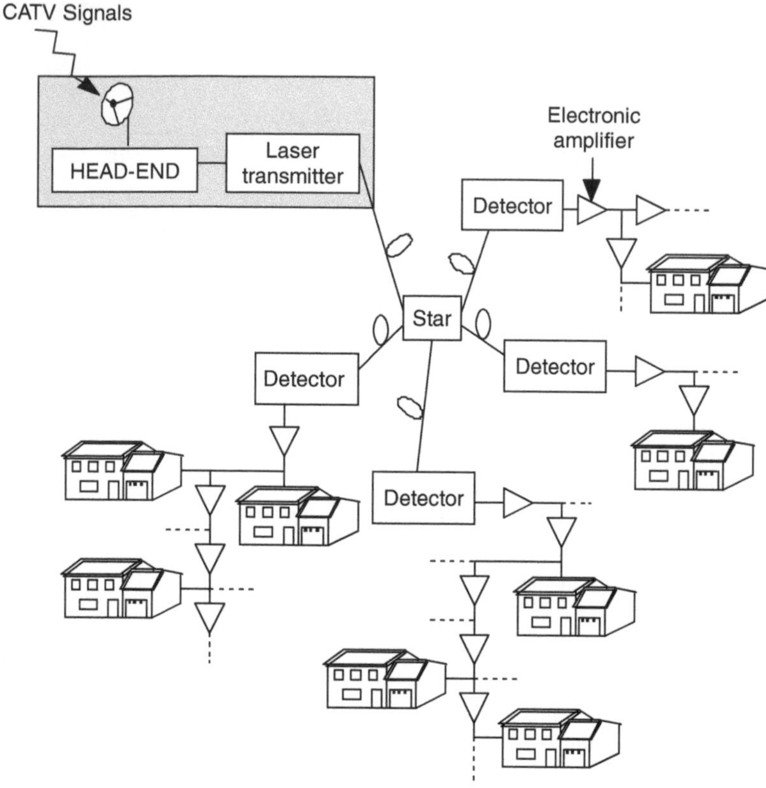

FIGURE 20.137 Cable television distribution system using optical fiber as the system backbone.

As digital video gains in popularity, it is commonly proposed to deliver compressed digital video in CATV systems with many digital subcarriers operating in the RF frequency range above the AM-VSB (e.g., 550–750 MHz) video subcarriers. Note that it is possible to transmit ten or more digital video signals in the same bandwidth required for a single analog video signal. This results in a dramatic increasing in channel capacity to several hundred or more in combination analog/digital systems. Hybird multichannel AM-VSB and M-ary quadrature amplitude modulation (M-QAM) subcarrier-multiplexed video lightwave transmission systems are currently deployed by both the telecom and cable industries as a promising technology for simultaneous delivery of both analog video and digital video/data services. Within the 50–750 MHz bandwidth, these systems can simultaneously carry up to 80 channels of traditional broadcast AM-VSB signals and more than 30 channels of M-QAM digital signals using a single laser transmitter. As an example, Figure 20.138 shows the spectrum of combining sixty-three AM/VSB analog channels and twenty-one 64-QAM digital subcarriers.[233] Due to the fact that digital signals are more robust with respect to noise and nonlinear distortions, the operating point of these hybrid systems is largely dictated by the stringent CNR requirement of the AM channels.[234]

20.5.2.4 Hybrid WDM/SCM

SCM can enable a single laser to transmit 20–50 channels, depending on the modulation format. If a system wishes to transmit more than this number of channels, or if the channels' origins are geographically separated by more than a few kilometers, then it may be quite advantageous to transmit the SCM channels on more than one laser. Future ultra-high-capacity systems may use many lasers to distribute hundreds of SCM channels, with each laser at a different wavelength. One encouraging experiment is shown in Figure 20.139 in which four separate wavelength lasers are each transmitting 50 AM-modulated SCM channels for a total system distribution capacity of 200 channels.[235] Optical filters

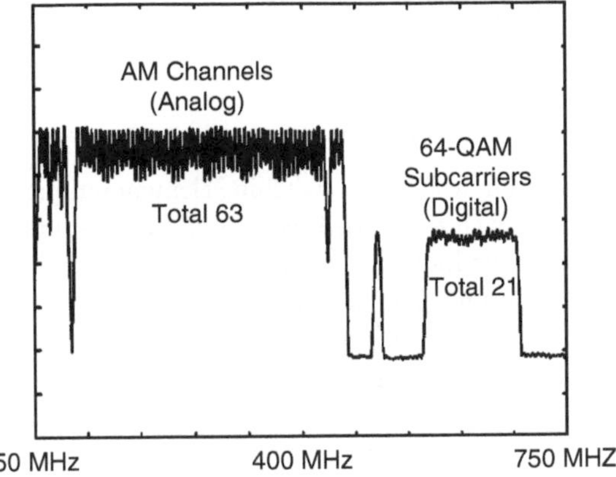

FIGURE 20.138 Spectrum of combined analog and digital subcarriers. Sixty-three AM subcarriers were spaced 6 MHz apart from 55.25 to 451.25 MHz. Twenty-one 64-QAM digital subcarriers were spaced 6 MHz apart from 564 MHz to 684 MHz. (Adapted from Joyce, G. R. and Olshansky, R. 1996. *IEEE J. Sel. Areas in Commun.*, 14(6), 1062–1065. With permission.)

are used to demultiplex the different wavelengths, and electrical filters are used to demultiplex each single-laser's 50 channels. One feature of this system is that it used 10 km of standard single-mode fiber, making this system compatible with the existing embedded fiber base in the U.S.; note that this was accomplished using 1.55-µm lasers and an EDFA power booster. This hybrid WDM/SCM system combines some of the best qualities of both types of multiplexing schemes, allowing (1) the modulation and demodulation to occur in the electrical domain, (2) compatibility with existing CATV transmission, and (3) the transmission of hundreds of channels thereby utilizing a larger part of the fiber bandwidth.

One more important point should be mentioned. In addition to send the multi-channel low-speed data, SCM subcarriers can be used in conjunction with high-speed digital baseband transmission for WDM optical performance monitoring, optical label switching (OLS) and routing. As an example, one

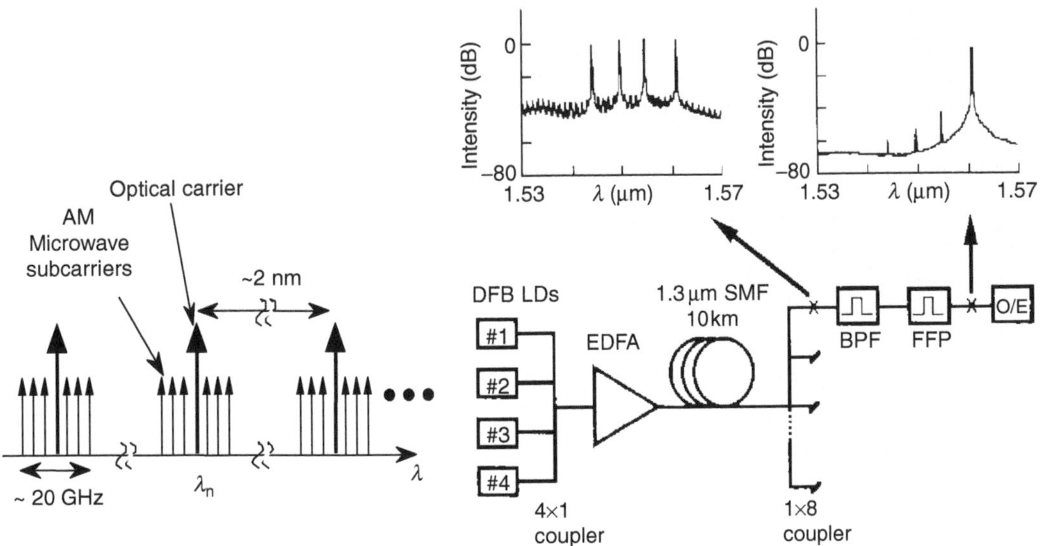

FIGURE 20.139 A hybrid WDM/SCM distribution system in which four lasers each transmit 50 AM-SCM channels.

optical network application is to use subcarrier signals to help control the network traffic and communication paths.[126]

20.5.3 Space-Division Multiplexing

A fairly mature technology is SDM in which each switching element uniquely defines an optical path and connection.[236] An optical signal will enter an SDM switch and can be routed from any input port to any output port by appropriately enabling individual space-switching elements. Therefore, the spatial path is unique and defines the connection link. The most basic element in an SDM switch is a 2×2 crossbar switch whose functionality is described in Figure 20.140. This switching element can route an optical signal from input 1 to either output 1 or output 2 depending whether the switch is biased in the bar or cross states, respectively. Many technologies are available for making optical crossbar switches, including electroabsorption in waveguides, lithium-niobate interferometers, and tunable directional couplers.[237] The concept of space multiplexing can be extended to large switches which incorporate these 2×2 switches as their basic building-block elements. For example, a 4×4 SDM switch can be composed of four 2×2 crossbar switches, with 2 switches at the input stage and 2 switches at the output stage. These switches must be properly set by an electronic controller in order to enable the routing of an incoming signal to the appropriate output port; note that optical control for such switches is still only a research topic. If two input ports request transmission to the same output port, only one signal will be allowed to reach its proper destination and the other signal must be deflected to an alternate output port. Output-port-contention resolution, deflection routing fairness and priority, and local control are all network management issues which must be considered by any fully-functioning switch.

As another illustrative example of an optical SDM switch, the 2-dimensional optical crossbar switch of Figure 20.141 is composed of N input fibers, N output fibers, and an $N \times N$ array of optical switching elements, such as microelectromechanical systems (MEMS) based switches. By making the mirrors stand up or lie down, the light from any inlet port may be switched to any desired outlet port. Since each spatial switching element (mirror) can be independently chosen to transmit or reflect the light, then a unique path connecting any input fiber to any output fiber can be established by allowing the appropriate spatial element (mirror) to reflect. Although SDM is a simple technique, it suffers from some disadvantages: (1) the number of elements scales as the square of the number of users, making this technology limited to a small number of users due to fabrication limitations; and (2) optical splitting and recombining when using passive optical $N \times N$ star couplers will each incur a $1/N$ optical splitting loss which limits operation. We can compensate for these losses by using external EDFAs or by using an array of SOAs as the switching elements themselves.[238] Since many switching technologies (i.e., liquid crystal) offer slow ms switching speeds, fast switching SOAs can be employed to obviate the speed bottleneck. An alternate method for implementing SDM is to use an array of electro-optic lithium-niobate switches arranged in a multi-stage planar orientation.[239] The switching speeds are high (GHz) but there is a practical fabrication-limited size associated with such switches.

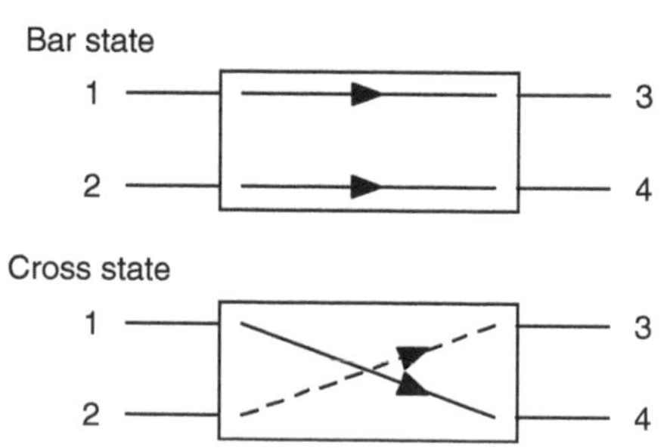

FIGURE 20.140 Functionality of a simple 2×2 crossbar switching element.

A desirable feature of SDM is that much of the technology is fairly mature and can be implemented in the short term, even though further technological advances are probable. It is quite possible that high-capacity multichannel systems will incorporate elements of SDM in conjunction with other multiplexing techniques.

Space-division multiplexing is different than the other multiplexing methods we have discussed in this chapter. All the other techniques have simultaneously multiplexed many signals onto a single optical fiber, thereby utilizing a larger fraction of the

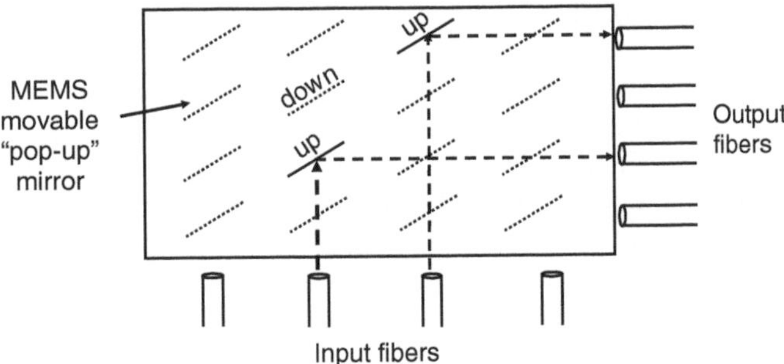

FIGURE 20.141 Space-division multiplexed optical switch.

enormous fiber bandwidth. However, SDM allows for transmitting only a single signal on an individual fiber. The usefulness of SDM is that it enables the efficient routing of many high-speed signals through a switching matrix. It is this switching matrix that has an extremely high capacity, but the fiber-optic transmission system outside the space switch is still using only a small fraction of the fiber bandwidth.

20.5.4 Code-Division Multiplexing

A method of encoding which predates optical communications is CDM, which establishes the unique communications link by encoding the destination inside of each bit time.[240,241] OCDMA is essentially a broadcast and select network with privacy. Each user sends a unique code to represent a "1" bit within the chip times, as shown in Figure 20.142. N users would require N codes that are orthogonal. Each destination has a receiver which will decode the intended data with the inverse of the transmitter coding hardware but will not decode data intended for another destination. (Note that a fixed transmitter code and a tunable receiver code is another alternative.) If a decoder has the correct inverse hardware as the transmitter, then the output will produce a peak in the optical signal called the autocorrelation. This autocorrelation signal will trigger a detected "1" bit in the receiver if it is above a certain threshold level. If the receiver does not have the correct decoder, then the cross-correlation signal, which appears as a noisy

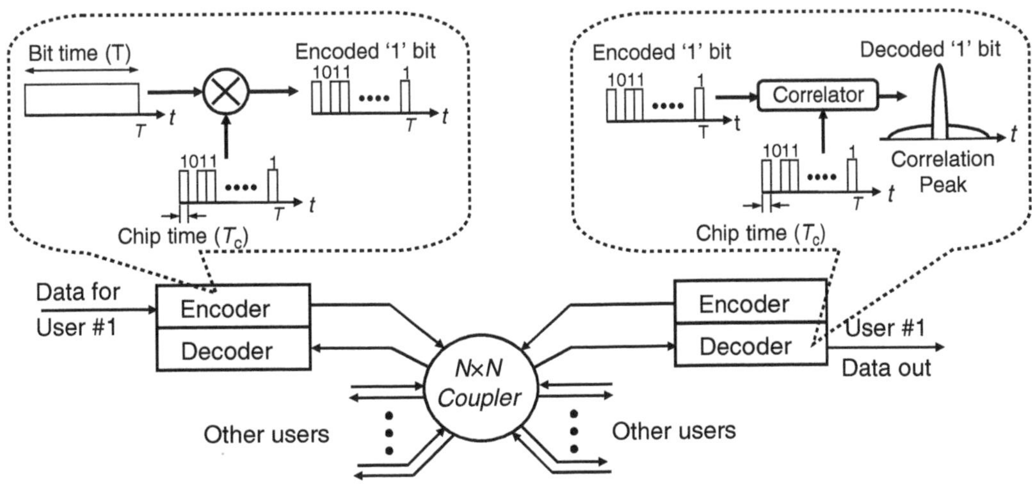

FIGURE 20.142 Principle of CDMA: Encoding and decoding of an optical bit with variable-length optical delay lines.

temporally-broad background signal will contain all the energy and the autocorrelation will be small, producing only "0" bits.

Figures-of-merit for operating of CDMA include (1) achieving a high auto-correlation signal when the receiver's decoder scheme inversely matches the encoder scheme, and (2) achieving a low cross-correlation signal when the decoding scheme does not inversely match the encoding scheme; cross-correlation is crosstalk due to other channels sharing the same fiber medium. These two factors will determine the SNR of the recovered signal when several channels are simultaneously transmitted over the fiber. We have mentioned that N users require N different codes. However, this is misleading in the sense that N distinct possible codes can be achieved naively with ($\log_2 N$) number of chips/bit. For example, 8 users could use only 3 chips/bit to uniquely identify each user. However, this means that one user would have a "111" code and another user would have a "110" code. If both these users are transmitting simultaneously, then 2 of the 3 coding chips will align in the decoder and produce a high cross-correlation crosstalk power, significantly reducing the recovered channel's SNR. This effect is increased if several channels are simultaneously transmitted. Therefore, many more than ($\log_2 N$) chips/bit are required in order to minimize the possible cross-correlation power from simultaneously-transmitted channels. This increased number of chips/bit will minimize the possibility that chips for any non-selected channel will align in the decoder. One performance gauge for a CDMA system is the achievable SNR given a certain number of users (N) and a certain number of chips/bit (K):[242]

$$\mathrm{SNR} \cong \frac{1}{0.29}\left[\frac{K}{(N-1)}\right]. \tag{20.73}$$

Furthermore, the receiver output, $r(t)$, can be given by[243]

$$r(t) = \int_{-\infty}^{+\infty} s(z) f(z-t)\,\mathrm{d}z \tag{20.74}$$

where each bit has been encoded into a waveform $s(t)$ and the receiver address is $f(t)$. The autocorrelation signal appears when $s(t)=f(t)$, and the cross-correlation signal appears when $s(t) \neq f(t)$. In order to minimize the cross-correlation, the different codes should be (pseudo)orthogonal.

One fundamental result of encoding bits into short chip times is that the bandwidth, or spectrum, of the transmitted signal has increased several times over. A 1 Gbit/sec signal would normally occupy ~1 GHz information bandwidth, but this same signal will occupy 10 times that bandwidth if 10 chips/bit are used. This effect is known as spread-spectrum transmission.[244] If several channels are simultaneously transmitted on the same fiber, the channels will each have a different code but their broad spectra will overlap. The decoding is temporal, not spectral, and has the effect of narrowing the transmitted spectrum upon decoding.

We have already mentioned that the decoding can be performed asynchronously. Another advantage of CDM is the increased transmission security because (1) the spectrum of the signal is so broad so that narrowband noises or jamming signals do not significantly affect the data transmission and recovery, and (2) the data can only be recovered by a unique and private decoding sequence, or key, which can be kept private. Without a decoder, the spectra of all the channels overlap and cannot be demultiplexed.

The disadvantages of CDM include (1) a limitation in the number of users for which a fixed number of M chip slots will produce a high autocorrelation and a low cross-correlation signal when the decoder and encoder match; (2) a limitation in the speed of the system since very short pulses are required for the chip times within each bit time, thereby limiting the bit rate for a finite-pulse-width transmitter; and (3) high optical splitting losses at the encoder/decoder. Note that disadvantage can be thought of as the most critical drawback at present since ~Gbit/s transmission, which is a standard speed for optical transmission systems, can be achieved in CDM only with great difficulty. Ultra-short-pulse laser transmitters can produce ultra-short chip times, but the complexity and cost may not be make this system achievable for practical use.

CDM received much attention in the mid-to-late 1980s, but has not been popular in the optical communications community due to the fundamental limitations in modulation speed, number of users, and optical encoding/decoding losses. However, CDM re-emerged recently as one promising candidate for next-generation broadband multiple access technique for local area networks due to the following facts: (1) novel tunable multi-wavelength chip scale devices enable compact and low-cost encoders and decoders recently; (2) photonic crystal-based delay lines make discrete fiber-based delays unnecessary; and (3) optical CDMA has many attractive characteristics, such as allowing nodes to transmit asynchronously without any media access delay, nearly perfect power control, fixed user positions and high signal to noise ratios.[245,246] There are several different OCDMA implementations that could be roughly classified according to operation principle as incoherent and coherent OCDMA. Coherent OCDMA using ultra-short optical pulse is receiving increasing attention with the progress of reliable and compact encoder/decoder devices, such as planar lightwave circuit (PLC) and superstructured fiber Bragg grating (SSFBG). In coherent OCDMA, encoding and decoding are based on optical field amplitude instead of power intensity. The coding can be either direct time-spreading the ultra-short optical pulse or spectral phase-encoded time spreading.

20.6 Optical Network Technologies

The chapter has already dealt extensively with the transmission and recovery of multichannel signals. Additionally, an extremely important topic is the switching or routing in optical networks as they relate to controlling a multichannel system. Some higher-level network issues include high-efficiency protocols, algorithms, and architectures. Optical networks offer enormous capacities, provide a common infrastructure over while a variety of services can be delivered, and are capable of delivering bandwidth in a flexible manner where and when needed. In this section, we will deal with the two fundamental types of optical network infrastructures: circuit-switching and packet switching. Furthermore, since we have concentrated mostly on WDM systems in this chapter, this section will discuss switching issues as they relate mostly to WDM, with some discussion applying to other forms of multiplexing as well.

20.6.1 Optical Circuit Switching

The two fundamental types of underlying telecommunication network infrastructures, based on how traffic is multiplexed and switched, are circuit-switched and packet switched. A circuit-switched network provides circuit-switched connections to its customers. Once the connection is established, a guaranteed amount of bandwidth is allocated to each connection and available to the connection all the time. The network is also transparent and the nodes seem to be directly connected. In addition, circuit switching requires a lower switching speed (>ms). Many types of communication links and distribution systems may satisfactorily be interconnected by circuit switching which is relatively simple to operate. The problems with circuit switching include: (1) it is not efficient at handling bursty data traffic (low utilization for traffic with changing intensity or short lived connections); (2) there is a delay before the connection can be used; (3) the resources are permanently allocated to a connection and cannot be used for any other users; and (4) circuit-switched networks are more sensitive to faults (for example, if a part of the connection fails, the whole transfer fails).

Optical networks based on WDM technology, called wavelength-routing networks, are now being deployed. Since optical circuit switching is the relatively mature technology today, the wavelength-routing network is generally circuit-switched. The basic mechanism of communication in a wavelength-routed network is a lightpath (corresponding to a circuit), which is an all optical connection (communication channel) linking multiple optical segments from a source node to a destination node over a wavelength on each intermediate link. At each intermediate node of the network, the lightpath is routed and switched from one link to another link. A lightpath can use either the same wavelength throughout the whole link or a concatenation of different wavelengths after undergoing wavelength conversion at intermediate optical nodes. In the absence of any wavelength conversion device, a lightpath is required to be on the same wavelength channel throughout its path in the network; this requirement is referred to as the wavelength continuity property of the lightpath. Once the setup of a lightpath is completed, the whole lightpath is available during the connection. Note that different lightpaths can use the same wavelength as long as they do not share any common links (i.e., same wavelength can be reused spatially in different parts of the network).

As shown in Figure 20.143, the key network elements in the wavelength-routing network are optical line terminal, optical add/drop multiplexers (OADM) (see Figure 20.82), and OXC (see Figure 20.87). There are many different OADM structures such as parallel or serial, fixed or reconfigurable. In general, an ideal OADM would add/drop any channel and any number of channels, and would be remotely controlled and reconfigured without disturbance to unaffected channels. The other requirements include low and fixed loss, independent of set of wavelengths dropped, and low cost. An OXC can switch the channels from input to output ports and input to output wavelengths. The functions in an OXC node includes providing lightpath, rerouting (switching protection), restoring failed lightpath, monitoring performances, accessing to test signals, wavelength conversion, and multiplexing and grooming. An OXC can be either electrical (performing O-E-O conversion for each WDM channels) or optical, transparent or opaque. To accomplish all these functions, the OXC needs three building blocks: (1) fiber switching, to route all of the wavelengths on an incoming fiber to a different outgoing fiber (optical space switches); (2) wavelength switching, to switch specific wavelengths from an incoming fiber to multiple outgoing fibers (multiplexing/demultiplexing); and (3) wavelength conversion, to take incoming wavelengths and convert them to another optical frequency on the outgoing port. This is essential to achieve strictly nonblocking architectures when using wavelength switching (wavelength conversion will be discussed in the later section).

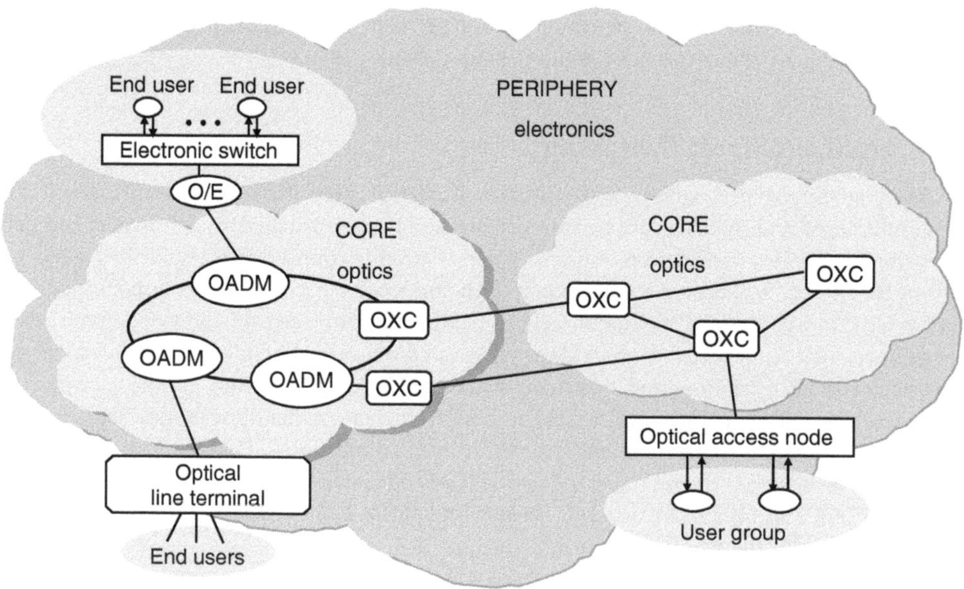

FIGURE 20.143 An optical network showing optical line terminal, OADM, and OXC.

20.6.2 Optical Packet Switching

In packet switched networks, the data stream is broken into small packets of data. These packets are multiplexed together with packets from other data stream inside the network. The packets are switched inside the network based on their destination. To facilitate this switching, a packet header is added to the payload in each packet. The header carries addressing information. The switching nodes read the header and then determine where to switch the packet. At the receiver end, packets belonging to a particular stream are put back together. Therefore, packet switching requires switching speeds of μs or less. For high-speed optical transmission, packet switching holds the promise for more efficient data transfer in which no long-distance handshaking is required and high-bandwidth links are used more efficiently.

As shown in Figure 20.144, an optical packet switching node is generally composed of three parts: control unit, switching unit, and input/output interfaces. The control unit retains information about network topology, the forwarding table, scheduling, and buffering. It decides the switching time and is in charge of resolving contentions at a node. The switching unit allows the data to remain in the optical domain during the routing process. It is especially important for optical packet switched networks that the switching speed be fast enough to minimize overhead. The input/output interface is where optical technologies are utilized to deal with contention problems. Optical buffers and wavelength converters are the building blocks of time and wavelength domain contention resolution modules, respectively, and are housed in the interface units. In addition, other physical layer functionalities required for an optical switching node such as synchronization are realized at the interface.

Note that network packet switching can be accomplished in a conceptually straightforward manner by requiring a node to optoelectronically detect and retransmit each and every incoming optical data packet. The control and routing information is contained in the newly-detected electronic packet, and all the switching functions can occur in the electrical domain prior to optical retransmission of the signal. Unfortunately, this approach implies that an optoelectronic speed bottleneck will eventually occur in this system. On the other hand, it is extremely difficult to accomplish the signal processing in the optical domain currently. Alternatively, much research is focused towards maintaining an all-optical data path and performing the switching functions all-optically with only some electronic control of the optical components. The reason is that the control unit detects and processes only the header of a packet (not the payload). Therefore it is possible to transmit the header at a lower bit-rate to facilitate the processing. As a result, the presence of electronics at the control unit does not necessarily pose a limitation on the data transmission rate. Figure 20.145 shows a generic solution that passively taps an incoming optical signal. Information about the signal is made known electrically to the node but the signal itself remains in the optical domain. The routing information may be contained in the packet header or in some other form (i.e., wavelength, etc.).

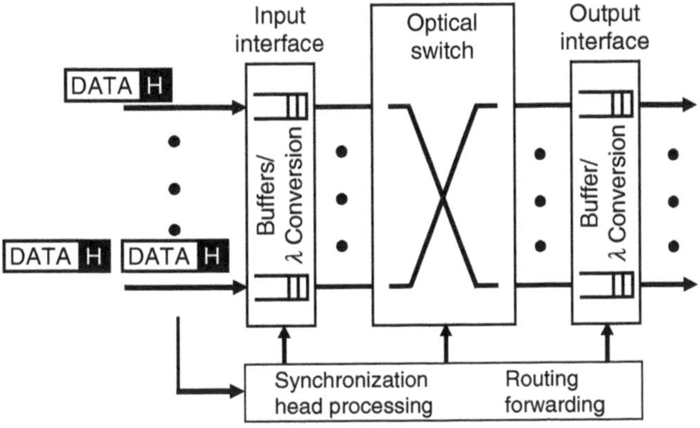

FIGURE 20.144 Schematic diagram of an optical packet switching node.

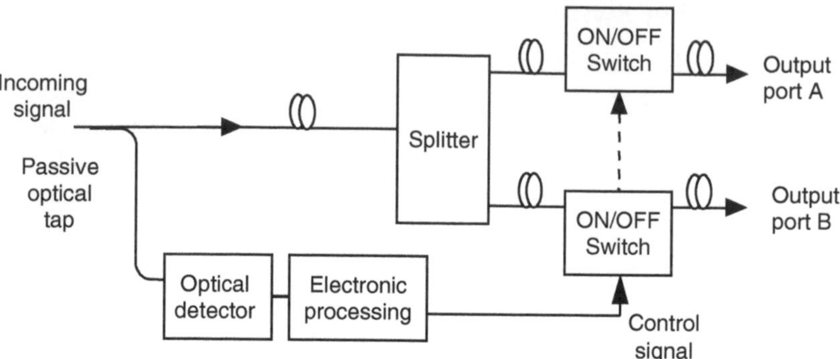

FIGURE 20.145 Passive optical tapping of an optical packet in order to determine routing information and allow a node to electronically control an optical switch.

The various functions that may be performed in an optical switching node include (1) address/label recognition to determine the intended output port, (2) header updating/label swapping to prepare the packet header for the next node, (3) bit and/or packet synchronization to the local node to time the switching process, (4) routing-table caching as a reference for routing decisions, (5) output-port contention resolution via buffering and/or wavelength conversion, (6) signal monitoring to assess the signal quality, (7) signal regeneration to combat the accumulated distortion of the signal, and (8) optical switching to direct packets to the appropriate output ports.

It is very important to mention that in the past several years, data traffic has been growing at a much faster rate than voice traffic. Data traffic is "bursty" in nature, and reserving bandwidth for bursty traffic could be very inefficient. Circuit switching networks are not optimized for this type of traffic. On the other hand, packet switching networks require a complex processing unit and fast optical switches. Therefore, optical burst switching (OBS) was introduced to reduce the processing required for switching at each node and to avoid optical buffering. OBS is somewhere between packet switching and circuit switching with the switching period on the order of many packets. OBS takes advantage of time-domain statistical multiplexing to utilize the large bandwidth of a single channel to transmit several lower-bandwidth bursty channels. At the network edge, packets are aggregated to generate bursts that are sent over the network core. In almost all OBS schemes, the header is sent separately from the payload with an offset time that is dependent on the scheme.[247,248] The header is sent to the switches, and the path is then reserved for the payload that follows. The loose coupling between the control packet and the burst alleviates the need for optical buffering and relaxes the requirement for switching speed.

20.6.3 Key Technologies for Optical Switching Nodes

Many challenging optical switching issues require solutions.[249] Most of these issues will relate to packet switching, although circuit switching will also require attention. We will be discussing packet switching issues with some discussion reserved for circuit switching. Due to the limited space, the specific issues that will be addressed here include routing control, contention resolution, and optical header processing.

20.6.3.1 Routing Control

In a network, control and routing of signals can be centralized or decentralized.[250] A centralized system requires that a central controller have access to all signal information and can command the individual nodes to connect a given link. In a decentralized system, each node has its own autonomy, connecting to any available link it wishes. Each scenario depends critically on the specific network.

Signal routing is a critical network issue. As a signal enters a node or switch, it is crucial that the node determines (1) if that signal it intended for it, (2) if the signal must be transmitted unaffected, or (3) if the signal requires active routing to a specific output port (see for example a WDM ring in Figure 20.146). In other words, how does a node know if the incoming packet is intended for it, and, if so, which wavelength should it tune its optical filter to given a WDM network? The straightforward approach of passive optical tapping and electronic circuit decision making is one approach, but it cannot easily accommodate a system which has many channels simultaneously transmitted on the same fiber. As with contention resolution, there are several approaches to attack this issue of multichannel routing, with no single favored solution. We will briefly mention three such approaches:

1. OLS: A node can handle an incoming packet by tuning its optical filter "on-the-fly" in a self-routing scheme. Each node passively taps the signal's on an incoming fiber. Each transmitter not only transmits data on its own wavelength, but it also transmits a subcarrier optical-label on a specific frequency. The optical label contains information pertaining to packet forwarding. The router forwards an optical packet based on the optical-label in the optical domain without converting the baseband data payload in the electrical domain. In this way, the packet forwarding is independent of the payload data format, bit rate, and underlying protocols. It pursues a natural form of IP over WDM with a simple optical-label shim layer, and provides true interoperability among circuit-, burst-, and packet-switching technologies at the optical layer.[251]
2. Common Wavelength Control: In a centralized WDM environment, each node can have its own signature wavelength and a common central controller can have its own wavelength. The nodes and the controller also each transmit a subcarrier signal at a specific subcarrier frequency. Each individual node monitors the common wavelength from the controller and deciphers the low-speed routing information in the controller's subcarrier signal. This routing information commands the node to tune its optical filter to a specific wavelength. Each node can communicate back to the controller on its own individual subcarrier frequency. Each node uses a dual-passband liquid-crystal optical filter for which one fixed wavelength monitors the controller and the other tunable wavelength recovers data from another node.[252]
3. Sequential Polling Using Optical Filters: A receiver with an optical filter can sequentially poll N distinct wavelengths in an N-user network.[125] As the filter is sequentially tuned to a given wavelength, the receiver listens momentarily in order to determine if that wavelength requests to speak with it. This method is useful in a circuit-switched environment, but is too slow (especially given large N) in a packet-switched network.

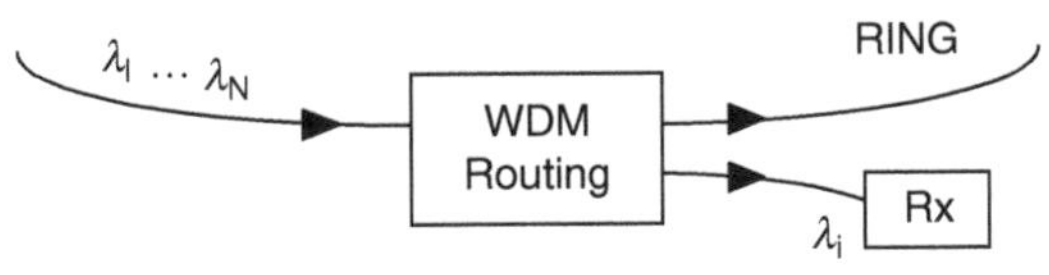

FIGURE 20.146 An example of routing concerns in a WDM ring network.

20.6.3.2 Contention Resolution

Potentially the most limiting issue for a high-speed network is a situation in which two or more input ports request a communications path with the same output port, known as output-port contention. In the optical regime, contention problems may be addressed in time, wavelength or space domains. Since we are dealing with a high-speed system, we require rapid contention resolution in which one signal is allowed to reach its destination and the other signal is delayed or rerouted in some fashion; this resolution can be based on any type of network fairness algorithm. When considering our multiplexing schemes, the issue of contention exists when signals from two input ports would request routing to the same output port and contain identical wavelengths (or codes or time slots). Several approaches exist for resolving contention. Figure 20.147 summarizes the three contention resolution schemes. We will briefly discuss three approaches in time, frequency, and space domain respectively.

Time Domain—Optical Buffering. Once it is determined that contention exists, the packet is to retain locally at the switching node and then switch it to the appropriate output port when that port is available. This requires local buffering of the packet, either in electrical or optical form. Electronic buffering is

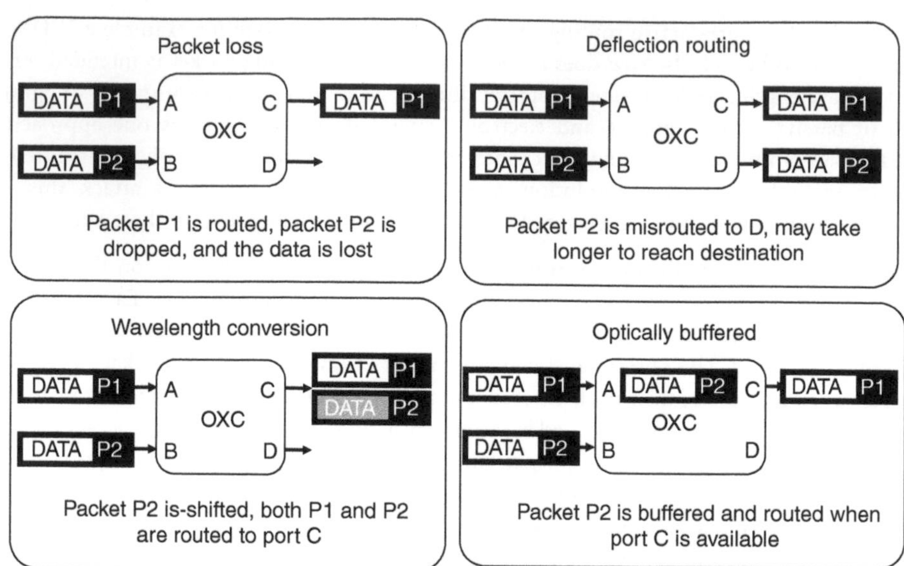

FIGURE 20.147 Contention and contention resolution techniques. Assuming all packets are destined to output port C, P, packet.

straightforward but requires undesirable optoelectronic conversions and may require very large buffers. On the other hand, optical buffering is difficult since (1) many buffering schemes require updating a "priority bit," and it is difficult to change a priority bit of an optical data stream; and (2) optical memory is not an advanced art, consisting mostly of using an optical delay line.[253,254] It is possible to construct a programmable delay line using a cascade of optical switches and fiber delay lines. This configuration can delay a signal by sending it through different lengths of fiber. There are several drawbacks in using fiber delay lines as optical buffers. First, generating even a small time delay requires a long length of fiber. For example, 200 m of fiber will generate only 1 μs delay! As a result, long delays should be avoided to make the buffer size reasonable. Secondly, fiber delay lines are first-in first-out (FIFO) memories with a predetermined delay time. Once an optical signal is sent into a fiber delay line, it will not be accessible until it comes out of the other end of the delay line. In addition, if one simply decides to delay a packet at one of the input ports by adding a fiber delay line to that path, all the subsequent packets coming from that port will go through an unnecessary delay as well. This would degrade the performance of a switching node. To circumvent this problem, fiber delay lines may be used in conjunction with optical switches. Various other techniques have been also presented to create optical delay lines including a fiber loop, Bragg gratings and free-space optics.[255–257] The major limitations these techniques suffer from include the ability to tune the delay, the tuning range (if any), and the repeatability.

It is important to mention that in the past few years, researchers have slowed light traveling through a semiconductor down to 6 miles a second,[258] 31,000 times slower than that light normally travels in a vacuum, based on the material's electromagnetically induced transparency (EIT). By slowing down light and developing chips that can handle slow light impulses would provide the ideal optical buffers and memory.

Note that contention resolution is not the only application for an optical buffer. Packets also need to be delayed before the optical switch to make sure that the control unit has enough time to process the header. Until the output port for a packet is determined, the packet cannot be switched.

Frequency Domain—Wavelength Conversion. Frequency domain contention resolution takes advantage of the additional dimension available in WDM systems: if two packets on the same wavelength are heading to the same output port at the same time, one of the packets is wavelength-converted to utilize any unused wavelength at that output port. Then both the packets will be routed to the same output port (at different wavelengths). This is perhaps the most efficient contention resolution technique as it does not delay signals, nor does it redirect the packets through a longer transmission distance.

An ideal wavelength converter would change the wavelength of an optical signal independent of its bit-rate, data format, polarization, wavelength or power. However, practical wavelength converters are far from ideal, and depending on the technique, are sensitive to some or all of the input signal parameters.[259] Other important performance metrics for wavelength converters are low NF and high output extinction ratio. The popular wavelength conversion techniques are based on one of three phenomena,[260] namely, gain saturation in SOA, interferometric effects, or nonlinear wave-mixing.

Perhaps the simplest technique for wavelength conversion is to use gain saturation (so-called XGM: Cross gain modulation) in SOAs (see Figure 20.99, Section 20.6.5). This technique can be used only for OOK data formats since the phase information is lost in this wavelength conversion scheme. The other drawbacks of this technique are SNR degradation due to ASE noise added by the SOA, low extinction ratio and chirp induced on the output signal. Slow gain recovery also limits the bit rate at which XGM in an SOA can be used for wavelength conversion.

Interferometric wavelength converters exploit the dependency of the refractive index on the optical intensity and carrier density in semiconductors. Due to XPM, the continuous optical wave will experience different phase shifts that depend on the intensity of the pump signal. This phase modulation then can be converted to amplitude modulation through an optical interferometer. The output signal can have a sharp rise/fall time as well as a high extinction ratio because of the sensitivity of the interferometer. However, this technique suffers from stability problems if the interferometer is not an integrated device. It also requires a very accurate adjustment of the SOA bias current and is sensitive to temperature changes. Other configurations of interferometric wavelength converters have also been reported using a NOLM.[261]

Nonlinear wave-mixing is the wavelength conversion technique that makes an exact wavelength-shifted copy of a signal (amplitude, frequency, conjugated phase). Wave-mixing can happen as a result of cascaded second-order nonlinearities, in which case it is referred to as difference frequency generation (DFG). $L_iN_bO_3$ is a good example of a material with a high second-order nonlinearity, high optical bandwidth, and high dynamic range.[262] Figure 20.148 shows the DFG mechanism. First, the high power pump signal initiates frequency doubling or second harmonic generation inside the material. Then the second harmonic component mixes with the probe signal to generate a wavelength-shifted copy of the probe. It is important to note that the frequency spacing between the pump and the probe dictates the output wavelength. In other words, the output signal's wavelength is a mirror image of the probe wavelength with respect to the pump. Consequently, DFG can be used to wavelength-shift several WDM channels simultaneously.

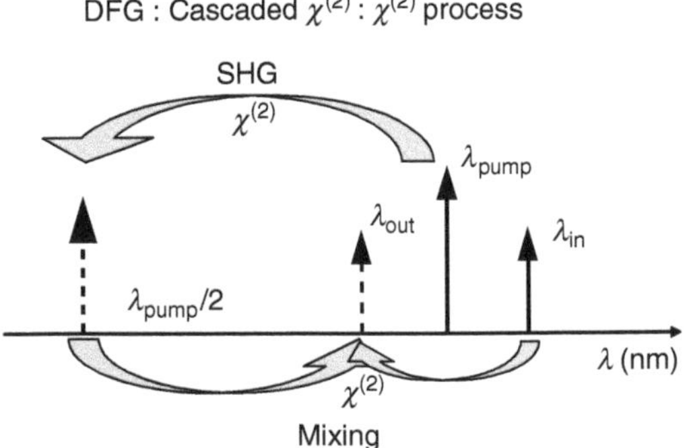

FIGURE 20.148 The mechanism of difference frequency Generation.

As shown in Figure 20.102, FWM that generates new wavelengths can also be used as the wavelength converter. FWM can be accomplished inside fiber as well as in SOAs, where the third-order nonlinearity is much higher than in fiber. In general, nonlinear wave-mixing techniques are polarization sensitive, though it is possible to reduce polarization sensitivity by wavelength-shifting each polarization component of the signal separately. Most wave-mixing techniques are capable of wavelength shifting many independent channels simultaneously in a single device, unlike XGM and XPM.

Cross absorption modulation (XAM) in an electro-absorption modulator has also been used for wavelength conversion.[263] The principle is similar to XGM but instead of the gain, the pump signal modulates the absorption seen by the probe signal. As a result, a copy of the data on the pump signal is imprinted on the probe wavelength.

Space Domain—Deflection Routing. Once it is determined that contention exists, one signal is routed to its desired output port and the other signal can be routed to another empty output port, known as deflection routing. The deflected packet is now traversing the network and will eventually be routed back to its desired destination. Increased delay will be incurred since the packet is being switched by more nodes than is the minimum number, adding extra "hops" to the deflected packet.[264] Deflection routing requires a complicated control algorithm to make sure packets are not lost inside the network. Deflection also adds overhead to the network by artificially increasing the network load.

20.6.3.3 Optical Header Processing

Each packet typically has its routing and destination information contained in the packet header. In a large network, it is quite possible that the routing of a packet may change as it traverses the network due to dynamic traffic patterns and link availabilities. If the routing is changed, then the packet header information itself must be changed. Moreover, if this rerouting is done in a WDM system, then active wavelength shifting may also be required. This can be done optoelectronically by detection and retransmission.[265] Alternatively, the header can also by recognized and replaced in the optical domain by using the all-optical processing techniques including SOAs, NOLM, and other nonlinear devices.[266,267]

Optical Header Recognition. Header information can be transmitted in several different ways. For example, the baseband header can be transmitted as a data field inside a packet either at the same bit-rate as the data, or at a lower rate to relax the speed requirement on the electronics for header detection. It can also be located out-of-band either on a subcarrier on the same wavelength or on a different wavelength, altogether.

For the first scheme, an optical baseband header recognition technique can be built based on optical matched filters (correlators) which determine the correlation between the received header and entries in a routing look-up table. An optical correlator adds weighted versions of the input signal in a tapped-delay-line structure. If the output of the correlator exceeds a threshold level, an address match is detected. In a report shown in Figure 20.149, an FBG array is served as 4-bit passive optical correlator. Different sections along the FBG array act as partial mirrors which reflect the incoming light with a predetermined delay. The weight of each delayed version of the signal is proportional to the strength of the grating at that point. In order to equalize the optical power reflected off each section of the grating, the reflectivity of each section increases as we go deeper inside the grating.[268] An optical address match can also be detected using active time-to-wavelength mapping.[269] In this technique, each address bit is converted to a unique

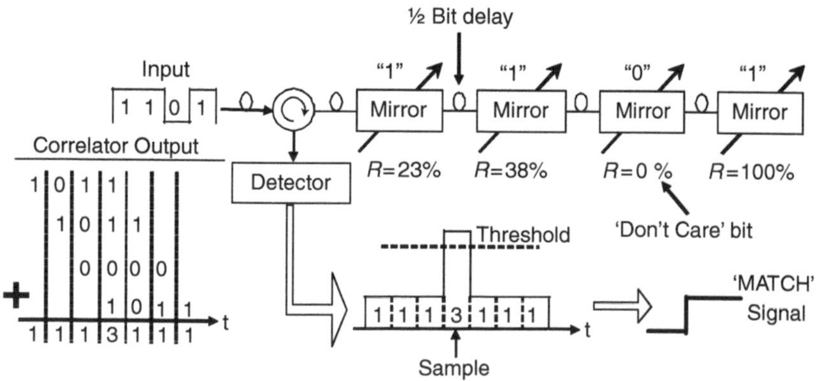

FIGURE 20.149 Architecture and operating principle of an FBG-array based optical correlator.

wavelength through a nonlinear wave mixing process (e.g., by using a Periodically-Poled Lithium Niobate waveguide—PPLN), and sent through a wavelength dependent delay. The output level will be proportional to the correlation between the incoming and the stored address. Though it is possible to use nonlinearities and/or wavelength conversion in other devices such as SOA, the PPLN waveguide offers the advantage of simultaneous processing of several WDM channels.

Note that the complexity of optical address recognition techniques increases drastically with the length of the address. Conventional internet routers process 32-bit destination addresses and match them to entries in a large routing table. This can be a significant source of latency in the electronic domain as routing tables in the core may contain $>$500,000 entries. On the other hand, most of the current optical techniques are capable of processing only a few address bits at a time. However, a recent study[268] shows that by using a software algorithm that searches and determines a small subset of addresses that correspond to the most popular network destinations, a significant amount of traffic can be correctly routed, all-optically, by checking only a few address bits. This "popular" look-up table can be implemented using a manageable number of optical correlators. The traffic that cannot be processed by optical means will be processed using conventional electronics. For example, if all incoming packets with the second, fourth and twentieth bits equal to "0", "1" and "1," respectively, should all go to the same output, then the optical correlator needs to be capable of matching only these 3 bits, in their respective positions, instead of all 32.[270]

Optical Label Swapping. As mentioned in the routing control, optical labels carrying routing information specific to the optical layer need to be updated (called label swapping) at each routing node. This involves erasure of the original label, determining the new label based on the information received from the incoming label and re-writing the new label into the original data stream. Wavelength conversion is also required to convert the packet and the new label to a different wavelength which is also determined by the information in the incoming label. The label is processed electronically to determine the output port, wavelength and the bits in the new label but the corresponding packet does not undergo any O-E-O conversion. Since the electronic processing modules operate only on the label, it is possible to maintain bit-rate and data format independence and transparency with respect to the data packet while using lower bit rate labels to accommodate the slower electronic speeds.

Various technologies have been explored to implement all-optical label swapping functions.[271] These usually involve selective conversion of the packet to another wavelength (thereby erasing the old label) and insertion of the new label by modulation of a tunable laser.[272] Since the label need not have the same bit rate or data format as the packet, it is possible to devise various schemes to exploit this flexibility in order to obtain highly efficient and high-speed label swapping techniques.

Synchronization. The switching scheme in optical packet switched networks can be implemented synchronously or asynchronously. Most of the research so far has been performed on fixed-length packets, time-slotted networks which require synchronization. Variable packet size switching networks, in general, are more susceptible to contention, and have a lower throughput as a result.[273]

In general, synchronization is realized between either two incoming packets or an incoming packet and a local clock signal. To synchronize two signals, their bits should be aligned in the time domain. Tunable delay lines can achieve this by accurately delaying one of the signals relative to the other. To achieve fine synchronization, the resolution of the tunable delay should be much smaller than the bit time. It is possible to get a continuous delay by exploiting fiber dispersion in conjunction with wavelength conversion techniques.[274]

All-optical synchronization techniques demonstrated hitherto have been fairly complex and expensive. Most optical synchronization approaches rely on nonlinear interaction between optical signals or the nonlinear response of a semiconductor device in order to generate a timing signal. In one approach, the fast-saturated gain of SOAs is exploited to detect the start of a packet.[275] A NOLM is then used to enhance the intensity of the synchronized pulse. In another approach, a terahertz optical asymmetric demultiplexer is utilized to extract the clock signal from an incoming packet, which can then be used for synchronization.[276]

Depending on the architecture of an optical packet switched network, it may be sufficient to only synchronize the time-slots of the optical packets rather than the actual bits. One approach for slot synchronization is called distributed slot synchronization.[277] In this technique, synchronization is achieved by adjusting the nodes' transmission time so that packets arrive synchronously at a chosen switching point in the network. Each node has to time its transmission by listening to the master synchronization signal and its own signal reflected from the switching node. This way the synchronization would be independent of transmission lengths. A packet arrival jitter of less than 13 ns has been demonstrated using this technique.[277]

20.6.4 Protocols

A standardized network protocol must be used to ensure that data packets are all formatted with recognizable routing information so that the packet can be switched through the network with full global compatibility. The two standards which show the most promise of full adoption for a global optical network is SONET[278,279](or its European equivalents is SDH: synchronous digital hierarchy) and asynchronous transfer mode (ATM);[280,281] these are shown in Figure 20.150 in very simple form. Data and header information is bunched into small 53-byte ATM packets. These packets arrive at a switching node at random times and are grouped together into a large 125-µs SONET frame which makes its way in predetermined synchronous time slots through the network. The lowest bit-rate for SONET is 51.84 Mbit/s and is called synchronous transport signal level-1 (STS-1) or optical carrier level-1 (OC-1). Higher bit-rates at STS-n are generated by interleaving n synchronized STS-1 signals. As an example, STS-192 or OC-192 corresponds to 9.95328 Gbit/s. The ATM packets are "unloaded" by the SONET frame when its direction is switched through the network and it can be placed into a different SONET frame. The analogy has been made that the ATM packets represent people randomly boarding a time-scheduled SONET train.[282]

Internet protocol (IP) is the most widely used protocol for the Internet. IP converts the data into variable-size packets and adds a header which includes the source and destination addresses and error control information. IP packets may be broken into several ATM cells which could be transferred on a SONET network. In general, IP offers no guaranteed quality-of-service (QoS). Rather, it uses a best-effort approach to transmit the packets to their destinations. In order to improve the switching efficiency in WDM networks and provide some QoS guarantees, multiprotocol label switching (MPLS) technology was introduced in 1998.[283] MPLS can provide a label-switched path (LSP) in the network that can be used to establish an end-to-end path between two stations. It also has some measures for protection and failure detection.

Ethernet networks are based on carrier sense multiple access with collision detection (CSMA/CD) and operates at 10 Mbit/s to connect users on a copper bus. Fast Ethernets operating at 100 Mbit/s and higher

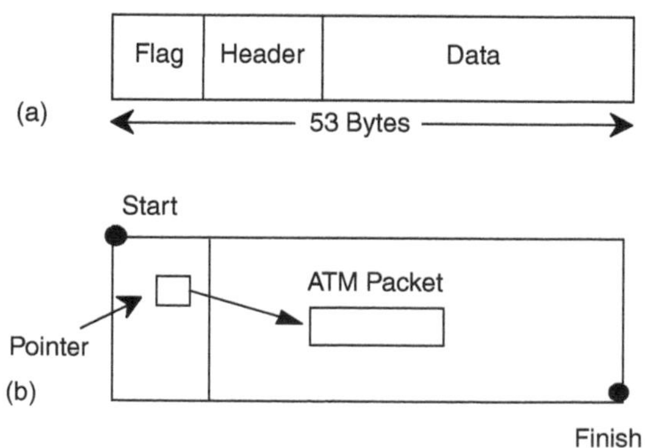

FIGURE 20.150 (a) An ATM packet, (b) a SONET frame containing many ATM packets and a header field.

use optical fibers. 100BASE-FX is a fiber-based fast Ethernet standard that supports a network span of up to 400 meters. Several standards have also been introduced to implement Gigabit Ethernet (e.g., 1000BASE-SX and 1000BASE-LX) and 10-Gbit/s Ethernet (10GBASE-S, 10GBASE-L, 10GBASE-E and 10GBASE-LX4) depending on the maximum reach and the operation wavelength (850, 1310 or 1550 nm), utilizing the large capacity of the optical fiber.[280]

20.6.5 Optical Access Networks

We have explored the use of optical networks for metro and long-haul backbone network applications. Here we will briefly discuss the optical access network through a few examples.

Access network is the last mile of the telecommunications network that runs from the service provider's facility to the home or business. Most residential connections to the Internet are still through dial-up modems operating at a low speed on twisted pairs. As the demand for access bandwidth increases with emerging high-bandwidth applications, such as high-speed Internet, distance learning, high-definition television (HDTV), and video on demand (VoD), video conferencing, the last-mile access networks have become a bandwidth bottleneck in today's telecommunications infrastructure.

To ease this bottleneck, it is imperative to provide sufficient bandwidth capacity in the access networks to open the bottleneck and thus present more opportunities for the provisioning of multiservices. Optical access solutions promise huge bandwidth to service providers and low-cost high-bandwidth services to end users and are therefore widely considered the technology of choice for next-generation access networks. To realize the vision of optical access networks, many key issues need to be addressed, such as network architectures, signaling protocols, and implementation standards. The major challenges lie in the fact that an optical solution must be not only robust, scalable, flexible, but also implemented at a low cost comparable to that of existing access solutions in order to increase the economic viability of many potential high-bandwidth applications.

A passive optical access network (PON) is an optical network that takes optical fiber right to the customer premises instead of copper. Figure 20.151 shows the general architecture of a PON. It does not use any active electronic components (devices consuming power) from the central office to the consumer's premises. The network carries a single strand of fiber, which undergoes a multiple split to serve many consumer installations. This splitting is achieved by means of passive splitters. On the side of the local exchange there is an optical line termination (OLT), on the user side there is an optical network unit (ONU). Using a passive point-to-multipoint fiber network, consisting of optical fibers and one or more splitters (in cascade), a number of ONUs are connected to an OLT in a tree topology. Bus and ring topologies are considered less suitable for user connections, as they run a higher risk of individual users causing disruptions for other users. An ONU can be combined with a network termination unit (NT). This produces an optical network termination (ONT). The OLT has the interfaces with the backbone network architectures that provide the services to the users. Hence a PON's passive part consists of splitters and fiber and is situated in the field. Reasonably complex active components are required in the local exchange (the OLT) and on the side of users (the ONU/ONTs).

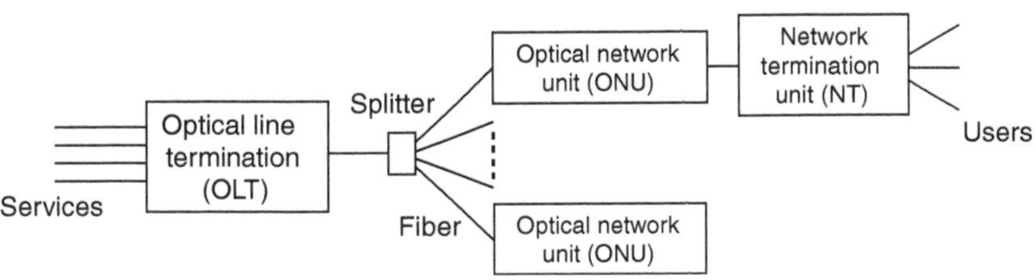

FIGURE 20.151 The architecture of a passive optical access network.

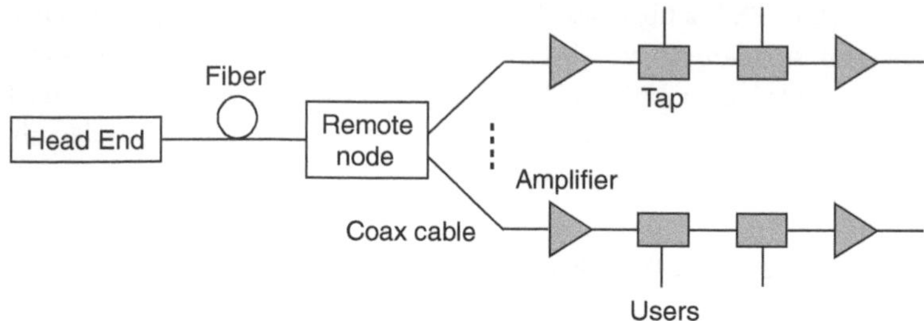

FIGURE 20.152 The hybrid fiber coax cable network. The head end broadcast signals over fiber to the remote node, which then distributes in to individual subscribers via coaxial cable drops.

The passive network constitutes a shared transmission room. The required connections can be made with the use of multiplexing technologies. For example, TDM at the transport level can be realized by alternating ATM cells on an ATM PON (APON) and Ethernet packets on an Ethernet PON (EPON). A broadcast star WDM topology with each network interface unit assigned a separate wavelength is called WPON. Note that there are some hurdles PON technology has to overcome, for example, the lack of availability of fiber in the access networks, distance limitations imposed by the lack of amplification or regeneration on the fiber. Active optical networks are the counterpart of the passive networks. Characteristically, electronics are set up in the last mile in street cabinets that also supply a traffic distribution function. Such architectures are usually called Fiber-to-the-Curb. It is intrinsically more error-prone, more sensitive to changes in temperature and sensitive to radio interference. An especially important factor is that active equipment introduces the additional risk of down time due to local power failures. The best-known architecture for an active optical last mile network is the hybrid fiber coaxial (HFC) network, as shown in Figure 20.152. This approach is attractive due to the fact that coaxial cable is already deployed to the home. It maintains compatibility with existing equipment, therefore is the logical evolution choice for cable companies. On the other hand, it has the disadvantages of a coax-based solution, such as limited upstream bandwidth, limited reliability, and powering needed for the many amplifiers in the path.

20.7 Future Trends

In this chapter, we have covered many different topics involving high-speed optical communication systems. We have endeavored to treat the most important topics that will impact optical systems for years to come. These include WDM for multichannel communications, and analog transmission for CATV or wireless systems. The "glue" binding and impacting all of these areas is the revolutionary introduction of the EDFA. Additionally, transmission limitations are being continually redefined. For example, by appropriately managing dispersion and nonlinearities, designing robust modulation formats, and using advanced DRAs, research results have demonstrated the ability to transmit spectral-efficient high-speed signals over much-longer distances than was previously thought possible.

One fascinating point is that the overwhelming amount of scientific and technological progress has occurred only over the past 30 years. In fact, an interesting trend observed by Dr. Tingye Li of AT&T Bell Laboratories is that optical communications technology has enabled the doubling of the bit rate-distance transmission product every year (see Figure 20.153)! This trend has been sustained by depending on different technologies such as the inventions of the single-mode fiber, the single-frequency laser, and the EDFA. It is hard to imagine how optical communications will continue to make such astounding progress, but the striving continues due to the enormous inherent potential of the optical fiber. The limits are continually being pushed back. Although the sky isn't the limit, the fiber certainly is!

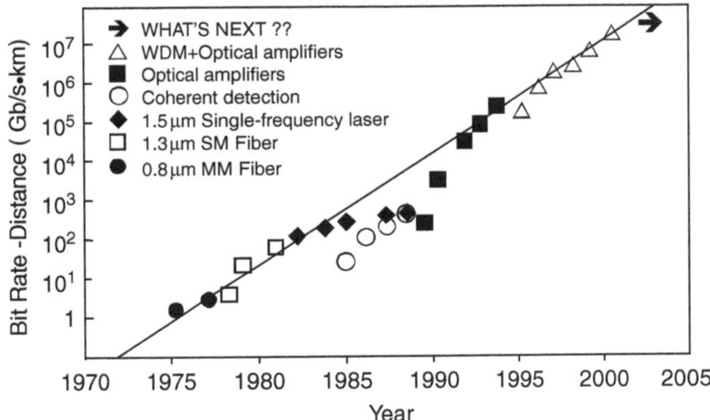

FIGURE 20.153 Bit rate-distance product as a function of year for different optical technologies.

The progress in deployed optical systems has already achieved an astounding rate of growth over the past 25 years. One should anticipate that the field of optical communications will continue to enjoy periods of explosive growth due to the enormous bandwidth demand for next-generation Internet services. This bandwidth demand may approaches and even goes beyond the limitations of today's fiber capacity. Therefore, breakthroughs in new materials and devices, spectral efficiency, the fundamental limitations of optical communications, and advanced technologies are needed to enable the optical networks of tomorrow. We would like to highlight a few ongoing research areas that may change the future optical communications significantly. Readers can find the more depth discussion in the related literatures.

Nano-photonic and photonic crystal devices. The concept of a photonic crystals (photonic bandgap structures) was first described by Yablonovitch[284] and John[285] in 1987. The readers can find the more depth discussion of nano-materials and photonic crystals in the early chapters.

Simply put, photonic crystals are artificial periodic dielectric structures with a dielectric constant that is spatially distributed according to certain symmetries. When a defect is introduced in an otherwise perfect photonic crystal, localized photonic states can be created in the gap, whose shapes and properties are dictated by the nature of the defect. A point defect can act like microcavity, a line defect like a waveguide, and a planar defect like a perfect mirror. This ability to manipulate an optical wave provides us with a new dimension in the ability to mould or control the properties of light. The easiest way to understand the behavior of light in a photonic crystal is to compare it to the movement of electrons and holes in a semiconductor such as silicon. The current explosion in information technology has been derived from our ability to control the flow of electrons in a semiconductor in the most intricate ways. But here photonic crystals control photons—rather than electrons. Photonic crystals provide a new mechanism for steering and localizing light that is different from that based on conventional optics. It has been generating much interest due to their possible applications in areas such as filters, waveguides, optical switches, design of more efficient lasers, and more. A number of basic applications of photonic crystals will likely start making an appearance in the market place, such as photonic-crystal waveguides, high-resolution spectral filtering, photonic-crystal lasers, and LEDs. These new devices may enable the all-optical signal processing so that to support new network functionalities. Given the impact that semiconductor materials have had on every sector of society, photonic crystals could play an even greater role in the 21st century, particularly in the optical-related industry.

Highly secure, reliable, and scalable optical networking. A critical challenge for the future implementation of optical network technologies is that they deliver superior performance while being cost effective and robust to deploy. An example is the issues associated with providing end-users with access to the core network shown in Figure 20.154. New architectures, protocols, management software, and transport

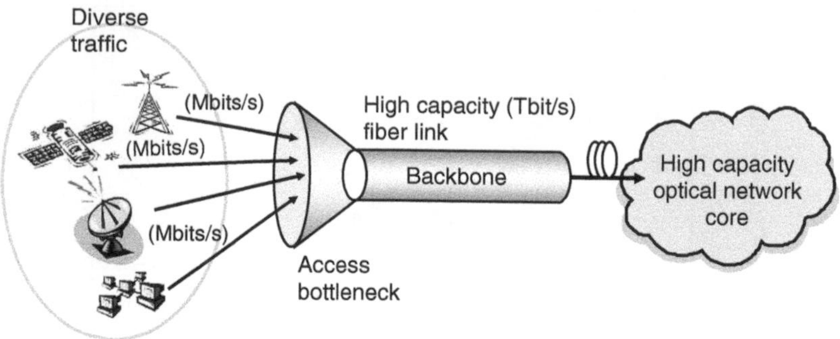

FIGURE 20.154 High-speed local area networks (LANs) are connected to the ultra-high-speed backbone through a lower-speed access-point bottleneck.

technologies are needed to fulfill the vision of providing innovative services to end-users with high security, reliability, and scalability. Security is the ability to prevent unwanted eavesdropping on end-to-end sessions. Reliability is the ability of the network to recover from one or more failures at any point throughout the network. Scalability is the ability to scale in functionality to provide richer services and/or the ability to achieve cost effective solutions that scale to a low cost/end-user. For example, security can be realized at the physical layer by using the quantum nature of the optical field (i.e., quantum cryptography: physics guarantees that an intruder cannot tap the line to intercept the key without being detected). At the protocol and network level, security may be provided by distributing the data along different fiber spans so that eavesdropping on any single link could not reconstruct the message. One may wish to speculate that total fiber bandwidth and the unlimited services will be available to each of us in the year 2010? 2015? 2020?

In summary, optics is far behind the electronics in terms of the intelligence. There are still many research challenges therefore opportunities. With the breakthrough in optical materials and devices, one may anticipate the new technical advances will be achieved in the near future and then will change optical communications area significantly.

Acknowledgments

The authors wish to thank the kind and critical help from all the members of USC's Optical Communications Laboratory family.

References

1. Marcuse, D. 1982. *Light Transmission Optics*. Van Nostrand Reinhold, New York.
2. Kressel, H. and Butler, J. K. 1978. *Semiconductor Lasers and Heterostructure LEDs*. Academic Press, New York.
3. Li, T. 1983. Advances in optical fiber communications: an historical perspective. In *Selected Areas in Communications*, edited by *IEEE J.*
4. Bergano, N. S., et al. 1995. 40 Gbit/s WDM Transmission of Eight 5 Gbit/s Data Channels Over Transoceanic Distances using the Conventional NRZ Modulation Format, Paper PD19 presented at Conference on Optical Fiber Communications, San Diego, CA.
5. MacMillan, J. 1994. *U.S. News and World Report*, Advanced Fiber Optics, May.
6. Henry, P. S., Linke, R. A., and Gnauck, A. H. 1988. Introduction to lightwave systems. In *Optical Fiber Telecommunications II*, S. E. Miller and I. P. Kaminow, eds., Academic Press, New York.
7. Agrawal, G. P. 2002. *Fiber-Optic Communication Systems*. Wiley, New York.

8. Nakazawa, M., Yamamoto, T., and Tamura, K. R. 2000. 1.28 Tbit/s-70 km OTDM transmission using third- and fourth-order simultaneous dispersion compensation with a phase modulator. *Electron. Lett.*, 36, 24, 2027–2029.
9. Vareille, G., Pitel, F., and Marcerou, J. F. 2001. 3 Tbit/s (300×11.6 Gbit/s) Transmission over 7380 km using C+L band with 25 GHz spacing and NRZ format, Paper PD22 presented at Optical Fiber Communication Conference and Exhibit, Anaheim, CA.
10. Cai, J. -X., et al. 2002. Transmission of thirty-eight 40 Gbit/s channels (>1.5 Tbit/s) over transoceanic distance, Paper PD FC-4 presented at Optical Fiber Communication Conference and Exhibit, Anaheim, CA.
11. Frignac, Y., et al. 2002. Transmission of 256 wavelength-division and polarization-division-multiplexed channels at 42.7 Gbit/s (10.2 Tbit/s capacity) over 3×100 km TeraLight™ fiber. Paper PD FC-5 presented at Optical Fiber Communication Conference and Exhibit, Anaheim, CA.
12. Born, M. and Wolf, E. 1999. *Principles of Optics, 7th Ed.*, Cambridge UP, Cambridge.
13. Klein, M. V. and Furtak, T. E. 1986. *Optics, 2nd Ed.*, Wiley, New York.
14. Gloge, D. 1971. Weakly guiding fibers. *Appl. Opt.*, 10, 2252–2258.
15. Gowar, J. 1984. *Optical Communication Systems.* Prentice Hall, New Jersey.
16. Keck, D. B. 1976. Optical fiber waveguides, In *Fundamentals of Optical Fiber Communications*, M. K. Barnoski, ed., Academic Press, New York.
17. Giallorenzi, T. G. 1978. Optical communications research and technology: fiber optics. *Proc. IEEE*, 66, 744–780.
18. Jones, W. B. Jr. 1988. *Introduction to Optical Fiber Communication Systems.* Holt, Rinehart, and Winston, New York.
19. Willner, A. E. and Hoanca, B. 2002. Fixed and tunable management of fiber chromatic dispersion, In *Optical Fiber Telecommunications IV*, Ivan P. Kaminow and Tingye Li, eds., Academic Press, New York.
20. Kaiser, P. and Keck, D. B. 1988. Fiber types and their status, In *Optical Fiber Telecommunications II*, S. E. Miller and I. P. Kaminow, eds., Academic Press, New York.
21. Marcuse, D., Chraplyvy, A. R., and Tkach, R. W. 1994. Dependence of cross-phase modulation on channel number in fiber WDM systems. *IEEE J. Lightwave Technol.*, 12, 5, 885–890.
22. Knudsen, S. N. and Veng, T. 2000. Large effective area dispersion compensating fiber for cabled compensation of standard single mode fiber. *Optical Fiber Communication Conference., Vol. 1*, pp. 98–100.
23. Eiselt, M., Garrett, L. D., and Tkach, R. W. 1999. Experimental comparison of WDM system capacity in conventional and nonzero dispersion-shifted fiber. *IEEE Photon. Tech. Lett.*, 11, 2, 281–283.
24. Poole, C. D. 1988. Statistical treatment of polarization dispersion in single-mode fiber. *Opt. Lett.*, 13, 687–689.
25. Poole, C. D. 1989. Measurement of polarization-mode dispersion in single-mode fibers with random mode coupling. *Opt. Lett.*, 14, 523–525.
26. Namihira, Y. and Wakabayashi, H. 1991. Fiber length dependence of polarization mode dispersion measurement in long-length optical fibers and installed optical submarine cables. *J. Opt. Commun.*, 12, 1–8.
27. Jones, K. A. 1987. *Introduction to Optical Electronics.* Harper and Row, Philadelphia.
28. Young, W. C. 1989. Optical fiber connectors, splices, and joining technology, In *Optoelectronic Technology and Lightwave Communication Systems*, C. Lin, ed., Van Nostrand Reinhold, New york.
29. Carlisle, A. W. 1985. Small-size high-performance lightguide connector for LANs, *Proceedings at Conference on Optical Fiber Communications*, San Diego, CA.
30. Siegman, A. E. 1986. *Lasers.* Univ. Science Books, Mill Valley, CA.
31. Yariv, A. 1989. *Quantum Electronics, 3rd Ed.*, Wiley, New York.

32. Yariv, A. 1976. *Introduction to Optical Electronics, 2nd Ed.*, Holt, Rinehart, and Winston, New York.
33. Fabry, C. and Perot, A. 1899. Theorie et applications d'une nouvelle de spectroscopie interferentielle. *Ann. Chim. Phys.*, 16, 115.
34. Casey, H. C. and Panish, M. B. 1978. *Heterostructure Lasers*. Academic Press, New York.
35. Agrawal, G. P. and Dutta, N. K. 1986. *Long-Wavelength Semiconductor Lasers*. Van Nostrand Reinhold, New York.
36. Saleh, B. E. A. and Teich, M. C. 1991. *Fundamentals of Photonics*. Wiley, New York.
37. Kogelnik, H. and Shank, C. V. 1972. Coupled wave theory of distributed feedback lasers. *J. Appl. Phys.*, 43, 2328.
38. Campbell, J. C. 1989. Photodetectors for long-wavelength lightwave systems. In *Optoelectronic Technology and Lightwave Communication Systems*, C. Lin, ed., Van Nostrand Reinhold, New York.
39. Lee, T. P. and Tingye, Li. 1979. Photodetectors. In *Optical Fiber Telecommunications I*, S.E. Miller and A.G. Chynoweth, eds., Academic Press, New York.
40. Palais, J. C. 1988. *Fiber Optic Communications, 2nd Ed.*, Prentice Hall, New Jersey.
41. Personick, S. D. 1977. Receiver design for optical fiber systems. *Proc. IEEE*, 65, 1670–1678.
42. Li, T. 1993. The impact of optical amplifiers on long-distance lightwave telecommunications. *IEEE*, 81, 1568–1579.
43. Yariv, A. 1991. *Optical Electronics, 4th Ed.*, Holt, Rinehart, and Winston, appendix D.
44. Green, P. E. Jr. 1993. *Fiber Optic Networks*. Prentice Hall, New Jersey.
45. Giles, C. R. 1992. System applications of optical amplifiers, Paper presented at conference on Optical Fiber Communications.
46. Hill, A. M., Wyatt, R., Massicott, J. F., Blyth, K. J., Forrester, D. S., Lobbett, R. A., Smith, P. J., and Payne, D. B. 1990. Million-Way WDM broadcast network employing two stages of Erbium-doped fibre amplifiers. *Electron. Lett.*, 26, 1882–1884.
47. Koester, C. J. and Snitzer, E. 1964. Amplification in a fiber laser. *Appl. Opt.*, 3, 1182.
48. Desurvire, E. 1991. Erbium-doped fiber amplifiers. Paper presented at Conference on Optical Fiber Communications.
49. OKI Corp. 1982. Product Brochure.
50. Giles, C. R. and Desurvire, E. 1991. Modeling Erbium-doped fiber amplifiers. *J. Lightwave Technol.*, 9, 271–283.
51. Miniscalco, W. J. 1991. Erbium-doped glasses for fiber amplifiers at 1500 nm. *J. Lightwave Tech.*, 9, 234–250.
52. Desurvire, E. 1990. Spectral noise figure of Er^{3+}-doped fiber amplifiers. *IEEE Photon. Technol. Lett.*, 2, 208–210.
53. Olshansky, R. 1988. Noise figure for Erbium-doped optical fibre amplifiers. *Electron. Lett.*, 24, 1363–1365.
54. Walker, G. R., Spirit, D. M., Williams, D. L., and Davey, S. T. 1991. Noise performance of distributed fibre amplifiers. *Electron. Lett.*, 27, 1390–1391.
55. Giles, C. R. and Di Giovanni, D. 1990. Spectral dependence of gain and noise in Erbium-doped fiber amplifiers. *IEEE Photon. Technol. Lett.*, 2, 797–800.
56. Giles, C. R., Desurvire, E., Zyskind, J. L., and Simpson, J. R. 1989. Noise performance of Erbium-doped fiber amplifier pumped at 1.49 μm, and application to signal preamplification at 1.8 Gbit/s. *IEEE Photon. Technol. Lett.*, 1, 367–369.
57. Way, W. I., Von Lehman, A. C., Andrejco, M. J., Saifi, M. A., Lin, C. 1991. Noise figure of a gain-saturated Erbium–doped fiber amplifier pumped at 980 nm, *Tech. Digest*.
58. Willner, A. E. and Desurvire, E. 1991. Effect of gain saturation on receiver sensitivity in 1 Gbit/s multichannel FSK direct-detection systems using Erbium-doped fiber preamplifiers. *IEEE Photon. Technol. Lett.*, 3, 259–261.

59. Sun, Y. et al. 1998. A gain-flattened ultra wide band EDFA for high capacity WDM optical communications systems. *24th Europ. Conf. Opt. Commun.*, 1, 53–54.
60. Gabla, P. M., Leclerc, E., and Coeurjolly, C. 1991. Practical implementation of a highly sensitive receiver using an Erbium-doped fiber preamplifier. *IEEE Photon. Technol. Lett.*, 3, 727–729.
61. Jacobs, I. 1990. Effect of optical amplifier bandwidth on receiver sensitivity. *IEEE Trans. Commun.*, 38, 1863–1864.
62. Laming, R. I., Gnauck, A. H., Giles, C. R., Zervas, M. N., and Payne, D. N. 1992. High sensitivity optical pre-amplifier at 10 Gbit/s employing a low noise composite EDFA with 46 dB gain. *Tech. Digest.*
63. Massicott, J. F., Wyatt, R., Ainslie, B. J., and Craig-Ryan, S. P. 1990. Efficient, high power, high gain Er^{3+} doped silica fibre amplifier. *Electron. Lett.*, 26, 1038–1039.
64. O'Mahony, M. J. 1988. Semiconductor laser optical amplifiers for use in future fiber systems. *J. Lightwave Technol.*, 6, 531–544.
65. Saitoh, T. et al. 1986. Recent progress in semiconductor laser amplifiers. *J. Lightwave Technol.*, 6, 1656–1664.
66. Saitoh, T. and Mukai, T. 1987. 1.5 μm GaInAsP traveling-wave semiconductor laser amplifier. *IEEE J. Quantum Electron.*, 23, 1010.
67. Simon, J. C. 1987. GaInAsP semiconductor laser amplifiers for single mode fiber communication. *J. Lightwave Technol.*, 5, 1286–1295.
68. Fye, D. M. 1984. Practical limitations on optical amplifier performance. *J. Lightwave Technol.*, 2, 403–406.
69. Stone, J. and Burrus, C. A. 1973. Neodymium doped silical lasers in end-pumped fiber geometry. *Appl. Phys. Lett.*, 23, 388.
70. Zah, C. E., Osinski, J. S., Caneau, C., Menocal, S. G., Reith, L. A., Salzman, J., Shokoohi, F. K., and Lee, T. P. 1987. Fabrication and performance of 1.5 mm InGaAsP traveling wave amplifiers with angled facets. *Electron. Lett.*, 23, 990–991.
71. Olsson, N. A., Kazarinov, R. F., Nordland, W. A., Henry, C. H., Oberg, M. G., White, H. G., Garbinski, P. A., and Savage, A. 1989. Polarization independent optical amplifier with buried facets. *Electron. Lett.*, 25, 1048.
72. Vassallo, C. 1989. Gain ripple minimization and higher-order modes in semiconductor optical amplifiers. *Electron. Lett.*, 25, 789–790.
73. Newkirk, M. A., Miller, B. I., Koren, U., Young, M. G., Chien, M., Jopson, R. M., and Burrus, C. A. 1993. 1.5 μm multiquantum-well semiconductor optical amplifier with tensile and compressively strained wells for polarization-independent gain. *IEEE Photon. Technol. Lett.*, 4, 406–408.
74. Suzuki, Y., Magari, K., Ueki, M., Amano, T., Mikami, O., and Yamamoto, M. 1993. High-gain, high-power 1.3 μm compressive strained MQW optical amplifier. *IEEE Photon. Technol. Lett.*, 4, 404–406.
75. Durhuus, T., Joergensen, C., Mikkelsen, B., Pedersen, R. J. S., and Stubkjaer, K. E. 1994. All optical wavelength conversion by SOAs in a Mach-Zehnder configuration. *IEEE Photon. Technol. Lett.*, 6, 1, 53–55.
76. Joergensen, C., Danielsen, S. L., Vaa, M., Mikkelsen, B., Stubkjaer, K. E., Doussiere, P., Pommerau, F., Goldstein, L., and Goi, M. 1996. 40 Gbit/s all-optical wavelength conversion by semiconductor optical amplifiers. *Electron. Lett.*, 32, 4, 367–368.
77. Lin, L. Y., Wiesenfeld, J. M., Perino, J. S., and Gnauck, A. H. 1998. Polarization-insensitive wavelength conversion up to 10 Gbit/s based on four-wave mixing in a semiconductor optical amplifier. *IEEE Photon. Technol. Lett.*, 10, 7, 955–957.
78. Wolfson, D., Kloch, A., Fjelde, T., Janz, C., Dagens, B., and Renau, M. 2000. 40 Gbit/s All-optical wavelength conversion, regeneration, and demultiplexing in a SOA-based all-optical Mach-Zehnder interferometer. *IEEE Photon. Technol. Lett.*, 12, 3, 332–334.

79. Diez, S., Schubert, C., Ludwig, R., Ehrke, H. J., Feiste, U., Schmidt, C., and Weber, G. 2000. 160 Gbit/s all-optical demultiplexer using hybrid gain-transparent SOA Mach-Zehnder interferometer. *Electron. Lett.*, 36, 17, 1484–1486.
80. Bromage, J. 2004. Raman amplification for fiber communications systems. *IEEE/OSA J. Lightwave Technol.*, 22, 1, 79–93.
81. Fludger, C. 2000. Paper FF2 presented at Optical Fiber Communication Conference and Exhibit.
82. Emori, Y., Tanaka, K., and Namiki, S. 1999. 100 nm bandwidth flat-gain Raman amplifiers pumped and gain-equalised by 12 wavelength-channel WDM laser diode unit. *Electron. Lett.*, 35, 1355–1356.
83. Mori, A., Masuda, H., Shikano, K., and Shimizu, M. 2003. Ultra-wide-band tellurite-based fiber Raman amplifier. *IEEE/OSA J. Lightwave Technol.*, 21, 5, 1300–1306.
84. Vasilyev, M. et al. 2001. Pump intensity noise and ASE spectrum of Raman amplification in nonzero dispersion shifted fibers, paper OMC3. In *Proceedings of Optical Amplifiers and Applications.*
85. Bromage, J. et al. 2001. Dependence of double Rayleigh scattering backscattered noise in Raman amplifiers on gain and pump depletion. In *Proceedings of Optical Amplifiers and applications.*
86. Fludger, C. R. S. et al. 2001. Pump to signal RIN transfer in Raman fiber amplifier. *IEEE J. Lightwave Technol.*, 19, 1140–1148.
87. Marhic, M. E. et al. 1996. Broadband fiber optical parametric amplifiers. *Opt. Lett.*, 21, 573–575.
88. Okuno, T. 2001. Nonlinear-fiber based discrete Raman amplifier with sufficiently suppressed degradation of WDM signal quality, OTuB5. In *Proceedings of Optical Amplifiers and applications.*
89. Willner, A. E. 1990. Simplified model of a FSK-to-ASK direct-detection system using a Fabry-Perot demodulator. *IEEE Photon. Technol. Lett.*, 2, 363–366.
90. Erdogan, T. 1998. Fiber grating spectra. *IEEE J. Lightwave Technol.*, 15, 1277–1294.
91. Achtenhagen, M., Mohrdiek, S., Pliska, T., Matuschek, N., Harder, C. S., and Hardy, A. 2001. L-I characteristics of fiber Bragg grating stabilized 980-nm pump lasers. *IEEE Photon. Technol. Lett.*, 13, 5, 415–417.
92. Nykolak, G., de Barros, M. R. X., Nielsen, T. N., and Eskildsen, L. 1997. All-fiber active add-drop wavelength router. *IEEE Photon. Technol. Lett.*, 9, 5, 605–607.
93. Feng, K. -M., Cai, J. -X., Grubsky, V., Starodubov, D. S., Hayee, M. I., Lee, S., Jiang, X., Willner, A. E., and Feinberg, J. 1999. Dynamic dispersion compensation in a 10-Gbit/s optical system using a novel voltage tuned nonlinearly chirped fiber Bragg grating. *IEEE Photon. Technol. Lett.*, 11, 3, 373–375.
94. Harurnoto, M., Shigehara, M., and Suganurna, H. 2002. Gain-flattening filter using long-period fiber gratings. *IEEE/OSA J. Lightwave Technol.*, 20, 6, 1027–1033.
95. Bernasconi, P., Doerr, C., Dragone, C., Cappuzzo, M., Laskowski, E., and Paunescu, A. 2000. Large N×N waveguide grating routers. *IEEE/OSA J. Lightwave Technol.*, 18, 7, 985–991.
96. Takada, K., Abe, M., Shibata, T., and Okamoto, K. 2002. A.2.GHz-spaced 1080-channel tandem multi/demultiplexer covering the S-, C-, and L-bands using an arrayed-waveguide grating with Gaussian passbands as a primary filter. *IEEE Photon. Technol. Lett.*, 14, 5, 648–650.
97. Tamir, T. 1979. Beam and waveguide couplers, In *Integrated Optics, Topics in Applied Physics, Vol. 7*, T. Tamir, ed., pp. 83–1039. Springer, New York.
98. Alferness, R. C. 1982. Switches and modulators, In *Integrated Optics: Theory and Technology*, G. Hunsperger, ed., Springer, New York.
99. Koch, T. L. and Bowers, J. E. 1984. Nature of wavelength chirping in directly modulated semiconductor lasers. *Electron. Lett.*, 20, 1038–1039.
100. Korotky, S. K. 1984. Three-space representation of phase-mismatch switching in coupled two-state optical systems. *IEEE J. Quantum Electron.*, QE-22, 952–958.
101. Schwartz, M. 1986. *Information Transmission, Modulation, and Noise, 3rd Ed.*, McGraw-Hill, New York.
102. Haykin, S. 1983. *Communication Systems, 2nd Ed.*, Wiley, New York.
103. Roden, M. S. 1995. *Analog and Digital Communication Systems, 4th Ed.*, Prentice Hall, New Jersey.

104. Couch II, L. W. 2001. *Digital and Analog Communication Systems.* Prentice Hall, New Jersey.
105. Killen, H. B. 1991. *Fiber Optic Communications.* Prentice Hall, New york.
106. Gnauck, A. H. and Winzer, P. J. 2005. Optical phase-shift-keyed transmission. *IEEE/OSA J. Lightwave Technol.*, 23, 1, 115–130.
107. Chinn, S. R., Boroson, D. M., and Livas, J. C. 1996. Sensitivity of optically preamplified DPSK receivers with Fabry-Perot filters. *J. Lightwave Technol.*, 14, 3, 370–376.
108. Xu, C., Liu, X., Mollenauer, L. 2003. Comparison of return-to-zero phase shift keying and on-off keying in long haul dispersion managed transmissions. Paper ThE3 presented in Proceedings OFC 2003, Atlanta, GA.
109. Senior, J. M. 1985. *Optical Fiber Communications.* Prentice Hall, New York.
110. Betti, S., de Marchis, G., and Iannone, E. 1995. *Coherent Optical Communication Systems.* Wiley, New York.
111. Stephens, T. D. and Nicholson, G. 1987. Optical homodyne receiver with a six-port fibre coupler. *Electron. Lett.*, 22, 1106–1107.
112. Kahn, J. M., Gnauck, A. H., Veselka, J. J., Korotky, S. K., and Kasper, B. L. 1990. 4 Gbit/s PSK homodyne transmission system using phase-locked semiconductor lasers. *IEEE Photon. Tech. Lett.*, 2, 285–287.
113. Sinsky, J. H., Adamiecki, A., Gnauck, A., Burrus, C., Leuthold, J., Wohlgemuth, O., and Umbach, A. 2003. A 42.7-Gbit/s integrated balanced optical front end with record sensitivity. Postdeadline paper PD39 presented in Proceedings OFC 2003, Atlanta, GA.
114. Winzer, P. J., Gnauck, A. H., Raybon, G., Chandrasekhar, S., Su, Y., and Leuthold, J. 2003. 40-Gbit/s return-to-zero alternate-mark-inversion (RZ–AMI) transmission over 2000 km. *IEEE Photon. Technol. Lett.*, 15, 766–768.
115. Linke, R. A. and Gnauck, A. H. 1988. High capacity coherent lightwave systems. *IEEE/OSA J. of Lightwave Technol.*, 6, 1750–1769.
116. Glance, B. 1987. Polarization independent coherent optical receiver. *IEEE/OSA J. Lightwave Technol.*, 5, 274–276.
117. Spiegel, M. R. 1968. *Mathematical Handbook.* Shaum' Outline Series in Mathematics. McGraw-Hill, New York.
118. Olsson, N. A. 1989. Lightwave systems with optical amplifiers. *IEEE/OSA J. Lightwave Technol.*, 7, 1071–1082.
119. Kaminow, I. and Koch, T. 1997. *Optical Fiber Telecommunications IIIB.* Academic Press, San Diego.
120. Personick, S. D. 1985. *Fiber Optics: Technology and Applications.* Plenum Press, New York.
121. Runge, P. K. and Bergano, N. S. 1988. Undersea cable transmission systems, In *Optical Fiber Telecommunications II*, S. E. Miller and I. P. Kaminow, eds., Academic Press, New York.
122. Brackett, C. A. 1990. Dense wavelength division multiplexing: principles and applications. *IEEE J. Sel. Areas in Commun.*, 8, 948.
123. Kaminow, I. P. 1990. FSK with direct detection in optical multiple-access FDM networks. *IEEE J. Sel. Areas in Commun.*, 8, 1005.
124. Willner, A. E., Kaminow, I. P., Kuznetsov, M., Stone, J., and Stulz, L. W. 1990. 1.2 Gbit/s closely-spaced FDMA–FSK direct-detection star network. *Photon. Technol. Lett.*, 2, 223–226.
125. Dono, N. R., Green, P. E., Liu, K., Ramaswami, R., and Tong, F. F. 1990. A wavelength division multiple access network for computer communication. *IEEE J. Sel. Areas in Commun.*, 8, 983–994.
126. Way, W. I., Smith, D. A., Johnson, J. J., and Izadpanah, H. 1992. A self-routing WDM high-capacity SONET ring network. *IEEE Photon. Technol. Lett.*, 4, 402–405.
127. WuFiber, T. -H. *Fiber Network Service Survivability*, Boston, Artech House.
128. Acampora, A. S., Karol, M. J., and Hluchyj, M. G. 1987. Terabit lightwave networks: the multihop approach. *AT&T Tech. J.*, 66, 6, 21–34.

129. Schwartz, M. 1987. Telecommunication networks. *Protocols, Modelling, and Analysis*. Addison Wesley, New York.
130. Chan, V. W. S., Hall, K. L., Modiano, E., and Rauschenbach, K. A. 1998. Architectures and technologies for high-speed optical data networks. *IEEE J. Lightwave Technol.*, 16, 2146–2168.
131. Aiki, K., Nakamura, M., Umeda, J., Yariv, A., Katzir, A., and Yen, H. W. 1975. GaAs–GaAlAs distributed feedback laser with separate optical and electrical confinement. *Appl. Phys. Lett.*, 27, 145.
132. Inoue, K., Toba, H., and Nosu, K. 1991. Multichannel amplification utilizing an Er^{3+}-doped fiber amplifier. *IEEE/OSA J. Lightwave Technol.*, 9, 368–374.
133. Coldren, L. A., Fish, G. A., Akulova, Y., Barton, J. S., Johansson, L., and Coldren, C. W. 2004. Tunable semiconductor lasers: a tutorial. *IEEE/OSA J. Lightwave Technol.*, 22, 1, 193–202.
134. Chang-Hasnain, C. J. 2000. Tunable VCSEL. *IEEE J. Selected Top. Quantum Electron.*, 6, 6, 978–987.
135. Riemenschneider, F., Maute, M., Halbritter, H., Boehm, G., Amann, M. C., and Meissner, P. 2004. Continuously tunable long-wavelength MEMS-VCSEL with over 40-nm tuning range. *IEEE Photon. Technol. Lett.*, 16, 10, 2212–2214.
136. Soole, J. B. D., et al. 1994. Paper ThB1 presented at Conference on Optical Fiber Communications. *Technical Digest*. San Jose, CA, Feb., 1994 (Optical Society of America, Washington, D.C., 1994).
137. Dragone, C. 1991. An $N \times N$ optical multiplexer using a planar arrangement of two star couplers. *IEEE Photon. Technol. Lett.*, 3, 812–815.
138. Chang, G. K., Liu, T. P., Gimlett, J. L., Shirokmann, H., Iqbal, M. Z., Hayes, J. R., and Wang, K. C. 1992. A direct-current coupled, all-differential optical receiver for high-bit-rate SONET systems. *IEEE Photon. Technol. Lett.*, 4, 339–342.
139. Ramaswami, R. and Sivarajan, K. N. 1996. Design of logical topologies for wavelength-routed optical networks. *IEEE J. Selected Areas of Commun.*, 14, 840–851.
140. Humblet, P. A. and Hamdy, W. M. 1990. Crosstalk analysis and filter optimization of single- and double-cavity Fabry-Perot filters. *IEEE J. Selected Areas of Commun.*, 8, 1095–1107.
141. Carlson, A. B., Crilly, P. B., and Rutledge, J. 2001. *Communication Systems, 4th Ed.*, McGraw Hill, New York.
142. Koch, T. L. and Bowers, J. E. 1984. Nature of wavelength chirping in directly modulated semiconductor lasers. *Electron. Lett.*, 20, 1038–1039.
143. Chung, Y. C., Pollock, K. J., Fitzgerald, P. J., Glance, B., Tkach, R. W., and Chraplyvy, A. R. 1988. *Electron. Lett.*, 24, 1313.
144. Kaminow, I. P., Iannone, P. P., Stone, J., and Stulz, L. W. 1988. FDMA–FSK star network with a tunable optical fiber demultiplexer. *J. Lightwave Technol.*, 6, 1406–1414.
145. Glance, B., Koch, T. L., Scaramucci, O., Reichmann, K. C., Koren, U., and Burrus, C. A. 1989. Densely spaced FDM coherent optical star network using monolithic widely frequency-tunable lasers. *Electron. Lett.*, 25, 672–673.
146. Keating, A. J. and Lowery, A. J. 1995. Wavelength stabilization in WDM packet-switched networks. Paper WR2 presented at Conference on Optical Fiber Communications, San Diego, CA.
147. Nasu, H., Takagi, T., Oike, M., Nomura, T., and Kasukawa, A. 2003. Ultrahigh wavelength stability through thermal compensation in wavelength-monitor integrated laser modules. *IEEE Photon. Technol. Lett.*, 15, 3, 380–382.
148. Goldstein, E. L., Elrefaie, A. F., Jackman, N., Zaidi, S. 1993. Multiwavelength fiber-amplifier cascades in unidirectional interoffice ring networks, Paper TuJ3 presented at conference on Optical Fiber Communications.
149. Blondel, J. P., Pitel, A., Marcerou, J. F. 1993. Gain-filtering stability in ultralong-distance links. Paper presented at conference on Optical Fiber Communications.

150. Taga, H., Edagawa, N., Yoshida, Y., Yamamoto, S., and Wakabayashi, H. 1993. IM-DD four-channel transmission experiment over 1500 km employing 22 cascaded optical amplifiers. *Electron. Lett.*, 29, 485.
151. Willner, A. E. and Hwang, S.-M. 1995. Transmission of many WDM channels through a cascade of EDFA's in long-distance link and ring networks. *IEEE J. Lightwave Technol.*, 13, 802–816.
152. Vengsarkar, A. M., Lemaire, P. J., Judkins, J. B., Bhatia, V., Erdogan, T., and Sipe, J. E. 1996. Long-period fiber gratings as band-rejection filters. *IEEE J. Lightwave Technol.*, 14, 58–65.
153. Sun, Y. et al. 1997. Transmission of 32-WDM 10-Gbit/s channels over 640 km using broad-band, gain-flattened Erbium-doped silica fiber amplifiers. *IEEE Photon. Technol. Lett.*, 9m, 1652–1654.
154. Wysocki, P. F., Judkins, J. B., Espindola, R. P., Andrejco, M., and Vengsarkar, A. M. 1997. Broad-band Erbium-doped fiber amplifier flattened beyond 40 nm using long-period grating filter. *IEEE Photon. Technol. Lett.*, 9, 1343–1345.
155. Pan, J. Y., Ali, M. A., Elrefaie, A. F., and Wagner, R. E. 1995. Multiwavelength fiber-amplifier cascades with equalization employing Mach-Zehnder optical filter. *IEEE Photon. Technol. Lett.*, 7, 1501–1503.
156. Yamada, M., Kanamori, T., Terunuma, Y., Oikawa, K., Shimizu, M., Sudo, S., and Sagawa, K. 1996. Fluoride-based Erbium-doped fiber amplifier with inherently flat gain spectrum. *IEEE Photon. Technol. Lett.*, 8, 882–884.
157. Desurvire, E., Giles, R., and Simpson, J. 1989. Gain saturation effects in high-speed, multi channel Erbium-doped fiber amplifiers at 1.53 μm. *IEEE J. Lightwave Technol.*, 7, 12, 2095–2104.
158. Zyskind, J., Sun, Y., Srivastava, A., Sulhoff, J., Lucero, A., Wolf, C., and Tkach, R. 1996. Fast power transients in optically amplified multiwavelength optical networks, Paper PD-331 presented at Conference on Optical Fiber Communications.
159. Hayee, M. and Willner, A. 1999. Fiber transmission penalties due to EDFA power transients resulting from fiber nonlinearity and ASE noise in add/drop multiplexed WDM networks. *IEEE Photon. Technol. Lett.*, 11, 889–891.
160. Cai, J. -X., Feng, K. -M., Willner, A. E. 1997. Simultaneous compensation of fast add/drop power-transients and equalization of inter-channel power differentials for robust WDM systems with EDFAs, Paper MC6 presented at Conference on Optical Amplifiers and Their Applications. Victoria, Canada.
161. Motoshima, K., Leba, L., Chen, D., Downs, M., Li, T., and Desurvire, E. 1993. Dynamic compensation of transient gain saturation in erbium-doped fiber amplifiers by pump feedback control. *IEEE Photon. Technol. Lett.*, 5, 1423–1426.
162. Srivastava, A., Zyskind, J., Sun, Y., Ellson, J., Newsome, G., Tkach, R., Chraplyvy, A., Sulhoff, J., Strasser, T., and Wolf, C. 1997. Fast-link control protection of surviving channels in multi-wavelength optical networks. *IEEE Photon. Technol. Lett.*, 9, 1667–1669.
163. Luo, G., Zyskind, J., Sun, Y., Srivastava, A., Sulhoff, J., and Ali, M. 1997. Performace degradation of all-optical gain-clamped EDFA's due to relaxation-oscillations and spectral-hole burning in amplified WDM networks. *IEEE Photon. Technol. Lett.*, 9, 1346–1348.
164. Cai, J., Feng, K., Chen, X., and Willner, A. 1997. Experimental demonstration of dynamic high-speed equalization of three WDM channels using acousto-optic modulators and a wavelength demultiplexer. *IEEE Photon. Technol. Lett.*, 9, 678–680.
165. Huang, S., Zou, X., Willner, A., Bao, Z., and Smith, D. 1997. Experimental demonstration of active equalization and ASE seppression of three 2.5 Gbit/s WDM-network channels over 2500 km using AOTF as transmission filters. *IEEE Photon. Technol. Lett.*, 9, 389–391.
166. Starodubov, D., Grubsky, V., Feinberg, J., Cai, J., Feng, K., Willner, A. 1998. Novel fiber amplitude modulators for dynamic channel power equalization in WDM systems, Paper PD-8 presented at Conference on Optical Fiber Communication.
167. Ford, J. and Walker, J. 1998. Dynamic spectral power equalization using micro-opto-mechanics. *IEEE Photon. Technol. Lett.*, 10, 1440–1442.

168. Doerr, C., Joyner, C., and Stulz, L. 1998. Integrated WDM dynamic power equalizer with low insertion loss. *IEEE Photon. Technol. Lett.*, 10, 1443–1445.
169. Jopson, R. M. and Darcie, T. E. 1988. Calculation of multicarrier intermodulation distortion in semiconductor optical amplifiers. *Electron. Lett.*, 24, 1372–1374.
170. Jopson, R. M. et al. 1987. Measurement of carrier density mediated intermodulation distortion in an optical amplifier. *Electron. Lett.*, 23, 1394–1395.
171. Agrawal, G. P. 1987. Four-wave mixing and phase conjunction in semiconductor laser media. *Opt. Lett.*, 12, 4, 260–262.
172. Wiesenfeld, J. M., Raybon, G., Koren, U., Eisenstein, G., and Burrus, C. A. 1991. Gain spectra and gain compression of strained-layer multiple quantum well optical amplifiers. *Appl. Phys. Lett.*, 58, 219–221.
173. Wiesenfeld, J. M., Gnauck, A. H., Raybon, G., and Koren, U. 1992. High-speed multiple-quantum-well optical power amplifier. *IEEE Photon. Technol. Lett.*, 4, 708–711.
174. Garrett, L. D. 2001. All about chromatic dispersion in dense WDM optical fiber transmission, Invited short course at Optical Fiber Communication Conference.
175. Antos, A. J. and Smith, D. K. 1994. Design and characterization of dispersion compensating fiber based on the LP_{01} mode. *IEEE J. Lightwave Technol.*, 1739–1745.
176. Gnauck, A. H., Garrett, L. D., Forghieri, F., Gusmeroli, V., and Scarano, D. 1998. 8×20 Gbit/s 315-km 8×10 Gbit/s 480-km WDM transmission over conventional fiber using multiple broadband fiber gratings. *IEEE Photon. Technol. Lett.*, 10, 10, 1495–1497.
177. Ibsen, M., Durkin, M. K., Cole, M. J., and Laming, R. I. 1998. Sinc-sampled fiber bragg gratings for identical multiple wavelength operation. *IEEE Photon. Technol. Lett.*, 10, 6, 842–845.
178. Eggleton, B. J., Rogers, J. A., Westbrook, P. S., and Strasser, T. A. 1999. Electrically tunable power efficient dispersion compensating fiber bragg grating 854. *IEEE Photon. Technol. Lett.*, 11, 7, 854–856.
179. Shirasaki, M. 1997. Chromatic dispersion compensation using virtually imaged phased array. *IEEE Photon. Technol. Lett.*, 9, 11, 1598–1600.
180. Hasegawa, A. and Tappert, F. 1973. Transmission of stationary nonlinear optical pulses in dispersive dielectric fibers. *Appl. Phys. Lett.*, 23, 142–144.
181. Mollenauer, L. F., Stolen, R. H., and Gordon, J. P. 1980. Experimental observation of picosecond pulse narrowing and solitons in optical fibers. *Phys. Rev. Lett.*, 45, 1095–1097.
182. Gordon, J. P. and Haus, H. A. 1986. Random walk of coherently amplified solitons in optical fiber transmission. *Opt. Lett.*, 11, 665–667.
183. Mollenauer, L. F., Gordon, J. P., and Islam, M. N. 1986. Soliton propagation in long fibers with periodically compensated loss. *IEEE J. Quantum Electron.*, QE-22, 157–173.
184. Evangelides, S. G. 1993. Private communication, in L. F. Mollenauer. Introduction to solitons. Conference on Optical Fiber Communications '94, Technical Digest, Tutorial TuF, San Jose, CA, Feb. (Optical Society of America, Washington, D.C., 1993).
185. Eisenstein, G., Tucker, R. S., Koren, U., and Korotky, S. K. 1986. Active mode-locking characteristics of InGaAsP single mode fiber composite cavity lasers. *IEEE J. Quantum Electron.*, QE-22, 142–148.
186. Mecozzi, A., Moores, J. D., Haus, H. A., and Lai, Y. 1991. Soliton transmission control. *Opt. Lett.*, 16, 1841.
187. Kodama, Y. and Hasegawa, A. 1992. Generation of asymptotically stable optical solitons and suppression of the Gordon–Haus effect. *Opt. Lett.*, 17, 31.
188. Mollenauer, L. F., Evangelides, S. G., and Gordon, J. P. 1991. Wavelength division multiplexing with solitons in ultra long distance transmission using lumped amplifiers. *IEEE J. Lightwave Technol.*, 9, 362.
189. Carter, G. M., Jacob, J. M., Menyuk, C. R., Golovchenko, E. A., and Pilipetskii, A. N. 1997. Timing-gitter reduction for a dispersion-managed soliton system: experimental evidence. *Opt. Lett.*, 22, 513–515.

190. Sato, K., Kuwahara, S., Miyamoto, Y., Murata, K., and Miyazawa, H. 2002. Carrier-suppressed return-to-zero pulse generation using mode-locked lasers for 40-Gbit/s transmission. *IEICE Trans. Commun.*, E85-B, 2, 410–415.
191. Yonenaga, K. and Kuwano, S. 1997. Dispersion-tolerant optical transmission system using duobinary transmitter and binary receiver. *IEEE/OSA J. Lightwave Technol.*, 15, 8, 1530–1537.
192. Yonenaga, K., Yoneyama, M., Miyamoto, Y., Hagimoto, K., and Noguchi, K. 1998. 160 Gbit/s WDM transmission experiment using four 40 Gbit/s optical duobinary channels. *Electron. Lett.*, 34, 1506–1507.
193. Ono, T., Yano, Y., Fukuchi, K., Ito, T., Yamazaki, H., Yamaguchi, M., and Emura, K. 1998. Characteristics of optical duobinary signals in terabit/s capacity, high-spectral efficiency WDM systems. *J. Lightwave Technol.*, 16, 788–797.
194. Mitra, P. P. and Stark, J. B. 2001. Nonlinear limits to the information capacity of optical fibre communications. *Nature*, 411, 1027–1030.
195. Willner, A. E., 2002. Polarization mode dispersion: playing Russian roulette with your network. *Lightwave Mag.*, April.
196. Karlsson, M. and Brentel, J. 1999. Autocorrelation f unction of the polarization-mode dispersion vector. *Opt. Lett.*, 24, 14, 939–941.
197. Khosravani, R. and Willner, A. E. 2001. System performance evaluation in terrestrial systems with high polarization mode dispersion and the effect of chirping. *IEEE Photon. Technol. Lett.*, 13, 296–298.
198. Winters, J. H. and Gitlin, R. D. 1990. Electrical signal processing techniques in long-haul fiber-optic systems. *IEEE Trans. Commun.*, 38, 9, 1439–1453.
199. Humblet, P. A. and Hamdy, W. M. 1990. Crosstalk analysis and filter optimization of single- and double-cavity Fabry-Perot filters. *IEEE J. Selected Areas of Commun.*, 8, 1095–1107.
200. Wagner, R. E., Alferness, R. C., Saleh, A. A. M., and Goodman, M. S. 1996. MONET: multiwavelength optical networking. *J. Lightwave Technol.*, 14, 1349–1355.
201. Legg, P. J., Hunter, D. K., Andonovic, I., and Barnsley, P. E. 1994. Inter-channel crosstalk phenomena in optical time division multiplexed switching networks. *IEEE Photon. Technol. Lett.*, 6, 661–663.
202. Goldstein, E. L. and Eskildsen, L. 1995. Scaling limitations in transparent optical networks due to low-level crosstalk. *IEEE Photon. Technol. Lett.*, 7, 93–94.
203. Goldstein, E. L., Eskildsen, L., and Elrefaie, A. F. 1994. Performance implications of component crosstalk in transparent lightwave networks. *IEEE Photon. Technol. Lett.*, 6, 657–660.
204. Kilper, D. C., Bach, R., Blumenthal, D. J., Einstein, D., Landolsi, T., Ostar, L., Preiss, M., and Willner, A. E. 2004. Optical performance monitoring. *IEEE/OSA J. Lightwave Technol.*, 22, 1, 294–304.
205. Yu, Q., Pan, Z., Yan, L.-S., and Willner, A. E. 2002. Chromatic dispersion monitoring technique using sideband optical filtering and clock phase-shift detection. *IEEE/OSA J. Lightwave Technol.*, 20, 12, 2267–2271.
206. Pan, Z., Yu, Q., Xie, Y., Havstad, S. A., Willner, A. E., Starodubov, D. S., and Feinberg, J. 2004. Real-time group-velocity dispersion monitoring and automated compensation without modifications of the transmitter. *Opt. Commun.*, 230, 1–3, 145–149.
207. Friskney, R., Warbrick, K., Poliakoff, S., Health, R. 2002. Link-based photonic path performance prediction and control. In *European Conference on Optical Communication (ECOC)*.
208. Oshima, K., Kitayama, T., Yamaki, M., Matsui, T., and Ito, K. 1985. Fiber-optic local area passive network using burst TDMA scheme. *J. Lightwave Technol.*, 3, 502–510.
209. Prucnal, P. R., Santoro, M. A., and Sehgal, S. K. 1986. Ultrafast all-optical synchronous multiple-access fiber networks. *IEEE J. Sel. Areas in Commun.*, 4, 1484–1493.
210. Tucker, R. S., Eisenstein, G., Korotky, S. K., Raybon, G., Veselka, J. J., Buhl, L. L., Kasper, B. L., and Alferness, R. C. 1987. 16 Gbit/s fibre transmission experiment using optical time-division multiplexing. *Electron. Lett.*, 23, 24.

211. Prucnal, P. R., Santoro, M. A., and Sehgal, S. K. 1986. Ultrafast all-optical synchronous multiple-access fiber networks. *IEEE J. Sel. Areas in Commun.*, 4, 1484–1493.
212. Alferness, R. C. 1989. Titanium-diffused lithium niobate waveguide devices, In *Guided-Wave Optoelectronics*, T. Tamir, ed., Springer, New York.
213. Doran, N. J. and Wood, D. 1988. Nonlinear optical loop mirror. *Opt. Lett.*, 13, 56–58.
214. Mestdagh, D. J. G. 1995. *Fundamentals of Multiaccess Optical Fiber Networks.* Artech House, Boston.
215. Kashima, N. 1993. *Optical Transmission for the Subscriber Loop.* Artech House, Boston.
216. Way, W. I. 1989. Subcarrier multiplexed lightwave system design considerations for subscriber loop applications. *IEEE J. Lightwave Technol.*, 7, 1806–1818.
217. Darcie, T. E. 1987. Subcarrier multiplexing for multiple-access lightwave networks. *IEEE J. Lightwave Technol.*, 5, 1103–1110.
218. Way, W. I., Maeda, M. W., Yan, A. Y., Andrejco, M. J., Choy, M. M., Saifi, M., and Lin, C. 1990. 160-Channel FM-video transmission using FM/FDM and subcarrier multiplexing and an Erbium-doped optical fiber amplifier. *Electron. Lett.*, 26, 139.
219. Sato, K. 1983. Intensity noise of semiconductor laser diodes in fiber optic analog video transmission. *IEEE J. Quantum Electron.*, QE-19, 1380.
220. Olshansky, R., Lanzisera, V. A., and Hill, P. M. 1989. Subcarrier multiplexed lightwave systems for broad-band distribution. *IEEE J. Lightwave Technol.*, 7, 1329–1342.
221. Stubkjaer, K. and Danielsen, M. 1980. Nonlinearity of GaAlAs lasers-harmonic distortion. *IEEE J. Quantum Electron.*, QE-16, 531.
222. Saleh, A. A. M. 1989. Fundamental limit on number of channels in subcarrier-multiplexed lightwave CATV system. *Electron. Lett.*, 25, 776–777.
223. Schmuck, H. 1995. Comparison of optical millimeter-wave system concepts with regard to chromatic dispersion. *Electron. Lett.*, 31, 1848–1849.
224. Yonenaga, K. and Takachio, N. 1993. A fiber chromatic dispersion technique with an optical SSB transmission in optical homodyne detection systems. *IEEE Photon. Technol. Lett.*, 5, 949–951.
225. Poole, C. D. and Darcie, T. E. 1993. Distortion related to polarization-modedispersion in analog lightwave systems. *J. Lightwave Technol.*, 11, 1749–1759.
226. Adamczyk, O. H., Sahin, A. B., Yu, Q., Lee, S., and Willner, A. E. 2001. Statistics of PMD-induced power fading for intensity-modulated double-sideband and single-sideband microwave and millimeter-wave signals. *IEEE Trans. Microwave Theory Tech.*, 49, 10, 1962–1967.
227. Gitlin, R. D., Hayes, J. F., and Weinstein, S. B. 1992. *Data Communications Principles.* Plenum Press, New York.
228. Darcie, T. E., Tucker, R. S., and Sullivan, G. J. 1985. *Electron. Lett.*, 21, 665.
229. National Cable Television Association 1993. 2nd ed. NCTA Recommended Practices for Measurements on Cable Television Systems. National Cable Television Association, Washington, DC.
230. Darcie, T. E., Lipson, J., Roxlo, C. B., and McGrath, C. J. 1990. Fiber optic device technology for broadband analog video systems. *IEEE LCS*, 1, 1, 46–52.
231. Kuo, C. Y. 1992. AM–VSB transmission using Erbium-doped fiber amplifiers. *IEEE J. Lightwave Technol.*, 10, 235.
232. Hill, A. M., Wyatt, R., Massicott, J. F., Blyth, K. J., Forrester, D. S., Lobbett, R. A., Smith, P. J., and Payne, B. 1990. 40-Million-Way WDM broadcast network employing two stages of Erbium-doped fibre amplifiers. *Electron. Lett.*, 26, 1882–1884.
233. Joyce, G. R. and Olshansky, R. 1996. Multichannel 64-QAM transmission in AM/digital and all-digital hybrid-fiber-coax based video dial-tone systems. *IEEE J Selected Areas in Commun.*, 14, 6, 1062–1065.
234. Dai, H., Ovadia, S., and Lin, C. 1996. Hybrid AM–VSB/M-QAM multichannel video transmission over 120 km of standard single-mode fiber with cascaded Erbium-doped fiber amplifiers. *IEEE Photon. Technol. Lett.*, 8, 12, 1713–1715.

235. Uno, T., Mitsuda, M., and Ohya, J. 1994. Low distortion characteristics in amplified 4×50-channel WDM AM SCM transmission, Paper WM2 presented at Conference on Optical Fiber Communications, San Jose, CA. Technical Digest, San Jose, CA, Feb (Optical Society of America, Washington, DC, 1994).
236. Diaz, A. R., Kalman, R. F., Goodman, J. W., Sawchuk, A. A. Fiber-optic crossbar switch with broadcast capability, *Opt. Eng.*, 27, 1087–1095.
237. Alferness, R. C. 1989. Titanium-diffused lithium niobate waveguide devices. In *Guided-Wave Optoelectronics*, T. Tamir, ed., pp. 174–194. Springer, New York.
238. Fujiwara, M., Nishimoto, H., Kajitani, T., Itoh, M., and Suzuki, S. 1991. Studies on semiconductor optical amplifiers for line capacity expansion in photonic space-division switching system. *J. Lightwave Technol.*, 9, 155–160.
239. Kondo, M., Takado, N., Komatsu, K., and Ohta, Y. 1987. 32 switch-elements integrated low-crosstalk Ti:LiNbO3 optical matrix switch, *IOOC-ECOC Tech. Digest*, 361–364.
240. Prucnal, P. R., Santoro, M. A., and Fran, T. R. 1986. Spread spectrum fiber-optic local area network using optical processing. *IEEE/OSA J. Lightwave Technol.*, 4, 547.
241. Salehi, J. A., Weiner, A. M., and Heritage, J. P. 1990. *IEEE/OSA J. Lightwave Technol.*, 8, 478.
242. Chung, F. R. K., Salehi, J. A., and Wei, V. K. 1989. Optical orthogonal codes: design, analysis, and applications. *IEEE Trans. Inf Theory*, 35, 595–604.
243. Cheng, R. S. 1992. Total capacities of maximum numbers of users in strictly band limited CDMA & FDMA gaussian channels. *MILCOM Proc.*, 247–251.
244. Lindsey, W. C. and Simon, M. K. 1991. *Telecommunications Systems Engineering*. Dover Publications, New York.
245. Kamath, P., Touch, J. D., and Bannister, J. A. 2004. The need for media access control in optical CDMA networks. *IEEE Infocom*.
246. Wang, X., Wada, N., Hamanaka, T., Kitayama, K., Nishiki, A. 2005. 10-user, truly-asynchronous OCDMA experiment with 511-chip SSFBG en/decoder and SC-based optical thresholde, Paper PDP33 presented at Conference on Optical Fiber Communications.
247. Jeong, M., Cankaya, H. C., and Qiao, C. 2002. On a new multicasting approach in optical burst switched networks. *IEEE Commun. Mag.*, 96.
248. Baldine, I. et al. 2002. JumpStart: A just-in-time signaling architecture for WDM burst-switched networks. *IEEE Commun. Mag.*, 82, .
249. Berthold, J. E. 1994. Networking fundamentals, Paper presented at Conference on Optical Fiber Communications, Tutorial TuK, 3-20, San Jose, CA.
250. Hui, J. Y. 1990. *Switching and Traffic Theory for Integrated Broadband Networks*. Kluwer Academic Publishers, Boston.
251. Yoo, S. J. B. and Chang, G. K. 1997. *High-throughput, Low-Latency Next Generation Internet Using Optical-tag Switching*, U.S. Patent, 6,111,673.
252. Maeda, M. W., Willner, A. E., Wullert, J. R., Patel, J., and Allersma, M. 1993. Wavelength-division multiple-access network based on centralized common-wavelength control. *IEEE Photon. Technol. Lett.*, 5, 83–86.
253. Goel, K. K. 1993. Nonrecirculating and recirculating delay line loop topologies of fiber-optic delay line filters. *IEEE Photon. Technol. Lett.*, 5, 1086–1088.
254. Chlamtac, I., Fumagalli, A., and Chang-Jin, S. 2000. Multibuffer delay line architectures for efficient contention resolution in optical switching nodes. *IEEE Trans. Commun.*, 48, 2089.
255. Agrawal, A., Wang, L., Su, Y., Kumar, P. 2001. All-optical erasable storage buffer based on parametric nonlinearity in fiber, Paper ThH5 presented at OFC.
256. Rader, A. and Anderson, B. L. 2003. Demonstration of a linear optical true-time delay device by use of a microelectromechanical mirror array. *Appl. Opt.*, 42, 8, 1409–1417.
257. Pape, D. R. and Goutzoulis, A. P. 1999. New wavelength division multiplexing tru-time-delay network for wideband phased array antennas. *J. Opt. A: Pure Appl. Opt.*, 1, 320–323.

258. Chang-Hasnain, C. J., Ku, P., Kim, J., and Chuang, S. 2003. Variable optical buffer using slow light in semiconductor nanostructures. *Proc. IEEE*, 91, 11, 1884–1897.
259. Rangarajan, S. et al. 2004. All-optical contention resolution with wavelength conversion for asynchronous variable-length 40 Gbit/s optical packets. *IEEE Photon. Technol. Lett.*, 16, 689.
260. Elmirghani, J. and Mouftah, H. 2000. All-optical wavelength conversion: Techniques and applications in DWDM networks. *IEEE Commun. Mag.*, 86–92.
261. Nesset, D., Kelly, T., and Marcenac, D. 1998. All-optical wavelength conversion using SOA nonlinearities. *IEEE Commun. Mag.*, 36, 12, 56–61.
262. Brener, I., Chou, M. H., Fejer, M. M. 1999. Efficient wideband wavelength conversion using cascaded second-order nonlinearities in $L_iN_bO_3$ waveguides, Paper FB6 presented at Proceedings Optical Fiber Communications Conference, pp. 39–41.
263. Hsu, A. and Chuang, S.L. 1999. Wavelength conversion by cross-absorption modulation using an integrated electroabsorption modulator/laser, Summaries of Papers Presented at the Conference on Lasers and Electo-Optics, p. 488.
264. Baresi, M. et al. 2003. Deflection routing effectiveness in full-optical IP packet switching networks. *Proc. IEEE Int. Conf. Commun.*, 2, 1360.
265. Choa, F. -S. et al. 2005. An optical packet switch based on WDM technologies. *IEEE/OSA J. Lightwave Technol.*, 23, 3, 994–1014.
266. Willner, A. E., Gurkan, D., Sahin, A. B., McGeehan, J. E., and Hauer, M. C. 2003. All-optical address recognition for optically-assisted routing in next-generation optical networks. *IEEE Commun. Mag.*, 41, S38–S44.
267. Calabretta, N., de Waardt, H., Khoe, G. D., and Dorren, H. J. 2004. Ultrafast asynchronous multioutput all-optical header processor. *IEEE Photon. Technol. Lett.*, 16, 4, 1182–1184.
268. Willner, A., Gurkan, D., Sahin, A., McGeehan, J., and Hauer, M. 2003. All-optical address recognition for optically-assisted routing in next-generation optical networks. *IEEE Commun. Mag.*, 41, 5, S38–S44.
269. Gurkan, D., Hauer, M, C., Sahin, A. B.,Lee, Z. S., Willner, A. E., Parameswaran, K. R., Fejer, M. M. 2001.Demonstration of multi-wavelength all-optical header recognition using a PPLN and optical correlators, *European Conference Optical Communication*.
270. Bannister, J., Touch, J., Kamath, P., Patel, A. 2001. An optical booster for internet routers, Paper presented at Proceedings of 8th International Conference High Performance Computing.
271. Blumenthal, D. J., Bowers, J. E., Rau, L., Chou, H.-F., Rangarajan, S., Wang, W., and Poulsen, K. N. 2003. Optical signal processing for optical packet switching network. *IEEE Commun. Mag.*, 41, 2, S23–S29.
272. Rau, L., Rangarajan, S., Blumenthal, D. J., Chou, H. -F., Chiu, Y. -J., Bowers, J. E. 2002. Two-hop all-optical label swapping with variable length 80 Gbit/s packets and 10 Gbit/s labels using nonlinear fiber wavelength converters, unicast/multicast output and a single EAM for 80–10 Gbit/s packet demultiplexing, Paper presented at Optical Fiber Communication Conference, pages FD2-1-FD2-3.
273. Yao, S., Mukherjee, B., and Dixit, S. 2000. Advances in photonic packet switching: An overview. *IEEE Commun. Mag.*84–94.
274. Blixt, P. and Bowers, J. E. 1994. An optical technique for bit and packet synchronization. Lasers and electro-optics society annual meeting. *LEOS*, 2, 103–104.
275. Xia, T. J., Kao, Y.-H., Liang, Y., Lou, J. W., Ahn, K. H., Boyraz, O., Nowak, G. A., Said, A. A., and Islam, M. N. 1999. Novel self-synchronization scheme for high-speed packet TDM networks. *IEEE Photon. Technol. Lett.*, 11, 269–271.
276. Wang, L. J., Shi, H. K., Lin, J. T., Guan, K. J., and Ye, P. 1998. Clock and frame synchronization recovery based on a terahertz optical asymmetric demultiplexer. *IEEE Photon. Technol. Lett.*, 10, 881–883.

277. Hofmeister, R. T., Lu, C. -L., Ho, M. -C., Poggiolini, P., and Kazovsky, L. G. 1998. Distributed slot synchronization (DSS): a network-wide slot synchronization technique for packet-switched optical networks. *IEEE J. Lightwave Technol.*, 16, 2109–2116.
278. SONET Telecommunications Standard Primer. 2001. Tektronix, application note.
279. Ching, Y.-C. 1993. SONET Implementation. *IEEE Commun. Mag.*34–40.
280. Stallings, W. 2004. *Data and Computer Communications. 7th Ed.* Pearson Prentice Hall, Upper Saddle River.
281. McEachern, J. A. 1992. Gigabit networking on the public transmission network. *IEEE Commun. Mag.*70–78.
282. Cisneros, A. 1992. Large scale ATM switching and optical technology, Paper TuJ3 presented at Conference on Optical Fiber Communications '93. Technical Digest, San Jose, CA.
283. Black, U. 2002. *MPLS and Label Switching Networks. 2nd Ed.* Prentice Hall, Upper Saddle River.
284. Yablonovitch, E. 1987. Inhibited spontaneous emission in solid-state physics and electronics. *Phys. Rev. Lett.*, 58, 2059.
285. John, S. 1987. Strong localization of photons in certain disordered dielectric superlattices. *Phys. Rev. Lett.*, 58, 2486.

Index

A

B

C

D

E

F

G

H

I

J

K

L

M

N

O

P

Q

R

S

T

U

V

W

X

Y

Z